WIND ENGINEERING

(IN TWO VOLUMES)

Volume 1

Also of Interest

CHEUNG
Finite Strip Method in Structural Analysis, 2nd Ed

FRANCIS
Introducing Structures

GIBSON
Thin Shells: Computing and Theory

HARRISON
Structural Analysis and Design: Some Minicomputer Applications

HORNE
Plastic Theory of Structures, 2nd Ed

LENCZNER
Elements of Loadbearing Brickwork

LENCZNER
Movements in Buildings, 2nd Ed

MILLER & SMITH
Mechanical Behaviour of Materials (3 volumes)

PIGGOTT
Load Bearing Fibre Composites

SACHS
Wind Forces in Engineering, 2nd Ed

TAPLIN
Advances in Research on the Strength and Fracture of Materials and Structures (6 volumes)

Pergamon Related Journals

Free Specimen Copies Gladly Sent on Request

CEMENT AND CONCRETE RESEARCH
COMPUTERS AND STRUCTURES
ENGINEERING FRACTURE MECHANICS
FATIGUE OF ENGINEERING MATERIALS AND STRUCTURES
INTERNATIONAL JOURNAL OF ENGINEERING SCIENCES
INTERNATIONAL JOURNAL OF SOLIDS AND STRUCTURES

WIND ENGINEERING

Proceedings of the Fifth International Conference, Fort Collins, Colorado, USA, July 1979

Edited by

J. E. CERMAK

Colorado State University, USA

Volume 1

PERGAMON PRESS

OXFORD · NEW YORK · TORONTO · SYDNEY · PARIS · FRANKFURT

U.K.	Pergamon Press Ltd., Headington Hill Hall, Oxford OX3 0BW, England
U.S.A.	Pergamon Press Inc., Maxwell House, Fairview Park, Elmsford, New York 10523, U.S.A.
CANADA	Pergamon of Canada, Suite 104, 150 Consumers Road, Willowdale, Ontario M2J 1P9, Canada
AUSTRALIA	Pergamon Press (Aust.) Pty. Ltd., P.O. Box 544, Potts Point, N.S.W. 2011, Australia
FRANCE	Pergamon Press SARL, 24 rue des Ecoles, 75240 Paris, Cedex 05, France
FEDERAL REPUBLIC OF GERMANY	Pergamon Press GmbH, 6242 Kronberg-Taunus, Hammerweg 6, Federal Republic of Germany

First edition 1980

British Library Cataloguing in Publication Data

International Conference on Wind Engineering,
5th, Colorado, 1979
Wind engineering.
1. Wind-pressure - Congresses
I. Title II. Cermak, J E
624'.175 TA654.5 80-40753

ISBN 0-08-024745-8

In order to make this volume available as economically and as rapidly as possible the author's typescript has been reproduced in its original form. This method has its typographical limitations but it is hoped that they in no way distract the reader.

Printed in Great Britain by A. Wheaton & Co. Ltd., Exeter

Foreword

The Fifth International Conference on Wind Engineering (ICWE) held at Colorado State University in Fort Collins, Colorado, USA, on 8-14 July 1979 marks a new phase in development of engineering related to wind and its effects. Whereas the four previous conferences --

first in Teddington, England (June 1963);
second in Ottawa, Canada (September 1967);
third in Tokyo, Japan (September 1971); and the
fourth in Heathrow, England (September 1975) --

were identified as International Conferences on Wind Effects on Buildings and Structures, the Fifth ICWE addressed a broader scope of effects and concerns associated with wind. In addition to the traditional theme of wind effects on buildings and structures, the Fifth ICWE included considerations of social and economic impact of wind storms and wind-engineering practice and an introduction to relatively new wind-engineering applications of growing significance. Movement of air through buildings, dispersion of heavy gases released from spills of chemicals and liquid natural gas, physical modeling of flow over complex terrain and transport of particles, and siting of wind turbines with respect to terrain features are among the most rapidly developing applications.

Substantial evidence is available that indicates an ever increasing interest and activity in wind engineering. Fifth ICWE statistics of 250 complete papers submitted for review and 350 participants from 25 countries reveal a strong growth trend. Another favorable sign is development of wind-engineering organizations in several countries. For example, the

Canadian Wind Engineering Association, the
Japan Association for Wind Engineering and the
Wind Engineering Research Council, Inc. (USA)

sponsor national wind-engineering conferences to promote exchange of information that compliments communication through the International Conferences and these Proceedings.

The Conference Organizing Committee

Foreword

Contents

Volume 1

Technical Session V: Wind Loading—Fluctuating External

Contents

Volume 2

Technical Session VII: Dynamic Response—Bridges, Transmission Lines and Roof Membranes

Technical Session VIII: Physical and Mathematical Modeling

Technical Session IX: Wind-Engineering Applications

Introduction

INTERNATIONAL ASSOCIATION FOR WIND ENGINEERING

The International Association for Wind Engineering (IAWE) came into being on 10 September 1975 through action of the Steering Committee for the International Study Group on Wind Effects on Buildings and Structures. Primary purposes of the IAWE are:

(1) organize and maintain continuity of international conferences on wind engineering that began with the International Conference on Wind Effects on Buildings and Structures held at Teddington, UK, in 1963 and

(2) establish liaison with national and international organizations working in related fields.

At the same meeting, held during the Fourth International Conference on Wind Effects on Buildings and Structures at Heathrow, UK, the Steering Committee chose to redesignate the series of quadrennial international conferences as the International Conferences on Wind Engineering (ICWE). This decision was prompted because of expanding needs and interests to consider transport of air pollutants, wind effects on people and plants and utilization of wind power. Furthermore, knowledge of wind effects on buildings and structures had advanced sufficiently to be accounted for through engineering design and analysis. The next conference, the Fifth International Conference on Wind Engineering, was scheduled for 8-14 July 1979 at Colorado State University (CSU) in Fort Collins, Colorado, USA.

Minutes of the Steering Committee meeting held during the Fifth ICWE are presented in Appendix 1 of these proceedings.

PURPOSE OF CONFERENCE

As in the four previous conferences, the Fifth ICWE was organized to facilitate international communication and exchange of new developments in knowledge and practice of wind engineering. One objective was to stimulate the exchange of research results recently obtained by engineers, fluid dynamicists and meteorologists related to wind effects on buildings, structures, people, agriculture, air pollution and energy production. Application of new findings to development of design methods and codification of information for use by practicing engineers, architects, meteorologists and urban planners was another important objective.

A new objective for the international conferences was to assess the social and economic impact of wind as a natural hazard. Equally significant was an attempt to realize the social and economic implications of codes and design practices developed to combat this hazard.

THEMES OF TECHNICAL SESSIONS

Subject matter of the Fifth ICWE was organized into ten consecutive technical sessions. These half-day sessions of approximately ten papers, following a paper on Recent Developments and Trends by each Chairman, were focused on the following themes:

I. Social and Economic Impact of Wind Storms

II. Wind--Characteristics and Description

III. Local Wind Environment

IV. Wind Loading--Mean and Fluctuating Internal and Mean External

V. Wind Loading--Fluctuating External

VI. Dynamic Response--Tall Buildings and Towers

VII. Dynamic Response--Bridges, Transmission Towers and Roof Membranes

VIII. Physical and Mathematical Modeling

IX. Wind-Engineering Applications

X. Wind-Engineering Practice

The theme of Session I, Social and Economic Impact of Wind Storms, was included for the first time on an ICWE program. Because of ever increasing economic losses through wind damage this topic was considered to be timely and appropriate to establish significance of the nine topics to follow. For example, a study by the J. T. Wiggins Company of California sponsored by the National Science Foundation concluded that wind damage to buildings in the United States during 1970 amounted to $3.5 billion (1978 dollars) and may reach $8 billion by year 2000.

TECHNICAL SESSION PAPERS AND DISCUSSIONS

Papers for the technical sessions were selected from 250 complete manuscripts submitted in response to the Call for Papers. Ten additional papers resulted from a paper on Recent Developments and Trends by the Chairman and Co-Chairman of each session. In an effort to provide adequate time for presentation and discussion, while keeping the tradition of non-parallel technical sessions, the Organizing Committee and a host of reviewers selected approximately 100 of the unsolicited papers for presentation. These selections were based on originality, quality and completeness of the paper with additional considerations given to coherence of subject matter for each session and geographical distribution of the authors. Authors of papers recommended for presentation that could not be placed on the program were invited to submit their papers to the Journal of Industrial Aerodynamics for further review and possible publication. Through courtesy of the U.S. Nuclear Regulatory Commission and assistance of Dr. R. F. Abbey, Jr. all papers, with exception of papers by the Chairmen, were presented to participants in a two-volume set of preprints at the time of registration.

Two discussion periods were scheduled for each technical session -- one preceding the mid-session recess for refreshments and the other immediately before closure of the session. The rapporteur of each session served to coordinate communication between discussants and edited the dialogue that ensued.

These proceedings contain all papers presented during the Conference and five papers selected for presentation but not presented. The papers of each session are preceded by the Chairman's paper on Recent Developments and Trends and are followed by the discussions.

PROGRAM SUMMARY

All Conference sessions were held in the Andrew G. Clark Building on the main campus of Colorado State University in Fort Collins, Colorado. The following program covered the period 8-14 July 1979:

Sunday, 8 July:

Afternoon	Registration
Evening	Reception

Monday, 9 July:

Morning	Opening Session Welcome by Dr. L.V. Baldwin, Dean of Engineering, CSU Address by Dr. A.R. Chamberlain, President, CSU Opening Remarks by Dr. J.E. Cermak, Chairman Technical Session I - Social and Economic Impact of Wind Storms
Noon	Luncheon and presentation by Dr. H.W. Teunissen -- Activities of Canadian Wind Engineering
Afternoon	Technical Session II - Wind: Characteristics and Description
Evening	Meeting and Dinner - ICWE Steering Committee

Tuesday, 10 July:

Morning	Technical Session III - Local Wind Environment
Noon	Luncheon and presentation by Dr. A.N.L. Chiu -- The Wind Engineering Research Digest
Afternoon	Technical Session IV - Wind Loading: Mean and Fluctuating Internal and Mean External
Evening	Conference dinner and program by R.L. Brown of Abraham Lincoln High School, Denver -- Ghost Towns of Colorado: Past and Present

Wednesday, 11 July:

Morning	Technical Session V - Wind Loading: Fluctuating External
Afternoon	Tour of Rocky Mountain National Park Meeting -- Wind Engineering Research Council, Inc.
Evening	Dinner at Lazy B Ranch, Estes Park

Thursday, 12 July:

Morning	Technical Session VI - Dynamic Response: Tall Buildings and Towers
Noon	Luncheon and presentation by Dr. H. Ishizaki -- Activities of Japan Association for Wind Engineering
Afternoon	Technical Session VII - Dynamic Response: Bridges, Transmission Towers and Roof Membranes
Evening	Picnic Tour of Fluid Dynamics and Diffusion Laboratory, CSU

Friday, 13 July:

Morning	Technical Session VIII - Physical and Mathematical Modeling
Noon	Luncheon and presentation by Dr. D.B. Simons -- Research in the College of Engineering at CSU
Afternoon	Technical Session IX - Wind-Engineering Applications

Saturday, 14 July:

Morning	Technical Session X - Wind-Engineering Practice Presentation by Dr. J.E. Cermak -- Plaque to Dr. Deliu Sfintesco on His 70th Birthday Closing remarks by Dr. W.H. Melbourne -- Plans for Sixth ICWE in Australia

The foregoing program was supplemented by a program for ladies and children. This program included a variety of tours, mountain hikes, luncheons and presentations on fine art.

Following the Saturday morning session many participants departed on Post-Conference Study Tours. Three basic tours were scheduled -- 1) northern travel to the University of Western Ontario, London, Ontario; Atmospheric Environment Service, University of Toronto and CN Tower, Toronto; and the National Research Council and National Aeronautical Establishment, Ottawa; 2) easterly travel to Center for Building Technology and George S. Vincent Memorial Wind Tunnel, Washington, D.C.; the World Trade Center Tower and Citicorp Building, New York; the Environmental Sciences Research Laboratory, Research Triangle Park, North Carolina; and the Wright Brothers Wind Tunnel and John Hancock Tower Boston, and 3) westerly travel to the Full-Scale Earthquake and Wind Engineering Laboratory, UCLA, Los Angeles; The California Institute of Technology, Pasadena; and the University of British Columbia and the Lions' Gate Bridge, Vancouver, British Columbia.

ACKNOWLEDGMENTS

The individual threads of this Conference and their integration into a fabric of distinct pattern and enduring value are the result of much effort by many contributors. Authors, chairmen, rapporteurs, delegates, IAWE Steering Committee and my associates of the Fluid Mechanics and Wind Engineering Program at Colorado State University provided major contributions for which we express our sincere thanks.

For extensive correspondence and arrangement of operational details before and during the Conference that resulted in an unusually enjoyable and problem-free week, we are especially grateful to our Conference Secretary, Edith L. Kruse. We are all indebted to Helen V. Sitzman, Fluid Mechanics and Wind Engineering Program Secretary, for many hours of painstaking effort dedicated to typing and assembling the manuscript for these Proceedings.

We thank the following Sponsors who provided necessary financial support and Co-Sponsors who published Conference announcements in their publications:

Sponsors

National Science Foundation
Libbey-Owens-Ford Company
AMAX Foundation, Inc.
Exxon Research and Engineering Company
Electric Power Research Institute
U.S. Nuclear Regulatory Commission
Fluid Mechanics & Wind Engineering Program, CSU

Co-Sponsors

American Academy of Mechanics
American Institute of Architects
American Meteorological Society
American Society of Civil Engineers
American Society of Mechanical Engineers
Canadian Wind Engineering Association
International Association for Bridge and Structural Engineering
Subsonic Aerodynamic Testing Association
Wind Engineering Research Council, Inc.

J. E. Cermak
Chairman

Opening Address

DR. A. R. CHAMBERLAIN

President Colorado State University

Let me add to the welcome given you by the Colorado State University Dean of Engineering, Dr. Lionel Baldwin. The governing Board of Colorado State University and the entire university community of 18,000 competitively selected students and 1,500 professional personnel take pride in having this opportunity to serve as host for this Fifth International Conference on Wind Engineering. It is my special privilege to have the opportunity to provide a brief opening address to you, a distinguished group of international scholars from many parts of the world.

Much of this conference will relate to the sharing of new knowledge from research. Very little is easily translated into public policy action by typical elective officials who control most of our economic and social behavior. Yet, the preprints for this conference makes it clear to me that the economic benefits of more wind engineering research and the application of it would provide a high rate of economic return, especially as construction costs continue to skyrocket.

I read recently that by the year 2,000, well within the anticipated life of most buildings we could build even this year of 1979, building damage caused by hurricane winds and storm surges may approximate $5 billion annually in constant dollars and will exceed that caused by any other natural hazard. This conveys the importance of increasing our collective research efforts on wind engineering. This conclusion was reinforced when further reading suggested that probably half of the estimated damage could be avoided by the application of current and probable near future knowledge. It also indicates a very high economic payoff from an increased investment in research.

To the above example we must add many, many other areas of our daily lives in which wind engineering can and should play an increasing role; roles which suggest the need for more graduate education and research in wind engineering to develop the human capital so necessary to conduct the research, design and planning functions.

As an increasing number of people occupy more land area and use more, larger, higher and longer structures, with greater efforts in their design to fulfill the aesthetic needs of people, the structural and environmental complexity will call clearly for an increasing frequency of use of wind engineering, planning and design talents. This will require enhanced knowledge by engineers, architects, lawyers, insurance and other types of professionals if we are to sustain a quality of living environment acceptable to our people and which is reasonably safe and yet economical to construct.

It seems that the central thrust of your profession has been around the influence of winds on structures. This has had some very high payoffs for the modest funds invested in research over the last 20 years. It is reassuring to note the increasing number of major structures being tested in wind tunnels before they are built. While this success will continue and should help encourage more funding for research, a number of recent events indicate that you have just raised the tip of the iceberg -- "windburst" -- for your profession.

Some examples come to mind. Renewed emphasis on wind engineering of the automobile as a means to conserve energy. The distress of the public about the potential wind dispersion of radioactive gases if the Three-Mile Island Nuclear plant problem got out of control. The economic, energy and ecologic problems created by a wind driven supertanker breaking up on a seashore due to power or steerage loss. What are the potential community problems of meteorological "downbursts", which while having caused fatal plane crashes, cannot even be tested in current wind tunnels as to their characteristics of downward vertical loading on structures such as those with large roofs? Lastly, how much have you done for the recreation field; wind engineering could contribute more to clothing design, both as to drag and insulating against wind chill?

There are many areas of research which have been going for decades that as you seek increased financial support for your efforts you will have to decide whether or not it is to your benefit before public policy people to claim the activity as wind engineering. As a clearly positive example, crop genetic research to develop wind resistant species of wheat, rye, millet, etc., has had a positive payoff. But it is less clear whether claiming weather modification as a part of wind engineering will help you!

My first professional experiences related to wind engineering were few; review of the Tacoma Narrows disaster, noting the winds around massive forest fires in the Pacific Northwest and observing the devastating effect of moderate velocity low humidity winds on agricultural plants in the Columbia River Basin. Current experience also suggests innumerable positive effects of winds on our life; winds solving for man problems we cannot or will not solve ourselves. An example is the Denver "Brown Cloud" of largely auto exhaust based pollution. When the winds come, this huge mass of pollution is dispersed. But can the atmospheric winds, the oceans and earth's biota continue to handle man's advertant and inadvertant effects on our environment? Only research and experience will tell us.

Now a few comments about Colorado State University. This University's role in Wind Engineering derived from the special interests of Dr. Jack Cermak, who has headed up much of our fluid dynamics work for three decades. His wind tunnel developments, supported by a legend of colleagues and graduate students now dispersed around the world, have placed this University in the forefront nationally of experimental studies of the effects of wind on buildings. I expect to see Dr. Cermak continue to provide leadership in this area for Colorado State University and his colleagues internationally. Parenthetically, Dr. Cermak was one of my best professors when I was a student under him; so, I clearly believe in him!

But such leadership, typified by many of you in the audience, needs support for a constant stream of new equipment. And each generation of equipment is more complex, leading to extremely rapid escalation in the cost of adequate equipment to do your job. Most state governments are increasingly reluctant to provide funds for such large equipment. The Federal government is recognizing the problem, but has not yet found adequate means to help the academic community keep its equipment inventory current, even for items much less costly than wind tunnels.

Many of the next generation of wind tunnels and their associated data processing hardware will have to be debt financed. Such debt financing plans will need the implicit and sometimes explicit endorsement of agencies such as the National Science Foundation. Further, the best bets for the next decade probably will be third-party lease financial arrangements for the achievement of joint ventured facilities by selected universities in cooperation with major industrial companies.

Lastly, in closing, let me extend to you the hope that you soon will search for ways to have your wind engineering skills contribute more aggressively to the development needs of the peoples of the Third-World and the Fourth-World.

While you are at Colorado State University, in addition to hoping you have a successful conference professionally, I hope you also take time to enjoy one another's company amidst the splendor of our Rocky Mountains. If you are lucky, Colorado summer weather will give you new inspiration and commitment to wind engineering by permitting you to share the awe inspiring and majestic experience of a supercell thunderstorm with all its visual, acoustical and high velocity characteristics, again making us as lowly humans have the ultimate in respect for nature and its massive forces and influence.

Good luck to each of you.

SESSION I

Social and Economic Impact of Wind Storms

CHAIRMAN

G. F. White

Director, Natural Hazards Research and
Applications Information Center
University of Colorado
Boulder, Colorado, USA

CO-CHAIRMAN

M. P. Gaus

Program Manager
National Science Foundation
Washington, D.C., USA

RAPPORTEUR

K. J. Eaton

Head Housing Section
Overseas Division
Building Research Establishment
Garston, United Kingdom

RECENT DEVELOPMENTS AND TRENDS: SOCIAL AND ECONOMIC IMPACT OF WIND STORMS

G. F. WHITE

Natural Hazards Research and Applications Information Center, University of Colorado, USA

During the past decade at least three major trends in the situation with respect to social and economic impact of windstorms may be distinguished. Basic information as to the nature and extent of losses in property and life as a result of windstorms has improved significantly. The variety of measures taken to cope with the consequences of windstorms has increased. At the same time there has been a refinement in modes of analyzing the decision as to what types of action should be taken by both individual and public managers of vulnerable property. All of these trends are reflected in the papers presented in Technical Session I.

TRENDS IN INFORMATION ON LOSSES

In a number of countries the rudimentary information collected by public agencies or available through press reports on the character of losses resulting from windstorms has improved through greater precision and coverage in government collection systems and through detailed studies of particular extreme events. In the United States the rough assemblage of data made by the National Weather Service has improved in detail and in degree of coverage and has been supplemented by a series of efforts to estimate damages. Ideally, one would like to see comprehensive recording of damages linked with detailed investigations of sample events: it obviously is impractical to make detailed studies of every tornado, hurricane, or down-slope wind.

In Australia and in reviews which have been made under auspices of the World Meteorological Organization there now is more accurate investigation of precisely what brought about failure of structures and loss of life.

We still lack a thoroughly insightful inquiry into the precise circumstances in which injuries occurred to humans and in which fatalities are caused. Some of the conventional wisdom as to what parts of a structure should be sought as places of refuge and what steps should be taken to open or close doors and windows has been questioned, and the sort of post-mortem examination of experience that occurred this spring at Wichita Falls, Texas is needed in other circumstances.

Reviewing the information for the United States as to the losses from wind, it seems clear that the property damage has been increasing and continues to increase from hurricane, tornado, and windstorm causes. The aggregate figures suggest that losses of life from hurricane and tornado have been decreasing while wind losses have tended to increase. There is some reason to think that the average figures for both hurricane and tornado may be reversed at the present time: recent events may turn the trend line upward.

Examination of particular sites such as the hurricane hazard in Miami, Florida and New Orleans, Louisiana indicates that there are large catastrophes in the offing. The prospect concerning wind damage is very similar to that from some other extreme atmospheric events. The average loss from high frequency events seems to be declining but the potential loss from the catastrophes that follow with major social dislocation in the trail of very rare, low frequency events increases. This is particularly the case for hurricanes where concentrations of population in areas subject to both high winds and storm surge make it likely that there will be larger losses of property and life in the future than we have experienced to date.

MITIGATING MEASURES

The two major and prevailing social responses to losses from high winds are structural and informational. Building design is, of course, the most confident measure to be adapted in dealing with high velocities and stress. Warning systems may assist in reducing property losses somewhat through provoking emergency measures and the evacuation or protection of people at risk. Less attention has been paid to the possibilities of land use planning as represented in microzonation, or of insurance as means of spreading losses and of stimulating improved structural design and construction. Across the field of natural hazards activity there is a broad trend toward increased emphasis upon mitigating measures in contrast to reliance upon relief and reconstruction measures in the trail of the disaster. In the case of windstorms this may have its principal emphasis upon understanding the conditions in which adequate building codes may be not only designed but enforced. It also may show in efforts to link insurance and public reconstruction measures with efforts to improve the adoption of building codes and to exercise more discrimination in the location of new structures.

An interesting aspect of the building code opportunities is that presented by provisions for earthquake stress. Here, it seems likely that greater collaboration between proponents of earthquake and windstorm safety would enhance the attention given to both.

DECISION MAKING

As more information is accumulated about losses and about the range of opportunities that are available in coping with them it has become easier to identify the components of any analysis of what would be an optimal decision on the part of either individual managers of property or public agencies faced with influencing the action of private operators. The need for improvements in decision making can be satisfied only by joining two extremes of analysis. One is the engineer-economic analysis of efficiency in the light of expected streams of benefits and costs for specified time periods and discount rates. The other is the social-political analysis of the conditions in which public agencies may have influence either by incentives or by restrictions upon the activities of private builders. The first has advanced more fully than the second. As a result, it is now possible to point out types of measures which might offer high confidence of minimal losses but which also might be inefficient in terms of either private or public investment. It is now possible, as several of the papers demonstrate, to sharpen the analysis for such purposes. At the same time there is still ambiguity as to what lines of public action in which social and economic circumstances may be counted upon to achieve what would be optimal public outcomes. This requires careful and sensitive assessment of the value of human life and discomfort. It also requires balancing the costs incurred by measures such as strengthened building construction for a long period over the short term benefits to those operating the property in the short run. It is not enough for the engineer to declare that a particular structure is the safest in a given situation; it is also necessary for the engineer with the help of others to think through the question of what might be in the best interest of the community in terms of allocation of capital and of administrative effort. The papers in this session point up some of these issues but show that we are still distant from arriving at satisfactory judgments in a variety of situations.

LOW-INCOME HOUSING AND HURRICANES

K. J. EATON

Overseas Division, Building Research Establishment, UK

ABSTRACT

About 750 million people in hurricane areas around the world live in poor houses, apartments, timber shelters or other types of construction that can be called 'home'. Most of these dwellings which have been built by the occupants, do not incorporate engineering principles and are not resistant to hurricanes. This paper describes some of the problems which exist for low-income housing in hurricane areas and describes research being carried out in the Caribbean to develop simple and economic techniques for hurricane resistance.

INTRODUCTION

The aim of this paper is to focus attention on some of the problems that exist when housing for low-income people in developing countries experiences the forces imposed by hurricanes, typhoons and cyclones*. A second aim is to outline the current research looking into these problems in the Eastern Caribbean area that is being undertaken by the Overseas Division of the UK Building Research Establishment.

The paper starts by discussing the main problems, which, with this sort of structure, involve socio-economic as well as engineering aspects; photographs are included in this section of hurricane damage to low-income housing.** Consideration is also given to the areas of the world where these non-engineered structures are vulnerable. The subsequent sections cover the current BRE research into low-income housing experiencing adverse environments and this leads on to the anticipated outcome of these studies including

---

* Throughout the rest of this paper, these terms and other names for the hurricane stage of tropical cyclones will be called 'hurricanes' for convenience.

**The term 'low-income' is used in preference to 'low-cost' and signifies housing for the very poorest sectors of the community concerned.

the provision of building codes and other specific application activities.

THE GENERAL PROBLEMS

In many tropical areas, and particularly in developing countries, it is unfortunate that sound principles of building that would help to resist hurricanes are rarely followed. This is particularly the case where there has been no record of severe wind activity for a number of years; and consequently people tend to ignore the need for adequate design and construction.

Every year, many hurricanes develop in six main tropical areas (1), as shown in Figure 1. Some of these only occur over the oceans and therefore cause no damage to buildings; some blow over uninhabited land and only cause damage to crops etc; but many cause structural damage, make people homeless and cause loss of life. The scale of the consequential effects varies considerably, depending on the intensity of the hurricane and its geographical location. For example, two of the recent hurricanes (November 1978) have been those in Sri Lanka and Belize. In Sri Lanka (population 13.8 million, area 66,000 km^2), thousands of people were killed, millions made homeless and half-a-million houses destroyed. Whereas in Belize (population 143,000, area 23,000 km^2) five people were killed, a few hundred people were made homeless, and about six hundred houses were damaged and very few totally destroyed. Belize is much less densley populated (2) - 6 persons/km^2 cf 209 persons/km^2 - and the majority of the hurricane damage was to crops (3,4).

Whatever the form of damage, enormous social and economic problems are created. Jobs, shelter, crops, animals, forms of transport, and supplies of food and fuel are all important to a community and can be seriously affected by hurricanes. To rectify the situation takes money - a commodity which is always in short supply in third world countries, both to the individual and to the central administration. To an individual the most important item listed above, after the supply of food, is his home. Climatic considerations in most countries mean that shelter and security are urgently needed after a hurricane has passed.

In some cases houses are only damaged, and can still be inhabited whilst being repaired. An example of this is shown in Figure 2, where the house had been a 'high-set' home on stilts, which lacked diagonal bracing in the substructure and end fixing of the columns. In other cases the houses are completely destroyed (Figure 3). This example had also been a 'high-set' house on stilts; there was good bracing forming a strong substructure, but the main superstructure was very weak at the joints.

Similar examples of damage in many countries have been cited in other papers (5 - 15); the writer unfortunately believes that this will continue to be the case, for as far as housing is concerned it takes a long time for the implications to be learnt and appropriate action taken. Following cyclone 'Tracy' in Darwin in December 1974, Dr G R Walker (10) wrote:

> 'The most significant consequence that has followed from the investigation of damage caused by cyclone 'Tracy' and the studies which have followed has been

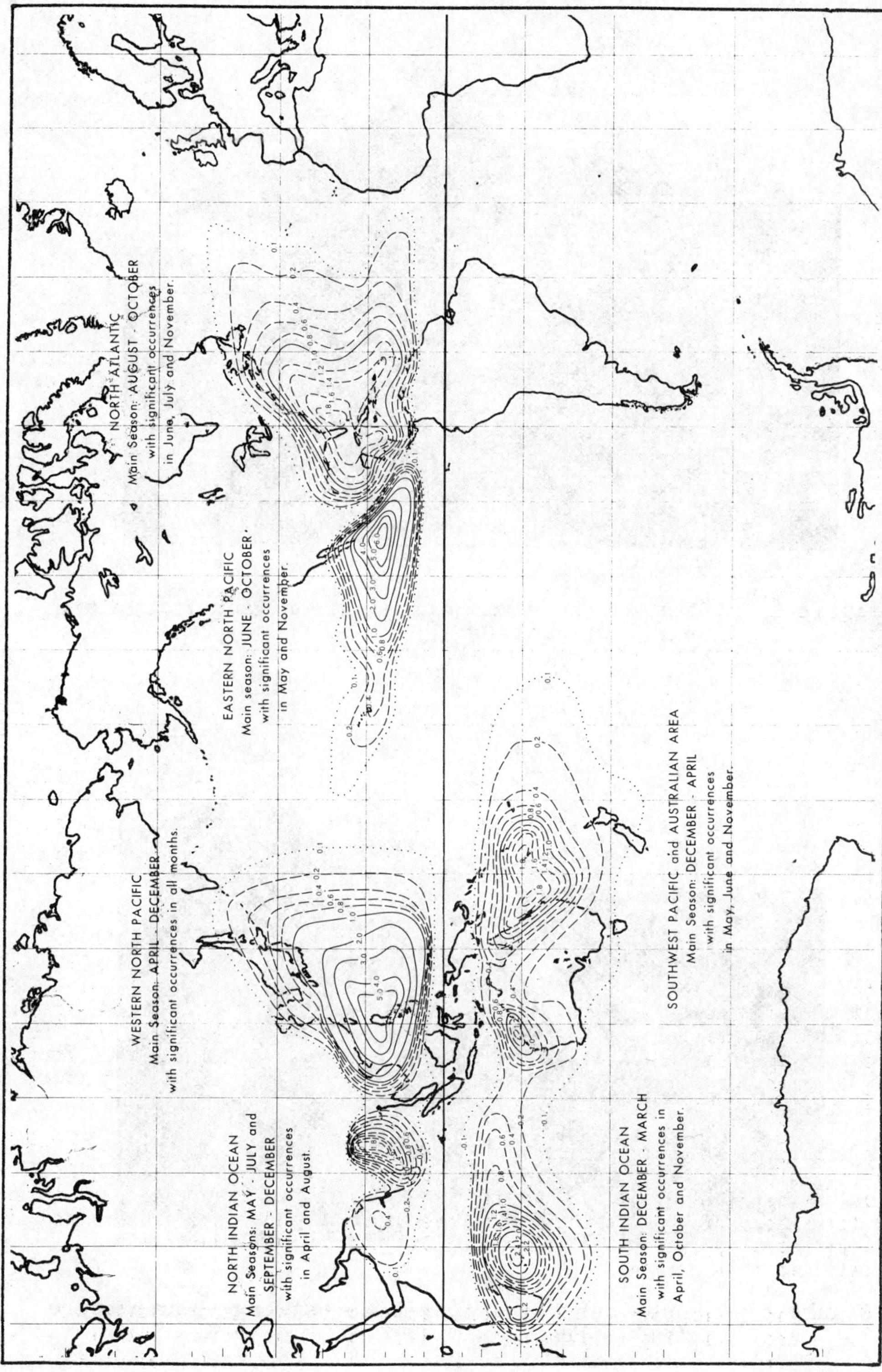

Figure 1 Areas of tropical cyclones and hurricanes

Figure 2 Damaged house in Belize: strong superstructure, weak substructure

Figure 3 House destroyed in Belize: strong substructure still standing

the recognition of the importance of incorporating structural engineering principles in the design of housing likely to be exposed to extreme winds such as those accompanying severe tropical cyclones.'

And whatever applies to housing in developed countries applies to an even greater extent in developing countries (16). There is often no building control, codes or regulations; inferior materials often have to be used, either through financial difficulties or the scarcity of better materials. With a lack of knowledge of engineering principles and the ability to adopt sound construction practices, the small house-builder cannot hope to build his home such that it will survive hurricane winds. The writer therefore firmly believes that more effort should be spent to overcome these problems, otherwise they will only get worse. By the turn of the century the population of the developing countries will have doubled; and about half of this population - more than 3,000 million people - will be living in very poor self-built housing. And it is estimated that about half of these people will be living in the hurricane areas indicated in Figure 1, particularly around the North Indian Ocean, in South East Asia, in some of the Pacific Islands, and in the North Atlantic/Caribbean Sea/Gulf of Mexico area.

THE RESEARCH BACKGROUND

Studies of wind loading problems have been carried out extensively over the past twenty years. Most of this research has been concerned with tall buildings and other engineered structures. Few studies have been undertaken into wind effects on low-rise structures such as housing. Notable exceptions are the research programmes carried out by the US National Bureau of Standards (7,17,18), by the UK Building Research Establishment (19-21) and by several organisations in Australia (22-26) following cyclone 'Tracy'. However, only the NBS study (7) specifically investigated hurricane effects on housing in developing countries.

The Overseas Division of BRE carries out building research sponsored by the UK Ministry of Overseas Development. In particular, a significant proportion of the studies relate to housing in developing countries. In 1978 a project was initiated to investigate the effects on such housing of 'adverse environments', meaning earthquakes, floods, hurricanes, and any other natural disasters that may occur. Many developing countries could experience any one type of disaster; few countries have a high risk of, say, both earthquakes and hurricanes happening. However, one such group of countries are those in the Eastern Caribbean where the design wind speed (1 in 50 year gust speed with 0.63 probability of exceedence) is between 50 m/s and 55 m/s (27), and where the design ground acceleration for earthquake loads (1 in 50 year bedrock value with 0.1 probability of exceedence) is between 0.15g and 0.4g (28-31). These values are as severe as those experienced in parts of Florida as far as hurricanes are concerned, and as severe as conditions in parts of California for earthquakes.

In the current study, reported for the first time in this paper, the aims are to assess the ability of the current low-income housing in some of the Eastern Caribbean islands to withstand the imposed forces, to modify the designs as necessary using engineering principles, to construct some full-scale trial units using these designs, to utilise more indigenous materials and thereby reduce the costs, and to report on the outcome of the research and get the results widely applied. At the time of writing this paper, the trial houses are being constructed; further information will be provided in the presentation at the Conference.

The writer has surveyed the low-income housing in six of the islands in the Caribbean. These countries all have a housing shortage; they all have a public sector housing authority; and these authorities are all currently undertaking a programme of 'low-income' house construction. Unfortunately, in common with a majority of low-income and middle income countries around the world, most of the housing that is constructed cannot be used for truly low-income people as construction costs are too high - typically up to US \$6,000 - US \$8,000 per unit for a 50 m^2 (500 ft^2) floor area. Some examples of these houses are shown in Figures 4 and 5.

In addition to the costs involved with these typical houses, there are many construction problems. For example, in St Vincent a complete structural RC framework is cast in-situ to resist the hurricane and earthquake forces - but this takes time to build the falsework, to cast, to cure, and to dismantle the falsework (Figure 6).

In most places, and particularly in St Lucia, the only land available for housing is on steep hillsides. This creates foundation problems and often means that the houses are sited in valleys where the hurricane forces can be increased due to funnelling. And houses with foundation problems - often overcome by long slender columns as in Figure 7 - are potentially unsafe. In some of the houses inspected by the writer, these problems are made worse by poor detailing of hurricane straps (Figure 8) - if there are any straps at all. And most houseowners want extra rooms and storage space; this is achieved by excavating the hillside, often undermining the foundations (Figure 9).

THE TRIAL HOUSES

To overcome most of the problems mentioned, to reduce the cost of imported cement and reinforcing steel, and at the same time to resist the likely forces imposed by hurricane winds, the houses detailed below have been designed (Figure 10). They are constructed at Barrouallie, St Vincent, a site owned by the Housing and Land Development Corporation. St Vincent was chosen for the project for several reasons, including the likelihood of hurricanes, the low-income level of the potential owners (St Vincent's GNP per capita is nearly the lowest in the Caribbean (32)), and the willingness of the government to explore all means of improving the structures and reducing the costs. The main features of the trial houses are:

1 Solid foundations produced by cut-and-fill, and retained in position by a reinforced hollow-concrete block wall.

2 Hardwood (greenheart) structural columns, rigidly cast into the floor slab, and connected by rigid brackets to -

3 Hardwood beams and rafters, which are in turn connected by straps to -

4 Hardwood purlins

Figure 4 Low-income house in St Lucia - timber construction

Figure 5 Low-income house in St Vincent - concrete blockwork

Figure 6 False work for R.C. ring-beam

Figure 7 Houses on hillside with long supporting columns

Figure 8 Inadequate hurricane straps

Figure 9 Excavations beneath foundation footing

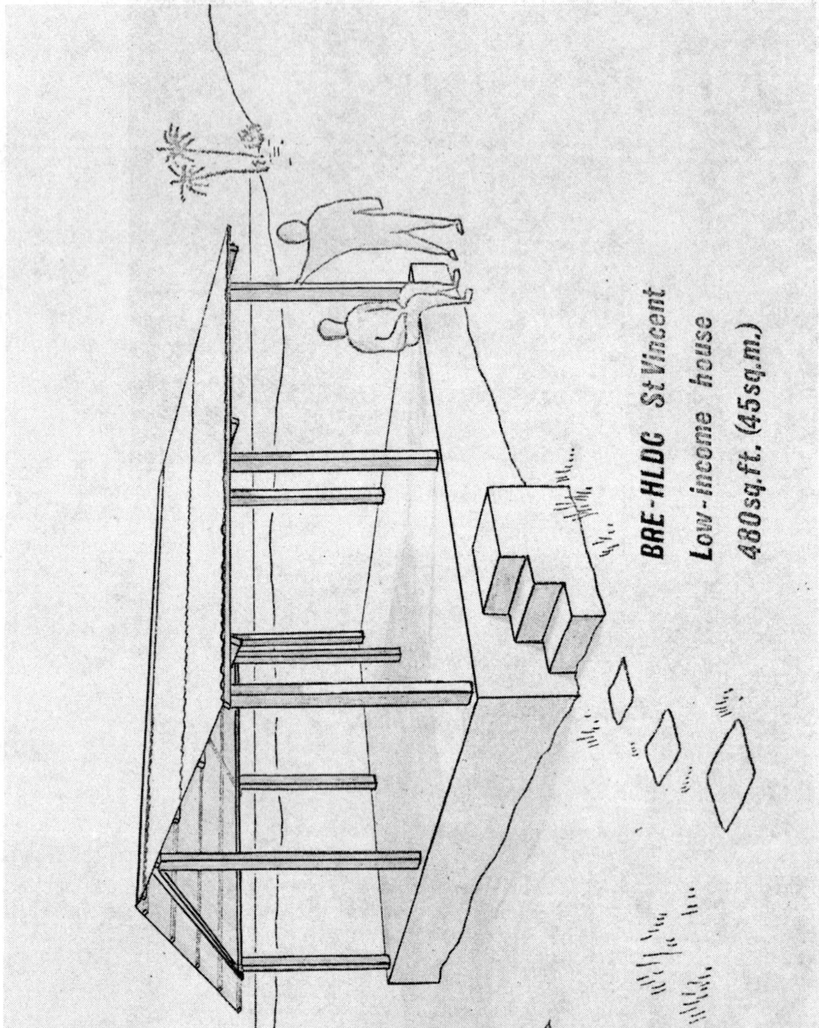

Figure 10 Diagram of a trial house

5 A corrugated bitumen/fibre sheeting, called Onduline, is used to clad the roof in preference to 'galvanised iron' as virtually no maintenance is required (other than checking nails and washers) for a period claimed to be 25 years.

6 Non-structural walls for a toilet/shower room; but no other walls are provided.

Having provided the infrastructure and a core-house, with all the structural elements in place, the walls are completed by the owner. In practice, with the assistance of friends and families, this is done very quickly - and even large extensions are added within a few months.

The important points in this design are the cost reducations due to the quicker and easier fabrication of the timber frame as opposed to an RC frame; the elimination of much of the mass previously incorporated at high levels in the gable end walls (which is important for earthquake considerations); the ability to pre-fabricate all the timber in a central workshop; and, most important of all, the set of rigid connecting brackets.

An engineering analysis for the house shown in Figure 10 indicates the need to resist a lateral force of about 8500 N (1900 lbf) and an uplift of about 50,000 N (11,000 lbf). Brackets to provide this resistance are shown in Figure 11. The brackets are galvanised to prevent corrosion.

A second trial house is being constructed with a permanently open, but covered ridge ventilator. It is known both theoretically (5) and by studying past hurricane damage, that such a ventilator reduces the forces across a roof structure during high winds, by tending to equalise the internal and external suctions (underpressures). By demonstrating this in a trial low-income house, the writer hopes to demonstrate social acceptance of such a feature by the occupants, and also to demonstrate the possibility of using smaller brackets and members with the reduced loads.

The site work should be completed by July 1979, including the installation of an anemometer on a 10 m mast. With this it is hoped to monitor the wind speeds during the hurricane season over the next few years and thereby relate, hopefully, the lack of damage to the trial houses with these speeds.

THE APPLICATION PHASE

The outcome of this research should be several house designs for low-income people in tropical countries, capable of withstanding the forces imposed by hurricanes and the other adverse environments listed above.

It is important to publicise these results; at present it is envisaged that this will be carried out by specific publications that will be made available to public sector housing agencies (not only in the Caribbean islands, but in the other vulnerable areas in Figure 1); by simple diagrammatic booklets for self-help housebuilders (as seen in References 33 and 34); by regional seminars; and possibly by incorporating the results into new building codes and regulations specifically dealing with housing. This last activity will be of a

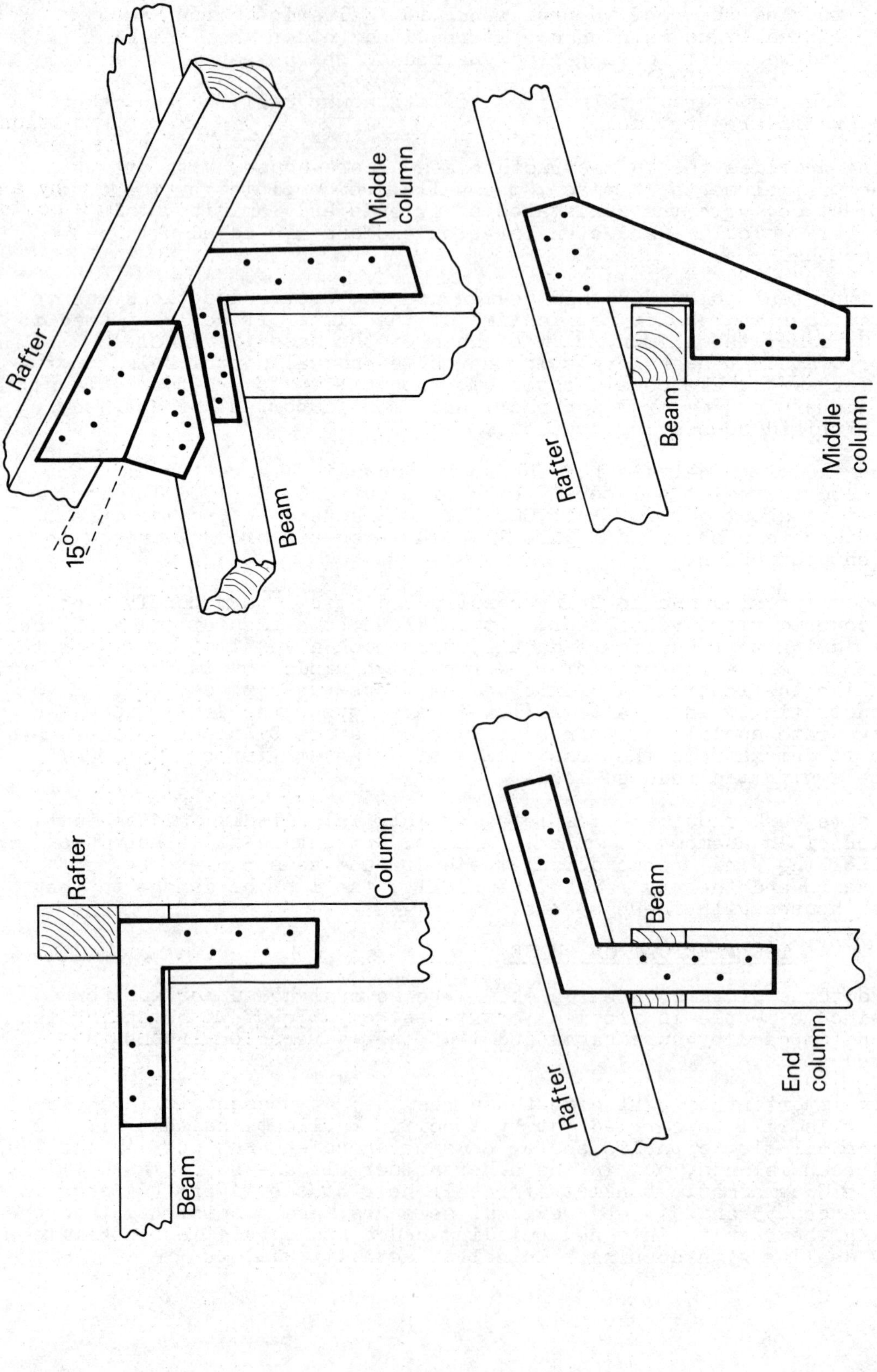

Figure 11 Rigid connecting brackets

longer-term nature, but the writer considers this most important; as far as is known this has only been done in Australia (35), and never for housing in developing countries.

CONCLUDING REMARKS

In developing countries subject to hurricanes, design and construction criteria should be developed for low-income housing as a guide to both public-sector and private organisations involved in the construction of these houses. Design criteria may be developed by a careful analysis of wind records, and criteria for construction could be incorporated in building regulations or codes. For low-income housing the criteria for construction should be simple and, in order to be economic, would need to give priority to key parts of the structure.

The research that is outlined in this paper may prove that the hurricane-resistance in low-income housing can be provided with very little extra cost. Indeed, with possible reductions in the material and/or the labour costs, the outcome could be both cheaper and stronger low-income houses than those currently built.

The project in St Vincent is not the answer to all the world's low-income housing that experiences hurricanes; but it is hoped that the outcome will go a long way from social, economic and engineering viewpoints towards assisting low-income people in developing countries to house themselves and shelter safely from hurricanes.

ACKNOWLEDGEMENTS

The writer wishes to acknowledge the valuable assistance provided by the Government of St Vincent, and in particular by the Housing and Land Development Corporation, in the research outlined in this paper.

The preparation of this paper has been carried out as part of the research programme of the Building Research Establishment of the Department of the Environment, and this paper is published by permission of the Director.

REFERENCES:

1. Crutcher, H.L., and Quayle, R.G., Mariners Worldwide Climatic Guide to Tropical Storms at Sea, US Naval Weather Service Command NAVAIR 50-1C-61, March 1974.

2. World Development Report 1978, The World Bank, August 1978.

3. Greta Lashes Belize, The Belize Times, No 3004, 24 September 1978.

4. Leslie, K.R., Meteorological Report on Hurricane Greta, September 1978, Belize Meteoroligal Service.

5. Newberry, C.W., and Eaton, K.J., Wind loading handbook, HMSO,1974.

6. Model Minimum Hurricane Resistant Building Standards for the Texas Gulf Coast, The Texas Coastal and Marine Council, September 1976.

7. Building to Resist the Effect of Wind, US National Bureau of Standards, Building Science Series 100, May 1977.

8. United Nations, Low-cost construction resistant to earthquakes and hurricanes, ST/ESA/23, United Nations, 1975.

9. Reps, W.F., and Simiu, E., Design Siting and Construction of low-cost housing and community buildings to better withstand earthquakes and windstorms, BSS 48, US National Bureau of Standards, 1974.

10. Walker, G.R., The design of buildings and their components for cyclone conditions, Institution of Engineers Australia, Queensland Division Technical Papers, Vol 18, No 31, October 1977.

11. Walker, G.R., Report of cyclone Tracy - Effect on buildings, Australian Department of Housing and Construction, Melborne, March, 1975.

12. Dickers, R.D., Marshall, R.D., and Thom, H.C.S., Hurricane Camille, National Bureau of Standards Technical Note 569, March 1971.

13. Leicester, R.H., and Reardon, G.F., Wind Damage in Australia, Special Report, Division of Building Research, CSIRO, 1976.

14. Ishizaki, H., Katsura, J., and Murota, T., The damage to structures caused by the second Miyakojima Typhoon, Bulletin of the Disaster Prevention Research Institute, Kyoto University, Vol 18, Part 1, August 1969.

15. Minor, J.E., Mehta, K.C., and McDonald, J.R., Failures of structures due to extreme winds, J Struct Div ASCE, Vol 98, No ST 11, Nov 1972.

16. Shankland Cox Partnership, Third World Urban Housing, UK Building Research Establishment, 1978.

17. Marshall, R.D., A study of wind pressures on a single-family dwelling - in model and full scale, Journal of Industrial Aerodynamics, Vol 1, No 2, October 1975.

18. Marshall, R.D., and Thom, H.C.S., Wind loads on buildings and structures, US National Bureau of Standards, 1970.

19. Eaton, K.J., and Mayne, J.R., Instrumentation and analysis of full-scale wind pressure measurements, Current Paper CP 1/69, UK Building Research Establishment, 1969.

20. Eaton, K.J., and Mayne, J.R., The measurement of wind pressures on two storey houses at Aylesbury, Current Paper CP70/74, UK Building Research Establishment, 1974.

21. Eaton, K.J., Mayne, J.R., and Cook, N.J., Wind loads on low-rise buildings - effects of roof geometry, Current Paper CP1/76, UK Building Research Establishment, 1976.

22. Parsons, A.A., Practical development from static and cyclic load testing of steel claddings, Proc Metal Structures Conference, 1976, Adelaide.

23. Walker, G.R., and Hughes, G.C., An investigation of the racking strength of asbestos cement clad timber framed walls, Dept of Civil and Systems Engineering, James Cook Univeristy of North Queensland, 1977.

24. Johnson, P.I., Impact testing of building products for cyclone prone areas, Workshop on guidelines for cyclone product testing and evaluation, Experimental Building Station, Sydney, 1977.

25. Armstrong, L.D., and Schuster, K.B., Plywood sheathed wood frames to resist wind forces - a technique of assessment, Proc 6th Australasian Conference on Mechanics of Structures and Materials, University of Canterbury, Christchurch, 1977.

26. Morgan, J., and Beck, V.R., Failure of sheet metal roofing under repeated wind loading, Proc Ann Eng Conf, 1976, Institution of Engineers Australia, Townsville.

27. Shellard, H.C., Extreme Wind Speeds in the Commonwealth Caribbean, J of the Barbados Association of Professional Engineers, December 1972.

28. Tomblin, J.F., Earthquake parameters for engineering design in the Caribbean, Proc 1st Caribbean Conf on Earthquake Engineering, Trinidad, January 1978.

29. Taylor, L.O., Aspinall, W., and Morris, P., A preliminary analysis of seismic risk in the Lesser Antilles and Trinidad and Tobago, Proc 1st Caribbean Conf on Earthquake Engineering Trinidad, January 1978.

30. Tomblin, J.F., Earthquake microzonation and prediction in the West Indies, OAS-IGP Seminar on Urban Seismic Microzonation, Peru, October 1978.

31. Tomblin, J.F., Annual Report for Academic Year 1977-78, Seismic Research Unit, University of the West Indies, Trinidad, October 1978.

32. World Bank, World Bank Atlas - population, per capita, product, and growth rates, World Bank, Washingston, 1977

33. Appropriate Reconstruction Training and Information Centre, How to make your house safer against strong winds, The Centre, India, 1978.

34. Kliment, S.A., Raufaste, N.J., and Marshall, R.D., How houses can better resist high wind, US National Bureau of Standards, NBSIR 77 - 1197, Washington, 1977.

35. Standards Association of Australia, Draft manual for guidance on construction of houses in high wind areas, DR 77109, SAA, Sydney, September 1977.

AN ASSESSMENT OF POTENTIAL CYCLONE DAMAGE TO DWELLINGS IN AUSTRALIA

R. H. LEICESTER*, C. T. J. BUBB**, C. DORMAN** and F. D. BERESFORD*

*Commonwealth Scientific and Industrial Research Organization, Melbourne, Australia

**Australian Department of Housing and Construction, Melbourne, Australia

SUMMARY

A risk assessment is made for potential cyclone damage to dwellings along the northern coastline of Australia. This is based on assessments of the wind regime and the strength of houses for nine major population centres considered to be at risk from cyclone attack. It is found that there is a wide range in the expected annual damage and that it is dangerous to base estimates solely on risks observed during the past few decades.

1. INTRODUCTION

In the wake of the extreme damage in 1974 casued by floods in the city of Brisbane, and by cyclone Tracy on the city of Darwin, the Australian Government proposed that a Natural Disaster Insurance Scheme be set up for Australia. This paper describes an assessment of potential cyclone damage to dwellings, undertaken by the Australian Department of Housing and Construction to provide technical information for the Insurance Scheme. It is believed that the method used is suitable for many other countries.

The tropical cyclone of Australia is synonymous with the hurricane and typhoon. Approximately 15 of the 75 severe cyclones that develop around the world each year occur near the north-east and north-west coasts of Australia (1). There are specific characteristics of cyclones that make them potentially more dangerous than other sources of strong winds such as tornadoes and thunderstorms. The long duration of a cyclone can cause failures through fatigue effects (2), and the large area of its influence can lead to very extensive damage. For example, out of approximately 8000 houses in Darwin, cyclone Tracy destroyed 5000 beyond repair (3). The city had to be evacuated and for the next two years about one-third of the Australian Government's construction budget was used solely for rebuilding that city. It is for these reasons that it is important to know the potential risk of damage by cyclones, irrespective of the risks that may exist from other sources of strong winds.

2. METHODOLOGY

The coastal areas of Australia north of 27° south and up to 50 km inland, are considered to be at serious risk from cyclone attacks (4). There is also some risk for coastal areas south of 27° south, particularly those areas along the west coast. Most of northern Australia is sparsely populated and so it is not possible to assess the risk of damage solely from previous experience. The method used in this study is to use a fairly detailed structural engineering theory to assess the risk at the most heavily populated centres of the northern coastal strip, and then to rely on rough interpolation procedures should risk assessments be required for other locations.

The centres chosen for analysis are shown in Fig. 1. These are all coastal cities except for (a) Roebourne Shire, which refers to a combination of the four small towns of Roebourne, Karratha, Wickham and Dampier, and (b) Hedland, which refers to the three small towns of Port Hedland, South Hedland and Finucane Island.

The two basic parts of the risk analysis of each centre are the following:

(a) an assessment of the regional wind regime, and

(b) an assessment of the damage that would be caused by various regional winds.

For the purposes of these assessments, the regional wind is specified in terms of a basic regional wind velocity, which will be defined as the peak three-second gust that occurs at a 10-metre-high anemometer located in an open field (4). It is to be emphasized that the basic regional wind velocity is used solely as a reference, and that there is no intention of implying that wind damage is caused only by peak gusts. Among other factors, there is strong evidence, as mentioned earlier, that fatigue effects may contribute significantly to the damage observed in cyclones (2).

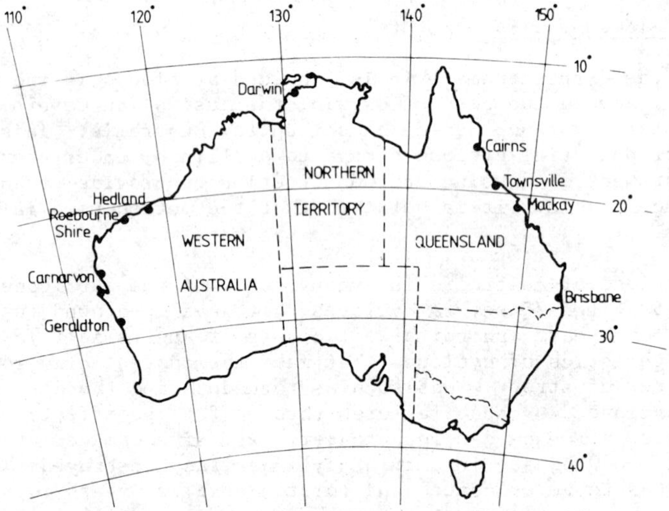

FIG. 1. LOCATIONS OF RISK SURVEYS.

3. REGIONAL WIND REGIMES

Because the record of winds due to cyclones on any one particular site is very sparse, the information available is aggregated and used in a Monte Carlo simulation technique. This technique has been applied to the coast of Texas by Russell (5) and to the northern Australian coast by Martin (6), Gomes and

Vickery (7), and Martin and Bubb (8). More recently the analyses for the Australian coast have been updated by Gomes (9) and Dorman (10).

To apply the method, a search area, typically a 5° or 10° square, as shown in Fig. 2, is chosen with the site or city of interest roughly centrally located. Then the real coastline is approximated by a set of straight-line segments, and also a suitable set of pseudo coastlines reaching out to sea, are chosen. These are denoted by C1-C4 and P1-P3 respectively in Fig. 2. Along the real and pseudo coastlines a fictitious 500-year record of cyclones is generated on a digital computer through a Monte Carlo simulation method. The statistical parameters of the fictitious cyclones are matched to the historical record, particularly with regard to their intensity and rate of occurrence. The effect of each of these cyclones on the basic regional wind at the city of interest is computed and thereby a complete wind regime due to cyclones for that city is derived.

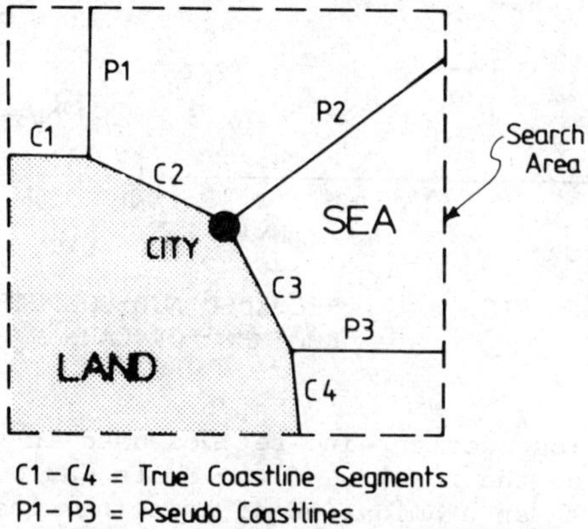

FIG. 2. TRUE AND PSEUDO COASTLINES FOR A TYPICAL SURVEY LOCATION.

The analyses for Australian coastal cities are based on data supplied by the Australian Bureau of Meteorology. This data is fairly detailed for the past 20 years, and contains information back to 60 years. However, it must be used with caution as all the important parameters of the data, such as the locations and the central pressures of cyclones, were not observed directly but were derived through indirect means.

In one assessment of the accuracy of this simulation technique, some 170 records of cyclone winds are being examined to ascertain the effectiveness of using the technique to predict winds at specific sites. Another check can be made by comparing the true wind record of a specific site, with the wind regimes for cyclones only derived through the Monte Carlo simulation technique. An example of this check applied to the city of Carnarvon is shown in Fig. 3. The wind record for this city is based on observations since 1953 and therefore should be fairly reliable for predicting the 20-year return wind (11). Obviously the 20-year return wind for cyclones only must be equal to or less than that for the total wind record. This is contradicted by the simulated cyclone wind record and hence the design wind was chosen to match the true wind record at the 20-year return period. Another point in the Gumbel line was taken to be the wind regime of the fictitious cyclones at the 1000 year return period. It was found necessary to apply this type of modification to the wind regimes generated by the Monte Carlo technique for all the cities except Darwin and Brisbane.

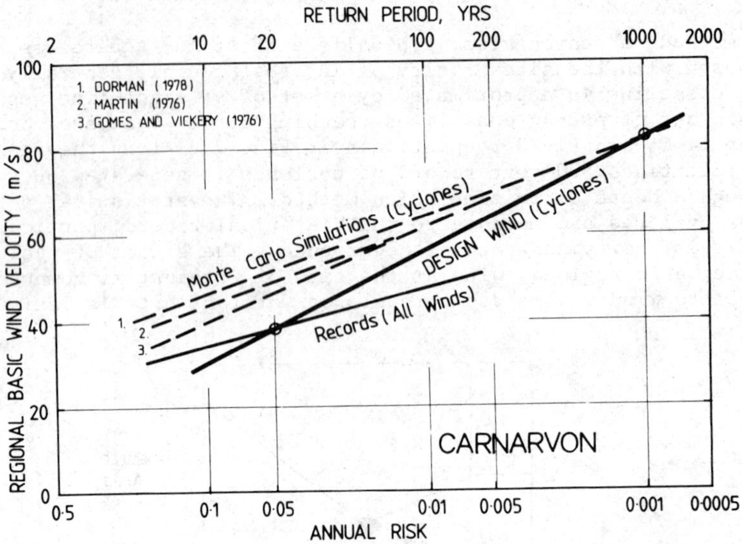

FIG. 3. RECORDED AND COMPUTED WIND REGIMES FOR CARNARVON.

Fig. 4 shows the cyclone design wind regimes used in this study. They were chosen after consideration of the results of the Monte Carlo simulation studies mentioned earlier, together with an examination of existing wind records and other historical data.

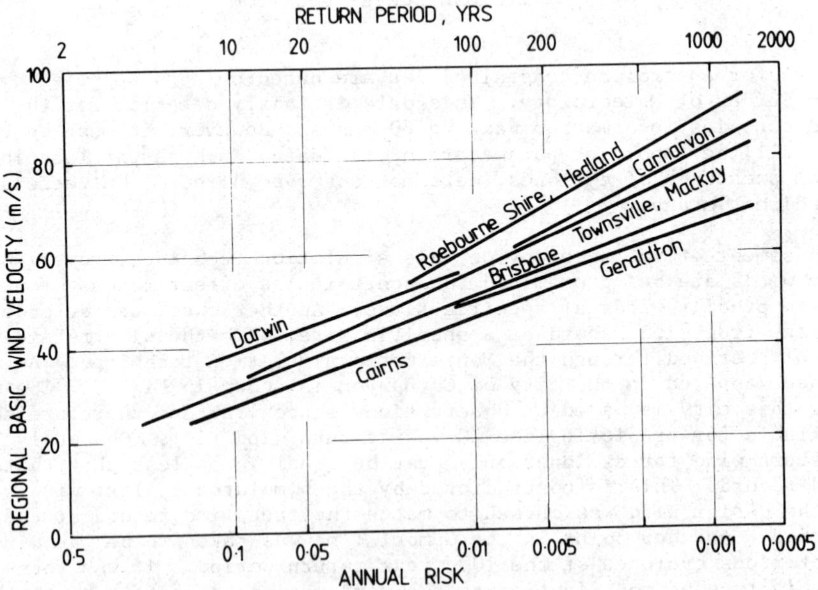

FIG. 4. DESIGN WIND REGIMES FOR ALL LOCATIONS.

4. DAMAGE CAUSED BY REGIONAL WINDS

4.1. The Damage Index

The damage to a house will be described in terms of an index that ranges from 0 to 1 and is defined as follows:

damage index = (repair cost)/(initial cost of building), where the costs are based on the assumption of stable building industry conditions, i.e. the effects of post-disaster inflation are not included.

Table 1 shows values of the damage index corresponding to specific degrees of damage for some typical forms of Australian housing (12).

TABLE 1

DEFINITION OF DAMAGE INDEX FOR A HOUSE

After Leicester and Reardon (12)

Worst damage feature	Damage index*		
	Elevated houses	Low-set houses	
		Brick walls	Asbestos cement walls
Negligible	0.00	0.00	0.00
Missile damage to cladding or windows	0.05	0.05	0.05
Loss of half roof sheeting	0.10	0.10	0.10
Loss of all roof sheeting	0.20	0.20	0.15
Loss of roof structure	0.25	0.25	0.20
Loss of half walls	0.50	0.65	0.60
Loss of all walls	0.75	0.90	0.90
Loss of half floor	0.85	-	-
Loss of all floor	0.95	-	-
Collapse of floor support piers	1.00	-	-

* damage index = (repair cost/initial cost of building).

4.2. Assessment of Building Strength

In order to assess building strength, surveys of dwellings were made over a total period of four weeks at the population centres shown in Fig. 1. For each building type the assessment of strength was based on the features related to design, approval, supervision and construction listed in Table 2, and on information obtained previously from field investigations of wind damage.

TABLE 2

FACTORS AFFECTING THE STRENGTH OF HOUSES

Aspect	Feature
Design	1. Wind sensitivity of the structure. 2. Quality of engineering design, if any. 3. Reliability of the application of current engineering theory to the particular house structure.
Approval	1. Structural engineering ability of the approval authority. 2. Degree to which critical aspects of the structure are shown on the plans.
Supervision	1. The understanding of structural engineering principles by the field inspector. 2. The number and variety of structures supervised within a given period. 3. The type of structure with respect to the effects of possible errors in supervision.
Construction	1. The quality of workmanship associated with the construction of critical structural features. 2. The quality of the structural materials

In particular, the data collated by Trollope (13), Reardon (14), Walker (3), Leicester and Reardon (12, 15), the Australian Bureau of Meteorology (16) and Holmes (17), were used to assess the strength of the older non-engineered houses, and the reports by Clarke (18, 19) and Wittenoom (20, 21) were used to assess the strength of the newer houses, which contain some direct engineering input by professional engineers. In addition, it was assumed that a portion of the non-engineered houses located in exposed situations have been strengthened above their usual condition in response to observations of weaknesses that had become apparent during earlier strong wind occurrences.

Most of the houses in Hedland, Roebourne Shire and Darwin have been built in the past few years and contain a high engineering input. The houses in the other cities can be broadly grouped into three categories. Those constructed prior to 1950 have useful wind resistant features such as their small size, steep hip roofs, and the use of vertical board linings to provide a structural tie between the rafters, wall top plates and floor joists. After 1950 sheet materials were used for internal linings and the method of attaching these to the structural frame was such that they did not provide a vertical structural tie. In addition, many of the houses were weakened because they had gable roofs which, unless properly engineered, are often inadequately braced. Moreover, the gable end walls of these houses are inadequately tied back to the ceiling. The damage caused by cyclones Althea and Tracy in 1971 and 1974, prompted investigations to uncover the weaknesses of this form of construction, and since then the structural integrity of houses in these areas has been progressively improved.

4.3 Specification of Building Strength

The description of structural strength of each particular type of house is stated in terms of five basic structures denoted by A, B, C, D and E. The relationships between wind gust speed and the damage indices of both the building and the contents of houses with these basic structures are defined in Table 3 and Fig. 5. The form of this relationship is based on the results of the analysis by Leicester and Reardon (12) of the damage by Tracy on Darwin. In general, more than one basic structure is used in specifying the strength of a particular type of house. This not only provides a means of interpolating between basic structural types, but also provides a method for modelling the fact that structural errors may occur in some of the houses.

TABLE 3

CLASSIFICATION OF BASIC HOUSE STRUCTURES

House Structure Type	Wind gust velocity (ms^{-1})	
	Minor damage	Major damage
A	35	45
B	40	50
C	45	60
D	50	70
E	∞	∞

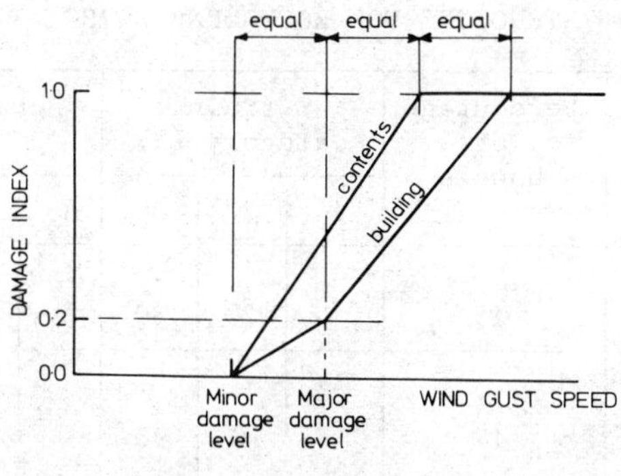

FIG. 5. POSTULATED RELATIONSHIP BETWEEN GUST SPEED AND DAMAGE

4.4 Shielding and Exposure

The damaging effects of the wind are influenced considerably by the shielding and exposure of the houses at risk. Table 4 shows the definition of terrain categories according to the Standards Association of Australia (4) and the corresponding effective wind speeds in these terrains.

In addition to these sources of shielding and exposure, the modifying effects of escarpments, hills, housing clusters and open spaces were also considered (4, 22, 23). The distributions of the effective terrain categories for all housing were noted during the course of the survey on house structures mentioned earlier.

TABLE 4

CHARACTERISTICS OF TERRAIN CATEGORY

Standards Association of Australia (4)

Terrain Category	Description of typical terrain	Effective wind speed*
1	Open sea coasts, flat tree-less plains.	1.02
2	Open fields.	0.93
3	Wooded areas, suburbs, industrial areas.	0.65

* Relative to the regional basic wind speed

4.5 Regional Damage Due to a Basic Regional Wind

To determine the expected damage due to a basic regional wind, an assessment was first made of the distribution of housing in terms of terrain category and structure type. An example of this for Roebourne Shire is shown in Table 5.

TABLE 5

DISTRIBUTION OF TERRAIN CATEGORY AND TYPE OF HOUSE STRUCTURE FOR ROEBOURNE SHIRE

House type	Percentage of Houses	Terrain category (%)			House structure type (%)				
		1	2	3	A	B	C	D	E
Dampier (mining company)	15		20	80			30	70	
Wickham (State Housing Commission and mining company)	17		10	90			40	60	
Roebourne (State Housing Commission and private)	8		70	30		20	30	50	
Karratha (mining company)	27	50	50					80	20
Karratha (State Housing Commission)	27	50	50				50	50	
Karratha (private)	6	50	50			20	40	40	

A summary of this information in condensed form is shown in Table 6 for all the locations surveyed.

TABLE 6

SUMMARY OF TERRAIN CATEGORY AND HOUSE STRUCTURE TYPE

Location	Terrain category	House structure type (%)				
		A	B	C	D	E
Geraldton	1	2	5	4	0	0
	2	2	6	5	0	0
	3	35	41	0	0	0
Carnarvon	1	0	2	2	1	0
	2	4	18	20	6	0
	3	12	23	10	2	0
Roebourne Shire	1	0	1	8	19	3
	2	0	2	11	25	3
	3	0	0	10	18	0
Hedland	1	0	0	3	8	0
	2	0	0	9	32	3
	3	0	0	10	32	3
Darwin	1	0	2	3	3	2
	2	0	4	6	7	3
	3	2	23	12	21	12
Cairns	1	1	2	1	1	0
	2	2	7	8	2	0
	3	33	40	2	1	0
Townsville	1	1	1	2	1	0
	2	2	5	7	1	0
	3	33	40	2	5	0
Mackay	1	1	2	2	1	0
	2	2	8	9	3	0
	3	26	38	6	2	0
Brisbane	1	1	1	0	0	0
	2	2	5	1	0	0
	3	66	24	0	0	0

Finally, from the use of the data in Fig. 5 and Tables 3, 4 and 6, the relationship between wind speed and expected damage for each survey location can be computed. These relationships are shown in Fig. 6. This figure also shows the magnitude of the cyclones that have caused recorded damage during the past decade, and accordingly have been used to calibrate the proposed models of structural resistance.

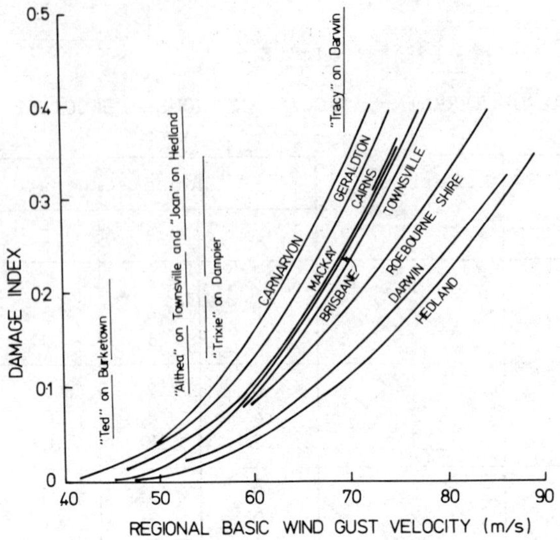

FIG. 6. COMPUTED RELATIONSHIP BETWEEN REGIONAL BASIC WIND SPEED AND REGIONAL DAMAGE INDICES.

5. COMPUTATION OF POTENTIAL DAMAGE

The simplest method of assessing the potential cyclone damage is to examine the damage computed through a Monte Carlo simulation of damage history. To do this, a regional peak wind gust V is chosen for each year and location through the equation

$$V = m + s \log_e (1/r)$$

where m and s are the modes and slopes of the Gumbel plots shown in Fig. 4, and r is a random number uniformly distributed between 0 and 1. Next, the damage caused by this wind gust to each particular type of house is computed and recorded for that year. These records are then analysed to determine the required damage statistics.

6. DAMAGE CHARACTERISTICS

The expected annual damage for buildings and contents, averaged over an infinite period of time, are shown in Table 7.

It is to be noted that there is quite a range in the expected risk, extending from the low value of 0.08 per cent for buildings in Cairns to 0.37 per cent for buildings in Carnarvon. Fig. 7 shows that while some of the damage occurs at extreme winds with return periods of more than 1000 years, most of the damage occurs at relatively frequent winds with return period of about 200 years. Another damage characteristic noted was the large proportion of the damage attributed to houses in exposed locations. For example, in Queensland the houses located in terrain categories 1 and 2 comprise only 20 per cent of the total houses but contribute 60 per cent of the total damage.

TABLE 7

AVERAGE ANNUAL DAMAGE INDEX

Locality	Annual damage index (%)	
	Buildings only	Contents only
Geraldton	0.15	0.30
Carnarvon	1.37	0.66
Roebourne Shire	0.31	0.60
Hedland	0.19	0.39
Darwin	0.15	0.32
Cairns	0.08	0.15
Townsville	0.23	0.43
Mackay	0.26	0.48
Brisbane	0.13	0.26

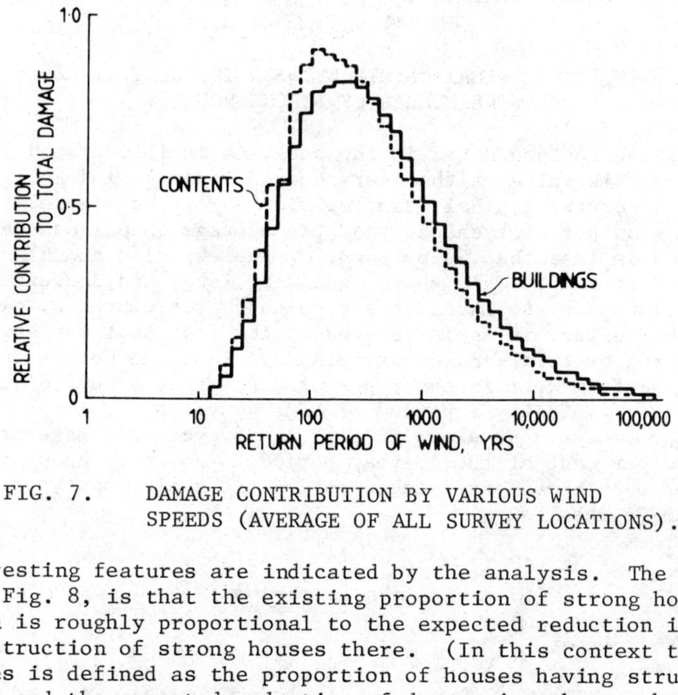

FIG. 7. DAMAGE CONTRIBUTION BY VARIOUS WIND SPEEDS (AVERAGE OF ALL SURVEY LOCATIONS).

Two other interesting features are indicated by the analysis. The first, illustrated in Fig. 8, is that the existing proportion of strong houses at each survey location is roughly proportional to the expected reduction in damage index due to the construction of strong houses there. (In this context the proportion of strong houses is defined as the proportion of houses having structures of type C, D or E, and the expected reduction of damage is taken to be the difference between the expected damage indices evaluated for two hypothetical cases in which the structures are assumed to be all of types A or C only.) This relationship between the number of existing strong houses and the expected reduction in damage reflects the fact that the structures of houses tend to develop through a trial-and-error process to resist the local wind regime. However, since the provision of a strong structure is estimated to add 5-15 per cent to the cost of a house, then a more rational approach to decisions related to the provision of strong structures could be for building regulations to specify a requirement that strong houses be built at all locations where the expected resulting reduction in damage index exceeds 15 per cent over a 30-year period.

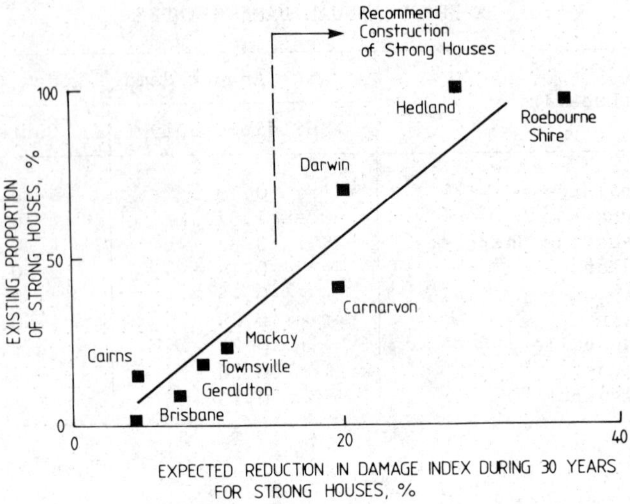

FIG. 8. RELATIONSHIP BETWEEN THE BENEFIT AND THE NUMBER OF STRONG HOUSES.

The other interesting feature noted in the analysis is illustrated in Fig. 9. This shows the probable value of the average annual damage index observed during a 40-year period. For the typical examples plotted in the figure, it is seen that there is roughly a 50 per cent chance that the average annual damage observed for any 40-year period is less than 30 per cent of the expected annual damage averaged over a long period of time. This shows that very large and dangerous errors can arise if damage risks are estimated solely on the basis of recent experiences. The reason for this effect is again related to the fact that the structures of houses are developed by trial-and-error methods. Thus the houses in any particular region will perform well in the types of winds that occurred during their evolutionary period, typically a period of some 40 years. This process leads to a threshold effect, wherein as shown in Fig. 9, no cyclone damage at all will occur in about 10 per cent of the 40-year periods. However, as wind gusts occur that are stronger than those experienced during the evolutionary period, a very rapid rise in damage results.

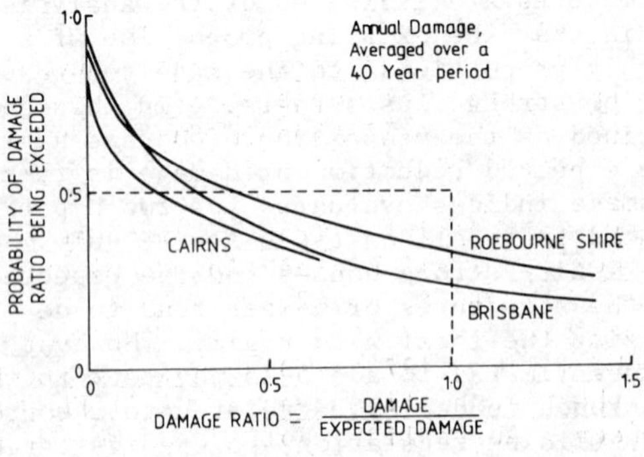

FIG. 9. PROBABLE VALUE OF MEAN ANNUAL DAMAGE OBSERVED DURING A 40-YEAR PERIOD.

7. CONCLUSIONS

A method, based on structural engineering concepts, has been described for assessing the risk of cyclone damage to houses.

An application of the method to the major population centres along the northern coastline of Australia, indicates that there is a wide range in the expected annual damage at the various locations surveyed, that much of the damage occurs to exposed houses by winds with return periods of around 200 years, and that it is dangerous to base estimates of risk solely on the damage observed during the past few decades. The analyses of expected damage also provide a means of assessing whether it is reasonable for building regulations to specify that strong structures be used in houses at particular locations.

8. ACKNOWLEDGEMENTS

The authors gratefully acknowledge the assistance of Dr. L. Gomes and the Australian Bureau of Meteorology for their contributions to the work related to assessing the cyclone wind regimes. Thanks are also due to the numerous structural engineers, academics, building regulatory authorities and building inspectors who provided invaluable information for assessing the strength of houses.

9. REFERENCES

1. Southern, R.L. and Scott, A.N. (1976) - The genesis and characteristics of tropical cyclones. (With special reference to north-west Australia.) Proceedings of Symposium on the Impact of Tropical Cyclones on Oil and Mineral Development in north-west Australia, Perth. Australian Government Publishing Service, Canberra. pp.1-27.

2. Morgan, J. and Beck, V.R. (1976) - Failure of sheet-metal roofing under repeated wind loading. Proceedings of Annual Engineering Conference, Institution of Engineers, Australia. Townsville, Australia, May, pp.290-294.

3. Walker, G.R. (1975) - Effect on buildings. Department of Housing and Construction. Report on Cyclone Tracy, Vol. 1, Melbourne, Australia, March.

4. Standards Association of Australia (1975) - AS:1170, Part 2 - 1975. SAA Loading Code. Part 2 - Wind Forces.

5. Russell, L.R. (1971) - Probability distributions for hurricane effects. Proc. of ASCE, Journal of Waterways, Harbours and Coastal Engineering Division Vol. 97, No. WW1, Feb., pp. 139-154.

6. Martin, G.S. (1974) - Probability distributions of hurricane wind gust speeds on the Australian coast. Proc. of Conference on Application of Probability Theory to Structural Design, Institution of Engineers, Australia, Melbourne, Nov., pp. 55-66.

7. Gomes, L. and Vickery, B.J. (1976) - Tropical cyclone gust speeds along the northern Australian coast. Civil Engineering Transactions, Institution of Engineers, Australia. Vol. CE18, No. 2, pp. 40-48.

8. Martin, G.S. and Bubb, C.T.J. (1976) - Discussion on tropical cyclone gust speeds along the northern Australian coast. Civil Engineering Transactions, Institution of Engineers, Australia. Vol. CE18, No. 2, pp. 48-49.

9. Gomes, L. (1978) - Tropical cyclone gust speeds. Private communication to Australian Department of Construction.

10. Dorman, C. (1978) - Tropical cyclone gust speeds. Report for Australian Department of Construction.

11. Whittingham, H.E. (1964) - Extreme wind gusts in Australia. Bulletin No. 46, Bureau of Meteorology, Melbourne, Australia.

12. Leicester, R.H. and Reardon, G.F. (1976) - A statistical analysis of the structural damage by Cyclone Tracy. Civil Engineering Transactions, Institution of Engineers, Australia, Vol. CE18, No. 2, pp. 50-54.

13. Trollope, D.T. (1972) - Cyclone "Althea". Part 1 - Buildings. Report by James Cook University of North Queensland, Townsville, Australia, March.

14. Reardon, G.F. (1975) - The Brisbane windstorm, November 1973. Report 33, CSIRO Division of Building Research, Melbourne, Australia.

15. Leicester, R.H. and Reardon, G.F. (1976) - Wind damage in Australia - A pictorial review with particular reference to domestic and other low rise structures. CSIRO Division of Building Research, Melbourne, Australia.

16. Australian Bureau of Meteorology (1976) - Summary of tropical cyclones and reported damage affecting the north-west coastline of Australia for the period 1950 to 1975. Private communication, Melbourne.

17. Holmes, J.D. (1977) - Report on structural damage at Burketown after Cyclone "Ted". Wind Engineering Report 1/77. Department of Civil and Systems Engineering, James Cook University of North Queensland, Townsville, Australia, April.

18. Clarke, A.G. (1975) - Cyclone Trixie. Effects on structures in north-west towns. Report by Department of Construction, Perth, Australia, February.

19. Clarke, A.G. (1976) - Cyclone Joan. Effects on structures in Port Hedland, 8th December, 1975. Report by Department of Construction, Perth, Australia.

20. Wittenoom, R.A. (1975) - The W.A. north-west cost - after "Trixie". Symposium on design of buildings for cyclones. Institution of Engineers, Perth, Australia.

21. Wittenoom, R.A. (1976) - The performance of Western Australian housing structures under cyclone loadings. Supplement to Annual Conference, Institution of Engineers, Australia, Townsville, Australia, May.

22. Vickery, B.J. (1976) - Wind loads on low rise buildings. Darwin Reconstruction Commission seminar, Darwin, March 30.

23. Reardon, (G.F. (1978) - Interim design criteria for houses to resit cyclone winds. CSIRO Division of Building Research, Melbourne, Australia.

THE ECONOMIC BENEFITS AND COSTS OF MANAGING WINDBLOWN SNOW IN THE NORTHERN PLAINS OF THE U.S.

J. P. OLIENYK, J. R. SNYDER, M. D. SKOLD and W. O. WILLIS

Department of Economics, Colorado State University and USDA-AR, Fort Collins, Colorado, USA

Among the dominant climatic characteristics of the North American Great Plains are its aridness and frequent winds. These factors combine to make snow management, the object of economic evaluation in this paper, a subject of importance to individual farmers and to society in general.

Farming in the Great Plains has long been recognized as a high risk operation. Precipitation in this region is only marginally adequate to sustain crop production, and farming practices have been tailored to accommodate the arid environment. Furthermore, a significant amount of the precipitation that does fall comes in the form of winter snow. This winter precipitation occurs outside of the crop-growing season, and its value for crop production depends on the extent to which water in the form of melted snow can be accumulated and stored in the soil for future use. This potential benefit from snow is limited, however, by the presence of frequent strong winds. Either the winds accompany the snowfall, resulting in blizzard conditions, or the winds occur after the snow has fallen, resulting in "ground blizzard" conditions. In either case much of the snow is blown off the fields and accumulates in roadside ditches, in farmyards and towns, and along fencerows and other manmade barriers and natural surface irregularities. The snow which does remain on the fields does not accumulate evenly, tending instead to form long narrow snowdrifts. Much of the water from these snowdrifts is lost to evaporation and to run-off as the drift melts. The water from melting drifts which does seep into the soil is not distributed evenly over the fields. Thus, the winter winds severely limit the potentially valuable contribution of snow water to crop production.

This loss of snow water to farmers is only one side of the coin. Significant social costs are also incurred. The snow blown from farmers' fields becomes at best a nuisance and at worst a hazard for people living in the Plains and, indirectly, for society in general. For example, blowing snow often results in hazardous road conditions, leading to traffic accidents, school closures, loss of commerce, etc. Substantial amounts of money are spent annually by state, county, and municipal governments for snow removal.

The purpose of this paper is to evaluate the economic benefits and costs of managing snow by trapping it on farmers' fields. The benefits and costs of snow management are of two types--private and social. The private or internal benefits and costs are those which accrue to the individual farmer exclusively, and their impact is on the farmer's net income. The social or external benefits and costs are those which are brought to bear on society in general as a result of actions taken by individual farmers. These external effects have their impact on the well-being of society. For example, if snow is trapped on fields for the purpose of enhancing crop production, society may enjoy an external benefit in the form of reduced expenditure for snow removal. The individual farmer, however, will normally not take the social benefits (and costs) into account, since he has no economic incentive to do so. His decision of whether or not to adopt snow management techniques will be based solely on the private benefits and costs.

SNOW MANAGEMENT

Precipitation in the Great Plains is sparse, irregular, and not always distributed over the growing season in a manner most favorable for plant growth. Wind often increases the evapotranspiration from soil-plant systems, and high water stress conditions are common among crops. Farmers have attempted to cope with this uncertain environment by the use of such practices as judicious selection of crops, use of soil-water-conserving tillage practices, summer fallowing, strip farming, and the use of stubble mulching in conjunction with summer fallowing. The contribution of snow to farming is recognized, but management of water from snow has not received sufficient emphasis.

Crop-fallow farming systems are the primary means through which soil water is accumulated. These systems typically involve leaving one-half of the farmland idle for one growing season so that the land can accumulate soil water for the benefit of the crop to be planted in the following season. These crop-fallow systems, however, are not highly efficient. The capacity of soils to hold water is limited. Evaporation losses are high and some water may percolate to a depth beyond the reach of plant root systems. Only 20 to 25 percent of the total precipitation is conserved during the fallow period (16).

Snow management may be a viable alternative to the highly inefficient crop-fallow systems. As one moves northwest in the plains, increasingly larger percentages of total precipitation fall as snow (see Figure 1). Most areas of the Northern and Central Plains receive more than ten centimeters of rainfall-equivalent precipitation in the form of snow. Since an additional four and one half centimeters of soil water on an annually cropped field of spring wheat at planting is equivalent to planting on fallowed land (9), capture of even half of the normal snowfall could easily match the soil water accumulated during summer fallowing. Thus, a workable snow management system could result in substantial benefits to farmers. It would reduce the need to summer fallow and would enable farmers to annually crop more of their cultivated land. Snow would be considered a resource to be managed for economic benefit rather than a climatic severity to be endured with life on the Great Plains.

Snow traps (or barriers) are constructed or planted in such a manner as to cause snow to drift evenly over the field. There are two general categories of snow barriers. Competitive barriers (trees, tall grass, and other plants) compete with crops for soil water, while noncompetitive barriers (fences, earth dikes, and small grain stubble) do not compete for soil water. Most experimental work has been done with tall grass barriers, and we will focus our attention on this type of barrier (7).

Current recommendations for grass barrier systems include: (a) a single row of tall wheatgrass, (b) distances between the barriers ranging from twelve to eighteen meters (the exact distance made to conform with machinery widths), and (c) barriers oriented perpendicular to prevailing winter winds and, if feasible, to topographical slope. Theoretically, optimal barrier placement requires consideration of factors such as soil type, slope, winter wind direction, and barrier height and porosity. These variables are not evaluated in this paper.

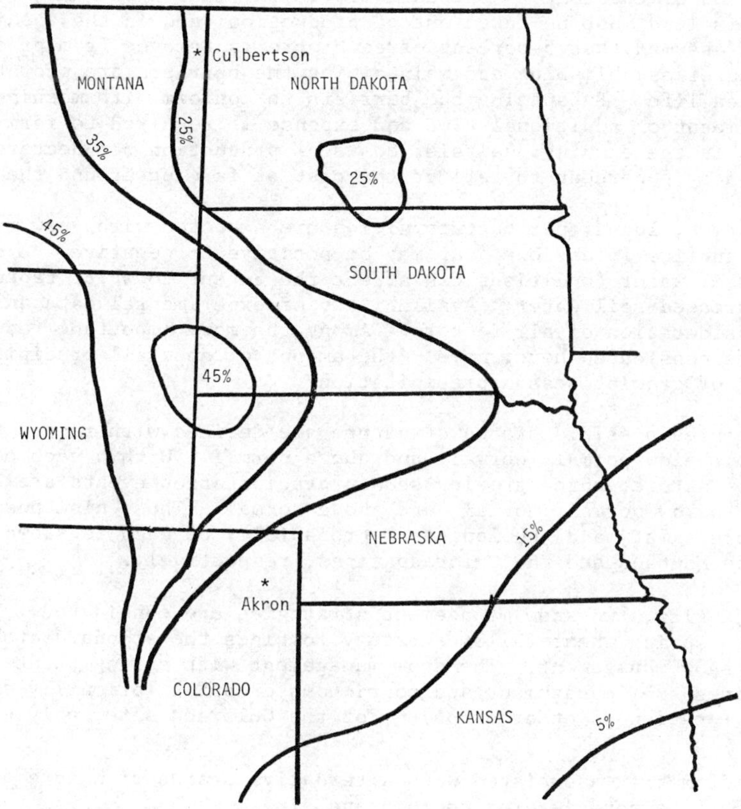

Figure 1. Snowfall precipitation as a percent of annual precipitation in the Northern Great Plains.

The economic analysis for this study is based on experiments with grass barriers conducted at two diverse locations. One site is near Culbertson, Montana, and the other is near Akron, Colorado. The locations differ in the amounts and percentages of precipitation which fall as snow, and in their potentials to utilize water from captured snow.

In Montana, Black and Siddoway found snow management using tall wheatgrass barriers resulted in enough additional water to increase winter wheat yields by 1.88 quintals/hectare/year on fallow and 3.77 quintals/hectare/year on continuous cropping (4). Spring wheat yields increased by 1.26 quintals/hectare/year on fallow and by 1.88 quintals/hectare/year on continuous cropping. The yield differences were statistically significant under both cropping systems. In a similar experiment near Akron, Colorado, an additional 3.8 centimeters of soil water accumulated inside the barrier area and the extra water resulted in increased wheat yields of about 2.52 quintals per hectare (8). Yield increases in Colorado were also statistically significant. The results of these experiments show the potential for snow management.

PRIVATE BENEFITS AND COSTS

The major portion of the cost of the barriers is the land upon which the barriers are grown. This land must be taken out of production, and in the ensuing analysis it is assumed that 5 percent of each acre is removed from production. The total costs of establishing and maintaining the barriers are prorated over a 20-year expected life. By spacing the barriers to conform with machinery widths, only a small amount of additional time and expense is required to farm around the barriers. In the ensuing analysis, costs of production per hectare were adjusted upward by 5 percent to reflect the cost of farming around the barriers.

Averaged over time, larger per hectare yields are expected with snow management; however, the benefits in any one year may be positive or negative. A number of climatic and soil water conditions can affect the extent to which captured snow results in increased soil water. Availability of experimental data and space do not permit consideration of all factors. Among the more important factors are those which are considered here; namely the amount of snowfall precipitation and the amount of growing season precipitation.

The following three possible states of nature are defined with regard to winter precipitation: below normal, normal, and above normal. Within each of these three states of nature, three growing season precipitation events are defined. These are also below normal, normal, and above normal. Thus, nine possible states of nature exist, and the long-term probability of each is shown in Tables 1 and 2 for the Montana and the Colorado sites, respectively.

For the Montana site, two snow management strategies are considered. The snow management with spring wheat-fallow strategy combines the standard crop-fallowing practice with snow management. The snow management with spring wheat-rotation strategy combines snow management and continuous cropping (primarily spring wheat with periodic rotation to other crops). For the Colorado site, only one snow

Table 1. Payoff matrix associated with alternative states of nature and farmers' strategies for the Montana site.

	States of Nature									E(V)
Snow	Below Normal (B)			Normal (N)			Above Normal (A)			
Growing season precipitation	B	N	A	B	N	A	B	N	A	
Probabilities	.125	.152	.099	.125	.152	.099	.083	.101	.066	
No snow management	(per hectare net returns, rounded to nearest U.S. $)									
Spring Wheat-Fallow	-44	33	83	-29	49	98	-14	63	113	34
Snow management Spring Wheat-Fallow	-45	45	95	-18	59	109	-10	67	117	42
Snow management Spring Wheat-Rotation	-78	54	138	-53	77	162	-35	97	182	52

Table 2. Payoff matrix associated with alternative states of nature and farmers' strategies for the Colorado site.

	States of Nature									E(V)
Winter precipitation	Below Normal (B)			Normal (N)			Above Normal (A)			
Growing season precipitation	B	N	A	B	N	A	B	N	A	
Probabilities	.124	.140	.113	.142	.160	.130	.062	.070	.057	
	(per hectare net returns, rounded to nearest U.S. $)									
No snow management, Forage Crop-Fallow	65	103	102	75	113	129	82	121	136	101
No snow management Continuous Forage Crops	-46	31	62	37	90	145	60	137	168	67
Snow management Continuous Forage Crops	27	100	130	65	139	168	85	170	200	114

management strategy, snow management with continuous forage cropping, is analyzed. No data was available for snow management with crop-fallow.

The elements in the last three rows in each table are the expected returns per hectare for each strategy under each state of nature, where expected return is equal to income minus variable costs and machine ownership costs. The values for normal precipitation were derived through the use of regression techniques for the Montana site (2, 3), and by mean values for the Colorado site (8). The above and below normal values were then adjusted to reflect diminishing returns to water by the application of a Spillman coefficient (10). Making use of the probabilities associated with the described states of nature and the net returns associated with those states, the expected value, E(V), for each strategy was computed as follows:

$$E(V) = \sum_{j=1}^{9} x_{ij} P_{ij}$$

where x_{ij} is the payoff associated with the i-th strategy and the j-th state of nature and P_{ij} is the probability associated with the corresponding state of nature.

It could be concluded a priori that holding growing season rainfall constant, the benefits from snow management would increase at a decreasing rate with the amount of winter precipitation. This conclusion is confirmed by the data. More important, however, is the fact that the data shows snow management to be the superior strategy for both sites. Crop-fallow always has a greater return than continuous cropping when snow management barriers are not present at the Colorado site, and continuous cropping as a higher expected value than crop fallow when barriers are present at the Montana site. At both sites, snow management with continuous cropping had the highest expected return of all strategies considered.

Tables 1 and 2 indicate, however, that while snow management may increase farmers' incomes in the long run, it may also increase the year-to-year variability of income. The data indicates that two consecutive years of below normal winter and growing season precipitation would cause much more hardship for farmers with snow management and continuous cropping as compared to the traditional practice of crop-fallowing. Crop-fallow is an attractive farming system to the extent that it reduces the variability of crop yields and farm incomes. The crop-fallow results in accumulation of a portion of the soil water and plant nutrients from two years and makes them available to one year's crop. It becomes a smoothing operation which reduces the effects of year-to-year variability in precipitation. The stabilizing effects of a crop-fallow system on yields is shown by statistical analysis of long-term crop experiments. Using 30 years of data from Akron, Colorado, the coefficient of variation (C_V) for wheat yields under continuous cropping is 1.3, while the C_V for wheat with the crop-fallow system is 0.59. During a similar 30-year period at Havre, Montana, the C_V for continuously cropped spring wheat is 0.84, while the C_V is only 0.64 under crop-fallow.

The coefficients of variation for the continuous cropping and crop-fallow experiments mentioned above apply to a situation in which snow management is absent. While crop-fallow lessens the variability of yields in the absence of snow management, insufficient data exist to make proper analysis of the long-term effect of snow management on yield variability, either with or without crop-fallow. However, at both the Montana and Colorado sites where the experiments with grass barriers were conducted, the data indicate increased variability of income with snow management. It is hypothesized that this variability would exist in general, and that there is some (unspecified) trade-off between the expected value of income and the variability of income. Individuals differ as to their preferences along this continuum, and high variability of expected income could be a significant deterrent to the adoption of snow management.

ADDITIONAL BENEFITS AND COSTS

Snow Removal

Snow management could yield social benefits resulting from a reduction in the amount of snow on roads and highways. These would include fewer school days lost, fewer man-days of labor lost, less reduction of trade and commerce in general, fewer traffic accidents (and therefore fewer injuries and deaths), and less expenditure for plowing and sanding roads. In addition, people living in the Northern Plains would enjoy some psychic income to the extent that they would feel that winter travel would be less hazardous.

Until more research is done on the physical relationships between such factors as wind velocity, the size, type, and location of barriers, the percentage of snow captured by barriers, the amount of snow which falls during a given period of time, etc., the benefits described in the preceding paragraph cannot be accurately measured. It is hypothesized, however, that where certain stretches of roads are consistently subject to severe drifting, it may be economical for highway departments to subsidize the adoption of snow management practices on fields upwind to the roads. A wide variety of subsidy schemes could be devised. For example, highway department subsidy payments could contribute to a reduction in the variability of expected income if the payments were made only in years of below normal precipitation.

Saline Seep

Saline seep is a problem which is caused by water percolating deep into the soil past the root zone. When the percolating soil water reaches an impervious geologic layer (like clay or shale), it moves horizontally along this layer through the so-called recharge area, dissolving and accumulating salts. Salinization occurs on the impervious layer and eventually surfaces at some point below the recharge area. As the water evaporates, the salt deposits are left on or near the surface of the soil (5).

Estimates of the amount of land already rendered useless are as follows: Alberta, Canada, 100,000 hectares; Montana, 55,000 hectares; and North Dakota, 40,000 hectares (13). No figures are available for Saskatchewan, Canada; South Dakota; Wyoming; or Nebraska. Thus, while we can safely say that there are at least 200,000 affected hectares north of the 45th parallel, the total amount of affected land is likely well in excess of that amount.* Furthermore, soil scientists at Montana State University estimate that the number of affected acres is increasing at a rate of ten percent per year (11).

The crop-fallow system of farming is a factor contributing to saline seep. During the 14 to 21 month period in which the land is kept free from vegetation, accumulating soil water will percolate down into the soil beyond the root zone and end up in a seep.

Concentrations of snow in large, uneven drifts in recharge areas have also been identified as a factor contributing to saline seep. In a study conducted in Alberta, Canada, it was determined that windbreaks such as shelterbelts around homesteads, ordinary fence rows, and snow fences designed to protect roads, highways, and railroads can cause excessive, localized accumulations of snow. As the snowdrifts melt, the soil under the drifts quickly becomes saturated, and when the water-holding capacity of the soil is reached, the excess water will find its way to a downslope discharge area. The study concludes that "a more desirable situation would be to have the snow spread in a uniform blanket over the fields and to retain the moisture derived from it on location, storing it in the soil where it would be available for plant use." (14)

In most cases, the seep problem can be virtually eliminated by converting the recharge areas to pasture, or by planting deep-rooted perennial crops like alfalfa, or by continuous cropping in the recharge areas (15). The discharge areas can then be reclaimed and returned to productive use.

For small grain farmers, however, conversion to pasture or planting deep-rooted perennial crops may not be suitable alternatives. These alternatives would require significant changes in the farmers' output mix, and the net return from land used for pasture or deep-rooted perennial crops may be substantially less than from land in small grain production. Thus, the third alternative, continuous cropping, would appear to be the most suitable alternative for many farmers. Since the expected value of net returns is greater under continuous cropping with snow management than continuous cropping without snow management, the former strategy would appear to be especially attractive where saline seep is a problem.

This conclusion is subject to an important qualification. In many cases the recharge area and the discharge area are on two separate farms. The recharge-area farmer would, in such a case, receive none of the benefits resulting from a reduction in the amount of land damaged by saline seep. He would, however,

---

*At $1,250 per hectare, the total loss would be at least $250,000,000.

bear all the additional costs and risks inherent in continuous cropping. If the recharge-area farmer should adopt snow management and continuous cropping, the discharge-area farmer would enjoy an external benefit in the form of increased land value.

Wind Erosion

Wind erosion is most likely to occur where the soil is smooth, fine, and dry, and fields are fallow, unsheltered, and wide. The threshold windspeed for erosion to occur is about 5.4 meters per second at a height of 30 centimeters. When this threshold is reached, surface soil is blown off fields and is deposited in drifts along fence rows, in gullies and ditches, and along other surface irregularities. The farmer whose land is being eroded incurs losses in the form of reduced land values and possible abrasion or "sandblasting" damage to young plants. Downwind farmers and society in general incur external costs in the form of dust storms and soil drifts. The United States Department of Agriculture has estimated that wind erosion is a serious problem on more than eight million hectares of cropland in the Northern Plains (12).

Tall wheatgrass barriers can substantially reduce wind erosion. The snow trapped by the barriers reduces soil exposure to the wind during the winter, and keeps the soil wetter longer (and therefore, less prone to erosion) in the spring. In addition, Aase, Black, and Siddoway (1) have shown that barriers very effectively reduce windspeed. With barriers planted nine meters apart, windspeed between the barriers ranged from 17 percent to 70 percent of open-field windspeed. (The percentage reduction in windspeed is a decreasing function of the distance downwind from the leeward side of the barrier.) The windspeed levels for an 18 meter barrier interval ranged from 19 percent to 84 percent of open-field velocity. These reductions are significant since eight times as much soil is eroded by a wind of ten meters per second as by a wind of five meters per second at a height of 30 centimeters. A wind of 15 meters per second at that height will erode almost 16 times as much soil as a wind of five meters per second (17). Thus, where wind erosion is a problem, the attractiveness of snow management is significantly enhanced.

Output Prices

The crop-fallow system of farming results in negative externalities in the form of blowing snow, wind erosion, and saline seep. These externalities are technological, i.e., they cause reductions in the productivity of publicly-owned and privately-owned resources. Adoption of snow management and continuous cropping would mean the elimination of the crop-fallow system and the negative externalities which result from it. Society at large, discharge-area farmers, and downwind farmers would receive benefits in the form of lower costs (increased resource productivity).

Widespread adoption of snow management with continuous cropping could, however, impose pecuniary externalities on farmers outside the Northern Plains. Pecuniary externalities have their impact on resource or product prices rather than resource productivity. All other things equal, continuous cropping in the Northern Plains would result in an increase in the annual output of wheat and other small grains. The increase in output could be substantial enough to cause market prices to fall. Farmers outside the Plains area would thus experience decreases in income, while consumers would enjoy at least marginally lower food costs.

The U.S. Department of Agriculture estimates that the price elasticity of demand for wheat, for example, is approximately -.24. This means that for each one percent increase in the production of wheat, price will fall by 4 percent. Currently, approximately 25 percent of U.S. wheat output is produced in the Northern Plains. Thus, a one percent increase in wheat output in the Plains would result in a .25 percent increase in total wheat output, and therefore a one percent decrease in price (6).

CONCLUSIONS

Capturing wind-blown snow on fields will increase soil water and enhance crop production, thus converting a sometimes undesirable climatic event into one with direct economic benefits. Research indicates that the expected value of net income per acre is greater with snow management and continuous cropping than with the traditional practice of crop fallow.

In addition to these private benefits, potential for social benefits also exists. If more snow is retained on fields, there should be less drifting of snow on roads and in cities and farmsteads. Furthermore, when snow management is accompanied by continuous cropping, economic losses due to saline seep and wind erosion will be reduced. Finally, the increased supplies of wheat and other small grains resulting from adoption of snow management and continuous cropping could depress grain prices to the detriment of farmers, but to the benefit of consumers.

Further research is needed to provide a more complete evaluation of the economics of snow management. Preliminary conclusions are, however, that this type of wind engineering holds much potential.

REFERENCES

(1) Aase, J.K., Siddoway, F.H., and Black, A.L., "Perennial Grass Barriers for Wind Erosion Control, Snow Management, and Crop Production," Proceedings of the Symposium on Shelterbelts on the Great Plains, Great Plains Agricultural Council Publication No. 78, pp. 69-78 (April, 1976).

(2) Bauer, Armand, "Effect of Water Supply and Seasonal Distribution on Spring Wheat Yields," North Dakota Agricultural Experiment Station Bulletin 490, (1971).

(3) Black, A.L., and Siddoway, F.H., "Dryland Cropping Sequences Within a Tall Wheatgrass Barrier System," Soil and Water Conservation, Vol. 31, pp. 101-105 (1976).

(4) Black, A.L., private communication (1977). (Northern Plains Soil and Water Research Center, Sidney, Montana.)

(5) Doering, E.J. and Sandoval, F.M., "Hydrology of Saline Seep Development on Upland Sites in the Northern Great Plains," Transactions of the American Society of Agricultural Engineers, Vol. 19, pp. 856-861 (1976).

(6) George, P.S., and King, G.A., Consumer Demand for Food Commodities in the United States with Projections for 1980, Giannini Foundation for Agricultural Economics, Monograph Number 26, p. 66 (March, 1971).

(7) Great Plains Agricultural Council, Snow Management in the Great Plains, Symposium, Great Plains Agricultural Publication No. 73, Lincoln, Nebraska, (1975).

(8) Greb, G.W., private communication (1977). (U.S. Central Great Plains Research Station, Akron, Colorado.)

(9) Haas, H.J., and Willis, W.O., "Moisture Storage and Use by Dryland Spring Wheat Cropping Systems," Soil Science Society of America Proceedings, Vol. 26, pp. 506-509 (1962).

(10) Heady, E.O. and Dillon, J.L., Agricultural Production Functions, Iowa State University Press, Ames, Iowa (1961).

(11) Jackson, Grant, private communication (1978). (Montana State University Saline Seep Program, Bozeman, Montana.)
(12) Lyles, Leon, "Wind Patterns and Soil Erosion on the Great Plains," Proceedings of the Symposium on Shelterbelts on the Great Plains, Great Plains Agricultural Council Publication No. 78, pp. 22-30 (April, 1976).
(13) Miller, M.R., Vander Pluym, H., Holm, H.M., Vasey, E.H., Adams, E.P., and Bahls, L.L., "An Overview of Saline Seep Programs in the States and Provinces of the Great Plains," Regional Saline Seep Control Symposium Proceedings, Montana State University Extension Service Bulletin No. 1132, pp. 4-17 (April , 1976).
(14) Sommerfeldt, Theron G., "Snow Trapping by Windbreaks," Regional Saline Seep Control Symposium Proceedings, Montana State University Cooperative Extension Service Bulletin No. 1132, pp. 87-90 (April, 1976).
(15) Thacker, Wendell, "An Overview of Saline Seep," Regional Saline Seep Control Symposium Proceedings, Montana State University Extension Service Bulletin No. 1132, pp. 1-3 (April, 1976).
(16) United States Department of Agriculture, Agricultural Research Service, "Summer Fallow in the Western United States," Conservation Research Report No. 17 (1974).
(17) "Wheatgrass Stops Erosion," Irrigation Age, Vol. 13, No. 7, p. 15 (April, 1979).

ECONOMIC IMPACT OF WINDSTORMS IN THE UNITED STATES

D. G. FRIEDMAN

Travelers Insurance Companies, Hartford, Connecticut, USA

ABSTRACT

A representative measure of the overall wind hazard to property and crops in the United States is not presently available. There are difficulties even in determining the relative importance of each of the four major windstorm types as damage producers. For many applications, information is needed on potential damaging effects of windstorms in addition to actual losses. Potential economic impact of wind is more realistically characterized by using two risk components: the average annual damage and the catastrophe producing potential of individual windstorms. Some of the problems inherent in estimating catastrophe potential for various types of windstorms are discussed.

INTRODUCTION

In a typical year, severe thunderstorms may produce as many as 800 tornadoes and 2000 strong "straight line" wind situations in the United States (1), about 800 of which will cause property or crop damage (2). In addition, a dozen or a more widespread winter or spring cyclones with accompanying high winds, two or three hurricanes, and several orographically related (Santa Ana or Boulder) windstorms will occur. Each year some of these storms produce losses that are of catastrophic proportions. There is a need for a more detailed understanding of the magnitude and other characteristics of economic impacts of windstorms (Refs. 3-5).

CURRENTLY AVAILABLE INFORMATION

A short description of each reported storm occurrence and its effect is listed in Storm Data (6) along with details on location, date, type of storm, casualties, and resulting property or crop damage. Although these damage tabulations are indispensible for many purposes, they are not complete or accurate enough to provide the sole basis for estimating either the overall magnitude of the wind hazard or the relative importance of each windstorm type as a damage producer in the United States. The data collection procedure, which is primarily based upon newspaper clippings, probably leads to a systematic underestimation of the damaging effects of some windstorm types. Many localized occurrences are missed. Because of the dependence upon these unverified reports of damage from throughout the United States, a standardized appraisal is not possible. As a result, damages listed in Storm Data are only approximations to actual amounts. Losses are coded in order-of-magnitude categories with a large range which limits their usefulness for estimating the relative damage production of various types of windstorms and the overall magnitude of the wind hazard when individual impacts by storm type are accumulated. Figure 1 illustrates this drawback by comparing the coded estimates of annual damages in the United States caused by tornadoes and hurricanes in each of the past ten years (7). Even though hurricane caused damages have a larger year-to-year fluctuation than tornado losses, the wide range of the coded damage categories makes it difficult to determine which of these hazards causes greater average annual damages.

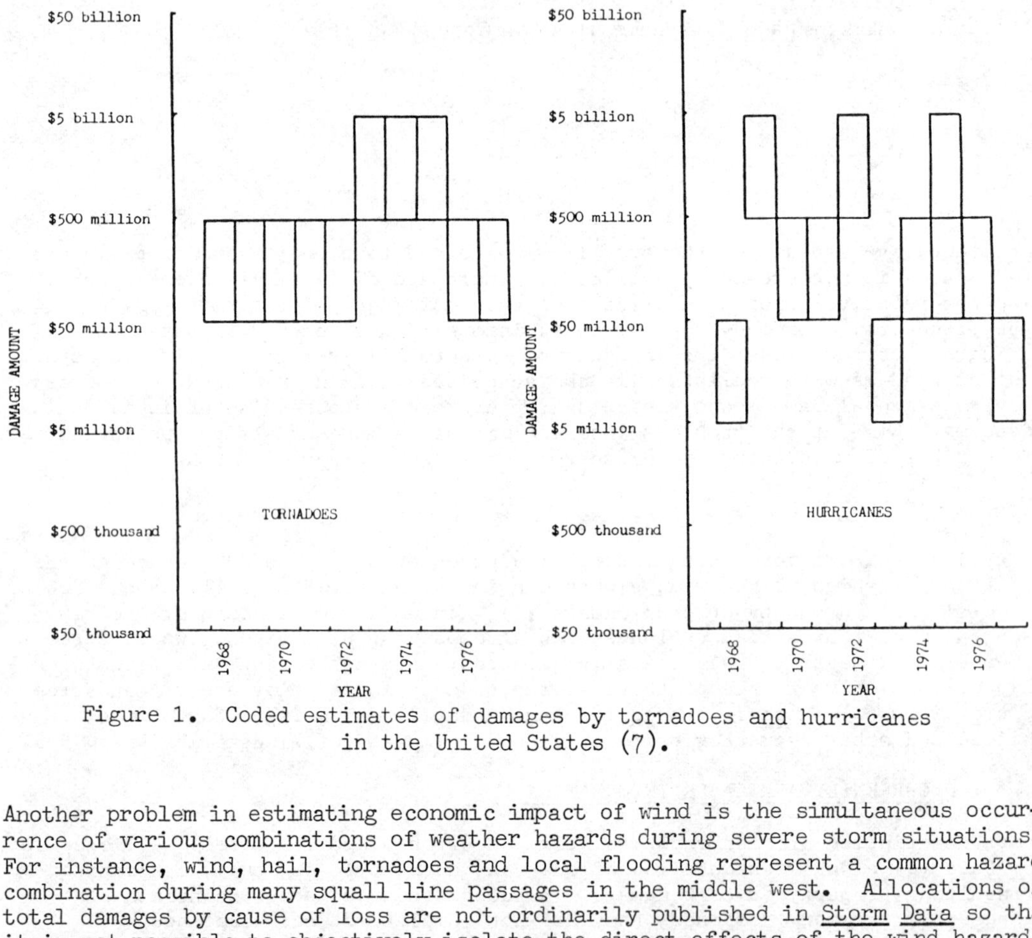

Figure 1. Coded estimates of damages by tornadoes and hurricanes in the United States (7).

Another problem in estimating economic impact of wind is the simultaneous occurrence of various combinations of weather hazards during severe storm situations. For instance, wind, hail, tornadoes and local flooding represent a common hazard combination during many squall line passages in the middle west. Allocations of total damages by cause of loss are not ordinarily published in Storm Data so that it is not possible to objectively isolate the direct effects of the wind hazard in many of the storm occurrences. An estimate of the overall impact of the wind hazard in the United States based directly upon Storm Data tabulations is not published no doubt because of these damage content differences and other data difficulties.

Over the past decade, a number of estimates of average annual damage caused by various types of windstorms, or combinations thereof, have been given which were expressed in terms of dollars and not damage categories. However, most of these estimates, which are listed in Table 1, are probably dependent to some degree upon the Storm Data listings. Four major windstorm types are used (8):

1. Tropical cyclones which include hurricanes and tropical storms (In subsequent discussions, this type is called HURRICANE).

2. Extratropical cyclones and anticyclones including widespread winter and spring windstorms plus orographic associated situations including Boulder and Santa Ana windstorms (WINTER WINDSTORM).

3. Thunderstorm spawned tornadoes (TORNADO).

4. Thunderstorm spawned "straight line" winds (THUNDERSTORM WIND).

Table 1 estimates are not directly comparable among themselves because of differing sets of assumptions regarding causes of loss and types of damage that are included. It is not known how much weight was given to Storm Data information when these estimates were made. The wide range of estimated average annual damages is due, in part, to these differing sets of assumptions.

Table 1. Estimates of average annual damages caused by various windstorm types.

Type of Storm	Hendricks & Friedman 1966 (Ref. 8)	Sugg 1967 (Ref. 9)	Kessler 1970 (Ref. 10)	Sanders 1971 (Ref. 11)	Ayre 1975 (Ref. 12)	Kessler & Pearson 1978 (Ref. 13)	Wiggins 1978 1970 conditions (Ref. 14)	Wiggins 1978 1980 conditions (Ref. 14)
Hurricane	$250 million	$310 million	$500 million	$530 million	$400 million	—	$1,200 million	$1,600 million
Winter Windstorm	25 million	—	—	280 million	—	—	—	—
Tornado	100 million	—	75 million	—	215 million	$300 million	1,500 million	$2,000 million
Thunderstorm Wind	125 million	—	—	—	—	—	—	—
Thunderstorm Wind and Tornado	—	—	—	570 million	—	—	—	—
Thunderstorm Wind and Winter Windstorm	—	—	—	—	100 million	—	at least $19 million	at least $19 million
Total all Windstorm types	$500 million	NC	NC	$1,380 million	$715 million	—	at least $2719 million	at least $3,619 million

NC—Not Complete

Another variable is the year in which the estimate was made. Recent rapid changes over time in property value and the cost of repair also prevents a direct comparison unless an adjustment in the estimates is made. Even when a common base is used - 1978 dollars, estimates of average annual damage vary by a factor of three for the hurricane category ranging from $500 million dollars for combined wind and coastal flooding damages to $1.6 billion for wind only damages to buildings under 1980 conditions. In the tornado category, the estimates vary from $140 million for all types of property damage to $2 billion for damages to only buildings using 1980 conditions. Damage estimates for the two other windstorm types range from (at least) $19 million to several hundreds of millions of dollars a year.

Estimates of the overall wind impact using a common base extend from about $900 million to (at least) $3.6 billion for only buildings under 1980 conditions. Even with an adjustment for the effect of inflation, there are still large differences in the estimates of the magnitude of the wind hazard to property in the United States. The impact of the wind hazard on agriculture may be even more difficult to estimate as evidenced by the lack of pertinent articles in the climatological literature.

It is possible that even though there is not much agreement about the overall level of the wind hazard to property, that there are consistencies in estimates of the relative importance of each windstorm type as a damage producer. Table 2 lists the percentage damage contributions by windstorm type, or combination of types, for each of the four sets of estimates that cover all of the storm categories. Among these sets, the hurricane contribution ranges from 39 percent to 56 percent. The tornado contribution to total wind damage varies from 20 percent for damage to insured property to (less than) 55 percent for buildings only. These large differences indicate that there is not much consistency even among estimates of the relative damage production to property of the various windstorm types.

Table 2. Relative importance of windstorm types as damage producers based upon sets of estimates in Table 1 that included all types. Expressed as a percentage of total damages.

Type of Storm	Hendricks & Friedman 1966	Sanders 1971	Ayre 1975	Wiggins 1978
Hurricane	50%	39%	56%	less than 44%
Winter Windstorm	5	20	—	—
Tornado	20	—	30	less than 55%
Thunderstorm Wind	25	—	—	—
Thunderstorm Wind and Tornado	—	41	—	—
Thunderstorm Wind and Winter Windstorm	—	—	14	greater than 1%
Total damage all Windstorm types	100%	100%	100%	100%

USE OF INSURANCE DAMAGE INFORMATION

Some additional information to use in estimating the overall magnitude of the wind hazard and the damage production of various windstorm types is available in insurance loss experience. Because of the current tendency to group various types of perils into a single package such as the homeowners policy, it is not possible to obtain overall insurance industry damage experience solely for the wind peril. However, information is available on insured losses incurred during weather catastrophes caused by wind and/or hail. The property insurance industry defines a "catastrophe" as a single event that produces at least $1.5 million (prior to 1975, $1 million) aggregate insured losses. Over four hundred of these catastrophes have been attributed to the insured weather perils since coding began in 1949. Insured property damages during these events amounted to about $11.5 billion (1978 dollars).

Although, insured damage amounts caused by wind and hail during non-catastrophe situations are not available, there are advantages to considering insured catastrophe losses in making estimates of the overall impact of the wind hazard. Insured damage represents a sizable, but unknown percentage of total wind caused property damage in the United States. These losses are based upon a fairly uniform appraisal system across the country and the damaging effects of other concurrent weather hazards have been minimized in these loss tabulations. In as much as insured damages during wind-caused catastrophes represent a portion of the total wind losses, they can be used to denote the minimum damage contribution of each storm type to the overall economic impact of the wind hazard.

For each coded catastrophe, the date of occurrence, location and amount of insured damages is listed (15). Over the years, the number of catastrophes caused by relatively infrequent winter windstorms and hurricanes has not changed appreciably. However, the number of thunderstorm - caused catastrophes increased significantly. One of the reasons for this increase is the effect of inflation upon property values and cost of repair. Currently, it is difficult for an intense squall line to sweep across a section of the United States without causing, at least, several million dollars in insured damages.

Figure 2 represents insured property damages incurred during weather catastrophes in each of the twenty-seven year period 1949-1975 and expressed in 1978 dollars. Of the three storm types that are represented, winter windstorms caused losses which, in most years, were relatively small. However, occasionally a catastrophic event does take place. A single storm that occurred in November of 1950 caused widespread serious wind damages over a large portion of the northeastern United States. The thunderstorm hazards were the second largest damage contributor with an occasional catastrophic event such as the swarm of tornadoes in 1974. Insured damages caused by winds associated with a hurricane's passage exhibit the greatest year-to-year variability ranging from near zero in some years to over a billion 1978 dollars in a single year. Refer to 1965 - the year of Hurricane Betsy.

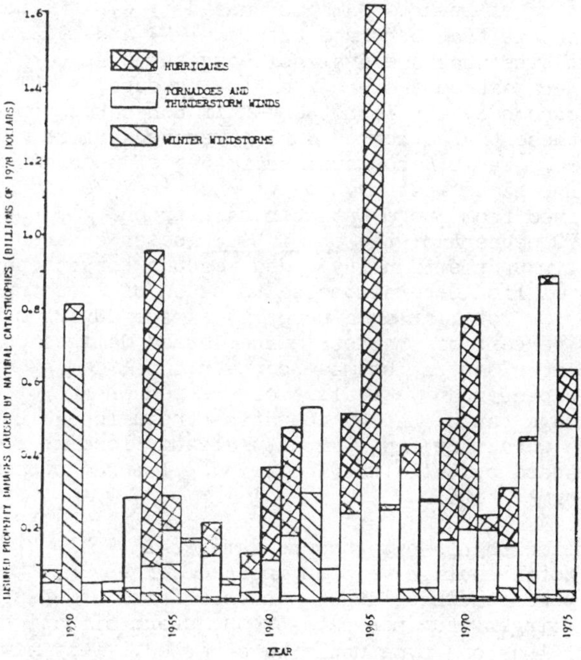

Figure 2. Insured property damages incurred during weather catastrophes.

To provide an estimate of the relative importance of the four windstorm types plus hail as insured damage producers, the date and location of the 411 weather catastrophes was used to obtain information on each storm situation from the Storm Data tabulations. This information provided the basis for allocating portions of the total insured damages of each catastrophe to the appropriate cause of loss category (16). With the large number of storms that were used, it was hoped that judgement errors in one direction or the other would average out. The entire 27-year period was taken in order to include as many as possible of the rare, but highly damaging, events that occur within each windstorm type. Results of the analysis are shown in Table 3. When the effect of hail is eliminated, the wind hazard accounted for about $375 million in average annual insured wind damages incurred during weather catastrophes.

Table 3. Estimated average annual property damage by storm type (16) incurred during 411 weather catastrophes in the period 1949-1975. Damage expressed in 1978 dollars.

Type of Storm	Amount of Damage	Average Annual damage	Relative importance of various combinations of storm types as damage producers. (percent of total damage)				
Hurricane	$4.69 billion	$174 million	41	45	—	—	61
Winter Windstorm	$1.48 billion	$ 55 million	13	15	27	—	—
Tornado	$3.00 billion	$111 million	26	30	55	76	39
Thunderstorm Wind	$0.97 billion	$ 36 million	8	10	18	24	—
Hail	$1.35 billion	$ 50 million	12	—	—	—	—
Total	$11.49 billion	$426 million	100	100	100	100	100

Unfortunately, a listing of insured damages caused by wind is not available for the non-catastrophe storms that occurred between 1949 and 1975. However, some loss experience statistics that are available suggest that total wind-and-hail damages caused by non-catastrophe storms, on the average, equal or exceed damages incurred during catastrophes. It is not known if non-catastrophe weather losses would have equalled those that occurred during weather catastrophes if the 27-year interval had been used. Table 4 compares estimates of insured property damages by type of storm during named weather catastrophes in 1976 and 1977 with property damages derived from Storm Data for catastrophe and non-catastrophe storm occurrences. About 70 percent of the Storm Data losses occured at the time and place denoted by the insurance industry coded weather catastrophes. The insured losses of about $450 million were allocated to cause of loss through the use of Storm Data information. Total insured property damage during catastrophes caused by wind was about 65 percent of the total wind caused damages during the same period. One other comparison can be made between total damage and the portion that is insured. A representative measure of the percentage of insured damage to total wind caused damage can be had by comparing these losses caused by Hurricane Celia in 1970. There were very few agricultural and flooding losses as a result most of the damage caused by Celia was from wind. Insured property damages were 70 percent of the storm's total.

If 65 percent is applied to non-catastrophe damages in Table 4, about $200 million of the $315 million total would have been insured. However, the actual insured amount would probably be somewhat larger because of the underestimation problem in the Storm Data listings. For non-catastrophe storm situations a different mixture of damage by windstorm type would be expected. Very few losses would come from the hurricane wind category. The largest percentage of the insured damages for non-catastrophe storms would no doubt be caused by the thunderstorm wind and tornado hazards.

Table 5 gives the estimated damage contribution from each of the four basic windstorm types based upon an allocation of the original combined windstorm groupings to individual categories. These percentages are compared with those estimated for insured damages during catastrophes and total insured wind damages.

Table 4. Comparison of insured property damage by type of storm during named weather catastrophes in 1976-77 with property damages derived from the 1976-77 Storm Data tabulations for catastrophe and non-catastrophe storm occurrences.

Type of Storm	Total Damage	Non-Catastrophe occurrences	Catastrophe occurrences	Insured damages incurred during catastrophes
Hurricane	$ 47 million (100)	0 (0)	$ 47 million (100)	$ 25 million
Winter Windstorm	$137 million (100)	$ 49 million (36)	$ 88 million (64)	$ 73 million
Tornado	$390 million (100)	$106 million (27)	$284 million (73)	$142 million
Thunderstorm Wind	$281 million (100)	$109 million (39)	$172 million (61)	$137 million
Hail	$158 million (100)	$ 51 million (32)	$107 million (68)	$ 75 million
Total	$1013 million (100)	$315 million (31)	$698 million (69)	$452 million

Several observations can be made on the basis of this cursory scanning of current estimates of the overall impact of the wind hazard and the relative importance of various windstorm types as property damage producers in the United States. There is not sufficiently accurate and detailed damage information available from which universally agreed upon estimates can be made. The information that can be used suggests that the overall economic impact of wind in the United States is currently averaging at least $1 billion a year. The greatest damage contribution comes from the hurricane wind category which exhibits a large year-to-year variability in damage impact. However, the winter windstorm and thunderstorm wind categories are apparently of much greater importance in producing damages than is generally assumed. Insurance loss statistics provide, at least, a lower bound to the damage production percentages of each windstorm type. Widespread minor wind damages caused by swaths of severe local winds can contribute significantly to annual damage totals because of the high frequency of these windstorms in any given year.

CATASTROPHE POTENTIAL

For many applications, including insurance, it is necessary to make prospective estimates of windstorm losses as well as damage surveys after a storm has occurred. Two measures of risk are needed to estimate these potential economic impacts. One is the average annual damage. A second measure, which for rare occurring severe windstorm situations is of greater importance, is the catastrophe producing potential of individual storms. The tendency for some types of windstorms to have large catastrophe potentials determines the degree of year-to-year variability in damages resulting from these storms.

The sole use of past damage experience is not a good measure of current catastrophe potential because of changes over time, in number, type, damage susceptibility, cost of repair, and geographical distribution of exposed property since the last major storm occurred. Fortunately, in any given coastline section, the probability of a direct hit by a hurricane is relatively low. However, from the stand-

Table 5. Estimated damage contribution (percentage of total damage) by windstorm type based upon Tables 2-4.

Type of Storm	Hendrick & Friedman 1966	Sanders 1971	Ayre 1975	Wiggins 1978	Insured catastrophe damages	Total insured damages
Hurricane	55	39	56	less than 44	45	40
Winter Windstorm	5	20	5**	greater than 1***	15	12
Tornado	22	24**	30	less than 55	30	32
Thunderstorm Wind	18*	17**	9**	greater than 1***	10	16
Total	100	100	100	100	100	100

* Hail damages excluded using Table 4 information.
** Allocation to individual windstorm class based on Table 4 information.
*** Winter wind and thunderstorm wind classes combined.

point of a risk analysis, this rareness of occurrence reduces the usefulness of damage experience resulting from past storms as measures of current damage potential.

One approach to estimating catastrophe potential is to use computerized simulation techniques (18). Computer generated wind speed patterns mathematically superimposed upon the geographical array of property provide a means of approximating the damage producing mechanism of the various windstorm types and a basis for calculating expected average annual damage and catastrophe potential of individual storms. For winter windstorms and hurricanes, a national network of 85 thousand grid areas are used to represent the spatial distribution of exposed properties in the contiguous forty-eight states. The damage production of a given storm depends not only upon its severity, but also upon its path relative to densely populated areas.

Figure 3 illustrates how the damage potential of a simulated storm of given intensity can vary depending upon the direction of its movement and landfall location. Relative wind damage production is indicated by percentages given at the end of each path arrow and is measured relative to the optimal path for producing maximum damage in the Houston area. One advantage of the simulation approach is that it provides an opportunity to utilize and translate pertinent information inherent in severe wind climatology into measures of potential economic impact of the wind hazard to the current geographical distribution of exposed property.

Much work needs to be done in developing simulation and other methods of estimating average annual damage and catastrophe potential of the large scale hurricane and winter windstorm hazards. Of critical importance is a more precise determination of basic relationships between physical characteristics of the geophysical event and its damaging effect upon a particular type of property, a better understandingofthe event itself, and finally development of a reliable method of estimating the number and type of properties-at-risk in any given geographic area.

Problems are even greater when one attempts to quantify catastrophe potential of the two other major windstorm types: thunderstorm wind and tornadoes. If computer simulation were to be used, the size of the unit grid area representing the geographical distribution of exposed property would have to be much less than

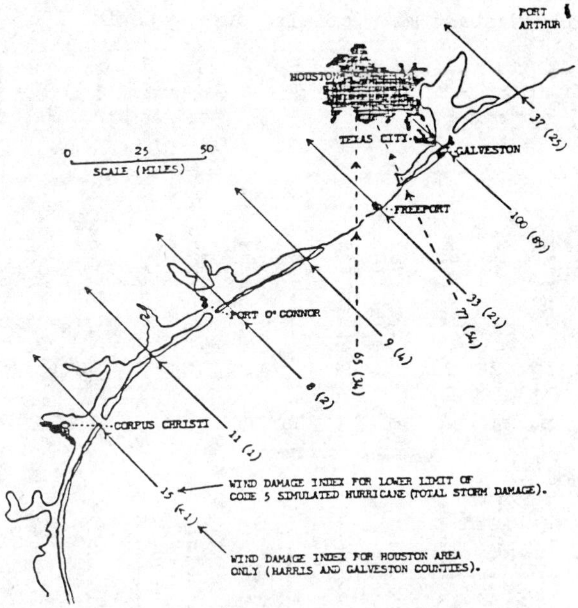

Figure 3. Index of aggregate damage production of a simulated hurricane of fixed intensity that follows different paths along the Texas coast.

for hurricanes or winter windstorms. In general, the size of the grid unit should be much smaller than the area of the severity pattern generated by the storm. The area of high winds swept out by the movement of a thunderstorm cell and the even smaller high wind severity area produced by a tornado could require a network at least an order of magnitude larger than the national grid system currently in use.

Of equal importance, is a need for development of a severe wind climatology especially with respect to the thunderstorm wind hazard. The climatology of tornadoes(frequency and severity) has received considerable attention in recent years (19). Very little has been done on specifying climatological characteristics of "straight line" thunderstorm winds or, for that matter, of other local damaging windstorms (2).

Each type of property such as buildings may have a different vulnerability to high winds and a different threshold wind speed for producing damage. As speed increases above this threshold value the probability of increased damage severity rises. It is likely that this damage threshold speed, for most types of buildings, is less than Thom's annual fastest mile of wind (20). Therefore, the frequency distribution of different high wind speeds at a location is needed. Table 6 is an example of a thunderstorm wind climatology that has been found to be useful for estimating average annual damage in the middle west. The occurrence of hail during a period of high thunderstorm winds greatly increases the insured damage potential because the stones are driven into wall and window surfaces that are not vulnerable when the wind is not blowing. With information given in Table 6, it is possible to select a threshold wind speed for a particular type of property and then determine the spatial pattern of frequencies above that speed.

Table 6. Frequency of daily fastest mile of wind during period 1967-1974 (21).

Wind Speed	Colorado Springs	Amarillo	Goodland	North Platte	Dodge City	Wichita Falls	Grand Island	Concordia	Oklahoma City	Wichita	Omaha	Tulsa	Topeka	Kansas City	Fort Smith	Springfield
DAYS WITH A THUNDERSTORM																
0- 4	-	-	-	-	-	-	-	-	-	-	-	-	-	-	-	-
5- 9	1	-	-	1	-	-	-	3	-	-	2	-	5	4	9	1
10-14	24	6	8	19	10	14	29	26	13	6	30	27	41	33	75	48
15-19	80	36	56	77	66	64	84	103	89	74	93	78	112	97	133	143
20-24	166	73	133	62	84	141	128	93	91	89	79	90	106	104	102	103
25-29	89	91	64	73	107	76	43	91	89	116	73	81	87	67	69	78
30-34	44	68	24	48	59	32	14	67	49	59	27	51	32	21	31	36
35-39	13	33	11	23	27	32	23	21	19	26	14	14	14	11	12	8
40-44	4	16	6	6	16	8	2	10	10	18	3	7	7	8	8	8
45-49	1	7	1	4	4	4	2	6	3	7	4	3	2	3	1	1
50-54	2	3	1	1	4	-	1	6	-	3	2	-	-	1	-	-
55-59	-	-	-	-	3	-	-	2	-	-	2	-	-	-	-	-
60-64	-	-	-	-	-	-	-	2	-	1	-	-	1	-	-	1
65-69	-	-	-	1	-	-	-	-	-	-	-	-	-	-	-	-
70-74	-	-	-	-	-	-	-	-	-	-	-	-	-	1	-	-
75-79	-	-	-	-	-	-	-	-	-	-	-	-	-	-	-	-
	424	333	304	315	380	371	326	430	363	399	329	351	407	350	440	427
DAYS WITH HAIL																
0- 4	-	-	-	-	-	-	-	-	-	-	-	-	-	-	-	-
5- 9	-	-	-	-	-	-	-	-	-	-	-	-	-	-	-	-
10-14	1	-	-	1	-	-	1	-	-	-	1	1	1	-	2	1
15-19	9	2	2	2	2	1	3	1	3	3	4	2	2	2	3	5
20-24	15	4	13	1	3	6	8	6	7	2	3	4	-	3	4	4
25-29	12	5	12	12	4	16	8	8	6	7	4	12	8	12	2	7
30-34	8	15	4	9	12	8	1	5	6	5	1	6	7	3	5	4
35-39	5	11	5	6	5	6	5	4	5	3	2	1	1	3	1	2
40-44	2	5	-	-	3	1	1	4	5	5	3	1	5	1	-	2
45-49	-	-	1	-	3	1	2	-	1	-	1	-	2	2	-	1
50-54	-	-	-	1	1	1	-	1	1	-	-	1	1	-	-	-
55-59	-	1	-	-	1	-	1	-	-	-	1	-	-	-	1	-
60-64	-	2	-	-	-	-	-	-	-	-	-	-	-	-	-	-
65-69	-	-	-	-	-	-	-	1	-	-	1	-	-	-	-	-
70-74	-	-	-	-	-	-	-	-	-	-	-	-	-	-	-	-
75-79	-	1	-	-	-	-	-	-	-	-	-	-	-	-	-	-
	52	46	37	32	34	40	30	30	34	25	21	28	27	26	18	26

To illustrate the use of information in Table 6 for estimating expected average annual damage and catastrophe potential of thunderstorm wind and hail refer to Figures 4-5. The number of times the fastest mile of wind on thunderstorm days exceeded a threshold value of 35 miles per hour (peak gust probably greater than 50 miles per hour) is shown in Figure 4. Locations of greatest damage potential implied from these frequencies covered the area stretching from the Texas panhandle across central Kansas. This frequency pattern used in conjunction with damage vulnerability relationships between wind speed, with and without hail, and damage severity provides a means of calculating average annual expected loss from the thunderstorm wind and hail hazards. For this eight year period, the greatest damage potential for each individual property was in western Oklahoma and central portions of Kansas. Calculation of the geographical pattern of expected damage is based upon the implicit assumption that the exposed property is uniformly distributed across the sampled area.

To illustrate the difficulty in estimating the catastrophe potential using only these climatological data, the location of significant damage from thunderstorm wind and hail during weather catastrophes in the eight year period is plotted in Figure 5. There is very little correspondence between the pattern of storm damage locations and the area of maximum storm frequency. This illustrates the need to incorporate the geographic distribution of the property-at-risk in attempting to define catastrophe potential. The pattern of storm damage locations depends upon the interaction of the climatological pattern of storm frequency with the population density distribution. Estimation of catastrophe potential must somehow take both factors into account. The damage producing potential of a windstorm is based upon a measure of the storms intensity. The percentage of this damage potential that is actually realized depends upon the interaction of the wind speed severity pattern with the exposed property-at-risk spatial distribution.

CONCLUSION

Wind is one of the major damage producers among the natural hazards. Much remains to be done to accurately define the magnitude and other characteristics of its economic impact upon property in the United States.

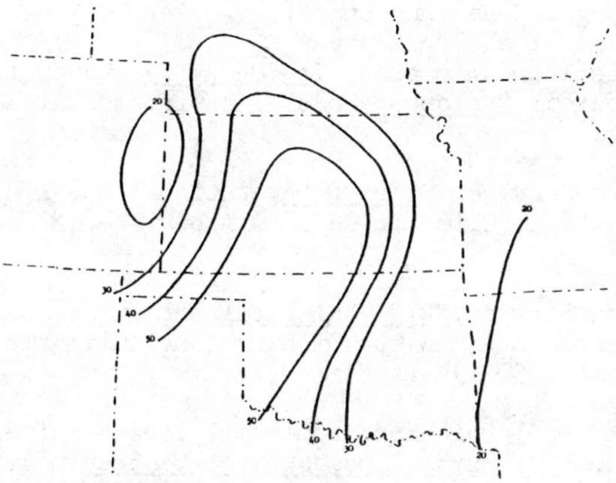

Figure 4. Frequency of thunderstorm day winds greater than 35 miles per hour during 1967-1974.

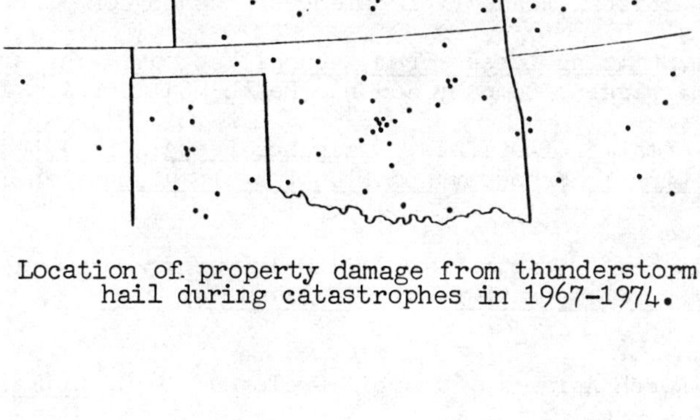

Figure 5. Location of property damage from thunderstorm and/or hail during catastrophes in 1967-1974.

REFERENCES

1. E. Kessler and A.D. Pearson, Severe Thunderstorm Systems, Geophysical Predictions, National Academy of Sciences, Washington, D.C., (1978) 130.

2. Brinkmann, W.A.R., Severe Local Storm Hazard, A Research Assessment, University of Colorado, Boulder, 1975.

3. Roshko, A., Wind Loads on Structures, California Institute of Technology, Pasadena, 1971.

4. White, G.F. and Haas, J.E., Assessment of Research on Natural Hazards, the MIT Press, Cambridge, Massachusetts, 1975.

5. World Meteorological Organization, The Quantitative Evaluation of the Risk of Disaster from Tropical Cyclones, Special Environmental Report number 8, Geneva Switzerland, 1976.

6. Environmental Data Service, Storm Data, National Oceanic and Atmospheric Administration, National Climatic Center, Asheville, N.C. monthly issues, 1950-1978.

7. Environmental Data Service, Climatological Data, National Summary, National Oceanic and Atmospheric Administration, National Climatic Center, Asheville, N.C. annual issues, 1950-1978.

8. R.L. Hendricks and D.G. Friedman, Potential Impacts of Storm Modification on the Insurance Industry, Human Dimensions of Weather Modifications, Department of Geography, Research Paper number 105, University of Chicago Press, Chicago, 229, (1966).

9. A.L. Sugg, Economic Aspects of Hurricanes, Monthly Weather Review, 95, 143-146 (1967).

10. E. Kessler Tornadoes, Bulletin of the American Meteorological Society, 51, 926 (1970).

11. F. Sanders, Toward defining human needs: how does the atmosphere hurt us? Bulletin of the American Meteorological Society 52, 448 (1971).

12. Ayre, R.S., Technological Adjustments to Natural Hazards, University of Colorado Boulder, 1975.

13. E. Kessler and A.D. Pearson, Severe Thunderstorm Systems, Geophysical Predictions, National Academy of Sciences, Washington, D.C., 131 (1978).

14. Wiggins, J.H., Building Losses From Natural Hazards: Yesterday, Today and Tomorrow, J.H. Wiggins Company Redondo Beach, California, 1978.

15. Insurance Information Institute, Insurance Facts: Selected Data of General Interest Relating to Property and Liability Insurance, New York, various issues 1949-1978.

16. Friedman, D.G., Hail Suppression Impact Upon Property Insurance, Technology Assessment of the Suppression of Hail, working paper 11, Illinois State Water Survey, Urbana, 1976.

17. C.R. Condon, North Atlantic Tropical Cyclones, 1970, Climatological Data

National Summary, Annual, National Climatic Center, Asheville, N. C., 21, 13, 66 (1971).

18. Friedman, D. G. Computer Simulation in Natural Hazard Assessment, University of Colorado, Boulder, 1975.

19. Abbey, R. F., Jr. and Mehta, K. C. Proceedings of the Symposium on Tornadoes: Assessment of Knowledge and Implications for Man, Texas Tech University, Lubbock, Texas, 1976.

20. H.C.S. Thom, New Distributions of Extreme Winds in the United States, Journal of the Structural Division, Proceedings of the American Society of Civil Engineers, 99, 1787-1800, (1968).

21. Environmental Data Service, Local Climatological Data, National Weather Records Center, Asheville, N. C. monthly issues, 1967-1974.

DAMAGE AND DECISION MAKING IN WIND ENGINEERING

W. PETAK* and G. C. HART**

*J. H. Wiggins Company, Redondo Beach, California, USA
**University of California, Los Angeles, California, USA

ABSTRACT

Damage due to wind natural hazards are compared with other natural hazard damage. The formulation of a decision making framework for proposing and obtaining optimal damage reduction decisions is presented. A comment is provided to indicate how wind microzonation is important in the decision making problem.

INTRODUCTION

One purpose of this paper is to report on part of the results of a study funded by the National Science Foundation (NSF). The study was a problem-oriented Technological Assessment aimed at identification of the potential impacts, problems, mitigation technologies and policy alternatives associated with the exposure of people and property to nine natural hazards in the United States. One group of hazards was the general category of wind. Specifically, tornadoes, hurricanes and general severe winds were examined. The results of this portion of the study are summarized in this paper.

The two basic components of the study were: (1) Problem Analysis, and (2) Policy Analysis. The former identifies the: (a) problem states of affairs, (b) networks of causes and effects to which these states are linked, (c) underlying values associated with the states, and (d) value-holders and problem-experiencing populations and areas. The latter identified: (a) alternative problem solving strategies and policy choices, (b) criteria for selection of policy alternatives, (c) decision influencing constraints, and (d) the payoffs, outcomes, and side effects which may be expected from each policy alternative.

This material is based upon research supported by the National Science Foundation under grant numbers ERS-75-09998-A01, and AEN-74-23992. Any opinions, findings, and conclusions or recommendations expressed in this publication are those of the author(s) and do not necessarily reflect the views of the National Science Foundation.

The overall approach to the study involved the application of risk analysis. Such an analysis is a specialized set of operations which, in our opinion, is the backbone of the problem analysis. Most Technology Assessments have failed to recognize the importance of a highly technical (or specialized) risk analysis. Therefore, this study made a fundamental contribution to wind engineering by incorporating a wind risk analysis in the wind portion of the technology assessment. This analysis recognized and accounted for the risk, associated with wind as a function of: (a) the probability that an event will occur and (b) the consequences of the event.

The wind risk analysis identified: (a) the probability that particular events at certain intensities will occur over a specified time frame, (b) the area or population which will be exposed to the events (e.g., population at risk), (c) the vulnerability of the area or exposed population to effects associated with the events; and (d) the consequences to the population at risk of exposure to the expected series and intensities of the events over a specified time frame. The output of the wind risk analysis results in a probabilistically-based annualized estimate of the damage to the building population-at-risk which is expected to arise from the wind hazard. Perhaps the most important contribution of the NSF study to wind engineering and this audience is the identification of where wind engineers involved in the many diverse aspects of the wind problem can provide expertise in order to improve the techniques and hence the numerical estimates of loss obtained in a risk analysis.

The second purpose of this paper is to show where the recently proposed concept of wind microzonation fits into a risk analysis. The division of relatively small geographic regions (e.g., the greater Los Angeles, California, area) into smaller zones of wind environment description is the goal of a wind microzonation study. Such a zonation enables the local meterological conditions to be more accurately modeled and the confidence band on any numerically derived wind speeds to be reduced.

WIND DAMAGE

The methodology used to evaluate and predict the future damage for nine natural hazards is summarized in reference (1). The wind methodology is discussed in detail in references (2) to (4). Tables 1 and 2 summarize the expected annual loss from natural hazard exposures for two time points-1970 and 2000.

The damage to buildings was a special focus of the study and it was revealed that a considerable difference existed as to the damage, type, and magnitude. Taking into account the specific geographic areas in which each type of hazard occurs and the vulnerability of buildings to the hazard, the computer models used in the study disclosed that light-to-moderate damage would be sustained by 62% of the buildings experiencing damage from earthquakes, by 27% of those damaged by storm surge, by 18% of those damaged by tsunami, and by approximately 16% of those damaged by either hurricanes or severe winds. There was only a small percentage of buildings damaged in this category due to tornadoes. However, structural collapse or severe damage is expected to be experienced by approximately 93% of the exposed buildings damaged by tornadoes, by 51% of those damaged by hurricanes and by 36% of those damaged by tsunami. Figure 1 summarizes this proportional distribution of damage.

Table 3 indicates for the time period 1961-1970 that major single event natural disasters resulted in over $6 billion dollars damage and 1,000 lost lives. Eight of the events were wind engineering related with two being tornadoes and six being hurricanes.

Table 4 indicates the magnitude of the exposure to both people and property to the nine natural hazards. For example, approximately 90% of the U.S. property and

Table 1. Expected annual losses from natural hazard exposures in the United States by type of hazard and type of loss, 1970

HAZARD	EXPECTED ANNUAL LOSSES								
	DOLLAR LOSSES RESULTING FROM INDICATED FACTOR (MILLIONS OF DOLLARS)					OTHER LOSSES			
	(1) BUILDING DAMAGE	(2) CONTENTS DAMAGE	(3) INCOME LOSS[a]	(4) SUPPLIER LOSS[b]	(5) TOTAL (1-4)	(6) NUMBER OF DEATHS	(7) HOUSING UNITS LOST	(8) PERSON YEARS OF HOMELESSNESS	(9) PERSON YEARS OF UNEMPLOY.
1. EARTHQUAKE	655.2	123.23	2.651	0.030	781.1	273	20,485	736	413.5
2. EXPANSIVE SOIL	798.1	-	-	-	798.1	-	-	-	-
3. HURRICANE	685.5	267.57	101.803	1.092	1056.0	62	31.885	34,505	21,003.7
4. LANDSLIDES	370.3	-	-	-	370.3	-	-	-	-
5. RIVERINE FLOODING	1901.0	847.02	10.166	0.120	2758.3	190	-	-	-
6. SEVERE WIND	11.4	4.47	2.090	0.022	18.0	5	547	852	373.1
7. STORM SURGE	441.6	197.24	2.367	0.028	641.2	37	24,521	7,290	369.7
8. TORNADO	879.8	469.93	302.821	3.451	1656.0	392	36,212	86,122	57,541.6
9. TSUNAMI	8.7	5.54	0.727	0.012	15.0	20	234	345	97.5
TOTALS	5751.6	1915.0	422.625	4.755	8094.0	979	113,884	129,850	79,799.1

[a] Total loss of worker earnings associated with hazard caused unemployment

[b] Total loss of income experienced by suppliers of businesses and industries experiencing hazard-induced shutdowns

Table 2. Expected annual losses from natural hazard exposures in the United States, by type of hazard and type of loss, 2000

HAZARD	EXPECTED ANNUAL LOSSES								
	DOLLAR LOSSES RESULTING FROM INDICATED FACTOR (MILLIONS OF DOLLARS)					OTHER LOSSES			
	(1) BUILDING DAMAGE	(2) CONTENTS DAMAGE	(3) INCOME LOSS[a]	(4) SUPPLIER LOSS[b]	(5) TOTAL (1-4)	(6) NUMBER OF DEATHS	(7) HOUSING UNITS LOST	(8) PERSON YEARS OF HOMELESSNESS	(9) PERSON YEARS OF UNEMPLOY.
1. EARTHQUAKE	1177.0	372.78	3.906	0.048	1553.7	400	22,868	648	634.9
2. EXPANSIVE SOIL	997.1	-	-	-	997.1	-	-	-	-
3. HURRICANE	1742.0	1504.98	276.191	3.095	3526.3	153	52,237	48,271	58,223.7
4. LANDSLIDE	871.2	-	-	-	871.2	-	-	-	-
5. RIVERINE FLOODING	1594.0	1572.54	8.68	.105	3175.33	159	-	-	-
6. SEVERE WIND	24.8	23.8	4.696	0.051	53.4	11	748	1,014	850.9
7. STORM SURGE	1176.0	1160.43	6.407	0.077	2342.9	103	43,757	10,330	1,018.3
8. TORNADO	2058.0	2401.32	750.780	9.042	5219.1	920	52,119	107,650	146,568.5
9. TSUNAMI	19.8	19.10	1.479	0.027	40.4	44	335	389	195.9
TOTALS	9659.9	7054.95	1052.139	12.445	17,779.43	1790	172,084	168,302	207,492.2

[a] Total loss of worker earnings associated with hazard caused disruption

[b] Total loss of income experienced by suppliers of business and industries experiencing hazard-induced shutdowns

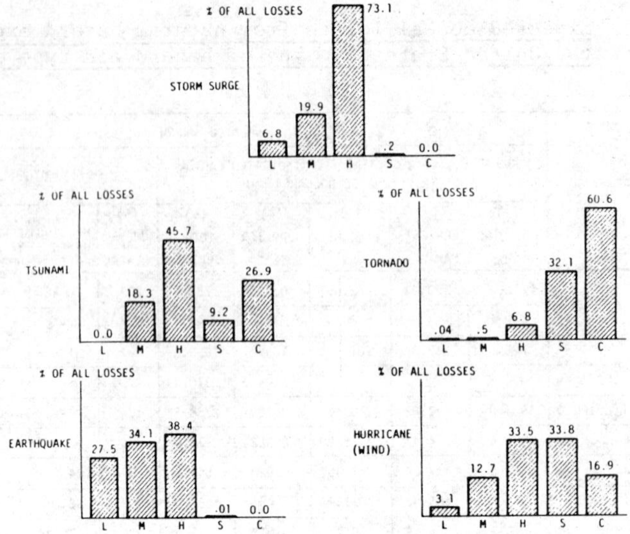

DAMAGE STATE	STRUCTURE CONDITION
None (0 - 0.5%) } N	No damage
Light (0.5 - 1.25%) } L	Minor ceiling tile or partition cracking; possible damage due to missiles.
Moderate (1.25 - 7.5%) } M	Many partitions cracked or ceiling tiles fallen down; a few structural members appear to be stressed beyond yield level.
Heavy (7.5 - 65%) } H	Significant number of structural members with structural damage, or damage to a structural system, roof having major damage.
Very Severe (65 - 100%) } S	Major damage; structure standing but will probably be taken down.
Collapse (100%) } C	Structure does not remain standing.

Fig. 1. Average proportionate distribution of building damage by selected hazard type in the United States and by level of severity of hazard effects

Table 3. Significant natural disasters, 1961-1970 (at least 100 million in damage and 10 lives lost) [Office of Emergency Preparedness, 1972]

MONTH/ YEAR	DISASTER	ESTIMATED DAMAGE ($M)	LIVES LOST
9/61	HURRICANE CARLA	408.3	46
3/64	ALASKA EARTHQUAKE	500.0	131
10/64	HURRICANE HILDA	100.0	38
12/64	CALIFORNIA, OREGON FLOODS	415.8	40
3-5/65	FLOODS - BASINS OF THE UPPER MISSISSIPPI, THE MISSOURI, AND THE RED RIVER OF THE NORTH	181.3	16
4/65	PALM SUNDAY TORNADO-MIDWEST	300.0	271
6/65	SOUTH PLATTE BASIN FLOODS	415.1	16
9/65	HURRICANE BETSY	1,440.0	75
9/67	HURRICANE BEULAH	200.0	15
1-2/69	CALIFORNIA FLOODS	399.2	60
8/69	HURRICANE CAMILLE	1,420.8	256
5/70	LUBBOCK, TEXAS TORNADO	135.0	26
8/70	HURRICANE CELIA	453.8	11
	TEN-YEAR TOTAL	6,369.3	1,001

Table 4. Exposure of building values and persons to natural hazards in the U.S. by type of hazard for 1970

HAZARD	TYPE OF AREA	HUMAN POPULATION EXPOSED		VALUE OF BUILDING EXPOSED TO HAZARD	
		N	PERCENT OF U. S. TOTAL	N ($ x 10^6)	PERCENT OF U. S. TOTAL
1. SEVERE WIND	COUNTIES	203,260,531	100.0	2,064,507.5	100.0
2. TORNADO	COUNTIES	181,198,749	89.1	1,788,989	89.1
3. EXPANSIVE SOILS	FRACTIONS OF COUNTIES			NOT AVAILABLE	
HIGH ZONES		17,730,021*	8.7**		
MEDIUM ZONES		23,710,910	11.7		
LOW ZONES		98,320,710	48.4		
4. EARTHQUAKE ANY INTENSITY	COUNTIES	143,169,495	70.4	1,494,293	72.4
5. EARTHQUAKE	FRACTIONS OF STATES				
ZONE #1		120,600,000	59.3	1,205,138	58.4
ZONE #2		33,340,000	16.4	340,210	16.5
ZONE #3		38,020,000	18.7	427,860	20.7
6. HURRICANE WIND	COUNTIES	62,741,264	30.9	682,476	30.9
7. LANDSLIDE	COUNTIES			NOT AVAILABLE	
HIGH		44,068,071	21.7		
MODERATE		39,426,341	19.4		
8. STORM SURGE	COUNTIES	38,387,247	18.9	438,733	21.3
9. STORM SURGE	COUNTIES SEGMENTS 0-20 FT. MEAN SEA LEVEL (ADJUSTED)	9,940,101	4.89	104,670.25	5.1
10. RIVERINE FLOOD	5,539 FLOOD PRONE CITIES:				
	ALL ZONES	24,112,000	11.86	731,977.34	35.5
	100 YR. FLOOD PL.	12,056,000	5.93	364,650.79	17.7
	50 YR. FLOOD PL.	8,776,416	4.32	265,471.01	12.9
	25 YR. FLOOD PL.	5,830,968	2.87	175,478.42	8.5
11. TSUNAMI	100 YR. INUNDATION AREA	109,400	0.054		
	500 YR. INUNDATION AREA	237,500	0.117		
	COUNTIES CONTAINING ABOVE AREAS	18,200,851	8.95	217,327.63	10.5

* Number of People in Single Family Residences

** % of Total Population

people are exposed to the tornado hazard with 31% exposed to hurricane winds. Tornado strikes, because of this high exposure and the severity of damage associated with each strike, are particularly hazardous. Table 5 summarizes the tornado strike frequency distribution. Column one of the table indicated eight different levels of tornado strike frequence for a 10° x 10° grid in the U.S. The total data time period is 1955-1967. These levels range from 75 or more strikes for the 13 years to 0 strikes. The 900 grids of 10° x 10° dimensions were each reviewed and compared with historical tornado statistics. Therefore, column two summarized the number of grids in the U.S. which experience the tornado strike frequencies indicated in column one. For example, 11 grids have 75 or more strikes, and 31 grids have 50 to 74 strikes. Column 3 and 4 present this same data in an alternate form which indicates for example that approximately one half of the grids (or approximately one half U.S. area) experiences 5 or more tornado strikes. Also approximately 40% of the grids experiences approximately one strike per year. The remaining column in the table show how the number of strikes are distributed according to grid tornado frequency. For example, from columns one and two of the table

Table 5. Tornado strike distribution within the United States (1955-1967)

NUMBER OF STRIKES PER GRID (1°X1°)	NUMBER OF GRIDS IN GROUP			NUMBER OF STRIKES		
	N	CUM	%, CUM	N	CUM	%, CUM
75+	11	11	1.22	917	917	7.66
50 - 74	31	42	4.66	1768	2685	22.43
25 - 49	134	176	19.56	4763	7448	62.23
10 - 24	192	368	40.89	3360	10808	90.30
5 - 9	94	462	51.33	658	11466	95.80
2 - 4	143	605	67.22	429	11895	99.38
1	74	679	75.44	74	11969	100.00
0	221	900	100.00	0	--	--
TOTAL	900	900	0	11969	11969	100

Source: Hart [1976] Figure 1-11.

it followed that 11 grids experienced 75 or more tornado strikes. These eleven grids accounted for 916 tornadoes or 7.66% of all tornadoes. Therefore, 90.3% of all tornadoes (3360 in number) occur in 40.89% of the grids (368 grids net of 900).

This grid containing the tornado strike distribution was then overlaid on a county level building and population exposure data base. When integrated with the damage algorithms, county, state and national level expected tornado wind damage rates were developed. The same procedure was applied to all the wind hazards considered in the study.

Table 6 provides a state by state summary of damage due to wind hazards and compares the damage with other earth and water related hazards. Figure 2 indicates the counties with high damage rates for the combined natural hazards based upon 1970 modeled losses. The damage rate is defined as the expected annual damage of an area divided by the total building wealth in that area. The damage rate for the U.S. for all natural hazard damage is 0.279%. High damage rate is defined herein to be a damage rate of 0.279% or greater. Table 6 can be reviewed to verify that twenty states recorded 1970 expected annual building damage rates for all hazards which were equal to or greater than the national rate. These twenty states, in 1970, represented a population of 82 million and a combined value of buildings of 785 billion dollars. Figure 3 shows the U.S. counties with a tornado induced damage rate that were two or more times the national damage rate.

Table 7 and Figure 4 summarize the natural hazard damage on a hazard comparative basis. Two of the five most severe hazards are wind related, with tornadoes accounting for approximately 15% of all the damage.

DECISION MAKING

What can be done to reduce this level of damage? Is the damage level acceptable? These two questions, and many others, can often be answered if the damage to

Table 6. Damage to buildings (in millions of dollars) by state for the year 1970

1970 BASELINE % DAMAGE	STATE	TOTAL STRUCTURE VALUE ($BILL.)	POPULATION (MILLIONS)	EARTH HAZARDS			WATER HAZARDS			WIND HAZARDS			ALL HAZARDS
				EQ	ES**	LS	FL*	SS	TS	TO	HU	SW	
.309	AL	26.7	3.44	.0	14.45	4.6	28.9	4.3	0	18.4	11.9	0	82.6
.164	AK	3.9	0.27	3.6	0	0	2.4	0	0.4	.0	0	0	6.4
.130	AZ	16.9	1.77	0.7	2.3	2.29	14.7	0	0	1.9	0	0	21.9
.299	AR	13.6	1.92	3.3	7.01	2.98	20.0	0	0	7.3	0	0	40.6
.375	CA	226.0	19.96	439.6	182.59	36.83	178.5	0	3.5	13.8	0	0.8	855.6
.319	CO	22.7	2.21	20.2	17.77	6.27	18.4	0	0	9.4	0	0.4	72.3
.284	CT	35.0	3.03	0.8	7.08	10.30	21.8	9.7	0	5.7	43.6	0.4	99.4
.184	DE	5.7	0.55	0.1	.64	1.59	4.5	0.5	0	1.4	1.8	0	10.5
.087	DC	12.6	0.76	0.1	.79	.82	6.3	0	0	1.1	1.9	0	11.0
.825	FL	61.1	6.79	1.0	14.60	3.91	56.9	195.0	0	41.8	190.4	0.2	503.8
.223	GA	41.5	4.59	0.5	17.17	5.12	38.5	4.4	0	18.6	8.4	0	92.7
.114	HI	10.0	0.77	0.3	0	0	6.9	0	4.1	.0	0	0.1	11.4
.179	ID	6.3	0.71	1.5	2.42	1.01	5.9	0	0	0.4	0	0.1	11.3
.274	IL	126.8	11.12	1.0	28.69	33.16	132.2	0	0	152.2	0	0.8	348.0
.252	IN	48.8	5.20	0.1	6.31	6.80	61.8	0	0	47.7	0	0.4	123.1
.291	IA	25.8	2.83	.0	14.69	5.21	36.4	0	0	18.6	0	0.3	75.2
.423	KS	21.5	2.25	0.2	25.88	3.41	28.9	0	0	32.3	0	0.2	90.9
.192	KY	25.6	3.22	1.3	8.58	5.96	25.4	0	0	8.0	0	0	49.2
.721	LA	30.3	3.64	1.9	26.29	5.95	37.8	84.5	0	13.3	48.6	0.1	218.4
.233	ME	8.3	0.99	.0	2.30	2.12	7.1	0.5	0	1.0	6.3	0	19.3
.155	MD	49.2	3.92	0.1	6.66	11.97	32.2	4.0	0	8.6	12.8	0.1	76.4
.272	MA	63.4	5.69	1.7	4.83	13.21	40.8	12.5	0	35.3	63.6	0.8	172.7
.230	MI	90.8	8.88	0.9	42.31	19.47	105.6	0	0	40.1	0	0.1	208.5
.215	MN	37.7	3.80	.0	5.87	5.52	48.8	0	0	20.3	0	0.4	80.9
.477	MS	14.6	2.22	0.4	13.19	2.76	23.1	13.2	0	8.4	8.7	0	69.7
.396	MO	45.1	4.68	15.3	37.22	10.75	60.1	0	0	55.2	0	0.2	178.8
.202	MT	6.4	0.69	1.4	4.02	1.18	5.8	0	0	0.4	0	0.1	12.9
.391	NE	14.3	1.49	0.2	21.33	2.64	19.2	0	0	12.3	0	0.2	55.9
.153	NV	5.8	0.49	2.7	1.12	.64	4.1	0	0	0.2	0	0.1	8.9
.220	NH	7.4	0.74	0.2	.70	1.47	5.3	0	0	2.8	5.8	0	16.3
.159	NJ	83.7	7.17	3.4	7.47	10.70	49.7	11.5	0	26.2	23.1	0.8	132.9
.182	NM	9.1	1.02	1.2	3.81	.82	8.5	0	0	2.2	0	0.1	16.6
.166	NY	229.6	18.24	20.2	14.76	29.36	126.5	50.0	0	10.7	129.4	1.0	381.9
.310	NC	40.0	5.08	0.1	18.92	9.76	41.8	7.7	0	12.7	32.	0.2	124.1
.277	ND	4.8	0.62	.0	1.88	1.13	8.0	0	0	2.2	0	0.1	13.3
.194	OH	106.3	10.66	1.0	14.57	21.87	126.8	0	0	41.9	0	0.3	206.4
.338	OK	23.1	2.56	0.5	4.86	2.16	26.6	0	0	43.5	0	0.4	78.0
.151	OR	19.9	2.09	1.7	4.97	4.19	18.7	0	0.1	0.4	0	0	30.1
.136	PA	116.1	11.80	0.4	14.84	24.91	81.8	0	0	20.1	15.7	0.4	158.2
.284	RI	9.3	0.95	0.1	.74	.60	6.8	7.1	0	0.5	10.4	0.2	26.4
.311	SC	19.6	2.59	1.9	8.66	3.21	21.7	7.0	0	7.9	10.5	0	60.9
.311	SD	5.3	0.67	0.1	3.49	1.31	8.6	0	0	2.9	0	0.1	16.5
.210	TN	30.8	3.92	15.1	6.67	2.00	30.9	0	0	10.0	0	0	64.7
.398	TX	103.7	11.20	0.8	143.69	11.36	116.5	15.7	0	92.6	30.6	1.0	412.3
.286	UT	10.6	1.06	12.2	5.53	3.13	8.8	0	0	0.5	0	0.1	30.3
.137	VT	3.8	0.44	.0	.44	.54	3.2	0	0	.8	0.2	0	5.2
.220	VA	48.8	4.65	0.4	8.87	8.94	38.2	13.5	0	8.5	28.8	0.3	107.5
.404	WA	35.6	3.41	96.9	4.71	10.41	30.5	0	0.7	.4	0	0.2	143.8
.166	WV	14.2	1.74	0.1	3.87	4.88	13.7	0	0	1.0	0	0	23.6
.213	WI	41.2	4.42	.0	6.28	10.43	52.6	0	0	18.2	0	0.2	87.7
.200	WY	3.2	0.33	.0	2.27	.71	2.8	0	0	0.5	0	0.1	6.4
.279	U.S.	2064.5	203.24	655.2	798.1	370.3	1901.0	441.1	8.8	880.2	685.4	11.3	5751.4

*Flood is estimated with a different model

**Only residential expansive soil losses

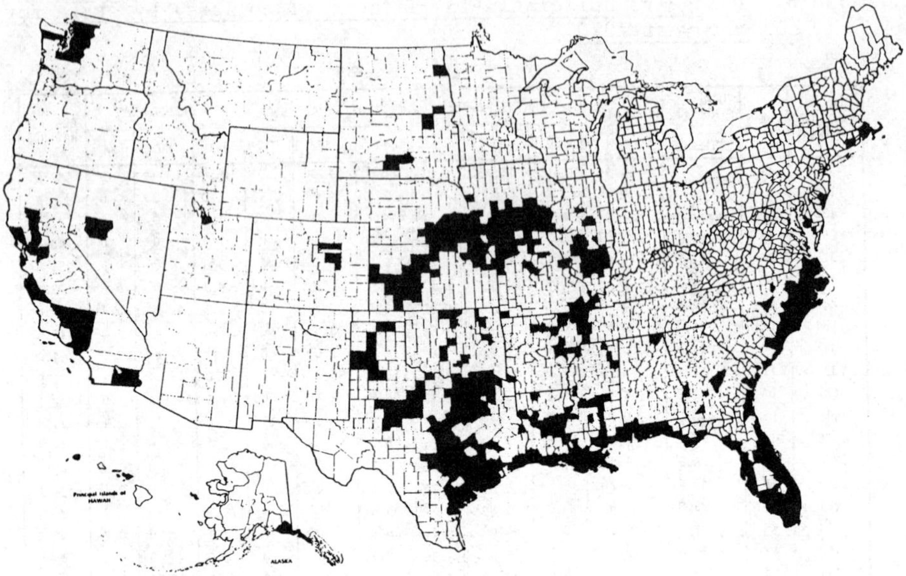

Fig. 2. Map of high damage rate counties for eight natural hazards, 1970

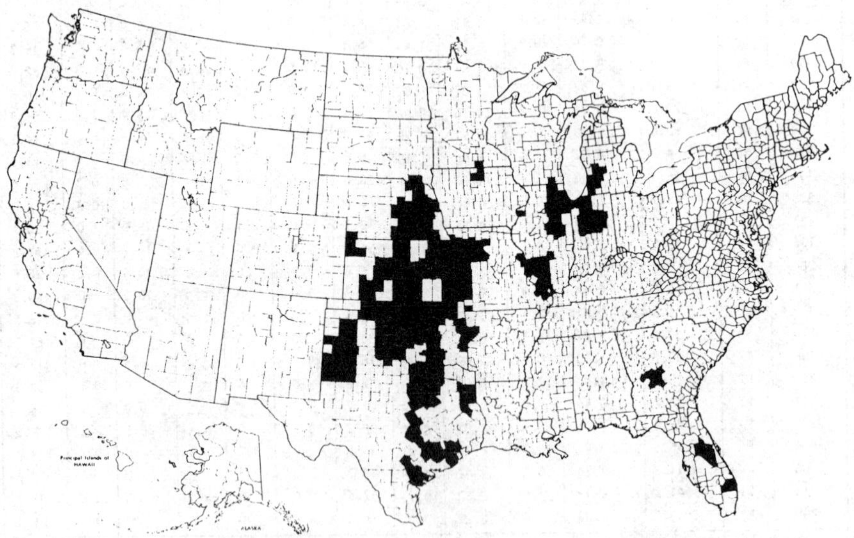

Fig. 3. Counties with a damage rate two or more time the NBDR for tornado, 1970

	HOUSING STRUCTURES	FARMING AND MINING STRUCTURES	CONSTRUCTION STRUCTURES	MANUFACTURING STRUCTURES	TRANSPORTATION & UTILITIES STRUCTURES	RETAIL AND WHOLESALE TRACE STRUCTURES	FINANCE STRUCTURES	SERVICES STRUCTURES	FEDERAL GOVERNMENT STRUCTURES	STATE AND LOCAL STRUCTURES	TOTAL VALUE OF STRUCTURES
Total Estimated Value Exposed in 10^6 1970$	1,106,073	21,026	27,058.1	109,897.4	235,109.4	85,075.9	104,911.4	93,039.0	128,561.01	153,753.8	2,064,507.5
Percent of Total	53.7	1.0	1.3	5.3	11.4	4.1	5.1	4.5	6.2	7.4	100.0
SEVERE WIND											
Exposure Total = 100	100.	100.	100.	100.	100.	100.	100.	100.	100.	100.	100.
Est. Loss 10^6 1970$	7.55	.07	.07	.26	.64	.60	.69	.64	.38	.42	11.32
Percent of Total	66.6	0.6	0.6	2.3	5.17	5.3	6.1	5.7	3.4	3.7	100.0
HURRICANE (Wind)											
Exposure Total = 100	32.6	19.2	33.3	28.9	33.5	32.6	39.5	36.5	36.2	31.2	33.1
Est. Loss 10^6 1970$	451.0	3.4	5.3	16.0	45.9	33.6	45.7	37.9	20.9	25.6	685.3
Percent of Total	65.9	0.5	0.8	2.3	6.7	4.9	6.7	5.5	3.0	3.7	100.0
STORM SURGE											
Exposure Total = 100	20.8	13.7	21.1	17.2	24.0	21.5	28.5	24.4	18.0	19.8	21.3
Est. Loss 10^6 1970$	333.	3.1	2.9	7.2	28.9	12.9	15.4	13.2	11.6	13.6	441.8
Percent of Total	75.4	0.7	0.7	1.6	6.5	2.9	3.5	3.0	2.6	3.1	100.0
TORNADO											
Exposure Total = 100	87.2	92.2	88.5	91.2	84.5	87.1	81.1	85.0	87.9	84.9	86.7
Est. Loss 10^6 1970$	502.	7.2	8.8	36.8	75.0	53.9	59.5	53.8	37.6	44.9	879.5
Percent of Total	57.1	0.8	1.0	4.2	8.5	6.1	6.8	6.1	4.3	5.1	100.0
TSUNAMI											
Exposure Total = 100	10.3	3.9	9.7	8.5	11.5	10.6	11.6	11.6	11.1	11.5	10.5
Est. Loss 10^6 1970$	4.52	.004	.02	.04	3.3	.17	.18	.24	.14	.11	8.72
Percent of Total	51.8	0.0	.2	.5		1.9	2.1	2.8	1.6	1.3	100.0
EARTHQUAKE											
Exposure Total = 100	71.9	72.8	71.3	68.5	72.7	71.8	75.1	74.6	75.9	72.7	72.4
Est. Loss 10^6 1970$	354.2	5.7	7.2	24.8	77.9	26.0	34.1	29.1	46.5	49.7	655.2
Percent of Total	54.0	0.9	1.1	3.8	11.9	4.0	5.2	4.4	7.1	7.6	100.0

Table 7. Distribution of building values and losses by type in 1970.

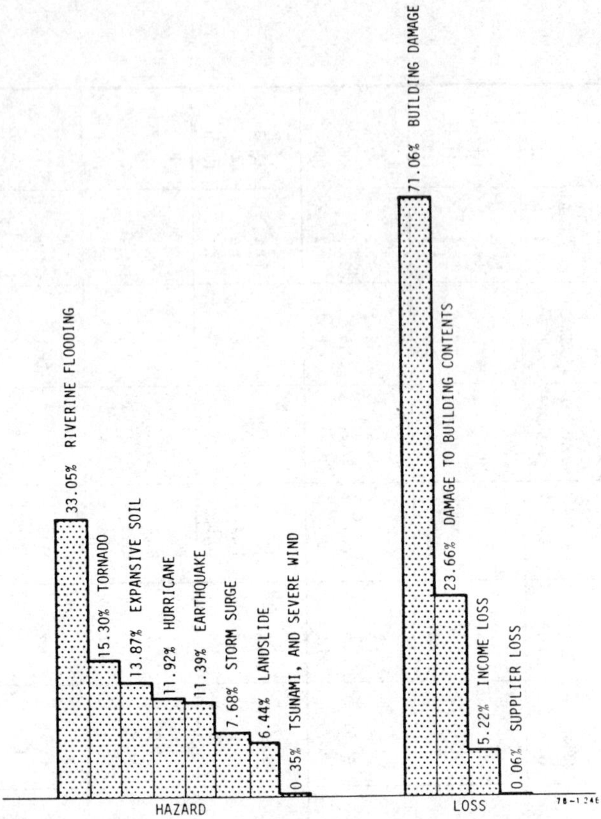

Fig. 4. Proportion of 1970 natural hazard annual expected dollar losses produced by each type of hazard and by each type of loss

structures is not viewed as a single entity, but as one part of a decision making problem. Figure 5 shows a schematic of the decision making process.

In this diagram, the point of decision is shown with various items of information, or "indications," feeding into it, and with various possible decision options emanating out from it. Since the outcome of a particular decision option is not known with certainty, the diagram shows a series of possible outcomes emanating from each option with each outcome assigned its own probability.

A given outcome will in general have many different impacts on people, property, economy, environment, and so on. These impacts are grouped into a linear list which is called in the diagram "The Impact Vector." The use of the term "vector" here emphasizes the idea that the "impact" of any decision action is a multidimensional, multiattribute quantity.

The decision maker, observing these impact vectors, will be able to select a set of impacts more desirable than another set, and so on. Thus, there is in the mind of the decision maker a notion of preference. We may express this by saying that there exists in the mind of the decision maker a "preference function" or a "utility function" which maps the multidimensional quantity, the impact vector, into a scalar quantity; i.e., a single number, called the "utility" of that vector.

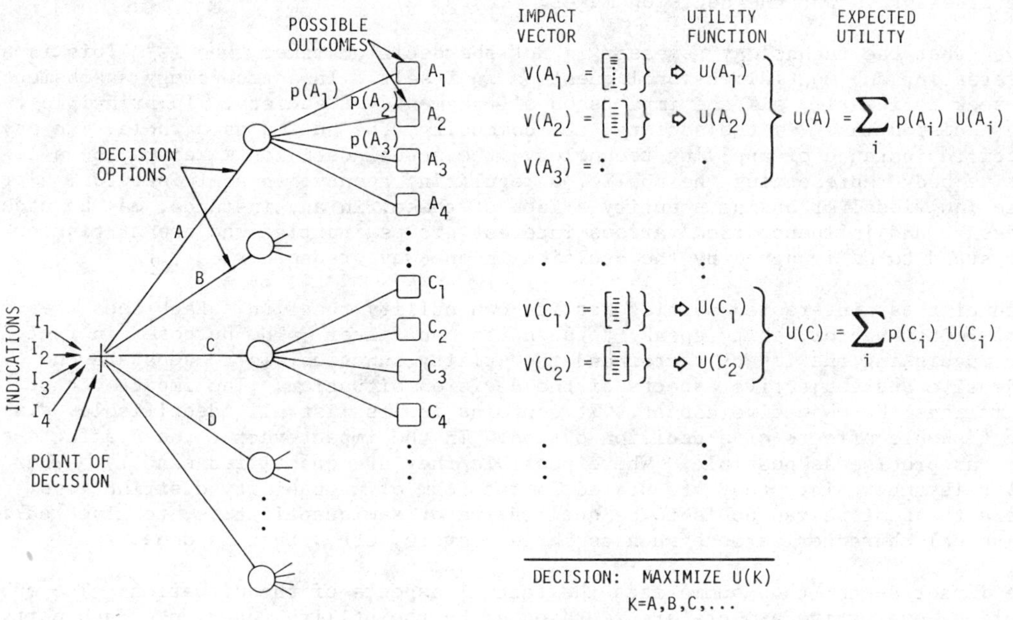

Fig. 5. Decision making schematic

The "expected" utility of a decision option then may be defined as the sum over the possible outcomes of the utility of each outcome weighted by the probability of that outcome. The optimum decision then, according to this model of the decision process, is to choose that option which has the maximum expected utility.

Thus, in terms of this model, the steps in the decision process are:

1. Identify options
2. Identify possible outcomes of each option
3. Assign probabilities to outcomes
4. Determine impact vectors for each outcome
5. Establish utility or preference function
6. Evaluate utility of impact vectors
7. Compute expected utilities
8. Identify maximum expected utility
9. Decide!

This idealized model of the decision process serves as a convenient framework in which to discuss the role of the technology assessor. Clearly the assessor may, and should, take full responsibility for items 1 through 4, the identification of decision options and the computation of outcomes, impacts and associated probabilities. Items 5 and 6, on the other hand, are entirely the responsibility of the decision maker. The assessor may assist in these steps by eliciting the preference function and by making it explicit and quantitative. The preference function is not his; however, it is the decision maker's. He may also assist in communicating, disseminating, and developing a consensus. If asked, he can recommend a decision but the assessor does not decide--the assessor "presents for" decision. The assessor is not the decision maker!

Given that the technology assessor is not the decision maker, who is? This is an interesting and multidimensional question in itself. In a technology assessment, we seek to identify all the impacts on all segments of society. In principle then, the decision maker is the society, the community, the nation as a whole. In any specific instance of applying technology, the actual decision maker may be a legislative body representing the public, a regulating agency, an administrator, a private individual or business entity. Each of these, in any instance, may be under pressure and influence from various interest groups, parties who are at risk or who stand to be impacted by the decision in one way or another.

Each of these interested parties has its own utility function. Each sees the same impact vector, so to speak, through its own filter. The purpose, in fact, of distinguishing the impact vector and the utility function, is to separate the objective and subjective aspects of the decision situation. The impact vector summarizes the objective aspect. It contains in its list all identifiable, distinguishable effects of a decision option. In the impact vector these effects are made as precise as possible. Where possible they are quantified exactly. Where there is uncertainty they are stated in the form of probability distributions. Where the best we can do is to be qualitative or semiquantitative, to give rankings or verbal characterizations such as large, severe, etc., that is done.

The impact vector thus summarizes the factual aspects of the situation. The subjective, evaluative aspects are represented by the utility function. Each party at interest thus maps the impact vectors through his own utility function into his own order of preference.

When the decision becomes a public one for which the legislative body has the task of assessing the "averaged" utility function of its total constituency. Each member of the body at the same time has his/her own personal utility function to consider. Superimposed on all this are the forces of the media and vocal special interest groups. This is the democratic process. Hopefully the utility function of the majority holds sway. In the course of this process, however, there is often enormous waste, delay, confusion, anger, litigation, miscommunication and dissipation of time and energies that could be more productively spent.

If the technology assessor does his job he can be of great value in reducing this waste and delay. If he does a good job of computing the impact vectors and probabilities attendant to each decision option, and if he does a good job of communicating these results, then everyone concerned will benefit by better understanding the facts of the case; the costs, the benefits, the risks, what is known and what is unknown. There will remain diverse points of view and different reactions to the facts, different economic/social positions, and different attitudes toward risk taking, but at least people will be more clear on what the differences are. The way is then open for arriving at agreement.

Our experience with decision theory is that much of the argument and destructive, right/wrong type, infighting that goes on is the result of miscommunication. It is hoped that the issues are clarified, and taken out of the verbal realm and put into the realm of equations and numbers, that people will tend to see that they are really on the same team concerned with the same problems, and thus more inclined to coordinate, compromise and cooperate.

MICROZONATION

Is the discussion of microzonation out of place in this paper? The authors believe not because it was apparent from the previous discussion that the probability of a natural wind hazard plays an important part in the solution of a decision making problem. Microzonation seeks to divide a geographic region or zone into smaller regions or zones which on a geographic scale are small enough to be characterized by a single meterological condition. Therefore, for such a micro-zone one is able to assign a single probability of occurence for a specific hazard.

An example study of wind microzonation has been done for the Los Angeles area of California (5). Figure 6 shows a microzonation map for severe wind speeds. Such a map must be recognized as a best estimate based on historical data, and must often be revised when uncertainties are clarified.

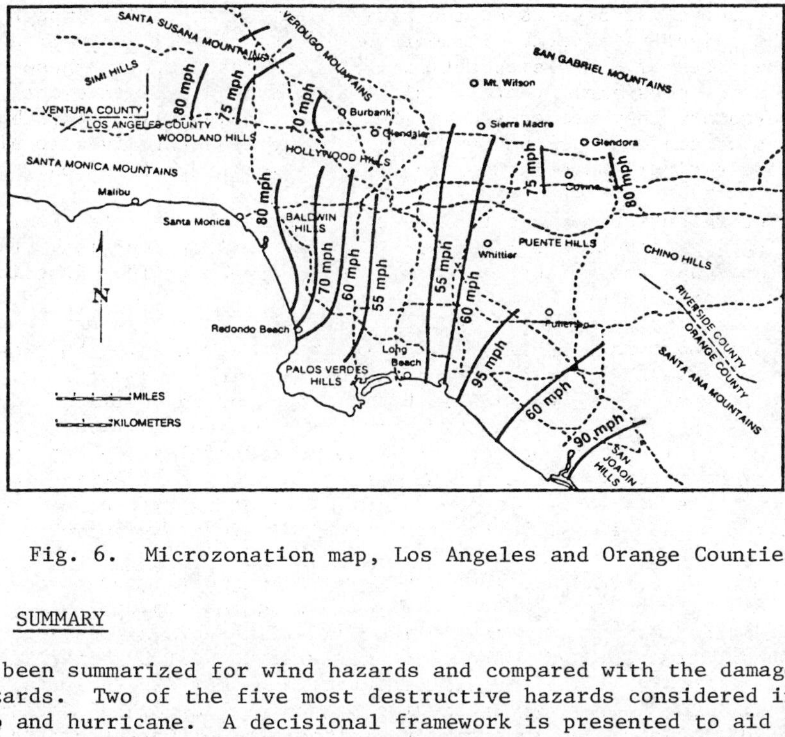

Fig. 6. Microzonation map, Los Angeles and Orange Counties

SUMMARY

Damage has been summarized for wind hazards and compared with the damage for other natural hazards. Two of the five most destructive hazards considered in the study are tornado and hurricane. A decisional framework is presented to aid in arriving at a solution to different problems related to wind engineering. Microzonation studies are valuable in quantifying accurate probabilities for the decision models.

REFERENCES

(1) W.J. Petak, A.A. Atkisson, P. Gleye, "Natural Hazards: A Public Policy Assessment," J.H. Wiggins Company, Redondo Beach, California 1978.

(2) G.C. Hart, "Natural Hazards: Hurricane, Tornado, Severe Wind Loss Models," J.H. Wiggins Company, Redondo Beach, California, 1976.

(3) L.T. Lee and J.D. Collins, "Engineering Risk Management for Structures," ASCE Journal of the Structural Division, Proceedings of the American Society of Civil Engineers, September 1977.

(4) L.T. Lee and J.D. Collins, "Engineering Risk Management of Natural and Man-Made Hazards," ASCE Journal of Structural Division, Proceedings of the American Society of Civil Engineers, September 1976.

(5) Lew, M. and G.C. Hart. "Microzonation." ASCE Journal of Structural Division, Proceedings of the American Society of Civil Engineers, 1979.

DISCUSSION ON SESSION I
Social and Economic Impact of Wind Storms

Rapporteur: K. J. Eaton
Building Research Establishment
Building Research Station, United Kingdom

Paper by K. J. Eaton--LOW-INCOME HOUSING AND HURRICANES

Question: T. G. Zambrano. Has there been any inclusion into your design recommendations for site selection or site modification to reduce the direct impact of severe storms?
Reply: K. J. Eaton. Much as I would like to do this, it is just not practical. Land available for housing in most developing countries is in very short supply, and by definition in agricultural-based communities, it is usually restricted to small areas or hillsides. I doubt if any efforts to incorporate shelter effects, other than by planting useful crop trees, would be economically viable with such low-income housing. Urban situations might be different.

Question: J. Hutson. What are the Units of the Contours in Figure 1?
Reply: K. J. Eaton. They are annual probability values for the expectation of a hurricane occurring in a given 5° square (latitude and longitude).

Question: L. H. Lamouria. Would you please expand on your statement regarding design to reduce differential pressure?
Reply: K. J. Eaton. As stated in reply to Dr. Hazelwood, I have not actually incorporated any ridge ventilators in the current housing. However, I believe that a reduction of differential pressures, by ridge ventilators or any other means, can significantly change the structural design and hence affect the economic considerations in the construction.

Question: R. A. Hazelwood. With reference to your use of ridge ventilators to reduce internal pressure, have you experienced any difficulty with consequential leakage of wind driven rain?
Reply: K. J. Eaton. Contrary to the statement in the paper (which was written in November 1978), ridge ventilators have not been incorporated in the trial houses constructed in St. Vincent. It is possible that this could be tried in future units.

Question: J. D. Holmes. I understand the author has structurally designed only a frame and roof for most of his proposed houses. The infill walls are left up to the unskilled and untrained house occupier. If these walls fail easily, as one would expect, will not the damage to life and property from direct wind loading, rain and debris be nearly as severe as if the houses were completely non-engineered?
Reply: K. J. Eaton. I believe this will not be the case. In the housing projects in St. Vincent, an integral part of the scheme is a large community development and training program. The potential house owners are trained in a central workshop, produce the material components for their own infill walls, and are supervised on site in the actual construction. In addition, certain minimum standards of construction are imposed, and simple pictorial advice is given to ensure that the walls are well tied to the structural frame.

Question: P. Siraprakasapillai. (1) Are the walls meant to be collapsible or are they able to withstand the wind? In the latter case, how were they tied to the columns? (2) Are the columns meant to act as cantilevers or what is the bracing method adopted?

Reply: K. J. Eaton. (1) The infill walls are meant to withstand the wind (and seismic) forces. The frame is designed to withstand the forces with or without the walls in place. The fixing details of the walls to the columns will depend on the materials used. In the case of reinforced hollow-concrete blocks, the horizontal reinforcement is taken around the columns; and the vertical reinforcement comes out of the foundations and is passed around the top structural beams and rafters. (2) The columns act as cantilevers from their rigid fixing in the foundation. With the rigid connecting brackets throughout, all the columns share the resistance to input loads.

Question: K. DeSilva. In the trial house (Figure 10), can the beams be avoided by transfering the roof load from the roof sheets to purlins, and from purlins to rafters and then straight to the columns? Lateral stability could be provided (a) by way of tension bracing or (b) walls. This idea is to reduce the cost.

Reply: K. J. Eaton. As the houses are designed to withstand both the hurricane wind speeds likely to be expected once in 50 years (50 m/s), and the corresponding seismic forces (0.2 g bedrock acceleration), any reduction in the members incorporated in the frame would lead to failure.
Any alternative resistance method would probably be too complicated for self-help builders, and infill walls for example might not always be in place in the future.

Paper by R. H. Leicester, C. T. J. Bubb, C. Dorman, F. D. Beresford--AN ASSESSMENT OF POTENTIAL CYCLONE DAMAGE TO DWELLINGS IN AUSTRALIA

Question: T. G. Zambrano. As tropical cyclones approach land the warm ocean waters which are their driving mechanism are cut off and the surface friction also changes. Given the great concern among designers of offshore structures for estimations of design storm winds, how can these results be interpreted? Does the Monte Carlo method as used apply out to say 100 k, from the Australia coast or must there be an adjustment for the different surface characteristics? Also results are defined in terms of peak gusts occurring at a 10 m height. How would this be interpolated to a building or structure of 50 m or 100 m height?

Reply: C. T. J. Bubb. The results reported in the paper are for typical Australian dwellings (houses) on land within say 50 km inland of the coast. They were adjusted downward in certain cases at the 20-year Return Period as described in the paper--Figure 3. These adjusted results should not be used for maritime structures or for design purposes in general.
The unadjusted Monte Carlo results are in general suitable for design. References 7 and 8 in the paper give the results for use on land but an upward adjustment would be needed for 100 km seaward (see Reference 5). Australian Wind Code SAA AS1170 Part 2 gives wind gust speed multipliers for any terrain category at any height.

Question: G. A. Berz. What is the reason for the steeper increase of the damage index in Figure 5 with wind gust speed for contents than for buildings?

Reply: C. T. J. Bubb. The contents damage is assumed to begin as soon as the minor damage level for the building structure is reached and then continue directly to a point of total damage at a wind speed less than that causing total destruction. Contents will be damaged substantially by water (or sand and water) penetration even before minor structural damage depending on type of structure and type of contents. Based on Australian experience for houses it is reasonable to take a higher damage index for contents for a given wind speed.

Question: C. F. Shieh. The damage indices strongly depend on the locations surveyed, Figure 6 in text. Is this due to the local terrain effect or something else?

Reply: C. T. J. Bubb. The damage indices computed and shown in Figure 6 depend both on the terrain category and the strength of the houses. Table 6 shows the percentage of each house structure type which is in each terrain category for each location. For example Brisbane houses are weak (Type A) but are well sheltered (Category 3). The new Darwin houses are more exposed and much stronger (Types B,C,D,E).

Question: J. E. Minor. In view of the zero percentage of "strong" houses reported for Brisbane (Figure 8), would you care to comment on the catastrophe potential for the city of Brisbane?

Reply: C. T. J. Bubb. Table 6 shows that all the Brisbane houses are "weak" (Type A and B) but also that they are all in Category 3 and therefore sheltered. This leads to a calculated annual damage index for Brisbane of 0.13 which is almost the lowest shown. However, the number of houses is by far the greatest so the catastrophe potential is high. The data on which the tropical cyclone risk for Brisbane is based is very sparse and poor in quality and the validity of the simulation method in this case is doubtful. There is no record of a hurricane strike or near miss in the 150 years of history of Brisbane. The best advice for Brisbane would be to enforce the quite high design wind speed of 50 m/sec which is in the Australian Code already and this would provide good protection against a large range of possible events.

Paper by J. R. Snyder, M. D. Skold, W. O. Willis--THE ECONOMIC BENEFITS AND COSTS OF MANAGING WINDBLOWN SNOW

Question: R. E. Whitbread. The economics of snow control using wheat grass seem to be dependent on the capital cost of the wheat grass planting. There is the additional inconvenience of having to avoid the wheat grass strips during harvesting. Could not a single deep furrow or series of furrows be ploughed instead so that the furrows would collect the snow and the area behind the ridges trap further deposits? This method would have the advantage that the furrows could readily be ploughed-in during the spring to give an unimpeded area for the working of farm machinery.

Reply: M. D. Skold. Grass barriers are among a number of techniques used as catchment barriers. Experimenters have used snow fences, crop residues cut to irregular heights and have applied each of these systems within the context of land forming (terracing). Grass barriers have proven to be the most effective. They are relatively easy to establish and can be planted with the same equipment which farmers use to plant their small grains. While the barriers go to seed and this seed sometimes germinates and grows in the space between the barriers intended for the cultivated crop, this is not a serious problem. Sometimes farmers have harvested the seed from the barriers for sale which represents another potential source of revenue. Furrows would likely work for snow catchment too; I expect that such furrows would occupy considerably more space than is required by grass barriers, however.

SESSION II

Wind - Characteristics and Description

CHAIRMAN

J. H. Golden
Program Development Scientist
Environmental Research Laboratory
National Oceanic & Atmospheric Administration
Boulder, Colorado, USA

CO-CHAIRMAN

R. F. Abbey, Jr.
Director, Meteorology/Hydrology Research Program
US Nuclear Regulatory Commission
Washington, D.C., USA

RAPPORTEUR

E. Simiu
Research Engineer
National Bureau of Standards
Washington, D.C., USA

RECENT DEVELOPMENTS AND TRENDS: WIND - CHARACTERISTICS AND DESCRIPTION

J. H. GOLDEN* and R. F. ABBEY, Jr**

*Environmental Research Laboratory, NOAA, USA
**U.S. Nuclear Regulatory Commission, USA

ABSTRACT

One should note that the papers in this session cover a wide spectrum of wind characteristics, ranging from tornadoes and tatsumakis, to boundary layer modeling and measurements in different regimes, and finally to terrain/roughness effects on atmospheric turbulence. In the brief space allotted to this review, therefore, the following important topics will be discussed: hurricanes, downslope mountain windstorms, tornadoes, and thunderstorm downbursts.

HURRICANES

Since the 4th International Wind Engineering Conference four years ago, the two new NOAA WP-3D instrumented aircraft of the Research Facilities Center in Miami, Florida have come on-line and demonstrated their great research potential in hurricanes. For many years meteorologists and engineers have needed high density in situ measurements of atmospheric and oceanographic parameters through hurricanes and other wind systems. To satisfy these requirements, new sensors and high-speed on-board recording techniques have been developed and have given impetus to the consideration of aircraft as research tools. The structural, flight characteristics, measurement capabilities and missions flown by the two NOAA WP-3D and C-130B Hercules turboprop aircraft are provided elsewhere.* Beginning with the intense Hurricane Anita in 1977, the NOAA P-3's have flown research and operational reconnaissance missions through tropical storms and hurricanes in the Atlantic, Caribbean, Gulf of Mexico, and off the West Coast of Mexico and Baja, California; during February-March, 1979, missions were flown off the coast of Australia. Sheets (1) has described the on-board measurement and data display systems of the NOAA P-3 aircraft. Figures 1 and 2 show the real-time displays of aircraft radar and measured wind and pressure profiles through Hurricane Anita in the Gulf of Mexico. Note the close correlation with the rapid rise in wind speeds and fall in sea-level pressure as the aircraft penetrated the intense ring of convection, the eyewall, which defines the hurricane (about 40 km diameter, for this case). Figure 3 shows a radial profile of 5-sec averaged aircraft winds through Cyclone "Kerry" off the northeast Australian coast on February 22, 1979. The gusty nature of the wind profiles in tropical cyclones

*Dr. C. B. Emmanuel, Director, Research Facilities Center, NOAA, P.O. Box 480197, Miami, Florida 33148

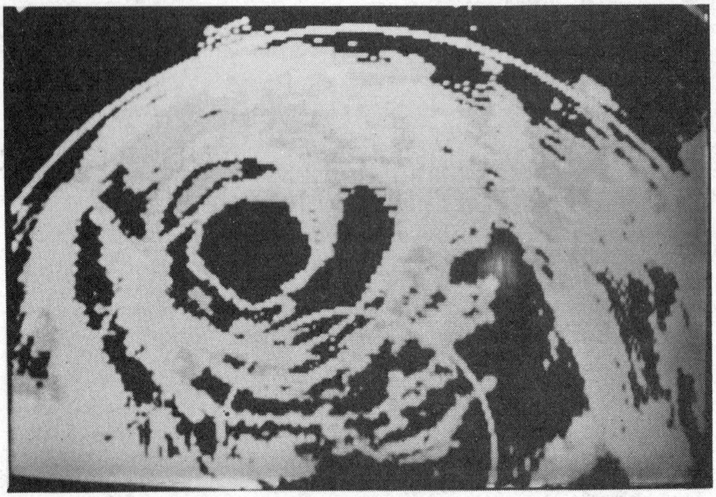

Fig. 1. On-board 5 cm PPI radar presentation from NOAA's WP-3D research aircraft showing spiral rainbands in Hurricane "Anita" on September 2, 1977. Note echoes are intensity-contoured and circular "eyewall" region in left center.

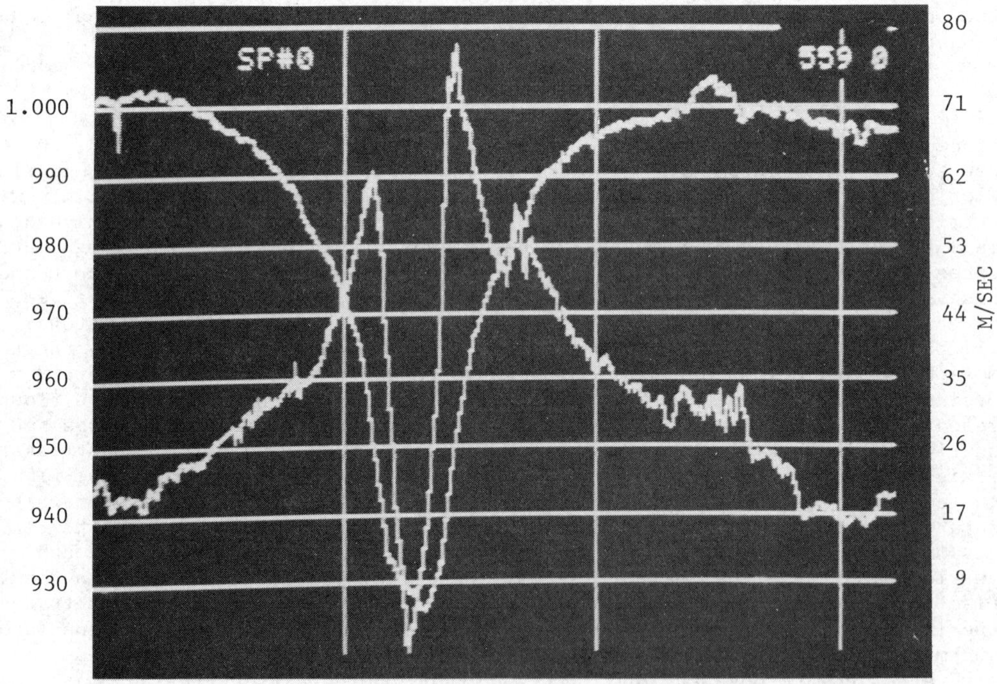

Fig. 2. Real-time plots of pressure and winds from NOAA's WP-3D research aircraft as it penetrated Hurricane "Anita" on September 2, 1977. Pressure scale in millibars on left, minimum pressure in hurricane eye was 926 mb and peak 5-sec averaged winds in right side of eyewall were 74 m/sec.

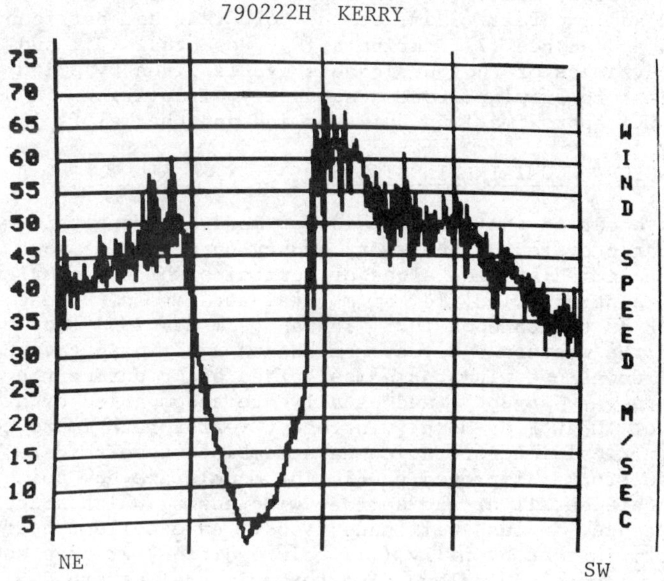

Fig. 3. Same as Fig. 2, except radial profile of NOAA's WP-3D aircraft-measured 5-sec winds through Australian Tropical Cyclone "Kerry" on February 22, 1979 when storm was off NE coast. Peak velocities in eyewall were 70 m/sec and the record for maximum wind speeds observed in Australian tropical cyclones.

penetrated by the NOAA aircraft in both hemispheres is quite evident by comparing Figures 2 and 3.

Recent studies by Neumann and Hill (2) have given detailed and computerized climatology of tropical cyclones in the Atlantic, Caribbean and Gulf of Mexico for the years 1886-1975. Gray (3) has recently completed a comprehensive global assessment of tropical cyclone frequency; moreover, these storms account for about 2% of the annual global rainfall and kinetic energy generation. They export kinetic energy in their upper tropospheric outflow levels; this is accomplished largely by horizontal eddy processes. A very useful reference on the deadliest, costliest and most intense U.S. hurricanes and other related statistics is provided by Hebert and Taylor (4). Sheets (6) presents an up-to-date summary of the present status and future plans of Project STORMFURY, NOAA's hurricane modification program. These hurricane experiments, which date back to the early 1960's, are designed to cause a reduction in the maximum wind speeds (and therefore surface damage) through an alteration in the location of energy released near the hurricane's center. Schwerdt et al. (5) provide criteria for determining wind fields along the gulf and east coasts of the United States for the most severe hurricane reasonably characteristic of a region; they also provide similar criteria for the hurricane that will produce the highest sustained wind that can probably occur at a specified coastal location.

We anticipate that the new aircraft and satellite measurement capabilities which continue to evolve at an ever-accelerating pace will greatly improve our knowledge of wind distributions in hurricanes, on smaller time and space scales with greater resolution. NOAA has proposed a new budget initiative called Hurricane STRIKE. It aims at using these new continuous monitoring capabilities in tandem with sophisticated 3-D numerical hurricane models at the National Hurricane and

Experimental Meteorology Laboratory to improve short-range forecasts of hurricane intensity, track and landfall. Some of the more relevant hurricane models are those by Diercks and Anthes (7), Kurihara (8), Rosenthal (9), and Jones (10). A new look at the dynamics of the hurricane's eye is given by Smith (11). Willoughby (12) has re-examined forced secondary circulations in hurricanes and deduced some useful approximations for the wind and thermal fields.

DOWNSLOPE MOUNTAIN WINDSTORMS

These windstorms occur to the leeward side of mountain ranges in most parts of the world and are characterized by their dry, gusty nature. The most severe downslope winds occur on the immediate lee slope of mountains which are oriented nearly perpendicular to a strong wind flow at mountaintop levels. These winds often have local names, such as the Chinook (U.S. Indian term meaning "snow eater"--aptly descriptive!) in the western U.S., or the "Santa Ana" in southern California. In these cases, the downslope windstorms most often occur during the fall and winter months; they begin very abruptly and usually are accompanied by a rapid rise in air temperature of 10-20°F or more. In recent years, both improved observational studies and theoretical interpretation have combined to elucidate both the basic dynamics of these storms; improved prediction models are now being tested. However, the fine-scale detail of the surface wind gusts, which account for most of the structural damage, is less well understood. An excellent overview of the mountain windstorm is that by Lilly (13). In addition, Fritsch and Chappell (14) illustrate the regions of downslope windstorms in the western U.S. and the windstorm climatology for Boulder, Colorado as shown in Figure 4. A good typical example of a conventional anemometer trace from a Boulder downslope windstorm is given in Figure 5. Note especially the abrupt onset of the strong winds at 8:45 p.m. with at least three peak gusts approaching 120 mph during the next 90 minutes.

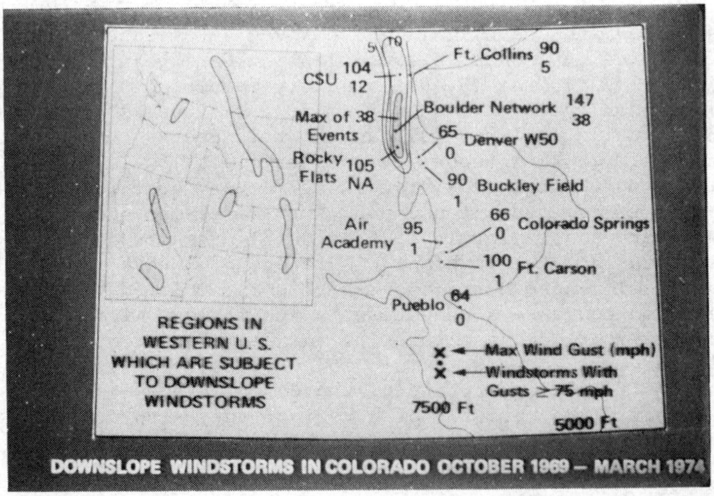

Fig. 4. Recent climatology of downslope mountain windstorms in western U.S. with enlargement of Colorado Rockies front range windstorm statistics at right after Fritsch and Chappell (14).

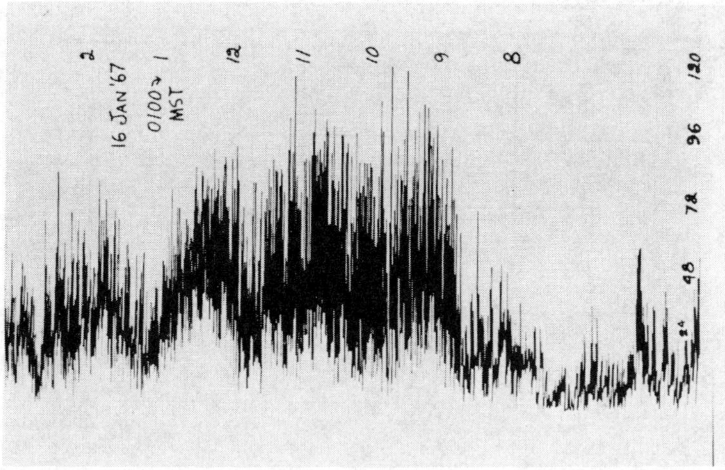

Fig. 5. Conventional cup-anemometer trace of intense Boulder, Colorado downslope windstorm on night of January 16, 1967. Time goes from right to left--note 3 peak gusts approaching 120 mph from 2130-2300 MST. From NCAR data.

In addition to the obvious threat these storms pose to buildings and trees, a second area of concern is aviation. Downslope windstorms are not only capable of destroying light aircraft on the ground, but they make takeoffs and landings extremely hazardous; they generate downdrafts which small aircraft may be unable to avoid in mountainous terrain. A cross-section of the potential temperature field along an east-west line through Boulder is given in Figure 6 , during the intense windstorm of January 11, 1972, after Lilly and Zipser (15). It is clear from this figure that the dominant feature is a long downdraft region from the peak of the mountaintops to well east of Boulder, an intense updraft just beyond that point, and rather chaotic-looking short waves downstream. More recently, Lilly (16) has found from a detailed analysis of NCAR aircraft and other data of this same case that: (a) optimal conditions for strong mountain wave generation are detectable from sounding data 12-24 hours in advance and about 1000 km upstream, and (b) the mesoscale structure is typified by a single quasi-hydrostatic wave of extreme amplitude and variable location with corresponding variations in the windstorm structure on the ground. The aircraft data in this and subsequent cases show that severe to extreme aircraft turbulence is observed in a deep boundary layer over the region of strong surface winds and also in a separate mid-tropospheric turbulence zone. Using these observations of downslope windstorm characteristics, Klemp and Lilly (17) have recently developed and tested a two-dimensional, hydrostatic simulation model for mountain waves. Finally, Bergen (18) has described the frequent occurrence of short-lived, rather intense vertical vortices, perhaps as strong as small tornadoes but without the visible funnel, which appear to be mechanically generated during some of the most intense downslope windstorms. Bergen has described these vortices using the local term "mountainadoes" and has ascribed many instances of severe and isolated damage to their occurrence.

As the population density continues to increase along the front range of the Rocky Mountains and in other mountainous areas, susceptibility of man-made structures to wind damage will increase unless sound engineering practice is applied and building codes strengthened and enforced. An experiment is about to begin in the

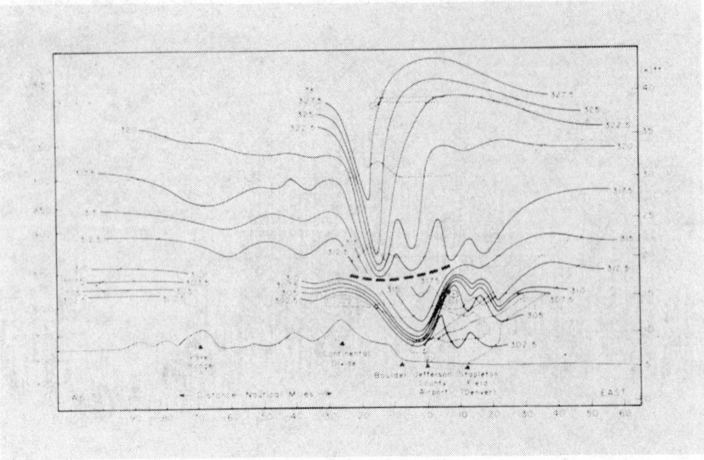

Fig. 6. Cross section of the potential temperature field in degrees K over the mountains and foothills as obtained from analysis of the Queen Air and Sabreliner data on January 11, 1972. Data above 500 mb are exclusively Sabreliner taken from 1700-2000. Data below 500 mb are primarily Queen Air taken from 1330-1500 MST. Flight tracks are indicated by the dashed lines, except by crosses in turbulent portions. Windstorm conditions on the ground extend eastward to the location where the isentropes rise sharply, a few miles east of the origin at the Jefferson County Airport. After Lilly and Zipser (15).

Boulder-Denver region to improve local weather services and forecasts known as the PROFS program (Prototype Regional Observing and Forecasting Service); the regional network of new remote and in situ wind sensors should provide data on downslope winds, on small, high resolution, time and space scales hitherto unavailable.**

TORNADOES

The most violent of atmospheric phenomena, tornadoes are probably the least understood. For wind engineering purposes, three aspects of tornadoes are pertinent: (1) magnitudes of tornado parameters (e.g., wind speeds and pressure deficits); (2) flow field characteristics for wind loads and missiles; and (3) tornado hazard probabilities and their regional variation. An excellent compilation of state-of-knowledge papers written by authorities on all aspects of tornado research appears in the proceedings of the Symposium on Tornadoes: Assessment of Knowledge and Implications for Man, held June 22-24, 1976, in Lubbock, Texas.*** Lilly (19) provides a well-written review of severe storms and storm systems.

The symposium was divided into six sessions, three each from the meteorological and the engineering viewpoints. The following topics were discussed in depth: (1) wind speeds in tornadoes; (2) tornado vortex models (theoretical and laboratory

**Further information is available from Dr. Don Beran, Chief, PROFS Program, ERL-NOAA (R45), Boulder, Colorado 80302

***Available from: Institute for Disaster Research, Texas Tech University, P.O. Box 4089, Lubbock, Texas 79409

simulations); (3) tornado prediction, detection, and assessment of risk; (4) tornado-structure interaction; (5) tornado-generated missiles and their effects; and (6) applications of tornado technology (in professional practice, in the nuclear industry, in the protection of people and essential facilities; and in an insurance operation).

Eight techniques for measuring or inferring tornado characteristics are currently being actively pursued. Six of these involved taking measurements while the tornado is actually occurring: (1) photogrammetry; (2) Doppler radar; (3) Doppler lidar; (4) acoustic recordings; (5) direct probing (of waterspouts) by aircraft, a North American T-28; and (6) firing instrumented rockets from an airborne platform (temperature and pressure measurements only). Two approaches back-calculate wind speeds necessary to produce (1) observed damage, preferably to engineered structures and missile behavior, and (2) cycloidal ground swaths.

There appears to be a consensus of opinion converging on 250-300 mph as being the maximum horizontal tornado wind speed near-ground level. Forbes (20) calculated the fastest measured velocity thus far, 275 mph at an elevation of 342 m in one of the suction vortices comprising the Parker, Indiana tornado of April 3, 1974. Vertical velocities of 160 mph have been measured in the Sayler Park, Ohio tornado of April 3, 1974. Unfortunately, this technique measures velocities only at the outer fringe of the dust cloud; probably higher wind speeds occur inside the dust sheath. Doppler radar and lidar technology may yield possible solutions to this dilemma, but the proximity and ground-clutter difficulties must be overcome. Direct penetration techniques may provide the necessary in-vortex measurements; the safety to the investigator and stress to the probe are the limiting factors. An interesting and potentially significant result from one aircraft penetration indicated that only 0.7 of the measured pressure deficit in a waterspout could be accounted for by the cyclostrophic equation and combined-Rankine vortex assumptions. In most of these investigations, significant vertical and horizontal accelerations have been documented. Asymmetric vortices appear to be the general condition rather than the exception. Correlations of wind speed measurements made by the different methods will enhance application of the historical tornado damage intensity reports to the determination of statistical distributions of tornado wind speeds.

To obtain a firmer handle on the complexities of vortex modeling and severe storm morphology, studies are being conducted which would analyze tornado dynamics with the goal of defining the bounds of the velocities and pressure gradients which may occur especially in the near-ground (lowest 100 m) boundary layer, e.g., Lewellen (21), Lilly and Rotunno (22), and Fujita (23). Since most efforts to date have assumed constant eddy viscosity, an attempt to account for turbulent diffusion of momentum by using sound-order closure of the Navier-Stokes equations in the computations of an axisymmetric vortex driven by buoyancy in the presence of the ground boundary has been proposed. Turbulence generated by the storms may provide the self-limiting mechanism that bounds the velocities attainable. Other efforts involve the computation of the three-dimensional tornado, including suction vortices. Fujita's (23) analytical model corroborates that of Lewellen (24,25) in that the maximum velocities occur just inside the height of the inflow layer. Laboratory-scale measurements of pressure loads on buildings arising from vortical flow will provide a comparison with similar pressure load measurements made in straight-line, laminar flow; Jischke (26) describes the preliminary results of this effort.

Several difficulties arise when one attempts to translate tornado incidence or touch-down frequency, into a realistic assessment of risk. The U.S. Nuclear Regulatory Commission Office of Research is supporting efforts at the University of Chicago, Texas Tech University, and the National Severe Storms Forecast Center to assess the adequacy of the tornado data record and the sensitivity of various probabilistic modeling approaches. One preliminary finding emerging from these efforts is that the area of greatest tornado hazard does not coincide with the tornado frequency maximum, but rather in the southeast where tornadoes are not as frequent as in Oklahoma and Texas, but when they do occur they tend to be much more intense than the midwestern variety. A map showing 45 years of historical occurrences of U.S. tornadoes with accurate portrayal of location, path length, and reassessed intensity assignments has been prepared by Fujita (27).

THUNDERSTORM WIND GUSTS AND DOWNBURSTS

Goff (28) has written a comprehensive treatise of cold-air outflow structure and its life-cycle stages from thunderstorms. From a multi-level 461 m tower in central Oklahoma, 20 outflows were sampled; the study was confined to the tower layer depth and focused on the characteristics of the outflow's leading edge. Goff did not consider outflow origin. Figures 7a and 7b illustrate thunderstorm circulations associated with cold-air outflow and a close-up schematic of the outflow's leading edge. The leading edge of the outflow is characterized by strong shears in horizontal and vertical winds. Updrafts occur in a 1.5 km wide band ahead of the outflow and often exceed 6 m/s. Maximum low-level horizontal wind gusts are found to be directly related to the translational speed of the front. In some other cases documented by the National Severe Storms Laboratory, wind gusts ranged upward to 75-100 mph (34-45 m/s). The windshift zone precedes the outflow thermal boundary by about 1 km and, in strong outflows, both coincide with a pressure jump. Outflow gust fronts may be observed up to 45 minutes before measurable precipitation commences and are frequently characterized by multiple surges (see Figure 7).

Two excellent summary articles which describe characteristics of severe thunderstorms, including idealized storm models, were prepared by Lilly (29) and by Newton et al. (30). The latter paper includes an account of the first scientific tornado prediction. The decade of the 1970's has been marked by great progress in our ability to elucidate the internal flow structure of severe thunderstorms with Doppler radar and other remote sensors. During the past four years, dual-Doppler studies have mapped the detailed three-dimensional winds in both tornadic storms, Brandes (31) and Ray (32), and hailstorms, Kropfli and Miller (33). The latter study also presents quantitative flux results. The first synthesis of three-dimensional winds in a thunderstorm complex from triple-Doppler observations was reported by Ray et al. (34).

During the past few years there has been a number of aircraft accidents during takeoffs and landings beneath thunderstorm cells (see e.g., Figures 8 and 9). Fujita and Caracena (34) analyzed three such incidents and found that the aircraft performance was degraded by a precipitous loss in air speed and a rapid rate of descent. In some of these cases, the winds were strong enough to cause damage about the immediate area of the airport. On the basis of these studies and the possible implication of severe downdrafts as destructive phenomena and aviation weather hazards, Fujita (36) coined the term "downburst," and defined it to be: "A strong downdraft inducing an outward burst of damaging winds on or near the ground." For aviation safety, Fujita (36) defines a downburst further as a downdraft that is strong enough to produce a surface divergence of 0.04 s^{-1} in the outflow, and provides a number of illustrative damage maps. During the late spring of 1978, Fujita (34) organized Project NIMROD (Northern Illinois Meteorological Research on Downbursts) in the western suburbs of Chicago to obtain more definitive data on downbursts and other severe thunderstorm phenomena. During 45 days of operation with three Doppler radars, a 27-station automatic surface mesonetwork and 30- to 60-minute interval rawinsoundings, Fujita and his collaborators observed: (1) downburst and tornado-inducing 'bow' radar echoes, (2) two mesocyclone echoes with caved-in tops, (3) isolated downbursts and

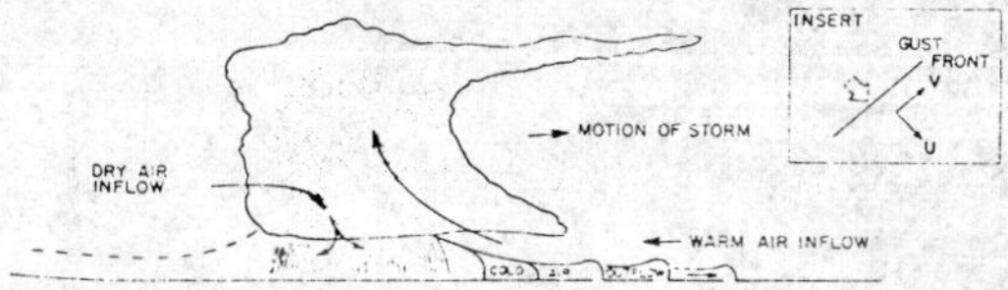

Fig. 7a. Thunderstorm circulations associated with cold-air outflow (schematic). Several outflow surges are shown. Stippled region is falling or suspended precipitation. Dashed line is upper boundary of the outflow left in the storm's wake. Insert shows horizontal wind coordinates relative to front. After Goff (28).

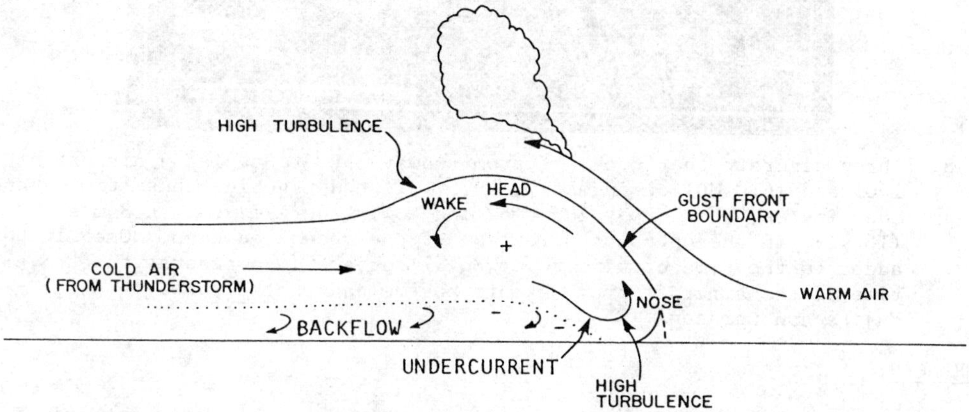

Fig. 7b. Schematic diagram of the outflow's leading edge. Cloud presence depends upon height of the lifting condensation level. Flow is relative to the gust front. Plus (+) and minus (-) refers to direct and indirect circulations, respectively. Dotted line indicates top of separation layer.

microbursts with up to 70 mph winds, and (4) a number of gust fronts. Two areas of unresolved controversy surround the downburst hypothesis: (1) the relative importance, in low-level aircraft accidents near thunderstorm outflows, of downbursts versus rapid variations of horizontal wind; (2) whether or not downbursts originate near the top of severe thunderstorms, and are in fact an avalanche of air descending from the cloud top to near the ground. McCarthy et al. (38) promulgate the idea that at least some of the aircraft accidents noted above were caused primarily by a loss of air speed due to a resonance effect between the aircraft's phugoidal frequency and horizontal wind-gust frequencies near the gust front.

ACKNOWLEDGEMENTS

The encouragement of Dr. Jack Cermak led to our synthesis of relevant and significant research accomplishments by the meteorological community that might be applied to engineering practices. Our continual contacts with our associates and colleagues made this summary possible, especially regarding future research plans and developments. The typing of the preliminary drafts was performed with the quick and able fingers of Delight Herman, NRC.

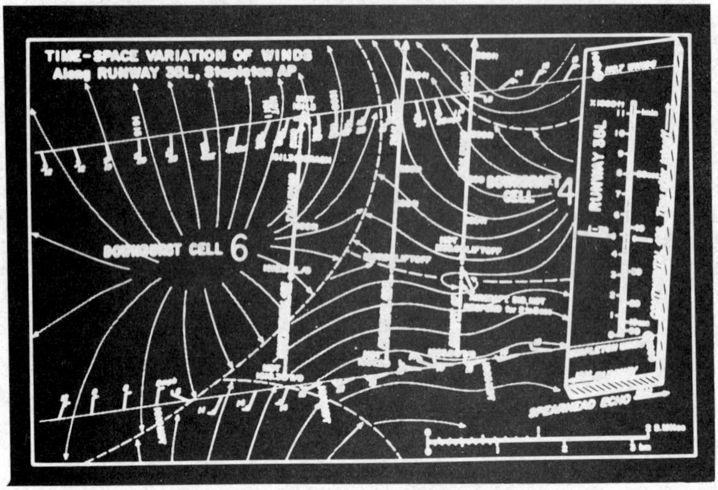

Fig. 8. Three aircraft that took off from runway 35L of Stapleton Airport between 1605 and 1610 MDT on August 7, 1975. In constructing these time-space coordinates, the runway and two wind towers were moved, at 8 m s^{-1} (16 kt), in the opposite direction of the spearhead echo. One minute was added to the time of wind tower No. 7 to take into account the travel time of the aircraft from takeoff to the end of the runway. After Fujita and Caracena (35).

REFERENCES:

(1) Sheets, R. C., Hurricane ANITA - a new era in airborne research, Mariner's Weather Log, Vol. 22(1), January 1978, pp. 1-8.

(2) Neumann, C. J. and D. A. Hill, Computerized tropical cyclone climatology, Mariner's Weather Log, Vol. 20(5), September 1976, pp. 257-262.

(3) Gray, W. M., Hurricanes: Their formation, structure and likely role in the tropical circulation, Proc., RMS/AMS Conf. on Meteorology over the Tropical Oceans, London, U.K., August 21-25, 1978, 117 p.

(4) Hebert, P. J. and G. Taylor, The deadliest, costliest, and most intense hurricanes of the century, NOAA Tech. Memo. NWS-NHC-7, August 1978, National Hurricane Center, Miami, 23 p.

(5) Schwerdt, R. W. et al., Meteorological criteria for standard project hurricane and probable maximum hurricane windfields, gulf and east coasts of the United States, NOAA Tech. Rept. NWS 23, Washington, D.C., September 1979.

(6) Sheets, R. C., Project STORMFURY: Present status - future plans, WMO Bulletin, World Met. Org., Vol. 28(1), January 1978, pp. 16-23.

(7) Diercks, J. W. and R. A. Anthes, Diagnostic studies of spiral rainbands in a nonlinear hurricane model, J. Atmos. Sci., Vol. 33, 1976, pp. 959-975.

(8) Kurihara, Y. and R. E. Tuleya, Structure of a tropical cyclone developed in a three-dimensional numerical simulation model, J. Atmos. Sci., Vol. 31, 1974, pp. 893-919.

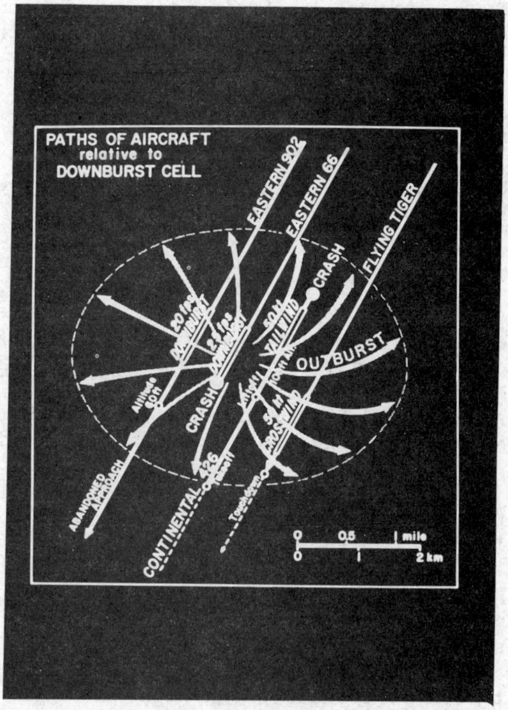

Fig. 9. Composite paths of aircraft relative to downburst cell for Flying Tiger, Eastern 66, and Eastern 92 at JFK on June 24, 1975, and Continental 426 at Stapleton on August 7, 1975. Allegheny 121 at Philadelphia on June 23, 1976 flew through the downburst center from right to left, experiencing a 25–30 m s^{-1} (50-60 kt) headwind, calm, and 3 m s^{-1} (6 kt) tailwind before it crashed on the runway. After Fujita and Caracena (35).

(9) Rosenthal, S. L., Numerical simulation of tropical cyclone development with latent heat release by the resolvable scales, J. Atmos. Sci., Vol. 35, 1978, pp. 258-271.

(10) Jones, R. W., A three-dimensional tropical cyclone model with release of latent heat by the resolvable scales, submitted to J. Atmos. Sci., March 1979.

(11) Smith, R. K., Tropical cyclone eye dynamics, paper presented at workshop on Tropical Cyclones, James Cook University, Townsville, Australia, February 1979, 15 p.

(12) Willoughby, H. E., Forced secondary circulations in hurricanes, J. Geophys. Res., Vol. 84(C6), June 20, 1979, pp. 3173-3183.

(13) Lilly, D. K., Chinooks and other strong downslope windstorms, paper presented at ASCE annual convention--Philadelphia, PA, September 27-October 1, 1976, 19 p.

(14) Fritsch, J. M. and C. F. Chappell, Severe downslope winds, preprints, 16th SAFE Symp. on Meteorological Hazards to Aviation, San Diego, CA, October 1978, 4 p.

(15) Lilly, D. K. and E. J. Zipser, The front range windstorm of January 11, 1972 - a meteorological narrative, Weatherwise, Vol. 25(2), April 1972, pp. 56-63.

(16) Lilly, D. K., A severe downslope windstorm and aircraft turbulence event induced by a mountain wave, J. Atmos. Sci., Vol. 35(1), 1978, pp. 59-77.

(17) Klemp, J. B. and D. K. Lilly, Numerical simulation of hydrostatic mountain waves, J. Atmos. Sci., Vol. 35(1), 1978, pp. 78-107.

(18) Bergen, W. R., Mountainadoes: A significant contribution to mountain windstorm damage, Weatherwise, Vol. 28, 1976, pp. 64-69.

(19) Lilly, D. K., Severe storms and storm systems: Scientific background, methods and critical questions, Pure Appl. Geophys., 113, pp. 713-734.

(20) Forbes, G. S., Three scales of motions associated with tornadoes, NUREG/CR-0363, U.S. Nuclear Regulatory Commission, Washington, D.C., September 1978.

(21) Lewellen, W. S. et al., Modeling tornado dynamics, NUREG/CR-0306, U.S. Nuclear Regulatory Commission, Washington, D.C., July 1978.

(22) Lilly, D. K. and R. Rotunno, A numerical study of atmospheric vortices, Annual Progress Report under Contract No. NRC-04-78-247 with U.S. Nuclear Regulatory Commission, Washington, D.C., 1979.

(23) Fujita, T. T., Workbook of tornadoes and high winds for engineering applications, SMRP Research Paper No. 165, University of Chicago, Illinois, September 1978.

(24) Lewellen, W. S. and M. E. Teske, Turbulent transport model of low-level winds in a tornado, Tenth Conf. on Severe Local Storms, October 18-21, 1977, Omaha, Nebraska, by American Meteorological Society, Boston, Mass., pp. 291-298.

(25) Lewellen, W. S. and Y. P. Sheng, Influence of surface conditions on tornado wind distributions, 11th Conf. on Severe Local Storms, October 2-5, 1979, Kansas City, Missouri, by American Meteorological Society, Boston, Mass.

(26) Jischke, M. C. and B. D. Light, Laboratory simulation of tornadic wind loads on a cyclindrical structure, Fifth International Conf. on Wind Engineering, July 8-14, 1979, Fort Collins, Colorado, VIII-7-1-11.

(27) Fujita, T. T., U.S. tornadoes, 1930-74, map available from the University of Chicago, Illinois, pub. in 1976.

(28) Goff, R. C., Vertical structure of thunderstorm outflows, Mon. Wea. Rev., 104, 1976, pp. 1429-1440.

(29) Lilly, D. K., The dynamical structure and evolution of thunderstorms and squall lines, Ann. Rev. Earth Planet. Sci., 7, 1979, pp. 117-161.

(30) Newton, C. W., R. C. Miller, E. R. Fosse, D. R. Booker, and P. McManamon, Severe thunderstorms: Their nature and effects on society, Interdisciplinary Science Reviews, Vol. 3(1), 1978, pp. 71-85.

(31) Brandes, E. A., Flow in severe thunderstorms observed by dual-Doppler radar, Mon. Wea. Rev., Vol. 105, 1977, pp. 113-120.

(32) Ray, P. S., Vorticity and divergence fields within tornadic storms from dual-Doppler observations, J. Appl. Meteor., Vol. 15, 1976, pp. 879-890.

(33) Kropfli, R. A. and L. J. Miller, Kinematic structure and flux quantities in a convective storm from dual-Doppler observations, J. Atmos. Sci., Vol. 33, 1976, pp. 520-529.

(34) Ray, P. S., K. K. Wagner, K. W. Johnson, J. J. Stephens, W. C. Bumgarner, and E. A. Mueller, Tripler-Doppler observations of a convective storm, J. Appl. Meteor., Vol. 17(8), 1978, pp. 1201-1212.

(35) Fujita, T. T. and F. Caracena, An analysis of three weather-related aircraft accidents, Bulletin of the American Meteorological Society, Vol. 58(11), 1977, 1164-1181.

(36) Fujita, T. T., Manual of downburst identification for project NIMROD, SMRP Res. Paper No. 156, Dept. of Geophys. Sci., University of Chicago, May 1978, 104 p.

(37) Fujita, T. T., Objectives, operation and results of project NIMROD, Preprints, 11th Conf. on severe local storms, by American Meteorological Society, Kansas City, Missouri, October 1979.

(38) McCarthy, J., E. F. Blick, and R. B. Bensch, Jet transport performance in thunderstorm wind shear conditions, NASA Contractor Rept., NASA Geo. C. Marshall Spaceflight Center, Scientific and Tech. Information Office, 1979 (submitted to J. Appl. Meteor.).

TORNADO TRACK CHARACTERISTICS AND HAZARD PROBABILITIES

J. T. SCHAEFER*, D. L. KELLY* and R. F. ABBEY, Jr**

*National Severe Storms Forecast Center, Kansas City, Missouri, USA
**U.S. Nuclear Regulatory Commission, Washington, D.C., USA

SUMMARY

A detailed examination of the statistics corresponding to every reported tornado over the United States from 1950 through 1977 is performed. Linear correlation between track length and path width does not explain more than 6% of the variance. It is further found that tornado area does not fit a log-normal distribution. The average tornado track length, path width and affected area increases with severity as classified by the F-scale. However, tornado intensity and damage area are not joint log-normally distributed.

Geographical distributions of the space-time area affected by the 1 mi^2 of tornado damage are presented for all reported tornadoes and for only the violent ones. The locations of "Tornado Alley" and "Dixie Alley" are quite apparent in the charts. A region of marked decrease in the tornado hazard occurs over the Ozark plateau. Violent tornadoes in the Texas Panhandle have the shortest mean return interval; in a typical year every 2,398 mi^2 include 1 mi^2 of devastation.

INTRODUCTION

During the 28 year period, 1950-1977, there have been 18,545 confirmed tornadoes over the coterminous United States as recorded in the National Severe Storms Forecast Center data log. The NSSFC, in conjunction with the Nuclear Regulatory Commission, has compiled data on the damage swath and occurrence times of these storms (1). The archived data include FPP indices (2) which categorize the intensity, track length and average path width of each tornado (Table 1). Also, the observed track length, average path width, and latitude/longitude of touchdown and retraction points are recorded.

Utilizing these data, it is possible to examine the interrelationships between the geometric characteristics of the tornado track and its severity. Further, it is possible to evaluate recurrence frequencies and strike probabilities for tornadoes of various intensities.

LENGTH, WIDTH AND AREA CORRELATIONS

Observations of both the actual length and width of the tornado track are available for 10,826 tornadoes. The correlation coefficient between the path length and path width is 0.25. With 95% confidence, the true correlation coefficient for

TABLE 1: Ranges of wind speed, path length and path width included within F, PL, PW and A indices (2).

Scale	F(mph)	PL(mi)	PW(yards)	A(mi^2)
0	<72	<1.0	<17	<0.001
1	73-112	1.0-3.1	18-55	0.001-0.009
2	113-157	3.2-9.9	56-175	0.010-0.099
3	158-206	10-31	176-556	0.100-0.999
4	207-260	32-99	0.34-0.9*	1.000<9.999
5	261-318	100-315	1.0-3.1*	10.00<99.99
6	319-380	316-999	3.2-9.9*	100.0<999.9

*miles

these data is within 0.03 of this value. A categorized scatter diagram of length versus width (Fig. 1) illustrates the dispersion in the data. It cannot be emphasized too strongly that only 6% of the variance can be explained by a linear correlation of tornado track width to the path length.

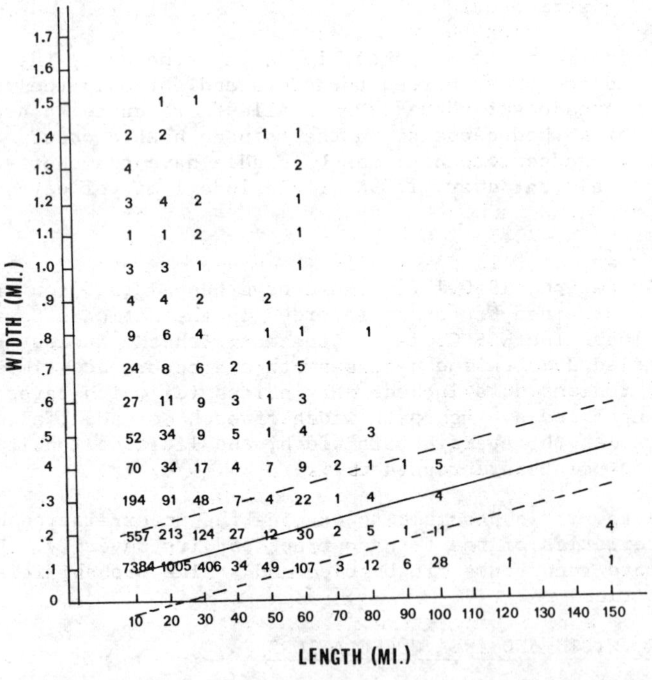

Fig. 1. Scatter diagram of track length to path width. Solid line is regression line; dashed lines depict one standard error of estimate from the regression.

Approximately 85% of the data fall into the lowest two categories of length and width in the figure. In order to compensate for this, the correlation between the natural logarithms of length and width was computed. Figure 2 shows the regression line in logarithmic space after this transformation. The correlation coefficient between these two parameters has increased to 0.53, a value much better than the 0.39 reported by Thom (3). However, it is in reasonable agreement with the 0.456 recently published by Garson, *et al.* (4).

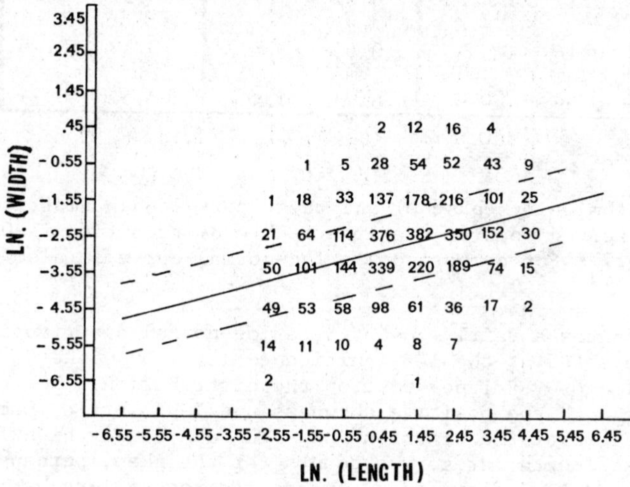

Fig. 2. Scatter diagram of natural logarithms of path length and track width.

Since the tornado damage area is the product of the path length and track width, it is possible for a stronger correlation to exist between the length and the area, or between the width and the area than for the length and width. Indeed, when these two correlations are computed, it is found that length correlates to area with a coefficient of 0.68 and width to area with a value of 0.57. When logarithms are used, these correlations become extremely strong, being 0.92 and 0.82 respectively. Note that the correlation between the actual data, not the logarithmic transformation of the data, is physically important. That is, when logarithms are considered, small numerical deviations lose significance.

These geometrical correlations can also be examined as a function of tornado intensity. Do stronger tornadoes show a closer correlation between length and width than weaker ones? To answer this, the data have been divided into three categories according to F-scale, (Table 2). Weak tornadoes have recorded F numbers of 0 and 1, strong tornadoes are F2 and 3's, and violent ones are those recorded as F4 or F5, following the convention established by Abbey and Fujita (5). As the intensity of the storm increases, there is a slight tendency for the length-width correlation to decrease. However, this trend is statistically insignificant at the 5% level. For completeness, the correlations of length to area and width to area as a function of intensity have also been computed. The variation of correlation coefficients with intensity is found to be insignificant at the 2.5% level. Thus, as a tornado becomes more intense, the relationships between its track parameters do not become better defined statistically.

DISTRIBUTION OF LENGTH, WIDTH AND AREA

For the 10,826 tornadoes for which an area could be computed, the average path length is 6.0 miles with a standard deviation of 12.2 miles. These data have been

TABLE 2: Sample Correlation Coefficients

	Length to Width		Length to Area		Width to Area		Sample Size
	Actual	*Log.*	*Actual*	*Log.*	*Actual*	*Log.*	
All Tornadoes	0.25	0.53	0.68	0.92	0.57	0.82	10,826
Weak Tornadoes	0.17	0.48	0.62	0.91	0.57	0.80	5,903
Strong	0.16	0.38	0.63	0.89	0.59	0.75	3,987
Violent	0.09	0.25	0.74	0.86	0.53	0.71	314

categorized logarithmically (base e) and their path length distribution is shown in Fig. 3. The observed mean of the logarithmic distribution is 0.46 with a standard deviation of 1.7. For comparison, a log-normal curve is also indicated on the figure.

While the eye indicates a fairly good fit, a log-normal distribution of tornado lengths must be rejected at the 1% significance level by either a χ^2 test on the frequency or a Kolmogorov-Smirnov test on the distribution. The cause of this rejection comes from both the positive skewness and leptokurtic (moment of kurtosis greater than 3) nature of the distribution. The abnormally high frequency of reports in the lowest length category accounts for the skew, perhaps due to an intrinsic bias in the data. There is a tendency to report lengths in decimal increments, with 0.1 mile generally being the shortest path record.

If this is the case, a sizable proportion of the lengths reported in the 0.05 to 0.14 category should really be in a smaller category. Shifting them would decrease the mean and reduce the skew. However, at the same time, the standard deviation would increase, aggravating the leptokurtic condition. Conversely, if long-track tornado reports can be dismissed as erroneous, the observed distribution approaches a true log-normal one.

Such an attack was followed by Thom (3) and Howe (6), who rejected all tornado tracks longer than 100 miles as "doubtful observations". This offhand rejection of data is not valid. Wilson and Morgan (7) have shown that a small but significant proportion of tornadoes have paths greater than 100 miles.

Path width (Fig. 4) also fails the statistical tests for a log-normal distribution. Again, to the eye it appears to be an adequate fit, but because of the large sample size (10,826) the slight platykurtic (moment of kurtosis less than 3) nature of the distribution makes it definitely not log-normal. The original papers on width distribution (4,6) again force a fit by ignoring width reports greater than 1,000 yds. The observed mean of tornado width is 140.8 yd. with a standard deviation of 221.8 yds. After a transformation of the width in miles to natural logarithms, the mean becomes -3.28 and the standard deviation is 1.19.

When the paired measurements of length and width are used to compute area, the resulting distribution also fails to be log-normal (Fig. 5). The average area is 0.85 mi^2 and its standard deviation is 3.18, while the logarithmic average is -2.82 with a standard deviation of 2.54. The failure of tornado length, width and area to be log-normally distributed substantiated the recent findings of Twisdale (8) who considered FPP indices for a considerably smaller sample.

Table 3 summarizes the mean path length, width and area data. It also presents the expected values that would be obtained if the data were actually distributed log-normally. Fcr comparative purpose, some recently reported values are also given in the table. The values computed from the NSSFC-NRC data are within the current range of estimates.

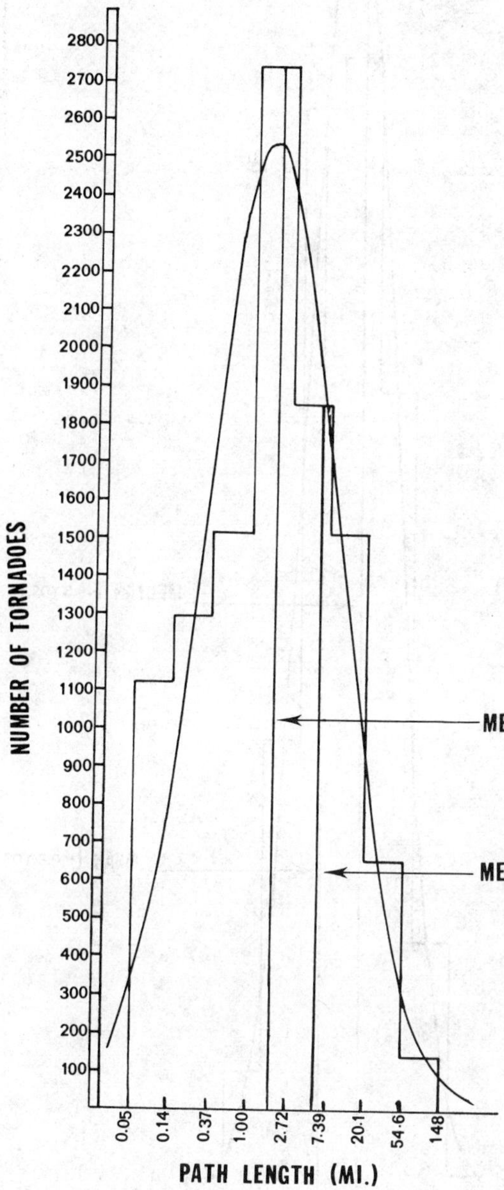

Fig. 3. Distribution of path length.

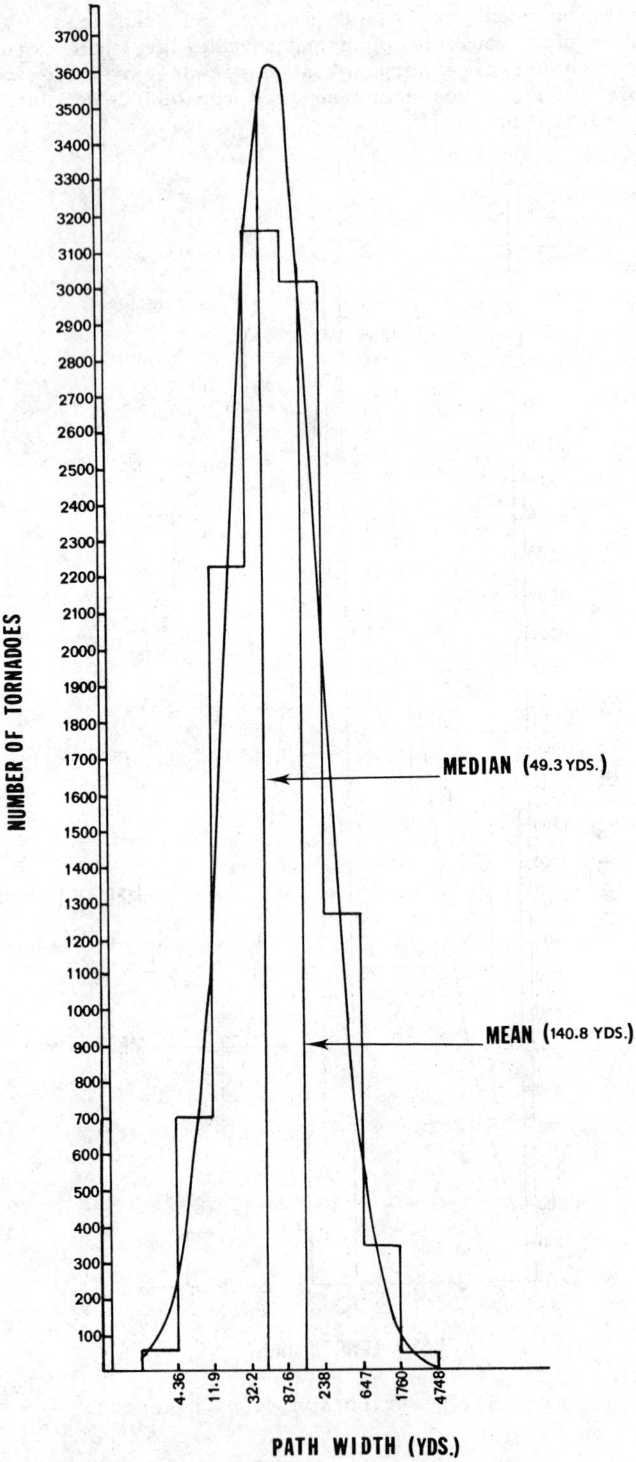

Fig. 4. Distribution of path width.

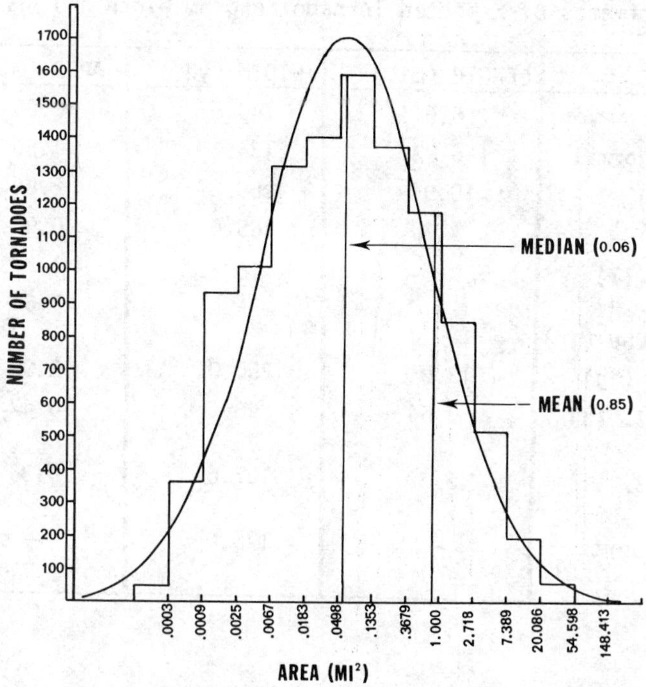

Fig. 5. Distribution of track area.

It should be noted that the mean area does not equal the product of the mean length and the mean width. This would only be true if there were no correlation ($\rho = 0$) between the length and width measurements. Further, since the geometric distributions are highly skewed (quasi-logarithmic) for characterizing the typical tornado, median values rather than means might be more appropriate. These values, 1.99 mi for length, 49.3 yds for width, and 0.06 mi^2 for area, are also included in Table 3.

TRACK PARAMETERS AND INTENSITY

The mean and standard deviation of the length, width and area as a function of tornado intensity are given in Table 4. There is a definite tendency for the averages to increase as the intensity increases. Even though the distributions are neither normal nor log-normal, the central limit theorem implies that the means are approximately normal. Thus, the statistical significance of these trends are examined and all are found to be significant at the 5% level. This implies that the mean violent (strong) tornado is longer, wider and affects more area than the mean strong (weak) one.

While the average track parameter for more intense tornadoes are larger than for weaker ones, this does not imply any relationship between the track parameters of any individual tornado and its severity. To examine this, the length, width and area indicies (indicated in Table 1) must be compared on a one-to-one basis with the intensity index. (Indices are needed so that the number of categories for each parameter is the same).

TABLE 3: Estimates of Expected Tornado Length, Width and Area

SOURCE	LENGTH (mi.)	WIDTH (yd.)	AREA (mi^2)
Mean	6.0	140.8	0.85
Expected Log-Normal	6.72	134.4	1.50
Thom (3)	10.94	259.87	2.82
Howe (6)	4.7	165.0	0.96
Skaggs (11)	-	-	1.82
Hart (12) (based upon Wen and Chu (18)	-	-	1.66
Sadowski (19)	13.0	250.0	2.00
Garson *et al.* (4) (Model)	3.64	98.6	0.47
Garson *et al.* (4) (histogram)	3.30	107.0	0.36
Brown and Roberts (20)	-	176.0	-

Table 4: Mean tornado parameters as related to storm intensity.

	Length (mi.)		Width (yd.)		Area (mi^2)	
INTENSITY	*Mean*	*Standard Deviation*	*Mean*	*Standard Deviation*	*Mean*	*Standard Deviation*
Weak	3.06	6.89	88	140.80	0.24	0.91
Strong	8.93	1.43	193.60	264.00	1.35	3.74
Violent	25.72	27.80	422.40	369.60	6.65	10.2

The relationship between path length and tornado intensity has been briefly examined by Kelly, *et al.* (9), and is shown in Table 5. Intensity and length indices are available for 14,409 tornadoes. No clear relationship between length and intensity is found. Only about one-fourth (28%) of the long-track (P_L 4&5) tornadoes are violent, while over half (59%) of them are strong. To complicate the relationship further, a majority (61%) of the violent storms have an intermediate length path (P_L 2&3). This is in general agreement with the findings of Fujita and Pearson (2).

Statistically, the P_L index has a 0.51 correlation with the intensity estimate. Thus, the track length only explains 26% of the variance of tornado intensity. The standard error of estimation for a linear regression of intensity upon length category is 0.75, i.e., a linear relationship based upon P_L selects the wrong intensity category almost one time out of three.

Table 5: Distribution of 14409 tornadoes by force (weak, strong, violent and length (short, intermediate, long) from (9).

F Scale	PL0-1	PL2-3	PL4-5	TOTAL
4-5	47	205	84	336
2-3	2921	2117	175	5213
0-1	7615	1206	39	8860
TOTAL	10583	3528	298	

The correlation between path width and tornado intensity is equally ambiguous (Table 6). A total of 13,384 tornadoes have both width and intensity estimates available. Less than one-fifth (18%) of the broad tornadoes (P_W4&5) are violent while over two-thirds of the violent storms have a moderate width. This is no great surprise when one considers the dramatic width changes that a single tornado can undergo (e.g., 10). The correlation between P_W and F is 0.50. Once again, a linear regression between path width and intensity does not appear to be significant.

TABLE 6: Distribution of 13384 tornadoes by force (weak, strong, violent) and width (narrow, intermediate, broad).

F Scale	PW 0-1	PW 2-3	PW 4-5	TOTAL
4-5	14	251	84	349
2-3	2082	2864	285	5231
0-1	5510	2193	101	7804
TOTAL	7606	5308	470	

It has been hypothesized recently that tornadoes follow a two-dimensional normal distribution between the A (area) and F (intensity) indices (11, 12). Accordingly, in making Table 7 relating these parameters, the indices were not sub-grouped. The mean A is found to be 2.78 and the mean F is 1.90, with standard deviations of $\sigma_A = 1.156$ and $\sigma_F = 0.997$. The correlation coefficient between A and F is 0.54.

TABLE 7: Distribution of 10204 tornadoes by force and area.

5	-	-	-	4	18	19
4	-	3	16	61	150	43
3	3	29	144	365	363	64
2	39	326	874	1208	527	45
1	196	998	1514	1086	286	9
0	320	827	452	177	38	-
F / A	0	1	2	3	4	5

With this information, it is possible to construct the theoretical distribution. The maximum normalized difference between observed cell frequency and expected cell frequency is less than 0.025. However, even with this close a fit, the distribution fails to be two-dimensional normal at the 5% level. Too many tornadoes occur when both A and F are low or high and too few occur at mid-values for the postulated joint log-normal distribution to hold.

GEOGRAPHIC ASSESSMENT OF TORNADO HAZARD

In order to assess the potential for tornado occurrence, it is necessary to know the geographic distribution of track parameters. For a crude estimate, Thom (3) suggested assuming geographic invariance of tornado-affected areas and applying the Iowa-Kansas expected value nationwide. Skaggs (11) and Howe (6) both have shown that this is not adequate. Further, since tornado area is not log-normally distributed, presentations based upon this assumed distribution (e.g., 11) are also questionable.

Accordingly, the geographical distribution of tornado-affected area has been constructed by summing the area affected by tornadoes within a 2° latitude/longitude cell and normalizing by year. Such values are computed at 1° latitude and longitude increments east of 107°W. This overlapping of 2° cells introduces a smoothing conceptually similar to that employed by Skaggs (13). It is equivalent (ignoring modest error caused by varying areas between adjacent squares) to smoothing 1° cell data with a running box-car average.

The ratio of the average annual area affected by tornadoes to the total area of each cell is computed. When this quotient is inverted, it represents the area-time space which will, on the average, contain one unit area of tornado damage. If a quasi-continuous distribution of this quantity is assumed, an isopleth chart (Fig. 6) can be drawn.

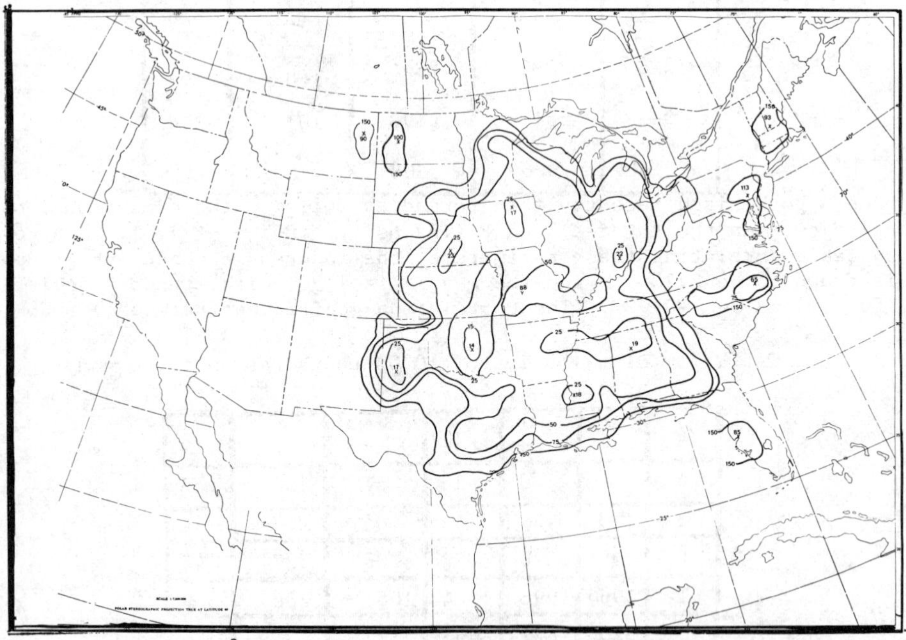

Fig. 6. Isopleths (mi^2-century) of the area-time product which will in the mean encompass 1 mi^2 of tornado damage.

Several interpretations of this area-time field are possible. If a one year time span and one square mile area is considered, this chart indicates the number of square miles which will sustain one square mile of tornado damage in an average year. This concept is similar to that suggested by Brown and Roberts (14).

By further assuming that the number of points affected by more than one tornado is negligible (i.e., tornado paths do not intersect or overlap), area is self-cancelling. The chart then gives the mean recurrence interval (MRI) of a tornado strike. This alternative interpretation is similar to that of Thom (3). However, it must be emphasized that we are not considering a true mathematical point. How small an area can be considered as a "point" is limited by both the precision of the measurements and the assumptions made in deriving the MRI. Since the area of each cell has been computed only to the nearest mile, the MRI is valid for a one square mile area (circle of radius 0.56 mi.) surrounding a point.

For sites larger than 1 mi^2, an MRI can be estimated from the area-time data. Consider the Kansas City metropolitan area (1,643 mi^2). The cell which contains it will have 1 mi^2 of tornado track for every 1,827 mi^2 in one year, or 1 mi^2 of tornado track for every 1,643 mi^2 in 1.1 years. Thus, there will be 1 mi^2 of tornado destruction in greater Kansas City on an average of once every 13 months.

Examination of Fig. 6 indicates that the regions which substain the most tornado damage annually do not exactly agree with the regions of most frequent tornadoes. Kelly *et al.* (9) found that "Tornado Alley", the region of most frequent tornadic activity, runs approximately between 97 and 98°W meridians. However, when damage area is considered, this Alley is rotated clockwise and runs in a broad band from the Texas Caprock escarpment through north central Iowa. While four relative maxima exist, perhaps the most striking feature is the protrusion of a minimum into the Oklahoma Panhandle and southwest Kansas. This feature is apparently anomalous and could possibly be related to the low-population density and the large proportion of undeveloped land.

The so-called "Dixie Alley" is seen to lie from east-central Texas to North Carolina. A minimum in this Alley occurs along the Appalachian Mountains. Other apparent areas of increased threat occur north of Chesapeake Bay, in southern New England, in eastern Montana/western North Dakota and over western Florida. The region in Florida is considerably smaller than one would expect from an examination of tornado frequency data alone. This arises from the decreased emphasis given to tornadic waterspouts when damage area is considered. A large area of decreased risk occurs over the Ozark Plateau. A survey of attempts to define the tornado hazard regionally may be found in (15) and (16).

An area-time chart is also presented for only violent (F4 & 5) tornadoes (Fig. 7). The pattern for these intense storms is much less smooth than the one for all tornadoes. Further, because of the smaller sample size, confidence in the pattern is decreased from the previous figure. The area of highest hazard for violent storms lies in west Texas between Lubbock and Amarillo. Abbey and Fujita (5) did not find this area to be one of relatively high hazard probability. Weak relative maxima are rather randomly scattered between the Rockies and Appalachians. Surprisingly, the area with the second greatest hazard is found in southwestern Mississippi. The "Tornado Alley" maximum is relegated to fourth place. Thus, when severity is considered, the Mississippi-Alabama region has a slightly larger area traversed by violent storms than the Oklahoma-Kansas region which agrees with the earlier findings by Abbey and Fujita (5).

The median tornado occurrence period can be found by multiplying the MRI by 0.69 (see e.g., 17). This period is the time during which there is a 50% chance of a given 1 mi^2 area being devastated by a tornado. Again, referring to Fig. 6, it

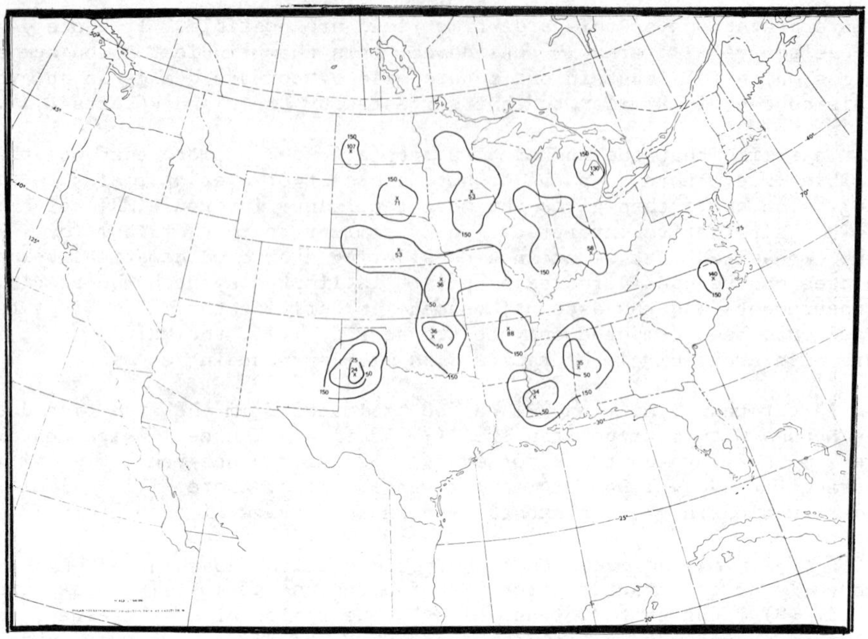

Fig. 7. Isopleths (mi^2-century) of the area-time product which will encompass 1 mi^2 of violent tornado damage.

can be seen that over central Oklahoma, there is an even chance of a tornado strike in approximately 966 years.

Also, from the MRI, the well-known formula (see e.g., 17)

$$P\{N\} = 1 - (1 - 1/MRI)^N$$

can be used to determine the probability of an event's occurrence over a span of N years. Figure 8 presents the probability that a 1 mi^2 "point" will be struck by a tornado in 50 years. Because of the exponential nature of the defining equation, this field is a filtered version of the area-time recurrence field. The highest probability of tornado occurrence over a 50 year span is 3.46% in central Oklahoma. Three relative minima occur: one over the Ozark Plateau, one over the Nebraska sand hills, and a very notable one in the mountains of West Virginia and Virginia.

For violent tornadoes, Fig. 9, the highest probability for destruction during a 50-year period is 2.06% and occurs in the latitude-longitude cell containing Lubbock, Texas. Because of data sparsity, the pattern is complex. Zones of enhanced probability occur in Mississippi-Alabama and in eastern Oklahoma-Kansas. Distinct minima occur from central Wisconsin, southward to west-central Illinois, and along the Ozark Plateau. An isolated zone of enhanced probabilities exists in west-central Florida.

In case the region of interest is larger than 1 mi^2, the appropriate MRI to be used in the probability formula can be estimated from Figs. 6 and 7.

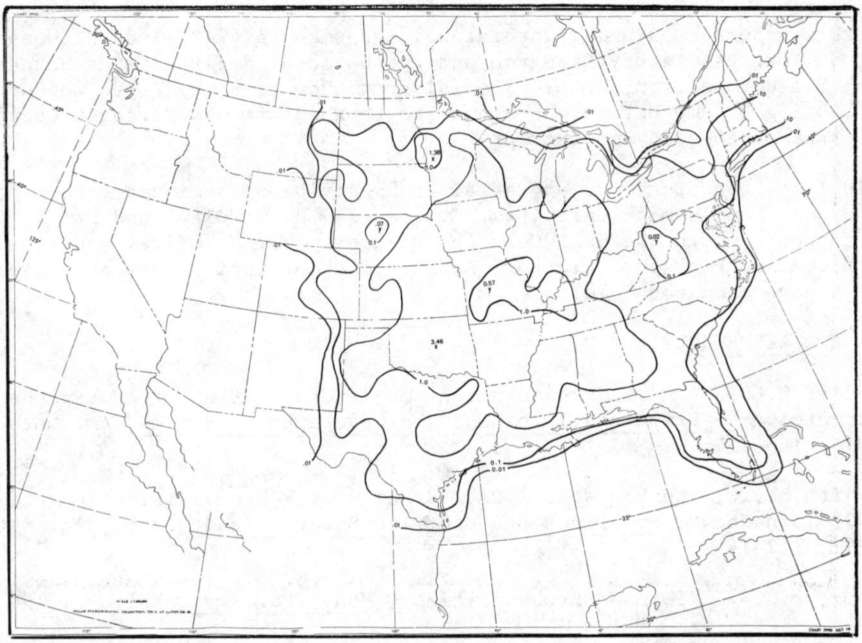

Fig. 8. 50-year probability (%) of 1 mi^2 of tornado damage.

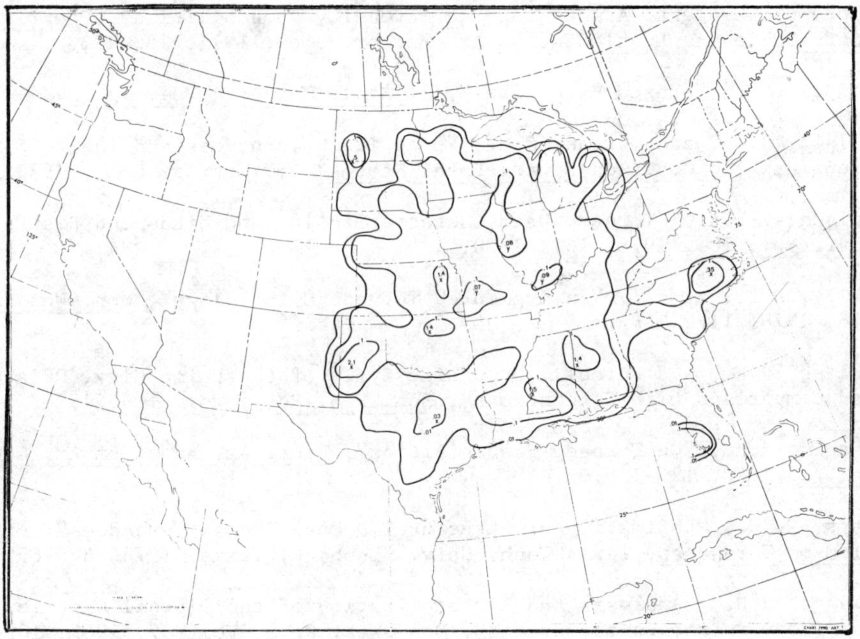

Fig. 9. 50-year probability (%) of 1 mi^2 of tornado damage for violent tornadoes.

ACKNOWLEDGEMENTS

This work was performed under Interagency Agreement AT49-25-1004 between the Office of Nuclear Regulatory Research and the National Severe Storms Forecast Center, National Weather Service, NOAA. Special gratitude to Sandy Boyse who drafted the figures, Delight Jump Herman who typed the rough draft and Margaret Coonfield who prepared the photo-ready manuscript.

We would like to express our appreciation by giving a lagniappe to our colleagues in Techniques Development Unit (Drs. C. Doswell, R. McNulty, and Mr. L. Lemon). To the forecasters in the SELS Unit (particularly Mr. J. Galway and R. Williams) and to Mr. Allen Pearson, Director, NSSFC. Without their assistance, this work would not have been possible.

REFERENCES:

1. Kelly, D.L., *et al.*, "Presentation and Interpretation of an Expanded Tornado Climatology", Preprints 10th Conf. on Severe Local Storms, Am. Meteor. Soc., 1977, 186-193.

2. Fujita, T.T., and Pearson, A.D., "Results of FPP Classification of 1971 and 1972 Tornadoes", Preprints 8th Conf. on Severe Local Storms, Am. Meteor. Soc., 1973, 142-145.

3. Thom, H.C.S., "Tornado Probabilities", Mon. Wea. Rev., Vol. 91, 1963, 730-736.

4. Garson, R.C., Catalan, J.M., and Cornell, C.A., "Tornado Design Winds Based on Risk", J. Struct. Div., Vol. 101, 1975, 1883-1897.

5. Abbey, R.F., Jr., and Fujita, T.T., "Use of Tornado Path Lengths and Gradations of Damage to Assess Tornado Intensity Probabilities", Preprints 9th Conf. on Severe Local Storms, Am. Meteor. Soc., 1975, 286-293.

6. Howe, G.M., "Tornado Path Sizes", J. Appl. Meteor., Vol. 13, 1974, 343-347.

7. Wilson, J.W., and Morgan, G.M., "Long-Track Tornadoes and Their Significance", Preprints 7th Conf. on Severe Local Storms, Am. Meteor. Soc., 1971, 183-186.

8. Twisdale, L.A., "Tornado Data Characterization and Windspeed Risk", J. Struct. Div., Vol. 104, 1978, 1611-1630.

9. Kelly, D.L., *et al.*, "An Augmented Tornado Climatology", Mon. Wea. Rev., Vol. 106, 1978, 1172-1183.

10. Golden, J.H., and Purcell, D., "Life Cycle of the Union City, Oklahoma Tornado and Comparison with Waterspouts", Mon. Wea. Rev., Vol. 106, 1978, 3-11.

11. Skaggs, R.H., "On Tornado Probabilities", Proc. Am. Assn. of Am. Geographers, Vol. 2, 1970, 123-126.

12. Hart, G. C., "Estimation of Structural Damage Due to Tornadoes", Proc. Symposium on Tornadoes, Texas Tech. Univ., Lubbock, Texas, 1976, 645-655.

13. Skaggs, R.H., "Analysis and Regionalization of the Diurnal Distribution of Tornadoes in the United States", Mon. Wea. Rev., Vol. 97, 1969, 103-115.

14. Brown, C.W., and Roberts, W.O.J., "The Distribution and Frequency of Tornadoes in the United States from 1880-1931", Trans. Am. Geophys. Union, Vol. 16, 1935, 151-158.

15. Abbey, R.F., Jr., "Risk Probabilities Associated with Tornado Windspeeds", Proc. Symposium on Tornadoes, Texas Tech Univ., Lubbock, Texas, 1976, 177-236.

16. Abbey, R.F., Jr., and Fujita, T.T., "Regionalization of the Tornado Hazard", Preprints 10th Conf. on Severe Local Storms, Am. Meteor. Soc., 1977, 526-533.

17. Court, A., "Some New Statistical Techniques in Geophysics", Adv. in Geophysics, Vol. 1, 1952, 45-85.

18. Wen, Y.K., and Chu, S.L., "Tornado Risks and Design Wind Speed", J. Struct. Div., Vol. 99, 1973, 2409-2421.

19. Sadowski, A., "Potential Casualties from Tornadoes", paper presented at 244th National Meeting of the Am. Meteor. Soc., Cloud Physics and Severe Local Storms, 1965, 11 pp.

20. Brown, C.W., and Roberts, W.O.J., "The Areal Frequency of Tornadoes in the United States by Counties, 1180-1931", Trans. Am. Geophys. Union, Vol. 18, 1937, 144-146.

A TATSUMAKI IN TOKYO ON FEBRUARY 28TH, 1978

H. ISHIZAKI*, Y. MITSUTA*, Y. TANIIKE* and Y. IWATANI**

*Disaster Prevention Research Institute, Kyoto University, Uji City, Kyoto, Japan
**College of Industrial Technology, Nihon University, 7-24-1 Narashinodai, Funabashi City, Japan

SUMMARY

In the evening of Feb. 28th, 1978, a strong wind caused severe damages in the area along the north-west coast of Tokyo Bay. The damage was distributed in a straight belt area beginning at Kawasaki, Kanagawa and ending at Kamagaya, Chiba. The length of the damaged area was about 40 km (25 miles) and the width 0.4 km (0.25 miles). Most outstanding damages of all were that a train (Tozai-line) was turned over on the bridge over the Arakawa River in Tokyo and that a prefabricated one-story school building (floor area 539 m^2) was lifted about 5 m up and carried over the next building and crashed over the third building.

1. INTRODUCTION

"Tatsumaki" is a Japanese term for tornado, "a violently rotating column of air, pendent from a cumulonimbus cloud and nearly always observable as a funnel cloud", which is seen almost all over the world. There have been several studies on Tatsumaki climatology in Japan (e.g. Mitsuta[1], Ibaraki and Tanaka[2], Shimada[3]). According to these studies most of Tatsumakis in Japan occur near the sea coast, especially on the southern part of the Pacific coast. The occurrence of Tatsumakis on main islands of Japan is about 7 per year on average and one person is killed, sixteen are wounded and 62 houses are damaged (Mitsuta[1]). The averaged characteristics and damage are shown in Table 1, which shows the amount of damage is not so large.

Table 1 Statistics of Tatsumaki in Japan

Tatsumakis 1955 - 1964			
Occurrence (per Year)	7.2	Path Width m	45(148 ft)
Number of Deaths		Length k	3(1.9 mile)
(per Year)	1.0	Forward Motion	
(per Tatsumaki)	0.14	Speed km/hr	45(28 mph)
Number of Wounded		Direction	to NE
(per Year)	15.9	Maximum Wind Speed m/sec	60(134 mph)
(per Tatsumaki)	2.2	Probability at any given point per year	10^{-6}
Number of Damaged Houses			
(per Year)	62.1		
(per Tatsumaki)	8.6		

The maximum wind speed of Tatsumaki has never been actually measured. We have only several reports showing anemograph records near the path of the Tatsumaki indicating maximum speed above 60 m/sec (134 mph), as a clue to the maximum wind speed of Tatsumakis. The estimated maximum wind speed obtained from damage is about 100 m/sec (224 mph) (Ishizaki[5] et al). In the United States, the maximum wind speed of tornadoes has also been a subject of study, and a maximum wind speed of 125 m/sec (280 mph) has been observed by a Doppler radar and 250 m/sec (560 mph) has been estimated from the analysis of damage to structures.

The occurrence probability of Tatsumaki at any one point in Japan is of the order of 10^{-6} a year (Mitsuta[1]), which means that the return period of a Tatsumaki wind is as long as a million years at a given point. But the occurrence probability of a tornado at any one point in the United States is estimated to be about 10^{-4} per year on the whole and 10^{-3} on the tornado belt in the central part (Thom[4]).

Clearly, the Japanese Tatsumaki is less severe and less frequent than the American tornado, hence less serious; but more detailed knowledge[5] of its characteristics, especially its maximum wind speed estimate, is required for the purpose of disaster prevention, because the structures which need great protection from damage such as an atomic power plant are increasing. A Tatsumaki hit the southeastern part of Tokyo near the coast of Tokyo Bay and caused severe damage including a train accident on the evening of Feb. 28, 1978, which indicates the undisregardful nature of Tatsumaki of highly populated areas.

2. A TATSUMAKI IN TOKYO

2.1 Damages

In the evening of Feb. 28, 1978, a Tatsumaki hit southeastern part of Tokyo and severe damage was caused along a line from the coast of Tokyo Bay near Haneda Airport to the northeast ward. The total amount of damage is indicated in Table 2. These statistics indicate that this Tatsumaki was not an extraordinarily severe one compared to the averaged Japanese Tatsumaki shown in Table 1. The distribution of the damage to the houses (including completely and partially destroyed and more than half of the roof tiles destroyed), and the damage to the electric facilities (the electric pole and power transmission line) and some other outstanding damage is indicated in Fig. 1. The damaged area around Kawasaki (southwest end of the figure) and that around Ichikawa seem to be connected approximately by one straight narrow belt. The length of the damaged area is about 40 km (25 miles) and the width is at most 0.4 km (0.25 miles). Even though the total damages were not so serious, we have to note that there were two cases which could have been serious disasters if they had occurred at slightly different times. This will be discussed again later.

Table 2 Statistics of damage caused by the Tatsumaki

Item	Quantity
Number of Wounded	34
Number of Damaged Houses	
Totally destroyed	27
Badly damaged	249
Slightly damaged	904
Number of Derailments	3
Number of Afflicted Houses	1515

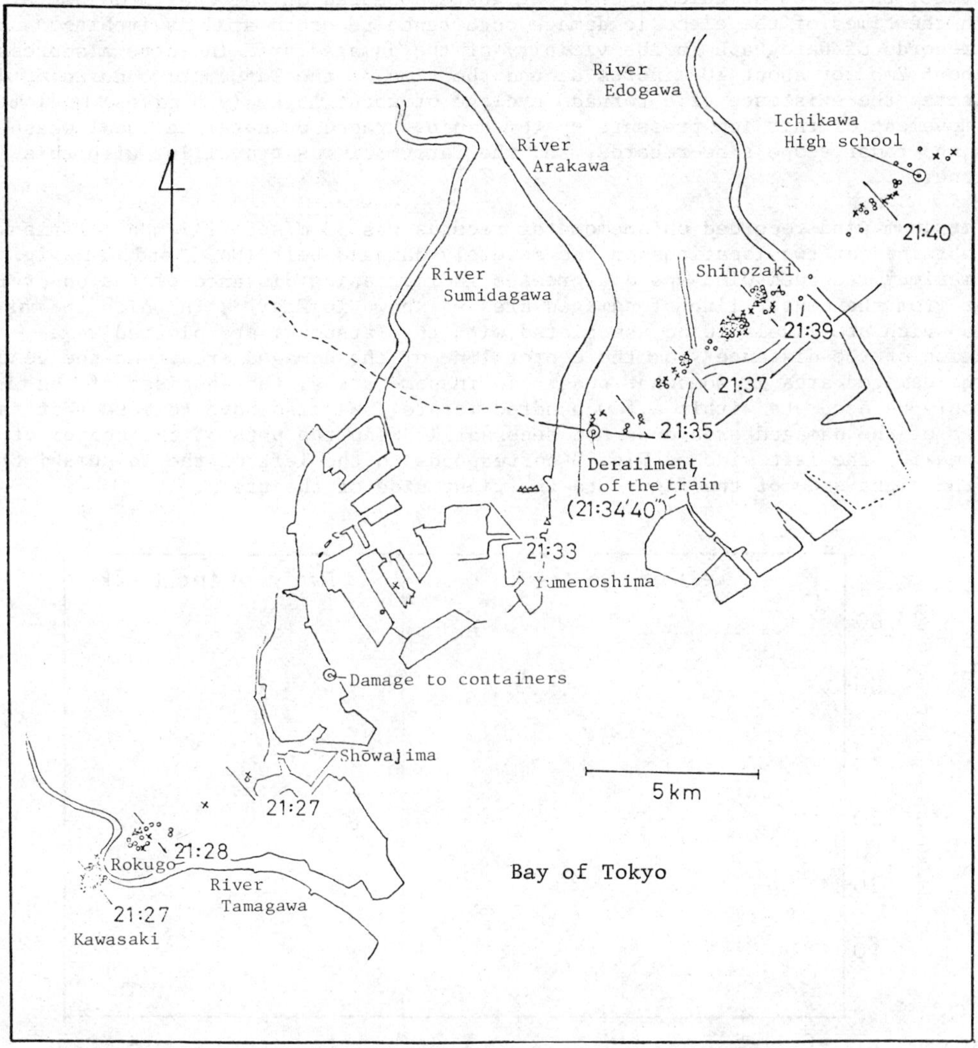

Fig. 1 The distribution of the damages to houses (○), the electric facilities (×) and other outstanding damages (◉). The numbers are the times of power failures and the time of the train derailment.

2.2 Meteorological conditions

On that day a developed extratropical cyclone was passing along the northern coast of the Main Island of Japan. The Tatsumaki appeared in the warm section between the warm and cold front. Fortunately, around the damaged area of this Tatsumaki, there distributed a number of meteorological instruments installed for the purpose of disaster prevention and environmental protection. Some of the wind records obtained at these stations in and around the damaged area are shown in Fig. 2. These records indicate a strong gust for a very short time with a sharp turning of the direction, associated with the Tatsumaki vortex passage.

Unfortunately, however, since the times on the strip chart were not sufficiently accurate, the times at which the strong gusts appeared on the charts do not always match the times of the electric damage occurrence recorded at the switching station. The records of barograph in the vicinity of the damaged area indicate a depression of about 2mb for about 10 minutes around the time of the Tatsumaki occurence, which indicates the existence of a tornado cyclone of about 15 km (9.3 miles) in diameter. The movement of this low pressure system can be traced on detailed local weather maps and radar scope cine-records, and the Tatsumaki was travelling with this cyclone.

The maximum wind recorded on anemometer records was 52 m/sec (117 mph). This value was obtained at two locations on the severely damaged belt (No. 7 and 15 in Fig. 2). The maximum recorded wind speed decreases as increasing distance of the observing point from the center line of damaged area as shown in Fig. 3, in which the maximum gusts which are likely to be associated with the Tatsumaki are plotted as a function of the distance from the center line of the damaged area. As the center of the damaged area is somewhat uncertain in some areas, the abscissa of the figure may only be accurate within a few hundred meters. We also have to note that the center of the damaged area does not necessarily mean the path of the center of the Tatsumaki. The left side of Fig. 3 corresponds to the left of the Tatsumaki track and the right side of the figure to the right side of the track.

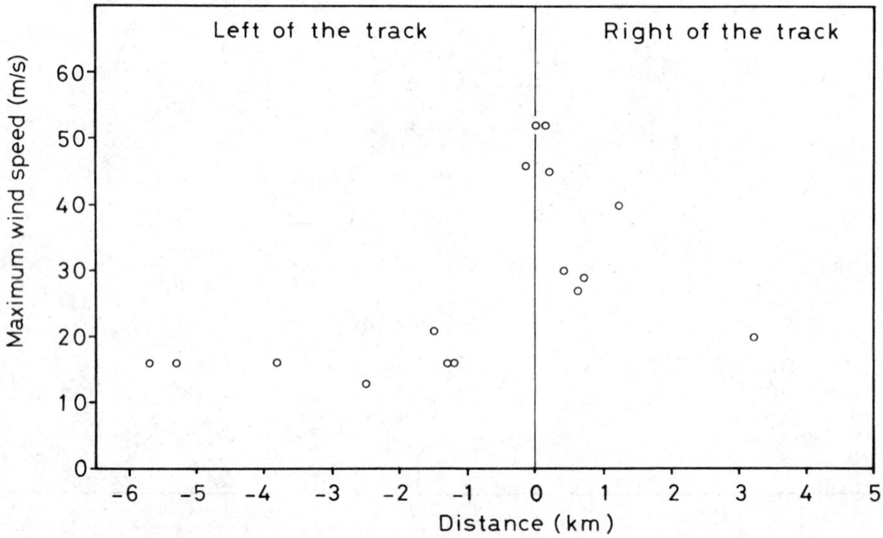

Fig. 3 The maximum wind speed as a function of the distance from the center of the damaged area. The left side of the figure corresponds to the left of the Tatsumaki track and the right side to the right of the track.

2.3 The time of the damage occurrence and the translation of the Tatsumaki

The most reliable determination of the time of the damage occurrence or Tatsumaki passage can be done by the times of the power failure, which are recorded at switching station with the accuracy of one minute and, the time of the train accident, which is probably accurate within an order of ten seconds. The times of the damage occurrence thus determined are indicated in Fig.1.

Although the path of the center of the Tatsumaki has not been finally determined yet, it is likely that the damage mentioned above was caused by a single Tatsumaki

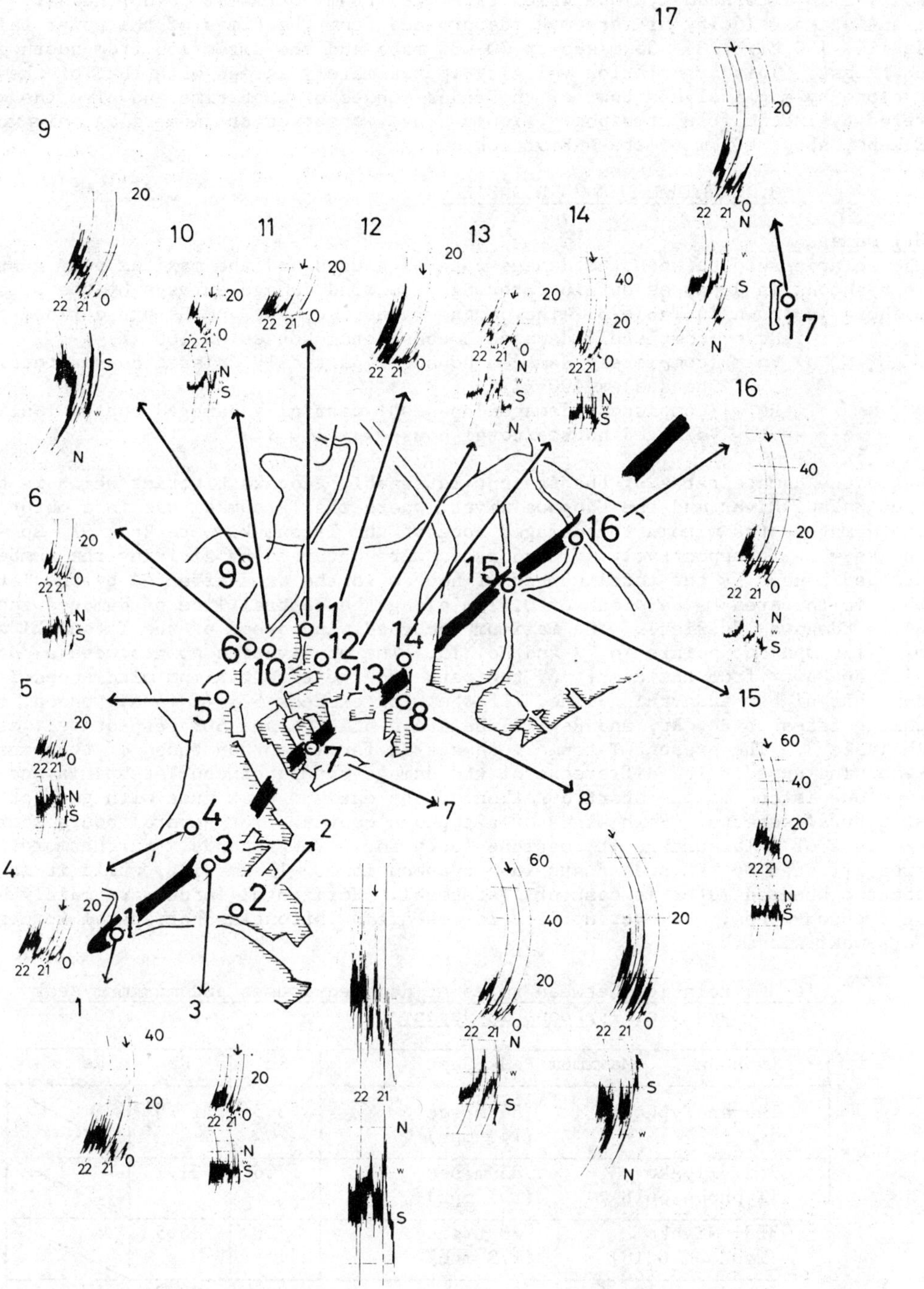

Fig.2 The records of wind speed and direction around the damaged area. The unit of the wind speed is m/sec. The abscissa indicates JST (Japan Standard Time). The peak gusts associated with the Tatsumaki are pointed by small arrows.

spawned in a tornado cyclone which travelled from southwest to northeast. The translation velocity of Tatsumaki determined from the times of the power failure is 110 - 130 km/h (30 - 35 m/sec or 70 - 80 mph) and the direction from southwest to northeast. This translation velocity approximately agrees with that of the tornado cyclone by map analysis, that of the radar echoes of that time and also the wind velocity in the free atmosphere, 700 mb above, observed at the aerological station at Takeno, about 60 km to the NNE of Tokyo.

3. CONSIDERATION OF DISASTERS

3.1 Damage rate

The relationship between the damage rate of houses and the maximum peak gust due to typhoons in Japan as obtained through the wind damage surveys by the present authors is shown in Table 3. The damage rates, Rc, Rp, and Rs are defined as

Rc = (completely damaged houses/total houses) x 100 (%)

Rp = (completely damaged houses + partially damaged houses/total houses) x 100 (%)

Rs = (completely damaged houses + partially damaged houses + slightly damaged houses/total houses) x 100 (%)

While the damage rates of the present Tatsumaki, Edogawa District which is between the Arakawa River and the Edogawa River, where the Tatsumaki was in a mature state, the damage rates within the damaged zone of the Tatsumaki were Rc = 0%, Rp = 2.5% and Rs = 5.2% respectively. These rates were obtained by dividing the number of damaged houses by the total number of houses in the area affected by the Tatsumaki, that is the area with a width of 0.4 km along the central line of damaged zone (5.7 km in Edogawa District). The maximum recorded wind speed of the Tatsumaki was 52 m/s (116 mph) at points No. 7 and No. 15. The observation points were a short distance away from the center of the path and the true maximum wind speed is likely somewhat higher than this value. If the wind exceeds 52 m/s in a typhoon, the damage rates of Rc, Rp, and Rs will be above 5%, 20% and 50% respectively as shown in Table 3. The present Tatsumaki damage is far less than those of typhoons of the same wind speed. The differecne of the damage rates between Tatsumakis and typhoons is due to the short duration of the maximum peak gust with respect to the structural aspects. High winds of a typhoon continue for several hours whereas in the case of a Tatsumakis it continues only for a few seconds. Furthermore, the area affected by the peak gusts in a typhoon is 10 - 15 km wide, while it is only about a hundred meter in case of a Tatsumaki. Horizontal forces are mainly dominant in typhoon winds. However up lift forces are as strong as horizontal forces in Tatsumaki winds.

Table 3 Relation between rates of damaged houses and maximum peak gusts of Typhoons in Japan.

Typhoon	Maximum Peak Gust	Rc	Rp	Rs
Isewan Typhoon	64 m/sec (143 mph)	3.5	17.5	
2nd. Miyako (Typhoon 6618)	85 m/sec (191 mph)	17.6	31.1	
3rd. Miyako (Typhoon 6816)	80 m/sec (179 mph)	7.0	33.5	
Typhoon 7513	68 m/sec (152 mph)	8.0	22.3	57
Typhoon 7709	60 m/sec (135 mph)	24.4	50.0	77.4

It is inappropriate to underestimate the force of Tatsumakis even if the damage rates of Tatsumakis are low compared to those of typhoons. In an open area such as a plain or a river, a train was turned over and a prefabricated one-story school building was lifted away by the Tatsumaki as described in the next section. This damage indicates the existence of strong up-drafts, because these objects were not fixed to the ground tightly compared with other structures. In a built-up urban area crowded with buildings, these structures are substituted for roughness blocks so that the boundary layer of the Tatsumaki increased in thickness and high wind does not reach down to near the ground. These results explain that the damage rates of low houses of one or two stories caused by the Tatsumaki are not so large as compared to the maximum peak gust.

3.2 Two special cases of disasters

The two last cars of a 10 car train travelling at the speed of 78 km/h (48 mph) were turned over on the rails of the other lane on the bridge over the Arakawa River at 21: 34' 40" by a strong wind blowing off from south (see Photo.1 and Figure 4). Weights of the last and second last cars were 27.5 tons and 36 tons respectively. Fortunately only 21 people were injured because there were not many passengers on the train. The driver in the lead car was not aware of anything unusual and did not slow the train down until he felt the train derail and lean to the right. He found the cars in the rear had turned over after the train stopped completely. Although the two derailed cars had a total of 8 wheels, only one trace of the wheel crossing on the rail was found in the subsequent investigation. Two cars are assumed to have been lifted up at first and turned over by the wind.

Photo.1 Derailment of a train

Photo.2 Damage to a prefablicated school building

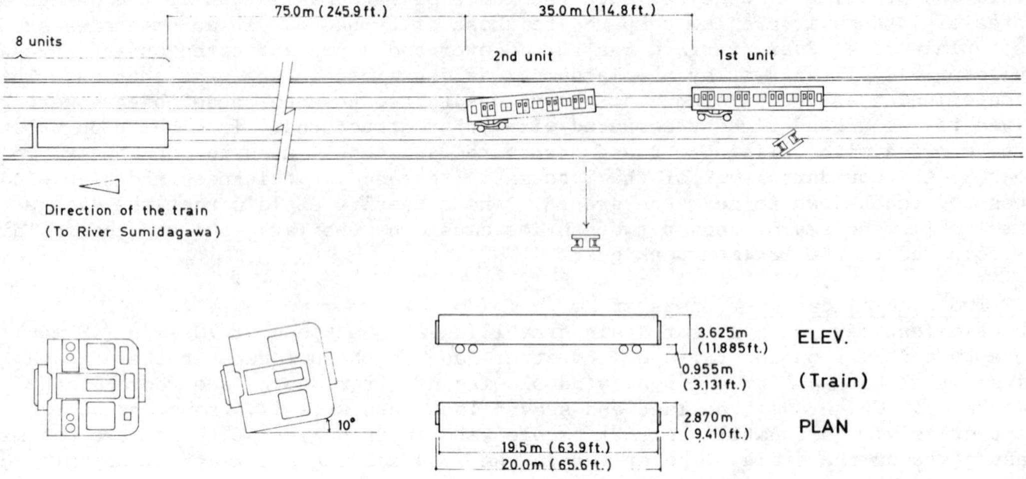

Fig.4 Derailment of a train on a bridge over the Arakawa River

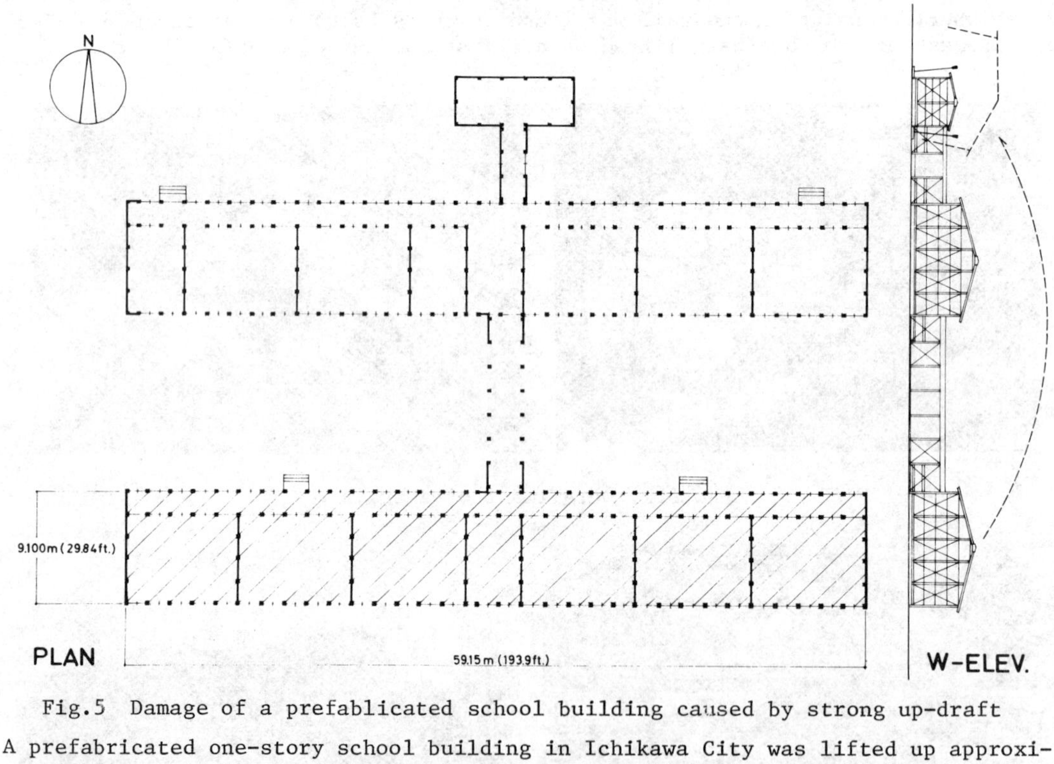

Fig.5 Damage of a prefablicated school building caused by strong up-draft

A prefabricated one-story school building in Ichikawa City was lifted up approximately 5m, carried over and crashed down on the other side of a neighbouring building which was 4.8 m (16 ft) in height (see Photo 2 and Figure 5). The school building had 7 rooms and a floor area of 9.1 m x 59.2 m (30 ft x 194 ft). The distance between the two buildings was 14.5 m (48 ft). The floor boards of the school building remained intact and the desks in the rooms also remained as they were. It was found that the side walls were not sufficiently fixed to the floor boards or to the foundation of the building. These two special cases indicate that the damage was caused by up-lift forces with short durations and therefore there was a strong up-draft.

3.3 Estimate of maximum peak gust

The measured maximum peak gust was 52 m/s (116 mph) and the true maximum peak gust was estimated to be about 75 m/s (168 mph). The maximum peak gust was not measured because the path of the Tatsumaki center was not so close to the observation point. The shortest distance between them is estimated to be about twice of the radious of the Tatsumaki. If we suppose the maximum peak gust is 75 m/s, we can obtain the maximum tangential velocity of 40 m/s (89 mph) by subtracting the translation velocity of 35 m/s given in Chapter 2. The maximum vertical velocity is supposed to be of the same order as the tangential velocity (C.A. Wan and C.C. Chang[6]). We take up two cases of the damages by the Tatsumaki, one is the damage to a prefabricated school building and the other is the derailment of a train.

The damage to the prefabricated school building: (see Photo.2 and Figure 5). The up-lift force is roughly approximated by the velocity pressure at the stagnation point at the maximum wind zone on the surface where wind is only vertical. If the vertical velocity is assumed to be 40 m/s (89 mph), the up-lift velocity pressure per a square meter is given as follows.

$$F_L = \frac{1}{2}\rho V^2 \quad (kg/m^2) \qquad (1)$$

putting ρ = 0.124 (kg-sec^2/m^4) (at 10°C, 984 mb) and V = 40 (m/s) in Eq. 1. We have

$$F_L = 99 \ (kg/m^2) > 82 \ (kg/m^2).$$

As the weight of the prefabricated building was only 82 kg/m^2, the up-lift force per a square meter could exceed the weight per a square meter of the prefabricated school building. It may safely be said that the assumptions of the maximum peak gust of 75 m/s is reasonable.

The damage to a train (see Photo.1 and Figure 4); In most cases of Tatsumakis, the increasing rate of wind velocity is not so small to be neglected. The force acting on the train was obtained by the summation of the resultant wind force and the accelerative force of the increasing velocity of winds. Thus the lift force per a square meter of the reference area of the train can be shown by

$$F_L = \frac{1}{2}\rho V^2 C_L + \rho l \frac{\partial V}{\partial t} C_V \qquad (2)$$

where the resultant wind velocity V equals the square root of the horizontal wind velocity of 75 m/s plus the vertical wind velocity of 40 m/s. The vertical lift coefficient C_L is assumed to be 1.3, the virtual mass coefficient $\frac{\partial V}{\partial t}$ is considered to be of the same order of C_L and the acceleration $\frac{\partial V}{\partial t}$ to be of the order of 10 m/sec^2 which was roughly calculated from the anemometer records. Putting ρ = 0.124 (kg-sec^2/m^4), $V = \sqrt{75^2 + 40^2}$ = 85(m/sec), C_L = 1.3, C_V = 1.0, $\frac{\partial V}{\partial t}$ = 10 (m/sec^2) and l = 3m (the approximate width of the train) into Eq. 2, we obtain

$$F_L = 586 \ (kg/m^2) \simeq 640 \ (kg/m^2),$$

where 640 kg/m^2 is the averaged weight of a car of the train and is almost the same as the estimated lift force of 586 kg/m^2. The train would be derailed by the joint action of the overturning moment due to the horizontal force and the lift force estimated above.

4. CONCLUSIONS

In the evening of Feb. 28th, 1978, a strong wind caused severe damage in the area along the northwest coast of Tokyo Bay. The damage was distributed in a straight belt area beginning at Kawasaki, Kanagawa ending at Kamagaya, Chiba. The length of the damaged area was about 40 km (25 miles) and the width was 0.4 km (0.25 miles). The most outstanding damages of all were that a train (Tozai-line) was turned over on the bridge over the Arakawa River in Tokyo, and that a prefabricated one-story school building (floor area 539 m^2) was lifted about 5 m up and carried over the next building and crashed over the third building. We assumed the maximum peak gust to be 75 m/s from the anemometer. From these damages it may safely be said that the assumption of the maximum peak gust of 75 m/s is reasonable.

The traces of barographs along the damaged area show a small scale depression of about 2 mb moved from SW to NE. The Tatsumaki is considered to have occurred in this small depression which may be a tornado cyclone. This small cyclone is first recognized in Yokohama at about 21:10, and can be traced up to Shonan-Cho at about 21:50. The translation velocity is estimated to be 126 km/hr (78 mph). This velocity approximately agrees with the upper wind velocity at 700 mb (SW 36 m/s or 130 km/hr). The translation velocity of the Tatsumaki estimated from the time of occurrence of the damages also coincides with this.

The Tokyo area has a dense meterological observation network, and the wind fluctuations corresponding to the Tatsumaki movement were recorded on several stations. This may be the first case in which wind patterns around the Tatsumaki (tornado) can be analysed in Japan.

REFERENCES

1. Mitsuta, Y., "Wind Disasters Caused by Small Scale Disturbances", Annuals of Disaster Prevention Research Institute, Kyoto Univ., 1968, No.11A, pp. 499 - 506, (in Japanese).

2. Ibaraki, T. and Tanaka, A., "Investigation on Spouts in Japan", J. of Meteorological Research, Vol. 13, No. 1, 1961, pp. 68 - 79, (in Japanese)

3. Shimada, M., "Tornadoes and Waterspouts in Japan During 1955 - 1964", J. of Meteorological Research, Vol. 19 No. 1. 1967, pp. 1 - 22, (in Japanese).

4. Tom, H.C.S., "Tornado Probabilities", Monthly Weather Review, Vol. 44, 1963, pp. 730 - 735.

5. Ishizaki, H., Mitsuta, Y., "Studies of a Tatsumaki at Toyohashi, Dec. 7th, 1969", Annuals of Disaster Prevention Research Institute, Kyoto Univ., 1971, No. 14A, pp. 481 - 500, (in Japanese).

6. Wan, C.A. and Chang, C.C., "Measurement of the Velocity Field in a Simulated Tornado-Like Vortex Using a Three-Dimensional Velocity Probe", J. of the Atmospheric Sciences 1972, Vol. 24, pp. 116 - 127.

7. Fujita, T.T., "Workbook of Tornadoes and High Winds", SMRP Research Paper 165, Sept. 1978.

ACKNOWLEDGEMENTS

The authors are greatly indebted to many persons and organizations that contributed to gathering data in field surveys of Tatsumaki damages. Special credit is due to Mr. U. Yoshikawa of Daiwa Housing Company and Dr. N. Monji of Kyoto University who have joined our survey and spent much time and effort in field surveys of Tatsumaki damages.

WIND AND PRESSURE DISTRIBUTIONS IN A TORNADO

W. S. LEWELLEN, M. E. TESKE and Y. P. SHENG

Aeronautical Research Associates of Princeton, Inc., Princeton, New Jersey, USA

ABSTRACT

A model based on turbulent transport theory is used to make predictions of wind and pressure distributions at low levels in a tornado. The influence of buoyancy, swirl ratio, surface roughness, updraft shape and inflow boundary layer thickness are discussed. In spite of the variation induced into the distributions by these variables, the horizontal wind distributions show considerable similarity when appropriately normalized. This similarity is used to present an engineering approximation to the wind distribution. Two different ways of estimating the maximum velocity occurring in a tornado are discussed.

INTRODUCTION

Detailed knowledge about the wind and pressure fields that may exist at low levels in a tornado is needed for the efficient design of critical buildings. The purpose of this paper is to present predictions of the distributions of these variables from a tornado model based on turbulent transport theory. The model formulation and some preliminary results were given by Lewellen and Teske (1). The present paper gives a very brief review of the model, presents some typical results, discusses the sensitivity of the model to the most significant variables, discusses estimates of V_{max} , and gives an engineering approximation which can be used for design studies.

BRIEF REVIEW OF MODEL

The tornado simulation model uses an axisymmetric, incompressible code which solves numerically for the three components of velocity, six components of Reynolds stress, three components of heat flux, temperature and temperature variance, turbulent scale, and pressure. The model equations given in Lewellen and Teske (1) and Teske and Lewellen (2) are based on the turbulent transport model described by Donaldson (3). The development and testing of the model has been summarized previously by Lewellen (4). The first of the above papers (1) contains a comparison between model predictions and laboratory observations of a contained vortex. This comparison was given to verify that the model can produce reasonable agreement with available laboratory data, not to imply that the laboratory vortex faithfully simulates a tornado flowfield. After discussing the verification tests, this paper also presents predictions for the wind and pressure distribution in a destructive tornado of the type which struck Xenia, Ohio in 1974.

Boundary conditions on all the model dependent variables are specified at the outer radius, r_+ (generally an inflow boundary); at the top of the domain, z_+, and at the surface. At the surface, the boundary conditions are specified in terms of an effective aerodynamic roughness z_o, and the surface heat flux $\overline{w'\theta'}|_o$. Use is made of the asymptotic solution given by Lewellen (5) to specify the relationship between the velocity gradients and turbulent correlations in the surface layer. At r_+, we specify the vertical distribution of the two horizontal components of wind, the gradient of the vertical wind, the potential temperature, and the turbulent kinetic energy and its length scale. The turbulent correlations are taken to be in equilibrium with the specified vertical gradients of the mean flow quantities.

The top of the domain is a major source of uncertainty in the model. If the top of the domain were placed above the thunderstorm, then pre-storm soundings and Doppler radar observations could be used to specify conditions along z_+. However, this would require modeling the complete storm. Due to the spectrum of scales involved, from the centimeters over which strong changes occur next to the ground to the kilometer scales in mid-troposphere, this restriction imposes formidable requirements on the computational grid. To permit better definition of the distributions next to the surface, most of our calculations have limited the computational domain to within a few hundred meters of the surface. At this open (generally outflow) boundary, we specify zero slope with respect to z on all the variables except ψ which is specified.

The major variables affecting the tornado model, coming from the real physical variations nature provides in the boundary conditions, may be parameterized as: a Froude number, $Fr = V_{max}/[g\, \overline{w'\theta'}|_o \ell/T_o]^{1/3}$; an incoming swirl ratio, $V/U|_{r+}$; a surface roughness, z_o/r_+ ; an incoming boundary layer thickness, δ_+/r_+ ; and a shape parameter for the updraft at altitude, $W(0,z_+)/\langle W(r,z_+)\rangle$.

MODEL RESULTS

We will not show any variation with Froude number in the present report. Calculations (6) show that a transition from a columnar, vortex type updraft to a buoyancy, plume-type flow occurs when Fr is reduced to order one. But the Froude number defined above, with ℓ limited to a few hundred meters, will remain much larger than 1 even with extremely strong buoyancy in the case of a severe tornado. Since we are primarily interested in the boundary layer flow, we will take $Fr = \infty$ in the calculations discussed herein. This effectively eliminates five variables from the system of equations: the temperature, temperature variance and three components of heat flux.

The major difficulty in limiting the height of the computational domain and essentially eliminating the influence of Froude number is the introduction of uncertainty as to what the top boundary condition should be on ψ. In Lewellen and Teske (1), we assumed a zero slope condition with respect to z on the radial velocity (equivalent to a zero curvature condition on ψ). However, as pointed out in that paper, this does not uniquely define the flowfield. That is, more than one steady wind distribution is possible for the same incoming flow conditions at large radius and the same surface conditions. These different wind distributions can be obtained by varying the radial pressure drop across the top of the domain. If everything else is held constant and a smaller pressure drop specified across the top of the domain, then a larger core size and smaller V_{max} is obtained. Since our model solves for stream function and vorticity with the pressure eliminated as an independent variable, we choose to solve the indirect problem of specifying ψ across the top of the domain and then determine the pressure distribution consistent with the resulting solution. We take the ratio of the vertical velocity on the centerline to the average vertical velocity across the top of the domain as a parameter to use for purposes of parameterizing this outflow variation.

The distributions of the incoming flow at large radius are parameterized here in terms of two variables: the average swirl ratio $V/U|_{r+}$, and the boundary layer thickness at this outer radius, δ_+/r_+. By far the more important of these is the swirl ratio. In fact, we find the distributions in the neighborhood of $\hat{r}$ (where V is maximum) to be quite insensitive to δ_+/r_+ as long as $r_+/\hat{r} \gtrsim 10$. This insensitivity is somewhat at odds with Rotunno's (7) conclusion that the entrant boundary layer thickness has a large effect on the maximum swirl attained for laminar calculations. This is probably a reflection of the fact that a turbulent boundary layer develops faster than a laminar one and thus forgets its history more quickly.

Typical results for the low-level distributions of the three components of mean wind, the mean pressure and the rms total velocity fluctuations are shown in Figures 1 through 5 for four different combinations of $V/U|_{r+}$ and $W(0,z_+)/\langle W(r,z_+)\rangle$. The surface roughness z_o/r_+ is approximately constant (10^{-3} for (a) and 0.9×10^{-4} for the others). To emphasize the similarities between the profiles, the lengths have all been normalized by the value of the radius at which V_{max} occurs, with velocities normalized by V_{max}. The ratio $r_+/\hat{r}$ varies from 2.5 to 13 for these simulations. The distributions given in case (a) are the same as those given in the example in our earlier paper (1).

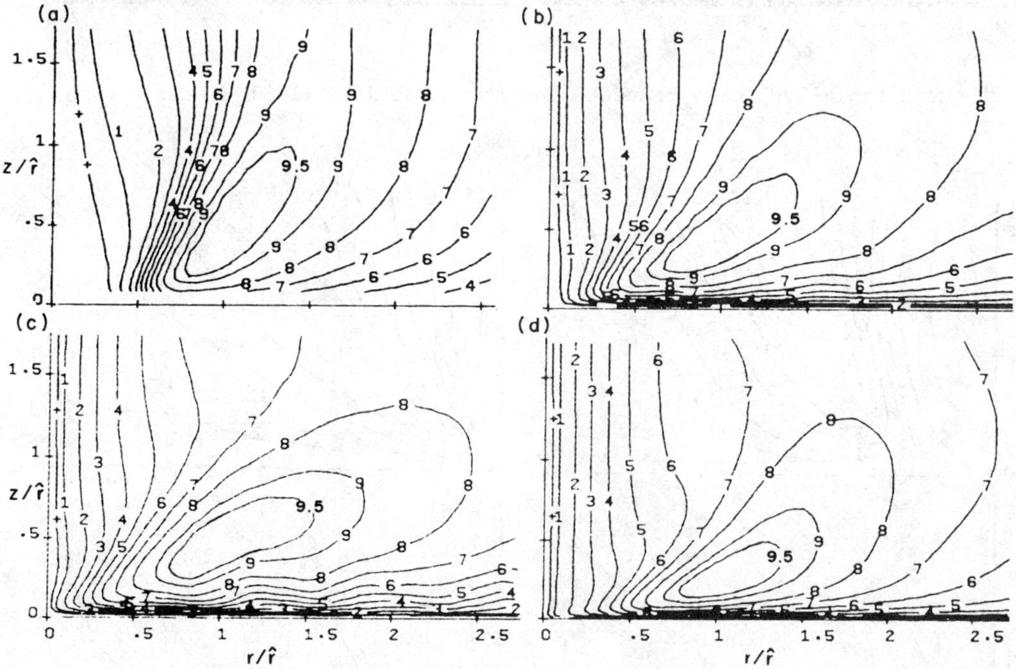

Fig. 1. Mean tangential velocity contours for four combinations of input parameters. Contours are labeled in tenths of V_{max} (+ = .01 V_{max}). Parameter values are given in the following table:

	$V/U\|_+$	z_o/r_+	$W(0,z_+/\langle W(r,z_+)\rangle$	δ_+/r_+
a)	4.0	10^{-3}	-0.1	0.2
b)	1.0	9×10^{-4}	1.0	0
c)	2.7	9×10^{-4}	1.0	0
d)	5.4	9×10^{-4}	1.0	0

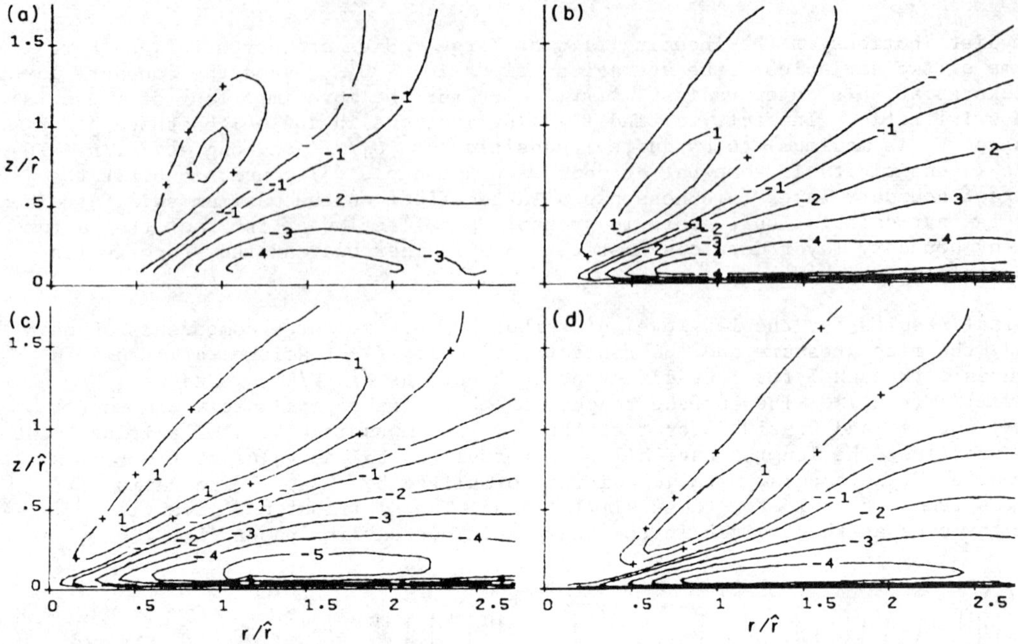

Fig. 2. Mean radial velocity contours for the conditions of Figure 1.

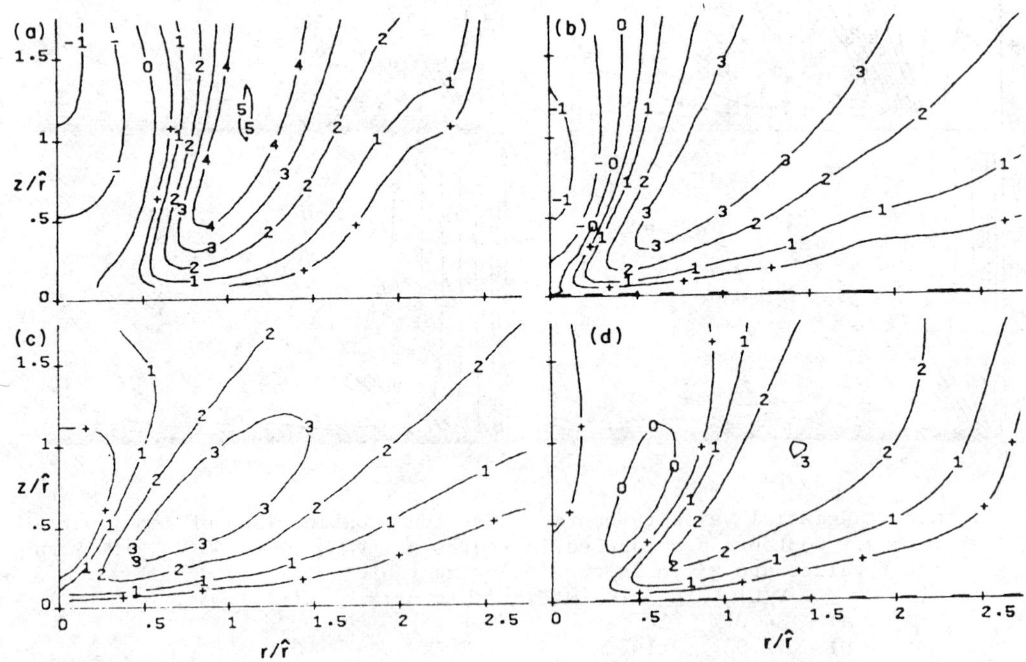

Fig. 3. Mean vertical velocity contours for the conditions of Figure 1.

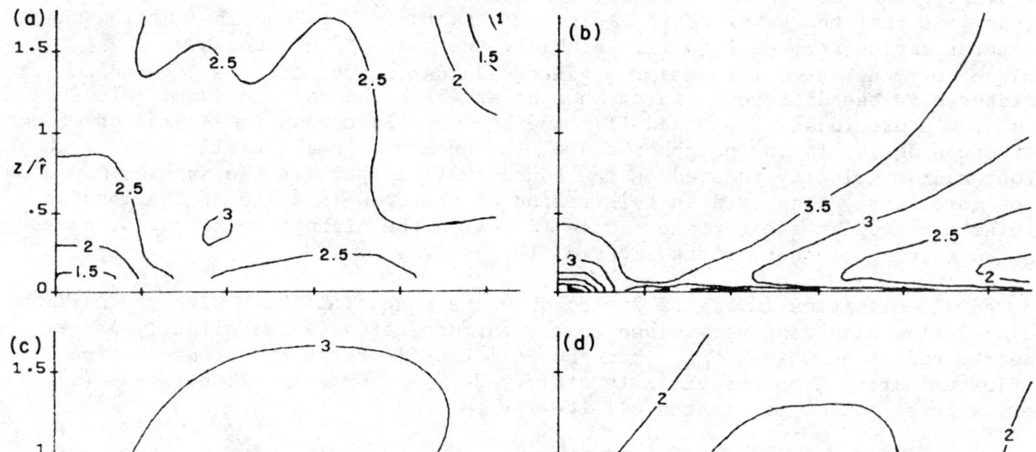

Fig. 4. Total velocity variance contours for the conditions of Figure 1.

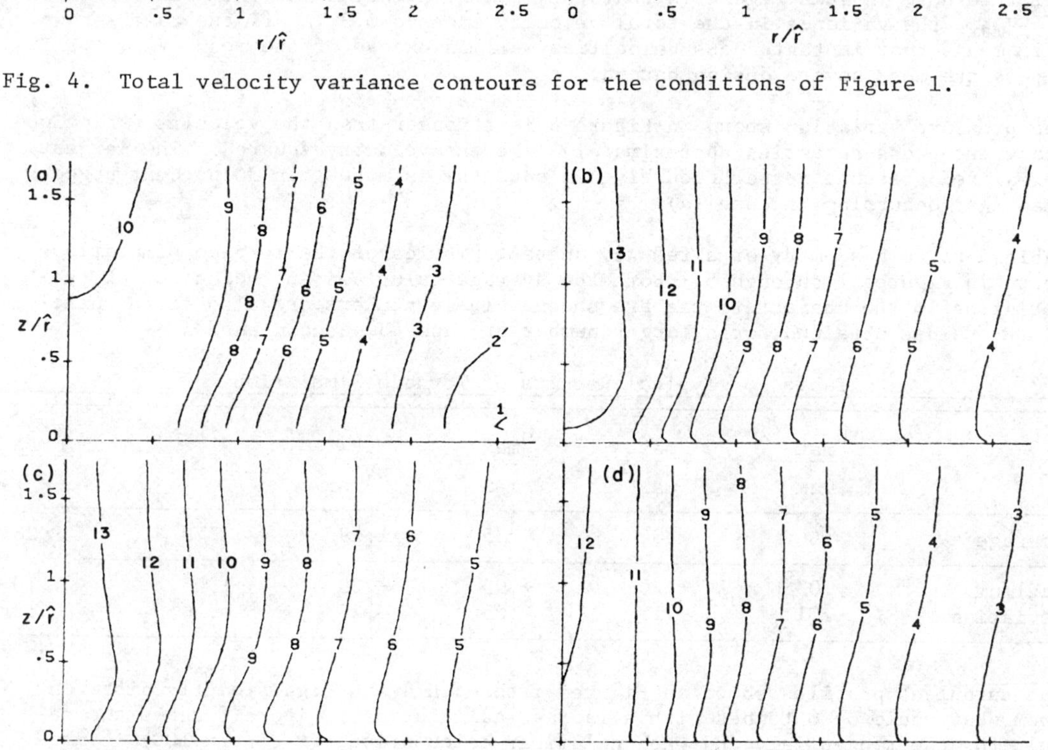

Figure 5. Mean pressure perturbation contours for the conditions of Figure 1. Contours labeled in tenths of $-\rho V_{max}^2$.

The swirl velocities shown in Figure 1 all show a remarkable similarity in spite of the fact that the swirl ratio varies by a factor of 5.4 and the outflow shape parameter varies from -0.1 to +1. Probably the only difference shown which would be of concern to the design engineer evaluating the effects of wind on structures is the difference in the height at which the maximum swirl velocity occurs. As previously explained (1), the boundary layer has the anomalous effect of increasing V in the neighborhood of the ground at small radii, because the strong radial velocity induced in the boundary layer permits the angular momentum to be more nearly conserved in this region of the flow in spite of the loss of angular momentum at the surface. Case (d), with the highest swirl ratio, shows V_{max} occurring closest to the surface (@ $z \approx 0.24\ \hat{r}$).

The radial velocities in Figure 2 also show a strong similarity with the maximum radial inflow occurring very close to the surface, at a radius slightly larger than the radius at which V_{max} occurs. All the figures show a conical type outflow emanating from the vicinity of $r = 0$, $z = 0$. In all but case (a), the low-level inflow penetrates all the way to $r = 0$.

The vertical velocities in Figure 3 show more variation than the two horizontal components. The downdraft along the axis extends to $r \approx 0.5\ \hat{r}$ in case (a), while in case (c) it is almost nonexistent. In case (d), it is more like a stagnation zone which extends out to $r \approx 0.8\ \hat{r}$.

The total velocity variance is shown in Figure 4. Perhaps the most important point to note in this figure is that the flow is highly turbulent. In the vicinity of V_{max} the variance in the total velocity exceeds 0.3 in all the cases shown. This means that instantaneous velocities well in excess of the mean V_{max} will impose the most severe design hazard.

The pressure variation shown in Figure 5 is stronger than the velocity variation since the pressure varies approximately like the velocity squared. The largest normalized pressure defect occurring in case (b) is more than 30 percent bigger than that occurring in case (a).

Table 1 gives a summary of a few key characteristics of the tornado simulations shown in Figures 1 through 5. Both the average value and an indication of the variation in the characteristic are shown. These numbers are also representative of the values obtained for a larger number of runs than shown herein.

Table 1. Typical Results of Tornado Simulation

	$\frac{U_{max}}{V_{max}}$	$\frac{W_{max}}{V_{max}}$	$\frac{q_{max}}{V_{max}}$	$\frac{\Delta p_+}{\rho V^2_{max}}$	$\frac{\hat{z}}{\hat{r}}$	$\frac{V_{max}}{V_{max_{z+}}}$
Average	.50	.36	.35	1.29	.40	1.28
Maximum deviation	+.03 -.01	+.20 -.19	+.05 -.05	+.28 -.23	+.06 -.16	+.10 -.18

The turbulent profiles shown in Figures 1 through 4 are significantly different from what would be obtained with a laminar calculation. Figure 6 shows the result for the tangential wind when a laminar calculation $(r_+U_+/\nu = 100)$ is made for the conditions of case (b). This value of ν was chosen because it leads to roughly the same value of V_{max} as obtained in the turbulent run. The most obvious difference between Figure 6 and Figure 1(b) is the much higher altitude at which V_{max} occurs $(\hat{z}/\hat{r} = 1.4)$. Another difference for this calculation is

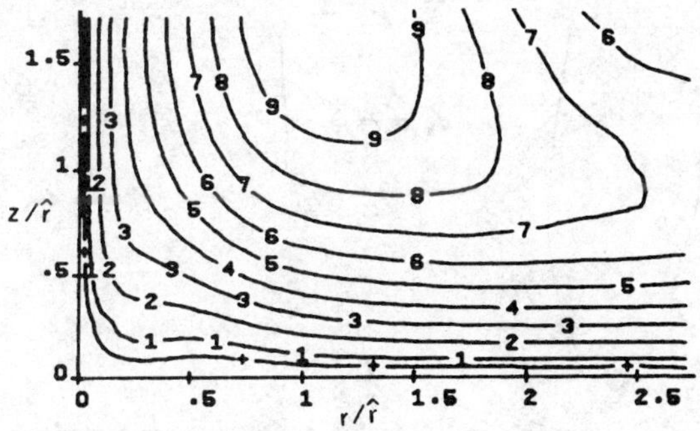

Fig. 6. Tangential velocity contours for the same conditions as Figure 1(c) except it is a laminar calculation ($r_+U_+/\nu = 100$).

that the vertical velocity is much more jet-like at $r = 0$ and small z. The maximum value of W is actually larger than V_{max} for this case.

The laminar profiles are naturally a strong function of the assumed Reynolds number when one attempts to calculate an atmospheric vortex using an effective eddy viscosity. The point we wish to emphasize here is that, even when an effective eddy viscosity is chosen to best match the results of the turbulent model, the velocity profiles are significantly different. It should not be surprising that the turbulent profile leads to higher values of velocity close to the surface.

MAXIMUM VELOCITY ESTIMATES

Two methods are available for estimating the maximum velocities likely to occur in a tornado. One method is to use available distributions of winds obtained from Doppler radar to set model boundary conditions. The other is to take advantage of the relative constancy of $V_{max}/\sqrt{\Delta p/\rho}$ with respect to variations in the major parameters to obtain an estimate of V_{max} based on an estimate of Δp.

As an example of the first method, we consider the Harrah, Oklahoma storm of 8 June 1974. Heymsfield (8) and Brandes (9) have published detailed distributions of winds obtained from Doppler radar. They cannot resolve velocities on sufficiently small scales to identify the maximum winds in the tornado, but their observations can be used to set model boundary conditions on a kilometer scale calculation domain. The largest vertical vorticity ω reported was $2.5\text{x}10^{-2}\text{sec}^{-1}$ over approximately 1 Km (see Figure 8, 1.3 Km altitude, Brandes (9)). The updraft velocity in the same region was 12 m/sec. For uniform flow in at r_+ and out at z_+, our swirl ratio V_+/U_+ is equal to $\omega\ell/W$. Thus, for this case $V_+/U_+ \approx 2.7$.

Figure 7 shows the wind distributions obtained from our model with the computational domain set for 1 Km radius and 1 Km altitude with the swirl ratio $V_+/U_+ = 2.7$, uniform U and V (= 12.5 m/sec) at 1 Km radius, and uniform W at z_+. The surface roughness was set to 0.1 m and the inflow turbulence level was moderate at $q = 0.18\ V_+$.

The maximum swirl velocity (59 m/sec) occurs at 53 m radius and 8.5 m altitude. This appears to be a reasonable value for a moderately strong tornado. The velocity variance (not shown) is predicted to have a maximum of 15.9 m/sec near the same position where V_{max} occurs. Thus, the maximum damage caused by the storm should be associated with those instances in time and space where the turbulent fluctuations add to the mean velocity.

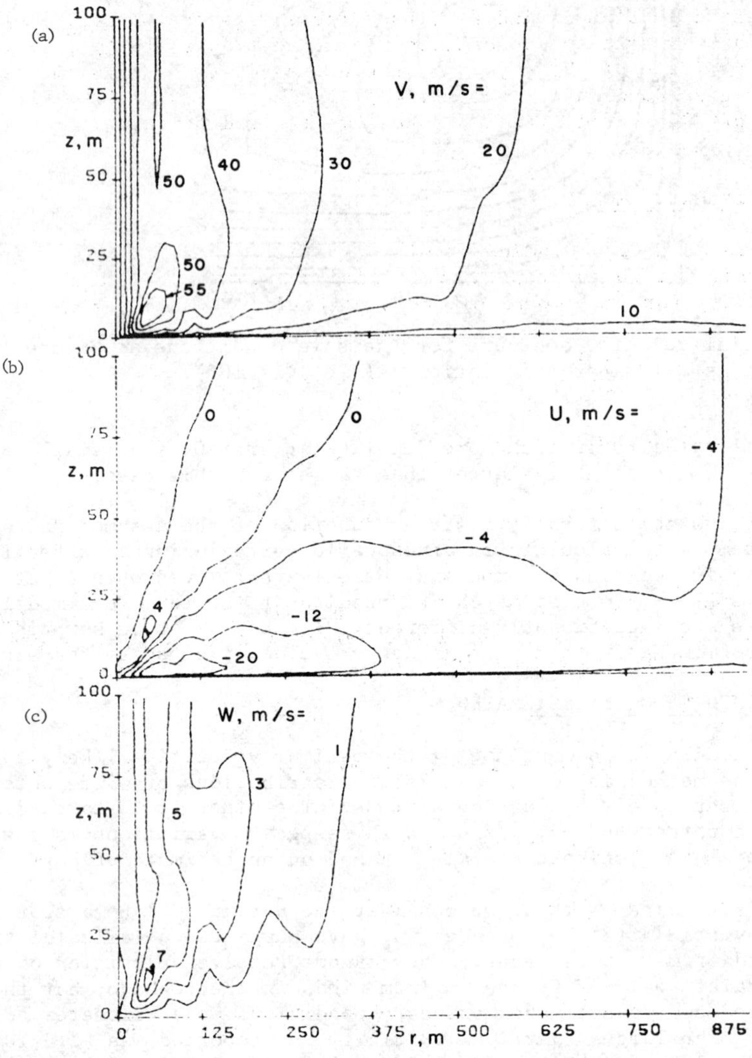

Fig. 7. Model predictions for the wind distributions in the 1974 Harrah, Oklahoma tornado: (a) tangential velocity, (b) radial velocity, (c) vertical velocity.

An alternate estimate of the maximum velocity in the tornado is available if a good estimate of the Δp supportable by the parent storm can be made. In a review article, Lewellen (10) pointed out three possible sources of error in estimating V_{max} from representative thermodynamic soundings of the air mass as proposed by several investigators (e.g., Lilly (11), Kessler (12) and Dergarabedian and Fendell (13)). The relative constancy of $\Delta p/\rho V_{max}^2$ means that two of the three can be bounded by using the number given in Table 1. We are still left with uncertainty in the hydrostatic approximation introduced by the large vertical velocities in the up (and down) drafts aloft in the storm. For the model to shed light on this problem, it will be necessary to increase the top domain of the calculation to heights where the internal Froude number is reduced to order one. If this nonhydrostatic problem is ignored, then Table 1 gives $\Delta p/\rho V_{max}^2 = 1.29$. When this estimate is used in conjunction with an estimate of Δp, it leads directly to an estimate of V_{max}, e.g., the estimate of $\Delta p = 151$ mb for the Enid tornado (Davies-Jones and Kessler (14)) leads to $V_{max} = 97.5$ m/sec. The variation in $\Delta p/\rho V_{max}^2$ indicated in Table 1 would lead to only a ±12 percent variation in this estimate.

ENGINEERING MODEL

For such purposes as design studies of wind-generated missiles, we have generated a rough parametric fit to the typical wind fields obtained from our model. The tangential velocity for this model may be specified as (with $\hat{r} = 1$)

$$\begin{aligned} V &= \frac{c}{r}\left[1 - e^{-\alpha_1 r^2}\right]\frac{\ln z/z_o}{\ln z_m/z_o} \quad ; \quad z_o \le z \le z_m \\ &= \frac{c}{r}\left[1 - e^{-\alpha_1 z_m^2 r^2/z^2}\right] \quad ; \quad z_m < z \le z_T \qquad (1) \\ &= \frac{c}{r}\left[1 - e^{-\alpha_1 z_m^2 r^2/z_T^2}\right] \quad ; \quad z > z_T \end{aligned}$$

with $c = [1 - e^{-\alpha_1}]^{-1}$; $z_m = \delta_1[1 - e^{-\alpha_2 r^2}]$ and α_1, α_2, δ_1 and z_T/z_m parameters to be specified. The radial and vertical mean velocities are specified in terms of a stream function ψ

$$U = -\frac{1}{r}\frac{\partial\psi}{\partial z} \quad , \quad W = \frac{1}{r}\frac{\partial\psi}{\partial r} \qquad (2)$$

with ψ related to V through two more parameters K_1 and K_2. This expression is

$$\psi = A[z \ln z/z_o - (z-z_o)] - K_2 V_m r(\ln z_m/z_o)(z^2-z_o^2) \quad ; \quad z_o \le z \le z_m \qquad (3)$$

$$\psi = \frac{z^2 K_1 \delta_1^2 c}{z_m^2 r}\left[1 - e^{-\alpha_1 z_m^2 r^2/z^2}\right]\left[1 - e^{-\alpha_2 z_m^2 r^2/z^2}\right]^2 \quad ; \quad z_m < z \le z_T$$

$$\psi = \frac{z_T^2 K_1 \delta_1^2 c}{z_m^2 r}\left[1 - e^{-\alpha_1 z_m^2 r^2/z_T^2}\right]\left[1 - e^{-\alpha_2 z_m^2 r^2/z_T^2}\right]^2 \quad ; \quad z > z_T$$

with

$$A = \frac{K_1 V_m z_m^2 + K_2 V_m r(\ln z_m/z_o)(z_m^2-z_o^2)}{z_m \ln z_m/z_o - (z_m-z_o)}$$

Plots of U, V, and W are given in Figure 8 for the following representative values of the parameters: $\alpha_1 = \alpha_2 = 1.26$, $\delta_1 = 1$, $z_T/z_m = 1.5$, $z_o = 0.002$, $K_1 = 0.2$ and $K_2 = 0.1$. These distributions are quite similar to those shown in Figures 1 through 3.

CONCLUDING REMARKS

Tornadic wind and pressure distributions have been predicted for a number of different input variables. The horizontal wind distributions show considerable similarity when normalized by V_{max} and $\hat{r}$. This similarity was used to generate an engineering approximation to the wind distribution which can be used for design studies. The maximum winds occurring in a tornado can be estimated either by using the wind distributions obtained by Doppler radar in storms to set model boundary conditions, or by using the similarity in $V_{max}/\sqrt{\Delta p/\rho}$ in the model output to obtain an estimate of V_{max} if an estimate of Δp is available from analysis of thermodynamics soundings.

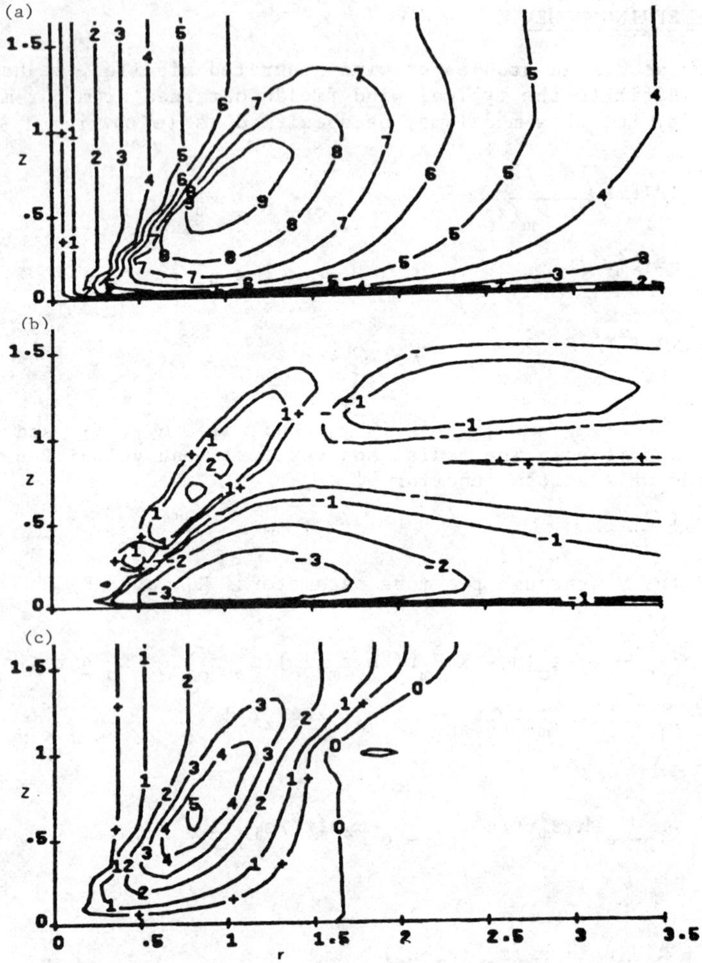

Fig. 8. Velocity distributions for the analytic engineering model as given by Eqs. (1)-(3): (a) tangential velocity, (b) radial velocity, (c) vertical velocity.

ACKNOWLEDGMENT

This research was supported by the Nuclear Regulatory Commission under Contract No. NRC 04-76-218 with R. F. Abbey as technical monitor.

REFERENCES

1. Lewellen, W.S. and Teske, M.E., Turbulent transport model of low-level winds in a tornado, Proceedings of AMS 10th Conference on Severe Local Storms, Oct. 18-21, 1977, Omaha, NE, 291-298 (1977).

2. Teske, M.E. and Lewellen, W.S., Turbulent transport model of a thunderstorm gust front, Proceedings of AMS 10th Conference on Severe Local Storms, Oct. 18-21, 1977, Omaha, NE, 143-150 (1977).

3. Donaldson, C.duP., Atmospheric turbulence and the dispersal of atmospheric pollutants, Proceedings of AMS Workshop on Micrometeorology (D.A. Haugen, ed.), Science Press, 313-390 (1973).

4. Lewellen, W.S., Use of invariant modeling, Handbook of Turbulence, Vol. 1 (W. Frost and T.H. Moulden, eds.), Plenum Publishing Corp., 237-290 (1977).

5. Lewellen, W.S., Influence of body forces on turbulent transport near a surface, Journal of Applied Mathematics and Physics (ZAMP), 28: 825-834 (1977).

6. Lewellen, W.S., Teske, M.E. and Sheng, Y.P., Modeling tornado dynamics, Annual Progress Report, Prepared for Office of Nuclear Regulatory Commission, A.R.A.P. Report No. 347 (1978).

7. Rotunno, R., A study in tornado-like vortex dynamics, Journal of the Atmospheric Sciences, to appear (1979).

8. Heymsfield, G.M., Kinematic and dynamic aspects of the Harrah tornadic storm analyzed from dual-Doppler radar data, Monthly Weather Review, 106: 233-254. (1978).

9. Brandes, E.A., Mesocyclone evolution and tornadogenesis: Some observations, Monthly Weather Review, 106: 995-1011 (1978).

10. Lewellen, W.S., Theoretical models of the tornado vortex, Proceedings of A Symposium on Tornadoes: Assessment of Knowledge and Implications for Man, June 22-24, 1976, Texas Tech University, Lubbock, TX (1976).

11. Lilly, D.K., Tornado dynamics, NCAR Manuscript 69-117, National Center for Atmospheric Research, Boulder, CO (1969).

12. Kessler, E., Tornadoes, Bulletin of the American Meteorological Society, 51: 926-936 (1970).

13. Dergarabedian, P. and Fendell, F.E., Estimation of maximum wind speeds in tornadoes, Tellus, 22: 511-515 (1970).

14. Davies-Jones, R.P. and Kessler, E., Tornadoes, Weather and Climate Modification (W.N. Hess, ed.), John Wiley, NY, 552-595 (1974).

AN EVALUATION OF DIAGNOSTIC MARINE BOUNDARY LAYER MODELS APPLIED TO TROPICAL CYCLONES

M. D. POWELL

National Hurricane and Experimental Meteorology Laboratory, NOAA, Miami, Florida, USA

Abstract

Four diagnostic marine boundary layer models were evaluated for applicability to the hurricane regime. The goal was to develope an operational method of estimating surface variables with research aircraft flight-level (500-m) data. To evaluate the models, we compared model wind speed with ground truth buoy and ship data from Hurricanes Eloise (9/22/75) and Anita (8/30-9/2/77) and with vertically stacked several-level aircraft data in Eloise (9/17/75) and Caroline (8/30/75). Three models estimated wind speed similarly, giving an error of 10%. Model results also included 10-m-level neutral drag coefficients which were examined and then compared to previous studies.

INTRODUCTION

Imminent landfall of a hurricane presents many problems. The likelihood of major disaster in low-lying heavily populated areas is high. Accurate warnings from detailed forecasts must be given to save life and property and to reduce the high costs of preparing relatively large coastal areas for hurricane occurrence. The National Hurricane and Experimental Meteorology Laboratory (NHEML) is investigating details of hurricane landfall in an effort to develop operational improvements of forecasts, warning procedures and damage potential estimates. Investigation of the hurricane boundary layer is a necessary part of the solution.

The object of this study is to evaluate and compare the capabilities of several boundary layer models for determining low-level wind structure in hurricanes. In addition, estimations of surface turbulent parameters, surface stresses, and neutral drag coefficients are evaluated. The goal of this evaluation is to develop an operational diagnostic model that is capable of operating on low-level (about 500 m) aircraft data to estimate data close to the surface (10 m).

The four models investigated in this study are the:

1) Deardorff (1) planetary boundary layer (PBL) parameterization scheme by Moss and Rosenthal (2)
2) Deardorff technique modified by the author,
3) Surface layer wind reduction model, Cardone I (3)
4) Cardone II model, which is based on Blackadar's two-layer neutral model (3,4)

All models depend upon the applicability of the Monin-Obukov (5) surface layer "similarity" theory. This similarity theory applies to that portion of the atmosphere in contact with the surface and within the planetary boundary layer (PBL). The PBL in tropical circulations, as defined by (6), pertains to the region adjacent to the earth's surface where small-scale, surface-induced turbulence occurs almost continuously in time, excluding convective turbulent regimes. The surface layer is defined as the region within which stress is approximately constant with height. Since direct measurements of stress are difficult to make at greater than 10% to 20% accuracy, the surface layer is determined to be the height at which the stress decreases to 80% of its surface value. This height ranges from 20 m to 200 m over land. Over water, on the periphery of the 1975 Hurricane Eloise, it was estimated to be 175 m.

MODELS

The four diagnostic models require low-level (500 m) research aircraft input data consisting of potential temperature, mixing ratio and wind speed, sea surface temperatures, sea-level pressure, and boundary layer height. These data are utilized in iterative schemes that involve surface layer similarity relationships. Wind speed is computed by

$$U = \frac{u_*}{k}\left(\ln \frac{Z}{z_o} - \Psi_m\right) \tag{1}$$

where u* is the frictional velocity, k is the Von-Karman constant, Z is the height, z_o is the roughness length, and Ψ_m is a stability function. The friction velocity provides for estimates of surface stress,

$$\tau = \rho u_*^2 \quad . \tag{2}$$

where ρ is the mean air density, L_v is the latent heat of vaporization and c_p is the specific heat at constant pressure.

The Moss-Rosenthal model includes relations for wind and temperature deficits and bulk momentum and heat transfer coefficients that are computed according to stability, as determined from the bulk Richardson number.

$$Ri_B = \frac{gZ_{AN}\Delta\theta}{\theta V^2} , \tag{3}$$

where g is the gravitational acceleration, $\Delta\theta$ is the air-sea potential temperature difference, θ is the bulk potential temperature, and V is the wind speed. The Powell model (6) employs the same deficit relationships for wind speed and temperature, but uses surface-layer similarity expressions as a method of parameterizing the unstable boundary layer, rather than bulk tansfer coefficient relationships. The remaining two diagnostic models were developed by Cardone (3); one is a surface boundary layer model that employs similarity relationships, but which assumes that the aircraft is at the top of the surface layer. The other Cardone model is based on a two-layer neutral baroclinic model by Blackadar (3). This model was used by Elsberry, et al. (7) in a semiempirical model of the hurricane boundary layer and is essentially a modified Ekman spiral. In addition to the four models mentioned, two wind speed estimation models were employed: Bates (8) developed hurricane wind profiles normalized by the 150- to 350-m level for over land and over water. These profiles are given in figure 1. A rough estimation of surface winds in hurricanes can be made by multiplying low-level aircraft winds (usually at 500 m) by 0.8. This estimation was used by Banner I. Miller to provide quick approximations of surface winds below flight level.

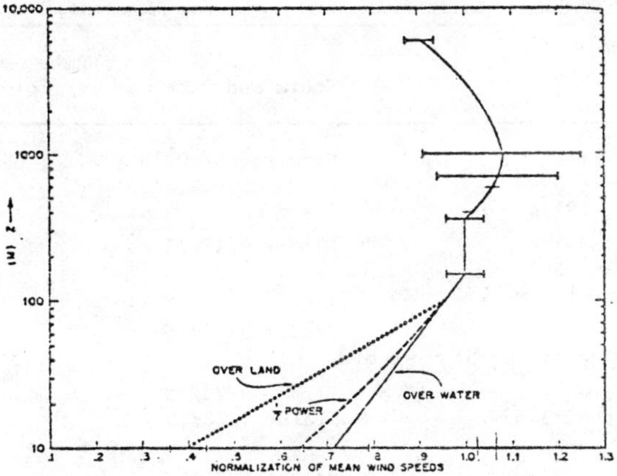

Fig. 1. Normalized profile of variation of mean wind speed with height in tropical cyclones. (Bates 1977).

PROCEDURE

Data from three storms were available for analysis: Hurricane Caroline on 30 August 1975, Hurricane Eloise on 17, 22 and 23 September 1975, and Hurricane Anita on 30 and 31 August and 2 September 1977. Complete details on these storms are available through the following works: Caroline - (10, 11, 12); Eloise - (6, 11, 12, 13).

Two types of data measuring and recording platforms were used for this study - one airborne and one at sea. The NOAA Data Buoy Office operates and maintains the "EB" series of instrumented buoys. Wind speed and direction, air and sea surface temperature, dew point, pressure and accumulated precipitation were measured in Hurricane Eloise, but dew point and precipitation were eliminated from the 1977 measurements in Anita. An instrumented NOAA DC-6 was employed in Hurricanes Caroline and Eloise in 1975. In 1977, Hurricane Anita was flown with a WC-130B (Hercules) and a WP3-D (Lockheed Orion). These aircraft were equipped with inertial navigation systems and sensors for measurement of temperature, pressure, wind speed and direction, and dew point.

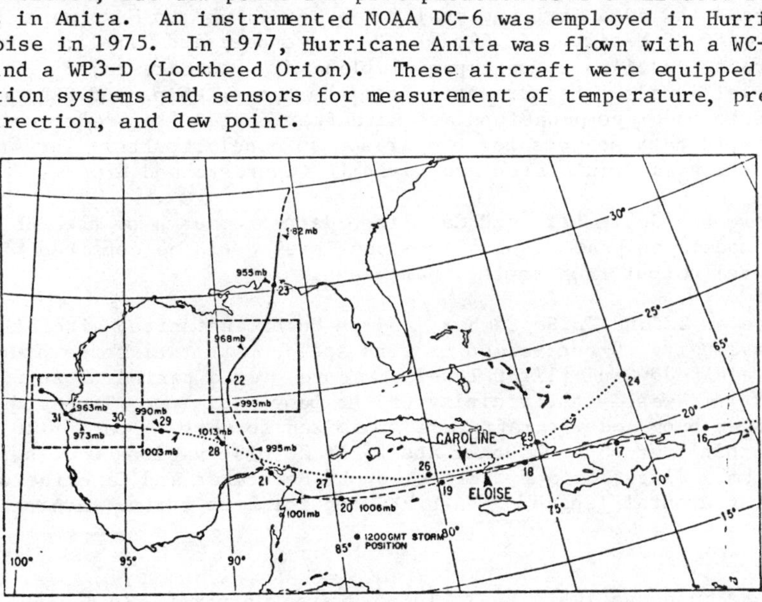

Fig. 2. Tracks of Hurricane Caroline and Eloise, including 12Z position minimum surface pressure and aircraft experiment locations (in boxes). Depression and tropical storm stages are indicated by dotted and dashed lines, respectively.

Table 1. Model Comparison Data Sources

Source	Storm and Date	Number of Observations for Comparison
NOAA Research Facilities Center aircraft DC-6, 39C Two levels: 267 m, 394 m	Caroline 8/30/75	2
NOAA Research Facilities Center aircraft DC-6, 39C Three levels: 86 m, 153 m, 362 m	Eloise 9/17/75	3
U. S. Navy aircraft carrier *Lexington* anemometer ht - 48 m	Eloise 9/23/75	1
Exxon tanker *San Francisco* Anemometer ht - 23 m	Eloise 9/22/75 Eloise 9/23/75	2
Environmental buoy EB-10; anemometer ht - 10 m	Eloise 9/22/75 Eloise 9/23/75	4
Environmental buoy EB-4; anemometer ht - 10 m	Eloise 9/23/75	2
Environmental buoy EB-4 anemometer ht - 5 m	Anita 8/30/77	5 aircraft obs to 3 buoy obs
	Anita 8/31/77	1 aircraft to 1 buoy
	Anita 9/02/77	1 aircraft to 1 buoy
Environmental buoy; EB-71 anemometer ht - 10 m	Anita 8/31/77	3 aircraft to 1 buoy
	Anita 9/02/77	1 aircraft to 1 buoy

Flight-level measurements of wind speed and other meteorological parameters from the NOAA Research Facilities Center aircraft were used as input data for the diagnostic models. Model computations were then compared to suface wind and other measurements made by NOAA environmental buoys and ships[1] in Hurricane Eloise and Anita in the Gulf of Mexico. In Eloise (17 September) and Caroline surface measurements were not available, but special planetary boundary layer (PBL) legs were flown at several levels. Measurements from the buoys, ships and PBL legs that were compared to model computations are hereafter called "reference data." Table 1 gives reference data sources for comparison to model results. The storm tracks and locations of data acquisition are given in figures 2 and 3.

For Eloise (17 September) and Caroline, data from each of the PBL levels were put into the models so that results from one level could be compared with results based upon model output from another data level.

In Eloise on 22 and 23 September, and in Hurricane Anita, aircraft data were prepared by averaging 10-sec values of wind speed, potential temperature, mixing ratio, radar altitude, and flight level pressure over a period (1 to 3 min) during which the aircraft was in the vicinity of the buoy position. Time and space differences between buoy and aircraft were minimized so that 16 out of 20 observations were within 1 hr of each other and 15 of 20 observations were separated by less than 20 km. Flight data for Eloise on 17 September and Caroline were averaged over the total length of each PBL leg for 2 to 4 min. Averaging times

[1] Ship data were made available by T. Inman, Exxon Corporation, for the Exxon Tanker *San Francisco*, and by Lt. Cmdr. W. Peterson, U.S. Navy, for the aircraft carrier *Lexington*.

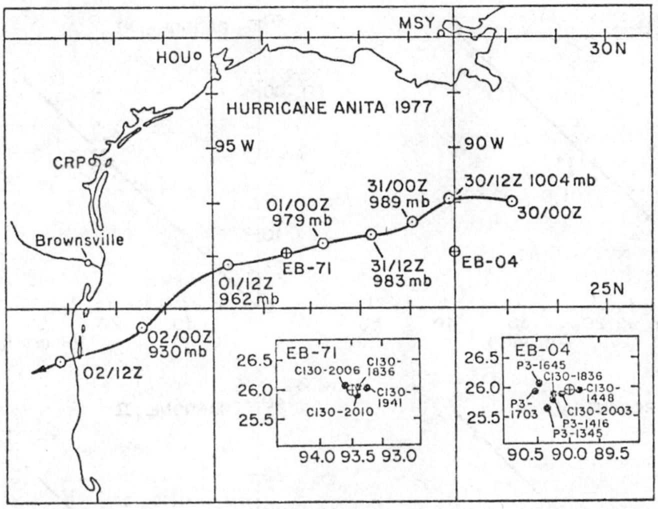

Fig. 3. Track of Anita including 12-hour positions, environmental buoy positions and minimum sea level pressures. Insets give aircraft positions used for comparisons with respect to the buoys.

for buoys EB-10 and EB-4 in Eloise (22 and 23 September 1975) were 15 and 40 min, respectively, and 8.5 min for buoys EB-4 and EB-71 in Anita. Ship data were available from the Exxon tanker *San Francisco* and the U. S. Navy aircraft carrier *Lexington*; these data were not averaged. Largest errors in low-level wind estimation usually occur in locations close to the inner core of the eye wall where wind velocity and other variables change greatly with radial distance. Here it would be difficult for aircraft averages to correspond to portions of the storm measured by the buoys and ships.

RESULTS

Wind Speed Estimations

Plots of reference wind speed, U_{ref}, versus model wind speeds, U_m, are given in figure 4 for the four diagnostic models and the Bates and "0.8 estimation" models. An idea of the accuracy of each model may be gained by evaluating the distance between the solid lines and the plotted points in figure 4. The Cardone II model and the Bates profile appear to underestimate winds, as evidenced by the large number of points above the line.

Table 2 lists average absolute, root mean square (RMS) and percentage wind speed errors for each of the models for all the comparison sources in Table 1 and for the buoy data only. These errors can be thought of as distance of points from the perfect-fit lines in figure 4, the distance being ΔU. The Cardone I model is slightly more accurate than the Moss-Rosenthal and Powell models, which achieved similar accuracy. The same trend (with decreased errors) was evident when only buoy comparisons were considered. The Cardone II model and the Bates profile give larger magnitude average absolute error because of wind speed underestimation.

The large bias in the Cardone II model may be caused by the specification of the surface layer height, the substitution of flight level wind speed for the geostrophic wind speed or the application of geostrophically based Ekman theory to a nongeostrophic flow.

The .8 estimation worked quite well for an absolute error of 1.8 m/s and a percentage error (comparable to the Moss-Rosenthal and Powell models) of 9%. The

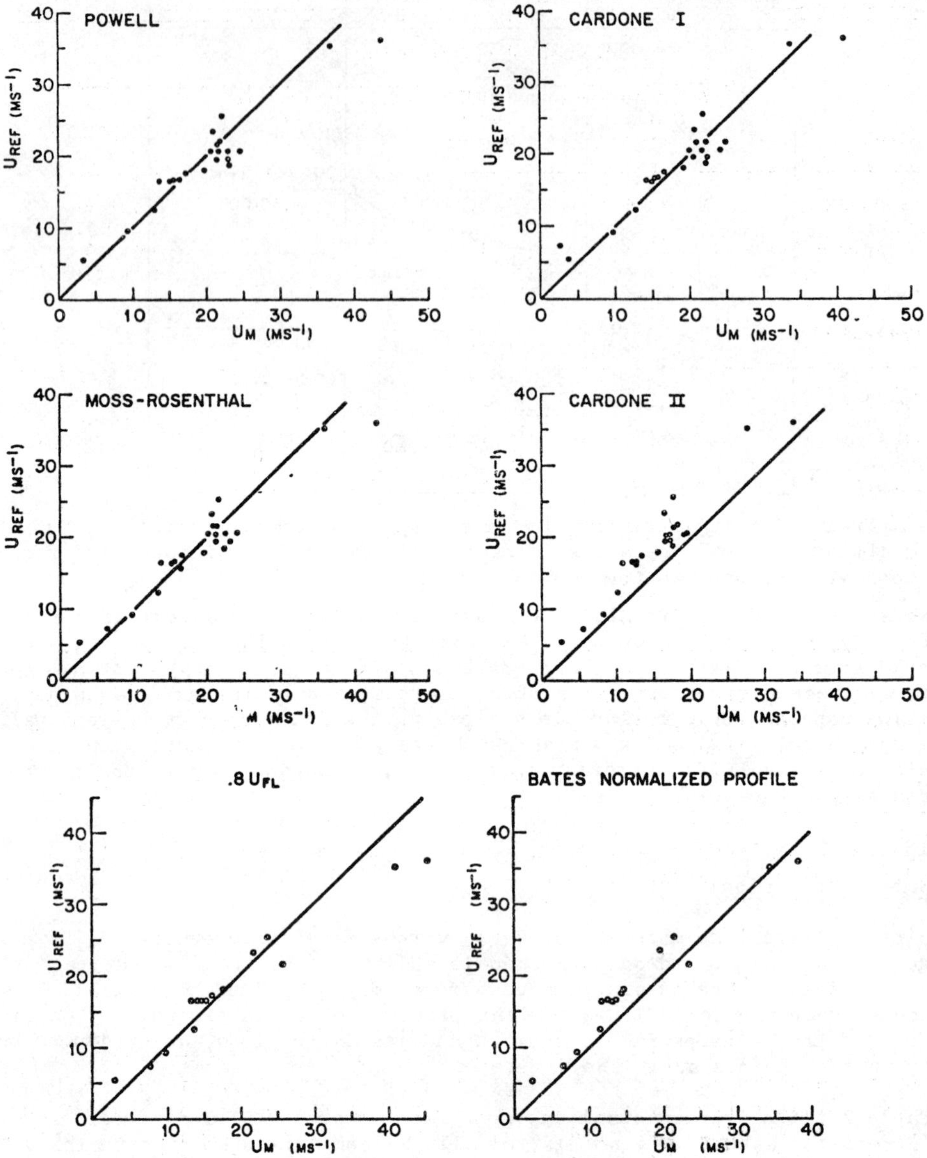

Fig. 4. Reference wind speed versus model wind speed. Solid line indicates perfect fit.

RMS error of ±2.5 m/s was large due to overestimates at high wind speeds (35 m/s), but this method appears to be a good way to make a quick estimate of surface wind from aircraft winds. Model errors (ΔU) were investigated for bias caused by wind speed (U_{ref}) or stability ($\Delta\theta_{va-s}$). Error increased with wind speed for the 0.8 estimation, since for high wind speeds, the fractional speed decrease from flight level was greater than 0.8. No stability bias was found.

When the natural variability of hurricanes and the time and space differences between reference observations and input data aircraft observations are considered,

Table 2. Model vs Reference Comparison Errors

Model	Avg absolute error (m/s)	RMS error ±(m/s)	Percentage error (%)	
		Wind speed comparison errors		
Cardone I	1.6	2.0	8.7	Based on 19 buoy, ship and A/C comparisons
Moss-Rosenthal	1.8	2.4	10.6	
Powell	1.9	2.6	10.3	
Cardone II	3.8	4.2	21.5	
Cardone I	1.5	1.8	7.8	Based on 12 buoy comparisons
Moss-Rosenthal	1.7	2.0	9.3	
Powell	1.7	2.0	8.5	
Cardone II	4.2	4.7	22.8	
0.8	1.8	2.5	9.1	
Bates	2.6	2.9	14.9	

a percentage error of 9% for the Cardone model and 10% for the Moss-Rosenthal, Powell and 0.8 models indicate that these models can estimate surface winds accurately from aircraft data.

Drag Coefficients Over Water

Neutral stability drag coefficient evaluated at anemometer level (10 m) is related to the friction velocity and is often used to calculate surface stress

$$\tau = \rho C_{D10} U_{10}^2 \tag{4}$$

where

$$C_{D10} = \left(\frac{u_*}{U_{10}}\right)^2 \tag{5}$$

by definition, (16). The behavior of the drag coefficient with increasing wind speed over the ocean is a controversial subject. Several investigators have attempted to relate the drag coefficient to the mean wind speed with varying results (18,19,20,21,22). Sheppard et al. found a linear increase of C_D with mean wind speed. Pond et al. reported a near-constant drag coefficient. Neumann measured a decrease in drag coefficient with increasing wind speed. Sethuraman and Raynor found a constancy of drag coefficient with mean wind speed for each of three roughness categories. Garratt combined results of 14 investigations and showed a linear dependence upon wind speed. More than 85% of his data were for speeds less than 10 m/s, however, and very little data are available for speeds over 15 m/s.

Investigators have calculated drag coefficients in hurricanes through several methods (23,24,25,26,27). Riehl and Malkus computed drag coefficients for Hurricane Daisy from heat and moisture budgets employing the bulk aerodynamic formulation. Miller, Hawkins and Rubsam, and Hawkins and Imbembo computed drag coefficients for Hurricanes Helene (1958), Hilda (1964), and Inez (1966) respectively by integrating the tangential equation of motion through the inflow layer,

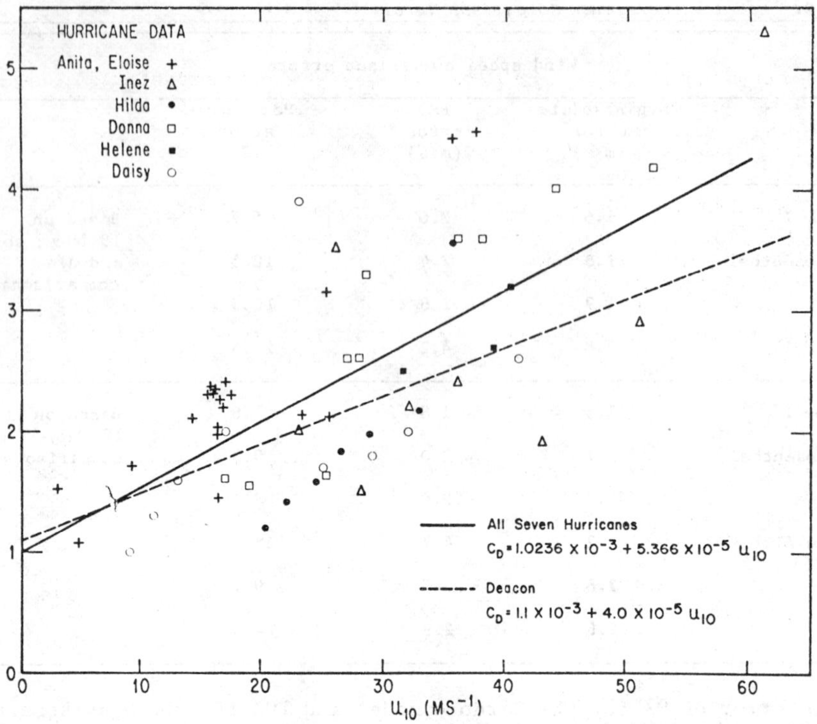

Fig. 5. Ten meter level neutral drag coefficient from several studies plotted versus wind speed.

$$\tau_r = \frac{1}{R^2} \frac{\partial}{\partial R} \int_{P_S}^{P_H} V_R \, RM \frac{dp}{g} + \frac{\omega_H M_H}{Rg}, \tag{6}$$

where τ_θ is the tangential surface stress component, R is radial distance, M is absolute angular momentum, ω is vertical velocity; the subscripts H and S pertain to the top of the atmosphere and the sea surface respectively. Miller made drag coefficient calculations from Hurricane Donna (1960) by integrating the tangential equation of motion in terms of vorticity

$$\tau_\theta \approx \int_{P_S}^{P_H} V_R \, \zeta_A \frac{dp}{g} \tag{7}$$

where V_R is the radial wind component and ζ_A is the absolute vorticity of the tangential wind. After computing surface stresses were computed, drag coefficients were calculated from equation 4.

Drag coefficients for Hurricanes Eloise (1975), Caroline (1975), and Anita (1977), were computed from model friction velocities and wind speeds for eight aircraft and ship reference cases and by the use of model friction velocities and 10 m level measured wind speeds for the remaining 12 buoy reference cases.[2] Figure 5 presents a scatter diagram of drag coefficient versus anemometer level wind speed for all of the storms mentioned above. Despite considerable scatter, figure 5 indicates that the drag coefficient increases with wind speed. A linear least-squares fit to the data of figure 5 gives the relation,

[2] Drag coefficient computations of the Powell, Moss-Rosenthal, and Cardone I models were averaged for each buoy wind speed observation.

$$C_{D_{10}}(\times 10^3) = 1.0236 + 5.366 \times 10^{-2} U_{10} \quad , \tag{8}$$

which is plotted on the figure together with the Deacon relationship used in NHEML numerical hurricane models (28),

$$C_{D_{10}}(\times 10^3) = 1.1 + 4 \times 10^{-2} U_{10} \quad . \tag{9}$$

Much of the scatter in figure 5 can be attributed to differences in method of computation, that is, from the momentum budget or heat budget, and to errors and differences in measurement systems. Previous hurricane studies involved aircraft wind data that were measured with a Doppler naviagation system while current studies make use of an inertial navigation system. Although it appears that the Deacon drag coefficients may be too low at speeds above 10 m/s, it is realized that more high quality data will be required to determine the accuracy of equation 8.

CONCLUSION

Four diagnostic boundary layer models and two simple estimation models were evaluated for their ability to estimate wind speed near the surface by a comparison to buoy, ship and aircraft data in Hurricanes Caroline, Eloise and Anita. The Cardone I model was most accurate (9% error) and the Moss-Rosenthal, Powell and 0.8 models followed with 10% errors. Drag coefficient was found to increase with wind speed to a greater extent than that shown by the Deacon relationship and ranges of turbulent parameters were established for the hurricane regime over water. It is concluded that diagnostic marine boundary layer models are able to estimate wind speed accurately in hurricanes. It is believed that the above models can be used on an operational basis to provide valuable low-level wind and surface layer information from research aircraft efforts in hurricanes and other high wind situations.

ACKNOWLEDGMENTS

I wish to thank Peter Black and Drs. Michael Moss and Frank Merceret of NHEML for providing data and helpful discussions. I greatly appreciate the valued suggestions and comments of Drs. Alfred K. Blackadar, Toby Carlson and John Lee of my M.S. thesis committee at the Pennsylvania State University. Figures were expertly drafted by Dale Martin and Antonio Mijares. The manuscript was typed by Lee Stirewalt and Lisa Strauss.

REFERENCES

1. Deardorff, J. W., "Parameterization of the Planetary Boundary Layer for Use in General Circulation Models," Mon. Wea. Rev., Vol. 100, Feb. 1972, pp. 93-106.

2. Moss, M. S., and Rosenthal S. L., "On the Estimation of Planetary Boundary Layer Variables in Mature Hurricanes," Mon. Wea. Rev., Vol. 103, Nov. 1975, pp. 980-988.

3. Cardone, V. J., "Specification of the Wind Distribution in the Marine Boundary Layer for Wave Forecasting," Geophysical Sciences Laboratory TR69-1, Dept. Meteor. and Oceanog., New York University, University Heights, New York, New York 1969.

4. Blackadar, A. K., "A Simplified Two-Layer Model of the Baroclinic Neutral Atmospheric Boundary Layer," Final Report: Flux of Heat and Momentum in the Planetary Boundary Layer of the Atmosphere, The Pennsylvania State University, University Park, Penn., 1965.

5. Monin, A. S., and Obukov, A. M., "Dimensionless Characteristics of Turbulence in the Surface Layer," Akad. Nauk. SSSR, Geofiz Inst., Tr. No. 24, 1954, pp. 163-187.

6. Moss, M., and Merceret, F., "A Note on Several Low-Level Features of Hurricane Eloise (1975)," Mon. Wea. Rev., Vol. 104, July, 1976, pp. 967-971.

7. Powell, M. D., "An Evaluation of Diagnostic Marine Boundary Layer Models Applied to Hurricanes," M.S. thesis, The Pennsylvania State University, University Park, Penn., 1978.

8. Elsberry, R. L., Pearson, N., and Corgnati, L., "A Quasi-Empirical Model of the Hurricane Boundary Layer," J. Geophys. Res., Vol. 79, No. 21, 1974, pp. 3033-3040.

9. Bates, J., "Vertical Shear of the Horizontal Wind Speed in Tropical Cyclones," NOAA Tech. Memo ERL WMPO-39, NOAA, NHEML, Coral Gables, Fla., 1977.

10. Merceret, F. J., "The Turbulent Microstructure of Hurricane Caroline (1975)," Mon. Wea. Rev., Vol. 104, October, 1976, pp. 1297-1307.

11. Herbert, P., "Atlantic Hurricane Season of 1975," Mon. Wea. Rev., Vol. 104, April, 1976, pp. 453-465.

12. Moss, M. S., "Low-Level Features of Two High Wind Speed Regimes," Ph.D. thesis, The Florida State University, Tallahassee, Fla., 1978.

13. Environmental Data Service, Data Buoy Office, "Buoy Observations During Hurricane Eloise," NOAA, National Oceanographic Data Center, Washington, D. C., 1975.

14. Sheets, R., "On the Evolution of Hurricane Anita." Papers, 11th Conference on Tropical Meteorology-AMS, Miami Beach, Fla., December, 1977, pp. 280-287.

15. Lawrence, M., "Atlantic Hurricane Season of 1977," Mon. Wea. Rev., Vol. 106, April 1978, pp. 534-545.

16. NOAA Data Buoy Office, "Data Report, Buoy Observations During Hurricanes Anita and Babe, August-September 1977," U. S. Dept. of Commerce Natl. Ocean. and Atmos. Admin., NSTL station, MS, August 14, 1978.

17. Kraus, E. B., "Atmosphere-Ocean Interaction," Oxford Monographs on Meteorology, Clarendon Press, Oxford, 1972.

18. Sheppard, P. A., D. T. Tribble, and J. R.-Garratt et al., "Studies of Turbulence in the Surface Layer Over Water I. Instrumenta tion, Programme, Profiles," Quart. J. Roy. Meteor. Soc., Vol. 98, July, 1972, p. 627.

19. Pond, S., Phelps, G., Paquino, et al., "Measurements of the Turbulent Fluxes of Momentum Moisture and Sensible Heat Over the Ocean," J. Atmos. Sci., Vol. 28, September, 1971, pp. 901-917.

20. Neumann, G., "Wind Stress on Water Surfaces," Bull. Amer. Meteor. Soc., Vol. 37, May, 1956, pp. 211-217.

21. Sethuraman, S., and Raynor, G., "Surface Drag Coefficient Dependence on the Aerodynamic Roughness of the Sea," J. Geophy. Res., Vol. 80, No. 36, 1975, pp. 4983-4988.

22. Garratt, J. R., "Review of Drag Coefficients Over Oceans and Continents," Mon. Wea. Rev., Vol. 105, July, 1977, pp. 915-927.

23. Hawkins, H. F., and Imbembo, S. M., "The Structure of a Small, Intense Hurricane - Inez 1966," Mon. Wea. Rev., Vol. 104, April, 1976, pp. 418-442.

24. Hawkins, H. F., and Rubsam, D. T., "Hurricane Hilda 1964: II. Structure and Budgets of the Hurricane on October 1 (1964)," Mon. Wea. Rev., Vol. 95, September, 1968, pp. 677-636.

25. Miller, B. I., "On the Momentum and Energy Balance of Hurricane Helene (1958)," NHRP Report No. 53, National Hurricane Res. Lab., Coral Gables, Fla., 1962.

26. Miller, B. I., "A Study of the Filling of Hurricane Donna (1960) over land," Mon. Wea. Rev., Vol. 92, September, 1964, pp. 389-406.

27. Riehl, H., and Malkus, J., "Some Aspects of Hurricane Daisy, 1958," NHRP Report No. 46, NOAA, NHEML, Coral Gables, Fla., 1961, pp. 1-26.

28. Rosenthal, S. L., "The Response of a Tropical Cyclone Model to Variations in Boundary Layer Parameters, Initial Conditions, Lateral Boundary Conditions and Domain Size," Mon. Wea. Rev., Vol. 99, October, 1971, pp. 767-777.

THE STRUCTURE OF MODERATELY STRONG WINDS AT A MID-ATLANTIC COASTAL SITE (BELOW 75m)

H. W. TIELEMAN and S. E. MULLINS

Virginia Polytechnic Institute and State University, Blacksburg, Virginia, USA

ABSTRACT

Velocity and temperature measurements were obtained from an instrumented 76-meter meteorological tower located near the Atlantic Ocean at Wallops Island, Virginia. The instrumentation consists of a slow-response cup-vane and resistance temperature (profile) system and a hot-film, thermocouple system for turbulence measurements. Results are presented for moderately strong winds from westerly directions (category I) and for on-shore winds from southerly directions (category II).

Results from category I winds indicate the presence of low-frequency velocity fluctuations affecting all turbulence parameters similar to the observations made at Kansas and Minnesota.

Winds of category II experience a change in surface roughness and surface temperature as they cross the coast line, resulting in the development of an internal boundary layer (IBL). The stable ocean air above the IBL shows extremely low turbulence levels, often less than 3%. Because of the lack of turbulent mixing with adjacent layers, Coriolis effects are important and velocity profiles with a maximum velocity (low-level jet) at heights between 60-200 m exist.

INTRODUCTION

Structures and structural elements are usually designed to support the maximum loads that are likely to be imposed upon them within the lifetime of the structure. Under the strongest wind conditions the assumption of a neutrally stable planetary boundary layer (PBL) is probably satisfied. However, problems of fatigue and serviceability of a structure are often overlooked in its design. These problems can become accute under much less severe or under moderate wind conditions. Consequently, in the field of wind loading on structures and their response, one needs to study the flow in the PBL not only under extreme wind conditions but also under more common moderate conditions. Flow characteristics of the PBL under these conditions, mostly corresponding to flow over relatively homogeneous terrain have been summarized in several review papers [1,2,3,4]. These results have formed the basis for empirical models of the wind structure which are also presented in these papers and others [5,6]. Most of the proposed models, with exception of the model proposed by Panofsky [5], are limited to the following conditions: 1. Strong, stationary winds, 2. Neutral thermal stability

3. Level terrain with uniform roughness.

Upon reading of the cited references one does not always get a clear picture when neutral conditions exist. Deaves and Harris [6] do not give any conditions for neutrally stable flow other then the mean velocity has to be strong enough, Counihan [4] considers only those data which specifically indicate adiabatic or near adiabatic conditions, or which have wind speeds in excess of 5 m/s at a heigh of z-10m. ESDU [2,3] considers neutral stability to exist when the hourly mean wind at a height of 10 m is greater than 10 m/s. Using the gradient Richardson number $Ri = (g/\theta)(d\theta/dz)/(dU/dz)^2$ as a measure for the degree of thermal stability, near neutral conditions are suggested to exist by Teunissen [1] for $|Ri| < 0.03$ and by Panofsky [5] for $|Ri| < 0.01$.

Daytime observations throughout the PBL under moderately strong wind conditions at Minnesota [7,8] indicate that the PBL consists of a surface layer, a mixed layer and an overlying inversion layer delineating the height of the PBL, z_i, above which the flow is relatively undisturbed by the presence of the earth's surface (free atmosphere). The height of the PBL increases during the day as the earth surface heats up. The large-scale convective turbulence structure in the PBL interacts with the air in the free atmosphere causing the formation of an inversion through which warmer air of higher velocity and with a clockwise directional shear is entrained into the mixed layer [7,8,9]. Full-scale measurements [8], in, as well as laboratory simulation [10] of the convective PBL seem to suggest the existence of an inverted thermal plume structure throughout the PBL. These large-scale structures contribute significantly to the variances of the turblence components down in the surface layer [8,11].

Even over the roughest surfaces and under the strongest wind conditions, the wind shear, dU/dz, decreases with height and the convective turbulence gains importance with height. The height above which the convective turbulence cannot be neglected varies with wind speed, surface roughness, time of day, and a variety of atmospheric conditions such as cloud cover, precipitation etc.

Also deviating from the neutrally stable model is an on-shore wind [12]. As the usually stable air passes a coast line, the flow is modified as a result of both a change in surface roughness and surface temperature, giving rise to the development of an internal boundary layer (IBL). Because of the lack of turbulent mixing with adjacent layers, the flow in the inversion layer above the IBL can manifest itself as a surface jet with a maximum mean velocity at a height varying between 60-200 m [12,13].

In this article mean velocity and turbulence measurements which were obtained under moderately strong wind conditions varying in direction from southwest to north (category I) and from southerly directions (category II) are described and compared with empirical relations for the PBL models as listed in the previously mentioned references.

EXPERIMENTAL DETAILS AND DATA HANDLING

Mean wind and turbulence measurements have been obtained with two types of instrumentation from the 76-meter (250 foot) meteorological tower at Wallops Island, Virginia. Wallops Island is one of the barrier islands protecting the so-called "Delmarva" peninsula from the Atlantic Ocean. The island, which consists of a narrow strip of dunes, approximately 3 meters above sea level is situated in a northeast-southwest direction and is separated from the peninsula by a tidal marsh about three kilometers wide (Fig. 1). The island is used by the National Aeronautics and Space Administration as a sounding rocket launch facility.

Profile Instrumentation

Measurements were obtained with cup-vane velocity instruments and aspirated temperature sensing probes at 5 levels mounted at 15 m (50 feet) intervals. This system is primarily used for profile measurements, the outputs are sampled once every two seconds. The data are analyzed in blocks of $2^8 = 256$ samples representing a data record of 512 seconds in duration. Inspection of block means leads to the selection of stationary data samples usually six to ten blocks in length (51 to 85 min). Sample averages of velocity, direction and temperature are obtained as well as variances and covariances of the horizontal components. In addition, a 256-point autocorrelation function of the u component is calculated from which the turbulence integral scale, L_u^x, is obtained. Data from this instrumentation system can only be analyzed in the frequency range of 0.00195 - 0.25 Hz.

Turbulence Instrumentation

Six three-dimensional split-film anemometers (TSI-1080 D) are used for turbulence measurements. These instruments are located at the same levels as the cup-vane instruments and also at the 9 m (30 foot) level. The electronics for the split-film anemometers as well as the data-acquisition system are located in an instrument trailer which is parked at the base of the tower. For a detailed description of the calibration procedures and the data-acquisition system consult references 14 and 15. The digitized data (sample rate 200 Hz) are recorded and consequently analyzed on an IBM-370 computer. Four main computer programs have been developed to calculate the following statistical parameters describing the turbulent flow: mean values, variances and covariances, spectra, cross spectra and integral scales [16].

The frequency range in which the data are analyzed is covered in three stages. The highest range (0.0244 - 100 Hz) is obtained by dividing each data record into 80 consecutive blocks of 8192 data points each. For the middle range (0.00153 - 6.25 Hz) and the lowest range (0.00031 - 1.25 Hz), the original data are subjected to a 16-point and an 80-point non-overlapping average respectively. The spectra computed from the averaged data inherently show some scatter in the low-frequency range but the agreement in the overlap regions is excellent.

RESULTS AND DISCUSSION

Terrain Description

Winds corresponding to category I (west-north) are usually encountered following the passage of a cold front and will have crossed in succession the mainland, the Chesapeake Bay (20-50 km), the "Delmarva" peninsula (20-50 km) and the tidal marsh (3-5 km). Depending on the direction, for the last 200-300 m the air travels over dry land before it arrives at the tower location (Fig. 1). The tidal marsh consists of shallow areas of water interchanged with swamp vegetation, mostly grass of a maximum height of 1 m. Some taller vegetation consisting of bushes and brush of a maximum height of 5 m exists in several upstream directions. For winds in the sector between 20 degrees south of west and west, a 6.5 m high rocket fuel storage bunker is approximately 85 m upstream. An elevated roadway 2 m above the surrounding terrain passes the tower on the west side with 200 m.

Southerly winds (category II) approach the tower from the Atlantic Ocean and cross the beach approximately 300 m upstream of the tower. In the southerly direciton a 6 m high rocket assembly building is about 100 m upstream possibly affecting the measurements at the two lower levels. Comparison of the measurements from another meteorological tower (91-m) at Wallops Island shows that the effects on the mean flow from a 30-m tall rocket launching tower about 300 m upstream in a direction 15 degrees west of south can be considered to be negligible.

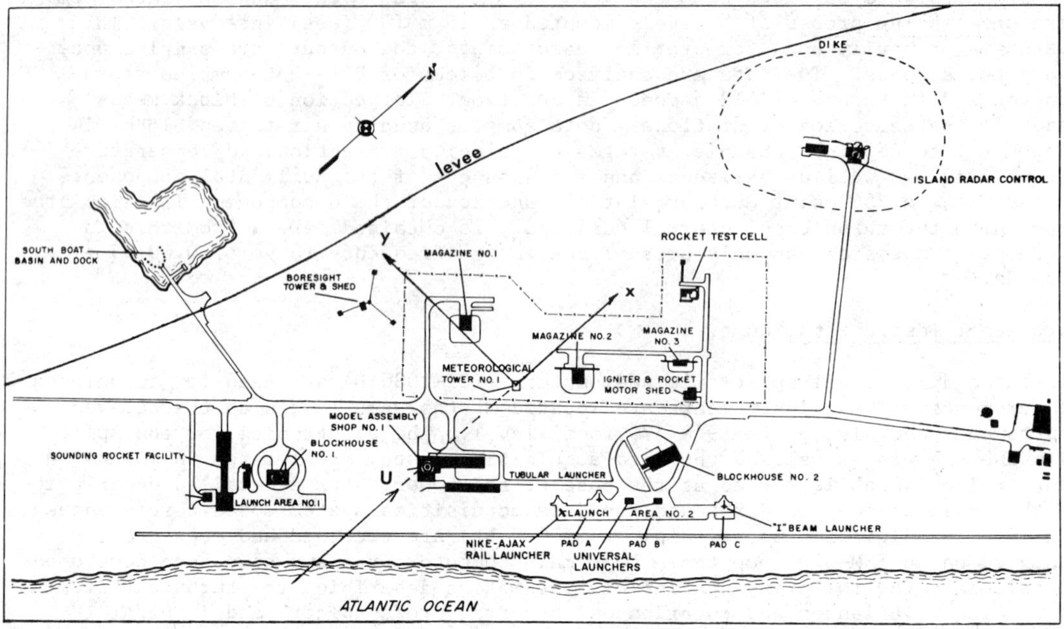

Fig. 1. Tower location on Wallops Island and coordinate system orientation

Category I Winds

Mean Wind and Mean Temperature Profiles

For flow over homogeneous terrain under near neutral conditions, the velocity profile should obey the logarithmic law: $U = (U_0^*/\kappa)\ \ln(z/z_0)$. In practical situations these conditions are seldom met and profile variations usually exist as a result of non-neutral conditions and/or non-uniform upstream terrain.

The logarithmic velocity profiles of category I winds over near-uniform terrain are either linear within the height of observation (76 m) or show two different linear segments (Fig. 2). All upper segments show a tendency towards a uniform profile. The heights of the kinks (intersection of the two linear segments) in the logarithmic velocity profiles seem to vary with the degree of thermal stability in the surface layer. However, values of the profile shear-velocity, U_0^*, obtained from the lower linear segments compare quite well with the shear velocity obtained from turbulence measurements in the surface layer. These observations were made in the stability range $Ri > -0.06$ with Ri evaluated at 15 m and with the shear velocity based on the horizontal shear stress according to $U^* = [(\overline{uw})^2 + (\overline{vw})^2]^{1/4}$ and analyzed in the high-frequency range (0.0244 - 100 Hz). The implications of these observations are that in this stability range, values of z_o can be obtained which correspond to those obtained under neutral conditions and that the low-frequency velocity fluctuation components do not affect the velocity profile in the surface layer.

Although the scatter of the data is appreciable, profile results indicate that for the stability range $Ri > -0.15$, the average value $z_0 = 0.034$ m corresponds

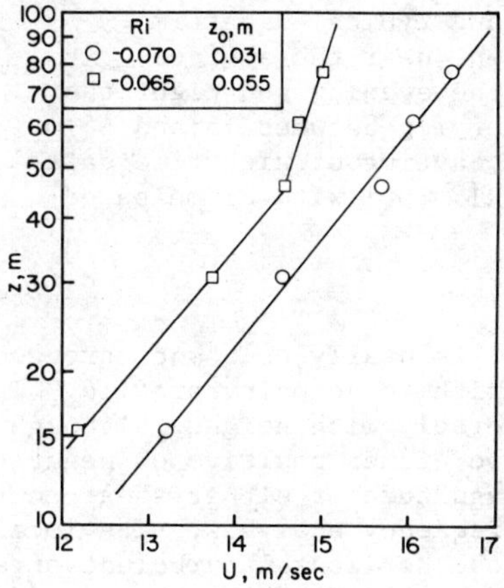

Fig. 2. Typical category I logarithmic wind profiles

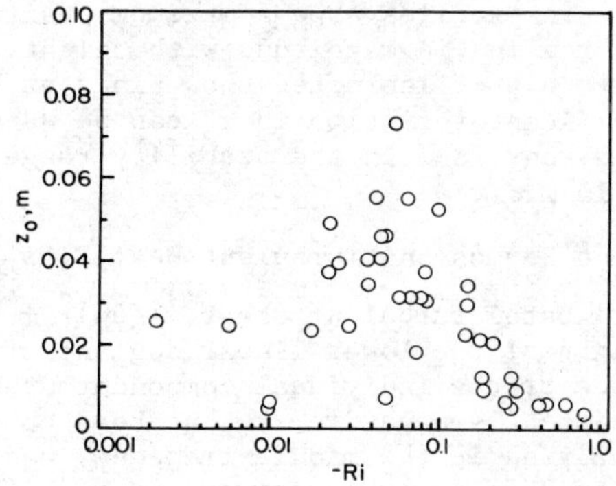

Fig. 3. The roughness parameter, z_o, versus Ri at z = 15.2 m for category I winds

quite well with predicted value from ESDU [2] for similar terrain (Fig. 3). For Ri < -0.15, z_0 decreases rapidly as the surface layer becomes more unstable. Figure 4 shows the variation of z_0 with wind direction, ϕ, for category I winds, clearly indicating the location of the bunker at $\phi = 90°$ and increased roughness for $\phi > 170°$.

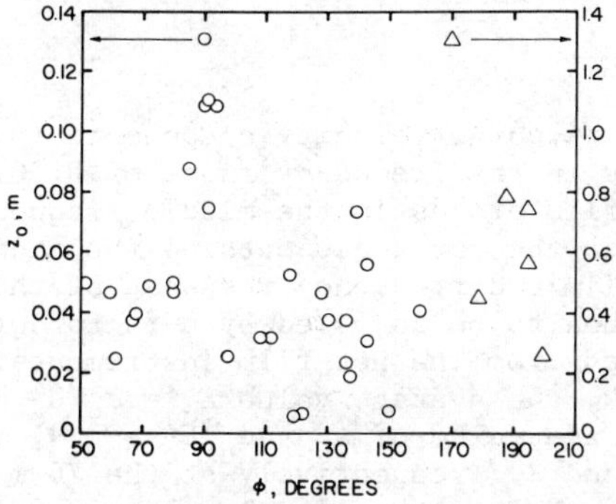

Fig. 4. The roughness parameter, z_o, versus wind direction for category I winds

For non-neutral conditions the velocity profile is given by $U=(U_0^*/k)[\ln(z/z_o)-\Psi]$, where ψ is a universal function depending on the stability parameter z/L which in turn depends on Ri [5]. Theoretically, if the velocity, U, is plotted as a function of $\ln z - \psi$, values of z_0 corresponding to neutral conditions could be obtained. This technique applied to the category I wind profiles at Wallops Island leads to large corrections resulting in unrealistic estimates of z_o.

Under moderately strong wind conditions, typical daytime temperature profiles have a maximum negative gradient near the surface and typical nighttime temperature profiles show a maximum positive gradient near the surface both decreasing in magnitude with height. During the evening and night the temperature profiles often show kinks at heights varying between 15 and 45 m. No significant variation in z_o can be detected for daytime or nighttime data provided they fall in the stability range $-0.15 < Ri < 0.1$ with Ri measured at z = 15 m.

Turbulent Stress and Turbulent Heat Flux

The horizontal turbulent shear, $[(\overline{uw})^2 + (\overline{vw})^2]^{1/2}$, is nearly constant throughout the height of the lower linear segment of the logarithmic velocity profile. The magnitude of the individual components vary considerably with height, although $\overline{uw}$ is consistently negative, $\overline{vw}$ has been observed to be either positive or negative. When analyzed in the middle-frequency range, the magnitude of either shear component is 25% larger in comparison with the high-frequency analysis. The local shear velocity, U*, defined as the square root of the horizontal turbulent stress obtained at the lowest three observation levels and analyzed in the high-frequency range compares well with the profile friction velocity, U_0^*.

Daytime observations show negative values for the vertical turbulent heat flux, $\overline{w\theta}$, while the horizontal components ($\overline{u\theta}$ and $\overline{v\theta}$) are either positive or negative. The simultaneous observations of a downward turbulent heat flux and a negative temperature profile lead to the problem that the gradient Richardson numbers (Ri) and flux Richardson numbers (R_f) have opposite signs. A possible explanation for this phenomena is that category I winds experience a discontinuity in surface heat flux while crossing the Chesapeake Bay and the tidal marsh. Consequently, the downward heat flux as the result of entrainment of warmer air from the free atmosphere can be expected to penetrate deeper into the PBL giving rise to the counter-gradient heat flux.

Variances

Standard deviations of the horizontal velocity components (σ_u, σ_v) were obtained with the cup-vane instruments in the frequency range of 0.00195-0.25 Hz, and of all components from the hot-film probes in the middle frequency range of 0.00153-6.25 Hz. The contributions to the variances outside the middle frequency range are negligible. Because of the limited frequency response of the cup anemometers, σ_u from the cup-vane system had to be adjusted by a factor of 1.17, which was based on comparison with σ_u obtained from the hot-film instruments. For logarithmic velocity profiles with a kink, the average value $A_u = \sigma_u/U_o^* = 2.5$ and without a kink in the profile, the average value of $A_u = \sigma_u/U_o^* = 2.9$, measured at the 15-m level and decreasing to 2.2 and 2.7 respectively at the 76-m level. Average values of $A_v = \sigma_v/U_o^*$ vary from 2.5 at the 15-m level to 2.3 at the 76-m level, while the average value of $A_w = \sigma_w/U_o^* = 1.2$ is uniform over the height of observation. These results obtained near the surface are in fair agreement with the average values of these coefficients obtained from several sources [1,4 and 5] of 2.5, 2.0 and 1.2 for A_u, A_v and A_w respectively. The variances of the horizontal velocity components show a great deal of variability, and on the average the ratio A_v of the lateral velocity (wind direction) at Wallops Island is larger than at other sites. Inspection of several years of strip-chart recordings of wind speed and direction obtained from a propeller-vane anemometer at Wallops Island indicates that for NW winds, the wind direction shows frequently large fluctuations towards the north, explaining the large values of A_v (Fig. 5).

These recordings indicate that the fluctuations in wind direction are larger than normally encountered and moreover, are not normally distributed which explains the non-zero turbulence stress, $\overline{vw}$.

The average turbulence intensities are in fair agreement with those suggested by ESDU [3], with the expection of the lateral turbulence intensities, σ_v/U, whose measured values are consistently higher (Fig. 6).

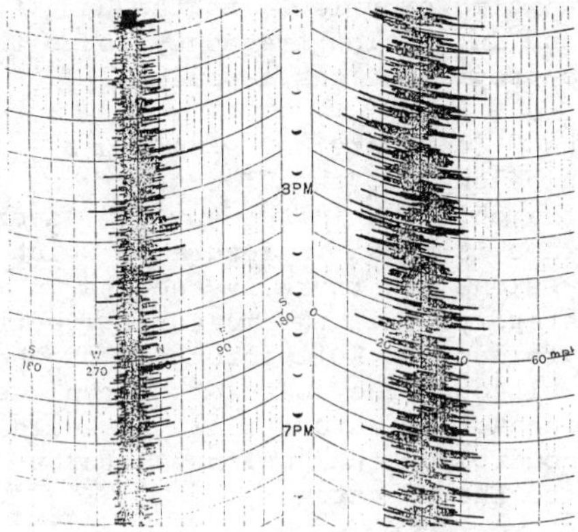

Fig. 5. Typical records of wind direction and wind speed for category I winds

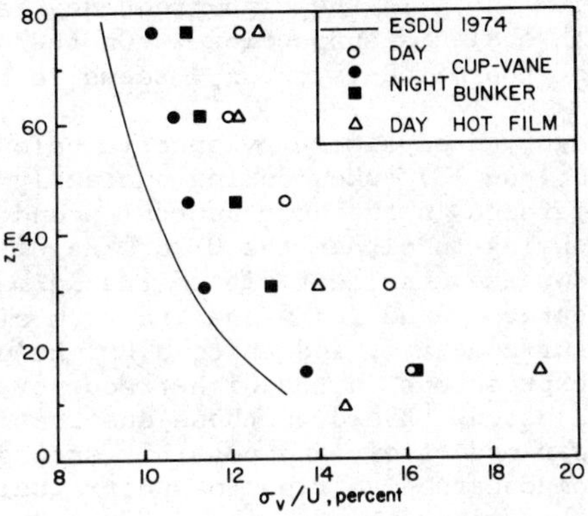

Fig. 6. Variation of turbulence intensity (lateral component) with height for category I winds

Velocity Spectra

For frequencies higher than $n = 3U/z$, measured logarithmic velocity spectra vary as $n^{-2/3}$ according to the Kolmogorov's law for the inertial subrange: $kS_u(k) = K_u\ \varepsilon^{2/3}\ k^{-2/3}$. Here k is the wave number in the streamwise direction ($k=2\pi n/U$ by Taylor's hypothesis) and ε is the dissipation rate. With the universal constants $K_u = 0.5$ and $K_{v,w} = 4/3\ K_u$, and introducing the dimensionless dissipation, $\phi_\varepsilon = \kappa\ z\,\varepsilon/(U^*_o)^3$, the inertial subrange logarithmic spectra normalized with the variances have the following form:

$$nS_u(n)/\sigma_u^2 = (0.27/A_u^2)\ \phi_\varepsilon^{2/3}(f)^{-2/3} \text{ and } nS_\alpha(n)/\sigma_\alpha^2 = (0.36/A_\alpha^2)\phi_\varepsilon^{\ 2/3}\ (f)^{-2/3} \text{ for } \alpha = v,w$$

where f is the dimensionless frequency $f = nz/U$. With $\phi_\varepsilon = 1$ (assuming near neutral conditions) and with values of A_α, based on either the local shear velocity, U^*, or the profile shear velocity, U^*_0, the Kolmogorov spectrum functions represent the inertial subrange u, v and w spectra quite well for all elevations up to 76m.

For frequencies lower than $n=3U/z$, the observed, u, v and w spectra vary a great deal so that no simple spectrum expression seem to exist for this frequency range. Kaimal et al [17] obtained empirical spectrum formulae based on measurements in the near-neutral surface layer over smooth and homogeneous terrain. These spectrum equations represent the spectra at near-neutral conditions approached from the stable regime (0^+). The Kansas results [17] show that the near neutral spectra of the horizontal components (u, v) approached from the unstable regime (0^-) have much higher spectral values at low frequencies than predicted by the Kaimal spectrum expressions. The discontinuity observed in the u, v spectra at near neutral conditions does not exist for the w component, and consequently the Kaimal spectrum expression should fit all w spectra at near neutral conditions.

Consequently, for near neutral conditions, u, v spectra as well as the variances show a great variability depending on the presence of low-frequency components. The latter seem to be the result of a large-scale turbulence structure observed in the mixed layer [8] during the daytime and penetrating down into the surface layer. The near neutral u, v spectra in the 0^- category do not follow a single similarity law. In the high-frequency range the Kolmogorov law with the height z as length scale is appropriate. On the other hand in the low-frequency range the height of the mixed layer, z_i, seems to be the appropriate length scale [11].

All category I daytime u, v spectra belong to the 0^- category while only one data run (run 16) taken during the early evening belongs to the 0^+ category. The difference in the near-neutral u and v spectra obtained in the surface layer and belonging to either the 0^- or the 0^+ categories is shown in figure 7, which also shows the excellent fit of the Kaimal expressions for the u, v spectra in the 0^+ category and for w spectra from either category. All the measured spectra in the surface layer and up to a height of 76m, follow the Kolmogorov law or the Kaimal expressions in the high-frequency range, although the constants often need to be adjusted. Based on these observations it cannot be automatically assumed that under moderately and possibly strong wind conditions the horizontal turbulence components give rise to spectra belonging to the 0^+ category.

Integral Scales

The longitudinal scales, L^x_α for α = u, v, w, in the direction of the mean flow are often calculated from the normalized autocorrelation function, assuming that Taylor's hypothesis is valid. A second method for obtaining these scales is using the best fit of measured velocity spectra to the corresponding empirical von Karman spectrum functions.

Figure 8 shows the measured scales, L^x_u, by these two methods for two different data runs. One data run (run 18) with, and a second run (run 16) without low-frequency velocity velocity fluctuations. In the case, low-frequency velocity fluctuations are present, the von Karman spectrum functions represent the data only in the high-frequency range. Consequently, integral scales obtained with the second method are based on the turbulence in the inertial subrange, excluding the contributions of the low-frequency fluctuations which are included when integral scales are obtained from the auto-correlation function. In the case the low-frequency fluctuations are absent the two methods predict almost equal scales (Fig. 8). The ESDU [3] predicted scales for corresponding terrain fall in between the measured scales for the two different data runs.

Probability Density Function

The probability density functions of each of the velocity components are usually considered to be Gaussian in the range of $\pm 3\sigma_\alpha$ [3,4]. Careful inspection of many hours of category I wind data show large asymmetrical fluctuations in direction towards the north (Fig. 5). Measured probability density functions indicate that the fluctuations of the velocity magnitude in contrast to the direction fluctuations are usually normally distributed. This was especially observed for daytime data, while for the evening run without low-frequency fluctuations (run 16) both magnitude and direction are almost normally distributed. Several probability density functions of direction fluctuations show a secondary peak or the nucleus of a secondary peak at -1.5σ (Fig. 9). These peaks could be caused by a downward flow of air with a more northerly direction probably as the result of entrainment.

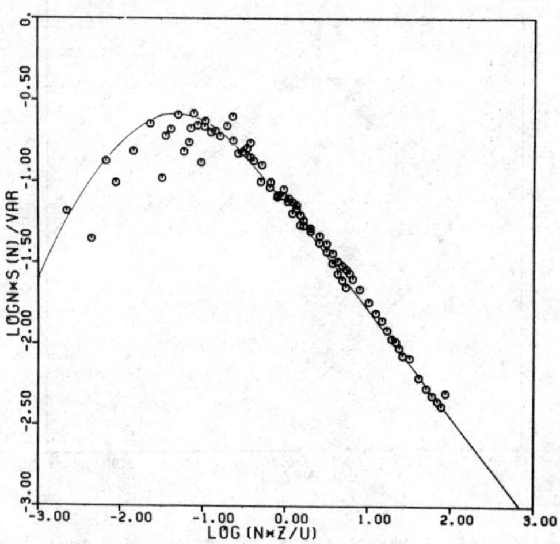

Fig. 7a. U- spectrum, run 16, z=9.1 m, U=6.23 m/sec σ_u=0.91 m/sec ——— Kaimal [17].

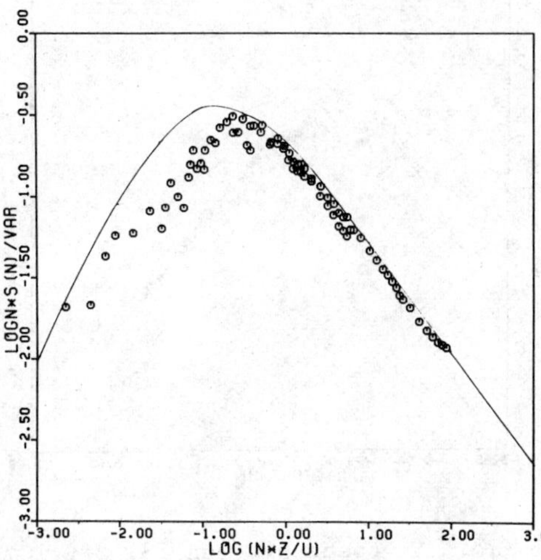

Fig. 7b. V- spectrum, run 16, z=9.1 m, U=6.23 m/sec σ_v =0.59 m/sec ——— Kaimal [17].

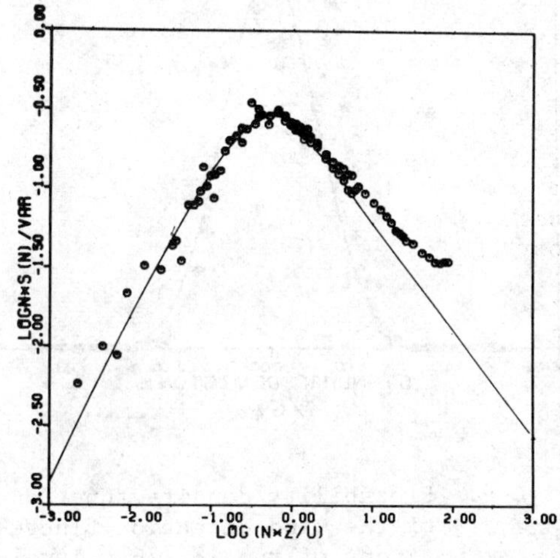

Fig. 7c. W- spectrum, run 16, z=9.1m, U=6.23 m/sec σ_w=0.53 m/sec ——— Kaimal [17].

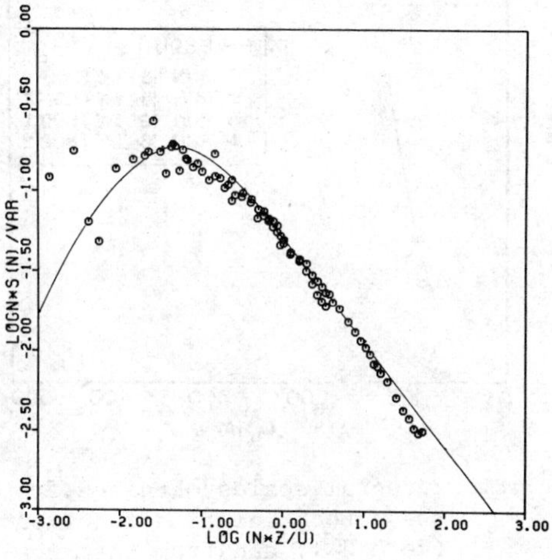

Fig. 7d. U- spectrum, run 18, z=9.1, U=10.14 m/sec σ_u=1.88 m/sec ——— Kaimal [17].

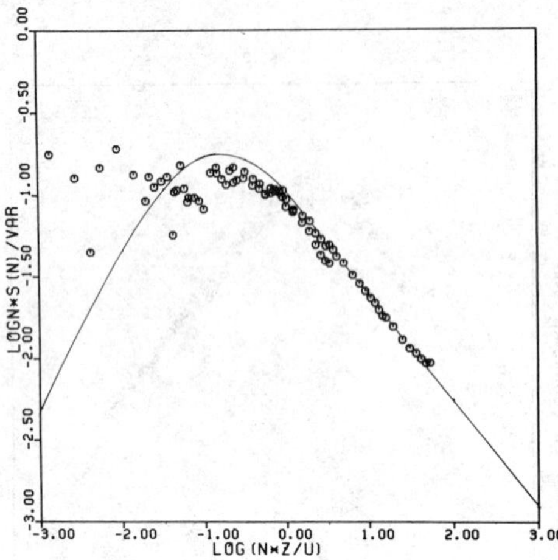

Fig. 7e. V- spectrum, run 18, z=9.1 m, U=10.14 m/sec σ_v=1.42 m/sec
_____ Kaimal [17].

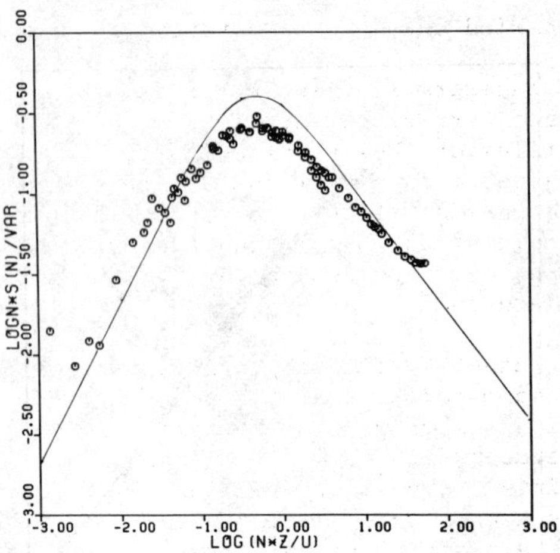

Fig. 7f. W- spectrum, run 18, z=9.1 m, U=10.14 m/sec σ_w=0.75 m/sec
_____ Kaimal [17].

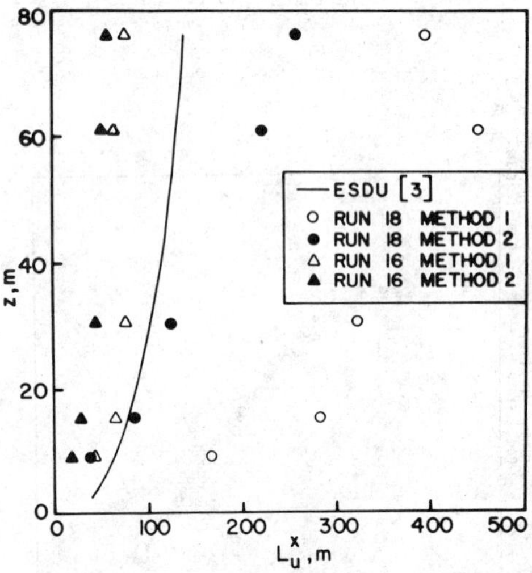

Fig. 8. Integral scales obtained from the autocorrelation function (method 1) and from the von Karman spectrum function (method 2) for two runs, one with low-frequency components (run 18) and one without low-frequency components (run 16).

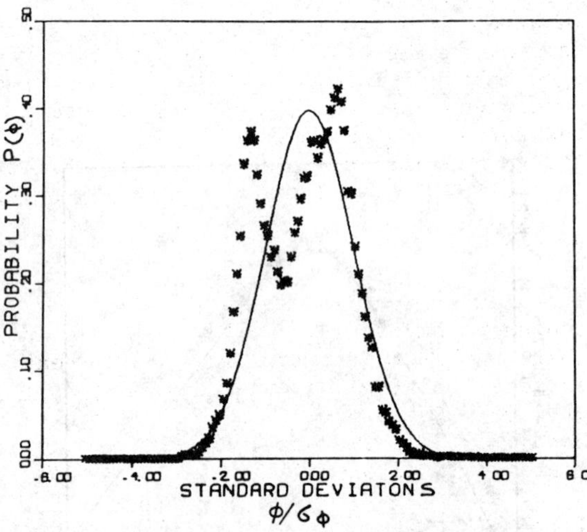

Fig. 9. Probability density function of the wind direction fluctuations of run 18, 76 m level, σ_ϕ=7.97°. Solid line represents the Gaussian distribution function.

Category II Winds

Mean Wind and Mean Temperature Profiles

The category II velocity profiles when plotted in the usual logarithmic fashion are either straight or curved opposite to the category I logarithmic profiles or show a maximum below the 76-m level (Fig. 10). The roughness lengths obtained from either the straight logarithmic profiles or from the lowest two observations of the curved profiles vary a great deal and are too high for the terrain at Wallops Island. Corrections of these profiles, in order to obtain values of z_o corresponding to the terrain, as suggested by Panofsky [5] are inadequate.

The mean-temperature profiles show some unusual kinks mostly between the 30 and 45 meter levels. The most interesting sequence of mean-temperature distributions is from three runs taken on one summer afternoon (Fig. 11). The early afternoon data show a decrease in temperature up to 30 m with a rather strong positive temperature gradient at higher elevations. As the afternoon progresses this temperature gradient decreases drastically. The majority of observations show a positive temperature gradient above 30 m of varying intensity and below 30 m the temperature gradient is either positive or negative depending on the time of day and the time of year.

Turbulent Stress and Turbulent Heat Flux

The measured turbulent stress $\overline{uw}$ has a maximum value at the 9 m level, and decreases rapidly with height. The vertical turbulent heat flux $\overline{w\theta}$ is positive (upward) below 30 m and is negative (downward) at this height and above, corresponding to the observed summer temperature profiles. These results indicate a dual type boundary layer within the height of observation (76 m), with unstable conditions near the surface and an overlying inversion. The unstable layer near the surface corresponds to the internal boundary layer which develops as the stable on-shore wind is heated from below after it crosses the beach. The depth of the IBL increases with distance from the shoreline.

Variances

The standard deviations of the velocity fluctuations, σ_α, are largest in the IBL and decrease rapidly in the overlying inversion. The ratios, $\sigma_\alpha/U*$, are quite sensitive to stability and increase for deviations from neutral conditions in both the stable and unstable regime [12]. For near-neutral conditions in the IBL, the average values of $\sigma_\alpha/U*$ are 1.75, 1.5 and 1.4 for α = u, v and w respectively. Turbulence intensities σ_u/U in the stable air above the IBL (25 m or higher) are about half of the ESDU [3] predicted values with z_o = 0.001 m (Fig. 12).

In the IBL with z_o = 0.01 m, the ESDU predictions [3] are comparable to the measurements. The hot-film data shown in Figure 12 were acquired during hot and sunny summer afternoons with mean velocities at the 76 m level varying from 6.7 m/sec to 12.7 m/sec. The cup-vane data were obtained during morning and early afternoon in the winter and early spring, with mean velocities at the 76 m level varying from 10.7 m/sec to 20.6 m/sec. For 15 of a total of 38 runs investigated, the streamwise turbulence intensity at the 76 m level is less than 3%. The lateral and vertical turbulence intensities vary in a similar fashion.

Included in Figure 12 are the turbulence intensities for run 7 obtained under extreme stable conditions (R_f = +25.5, U = 10.7 m/sec at z=61 m) during a sunny

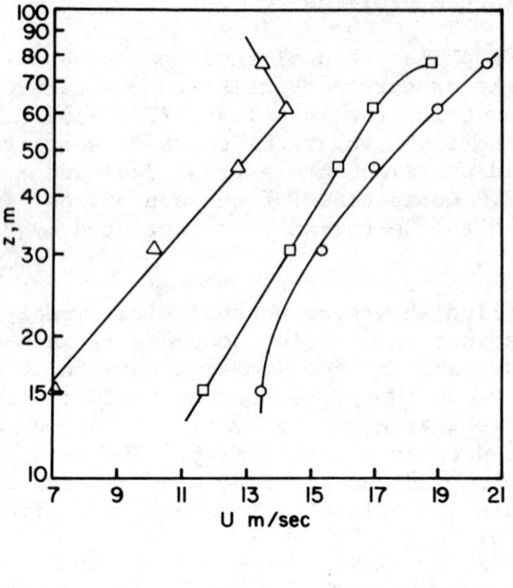

Fig. 10. Typical category II logarithmic wind profiles.

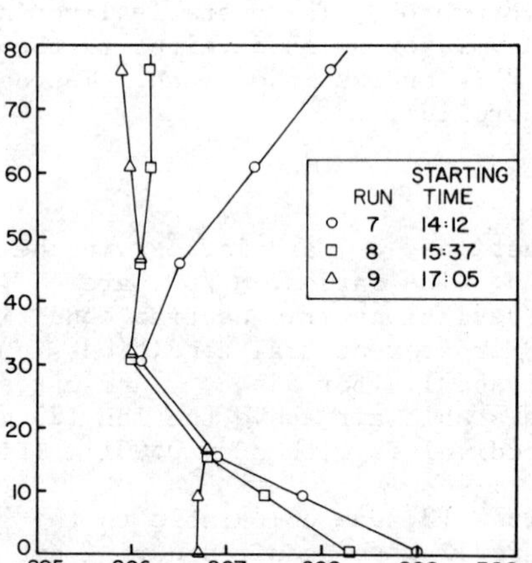

Fig. 11. Temperature profiles for runs 7.8 and 9.

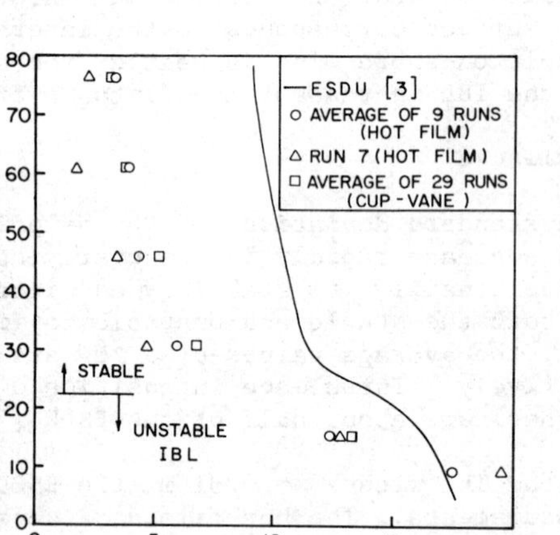

Fig. 12. Variation of turbulence intensity (streamwise component) with height for category II winds.

warm afternoon in August. Under these conditions the turbulence almost disappears. During a trial run with the hot-film system under similar conditions, the turbulence disappeared completely at the two highest observation levels for at least 20 minutes. For many of the runs, a distinct clockwise rotation of a maximum of 20 degrees of the mean velocity vector with height is observed. With stable air coming from the ocean, maximum velocities at elevations below 200 m have been observed visually from exhaust trails of launched rockets and occasionally measured with the tower instruments below 76 m. (Fig. 10).

Based on these observations it can be concluded that the stable air above the IBL with low turbulence intensities and small turbulence integral scales, loses dynamic contact with the air above and below it. Little mixing and little momentum transfer takes place between the stable air and the adjacent air so that Coriolis forces become important and the stable air encounters little resistance resulting in the low-level jet.

Velocity Spectra

Measured logarithmic velocity spectra of category II winds vary as $n^{-2/3}$ in the inertial subrange and those spectra originating in the IBL belong to the 0^+ category, varying as n^{+1} in the low frequency range. However, for elevations above the IBL, low-frequency components are present in all three components. Under stable conditions, wave phenomena have been observed near the surface [13, 18] and can be recognized from the turbulence for those flows where the turbulence levels are low. Waves are not random and usually occur at a different frequency band so that a distinct spectral gap exists between the spectral peak due to the turbulence and the spectral peak at lower frequencies due to the waves.

The part of the spectra due to the turbulence in both the IBL and overlying inversion are well represented by the Kaimal expressions, although the constants in these expressions vary with stability [12]. The presence of the low-frequency waves varies a great deal with atmospheric conditions and their effect on the spectra seems to grow with height, as the fluctuations due to turbulence and wave motion fall in similar frequency bands.

Integral Scales

Integral scales obtained from the integration of the normalized autocorrelation functions vary a great deal with the presence of the low-frequency components. Values of the integral scales obtained with this method for the consecutive afternoon runs 7, 8 and 9, measured at the 76-m level, vary from 406, 162 to 56 m respectively. The wavelengths corresponding to the peak of the logarithmic spectra vary substantially with stability and height [12]. For near-neutral conditions in the IBL, the peak-wave length λ_m^α equals z/0.15, z/0.35 and z/0.15 for α = u,v and w respectively. Normalized spectral values $nS_\alpha(n)/U^{*2}$ evaluated at f=10, vary a great deal with stability [12]. Consequently, integral scales obtained from fitting of the von Karman functions vary with stability as well.

CONCLUDING REMARKS

The wind measurements made at this eastern coastal region under sustained moderately strong wind conditions show marked deviations from observations obtained in purely shear-generated turbulent flow. For near-neutral conditions of cateogry I (west-north) winds, low-frequency velocity and direction fluctuations influence all parameters describing the turbulence. The large direction fluctuations can have a pronounced effect on the response of tall buildings especially in the case the mean wind blows diagonally onto the building [19,20]. In wind tunnel model tests the non-mechanically produced turbulence cannot be simulated adequately resulting in a discrepancy between model and full-scale tests [20].

For moderately on-shore winds from southerly directions, a surface based inversion is present even with the hourly mean wind speed over 20 m/sec and a maximum velocity (low-level jet) may exist at heights between 60-200 m. The turbulence intensities in the stable layer are often less than 3% and under certain atmospheric conditions the turbulence vanishes completely. Such flow could affect the response of suspension bridges located near coastal areas especially when the bridge decks are at the same height as the almost turbulence-free low-level jet.

Based on the observations made at Wallops Island under moderately strong wind conditions, the turbulence is not of purely mechanical origin. Moreover, as a result of non-homogeneous upstream surface conditions at this site, similarity theories are not applicable.

ACKNOWLEDGEMENTS

The research reported in this article is supported by the NASA Grant NGL 47-004-067. Continuous support from the grant monitor, Mr. R. E. Carr of NASA Wallops Flight Center is gratefully acknowledged.

REFERENCES

1. Teunissen, H. W., Characteristics of the Mean Wind and Turbulence in the Planetary Boundary Layer, Institute for Aerospace Studies, University of Toronto, UTIAS Rep. No. 32, October 1970.

2. Engineering Sciences Data Unit, Characteristics of Wind Speed in the Lower Layers of the Atmosphere near the Ground: Strong Winds (Neutral Atmosphere), ESD Item No. 72026, November 1972, with Amendments A and B October 1974.

3. Engineering Sciences Data Unit, Characteristics of Atmospheric Turbulence near the Ground, Part II: Single Point Data for Strong Winds (Neutral Atmosphere), ESD Item No. 74031 October 1974, with Amendment A March 1975.

4. Counihan, J., Adiabatic Atmospheric Boundary Layers: A Review and Analysis of Data from the Period 1880-1972, Atmospheric Environment, 9, 871 (1975).

5. Panofsky, H. A., Wind Structure in Strong Winds Below 150 m, Wind Engineering, 1, 91 (1977).

6. Deaves, D. M. and Harris, R. I., A Mathematical Model of the Structure of Strong Winds, Environmental Sciences Research Unit, Cranfield, England, ESRU Report No. 24, Dec. 1976.

7. Izumi, Y. and Caughey, J.S., Minnesota 1973 Atmospheric Boundary Layer Experiment Data Report, Air Force Cambridge Research Laboratories, Hanscom AFB, Mass., AFCRL-TR-76-0038, Jan. 1976.

8. Kaimal, J.C. et al., Turbulence Structure in the Convective Boundary Layer, Journal of the Atmospheric Sciences, 33, 2152 (1976).

9. Businger, J.A., The Need for Boundary-Layer Measurements and Their Significance for Solving Atmospheric Problems, Atmospheric Technology, 7, 2 (1975).

10. Willis, G. E. and Deardorff, J. W., Laboratory Simulation of the Convective Planetary Boundary Layer, Atmospheric Technology, 7, 80 (1975).

11. Kaimal, J.C., Horizontal Velocity Spectra in an Unstable Surface Layer, Journal of the Atmospheric Sciences, 35, 18 (1978).

12. Tieleman, H. W. and Derrington, D. B., Jr., An Experimental Study of the Atmospheric Boundary Layer Modified by a Change in Surface Roughness and Surface Temperature, Department of Engineering Science and Mechanics, Virginia Polytechnic Institute and State University, Blacksburg, VPI-E-77-17, May 1977.

13. Sethuraman, S., The Observed Generation and Breaking of Atmospheric Internal Gravity Waves over the Ocean, Boundary-Layer Meteorology,12, 331 (1977).

14. Tieleman, H. W. and Tavoularis, S. C., A Method for the Measurement and the Statistical Analysis of Atmospheric Turbulence, Department of Engineering Science and Mechanics, Virginia Polytechnic Institute and State University, Blacksburg, VPI-E-74-26, October 1974.

15. Tieleman, H. W. and Tavoularis, S. C., An Instrumentation System for the Measurement of Atmospheric Turbulence, Journal of Industrial Aerodynamics, 2, 49 (1977).

16. Tieleman, H. W. and Chen, W. W. L., Statistical Analysis of Low Level Atmospheric Turbulence, Department of Engineering Science and Mechanics, Virginia Polytechnic Institute and State University, Blacksburg, VPI-E-74-3, January 1974.

17. Kaimal, J. C. et al., Spectral Characteristics of Surface Layer Turbulence, Journal of the Royal Meteorological Society, 98, 563 (1972).

18. Caughey, S. J., Boundary-Layer Turbulence Spectra in Stable Conditions, Boundary Layer Meteorology, 11, 3 (1977).

19. Reinhold, T. A., et al., The Effect of Wind Direction on the Static and Dynamic Wind Loads on a Square-Section Tall Building, 3rd Colloquium on Industrial Aerodynamics, Aachen, Germany, June 14-16, 1978, Proceedings Part 1, pp. 263-279.

20. Jeary, A. P. (Ed.), The Dynamic Behaviour of the Arts Tower, University of Sheffield, and its Implications to Wind Loading and Occupant Reaction, Building Research Establishment, Building Research Station, Garston, Watford, England, Current Paper CP 48/78 June 1978.

FULL SCALE MEASUREMENTS OF THE ATMOSPHERIC TURBULENCE OVER TWO ESCARPMENTS

A. J. BOWEN

University of Canterbury, Christchurch, New Zealand

ABSTRACT

The results from recent field measurements of the atmospheric turbulence over a low two dimensional 2:1 slope and cliff escarpment are presented as part of an investigation into the effects of complex terrain on the wind flow close to the ground. The variation of mean wind speed, turbulence intensity, longitudinal power spectral density and gust factors are discussed in detail and compared with model tests. The effects of the escarpments are varied and complex and the measurements clearly indicate that the standard conditions applicable to flat homogeneous terrain should not be applied to a hilly site.

INTRODUCTION

The need for accurate information on the atmospheric boundary layer for wind engineering purposes is well known as it is from this information that the structural wind loading is estimated. Although the flow over homogeneous terrain is well understood, the influence of local topography on the wind is still far from clear. There is therefore a real need for reliable guidelines for the wind loading of structures intended for exposed hilly sites. Full scale investigations are notoriously difficult to accomplish due to the many practical problems that have to be overcome, so it is natural that model tests are the preferred means of investigation for a wind engineering problem. However the accuracy and limitations of the modelling techniques employed can only be verified by direct comparison with field tests of the full scale situation so that subsequent wind tunnel investigations may be carried out with confidence.

The current research programme into the effects of complex terrain on the wind flow close to the ground was initiated with these thoughts in mind. The two escarpments under scrutiny were chosen as simple two dimensional hills in order to find the individual effects of slope and wind direction on the wind flow. The cliff represented the most abrupt case involving separated flows whilst the 2:1 slope represented a substantial obstruction, but with attached flows. Preliminary field measurements were accomplished with cup anemometers followed later by a wind tunnel investigation involving a series of plain two dimensional escarpments normal to the flow and have been reported by Bowen and Lindley (1,2). The development of a field measurement facility made the measurement of atmospheric turbulence possible and the results from the use of this facility on the same escarpments forms the basis of this paper. The field results are compared with recent model tests which were carried out using accurate models of the two escarpments at the same angles of incidence as in the field and using the same modelling techniques and equipment as reported in the earlier work (2). The work programme is continuing with a study of all parameters that may affect the flow over two dimensional hills of various shapes. Model and field investigations of the wind flow over unique complex terrain situations are in progress.

DATA ACQUISITION SYSTEM

Anemometers

The propeller anemometers that were used in the field tests were developed by the University of Canterbury and used a four bladed, 19 cm (7½") diameter helicoid propeller similar to that of the Gill anemometer. The propeller drove a light-chopping disc which generated two out-of-phase pulse trains from which the direction and number of turns of each instrument could be recorded on magnetic tape every 1/16th second. The distance constant of the anemometers aligned into the wind was found to be 0.85 m (2.8') from wind tunnel acceleration tests, but was almost 1 m (3.3') for wind angles close to 45° to the shaft axis which was considered more relevant to field conditions. The effect of their first order dynamic response on the power spectra may be seen in Fig.8 from which it may be assumed that frequencies, n below that represented by $n/\bar{V} \approx 3x10^{-2}$ would be accurately recorded. For a typical wind speed of 8 m/s (18 m.p.h.) say, the upper frequency limit would be 0.25 Hz for a pure sinusoidal velocity fluctuation.

Data processing

The recorded pulse counts were converted to wind speed for each scan, a running mean calculated for each instrument to show trends and consecutive data checked for large differences that signified recording errors. After correcting for non-cosine directional response, the velocity readings were recorded on a library tape for further analysis. Subsequent processing involved the conversion of the reference axes to align with the mean wind direction at each instrument position, accumulating turbulence data, peak gust occurrences and calculating power spectral densities using a Fourier transformation program.

SLOPING ESCARPMENT FIELD TEST

Site description

The sloping escarpment was located at Amberley, 40 km (25 miles) north of Christchurch and was separated from the open sea by 1:3 km (0.8 miles) of ploughed land, low scrub and the occasional copse of trees. A map of the surrounding terrain may be seen in Fig.1 and a general view looking NE showing the caravan and #2 mast seen in Fig.2. Apart from the two eroded gullies some distance from the tower sites, the crest lay in a straight line with a height of H = 13 m (42.6') above the seaward ground level with an angle of inclination of approximately 26°, giving it a consistent 2:1 slope. The top of the escarpment was flat and smooth and together with the slope, was covered in well cropped grass. The recorded winds came in off the sea at a mean angle of 30±5 degrees to the crest.

Three towers were erected to support the orthogonal arrays of propeller anemometers. The 10 m high (33') #1 tower was placed on the lower level at a distance of 10 H from the escarpment foot to monitor the oncoming wind conditions with instruments at four levels. The 20 m (66') high #2 tower shown in Fig.2 was placed as close as possible to the crest on the upper level which was conveniently 0.5 H from the crest, with instruments at six levels. A further 10 m (33') #3 tower was placed on the upper level at a distance of 4 H behind the crest, with instruments at four

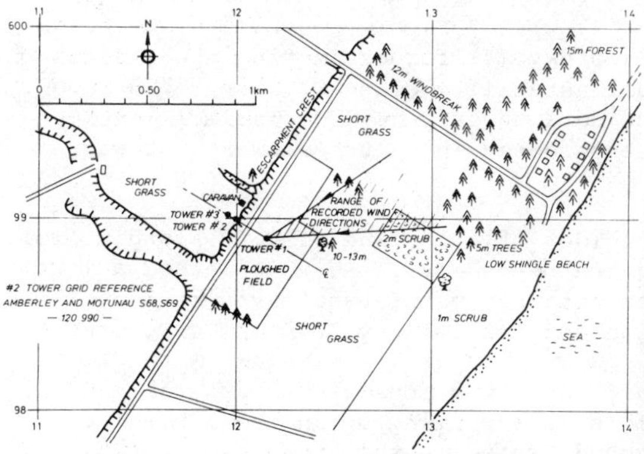

Fig.1 A map of the surrounding terrain at the Amberley site.

Fig.2. The 2:1 escarpment at Amberley looking NE.

levels. The lower four instrument levels on all three masts were the same at Z = 9.87 m (32.4'), 4.90 m (16.1'), 2.88 m (9.4'), and 1.56 m (5.1') with the two extra levels on the #2 tower at 18.78 m (61.6') and 14.87 m (48.8'). Three propeller anemometers were set up in an orthogonal array for each instrument level on the #2 tower and the top level on the other two towers. The other levels were each fitted with a pair of anemometers with the vertical instrument omitted. In this way the 36 anemometers were distributed over fourteen positions. All the instruments were permanently wired to the data acquisition unit in the caravan and recordings were initiated manually when the weather conditions were appropriate.

Site recordings

It was intended to take recordings during conditions of neutral atmospheric stability which normally could be expected to occur during strong winds and cloudy weather and no temperature profile measurements were attempted. Unfortunately, the summer during which the recordings were taken was unusual for its lack of wind and cloud. Over a period of six weeks, four half hour and one 15 minute recordings were taken with the mean upstream wind speed at 10 m (33') between 7.1 and 9.7 m/s (15.9 and 21.7 m.p.h.) under similar sunny conditions. Non-stationary trends in the mean wind speed were less than 0.5 m/s (1.1 m.p.h.) and were not removed from the data during processing. The mean wind and turbulence data from the upstream tower #1 at 10 m (33') height above ground are shown in Table 1 alongside the data given by Counihan (3) for a neutrally stable 'rural' boundary layer.

Atmospheric stability

An indication of the average atmospheric stability existing during the tests may be drawn by comparing the effect of the stability term $\psi(\frac{Z}{L})$ in equation 1 on the turbulence data normalised by the friction velocity, V_* (L = Monin-Obukhov Length).

$$\bar{V}_Z = \frac{V_*}{k}\left[\log_e\left[\frac{z-d}{Z_o}\right] - \psi\left(\frac{Z}{L}\right)\right] \qquad (1)$$

Case 1:- If neutrally stable conditions were assumed and $\psi(\frac{Z}{L})$ ignored, then the recorded turbulence data may be presented as in Table 1, column #1. The intensity of turbulence and Reynolds stress values were close to that expected from a rural boundary layer but the values of the surface roughness parameter, Z_o and $V_*/\bar{V}_{10}$ were too low implying that the assumption of neutral stability was incorrect. The low value of $V_*/\bar{V}_{10}$ raised the turbulence parameters $\sigma u/V_*$ etc. well above those of

Counihan (3).

Case 2:- By using the method set out by Panofsky (4) for estimating the effects of atmospheric stability and assuming a Pasquill stability class C/D for the strong sun and medium wind speeds encountered, the data under column #2, Table 1 were generated. The turbulence parameters have been reduced in value somewhat but were still higher than that expected.

Case 3:- It was also possible to obtain an idea of the value of $V_*/\bar{V}_{10}$ and Z_o from the measured Reynolds stresses by assuming that $\sqrt{-\overline{uw}} = V_*$. By this method a more appropriate value for Z_o of 0.05 m (2") was obtained using Panofsky's method, which also indicated a class C stability condition and realistic values of $\sigma u/V_*$ etc.

It may be concluded that the upwind boundary layer was consistently unstable which may have been caused by the cold wind from the sea warming up in the lower levels of the atmosphere as it passed over the ground heated by the sun. The instability caused the unusually full velocity-height profile (α = 0.13) compared with that expected from a rural boundary layer (α = 0.18) and which was modelled in the wind tunnel for the model tests.

Table 1. Upstream turbulence field data showing the effect of various atmospheric stability assumptions

PARAMETER	STANDARD RURAL DATA COUNIHAN (3)	AMBERLEY 2:1			RAKAIA CLIFF			
		ASSUME			ASSUME			
		1 NEUTRAL STABILITY	2 CLASS C/D STABILITY	3 $V_*=\sqrt{-\overline{uw}}$	1 NEUTRAL STABILITY	2 CLASS C/D STABILITY	3 $V_*=\sqrt{-\overline{uw}}$	4 CLASS D/E STABILITY
α	0.143-0.167	0.13[a]	0.17[b]	0.19[b]	0.28[a]	0.29[b]	0.16[b]	0.20[b]
Z_o m	0.01-0.15	0.003[a]	0.025	0.050	0.10[a]	0.25	0.013	0.067
$V_*/\bar{V}_{10}$		0.049	0.072	0.087	0.090	0.112	0.053	0.073
$V_*/\bar{V}_G$		0.029[e]	0.042[e]	0.050[e]	0.028[e]	0.035[e]	0.017[e]	0.023[e]
$\sigma u/\bar{V}_{10}$		0.22[a]	0.22[a]	0.22[a]	0.17[a]	0.17[a]	0.17[a]	0.17[a]
$\sigma v/\bar{V}_{10}$		0.16[a]	0.16[a]	0.16[a]	0.16[a]	0.16[a]	0.16[a]	0.16[a]
$\sigma w/\bar{V}_{10}$		0.11[a]	0.11[a]	0.11[a]	0.09[a]	0.09[a]	0.09[a]	0.09[a]
$\overline{uw}/\sigma u.\sigma w$		-0.31[a]	-0.31[a]	-0.31[a]	-0.18[a]	-0.18[a]	-0.18[a]	-0.18[a]
$\sqrt{-\overline{uw}}/V_*$		1.77	1.20	1[c]	0.58	0.47	1[c]	0.72
$-\overline{uw}/\bar{V}_G^2$	0.002 - 0.003	0.0026	0.0026	0.0026	0.00027	0.00027	0.00027	0.00027
$A = \sigma u/V_*$	2.50	4.49	3.06	2.53	1.89	1.52	3.21	2.33
$B = \sigma v/V_*$	1.875	3.26	2.22	1.84	1.78	1.43	3.02	2.19
$C = \sigma w/V_*$	1.25	2.24	1.53	1.26	1.00	0.80	1.70	1.23
$\sigma v/\sigma u$	0.75	0.73[a]	0.73[a]	0.73[a]	0.94[a]	0.94[a]	0.94[a]	0.94[a]
$\sigma w/\sigma u$	0.50	0.50[a]	0.50[a]	0.50[a]	0.53[a]	0.53[a]	0.53[a]	0.53[a]
STABILITY CLASS	D	D	C/D	C[d]	D	C/D	E	D/E[d]

NOTES:-

a MEASURED VALUE (d = 0 FOR AMBERLEY 2:1 TESTS; d = 0.5m(20") FOR RAKAIA CLIFF TESTS)

b ESTIMATED VALUE FOR NEUTRALLY STABLE CONDITIONS

c THEORETICAL ASSUMPTION

d PREFERRED OPTION

e ASSUME $Z_G \approx$ 600m (2000ft)

Test results

The variations of $\bar{V}_Z$ and σu over the escarpment are shown in Fig.3 alongside those obtained from recent wind tunnel tests using an accurately made model in a rural boundary layer and the same angle of incidence. Good agreement is evident between the full scale and model test results except for the unusual upstream mean velocity -height profile in the field results. The scatter in the test results was a feature of the field measurements and a result of the different daily atmospheric conditions. No evidence of flow separation was encountered in either tests.

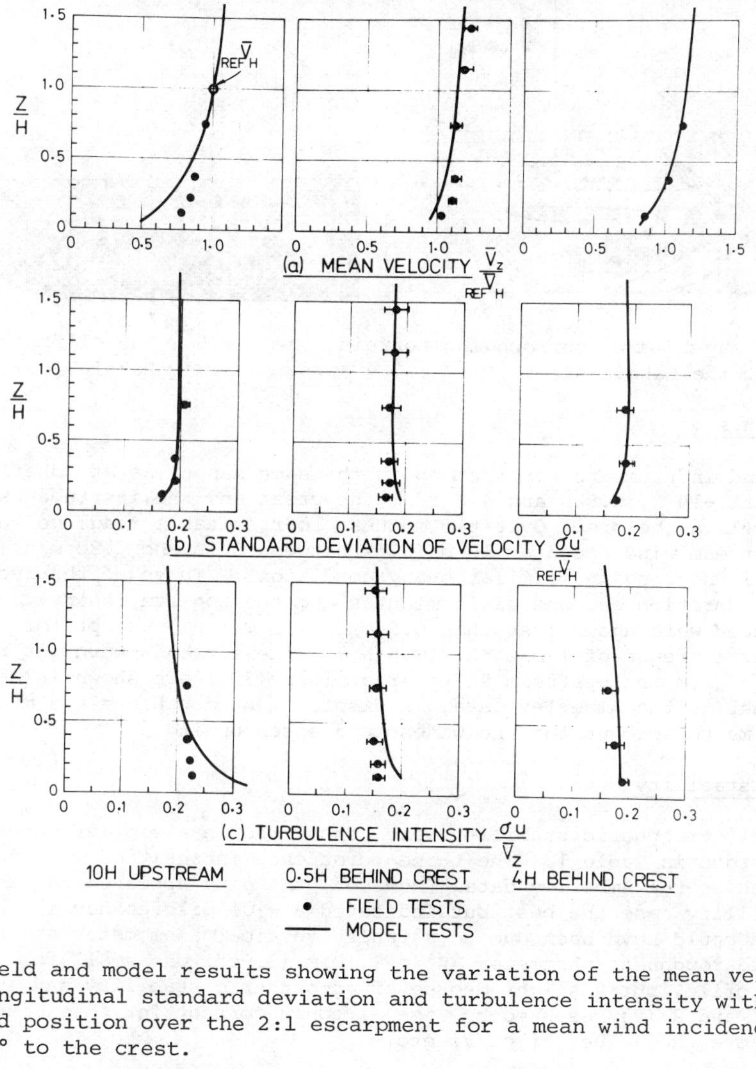

Fig.3. Field and model results showing the variation of the mean velocity, longitudinal standard deviation and turbulence intensity with height and position over the 2:1 escarpment for a mean wind incidence of 30° to the crest.

CLIFF ESCARPMENT FIELD TEST

Site description

The cliff site was located on the south bank of the Rakaia River 70 km (44 miles) south of Christchurch, and faced NE over a 2 km (1¼ miles) wide flat shingle riverbed with occasional patches of 1 m (40") high lupin scrub. A map of the surrounding terrain may be seen in Fig.4 and a general view looking SW showing the three masts seen in Fig.5. The countryside on the opposite side of the river and upwind of the site was flat open farmland with the occasional row of trees. The cliff was 11.56 m (37.93') high with a straight, sharply defined crest and a flat downwind terrain covered with short grass. The recorded winds came across the riverbed and onto the escarpment at a mean angle of 65±3 degrees to the crest.

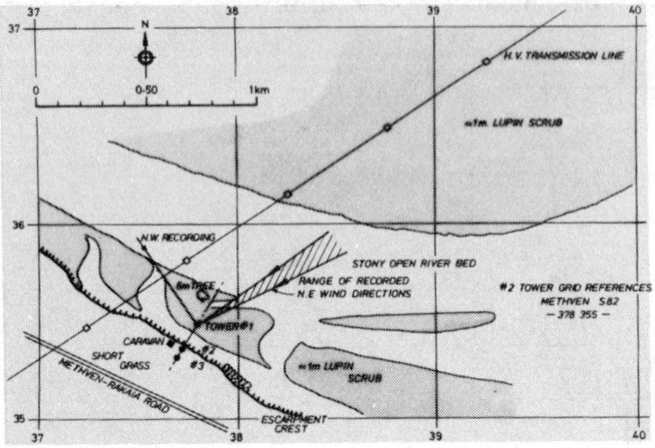

Fig.4. A map of the surrounding terrain at the Rakaia site.

Fig.5. The cliff escarpment at Rakaia looking SW.

Site recordings

The towers and instruments were set up in the same manner as at Amberley with the three masts at -10 H, 0.5 H and 4 H from the crest and the instruments close to the same levels as before. Over two months, four suitable readings were made with the mean upstream wind speed at 10 m (33') between 6.76 and 7.80 m/s (15.12 and 17.45 m.p.h.) under again, similar sunny conditions. Three of the recordings were of half hour duration and one of 15 minutes and the non-stationary trends in the mean wind speed were again less than 0.5 m/s (1.1 m.p.h.) except for one recording with a linear increase of 1 m/s (2.2 m.p.h.). These trends were not removed. The averaged data from the upstream #1 tower at 10 m (33') are shown in Table 1 in a similar manner as the Amberley data. A displacement depth, d = 0.5 m (20") was used this time to account for the widespread scrub upwind.

Atmospheric stability

Four cases of atmospheric stability were considered in a similar manner as before and are recorded in Table 1. The three turbulence intensities were not as close to that expected as the Amberley data. However, the data appeared to fit the near neutral stability case the best but indicated a wide discrepancy in the Reynolds stress which could have been due to a faulty vertical anemometer at that position. The indicated roughness parameter, Z_o = 0.10 m (3.94") was small for the measured power law profile, but a slight degree of atmospheric stability say somewhere between Case 1 and 2 (represented by Case 4) could correct for this discrepancy as well as improve the values of $\sigma u/V_*$ etc.

Test results

The variations of $\bar{V}_Z$ and σu over the cliff are shown in Fig.6 and compared with the recent wind tunnel data. Agreement between model and field results were confined to general trends only, except for positions outside the strong wake generated behind the crest where agreement was good. In the wake region, a strong reduction in mean velocity and a very high turbulence in all three directions were evident in

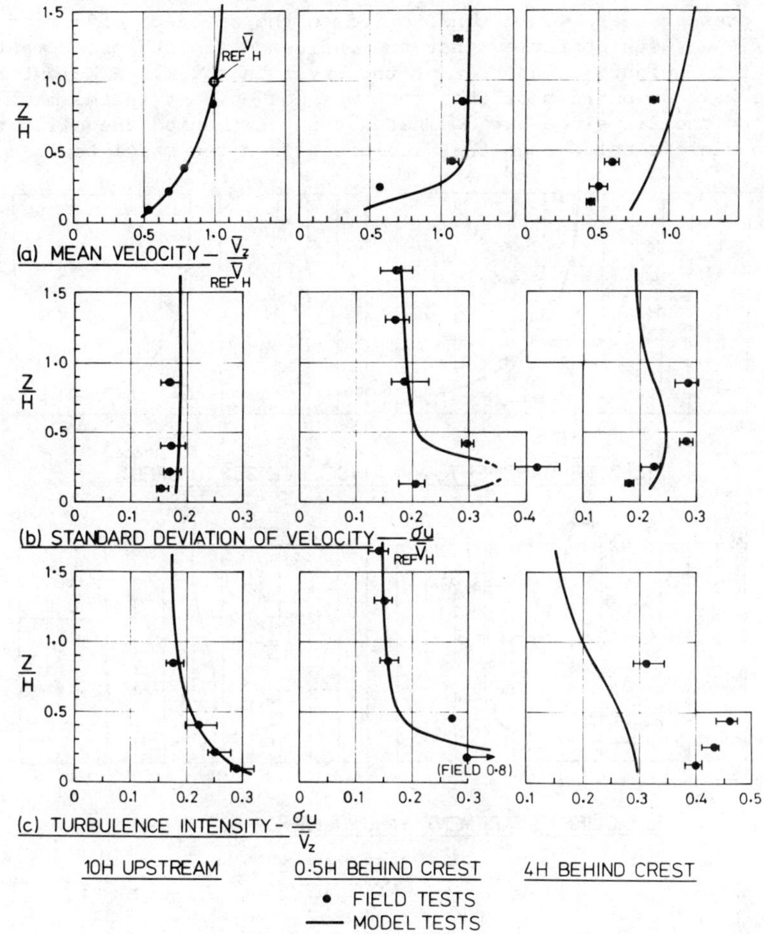

Fig.6. Field and model test results showing the variation of the mean velocity, longitudinal standard deviation and turbulence intensity with height and position over the cliff escarpment for a mean wind incidence of 66° to the crest.

both the model and field tests, which would make the accuracy of the data taken by hot wire and propeller anemometers increasingly suspect. However, the reverse flow was only found at the lowest level on tower #2 during the field tests. This restricted region of separated flow was probably due to the oblique angle of incidence of the wind.

DISCUSSION

Mean velocity

The variation in mean velocity over the two escarpments showed similar trends to those observed in earlier model tests by Bowen and Lindley (2). However, the difference in the angle of incidence and the surface roughness of the models made close comparison difficult with the earlier tests and other research work such as Jackson (5), Meroney et.al.(6) and Freeston (7). It was expected that the smooth surface of the present models would have caused higher flow velocities close to the ground than with the previous rough models, whilst the more oblique angles of incidence of the present tests would tend to reduce the speed-up effect of the hill. The present field tests confirmed that the influence of both escarpments was still strong after four heights downstream of the crest in the 1.5 H height range considered. The wake downstream of the crest was not evident in the mean velocity profiles behind the 2:1 slope but dominated the flow behind the cliff to a greater extent in the field results than that indicated from the model tests.

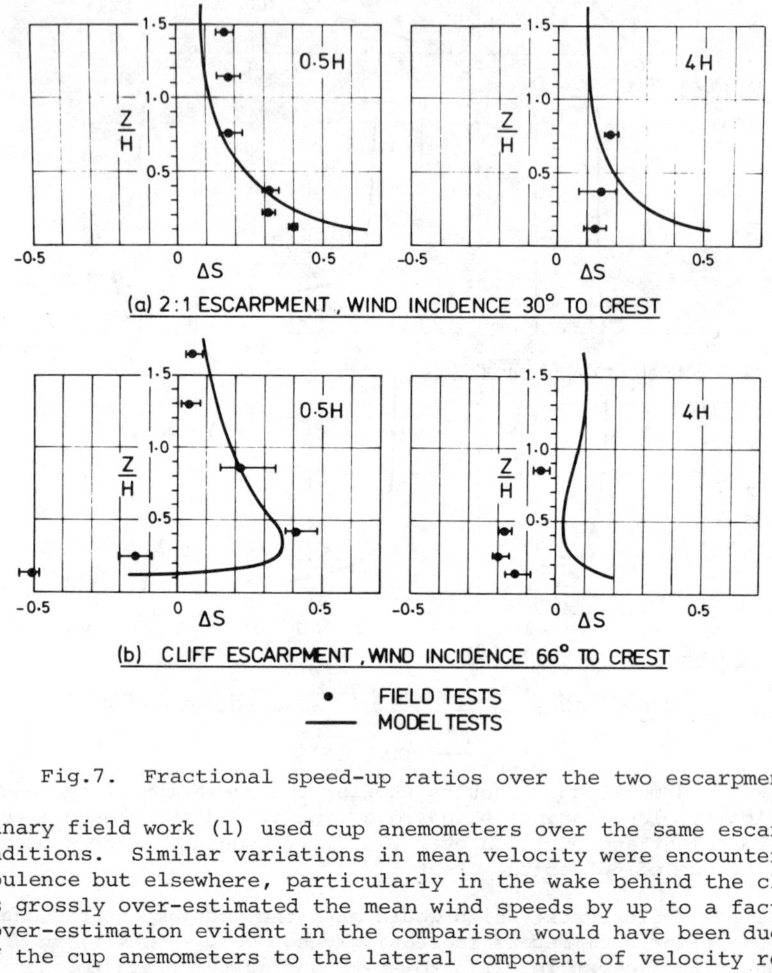

Fig.7. Fractional speed-up ratios over the two escarpments

The preliminary field work (1) used cup anemometers over the same escarpments in similar conditions. Similar variations in mean velocity were encountered in areas of low turbulence but elsewhere, particularly in the wake behind the cliff, the cup anemometers grossly over-estimated the mean wind speeds by up to a factor of two. The large over-estimation evident in the comparison would have been due to the sensitivity of the cup anemometers to the lateral component of velocity regardless of sign and their poor dynamic response characteristics leading to large over-estimation in regions of high turbulence.

Agreement between the present model and field results was satisfactory in flow regions where the turbulence intensity was less than 0.2. However, major differences occurred in the highly turbulent wake region behind the cliff which would be mainly attributable to the inaccuracy of the single horizontal hot wire probe in such flow conditions. In many cases the propeller anemometers recorded turbulence intensities well over 0.3 in both longitudinal and lateral directions. Changes in flow direction over the crest could also account for a significant error from the hot wire which was aligned normal to the upstream air flow.

The variation in the fractional speed-up ratio, ΔS is shown in Fig.7 for the two escarpments. Larger discrepancies between model and field tests were evident here as the ratio incorporated the instrumentation errors as well as the differences in the upstream velocity profiles. The large fluctuations between the individual field readings suggested that precise values of ΔS were meaningless in the full scale situation where the atmospheric stability was constantly changing. However, it may be seen that the separated flow behind the cliff was sufficiently strong to reduce the values of ΔS close to the ground to such an extent that they were less than that for the 2:1 slope. Maximum values of $\Delta S \approx 0.4$ occurred close behind the crests of both escarpment field recordings but were exceeded consistently by the model test results. The results suggested that a maximum ΔS would be created close to the ground behind a rounded crest, whose upstream slope was as steep as possible without separation occurring.

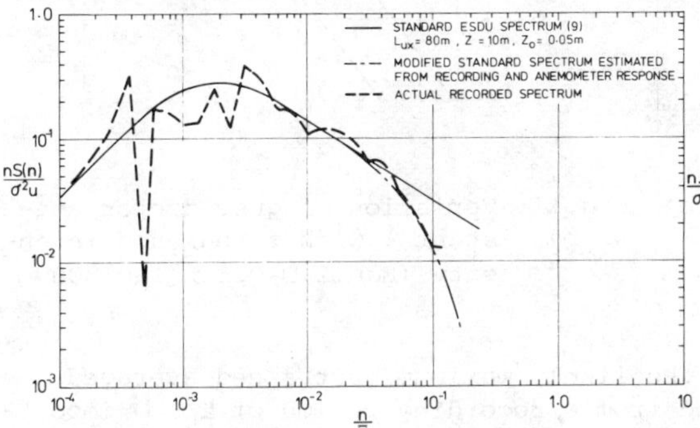

Fig.8. A recorded upstream longitudinal velocity power spectrum at Z=10 m and standard data.

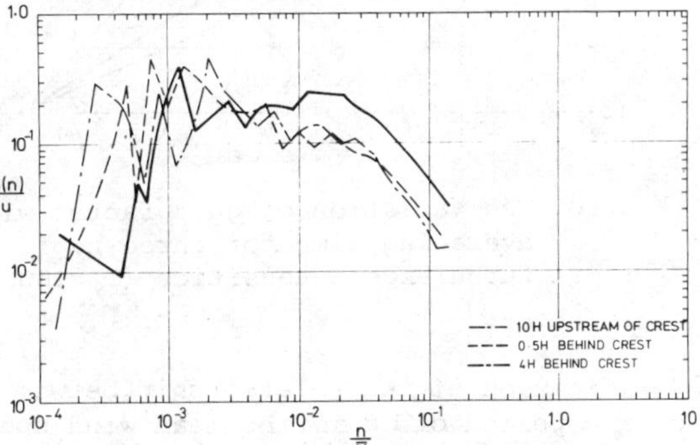

Fig.9. Change in longitudinal velocity power spectrum at Z=10 m with position over cliff.

Turbulence intensity

The model and field results showed good agreement for the turbulence intensity behind the 2:1 slope. Values of σu increased slightly at low levels close behind the crest, but a small decay in magnitude was evident at 4 H in contrast to the earlier tests using a rough model, when σu remained higher for some distance downstream. The wake behind the cliff crest exhibited very high levels of turbulence which showed signs of weakening at the 4 H downstream position and diffusing upwards to a depth of about one cliff height.

Power spectra

The power spectral density of longitudinal turbulence such as in Fig.8, which were recorded at a height of 10 m (33') on the upstream tower, were in all cases close to that expected if due regard for the frequency response of the instrument was taken. The spectra recorded at positions behind the crest exhibited a flatter top with increased energy at the higher frequencies as can be seen in Fig.9, a trend which was commented on previously (2). This trend was more definite behind the cliff escarpment for positions submerged in the highly turbulent wake but was hardly noticeable at the 10 m (33') height behind the 2:1 slope.

Velocity probability distribution

The frequency of occurrence of the longitudinal velocity component was analysed for all field test positions over the two escarpments. It was found that the probability distributions were very close to Gaussian for all positions with the only difference being in the value of peak gust recorded.

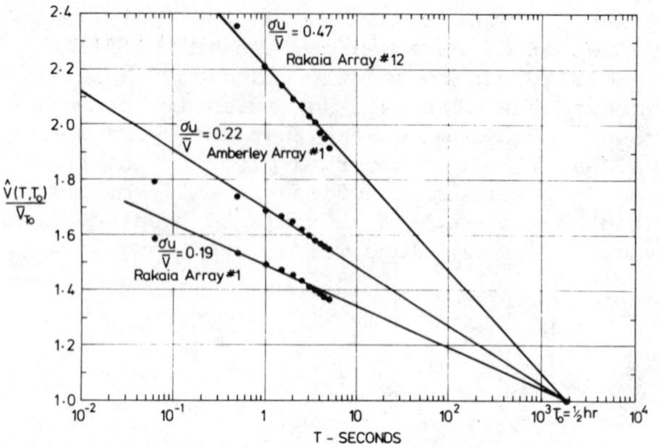

Fig.10. The variation of gust factor with averaging time for three typical turbulence intensities (T_o=1800 s).

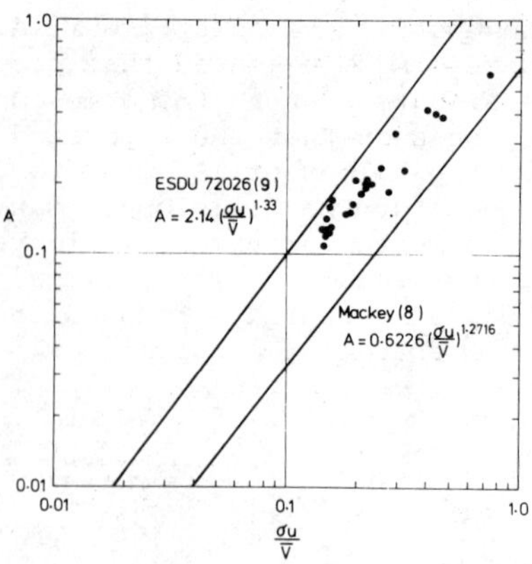

Fig.11. Variation of gust factor constant A with turbulence intensity (Equ.2, T=0.5 s, T_o=1800 s).

Gust factors

Verification of the relationship between the likely maximum gust speed averaged over a period of T s and the mean wind speed from a recording period of T_os is important for a number of reasons. Simple wind loading calculations are often carried out using a T=3,5 or 15 s gust speed and ignoring the dynamic response of the structure. Field data providing these gust speeds are fragmented and furthermore, the anemometers used in the collection of these design wind speeds have differing response characteristics and would only provide accurate gust speeds if the so-called gust averaging time, T is correctly matched to the dynamic response of the instrument. A knowledge of the variation of maximum gust speed with averaging time would thus allow a means of standardising peak gust speeds obtained from different types of anemometers and in flows of different turbulence intensities. Conversely, the turbulence intensity could be estimated directly from the peak gust speeds recorded by conventional anemometers which otherwise would require more elaborate fast response equipment.

It was considered worthwhile then to check the gust factors obtained in the present tests with current understanding and to see if the terrain had affected the situation significantly. The longitudinal velocity recorded every 1/16th second at each instrument position was scanned over the whole length of the half-hour recordings and the peak velocity reading located. The data on either side of this reading was then averaged over different periods to produce peak velocities with T=0.06,0.5, 1, 1.5 etc. up to 5 seconds averaging time. The peak velocities were averaged over the number of recordings and three examples have been plotted in Fig.10 as gust factor, $\hat{V}/\bar{V}$ against averaging time, T representing conditions of different turbulence intensity. These plots confirmed the conclusions of Mackey (8) and ESDU 72026 (9) that

$$\frac{\hat{V}}{\bar{V}}(T,T_o) = 1 - A\log_{10}\left[\frac{T}{T_o}\right] \qquad (2)$$

It also became clear that the gust factor averaged over 0.5 s behaved in the same manner as those averaged over longer periods despite the upper frequency limit of 0.25 Hz for adequate measurement of a pure sinusoidal velocity fluctuation by the propeller anemometers. The average 0.5 s gust factors obtained from each position over the two escarpments, represented data from a wide range of turbulence intensities and were used in Fig.11 to define the factor A in equation 2. When considering the scatter in the results it must be remembered that the peak velocity has a certain probability of occurrence with its own mean and standard deviation and also were obtained from the disturbed flow over an obstruction. However no consistent trend was evident that could be attributed to the effect of the escarpments. The results showed a compromise between the two estimates of Mackey and ESDU being closer to the ESDU estimate namely :

$$\frac{\hat{V}(T,T_o)}{\bar{\bar{V}}} = 1 - 2.14\left[\frac{\sigma u}{\bar{\bar{V}}}\right]^{1.33} \log_{10}\left[\frac{T}{T_o}\right] \quad (3)$$

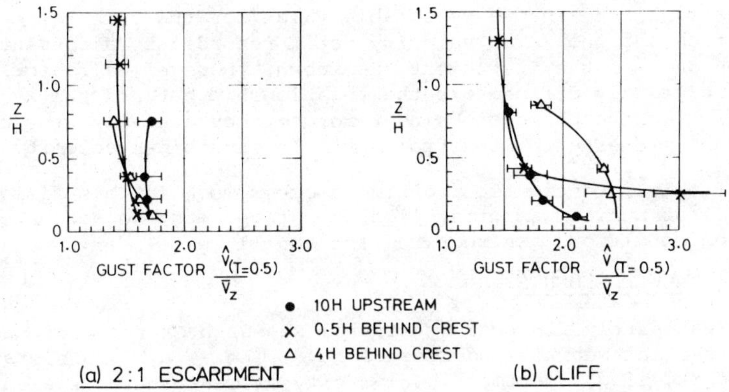

Fig.12. Variation of gust factor over the escarpments (T=0.5 s, T_o = 1800 s)

An alternative approach using the longitudinal velocity power spectra and extreme value analysis has been suggested by Davenport (10) when he derived expressions for the average gust velocity and its standard deviation as

$$\text{Mean} \quad \hat{\eta} = (2 \log_e \nu T_o)^{\frac{1}{2}} + 0.577(2 \log_e \nu T_o)^{-\frac{1}{2}} \quad \text{and} \quad \sigma(\hat{\eta}) = \frac{\pi}{\sqrt{6}}(2 \log_e \nu T_o)^{-\frac{1}{2}} \quad (4),(5)$$

$$\text{where } \nu^2 = \frac{\int_o^\infty n^2.nS(n).d \log n}{\int_o^\infty nS(n).d \log n} \quad \text{and} \quad \eta = \frac{V - \bar{V}}{\sigma u}$$

Using the modified standard longitudinal power spectrum shown in Fig.8, a value of νT_o = 210 was obtained using $\bar{V}$ = 7.88 m/s and T_o = 1800 s. This then yielded a satisfactory mean gust factor of 1.74 ± 0.08 in comparison with the particular field data shown in Fig.10 (Amberley array #1, upstream at Z = 10 m).

The variation of gust factor over the two escarpments is shown in Fig.12 and exhibited the same trends as the turbulence intensity. Typical values between 1.4 and 1.7 occurred except in areas of high turbulence where values of up to 3 were encountered.

CONCLUSIONS

The wide scatter of results from the field tests during apparently similar weather conditions indicated that small changes in atmospheric stability would be significant when estimating the wind conditions over a hilly site. Wind tunnel model tests would normally only yield data applicable to neutral conditions which may not represent the worst conditions, such as during squally weather or thunder-storms. Model studies would also be suspect in areas of high turbulence or separated flows where conventional measuring techniques become inaccurate.

Typical increases in velocity close to the ground behind the crest for both slopes were close to 40% but reduced rapidly with height above the ground. Earlier model tests indicated that larger values would be expected with winds normal to the crest. The influence of the escarpments extended well beyond 4 H downstream of the crest.

The wake generated by the sharp crest created high turbulence levels which systematically decayed downstream whilst the wake grew in height. The turbulence levels close to the ground depended on the surface roughness and detailed geometry of the hill and crest. Outside the wake, turbulence levels remained almost unchanged and behaved as if convected along the streamlines. The power spectral densities remained unchanged except close to the ground in the wake region, where an increase in high frequency energy was evident.

The field data confirmed that a reasonable value for the turbulence intensity could be estimated from the peak/mean velocity ratio recorded by a conventional anemometer if the response of the instrument was known. However, a factor which still remained to be accurately defined was the relationship between the averaging time of the highest peak gust capable of being recorded accurately and the upper frequency limit of the anemometer estimated from a single sinusoidal velocity input.

The results of this investigation confirmed other work in this field that the modifying influence of a hill was often significant and must be accounted for if accurate design conditions are required for a hilly site.

ACKNOWLEDGEMENTS

This project was partly financed by several Research Grants from the University Grants Committee and New Zealand Electricity. The valuable assistance from the Mechanical Engineering Department, University of Canterbury is also gratefully acknowledged.

REFERENCES

1. Bowen, A.J. & Lindley, D., "Measurements of the Mean Wind Flow over Various Escarpment Shapes", Proc.Fifth Australasian Conf.on Hydraulics and Fluid Mechanics, University of Canterbury, New Zealand, 211-219 (1974).
2. Bowen, A.J. & Lindley D., "A Wind Tunnel Investigation of the Wind Speed and Turbulence Characteristics Close to the Ground over Various Escarpment Shapes", Boundary Layer Meteorology, 12, 259-271 (1977).
3. Counihan, J., "Adiabatic Atmospheric Boundary Layers: A Review and Analysis of Data from the period 1880-1972", Atmospheric Environment, 9, 871-905 (1975).
4. Panofsky, H.A., "Wind Structure in Strong Winds Below 150 m", Wind Engineering, 1, No.2, 91-103 (1977).
5. Jackson, P.S., "A Theory for Flow over Escarpments", Proc.Fourth International Conf.on Wind Effects on Buildings and Structures, Heathrow, U.K.,33-40 (1975).
6. Meroney, R.N., Sandborn, V.A., Bouwmeester, R.J.B. & Rider, M.A., "Sites for Wind Power Installations", 1976 Annual Report ERDA/NSF-00702/75/Tl; 1976 Progress Report ERDA/Y-76-S-06-2438/7611 and 1977 Annual Report Dept.of Energy RLO/2438-77/1, Colorado State University, Fort Collins, U.S.A.
7. Freeston, D.H., "Atmospheric Shear Flow over Ramps and Escarpments", Presented to Fifth Australasian Conf.on Hydraulics and Fluid Mechanics, Univ. of Canterbury, N.Z., Dec.1974. Available from Mechanical Engineering Department, University of Auckland.
8. Mackey, S., "Wind Studies in Hong-Kong - Some Preliminary Results", Industrial Aerodynamics Abstracts, 5, 1-16 (1970).
9. ESDU 72026, "Characteristics of Wind Speed in the Lower Layers of the Atmosphere Near the Ground: Strong Winds (Neutral Atmosphere)", Engineering Sciences Data Unit, 251-259 Regent Street, London, W1R 7 AD (1972).
10. Davenport, A.G., "Note on the Distribution of the Largest Value of a Random Function with Application to Gust Loading", Proceedings I.C.E. 28, 187-196 (1964).

TERRAIN EFFECTS ON WIND FLUCTUATIONS

H. A. PANOFSKY, M. VILARDO and R. LIPSCHUTZ

The Pennsylvania State University, University Park, Pennsylvania, USA

ABSTRACT

The relative intensity of turbulence has been measured over complex terrain near State College, PA. The results show that the roughness length alone accounts for the effect of terrain on fluctuations of vertical velocity. On the other hand, also terrain features on a larger scale increase fluctuation of the horizontal wind components.

INTRODUCTION

The destructive force of winds not only depends on mean winds, but also on turbulent wind fluctuations. Given the mean winds and the standard deviation of wind speed fluctuations, it is possible to calculate peak wind statistics.

The theory of mean wind profiles is well understood; but the properties of the wind fluctuations still present problems. The purpose of this note is to report on results from a field experiment in central Pennsylvania, which casts some light on the effect of terrain variations on wind speed fluctuations.

THEORY

Panofsky et al. (1977) have shown that Monin-Obukhov similarity is unable to explain the properties of the variations of the horizontal wind components. Instead, these authors suggest that, in unstable and neutral stability,

$$\frac{\sigma_u}{u_*} = f\left(\frac{z_i}{L}\right) \quad . \tag{1}$$

Here, σ_u is the standard deviation of the wind component parallel to the mean wind speed; very nearly σ_u is the standard deviation of the wind speed itself. u_* is the friction velocity, L the Monin-Obukhov length and z_i the height of the lowest inversion, approximately the thickness of the mixed layer. f is a function, presumably universal, for which the authors quoted give some empirical expressions.

For destructive winds, L becomes very large, so that

$$\frac{\sigma_u}{u_*} = f(0) = A \quad . \tag{2}$$

We are therefore particularly interested in f(0)=A. The friction velocity u_* can be eliminated from equations (1) or (2) through the equation for the wind profile:

$$V = \frac{u_*}{k}\left[\ln\frac{z}{z_o} - \psi\left(\frac{z}{L}\right)\right] \tag{3}$$

where V is the wind speed, k the von Karman constant, often taken as 0.4, z the height, z_o the roughness length, and ψ is another universal function, the properties of which are quite well known (see, e.g. Paulson, 1970).

Combining equations (1) and (3), we obtain:

$$\frac{\sigma_u}{V} = \frac{k f\left(\frac{z_i}{L}\right)}{\ln\frac{z}{z_o} - \psi\left(\frac{z}{L}\right)} \tag{4}$$

which gives the theoretical behavior of σ_u/V in terms of roughness and other basic variables.

For destructive winds, we can put ($|L| \to \infty$)

$$\frac{\sigma_u}{V} = \frac{kA}{\ln\frac{z}{z_o}} \tag{5}$$

Panofsky et al. (1977) had pointed out that $f(\frac{z_i}{L})$ and A may not be universal, but depend on terrain features not described by z_o, such as topography on a horizontal scale of order of 1 km or more. For this reason the variables in equation (4) were measured in undulating farm land, near a low mountain ridge, in central Pennsylvania. Particular emphasis was placed on determining the characteristics, at this site, of the quantity $A \equiv f(0)$. Over flat terrain, and in wind tunnels, $A \sim 2.5$ (see, e.g. Ariel and Nadezhina, 1976) so that we have the simple result:

$$\frac{\sigma_u}{V} = \frac{1}{\ln\frac{z}{z_o}} \quad . \tag{6}$$

However, even in the sixties (see Lumley and Panofsky, 1964) it had become clear that A is not a constant and equation 6 not always valid.

SITE AND INSTRUMENTATION

Under the direction of John Norman, a field site has been established at Rock Springs, eight miles SW of State College, PA. The site and instrumentation have been described in detail by Perry et al. (1978). Here, only a brief summary is given. The site is located in a field covered with different crops at different times, located in rolling farm land. To the SE of the site, an almost two-dimensional wooded ridge is located rising 300 m above the valley, oriented from 240° to 60°. The distance from the site to the first trees of the ridge is about 400 m.

At first, measurements of mean winds were made by Thornthwaite anemometer at three levels, allowing the roughness length z_o to be computed for various crops. Fast-response measurements were confined to two meters. These were continued after the profile measurements had stopped.

Wind fluctuations were measured with Norman drag anemometers, temperature fluctuations with platinum-wire anemometers, and mixing depths for the later experiments, with an acoustic sounder. From these measurements, standard deviations of the velocity components, Reynolds stress, vertical heat flux and Monin-Obukhov L were computed.

RESULTS

Fig. 1 shows, first of all, values of σ_u/V observed at O'Neill, Nebraska, plotted as functions of z/L. The observations were taken from Haugen (1959). z/L was computed from Richardson numbers (obtained by Allen Weber for 2 m height) by the formulae: z/L = Ri (unstable) and z/L = Ri/(1-5 Ri), stable. A line is fitted to these points by eye. The parameter A (see equations 2 and 5) computed for O'Neill, where $z_o \approx 0.8$ cm, is 2.4. This is close to the usual value for A in flat terrain.

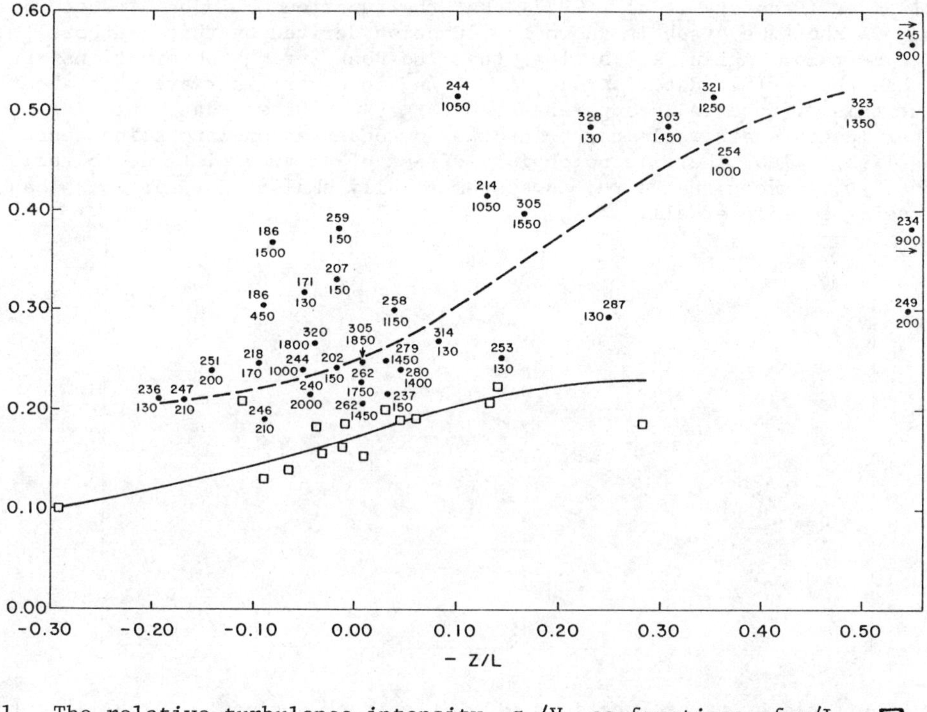

Fig. 1. The relative turbulence intensity, σ_u/V, as functions of z/L. □ and solid line, O'Neill; circles and dashed line, Rock Springs. Upper numbers, wind direction; lower numbers, mixing depth in meters.

Also shown in Fig. 1 are observations of σ_u/V for Rock Springs, multiplied by $\ln z/0.8/\ln^z/z_o$ where z is in centimeters, to "reduce" the Rock Springs data to the O'Neill roughness.

Next to each dot denoting Rock Springs observations are given the wind direction (above) and mixing depth in meters (below).

The most outstanding feature of the Rock Springs values is that σ_u/V there is systematically larger than at O'Neill. A dashed line has been drawn through all points with wind direction greater than 240°. The scatter about this line is only moderate. The value of σ_u/V interpolated for $z/L = 0$ is about .25. Solving equation 5 for A then yields $A \simeq 3.5$, roughly 50% larger than at O'Neill.

The values of σ_u/V for directions 240° or less are, on the average, larger, and widely scattered. Part of the reason for this is the wake effect of the ridge. For example, wind tunnel studies by Plate and Lin (1965) have shown that σ_u/V is much larger in the wake of two-dimensional hills than 2.5. For our data, A for these cases averages about 4.5, with large scatter.

If the wind is just parallel to the ridge (wind direction around 240°) the reason for the large A-values is presumably the creation of quasi-horizontal eddies by the shear created between the valley floor and the stationary ridge.

Fig. 2 shows σ_u/u_* plotted as functions of z_i/L, which, according to equation 1, is expected to be a universal function. Actually, it was suggested in the original paper (Panofsky <u>et al</u>., 1977) that the function would be affected by terrain. On the same graph is shown the function derived by these authors over flat terrain. Again, it is clear that the Rock Springs observations are larger than the earlier data. Again, $A = \sigma_u/u_*$ in neutral air averages about 3.5. But this difference is apparent even for $z_i/L \to 0$, so that replacing z_i by another length scale would not bring the new observation into coincidence with the line. Also, there is no obvious effect of varying distance to the ridge on σ_u/u_*, except that σ_u/u_* is systematically smaller for air which has not crossed the ridge at all.

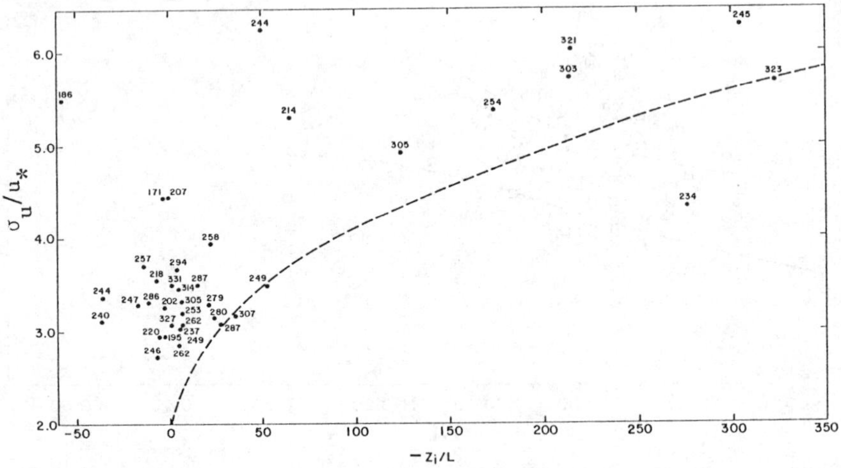

Fig. 2. The ratio σ_u/u_* as functions of z_i/L at Rock Springs. Solid line: the equation

$$\frac{\sigma_u}{u_*} = \left(4 + 0.6 \left\{\frac{z_i}{-L}\right\}^{2/3}\right)^{1/2}$$

Valid over uniform terrain. Numbers represent wind direction.

Thus, both Figures (1) and (2) suggest that the effect of mesoscale terrain is to increase σ_u/V and σ_u/u_* at all z/L or z_i/L by about the same amount.

Additional increases seem likely if the air has crossed a ridge or blows parallel to a ridge.

It has been shown in another paper (Panofsky, et al., 1978) that the fluctuations of vertical motion are insensitive to mesoscale terrain, in agreement with a statement by Egan (1975) that vertical diffusion over complex terrain is the same as over uniform terrain. Therefore, a good indicator of the effect of terrain on turbulence is the ratio σ_u/σ_w, where σ_w is the standard deviation of vertical velocity. Over uniform terrain near neutral conditions, $\sigma_u/\sigma_w \sim 1.9$. Near Rock Springs this ratio averages about 2.8. Again, larger values dominate in air that has crossed the ridge or moved parallel to it.

The results given in this note clearly do not apply to all types of mesoscale terrain features, but only to gently rolling terrain, and to the wake of a small hill.

In particular, conditions are quite different on top of hills and escarpments. Bradley (1979) has recently summarized theory and observations relating to this subject. In particular, $\sigma_u/\overline{V}$ is close to its flat-land value right at the surface and then rapidly decreases upward. In contrast, $\sigma_w/\overline{V}$ becomes larger. u_* also decreases with height near the surface so that σ_u/u_* stays relatively close to its flat-land value in the region below a level of maximum wind speed.

REFERENCES

Ariel, N. A., and Y. D. Nadezhina, 1976: Dimensionless turbulence characteristics under various stratifications. Izv., Atmos. and Ocean Physics, 12, pp. 802-809.

Bradley, E. F., 1979: An experimental study of the profile of wind speed, heating, stress and turbulence at a crest of a large hill. Submitted to Quarterly Journal Roy. Met. Soc.

Egan, B. A. (Ed.), 1959: Project Prairie Grass, a field program in diffusion, Geophysical Res. Paper, AFCRC, No. 59.

Lumley, J. L., and H. A. Panofsky, 1964: Structure of Atmospheric Turbulence, Wiley, New York.

Merry, M. and H. A. Panofsky, 1976: Statistics of vertical motion over land and water, Quart. Roy. Meteor. Soc., 102, pp. 255-260.

Panofsky, H. A., H. Tennekes, D. H. Lenschow, and J. C. Wyngaard, 1977: The characteristics of turbulent velocity components in the surface layer under convective conditions, Bound. Layer Meteor., 11, 355-359.

Panofsky, H. A., J. M. Norman, C. A. Egolf, and R. Lipschutz, 1978: On characteristics of wind direction fluctuations in the surface layer, Bound. Layer Meteor., 15, 439,446.

Paulson, C. A., 1970: The mathematical representation of wind speed and the temperature profiles in the unstable atmospheric surface layer, J. Appl. Meteor., 9, 857-861.

Perry, S. G., J. M. Norman, H. A. Panofsky, and J. D. Martsolf, 1978: Horizontal coherence decay near large mesoscale variations of topography, J. Atmos. Sci., 35, 1884-1889.

Plate, E. J. and C. W. Lin, 1965: The velocity field downstream of a two-dimensional model. Final Report U.S. Army OA AMC-36-039-63-G7, Fluid Dynamics and Diffusion Laboratory, Colorado State University.

EFFECT OF HIGH ROUGHNESS ON THE CHARACTERISTICS OF TURBULENCE IN CASES OF STRONG WINDS

Ph. DUCHÊNE-MARULLAZ

Centre Scientifique et Technique du Bâtiment, Nantes, France

ABSTRACT

This paper presents the results of an investigation of the structure of strong winds over a rough site. Owing to the heterogeneity of the site, the examination of the mean wind profiles allowed the rapid variation of the roughness parameter to be shown as a function of wind direction. Also evidenced was the important influence of the height and roughness on the principal turbulence parameters. The experimental results are in good agreement with some of the recently proposed theoretical and empirical models of the surface boundary layer.

INTRODUCTION

In the last several years experimental studies of wind structure in the surface boundary layer have been more and more oriented towards an examination of the phenomena observed in turbulent zones, for example where there is a high roughness (1,2,3), where there are the effects of obstacles (4) or where there is a change in roughness (5). At the same time, models of the atmospheric boundary layer, recently proposed in the instances of strong winds and neutral stability (6,7,8), take into account wind structure modifications in high roughness zone, more so than previous models which were only the extrapolation of surface boundary layer theories, mainly verified in the cases of smooth roughness (z_o < 0.1 m).

Nevertheless, despite the increase of experimental studies in the field of high roughness, there is still a need of measurements to complete the already existing results (as verified by Counihan (7) and Teunissen (3)), in order to verify and validate the theoretical models, or in order to propose other ones.

The following paper presents the results of the measurements of strong winds, carried out between 1974 and 1977 in a suburban area of Nantes. In two previous papers (2,9) some partial results of these measurements were given. All the runs that have been studied are brought together here. The characteristics of the mean flow and of the turbulent parameters for different wind directions will be successively examined. Greater consideration will be given to the longitudinal component of the wind, the component the most readily used in its applications related to wind effects on buildings. The results obtained will be analyzed by comparing them to already existing models and to measurements carried out on other sites.

THE SITE AND THE MEASURING EQUIPMENT

Few details on the site and the measuring instruments will be presented here. Further information is given in papers (2) and (10). The experiment is set up on the grounds of C.S.T.B. which is situated northeast of Nantes (a city of about 400,000 inhabitants) at the city limits. The prevailing strong winds (S-SW - NW) cross a large part of the city before reaching C.S.T.B.

The environment around the measuring site, within a 1000 meter radius, is very heterogeneous. It was divided into four sectors (A,B,C,D) (Fig. 1) which correspond to the zones where the vertical profiles of the mean wind speed are similar.

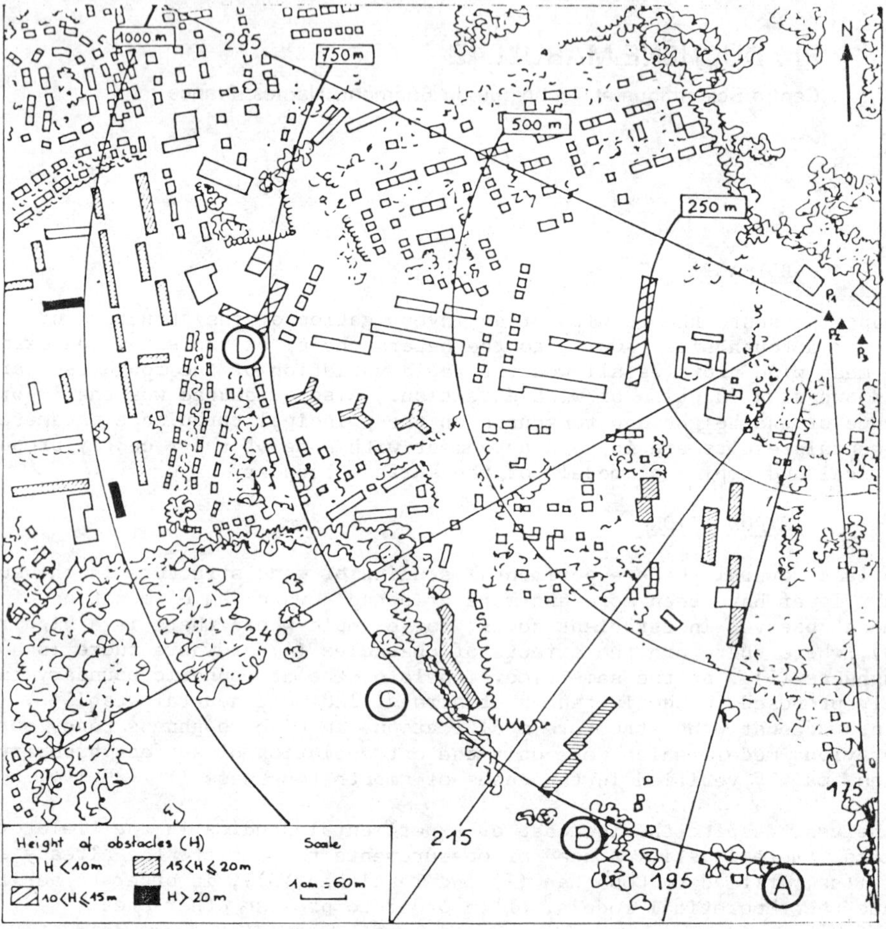

Fig. 1. Map of the site.

Sector A is mainly vegetation and low-level trees (< 5 m high); sector B is influenced by a group of buildings built side by side, about 18 m high and situated less than 500 m from the site; sector C is fairly heterogeneous (a flat zone with tall trees 15 to 20 m high) with some isolated buildings and groups of houses; finally, sector D is more homogeneous (groups of houses and small trees, < 8 m) and up until 850 m from the site there are well-spaced buildings less than 15 m high. Behind the masts, in the opposite direction of the strong prevailing winds, the roughness is made up of meadows and small trees about 4 to 5 m high.

The measuring equipment includes three 60 m high masts (P_1,P_2,P_3) situated on a line perpendicular to the sector of the prevailing strong winds and set up respectively at 20 m and 60 m from P_1 (Fig. 1). They are each equipped with Gill Propeller bi-directional anemometers (tri-directional for P_1) at four principal levels (10 m, 20 m, 40 m and 60 m above the ground); P_1 is also equipped at four supplementary levels (15 m, 30 m, 50 m and 55 m above the ground).

The temperature vertical gradient is measured as well at P_2 at six different levels (5, 10, 20, 30, 40 and 60 m above the ground). This is done with the help of platinum resistance sensors installed in aspirated radiation shields.

The analogous signals provided by the anemometers are transmitted to a data acquisition system which digitalizes them at the frequency of 2 Hz for each channel.

MEAN WIND PROFILES

Choice of the runs

Sixty-four runs have been selected of which the lengths vary from 10 to 20 minutes. They correspond to strong winds ($\overline{U}$(10 > 5 m/s)) coming from the sector 175-295°. The stationarity of each of these runs has been verified; the atmospheric stability has also been examined by calculating the Richardson number Ri. For these runs, as the values of Ri were between 0.025 and 0.015 inclusive, the hypothesis of a neutral stability (or nearly neutral) was respected.

Roughness parameters

To study the vertical profiles of the mean wind speed, the data of the three masts has been used, and a unique mean profile characteristic of the site for the direction in consideration has been determined.

The examination for each run of the vertical profile of the mean wind speed $\overline{U}(z)$ showed that the logarithmic law

$$\overline{U}(z) = \frac{U_*}{k} \,\mathrm{Ln}\,(z/z_o)$$

where U_* is the friction velocity, z_o is the roughness length and k is the von Karman constant (k = 0.4) suitable to these profiles, provided that for sectors B, C and D the value at 10 m, and sometimes at 15 m (sectors B and C), is not taken into account. It is found that in all the cases, at these levels, there is an acceleration effect which shows up a kink on a graph with the coordinates LnZ, $\overline{U}(z)$ (between 15 m and 20 m). This modification in the profile is perhaps due to a change in roughness: there is, in fact, between the masts and the built-up area, an area without obstacles about 120 m long. Using the model proposed by Rao et al. (11), for a change in roughness, from rough to smooth, the internal boundary layer at the masts would have a height δ of 12 m ($\delta/x \simeq 1/10$, x being the distance between the roughness change and the point in consideration).

Consequently, it was possible to determine, for each run, a value of the roughness length z_o and a value of the friction velocity U_*. But, owing to the scattering of these values for a single direction, or for neighboring directions where the roughness is not modified, it seemed necessary to regroup the profiles in order to determine the characteristic values of the roughness length z_o; lastly, after having regrouped the profiles for every 5 degrees, the site was divided into four sectors: A: 175°-195°, B: 195°-215°, C: 215°-240°, and D: 240°-295°. Each sector corresponds to a zone where the mean wind speed profiles can be considered as similar and where the roughness presents the same characteristics. For each sector a mean profile has been determined and the logarithmic law has been adjusted to the experimental points by using the least squared method. The values of z_o are shown in Table 1.

TABLE 1. The roughness parameter.

	Sector A	Sector B	Sector C	Sector D
z_o (m)	0.55	2.1	1.3	0.70

If for sector B, the value of 2.1 m seems high (but this sector is very disturbed), the values for sectors A, C and D correspond well to the value which can be expected for an urban site of this type. These last values are readily comparable to the values obtained by Ramsdell (1) in a Seattle suburb with a neutral stability (0.4 m to 1.5 m); his profiles present the same "kink" as the ones determined on the site of C.S.T.B. at a height between 10 and 20 m; and the Seattle values are equally affected by a flat stretch without obstacles which is 100 to 200 m long and situated between the rough zone and the measuring equipment.

The Reynolds shear stress

The Reynolds shear stress $\tau = -\overline{u'w'}$, per mass unit, was calculated (u' and w' are the fluctuations of the longitudinal and vertical wind velocity components) at the levels 10, 20, 40 and 60 m, for all the runs. For each sector, the vertical variation has been examined by calculating the ratios τ_i/τ_{60} where τ_i is the Reynolds shear stress at 10, 20 and 40 m and where τ_{60} is the Reynolds shear stress at 60 m (Table 2). Theoretically, where the logarithmic profile applies τ should be constant in the surface boundary layer. On examining Table 2, where the values of τ_i/τ_{60} are presented, it is noticeable that this condition is approximately verified between 20 and 60 m for all the sectors.

TABLE 2. Mean value and standard deviation (S.D.) of τ_i/τ_{60} (N is the number of runs).

		SECTOR A		SECTOR B		SECTOR C		SECTOR D	
	z	MEAN VALUE ± S.D.	N	MEAN VALUE ± S.D.	N	MEAN VALUE ± S.D.	N	MEAN VALUE ± S.D.	N
τ_i/τ_{60}	10	0.96 ± 0.37	7	0.78 ± 0.29	11	0.58 ± 0.15	16	0.72 ± 0.28	23
	20	1.09 ± 0.19	7	1.09 ± 0.37	11	0.78 ± 0.19	16	0.92 ± 0.36	23
	40	1.03 ± 0.15	7	1.09 ± 0.18	11	0.98 ± 0.10	16	0.97 ± 0.21	23

The values of τ_{10}/τ_{60} are less than the other values of τ_i/τ_{60} and clearly inferior to 1 for sectors B, C and D; this could confirm the hypothesis that a roughness change influences the layer between 0 to 20 m.

The ratio between $\sqrt{\tau}$ and the friction velocity U_* has also been studied; by definition according to the surface boundary layer theory $\sqrt{\tau} = U_*$. This ratio has been determined for each run, based upon the friction velocity deduced from the vertical profiles of the mean wind velocity and from the mean Reynolds shear stress derived from τ_{20}, τ_{40} and τ_{60} (Table 3).

TABLE 3. Mean value and standard deviation (S.D.) of $\sqrt{\tau}/U_*$, $\mathrm{Tan}^{-1}(\overline{W}/\overline{U})$ (for each run the mean of the values at 20, 40 and 60 m expressed in degrees), and $(U_*/\overline{U}(10)^2)$ (N is the number of runs).

	SECTOR A		SECTOR B		SECTOR C		SECTOR D	
	MEAN VALUE ± S.D.	N	MEAN VALUE ± S.D.	N	MEAN VALUE ± S.D.	N	MEAN VALUE ± S.D.	N
$\sqrt{\tau}/U_*$	0.89 ± 0.12	7	0.75 ± 0.06	11	0.88 ± 0.10	16	0.93 ± 0.11	23
$\mathrm{Tan}^{-1}(\overline{W}/\overline{U})$	2.9 ± 0.6	20	2.7 ± 0.9	32	2.2 ± 0.6	40	0.6 ± 0.7	69
$(U_*/\overline{U}(10)^2)$	0.022±0.005	9	0.064±0.010	14	0.045±0.008	16	0.027±0.004	25

For sectors A, C and D this ratio conserves a value close to 1 (with about a 10% variation); on the other hand, for B the value is notably inferior to 1: B is the sector the most highly disturbed; it is the sector where the results deviate the most from the theoretical values readily accepted. However, any interpretation of the results of the Reynolds shear stress measurements should be made with a certain caution: the behavior of the anemometer used for the measurement of the vertical velocity can bring about an underestimation of τ (which amounts to few percent at lower heights), and an inexact leveling (or what amounts to the same thing, a flow not perfectly horizontal) can bring about, in the calculation of τ, an error greater than 10% (12). The accuracy of the leveling was estimated at about 2° for our measurements and it was possible to verify that the angle of inclination of the flow in relation to the horizontal plane ($\mathrm{Tan}^{-1}(\overline{W}/\overline{U})$) was not negligible and varied in a systematic way from one sector to another (Table 3); this angle is about 2.5° to 3° for sectors A and B, 2° for sector C and has a very small value for sector D (0.5°).

Lastly, to complete the characterization of the different sectors studied, the drag coefficient at 10 m, which is expressed as $C_D = (U_*/\overline{U}(10)^2)$, has been calculated (Table 3). The values found for sectors A and D are readily comparable to the values proposed by Davenport (13) (C_D = 0.015 to 0.030 in a suburban or wooden area); on the other hand, the values for sectors B and C are higher and correspond better to the values found, for the center of towns (C_D = 0.030 to 0.050).

Wind direction

Also examined was the vertical variation of the mean wind direction $\Delta\overline{\theta}$: noted was a fairly clear variation of this in a clockwise direction between 10 m and 60 m (Table 4). Under these circumstances the sectors are also very differentiated: the variation for D, between 20 and 60 m, is small (1.5°), and it amounts to the same as the error of the azimuth position of the anemometers (evaluated at ± 1.5°); on the other hand, for B and C this variation is clearly greater than this error. These results are contrary to those obtained by Harris (14) and Teunissen (3) who did not remark any noticeable variation of the mean wind direction for a comparable height above the ground.

TABLE 4. Mean value and standard deviation (S.D.) of the difference between the mean wind direction at 60 m and at z_i m ($\Delta\bar{\theta}_i^{60}$ is in degrees) (N is the number of runs).

		SECTOR A		SECTOR B		SECTOR C		SECTOR D	
	Z	MEAN VALUE ± S.D.	N	MEAN VALUE ± S.D.	N	MEAN VALUE ± S.D.	N	MEAN VALUE ± S.D.	N
$\Delta\bar{\theta}_i^{60}$	10	3.5 ± 2.5	9	5.0 ± 2.0	14	6.0 ± 1.5	16	1.0 ± 1.5	25
	20	3.0 ± 2.0	9	4.5 ± 1.75	14	4.5 ± 1.0	16	1.5 ± 1.5	25
	40	1.75 ± 0.75	9	2.5 ± 0.5	14	2.5 ± 0.75	16	1.5 ± 0.75	25

TURBULENCE PARAMETERS

The standard deviation of the longitudinal (σ_u), lateral (σ_v) and vertical (σ_w) wind components

The standard deviations σ_u, σ_v and σ_w vary in the same way as the wind speed in neutral stability; on the other hand, the ratios σ_u/U_*, σ_v/U_* and σ_w/U_* are theoretically considered as constant in the constant shear stress layer and have respectively as mean values 2.5, 1.9 and 1.25 according to Lumley and Panofsky (15). The experimental results obtained on various sites show a large scattering around these mean values.

These ratios have been calculated for all runs and the mean of the values of all the levels has been calculated for each sector; the results are shown in Table 5.

TABLE 5. Mean values of σ_u/U_*, σ_v/U_*, and σ_w/U_*.

	SECTOR A	SECTOR B	SECTOR C	SECTOR D
σ_u/U_*	2.07 ± 0.04	1.65 ± 0.09	1.97 ± 0.11	2.20 ± 0.11
σ_v/U_*	1.65 ± 0.05	1.33 ± 0.02	1.48 ± 0.04	1.66 ± 0.06
σ_w/U_*	1.02 ± 0.12	0.82 ± 0.12	0.97 ± 0.15	1.07 ± 0.13
z_o(m)	0.55	2.1	1.3	0.7

It has been noted that the higher the roughness the more clearly inferior are these values in relation to the theoretical ones. A close examination of Table 6 and Fig. 2 (values for sector D) reveals a vertical variation of these ratios greater for σ_u/U_* and σ_w/U_* than for σ_v/U_*; this variation is similar for the other sectors.

The vertical profiles of these ratios show that the greatest variation and the smallest values are found below 30 m; above that height they stay practically constant. Also indicated in Fig. 2 are the values of the ratio σ_u/U_*, based upon the theoretical model recently proposed by Deaves and Harris (8): for sectors C and D their model fits the experimental datas fairly well and deviates notably from the results obtained for sectors A and B.

TABLE 6. Mean value and standard deviation of σ_u/U_*, σ_v/U_*, σ_w/U_*, $\sigma_u/\overline{U}(z)$ and L_u^x (mast P_1).

SECTOR D (25 RUNS)					
Z	σ_u/U_*	σ_v/U_*	σ_w/U_*	$\sigma_u/\overline{U}(z)$	L_u^x
10.5	1.97 ± 0.16	1.56 ± 0.18	0.91 ± 0.08	0.294 ± 0.022	87 ± 28
15.5	2.06 ± 0.16	1.62 ± 0.18		0.285 ± 0.023	98 ± 36
20.5	2.15 ± 0.19	1.63 ± 0.19	1.04 ± 0.09	0.265 ± 0.025	116 ± 52
30.5	2.21 ± 0.20	1.69 ± 0.20		0.253 ± 0.023	146 ± 64
40.5	2.23 ± 0.20	1.67 ± 0.21	1.14 ± 0.10	0.236 ± 0.022	166 ± 73
50.5	2.28 ± 0.23	1.72 ± 0.24		0.224 ± 0.023	184 ± 78
55.5	2.30 ± 0.25	1.72 ± 0.24		0.221 ± 0.024	196 ± 90
60.5	2.28 ± 0.24	1.70 ± 0.24	1.20 ± 0.14	0.213 ± 0.023	207 ± 73

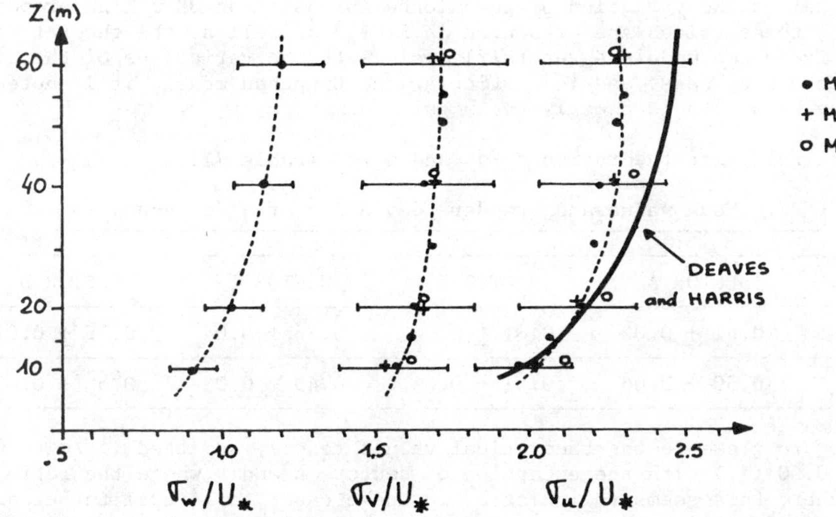

Fig. 2. Vertical variation of σ_u/U_*, σ_v/U_* and σ_w/U_* for sector D.

The pronounced variation of the mean values of these ratios with the sectors leads to an examination of the influence of the roughness.

In order to do that, the mean values of σ_u/U_* at 30 m, are presented in Fig. 3 for all the sectors.

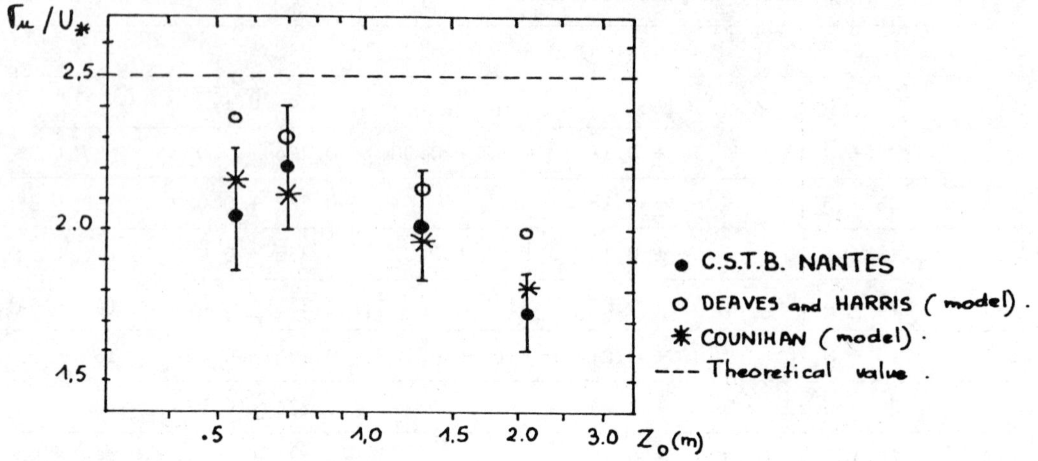

Fig. 3. Variation of σ_u/U_* with roughness length z_o at 30 m.

In Fig. 3 it appears clearly that σ_u/U_* decreases when z_o increases. The values corresponding to the mean value at 30 m high have also been calculated with the help of the Deaves and Harris model (8), as well as those values deduced from the empirical model of the variation of turbulence intensity at 30 m high proposed by Counihan (7); these values are presented in Fig. 3 as well as the theoretical value 2.5. These two models agree fairly well with the variations of the data in function of the roughness, and in particular the Counihan model; it is noted as well that the value 2.5 is clearly too high.

Calculated as well were the ratios σ_v/σ_u and σ_w/σ_u (Table 7).

TABLE 7. Mean value and standard deviation of σ_v/σ_u and σ_w/σ_u.

	SECTOR A	SECTOR B	SECTOR C	SECTOR D
σ_v/σ_u	0.81 ± 0.02	0.81 ± 0.04	0.76 ± 0.04	0.76 ± 0.02
σ_w/σ_u	0.50 ± 0.06	0.51 ± 0.06	0.49 ± 0.03	0.50 ± 0.03

These values are close to the theoretical values readily admitted (σ_v/σ_u = 0.75 and σ_w/σ_u = 0.50 (7)) with the exception of sectors A and B where the ratio σ_v/σ_u is higher; this seems to indicate, in these cases, a lateral turbulence systematically higher for these two sectors.

Turbulence intensity

The values of the turbulence intensity $\sigma_u/\overline{U}(z)$ have been calculated for each level and for each direction and the values for sector D are presented in Table 6.

The vertical variation of the turbulence intensity has been examined first of all. The results are compared to the theoretical model proposed by in particular Teunissen (16) ($\sigma_u/\overline{U} = 1/Ln(z/z_o)$), the model ESDU 1974 (6) and the Deaves and Harris theoretical model (8) (Fig. 4).

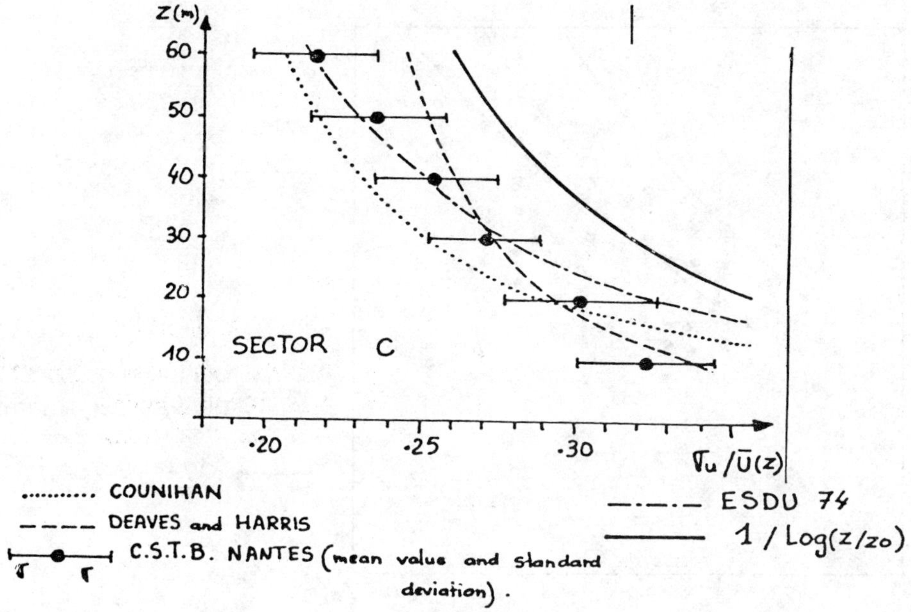

Fig. 4. Vertical variation of $\sigma_u/\bar{U}(z)$ (sector C).

Except for sector D, the theoretical model $1/\mathrm{Ln}(z/z_o)$ leads to excessive turbulence intensity values in relation to the experimental values found. It was also noticed that for sector B none of the above mentioned models were suitable. The model ESDU 1974 (6) (apart from the values at 10 m) and the one proposed by Deaves and Harris (8) are within the range of scattering of the values at each level, for the sectors A, C and D.

The influence of the roughness was also examined in greater detail by studying the variation of the turbulence intensity, at 30 m, in function of the roughness length z_o. In regard to this, there is presented in Fig. 5 the 64 values of the turbulence intensity, measured at 30 m high on mast P_1, in function of the roughness length of each of the corresponding profiles; the scattering is large; however, an augmentation of the turbulence intensity when z_o increases shows up fairly clearly on this figure.

On Fig. 5 the theoretical model $1/\mathrm{Ln}(z/z_o)$ has also been represented as well as the model ESDU 1974 (6), the Deaves and Harris theoretical model (8), and the empirical model of Counihan (9): it is the latter which fits best the experimental data.

On the other hand, the theoretical model $1/\mathrm{Log}(z/z_o)$ very clearly overestimates the values of the turbulence intensity for the roughness encountered here ($z_o \geq 0.4$ m).

TURBULENCE SCALE

In order to complete the description of the characteristics of the turbulence observed at the C.S.T.B. site, the longitudinal (L_u^x) and the lateral (L_u^y) scales of the longitudinal wind component have been determined.

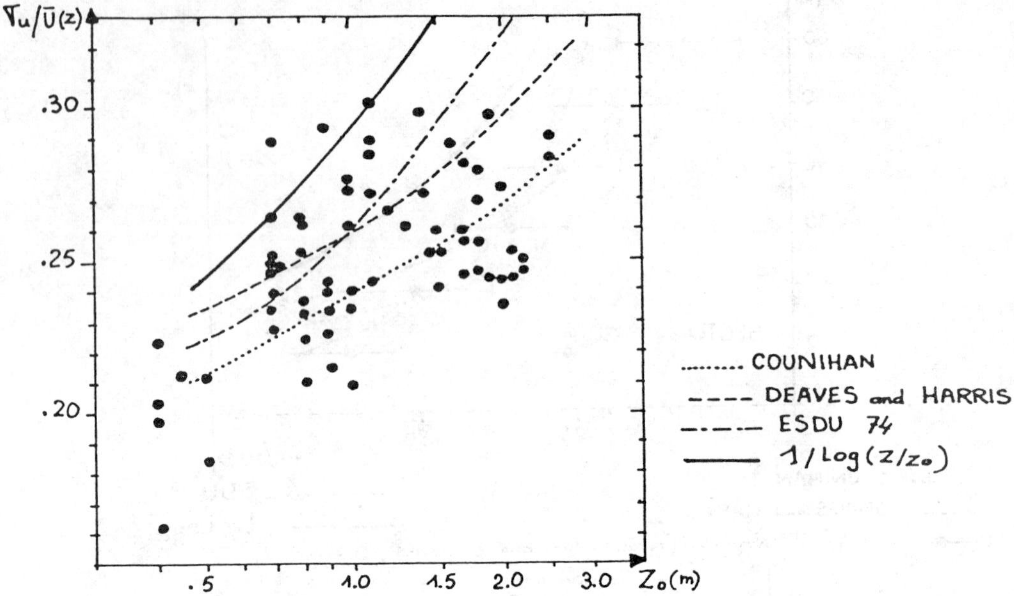

Fig. 5. Variation of $\sigma_u/\bar{U}(z)$ with roughness length at 30 m.

Longitudinal scale L_u^x

Assuming that Taylor's hypothesis is true, the calculation of the longitudinal scale L_u^x can be based upon the autocorrelation function $R_{uu}(\tau)$ using the expression:

$$L_u^x = \bar{U}(z) \int_0^\infty R_{uu}(\tau)\, d\tau \qquad (\bar{U}(z) \text{ mean wind speed at the height } z)$$

The autocorrelation function which decreases towards zero when τ increases, often involves some oscillations of a small amplitude around zero after it crosses the τ axis; it was assumed that these oscillations compensated one another and for the calculation of L_u^x only the part of $R_{uu}(\tau)$ occurring before the first zero crossing, was considered. The mean values of L_u^x and the corresponding scattering for each level are presented in Table 6 for sector D. An examination of this table shows that the dispersion of the values of L_u^x is very large and that it can reach 40% of the mean value.

The scattering of measurements gathered from different sites is as large as the one noted in the preceding paragraph. Besides the variations linked to the height above the ground, the nature of the terrain, the atmospheric stability and the measuring instruments, the methods of calculation used to determine L_u^x can themselves bring about very scattered results. At 76 m, for a roughness varying from 0.02 to 0.07 m, the mean of the values presented by Tieleman and Mullins (17), is 288 m; extrapolating up to 76 m, the values found in Nantes for sector D gives $L_u^x \simeq 240$ m. Shiotani (18) at 40 m, finds 154 m for a land wind; the value at the same height is 145 m for the sector A. Lastly, Ramsdell (1) finds 138 m at 48 m high for a roughness close to 1 m; for sectors C and D, where the roughness is comparable at 50 m high $L_u^x = 166$ m for sector C and $L_u^x = 184$ m for sector D. Taking into account the scattering of L_u^x, these values although they cannot be considered as identical are in the same range.

Table 6 also shows that the longitudinal scale increases in a considerable way with height: the value at 60 m is practically double of the one at 20 m. Numerous models of the vertical variation of L_u^x exist: Ramsdell (1) has compared six of them with his data, only examined here are the models recently proposed: ESDU 1974 (6), Deaves and Harris (8), Ramsdell (1), Counihan (7) and the oldest model (1970) proposed by Teunissen (16).

The values of sector C and the models used are presented in Fig. 6.

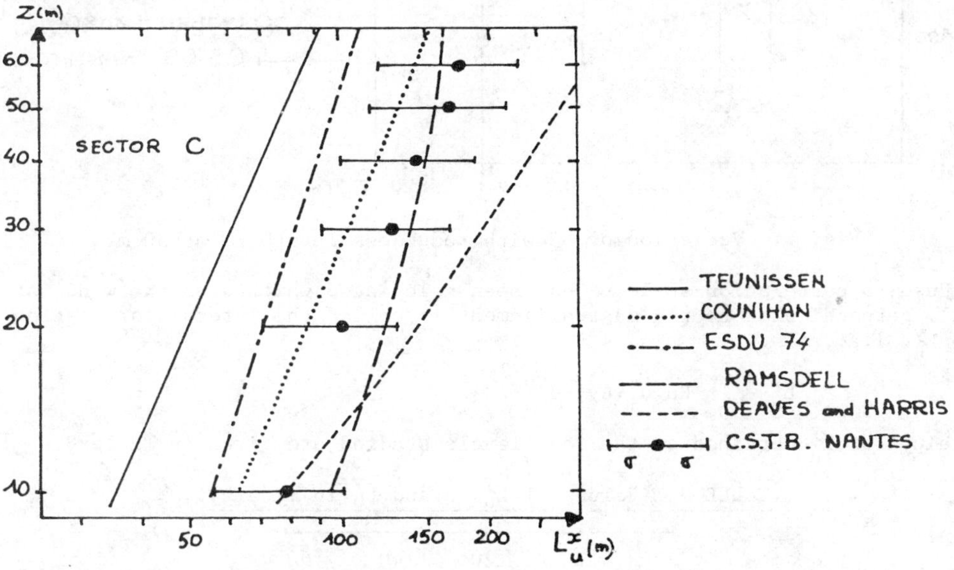

Fig. 6. Vertical variation of the longitudinal scale L_u^x.

For all the sectors, the Teunissen (16) and ESDU 1974 (6) models present values that are too small, while on the other hand, the Deaves and Harris model (8) presents values that are too high; it is the models proposed by Ramsdell (1) and in particular by Counihan (7) which fit the data best.

All the models (except Teunissen (16)) take into account the roughness. To examine the influence of the roughness the mean values of L_u^x obtained at 30 m for the different sectors are given in Fig 7; this figure shows a pronounced decrease, when the roughness increases. This trend is the same as the one observed by Counihan who proposed for a height of 30 m above the ground, the expression $L_u^x = 108(1/z_o)^{1/6}$ (Fig. 7).

Lateral scale L_u^y

The intercorrelation function $Ru_iu_j(\Delta y)$ (lag $\tau = 0$) at the four main levels (10, 20, 40 and 60 m) has been calculated for 54 runs; u_i and u_j are the longitudinal components of the wind speed for two anemometers placed at the same height z above the ground and separated by a distance Δy.

At a fixed height z, the function $Ru_iu_j(\Delta y)$ decreases when the interval between two sensors Δy increases; this function also decreases for a given height z and for a given range Δy, when the angle α between the mean wind direction and the alignment of the masts increases from 0 to 90°; finally, for a given angle α and a given interval Δy, $Ru_iu_j(\Delta y)$ increases with height z (9).

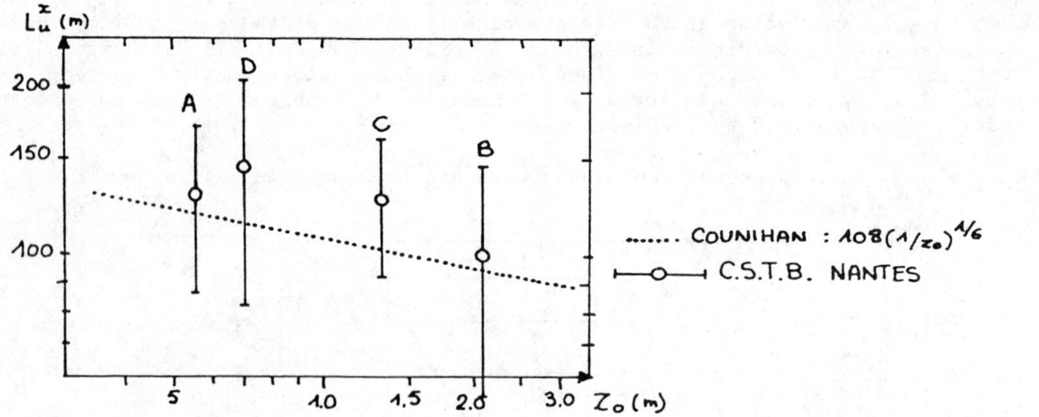

Fig. 7. Variation of L_u^x with roughness length z_o at 30 m.

From Ru_iu_j a correlation scale L_u has been calculated which when the wind is exactly perpendicular to the mast alignment is called the lateral correlation scale L_u^y, i.e.,

$$L_u^y = \int_0^\infty Ru_iu_j(\Delta y)\, dy$$

The values of L_u^y obtained at the four levels studied are given in Table 8.

TABLE 8. Values of L_u^y (z and L_u^y in meters).

z	10	20	40	60
L_u^y	18	22	33	38

In the preceding calculations the runs have not been differentiated by roughness; the main part of the 54 runs corresponds to a roughness ranging from 0.7 m to 1.3 m.

To the author's knowledge there exists few direct measurements of the lateral scale L_u^y. The values found by Harris (14) for a smooth roughness (35 m at 10 m high and 52 m at 55 m high) are much higher than those found here; this holds true for Shiotani's values (18) which vary from 54 m for a sea wind to 65 m for a land wind at a height of 40 m. These last two values, which are the averages calculated from the values obtained for wind directions not all exactly perpendicular to the mast alignment, may be slightly overestimated.

In Fig. 8 the experimental values of Shiotani (18) and Harris (14) are given and the curves, calculated from the ESDU 1974 model (6) and the model proposed by Deaves and Harris (8), are represented; these models vary with roughness and the roughness of sector D (z_o = 0.7 m and U_* = 1.25 m/s), was chosen for these calculations. The Deaves and Harris model presents a very large vertical variation of L_u^y; the ESDU model seems to adjust better to the experimental values presented on this graph.

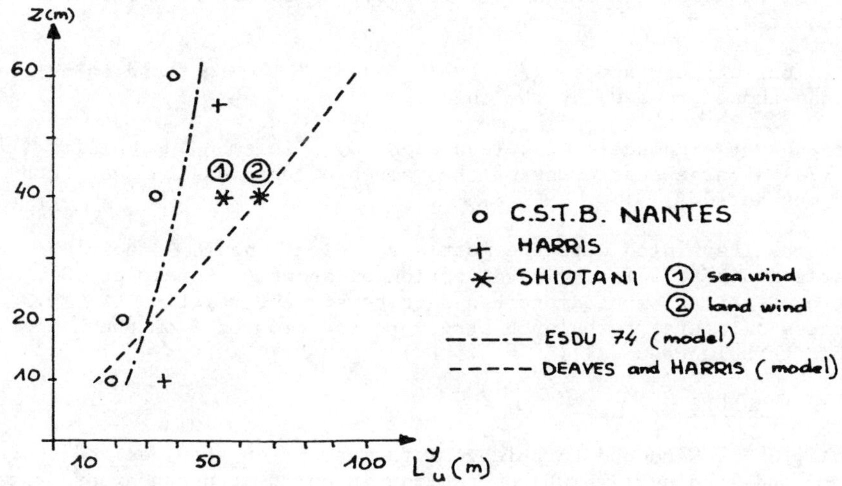

Fig. 8. Vertical variation of the lateral scale L^y_u.

Finally, in comparing the values of L^y_u with the values of L^x_u relative to sector D, the following relation was found:

$$L^y_u \simeq 0.2\ L^x_u$$

This relation is far removed from the theoretical one determined for a homogeneous and isotropic turbulence $L^y_u = 0.5\ L^x_u$; it is, on the other hand, close to the relation proposed by Counihan (7) ($L^y_u \simeq 0.3\ L^x_u$).

CONCLUSION

The detailed examination of the values of the roughness and turbulence parameters obtained at the C.S.T.B. site in Nantes, show that most of these results are comparable to those obtained at similar experimental sites and that they are in good agreement with the models recently proposed for the surface boundary layer in the case of neutral stability and strong winds. This is especially true for sectors D and C, and to a lesser extent, for Sector A; on the other hand, for sector B which is very influenced by large obstacles situated at a distance between 300 and 500 m from the site, the obtained results diverge noticeably from the values foreseen by these models.

With the strong prevailing wind arriving at the measuring site in a wide sector and with the roughness presenting values fairly different at the interior of this sector, it was possible to examine the influence of the roughness on the main turbulence parameters for values of z_o varying from 0.5 m to 2 m. In particular, it was found that the ratio σ_u/U_* and the longitudinal turbulence scale L^x_u decreased when z_o increased and that on the other hand, the turbulence intensity ($\sigma_u/\overline{U}(z)$), increased but much less rapidly than is foreseen by the theoretical model $1/\text{Ln}\ (z/z_o)$. This variation with roughness, observed here, is readily comparable with the ones proposed by the Counihan (7) and Deaves and Harris (8) models.

Besides the influence of the roughness, it seems necessary to stress the following results:

- the mean values of the ratios σ_u/U_*, σ_v/U_* and σ_w/U_* were found to be clearly smaller than those proposed by the theory.

- like other authors (Ramsdell (1), Teunissen (3), Tieleman and Mullins (17) and Counihan (7)) a large scattering of the values of the longitudinal turbulence component was noted.

Finally, two results, which were the most surprising, should be pointed out: the vertical variation of the mean wind direction of about 4.5° between 20 m and 60 m above the ground for the most disturbed sectors and the relation $L_u^y \simeq 0.2\ L_u^x$ which indicates a lateral turbulence structure far removed from the homogeneous and isotropic turbulence model ($L_u^y = 0.5\ L_u^x$).

REFERENCES

1. Ramsdell, J.V., Wind and turbulence information for vertical and short take-off and landing (V/STOL) operations in built-up urban areas - Results of Meteorological Survey Battelle, Pacific Northwest Laboratories, Richland, Washington, June, 1975.

2. Duchêne-Marullaz, Ph., Full-scale measurements of atmospheric turbulence in a suburban area, Proceedings of the Fourth International Conference on Wind Effects on Buildings and Structures, Heathrow, 1975.

3. Teunissen, H.W., Measurements of planetary boundary layer wind and turbulence characteristics over a small suburban airport, Journal of Industrial Aerodynamics 4, 1979, pp. 1-34.

4. Sacré, C., Influence d'un obstacle topographique sur l'écoulement de l'air dans les basses couches de l'atmosphère, Internal report EN-CLI 77.2, Nantes, 1977.

5. Jensen, N.O. and Peterson, E.W., On the escarpment wind profile, Quarterly Journal of the Royal Meteorological Society, Vol. 104, 1978, pp. 719-728.

6. Engineering Sciences Data Unit, Characteristics of atmospheric turbulence near the ground, ESDU Item number 74031 and 75001, Regent Street, London, 1974-1975.

7. Counihan, J., Adiabatic atmospheric boundary layers: a review and analysis of data from the period 1880-1972, Atmospheric Environment, Vol. 9, 1975, pp. 871-905.

8. Deaves, D.M. and Harris, R.I., A mathematical model of the structure of strong winds, Report 76, May, 1978, CIRIA, London.

9. Duchêne-Marullaz, Ph., Structure du vent en zone suburbaine, Rugosité et corrélation latérale, Internal report EN-CLI 76.2, Nantes, 1976.

10. Duchêne-Marullaz, Ph., Installation de mesures pour l'étude de la structure fine du vent, Internal report EN-CLI 73.3, Nantes, 1973.

11. Rao, K.S., Wyngaard, J.C., and Cote, O.R., The structure of the two-dimensional internal boundary layer over a sudden change of surface roughness, Journal of the Atmospheric Sciences, Vol. 31, n° 3, April, 1974.

12. Dyer, A.J. and Hicks, B.B., The spatial variability of eddy fluxes in the constant flux layer, Quarterly Journal of the Royal Meteorological Society, 1972, pp. 206-212.

13. Davenport, A.G., The spectrum of horizontal gustiness near the ground in high winds, Quarterly Journal of the Royal Meteorological Society, 1961, pp. 191-211.

14. Harris, R.I., Measurements of wind structure, Symposium on External Flows, July 4-6, 1972, University of Bristol.

15. Lumley, J.L. and Panofsky, H.A., The structure of atmospheric turbulence, Interscience Publishers, New York, 1964.

16. Teunissen, H.W., Characteristics of the mean wind and turbulence in the planetary boundary layer, UTIAS Review n° 32, University of Toronto, October, 1970.

17. Tieleman, H.W. and Mullins, S.E., Atmospheric turbulence below 75 m in the convective boundary layer (strong wind conditions), Proceedings of the 3rd Colloquium on Industrial Aerodynamics, Aachen, June 14-16, 1978.

18. Shiotani, M. and Iwatani, Y., Horizontal space correlations of velocity fluctuations during strong winds, Journal of the Meteorological Society of Japan, Vol. 54, n° 1, February 28, 1976.

VERTICAL COHERENCE IN THE ATMOSPHERIC BOUNDARY LAYER

G. R. STEGEN* and R. L. THORPE**

*Science Applications, Inc., Bellevue, Washington, USA
**Aeronautical Research Associates of Princeton, Princeton, New Jersey, USA

INTRODUCTION

For a number of years, atmospheric scientists have been concerned with describing the three-dimensional characteristics of the atmospheric boundary layer. Attempts to describe the small-scale turbulent-like motions have relied upon simple physical models of the statistical properties of the flow field. It has been almost two decades since Davenport described the spectral characteristics of the fluctuations of the horizontal wind field (2). In that paper he proposed a simple scaling law to describe the coherence of fluctuations of the horizontal winds at different vertical heights. This proposed model was only strictly true for stabilities near neutral ($Ri \to 0$). Pielke and Panofsky (6) extended Davenport's model to the general case of coherence between winds in any of the three directions with separations in any of the three directions. Subsequent papers by Panofsky and coworkers (7, 4, 5) have principally addressed the coherence between horizontal wind field fluctuations at different horizontal locations. As a result of these studies we have an excellent understanding of the various parameters which influence the decay of "horizontal coherence".

In the present paper we will re-examine Davenport's original hypothesis of the decay of "vertical coherence" in light of the present understanding of the decay of horizontal coherence. The experiment and analysis described herein were designed to examine the scaling laws, and hence do not directly address the parametric dependence on environmental parameters.

EXPERIMENTAL DESIGN

The measurements were made on November 2, 1973, at the Tuckerton Experimental Station of Rutgers University. The Experimental Station is located in a state protected salt water marsh at the entrance to Little Egg Harbour, New Jersey, approximately 20 miles north of Atlantic City. The data was collected in the afternoon under the following conditions: air temperature 61 °F; water temperature 57 °F; offshore wind steady 15 mph SW. It was a clear day with few clouds, the atmospheric stability being very near neutral.

The horizontal velocity long the direction of the wind was measured using a hot-film anemometer (Thermo-Systems, Inc., VT 161). The sensing element was cylindrical film 0.0025 cm diameter, 0.05 cm long, which was oriented vertically so as to be sensitive to the total wind speed and insensitive to changes in wind direction.

The analog voltage output of the anemometers was connected to the recording trailer via shield twist pair cables. At the recording trailer the data was preconditioned by buffering, amplifying, and lowpass-filtering (f_o = 100 Hz, 4-pole Butterworth). The data was then digitized with a 12-bit A/D converter and recorded on an IBM-compatible nine-track digital magnetic tape.

The two anemometers were mounted on one-meter-long horizontal spars which were attached to a 7.6-cm diameter vertical mast three meters long. The vertical mast was part of an erectable meteorological tower. The vertical aperture of the array was varied by changing the separation of the horizontal spars, while the height of the array was controlled by erecting the tower. The tower was situated approximately 300 meters from the water's edge, a location which ensured that all measurements would be made within a fully developed boundary layer.

Sampling Strategy

Previous investigations of the decay of coherence have been concerned with the influence of a variety of environmental parameters including stability, surface roughness, terrain, and wind veering. The present experiment was principally concerned with physical scaling of the vertical coherence, and hence sought to minimize the variability of the various environmental parameters. To ensure that the meteorological parameters would have a minimum variability, the experiment was performed during a single three-hour period. In order to obtain a range of array apertures and heights, it was necessary to limit the duration of each set of measurements to 10 minutes. The measurement parameters for the nine runs are summarized in Table 1 where ΔZ is the probe separation, Z_{max} is the height of the uppermost probe, $\overline{U}$ is the average horizontal wind speed measured by the two anemometers, and $\overline{\sigma}$ is the standard deviation of the wind speed. The variations in $\overline{U}$ and $\overline{\sigma}$ are primarily a consequence of the short averaging period and do not represent any significant large-scale unsteadiness in the wind field.

Table 1. Summary of run parameters

ΔZ (m)	Z_{max} (m)	$\overline{U}$ (m/sec)	$\overline{\sigma}$ (m/sec)
0.30	4.14	7.03	0.99
0.30	6.20	7.78	0.93
0.30	8.25	7.74	0.71
0.61	4.19	6.32	1.12
0.61	6.20	8.01	1.06
0.61	8.25	7.89	0.80
1.22	4.80	7.09	0.91
1.22	6.20	6.72	0.80
1.22	8.25	7.89	0.72

ANALYSIS

The coherence $R^2(f)$ between two time series plays the role of a correlation coefficient defined at each frequency. The coherence is estimated by

$$R^2(f) = S_{xy}(f)S^*_{xy}(f)/E_x(f)E_Y(f) \quad (1)$$

where $E_X(f)$ and $E_Y(f)$ are the energy spectra of the two time series, $S_{xy}(f)$ is the cross-spectrum, and $S^*_{xy}(f)$ is the complex conjugate of the cross-spectrum. By Schwarz's inequality the coherence at each frequency is bounded in the region $0 \leq R^2(f) \leq 1$, where the lower limit corresponds to no correlation, and the upper limit corresponds to perfect correlation. Davenport's original proposal for the physical scaling of coherence can be expressed as

$$R^2(\Delta Z/\lambda_H) = \exp\{-a\Delta Z/\lambda_H\} \quad (2)$$

where a is a decay parameter and λ_H is the horizontal wavelength, $f/\overline{U}$. Note: some authors refer to $\Delta Zf/\overline{U}$ as ΔF, a nondimensional frequency.

The spectral estimates were made using the discrete Fourier transform method. The 10-minute-long time series were divided into 25 equal sections. The spectra were estimated for each section using an FFT algorithm and then were ensemble-averaged. Figures 1 and 2 present a typical set of coherence estimates on linear-linear and log-linear scales, respectively. Together they illustrate that R^2 is reasonably

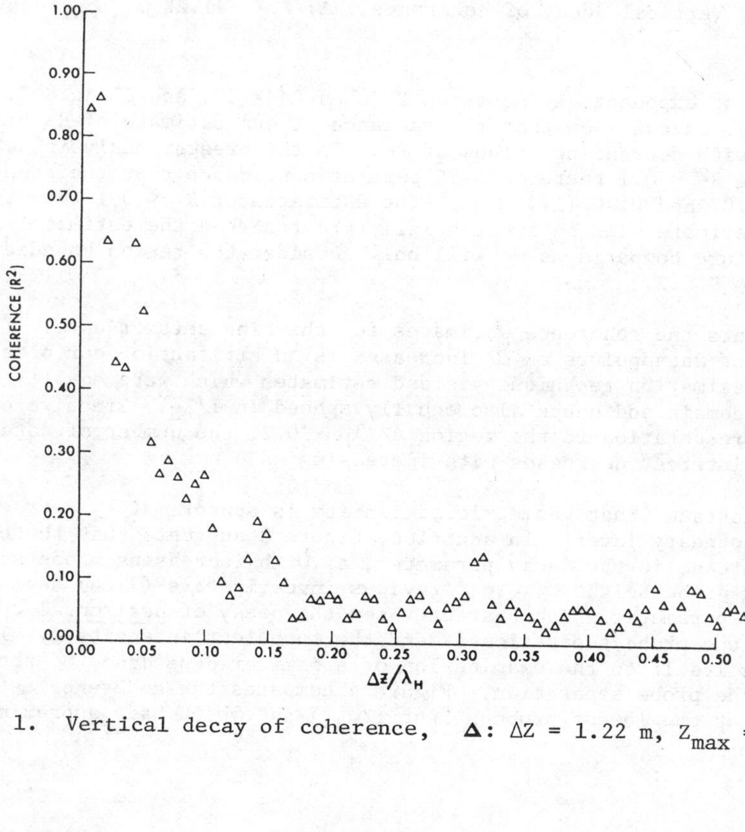

Figure 1. Vertical decay of coherence, Δ: $\Delta Z = 1.22$ m, $Z_{max} = 4.80$ m

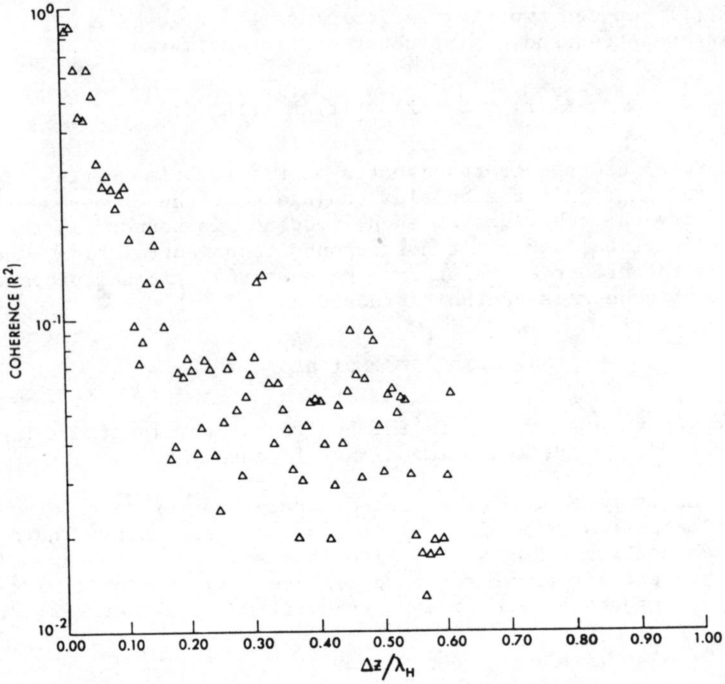

Figure 2. Vertical decay of coherence, **Δ:** $\Delta Z = 1.22$ m, $Z_{max} = 4.80$ m

represented by an exponential (Equation 2) when $R^2 > 0.1$ and $\Delta Z/\lambda_H < 0.2$. Following Goodman (3), we can show that the variance of our estimate of R^2 will increase monotomically with decreasing values of R^2. In the present analysis, when the estimated coherence $R^2 = 0.1$ there is a 90 percent confidence that the true value of R^2 lies between 0.02 and 0.20 (1). Thus, for estimates of $R^2 \leq 0.1$ the standard deviation of the estimate will be substantially greater than the estimate. For these reasons, in future comparisons we will only consider the region bounded by $R^2 > 0.1$, $\Delta Z/\lambda_H < 0.2$.

Figure 3 presents the coherence estimates for the nine sets of data. The decrease in the number of data points as ΔZ increases is an artifact of our presentation. The spectral estimation technique yielded estimates which were equally spaced in the frequency domain and hence also equally spaced in $1/\lambda_H$. Since we have restricted our presentation to the region $\Delta Z/\lambda_H < 0.2$, the number of data points in the domain of interest decreases with increasing ΔZ.

Figure 3 demonstrates that geometric similarity is approximately correct for the near surface boundary layer. In addition, Figure 3 suggests that there exists a systematic increase in the decay parameter, a, with increasing probe separation ΔZ and decreasing probe height Z_{max}. Previous investigators (7, 4) have indicated that the decay parameter which characterizes the decay of horizontal coherence depends on both the probe separation ΔX and the turbulent intensity $\bar{\sigma}$. The present data set lends itself to the examination of a similar dependence of the vertical coherence on the probe separation. Figure 4 compares the coherence estimate for two runs made at the lowest probe height (Z_{max}) and which have approximately the

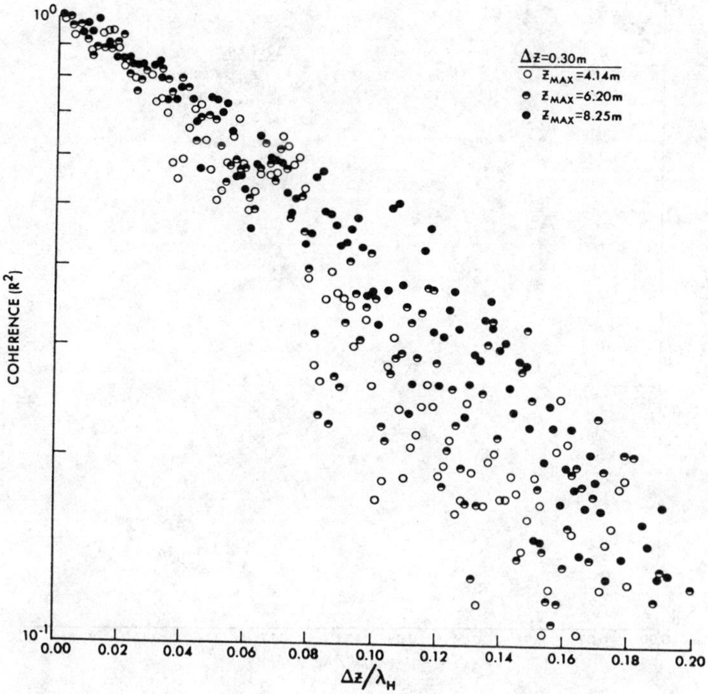

Figure 3. (a) Vertical decay of coherence, ΔZ = 0.30 m

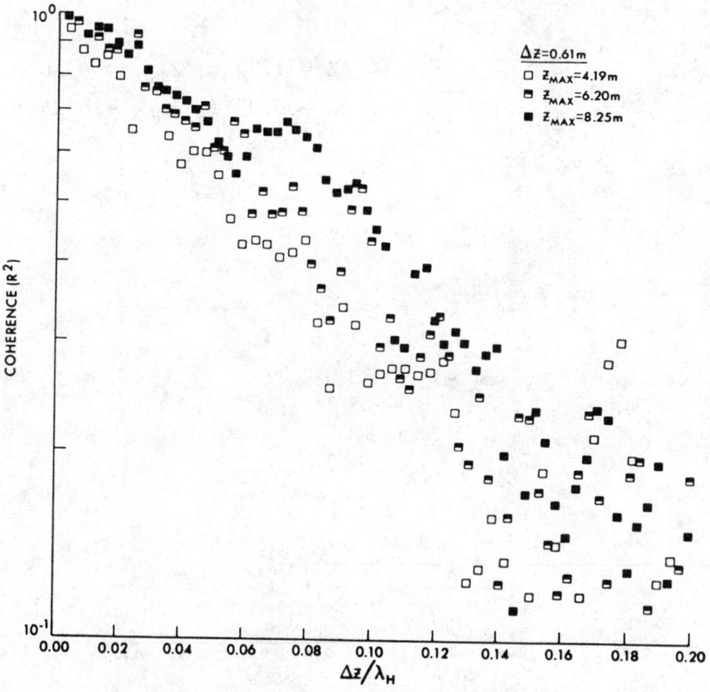

Figure 3. (b) Vertical decay of coherence, ΔZ = 0.61 m

WE H

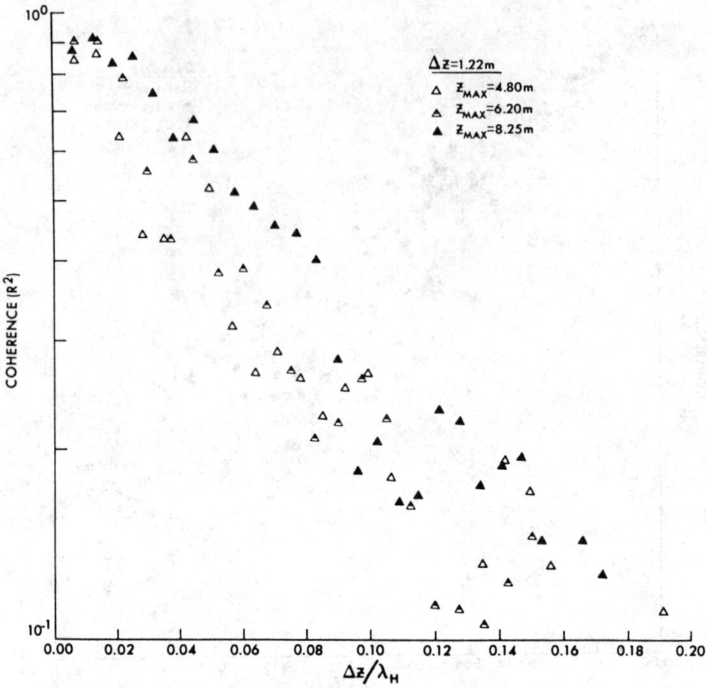

Figure 3. (c) Vertical decay of coherence, ΔZ = 1.22 m

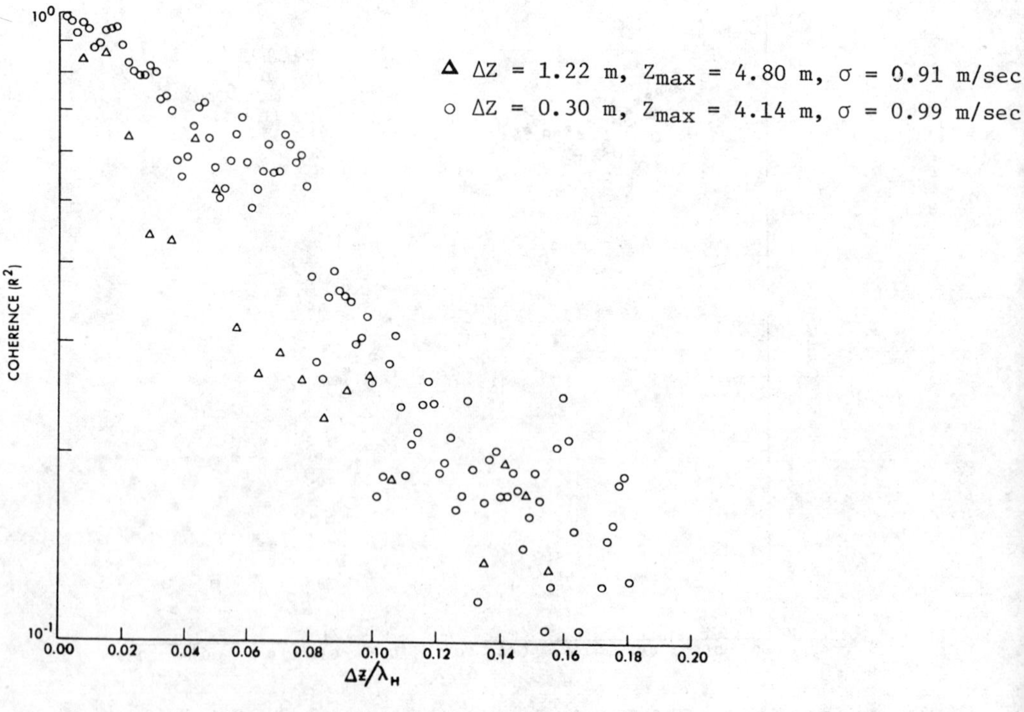

Figure 4. Vertical decay of coherence

same turbulence intensity ($\overline{\sigma}$), but have significantly different vertical probe separations (1.22 m vs. 0.30 m). The coherence decays approximately 20 percent faster for the run with the greater probe separation. Figure 5 presents similar data taken at the maximum probe height, which also have comparable values of turbulence intensity. Once again, the coherence decays faster for the run with the greater probe separation. Comparing Figures 4 and 5, it is apparent that the decay parameter must also depend upon the probe height and the turbulence intensity. However, the present data set is inadequate to determine the dependence of the decay parameter, a, on the identified parameters ΔZ, Z_{max}, and $\overline{\sigma}$.

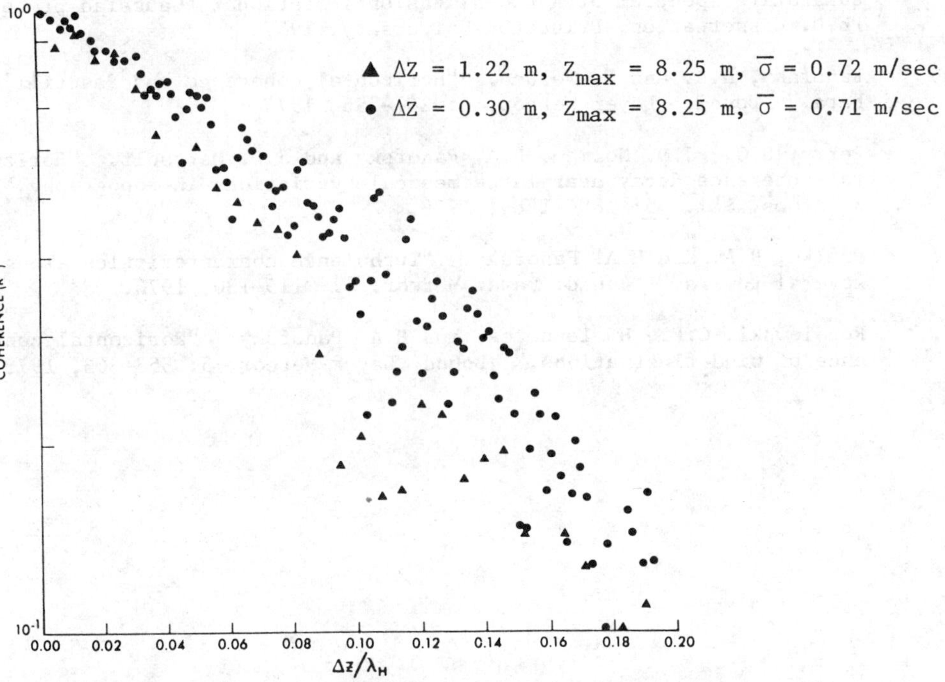

Figure 5. Vertical decay of coherence

FINAL REMARKS

This study has shown that Davenport's geometric similarity is approximately correct for the decay of vertical coherence of horizontal wind speed fluctuations. It further demonstrates that the decay parameter depends on the probe separation, probe height, and the turbulent intensity. Clearly, with additional data we could develop an empirical model suitable for parameterization of the decay characteristics. However, we must ask ourselves if in fact we shouldn't be seeking a more fundamental description of the loss of vertical coherence. In this regard, it is recommended that future efforts examine the role of vertical length scales on the decay of vertical coherence, particularly very near the surface.

The authors would like to thank Professor R. L. Peskin, Rutgers University, for his cooperation in conducting these experiments. The design and analysis of these experiments were performed at Princeton University and supported under NSF Grant GK35885 and NOAA/GFDL Grant E22-21-70 (G).

REFERENCES

1. Amos, D.E. and L.H. Koopmans. "Tables of the distribution of the coefficient of coherence for stationary bivariant Gaussian processes." Sandia Monograph SCR-483, 1963.

2. Davenport, A.G. "The spectrum of horizontal gustiness near the ground in high winds." Quart. J. Roy. Meteor. Soc. 87: 194-211, 1961

3. Goodman, N.R. "On the joint estimation of the spectra cospectrum and quadrative spectrum of a two-dimensional stationary Gaussian process." Ph.D. dissertation, Princeton University, 1957.

4. Panofsky, H.A. and T. Mezuno. "Horizontal coherence and Pasquill's beta." Bound. Layer Meteor. 9: 247-256, 1975.

5. Perry, S.G., J.M. Norman, H.A. Panofsky and J.D. Martsolf. "Horizontal coherence decay near large mesocale variations in topography." J. Atmos. Sci. 35: 1884-1889, 1978.

6. Pielke, R.A. and H.A. Panofsky. "Turbulence characteristics along several towers." Bound. Layer Meteor. 1: 115-130, 1970.

7. Ropelewski, C.F., H. Tennekes and H.A. Panofsky. "Horizontal coherence of wind fluctuations." Bound. Layer Meteor. 5: 353-363, 1973.

GUST STRUCTURES OVER FLAT TERRAINS AND THEIR MODIFICATION BY A BARRIER

M. SHIOTANI and Y. IWATANI

The Physical Science Laboratories, Nihon University at Narashino, Funabashi, Chiba, Japan

ABSTRACT

The measurement of wind velocities and turbulence was made at the sea coast during high winds. Turbulent intensity is influenced greatly by the roughness of the ground and its estimate becomes large with increasing terrain roughness. But, turblent length scale and two-point correlation are hardly influenced by the roughness. The dependence of turbulent intensity and length scale on wind velocity is not clearly observable. The turbulent length scale increases in proportion to the power of height whose exponent is in the range from 0.2 - 0.4. It was found that the decay parameter of narrow-band cross-correlation coefficient is not constant but it is expressed as function of the ratio of distance to height. As a case study of topographical influence on wind velocities and turbulence, the wind flow over a slender peninsula like a barrier was examined. Though the air flow is accelerated over the peninsula, marked variations on turbulent intensities and length scales are not noticed in longitudinal velocity components compared with the wind over undisturbed open sea.

INTRODUCTION

The turbulent properties of the wind over a flat terrain with uniform roughness are influenced by the stability of the air layer and the roughness of the ground. In strong winds, the effect of the stability may be ignored, and the turbulent properties mainly depend on the roughness length z_0. A number of researches are made on the variation of the intensity of turbulence of an alongwind velocity $\sqrt{\overline{u^2}}/U$ with roughness length z_0. $\sqrt{\overline{u^2}}/U$ always increases with increasing z_0. Counihan (1) reviews these results and expresses them that $\sqrt{\overline{u^2}}\ U$ is a quadratic equation of $\log z_0$.

The turbulent length scale L_x in a longitudinal direction may be influenced by the roughness of the ground. However, the relation between L_x and z_0 is not so definite as that between $\sqrt{\overline{u^2}}/U$ and z_0. According to the review of Teunissen (2), the turbulent length scale has a definite trend to increase with the height, but the roughness length has little effect on the length scale, while large non-uniformities on the surface tend to decrease it. According to aforementioned review by Counihan, L_x would be nearly independent of terrain roughness in the air layer above the height, which is assumed to be the order of 1/3-rd of the boundary layer thickness. Below this height, L_x decreases with height and the rate of decrease is greater with increasing terrain roughness, so that within a few tens of meters above the ground L_x becomes greater over smoother terrain.

The length scale L_x is usually obtained from autocorrelation function or power spectrum of wind velocities measured at one point by the help of Taylor's hypothesis. Properly speaking, the turbulent length scale should be obtained by the integration of two-point correlation of wind velocities by distances. The relation between two-point correlation and roughness length has not been conclusively reported within the scope of the author's knowledge. According to wind and turbulent data of ESDU (3,4), the length scale would be independent of terrain roughness at the top of the boundary layer, from which they decrease downwards at a definite rate irrespective to the terrain roughness. While, the thickness of the boundary layer is defined as the function of the roughness length as it becomes larger with increasing roughness. Accordingly, in the lower part of the boundary layer turbulent length scales would be larger over smoother terrain.

We had made the measurements of wind velocities at the sea coast during strong winds for quite a long time, and we obtained wind profiles, turbulent intensities and length scales, two-point correlations, and spectra. Some results on the two-point correlation are recently reported by present authors (5). This paper deals with the dependence of aforementioned turbulent properties on terrain roughness in the winds over the sea and those over the flat land. In particular, two-point correlations both in lateral and in vertical directions are discussed and their relations with respect to heights and distances are investigated when the wind from an open sea blows over a slender peninsula like a barrier.

THE EFFECT OF TERRAIN ROUGHNESS ON TURBULENT PROPERTIES OVER A FLAT AND UNIFORM BOUNDARY

Turbulent intensities

Velocity profiles and turbulent properties were measured in the air layer from 30 m to 150 m above the ground at the coast of Nakagawa. (see Fig. 1 (a)). The tower is situated in a loose pine forest apart 130 - 150 m from the beach. The wind from the sea is modified by the land trajectory in the air layer up to 30 m above the ground. In the air layer ranging from 50 m to 150 m, turbulent intensity is expressed as follows:

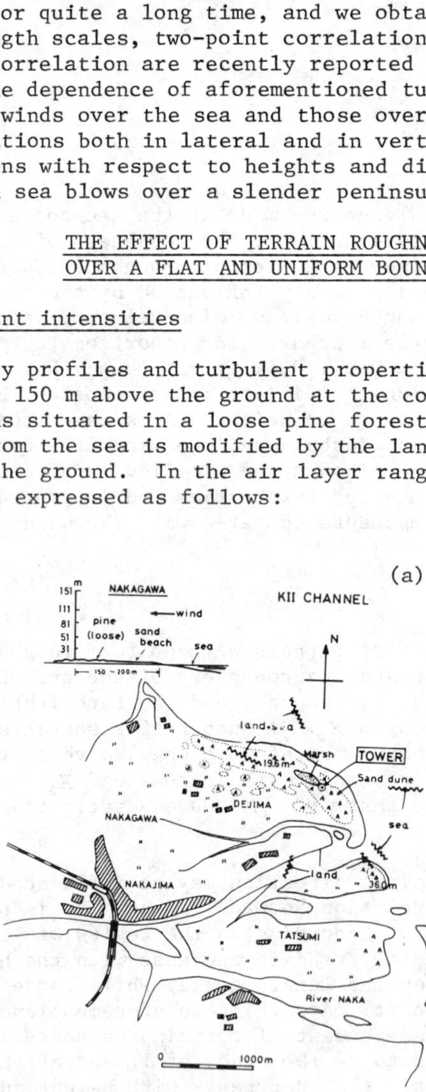

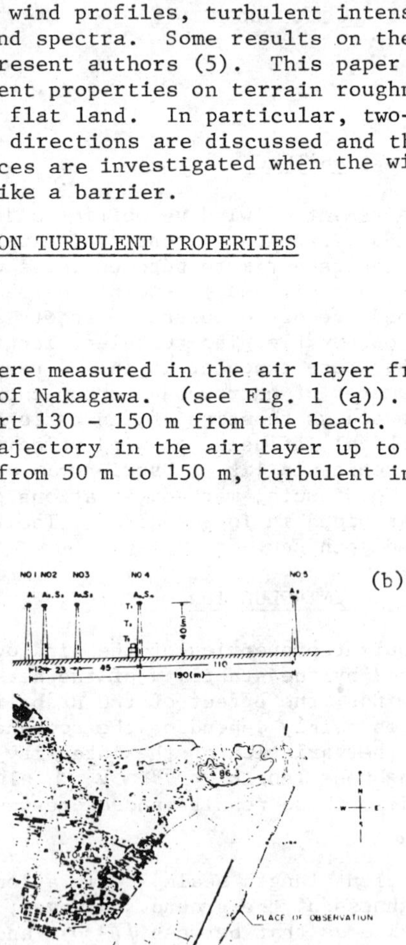

Fig. 1. Observation sites. (a) Nakagawa, (b) Satoura

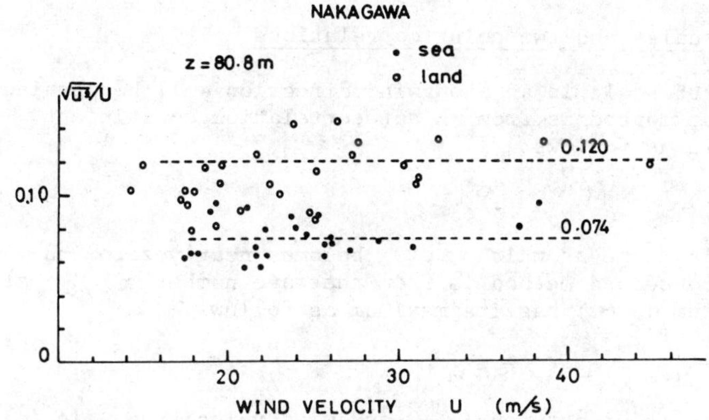

Fig. 2. The variation of turbulent intensities $\sqrt{\overline{u^2}}/U$ with wind velocities U in land and sea winds (Nakagawa)

$$\sqrt{\overline{u^2}}/U = 0.185\, z^{-0.21} \quad \text{in the sea wind}$$
$$\sqrt{\overline{u^2}}/U = 0.125\, z^{-0.01} \quad \text{in the land wind} \qquad (1)$$

where z_0 is meter.

According to the logarithmic law of wind profile, it is estimated that the roughness length z_0 is in the range from 0.04 m - 0.08 m in the sea wind and from 0.2 m - 0.8 m in the land wind. $\sqrt{\overline{u^2}}/U$ at 80.8 m is shown in Fig. 2 as function of mean wind velocity U. The magnitude of $\sqrt{\overline{u^2}}/U$ becomes large with increasing terrain roughness. $\sqrt{\overline{u^2}}/U$ seems to be unchangeable by the change of wind velocity.

Turbulent intensities and horizontal space correlation were obtained at Satoura, where five towers, 40 m in height, were arranged along the coast over 190 m distance as shown in Fig. 1 (b) and the wind was measured at the top of each tower. Fig. 3 shows the relation between the turbulent intensity $\sqrt{\overline{u^2}}/U$ and wind velocity U for different wind directions. The magnitude of $\sqrt{\overline{u^2}}/U$ becomes large with increasing terrain roughness. Particularly in the directions ranging from WNW - NNW, $\sqrt{\overline{u^2}}/U$ scatters in a wide range and its magnitude is too large. It might be influenced by the hills of 80 m in height, situated about 1 km in the windward direction. $\sqrt{\overline{u^2}}/U$ seems to be not influenced by the change of wind velocity as in the case of Nakagawa. However, in the sea wind from SSE, $\sqrt{\overline{u^2}}/U$ is 0.07 in the range of U from 8 - 25 m/s, and it becomes large as 0.10 in the range of U from 40 - 60 m/s.

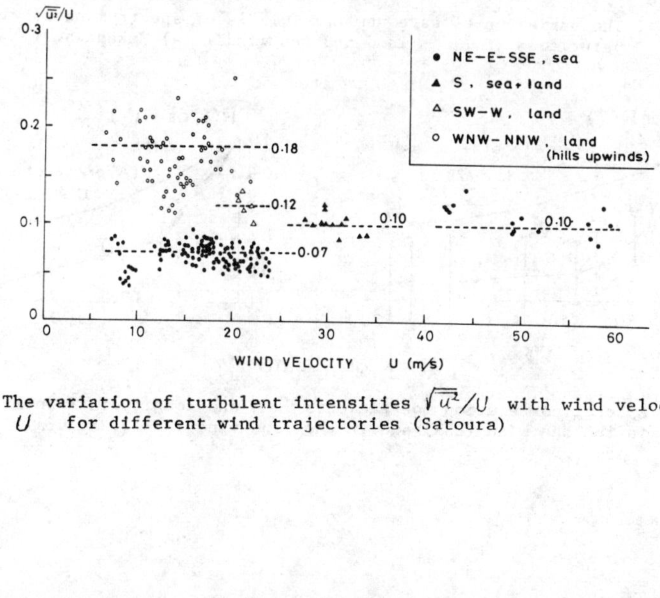

Fig. 3. The variation of turbulent intensities $\sqrt{\overline{u^2}}/U$ with wind velocities U for different wind trajectories (Satoura)

Turbulent length scales and two-point correlations

The turbulent length scale in an alongwind direction will be obtained in various methods. The first method is from an autocorrelation coefficient $R(\tau)$ by use of Taylor's hypothesis as follows:

$$L_x = U\int_0^{\tau_0} R(\tau)\,d\tau \tag{2}$$

where τ_0 is the lag time at which $R(\tau)$ becomes nearly zero and U is the mean wind velocity. The second method is from the wave number n_m/U at which logarithmic power spectrum $nS(n)$ has its maximum as follows:

$$L_x = U/2\pi n_m \tag{3}$$

Fig. 4 shows the relation between n_m/U and wind velocity U in (a) Nakagawa's data and in (b) Satoura's ones. As the points in the figure scatter in a wide range, it is hard to distinguish the data of the sea wind from those of the land wind. It seems n_m/U is not influenced by the roughness of the ground. The estimate of n_m/U is almost constant or somewhat decreases with the increase of wind velocities. From the estimates for sea wind and land wind at both observation sites, the relation between non-dimensional frequency $f(=n_m z/U)$ and the height z is obtained in the wind velocity of about 20 m/s as follows:

$$f_m = 0.006\, z^{0.58} \qquad 20 \leq z\ (m) \leq 150 \tag{4}$$

where z is in meter. From (4), the turbulent scale L_x is obtained by the second method mentioned above as follows:

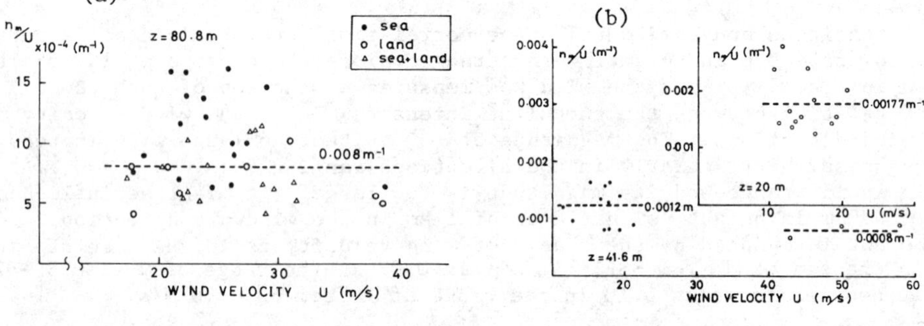

Fig. 4. The variation of wave numbers n_m/U of spectrum peaks with wind velocities U, in land and sea winds. (a) Nakagawa, (b) Satoura

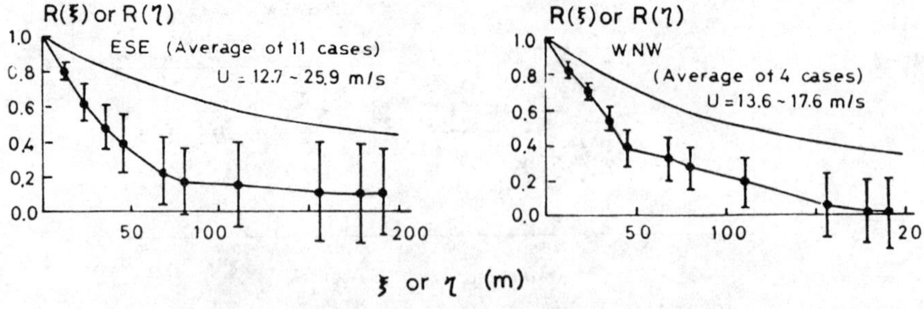

Fig. 5. Lateral correlation coefficients $R(\eta)$ and longitudinal ones $R(\xi)$ in the sea wind (ESE) and in the land wind (WNW), (Satoura)

$$L_x = 53 z^{0.42}, \qquad 20 \leq z \text{ (m)} \leq 150 \quad (5)$$

where z is in meter. Shiotani (6) has obtained L_x by the first method by using Nakagawa's data and showed the relation between L_x and z at higher air layer than 50 m as follows:

$$L_x = 90 z^{0.19} \qquad 50 \leq z \text{ (m)} \leq 150 \quad (6)$$

where L_x is averaged in the range of wind velocities from 15 - 45 m/s. It seems the rate of increase of L_x with height becomes slower as the height becomes larger. Anyway, L_x increases with height with the power of z, whose exponents are in the range from 0.2 - 0.4.

Next, we shall examine the space correlation coefficient at lag time zero. In a horizontal direction, the correlation coefficient depends only on the separation distance between two points. The estimates of correlation coefficient $R(\eta)$ from Satoura's wind data are shown in Fig. 5 by solid symbols with their standard deviations, where η is the distance in a lateral direction. A longitudinal correlation coefficient $R(\xi)$, where ξ is the distance in an alongwind direction, may be obtained by the help of Taylor's hypothesis from autocorrelation $R(\tau)$, and its estimates are shown in the same figure with smooth curves. The wind directions are perpendicular to the coast. The wind of WNW direction comes from the land and that of ESE direction comes from the sea. We cannot notice the difference between the estimates of the sea wind and those of the land wind. In Table 1, the estimates of $R(\eta)$ in the sea wind and those of land wind are compared for five different η. In this case, we do not limit to the wind in only perpendicular direction to the coast, but include the wind ranging from it to 45° direction. Both estimates are nearly equal within the accuracy of estimation.

Turbulent length scales L_x and L_y in longitudinal and lateral directions at Satoura are eatimated respectively from Eq. (2) and the following equation,

$$L_y = \int_0^{\eta_m} R(\eta)\, d\eta \quad (7)$$

where η is the maximum separation distance in observation, that is, 190 m. In the sea wind of 11 cases, L_x= 195 ± 55 m and L_y = 53 ± 35 m. In the land wind of 4 cases, L_x= 135 ± 29 m and L_y = 55 ± 20 m. In spite of large standard deviations of the estimates, it seems turbulent length scales in longitudinal directions are larger in the sea wind. This trend is in agreement with Counihan's and ESDU's results that a longitudinal scale becomes larger over smoother terrain. The difference between lateral length scales of sea wind and those of land wind is not significant.

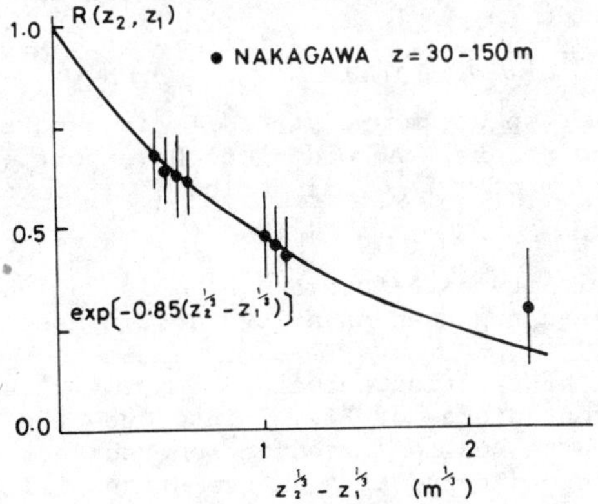

Fig. 6. Vertical correlation coefficients as function of $z_2^{1/3} - z_1^{1/3}$, (Nakagawa)

Table 1. Horizontal correlation coefficients for allover frequency-range. (Satoura)

η (m)	12		23		35		45		68	
Wind trajectory	sea	land	sea	land	sea	land	sea	land	sea	land
Estimate $R(\eta)$	0.81	0.81	0.61	0.63	0.47	0.49	0.36	0.37	0.23	0.26
Standard deviation	0.05	0.06	0.12	0.08	0.15	0.12	0.17	0.13	0.16	0.15
Number of measurement	19	10	19	10	19	10	19	10	19	10

Table 2. Vertical Correlation coefficients for allover frequency-range. (Nakagawa)

$z_2 - z_1$ (m)	80 — 50		50 — 30		80 — 30		110 — 80	
Wind trajectory	sea	land	sea	land	sea	land	sea	land
Estimate $R(z_2,z_1)$	0.61	0.62	0.61	0.64	0.44	0.45	0.67	0.69
Stndard deviation	0.08	0.09	0.10	0.07	0.09	0.07	0.05	0.09
Number of measurement	21	15	13	9	12	8	8	11

In a vertical direction, the correlation coefficient cannot be expressed by the function of only the distance between two points, but it depends on the positions of both points. The estimates of correlation coefficient $R(z_2, z_1)$ between two points z_2 and z_1 on a vertical line are shown in Table 2 in the sea wind and in the land wind at Nakagawa. Both eatimates are nearly equal within the accuracy of estimation as shown in Table 1. From the results in Figs. 4 and 5 and Tables 1 and 2, we may conclude that the two-point correlation is not definitely influenced by the roughness of the boundary surface.

According to Panofsky and Singer (7), $R(z_2, z_1)$ obtained in both sea winds and land winds can be expressed as an exponential function of $z_2^{1/3} - z_1^{1/3}$ as shown in Fig. 6 (6). A smooth line in a figure is,

$$R(z_2, z_1) = \exp\{-0.85(z_2^{1/3} - z_1^{1/3})\} \tag{8}$$

NARROW-BAND CROSS-CORRELATION COEFFICIENTS AND DECAY PARAMETERS

The lateral correlation coefficient $R(\eta)$ as shown in Fig. 5 can be expressed as follows:

$$R(\eta) = \int_0^\infty \sqrt{Coh(\eta, n)}\, S(n)\, dn \Big/ \int_0^\infty S(n)\, dn \tag{9}$$

where, $S(n)$ is a power spectrum and $\sqrt{Coh(\eta, n)}$ is a narrow-band cross-correlation coefficient between two points. According to Davenport (8), the latter may be expressed as the function of $n\eta/U$ as follows:

$$\sqrt{Coh(\eta, n)} = \exp(-k_\eta n\eta/U) \tag{10}$$

where U is a mean velocity as before and k_η is called a decay parameter or a decay factor and has been regarded as a constant in most bibliographies.

In an illustration of the relation between $\sqrt{Coh(\eta, n)}$ and $n\eta/U$ which would be drawn by overlapping four pieces of Fig. 7 into one, the points would scatter so widely that we allow ourselves to introduce some subjective methods in the estimation of the k_η. Accordingly, at a glance of the illustration, we think there would be no difference whether we adopt a constnt k_η or variable k_η for each η/x. However, examining the four pieces of Fig. 7 separately, we find that

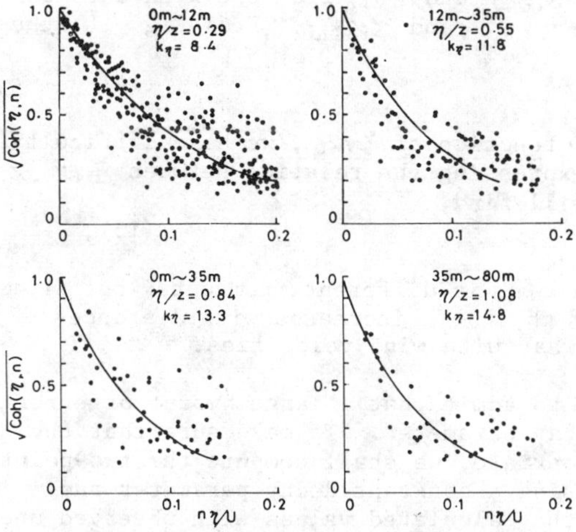

Fig. 7. Variation of lateral (narrow-band) cross-correlation coefficients with $n\eta/U$

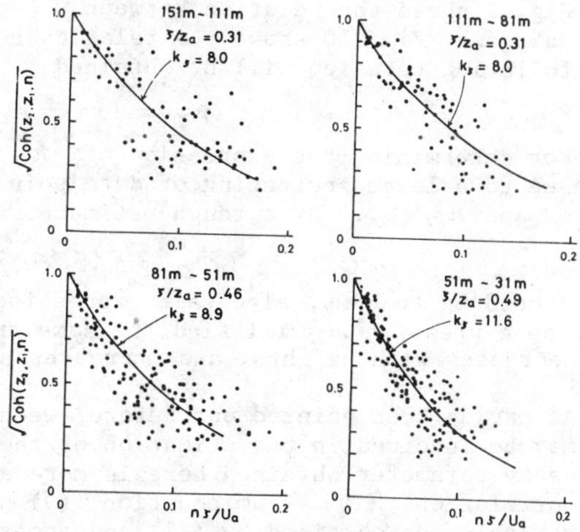

Fig. 9. Variations of vertical (narrow-band) cross-correlation coefficients with $n\zeta/U_a$

k_η is not constant but is increasing with η/z, as having been pointed out by Naito snd Kondo (9) and Iwatani (10). The latter has obtained the following relation between k_η and η/z. (See Fig. 8)

$$k_\eta = 14\,(\eta/z)^{0.45} \tag{11}$$

In a vertical direction, the correlation coefficient $R(z_2, z_1)$ can be expressed as follows:

$$R(z_2, z_1) = \int_0^\infty \sqrt{Coh\,(z_2, z_1, n)}\,\sqrt{S(z_2, n)\,S(z_1, n)}\,\cos\{\Phi(z_2, z_1, n)\}\,dn \Big/ \sqrt{\int_0^\infty S(z_2, n)\,dn}\cdot\sqrt{\int_0^\infty S(z_1, n)\,dn} \tag{12}$$

where $S(z_2, n)$ and $S(z_1, n)$ are power spectra at z_2 and z_1, respectivly. $\sqrt{Coh(z_2, z_1, n)}$ and $\Phi(z_2, z_1, n)$ are narrow-band cross-correlation coefficient and phase between z_2 and z_1, and they are respectively expressed as follows:

$$\sqrt{Coh\,(z_2, z_1, n)} = \exp(-k_\zeta n\zeta/U_a)$$
$$\Phi(z_2, z_1, n) = k'_\zeta n\zeta/U_a \tag{13}$$

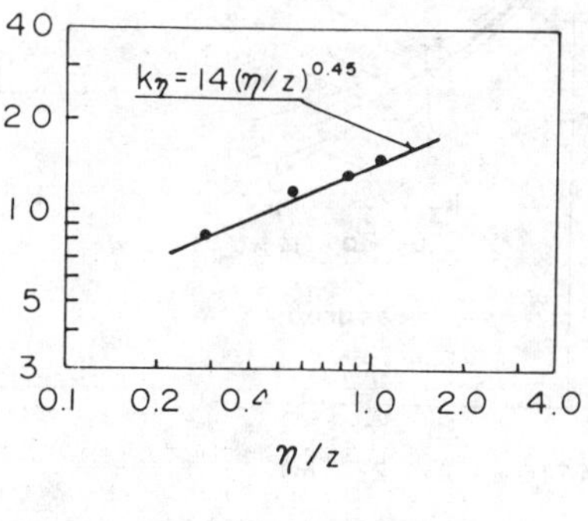

Fig. 8. The relation between k_η and η/z

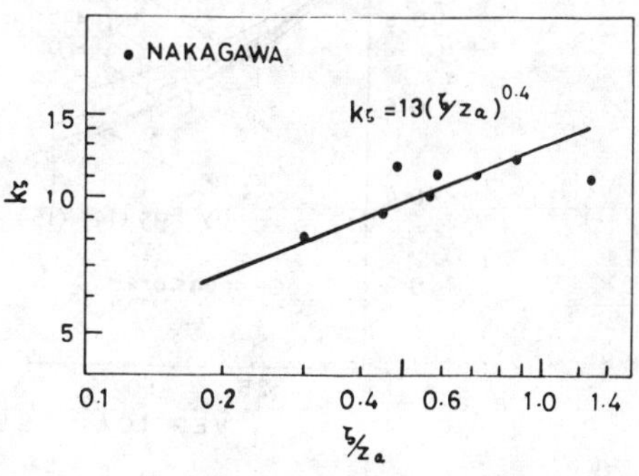

Fig. 10. The relation between k_ζ and ζ/z_a

where $\zeta = |z_2 - z_1|$ and U_a is the average of wind velocities at z_2 and z_1. Fig. 9 shows the relation between $\sqrt{Coh(z_2, z_1, n)}$ and $n\zeta/U_a$ from the wind data of Nakagawa. Fig. 10 shows the relation between k_ζ and ζ/z_a. From Fig. 10, the following relation will be obtained,

$$k_\zeta = 13(\zeta/z_a)^{0.4} \tag{14}$$

For determining the expression of k'_ζ as function of ζ/z_a, we have difficulty due to a large scattering of data points expressing the relation between $\Phi(z_2, z_1, n)$ and $n\zeta/U_a$. By a rough eatimate, we will find,

$$k'_\zeta = 9(\zeta/z_a)^{0.4} \tag{15}$$

According to ESDU, algebraic expression of k_η is different from ours, but from the expression we will find the same trend that k_η increases with distances, decreases with heights, and increases somewhat with wind velocities.

As having been pointed out before, we have to admit that a large amount of error may be involved in the etimation of the decay parameter. To make sure that the decay parameter obtained here is more appropriate, we shall compute the two-point correlation $R(\eta)$ from equation (9) by using a constant decay parameter and variable one obtained by (11) and compare the calculated values with observed ones. We use Karman-type spectrum (11) for $S(n)$, as follows:

$$\frac{nS(n)}{\overline{u^2}} = \frac{0.475x}{(1+x^2)^{5/6}} \qquad x = \frac{Ln}{U} \quad \text{and} \quad L = \frac{3}{2}\frac{z}{f_m} \tag{16}$$

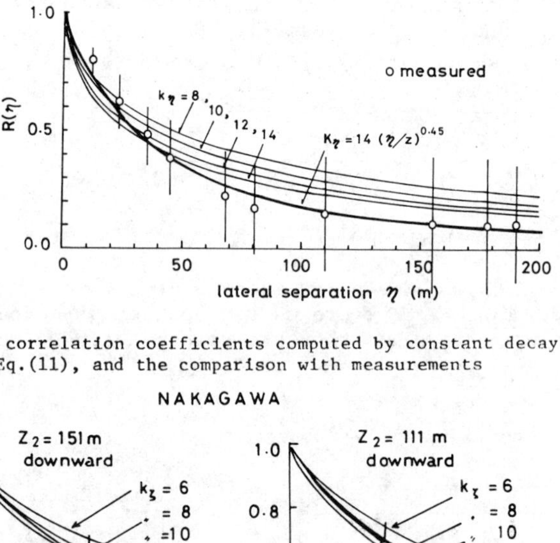

Fig. 11. Lateral correlation coefficients computed by constant decay parameters and by Eq.(11), and the comparison with measurements

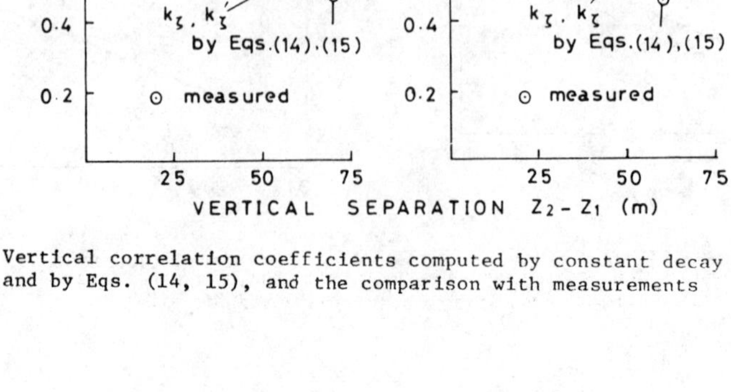

Fig. 12. Vertical correlation coefficients computed by constant decay parameters and by Eqs. (14, 15), and the comparison with measurements

and f_m is computed from (4).

Fig. 11 shows the results of computation. When $k_\eta = 14(\eta/z)^{0.45}$, computed values of $R(\eta)$ agree surprisingly well with the observed values. When k_η = 8 - 14, computed values of $R(\eta)$ are always less than observed ones in small range of η and they become larger than observed ones in large range of η.

We shall compute the vertical correlation $R(z_2, z_1)$ by using the Eq. (12) in the same manner as the lateral correlation. The results are shown in Fig. 12. We will find that computed values from k_ζ and k'_ζ by Eqs. (14) and (15) respectively agree better with observed values than computed values from constant k_ζ = 6 - 10 and k'_ζ = 0, though the difference among computed curves is not so distinctly observable as the case of the lateral correlation. From the computed values of $R(z_2, z_1, \tau_m)$, where τ_m is the lag time at which $R(z_1, z_2, \tau)$ has its maximum, the same trend is noticed though the results are not shown here. Thus, we may conclude that the decay parameter does not keep constant values, but it should vary as the function of the ratio of distance to height.

As already pointed out by present authors (12) and others (4), the decay parameter has a trend to increase with wind velocity. However, at present, we have not an appropriate algebraic expression to represent the trend.

A CASE STUDY OF THE EFFECT OF SMALL TOPOGRAPHIES ON WIND VELOCITY AND TURBULENCE -- THE WIND FLOW OVER A BARRIER

The wind was measured near the head (called Tozaki) of slender peninsula which projects about 1.5 km into Naruto Straits from the side of Awaji Island. (see Fig. 13 (a)) When the wind with a considerable fetch over the sea impinges perpendicularly to the peninsula, the wind profile and turbulent properties of the sea wind may be modified by the topographies. The cross section of the peninsula is horseback-shape as shown in Fig. 13 (b) in the direction of the flow. The height of peninsula H is 42 m at the base of the tower, and both sides fall into the sea

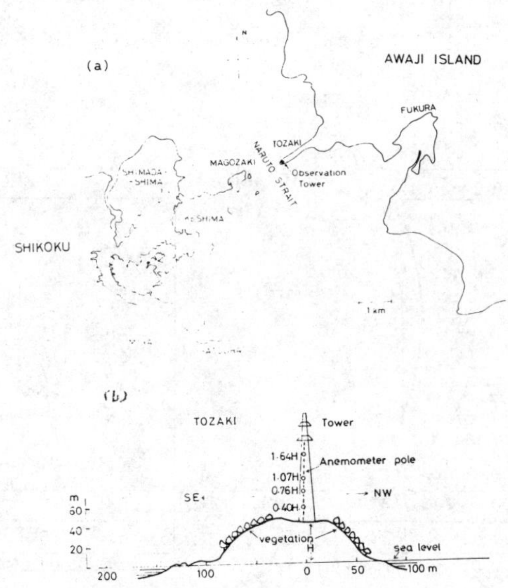

Fig. 13. The map of Naruto Straits and the cross section of the peninsula at the measuring site

with a steep slope of nearly 45 degrees. The wind was measured at 0.40H, 0.76H, 1.7H, and 1.64H above the peninsula.

According to Mitsuta (13), the wind velocity at the highest measuring point is hardly influenced by the topography. Accordingly, we regard mean and turbulent velocities at this point as the reference for convenience' sake and denote them U_s and $\sqrt{\overline{u^2}}_s$, respectively. Fig. 14 shows the vertical profiles of U/U_s and $\sqrt{\overline{u^2}}/\sqrt{\overline{u^2}}_s$ together with those of turbulent intensity $\sqrt{\overline{u^2}}/U$, length scale L_x and the vertical angle of the wind $\bar{\alpha}$. The estimate of $\bar{\alpha}$ is 8 deg. at 0.76H and it decreases 5 deg. at 1.64H in the wind of NW-direction. In the wind of SSE-direction $\bar{\alpha}$ become 12 deg. and 5 deg. at respective heights. We can imagine that the wind is converged vertically when it blows over the peninsula.

The effect of the convergence is clearly noticeable in the profile of wind velocity U/U_s. The magnitude of wind velocity is largest at 0.40H in NW wind but, in SSE wind, the level of 0.40H is just inside the wake which lies under separated flow at the top edge (see Fig. 13 (b)) of the windward slope. In this wind direction, an amplification of the wind velocity is clearly observable at the level of 0.76H. Inside the wake, the mean velocity is small and the turbulent intensity $\sqrt{\overline{u^2}}/U$ is considerably large. The estimate of vertical correlation coefficient between 0.76H and 0.40H is very small as 0.11 in spite of small separation in SSE wind, but, in NW wind, it is large as 0.72. Accordingly, turbulent properties in the wake is completely uncorrelated with those in the accelerated flow over it.

The estimate of turbulent intensity $\sqrt{\overline{u^2}}/U$ is in the range from 0.06 - 0.07 at 87 m above the sea in the accelerated flow and that of turbulent length scale is 250 m on the average. Estimates at Nakagawa (in the sea wind without topographical influence) at the same height are respectively $\sqrt{\overline{u^2}}/U = 0.075$ and L_x = 210 m on the average. It seems in an accerelated flow the turbulent intensity becomes smaller and the turbulent length scale becomes larger in a longitudinal direction.

The vertical correlation coefficient in the accelerated flow over the peninsula is expressed in terms of Eq. (8) as follows:

$$R(z_2, z_1) = \exp\{-0.76(z_2^{1/3} - z_1^{1/3})\}$$

$$59 \leq z_2, z_1 \leq 111, \text{ in NW direction} \qquad (17)$$

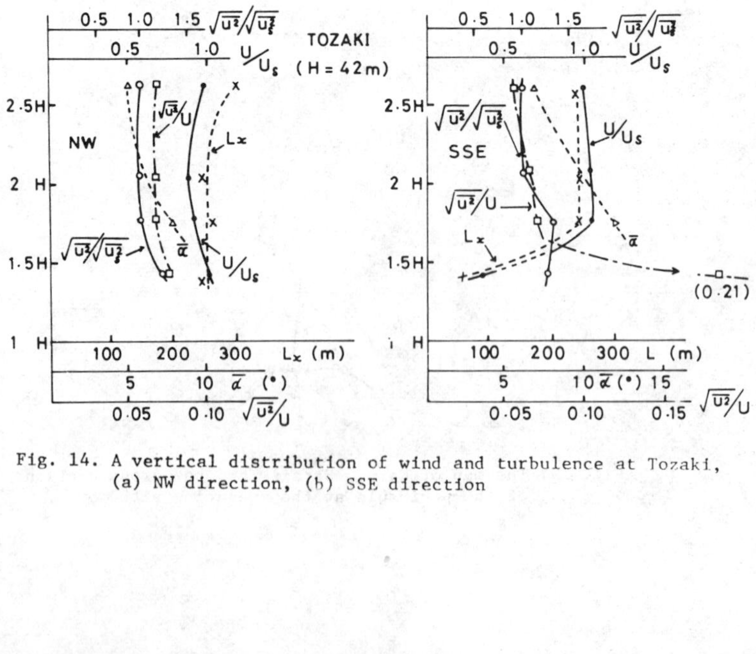

Fig. 14. A vertical distribution of wind and turbulence at Tozaki, (a) NW direction, (b) SSE direction

$$R(z_2, z_1) = \exp\{-1.15(z_2^{1/3} - z_1^{1/3})\}$$

$$74 \leq z_2,\ z_1 \leq 111 \text{ , in SSE direction} \quad (18)$$

where heights z_2 and z_1 are measured from the sea surface in meter. In the wind of NW direction estimates of $R(z_2, z_1)$ are nearly the same as those of undisturbed sea wind at Nakagawa. (Compare with Eq. (8).) But, in the wind of SSE direction, the estimates of $R(z_2, z_1)$ are smaller than those of Nakagawa. We cannot understand the reason why such a marked difference has occurred in the vertical correlation between the accelerated flow from NW direction and that from SSE direction, though the detailed features of windward terrain may be somewhat different as shown in Fig. 13. In the longitudinal scales L_x , we can not find any difference between both wind directions, though, in the estimate of L_x , standard deviation is much greater than that of $R(z_2, z_1)$.

CONCLUSION

The measurements of wind velocities and turbulence are made at the sea coasts during high winds, and the wind data are accumulated for different wind trajectories. Turbulent intensities are greatly influenced by terrain roughness. But, in two-point correlations in both lateral and vertical directions, we can not find a marked variation by the change of terrain roughness, though the accuracies of the estimation for these properties are not good. It seems turbulent scales may be very slightly influenced by the roughness of the boundary surface as compared with turbulent intensities. As an approximation, it may be allowable to mix up all data irrespective of their wind trajectories for obtaining the variations of turbulent scales with heights.

Turbulent length scale increases in proportion to the power of height whose exponents are in the range from 0.2 - 0.4. The estimate of L_y is about 50 m at 41 m. The vertical correlation coefficient for allover frequency range $R(z_2, z_1)$ may be expressed by the form suggested by Panofsky and Singer as follows:

$$R(z_2, z_1) = \exp\{-C(z_2^{1/3} - z_1^{1/3})\}$$

The estimate of C is 0.85 $m^{-1/3}$.

It is necessary to express the two-point correlation in narrow-band frequency range for applying it for the calculation of the gust response of the structure. When the narrow-band cross-correlation coefficients both in lateral direction and in vertical one are approximated by exponential functions as having been widely used, the decay parameters are not constant, but should be expressed as the function of the ratio of distance to height. Their algebraic expressions are tentatively shown. Numerical constants contained in both lateral and vertical decay parameters are almost the same. It is shown that correlation coefficients for allover frequency range computed by use of power spectra and cross-correlation coefficients with the decay parameters obtained here agree better with observed results than those with constant decay parameters do, particularly for a lateral direction.

When the wind from a wide open sea comes into the land, the wind velocities and the turbulence may be modified by the topographies of the land. As a case study, we examine the wind flow over a slender peninsula like a barrier. The air flow is accelerated above the peninsula and the velocity profile changes considerably. In an accelerated flow, it seems turbulent intensities decrease a little and turbulent length scales increase a little in alongwind velocity components. But, within the accuracy of the measurements, they are about the same as those in an undisturbed sea wind.

It was noticed from the measurements of fluctuations of vertical wind directions here and at other places that a vertical velocity component seems to be more sensible than an alongwind one to the protuberance from the flat surface, and it seems that the topographical influence is three-dimensional in nature. It will be necessary to measure three components of the wind velocity to understand the effect of topography.

The authors wish their hearty thanks to Mr. S. Nishino, Honshu-Shikoku Bridge Authoritees for making the measurements at Tozaki.

REFERENCES:

1. Counihan, J., Adiabatic atmospheric boundary layer; A review and analysis of data from period 1880 - 1972, Atmos. Environ. 9: 871-905 (1975)

2. Teunissin, H.W., Characteristics of the mean wind and turbulence in the planerary boundary layers, UTIAS Rev. No. 32, Inst. for Aero. Space Studies, Univ. of Toronto, 1-46 (1970)

3. ESDU, Characteristics of atmospheric turbulence near the ground, Part 2: single point data for strong wind (neutral atmosphere), Item 74031: 1-29 (1975)

4. __________, ______________________________, Part 3: variation in space and time for strong winds (neutral atmosphere), Item 75001: 1-27 (1976)

5. Shiotani, M. and Iwatani, Y., A study of two-point correlation during strong winds, Proc. 3rd Colloquium on Indust. Aerodynamics Part 1: Aachen, 43-54 (1978)

6. Shiotani, M., Turbulence measurements at the sea coast during high winds, J. Meteor. Soc. Japan 53: 340-354 (1975)

7. Panofsky, H.A. and Singer, I.A., Vertical structures of turbulence, Quart. J. Roy. Meteor. Soc. 91: 339-344 (1965)

8. Davenport, A.G., The spectrum of horizontal gustiness near the ground in high winds, Quart. J. Roy. Meteor. Soc. 87: 194-211 (1961)

9. Naito, G. and Kondo, J., Spatial structures of fluctuating components of horizontal wind speed above the ocean, J. Meteor. Soc. Japan 52: 391-399 (1974)

10. Iwatani, Y., Some features of spatial structures of the surface layer turbulence in high wind conditions, J. Meteor. Soc. Japan 55: 130-138 (1977)

11. Hino, M., Spectrum of gusty wind, Proc. 3rd Intern. Conf. Wind Effects on Buildings and Structures, Tokyo: 69-77 (1971)

12. Shiotani, M. and Iwatani, Y., Velocity correlation in relation to gust loadings, Proc. 3rd Intern. Conf., Wind Effects on Buildings and Structures, Tokyo: 57-68 (1971)

13. Mitsuta, Y., Characteristics of air flow over the barrier in the storm, Proc. 3rd Intern. Conf. Wind Effects on Buildings and Structures, Tokyo: 33-44 (1971)

DISCUSSION ON SESSION II
Wind--Characteristics and Description

Rapporteur: E. Simiu
National Bureau of Standards
USA

Paper by H. Ishizaki, Y. Mitsuta, Y. Taniike--A TATSUMAKI IN TOKYO ON FEBRUARY 28, 1978

Question: M. C. Jischke. How can a lift force (e.g., a vertical force) be developed on the train which is on the ground? I would assume the velocity within a few meters of the ground would be essentially horizontal. Thus, only a horizontal force--which could tip the train rather than lift it--would be developed.

Reply: H. Ishizaki, Y. Mitsuta, Y. Taniike. The train was not on the ground but on the lattice bridge over the river which was about 20 m high.
As written in our paper, the train would be derailed by the joint action of the overturning moment due to the horizontal force and the lift force estimated in Equation (2).

Comment: R. K. Cooper. If we suppose that the aerodynamic side force acts at the centroid of the side area of a vehicle, then the wind speed necessary to rotate the vehicle about the lee rail is

$$V > V_c = \sqrt{\frac{mgs}{C_S \tfrac{1}{2}\rho \ AH_c}}$$

where s = 0.56 m is the semi-track. Using h = 2.3 m as the centroid height, M^c = 27500 kg (Taniike), C_S = 1.3 and other values given in the paper we obtain

$$V_c = 40 \text{ m/s}$$

much less than the measured speeds. This suggests that the vehicles were overturned by rotation about the lee rail, and that the lift force was not a major factor in the incident. The tatsumaki won very easily.

Question: J. C. R. Hunt. These authors on p. 9 of their paper calculate the lift force on the train as the sum of the steady lift force ($\tfrac{1}{2}\rho\ V^2C_L$) and the added mass effect ($\rho \ell \partial V/\partial t C_V$). They find that the latter component is negligible. Computation such as those of Wen (1975) show that the added mass effect can be as large or larger than the steady lift or drag force for large structures. Is this fact used in current comparisons between damage and wind speeds?

Reply: H. Ishizaki, Y. Mitsuta, Y. Taniike. We write in our paper on p. 9 that "the increasing rate of wind velocity is not so small to be neglected." As the reference length of "ℓ" is not large in the case of the force acting on the train, the accelerative force is small as compared to the steady force.

Comment: E. Simiu. The added mass coefficient is generally not used in engineering calculations to estimate drag forces associated with local flow fluctuations. An experimental justification for this appears to have been provided by Vickery (J. of the Structural Division, ASCE, 1972).

Paper by Dr. W. S. Lewellen, M. E. Teske, Y. P. Sheng--WIND AND PRESSURE DISTRIBUTIONS IN A TORNADO

Question: S. Ayad. What was the inward (radial) velocity profile used as a boundary condition? Was the value U^+ the average of that profile?

Reply: W. S. Lewellen, M. E. Teske, Y. P. Sheng. Yes. U^+ is the average radial velocity at the large radius boundary. The input radial velocity is assumed constant for cases b, c and d in Figures 1-5. The detailed profile for case a is given in Reference 1.

Question: B. M. Leadon. If I understood your reference to the multivortex tornadoes as highly turbulent flows, you found that your turbulence levels calculated were also very high in some cases, seeming to confirm the real observations--to some extent. My comment is that multiple vortices' fluctuations are not themselves evidence of turbulence, though they may help to intensify turbulence, but the vortices are organized flows. Have you studied the stability of your axial flows?

Reply: W. S. Lewellen, M. E. Teske, Y. P. Sheng. In a few exceptional cases multiple vortices persisting for a fairly long time have been observed in tornadoes. However, the much more common observation is that the secondary vortices have lifetimes less than one rotation about the axis of the tornado. Thus, it should be quite consistent to treat them as turbulent eddies. The relative stability of the flow is indicated by how steady the long time numerical solution is. We do see some quasi-periodicity of frequency approximately $2V_{max}/\hat{r}$ in the long time solution.

Paper by M. D. Powell--AN EVALUATION OF DIAGNOSTIC MARINE B.L. MODELS--CYCLONES

Question: D. J. Norton. At hurricane wind speeds the ocean surface will have large waves and sea spray. What meaning will the U_{10} wind speed have if waves ($H \simeq 20$ m) exist? What effect will sea spray have on the interpretation of the surface shear stress coefficients reported in Figure 5.

Reply: M. D. Powell. The IDM wind was used in order to compute model results to buoy measurements that were taken at the IDM height. The models do not take waves and sea spray into account; the buoy averages were taken over a time period of 8.5 min., 40 min., or 15 min. so that extreme effects of spray and waves may have a smaller effect. The effects of waves and sea spray are difficult to parameterize in the models at present but it is thought that these effects would act to increase drag coefficients. Future work at NHEML with vertical camera data will be employed to help determine the effects of spray and waves.

Question: E. Simiu. The paper indicates that the wind speed is calculated by the logarithmic law (Equation 2). Up to what height was the log law found to be valid? Was its range of validity comparable to Figure 1?

Reply: M. D. Powell. The range of validity of the log law was comparable to Figure 1 in the periphery of hurricanes Caroline and Eloise. These vertically stacked profiles indicated that the log worked well below the estimated height of the surface layer of 175 m. However, no data is available for a high vertical resolution profile over water in winds over 23 MS^{-1}.

Question: F. H. Durgin. Do you have any explanation as to why the drag coefficient apparently increases with velocity? Is the entrainment of seawater by air a part of this?

Reply: M. D. Powell. Since the model expressions for roughness length depend on the magnitude of the friction velocity, the model drag coefficients will show an increase with wind speed. The budget study computations have no such dependence but also show an increase with wind speed. The agreement indicates that Charnock's relationship is reasonable in the hurricane regime. If this is true, then, as wind increases, the surface roughness increases due to capillary and gravity waves is included in Charnock's expression to some extent. The effect of sea spray would presumably be to increase the drag coefficient but this is still unknown.

Question: H. A. Panofsky. Did you check on Charnock's relations for roughness length as functions of U_*^2/g^2?

Reply: M. D. Powell. The Charnock relation for roughness length with $\alpha = .035$ was used by the Powell Model and Moss-Rosenthal Model. The Cardonel Model used a regression employing U_* that gave similar results to the Charnock relation. The three models were averaged to give AU_* that was used to compute drag coefficients. The agreement of the drag coefficient computations with earlier hurricane budget studies indicates that Charnock's relationship can provide roughness length estimates that can produce reasonable hurricane drag coefficients. Large scatter in the data base indicates that further work is needed to determine a correct value of α.

Paper by H. W. Tieleman and S. E. Mullins--THE STRUCTURE OF MODERATELY STRONG WINDS AT A MID-ATLANTIC COASTAL SITE (BELOW 75 m)

Question: H. W. Teunissen. For winds approaching the measurement tower from the northwest, there appear to be a number of changes in the nature of the upstream fetch. Is there any possibility that the development of internal boundary layers could account for some of the kinks in the observed mean velocity profiles?

Reply: H. W. Tieleman, S. E. Mullins. The kink in the mean velocity profile of Figure 2 is the result of the transition of the logarithmic profile in the surface layer to the uniform mixed layer profile. Many of the other profiles under either stronger wind conditions (Figure 2) or more moderate conditions ($Ri < -0.15$) are linear when presented in semi-logarithmic coordinates, not giving any indication of kinks which could occur as the result of a change in roughness (possible smooth-rough, see left-hand profile, Figure 2). However, there are indications that one internal thermal boundary layer is present as a result of transition from wet surface conditions in the upstream tidal swamp to the warmer and dryer island surface. However, the change in surface temperature does not seem large enough and occurs far enough upstream to affect the flow above a height of 15 m when measured at the tower.

Question: H. W. Teunissen. The average values of σ_α/U^* for near-neutral conditions in the IBL for Category 2 winds from the south (i.e., 1.75, 1.5 and 1.4) seem rather low for neutral conditions. Is there a particular reason for these low values in the IBL?

<u>Reply</u>: H. W. Tieleman, S. E. Mullins. South winds passing the beach are modified giving rise to the development of the IBL. Shear velocities cannot be obtained from the mean velocity profile, but must be obtained from the covariance $\overline{u}$. Consequently, the values of $\sigma\alpha/U*$ depend on turbulence measurements in the IBL. Since there is a drastic change in stability and surface conditions from the upstream surface flow over the ocean to the developing IBL, it is quite possible that the turbulence measured at the lower locations has not had sufficient time to adapt itself to the new terrain and stability conditions. Consequently, measured values of $\sigma\alpha/U*$ in the IBL cannot be expected to be the same as those measured over uniform terrain. A comprehensive experimental study of the modification of the mean flow and turbulence after a change in surface conditions (including oblique angles) is long overdue.

<u>Question</u>: H. W. Teunissen. Is there a seasonal pattern to the stable, on-shore flows approaching from the south (i.e., Category 2)? How often do such flows occur, and are there any particular local water features, such as local cool patches, which might be reasonable for creating this?

<u>Reply</u>: H. W. Tieleman, S. E. Mullins. The on-shore southern winds occur more frequently during the summer, but also occur during the other seasons as well. Based on observations made at Wallops Island, as well as those near Long Island by Sethuraman (Reference 13), there is no evidence of the existence of locally cooler patches of water affecting the flow over the ocean. However, the height of the surface-based inversion and its degree of thermal stability varies a great deal throughout the course of a sunny day (Figure 11).

<u>Question</u>: M. D. Powell. The leftmost profile of Figure 10 is very similar to data taken from a 500 m tower on the N.W. Cape of Australia by Ken Wilson of the Australian Bureau of Meteorology. He suggested that it occurs below a stable region below cloud base produced by between cloud subsidence. This may occur often for moderate or strong on-shore flows in a connective regime or stable regime that may have been missed previously due to sparsity of observations. Do you know of any other studies which show a similar profile?

<u>Reply</u>: H. W. Tieleman, S. E. Mullins. Low-level jets have been observed for southerly winds at Wallops Island, with maximum velocities above a height of 70 m, by following rocket exhaust tails on clear days as well as on days with cloud cover. Work done by Dr. S. Sethuraman at the Brookhaven National Laboratories, Long Island, New York, support these observations. Wind speed profiles obtained from pilot-balloon soundings at the beach on Long Island show clearly the low-level jet. See Figure 3 of the paper by S. Sethuraman, "The Observed Generation and Breaking of Atmospheric Internal Gravity Waves over the Ocean," Boundary Layer Meteorology, Vol. 12, No. 3, October 1977, pp. 331-349.

Paper by P. Duchene-Marullaz--EFFECT OF HIGH ROUGHNESS ON THE CHARACTERISTICS OF TURBULENCE IN CASES OF STRONG WINDS

<u>Question</u>: P. Arya. The increase of σ_u/U_* and σ_v/U_* with height is likely to be caused either by convection or by the limited response of instruments to high frequency fluctuations at the lower level. The author may comment on the effect of these factors on the distributions of σ_u and σ_v/U_*.

<u>Reply</u>: P. Duchene-Marullaz. For all the runs, we have verified the atmospheric stability between 5 m and 60 m (Ri $\simeq$ 0.00), so I do not think that the increase of σ_u/U_* and σ_v/U_* is due to convection. The response of the Gill anemometer is correct up to 0.5 Hz, so it is not the cause of the variation with height. I do not have a very good explanation for that increase but I think that it is due to the probable presence of an internal boundary layer from 0 to 20 m above ground at the measuring site.

Question: H. A. Panofsky. How did you determine Z_o and U_*? Did you allow for influence of stratification?

Reply: P. Duchene-Marullaz. Z_o and U_* were determined by using the profile above 20 m. We have also calculated τ by measuring the vertical component; above 20 m the ratio $\sqrt{\frac{\tau}{\rho}}/U_*$ was nearly equal to unity.
For all the runs we have verified that the atmospheric stratification was nearly neutral ($Ri \simeq 0.00$).

Paper by G. R. Stegen, R. L. Thorpe--VERTICAL COHERENCE IN THE ATMOSPHERIC BOUNDARY LAYER

Question: H. P. Irwin. A parameter that is often neglected in coherence measurements is the ratio of separation, ΔZ, to the turbulence integral scale, L. Isotropic turbulence theory shows that the coherence is unlikely to collapse onto a single curve when plotted against $\Delta Z/\lambda H$ unless $\Delta Z/L$ is small. As $\Delta Z/L$ increases the coherence decreases at a given value of $\Delta Z/\lambda H$. Information on the predictions for isotropic turbulence has been given by several authors (see Irwin, H.P., J. Boundary Layer Met., 1979). Have the authors considered the effect of $\Delta Z/L$?

Reply: G. R. Stegen, R. L. Thorpe. The authors did not explicitly consider the influence of the integral scale on the decay rate (a). Although it was not computed in this study, it is estimated that the integral scale is several meters. It is possible that the ratio $\Delta Z/L$ is large enough to influence the decay during some of the runs, particularly at the lower heights where the integral scale may be smaller. However, the decay rate is also known to depend upon the Richardson number (Ri). As pointed out by Prof. Panofsky, Ri is proportional to height near the ground, which may also explain the variation of decay rate with height.

Paper by H. A. Panofsky, M. Vilardo, R. Lipschutz--TERRAIN EFFECTS ON WIND FLUCTUATIONS

Question: H. C. Chien. In the spectra you showed, what is the height at which you made the measurements? If you move to a higher position, what would happen? Would a single length scale be identified?

Reply: H. A. Panofsky, M. Vilardo, R. Lipschutz. The graphs I showed were models. They fit data up to several hundred meters over uniform terrain and slightly rolling terrains.

Question: H. W. Tieleman. (1) How far is the mountain upstream from the location where measurements were made? (2) What length scale is used during nighttime, when the mixed layer is difficult to recognize, and the mixing height Z_i is not well-defined?

Reply: H. A. Panofsky, M. Vilardo, R. Lipschutz. (1) The distance of the mountain edge was 400 m at right angles. But the fetch was variable, depending on wind directions. For many runs, the fetch was ∞. (2) We measured mixing depth at night by acoustic sounder, but found no clear relations between it and the σ_u/V data.

Comments: J. C. R. Hunt. I should like to emphasize the field results of Bowen and Shiotani over a hillcrest or escarpment that σ_u decreases well above the surface and increases near the surface. Bradley (to appear in Quart. J. Roy. Met. Soc., 1979) finds that σ_w increases well above the surface and near the surface. As Shiotani indicated in his diagram, the explanation is that well above the surface the accelerating streamlines reduce σ_u and significantly amplify σ_w while close to the surface in an inner layer both σ_u and σ_w are amplified. Panofsky's remark that σ_w is generally unaffected does not hold to hillcrests (Brilto, Hunt and Richards, submitted to Quart. J. Roy. Met. Soc., 1979). I would like to add that the fluctuations of the vertical angle α were measured over the peninsula. The standard deviations $\sigma\alpha$ was estimated. By the rough approximation, $\sigma_w = U\sigma\alpha$. The ratio of σ_u to σ_w was about 1.4 over the peninsula. On the other hand, over a flat terrain the ratio was about 2. Though the measurement of the fluctuations α was not accurate due to the response characteristics of bi-directional vane, it seems that σ_w was somewhat amplified over the terrain.

Comment: H. A. Panofsky. Our observations in rolling farmland did show smaller values of σ_w/σ_u than in uniform terrain. This is in agreement with Bruce Egan's observations that, in complex terrain, he found stronger lateral dispersion but equal vertical diffusion, compared to uniform terrain.

Paper by M. Shiotani, Y. Iwatani--GUST STRUCTURES OVER FLAT TERRAINS AND THEIR MODIFICATION BY A BARRIER

Question: H. W. Teunissen. Equation 5 for the variation of length scale with Z (i.e., $L_x = 53Z^{0.42}$) is obtained from Equations 3 and 4, according to the authors. But combining Equations 3 and 4 actually yields $L_x = 26.5Z^{0.42}$) which seems to yield more reasonable values for L_x. Is the latter expression actually the correct one, or is there a missing factor of two in Equation 4?

Reply: M. Shiotani, Y. Iwatani. Equation 5 is in error. It should be read as $L_x = 27X^{0.42}$. Equation 4 is correct.

Question: H. W. Teunissen. For the data of Figure 14, how many data sets are involved, and what was their length? What mean speeds were involved (i.e., what was U_S typically)? Were any air or water temperatures measured to check atmospheric stability conditions? What were the general climatic conditions?

Reply: M. Shiotani, Y. Iwatani. The number of wind data in Figure 14 are about 10 ~ 15. The range of wind velocities is from 10 to 30 m/s. The length of the data was 15 minutes. We did not check the stability of the atmosphere, or the change of water temperature. Most of the strong wind data were taken in the passage of typhoon. In that time, the wind was usually accompanied by heavy rain. Some data were taken in winter monsoon, where the weather was cold and dry. Other data were taken in the passage of cyclone where the weather was dry or rainy according to the path of cyclone. Except for the typhoon, wind velocity was less than 20-25 m/s.

Question: H. W. Teunissen. For the sea-fetch turbulence intensities of Figures 2 and 3, was air or water temperature data measured to assess atmospheric stability conditions? Is it possible that stability differences might cause some of the scatter in the observed data, and in particular that stable on-shore flows might be the cause of some of the lower turbulence intensity levels?

Reply: M. Shiotani, Y. Iwatani. The temperature differences were measured in both cases of Figures 2 and 3 in early stage of measurements. These measurements were not very good, so that we could not obtain correct values of stability paramete (for example, Richardson number of Z/L). But, it seemed that the atmosphere was under neutral or slightly unstable conditions during strong winds. In the wind of sea-fetch, the turbulent intensities sometimes decreased down to 0.04 ~ 0.03 at Satoura and Tozaki, when the wind velocity was about 20 m/s. At these sites sea current is rapid near the shore. It might be possible to consider that such a small turbulent intensity may be caused by a cold sea current. But we now have no means to certify the phenomena. The main reason of the scatter in the observed data is that there were frequently large scale fluctuations involved in the observed wind series. Whether these fluctuations are caused by the stability of the air layer or horizontal inhomogeneity, I cannot prove.

SESSION III

Local Wind Environment

CHAIRMAN

H. Ishizaki

Disaster Prevention Research Institute
Tokyo, Japan

CO-CHAIRMAN

J. Gandemer

Ingénieur Aérodynamicien
Centre Scientifique et Technique
du Bâtiment
Nantes, France

RAPPORTEUR

J. L. Way

Illinois Institute of Technology
Chicago, Illinois, USA

RECENT DEVELOPMENTS AND TRENDS: LOCAL WIND ENVIRONMENT

H. ISHIZAKI

Disaster Prevention Research Institute, Kyoto University, Japan

"Local Wind Environment" includes broad and various problems. In a summary paper for the Fourth International Conference on Wind Effects in 1975, A.F.E. Wise mentioned that the environmental aspect of the subject itself formed a wide and complex area of research. Accordingly, it is difficult to summarize and discuss the progress of the research in this special field, and the following description of the subject matter may not cover the whole area of research.

Environmental winds were used to be discussed mainly in relation to comfort for outdoor pedestrians, and a lot of papers on the assessment of ground level winds and the criteria have been published. In order to assess the environmental wind, wind tunnel experiments are usually performed. However, the experiment is ocassionally an elaborate work and a quick or expeditious experimental method is demanded. For this purpose, two papers on the experimental method using the sand erosion technique by Beranek and Borges have been submitted for Session III.

Concerning the criteria for pedestrians, several papers were presented at the Third and Fourth International Conferences, and a paper by Murakami, et al., has been submitted to the Session. Considering the achievements of these papers, we can set up a standard for the criteria although our knowledge is not sufficient for design purposes in some cases. On the other hand, the criteria for low-rise buildings around tall buildings have not been discussed. Most low-rise buildings are residential houses having non-engineering structures and they are vulnerable to some extent by severe winds. If a tall building causes the wind speed increase in the surrounding area and there are some low-rise buildings in the area, the damages of these buildings will be increased in the event of severe winds. Actually, the more residential houses around a tall reinforced concrete building were damaged than the houses afar from the tall building by typhoon winds. Even if the wind speed increase caused by the tall building is very small, the damage increase may not be small. If the wind speed exceeds the endurable speed for the structure, it will be collapsed and severe damages will occur. Environmental winds must be considered not only for pedestrians but also for structures, and any criterion for structures cannot be found under violent winds.

In two papers submitted to the Session, wind environments are discussed in relation to the indoor climate or comfort in buildings. Aynsley performed some wind tunnel tests on pressure distributions on various housing types for estimating air flow necessary to restore thermal comfort in hot humid housing. Burns describes a series of model and full scale tests to provide an accurate method of measuring the convective heat transfer coefficient from a building surface. A paper by Paciuk, et al., deals with the wind in a stadium based on the wind observation at the site and a wind tunnel investigation.

The wind characteristics in wakes behind buildings or obstacles are discussed in three papers for Session III. Davies, et al., made the measurements of the velocity field in the wake of a three dimensional square-section building model. In the papers by Castro and Ogawa, et al., the wind characteristics in the wakes behind the two dimensional models are described. In the final paper for the Session by Snyder, et al., the flow structures around simple three dimensional obstacles are investigated experimentally in a towing tank. These papers are concerned with the aerodynamic aspect of wind environment.

The fundamental aerodynamic aspect of local wind environment will be wind speed distributions or mean wind velocity fields around buildings and structures. It is feasible to obtain mean velocity fields experimentally at the ground level and to consider wind speed effects on people and structures as mentioned above. However, it is difficult to know the flow structure in detail or turbulent characteristics of the wake behind obstacles. The wind structure is susceptible to the geometrical changes of obstacles, the turbulent characters of oncoming flow and the meteorological conditions. The flow separations from obstacles and the reattachment of flows also depend on the various factors.

In spite of the complexity of the phenomena, the turbulence structures around obstacles have become to be better understood than before due to the efforts of many researchers. Although it is doubtful whether there is strict two-dimensional turbulent flow, the wakes behind two-dimensional blocks in boundary layers have been studied. On the other hand some papers on the air flows around simple three dimensional models in wind tunnels have been published. These are important as the fundamental studies on environmental winds but the actual wind flow must also be studied.

It is difficult to measure the turbulent wind flow around actual structures. The wind at the ground level can be measured but the wind observation at higher levels around a tall building is almost impossible because of the difficulties in instrumentation. Accordingly, the experimental techniques on actual winds including the improvements of instruments must be developed. In order to understand the complicated wind characteristics in urban areas which are crowded with various buildings and structures, a lot of wind data should be accumulated.

At present the conclusions of most studies on environmental winds are drawn only from individual experimental results. Further studies are needed in order to obtain the conclusions usefully generalized for applications. These studies are essential to the clarification of the diffusion problems on air pollution and to the determination of wind loading on structures. They are relevant to the safety rather than the comfort of human beings.

GENERAL RULES FOR THE DETERMINATION OF WIND ENVIRONMENT

W. J. BERANEK

Institute TNO for Building Materials and Building Structures (IBBC), Delft, The Netherlands

SUMMARY

Sand erosion tests and flow visualization tests have been carried out on a large amount of rectangular buildings. Global rules have been formulated to determine the wind environment. Data are supplied to determine the influence area around a building and to obtain numerical values for a supplementary criterion: the so-called wind environment area. An example has been worked out and some test results of the interaction between buildings are given.

1. INTRODUCTION

For architects and town-planners some practical rules for the determination of wind environment would be most helpful. With the help of these rules numerical values should be obtained for the improvement or worsening of the environment, due to changes in the built-up area. To obtain more insight into these phenomena, extensive windtunnel tests have been carried out in the small boundary-layer wind-tunnel of IBBC-TNO, on behalf of the Foundation Building Research (SBR).

A new criterion was developed, based on allowable 3s-gustspeeds with stated frequency of occurrence [1,3]. To obtain full-field information, two overall measuring techniques were employed [1,2]. The sand-erosion technique provides a pattern of areas where the discomfort equals or exceeds given values. The sand-erosion proves to be mainly dependent on the maximum gustspeed, and partly dependent on some other factors, which are also of great importance for the wind environment. So a discomfort parameter γ was defined [2], which should be considered primarily as a magnificationfactor for the gustspeed. The usual flow-visualization technique (oil-film) provides data about the average wind direction on walking level. An optical dynamometer was employed to determine magnitude and direction of the gustspeed in points of special interest [1,2].

From numerous tests a number of global rules was detected, which were kept as simple as possible. These will give the architect and building engineer a practical device to judge the environment without the necessity of using windtunnel tests for each occasion. This paper describes the results which were obtained with single rectangular buildings [4]. Tests about the interaction of buildings are in full operation, a few examples are given.

Extensive results are supplied in the SBR-publication [4] (in Dutch). The first volume consist of a theoretical part and a practical part with an enclosure of fourty pages discomfort maps and streamline patterns regarding rectangular buildings. The phenomena are explained and sufficient numerical data are supplied to determine magnitude and position of the discomfort areas (English translation available). In the second volume the interaction between buildings will be treated.

2. TESTS ON SINGLE BUILDINGS

An important part of the investigation has been concentrated on tests of single rectangular buildings for several reasons:

a. Most buildings have a rectangular shape.
b. The wind profile and turbulence intensity are caused by the vegetation and built-up area all together, but around each building the airflow is mainly influenced by the building itself. So first of all the phenomena around a single building should be fully understood, before the interaction of buildings is investigated.
c. In many cases high buildings are erected far apart from other buildings or in the vicinity of much lower buildings. The influence of the high building is then predominant and the worsening effect of the high building is practically the same as if the lower buildings would not be present. An example is given in fig. 1 where a tower-shaped building is situated amidst a regular configuration of much lower buildings.
d. In a lot of cases the interaction of high buildings is of minor importance and the situation can be regarded more or less as a superposition of the situations which would be valid for each of the buildings. A striking example is given in fig. 9, where two buildings in a rectangular configuration were tested. There are important exceptions however, especially due to the occurrence of transverse airstreams between buildings. Examples are given in fig. 10 and 11. All these examples are treated more in detail in paragraph 8.

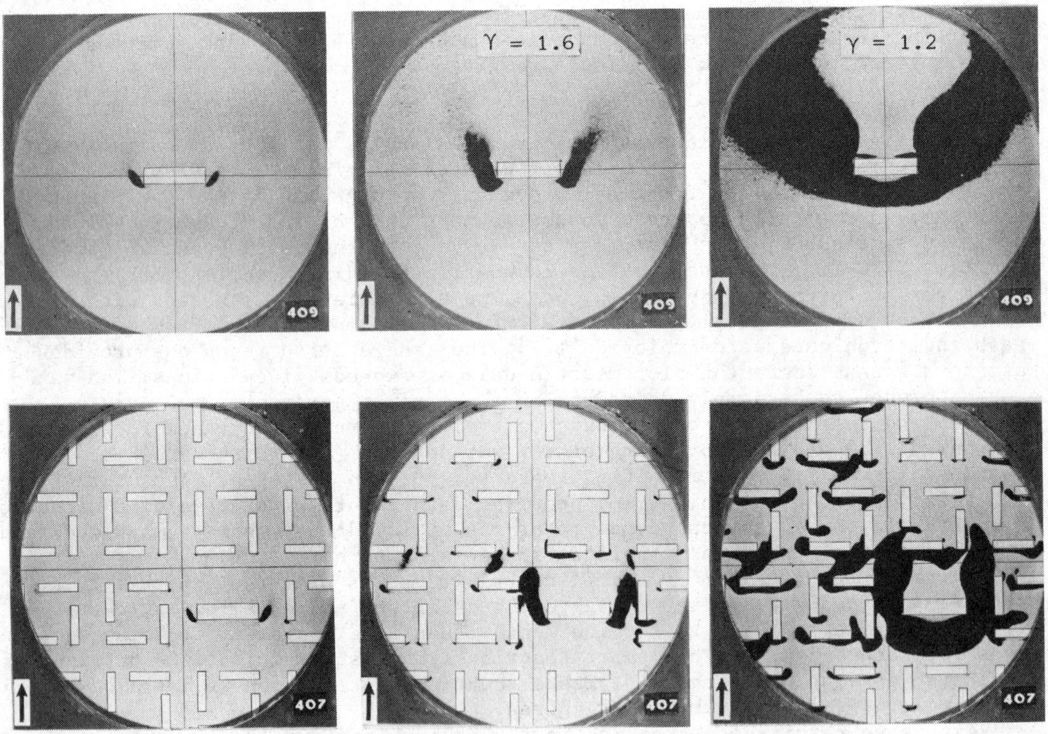

a.

b.

Fig. 1. Discomfort patterns for the values γ = 2.0 - 1.6 - 1.2.
a. Patterns for the high building alone (b × l × h = 20 × 70 × 70 m).
b. Patterns when the high building is surrounded by a regular configuration of much lower buildings (b × l × h = 10 × 50 × 12 m).

For γ = 2.0 and 1.6 the areas in fig. 1a and 1b are practically the same, only for γ = 1.2 the sheltering effect of the lower buildings is clearly visible.

3. COMPLEMENTARY CRITERION FOR WIND ENVIRONMENT

All modern criteria for wind environment are based on allowable windspeeds with stated frequency of occurrence [3]. In most investigations the disturbance due to the wind is given as a dimensionless coefficient. By this coefficient some reference windspeed in the new situation is compared with the reference windspeed at the same point in the previous situation. Sometimes mean windspeeds are compared, sometimes reference windspeeds like $\hat{u} = \bar{u} + 3\sigma$, or $\hat{u} = \bar{u} + \sigma$, as has been done by Gandemer [5,6], introducing the so-called comfort parameter.
In principle, gustspeeds are compared in the sand erosion tests and the discomfort parameter should primarily be considered as a magnification in gustspeed.

All these approaches are more or less similar and the results are in outline comparable. But if measurements are carried out on a point by point basis, one must hope to have applied sufficient measuring points to detect the worst spots. The results of the sand erosion tests have shown that practically always - even with medium high buildings - somewhere a discomfort parameter with magnitude 2 will be reached, sometimes over very small areas. This will be quite obvious if it is borne in mind that even in a laminar two-dimensional airflow around a cylinder the maximum windspeed will be twice the mean value [7].

Tests have shown that the maximum value of the discomfort parameter increases very slowly with increasing height of the building. As is shown in [3] this maximum value can be connected with the criterion if the statistical distribution of the mean hourly windspeed is known. The area however in which the discomfort parameter exceeds given values, increases very rapidly with increasing height of the building. So it was decided that the magnitude of the area in which the factor γ exceeds given values, is of more importance for the environment than the value of γ_{max} itself. As a measure for the environment, the area has been chosen in which the discomfort parameter exceeds the value $\gamma = 1.6$. This magnification of the gustspeed is large enough to be significant and for single buildings this area always belongs to the two jetstreams starting from the windward corners. Furthermore, the regarded areas are large enough to draw some conclusions with confidence.

4. AIRFLOW AROUND RECTANGULAR BUILDINGS

The airflow around rectangular buildings is well-known if the wind is parallel to one of the main axes of the building. Clearly three wind régimes can be distinguished:

a. Tall slender buildings, see fig. 2a.
The airflow is mainly transported alongside the building and reverse airflows on the windward and leeward side of the building occur only over small areas.

b. Tall building of the transition type, see fig. 2b.
The airflow on the windward face of the building is to a great extent transported downwards, causing a strong vortex flow and important corner streams. On the leeward side of the building a large area with reverse airflow occurs.

c. Long buildings, see fig. 2c.
The airflow is mainly transported over the building. The vortex flow on the windward side of the building and the cornerstreams are less important than in the previous case. There is a large area on the leeward side of the building with reverse airstreams. When the wind is incident in an angle between 30^0 and 60^0 with the axes, a vortex is "falling over the building", see fig. 2d.

The stylized airflow patterns of fig. 2 are only dependent on the ratio between the main dimensions of the building (b : l : h), i.e. as long as the velocity profile is of the exponential type: $\bar{u}_1 = (h_1/h_o)^\alpha \bar{u}_o$.
A linear magnification of all the dimensions of the building will also cause a linear magnification of the streamline pattern. The discomfort pattern will show a similar shape as well. However the magnitude of the discomfort parameter will grow with increasing dimensions of the building.

This means that especially the visualization of the airflow pattern on ground-level is suited to determine the area in which the airflow is mainly influenced by the building itself. As stated in [2], this influence area is chosen equal to the horse-shoe shaped pattern which is clearly shown on all flow visualization photographs. In this influence area the turbulence is higher than outside this area and reverse airstreams may occur.

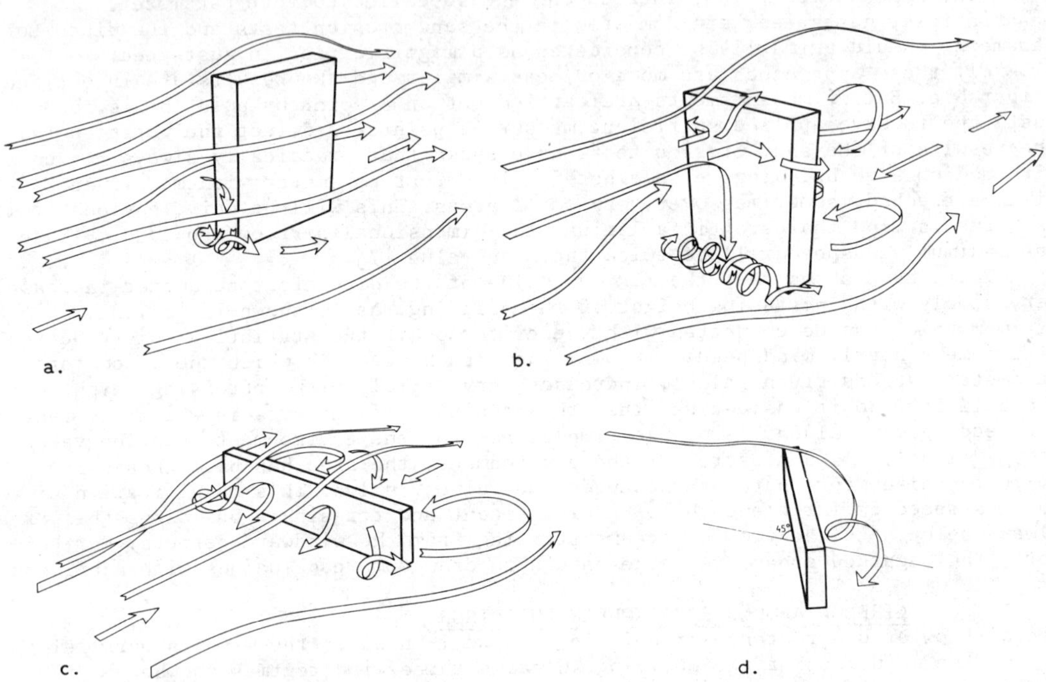

Fig. 2. Airflow pattern around three types of single buildings.
a. Tall slender building. b. Tall building of the transition type
c. Long building. d. Long building, winddirection $\phi = 45^0$.

5. RULES FOR THE DETERMINATION OF THE INFLUENCE AREA

For buildings of the transition type, the influence area can be determined by drawing a circle through the front and back stagnation points [2]. The radius R is practically only dependent on the dimensions of the windward face of the building. The same is valid for the distance e between the center of the circle and the windward face of the building.

$$R = 1.6\sqrt{ah} \qquad (1)$$

$$e = 0.9\sqrt{ah} \qquad (2)$$

These rules are valid for $1.25 > h/a > 0.33$, see fig. 3b.

For tall slender buildings the radius R is hardly growing with increasing height. So if $h > 1.25a$ a fictitious height $h' = 1.25a$ is substituted in formula (1). The value for e proves to be somewhat too high, so in this case another coefficient was chosen, see fig. 3a.

For long buildings the radius R is practically only dependent on the height of the building and the effect of an increasing width may be neglected. So if $a > 3h$, the middle part of the building is neglected and the influence areas are sketched as for two buildings, each with a fictitious width $a' = 3h$, see fig. 3c.

If the dimensions of the building parallel to the wind are rather large, then the influence area has to be adapted somewhat. If the wind is incident under a variable angle with the axes of the building, also some indication of the influence area can be given [4].

Knowledge of the influence area around a building is important for two reasons:

a. If a high building has to be erected in the vicinity of lower buildings, the situation of the new building should be such that the lower buildings are outside the influence area of the new building.
b. When the influence areas of various buildings are overlapping there is a possibility of serious interference between the buildings.

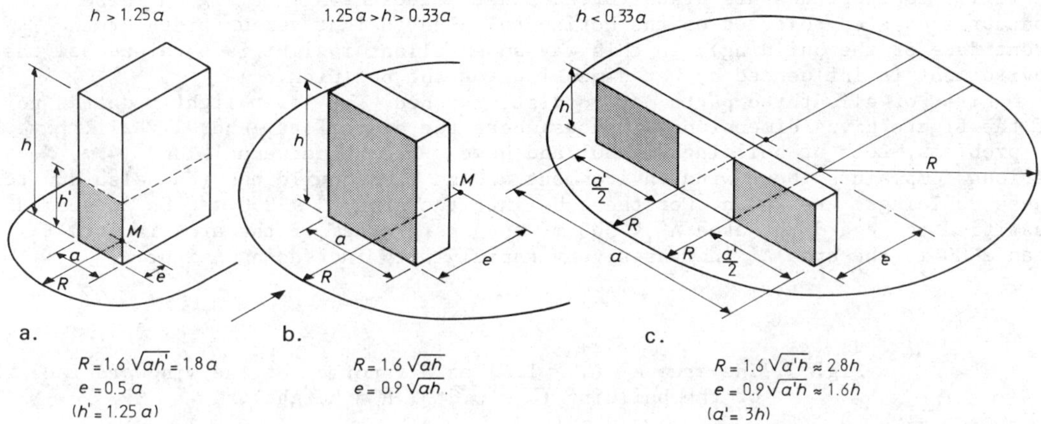

Fig. 3. Determination of the influence area for the three wind régimes.
a. Tall slender building. b. Transition type. c. Long building.
The effective areas of the front faces of the buildings have been shaded.

6. RULES FOR THE DETERMINATION OF THE DISCOMFORT

In fig. 4 for two single buildings the area A is given where the discomfort parameter γ exceeds the values $\gamma = 1.6 - 1.8 - 2.0$. The wind direction has been changed in steps of 15^0. Especially in the case of the tower-shaped building the effect of the "streamline shape" for $\phi \approx 45^0$ is obvious. In the flat-shaped building this effect is hardly recognized. But in both cases the whole situation is rather well defined when the effects are known for wind parallel to the axes.

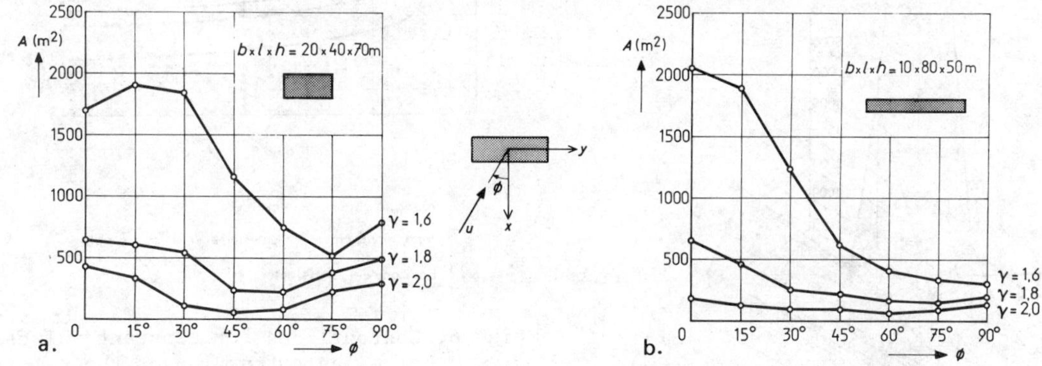

Fig. 4. Magnitude of the area where the discomfort parameter γ exceeds the values $\gamma = 1.6 - 1.8 - 2.0$.
a. Tower shaped building. b. Flat-shaped building.

Considering the areas A_x^* and A_y^* for $\gamma = 1.6$, the areas for other wind directions can roughly be approximated with the formula (see also fig. 7c):

$$A_\phi = \frac{A_x^* + A_y^*}{2} + \frac{A_x^* - A_y^*}{2} \cos 2\phi \tag{3}$$

For a large number of buildings the area A^* has been determined for wind parallel to the axes. It follows from the tests that only the dimensions of the windward face of the building are of importance. The results are shown in fig. 5. It is quite obvious that the area A^* grows very rapidly with increasing height, provided that the width of the building has more or less the same magnitude as the height. In fig. 5 contourlines are drawn for constant values of A^*. In fig. 6 these contourlines are projected on the horizontal plan (height versus width of the front face of the building). In this way an excellent insight is obtained how the environment is influenced by het dimensions of the building.

First of all, three parts may be distinguished. The lower lightly dotted part of the figure gives dimensions of faces where the discomfort generally will give no problems. Lots of buildings in Holland have a height between 12 and 14 m. No serious complaints about wind environment around these buildings (and also due to these buildings) are known. For these heights the area $A^* \simeq 250$ m^2. So it seemed justified to regard the area $A^* < 500$ m^2 as a safe area. If the area is larger than 2000 m^2 the area A^* increases very rapidly with increasing height.

Fig. 5. The area A^* ($\gamma > 1.6$) as a function of the windward face of the building (a = width, h = height).

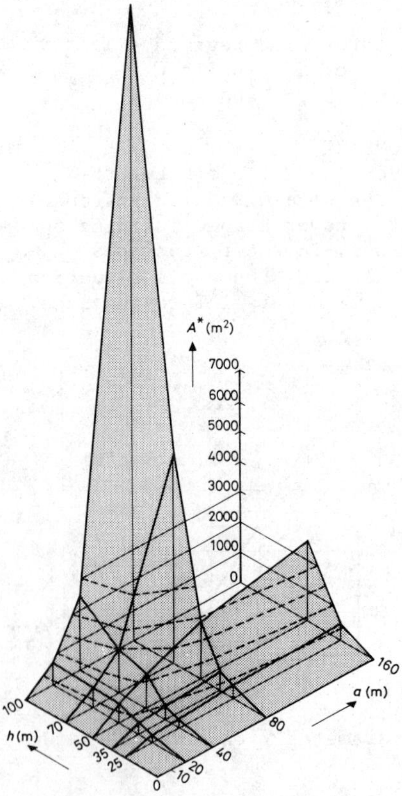

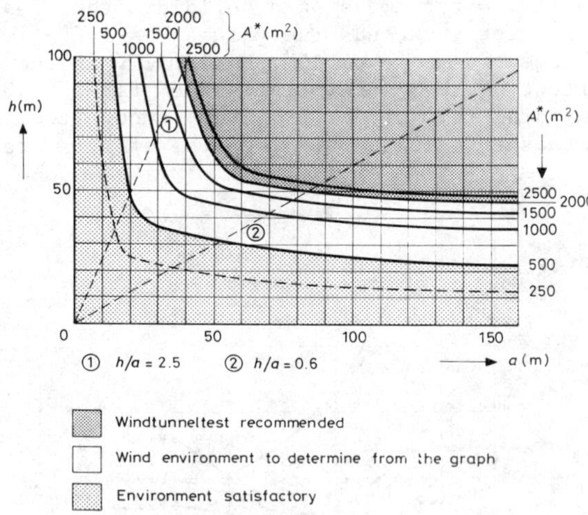

Fig. 6. Contourlines for constant values of A^* as a function of the dimensions of the windward face of the building (a = width, h = height).

For long buildings this means a height of 50 m or more. So if a front of a building belongs to the heavily dotted area, a windtunneltest is generally recommended. The blank area between the two dotted areas contains all the necessary information to judge the wind environment.

The three régimes of the airflow are also recognised in fig. 6. The given boundaries however are not quite the same as used in the determination of the influence area in paragraph 5.

a. Tall slender buildings ($h > 2.5a$), fig. 2a.
   The wind environment is practically independent of an increasing height and depends mainly on the width of the windward face of the building.
b. Tall buildings of the transition type ($2.5a < h < 0.6a$), fig. 2b
   The wind environment depends both on the width and on the height of the windward face of the building.
c. Long buildings ($h < 0.6a$), fig. 2c.
   The wind environment is mainly dependent on the height of the windward face of the building and practically independent of an increasing width.

The just mentioned boundaries are only giving a rough indication where the airflow has changed from one régime to another, but this is hardly a disadvantage as more precise information can be taken from fig. 6.

7. EXAMPLE

To demonstrate the possibilities of a numerical evaluation of the wind environment, a simple example has been worked out in fig. 7 and 8. The dimensions of the building and the situation are used as a starting point, see fig. 7a. For the determination of the influence area the formulas of fig. 3 are used. For wind in the x-direction one finds: $h/a = h/l = 45/75 = 0.6$, so this face belongs to a building of the transition type. $R = 1.6\sqrt{ah} = 1.6\sqrt{75 \times 45} = 93$ m; $e = 0.9\sqrt{ah} = 52$ m. The influence areas for wind from both sides are also shown in fig. 7a; the solid lines belong to the windward faces of the building, the dotted lines to the leeward faces. For wind in the y-direction one finds: $h/a = h/b = 45/15 = 3.0$, so this face belongs to a tall slender building. $R = 1.8a = 1.8 \times 15 = 27$ m; $e = 0.5 \times 15 = 7.5$ m. For the sake of completeness this much smaller area is shown as well in fig. 7a.

For the determination of the wind environment the areas A_x^* and A_y^* will serve as a starting point. In fig. 7b a part of fig. 6 has been repeated and the two faces of the building are sketched in the figure (shaded area). One finds: $A_x^* \simeq 1300$ m^2 and $A_y^* \simeq 300$ m^2. The magnitude A_ϕ^* for other winddirections can be determined with the help of formula (3), see fig. 7c.

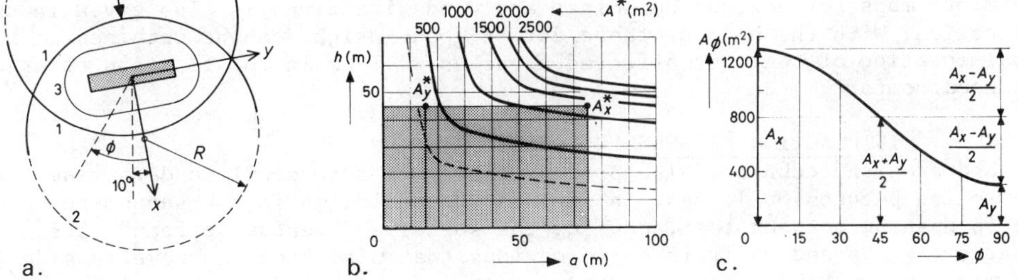

Fig. 7. Determination of the wind environment around a building.
a. Situation of the building ($b \times l \times h = 15 \times 75 \times 45$ m, $\alpha = 10^0$) and position of the influence areas (1 - 2 = wind in x-direction, 3 = wind in y-direction).
b. Determination of A_x^* and A_y^*. c. Determination of A_ϕ^* according to formula (3)

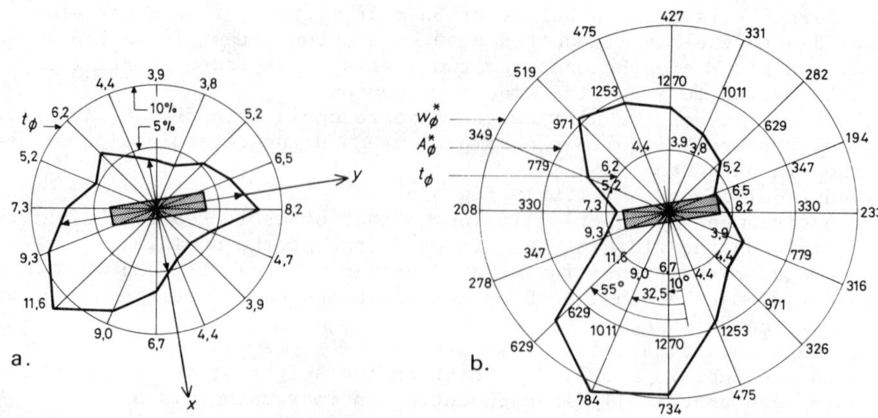

Fig. 8. Determination of the wind climate around a building.
a. Countours of the hourly wind velocity (t_ϕ = percentage of the time that the wind is blowing in the given direction; sectors of 22.5^0).
b. Contours of the wind environment area W^*_ϕ according to formula (4).

In several cases the determination of A^*_x and A^*_y alone will be sufficient to determine the quality of the building in respect to wind environment. The effect of changes in the dimensions of the building are immediately recognised. In other cases however more detailed information might be wanted, as the actual wind environment depends to a large extent on the situation of the building towards the prevailing winds.

In such cases the contours of the hourly wind velocity should be taken into account, taken from a wind registration station in the vicinity, see fig. 8a. The numbers in the figure give an indication of the percentage of the time that the wind is blowing in the given direction. For each sector of 22.5^0 the total time t_ϕ that the wind is blowing in such a direction is compared with the time t_o that the wind is blowing in the prevailing direction. Finally the wind environment is judged by the so-called Wind Environment Area W^* which is defined as:

$$W^*_\phi = (t_\phi/t_o)\, A^*_\phi \tag{4}$$

In this way the properties of the building itself are connected with the properties of the wind itself. The results of the determination of the Wind Environment Area W^* are given in fig. 8b. It is easily seen that the southern winddirections have become more important due to the situation of the building. In the described way influences of changes in the dimensions of the building and changes in the situation are immediately recognised. The general procedure is described more in detail in [4]. A large amount of streamline patterns and discomfort maps for various buildings and wind directions is also given in this publication. With the help of these data a good insight can be obtained in the actual location of the most affected areas as well as in the location of areas with good comfort.

8. INTERACTION BETWEEN BUILDINGS

The sand erosion technique also proves to be very succesfull to determine the interaction between buildings. In fig. 9a two buildings in a V-shape are shown with an opening between them, causing the so-called "venturi effect" [5,6]. If one compares fig. 9a and 9b it is quite obvious that wind from the reverse side is even more unfavourable. The windward corners of the buildings show in both cases the highest discomfort. Comparison with the situation for the similar single building [2] shows that the main effects are only slightly changed. This means that the pattern can more or less be regarded as a "superposition" of the patterns for the two single buildings.

In the figures 10 and 11 two situations are shown where the buildings are parallel. Now for various winddirections the "superposition" of the basic cases is no longer valid. In fig. 10a two buildings are more or less in the "wind shadow" of the windward building. For $\phi = 45^0$ the discomfort between and behind the buildings has increased considerably, see fig. 10b. Only when the wind is parallel with the length of the buildings there is no interference at all, see fig. 10c. In fig. 11 the three blocks are shifted towards each other. Now the worst effects occur when the wind is perpendicular to the length of the buildings, see fig. 11a. The streamline pattern in fig. 11b gives a good indication of the variation in wind direction, which might hardly be expected by pedestrians. The situation is improving for $\phi = 45^0$, see fig. 11c. Again there is no interaction at all for wind parallel to the length of the buildings.

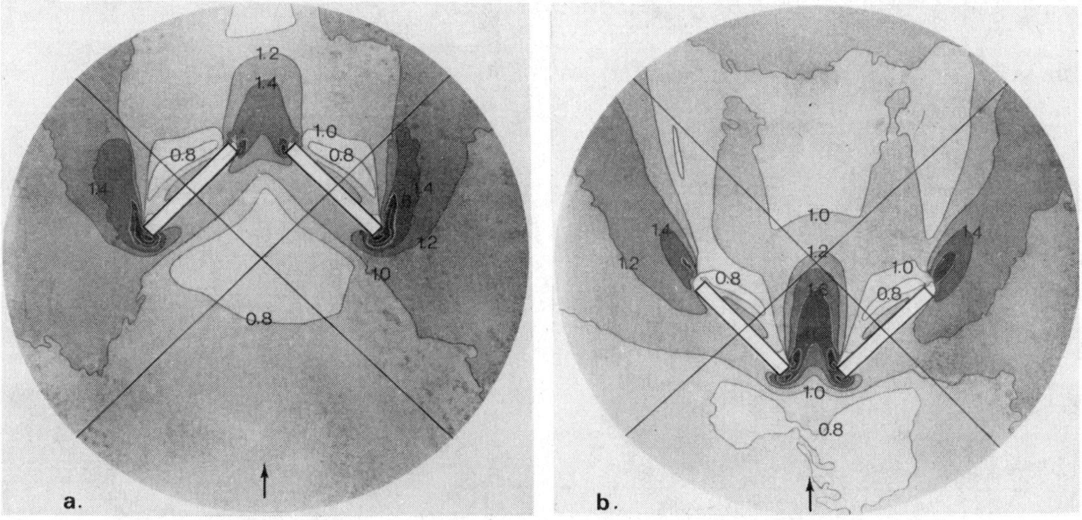

Fig. 9. Two buildings (10 × 80 × 50 m) with an opening between them.
a. The wind causes the "venturi effect".
b. The wind is blowing in the opposite direction.

REFERENCES

[1] Beranek, W.J. and Koten, H. van "Wind environment around buildings" Proceedings of the third U.S. National Conference Wind Engineering Research, Gainesville, 1978, pp. 193-197.

[2] Beranek, W.J. and Koten, H. van "Visual techniques for the determination of wind environment" Proceedings of the third Collequium on Industrial Aerodynamics, Aachen 1978, Part 3, pp. 1-14.

[3] Koten, H. van and Beranek, W.J. "Criteria for windspeeds around buildings (values and applications)" Proceedings of the third Colloquium on Industrial Aerodynamics, Aachen 1978, Part 3, pp. 15-22.

[4] Beranek, W.J. and Koten, H. van "Beperken van windhinder rond gebouwen, deel 1" Stichting Bouw Research, 1979.

[5] Gandemer, J. "Wind environment around buildings, aerodynamic concepts" Proceedings of the fourth International Conference Wind Effects on Buildings and Structures, Heathrow, 1975, pp. 423-432.

[6] Gandemer, J. and Guyot, A. "Intégration du phénomene vent dans la conception du milieu bati". 1976 La documentation francaise, Paris.

[7] Prandtl, L. and Tietjens, O.G. "Applied Hydro- and Aeromechanics" Dover Publications New York (103. Superposition of a Parallel Flow and a Circulation Flow).

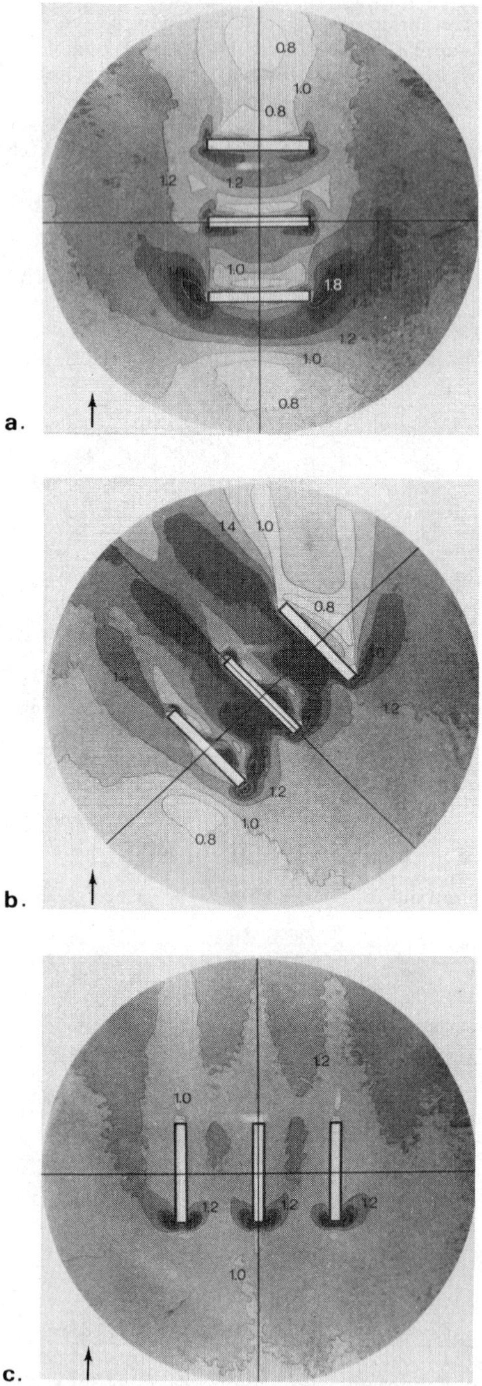

Fig.10. Three parallel buildings in line. Dimensions b×l×h = 10×80×50 m.
a,b,c. Discomfort maps for $\phi = 0^0$-45^0-90^0.

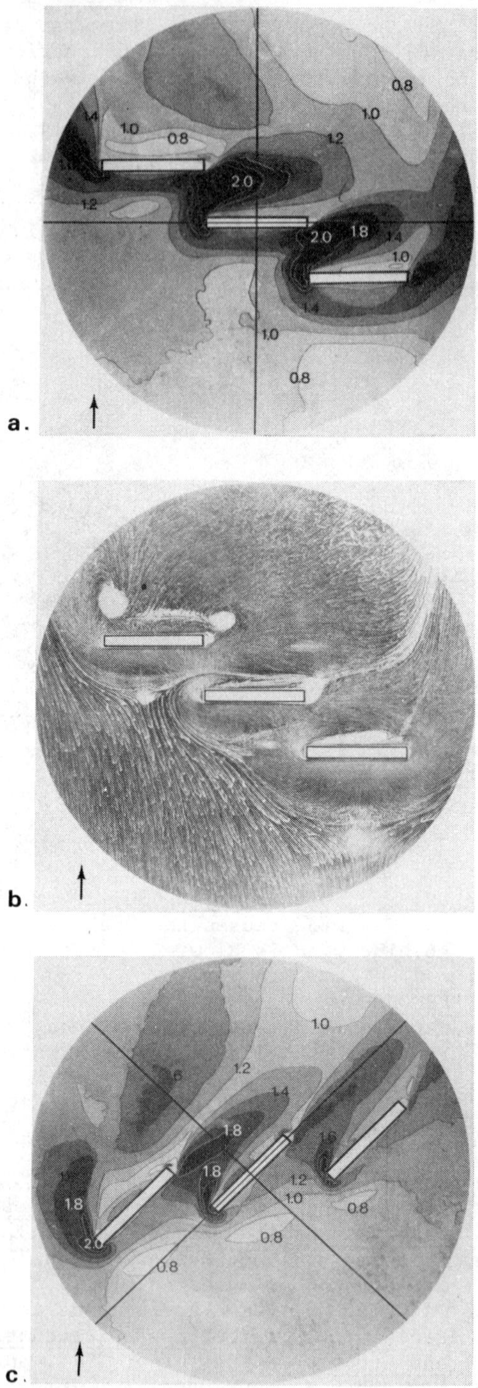

Fig.11. Three parallel buildings, shifted towards each other (10×80×50 m).
a.c. Discomfort maps for $\phi = 0^0$-45^0.
b. Streamline pattern for $\phi = 0^0$.

AN EROSION TECHNIQUE FOR ASSESSING GROUND LEVEL WINDS

A. R. JANEIRO BORGES* and J. A. GIL SARAIVA**

*Universidade Nova de Lisboa, Lisbon, Portugal
**Laboratório Nacional de Engenharia Civil, Lisbon, Portugal

ABSTRACT

Previous research as well as practical evidence has shown that due attention should be given to ground level wind effects produced in the vicinity of tall buildings. An experimental method suitable for use as a wind tunnel technique is presented and some typical results shown. The method is based on the erosion patterns formed by the saltation of fine grains of sand uniformly spread on a smooth ground plate.

SYMBOLS

- b distance from center of nozzle outlet to the plate
- d diameter of sand grains
- D nozzle outlet diameter
- F Froude number for saltation
- g acceleration of gravity
- h height of lower building
- H height of taller building
- L gap between buildings
- r radial distance
- R Reynolds number of roughness
- R_o Reynolds number of jet
- u_τ saltation friction velocity at the contour of erosion pattern
- $u_{\tau o}$ saltation friction velocity upstream
- T depth of buildings
- U mean velocity at level z
- U_H mean velocity at level H
- W width of buildings
- x streamwise distance from nozzle outlet
- ρ air density
- ρ_s sand density
- ν air viscosity

INTRODUCTION

High rise construction made common in the past twenty years due to the general tendency towards demographic concentration and the concomitant development and acceptance of new concepts and criteria for urban planning strongly changed the profile of big cities and smaller towns. In the process, in many cases, the presence of tall buildings has created inhospitable, inconvenient and even dangerous ground level wind situations. As a result questions related to the prediction of wind environment in the vicinity of tall buildings evolved to its present status as a major design consideration.

The problem is mainly one of interaction between adjacent buildings whose relative sizes and group configuration are wildely variable. The inherent complexity of the flow fields makes wind tunnel investigations the only available method of precise and detailed study of such problems. The experimental methodology to use should no doubt provide detailed description and precision of the measurements but combined with promptitude of the results and overall versatility because the typical remedy for a poor situation usually entails considerable changing of the model.

Most frequentely the wind tunnel engineer is faced with the primary but no lesser problems of convincing the architect and the planner that their designs or plans, although satisfying the general specifications, are defective from an environmental point of view and the alterations suggested merit to be considered. In the course of so doing no argument is more persuasive than some form of visualization of the flow details under discussion.

The present paper introduces an erosion wind tunnel technique that can be used for the assessment of ground level winds in urban areas.

The technique is illustrated by a short presentation of two studies: One is related to the interaction of a turbulent circular jet with a flat plate and the other is the typical troublesome situation of a tall building erected parallel and in the vicinity of a low building, proviously described by Wise (1) and others.

THE EROSION TECHNIQUE

When the wind blows over a layer of loose sand spread on a horizontal, solid, flat surface a relatively well defined velocity is required to initiate saltation of the grains. If they are almost uniform in size and are small enough for the Reynolds number

$$R = \frac{u_\tau d}{\nu}, \tag{1}$$

based on average grain diameter d and friction velocity u_τ, to have a value under about 5 the sand layer deposited on the surface will be thinner than the viscous sub-layer of the flow and the solid boundary will behave as aerodynamically smooth. Under these conditions we expect the saltation velocity threshold not to depend on such characteristics of the flow as velocity profile and boundary layer thickness.

Hence, for grains of a given shape, spread with a certain concentration (weight of sand per unit area) on the same surface, the saltation Froude number

$$F = \frac{\rho u_\tau^2}{\rho_s gd} \tag{2}$$

will be a function of R.

Furthermore, since according to Owen (2) and others during saltation the trajectories of the grains initiate on the overage vertically, a strong dependence of this phenomenon on the exact properties of the supporting surface (smoothness, angle of friction, etc.) is not to be expected.

For sufficiently high free-stream velocity U saltation takes place. For tridimensional flow, after keeping U constant a certain time, a definite scour pattern is formed. The contours of the picture so obtained are lines where the friction velocity assumes the threshold value for saltation. By varying the velocity of the flow one can easily obtain a family of such lines. Each member of the family is related to a particular value of the free-stream velocity and the same same value of u_τ.

The form of the above mentioned relation between F and R can be deduced from the experimental data shown in fig. 1. It was obtained using sieved sea-beach sand of six different sizes .15 mm (0.059 in) < d < .60 mm (.023 in) sparsely spread (superficial concentration varying from 5 mg/cm^2 (.5 gr/in^2) to 9 mg/cm^2 (.9 gr/in^2) on a plate of "Laminite" of the type currently used for the manufacture of blackboards. The flow was produced by a turbulent circular jet ($3,5 \times 10^3 < R_o < 10^4$) impinging at right angles to the plate or being parallel to it.

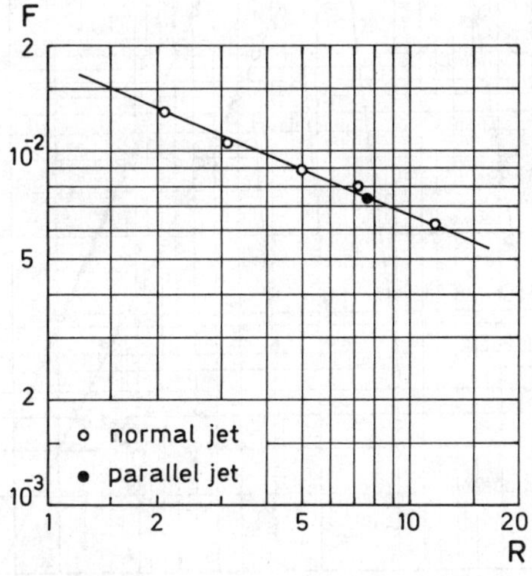

Fig. 1 - Relationship between the Froude number and the Reynolds number

Sample results of this investigation, to be published later on elsewhere (3), are presented on fig. 2. They were obtained by measuring the diameter (normal jet) or the lenght along the longitudinal direction (parallel jet) of the scour patterns formed. Present results agree quite well with the available data obtained by other workers (5,6,7) using completely different techniques. The fitting so obtained made it possible for us to determine the absolute values of u_τ and the points on fig. 1. Within the range of the present tests the relationship between F and R is:

$$F = 0.0180 \quad R^{-0,432} \tag{3}$$

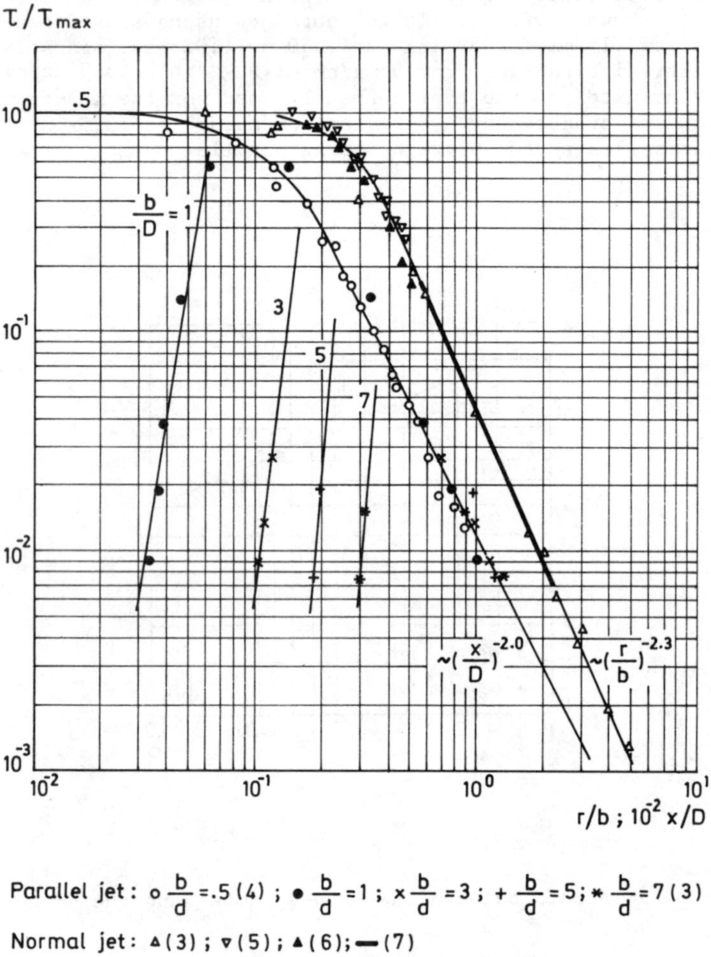

Fig. 2 - Wall shear stresses produced by a circular jet interacting with a flat plate.

GROUND LEVEL WINDS

The erosion technique described above is currently used in our laboratory as a straightforward means of predicting the ocurrence of high winds at ground level in complex urban situations. However, for the sake of simplicity, only some results in reference to the classical building arrangement discussed by Wise (1) and others will be briefly presented. The model arrangement and dimensions are shown in fig. 3. As can be seen the model buildings are prismatic, have the same width (W) and were fixed in position on the floor of the wind tunnel working section, the taller model behind the lower one and both facing the wind at right angles. The wind tunnel used is a closed return, closed working section 1x1.25 m^2 3 m long, fully described in (8). The shear of the wind aproaching the model is typical of open country sites. Also shown in the figure is the velocity profile at mid distance between the models in the plane of symmetry of the flow. This clearly shows the vertical development and the ground level region of high velocity reversed flow associated with the horse-shoe shaped strong vortex at the base of the tall building.

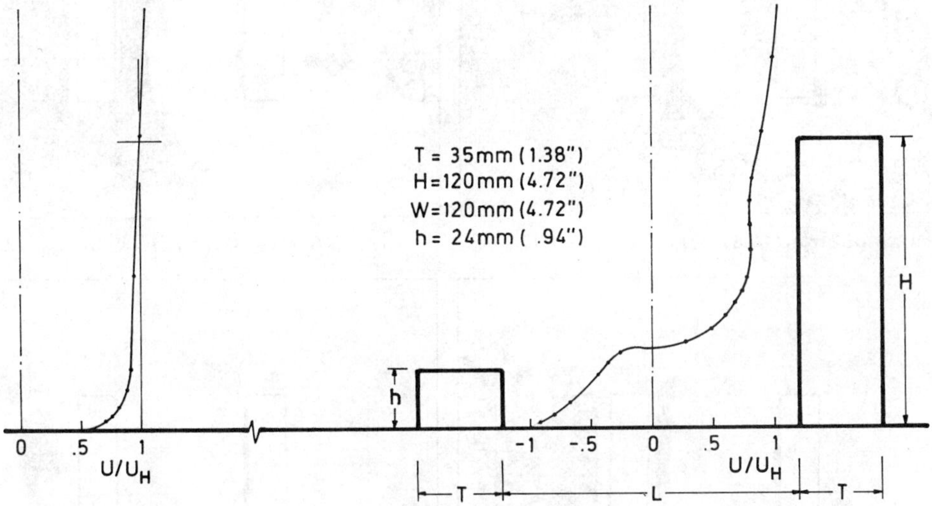

Fig. 3 - Model dimensions and velocity profiles.

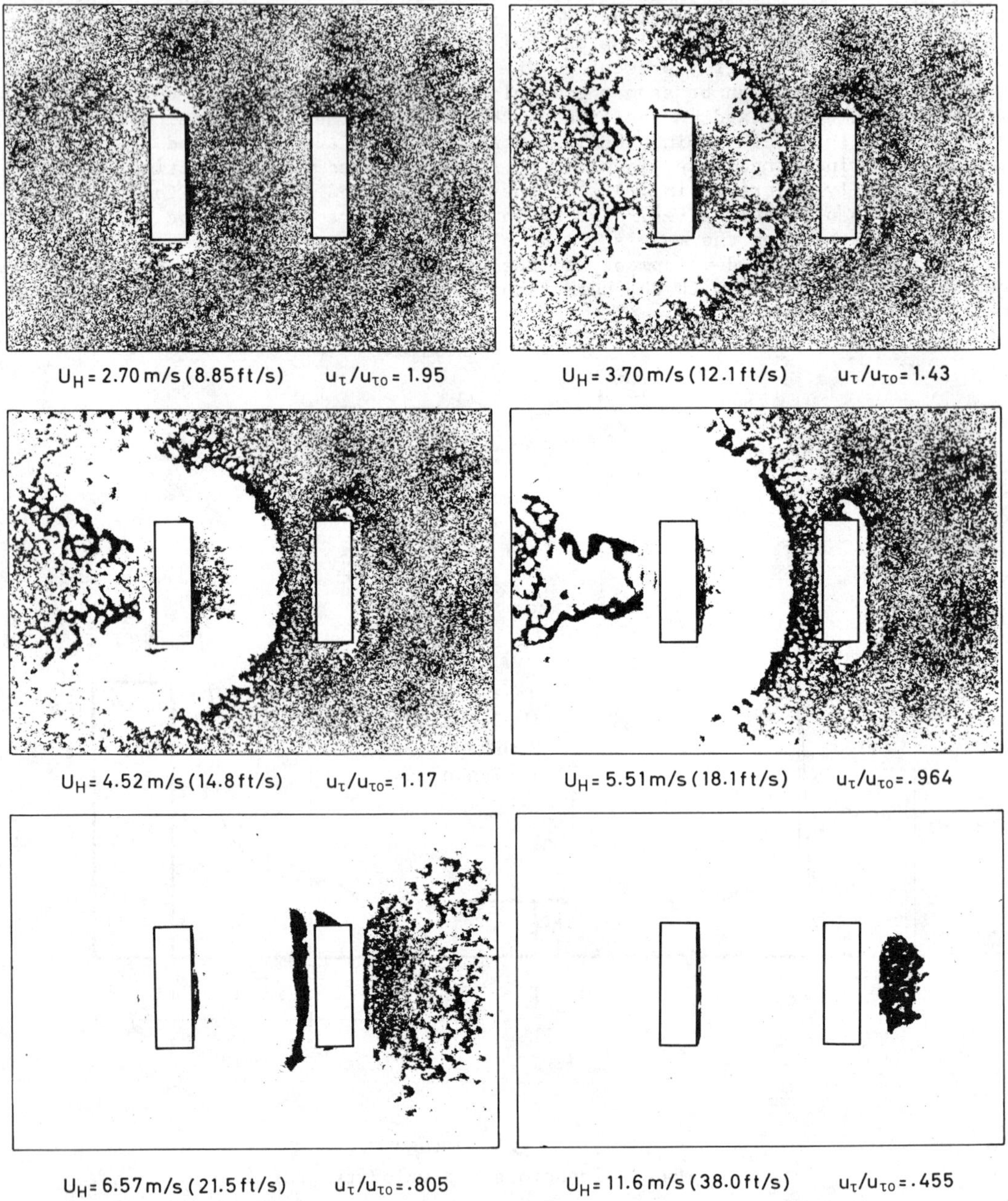

Fig. 4 - Photographic sequence of erosion pictures

Fig. 4 is a photographic sequence of scour patterns obtained for different wind speeds. The reference velocity U_H is taken upstream at height H. The sand used had an average grain size of d = .40 mm (.016 in), 2.62 specific gravity and an initial superficial concentration of about 8.6 mg/cm^2 (.86 gr/in^2).

The scouring action starts close to the front corners of the tall building where the friction velocity u_τ is about twice as high as the upstream friction velocity $u_{\tau o}$. The erosion develops extensively around the taller building far before it starts at the proximity of the lower one. In front of this building there is a region of quite considerable calm.

Fig. 5 is a detailed mapping of friction velocities around the buildings referred to their upstream value. It is based on a photographic sequence of sixteen pictures including those shown in the preceding figure.

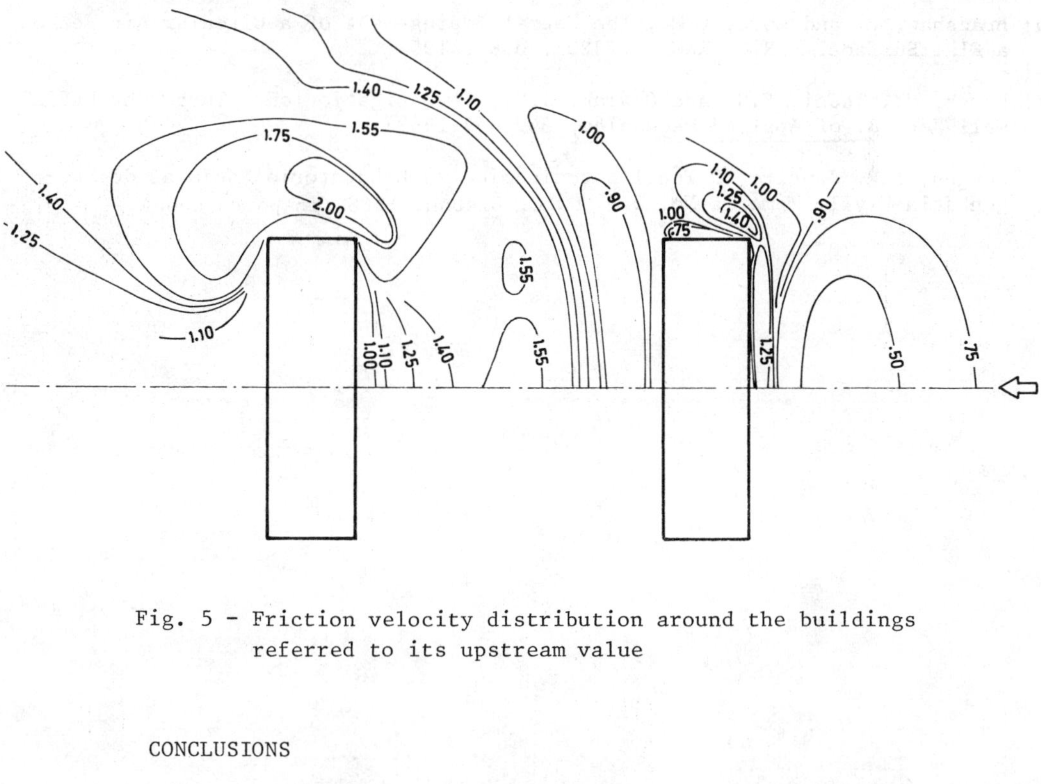

Fig. 5 - Friction velocity distribution around the buildings referred to its upstream value

CONCLUSIONS

A simple expeditious experimental method for determining the friction velocity distribution on the ground around buildings was briefly presented.

The method associates quickness and sufficient accuracy for most practical purposes. The visualization of the flow at ground level offered by the saltation process is thought to be a major advantage. If complemented with a few velocity measurements above important points, for instance with a Pitot-static tube, a complete and comprehensive assessment of the flow at low-levels is obtained.

Further study will be needed to fully understand the mechanism of saltation in the transition from smooth to rough surface regime, namely the exact significance of equation (3).

REFERENCES

(1) Wise, A.F.E., Wind Effects due to Groups of Buildings, CP 23/70, Building Research Station, Watford, U.K., Jul. 1970.

(2) Owen, P.R., Saltation of Uniform Grains in Air, J. Fluid Mechanics, 20, 225 (1964)

(3) Borges, A.R. Janeiro and Viegas, D.X., to be published.

(4) Rajaratnam, N. and Pani, B.S., Three Dimensional Turbulent Wall Jets, J. Hyd. Div., Proc. ASCE, 100, HY1, 69(1974)

(5) Beltaos, S. and Rajaratnam, N., Circular Turbulent Jets, J. Hyd. Div., Proc. ASCE, 100, HY 10, 1313(1974)

(6) Bradshaw, P. and Love, E.M., The Normal Impingement of a Circular Air Jet on a Flat Surface, A.R.C, R&M No. 3205, U.K., 1959

(7) Poreh, M., Tsuei, Y.G. and Cermak, J.E., Investigation of a Turbulent Radial Wall Jet, J. of Applied Mechanics, 34, 457(1967)

(8) Borges, A.R. Janeiro, O Túnel Aerodinâmico do Laboratório Nacional de Engenharia Civil, Memória No. 319, LNEC, Lisbon, 1968 (in portuguese).

WIND-GENERATED NATURAL VENTILATION OF HOUSING FOR THERMAL COMFORT IN HOT HUMID CLIMATES

R. M. AYNSLEY

University of Sydney, Sydney, Australia

ABSTRACT

This paper investigates a rational method of utilizing recent improvements in wind tunnel techniques and meteorological data, to estimate potential wind-generated airflow through housing in hot humid climates. The method uses mean pressure differences obtained from solid wind tunnel models, together with appropriate discharge coefficients for rectangular openings. Changes in pressure distributions, due to typical openings through models, indicated that the use of pressure data from solid models results in errors similar to those associated with the local wind data. Pressure distribution characteristics are provided for six model types, both isolated and in two parallel rows, for a range of wind incidences.

Orientation of long walls of houses to the north and south is recommended to avoid sun control devices covering wall openings. A simple graph provides estimates of airspeed necessary to restore thermal comfort inside well insulated, and naturally ventilated houses in hot humid climates, using input from readily available climatic data.

INTRODUCTION

Natural ventilation has been a traditional feature of much of the housing built in hot humid tropical regions. Even today many residents in such regions prefer natural ventilation to air conditioning or mechanical ventilation systems which can be expensive to operate and are often difficult to maintain in remote locations. Where mechanical equipment is used there is a preference toward simpler units such as low speed ceiling fans or evaporative coolers (1).

Almost three million square kilometres of the Australian continent, 39%, lie within the tropics, with major cities located in hot humid areas along the coastline. Design of housing in these cities demands careful attention to thermal characteristics of the building envelope, sun control, as well as provision of adequate air conditioning or air movement for comfort.

In contrast with the common use of long-established techniques for estimating thermal characteristics, sun shading and air conditioning loads, there are no established techniques in use for estimating potential natural airflow through tropical housing. This paper investigates rational methods of utilizing recent improvements in wind tunnel techniques and meteorological data, to estimate potential wind-gener-

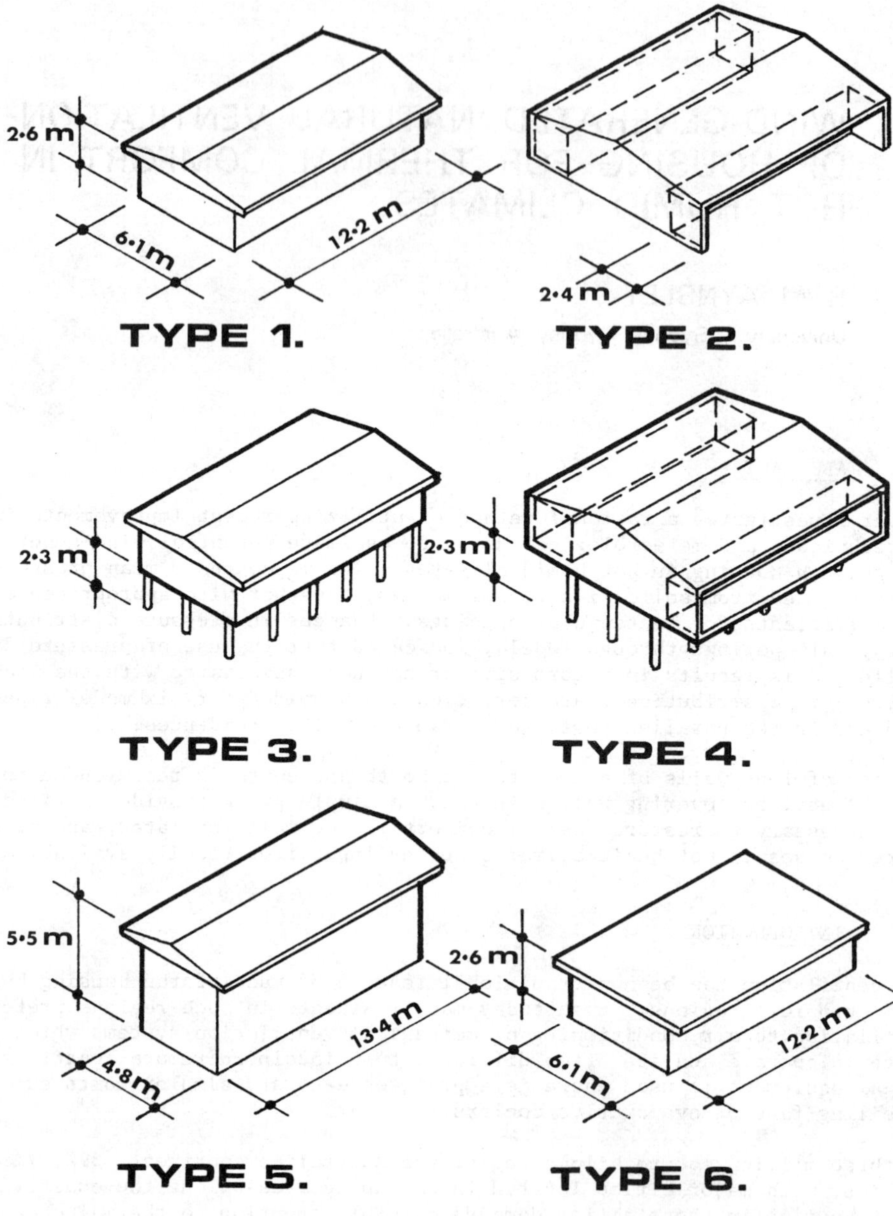

Figure 1. Solid model types used in mid-wall height pressure distribution studies

ated airflow through tropical housing.

Outline of Wind Tunnel Wind Pressure Distribution Studies

Initial wind tunnel studies investigated pressure distributions on the walls of three dimensional solid models both singly and in groups. Six model types were used representing symmetrical housing forms, rectangular in plan, common to Australia's hot humid tropics: single storey on-grade models; elevated single storey models with and without extended eaves and end walls; and, a two storey model. All models had low pitched gable roofs, except one on-grade model which had a flat roof (Fig. 1).

The wind tunnel used was of open circuit blower design with test section 450 mm in width by 300 mm high at its outlet. The lower portion of the mean windspeed profile was modelled with an exponent of 0.28 and a gradient height of 396 metres at a scale of 1 : 150. Maximum model blockage was limited to 2.4% of the test section. Mean windspeed at a height equivalent to 10 metres above ground was 7.8 m/s. All pressure coefficients were referenced to the dynamic pressure upwind of the models at this height to allow direct use of local 10 m meteorological mean windspeed data.

Basis for Pressure Measurements at Mid-Height of Walls

In low, residential buildings most openings for natural ventilation are near the mid-height of walls. A study of the vertical distribution of wind pressure, at a number of points on walls of a range of model buildings, with projecting eaves at various incidences to the wind, indicated wind pressure at a mid-height point on a wall provided a good estimate of the average of wind pressures on a vertical line through that point, over the full height of the wall. Where wind pressures increased near the top of a wall they were generally offset by similar decreases in pressures near the base of the wall. The reduction in the number of measurements achieved by restricting measurements to mid-wall height observations allowed more types and arrangements of models to be studied in the time available.

Influence of Architectural Features on Isolated Solid Model Pressure Distributions

Architectural features such as projecting eaves, projecting end walls and balconies, as well as elevation of single storey buildings raised above ground level, were found to have a significant effect on the magnitude and distribution of wind pressures on wall surfaces of solid, isolated model buildings in wind tunnel tests.

Comparison of average pressure difference coefficients between the mid-height of the long walls on six types of isolated solid model houses with wind at normal, 30^0 and 45^0 incidence to a long wall is indicated in Table 1.

Orientation of long walls of houses to the north to minimize solar heat gain and obstruction to airflow by sun control devices at locations along the north Queensland coast, often results in a prevailing summer wind incidence of 45^0 to a long wall (Fig. 2).

Ranking by average pressure difference coefficient between the long walls of the six model types at 45^0 wind incidence to a long wall is indicated in Table 2.

Generally the average pressure difference between the long walls of solid isolated models tested (indicating average potential for airflow between long walls) was maximum with wind incidence normal to a long wall. The exception was the single storey on-grade model type 2 with extended eaves and end walls. On this model the average pressure difference coefficient between the long walls was greatest (0.71) with wind at inclined incidence of 30^0 to a long wall (see Table 1).

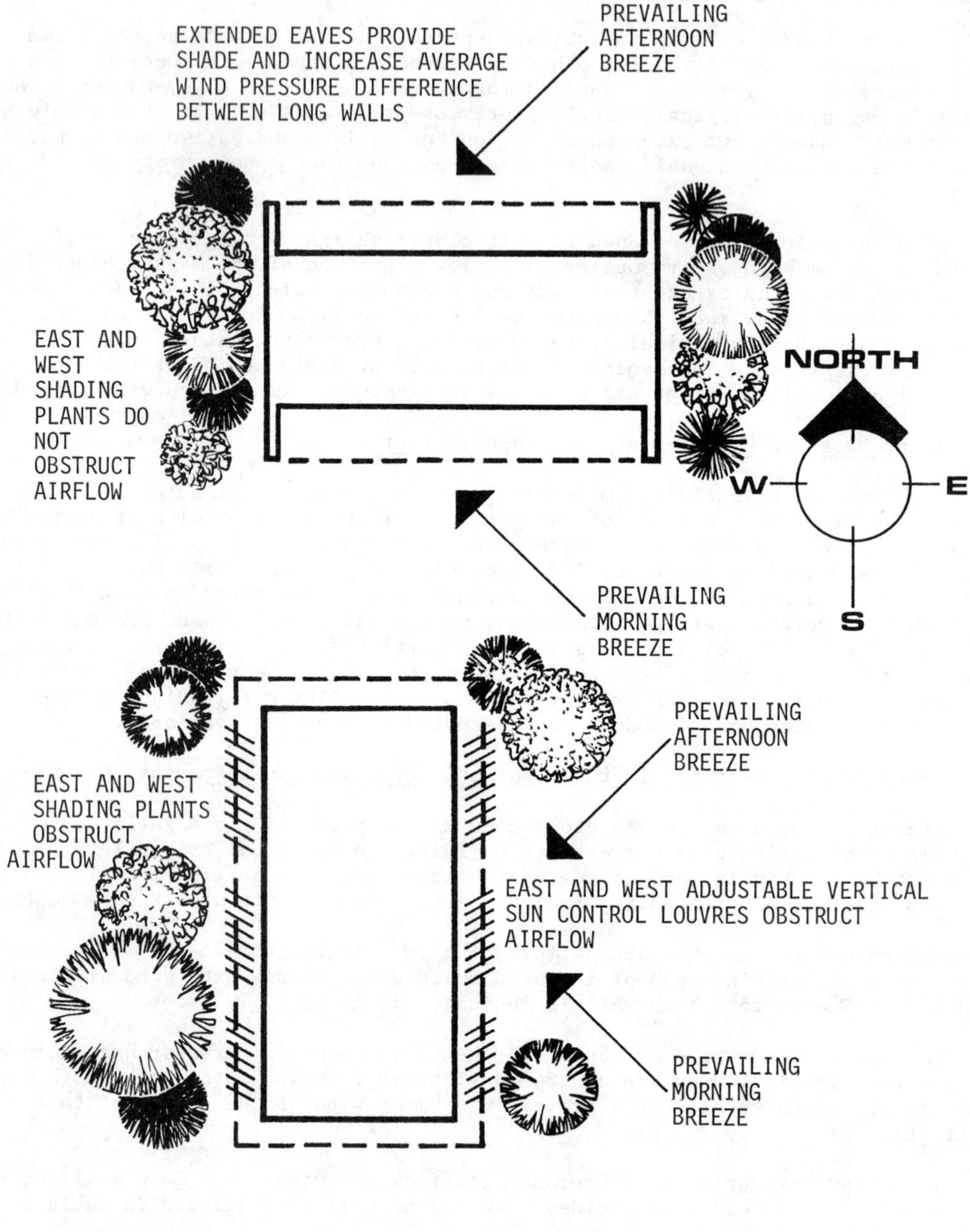

Figure 2. Orientation of long walls to minimise obstruction of openings by shading devices

TABLE 1. Summary of Wind Pressure Difference Characteristics at Mid-Height of Walls on Six Types of Isolated Models at Wind Incidences, Normal, 30^0 and 45^0 to Long Walls

Model Features	Model Type	Overall Ranking Based on Average Pressure Diff. Coefft. Between Long Walls	Incidence to Long Wall	Average Pressure Diff. Coefft. Between Long Walls	Maximum Pressure Diff. Coefft. Between Opposite Points on Long Walls	Minimum Pressure Diff. Coefft. Between Opposite Points on Long Walls
Single storey on-grade	1	Worst	0°	0.49	0.56 C	0.42 BE
			30°	0.41	0.67 WE	0.10 LE
			45°	0.34	0.71 WE	-0.02 LE
			Average	0.41		
Single storey on-grade with extended eaves and end walls	2	2nd best	0°	0.51	0.56 C	0.46 BE
			30°	0.66	0.71 WE	0.58 LE
			45°	0.58	0.63 C	0.52 C
			Average	0.58		
Single storey elevated above ground	3	3rd best	0°	0.65	0.71 C	0.60 BE
			30°	0.58	0.79 WE	0.33 LE
			45°	0.43	0.81 WE	0.00 LE
			Average	0.55		
Single storey elevated above ground with extended verandahs and end walls	4	Best	0°	0.81	0.85 BE	0.77 C
			30°	0.67	0.75 LE	0.50 WE
			45°	0.75	0.81 LE	0.73 BE
			Average	0.74		
Two storey	5	5th best	0°	0.61	0.69 C	0.52 BE
			30°	0.33	0.37 C	0.24 WE
			45°	0.31	0.42 C	0.12 LE
			Average	0.42		
Single storey on-grade with flat roof	6	4th best	0°	0.51	0.58 C	0.44 BE
			30°	0.42	0.69 WE	0.08 LE
			45°	0.35	0.71 WE	0.00 LE
			Average	0.43		

C = centre one third of long walls
BE = both ends of long walls
WE = windward ends of long walls
LE = leeward ends of long walls

TABLE 2. Wind at 45^0 Incidence to a Long Wall

Model Features	Average Pressure Diff. Coefft. Between Mid-height of Long Walls for Wind at 45^0 Incidence to a Long Wall (Ref. Press.-Dynamic 10 m above Ground in Wooded Suburban Terrain)
Single storey elevated above ground with extended eaves and end walls (type 4) *	0.75
Single storey on-grade with extended eaves and end walls (type 2) *	0.58
Single storey elevated above ground (type 3) *	0.43
Single storey on-grade with flat roof (type 6)	0.35
Single storey on-grade with 10^0 pitch roof (type 1)	0.34
Two storey (type 5)	0.31
Note that the three highest ranking model types *, 4, 2 and 3 at 45^0 incidence were also highest ranking averaged over wind incidences of 0^0, 30^0 and 45^0.	

Maximum pressure difference between opposite points on the long walls of solid isolated models (indicating potential for airflow at a particular location between the long walls) was generally greater at inclined wind incidence than normal incidence to a long wall. The exceptions were the two storey model (type 5) and the single storey elevated above ground with extended verandahs and walls (type 4), on which maximum pressure difference coefficient between opposite points on long walls occurred at wind incidence normal to a long wall.

Influence of Building Groups on Wind Pressure Distributions on Walls

As the principal arrangement of houses in suburban development is in rows along roadways, a series of tests were performed to determine the effects of combinations of downwind and crosswind spacing on pressure distributions. The practice of erecting houses at the minimum legal set back from the street boundary required by local government authorities is the main influence on the establishment of such a pattern. This arrangement together with minimum legal allotment size results in spacings between the rear walls of houses in the order of 30.5 - 36.6 m (100 - 120 ft.). The upwind terrain roughness and its associated mean windspeed profile in the test section accounted for upwind buildings.

Spacings between walls facing the roadway depend on the width of the streets which result in typical wall spacing from 24.4 to 32.3 m (80 - 106 ft.). Spacing of walls parallel to side boundaries depends on the width of allotments and side set backs required by authorities or provision for vehicular access to the rear of the site. Minimum spacing would be 3.0 m (10 ft.) as required by local government regulations with a typical maximum of around 9.1 m (30 ft.).

In practice, crosswind and downwind spacings would be somewhat random. Because of the considerable time required to perform wind tunnel studies, spacings were rationalised to a series of equal downwind and crosswind spacings.

Pressure distribution studies of solid model on-grade or 'low set' type 1, elevated or 'high set' type 3 and two storey type 5, in parallel rows with wind normal to a long wall, with 3.8 m (12.5 ft.) crosswind spacing and 34.3 m (112.5 ft.) downwind spacings similar to typical subdivisions with 20 m (1 chain) width roadways (Table 3), indicated high set and low set models had similar overall average pressure differences in windward and leeward rows.

TABLE 3. Wind Pressure Differences at Normal Incidence

Model Type	Average Pressure Difference Coefft. Between Long Walls of Windward Models	Average Pressure Difference Coefft. Between Long Walls of Leeward Models
Low set (type 1)	0.58	0.45
High set (type 3)	0.60	0.43
Two storey (type 5)	0.68	0.29

These average pressure difference coefficients between the long walls indicate that the average potentials for airflow between long walls on low set models, type 1, and high set models, type 3, were similar in both windward and leeward rows at this typical suburban spacing with wind normal to long walls.

Although the two storey models, type 5, had greater average potential for airflow between long walls than low set and high set model, types 1 and 3, in the windward row, average potential for airflow between long walls on two storey models, type 5, in the leeward row was much lower than corresponding potential on model types 1 and 3.

TABLE 4. Wind Pressure Differences at 30^0 and 45^0 Wind Incidence

Model Type	Crosswind Spacing	Downwind Spacing	Average Pressure Diff. Coefft. Between Mid-height of Long Walls of Windward Models		Average Pressure Diff. Coefft. Between Mid-height of Long Walls of Leeward Models	
			30^0	45^0	30^0	45^0
High Set (type 3)	5.7 m (18.75 ft)	26.7 m (87.5 ft)	0.63	0.53	0.48	0.40
Low Set (type 1)	5.7 m (18.75 ft)	26.7 m (87.5 ft)	0.30	0.20	0.46	0.45
Two Storey (type 5)	7.6 m (25.0 ft)	26.7 m (87.5 ft)	0.38	0.26	0.32	0.29

Pressure Distributions on Solid Models in Two Parallel Rows with Wind Incidence of 45^0 to Long Walls

Average pressure difference coefficients between mid-height of long walls for model types 1, 3 and 5 at wind incidences of 30^0 and 45^0 to the long walls of models at a typical downwind spacing of 26.7 m (87.5 ft.) are indicated in Table 4.

The effects of increasing wind incidence to long walls from 0^0 to 45^0 on average pressure difference coefficient between the long walls varied with each model type and windward or leeward positions.

Average pressure difference coefficients between the mid-height of long walls on low set model type 1 at 45^0 incidences increased on models in the leeward row and decreased on models in the windward row compared to corresponding average pressure differences with wind normal to long walls.

Average pressure difference coefficients between mid-height of long walls on high set and two storey model types 3 and 5 were lower in both windward and leeward rows at 45^0 incidence compared to corresponding average pressure difference coefficients with wind normal to long walls.

Effects of Wall Openings

It can be shown (2) that the pressure difference responsible for mean airflow through openings in buildings is the difference between the mean total pressure across the windward opening and the static pressure outside the leeward opening.

Static pressure outside leeward openings will be similar to the air pressure on the leeward surface, and readily available from existing data.

Mean total pressures across windward openings are not so easily determined. They can be obtained from wind tunnel studies but this is tedious. A reasonable approximation of the mean total pressure across windward openings in the case of low buildings with projecting eaves, is the average of surface pressures at stagnation points on each side at mid-height of the windward opening. Wind tunnel measurements are required to measure this pressure.

Because of the lack of published pressure distribution data for residential buildings with wall openings, a wind tunnel study was performed to determine the relationship between this mean total windward pressure and leeward static pressure across openings through residential buildings, and the readily available pressure distribution data for similar solid building forms.

Two building types were tested with wind normal to a long wall, the single storey on-grade design and a similar design elevated 2.3 m above ground level. Equal slot openings from floor to ceiling were formed in the centre of three equal dimensions on each long wall and incrementally increased in size (Fig. 3). Wind was normal to a long wall and all observations were made on the centre model in the leeward row of two rows with downwind spacing of 34.4 m (112.6 ft.) and a crosswind spacing of 5.7 m (18.75 ft.).

It was noted that typical percentage wall opening porosities of houses in Townsville tended to follow the type of windows used. Pressure differences between the windward and leeward walls (Fig. 4) were found to increase with increases in porosity up to between 10% and 20% and then decrease with further increases in porosity up to porosities of between 50% and 55%. With further increases in porosity from 55% to 80%, pressure differences between windward and leeward walls remained constant. These changes in pressure differences indicated that, for wind at normal incidence

Figure 3. Porous low set model with three equal width channels between long walls centred between pressure measurement point. High set model similar but raised equivalent of 2.3 m above ground.

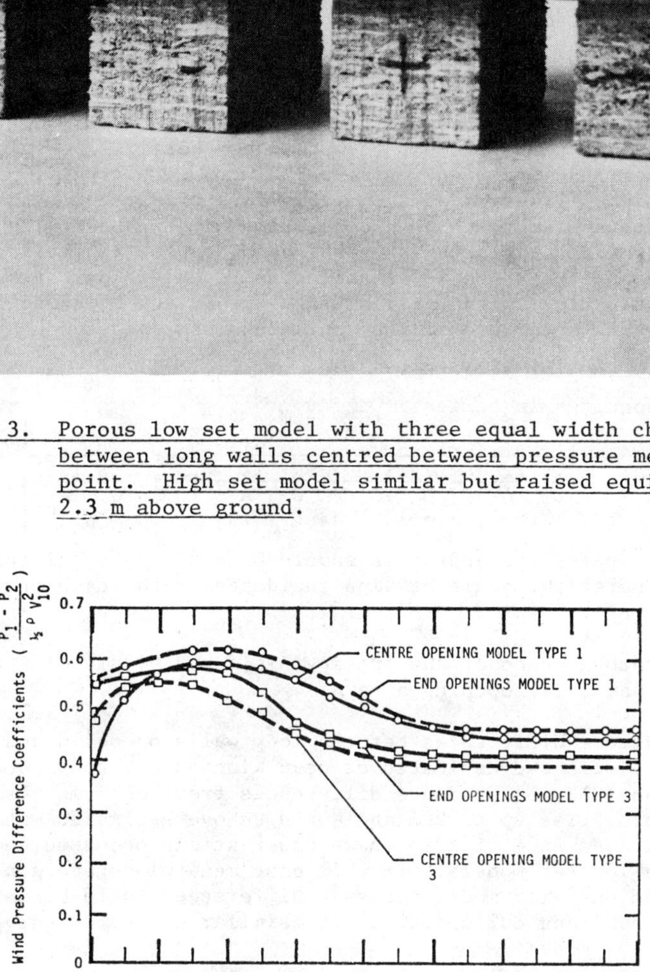

Figure 4. Percentage area of 3 openings between windward and leeward walls

to a long wall, maximum potential for mean windspeed through low set models of the type studied would be achieved with a porosity of 20%, and through high set models with a porosity of 10%.

Estimation of Flow Through Openings

An equation (1) can be written for the mean windspeed V_o through an opening using appropriate discharge coefficients, C_d, from Table 5 and pressure coefficients for total windward pressure C_{p1}, and leeward static pressure C_{p2}, referenced to dynamic pressure of the mean windspeed, V_z (m/s), Z being the height (metres) above ground (usually 10 m) for which long term mean windspeed data is available.

$$V_o = C_d \, [(C_{p1} - C_{p2}) \, V_z^{\,2}]^{\frac{1}{2}} \text{ m/s (ft/sec)} \tag{1}$$

or in terms of volumetric flow rate Q:

$$Q = C_d A \, [(C_{p1} - C_{p2}) \, V_z^{\,2}]^{\frac{1}{2}} \text{ m}^3\text{/s (ft}^3\text{/sec)} \tag{2}$$

These equations for airflow through a single opening in a building due to differences in wind pressure are difficult to apply, as there is little detailed data available on pressure due to wind inside buildings, with significant openings in windward and leeward walls. To overcome this lack of data, in cases where internal airflow follows a simple path through a building without branching, it is convenient to combine the discharge characteristics of each opening in series and use the wind pressure difference between the inlet and outlet openings on external walls. The condition of no branching in the airflow path within the building is not a major restriction, as it is a feature of most buildings designed to encourage natural airflow.

For a number of openings in series

$$Q = \left| \frac{(C_{p1} - C_{p_{n+1}}) V_z^{\,2}}{\dfrac{1}{C_{d_1}^{\,2} A_1^{\,1}} + \dfrac{1}{C_{d_2}^{\,2} A_2^{\,2}} + \dfrac{1}{C_{d_3}^{\,2} A_3^{\,2}} + \dots \dfrac{1}{C_{d_n}^{\,2} A_n^{\,2}}} \right|^{\frac{1}{2}} \text{ m}^3\text{/s (ft}^3\text{/sec)} \tag{3}$$

The area of openings in windward walls should be reduced by multiplying by a factor equal to the cosine of the angle of wind incidence in the case of inclined wind incidence to openings.

Mean windspeeds through each of the openings is found by dividing the common discharge rate Q by the local opening area A.

Comparison of pressure differences between long walls on solid and porous models at normal incidence, indicated estimates of mean windspeed through buildings using the discharge equation (3) with pressure differences from solid models would be underestimated with porosities up to 20% and would be overestimated with porosities over approximately 35%. An exception to these observations occurred near the ends of the long walls of the low set models. In this case mean windspeed estimates using equation (3) based on solid model pressure differences would be underestimated for porosities between 5% and 80% compared with similar estimates using pressure differences from porous models.

For buildings similar to model types 1 and 3, pressure difference coefficients can be taken from Fig. 4 for appropriate long wall porosities. Further wind tunnel work is planned for other model types and inclined wind incidences.

Errors in estimated mean windspeed resulting from the use of pressure differences on solid high set models, being proportional to the square root of the pressure difference, would be +2% to -13% near the ends of the long walls and +7% to -10% near

TABLE 5. Typical Discharge Coefficients for Single Inlet or Intermediate Openings in Buildings

Description of Opening	Typical Range of Discharge Coefficients C_d for Normal Incidence	Jet Characteristics
Small openings in thin walls less than 10% of wall area near the centre of the wall	0.50-0.65	Small inertia due to small mass of air in jet
Openings 10-20% near the centre of a wall with aspect ratio similar to the cross-section of the downwind space	0.65-0.70	Significant inertia due to increased mass of air in jet
Openings 10-20% of a wall with one edge common with the downwind space such as a doorway	0.70-0.80	Wall effect reduces energy losses on one side of jet
Openings similar in size to the cross-section of the downstream space	0.80-0.90	Wall effect around the perimeter of the jet significantly reduces turbulent energy losses

Typical Discharge Coefficients for Leeward Outlet Openings

A_0/A_1	C_d
Approaching	
0.0	0.63
0.2	0.64
0.4	0.67
0.6	0.71
0.8	0.81
1.0	1.00

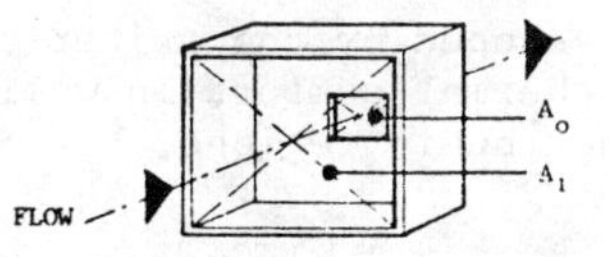

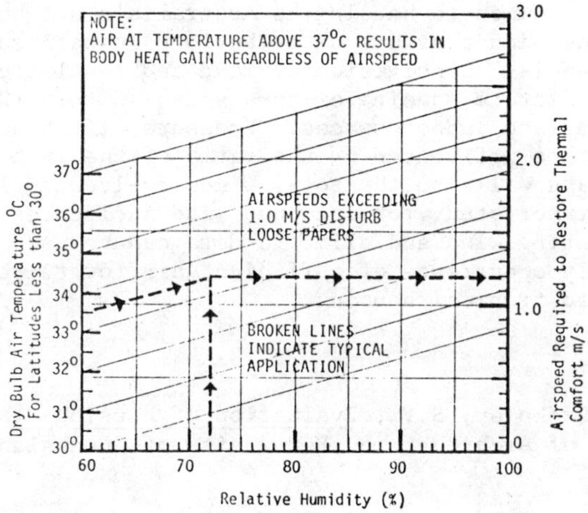

Graph for Estimating Minimum Airspeed Required to Restore Thermal Comfort for a Range of Dry Bulb Temperatures and Relative Humidity

Figure 5. Graph for estimating minimum airspeed required to restore thermal comfort for a range of dry bulb temperatures and relative humidity the centre of the long walls.

When dealing with variables such as estimates of the speed and direction of the natural wind, variations of say 15% must be expected. This would suggest that the use of pressure differences from similar groups of solid models would be of at least the same order of accuracy as short term estimates of local windspeeds and could be justified for estimating airflow through high set houses with wind normal to a long wall.

Estimates of Indoor Thermal Comfort

To estimate the percentage of time that indoor thermal comfort will occur at a particular time of day and month of year, estimates of windspeeds through openings, (using local windspeed, direction and frequency data), relative humidity, 86 percentile daily maximum dry bulb temperature and thermal comfort criteria, such as those suggested by Macfarlane (3) are required. As dry bulb temperatures and relative humidity are relatively constant for each hour of the day during monsoonal months, the writer suggests the percentage of time thermal comfort is achieved will follow directly the percentage of time that airspeeds occur sufficient to restore thermal comfort near wall openings in buildings [examples given in (4)].

Estimates of indoor comfort are limited to 9 am and 3 pm, as these are the only times of day for which reliable, long term, relative humidity and dry bulb air temperature data are readily available in Australia.

A graph (Fig. 5) was developed by the writer to simplify estimation of speed necessary to restore indoor thermal comfort in well shaded and well insulated, naturally ventilated housing in hot humid regions, based on Macfarlane's thermal comfort criteria.

CONCLUSIONS

Architectural features common to housing in Australia's hot humid tropics have significant effects on the wind pressure distributions on wall surfaces and need further study in wind tunnels. Subdivision of land and regulations on building setbacks from boundaries dictate spacing of houses and in turn the wind energy available to naturally ventilate indoor spaces. Pressure differences across houses at normal wind incidence are influenced by the ratio of the area of openings between the windward and leeward walls to the total windward-leeward wall areas. These influences require further study for inclined wind incidences and a wider range of building shapes. Existing wind and climatic data can be used to make preliminary estimates of the likely occurrence of sufficient airflow to restore thermal comfort in naturally ventilated tropical housing.

REFERENCES

(1) Saini, B.S. and Szokolay, S.V., Evaluation of Housing Standards in Tropical Australia, Dept. of Architecture, University of Queensland, St. Lucia, Queensland, 1975.

(2) Aynsley, R.M., A Study of Airflow Through and Around Buildings: With Particular Reference to Airflow for Thermal Comfort in Hot Humid Tropical Housing, Ph.D. Thesis, School of Building, University of New South Wales, Kensington, N.S.W., 1977.

(3) Macfarlane, W.V., Thermal Comfort Zones, Architectural Science Review, 1, 1-13, (1958).

(4) Aynsley, R.M., Melbourne, W. and Vickery, B.J., Architectural Aerodynamics, Applied Science Publishers, London, 1977, pp. 212-217.

WIND-INDUCED HEAT LOSS FROM BUILDINGS

A. P. BURNS

BHRA Fluid Engineering, Cranfield, Bedford, UK

SUMMARY

To provide an accurate method of measuring the convective heat transfer coefficient from a building surface, a series of model and full scale tests is described. The tests rely on the measurement of local flow conditions (i.e. at the edge of the local boundary layer), thereby accounting automatically for the influence of body geometry on the flow and resulting heat transfer. Empirical relationships are presented from wind tunnel tests on a flat plate and a cube; these support the reasoning behind local flow measurement. A full scale test highlights the variation of the local heat transfer but agrees less well with the empirical correlations.

1. GLOSSARY OF TERMS

h Heat transfer coefficient, W/m^2 °K(Btu/h ft^2 °F)

I Percentage turbulence intensity

L Non-dimensional turbulence scale (=λ/l)

l Characteristic body length, mm (inch)

Nu Nusselt number

Pr Prandtl number

Re Reynolds number

V Mean wind velocity, m/s (ft/sec)

ϕ Incidence of body or face to the flow

λ Scale of turbulence, mm (inch)

Subscript

ℓ Local condition i.e. at the edge of the boundary layer or sub-layer

2. INTRODUCTION

While heat gains and losses from buildings are currently the subject of considerable interest, it is ironic that the thermal performance of buildings has declined steadily from pre-industrial revolution times to the present day. For heat transfer purposes buildings have been treated as single whole units - local effects being obscured. These local effects have become much more noticeable with the development of multistorey structures. It has been found that different storeys require widely differing heat inputs to maintain similar temperatures. The same is true of different locations on the same level. This paper presents measurements of the extent to which these local variations are dependent on air flow conditions as affected by building shape, orientation to the wind and architectural detail.

The visual appearance of a building very often tends to be the overriding design criterion. Any internal environmental deficiencies are modified by electrical or mechanical services. This leads to a situation in which 80 per cent of the buildings constructed in Europe between 1955 and 1970 use 3-4 times the energy that they would have used had a more technological approach to design been adopted (1). Given the exact nature of wind cooling, as affected by site, orientation, shape and design detail, considerations such as the extent of insulation and glazed areas may be selectively applied.

Convective heat loss to the air from the walls of a building has always proved more difficult to measure than the conductivity of the walls themselves.

By approximating the effect of orientation and exposure, convective coefficients for buildings have been produced in (2). Effects of surface finish have been studied in (3), where tables were presented showing 'effective transmittance',

$\left(= \dfrac{1}{\text{effective convection coefficient}} \right)$, for different locations, times of year, orientations and surface properties. The Institute of Heating and Ventilating Engineers proposes an empirical formula,

$$h = 1 + 0.22\,V \qquad \ldots\ldots \quad (1)$$

where h is the convection coefficient in Btu/ft^{2}h$^\circ$F and V is the wind speed in ft/sec. The basis of the above results is still the overall heat transfer from an isothermal flat plate whereas local or point measurements on a bluff body are more representative of heat loss from different areas of a building.

Even the more recent studies in (4) and (5) neglected some of the important parameters of the physical problem. In (4) no account was taken of wind direction nor was turbulence quantified, while in (5) overall cooling was measured and wind direction neglected. Most correlations between heat loss and wind effects are based on flat plate heat transfer and free-stream flow conditions. Buildings distort flow conditions to such an extent that free-stream measurements do not have much relevance to the flow close to the structure, where the heat transfer takes place. When the presence of other buildings upstream, different surfaces finishes and a three-dimensional, non-homogeneous flow are also considered, any hope of correlating the forced convection and the parameters of the atmospheric boundary layer seems remote. From the results presented here an attempt is made to relate local convective heat transfer to local flow conditions at the edge of the boundary layer. By doing this a knowledge of the flow close to a building - measured with an anemometer - can then be related to the local heat transfer.

3. WIND TUNNEL TESTS

Dimensional analysis shows that a turbulent flow field is completely defined at any point when the parameters of Reynolds number {Re}, turbulence intensity {I}, and turbulence scale {L} are known. Similarly, forced convection at any point in a turbulent flow can be written as

$$Nu = f(Re, I, L, Pr) \quad \ldots\ldots (2)$$

where Nu is the Nusselt number and Pr is the Prandtl number. The use of the free-stream value of these parameters has been found to be convenient in most research to date largely because of their independence of the shape of the body in the flow. The nature of the function in equation 2 can easily be determined for geometrically simple bodies such as cylinders and flat plates. Here, because of the strong influence of building shape on the flow field and hence on the transport properties of the flow, flow parameters are measured close to the body surface, i.e. at the edge of the boundary layer, or sub-layer in the case of separated flow. Considering the Prandtl number {Pr} as a constant over the air temperature range of interest, equation 2 is then written in terms of these local values.

$$Nu = f(Re_\ell, I_\ell, L_\ell) \quad \ldots\ldots (3)$$

3.1. Flat Plate Tests

The measurement of heat transfer from a flat plate and the relevant flow parameters at the edge of the boundary layer served two purposes:

(i) validation of the experimental technique by duplicating 'classical' flat plate heat transfer results and,

(ii) establishment of empirical correlations which, it was hypothesized, would be independent of body shape since local flow measurement automatically accounts for the influence of shape.

Within the range of available adjustment the turbulence scale was always of the same order of size as the heat transfer measuring device; its variation in this range made no measurable difference to the heat transfer and so it was held constant, reducing equation 3 to

$$Nu = f(Re_\ell , I_\ell) \quad \ldots\ldots (4)$$

Purpose (i) above was accomplished and heat transfers, in a variety of free-stream speeds and turbulence, were measured from the flat plate over the incidence range, $\phi = -90°$ (plate facing downstream) to $+90°$ (plate facing into flow). Typical results for the parameters of interest are shown in figure 1. The flow conditions and heat transfer divide into three distinct incidence ranges as follows:

(i) $+90° > \phi > +5°$, Laminar, boundary layer flow with the intensity of free stream turbulence reflected in the measurements of heat transfer and local turbulence intensity.

(ii) $+5° > \phi > -20°$, Complex transition flow with laminar boundary layer, turbulent boundary layer and separated flow present. Free-stream turbulence plays a secondary role to that which is body-generated.

(iii) $-20° > \phi > -90°$, Fully separated flow with a boundary sub-layer. The characteristics of both heat transfer and flow are dictated by the disturbing effect of the plate.

If local Nusselt number {Nu} is plotted against local Reynolds number {Re_ℓ} (based on local mean velocity) for each of the above flow types, then curves of equal local turbulence intensity can be drawn as shown in figures 2 - 4. Because of the entirely different flow regimes operating in each case, it was not possible to combine the three sets of curves.

Wherever the present work coincided with that of previous research, comparison was made and, generally, agreement was found to be reasonable. For example: heat transfer from a flat plate at positive incidence, (6) and (7); heat transfer in separated flow, (8) and (9).

Although not representative of a general bluff body shape, such as a building, the above results are indicative of the flow and heat transfer from a sharp-edged, flat-faced solid. A qualitative application of the results to a building would show the areas that are sensitive to wind-induced heat loss. For example, on a site of prevailing westerly winds, if the sides of a building face the cardinal points of the compass, then the ranking of the sides suffering convective heat loss would be (in descending order): north and south, west, east.

3.2. Cube Test

The use of a cube model in the wind tunnel served, firstly, to study the convective heat transfer from a shape more representative of a free standing building and, secondly, to test the hypothesis that local measurements account automatically for differences in body shape so that a wider interpretation can be given to the correlations of figures 2 - 4. For measurement purposes the sides of the cube were treated as 'flat plates' and there were nine measuring positions per face.

Figure 5 shows the distributions of heat transfer coefficient, local turbulence intensity and local mean speed with changing incidence {ϕ}. ($\phi = 0°$ corresponds to a face being parallel to the mean flow direction in the free-stream). While on comparison with figure 1 (flat plate), there is general similarity in the distributions, there is a notable difference in the incidence ranges corresponding to the various flow regimes, i.e. $+90° > \phi > +45°$ laminar, $+45° > \phi > -10°$ transition, $-10° > \phi > -90°$ separated. The shift in the characteristics of the curves to higher incidences can be attributed to body geometry whose influence cannot be detected in the free-stream but is picked up by the local measurements. For the heat transfer from a building the widening of the transition flow incidence range is significant, as here the convection is maximum.

The results from the cube model tests are plotted with the correlation curves derived from the flat plate tests in figures 6-8 and within acceptable limits the agreement is good. The use of local values of the flow parameters does seem to show an independence of the correlations from body shape.

4. FULL SCALE TEST

The full scale test was carried out on a small, glass-clad building (4.1 m (13.4 ft) x 3.6 m (11.8 ft) x 2.0 m (6.6 ft) high) situated on the flat roof of a 4 storey building and exposed to the prevailing winds. As with the cube model test, the heat transfer was measured at various points on each face which was considered to be a flat plate for purposes of comparison with wind tunnel results. The flow conditions were again measured at the edge of the local boundary layer.

The heat transfer coefficients measured for three different wind conditions are shown in figures 9 - 11. A formula such as equation 1 gives the values of h of 26.2 $W/m^2°K$ (4.61 $Btu/ft^2h°F$), 37.6 $W/m^2°K$ (6.62 $Btu/ft^2h°F$) and 32.8 $W/m^2°K$ (5.77 $Btu/ft^2h°F$) corresponding to figures 9, 10 and 11 respectively. These can be up to 500 per cent higher or 25 per cent lower than the measured values. Comparative results from the correlations of figures 2 - 4 are also given in figures 9 - 11 and do not provide the good agreement achieved in the wind tunnel. The empirical correlations for laminar and transition flow give results lower than those measured, while the separated flow results are not consistent. However, against the poor agreement should be set the difficulties encountered in these full scale measurements, i.e. large fluctuations in flow direction, edge of boundary layer less clearly defined than in the wind tunnel, scale of turbulence now very much a variable.

5. CONCLUSION

Both model and full scale studies showed the dependence of local turbulent convection on local flow conditions. Within the scope of the wind tunnel tests the hypothesis of using local measurement to provide independence from body shape was supported. Extension of this result to the full scale is not justified on the basis of the results presented but, given improved experimental techniques, such an extension seems a possibility. However, the empirical correlations do indicate the relative magnitudes of turbulent convection in the different flow conditions encountered.

6. ACKNOWLEDGEMENTS

The work presented was carried out by the author at the University of Glasgow as part of a PhD thesis and was funded by the British Science Research Council. This paper was prepared under the sponsorship of the National Building Research Institute of the South African Council for Scientific and Industrial Research and BHRA Fluid Engineering, Cranfield, England.

REFERENCES

(1) Wyatt, T.J. "A description of integrated design involving thermal balance", Parsons Brown Report, May 1971

(2) Dufton, A.E. Journal of the Institute of Heating and Ventilating Engineers, Vol 8, 1940

(3) Loudon, A.G. Journal of the Institute of Heating and Ventilating Engineers, Vol 36, 1968

(4) Sturrock, N.S. "Localized boundary layer heat transfer from external building surfaces", PhD Thesis, Liverpool University, 1972

(5) Jackman, P.J. "Heat loss in buildings as a result of infiltration", Building Services Engineer, Vol 42, 1974

(6) Drake, R.M. "Investigation of the variation of point unit heat transfer coefficients for laminar flow over an inclined plate" ASME Vol 71, 1949

(7) Eckert, E. "Die Berechnung des Wärmeübergangs in der laminaren Grenzschicht umstromter Kärper", V D I Forschungsheft, Vol 416, 1942

(8) Sogin, H.H. "A summary of experiments on local heat transfer from the rear of bluff obstacles to a low speed airstream", Journal of Heat Transfer Vol 86, 1964

(9) Petrie, A.M., and Simpson, H.C. "An experimental study of the sensivity to free stream turbulence of heat transfer in wakes of cylinders in crossflow", International Journal of Heat and Mass Transfer, Vol 15, 1972

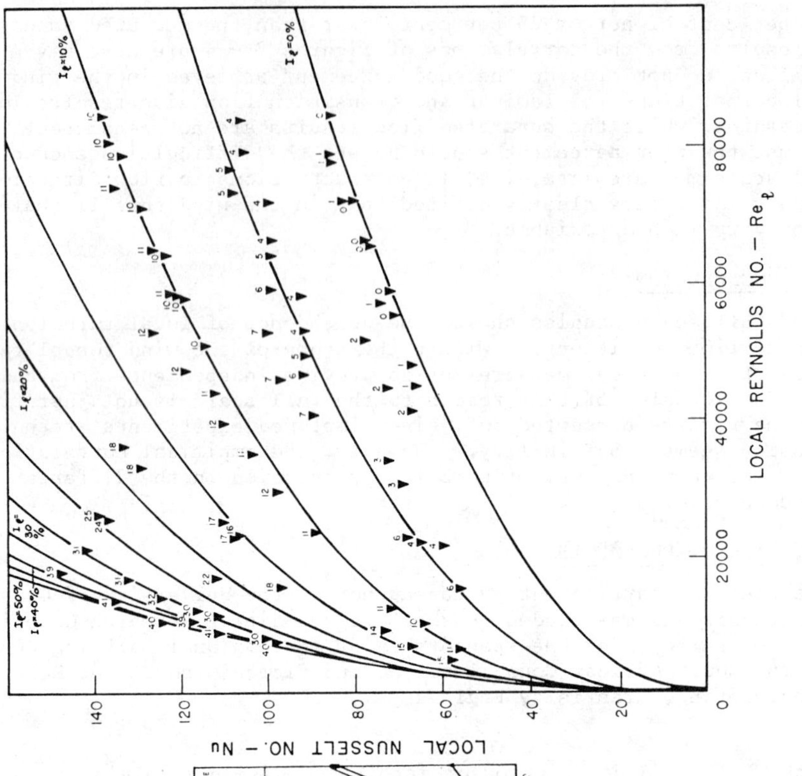

FIGURE 2 *Correlation of laminar heat transfer with local intensity and Reynolds No. – Flat Plate*

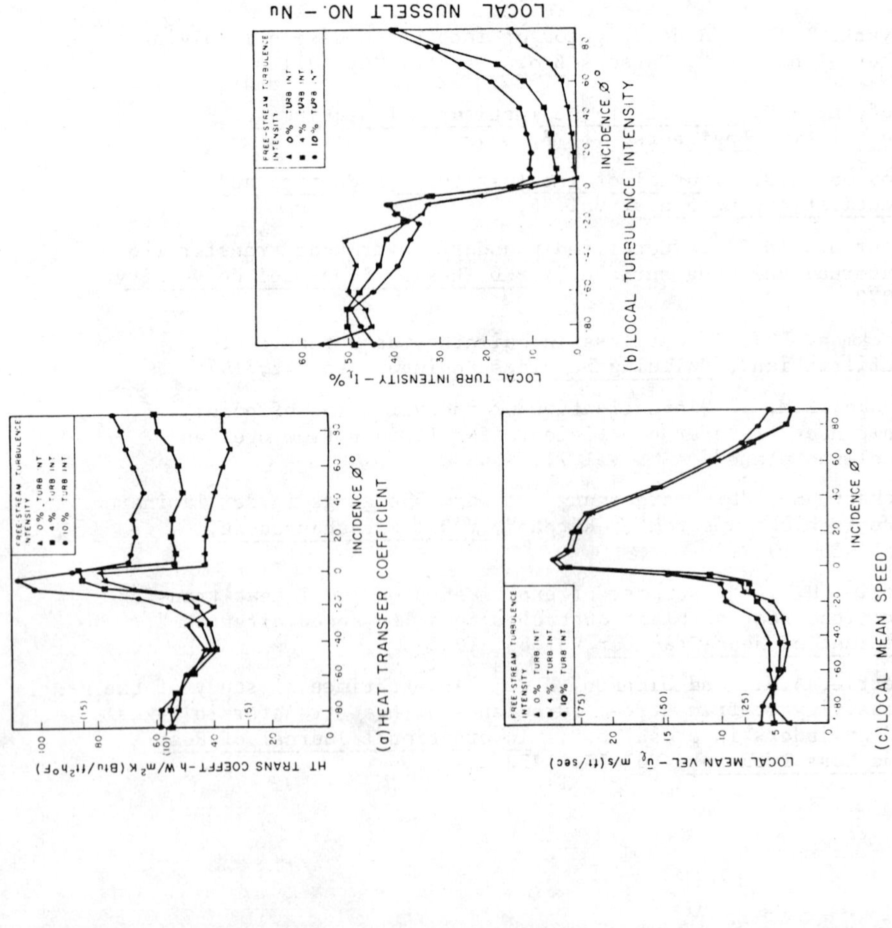

FIGURE I *Variation of heat transfer and flow conditions with changing incidence for a flat plate*

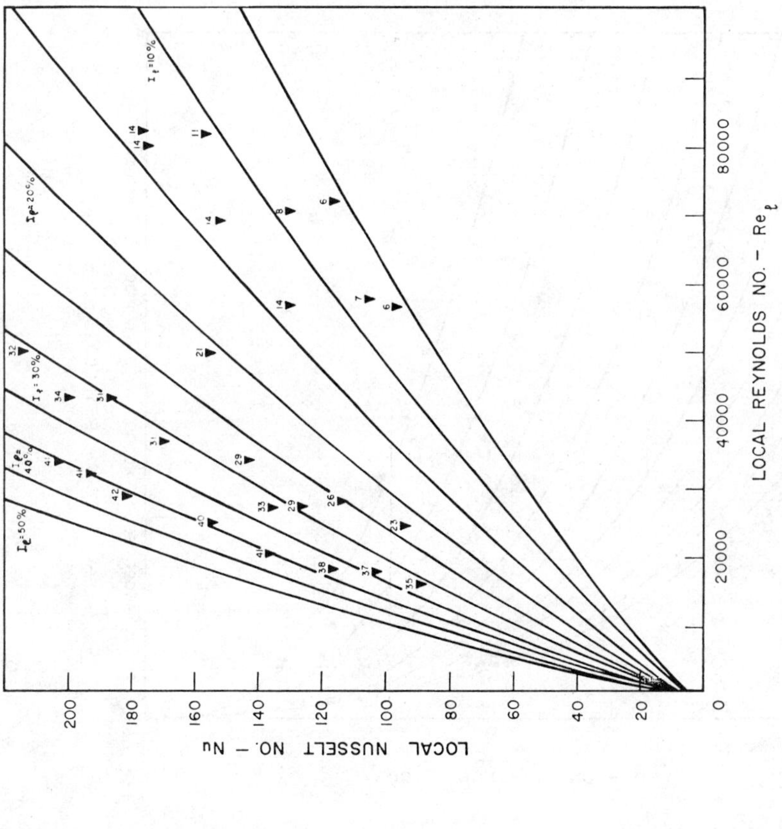

FIGURE 3 Correlation of transition heat transfer with local intensity and Reynolds No. – Flat Plate

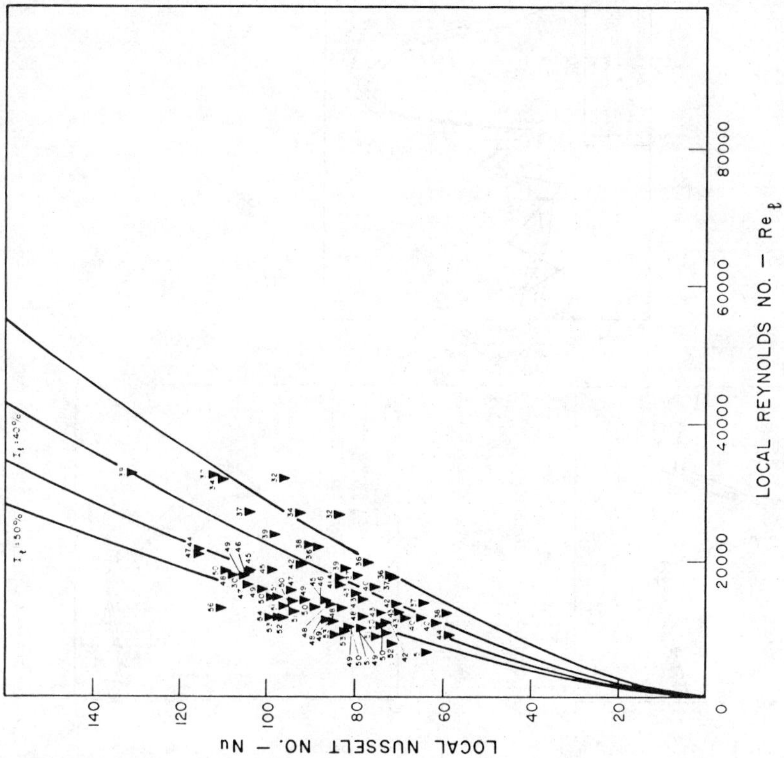

FIGURE 4 Correlation of separated heat tranfer with local intensity and Reynolds No. – Flat Plate

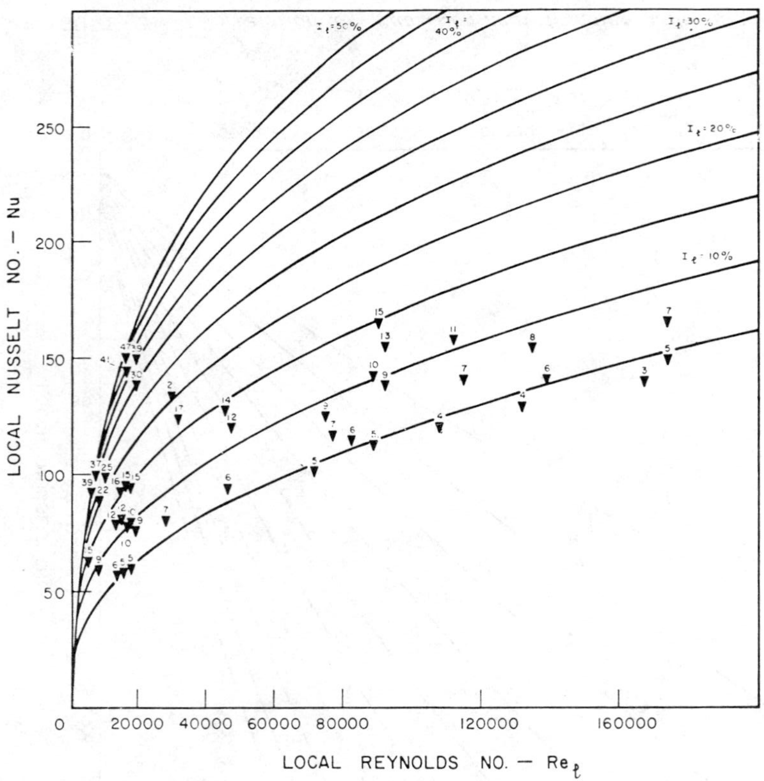

FIGURE 6 *Laminar heat transfer correlation curves (from flat plate) and experimental points from cube test*

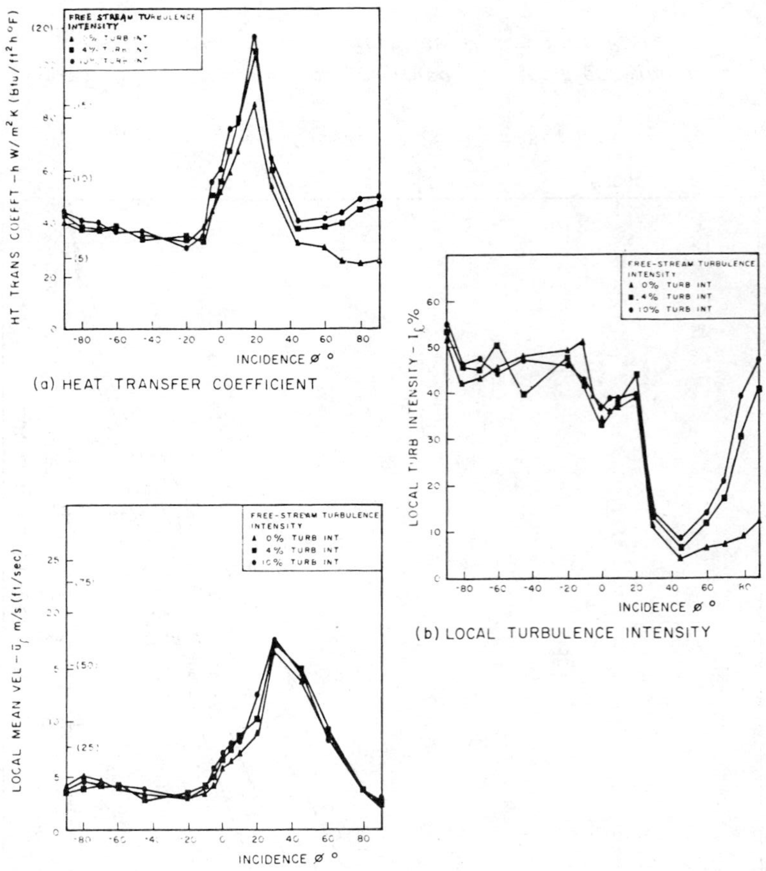

FIGURE 5 *Variation of heat transfer and flow conditions with changing incidence for the mid-point on the face of a cube*

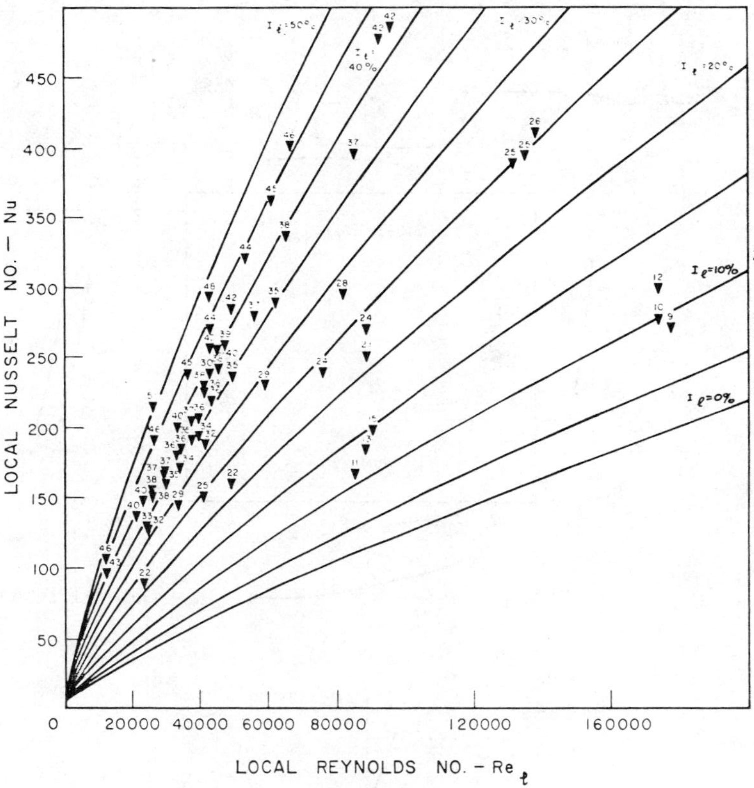

FIGURE 7 *Transitional heat transfer correlation curves (from flat plate) and experimental points from cube test*

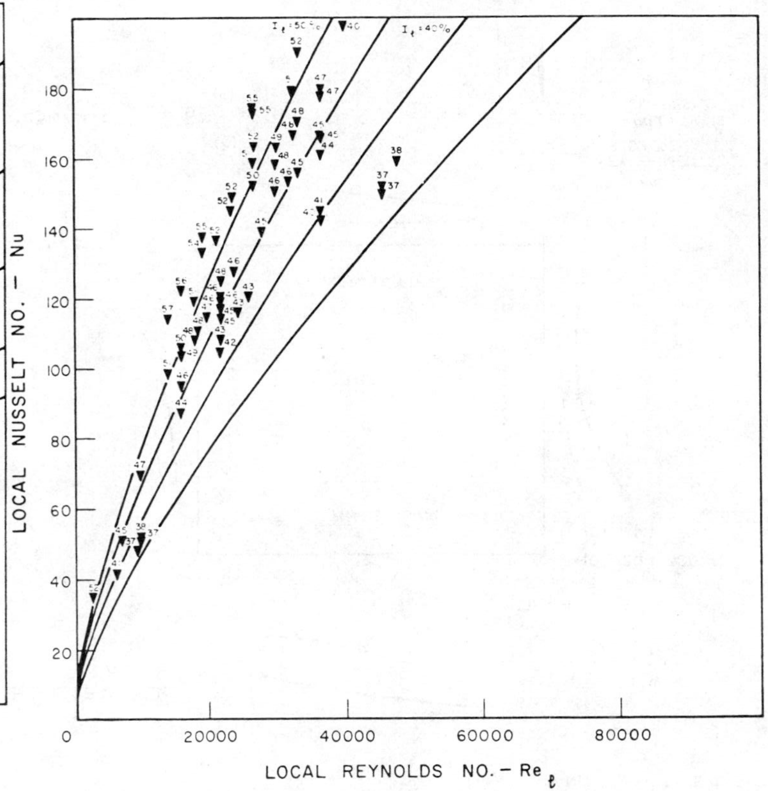

FIGURE 8 *Separated heat transfer correlation curves (from flat plate) and experimental points from cube test*

WE K

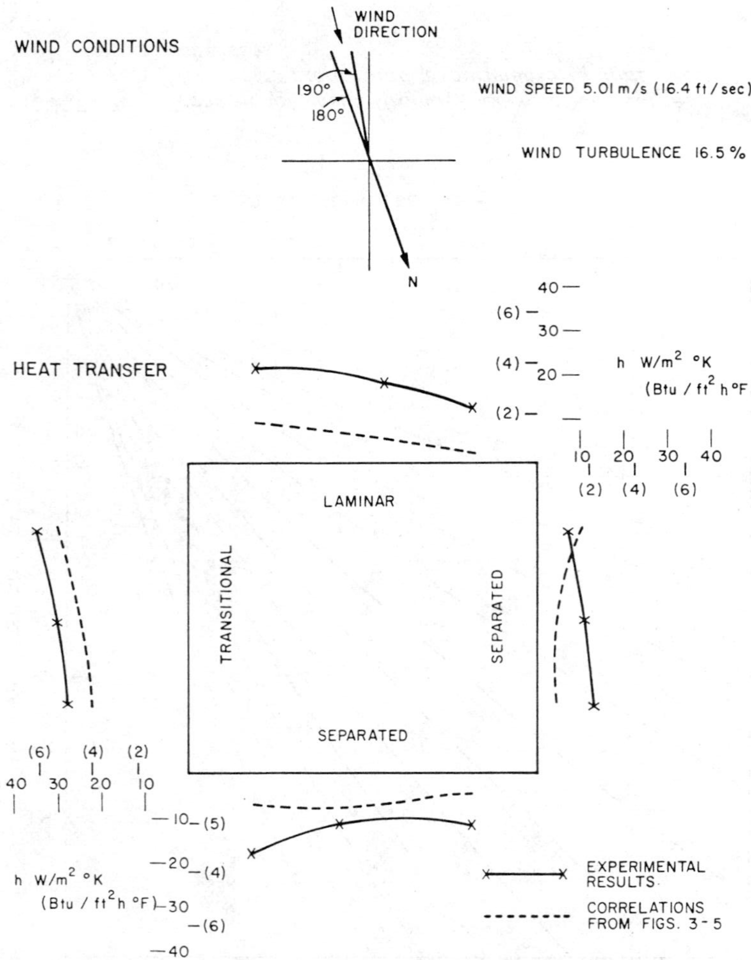

FIGURE 9 *Variation of heat transfer round the test building and comparison with empirical correlations*

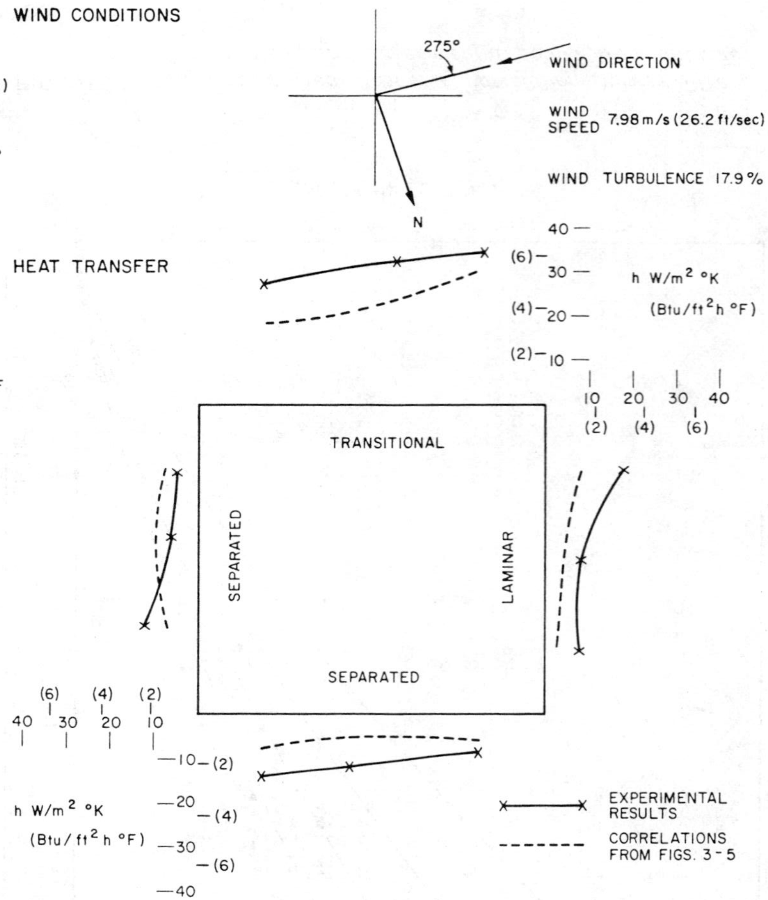

FIGURE 10 *Variation of heat transfer round the test building and comparison with empirical correlations*

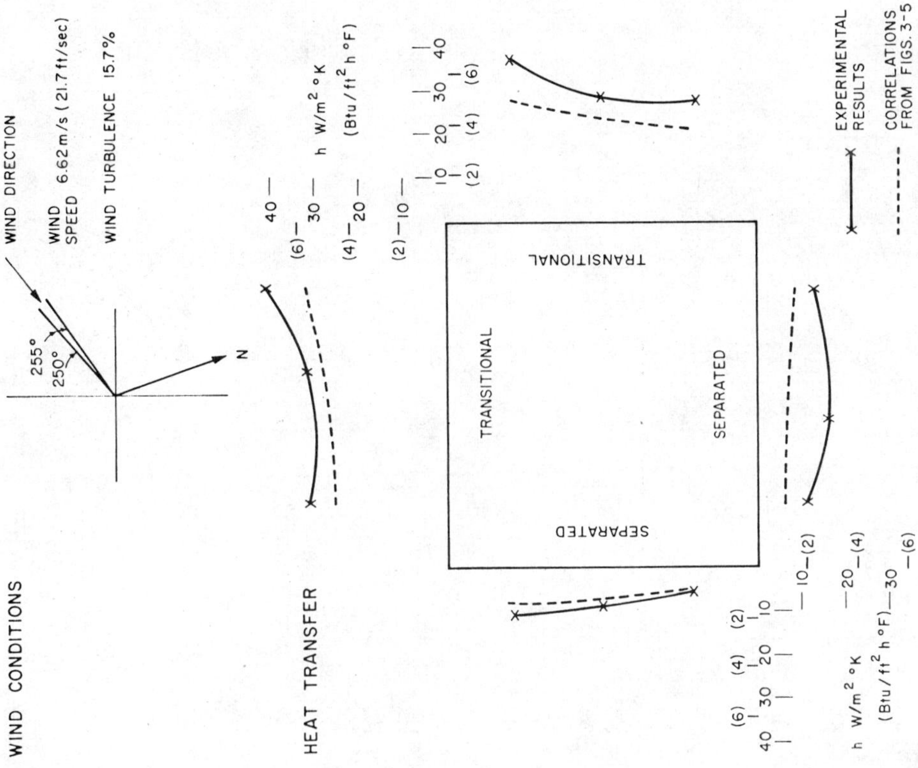

FIGURE 11 *Variation of heat transfer round the test building and comparison with empirical correlations*

A WIND TUNNEL INVESTIGATION ON THE GROUND LEVEL WIND ENVIRONMENT IN A PROPOSED OPEN STADIUM

M. PACIUK and M. POREH

Technion, Israel Institute of Technology, Haifa, Israel

INTRODUCTION

During the preliminary stages of design of a Center for Sport and Recreation, to be located at the top of one of the highest mountain ridges surrounding the city of Jerusalem, both client* and architect** expressed concern regarding the quality of the wind environment which would prevail within and around the center. In particular, the question was raised whether problems of discomfort, annoyance, or even danger, would arise in the large, open stadium, as a result of the mechanical effects of wind.

Mean hourly wind speed and direction at the site were measured at a height of 3.5m above ground during the year 1974/75, and results were correlated (1) with data from two permanent meteorological stations in the Jerusalem area. The resulting estimation of wind conditions at the site revealed that strong winds (Beaufort 5 and above) are expected to prevail in the mornings and early afternoons during at least 36% of the winter season.

These findings prompted the client to sponsor a wind tunnel model study at the Technion, Israel Institute of Technology, in order to obtain quantitative estimates on wind comfort levels in the proposed stadium, and to suggest, if necessary, design modifications which could improve the ground level wind environment. Within the scope of the study only broad design concepts were investigated from an aerodynamic, rather than an economic or engineering, point of view.

The investigation, which is described in the following, showed that difficult conditions would prevail within the stadium during a high percentage of the football season. Several modifications to the proposed design were tested and evaluated, and it was found that the construction of a peripheral, sloped roof, partially extending over the upper stands and deflecting the oncoming wind, could reduce the occurrence of unfavorable conditions by more than half.

---

* The Jerusalem Sports and Recreation Center Ltd.

** Pascual Broid and Assoc., Jerusalem.

WIND TUNNEL MODEL STUDY

The investigation was carried out in the newly built Environmental Wind Tunnel of the Technion. The tunnel has a 2m wide, 15m long working section. An adjustable ceiling, varying in height between 1.50m and 1.95m, makes it possible to maintain a zero longitudinal pressure gradient along the tunnel. Air enters the tunnel through a 3.2m x 2.8m honeycomb, followed by a smooth transition to the working section, where Counihan type vortex generators are installed. These vortex generators and a long fetch of floor roughness generate a turbulent boundary layer whose thickness and characteristics can be controlled (2).

Wind flow patterns at the site, and the general features of the approaching flow, were first observed in a 1:1250 scale model of the area (Fig. 1).

Fig. 1 - The 1:1250 topography-stadium model in the wind tunnel

In the absence of upper level data on wind speeds over the site, it was not possible to undertake a full fledged simulation of the local boundary layer. The thickness of this layer was, however, estimated to be relatively small because of the steady increase in altitude of the terrain from the Mediterranean Coast toward the mountain range. It was, therefore, decided to include in the model a considerable fetch (approx. 6 km) of the upstream topography, to ensure that significant features in the approaching flow at ground level would be simulated to a large degree.

Visualization of the ground level flow pattern was accomplished by means of an extensive grid of tiny "flags" fixed on the model (see Fig. 1), and by means of smoke released at several locations.

The observation of flow patterns clearly indicated that the direction of the ground level flow near the top of the ridge on which the stadium is to be located, was identical to that of the upper, unmodified flow in the tunnel, while at other locations of the model, on the northern and southern slopes of the ridge and in the surrounding valleys, a strong influence of topographic features on the local wind direction was felt. No separation of the flow passing over the upstream hills was detected. Moreover, it was found that the flow above, around and within the stadium, was practically independent of the exact nature of the upstream flow, and that changes in roughness or turbulence in the upstream boundary layer, or of its thickness, could hardly be detected within the stadium.

The variation of mean wind velocity with height at the base and at the top of the ridge was also recorded. A comparison between these velocity profiles showed the well known phenomenon of a shallower boundary layer at the top of the ridge, where the wind velocity attained the free stream value at approximately 30m above ground level.

From these observations, it was concluded that a 1:250 scale model of the stadium (Fig. 3) which includes the major part of the ridge and its slopes, and is placed at the downstream end of the tunnel, in a 300m (full scale) thick boundary layer flow developed over a rough surface, would give a satisfactory simulation of wind patterns in the stadium. When introduced into the tunnel, the top of the ridge in the 1:250 model reached a height of 40cm above the tunnel floor level. The adjustable tunnel's ceiling was therefore raised in that region, to obtain a zero pressure gradient flow high above the model.

Measurements were made for four different wind azimuths (90°, 180°, 250°, 270°) which covered most of the strong wind events occurring at the site, by rotating the model in the tunnel. At first, the topographic model (no stadium present) was tested in order to determine the velocity profile at the top of the ridge, at a point where the anemometer station used in (1) had been located. It was found that the wind speed at a height of 3.5m above ground ($U_{3.5}$) was 20% lower than the free stream speed over the ridge (U_{ref}), which was consistent with the measurements made in the 1:1250 model. The Reynolds number of the model being sufficiently large, it was assumed that the ratio of the wind speed U at any point of the stadium to the wind speed $U_{3.5}$ measured at the site, can be approximated by the following expression

$$\left.\frac{U}{U_{3.5}}\right|_{\substack{\text{in}\\ \text{prototype}}} = \left.\frac{U}{U_{ref}}\right|_{\substack{\text{in model}\\ \text{with stadium}}} \times \left.\frac{U_{ref}}{U_{3.5}}\right|_{\substack{\text{in model}\\ \text{without stadium}}} \qquad (1)$$

and that the local turbulence level (u'/U) was the same for both the prototype and the model.

Values of mean velocity and turbulence were measured using a linearized hot wire anemometer at various points of the model, and particularly at a height of 1.5m (full scale) above the ground, with the reference velocity kept at 10m/sec. To measure the ground level wind characteristics in the stadium, the sensing element was always placed horizontally and normal to the mean wind direction, which had been determined earlier.

It was realized that the recorded values of the longitudinal component of both the mean wind speed and turbulence, were affected by the high turbulence level, the wire indicating higher mean velocities and lower turbulence values than in reality. Using Bradbury's model (3) it was, however, estimated that, up to a turbulence level of 50%, and for the above mentioned orientation of the wire, the error was negligible. Moreover, since the errors in reading the mean velocity and turbulence level were of opposite sign, while both affect human comfort in the same manner (as will be discussed further on) it was decided not to correct the hot wire readings, when used to evaluate the stadium's ground level wind environment.

CRITERIA FOR THE ACCEPTABILITY OF WIND CONDITIONS

Several criteria for evaluating the acceptability of given wind climates has been suggested lately (4)(5)(6)(7). The different criteria usually suggest particular values related to thresholds of discomfort, annoyance, and danger levels, as well as acceptable frequencies of occurrence of these levels. None of the criteria found in the current literature refers specifically to sport arenas.

Recent studies have clearly indicated that the effects of wind on man are related to both mean and fluctuating speeds. Based on limited wind tunnel experiments, Hunt (5) stresses that a strongly turbulent wind is equal, in its effects on man's comfort and activities, to a much higher mean wind speed with a lower degree of turbulence, and proposes to account for both effects by using an equivalent wind speed U_{eq} defined as

$$U_{eq} = U\,(1 + 0.03\,I) \qquad (2)$$

where U is the mean wind speed and I is the intensity of turbulence in percent. Hunt (5) also suggests approximate threshold values which define different reactions of man to the wind environment and which are described in Table 1.

Table 1 - Evaluation of wind effects on people (after Hunt (5))

U_{eq} (m/sec)	Assessment	Description of effect	Beaufort scale range*
<6	comfortable	little effect on performance	<3
6-9	uncomfortable	most performance unaffected	3-4
9-15	difficult	force of wind felt on body	4-5
15-20	intolerable	difficult to walk steadily	6-7
>20	dangerous	progress impeded people blown over by gusts	7-8

*Measured at h = 10m above ground in an open site (assuming 18% turbulence level).

As pointed out earlier (4)(7), man's feelings and reactions to wind are governed, among other parameters, by his size, health, physical condition, degree of acclimatization, expectations, etc. Thus, it would be sensible to expect that young people or trained sportsmen will be less sensitive to wind effects than the general public. No concrete information on the subject was, however, available and it was therefore decided to examine the reaction of Israeli players following two different football games.

The first game was held at the YMCA Stadium in Jerusalem, a location surrounded by a dense urban development. The mean ground level wind speed recorded during the game was in the order of 3m/sec, with occasional gusts climbing to twice this speed. The turbulence level in the field, located within the wake of other buildings and of a high impermeable wall, was estimated to be as high as 65%. The equivalent wind speed (Eq. (2)) was in the order of 9m/sec, a value at which, according to Table 1, disturbance from wind is expected to occur. And indeed, most players complained, and described the wind conditions at the field as annoying. However, no one felt that the game had to be cancelled.

The second game was held at Kiryat Haim, a suburban area near Haifa, in an open stadium surrounded by a rather low fence and scattered groups of trees. The measured mean wind speed within the field was, this time, higher, about 4m/sec, but with a rather low turbulence level, giving a value of $U_{eq} < 8$m/sec. The interviewed players reported that the wind, although felt during the game, was not a truly disturbing factor, which again, was consistent with the criteria given above. It was, therefore, decided to accept Hunt's criteria for the evaluation of acceptable wind conditions within the stadium, by giving it the following interpretation: Whenever local equivalent wind speeds rise above 9m/sec, increasing disturbance from wind will be felt and complaints from both players and spectators will be heard. When reaching 15m/sec, conditions will become extremely difficult and, at higher speeds, intolerable, and most certainly games will then have to be cancelled.

The effect of the wind on the ball was not considered to be a dominant reason for cancelling games, since it was learned that experienced players adapt their playing style to the existing wind conditions, for example, by refraining from high kicks or long distance passes.

EVALUATION OF THE ORIGINAL STADIUM DESIGN AND PROPOSED MODIFICATIONS

The original design of the stadium, as proposed by the architect, was first tested for four principal wind directions. Flow visualization around and within the model clearly showed that, for westerly wind directions (and similarly for the easterly wind direction), a large vortex was formed in the wake of the stadium's wall (see Fig. 2), with downstream reattachment taking place somewhere near the last third of the playing field. Ground level flow conditions within the wake were highly turbulent. Also, the varying height of the circumscribing stadium's wall produced horizontal vortices, which were noticed in the sportsfield.

Ground level wind velocity and turbulence were measured in the model at points of interest, for various locations (sportsfield, racing track, stands, plaza). A typical flow pattern distribution is illustrated in Fig. 3 for a westerly wind direction ($\theta = 270°$) and for a free stream velocity $U_{ref} = 10$m/sec.

The calculation of equivalent wind speed values for each measuring point was a straightforward task. However, since it is clear that these values vary from one point to the other, a simple criteria, based on some average value, had to be adopted in order to evaluate the wind climate for any particular section of the stadium. It was decided to choose, as a representative value, the equivalent wind speed averaged over that part (generally one third to one fourth) of the section which showed the worst wind conditions. This procedure was backed by the argument that, if wind conditions in a representative portion of the field would be intolerable, the game would have to be cancelled, even if conditions at the rest of the field happened to be slightly milder.

The averaged values, in combination with frequency of wind speed and direction data, were used in the computation of the relative frequency of occurrence of different conditions in the stadium's wind environment. The results were arranged into tables giving the expected wind climate for different hours and seasons of the year. Table 2, for example, shows the expected distribution of wind conditions in and around the stadium during the early afternoon hours (2.00 p.m.) of the winter season. Table 3 summarizes the expected occurrence, at the sportsfield, of the conditions defined as difficult or worse.

As can be observed, during 30% of the winter season, conditions at the original stadium's sportsfield are expected to be at least difficult, and sometimes intolerable or even dangerous. During the spring season, conditions are expected to be slightly better. According to the criteria presented before, difficult wind conditions do not necessarily imply cancellation of games, but undoubtedly such conditions will cause annoyance to both players and spectators and give the stadium a reputation of a "windy place", discouraging people from frequenting it. Since the football season in Israel extends over the winter and spring months, most of the games being played on Saturday's early afternoon hours, the above given figures approximately represent an average of every fourth Saturday with difficult or intolerable wind conditions at the sportsfield.

Table 2 - Estimated distribution of wind conditions in original and modified stadium - Winter season, 2.00 p.m.

	Comfortable		Uncomfortable		Difficult		Intolerable		Dangerous	
	Orig.	Mod.	Orig.	Mod.	Orig.	Mod.	Orig.	Mod.	Orig.	Mod.
Location										
Sportsfield	50	70	19	17	25	13	5	0.2	0.3	--
West.stands	64	67	17	18	18	14	0.5	1	--	--
East.stands	56	67	14	18	24	14	6	1	--	--
Plaza	44	44	16	16	26	26	10	10	4	4

Table 3 - Summary of relative frequency of occurrence of difficult or worse wind conditions at the unbuilt site and at the sportsfield in the original and modified stadium.

Season	Hour	Unbuilt Site	Original Stadium	Modified Stadium
Winter	11.00	36	20	7
	14.00	40	30	13
Spring	11.00	24	10	2
	14.00	32	21	5

In an attempt to alleviate the unsatisfactory wind conditions created within the stadium and eliminate the objectionable features of the existing flow patterns within the wake, several modifications to the stadium's original design were considered and evaluated. The proposed modifications did not alter the basic conception of the stadium's design and were elaborated in close cooperation with the architect.

The most effective of the considered modifications consisted in changing the shape of the western overhang so that it could now trap part of the oncoming wind and divert it downward into the field, in a direction opposite to that of the returning flow in the wake. The overhang (see Figs. 4 and 5), which was also extended to circumscribe the entire stadium, was designed as a corrugated roof, 15m wide, and slightly tilted inwards, almost following the slope of the stands. Between the overhang and the wall a 2.5m high gap was left, and the roof was built in such a way that it extended outwards, some 3m beyond the wall's edge.

The modified flow pattern distribution is illustrated in Fig. 5 for a wind azimuth angle $\theta = 270°$ and $U_{ref} = 10m/sec$. A comparison between figures 3 and 5 clearly shows the general reduction in both wind speeds and turbulence intensity achieved within the stadium as a result of the proposed design changes.

The relief in conditions resulting from the addition of the peripheral roof cover can be clearly appreciated from Tables 2 and 3. The percentage of time during which conditions at the sportsfield are expected to be intolerable is almost non-existent. Difficult conditions are still likely to appear during the football season, their frequency of occurrence, however, being reduced by half.

The proposed overhang also ameliorates conditions at the stands, especially at the eastern stands. For the original design, the probability exists that every third Saturday of the winter, the use of the eastern stands will have to be restricted because spectators would find it an annoying or even intolerable experience to endure existing wind conditions for any prolonged lapse of time. The addition of the peripheral overhang reduces this probability to two games during the entire season. Attention should be paid, however, to the fact that a localized problem of high wind speeds was observed near the upper rows receiving the full impact of the flow, trapped by the roof and redirected downwards toward the field. This local problem can be solved with a proper redesign of the upper stands area.

From Table 2, it is possible to conclude that, by and large, the most unpleasant and sometimes even dangerous conditions, are likely to originate at the plaza surrounding the stadium. This estimation is not surprising, since the open plaza offers no protection at all from the oncoming winds, conditions there being even slightly worse than at the unbuilt site. These conditions could be alleviated, to a great extent, by the use of local artificial and natural windbreaks, such as rows of trees, fences, walls and other attachments, intended to break down the flow at ground level and create a more sheltered environment.

CONCLUSIONS

The wind tunnel model study and subsequent analysis of data, in combination with the statistics of the local wind climate, showed that rather frequent severe and even intolerable wind conditions are expected to prevail in the proposed open stadium. These conditions will create discomfort and annoyance to both spectators and players, and will sometimes lead to a suspension of games.

As a result of this appraisal, several design modifications were proposed and evaluated for effectiveness. Results showed that objectionable features in the flow pattern could be controlled and a considerable reduction in wind speed and turbulence within the stadium could be achieved with a modified stadium's design, featuring a specially shaped, peripheral overhang extending over the outer wall and upper stands.

No attempt was made by the investigators to draw conclusions concerning the suitability of the assigned location or the quality of design of the proposed stadium. Other solutions, such as an alternative site or a completely different design concept, are also open for consideration. To arrive at any decision, questions of communal interest, financial resources, general urban policy, and of course, the client's previous expectations regarding the quality of the stadium's environment, have to be considered and weighed.

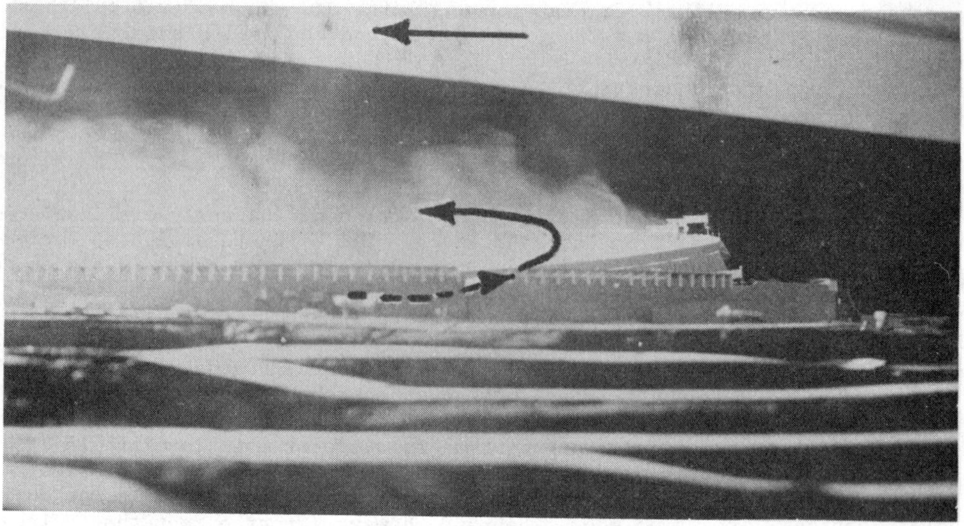

Fig. 2 - Smoke visualization of the vortex created in the stadium by westerly winds. Arrows designate the direction of the flow

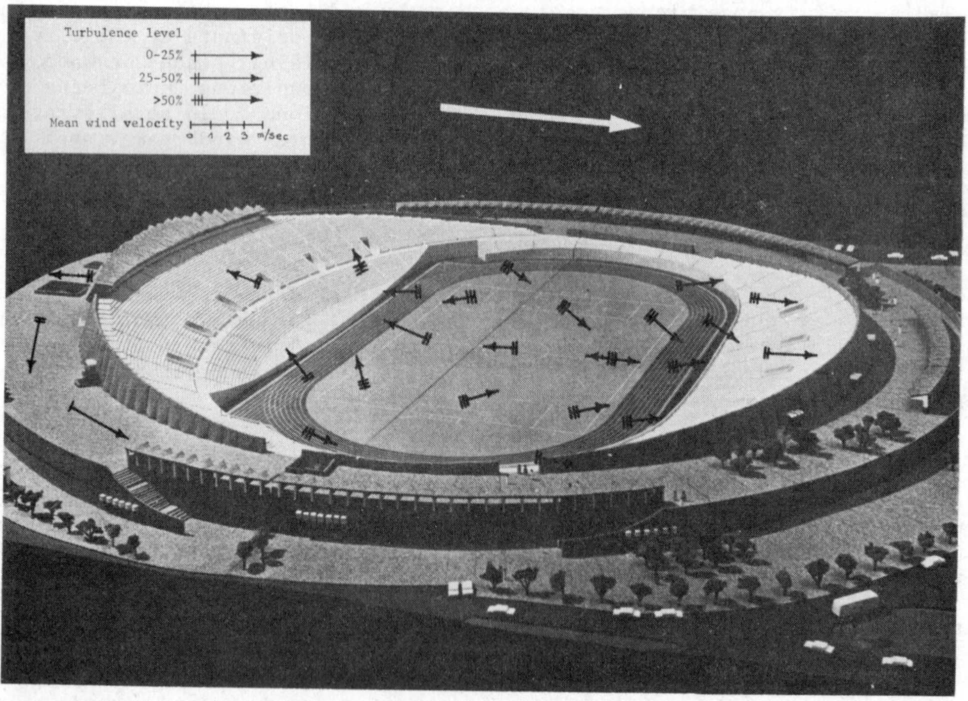

Fig. 3 - Flow pattern distribution in the original stadium for a westerly wind direction (θ = 270°) and a free stream velocity U_{ref} = 10m/sec.

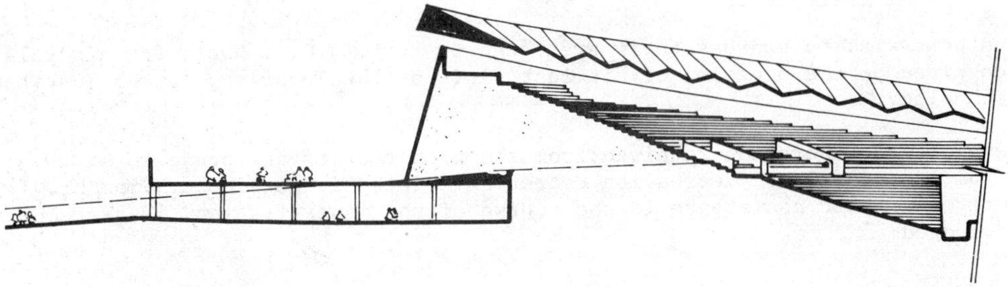

Fig. 4 - Proposed peripheral overhang for the modified stadium design

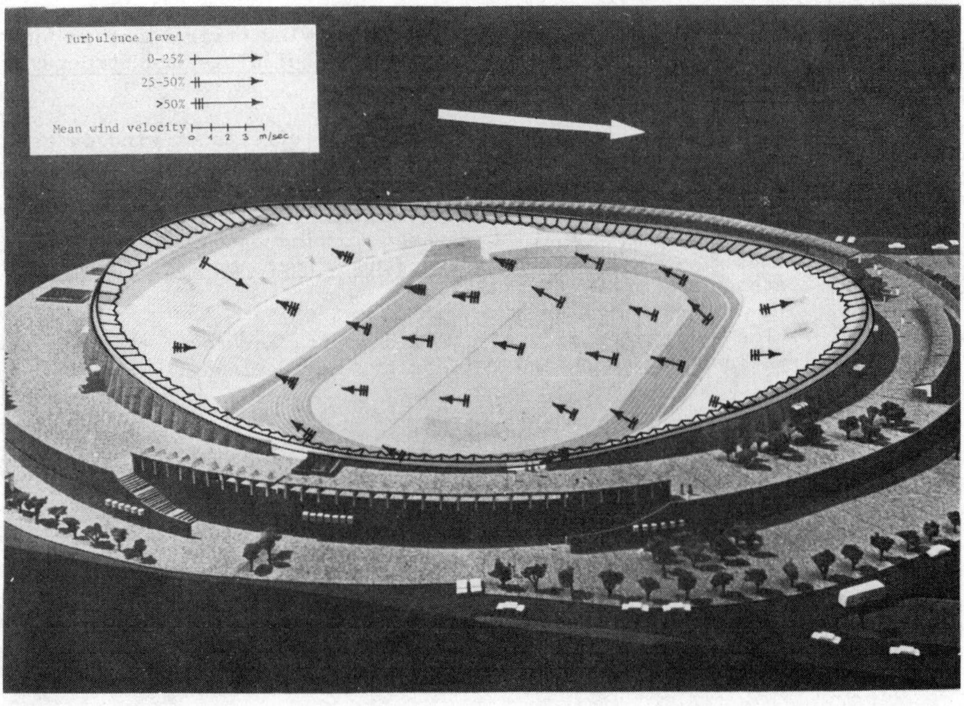

Fig. 5 - Flow pattern distribution in the modified stadium for a westerly wind direction (θ = 270°) and a free stream velocity U_{ref} = 10m/sec.

It is believed that this investigation supplies sufficient information regarding the expected ground level wind climate within and around the stadium to enable such a decision, and also offers a simple design alternative intended to improve the wind environment in the originally proposed stadium.

ACKNOWLEDGEMENTS

The authors wish to express their gratitude to Prof. J.C.R. Hunt, for his valuable advice given on the occasion of his last visit at the Technion, Israel Institute of Technology.

The excellent cooperation received from the architect of the project, Mr. P. Broid, and from the Sports and Recreation Center officials, in particular from Mr. I. Samberg, and their assistance in the course of the investigation, is gratefully acknowledged.

REFERENCES

(1) Sharon, D. and Nachmanovitz, A., The local climate at the Shoa'fat Site, Department of Geography, The Hebrew University, Jerusalem (1977).

(2) Counihan, J., An improved method of simulating a neutral atmospheric boundary layer in a wind tunnel, Atmospheric Environment 3, 197-214 (1969).

(3) Bradbury, L.J.S., Measurements with a pulsed-wire and hot-wire anemometer in the highly turbulent wake of a normal flat plate, J. Fluid Mech., 77, 473-497 (1976).

(4) Isyumov, N. and Davenport, A.G., The ground level wind environment in built-up areas, Proc. 4th Int. Conf. on Wind Effects on Buildings and Structures, Heathrow 1975, 403-422 (1977).

(5) Hunt, J.C.R., Poulton, E.C. and Mumford, J.C., The effects of wind on people; new criteria based on wind tunnel experiments, Build. and Environment, 11, 15-28 (1976).

(6) Lawson, T.V. and Penwarden, A.D., The effects of wind on people in the vicinity of buildings, Proc. 4th Int. Conf. on Wind Effects on Buildings and Structures, Heathrow 1975, 605-622 (1977).

(7) Jackson, P.S., The evaluation of windy environments, Report of the Department of Mechanical Engineering, Auckland University, New Zealand (1978).

WIND EFFECTS ON PEDESTRIANS: NEW CRITERIA BASED ON OUTDOOR OBSERVATION OF OVER 2000 PERSONS

S. MURAKAMI*, K. UEHARA** and K. DEGUCHI***

*University of Tokyo, Japan
**National Institute for Environmental Study, Japan
***University of Hosei, Japan

SUMMARY

Experiments concerned with wind effects on pedestrians are described. The experiments consisted of three parts. In Part I (1975) walking tests were conducted in a large wind tunnel. In Part II (1976) walking tests were held in a precinct at the base of a high-rise building. Part III (1978) again consisted of observation of pedestrians in a precinct at the base of a high-rise building. In Part I and Part II, 5 to 10 persons were used as subjects and movie records of their footsteps were analysed. In Part III, movie records taken by video camera of over 2000 pedestrians were analysed as to footstep-irregularity and body-balance.
Based on these three-part experiments, the following criteria are proposed for evaluating wind effects on pedestrians:

$u<5$ m/s	no effect
$5<u<10$	some effect
$10<u<15$	serious effect
$15<u$	very serious effect

here u = instantaneous wind speed averaged over 3 seconds.
These criteria are the same as or a little more severe than the ones proposed by Hunt et al. (6) and are considerably more severe than others.

INTRODUCTION

In the assessment of wind-induced discomfort around a high-rise building, the choice of suitable criteria for evaluating its acceptability is one major problem. Various criteria concerned with this problem have been suggested already by W.H. Melbourne and P.N. Joubert (1); A.D. Penwarden (2); N. Isyumov and A.G. Davenport (3); T.V. Lawson and A.D. Penwarden (4); J. Gandemer (5); J.C.R. Hunt, E.C. Poulton, and J.C. Mumford (6); S. Murakami and K. Uehara (7,8); and I. Kamei and E. Maruta (9) etc. Especially admirable are the criteria suggested by Hunt which are based on extensive wind tunnel experiments using 40 persons.
The authors of this study have, in the last five years, completed three stages of experiments concerned with this problem. In these experiments over 2000 pedestrians walking in an outdoor passageway were observed and the wind effects on them analysed. The criteria proposed here are more reliable than previous ones, as they are based on the outdoor observation of a very large number of people.

PART I WALKING TEST IN LARGE WIND TUNNEL

1.1 Symbols used in Part I

U_∞ = wind speed in tunnel, m/s
U_s = wind speed in street, m/s
$\overline{U}_s$ = mean of U_s, m/s
X_5 = length of one step taken against-wind with $U_\infty = 5$; refer to Fig. 1.2, m
$\overline{X}_5$ = mean of X_5, m
X_i = same as X_5, but with $U_\infty = i$, m
Y_5 = distance of step from walking line under cross-wind with $U_\infty = 5$, m
$\overline{Y}_5$ = mean of Y_5, m
Y_i = same as Y_5, but with $U_\infty = i$, m
D_i(against-wind) = deviation of X_i; refer to equation (1.1)
D_i(cross-wind) = deviation of Y_i

1.2 Outline of experiments

1) The experiments took place in July and August, 1975.
2) Methodology used for walking tests
Walking of subjects on the wind tunnel test floor was observed and the wind effects were analysed. The subjects were directed to walk on lines as shown in Fig. 1.1, under various conditions, namely, walking with baggage; with umbrella (refer to Photo 1.1); etc.

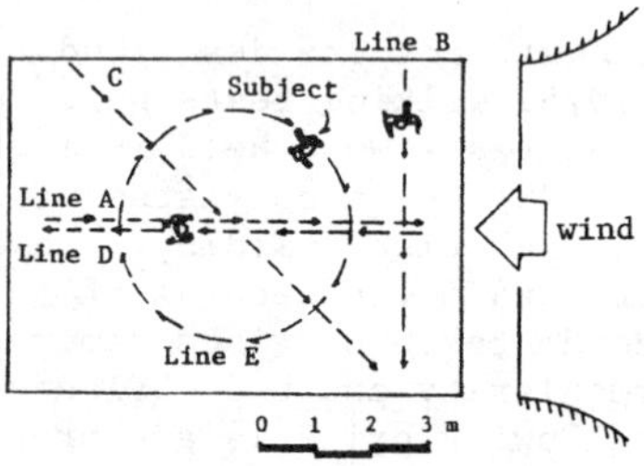

Fig. 1.1 Plan of working section of wind tunnel, showing walking lines used by subjects

Photo 1.1 Walking with umbrella, U_∞=12.5 m/s

Three methods were used for observation and analysis of walking:
a) Mechanical recording and analysis of footsteps;
(Each subject's walking was recorded by a 16 mm movie camera. The footsteps were analysed in detail by a motion-analyser.)
b) Visual assessments by the experimenters on walking conditions, disturbance of hair and clothes, etc;
c) Verbal assessments by the subjects on the relative ease or difficulty of walking.

3) The large wind tunnel used was of the open-jet type. It belongs to the Japan Automobile Research Institute in Ibaragi Prefecture, Japan. The size of the working section measures 6 m in width, 13 m in length, and 5 m in height.
4) Wind profile at working section
Two types of wind profile were used in the test:
a) uniform-flow
b) non-uniform flow with fences.
The height of the fences used was 1.8 m and the lengths were 1.5 m and 1.0 m. The distance between the two fences and the opening ratio of the fences were changed so as to produce various wind profiles.
5) A total of 8 subjects was used, 5 male and 3 female, all of whom were between the ages of 20 and 22. Their heights and weights were average.
6) Analysis of footstep-traces
The sequential-path of footsteps was reproduced from the 16 mm movie film by means of a motion-analyser. Subjects were directed beforehand to walk on the walking lines, as is shown in Fig. 1.1. Their footsteps at $U_\infty = 5$ m/s were used as the standard because wind had no effect on the subjects at this wind speed. For cal-

culating the deviation of footsteps from the walking line, certain equations were defined. Different methods of calculation were used for against-wind and cross-wind conditions, as is shown in Fig. 1.2.
D_i(cross-wind) includes not only the irregularity of the steps themselves but also their lateral displacement.

a) Deviation under against-wind condition:

$$D_i(\text{against-wind}) = \frac{|\bar{X}_5 - X_i|}{|\bar{X}_5 - X_5|} \qquad (1.1)$$

b) Deviation under cross-wind condition:

$$D_i(\text{cross-wind}) = \frac{|(\bar{Y}_5 - Y_i)_{right} + (\bar{Y}_5 - Y_i)_{left}|}{|(\bar{Y}_5 - Y_5)_{right} + (\bar{Y}_5 - Y_5)_{left}|} \qquad (1.2)$$

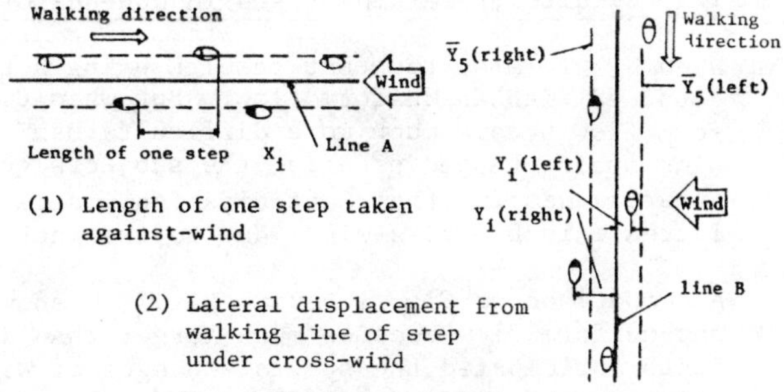

(1) Length of one step taken against-wind

(2) Lateral displacement from walking line of step under cross-wind

Fig. 1.2 Analysis of footstep-traces

1.3 Results of walking tests in uniform flow

Verbal and visual assessments by subjects and experimenters are shown in Table 1.1. In these tests male subjects dressed in shirts and trousers and female subjects wore skirts. When the wind speed exceeded 10 m/s, some subjects felt it was not easy to walk. As the wind speed passed 15 m/s, most subjects felt it was difficult to control their walking. Wind effects on females were more severe than those on males because of the turbulence created by their longer hair and skirts and because of their inferiority of strength. The eyes proved to be the point most damageable. When the wind speed exceeded 15 m/s, it became impossible to keep the eyes open continuously. Damage to eyes was most severe when the wind hit the face from an angle of 45° rather than frontally. It was impossible to hold an umbrella when the wind speed exceeded 12.5 m/s (refer to Photo 1.1). The footsteps of a female subject under various wind speeds are diagrammed in Fig. 1.3. As the wind speed increased, female subjects were blown laterally to an ever-increasing degree despite

Table 1.1 Criteria based on tests in Part I: results of verbal and visual assessments (for non-uniform flow, downward revision is required)

based on U_∞ (can be assumed $= U_{s\,max}$)		5	10	15	20 m/s
(based on $\bar{U}_s$ (estimated as G.F.=2.0)		2.5	5	7.5	10)
Uniform flow	effects on walking		walking not easy footsteps irregular body-balance lost	walking difficult to control upper body bends to windward	walking very difficult whole body bends to windward
	on hair	minor disturbance	disturbed	violently disturbed	
	on face eyes ears	wind felt on face	frequent blinking wind noisy	impossible to open eyes continuously falling tears	impossible to face wind, facial pain ear-ache, headache breathing difficult
	on clothes and umbrella	minor disturbance	fluttering of clothes difficult to hold umbrella	impossible to hold umbrella	violent fluttering of clothes
Non-uniform flow	effects on walking		walking not easy	walking difficult stumbling	dangerous possible falling

their efforts to keep their body-balance and to walk on the line by bending their bodies to windward. Since wind turbulence is much larger in the street, it is very difficult to keep one's body continuously in balance. The increase of D_i(cross-wind) is shown in Fig. 1.4.

1.4 Results of walking tests in non-uniform flow

When $U_\infty \leqq 10$ m/s, the test results using a non-uniform flow were similar to those obtained with an uniform flow. But when U_∞ exceeded 15m/s, walking in a non-uniform flow became much more difficult than walking in an unifrom flow. Where the wind speed changed greatly, the subjects could not keep their balance and staggered severely despite their anticipation, as is shown in Fig. 1.5 and Photo 1.2. The increase in D_i(cross-wind) due to the increase of U_∞ is shown in Fig. 1.6.

A comparison of Fig. 1.6 with Fig. 1.4 shows that the value of D_i(cross-wind) in a non-uniform flow becomes much larger than in an uniform flow. Although the subjects anticipated the spatial changes of wind speed, the sudden changes in wind velocity made it very difficult for them to keep their balance.

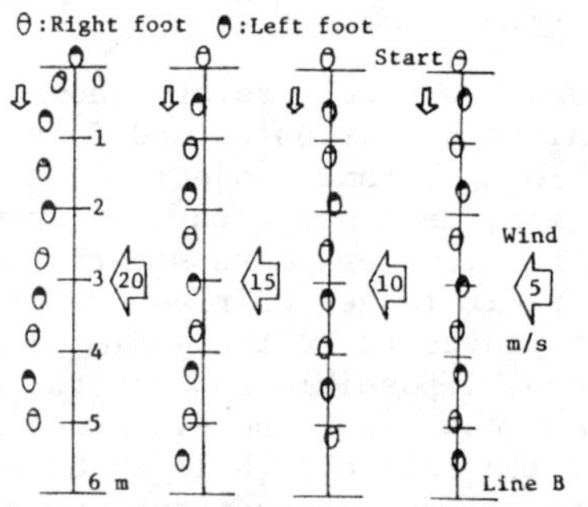

Fig. 1.3 Footsteps of female subject taken in uniform cross-wind (as indicated by walking line B in Fig. 1.1)

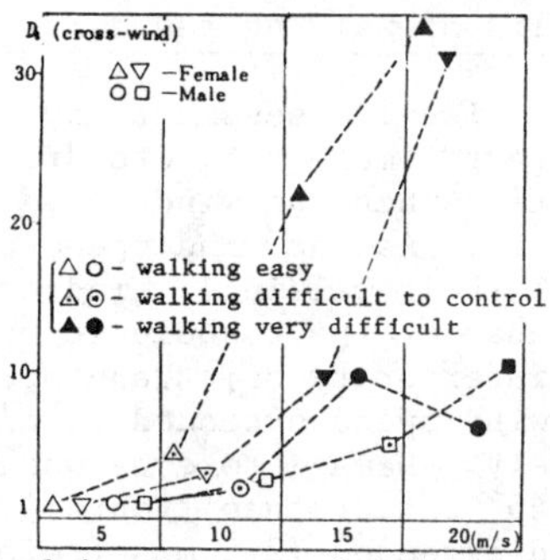

Fig. 1.4. Increase of D_i(cross-wind) due to increase of U_∞

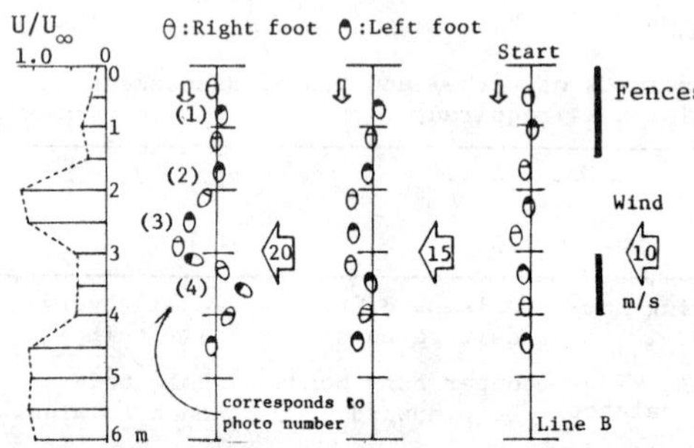

Fig. 1.5. As Fig. 1.3, but non-uniform flow behind fences

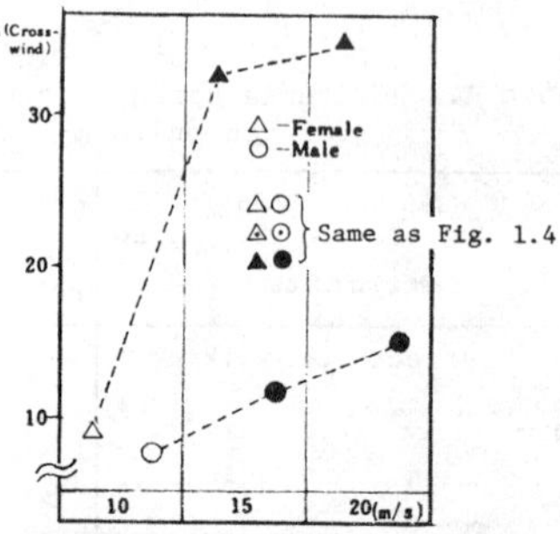

Fig. 1.6 As Fig. 1.4, but with non-uniform flow behind fences

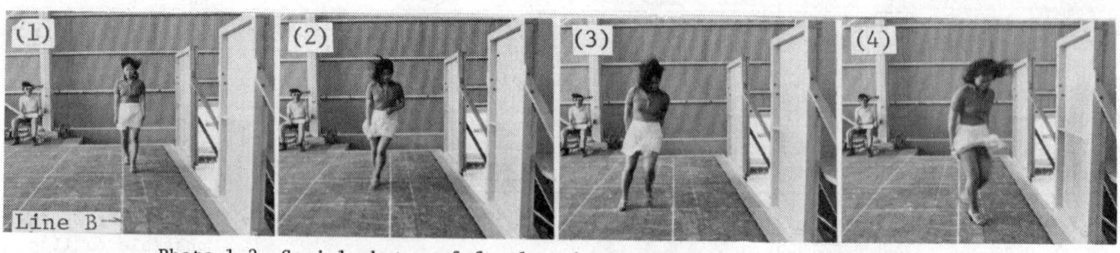

Photo 1.2 Serial photos of female subject taken at 1 second intervals in non-uniform flow behind fences, with U_∞=20 m/s

1.5 Comparison of walking outdoors with walking in wind tunnel

Walking in an outdoor area is much more difficult than in the wind tunnel because:

a) The change of wind speed in both time and space is very large and difficult to anticipate;

b) Eyes are damageable by the dust in the outdoor air.

These remarks are based on the experiments of Part I described above. U_∞, which is constant in the wind tunnel, may be assumed to correspond to the peak of U_s in the street. If $U_{s\,max} / \overline{U}_s$ is assumed to be 2.0, the criteria based on $\overline{U}_s$ may be shown below as an example.

walking difficult to control ------- $U_\infty > 15$ m/s or $\overline{U}_s > 7.5$ m/s
walking difficult ------------------ $U_\infty > 20$ m/s or $\overline{U}_s > 10$ m/s

1.6 Conclusions of Part I

1) The results of the tests conducted in Part I are summarized in Table 1.1. In cases of non-uniform flow with fences, these criteria need to be rivised downward.
2) The eyes are the point most damageable by wind.
3) Walking in non-uniform wind is much difficult than in uniform flow.

PART II WALKING TESTS IN PRECINCT AT BASE OF HIGH-RISE BUILDING

2.1 Symbols used in Part II

U = instantaneous wind speed in precinct at height of 1 m (horizontal component measured by sonic-anemometer), m/s

$\overline{U}_{10\text{-}sec}$ = mean value of U, as averaged over a period of about 10 seconds, during which time one subject finished one trial of walking, m/s

X = length of one step; refer to Fig. 2.2, m

$\sqrt{\overline{X^2}}$ = deviation of X, m

Y = distance between right and left foot, m

$\sqrt{\overline{Y^2}}$ = deviation of Y, m

2.2 Outline of experiments

1) Wind velocity on the date of tests

Date of test	Daily maximum wind speed at Tokyo Met.Observ.	Mean wind speed $\overline{U}_{10\text{-}sec}$ at walking line	Intensity
No.1 Oct. 2, 1976	6.9 m/s	1.56 m/s	about25 %
No.2 Oct.25, 1976	23.4	3.80	53
No.3 Dec. 9, 1976	13.5	3.92	10
No.4 Mar. 9, 1977	25.4	7.62	15

2) The high-rise building and the pedestrian precinct are located at Shinjuku, Tokyo. The plan of the precinct is shown in Fig. 2.1.

3) Methodology used for walking tests

The walking of the subjects was recorded by a 16 mm movie camera. The subjects were directed beforehand to walk on the walking line. The walking line was set at the most windy space in the precinct, which changed according to the prevailing wind direction, as shown in Fig. 2.1. Its length was about 7 to 8 m. One walking trial along the line took about 10 seconds per subject, and the trials were repeated many times. After each walking trial, verbal assessments by

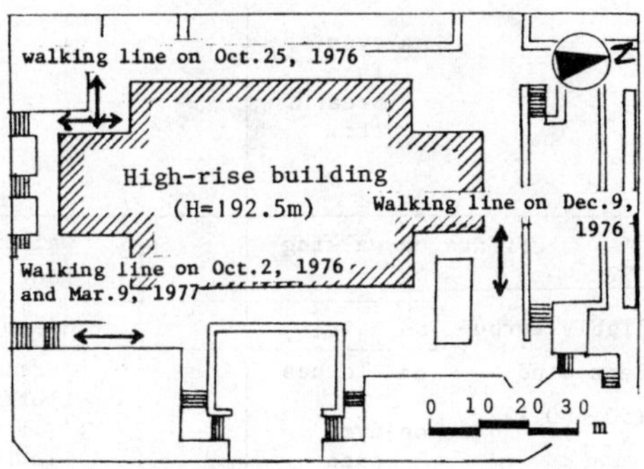

Fig. 2.1. Plan of precinct and walking lines

the subjects were recorded, as well as visual assessments by the experimenters. The wind speed at the walking line was measured by a sonic-anemometer set at a height of 1 m. Footsteps of the subjects were analysed by a motion-analyser using the film record of the walking tests.

4) Subjects were all female aged 20 to 22.

5) Analysis of footstep-traces (Fig. 2.2)

To analyse the walking conditions, the following five measures were selected:

X; $\sqrt{\overline{X}^2}$; Y; $\sqrt{\overline{Y}^2}$; and walking speed.

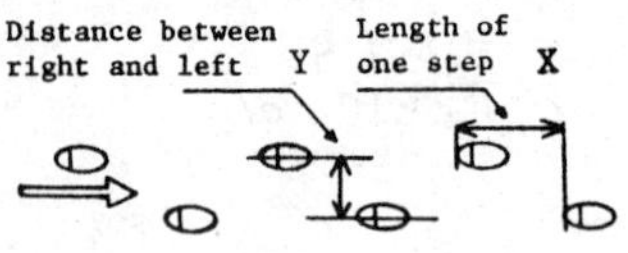

Fig. 2.2 Analysis of footstep-traces

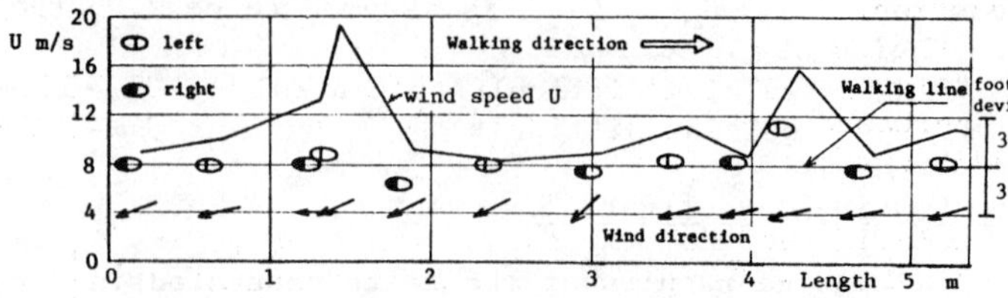

Fig. 2.3. Relation between wind speed and footsteps for female subject (walking speed is about 1 m/s)

2.3 Verbal and visual assessments by subjects and experimenters

The results of the verbal and visual assessments are shown in Table 2.1. It is clear from these results that walking in highly turbulent wind is much more difficult than in non-highly turbulent wind.

On Oct. 25, the wind was highly turbulent, while on Oct. 2nd, Dec. 9, 1976 and Mar. 9, 1977, it was not highly turbulent. It is notable that the assessments "footsteps irregular" or "walking difficult to control" occurred at wind speed 3 m/s when wind turbulence was high but they occurred at wind speed 7 to 9 m/s when wind turbulence was not high. When wind was highly turbulent the instantaneous maximum wind speed reached 8 to 9 m/s although the mean value stayed about 2 to 3 m/s. The results in Table 2.1 agree with the criteria based on street wind $\overline{U}_S$ given in Table 1.1, allowing for the difference in turbulence characteristics in each case.

Table 2.1 Criteria based on tests in Part II : results of verbal and visual assessments

$\overline{U}_{10\text{-sec}}$ (at height of 1 m averaged over 10 sec)		1 2	3 4	5 6	7 8	9 10 m/s
non-highly turbulent wind (around 20 %)	effects on walking			little effect	footsteps irregular	walking difficult to control
	on hair			hair disturbed	hair blown strongly	hair violently disturbed
	on clothes			partial fluttering	clothes violently fluttered	
	on eyes face breathing etc.				skirt flew up wind felt strong	eyes felt dry difficult to face wind wind felt violent
highly turbulent wind (50 ~ 60 %)	effects on walking		walking difficult to control footsteps irregular			
	on hair		disturbed	hair blown strongly		
	on clothes		partial fluttering	clothes violently fluttered skirt flew up		
	on eyes face breathing etc.		wind felt strong	difficult to face wind		

2.4 Changes in footsteps due to increase of wind speed

1) Footstep irregularity caused by change of wind speed U (Fig. 2.3)
It is clear that a change of U is greatly influential on walking.

2) Relation between $\overline{X}$ (length of one step) and $\overline{U}_{10\text{-sec}}$ (Fig. 2.4)
$\overline{X}$ decreases as $\overline{U}_{10\text{-sec}}$ increases. This tendency is clear in cases of against-wind and highly turbulent wind conditions, but is not clear in the case of follow-wind condition.

3) Relation between $\sqrt{\overline{X^2}}$ (deviation of X) and $\overline{U}_{10\text{-sec}}$ (Fig. 2.5)
$\sqrt{\overline{X^2}}$ increases as $\overline{U}_{10\text{-sec}}$ increases in cases of against-wind and highly turbulent wind conditions.

4) Relation between $\overline{Y}$ (distance between right and left foot) and $\overline{U}_{10\text{-sec}}$ (Fig. 2.6)
$\overline{Y}$ increases as $\overline{U}_{10\text{-sec}}$ increases, but in this case, there is no difference in this tendency between against-wind, follow-wind, and highly turbulent wind conditions.

5) Relation between walking speed and $\overline{U}_{10\text{-sec}}$ (Fig. 2.7)
Walking speed decreases as $\overline{U}_{10\text{-sec}}$ increases. This tendency is clear in cases of against-wind and highly turbulent wind conditions, but is not clear in the case of follow-wind condition.

The results diagrammed in these five figures (Fig. 2.3 to 2.7) are based on the tests of one subject who is rather sensitive to wind, but the tests of other subjects show similar results.

Date of tests	Oct.2	Oct.25	Dec.9	Mar.9
Highly turbulent	—	×	—	—
Non-highly, follow	●	—	■	▲
Non-highly, against	○	—	□	△

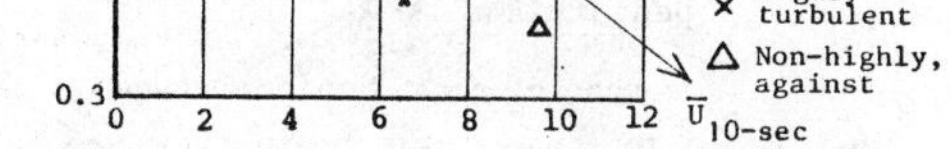

Fig. 2.4 Relation between $\overline{X}$ and $\overline{U}_{10\text{-sec}}$

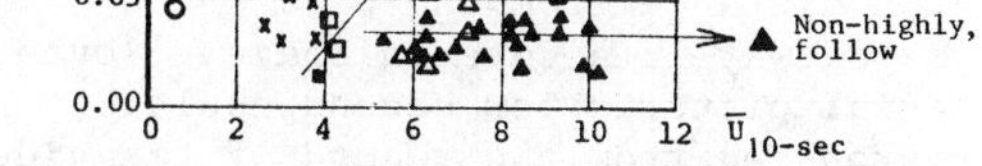

Fig. 2.5 Relation between $\sqrt{\overline{X^2}}$ and $\overline{U}_{10\text{-sec}}$

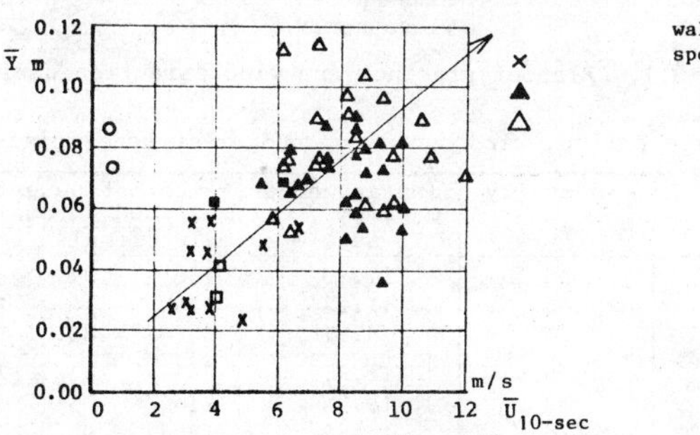

Fig. 2.6 Relation between $\overline{Y}$ and $\overline{U}_{10\text{-sec}}$

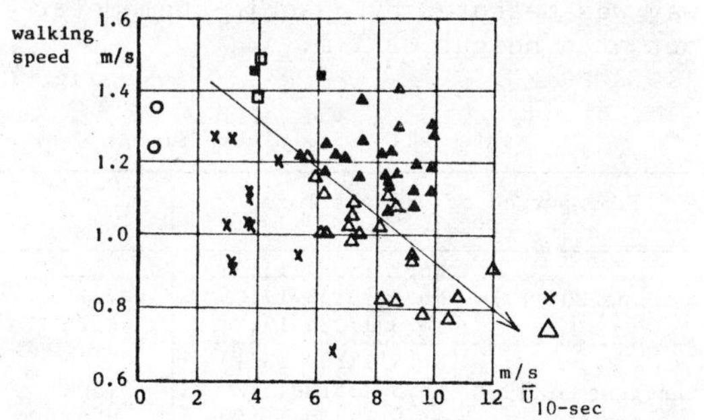

Fig. 2.7 Relation between walking speed and $\overline{U}_{10\text{-sec}}$

2.5 Conclusions of Part II

1) Wind effects on pedestrians examined in Part II are summarized in Table 2.1.
2) The footstep pattern is influenced by wind turbulence, wind speed, and wind direction. Highly turbulent wind and against-wind conditions are especially influential.
3) When the wind is highly turbulent, the assessments "footsteps irregular" or "walking difficult to control" may occur at wind speed 3 m/s, while they occur at wind speed 7 to 9 m/s when it is not highly turbulent.
4) The criteria in Table 2.1 agree with the criteria based on street wind speed $\overline{U}_s$ given in Table 1.1, allowing for the difference in turbulence characteristics and averaging-time in each case.

PART III OBSERVATION OF PEDESTRIANS IN PRECINCT AT BASE OF HIGH-RISE BUILDING

In the experiments of Part I and Part II, tests were conducted with pre-selected subjects. For Part III, random pedestrians within the precinct were recorded by a video TV camera. Using these records the wind effects were analysed.

3.1 Symbols used in Part III

U = instantaneous wind speed in precinct at height of 1 m, (horizontal component measured by sonic-anemometer) m/s

$U_{3\text{-sec}}$ = wind speed averaged over 3 seconds, m/s

$U_{10\text{-min}}$ = mean value of U, averaged over 10 minutes, m/s

3.2 Outline of experiments

1) The high-rise building and the pedestrian precinct are located at Shinjuku, Tokyo. The plan of the precinct is shown in Fig. 3.1.

2) Methodology used for walking tests
The walking conditions of pedestrians were recorded by a video TV camera set at a distance from the pedestrian passageway in a covered space. Seven serial pictures per second could be reproduced from the record of the video camera. Wind velocity at the passageway was measured by a sonic-anemometer set at a height of 1 m.

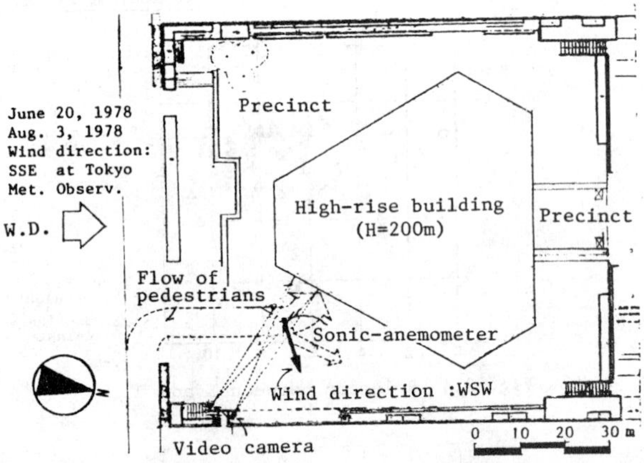

Fig. 3.1. Plan of precinct and wind direction during tests

Table 3.1 Date of experiments, time period, wind speed and number of pedestrians

Time period of video filming used in analysis		Mean $\overline{U}$ (m/s)	Intensity $\sqrt{\overline{u^2}}/\overline{U}$ (%)	Instantaneous maximum $U_{3\text{-sec}}$ (m/s)	Number of pedestrians included in analysis: male	female	total
June 20, 1978	No.1 15:44-15:54	7.22	28	11.90	138	91	229
	No.2 16:54-17:04	7.72	26	11.21	179	118	297
	No.3 17:24-17:34	4.26	36	8.91	336	285	621
August 3, 1978	No.4 13:00-13:10	9.65	38	17.17	261	153	414
	No.5 13:10-13:20	11.02	30	19.34	214	105	319
	No.6 13:30-13:40	9.67	28	14.82	198	98	296
total pedestrians used in analysis					1326	850	2176

3.3 Wind speed and number of pedestrians

Filming lasted for about 4 to 5 hours on both days. For the analysis of walking conditions, 6 time periods (No.1 to No.6) were selected, as is shown in Table 3.1. The number of pedestrians included in the analysis totaled 2176 persons. Wind speed and turbulence intensity for each time period is also shown in Table 3.1. The value of $U_{3\text{-sec}} / \bar{U}_{10\text{-sec}}$ as measured in the precinct ranges from about 1.5 to 2.5.

3.4 Grading system for footstep-irregularity and walking-balance

The grades for footstep-irregularity and walking-balance are shown in Tables 3.2 (1), (2). Photo 3.1 is an example of a pedestrian walking under footstep-irregularity grade 3. (The subject of the analysis in each photo is indicated by an arrow; the number of each photo corresponds to the same number on the following record of wind velocity.)

Emerging from building — Footsteps irregular — Footstep-path recovered

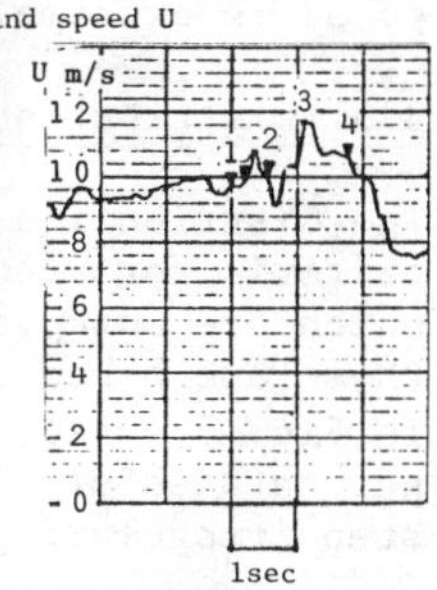

Photo 3.1 Footstep-irregularity "Grade 3"

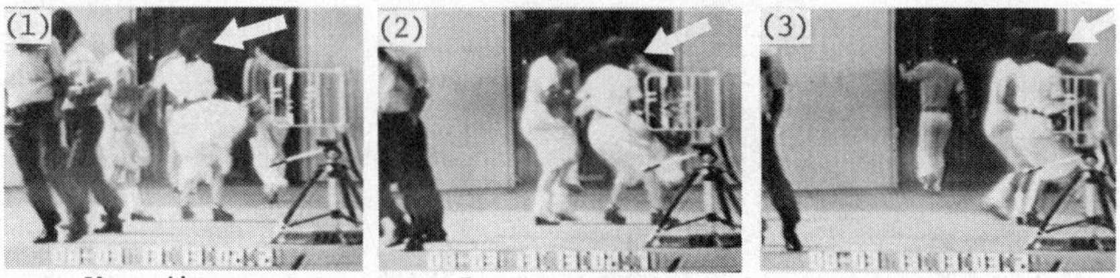

Blown sideways — Turned to leeward — Blown sideways again

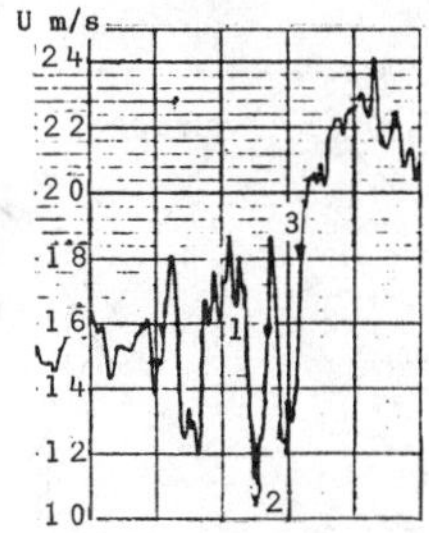

Photo 3.2 Footstep-irregularity "Grade 5"

Bending upper body to windward, despite being a large full-grown man with relatively great strength to resist the wind

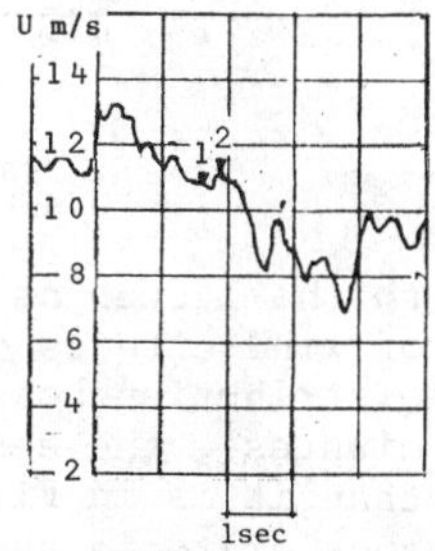

Photo 3.3 Walking-balance "Grade C"

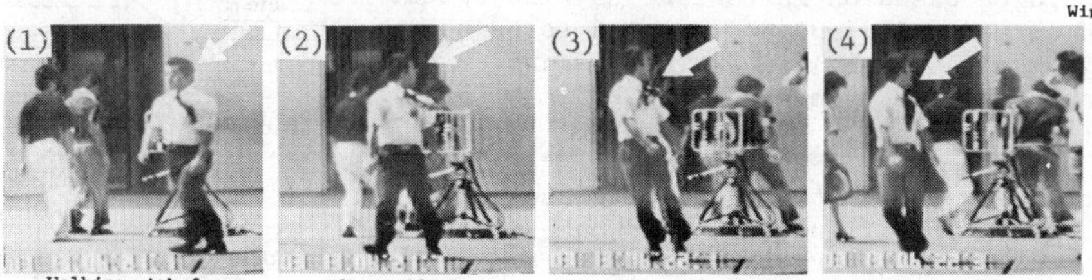

Walking with face to wind — Blown sideways — Walking to windward but facing to leeward

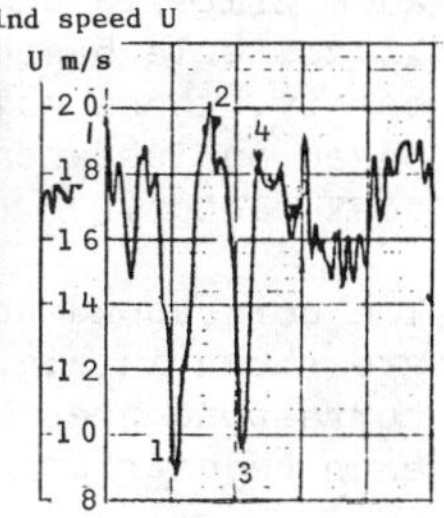

Photo 3.4 Walking-balance "Grade D"

Table 3.2(1) Grades of footstep-irregularity

Grade 1	no effect
Grade 2	minor irregularity of footsteps
Grade 3	footsteps irregular
Grade 4	walking line greatly irregular (walking difficult to control, body sometimes blown sideways)
Grade 5	body blown sideways or to leeward (walking impossible to control, walking difficult, walking stop)

Table 3.2(2) Grades of walking-balance

Grade A	no effect
Grade B	sensitive to wind (face turns sideways to avoid gust, face covered with hand to break gust)
Grade C	upper-half of body bends to windward
Grade D	whole body bends to windward, whole body swings

Photo 3.2 is an example of a pedestrian walking under footstep-irregularity grade 5. Photo 3.3 is an example of a pedestrian walking under walking-balance grade C. Photo 3.4 illustrates an example of walking-balance grade D. The change of wind speed is extremely large.

3.5 Distributions for each grade

The distribution for each grade is shown in Figs. 3.2 (1), (2). Acceptance criteria for wind effects on people can be easily proposed from these figures, which are more valid than those previously suggested by other researchers because these are based on a far greater number of observations of pedestrians.
In Fig. 3.2 (2), grade B for females and grade C for males and females occupy large percentages of the total. The separation of each grade is much clearer for footstep-irregularity than it is for walking-balance. Thus the former seems preferable

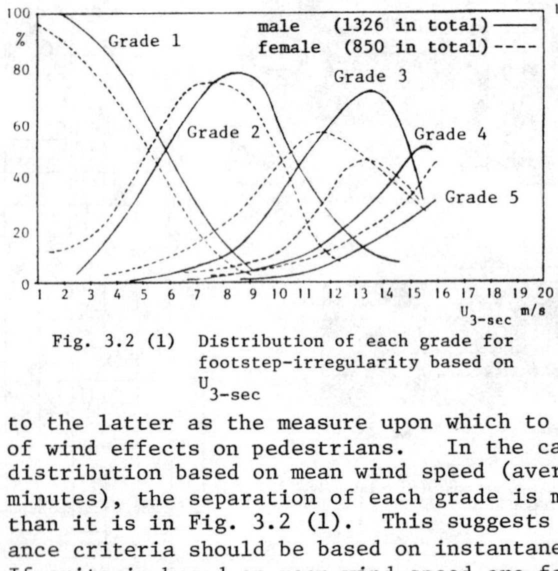

Fig. 3.2 (1) Distribution of each grade for footstep-irregularity based on U_{3-sec}

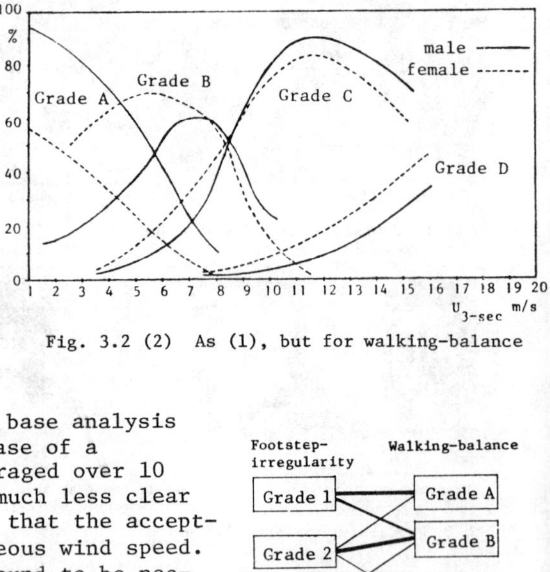

Fig. 3.2 (2) As (1), but for walking-balance

to the latter as the measure upon which to base analysis of wind effects on pedestrians. In the case of a distribution based on mean wind speed (averaged over 10 minutes), the separation of each grade is much less clear than it is in Fig. 3.2 (1). This suggests that the acceptance criteria should be based on instantaneous wind speed. If criteria based on mean wind speed are found to be necessary, these may be obtained by converting the criteria based on instantaneous value by means of peak factor or gust factor.

The correspondences between the two types of grades are shown in Fig. 3.3. This figure is based on correspondences between the two types of grades for each pedestrian.

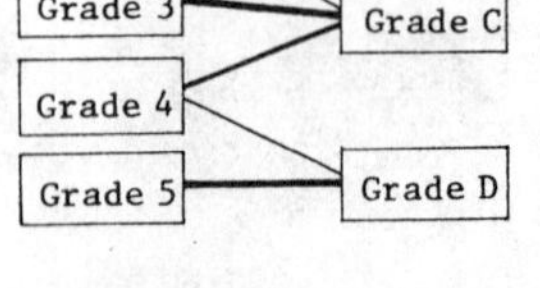

Fig. 3.3 Correspondence between footstep-irregularity and walking-balance

3.6 Conclusions of Part III; comparison between proposed and previous criteria

The acceptance criteria based on instantaneous wind speed that are proposed here are compared to previous ones in Fig. 3.4. The difference in averaging-time used in each criteria should be considered with care. The vertical screen-tone lines show the velocity values where neighboring distribution curves for female cases cross. (As, for example, at 5 m/s, where the percentage of grade 2 becomes higher than that of grade 1.) The values of 5, 10, and 15 m/s --which correspond well with the grades proposed by Hunt et al.(6)-- are:

onset of some effect ---------- $U_{3\text{-sec}} > 5$ m/s
onset of irregular footsteps or walking difficult to control ---- $U_{3\text{-sec}} > 10$ m/s
onset of walking difficult or dangerous for elderly person ------ $U_{3\text{-sec}} > 15$ m/s

These results are virtually the same as those of Table 1.1 in Part I and Table 2.1 in Part II, allowing for the difference in turbulence characteristics and averaging-time in each case.

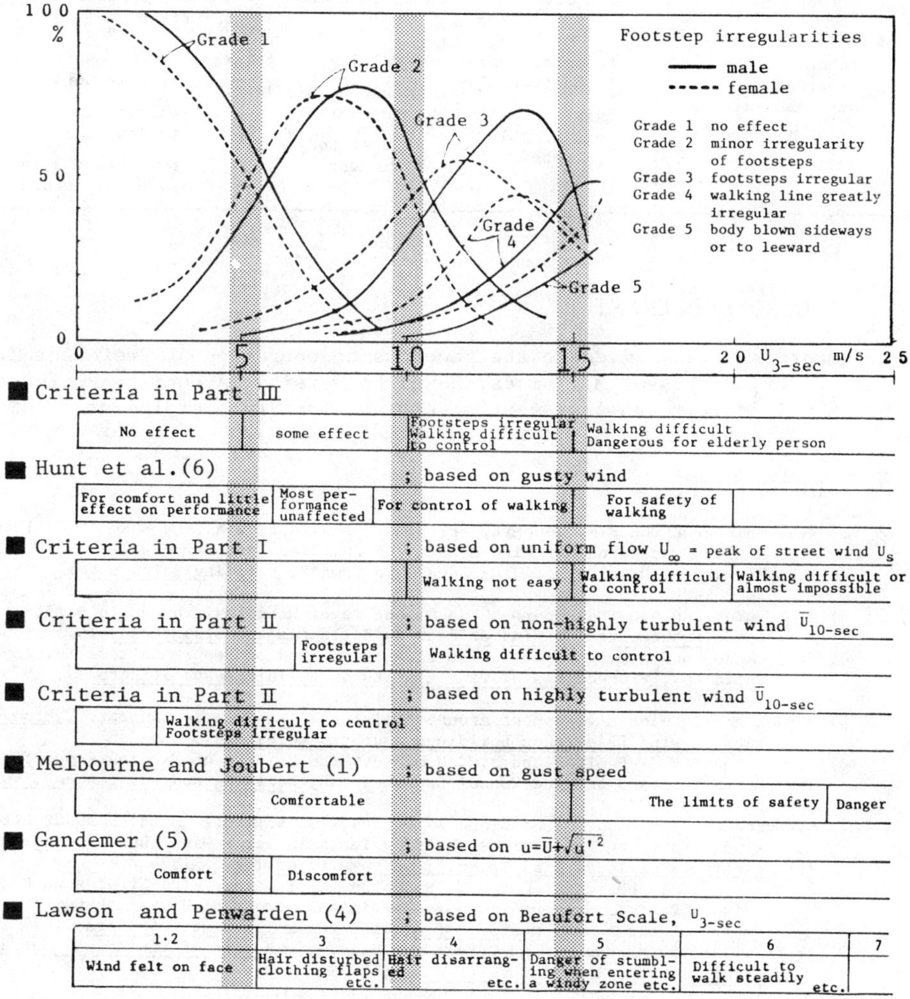

Fig. 3.4 Comparison between proposed and previous criteria based on instantaneous wind speed (difference of averaging-time in each criteria should be considered with care)

CONCLUDING REMARKS

1) Acceptance criteria based on the tests in Parts I, II and III are proposed in Table 4.1. When wind is highly turbulent in time and space, downward revision is required.
2) Non-uniformity of wind speed in time and space greatly affects walking. In highly turbulent wind walking is extremely difficult and assessment "walking difficult to control" may occur at wind speed 3 m/s.
3) Eyes are the point most damageable by wind.

Table 4.1 Acceptance criteria for pedestrians;
based on instantaneous wind speed $U_{3\text{-sec}}$
(for highly turbulent wind, downward revision is required)

$U_{3\text{-sec}} < 5$ (averaged over 3-sec)	5 ~ 10	10 ~ 15	$U_{3\text{-sec}} > 15$ m/s
no effect	some effect	serious effect	very serious effect
in case of female, minor effect on hair and skirt	footsteps sometimes irregular hair and skirt considerably disturbed	walking irregular walking difficult to control upper body bends windward	dangerous for elderly person walking impossible to control body blown sideways or leeward

ACKNOWLEDGEMENT

Acknowledgment is hereby given to the generous co-operation pf Professor T. Shoda, Nihon University; Professor M. Yamada, Hosei University; Associate Professor T. Goto, Hosei University; and Assistant H. Komine, Institute of Industrial Science, University of Tokyo.

REFERENCES

1) W.H. Melbourne and P.N. Joubert, "Problems of Wind Flow at the Base of Tall Buildings", Proceedings of Wind Effects on Buildings and Structures, 1971 Tokyo
2) A.D. Penwarden, "Acceptable Wind Speeds in Towns", Building Science Vol.8,1973, pp 259-267
3) N. Isyumov and A.G. Davenport, "The Ground Level Wind Environment in Built-up Areas", Proceedings of Wind Effects on Buildings and Structures, 1975 London
4) T.V. Lawson and A.D.Penwarden, "The Effects of Wind on People in the Vicinity of Building", Proceedings of Wind Effects on Buildings and Structures, 1975 London
5) J. Gandemer, "Wind Environment around Buildings: Aerodynamic Concepts", Proceedings of Wind Effects on Buildings and Structures, 1975 London
6) J.C.R. Hunt, E.C. Poulton and J.C. Mumford, "The Effects of Wind on People; New Criteria Based on Wind Tunnel Experiments", Building and Environment, Vol.11 pp 15-28, 1976
7) S. Murakami, T. Shoda, T. Goto and K. Uehara, "Experimental Studies about Strong Wind Effects on Pedestrians; Walking Tests in Large Wind Tunnel", Journal of Institute of Industrial Science, University of Tokyo, 1976
8) S. Murakami, K. Uehara, T. Goto, T. Shoda and M. Yamada, "Wind Effects on Pedestrians; Part II Walking Tests in Precinct in Front of High-rise Building", Report of the Annual Meeting of Architectural Institute of Japan, 1977
9) I. Kamei and E. Maruta, "Wind Effects on Walking", Report of the Annual Meeting of Architectural Institute of Japan, 1978

THE NEAR-WAKE OF A TALL BUILDING BLOCK IN UNIFORM AND TURBULENT FLOWS

M. E. DAVIES, V. G. QUINCEY and S. J. TINDALL

National Maritime Institute, Teddington, Middlesex, UK

1. SUMMARY

Measurements have been made of the velocity field in the near-wake of a square-section building model of height to side ratio of six. In the two cases explored fully the complete mean velocity fields have been represented by three dimensional streamlines and the resulting complex topological patterns examined. A contraction of the wake recirculation region is evident when the body is subject to an oncoming turbulent shear flow and attention is focussed on the modification of the other major features or "modules" that characterise these flows. The experiments were intended to aid the understanding of environmental flows of this sort and also to provide more information on the relevance of atmospheric boundary layer simulation. A movie film of computer generated streamline development will accompany the presentation.

2. INTRODUCTION

Recent work at the National Maritime Institute at Teddington (formerly a division of the National Physical Laboratory) has included the implementation of a "counter-jet" atmospheric boundary layer simulation method in a large, newly modified wind tunnel designed for industrial aerodynamic applications. Following the lead of the work at IIT (1),(2) an extensive study has been made of the characteristics of the turbulent shear flows produced by counter-jets. The experiments were conducted in a 1.2m (4ft) by 0.6m (2ft) blower tunnel to determine a viable system for the new facility which has a cross-section of 4.8m (16ft) by 2.4m (8ft). The resulting flexible generation methods are described in (3), and include descriptions of simulations covering smooth-sea to city-centre terrains, together with results showing a good range of attainable scales. Complementary to investigations of simulation technique, there exists the continuing need to understand the effects of deficient modelling of the oncoming flow and also, in sensitive situations, to have available as much full scale information as possible. These elements are part of the current NMI programme; they include the work reported here and measurements of the wind over the ocean in different sea states, (4).

It has been demonstrated in various experiments that turbulence and shear in the approach flow can have significant effects on the nature of the wake of both two and three-dimensional bluff obstacles and on the pressures and forces experienced by such bodies, (e.g. (5),(6),(7),(8),(9),(10)). However, much remains to be understood about the details of the near-wake of surface-mounted bluff bodies, there being few measurements in the difficult reversed flow regions with the notable exception of the work of Castro and Robins (11), who examined the flow around a cube.

The present experiments concentrate on the details of the wake of a square-section prism of height to side ratio of six. The investigation was intended to throw more light on the influence of the approach flow on a tall building model, and also to attempt to reveal the nature of the flow structure in such cases. An interesting characterisation of three dimensional bluff body flows has been given by Morkovin (12) and these experiments present an opportunity to appraise some of his conjectures. He reasonably suggests that through an appreciation of the constituent elements of a flow, it may then be possible to construct satisfactory wind environments by the sensible use of these "modules". Similarly, the same flow modules can be used as vehicles to discuss the changes in wake pattern due to varying upstream input and further serve to order one's picture of an otherwise highly complex three-dimensional flow.

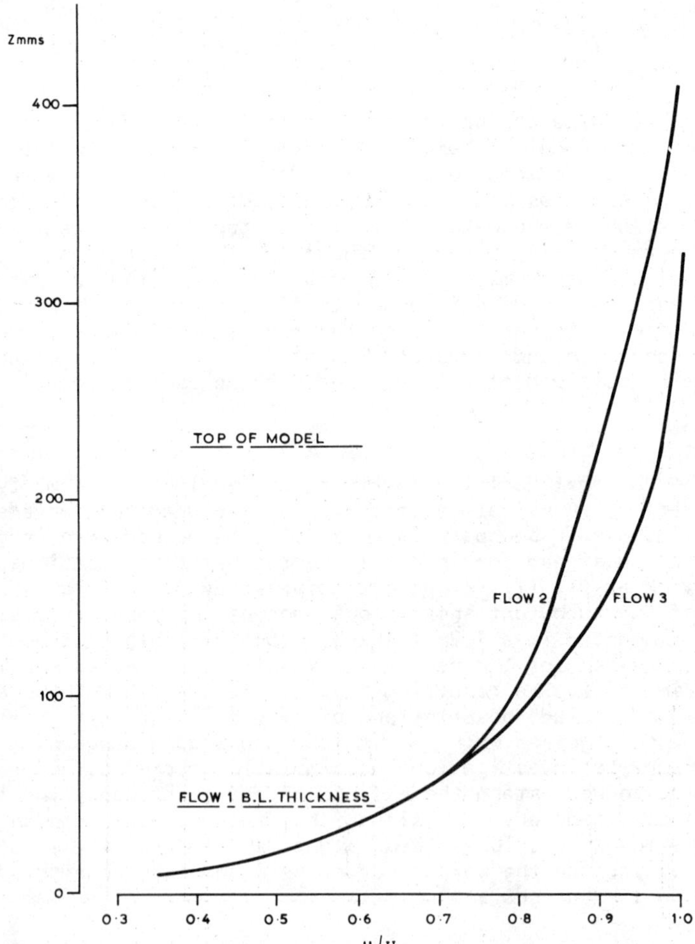

FIG. 1. MEAN VELOCITY PROFILES OF APPROACH FLOWS

3. EXPERIMENTAL DETAILS

The flow around a 229mm (9in) high by 38mm (1.5in) wide square section block was investigated in a wind tunnel 1200mm (4ft) wide and 600mm (2ft) high. Three approach flows were considered, two generated by the counter-jet system over a rough surface and the other developed naturally in an empty tunnel. The "smooth uniform" stream (flow 1), grew a floor boundary layer approximately 0.15h to 0.2h thick, where h is the block height. The rough matting surface had an effective roughness length, zo, of order 0.009h and the boundary layer thicknesses were in the range of 1.5h to 2h, as shown by the profiles of Figure 1.

Explorations of the velocity field in the block's wake were made with a pulsed-wire anemometer. For the full traverses all three components were measured and the extensive survey was made possible by an automatic system of data collection. The stepper-motor driven traverse gear was interfaced to a microcomputer, as was the output from the pulsed-wire. Vertical planes (y,z) normal to the stream direction were covered by an array of 13 x 13 points, extending from the streamwise centre-line of the body out to one side. The vertical plane (x,z) containing this line was checked as being a plane of symmetry in the mean. Mean and root mean square velocities were computed over a variable number of samples depending on an initial estimate of the fluctuations at any point. On a purely statistical basis, errors in the mean and rms were kept within ±1% and ±2 to 3%, of their respective maximum values.

4. VELOCITY MEASUREMENTS ON THE STREAMWISE CENTRE-LINE

Measurements of streamwise (u) and vertical (w) velocities were made on the wake centre-plane for the three approach flows considered. Figure 2 shows the variation of the u-intensity with downstream distance and reveals the marked effect of the character of the approach flow. In all cases the turbulence level rises in the near-wake and increases downstream for a short distance. In the case of the uniform smooth flow (flow 1) a maximum is reached approximately 4b (0.66h) behind the rear face where b is the breadth of the model. For flows 2 and 3 the distance to the maximum is 1.5b (0.25h) and the u-intensity at this position is 30% of the free stream velocity over a large proportion of the height of the model. For flow 1 the maximum intensity is of order 27%, only slightly less than for the turbulent streams in spite of the marked initial

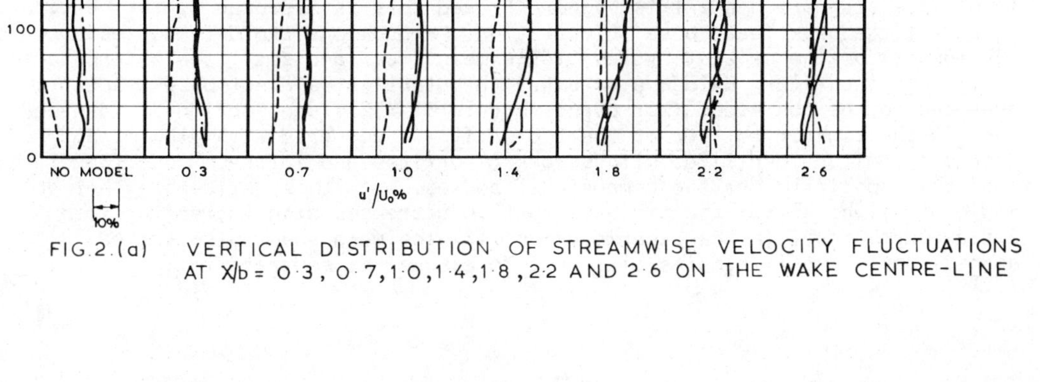

FIG.2.(a) VERTICAL DISTRIBUTION OF STREAMWISE VELOCITY FLUCTUATIONS AT X/b = 0·3, 0·7, 1·0, 1·4, 1·8, 2·2 AND 2·6 ON THE WAKE CENTRE-LINE

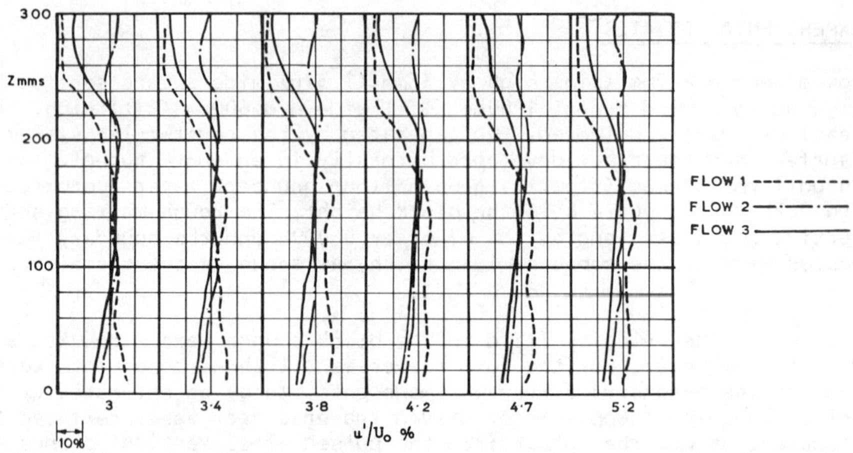

FIG. 2. (b). VERTICAL DISTRIBUTION OF STREAMWISE VELOCITY FLUCTUATIONS AT X/b = 3, 3·4, 3·8, 4·2, 4·7, 5·2 ON THE WAKE CENTRE-LINE

difference. In the region between 3b and 3.5b downstream the three flows give similar distributions of the u-fluctuations, but it is clear that the overall flow field is vastly different, at least between the smooth and turbulent flows. Flow 1 is still in near-wake "development", whereas 2 and 3 are relaxing as the wake decays. The decay region has been well explored (13),(14) and it was shown in (11) how much more quickly the wakes in a turbulent stream decay. The major difference in the near-wake, implied by these measurements, is even more clearly demonstrated by examining the mean velocity field. Only the wakes due to flows 1 and 3 have been thoroughly traversed, but mean velocity vectors on the centre-line are available for all three cases, (assuming the lateral mean flow to be zero). The mean streamlines in this plane have been computed as the paths particles would take if a steady flow existed, since the meaning of a streamline is simply that its tangent at any point is along the local flow direction. Figures 3, 4 and 5 demonstrate the combined effect of increasing turbulence and changing mean shear. The distances to maximum fluctuation deduced from Figure 2 are marked on these streamline plots and are seen to coincide with the "closure" of the wake recirculation. In the presence of a turbulent shear flow it is noticeable that not only is the recirculation region shortened, but also the vertical character is changed: in particular the "free stagnation point" (marked S) is elevated. Clearly, one must be wary of reading too much into a structure implied by mean measurements in a highly turbulent flow, but these features must be some indication of the type of wake flow that exists instantaneously. For flow 1 the streamlines of Figure 3 show a vortical structure near the ground at the rear base of the model. Its vertical extent is around 0.2h which is the same order as the thickness of the approach flow boundary layer, and therefore, the scale of the front face horse-shoe vortex. When the approach flow has more significant shear (flows 2 and 3), the resulting front face stagnation point is higher (8) and this is mirrored by the vertical extent of the back face "base" circulation. Perhaps one can only conjecture, for the moment, on the separate roles of the mean shear and small scale turbulence in the approach flow. Evidence of extra turbulent entrainment by the shear layer over the top of the model (Morkovin's module 3, (12), see Section 5), can be seen in the tighter roll-up at the tip in flow 2 compared with flow 3. Such trends are often identified with stronger vortices and increased suction at the surface, especially for two dimensional bodies, but it is noticeable that on a model of height 4b (8) the tip base suction decreases with increasing free stream turbulence. In the experiments of (8) the mean shear also varies significantly, so it remains difficult to assess the separate influences.

Figure 3. Streamlines on the wake centre-line for flow 1.

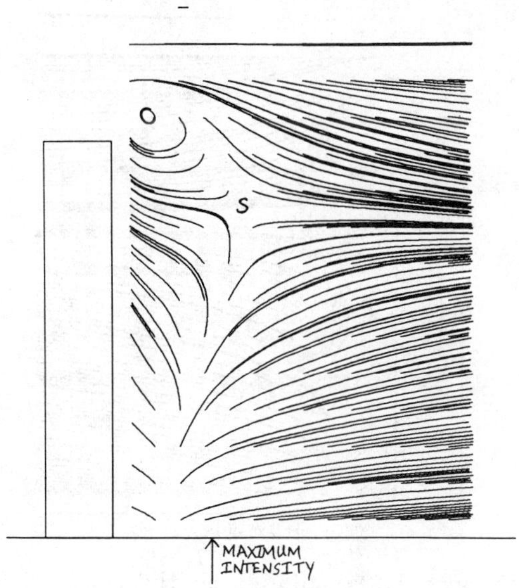

Figure 4. Streamlines on the wake centre-line for flow 3.

Experiments which vary one parameter at a time are called for to assist this understanding and the counter-jet technique can play a role here, since it is possible to contrive a variation of turbulence distribution with little change in mean velocity, (3). Measurements in low turbulence shear flows would also be of value in determining how much the changes in wake "bubble" length, wake "stagnation" position and length of tip influence, (cf. figures 3, 4 and 5), can be attributed to the mean flow alone.

5. A THREE-DIMENSIONAL VIEW OF THE VELOCITY FIELD DUE TO

(a) A Uniform Smooth Flow (flow 1)

We have examined the wake flows in Section 4 by streamlines in the plane of symmetry through the centre-line. Clearly, there are closed circulation zones in Figure 3, 4 and 5, which are apparently not fed from any outside region. In reality, a considerable flow around the block exists and the centre-plane is fed, in part, laterally. A clearer picture of these average features is obtained by considering the complex flow field in a three-dimensional rather than planar fashion.

To aid discussion of this presentation it is convenient to consider the modular simplification of such a flow presented by Morkovin (12). He suggested four major elements as, [1] the horse-shoe vortex of the front-face, resulting from shear in the approach flow; [2] the side shear layers, which may produce "two-dimensional" vortex shedding if vertical flows along the body are small; [3] the separated shear layer over the top of the block; [4] the "flow-through" module, i.e. the recirculation process of the near-wake.

No measurements have been made upstream of the model for the experiment described here, but some indication of the vertical scale of the horse-shoe vortex has been inferred in Section 4. To provide more information on module 2,

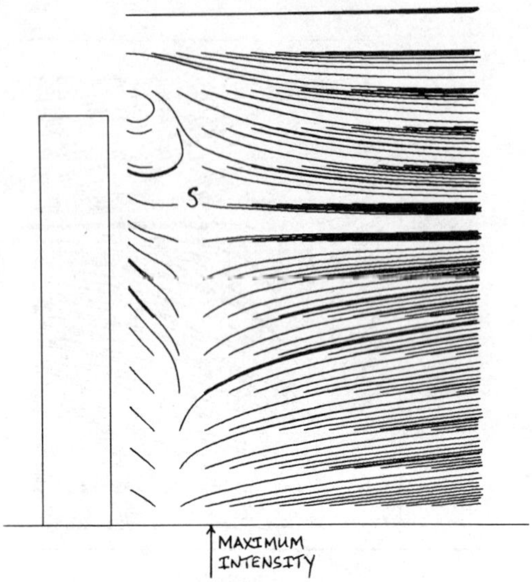

Figure 5.Streamlines on the wake centre-line for flow2.

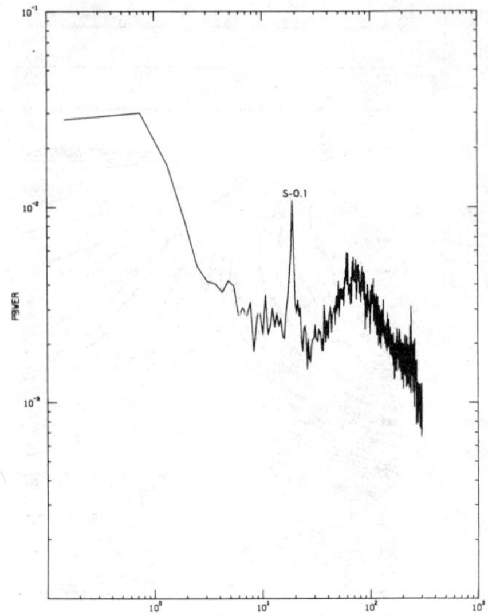

Figure 6. U-velocity spectrum for flow 1, x/b=0; y/b=1.3; z/h=0.55

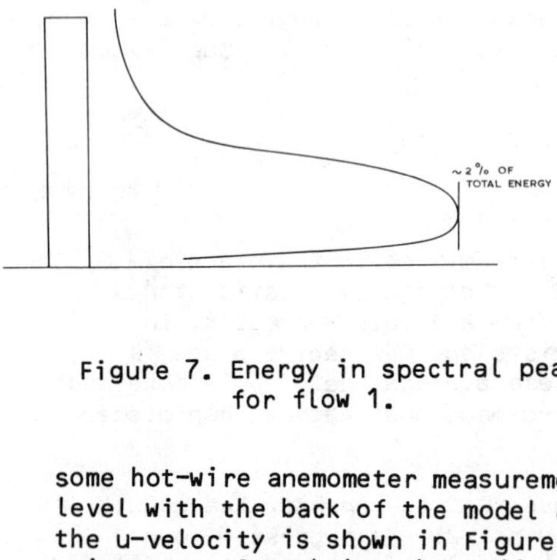

Figure 7. Energy in spectral peak for flow 1.

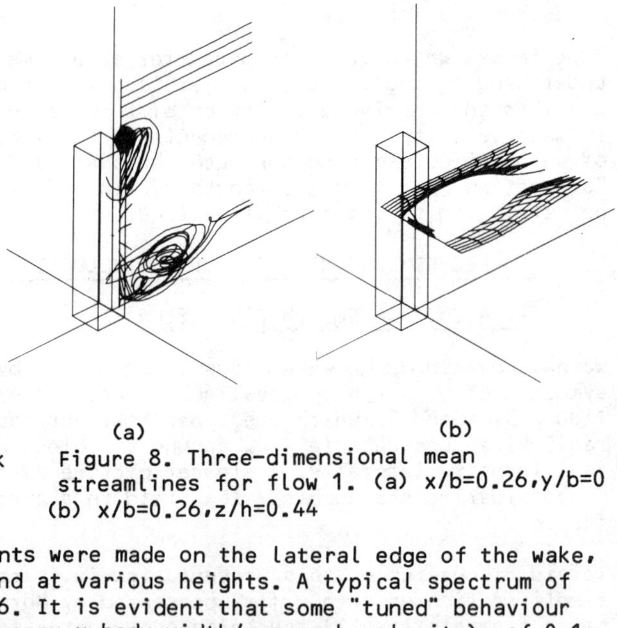

(a) (b)

Figure 8. Three-dimensional mean streamlines for flow 1. (a) x/b=0.26,y/b=0 (b) x/b=0.26,z/h=0.44

some hot-wire anemometer measurements were made on the lateral edge of the wake, level with the back of the model and at various heights. A typical spectrum of the u-velocity is shown in Figure 6. It is evident that some "tuned" behaviour exists at a Strouhal number, (frequency x body width/approach velocity), of 0.1, and this could represent a vortex-shedding cross-wake oscillation. The maximum energy at this frequency (see Figure 7), occurs just above the base vortex in Figure 3, where the flow at the centre-plane is reasonably horizontal. Figure 8 provides more information on the vertical distribution of the flow features, which is relevant to this aspect. Steady flow particle paths describing the mean streamlines are shown in Figure 8(a) starting just behind the model on a vertical line close to the streamwise centre-line. A region exists between the complex base region and the tip vortex, where the mean velocity is fairly horizontal and directed upstream and out from the centre-line. In this zone a horizontal line originating a little below mid-height produces paths with only a little vertical displacement. The isometric projection of Figure 8(b) indicates a tendency to downward movement at the "neck" of the wake closure and slight upward displacement close to the back face. Within the recirculation area at this level streamlines must enter from just above or below. (In Figure 8(b) the cross-lines joining the streamlines are through points of equal time in the "path" generation and give an indication of velocity magnitude along the streamlines).

There is some indication, then, of a degree of horizontal flow in the wake, which might result in the lateral vortex sheets rolling up alternately into

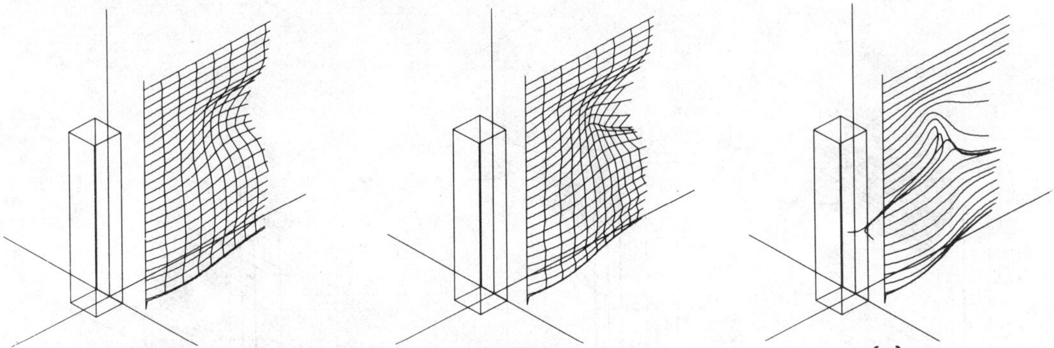

(a) (b) (c)

Figure 9. Three-dimensional mean streamlines for flow 1.
(a) x/b=0.26,y/b=1.15; (b) x/b=0.26,y/b=1.02; (c) x/b=0.26,y/b=0.89

vortices, but the evidence for this is weak. The energy in the frequency band around S = 0.1 is not great, (a maximum of 2% of the total signal in Figure 7) and it may be that this feature plays only a minor role in forming the wake character. Certainly there is more energy in the low frequency part of the spectrum and, by analogy with two-dimensional bluff body flows (e.g. (15)), this is likely to be the result of an in-phase (across the wake) pulsing of the near-wake, with little asymmetric oscillation. If this is so, then the mean structure, as deduced from the pulsed-wire measurements, is more likely to resemble an instantaneous view.

A further impression of the distortion of the side sheet streamlines is given in Figure 9. The vertical line of origin of the streamlines moves closer to the body from (a) to (c). The drawing-in of the paths toward the centre-line is seen to increase as the top of the block is approached. This is the result of a contracting recirculation region due to the high suction at the tip of the model. Towards the tip the streamlines turn down as they move inwards, the side sheets being influenced by the energetic top vortex sheet (module 3). The dramatic role of this shear layer is demonstrated in Figure 10, where the sequence shows streamlines emanating from a horizontal line which is lowered to just below the level of the model roof. This vortex sheet produces the high suction tip vortex, but also feeds much of the low velocity near-wake region together with input from the side sheets. It remains difficult to pinpoint an entry to the "flow-through" module of the wake, but Figures 9(c), 10(d) and 10(e) show paths entering the body of the wake from a region above the stagnation area of Figure 3. Figure 8(a) suggests the base vortex region to be somewhat separate from the remainder of the wake: low level paths close to the body become elevated and very contorted, but leave the region downstream just below the stagnation area.

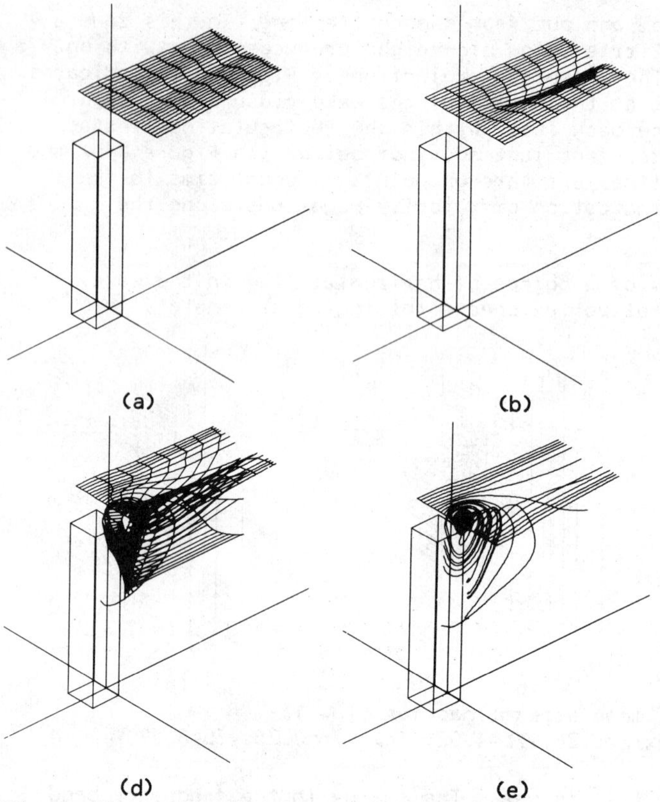

Figure 10. Three-dimensional streamlines for flow 1. x/b=0.26, (a) z/h=1.09; (b) z/h=1.07; (c) z/h=1.05; (d) z/h=1.01; (e) z/h=0.96

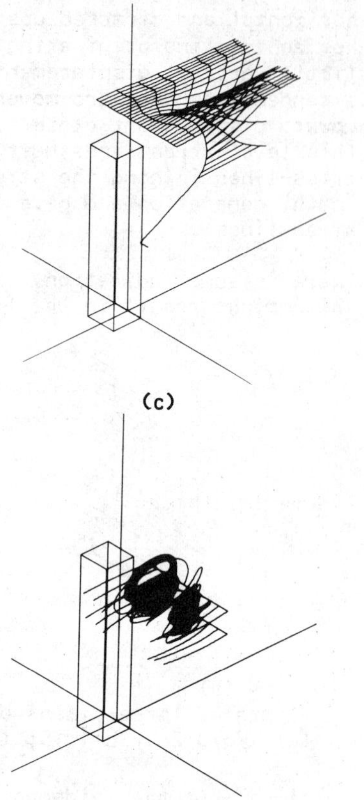

Figure 11. Three-dimensional streamlines for flow1 generated in reverse direction from x/b=2.6, z/h=0.44 and z/h=0.57

Finally to try to examine the conjecture of an exit funnel in module 4, streamlines were computed "backwards in time" from certain downstream positions. Figure 11 shows paths which arrive at horizontal lines close to the mid-height position, having spiralled down the two symmetric inclined "vortices" shown. This is part of the downward entry into the rear of the pattern in Figure 8(b). The figures appear consistent with the idea of an upward movement, close to the rear face, toward the tip suction and a recirculating downward spiral downstream. It is not possible without further analysis to characterise the exit flows any more clearly, but it appears that the streamlines entering the near-wake in Figure 9(c) do so from a level higher than the exit lines in Figure 11.

(b) A Turbulent Shear Flow (flow 3)

A preliminary examination of the full wake measurements for flow 3 has confirmed the general observations made in section 4. Some of the differences in wake structure compared with flow 1 are shown in the streamline patterns of figure 12. The impression of a contracted "wake bubble" illustrated in figures 3, 4 and 5, is maintained in these results. In figure 12(a) the side sheets are tightly drawn in behind the body, but the streamlines do not recirculate in the manner of figure 9(c). This suggests that the wake suction is stronger and correspondingly the reverse-flow region is more compact. The spectra of velocity fluctuations just outside this region show no predominant frequency, implying that the stronger base suction is not accompanied by a strengthening of regular vortex shedding.

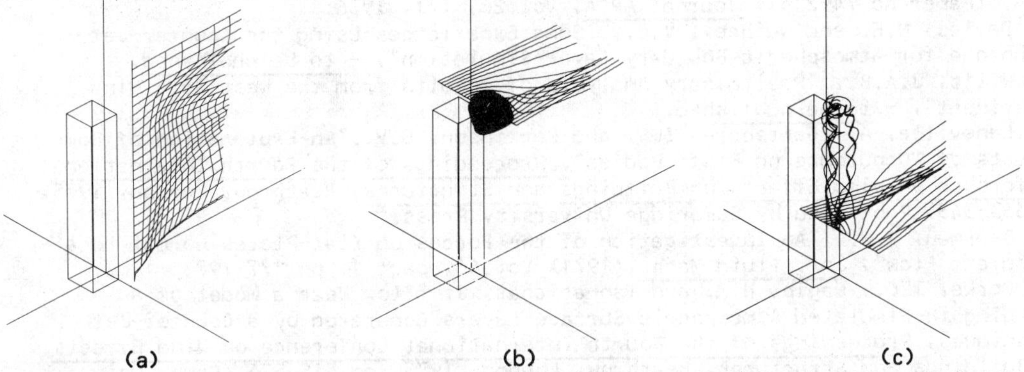

Figure 12. Three-dimensional mean stremlines for flow 3. (a) x/b=0.26,y/b=0.89; (b) x/b=0.26,z/h=0.96; (c) x/b=0.52,z/h=0.18

Figure 12(b) pictures the top shear layer rolling up into a tight vortex at the top of the model. A comparison of this pattern with that of figure 10(e) demonstrates the changed tip influence. More elevated streamlines are strongly "bent" by the tip vortex but the region of "capture" of streamlines is not extended upwards.

Figure 12(c) shows entry into the wake from a low level, the flow being subsequently drawn up the back face by the low pressure at the top of the block.

6. CONCLUSIONS

The pulsed-wire velocity measurements in the near-wake of a building block model have enabled an average description of the wake structure to be formed. It has been possible to make general statements about the length scales and positions of the major flow elements. The top-vortex sheet is seen to play a strong role in determining the character of the wake, but it does so in combination with a ground level structure which is related to the shear in the oncoming flow. Increased shear in the approaching stream causes a higher front stagnation point on the body and strongly influences the wake "free stagnation" zone. There are unresolved questions as to the separate effects of shear and turbulence, but it is clear that these results lend yet more weight to arguments for the use of some representation of the atmospheric boundary layer in model testing. The modular concept of Morkovin has been found extremely useful in examining these flows, though it remains difficult to understand the recirculation process fully, and to be sure that all can be revealed from mean measurements.

ACKNOWLEDGEMENTS

This work was supported, in part, by the Department of the Environment through the Building Research Establishment, Garston, England.

REFERENCES

(1) Nagib, H.M., Morkovin, M.V., Yung, J.T., and Tan-atichat, J., "On modelling of Atmospheric Surface Layers by the Counter-Jet Technique", AIAA Journal,

Vol.14., (2), 1976, pp 185-190

(2) Tan-atichat, J. and Nagib, H.M., "Wind Tunnel Simulation of Neutral Atmospheric Surface Layers by the Counter-Jet Technique", Proceedings of the 67th Annual Meeting of the Air Pollution Control Association, Denver, Colorado, 1974, (Paper No 74-214), Journal APCA, Vol.26, (7), 1976.

(3) Davies, M.E. and Quincey, V.G., "Some Experiences Using the Counter-Jet Technique for Atmospheric Boundary Layer Simulation", - to be published.

(4) Wills, J.A.B., "Preliminary Analysis of Results from the West Sole Wind Experiment", - to be published.

(5) Laneville, A., Gartshore, I.S. and Parkinson, G.V.,"An Explanation of Some Effects of Turbulence on Bluff Bodies", Proceedings of the Fourth International Conference on Wind Effects on Buildings and Structures, Heathrow, London, 1975, pp 333-343, Published by Cambridge University Press.

(6) Bearman, P.W., "An Investigation of the Forces on Flat Plates Normal to a Turbulent Flow", Jnl. Fluid Mech. (1971) Vol.46, part 1, pp 177-198.

(7) Corke, T.C., Nagib, H.M. and Tan-atichat, J, "Flow Near a Model of a Building in Simulated Atmospheric Surface Layers Generated by a Counter-Jet Technique", Proceedings of the Fourth International Conference on Wind Effects on Buildings and Structures, Heathrow, London, 1975, pp 515-526 (Camb. Univ. Press).

(8) Corke, T.C. and Nagib, H.M., "Sensitivity of Flow Around and Pressures on a Building Model to Changes in Simulated Atmospheric Surface Layer Characteristics", IIT Fluids and Heat Transfer Report R76-1, May 1976.

(9) Lee, B.E., "Some Effects of Turbulence on the Mean Forces on a Bluff Body", Jnl. of Insutrial Aerodynamics, 1 (1975/1976), pp 3161-370.

(10) Melbourne, W.H.,"Comparison of Measurements on the CAARC Standard Tall Building Model in Simulated Model Wind Flows", to be published.

(11) Castro, I.P. and Robins, A.G., "The Flow Around a Surface-Mounted Cube in Uniform and Turbulent Streams", Jnl. Fluid Mech. (1977), Vol.79, part 2, pp 307-335.

(12) Morkovin, M.V., "An Approach to Flow Engineering via Functional Flow Modules", Paper No 72-27, Beitrage zur Stromungsmechanik, A. Walz 65th Anniversary Volume, Deutsch Luft und Raumfahrt Forschungsvericht, 1972.

(13) Lemberg, R., "On the Wake Behind Bluff Bodies in a Turbulent Boundary Layer", University of Western Ontario, BLWT-3-73, 1973.

(14) Peterka, J.A. and Cermak, J.E., "Turbulence in Building Wakes", Proceedings of the Fourth International Conference on Wind Effects on Buildings and Structures, Heathrow, London, 1975, pp 447-464, (Camb. Univ. Press).

(15) Davies, M.E., "Wakes of Oscillating Bluff Bodies", Ph.D Thesis, University of London, 1975.

THE RELAXATION AND STEADINESS OF WAKES BEHIND OBSTACLES IN BOUNDARY LAYERS

I. P. CASTRO

University of Surrey, Guildford, UK

ABSTRACT

Detailed measurements in the wakes behind two-dimensional square section blocks (height h) mounted in thick rough wall boundary layers (height δ) are presented for cases in which $h/\delta << 1$. The data provides some insight into the relaxing flow downstream of reattachment, confirming the conclusion of Bradshaw and Wong (1) that reattaching flows are surprisingly complicated and involve considerable distortion of the separated shear layer. In particular, the measurements show that eddy length scales are considerably reduced so that, since the flow eventually relaxes to a boundary layer similar to that upstream, turbulence models based on eddy viscosity concepts cannot, in principle, be expected to be satisfactory.

It has been suggested in the literature that such wake flows contain significant periodicity, particularly in the separated shear layer near the body in cases where $h/\delta > 1$. This has obvious implications for those workers attempting to calculate the flow using numerical methods and physical models based on the steady flow equations. However, the present measurements demonstrate that such periodicity, if present in the flow at all, is extremely weak.

INTRODUCTION

'At present little is known about the wake behind surface obstacles in turbulent boundary layers'. So wrote Counihan et al (2) in the introduction to their theoretical work on such wakes behind two-dimensional bodies. Despite their rather limited experimental work, undertaken as a test of the theory, it seems that the truth of their statement has altered only marginally for two-dimensional bodies, although some useful measurements behind three-dimensional bodies have since been reported (eg: Peterka and Cermak, 3, Castro and Robins, 4). The theory developed by Counihan et al (2), hereafter known as 'CHJ', was recognised to be fairly primitive but, partly in view of the lack of detailed experimental data, it has not seemed worthwhile attempting any major improvements (Hunt, private communication). The present work was originally undertaken as a more detailed test of the theory and, perhaps more importantly, as an attempt to investigate directly the effects, on the decay and turbulence structure of the wake, of varying parameters like h/δ (where $h/\delta << 1$) and the roughness of the boundary.

Some workers are attempting to calculate recirculating flows like those investigated here, usually with the assumption that the flow contains no strong periodicity.

Whilst vortex shedding, in the usual sense, certainly cannot occur, there is some uncertainty regarding the presence of other forms of large scale unsteadiness. The present investigation has therefore been recently extended to study this aspect of the flow in more detail, particularly for cases where $h/\delta = O(1)$ in which, arguably, such unsteadiness may be rather more likely than where $h/\delta << 1$. Preliminary results are presented in this paper.

EXPERIMENTAL ARRANGEMENTS

General Details

The wake decay experiments were undertaken in the 6 x 0.9 x 0.75 m suck-down wind tunnel at Marchwood Engineering Laboratories. Figure 1 shows the general experimental arrangements. In the empty working section mean velocity variations were less than ± 0.25% and the longitudinal turbulence intensity was less than 0.05%. A false floor spanning the 0.9 m horizontal width of the tunnel and extending down most of the working section length was installed about 100 mm above the floor. It was equipped with an adjustable flap at the rear and calibrated yaw tubes at the nose to enable zero circulation to be maintained in all the flows studied. A castellated barrier wall, vorticity generators and distributed surface roughness were used to generate thick turbulent boundary layers on the false floor. The general details of such a system have been discussed elsewhere (Counihan, 5, Robins, 6) and the two boundary layers used in the present study are described in detail by Castro (7). In the present paper, the smoother of the two layers is designated BL1 and the rougher BL2; salient details are given in Table 1, in which z_o is the surface roughness length, d is the zero plane displacement, U_τ is the wall friction velocity and U_o the free stream velocity.

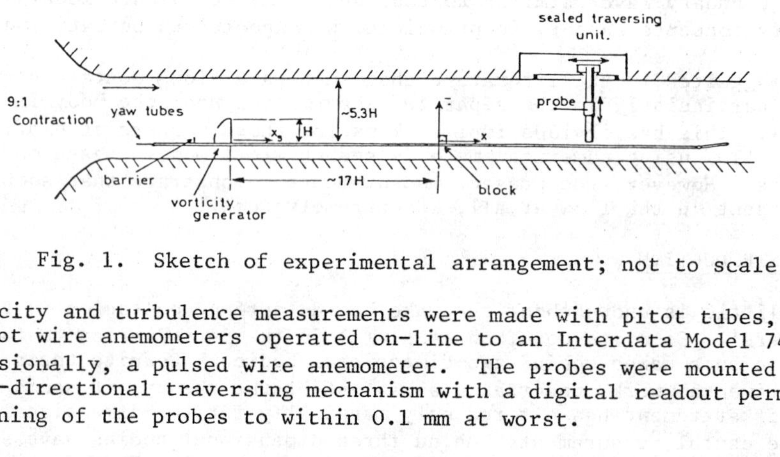

Fig. 1. Sketch of experimental arrangement; not to scale

Mean velocity and turbulence measurements were made with pitot tubes, single and crossed hot wire anemometers operated on-line to an Interdata Model 74 Minicomputer and, occasionally, a pulsed wire anemometer. The probes were mounted on an automated two-directional traversing mechanism with a digital readout permitting repositioning of the probes to within 0.1 mm at worst.

TABLE 1

Flow	d (mm)	z_o (mm)	(U_τ/U_o)	n
BL 1	1.8	0.071	0.048	0.19
BL 2	4.0	0.320	0.055	0.23

n in Table 1 is the exponent giving the most reasonable fit of a mean velocity power law profile to the experimental data

Three flows, designated F1L, F1S and F2L were investigated; the appropriate values of h/z_o, h/δ and $(U/U_o)_h$ are given in Table 2 along with some other basic flow parameters introduced later. In Table 2 and all subsequent figures the appropriate zero plane displacement, d, has been subtracted from h and δ (and all vertical heights).

Measurements of vertical profiles of mean velocity, shear stress and all three components of normal stress were made at a minimum of ten downstream locations between x/h of 10 and 60, where x is measured from the front face of the block. Less extensive measurements of the turbulence dissipation rate and some velocity probability density distributions were also made.

TABLE 2. Various parameters for the three flows

Flow	Upstream Boundary Layer	Nominal h,mm	h/z_o	h/δ	$(U/U_o)_h$	$K \equiv \frac{2\nu o(h)}{hU(h)}$
F1L	BL1	25	327	0.157	0.685	0.045
F2L	BL2	25	78	0.169	0.625	0.049
F1S	BL1	12	144	0.068	0.580	0.068

To investigate the presence, or otherwise, of strong periodicity within the flow, spectra, auto-correlations and probability density distributions of the longitudinal velocity component were measured both in the flows described above and, more recently, in similar flows set up in the 0.6 x 1.0 x 4.0 m wind tunnel in the Department of Mechanical Engineering at the University of Surrey. In the latter case, a similar false floor arrangement was used and attention was concentrated on the case $\delta/h < 1$. The smooth wall boundary layer was tripped just downstream of the nose of the false floor, and measurements of the growth of the boundary layer were used to determine the positions at which to place the 26 mm and 52 mm blocks used; the range covered was $0.34 < \delta/h < 0.8$. All hot wire calibrations and measurements were performed digitally on-line with a Commodore PET 2001 computer.

EXPERIMENTAL RESULTS: WAKE DECAY

A considerable amount of mean velocity and turbulence data was accumulated during the course of the work and a complete presentation of all the basic data would be inappropriate. Typical examples of the relaxation of the wake flow are given in Figure 2, which shows the way in which the profiles of mean velocity and shear stress gradually recover to the basic boundary layer profiles. For clarity results

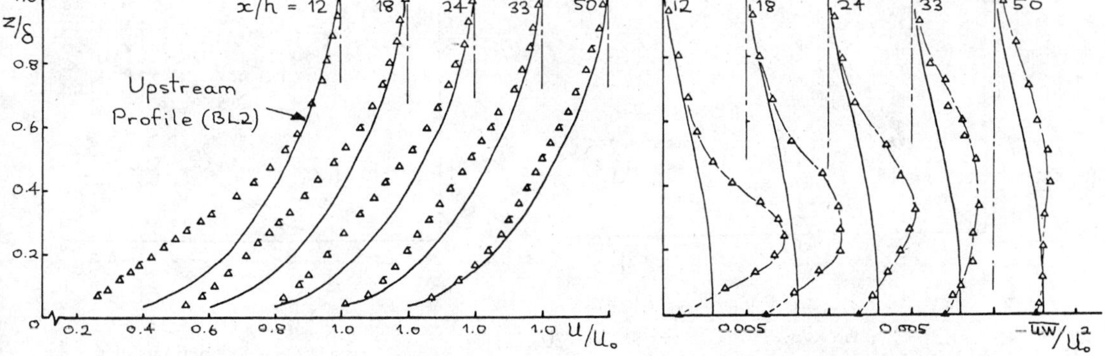

Fig. 2. Decay of mean velocity and shear stress perturbations; F2L

at only five x/h stations are presented, though measurements were made at ten. It is clear that the wake is evidenced by a velocity deficit and increases in shear stress and turbulence intensity. Near the wall, however, the shear stress is actually lower than its undisturbed value; it must, of course, be zero at or near reattachment (around x/h = 10 in this case). The recovery process appears to be most rapid at the lower velocity edge of the wake - the near wall region. At x/h = 50 the measurements below $y/\delta_o = 0.15$ are virtually indistinguishable from the undisturbed boundary layer profiles, but elsewhere there is still a clear wake region which has, in fact, grown at its outer edge to fill the entire boundary layer. A series of careful single hot-wire measurements in the near-wall region at each axial station was undertaken in an attempt to obtain an independent measure of the wall shear stress. With the assumption that the zero plane displacement, d, was the same everywhere as it was in the basic boundary layer (BL2), straight line fits to the data plotted in log-law form led to estimates of the wall shear stress, and these are included in Figure 2. These estimated wall shear stresses, whilst probably not very accurate, are not inconsistent with the cross-wire measurements.

Similar general features of the wake flow were evident for F1L and F1S although the magnitudes of the perturbations and the rates at which they decayed were different. With the raw results for all three cases it is possible to investigate the perturbation flows in some detail, and these are now briefly compared with the theoretical results of CHJ.

COMPARISONS WITH THEORY

The Mean Velocity Perturbations

One of the basic results of the theory of CHJ is that the maximum value of the velocity perturbation should be inversely proportional to $\overline{x}$, ((x-a)/h), and in Figure 3 $(u/U_h)^{-1}_{max}$ is plotted against x/h for all three cases. Using the resulting value for a and the appropriate value for K (Table 2) the profile for F1S is plotted in the self-similar form in Figure 4. Theoretical self-preserving profiles can be obtained by fixing the disposable constant occuring in the theory so that the maximum velocity deficit is equal to the measured value; figure 4 includes the resulting profile. There is a considerable region of the wake which does exhibit

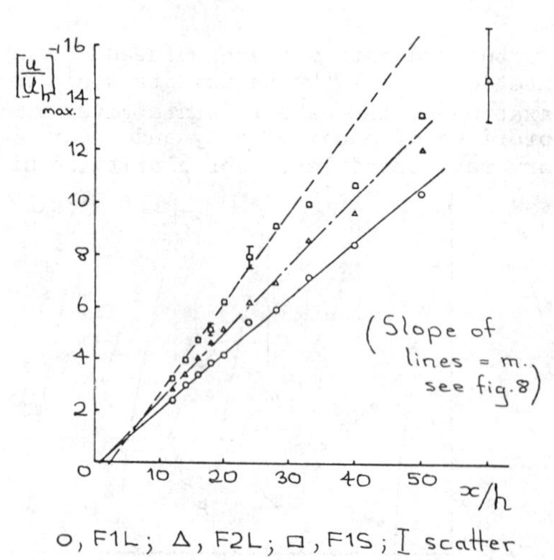

Fig. 3. Decay of Maximum Velocity Deficit

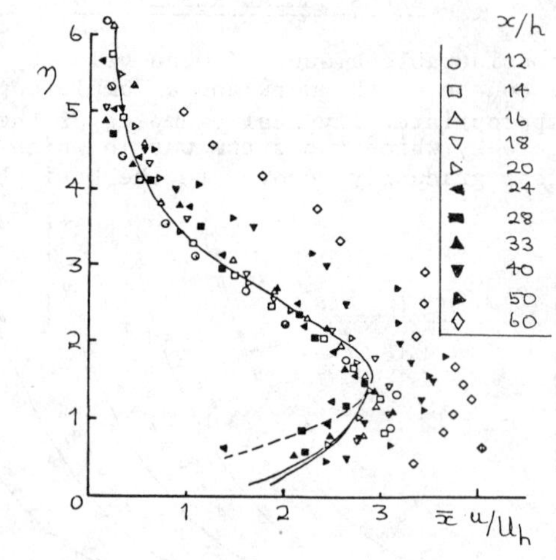

Fig. 4. Velocity Deficit Profiles for F1S

the expected self-similar behaviour away from the wall region but clearly the far wake is not well predicted, although for F1L and F2L agreement is rather better, as might be anticipated from Figure 3. As K, a 'turbulence' Reynolds number of the body defined by CHJ (See Table 2), increases it seems that the rate of decay of the velocity deficit begins to fall below the $\overline{x}^{-1}$ rate at smaller and smaller $\overline{x}$. Even for F1S, however, the predictions seem satisfactory up to x/h = 30, which is further downstream than the extent of the only previously reported experiments of similar type (CHJ).

In the wall region of the wake flow, the perturbation velocity is not expected to have the self-similar form assumed in the mixing region, mainly because the assumption of a constant perturbation eddy viscosity (used in the derivation of equation (1)) was not thought to be adequate in such a region. CHJ therefore derived a separate solution for this wall region by assuming it to be in local equilibrium.

An implication of their theoretical result for the wall region is that the velocity deficit decays more slowly than $\overline{x}^{-1}$ and the theoretical profiles are included in Figure 4 for x/h = 12 and 50, demonstrating this behaviour. Now, by contrast, it is clear that the data, although somewhat scattered in this region because the experimental errors are rather higher than elsewhere, exhibits an entirely opposite trend, at least in the axial region where the self-preserving solution seems adequate. The velocity deficit actually decays more rapidly in this wall region than elsewhere.

Perturbations of the Turbulence Field

The perturbation shear stress data is plotted in the expected self-preserving form in Figure 5 for F2L compared with the theoretical profile obtained by using the same value for the disposable constant used in the corresponding mean velocity comparisons.

The maximum shear stress increase has about the expected magnitude just downstream of the separated region (x/h = 12) but clearly decays significantly less rapidly than the theory suggests. Furthermore, a plot of $\Delta(-\overline{uw})/U_h^2{}_{max}$ vs. x/h indicates that the rate of decay decreases with increasing K, Figure 6, in line with the trends described in the previous section.

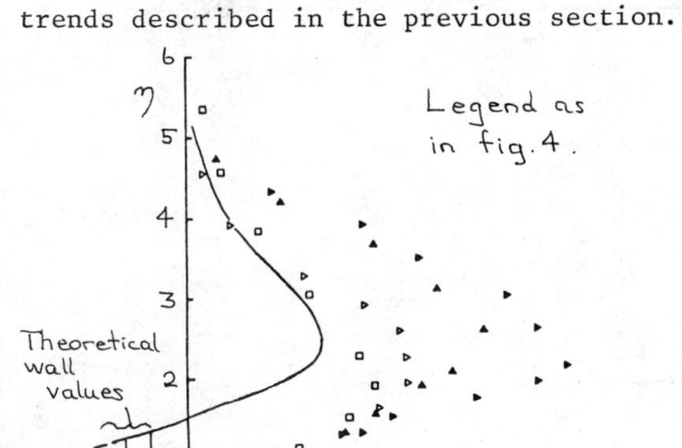

Fig. 5. Shear Stress Perturbation Profiles for F2L

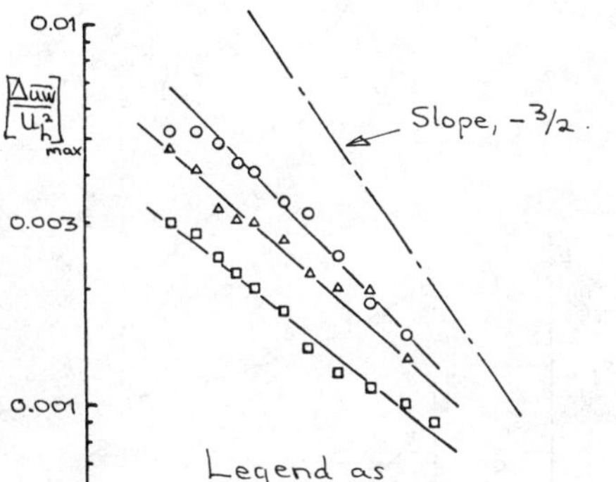

Fig. 6. Decay of Maximum Shear Stress Excess

WE L*

In the wall region the assumption of equilibrium with a matching of solutions in this and the mixing region leads to a constant perturbation shear stress and the theoretical result is included for typical values of x/h in Figure 5. As in the case of the theoretical velocity perturbation, the implication is that the perturbation shear stress decays less rapidly in the wall region whereas, in fact, the data indicates that it decays rather more rapidly.

Although the turbulent shear stress is reduced below the boundary layer value in the near wall region the normal turbulent stresses all increase so that the turbulence structure is rather different from an equilibrium flow. However, away from the wall the increases in turbulence intensities are very similar in magnitude to the increases in shear stress. For F1S, for example, the maximum increase in turbulence energy expressed as a fraction of the upstream value, $|\Delta\overline{q^2}/\overline{q^2}|_{max}$, is about 0.9 at x/h = 14, compared with a corresponding increase in shear stress of about 1.3. For F1L the corresponding figures are about 2.1 and 2.3. In addition, the rate at which the perturbation intensities decay is very close to the rate of decay of the shear stress perturbations, so that for F1L, for example, $|\Delta\overline{q^2}/\overline{q^2}|_{max}$ and $|\Delta\text{-}\overline{uw}/\overline{uw}|_{max}$ are about 0.8 and 1.0 respectively, by x/h = 50.

Implications

Discussion of the decaying, perturbed flow downstream of reattachment begins with a consideration of the flow generated by the smaller of the two bodies (Flow F1S), largely because in this case the assumptions used by CHJ are likely to be most closely satisfied. Using the present data for F1S, the measured eddy viscosity, $\nu_T \equiv \overline{uw}/(\partial U/\partial z)$, at x/h = 12, 28 and 60 is shown in Figure 7a. It is obvious that

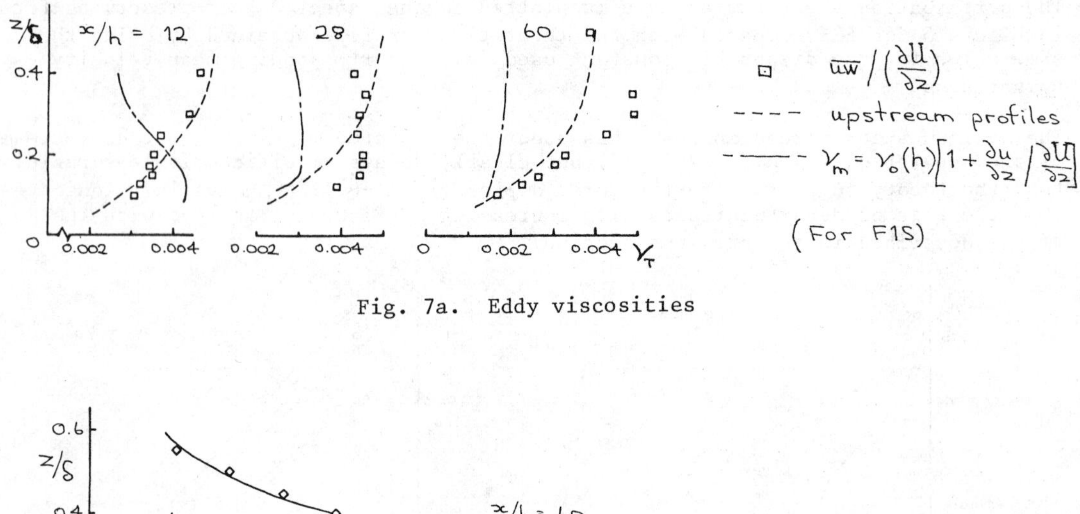

Fig. 7a. Eddy viscosities

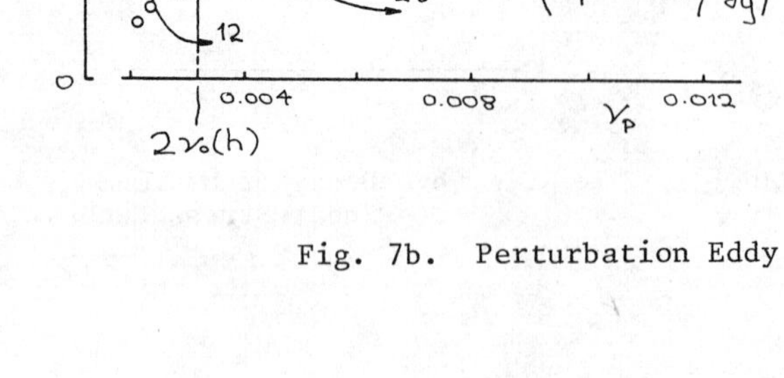

Fig. 7b. Perturbation Eddy Viscosities; F1S

the variations of ν_T could not be predicted by simple formulae of the type used by CHJ - which is included in the Figure. In addition further experiments showed that there were significant variations in the ratio of the mixing length to the dissipation length scale, so that shear stress transport terms must vary with x; it therefore seems likely that any eddy viscosity relation will be unsuccessful in predicting the total flow. With simple assumptions about the response of the eddy viscosity to extra strain rates CHJ derived a perturbation eddy viscosity $\nu_p = 2\nu_o(h)$ and this was the value used to link the perturbation mean velocity and shear stress fields in their prediction of the perturbed flow. It is compared with the experimental data in Figure 7b. Since the points of zero $\Delta(-\overline{uw})$ and $\partial u/\partial y$ do not coincide, there are discontinuities in ν_p at the point of maximum velocity deficit, but nevertheless it is clear that a constant perturbation eddy viscosity is not an adequate assumption. Indeed, since the data of Figure 7 clearly demonstrate that the perturbation mean velocity decays more rapidly than the perturbation shear stress, it seems probable that no eddy viscosity relation for the perturbation flow would be satisfactory, any more than it would for the total flow.

Now the predicted mean velocity is generally less sensitive than the turbulence quantities to the exact nature of the turbulence model and since, in addition, it so happens that $2\nu_o(h)$ is not too far in magnitude from the measured ν_p in the early part of the wake ($x/h < 28$, say), the mean velocity perturbations are reasonably well predicted in this region (Figure 4 and Section 4), whilst the shear stress perturbations are, of course, less well predicted.

Consider now flows F1L and F2L, generated by the larger of the two blocks. It is at first sight surprising that, at least as far as the mean velocity perturbations are concerned, the wake decay is apparently better predicted than it is for F1S (Figure 3), despite the fact that results not presented here showed that the shear layer turbulence after separation is more highly distorted. The maximum velocity deficit decay rate does not decrease as $\overline{x}$ increases, as it does for F1S. However, for F1L and F2L the wake practically fills the boundary layer by about $x/h = 40$, so that the CHJ assumption that δ is not a relevant length scale cannot be very satisfactory. Indeed, the fact that the x^{-1} decay rate is actually maintained over the whole measured extent of the wake is likely to be a result of the opposing effect of two inadequate assumptions in the theory. Convincing substantiation of these conclusions would necessitate a closer investigation of the turbulence structure than was possible in the present study.

However, what is clear is that the wake flow is sufficiently complicated to ensure that theories or, for that matter, numerical predictions which use eddy viscosity turbulence models are not likely to be successful in predicting the total flow, although they might give seemingly adequate results in limited regions.

One result of the theory which should not, in principle, depend on the assumed turbulence model, is that the rate of decay of the wake is inversely proportional to a ratio ($1/K$) of inertial stresses to turbulence shear stresses in the upstream flow at, say, the height of the body. The experimental data do seem to confirm this (Figure 8) and it appears that the primary effect of this ratio is on the distortion of the separated shear layer up to reattachment. Downstream of this, the wake relaxation is slowest in the case of the greatest distortion (F1L). On the other hand, in the wall region the velocity perturbations decay more rapidly for F2L than for F1L but this is almost certainly a result of the considerably rougher surface in the former case. The thickness of the region obeying the log-law grows at a similar rate in the two cases but since, at a given z, the velocity in the undisturbed boundary layer is less for the rougher surface, the perturbation velocity will in the equilibrium wall layer downstream of reattachment also be less.

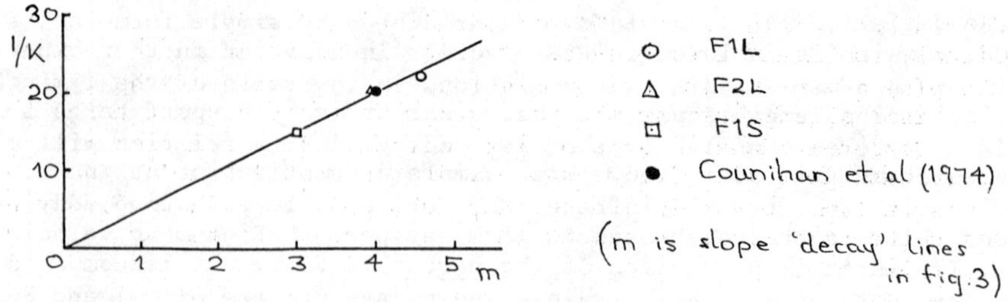

Fig. 8. Decay rate (m) vs. 'Turbulence' Reynolds Number (1/K)

What is, perhaps, more interesting is that the log-law region grows at about the same rate in all three cases, which agrees with the suggestion of Bradshaw and Wong (1) that a local-equilibrium region re-establishes itself at a rate almost independent of the turbulent flow that provides its outer boundary condition.

WAKE UNSTEADINESS

If wake unsteadiness were to occur it would presumably be most noticeable either in the separated shear layer near the body itself or around the region of reattachment. Crabb, Durao and Whitelaw (8) using laser anemometry, measured longitudinal velocity probability density distributions (p(u)) and apparently found distinctly bimodal behaviour symptomatic of considerable energy at a particular frequency. Figure 9 shows, for a range of δ/h, single hot wire measurements of p(u) from the present work, taken at x/h = 1.5 and near the centre of the shear layer - where the local turbulent intensity was about 25%. Pulsed wire measurements of p(u) at the same location agreed closely with the hot wire results which were themselves very similar to corresponding distributions in a classical plane mixing layer. There is clearly no evidence of bimodal behaviour and measurements made at other positions within the shear layer and also near reattachment gave similar results.

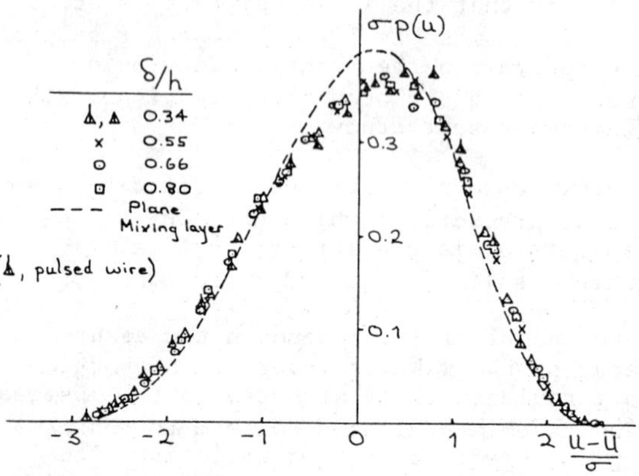

Fig. 9. p(u) on Shear Layer Centreline at x/h = 1.5

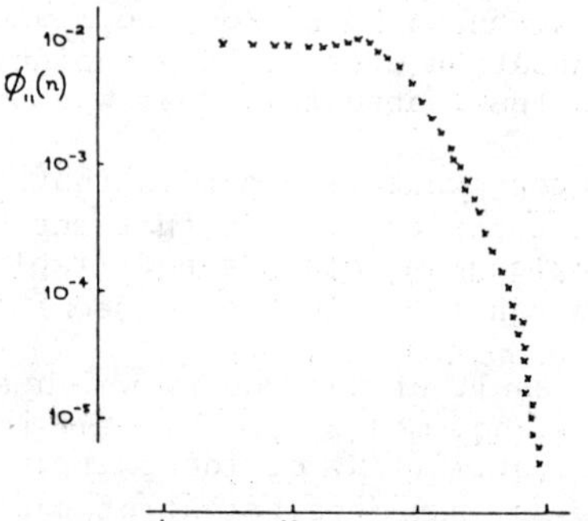

Fig. 10. Energy Spectrum x/h=1.5; $\delta/h=0.34$

However, it must be emphasised that the addition to a pure sine wave of white noise of just sufficient amplitude to eradicate the bimodal shape of the probability distribution would still leave a very substantial peak in the frequency spectrum. Lack of bimodality in the p(u) results like those in Figure 9 does not therefore necessarily imply lack of strong periodicity in the flow. A Hewlett-Packard 3721A digital correlator was therefore used to measure autocorrelations of the hot wire signal, and hence frequency spectra, at the same positions in the flow used for the p(u) measurements. Figure 10 shows a typical spectrum for $\delta/h = 0.34$. Whilst there is a slight hump in the spectrum at a frequency around 30 Hz (corresponding to a Strouhal number, based on free stream velocity and body height, of about 0.1) there was clearly no strong periodicity in the flow at that position. 'Ringing' at large time delays in the corresponding autocorrelation was scarcely discernable above the noise level although there was, as might be expected, a noticeable negative loop.

The result was again typical of measurements made elsewhere in the various flows and, taken with the probability density measurements, is clear evidence of the lack of strong periodic components in the flow over a two-dimensional surface mounted block, whatever the thickness and characteristics of the upstream boundary layer.

CONCLUSIONS

Two major conclusions of the work have emerged. Firstly, for $h/\delta \ll 1$, although the velocity and turbulence characteristics in the region of the wake not too far downstream of reattachment decay in a manner governed largely by the characteristics of the upstream flow, turbulence models based on eddy viscosity concepts, even for the 'perturbed' flow as in the theory of Counihan et al, are not likely to be sufficiently realistic to give good predictions of the flow even downstream of reattachment.

Secondly, however, there seems to be little evidence of large scale unsteadiness in the wake, even in cases where the upstream boundary layer is small compared with the body height. There is therefore some hope that good numerical predictions based on steady flow equations may be feasible, although the serious problems concerning numerical accuracy and adequate physical modelling have yet to be overcome.

ACKNOWLEDGEMENT

The author acknowledges with thanks the valuable assistance of Professor H.H. Fernholz during the more recent part of the investigation whilst the latter was on a short study visit at the University of Surrey.

REFERENCES

1. Bradshaw, P. and Wong, F.W.F., J. Fluid Mech., 52: 113 (1972)

2. Counihan, J., Hunt, J.C.T. and Jackson, P.S., J. Fluid Mech., 64: 529 (1974)

3. Peterka, J.A. and Cermak, J.E., Turbulence in building wakes, Proc. of 4th Int. Conf. on Wind Effects on Buildings and Structures, Heathrow, CUP, (1975)

4. Castro, I.P. and Robins, A.G., J. Fluid Mech., 79: 307 (1977)

5. Counihan, J., Atmos. Environ., 3: 197 (1969)

6. Robins, A.G., CEGB Report R/M/R246 (1977)

7. Castro, I.P., CEGB Report R/M/N1006 (1978)

8. Crabb, D., Durao, D.F.G. and Whitelaw, J.H., Paper presented at 4th Brazilian Congress of Mechanical Engineering (1977)

THE EFFECTS OF THE CHARACTERISTICS OF THE ONCOMING WIND ON THE FLOW BEHIND A TWO-DIMENSIONAL MODEL BUILDING

Y. OGAWA*, P. G. DIOSEY* and S. OIKAWA**

*National Institute for Environmental Studies, Japan

**Shimizu Construction Co., Ltd., Tokyo, Japan

SUMMARY

Scale-model field and wind tunnel experiments were conducted to examine the inflow and similarity considerations for flow behind a building. In both cases the model was simplified to a two-dimensional fence placed perpendicular to the oncoming wind, and the nondimensionalized cavity wake length, L_c/H, was the flow characteristic examined. In the field, the flow downwind of the fence was visualized with a grid-like arrangement of wind vanes. It was found that L_c/H increases with increasing Re_{turb} but decreases with increasing atmospheric stability. In the wind tunnel, upwind surface roughness and thermal stability were varied. Over a range of stabilities from unstable to stable, it was found that in the stable region L_c/H decreases rapidly as stablity increases while in the unstable region L_c/H decreases only slightly as the flow becomes more unstable. By changing the surface roughness, the turbulent Reynolds number was varied from approximately 15 to 60(keeping the model well within the surface boundary layer, $H/\delta \le 0.2$). As Re_{turb} increases, L_c/H increases. A comparison of the field and wind tunnel data shows that in both, L_c/H increases with increasing Re_{turb}. This relationship is not linear, but seems to approach an upper limit, perhaps the $L_c/H \simeq 17$ found by Arie and Rouse. Additionally, it was found that in both studies that as the local turbulent intensities increase, L_c/H decreases.

LIST OF SYMBOLS

g	acceleration due to gravity
H	model height
k	Karman constant
K	turbulent diffusion coefficient
K_m	eddy diffusivity for momentum, $-\overline{uw}/(\partial U/\partial z)$
L	Monin-Obukhov stability length
L_c	cavity wake length
Re_{turb}	turbulent Reynolds number
Ri	gradient Richardson number, $[(g/T)(dT/dz)]/[d\overline{u}/dz]^2$
Ri_f	flux Richardson number, $[-(g/T)\overline{w\theta}]/[\overline{uw}(\partial u/\partial z)]$
T	temperature
u,v,w	turbulent components of velocity in the x, y, and z directions
U,V,W	mean components of velocity in the x, y, and z directions
u_*	friction velocity
x,y,z	spatial coordinates in the windward, crosswind, and vertical directions

z_o	roughness length
θ	potential temperature
$\sigma_u, \sigma_v, \sigma_w$	standard deviations of u, v, and w
$( )_a$	ambient
$( )_f$	floor
$( )_\infty$	freestream
$\overline{( )}$	mean

1. INTRODUCTION

When studying the interaction of buildings and airflow(and the resulting complex flow) in the laboratory, e.g., wind tunnel and water channel, certain similarity criteria must be satisfied in the scale-model study in order to accurately simulate the atmospheric phenomena involved. Inoue(1) suggested the equality of turbulent Reynolds number, $Re_{turb} = UH/K$, where U is the mean wind velocity at model height H and K is the turbulent diffusion coefficient. Jensen(2) suggested the equality of the ratio of roughness length to model height, z_o/H. Tani(3) used the H/z_o equality for a model shelter-hedge wind tunnel study and suggested that if logarithmic wind profile conditions apply, the equality of H/z_o is equivalent to the Re_{turb} equality. Plate(4) found identical drag coefficients on a two-dimensional shelter in the wind tunnel when the H/z_o ratio of the approach flow was kept constant.

If, then, Re_{turb}(or H/z_o) is preserved, does the laboratory experiment simulate the atmospheric phenomena? And how will changes in the inflow conditions, such as surface roughness and thermal stratification, change the flow behind the model? This paper reports the results of a combined field and wind tunnel investigation of both the similarity and inflow considerations for a simple two-dimensional fence placed perpendicular to the oncoming wind. In both, the cavity wake length(the distance from the fence where the flow separates to the reattachment point downwind) is the flow characteristic under consideration.

2. FIELD STUDY

2-1. EXPERIMENTAL PROCEDURE

The field experiment was conducted in a large, flat, open field covered with short grass. The model is shown in Fig. 1. Note that to make the model near two-dimensional, two high canvas fences were positioned at either end of the model extending far downwind.

The oncoming mean velocity and temperature profiles(10min averages) were measured along an 8m meteorological pole equipped with five cup anemometers and shielded thermocouples. Turbulence characteristics were measured with ultrasonic anemometers. Lightweight wind vanes were manufactured from thin aluminum poles with styrofoam tails and lead counterweight heads and were used to detect and visualize the wind direction behind the model. Sixty vanes were attached to ten poles positioned downwind of the fence in multiples of x/H(from 1 to 10) so that they could rotate freely in the xz-plane. Over each 10min test run, slide photographs were taken at 5sec intervals with a 35mm OMI equipped with motor drive and wide angle lens. The slides were later projected onto a special chart reader desk and the head and tail points were marked and automatically punched onto paper tape. This data was computer analyzed and averaged for each vane's wind direction over the test run. In seven of the fourteen tests, the wind direction shifted so that only the remaining seven tests plus one non-perpendicular case(Test No. NP, wind direction 31.9°C normal to the fence) were analyzed. Test No. NP was analyzed for the effect of wind direction on the flow behind the model.

The upwind anemometer measured mean and turbulent velocities, Reynolds stress, and heat flux. From these, the velocity and temperature gradients, flux Richardson number(Ri_f), Monin-Obukhov stability length(L), and turbulent Reynolds number were calculated. Table 1 is a summary of these upwind flow characteristics. Testing was conducted from morning until sunset so that the range of atmospheric stability went from slightly unstable to slightly stable.

2-2. FIELD RESULTS

Figure 2 shows the average mean wind directions behind the model determined by the small vanes. The mean vane directions show a clear cavity wake behind the model. The size of the cavity differs slightly among the tests. Test No. NP, especially, shows a small cavity due to the non-perpendicular direction of the wind.

Figure 3 shows L_c/H versus Re_{turb}. The cavity wake length increased with increasing Re_{turb}. Although the range of flux Richardson number was not large, a comparison of L_c/H and Ri_f in Fig. 4 indicates that cavity wake length decreased as stability increased.

3. WIND TUNNEL STUDY

3-1. EXPERIMENTAL APPARATUS

The new 2m x 3m x 24m(test section height, width, and length) wind tunnel at NIES was used for this study. The wind tunnel is a vertical closed-return type and a schematic is shown in Fig. 5.

This wind tunnel was designed for the simulation of atmospheric diffusion processes with special focus on the effects of thermal stratification. For this purpose, the tunnel is equipped with three temperature control systems permitting the simulation of a wide range of atmospheric conditions. The first system is an ambient air system which controls the temperature in the wind tunnel from 15°C to 87°C. The second is a surface temperature control system. The test section is equipped with eight 3m x 3m floor panels whose individual surface temperatures can be controlled from 7°C to 112°C. These panels may be arranged in any combination of temperatures. The third system is a temperature cart which can be inserted at the entrance to the test section. This cart is divided vertically into twenty sections; each section contains 43 horizontal rod heaters, 9mm in diameter, and 12 vertical rod heaters to compensate for crosswind temperature differences. At the freestream velocity of 5m/sec, the cart can increase the temperature of each of the 20 sections to as much as 30°C above the ambient.

The freestream velocity in the tunnel ranges from 0.20 to 10m/sec. When a specific velocity profile is desired, a velocity profile-generating cart can be installed before the test section(immediately downwind of the temperature cart if both are used). The velocity cart is divided vertically into twenty sections whose free-flow areas may be varied from 0% to 66%. Each section is independently controlled so that profiles such as linear, logarithmic, exponential, and jet can be created. Additionally, with the use of special floor panels, the size and arrangement of roughness elements can be changed to create various surface roughness conditions.

The model was a thin aluminum plate spanning the entire width of the test section; it was placed perpendicular to the flow. The upper leeward edge of the plate was sharply tapered to prevent blunt edge effects. Two models, 4cm and 8cm, were tested. The model was placed at a position 10m downwind of the test section entrance.

A two-color laser-Doppler anemometer(DISA) with counter processor was used in the backscatter mode to measure the flow behind the model. This enables the measurement of the u and w components simultaneously, and because of the frequency shift, reverse flow can also be measured. By using the LDA technique, the insertion of flow-disturbing probes into the highly sensitive flow field behind the model was avoided.

Oncoming flow velocity was measured with a pitot-static tube and MKS Baratron and oncoming temperature was measured with a thermocouple. In addition, a DISA single-wire hot wire anemometer was used to measure turbulence and stress.

3-2. EXPERIMENTAL CONDITIONS

The model was investigated for the separate effects of two inflow conditions: thermal stratification and surface roughness. Thermal stratification was simulated by creating a temperature difference between the wind tunnel floor and ambient temperatures. Five stability cases(five different Richardson numbers) including the neutral case were examined. Two stable stratifications were created by cooling the floor of the test section and heating the ambient air($T_f - T_a = -69.2$°C and $T_f - T_a = -31.0$°C) at an ambient velocity of $U_\infty \simeq 2$m/sec. Two unstable cases were created by heating the floor and cooling the ambient air($T_f - T_a = 65.8$°C and $T_f - T_a = 32.6$°C) at the same wind velocity.

The second parameter was surface roughness, leading to changes in turbulent Reynolds number. Four different conditions including a smooth upwind surface were created by changing the size and arrangement of the surface roughness elements. These conditons are described in Table 2.

3-3. WIND TUNNEL RESULTS

The first set of experiments was conducted with the 4cm model, mainly to study the effects of thermal stratification. The test section was kept smooth and the ambient air and floor temperatures were changed as follows:

Unstable: $T_a = 32.6$°C, $T_f = 98.4$°C and $T_a = 31.6$°C, $T_f = 64.2$°C
Stable: $T_a = 80.4$°C, $T_f = 11.2$°C and $T_a = 51.7$°C, $T_f = 20.7$°C

Figure 6 shows the oncoming flow mean velocity profiles for the five different stabilities. Due to the effects of the stratification, the stable cases have decreased wind velocity near the surface, while the unstable cases have increased wind velocity when compared with the neutral case. Figure 7 shows the oncoming flow mean excess temperature profiles(temperature minus ambient temperature). This excess temperature $\Delta\theta = \theta - \theta_a$ has a positive value for unstable stability. From these figures, the gradient Richardson number can be calculated.

Figures 8a, b, and c show the U velocity profiles behind the model as measured by the LDA. Mean velocity was digitally averaged over 20,000 data for each point for accuracy. Note that for the stable case, the cavity wake length, Lc/H was 7.0, a drastic decrease from the neutral case whose cavity wake length was 13.0 and from the unstable case with a wake length of 12.5.

The cavity wake length versus gradient Richardson number at model height is shown in Fig. 9. For the stable cases, L_c/H decreases as stability increases. For the unstable cases, the cavity wake length decreases slightly as Ri decreases.

In the atmospheric boundary layer, surface roughness is an important parameter and thus should be considered in wind tunnel simulation. At the same time, the model in the wind tunnel should represent its prototype by being well within the thick surface boundary layer, that is, $H/\delta \leq 0.2$. For each of the cases considered here, mean velocity, turbulence, and shear stress profiles were measured; K_m values were calculated and Re_{turb} values at model height were estimated. These values are shown in Table 2.

Turbulent components were measured by rotating a single hot wire. The signals were linearized and fed to a mini-computer to get the means and standard deviations(two sets of 100sec sampling duration, 10Hz sampling time). The signals were also fed into an RMS-meter(DISA), digital voltmeter, and onto a chart recorder for comparison with the computer output. Figure 10 shows examples of the measured mean velocity and turbulent intensity profiles at the model position for Case 1(smooth) and Case 4(very rough). Figures 11a and b show the mean velocity profiles(LDA) behind the model for Cases 3 and 4. As surface roughness increases, the cavity wake length decreases.

To vary Re_{turb} from 62.4 for the smooth case, roughness elements were put upwind of the model. However, it is generally not realistic for roughness to be larger than or even identical to model height, thus imposing a lower limit on Re_{turb}. In Case 4, the roughness was 0.4H high and a tripping fence of height H was put 10m upwind of the model. With such a configuration, Re_{turb} was 15.2, the smallest tested in this study and quite close to the value found in the field study. Changing Re_{turb} beyond these points with reasonable roughness and within the thick surface boundary ($H/\delta \leq 0.2$) appears difficult.

4. COMPARISON OF FIELD AND WIND TUNNEL STUDIES

Figure 12 shows a summary of L_c/H versus Re_{turb} for the field and wind tunnel. For additional comparison, several other wind tunnel data for similar models are shown. The value of Re_{turb} was not measured for most; however, by assuming that there was a logarithmic velocity profile and using Tani's(3) suggestion, it could be found from:

$$Re_{turb} = UH/K_m = [u_*/k(\ln(H/z_o))]/ku_*H = 1/k^2[\ln(H/z_o)].$$

This calculation is valid only when the model is in the boundary layer; if it is above the boundary layer, Re_{turb} may be larger than calculated.

Also shown in this figure is the cavity wake length found by Arie and Rouse(5) in careful measurements of flow behind a normal plate in uniform flow; they found that $L_c/H \simeq 17$. If one can define a turbulent Reynolds number for uniform flow, since K_m is small here, Re_{turb} would be very large; this then may be the upper limit for cavity wake length for this model. If this is the case, then the relationship between L_c/H and Re_{turb} is not linear and the cavity wake length approaches the upper limit as Re_{turb} increases.

Note that while in both the field and wind tunnel experiments L_c/H increases as Re_{turb} increases, the data points do not overlap. It must be mentioned here that the calculation of Re_{turb} requires the measurement of shear stress and mean velocity gradients, both of which remain difficult to accurately measure.

Figure 13 shows L_c/H versus the local turbulent intensity. Note that the value of L_c/H decreases as the turbulent intensity increases. This was also true for $\sqrt{\overline{u}^2}/U$ and $\sqrt{\overline{w}^2}/U$, and while the wind tunnel intensities do not match those found in the field, this tendency does match that found in the field.

5. CONCLUSIONS

To better understand the complex flow behind a building and to find proper similarity criteria for its wind tunnel simulation, related field and wind tunnel studies were conducted on a simple two-dimensional model fence placed in a perpendicular flow field. Using the cavity wake length(nondimensionalized by model height) as flow characteristic, it was found in the field study that L_c/H increased as Re_{turb} increased and decreased as atmospheric stability increased. In the wind tunnel study, the effects of surface roughness and thermal stratification were examined. For stable stratification, L_c/H decreases rapidly as stability increases, while for the unstable region, L_c/H decreases only slightly as the flow becomes more unstable. Changing the size, or the configuration, or the arrangement of the surface roughness leads to changes in Re_{turb} in the wind tunnel, and in this study it ranged from approximately 15 to 60(very rough upwind surface to smooth upwind surface) when the model was kept well within the surface boundary layer. As Re_{turb} increases, L_c/H increases. This increase is not linear but appears to approach an upper limit. This upper limit may be near the value of $L_c/H \simeq 17$ found by Arie and Rouse for a normal plate in uniform flow.

In both the field and wind tunnel studies, cavity wake length increases as Re_{turb} increases; however, the data do not overlap and this may be due to inaccurate measurement of Re_{turb}. Both studies also show a tendency for L_c/H to decreases as stability increases. The field and wind tunnel data also indicate that as the local turbulent intensity increases, the cavity wake length decreases. While the turbulent intensities found in the field could not be matched in the wind tunnel, their similar trends suggest that if u_*/U, $\sqrt{w^2}/U$, and $\sqrt{u^2}/U$ can be simulated, similar L_c/H might be found in the wind tunnel.

REFERENCES

(1) Inoue, E., "The structure of the surface wind," Bull. Inst. Agri. Sci., No. 2, 1952(in Japanese).
(2) Jensen, M., "The model law for phenomena in a natural wind," Ingenioren-Int. Ed. 2, No. 4, 1958.
(3) Tani, N., "On the wind tunnel test of a model shelter-hedge," Bull. Inst. Agri. Sci., Series A, Vol. 6, 1958(in Japanese).
(4) Plate, E., Aerodynamic Characteristics of the Atmospheric Boundary Layer, NTIS-TID-25465, 1971.
(5) Arie, M. and Rouse, H., "Experiments on two-dimensional flow over a normal wall," J. Fluid Mech., Vol. 1, pp. 129-141, 1956.
(6) Good, M.C. and Joubert, P.N., "Form drag of two-dimensional bluff plates immersed in turbulent boundary layers," J. Fluid Mech., Vol. 31, 1968.
(7) Shoda, R., "Experimental study of flow characteristics in the environment of buildings," MS Thesis, Hokkaido Univ., 1967(in Japanese).
(8) Kamata, Y., "A wind tunnel study of the effects of buildings," 29th Annual Civil Engineers of Japan Meeting, 1974(in Japanese).

TABLE 1. Summary of Field Upwind Flow Characteristics at z/H = 1

Test No.	U $\frac{cm}{sec}$	V $\frac{cm}{sec}$	σ_u $\frac{cm}{sec}$	σ_v $\frac{cm}{sec}$	σ_w $\frac{cm}{sec}$	$-\overline{uw}$ $[\frac{cm}{sec}]^2$	$\overline{w\theta}$ $\frac{cm}{sec}$ °C	Ri_f	H/L	z_o cm	K_m $\frac{cm^2}{sec}$	Re_{turb}	L_c/H	u_*/U	$\sqrt{\overline{w}^2}/U$
1	327	-34	101	91	46	1318	3.1	-0.013	-0.016	2.0	2501	23.8	7.2	0.111	0.141
2	395	79	120	111	64	2593	6.5	-0.014	-0.012	3.8	3507	20.5	7.0	0.129	0.162
3	301	51	95	89	47	1436	3.1	-0.015	-0.014	2.5	2872	19.1	6.8	0.126	0.156
4	311	75	88	83	48	1467	7.0	-0.028	-0.031	4.5	2516	22.5	7.0	0.123	0.154
5	288	-57	78	87	44	1163	-0.3	0.002	0.002	2.0	2438	21.5	7.2	0.118	0.153
6	217	6	63	50	33	635	-2.0	0.026	0.030	2.5	1580	25.0	7.4	0.116	0.152
7	164	30	53	44	27	462	-4.6	0.127	0.115	3.0	1731	17.2	6.3	0.131	0.165
NP	259	161	73	77	41	992	0.6	-0.007	-0.008	2.8	1997	23.6	4.2	0.122	0.158

TABLE 2. Summary of Wind Tunnel Roughness Conditions and Flow Characteristics

Case	Roughness	U $\frac{cm}{sec}$	Hcm	H/δ	$U_{z=H}$	$\sqrt{\overline{u}^2}/U$	$\sqrt{\overline{w}^2}/U$	u_*/U	K_m	Re_{turb}	L_c/H
Case 1	No roughness	200.3	4	0.20	146.2	0.089	0.052	0.051	9.38	62.4	13.0
Case 2	1cm L-angles, 150cm apart	200.0	4	0.167	134.9	0.110	0.068	0.062	16.1	33.5	11.2
Case 3	1cm and 2cm L-angles, alternating, 7.5cm apart	201.5	4	0.129	123.3	0.144	0.088	0.072	19.4	25.4	10.0
Case 4	3.2 x 3.7 x 25cm(H x W x L) wooden blocks, lat. and long. spacing = 25cm; 8cm fence, 10m upwind of model	220.8	8	0.160	127.6	0.242	0.118	0.107	67.2	15.2	7.5

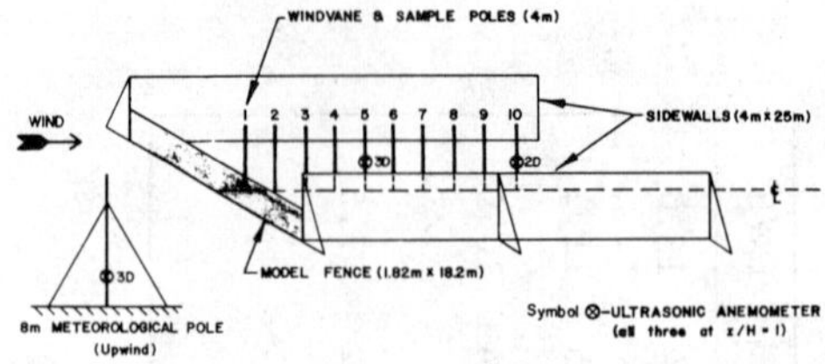

Fig. 1. Schematic of field model

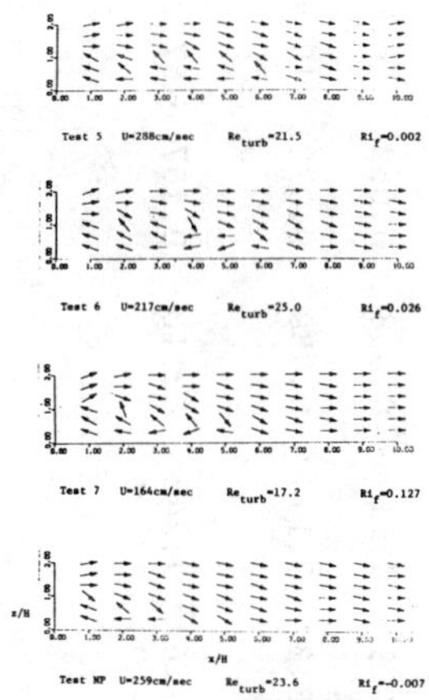

Fig. 2. Averaged vane directions behind the model

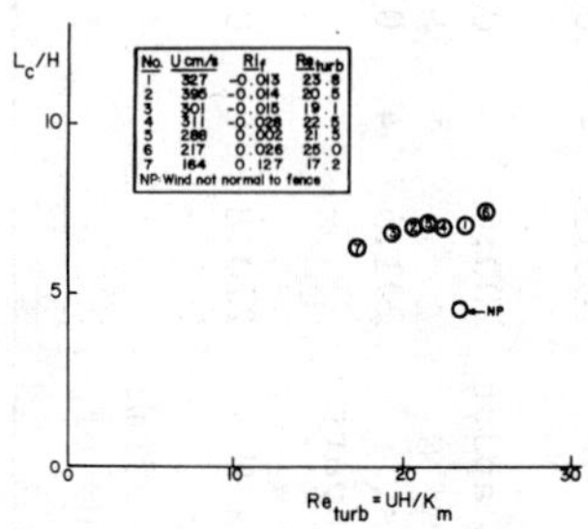

Fig. 3. L_c/H vs. Re_{turb}

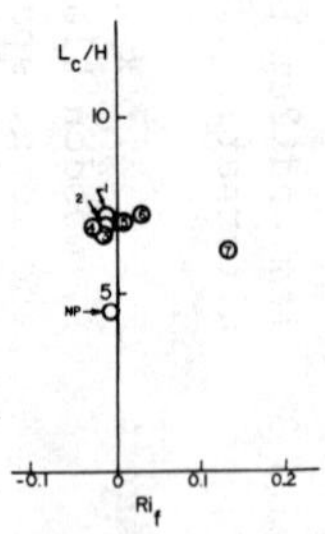

Fig. 4. L_C/H vs. Ri

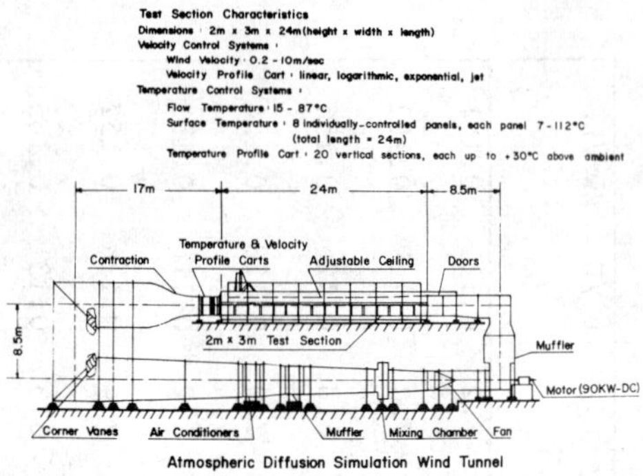

Fig. 5. Schematic of NIES wind tunnel

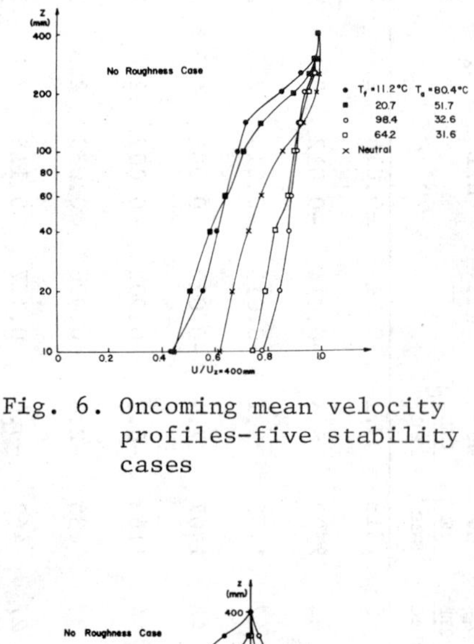

Fig. 6. Oncoming mean velocity profiles-five stability cases

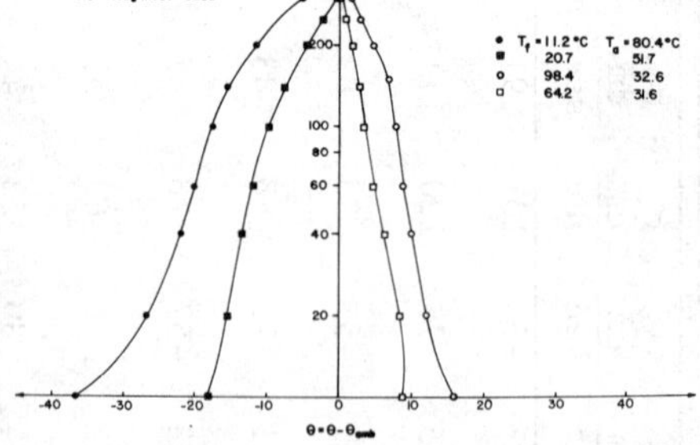

Fig. 7. Oncoming mean excess temperature profiles-five stability cases

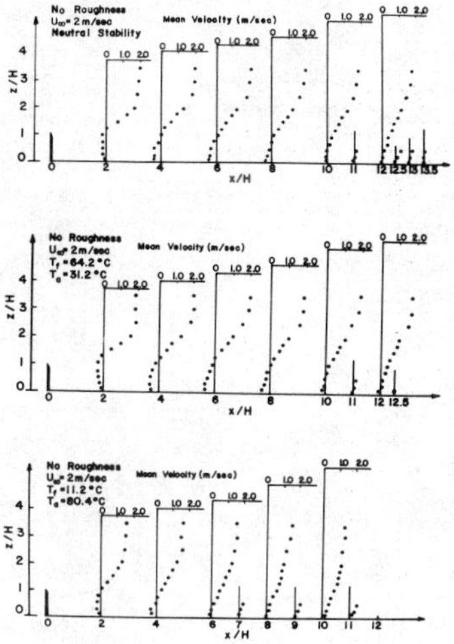

a. Neutral

b. Unstable

c. Stable

Fig. 8. Mean velocity profiles behind the model

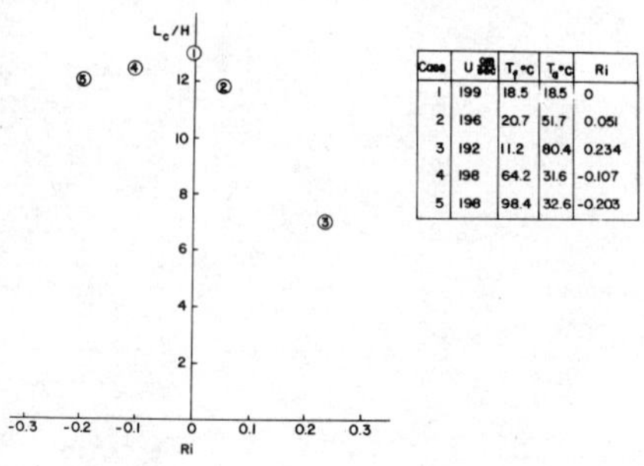

Fig. 9. Cavity wake length vs. $Ri_{gradient}$

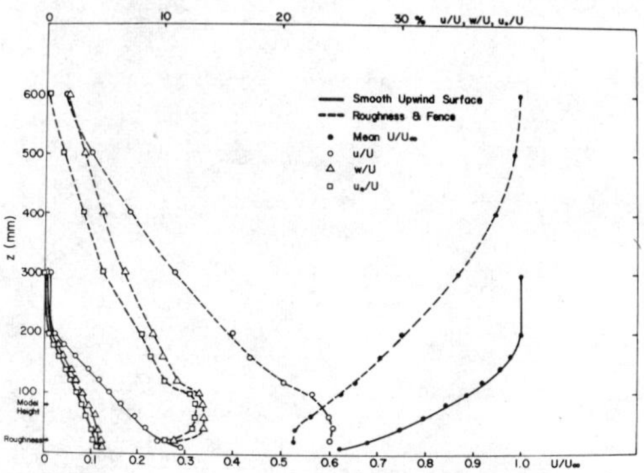

Fig. 10. Mean velocity and turbulent intensity profiles at model position-Cases 1 and 4

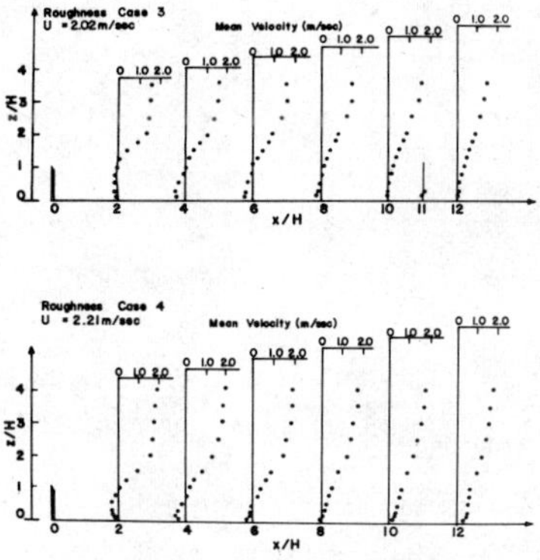

a. Case 3

b. Case 4

Fig. 11. Mean velocity profiles behind the model

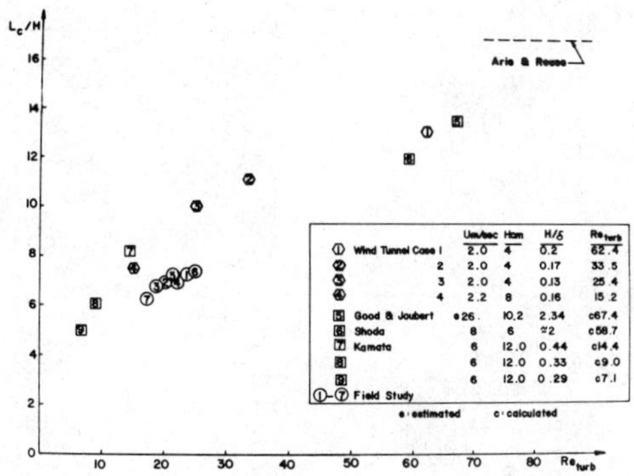

Fig. 12. L_c/H vs. Re_{turb} for the field wind tunnel studies

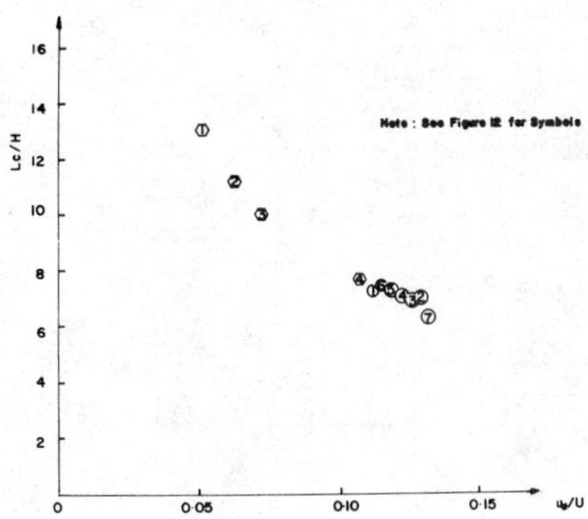

Fig. 13. L_c/H vs. local turbulent intensity

A FLUID MODELING STUDY OF THE FLOW STRUCTURE AND PLUME IMPINGEMENT ON A THREE-DIMENSIONAL HILL IN STABLY STRATIFIED FLOW

W. H. SNYDER§*, R. BRITTER§§** and J. C. R. HUNT**
*Environmental Protection Agency, Research Triangle Park, North Carolina, USA
**University of Cambridge, Cambridge, UK

ABSTRACT

The flow structure around simple three-dimensional obstacles in stably stratified flow is experimentally studied using a large towing tank. Under certain conditions, plumes may impinge directly on an obstacle giving far higher concentrations in comparison with plumes diffusing to the surface. The conditions for plume impingement and resulting concentrations are quantified and interpreted. Requirements for a strong downslope wind in the lee of the obstacle are noted. The results are relevant to winds around hills in a stable atmosphere and local transport and dispersion of pollutants.

INTRODUCTION

Far larger surface concentrations are expected when a pollutant plume from an elevated source impinges on rather than diffuses to a particular surface. We define impingement as a pollutant source on a streamline that stagnates on a surface. For example, winds occurring with a stable density profile blowing over complex terrain can result in plume impingement on isolated hills.

This paper deals with stable flows near simple isolated obstacles and forms part of a larger program to understand transport and diffusion of pollutants in complex terrain, under neutral and stably stratified flow conditions. The objective of this project was towards generic, instead of specific case studies, providing increased comprehension of the flow structure for the interpretation of concentration measurements. Concurrently, the experimental results provided a data base for the testing and validation of analytical and numerical models for such flows.

Use was made of a meteorological wind tunnel (2.1x3.7x18.3m) and a salt water stratified towing tank (1.2x2.4x25m). Three axisymmetric hill shapes were used; bell-shaped (polynomial, $1/(1+r^4)$), conical (26.5^o slope), and hemispherical, all with a height of 23 cm. Flow visualization was extensively used to obtain both qualitative and quantitative information on the flow structure around the hills.

---

§On assignment from the National Oceanic and Atmospheric Administration, U.S. Department of Commerce.

§§Financial support provided through an appointment at the North Carolina State University through EPA Grant 805595.

Tracers were released from stacks upwind and resulting (primarily surface) concentration patterns were measured.

Four major topics are discussed:

- The effect of changes in stratification, from highly stable to neutral, on the flow structure.
- The effects of hill shape on the flow structure.
- Simple rules for determining whether a plume from an upwind source will impinge on the hill or go over the top or around the sides for both continuous linear stratification and elevated step inversions.
- Rules for predicting maximum surface concentrations in the event of plume impingement.

Although the material covered concerns flow around hills, the results should be applicable to flow around isolated obstacles in general.

FLOW STRUCTURE

Consider inviscid flow around a simple, three-dimensional shape, e.g., a hemisphere on a flat horizontal surface. If the fluid is neutrally stratified the velocity field is, in the absence of separation, known from potential flow theory. Of particular interest here are those streamlines well upstream which are in the plane of symmetry. All streamlines in this centerplane pass over the crest of the obstacle, even branches which were on the surface well upstream.

When the fluid is stably stratified, the kinetic energy in the approach flow may be less than the potential energy required to raise a fluid element from its upstream elevation to the top of the hill through the density gradient. For a uniform velocity profile and a linear density gradient, this energy ratio is characterized by:

$$\rho U^2/(gh^2 d\rho/dz) = U^2/N^2h^2 = F^2,$$

where U is the approach velocity, N is the Brunt-Väisäla frequency $(=(g/\rho)d\rho/dz)^{\frac{1}{2}}$, h is the height of the hill and F is the Froude number. A small Froude number, indicating strongly stable stratification, implies insufficient kinetic energy in the approach flow to raise fluid from the base to the top of the hill. For a two-dimensional hill, this may result in complete upstream blocking of the flow below the top (1). For a three-dimensional hill, the fluid, rather than being blocked, can go <u>round</u> the sides. A fundamental question which is yet unsolved concerns the flow of a stably stratified fluid near an obstacle of arbitrary aspect ratio, and when it has a two- versus a three-dimensional character.

Existing theory (2), model experiments (3,4), and field observations (5) of air flow around three-dimensional hills indicate that during strong stratification ($F \rightarrow 0$) the fluid is constrained to move in essentially horizontal planes round the topography. An exception is near the summit of the hill where a fluid element requires only a small vertical displacement to surmount the obstacle. In fact, Drazin's (2) analysis breaks down within a distance of order Fh of the summit, where the flow is three-dimensional.

The basic flow patterns expected are sketched in Figure 1 for (a) neutral and (b) very stably stratified flow around a three-dimensional hill. Apart from the differences between the flows already described, the separated flow region in the lee of the hill has also been sketched. Under very stable conditions, the flow separates from the surface as it goes round the sides, much as it would from a circular cylinder, and the wake adopts a two-dimensional horizontal nature.

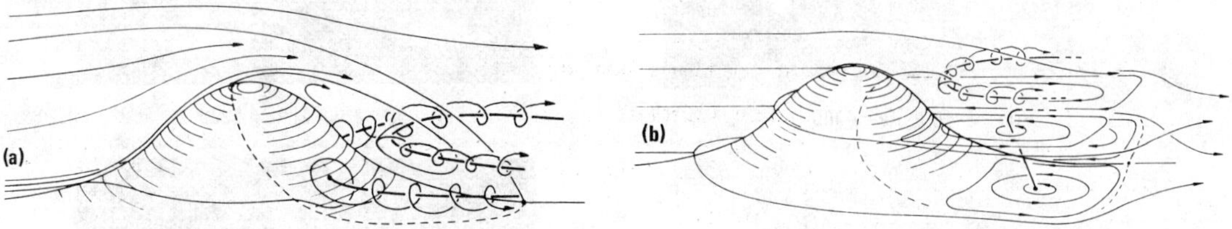

Figure 1. Sketches of (a) neutral and (b) highly stratified flows over three-dimensional hills.

Although extensive flow visualization results were obtained, only a few typical and pertinent photographs were chosen to constitute Figures 2 and 3. Figures 2a-f illustrate dye releases in the center plane upstream of the hemisphere at Froude numbers from 0.2 to ∞, while Figures 3a-f present shadowgraphs of the flow around the cone for Froude numbers of 0.2 to 3.8. In most respects, the main features of the flow were similar for all three shapes studied at any particular Froude number. For additional photographs and a more extensive discussion of the flow structure over the bell-shaped hill, the reader is referred to the EPA report (6).

Upstream of the crest, the flow is essentially horizontal at a Froude number of 0.2 and effectively neutral (fully three-dimensional) at a Froude number of order 2, with a transition between. Downstream from the crest, separation was progressively delayed as the Froude number was increased above 0.2 and, at a Froude number of order unity, separation did not occur at all on the lee slopes of either the cone or polynomial hill. Due to the steep slope near the base of the hemisphere, separation on its lee surface was never completely suppressed. A further increase in Froude number induced the separation point to move rapidly back to the top of the lee slope.

To characterize the flow structure in the immediate vicinity of the hill (excluding, in particular, the far wake), we have chosen two obvious and relevant parameters:

(1) The upstream height H_s of the dividing streamline, above which the flow goes over the hill top and below which the flow goes round the sides.

(2) The distance x_{sep} from the crest to the point of separation on the centerline in the lee.

A criterion for predicting the upstream height of the dividing streamline was suggested by Hunt and Snyder (7) on the basis of the low Froude number theory of Drazin as $H_s/h=1-F$. A comparison of this formula with data for the hemisphere, polynomial hill and cone (Figure 4) shows good agreement and validates the prediction. It is clear that all streamlines pass over the top if the Froude number is larger than unity, although they are more closely spaced when compared to those in neutral flow.

Suppression of separation on lee slopes has been commented upon by Scorer (8) and others for two-dimensional flows and is best interpreted in terms of lee wave formation. Assuming the fluid is infinite* in its vertical extent, the fundamental lee wave mode will have a wavelength $\lambda = 2\pi U/N$ independent of hill shape (9).

*If the vertical extent of the fluid is D (as in a tank of finite depth), then $\lambda=2\pi U/N(1-\pi^2U^2/N^2D^2)^{\frac{1}{2}}$. There is no possibility of lee waves when $ND/U<\pi$.

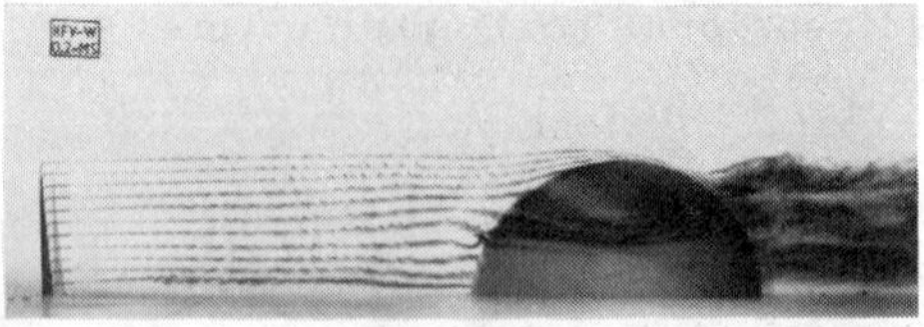

(a) F = 0.2, Re = 4300

(b) F = 0.4, Re = 8700

(c) F = 0.6, Re = 13000

(d) F = 1.0, Re = 21000

(e) F = 1.7, Re = 36000

(f) F = ∞, Re = 55000

Figure 2. Centerplane streamlines near hemisphere.

(a) F = 0.2, Re = 5700

(b) F = 0.4, Re = 12000

(c) F = 0.6, Re = 18000

(d) F = 1.0, Re = 29000

(e) F = 1.2, Re = 35000

(f) F = 3.8, Re = 110,000

Figure 3. Shadowgraphs of flow near cone.

When this wavelength is comparable to the overall length of the hill, large amplitude (of the order of the hill height) lee waves are to be expected. The pressure field of the wave acts to suppress separation on the lee face, but it may also promote separation downstream, possibly forming a rotor. We define the length scale of the hill as the semi-length between the half-height positions. Typically, the overall length of the hill is about 4L. It is important to note that, in discussing the influence of density stratification on separation, it is the parameter $F_L=U/NL$ that is relevant, not $F=U/Nh$. Accordingly, we plotted x_{sep}/L as a function of F_L (Figure 5) for the three shapes studied. Data from the experiments of Davis (10) for a two-dimensional triangular ridge and computations from Sykes (11) for a steep, two-dimensional smooth ridge are included in the figure.

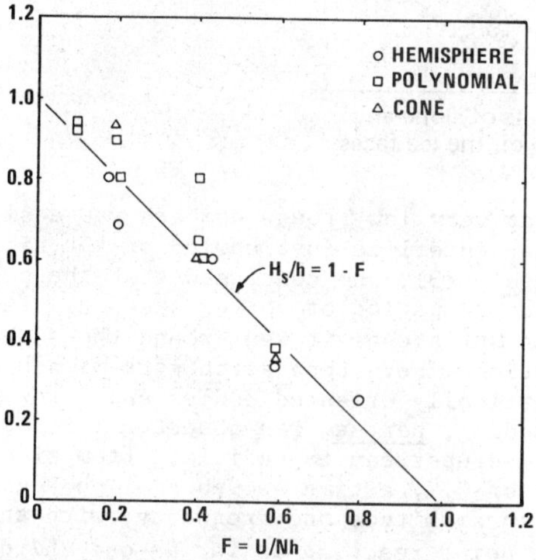

Figure 4. Upstream height of the dividing streamline as a function of Froude number.

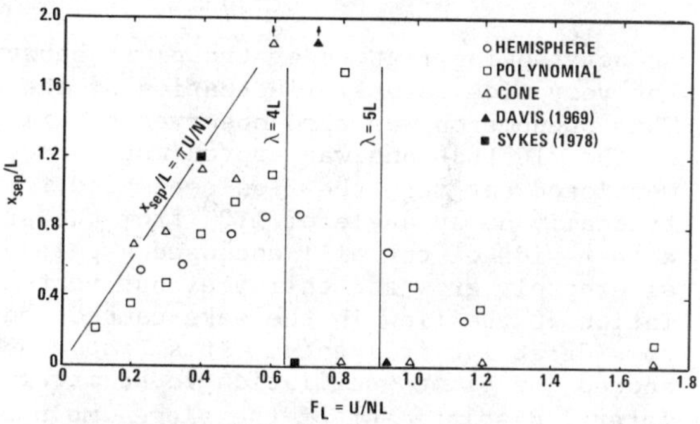

Figure 5. Position of separation on the lee slope as a function of F_L.

We expect that separation will occur at the first trough of the fundamental lee wave, that is $x_{sep}=\lambda/2$ as $x_{sep}/L=\pi U/NL=\pi F_L$. However, if $\lambda>>L$, the pressure field induced by the lee wave will not maintain the flow on the lee face and separation will return to near the crest. Thus, for λ much greater than the overall length of the hill, say 4L, we anticipate $x_{sep}/L\simeq 0$. Therefore, $x_{sep}\simeq 0$ for $U/NL=F_L\gtrsim 2/\pi$. The data in Figure 5 support the two expressions $x_{sep}=\lambda/2$ and $x_{sep}=0$. Separation from the hemisphere is influenced by the large local slopes in the lee. Although we have no documentation, it was apparent that the maximum velocity of the downslope flow occurred when x_{sep}/L was a maximum.

When the gross slope (characterized by h/L) is small, there will never be separation on the lee face, but if it is large, separation will always occur near the crest. Figure 6 summarizes the anticipated effect of both slope and Froude number on separation on the lee face of three- and two-dimensional hills. The data reported here fall in the moderate slope category (h/L=1) with the Froude number varying from small to large.

SURFACE CONCENTRATIONS UNDER PLUME IMPINGEMENT

From our previous discussion, it is evident that plumes below the dividing streamline will impinge directly on the windward slope, split, then travel round the sides of the hill. It is also evident that this type of situation will result in the highest short-term ground level concentrations (excluding wind meander factors) due to limited vertical diffusion of the plume and direct impingement on the hill surface.

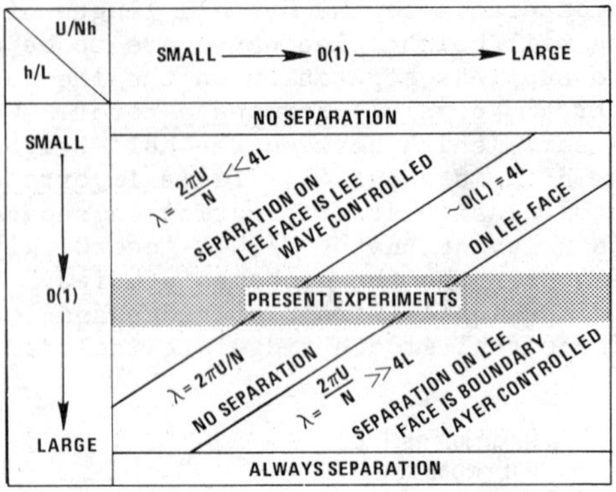

Figure 6. Anticipated effects of slope and stratification on separation on the lee faces of hills.

An outstanding feature of the plume behavior at very low Froude numbers was a slow but very wide lateral fluctuation of the plume centerline just upwind of the hill. This phenomenon was also observed by Riley et al. (3), who concluded that the period of the fluctuations was approximately equal to the period of vortex shedding. As mentioned earlier, the flow separated from the surface as it went round the sides, typically at an angle of 110° from the stagnation line. Separation streamlines on either side of the hill enclosed a pair of vertically oriented eddies appearing to alternately grow and collapse, but vortex shedding, per se, was absent. This oscillation of the flow in the wake caused the plumes upstream to vacillate from side to side at low frequency. At a Froude number of 0.2, a time sequence of photographs showed the plume oscillation to be irregular in amplitude and frequency, with the lateral displacement of the plume impingement point reaching as far as one-third the local hill radius on either side of the centerline. At Froude numbers greater than 0.4, oscillations of the plume position progressively diminished. At lower Froude numbers ($F<0.1$), Brighton (4) observed coherent vortex shedding.

Hunt et al.(12) described a method for computing concentrations around the impingement point on the surface. Using fairly nonrestrictive assumptions, they reduced the problem to calculating the two-dimensional diffusion from a line source around a cylinder with axis parallel to the line source axis, the cylinder having the same radius as the hill at the height of the plume. Their results predicted, for the particular parameters of the present laboratory model, that the ratio of the maximum surface concentration χ_{mx} to the concentration in the center of the plume in the absence of the hill χ_o would be 1.12.

For neutral turbulent flow round a cylinder or sphere, calculations and wind tunnel observations of Britter et al. (13) demonstrate that $\chi_{mx}/\chi_o=1$ for line or point sources situated on the centerline more than two diameters upstream. The EPA valley model (14) assumes $\chi_{mx}/\chi_o=2$, attributable to plume reflection at the surface. Note, however, that the average concentration predicted may depend more upon the assumptions of wind meander (sector averaging) than upon the precise value of the ratio χ_{mx}/χ_o.

At this point it will be enlightening to briefly describe the concentration measurement program. Over one hundred sampling ports were fixed on the polynomial hill surface at the radial planes $\Theta=180^o$, -165°, -90°, and 0°. (See Figure 7.)

Blue food dye mixed with salt water constituted the neutrally buoyant effluent. It was isokinetically injected from the source and used for both flow visualization and quantitative measurements of concentration. A vacuum system withdrew samples

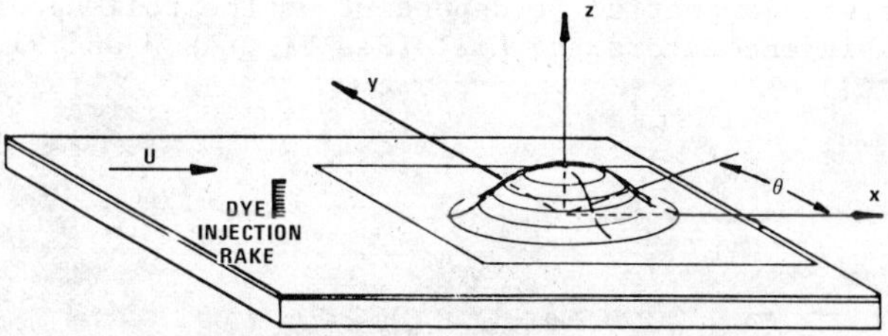

Figure 7. Sketch of polynomial hill model.

through the surface ports into individual collection bottles for quantitative concentration analysis via a colorimeter, calibrated with samples of known concentration. For each run, concentration profiles were obtained along each of the four radial lines. Variables were the source height h_s, the lateral offset of the stack from the hill centerline y_s, and the Froude number. Since it was desired to compare concentrations measured on the hill surface with those on the centerline of the plume in the absence of the hill, baseline measurements were essential. These were accomplished using vertical and lateral rakes of sampling tubes at a distance downstream of the source corresponding to the center of the hill.

Figure 8 is a graph of the maximum surface concentration determined from each profile on the hill surface using source offset as a parameter. The ratios of maximum concentration observed on the hill surface to those observed at the plume centerline in the absence of the hill were obtained from similar graphs (Figure 8) for varying source heights and Froude numbers (Table 1).

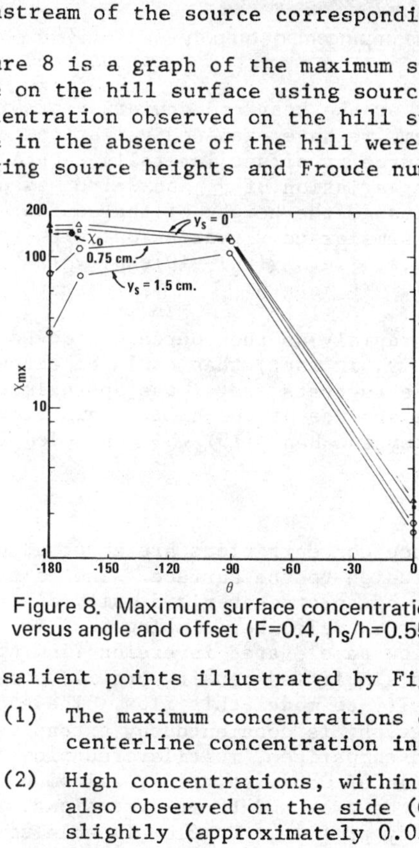

Figure 8. Maximum surface concentration versus angle and offset (F=0.4, h_s/h=0.55).

Table 1. χ_{mx}/χ_o ON POLYNOMIAL HILL.

h_s/h	0.39	0.55
F		
0.2	0.53	0.63
0.4	0.83	1.12

The salient points illustrated by Figure 8 and Table 1 are:

(1) The maximum concentrations observed range from 0.5 to 1.2 times the centerline concentration in the plume in the absence of the hill.

(2) High concentrations, within 80% of the largest observed on $\Theta=180^o$, are also observed on the side ($\Theta=90^o$) of the hill when the source is only slightly (approximately 0.03 radius) displaced from the centerplane.

The repeatability of these experiments was not good. Oscillations in the wake, the limited length of the runs, and difficulties with the colorimeter combined to yield repeatabilities generally within a factor of 2. Hence, differences in the concentration ratios listed in Table 1 are not regarded as very significant, although the lower values at lower Froude numbers appear to be caused by oscillations of the plume caused, in turn, by oscillations in the wake. Similarly, larger values at larger source heights appear to be dependent on the roll-up of the plume into a horseshoe vortex as it encounters the hill (See Figures 3 and 9).

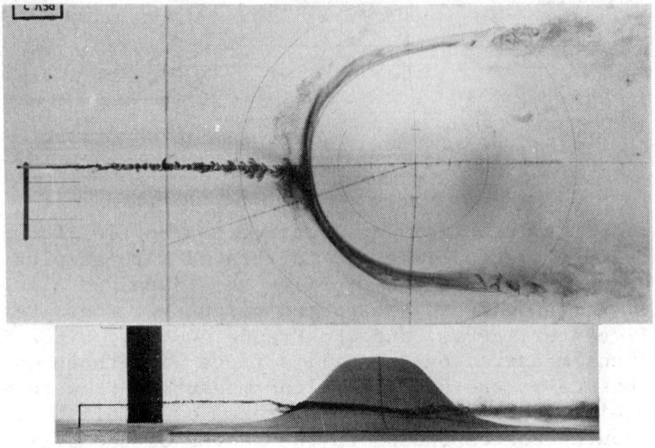

Figure 9. Top and side views of plume impingement on polynomial hill (h_s/h=0.4, F=0.2).

An important atmospheric factor is wind meander or the standard deviation of wind direction fluctuations. To address this problem, we have examined variations of concentration on the hill centerplane as the source is offset laterally. Data from Figure 8 are replotted in Figure 10 as the variation of χ_{mx} on $\Theta=180^o$ as a function of $(y_s/R_s)^2$, where R_s is the hill radius at the height of the source. An initial hypothesis is that χ_{mx} varies with y_s, similar to χ_o variations with the lateral source displacement in the hill's absence, i.e., $\chi(y_s)=\chi(0)\exp(-y_s^2/2\ell^2)$, where ℓ is the plume width (standard deviation).

Concentrations on the 180° line decreased very rapidly as the source was moved laterally from the centerplane, much more rapidly, in fact, than could be accounted for by our hypothesis. A better fit to the data suggests that ℓ was approximately 0.7 cm, only half the σ_{yo} value measured in the absence of the hill. This finding is contrary to the wind tunnel results of Britter and Hunt (15) who concluded $\ell \approx \sigma_{yo}$.

ELEVATED INVERSION

The previous section indicated that large surface concentrations are observed upon plume impingement in contrast to pollutant diffusion to the surface. The continuous linear density distribution is an approximation of an atmospheric density distribution making plume impingement possible. Another example where impingement is possible is a plume in a "well mixed" layer below an elevated inversion flowing toward a three-dimensional obstacle. We used a simplified density profile consisting of two homogeneous layers of different density to model this flow. The study of multilayered flow over two-dimensional topography is documented by extensive experimental and analytical literature (1). In comparison, multilayered flow around isolated obstacles has, with the exception of Oobayashi (16), seldom been investigated. An important difference between the two- and three-dimensional flows is that, similar to the continuous linear density profile, the flow is able to go round rather than over an isolated obstacle.

In the experiments, the obstacle was the same conical hill used in Section 2. The towing tank was filled to a total depth D with two layers of density ρ_1 and ρ_2 and the hill was towed (inverted) at a velocity U along the tank. The density interface was h_o above the hill base and the ratio h/D was fixed at 0.22. By invoking the Boussinesq approximation (valid as $\rho_2/\rho_1 \to 1$), the flow structure is a function of two dimensionless parameters, h/h_o and $U/(g'h_o)^{\frac{1}{2}}$, where $g'=g(\rho_2-\rho_1)/\rho_1$. Flow visualization studies indicated many of the same features (large amplitude waves in the lee, strong downslope flow, hydraulic jumps, etc.) previously discussed in Section 2.

Presented here are those results bearing on the possibility of plume impingement, specifically, a splitting of the plume into two sections traveling round the sides instead of a single plume passing over the top. As in the case of the linear density profile, the maximum concentration on the surface will be almost identical to the maximum concentration in the absence of the hill. A vertical rake of dye sources (h_s/h=0.1 to 1) was placed on the centerline upstream of the hill and plume trajectories (streamlines) were photographed. Figure 11 summarizes the data for the existence of plume impingement as a function of $U/(g'h_o)^{\frac{1}{2}}$ and the ratio of hill height to inversion height h/h_o.

It is obvious that impingement must occur when the source is in the fluid near the hill base and the interface does not pass over the hill; the fluid layer near the base is thus doubly-connected while the layer above the interface is singly connected. In the limits $U/(g'h_o)^{\frac{1}{2}} \rightarrow 0$ and $h/h_o \rightarrow \infty$, the fluid layer near the base may be treated as neutral flow around a two-dimensional cylinder while the upper layer may be treated as neutral flow around a cone of height $(h-h_o)$.

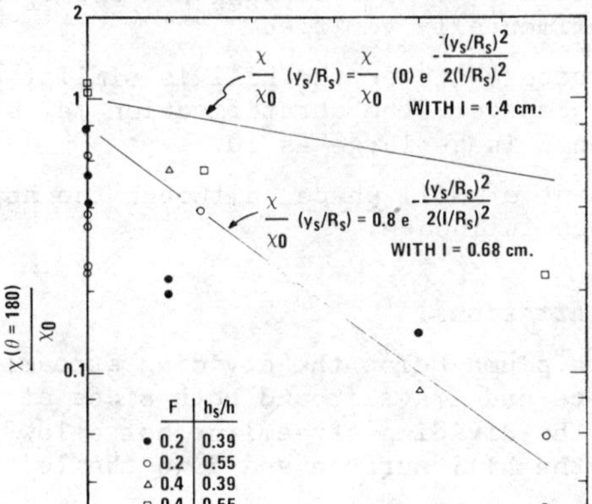

Figure 10. Maximum surface concentration on stagnation line (θ=180°) as a function of source offset.

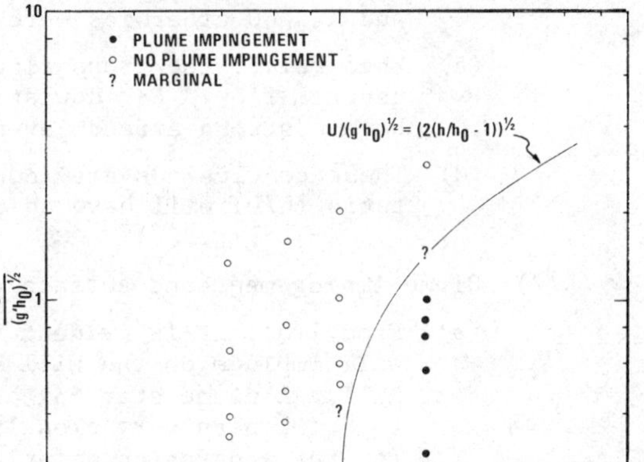

Figure 11. Existence of plume impingement under an elevated inversion as a function of $U/(g'h_o)^{1/2}$ and h/h_o.

A simple energy argument (Bernoulli's equation) shows that the interface cannot surmount the hill when $U/(g'h_o)^{\frac{1}{2}} < (2(h/h_o-1))^{\frac{1}{2}}$. This curve is included in Figure 11, indicating good agreement with our limited experimental results. We conclude that plume impingement occurs when this inequality is met and the source height h_s is beneath the inversion, i.e., $h_s < h_o$. Unlike the continuous linear density profile where the flow was stably stratified everywhere, the model of the elevated inversion has neutral stratification above and below the interface. Diffusion of the plume, per se, will be markedly different within the two flows.

CONCLUSIONS

The structure describing the stably stratified flow in the vicinity of isolated three-dimensional hills was derived from elementary physical arguments and experiments. Flow visualization was extensively used as an aid to understanding plume behavior under both constant density gradient and elevated inversion conditions.

The salient features of the flow structure and surface concentration patterns are summarized below.

(1) Flow structure under constant density gradient:

(a) When $F \ll 1$ (i.e., $F < 0.3$ for hills with moderate slope), the flow is essentially horizontal except in a narrow region near the top having a thickness of Fh. Thus, there is a dividing streamline with an upstream height H_s, above which the flow goes over the top and below which the flow goes round the sides. A formula for determining the height of the dividing streamline is $H_s/h = 1-F$, which is experimentally verified for the range $0<F<1$. A small lee wave or internal hydraulic jump may be found in the lee of the hill above H_s; below H_s, the flow separates from the sides of the hill, with separation streamlines enclosing a primarily horizontal recirculating wake.

(b) When $F \simeq 1$, it is fruitful to think in terms of the response of the flow over the hill to the wavelength λ of the lee wave pattern set up by the hill. The most important parameter is the Froude number based on the hill length. Physical arguments suggesting the location of the separation point on the lee side of the hill as $x_{sep} = \lambda/2$ for $F_L < 2/\pi$ and $x_{sep} = 0$ otherwise were experimentally verified.

(c) When $F \gtrsim 1.5$, the gross flow structure around the hill is similar to neutral flow. Far downstream, however, the stratification may still have a strong effect, even when F is as large as 10.

(d) These conclusions are independent of hill shape, although the aspect ratio (h/L) will have an obvious influence.

(2) Plume impingement and surface concentrations:

(a) From 1(a), it is evident that a plume below the dividing streamline will impinge on the hill surface and travel round both sides of the hill. A plume starting above the dividing streamline but below the top will pass very closely to the hill surface and down the lee slope (to the separation point).

(b) Under plume impingement conditions, maximum surface concentrations observed were between 0.5 and 1.2 times the plume centerline concentration in the absence of the hill. High concentrations, within 80% of the largest observed on the stagnation line, were also observed on the side of the hill when the source was only slightly displaced from the centerline.

(3) Elevated inversion:

(a) A simple energy argument suggesting that the interface between two fluids cannot surmount a hill if $U^2/(g'h_o) < 2(h/h_o-1)$ was experimentally verified. Under these conditions, a plume originating in the lower layer of fluid will impinge on the hill surface.

(b) Surface concentrations resulting from the above plume may be predicted by calculating the diffusion under neutral flow conditions (with appropriate adjustments for limited mixing depth, etc.).

REFERENCES:

1. Long, R.R., "Finite Amplitude Disturbances in the Flow of Inviscid Rotating and Stratified Fluids Over Obstacles", Annual Review of Fluid Mechanics, Vol. 4, Annual Reviews, Palo Alto, CA, 1972, pp. 69-92.

2. Drazin, P.G., "On The Steady Flow of a Fluid of Variable Density Past an Obstacle", Tellus, Vol. XIII, No. 2, 1961, pp. 239-251.

3. Riley, J.J., Liu, H.T. and Geller, E.W., "A Numerical and Experimental Study of Stably Stratified Flow Around Complex Terrain", U.S. Envir. Prot. Agcy. Rpt. No. EPA-600/4-76-021, Res. Tri. Pk., NC, May, 1976.

4. Brighton, P.W.M., "Strongly Stratified Flow Past Three-Dimensional Obstacles and Mesoscale Vortex-Streets in the Atmosphere", Q. J. Roy. Meteorol. Soc., Vol. 104, 1978, pp. 289-307.

5. Queney, P., Corby, G.A., Gerbier, N., Koschmieder, H. and Zierep, J., "The Airflow Over Mountains", World Meteorol. Organ. Tech. Note No. 34, Geneva, Switz., 1960.

6. Hunt, J.C.R., Snyder, W.H. and Lawson, R.E., Jr., "Flow Structure and Turbulent Diffusion Around a Three-Dimensional Hill: Fluid Modeling Study on Effects of Stratification: Pt I: Flow Structure", U. S. Envir. Prot. Agcy. Rpt. No. EPA-600/4-78-041, Res. Tri. Pk., NC, July, 1978.

7. Hunt, J.C.R. and Snyder, W.H., "Experiments on Stably and Neutrally Stratified Flow Over a Model Three-Dimensional Hill", To appear in J. Fluid Mech., 1979.

8. Scorer, R.S., Environmental Aerodynamics, Ellis Horwood Ltd., England, 1978.

9. Crapper, G.D., "A Three-Dimensional Solution for Waves in the Lee of Mountains", J. Fluid Mech., Vol. 6, 1959, p. 51.

10. Davis, R.E., "The Two-Dimensional Flow of a Fluid of Variable Density Over an Obstacle", J. Fluid Mech., Vol. 36, 1969, p. 127.

11. Sykes, R.I., "Stratification Effects in Boundary Layer Flow Over Hills", Proc. Roy. Soc. Lond., Ser. A, Vol. 361, 1978, p. 225-243.

12. Hunt, J.C.R., Puttock, J.S. and Snyder, W.H., "Turbulent Diffusion from a Point Source in Stratified and Neutral Flow Around a Three-Dimensional Hill: Part I: Diffusion Equation Analysis", Atmos. Envir., 1979 (in press).

13. Britter, R.E., Hunt, J.C.R. and Puttock, J.S., "Predicting Pollution Concentrations Near Buildings and Hills", Proc. Conf. Systems and Models in Air and Water Pollution, Inst. Meas. and Control, London, 1976.

14. Burt, E.W. and Slater, H.H., "Evaluation of the Valley Model", AMS-APCA Joint Conf. on Appl. of Air Poll. Meteorology, Salt Lake City, UT, Nov. 29-Dec. 2, 1977.

15. Britter, R.E. and Hunt, J.C.R., "Dispersion from Sources Upwind of Bluff Bodies in Grid Turbulence", Submitted to Atmos. Envir., 1979.

16. Oobayashi, T., "Numerical Study of Two-Dimensional Airflow Over an Isolated Mountain", J. Meteorol. Soc. of Japan, Vol. 48, 1970, p. 118-127.

DISCUSSION ON SESSION III
Local Wind Environment

Rapporteur: J. L. Way
Illinois Institute of Technology
USA

Paper by R. M. Aynsley--WIND GENERATED NATURAL VENTILATION OF HOUSING FOR THERMAL COMFORT IN HOT HUMID CLIMATES

Question: R. A. Hazelwood. Since in very low wind speeds, stack pressures (due to temperature and humidity differences) will be very important, do you have any information on where these effects are masked by wind pressures, and have you any views on how to incorporate this in architectural design, particularly in two story types?

Reply: R. M. Aynsley. Typical two story housing in Australia's tropics have little space penetration between floors, usually a single stair which has a small open area compared to window and door openings for horizontal flow. With the stack effect being proportional to the square root of the difference in height of inlet and outlet openings and the temperature difference between internal and external air, residential buildings only two storys in height would require significant internal to external temperature differences to develop a significant stack effect. In naturally ventilated buildings in hot climates with large wall openings and good prevailing breezes, the airflow rate through such buildings is such that with shaded walls, and suitably insulated roofs, internal dry bulb air temperatures are the same as external shade dry bulb temperatures. If high temperatures of roof sheeting is used to obtain a temperature difference the underside should be well insulated to prevent radiant heat gain to building occupants. CSIRO has done some experiments along this line. In cold climates buildings with small openings will experience significant stack effect when the interior is heated. These effects were well described by Dick (BRE) 1950.

Paper by W. J. Beranek--GENERAL RULES FOR THE DETERMINATION OF WIND ENVIRONMENT

Question: V. R. Cartelli. These studies apparently show interesting results for certain configurations of buildings. Can any correlation be derived for buildings and structures which are located in large cities with varying shielding effects? Also, do you consider our present design parameter coefficients of 0.7 windward and 0.3 leeward reasonable?

Reply: W. J. Beranek. In many cases the rules for a single building remain valid if the high single building is situated among lower buildings, see Figure 1. Between high buildings sometimes serious interaction will occur, see same figure.
Various configurations are now under investigation.
A distribution of the pressure over windward and leeward side of the building in the ratio 0.7 and 0.3 is a good start for a calculation.

Question: C. F. Shieh. The results indicate the flow patterns around obstructions strongly depend on the gust speed. This can be caused by the different approaching wind profile. Does the vertical velocity profile remain the same in all the cases investigated? Has the effect of the undisturbed wind profile been considered? Is the undisturbed wind speed uniform?

Reply: W. J. Beranek. The incoming wind is turbulent (boundary layer $\alpha = 0.28$) with appropriate horizontal and vertical turbulence intensities, the incoming wind remaining the same in all situations. Between the buildings the airstream is modified by the buildings.

Question: N. Feis. How can you state that a discomfort factor of 2, which should always be reached even with medium high buildings, is quite obvious if one looks to the maximum speed in laminar two-dimensional flow around a circular cylinder? It seems impossible to compare the simple laminar two-dimensional flow around a front-streamlined body with the very complicated three-dimensional highly turbulent flow around a bluff sharp-edged body in a boundary layer; the flow patterns are completely different. As high energy flow particles are brought down from almost the top of the building to the corner streams it seems reasonable that the maximum speed is g0-g5% of the speed in the oncoming flow at the top of the building as found by research at the BRE and our laboratory.
Reply: W. J. Beranek. Even a laminar two-dimensional flow around a circular gives a magnification 2 in respect to the mean speed. Flow around rectangular bluff bodies creates a more unfavorable situation, so a magnification of 2 can most of the time be expected at corners of rectangular buildings.

Question: N. Feis. Why do you talk about a principlely impossible superposition (= vectorial addition) of flow patterns while even joining of erosion patterns does not occur? How can you state that surrounding lower buildings have not much influence on the discomfort around a higher building while we all know the strong influence of a low building placed in front of a high building on the strength of the horseshoe vortex around the latter? From our own experience of wind velocity measurements around many holes of conglomerates of building in a broad variety of built surroundings, it is clear that one has to simulate all the low buildings and higher vegetation within a radius of at least 200 to 300 m around the high building to get reliable results.
Reply: W. J. Beranek. The discomfort maps give a good indication of the flow around a building. If in a certain situation two single buildings give the same pattern of each building alone, a very practical rule for a building engineer is achieved. It is stated in the paper, however, that in certain building configurations such a "superposition" is not possible.

Question: N. Feis. How can one rely on sand erosion or oil-streak patterns for the prediction of wind nuisance to persons or traffic? For, in the altering boundary layer of the flow around a building there is no fixed relation between the turbulent wall shear stress governing erosion phenomena, and the mean velocity nor the turbulence intensity nor the maximum gust velocity at a certain distance from the ground. Did you measure the vertical profiles of the velocity and the turbulence along the found erosion lines or boundaries and what were the results? How did you test and calibrate the mentioned optical dynamometer in different boundary layers with known velocity and turbulence profiles, included the very deformed profiles near the ground around buildings, and what are the results?
Reply: W. J. Beranek. The sand erosion patterns and streamline patterns are compared with the readings of the optical dynamometer (which is comparable with a stylized human body) and the results were very satisfactory (see also Wind Environment around Buildings by Beranek and Van Koten, Gainsville, February 1978).

Question: N. Feis. On what experience is your statement based that in Holland we have no serious complaints about the wind environment around buildings lower than 12 to 14 m? At least 10% of all the models tested in our wind tunnels are low-building configurations not higher than 10 m, for which we have to find solutions for existing wind nuisance problems. Why do you choose a magnification factor of 1.6 as a criterion for wind nuisance? From our experience we have learned that a factor 1, i.e., the same wind conditions as in the free field, is in Holland often unexceptable especially in living and recreation areas and at shopping centers.

Reply: The rules are meant for normal shelter (see Wind Environment around Buildings by Beranek and Van Koten). Large parts of Dutch towns and cities consist of blocks with a height of 4 or 5 stories so the environment around these buildings has to be regarded as the normal case. Our experience is further based on interviews carried out in areas with apartment buildings of varying heights. It is certainly possible that in certain circumstances when increased comfort is required a wind-tunnel investigation is very useful (shopping precincts on the edge of cities near open fields will often have a serious lack of shelter).

Question: N. Isyumov. I have two general questions, one on experimental technique and the other on the applicability of these recommendations to real environments. (1) The use of sand on a smooth model base plate results in different erosion rates and thus different apparent patterns of flow. That is, particles are more susceptible to motion at the "bare" spots. For example, the corner streams where sand particle motion first starts will tend to become accentuated. One variation of this technique attempted at the University of Western Ontario is to use a rough (sand roughness) base plate. Using this improvement we have obtained results within say 30% of hot-wire measurements at the hot spots at corners. Has this author attempted a similar approach? (2) Our experience is that the building height relative to the heights of the surroundings is one of the most important parameters determining actual local speed amplifications in real building environments. Would the author clarify his comments on the reduced effect of h for $\frac{h}{a} > 2.5$? Also, how do his recommendations for single buildings apply to real city environments?

Reply: W. J. Beranek. (1) We did not try out a rough surface and it seems a good suggestion to compare the results of both techniques. (2) This might depend on the ratio between the heights of the lower and higher buildings. In our case ($h_1 > 35$ m and $h_2 \leq 12.5$ m) the lower buildings had practically no influence on the discomfort pattern of the high building (see Figure 1). For $h \geq 2.5$ the first wind regime is valid, the discomfort patterns remain practically unchanged for increasing height. In city environment it should be checked if the interaction between buildings is not too important.

Paper by A. R. J. Borges, J. A. G. Saraiva--AN EROSION TECHNIQUE FOR ASSESSING GROUND LEVEL WINDS

Question: C. F. Shieh. Figure 4 shows different friction velocity ratio $U\tau/Uo$ associated with different reference velocity U_H. If the undisturbed wind profiles of the cases studied remain the same (Figure 3), the ratio $U\tau o/U_H$ can approach a constant (e.g., a logarithmic wind profile). Thus, the magnitude of $U\tau$ approaches a constant as one can calculate from the data given in Figure 4. It is reasonable that the magnitude of $U\tau$ is independent of the $U\tau o$ magnitude?

Reply: A. R. J. Borges, J. A. G. Saraiva. We think of this particular result as an approximation.

Question: R. E. Britter. How were the velocity measurements made in the highly turbulent flow between the buildings? Are the turbulence measurements also available?

Reply: A. R. J. Borges, J. A. G. Saraiva. We used a small Pitot-static tube. The only purpose of these measurements were to give a rough indication of the velocity profile in the gap between the building models. We made no turbulence measurements.

Question: R. E. Britter. Both papers discuss the sand erosion technique; Borges and Saraiva imply mean parameters from the technique while Beranek implies peak parameters. Is there a conflict here?

Reply: A. R. J. Borges, J. A. G. Saraiva. No. While Dr. Beranek's main interest seems to be the development of design criteria, I am interested in understanding the fundamentals underlying the erosion technique.

Question: R. E. Britter. The calibration technique used has a small ratio of fluctuating to mean wall stress compared with that expected in the experiment. Does this difference influence the applicability of the Sand Erosion Technique?
Reply: A. R. J. Borges, J. A. G. Saraiva. Both the calibration and the experiment with building models seem to give consistent results. However, as pointed out in the paper, further study will be needed to fully understand the mechanism of erosion.

Question: R. J. Kind. The Froude number (F) vs. Reynolds number (R) relationship shown in Figure 1 is substantially different from that obtained by BAGNOLD (1) and others for a two-dimensional flow with a well-developed logarithmic velocity profile region above the sand surface. This indicates that the F vs. R relationship is not unique except for equilibrium shear flows and that caution should therefore be exercised in assigning numerical values, as in Figure 5, on the basis of sand erosion tests. Reference 1--Bagnold, R. A., "Wind Blown Sand and Desert Dunes," Methuen, London, 1941 (book).
Reply: A. R. J. Borges, J. A. G. Saraiva. As far as I can remember Bagnold's results relate to flow over thick layers of sand while present results relate to a monolayer of sand over a solid surface of given roughness. Nevertheless, the values of F I arrived at are of the same order of magnitude as those in Bagnold's work, i.e., 10^{-2}.

Question: C. F. Shieh. Figure 4 shows different friction velocity associated with different reference velocity U_H. Does this cause the wind profile different from the one shown in Figure 3?
Reply. A. R. J. Borges, J. A. G. Saraiva. No.

Paper by S. Murakami, K. Uehara, K. Deguchi--WIND EFFECTS ON PEDESTRIANS: NEW CRITERIA BASED ON OUTDOOR OBSERVATION OF OVER 2000 PERSONS

Question: R. L. Wardlaw. One of the more demanding criterions that the pedestrian is exposed to is the abrupt changes in wind speed encountered when passing building corners or exiting from doors. Therefore, it is of interest to see the observations made for such circumstances. The observations appear to have been made for crosswinds. Walking downwind or changing direction can also be difficult. Were any observations made to determine if walking downwind was a less severe limitation than walking crosswind?
Reply: S. Murakami, K. Uehara, K. Deguchi. To observe the effects of the abrupt changes of wind speed such as at a building corner, the experiments are conducted under the condition of non-uniform flow with fences. Observations are made under the wind conditions of crosswind, follow-wind and against-wind. Walking downwind (follow-wind) is less severe than walking against-wind or crosswind. The examples of the results of observations are shown in Figures 2.4, 2.5 and 2.7.

Question: J. C. R. Hunt. My congratulations to the authors on some fine experiments. (1) Am I correct in thinking that the turbulence was always present in conjunction with a non-uniform mean flow? If so, it is difficult to distinguish between these two effects. (2) The authors final conclusion is that criteria should be based on a 3-second average gust. In practice, this is similar to criteria of Hunt, Mumford and Poulton and Gardemer using a mean velocity and a turbulence intensity, but there is a difference in principle because we found that people were affected by quite low turbulence levels which would not be represented by a 3-second gust. (3) Your pictures of people in

Tokyo indicates large turbulence and shear. Do they have handrails or ropes to help pedestrians as they do in Chicago?

Reply: S. Murakami, K. Uehara, K. Deguchi. (1) In the experiments of Part I, which are the walking tests in the large wind tunnel, mean flow and turbulence are not distinguished. The effects of turbulence are investigated in Part II and Part III which are the outdoor observations. In the experiments of non-uniform flow with fences in Part I, the aim is to observe the effects of abrupt changes of wind velocity by space, not by time. (2) Of course, people are affected by quite low turbulence wind. But, the degree of effect is much lower, compared to that by gusty wind. Therefore, in the case of a steady wind tunnel flow, the mean value of wind speed should be regarded to correspond to the maximum value of a gusty wind. (3) On windy days, watchers appear and set the ropes to guide the pedestrians, at the precinct of some high-rise buildings in Tokyo.

Paper by I. P. Castro--THE RELAXATION AND STEADINESS OF WAKES BEHIND OBSTACLES IN BOUNDARY LAYERS

Question: S. P. S. Arya. Referring to Figures 3 and 8, it needs to be pointed out that the magnitude of the normalized perturbation at a given position (X/h) in the wake and, hence, the slope (m) of the lines in Figure 3 are more strongly dependent on the hill shape or the extent of the cavity region than they are on the parameter K. Our measurements (Arya and Shipman, 1979) in the wakes of a steep ridge of slope 2:1 and a 'gentle' hill of slopes $\leq$ 0.1 show m to be 6 times as large in the former even though K and other boundary layer parameters were kept the same in the two cases. Interestingly, for both the hill shapes, normalized perturbations in Reynolds stress and variances in the wake are found to decay inversely proportional to X/h.

Reply: I. P. Castro. It is, certainly, true that the decay of the wake perturbations is strongly influenced by body geometry. I think that this confirms my belief that the separated shear layer distortion and the recirculating region dynamics are critical in determining the wake decay. Body geometry changes lead to particularly significant changes in these areas of the flow, although I do not think that changes in upstream conditions are without significance. We need much more information about the 'near' wake dynamics: They provide the initial condition for the far wake decay region, whilst the upstream flow provides an outer boundary condition.

Paper by Y. Ogawa, P. G. Diosey, S. Oikawa--THE EFFECTS OF THE CHARACTERISTICS OF THE ONCOMING WIND ON THE FLOW BEHIND A TWO-DIMENSIONAL MODEL BUILDING

Question: J. C. R. Hunt. I am doubtful about the author's conclusion that stable stratification always reduced the length Lc of the separation bubble. There are two effects of stable stratification; it reduces the turbulence (which can increase Lc as has been noted by other experiments in the laboratory and field) and by inducing a leewave with wavelength less than Lc, it can reduce Lc. I believe that Ogawa's observation that Lc is reduced by stable stratification is explained by the low turbulence level. Had the upwind turbulence been greater, the damping effect of stratification may have increased Lc. Does the author have any evidence to support this conjecture?

Reply: Y. Ogawa, P. G. Diosey, S. Oikawa. I agree with Dr. Hunt that there are two effects caused by stable stratification: one effect is to reduce the oncoming flow turbulence (or increase return) which leads to large Lc, the other effect is to inhibit the change of flow caused by the fence leading to smaller Lc. I believe that in our smooth surface case the latter effect dominates thus causing Lc to decrease. We did not try a rough surface case with stratification change in this study, but we do plan such a study.

Paper by W. H. Snyder, R. E. Britter, J. C. R. Hunt--A FLUID MODELING STUDY OF THE FLOW STRUCTURE AND PLUME IMPINGEMENT ON A THREE-DIMENSIONAL HILL IN STABLY STRATIFIED FLOW

<u>Question</u>: R. M. Aynsley. In the top view of plume impingement on polynomial hill (Figure 9) there is no indication of the side boundaries of the towing tank. What was the ratio of the hill diameter to the tank width? Was there any possibility of flow pattern distortions due to tank blockage?

<u>Reply</u>: W. H. Snyder, R. E. Britter, J. C. R. Hunt. The ratio of hill diameter to tank width was approximately one-sixth. Our belief is that flow pattern distortions due to tank blockage were negligible.

<u>Question</u>: R. L. Petersen. (1) What effect will changing aspect ratio have upon the results presented? (2) What effect upon the results would occur if a turbulent boundary layer were simulated? (3) How will results be incorporated into EPA modeling guidelines?

<u>Reply</u>: W. H. Snyder, R. E. Britter, J. C. R. Hunt. (1) Under very strong stratification, we believe the aspect ratio h/L will have negligible effects; the fundamental question is whether or not a fluid particle approaching the hill has sufficient kinetic energy to overcome the potential energy required to lift it to the top of the hill and time is totally independent of the shape of the hill. (2) Under strong stable stratification, surface turbulence will be confined to very thin layers (n 10 m) and possibly totally absent. Since the plume is elevated, we believe simulation of the turbulent surface layer is unimportant. Under moderate stratification $10 \gtrsim U/Nh \gtrsim 1$, we speculate that the mean flow pattern over the hill (but not in the wake) is independent of the upwind turbulent boundary layer structure, just as the change of wind speed in neutral flow over hills appears to be rather insensitive to upwind shear, provided U is chosen in the scaling as U (z = L). (3) The data are being used to develop mathematical models for treating flow and diffusion processes in complex terrain (Hunt, Puttock and Snyder, Atmos. Env., 1979).

<u>Question</u>: K. M. Kothari. Up to what downstream distance the effect of the hill was observed on the concentration field? Did you always observe the lower concentration with hill present?

<u>Reply</u>: W. H. Snyder, R. E. Britter, J. C. R. Hunt. Whereas we did not make any measurements farther than one hill diameter downstream, it was quite evident that if plume impingement occurred, hill effects were felt for many diameters downstream. Maximum hill surface concentrations were observed to range from 50% to 120% of what would have been observed at the centerline of the plume in the absence of the hill, which is, of course, very much larger than would have been observed on the ground surface in the absence of the hill.

SESSION IV

Wind Loading - Mean External and Internal and Fluctuating Internal

CHAIRMAN

W. H. Melbourne

Professor, Monash University
Clayton, Australia

CO-CHAIRMAN

R. D. Marshall

Leader, Structural Engineering Group
National Bureau of Standards
Washington, D.C., USA

RAPPORTEUR

B. M. Leadon

Professor, University of Florida
Gainesville, Florida, USA

RECENT DEVELOPMENTS AND TRENDS: WIND LOADING - MEAN EXTERNAL AND INTERNAL AND FLUCTUATING INTERNAL

W. H. MELBOURNE

Monash University, Australia

The state of the art in wind engineering is now such that we do not admit to the existence of only mean loads; even when using quasi-steady approaches to design. Whilst several papers in this session do address themselves to mean pressures and loads, they are all concerned with real dynamic loading phenomena, and as such the subject matter in this session blurs with that of the next session on unsteady wind loads.

The papers in the session do not have a single theme, but can be identified under several headings -

(i) Full scale measurements on buildings.
(ii) Pressure distributions on rectangular bodies.
(iii) Loading on frames and shielding effects.
(iv) Internal pressures in buildings.
(v) Interaction effects of neighbouring tall buildings.

Since the meeting on "Full Scale Measurements of Wind Effects on Tall Buildings and other Structures" at the University of Western Ontario in 1974, considerable progress has been made with a number of full scale measurement programmes, along with comparisons with model measurements. Whilst the enormous problems associated with obtaining useable full scale data are still only appreciated by a few, it can be fairly be said that evidence being assembled is beginning to provide the validation of model measurements that is so urgently needed. In respect of measurements on buildings, the work of Dalgliesh and his colleagues at the N.R.C. in Canada on the Commerce Court Tower, and the work of Eaton and Mayne at B.R.S. in England on the Aylesbury houses, stand out at two ends of the size spectrum. The N.R.C. group have brought their work through a series of publications in recent years to a paper in this session on the critical evaluation of model/full scale comparison. In particular, this paper focuses on the complex problem of peak pressures and descriptions of same, which I venture to suggest is a subject that will receive considerable attention at this meeting. Whilst much of the full scale data from Aylesbury study has not been published, there have been extensive comparative model studies on the experimental building configuration carried out in boundary layer wind tunnels in Australia, Canada and England, which have been published in internal publications and which hopefully will receive some discussion at this meeting. There will be other papers at this meeting highlighting full scale measurements on other structures;

these, plus papers presented at the recent International Conference on Environmental Forces on Engineering Structures at Imperial College, London, and those to be presented at the meeting of the International Symposium on Practical Experiences with Flow Induced Vibrations at Karlsruhe in September, will provide a quite extensive coverage of current full scale measurements in the field of wind engineering in 1979.

Papers to be presented by Kim & Mehta and Kavanagh in this session describe two semi-full scale projects aimed at measuring total wind loads on panels and wind loads which finally get transferred to the structure of a light steel framed building. Kim and Mehta are presenting some data already collected on peak total loads on the flat roof of their facility. Few model studies to date have tackled this difficult measuring problem and it is very encouraging to see this type of work being undertaken in the even more difficult full scale arena.

The intensification of model pressure measurement studies on a variety of shapes will be evident throughout this meeting. Apart from the volume of work undertaken in the past four years, in which low-rise building studies have figured prominently, there is some controversy at a fundamental level concerning the effects of turbulence on surface pressures, the measurement and interpretation of peak pressure measurements and the integration of peak pressures over finite surface areas. In this session one paper by Akins, Peterka and Cermak discusses the problem of integrating pressures to arrive at cladding and support structure design loads, and some comparison with wind loading codes is also made. Unfortunately, the method used only permitted the integration of mean pressures and not instantaneous peak pressures, hence not permitting integrated peak loads on a finite area to be compared with the design loads estimated using quasi-steady assumptions and a peak design wind speed with integrated mean pressure coefficients. Hopefully, this work might be extended to make this comparison. Laneville and Williams are presenting a paper which discusses another of the fundamental problems, namely the effect of intensity and large scale turbulence on mean pressure and drag on two-dimensional rectangular cylinders. This work has confirmed much of the earlier indications about the effect of turbulence intensity and in addition has shown that the effect of changes in the large scale turbulence is not significant. The writer is presenting a paper in the next session which looks at the effect of the fine scale turbulence on surface pressures and which concludes that surface pressures are significantly dependent on the amount of fine scale turbulence in the incident flow and that effect of the large scales, other than the generation of the smaller scales, is not significant.

In any review of work on the wind loading of lattice frames and high porosity fabricated structures, including the effect of member shielding, it would have to be noted that until recently very little work has been done over the past two decades, and virtually no data is available with respect to drag measurements in turbulent flows. Hence, it is most significant that in the last two years we have seen renewed interest in the problem, resulting in four papers at the recent Colloquium on Industrial Aerodynamics at Aachen in 1978 and not two papers to be presented in this session. The paper by Georgiou and Vickery presents the first part of a much larger programme and is addressed mainly to frame shielding effects in low turbulence flow. The effects of turbulence on drag of upstream frames and subsequent shielding effects, and the effects of longitudinal edge beams from part of the larger programme. Despite some early indications that turbulence effects are small, it will be surprising if in the long run they turn out to be insignificant in the knowledge of the effects of turbulence on the drag of isolated bluff bodies. It can be expected that the effects of turbulence will be complicated in that upstream members with significant streamwise depth will have reduced drag with increased turbulence, and vice versa for flat members with shallow streamwise depth, in both cases increased turbulence would be expected to reduce wake width and hence the shielding effect may be reduced, at least on the adjacent downstream frame. The paper by Whitbread reports on an extensive wind tunnel testing programme aimed at providing data for future codification in the U.K. Both papers make extensive comparisons with existing design methods and code recommendations, in particular, highlighting a number of deficiencies in these.

A by-product of the recent intensive activity in measuring wind loads on low rise structures are several studies on internal pressures in buildings due to various openings and degree of wall porosity. Holmes has presented some experimental data supported by a theoretical analysis which provides some understanding of why inertia and resonance effects are usually small. The paper by Stathopoulos, Surry and Davenport provides model measured data for internal pressures for a variety of building and porosity configurations; in particular, the high correlation found between external and internal pressures further highlights the need to account for internal pressures at the design stage and, or, taking steps to provide adequate venting or orientation of dominant openings to reduce the internal pressure.

The remaining paper by Blessman and Riera studies the effects of interference interaction of tall buildings on each other. Whilst the model data presented confirms the possible magnitude of these effects on along-wind and cross-wind response, the paper highlights the possibility of even more significant amplifications of torsional moments and consequent response.

Possibly because of personal bias, I anticipate the trend of future research both in full scale and model scale studies to be even more involved with the fundamental aspects of turbulence and its effect on the loading of structures and the estimation of peak (design) loads. In the latter area, I think the peak response, or equivalent loads, resulting from essentially resonant processes, are fairly well understood and documented, but it is in the area of wind loading on cladding elements and the loads transferred to the structure of small buildings where the knowledge of peak loads for design purposes is least understood.

A FIELD MEASUREMENT FACILITY BASED UPON A WIND-SENSITIVE LOW-RISE STRUCTURE

K. T. KAVANAGH
University of Western Australia, Nedlands, Western Australia

SYNOPSIS

A field measurement system is described in which a flexible steel structure is employed to monitor wind forces. The system is transportable, so that a comparison of measurements with building code requirements can be obtained for a variety of urban and sub-urban environments. Design forces, including leeward and windward wall and roof forces, net building uplift, and net building base shear, are derived from dynamic building displacements using a calibrated structural model. Laboratory calibration is achieved through a least-squares parameter estimation process.

Initial tests on the prototype facility in an open terrain are described. Measurements indicate that peak forces occur over very short time intervals, and that individual panel forces are not in phase at the individual peak values. Results of the initial prototype tests are used to examine further modifications to the measurement system, and to describe a test programme relevant to the Australian wind loading code, AS1170/2.

INTRODUCTION

Criticism of the wind force provisions of the current Australian Standard, AS1170, has arisen in Western Australia with respect to the design of low rise structures in semi-urban environments. The design of light steel structures, a prevalent building form in Australian light industry, is particularly sensitive to wind forces in terms of bracing and foundation cost. Designers are currently reluctant to adopt the higher wind forces in AS1170 (Category 2) when less conservative building designs, based on earlier wind force provisions, have behaved satisfactorily.

In March, 1978, the University of Western Australia proposed a research programme to obtain field measurements of wind forces applicable to both cyclonic and non-cyclonic areas. The basis of the proposal was contained in the design of a transportable steel building, instrumented for deflection, and connected to an Hewlett Packard 21MX computer via an FM tape recorder and an A/D converter. Design of the complete system utilised a maximum of existing data acquisition equipment, with a minimum commitment to the purchase of commercially produced pressure transducers.

The following description of the experimental facility includes the design basis for structure, the selection of measurement points, and the problems in data reduction. Limited field experiments have been conducted which indicate a highly successful behaviour of the system in light to moderate winds.

CONSIDERATIONS IN DESIGN

The design of a wind sensitive building is, in its simplest terms, the design of a mechanical transducer. The complexity in the design lies in the multiple degree of freedom behaviour of the transducer. The design must be constrained to meet three conflicting service requirements: 1) It must have high strength to prevent local or global failure 2) It must be flexible (or responsive) in terms of the loads and the accuracy of the monitoring devices, and 3) It must be of relatively high frequency, to minimise the occurrence of resonances. Criteria (1) and (3) are natural consequences of one another, but criterion (2) is in direct opposition. The design, therefore, is necessarily a compromise, and in this instance, is a compromise which leads to a sophisticated data reduction exercise.

Utilisation of existing measurement devices (LVDT'S) with a ± 20 mm displacement range set the initial criterion for full scale deflection under a 30 knot wind. The 20 mm deflection led to an initial design based on a portal frame structure, which had an undesirably low frequency. The initial portal frame concept was subsequently altered to produce a building with a minimum of mass and a resultant higher frequency. Reduction of the building mass was accomplished by dividing the structure into two independent sub-systems. (See Fig. 1). The two sub-systems isolated vertical and horizontal motions, which had the desirable side effect of simplifying the displacement measurements.

In Fig. 1, the horizontal sub-system is shown to be comprised of two welded portal frames, with individual wall panels mounted on the front and rear. The portal frames are pin-based, with a roller bearing to provide the pin support. The vertical sub-system is comprised of two trusses supporting individual roof panels on a 20° pitch. The trusses are mounted on a pair of simply supported beams with cantilevered ends. Roller bearings are employed in the second sub-system to simulate the simple beam supports.

Measurements of deflection for each of the vertical and horizontal sub-systems are taken at the centre of the building. The use of a centre deflection on a symmetric structure automatically isolates the symmetric component of the wind forces, and automatically eliminates any antisymmetric influences. The symmetry argument is used in the subsequent data reduction to define an average uniform panel force, similar to the equivalent static load defined in the loading code.

THE MATHEMATICAL MODEL

The most fundamental consideration in the establishment of a transducer model is concerned with the choice of an appropriate cut-off frequency. For structural design, it was decided to establish an upper cut-off value between ten (10) and fifteen (15) Herz, since frequencies above this magnitude are not significant in their contributions to global structural response. In this sense, the measured results pertain to 'average' pressure values, integrated over significant areas of the structure, and not to local effects, such as connections. The cut-off value is explicitly contained in the data reduction through a low pass digital filter. The transducer, therefore, is a low frequency transducer, rather than a high frequency device, which is designed to respond dynamically to wind gusts.

In the first prototype, the wall and roof panels of the building were constructed of identical dimensions and materials: 1) a pair of 75 x 32 x 1.2 mm cold formed steel channels, and 2) a 'V-crimp'* steel cladding, a proprietary sheeting of Osborne Metal Industries, Osborne Park, W.A. As shown in Fig. 2, the cladding and support channels act in separate directions, creating a simple two-way load carrying system. The fundamental frequency of the panel system was recorded experimentally, and was found to be slightly in excess of five Hz. The second fundamental frequency was in excess of 12 Hz.

The support frames are different for the vertical and horizontal sub-systems, and consequently, produce different fundamental mode frequencies. The wall panels and portal frames exhibit a fundamental mode frequency of 3.8 cycles per second, while the roof panels and support beams are slightly lower, at 3.4 Hz. The response times of the transducer, therefore, are not low enough to be treated independently of the dynamic component of deflection. This dynamic component, however, can be accurately modelled, and the errors for meaningful wind gusts are acceptably small.

The wide spread between the first and second mode frequencies of the system is important, since a simple model can be adopted for the structural dynamics. The 'equivalent' spring mass model for each sub-system is shown in Fig. 3. In these models, only one frequency is included for each panel, and one frequency for each frame. The resulting combination of two panels and support frame is a three degree of freedom system with three vibration frequencies below the ten Hz cut-off value. The mass matrix for the individual panels is a 'consistent' mass matrix, calculated from measurements of the panel mode shape. The stiffness and mass matrices for the sub-systems are given below:

Horizontal Frame

$$[M] = \begin{bmatrix} 13.5 & 3.34 & 0 \\ 3.34 & 41.1 & 3.34 \\ 0 & 3.34 & 13.5 \end{bmatrix} \text{(kg)} \qquad [K] = 10^4 \begin{bmatrix} 2.40 & -1.20 & 0 \\ -1.20 & 4.07 & -1.70 \\ 0 & -1.20 & 2.40 \end{bmatrix} \text{(N/M)}$$

Vertical Frame

$$[M] = \begin{bmatrix} 13.5 & 3.8 & 0 \\ 3.8 & 35.6 & 3.8 \\ 0 & 3.8 & 13.5 \end{bmatrix} \text{(kg)} \qquad [K] = 10^4 \begin{bmatrix} 2.4 & -2.4 & 0 \\ -2.4 & 6.1 & -2.4 \\ 0 & -2.4 & 2.4 \end{bmatrix} \text{(N/M)}$$

The system mass and stiffness matrices (above) represent initial estimates of the system parameters, based upon the computed section properties. These initial estimates can be refined by a comparison of the computed values with static and dynamic laboratory tests. Tests were carried out on the individual panels, as well as on the assembled structure. Variations in the panel stiffnesses were found to be within ± 3% of the average value, and the assembled building exhibited a fundamental mode frequence 2.5% higher than the predicted by the initial model.

To fit the properties of the structure to the existing data, the total mass of the system, and the mass of each individual panel were assumed error free. The deviations from computed values, therefore, could be attributed to two parameters: a) the moment of inertia of the V-crimp sheet, (I_1) and b) the effective moment of inertia of the channels and attached sheet (I_2). The theoretical frame stiffnesses were compared to experiment and found to be accurate to within 1% of the experimental values. The portal frame properties were assumed to be exact.

The two parameters, I_1 and I_2, affect the mass and stiffness matrices by altering both the fundamental frequency and the mode shape of the individual panels. Since no shake table or servo motor was available to test the system, the properties were 'fit' to data obtained from free vibration tests. The free vibration tests used markedly different initial conditions in order to excite the maximum number of modes. Damping values were required, and these were estimated from a first mode logarithmic decay. The damping was small, 0.75%, and was assumed constant for each mode.

A special parameter estimation programme was written based upon 50 data points from each of the three tests. I_1 and I_2 were assumed to vary within ± 5% so that nine trial points could give a quadratic error function approximation. The trial space is shown in Fig. 4. For each point, the mode shape, mass and stiffness matrices were computed, the mode shape calculation requiring an independent panel analysis. Successive refinements of the trial space were used, until the minimum least squares error converged to within 0.5%.

Using the results of the fitting programme, a second Kalman Filter programme was written, based upon a sequential least squares algorithm.(1) For the Kalman filter method, the mode shape was fixed to simplify the algebra. Results of the Kalman Filter were not significantly different from the results of the specialised least squares procedure described above.

Additional trials were used to refine the estimated damping properties. An attempt to fit both mass and stiffness proportional damping:

$$[C] = \alpha[M] + B[K] \qquad \alpha,\ B = \text{Damping Parameters}$$

proved futile. The difficulty encountered in refining the damping indicates an overall model error of about 1% in the building model. Any error in the damping term is insignificant for long-term forced response, but this error can be significant for excitations near resonance.

The final mass and stiffness matrices are shown below:

Horizontal Frame

$$[M] = \begin{bmatrix} 13.5 & 3.34 & 0 \\ 3.34 & 41.1 & 3.34 \\ 0 & 3.34 & 13.5 \end{bmatrix} \text{(kg)} \qquad [K] = 10^4 \times \begin{bmatrix} 2.52 & -1.26 & 0 \\ -1.26 & 4.03 & -1.13 \\ 0 & -1.13 & 2.26 \end{bmatrix} \text{(N/M)}$$

Vertical Frame

$$[M] = \begin{bmatrix} 13.5 & 3.8 & 0 \\ 3.8 & 35.6 & 3.8 \\ 0 & 3.8 & 13.5 \end{bmatrix} \text{(kg)} \qquad [K] = 10^4 \times \begin{bmatrix} 2.2 & -2.1 & 0 \\ -2.1 & 7.3 & -2.2 \\ 0 & -2.2 & 2.25 \end{bmatrix} \text{(N/M)}$$

DATA REDUCTION

Using the accurate model of the dynamic system, it is a relatively straight forward procedure to derive wind forces from the dynamic motion. The equations of motion:

$$\begin{Bmatrix} P_1 \\ P_2 \\ P_3 \end{Bmatrix} = \begin{bmatrix} M_{11} & M_{12} & M_{13} \\ M_{21} & M_{22} & M_{23} \\ M_{31} & M_{32} & M_{33} \end{bmatrix} \begin{Bmatrix} x_1 \\ x_2 \\ x_3 \end{Bmatrix} + \begin{bmatrix} C_{11} & C_{12} & C_{13} \\ C_{21} & C_{22} & C_{23} \\ C_{31} & C_{32} & C_{33} \end{bmatrix} \begin{Bmatrix} x_1 \\ x_2 \\ x_3 \end{Bmatrix} + \begin{bmatrix} K_{11} & K_{12} & K_{13} \\ K_{21} & K_{22} & K_{23} \\ K_{31} & K_{32} & K_{33} \end{bmatrix} \begin{Bmatrix} x_1 \\ x_2 \\ x_3 \end{Bmatrix}$$

or:

$$P = M\ x + C\ x + K\ x\ .$$

can be converted to a discrete time increment form by the use of a finite difference (or equivalent) operator,

$$\begin{Bmatrix} x_1 \\ x_2 \\ x_3 \end{Bmatrix}^{(i)} = (1/\Delta t)^2 \left(\begin{Bmatrix} x_1 \\ x_2 \\ x_3 \end{Bmatrix}^{(i+1)} - 2 \begin{Bmatrix} x_1 \\ x_2 \\ x_3 \end{Bmatrix}^{(i)} + \begin{Bmatrix} x_1 \\ x_2 \\ x_3 \end{Bmatrix}^{(i-1)} \right)$$

$$\begin{Bmatrix} x_1 \\ x_2 \\ x_3 \end{Bmatrix}^{(i)} = (1/\Delta t) \left(\begin{Bmatrix} x_1 \\ x_2 \\ x_3 \end{Bmatrix}^{(i)} - \begin{Bmatrix} x_1 \\ x_2 \\ x_3 \end{Bmatrix}^{(i-1)} \right)$$

to produce a discrete force time relationship:

$$p^{(i)} = [(1/\Delta t)^2 \ M]x^{(i+1)} + [(1/\Delta t) \ C - (2/\Delta t)^2 \ M + K] \ x^{(i)} +$$

$$[(1/\Delta t)^2 \ M - (1/\Delta t) \ C] \ x^{(i-1)}$$

The finite difference procedure leads to the computation of wind forces directly, except for the presence of experimental noise. The finite difference expressions amplify the high frequency components of the noise, so that a displacement record cannot accurately define accelerations. The displacement, velocity, and acceleration records in Fig. 5 illustrate this artificial noise generation.

Elimination of the high frequency noise is accomplished by the use of a high frequency filter. The filter can be either analogue or digital, dependent upon the nature of the instrumentation. In this prototype, the signal was filtered digitally, after passing through an analogue-to-digital converter. The filter was an optimising filter based upon the use of a 16th order Chebyshev polynomial[2]. Characteristics of the filter are shown in Fig. 6, and the effect of the filter on the calculation of force is shown in Fig. 7.

The calculation of time-dependent generalised forces (P_1, P_2, and P_3) must be converted mathematically to average wind pressures on the building faces. The conversion is obtained from the expression for the work done by an equivalent uniform load moving through the displacement pattern shown in Fig. 8. The displacement pattern is a function of the panel mode shape and of the rigid body motion of the panel on the building frame.

FIELD MEASUREMENT

The first prototype structure was placed in the field in September of 1978. The facility is shown in the photographs of Fig. 9. The terrain was selected to provide for a stretch of unobstructed land between the building and the seashore. The position was ideal for monitoring the prevailing (westerly) sea breezes.

Initial tests of the structure were conducted on single sub-systems, so that a single four channel tape recorder could be used to record the analogue displacement signals. The wind speed could not be directly correlated to the time-dependent force calculations, since a separate chart recorder was used to record wind speed. The anemometer was a simple cup-type anemometer, which provided measurements of mean-hourly wind speeds. Additional wind data was obtained from a separate station several miles to the south, but the lack of a common time frame made this information of little value.

Wind measurements were taken for several days in which the mean hourly wind speeds were less than 10 knots. For these quiet days, the wind force calculations correlated reasonably well with the Australian design code values. The distribution of windward and leeward panel forces were roughly in the ratio of 1.5 to 1 at the point of peak response.

Measurements taken for a series of seasonably windy days (30 knots) produced the records shown in Fig. 10. Although gusting conditions prevailed for several hours of recording, peak building responses were observed over a very small fraction of that time (1.5%). At maximum response, the individual panels appeared to respond at a higher frequency than the frame. Peak panel forces and peak building forces did not occur simultaneously, with the leeward panels showing the greatest discrepancies. Overall building forces were less than predicted by the Australian code, which is attributed to generally lower leeward values and to a lack of consistent phase.

Expansion of the facility is currently being tested for additional input channels. The expanded facility will provide for a common time frame between the two subsystems, as well as for a higher frequency wind measurement.

IMPROVED PROTOTYPE

A second prototype structure has been developed since the placement of the first prototype in the field. To be fully transportable, the new structure is designed in 'splits', i.e. each half of the building is a self-contained unit, with individual trailer mounts and supports. The building faces are welded steel frames with cold-formed steel sheet inserts to act as a skin without contributing to the overall stiffness. The panels are designed to provide a sensitivity to edge loads, a noticeable deficiency in the first prototype. The modified system model is shown in Fig. 11.

Instrumentation for the second prototype is built around a pair of 30 channel multiplexers, providing analogue input to the four channel tape recorder. Decoding of the multiplexed signals is achieved by a dummy 'timing' channel, which records a high voltage spike at the end of each multiplexer sweep. Automatic (remote) off/on switching is being provided to eliminate the recording of low intensity data.

To measure forces on 'portions' of the structure, several low-frequency strip transducers are being constructed. The strip transducers operate on a similar principal to the prototype structure, translating displacement and tilt into force and center of pressure. The response time of the transducer is approximately 1/5 of the response of the structure. (See Fig. 12).

WIND CODE STUDIES

The two prototypes, the first now in service and the second under construction, are to be used to monitor building forces in the Perth metropolitan area. The first prototype will remain at the open seafront to experiment with artificial obstructions, while the second will be moved to a variety of residential sites. Sites have been chosen to monitor wind forces in a suburban housing tract adjacent to open terrain. An attempt will be made to correlate wind forces with the position in the tract, as a function of distance from the open terrain.

Both prototypes will be studies to determine the most significant gust duration. Initial measurements have indicated that peak responses occur for wind gusts of 0.75 to 1.5 seconds duration. The building responds dynamically to the approach of the gust, but the magnitude of the dynamic is far less than would be expected if the gust were assumed to act instantaneously. Thus, for short duration gusts, the structure must filter out the very sharp pressure fluctuations due to its finite dimension in the wind stream.

The monitoring of structural reactions, notably base shear and overturning, will be used to recommend possible reductions in the current static force requirements. The current proposal is to examine the effect of four parameters: 1) effective terrain category, 2) gust duration, 3) building frequency, and 4) skin frequency.

CONCLUSIONS

The construction of a wind-sensitive structure has demonstrated the feasibility of obtaining useful wind force data from such a facility in the field. Data reduction requires a recognition of the relatively low frequency response of the system, and requires accurate mathematical modelling in order to calculate the magnitude of time-varying forces. The advantage of the structural facility is its ability to produce global forces directly, without requiring averaging procedures for local point pressures or point forces.

Data reduction procedures for the field measurements must include a high frequency filter, to account for artificial generation of 'noise' in the analytical procedure. Average wind pressures can be deduced from the analytical model by standard matrix energy techniques. The field station, therefore, produces forces directly, and wind pressures indirectly. The approach stands in contrast to standard field monitoring stations.

Procedures utilised in the prototype experiment are general enough to imply that wind forces can be obtained from any instrumented structure. Tests on the prototype, however, indicate that there is considerable difficulty in calibrating the mathematical model, and that sophisticated 'parameter estimation' techniques would have to be developed. For simple structures, as exemplified by the prototype, number of identifiable parameters is small, and the mathematical model is manageable.

REFERENCES:

(1) Jazwinski, A.H., Stochastic Processes and Filtering Theory, Academic Press, Newport, 1970

(2) Lynn, P.A., An Introduction to the Analysis and Processing of Signals, MacMillan, London, 1973

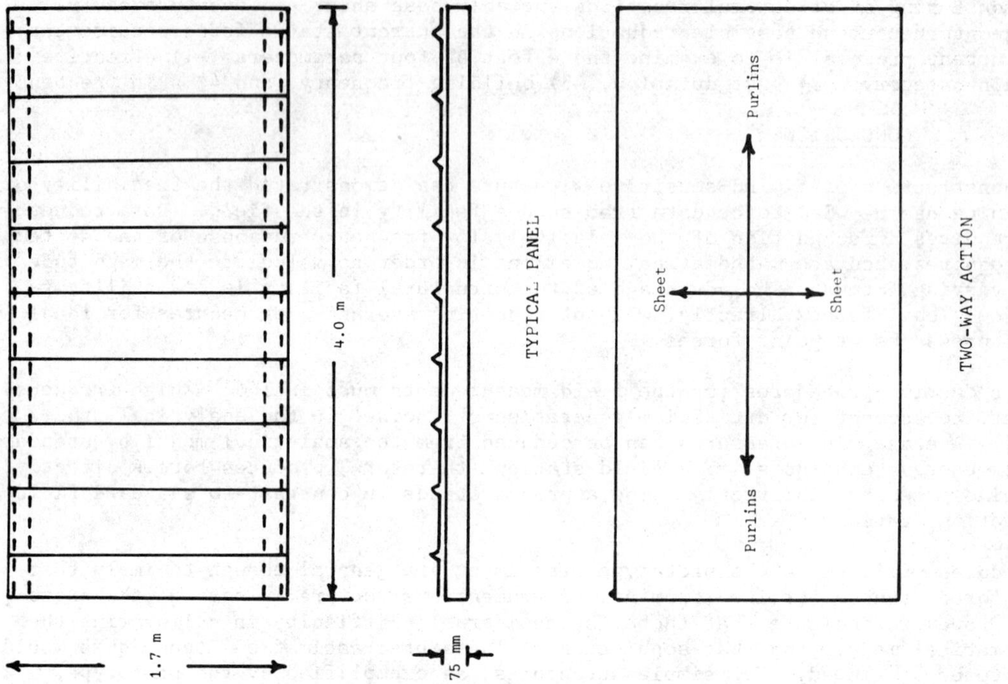
1.7 m
4.0 m
75 mm
TYPICAL PANEL
Purlins
Purlins
Sheet
Sheet
TWO-WAY ACTION

FIG. 2

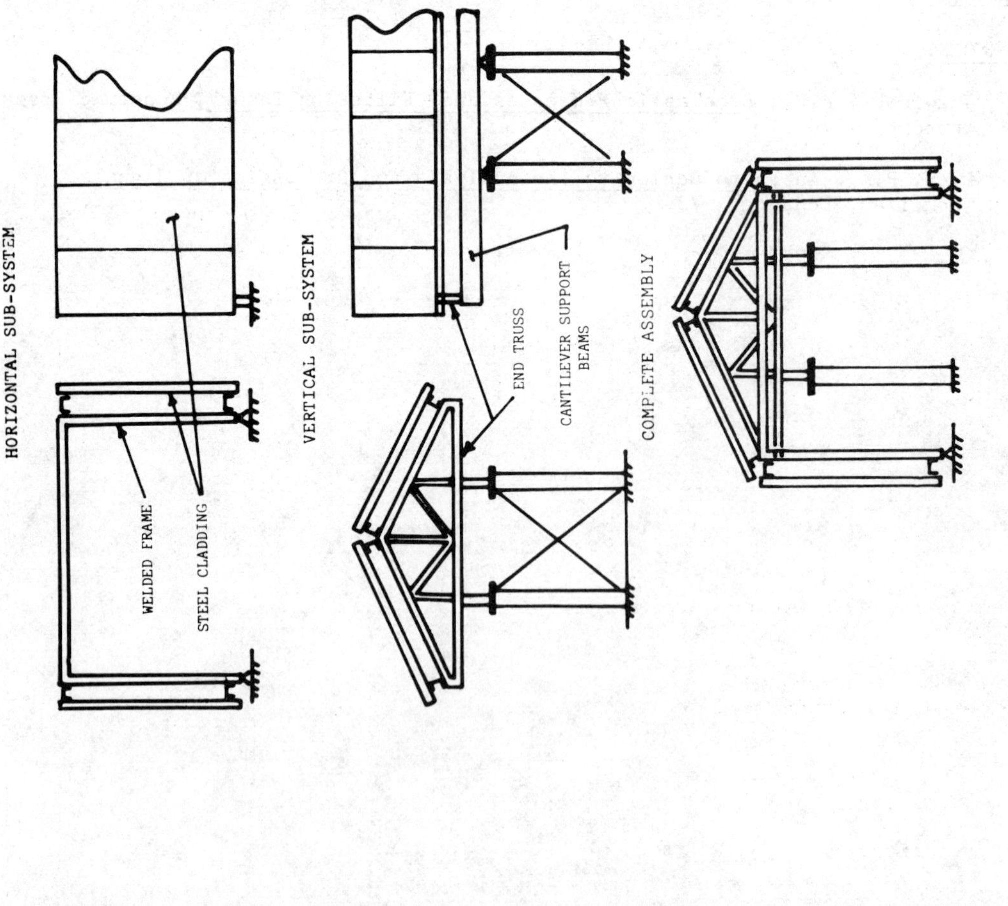
HORIZONTAL SUB-SYSTEM
WELDED FRAME
STEEL CLADDING
VERTICAL SUB-SYSTEM
END TRUSS
CANTILEVER SUPPORT BEAMS
COMPLETE ASSEMBLY

FIG. 1

ERROR FUNCTION MINIMIZATION

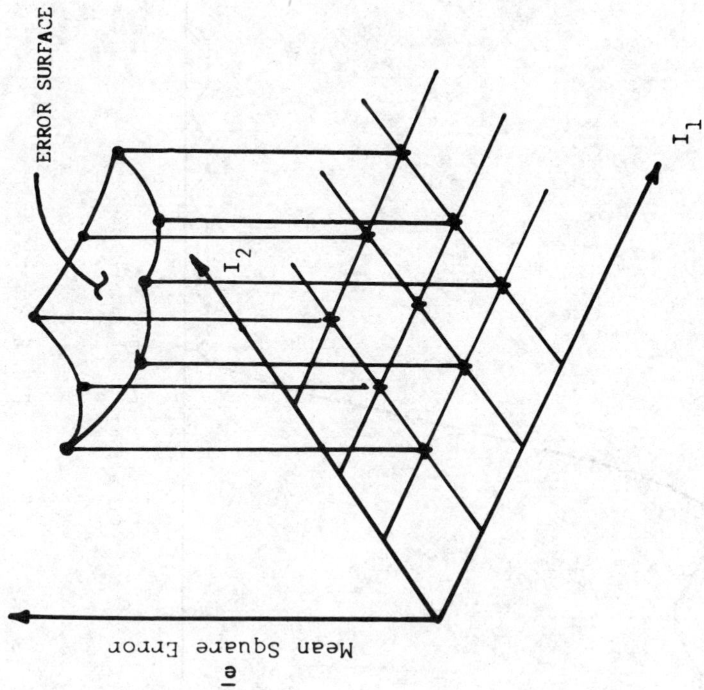

FIG. 4

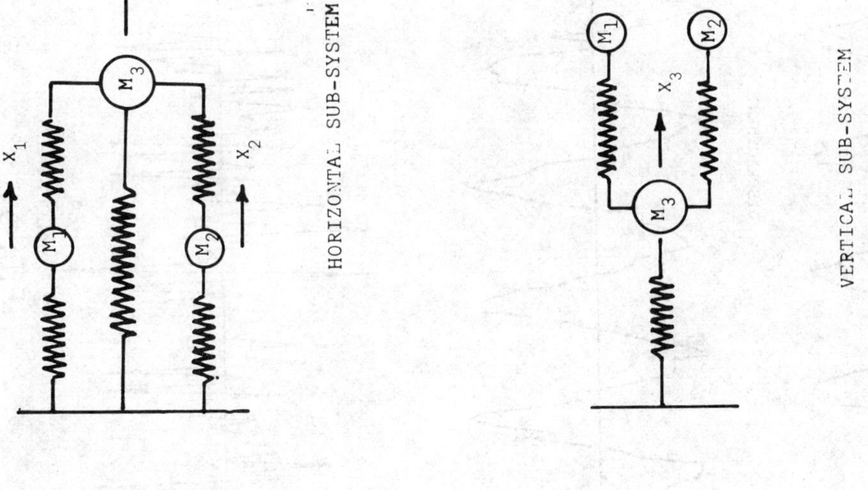

FIG. 3

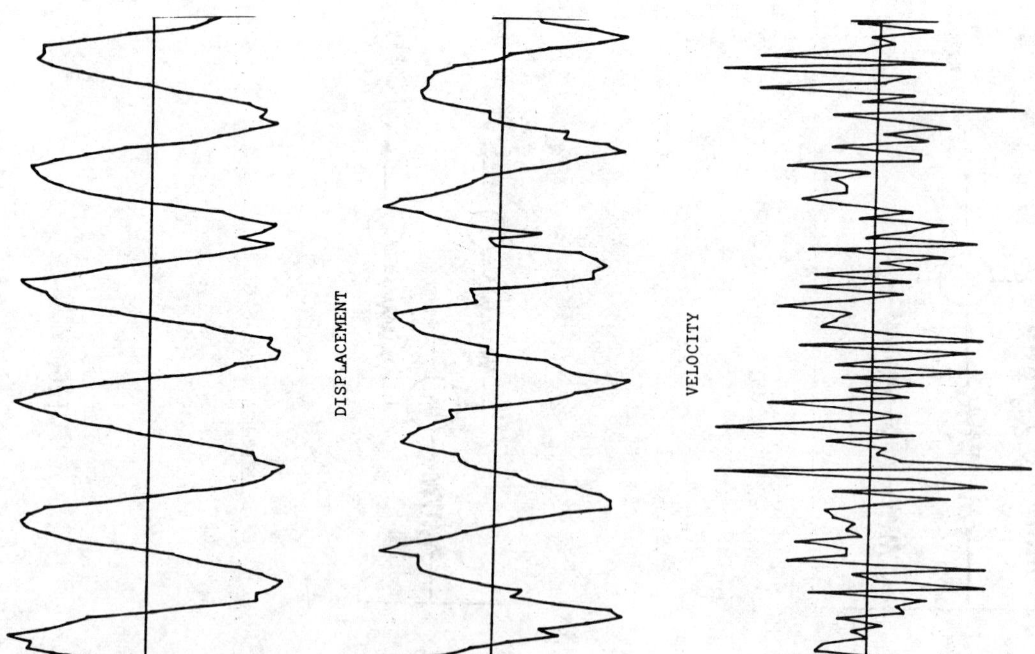
DISPLACEMENT
VELOCITY
ACCELERATION

FIG. 5

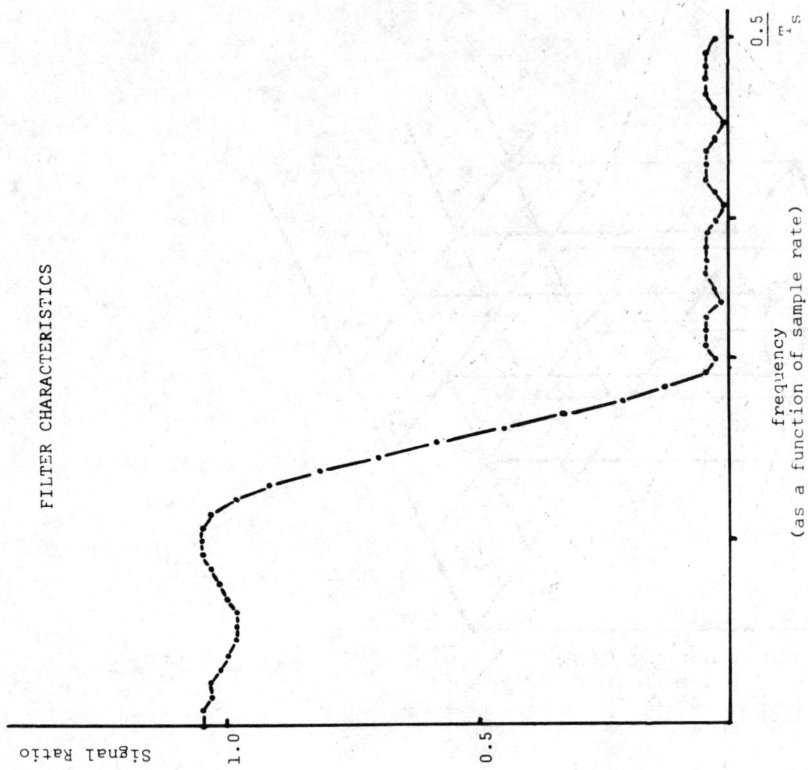
FILTER CHARACTERISTICS
Signal Ratio
1.0
0.5
frequency
(as a function of sample rate)
0.5
T
s

FIG. 6

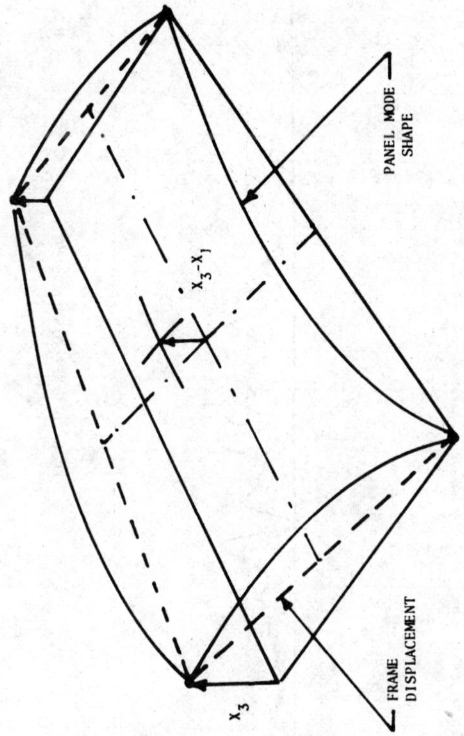
PANEL MODE SHAPE
$X_3 - X_1$
X_3
FRAME DISPLACEMENT

FIG. 8

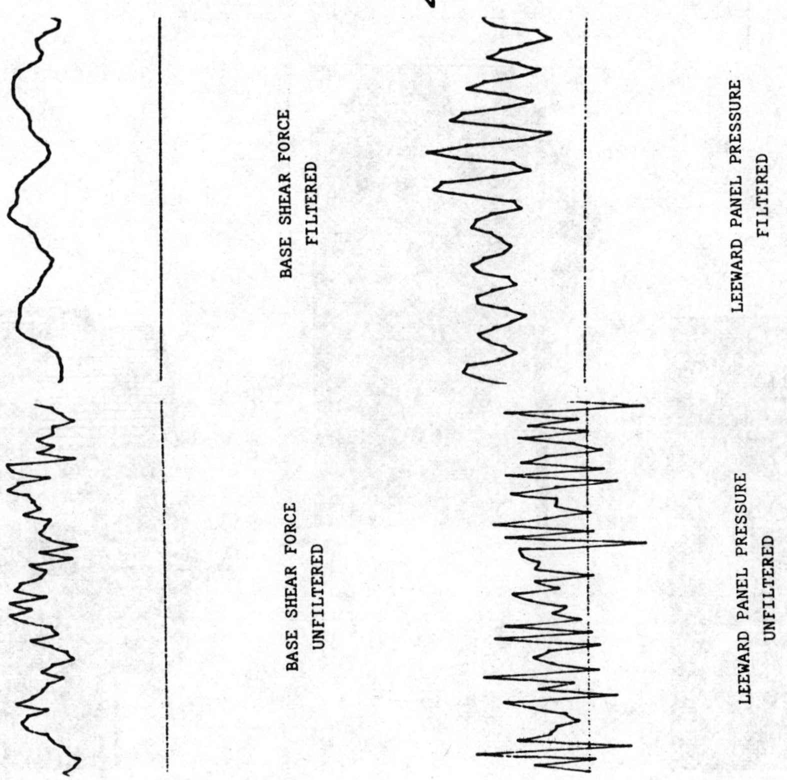
BASE SHEAR FORCE FILTERED
BASE SHEAR FORCE UNFILTERED
LEEWARD PANEL PRESSURE FILTERED
LEEWARD PANEL PRESSURE UNFILTERED

FIG. 7

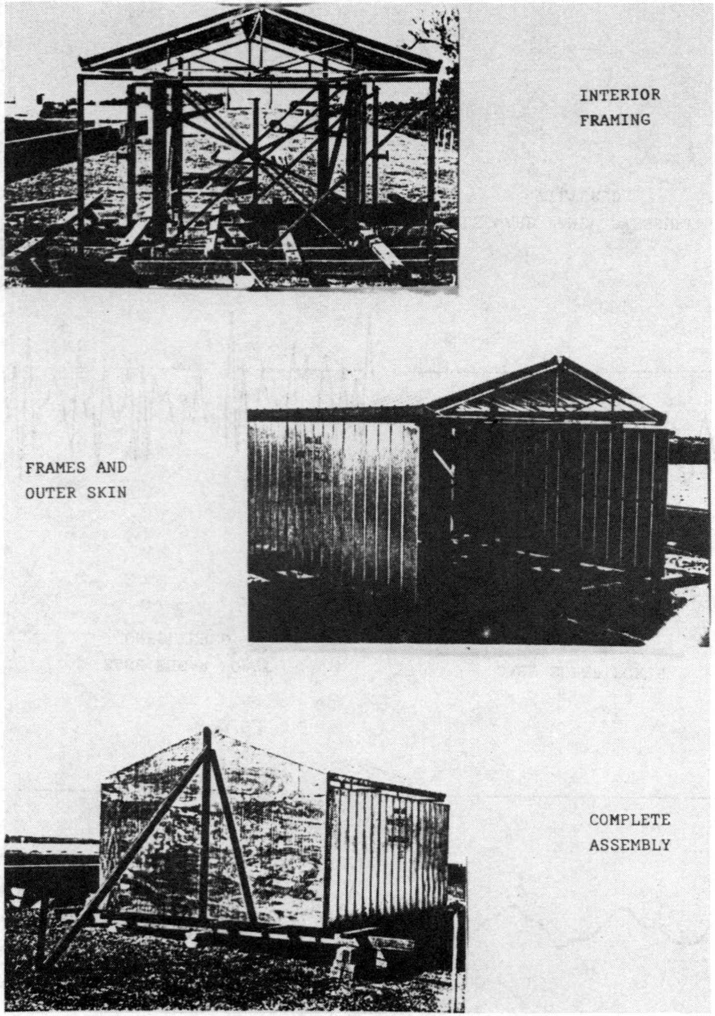
INTERIOR
FRAMING
FRAMES AND
OUTER SKIN
COMPLETE
ASSEMBLY

FIG. 9

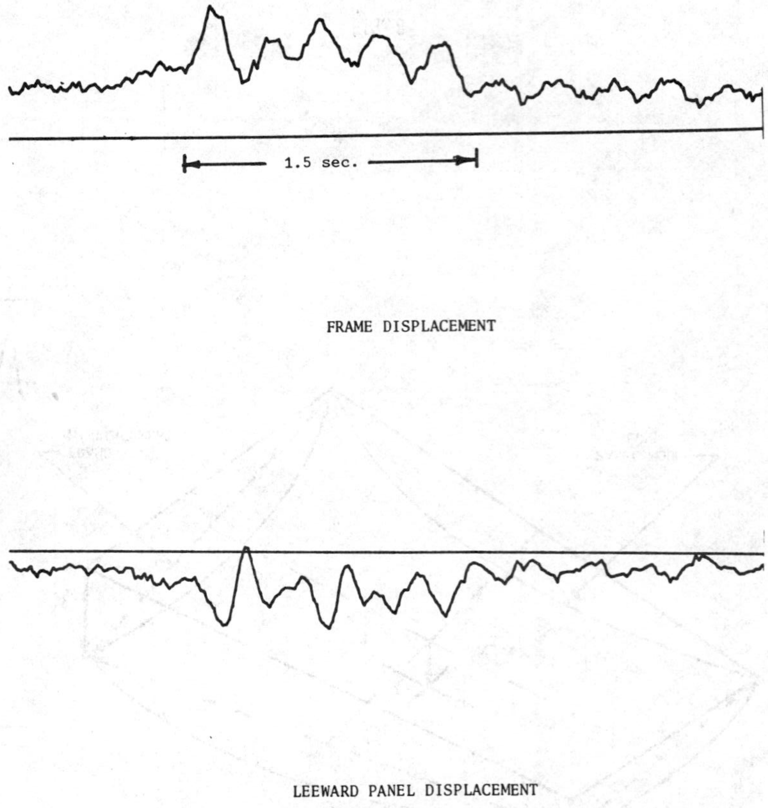
1.5 sec.
FRAME DISPLACEMENT
LEEWARD PANEL DISPLACEMENT

FIG. 10

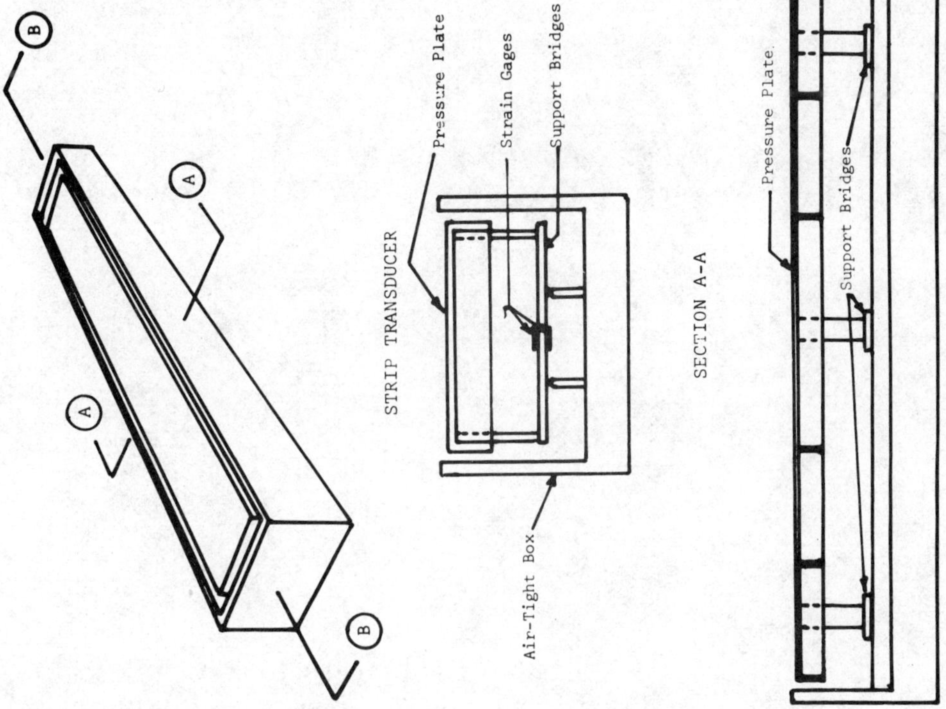
A
B
STRIP TRANSDUCER
Pressure Plate
Strain Gages
Support Bridges
Air-Tight Box
SECTION A-A
Pressure Plate
Support Bridges
SECTION B-B

FIG. 12

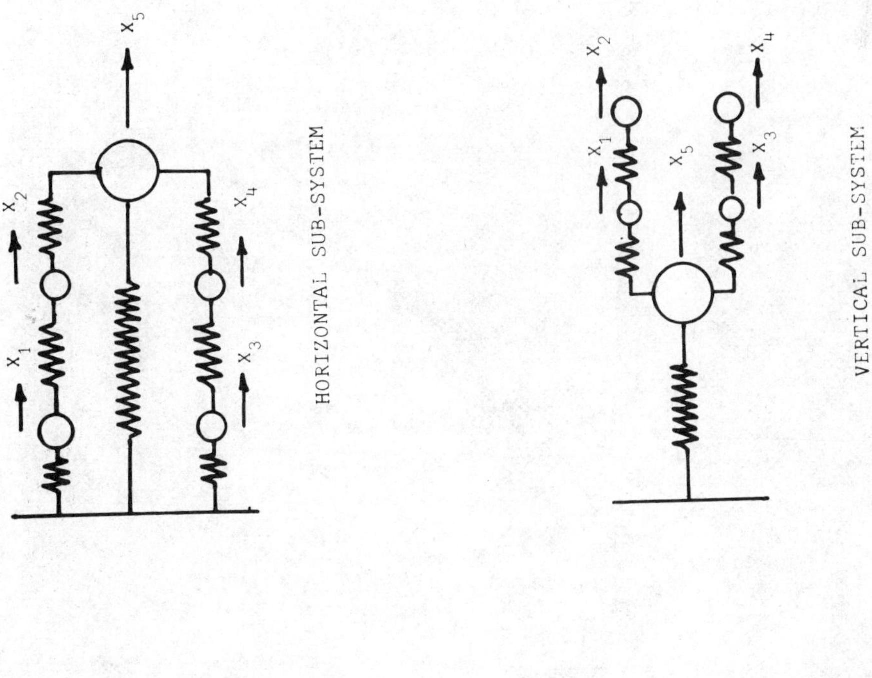
X_1
X_2
X_3
X_4
X_5
HORIZONTAL SUB-SYSTEM
VERTICAL SUB-SYSTEM

FIG. 11

FULL-SCALE MEASUREMENTS ON A FLAT ROOF AREA

SOO-II KIM* and K. C. MEHTA**
*Yonsei University, Seoul, Korea
**Texas Tech University, Lubbock, Texas, USA

ABSTRACT

The paper presents experiment, data acquisition, and results of roof uplift load measurements in a natural wind environment. In addition, development of a statistical model to predict probabilistic uplift load on a flat roof and comparison of the predicted loads with the current standards of practice are discussed. The measurements were made on a building with dimensions of 4.9 m x 4.9 m x 3 m (16 ft x 16 ft x 10 ft). The experiment is unique in the sense that average loads over a surface area are measured instead of the popularly employed technique of measurement of pressures at discrete points. The load measurements and the analysis of the data yield results of loads over a roof area.

INTRODUCTION

Current design wind loads suggested in standards and codes (1, 2) are based on pressure coefficients which have been obtained through measurements in aeronautical wind tunnels. The actual wind regime in boundary layer in the field is quite different from the air flow characteristics in aeronautical wind tunnels. Even though, in recent years, boundary layer wind tunnels which simulate wind characteristics closely to the actual wind regime have furnished improved pressure coefficient data, it is essential to continue to measure wind loads in the field on a limited basis.

Pressure measurements on roofs of low-rise model buildings in boundary layer wind tunnels have been accomplished by Cermak (3), Surry (4), Stathopoulos (5), and Davenport (6). On a full-scale basis Eaton (7) and Marshall (8) (9) (10) have measured wind pressures on roofs of low-rise buildings in natural wind environment. Except for the Stathopoulos (8) measurements in wind tunnel, the wind pressures on roof by these investigators were measured using pressure taps at discrete points on the roof. These pressure data fail to reflect design criteria of a structural component which should be design wind pressures averaged over an influence area.

The full-scale measurements described in this paper measured uplift loads over the entire roof of the building. This type of measurement provides average pressure over an area of the roof. The experiment, the data acquisition and analysis, the pertinent results, and implications of the results are presented in this paper.

EXPERIMENT

The building was designed to measure wind loads on a flat roof. The building had dimensions of 4.9 m x 4.9 m (16 ft x 16 ft) in plan. Its roof was 3 m (10 ft) above the ground. A general view of the building and accompanying wind measuring station is shown in Figure 1.

The roof of the experimental building was supported on rigid steel frame. The steel pipe columns of the frame were capped with one inch thick steel plates to facilitate mounting of the load cells. The roof structure consisted of two steel beams (M 8 x 8.5) and five wood joists 51 x 152 mm (2 in x 6 in). The wood joists were supported on the two steel beams. The roof decking, which was nailed to the wood joists, was 13 mm (0.5 in) thick plywood. The cross section of the roof structure is shown in Figure 2. The entire roof was supported at four corners by supporting the steel beams on load cells. The four load cells were connected to the cap plates of the pipe columns. This support arrangement permitted measurement of the fluctuation of load on the roof through monitoring of the load cells. In order to minimize bending and lateral shear effect on the load cells, the steel beams were supported on 25mm (1 in) diameter bolts, creating a 203 mm (8 in) space between the load cells and the beams. A closeup view of the load cell and the steel beam connection is shown in Figure 3. The natural frequency of vibration of the roof structure was 12.5 Hertz, which made dynamic amplification of the roof response negligible.

The top edge of the wall of the building was 25 mm (1 in) below the roof decking, thus providing a separation between the walls and the roof. The gap between the walls and the roof was covered with a flexible, loose plastic membrane all around the perimeter to reduce leakage of air into the building and to eliminate transfer of wall vibrations to the roof structure. Several small diameter holes were drilled in the walls to equalize the inside and outside static pressures of the building.

To measure windspeeds and wind directions at the building site a wind measuring station was established 12 m (40 ft) west of the building. Three lightweight cup anemometers were placed at heights of 3, 5.5, and 9.1 m (10, 18, and 30 ft) and a wind vane was placed at the height of 10 m (33 ft) above the ground. The anemometers were mounted on arms which extended 1.2 m (4 ft) to the west from the supporting pole. Electrical cables were run from the anemometers and the wind vane to the building so that all instrumentation could be placed inside the building.

The experiment site is located on the outskirts of Lubbock, Texas, which is a city with a population of 170,000. The natural terrain is extremely flat in this part of the country. As depicted in Figure 4, there are no trees in the vicinity of the experiment building and there are only a few residences with the nearest one being 91 m (300 ft) from the building. During the measurement period, the major wind directions were from the south and north quadrants.

DATA ACQUISITION

Wind load on roof, windspeed and wind direction were recorded simultaneously on an analog recorder. During high winds a set of data of these parameters for a continuous 15 minute period was recorded. A total of 28 sets of data during a three month period were obtained. Wind directions for each set of data are shown in Figure 5. Mean windspeeds for the recorded sets of data are between 7.2 mps (16 mph) and 18.4 mps (30 mph). Peak gust during the recording period was as high as 19.7 mps (44 mph).

Fig. 1. General view of the building and wind measuring station

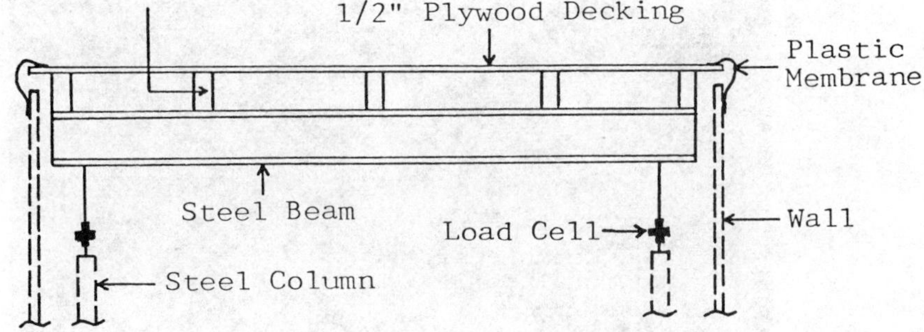

Fig. 2 Cross section of the roof structure

Fig. 3 Details of mounting of load cell

Fig. 4 Experimental site looking north

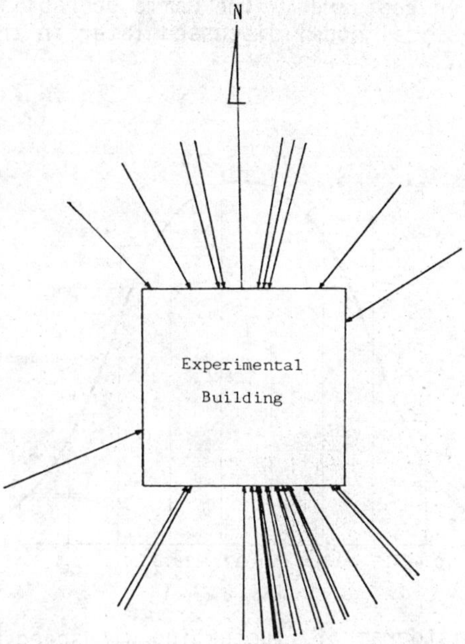

Fig. 5 Wind direction for each set of recording

The analog data were digitized using 25 points per second for wind load and one point per second for windspeed. Digitizing rate for wind load data was sufficient to indicate extent of dynamic amplification of the roof structure response at the resonance frequency. The digitized data were analyzed utilizing IBM 370/145 computer at Texas Tech University.

DATA ANALYSIS AND RESULTS

For the purposes of analysis, each set of data was trimmed to 800 seconds of continuous recording by eliminating front or tail end of the data set. Reference windspeed used in the analysis is the one measured at 3 m (10 ft) height; same height as that of the roof structure. Stationarity for windspeed data of each set were checked by dividing each record into several equal time intervals. Details of the stationarity check are given in the dissertation by the first author (11).

Probability distribution of the fluctuating component of windspeed was found to be Gaussian in most of the cases. A typical histogram of a windspeed record and associated Gaussian probability density function are shown in Figure 6. This result has been observed previously by other researchers (12, 13).

Fluctuating component of wind loads on roof was found not to follow Gaussian probability density function in most of the records. Cermak (3), Peterka (14), and Surry (15) found this same characteristic for fluctuating component of wind pressures. When Weibull and Gamma probability density functions were attempted, it was found that wind loads followed Gamma distribution more often than the Weibull distribution. A typical histogram of a recorded wind load and associated Gamma density function are shown in Figure 7. This correlation of measured data with Gamma probability density function is significant because it permits determination of peak loads with accuracy and confidence. If Gaussian probability density function is assumed for wind load instead of Gamma function, it would drastically underestimate peak loads on roof area. The Gamma probability density function is used to develop the statistical model discussed later in the paper.

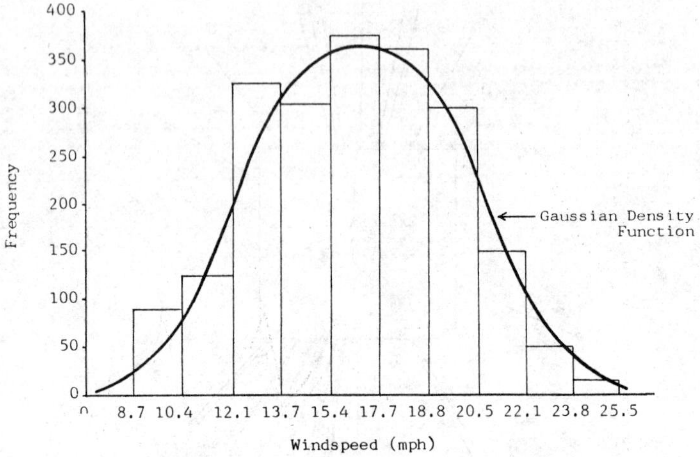

Fig. 6 Histogram of fluctuating windspeed data and associated Gaussian density function

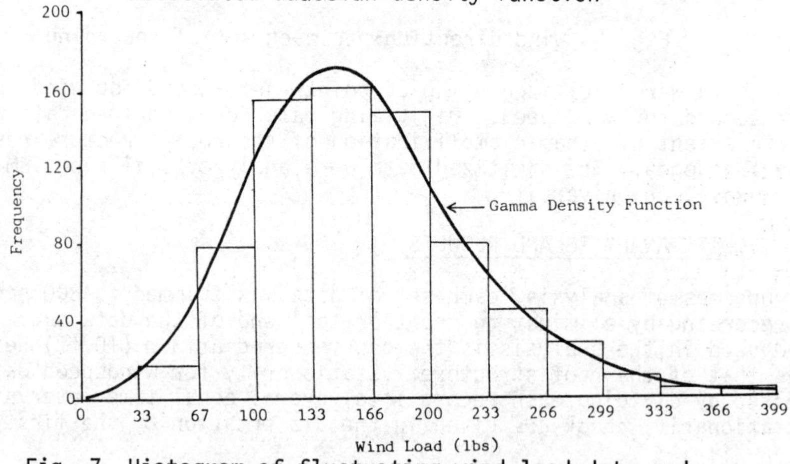

Fig. 7 Histogram of fluctuating wind load data and associated Gamma density function

In addition to the probability distributions of the recorded data, spectra for windspeed and wind load data are determined. Windspeed spectra and associated wind load spectra are plotted for each set of data. By comparing windspeed and wind load spectra in graphical form, it is observed that the spectra are relatively similar and major peaks of load spectrum occur at almost the same frequencies as the corresponding windspeed spectrum peaks. A typical load spectrum and the corresponding windspeed spectrum are shown in Figure 8. The similarities of the spectra demonstrate that, even though the free stream wind is separated over the roof, the transfer of energy from wind to load over an area of the roof takes place at corresponding frequencies. The closeness of the frequency ranges of the spectral peaks indicates that the small scale turbulence over the roof plays a minor role if energy is summed up over an area.

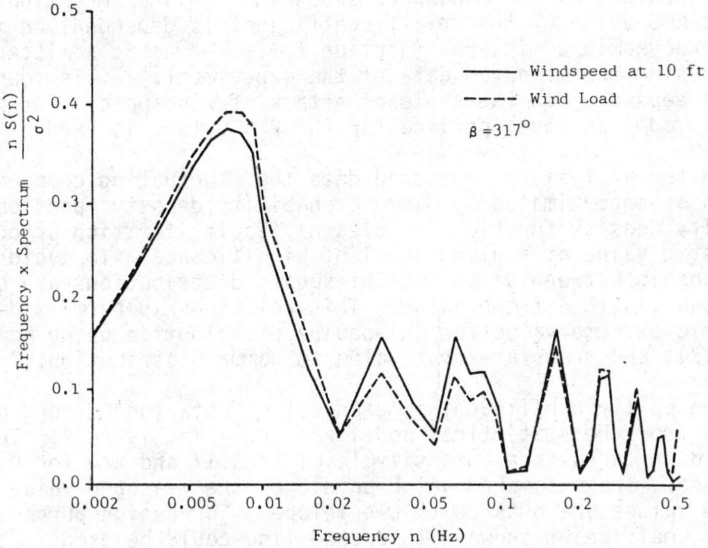

Fig. 8 Windspeed spectrum and associated wind load spectrum showing similar peaks

In full-scale experiment, Eaton and Mayne (7) found that spectra of the free field wind and the roof pressure at a point were not close to each other. In their results, the peak of the point pressure spectrum was shifted by an order of magnitude in frequency from the peak of the windspeed spectrum, although the general shapes of the two spectra remained similar. This shift of the spectrum peak (pressure frequency higher than free field wind turbulence frequency) may be explained by the introduction of small scale turbulences creating higher pressure frequencies at a point. However, if pressures are measured over a surface area rather than at a point, as accomplished in the experiment presented in this paper, the small scale turbulences seem to have negligible effects on the roof load.

STATISTICAL MODEL

A statistical model to predict probabilistic uplift wind load on a flat roof of a building of cubical configuration is developed by the first author and is presented in the dissertation (11). The model is based on the analysis and results of the measured data. Gamma probability density function for the fluctuating component of the wind load is employed in the model. Space limitations prevent presentation of the details of the development of the statistical model; details are contained in the reference 11. Major points of the development of the statistical model are presented below.

Probabilistic wind load on a roof can be predicted with the knowledge of mean uplift coefficient, roof mean square (RMS) value of uplift coefficient and probability density function. The mean uplift coefficient values are obtained for each set of measured data of 800-second duration. A function for mean uplift coefficient with angle of wind attack as a variable is established for use in the statistical model. The RMS values of the uplift coefficient are determined utilizing results observed in the experiment and the concept advanced by Davenport (16). Similarity of wind-speed and wind load spectra shown in Figure 8 suggests that aerodynamic admittance function is independent of the windspeed spectrum. This observation leads to a conclusion that RMS value of the uplift coefficient is dependent on mean uplift coefficient, aerodynamic admittance function for aerodynamic admittance function is established using the measured data of the experiment. It is found that the RMS function is dependent on the angle of attack of windspeed. For purposes of the statistical model an envelope line for the RMS values is used.

As indicated in the analysis of measured data the fluctuating component of uplift coefficient can be approximated by Gamma probability density function. However, Gamma probability density function invloves difficult iteration process to determine probabilistic value at a given level of significance. To avoid interative process relationship between Gamma and Chi-square distributions are utilized to obtain the probabilistic extreme value. This relationship permits determination of probabiliistic extreme value for Chi-square distribution using nomogram developed by Boyd (17), and translate that value to Gamma Distribution.

Measured extreme uplift coefficient in each set of data (one second average) and estimated value from the statistical model are shown in Figure 9. The estimated values are based on turbulence intensity level of 0.17 and are for 0.001 probability of exceedance. The estimated value envelopes the measured value, though two of the measured values are outside this envelope. For design purposes the maximum value of uplift coefficeint shown with dashed line could be used. It should be noted that the concepts employed in the development of this statistical model are general but the numerical values are restricted to a cubical configuration with specific relative dimensions of the building.

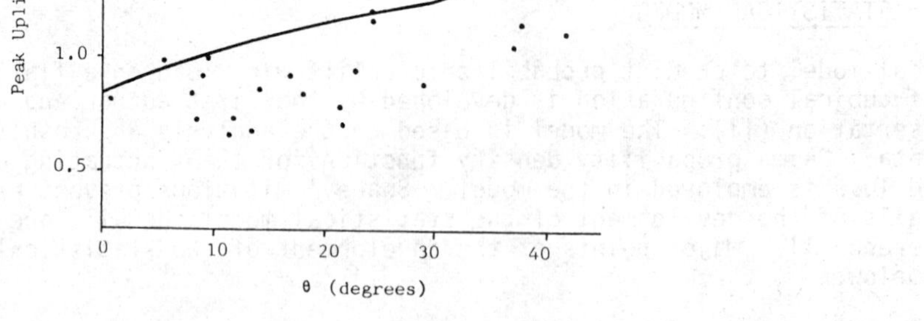

Fig. 9 Measured and estimated peak uplift coefficients

COMPARISON OF UPLIFT LOADS

Probabilistic load estimated by the statistical model is compared with design load recommended in the ANSI A58.1-1972 (1) of the United States, the British Standards Institute CP 3:1972 (18), the National Building Code of Canada-1975 (19), and the Australian Standard SAA 1170:1975 (20). Since each standard of practice utilizes different averaging time for windspeed, the comparison is made between design pressures and not between pressure coefficients. Shape of the building and relative dimensions of the building are assumed to be the same as the experiment building.

Reference fastest mile windspeed is assumed to be 44.7 mps (100 mph). Equivalent mean hourly windspeed, average 800 second windspeed, and two second gust speed are 34.9, 35.8, and 52.8 mps (78, 80, and 118 mph) respectively. The reference windspeeds, effective velocity pressures, pressure coefficients, and design pressures as recommended by various standards are shown in Table I.

TABLE 1. Comparison of roof uplift load

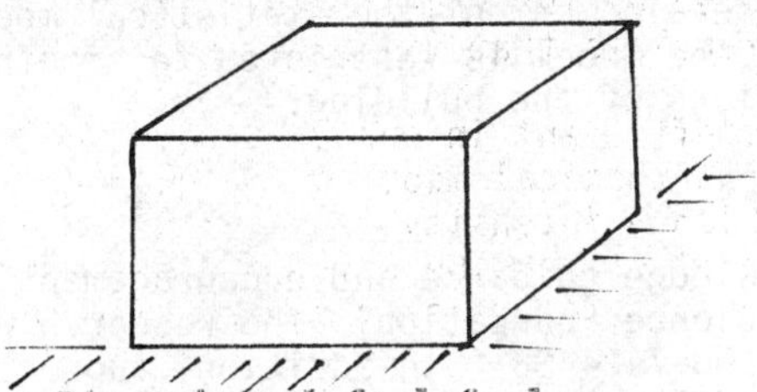

Dimensions 1.6: 1.6: 1

Design Standard	Reference Windspeed Duration	Reference* Windspeed mps (mph)	Uplift Load Pa (psf)
Australian Standard (20)	2-second gust	52.8 (118)	1338 (29)
British Code (18)	2-second gust	52.8 (118)	1197 (25)
Canadian Code (20)	Mean hourly	34.9 (78)	1484 (31)
ANSI Standard (1)	Fastest mile	44.7 (100)	1101 (23)
Statistical Model	800-second average	35.8 (80)	1149 (24)

*Windspeeds shown are equivalent to Fastest Mile Wind of 100 mph

The uplift coefficient for the statistical model is based on the worst angle of attack and 0.001 probability of exceedance for the extreme load. As evident in Table 1 the statistical model suggests design pressure which is in line with the various national standards.

CONCLUSIONS

Measurements of wind loads on flat roof of the experimental building and the analysis of the data lead to following observations:

1. Windspeed and wind load spectra are similar with spectral peaks occurring at virtually the same frequencies. This observation indicates that the wind load averaged over an influence area is not significantly affected by the localized turbulence generated due to presence of bluff body in the wind field.

2. Fluctuating component of the roof uplift load is best represented by the Gamma probability density function.

These two observations, combined with the analysis of measured data, lead to development of a statistical model. The model predicts probabilistic extreme loads which compare reasonably well with the design loads of the current standards. The concepts employed in the development of the statistical model are general, but the numerical formulation of the model is restricted to smooth terrain and specific shape and relative dimensions of the building.

ACKNOWLEDGMENTS

The authors wish to acknowledge guidance and encouragement provided by Dr. Michael P. Gaus of the National Science Foundation. The research was conducted under the sponsorship of NSF Grant ENG-7424406. In addition, support of the Institute for Disaster Research at Texas Tech University and the State of Texas are acknowledged.

REFERENCES:

1. American National Standard Building Code Requirements for Minimum Design Loads in Buildings and Other Structures, ANSI A58.1-1972, American National Standards Institute, New York.

2. Uniform Building Code, International Conference of Building Officials, 5360 South Workman Mill Road, Whittier, California, 1976.

3. Cermak, J.E., Peterka, J., and Dreher, K.J., "Wind Pressures on a House Roof," Proceedings of the Second U.S.-Japan Research Seminar on Wind Effects on Structures, Edited by H. Ishizaki and A.N.L. Chiu, University of Tokyo Press, 1976.

4. Surry, D., Stathopoulos, T., and Davenport, A.G., "Wind Loading of Low Rise Buildings," a paper presented at the Canadian Structural Engineering Conference, 1978.

5. Stathopoulos, T., "Wind Pressure Loads on Flat Roofs," Master thesis, The University of Western Ontario, London, Canada, 1975.

6. Davenport, A.G., and Surry, D., "The Pressures on Low Rise Structures in Turbulent Wind," Canadian Structural Engineering Conference, 1974.

7. Eaton, K.J., and Mayne, J.R., The Measurement of Wind Pressures on Two-story Houses at Aylesbury, Building Research Establishment, Dept. of Environment, England, July 1974.

8. Marshall, R. D., and Raufaste, N.J., Building to Resist the Effect of Wind NBS Building Series 100-1, Vol. 1, Institute for Applied Technology, National Bureau of Standards, May 1977.

9. Marshall, R.D., A Study of Wind Pressures on a Single-Family Dwelling Model and Full Scale, NBS Technical Note 852, Institue for Applied Technology, National Bureau of Standards, October 1974.

10. Marshall, R.D. The Measurement of Wind Loads on a Full-Scale Mobile Home, NBSIR 77-1289, Center for Building Technology, National Bureau of Standards, 1977.

11. Kim, S., Wind Loads on Flat Roof Area through Full-Scale Experiment, a Ph.D. Dissertation, Dept. of Civil Engineering, Texas Tech University, Lubbock, Texas, August 1977.

12. Shiotani, M., Arai, H., and Hayashida, C., "Vertical Distribution of Wind Velocities and Vertical Structures of Gusts in Storm Winds", Railway Technical Research Institute, Japanese National Railways, Report No. 514, December 1965.

13. Singer, I.A., "Wind Gust Spectra," Proceedings of Conference on Large Steerable Radio Antennas-Climatological and Aerodynamic Considerations, Annals of the New York Academy of Science, Vol. 116, p. 116.

14. Peterka, J.A., and Cermak, J.E., "Probability Distribution of Wind Pressure Fluctuations on Buildings", Probabilistic Methods in Engineering, Abstracts for the ASCE-EMD Specialty Conference at Stanford University, June 24-26, 1974, p. 90.

15. Surry, D., and Truggvason, B.V., "Fluctuating Pressures on Tall Buildings," Proceedings of the Second U.S. National Conference on Wind Engineering Research, Colorado State University, Fort Collins, Colorado, 1975.

16. Davenport, A.G., "The Application of Statistical Concepts to the Wind Loading of Structures", Proceedings of Institute of Civil Engineers, Vol. 19, August 1961, p. 449-471.

17. Boyd, W.C., "A Nomogram for Chi-square," Journal of the American Statistical Association, Vol. 60, 1965, p. 344-346.

18. Code of Basic Data for the Design of Buildings, CP3: Chapter V: Part 2: Wind Loads, British Standards Institution, England, 1972.

19. National Building Code of Canada 1975, Issued by Associate Committee on the National Building Code, National Research Council of Canada, NRC No. 13982, p. 156-157.

20. Australian Standard AS 1170 Part 2-1973, Standards Association of Australia, Standards House, 80 Arthur Street, North Sydney, 1975.

AVERAGED PRESSURE COEFFICIENTS FOR RECTANGULAR BUILDINGS

R. E. AKINS*, J. A. PETERKA** and J. E. CERMAK**
*Virginia Polytechnic Institute and State University, Blacksburg, Virginia, USA
**Colorado State University, Fort Collins, Colorado, USA

SUMMARY

Mean pressures obtained from an extensive series of wind-tunnel tests have been averaged over an entire side of a building to obtain a single mean pressure coefficient for a given side and wind direction. Coefficients based on either a local reference wind speed or a reference wind speed measured at the level of the roof of the building have been computed. An example and a comparison with existing building codes and standards are included to illustrate the use of these coefficients.

INTRODUCTION

Engineers who design structures for wind loading are faced with two distinct problems, the selection of an appropriate wind speed for design and the selection of a force or pressure coefficient which allows the design wind speed to be expressed in terms of a design load. The choice of a design wind speed is a function of the projected life of the structure and the location of the structure. In some cases a design wind speed is dictated by a local building code. The choice of a force or pressure coefficient is usually determined by data available in the technical literature or incorporated into standards or building codes. These coefficients will not necessarily correspond to the structure being designed and in some instances sufficient information about how the coefficients were obtained does not exist to adequately judge the values reported.

A considerable body of data is available describing mean pressures on structures of various shapes measured in wind tunnels with a uniform approach velocity (no variation of wind speed with height) and very low incident turbulence intensity. Such data have been incorporated into most modern building codes and standards. However, Jenson (1) has shown that a wind-tunnel model of a structure should be tested in a turbulent shear flow that simulates the natural wind if accurate assessments of the mean pressures caused by wind loading of buildings are to be obtained. The techniques of modeling wind forces on buildings in boundary-layer wind tunnels have advanced sufficiently to allow a more detailed examination of the effects of building geometry and incident flow properties on mean wind pressures than are currently available in modern building codes and standards. The history and present status of wind-tunnel modeling for wind loading of structures has recently been reviewed by Cermak (2) (3).

Despite these advances, very little organized information is available to describe the mean pressures on structures in properly simulated natural winds. In this paper, mean pressure coefficients which have been averaged over an entire side of a building are presented and discussed. These pressure coefficients are for rectangular flat-roofed buildings and were obtained with the structures immersed in thick turbulent boundary layers with properties which simulate neutrally-stable atmospheric flows.

The techniques used in the determination of these pressure coefficients are described briefly. A detailed discussion of the results including examples of applications is presented. These measurements were conducted using isolated, sharp-edged, smooth rectangular models. The results presented do not include information concerning the effects of corner geometry, surface roughness of the building, or adjacent structures. Because many of these effects, but certainly not all, result in reduced mean loads, these data provide a useful, and in most cases, slightly conservative estimate of surface pressure for rectangular buildings.

EXPERIMENTAL TECHNIQUES

Buildings and Pressure Measurements

A series of thirteen flat-roofed rectangular buildings were used in the study. The building geometry is described by two ratios--the side ratio, defined as the ratio of the width of the small side to the width of the large side, and the aspect ratio, defined as the ratio of the building height to the width of the small side. Side ratios of 0.25, 0.50 and 1.0 were examined over a range of aspect ratios from 1.0 to 8.0. The models were made of plexiglass and instrumented in most cases at 272 locations with pressure taps (60 on each vertical face and 32 on the roof). The buildings were mounted on a turntable at the downstream end of 16.7-m (55 ft) long test section of the industrial aerodynamics wind tunnel of the Fluid Dynamics and Diffusion Laboratory, Colorado State University.

Pressure measurements were made using a system consisting of a 72-channel pressure-selector switch, strain-gauge pressure transducers, and a digital data acquisition system. The pressures measured were the difference between the instantaneous local pressure at a location on a building and the static pressure in the ambient flow above the model building. A more detalied description of the buildings and the pressure measurement system is available in references 4 or 5.

The Boundary Layers

The four boundary layers used in the study were developed over the length of the test section using various-sized elements on the floor and spires at the entrance to the test section. The spires were used to artificially stimulate the growth of the boundary layer and in addition provided a constant-depth boundary layer for all of the roughness configurations. Table 1 is a summary of the properties of these boundary layers. The properties included in the table are: δ, the boundary layer thickness; p the exponent the power law representation of the mean velocity profile; z_o, the surface roughness length; u_*/U_δ, the ratio of the friction velocity to the free-stream velocity; L_x, the longitudinal integral scale of the turbulence; and T_{ux}, the longitudinal turbulence intensity. Further details concerning the boundary layers including velocity spectra, cross-correlations, and profiles of turbulence intensity in three dimensions are available in reference 5.

TABLE 1. Boundary Layer Characteristics

Boundary Layer	δ (m)	p	z_o (m)	u_*/U_δ	L_x @ z/δ=0.2 (m)	T_{ux} @ z/δ=0.2 (%)
1	1.27	0.12	1.2×10^{-5}	0.028	0.40	7.2
2	1.27	0.27	2.8×10^{-3}	0.052	0.54	12.5
3	1.27	0.34	5.0×10^{-3}	0.051	0.43	15.0
4	1.27	0.38	1.1×10^{-2}	0.062	0.34	17.3

Data Reduction

Mean pressure coefficients were averaged over an entire side of a building. Through the use of a more sophisticated integration procedure it was established that an accurate averaging was obtained if an appropriate area was assigned to each pressure-tap location and a vector summing of pressure times a representative area was used. The pressure coefficients used were defined by two equations,

$$\overline{C}_{PL} = \overline{\left(\frac{P-P_o}{0.5\rho V(z)^2}\right)} \tag{1}$$

$$\overline{C}_{PR} = \left(\frac{\overline{P-P_o}}{0.5\rho V(H)^2}\right). \tag{2}$$

$\overline{C}_{PL}$ is a local pressure coefficient averaged over an entire side of a building. It is the nondimensional ratio of $P-P_o$, the difference between the pressure at a location on the building and the local static pressure, to $0.5\rho V^2(z)$, the dynamic pressure based on the wind speed in the approach flow at the height of the pressure measurements, V(z). This value was determined at each point on the side of a building and averaged as a pressure coefficient. The overbar in both equations is used to denote spatial averaging over an entire side. $\overline{C}_{PR}$ is a pressure coefficient based upon a fixed reference wind speed measured at the level of the roof of the building. This coefficient was also averaged over an entire side of a building. V(H), used in equation (2), denotes the wind speed at the height of the building, H.

Based on the results of previous work (5), the buildings used were grouped by side ratio for all aspect ratios and boundary layers tested. The values of both $\overline{C}_{PR}$ and $\overline{C}_{PL}$ for all cases of corresponding side ratio were averaged to obtain one set of values. In most cases from eight to ten separate cases were averaged to obtain a single result.

The coordinate system used to describe the averaged pressure coefficients is shown in Fig. 1. The pressure coefficients were obtained for each of four sides and the roof. Measurements were obtained at eleven wind directions from 0 to 90°. Only five of the directions are shown in the figure. The sides were numbered such that side 2 was upwind for a wind direction of 0° and side 3 was upwind for a wind direction of 90°. All of the buildings studied were placed in an isolated environment with no adjacent structures present. A 90° variation in wind direction was therefore adequate to define the pressures acting on the structure for any wind direction. No corrections for tunnel blockage were applied because the blockage was small (less than 7 percent) and the flexible roof was adjusted to remove the longitudinal pressure gradient in the tunnel.

RESULTS AND DISCUSSION

The consideration of these two types of averaged pressure coefficients was motivated by interest expressed in more complicated pressure distributions reported in reference 5. These distribtuions were reported in terms of local pressure coefficients defined using a local wind speed at the height of the pressure measurement in the approach flow. This choice of coefficient allowed the pressure distributions to be expressed as a function of side ratio only with very little dependence on aspect ratio or approach flow properties. The use of such a distribution allows an accurate estimate of the distribution of pressures on a building. Such a procedure may be too detailed or complicated for some applications. In certain instances a single pressure coefficient over the side of a structure may be adequate. If this coefficient is obtained by averaging, there will naturally be some locations which will experience pressures above the average and other locations which will experience pressures below the average. If such an approximation is appropriate in the design of a structure, then the values of averaged pressure coefficients reported in this paper should be useful.

In order to provide a summary of the current methods of using pressure coefficients, the techniques suggested in the ANSI A58.1-1972 (6), "Building Code Requirements for Minimum Design Loads in Buildings and Other Structres," will be briefly reviewed. Techniques suggested in other codes and standards for the determination of mean pressures on a building are quite similar to those used in ASNI A58.1-1972. Table 7 of the standard provides values of external pressure coefficients for walls. These are reported for windward walls, side walls, and leeward walls, corresponding to wind directions of 0° or 90° from Fig. 1 of this paper. In order to obtain a pressure, these coefficients are multiplied by a reference dynamic pressure which increases with height. This corresponds to the reference pressure used in the definition of a local pressure coeeficient, Eq. 1. While this may indicate that the pressure coefficients reported in ANSI A58.1-1972 are local pressure coefficients (based on a reference wind speed measdured at the height of the pressure measurement in the approach flow), such is not the case. The coefficients provided in the standard were obtained in a uniform flow, no variation of wind speed with height, and for this case both definitions of pressure coeffient reduce to the same value, i.e. $V(z) = V(H)$ for all z. The choice of a height dependent reference pressure (proportional to the square of the wind speed) is interpreted as an effort to make the coefficients obtained in a uniform flow useable in a boundary layer flow. It should be noted that these coefficients were obtained prior to recent advances in techniques in wind-tunnel modeling. The use of a pressure coefficient which is constant over an entire side of a building with a reference pressure which increases with height provides a design pressure which increases with height but does not vary in the horizontal across a side of a building.

The averaged local pressure coefficients, $\overline{C}_{PL}$, for side ratios of 1.0, 0.5 and 0.25 are presented graphically in Figures 2-4 respectively. The curves shown in Figure 2 for a side ratio of 1.0 are symmetrical about 45° for sides 2 and 3 and sides 4 and 1. This symmetry is to be expected. If the values for a wind direction of 0° are considered the maximum magnitudes of the coefficients are obtained. For a side ratio of 0.5, Figure 3, the values for a wind direction of 90° (normal to the wider-side) must be used to obtain the maximum magnitudes. For a side ratio of 0.25, the maximum magnitude of the pressure coefficient for side 2 occurs at an approach wind of 70°. For side ratios of both 0.5 and 0.25, the coefficients

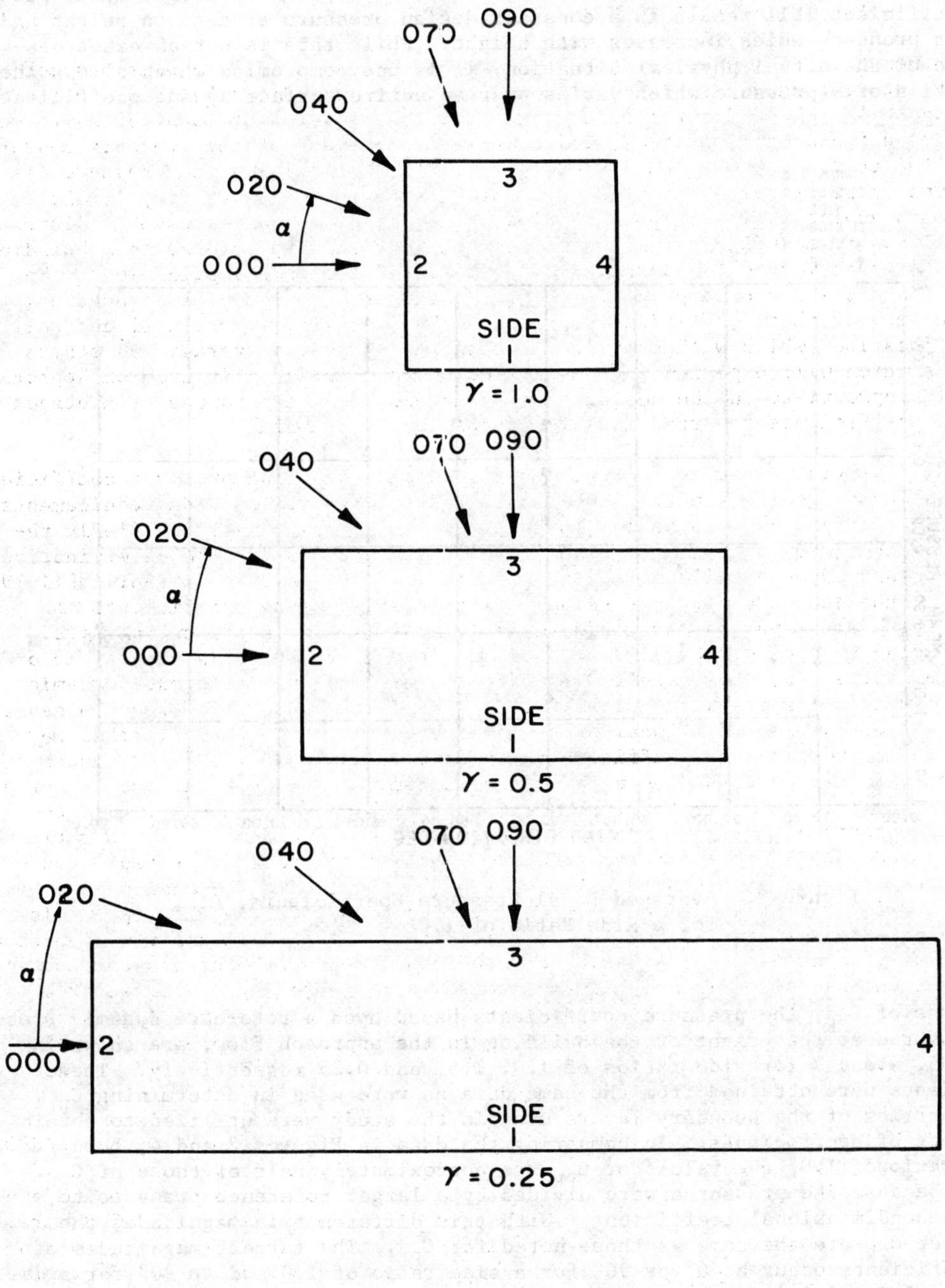

Figure 1. Wind Directions and Nomenclature for the Sides

for a 0° wind (normal to the narrow side), are of a smaller magnitude on the sides and the rear surface than for the case of 90° wind. The format of an averaged local pressure coefficient incorporates the effects of different approach boundary layers and different height buildings in the same boundary layer. The use of a variable reference pressure, $0.5\rho V(z)^2$, with a constant averaged local pressure coefficient will result in a constant design pressure at a given height and a design pressure which increases with height. While this is not an exact description of the actual physical situation, it is one compromise which allows the description of a pressure which varies over an entire surface by one coefficient.

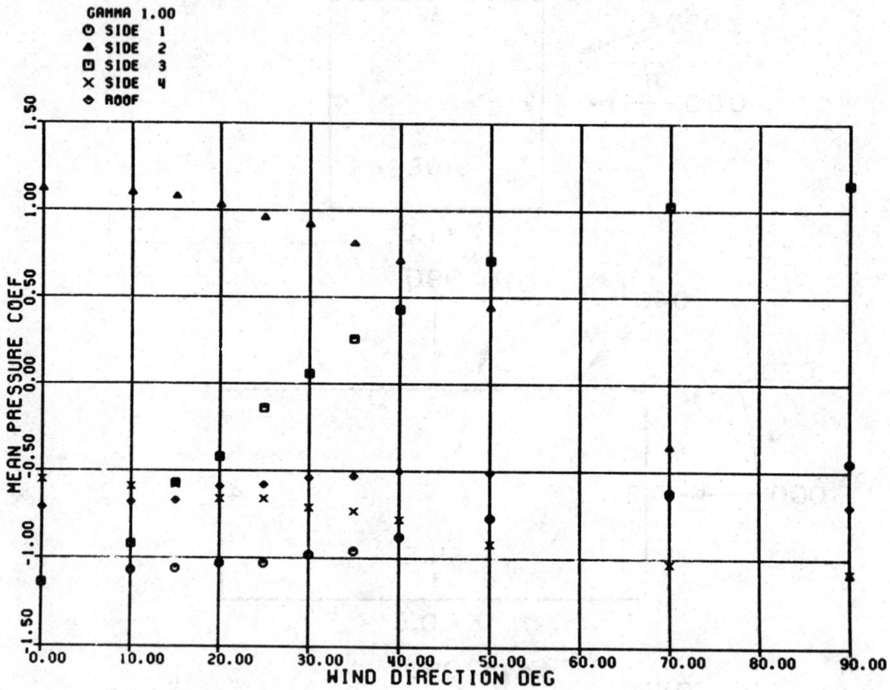

Figure 2. Averaged Local Pressure Coefficients, $\overline{C}_{PL}$, for a Side Ratio of 1.0

The values of $\overline{C}_{PR}$, the pressure coefficients based upon a reference dynamic pressure measured at the height of the building in the approach flow, are shown in Figures 5, 6 and 7 for side ratios of 1.0, 0.5, and 0.25 respectively. These coefficients were obtained from the same data as were used in determining $\overline{C}_{PL}$. The properties of the boundary layers used in the study were utilized to obtain both types of coefficients. In comparing the data in Figures 2 and 4, both for a side ratio of 1.0, the values of $\overline{C}_{PR}$ are approximately half of those of $\overline{C}_{PL}$. This is because the pressures were divided by a larger reference pressure to create a nondimensional coefficient. With this difference in magnitude, the trends for $\overline{C}_{PR}$ are the same as those noted for $\overline{C}_{PL}$. The largest magnitudes of the coefficients occur at 0° or 90° for a side ratio of 1.0 and at 90° for side ratios of 0.5 and 0.25. Again the maximum magnitude of the coefficients for a side ratio of 0.25 occurs at 70° for side 2. The coefficients based upon a

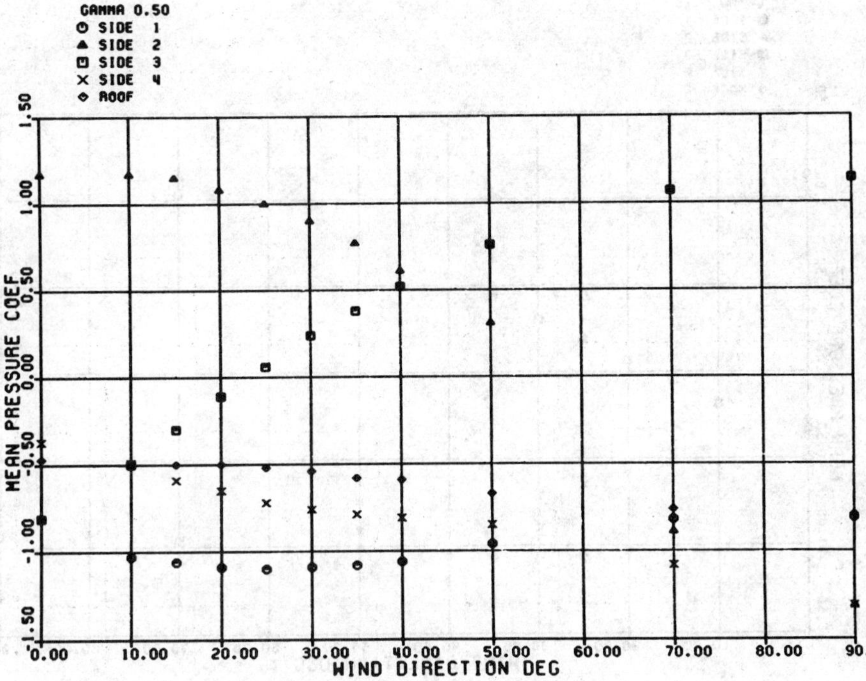

Figure 3. Averaged Local Pressure Coefficients, $\overline{C}_{PL}$, for a Side Ratio of 0.5

reference pressure of $0.5\rho V(H)^2$ will only require a single value of reference wind speed, V(H). The use of these coefficients will result in a constant design pressure over an entire side of a building.

At this point it should be noted that no matter how one chooses to define a pressure coefficient the decision to average the coefficients over an entire side of a building will result in some differencies between the actual pressure a building experiences and the pressure predicted using an average coefficient. To obtain a feel for the magnitude of the differences which could be expected, the averaged local pressure coefficients, $\overline{C}_{PL}$ may be compared with the distributions of local pressure coefficients over a surface reported in reference 5. Table 2 is a comparison for the case of a wind direction of 0° and a side ratio of 1.0. The three columns of data show the averaged local pressure coefficient for each side (corresponding to Fig. 2) and the maximum and minimum mean local pressure coefficient on a side. These data show that the use of an averaged pressure coefficient can result in both over and under prediction of pressures by up to 100%. Some design situations such as cladding design for large buildings may not be appropriate application of a spatially averaged pressure coefficient. In other instances such as low rise buildings or overall loadings, the concept should be useful.

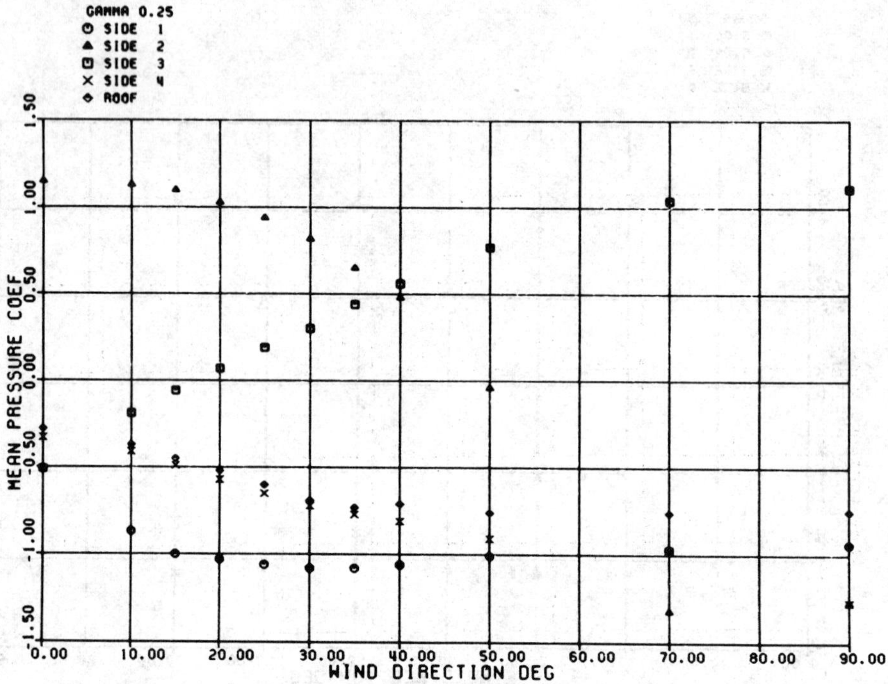

Figure 4. Averaged Local Pressure Coefficients, $\overline{C}_{PL}$, for a Side Ratio of 0.25

TABLE 2. Variation of local pressure coefficient side ratio 1.0, wind direction 0°

Side	$\overline{C}_{PL}$	Maximum C_{PL}	Minimum C_{PL}
1	-1.1	-0.5	-2.6
2	1.1	1.6	0.6
3	-1.1	-0.5	-2.6
4	-0.6	-0.4	-0.9
Roof	-0.7	-0.5	-1.0

In order to obtain a more qualitative idea of the differences between the two types of coefficients presented in this paper, an example of the calculation of the mean pressures on a 70 m (200 ft) building will be examined. This comparison will only consider the determination of mean pressures. Design techniques for

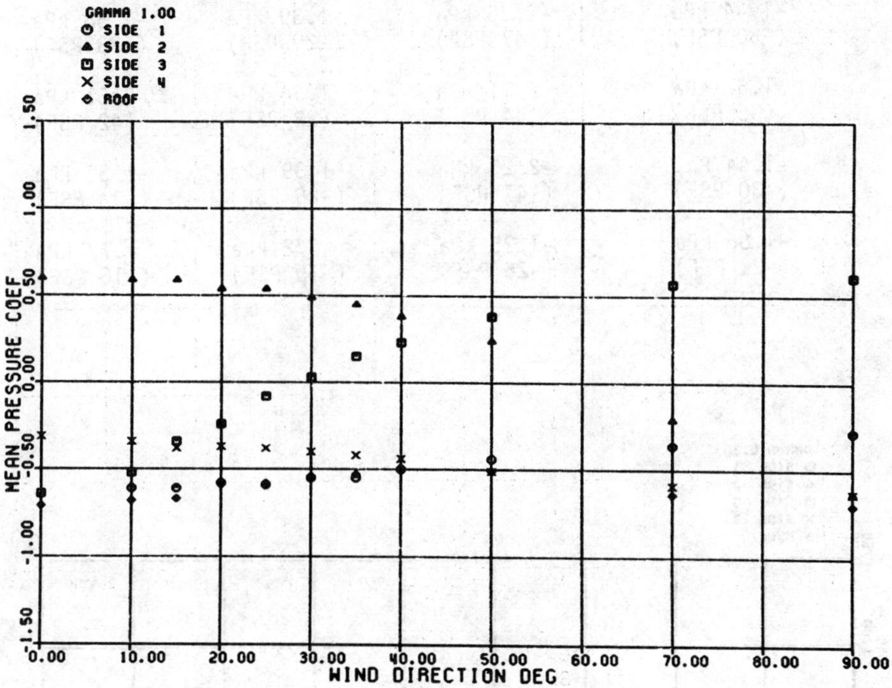

Figure 5. Averaged Presure Coefficients Based on a Reference Velocity Measured at the Roof, $\overline{C}_{PR}$, for a Side Ratio of 1.0

peak pressures are beyond the scope of this paper. The example structure is assumed to be located in an exposure with a power-law exponent of 0.22 (exposure B of Reference 6) and to experience a wind of 40 m/s (90 mph) at an elevation of 9 m (30 ft). Four separate pressures were computed for each side for a building with a side ratio of 1.0 and an incident wind normal to side 2. These four cases are: (1) using pressure coefficients from Ref. 6, (2) using $\overline{C}_{PL}$, (3) using $\overline{C}_{PR}$ and (4) using data from Ref. 5. For the purpose of this comparison, the data from Ref. 5 will be assumed to be the actual pressure at a selected location. The comparison was conducted at eight locations, in the middle of each side horizontally and at elevations representing 20% and 80% of the height of the building. The results as shown in Tables 3 and 4. At the 80% of building height level, the ANSI values have the best agreement, the $\overline{C}_{PR}$ values are low and the $\overline{C}_{PL}$ are high when compared with the actual values. At the 20% of building height level all predicted values are less than the actual values. The values predicted using $\overline{C}_{PL}$ and $\overline{C}_{PR}$ are closer to the actual values than those predicted using the coefficients from the ANSI standard.

TABLE 3. Comparison of predicted mean pressures, side ratio 1.0, elevation equal to 80% of building height

SIDE	ANSI A58.1[6]	$\overline{C}_{PL}$	$\overline{C}_{PR}$	ACTUAL VALUE[5]
1	-1.44 kPa (-30 PSF)	-2.25 kPa (-47 PSF)	-1.39 kPa (-29 PSF)	-1.58 kPa (-33 PSF)
2	1.67 kPa (35 PSF)	2.11 kPa (44 PSF)	1.34 kPa (28 PSF)	2.01 kPa (42 PSF)
3	-1.44 Pa (-30 PSF)	-2.25 kPa (-47 PSF)	-1.39 kPa (-29 PSF)	-1.58 kPa (-33 PSF)
4	-1.06 kPa (-22 PSF)	-1.25 kPa (-26 PSF)	-.72 kPa (-15 PSF)	-.77 kPa (-16 PSF)

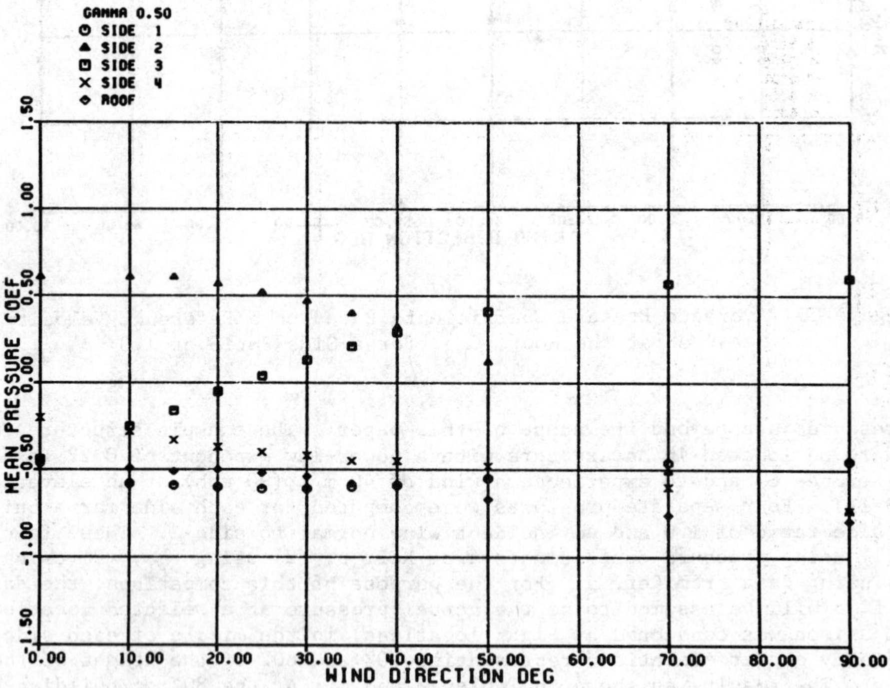

Figure 6. Averaged Pressure Coefficients Based on a Reference Velocity Measured at the Roof, $\overline{C}_{PR}$, for a Side Ratio of 0.5

TABLE 4. Comparison of predicted mean pressures, side ratio 1.0, elevation equal to 20% of building height

SIDE	ANSI A58.1 (6)	$\overline{C}_{PL}$	$\overline{C}_{PR}$	ACTUAL VALUE (5)
1	-.77 kPa (-16 PSF)	-1.24 kPa (-26 PSF)	-1.39 kPa (-29 PSF)	-1.58 kPa (-33 PSF)
2	.91 kPa (19 PSF)	1.19 kPa (25 PSF)	1.34 kPa (28 PSF)	1.53 kPa (32 PSF)
3	-.77 kPa (-16 PSF)	-1.24 kPa (-26 PSF)	-1.39 kPa (-29 PSF)	-1.58 kPa (-33 PSF)
4	-.53 kPa (-11 PSF)	-.67 kPa (-14 PSF)	-.72 kPa (-15 PSF)	-.81 kPa (-17 PSF)

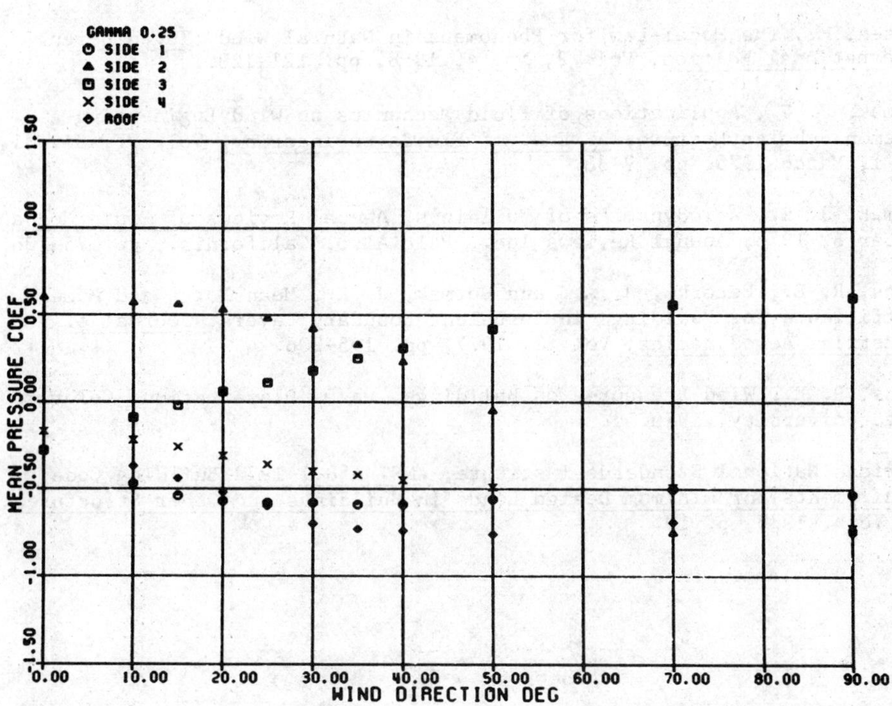

Figure 7. Averaged Pressure Coefficients Based on a Reference Velocity Measured at the Roof, $\overline{C}_{PR}$, for a Side Ratio of 0.25

SUMMARY

Mean pressure coefficients spatially averaged over an entire side of a building have been computed from a large volume of data obtained in wind-tunnel tests using a boundary-layer wind tunnel. These coefficients were determined in terms of a local pressure coefficient, $\overline{C}_{PL}$, based upon a reference wind speed measured in the approach flow at the height of the pressure measurement and in terms of a pressure coefficient, $\overline{C}_{PR}$, based upon the wind speed measured in the approach flow at a height equal to the height of the building. These coefficients were obtained for a range of wind directions from 0° to 90° for a series of flat-roofed rectangular buildings. Based on a limited comparison the errors involved in using either type of averaged coefficient seem comparable and there may be some justification for selectng $\overline{C}_{PR}$ for use in this context due to the need to only select one reference wind speed. Such a choice is appropriate only for spatially averaged mean pressure coefficients. When looking at actual distributions of pressure on a face, the use of a local pressure coefficient is a more useful concept.

ACKNOWLEDGEMENTS

The experimental work used to obtain the results presented was supported by National Science Foundation Grant ENG-76-03135. Support for preparation of the paper was provided by the Department of Engineering Science and Mechanics, Virginia Polytechnic Institute and State University.

REFERENCES

L. Jensen, M., The Model-Law for Phenomena in Natural Wind, Ingenioeren, International Edition, Vol. 2, No. 4, 1958, pp. 121-128..

2. Cermak, J. E., Applications of Fluid Mechanics to Wind Engineering-A Freeman Scholar Lecture, ASME J. of Fluids Engineering, Vol. 97, Ser. 1, No. 1, March 1975, pp. 9-38.

3. Cermak, J. E., Aerodynamics of Buildings, Annual Reviews of Fluid Mechanics, Number 8, 1976, Annual Reviews Inc., Palo Alto, California., pp. 75-106.

4. Akins, R. E., Peterka, J. A., and Cermak, J. E., Mean Force and Moment Coefficients for Buildings in Turbulent Boundary Layers, Journal of Industrial Aerodynamics, Vol. 2, 1977, pp. 195-206.

5. Akins, R. E., Wind Pressures on Buildings, Ph.D. Dissertation, Colorado State University, 1976.

6. American National Standards Institute, ANSI A58.1-1972 Building Code Requirements for Minimum Design Loads in Buildings and Other Structures, New York, 1972, p. 19.

INTERACTION EFFECTS IN NEIGHBOURING TALL BUILDINGS

J. BLESSMANN and J. D. RIERA
Pós-Graduacâo, Eng. Civil - UFRGS, Porto Alegre-RS, Brazil

ABSTRACT

Wind forces used in tall buildings structural design are usually determined on the basis of experimental results obtained for isolated models. The influence of urban environment is then taken into account in the selection of parameters for the vertical wind profile and longitudinal wind turbulence.

Neighbouring buildings, however, may exhibit significant interaction, especially with regards to torsional effects. Taking into account that the problem has not yet been considered in the specification of aerodynamic coefficients for current building design, an extensive interaction study between two square prisms was conducted at the LAC, UFRGS. Fifteen relative prism positions were studied, for 24 different orientations of wind incidence, both under uniform as well as shear wind. Mean force, torsional moment coefficients and point of application of resultant mean forces are given. The practical implications of the results so far obtained are also discussed.

1. INTRODUCTION

The wind forces used in the structural analysis & design of tall buildings are determined, in normal engineering practice, on the basis of aerodynamic coefficients recommended by wind Codes, which are generally the result of experimental studies conducted on isolated models. The influence of a urban environment is taken into account in the selection of both the vertical profile for the mean wind velocity and the parameters that define the wind turbulence.

Nevertheless, neighbouring buildings may exhibit significant interaction phenomena, especially with regards to torsional effects. Back in 1934, Harris[1] found by means of a wind tunnel study that two blocks planned to be built in the vicinity of the Empire State Building in New York City, would practically double the torsional wind loading in the latter, throughout the height of the proposed new buildings. Several authors, such as Chien 'et al'[2], Hamilton[3], Scruton & Newberry[4], Blessmann[5] and Leutheuser[6] observed that in two prismatic buildings located close to each other, suctions may attain values two to three times higher than those observed in the isolated edifices.

Kelnhofer[7] studied the influence of a neighbouring construction on the suction on the flat top of a prismatic building. Ishizaki e Sung[8] and Wiren[9] measured the air velocity between opposing faces of closely separated buildings. It was found that, under certain conditions, the velocity can be significantly higher than the wind velocity in the free-field because of Venturi-type effects.

Reinhold, Tieleman and Maher[10], in a more recent paper, studied the interaction between two square prisms in bidimensional flow, both for uniform as well as turbulent wind. Coefficients for the mean and fluctuating drag and lateral forces as well as for the torsional moment are given for 4 relative positions of the two prisms and two orientations of wind incidence in each case. It should be emphasized that the contribution by Reinhold 'el al' constitutes the only paper in which information on force and moment coefficients could be found, in spite of the great significance of the subject in actual practice. Unfortunately, the limited number of cases reported by the authors of Ref.(10), and the fact that their results are based on pressure readings at only three points on each face of the model, do not permit drawing general conclusions from their work.

The experimental evidence already available, however, strongly suggests that aerodynamic coefficients for tall building design in an urban environment should somehow incorporate the potential influence of neighbouring constructions. Of particular concern are torsional effects on buildings within the 30 to 40 stories high range --- for which complete wind tunnel studies are seldom made --- located along a water front or an open area, such as a park.

With the purpose of contributing to clarify this problem, an extensive and systematic interaction study between two square prisms both in uniform and turbulent shear flow was undertaken at the LAC (Laboratório de Aerodinâmica das Construções) of the Federal University of Rio Grande do Sul, Brazil (UFRGS). Fifteen relative prism positions were studied, with the orientation of the incident wind varying at 15° intervals. Pressure coefficients for uniform wind were read at about 67000 points. Studies under turbulent shear flow are still in progress. The test data was processed at the Computer Center of the UFRGS, with the objective of calculating (a) force coefficients for each face of the building, (b) force and torsional moment coefficients for the entire structure and (c) point of application of the resultant forces.

Preliminary results for in-line buildings under uniform flow were presented by the authors in Ref. (11). Complete results for uniform flow and partial ones for turbulent shear flow are reported in this paper. The practical significance of those results is also discussed. The study is actually being continued with a series of tests using aeroelastic models.

2. DESCRIPTION OF THE TESTS

The tests corresponding to uniform flow were conducted in the TV-1 Wind Tunnel of the LAC, with a boundary layer 50 mm (2.0 in) high in the area where the models were located. The intensity of turbulence in this region, outside the boundary layer, is about 0.6%, and Reynolds Number, referred to side length of model, equals 140000.

The tests corresponding to turbulent shear flow, on the other hand, were conducted in the modified TV-2 LAC Wind Tunnel, which is partially shown in Fig. 2.1. Profiles of the mean wind velocity (power law exponent 0.34), intensity and integral scale of turbulence along the vertical axis of the TV-2 turntable are given in Fig. 2.2. For turbulent shear flow, Re corresponding to the top of the model was 127000.

Fig.2.1. Partial view of LAC TV-2 Wind Tunnel

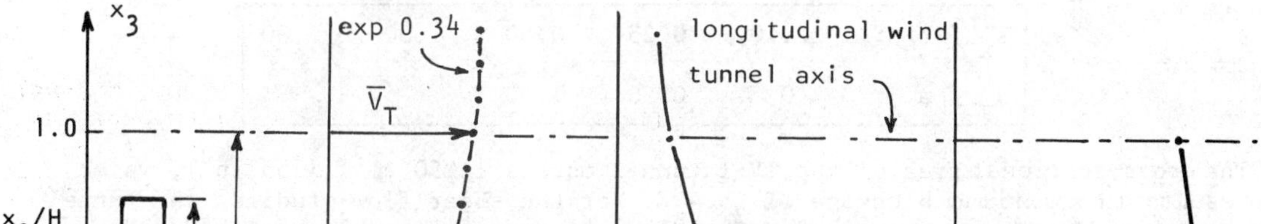

Fig. 2.2. Properties of wind at the test section

Two square prisms of dimension 60x60x360 mm (2.36x2.36x14.17 in) were used. The side to height ratio was therefore 1 x 6. One of the models had two adjoining faces instrumented with forty pressure taps distributed in rectangular grids with 8 lines and 5 columns. Plastic tubes with a lenght of about 120 cm (4.7 in) and an internal diameter of 1 mm (0.039 in) are directly attached to the walls of the model, resulting in pressure holes with a diameter slightly smaller than 1 mm (0.039 in). The former are connected to the manometer glass tubes ($\emptyset_i$ = 5mm --- 0.20 mm) by means of about 500 cm (200 in) long tubes with $\emptyset_i$ = 2 mm (0.079 in). The models were located in plan as shown in Fig.2.3. The separation distances in the x and y directions, denoted by s_x and s_y, were varied according to all possible combinations of the values given in Table 2.1.

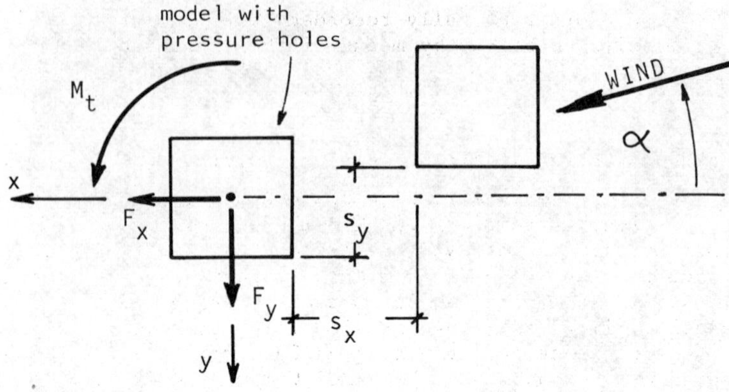

Fig. 2.3. Notations and sign conventions

Table 2.1 Relative positions studied

s_x / a :	0.10	0.25	0.50	1.00	2.00
s_y / a :	0	0.50	1.00		

The cross-sectional area of the TV-1 tunnel equals 11250 cm^2 (1755 in^2), which results in a maximum blockage of 5.4 %. For the shear flow studies, the tunnel cross-sectional area was 11700 cm^2 (1813 in^2) and maximum blockage 5.2%. No correction for blockage was made, which results in an overestimation of the pressure coefficients that varies between 1% and 3%, depending on the models relative position and the angle of wind incidence.

Fig. 2.4 shows the models in the wind tunnel. The blocks and the spires used to obtain the turbulent shear flow can also be seen.

Fig. 2.4. View of model in wind tunnel

3. PROCEDURE FOR DATA RECORDING AND PROLESSING

The test results were photographically recorded. The pressure coefficients were then obtained from enlarged prints by means of the expression:

$$c_{pe} = \Delta p_e \, / \, q \tag{3.1}$$

in which Δp_e = effective static pressure at point under consideration;
q = reference dynamic pressure, measured with the aid of the piezometric rings on the wind tunnel.

The force and moment coefficients were calculated by numerically integrating the pressures recorded on each face of the model. To that end, an integration algorithm that uses interpolating cubic parabolas was developed and programmed in FORTRAN. The program allows the consideration of pressure readings at arbitrarily spaced points in both directions of a rectangular grid. The data were processed in the Burroughs B 6700 computer of the UFRGS Computer Center, to obtain the following coefficients (see Fig. 2.3 for sign convention):

(a) force coefficients, defined by

$$C_x = \frac{F_x}{qA} \qquad C_y = \frac{F_y}{qA} \tag{3.2}$$

F_x = total wind force in x-direction;
F_y = total wind force in y-direction;
q = incident wind dynamic pressure (in free stream);
A = area of building face = ah.

(b) torsional moment coefficient, defined by

$$C_{Mt} = \frac{M_t}{q\ A\ a} \tag{3.3}$$

M_t = total torsional moment produced in building by the wind pressure;
a = lenght of building base.

The coefficients defined above are given in Tables A.1-3 for uniform flow and in Table A.4 for turbulent shear flow. Additionally, Figs. A.1 - 6 show the variation of the force coefficients with angle of incidence.

It seems germaine to comment upon an interesting effect observed in the tests with aligned models ($s_y = 0$) and wind incidence parallel to the facing walls (Figs. A.2 & A.3). From s_x = 0.25a up to s_x = a the readings in all pressure taps of the instrumented fãcing wall presentẽd, at approximately constant time intervals --- about four seconds --- sudden and uniform decreases, which lasted for about one second. The Figs. A.2 & A.3 indicate the limiting values of the oscillating forces.

4. ANALYSIS AND DISCUSSION OF RESULTS

The horizontal components of the total wind force as well as the torsional moment on the instrumented model were significantly influenced by the presence of an upstream model of identical dimensions even for the largest separation tested, in which the distance between vertical axes equals three times the side lenght.

For the downstream building, the total horizontal force, given by

$$F = q\,A\sqrt{C_x^2 + C_y^2} = q\,A\,C \qquad (4.1)$$

is usually smaller --- and even of opposite sign --- than that corresponding to the isolated model. For aligned models ($s_y = 0$), Fig. 4.1 shows the variation of C_x --- which in this case equals C --- with model separation s_x. Also shown are coefficients obtained by Reinhold 'et al'[10] for bidimensional flow as well as values corresponding to a section of the building located at a height equal to 0.4 h. It may be seen that the coefficients for the bidimensional model differ considerably from those obtained for the entire model. As might be expected, the approximation fares better when compared with the local aerodynamic coefficient c_x, calculated at 0.4 h.

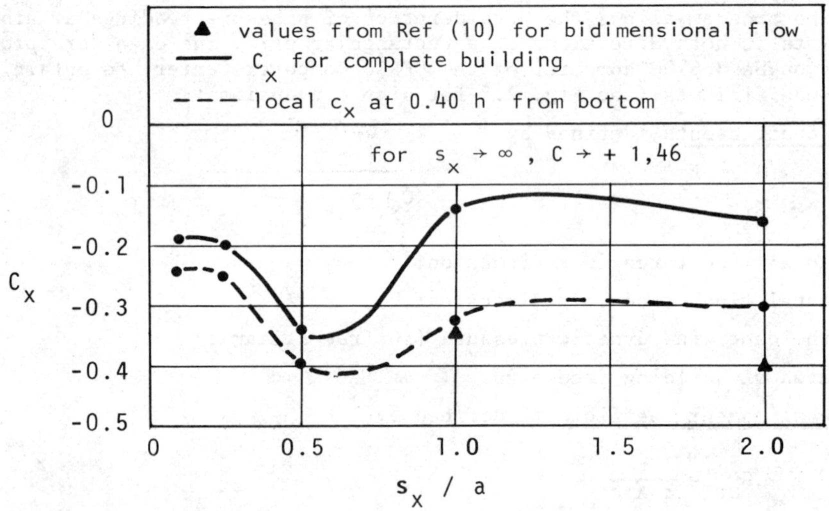

Fig. 4.1. Variation of C_x with model separation s_x, for $\alpha = 0^o$

The drag coefficient for the isolated model in uniform flow was found to be $C_x = 1.46$, which differs by only 2% from the value recommended by the Brazilian Wind Code[12] ($C_x = 1.43$). For turbulent flow, however, a coefficient $C_x = 0.88$ (referred to the wind velocity at the top of the model) was measured. Refs. (10) and (12), on the other hand, recommend, for highly turbulent flow, the value 1.08, but it is not clear whether it applies to the velocity at the top of the building or to the actual vertical profile. It is interesting to observe that if a simplified version of Baines' criterion is applied to the force coefficients determined for uniform flow, rough estimates of the forces corresponding to turbulent shear flow are obtained. The torsional moments, on the other hand, can not be estimated in a similar fashion. In fact, for the cases already studied, the turbulent shear flow did not seem to exert a significant overall influence on the torsional moments determined for uniform flow, thus confirming results from Akins 'et al'[15] for prismatic isolated buildings.

The total force on the upstream model may increase substantially in cases when the downstream model is not within its wake. For instance, for $s_x = 0.1a$ and $s_y = a$, the maximum value of the C coefficient increases 40% with respect to the isolated model.

Perhaps the most important aspect of the interaction problem are the substantial increases observed in the maximum torsional moment, which may be more than three times the values measured on the isolated model, as may be appreciated in Fig. 4.2. This figure indicates the maximum absolute torsional moment coefficients measured, regardless of angle of wind incidence, as a function of building separation in the x direction.

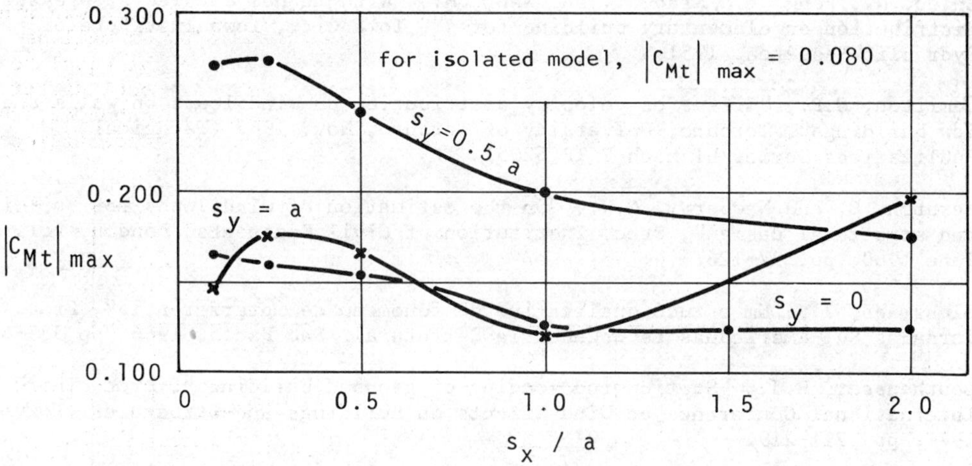

Fig. 4.2. Variation of maximum absolute torsional moment coefficient with separation s_x between models.

No explanation has yet been attempted for the cyclic variations of pressure between two models facing each other with wind incidence parallel to the orientation of the gap. It is expected that additional data obtained with aeroelastic models may help to explain this effect.

5. CONCLUSIONS

Results of an extensive interaction study between two squares prisms subjected to both uniform and turbulent shear flow have been presented. They show that the presence of neighbouring buildings may cause increases in the horizontal wind forces of up to 40%. Even more important is the increase observed in the total torsional moments, which may exceed three times the maximum values corresponding to the isolated construction. These interaction effects, which do not seem to be greatly influenced by the turbulence of the incident wind, ought to be taken into account in the selection of aerodynamic coefficients for tall building design. Of particular concern are torsional effects on buildings within the 30 to 40 stories high range -- for which complete wind tunnel studies are seldom made-- located along a water front or an open area, such as a park.

ACKNOWLEDGMENTS

The work described in this paper was partially supported by the Conselho Nacional de Desenvolvimento Científico e Tecnológico do Brasil (Proc. 1111.2551/76). The authors wish to acknowledge the cooperation of Mr. Paulo F. Bueno during the wind tunnel testing and the general support of the UFRGS Post-Graduate Program in Civil Engineering.

REFERENCES:

1. Harris, C.L., "Influence of neighboring structures on the wind pressure on tall buildings", Bureau of Standards Journal of Research, Vol. 12, Jan 1934 (Research Paper RP 637).

2. Chien, N., Feng, Y. ,Siao, T. and Wang, H., "Wind-tunnel studies of pressure distribution on elementary building forms", Iowa City, Iowa Institute of Hydraulic Research, 1951.

3. Hamilton, G.F., "Effect of velocity distribution on wind loads on walls and low buildings", Toronto, University of Toronto, Nov. 1962 (Technical Publications Series UT Mech E TP 6205).

4. Scruton, C. and Newberry, C.W., "On the estimation of wind loads for building and structural design", Proc. Institution of Civil Engineers, London, Vol. 25, June 1963, pp. 97-126.

5. Blessmann, J., "Um estudo qualitativo do fenômeno de interferência", Proc. XI Jornadas Sul Americanas de Engenharia Estrutural, São Paulo, 1966, pp.585-600.

6. Leutheusser, H.J., "Static wind loading of grouped buildings", Proc. Third International Conference on Wind Effects on Buildings and Structures, Tokyo, 1971, pp. 211-219.

7. Kelnhofer, W.J., "Influence of a neighboring building on flat roof wind loading", Proc. Third International Conference on Wind Effects on Buildings and Structures, Tokyo, 1971, pp. 221-230.

8. Ishizaki, H. and Sung, I.W., "Influence of adjacent buildings to wind", Proc. Third International Conference on Wind Effects on Buildings and Structures, Tokyo, 1971, pp. 145-152.

9. Wirén, B.G., "A wind tunnel study of wind velocities in passages between and through buildings", Proc. Fourth International Conference on Wind Effects on Buildings and Structures, London, 1975, pp. 465-475.

10. Reinhold, T.A., Tieleman, H.W. and Maher, F.J., "Interaction of square prisms in two flow fields", Journal of Industrial Aerodynamics, Vol. 2, Nº 3, Nov. 1977, pp. 223-241.

11. Blessmann, J. and Riera, J.D., "Interação em edifícios altos vizinhos submetidos a ação do vento", Proc. XIX Jornadas Sul Americanas de Engenharia Estrutural, Santiago, Chile, April 1978, Vol. 1, Paper A-9.

12. Associação Brasileira de Normas Técnicas (ABNT): "Forças devidas ao vento em edificações--- NB-599", 1977 (Brazilian Wind Code).

13. Newberry, C.W. and Eaton, K.J., "Wind loading handbook", Building Research Establishment Report, London, 1974, pp.42-43.

14. Baines, W.D., "Effects of velocity distribution on wind loads and flow patterns on buildings", Proc. Firth International Conference on Wind Effects on Buildings and Structures, National Physical Laboratory (Symposium Nº 16), 1963, Paper 6, pp. 197-223.

15. Akins, R.E., Peterka, J.A. and Cermak, J.E., "Mean force and moment coefficients for buildings in turbulent boundary layers", Journal of Industrial Aerodynamics, Vol. 2, Nº 3, Nov. 1977, pp. 195-209.

APPENDIX A

Table A.1 Force and torsional moment coefficients for uniform flow, $s_y = 0$.

α^o	Isolated model			$s_x = 2.0a$; $s_y = 0$			$s_x = 1.0a$; $s_y = 0$		
	C_x	C_y	C_{Mt}	C_x	C_y	C_{Mt}	C_x	C_y	C_{Mt}
0	+1.46	0.00	0.000	-0.16	0.00	0.000	-0.14	0.00	0.000
15	+1.39	-0.38	+0.080	+0.72	+0.45	+0.126	+0.32	+0.46	+0.078
30	+1.25	+0.53	+0.064	+0.92	+0.58	+0.077	+0.54	+0.66	+0.075
45	+1.00	+1.00	0.000	+0.80	+0.90	+0.018	+0.54	+0.88	+0.035
60	+0.53	+1.25	-0.064	+0.48	+1.17	-0.045	+0.34	+1.18	-0.028
75	-0.38	+1.39	-0.080	-0.14	+1.36	-0.126	-0.06	+1.28	-0.058
90	0.00	+1.46	0.000	-0.19	+1.54	-0.048	-0.20	+1.51	-0.126
105	+0.38	+1.39	+0.080	+0.25	+1.64	+0.034	+0.08	+1.63	+0.013
120	-0.53	+1.25	+0.064	-0.49	+1.48	+0.090	-0.44	+1.46	+0.095
135	-1.00	+1.00	0.000	-1.04	+1.17	+0.004	-1.05	+1.18	-0.004
150	-1.25	+0.53	-0.064	-1.26	+0.62	-0.048	-1.28	+0.66	-0.048
165	-1.39	-0.38	-0.080	-1.28	-0.34	-0.069	-1.35	-0.43	-0.079
180	-1.46	0.00	0.000	-1.42	0.00	0.000	-1.54	0.00	0.000

α^o	$s_x=0.5a$; $s_y = 0$			$s_x = 0.25a$; $s_y = 0$			$s_x = 0.1a$; $s_y = 0$		
	C_x	C_y	C_{Mt}	C_x	C_y	C_{Mt}	C_x	C_y	C_{Mt}
0	-0.34	0.00	0.000	-0.20	0.00	0.000	-0.19	0.00	0.000
15	+0.08	+0.46	+0.010	-0.12	+0.47	-0.062	-0.07	+0.44	-0.058
30	+0.18	+0.60	+0.031	-0.27	+0.62	-0.046	-0.31	+0.58	-0.144
45	+0.14	+0.86	+0.008	-0.40	+0.88	-0.077	-0.34	+0.84	-0.165
60	-0.05	+1.10	-0.035	-0.36	+1.10	-0.087	-0.24	+1.08	-0.139
75	-0.40	+1.26	-0.114	-0.48	+1.26	-0.160	-0.17	+1.28	-0.145
90	-0.44*	+1.56	-0.154	-0.42	+1.42	-0.148	-0.16	+1.42	-0.098
105	-0.34	+1.54	-0.040	-0.24	+1.55	-0.116	-0.11	+1.59	-0.105
120	-0.64	+1.45	+0.065	-0.78	+1.44	-0.022	-0.46	+1.48	-0.032
135	-1.16	+1.18	-0.012	-1.44	+1.22	-0.099	-1.27	+1.28	-0.152
150	-1.38	+0.67	-0.049	-1.63	+0.66	-0.112	-1.52	+0.75	-0.166
165	-1.40	-0.48	-0.088	-1.46	-0.46	-0.099	-1.32	-0.28	-0.104
180	-1.57	0.00	0.000	-1.55	0.00	0.000	-1.52	0.00	0.000

* Average of extreme values. Pressure readings oscilating as described in Section 4.

Table A.2 Force and torsional moment coefficients for uniform flow, $s_y = 0.5$ a

α^o	Isolated model			$s_x = 2.0a; s_y = 0.5a$			$s_x = 1.0a; s_y = 0.5a$		
	C_x	C_y	C_{Mt}	C_x	C_y	C_{Mt}	C_x	C_y	C_{Mt}
0	+1.46	0.00	0.000	-0.06	-0.23	-0.029	-0.15	-0.15	-0.061
15	+1.39	-0.38	+0.080	+0.06	+0.69	-0.041	-0.23	+0.46	-0.015
30	+1.25	+0.53	+0.064	+0.50	+0.77	+0.034	-0.18	+0.89	-0.118
45	+1.00	+1.00	0.000	+0.60	+0.97	0.000	-0.01	+1.00	-0.077
60	+0.53	+1.25	-0.064	+0.31	+1.19	-0.073	-0.04	+1.18	-0.111
75	-0.38	+1.39	-0.080	-0.34	+1.34	-0.136	-0.35	+1.28	-0.078
90	0.00	+1.46	0.000	-0.01	+1.46	-0.009	-0.06	+1.34	-0.021
105	+0.38	+1.39	+0.080	+0.19	+1.64	+0.024	+0.03	+1.65	-0.079
120	-0.53	+1.25	+0.064	-0.43	+1.52	+0.098	-0.34	+1.55	+0.095
135	-1.00	+1.00	0.000	-1.04	+1.25	0.000	-1.01	+1.29	-0.004
150	-1.25	+0.53	-0.064	-1.25	+0.69	-0.052	-1.25	+0.79	-0.046
165	-1.39	-0.38	-0.080	-1.28	-0.33	-0.080	-1.25	-0.33	-0.080
180	-1.46	0.00	0.000	-1.42	-0.07	-0.007	-1.46	-0.19	-0.019
195	-1.39	+0.38	+0.080	-1.33	+0.38	+0.075	-1.36	+0.42	+0.078
210	-1.25	-0.53	+0.064	-1.23	-0.55	+0.046	-1.36	-0.60	+0.051
225	-1.00	-1.00	0.000	-1.03	-1.10	+0.011	-1.09	-1.13	-0.002
240	-0.53	-1.25	-0.064	-0.51	-1.41	-0.088	-0.53	-1.41	-0.086
255	+0.38	-1.39	-0.080	+0.30	-1.60	-0.045	-0.01	-1.45	-0.011
270	0.00	-1.46	0.000	-0.40	-1.61	+0.046	+0.32	-1.67	+0.094
285	-0.38	-1.39	+0.080	+0.07	-1.38	+0.115	+0.30	-1.29	+0.063
300	+0.53	-1.25	+0.064	+0.61	-1.16	+0.033	+0.73	-1.13	-0.006
315	+1.00	-1.00	0.000	+0.93	-0.85	-0.020	+0.90	-0.82	-0.050
330	+1.25	-0.53	-0.064	+1.08	-0.42	-0.094	+1.02	-0.33	-0.141
345	+1.39	+0.38	-0.080	+1.13	+0.10	-0.174	+0.03	+0.27	-0.198

α^o	$s_x = 0.5a; s_y = 0.5a$			$s_x = 0.25a; s_y = 0.5a$			$s_x = 0.1a; s_y = 0.5a$		
	C_x	C_y	C_{Mt}	C_x	C_y	C_{Mt}	C_x	C_y	C_{Mt}
0	-0.07	+0.06	-0.084	-0.05	+0.17	-0.085	0.00	+0.24	-0.118
15	-0.30	+0.13	+0.037	-0.37	-0.03	+0.030	-0.40	-0.10	+0.022
30	-0.07	+0.25	+0.071	-0.08	+0.07	+0.029	-0.12	-0.04	+0.008
45	-0.15	+1.06	-0.101	-0.04	+0.72	+0.067	-0.06	+0.37	+0.118
60	-0.22	+1.17	-0.103	-0.33	+1.14	-0.147	-0.20	+0.92	0.000
75	-0.26	+1.27	-0.067	-0.44	+1.22	-0.147	-0.42	+1.24	-0.148
90	-0.29	+1.59	-0.112	-0.44	+1.29	-0.136	-0.39	+1.36	-0.155
105	-0.28	+1.56	-0.072	+0.05	+1.61	-0.113	-0.12	+1.59	-0.130
120	-0.57	+1.44	+0.049	-0.08	+1.55	0.000	-0.08	+1.59	-0.036
135	-1.25	+1.25	-0.035	-0.90	+1.46	+0.060	-0.98	+1.56	-0.136
150	-1.42	+0.83	-0.070	-1.28	+1.10	+0.001	-1.37	+1.33	-0.183
165	-1.37	-0.29	-0.103	-1.46	+0.01	-0.160	-1.34	+0.52	-0.095
180	-1.45	-0.20	-0.011	-1.38	-0.26	-0.010	-1.37	-0.30	-0.012
195	-1.45	+0.38	+0.084	-1.48	+0.39	+0.088	-1.47	+0.33	+0.084
210	-1.27	-0.46	+0.061	-1.41	-0.52	+0.058	-1.42	-0.48	+0.066
225	-0.81	-1.13	+0.001	-0.88	-1.06	-0.015	-0.95	-0.98	+0.004
240	-0.34	-1.36	-0.008	-0.18	-1.43	-0.126	-0.08	-1.37	-0.112
255	+0.53	-1.58	+0.040	+0.50	-1.57	-0.066	+0.52	-1.52	-0.108
270	+0.12	-1.31	+0.029	+0.47	-1.54	-0.072	+0.49	-1.55	-0.117
285	+0.38	-1.31	-0.028	+0.49	-1.26	-0.089	+0.60	-1.32	-0.118
300	+0.63	-1.10	-0.065	+0.65	-1.12	-0.103	+0.66	-1.08	-0.123
315	+0.75	-0.73	-0.118	+0.71	-0.69	-0.159	+0.63	-0.63	-0.185
330	+0.85	-0.20	-0.208	+0.73	-0.02	-0.250	+0.64	+0.08	-0.261
345	+0.86	+0.31	-0.244	+0.66	+0.44	-0.272	+0.59	+0.50	-0.270

Table A.3 Force and torsional moment coefficients for uniform flow, $s_y = a$

α^o	Isolated model			$s_x=2.0a$; $s_y = 1.0a$			$s_x = 1.0a$; $s_y = 1.0a$		
	C_x	C_y	C_{Mt}	C_x	C_y	C_{Mt}	C_x	C_y	C_{Mt}
0	+1.46	0.00	0.000	+0.80	+0.31	-0.198	+1.11	-0.09	-0.100
15	+1.39	-0.38	+0.080	-0.19	+0.31	-0.022	-0.14	+0.06	-0.057
30	+1.25	+0.53	+0.064	-0.12	+0.94	-0.143	-0.13	+0.13	+0.034
45	+1.00	+1.00	0.000	+0.21	+1.03	-0.060	-0.13	+1.03	-0.020
60	+0.53	+1.25	-0.064	+0.08	+1.21	-0.117	-0.49	+1.22	-0.121
75	-0.38	+1.39	-0.080	-0.45	+1.39	-0.104	-0.12	+1.32	-0.051
90	0.00	+1.46	0.000	0.00	+1.45	-0.006	+0.02	+1.38	-0.015
105	+0.38	+1.39	+0.080	+0.20	+1.58	+0.027	+0.09	+1.41	+0.007
120	-0.53	+1.25	+0.064	-0.42	+1.54	+0.100	-0.25	+1.62	+0.107
135	-1.00	+1.00	0.000	-1.06	+1.30	-0.008	-1.01	+1.44	+0.008
150	-1.25	+0.53	-0.064	-1.31	+0.80	-0.046	-1.27	+1.03	-0.041
165	-1.39	-0.38	-0.080	-1.39	-0.23	-0.106	-1.41	+0.14	-0.104
180	-1.46	0.00	0.000	-1.36	-0.05	-0.007	-1.37	-0.12	-0.018
195	-1.39	+0.38	+0.080	-1.36	+0.42	+0.081	-1.35	+0.27	+0.084
210	-1.25	-0.53	+0.064	-1.28	-0.57	+0.053	-1.37	-0.59	+0.061
225	-1.00	-1.00	0.000	-1.03	-1.06	-0.005	-0.87	-1.01	+0.016
240	-0.53	-1.25	-0.064	-0.51	-1.39	-0.087	-0.58	-1.42	-0.081
255	+0.38	-1.39	-0.080	+0.26	-1.56	-0.041	+0.05	-1.41	-0.031
270	0.00	-1.46	0.000	-0.33	-1.58	+0.038	+0.22	-1.58	+0.093
285	-0.38	-1.39	+0.080	+0.22	-1.49	+0.002	+0.69	-1.61	+0.052
300	+0.53	-1.25	+0.064	+0.71	-1.20	+0.027	+0.84	-1.13	0.000
315	+1.00	-1.00	0.000	+1.00	-0.88	-0.020	+1.08	-0.76	-0.040
330	+1.25	-0.53	-0.064	+1.17	-0.31	-0.113	+1.19	-0.16	-0.121
345	+1.39	+0.38	-0.080	+1.23	+0.29	-0.128	+1.36	+0.25	-0.080

α^o	$s_x = 0.5a$; $s_y = 1.0a$			$s_x = 0.25a$; $s_y = 1.0a$			$s_x = 0.1a$ $s_y = 1.0a$		
	C_x	C_y	C_{Mt}	C_x	C_y	C_{Mt}	C_x	C_y	C_{Mt}
0	+1.21	-0.01	-0.065	+1.22	-0.01	-0.073	+1.23	+0.06	-0.065
15	+0.45	0.00	-0.167	+0.64	0.00	-0.163	+0.62	+0.08	-0.133
30	-0.18	-0.06	+0.002	-0.20	-0.11	-0.004	-0.20	-0.08	-0.013
45	-0.05	+0.05	+0.034	-0.12	-0.14	0.000	-0.10	-0.13	-0.004
60	-0.02	+0.96	+0.073	+0.02	+0.39	+0.124	-0.03	+0.07	+0.077
75	-0.08	+1.33	-0.035	-0.09	+1.14	+0.050	-0.18	+0.14	+0.042
90	-0.05	+1.33	-0.020	-0.04	+1.34	-0.022	-0.26	+1.29	-0.063
105	-0.29	+1.61	-0.044	0.00	+1.61	-0.110	-0.01	+1.52	-0.088
120	-0.14	+1.59	+0.092	0.00	+1.56	-0.157	+0.31	+1.57	-0.004
135	-0.99	+1.52	-0.004	-1.00	+1.47	-0.074	-0.69	+1.59	-0.064
150	-1.32	+1.25	+0.124	-1.43	+1.32	-0.103	-1.17	+1.52	-0.119
165	-1.50	+0.72	-0.096	-1.61	+0.93	-0.175	-1.42	+1.24	-0.148
180	-1.28	-0.34	-0.017	-1.47	-0.06	-0.107	-1.49	+0.57	-0.072
195	-1.14	-0.16	+0.072	-1.12	-0.40	+0.043	-1.32	+0.06	+0.020
210	-1.41	-0.68	+0.062	-1.33	-0.70	+0.058	-1.24	-0.74	+0.058
225	-1.14	-1.02	0.000	-1.18	-1.09	-0.001	-1.11	-1.06	+0.002
240	-0.16	-1.29	-0.031	-0.34	-1.23	-0.051	-0.60	-1.25	-0.061
255	+0.26	-1.35	+0.025	+0.63	-1.46	+0.048	+0.79	-1.48	+0.020
270	+0.76	-1.58	+0.076	+0.98	-1.57	+0.067	+0.97	-1.52	+0.063
285	+0.77	-1.27	+0.045	+1.07	-1.29	+0.050	+1.26	-1.32	+0.069
300	+1.07	-1.08	+0.004	+1.27	-1.13	+0.018	+1.42	-1.11	+0.051
315	+1.20	-0.68	-0.043	+1.35	-0.60	-0.039	+1.48	-0.50	-0.015
330	+1.25	+0.14	-0.121	+1.33	+0.31	-0.095	+1.45	+0.40	-0.055
345	+1.33	+0.08	-0.043	+1.36	+0.03	-0.025	+1.45	+0.06	+0.013

Table A.4 Force and torsional moment coefficients for turbulent shear flow

α^o	Isolated model			$s_x = 0.1a$; $s_y = 0.5a$		
	C_x	C_y	C_{Mt}	C_x	C_y	C_{Mt}
0	+0.90	0.00	0.000	+0.06	-0.02	-0.099
15	+0.89	+0.02	+0.081	-0.26	0.00	+0.011
30	+0.84	+0.45	+0.036	-0.14	+0.11	+0.035
45	+0.72	+0.72	0.000	-0.07	+0.38	+0.058
60	+0.45	+0.84	-0.036	-0.15	+0.67	-0.022
75	-0.02	+0.89	-0.081	-0.21	+0.87	-0.102
90	0.00	+0.90	0.000	-0.20	+0.98	-0.121
105	-0.02	+0.89	+0.081	+0.23	+0.61	-0.060
120	-0.45	+0.84	+0.036	-0.23	+1.10	-0.041
135	-0.72	+0.72	0.000	-0.72	+1.05	-0.110
150	-0.84	+0.45	-0.036	-0.94	+0.93	-0.136
165	-0.89	+0.02	-0.081	-0.95	+0.43	-0.109
180	-0.90	0.00	0.000	-0.86	+0.03	-0.025
195	-0.89	-0.02	-0.081	-0.89	-0.14	+0.060
210	-0.84	-0.45	-0.036	-0.90	-0.44	+0.040
225	-0.72	-0.72	0.000	-0.61	-0.70	-0.002
240	-0.45	-0.84	+0.036	-0.18	-0.93	-0.063
255	-0.02	-0.89	+0.081	+0.51	-1.04	-0.116
270	0.00	-0.90	0.000	+0.45	-1.10	-0.065
285	+0.02	-0.89	-0.081	+0.38	-0.90	-0.057
300	+0.45	-0.84	-0.036	+0.42	-0.76	-0.074
315	+0.72	-0.72	0.000	+0.39	-0.50	-0.110
330	+0.84	-0.45	+0.036	+0.38	-0.09	-0.166
345	+0.89	-0.02	+0.081	+0.35	+0.11	-0.175

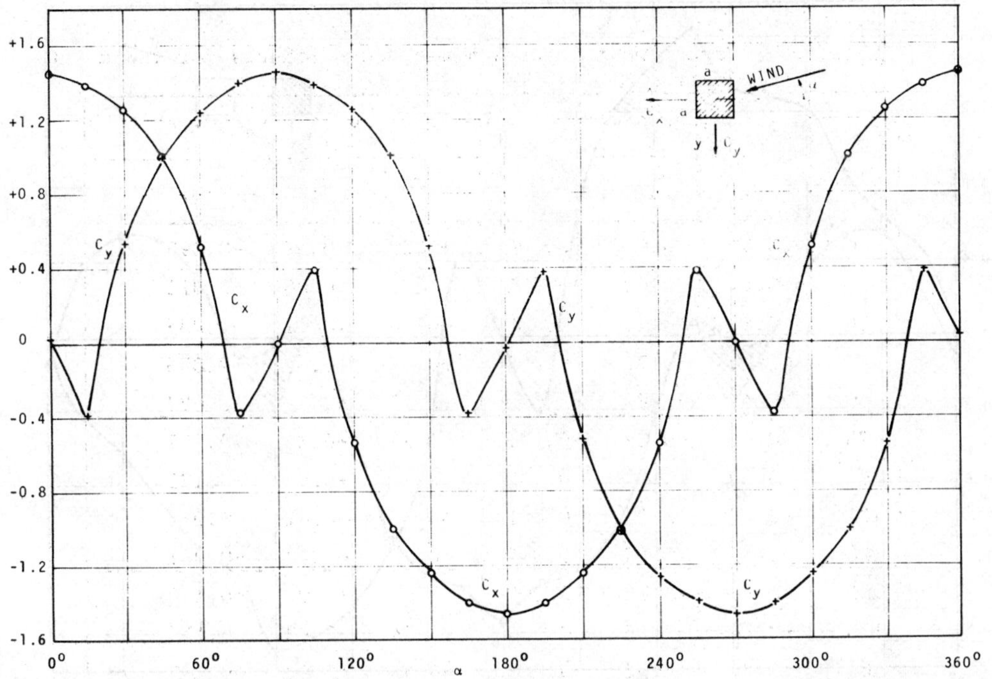

Fig. A.1. Variation of force coefficient with angle of incidence. Uniform wind, isolated model.

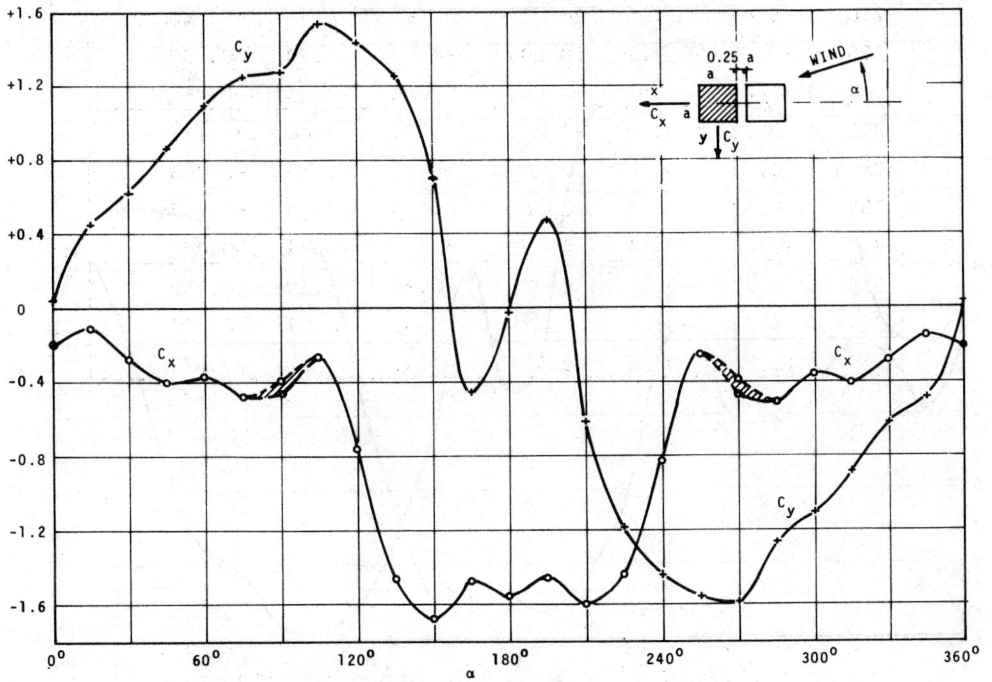

Fig. A.2. Variation of force coefficient with angle of incidence. Uniform wind, $s_x = 0$, $s_y = 0.25$ a.

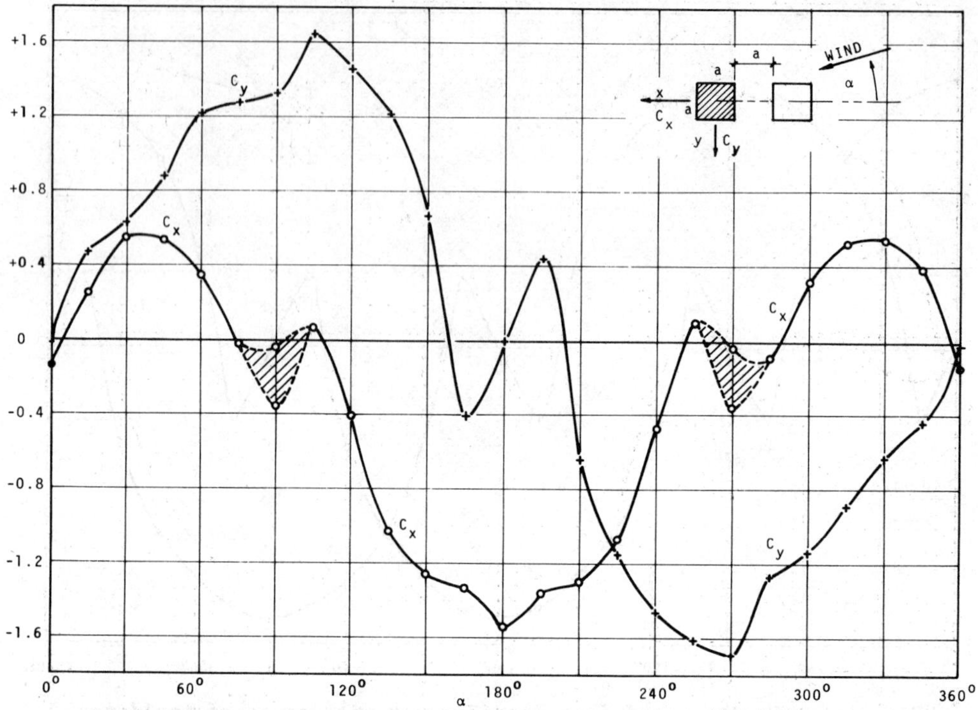

Fig. A.3. Variation of force coefficient with angle of incidence. Uniform wind, $s_x = 0$, $s_y = a$.

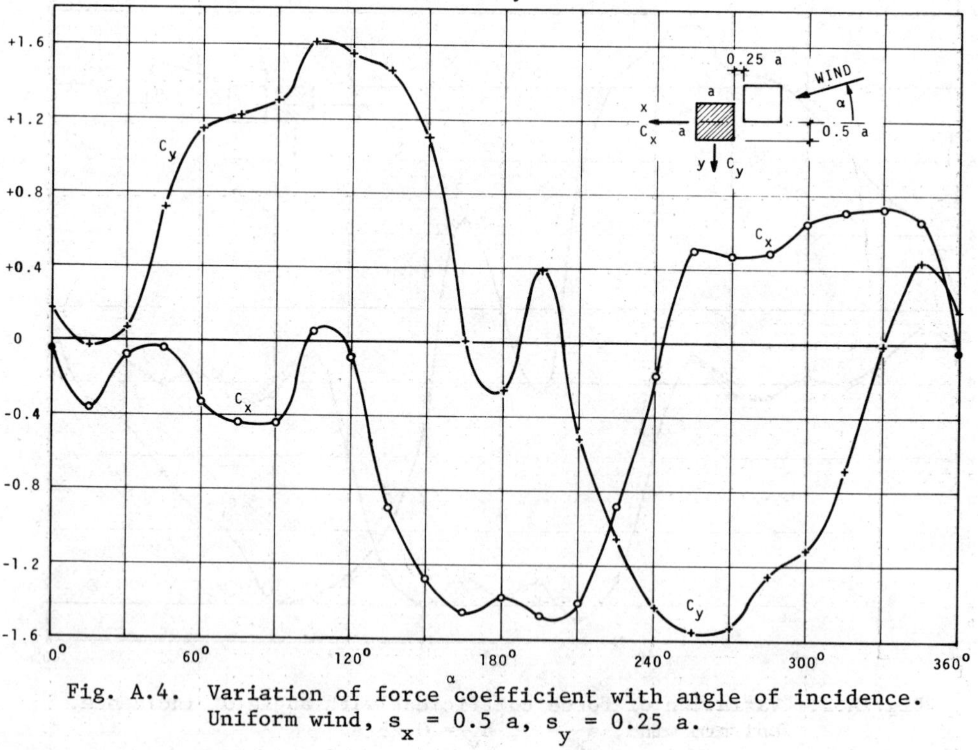

Fig. A.4. Variation of force coefficient with angle of incidence. Uniform wind, $s_x = 0.5\ a$, $s_y = 0.25\ a$.

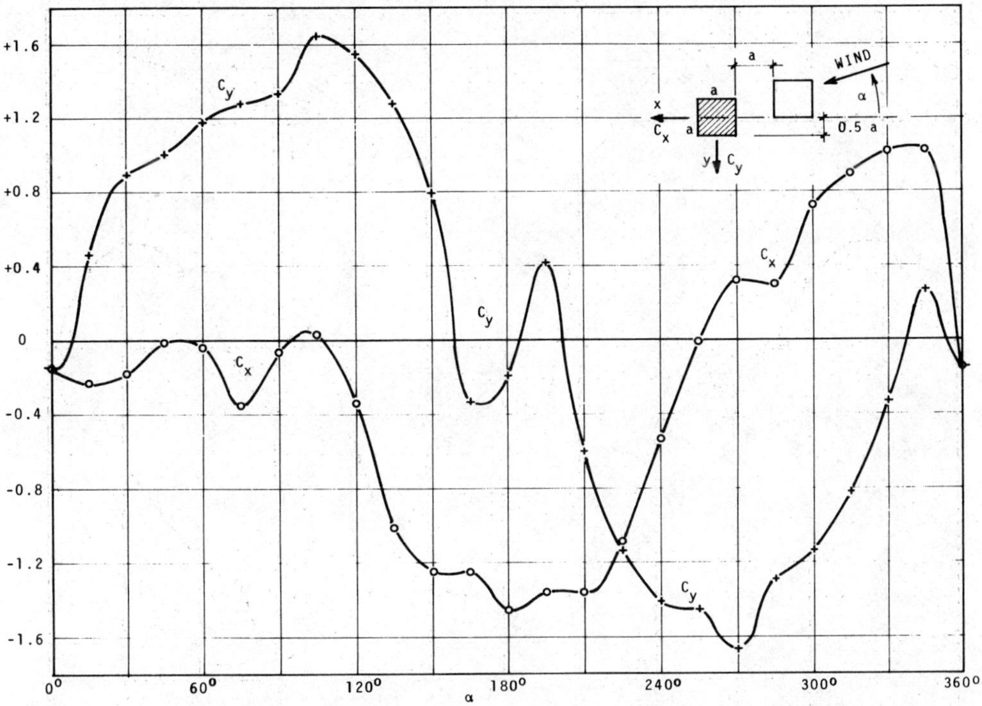

Fig. A.5. Variation of force coefficient with angle of incidence. Uniform wind, s_x = a, s_y = 0.5 a.

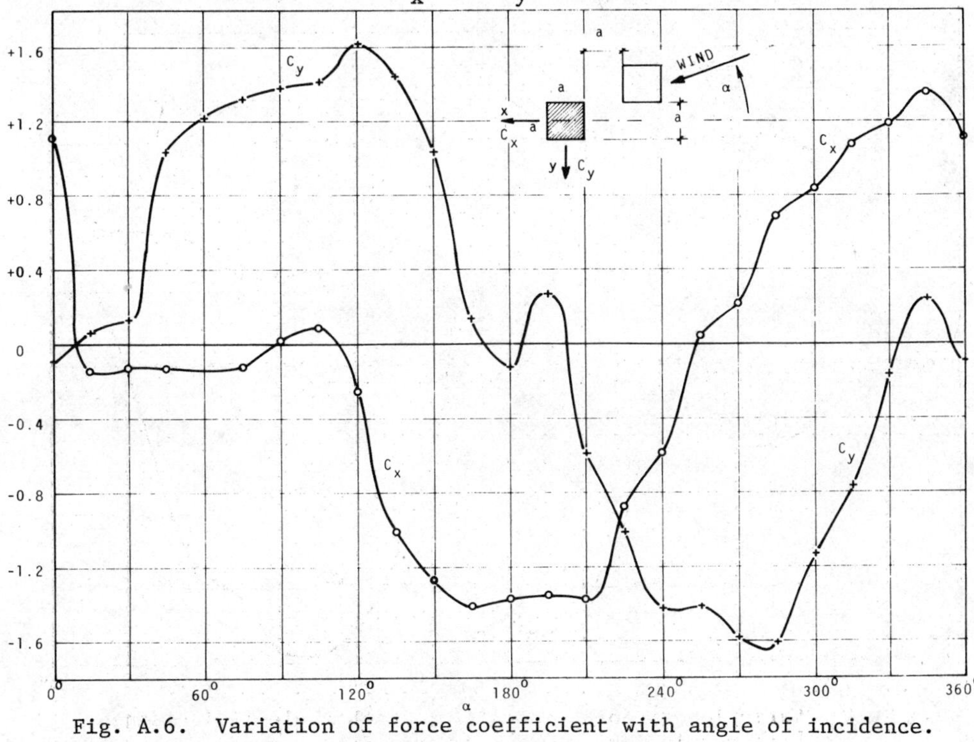

Fig. A.6. Variation of force coefficient with angle of incidence. Uniform wind, s_x = a, s_y = a.

THE EFFECT OF INTENSITY AND LARGE-SCALE TURBULENCE ON THE MEAN PRESSURE AND DRAG COEFFICIENTS OF 2D RECTANGULAR CYLINDERS

A. LANEVILLE* and C. D. WILLIAMS**

*University of Sherbrooke, Sherbrooke, Quebec, Canada

**National Research Council of Canada, Ottawa, Ontario, Canada

SUMMARY

Tests were performed to investigate the effect of intensity and large scale turbulence on the drag coefficient of two-dimensional rectangular cylinders. The results show that the effect of the scale of turbulence is not significant while the effect of turbulence intensity is important.

NOMENCLATURE

D	frontal dimension of the cylinder
H	longitudinal dimension of the cylinder
C_p	measured mean pressure coefficient
C_{pb}	mean base pressure coefficient
C_D	measured mean drag coefficient
C_{pbc}	corrected base pressure coefficient
C_{Dc}	drag coefficient corrected for tunnel blockage
L_x	longitudinal integral scale of turbulence
I	longitudinal intensity of turbulence $(u'/\overline{U})$
u'	rms longitudinal wind speed
$\overline{U}$	mean longitudinal wind speed
C_{ps}	corrected side pressure coefficient near the separation point.

INTRODUCTION

A useful conventional measure of the size of the turbulence is the integral turbulence scale L_x and, for atmospheric flows, its value, while increasing on the average with height above ground, exhibits considerable variations from site to site. For an open exposure, measured integral scales were reported (1) to vary between 120 m (396 feet) and 630 m (2000 feet), at a 151 m (500 feet) elevation. Under such conditions, the ratio of the integral scale L_x to a structure sectional characteristic dimension D reaches values of the order of 100 and, this is a source of uncertainty in the aerodynamic laboratory testing of structures, where the same ratio is of the order of unity. To investigate this uncertainty, a series of wind tunnel tests were performed on 2D rectangular cylinders where the only variable L_x/D, the scale ratio, ranged from 6 to 18.

PREVIOUS INVESTIGATIONS

Most of the previous determinations of the integral scale effect were limited to ratios of L_x/D smaller than 5. Laneville et al (2), in the range $0.5 < L_x/D < 5$, observed no effect of the scale of turbulence on the mean drag coefficients (and on the galloping oscillations) for two-dimensional rectangular cylinders at zero incidence with $H/D = 0.5$, 1 and 2. Nakamura et al (3), in a wider range of longitudinal lengths ($0.1 < H/D < 3$), also observed no integral scale effect for a similar range of scale ratios. On the other hand, Lee (4) did observe a strong integral effect (20% increase) on the mean drag coefficient of square prisms, in the neighbourhood of $L_x/D = 1$. Petty (5), using Lee's results and Allen and Vincenti's blockage correction, showed the scale effect to be of the order of 5%. Courchesne et al (6) investigated closely the range $0.3 < L_x/D < 1.5$ and Figures 1 and 2 summarize their results for the effects of the intensity and integral scale of turbulence respectively, on the mean drag coefficient for rectangular cylinders.

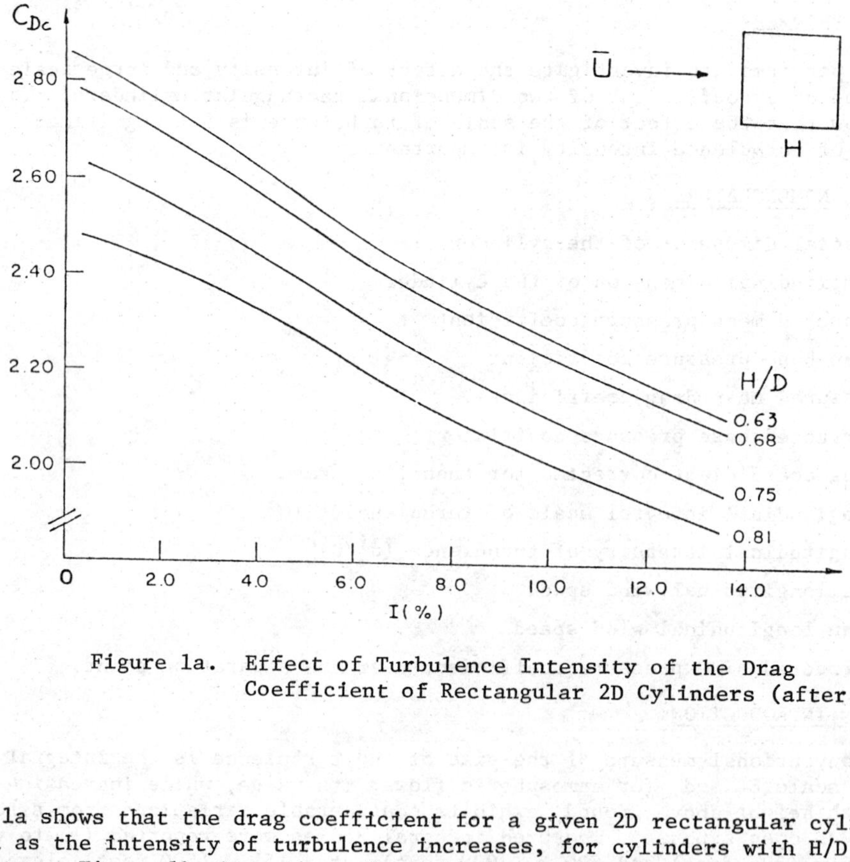

Figure 1a. Effect of Turbulence Intensity of the Drag Coefficient of Rectangular 2D Cylinders (after Ref. 6)

Figure 1a shows that the drag coefficient for a given 2D rectangular cylinder is reduced as the intensity of turbulence increases, for cylinders with H/D of 0.63 or greater. Figure 1b shows that, for cylinders with $0.5 < H/D < 0.63$, the drag coefficient for a given cylinder increases to a relative maximum as the intensity of turbulence increases, and then decreases. For cylinders with $H/D \leq 0.5$, the trend is similar, but the maxima are not so pronounced.

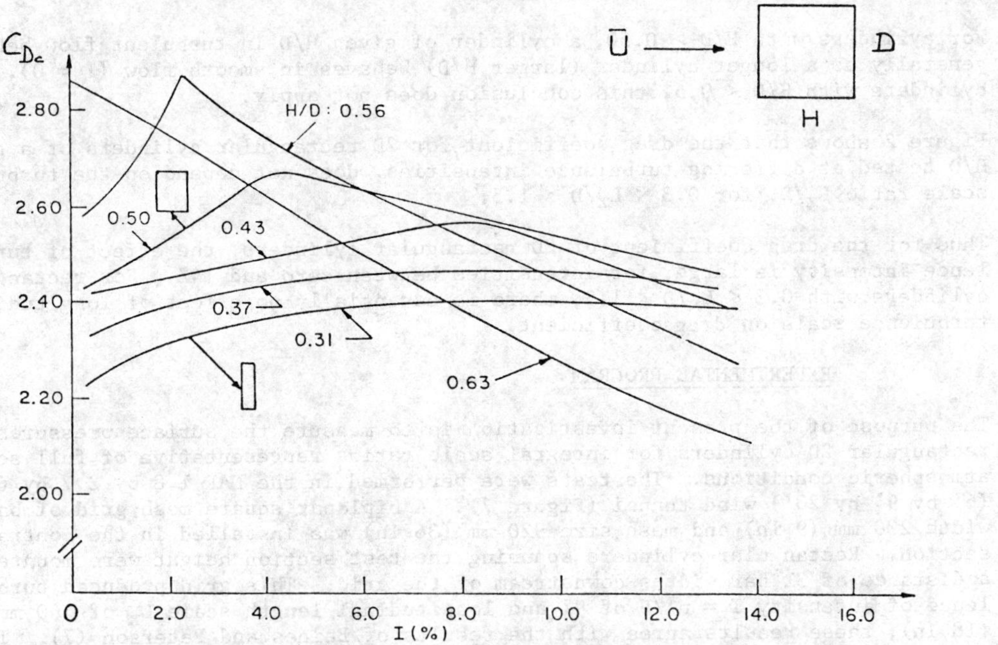

Figure 1b. Effect of Turbulence Intensity of the Drag Coefficient of Rectangular 2D Cylinders (after Ref. 6)

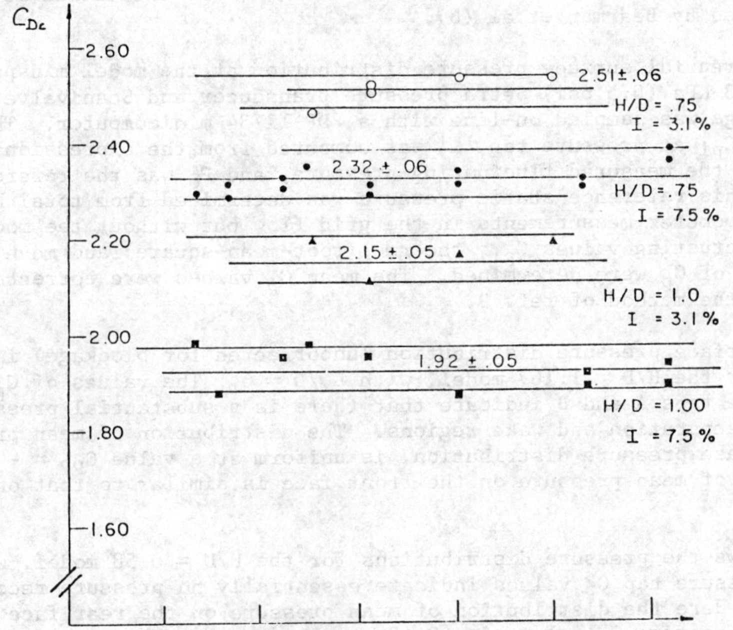

Figure 2. Effect of Longitudinal Scale of Turbulence on the Drag Coefficient of Rectangular 2D Cylinders (after Ref. 6)

WE O*

For cylinders with $H/D \geq 0.50$, a cylinder of given H/D in turbulent flow behaves generally as a longer cylinder (larger H/D) behaves in smooth flow ($I = 0$). For cylinders with $H/D \leq 0.5$, this conclusion does not apply.

Figure 2 shows that the drag coefficient for 2D rectangular cylinders of a given H/D tested at differing turbulence intensities, does not depend on the turbulence scale ratio L_x/D, for $0.3 < L_x/D < 1.5$.

Thus for the drag coefficient of 2D rectangular cylinders, the effect of turbulence intensity is large, for intensities between zero and 14%. For rectangular cylinders with $0.3 < L_x/D < 1.5$, there is essentially no effect of longitudinal turbulence scale on drag coefficient.

EXPERIMENTAL PROGRAM

The purpose of the present investigation is to measure the surface pressures on rectangular 2D cylinders for integral scale ratios representative of full scale atmospheric conditions. The tests were performed in the NAE 1.8 by 2.7 by 6 m (6' by 9' by 20') wind tunnel (Figure 7). A biplanar square mesh grid of bar width 230 mm (9 in) and mesh size 920 mm (36 in) was installed in the contracting section. Rectangular cylinders spanning the test section height were mounted at a distance of 31 bar widths downstream of the grid. This grid produced turbulence of intensity $I = u'/U$ of 9% and longitudinal length scale L_x of 460 mm (18 in); these results agree with the results of Baines and Peterson (7). The width D of the four models was 25, 38, 51 and 76 mm (1, 1.5, 2 and 3 in). Full-width spacers were used on each model to vary the longitudinal length H, so H/D could vary from a value of 0.15 to 2.75. The resulting integral scale ratio L_x/D ranged from 6 to 18. Reynolds number, based on D, was maintained at 60×10^3, as recommended by Bearman et al (8).

The circumferential surface pressure distribution at the model midspan was measured with a 3 KPa (0.5 psi) Setra pressure transducer and Scanivalve system. The output voltage was sampled on-line with a PDP 11/34 minicomputer. The pressure coefficient C_{pi} at pressure tap 'i' was computed from the expression $(P_i - P_o)/\frac{1}{2}\rho U^2$, where P_i was the measured fluctuating pressure, and P_o was the reference static pressure. This reference static pressure was determined from total head tube and hot wire anemometer measurements in the grid flow but without the model in place. From the fluctuating values C_{pi}, the rms (root-mean-square) and mean (time averaged) values of C_p were determined. The mean C_p values were corrected for tunnel blockage by the method of ref. 9.

A typical surface pressure distribution (uncorrected for blockage) is shown in Figure 3, for the H/D = 1.167 model, with $L_x/D = 6$. The values of C_p at the two side pressure taps A and B indicate that there is a substantial pressure recovery between the separation and wake regions. The distribution of mean pressure on the rear face (base pressure distribution) is uniform at a value $C_{pb} = -1.15$. The distribution of mean pressure on the front face is similar to that on a normal flat plate.

Figure 4 shows the pressure distributions for the H/D = 0.58 model, with $L_x/D = 6$. The side pressure tap C_p values indicate essentially no pressure recovery on the side faces. Here the distribution of mean pressure on the rear face is curved, with a peak negative C_p value of -2.45 at the center. The pressure distributions on the front faces of the H/D = 0.58 and H/D = 1.167 models are essentially identical. The equivalent blockage corrections on tunnel dynamic pressure for these two models were both 3%.

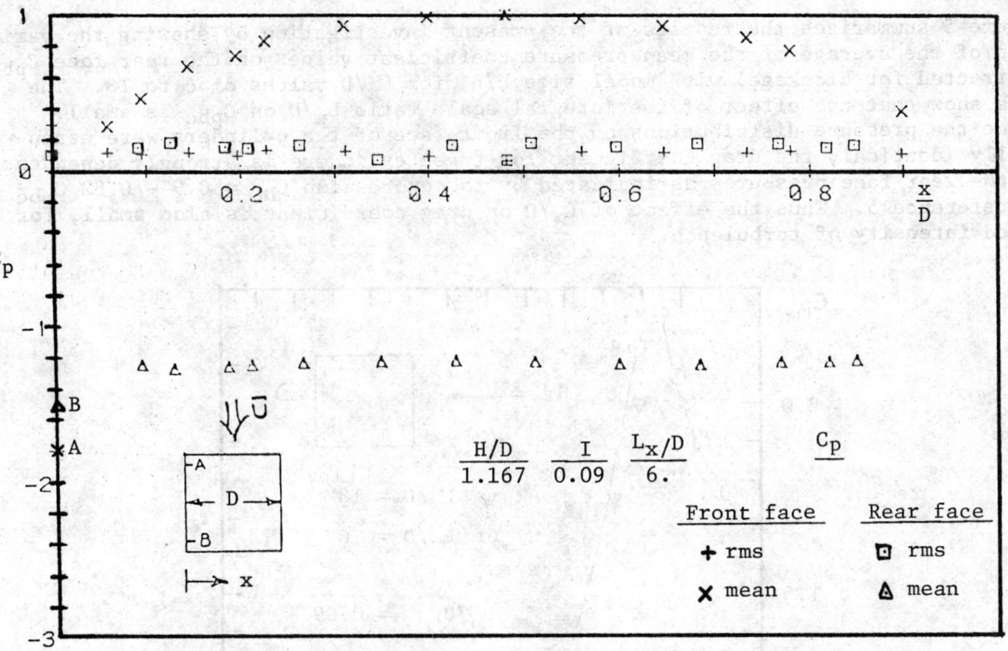

Figure 3. Pressure Distribution on the H/D = 1.167 Cylinder (uncorrected)

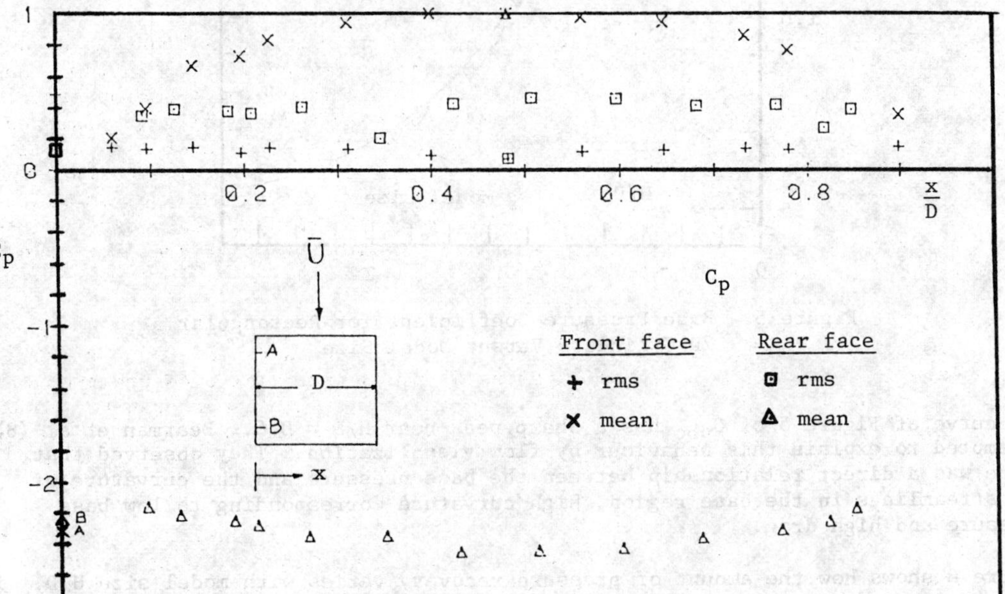

Figure 4. Pressure Distribution on the H/D = 0.58 Cylinder (uncorrected)

Figure 5 summarizes the results of the present investigation by showing the variation of the average of the mean pressure coefficient values on the rear face C_{pbc} (corrected for blockage) with model size H/D, for L_x/D values of 6 to 18. The data show that the effect of the integral scale ratio L_x/D on C_{pbc} is small. Since the pressure distributions on the front face of the cylinders were essentially identical, the drag coefficient for these cylinders is strongly dependent on the rear face pressures as indicated by the expression $C_{DC} = 0.9 - 0.82\ C_{pbc}$ of reference 5. Thus the effect of L_x/D on drag coefficient is also small, for fixed intensity of turbulence.

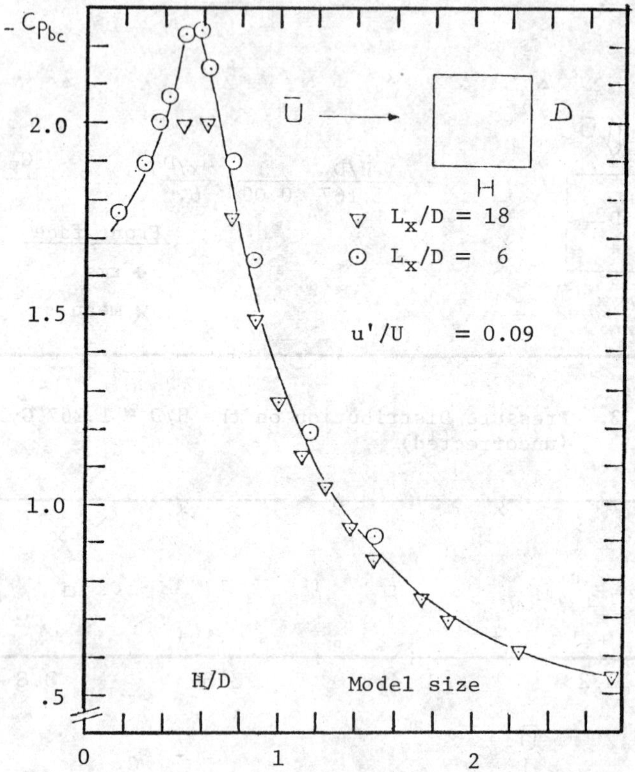

Figure 5. Base Pressure Coefficient for Rectangular 2D Cylinders Versus Model Size

The curve of Figure 5 of C_{pbc} has a sharp peak near H/D = 0.5. Bearman et al (8) attempted to explain this behaviour by flow visualization. They observed that there was a direct relationship between the base pressure and the curvature of the streamlines in the base region, high curvature corresponding to low base pressure and high drag.

Figure 6 shows how the amount of pressure recovery varies with model size H/D. It appears that a pressure recovery begins only when a value of H/D $\simeq$ 0.6 is reached and that the recovery, for values of H/D > 2.8, tends to an asymptotical value of the order of 0.8. This last behaviour is certainly related to the reattachment of the shear layers. The scatter of the data points is due to the location of the tap on the 1 inch model.

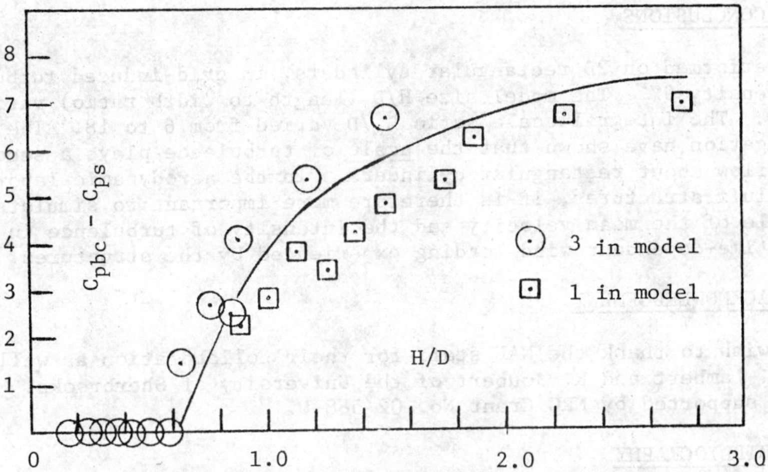

Figure 6. Side Face Pressure Recovery for Rectangular 2D Cylinders versus Model Size

Figure 7. A Rectangular Cylinder Model and Turbulence Grid in the Wind Tunnel

CONCLUSIONS

Tests were performed on 2D rectangular cylinders, in grid-induced turbulence of constant intensity 9%. The model size H/D (length to width ratio) varied from 0.15 to 2.75. The integral scale ratio L_x/D varied from 6 to 18. The results of this investigation have shown that the scale of turbulence plays a secondary role in the mean flow about rectangular cylinders. In the aerodynamic laboratory testing of bluff structures, it is therefore more important to simulate adequately the profile of the mean velocity and the intensity of turbulence in order to predict the time-dependent wind loading experienced by the structures.

ACKNOWLEDGEMENT

The authors wish to thank the NAE staff for their collaboration as well as P. Jutras, D. Lambert and R. Joubert of the University of Sherbrooke. This research was supported by NRC Grant No. 02-588-0.

BIBLIOGRAPHY

(1) M. Shiotani, "Structure of Gusts in High Winds", Interim Reports Parts I-IV, The Physical Sciences Laboratory, Nihon Univ. Furabashi, Chiba, Japan, 1967-71.

(2) A. Laneville, I.S. Gartshore and G.V. Parkinson, "An Explanation of Some Effect of Turbulence on Bluff Bodies", *Proceedings of the 4th Int. Conf. on Winds Effects on Buildings and Structures*, Heathrow, England, 1975.

(3) Y. Nakamura and Tomonari, "The Effect of Turbulence on the Drag of Rectangular Prisms", *Trans. Japan Soc. Aero. and Space Science*, Vol 19, pp 81-86.

(4) B. Lee, "Some Effects of Turbulence Scale on a Bluff Body", *Journal of Industrial Aerodynamics*, Vol 1, no 4, pp 361-371.

(5) D.G. Petty, "The Effect of Turbulence Intensity and Scale on the Flow Past Square Prisms", *Proceedings of the 3rd Colloquium on Industrial Aerodynamics*, Part 2, Aachen, Germany, June 1978, p 281.

(6) J. Courchesne et A. Laneville, "Les Cylindres à Section Rectangulaire soumis à des Ecoulements bi-dimensionnels peu Turbulent et Turbulent", *Comptes Rendus du Second Atelier Canadien sur le Génie Eolien*, I.R.E.Q., Varennes, Québec, Sept. 1978.

(7) W.D. Baines and E.G. Peterson, "An Investigation of the Flow Through Screens", *Trans. A.S.M.E.*, July 1951, pp 467-480.

(8) P.W. Bearman and D.M. Trueman, "An Investigation of the Flow around Rectangular Cylinders", *Aeronautical Quat.*, Vol XXIII, Part 3, 1972, pp 229-237.

(9) J. Courchesne and A. Laneville, "Experimental Evaluation of Drag Coefficients for Rectangular Cylinders", to be published in the *Journal of Fluids Engineering*, A.S.M.E.

THE INFLUENCE OF SHIELDING ON THE WIND FORCES EXPERIENCED BY ARRAYS OF LATTICE FRAMES

R. E. WHITBREAD
National Maritime Institute, Teddington, UK

SUMMARY

Based on a systematic series of measurements of the influence of solidity, spacing ratio, number of frames and type of member, it is shown that existing methods for the estimation of the wind-force coefficients for lattice frames can prove very unreliable. An empirical model for the prediction of shielding factors is proposed which greatly improves the accuracy of wind-force estimates. The same basic model has been extended to allow the wind force on an array of up to five frames to be predicted with reasonable accuracy.

List of Symbols

A_m	area of all the members of a frame projected on to a plane normal to the direction of the wind.
A_o	area of the outline of a frame projected on to a plane normal to the direction of the wind.
C_F	wind-force coefficient, defined as the wind force divided by $0.5\rho V^2 A_m$.
C_{F_o}	wind-force coefficient for a single isolated frame.
V	mean wind speed.
η	shielding factor ($=C_F/C_{F_o}$)
η_1, η_2	shielding factors for the upwind and downwind frames of a pair.
η_e	equivalent shielding factor, including the influence of all downwind frames where η_1 is assumed to be equal to 1.0.
λ	spacing ratio (= distance between consecutive frames divided by the depth of the frame, as defined in Fig. 5).
ρ	air density.
ϕ	solidity ratio ($= A_m/A_o$).
ϕ_e	an equivalent solidity for use with frames composed of round-section members.

1. INTRODUCTION

The preparation of a new BSI draft Code of Practice for Lattice Towers(1) and consideration of possible revision to the BSI Code CP3 Chapter V(2) have stimulated a particular interest in the UK in the more accurate prediction of the wind loading experienced by lattice structures. In recent years there have been a number of publications in this general field notably by CONSTRADO(3) and by the Engineering Science Data Unit (ESDU)(4). The former is based on the work of Schulz(5) and Gould(6), whilst the latter is essentially an evaluation of all available published data. In this respect, it is of interest to study the bibliography given in the ESDU publication for it reveals that, apart from the work of Schulz and Gould, there appear to have been surprisingly few systematic studies in this field since the early work of Flachsbart, first published in 1932 and now available as an English translation(7). It seems likely therefore that many of the current national codes of practice have been based for the most part on this original work of Flachsbart.

Over the period 1976-78 the National Maritime Institute has undertaken a number of wind-tunnel programmes, sponsored by the Building Research Establishment, to provide a more substantial volume of data on which to base future codification. The initial work involved systematic measurements for typical tower sections and included the influence of ancillaries in idealised forms. Subsequently however the work has developed in two different directions. Firstly, to test the accuracy of the prediction procedures incorporated in Reference 1, a series of measurements has been made on a model of a complete tower including the influence of accurately modelled dish aerials, ladders, feeders etc. This work is to be published separately. The second direction taken was a more careful investigation of the combined influence of frame solidity and spacing in determining wind forces experienced by arrays of lattice frames(8). This led directly into an investigation of shielding factors which is the subject of this paper.

A programme of wind-tunnel testing was undertaken with the following objectives:-

(i) To obtain wind-force coefficients for single frames composed of both angle-section and round-section members. The frames were to be normal to the direction of the wind and the programme was to include a wide range of solidity.

(ii) To extend the data obtained under (i) to include the influence of a second identical frame at various distances downstream. Separate measurements were to be made on each frame.

(iii) To further extend the data obtained under (i) and (ii) to include the influence of additional downstream frames.

Due to escalation in the parametric requirements for (iii) above, it was decided to limit this section of the programme to two spacings namely, $\lambda = 0.5$ and 1.0 and to place most emphasis on arrays of frames composed of angle-section members since these were unlikely to be significantly influenced by Reynolds number. It was assumed that it would be possible to deduce the effects of round-section members in the light of the data obtained from the earlier sections of the programme although it was recognised that the relatively low Reynolds numbers achieved in the tests would cause the results to be somewhat unrepresentative of full-scale conditions.

2. WIND-TUNNEL TESTING

2.1 The models

Details of the models used for the test programme are given in Fig. 1. For the frames composed of round-section members, separate models were built to represent the three different solidities. A different approach was adopted however in the case of the angle-section frames where a greater range of solidity was required. Only a single set of five basic models was built. To each frame were then added flat strips, evenly distributed over the face of the frame to give the required range of solidities. This technique had been used in the earlier programme of tests with the complete tower sections and the results appeared to justify this simplification of the representation of solidity.

SOLIDITY RATIO (ϕ)	NUMBER OF 12·7mm STRIPS ADDED TO BASIC MODEL
0·163	NONE
0·352	4
0·447	6
0·543	8
0·635	10
0 733	12

DIMENSIONS IN mm

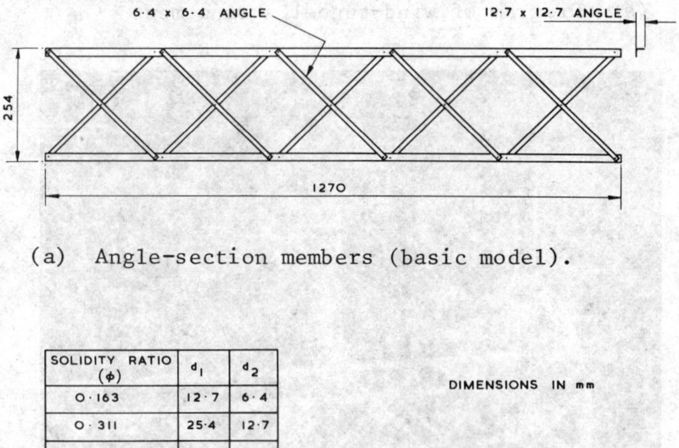

(a) Angle-section members (basic model).

SOLIDITY RATIO (ϕ)	d_1	d_2
0·163	12·7	6·4
0·311	25·4	12·7
0·443	38·1	19·1

DIMENSIONS IN mm

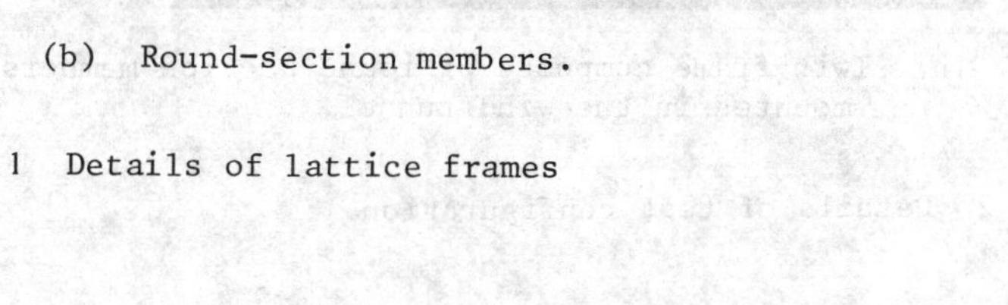

(b) Round-section members.

Fig. 1 Details of lattice frames

2.2 The measurement programme

The measurements were undertaken in the NMI No. 6 Wind Tunnel which has a test section 2.74m by 2.14m. A single active frame was attached at either end to balance struts concealed within a pair of false walls. This arrangement is illustrated in Fig. 2.

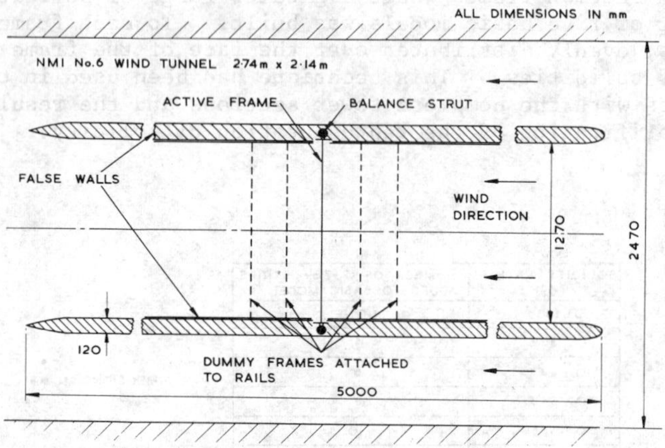

(a) Diagram of wind-tunnel arrangement

(b) Twin-frame composed of round-section members mounted in the wind tunnel.

Fig. 2 Details of test configuration.

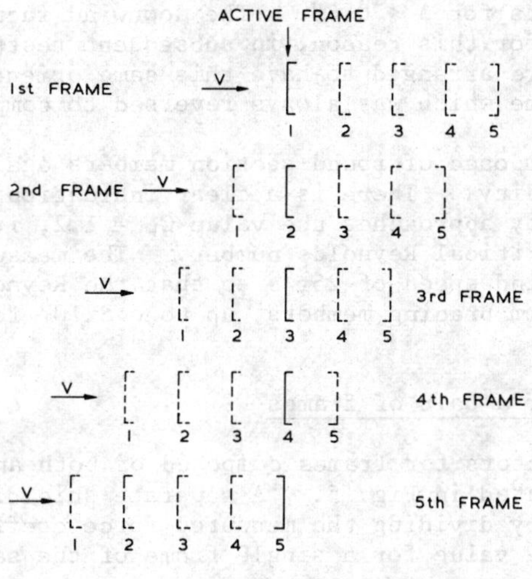

Fig. 3 Schematic procedure for the testing of arrays of multiple frames

The position of the active frame relative to others in an array was varied by placing a dummy frame or frames upwind or downwind according to the requirements of the configuration under test. A set of rails extending up and downwind from the active frame permitted changes in the configuration to be made quite readily. A typical set of test configurations is illustrated in Fig. 3.

The programme of measurements involved the use of a single isolated frame, two frames separated by up to four frame widths and arrays of up to five frames at spacings ratios of $\lambda = 0.5$ and 1.0.

The wind forces were recorded using a semi-automatic overhead wind-tunnel balance and corrections were applied both for blockage and for the forward influence of the model on the indicated wind speed.

3. WIND FORCES AND SHIELDING EFFECTS

3.1 Wind forces on a single frame

A plot of wind-force coefficient versus solidity is shown in Fig. 4 and, for comparison, the results of Flachsbart(7) are included as the hatched area together with three specific points obtained from a paper by Ower(9).

The NMI results for angle-section members are shown to be significantly higher than those of Flachsbart over the range $0.6 > \phi > 0.25$. For the lowest solidities, the Flachsbart results appear to approach the value 2.0 whereas the lowest solidity achieved in the NMI tests ($\phi = 0.163$) gave a force coefficient of 1.72 with virtually no variation up to $\phi = 0.45$. Ower's result for a solidity of $\phi = 0.27$ is in good agreement with the present data although for his two higher solidities there is a marked reduction in force coefficient with solidity. This reduction may possibly be explained by the fact that Ower's model consisted of a complete bridge girder of finite length. At the higher solidities therefore, the influence of aspect ratio will become increasingly apparent and a steady reduction in the force coefficient with increasing solidity would be expected.

The direction of the flange on the edge member had some influence on the observed force coefficients for $\phi < 0.45$. The downwind facing flange gave the highest coefficient and, for this reason, in subsequent tests with multiple frames, all the frames were arranged to have this same orientation with the exception of the last frame which was always reversed to complete the array.

The results for frames composed of round-section members again showed relatively little dependence on solidity. There is a clear indication however that the coefficient at low solidity approaches the value $C_F = 1.2$, i.e. the value for a circular cylinder at subcritical Reynolds number. The measurements were conducted at a constant wind speed of 25m/s so that the Reynolds numbers ranged from 1.14×10^4 for the 6.4mm bracing members, up to 6.8×10^4 for the 38.1mm main members.

3.2 Shielding effects for a pair of frames

The measured shielding factors for frames composed of both angle and round-section members are presented in Fig. 5. A separate shielding factor has been obtained for both frames by dividing the measured force coefficient for each frame by the corresponding value for a single frame of the same solidity.

The results for the downwind frame (see Fig. 5(a)) show a complicated pattern of behaviour with a reversal of the anticipated dependence of shielding factor on spacing ratio for $\phi > 0.35$. The proximity of the downwind frame is also shown to have an appreciable influence on the wind forces experienced by the upwind frame.

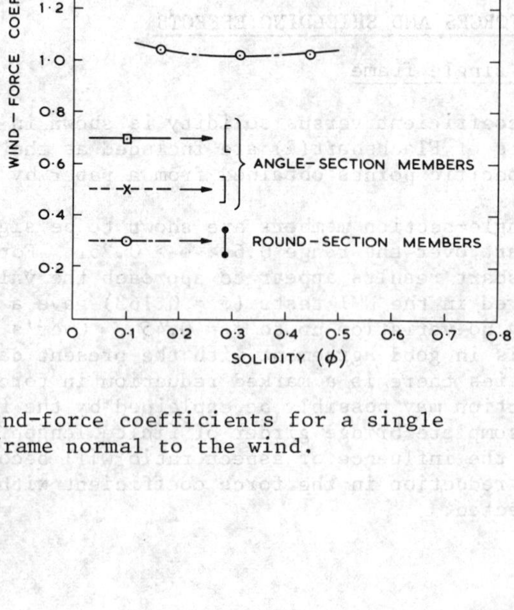

Fig. 4 Wind-force coefficients for a single frame normal to the wind.

Fig. 5(b) indicates a better ordered situation for the frames composed of round-section members. Here the anticipated dependence of shielding factor on spacing ratio gives no indication of a reversal for any practicable solidity.

Reference 10 shows that the shielding factors for arrays of tubular members are dependent not only on solidity and spacing ratios but also on Reynolds number. It is further suggested that shielding factors obtained from measurements at subcritical Reynolds numbers may underestimate the influence of a downstream frame at Reynolds numbers above the critical.

3.3 Shielding effects for multiple arrays

Figs.6(a) and (b) present the corresponding shielding factors for a progressive increase in the number of frames up to a maximum of 5, for the two spacing ratios $\lambda = 0.5$ and 1.0. The addition of frames at the downwind end of the array has an influence on all the upwind frames such that there is a slight but systematic readjustment in the distribution of loading throughout the array in each case. For a solidity of $\phi = 0.5$ this mutual interaction appears to have virtually disappeared.

On the basis of these observations, it would seem inadvisable to attempt to represent the influence of multiple frames in terms of separate shielding factors for each frame. Rather, it is the joint effect of all the frames acting together in a particular array that is important.

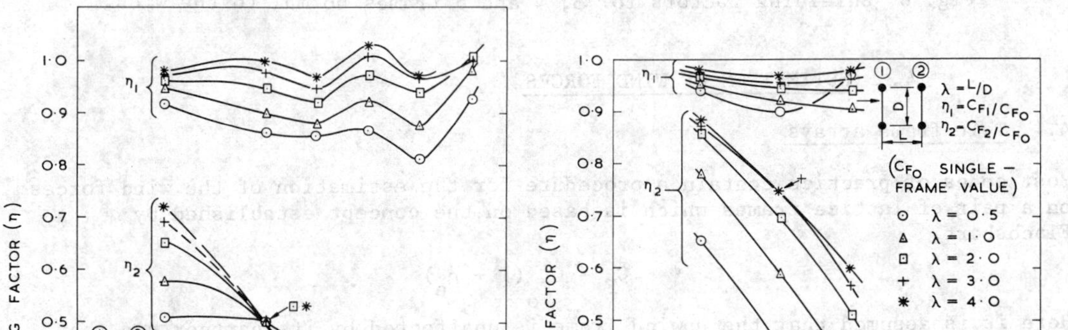

(a) Angle-section members (b) Round-section members

Fig. 5 Shielding factors for a pair of lattice frames normal to the wind.

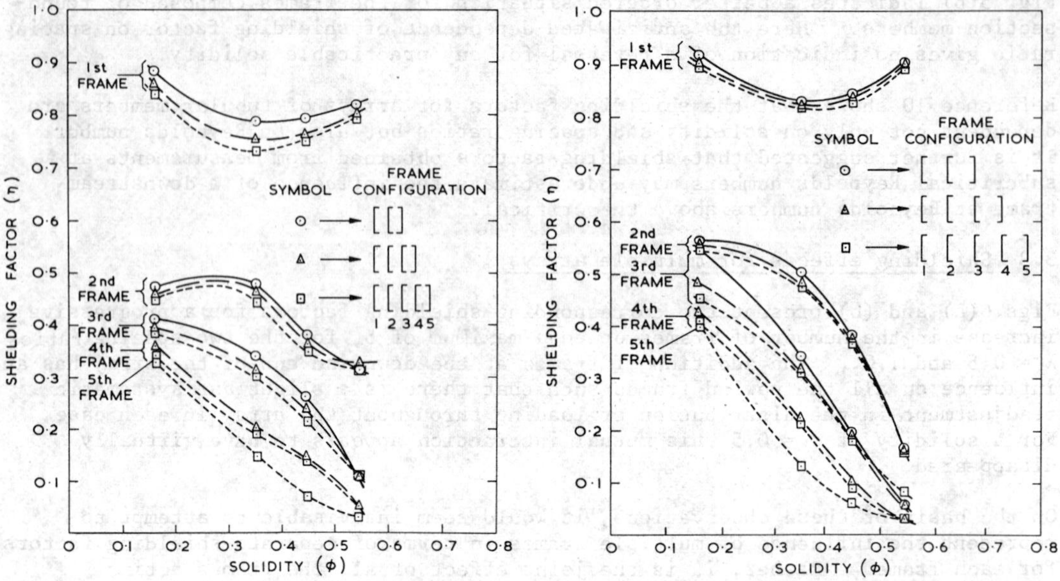

(a) λ = 0.5, angle-section members. (b) λ = 1.0, angle-section members

Fig. 6 Shielding factors for 3, 4 and 5 frames normal to the wind.

4. ESTIMATION OF WIND FORCES

4.1 Twin frame arrays

Most codes of practice contain a procedure for the estimation of the wind forces on a pair of lattice frames which is based on the concept established by Flachsbart.

$$C_F = C_{F_o} (1 + \eta_e)$$

Here it is assumed that the upwind frame is unaffected by its partner and that all the shielding is confined to the downwind frame (i.e. $\eta_1 = 1$). Figs 5 and 6 show that this is not a particularly realistic assumption but it does have the virtue of simplifying the problem.

The expression proposed by Flachsbart for the shielding factor is

$$\eta_e = 1.15 \left[1-1.45\phi\lambda^{-0.25}\right]$$

(In the original paper the exponent of λ is printed as 3/4 but Ower(9) has shown that it was intended to be - 1/4). This expression is shown plotted against solidity (φ) for a range of spacing ratio (λ).

CP3 Chapter V(2) contains a tabulated set of shielding factors and the ESDU Data Sheet(4), a graphical presentation. Both appear to be based on the Flachsbart model. A somewhat different approach has been adopted for the new UK Code for Lattice Towers(1). Here an empirical curve is given for the basic shielding factor covering the full range of solidity 1.0 > φ > 0 for the single spacing ratio of λ = 1.0. Empirical formulae are then given which allow for other spacing ratios and for specific configurations such as triangular sections.

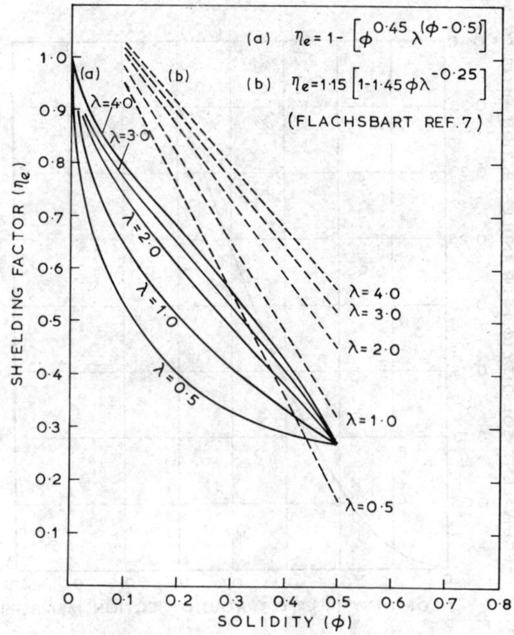

Fig. 7 Predicted shielding factors for a pair of lattice frames normal to the wind.

Predicted force coefficients obtained by the use of Refs. 1 and 2 are presented in Figs. 9(a)-(j) for comparison with the data measured in the present programme. Since this measured data related essentially to two separated frames, some further points have been added which include the influence of bracing in the plane of the wind. Such bracing can be seen to have had relatively little influence on the measured wind-force coefficients.

The predictions obtained using Refs. 1 and 2 are shown to overestimate the measured force coefficients in most cases. Ref. 1 consistently overestimates by a large margin; the error in the shielding factor can be as large as 100%. Ref. 2 provides a better prediction although the force coefficients are shown to vary from a significant overestimate for angle-section frames of low solidity through to an underestimation for frames of round-section members where $\lambda > 1.0$. It is clear that even the more recently advocated procedures cannot be relied on for more than a limited range of solidity and spacing ratio. Using data obtained from the present programme it has been possible to devise an improved empirical model which more accurately reflects the influence of the principal parameters namely, solidity and spacing ratio. The following expression is now proposed for the estimation of the shielding factor

$$\eta_e = 1 - \left[\phi^{0.45}\lambda^{(\phi-0.5)}\right] \quad \text{for } 0.5 > \phi > 0$$

It is shown plotted in Fig. 7 for comparison with the original Flachsbart model. The differences are quite significant especially in the solidity range of practical interest $0.3 > \phi > 0.1$. It is considered inadvisable to extend the use of the model beyond $\phi = 0.5$ although, since the curves for all spacing ratios converge on to a single shielding factor at this solidity, it is convenient to assume this particular value for all higher values of solidity.

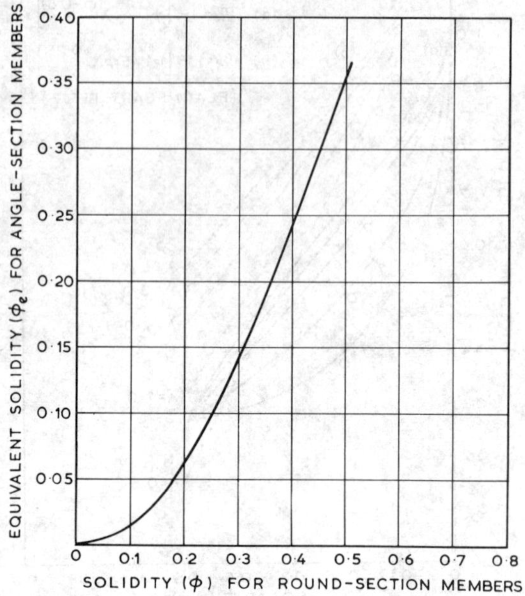

Fig. 8 Equivalent solidity for lattice frames composed of round-section members (subcritical Reynolds number).

Wind-force coefficients obtained with the new model are shown plotted in the Figs. 9(a)-(j); in most cases the error is within 10%. Although derived entirely on the basis of angle-section frames, it has been found that the same model can be used for frames of round-section members by the simple expedient of employing an equivalent solidity (ϕ_e). Values of ϕ_e have been deduced as that value of solidity necessary to obtain the measured shielding factors for round-section frames of the same solidity. Inevitably, the results showed a fair degree of scatter and it was necessary to select a minimum value for ϕ_e in each case such that the shielding factor was never underestimated. The result of this exercise is shown in Fig. 8. The curve shown will apply strictly only to Reynolds numbers well below the critical. However further curves of equivalent solidity for higher Reynolds numbers could readily be constructed from relevant experimental data.

The comparisons shown in Figs. 9(a)-(j) indicate a very satisfactory agreement for angle-section members. Some degree of overestimation is apparent for the round-section frames. This is inevitable in view of the basis for the choice of the values for ϕ_e discussed above. There is of course, also the fundamental difference between the aerodynamic behaviour of angle- and round-section frames as is shown in Fig. 5.

4.2 Multiple-frame arrays

There appears to have been no serious attempt to date, certainly to the knowledge of the author, to deal with the case of multiple frames. Reference 2 for example, suggests that all additional frames should be assumed to have the same shielding factor as that deduced for the second frame of a pair. Fig. 6 demonstrates how conservative such an assumption would be.

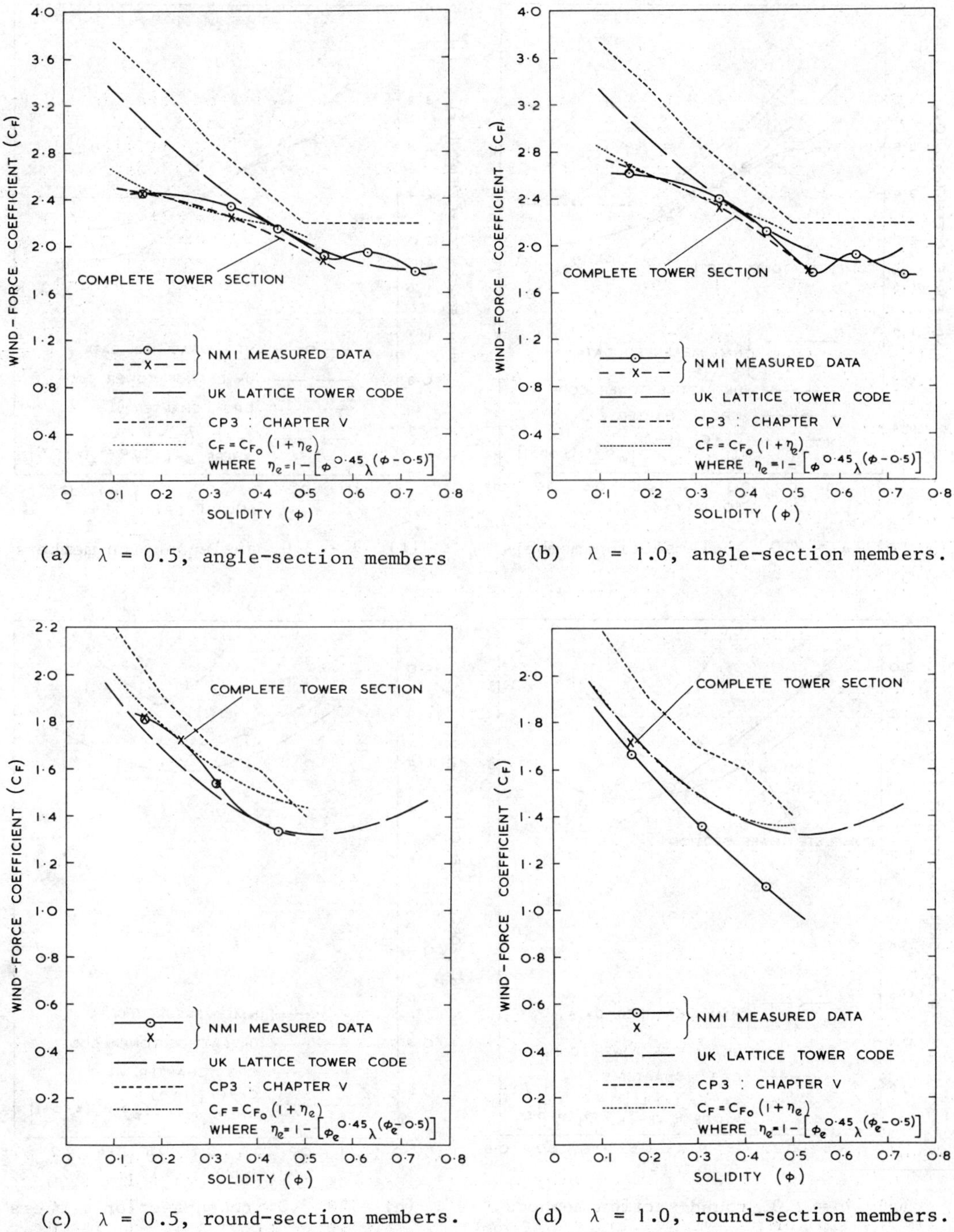

(a) λ = 0.5, angle-section members

(b) λ = 1.0, angle-section members.

(c) λ = 0.5, round-section members.

(d) λ = 1.0, round-section members.

Fig. 9 Measured and predicted wind-force coefficients for a pair of lattice frames normal to the wind.

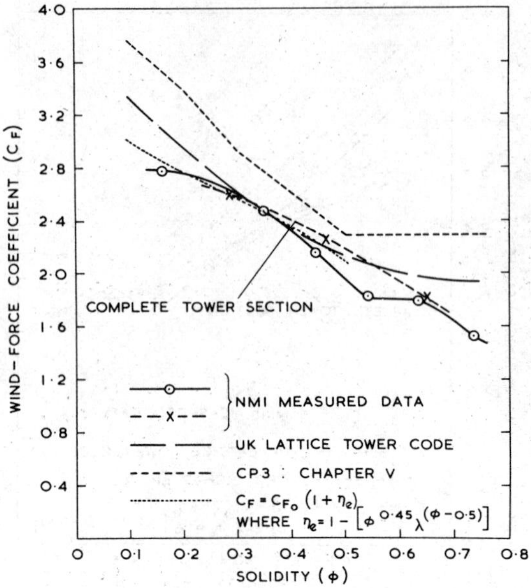

(e) λ = 2.0, angle-section members

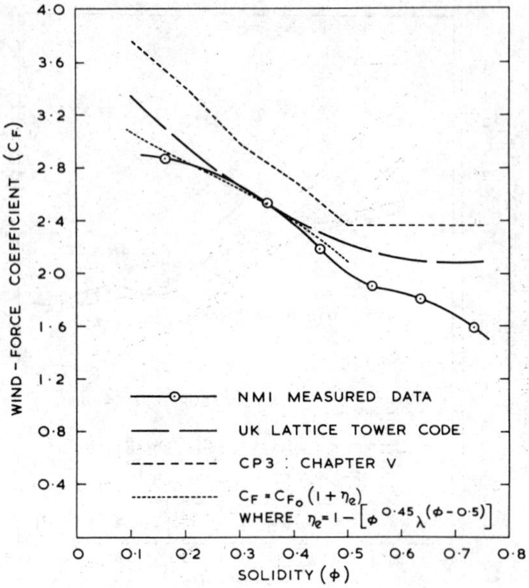

(f) λ = 3.0, angle-section members

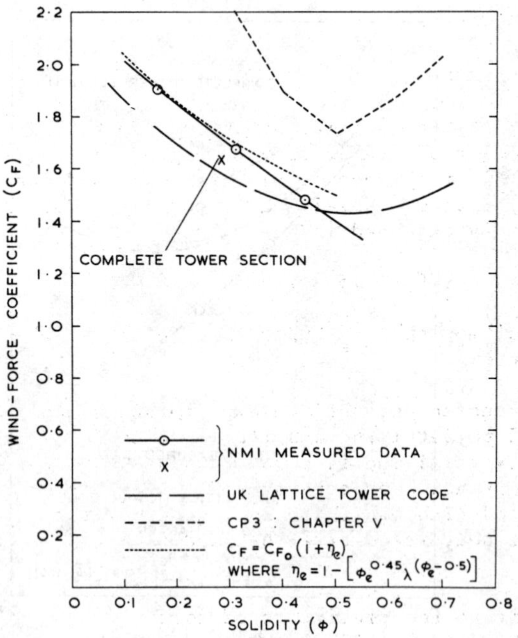

(g) λ = 2.0, round-section members.

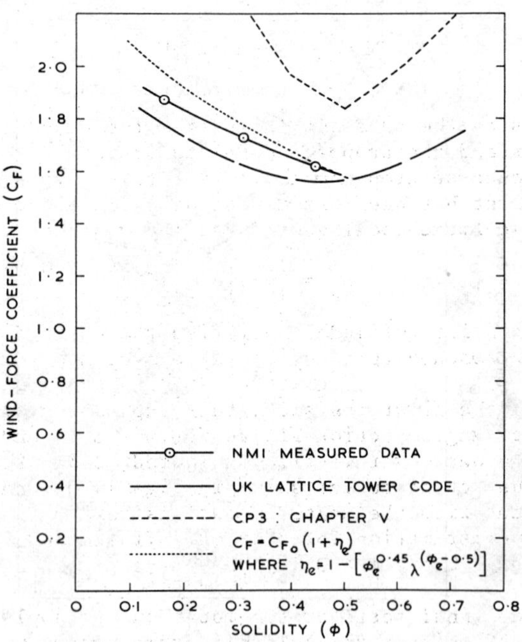

(h) λ = 3.0, round-section members

Fig. 9 (continued)

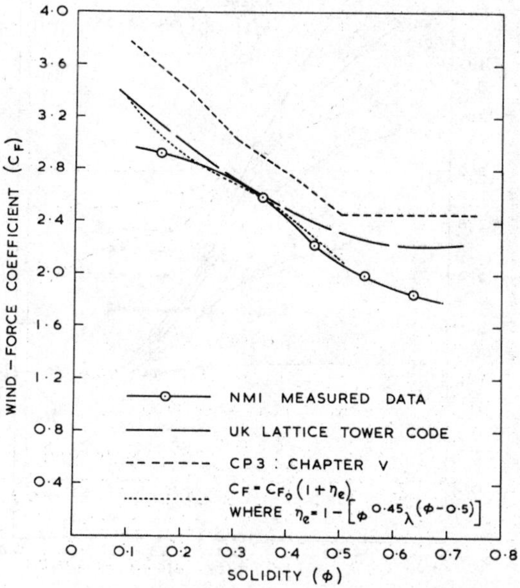

(i) $\lambda = 4.0$, angle-section members

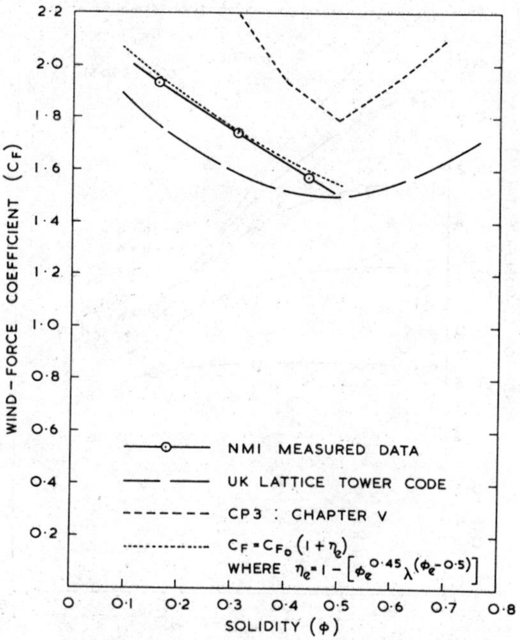

(j) $\lambda = 4.0$, round-section members.

Fig. 9 (continued)

Using the data available from the present tests, an attempt has been made to extend the proposed model to include any number of frames of both angle and round-section members. For this purpose, the influence of all frames beyond the first has been assumed to be represented by a single shielding factor η_e. The two-frame model quoted earlier was then simply extended in the form

$$\eta_e = \left[n-1\right] - \left[\phi^{0.45}\lambda^{(\phi-0.5)}(n-1)^{\gamma}\right] \quad \text{for } 0.5 > \phi > 0$$

This form of model satisfies the basic condition that η_e approaches (n-1) as (ϕ) approaches zero and as (λ) approaches infinity.

In the first instance, this expression was tested against the experimental data for angle-section frames for n = 3, 4 and 5 to determine a suitable value for γ. The best overall fit was judged to be with $\gamma = 1.1$ and it is on this basis that the set of curves shown in Figs. 10(a) and (b) has been constructed. The agreement with the measured data is generally good although there is a tendency for overestimation for 5 or more frames and for solidities at either end of the range $0.5 > \phi > 0.1$.

The final test for the model was to apply it to the prediction of force coefficients for multiple frame arrays with round-section members. For this purpose the equivalent solidity values plotted in Fig. 8 were again used. The results are shown plotted in Figs. 10(c) and (d) for comparison with the somewhat limited experimental data.

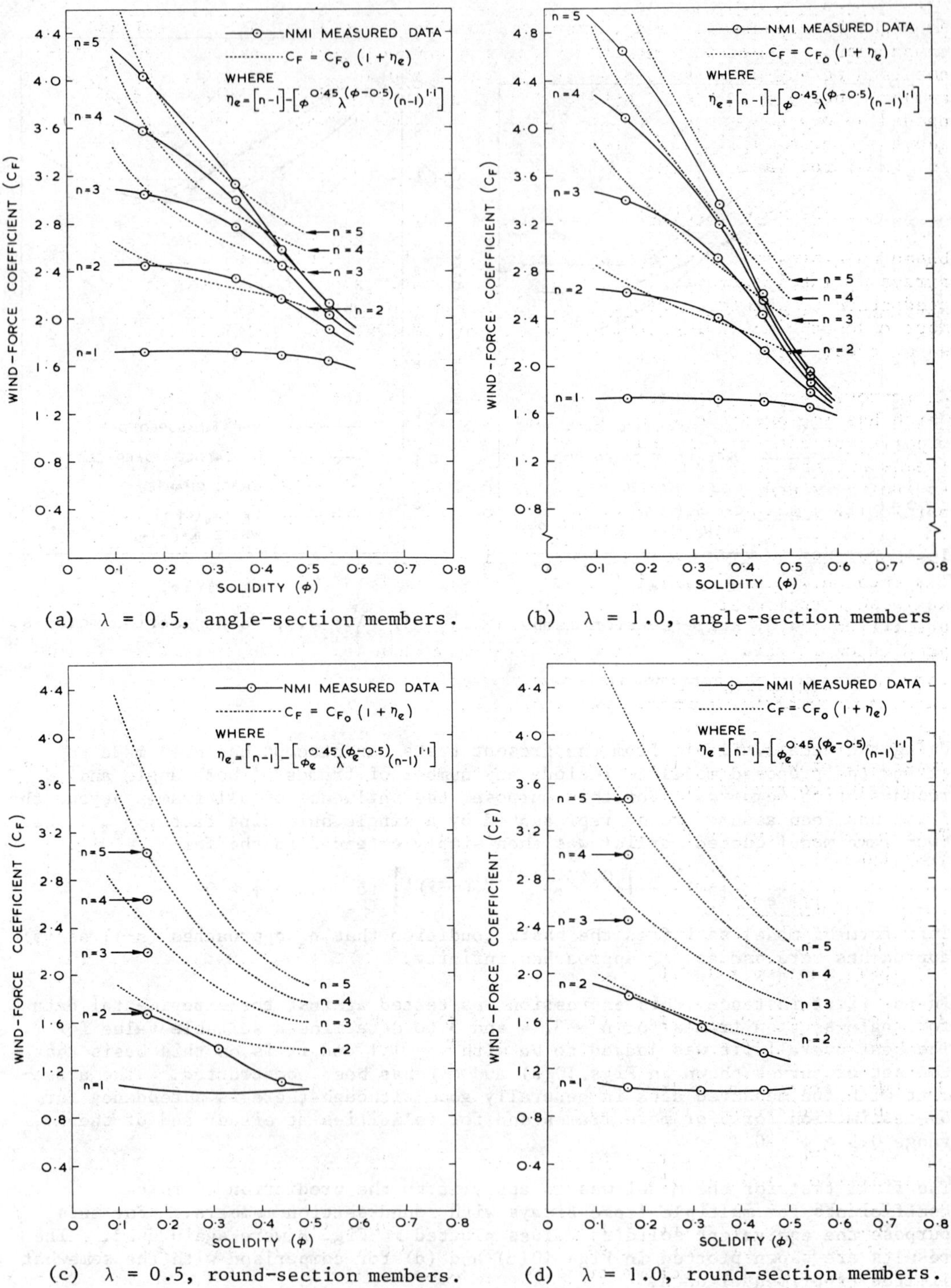

(a) λ = 0.5, angle-section members.

(b) λ = 1.0, angle-section members

(c) λ = 0.5, round-section members.

(d) λ = 1.0, round-section members.

Fig. 10 Wind-force coefficients for multiple arrays of frames normal to the wind.

As might be expected, the agreement is less satisfactory than in previous comparisons. Nevertheless, the estimates are certainly realistic and should not lead to any underestimation. It must again be emphasised that the Reynolds numbers obtained in these tests were relatively low and that further data are needed before the procedure can be considered for use at Reynolds number at or above the critical. Such data would be equivalent to a range of curves of η_e in Fig. 8 for various Reynolds numbers.

5. CONCLUSIONS

Based on a systematic series of measurements of the wind forces experienced by arrays of lattice frames, it has been shown that existing methods for the prediction of shielding factors are deficient in a number of respects. In most cases the shielding factors are overestimated; by as much as 100% in the worst cases.

An improved empirical model has been proposed for the twin-frame configuration which has improved the level of accuracy of prediction to around 10% in the majority of cases. The basic model was developed for use with angle-section frames but has been successfully extended, by the use of an equivalent solidity, to include round-section members without undue loss of accuracy for the particular case of Reynolds numbers below the critical.

The model has been further extended to include the effects of multiple frames and has been shown to give satisfactory predictions for up to 5 frames composed of angle-section members. The use of the equivalent solidity concept has enabled predictions to be made for round-section members although the accuracy of prediction in this case is less satisfactory than for the other configurations.

ACKNOWLEDGEMENTS

The experimental data on which this paper is based was obtained as part of a programme of wind-tunnel testing sponsored by the Building Research Establishment. Permission to publish the data is hereby acknowledged.

This paper is published with the approval of the Director, National Maritime Institute.

REFERENCES

1. Draft Code of Practice - Lattice Towers and Masts - Loading. BSI, September (1978).

2. Code of Basic Data for the Design of Buildings. BSI CP3, Chapter V, Part 2: Wind Loads, London (1972).

3. Wind Forces on Unclad Tubular Structures. Constructional Steel Research and Development Organisation, (1975).

4. Fluid Forces on Lattice Structures. Engineering Sciences Data Unit, Item No 75011, (1975).

5. Schulz, G. I - The drag of lattice structures constructed from cylindrical members (tubes) and its calculation. II - International comparison of standards on the wind loading on structures. CIDECT Report No 69/29, (1969).

6. Gould, R. W. F. and Raymer, W G. Measurements over a wide range of Reynolds numbers of the wind forces on models of lattice frameworks with tubular members. NPL Mar Sci Report 5-72, (1972).

7. Flachsbart, O. Wind loadings on lattice girder structures. BRE Library Translation 2202, (1977).

8. Whitbread, R. E. Wind-force measurements on single and multiframe arrays of lattice frames. To be published as an NMI Report.

9. Ower, E. The wind resistance of lattice girder bridges. The Institution of Civil Engineers and the Institution of Structural Engineers, London, (1948).

10. Pearcey, H. H. Review of Progress - Part I - Research Related Primarily to Jacket-Type Structures (Viscous Drag Regime). Proc. Seminar on Fluid Loading of Offshore Structures, National Physical Laboratory, (1975).

WIND LOADS ON BUILDING FRAMES

P. N. GEORGIOU and B. J. VICKERY
The University of Western Ontario, London, Ontario, Canada

ABSTRACT

The drag wind loads imposed upon single, square or triangular-framed lattice structures have been studied extensively in wind tunnel testing programs over a considerable period of time. However, the shielding effects which are present in multiple building frame configurations and are recognized by most current Wind Loading Codes are not nearly so well documented.

The lack of any comprehensive study into frame shielding to date, has prompted the present undertaking - a parametric investigation including the influence of wind direction, frame solidity, frame spacing, frame aspect ratio and the number of frames.

The shielding coefficients derived from the testing program have shown considerable divergence, both on the conservative and non-conservative side, from values quoted in present codes. The results presented herein, applying only to uniform, smooth flow, are part of a total program which includes the effects of fluctuating loads due to turbulent flow, and the influence of longitudinal edge beams and floor slabs.

NOTATION

- A_S = Effective solid area of a single frame
- A_E = Total area enclosed by a single frame
- ϕ = Frame solidity ratio A_S/A_E
- H = Frame height
- B = Frame width
- S = Frame spacing, centreline to centreline
- λ = Frame aspect ratio
- ρ = Density of air
- V = Free stream velocity
- α = Angle of attack of wind
- $F_D^N(\alpha)$ = Force on a group of N frames, at angle of attack α $= C_D^N(\alpha) \cdot \frac{1}{2}\rho V^2 \cdot A_S$
- $C_D^N(\alpha)$ = Force coefficient, for a group of N frames, at angle of attack α
- α_{max} = α giving a maximum value of $F_D^N(\alpha)$.
- $\eta_N(\alpha)$ = Directional shielding coefficient, defined so that

$$C_D^N(\alpha) = C_D^1(\alpha)\ \{1 + [N-1]\eta_N(\alpha)\}$$

$\overline{\eta}_N$ = Effective shielding coefficient, defined so that

$$C_D^N(\alpha_{max}) = C_D^1(\alpha_{max})\ \{1 + [N-1]\overline{\eta}_N\}$$
$$= C_D^1(0)\ \{1 + [N-1]\overline{\eta}_N\}$$

NOTE: In all cases the force measured was the component of the total force acting on a group of frames along a line normal to the plane of the frames.

SURVEY OF PREVIOUS INVESTIGATIONS

The drag loads on a single truss composed of flat sharp-edged members have been investigated by Flachsbart and Winter (1), Biggs (2), Pagon (3), and the drag coefficients quoted in Wind Loading Codes are consistent with their results.

The critical variables are only two - the frame solidity and frame aspect-ratio, since the maximum drag force always occurs at zero angle of attack. Flachsbart and Winter (1), Flachsbart (4), Pagon (3,5) and others (2,6,7) looked at various configurations of trusses and various member breadths and depths, and found the resulting drag coefficients to be consistently independent of these factors. The drag coefficients associated with trusses comprised of sharp-edged members are also independent of Reynold's Number effects, unlike those which include cylindrical members (8).

Normally the results are compiled in composite form. Firstly an infinite aspect-ratio drag coefficient, depending solely upon solidity ratio, is graphed or tabulated. Figure 1 shows Flachsbart's test values (4) and those found in various Wind Codes (9). It can be seen that all the Codes are slightly conservative. Secondly, a reduction factor which takes into account the effect of aspect-ratio is introduced. Figure 2 shows Flachsbart's results (1,4). Only the Canadian, Swiss and New Zealand Codes include a reduction factor which is dependent upon solidity ratio.

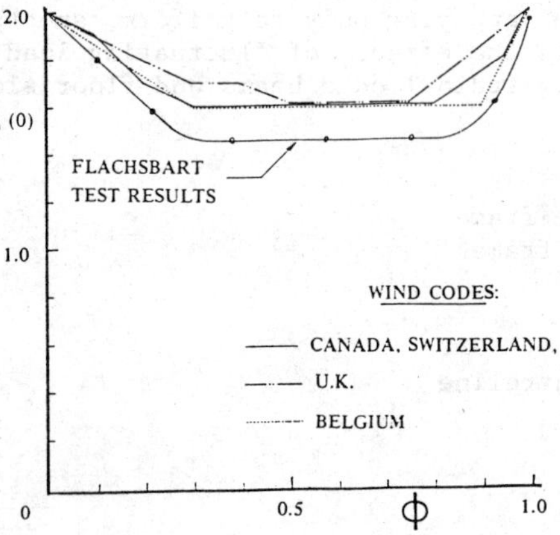

Fig. 1. Single frame drag coefficient

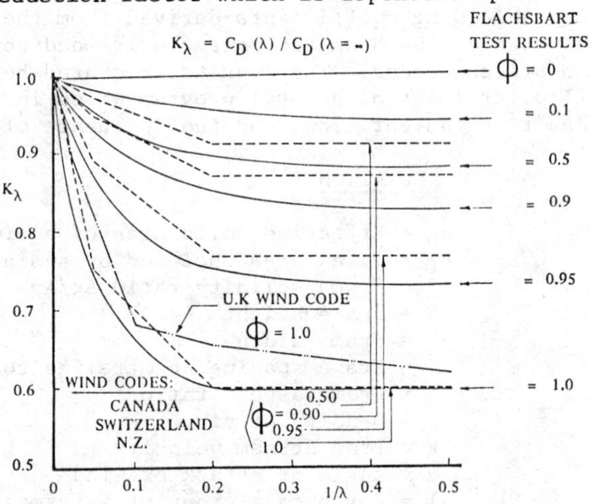

Fig. 2. Aspect ratio reduction factor

A truss of very small solidity ratio behaves like a group of connected but independent solid plates, while a truss of very high solidity ratio takes on the appearance of, and indeed the behaviour of a solid, flat plate. Hence the two extremes give similar drag coefficients. Intermediate values of solidity introduce interference effects between adjoining members, causing the observed drop in drag coefficient. While low and high solidity ratio trusses have numerically similar drag coefficients, their interactions with the flow as a whole are vastly dissimilar. Hence any associated frame shielding reflects this difference.

In the first detailed study of shielding, Flachsbart (4) determined drag loads on truss pairs, aligned with each other and displaced one half-panel width with respect to each other. He then proposed the following formula for the shielding coefficient, for a spacing ratio of about unity -

$$\eta_2(0) = K(1-\phi)^2 \qquad K = 1.0 \text{ aligned trusses}$$
$$\qquad\qquad\qquad\qquad 1.2 \text{ displaced trusses.}$$

Since this endeavour, no comprehensive study has been made, involving all the critical variables, to establish shielding coefficients for two or more frames. Instead, whatever experimental work has been carried out since, has been concerned with particular values of either frame solidity, spacing ratio or aspect-ratio. Understandably particular attention has been paid to tower structures (10,11), both of square and triangular planform. Such tests, and the limited data available on frame shielding, have shown considerable divergence from the values found in most Wind Loading Codes.

Shielding coefficients found in present Wind Codes are tabulated for various values of solidity ratio and spacing ratio. Figures 3, 4, 5 and 6 show a representative spectrum of data found in the Codes. Discrepancies are immediately apparent at high solidity ratios for most spacing ratios, and for low solidity ratios at small spacing ratios.

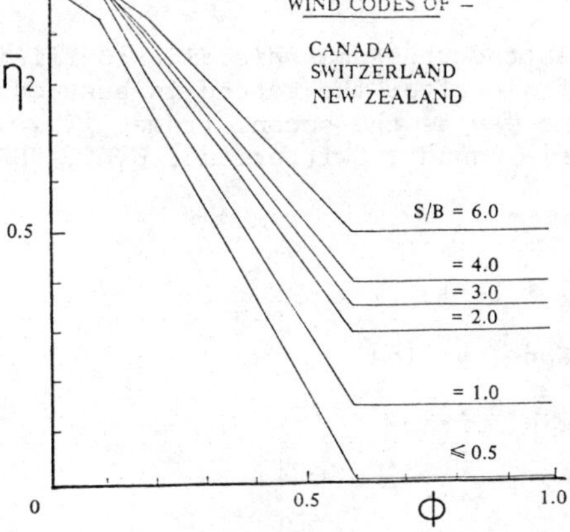

Fig. 3. $\bar{\eta}_2$ vs ϕ

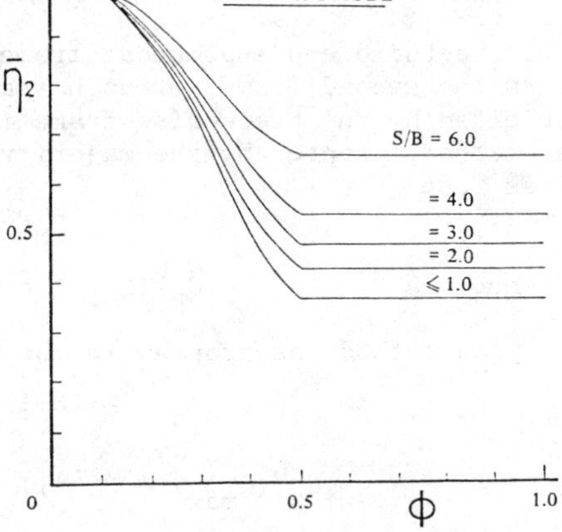

Fig. 4. $\bar{\eta}_2$ vs ϕ

While investigating the wind loads on a Power Station Boiler House during erection, Walshe (12) found that the maximum force imposed on the structure occurred with the least cladding. Moreover for the particular case studied (ϕ = 0.29 and S/B = 0.33), the values determined for multiple frame shielding coefficients (N > 2) indicated a decreased average load on each succeeding frame in the groups studied. The present test results indicate that this is not always the case.

Wind Tunnel tests done on a high rack-storage structure by Möll and Thiele (13) showed considerable disagreement with the German Wind Code, DIN1055, for the drag loads on an unclad structure. The tests illustrated the importance of angle of attack, and for the parameters studied (ϕ = 0.114 and S/B = 0.05), up to 40 frames in an unclad group configuration were looked at, once again revealing a decreased shielding coefficient with increasing number of frames.

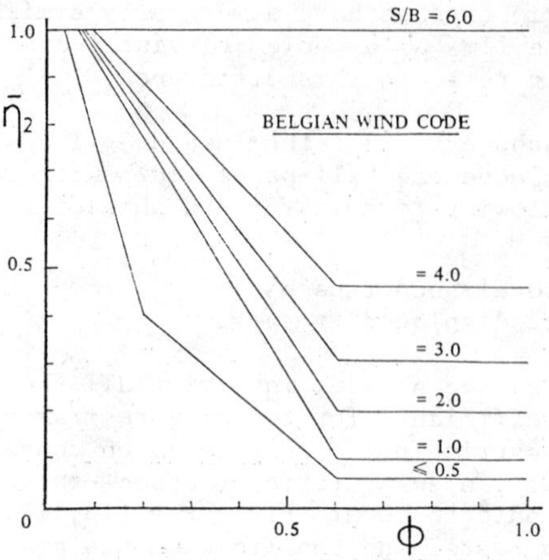

Fig. 5. $\bar{\eta}_2$ vs ϕ

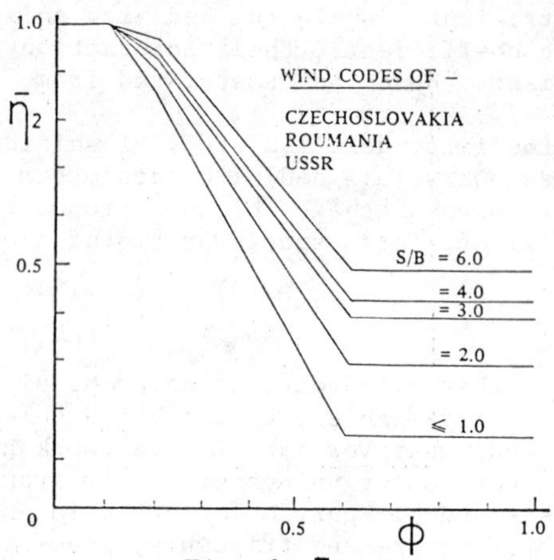

Fig. 6. $\bar{\eta}_2$ vs ϕ

Most recently, a series of studies on open framework models commissioned by the British Post Office (14,15,16,17) has indicated a wide disparity between experimentally measured force coefficients and the relevant U.K. Code Values. Differences of up to 100% were noted for the square towers tested.

The method by which multiple shielding coefficients for groups of 3 or more frames are handled by the Codes brings about an even wider variation in resulting total frame loads. They fall broadly into two categories.

Either the third and subsequent frames are allotted the same shielding coefficients as the second frame, or each succeeding frame after the second is assumed to be shielded by the preceeding frame in the same way as the second frame. The first method, adopted by the majority of Codes - Canada, Switzerland, U.K., USSR, implies that

$$\bar{\eta}_N = \bar{\eta}_2 \qquad N > 2$$

and hence

$$C_D^N(\alpha_{max}) = C_D^1(0)\ \{1 + [N-1]\bar{\eta}_2\}$$

The second method, as adopted in the Belgian Code, yields

$$\bar{\eta} = \bar{\eta}_2^{(N-1)} \quad \text{for the } N^{th} \text{ frame}$$

so that

$$C_D^N(\alpha_{max}) = C_D^1\ \{1 + \bar{\eta}_2 + \bar{\eta}_2^{\,2} + \ldots + \bar{\eta}_2^{\,N-1}\}$$

$$= C_D^1\ (0)\ (1-\eta^N)/(1-\eta) \qquad \eta = \bar{\eta}_2.$$

The latter approach significantly reduces the total frame load. Both methods appear to be based on very limited data. Indeed the results from the German series of tests (13) seriously undermine any confidence in the suggested Code values. Figure 7 contrasts these test results and the corresponding values to be found in the Canadian, U.K., German and Belgian Codes. A similar comparison with Walshe's results (12) shown in Figure 8, and the Canadian, U.K. and Belgian Codes reinforces these apparent inadequacies.

TEST PROGRAM

Models

All model frames were constructed from rectangular members with a breadth exceed-

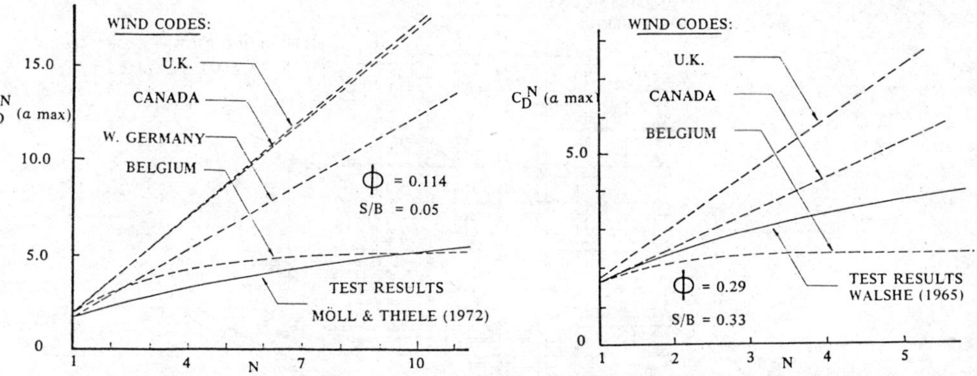

Figs. 7 & 8. Comparison of multiple frame force coefficients

ing twice the depth and were fabricated as biplanar lattices. The frame dimensions are given in Table I.

Table I. Model frame dimensions

FRAME TYPE	H/B	B	MESH	ϕ
SOLID	4	2.0	-	1.0
1	4, 2, 1	2.2	0.4	.773, .770, .778
2	4, 2, 1	2.625	0.5	.464, .466, .468
3	4, 2, 1	3.6	0.7	.286, .286, .295
4	4, 2, 1	4.063	1.0	.136, .131, .142

Test Procedure

The loads on the model frames were measured by a rotating strain-gauge balance as seen in Figure 9. The frames were held vertically in place by a supporting rig which fastened the sides of the frames. The base of the frames could be positioned in various configurations on a mounting plate provided with quarter-inch parallel grooves. The strain-gauge balance yielded a measure of the total load on the frame group in a direction normal to the plane of the frames. Force measurements were made at values of angle of attack ranging from 0° to 90°, in increments of 15°.

The initial tests were performed in a wind tunnel with a working section of 18 ins. by 18 ins., in smooth uniform flow.

Data Reduction

The measured loads were corrected to account for 3 effects. Firstly an adjustment was made to account for a thin boundary layer existing near the floor of the tunnel. Secondly, a rig interference correction was applied. This correction was obtained by performing a second set of measurements with the frames positioned but disconnected from the balance. Not all frame configurations were studied but the interference effects were found to be less than 2-3% of the total measured load.

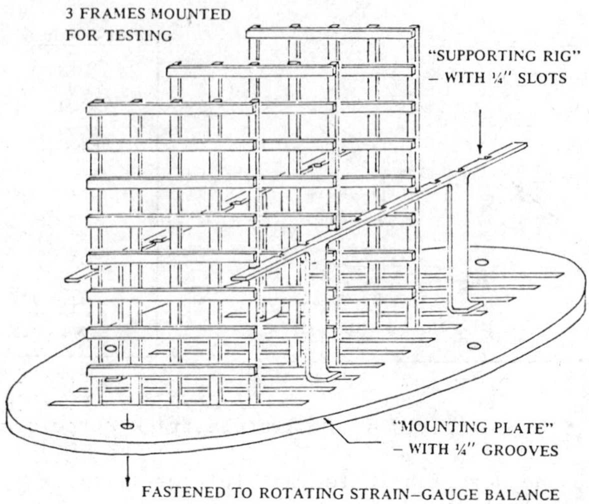

Fig. 9. Support structure for testing frames

Finally blockage corrections for all the tests were applied. Blockage effects were handled by performing a third series of tests, in which a particular configuration was retested in a wind tunnel with an 8 ft. by 8 ft. working section, in which it was presumed that the blockage effect would be zero. Corrections for those groups not retested were obtained by interpolation. The greatest blockage corrections were about 25%, but the average correction was around 10%.

The level of accuracy achieved for independent load measurements was better than 2%. After applying the rig interference and blockage corrections and manipulating the measurements to obtain the values of shielding coefficient, this level was about 10%.

TEST RESULTS

For each configuration, the measured force coefficients were plotted as shown in Figures 10 and 11. From figures of this type values of $\overline{\eta}_N$ and $\eta_N(0)$ were deduced.

Both figures are representative of the shielding behaviour for the particular solidity ratios shown. Figure 10 exhibits a radical change as the angle of attack, α, varies. Maximum drag values occur uniformly at around $\alpha = 55°$. Increases of load up to 100% and more occur from $\alpha = 0°$ to the value at α_{max}. On the other hand, in Figure 11, for this low solidity group ($\phi = 0.136$) maximum drag coefficients are located at around $\alpha = 15°$. The shielding at $\alpha = 0°$ is only slightly less than at values of α_{max}, and maximum loads are never more than 20% higher than those at $\alpha = 0°$. These same effects are apparent when one compares the shielding at $\alpha = 0°$ (Figure 12) with the shielding at α_{max} (Figure 13). For example, the curves of $\eta_2(0)$ and $\overline{\eta}_2$ for the lowest solidity ratio models, $\phi = 0.136$, are not markedly different, but for the curves at high solidity there is a substantial difference.

In the same manner as Figure 12 and 13 were compiled, graphs for the corresponding shielding coefficients for up to 10 frames were constructed. These showed the same trends in relation to the influence of angle of attack as was seen for the two-frame groups. Figures 14 and 15 are two such examples, for $\eta_5(0)$ and $\overline{\eta}_{10}$ respectively.

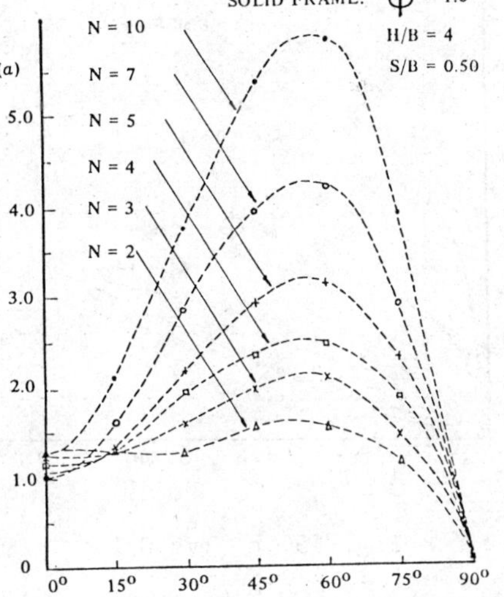

Fig. 10. $C_D^N(\alpha)$ vs α (solid frame)

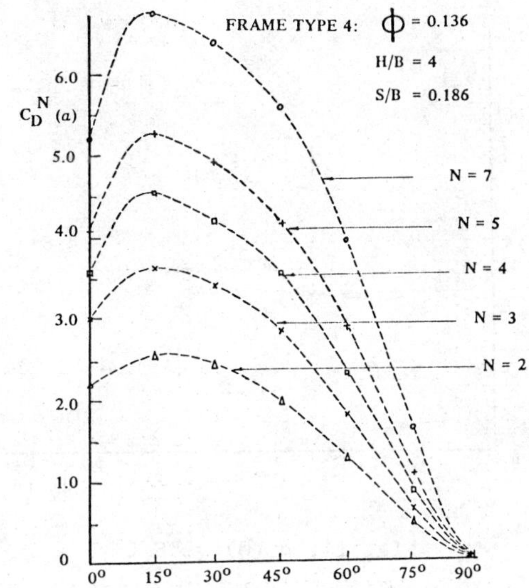

Fig. 11. $C_D^N(\alpha)$ vs α (frame type 4)

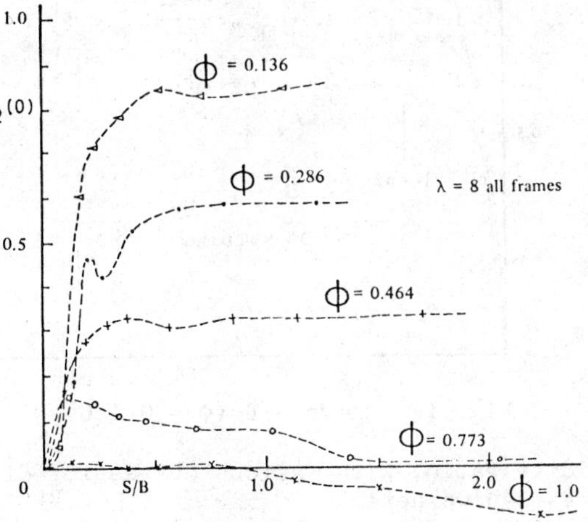

Fig. 12. $\eta_2(0)$ vs S/B

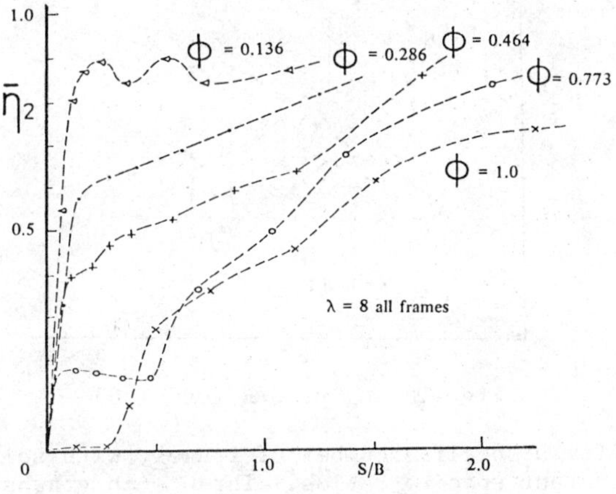

Fig. 13. $\bar{\eta}_2$ vs S/B

Of major importance to the present study was the investigation into the change in shielding behaviour due to increasing numbers of frames. A more illuminating picture of this phenomenon is given in Figures 16 and 17. Figure 16 shows an increase in effective shielding coefficient for $\phi = 1.0$, while the opposite occurs at low solidity, $\phi = 0.136$, as seen in Figure 17. Intermediate values of solidity yielded approximately constant values of shielding coefficient for up to 10 frames.

The effect of Aspect Ratio on the values of effective shielding coefficient was somewhat less well-defined. Figure 18 shows a representative example for values of $\bar{\eta}_{10}$. For the three Aspect Ratios investigated (λ = 2, 4 and 8), the curves are almost coincident.

The most direct formulation for shielding coefficients however coincides with the presentation found in Wind Codes, giving the variation of shielding coefficient,

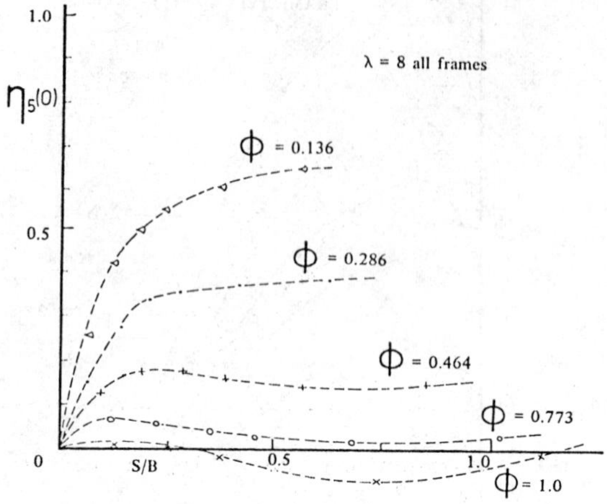

Fig. 14. $\eta_5(0)$ vs S/B

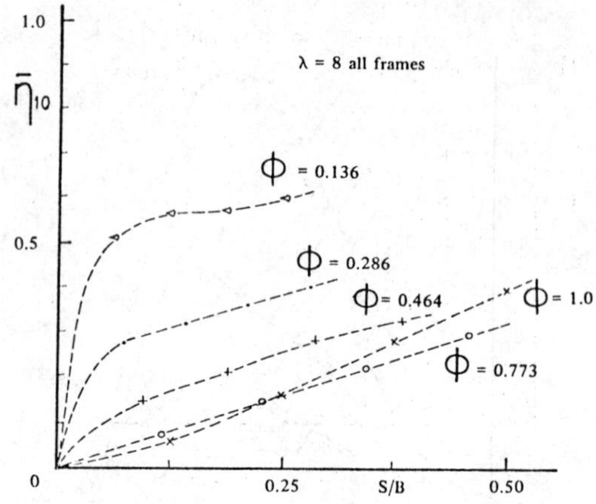

Fig. 15. $\bar{\eta}_{10}$ vs S/B

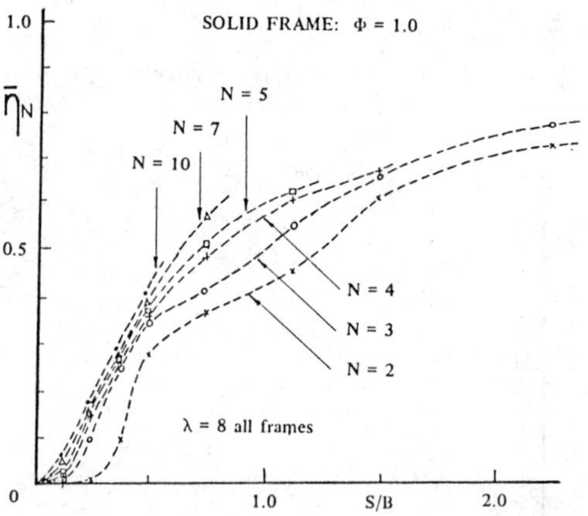

Fig. 16. $\bar{\eta}_N$ vs S/B (ϕ = 1.0)

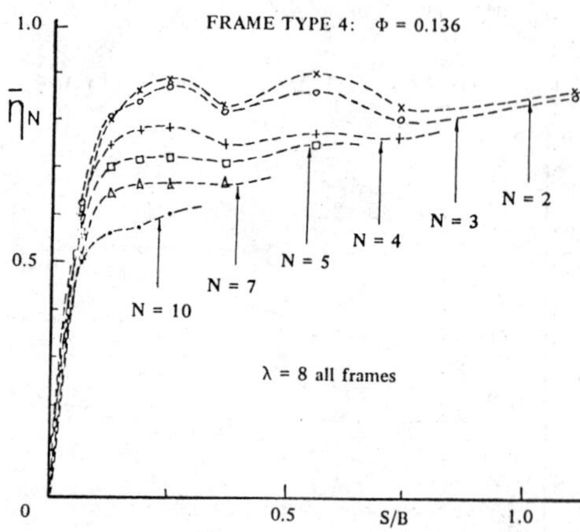

Fig. 17. $\bar{\eta}_N$ vs S/B (ϕ = 0.136)

for a specific number of frames, with solidity ratio, with various curves for different spacing ratios. Three such graphs are shown here.

Fig. 19 depicts the variation of $\eta_2(0)$ with ϕ, and compares the test results to Flachsbart's proposed formula - i.e. $\eta_2(0) = (1-\phi)^2$ for S/B $\sim$ 1.0. The agreement is within the experimental error bounds and the present test results indicate a small but definite dependence on spacing ratio.

Of greater practical importance are the graphs of effective shielding coefficient, shown in Figures 20 and 21, for $\bar{\eta}_2$ and $\bar{\eta}_5$ respectively. The behaviour illustrated here is consistent in form with most Wind Codes, but there are marked differences in the actual magnitudes involved. The discrepancies between these results and Codes is examined in the next section.

An attempt was made to link the shielding coefficients derived for three frames or more to those for two frames, by constructing graphs of $\bar{\eta}_N/\bar{\eta}_2$ (N > 3). Two such graphs are shown in Figures 22 and 23, for spacing ratios of 0.25 and 0.50 respectively. As can be seen from the graphs, the variation with spacing ratio and

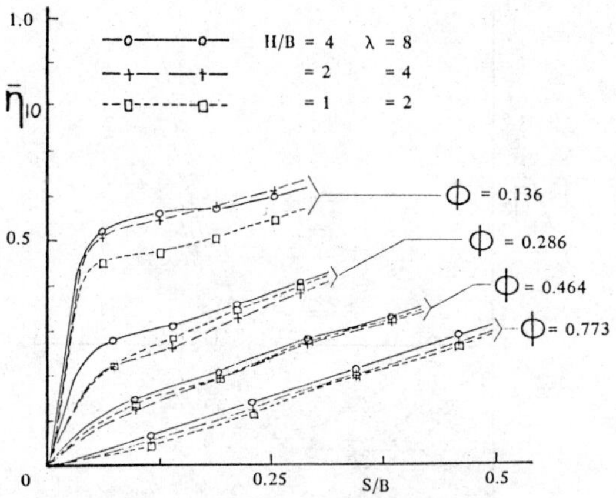

Fig. 18. Effect of aspect ratio (N = 10 frames)

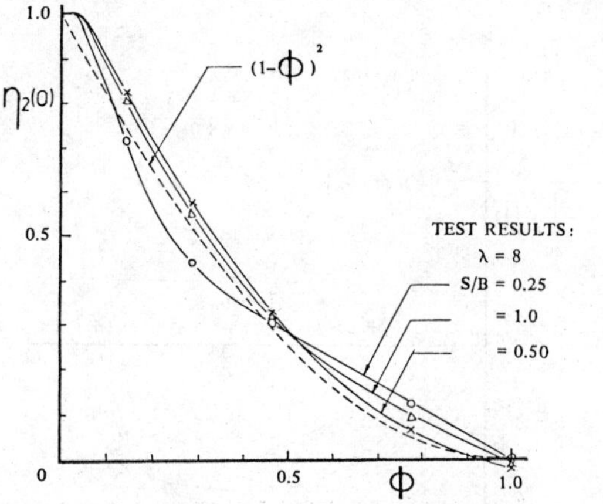

Fig. 19. Directional shielding coefficient N = 2, α = 0°

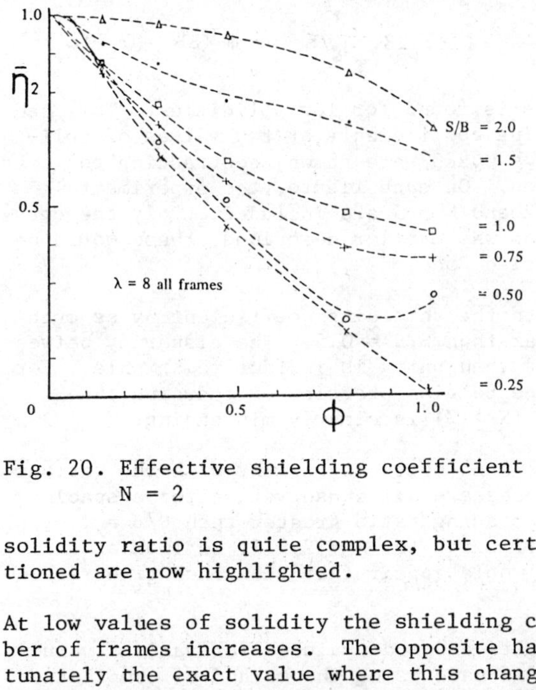

Fig. 20. Effective shielding coefficient N = 2

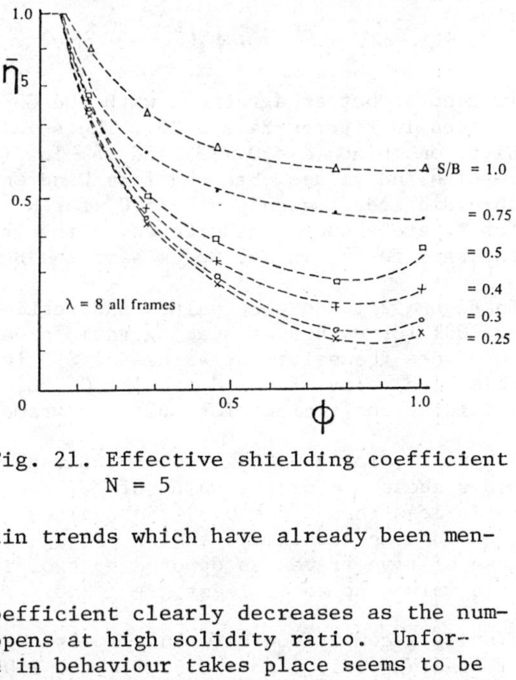

Fig. 21. Effective shielding coefficient N = 5

solidity ratio is quite complex, but certain trends which have already been mentioned are now highlighted.

At low values of solidity the shielding coefficient clearly decreases as the number of frames increases. The opposite happens at high solidity ratio. Unfortunately the exact value where this change in behaviour takes place seems to be a function of the particular spacing ratio.

It would appear that the most compact presentation of the results is a series of graphs, as in Figures 20 and 21, for various values of N, say 2, 3, 4, 5, 7 and 10 (the test values), where the shielding coefficients for intermediate values of N would be evaluated by interpolation.

COMPARISON WITH CODE PROVISIONS

The method of determining shielding coefficients as treated by Wind Codes was discussed in the literature survey. Comparisons with the presently derived test results are now made, and the results show discrepancies which in some instances exceed 100%.

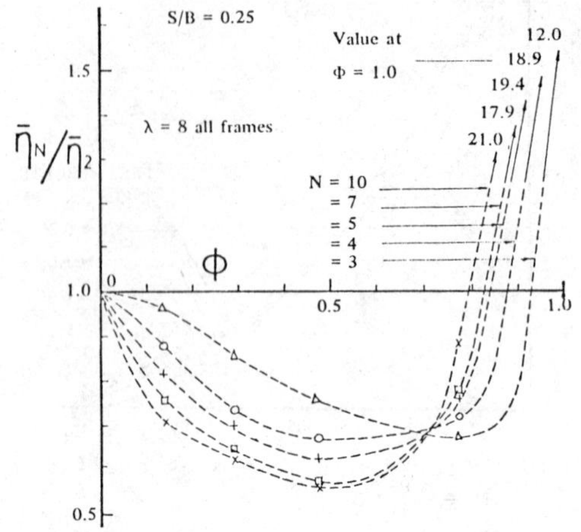

Fig. 22. $\bar{\eta}_N/\bar{\eta}_2$ vs ϕ (SB = 0.25)

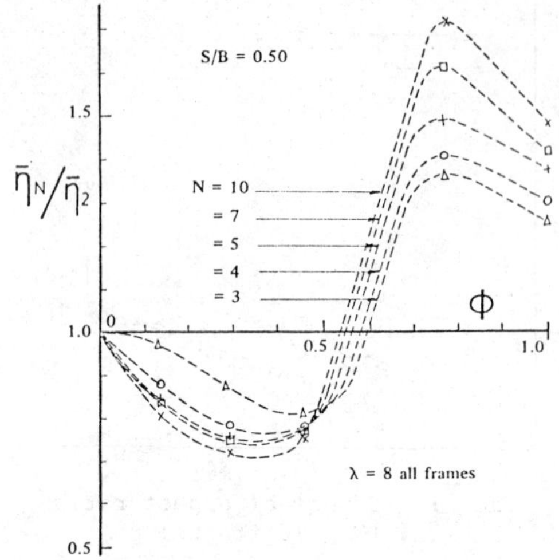

Fig. 23. $\bar{\eta}_N/\bar{\eta}_2$ vs ϕ (SB = 0.50)

In general better agreement with the Codes is found for low solidities. This can be seen in Figures 24 and 25. The shielding coefficients at two values of solidity, one high (ϕ = 0.773) and one low (ϕ = 0.286) are shown, contrasting the test results and values obtained from Wind Codes. On each figure, the experimental shielding coefficients at $\alpha = 0°$ for N = 2 and N = 5 are included. Only the curves for $\bar{\eta}_2$ are shown for the Codes, although as was earlier described, these would be the same for $\bar{\eta}_5$ in all Codes save the Belgian Code.

In Figure 24, the Code values underestimate the shielding coefficient by as much as 100% for values of spacing ratio greater than S/B = 0.5. The disparity between the Codes themselves at values of S/B less than unity is in fact remarkable. For this high value of solidity, ϕ = 0.773, the Belgian practice of reducing the shielding coefficient for multiple frames (N > 3) is clearly misleading.

Figure 25, for a value of ϕ = 0.286, shows much better agreement with the various Codes chosen. For the value of $\bar{\eta}_2$, the Codes are all conservative for a spacing ratio less than S/B = 0.5. For values of spacing ratio greater than S/B = 1.0, all the Codes underestimate the shielding coefficient for N = 2. If we look at the case of five frames as opposed to two, it would appear that the codes give reasonable values up to at least S/B = 2.0.

Finally a comparison between the test results and code values, for various values of spacing ratio (S/B = 0.5, 1.0 and 2.0) is made using the method of compilation found in the codes. The comparisons are shown in Figures 26, 27 and 28 respectively.

The areas in which the worst discrepancies occur are clearly revealed as those frame groups with high solidity ratio and spacing ratio greater than unity. In these cases the shielding coefficient is severely underestimated by the Codes.

Although Figure 28 only shows results for $\bar{\eta}_2$, extrapolation of our data indicates that the curves for N > 2 would be very near or above the $\bar{\eta}_2$ curve.

At low values of solidity however, the Codes appear to be slightly conservative at spacing ratios less than unity. There is also evidence for a further reduction in the shielding coefficient for frame groups with N > 2.

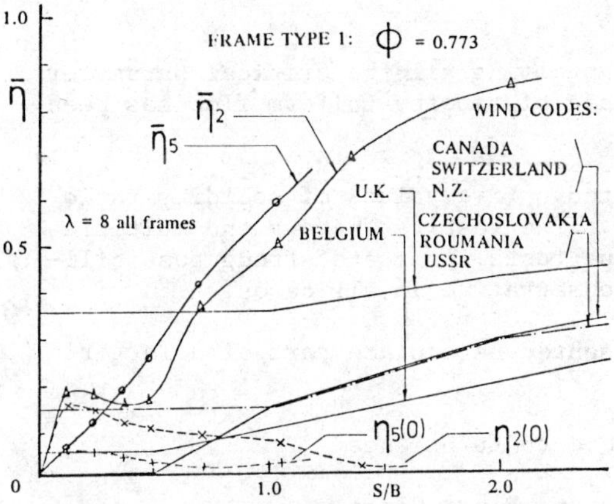

Fig. 24. Comparison of test results with code values

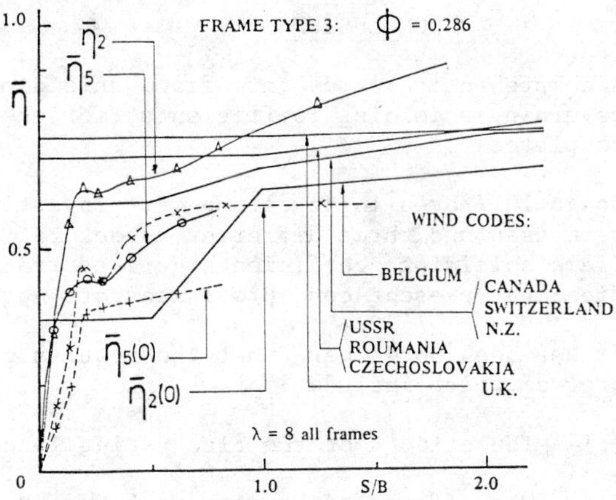

Fig. 25. Comparison of test results with code values

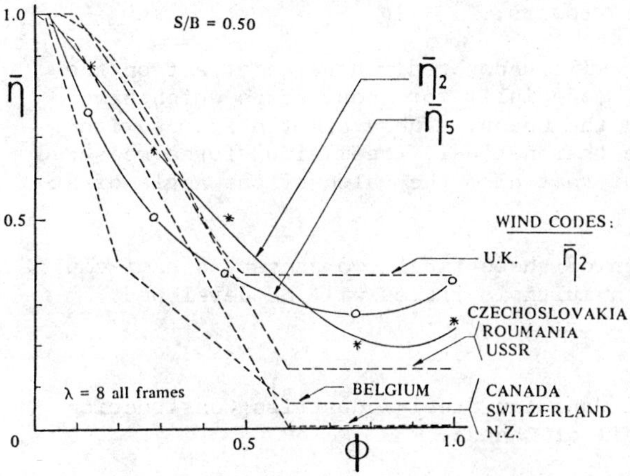

Fig. 26. $\bar{\eta}$ vs ϕ (S/B = 0.50)

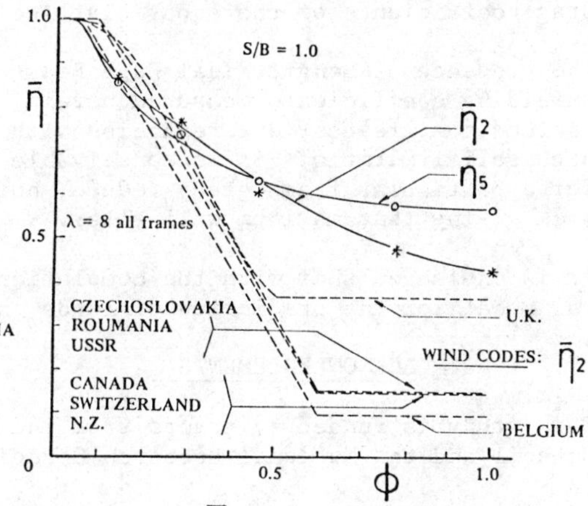

Fig. 27. $\bar{\eta}$ vs ϕ (S/B = 1.0)

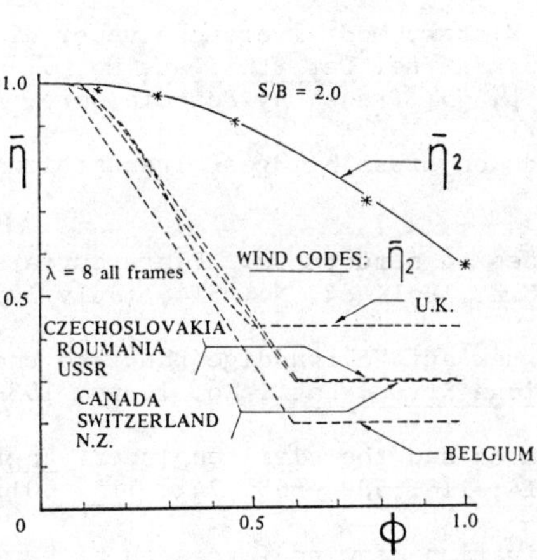

Fig. 28. $\bar{\eta}$ vs ϕ (S/B = 2.0)

CONCLUSIONS

A comprehensive study into frame shielding involving all the critical parameters governing shielding coefficients, for the case of smooth, uniform flow has been completed.

Up to 10 frames in any group were investigated. Five values of solidity ratio were used and three different aspect ratios. The results, namely the multiple frame shielding coefficients derived from the tests, indicate strong possibilities that present code provisions are not conservative in all cases.

It has been noted here that the results presented herein are part of a larger program which include:

(i) The effects of the fluctuating loads due to turbulence.

(ii) The effects of Longitudinal Edge Beams and Floor Slabs.

The work of Sykes (15) indicated negligible influence of turbulence on the mean drag coefficients of the square latticed towers.

The presence of Longitudinal Edge Beams will undoubtedly have an effect on the shielding coefficients measured herein, especially for those cases which have yielded the greatest discrepancies with the Codes. These are the frames with high solidity ratio. It is conceivable that not only the maximum force measured for a particular frame group reduce, but that also the value of the angle of attack giving that maximum will alter.

It is envisaged that with the completion of these final two stages, a semi-empirical model for the prediction of loads on multiple frames will be developed.

ACKNOWLEDGEMENTS

This study is funded by grants from the Canadian Steel Industries Construction Council and the National Research Council of Canada.

REFERENCES

(1) O. Flachsbart and H. Winter, Modellversuche ueber die Belastung von Gitterfachewerken durch Windkraefte, Der Stahblau, No. 9, p. 65, No. 10 (1934), p. 73; No. 8 (1935), p. 65; Trans. by Scandia Corp. AFSWP-464.

(2) J.M. Biggs, Wind loads on truss bridges, Transactions, A.S.C.E., Vol. 119, p. 879 (1954).

(3) W.W. Pagon, Wind forces on structures: Plate curders and trusses, Proc. Paper No. 1711, A.S.C.E., Vol. 84, No. ST4, (July 1958).

(4) O. Flachsbart, Winddruck auf Vollwandige Bauwerke und Gitterfachewerke, Publ. Int. Assoc. Bridge Struc. Eng., No. 1, pp. 153-172.

(5) W.W. Pagon, Aerodynamics and the civil engineer, Engineering News Record, 112, 41, 348, 456, 814; 114, 582, 665, 742; 115, 601.

(6) G.S. Vincent, Investigation of wind forces on highway bridges, Highway Res. Board, Spec. Rep. 10, Publ. 272, Washington, D.C., N.R.C.

(7) G. Ower, Wind resistance of lattice girder bridges, Inst. of Civil Eng. and Inst. of Struct. Eng., London, England, 58 pp. (1948).

(8) R.W.F. Gould and W.G. Raymer, Measurements over a wide range of Reynold's numbers of the wind forces on models of lattice frameworks with tubular members, N.P.L. Maritime Services, Report No. 5-72 (May 1972).

(9) Wind Resistant Design Regulations: A World List 1975, complied by Jap. Org. Comm. of the 3rd Int. Conf. on Wind Effects on Buildings and Structures and Comm. of Wind Loading, Soc. of Steel Construction of Japan (July 1975).

(10) W.B. Woodhouse, Wind pressure on latticed towers, J. Inst. Elect. Eng., 77, pp. 189-196.

(11) E. Cohen and H. Perrin, Design of multi-level guyed towers; wind loading, Proc. Paper No. 1355, A.S.C.E., Vol. 83, No. ST5, (September 1957).

(12) D.G. Walshe, Measurements of wind force on a model of a power station boiler house at various stages of erection, N.P.L. Aero. Report 1165 (September 1965).

(13) R. Möll and F. Thiele, Windkanalversuche am Modell eines Stanlskelett-Hochregals zur Bestimmung des Widerstaudsbiewertes nach DIN 1055 für ein filigranartiges Bauwerk, Der Stahblau, No. 3 (March 1972).

(14) R.J. Melling, Loads on open lattice structures in nominally smooth flow, Proc. 3rd Coll. on Ind. Aerodynamics, Aachen, pp. 107-127 (1978).

(15) D.M. Sykes, Wind loads on lattice structures in turbulent airflow, Proc. 3rd Coll. on Ind. Aerodynamics, Aachen, pp. 129-148 (1978).

(16) B.E.A. Jacobs, Determination of shielding factors for multiple frame structures, Proc. 3rd Coll. on Ind. Aerodynamics, Aachen, pp. 149-164 (1972).

(17) D.G. Chow, Loads on open lattice ctructures - a comparative study, Proc. 3rd Coll. on Ind. Aerodynamics, Aachen, pp. 165-177 (1978).

MEAN AND FLUCTUATING INTERNAL PRESSURES INDUCED BY WIND

J. D. HOLMES

James Cook University of North Queensland, Townsville, Australia

ABSTRACT

Mean and fluctuating pressures inside buildings, induced by wind, have been investigated using boundary layer wind tunnel and computer simulation techniques. Mean and r.m.s. fluctuating internal pressure coefficients were both found to be monotonic functions of the ratio of windward to leeward opening areas.

The case of a single windward opening was treated as a damped Helmholtz resonator. Inertia effects were found to produce resonance amplification in the response of the internal pressure to turbulent external pressures and to a step change in external pressure. However, these effects are unlikely to be of much practical significance except for the case of a sudden large opening occurring in a relatively rigid building.

For correct scaling of fluctuating internal pressures, wind tunnel testing should be tested at full scale wind velocities. However, it is shown that similarity can be maintained at lower wind tunnel velocities, by distorting the internal volume by a factor equal to the square of the velocity ratio.

NOMENCLATURE

- A - opening area
- b - breadth of opening
- c - speed of sound
- C_p - pressure coefficient
- C_y, C_z - decay constants for external pressure co-spectra (Eqn. 10)
- F_u - empirical factor in E.S.D.U. expression for turbulence intensity (Fig. 1)
- h - (1) eaves height of building (2) height of opening (Eqn.10)
- I_u - longitudinal turbulence intensity
- k - orifice discharge coefficient
- ℓ_e - effective length of air "slug"
- n - (1) polytropic exponent (2) frequency
- n_0 - Helmholtz resonance frequency
- p - pressure
- p_0 - atmospheric (static) pressure
- q - dynamic pressure
- Q - volume rate of flow
- r - $\ell_e/\sqrt{A}$
- $S(n)$ - auto-spectral density
- t - time
- u - longitudinal velocity component
- V_0 - internal volume
- x - displacement of air slug
- y - lateral displacement
- z - height
- z_0 - roughness length (Fig. 1)
- z_r - reference height (Fig. 1)
- Δ - change
- λ_u - peak wave length of longitudinal velocity spectrum

μ - dynamic viscosity
π_j - non-dimensional group
ρ - air density
τ - equilibrium time
χ^2 - aerodynamic admittance
Subscripts:
e - external
i - internal
L - leeward
W - windward
m - model scale
f - full scale
Superscripts:
— - mean value
' - fluctuating value
* - non-dimensional

1. INTRODUCTION

Internal pressures induced by wind on buildings have traditionally received much less attention than have external pressures. This is the case despite the fact that, for low-rise buildings in particular, the internal pressure loading may form a high proportion of total design wind loading on both structure and cladding. During Cyclone Tracy in Darwin, Australia, for example, a high proportion of roof failures on domestic buildings were believed to have been preceded by window failure on the windward side of the building, resulting in high internal positive pressures combining with roof suctions [Walker(1)] .

The relationship between *mean* internal pressures and the relative size of windward and leeward wall openings has been understood for some time [Euteneuer(2), Van Koten(3), Newberry and Eaton(4), Liu(5)], and is reflected in the coefficients of internal pressure specified in wind loading codes [e.g. S.A.A.(6)].

Euteneuer(7) derived an expression for the "response time" of the pressure inside a low-rise building to a step change in external pressure such as that caused by the sudden failure of a window; Vickery(8) and Liu(9) later derived similar expressions. Jancauskas and Sharp(10) carried out some boundary layer wind tunnel measurements in which a window failure was simulated, and the change in lift force on the roof of a 1/100 scale house model was measured. Jancauskas and Sharp also measured mean and fluctuating internal pressures for a range of windward to leeward opening ratios. Kramer, Gerhardt and Scherer(11) carried out wind tunnel measurements of response time and attributed differences from theoretical values to the effect of inertia forces.

In the present paper, a study of internal pressures in low-rise buildings, with both windward and leeward wall openings, using boundary layer wind tunnel and numerical simulation techniques, is described. The case of a single windward wall opening is treated in some detail: the effect of inertia forces, previously neglected, on the "response time" to a step change in external pressure is examined by means of numerical simulation. The fluctuating internal pressure resulting from external pressure fluctuations, due to wind turbulence, is studied using both wind tunnel and numerical simulation methods.

Although many of the results are applicable to all types of buildings, the work is mainly directed towards domestic houses and other low-rise buildings, and forms part of a research study sponsored by the Australian Housing Research Council.

This paper is a shortened version of a report [Holmes(12)] , in which more detail of the theory and experimental methods is included; design applications are also discussed.

2. THEORY

2.1 Mean Pressures

The mean internal pressure coefficient inside a building with total areas of windward openings, A_W, and leeward openings A_L, can readily be derived by assuming standard orifice flow relationships in and out of the openings and considering mass conservation(3), (5), (12):

$$\overline{Cp_i} = \frac{\bar{p}_i - p_0}{q} = \frac{\bar{C}p_W}{1 + (\frac{A_L}{A_W})^2} + \frac{\bar{C}p_L}{1 + (\frac{A_W}{A_L})^2} \qquad (1)$$

where p_0 is the reference static (atmospheric) pressure and $q = \frac{1}{2}\rho\bar{u}^2$, is the reference dynamic pressure. $\bar{C}p_W$, $\bar{C}p_L$ are the external pressure coefficients at the windward and leeward openings, respectively.

2.2 Single Windward Opening

The case of a single windward opening is normally the critical design case for low-rise buildings and deserves special attention. The mean internal pressure coefficient, from equation (1), is equal to the mean external pressure coefficient. However, when a building is immersed in a turbulent boundary layer flow such as that characteristic of the earth's boundary layer in strong winds, the external pressure will be fluctuating or turbulent, especially for low-rise buildings, and the internal pressure will respond in some way to these fluctuations. Since there is only a single opening, flow into the building resulting from an increase in external pressure will cause an increase in density of the air within the internal volume; this, in turn, causes an increase in internal pressure. The pressure changes due to wind *relative to atmospheric pressure* are small, and the *relative* density changes are of the same order. These density changes can be maintained by only relatively small mass flows in and out of the building and the internal pressure can therefore be expected to respond fairly rapidly to external pressure changes, except for the case of a very small opening size.

It is useful, first, to carry out a dimensional analysis to establish the non-dimensional groups involved.

2.2.1 Dimensional analysis

From dimensional analysis the fluctuating internal pressure coefficient, Cp_i' can be written:

$$Cp_i' = \frac{p_i - p_0}{\frac{1}{2}\rho\bar{u}^2} = F(\pi_1, \pi_2, \pi_3, \pi_4, \pi_5) \qquad (2)$$

where $\pi_1 = \frac{A^{3/2}}{V_0}$; $\pi_2 = \frac{p_0}{\frac{1}{2}\rho\bar{u}^2}$; $\pi_3 = \frac{\rho\bar{u}\sqrt{A}}{\mu}$ (Reynolds Number)

$\pi_4 = \frac{\sqrt{\overline{u'^2}}}{\bar{u}}$ (turbulence intensity); $\pi_5 = \frac{\lambda_u}{\sqrt{A}}$

where μ is viscosity

$\bar{u}$ is the mean velocity of the stagnation flow

$\sqrt{\overline{u'^2}}$ is the root-mean-square longitudinal turbulence component

λ_u is the peak wave length of the longitudinal turbulence spectrum i.e. a measure of turbulence scale

π_2, the ratio of atmospheric pressure to the reference dynamic pressure, is a parameter closely related to Mach Number. The significance of this parameter remains to be established, but establishing its importance is relevant to wind tunnel testing since equality with values for full scale design cases is normally not possible in conventional boundary layer wind tunnels. The Reynolds Number, π_3, is not likely to be a significant parameter except for small opening areas, A.

2.2.2 Response time

The case of the response of the internal pressure to a sudden external pressure increase, such as that caused by a sudden window failure, has been considered previously [Euteneur(7), Vickery(8) and Liu(9)].

The conservation of mass equation can be written in this case:

Rate of mass flow in = Increase in internal mass of air

$$\rho_i Q = \left(\frac{d\rho_i}{dt}\right) V_0 \qquad (3)$$

Assuming a polytropic law relating pressure and density:

$$\text{i.e.} \quad \frac{p_i}{\rho_i^{\,n}} = \text{constant} \qquad (4)$$

and for flow through the orifice:

$$Q = k\,A\sqrt{\frac{2(p_e - p_i)}{\rho}}, \qquad (5)$$

it can be shown that the equilibrium time when the internal pressure becomes equal to the internal pressure is given by:

$$\tau = \frac{\rho V_0 \bar{u}}{nkAp_0} \sqrt{Cp_e - Cp_{i_0}} \qquad (6)$$

where Cp_{i_0} is the initial value of Cp_i

2.2.3 Helmholtz resonator model

The theoretical basis of the computer simulation models used in this paper is now described. The orifice flow and polytropic pressure-density relationship used in the previous section are retained, but in addition, the inertia effects of the air moving in and out of the opening are included. The case is, in fact, a special case of a "Helmholtz resonator" well known in acoustics [e.g. Rayleigh(13), Malecki(14)].

A "slug" of air can be considered to be moving in and out of the opening in response to the external pressure changes.

A differential equation for the motion of this mass of air can be written:

$$\rho A \ell_e \ddot{x} + \frac{\rho A}{2k^2}\dot{x}|\dot{x}| + \frac{np_0 A^2}{V_0}x = \Delta p_e A \qquad (7)$$

This is the differential equation of a single degree of freedom dynamic system with non-linear (square law) damping. The mass of the "slug" of air is $\rho A \ell_e$, where ℓ_e is the effective length of the slug. ℓ_e can be taken to be $\sqrt{\pi A/4}$; this is strictly correct for circular openings only, but is a good approximation for rectangular openings of low aspect ratio [Malecki(14)].

The "stiffness" is the resistance of the internal pressure to unit deflection of the "slug". The damping force follows from the turbulent flow equation(5).

The undamped resonant frequency is:

$$n_0 = \frac{\omega_0}{2\pi} = \frac{1}{2\pi}\sqrt{\frac{nAp_0}{\rho \ell_e V_0}}$$

$$= \frac{cA^{1/4}}{2^{1/2}\pi^{5/4}V_0^{1/2}} \qquad (8)$$

where $c = \sqrt{\frac{np_0}{\rho}}$ is the speed of sound

Equation (8) is a well-known formula for the Helmholtz resonance frequency [Rayleigh(13), p. 187].

Equation (7) can be written with C_{p_i} as the dependent variable:

$$\frac{\rho \ell_e V_0}{np_0} \ddot{C}_{p_i} + \frac{\rho V_0^{\,2} q}{2k^2 n^2 A p_0^{\,2}} \dot{C}_{p_i} |\dot{C}_{p_i}| + A.C_{p_i} = A.C_{p_e} \qquad (9)$$

Equation (9) should be valid for the case of a fluctuating external pressure coefficient Cp_e induced by wind turbulence, provided that the parameter $\lambda_u / \sqrt{A}$ (π_5 in section 2.2.1) is large, i.e. if the opening dimensions are small in comparison with the scale of turbulence. Equation (9) was used as the basis of the computer simulation method described in this paper. The equation is solved for the time-varying internal pressure coefficient $C_{p_i}(t)$ for a specified external pressure variation $C_{p_e}(t)$.

3. INVESTIGATION TECHNIQUES

3.1 Boundary Layer Wind Tunnel Measurements

The wind tunnel measurements described in this paper were carried out on a 1/50 scale model of a two storey house in the James Cook University boundary layer wind tunnel; the latter has been described in detail elsewhere [Holmes(15)]. A rural atmospheric boundary layer at the scale of 1/50 was simulated by placing a 300 mm high plain fence or barrier at the start of the 13.5 m long test section followed by floor roughness consisting of carpet. This simulation method has been used for a number of studies of external wind pressures on house models carried out in this tunnel [Holmes and Best(16,17,18), and Best and Holmes(19)]. Profiles of mean velocity and turbulence intensity are shown in Figure 1 and compared with E.S.D.U.(20) data. The turbulence scale as indicated by the longitudinal turbulence spectrum is approximately half that suggested by the E.S.D.U. data for the scale of 1/50(12).

The house model is shown in Figure 2, and was specially constructed to enable pressures at taps vented to the internal volume, to be measured, as well as at external taps. Pressure tubing connecting the external taps to the pressure transducer beneath the wind tunnel turntable, was brought in between double side walls giving a clear internal space. A previous study [Munarin(21)] had shown that the internal pressures were substantially uniform throughout the internal volume and for most of the tests described in this paper, a pressure transducer (Setra 237) was connected by a 6 mm hole to a single pressure tap on the floor of the model. The model was made from "perspex", and the external pressure taps consisted of short lengths of stainless steel tubing of about 1 mm internal diameter. A number of interchangeable panels with openings of various sizes were made to fit open sections located centrally in the front and rear walls. This enabled the windward and leeward open areas A_W, A_L to be varied. The maximum open area obtainable was approximately 22% of the total wall surface area.

The reference dynamic pressure was obtained by measuring the free stream mean velocity $\bar{u}_h$, at a height equal to the eaves height of a building using a linearised hot-film anemometer system (T.S.I. 1054B). The reference position was about 500 mm to the side of the model centre and slightly upwind. The static pressure reference, p_0, was derived from a pitot-static tube mounted near the top of the wind tunnel. In a separate test, the static pressure at the reference (eaves) height was obtained and a small correction was applied to the results.

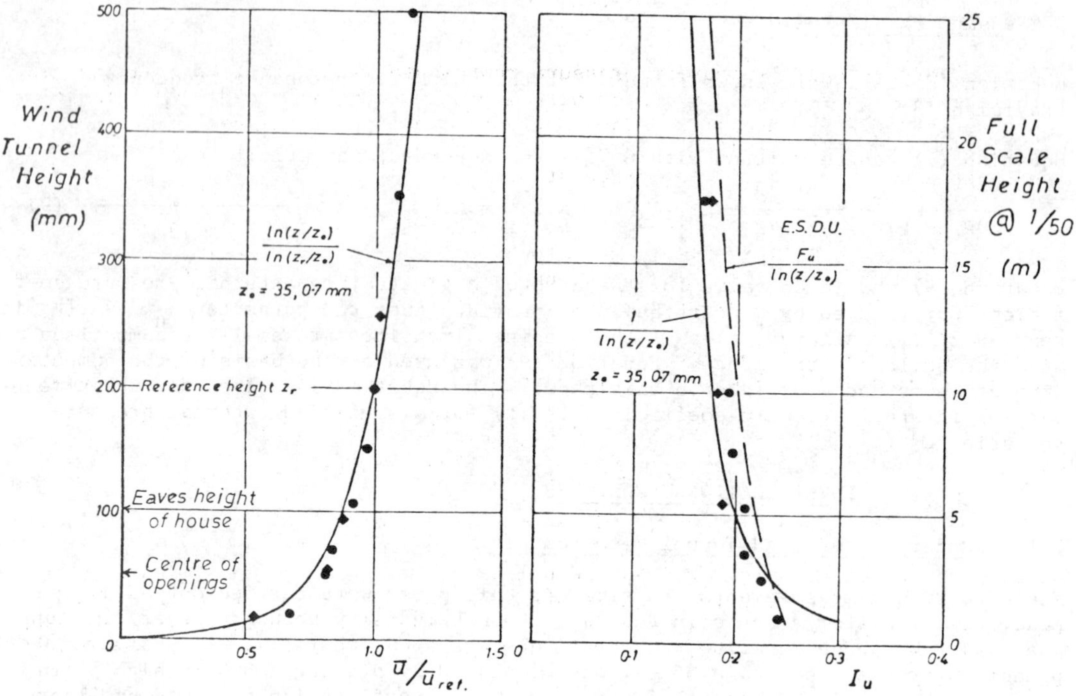

Figure 1. Wind tunnel mean velocity and turbulence intensity profiles

Figure 2. 1/50 scale house model

3.2 Computer Simulation

"Records" of external stagnation pressure representative of that on buildings in a turbulent atmospheric boundary layer were computed from simulated wind velocity records generated by the Inverse Fast Fourier transform technique [Holmes(22)]. The generation procedure incorporated a von Karman-Harris spectral model in which the length scale, λ_u, could be varied at will.

The records so generated became the "forcing function" on the right hand side of equation (9) for the single windward opening case. The equation was solved for the internal pressure variation using the lumped-impulse or central-difference method commonly used in structural dynamics [e.g. Biggs(23)].

An optional filter was incorporated to allow for the effective pressure reduction (aerodynamic admittance) over the windward opening. This filter took the form:

$$\chi^2(n) = \int_0^b\int_0^b\int_0^h\int_0^h \exp\{-\frac{n}{\bar{u}}[C_y{}^2(y_1 - y_2)^2 + C_z{}^2(z_1 - z_2)^2]^{\frac{1}{2}}\}dy_1 dy_2 dz_1 dz_2$$

$$= \frac{4}{b^2h^2}\int_0^b\int_0^h (b-y)(h-z)\exp\{\frac{-n}{\bar{u}}[C_y{}^2y^2 + C_z{}^2z^2]^{\frac{1}{2}}\}\,dy\,dz$$

$$= 4\int_0^1\int_0^1 (1-\frac{y}{b})(1-\frac{z}{h})\exp\{\frac{-n}{\bar{u}}[C_y{}^2y^2 + C_z{}^2z^2]^{\frac{1}{2}}\}\,d(\frac{y}{b})\,d(\frac{z}{h}) \qquad (10)$$

where b, h are the opening breadth and height respectively. The reduction to a double integral is after Bearman(24). This filter attenuated the external pressure fluctuations, the attenuation being larger for high frequencies and for large opening sizes. The constants C_y, C_z were derived from wind tunnel tests.

4. RESULTS

4.1 Mean Internal Pressures

The mean internal pressure coefficient for a range of windward and leeward opening sizes obtained from the wind tunnel experiments is shown in Figure 3. Values obtained previously by Munarin(21) by averaging 20 second (wind tunnel time) mean values from each of the 19 individual internal pressure taps on the model are shown, as well as those from the present study, in which an ensemble average of 27 separate runs from the single floor pressure tap was calculated.

Equation (1), with appropriate values of external mean pressure coefficients, agrees well with measured data for windward opening areas greater than about 5% of the total windward wall surface area. However for the single windward opening case, there is a reduction in mean internal pressure with decreasing area, for smaller opening areas. This may be due to the effects of leakage on the leeward sides of the model becoming important.

4.2 Internal Pressure Fluctuations

Figure 4 is the corresponding plot to Figure 3 for the r.m.s. fluctuating internal pressure coefficient. Again a steady increase occurs with increasing A_W/A_L ratio, and for opening areas over about 5% of the windward surface it is insensitive to changes in opening area. For non-zero leeward opening areas, the internal fluctuating pressure depends on the fluctuating external pressure on the leeward wall, as well as that on the windward side. The leeward external pressure fluctuations are markedly less in magnitude than the windward pressure fluctuations, and this adequately explains the reduction in internal pressure fluctuations with increasing leeward area, A_L.

For the single windward opening case, as is clear from sections 2.2.1 and 2.2.3,

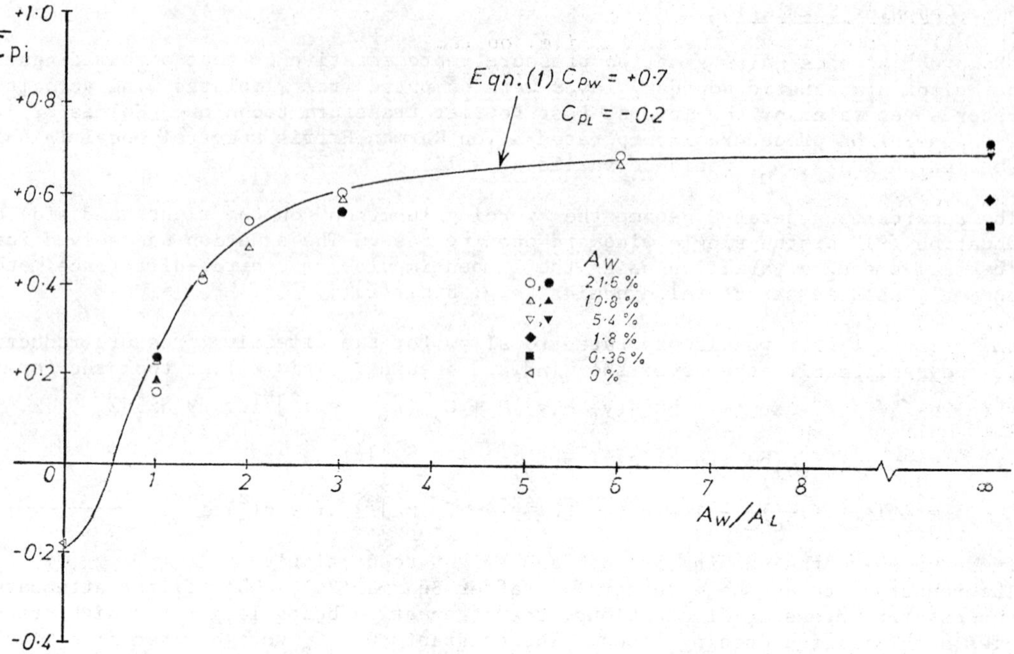

Figure 3. Mean internal pressure coefficient versus A_W/A_L
(Open symbols - averages over pressure taps
Solid symbols - ensemble averages for single floor tap).

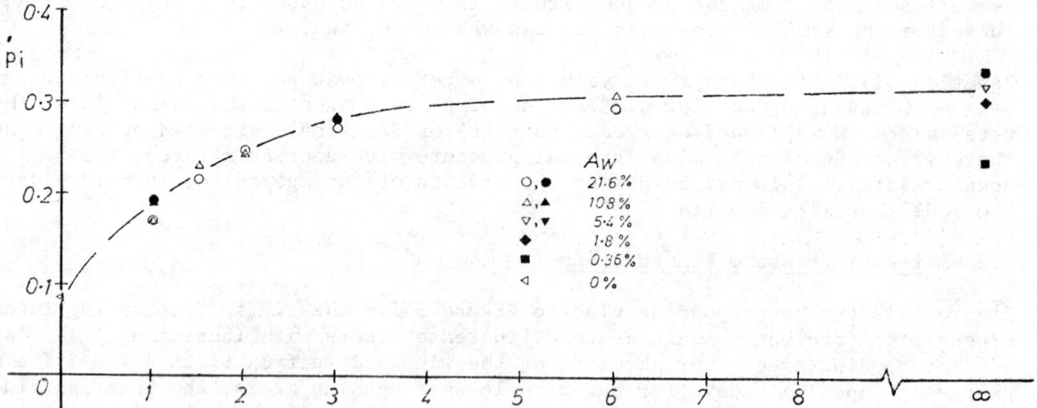

Figure 4. R.m.s. fluctuating internal pressure coefficient versus A_W/A_L
(Open symbols - averages over pressure taps
Solid symbols - ensemble averages for single floor tap)

the internal fluctuating pressure coefficient depends on the internal volume, V_0, as well as the windward area, A_W, i.e. on the ratio $A_W^{3/2}/V_0$. It is clear from Figure 4, for the house model, that a reduction in A_W, for A_W/A_L equal to infinity, generally causes a reduction in the internal r.m.s. pressure coefficient.

Figures 5 to 8 show spectra estimates of the internal pressure fluctuations from the house model, for the single windward opening case, for four different values of the opening area, A_W. Also shown are the corresponding spectra estimates obtained from the computer simulated data. Parts of the recorded and simulated data, corresponding to Figure 6, are shown in Figure 9.

The measured and simulated data clearly show resonant peaks occurring at frequencies close to those calculated from equation (8). In agreement with this equation, the resonant frequencies increase with the opening area. In addition, the height of the peaks increases and the width decreases with increasing opening area, indicating lower damping; this again is consistent with the damping term in equation (9). The small resonant effects are apparent in the recorded and simulated data (Figure 9).

For the computer simulations, the aerodynamic admittance filter (equation (10)) was applied, using values for the constants C_y and C_z of 7 and 3 respectively. These were obtained by measuring the co-spectra for external pressure fluctuations on the front wall of the model with the windward opening sealed and fitting exponential decay functions, appropriate to equation (10). However, no account of the turbulence distortion effects which tend to attenuate the pressure fluctuations at high frequencies [e.g. Bearman(25), Holmes(26)] was taken. This probably partly explains the overestimates in the simulated data of the internal pressure fluctuations at frequencies below the resonant frequency for the lower windward areas (Figures 5 and 6). The leeward leakage mentioned previously may also contribute to this.

The polytropic exponent, n, is 1.0 for isothermal and 1.4 for adiabatic conditions; an intermediate value of 1.2 was used in the simulations. This was vindicated by the good agreement between the computed and measured resonant frequencies. The orifice discharge coefficient, k, is about 0.6 under steady flow conditions [Pankhurst and Holder(27)], but is likely to differ considerably from this under highly fluctuating and reversed flow conditions, as in the present case. A value of 0.15 gave resonant peaks on the simulated spectra in quite good agreement with the measured data, although a larger value may be more appropriate for smaller openings.

Although the resonant peaks on the spectra are quite marked, their contribution to the total variance is small as the frequencies are high in comparison with those of the external 'forcing' pressure fluctuations. More significant effects appear to be those of the aerodynamic admittance and turbulence distortion effects described above. Values of r.m.s. fluctuating pressure coefficients corresponding to the spectra in Figures 5 to 8 are given in Table 1.

All the results discussed previously were for a mean velocity of about 10 m/s at the eaves height of the building, and this velocity applies to both full-and model-scale. Thus the value of the $p_0/\frac{1}{2}\rho\bar{u}^2$ (π_2 in section 2.2.1) was about 1670. The effect of increasing the mean velocity is to move the resonance wave number downwards towards the main peak of the pressure spectrum, but the damping is also increased as is clear from equation (9). (Figure 7)

4.3 Response to a Step Change in External Pressure

The same computer program used to solve equation (9), for the simulated turbulent pressures, was also used to illustrate the effect of inertia terms on the response

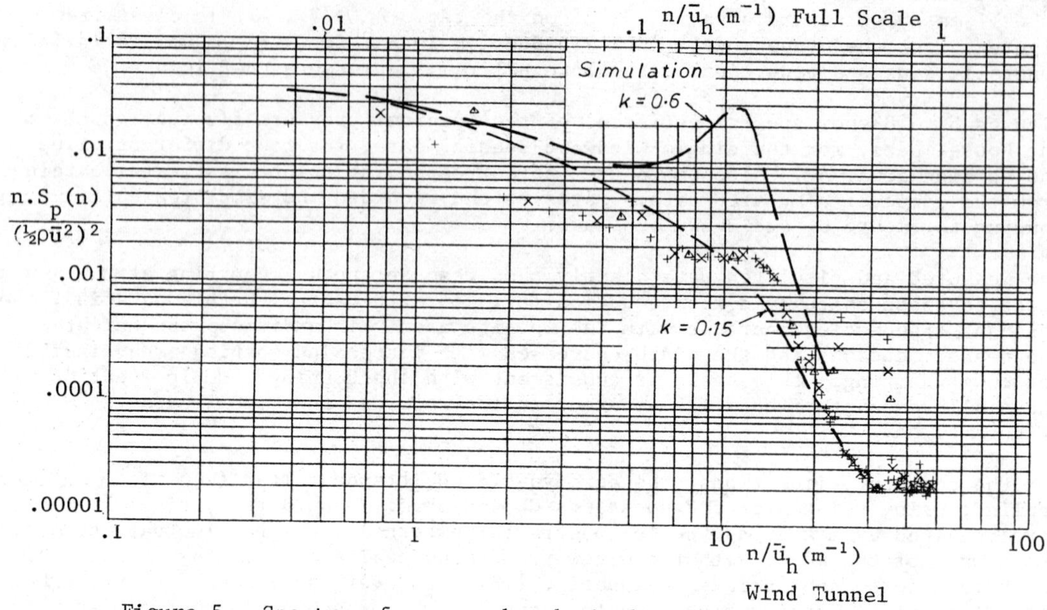

Figure 5. Spectra of measured and simulated internal pressure fluctuations. $A_w^{3/2}/V_0 = .00044$. $\bar{u}_h \simeq 10$ m/s.

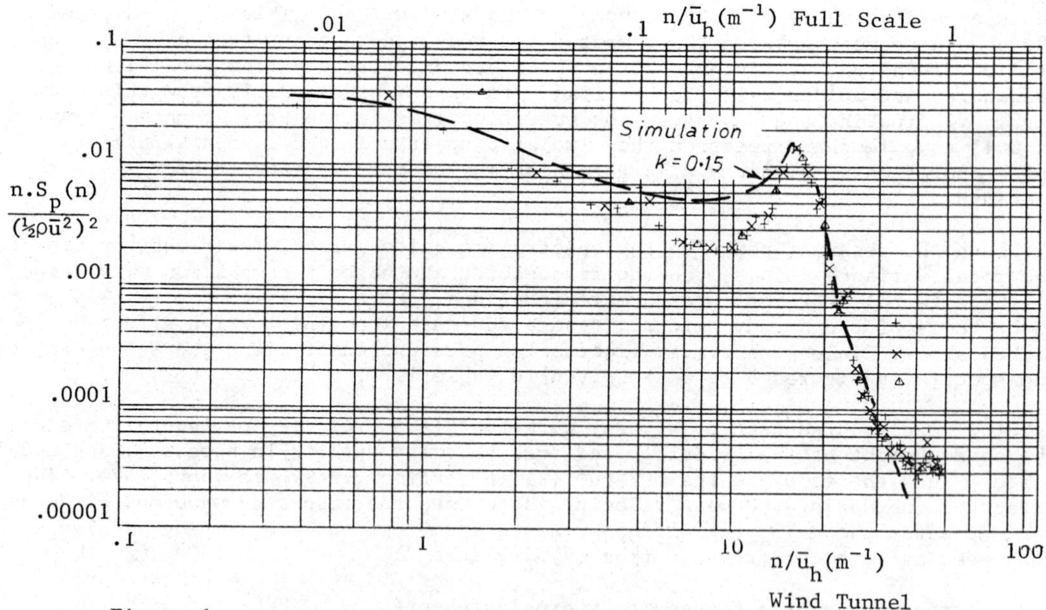

Figure 6. Spectra of measured and simulated internal pressure fluctuations. $A_w^{3/2}/V_0 = .0050$. $\bar{u}_h \simeq 10$ m/s.

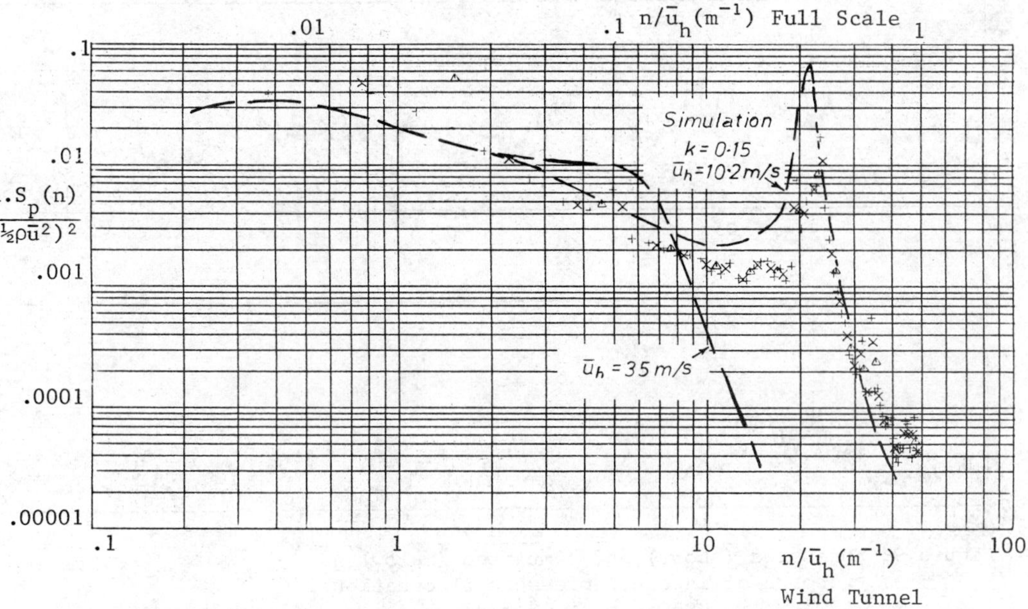

Figure 7. Spectra of measured and simulated internal pressure fluctuations. $A_W^{3/2}/V_0 = .026$.

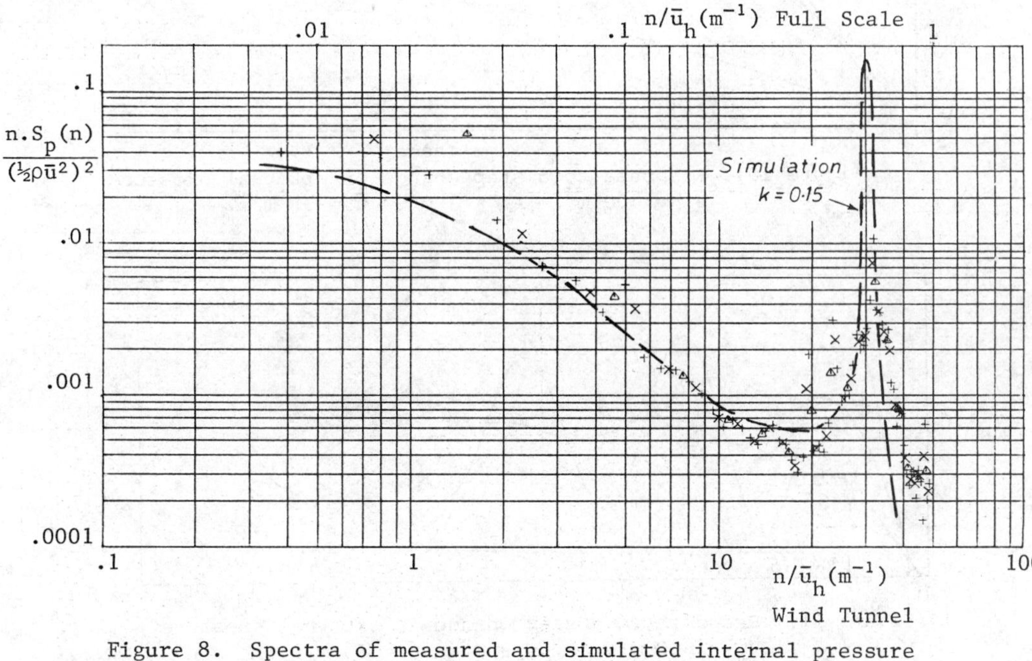

Figure 8. Spectra of measured and simulated internal pressure fluctuations. $A_W^{3/2}/V_0 = .21$. $\bar{u}_h \simeq 10$ m/s.

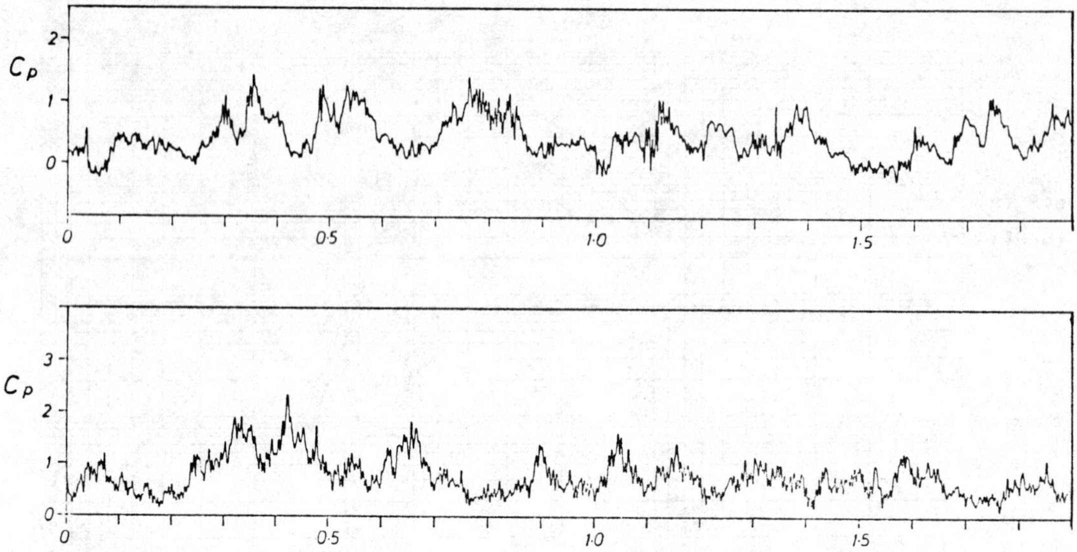

Figure 9. Measured (above) and simulated (below) records of internal pressure fluctuations. $A_W^{3/2}/V_0$ = .0050. $\bar{u}_h \simeq 10$ m/s.

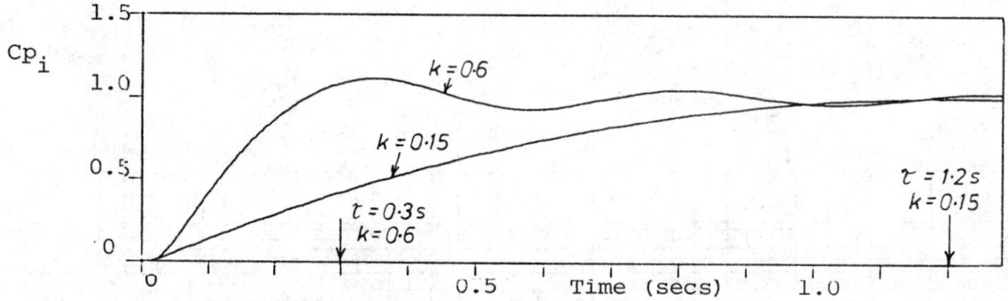

Figure 10. Response to a step change in external pressure V_0 = 600 m^3. A_W = 1 m^2. $\bar{u}$ = 30 m/s.

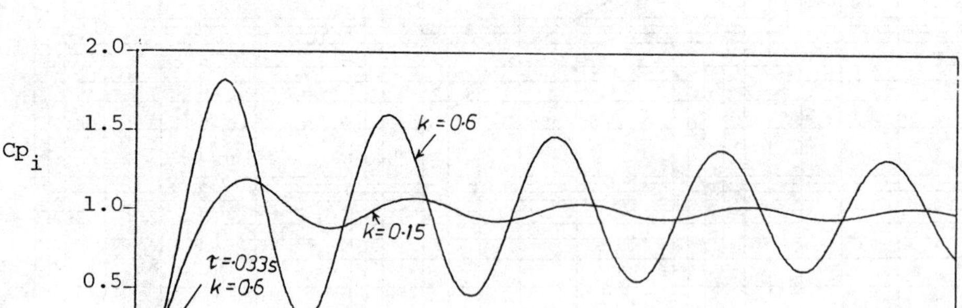

Figure 11. Response to a step change in external pressure V_0 = 600 m^3. A_W = 9 m^2. $\bar{u}$ = 30 m/s.

TABLE 1. $C\acute{p}_i$ from wind tunnel measurements and simulated data for $A_W/A_L = \infty$

$A_W^{3/2}/V_0$	Cp_i	
	Measured*	Simulated (k = 0.15)
.00044	.238	.326
.0050	.303	.336
.026	.338	.339
.21	.337	.330

* ensemble average for 27 separate runs

of an internal volume to a sudden change in external pressure.

Figures 10 and 11 show the response of an internal volume of 600 m^3 to a step increase in external pressure coefficient from zero to 1.0; a velocity of 30 m/s, and opening areas of 1 m^2 and 9 m^2 were considered. These could represent window and door failures in a small industrial building [the 1 m^2 area case was used as an example by Liu(9)]. A value of orifice coefficient, k, of 0.6 might be appropriate in the initial stages when the flow is all in one direction, but if oscillations occur then, from the previous section, it appears that a lower value is applicable. Calculations were done with values of 0.6 and 0.15 for k, and the results of both cases are shown in Figures 10 and 11. Values of the equilibrium time, τ, calculated from equation (6), in which inertia effects are ignored, are also shown.

Despite the doubt about the orifice coefficient, it is apparent that inertia effects are significant in the case of the larger opening when the damping term in equation (9) is much lower. This is, at least qualitatively, in agreement with the wind tunnel measurements of Kramer, Gerhardt and Scherer(11). It should be noted, however, that the effect of building flexibility has not been considered in the analysis; this is likely, in real buildings, to have a significant effect by increasing damping and reducing 'overshoot' and oscillation effects. Vickery(8), in fact, treated this effect in his consideration of response time by assuming an effective atmospheric pressure p_0' given by the actual pressure p_0 minus a bulk modulus for the building. This procedure could easily be incorporated in the computation procedure described above, provided realistic values of bulk modulus could be established.

5. INTERNAL VOLUME DISTORTION

As discussed in section 2.2.1, the non-dimensional parameter π_2 $(= p_0/\frac{1}{2}\rho\bar{u}^2)$ will not have the same value in full and model scales, unless the wind tunnel tests are carried out at full scale design wind velocities. Normally this is not possible and, typically, wind tunnel velocities are about one third of full scale values. The effect of this incorrect scaling was mentioned in section 4.2, and illustrated in Figure 7. For small internal volumes typical of houses and smaller industrial buildings, the effect on the r.m.s. internal pressure due to incorrect velocity scaling is small. For the case shown in Figure 7, the simulation results indicated a reduction in $C\acute{p}_i$ of only about 4% due to the increase in mean velocity. However for internal volumes characteristic of larger industrial buildings (5-10,000 m^3 in full scale), the computer simulation indicated somewhat larger reductions of 15-20% in r.m.s. internal pressure when the mean velocity was increased from 10 to 30 m/s.

It appears to be possible to simulate the effect of higher wind velocities in wind tunnel tests by introducing internal volume distortion, at least for the single windward opening case. This may be seen by writing equation (9) in the following non-dimensional form:

$$\left(\frac{r}{n}\right)\frac{2}{\pi_1\pi_2\pi_5{}^2}\frac{d^2Cp_i}{dt^{*2}} + \left(\frac{1}{k^2n^2}\right)\frac{1}{\pi_1{}^2\pi_2{}^2\pi_5{}^2}\frac{dCp_i}{dt^*}\left|\frac{dCp_i}{dt^*}\right| + Cp_i = Cp_e \qquad (11)$$

where ℓ_e has been written as $r\sqrt{A}$ and t^* is a non-dimensional time equal to $\frac{t\bar{u}}{\lambda_u}$

In this equation, the product $\pi_1\pi_2$ is equal to, and can be replaced by a single non-dimensional parameter, π_{12} equal to: $\frac{2A^{3/2}p_0}{\rho(\bar{u}^2V_0)}$

It should be noted that $\bar{u}$ and V_0 only occur together in equations (9) or (11) as a product, $\bar{u}^2V_0$. Thus, equality of π_{12} can be maintained by ensuring that:

$$\frac{[\bar{u}^2V_0]_m}{[\bar{u}^2V_0]_f} = \frac{[A]^{3/2}{}_m}{[A]^{3/2}{}_f} \times \frac{[p_0]_m}{[p_0]_f} \times \frac{[\rho]_f}{[\rho]_m}$$

If the atmospheric pressure and density are the same in full-and model-scale, this gives: $\frac{[V_0]_m}{[V_0]_f} = \frac{[L]^3{}_m}{[L]^3{}_f} \times \frac{[\bar{u}]^2{}_f}{[\bar{u}]^2{}_m}$ where $[L]$ is a characteristic length scale

Thus similarity can be maintained by increasing the internal volume in the wind tunnel test by the square of the full scale/model velocity ratio. It is interesting and significant that the same scaling relationship has been arrived at by Tryggvason (28) in considering the modelling of inflatable structures. In practice, it is not difficult to provide additional volume beneath the wind tunnel floor to provide the distorted internal volume(28).

6. CONCLUSIONS

A combination of theoretical, numerical simulation and boundary layer wind tunnel techniques have been used to investigate internal pressures induced by wind, allowing the following conclusions to be drawn:

(i) Mean and root-mean-square fluctuating internal pressure coefficients increase monotonically with increasing windward/leeward open area ratio, in the former case, in good agreement with a theory based on turbulent flow and mass continuity.

(ii) For the case of a single windward opening, wind tunnel measurements and computer simulated data showed resonance effects on the fluctuating internal pressures. The resonant frequency increases and damping decreases with increasing open area.

(iii) Reasonable agreement between the simulated and measured spectra was achieved when an orifice coefficient of 0.15 was used for the highly unsteady and reversing flow in and out of the orifice.

(iv) Because the resonant frequencies are high in comparison with those of the external wind velocity and stagnation pressures, the resonant effects do not contribute greatly to the total r.m.s. pressure fluctuations.

(v) Inertia effects may, however, be significant if a sudden change in external pressure, such as that due to a window failure, occurs for large openings.

(vi) The effects of not correctly satisfying the scaling requirements for fluctuating internal pressures by not wind tunnel testing at full scale wind velocities, are probably small in most cases. However, it appears that the effect of higher wind velocities can be simulated by distorting the internal volume by a factor equal to the square of the velocity ratio.

ACKNOWLEDGEMENTS

The author would like to acknowledge the considerable contribution of Joe Munarin, Gordon McNealy and Ross Best towards the work described herein. Discussions with Bjarni Tryggvason also proved extremely useful. The work was sponsored by the Australian Housing Research Council.

REFERENCES:

(1) Walker, G.R., "Report on Cyclone 'Tracy' - Effect on Buildings" Australian Department of Housing and Construction (1965).

(2) Euteneuer, G.A., "Einfluss des Windeinfalls auf Innendruck und Zugluft-Erscheinung in teilweise offenen Bauwerken". Der Bauingenieur 46: 355-360(1971)

(3) Van Koten, H., "The Wind Pressures on Windows" Symposium on Full Scale Measurements of Wind Effects London, Ontario (1974).

(4) Newberry, C.W. and Eaton, J.J., "Wind Loading Handbook" Building Research Establishment (U.K.) (1974).

(5) Liu, H., "Wind Pressure inside Buildings" 2nd U.S. National Conference on Wind Engineering Research Fort Collins, Colorado (1975).

(6) S.A.A. "Loading Code. Part 2 - Wind Forces" A.S. 1170 Part 2 Standards Association of Australia (1975).

(7) Euteneuer, G.A., "Druckansteig im Inneren von Gebaüden bei Windeinfall" Der Bauingenieur 45: 214-216 (1970).

(8) Vickery, B.J., "Internal Pressures", Section 10.8 in The Structural and Environmental Effects of Wind on Buildings and Structures Course Notes, Monash University (1975).

(9) Liu, H., "Building Code Requirements on Internal Pressure", 3rd U.S. National Conference on Wind Engineering Research Gainesville, Florida (1978).

(10) Jancauskas, E.D. and Sharp, D.B., "Wind Loading on the Roof of a Low-Rise House" 6th Australasian Hydraulics and Fluid Mechanics Conference Adelaide(1977)

(11) Kramer, C., Gerhardt, H.J. and Scherer, S., "Wind Pressure on Block-type Buildings" 3rd Colloquium on Industrial Aerodynamics Aachen (1978).

(12) Holmes, J.D., "Mean and Fluctuating Internal Pressures Induced by Wind" James Cook University of North Queensland. Wind Engineering Report 5/78 (1978).

(13) Rayleigh, Lord "Theory of Sound - Volume 2" Macmillan 1896. Reprinted by Dover Publications (1945).

(14) Malecki, I., Physical Foundations of Technical Acoustics Pergamon Press (1969).

(15) Holmes, J.D., "Design and Performance of a Wind Tunnel for Modelling the Atmospheric Boundary Layer in Strong Winds" James Cook University of North Queensland. Wind Engineering Report 2/77 (1977).

(16) Holmes, J.D. and Best, R.J., "Wind Tunnel Measurements of Mean Pressures on House Models and Comparison with Full Scale Data" 6th Australasian Hydraulics and Fluid Mechanics Conference Adelaide (1977).

(17) Holmes, J.D. and Best, R.J., "Wind Pressures on an Isolated High-Set House" James Cook University. Wind Engineering Report 1/78 (1978).

(18) Holmes, J.D. and Best, R.J., "Further Measurements on Wind Tunnel Models of the Aylesbury Experimental House" James Cook University of North Queensland. Wind Engineering Report 4/78 (1978).

(19) Best, R.J. and Holmes, J.D., "Model Study of Wind Pressures on an Isolated Single-Storey House" James Cook University. Wind Engineering Report 3/78 (1978).

(20) E.S.D.U. "Characteristics of Atmospheric Turbulence Near the Ground - Part II: Single Point Data for Strong Winds (neutral atmosphere)" Engineering Sciences Data Unit (U.K.) Data Item 74031 (1974).

(21) Munarin, J., "Wind Tunnel Measurements of Internal Pressures in a Two Storey House" B.E. Thesis Department of Civil and Systems Engineering, James Cook University (1978).

(22) Holmes, J.D., "Computer Simulation of Multiple Correlated Wind Records Using the Inverse Fast Fourier Transform" Civ. Engg. Trans. Inst. Eng. Austr. CE20: 67-74 (1978).

(23) Biggs, J.M., "Introduction to Structural Dynamics" McGraw-Hill (1964).

(24) Bearman, P.W., "An Investigation of the Forces on Flat Plates in Turbulent Flow" N.P.L. Aero. Report 1296 (1969).

(25) Bearman, P.W., "Some Measurements of the Distortion of Turbulence Approaching a Two-Dimensional Body" A.G.A.R.D Conference on Turbulent Shear Flows (1971).

(26) Holmes, J.D., "Pressure Fluctuations on a Large Building and Along-Wind Structural Loading" J. Indust. Aero. 1: 249-278 (1976).

(27) Pankhurst, R.C. and Holder, D.W., Wind Tunnel Techniques Pitman, London (1952).

(28) Tryggvason, B.V. "Aeroelastic Modelling of Pneumatic and Tensioned Fabric Structures" 5th International Conference on Wind Engineering Ft. Collins(1979).

INTERNAL PRESSURE CHARACTERISTICS OF LOW-RISE BUILDINGS DUE TO WIND ACTION

T. STATHOPOULOS*, D. SURRY** and
A. G. DAVENPORT**

*Concordia University, Montreal, Quebec, Canada
**The University of Western Ontario, London, Ontario, Canada

SUMMARY

Wind-induced internal pressures have been experimentally determined using models of low-rise buildings of different geometry and internal volume. Three different uniform porosities (0.0, 0.5 and 3.0% of the total surface area) have been examined in combination with openings in a wall ranging from 0 to 100% of that wall's area. Tests have been carried out in a simulated atmospheric boundary layer over two terrain roughnesses corresponding to open country and suburban regions. Results show that internal pressures are dynamic but their magnitudes are generally lower than those of local external pressures. For windward openings, internal pressure coefficients are generally positive with the exception of cases with high background porosity combined with small openings, when they become zero or slightly negative. The lower the background porosity, the smaller the size of the wall opening necessary to make the internal pressures insensitive to further increases of the wall opening. For all cases, high correlation has been found between those components of external and internal pressures which participate in loading the primary structure.

INTRODUCTION

Most full-scale and wind tunnel studies dealing with wind loads on buildings are designed to provide external pressure coefficients only. As such they define the external surface loading relative to a reference static pressure well away from the building. The total wind loading, even for a building which has minimal leakage and hence a steady uniform internal pressure, must include the contribution of the internal pressure to both local and distributed loads. In reality, a building may have considerable inherent leakage as well as the potentiality for large openings due to the presence of doors or windows. The resulting internal pressures can be a significant proportion of the total loading. Thus, it is important for the structural designer to account for the internal pressures for all likely conditions of permeability and wall openings. Note that a precise description of the net load would require knowledge of the correlation between external and internal pressures.

Although the significance of the internal pressure is well recognized, very few studies have dealt with it. Among those that have, only one or two are either full-scale cases or model cases in appropriate atmospheric simulations. No study, to the writers' knowledge, exists that deals with the determination of internal pressures in buildings with various uniform porosities combined with different wall openings. As a result only limited comparisons of the results of this study with others' findings can be made.

METHODOLOGY AND MODELS

The Boundary Layer Wind Tunnel at the University of Western Ontario has a working section about 24.5 m (80 ft.) long, 2.4 m (8 ft.) wide and about 2.1 m (7 ft.) high. Most of this fetch is required for the natural production of a boundary layer which grows in a manner paralleling the atmospheric process under neutral conditions. The surfaces in the wind tunnel can be changed to represent different terrains. Boundary layer depths ranging from about 0.5 to 1 m (2 to 4 ft.) are obtained at the test section with different surface roughnesses. In relation to the atmospheric boundary layer this implies that geometric scales of between 1:400 and 1:500 are most appropriate for studies of wind effects on buildings and structures; however, in the present study, models at this scale posed considerable difficulties because of their small size. Instead, most of the tests were carried out at a somewhat relaxed scale of 1:250, after an intensive set of experiments had been completed to verify that such a relaxation in scale did not materially affect the results (1). For the tests reported here, two terrain models have been used. The 'smooth' exposure represents open country conditions whereas the 'built-up' exposure corresponds to suburban conditions.

Three basic models were constructed, all providing variable side-wall and end-wall openings and three background porosities of 0%, 0.5% and 3.0% of the total surface area. The models shown diagrammatically in Figures 1 and 2, are sealed underneath so that the pressure inside the model is determined completely by the flow conditions within the tunnel. The two 'small' models are 1:250 scale models of 24.4 x 38.1 m (80 x 125 ft.) buildings with 1:12 and 4:12 roof slopes. The 'large' model is a 1:250 scale model of a building geometrically similar in plan, and having 2.5 times the plan dimensions. The small models could be tested at 4.9, 7.3 and 9.8 m (16, 24 and 32 ft.) eave heights whereas the large model could be tested only for a 9.8 m (32 ft.) eave height. It was found to be unnecessary to test all possible configurations. The background porosity or permeability of the models was achieved by a series of holes of various sizes which could be left open or closed. Although this simulates porosity in a manner somewhat different to reality, local effects were found to be small and hence for pressure measurements at locations a reasonable distance from the holes, the difference is not likely to be significant.

The chosen background porosity ratios (0.0, 0.5 and 3.0% of the total surface area) arose from the following considerations. First, it is very difficult to measure or even to estimate actual building permeabilities since they vary significantly even for the same type of buildings. Tamura (2) measured air leakages through six different one or two-storey houses in Canada. From his reported results, overall porosity values up to 0.09% can be calculated, although this permeability ratio does not include leakages through doors and windows. Newberry and Eaton (3) suggest porosity ratios of the order of 0.01 to 0.1% or more, as for instance 0.2% for the study of Royex House (4) in England. An ad hoc survey among several members of the Metal Building Manufacturers Association suggested typical values of about 0.5% for metal buildings without windows. For buildings with windows, it was estimated that the porosity could be as large as 1.0 to 3.0%, although these upper bounds are likely to significantly decrease with increasing awareness of energy losses. Furthermore, it is obvious that the various climatic conditions and construction techniques in different countries naturally yield a variety of building permeability ratios. Thus the range of 0.0 – 3.0% appears to include most cases of practical interest.

All models are equipped with pressure taps on both the internal and external surfaces. Each tap is connected to a pressure transducer through a length of plastic tubing of 1.6 mm (1/16″) I.D. which contains a restrictor to optimize its frequency response characteristics (6). A 610 mm (2 ft.) length of tubing provides a frequency response flat to about ± 6% up to 100 Hz. Such tubes were used for the larger model. The small model used tubes of half this length which improved the frequency response to 120 Hz. Comparisons of external pressure measurements made using both tubing configurations show no significant differences (1,7).

Both local pressure and 'generalized area loads' were measured. Local loads were determined using conventional techniques (1,6). For area loads, differences of external and internal pressures were measured simultaneously at ten locations. These were integrated together with appropriate influence factors to provide sets of structural 'outputs' or generalized loads, corresponding typically to horizontal and vertical forces or bending moments acting on building frames. Since this sampling and multiplication is done on-line at a high rate (200 times/sec) the result is the time-varying behaviour of the total wind-induced structural reactions which are of primary design interest (1,6,8). For both local and generalized loads, sampling is continued for a period of up to about a minute in real time during which the computer records, for each output, the maximum (max) and minimum (min) values that occur, and computes the mean and root mean square (rms) values. The reference dynamic pressure is monitored similarly. At the end of the sampling period the resultant measurements, consisting of the max, min, mean and rms pressure or load for each channel are put into coefficient form by dividing each by the mean dynamic pressure at eave height.

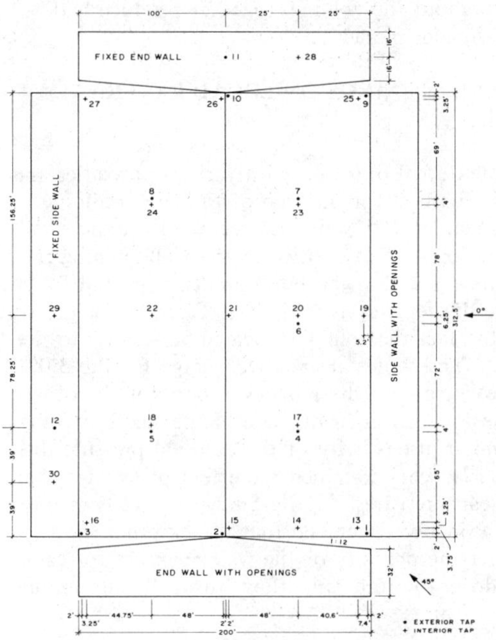

Fig. 1 Exploded plan view of the large model for internal pressure measurements showing the positions of pressure taps

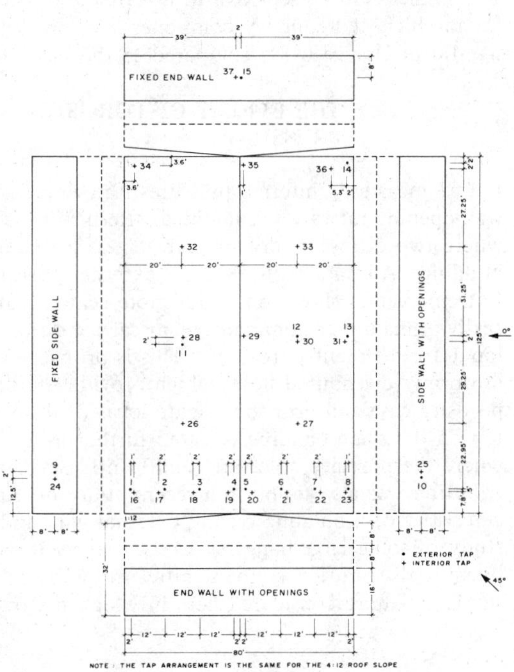

Fig. 2 Exploded plan view of the small model (3 heights, 2 roof slopes)

Scaling and similarity requirements for models designed for wind tunnel determination of external loads have been thoroughly discussed elsewhere (see (9) for example). Such similarity considerations lead to the determination of velocity and time scales of the order of 3:10 and 1:100 respectively for typical full-scale design speeds. Consequently, the use of a 30 second or one minute wind tunnel sample under flow conditions where the speed above the boundary layer is about 14 m/sec provides a statistically stable estimate of mean and rms pressures and an assessment of peak values associated with a full-scale period of about an hour.

Additional scaling requirements related to modelling internal pressures for flexible buildings have been discussed in (10). For rigid buildings of low height where gravitational terms are unimportant, the unsteady pressure change induced in the building by the wind essentially requires the Mach number to be simulated. This ensures that propagation of pressure fluctuations occurs similarly in model and full scale, and hence internal pressure fluctuations maintain a correct phase relationship with external pressures. Essentially, this reduces to maintaining the same velocities for model and full scale. For the cases examined here, a Mach number discrepancy equal to the velocity scale has been introduced;

however, the consequences are likely to be small since a typical propagation time for a full-scale pressure wave would be about one-fifth of a second in the larger building, and about one-thousandth of a second in model scale. These periods correspond to turbulent fluctuations within the wind which have very small spatial correlation. Hence, for larger gusts likely to have significant impact on the internal pressures, these pressures will respond essentially instantaneously; the fact that the model's interior responds three times too quickly for the high frequency components of small spatial correlation will be of little importance. The above is supported by Euteneur's work (11) on the estimation of time required for the transmission of pressures inside volumes not totally enclosed. Also the results reported below for the two model sizes verify that the internal pressures reflect gusts which essentially envelop the buildings. Furthermore, the results for the larger model indicate that the combined effect of a larger open frontal area (which will be sensitive to less of the gust content within the wind) and the longer times required for the highest frequency components to propagate throughout the volume, together produce little significant change when compared to the data from the smaller model.

THE EFFECT OF POROSITY AND WALL OPENINGS ON LOCAL EXTERNAL PRESSURES

Before measuring internal pressures, the variation of the external pressures with various porosities and/or wall-opening ratios was examined. Results are shown in Figure 3 for the case of the large building, where two different roof pressure taps are examined. Similar data were also recorded for the small building. All data show that the external pressure coefficients are insensitive to the wall-opening ratios. Extreme values show somewhat more scatter but the data are within the repeatability expected for successive runs under nominally identical conditions. Some discrepancy is evident in Figure 3 for pressure tap 1 for different porosities. This is probably due to the fact that tap 1 is located very close to the uniformly distributed holes which provide the porosity. These holes are relatively large for the 3.0% porosity case and probably create local flow disturbances. No such discrepancy occurs for the other tap on the same building (located further away from these holes) or in the case of the small building where the porosity holes are significantly smaller. A similar insensitivity of the external pressure distribution was also found in a recent study by Kandola (12), who examined the effect of leakage characteristics of a building on the external wind-induced mean pressures, and by Surry et al (13) in the study of some large hangars. Consequently it can be reasonably concluded that the external wind pressure distribution is not significantly affected by either the porosity or the wall openings for cases similar to those considered here in which the openings do not permit large flows through the buildings.

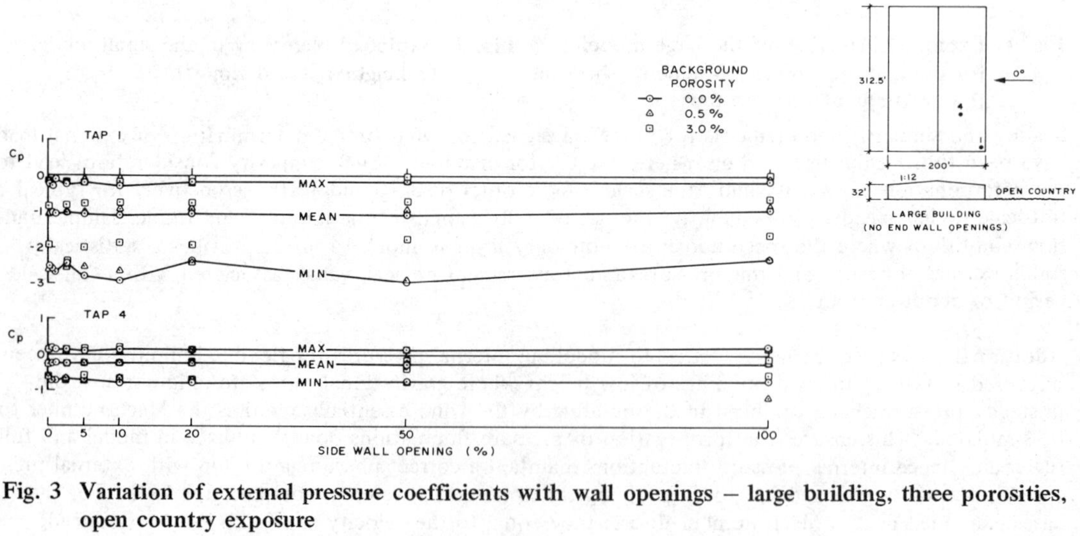

Fig. 3 Variation of external pressure coefficients with wall openings – large building, three porosities, open country exposure

LOCAL INTERNAL PRESSURES

Figure 4 illustrates the high degree of correlation of the internal pressures. Typical time traces of pressures from two internal taps located at opposite ends of the large building are shown. The traces are virtually identical. This characteristic is quite general for all the various configurations examined and leads to the conclusion that the spatial variation of the internal pressures is minimal. This was also observed in the substantially larger buildings studied in (13). Hence characteristics of the internal pressure can be given below without reference to the particular location within the building. The only exception to the above was found for points inside the building located in regions very close to openings or other architectural details of odd geometry. However, the uniformity of internal pressures is likely to be somewhat reduced in buildings with significant openings on more than one wall.

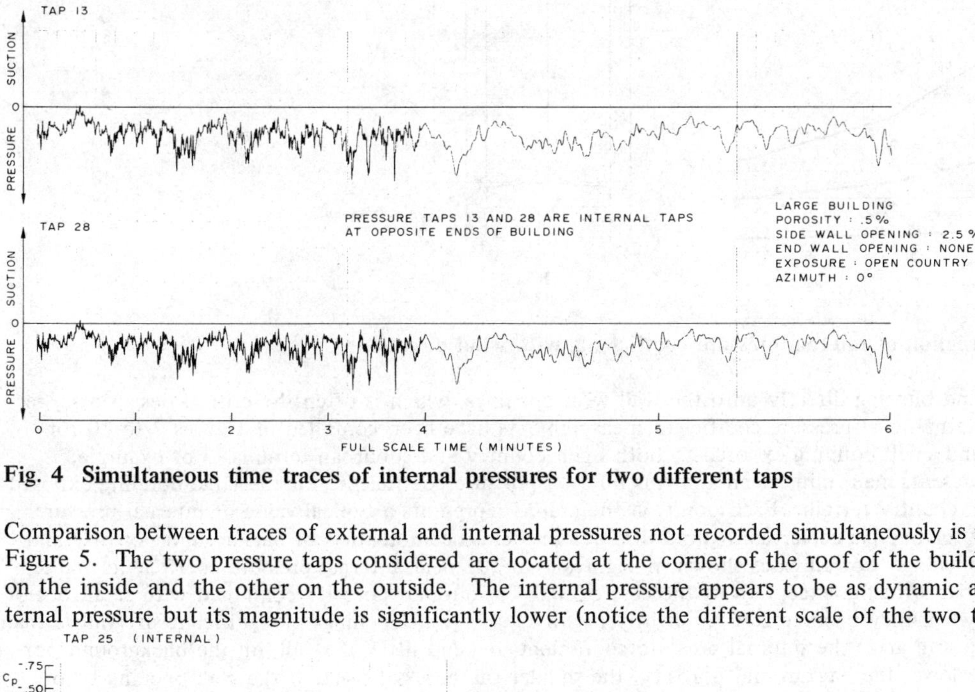

Fig. 4 Simultaneous time traces of internal pressures for two different taps

Comparison between traces of external and internal pressures not recorded simultaneously is made in Figure 5. The two pressure taps considered are located at the corner of the roof of the building, one on the inside and the other on the outside. The internal pressure appears to be as dynamic as the external pressure but its magnitude is significantly lower (notice the different scale of the two traces).

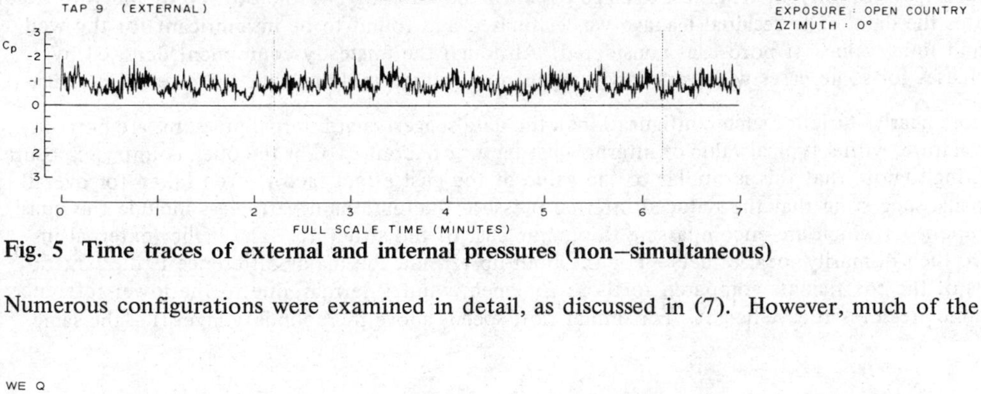

Fig. 5 Time traces of external and internal pressures (non–simultaneous)

Numerous configurations were examined in detail, as discussed in (7). However, much of the data were made redundant by the high spatial correlation of the internal pressures. Figure 6 shows a typical example of the variation of extreme and mean values of internal pressure coefficients with azimuth. As expected, the maximum values occur when openings are on the windward wall. These and the data for other cases indicate that the influence of the wind direction is greater for the extreme values than for the mean values; Furthermore, little difference was found in either the mean or peak internal suctions between wind directions of 90° and 180°. This differs from results found in (13) for a substantially higher hangar building. In those tests, peak internal suctions were nearly twice as large for 90° as compared to 180°. This presumably reflects the more severe suctions induced near the wall corners for taller buildings.

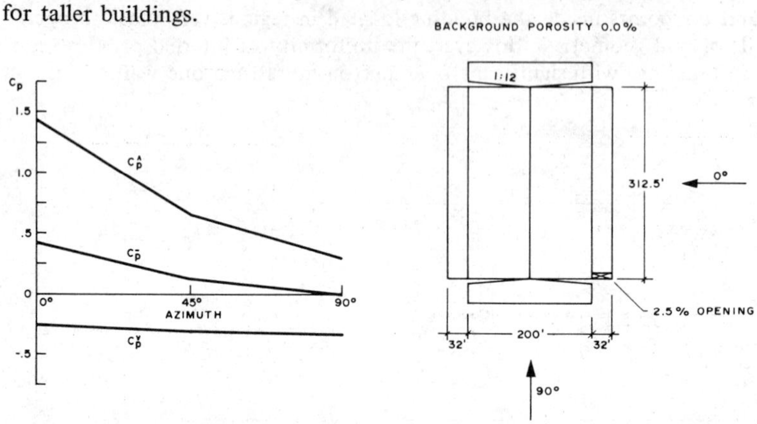

Fig. 6 Variation of internal pressure coefficients with wind direction

For the wind blowing directly onto the wall with openings, which is often the critical design case, results of internal local pressure coefficient measurements have been compiled in Figures 7 to 10 for the large and small building exposed to both open country and suburban terrains. For example, Figure 7 presents max, min, mean and rms internal pressure coefficients for the large building exposed to an open country terrain. Each point on the graphs represents a typical value of internal pressure coefficient for one particular configuration, i.e. a particular combination of background porosity and side wall opening. For all cases shown in Figures 7 to 10, mean internal pressure coefficients are generally positive with the exception of the case of high background porosity combined with small or no openings, when they become zero or slightly negative. For wall openings of significant size (more than 20% of the wall area) the internal pressure coefficients depend little, if at all, on the background porosity. The lower the background porosity, the smaller the necessary size of the wall opening before the internal pressure coefficients become insensitive to further increases of the openings. Some increase of the maximum pressure coefficient values appears for small openings, particularly for the larger building in a built-up exposure. This is attributed to the effects of small scale gusts whose integrated impact becomes smaller as the wall opening becomes larger. For very small openings ($<1\%$), large changes occur in the data for the zero porosity case. These changes are not significant since under conditions approaching nominal impermeability the slightest leakage between the building and the outside of the wind tunnel contaminates the data. The residual leakage was estimated and found to be insignificant for the wall openings and finite values of porosities considered. Although the unsteady component denoted by $C_{p_{rms}}$ decreases for some cases with very low wall-opening ratios (implying that the internal pressure becomes more nearly static for such configurations), the data suggest that internal pressures are normally dynamic in nature, with a typical value of internal gust factor of about two for the open country exposure. It is interesting to note that this is similar to the value of the gust effect factor often taken for overall external loads, suggesting that the reduced internal pressure fluctuations nevertheless include the quasi-steady components which are encompassing the major part of the structure. The higher external intensities are then primarily local effects. For the built-up terrain, the major difference is the larger magnitudes of the coefficients compared to those for open country terrain, due to the lower reference mean dynamic pressures at eave height. For similar flow speeds above the boundary layer (i.e. the same full-scale storm system) the actual pressures induced are lower. Comparison of the results for the small and large buildings indicate little significant difference in the general trends of the data for the two model sizes. The internal mean pressure coefficients are similar (a little bit higher for the small building) whereas the peak and rms values are somewhat higher for the small building, particularly for the lower background porosities. Some differences in the model instrumentation, as described earlier, might have led to shorter response times for the peaks, together with the effect of the smaller volume itself; however, the effect does not appear significant over the range of typical low-rise building geometries investigated.

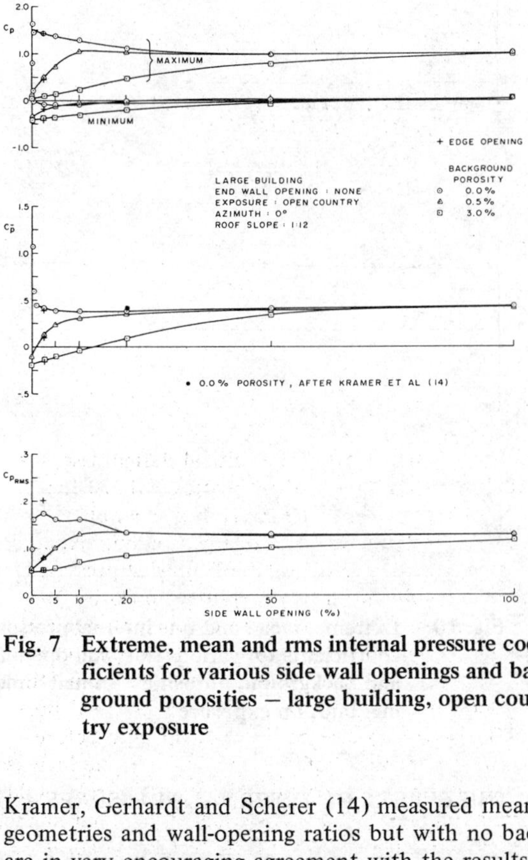

Fig. 7 Extreme, mean and rms internal pressure coefficients for various side wall openings and background porosities – large building, open country exposure

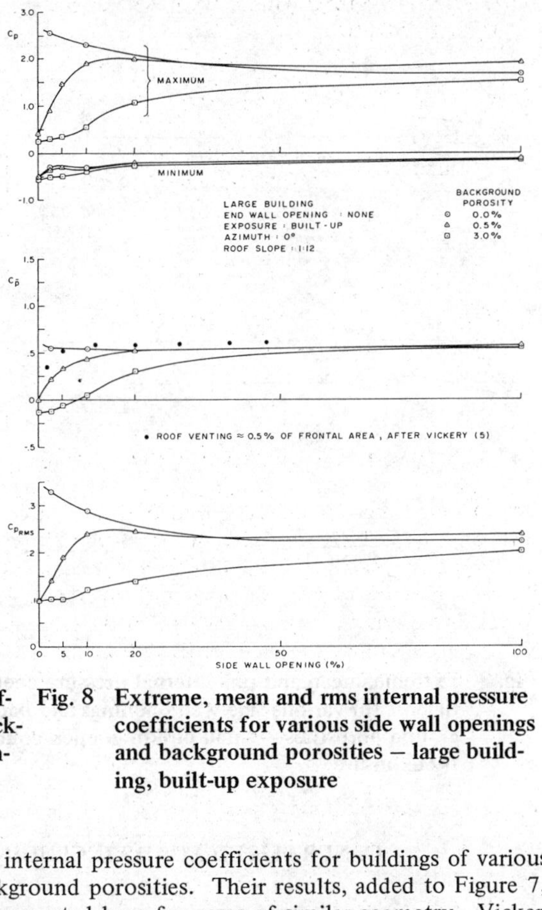

Fig. 8 Extreme, mean and rms internal pressure coefficients for various side wall openings and background porosities – large building, built-up exposure

Kramer, Gerhardt and Scherer (14) measured mean internal pressure coefficients for buildings of various geometries and wall-opening ratios but with no background porosities. Their results, added to Figure 7, are in very encouraging agreement with the results presented here for cases of similar geometry. Vickery's results (5) for internal pressure coefficients measured on a large low-rise hangar also show a fair agreement with the present data as shown in Figure 8. In Vickery's work the hangar roof had a permeability estimated at 0.3 to 1.0% of the frontal area due to instrumentation requirements, and was tested for various wall opening ratios in flow over a rough terrain.

Several internal pressure measurements were made for wind acting perpendicularly to the closed wall (the wall with openings being in the wake). In these cases the variation of internal pressure coefficients versus background porosity or size of openings is not large. Typical values of pressure coefficients referenced to eave height for this case are as follows:

Exposure	$C_{p_{max}}$	$C_{p_{min}}$	$C_{\bar{p}}$	$C_{p_{rms}}$
Open Country	0.10 to 0.20	-.30 to -.45	-.05 to -.15	0.05 to 0.10
Built-up	0.30 to 0.45	-.40 to -.55	0.00	0.10

These values are also representative of different size buildings (different internal volumes).

Spectra of internal pressures were also measured (7). Their characteristics are very similar to the spectral characteristics of external pressures, which indicate the rapidity with which pressure fluctuations are transferred to the interior of the building. This was also noticed by Eaton and Mayne (15) in their full scale experiments in England.

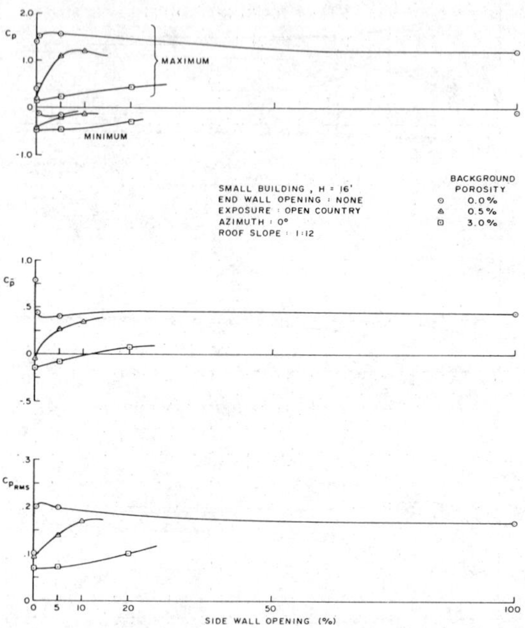

Fig. 9 Extreme, mean and rms internal pressure coefficients for various side wall openings and background porosities – small building, open country exposure

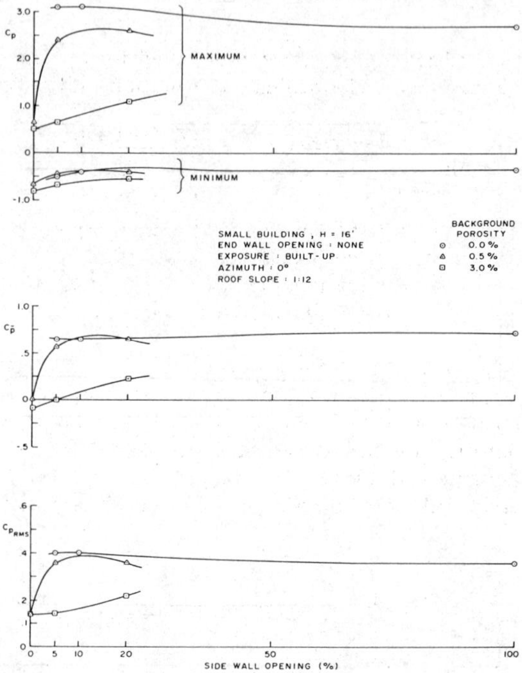

Fig. 10 Extreme, mean and rms internal pressure coefficients for various side wall openings and background porosities – small building, built-up exposure

GENERALIZED LOADS INCLUDING THE EFFECT OF INTERNAL PRESSURES

All of the above data are for local internal pressure coefficients. Measurements of generalized loads, including the simultaneous action of external and internal pressures, have also been made for the end bay of the small building. These generalized bay loads refer to forces and moments applicable to frame design; however, characteristics of the component loads which are appropriate for the design of structural elements such as wall panels, purlins and girts, were also measured. Similar results for external loads only have been obtained and described in detail elsewhere (1,7,8). In this previous work, the generalized loads were derived by measuring eight roof loads and two wall loads simultaneously, and combining them on-line using a computer to form generalized bay loads, which reflected the interaction of the wind-induced forces and the structural action. Each of the ten component loads was formed by pneumatically averaging (6) four or five pressure taps distributed over the component area of interest. In the present case, instrumentation and time constraints led to the use of a simplified approach for the determination of external purlin and wall loads by using one instead of multiple pressure taps to represent the component loads. The simplified approach was validated through successful comparisons of external pressure data obtained by both methods (7). In all cases the simplified approach appears slightly conservative. In order to obtain the total purlin load (external – internal), the external and internal pressures at each purlin location were differenced directly across each pressure transducer. The resultant pressure load was sampled by the on-line data system and used directly in the generalized load program. The use of the pressure transducer in this way led to a slight reduction of the frequency response to internal pressures; however, this is unlikely to have any significant effect on the integrated loads.

Generalized load data were measured for the critically-loaded end bay of the small building for various porosity and wall-opening configurations. Results appear in a non-dimensional form defined as follows:

$$\text{Force Coefficient, } C_F = \frac{\text{Force per unit area}}{q}\ ; \quad \text{Moment Coefficient, } C_M = \frac{M}{q\,w^2 b}$$

where M is the bending moment at various cross-sections of the frame
w is the building width
b is the longitudinal extent of the bay [taken as 7.6 m (25 ft)]
q is the dynamic wind pressure at eave height ($\frac{1}{2}\,\rho\,\overline{V}_e^2$)

Typical results appear in Figures 11 and 12 which show max, mean and min bay force and bending moment coefficients respectively, for a building with each of the three background porosities in combination with a 5% end-wall opening. Again, as with the local loads, the higher the building background porosity, the lower the critical bay loads. Both vertical force and bending moment coefficients of the building with no background porosity are significantly greater (for critical wind directions) than for cases with porosity, whereas the horizontal thrust coefficients are essentially unaffected.

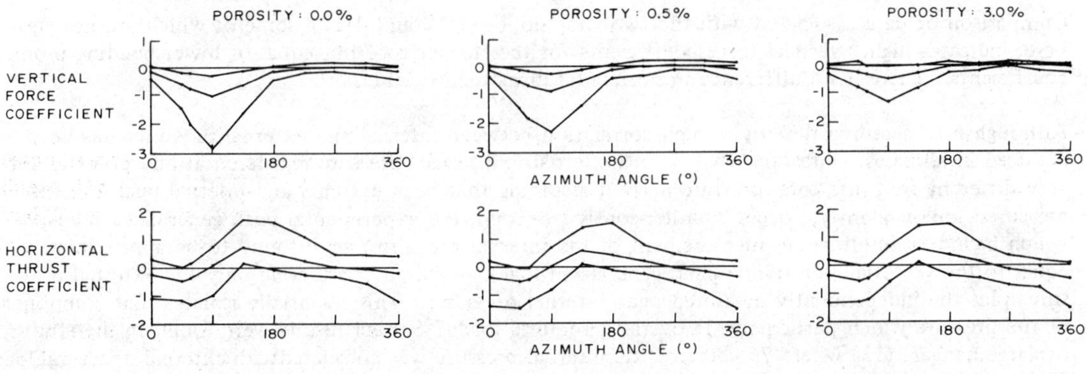

Fig. 11 End bay force coefficients for total (external – internal) load conditions – 3 porosities, 5% end wall opening (roof slope = 1:12, eave height = 16′, exposure = open country)

Comparison of the data for the building with no background porosity with results of the generalized loads using external pressures only, as obtained previously (1), show the true effect of internal pressures on the loading mechanism of the structural members. In fact, this effect is crucial for some loads, as for instance the vertical forces, and not critical at all for others, as for instance the horizontal thrusts. The latter is predictable since the internal pressure uniformity cancels the overall effect on the vertical walls. In general, vertical force and bending moment coefficients for the building with no background porosity and for critical wind directions are 50 – 100% higher when the internal pressures are considered. These values are reduced for increasing porosity. Similar data obtained in suburban exposure (presented in (7)) show the same trends except that the background porosity does not appear as strong a factor for the determination of loads. All results for low background porosities are relatively insensitive to wall openings greater than about 5% of the wall area. For high background porosity the results indicate some dependence on the size of the wall opening, as did the local loads; in fact, all the coefficients are significantly increased in the case of a missing wall.

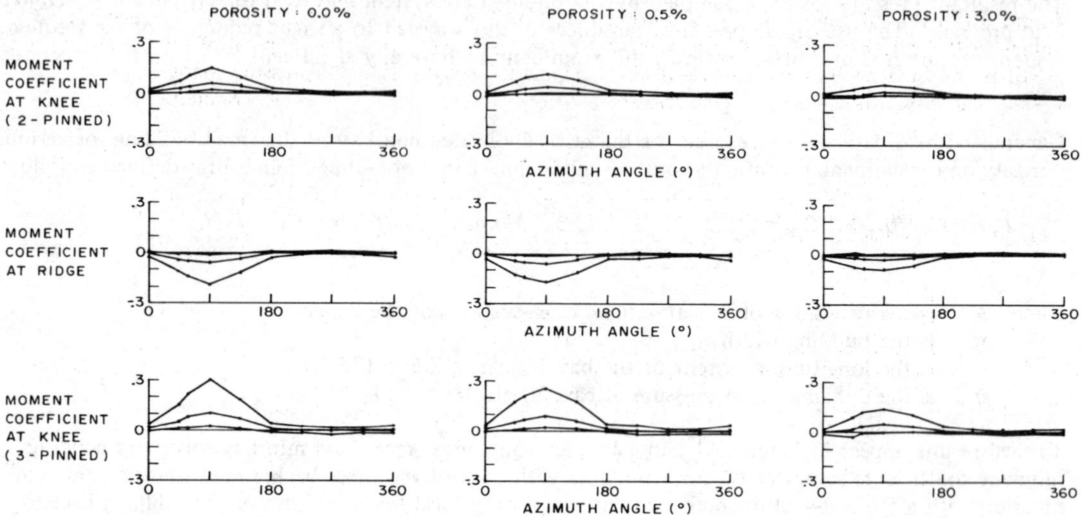

Fig. 12 End bay bending moment coefficients for total (external – internal) load conditions – 3 porosities, 5% end wall opening (roof slope = 1:12, eave height = 16′, exposure = open country)

Comparison of data associated with the two roof slopes (1:12 and 4:12), some of which are not shown here, indicates higher vertical force coefficients for the steeper roof but equal or lower bending moment coefficients. There is no difference in horizontal thrust coefficients.

Although not measured directly, a high correlation between internal and external pressures has been deduced in all cases. First, the peak total (external – internal) pressure registered at any point is not very different from the total pressure derived assuming that both external and internal peak pressures, measured independently, occur simultaneously. Second, the experimental peak generalized loads, which included simultaneous measurement of the internal pressures, were found to be approximately equal to the vectorial sum of the peak generalized loads measured considering only the external pressures plus the independently measured peak internal pressures. This essentially isolates that component of the pressure which participates in overall structural loads. Similar results were found in the study on large hangars (13) where 75–80% of the internal pressure was correlated with external structural loads. These characteristics are consistent with the physical picture of the building responding essentially quasi-statically to the larger-scale wind gusts. Smaller gusts and local effects introduce contributions to local peaks which are somewhat less spatially-correlated.

COMPARISON OF INTERNAL PRESSURE DATA WITH CURRENT CODES AND STANDARDS

To provide a better perspective on the experimental internal pressure data, comparisons have been made with pressures suggested by ANSI A58.1 (16) and the National Building Code of Canada (17,18). Since both documents do not recognize the effect of building background porosity, a separate comparison for each porosity tested is shown. Since, in general, pressure coefficients are referenced to different heights and speeds, all internal pressure coefficients have been transformed to internal building pressures assuming a particular storm system which produces a fastest mile speed of 100 mph at 10 m (30 ft) and consequently $\overline{V}_{10}$ = 34.4 m/sec (77 mph) and $\overline{q}_{4.9}$ = 608 N/m^2 (12.7 psf) in open country exposure (ANSI: type C, NBCC: type A). The building considered has a 1:12 roof slope and is 4.9 m (16 ft) high. Four cases of different side-wall openings have been considered. The calculations are shown in Table 1 to indicate the pressure coefficients used and the appropriate gust factors, if any. In all cases the internal pressure coefficients drawn from the current results are the worst peak values found for any wind angle. The final two columns in Table 1 indicate the ratios of the recommended values to the current results. It can be seen for this example, using data typical of all the results found, ANSI is conservative in some cases but also underestimates the internal pressure loads in some other cases. The NBCC is always conservative if, as shown, a gust factor of two is assumed.

TABLE 1

SIDE–WALL OPENING	BACK-GROUND POROSITY	Internal Pressures (psf) (1 psf = 47.88 N/m^2) BLWT	ANSI	NBCC	ANSI / BLWT	NBCC / BLWT
windward 50%	any	1.0 x 12.7 = 12.7	0.8 x 26 = 21.0	0.7 x 2.0* x 16 = 22.4	1.65	1.76
windward 2.5%	0.0%	1.5 x 12.7 = 19.0			0.46	1.18
	0.5%	0.7 x 12.7 = 8.9	(0.3 + 1.67 x .025) x 26 = 8.9	0.7 x 2.0* x 16 = 22.4	1.00	2.52
	3.0%	0.2 x 12.7 = 2.5			3.56	8.96
leeward opening of any size	any	-0.45 x 12.7 = -5.7	(-0.3 - .025) x 26 = -7.1	-0.5 x 2.0* x 16 = -16.0	1.25	2.81
			(-0.6 x 26 = -15.6 for > 30% opening)		2.74	
0	3%	-0.5 x 12.7 = -6.3	± 0.3 x 26 = ± 7.8	-0.3 x 2.0* x 16 = -9.6	1.24	1.52

* left to the designer to decide whether this value will be 2.0 or 2.5 or whether a gust factor should apply at all.

CONCLUSIONS

The following conclusions may be drawn from the study of wind-induced internal pressures in low-rise buildings:

i) Internal pressures fluctuate significantly; but their overall magnitudes are generally less than that of the local external pressures. The overall gust factor – the ratio of the peak pressure to the mean – is roughly two in open country.

ii) The fluctuations in internal pressure show little or no spatial variation except in regions close to dominant openings.

iii) A high correlation between external and internal pressures has been found in all cases, particularly for the component of pressure participating in the overall structural loading.

iv) The largest internal pressures occur when the wind direction is perpendicular to the wall with dominant openings.

v) When the dominant openings are to the lee of the structure and the windward wall is closed the internal pressures are generally negative and are not very sensitive to the size of wall openings or to the background porosity.

vi) For windward openings, although internal pressure coefficients are generally positive, cases with high background porosity combined with small openings produce zero or slightly negative coefficients. For wall openings of significant size (more than 20% of the wall area) the internal pressure coefficients become essentially independent of the background porosity. The lower the background porosity, the smaller the necessary size of the wall opening needed to make the internal pressure coefficients insensitive to further increases of the wall opening.

vii) Large-scale structural loads, represented here by bay loads, are also affected by the background porosity. For constant wall openings the bay loads tend to be higher for lower background porosity.

ACKNOWLEDGEMENTS

The work described herein is a part of a major study on wind loads of low-rise buildings, which has been financially supported by the Metal Building Manufacturers Association, The American Iron and Steel Institute, the Canadian Steel Industries Construction Council, the National Research Council of Canada and the Boundary Layer Wind Tunnel Laboratory.

REFERENCES

1. Davenport, A. G., Surry, D and Stathopoulos, T., Wind Loads on Low-Rise Buildings: Final Report of Phase I and II, Parts 1 and 2, Boundary Layer Wind Tunnel Laboratory, The University of Western Ontario, London, Canada, BLWT–SS8–1977, November 1977.

2. Tamura, G. T., Measurement of Air Leakage Characteristics of House Enclosures, ASHRAE Transactions, Vol. 81, Part 1, 1975.

3. Newberry, C. W. and Eaton, K. J., Wind Loading Handbook, Building Research Establishment Report, Department of the Environment, London, England, 1974.

4. Newberry, C. W., Eaton, K. J. and Mayne, J. R., Wind Loading on Tall Buildings – Further Results from Royex House, Industrial Aerodynamics Abstracts 4, July – August, 1973.

5. Vickery, B. J., Wind Loads on Low-Rise Buildings, presented at D.R.C. Seminar, Darwin, Australia, March 30, 1976, (Unpublished, 11 pages, 19 figs).

6. Surry, D., Stathopoulos, T., An Experimental Approach to the Economical Measurement of Spatially –Averaged Wind Loads, Journal of Industrial Aerodynamics, Vol. 2 (1977/1978), pp. 385–397.

7. Davenport, A. G., Surry, D. and Stathopoulos, T., Wind Loads on Low-Rise Buildings: Final Report of Phase III, Parts 1 and 2, Boundary Layer Wind Tunnel Laboratory, The University of Western Ontario, London, Canada, BLWT–SS4–1978, July 1978.

8. Surry, D., Stathopoulos, T. and Davenport, A. G., The Wind Loading of Low-Rise Buildings, Proceedings of Canadian Structural Engineering Conference 1978, The Canadian Steel Industries Construction Council.

9. Surry, D. and Isyumov, N., Model Studies of Wind Effects – A Perspective on the Problems of Experimental Technique and Instrumentation, Proceedings of the Sixth International Congress on Instrumentation in Aerospace Simulation Facilities, Ottawa, Canada, Sept. 22–24, 1975.

10. Tryggvason, B. V. and Isyumov, N., Similarity Requirements for Inflatable Structures, Proceedings

of the Third U.S. National Conference Wind Engineering Research, Gainesville, Florida, February 26–March 1, 1978.

11. Euteneuer, G. A., Druckanstieg Im Inneren von Gebauden Bei Windeinfall, Der Bauingenieur, Vol. 45 (1970), pp. 214–216.

12. Kandola, B. S., Wind Effects on Buildings with Varying Leakage Characteristics – Wind–Tunnel Investigation, Journal of Industrial Aerodynamics, Vol. 3 (1978), pp. 267–284.

13. Surry, D., Kitchen, R. B. and Davenport, A. G., Wind Loading of Two Hangars Proposed for Jeddah Airport, Saudi Arabia – A Rigid Model Study, Report to be published, Boundary Layer Wind Tunnel Laboratory, The University of Western Ontario, London, Canada.

14. Kramer, C., Gerhardt, H. J. and Scherer, S., Wind Pressure on Block Type Buildings, Proceedings of the 3rd Colloquium on Industrial Aerodynamics, Part 1, Aachen, W. Germany, June 14–16, 1978.

15. Eaton, K. J. and Mayne, J. R., The Measurement of Wind Pressures on Two-Storey Houses at Aylesbury, Journal of Industrial Aerodynamics, Vol. 1 (1975), pp. 67–109.

16. American National Standard Building Code Requirements for Minimum Design Loads in Buildings and Other Structures, ANSI A58.1 – 1972, Amer. Nat. Standards Inst., New York, 1972.

17. National Building Code of Canada 1975, NRCC No. 13982, Associate Committee on the National Building Code, National Research Council of Canada, Ottawa, 1975.

18. Commentaries on Part 4 of the National Building Code of Canada 1975, Supplement No. 4, NRCC No. 13989, Associate Committee on the National Building Code, National Research Council of Canada, Ottawa, 1975.

WIND PRESSURES ON COOLING TOWER SHELLS

H. PRÖPPER and J. WELSCH
Ruhr-Universität, Bochum, Federal Republic of Germany

ABSTRACT

During a great research program the nature of wind pressure acting on hyperbolic cooling tower shells was thoroughly investigated. Numerous wind tunnel tests with rigid cooling tower models were carried out in smooth and in turbulent flow, with simulating the earth's boundary layer by Counihan's method for different types of terrain. In order to check the influence of some model law parameters especially with regard to the Reynolds number the results of the measurements on three full-scale cooling towers (Weisweiler, Schmehausen, both W.-Germany; Martin's Creek, USA) were analyzed, too.

Corresponding to the random nature of the wind the data analysis was carried out by using statistical methods. The results give a detailed picture about the vertical and circumferential distribution of the mean pressure, the rms values, and the correlation coefficients. The form and the frequency distribution of the pressure spectra is also presented. The influence of some important parameters as the structure of the natural wind (i.e., wind profile, turbulence intensity, length scale), the dimensions of the cooling tower, and the rib roughness of the shell surface is discussed.

INTRODUCTION

For the design of cooling tower shells the load of the natural wind prevails. On the one hand cooling towers offer because of their dimensions a particular large working surface to the flow, on the other hand these buildings are rather thin-walled and light, and the higher such a tower is the lower the eigenfrequencies are.

During a storm the wind pressures acting on the surface of the shell occur spatially and temporarily completely irregular. They are random phenomena which can only be interpreted by means of statistic methods. For the first time Abu-Sitta and Hashish (1) and Hashish and Abu-Sitta (4) showed how to apply this theory of random vibrations to the calculation of cooling towers. But as they stated by themselves, there are still up today some uncertainties about the correct aerodynamic statements which are to introduce into the calculation because test results were not available especially those which were proved by full-scale measurements.

Therefore in an extensive research program, sponsored by the Deutsche Forschungsgemeinschaft, the aerodynamic load of a single detached cooling tower was investigated.

In doing so, all the values which are necessary to make a calculation of a cooling tower were taken into consideration, i.e., the distribution of the mean pressure and rms values as well as the correlation coefficients and the pressure spectra. Meanwhile a detailed description of the test results is published (9).

TEST FACILITIES

Model Tests

Most of the represented results are based on pressure measurements at rigid cooling tower models put through in the wind tunnel of VFW-Fokker, Bremen. Using vorticity generators and roughness fields according to Counihan's method the temporal and spatial nature of the wind in the earth's boundary layer was modeled for rural terrain (characterized by the power-law exponent $\alpha = 0.18$) and for suburban terrain ($\alpha = 0.27$).

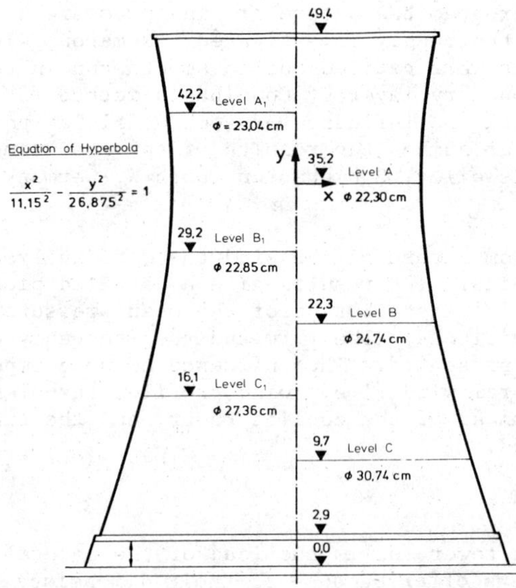

Fig. 1. Measurement levels of model cooling tower.

The 1/333 models of a 170 m high cooling tower with typical dimensions (Fig. 1) were equipped with 36 pressure measurement points arranged in three levels. They differed essentially from one another in the variation of number and height of the meridional wind ribs.

FULL-SCALE MEASUREMENTS

To prove the influence and the correct use of the non-dimensional scaling parameters, additional measurements were carried out at the 120 m full-scale cooling tower in Schmehausen, Westfalen (Fig. 2). For the evaluation those wind directions were selected where the wind approached from the west over flat open country without any disturbance by buildings (Fig. 3). Even the adjacent cable

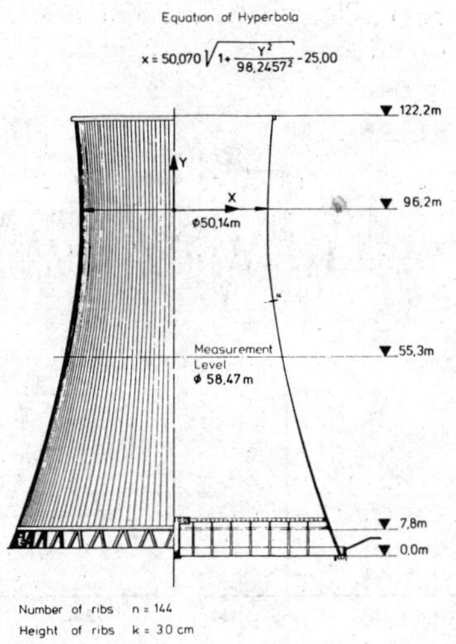

Fig. 2. Full-scale cooling tower at Schmehausen/Ger.

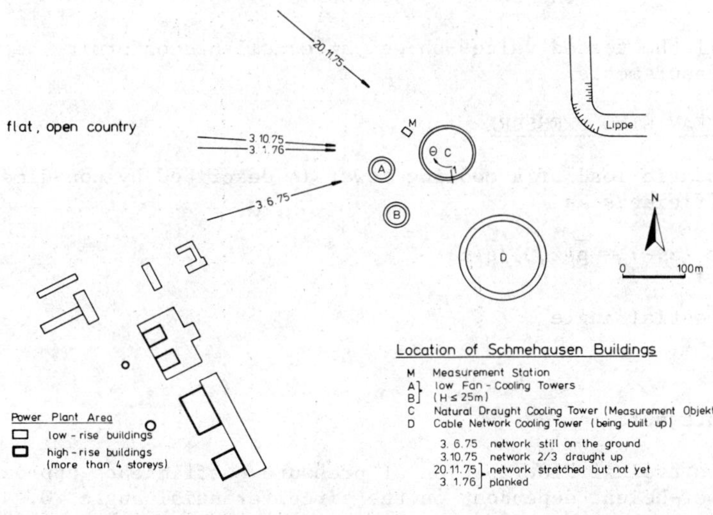

Fig. 3. Ground plan of Schmehausen power plant.

network cooling tower did not affect the pressure distribution because it was not yet planked during the measurement period. Therefore the full-scale results are characteristic for a detached single tower.

For example, Fig. 4 shows the time record of wind velocity measured during the storm of January 3, 1976 which was the strongest one of the last 20 years in this area.

Moreover minutes of pressure measurements at Martin's Creek Cooling Tower, Pennsylvania, were available to make statistical evaluations.

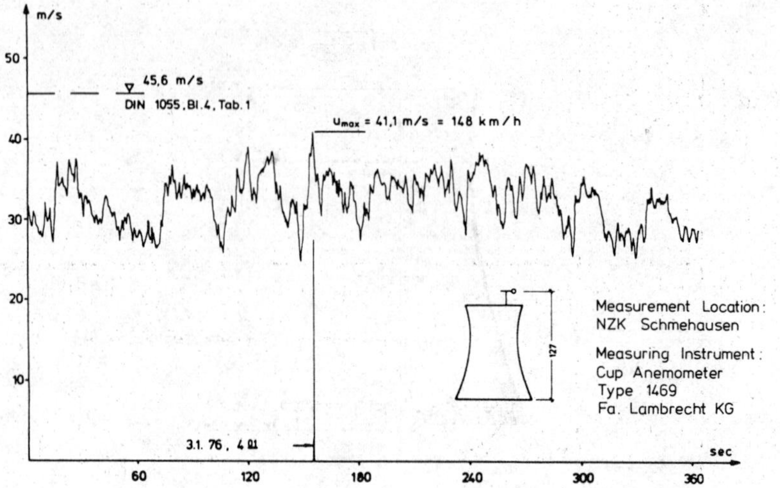

Fig. 4. Time record of wind velocity measured during the storm of January 3, 1976.

In general all the tested values showed an excellent conformity between model and full-scale measurements.

MEAN WIND PRESSURE

Usually the static load of a cooling tower is described by non-dimensional pressure coefficients as

$$c_p(z,\Theta) = \overline{p}(z,\Theta)/\overline{q}(z) \tag{1}$$

Θ = circumferential angle

z = height

q = dynamic pressure

Fig. 5 shows a typical distribution of pressure coefficients approximately in the middle of tower-height dependent on the circumferential angle Θ. This distribution can be described by

$$\begin{aligned} c_p(z,\Theta) &= A - B \cdot \sin^c(D \cdot \Theta) \qquad \Theta \leq \Theta_N \\ c_p(z,\Theta) &= \text{const.} \qquad \Theta > \Theta_N \end{aligned} \tag{2}$$

As Fig. 5 shows the circumferential distribution is characterized by six typical values which are the maximum pressure max c_p at stagnation point, the angle Θ_o of zero crossing, the maximum suction (≙ minimum pressure) min c_p and its position Θ_1 and the constant value c_{pN} of base pressure with the position Θ_N of its beginning. By means of this the unknown values of Eq. (2) can be replaced by

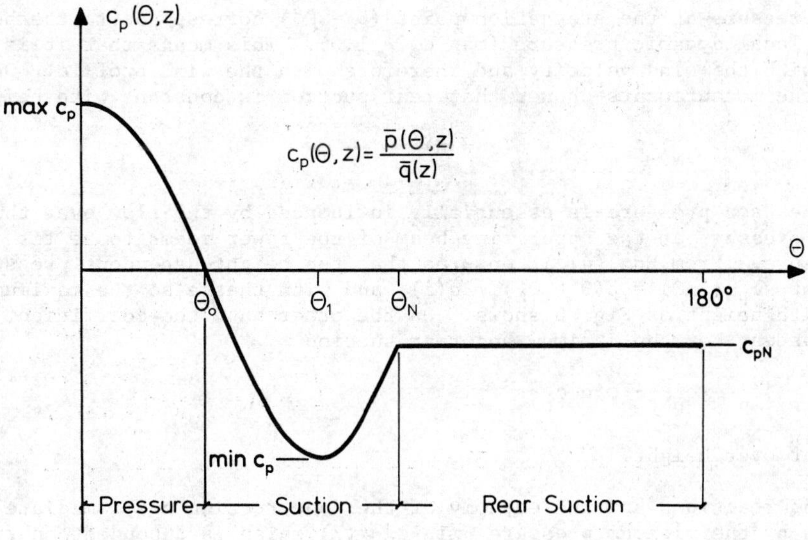

Fig. 5. Circumferential distribution of mean pressure coefficients.

$$
\begin{aligned}
A &= \max c_p \\
B &= A - \min c_p \\
D &= 90/\Theta_1 \\
C &= \ln (A/B) \,/\, \ln (\sin (D \cdot \Theta_o))
\end{aligned}
\qquad (3)
$$

The test results in (8) show a considerable influence of the wind profile on the pressure distribution (Fig. 6).

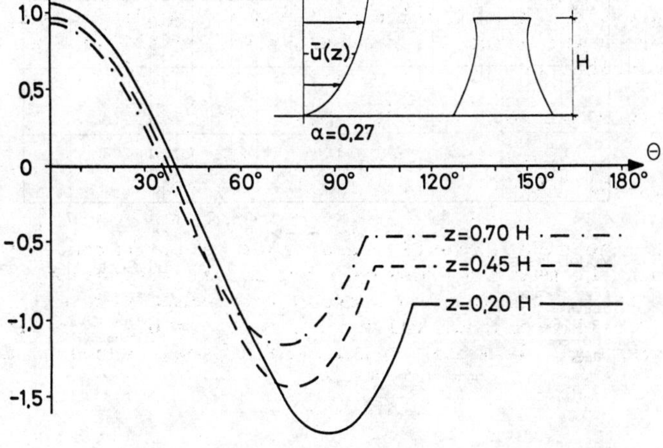

Fig. 6. Height dependence of mean pressure.

The mean pressure at the stagnation point ($\Theta = 0°$) corresponds to the height dependent local dynamic pressure (max $c_p \sim 1.0$). This means that it is connected directly with the wind velocity and therefore with the wind profile. On the contrary the measurements showed that rear suction is constant with height:

$$\overline{p}(\Theta > \Theta_N) \simeq -0.4 \cdot \overline{q}(z = H) \tag{4}$$

Because the base pressure is essentially influenced by the flow over the cooling tower the pressure at the upper ring beam of the tower seems to be fit as reference pressure. From Eq. (4) it ensures that the height-dependent pressure coefficient $c_{pN}(z,\Theta) = \overline{p}(\Theta > \Theta_N) / \overline{q}(z)$ and with that also the maximum suction ascends with height as Fig. 6 shows. On the other hand the coefficient of the pressure ascent between maximum and rear suction

$$\Delta c_p = c_{pN} - \min c_p \tag{5}$$

is constant over height.

The cutting reactions and the economy of the construction are immediately influenced by the minimum pressure min c_p (7) which is dependent on the shell's surface roughness which practically is controlled by the distribution of meridional wind ribs. Instead of min c_p in (9) the coefficient Δc_p which is constant over height is established as the characteristic value of the mean pressure distribution. The influence of the surface roughness is described by the parameter k/a (k = height of the ribs, a = distance of the ribs) which is substantially a product of the number of ribs n and the parameter k/D which is usual in fluid mechanics (8). Figure 7 shows the interaction between Δc_p and k/a which yielded by wind tunnel tests in smooth flow (Nieman (6), Farell et al. (3)) as well as in turbulent flow with boundary layer simulation (from (9)) and some full-scale measurements. It is easy to see that the Δc_p values continuously descends with ascending surface roughness. In the range $k/a > 8 \cdot 10^{-2}$ however the coefficients grow a little bit because the flow does not adjoin the surface between the too broad or too large wind ribs.

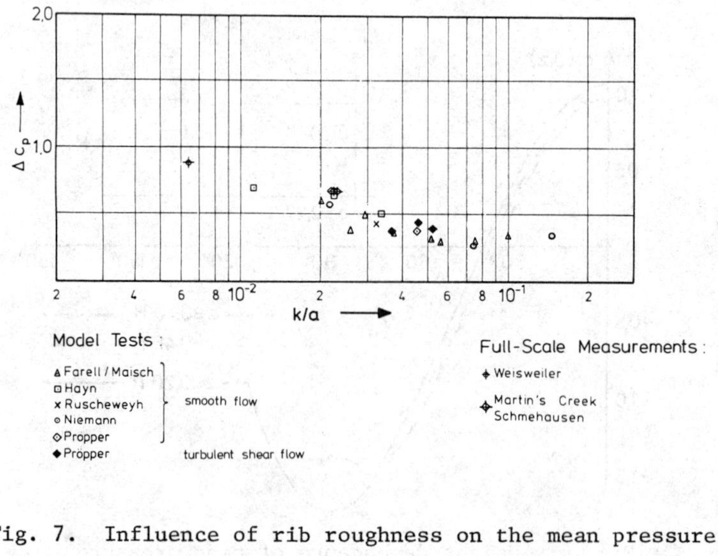

Fig. 7. Influence of rib roughness on the mean pressure.

With the aid of Eq. (4) and Δc_p from Fig. 7 min c_p can be determined for every height z.

RMS VALUES AND CORRELATION COEFFICIENTS

The rms values σ_p of the wind pressure give a direct and vivid view of the value of the pressure fluctuations. The ratio between σ_p and the mean pressure at the stagnation point $\overline{p}(\Theta = 0, z) \sim \overline{q}(z)$ is directly connected with the turbulence intensity of the natural wind:

$$\sigma_p(\Theta = 0, z)/\overline{q}(z) = 1.8 \cdot \sigma_u/\overline{u}(z) \tag{6}$$

Equation (6) shows that the fluctuations of the pressure time record are relatively greater than those of the oncoming flow.

Normalizing the circumferential distribution of the rms values $\sigma_p(\Theta, z)$ by the value $\sigma_p(\Theta = 0, z)$ at the stagnation point you come to a representation which is independent from the turbulence of the flow. Figure 8 shows some distributions of rms values $\sigma_p(\Theta, z)$. It was not possible to find an analytical term comparable to Eq. (2).

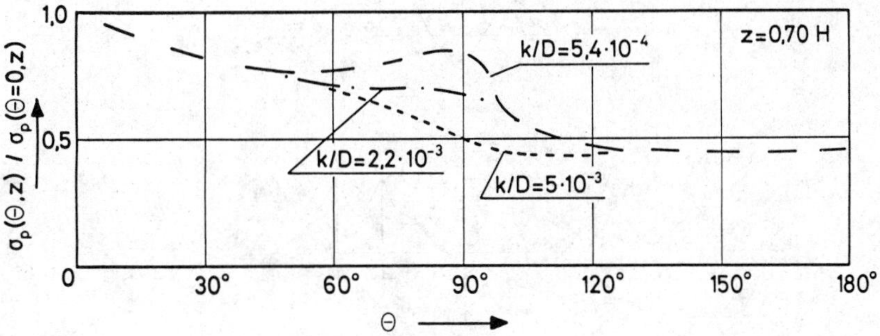

Fig. 8. Circumferential distribution of rms fluctuating pressure coefficients.

Proceeding from the stagnation point the curves in Fig. 8 descend slowly and after that they rise again in the reach of the pressure ascent Δc_p between Θ_1 and Θ_N. It is easy to see that here the values are strongly influenced by the surface roughness. Similar to the mean pressure distribution the values of the roughest tower are the best ones, but here k/D and not k/a is the scaling parameter which means that the number of ribs is insignificant. In the reach of the rear suction the curves uniformly sink down to half the values at the stagnation point.

The correlation coefficients, which are necessary for the calculation of the quasi-static response of the tower (1) give a proportion of the similarity of pressure time records at two different points on the shell surface and by this a glimpse into the spatial connections. With that, the fact is taken into consideration, that a wind gust has only a limited spatial spread and consequently, the maximum load is not able to appear everywhere at the same time. Again the vertical distribution of the correlation coefficients is directly connected with the structure of the wind turbulence by the ratio of the integral length scale $L_{u,z}/L_p$.

Figure 9 shows the circumferential distribution of the correlation function $\rho_p(\Theta, \Delta\Theta)$ for different angles $0 \leq \Theta \leq 180°$ found on a cooling tower model. It is well to see that the simplification established by Abu-Sitta and Hashish (1) that the rear suction is not correlated with the reach $\Theta < \Theta_N$ is not verified. Above all the zones of the maximum suction min c_p and the pressure ascent Δc_p is very much positively correlated because of the nearly periodical shedding of vortices in this reach. Furthermore it is conspicuous that the distribution of the correlation coefficients depends on the reference angle Θ as well as on the direction of $\Delta\Theta$. If $\Delta\Theta$ runs upstream the zones of pressure and suction have a strong negative correlation, on the contrary they are only slightly correlated if $\Delta\Theta$ runs downstream.

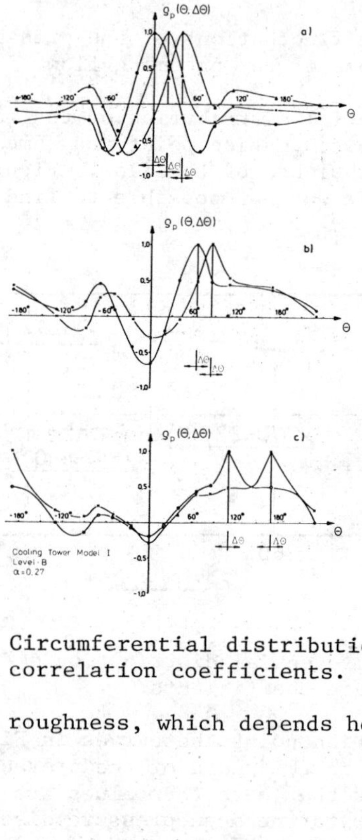

Fig. 9. Circumferential distribution of correlation coefficients.

The influence of the surface roughness, which depends here again on k/a, is relatively unimportant.

PRESSURE SPECTRA

Wide Band Excitation by Wind Turbulence

The represented pressure spectra are calculated from AD-transduced time records by means of the standard method (full-scale) and FFT method (model). Using the full capacity of the computer each time record of the model measurements for instance was divided into 153,600 discrete values in order to make a spectral analysis of them. In doing so it was possible to reduce the error of statistical uncertainty to 7 percent by using special smoothing procedures (2).

It was already shown that the fluctuations of the pressure immediately follow those of the wind velocity. Therefore it is obvious to describe the pressure spectra by similar empirical terms as the turbulence spectrum. In the range of higher frequencies where you can find the eigenfrequencies of the most of vibration endangered buildings the following statement is sufficient.

$$\frac{f \cdot S_p(\Theta, f)}{\sigma_p^2(\Theta, z)} = a_o(\Theta) \cdot X^{a_1(\Theta)} \tag{7}$$

X = any non-dimensional frequency according to Eqs. (11), (12) and (14).

But if you wish to describe the pressure spectra within the total frequency range a generalized form of the Karman-Spectrum is used:

$$\frac{f \cdot S_p(\Theta, f)}{\sigma_p^2(\Theta, z)} = \frac{\beta_p(\Theta) \cdot X}{(1 + \gamma_p(\Theta) \cdot X^2)^{\delta(\Theta)}} \tag{8}$$

In the higher frequency range, which now can be defined by $\gamma_p \cdot X^2 >> 1$, the following connection between the parameters of Eqs. (7) and (8) can be found:

$$\delta(\Theta) = (1 - a_1(\Theta))/2$$

$$a_o(\Theta) = \beta_p(\Theta)/\gamma_p(\Theta)^{\delta(\Theta)} \tag{9}$$

Doing the spectral analysis at first the parameters a_o and a_1 of Eq. (7) were evaluated by means of geometrical regression. After that with the aid of the method of least squares the optimum values of γ_p and β_p connected with a_o (Eq. (9)) were found.

The Eq. (8) renders possible to describe completely the pressure spectra over the whole interesting frequency range by only three parameters. The deviations from the measured spectra values are utmostly unimportant, in particular in the range of higher frequencies and the resultant errors from this are mostly less than those from the spectra calculation.

Figure 10 shows some pressure spectra at the stagnation point ($\Theta \approx 0°$) of cooling towers in the form of Eq. (8) for different flow structures. Compared with them the Karman-Turbulence-Spectrum

$$\frac{f \cdot S_u(f)}{\sigma_u^2} = \frac{4X_u}{(1 + 70.78\, X_u^2)^{5/6}} \tag{10}$$

is likewise drawn in. Therefore the plotting took place over the non-dimensional frequency X_u

$$X_u = f \cdot L(z)/\bar{u}(z) \tag{11}$$

with the integral length scale $L = L_{u,x}(z)$ of the turbulence as reference length.

All the pressure spectra show a remarkable stronger declination in the higher frequency range than the Karman-Spectrum. It is noticeable that all the spectra have the same declination and consequently they would fall into one another by adequate movement. Hence it follows that the declination fixing parameters a_1 (Eq. (7)) respectively, δ (Eq. (8)) are independent from the structure of the oncoming flow.

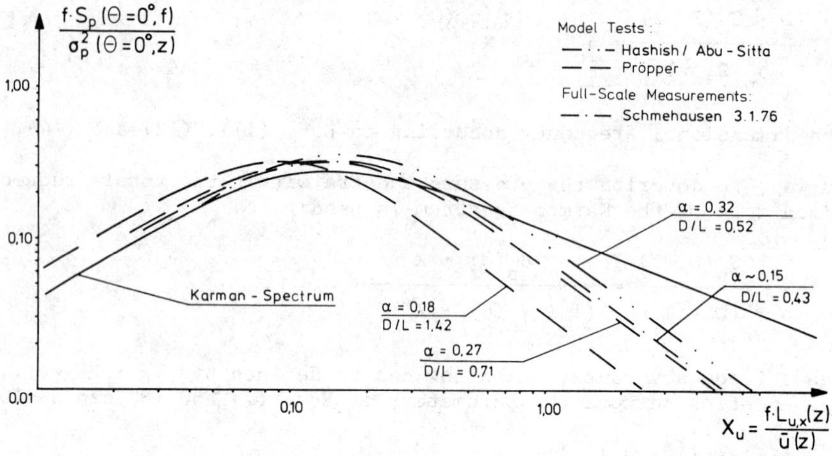

Fig. 10. Comparison between the wind spectrum and some pressure spectra at the stagnation point.

Figure 11 shows the aerodynamic admittance $X^2(\Theta, f)$, thus the ratio of the normalized pressure and turbulence spectra. The influence of the parameter D/L is eliminated if the plotting takes place over the non-dimensional frequency X_D

$$X_D = f \cdot D/\bar{u}(z) \tag{12}$$

with the tower's diameter D as building reference length. Therefore the results of the full-scale measurements ($\alpha \approx 0.15$) compared with those of the boundary layer simulation ($\alpha = 0.18$) are nearly identical, although the belonging D/L ratios are considerably different.

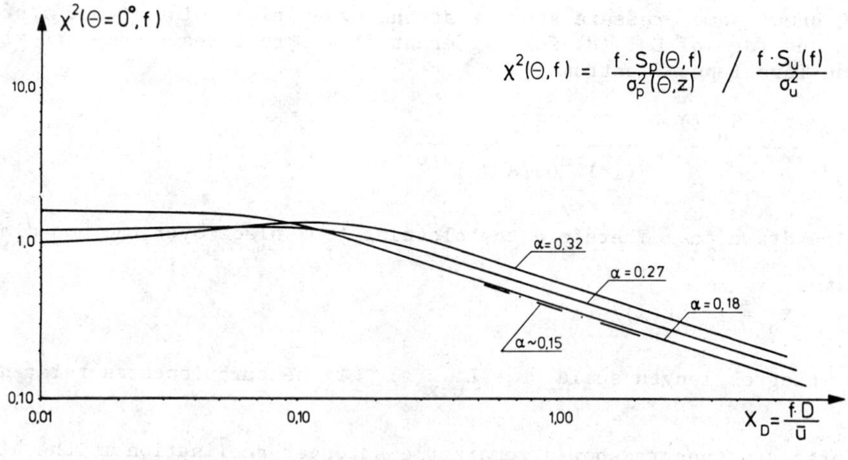

Fig. 11. Aerodynamic admittance.

Similar results were found by Hunt (5) and Vickery (11). However, both of them did not consider the influence of the wind profile which is well to see in Fig. 11.

The higher frequency range of the aerodynamic admittance function can be described by:

$$X^2(\Theta, f) = b_o^{1/\alpha}(\Theta) \cdot X_D^{b_1(\Theta)} \qquad (13)$$

$b_o^{1/\alpha}$ = order of set of curves.

The calculation of the pressure spectrum out of X^2 gives

$$\frac{f \cdot S_p(\Theta, f)}{\sigma_p^2(\Theta, z)} = 0.115 \cdot b_o^{1/\alpha} \cdot X_D^{b_1} \cdot X_u^{2/3}$$

and comparing this with Eq. (7) it conducts with $a_o = 0.115$ to a new non-dimensional frequency:

$$X_p(\Theta) = (b_o^{1/\alpha}(\Theta) \cdot (D/L)^{2/3})^{1/a_1} \cdot X_D \qquad (14)$$

Thus, it is possible to represent the pressure spectra uniformly and independent of the structure of the oncoming flow as well as of the size of the building.

Figure 12 gives a general view of the analytical equations of the wind pressure spectra. Still you only need to know the value of the parameters a_1, b_o and β_p. The circumferential distribution of them is shown in Fig. 13. They are nearly independent of height.

simple statement: $\frac{f \cdot S_p(\Theta, f)}{\sigma_p^2(\Theta, z)} = 0{,}115 \cdot X_p^{a_1(\Theta)}$

complete statement: $\frac{f \cdot S_p(\Theta, f)}{\sigma_p^2(\Theta, z)} = \frac{\beta_p(\Theta) \cdot X_p}{[1 + \gamma_p(\Theta) \cdot X_p^2]^{\delta(\Theta)}}$

dependence: $0{,}115 = \beta_p(\Theta) / \gamma_p(\Theta)^{\delta(\Theta)}$

$\delta(\Theta) = [1 - a_1(\Theta)] / 2$

normalizing of frequency: $X_p(\Theta) = [b_o^{1/\alpha}(\Theta) \cdot (D/L)^{2/3}]^{1/a_1(\Theta)} \cdot \frac{f \cdot D}{\bar{u}(z)}$

parameters from fig. 13 a ÷ c: a_1, b_o, β_p

Fig. 12. Analytical description of pressure spectra by empirical parameters.

Finally in Fig. 14 some pressure spectra at different points of the shell surface are plotted over X_p. The test results of a cooling tower model as well as the full-scale measurements of Schmehausen are compared. Both of the towers have the same roughness parameters k/a respectively k/D, but the parameters α and D/L differ very much. The generally good consistency of measuring values makes clear that the normalizing of the frequencies is right and that it is possible to scale the measured results to nature.

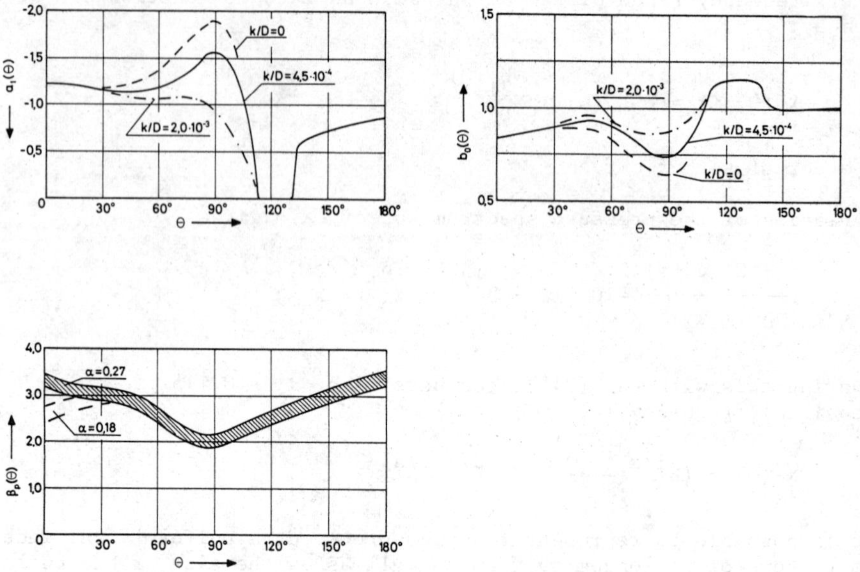

Fig. 13(a-c). Circumferential distribution of the spectra describing parameters.

NARROW BAND EXCITATION BY VORTEX SHEDDING

Since a long time the vortex shedding phenomena are intensively researched in fluid mechanics. The behavior of the flow depends strongly on the Re-number. While in the subcritical range the shedding of vortices is completely irregular without any dominating frequency, in the range of transcritical Re-numbers, which is particularly interesting in this context, a nearly periodical vortex shedding with predominant frequency appears.

The separation frequency f_{St} is defined by the Strouhal number Str

$$Str = f_{St} \cdot D/\bar{u}_m \tag{15}$$

which for cooling towers universally amounts to $Str = 0.18$. $\bar{u}_m$ is the mean value of the wind velocity over height.

Only then the non-dimensional frequency Str follows from X_p when in Eq. (14) the value in brackets is equal one, i.e., that the vortex shedding is independent of the parameters α and D/L. That is why the peaks on Fig. 14 resulting from separation frequencies do not have the same position.

Furthermore the separation frequency f_{St} itself is independent of the profile of the oncoming flow and constant over height as well as flowing around in the reach of separation, which is used by Roshko in (10) as reference velocity for the Strouhal number.

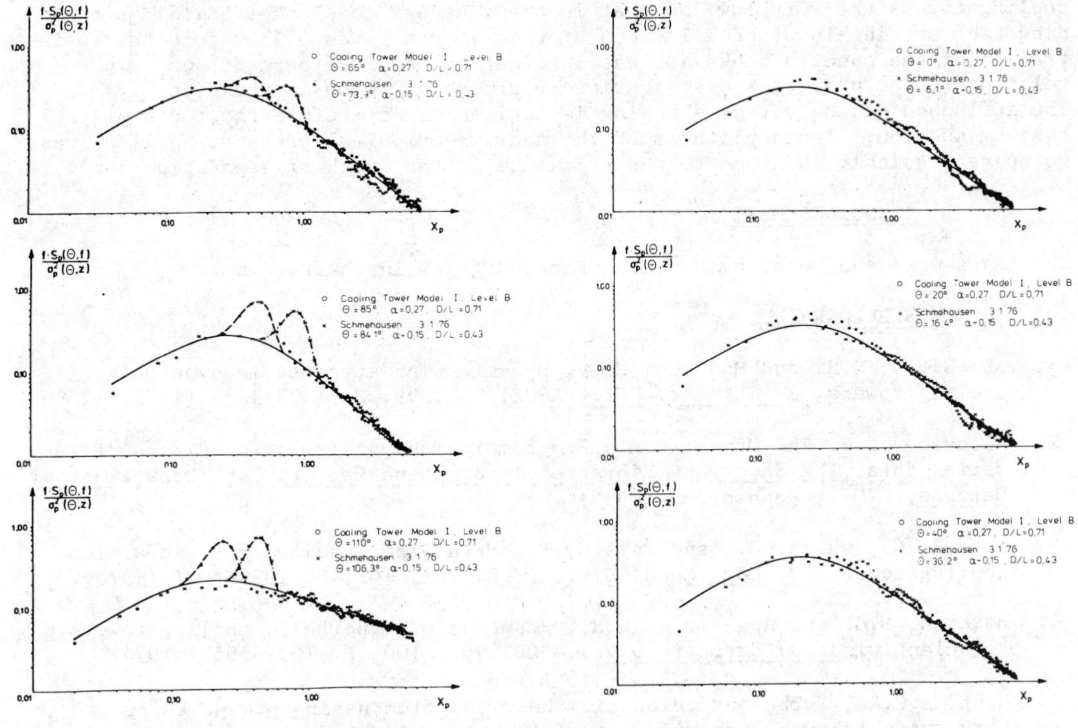

Fig. 14. Comparison between the pressure spectra of the model and full-scale measurements.

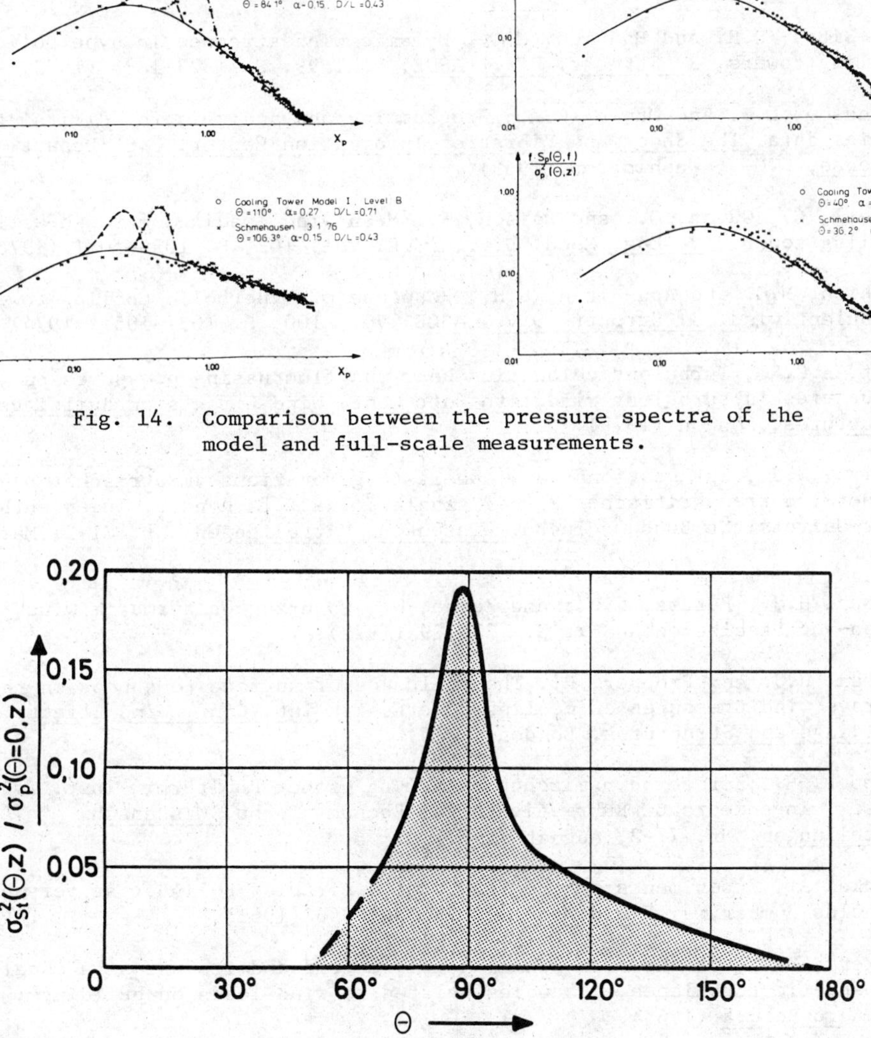

Fig. 15. Influence of vortex shedding on the pressure fluctuation.

Because of the very large tower diameters the dominating vortex frequencies of cooling towers are very low, so that in the contrary to slender buildings the danger of setting-in of oscillations by them is not given. Therefore the registration of the spectra banks is less interesting than the part of σ^2_{St} in relation to the whole rms distribution of pressure, which is shown in Fig. 15. The influence of the vortex shedding is noticeable not before the reach of the maximum suction. The amplitudes of the nearly periodical pressure oscillations go up very quickly and come to their maximum in the reach of separation zone.

ACKNOWLEDGMENT

The tests were sponsored by Deutsche Forschungsgemeinschaft, Bonn.

BIBLIOGRAPHY

1. Abu-Sitta, S.H. and Hashish, M.G., Dynamic wind stresses in hyperbolic cooling towers, J. Structr. Div., ASCE, Vol. 99, S. 1823-1835 (1973).

2. Enochson, L.D. and Otnes, R.K., Programming and analysis for digital time series data, The Shock and Vibration Information Center, U.S. Department of Defense, SVM-3, Washington (1968).

3. Farell, C., Güven, O., and Maisch, F., Mean wind loading on rough-walled cooling towers, J. Eng. Mech. Div., ASCE, Vol. 102, S. 1059-1081 (1976).

4. Hashish, M.G. and Abu-Sitta, S.H., Response of hyperbolic cooling towers to turbulent wind, J. Structr. Div., ASCE, Vol. 100, S. 1037-1051 (1974).

5. Hunt, J.C.R., Turbulent velocities near and fluctuating pressures on structures in turbulent wind, 4th Int. Conf. Wind Effects on Buildings and Structures, London (1975).

6. Nieman, H.J., Zur stationären windbelastung rotationssymmetrischer bauwerke im bereich transkritischer Reynoldszahlen, Inst. f. Konstr. Ingenieurbau, Ruhr-Universität Bochum, Techn. Wissensch. Mitteilungen, Nr. 71-2, März, (1971).

7. Nieman, H.J., Peters, H.L., and Zerna, W., Naturzugkühltürme im wind, Beton- u. Stahlbetonbau 67, S. 121-129 (1972).

8. Nieman, H.J. and Pröpper, H., The influence of surface roughness on the average wind pressures on cooling towers, 4th Int. Conf. Wind Effects on Buildings and Structures, London (1975).

9. Pröpper, H., Zur aerodynamischen belastung grober kühltürme, Inst. f. Konstr. Ingenieurbau, Ruhr-Universität Bochum, Techn. Wissensch. Mitteilungen, Nr. 77-3, August (1977).

10. Roshko, A., Experiments on the flow past a circular cylinder at very high Reynolds number, J. Fluid Mech. 10, S. 345-356 (1961).

11. Vickery, B.J., On the flow behind a coarse grid and its use as a model of atmospheric turbulence in studies related to wind loads on buildings, NPL Aero Report (1965).

DISCUSSION ON SESSION IV
Wind Loading--Mean External and Internal
and Fluctuating Internal

Rapporteur: B. M. Leadon
University of Florida
USA

Papers by K. T. Kavanagh--A FIELD MEASUREMENT FACILITY BASED UPON A WIND-SENSITIVE LOW-RISE STRUCTURE
K. T. Mehta--FULL SCALE MEASUREMENTS ON A FLAT ROOF AREA

Question: J. D. Holmes. I would like authors to comment on the influence of fluctuating internal pressures on their measurements. What attention, if any, has been paid to the sealing of the building, and are (were) measurements of internal pressure being made during the experiments?

Reply: K. T. Kavanagh. The internal pressure and external pressure were equalized by providing several 1-inch diameter holes in appropriate walls. The holes were opened or covered as necessary to equalize the pressures. The internal and external static pressures were monitored using a static pressure probe. A U-tube located inside the building was connected to the static probe outside. Even though the liquid in two sides of the tube fluctuated, the fluctuation was kept within 1/16-inch during the period of data acquisition. The original experimental program was designed to test the effects of terrain on low-rise structures, and not to test the magnitude of internal pressures. The building, therefore, was completely sealed with tape and polyethylene sheet before taking measurements (as shown in the last two slides of the presentation). It should be emphasized, however, that there is a redundancy of data which could be used to determine internal pressure. Since panel displacements define average pressures, only two displacements represent independent data. The additional measurement of frame displacements (in each subsystem) can be used to calculate a common internal pressure on each panel. The assumption of common internal pressures would appear to be consistent with the data shown by Stathopoulos et al. (paper IV-9-1).

The field facility described in this paper contains three resonant frequencies (the 'model' is a three degree of freedom system). As described in the paper, a lengthy systems analysis was performed to calibrate the dynamic model, so that the pressure could be deduced from the dynamic response (through resonant peaks, if necessary). High frequency noise was filtered above the model response range by digital filtering to eliminate errors from higher mode resonances.

Perth is fortunate to have summer sea breezes in the 10-25 knot range. The 'breezes' generally occur over a period of 1 to 2 hours with fluctuating intensity, but consistent direction. It was initially intended to use 30-minute segments of data in each significant record. Present results indicate that the calculations of pressure are non-stationary over these periods. It is too early to draw conclusions from these records, since the analytical procedures are being refined during the early field tests.

Question: A. P. Jeary. When performing spectral analysis it is necessary to reduce bias and variance errors by using sufficiently long data records. In order to obtain spectral estimates of which approximately 11% of estimates are within one standard deviation of the true point it is necessary to have a record length of 200 $[\xi_r f_r]^{-1}$ seconds (where f_r is a resonance frequency and ξ_r is its' associated damping value). Would the author comment on what precautions he took in this respect.

Reply: K. C. Mehta. The continuous recording period of each data set is a significant factor for the analysis of the random wind load. Davenport has suggested in the literature that the period be in the range of 5 to 30 minutes. This suggestion is based on the consideration of spectral gap in natural wind data, periods of thunderstorms, and non-stationarity of wind speed. Researchers e.g., Marshall, Eaton and Mayne, have used recording periods in the range of 15-17 minutes in field measurements. In the experiment described in the paper, the recording period of 15 minutes (900 seconds) was used. The analysis was performed using either the front or back 800 seconds of the total period. Resonance frequency of the load measuring system was estimated to be 50 Hz, while that of the roof system was measured to be 12.5 Hz. Damping of the system was not obtained. It was observed in 4 out of 28 data sets that the wind speed data was non-stationary in the 800 second period. This indicates that longer periods of recording may cause concern for stationarity.

Question: J. R. Mayne. I am doubtful of the accuracy of the spectra presented and I think that the similarity of the speed and load spectra (Figure 8) may be due to the data handling techniques. The standard error is probably high. How many 800 second data sets were used to prepare each spectrum? What analog filtering and anti-aliasing techniques were used? What was the frequency response of the load-cell system?

Reply: S. Kim, K. C. Mehta. The frequency response of the load measuring system with roof weight resting on the load cells was estimated to be 50 Hz. Without the roof weight, the load cells would respond to frequency of about 300 Hz. The speed and load spectra, such as the one shown in Figure 8 of the paper, are based on one 800 second data set. Spectra were obtained for each 800 second data set (28 in all) to observe shape and peaks of a speed spectrum and the corresponding load spectrum. Since each spectrum is based on one continuous 800 second period, the comment concerning standard error is not applicable. The data were recorded on an analog tape recorder. The noise level was kept to a minimum by using a tape speed of 15 in. per second. The records were digitized without filtering. Admittedly, subsequent data processing introduced some averaging and smoothing effects, but these effects would not be significant enough to alter overall results. Details of the data analysis are given in Reference 11 of the paper.

Question: C. Kramer. The shape of the roof edges in your experiment is not well-defined by the plastic membrane. This may be of significant influence on the peak pressures. Did you investigate the influence of the edge shape on your wind loading measurements?

Reply: S. Kim, K. C. Mehta. The plastic membrane was used at the edges of the roof to seal the building and to mitigate transmission of wall vibrations to the roof system. We did not change this edge condition throughout the experiment.

Question: C. Kramer. The higher frequency pressure fluctuations on a flat roof are not caused by "small scale turbulence" but by a shifting of flow reattachment zones or lines. These regions grow with increasing roof size. Neglecting the higher frequencies the loading change, for example, of a fixation element during the lifetime of a building will be underestimated by some orders of magnitude.

Reply: S. Kim, K. C. Mehta. Fixation elements or anchorages such as bolts, screws, welds, etc. are likely to experience higher frequency fluctuations. Tributary areas for these anchorages are relatively small. The results presented in the paper are for much larger areas and may be applicable to larger roof components.

Comment: H. A. Panofsky. In my opinion the "spectra" in this paper represented just a mathematical transformation of the data in this particular sample. There was no question about representing the "population." The similar appearance of the spectra yielded a valid comparison. Coherence would have been better because it also tests the phase relationships.

Paper by R. E. Akins, J. A. Peterka, J. E. Cermak--AVERAGED PRESSURE COEFFICIENTS FOR RECTANGULAR BUILDINGS

Question: V. R. Cartelli. In Tables 3 and 4, I assume the coefficients and mean pressures are for each face. In designing buildings for total wind pressure should we add the values for sides 1 and 3 and 2 and 4 respectively? Please clarify.

Reply: R. E. Akins, J. A. Peterka, J. E. Cermak. Yes, the values for sides 1 and 3 and 2 and 4 should be added for an overall pressure. This has been done in Reference 4 of the paper. For the design of local cladding it would be appropriate to use the values for each side individually.

Paper by A. Laneville, C. D. Williams--THE EFFECT OF INTENSITY AND LARGE SCALE TURBULENCE ON THE MEAN PRESSURE AND DRAG COEFFICIENTS OF 2D RECTANGULAR CYLINDERS

Question: I. P. Castro. The overall spanwise aspect ratio of the models was apparently large enough to suggest that three-dimensional end effects were small. However, did the authors check the spanwise uniformity of the upstream mean velocity, turbulence intensity and scale? It is very difficult, with the mesh size tunnel size ratios apparently used, to ensure adequate flow uniformity at any cross section.

Question: A. Kareem. Was the intensity of turbulence uniform along the span of the prism? It appeared from the slide that the turbulence grid bars were very wide and larger than the width of the prism. This might result in nonuniform turbulence intensity depending how far downstream the prism is.

Question: F. H. Durgin. How many mesh lengths as opposed to bar lengths between the grid and model? Did you check for homogeneity of flow?

Reply: A. Laneville, C. D. Williams. The bar size was 9 inches and the mesh size 36 inches. The distance between the grid and the model was 279 inches. The mean velocity distribution was uniform within 5% for the core of the flow. The turbulence intensity was 9% measured on the tunnel centerline. Since these results agree with those of Reference 7, the distributions of the turbulence intensity and of the integral scale were not measured.

Question: R. E. Britter. Does the limited experimental range of L_x/D justify the author's conclusion that turbulence scale effects are not important particularly when the author states that cases with $L_x/d \sim 100$ are of interest. Although Figure 5 was shown to indicate that turbulence scale effects are unimportant, the peak of the curves in Figure 5 differ by more than 10% over $6 < L_x/D < 18$.

Reply: A. Laneville, C. D. Williams. In fact, the effect of L_x/D is negligible for the range $0.3 < L_x/D < 18$ as was shown by the experiments. Figure 5 shows the base pressure coefficient values only for $L_x/D = 6$ and 18. However, the values for all other cases were clustered about the curve shown. The maximum differences of $\pm 5\%$ between data points shown are within experimental error. The curve was drawn only to indicate the trend of the data.

Question: W. Z. Sadeh. It is questionable whether the turbulence intensity is the parameter of significance with regard to the body's C_D. Change in the C_D can be affected by supplying turbulent energy at scale susceptible to interaction with the body boundary layer. Consequently, the distribution of turbulent energy is to be considered and not an overall parameter such as turbulence intensity.

Reply: A. Laneville, C. D. Williams. There is some evidence that it is the small scale turbulence which is entrained into the shear layer which can alter the characteristics of the shear layer and thus alter the drag force. In the case of grid turbulence, the spectrum is well-defined once the integral scale and the intensity are known. At this point it is the overall intensity which is used as a parameter in categorizing the effect of turbulence on drag force. Perhaps an intensity factor can be devised which is a measure of the contribution of the small scales to the overall intensity which will be useful in explaining the effect of intensity on drag force. In grid flow, this intensity factor might be simply related to the overall intensity.

Question: I. S. Gartshore. There must be an effect of scale at very large scales. If, as you have suggested, the very large scales introduce a "quasi-steady" effect since, when this is the case, only a part of the intensity is effective. So, we need to know what fraction of the total intensity to use as the scale goes up since clearly the "effective" intensity is a very important parameter in assessing shear layer behavior, drag, fluctuating lift, etc. Can you suggest what scale ratio introduces quasi-steady effects?

Reply: A. Laneville, C. D. Williams. Large scale turbulence can be visualized as large eddies imposing a fluctuating angle of incidence upon the cylinder. On a time averaged basis, because of the randomness of the motion of those eddies, it should have no net effect on the time-averaged drag coefficient. But, since in turbulent flow diffusion is always present, the parameter should be an intensity of turbulence, either the overall intensity or an intensity defined from a sub-range of the spectrum. Clearly some more work is required to explain the phenomenon. An important parameter that should come into the picture is the ratio of the energy contained in the shear layer to that contained in the incident turbulence eddies.

Question: Y. Nakamura. Regarding Figure 6, the results indicate that the side face pressure recovery for the subcritical cylinders ($H/D < 0.63$) is nearly zero. This is not in agreement with our recent measurements (Y. Nakamura and Y. Tomonari, Pressure Distributions on Rectangular Prisms at Small Incidences, Trans. Japan Soc. Aero. and Space Sciences, February, 1979) where it takes a considerably negative value as H/D approaches the critical, and then turns abruptly to be a positive one after the critical is reached. I think that our trend is consistent to the closest proximity of the shear layer toward the rear corner at the critical depth.

Reply: A. Laneville, C. D. Williams. The work referred to by Prof. Nakamura has been in smooth flow while here the flow was turbulent. Also in Figure 6, only positive pressure recovery was considered in order to explain the drop in C_D with increasing H/D. Only two sidewall pressure taps were available to determine the amount of pressure recovery. Clearly, more sidewall pressure taps are required to investigate the relationship between the amount of pressure recovery as a function of H/D.

Paper by R. E. Whitbread--THE INFLUENCE OF SHIELDING ON THE WIND FORCES EXPERIENCED BY ARRAYS OF LATTICE FRAMES

Question: B. J. Vickery. Why was it chosen to examine shielding at $\alpha = 0°$ rather than at the angle yielding maximum load?

Reply: R. E. Whitbread. The data presented in my paper is just part of a continuing program on lattice structures. At the onset, the work was concerned with complete square, triangular and rectangular sections relating to transmission towers. These sections were tested over a full range of wind direction and many of the features observed by Georgiou and Vickery were reproduced. Tests were also made over a full range of wind direction for a complete model of a typical transmission tower as a calibration of the wind resistance clauses of the proposed UK Code of Practice for Lattice Towers.
It was only at a late stage in the program that doubts were raised concerning the validity of existing published shielding factors. For this reason, it was decided to restrict the scope of the parameters to the zero degree wind direction.

Paper by N. Georgiou, B. J. Vickery--WIND LOADS ON BUILDING FRAMES

Comments: F. H. Durgin. It is shown that as the porosity increases the drag of the second rectangular frame begins to increase at a porosity of about 0.3. In creating boundary layers in a wind tunnel using spires I have observed that as one moves downstream from the spires the velocity defect became an excess, etc. See work on spires at NRC. I believe the effect shown is related.
Reply: N. Georgiou, B. J. Vickery. The fact that the anticipated behavior of the shielding factors for angle-section frames was reversed above a solidity of 0.35 did not become apparent until the results had been analyzed. No further measurements were possible at that stage. It will be necessary therefore to look to published results to seek for this unexpected effect. This comment is helpful in this respect.

Papers by J. D. Holmes--MEAN AND FLUCTUATING INTERNAL PRESSURES INDUCED BY WIND
T. Stathopoulos, D. Surry, A. G. Davenport--INTERNAL PRESSURE CHARACTERISTICS OF LOW-RISE BUILDINGS DUE TO WIND ACTION

Question: R. A. Hazelwood. The evidence of relatively high frequency pulsations due to cavity resonance suggests that this may excite building components with similar natural vibration frequencies. Has either author evidence for the increase in damping that may be expected with the introduction of porous walls?
Reply: J. D. Holmes. There seems to be a possibility that internal cladding panel vibrations may be excited by internal pressure resonance, but since the excitation as well as the response is narrow-band, exact coincidence of the resonant frequencies would be required. Damping due to porous walls may be estimated in the same way that the damping term in Equation 9 of the paper was derived.
Reply: D. Surry. We have no evidence of any increase in damping or its potential values for building with porous walls. We have found evidence suggesting that the highest frequency components, unlike the overall structural loads, are not well correlated between internal and external loads, especially for the high intensity external local pressures occurring on the roof.

Question: B. M. Leadon. Surely there may be a very small external compressibility involved. (1) Why not modify the Helmholtz resonator formula for long buildings to $n \sim a/4L$ which is the organ-pipe frequency for one end open? An L/D of only about 10 is required. (2) Unsteady flow orifice coefficients may be found in the shock-tube literature of the late 1940's: RUDINGER, "Wave diagrams for unsteady flow in ducts," Van Nostrand, for example.
Reply: J. D. Holmes. (1) An organ pipe model may be more valid for very long buildings, perhaps 200 m or greater in length. However, for most buildings and certainly for the houses, I was mainly concerned with this research, the Helmholtz resonator concept is quite adequate. (2) Thank you for this reference. I was not aware of it and will investigate it.

Question: H. P. Irwin. The model tests in both these experiments appear to be essentially on an empty building. The presence of interior partitions would presumably reduce the interior pressure fluctuations. Do the authors have any comments on this?

Reply: T. Stathopoulos, D. Surry, A. G. Davenport. It is correct that the buildings considered did not have any partitions, although models did have a significant amount of pressure tubing inside, which was different for the small and large models. This did not appear to interfere with the good correlation of internal pressures. Partitions, depending on their degree of sealing, could interfere with this correlation and could reduce the fluctuations in reasonably well-sealed compartments away from any openings to the outside. This would certainly be a factor when extending the results to large structures.

Reply: J. D. Holmes. My tests were conducted with a completely clear internal space inside the model. Internal partitions clearly will affect the internal pressure fluctuations and further work on this is required. The internal rms pressure coefficient is a function of the parameter $A_w^{3/2}/V_o$ in which V_o is the internal volume as indicated in the paper.

Question: N. J. Cook. You stated in your presentation that the difference between internal and external pressures of the model was measured by referencing the transducer backing pressure to the internal pressure. What transducer did you use? I ask because some transducers, including the SETRA and DISA transducers, appear to respond nonlinearly to backing-pressure fluctuations resulting in an offset mean.

Reply: J. D. Holmes. A SETRA 237 pressure transducer was used for the measurements. If there is a problem with the transducer I would like to talk to Dr. Cook about it. However, the comparison of the mean pressure differences measured simultaneously and non-simultaneously shown on the slide would not appear to indicate an offset.

Question: B. J. Vickery. It appears that the theory does not take account of the flexibility of the building. This is not significant for the wind tunnel model but for prototype structures the volume increase due to wall and roof deflections is comparable with that associated with the compression of the enclosed air. Has the author considered this point?

Reply: J. D. Holmes. Professor Vickery is correct in saying that building wall flexibility was not included in the computer model results presented in the paper, which were intended for comparison with the wind tunnel model data; for the latter, of course, the walls were essentially rigid.
Wall flexibility was, in fact, mentioned briefly in the paper at the end of Section 4.3. This effect can be very easily incorporated in the computer model as an effective reduced atmospheric pressure, as suggested by Professor Vickery himself in Reference 8. The influence of this effect will be investigated in future work.

Question: J. L. Wood. Have the authors considered the flow conditions required for ventilation to control condensation in wall cavities and roof voids? Condensation is becoming a major problem in metal clad heavily insulated buildings in cool humid climates.

Reply: J. D. Holmes. Not at this time.

Reply: T. Stathopoulos, D. Surry, A. G. Davenport. No, we have not although we understand this problem is receiving some attention by building research groups in a number of countries, e.g., Canada and the United Kingdom.

Question: R. L. Wardlaw. The porosity of buildings is Reynolds number sensitive and also nonlinear with pressure differential. These effects are not well established, partly due to the use of porosity coefficients that are not in dimensionless form. Consequently, it is difficult to know how to model porosity at small scale. Any full scale porosity coefficients that I have seen relate volume flow and pressure differential. The authors have used porosities in terms of area ratios. I would like to know how they have inferred these area ratios from the other form of coefficients and have they comments on scaling.

Reply: T. Stathopoulos, D. Surry, A. G. Davenport. The porosity values used cover a range of values which were assumed to be representative of full-scale practice. The climate, the building material, construction techniques, etc. all lead to a good deal of scatter in actual porosities. Energy conservation in cold climates leads to very small values, for instance, whereas deliberate ventilation produces the reverse in hot climates. Our values of up to 3% were derived from estimates made by manufacturers of metal buildings. In North America we suspect the most likely values are of the order of 0.5% or less, not counting deliberate openings such as windows or doors.
The Reynolds number effect is a real problem, both in ascertaining porosity values from full-scale infiltration measurements and in assigning effective porosity values to the model scale representations, which are described only in terms of geometrical area ratios. More work is required in this area.

Question: G. R. Walker. Internal pressure changes are presumably transmitted at sonic velocity. In small scale models this will tend to cause fluctuations to be apparently transmitted faster than in the corresponding full-scale building. How much value can therefore be placed on the information presented in Figure 4?

Reply: T. Stathopoulos, D. Surry, A. G. Davenport. This effect is mentioned in the paper and essentially requires Mach number scaling to be correctly maintained. However, it is also argued there that the effect is likely to be small except possibly for models of very large buildings where the pressure transmission time (the dimension divided by the sound speed) could approach periods of interest for the structural response of the building. The Mach number omission for model testing is also required by the simplified distorted geometrical scaling approach suggested to correctly maintain the Helmholtz resonator frequency in the previous paper by Holmes. Both of these effects are likely to be small for typical low-rise buildings but should be checked for larger structures.

N.B.: Comment on the question of Mach number scaling appears in the Discussion of Session VII.

Paper by J. Blessmann, J. D. Riera--INTERACTION EFFECTS IN NEIGHBORING TALL BUILDINGS

Question: H. Ruscheweyh. Two cylinders close together have a distinct aeroelastic instability at a small angle of attack depending on distance ratio. Because of the abrupt change of the flow through the gap at this angle of attack the static side force jumps from one value to another large one. If the cylinder can vibrate, hysteresis effects occur depending on the Strouhal number. This hysteresis leads to a dangerous aeroelastic instability similar to a galloping phenomenon (interference galloping). Have you looked for this effect?

Reply: J. Blessman, J. D. Riera. It has been observed, and it is pointed out in the paper, that for ratios S_x/a between 0.25 and 1 and wind incidence approximately parallel to the orientation of the gap ($\alpha = 90°$, $S_y = 0$), the pressure within the gap presented cyclic variations (see Figures A.2 and A.3). It appears that Dr. Ruscheweyh is referring to this effect, which will be the subject of special attention during the tests with aeroelastic models to be conducted shortly at the LAC. However, it is anticipated that on account of the low frequency of the pressure oscillations, they should not seriously affect building structures.

SESSION V

Wind Loading - Fluctuating External

CHAIRMAN
G. V. Parkinson
Professor of Mechanical Engineering
University of British Columbia
Vancouver, Canada

CO-CHAIRMAN
A. Roshko
Professor of Aeronautics
California Institute of Technology
Pasadena, California, USA

RAPPORTEUR
J. A. Peterka
Professor of Civil Engineering
Colorado State University
Fort Collins, Colorado, USA

RECENT DEVELOPMENTS AND TRENDS: WIND LOADING - FLUCTUATING EXTERNAL

G. V. PARKINSON

University of British Columbia, Canada

Four years ago, at the fourth of these International Conferences on Wind Engineering, Scruton, in introducing the session which corresponds to the present Session IV and V, made some general observations about the nature of the problems of wind loading, and our progress in understanding and dealing with them. Those observations remain largely valid today, despite the further progress that has occurred. I will confine my remarks to brief comments on some evidence of that further progress, from papers to follow in this Session.

As we contemplate the general problem of fluctuating external wind loads on buildings and structures, we should feel no chagrin at the relatively slow pace of progress in dealing with it. In dealing with turbulent separated flows, progress is bound to be slow - indeed, we might well feel some satisfaction at the amount that has been achieved. Most of the themes of this Session had already been sounded at the previous Conference, or at still earlier ones, but there have been changes in emphasis. For example, papers by Sockel, by Dalgliesh, Templin, and Cooper, and by others in other Sessions consider different statistical models for predicting extreme values of fluctuating surface pressures, obviously desirable for cladding design. In a continuing, solitary theme Hunt and Durbin present further developments in Hunt's theory of the effects of weak incident turbulence near the stagnation streamline on fluctuating pressures on the parts of body surfaces exposed to unseparated flow.

Unlike the purely statistical models, which do not attempt to deal with the physics of the flow fields, Hunt's theory involves both flow field and statistical turbulence concepts, and its quasi-steady and small-scale turbulence limits appear to straddle relevant experimental data, and produce realistic trends.

Another paper, by Kawai, Katsura, and Ishizaki deals with the experimental correlation between incident turbulence and fluctuating pressures on windward surfaces, in this case on rectangular building forms, both model and full-scale, and the authors infer larger scales for the pressure than for the velocity fluctuations. This effect is also noted by Lam, Lam, and Mackey. Model and full-scale comparisons of building surface pressures are also a part, indeed the main part, of the paper by Dalgliesh et al., and the close agreement shown gives increased confidence in the techniques of boundary-layer wind tunnel simulations, at least for strong wind conditions. Their tests also make evident the need for a better reference pressure for full-scale tests.

Saunders and Melbourne, in the last of the Session's papers dealing primarily with wind loading, present useful results from boundary-layer wind tunnel tests simulating buffeting of a rectangular tall building by the wakes of single or paired upstream buildings of similar size and shape. They provide data for wind loading buffeting factors, which indicate approximately doubled fluctuating base bending moments in the worst cases tested.

Other papers in the Session deal either with wind-induced oscillations, or with fundamental phenomena of separated flow, or both. A theme common to papers by Melbourne and by Miyata and Miyazaki in this Session (and by Laneville and Williams in Session IV) is that of the effects of incident turbulence on the separating and reattaching shear layers from rectangular shapes. This theme was sounded by Laneville and Parkinson at the 1971 Tokyo Conference, and developed by Laneville, Gartshore, and Parkinson at the 1975 Heathrow Conference. Melbourne's concern is with evidence for his instability hypothesis as an explanation of high negative pressures near the leading edge of the surface under such a separating and reattaching shear layer. He finds himself in agreement with Gartshore on the importance of small-scale turbulence in enhancing this effect.

Miyata and Miyazaki, in contradiction to previous tentative findings by Laneville and Parkinson, and current conclusions by Laneville and Williams, find significant effects of turbulence integral scale on the transverse wind force and consequent galloping oscillations for rectangular-section cylinders in wind tunnel tests.

In other papers in this Session dealing with organized forms of wind-induced oscillations, Howell and Novak present data on vortex shedding effects in turbulent flows for circular cylinders both oscillated mechanically and allowed to self-oscillate. The turbulent flows are of both grid-produced and boundary-layer types. They find the vortex-induced response insensitive to the turbulence if the system damping is low.

Finally, Griffin presents further evidence for his version of that hardy perennial, the universal Strouhal number, first proposed by Roshko about a quarter century ago. Griffin's model uses a measured wake width as the characteristic length, and the data from a variety of sources collapse well onto a single curve as a function of a universal wake Reynolds number.

FLUCTUATING SURFACE PRESSURES ON BLUFF STRUCTURES IN TURBULENT WINDS: FURTHER THEORY AND COMPARISON WITH EXPERIMENT

P. A. DURBIN and J. C. R. HUNT
University of Cambridge, Cambridge, UK

SUMMARY

The theory for the distortion of weak turbulence impinging onto bluff obstacles developed by Hunt (1973, 1974) has been extended to predict the surface pressure fluctuations, p*, around the obstacle in terms of the fluid density and upwind mean and r. m. s. velocities, u'_∞ and $\overline{u}_\infty$, when the scales of turbulence, L_x, are small compared with a typical cross-sectional dimension of the structure (say a radius, a), or when the frequencies are large. It is found: that the surface pressure spectra around cylindrical or spherical structures have an effective 'cut off' at frequency of order $\overline{u}_\infty \Pi / a$ after which the spectrum decreases typically as $n^{-16/3}$; that when $a/L_x >> 1$, p^{*2} is proportional to $(\rho \overline{u}_\infty u'_\infty)^2 (a/L_x)$ and decreases around the surface from the stagnation point (in contrast to the case $a/L_x << 1$, where it increases); that, even if $a/L_x >> 1$, p* is correlated in the streamwise direction over a distance of the order of a over the surface of a bluff obstacle. On a cylindrical structure p* is correlated in the axial direction over a distance greater than, but of the order of L_x. Consequently, when $a/L_x >> 1$, on a spherical structure the mean square drag is proportional to $(\rho \overline{u}_\infty u'_\infty)^2 (a/L_x)^{-1}$, and on a cylindrical structure is proportional to $(\rho \overline{u}_\infty u'_\infty)^2 (a/L_x)^{-2}$.

Our results explain why the aerodynamic admittance, based on integrating the effect of upstream turbulence across an area equal to the obstacle's (the porous or lattice network approach) is too small for solid structures. The theory is compared with field and laboratory measurements and shows that most measurements lie between the well known quasi-steady predictions and these new small scale turbulence predictions.

INTRODUCTION

Fluctuating pressures on bluff obstacles in turbulent flows are caused directly by upstream turbulence and the unsteady velocity field produced by one or more separations of the flow from the surface of the obstacle. Although the mean pressure field and the intensity of fluctuating pressures produced by the separated regions of the flow are affected by the intensity and scale of the incident turbulence (by means that remain still rather obscure), the correlation between those fluctuating pressures and velocities produced by the separated flows and those produced by the upstream turbulence is generally thought to be small, on the basis of measurement of surface pressure correlation inside and outside the wake and hot wire velocity measurements. (See for example Batham, 1973 and

Britter, Hunt & Mumford, 1979). Given this assumption and given the mean velocity field, and the turbulent velocity field near the obstacle and fluctuation of surface pressure, p*, outside separated flow regions, can be calculated in terms of the upstream turbulent velocity field, provided its intensity is small (say less than 15%). (Hunt, 1973, 1974, 1976).

As regards pressure fluctuation, this theory has been used to show: (i) that the quasi-steady limit for p* is the same as the exact theory in the limit $(a/L_x) \to 0$; (ii) the correction for the low frequency pressure spectrum $Ⓗ^{(P)}(K_1)$ when $a/L_x \ll 1$; (iii) that the high frequency spectrum $Ⓗ^{(P)}(K_1)$ at the stagnation point decreases very much more rapidly with K_1 (like $e^{-K_1/(a/L_x)}$) than the upwind velocity spectrum; (iv) how the drag on a porous plate varies with the scale of the incident turbulence and the porosity of the plate. (Graham, 1976).

In this paper we show how the theory can be applied to calculate the spectra, the mean square value, the form of the cross correlation around the cylinder, and the axial length scale of the fluctuating pressure around circular cylinders and thence other obstacles. We compare our predictions with laboratory and field measurements, particularly aiming to see if the two theoretical asymptotic limits of large ($a/L_x \ll 1$) and small scale turbulence ($a/L_x \gg 1$) straddle the observations and explain their trends.

We conclude with some comments on the next steps to be tackled to understand how turbulence affects wind loads.

2. Rapid distortion theory for fluctuating pressures

Assumptions As explained and justified in more detail by Hunt (1974, 1973), given certain assumptions, the fluctuating velocity near a bluff body and the surface fluctuating pressures can be calculated in statistical terms as a function of the spectra or correlations of the upwind turbulence. These assumptions are: (A1) The upwind turbulence intensity is weak, i.e. $u'_\infty/\overline{u}_\infty \ll 1$. Atmospheric turbulence intensities of 0.1 to 0.2 are just about small enough to qualify; (A2) The time taken for turbulent eddies to pass round the structure is small compared with the time scale of the non-linear interaction between the eddies i.e.

$$(u'_\infty / \overline{u}_\infty)\ (a/L_x) \ll 1$$

a being the radius of the structure and L_x is the upwind integral scale. (If $a/L_x \gg 1$, then the theory is strictly only valid if the turbulence intensity is very weak indeed); (A3) The Reynolds number of the turbulence is large

$$u'_\infty L_x/\nu \gg 1.$$

(A4) The scale of turbulence is large compared with the thickness (δ) of the boundary layer on the structure; (A5) The fluctuating velocity and pressure field outside the wake can be divided up into 2 components, one of which is caused by the upwind turbulence and the other by the wake induced fluctuations; (A6) The upwind turbulence is assumed to be homogeneous†.

† As we shall see later, for calculating surface pressure fluctuations it need only be homogeneous over a distance of $(O2L_x)$ either side of the stagnation line.

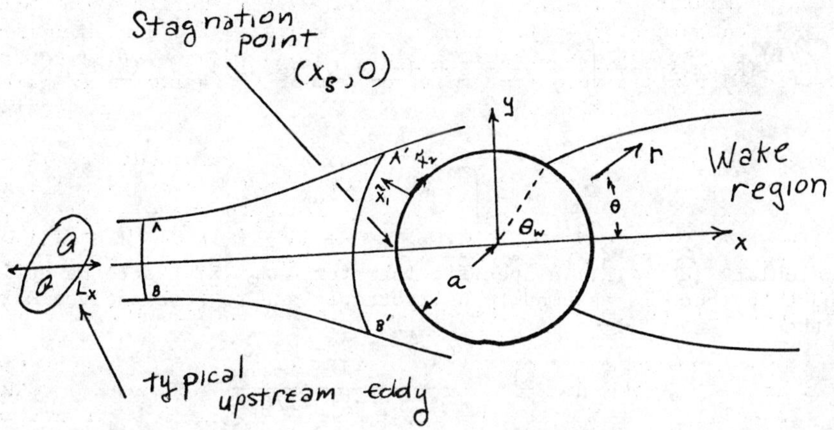

Fig.1 Defining sketch showing (x,y), (r,θ), (x_1,x_2), coordinates, wake region and stretching of vortex line AB as it approaches the cylinder.

Normalisation The local mean and fluctuating velocity $\overline{u}_i*$, $u_i*(\underline{x}*,t*)$ and local mean and fluctuating pressure $\overline{p}*$, $p*(\underline{x}*,t*)$ at a location $\underline{x}*$, at time t*, are normalised on the upwind r.m.s. velocity $u^1{}_\infty$, mean velocity $\overline{u}_\infty$ and density ρ. $\underline{x}*$, t* are normalised on the structure's radius a and $\overline{u}_\infty$. Thus

$$u_i = u_i*/u'_\infty, \; U_i = \overline{u}_i*/\overline{u}_\infty, \; P = (C_p/2) = \overline{p}*/\rho\overline{u}_\infty{}^2$$

$$p = p*/(\rho\overline{u}_\infty u'_\infty), \; x_i = x_i*/a, \; t = t*/(a/\overline{u}_\infty) \qquad (2.1)$$

Note the different normalisation of p* and $\overline{p}*$. C_p is the conventional pressure coefficient.

Spectral analysis The object of our analysis is to calculate the following:

(i) local mean square fluctuating pressure $\overline{p^2}$ or its r.m.s. value p′. (ii) the one-dimensional cross spectrum at two points $\underline{x}*$, $\underline{x}*^1$ (or the power spectrum at one point $\underline{x}*$) which, when normalised on the upwind integral scale L_x is written as

$$\frac{Ⓗ*^{(p)}(K_1*)\; \underline{x}*,\underline{x}*^1)}{(\rho\overline{u}_\infty u^1{}_\infty)^2\, L_x} = Ⓗ^{(P)}(K_1,\underline{x},\underline{x}^1) = \frac{1}{2\Pi}\int_{\infty}^{\infty} \overline{p(\underline{x},t)p(\underline{x}^1,t+\tau)}\; e^{iK_1/(a/L_x)}\,d\tau \qquad (2.2)$$

here $K_1 = K_1*\, L_x = 2\Pi n\, L_x/\overline{u}_\infty$ and n is a frequency. It is also convenient to define a wave number

$$K_1 = K_1*\; a* = \hat{K}_1\; a/L_x$$

This spectrum can be expressed in terms of the spectrum of the upwind velocity by means of the transfer function, $Q_n(x_1,x_2;\hat{\underline{K}})$ which relates the Fourier transform (or Fourier coefficient) of the normalised pressure to the Fourier transform of the normalised upwind velocity. If the upwind turbulence is homogeneous a normalised three dimensional energy spectrum tensor $\hat{\Phi}_{\infty nm}$ $(\hat{\underline{K}})$ can be defined in terms of which the local pressure spectrum is calculated:

$$\hat{H}^{(P)}(\hat{K}_1,\underline{x},\underline{x}^1) = \int_{-\infty}^{\infty}\int_{-\infty}^{\infty} \sum_n \sum_m Q_n(\underline{x},\hat{\underline{K}})\; Q_m{}^*(\underline{x}^1,\hat{\underline{K}})\hat{\Phi}_{\infty\, nm}\;(\hat{\underline{K}})\,d\hat{K}_2 d\hat{K}_3 \tag{2.3}$$

and thence

$$\overline{p^2} = \int_{-\infty}^{\infty} \hat{H}^{(P)}(\hat{K}_1)\; d\hat{K}_1 \tag{2.4}$$

To calculate $\hat{H}^{(P)}(\hat{K}_1)$ a specific form for $\hat{\Phi}_{\infty nm}$ $(\hat{\underline{K}})$ has to be defined. If the upwind turbulence is assumed to be isotropic and to have the von-Karman spectrum then

$$\hat{\Phi}_{\infty nm}\;(\hat{\underline{K}}) = \frac{c_3\, \hat{k}^2\, \delta_{ij} - \hat{K}_i\, \hat{K}_j}{c_2 + \hat{k}^2 \;\; 17/6} \tag{2.5}$$

where

$$c_3 = (55/36\;\Pi^2)\,c_2^{\;5/6},\; c_2 = 0.558,\; \hat{k}^2 = \sum_{i=1}^{3} \hat{K}_i^{\;2}$$

Using the dynamical equations of motion we now show how Q_n can be expressed in terms of the velocity transfer function $M_{in}(x_1,x_2;\;\hat{\underline{K}})$ which relates the Fourier coefficient of the ith velocity component at (x_1,x_2) to the Fourier coefficient of the nth upwind velocity component.

Finding the pressure transfer function Q_n The fluctuating pressure is determined by the linearised inviscid Navier-Stokes equations using assumptions A1-4 (Hunt, 1974). To calculate the pressure spectrum it is convenient to write these equations in terms of the pressure transfer function Q_n and the velocity transfer fraction M_{in}, so that

$$-\partial\, Q_n/\partial\, x_i = -\, iK_1\, M_{in} + \partial\,(U_k M_{jn} + U_j M_{in})\,/\partial\, xj \tag{2.6}$$

where $\partial/\partial x_3 \equiv iK_3$. In the approaching stream, (where $U_1 = 1$) the turbulence is convected in a 'frozen' pattern, so that as

$$x \to -\infty,\; -iK_1\, M_{in} + \partial\, M_{in}/\partial\, x_1 = -\partial\, Q_n/\partial\, x_i = 0 \tag{2.7}$$

It is convenient to define $M_{in}(\infty)$ $(x_1,x_2;\underline{K})$ as the value of M_{in} in the absence of the obstacle, where

$$M_{in}^{(\infty)} = \delta_{in}\, e^{i(K_1x_1 + K_2x_2)} = \delta_{in}\, e^{i(K_1x + K_2y)} \tag{2.8}$$

Low frequency turbulence For large scale turbulence and for the low frequency part of the spectrum, given M_{in}, equation (2.6) is the most convenient means of calculating Q_n. Thus at the stagnation point ($x = x_s$, $y = 0$) of a symmetrical cylindrical obstacle, by integrating (2.6) along the stagnation line and using (2.7), (2.8), Q_n is given by

$$Q_n(x_s,0;\underline{K}) = iK_1 \int_{-\infty}^{x_s} (M_{in}(x,0) - M_{in}^{(\infty)}(x,0)\; d_x + M_{in}^{\;\infty}\;(x_s,0) \tag{2.9}$$

Thus when

$$K_1 \rightarrow 0, \; Q_n(x_s,0;K_2,K_3) = \delta_{in} \qquad (2.10)$$

Around the surface of the structure, if $\tilde{x}_2$ is the distance from the stagnation point and $\tilde{U}_2$, $\tilde{u}_2$ the mean and turbulent velocity components parallel to the surface, with $\tilde{M}_{2n}$ being the transfer function relating $\tilde{u}_2$ to the upstream velocity components u_n, $(n = 1,2,3)$, then

$$Q_n(\tilde{x}_2,\underline{K}) = Q_n(x_s,0;\underline{K}) + iK_1 \int_0^{\tilde{x}_2} \tilde{M}_{2n}(x_2{}^1,\underline{K})\, d\tilde{x}_2{}^1 - \tilde{U}_2 \tilde{M}_{2n}(\tilde{x}_2;\underline{K}) \qquad (2.11)$$

In the limit of large scale turbulence ($|\underline{K}_1| \rightarrow 0$) (2.11) becomes

$$Q_n(\tilde{x}_2,\underline{K}) = 1 - \tilde{U}_2 \tilde{M}_{2n}(\tilde{x}_2,\underline{K}), \qquad (2.12)$$

which for a potential flow around a circular cylinder is

$$Q_n(s,\underline{K}) = (1 - 4 \sin^2 \;, \; 2 \sin 2 \;, \; 0). \qquad (2.13)$$

Quasi-steady analysis This result can be obtained quite independently by considering the fluctuating pressure caused by very slow fluctuations in u_1, u_2, u_3 in the free stream, the so-called 'quasi-steady approach'. A change in u_3 has no effect on a cylinder. A change in u_1 is effectively a change in the mean velocity $\delta\overline{u}_\infty$, and a change in u_2 is effectively a change, $\delta\alpha$, in the orientation of the structure to the free stream, so that the change in pressure at a point on the obstacle is

$$\delta_p{}^* = C_p \tfrac{1}{2} \rho\, 2\overline{u}_\infty \delta\overline{u}_\infty + \frac{\partial C_p}{\partial \alpha} \tfrac{1}{2} \rho (\overline{u}_\infty + \delta\overline{u}_\infty)^2 \delta\alpha$$

where $C_p = 2P$ is the mean pressure coefficient. Then since $\delta\overline{u}_\infty = u_1 u'_\infty, \delta\alpha = u_2 u'_\infty/\overline{u}_\infty$

$$Q_n = \left[C_p, \tfrac{1}{2} \partial C_p/\partial \alpha, 0\right] \qquad (2.14)$$

In the limit of very small scale turbulence around a circular cylinder (2.11) can be evaluated in the limit $K_1 = 0$, $|K_2|$, $|K_3| \rightarrow \infty$. Using the expression for $\tilde{M}_{2n}$ in (5.47) of Hunt (1973), it follows that

$$Q_n(K_1 = 0,K_2,K_3) = \left[1 - \frac{K_3(1+\cos\theta) - i2K_2\sin\theta}{K_3 - i2K_2\sin\theta}, 0, 0\right] \qquad (2.15)$$

Small scale turbulence When $|\underline{K}| \gg 1$ the integrals in (2.9) and (2.11) are difficult to perform analytically. It is more straightforward to calculate Q_n from the divergence of (2.6) namely

$$\partial^2 Q_n/\partial x^2 + \partial^2 Q_n/\partial y^2 - K_3{}^2 Q_n = -2 \sum_{i=1}^{2} \sum_{j=1}^{2} \frac{\partial U_i}{\partial x_j} \frac{\partial M_{jn}}{\partial x_i} \qquad (2.16)$$

When $|\underline{K}| \gg 1$, the first stage in solving (2.16) for Q_n near the surface is to express the right hand side in terms of the velocity components normal and parallel to the surface, $\tilde{U}_i$ and $\tilde{M}_{jn}$, and the local co-ordinates $(\tilde{x}_1,\tilde{x}_2,\tilde{x}_3)$.

Thus, invoking the incompressibility and irrotationality of $\tilde{U}_i$;

$$\frac{\partial^2 Q_n}{\partial \tilde{x}_1^2} + \frac{\partial^2 Q_n}{\partial \tilde{x}_2^2} - K_3^2 Q_n = -2 \frac{\partial \tilde{U}_2}{\partial \tilde{x}_2} \left(\frac{\partial \tilde{M}_{2n}}{\partial \tilde{x}_2} - \frac{\partial \tilde{M}_{1n}}{\partial \tilde{x}_1} \right) + \frac{\partial \tilde{U}_2}{\partial \tilde{x}_1} \left(\frac{\partial \tilde{M}_{1n}}{\partial \tilde{x}_2} + \frac{\partial \tilde{M}_{2n}}{\partial \tilde{x}_1} \right) \tag{2.17}$$

Local solutions to (2.17) are obtained by introducing a 'local wave number', $\tilde{\chi}_2$, and writing Q_n in the form

$$Q_n = F_n(\tilde{x}_1, \underline{K})\, e^{i(\tilde{\chi}_2 \tilde{x}_2)}$$

(Hunt, 1973, §5.2).

On substitution into (2.17), F_n satisfies (in the case of a circular cylinder):

$$\frac{\partial^2 F_n}{\partial \tilde{x}_1^2} - (\chi_2^2 + K_3^2) F_n = 4 \left[\cos\theta \, (i\tilde{\chi}_2 \tilde{f}_{2n} - \partial \tilde{f}_{1n}/\partial \tilde{x}_1) + \sin\theta \, (i\tilde{\chi}_2 f_{1n} + \partial f_{2n}/\partial \tilde{x}_2) \right] \tag{2.18}$$

For details of the solution of (2.18) see Durbin (1979).

<u>Pressure spectra</u> On substituting the solutions for Q_n into (2.3), and using (2.5), the one-dimensional spectrum $\textcircled{H}^{(P)}(\hat{K}_1)$ can be computed. For large scales of turbulence, no computing is required because Q_n is independent of ($\underline{K}$), so that as $a/L_x \to 0$, from (2.14),

$$\hat{\textcircled{H}}^{(P)}(\hat{K}_1) = C_p^2 \, \hat{\textcircled{H}}_{\infty 11}(\hat{K}_1) + \frac{1}{4}\left(\frac{\partial C_p}{\partial x}\right)^2 \hat{\textcircled{H}}_{\infty 22}(\hat{K}_1) \tag{2.19}$$

which for a circular cylinder in isotropic turbulence becomes

$$\hat{\textcircled{H}}^{(P)}(\hat{K}_1) = \hat{\textcircled{H}}_{\infty 11}(\hat{K}_1)\,(1 - 4\sin^2\theta)^2 + 4\sin^2 2\theta \, \hat{\textcircled{H}}_{\infty 22}(\hat{K}_1) \tag{2.20}$$

The small correction to this result for K_1, or $\hat{K}_1(a/L_x) \ll 1$ and $a/L_x < 1$ was given by Hunt (1974). Even if $a/L_x \ll 1$, (2.20) becomes a poor approximation when K_1 or $\hat{K}_1(a/L_x) \gtrsim \frac{1}{2}$, as shown later.

For high wave numbers or small scales of turbulence a similarly general analysis is not possible. But we can obtain useful analytical results for high and low frequency limits. For a circular cylinder from the solution (2.18) or from (2.15), (2.3), and (2.5)

$$\hat{\textcircled{H}}^{(P)}(\hat{K}_1 \to 0) = \cos^2\theta \,/\, \left[\Pi(1 + 2|\sin\theta|)\right], \tag{2.21a}$$

and from (2.18)

$$\hat{\textcircled{H}}^{(P)}(\hat{K}_1 \to \infty) = 9.58\, \hat{K}_1^{-16/3} / 2 \,|\sin\theta|^{10/3} / (a/L_x)^{11/3}, \tag{2.21b}$$

or if the frequency and spectrum are normalised on the radius a

$$\hat{H}^{(P)}(K_1 \to \infty) \sim 9.58\,(a/L_x)^{2/3} K_1^{-(16/3)} \,/2 \sin\theta\,/^{10/3} \quad (2.21c)$$

In the limit of θ tending to Π (2.21a) remains valid but the expression in (2.21b) tends to zero, which is consistent with the theoretical result that at the stagnation point $H^{(P)}$ decreases exponentially as $K_1 \to \infty$. It is shown in detail in Durbin's thesis that on $\theta = \Pi$, as $K_1 \to \infty$, for a circular cylinder

$$H^{*(P)}(K_1^* a)\ \alpha\ (\rho\,\overline{u}_\infty u'_\infty)^2\, a(a/L_x)^{2/3}(K_1^* a)^{-7/3} e^{-\Pi/2K_1^* a} \quad (2.22)$$

(Regrettably there was an algebraic error in the analysis of Hunt (1976), but the similarity plot of

$$H^{*(P)}(K_1^* a) \,/ \left[(\rho\,\overline{u}_\infty u'_\infty)^2 a\,(a/L_x)^{2/3}\right]$$

which tended to collapse the high frequency measurements is still valid).

In fig. 2a computations of $\hat{H}^{(P)}(\hat{K}_1)$ at $\theta = 5\pi/6$ are plotted for various values of a/L_x, which are all large. Note that at low frequency the spectra tend to the same limit for all scales but at high frequency, for given K_1, $H^{(P)}$ decreases as a/L_x increases.

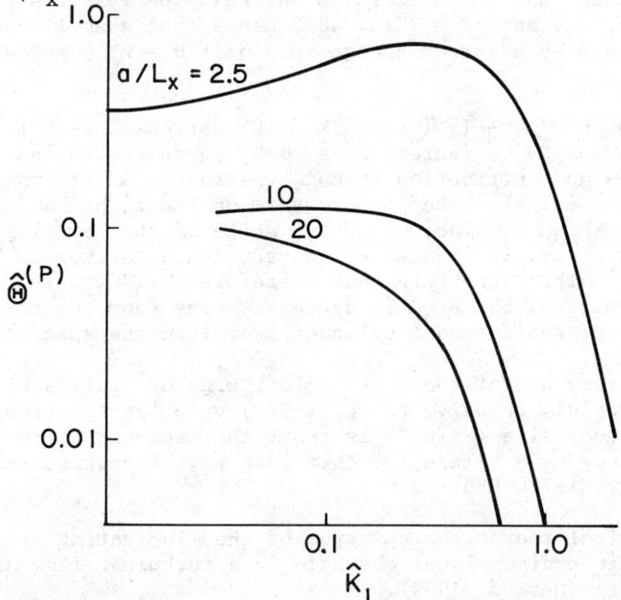

Fig. 2a Pressure spectra at $\theta = 5\Pi/6$ as a function of integral scale - (Theory assuming $a/L_x >> 1$).

In fig. 2b we present a schematic diagram to summarise the general form of the pressure spectra around a bluff cylinder for large and small scales. The main differences near $\theta = \Pi$ can be explained as follows: at low frequency, when

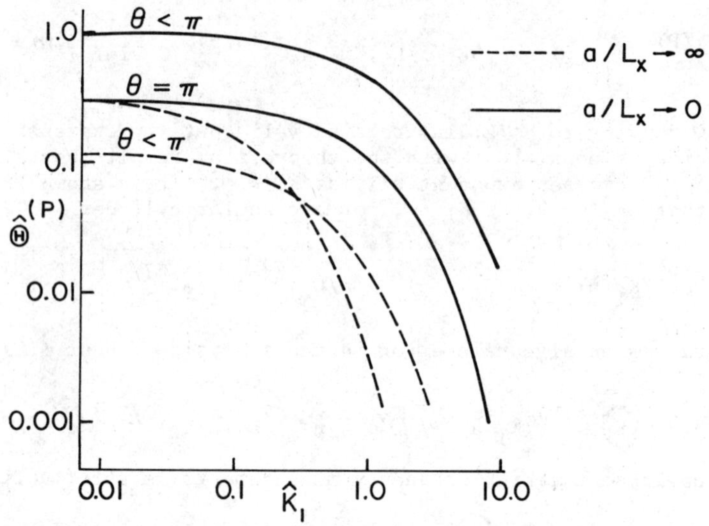

Fig. 2b Schematic diagram of pressure spectra around circular cylinder - showing the difference between large and small scale turbulence and the 'cut off' of the spectra at $K_1 \approx \Pi$.

$a/L_x << 1$, $\textcircled{H}^{(P)}$ increases as $|\theta-\Pi|$ increases basically because the rapid variation of the mean pressure with θ near $\theta = 2\Pi/3$, $4\Pi/3$ means that a small change in flow direction (due to u_2) causes a larger change in p near $\theta = 2\Pi/3$ and $4\Pi/3$ than near $\theta = \Pi$.

At small scale, when $a/L_x >> 1$, $\hat{\textcircled{H}}^{(P)}$ $(K_1 \to 0)$ decreases as θ increases because at low frequencies $\tilde{U}_2\ \tilde{u}_2$ increases as $|\theta-\Pi|$ increases so that $(1- \tilde{U}_2\ \tilde{M}_{21})$ decreases, and there is no contribution from u_2 upstream. An alternative explanation may be sought from (2.17b). The largest term on the right hand side is effectively $2\,\partial\tilde{U}_2/\partial\tilde{x}_2\ \ \partial\tilde{M}_{1n}/\partial\tilde{x}_1$ since $\partial\tilde{M}_{1n}/\partial\tilde{x}_1$, which is the normal gradient of the normal component of velocity, is induced by the stretching of vortex lines around the structure (fig. 1). This stretching, and therefore $\partial\tilde{M}_{1n}/\partial\tilde{x}_1$, is roughly constant around a bluff structure, but $\partial\tilde{U}_2/\partial\tilde{x}_2$ decreases away from the stagnation point, which is why $\textcircled{H}^{(P)}$ decreases around a cylinder away from the stagnation point.

At the high frequency end of the spectra for large or small scale turbulence, $\hat{\textcircled{H}}^{(P)}$ $(\hat{K}_1)$ decreases rapidly relative to the (-5/3) velocity spectrum. This 'cut off' occurs where the eddy size $(2\pi/K_1{}^*)$ is about the diameter of the structure (2a), i.e. when $K_1 = \Pi$ or $K_1 = \Pi/(a/L_x)$. This idea is, of course, well known (Davenport, 1961; Surry, 1972; Hunt 1976).

For the same physical reason the spectrum of the fluctuating stagnation pressure of an axisymmetric 3-dimensional obstacle in a turbulent flow also has a high frequency 'cut off' (Durbin, 1979).

Cross spectra Using our expressions for Q_n in (2.13) or (2.15) for turbulent flow round a circular cylinder, where the mean flow is potential flow, we find that the co-spectrum of pressure between the stagnation point and a point $(1,\theta)$ is:

$$\textcircled{H}^{(P)}(K_1;\ \theta,\Pi) = (2\cos 2\theta - 1)\,\textcircled{H}_{\infty 11}(K_1) \qquad (2.23)$$

when $(a/L_x) \to 0$. When $a/L_x >> 1$ we can only derive a simple expression for the case when $K_1 \to 0$:

$$\textcircled{H}^{(P)}(K_1 = 0; \theta, \Pi) = \frac{-\cos\theta}{\Pi(1+2/\sin\theta/)} \quad (2.24)$$

Variances and cross correlations of pressure fluctuations From the pressure spectrum $\textcircled{H}^{(P)}(K_1)$ or co-spectrum the variance p^2 and cross correlations

$$R^{(P)}(\tilde{x}_2, \tilde{x}_2{}^1) = \overline{p(\tilde{x}_2)p(\tilde{x}_2{}^1)} \Big/ \left[\overline{p^2}(\tilde{x}_2)\overline{p^2}(\tilde{x}_2{}^1)\right]^{\frac{1}{2}} \quad (2.25)$$

can be calculated. We present here normalised results for very large and very small scales of turbulence. To translate these into dimensional results refer to (2.1).

When $a/L_x \rightarrow 0$, from (2.20a) it follows that

$$\overline{p^2}(\tilde{x}_2) = \left[C_p{}^2 \overline{(u_1{}^2)}^{(\infty)} + \left(\frac{\partial C_p}{\partial \alpha}\right)^2 \overline{u_2{}^2}^{(\infty)} + C_p \frac{\partial C_p}{\partial \alpha} \overline{u_1 u_2}^{(\infty)}\right] \quad (2.26)$$

where C_p and $\partial C_p/\partial\alpha$ are defined at $\tilde{x}_2$. Note that $\overline{p^2}$ and $R^{(P)}$ can be affected by the upstream cross correlation of u_1 and u_2, which may not be negligible e.g. near a building lying in the wake of another. For the case of a circular cylinder in a turbulent flow where $(\overline{u_1 u_2})^{(\infty)} = 0$ (2.20b) leads to

$$\overline{p^2} = 5-4\cos 2\theta . \quad (2.27)$$

For the interesting case of a cross correlation between the pressure at the stagnation point ($\theta = \Pi$) and at an angle θ round the cylinder

$$R(\theta,\Pi) = \frac{2\cos 2\theta - 1}{\sqrt{5-4\cos 2\theta}} \quad (2.28)$$

as is plotted in fig. 3. Note how $R(\theta, \Pi)$ is zero when $|\theta-\Pi| = \Pi/6$ and is -1 when $|\theta-\Pi| = \Pi/2$.

To estimate the variation of $\overline{p^2}$ at the stagnation point with integral scale when $(a/L_x) \ll 1$, one can use the fact that the pressure spectrum $\textcircled{H}^{(P)}(K_1)$ tends to be 'cut off' when $K_1 \approx \Pi/(a/L_x)$, Consequently, since $\textcircled{H}^{(P)}(K_1) \approx \textcircled{H}_{\infty 11}(K_1)$ at the stagnation point when

$$K_1 \lesssim \Pi/(a/L_x)$$

$$\overline{p^2} \simeq 2\int_0^{\gamma\,\Pi/(a/L_x)} \textcircled{H}_{\infty 11}(K_1)\,dK_1 \quad (2.29)$$

$$\simeq 1 - \lambda\,(a/L_x)^{2/3} \quad (2.30)$$

where γ and λ are constants. Thus the difference between $\overline{p^2}$ and its quasi steady value is the order of $(a/L_x)^{2/3}$. It is found that when $a/L_x \gg 1$ for a circular cylinder

$$\overline{p^2} = \cos^2\theta / (a/L_x) \quad (2.31)$$

or, at the stagnation point,

$$\overline{p^2} = L_x/a \qquad (2.32)$$

Since (2.32) can be obtained from (2.29) with $\gamma = .5$, it suggests that an estimate for p^2 on $\theta = \Pi$ is to use (2.29) with $\gamma = .5$ for the whole range of (a/L_x). This is the estimate plotted in fig. 4.

Since $\hat{Ⓗ}^{(P)}(\hat{K}_1)$ on a 3-dimensional obstacle has the same 'cut off' as for a cylindrical obstacle, when $(a/L_x) \gg 1$ at the stagnation point on 3-dimensional obstacles

$$\overline{p^2} \; \alpha \; (\overline{u}_\infty u'_\infty)^2 (a/L_x)^{-1}. \qquad (2.33)$$

The cross correlation $R^{(P)}(\theta,\theta')$ has not yet been computed, when $a/L_x \gg 1$, but it can be estimated from $Ⓗ^{(P)}(\hat{K}_1 = 0;\theta,\theta')$ using the argument of (2.33).

$$R^{(P)} \simeq \hat{Ⓗ}^{(P)}(0;\theta,\theta^1) \Big/ \sqrt{\hat{Ⓗ}^{(P)}(0,\theta)\ \hat{Ⓗ}^{(P)}(0,\theta')} \qquad (2.34)$$

Thus,

$$R^{(P)}(\theta,\Pi) = \frac{1}{\sqrt{1+2/\sin\theta/}}$$

as shown in fig. 5. This result means that the fluctuating pressure is always correlated around the cylinder even if the upwind turbulence scale is much smaller than the radius of the cylinder.

Axial length scale To estimate the loads on a tall structure, it is also necessary to know the length scale $L_3^{(P)}$ of pressure fluctuations in a direction parallel to the axis of the structure. It is of particular interest to relate $L_3{}^{(P)}$ to L_3, the axial length scale of the upstream velocity fluctuations. For isotropic turbulence $L_3 = L_x/2$. The definition $L_3{}^{(P)}$ is as follows (for two points with the same location around the structure $\tilde{x}_2$, but different axial locations $x_3, x_3{}^1$):

$$L_3{}^{(P)}(\tilde{x}_2)/a = \int_0^\infty \overline{p(\tilde{x}_2,\tilde{x}_3)p(\tilde{x}_2,\tilde{x}_3{}^1)}/\overline{p^2}(\tilde{x}_2)\,dx_3{}^1 \qquad (2.35)$$

For small scales of turbulence, it is shown in Durbin (1979) that

$$\overline{p^2}(L_3{}^{(P)}/a) = (0.935/(a/L_x)^2)\left[\left(\frac{\partial \tilde{U}_2}{\partial \tilde{x}_2}\right)^2 + \left(\frac{\partial \tilde{U}_2}{\partial \tilde{x}_1}\right)^2\right] \qquad (2.36)$$

Thus around a circular cylinder where $(\partial \tilde{U}_2/\partial \tilde{x}_2) = 2\cos\theta$ and

$$\partial \tilde{U}_2/\partial \tilde{x}_1 = 2\sin\theta, \qquad (2.37)$$

when $a/L_x \gg 1$

$$\overline{p^2}\, L_3{}^{(P)}/a = 3.74/(a/L_x)^2:$$

in other words (for the particular case of a circular cylinder, but not for other shapes) $p^2\, L_3{}^{(P)}$ is independent of distance around the surface.

With large scales of turbulence, around a circular cylinder, as $(a/L_x) \to 0$

$$\overline{p^2}(L_3{}^{(P)}/a) = (5-4\cos 2\theta)/\,2(a/L_x) \qquad (2.38)$$

which implies that when $a/L_x \to 0$, and $\theta = 0$

$$L_3^{(P)} = \frac{1}{2(a/L_x)} \qquad (2.39)$$

From (2.37) and (2.31) the axial pressure integral scale, when $a/L_x \gg 1$, is

$$L_3^{(P)}/L_3 = 7.48/\cos^2\theta \qquad (2.40)$$

which is plotted in fig. 6. (The singularity at $\theta = \Pi/2$ is not real; it means that, since $\overline{p^2}$ is $O((a/L_x)^{-2})$ at $\theta = \Pi/2$, $L^{(P)}$ is $O(a)$).

The results in (2.40) and (2.34) have a straightforward physical explanation. (see fig. 6b). Let the scales of eddies of the upstream turbulence that contribute most to the fluctuating pressure, $\overline{p^2}$, be l_x, l_y, l_z in the x y z direction respectively. We have seen in fig. 2b that when $a/L_x \ll 1$ the only eddies that contributed to $\overline{p^2}$ are those for which $l_x > 2a$ (or $K_1 > \Pi$). What are the scales l_y, l_z?

In the following figure we illustrate 2 eddies (B), (C) where $l_x \sim 2a$:

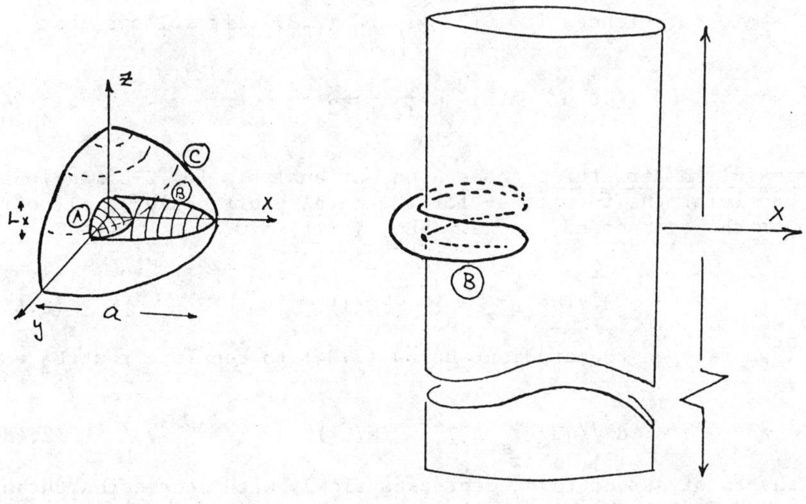

Quadrants of some eddies impinging on a cylinder.

for the eddy (B) $l_y \sim l_z \sim L_x$; whereas, for (C) $l_y \sim l_z \sim l_x \sim 2a$. The equivalent wave number range for (B) is $(1/L_x)$ and for (C) is $(\frac{1}{2}a)$. Since the energy spectrum is a maximum when $k \sim 1/L_x$; and since $(1/L_x) \gg (\frac{1}{2}a)$, the energy of eddy (B) is much greater than that of (C). Consequently eddies such as (B) contribute most to $\overline{p^2}$. This is the explanation why the axial length scale of pressure $L_3^{(P)}$ is of the order of L_x or L_3 and not of order a. (i.e. the result (2.40)). Eddy (A) has also been depicted, which has slightly greater energy than (B), but does not contribute much to $\overline{p^2}$ since for (A) $l_x \sim L_x \ll a$; hence, its contribution is suppressed by 'piling-up'.

Fluctuating drag From the results for the variance of the fluctuating pressure, p^2, the cross correlation $R^{(P)}(\theta,\theta^1)$ and the axial length scale $L_3^{(P)}$, it is possible to see how the variance of the drag $\overline{d^2}$ of a cylinder varies with the integral scale of the oncoming turbulence. By hypothesis the upwind turbulence only affects the fluctuating pressure outside the wake i.e. $(2\Pi-\theta_w) < \theta < \theta_w$, and also by hypothesis the length of the cylinder H is very much greater than L_x, so

$$\overline{d^2} = (\rho\overline{u}_\infty u'_\infty)^2\, a^2 \int_0^H\!\!\int_0^\infty \{ \int_{\theta_w}^{2\Pi-\theta_w}\!\!\int_{\theta_w}^{2\Pi-\theta_w} \overline{p(\theta)p(\theta^1)\cos\theta\cos\theta^1}\, d\theta d\theta^1 \}\, dz\, dz^1 \qquad (2.42)$$

Thus for large scale turbulence, from (2.28), (2.35), (2.38), when $a/L_x \ll 1$,

$$\overline{d^2} \propto (\rho\overline{u}_\infty u'_\infty)^2\, a^2\, L_x H \qquad (2.43a)$$

or if $\overline{d^2}$ is normalised on the frontal area,

$$\overline{d^2} / (\rho\overline{u}_\infty u'_\infty\, a\, H)^2 \propto L_x/H \qquad (2.43b)$$

For small scale turbulence, from (2.34) and (2.37), it follows that

$$\overline{d^2}/(\rho\overline{u}_\infty u'_\infty\, a\, H)^2 \propto \frac{1}{(a/L_x)^2}\,\frac{a}{H} \propto \frac{L_x^2}{aH} \qquad (2.44)$$

Cube/hemispherical-like structures Based on our analysis for 2-dimensional obstacles, we can infer the form of d^2 for a general bluff obstacle all of whose dimensions are of the same order, say a. When $(a/L_x) \to 0$

$$\overline{d^2}/(\rho\overline{u}_\infty u'_\infty\, a^2)^2 = O(1) \qquad (2.45a)$$

When $(a/L_x) \gg 1$ we rely on the argument below (2.34) to conclude that $L_3 = O(a)$. Thus,

$$\overline{d^2}/(\rho\overline{u}_\infty u'_\infty\, a^2)^2 \propto (a/L_x)^{-1} \qquad (2.45b)$$

For most shapes of structure $\overline{d^2}$ decreases slowly with decreasing turbulence scale or increasing (a/L_x); but the rate of decrease is slower on 3-dimensional bodies than on 2-dimensional ones.

Comparison with porous structures It is instructive to compare these variations of d^2 with (a/L_x) with those for porous or permeable structures, because calculations of d^2 for such structures have been used to estimate d^2 for non-porous structures (Vickery (1965), Graham (1976). For porous structures $p^2 \propto (\rho\overline{u}_\infty u'_\infty)^2$ for all values of a/L_x, and the cross correlation of p at two points on the surface is the same as that of the upwind turbulent velocity component u_1 at the same points. Then for a tall porous structure if $a/L_x \ll 1$ and $H \gg L_x$

$$\overline{d^2}/(\rho\overline{u}_\infty u'_\infty\, a\, H)^2 \propto L_x/H \qquad (2.46a)$$

if $a/L_x \gg 1$,

$$\overline{d^2}/(\rho \overline{u}_\infty u'_\infty \ a H)^2 \ \alpha \ L_x^{\ 2}/(a/H) \qquad (2.46b)$$

On the other hand for a cubical or spherical type of porous structure, if

$$a/L_x << 1 \quad \overline{d^2}/(\rho \overline{u}_\infty u'_\infty \ a^2)^2 = O(1) \qquad (2.47a)$$

and if

$$a/L_x >> 1 \quad \overline{d^2}/(\rho \overline{u}_\infty u'_\infty \ a^2)^2 \ \alpha \ \frac{1}{(a/L_x)^2} \qquad (2.47b)$$

Comparing (2.46a) and 2.47a) with (2.43b) and (2.45a) shows that when $L_x >> a$ the variation of d^2 with a/L_x is, for obvious physical reasons, similar for porous and solid structures. Where $L_x << a$ any similarity of the results between porous and non-porous structures is fortuitous because p^2 is greater for the former than the latter, but the correlation is greater for the latter than the former. For cylindrical structures these two effects balance each other (c.f. (2.46b) and (2.44), whereas for cubical or spherical structures the greater correlation on solid structures outweighs the lower value of p^2. (c.f. (2.45)b and (2.47b)).

It is convenient for later discussion to define a ratio of the spectrum of the drag of solid to porous structures as Λ = Ⓗ$^{(d)}$ ($\hat{K}_1$) (solid)/ Ⓗ$^{(1)}$($\hat{K}_1$) porous).

3. Comparison with experiment

Pressure spectra The most detailed fluctuating pressure measurements when a/L_x <<1 are those of tunstall (1974) made on Fawley chimney (diam. 19.7m., height 192 m. and $a/L_x \approx 0.066$). His measurements of the pressure spectrum show that outside the wake region Ⓗ$^{(P)}$ and also p^2 are approximately described (when $K_1 < 1.0$) by the quasi-steady prediction (2.20). In particular the data confirm the prediction that Ⓗ$^{(P)}$($\hat{K}_1 << 1$) increases when $|\theta - \Pi|$ increases, reaching a maximum near $\theta = \Pi/2$. Also at high frequencies (defined by $\hat{K}_1 \gtrsim 1$) the slope of log Ⓗ$^{(P)}$($\hat{K}_1$) against log $\hat{K}_1$ increases steadily from -1.7 to -3 at $K_1 \approx 4$ or $\hat{K}_1 = 61$. This is broadly consistent with the predicted form of Ⓗ$^{(P)}$($\hat{K}_1$) in fig. 2a when $\hat{K}_1$ >>1 which shows that when $K_1 = 4$, Ⓗ$^{(P)}$($\hat{K}_1$) typically has a slope of -3 to -4 and does not reach its asymptotic, $\hat{K}_1^{-16/3}$, rate of decay until $K_1 = 10$.

Laboratory measurements by Batham (1973), Bruun & Davies (1975) of pressures on circular cylinders in grid turbulence over the range $0.9 < a/L_x < 2.6$ provide a test of the theory when $K_1 >> 1$, and to some extent when $a/L_x >> 1$. In their experiments the Reynolds numbers were about 10^5 and turbulence intensities about 0.1 to 0.12.

Bruun & Davies' measurements when $a/L_x = 2.6$ show that $\hat{Ⓗ}^{(P)}$ ($\hat{K}_1 \to 0$) decreases as $|\theta - \pi|$ increases away from the stagnation point. For example they found the ratio

$$\frac{Ⓗ^{(P)}(K_1 \to 0;\ \theta = \Pi)}{Ⓗ^{(P)}(K_1 \to 0;\ \theta = 5\Pi/6)}$$

=	3.5, 1.7	
=	5.5, 4.0	(corrected)
=	2.67	(Theoretical) $a/L_x >> 1$

for the cases a/L_x = 0.9, 2.5; Re = 4.4 x 10^5 and 2.3 x 10^5. If these measurements have the value of $\textcircled{H}^{(P)}$, when there is no turbulence, subtracted from them to correct for the effects of vortex shedding, then this ratio becomes the 'corrected' value. These values are greater than the theoretical value; but they are greater than 1.0 in agreement with the prediction of the small scale turbulence theory.

There is little data at higher frequencies than $K_1 \simeq 6$, when it is found that

$$\textcircled{H}^{(P)} (K_1) \propto K_1^{-2.5}$$

so the high frequency limit of the theory has not really been tested. Measurements of pressure spectra at the stagnation point in the range $1<K_1<4$ were discussed by Hunt (1976).

R.m.s. pressure fluctuations Tunstall (1964) has shown quite definitively that when a/L_x is large enough the variation of p^2 with θ is given by the quasi-steady prediction (2.27) in which the measured values of $u^2_1(\infty)$, $u_2^2(\infty)$, ∂C_p, $\partial C_p/\partial\theta$ are used. His measured and predicted values of $\sqrt{p^2}$ ($=\sqrt{p^{*2}}/(\rho\overline{u}_\infty u'_\infty)$) are shown in fig. 3 and compared with the values of (2.27) calculated assuming potential flow

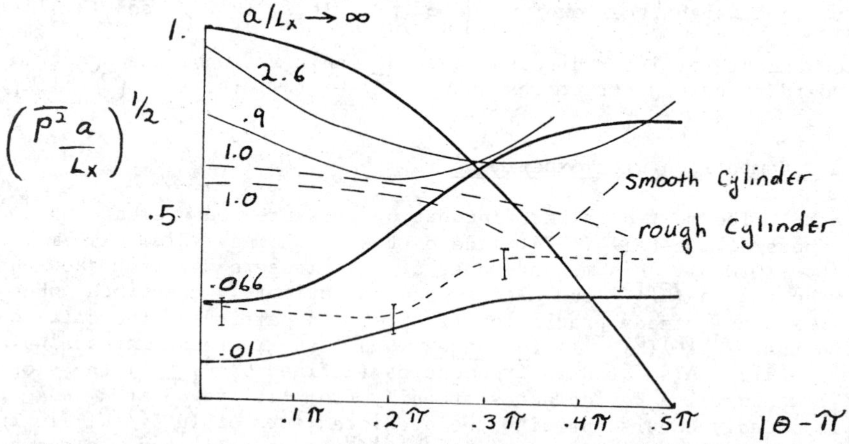

Fig. 3. Normalised r.m.s. pressure fluctuations as a function of angular distance from the stagnation point. Comparison of theory with experiment for various a/L_x. ▬ Theory; — Bruun & Davies; - - Batham; -I- Tunstall.

around the cylinder and isotropic upstream turbulence. The observed C_p only decreased to -1.6 rather than to the potential theory value of -3; which explains why $\sqrt{p^2}$ ($\theta = \Pi/2$)/ $\sqrt{p^2}$ ($\theta = \Pi$) was only 1.6 rather than 3.0. Tunstall found $u_2^2(\infty)$ to be 75% of $u_1^2(\infty)$; and this is why $\sqrt{p^2}$ slightly decreases as $|\theta-\Pi|$ increases, before increasing near $|\theta-\Pi| = \Pi/4$.

The theoretical limit of $\sqrt{p^2}(a/L_x)$ when $a/L_x >>1$ is also plotted on fig. 3 along with the measurements of Bruun & Davies (.975) and Batham (1973). The measured values have been corrected for the pressure fluctuations in the absence of upstream turbulence. Note how as (a/L_x) increases the observed values of $\sqrt{p^2}\,(a/L_x)^{\frac{1}{2}}$ increase and that the decrease of p^2 with θ becomes more marked as a/L_x increases.

In figure 4 we have plotted the variation of the r.m.s. fluctuating stagnation

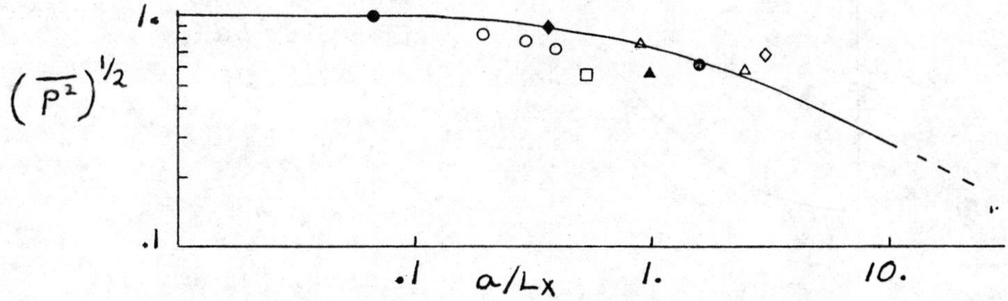

Fig. 4 Normalised r.m.s. fluctuating pressure as a function of relative turbulence scale. ▲ Batham; O Bearman; △Bruun & Davies; □ Lee; ◊ Marshall; ♦ Propper; • Tunstall

pressure $\sqrt{\overline{p^2}}$ with a/L_x not only for turbulent flow around circular cylinders, but also around other bluff bodies, along with the asymptotic estimates for $\sqrt{p^2}$ derived in (2.31b) and (2.33); which broadly follow the observed tendency of $\sqrt{p^2}$ to decrease below 1.0 as a/L_x increases.

Cross correlation of fluctuating pressure In figure 5 the measurements by Batham

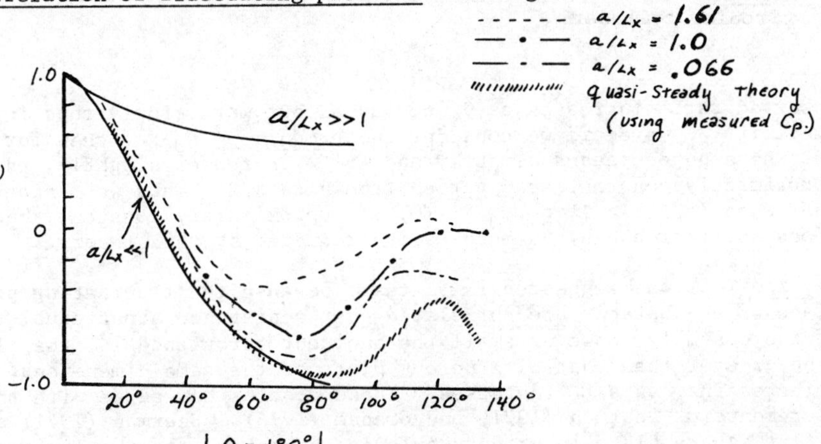

Fig. 5 Measurements of Batham (1973) and Tunstall (1977) of the cross correlation of fluctuating pressure between the stagnation point (θ=180°) and other points around the cylinder. Also shown are theoretical limits for large and small scale turbulence.

(wind tunnel) and Tunstall (chimney) of $R^{(P)}(\Pi,\theta)$ around circular cylinders are compared with the theoretical limits (2.30) and (2.34b). We observe that as a/L_x increases the measured curves of $R^{(P)}$ tend to move from one theoretical limit to the other, much as we found with the variation of p^2 with a/L_x.

Axial length scales of pressure In figure 6 the measurements of the ratio of the axial length scale of the pressure to that of the turbulence, $L_3^{(P)}/L_3$ as a function of for various values of (a/L_x) are compared with the theoretical limits of $a/L_x \to 0$ and $a/L_x \gg 1$. We note that when $\theta < \Pi/4$, $L_3^{(P)}/L_3$ lies between 1.5 and about 5.0 whereas the theoretical asymptotic values are between 1.0 and 15.

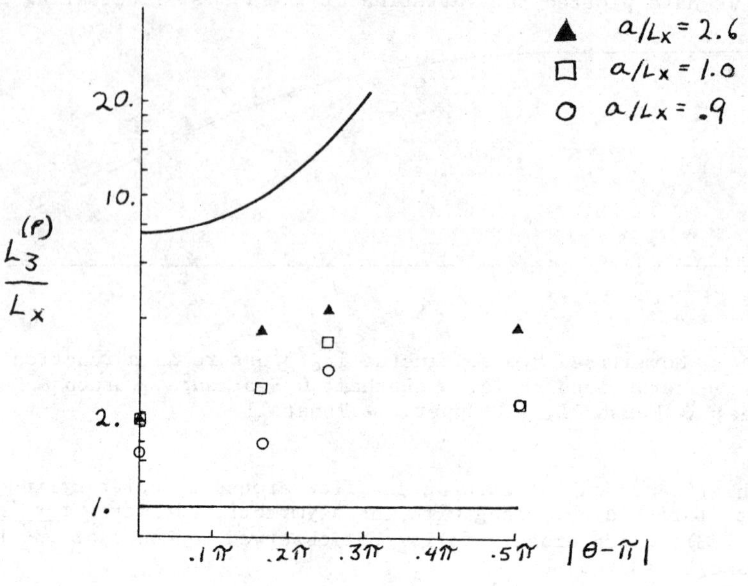

Fig. 6 Integral scales, measured and predicted, around a circular cylinder.

The observed ratio increases as θ increases, but more slowly than in the theoretical limit. However if we consider the product $p^2\ L_3^{(P)}$, then for all these observations where p^2 decreases slightly and $L_3^{(P)}$ increases slightly, the product remains approximately constant as θ varies from Π to Π/2. Thus we tentatively conclude that when $(a/L_x) \gtrsim 1$, $p^2\ L_3^{(P)}\ (\theta)$ is approximately constant, but that $L_3^{(P)}\ (\theta)$ does not increase by as much as is predicted by this theory.

Fluctuating drag It was suggested in §2 that: because the fluctuating pressures on the surface of an obstacle are correlated over a distance about equal to the diameter of the obstacle, however small the incident turbulence is, the fluctuating drag would be greater than that of a porous plate of the same dimensions; the ratio of the two increasing, as a/L_x increases. These conclusions agree with the wind tunnel measurements of Bearman (1971) and Graham (1976). Bearman (1971) showed that for a flat square plate in grid turbulence the ratio Λ of the low frequency spectrum of the drag Ⓗ$^{(d)}(K_1)$ to that calculated by integrating the quasi-steady (or porous plate) pressure across the plate was about 1.2 when $a/L_x = 0.33$, and was about 3.0 when $a/L_x = 1.33$. At higher frequencies the ratio became even greater. Graham's (1976) measurements of the fluctuating drag on a porous plate also confirmed this trend, by showing that as the turbulence scale decreased and the solidity increased, the ratio Λ increased. Graham also applied the rapid distortion theory of Hunt (1973) to the calculation of the fluctuating drag and showed how the distortion of the turbulence by the flow around and through the plate tended to increase d^2 and Ⓗ$^{(d)}(K_1)$ as the solidity increased.

An important corollary of our formula (2.34b) is that the fluctuating pressure in this case is produced only by the turbulent eddies on the stagnation line. Consequently it should be no surprise that Bearman (1971) found the very high correlation of 0.8 between the upwind low frequency velocity fluctuation on the stagnation line and the fluctuating drag when $a/L_x = 0.67$. Nor therefore should it be a surprise, though it certainly was when first presented by Gartshore et al (1976), that if the only upwind turbulence is restricted to a narrow region of about $2L_x$ either side of the stagnation line, then the effect on the surface pressure fluctuations and probably on the wake is the same as if the upwind turbulence was homogeneous with the same intensity and scale. (The effect on surface pressure fluctuation can be proved formally, using the techniques of Durbin (1979) for analysing turbulence in a restricted region of space).

REFERENCES

1. Batham, J.P. (1973). Pressure distributions on circular cylinders at critical Reynolds numbers. J. Fluid Mech. 57, 209.
2. Bearman, P.W. (1971). An investigation of the forces on flat plates normal to a turbulent flow. J. Fluid Mech. 46, 177.
3. Bearman, P.W. (1972). Some measurements of the distortion of turbulence approaching a two-dimensional bluff body. J. Fluid Mech. 53, 451.
4. Bruun, H.H., & Davies, P.O.A.L. (1975). An experimental investigation of the unsteady pressure forces on a circular cylinder in a turbulent cross flow. J. Sound Vibr. 40, 535.
5. Davenport, A.G. (1961). The application of statistical concepts to the wind loading of structures. Proc. Inst'n Cit. Engrs. 19, 449.
6. Durbin, P. (1979). Rapid distortion theory of turbulent flows, Ph.D. Dissertation, University of Cambridge. -------(1979b). Distortion of turbulence by a constant shear layer adjacent to a wall. (To be published in J. Fluid Mech.)
7. Graham, J.M.R. (1976). Turbulent flow past a porous flat plate. J. Fluid Mech. 73, 565.
8. Hunt, J.C.R. (1973). A theory of turbulent flow around two-dimensional bluff bodies. J. Fluid Mech. 65, 625.
   ----------- (1974). A theory for fluctuating pressures on bluff bodies in turbulent flows. IVTAM-AHR Symp. on Flow Induced Vib.; Karlsruhe Aug. 1971.
   ----------- (1976). Turbulent velocities near and fluctuating surface pressures on structures in turbulent winds. Proc. 4th Int. Conf. on Wind Effects on Buildings and Structures. Camb. Univ. Press.
9. Lee, B.E. (1975). The effect of turbulence on the surface pressure field of a square prism. J. Fluid Mech. 69, 263.
10. Marshall, (1968). Pressure fluctuation correlations near an axisymmetric stagnation point. Colo. State Ph.D. diss.
11. Propper, H. (1977). Zur aerodynamischen Belastung gro Ber Kuhlturme Tech. Rep. 77-3. Inst. Komst. Ing. Rhur-Univ. Bochum, W. Germany.
12. Robertson, J.M., Wedding, J.B., Peterka, J.A., Cermak, J.E., (1977). Wall pressures of separation - reattachment flow on a square prism in uniform flow. J. Ind. Aero. 2, 345.
13. Surry, D. (1972). Some effects of intense turbulence on the aerodynamics of a circular cylinder at subcritical Reynolds number. J. Fluid Mech. 52, 543.
14. Tunstall, M.J. (1974). Some measurements of the wind loading on Fawley Generating Station Chimney. Proc. Sympt. on Full Scale Fluid Dynamic Measurements p. 26. Leicester Univ. Eng. Dept., England.
15. Vickery, B.J. (1965). On the flow behind a coarse grid and its use as a model of a transpheric turbulence in studies related to wind loads on buildings. NPL Aero. Rep. 1143.

LOCAL PRESSURE FLUCTUATIONS

H. SOCKEL

Institute for Fluid-Dynamics, Technical University Vienna, Vienna, Austria

ABSTRACT

The probability density distributions of the pressure fluctuations are not symmetrical, especially for negative mean values of the pressure coefficient. A kth order Weibull distribution function gives a better agreement with the experimental data than a Weibull distribution of order one. The extreme value distribution of the pressure coefficient calculated with the aid of the initial distribution fits the extreme value data well.

1. INTRODUCTION

In building aerodynamics we are interested in the local peaks of pressures, because they are essential for the design of cladding and of glass panels. High peaks are caused mainly by the turbulence generated by the structure but depend on the turbulence of the approaching flow too. The pressure fluctuations are in a random manner and therefore we have to apply statistical methods.

To predict peak wind loading on a structure Davenport (6) and others provided peak value theories for wind loads based on the assumption of a Gaussian pressure distribution. But recent work on the nature of pressure fluctuations has shown, that the concept of Gaussian pressure fluctuations is at least valid on the windward face. Very extensive model experiments in a wind tunnel by Peterka and Cermak (3) have shown, that the probability densities of pressure fluctuations fall into two basic classis, one for $c_{p\ mean} > -0,1$ and another for $c_{p\ mean} < -0,25$. The densities in the class for $c_{p\ mean} > -0,1$ are nearly Gaussian, but the densites in the class of $c_{p\ mean} < -0,25$ are skewed such, that the probability for large negative pressure differences is much greater than for a Gaussian distribution. Full scale experiments in Aylesbury (4) have shown that there is an enormous deviation from the Gaussian distribution in the tail regions of separated flows. But it is very important to have a good fit of the data to the distribution curve in the region of the tail, because this part has an important influence on the extreme values. In all cases of the results of Tieleman and Reinhold (5) the negative pressure coefficents follow a Weibull probability distribution. But the fit of the data to this distribution is only good on one side of the maximum of the curve, as will be shown later.

A reason for the selection of a Weibull distribution is, that its extreme value distribution is of type I, which gives a very good agreement with experimental data (3, 4). Mayne and Cook (4) mentioned: "...whichever distribution is a correct description of the parent population, the extreme value distribution will be type I," which is verified by our own experimental results.

In this paper it is shown, that the kth order Weibull distribution gives a better fit to the experimental data than a Weibull distribution. The extreme value distribution of this new parent distribution is again of type I and so the above demand by Mayne and Cook is satisfied. As the proposed distribution is of kth order, I will talk about order statistics first.

2. ORDER STATISTICS

In building aerodynamics we are interested in the minimal and maximal pressure coefficients and their statistics. This is similar to other engineering problems. For example, the largest value of loads determines the specification of a structure and the highest stress on a surface may initiate a failure. If we make n measurements and get the results $x_{(i)}$ we can rearrange this given set of measurements into an ordered set

$$x_{(1)} \le x_{(2)} \le x_{(3)} \le \ldots\ldots \le x_{(r)} \le \ldots\ldots x_{(n)}$$

Thus the given results are relabeled by a subscript to indicate their order in magnitude. $x_{(1)}$ is the smallest measurement and $x_{(n)}$ the largest value of the sample. The order statistic $x_{(r)}$ is again a random variable with a probability density function f_r and a cumulative probability F_r. For the parent distribution the probability density $f(x)$ and the cumulative probability $F(x)$ are connected by

$$F(x) = \int_{-\infty}^{x} f(x)dx \tag{1}$$

If we have n independent samples, the cumulative probability F_n is given by (1)

$$F_n(x) = [F(x)]^n \tag{2}$$

and its density function is

$$f_n(x) = \frac{d}{dx}F_n(x) = n[F(x)]^{n-1} f(x) \tag{3}$$

The probability to get in one trial the result $X > x$ is

$$P\{X > x\} = \int_{x}^{\infty} f(x)dx = 1 - F(x) \tag{4}$$

For the cumulative probability for the smallest value of n independent trials we get (1)

$$1 - F_1(x) = [1 - F(x)]^n \tag{5}$$

and the probability density function

$$f_1(x) = n \cdot [1 - F(x)]^{n-1} f(x) \tag{6}$$

Equations 2, 3 and 5, 6, respectively, give the exact order statistics of n independent observations of x. When the sample size n is large enough, asymptotic distributions can be used instead of these equations to model the extreme value of interest. Since an extreme value occurs less frequently in repeated samples than the order values of the underlying variable x, the nature of the tail of f(x) of the parent distribution influences the distribution of the extreme value more stronlgy than the bulk of f(x). The so-called type I extreme value model arises from initial distributions that are unbounded in the direction of the extreme value, provided the relevant tail of the parent probability density function decreases at least as rapidly as an exponential function (1, 2).

$$F_{I\,max} = \lim_{n\to\infty} F_n(x) = \lim_{n\to\infty} [F(x)]^n = \exp\{-\exp[-\alpha_n(x-u_n)]\} \tag{7}$$

R. v. Mises has derived this asymptote under the condition that

$$\lim_{x\to\infty}\left\{\frac{d}{dx}\left[\frac{1-F(x)}{f(x)}\right]\right\} = 0 \tag{8}$$

u_n is the so-called characteristic largest value that has on the average only one exceedance in the sample of size n. It is defined by

$$F(u_n) = 1 - \frac{1}{n} \tag{9}$$

u_n is a quantile of the order $q = 1-\frac{1}{n}$. The quantity α_n is defined by (2)

$$\alpha_n = n \cdot f(u_n) \tag{10}$$

According to the above equations u_n and α_n are functions of the sample size n. The quantity u_n is a location factor and α_n is a scale factor. Expressing these two quantities in the normal notation we get

$$F_{I\,max} = \exp\{[-\exp(-\frac{x-\mu_e}{\sigma_e})]\}\,; \quad \frac{1}{\sigma_e} = \alpha_n\,; \quad u_n = \mu_e \tag{11}$$

$$f_{I\,max} = \frac{dF_{I\,max}}{dx} = \frac{1}{\sigma_e}\exp\left\{\frac{x-\mu_e}{\sigma_e} - \exp\left[-\frac{x-\mu_e}{\sigma_e}\right]\right\}$$

The index e denotes that we have values of an extreme distribution. This is the asymptote for the largest extremes. For the smallest value we have according to equation 5 the exact cumulative probability. For the extreme value distribution we get

$$1-F_{I\,min}(x) = \lim_{n\to\infty}[1-F(x)]^n = \exp\{-\exp[\alpha_1(x-u_1)]\}$$

$$F_{I\,min}(x) = 1-\exp\{-\exp[\alpha_1(x-u_1)]\} \tag{12}$$

Herein u_1 is the expected smallest value, the value that has on the average only one smaller observation in a sample of size n.

$$F(u_1) = \frac{1}{n} \tag{13}$$

α_1 is analogous to equation 10 defined by

$$\alpha_1 = n f(u_1) \qquad (14)$$

According to the equations 13 and 14 x_1 and u_1 depend on the sample size n. If we write the location factor u_1 and the scale factor α_1 in the normal notation we get for the distribution of the smallest value (2)

$$F_{I\,min} = 1 - \exp\left\{-\exp\left[\frac{x-\mu_e}{\sigma_e}\right]\right\} ; \quad \frac{1}{\sigma_e} = \alpha_1 ; \quad u_1 = \mu_e$$

$$f_{I\,min} = \frac{dF_{I\,min}}{dx} = \frac{1}{\sigma_e} \exp\left\{ \frac{x-\mu_e}{\sigma_e} - \exp\left[\frac{x-\mu_e}{\sigma_e}\right]\right\} \qquad (15)$$

In the next paragraph I will show that the kth order distribution of a Weibull distribution gives a good fit to the experimental values of a pressure coefficient distribution.

$$F(x) = \left\{1 - \exp\left[-\left(\frac{x}{\sigma}\right)^{\lambda}\right]\right\}^{K} \qquad x \geq 0$$

$$f(x) = \frac{dF(x)}{dx} = K \frac{\lambda}{\sigma}\left(\frac{x}{\sigma}\right)^{\lambda-1} \left\{1 - \exp\left[-\left(\frac{x}{\sigma}\right)^{\lambda}\right]\right\}^{K-1} \cdot \exp\left[-\left(\frac{x}{\sigma}\right)^{\lambda}\right] \qquad (16)$$

The extreme value distribution of equation 16 is again a type I distribution. The exact nth order distribution for n independent samples is

$$F_n(x) = \left\{1 - \exp\left[-\left(\frac{x}{\sigma}\right)^{\lambda}\right]\right\}^{Kn} \qquad (17)$$

It can be shown, that condition 8 by Mises is valid for the distribution 17 if $\lambda > 0$ and so the extreme value distribution is given by equation 11. For the location factor μ_e and the scale factor α_e we get according to equations 9 and 10.

$$F(u_n) = \left\{1 - \exp\left[-\left(\frac{u_n}{\sigma}\right)^{\lambda}\right]\right\}^{K} = 1 - \frac{1}{n}$$

$$\mu_e = u_n = \sigma\left\{-\ln\left[1 - \left(1 - \frac{1}{n}\right)^{\frac{1}{K}}\right]\right\}^{\frac{1}{\lambda}}$$

$$\frac{1}{\sigma_e} = \alpha_n = n f(u_n) = \frac{nK\lambda}{\sigma}\left[1 - \left(1-\frac{1}{n}\right)^{\frac{1}{K}}\right]\left\{-\ln\left[1 - \left(1-\frac{1}{n}\right)^{\frac{1}{K}}\right]\right\}^{\frac{\lambda-1}{\lambda}} \left[\left(1-\frac{1}{n}\right)^{\frac{K-1}{K}}\right] \qquad (18)$$

For very large n we can simplify these expressions

$$n \gg 1 ; \ \lambda \gg 1 : \quad \mu_e = u_n = \sigma\left[\ln(Kn)\right]^{\frac{1}{\lambda}}, \quad \frac{1}{\sigma_e} = \alpha_n = \frac{\lambda}{\sigma}\left[\ln(Kn)\right]^{\frac{\lambda-1}{\lambda}} \qquad (19)$$

This means, that the extreme value distribution of order n of the initial distribution 16 is the same as the extreme value distribution of a Weibull distribution of the order nk. This approximation is only reasonable for $\lambda \gg 1$ (1). For $x \ll 0$ we get similar results

$$F(x) = 1 - \left\{1 - \exp\left[-\left(-\frac{x}{\sigma}\right)^{\lambda}\right]\right\}^{K} \qquad x \leq 0$$

$$f(x) = \frac{K\lambda}{\sigma}\left(-\frac{x}{\sigma}\right)^{\lambda-1}\left\{1 - \exp\left[-\left(-\frac{x}{\sigma}\right)^{\lambda}\right]\right\}^{K-1} \exp\left[-\left(-\frac{x}{\sigma}\right)^{\lambda}\right] \qquad (20)$$

For the exact nth order distribution we get according to equation 5

$$F_1(x) = 1 - \left\{1 - \exp\left[-\left(-\frac{x}{\sigma}\right)^{\lambda}\right]\right\}^{Kn} \qquad (21)$$

For the location factor μ_1 and the scale factor α_1 we get with the equations 13, 14, 15

$$F(u_1) = 1-\left\{1-\exp\left[-\left(-\frac{u_1}{\sigma}\right)^{\lambda}\right]\right\}^{k} = \frac{1}{n}$$

$$\mu_e = u_n = -\sigma\left\{-\ln\left[1-\left(1-\frac{1}{n}\right)^{\frac{1}{k}}\right]\right\}^{\frac{1}{\lambda}} \tag{22}$$

$$\frac{1}{\sigma_e} = \alpha_1 = n\cdot f(u_1) = \frac{nk\lambda}{\sigma}\left[1-\left(1-\frac{1}{n}\right)^{\frac{1}{k}}\right]\left\{-\ln\left[1-\left(1-\frac{1}{n}\right)^{\frac{1}{k}}\right]\right\}^{\frac{\lambda-1}{\lambda}}\left[\left(1-\frac{1}{n}\right)^{\frac{k-1}{k}}\right]$$

If we compare the equations 18 and 22 we see that $\alpha_n = \alpha_1$ and $u_1 = -u_n$ (2). For large values of n we get again simplified expressions

$$n \gg 1;\ \lambda \geq 1: \quad \mu_e = u_1 = -\sigma\left[\ln(kn)\right]^{\frac{1}{\lambda}}, \quad \frac{1}{\sigma_e} = \alpha_1 = \frac{\lambda}{\sigma}\left[\ln(kn)\right]^{\frac{\lambda-1}{\lambda}} \tag{23}$$

The distributions of the extreme values (smallest and largest) of a kth order Weibull distribution are therefore type I distributiones (equ. 11 and 15) and the parameters u_n, α_n and u_1 and α_1 are given by the equations 19 and 23, respectively.

3. EXPERIMENTAL RESULTS

In the boundary layer wind tunnel of the Bundesversuchsanstalt Arsenal in Vienna we measured the pressure fluctuations in several points on the top of a cube (length 12 cm) in the separation region. In a quarter of the roof plain there were 89 pressure holes connected by short length of plastic tubes to a scanivalve fitted with a pressure transducer. The measurement was accomplished by processing the fluctuation signal from the transducer to a computer. Each pressure port was sampled at 1000 samples/s for a total of 13920 samples (about 14 s). All data were converted to a nondimensional pressure coefficient

$$c_p = \frac{p-p_A}{\frac{1}{2}\rho U^2}$$

in which p_A is the static pressure of the undisturbed flow and $\frac{1}{2}\rho U^2$ a reference dynamic pressure (height 60 cm). We computed further the average c_{pm} of the 13920 data, the largest value $c_{p\,max}$ and the smallest value $c_{p\,min}$ of each sample. Now I want to show two examples of initial distributions. In fig. 1 there is plotted the density multiplied with a factor over $c_p - c_{pm}$. The measurements were taken in point 5 near the edge of the flat roof of the cube, the flow direction was 45° to the front face. The full line is a curve due to equation 20 with the special values $x = c_p - c_{pm} - 4$; $\sigma = 2{,}9$; $\lambda = 3{,}5$; $k = 15$. For the calculation of the extreme values from the initial distribution a good fit in the tail is very important, as already mentioned. Therefore the same distribution is plotted in a logarithmic scale in fig. 2. Here we can see a rather good fit of the curve to the data. In fig. 3 we see the pressure distribution of point 7, again a point near the edge, a narrower distribution than we have in point 5. In fig. 4 the same distribution is plotted in a logarithmic scale. The agreement between the full line (equation 20) and the data is quite good. If one tries to fit a Weibull distribution of first order (k =1) to the data, one can get a good agreement only for one side.

We tried to do this with the results given by Peterka and Cermak (3) too. The measured points (fig. 6 of their paper) are marked in fig. 5 by a broad band. If we take $-\frac{x}{\sigma} = \frac{c_p - c_{pm}}{c_{prms}} - \frac{x_0}{\sigma}$, where x_o is a free parameter, we get a good approximation with a Weibull distribution (k = 1), but a very steep gradient on the right side of the maximum. Well, there are no experimental data on this side, but due to my experience, the gradient is to steep. If we try to fit the data with equation 20 we get a good agreement of the left side too, but a less steep gradient on the right. To have one parameter more in a function is always an advantage in fitting it to data.

From several points we made records of 111 360 data and collected sets of 1000, 2000, 3000, 4000 and 5000 values. Of each set there was taken the smallest value to get distribution of extreme values. In fig. 6 there is plotted the cumulative distribution of extreme values for point 7 for a set of 1000 data. The horizontal coordinate Y is given by

$$Y = -\ln\{-\ln[1-F_1(x)]\}$$

From the data there were computed the mean value μ'_e and the variance σ'^2_e. With these two values one can calculate the parameters μ_e and σ_e to fit the distribution function 15 to the data (1).

$$\mu_e = \mu'_e + 0{,}57722\,\sigma_e\,; \quad \sigma_e^2 = \frac{\sigma_e'^2}{1{,}64493}$$

The result is shown by a full line in fig. 6.

Otherwise we can calculate μ_e and σ_e from the initial distribution curve based on 13920 data (fig. 3, 4) with equation 23 for n = 1000. The dotted line in fig. 6 represents the results. In fig. 7 and 8 we see the extreme value distributions for the same point for a set of 3000 and 5000 values, respectively. There is a very good fit of the curve calculated with the parameters of the initial distribution to the experimental data. But one has to be careful with the parameter λ of initial distribution, because the extreme value distribution is rather sensitive to a change of λ . But if we have a good fit between the initial distribution curve (equ. 20) and the data, expecially in the tail, we can calculate the extreme value distribution, it is not necessary to measure a lot of data.

4. SUMMARY

It is shown, that the kth order Weibull distribution gives a better approximation of the parent distribution of the pressure coefficient c_p than the Weibull distribution. With the parameters of this initial distribution the parameters of the extreme value distribution of type I can be calculated. The curve calculated in this way is in good agreement with experimental data.

5. ACKNOWLEDGEMENTS

I am indepted to Dipl.-Ing. H. Schnattinger, who made the measurements and tried to fit the curves to the data and to Dipl.-Ing. M. Waclawiczek for their help.

6. REFERENCES

1. Bury, K.V., Statistical Models in Applied Science, John Wiley & Sons, New York, London, Sydney, Toronto 1975

2. Gumbel, E.J., Statistical Theory of Extreme Values and Some Practical Applications, A Series of Lectures, Nat. Bureau of Standards; Appl. Math. Series 33 (1954)

3. Peterka, J.A., Cermak, J.E., Wind Pressure on Buildings-Probability Densities, J. of the Struct. Div., Proc. of the ASCE Vol. 101, No.ST 6, p. 1225 - 1267 (1975)

4. Mayne, J.R., Cook, N.J., On Design Procedures for Wind Loading, Building Res. Est. CP 25/78, 29p. (1978)

5. Tielemann, H.W., Reinhold, T.A., Wind Tunnel Investigation for Basic Dwelling Geometries, College of Eng. Virginia Polytechnic Inst. & State Univ., Blacksburg, Dep. of Eng. Science & Mech., VPI -E-76-8, 157 p. (1976)

6. Davenport, A.G., Note on the Distribution of the Largest Value of a Random Function with Application to Gust Loading, Proc. of the Inst. of Civ. Engrs., Vol. 28, pp. 187 - 196 (1964)

7. FIGURES

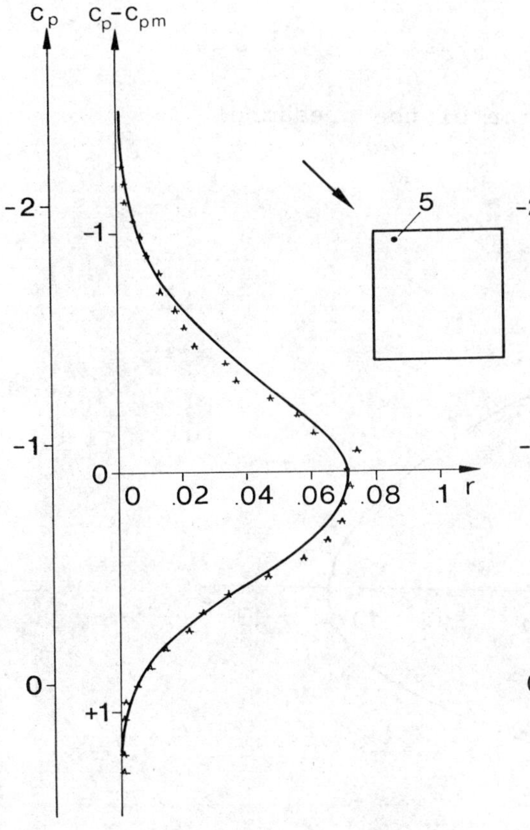

Fig.1: Initial distribution of the pressure coefficient in point 5

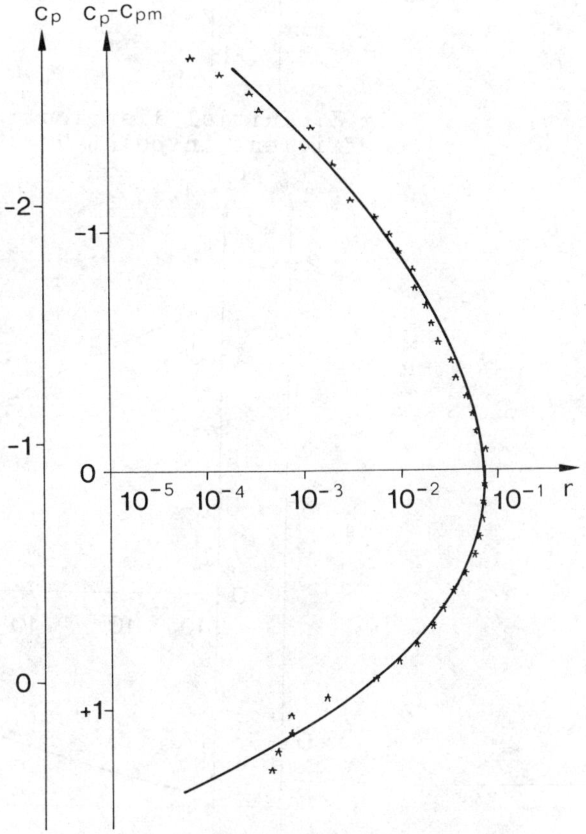

Fig.2: Initial distribution of the pressure coefficient in point 5 in a logarithmic scale

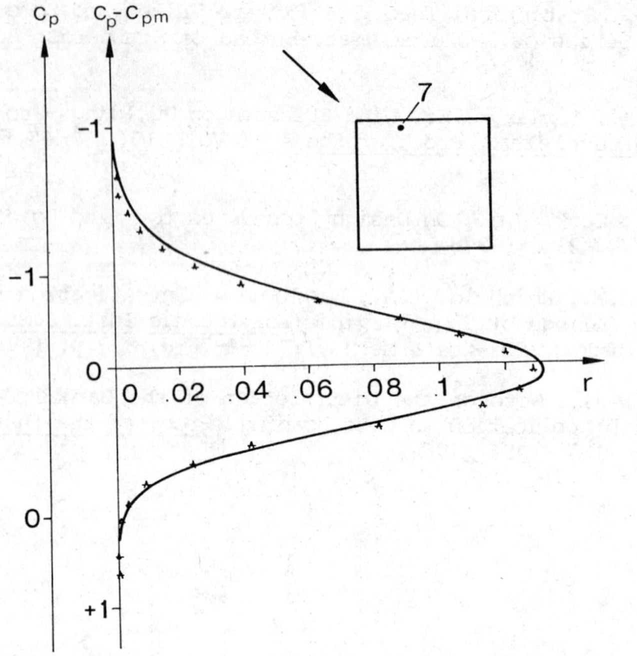

Fig.3: Initial distribution of the pressure coefficient in point 7

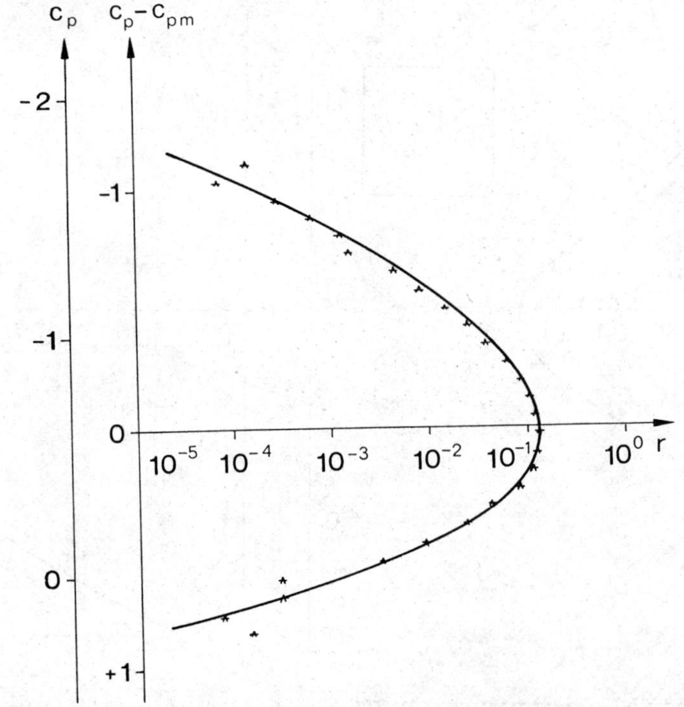

Fig.4: Initial distribution of the pressure coefficient in point 7 in a logarithmic scale

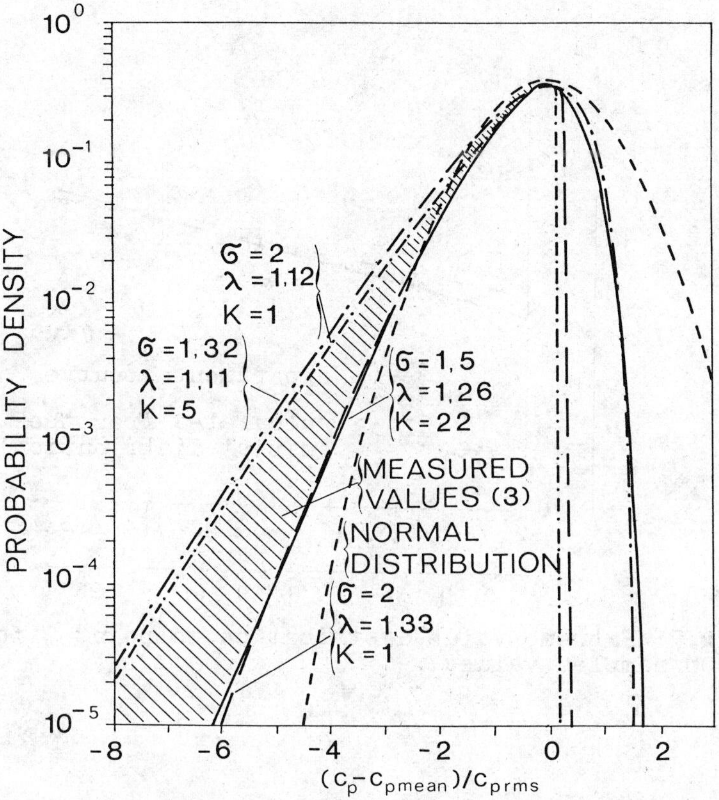

Fig.5: Initial distribution functions for the pressure coefficient and measured values(3)

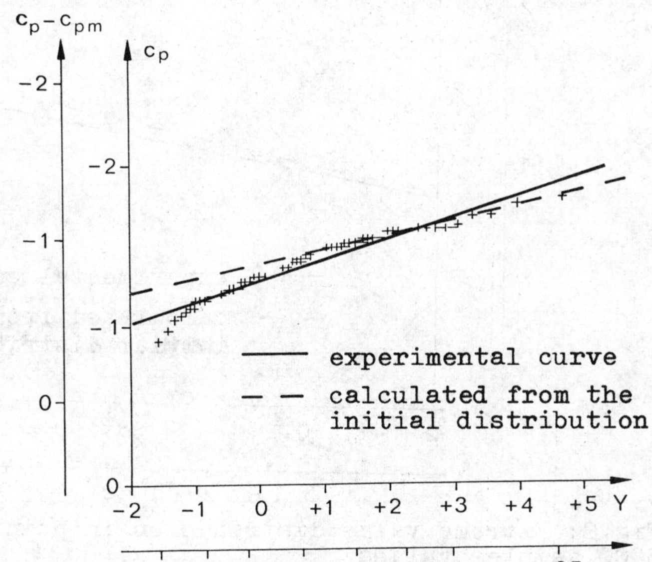

Fig.6: Extreme Value distribution in point 7 for 1000 sample values

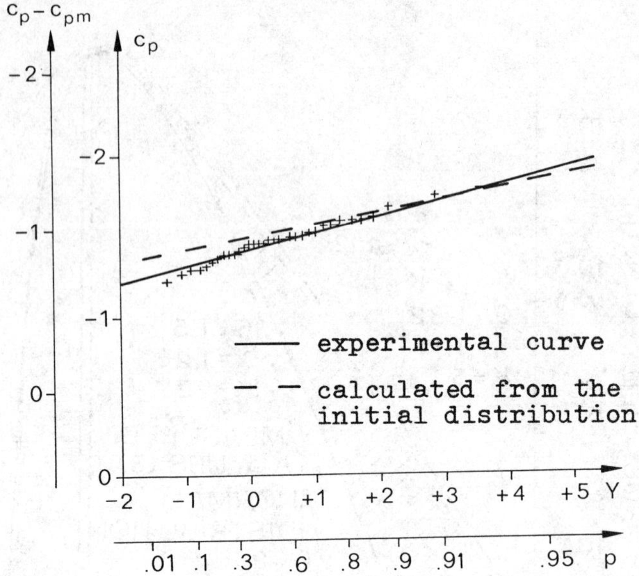

Fig.7: Extreme value distribution in point 7 for 3000 samples values

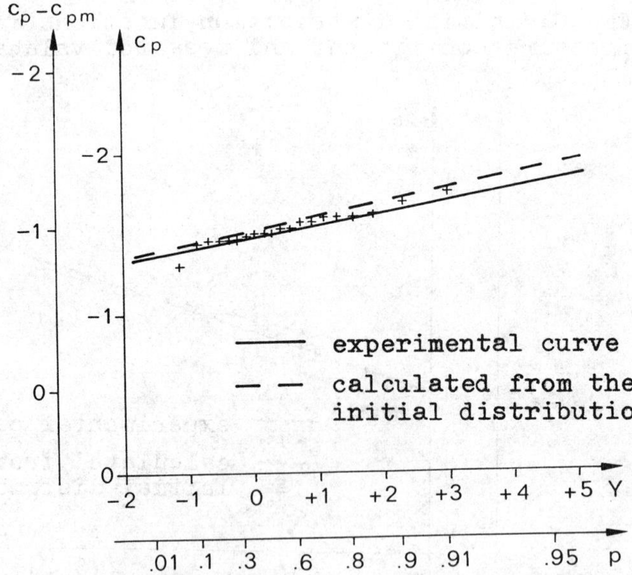

Fig.8: Extreme value distribution in point 7 for 5000 samples values

CHARACTERISTICS OF PRESSURE FLUCTUATIONS ON THE WINDWARD WALL OF A TALL BUILDING

H. KAWAI*, J. KATSURA** and H. ISHIZAKI**

*Tokyo Denki University, Hatoyama, Japan

**Kyoto University, Gokasho, Japan

ABSTRACT

Characteristics of pressure fluctuations on the windward wall of a tall building were investigated by wind tunnel tests and compared with the results of full scale measurements. From cross-correlation and co-spectra measurements between the upstream velocity fluctuation and the pressure fluctuation, it appears that the pressure fluctuation on the windward face can be expressed by a linear transformation of the velocity fluctuation. This linear transformer of the velocity to pressure, which is called "pressure/velocity admittance" by the authors, can be expressed as a function of only the reduced frequency nB/U for any B/Lx; where n is the frequency, B the breadth of the buildings, U the mean velocity and Lx the longitudinal scale of turbulence. On the other hand, it was also found that the scales of the pressure fluctuations were fairly large compared to those of the velocity fluctuations.

Using these findings, the admittance of a flat plate was computed and compared to experimental results and to values computed by the lattice theory. A comparison of gust factors for a tall building computed by the method of A.G.Davenport with those computed using these findings is also presented.

INTRODUCTION

The alongwind vibration of structures may be induced by the turbulence in the approaching flow. A.G.Davenport estimated the along wind response of structures by a statistical approach assuming that the fluctuating drag could be obtained through a linear transformation of the upstream velocity fluctuations. He called this linear transformation, which was applied in the frequency domain, the "aerodynamic admittance" and he measured the admittance of flat plates (1). As the admittance was the key parameter required to estimate the response, many scientists investigated the admittance by experimental and theoretical approaches (2),(3),(4). The theoretical approaches are based on the assumption that the pressure fluctuations on the windward face are directly related to the upstream velocity fluctuations, so that the spectrum and the co-spectrum of the pressure fluctuations are of the same form as those of the velocity fluctuations. However, many experimental results on full scale structures and in wind tunnel tests have shown that the spectra and the co-spectra are slightly different from those of the velocity fluctuations (5),(6). This fact was also found by J.C.R.Hunt using a new theoretical approach (7).

The purpose of the present study is to examine the assumption for the pressure fluctuations in light of the experimental data. From this study, we have also found that the pressure fluctuations on the windward face can be expressed by a linear transformation of the upstream velocity fluctuations. However, this linear transformer which is called the "pressure/velocity admittance" by the authors, is not constant as assumed by Davenport, B.J.Vickery and others, but decreased in high frequency range, and it can be expressed by a function of only nB/U for any sized building, where n is the frequency, B the breadth of a building and U the mean velocity. We have also found that the scale of the pressure fluctuations is not equal to that of the velocity fluctuations in either vertical or horizontal direction, but is somewhat larger than that of the turbulence even in the case of two dimensional flow.

To examin these findings, the aerodynamic admittance for a square flat plate was computed using the pressure/velocity admittance and the co-spectra of the pressure fluctuations, and compared with experimental data and with a value computed by the lattice theory of Vickery (3). The value computed on the basis of the present study was in better agreement with the experimental data than the value based on the lattice theory. A comparison of gust factors computed by the method proposed by Davenport with those computed using the pressure/velocity admittance and the co-spectra of the pressure is presented in the last section.

EXPERIMENTAL PROCEDURE

The experiment was performed in a wind tunnel at the Disaster Prevention Research Institute of Kyoto University. The wind tunnel is of a closed circuit type with an open octagonal test section of 100cm diameter. A highly turbulent flow was generated by the installation of square-mesh grids at the nozzel outlet. Details of the grids, which were constructed of bars of rectangular cross-section, are shown in TABLE 1.

TABLE 1. Details of grids

Grid	Mesh size (cm)	Bar size bxd(cm)	Turbulent intensity $(u^2)^{\frac{1}{2}}/U$	Longitudinal scale Lx(cm)
1	22.5	4.5x4.5	0.145	6.7
2	10.0	2.3x3.5	0.070	5.1

Square prisms were mouted normal to the flow at 115cm downstream from the grids at which point the mean velocity distributions were nearly uniform and the turbulence intensities were 14.5% for Grid 1 and 7% for Grid 2, and the longitudinal scales of turbulence were 6.7cm for Grid 1 and 5.1cm for Grid 2.

In order to measure fluctuating pressure on the windward face of prisms, semiconductor pressure transducers were connected to the 1mm diameter surface hole using 2cm long tubing. The frequency response for the tubing was acceptably flat from D.C. to 300 Hz. Turbulence measurement in front of the prisms were made with constant-temperature linearized hot wire anemometers. The fluctuating pressure and the velocity signals were simultaniously recorded on a 14ch tape recorder for later digitization and analysis by a computer (8).

EXPERIMENTAL RESULTS

Power spectra of pressure fluctuations

One of the most important quantity to evaluate the alongwind response is the power spectrum of the pressure fluctuations on the windward face. Fig.1 shows the power spectra of the pressure fluctuations of the mid points of the windward faces of the various sized prisms. The spectra are plotted as the non-dimensional quantity

$Sp(n)/\rho^2 Lx\overline{u}^2$ against the reduced frequency nLx/U where Sp(n) is the power spectrum, ρ the air density, Lx the longitudinal scale of turbulence, U the mean velocity and u the longitudinal component of turbulence. In the low frequency range, the spectra of the pressure fluctuations showed similar features to that of turbulence plotted as a solid line in Fig.1, but in the high frequency range they falled off more rapidly than that of the turbulence. The spectra of the pressure fluctuations in the high frequency range were smaller as B/Lx increased.

In order to examine the effect of the position of the pressure measurement on the power spectra, the spectra at various points of the two-dimensional prism with 10cm breadth were measured and are shown in Fig.2. Near the edge, the pressure fluctuations were affected by the wake and a slight amplification of the spectra occured at the reduced frequency nB/U=0.1. On the other hand, near the middle ponit the spectra falled off at a rate greater than -5/3 due to distortion of turbulence as mentioned by Hunt (7). The rate of the spectral decay in the high frequency range were from -3 at the middle point to -5/3 at the edge point, and they agreed with that of full scale measurements at Osaka Hotel Plaza shown in Fig.2 (5).

Scales of pressure fluctuations

Another important quantity required to evaluate the response is the scale of the pressure fluctuations. Fig.3 shows the horizontal cross-correlation of the pressure fluctuations on the windward face for various prisms. The solid lines show the cross-correlations of the velocity fluctuations in front of the two-dimensional prism with 5cm breadth. The horizontal cross-correlations of the pressure fluctuations were little affected by B/Lx or the size of the prism, and they were somewhat larger than that of the velocity fluctuation. The mean scale of the pressure fluctuations was about 2 times larger than that of the free stream turbulence. On the other hand, the vertical cross-correlations of the pressure fluctuations are shown in Fig.4 with those of the velocity fluctuations. The cross-correlation

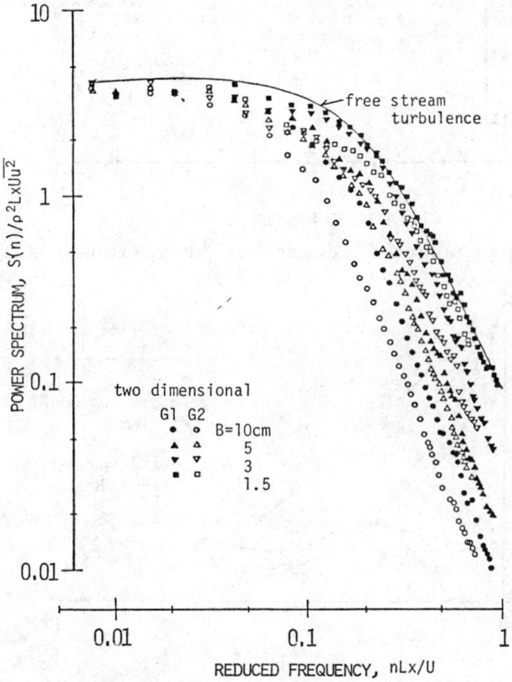

Fig. 1 Power spectra of pressure fluctuations on windward faces of square prisms

of the velocity fluctuations rapidly decreased near the prism. The vertical cross-correlation of the pressure fluctuations was again larger than that of the free stream velocity fluctuations even in two-dimensional flow. The vertical scales for the pressure fluctuations were about 1.5 times larger than that of the free stream turbulence, and similar to the horizontal scales they were not affected much by the size of the prism and the type of the flow.

Fig.5 and Fig.6 show the vertical and horizontal normalized co-spectra of the pressure fluctuations. The shaded area in Fig.5 shows the full scale results at Osaka Hotel Plaza (5). The vertical pressure co-spectra were affected by $\Delta z/Lx$ as well as the velocity co-spectra. On the other hand, the horizontal pressure co-spectra were little affected by $\Delta y/Lx$, and agreed well with the full scale results. If the

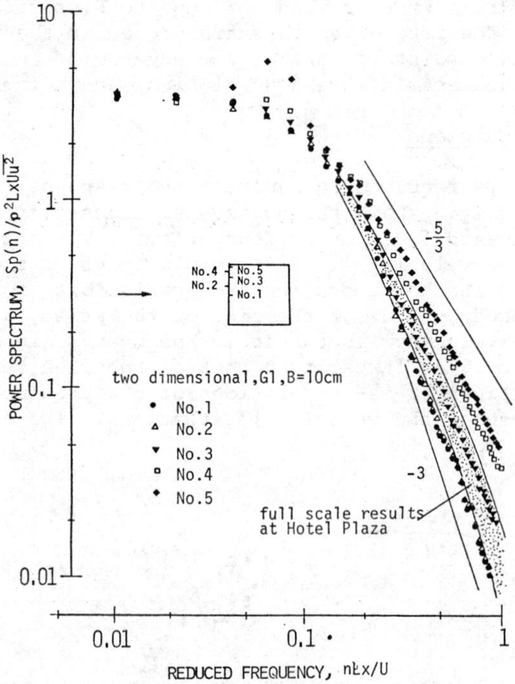

Fig. 2 Power spectra of pressure fluctuations at various points on a square prism

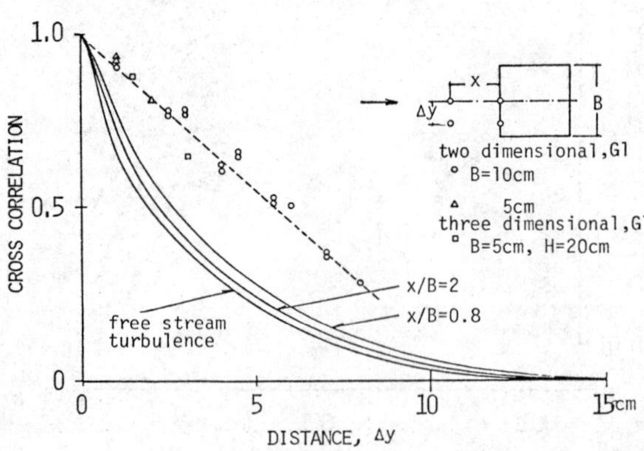

Fig. 3 Cross-correlations of pressure fluctuations and upstream velocity fluctuations in the horizontal direction

co-spectra were made an approximation with the exponential function of the reduced frequency, they could be expressed as equation 1 and 2. In the horizontal direction

$$F_{p_1p_2}(y_1,y_2,n)=S_{p_1p_2}(y_1,y_2,n)/\{S_{p_1}(y_1,n)S_{p_2}(y_2,n)\}^{\frac{1}{2}}=\exp(-4|y_1-y_2|n/U) \quad (1)$$

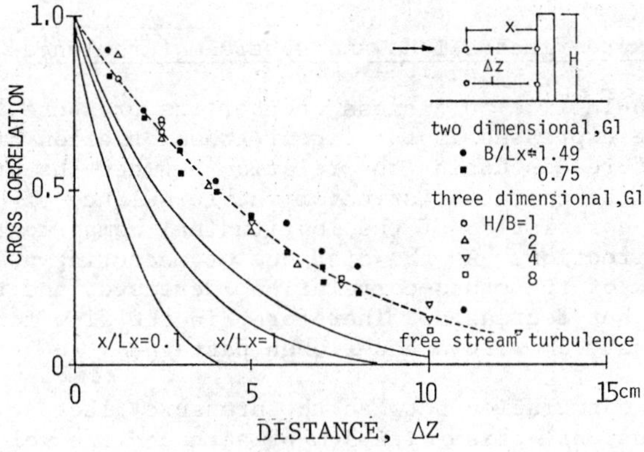

Fig. 4 Cross-correlation of pressure fluctuations and upstream velocity fluctuations in the vertical direction

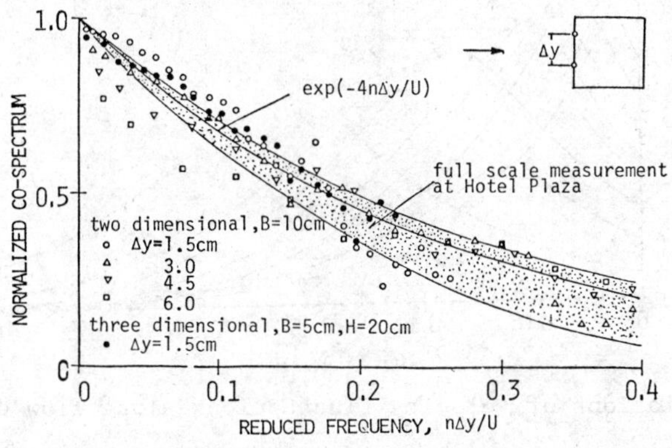

Fig. 5 Normalized co-spectra of pressure fluctuations in the horizontal direction

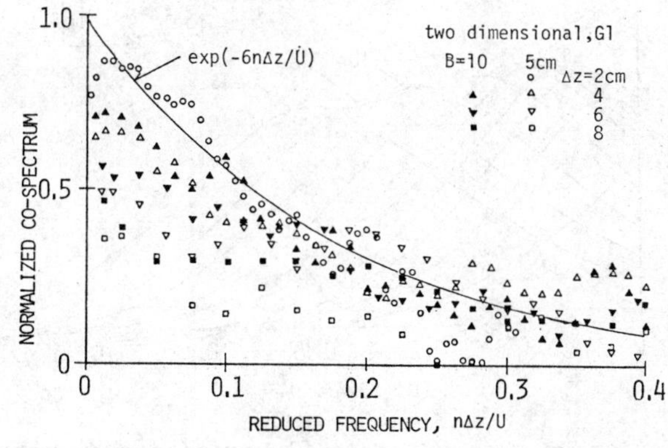

Fig. 6 Normalized co-spectra of pressure fluctuations in the vertical direction

and in the vertical direction

$$Fp_1p_2(z_1,z_2,n)=Sp_1p_2(z_1,z_2,n)/\{Sp_1(z_1,n)Sp_2(z_2,n)\}^{\frac{1}{2}}=\exp(-6|z_1-z_2|n/U) \qquad (2)$$

where Fp_1p_2 is the normalized co-spectrum, Sp_1p_2 the co-spectrum, Sp the power spectrum and z_1,z_2,y_1,y_2 the co-ordinate of points on the windward face of the prism.

Relation between pressure fluctuations and upstream turbulence

In this section, we would like to discuss whether the pressure fluctuations on the windward face could be expressed as the linear transformation of the upstream velocity fluctuation. Before discussing the relation between the pressure and the velocity, we describe the characteristics of the turbulence of Grid 1 in detail. Fig.7 shows the cross-correlation of the longitudinal compornent of the turbulence along the mean wind direction. As the distance of measuring points increased, the time delay of the peak of the cross-correlation increased, and the value of the peak was not constant but decreased. Therefore, in the flow for Grid 1, the turbulence convected downstream with a non-flozen pattern.

Fig.8 shows the cross-correlation between the pressure fluctuation on the middle point of the two-dimensional prism with 5cm breadth and the velocity fluctuation

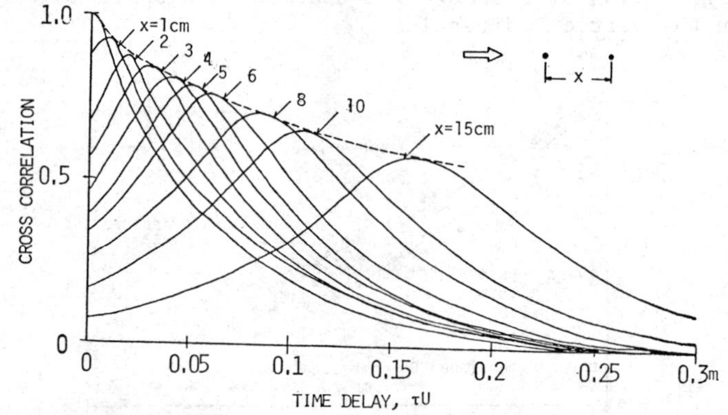

Fig. 7 Cross-correlations of velocity fluctuations along flow direction

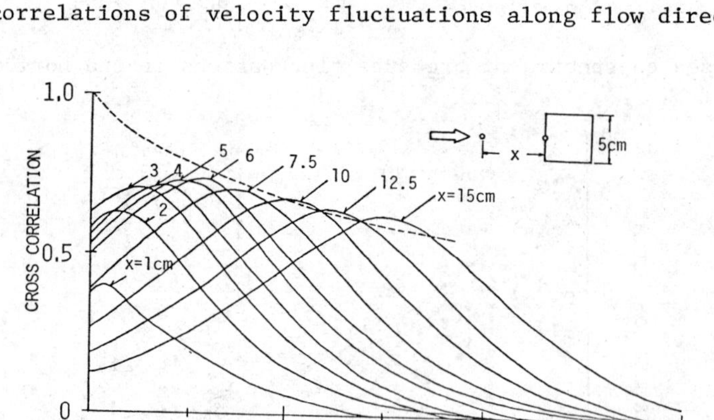

Fig. 8 Cross-correlation between pressure fluctuation and upstream velocity fluctuation

in front of the prism in the flow for Grid 1. The broken line shows the envelope of the cross-correlation of the free stream turbulence shown in Fig.7. The peak of the cross-correlation had the largest value at x/B=1.2, which is the ratio of the distance between the measuring points to the breadth of the prism. At x/B<0.4, the peak value of the correlation rapidly decreased as the measuring point of the velocity approaching the prism, so the pressure fluctuations might not be caused by the turbulence very close to the prism. The peak values of the correlations at x/B>2 were a little larger than that of the velocity correlations in Fig.7, so we can expect that the pressure fluctuations would be the transformation of the veloctiy fluctuation.

The convection velocity of the upstream velocity fluctuations to the pressure fluctuations was computed from the time delay of the peaks of the cross-correlation and is shown in Fig.10 along with the local mean velocity in front of the prism. If the turbulence is convected with the local mean velocity, this convection velocity should be the mean value of the local mean velocities from the prism to the velocity measuring point. However, the convection velocities were in general larger than the mean velocity and they increased as the measuring point of the velocity fluctuation approaching the prism, whereas the local mean velocity decreased.

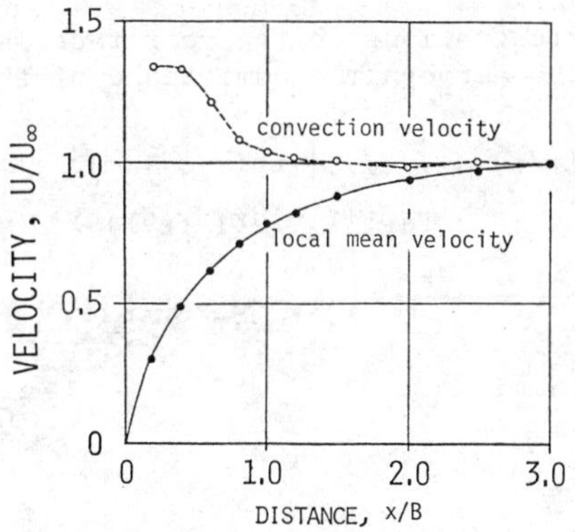

Fig. 9 Convection velocity of upstream velocity fluctuation to pressure fluctuation

DISCUSSION OF RESULTS

Pressure/velocity admittance

In the previous sections, we showed that the pressure fluctuations could be expressed as a linear transformation of the upstream velocity fluctuations. This transformer in the frequency domain which is defined as the "pressure/velocity admittance" by the authors, is given by the ratio of the spectrum of the pressure fluctuations to that of the velocity fluctuations or equation 3.

$$|\chi up(n)|^2 = Sp(n)/\rho^2 U^2 Su(n) \qquad (3)$$

where $|\chi up(n)|^2$ is the pressure/velocity admittance, Sp(n) the spectrum of the pressure and Su(n) the spectrum of the velocity.

Fig.10 shows the pressure/velocity admittance at middle point for various prisms, plotted against the reduced frequency nB/U. Fig.10 shows that the admittance was not affected by both the size of the prism and the turbulence intensity so that it can be taken as a function of only the reduced frequency nB/U. Fig.11 shows the pressure/velocity admittance for various points of the two-dimensional prism with 10cm breadth in the flow for Grid 1. The rate of the decay of the admittance function in the high frequency range was largest at the mid-point of the prism.

From the pressure/velocity admittance shown in Fig.10 and Fig.11, we formulated equation 4 which was a little different from the previous form given by the authors (8).

$$|\chi up(y,n)|^2=1/\{1+20(nB/U)^2\}^{\alpha} \tag{4}$$

where

$$\alpha=\frac{2}{3}\{1-(2y/B)^2\}$$

Equation 4 is shown as the solid lines in Fig.10 and Fig.11.

Aerodynamic admittance of a plate

The power spectrum of drag fluctuation of a flat plate can be given by equation 5 using the power spectrum and the normalized co-spectrum.

$$S_{FD}(n)=\iint_A\iint_A\{Sp_1(n)Sp_2(n)\}^{\frac{1}{2}}Fp_1p_2(r,n)dx_1dx_2dy_1dy_2 \tag{5}$$

Where $S_{FD}(n)$ is the power spectrum of the drag fluctuation and $Fp_1p_2(r,n)$ the normalized co-spectrum of the pressure fluctuations. x_1,x_2 and y_1,y_2 are the co-ordinate of two points on the plate surface. A is the area of the plate and

$$r=\{(x_1-x_2)^2+(y_1-y_2)^2\}^{\frac{1}{2}}$$

Assuming that the pressure fluctuations on the rear face may be directly related to that on the windward face, the aerodynamic admittance of the plate can be expressed as equation 6.

$$|\chi(n)|^2=S_{FD}(n)/A^2Cp^2\rho^2U^2Su(n)=\frac{1}{A^2}\iint_A\iint_A|\chi up(x_1,y_1,n)||\chi up(x_2,y_2,n)| \cdot Fp_1p_2(r,n)dx_1dx_2dy_1dy_2 \tag{6}$$

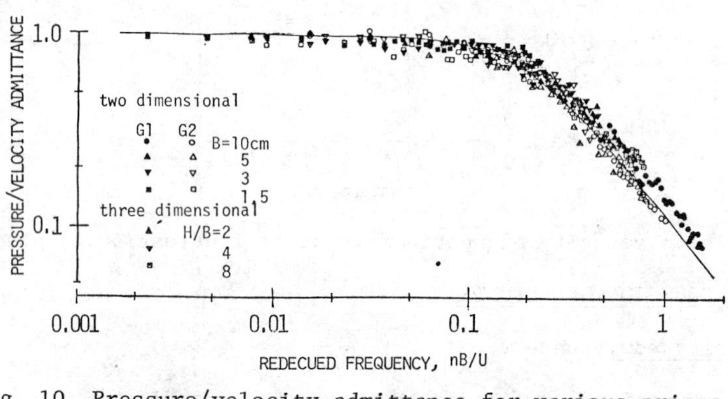

Fig. 10 Pressure/velocity admittance for various prisms

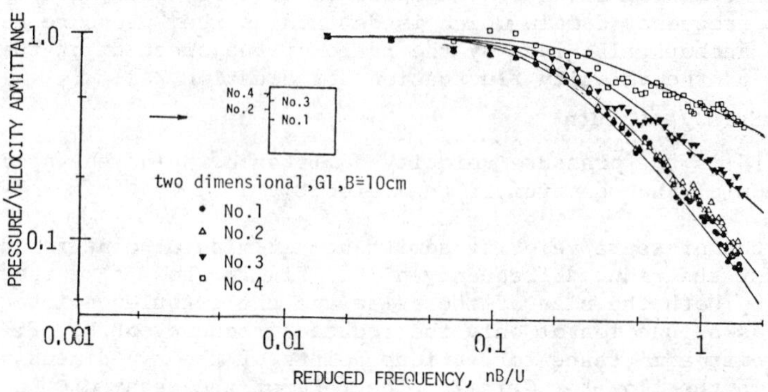

Fig. 11 Pressure/velocity admittance at various points

where $|\chi(n)|^2$ is the aerodynamic admittance, Cp the pressure coefficient and $|\chi up(x,y,n)|$ the pressure/velocity admittance in the previous section.

On the other hand, Vickery finds that the aerodynamic admittance for a lattice plate is given by equation 7 (3),(4).

$$|\chi(n)|^2=\frac{1}{A^2}\iint_A\iint_A Fu_1u_2(r,n)\,dx_1dx_2dy_1dy_2 \qquad (7)$$

where $Fu_1u_2(r,n)$ is the normalized co-spectrum of the velocity fluctuations. The form of this co-spectrum, suggested by Davenport, is given by equation 8.

$$Fu_1u_2(r,n)=\exp(-8nr/U) \qquad (8)$$

Aerodynamic admittance for a square plate were computed using 6 and 7, and are presendted as solid line and the broken line in Fig.12. The shaded area in Fig.12 shows the experimental results by Davenport, Vickery and P.W.Bearman (2),(3),(4). In the case of computing the aerodynamic admittance using equation 6, it was assumed that the normalized co-spectrum of the pressure fluctuations was given by the equation 1 and the pressure/velocity admittance did not change from point to point, so it was given by equation 4 at 2y/B=0.6. The aerodynamic admittance computed from equation 6 shows a better agreement with the experimental results at all values of nB/U than that computed from equation 7.

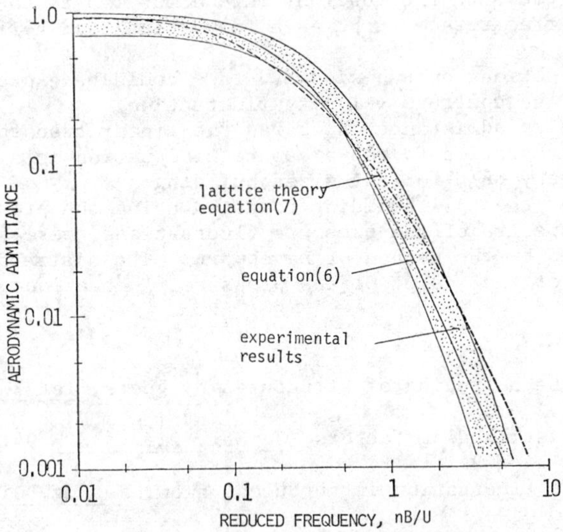

Fig. 13 Aerodynamic admittance of square flat plates

Comparison of gust factors

Gust factors for a tall building in a city and a suburb were computed. The properties of the building and the flow conditions are presented in TABLE 2 with the gust factors. The gust factors computed using the pressure/velocity admittance and the pressure co-spectra are about 17% higher than those computed by the method of Davenport in the case of Cz=8.0 and Ch=20.0, and are nearly equal to that in the case of Cz=8.0 and Ch=8.0.

The gust factors in TABLE 2 were computed on the assumption that the power spectrum of the velocity fluctuation did not vary with height and the pressure fluctuation on the rear face was directly related to that on the windward face, so that they would be overestimated. If the pressure fluctuation on the rear face was little related to that on the windward face, the gust factors must be reduced about 15%.

TABLE 2. Comparison of gust factors

		A.G.Davenport				present method	
hight (m)	H	200		200		200	
breadth (m)	B	50		50		50	
natural frequency (Hz)	n	0.2		0.2		0.2	
critical damping ratio	β	0.01		0.01		0.01	
		city	suburb	city	suburb	city	suburb
mean wind velocity at 300 m (m/sec)	U	60	60	60	60	60	60
surface drag coefficient	κ	0.05	0.015	0.05	0.015	0.05	0.015
wind profile $(Uz/Uref)=(z/z_{ref})^{\alpha}$	α	0.4	0.25	0.4	0.25	0.4	0.25
vertical co-spectrum exp(-Cz·n/Uz)	Cz	8.0	8.0	8.0	8.0	6.0	6.0
horizontal co-spectrum exp(-Ch·n/Uz)	Ch	20.0	20.0	8.0	8.0	4.0	4.0
pressure/velocity admittance	$\|\chi_{up}\|^2$	1.0	1.0	1.0	1.0	eq.(4)	eq.(4)
gust loading factor, Ymax/Ymean	G	2.45	2.42	2.77	2.76	2.83	2.83

CONCLUSION

(1) The spectra of the pressure fluctuations on the windward face decayed more rapidly in the high frequency range than that of the free stream turbulence. The rate of the spectral decay of the pressure fluctuations was from -3 at the middle point to -5/3 at the edge point.

(2) The scales of the pressure fluctuations were about 1.5 times and 2.0 times larger than those of the free stream turbulence in vertical and horizontal directions respectively.

(3) The pressure fluctuations on the windward face could be expressed by a linear transformation of the upstream velocity fluctuation.

(4) The pressure/velocity admittance which was the linear transformer of the upstream velocity fluctuation to the pressure fluctuation is a function of only the reduced frequency nB/U for any sized building.

(5) The gust factors of the tall building computed using the pressure/velocity admittance and the co-spectra of the pressure fluctuations, were about 17% higher than those computed by the method of Davenport. The discrepancy must be caused by the difference of the scales of the pressure fluctuations.

REFERENCES

(1) A. G. Davenport, The buffeting of structures by gusts, 1st IWCE, 358-391 (1963).

(2) A. G. Davenport, Gust Loading factors, J. Str. Div., ASCE 94, ST3, 11-33 (1967).

(3) B. J. Vickery, Load fluctuation in turbulent flow, J. Eng. Div., ASCE 94, EM1, 31-47 (1968).

(4) P. W. Bearman, An investigation of the forces on flat plates normal to a turbulent flow, J. Fluid Mech. 46, 177-198 (1971).

(5) H. Kawai and H. Ishizaki, Local wind pressure characteristics on a full-scale tall building, Proc. the Regional Conferece on Tall Buildings, Hongkong, 161-173 (1976).

(6) D. J. Holmes, Pressure fluctuations on a large building and along-wind structural loading, J. Industrial Aerodynamics 1, 249-278 (1976).

(7) J. C. R. Hunt, Turbulent velocities near and fluctuating surface pressure on structures in turbulent wind, Proc. 4th IWCE, 309-320 (1975).

(8) H. Kawai and H. Ishizaki, On the relation between the fluctuations of wind speed and pressure on the windward surface, Proc. Pan-Pacific Tall Building Conference, Hawaii, 267-278 (1975).

SOME EXPERIENCES WITH FLUCTUATIONS MEASUREMENTS ON MODELS

M. PIRNER

Technical and Research Testing Institute in Civil Engineering, Prague, Czechoslovakia

Summary

Measurements of the fluctuating components of pressures affecting the surfaces of models of tall buildings (Fig. 1.1) were made using an airstream and a boundary layer produced by a turbulence generator. The characteristics of the airstream are given in Ref. (1). The purpose of the measurements was the determination of power spectral densities of pressures and the coherence functions of pressures to enable the result to be used for the calculation of forced vibrations and/or for an analysis of local wind effects.

The measurements included particularly those concerning the power spectral densities of pressure on parallel bypassed surfaces and the coherence functions in dependence on the angle of incidence α. The pressures were measured by 1/4 in. Brüel-Kjaer microphones.

1. Introduction

The variable component of wind pressure affecting the surfaces of actual buildings was measured by several authors, particularly by Newberry (2) on several tall buildings in London. The descriptions of these experiments have revealed the high requirements imposed on the reliability of the pick-up sensors, and the processing of the record data as well as the high costs of such measurements. For this reason model research of the variable components of pressures affecting the surfaces of tall buildings of customary shapes was undertaken (Fig. 1.1).

The response of a structure to wind load is expressed in matrix form by the equation

$$G_y(f) = H(f) \cdot H^*(f) \cdot G_p(f) \qquad (1)$$

where H is the matrix of transfer function and H^* its complex adjoint matrix. G_p is the load matrix whose terms are mutual spectral densities referred to the individual points of the surface of the structure. The points are situated in the centers of the surfaces of elements of which the dynamic model of the structure consists. The terms of the load matrix situated outside the diagonal can be expressed by means of coherence function.

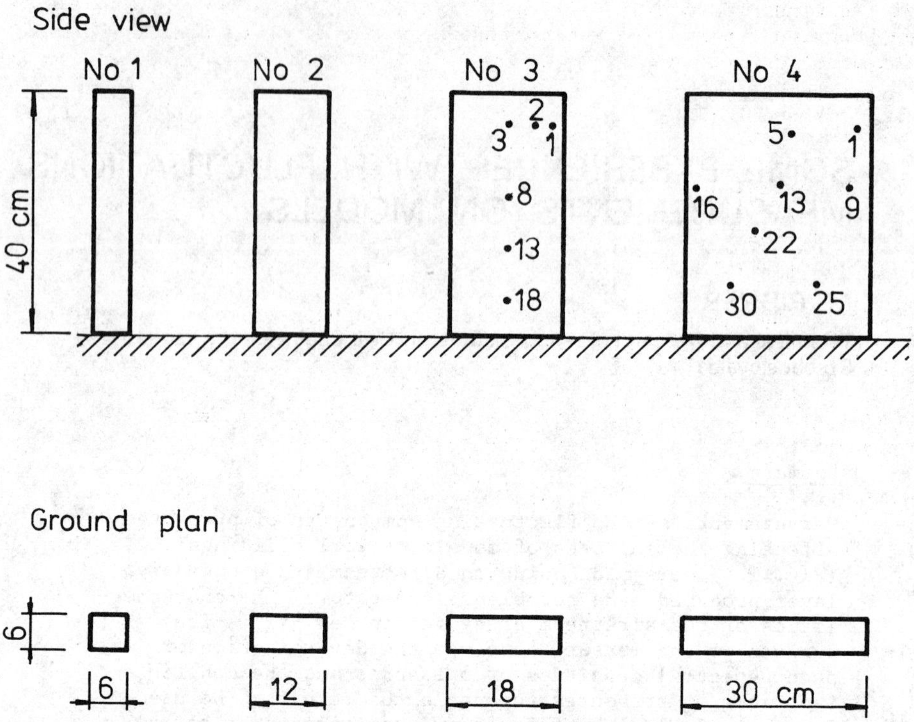

Fig. 1.1. Geometry of models.

The terms of the load matrix were ascertained by a pair of 1/4 in. Brüel-Kjaer microphones, a tape recorder, an analyzer and a computer; for detailed descriptions see Ref. (1).

2. Power Spectral Densities of Pressure

In the proximity of a sharp-edged rectangular body submerged into a turbulent stream four regions originate, viz.:

a) on the windward side of the body there are pressures corresponding with longitudinal turbulence of uninfluenced stream (this correspondence is of limited validity only);

b) on parallel bypassed surfaces there are largest fluctuation effects of the boundary layer, regardless of whether it is or is not going to reattachment;

c) on the leeward surfaces there is a wake;

d) on the roof surface the phenomena are similar to those arising on parallel bypassed surfaces; they can be affected by whirling due to vortices tearing off the corners of the roof surface.

2.1 Let us afford particular attention to the parallel bypassed surfaces ($\alpha = 90°, 270°$) which are subjected to higher dynamic stresses than other surfaces. From the positions of the maxima of ascertained power spectral densities the region of Strouhal number was derived (Table 1) for the reference width of the windward surface. Figure 2.1 shows the frequency of the fluctuation component, corresponding with the maximum ordinate of power spectral density, plotted against the distance d of the measured point from the vertical incidence edge.

Table 1. Strouhal number.

Model No. 1	Str = 0.070
Model No. 2	Str = 0.120 - 0.140
Model No. 3	Str = 0.050 - 0.165
Model No. 4	Str = 0.050 - 0.200

The regions are in agreement with the results of the works described in Refs. (3) and (4). Low values of Strouhal number (Str = 0.02) were ascertained by Naudascher (5) for the flow around a prism; a detailed explanation is given in Ref. (6). The peakness and the position of the peal of the power spectral densities of pressures, consequently, is influenced by the geometric shape of the building model and by the position of the investigated point.

The powers of the spectra of parallel bypassed surfaces according to the authors of Ref. (7), (8), and (9) do not accentuate the region corresponding with Strouhal frequency; for this reason the power spectral densities, by the equation which is a modified equation (Ref. (10)) for the points in the wake, viz.,

$$S_p(f) = \frac{(\frac{1}{2}\,\rho\,\bar{V}_\infty^2)^2}{\pi^2\,\bar{V}_\infty}\, C \exp - \left[(1 - \frac{f}{f_{Str}})/B\right] \qquad (2)$$

The first term of the right-hand side for the given tests is

$$2533\ N^2\ s/m^5$$

The mean values of the coefficient B and the quantity C for the given models are given in Table 2 below:

Table 2. Coefficient B and the quantity C.

Model	B[-]	C[m]
No. 1	0.58	0.005
No. 3	1.60	0.015
No. 4	1.09	0.023

Naturally, the above mean values do not express the effects corresponding with the particular positions of the investigated points; in such cases it is necessary to use the values of B and C accurately corresponding with them. Figure 2.2 shows the power spectral densities of pressures on the model No. 3. Figure 2.3 shows the power spectral densities of pressures on the model No. 4. The overall energy of the flucutating component expressed as m.s. depends on the distance of the measured point from the vertical incidence edge and on the turbulence of an uninfluenced airstream (see Fig. 2.1.b).

2.2 Let us further note the power spectral densities on the leeward surfaces (α = 180°). The energy ascertained on this surface on the model No. 4 is approximately of one order lower than on the windward surface. In the case of other models, Nos. 1, 2 and 3 are more or less smooth (Fig. 2.4), in the case of the model No. 4 it contains a number of peaks, the greatest energy being contained in the peak corresponding with Strouhal frequency. While in the tests of a

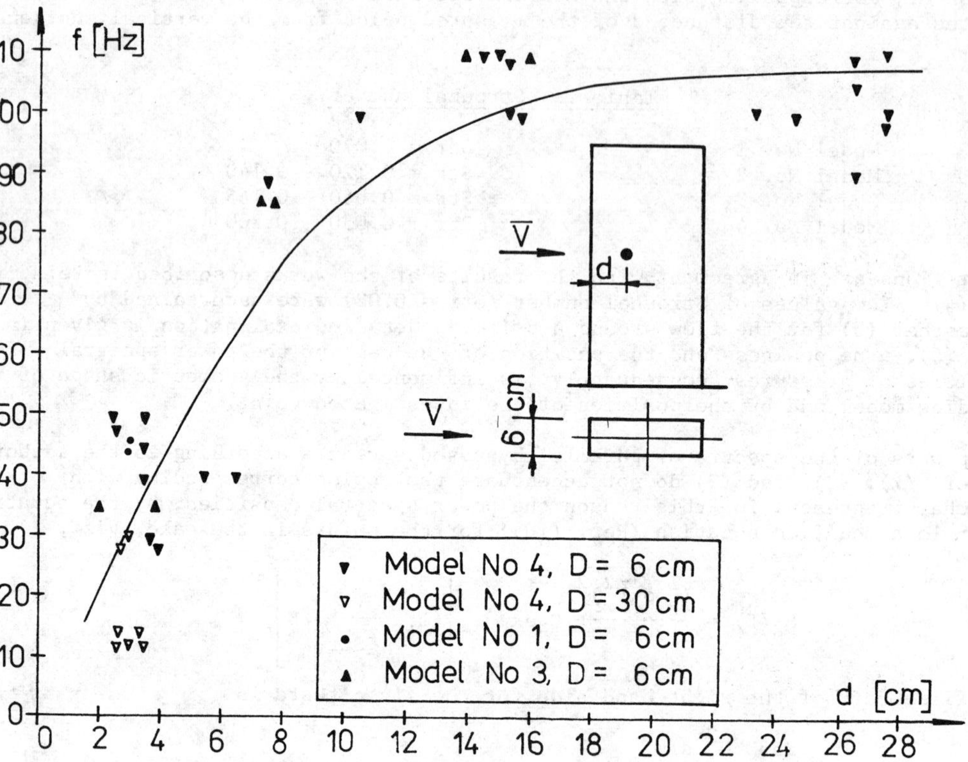

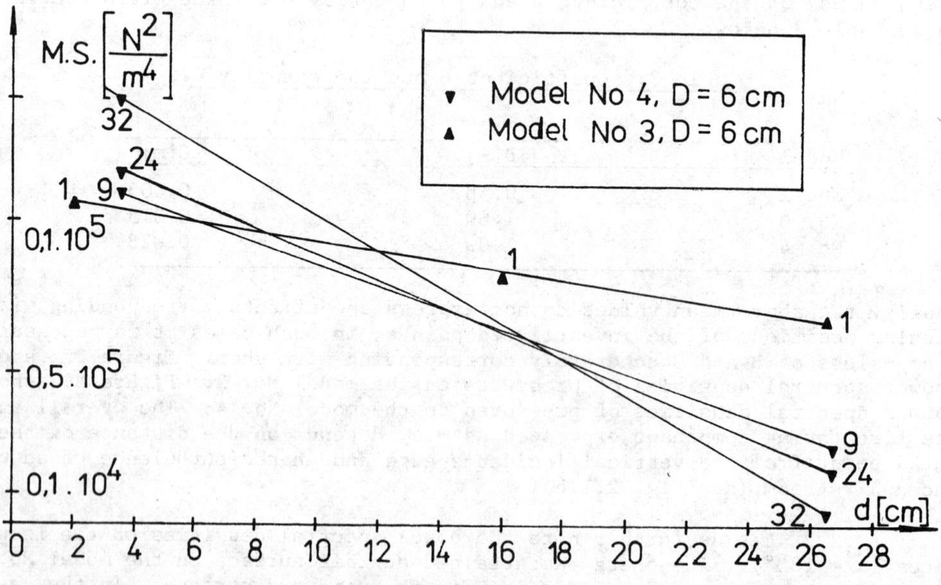

Fig. 2.1.a Relation between dominant frequency and distance d.

Fig. 2.1.b Relation between m.s. fluctuating pressure and distance d.

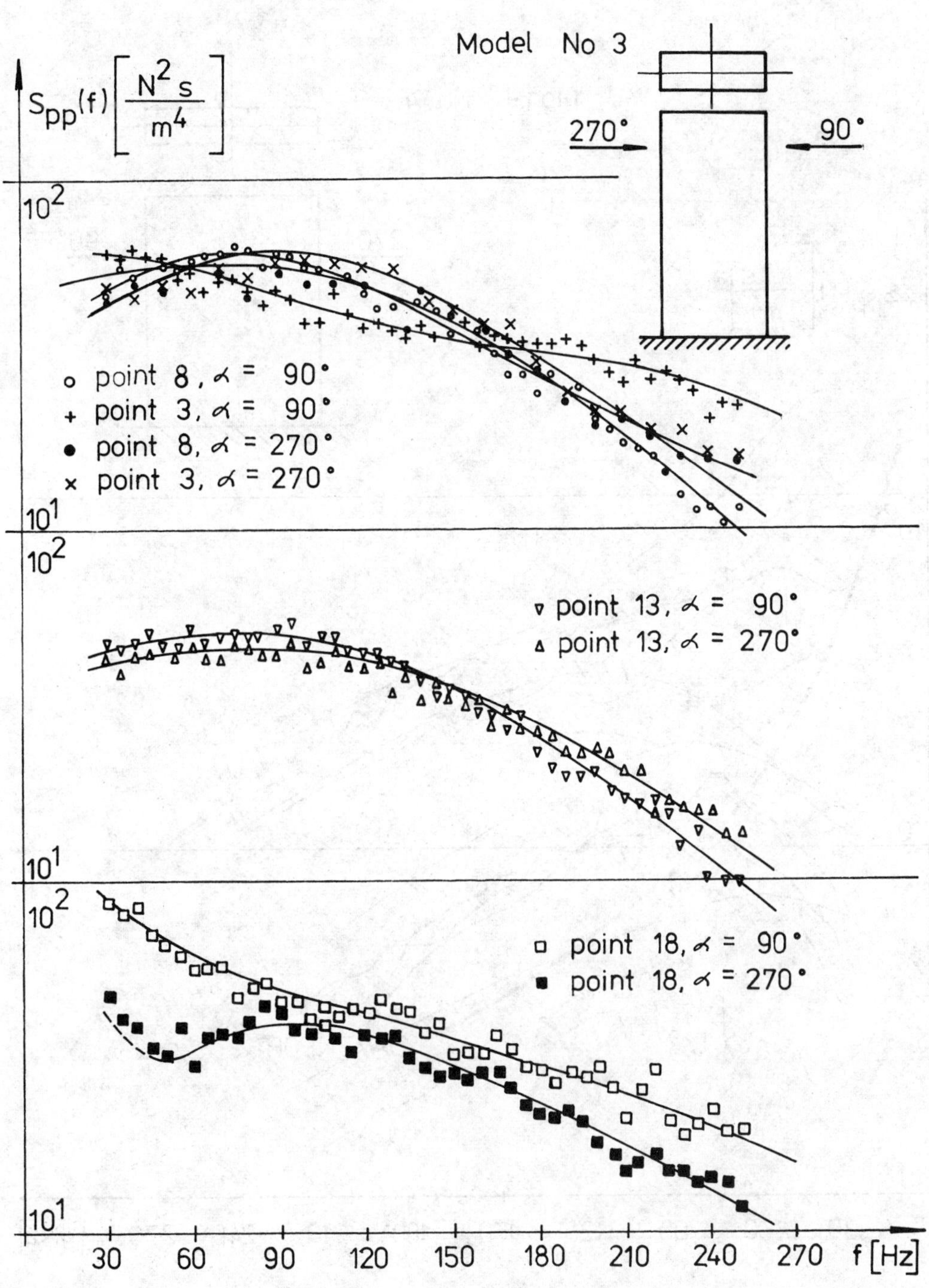

Fig. 2.2. Spectral densities on model No. 3 for α = 90° and 270°.

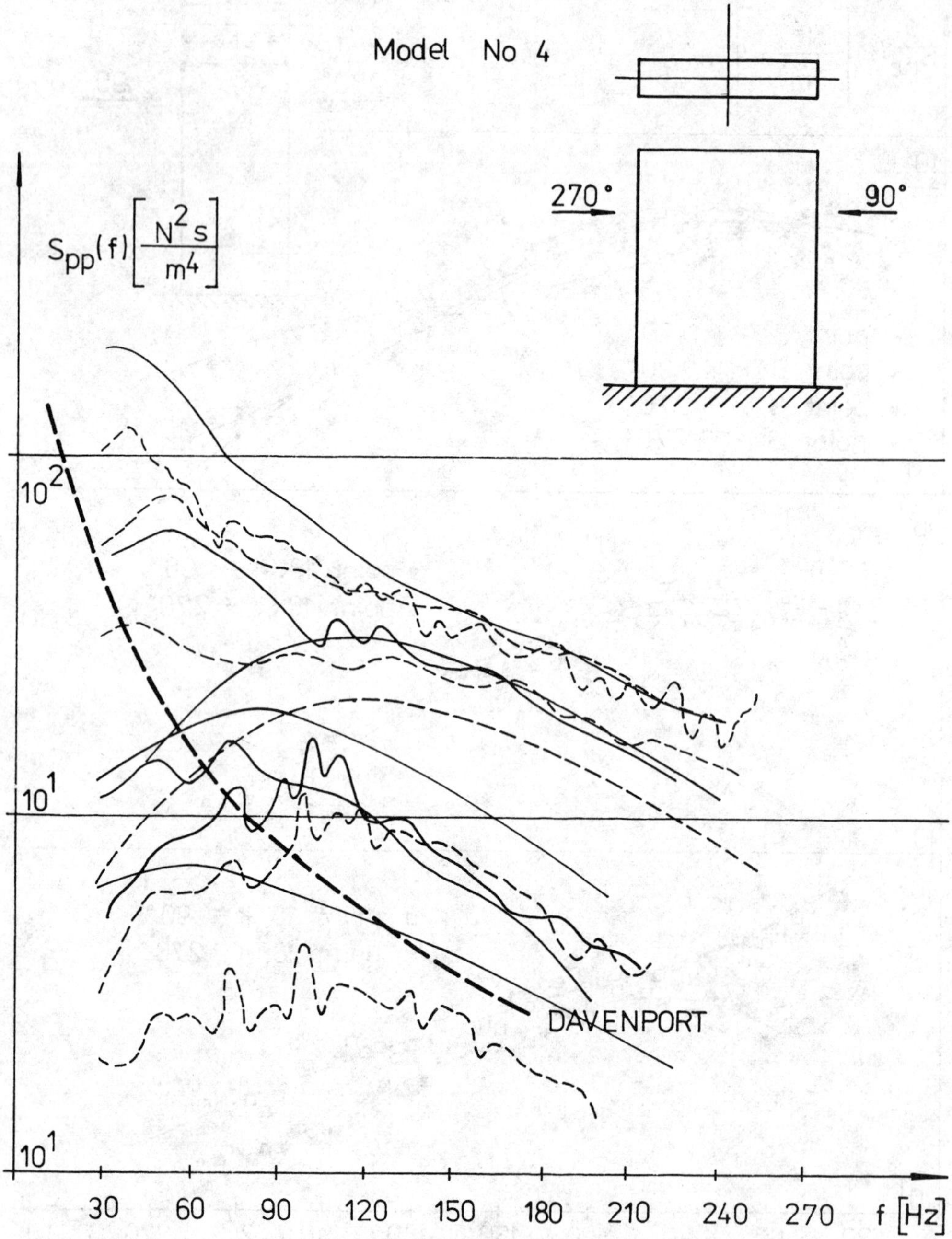

Fig. 2.3. Spectral densities on model No. 4 for α = 90° and 270°.

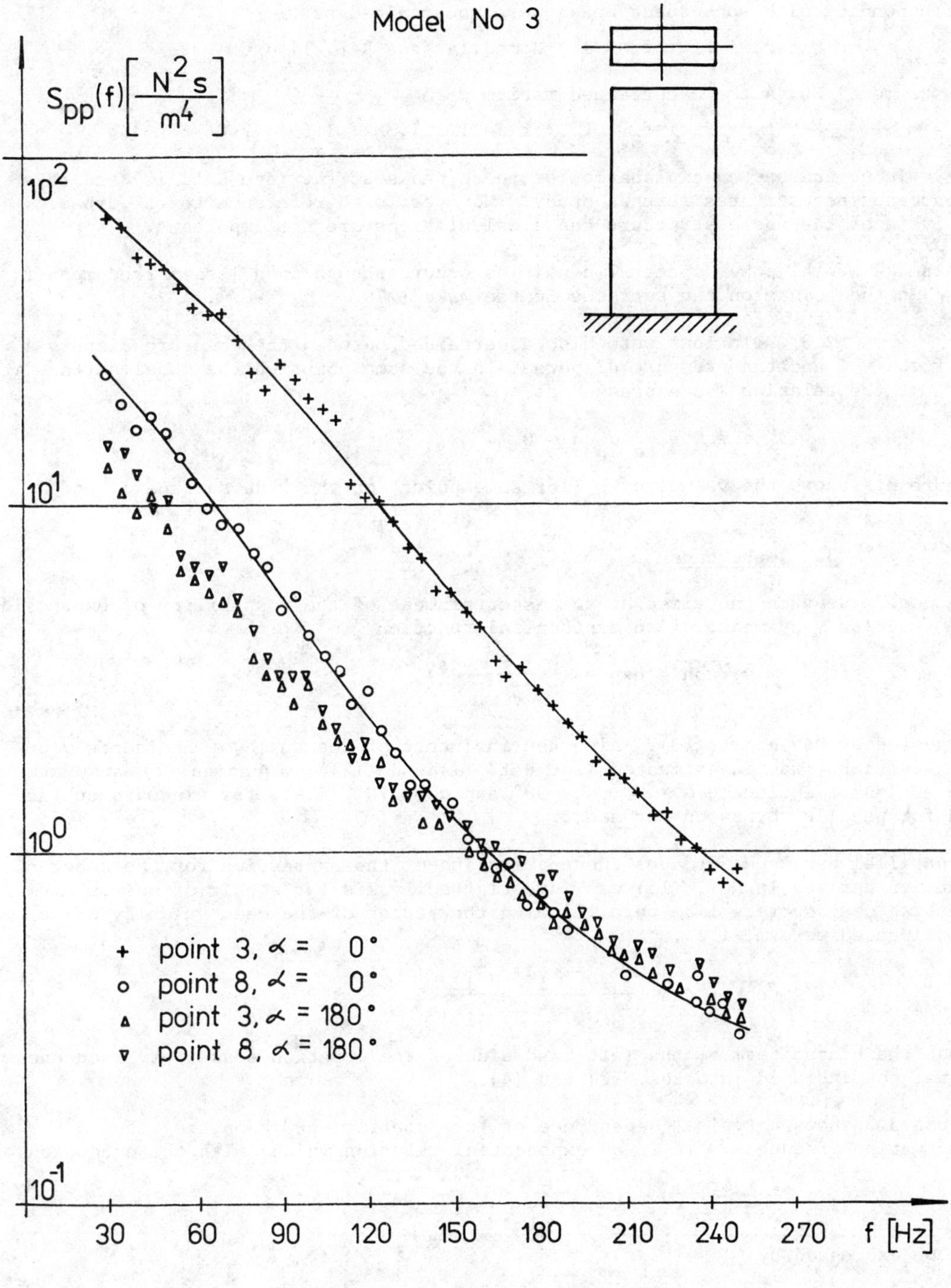

Fig. 2.4. Spectral densities on model No. 3 for α = 0° and 180°.

cylindrical surface of a circular cross section the frequency ratios corresponding with subsequent peaks were ascertained as

$$f_1 : f_2 : f_3 = 1 : 2 : 3.14 \text{ (see Ref. 11)}$$

On the model No. 4 the ascertained ratios were

$$f_1 : f_2 : f_3 : f_4 : f_5 = 1 : 1.38 : 1.69 : 2.06 : 2.44$$

There is no undisputed explanation of these peaks so far (see Ref. 12) and this phenomenon necessitates further study. However, with reference to the dynamic analysis of the whole structure these calculations are not important.

The peaks of the power spectral densities occur, though in a less marked manner, also in the points on the surfaces in the wake (270° > α > 90°).

2.3 Relations were also ascertained between the pressure dispersal σ_p^2 for $\alpha = 0$ and the pressure dispersal in the same point on the model surface for $\alpha \neq 0$: the relation was expressed as

$$\sigma_p^2 \; \alpha \neq 0 = k_\alpha \, \sigma_p^2 \; \alpha = 0 \tag{3}$$

Figure 3.2 shows the values of k_α for some points on the wide surface of model No. 4.

3. Coherence

The model research was aimed at the ascertainment of the possibility of expressing the function γ by means of an exponential function

$$\gamma = \sqrt{coh} = \exp - \left(\frac{k \cdot f \cdot \Delta r}{\overline{V}_{10}}\right) \tag{4}$$

suggested by Davenport (13). Also the influence of the angle of incidence α on the function γ was investigated (see Ref. (1) and (12)). Equation (2) was used for the angles of incidence $\alpha = 0°$, $\alpha = 90°$ and $\alpha = 270°$, i.e., for windward surface and for parallel bypassed surfaces.

On parallel bypassed surfaces the application of the expression for the coherence function derived in Ref. (15) was investigated, valid for a pair of points situated on a surface line parallel with the vector of the mean velocity of an uninfluenced stream, i.e.,

$$\sqrt{coh} \cdot \cos \frac{w(x_2 - x_1)}{\overline{V}} = \frac{G_{12}}{G_{11}} \tag{5}$$

Since the second term of the left-hand side of the equation nears very much one, it was not included into Eqs. (2) and (4).

Figure 3.3 shows a typical dependence of the "coefficient" k on the reduced fluctuation frequency. It is an exponential relation which, with the exception of

$$\frac{f \cdot \Delta r}{\overline{V}} \leqq 0$$

can be expressed by

$$k \cong - \frac{\lg \frac{f \cdot \Delta r}{\overline{V}}}{a} \tag{6}$$

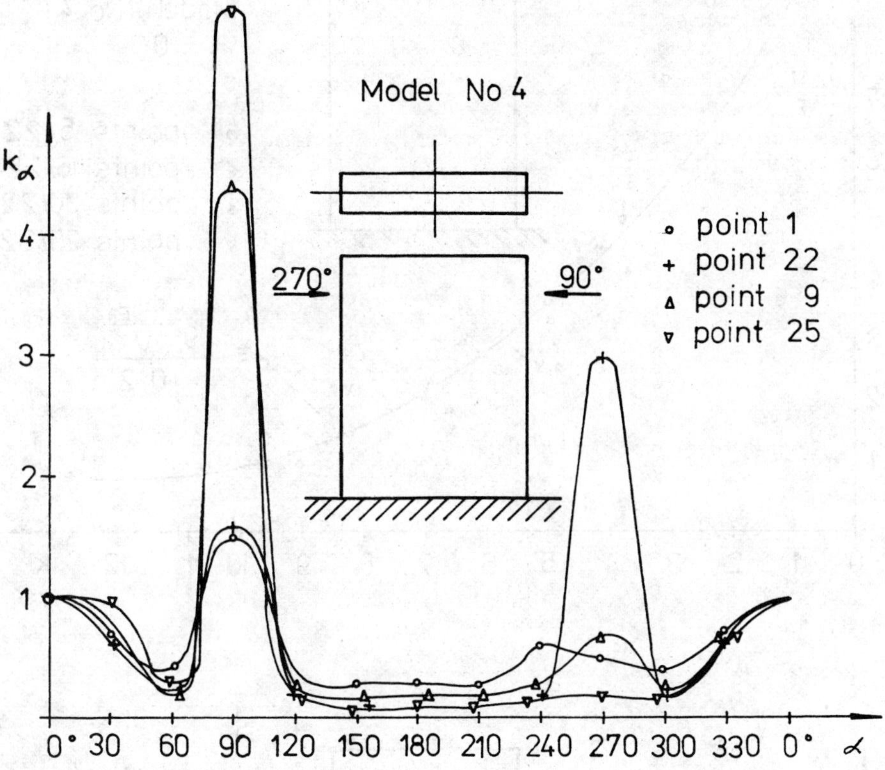

Fig. 3.2. Relation between k_α and α on model No. 4.

so that the function γ can be expressed once again with the exception of the = 0 by

$$\gamma = \exp - \left[\frac{\lg \frac{f \cdot \Delta r}{\overline{V}}}{a} \; \frac{f \cdot \Delta r}{\overline{V}} \right] \tag{7}$$

The constant a is shown in Fig. 3.3 for the model No. 4 for the angles of incidence $\alpha = 0°$ and $\alpha = 90°$.

In this way we have expressed the reduction of k with the growth of the frequency f ascertained, for example, also by Brook (14); thus, we have eliminated the underrating of the mutuality at higher frequencies which need not be negligible in some cases.

A number of peaks ascertained in some power spectral densities manifests itself also in the shape of the function γ; in the cases when they cannot be neglected the shape of the function γ is not determined with sufficient accuracy by Eq. (4). The procedure in such cases is described in Ref. (1).

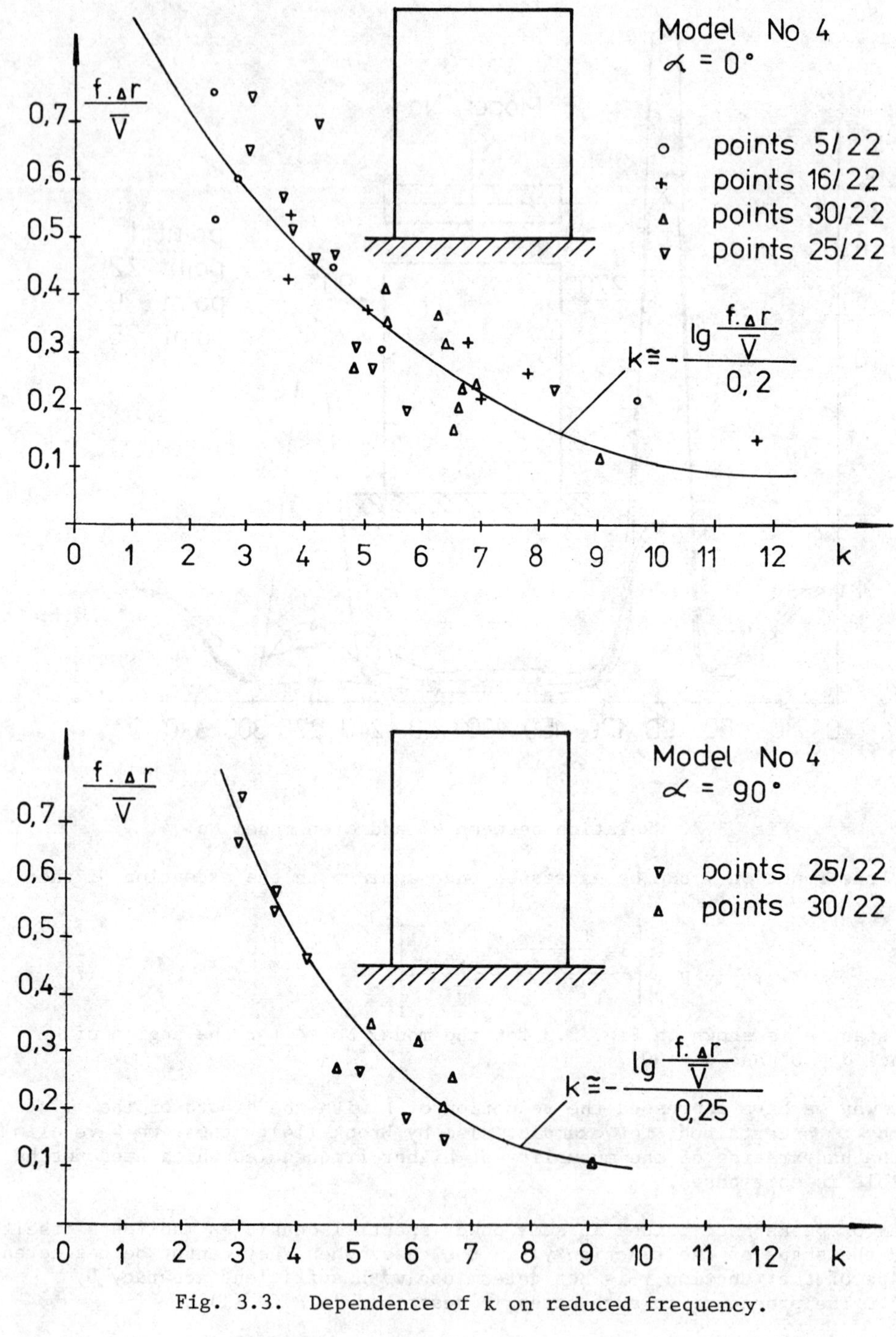

Fig. 3.3. Dependence of k on reduced frequency.

References

(1) Pirner, M., Some results of model measurements on tall buildings, Journal of Industrial Aerodynamics, ASCE, 3, 243, 1977.

(2) Newberry, C.W., Eaton, K.J., and Mayne, J.R., Wind loading on tall buildings - further results from Royex-House, Building Research Establishment Current Papers, report CP 29/73.

(3) Vickery, B.J., Wind loads on buildings, A Dissertation for Doctor of Philosophy, Department of Civil Engineering, University of Sydney, 1968.

(4) Saunders, J.W. and Melbourne, W.H., Tall rectangular building response to cross-wind excitation, Fourth Int. Conf. on Wind Effects on Buildings and Structures, Heathrow, 1975.

(5) Locher, A. and Naudascher, E., Some characteristics of macro-turbulence in flow past a normal wall, International Association for Hydraulic Research, 1974.

(6) Ermshaus, R., Morkovin, M.V., and Tödten, H., Oscillations induced unstable-reattachment, University of Karlsruhe, 1978.

(7) Bull, M.K., Wall-pressure fluctuations associated with subsonic turbulent boundary layer flow, J. Fluid Mech., 28, 719, 1967.

(8) Elashkar, J., Dynamic response of tension roofs to turbulent wind, The University of Western Ontario, Report BLWT-3-74, 1974.

(9) Maestrello, L., Measurement and analysis of the response field of turbulent boundary layer excited panels, J. Sound Vib., 2, 270, 1965.

(10) Vaicaitis, R., Nonlinear panel response to nonstationary wind forces, Journal of the Engineering Mechanical Division, ASCE, 12, 333, 1975.

(11) Johns, D.J. and Nataraja, R., Response of thin-walled cylinders to aerodynamic excitation, Department of Transport Technology, Lougborough University of Technology, Report TT 7413, 1974.

(12) Pirner, M., Dynamic effects of the fluctuating wind pressure in civil engineering, Academia, Prague, printed.

(13) Davenport, A.G., The response of slender line-like structures to a gusty wind, Proc. Inst. Civ. Eng. 6610, 23, 1967.

(14) Brook, R.R., A note on vertical coherence of wind measured in an urban boundary layer, Boundary Layer Meteorology, 9, 1975.

(15) Handa, K., Response of tall structures to atmospheric turbulence, University of Southampton, Report No. 4, 1967.

TURBULENCE EFFECTS ON MAXIMUM SURFACE PRESSURES - A MECHANISM AND POSSIBILITY OF REDUCTION

W. H. MELBOURNE

Monash University, Melbourne, Australia

ABSTRACT

Some experimental evidence is presented which provides further support for the hypotheses of Gartshore (1) and Melbourne (2) concerning the dominance of the incident small scale turbulence on separated shear layer behaviour and the instability mechanism by which high negative pressures are created under re-attaching shear layers. In particular it is shown that increase of small scale turbulence significantly increases the magnitude of high negative surface pressures under re-attaching shear layers, but that venting of this region can substantially reduce the magnitude of these negative pressures by causing the instability process to break down. The practical applications of these observations to the reduction of loads on roofs, etc., and some implications of errors in model scaling, if the small scale turbulence is incorrectly scaled, are discussed.

NOMENCLATURE

A	frontal area of upstream face
b	width of upstream face normal to flow
C_p	pressure coefficient $\left\| = \frac{p_1 - p_o}{\frac{1}{2}\rho \bar{u}_o^2} \right\|$
C_{σ_p}	standard deviation (or rms of fluctuating component) pressure coefficient $\left\| = \frac{\sigma_p}{\frac{1}{2} \bar{u}_o^2} \right\|$
$C\check{p}$	peak negative going pressure coefficient $\left\| \frac{\check{p} - p_o}{\frac{1}{2}\rho \bar{u}_o^2} \right\|$
d	streamwise depth of building
g	peak factor, defined by Equation 3
G	gust factor
h	depth of upstream face normal to flow, or height of building to eave
L_x	integral scale, defined by Equation 1
n	frequency
p_b	base pressure
p_1	pressure at subscripted location
p_o	freestream static pressure
$\check{p}$	peak negative pressure (defined as having a cumulative probability of exceedance of 10^{-4})
S	Small Scale Spectral Density parameter, defined by Equation 2

$S_u(n)$	Spectra of longitudinal component of velocity	σ_p	standard deviation (or rms of fluctuating component) of pressure
u	longitudinal velocity component	σ_u	standard deviation (or rms of fluctuating component) of longitudinal velocity
$\bar{u}_o$	freestream mean velocity		
β	wind direction, (zero when normal to upstream face)		

1. INTRODUCTION

In recent years there has been increasing evidence of the occurrence of very high negative pressures on streamwise surfaces near a leading edge from which a shear layer is separating. This phenomenon has led to failure of flat roofs on low rise structures and window panels on tall buildings. In both cases, full scale and model measurements have confirmed the existence of very high negative pressures occurring in the region near the leading edge under the separating shear layer. Melbourne (2) advanced an hypothesis suggesting that the high negative pressures, which are often of an intermittent nature, could be due to an instability under certain conditions as follows:

> "The shear layers and pressure underneath fluctuate. For certain conditions of freestream turbulence, angle of attack, (and radius of leading edge curvature) the shear layer commences to occasionally reattach onto the wall. As this occurs the cavity region under the shear layer is no longer vented and any decrease in pressure causes the initial radius of curvature of the shear layer to decrease further which in turn increases the local freestream velocity outside the shear layer and further decreases pressure. This is an unstable process which proceeds until the shear layer breaks up into a complete separation again and the cavity becomes vented. During this unstable process, very low, intermittent pressures can occur on the surface under the shear layer near the leading edge."

The development of this hypothesis, shown diagramatically in Figure 1, was influenced by the work of Gartshore (1) who, when working on the paradoxical effects of turbulence on the drag and base pressure of prisms, concluded as follows:

> "Small scale free stream turbulence can influence the drag, fluctuating lift, base pressure coefficient and galloping behaviour of bluff bodies by increasing the rate of entrainment of the shear layers shed from them. The free stream turbulence, after severe distortion by the mean flow field, produces the increased mixing and entrainment by its presence in the shed shear layers, probably in the form of local concentrations of streamwise vorticity.
>
> Only turbulence approaching a bluff body along its front stagnation line is required to produce the major effects of free stream turbulence on the flow around the body, so that a narrow wake effectively replaces grid-produced turbulence for experimental purposes. This observation also emphasises the importance of nearby upstream structures when considering the flow around tall buildings in the natural wind.
>
> Turbulence can produce increased drag and unstable galloping oscillations in shallow rectangular bodies ($l/h<0\cdot5$) which are stable in the absence of turbulence. It can also decrease the drag and stabilize bodies of deeper section ($l/h>0\cdot6$) which are unstable without turbulence. These effects are explicable qualitatively in terms of the increased entrainment resulting from the turbulence.

The effect of turbulence scale is small provided this scale is significantly smaller than the body dimension."

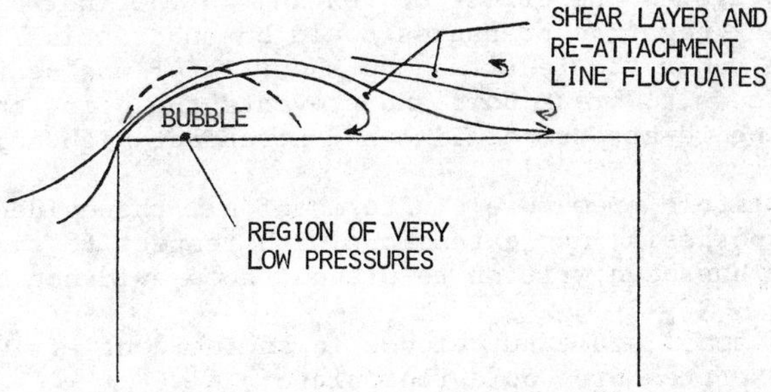

1. SHEAR LAYER RE-ATTACHMENT EFFECTIVELY SEALS OFF BUBBLE FROM DIRECT VENTING TO FREESTREAM.
2. FREESTREAM FLOW OVER THE FRONT OF THE SHEAR LAYER ACCELERATES, THE INCREASED VELOCITY IS ACCOMPANIED BY A DECREASE IN PRESSURE.
3. THE COMBINED EFFECTS OF ENTRAINMENT INTO THE SHEAR LAYER FROM INSIDE THE BUBBLE AND THE DECREASE OF PRESSURE AT THE BOUNDARY CAUSES THE SHEAR LAYER TO MOVE TO REDUCE BUBBLE VOLUME AND INTERNAL PRESSURE (DOTTED LINE).
4. THIS PROCESS CONTINUES IN AN UNSTABLE WAY TO REDUCE SURFACE PRESSURES UNDER THE BUBBLE UNTIL IT IS VENTED BY BURSTING OR BY ARTIFICIAL MEANS.

Fig. 1. Diagrammatic representation of an hypothesis on the cause of high negative pressures occurring near the leading edge under a separating and re-attaching shear layer.

This leads to the observation that if the freestream turbulence is increased, and in particular if the small scale turbulent energy is increased, the increased rate of entrainment into the shear layer would cause earlier re-attachment; the smaller radius of curvature of the shear layer further increases local freestream velocity outside the shear layer which decreases the pressure on the surface under the shear layer separation bubble. The combined effects of high freestream turbulence (in particular the small scale turbulence) and the instability process hypothesised are suggested as being the major causes of the very high negative pressures being experienced under shear layers which are re-attaching, and which have led to roof and window glass failures.

Bearman (3), in a study of the effects of turbulence on pressures and forces on flat plates, whilst noting "that the rate at which turbulence can entrain fluid from adjoining undisturbed fluid is set by the scale and intensity of its energy containing eddies", went on to show that the base pressure became more negative with increase of integral scale. Increase of integral scale implies a decrease in the small scale energy for a given turbulence intensity. Gartshore's observations that the base pressures for shallow (near flat) prisms become more negative with increased small scale turbulence in the incident stream are somewhat different to Bearman's conclusions. Subsequent work by Kwok and Melbourne (4) supports Gartshore's conclusions, and some further work will be included in this study to help resolve these differences.

A step towards validating Melbourne's hypothesis has been made by Sharp (5), who has shown that the peak negative pressures under a re-attaching shear layer occur near the leading edge, when the oscillating re-attachment line of the shear layer is closest to the leading edge from which the separating shear layer originated. Sharp further investigated the effect of venting on the instability process by showing that the peak negative pressures could be substantially reduced (in magnitude) if the bubble under the re-attaching shear layer was vented, which causes the instability process to break down and prevents the occurrence of the extreme forward travel of the re-attachment line and attendant high negative pressures.

Whilst it may be a little premature, in terms of all the evidence needed to validate Melbourne's hypothesis, and extensions with respect to the effect of turbulence, this paper has been written to present some evidence that,

(i) increase of small scale turbulence in an incident air stream causes more negative base pressures on a flat plate,

(ii) increase of small scale turbulence in an incident air stream increases both the mean and fluctuating (and hence peak) pressures on a surface under a re-attaching shear layer,

(iii) by venting the bubble under a re-attaching shear layer on a model of a flat roofed building, by slotting the eave, a substantial reduction in the maximum pressures on the roof can be achieved, and

(iv) that careful attention to the magnitude of the small scale end of the incident turbulence spectrum is necessary when making model measurements of maximum surface pressures on bluff bodies.

2. EXPERIMENTAL PROCEDURE

An experimental procedure was required which would facilitate a study of the effect of turbulence scale and turbulence intensity on the base pressure on a disk and on the pressure on the streamwise surface of a rectangular prism under a separating and re-attaching shear layer. To this end, several turbulent flows were developed in a 2 m square wind tunnel working section with different turbulence intensities and distributions of longitudinal spectra, as shown in Figure 2 and described as follows:

Flow A. All screens and honeycombs were taken out of the wind tunnel circuit, which produced a moderately turbulent flow with a relatively large integral scale (mostly generated by the wind tunnel turning vanes).

Flow B. A more turbulent flow was generated by a grid of slats, 30 mm wide on 150 mm centres, positioned 2·5m upstream of the model.

Flow C. Increased small scale turbulence was superimposed on Flow A by placing two 12·7 mm diameter rods, 40 mm apart, 200 Diameters (2·54 m) upstream of the model.

Flow D. A further increase of small scale turbulence was obtained by moving the rods in Flow C to 100 Diameters (127 m) upstream of the model.

Flow E. A turbulent boundary layer flow with a power law exponent of approximately 0·1 was developed on the floor of the working section. The spectrum and turbulence intensities given for this flow in Figure 1 were measured at a height of 28·5 mm above the floor, which is the height of the flat roof of the model tested in this flow.

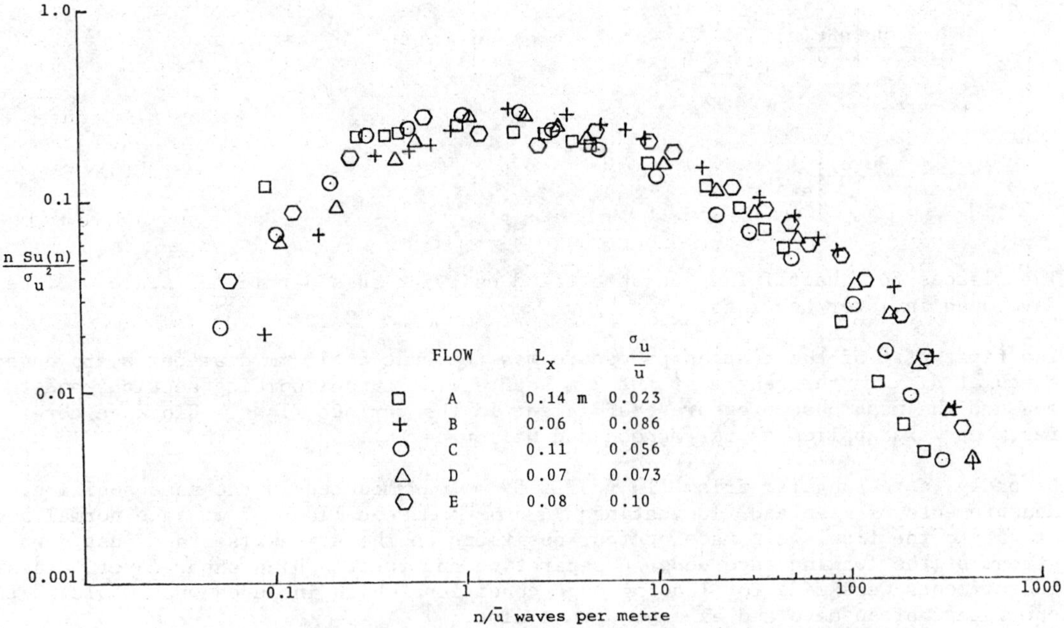

Fig. 2. Normalised spectral density of the longitudinal velocity component for the five wind tunnel flows.

The integral scale, L_x, quoted for these flows was obtained by centrally fitting a von Kármán spectrum,

$$\frac{Su(n)}{\sigma_n^2} = \frac{4\left(\frac{nLx}{\bar{u}_z}\right)}{\left|1 + 70{\cdot}7\left(\frac{nLx}{\bar{u}_z}\right)^2\right|^{5/6}} \qquad (1)$$

to the spectra shown in Figure 2. For such distorted flows the integral scale does not afford a very precise description, but it is adequate to discuss trends. These flows gave a factor of about 2½ variation in integral scale and 3½ variation in turbulence intensity. In addition, these flows gave a large variation in the amount of small scale energy available.

To quantify the amount of small scale energy available in the freestream flow, which is thought to influence the shear layer behaviour, a parameter called the "Small Scale Spectral Density Parameter", (S) is introduced. This parameter is defined as being the spectral density of the longitudinal velocity component normalised on mean velocity at a value of $n/\bar{u} = 1/0.1\sqrt{A}$, where A is the frontal area of the body from which the shear layer is being shed. The spectral density has been normalised by the mean velocity, rather than the variance, so that the effects of turbulence intensity can be included, and the value of wave number has been selected so that it is the energy at scales an order of magnitude lower than the typical frontal body dimension which are being defined. Hence it can be seen that the Small Scale Spectral Density Parameter,

$$S = \frac{\sigma_u^2}{\bar{u}^2} \left| \frac{nSu(n)}{\sigma_u^2} \right|_{\frac{n\sqrt{A}}{\bar{u}} = 10}$$

$$= \left| \frac{nSu(n)}{\bar{u}^2} \right|_{\frac{n\sqrt{A}}{\bar{u}} = 10} \qquad (2)$$

The reasons for quantifying the spectral density at this particular scale will be developed in later sections.

The first part of the testing procedure was to mount a 114 mm diameter sharp edged circular disk in the centre of the 2 m square wind tunnel working section, and to measure the mean base pressure coefficient in the various flows. Blockage corrections were applied as per McKeon and Melbourne (6).

Secondly, a rectangular prism 114 x 57 x 57 mm was mounted in the same position. Measurements of mean and fluctuating pressures, with one 114 x 57 mm face normal and at 45° to the flow, were made at four positions on the streamwise face just downstream of the leading edge under a separating and re-attaching shear layer. These measurements were made for leading edge conditions which included what would be the equivalent of an eave and an eave with a slot.

The same rectangular prism was then mounted on the floor of the tunnel as a 114 x 57 x 28·5 mm (high) prism and the pressure measurements on the same configurations as tested in the various flows in the centre of the tunnel were measured in the turbulent boundary layer, Flow E.

The spectral measurements of the longitudinal velocity component were made using a linearised hot wire system and both digital and band pass filter techniques were used to reduce the spectral values. The pressure measurements were made using Setra transducers mounted with minimum volume conditions, with less than 15 mm of 1 mm diameter tubing to ensure a flat frequency response up to 500 Hz.

3. MEASUREMENTS ON THE CIRCULAR DISK

A starting point to the exploration of the effects of turbulence scale on the behaviour of flow fields, shear layers and consequent surface pressures on bluff bodies appeared to lie in a repetition of Bearman's measurements of base pressure behind a circular disk. The results of the measurements of base pressure coefficient, C_{p_b}, behind the 114 mm diameter, sharp edged circular disk in Flows A to D are given in Table 1. In each case the mean pressure coefficient at the centre of the upstream face was measured to be within ±1% of unity, when based on the measurement of the velocity at the model position with a vertical hot wire relative to a reference pitot static tube measurement of dynamic pressure.

In Flow A, where the integral length scale is 0·14 m, and Bearman's parameter,

$$\frac{\sigma_u}{\bar{u}} \frac{L_x^2}{A} = 0{\cdot}045 ,$$

the value of the base pressure coefficient, $C_{p_b} = -0{\cdot}46$ compares well with that measured by Bearman (3). However, here the similarity ends, as in Flow B, where the length scale is halved and the turbulence intensity nearly quadrupled to give a value of $\frac{\sigma_u}{\bar{u}} \frac{L_x^2}{a} = 0{\cdot}032,$ similar to that in Flow A, the base pressure coefficient,

C_{p_b} = - 0·52. In Flows C and D where small scale turbulence from 12·7 mm diameter rods have been superimposed on Flow A a similar trend is shown. Whilst it can be shown that a 17% increase in negative base pressure can be obtained for very similar values of Bearman's parameter, it would only be reasonable at this stage to conclude that the change in base pressure is being influenced by variables other than specified by the grouping in this parameter.

Table 1. Base pressure coefficient measured behind a 115 mm diameter sharp edged circular disk in flows with various turbulence characteristics

Flow	Lx m	$\frac{\sigma_u}{\bar{u}}$	Bearman's Parameter $\frac{\sigma_u}{\bar{u}}\frac{L_x^2}{A}$	Small Scale Spectral Density Parameter $\left.\frac{n\,Su(n)}{\bar{u}^2}\right\rvert_{\frac{n\sqrt{A}}{\bar{u}}=10}$	C_{p_b}
Flow A	0·14	0·023	0·045	10×10^{-6}	-0·46
Flow B	0·063	0·086	0·032	340×10^{-6}	-0·52
Flow C	0·112	0·056	0·070	90×10^{-6}	-0·52
Flow D	0·073	0·073	0·039	200×10^{-6}	-0·54

The major clue seems to be in the observations of Gartshore (1) and later Kwok and Melbourne (4), that it is the small scale turbulence which is primarily responsible. Some insight to a possible mechanism has been given by Sadeh (7) who filmed filaments of small scale turbulent flow undergoing vortex stretching and vorticity amplification on impact with the surface of a cylinder; these vortices stretched to be axial with the local flow at scales similar to the boundary shear layer thickness. Such vortices could have the potential to enhance the entrainmnet of flow into the shear layer in much the same way as line vortices behind surface vortex generators. It seems possible that turbulent scales of the order of the shear layer thickness might be responsible for the behaviour observed by Bearman and Gartshore.

If one now looks at the amount of small scale energy available in the various flows it is immediately apparent from Figure 2 that Flows B, C and D have much more energy available than Flow A at length scales on order of magnitude less than the disk diameter, i.e. at $n/\bar{u} \simeq 100$ waves per metre. In these flows these scales are at the end of the inertial subrange, at about the region where the effects of viscous dissipation become apparent. The amount of small scale energy available in these experimental flows, including the effect of turbulence intensity, can be quantified by the previously introduced Small Scale Spectral Density Parameter, which are included in Table 1.

It is immediately evident from Table 1 that Flows B, C and D, which gave base pressure coefficients C_{p_b} = -0·52 to -0·54, have values of the Small Scale Spectral Density Parameter an order of magnitude greater than Flow A which gave the base pressure coefficient of C_{p_b} = -0·46. These measurements add some further evidence to support the conclusion that it is the increase of small scale turbulence in an incident air stream which causes the more negative base pressures on a flat plate via the mechanism hypothesised by Gartshore.

4. MEASUREMENTS ON A RECTANGULAR PRISM

To illustrate the effect of freestream turbulence on surface pressures near the leading edge under a re-attaching shear layer the very simple case of flow around a rectangular prism 114 x 57 x 57 mm has been used. In particular, two pressure tapping points have been chosen, as shown in Table 2. Pressure tapping 1 is centrally located a distance of 0·05h (or 0·1h) back from the leading edge, and pressure tapping 2 is locatated the same distance from the leading edge, but 0·2b from the end, in a region known to have particularly high negative mean and fluctuating pressures for a flow direction at 45° to the upsteam face.

Pressure measurements at these two points are given in Table 2 for Flows A and D. It can be seen that relative to Flow A the mean pressures change little, whilst the fluctuating pressures as given by the standard deviation pressure coefficient have doubled in the presence of the Flow D which has an order of magnitude greater small scale energy than Flow A .

TABLE 2. Pressure coefficients measured on the streamwise surface of a 114 x 57 x 57 mm rectangular prism with and without an eave in two turbulent flows

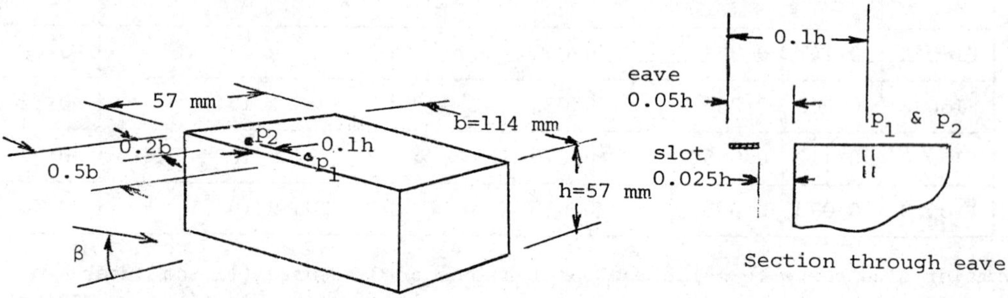

	L_x	$\frac{\sigma_u}{\bar{u}}$	$\left.\frac{n\,Su(n)}{\bar{u}^2}\right\vert_{\frac{n\sqrt{A}}{\bar{u}}=10}$	β	Pressure 1 $C_{\bar{p}_1}$	Pressure 1 $C_{\sigma_{p_1}}$	Pressure 2 $C_{\bar{p}_2}$	Pressure 2 $C_{\sigma_{p_2}}$
Basic Prism				0	-0.55	0.07	-0.56	0.09
Flow A	0.14	0.023	7×10^{-6}	45°	-1.23	0.08	-1.81	0.34
				0	-0.79	0.15	-0.87	0.21
Flow D	0.073	0.073	160×10^{-6}	45°	-1.31	0.16	-1.77	0.60
Prism with eave sealed				0	-0.57	0.08	-0.60	0.11
Flow A				45°	-1.30	0.13	-1.76	0.43
				0	-0.79	0.17	-0.88	0.24
Flow D				45°	-1.31	0.21	-1.50	0.56
Prism with slotted eave				0	-0.52	0.08	-0.52	0.09
Flow A				45°	-0.99	0.07	-1.85	0.12
				0	-0.69	0.15	-0.74	0.18
Flow D				45°	-1.05	0.13	-1.89	0.25

The effect of venting the bubble was investigated initially by adding an eave of 0·05h to the leading edge of the prism and moving the pressure tappings forward so that they are still 0·01h from the leading edge. The pressure measurements with the eave and eave with a 0·025h slot in Flows A and D are given in Table 2. The addition of the sealed eave has made little difference to the pressures as measured on the basic prism with the exception of the fluctuating pressures with the flow at 45° which have increased significantly. This effect could be related to the turbulence generated under the eave, part of which would feed into the separating shear layer and could have much the same effect as increased freestream turbulence. Opening the slot on the eave and in so doing venting the bubble, has had a small effect on the mean pressures, but has more than halved the maximum fluctuating pressures. A preliminary conclusion which could be drawn at this stage is to observe that the effect of venting the bubble under the re-attaching shear layer has had a significant effect in reducing the larger fluctuating pressures. Much more work would be required to determine how effective this reduction might be in more turbulent flows and for other bluff body shapes, venting configurations and location of maximum pressures.

5. MEASUREMENTS ON THE PRISM WITH AN EAVE IN A TURBULENT BOUNDARY LAYER FLOW

To extend the application of the hypothesis on the cause of the high negative pressures and a possible way of reducing these pressures to a flat roofed house, the model with the slotted eave tested in Flows A and D was tested in boundary layer Flow E, which had turbulence characteristics similar to Flow D as shown in Figure 2. The model was mounted with the horizontal plane of symmetry on the floor of the wind tunnel giving h = 28·5 mm. Mean and standard deviation pressure coefficients are given as a function of wind direction in Figure 3.

The peak pressure coefficients are determined from a probability distribution of each pressure record where the peak pressure is defined as the pressure with a cumulative probability of exceedance of 10^{-4}; in full scale terms this cumulative probability of exceedance is approximately 1/3 second per hour and normally there would be several upcrossings of this level per hour. In this case the probability distributions of pressures were near normally distributed both with and without the eave sealed, and so a peak factor of g = 4 was used to obtain the peak pressures through the formula,

$$C_{\check{p}} = C_{\bar{p}} (1 + g C_{\sigma_p}) \qquad (3)$$

The fact that the measured pressures were near normally distributed, and the peak negative pressures relatively small in magnitude for the eave sealed case, indicates that the freestream turbulence intensity was not high enough to produce the extreme peak pressures associated with flat roofs on low rise buildings.

Even though these tests do not illustrate very intermittent or extreme behaviour of the pressure fluctuations, the reductions in peak pressures occasioned by opening a slot in the eave are quite substantial; for the maximum peak pressure coefficients at 2 for β = 40° a reduction of 33%, from $C_{\check{p}}$ = 3.3 to 2.2, is shown.

6. CONCLUSIONS

Some experimental evidence has been presented to indicate that,

(i) increase of small scale turbulence in an incident air stream on a flat circular disk causes the mean base pressure to become more negative (there is no evidence that increased integral scale has the same effect);

(ii) increase of small scale turbulence in an incident air stream increases the magnitude of high negative pressures occurring near the leading edge under a re-attaching shear layer;

(iii) venting of the bubble under a re-attaching shear layer by using a slotted eave substantially reduces the magnitude of the maximum negative pressures in this region.

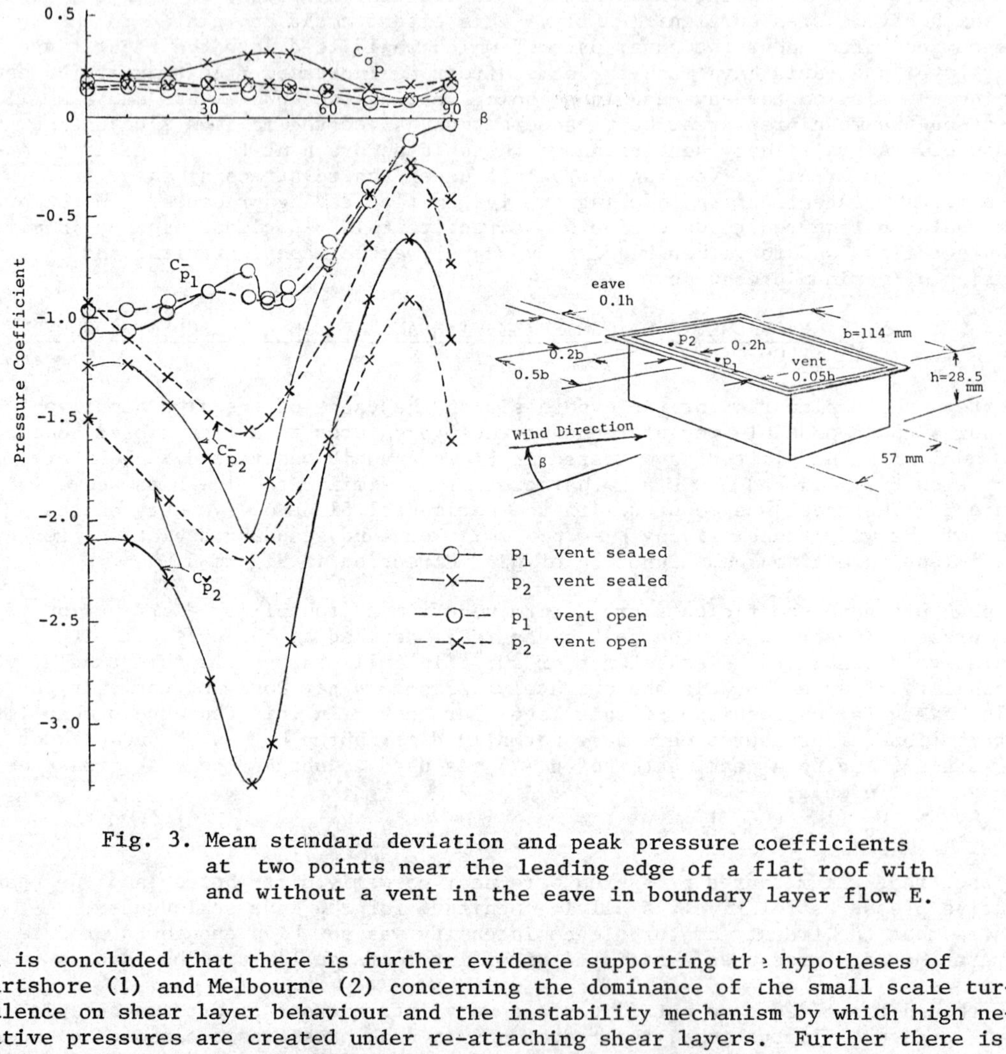

Fig. 3. Mean standard deviation and peak pressure coefficients at two points near the leading edge of a flat roof with and without a vent in the eave in boundary layer flow E.

It is concluded that there is further evidence supporting the hypotheses of Gartshore (1) and Melbourne (2) concerning the dominance of the small scale turbulence on shear layer behaviour and the instability mechanism by which high negative pressures are created under re-attaching shear layers. Further there is evidence that venting of the bubble under the re-attaching shear layer, thereby breaking down the instability mechanism, results in substantial reductions of the maximum negative pressures. The venting of leading edge regions, with devices such as a slotted eave, has practical applications in respect of reducing loads on low pitched roofs of houses and industrial buildings, and cantilevered grandstand roofs; there are also possible applications in reducing loads on glass panels in tall buildings.

A peripheral conclusion of these observations is to note that in model studies of loads on bluff bodies in turbulent flow it is important to scale the small scale turbulence correctly as well as the larger scale on which resonance phenomena depend. In particular, use of too small a scale model might cause premature reduction of important turbulence scales by the action of viscosity, and conversely the practice of using model scales larger than flow scales provides a situation where the model is exposed to relatively more small scale turbulence which would cause over prediction of maximum negative pressures under re-attaching shear layers.

Lee (8) in a study of turbulence effects on two dimensional square prisms showed large variations in base pressure (30%) and drag for the same value of turbulence intensity, and he noted, "there would appear to be some mechanism in addition to that associated with incident turbulence intensity which is capable of affecting the mean forces on a bluff body". It can be observed in Lee's results that the base pressures associated with the small bar grid are much less negative than those associated with the large bar grid for the same turbulence intensity. Whilst there are no spectra available from which to determine the Small Scale Spectral Density Parameter, it can be inferred that for the same turbulence intensity there would be relatively more small scale energy available behind the small bar grid, and hence the observed differences in base pressure are consistent with the hypothesis that it is the small scale turbulence which is primarily responsible for increasing the rate of entrainment into the shear layer, which is followed by earlier re-attachment and for the square prism less negative base pressures.

ACKNOWLEDGEMENTS

The work of Mr. D. Sharp in studying the correlation between peak pressures and shear layer re-attachment position and venting, currently being written up in a Master's Thesis, is gratefully acknowledged. The continued support of the Australian Research Grants Commission for this work generally is also acknowledged.

REFERENCES

(1) Gartshore, I.S. The effects of freestream turbulence on the drag of rectangular two dimensional prisms. Boundary Layer Wind Tunnel Report-4-73, University of Western Ontario, Canada (1973).

(2) Melbourne, W.H. The relevance of codification to design. Proc. 4th Int.Conf. on Wind Effects on Buildings and Structures, London, Univ. of Cambridge Press, pp 785-790, (1975).

(3) Bearman, P.W. An investigation of the forces on flat plates normal to a turbulent flow. J. Fluid Mech. vol. 46, part 1, pp 177-198, (1971).

(4) Kwok, K.C.S. & Melbourne, W.H. The effects of freestream turbulence on a galloping square tower. Proc. 6th Aust. Conf. on Hydraulics & Fluid Mechanics, Inst. of Engineers Aust. Conf. Publ. 77/12, pp 447-450, (1977).

(5) Sharp, D.B. Master's Thesis in preparation, (1979).

(6) McKeon, R.J. & Melbourne, W.H. Wind tunnel blockage effects and drag on bluff bodies in a rough wall boundary layer. Proc. 3rd Int. Conf. on Wind Effects on Buildings and Structures, Tokyo, Publ. Saikan, (1971).

(7) Sadeh, W.Z., Brauer, H.J. & Garrison, J.A. Visualization study of vorticity amplification in stagnant flow. Project Squid Technical Report CSU-1-PU, Colorado State University, Fort Collins, (1977).

(8) Lee, B.E. Some effects of turbulence scale on the mean forces on a bluff body. J. of Industrial Aerodynamics, vol.1, No. 4, pp 361-370, (1976).

COMPARISONS OF WIND TUNNEL AND FULL-SCALE BUILDING SURFACE PRESSURES WITH EMPHASIS ON PEAKS

W. A. DALGLIESH, J. T. TEMPLIN and
K. R. COOPER
National Research Council of Canada, Ottawa, Ontario, Canada

SUMMARY

Full-scale pressure coefficients obtained from a 57-storey building in Toronto and wind tunnel results from tests in the 9 m by 9 m wind tunnel at the National Research Council of Canada are compared and demonstrate good agreement where sufficient full-scale data exist. A method of treating peak pressures is proposed based on the fit of an exponential distribution to a population of "significant independent events," called pressure spikes. This distribution provides a good fit to both full-scale and wind tunnel results, which generally agree.

INTRODUCTION

An important application of wind tunnel testing is the provision of the surface pressure loads that are required for the design of cladding of tall buildings. There is a growing confidence that careful simulation of the flow-field and nearby structures will produce information that accurately reflects the full-scale situation. Recent comparisons of model and full-scale pressure measurements have shown that this can be said about the mean value and the standard deviation of surface pressure fluctuations(1, 2, 3), but not with equal confidence about the peak pressures because it is difficult to find a representative value(4) and because unusually high peaks, occasionally found in full scale(5), might not be reproduced or detected at model scale.

This paper explores two subjects:

(1) The difficulties in obtaining full-scale pressure measurements for the comparison with wind tunnel findings are discussed based on recent full-scale observations and the results of a new wind tunnel test. The full-scale measurements were made on the Commerce Court building in downtown Toronto; the model experiments were performed on a 1:200 scale model mounted in the 9 m by 9 m wind tunnel of the National Aeronautical Establishment in Ottawa. The discussion emphasizes the variability in full-scale pressure measurements that does not arise in the wind tunnel.

(2) An approach to the study of peak pressures is introduced based on fitting an exponential distribution to a population of "significant independent events" called spikes. A discussion of time scaling is presented which enables comparison of model and full-scale observations.

Figure 1 Commerce Court Tower (arrow) in downtown Toronto viewed from $\alpha=330°$

Figure 2 Commerce Court Model in NRC 9 m × 9 m Wind Tunnel viewed from $\alpha=10°$

MODEL/FULL-SCALE BUILDING GEOMETRY

The 57-storey (239 m) Commerce Court Tower is situated in the core of Toronto and is surrounded on three sides by buildings of comparable height (Figure 1). The nearby tall buildings cover an area of 1 km^2; a strip of tall buildings, ½ km wide, extends several kilometres to the north. Surrounding these areas are several kilometres of relatively low buildings. Lake Ontario provides a smooth fetch beyond one kilometre to the south.

A 1:200 scale model of the Tower and its environs, including all adjacent tall buildings, was constructed (Figure 2). A power law profile of 0.33 was established outside the central core for approaches over the city from the west, north and east while a profile exponent of 0.15 was used for the lake exposure to the south. A third profile was provided for winds coming from near north by adding larger blocks to the urban roughness to simulate the strip of tall buildings. Each boundary layer was developed using spires at the inlet to the test section followed by roughness blocks. The detailed proximity model was mounted 18.3 m downstream of the spires on a 7.3 m diameter turntable.

MODEL-SCALE PRESSURE MEASUREMENTS

The mean and the root-mean-square-about-the-mean (rmsm) of the surface pressure fluctuations of the model as well as the associated minima and maxima were measured through a 360° wind direction range using increments of no greater than 10°. The Commerce Court pressure model had 96 surface pressure taps on four levels, 32 of which corresponded to locations on the full-scale building. Measurements were made with a calibrated pressure-tube-scanivalve-transducer system developed at the National Aeronautical Establishment(6), with a flat frequency response to 135 Hz. Each tap was sequentially sampled at 200 samples per second (200 s^{-1}) for 20 s.

The mean pressure coefficient is defined as

$$\overline{C_p} = \overline{(p - p_{ref})}/q = \overline{p}/q \qquad (1)$$

where the bar indicates an average over time, p is the measured pressure, p_{ref} is the reference static pressure and $q = \frac{1}{2}\rho\overline{v}^2$ is the reference dynamic pressure at the height of the anemometer location. (ρ is the density of air corrected for atmospheric temperature and pressure and $\overline{v}$ is the mean speed from the anemometer.) The rmsm surface pressure coefficient is

$$C'_p = \left[\overline{(p - \overline{p})^2}\right]^{\frac{1}{2}}/q = p'/q \tag{2}$$

Time-series digital data from selected taps were recorded for several wind directions, at 500 s^{-1} for 200 s. These were used for the detailed studies of peak pressures.

FULL-SCALE PRESSURE MEASUREMENTS

The pressures were measured at 32 locations around the exterior of the building on four levels (Figure 3). Measurements were made relative to a common internal reference pressure. An anemometer and wind vane mounted on a mast on the roof 285 m above ground were used for the reference wind velocity measurements. The mean and rmsm of the full-scale pressure fluctuations, wind speeds and directions have been recorded for 5-min periods every hour for 6 years. Only the most recent 2 years of data (February 1976 to June 1978) are analyzed in this paper because new buildings were being erected in the vicinity prior to this period. Beginning in April 1977, maximum and minimum pressures were also recorded for each 5-min sampling interval. The signals were filtered with a 2-pole, 15 Hz low-pass filter and then sampled digitally at 20 s^{-1}. In addition to these summary data, time-series data sampled at 20 s^{-1} were recorded for longer periods during high winds.

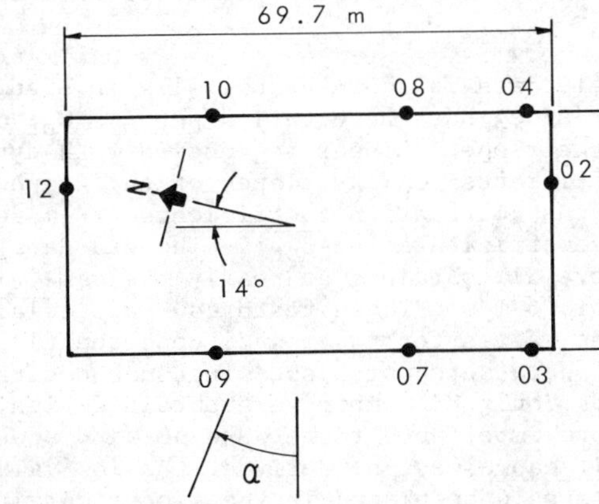

LEVEL	ELEV. m
ANEMOMETER	285
ROOF	239
3	202
4	166
5	138
7	102

e.g., TAP 304
3 - LEVEL
04 - TAP NUMBER

Figure 3 Location of pressure taps common to model and full-scale building

COMPARISON OF MEAN AND RMSM PRESSURES

Because full-scale pressures are affected not only by wind speed and direction, but also by the internal pressure due to building permeability, ventilation and temperature effects, a multiple linear regression must be performed to obtain mean pressure coefficients. The measured pressures are assumed to be described by

$$\overline{p} = \overline{p_o - p_i} = A + \overline{C_{pt}}\, q + \overline{C_T}\, p_b \left(\frac{1}{T_o} - \frac{1}{T_i}\right) \qquad (3)$$

where p_o and p_i are outside and inside pressures respectively, p_b is barometric pressure, T_o and T_i are outside and inside absolute temperatures respectively and A, $\overline{C_{pt}}$ and $\overline{C_T}$ are coefficients determined by the least squares fit. The coefficient A represents the fixed pressure difference across the wall resulting from pressurization of the building with no wind and no temperature difference. The last term in the equation represents the pressure across the wall due to a temperature difference between inside and outside (stack effect). The wind pressure coefficient ($\overline{C_{pt}}$) represents the pressure difference that changes with wind velocity due to aerodynamic effects and can be represented by

$$\overline{C_{pt}} = \overline{C_p} - \overline{C_{pi}} \qquad (4)$$

where $\overline{C_p}$ is the external pressure coefficient and $\overline{C_{pi}}$ is the mean internal pressure coefficient representing wind-induced variations in inside pressure due to building permeability.

A multiple linear regression performed on the mean pressure data reveals that the pressure is highly correlated with the stack effect so $\overline{C_T}$ has a large partial correlation coefficient ($r_{pT/q}$) typically greater than 0.85. On the other hand, the partial correlation coefficient ($r_{pq/T}$) of pressure with respect to dynamic reference pressure, q, ranges from 0.85 to 0.15 depending on wind direction. Large values of $r_{pq/T}$ are associated with relatively well-defined values of $\overline{C_{pt}}$; small values of $r_{pq/T}$ should be regarded as a warning that $\overline{C_{pt}}$ derived from that particular data set is unreliable.

The process of deriving $\overline{C_{pt}}$ from full-scale data is shown graphically in Figure 4. The line on each mean pressure plot (left side) has the equation $\overline{p} = A + \overline{C_{pt}}\, q$ so that A is the y-intercept and $\overline{C_{pt}}$ is the slope. Linear regression with dynamic pressure only provides rmsm pressure coefficients (C_p') as slopes of the graphs on the right side of Figure 4. In general, the correlation coefficients (r_{pq}) for C_p' are greater than the partial correlation coefficients for $\overline{C_{pt}}$. The 225 data points plotted in Figures 4(a) and (b) were all obtained at nearly the same wind direction and extend to a maximum speed of 26 m/s. These features, particularly the wide wind-speed range, result in large values for $r_{pq/T}$ and, consequently, reliable estimates for $\overline{C_{pt}}$. Examples of data sets not possessing such desirable features are given in Figures 4(c) and (d); only 83 points were obtained within a 10° sector (253° to 263°) and, what is more important, there were no wind speeds over 12 m/s. Although a least-squares fit can always be derived, the low values of $r_{pq/T}$ indicate that little confidence should be placed in the slopes obtained. Another problem associated with low wind speeds in full-scale measurements is that thermal effects are more likely to produce a different flow than the neutral boundary layer simulated in the wind tunnel.

The difficulty in generating full-scale pressure coefficients to compare with wind tunnel results is illustrated by the scatter in Figure 4, for which there is no wind tunnel equivalent. In the wind tunnel, a mean pressure coefficient can be obtained at a fixed value of dynamic pressure and wind direction, with a repeatability of ±0.04 for $\overline{C_p}$ and ±0.01 for C_p' , even when the tests are performed one year apart. Much of this variation is due to minor differences in the model and boundary layer setup. If two runs of 20 s are made one after the other, the repeatability improves to ±0.01 and 0.005 for $\overline{C_p}$ and C_p' respectively.

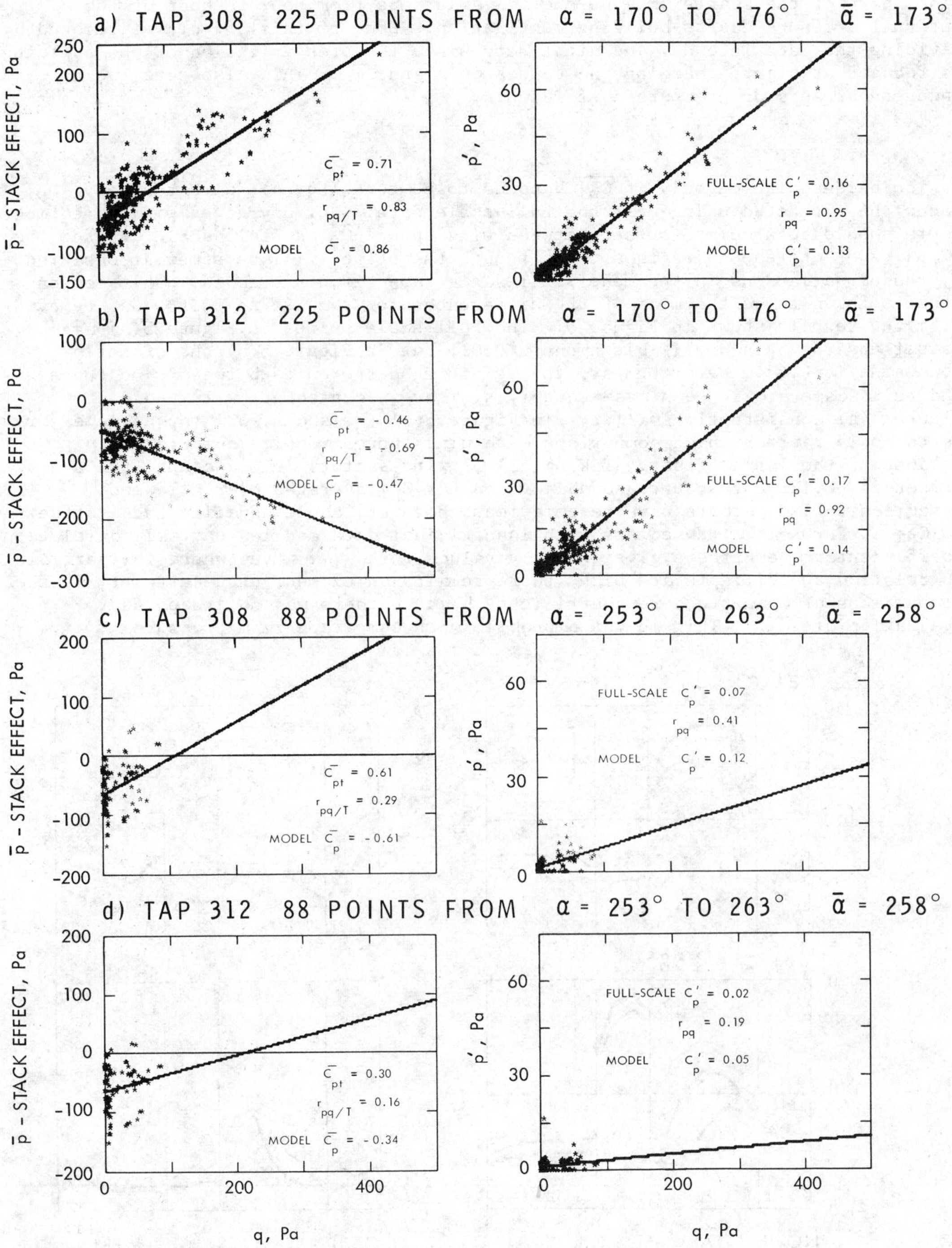

Figure 4 Measured full-scale surface pressures vs dynamic reference pressure

Since the pressures measured in full scale are differences between inside and outside, part of the scatter may be due to variations from hour to hour of the internal pressure due to building ventilation. Some of this effect was removed by dividing the data into day and night sets since the pressurization of the building is reduced at night, changing the values of A and $\overline{C_T}$. This also gave two independent sets of pressure coefficients.

$\overline{C_{pt}}$ coincides with $\overline{C_p}$ only if $\overline{C_{pi}}$ happens to be zero (Eq. 4); unfortunately, $\overline{C_{pi}}$ cannot be extracted using only the full-scale data. For any given wind direction, there should be a common offset between wind tunnel $\overline{C_p}$ and full-scale $\overline{C_{pt}}$ for the 32 pairs of corresponding taps, even though the offset will be somewhat obscured by random variations in the model/full-scale tap-pair differences. This common offset, taken to be the mean of the differences for the 32 tap-pairs, was removed from the results shown in Figure 5. The full-scale points in Figure 5(a) are offset upwards by a negligible amount, 0.01, but in Figure 5(b) the offset downwards is 0.67. Nevertheless, the remaining differences between wind tunnel and full scale are in each case roughly similar in character, with those of Figure 5(a) consistently smaller. In Figure 5(b) the south wall tap on level seven is the only tap for which comparison with wind tunnel results suffers badly because of the large offset. Taking all 51 wind sectors into account, the average offset is -0.12, not an unreasonable value for $\overline{C_{pi}}$ averaged over all wind directions. The authors have reservations, however, about equating these offsets to $\overline{C_{pi}}$. For some wind sectors with insufficient data and low partial correlation coefficients, the offsets give unlikely values for $\overline{C_{pi}}$. Even where the partial correlation coefficients are high, there remains an element of uncertainty in the comparisons of mean pressure coefficients because there was no independent evaluation of $\overline{C_{pi}}$. This is not a problem when comparing rmsm pressure

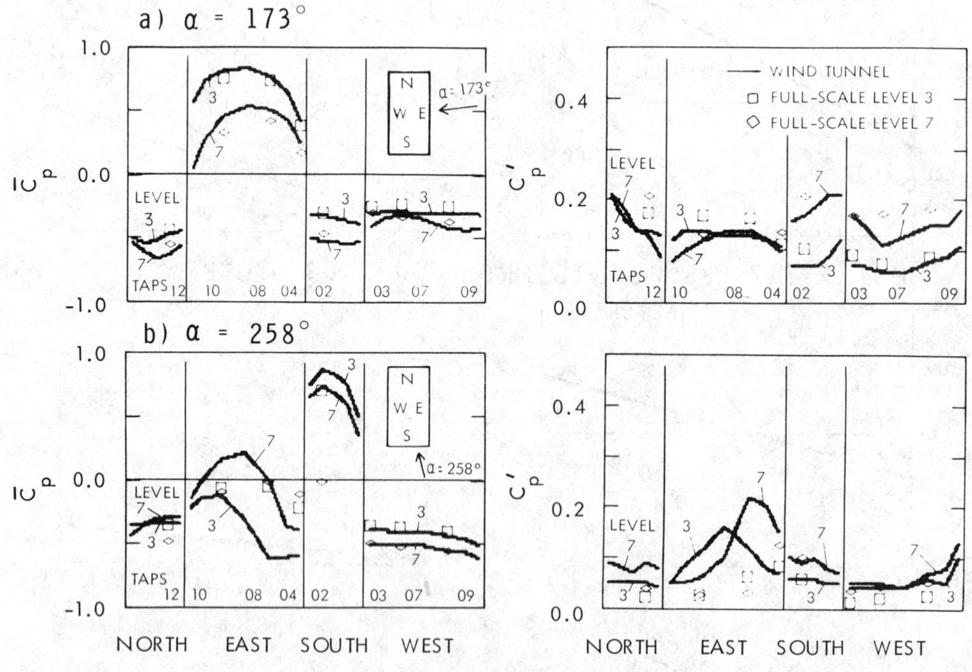

Figure 5 Level plots comparing model and full-scale pressures for two levels at 2 wind directions

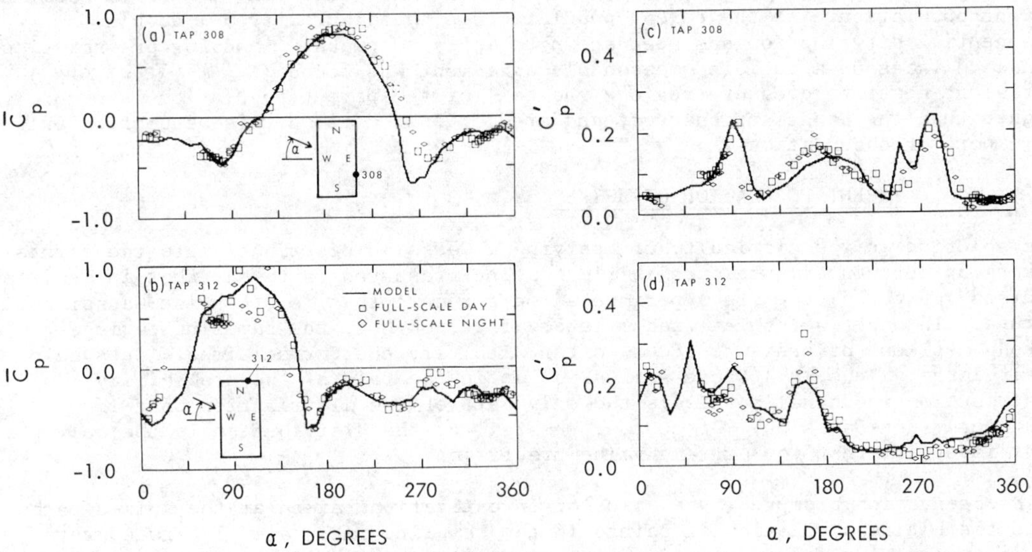

Figure 6 Pressure plots comparing model and full-scale results for two taps at all wind directions

coefficients because they are calculated on the same basis in wind tunnel and full-scale . Figure 6 shows general agreement between wind tunnel and full-scale pressure coefficients for two typical taps at all wind directions. The full-scale mean pressure coefficients (left) have the offset removed but the rmsm pressure coefficients (right) are directly from the linear regression.

PEAK PRESSURES

The peak pressure coefficient required for cladding design can be related to the mean and rmsm pressure coefficients by introducing a design peak factor, g_d:

$$C_{\hat{p}} = \overline{C_p} + g_d C'_p \tag{5}$$

The symbol ^ denotes a positive extreme value; to simplify the discussion all peak factors will be treated as if positive. The assignment of a value for g_d is not as straightforward as for $\overline{C_p}$ and C'_p because the peak factor must account for the randomness inherent in pressure fluctuations. Although, for a given flow situation and building geometry, $\overline{C_p}$ and C'_p are nearly constant from one sampling interval to another, the maximum pressure observed will vary in a random fashion. A reasonable way of choosing a representative value for g_d might be to pick one of the following: the value that occurs most frequently (the mode), the average value (the mean), or the value with an acceptably small probability of being exceeded. Whatever the choice, information is required about the extreme value distribution of peak factors.

Peak factor parameters do not apply to the collection of values making up the sampling interval. Each sampling interval provides only one maximum; the extreme value distribution parameters apply to a collection of maxima derived from many such intervals. This does not mean that the statistics of the parent population of points in the sampling interval are unrelated to those of the extremes; on the contrary, an expression for the extreme value distribution can be derived from the parent population. If the parent population is Gaussian, a type I double exponential distribution has been shown to apply to peaks[8] and for pressures on windward faces of a building reasonable agreement was found[4, 7]. In strongly turbulent negative pressure regions the tail of the parent population is much higher than the Gaussian, however, and predictions based on this assumption fall far short of observations.

PARENT POPULATION OF SPIKES

One solution to the difficulty of applying a Gaussian distribution to the points making up certain time records of fluctuating pressures is to say that intermittent "bursts" of turbulence are superimposed on a record that is otherwise Gaussian in nature. In other words, at random intervals throughout the sampling period, strong "spikes" of pressure occur, often with very short rise times. Because of their large values, all these spikes end up in the tail of the probability distribution and usually this is the only part of the distribution that is distinctly non-Gaussian. Of course, the tail of the distribution is the only part that is of interest for extreme value prediction.

This observation prompted a new line of investigation, aimed at the spikes rather than the entire population of points in the sampling interval. Two rules were devised, one to restrict consideration to events that would be significant from the standpoint of locating the extreme, and the second to ensure that the events were independent of one another. Each event is described by a single number called the spike value:

1) A spike value must be away from the mean by a threshold value at least twice the rmsm. Both mean and rmsm are measured over the entire sampling interval.

2) A spike must be separated from adjacent spikes by returns toward the mean that extend at least the threshold value from the spike value.

Figure 7 shows a short segment of record containing three spikes. The readings have been normalized by subtracting the mean value, then dividing by the rmsm value. Note that the second rule makes smaller "jiggles" in the record ineligible as spikes, even though they may be beyond the threshold level.

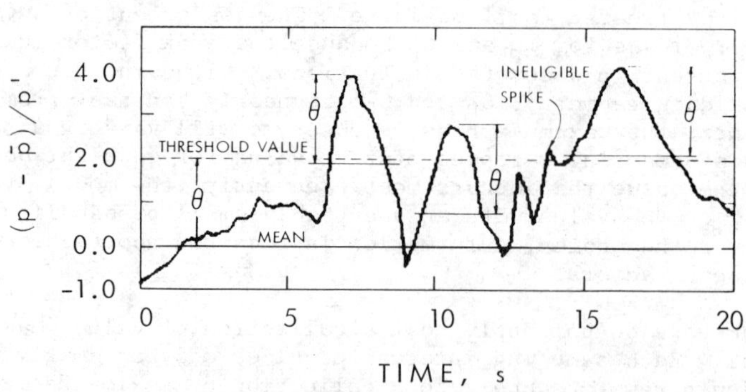

Figure 7 Typical time history with three spikes

The simplest distribution to apply to this new parent population, that of the spikes, is the negative exponential[9]. The cumulative distribution is:

$$F(g) = 1 - \exp\left[-(g - \theta)/(\overline{g} - \theta)\right] \qquad (6)$$

The spike value g is normalized as $|p - \overline{p}|/p'$. θ is the threshold value, chosen to be 2, mentioned in the first rule, and $\overline{g}$ is the mean of the spike values in the interval. The extreme value distribution of peaks, $\hat{g}$, where each peak in the distribution is the maximum spike value from a sampling interval, is closely related to the above expression and requires only one more parameter:

$$F_n(\hat{g}) = \left\{1 - \exp\left[-(\hat{g} - \theta)/(\overline{g} - \theta)\right]\right\}^n \qquad (7)$$

The new parameter n is the average number of spikes per interval. As one might expect, the longer the sampling interval, the larger the maximum spike value (peak). The value of the most probable peak (the mode $\tilde{\hat{g}}$) increases linearly with the natural logarithm of n as follows:

$$\tilde{\hat{g}} = (\overline{g} - \theta) \log_e n + \theta \qquad (8)$$

For a proper comparison of peak pressures, the model sampling period must be the properly scaled equivalent of the full-scale period. The time scaling parameter is $\tau = t\,\overline{v}/\ell$, where t and ℓ are a characteristic time and length. If t is chosen to be the sampling period (t_s) and ℓ is thought of as an average "spacing between spikes" (ℓ_s), then the average number of spikes per interval would be given by:

$$n = t_s\,\overline{v}/\ell_s \qquad (9)$$

INVESTIGATION OF PRESSURE SPIKES AND PEAK FACTORS

Nine selected wind tunnel taps were sampled at 500 s^{-1} at a reference speed of $\overline{v}$ = 15 m/s to provide one hundred 2 s sampling intervals. Four histograms of pressure spikes were developed from groupings of two or three taps at a time using the two rules previously defined. The grouping was necessary to provide an acceptably large sample of peak factors.

One of the full-scale histograms was obtained by sampling eight pressure taps on the windward face of the building at 20 s^{-1} for 160 min when the wind had a direction of 350° and a mean speed of 23 m/s. This provided 256 5-min sampling intervals from which n and $\overline{g}$ could be established. The other three full-scale distributions were obtained from the peak values recorded for the hourly observations of 5-min periods sampled at 20 s^{-1} when the mean speed was, on average, 17 m/s. Histograms of peaks were fitted by extreme value distributions in order to obtain n and $\overline{g}$ for comparison with model scale. This method of "back-calculating" to determine n and $\overline{g}$ gave reasonable agreement when compared with directly measured values of n and $\overline{g}$.

Table 1 summarizes the exponential parameters found for both model and full scale. The histograms and the corresponding exponential curves are shown in Figures 8(a) to 8(d). Except for the three full-scale cases just mentioned, the exponential curves are defined by the observed values of n and $\overline{g}$ obtained from the time series

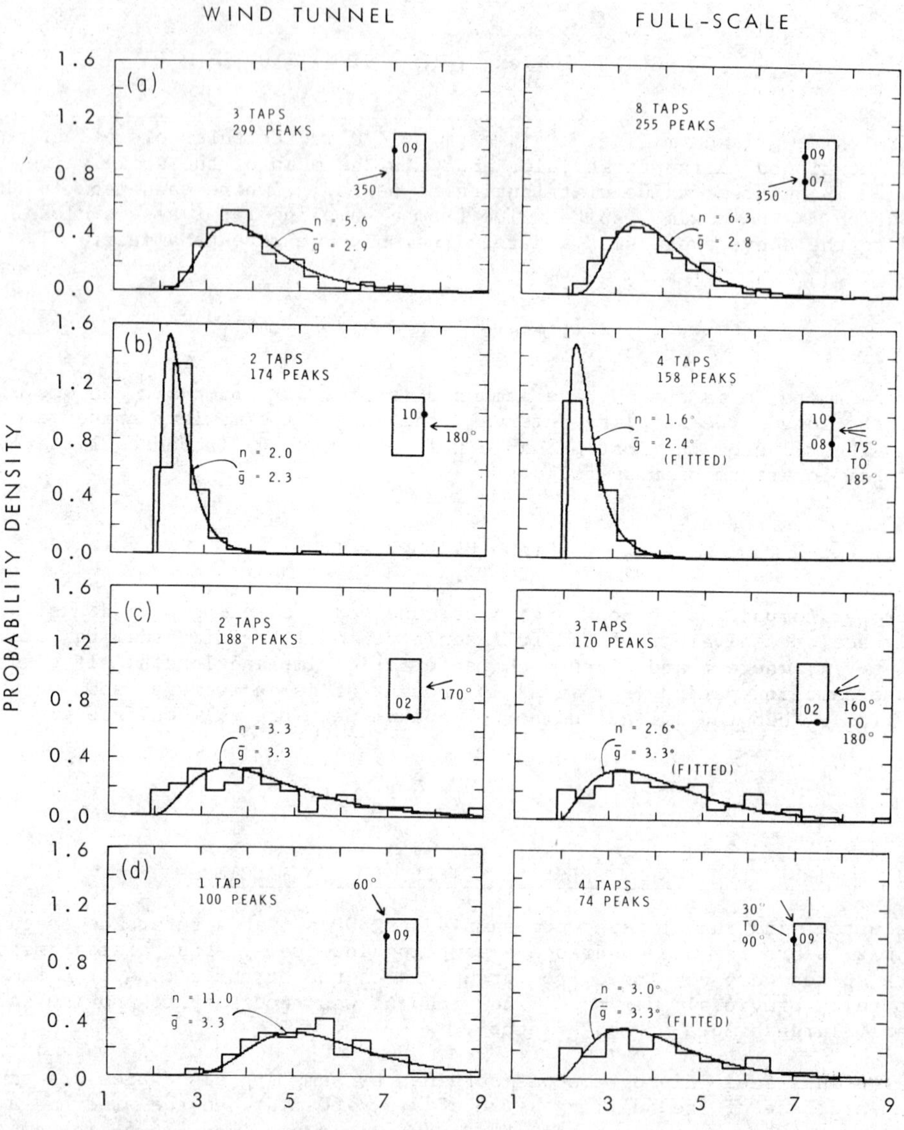

Figure 8 Peak pressure histograms and exponential curves

records defining the parent population of pressure spikes, and are not "fits" to the histograms. The fact that agreement is good even though most of the curves are derived from the parent population and not directly from the histograms of peaks indicates the usefulness of the concept of a parent population of spikes. Further, comparison of the model and full-scale distribution shows them to be in reasonable agreement, except for Figure 8(d) for which the model pressures were measured at a fixed wind direction of 60°, in the wake of a building 285 m high. To obtain sufficient data points at full scale for a histogram, data from a band between 30 and 90° had to be used, and even then only 74 points were found. Because of the localized variations of flow in a wake it is anticipated that the

Table 1. Properties of pressure spikes and peaks from wind tunnel and full scale

Item	Sign of Peak	F.S. Speed m/s	Parent Population (spikes)				Extreme Values (peaks)					
			n		$\bar{g}$		$\tilde{g}$		$\bar{\bar{g}}$		$\hat{g}'$	
			W.T.	F.S.	W.T.	F.S.	W.T.	F.S.	W.T.	F.S.	W.T.	F.S.
a	pos	23	5.6(6.4)	6.3	2.9	2.8	3.5	3.4	3.9	3.8	0.9	1.0
a'	pos	23		4.8		2.7		3.2		3.5		0.9
b	pos	17	2.0(1.7)	1.6*	2.3	2.4*	2.2	2.2	2.5	2.5	0.4	0.4
c	neg	17	3.3(2.8)	2.6*	3.3	3.3*	3.6	3.3	4.5	4.2	1.9	1.5
d	neg	17	11.0(9.4)	3.0*	3.3	3.3*	5.1	3.4	5.4	4.3	1.1	1.5

* no parent distribution available: n and $\bar{g}$ obtained from extreme value distribution fitted to match the mean and standard deviation, $\bar{\bar{g}}$ and $\hat{g}'$

( ) parentheses indicate wind tunnel values of n adjusted to represent a model sampling period equivalent to 5 min in full scale

unusual data at 60° in the wind tunnel would be masked in full scale by data from the rest of the sector.

Time scaling must be taken into account in a proper comparison of model and full-scale results. According to Eq. (9), values of n obtained from the 2 s records in the wind tunnel at a reference speed of 15 m/s correspond to those from the 5-min records in full scale only if the full-scale reference speed is 20 m/s, assuming the scale of mean "spike spacing" ℓ_s is the same as the geometric scale. Values of n adjusted to correspond to the full-scale 5-min record are shown in parentheses beside the measured wind tunnel values in Table 1.

Further investigation of the effects of time scaling indicates that sampling at 500 s^{-1} in the wind tunnel is equivalent to sampling at 3.8 s^{-1} for Item a (Table 1) and 2.8 s^{-1} for Items b to d. One might expect lower values at these lower model sampling rates compared to 20 s^{-1} which was used in full scale, but in fact the wind tunnel results are equal to or greater than the full scale in all but one case. The continuous record from full scale used for Item a was digitally filtered to simulate a single pole filter at 1.5 Hz and sampled at 4 s^{-1}. The properties of the filtered record, Item a′, range from 4 to 10 percent below those for Item a except for n which dropped from 6.3 to 4.8. The filter characteristics of the model sampling system were not well represented by a single pole filter, however, and the half-power point was closer to 5 Hz (full scale) than to 1.5 Hz. In any case, filter characteristics and sampling rates of the measuring system evidently influence the magnitude of the pressure spikes recorded, and even larger effects may be seen in their shapes.

SLOPES OF PRESSURE SPIKES

Spikes come in a variety of shapes. Although on a compressed time scale they may look like vertical lines, on an expanded scale like that in Figure 7 they can often be roughly described as a triangular pulse with a base width (extended back to the mean level) varying from 1 to 10 s. The slopes of 29 of the largest spikes of the full-scale records for Item a of Table 1 were found to average 7.3 C_p'/s with a coefficient of variation (c.o.v. = standard deviation ÷ mean value) of 0.7. On the other hand, 56 of the largest spikes from the wind tunnel study gave an average of 3.0 C_p'/s (c.o.v. = 0.6). The large discrepancy may be due to the differences in high frequency filtering and sampling rates. When the filtered full-scale record sampled at 4 s^{-1} was examined, the rate was found to be 2.7 C_p'/s (c.o.v. = 0.7). The slopes discussed so far have been for positive peaks; the average slope for 98 negative slopes from 19 tap locations sampled at 20 s^{-1} either in the wake or in glancing side-face flow was nearly double (14.2 C_p'/s with c.o.v. = 0.9). The full-scale slopes refer to a mean reference speed of 23 m/s, and would presumably be even steeper for higher speeds.

CONCLUDING REMARKS

The wind tunnel and full-scale mean and rmsm pressure coefficients agree well where there are sufficient full-scale data at high wind speeds. Wind speeds of at least 20 m/s are required to obtain reliable estimates of pressure coefficients for a particular wind direction at full scale. Otherwise, the full-scale data have too much random variation to be of use. Part of the pressure variation could be eliminated by obtaining a reference pressure independent of internal building pressure, as would be provided by a static pressure tap away from the influence of the building.

A scheme has been proposed for identifying significant independent peak events, pressure spikes, as they occur at random intervals in a time series, the statistical properties of which can be adequately described by an exponential distribution. Two parameters, the number of spikes in the interval and the average spike value, describe the extreme value distribution of peaks at a particular pressure tap or group of taps, allowing the peak at any desired probability level to be predicted for any storm duration. Positive pressure regions, for which positive peaks are of interest, typically have narrow probability density functions centred at peak factors lower than regions of positive pressure that are influenced by wakes. Negative pressure regions, particularly on side walls near upstream corners, exhibit broader distributions extending to larger peak factors than for positive pressures.

Considering the differences in the model and full-scale measuring systems, the agreement found between peak distribution parameters is encouraging. Additional experiments are planned in which greater attention will be paid to a proper matching of sampling rates and filter characteristics of wind tunnel and full-scale measuring systems. Changing the filter and sample rate for one full-scale record gave higher peaks for the larger rate, a trend that may continue even beyond 20 s^{-1}. In view of the difficulty and expense of raising sampling capabilities, it is worth examining the response of the cladding to peaks.

Annealed float glass office windows provide a particularly interesting example because their capacity to withstand peak pressures is thought to increase approximately as the 1/13th power of the peak slope[10]. This would mean that the 9 percent increase in peak magnitude ($\bar{g}$) at the higher sampling rate is almost matched by an 8 percent increase in capacity resulting from the increase in peak slope from 2.7 C_p' s^{-1} to 7.3 C_p' s^{-1}. This example, although over-simplified and based on insufficient data, illustrates some of the considerations in assessing the adequacy of design data. Each cladding application may have somewhat different requirements, depending on material properties, size, and natural frequency. The need for more information on the response of cladding as well as the characteristics of the loading is apparent. Nevertheless, the authors suggest that wind tunnel procedures of the sort described can provide adequate data for cladding design for high-rise buildings. Unless the effects of scale (sampling rate, sample length, reference speed) are accounted for, however, systematic peak prediction errors will result, and the practice of observing only a single peak to determine the design pressure coefficient will occasionally yield unrepresentative values. The authors recommend that peaks be selected from at least 10 subsets of an extended sampling interval, to permit "back-calculation" of n and $\overline{g}$ and a more soundly based prediction of peak pressure coefficients.

ACKNOWLEDGEMENTS

The authors wish to express their sincere appreciation to the following organizations and individuals for their contributions to this paper: Canadian Imperial Bank of Commerce, owners of the building; members of the Low Speed Aerodynamics Section, National Aeronautical Establishment; and colleagues, particularly P.J. Daly and F.W.K. Hummel, of the Division of Building Research. This paper is a joint contribution of two Divisions of the National Research Council of Canada: the National Aeronautical Establishment and the Division of Building Research. It is published with the approval of the Directors of these Divisions.

REFERENCES:

1. K.J. Eaton and J.R. Mayne, The measurement of wind pressures on two-storey houses at Aylesbury, J. Ind. Aerodynamics, 1, 67 (1975).

2. R.D. Marshall, A study of wind pressures on a single-family dwelling in model and full-scale, J. Ind. Aerodynamics, 1, 177 (1975).

3. W.A. Dalgliesh, Comparison of model/full-scale wind pressures on a high-rise building, J. Ind. Aerodynamics, 1, 55 (1975).

4. J.A. Peterka and J.E. Cermak, Wind pressures on buildings - probability densities, Proc. ASCE, Vol. 101, No. ST6, p. 1255-1267 (1975).

5. K.J. Eaton, Cladding and the wind, Proc. ASCE, Vol. 102, No. ST5, p. 1043-1053 (1976).

6. H.P.A.H. Irwin and K.R. Cooper, Correction of distortion effects caused by tubing systems in measurement of fluctuating pressures, National Research Council of Canada, LTR-LA-222 (1979).

7. W.A. Dalgliesh, Statistical treatment of peak gusts on cladding, Proc. ASCE, Vol. 97, No. ST9, September 1971, p. 2173-2187 (1971).

8. A.G. Davenport, Note on the distribution of the largest value of a random function with application to gust loading, Proc. Inst. Civ. Eng. 28, 187 (1964).

9. E.J. Gumbel, Statistics of extremes, Columbia University Press, New York, 1958, p. 113-118.

10. W.G. Brown, A load duration theory for glass design, National Research Council of Canada, Division of Building Research, Res. Paper 508 (1972).

FULL-SCALE MEASUREMENT OF WIND PRESSURE FLUCTUATIONS ON A BLUFF BUILDING

L. C. H. LAM, R. P. LAM and S. MACKEY
University of Hong Kong, Hong Kong

SUMMARY

This paper presents some properties of the fluctuating component of wind pressure on a full-scale multistorey building, with special emphasis on the 'air cushion' phenomenon, which has been scarcely described by wind-tunnel researchers. The aim is to stimulate more discussions and contributions in this area, which is likely to have significant influence on the dynamic design of certain types of structures.

INTRODUCTION

The current approach to design of tall buildings and structures which are susceptible to wind induced vibrations is to regard the wind forces as random loadings. This procedure consists of a series of calculations involving two groups of parameters: (i) the wind characteristics for the site as a function of height and of the type of surrounding terrain, and (ii) structural properties such as dimensions, natural modes and frequencies of vibration, and damping. The first group includes basic wind data such as design speed, exposure factor, velocity profile and intensity of turbulence. These parameters are sufficient to define the general wind climate at the site, but are inadequate to describe the fluctuating wind pressures acting on the structure. No satisfactory theory has yet been evolved to evaluate analytically the magnitude and distribution of fluctuating wind forces. Assessment of these is still largely dependent upon experimental data supplied by wind-tunnel model studies and full-scale measurements. Wind-tunnel model studies which include simulation of atmospheric turbulence characteristics and reproduction of adjacent topographic features around the structure under consideration have the advantage of providing direct insight into the particular problem. However, owing to incomplete dynamic similitude, scale effects cannot be avoided. On the other hand, although application of the results of full-scale measurements should be restricted to similar structures situated in a similar environment, full-scale experimental data are invaluable in providing a yardstick for wind-tunnel model simulations.

This paper presents some properties of the fluctuating component of wind pressure on a rectangular multistorey building in strong winds blowing in a direction normal to the major face of the building. The investigation has revealed some interesting facts relating to the "air cushion" phenomenon first reported by Mackey (1). The authors hope that the added information presented in this paper will be useful to research workers wishing to improve the reliability of their wind tunnel model investigations and encourage further full-scale study of the effect of varying degrees of permeability on wind pressure distribution on bluff buildings.

THE FULL SCALE EXPERIMENT

The full-scale wind pressure measurements were carried out at the ten-storey, 30.5 m (100 ft) tall, steel-framed, experimental building of the University of Hong Kong. The building, which has been described elsewhere (2) has dimensions 18.3 m (60 ft) by 9.1 m (30 ft) in plan and is located on the low-lying open foreshore shown in Fig. 1 with one of its major faces facing northeast.

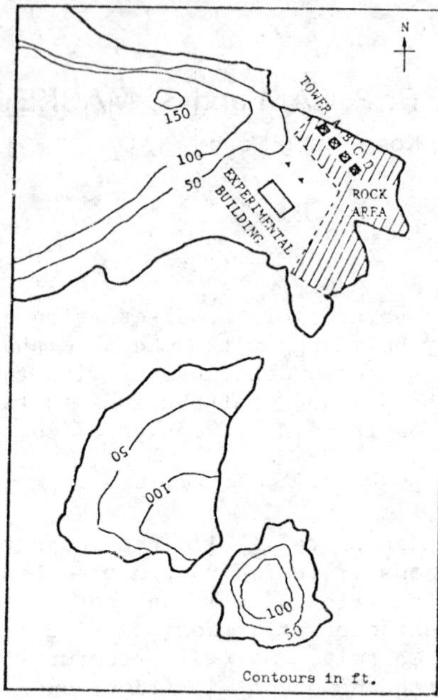

Fig. 1

Map of Cape D'aguilar showing location of the experimental building and the steel towers

RESULTS

Owing to their random nature, the characteristics of fluctuating wind pressures are best revealed by the correlation and spectral density functions. These functions for wind pressures measured on the four building faces under various strong monsoon winds and typhoons have been computed and are described below.

(a) Monsoon Wind

Typical auto-correlation curves for windward and leeward pressures under a face-on attack by the north easterly winter monsoon, having a mean wind speed of approximately 10 m/s, are plotted in Figure 2. Fluctuations having a period of about 10 s are apparent, especially for pressures measured near mid-height of the building, i.e. at the 5th and 7th floor levels. Corresponding spectra in Figure 3 show prominent peaks (at frequencies of around 0.07 and 0.14 Hz. respectively) which are absent from the horizontal wind speed spectrum. It is noted that these peaks are more pronounced for the leeward pressure spectra.

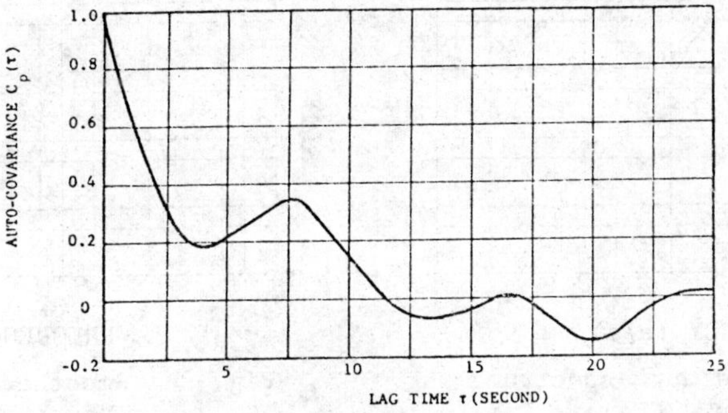

Fig. 2a
Typical auto-covariance function on windward face

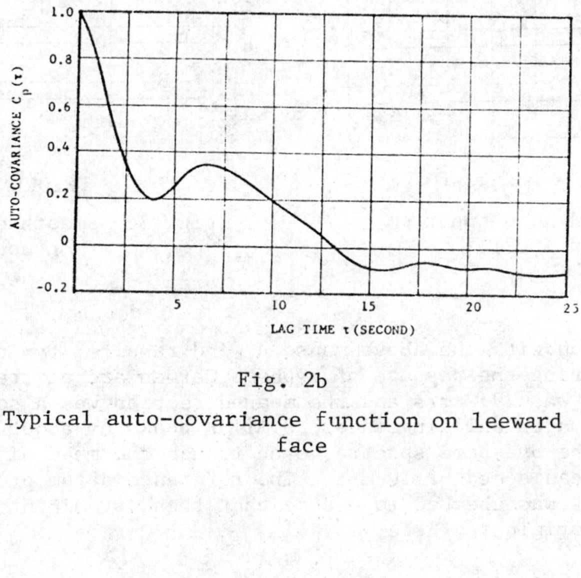

Fig. 2b
Typical auto-covariance function on leeward face

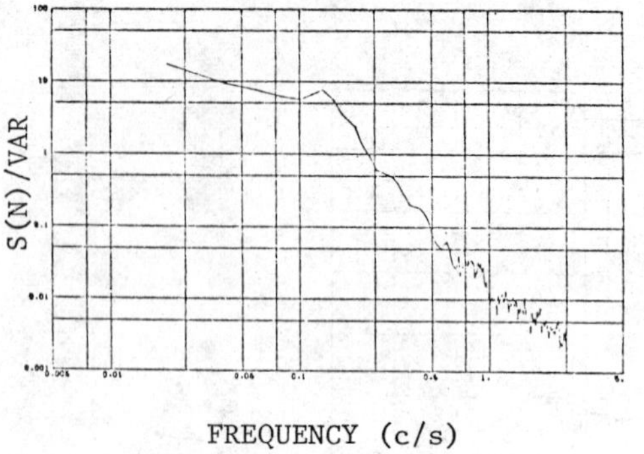

Fig. 3a Smoothed auto-spectrum (Monsoon, 7NE4)

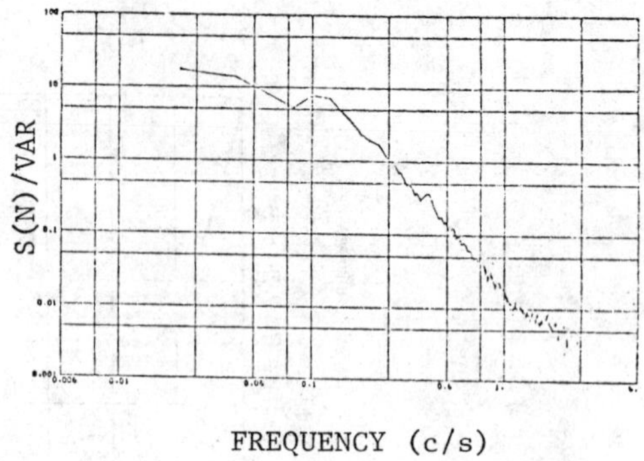

Fig. 3b Smoothed auto-spectrum (Monsoon, 5NW2)

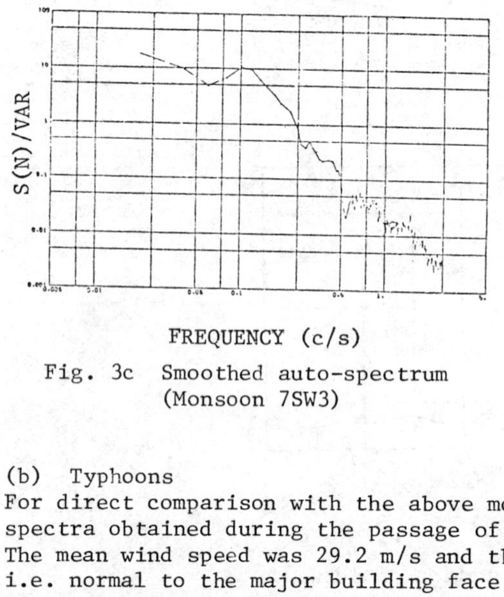

Fig. 3c Smoothed auto-spectrum (Monsoon 7SW3)

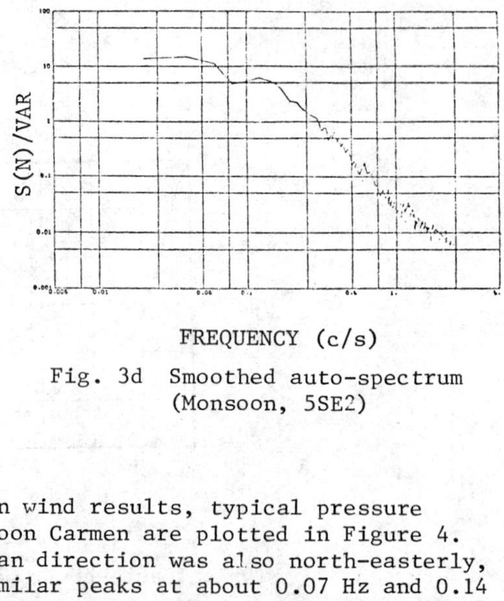

Fig. 3d Smoothed auto-spectrum (Monsoon, 5SE2)

(b) Typhoons

For direct comparison with the above monsoon wind results, typical pressure spectra obtained during the passage of Typhoon Carmen are plotted in Figure 4. The mean wind speed was 29.2 m/s and the mean direction was also north-easterly, i.e. normal to the major building face. Similar peaks at about 0.07 Hz and 0.14 Hz can be seen in the pressure spectra, even though the mean wind speed was three times the monsoon mean speed of 10 m/s. The reference datum pressure of the pressure transducers was checked to ensure that these prominent fluctuation components did not originate there.

DISCUSSION

The occurence of the fluctuating components within the frequency range, 0.07 Hz to 0.15 Hz cannot be predicted from the wind velocity spectrum obtained directly in the free-wind stream at the experimental site. None of the wind velocity spectra shows a peak within that frequency range. Existing knowledge of building aerodynamics has been examined in order to find an explanation for this phenomenon. However, none of the known factors nor established flow characteristics in wind-tunnel study appear to provide a satisfactory solution to the problem. Some unknown flow characteristics are therefore suspected and the hypothesis described below is suggested in order to explain this phenomenon which is found only in full-scale measurement.

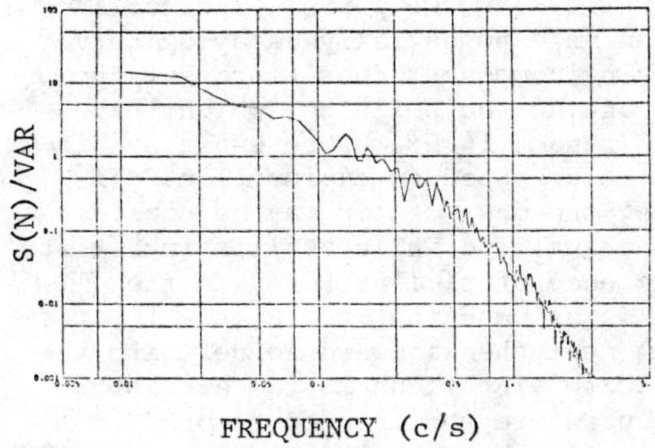

Fig. 4a Smoothed auto-spectrum (Carmen-4, 9NE3)

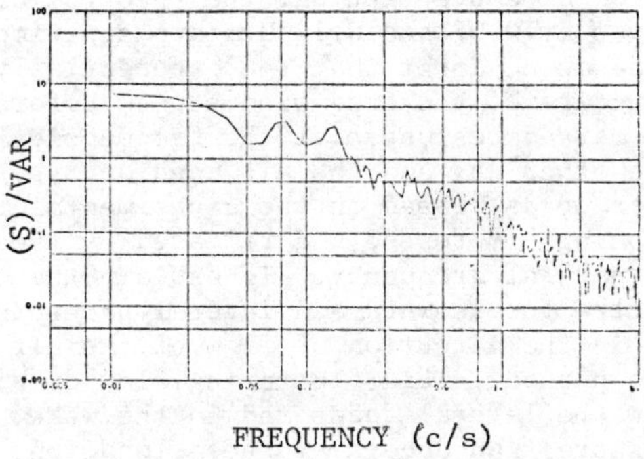

Fig. 4b Smoothed auto-spectrum (Carmen-4, 7NW2)

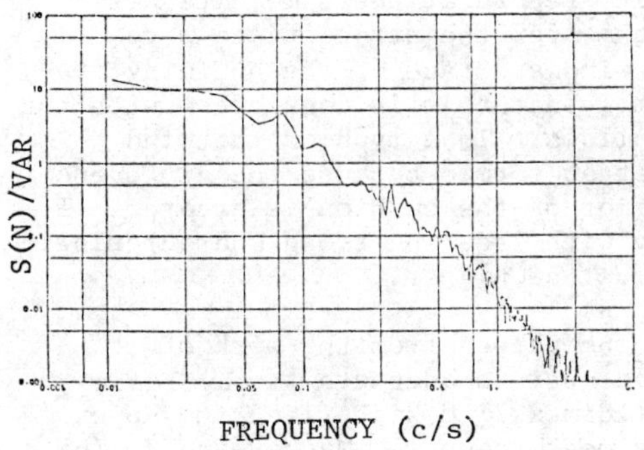

Fig. 4c Smoothed auto-spectrum (Carmen-4, 7SW3)

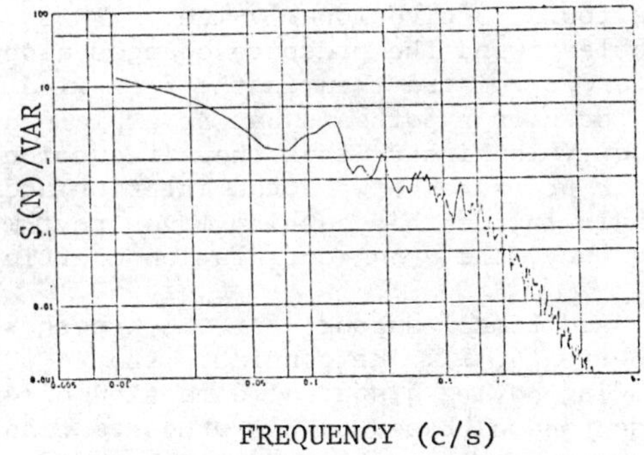

Fig. 4d Smoothed auto-spectrum (Carmen-4, 7SE2)

The free stream of air, as it approaches a building face, reduces its velocity. The energy is converted from the kinetic form to the pressure form. As a result, there is a mass of air in front of the frontal face of the building the average relative velocity of which is small. This mass, which has been previously defined as an "air cushion", will be most prominent on buildings which have large unbroken frontal areas facing into the wind. Once this air cushion is formed, it possesses certain characteristics of its own which may effect the dynamic motion of the building. The cushion will have periodic changes in size and shape with a period which is its own natural frequency and such vibrations might be expected to add to the oscillatory pattern of the approaching wind. Thus, inside this region the air is not completely stagnent, some fluctuations of shorter periods and small size eddies may be present.

The above hypothesis is advanced to explain some of the following unusual phenomena observed in the experimental results.

(1) The raw auto-spectrum computed from pressure transducer records is found to have either one or two peaks at the frequencies of 0.07 and 0.14 Hz. These peaks are shown by the windward, lateral and leeward faces. By non-linear regression method of curve fitting, the spectrum can be split into two parts. The swelling bulges at 0.07 and 0.14 Hz. are superimposed on an exponentially decaying curve. There is no doubt that the exponentially decaying component does represent the structure of the free wind stream before approaching the building, which leaves the marked responses at the frequencies of 0.07 and 0.14 Hz. to be explained in some other way and the air cushion hypothesis is proposed in explanation. In other words, based on the experimental results, the mass of air may be considered as being compressed and released by the approaching wind while oscillating at its own natural frequency. To explain the co-existence of similar terms in the spectra for leeward and lateral pressures, it is assumed that there is a fluttering in the direction of flow of the air stream near the windward corner with the same period. This fluttering flow of air produces vibrations in the air both near the lateral faces and in the wake, hence pressure spectra for records from pressure transducers at these locations also show such peaks.

(2) In comparing the cross correlation curves for the free wind records with those obtained from the pressure transducers, Lam (3) has reported that the cross-correlation for a given spatial separation in pressure records is higher for typhoons than monsoons although the velocity record at the same site has been found to give lower values. This shows that for the stronger and more turbulent wind the pressure changes along the windward face of the building are better correlated than in the free wind stream. That this is consistent with the air cushion hypothesis can be explained as follows. In a typhoon the wind velocity is higher, thus the air cushion is better formed because the difference in air pressure between the interior and exterior of the cushion is greater. If the air inside this cushion moves rhythmically with its own strong characteristic frequency, the cross correlation coefficient must be higher.

(3) Additional support for the hypothesis can be gained from the work of Newberry (4, 5). He considers that a 10-second gust is adequate to envelop a building having a spatial dimension of 43 m x 18 m x 71 m high (i.e. the Royex House London) based on his pressure transducer measurements. Mackey and Ko (6), considered this to be a reasonable deduction from measurements made on building faces. But on the other hand, when working on correlation studies of the free wind records they did not find such large envelopment. Further comparative studies on the values of horizontal and vertical cross-correlation coefficients show that for the same spatial separation, the correlation coefficient for wind pressure on the building face are higher than those for the free wind stream (7). Thus, in general, the wind pressures at a given spatial separation are better correlated than in the free wind. The existence of an air cushion in front of the building may explain these experimental findings. Since the air cushion is a mass of air acting with its own characteristic mechanism, the spatial correlation between pressure measurements at different points on the frontal face of the building must be higher and a gust of shorter duration than that predicted from free wind characteristics will appear to envelop the building.

(4) As a consequence of the presence of the peaks at 0.07 and 0.14 Hz in the pressure spectrum, the response spectrum of the building should also contain such peaks. This is verified by the building response spectrum obtained at low and moderate wind speeds. Prominent peaks at 0.14 Hz. are observed. However, at certain wind speeds and directions, the occurrence of this peak is less prominent. Cross-correlation between floor load time series in this particular case reveals that the fluctuations on windward and leeward faces at such a frequency are almost opposite in phase (8) and hence the peak at this frequency in the response spectrum is less distinct.

(5) In order to check the existence of the air cushion and to carry out further investigations, two guyed masts situated between the free standing masts and the building were erected and anemometers installed at the top of these masts. The auto-spectrum of wind direction obtained at the top of the guyed mast standing on the centre line of the building shows a distinct peak at 0.14 Hz. (Fig. 5) and both the wind velocity and direction spectra obtained at the guyed mast standing at a distance 12.2 m to the north-west from the centre line show very distinct peaks at the frequency of 0.14 Hz. (Figure 6 & 7). Simultaneous records of wind speed obtained at the free standing towers do not exhibit this peak. This result clearly shows that the periodic movement of air occurs in front of the building and gives additional support to the air cushion hypothesis.

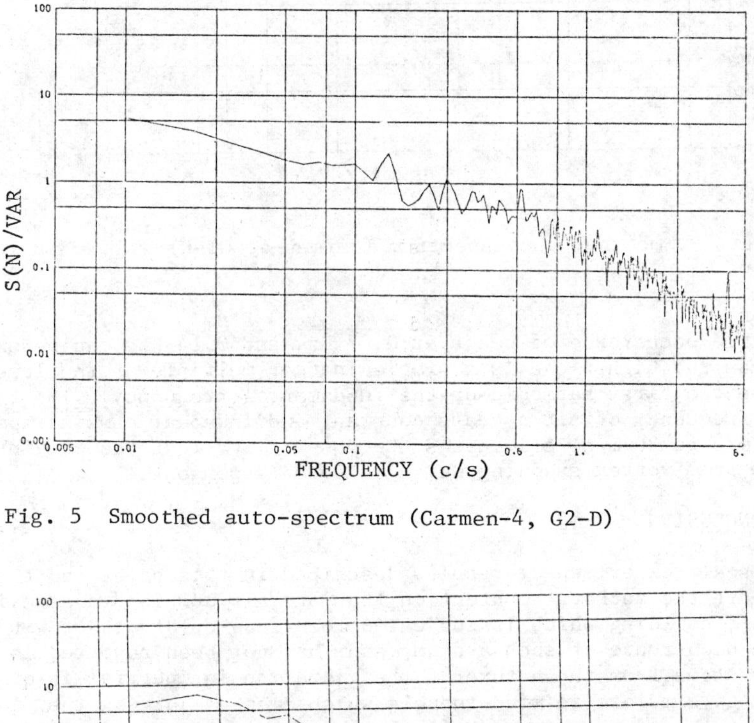

Fig. 5 Smoothed auto-spectrum (Carmen-4, G2-D)

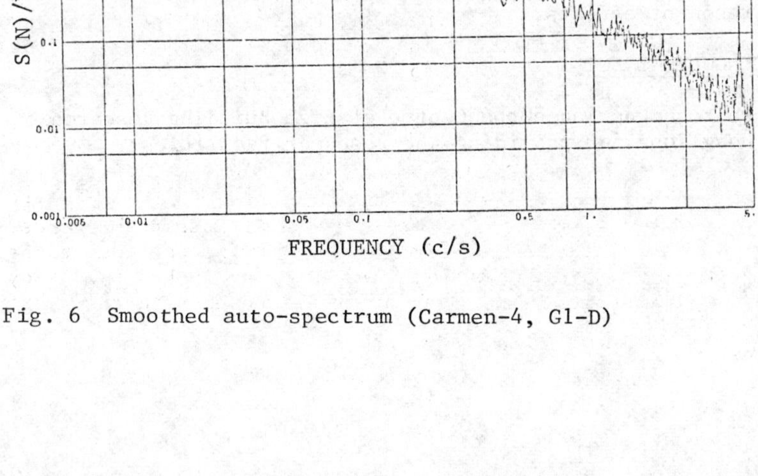

Fig. 6 Smoothed auto-spectrum (Carmen-4, G1-D)

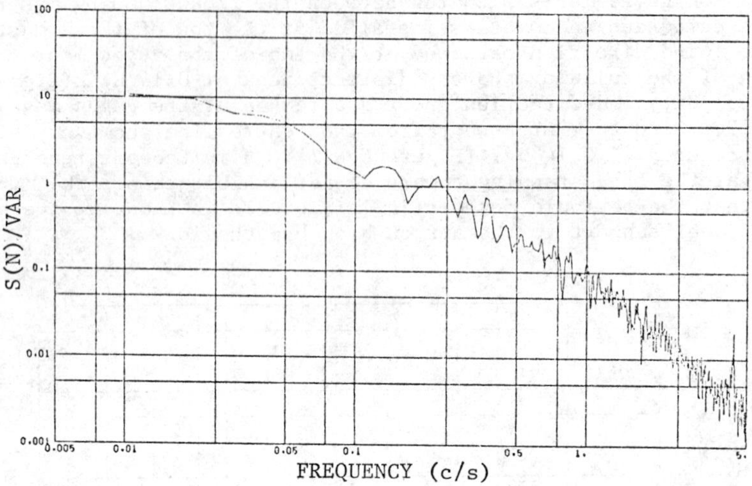

Fig. 7 Smoothed auto-spectrum (Carmen-4, G1-R)

(6) The possible occurrence of peaks at 0.07 Hz. and 0.14 Hz. can be explained by the fact that as the higher one is a simple integer multiple of the lower one it may possibly be the first harmonic of the fundamental frequency. The fact that the vibrating frequency of air appears constant and indepent of wind speed (at least from 10 m/s to 30 m/s) eliminates the possibility that the origin of such fluctuation is from vortex shedding which is speed dependent.

CONCLUSION

Both the full-scale experimental results described in this paper and the above reasoning support the authors' contention that an "air cushion" does exist in front of a bluff building which is subjected to a face-on attack of wind. The reason why the occurrence of such a phenomenum has not been reported in wind-tunnel studies stems from the nature of the vibration is inertial in origin and thus difficult to simulate in wind-tunnels which also use air as the fluid medium. Due to the existence of the air cushion theoretical approaches to predict building response based on the background excitation obtained from the characteristics of free wind and the dynamic amplification of the structure may need further improvement so as to incorporate the effect of the air cushion, and to avoid any risk of unanticipated resonant effects.

ACKNOWLEDGEMENT

The authors wish to acknowledge the Centre of High Building Research of the University of Hong Kong for providing the research facilities.

REFERENCES

(1) Mackey, S., "Wind Loading and Wind Effects", Summary Report, Proceedings of International ASCE - IABSE Conference on Planning and Design of Tall Buildings, Lehigh University, Pa., Technical Committee &, Ib, Aug., 1972, pp. 445 - 458.

(2) Lam, L.C.H., Lam, R.P. and Mackey, S., "Mean Wind Pressure Distribution Pattern on a Multistorey Building - Comparison of Full-scale and Wind-tunnel Test Results", Paper submitted to the Fifth International Conference on Wind Engineering.

(3) Lam, R.P., "Dynamic Response of a Tall Building to Random Wind Loads", Proceedings of the Third International Conference on Wind Effects on Buildings and Structures, Tokyo, Japan, Sept. 1971.

(4) Newberry, C.W., Eaton, K.J. and Mayne, J.R., "The Nature of Gust Loading on Tall Buildings", International Seminar on Wind Effects on Buildings and Structures, Ottawa, 1967; also BRS Current Paper CP 66/68.

(5) Newberry, C.W., Eaton, K.J. and Mayne, J.R., "Wind Loading on Tall Buildings - Further Results from Royex House", Industrial Aerodynamics Absts., July - Aug., 1973.

(6) Mackey, S. and Ko, P.K.L., "Spatial Configuration of Gusts", Proceedings of the Fourth International Conference on Wind Effects on Buildings and Structures, Heathrow, England, Sept., 1975, pp. 41 - 52.

(7) Lam, L.C.H., "Investigation of the Dynamic Response of a Multi-storey Building under Wind Loading", Ph.D. Thesis, University of Hong Kong, 1972.

(8) Wong, K.Y., "Some aspects of Wind Loading Effects on the Design of Multi-storey Buildings", Ph.D. Thesis, University of Hong Kong, 1976.

THE DETERMINATION OF MODAL WIND LOADS FROM FULL-SCALE BUILDING RESPONSE MEASUREMENTS

A. P. JEARY*, B. E. LEE** and P. R. SPARKS***

*Building Research Establishment, UK
**University of Sheffield, Sheffield, UK
***Virginia Polytechnic Institute and State University, Blacksburg, Virginia, USA

SUMMARY

As part of a research programme set up to study the response of tall buildings to wind loading, the 80 m high Arts Tower at the University of Sheffield has been subjected to an extensive investigation into its dynamic characteristics and its response to the wind. The work has been carried out by the Building Research Establishment in conjunction with the Department of Building Science of the University of Sheffield and the Centre Experimental de Recherches et d'Etudes du Batiment et des Travaux Publics, Paris (CEBTP).

The programme of research was as follows. The building was first subjected to a forced vibration test using an eccentric-mass vibrator supplied and operated by CEBTP. In this test the frequencies of the fundamental translational and torsional modes and a second translational mode were found, together with the associated building properties. Then over a period of time, extending from early 1976 to the present, the response of the building in windy conditions was recorded.

This paper describes the experimental procedure and the analysis of the results to determine the natural frequencies, mode shapes and associated stiffnesses, masses and damping. In the case of the wind response, the building is used as a large transducer and the modal wind loads are obtained from the modal responses. Comparisons are made between the design values of the building's properties and wind loads and those obtained experimentally. Finally, the wind climate of Sheffield is examined to determine the expected frequency of occurrence of perceptible levels of vibration in the building.

The torsional response of this building, which is essentially symmetrical, has been found to be of the same order as that of the translational modes. This would not be expected from normal design considerations.

Modal forces associated with oblique directions have been shown to be larger than the in-line directions. This is a result which is surprising and disturbing since no method of response prediction available at present covers anything other than in-line forces.

INTRODUCTION

Over the past 10 to 15 years, economic pressures have encouraged the structural engineer to use building materials more efficiently and this has resulted in a trend towards buildings of lighter and more slender construction.

Building owners often require their buildings to have a light and slender appearance and to contain large clear floor spaces which allow flexibility of use. Unfortunately some buildings designed to meet requirements of this nature are deemed to have an unacceptable performance when buffeted by the wind. Despite the provision of sufficient strength in such a slender construction, this design philosophy can result in a structure which is inadequately stiff and hence serviceability problems associated with excessive movement can arise. Occupants of these buildings, able to perceive the building movement, become disturbed and in extreme cases suffer from motion sickness. Further disturbance may arise from noise generated by the relative motion of building elements, cracking of partitions and leakage of rainwater past moving cladding.

Wind loads are dynamic, and buildings which have low levels of damping are dynamically sensitive structures. Hence, it is no longer possible to produce an adequate design for tall buildings from a consideration of the static loads alone. There are a number of methods available which permit the designer to predict the dynamic response of a building but each method requires a knowledge of the building's stiffness, mass, natural frequencies, mode shapes and damping parameters, together with an accurate estimate of the dynamic loads caused by the wind.

A research programme has been initiated by the Building Research Establishment (BRE), in collaboration with a number of other organisations, in order to provide realistic information on the dynamic properties of slender buildings and the wind loads acting upon them. This paper describes tests carried out on the Arts Tower of the University of Sheffield in collaboration with the Department of Building Science of the University. Further assistance came from the Centre Experimental de Recherches et d'Etudes du Batiment et des Travaux Publics, Paris (CEBTP) who, as part of a joint research programme with BRE, carried out a series of forced vibration tests on the building.

A brief description of this research study has already been reported, (1) and (2), in which some of the initial results from the wind response measurements have been presented and discussed. The purpose of the present paper is to make a far more detailed and reliable data collection available to wind engineering research workers with the aim of producing a more definitive series of conclusions.

DESCRIPTION OF THE BUILDING

The Arts Tower is situated within the University site near the centre of the city of Sheffield. It stands with its north-south axis at an angle of -20° to true north, on a sloping site. It is exposed to the north and east and is partially sheltered to the south and west by other low rise University buildings.

The building is approximately 80 m high, 36 m wide and 20 m deep. Structurally it consists of a cast in situ reinforced concrete core with deep floor slabs spanning between the core and the external reinforced concrete columns. Above the first floor there are 88 columns evenly spaced around the periphery of the building, each 203 mm x 406 mm. Below first floor level the number of columns is reduced to 16, each 965 mm x 965 mm.

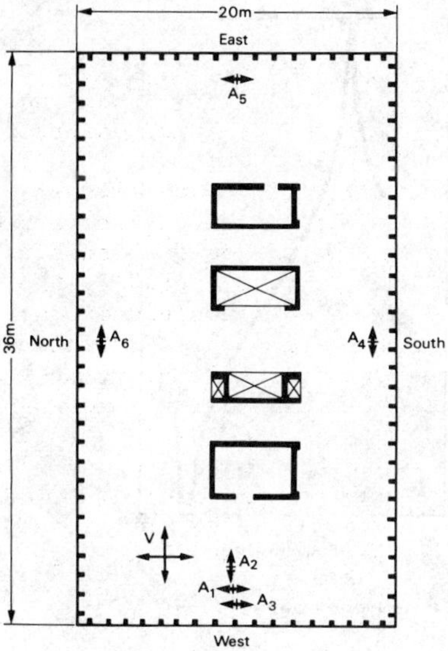

Figure 1 Structural section above first floor

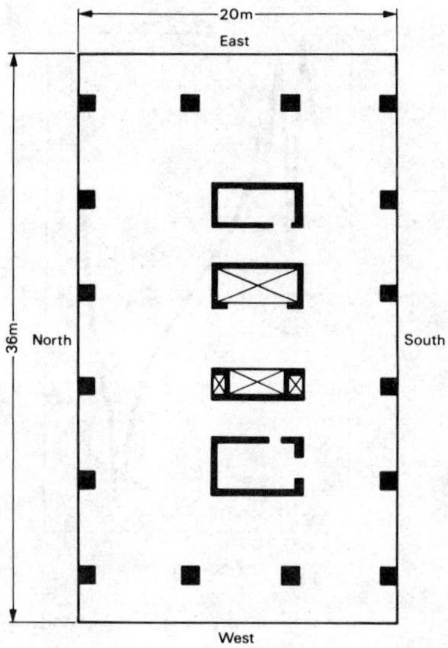

Figure 2 Structural section below first floor

Figure 1 shows a typical section above the first-floor level and Figure 2 a section below first-floor level. Non-structural blockwork partitions which vary in arrangement from floor to floor have been omitted from these sections. The cladding is of both glass and lightweight panelling. The building has a link to an adjacent building on the west side at mezzanine level, between the ground and first floors. There is a deep basement and the building is founded on piles driven into shale.

FORCED VIBRATION TESTS

To determine the dynamic properties of the building and to calibrate it in preparation for the wind response measurements, a series of forced vibration tests were carried out. The CEBTP eccentric-mass vibrator was mounted on the 19th floor of the building at position V, shown in Figure 1, and the response was monitored by accelerometers in positions A_1 and A_2. The vibrator was used to generate discrete-frequency forces, in which the excitation frequency was sequentially stepped up, first in the north-south direction and then in the east-west direction. Figure 3 shows the response of the building in the north-south direction measured at position A_1, and Figure 4 shows the response in the east-west direction with the accelerometer in position A_2.

The north-south response as shown in Figure 3, has four peaks: the first at 0.69 Hz is associated with the fundamental translational mode ($N\text{-}S_1$), the second at 0.79 Hz with the fundamental torsional mode (θ_1), the third at 0.87 Hz with the torsional component of the fundamental east-west translational mode ($E\text{-}W_1$) and the fourth at 2.46 Hz with the second north-south translational mode ($N\text{-}S_2$).

The east-west response shown in Figure 4 has only one major peak, this indicating the translational component of the east-west fundamental mode ($E\text{-}W_1$). At the upper end of the frequency sweep the small coupled response from the $N\text{-}S_2$ translational mode is observed.

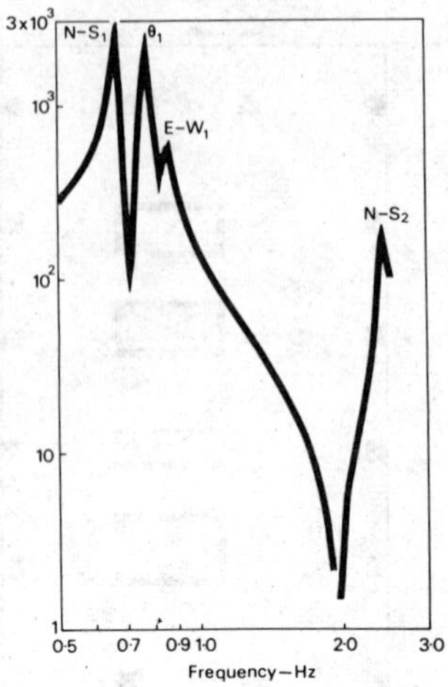

Figure 3 Frequency response north-south direction

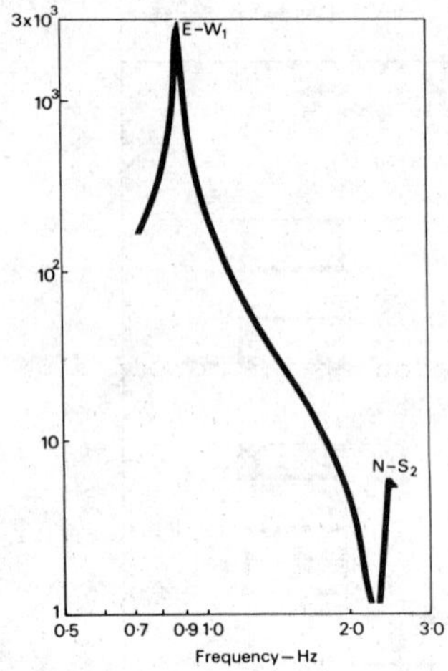

Figure 4 Frequency response east-west direction

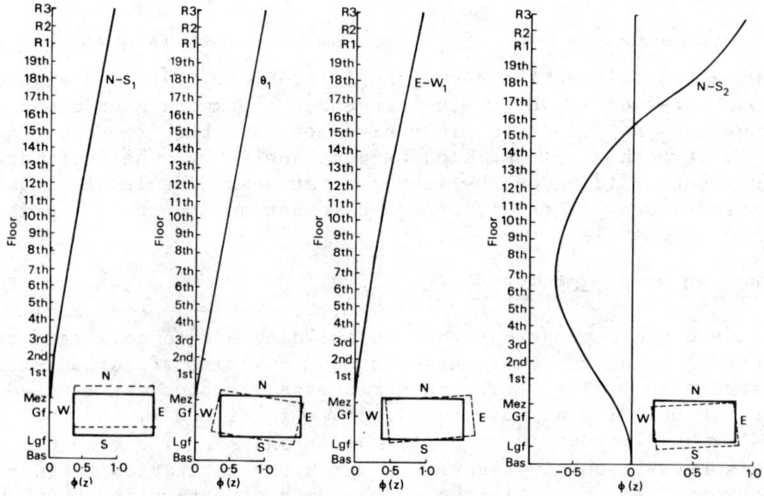

Figure 5 Plan and elevation mode shapes

Having identified the various natural frequencies, each mode was then excited in turn. A moveable accelerometer was used to compare the response of various parts of the building with reference to the accelerometers mounted on the 19th floor. Figure 5 shows the plan and elevation mode shapes of each of the modes obtained in this way. This technique, which involves the use of long lengths of cable, is only practicable, at present, when servo-type accelerometers are used. These do not suffer from the drift or reactance effects normally associated with devices using bridge networks (e.g. inductive accelerometers). The use of servo-accelerometers, which operate down to 0 Hz, also allows a direct calibration to be made by measuring their response due to a tilt. This can be performed on site with the system connected as in the experiment, and gives a high degree of confidence as any problems associated with electronic mis-matching are incorporated into the calibration figure. A consequence of using accelerometers at very low frequencies is that the signals produced are small, and stray signals from higher frequencies can swamp the useful signals.

For this reason, 10 Hz low pass filters are used. After filtering, the accelerometer signals are usually amplified 1000 times, before recording on an FM tape recorder, for subsequent analysis at the laboratory.

At the termination of the investigation of each mode shape the vibrator is stopped suddenly and the ensuing decay of oscillation is used to give an estimate of the structural damping. These tests were carried out in calm conditions to avoid the unwanted response which would result from additional excitation due to the wind.

The interpretation of the forced vibration test results is based upon the assumption that the response in each mode is given by the solution to the differential equation

$$\ddot{X}_r + 4\pi\zeta_r f_r \dot{X}_r + 4\pi^2 f_r^2 X_r = \frac{F_r}{M_r} \tag{1}$$

where X_r is the modal response
f_r is the modal frequency
M_r is the modal mass
ζ_r is the modal damping
and F_r is the modal force

It can be shown, (3), that a solution to equation (1) in the frequency domain is

$$X_r(f) = |H_r(f)|\,F_r(f) \tag{2}$$

where

$$\frac{1}{|H_r(f)|} = 4\pi^2 f_r^2 M_r \left\{\left[1-\left(\frac{f}{f_r}\right)^2\right]^2 + \left[2\zeta_r \frac{f}{f_r}\right]^2\right\}^{1/2} \tag{3}$$

At resonance, in the vibrator tests, the force is sinusoidal, and is applied at the frequency f_r with a modal amplitude of F_r. The maximum modal displacement, $\hat{X}_r$, is given by

$$\hat{X}_r = \frac{F_r}{8\zeta_r \pi^2 f_r^2 M_r} \tag{4}$$

with a maximum acceleration response of

$$\hat{\ddot{X}}_r = \frac{F_r}{2\zeta_r M_r} \tag{5}$$

Equation (5) is used to determine the modal mass of the structure and from it, using the following relationship, the modal stiffness is determined

$$K_r = 4\pi^2 f_r^2 M_r \tag{6}$$

In the vibrator tests, only the modal damping, ζ_r can be measured directly. Modal acceleration and forces are calculated from measured values, using the measured mode shapes. These mode shapes, $\phi_r(z)$, are obtained from the relative values of accelerations in various parts of the building. For convenience the mode shape values are normalised to 1 at the top of the building.

If the maximum measured acceleration at a height Z1 up the building is $\hat{\ddot{x}}_r(Z_1)$, then the maximum modal acceleration is given by

$$\hat{\ddot{X}}_r = \frac{\hat{\ddot{x}}_r}{\phi_r(Z_1)} \tag{7}$$

If the amplitude of the sinusoidal force applied at a height Z_2 up the building is A, the maximum modal force is given by

$$\hat{F}_r = A\,\phi_r(Z_2) \qquad (8)$$

Apart from permitting the modal stiffnesses to be determined, a knowledge of the modal masses has a number of other uses. The modal mass of the building, height H, is given by

$$M_r = \int_0^H m(Z)\,\phi_r^2(Z)\,dZ \qquad (9)$$

where m(Z) is the mass per unit height of the building. If m(Z) is assumed constant then the total mass M_t of the building can be determined from

$$M_t = \frac{M_r}{\int_0^H \phi_r^2(Z)\,dZ} \qquad (10)$$

and hence the average density of the building can be calculated. The average densities of buildings are known to lie within fairly close limits and can be checked against the working drawings. Average density values therefore provide a valuable check on the accuracy of values of stiffness obtained in a particular test.

The derivations from the modal masses of the building have another use. This is when the total mass determined from one fundamental translational mode differs from that determined in the orthogonal fundamental mode. If one of these is known to be a pure mode and the other is known to contain some torsion, then the difference between the masses can be used to determine the shear centre of the building. The difference between the two calculated masses is due to the fact that in the complicated mode some of the motion accredited to translation is actually due to rotation.

By analogy the torsional responses may be dealt with in a similar way to the translational responses. The maximum modal torsional acceleration is given by

$$\hat{\ddot{\theta}}_r = \frac{\hat{T}_r}{2\zeta_r I_r} \qquad (11)$$

where $\hat{T}_r$ is the peak modal torque applied and I_r is the modal polar moment of inertia.

The torsional stiffness is given by

$$K_{\theta r} = 4\pi^2 f_r^2 I_r \qquad (12)$$

Finally, it is worth considering the case when static loads are applied to a structure , i.e. when the frequency of excitation is zero. Substituting equation (6) and equation (3) into equation (2) the modal response at zero frequency is given by

$$X_r(0) = \frac{F_r}{K_r} \qquad (13)$$

and the total response due to all the modes is

$$\sum X_r(0) = \sum \frac{F_r}{K_r} \qquad (14)$$

In practice the second and higher modes normally contribute only about 2 per cent to the total static displacement and so the fundamental modal stiffness is a good estimate of the conventional static stiffness of a structure.

The results of the forced vibration tests are summarised in Table 1.

Table 1 Building Properties

(measured April 1976)

	N-S	E-W	
Fundamental natural frequency (Hz)	0.678	0.863	0.787
Modal mass (kg)	6.35×10^6	5.84×10^6	-
Modal inertia (kg/m^2)	-	-	8.14×10^8
Modal lateral stiffness (N/m)	1.15×10^8	1.70×10^8	-
Modal torsional stiffness (Nm/radian)	-	-	1.99×10^{10}
Damping (% critical)	0.86	0.86	0.90
Second mode natural frequency (Hz)	2.46	-	-

An inspection of the original design data for the building showed that the central core was intended to take all the wind load. Despite this, it contributes surprisingly little to the total stiffness of the building, 2 per cent in the north-south direction and 17 per cent in the east-west direction. Even when an allowance for the stiffness of the external columns is included, this only accounts for 25 per cent of the stiffness in the north-south direction and 60 per cent in the east-west direction. The additional stiffness is presumed to be provided by the internal partition walls and the external cladding panels.

The plan mode shapes show that the shear centre of the building does not coincide with the geometrical centre but that it lies to the north of the centre on the north-south axis. Thus the $E\text{-}W_1$ mode shape contains a torsion component and the θ_1 mode shape contains a translation component. This means that whereas the motion of the centre of the building in the north-south direction is due to pure translation, the motion of the centre of the building in the east-west direction contains an element of torsion. This coupling effect accounts for the difference between the $E\text{-}W_1$ modal mass and the $N\text{-}S_1$ modal mass. The difference in the modal masses suggests that the $E\text{-}W_1$ torsional component amounts to approximately 9 per cent of the apparent translation of the centre. From this it can be deduced that the shear centre of the building must lie approximately 3 m north of the centre of the building on the north-south axis.

A simple calculation based on the areas of the columns and of the core shows that the shear centre ought to lie about 0.7 m south of the geometrical centre of the building. In the previous paragraph it was reasoned that the blockwork partitions and cladding must make a significant contribution to the building stiffness. It would appear to be indicated that they are also capable of shifting the overall shear centre of the building away from the geometrical centre since the columns and core apparently contribute only approximately 17 per cent of the total torsional stiffness.

WIND RESPONSE MEASUREMENTS

After the forced vibration tests, the response of the building to the wind was measured using the four fixed accelerometers on the 19th floor in positions A3, A4, A5 and A6 (Figure 1). The signals from these accelerometers were filtered and amplified prior to being recorded on an FM tape recorder. These tapes were then analysed to determine the spectral content of the acceleration response.

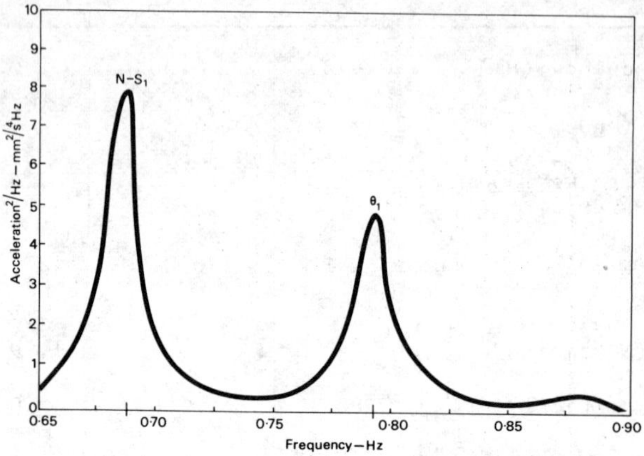

Figure 6 Acceleration spectrum – west accelerometer

Figure 6 shows the acceleration spectrum of the west accelerometer responding to motion in the north-south direction at frequencies in the region of the fundamental frequencies of the building. Wind-induced responses in the higher modes were found to be negligible. Figure 7 shows a similar spectrum from the south accelerometer recording motion in the east-west direction.

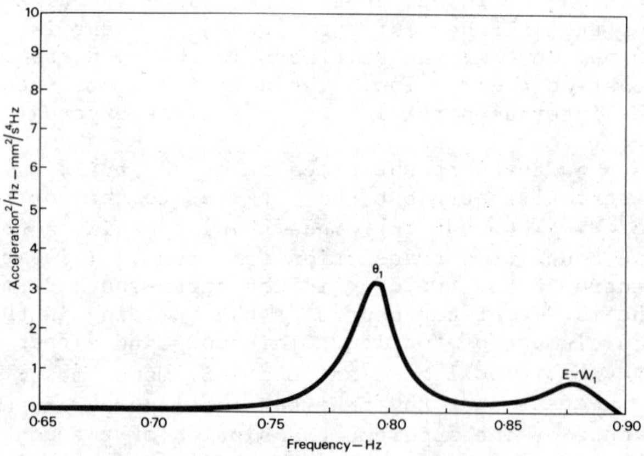

Figure 7 Acceleration spectrum – south accelerometer

From such spectra the mean square accelerations in each mode, taken to be the value of the spectral peaks at the appropriate modal frequency, were obtained ($\overline{\ddot{x}_r^2}$). It can be shown that the mean square acceleration in a particular mode is given approximately by:

$$\overline{\ddot{x}_r^2}(Z,t) = \phi_r^2(Z)\,\overline{F_r^2}\,\pi f_r / 4 M_r^2 \zeta_r \qquad (15)$$

M_r, ϕ_r, f_r and ζ_r, defined earlier, have been evaluated from the forced

Table 2 - Wind Conditions, Forces and Building Accelerations, (1976-1977)

	Mean Wind Direction	Mean Wind Speed (m/s)	$\bar{F}_r^2$ N-S_1 (KN^2/Hz)	$\bar{F}_r^2$ E-W_1 (KN^2/Hz)	$\bar{F}_r^2$ θ_1 (KN^2m^2/Hz)	R.M.S. Accelerations N-S_1 (mm/s^2)	R.M.S. Accelerations E-W_1 (mm/s^2)	R.M.S. Accelerations θ_1 (mm/s^2)
WIND EXCITATION RESULTS	254	9.3	6.3	2.2	427	2.8	1.8	3.0
	254	11.1	8.5	1.7	1012	3.2	1.6	4.7
	275	8.5	3.4	1.8	324	2.0	1.7	2.6
	275	11.8	35.2	6.5	3149	6.4	3.2	8.1
	295	9.1	8.3	2.8	349	3.1	2.1	2.7
	295	11.8	24.1	6.5	1671	5.3	3.2	5.9
	305	7.0	3.4	1.5	174	2.0	1.5	2.0
	305	9.6	15.2	4.3	920	4.2	2.5	4.5
	305	11.4	23.4	6.1	1472	5.3	3.0	5.7
	305	13.3	27.0	6.7	1564	5.7	3.1	5.8
	338	10.0	3.4	0.3	133	2.0	0.7	1.7
	338	11.9	6.7	1.1	216	2.8	1.3	2.2
	338	13.3	8.4	1.7	400	3.2	1.6	2.9
	350	8.1	5.8	0.6	156	2.6	1.0	1.8
ESDU 76001 PREDICTIONS	270	10.0	--	1.2	--	--	1.7	--
	270	11.9	--	3.2	--	--	2.8	--
	270	13.3	--	5.9	--	--	3.8	--
	360	10.0	3.5	--	171	2.4	--	2.2
	360	11.9	9.3	--	452	3.9	--	3.6
	360	13.3	17.2	--	844	5.4	--	5.0

vibration tests. Hence the mean square modal force produced by the wind at each of the fundamental frequencies ($\overline{F_*^2}$) can be determined. In Table 2, these forces are given for a number of wind conditions together with the measured rms accelerations. The wind speeds shown were obtained from anemometers positioned on top of the building and in the vicinity of the building, and have been reduced to values corresponding to a standard height of 10 m in open level country. The wind direction is given relative to the axes of the building; thus 360° is a wind blowing normal to the north face of the building, etc. The rms accelerations are given for a point in the centre of the east or west faces of the building on the 19th floor. The N-S1 and θ_1 mode accelerations are the average of the relevant components measured by accelerometers A3 and A5 and the E-W1 mode accelerations are the average of the relevant components measured by A4 and A6.

Included in Table 2 are the mean square modal forces and rms accelerations predicted by the method given in the Engineering Sciences Data Unit data item 76001, (4), for wind blowing onto the faces of the building. The method predicts only the in-wind response and considers the worst case to be when the wind is blowing normal to a face.

The data from the wind induced response measurements have been analysed for two separate periods of time, for Spring 1976 to Autumn 1977, in Table 2, and for Spring to Autumn 1978. The reasons for this are explained in the discussion.

1976-1977 Records

Considering first the information in Table 2, taken from an analysis of more than 200 hours of data, any conclusions drawn should be treated with some caution since many of the spectral estimates have confidence limits as wide as 30 per cent.

From the data in Table 2 the relationships between the wind speed and the rms modal forces were calculated. After averaging the relationships for all wind directions it was found that the N-S, the E-W and the θ modal forces are proportional to the velocity to the power 1.96, 2.00 and 2.61 respectively. The corresponding relationship given by ESDU for all the modes, is velocity to the power 2.8. Using the calculated relationship for each wind direction, the equivalent modal forces for a nominal wind speed of 10 m/s were determined. Figure 8 shows the variation of modal forces with wind direction for this wind speed, where the wind directions are now shown relative to the building such that 360° represents a wind blowing normal to the north face.

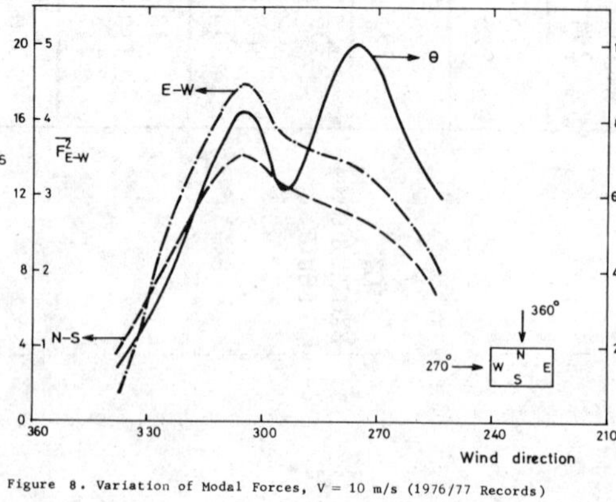

Figure 8. Variation of Modal Forces, V = 10 m/s (1976/77 Records)

A number of conclusions are apparent from this figure:

(a) The variation of the N-S and E-W modal forces with wind direction are almost identical, where the magnitude of the N-S modal forces is four times that of the E-W modal forces. These translational modal forces exhibit a maximum when the wind direction is onto the corner of the building.

(b) The variation of the torsional modal force shows two maxima, one when the wind is directed onto the narrow face of the building (270°), and one when the wind direction is onto the corner.

(c) In terms of rms accelerations, or displacements, the torsional response component of the north-south movement, measured on the east and west faces, is of the same order as the $N\text{-}S_1$ translational response for all the measured wind directions.

(d) The modal forces are greater when the wind is normal to the narrow face than when it is normal to the wide face of the building.

(e) As far as it can be judged, the value of the ESDU predictions is dependent on the wind direction. Whilst the ESDU method ignores the large forces which occur when the wind blows diagonally onto the building, it tends to underestimate the E-W translational modal force for a wind direction of 270° and to overestimate the N-S and modal forces particularly at higher wind speeds for a wind direction of 360°. For instance, at a wind speed of 13.3 m/s, the worst case predicted by the ESDU method shows a response rms acceleration of 5.4 mm/s^2 in the $N\text{-}S_1$ mode. The worst response measured is at first sight apparently in close agreement with this figure, being 5.7 mm/s^2 for the same wind-speed value. However, since the wind angle in this case was 305°, the $N\text{-}S_1$ translational component is not the total response in the north-south direction and must be combined with the north-south component in the θ_1 mode, 5.8 mm/s^2, to give an effective rms value of 8.3 mm/s^2. This coincides with an orthogonal acceleration of 3.1 mm/s^2 in the east-west direction ignored by the ESDU method.

1978 Records

Because of a malfunction in the data storage equipment no records were obtained during the winter months of 1977. When the first records available from early 1978 were analysed it was suspected that the natural frequencies of the building had all changed slightly. From an analysis of these wind records it was possible to arrive at an approximate estimate of the three fundamental modes, which were considered then to occur at 0.683 Hz ($N\text{-}S_1$), 0.798 Hz (θ_1) and 0.876 Hz ($E\text{-}W_1$).

In order to investigate this apparent change in modal frequencies a further series of forced vibration tests were conducted on the building in February 1979. It is not the authors' intention to report fully on the outcome of these tests here, but to publish full details at a later date.

In the modes where comparison is possible with the tests conducted in April 1976 it has been found that all the natural frequencies have fallen, although in the case of the fundamental $N\text{-}S1$ mode this has been by only a small amount. The drop in frequency for each mode, at similar force levels in the two test series, is as follows:

Mode	Frequency change
N-S_1	-0.004 Hz
E-W_1	-0.008 Hz
θ_1	-0.008 Hz
N-S_2	-0.060 Hz

These results are contrary to the indications of the frequency estimates from the 1978 wind records which appear to indicate an increase in modal frequencies, though no satisfactory explanation is presently available to explain the difference. The modal frequencies were also found to vary with the forcing amplitude during the tests, though only at force input levels considerably in excess of those which might be produced by wind action. Variation of damping with amplitude was also demonstrated, though the indications are that the most recent results are consistent with those obtained in the April 1976 tests, Table 1

The implications of the apparent change in natural frequency will require further attention before the 1978 data can be reliably presented.

PERCEPTION OF MOTION IN THE BUILDING

Shortly after the Arts Tower was first occupied in 1966, a number of the occupants of the upper floors of the building claimed that they could clearly perceive movement of the building. Since then there have been few complaints until recently when, perhaps aware of these measurements being made, one or two people have said that they thought that they had detected some movement.

It is believed that at the fundamental frequencies of this building, peak accelerations in the range 30-50 mm/s^2 represent the threshold of perception, (5). These are likely to be contained in periods which have rms accelerations in the range 9-15 mm/s^2. The results presented in Table 2 are based on ensemble averages of several records, each of which has a length of 1024 seconds, approximately 17 minutes. In one storm included in the 305°/13.3 m/s group there was a 17 minute period in which the mean wind speed was approximately 18 m/s. Based on the information from these tests one would expect a combined rms response in the N-S_1 and θ_1 modes of approximately 15 mm/s^2 over that 17-minute period,with a high probability of peak acceleration reaching the perceptible level.
It seems likely that in the past, perceptible movement of the building would have been restricted to storms in which the mean wind speed averaged over the worst 20 minutes of the storm exceeded 15 m/s. Table 3 lists all storms of this magnitude which occurred in Sheffield between 1962 and 1977.

In 1967, the year following the occupation of the building, there was only one storm in excess of 15 m/s and that occurred on the west face of the building. It is considered unlikely, on the evidence of the building's behaviour at present, that this storm would have produced perceptible motion. In 1968 there was another storm of 15 m/s, this time blowing onto the north face of the building, again unlikely on present evidence to have produced perceptible motion.

However a second storm did come from the south west later that year and this could possibly have produced preceptible motion, as might the more severe storms of 1974 and 1976 coming as they did from the north west.

At this stage it is interesting to consider the implications of a scenario based on an occupation of the building before the famous Sheffield gales of February 1962. The highest mean hourly wind speed recorded at a height of 10 m within the city, during this period of continual gales which lasted for some 4 days,

Table 3 Wind storms in Sheffield with 20-minute means in excess of 15 m/s at 10 m

Year	Wind speed (m/s)	Wind direction (relative to Arts Tower)
1977	15.0	290°
1976	16.9	310°
1974	17.5	290°
1974	15.5	300°
1973	19.0	310°
1968	15.9	250°
1968	15.0	010°
1967	15.4	280°
1962	21.0	290°
1962	17.5	280°
1962	15.0	290°

was 21 m/sec, (6).

One of the first considerations here is that the design wind speed would have been lower because this figure for the Sheffield area was substantially increased after the damaging gales of 1962. A structure designed before 1962 is therefore likely to be less stiff than a building of similar structural form designed after 1962.

Most certainly the storms of 1962 would have produced perceptible and perhaps alarming movement of the building. It is also possible that damage would have been done to partitions and cladding. This would have resulted in a public condemnation of the building as being badly designed and yet it is unlikely to have produced perceptible motion again until 1973.

On the other hand, an identical building constructed in 1969 would not have been subjected to a severe storm until 1973 and even if perceptible motion occurred then its reputation as a solid building would possibly have been established, and the storm response dismissed as a freak occurrence - unless of course the two storms of 1974 also produced perceptible motion.

Apart from the perception of motion in the building the wind storms which have occurred over the last twelve years have had other, more identifiable, effects. It has been pointed out earlier that a significant contribution to the stiffness of the building results from its so-called non-loadbearing elements. The blockwork internal partition walls were not designed for the shear loads which they have had to resist in their role as stiffeners and in the upper floors of the building diagonal shear cracks are apparent in these walls. The surface finish of these walls is regularly re-decorated to cover the cracks and yet they continue to appear. One particular instance of this was observed by one of the authors in January 1977 on a day during which a severe storm passed in the space of approximately 12 hours. A few days prior to this storm the room containing the recording apparatus had been re-decorated and the walls were covered in a fresh, uniform plaster surface, which had been painted. Within 6 hours of the start of the storm the diagonal cracks had re-appeared in the plaster, opening up to 3 mm wide at their centres. Cracks such as these are not readily apparent below the 12th floor of the building.

CONCLUSIONS

When the Arts Tower of the University of Sheffield was designed it easily satisfied the then current wind-loading requirements. It gives the appearance of being a well proportioned building of solid construction. Yet were it not for the significant contribution of the so-called non-loadbearing elements of the building, the building would undoubtedly have been excessively flexible and would have had an unsatisfactory performance in the wind.

These 'non-structural' elements also have a surprisingly large effect on the position of the shear centre of the building. A wise designer will try to ensure that his building's structural form is symmetrical about both axes. This reduces the likelihood of coupling between modes and of significant torsional movement in translational modes. The structural form of this building suggested that the shear centre would be approximately 0.7 m to the south of the geometrical centre, whereas in fact it lies approximately 3 m to the north of the centre. Consequently there is a large torsional component associated with the $E\text{-}W_1$ translational mode. Clearly our current understanding of the real stiffness of buildings is inadequate.

This paper also underlies the inadequacy of our knowledge about the wind forces on buildings. The most surprising fact to emerge from the wind response measurements presented here is the relative magnitude of the torsional and translational responses. On a building of this type it would be normal practice to expect that the torsional responses, and hence the torsional loads, would be small in comparison with the translational responses. Yet for all the wind directions recorded, the torsional response measured on the east or west faces is of the same order as the north-south translational responses. Additionally, the large modal forces associated with oblique wind directions is surprising. Of the three methods commonly in use to predict the response of tall buildings to the wind, two completely ignore the torsional loads and responses, (7), (8), and the third sadly underestimates the extent of the loading, (4). Also, none of the available methods deals with the seemingly critical case of oblique loading.

ACKNOWLEDGEMENTS

The authors wish to thank the University of Sheffield for permitting tests to be carried out on their building and also for the provision of services and laboratory accommodation.

The authors would like to thank Mr. B. Ellis of the Building Research Station and Dr. R. Evans of Sheffield University for their assistance with the forced vibration tests, February 1979.

The authors are also grateful for the contributions to this work made by V.C.M. de Souza, supported by the National Research Council of Brazil, and by J. Plaquet of CEBTP, Paris.

BIBLIOGRAPHY

1. Jeary, A.P., Lee, B.E., Plaquet, J., de Souza, V.C.M. and Sparks, P.R., The Dynamic Behaviour of the Arts Tower, University of Sheffield, Proc. Int. Conf. Behaviour of Slender Structures. City University, London, U.K. Sept. 1977.

2. Jeary, A.P., Lee, B.E. and Sparks, P.R., The Determination of Wind Induced Modal Forces, Proc. 3rd U.S. National Congress on Wind Engineering,

University of Florida, Gainesville, Florida. Feb. 1978.

3. Hurty, W.C. and Rubenstein, M.F., Dynamics of Structures, Prentice Hall, Englewood Cliffs N.J., 1964.

4. Engineering Services Data Unit, Response of Flexible Structures to Atmospheric Turbulence, Data Item 76001, E.S.D.U., London, 1976.

5. Steffens, R.J., Structural Vibration and Damage, Building Research Establishment Report. London, HMSO, 1974.

6. Lee, B.E., Extreme Wind Data and Building Design, Weather, Vol. 33, No. 11, Nov. 1978. pp. 416-423.

7. National Research Council of Canada, Commentaries on Part 4 of the National Building Code of Canada 1975, Ottawa, 1975.

8. Simiu, E. and Lozier, D.W., The Buffeting of Tall Structures by Strong Winds, U.S. Department of Commerce, National Bureau of Standards, Washington D.C., 1975.

BUFFETING EFFECTS OF UPSTREAM BUILDINGS

J. W. SAUNDERS* and W. H. MELBOURNE**

*Royal Melbourne Institute of Technology, Melbourne, Australia
**Monash University, Melbourne, Australia

ABSTRACT

This study systematically modelled the buffeting effects on a 150 x 37 x 37 m building of upstream single and twin buildings having a clear upstream fetch (e.g. water, desert). The results show that the presence of an upstream building of the same size can increase the dynamic loads by a factor of 1.65. A 220 m building upstream will increase the factor to 1.9. Twin 220 m buildings upstream can further increase this factor to 2.15. Buffeting effects can be significant for distances of 1 km. The maximum buffeting loads would have serious effects on a building designed to usual wind codes.

INTRODUCTION

The buffeting of small, medium and large buildings by upstream buildings has become an increasing problem because:

(a) Tall buildings have been built upstream of existing structures, particularly on re-claimed land.

(b) Buffeting of downstream buildings by upstream buildings can and has produced undesirable dynamic motion and loading of the downstream buildings. Habitation has become unpleasant in a number of instances.

(c) If the upstream building is built after the downstream building has been built, the owners of the downstream building can take legal action.

(d) Whilst there is proprietary information on buffeting effects from specific projects, little is readily available to the designer of a more general nature.

It is hoped that this initial study will promote discussion of this problem and also provide some more generally useful information to planners and designers.

There are a large number of variables present when upstream structures are considered. Therefore the experimental approach was to keep the model building configuration as simple as possible and search for limit conditions.

High winds can occur at low levels off waterfronts or near deserts. A designer of a medium-size building in using standard wind codes would normally only design for low intensity winds. Standard codes offer little guidance for the effects of buffeting if a large building is eventually built upstream of the medium-size building.

The effect of an upstream building will be to increase the turbulence intensity. This will broaden the band-width of the crosswind force spectrum (1,2). If the upstream building is of rectangular section and facing the mean wind, then the pressure distribution of the front face will also tend to induce higher velocity wind to lower levels. Both of these effects will increase the amount of energy available in the high-frequency side-band of the vortex-shedding mechanism. This will substantially increase the crosswind vibration of the downstream building. This effect may be further amplified by the presence of twin upstream buildings rather than a single upstream building.

Therefore, if the downstream medium-size buildings have been designed fairly economically, then the presence of upstream buildings may present a potentially dangerous (or at least expensive) configuration, causing excessive dynamic motion, creaking and groaning, and motion-perception by the occupants of the downstream building.

Melbourne and Sharp (3) systematically studied the buffeting effects on a model building of proportions hxbxd = 6x1x1 in a suburban and urban environment with a single upstream building of the same size (h=height, b=breadth, d=depth). They found that the rms crosswind (CW) response was increased by a maximum of 75% in the suburban environment, while in the urban environment the increase was a lesser value of 25%. In the alongwind (AW) direction the respective increases were 25% and 15%. These figures show the trend discussed above. Melbourne and Sharp used a higher intensity profile than was used in this study.

Therefore the possible limiting conditions for maximum buffeting of a medium-size building have assessed as:

(a) a clear upstream fetch, such as open water, or desert, or open plants;

(b) an upstream building having the same, larger and smaller height, but with the same plan area; and alternatively

(c) twin upstream buildings.

WIND MODEL

A 1/400 wind model was developed by Melbourne (4) and the tests were conducted using this wind model in the Boundary Layer Wind Tunnel at Monash University. Its characteristics are shown in Fig. 1. The power law exponent for the mean velocity

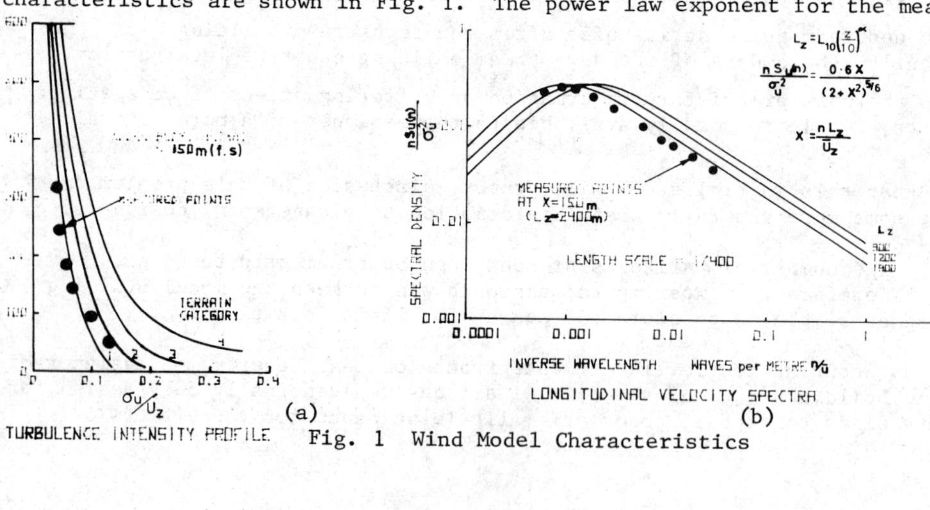

(a) (b)

Fig. 1 Wind Model Characteristics

profile is 0.13 based on a gradient height of 350 m (1150 ft). This wind model compares quite closely with the Australian Wind Code recommendations for the wind characteristics over sea or desert (5).

DYNAMIC MODEL BUILDING

The characteristics of the building modelled (to a 1/400 scale) were:- proportions hxbxd = 4x1x1; height,h = 150 m (500 ft); breadth,b = 37 m (120 ft); density = 175 kg/m^3 (11 lb/cu.ft.); and percentage critical damping = 1%. The mode shape of the model buioding was linear and the overturning moment was measured at the base of the model using strain gauges.

REDUCED VELOCITY RANGE

For the purposes of this paper, the reduced velocity, U_r is defined as,

$$U_r = \frac{\overline{U}_h}{n_o b} \qquad \ldots (1)$$

where $\overline{U}_h$ = hourly mean wind velocity at the top of the building;

n_o = natural frequency of building; and b = building breadth.

The prime wind design criterion of a building being subjected to buffeting from an upstream building will normally be motion perception by the occupants. A wind return period of less than 5 years will normally be appropriate. At a 5 year return period, a medium-size building will normally operate at reduced velocities up to about 6. At lower reduced velocities the level of response will be more important as the frequency of occurrence will be more often. Therefore U_r was varied from 2 to 6.

ISOLATED MODEL BUILDING CONDITIONS

Fig. 2 shows the dynamic characteristics of the model tested in isolation which were used as the reference conditions. Moment coefficient is defined as $m/\frac{1}{2}\rho\overline{U}_h^2 h^2 b$ where m = overturning moment about ground level and measured at the base of the building and ρ = air density.

BUFFETING FACTOR

A Buffeting Factor (BF) has been defined which is the ratio of the overturning moment about the base when there are upstream structures present to the overturning

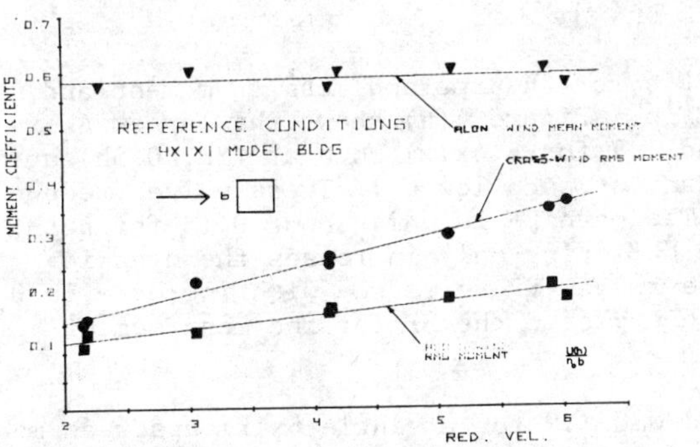

Fig. 2 Isolated Model Building Reference

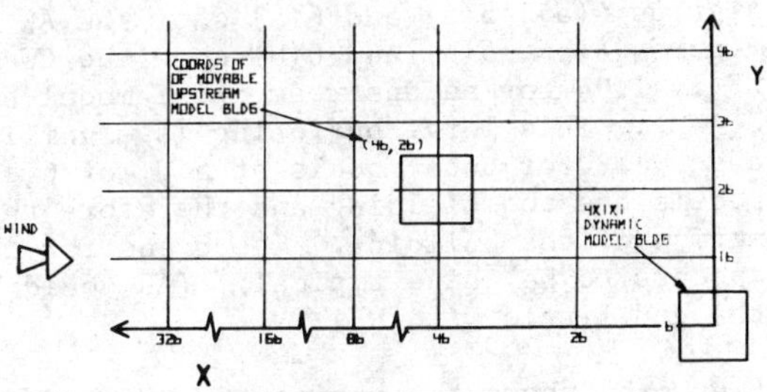

Fig. 3 Moveable Upstream Model Building Grid System

moment on the isolated building. Therefore,

$$BF = \frac{\text{Moment (upstream structures present)}}{\text{Moment (isolated building)}} \qquad \ldots (2)$$

The use of a BF concept allows convenient use with existing procedures for evaluating loads on buildings. In particular, the approach developed by Saunders (1,2,6) for evaluating the crosswind dynamic response of an isolated building can be easily adjusted for the presence of an upstream building using a BF.

UPSTREAM MODEL BUILDING TYPES

The types of upstream building types considered were:- 'Tall' building : hxbxd = 6x1x1; 'Medium-size' building : hxbxd = 4x1x1 (same size as the dynamic model); and 'Smaller' building : hxbxd = 3x1x1. All models had the same plan area. The upstream models were placed on a grid on the tunnel floor as shown in Fig. 3.

ORIENTATION OF DOWNSTREAM DYNAMIC BUILDING

Extensive systematic tests by Saunders (2) on various rectangular section model buildings showed that rectangular buildings were more sensitive to crosswind motion when one face is perpendicular to the mean wind velocity than at any other orientation.

A model building of same size was placed upstream of the dynamic model and initial tests showed that at about X = 8b, the crosswind response was significant. The upstream building was then progressively moved across the front of the dynamic model from Y = 0 to 4b. The dynamic model was orientated with one face normal to the flow and also at 45° to the flow. The results consistently showed the model orientated with zero angle of attack to have a response greater than with an angle of attack of 45°. This confirmed the expected trend.

BUFFETING FACTORS FOR SINGLE UPSTREAM MODEL BUILDINGS

Figs. 4,5 and 6 show the buffeting factors for the along-wind (AW) and cross-wind (CW) buffeting at reduced velocities, U_r of 2,4 and 6, when a single upstream model building of the various sizes was located at the grid positions shown. (One test was conducted with a 4x1x1 model outside of the grid range at X=0, Y=2b. No significant buffeting effect was evident.)

Each of these figures is a build-up of 6 graphs, each graph with the Y axis grid position on the vertical axis and the buffeting factors (BF's) along the horizontal scale from 0 to 2. The BF's for the 4x1x1 dynamic model are plotted for the upstream models located on the various grid positions. For example in Fig. 4b for a 6x1x1 model placed at X=8b and between Y=3b - 4b, the BF's were measured to be between 1.4 - 1.5.

Figures 4(a), 5(a) and 6(a) show the AW BF's for the rms and mean AW moment and Figures 4(b), 5(b) and 6(b) show the CW BF's. Figure 4(a) shows the largest AW rms BF is 1.94 for an upstream 6x1x1 model and 1.7 for a 4x1x1 both at (2b, 0.5b) at U_r = 2. This large buffeting is significant at such low U_r. These values reduce to about 1.2 for both models at a U_r of 6. The mean BF is only about 0.15 for both models due to shielding and therefore the buffeting only increases the dynamic motion of the building. The trend for the rms AW BF to reduce with increase in U_r appears to be quite typical. (As could be expected the BF for the mean moment changed little with U_r).

Fig. 6(b) shows the maximum CW BF measured was 1.9 for a single 6x1x1 upstream model at X = 2b directly upstream and at U_r = 6. The CW BF reduced to 1.02 at U_r = 2.

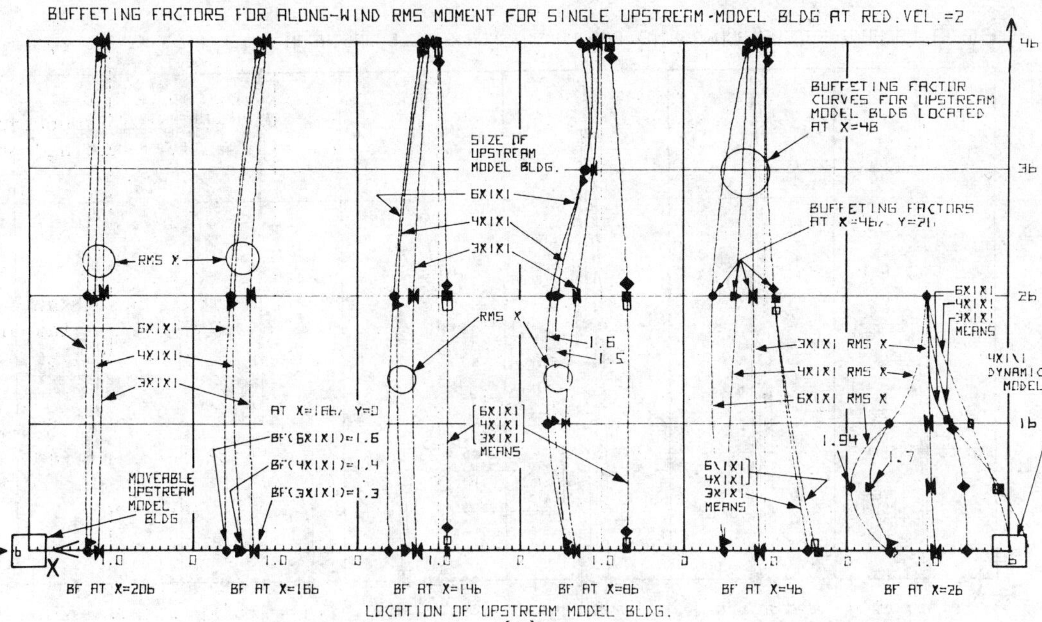

(a)

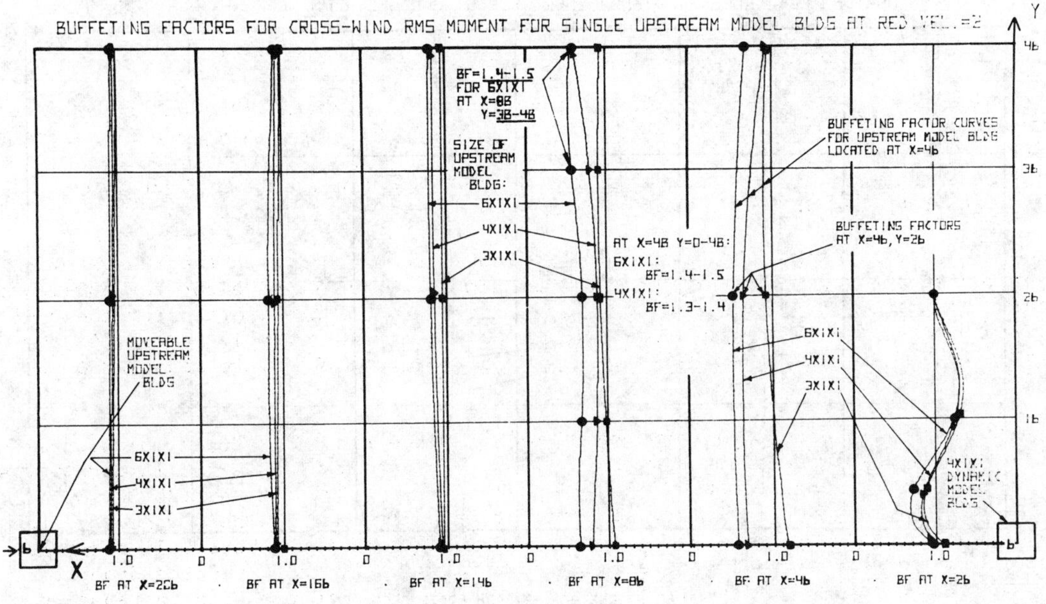

(b)

Fig. 4 Buffeting Factors for Single Upstream Building at Reduced Velocity = 2

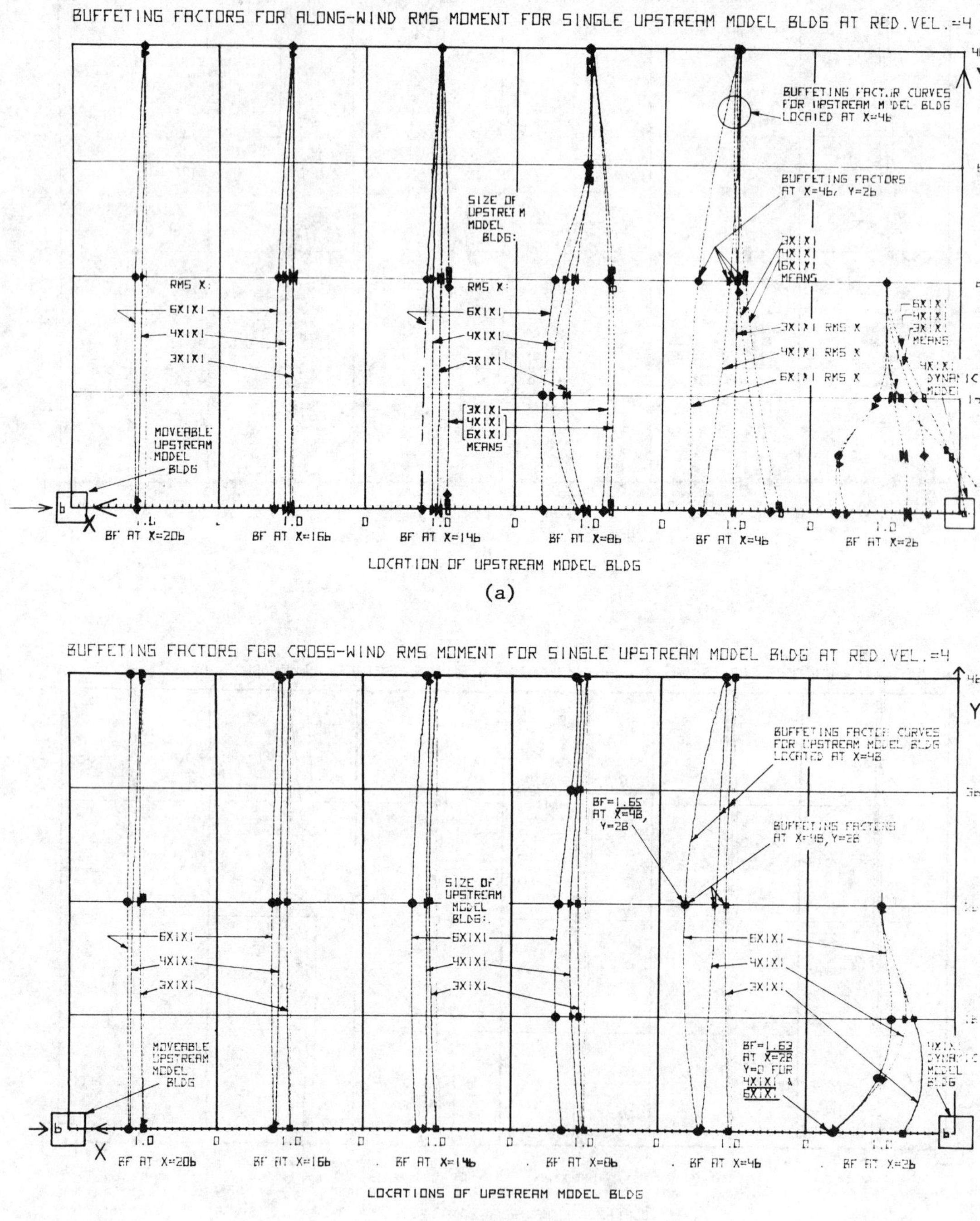

Fig. 5 Buffeting Factors for Single Upstream Building at Reduced Velocity = 4

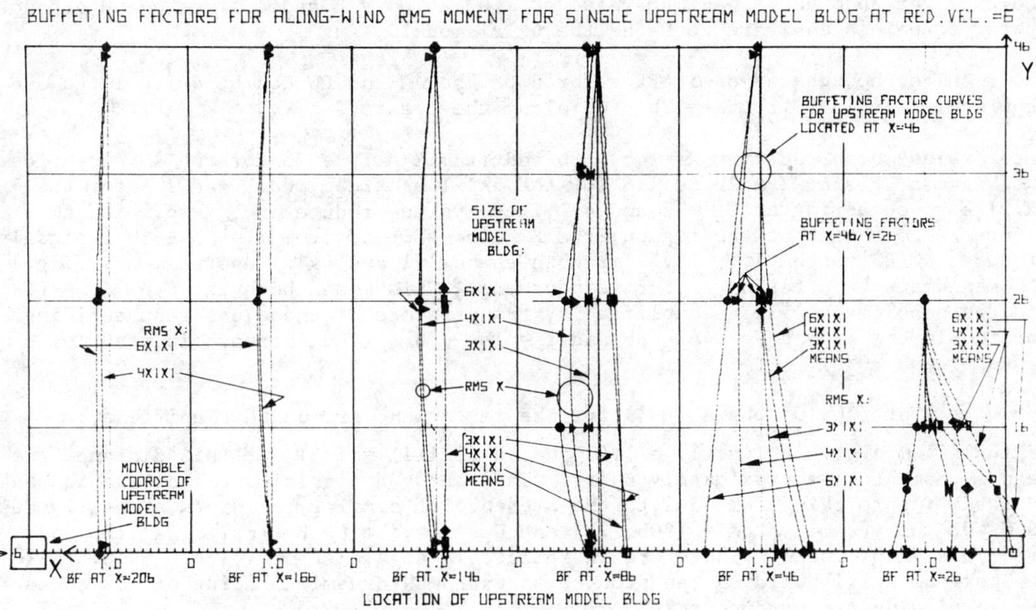

(a)

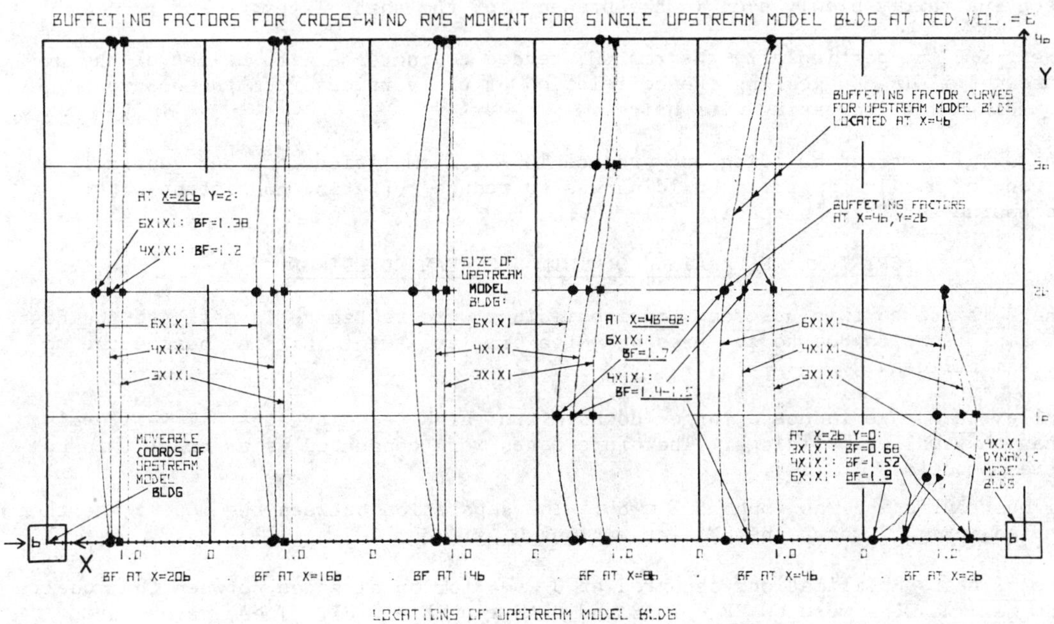

(b)

Buffeting Factors for Single Upstream Building at Reduced Velocity = 6

The CW BF typically showed an increase with U_r. For the 4x1x1 upstream building, the CW BF at 2b,0 had a maximum value of 1.63 at U_r = 4 and 1.52 at 6 and did not appear to be as sensitive to U_r as the 6x1x1 model.

The results show that dynamic BF's for both the AW and CW can be quite large and show maximum values from X = 2b - 4b along the Y axis.

Large values occurred from X = 4b - 8b and from Y = 1b - 3b for the AW direction. The highest BF's at Y = 2b is 1.6 for the 6x1x1 upstream model and 1.5 for the 4x1x1 both occurring at U_r = 2 and 4. These values reduce at U_r = 6. The CW figures also show that the region from X = 4b - 8b and from Y = 1b - 3b typically produces CW BF's from 1.4 - 1.7 for both the 6x1x1 and 4x1x1 upstream building. The mean moment buffeting factor is near one. This means both the dynamic effectc and the peak loads will be amplified by the presence of these upstream buildings located in the region X = 4b - 8b and Y = 1b - 3b. (The increase in peak loads can approach 20%.)

For U_r = 2 at (16b,0) the AW BF's for the rms moment are quite significant: BF (6x1x1) = 1.6; BF (4x1x1) = 1.4 and BF (3x1x1) = 1.3. At this distance the mean AW moment is approximately unity. This shows a significant increase in peak AW loads (approaching 30%) due to the presence of similar buildings at nearly 600 m (650 yds) upstream. At X = (20b, 2b) and U_r = 6, the CW BF (6x1x1) = 1.38 and 1.2 for a 4x1x1 model. Therefore in full scale at a distance of 750 m (0.5 mile), an upstream 6x1x1 building can produce an extra CW dynamic loading of 40% and a 4x1x1 building can produce a 20%.

Both these results showed that the distance which the vortices carry energy downstream is apparently quite substantial and the energy in the vortices is therefore decaying fairly slowly even in the presence of the shear layer.

The rms BF's, particularly the rms CW, tended to increase with height of the upstream building suggesting the correlation of the vortices from the upstream buildings was increasing with height.

The 3x1x1 upstream building can produce 10 - 15% amplification, but generally the effect of a 3x1x1 upstream building was to reduce buffeting consistent with Melbourne and Sharp's results (3).

'CRITICAL' SEPARATION OF TWIN UPSTREAM BUILDINGS

The BF's due to twin upstream model buildings were tested to investigate the degree of amplification due to the presence of a similar 'twin' building beside the upstream building.

Taller buildings induce a larger downdraft of high velocity air to lower levels than the smaller buildings. Therefore tests were conducted using 6x1x1 twin upstream models.

Highest buffeting was found at X = 8b. The separation between the models was then varied. Fig. 7 shows the BF's at various U_r's.

The highest amplifications occurred at U_r = 4 for 5b airspace between the models. The maximum BF's were CW BF = 2.13 and AW rms BF = 1.87. These values are substantially higher than for a single 6x1x1 model at 8b, 3b (corresponding to one half of the pair): CW BF = 1.2 and AW rms BF = 1.0. The presence of the additional building has produced a 90% amplification of CW rms BF and 80% amplification of the AW rms BF.

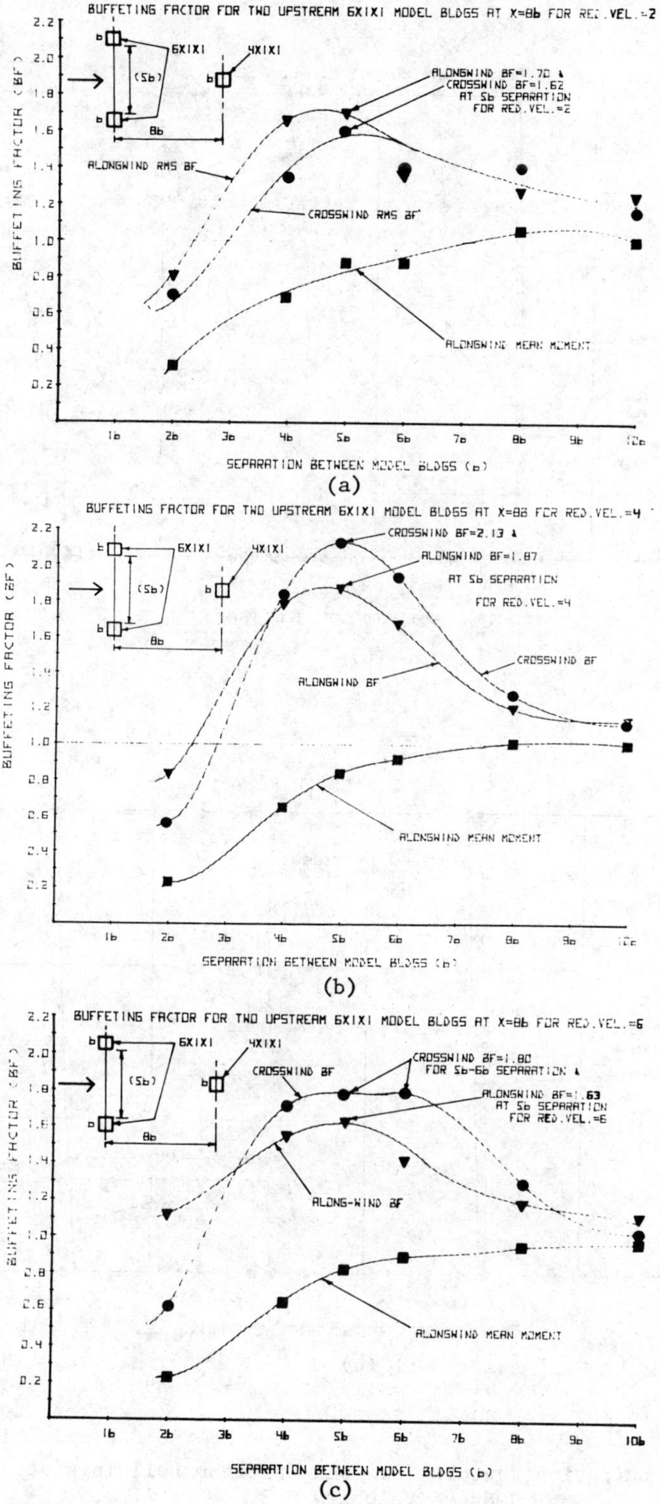

Fig. 7 Buffeting Factors at Varying Separation

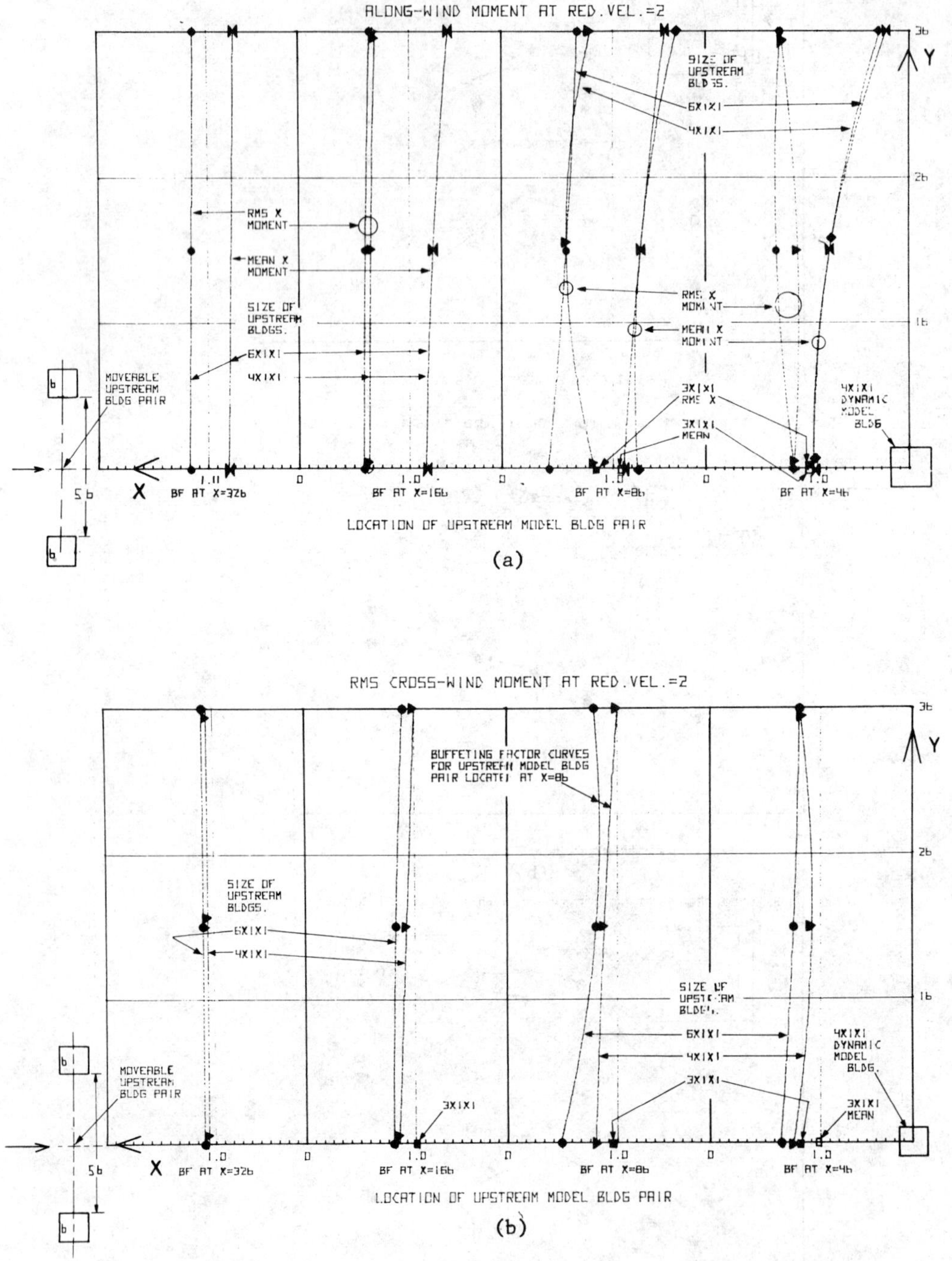

Fig. 8 Buffeting Factors for Twin Upstream Buildings at Reduced Velocity = 2

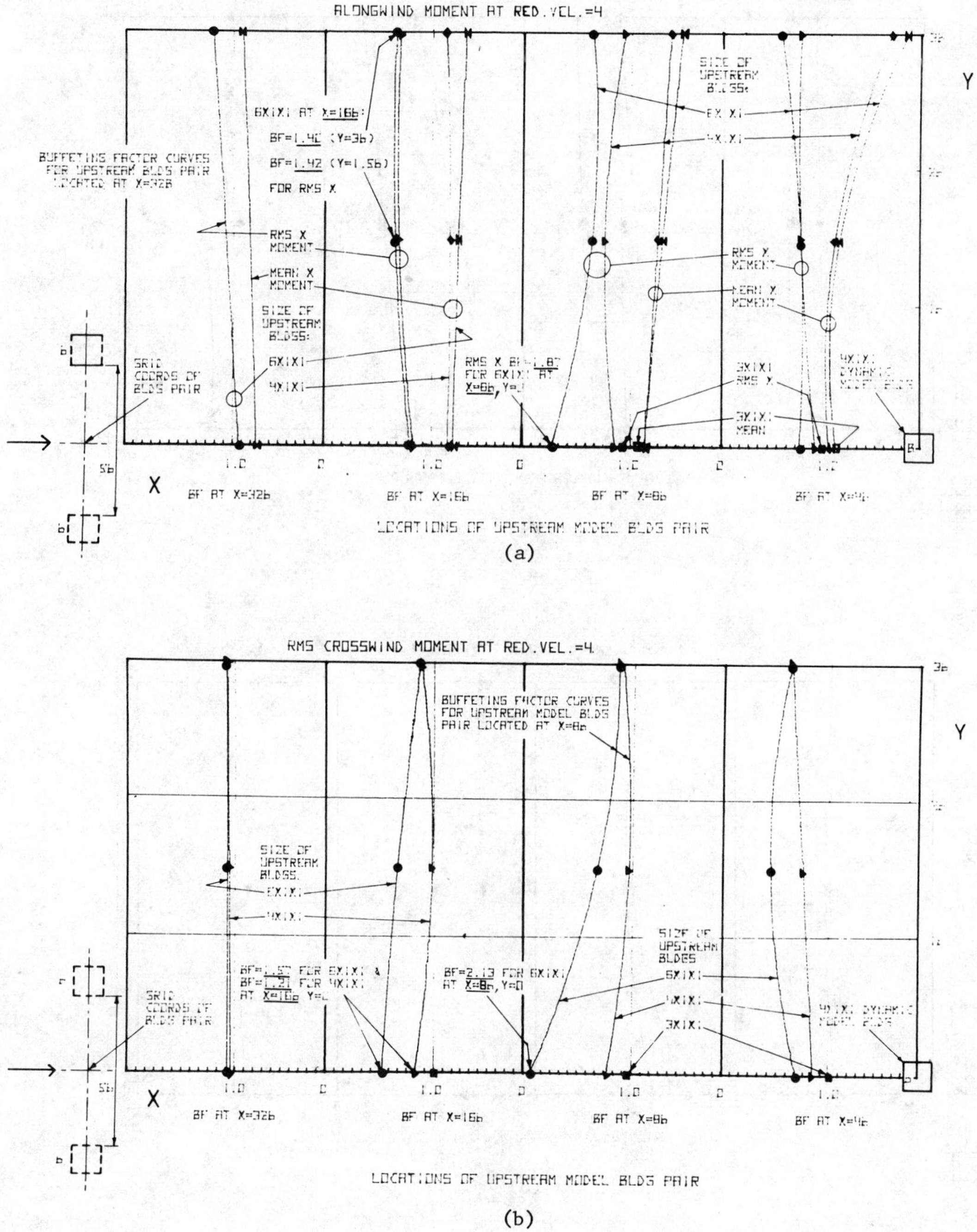

Fig. 9 Buffeting Factors for Twin Upstream Buildings at Reduced Velocity = 4

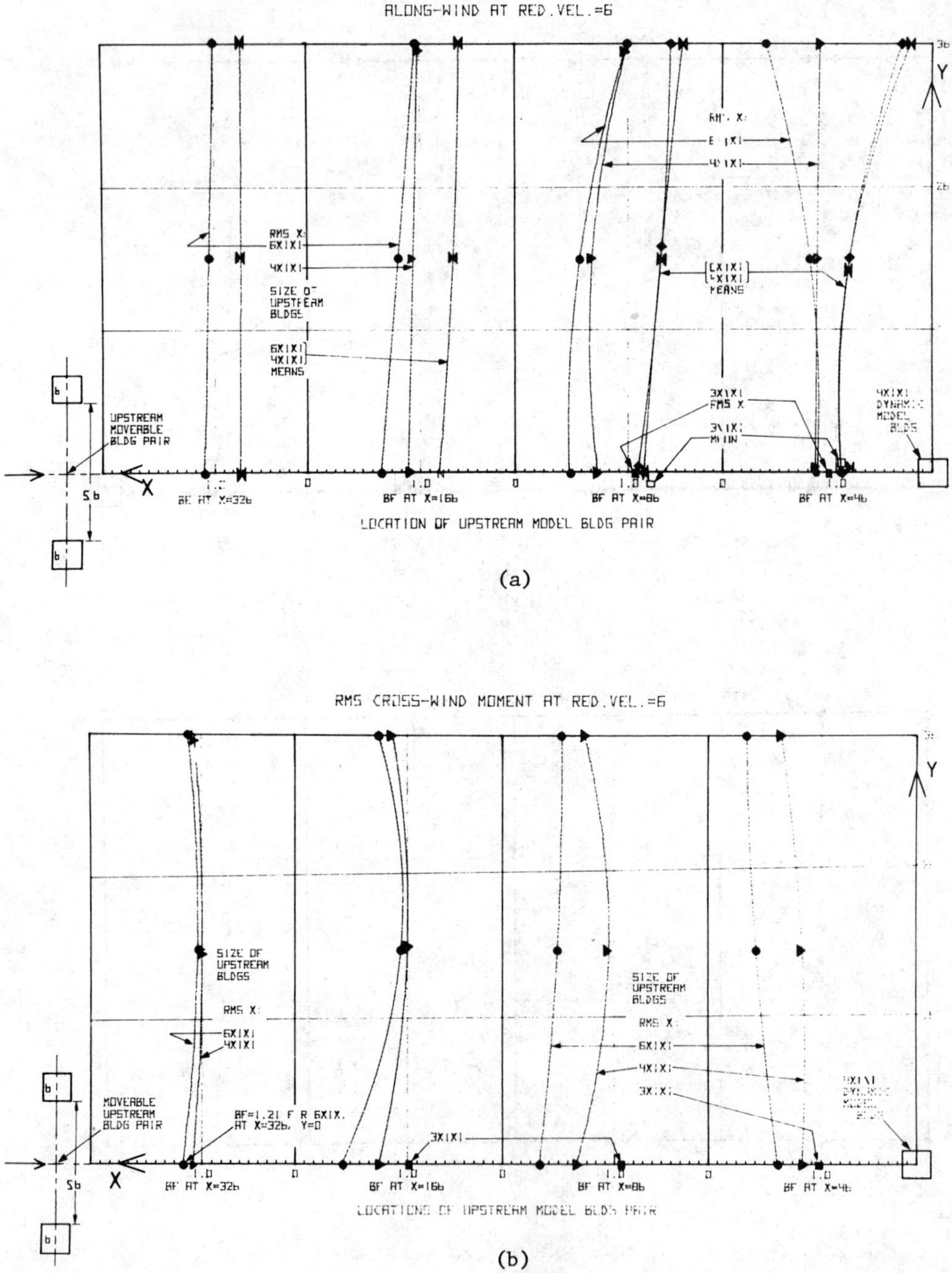

Fig. 10 Buffeting Factors for Twin Upstream Buildings at Reduced Velocity = 6

An airspace of 4b - 6b produces the most buffeting at each U_r and the rms buffeting factors are generally above 1.6 for both the AW and CW direction.

Tests were also conducted at the same upstream location with the orientation of the axis between the two buildings set at an angle of attack of 45° and 90° , and also with the separation between the models increased and also decreased. In all cases the buffeting was lower. It was then assumed that for twin upstream buildings that an airspace of about 5b produced the maximum buffeting effects as the pair were moved around the BF grid.

The tests were repeated on twin upstream buildings with each body axis set at 45 (model corners CW). These tests resulted in an airspace of 5b also producing the maximum buffeting but with lower buffeting factors.

BUFFETING FACTORS AT 5b AIRSPACE

The twin model buildings were set at 5b airspace and the pair were moved around a grid bounded by X = 4b - 32b and Y = 0 - 3b. The model pair centreline at Y = 3b corresponded to one model in line with the downstream 4x1x1 model.

Figs. 8, 9 and 10 show the AW and CW results for U_r = 2,4 and 6 for the 6x1x1, 4x1x1 and 3x1x1 upstream twin models.

The maximum buffeting factors occur at 8b directly downstream at U_r = 4. The maximum BF (6x1x1 pair) for the CW rms is approximately 2.15 and the maximum AW rms BF is approximately 1.9. The mean AW BF = 0.9 and therefore there will be a substantial increase in peak AW response (40% increase). These BF's produce larger dynamic and peak deflections than for those occurring under the maximum conditions for a single building near (2,0). A test of an 8x1x1 pair at X = 8b, Y = 0 did not increase the BF's above those of the 6x1x1 pair. Alternatively, the 3x1x1 pair produced minimal amplification.

In general X = 4b - 16b produces the highest BF's for both the 6x1x1 and the 4x1x1 twin models where mean BF = 0.8-1.0; AW rms BF = 1.2 - 1.9; and CW rms BF = 1.2 - 2.15.

When one model of the pair was in line with the downstream dynamic model (along X axis), the rms BF's were consistently above the values for a single upstream building located on the X axis.

At (32b,0), which is 1.2 km (0.75 mile) upstream in full scale, a 6x1x1 upstream pair produced the alongwind BF = 1.2 for low reduced velocities and the crosswind value was 1.2 at U_r = 6. This again shows that at large distances downstream the energy in the wake of large buildings is quite significant.

Tests were also conducted on twin upstream buildings with each body axis set at 45° (corners crosswind). The results were similar but with generally lower buffeting factors.

PEAK FACTORS

The peak factors required to calculate the peak moments were measured for the configurations producing the maximum BF's. The response was found to approximately Normal and a peak factor of 3.7 is appropriate for 1000 cycles/hour full scale.

ACCURACY OF BUFFETING FACTORS

The accuracy of the BF's factors measured at reduced velocities of 4 and 6 was assessed to be within 10% and at the lowest reduced velocity of 2 to be within 15% due to lower incident velocity and some background noise from tunnel vibration.

CONCLUSIONS

The various detailed conclusions in the test will not be repeated.

The general conclusions are that a medium-size (4x1x1) or a tall (6x1x1) building can produce substantial buffeting of similar size and smaller downstream buildings. The presence of an additional building forming a building pair of 5b airspace will further amplify the buffeting effects of the downstream building.

These buffeting factor results can be used with existing wind codes to make an allowance for ex isting or future buildings. In particular, the buffeting factors can be used in conjunction with the procedure proposed by Saunders (1,2,6) for evaluating the design sway stiffness of a building.

It can be seen that from the level of the buffeting found in this study, if a medium-size building near a waterfront or other clear fetch has similar or taller buildings built upstream at the critical locations, then the downstream buildings will suffer significant increase in peak loads and serious dynamic effects.

ACKNOWLEDGEMENTS

The senior author is indebted to the Mechanical Engineering Department at Monash University for making the Boundary Layer Wind Tunnel available and to the staff of RMIT, in particular for Peter Kilpatrick's assistance.

REFERENCES

1. Saunders, J.W., Wind Excitation of Tall Buildings. PhD thesis, Monash University, Clayton, Victoria, Australia, 1974.

2. Saunders, J.W. and Melbourne, W.H., Tall Rectangular Building Response to Cross-Wind Excitation, Proc. Fourth International Conference on Wind Effects on Buildings and Structures ed. K.J. Eaton, Cambridge University Press 1975, pp. 369-379.

3. Melbourne & Sharp, D.B., Effects of Upwind Buildings on the Response of Tall Buildings. Proceedings of the Regional Conference on Tall Buildings, Hong Kong, 20-22 Sept., 1976, pp. 174-191.

4. Melbourne, W.H., Development of Natural Wind Models at Monash University. Proc. 6th Australasian Hydraulics and Fluid Mechanics Conference, Adelaide, 1977, Institution of Engineers, Australia National Conference Publication No. 77/12, Canberra, pp. 190-194.

5. SAA Loading Code Pt. 2 - Wind Forces, AS 1170 - Pt. 2 - 1975, Standards Association of Australia, Sydney.

6. Saunders, J.W. and Melbourne, W.H., Wind-Excited Buildings : Design Sway Stiffness, submitted to ASCE for possible publication Nov. 1977, under review.

UNIVERSAL SIMILARITY IN THE WAKES OF STATIONARY AND VIBRATING BLUFF STRUCTURES

O. M. GRIFFIN

Naval Research Laboratory, Washington, D.C., USA

ABSTRACT

A universal wake Strouhal number, $St^* = f_s d'/U_b$, has been proposed and is based upon the Strouhal frequency f_s of the incident flow, the *measured* wake width d' at the end of the vortex formation region, and the mean velocity U_b at the edge of the separated boundary layer. This universal parameter collapses the characteristic wake scales for bluff bodies onto a single curve over the entire range of subcritical and transcritical wake Reynolds numbers Re*. The pressure drag, vortex shedding frequency and base pressure are related through a dependence between St C_D, the product of the Strouhal number and drag coefficient, and the base pressure parameter K. There is a general collapse of the data up to K = 2, which is the upper limit thus far for bluff body flows.

INTRODUCTION

It is often found that bluff, or unstreamlined, structures display damaging oscillatory instabilities due to wind action. A common mechanism for causing the wind-excited oscillations of structures is the organized and periodic shedding of vortices as the flow separates alternately from the structure. The flow field exhibits a dominant periodicity and the body is acted upon by time-dependent pressure loads which result in steady and unsteady drag forces in-line with the flow and unsteady lift or side forces in the cross flow direction. If the structure is flexible and lightly damped, then resonant oscillations can be excited normal or parallel to the incident flow. For the more common cross flow oscillations the body and the wake have the same frequency near one of the characteristic frequencies of the structure. The shedding then is shifted away from the natural, or Strouhal, frequency at which pairs of vortices would be shed if the structure were stationary.

Bluff body wakes have been the subject of constant study since the 1870's because of the intrinsic complexities and the practical importance of these flows. Just recently McCroskey (1) wryly noted that the problem of a bluff cylinder in a cross flow may have been invented "to insure a perpetual livelihood for fluid dynamicists and flow-visualization photographers."

This paper investigates the question of fluid dynamic similarity in the wakes of bluff bodies. Griffin (2) and Simmons (3) recently have introduced a universal Strouhal number which is valid at subcritical and transcritical Reynolds numbers for stationary two-dimensional bluff bodies with fixed and free separation points, vibrating bluff structures, and cylinders in confined flow and at large yaw angles to the incident flow. This Strouhal number $St^* = f_s d'/U_b$ is based upon the natural Strouhal frequency of the wake, f_s; the wake width d' at the end of the vortex formation region; and the mean velocity U_b at the edge of the boundary layer. The pressure drag, vortex shedding frequency and base pressure are also related by means of a correlation between St C_D, the product of the Strouhal number and the drag coefficient, and the base pressure parameter $K = (1-C_{pb})^{1/2}$.

RELATED INVESTIGATIONS

Some years ago Roshko (4) proposed a universal Strouhal number to scale the flow separation and vortex formation which take place in the near wakes of stationary bluff bodies of various cross sections. More recently Bearman (5) expanded on Roshko's original ideas and proposed a somewhat different semi-empirical model for unifying the various parameters which affect the vortex formation and the wake dynamics. Both formulations were similar in that the wake width—the principal length scale in the two models—was determined analytically. In Roshko's model the wake width was determined from a "notched hodograph" theory, while in Bearman's formulation the Kronauer stability criterion was used as a basis for computing the wake width. The vortex shedding frequency and the base pressure or, equivalently, the mean velocity on the free streamline at separation were obtained from experimental measurements in both cases.

Calvert (6) introduced a universal Strouhal number in his measurements of the wake formation behind cones with different vertex angles, and defined d' to be the *measured* transverse distance between the peaks of the velocity fluctuations, at the point of minimum mean pressure on the downstream axis of the wake. The Strouhal number St^* obtained in this way was found to be constant at about $St^* = 0.19$ for cone angles between $\alpha = 0$ and $90°$.

The measured wake width d' received no attention until Simmons (3) and Griffin (2) employed such a length parameter to scale the vortex formation processes in the wake of two-dimensional bluff bodies. Physically d' is the wake width at the end of the vortex formation region, where a vortex is fully formed and moves away from the base region of a bluff body.

Simmons introduced a universal Strouhal number St^*, based on the measured wake width d', for the wakes of two-dimensional bluff bodies with fixed separation angles between $\alpha = 0°$ and $90°$ and found a constant value of about $St^* = 0.163$. Griffin at the same time introduced the same St^* as part of an investigation of similarity in the wakes of circular cylinders vibrating in the regime of locking-on between the vortex and vibration frequencies. An average $St^* = 0.178$ was found for wake Reynolds numbers between 700 to $5(10^4)$ for vibrating cylinders.

Ramberg (7) has shown that the concept of universal wake similarity is valid for circular cylinders at yawed incidence. Measurements of the vortex shedding frequency f_s, the wake width d' and the base pressure coefficient C_{pb} were made at yaw angles up to $60°$ away from the normal incidence.

THE 'UNIVERSAL' STROUHAL NUMBER

Roshko (4) demonstrated that a characteristic non-dimensional parameter for scaling of the wakes of bluff bodies could be derived by applying relatively simple physical arguments. The result, after several improvements and refinements (2,3,5,6), is a universal wake Strouhal number St^* for vortex shedding from bluff bodies based upon measured parameters of the wake.

If one considers two shear layers a distance d' apart with the velocity just outside the layers equal to U_b, the mean velocity at separation, then a wake Strouhal number can be defined as

$$St^* = \frac{f_s d'}{U_b} = St\left(\frac{U}{U_b}\right)\left(\frac{d'}{d}\right) . \quad (1)$$

The characteristic frequency f_s associated with the configuration is assumed to be proportional to the ratio U_b/d', as sketched in Figure 1. Here the usual Strouhal number of the cylinder is

$$St = \frac{f_s d}{U} \quad (2)$$

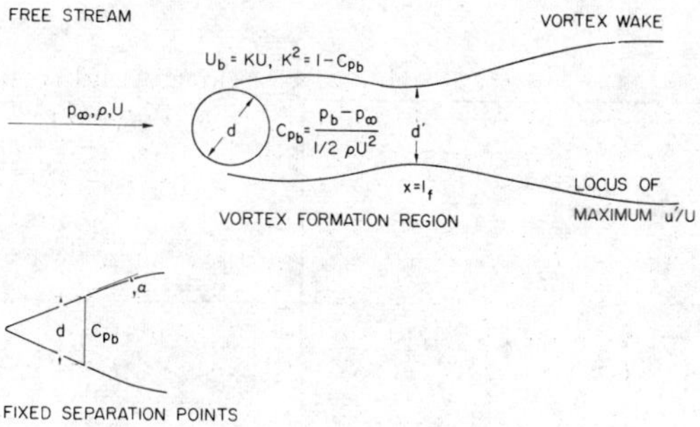

Fig. 1. Characteristic length, velocity, and pressure scales for the wakes of two-dimensional bluff bodies.

where d is the cylinder diameter and U is the incident flow speed. When Bernoulli's equation is applied to the flow just outside the boundary layer at separation, the base pressure coefficient is

$$C_{pb} = \frac{2\left(p_b - p_\infty\right)}{\rho U^2} = 1 - \left(\frac{U_b}{U}\right)^2 . \tag{3}$$

If the base pressure parameter $K = U_b/U$ is introduced, then

$$K^2 = 1 - C_{pb} \tag{4}$$

and

$$St^* = \frac{St}{K}\left(\frac{d'}{d}\right) . \tag{1a}$$

A wake Reynolds number Re* is defined in a corresponding way as

$$Re^* = \frac{U_b d'}{\nu} = Re\, K\left(\frac{d'}{d}\right). \tag{5}$$

where $Re = Ud/\nu$ is the usual free-stream Reynolds number.

THE WAKE WIDTH FOR BLUFF BODY FLOWS

Stationary circular cylinders.

The width d' of the wake behind a stationary circular cylinder is plotted as a function of the Reynolds number in Figure 2, and the results encompass the entire range of subcritical Reynolds numbers over which regular vortex shedding takes place. The legend for the data points is given in Table 1. Such a definition of the wake width has been shown to be an important scaling parameter for the wakes of both stationary and vibrating circular cylinders (2,8,9) and stationary, two-dimensional bluff cylinders with fixed separation points (3).

With the exception of Reynolds number below Re = 500, there is very little dependence of the wake width upon the Reynolds number. This is true for both smooth cylinders in flows with very low turbulence levels (12, 13) and moderate, or typical, wind tunnel turbulence levels (14, 15, 16). For several of the cases plotted in the figure d' is given by the free-streamline equation (10, 11)

$$\frac{d'}{d} = \frac{C_D}{-C_{pb}} \tag{6}$$

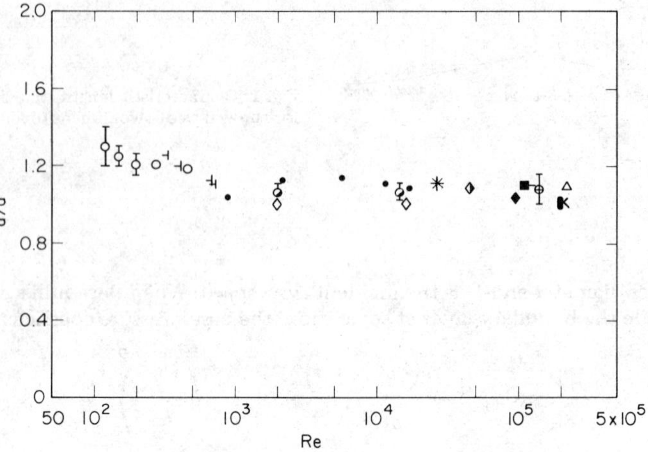

Fig. 2. Wake width d′/d for circular cylinders plotted against the Reynolds number Re; legend as in Table 1, except for the data (◉) of Ballou (14).

Table 1. Universal wake similarity for two-dimensional bluff bodies

Legend for experimental data employed in the St* vs. Re* and St C_D vs. K correlations

Symbol	Bluff body type	Medium	Investigators
○	Smooth circular cylinder (stationary and vibrating)	Air and water	Griffin (1978)+
⊣	Smooth circular cylinder at yawed incidence	Air	Ramberg (1978)+
+	Fixed separation angles between $\alpha = 0°$ (D-shaped) and $\theta = 90°$ (Flat Plate)	Air	Simmons (1977)+
◀	Blunt-based hydrofoil models	Air and water	Blake (1977)+
⊕x	Flat Plate, half cylinder	Air	M.E. Davies (1977 private communication)+; Bearman & Davies (1975)
◆	Smooth circular cylinder in confined flow	Air	Modi & El-Sherbiny (1977)++
⊕	Smooth circular cylinder	Air	Cantwell (1976)+, ++
△	Smooth circular cylinder	Air	Brüun & Davies (1975)++
*	Smooth circular cylinder with and without splitter plate; flat plate	Air and water	Apelt & West (1975); Apelt, West & Szewczyk (1973)+, ++
■	Smooth circular cylinder	Air	Batham (1973)++
◀	Triangular cylinder ($\alpha = 30°$)	Air	Twigge-Molecey & Baines (1973); reference (32)
◆	Smooth circular cylinder in turbulent flow	Air	Surry (1972)
◒	Rectangular cylinders	Air	Bearman & Trueman (1972)
●<	Smooth circular cylinder	Air	Bearman (1969)++
◇	Smooth circular	Air	Gerrard (1965, 1966)++
▽ ▽	D-shaped cylinder with and w/o splitter plates	Air	Bearman (1964)+
◒	Smooth circular cylinder	Air	Roshko (1961)++
●	Smooth circular cylinder	Air	Roshko (1954, 1955)++

+Wake width d′ measured at the end of the vortex formation region.
++Wake width d′ from equation (6).

based upon measured values of C_D and C_{pb}. This equation is valid for the limited conditions of small separation angle α and uniform base pressure C_{pb}. The approximation given by equation (6) holds for stationary circular cylinders with no separation bubbles, roughness effects on transition, etc. as in the critical Re range.

The excellent agreement between this equation and the wake width d′ measured from velocity traverses is typified by the plotted results from Cantwell (16). The vortex formation length ℓ_F, the wake width d′ from hot-wire traverses of the wake, the drag coefficient C_D and the base pressure C_{pb} were measured at a Reynolds number of Re = 1.4 (10^5). The wake widths at formation deduced from equation (6) and from rms profiles of the coherent vortex structure of the turbulent wake are virtually identical and are plotted as a single point in Figure 2. Apelt, West & Szewczyk (17) also measured d′, C_{pb} and C_D, and their result at Re = 2.5 (10^4) also is in good agreement with equation (6).

The results obtained with equation (6) from the experiments of Bruun & Davies (15), Surry (18) and Batham (19) are in excellent agreement with the wake width d′ as measured directly by Ballou (14), Bloor & Gerrard (13) and Ramberg (7) over the Reynolds number range from Re = 700 to 2 (10^4). The wake width behind a circular cylinder increases somewhat at the lowest Reynolds numbers in Figure 2, but there is only a slight Reynolds number dependence from Re = 700 to 2 (10^5) for smooth cylinders in uniform flow.

Vibrating circular cylinders.

When a bluff structure is self-excited by vortex shedding or is forced to vibrate laterally relative to an incident flow in the regime of locking-on or wake capture, there is a complex nonlinear interaction between the wake and the body. The vortex circulation and the base section are strongly enhanced, as are the unsteady lift and the steady drag forces. The response of the wake is dependent upon both the displacement amplitude and the frequency of an oscillation as shown, for example, by Griffin & Ramberg (2,8,9), Bearman & Davies (20) and Stansby (21).

The wake width measured downstream from a vibrating circular cylinder (both self-excited and forced) is plotted in Figure 3 against the vibration frequency (normalized by the corresponding Strouhal frequency f_s) for a typical lateral displacement of 30% of a diameter. Also plotted in the figure are measurements of the base pressure coefficient C_{pb} for a forced vibrating cylinder (21). The similarities of the wake response to the vibrations in terms of the wake width d′ at formation and the base pressure coefficient C_{pb} are evident for this case of locking-on between the vortex and vibration frequencies. All of the data in the figure are divided by corresponding stationary cylinder measurements in order to minimize any Reynolds number effects. Only the frequency dependence of the wake is shown in Figure 3; both the wake width and base suction increase monotonically with the cylinder displacement at a constant frequency in the regime of locking-on, as shown by the results of Griffin (27) and Stansby (21).

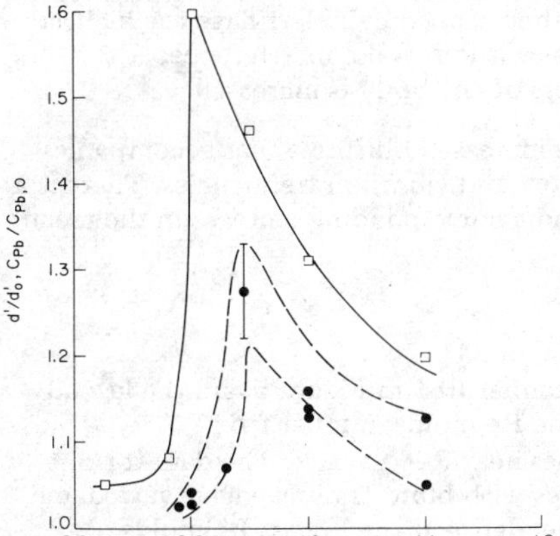

Fig. 3. Wake width d′ and base pressure coefficient C_{pb} measured in the wake of a vibrating circular cylinder, in the regime of locking-on between the wake and cylinder frequencies. Subscripts "O" denote corresponding measurements in the wake of a stationary cylinder. Cylinder displacement amplitude (peak-to-peak) = 0.30 diameter. Present wake width measurements, ●, Reynolds number = 150-700; base pressure measurements, Stansby (21), □, Reynolds number = 8600.

Other two-dimensional bluff bodies.

The wake widths measured at the end of the vortex formation region for these and other bluff bodies are plotted against the base pressure parameter K in Figure 4. The body shapes for which results are included in the figure range from stationary, blunt-based D-shaped cylinders of 6:1 aspect ratio with turbulent boundary layers (22) and of 1.5:1 aspect ratio with laminar boundary layers (3); to stationary circular cylinders at Reynolds numbers between Re = 120 and 2 (10^5); and to wedges, flat plates and half-cylinders normal to the incident flow. Also included in the plot are results for vibrating circular cylinders in the regime of locking-on between the vortex and vibration frequencies (2).

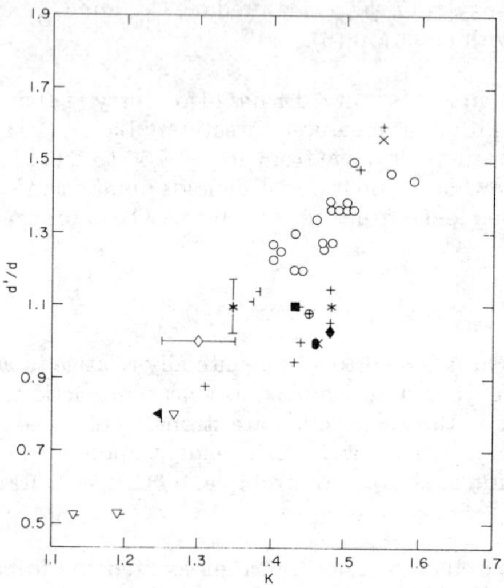

Fig. 4. Wake width d′/d plotted against the base pressure parameter K; legend as in Table 1.

The smallest measured wake widths at formation, for K = 1.1 to 1.2, were obtained by Bearman (22) when the vortex formation was delayed by wake splitter plates. The data in Figure 4 correspond to splitter plate lengths of up to one base height, since under these conditions a regular vortex street was formed. The largest wake widths are those measured downstream from a flat plate, from a half-cylinder and in the wakes of vibrating cylinders.

The results in Figure 4 demonstrate in general the relationship between bluffness (as denoted by d′) and base suction or drag (as denoted by K). An overall dependence between d′ and the separation angle α is also evident. Simmons (3) employed seven models with separation angles between $\alpha = 0°$ (D-shaped cylinder) and $\alpha = 90°$ (flat plate); see the sketch in Figure 1. Both D-shaped cylinders have the highest base pressures (lowest drag) and the narrowest wakes of the bodies without wake interference; as the separation angle increases, the wake becomes wider as the bluffness of the body is increased.

The vibrating cylinder measurements in Figure 4 demonstrate the increased bluffness that accompanies the lateral vibrations in the regime of locking-on between the vortex and vibration frequencies. There is a decrease in base pressure as the cylinder vibrates (see Figure 3) and a corresponding increase in the mean separation velocity U_b as the wake width increases.

EXPERIMENTAL MEASUREMENTS OF St*

The Strouhal number St* is plotted against the wake Reynolds number Re* in Figure 5 and the legend for the data is given in Table 1. The results span five decades of the Reynolds number from $Re^* = 10^2$ to 10^7, except for the critical and supercritical regimes near $Re^* = 10^6$. The data for St* encompass both stationary and vibrating bluff cylinders and bodies with blunt trailing edges, yawed and normal to the incident flow. There is some Reynolds number dependence at the lowest Reynolds numbers, as is the case for the classical St vs. Re relationship, but only a slight dependence of St* upon Re* over the remainder of the subcritical range.

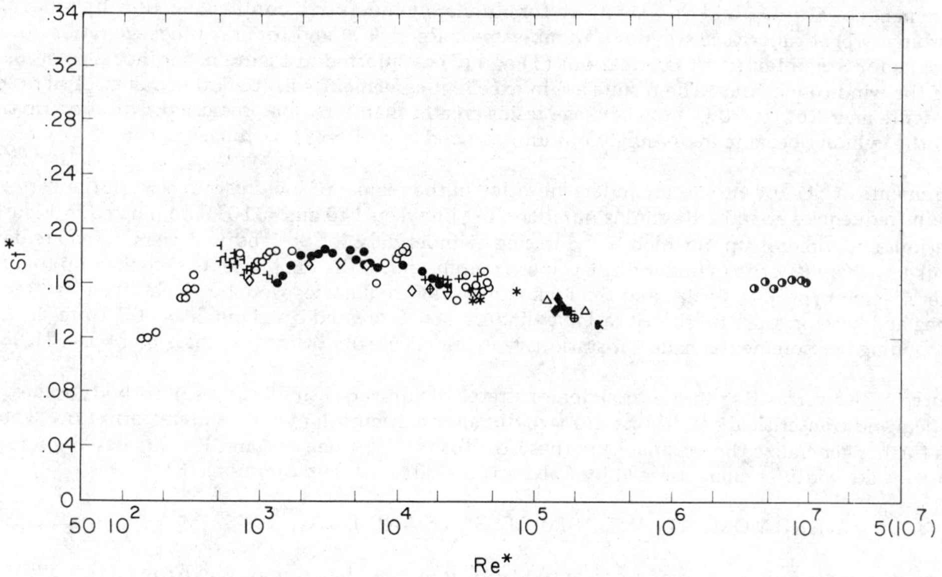

Fig. 5. Universal Strouhal number St* plotted against the wake Reynolds number Re*; legend as in Table 1

A number of results for yawed cylinders are included in the correlation (7). The yaw angle θ of a stationary circular cylinder was increased to as much as 60° from normal incidence and simultaneous measurements were made of the wake structure, including the base pressure coefficient C_{pb}, the wake width d' at formation and the shedding frequency f_s. The flow deviated substantially from the so-called Independence Principle, or Cosine Law, where the wake flow and fluid forces are assumed to be governed by the normal component of the incident velocity to the cylinder, but the wake parameters C_{pb}, d' and f_s varied in such way as to maintain a nearly constant value of St*. The base suction C_{pb} at yaw angles of θ = 40° to 60° was 20% and 30% higher than would be predicted by the Independence Principle.

Data for St* derived from nominally smooth circular cylinders in uniform flow reach a limiting range of St* = 0.14 to 0.15 as the critical Reynolds number, Re* = 2 to 3 (10^5), is approached. All of the St* values obtained in this range from the results of Batham (19), Bearman (12), Bruun & Davies (15) and Cantwell (16) are in good agreement. The flow was subcritical in all cases, at Reynolds numbers just below the critical value. The measurements of St* made at transcritical Reynolds numbers by Roshko (11) are also plotted in Figure 5. These results were based upon measurements of the shedding frequency, base pressure and drag and confirm the universality principle not only at subcritical Reynolds numbers but also at transcritical Reynolds numbers at least to Re* = 10^7.

There is considerable diversity in the results for St* over the Reynolds number range from Re* = 10^4 to 10^5. The body cross-sections included there (see Table 1) consist of stationary and vibrating circular cylinders, and a variety of bluff bodies with fixed separation angles between α = 0° (D-shaped cylinder) and 90° (flat plate). Several of the models (22,23) also include wake interference with splitter plates; Bearman's experiments with a blunt-based, D-shaped cylinder included direct measurement of the wake width d' at formation in addition to the shedding frequency f_s and the base pressure coefficient C_{pb}. The experiments reported by Blake (24), Apelt, West & Szewczyk (17) and Simmons (3) also included the simultaneous measurements of the parameters that comprise the universal Strouhal number St*.

The values of St* for stationary circular cylinders obtained from Roshko's and Gerrard's experiments (13,23,25,26), and based on the wake width d' as shown in Figure 2, are plotted in Figure 5. There is a slight Reynolds number dependence over the range Re* = 10^3 to 10^4, but in general these results for St* also are in good agreement with the concept of universal wake similarity.

Measurements of C_{pb}, C_D and St were made for circular cylinders in a confined air flow by Modi & El-Sherbiny (27) at subcritical Reynolds numbers near Re = 10^5 and for area blockage ratios up to 36%. The results for St* obtained from equations (1) and (6) are plotted in Figure 5, for blockage ratios up to 20% of the wind tunnel area. The results are in excellent agreement with the other values of St* plotted in the figure near Re* = 10^5. For blockage ratios greater than 20% the pressure distribution on the rear face of the cylinder became increasingly nonuniform and equation (6) no longer is valid.

Measurements of St* for circular cylinders vibrating in the regime of locking-on between the vortex and vibration frequencies at wake Reynolds numbers Re* between 140 and $5(10^4)$ are plotted in Figure 5. The various experiments upon which the vibrating-cylinder data for St* and Re* were based are described in detail elsewhere (2); the cylinder displacement amplitudes in the experiments extended up to a full cylinder diameter (peak-to-peak), and the locked-on wake frequencies were between 70 and 120% of f_s. The drag and base pressure on the vibrating cylinders were increased by as much as 80% from the corresponding measurements made for stationary cylinders at subcritical Reynolds numbers (21,28).

The curve of St* versus Re* in Figure 5 demonstrates the universal similarity of bluff body wakes under subcritical and transcritical conditions, for two-dimensional smooth cylinders in uniform flow. These results further generalize the original hypothesis of Roshko (4) when St* and Re* are based upon the measured wake width as characterized by Calvert (6), Griffin (2) and Simmons (3).

WAKE MOMENTUM, BASE PRESSURE AND DRAG

Bearman (5), in his formulation of a universal Strouhal number based upon the Kronauer stability criterion, showed that St* was dependent only upon the pressure drag coefficient C_D, the base pressure coefficient C_{pb} (or K) and the Strouhal number St. Simmons (29) reconsidered Bearman's approach and found that the Kronauer criterion was not applicable for K greater than 1.5. The Kronauer criterion is no longer considered to be universally applicable, since its validity extends only to K = 1.5 and the ranges of C_{pb} and K over which regular vortex takes place from bluff extend to K = 2. However, as Bearman recently noted (Private communication, 1978), the breakdown of Kronauer's hypothesis does not invalidate the basic argument that a universal correlation exists between St, C_D and K. This point is now considered.

Goldburg & Florsheim (30) proposed a wake Strouhal number of the form

$$St_\theta = \frac{f_s \theta'}{U} \quad ,$$

where θ' is called the "total wake momentum thickness". It was shown by dimensional arguments for two-dimensional bluff bodies that

$$\frac{\theta'}{d} = \frac{1}{2} C_D \tag{7}$$

or equivalently that

$$St\, C_D = 2St_{\theta'} .$$

Since Bearman showed that St C_D is dependent only upon K, the vortex shedding frequency, the drag and the wake momentum are related to the base pressure (at least in principle) by means of this equation.

The product St C_D is plotted against the base pressure parameter K in Figure 6, and the data encompass the entire range of conditions over which vortex shedding takes place from bluff bodies. The legend for the data points in the figure is given once again by Table 1. The shaded area at low values of K refers to experiments reported by Bearman (5) for vortex shedding from blunt-based, D-shaped cylinders. These experiments included interference with the vortex formation by means of base bleeding of low velocity air and wake splitter plates. A single point from the experiments of Batham (19) with circular cylinders in the critical range of Reynolds numbers falls within this range of the correlation.

At the higher values of K the results encompass a wide range of conditions that includes smooth circular cylinders in uniform flow at subcritical, critical and transcritical Reynolds numbers; and smooth circular cylinders in highly turbulent incident flows (18) at subcritical Reynolds numbers. Rough cylinders in uniform and turbulent flows in the critical and supercritical ranges of Re* (19) also fit nicely into the correlation.

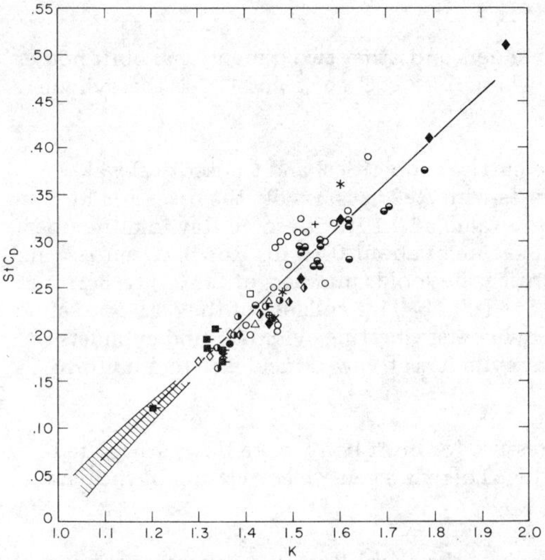

Fig. 6. The product of drag coefficient and Strouhal number, St C_D, plotted against the base pressure parameter K; legend as in Table 1. A least-squares fit to the data is given by the solid line, equation (8).

The results obtained for vortex shedding from flat plates cover the same range of wake momentum or bluffness as vibrating circular cylinders (2) and circular cylinders in confined flow (27) at the largest values of K. The base pressure parameter is a measure of the bluffness of a cylindrical body, and the data in Figure 6 confirm once again that flow confinement, rectangular cylinders and vibration in the locking-on regime result in the most extreme conditions of wake momentum and bluffness. In fact, the measurement by Modi & El-Sherbiny of vortex shedding from a circular cylinder at a blockage of 36% resulted in the largest value of the base pressure parameter, K = 1.95, observed by any of the host of investigators listed in Table 1. Previous measurements of C_D, St and K for a rectangular cylinder with a critical aspect ratio (depth/width) of t/d = 0.6 (31) were thought to be the limiting bluffness condition.

The drag and base pressure measurements made by Bearman & Trueman (31) showed a slight dependence on the cylinder aspect ratio t/d; this is considered by them to be an effect of increasing diffusion length, or shear layer thickness, on the vortex formation, as the afterbody length is increased from the case of a flat plate (t/d = 0) to a square cylinder (t/d = 1). Near the critical aspect ratio the shear layers never became parallel and the pressure coefficient C_{pb} was very non-uniform over the downstream face of the cylinder.

To give a quantitative idea of the correlation between drag, base pressure and shedding frequency, a least-squares fit to the St C_D versus K correlation is plotted in Figure 6. A linear fit to the data

$$a = -0.48 \pm 0.03$$

$$\text{St}\, C_D = a + b\, K \qquad (8)$$

$$b = +0.50 \pm 0.02$$

yielded the minimum mean-square deviation for all polynominal functions up to order 10.

The St C_D versus K correlation was originally suggested from considerations of universal wake similarity (5). All of the data that follow the St* versus Re* correlation in Figure 5 are included in the correlation. Additional data for vortex shedding from smooth and rough cylinders in turbulent flow at critical and subcritical Reynolds numbers are also included in Figure 6.

SUMMARY AND CONCLUDING REMARKS

A universal wake Strouhal number, $St^* = f_s d'/U_b$, has been proposed and is based upon the natural Strouhal frequency f_s of the incident flow to a bluff cylindrical strucutre, the *measured* wake width d′ at the end of the vortex formation region, and the velocity U_b at the edge of the separated boundary layer.

The wake width d′ for stationary and vibrating circular cylinders and other two-dimensional bluff bodies is correlated with the base pressure parameter $K = U_b/U$, where $K = (1-C_{pb})^{1/2}$ and C_{pb} is the cylinder's base pressure coefficient.

St* collapses these characteristic wake scales onto a single curve at subcritical and transcritical wake Reynolds numbers Re*. For Re* less than 500, St* increases with Re* as is usually the case. For Re* between 500 to 10^4 the data for St* are distributed about a value of 0.17, with some Reynolds number dependence. The data for St* are distributed about a mean value of about 0.16 for Re* between 10^4 and 10^5, and a range of St* = 0.14 to 0.15 is approached at critical Reynolds numbers of $2(10^5)$ to $3(10^5)$. For transcritical Reynolds numbers the Strouhal number St* is 0.16. This collapse of the data onto a single curve is valid for stationary two-dimensional bluff cylinders, vibrating cylinders, and cylinders in confined flow. Also included in the correlation are circular cylinders at yawed incidence to a uniform flow.

The pressure drag, vortex shedding frequency and base pressure for bluff body wake flows are related by means of a correlation between the product St C_D and K. There is a general correlation of the data up to K = 2, which is the upper limit achieved thus far.

This paper has been prepared as part of the research program of the Naval Research Laboratory and is published by permission. The author is indebted to Professor Anatol Roshko and to Professor Peter Bearman for many helpful comments and criticisms on the subject of wake similarity.

REFERENCES

1. McCroskey, W.J., "Some Current Research in Unsteady Fluid Dynamics," Trans. ASME, J. Fluids Eng., Vol. 99, 1977, pp. 8-39.
2. Griffin, O.M., "A universal Strouhal number for the 'locking-on' of vortex shedding to the vibrations of bluff cylinders," J. Fluid Mech., Vol. 85, 1978, pp. 591-606.
3. Simmons, J.E.L., "Similarities Between Two-Dimensional and Axisymmetric Vortex Wakes," Aero. Quarterly, Vol. 26, 1977, pp. 15-20.
4. Roshko, A., "On the Wake and Drag of Bluff Bodies," J. Aero. Sci., Vol. 22, 1955, pp. 124-132.
5. Bearman, P.W., " On vortex street wakes," J. Fluid Mech., Vol. 28, 1967, pp. 625-641.
6. Calvert, J.R., "Experiments on low speed flow past cones," J. Fluid Mech., Vol. 27, 1967, pp. 273-279.
7. Ramberg, S.E., "The Influence of Yaw Angle Upon the Vortex Wakes of Stationary and Vibrating Cylinders," Naval Research Laboratory, Washington, D.C., Memorandum Report 3822, 1978.
8. Griffin, O.M. and Ramberg, S.E., "The vortex street wakes of vibrating cylinders," J. Fluid Mech., Vol. 66, 1974, pp. 553-576.
9. Griffin, O.M. and Ramberg, S.E., "On vortex strength and drag in bluff body wakes," J. Fluid Mech., Vol. 69. 1975, pp. 721-728.
10. Eppler, H., "Beiträge zu Theorie und Anwendung der unstetigen Strömungen," J. Rational Tech. Anal., Vol. 3, 1954, pp. 591-644.
11. Roshko, A., "Experiments on the flow past a circular cylinder at very high Reynolds numbers," J. Fluid Mech., Vol. 10, 1961, pp. 345-356.

12. Bearman, P.W., "On vortex shedding from a circular cylinder in the critical Reynolds number regime," J. Fluid Mech., Vol. 37, 1969, pp. 577-586.

13. Bloor, M.S. and Gerrard, J.H., "Measurements on turbulent vortices in a cylinder wake," Proc. Royal Soc. London Ser. A, Vol. 294, 1966, pp. 319-342.

14. Ballou, C.L., "Investigation of the Wake Behind A Cylinder at Coincidence of a Natural Shedding Frequency of Vibration of the Cylinder and the Vortex Shedding Frequency," Massachusetts Institute of Technology, Cambridge, MA, Report 76028-2, 1967.

15. Bruun, H.H. and Davies, P.O.A.L., "An experimental investigation of the unsteady pressure forces on a circular cylinder in a turbulent cross flow," J. Sound Vib., Vol. 40, 1975, pp. 535-560.

16. Cantwell, B.J., "A Flying Hot-Wire Study of the Wake of a Circular Cylinder," California Institute of Technology, Pasadena, CA, Ph.D. Thesis (1976).

17. Apelt, C.J., West, G.S., and Szewczyk, A.A., "The effects of wake splitter plates on the flow past a circular cylinder in the range $10^4 < R < 5\times10^4$," J. Fluid Mech., Vol. 61, 1973, pp. 187-198.

18. Surry, D., "Some effects of intense turbulence on the aerodynamics of a circular cylinder at subcritical Reynolds number," J. Fluid Mech., Vol. 52, 1972, pp. 543-564.

19. Batham, J.P., "Pressure distributions on circular cylinders at critical Reynolds numbers," J. Fluid Mech., Vol. 57, 1973, pp. 209, 228.

20. Bearman, P.W. and Davies, M.E., "The flow about oscillating bluff structures," *Proceedings of the Fourth International Conference on Wind Effects on Buildings and Structures*, Cambridge University Press, 1975, pp. 285-296.

21. Stansby, P.K., "Base Pressure of Oscillating Circular Cylinders," Proc. ASCE, J. Eng. Mech., Vol. 104, 1976, pp. 591-600.

22. Bearman, P.W., "Investigation of the flow behind a two dimensional model with a blunt trailing edge and fitted with splitter plates," J. Fluid Mech., Vol. 21, 1965, pp. 241-255.

23. Roshko, A., "On the drag and shedding frequency of two-dimensional bluff bodies," National Advisory Committee for Aeronautics, Washington, D.C. Technical Note 3169, 1954.

24. Blake, W.K., "Hydroelastic variables influencing propeller and hydrofoil singing," *Noise and Fluids Engineering*, ASME, 1977, pp. 191-199.

25. Gerrard, J.H., "The mechanics of the formation region of vortices behind bluff bodies," J. Fluid Mech., Vol. 25, 1966, pp. 401-413.

26. Gerrard, J.H., "A disturbance-sensitive Reynolds number range for the flow past a circular cylinder," J. Fluid Mech., Vol. 22, 1965, pp. 187-196.

27. Modi, V.J. and El-Sherbiny, S., "A Free Streamline Model for Bluff Bodies in Confined Flow," Trans. ASME, J. Fluids Eng., Vol. 99, 1977, pp. 585-592.

28. Griffin, O.M., Skop, R.A. and Koopmann, G.H., "The Vortex-Excited Resonant Vibrations of Circular Cylinders," J. Sound Vib., Vol. 31, 1973, pp. 235-249.

29. Simmons, J.E.L., "Effect of Separation Angle on Vortex Streets," Proc. ASCE J. Eng. Mech., Vol. 101, 1975, pp. 649-661.

30. Goldburg, A. and Florsheim, B.H., "Transition and Strouhal Number for the Incompressible Wakes of Bluff Bodies," Phys. Fluids, Vol. 9, 1966, pp. 45-50.

31. Bearman, P.W. and Trueman, D.M., "An Investigation of the Flow Around Rectangular Cylinders," Aero. Quarterly, Vol. 23, 1972, pp. 229-237.

32. Twigge-Molecey, C. and Baines, W.D., "Aerodynamic Forces on a Triangular Cylinder," Proc. ASCE J. Eng. Mech., Vol. 99, 1973, pp. 803-818.

VORTEX SHEDDING FROM CIRCULAR CYLINDERS IN TURBULENT FLOW

J. F. HOWELL* and M. NOVAK**

*University of Bath, Bath, UK
**The University of Western Ontario, London, Ontario, Canada

SUMMARY

Vortex shedding from vibrating circular cylinders in turbulent flow is investigated experimentally in a wind tunnel at subcritical Reynolds numbers. The cylinders are freely vibrating or driven. Two types of turbulent flow are used: grid generated homogeneous turbulence and turbulent boundary layers. Pressures and their correlation, lift coefficient and response are studied. The response observed is compared with some of the mathematical models for the prediction of vortex induced oscillation.

INTRODUCTION

Vortex shedding from circular cylinders and vibration of various structures resulting from it have been the subject of many investigations. The extent of these studies may be gauged from the extensive bibliographies given by Morkovin (1), Mulcahy and Chen (2), Roshko (3) and others. In addition to experimental studies, a number of mathematical models for theoretical prediction of vortex induced oscillation has been presented (4 to 10). Some of these models were evaluated by Sarpkaya (11). However, in spite of the vast body of information available and a good qualitative understanding of the phenomenon achieved, the quantitative prediction of the vortex induced vibration of real structures in natural wind remains very uncertain. One of the main reasons for this is that the effect of turbulence on vortex shedding is not sufficiently understood. For example, pressure measurements indicate that turbulence broadens the power spectrum of the lift and shortens the correlation length (12). This would seem to suggest that the vortex induced response of a cylinder should be greater in smooth flow than in turbulent flow. Yet, response measurements have shown that the opposite can also occur.

This paper describes a wind tunnel study of vortex shedding from circular cylinders vibrating in turbulent flow. Freely vibrating and driven models are used. Some of the theories for the prediction of vortex induced oscillation are examined. The experiments are limited to the subcritical region but there is good reason to assume that such data may be at least partly relevant for the transcritical region and particularly for the case of turbulent flow (3,13).

HORIZONTAL DRIVEN MODELS

Many experiments have been devoted to the determination of the lift coefficient, C_L, under different flow conditions. In order to reduce the effect of end disturbances the models are usually a number of diameters in length. This, however, makes the total life force measured dependent on the spanwise correlation of the local lift. To account for this factor, the following approach was used.

Relationship between total lift and local lift. - The local force per unit length

$$F_L = \frac{1}{2} \rho D\bar{V}^2 C_L(t) \qquad (1)$$

in which ρ = air density, D = diameter of the cylinder, $\bar{V}$ = mean wind velocity and C_L = the time dependent coefficient of local lift assumed to be random. The power spectrum of this local lift force is

$$S_L(f) = (\frac{1}{2} \rho D\bar{V}^2)^2 S_{CL}(f) \qquad (2)$$

where S_{CL} = the power spectrum of the local lift coefficient, C_L, and f=frequency.

For a horizontal cylinder, the local vortex force may be considered to be homogeneous and thus, the spectrum of the local lift may be assumed to be constant along the span. Then, the spectrum of the total lift force, S_F, acting on length, H, can be written as

$$S_F(f) = S_L(f) \int_o^H \int_o^H R(z_1, z_2, f)\, dz_1 dz_2 \qquad (3)$$

where $R(z_1, z_2, f)$ = the normalized cross-spectrum. For a narrow band process such as vortex shedding, this function depends only on separation $r = |z_2 - z_1|$ and reduces to the correlation coefficient. Experiments indicated (11) that the correlation coefficient can be described as

$$R(\frac{r}{D}) = e^{-\frac{cr}{D}} \qquad (4)$$

in which c = dimensionless coefficient. Substitution of Eq. 4 into Eq. 3 yields for the power spectrum of the total lift force

$$S_F(f) = S_L H^2 \varepsilon \qquad (5)$$

where the dimensionless coefficient $\varepsilon(\leq 1)$ is

$$\varepsilon = 2 \frac{D}{cH} [1 - \frac{D}{cH} (1 - e^{-\frac{cH}{D}})] \qquad (6)$$

The coefficient, c, is a function of amplitude of motion and exposure and can be conveniently described in terms of the correlation length

$$L = \int_o^\infty R(r)\, dr \qquad (7)$$

Substitution of Eq. 4 into Eq. 7 yields c = D/L. Introducing the normalized power spectrum of the local lift coefficient, S'_{CL}, the spectrum of the local lift coefficient is $S_{CL}(f) = \sigma_L^2 S'_{CL}(f)$ where $\sigma_L^2 = \overline{C_L^2(t)}$ is the variance of the lift coefficient. Then, the relation between the spectrum of the total lift force, S_F, and the spectrum of the local lift becomes

$$S_F(f) = (\frac{1}{2} \rho D\bar{V}^2)^2 H^2 \varepsilon \sigma_L^2 S'_{CL}(f) \qquad (8)$$

Integrating Eq. 8, the variance of the total lift force is

$$\sigma_F^2 = (\frac{1}{2} \rho D\bar{V}^2 H)^2 \varepsilon \sigma_L^2 \qquad (9)$$

From Eq. 9, the r.m.s. lift coefficient

$$\sigma_L = \frac{\sigma_F}{(\frac{1}{2}\rho D\bar{V}^2 H)\sqrt{\varepsilon}} \quad (10)$$

From Eq. 10, the local r.m.s. lift coefficient, σ_L, can be calculated using the r.m.s. value of the total lift established experimentally. With vertical cylinders in a boundary layer, the same procedure can be applied with the exception that the mean wind profile and the vibration mode of the cylinder have to be accounted for.

Experiments with a horizontal cylinder. - The above approach was utilized in experiments conducted in the Boundary Layer Wind Tunnel Laboratory at The University of Western Ontario. The main purpose of the experiments was to evaluate the lift force as a function of amplitude and exposure.

The experimental apparatus consisted of a specially constructed mechanical oscillator which produced sinusoidal oscillations in the vertical plane which could be varied in amplitude and frequency over a large range. The mounting arms were provided with large streamlined fairings and the model was mounted horizontally on the oscillator. The total force on the cylinder was measured using a strain gauge bridge on the four pin-ended support arms of the cylinder. For the stationary cylinder, this records the lift force on the cylinder only, but when the cylinder oscillates, the strain gauges record the sum of the inertial forces due to motion plus the lift forces. In order to minimize inertial forces on the driven model, the cylinder was made as light as possible, consisting of a thin veneer of birch over balsa diaphragms. The assembled model and support bearings have the following properties: generalized mass = 0.0267 slug (0.39 kg), length = 34.75 inches (0.883 m) and diameter = 6.0 inches (0.152 m).

The inertial forces were electrically subtracted by means of a strain gauge accelerometer mounted on the support arm of the cylinder. (This technique was also used by Yano and Takahara, Ref. 14.) This produced a signal in phase with the inertial force and its amplitude was adjusted so that the electrical sum of the signals from the support strain gauges and the accelerometer was zero for no wind flow. Yet, the residual noise was of the order of 10% of the total lift force recorded with the vibrating model. The wind flow was either "smooth" with minimal turbulence or the almost isotropic turbulent flow produced by a mesh grid. This had an intensity of 11%, longitudinal scale ℓ = 4.75 in (0.12 m) and lateral scale of 2.5 in (0.064 m). This turbulence is referred to as "grid turbulence" in the figures to distinguish it from the turbulent flow generated in the boundary layer.

A preliminary investigation of the flow conditions around the model was conducted using a hot-wire probe in the wake of the stationary cylinder about 4 diameters downstream of the model in smooth flow. The spectra of wake velocity fluctuations showed that the Strouhal number is almost independent of flow velocity and is equal to 0.19 in the Reynolds number range available in the wind tunnel.

An example of the results obtained in both smooth and turbulent flow is shown in Figure 1. This figure presents the normalized spectrum of total lift and the r.m.s. of the total lift coefficient, C_L, which is evaluated from Eq. 1 as

$$\sqrt{\overline{C_L^2}} = \sigma_F / \tfrac{1}{2}\rho\bar{V}^2 DH.$$

For a stationary cylinder, the effect of isotropic turbulence is to increase the total lift slightly and to broaden its spectrum.

The effect of amplitude 2a/D on total lift presented in Figure 1, shows the expected result of the broadening of the spectrum with turbulence and the narrowing of

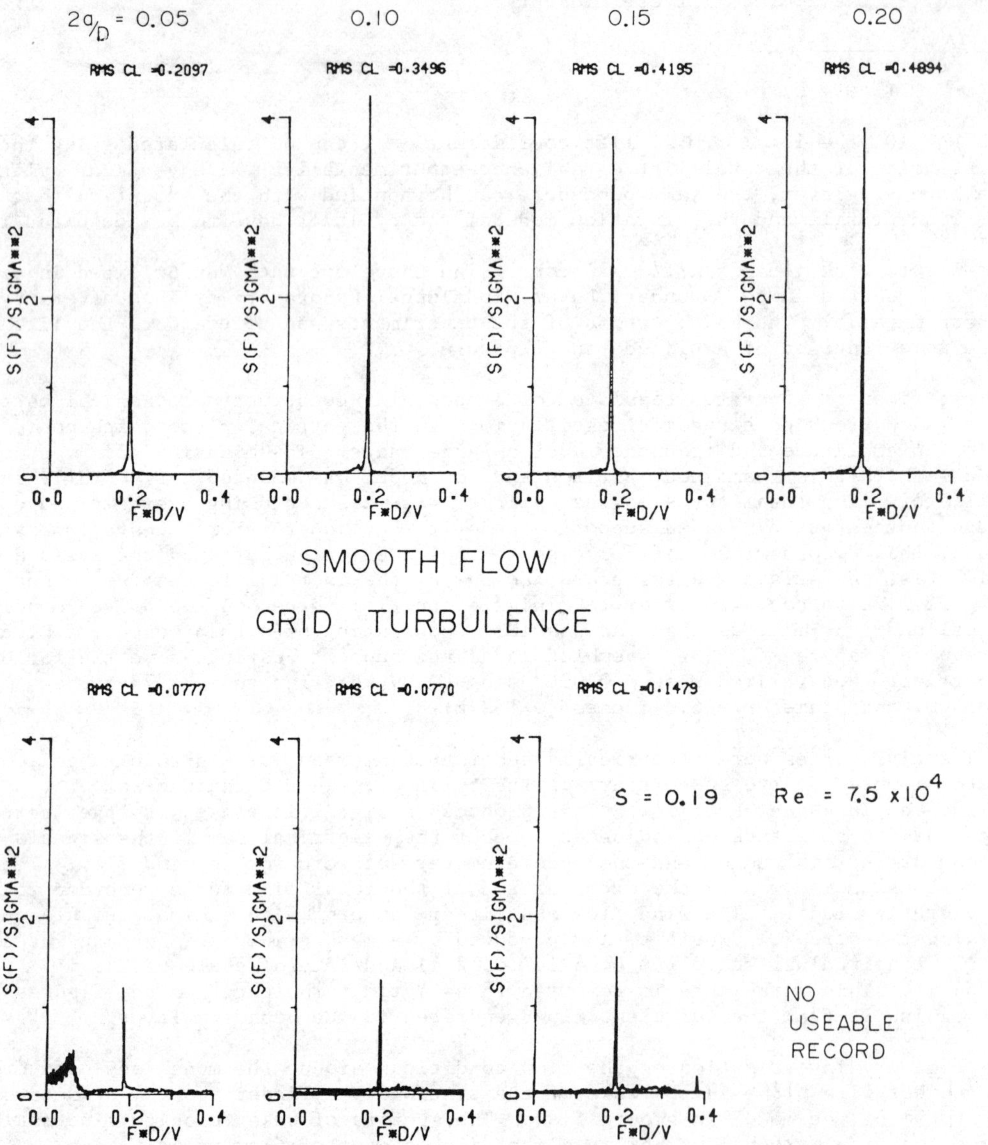

Fig. 1. Effect of amplitude 2a/D on the normalized spectrum of total lift C_L for the horizontal cylinder

it with amplitude accompanied by an increase in the r.m.s. of total lift. From these values for total lift, the local lift σ_L is evaluated from Eq. 10 and shown in Table 1. The correlation length L/D used is that established previously by Novak and Tanaka (12).

For smooth flow, the values of σ_L show a continuous increase with amplitude and on the whole, are consistent with other data reported for the given Reynold's number. The values found for grid turbulence, however, are much lower and do not appear sufficiently reliable. It may be noted, though, that the effects of turbulence may be strongly Reynolds number dependent (15). The data are being further analyzed in order to find out whether the in-phase (driving) component of the lift could be separated from its out-of-phase (damping) component.

Table 1. Local lift coefficient σ_L evaluated from experiments ($R_e = 7.5 \times 10^4$)

Flow	Amplitude 2a/D	L/D	ε	C_L	σ_L
Smooth	0	3.09	0.652	0.1	0.124
	0.05	4.65	0.767	0.210	0.240
	0.1	7.40	0.873	0.350	0.375
	0.15	14.3	0.979	0.420	0.425
	0.2	24.6	1.0	0.489	0.489
Turbulent	0	2.22	0.551	0.070	0.094
int. = 11%	0.05	2.85	0.628	0.077	0.097
$\ell_x/D = 0.8$	0.1	3.47	0.686	0.077	0.093
	0.15	6.13	0.833	0.150	0.164

EXPERIMENTS WITH A VERTICAL CYLINDER

The grid-generated turbulence, while having an intensity similar to that of the natural wind has length scales which are substantially different. Flow conditions simulating the natural wind are generated in the wind tunnel by roughening elements on the tunnel floor and a low grid located at the entrance. The longitudinal scale of turbulence, ℓ_x, is much greater than for grid-generated turbulence (up to about ten feet (3.05 m)) so that the ℓ_x/D ratio is more realistic. Some of the available mean velocity profiles are shown in Fig. 2 which also records the longitudinal turbulence intensity. The velocity profile may be represented by the power law. Three boundary layers were used with exponents, α, equal to 0.186 for the shallow boundary layer, 0.264 for the medium boundary layer and 0.355 for the deep boundary layer.

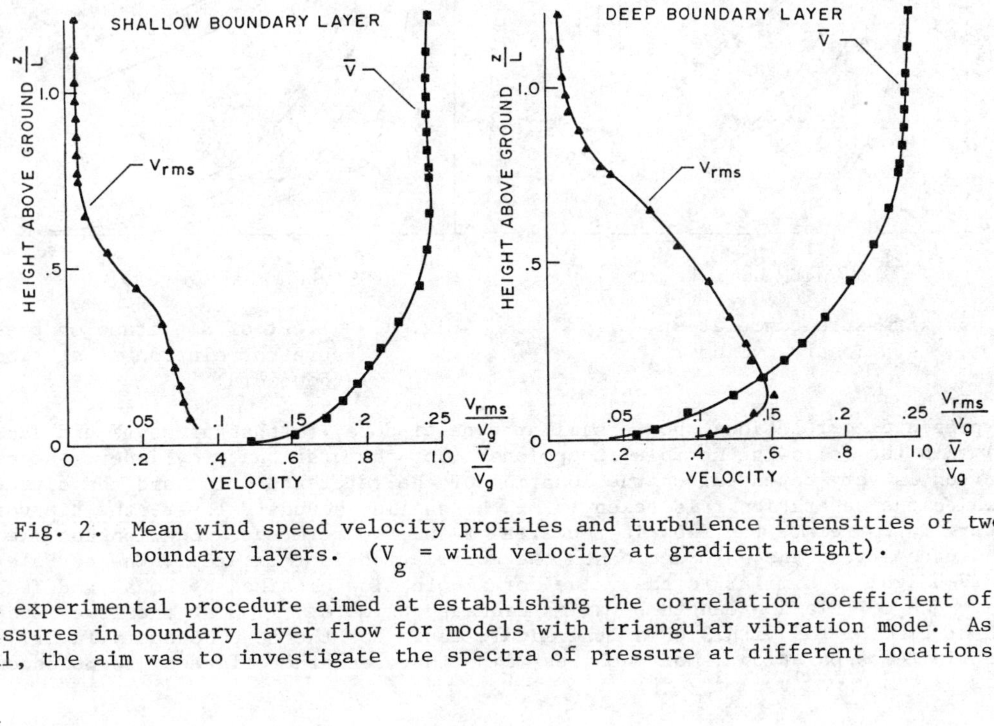

Fig. 2. Mean wind speed velocity profiles and turbulence intensities of two boundary layers. (V_g = wind velocity at gradient height).

The experimental procedure aimed at establishing the correlation coefficient of pressures in boundary layer flow for models with triangular vibration mode. As well, the aim was to investigate the spectra of pressure at different locations in the flows for different amplitudes of motion. The lift force was not measured but the response of the structure was investigated to permit the comparison of predicted response with experimental values.

The model studies were separated into two investigations. In the first study, a rigid heavy plexiglass model was used which could be driven at controlled amplitudes. The second part of the study employed a light balsa model of the same internal dimensions which was elastically mounted to permit oscillation transverse to the flow only. Amplitude of response was controlled by varying the damping of the model. The structural damping ratios were 1%, 3%, 9% and 11%. Amplitude of motion for the pressure model was controlled by a light inextensible wire connected to a mechanical oscillator. Tip amplitude 2a/D was variable up to 0.25. All pressure measurements were performed at an angle of attack to the flow of 60°. The light balsa model was also provided with pressure taps at the location of the reference points of the plexiglass pressure model. This model was subsequently shortened for some experiments. The pressure model had a diameter D = 4.0 in (0.1 m) and the aspect ratio H/D = 10. The elastically mounted model had D = 4.125 in (0.105 m), H/D = 10, generalized mass = 8.01 x 10^{-3} slug (0.117 kg) and the natural frequency = 9.7 Hz.

The correlation coefficient of pressures, $R_p(r/D)$, was investigated as a function of location of reference point. For a stationary cylinder in the shallow boundary layer, the results are shown in Fig. 3. The effect of motion amplitude is shown in Fig. 4 for the deep boundary layer.

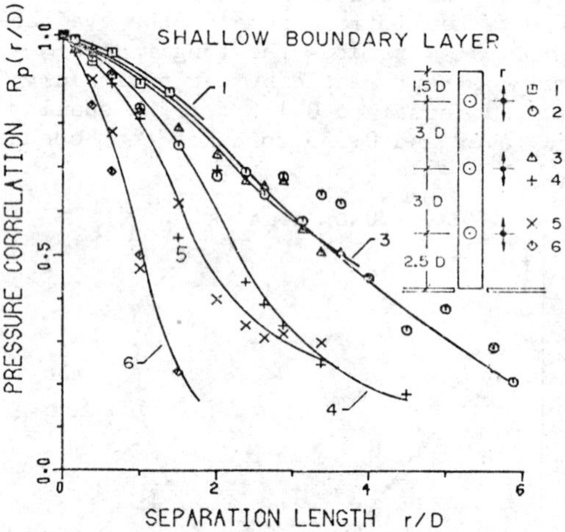

Fig. 3. Pressure correlation for stationary cylinder

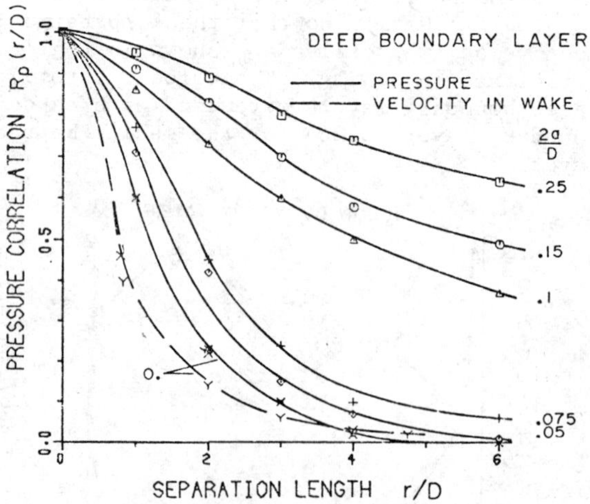

Fig. 4. Effect of amplitude on pressure correlation at station 2 (downward)

The pressure correlations show a similar type of decay to that of Novak and Tanaka (12) with the following notable exceptions. For the stationary cylinder, the correlation is very dependent on the location of the reference point and the direction in which the separation r is taken. For the shallow boundary layer, the tip vortex appears to correlate the flow in the area of the top reference location and the turbulence around the base decreases the correlation in this area. The correlation at mid-height is similar to the value for smooth flow obtained by Novak and Tanaka (12). The correlation length L varies from approximately 1.8D at the lowest tap to 4.1D at the top tap with 3.4D at the middle tap. For the deep boundary layer, the direction of separation is of much less importance and the effect of turbulence is significant, reducing the correlation. For this case, the average correlation length L = 1.7D. The correlation lengths were determined from a least squares fit of the data points.

For the oscillating cylinder, the general trend is as expected with amplitude of motion increasing the correlation and turbulence decreasing it. Because of the triangular mode shape, the amplitude decreases towards the base and it would be expected that the increase in correlation due to motion would not be as pronounced as for horizontal cylinders. The correlations of velocities in the wake compare with those of Toebes (16) but do not show as large an increase with amplitude as do the pressure correlations.

Numerous response curves were measured under different conditions such as exposure and damping. A few typical examples are shown in Figs. 5 and 6. The response curves may be summarized by noting that at low values of damping, the maximum response was almost independent of exposure, with the shallow boundary layer producing the highest response. As the damping was increased, grid turbulence produced a greater response than the other exposures. It would be expected that boundary layer flow would produce smaller response than constant mean velocity flow as the vortex shedding, at least for small amplitudes below "locking in", is at the Strouhal frequency only over a part of the body. At higher amplitudes, locking in of the wake to the oscillation frequency results in this response being insensitive to exposure. While grid turbulence produces the maximum amplitude, the response in smooth flow is consistently smaller. The difference between these two extremes may be up to 100%. Some of the response curves exhibit slight hysteresis. Strong hysteresis such as that found by Ferguson and Parkinson (17) at very small damping was not observed.

As well as the response curves, normalized spectra of pressures at three locations and of response were established for the elastically mounted model. Examples of these spectra are shown in Fig. 7 as a function of tip amplitude 2a/D. Vibration

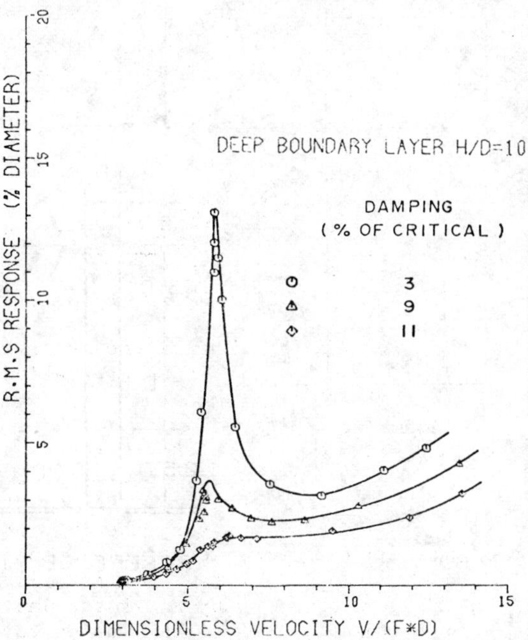

Fig. 5. Tip response in deep boundary layer(Damping 3,9,11%)

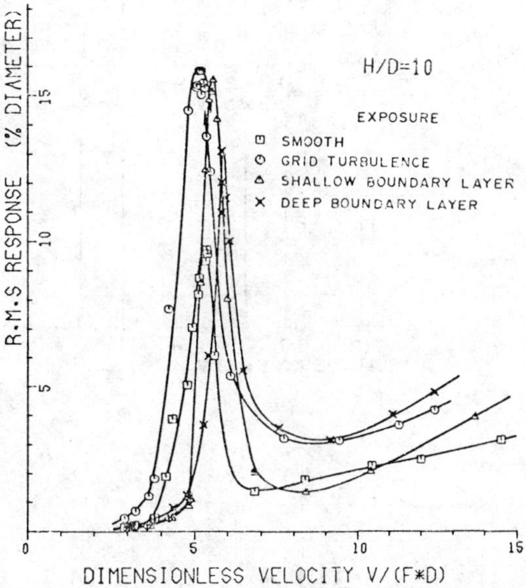

Fig. 6. Tip response for damping ratio 3% in different flows

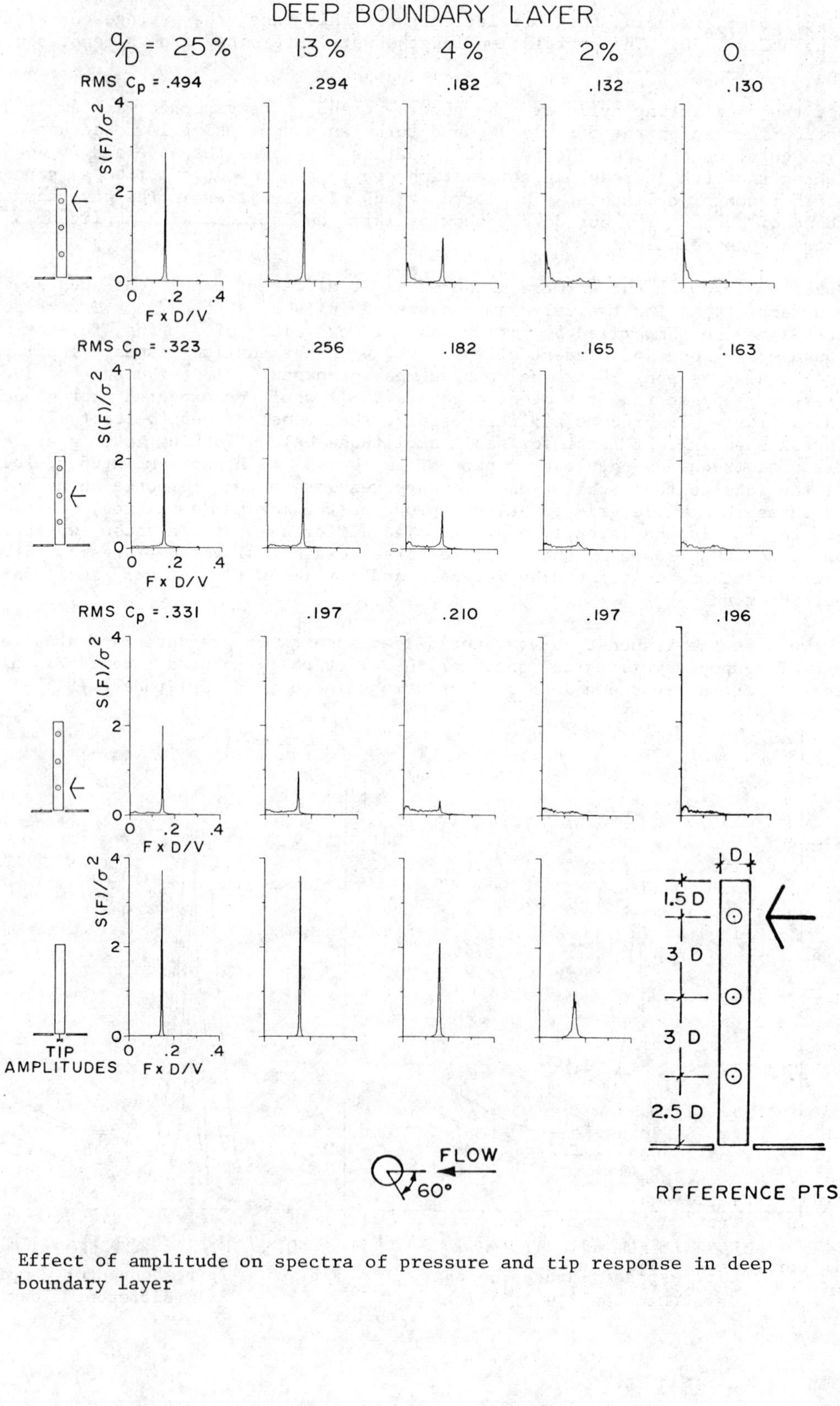

Fig. 7. Effect of amplitude on spectra of pressure and tip response in deep boundary layer

amplitude increases the r.m.s. of the pressures and narrows the pressure and response spectra, resulting in a sharper peak. Turbulence decreases the r.m.s. of the pressure and broadens the spectra. Several other observations are of interest too. The spectrum of pressure at some locations show two well defined peaks fairly close together, indicating that the shedding frequency and vibration frequency at that location were different. Occasionally, the second harmonic is apparent at approximately twice the Strouhal frequency. For some of the spectra on the stationary cylinder, the pressure spectrum is quite broad with possibly no peak at the Strouhal frequency. Fluctuating pressures on stationary cylinders usually have a well defined peak at the Strouhal frequency and it appears that the pressure tap was in a location where the oncoming flow was not yet affected by the vortex shedding for these cases. This phenomenon was not observed for the grid turbulence. It should be noted that the resolution for all spectra presented was better than 0.23 Hz.

PREDICTION OF THE RESPONSE

The response curves measured under different conditions can be used to verify the many theoretical models proposed. Two approaches are examined below. The forced vibration model is one of the most often used for practical predictions. Assuming narrow band random excitation, the r.m.s. tip displacement can be calculated from the expression (9,18)

$$\sqrt{\overline{a^2}} = \frac{\pi^{\frac{1}{4}}\, \sigma_L\, \rho\, D^4}{\sqrt{B}\, D_s\, (4\pi S)^2 M}\, C \tag{11}$$

where for the triangular mode shape and wind profile exponent α

$$C^2 = \frac{(H/D)^2}{1 + (H/2\ L\ D)}\ \frac{1}{3\alpha + 3} \tag{12}$$

in which B = bandwidth of the spectrum of lift, D_s = structural damping ratio and L = correlation length. For typical values of B = 0.1, L = 2.5 and σ_L = 0.5, the predicted r.m.s. tip response is compared with the experimental results in Table 2. It may be seen that the use of constant values for B, L and σ_L results in overestimation at low amplitudes and underestimation at high amplitudes. A better prediction may be made if L, B and σ_L reflect the described effects of vibration amplitude on the bandwidth, correlation length and lift coefficient. This may be done using an iterative procedure, first predicting the amplitude using estimated values of L, B and σ_L and then selecting more appropriate values indicated by Table 1 and recomputing the amplitude. This prediction is shown in Table 2 also. Another way of improving the forced vibration model is to introduce amplitude dependent aerodynamic damping as is done by Vickery (10).

Iwan (8) developed another model based on energy consideration in the wake and presented a universal response curve for structural elements including the pivoted rod used for the experiments here. The prediction was for smooth flow and plotted maximum amplitude in the mode considered, Y_{max}, normalized by a mode shape coefficient γ_n versus mass ratio-damping parameter which was approximately $2\ m\ \delta_s/10\rho D^2$ where m = mass per unit length and δ_s = logarithmic decrement of structural damping. For the model, the peak response was estimated from the r.m.s. using a peak factor of 3. The results are shown in Fig. 8 as solid symbols with other experimental results summarized by Iwan (8). The agreement between theory and prediction is good. It is interesting to note the scatter of the experimental results. This suggests that any theory is likely to provide only an approximate estimate of the response to vortex shedding. The model used by Iwan does not include turbulence and this may be a drawback for the prediction of response in turbulent flow.

Table 2. Comparison of experimental and predicted response to vortex shedding in boundary layer flow

Velocity Exponent α	Structural Damping D_s %	Expts 100 a/D %	Theory 100 a/D % Constant B, L, σ_L	Theory 100 a/D % Variable B, L, σ_L	
0.355	11	2.1	3.8	1.8	L = 1.8
0.264	11	1.7	3.9	1.9	B = .15
0.186	11	1.1	4.0	2.0	σ_L = .35
0.355	9	3.2	4.2	3.9	L = 2.
0.264	9	4.1	4.3	4.0	B = .1
0.186	9	4.8	4.4	4.1	σ_L = .5
0.355	3	13.5	7.4	9.8	L = 3.5
0.264	3	14.1	7.5	10.2	B = .08
0.186	3	15.4	7.7	10.6	σ_L = .55
0.355	1	26.5	12.5	22.1	L =10.
0.264	1	25.1	12.9	22.5	B = .08
0.186	1	-	13.3	23.2	σ_L = .55

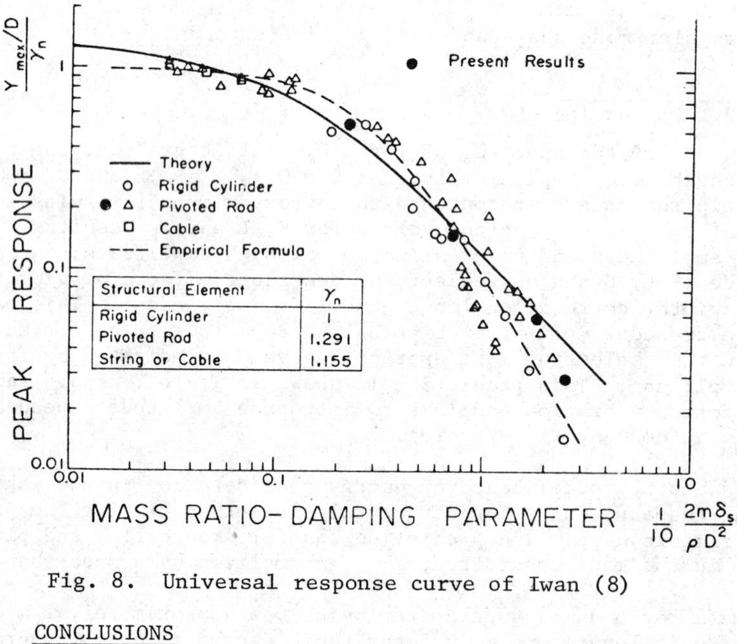

Fig. 8. Universal response curve of Iwan (8)

CONCLUSIONS

Wind tunnel experiments on circular cylinders have been described. The effects of various types of turbulence and the motion on vortex shedding were examined. In general, turbulence weakens the spanwise pressure correlations and broadens the pressure spectra while amplitude of motion increases the correlations and narrows the spectra. The pressure correlations in boundary layers are similar to those observed in grid generated turbulence.

The response of cylinders is insensitive to exposure for small damping but increases with turbulence for larger damping.

The forced vibration model does not predict the response over a broad range of amplitudes if only one set of input constants is used. The response prediction can be improved considerably if the correlation length, lift coefficient and bandwidth are chosen according to the amplitude of the motion and the type of turbulence.

ACKNOWLEDGEMENT

The study was supported by a grant in aid of research from the National Research Council of Canada.

REFERENCES

(1) M.V. Morkovin, Flow around circular cylinder - a kaleidoscope of challenging fluid phenomena, Proceedings Symposium on Fully Separated Flows, ASCE, (1964).
(2) T.M. Mulcahy and S.S. Chen, Annotated bibliography on flow induced vibrations, ARGONNE National Lab. Techil. Memorandy ANL-CT-74-05 (Jan. 1974).
(3) A. Roshko, A review of concepts in separated flow, Canadian Congress of Appl. Mechanics, CANCAM 67, Laval, Quebec, pp. 3-081 - 3-115 (1967).
(4) Y. Nakamura, Vortex excitation of a circular cylinder treated as a binary flutter, Rept. of Res. Institute for Applied Mechanics, Kyushu University, Vol. XVII, No. 59, pp. 217-234 (1969).
(5) R.T. Hartlen and I.G. Currie, Lift oscillator model of vortex-induced vibration, Proc. ASCE, Vol. 96, No. EM5, pp. 577-591 (1970).
(6) R.A. Skop and D.M. Griffin, A model for the vortex-excited resonant vibrations of bluff bodies, J. of Sound and Vibration, Vol. 27, pp. 225-233 (1973).
(7) W.D. Iwan and R.D. Blevins, A model for vortex induced oscillation of structures, J. of Applied Mechanics, Vol. 41, No. 3, pp. 581-586 (1974).
(8) W.D. Iwan, The vortex induced oscillations of elastic structural elements, J. of Engineering for Industry, ASME, pp. 1378-1382 (November 1975).
(9) A.G. Davenport and M. Novak, Vibration of structures induced by wind, Chapter 29-II in Shock and Vibration Handbook, 2nd edition,edited by C.M. Harris and C.E. Crede, McGraw-Hill (1976).
(10) B.J. Vickery, A model for the prediction of the response of chimneys to vortex shedding, 3rd Int. Symp. on Chimneys, Munich, 13 (1978).
(11) T. Sarpkaya, Notes for the 4-5 March 1976 meeting on cable strumming held at the California Institute of Technology between the sponsor, consultants and the investigators, Naval Postgraduate School, Monterey, California (1976).
(12) M. Novak and H. Tanaka, Pressure correlations on a vibrating cylinder, Proc. 4th Int. Conf. Wind Effects on Buildings and Structures, Cambridge University Press, pp. 227-232 (1975).
(13) A. Naumann and H. Quadfleig, Vortex generation on cylindrical buildings and its simulation in wind tunnels, IUTAM-IAHR Symposium on Flow-Induced Structural Vibrations, Karlsruhe, (August 1972).
(14) T. Yano and S. Takahara, Study on unsteady aerodynamic forces acting on an oscillating cylinder, Proc. Wind Effects on Buildings and Structures, Tokyo, pp. 737-746 (1971).
(15) J.H. Gerrard, A disturbance-sensitive Reynolds number range of the flow past a circular cylinder, J. of Fluid Mechanics, Vol. 22, pp. 187-196 (1965).
(16) G.H. Toebes, Fluid elastic features of flow around cylinders, Int. Research Seminar, Wind Effects on Buildings and Structures, Ottawa, Canada (1967).
(17) N. Ferguson and G.V. Parkinson, Surface and wake flow phenomena of the vortex-excited oscillations of circular cylinder, Trans. ASME, J. Engineering for Industry Series B89, No. 4, pp. 831-838 (1967).
(18) B.J. Vickery and A.W. Clark, Lift or across-wind response of tapered stacks, J. of the Structural Division, ASCE, 98, pp. 1-20 (1972).

TURBULENCE EFFECTS ON AERODYNAMIC RESPONSE OF RECTANGULAR BLUFF CYLINDERS

T. MIYATA and M. MIYAZAKI
University of Tokyo, Tokyo, Japan

In evaluating the effects of turbulence on the wind loads and the response behaviors of structures in natural wind, it has been supposed that the sensitivities to turbulence intensity is important and there is found to be little influence of the scale. In this paper, the effect of turbulence scale is studyed by analysing the results of wind tunnel experiments carried out on the rectangular bluff cylinders for the drag and lift forces, heaving oscillation and vortex shedding. It can be seen that scales of turbulence of the order of the body size affect the lift and the response in the lift direction.

List of Symbols

- d depth, dimension normal to flow
- b width, streamwise dimension
- ℓ length, spanwise dimension
- a amplitude of oscillation normal to flow
- U free stream velocity
- $Ur = U/Nd$ reduced velocity
- Iu turbulence intensity
- L_x^u longitudinal integral turbulence scale
- m mass of cylinder
- $\Gamma = \frac{2m}{\rho b d \ell} \cdot \delta s$ mass ratio . logarithmic structural damping
- α angle of attack
- N natural frequency of cylinder
- L lift force on cylinder
- D drag force on cylinder
- $C_L = L/\frac{1}{2}\rho U^2 d\ell$ lift coefficient
- $C_D = D/\frac{1}{2}\rho U^2 d\ell$ drag coefficient
- C_{Fy} lateral force coefficient
- $C_P = \overline{P-P_o}/\frac{1}{2}\rho U^2$ mean pressure coefficient
- C_{Prms} rms pressure fluctuation
- R_P spanwise correlation of pressure
- ρ air density
- r/d nondimensional spanwise distance from center of cylinder

1. Introduction

Atmospheric turbulence effects on the wind loads and the various response behaviors experienced by buildings and structures can be in the form of a simple load correlation connected directly with the approaching turbulence, but also a complex interference between the turbulence and the body which results in changing the flow around the body. These effects depend on the shape of the afterbody and how the body responses to vortex shedding, galloping and torsional flutter.

Wind loads and responses of structures are usually estimated by using of data from wind tunnel experiments in which the properties of the natural wind and the structure are correctly modelled. However, knowledge on the turbulence effects is not complete, and it is still not known how precise the simulation needs to be. Bearman[1] has analysed the influences of turbulence on the drag of bluff bodies and suggested that it is more important to reproduce the correct intensity than the scale in order to simulate the mean flow about a body in turbulent flow. As far as square section cylinder is concerned, for instance, the drag is sensitive to the turbulence intensity but no influence of scale is found[1, 2]. In this paper another aspect of the simulation will be discussed, that is the effect of scale on the lift force and response due to vortex shedding and galloping for some rectangular cylinders.

2. Outline of Experiments

In order to examine some of these effects of intensity and scale on wind loads and aerodynamic responses, wind tunnel experiments were carried out on three different rectangular two-dimensional cylinders with the dimension ratio $d/b = 1/1$, $2/3$ and $1/2$. The turbulent flow experienced by these cylinders was generated by a grid, featuring intensity of $Iu = 11$ % and longitudinal scale of $L_x^u = 73$ mm at the test position. The streamwise dimensions of $b = 50$, 75 and 150 mm were chosen so as to alter the scale ratio of $L_x^u/b = 1.46$, 0.97 and 0.49. Cross section of wind tunnel and model installation are such as shown in Fig. 1. The experiments in smooth flow were also conducted to see the influence of aspect ratio of $\ell/b = 21$, 14 and 7 as well as for comparison with those in turbulent flow. Force measurements and response observations were made with the same model and pressure measurements with the other, where $Re = 10^4 \sim 10^5$.

3. Measurement of Mean Forces and Pressure Distributions at Rest

The measurements for the drag and lift forces on the cylinders at rest were made by using of a dynamic balance. Mean drag coefficients C_D of the cylinders, with one face normal to the flow, are illustrated in Fig. 2, in which the abscissa is the ratio d/H of cylinder depth to wind tunnel height to see how is the blockage effect of the body. The drag coefficient measured in smooth flow for flat plates[3] with high aspect ratio, set normal to the flow, is also shown in Fig. 2 and it is found that it depends on the ratio d/H. It appears, however, that the drag for present rectangular cylinders has low dependence on the ratio d/H and aspect ratio b/ℓ (or d/ℓ) as well as scale ratio L_x^u/b. In all cylinders a striking decrease can be seen in turbulent flow. This fact affirms again that the drag of bluff bodies is sensitive to turbulence intensity but no influence of scale is found.

Figs. 3 and 4 show how is the dependence of lift and drag forces on moderate angles of attack of the flow. In the case of drag coefficient transition, there is found to be remarkable difference between smooth and turbulent flow in both 1/1 and 1/2 cylinders. However, little or no dependence of scale is found on this difference, which seems due to high response to turbulence intensity in the same way as the cylinders with one face normal to the flow.

On the other hand, aiming at the negative inclination $dC_L/d\alpha$ of lift in the vicinity of $\alpha = 0°$, there is found to be no difference between smooth and turbulent flow in 1/1 cylinders, that is little or no effect of turbulence, while the difference is pronounced in 1/2 cylinders in the form of high reduction in turbulent flow. The majority depends on turbulence scale as well as intensity, that is L_x^u/b the larger, the lower becomes the inclination. The quasi-steady theory on galloping describes that the inception of oscillation is highly dependent on the lateral force coefficient $dC_{Fy}/d\alpha = dC_L/d\alpha + C_D$ which acts in the direction of heaving oscillation. The experimental results make us suppose the onset to be influenced by turbulence scale. One would expect that there is almost no difference in the galloping of 1/1 cylinders between smooth and turbulent flow and there exists a shift of starting velocity to higher range in 1/2 cylinders, depending on turbulence intensity and greater scale.

The critical angle of attack, at which a shear layer separated from a front corner of the cylinder reattaches on the side surface downstream, becomes lower in turbulent flow in both 1/1 and 1/2 cylinders. It appears that the reattachment occurs at the angle of attack where the lift coefficient is minimum[4]. The critical angle also depends on turbulence scale and slightly shifts to lower value with increase of L_x^u/b. This means that turbulence modifies curved free shear layer development and greater scale causes more highly curved shear layer and more rapid reattachment.

The characteristics of lift and drag on the cylinder relates in detail with those of pressure distribution about the body. The pressure was measured as the difference from reference pressure by a gauge type transducer at the mid section of the cylinders. The results are shown in Fig. 5 for 1/1 cylinders at angles of attack of $\alpha = 0°$ and $9°$, and in Fig. 6 for 1/2 cylinders at $\alpha = 0°$, $4°$ and $8°$. The characteristics common in both cylinders are as follows. The frontal distribution is almost the same in smooth and turbulent flows. The presence of turbulence decreases the base pressure, depending on the angle of attack. The pressure recovery on the upper surface is found with increase of angle of attack, resulting in the reduction to uniform distribution and in the decrease of the effect of turbulence. The variation of negative pressure distribution on the lower surface is highly related with the critical angle for the reattachment of a separated shear layer. The pressure recovery developed with reattachment shifts upstream in order from the nearest pressure hole to the rear corner. That is, the recovery occurs at lower angle of attack in turbulent flow and with increase of L_x^u/b, slightly depending on turbulence scale.

As a result of pressure measurements, it is found that the influence of turbulence scale is not so distinct on the pressure distribution on the upper and lower side surfaces. The variation of pressure distribution needs to be treated with some caution, however, since the lift calculated from the difference of pressure distributions on both side surfaces is influenced by turbulence scale.

4. Observation of Heaving Oscillation due to Vortex Shedding and Galloping

The heaving oscillations were observed for the elastically mounted cylinders with one face normal to the flow. Fig. 7 shows a plot of nondimensional amplitude a/d of 1/2 cylinders versus reduced velocity in a specified mass damping value, where a is $\sqrt{2}$ times rms value of fluctuating oscillation with almost constant frequency. It appears that the responses in turbulent flow are different from those in smooth flow and the former clearly depends on turbulence scale, that is the response shifts to higher velocity with increase of L_x^u/b. Fig. 7 includes the experimental works by Laneville and Parkinson[5), where it has been concluded that the galloping responses are sensitive to the turbulence intensity, but there is no appreciable effect on the scale.

The diagrams to exmine all the present exprimental results for respective cylinders are Figs. 8, 9 and 10, in which the response characteristics at a specified nondimensional amplitude $a/d = 0.2$ are shown in the form of reduced velocity versus mass damping value. The scale ratio L_x^u/b and aspect ratio ℓ/b are parameters. This expression originates in the description ($a/d = 0.1$) for a square cylinder in smooth flow studied by Scruton[6]. In 1/1 cylinders in Fig. 8, there is found to be little or no effect of turbulence scale as well as intensity, that is little or no difference between smooth and turbulent flow. In 2/3 and 1/2 cylinders, however, it is distinctly found for the onset of heaving oscillations to shift to higher velocity range with increase of L_x^u/b. This indicates that the turbulence scale as well as intensity affects significantly the occurrence of galloping.

As far as the description of galloping of such bluff cylinders is concerned, the quasi-steady approach is relatively effective, where the response is described by using the lateral force with coefficient $C_{Fy} = -(C_L + C_D \tan\alpha)\sec\alpha$ which acts in oscillation direction. Figs. 11, 12 and 13 show the coefficient C_{Fy} versus angle of attack for respective cylinders. In all the figures, the peak value of C_{Fy} decreases and shifts to lower angle in turbulent flow, compared with that in smooth flow. These effects of turbulence have been so far indicated[5),7),8)]. Aiming at the effect of turbulence scale, there is found to be a trend of lower value of C_{Fy} for larger value of L_x^u/b in the range below a moderate angle of attack $\approx 10°$ for all cylinders. The value of inclination $dC_{Fy}/d\alpha$ in the vicinity of $\alpha = 0$ of the lateral force, which is inversely proportional to the galloping inception, is seen to be almost constant in 1/1 cylinders, while is found to be influenced by turbulence scale as well as intensity in 2/3 and 1/2 cylinders. The 1/2 cylinders especially respond to the effect of scale. This is because the variation in $dC_{Fy}/d\alpha = dC_L/d\alpha + C_D$ at $\alpha \approx 0$ highly depends on the lift inclination rather than the drag, where the effect of scale, as mentioned in section 3, is pronounced on the lift but not on the drag. These results can well explain the response characteristics of galloping mentioned above.

5. Vortex Shedding and Spanwise Pressure Correlation

It is argued that the high drag of large aspect ratio bluff bodies is associated with the regular formation and shedding of vortices. For low aspect ratio bluff bodies, there is no such shedding and the base pressure is comparatively high. The interference of the rear corners of the rectangular section with the highly curved free shear layers weakens the strength of the shed vortices and the drag decreases. With the introduction of turbulence the shear layers is to be expected to grow more rapidly. How is the effect of the rapid growth rate of the shear layers on the lift in turbulent flow.

To examine the effect of turbulence on the unsteady lift force associated with vortex shedding, the spanwise distribution and correlation of the pressure fluctuation were measured along the center line of the upper side surface of 1/1 and 1/2 cylinders, with one face normal to flow, at rest and in steady heaving motion with appropriate nondimensional amplitudes. The motion for the cylinders was forced almost at the reduced velocity of the inverse of Strouhal number, that is at $Ur \approx 8$ and $Ur \approx 12$ respectively. The spanwise distributions of rms pressure fluctuation are such as in Figs. 14 and 15. For both cylinders it is found that the spanwise distribution is comapratively uniform and the fluctuation in heaving motion is larger than that at rest, probably due to the vortex shedding locked in the motion. It seems that the turbulence reduces the fluctuation for 1/1 cylinders, but not so for 1/2 cylinders. However, it is difficult to isolate and identify the influences of various parameters of the aspect ratio, the presence of turbulence and the scale ratio on respective differences in the magnitude of pressure fluctuation due to vortex shedding.

On the contrary, there is found to be a clear distinction in the spanwise correlation for vortex shedding. The results are shown in Figs. 16 and 17, where the abscissa is the nondimensional distance from the center of the cylinder. Generally, the correlation in heaving motion is higher than that at rest in both 1/1 and 1/2 cylinders, probably attributable to the fact that the vortex-shedding is controlled by the oscillating motion, that is the locking-in phenomenon. The presence of turbulence decreases the correlation. The extent of decrease is low in 1/1 cylinders, while considerable in 1/2 cylinders. For the latter the effect of turbulence scale is also found, there being higher correlation with increase of L_x^u/b. The oscillation amplitude is highly related to the spanwise correlation and the improvement of correlation rises up to a saturated value with increase of amplitude. The 1/2 cylinders in turbulent flow are significantly affected by the turbulence scale in the above-mentioned improvement depending on the amplitude. The scale ratio of L_x^u/b is the larger, the lower is the effect. Novak and Tanaka[9] have also presented almost the same effect of turbulence and amplitude on the spanwise correlation for vortex shedding in fluctuating pressure on the surface of a circular cylinder, although no scale effect was referred to. As far as the effect of aspect ratio on the correlation is concerned for the present results, it seems that there is found to be little or no effect in smooth flow except the neighborhood of end plates.

6. Concluding Remarks

It has been argued that the main effect of turbulence is to increase the growth rate of the shear layers for sharp-edged bluff bodies. As a result of this increased growth rate, there is found a significant dependence on intensity and little or no effect of scale for the drag on such rectangular cylinders as in the present study. However, an analysis of experimental results indicates that not only the intensity of turbulence but also the scale affect the unsteady lift forces and heaving responses due to vortex shedding and galloping. There may be found to be some influences of turbulence scale on the mean and unsteady flow about bluff bodies. To simulate the forces and responses depending on the mean and unsteady flow near the side surfaces of a body in turbulent flow, it may be inevitable to consider the similarity of the scale as well as the intensity.

7. Acknowledgements

A part of experiments was supported by a grant-in-aid-of-research from Ministry of Education of Japan. The encouragement by Prof. M. Ito, University of Tokyo and the discussion with Prof. Y. Nakamura, Kyushu University are gratefully acknowledged.

8. References

1) Bearman, P.W., "Turbulence Effects on Bluff Body Mean Flow", Proc. 3rd U.S. Nat. Conf. Wind Eng. Res., Florida, 1978, pp. IV-1~8.
2) Laneville, A., Gartshore, I.S. and Parkinson, G.V., "An Explanation of Some Effects of Turbulence on Bluff Bodies", Proc. 4th Int. Conf. Wind Effects on Buildings and Structures, London, 1975.
3) Okamoto, T. and Takeuchi, M., "Effects of Wind Tunnel Side Walls on Wake of Flat Plates", Proc. Ann. Conv. JSME, Mar. 1973, 730-16 (in Japanese).
4) Nakamura, Y., Mizota, T. and Yoshimura, T., "Three Components of Steady Force on Rectangular and H-Shaped Cylinders", Rep. R.I.A.M., Kyushu Univ., No.40, 1973, pp. 245~255 (in Japanese).
5) Laneville, A. and Parkinson, G.V., "Effects of Turbulence on Galloping of Bluff Cylinders", Proc. 3rd Int. Conf. Wind Effects on Buildings and Structures, Tokyo, 1971.
6) Scruton, C., "On the Wind-excited Oscillations of Stacks, Towers and Masts", Proc. 1st Int. Conf. Wind Effects on Buildings and Structures, Teddington, 1963.
7) Novak, M., "Galloping Oscillations of Prismatic Structures", Proc. ASCE, Vol.98, EM1, Feb. 1972, pp.27~46.
8) Novak, M., "Effect of Turbulence on Galloping Instability", Proc. ASCE,

Vol.100, EM1, Feb. 1974, pp.27~47.
9) Novak, M. and Tanaka, H., "Pressure Correlation on a Vibrating Cylinder", Proc. 4th Int. Conf. Wind Effects on Buildings and Structures, London, 1975.

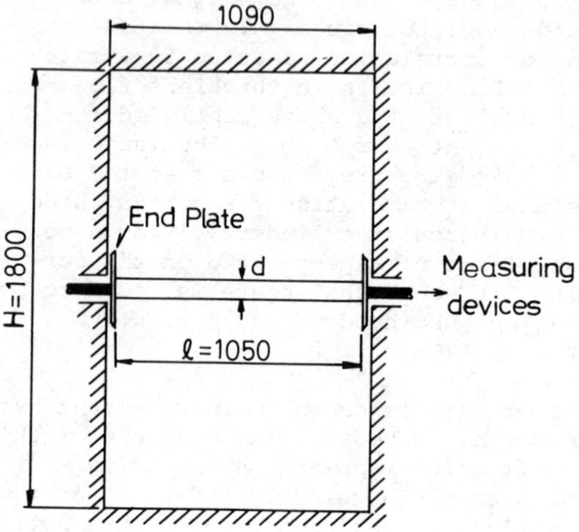

Fig. 1

Cross section of wind tunnel and model installation (Unit: mm)

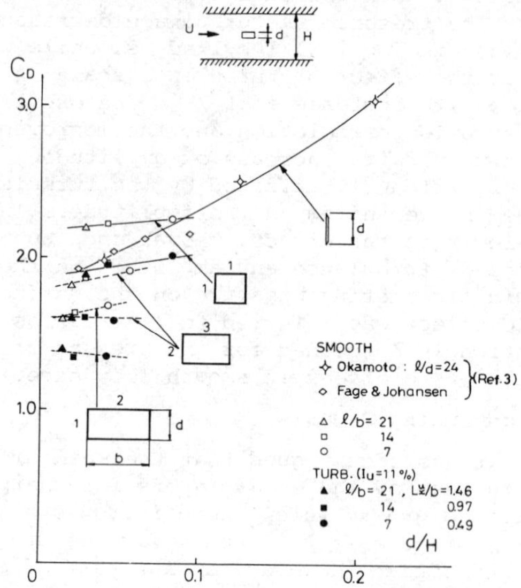

Fig. 2

Drag coefficient for respective cylinders versus ratio of depth to wind tunnel height

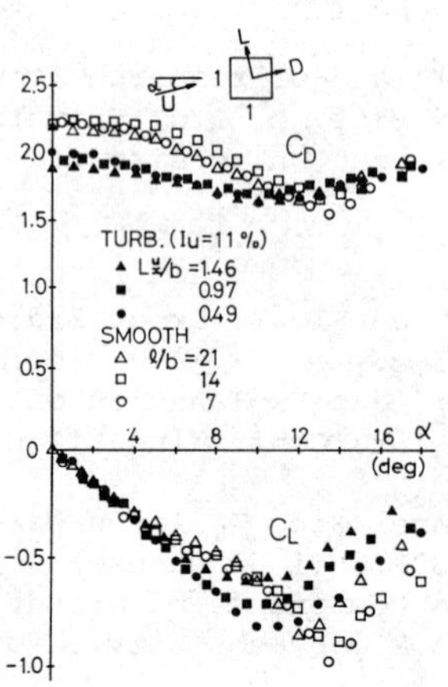

Fig. 3

Mean lift and drag coefficients on 1/1 cylinders versus angle of attack of flow

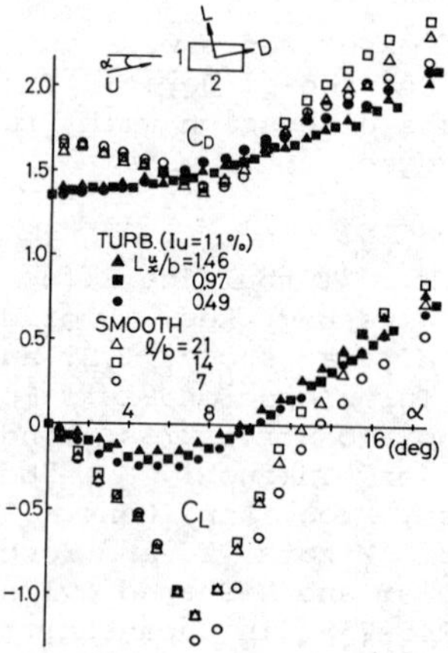

Fig. 4

Mean lift and drag coefficients on 1/2 cylinders versus angle of attack of flow

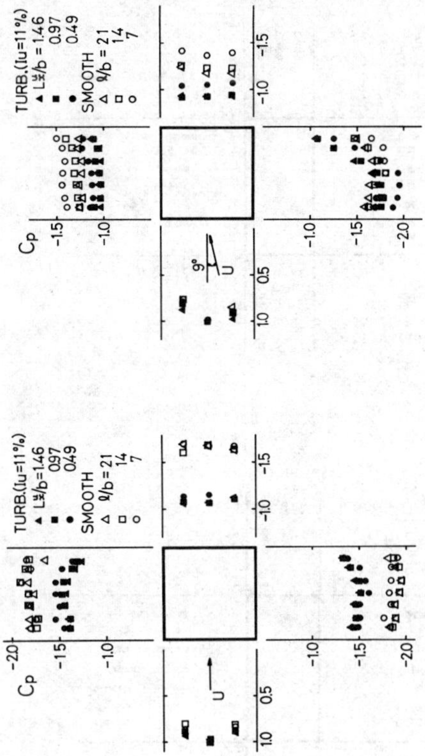

(a) α = 0° (b) α = 9°

Fig. 5 Mean pressure distribution around 1/1 cylinders

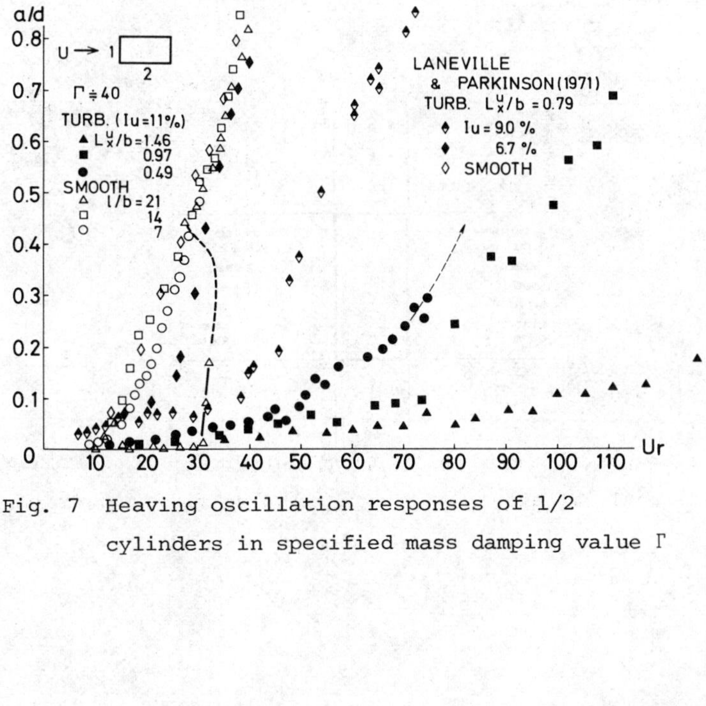

Fig. 7 Heaving oscillation responses of 1/2 cylinders in specified mass damping value Γ

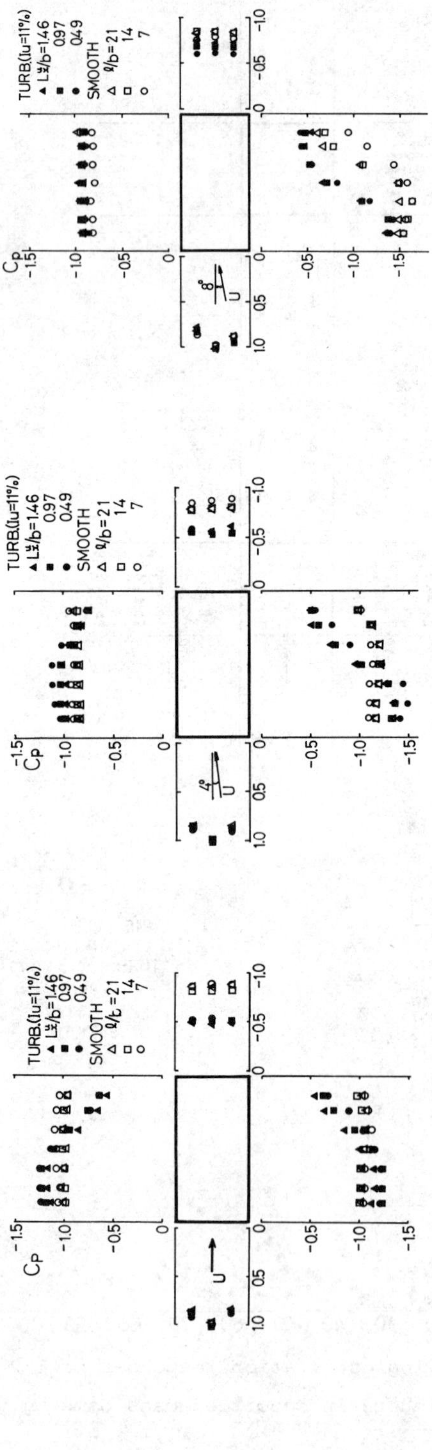

(a) α = 0°

(b) α = 4°

(c) α = 8°

Fig. 6 Mean pressure distribution around 1/2 cylinders

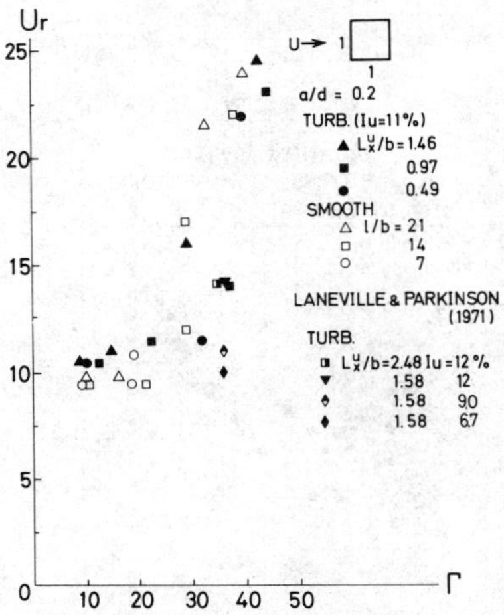

Fig. 8

Heaving oscillation responses of 1/1 cylinders at specified nondimensional amplitude

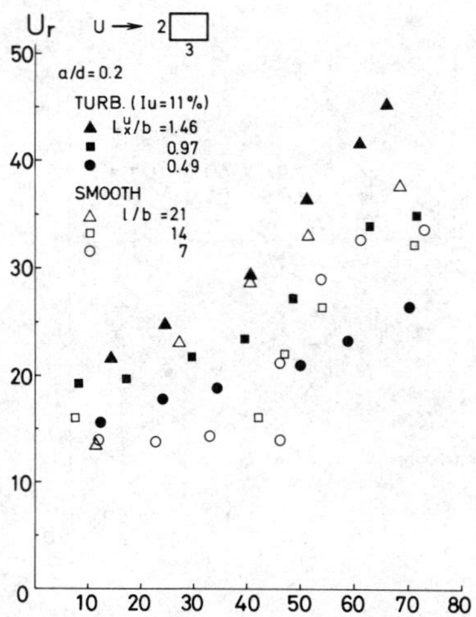

Fig. 9

Heaving oscillation responses of 2/3 cylinders at specified nondimensional amplitude

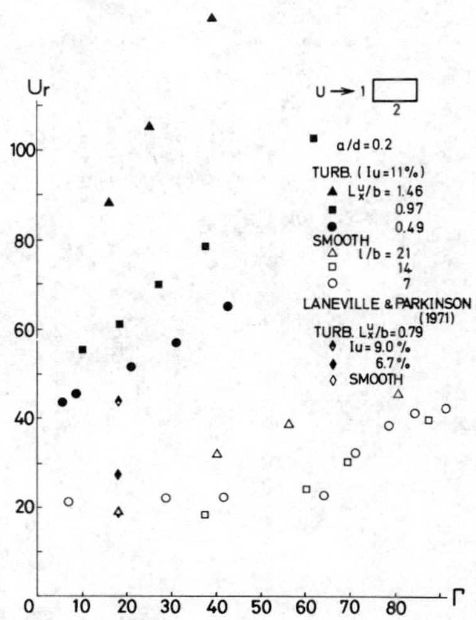

Fig. 10

Heaving oscillation responses of 1/2 cylinders at specified nondimensional amplitude

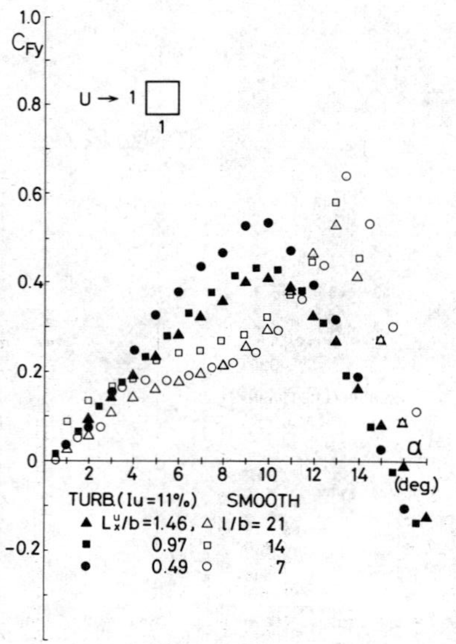

Fig. 11

Lateral force coefficient on 1/1 cylinders versus angle of attack

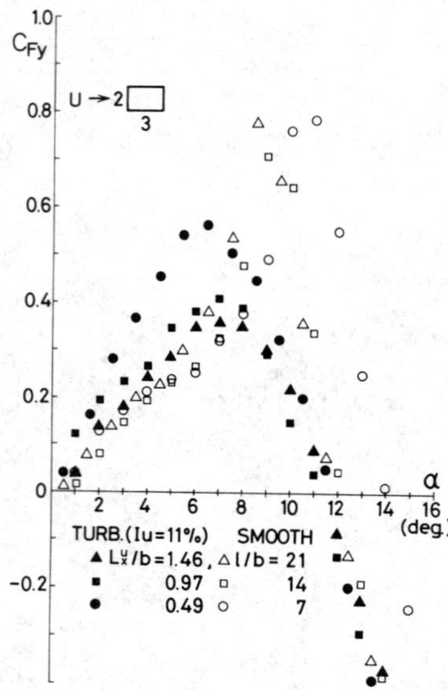

Fig. 12

Lateral force coefficient on 2/3 cylinders versus angle of attack

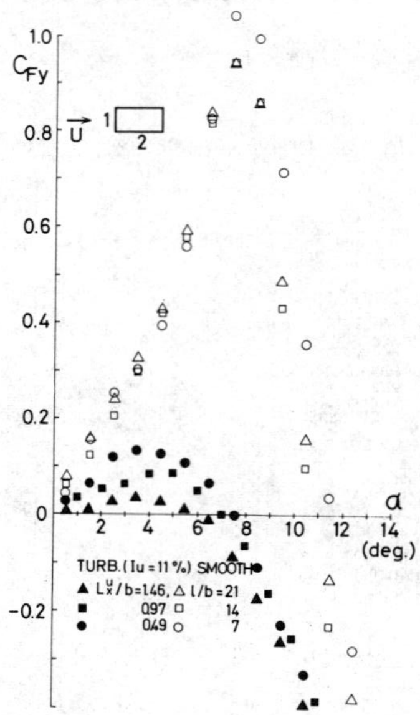

Fig. 13

Lateral force coefficient on 1/2 cylinders versus angle of attack

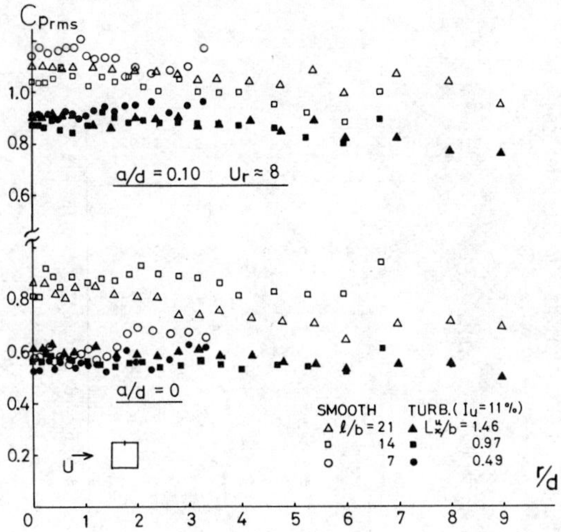

Fig. 14 Spanwise distribution of rms pressure fluctuation along center line of side surface of 1/1 cylinders

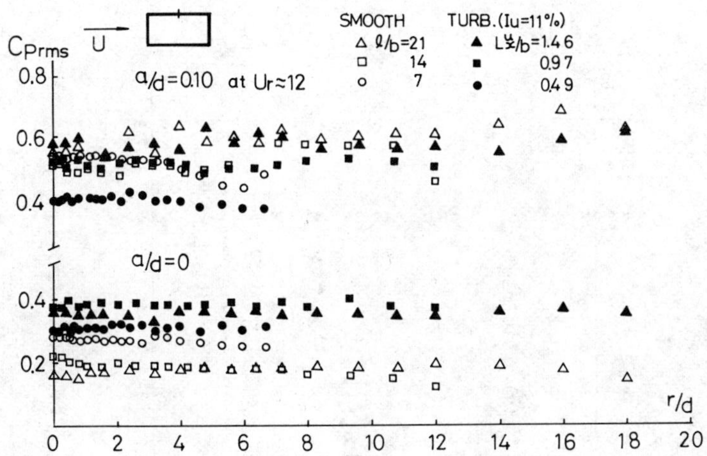

Fig. 15 Spanwise distribution of rms pressure fluctuation along center line of side surface of 1/2 cylinders

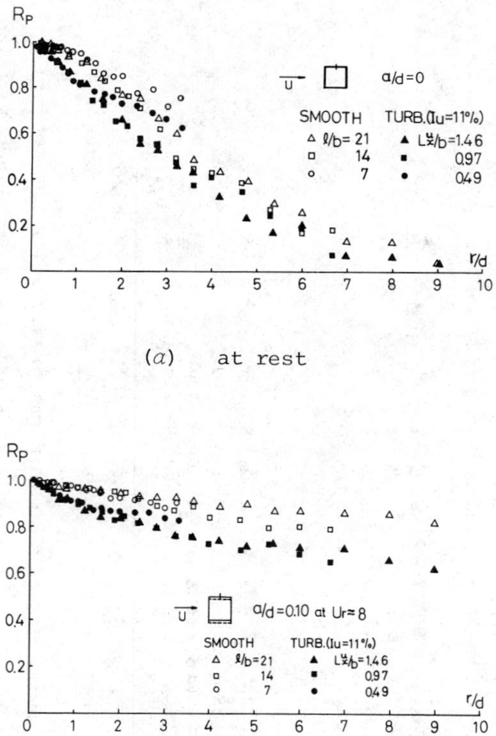

(a) at rest

(b) in heaving motion

Fig. 16 Spanwise pressure correlation along center line of side surface of 1/1 cylinders

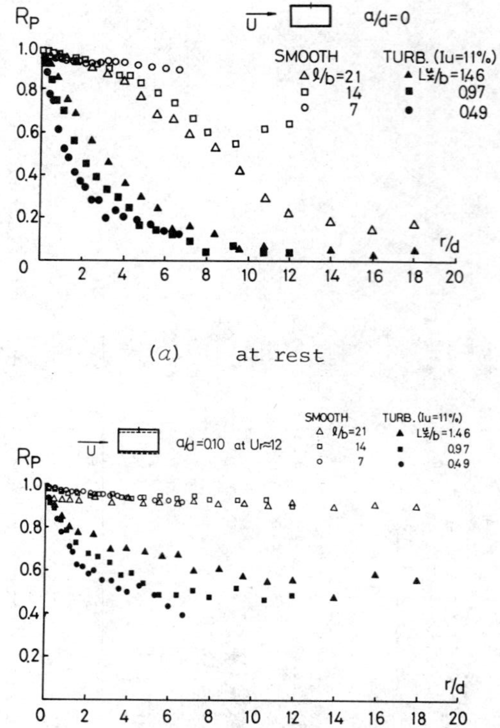

(a) at rest

(b) in heaving motion

Fig. 17 Spanwise pressure correlation along center line of side surface of 1/2 cylinders

DISCUSSION ON SESSION V
Wind Loading--Fluctuating External

Rapporteur: J. A. Peterka
Colorado State University
USA

Paper by P. A. Durbin, J. C. R. Hunt--FLUCTUATING SURFACE PRESSURES ON BLUFF STRUCTURES IN TURBULENT WINDS: FURTHER THEORY AND COMPARISON WITH EXPERIMENT

Question: H. A. Panofsky. In many applications the longitudinal integral scale, L_x, is required. In practice this depends on the method chosen for separation of turbulence from mean flows. Hence, in practice, it is rather arbitrary. Would you suggest a recipe for designing an unambiguous L_x?

Reply: P. A. Durbin, J. C. R. Hunt. When the frequency is low enough that eddies are large compared with the obstacle radius $(na/\overline{U}_\infty) \ll 1$, then a structure responds quasi-steadily so that the pressure spectrum $H^{(p)}$ is proportional to the spectra of the U, V component of the incident turbulence $H^{(u)}$, $H^{(v)}$. (Thus, the mean and fluctuating pressure can be distinguished in terms of the same time scale as the velocity.)
But when $na/\overline{U}_\infty \gtrsim 1$, then a turbulence scale must be defined because the pressure spectrum for a given $na/\overline{U}_\infty$ is obtained by integrating over wave numbers in the transverse and vertical directions. This scale can be defined from cross-correlation measurements of u, v or w. This can be reasonably unambiguous because such cross-correlations do not, I understand, depend sensitively on the averaging time (greater than say 15 minutes).

Question: J. L. Wood. Is your treatment applicable to a structure with high aspect ratio and small cross section, e.g., a flagpole or street lighting column? Can the method be extended to estimate the amount of energy transfer to a vibrating pole?

Reply: P. A. Durbin, J. C. R. Hunt. The detailed treatment in our paper is applicable to tall slender structures, though Durbin in his thesis has extended the analysis to axisymmetric bodies. The new feature of the analysis is that it is applicable to structures that are large compared with the turbulence scale, not small which you seem to refer to. For such structures quasi-steady analysis works well, and would for example be applied to a vibrating pole, but you might have to include the added mass due to the poles acceleration (see Goldstein and Atasi, J. Fluid Mech. 74, 741, 1976).

Paper by H. Sockel--LOCAL PRESSURE FLUCTUATIONS

Question: J. A. Peterka. Our laboratory investigations have shown that the probability distributions of pressures in separated regions with large negative mean pressures can fall close to a Gaussian form when vortices are present in the flow near the surface. Use of your technique could experience difficulty in those cases. Would you comment on the applicability of your technique for these cases.

Reply: H. Sockel. In our distribution function there are three free parameters and one can get a rather good fit to experimental data following a nearly Gaussian distribution too.

Question: R. D. Marshall. It is my understanding that the probability analysis is based upon samples taken every 1/1000 sec. Therefore, the probability densities relate to percent of time within a specified amplitude band. Ideally, one is interested in discrete fluctuations. Have you carried out any analysis on duration times of fluctuations or peak values associated with upcrossings of the mean?

Reply: H. Sockel. I agree with you, one is interested in discrete fluctuations, especially in peak values. But, one can calculate the probability of the peak value for a given time interval, that equals a given set of values, applying nth-order statistics, as I have shown.

Paper by H. Kawai, J. Katsura, H. Ishizaki--CHARACTERISTICS OF PRESSURE FLUCTUATIONS ON THE WINDWARD WALL OF A TALL BUILDING

Question: A. Laneville. In the case of your measurement of the correlation between the velocity and pressure fluctuations, how was the hot wire mounted? And, are you confident that there was no interference on the pressure measurement when the distance between the wire and the model was small? Finally, what was the tubing length between the pressure tap and the pressure transducer?

Reply: H. Kawai, J. Katsura, H. Ishizaki. We mounted a hot-wire anemometer laterally in front of the prism. The size of the hot wire was 2 mm and the supporting rod was 5 mm. Therefore, the hot wire did not have a significant effect on pressure and velocity measurements. The power spectra and other statistical quantities of the pressure fluctuations with the hot wire were the same as those without the hot wire.

Comment: J. C. R. Hunt. I was interested to see some strong similarities between the observations of Kawai et al. and the theoretical predictions of Durbin and Hunt. In particular we note: (1) in Figure 1 the high frequency pressure spectrum decreased as B/Lx increased (c.f. Figure 2b of Durbin and Hunt), (2) in Figure 2 the high frequency pressure spectrum is greater at the edge than at the stagnation point (c.f. Figure 2b of Durbin and Hunt); (3) the cross-correlations in the y and z directions are both larger when B/Lx increases (c.f. Figures 5 and 6 of Durbin and Hunt); (4) the inapplicability of the lattice network approach to calculating the fluctuating load and the need to use or calculate the pressure correlations (c.f. p. 13, Durbin and Hunt).

Question: A. Kareem. I am a little curious about your tube length of 2 cm with frequency response flat from D.C. to 300 Hz. For this small length I expect a peak below 30 Hz. Your experiment was performed in two-dimensional flow. I feel the aerodynamic admittance function as such cannot be applied to boundary layer flows. I have performed similar experiments in boundary layer flows and cannot come up with a general expression similar to yours given in Equation 4. I have also observed such admittance functions for the (leeward) wake face and they exhibit a peak at the Strouhal frequency for the locations near the edges on the leeward face. I would like to know if you have had such an observation.

Reply: H. Kawai, J. Katsura, H. Ishizaki. The natural frequency of our measuring system was about 1400 Hz. Dr. Katsura who is one of the authors of the paper measured the frequency response of this system from D.C. to 2000 Fz. The result of measurements showed that the frequency response was almost flat from D.C. to 300 Hz. Our experiments were carried out in the uniform turbulent flow in the case of the three-dimensional prism. We did not have any data in a boundary layer wind tunnel. However, many experimental data on full-scale structures showed that the power spectra and the co-spectra are of the same form as those presented in the papers. Therefore, I believe that even in turbulent boundary flow the modified universal relation of the spectra including the effects of the shear will exist.

Question: H. Ruscheweyh. You have calculated the gust factor with the assumption that the rear side pressure fluctuation is directly related to that of the windward face. Because of the vortex separation on the building itself the fluctuation pressure intensity on the rear side is much higher than on the front side. Have you made cross-correlation measurements between rear and front sides of your model to clarify this point?

Reply: H. Kawai, J. Katsura, H. Ishizaki. We did not measure the cross-correlation between the pressure fluctuation on the windward face and that on the leeward face. If the pressure fluctuations on the windward face are not correlated to that on the leeward face, the gust factors presented in our paper may be reduced more than 15%. I think that the pressure fluctuations on the leeward face are strongly affected by the turbulent property and the aspect ratio. Many experimental data on full-scale structures and in wind tunnel tests showed that the pressure fluctuations might not be as high as the pressure fluctuations on the windward face in the strong turbulent shear layer as like the atmospheric boundary layer. Therefore, it is reasonable to assume that the ratio of the magnitude of the fluctuating pressure on the windward face and on the leeward face may be the same as that of the mean pressure.

Paper by W. A. Dalgliesh, J. T. Templin, K. R. Cooper--COMPARISONS OF WIND TUNNEL AND FULL-SCALE BUILDING SURFACE PRESSURES WITH EMPHASIS ON PEAKS

Question: T. Miyata. Can we consider some discrepancy of Cp between full-scale and experimental measurements to be attributed to: impossibility of generation of long-term fluctuations in approaching flow in the wind tunnel as in the natural wind, though depending on the similitude scale, and difficulties in determination of time interval for Cp to be evaluated, depending on the above long-term fluctuation.

Reply: W. A. Dalgliesh, J. T. Templin, K. R. Cooper. We have purposely avoided long-term fluctuations in full-scale by averaging over relatively short-time intervals, i.e., 5 minutes.

In the wind tunnel, the 20 second averaging time was chosen to provide stable mean and rms values during routine data acquisition. In a special test, shorter periods down to 2 seconds were examined but the main result was to increase the variability of the mean and rms values without systematic changes in the values themselves. We concluded that the 20-second averages could be related to the 5 minute full-scale mean and rms values. Of course, this is not true for peak values because the average number of "spikes," n, increases linearly with length of sample interval.

Some instances of disagreement between rms pressure coefficients, Cp^1, may be related to fluctuations in wind direction in full-scale that may not have been simulated well in the wind tunnel. An example of the effect may be seen in Figure 6b and 6d for tap #312 at a wind direction of approximately 160° where $\overline{Cp}$ is very sensitive to small variations in wind direction. The corresponding full-scale Cp' is much larger than the wind tunnel value.

Paper by W. H. Melbourne--TURBULENCE EFFECTS ON MAXIMUM SURFACE PRESSURES - A MECHANISM AND POSSIBILITY OF REDUCTION

Question: A. Kareem. I would like to add some evidence to your hypothesis. I measured pressure fluctuations along the height of a building model very near the separation line. The flow was normal to the surface. Looking at the simultaneously measured time histories one could point out very large intermittent spikes occurring along the height of the building in the bubble. This provides evidence that this pressure fluctuation phenomenon is highly correlated spanwise when the approach shear flow separates from bluff bodies like a building. I would like to have your comment on this.

Reply: W. H. Melbourne. Thank you for your observation. I can add that we have observed the same phenomena and it is proposed that the mechanism is the same for the pressures under the bubble on the side of buildings and where the flow is angled to the leading edge.

Question: I. P. Castro. Bearman has shown that the effects of changes in the separated shear layer caused by different upstream conditions depends strongly on the geometry of the body downstream of separation. Does Prof. Melbourne see his stability mechanism as able to explain the fluctuating reattachment length on bodies for which the shear layer is permanently reattached onto the top surface, as well as the case, for short bodies, where surface reattachment is only intermittent so that the variations in 'cavity' volume are an order of magnitude greater than in the former case?

Reply: W. H. Melbourne. The high intermittent suction peaks still occur when the shear layer is permanently attached, particularly for the flow at an angle to the leading edge. Often the standard deviation of the pressure is less but the intermittency greater (i.e., a higher peak factor is measured). The reattachment region still fluctuates considerably with turbulent incicent flow, and it is suggested that the mechanism is the same in this case.

Question: F. H. Durgin. The fundamental causes of the extreme low pressures on a building are the vortices formed at three-dimensional corners whether the corners are aerodynamic or physical. Further, the most extreme peaks frequently are not at the leading edge but further back. Any explanation of these peaks must involve the strength and motion of these vortices as well as their point of origin.

Reply: W. H. Melbourne. I agree, and of course pressure further back from the leading edge under the bubble and along the line of the vortex shed in angled flow or from a discontinuity on the corner of a building are similarly highly negative.

Question: C. Kramer. I appreciate very much your explanation of the small scale turbulence influence on the separated boundary layer. As we found in wind tunnel tests with block type buildings this small scale turbulence will be affected by modifications of the shape of the leading edge where the flow separates. This may be an essential reason why peak pressure measurement on block type models in different wind tunnels for models with the same relative geometry often disagree greatly.

Reply: W. H. Melbourne. I believe much of the reason for the range of pressures measured in these situations in various wind tunnels can be explained by the difference in the fine scale energy present. It would help if experimenters published velocity spectra with all wind measurements. The amount of acoustic energy is also known to have a significant effect on shear layer behaviors.

Question: C. Kramer. The vent in the eave shows the greatest effect for wind direction diagonal to the roof plan form. Then, as mentioned in the previous comment too, the delta-wing type vortices occur. As reported earlier (Gainesville, 1978) we would reduce the suction peaks on flat roofs by plough shaped devices, fitted only at the roof corners and creating a disturbance of these vortices. Don't you think that the effect of the vent is mainly a disturbance of the vortices and less an influence of filling up a separation bubble?

Reply: W. H. Melbourne. Our observations are that the venting of the bubble both in normal and angled flows to the leading edge reduces the amount of intermittent forward travel to the reattachment and the mean reattachment is further back. The process is very complex. We have found a variety of behavior with leading edge conditions but no such clear picture emerges. In most cases, I think these variations cause changes in entrainment rate into the shear layer in the same way as changing the fine scale energy in the incident flow.

Question: G. W. Tieleman. Full-scale measurements of pressures on solar panels taken recently at Virginia Polytechnic Institute and State University, seem to indicate that local low pressure peaks correlate well with peak gusts. Is it possible that with an increase in approach velocity, the collapse of the separation bubble is increased? From recent wind tunnel tests of pressures on roofs of small houses, I have observed that the time interval between low pressure peaks varies a great deal with the orientations of the roof with respect to the mean wind direction. Does this phenomenon fit your collapsing separation bubble hypothesis?

Reply: W. H. Melbourne. I believe the correlation you observe is due to the higher peak velocities in the larger scales causing lower pressures outside the shear layer. Increased velocity causing lower pressure outside the shear layer is the mechanism by which low pressures in the wake occurs. The hypothesis is intended to cover the very intermittent pressure you observe when the flow is angled to the leading edge.

Question: D. Surry. It has been our experience that high peak suctions may also be observed well away from the edge from which the flow is separating for isolated building models as well as near that edge. The trace of such high peak suctions appear to follow the line where one might expect the shear layer to reattach or along which the base vortex might be. It is important to comment then that high peak suctions do not just occur near separating edges and to ask how these secondary peaks fit into the suggested mechanism.

Reply: W. H. Melbourne. Yes, I agree that the high suction peaks do not just occur near the leading edge and are similar further back in the bubble and along the bubble under angled vortices. It has not been our experience that the highest peak pressures occur near the reattachment region although the highest standard deviation of pressures often occur in this region.

Question: E. Simiu. Do you feel that the Reynolds number may affect the phenomena associated with small scale turbulence described in your paper?

Reply: W. H. Melbourne. Our evidence to date is that the Reynolds number effects on pressures under a reattaching shear layer are not evident in the range of 10^5 to 10^6 for which we typically make these measurements. However, I believe that at lower Reynolds numbers,when the viscous dissipation region closes up the inertial subrange,that eventually the amount of fine scale energy on which shear layer entrainment rate is dependent, would get to the stage where Reynolds number effects would become significant, and pressure start to look more like those in low turbulence flow.

Question: J. L. Wood. European architects are currently styling buildings with curved eaves, i.e., a radiused junction between wall and roof. What effect will this have on "bubble" behavior? Can any guidance be given on relation of radius and peak coefficient values?

Reply: W. H. Melbourne. I cannot give any specific guidance in this respect. Limited experience suggests that radiused edges cause the separation line to move backward and forward and then probably has the same effect as increasing fine scale turbulence in the incident flow and in that it causes increased entrainment rates into the shear layer and consequent increases of high suction peaks under the shear layer. Certainly some very high pressures have been measured in model and full-scale behind radiused corners.

Question: R. E. Whitbread. A grandstand roof was quoted in the paper as an extreme example of peak fluctuating pressures at the leading edge. What general guidance can you give on the venting requirements for this type of structure?

Reply: W. H. Melbourne. As yet we have not made any measurements on the grandstand roof dynamic model with vents. This will be done shortly and I am confident that very significant reduction in roof loading will be realized. The vent can be put further back from the leading edge in this situation which helps as we have observed in several cases on the block model, and the vented eave model. If the vent is too near the leading edge, the flow through the vent is entrained into the shear layer, resulting in higher overall shear layer entrainment rates and higher negative pressure peaks under the bubble.

Paper by J. W. Saunders, W. H. Melbourne--BUFFETING EFFECTS OF UPSTREAM BUILDINGS

Question: T. Reinhold. Your tests were carried out in an open terrain environment and you suggested that the interference effects may be attributed, at least in part, to increases in turbulence caused by the upstream building. Would you expect to find interference effects which are as large for tests conducted in an urban type boundary layer where the incident turbulence level is already high? Would you expect the influence to be felt for as large separation distances as found for the open terrain tests?

Reply: J. W. Saunders, W. H. Melbourne. Melbourne and Sharp (3) found the buffeting in urban terrain to be less than the values presented in this paper. A design of a building downstream of a lake using a gust factor approach will result in a design which will be unconservative if a tall building is located upstream at a later stage. The placement of an upstream building will have a more dramatic effect than in the urban case because the proportional increase in the force spectrum will be greater due to the increased turbulence. In the urban case the increased shear will reduce the distance the wake energy carries (Peterka and Cermak, 1975 conference).

Question: R. E. Whitbread. The paper mentions a simple damping ratio of 1% of critical. Were other damping levels used or was it assumed that the so-called buffeting factors are independent of damping?

Reply: J. W. Saunders, W. H. Melbourne. Only one typical value of damping was used because, based on the work of Saunders (1,2) and recent information presented by Reinhold and Sparks; Kareem and also Kwok presented at this conference, these four independent sets of results have again shown that the influence of aerodynamic damping is not significant up to a reduced velocity of 10 for habitable buildings. Therefore, the aerodynamic input into the downstream building will not be sensitive to its structural damping magnitude. The buffeting factor approach will therefore cancel the particular structural properties.

Question: J. C. R. Hunt. (1) Did you consider the effects of the downwind building on the upwind building? It is well known that even when the spacing is about one diameter this may be a strong effect. I have found it helpful to think in terms of a strong or a weak interaction depending on whether there is or is not a two-way interaction. (2) When there is a weak interaction, have you considered using wake data (e.g., Peterka and Cermak, 1975) to estimate the forces on the downwind structure? (3) Field and laboratory measurements show how the velocity deficit (and turbulence) decays more rapidly in strong external turbulence.

Reply: J. W. Saunders, W. H. Melbourne. (1) Thank you for this comment. (2) No, the amount of data required is substantial; however, we shall investigate this further. (3) Your point is well made and is highlighting that the wake energy will dissipate faster in an urban environment than the wind environment considered in this paper.

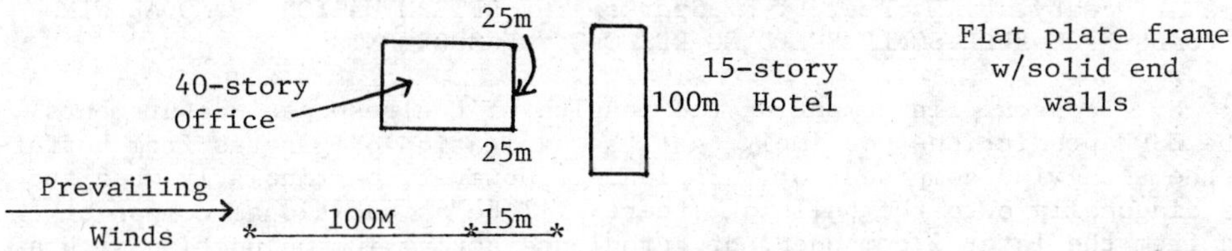

Question: D. W. Ebeling. Under 50 km/hr winds for periods of 1 hour or more the hotel would exhibit considerable motion of the upper floors in a torsional mode. Extrapolating from chimney reports we calculated that vortex shedding at that wind speed was very close to resonant vibration of the hotel causing partition damage. Does this seem reasonable? 20 years ago.

Reply: J. W. Saunders, W. H. Melbourne. Equating a reduced velocity of 10 for both buildings shows that the torsional natural frequency of the hotel must have been extremely low for resonance to occur. However, buildings with a breadth to depth ratio of 3 to 1 or higher such as you have presented are very sensitive to torsional aerodynamic input due to vortex action even though there is a large frequency difference between the frequency of the vortices and the natural frequency in torsion of the building.

Paper by L. C. H. Lam, R. P. Lam, S. Mackey--FULL-SCALE MEASUREMENT OF WIND PRESSURE FLUCTUATIONS ON A BLUFF BUILDING

Question: D. M. Sykes. The smoothed auto spectra shown in Figures 5, 6 and 7 contain large peaks at about 4 Hz. What is the cause of these high frequency peaks?

Reply: L. C. H. Lam, R. P. Lam, S. Mackey. The peaks are not high, relatively speaking, if you would note that the spectra are plotted on log scale. The authors would consider that this may have been the natural frequency of the guyed masts on which the anemometers were mounted.

Question: D. Surry. I must admit to being skeptical over the proposed 'air cushion' effect. First, the spectral peaks do not appear particularly prominent compared to the expected variability of the spectra, making their precise definition apparently difficult. Secondly, from the numbers given in the paper, it appears that these peaks occur roughly where they might be associated with vortex shedding - i.e. around a Strouhal number of 0.15. Has the author considered this as a possible mechanism and, if so, why has he decided it is not applicable?

Reply: L. C. H. Lam, R. P. Lam, S. Mackey. The auto-spectra showing the existence of the spectral peaks were plotted on logarithm-scale grids so as to cover a wider frequency range variation. Had the region near the mentioned spectral peaks been plotted on natural-scale grids, they would be much more pronounced. The answer to the second part of the question concerning vortex shedding effect has already been discussed in the last paragraph of the Discussion Section of the paper presented.

Paper by A. P. Jeary, B. E. Lee, P. R. Sparks--THE DETERMINATION OF MODAL WIND LOADS FROM FULL SCALE BUILDING RESPONSE MEASUREMENTS

Question: N. Thompson. In comparing the results of the response measurements with the ESDU predictions you imply that the excitation originates from buffeting by the alongwind component of turbulence. However, particularly when the wind is diagonally onto the building, there will be an additional component arising from the lateral component of turbulence acting in conjunction with a large rate of change of side force. Thus, the fluctuating wind incidence could produce a very significant component of the overall response. The ESDU item does not pretend to cover this component and so the only comparisons that are valid are when the wind is normal to one face when better agreement is obtained.

Reply: A. P. Jeary, B. E. Lee, P. R. Sparks. The authors accept Mr. Thompson's comment and would like to make it clear that the predictions of modal forces using the ESDU method, referred to in Table 2 of the paper, only apply to wind directions of 270° and 360°, i.e., when the wind is normal to the narrow and wind faces of the building, respectively.

Paper by J. F. Howell, M. Novak--VORTEX SHEDDING FROM CIRCULAR CYLINDERS IN TURBULENT FLOW

Question: H. Tanaka. (1) Any comments on the relation between pressure and velocity correlations under various flow conditions? (2) Have any attempts been made to extend the idea of universal response curve based on Iwan's model to the results in turbulent flow considering your results?

Reply: J. F. Howell, M. Novak. (1) Some measurements were made of wake velocity correlations but only for the driven vertical cylinder. These showed the same trends as those reported by Tobes (16); however, the correlation length was smaller and less sensitive to amplitude than those of pressures. One set of wake velocity correlations is shown in Figure 4. (2) The universal curve of Iwan only considers smooth flow. The present results for two turbulent flows have been plotted in Figure 9. The smooth flow results show good agreement as do those for turbulent flows at low damping. With increasing damping both the theoretical and empirical curves significantly underestimate the response. It would be of value if Iwan's universal curve could be extended to include turbulent flow. This may be done for the empirical curve but requires a better understanding of the problem than at present.

Paper by T. Miyata, M. Miyazaki--TURBULENCE EFFECTS ON AERODYNAMIC RESPONSE OF RECTANGULAR BLUFF CYLINDERS

Question: A. Laneville. Would you comment on the fact that only lift is influenced by L_x/d and not drag as indicated by your Figure 4? Was your damping kept constant (including the drag force induced in the motion)?

Reply: T. Miyata, M. Miyazaki. Greater scale causes more highly curved shear layer and more rapid reattachment which affects on the lift more highly than the drag in the case at $\alpha \neq 0$. The extent of influences of shear layer are judged to be extremely delicate. The structural damping (including the drag induced by the motion) was assumed to be kept almost constant, judging from the response results in Figure 10.

Question: A. Laneville. First I must say that, from our previous observations, the effect of integral scale was secondary in the galloping of bluff cylinders and that the quasi-steady theory worked adequately. Now, referring to Figure 4, it appears that the drag and life coefficient curves are not that much different for all the L_x/d values considered. Thus, the quasi-steady theory should predict displacement curves fairly similar for all the L_x/d considered, which is not the case as shown in your Figure 7. Do you expect a range of L_x/d that would invalidate the quasi-steady theory?

Reply: T. Miyata, M. Miyazaki. According to the quasi-steady theory, the onset velocity is inversely proportional to the inclination $dC_{Fy}/d\alpha$ of lateral force coefficient, where $dC_{Fy}/d\alpha \simeq dC_L/d\alpha + C_D$. The coefficient C_{F_y} for 1:2 section in Figure 13 are correctly derived from the drag and life coefficients in Figure 4, where we can see the difference for the L_x^u/b values considered in the inclination $dC_L/d\alpha$ of life coefficient at $\alpha \simeq 0$. The values of inclination $dC_{Fy}/d\alpha$ in Figure 13 are obviously different for the L_x^u/b values considered, which are reflected in the universal response diagram in Figure 10, including the case shown in Figure 7.

We believe that the quasi-steady theory is valid even in turbulent flows.

WIND ENGINEERING

(IN TWO VOLUMES)

Volume 2

Also of Interest

CHEUNG
Finite Strip Method in Structural Analysis, 2nd Ed

FRANCIS
Introducing Structures

GIBSON
Thin Shells: Computing and Theory

HARRISON
Structural Analysis and Design: Some Minicomputer Applications

HORNE
Plastic Theory of Structures, 2nd Ed

LENCZNER
Elements of Loadbearing Brickwork

LENCZNER
Movements in Buildings, 2nd Ed

MILLER & SMITH
Mechanical Behaviour of Materials (3 volumes)

PIGGOTT
Load Bearing Fibre Composites

SACHS
Wind Forces in Engineering, 2nd Ed

TAPLIN
Advances in Research on the Strength and Fracture of Materials and Structures (6 volumes)

Pergamon Related Journals

Free Specimen Copies Gladly Sent on Request

CEMENT AND CONCRETE RESEARCH
COMPUTERS AND STRUCTURES
ENGINEERING FRACTURE MECHANICS
FATIGUE OF ENGINEERING MATERIALS AND STRUCTURES
INTERNATIONAL JOURNAL OF ENGINEERING SCIENCES
INTERNATIONAL JOURNAL OF SOLIDS AND STRUCTURES

WIND ENGINEERING

Proceedings of the Fifth International Conference, Fort Collins, Colorado, USA, July 1979

Edited by

J. E. CERMAK

Colorado State University, USA

Volume 2

PERGAMON PRESS

OXFORD · NEW YORK · TORONTO · SYDNEY · PARIS · FRANKFURT

U.K.	Pergamon Press Ltd., Headington Hill Hall, Oxford OX3 0BW, England
U.S.A.	Pergamon Press Inc., Maxwell House, Fairview Park, Elmsford, New York 10523, U.S.A.
CANADA	Pergamon of Canada, Suite 104, 150 Consumers Road, Willowdale, Ontario M2J 1P9, Canada
AUSTRALIA	Pergamon Press (Aust.) Pty. Ltd., P.O. Box 544, Potts Point, N.S.W. 2011, Australia
FRANCE	Pergamon Press SARL, 24 rue des Ecoles, 75240 Paris, Cedex 05, France
FEDERAL REPUBLIC OF GERMANY	Pergamon Press GmbH, 6242 Kronberg-Taunus, Hammerweg 6, Federal Republic of Germany

First edition 1980

British Library Cataloguing in Publication Data

International Conference on Wind Engineering,
5th, Colorado, 1979
Wind engineering.
1. Wind-pressure - Congresses
I. Title II. Cermak, J E
624'.175 TA654.5 80-40753

ISBN 0-08-024745-8

In order to make this volume available as economically and as rapidly as possible the author's typescript has been reproduced in its original form. This method has its typographical limitations but it is hoped that they in no way distract the reader.

Printed in Great Britain by A. Wheaton & Co. Ltd., Exeter

Contents

Volume 2

Technical Session VI: Dynamic Response—Tall Buildings and Towers

Technical Session VII: Dynamic Response—Bridges, Transmission Lines and Roof Membranes

Technical Session VIII: Physical and Mathematical Modeling

Technical Session IX: Wind-Engineering Applications

Contents

Volume 1

Technical Session V: Wind Loading—Fluctuating External

SESSION VI

Dynamic Response - Tall Buildings and Towers

CHAIRMAN

A. G. Davenport
Director and Professor
Boundary Layer Wind Tunnel Laboratory
The University of Western Ontario
London, Canada

CO-CHAIRMAN

D. Sfintesco
Director of Research
Centre Technique Industriel de la Construction Métallique
Puteaux, France

RAPPORTEUR

A. N. L. Chiu
Professor of Civil Engineering
University of Hawaii at Manoa
Honolulu, Hawaii, USA

RECENT DEVELOPMENTS AND TRENDS: DYNAMIC RESPONSE - TALL BUILDINGS AND TOWERS

A. G. DAVENPORT
Boundary Layer Wind Tunnel Laboratory, University of Western Ontario, Canada

The various chairmen of the sessions have been charged with the task of reviewing the papers in the sessions and predicting future trends and developments. This is never a particularly easy task. In view of my misgivings, I decided to ask for the assistance of three acquaintances, each of whom had some knowledge and experience in wind engineering but with somewhat different viewpoints. They had helped me out on a similar occasion[1], and once again agreed to an informal discussion on the subject at hand. I would like to report briefly on this rather unusual meeting and the comments they made.

I should first introduce you to these experts. The most senior of them was a very distinguished French engineer - a member of the Legion d'Honneur in fact. He was a prominent and successful tower designer - in fact the only one of my acquaintances whose structure was named after him: his name, Monsieur Gustaf Eiffel. His inclusion in the group was appropriate. His extensive professional experience and skill was matched only by his pioneer research achievements in meteorology (in his Laboratory on top of the tower) and in aerodynamics - propellors, wing sections and wind loads - in his new wind tunnel in a suburb of Paris. He was in fact, if we care to look back, one of the oldest and most illustrious patrons of wind engineering.

Next attending this meeting was a much younger man, a recent doctoral candidate at one of the good universities where he defended his thesis on the "Random vibration of random structures in random wind". He is now a junior professor at an even better university. His excellent training in the theoretical and experimental aspects of meteorology, aerodynamics and structural mechanics have given him a keen insight into the difficult problems of wind engineering. He spends much of his year writing papers and attending conferences. He consented to attend our meeting on condition we paid his travel expenses. His name is Dr. Ventus Explorator.

[1]Canadian Workshop on Wind Engineering, Sept 23-24, 1976. Summary report issued by National Research Council, Canada, Associate Committee on Aerodynamics/Canadian Wind Engineering Association, dated Dec. 1976.

The third member of this group was a prominent professional man whose consulting firm now has offices in many major cities in the world. His relations with the architectural profession were good, and his expert team of designers could be relied on to produce the structural drawings on time even at the expense of their weekends. As well as his good business sense, he had an impeccable knowledge of the codes of practice, standards and various handbooks of design. His well tailored suits usually bulged with a calculator having 3 decimal places. The recent entry of his firm into design of tall buildings and towers in the Middle East had given him a keen interest in the subject. His name was Mr. Ingenius Constructus of the firm which bore his name.

This meeting between Monsieur Eiffel, Dr. Explorator and Mr. Constructus was held at Cape Race, Newfoundland, a notoriously windy promontory where the strength of the westerlies is exceeded only by the strength of the easterlies as they roll in over thousands of miles of stormy Atlantic fetch. This exposed locality, where such trees as there are cling desperately to the rock and grow horizontally, had the right atmosphere for this discussion on wind and vibration.

The three participants had received copies of the preprints in good time from the conference organizers and had been asked in particular to address the papers in session six. I tape recorded the conversation - when the howl of the wind permitted - and I first asked them for a general statement. Monsieur Eiffel began.

"Mesdames et messieurs (he was very formal) I have as you say more time on my hands these days so have read zees papers on recherche. In fact I have read all the previous proceedings of the other four conferences. But tell me this, why is it the volumes of the papers increase in number all the time. C'est formidable! If you are finding solutions to the problems of wind the size should decrease. Before long even I will have no time to read it all. Have you two gentlemen read the papers?"

It was Dr. Explorator's turn.

Dr. Explorator: "I think you are sadly out of touch with the problems Mr. Eiffel, the difficulties are increasing all the time and so must the papers. But to answer your question; I have only had time to read the abstracts, but there are obviously several theories I strongly disagree with and there are some mistakes in some of the experiments. I plan to correct these when I receive my next research grant."

Mr. Eiffel: "S'il vous m'excuse, you are an impetuous young man. I have some experience. How do the problems today differ from when I built La tour Eiffel? Are the structures much higher? Is the wind any stronger? At that time we had little information - no codes of practice and just some simple rules, Mr. Newton's for example. Of course we were conservative and a wind pressure of 350 kg/m^2 might be a little higher than you would use today. And I did not use, how you say, the "shielding factor". But Dr. Explorator, if you visit the beautiful city of Paris, voila, my tower is still there!"

Dr. Explorator: "Your structure is unfortunately not economic. It has three times more material than you need. We consider rivets impractical and the flat members you used are inefficient. Round members are much better. We can build much higher, more efficient structure today. Of course some of these light structures seem to vibrate and the damping is much less because of the bolts and welding. But we have some excellent statistical theories for predicting these vibrations so they should present no problems to speak of."

Mr. Constructus chimed in at this point:

Mr. Constructus: "I read a few of the papers on the plane; I found them hard to swallow and couldn't see the point of some of them. In my business I don't have the time to follow all that theoretical stuff. I have to get the job done. Of course I wouldn't like to use the loads you used Mr. Eiffel - or my clients wouldn't. We have to keep the cost down you know. But your theories are far too complicated Dr. Explorator. We want something simple, safe and economical. All this talk of statistics and probability - you can't use that sort of thing in a code. We want clear cut rules,not a sort of lottery."

Mr. Eiffel: "S'il vous me permittez, both you gentlemen have made some good points. Dr. Explorator I agree that if I was building the tower today it would be much lighter. I understand about the vibrations: I was worried our visitors would have, how you say, mal-de-mer, sea sickness. But I concluded it would only happen on windy days and on those days the visitors would surely seek the other pleasures of Paris. Besides our structure was so heavy; even so when I made my measurements I observed its vibration. These vortices of Monsieur Strouhal and the forces perpendicular to the wind are very curious - I remember reading his paper written in Prague - exactly 100 years ago. I hope you find a solution to these. But I agree with Mr. Constructus the engineer must have simple rules; your codes are a help if not too unflexible. I know the practical problems: the Paris Exhibition was not going to wait for the ink on my recherche memoires to dry. Research into the mysteries of nature is fascinating but you must, how you say in English, see the wood for the trees."

At this point a huge gust of wind obliterated all the conversation from the tape.

The discussion went on late into the evening. Mr. Constructus emphasized the practical questions. Dr. Explorator emphasized the importance of the verification of scientific theories and the many variables of the wind loading problem and how urgently research funds were needed to resolve the questions. Monsieur Eiffel always the eloquent compromiser, took an interest in the new tools; he digressed at length on the role of wind tunnels, the use of which he fully endorsed. But he reminded them of the need always to observe how full scale structures withstood the theories and the experiments.

In spite of a few critical comments the three gentlemen all found something of interest in the papers of session VI which illustrated their points of view.

As far as future trends and developments, I am not sure if a true consensus was expressed. Perhaps there wasn't time to work it out. Dr. Explorator had to return to his laboratory to calibrate some instruments and review some computer printout. Mr. Constructus had a busy schedule ahead meeting clients and Monsieur Eiffel wanted to return to his haunts in Paris. They agreed if necessary to meet again in four years time.

CROSSWIND RESPONSE OF HIGH-RISE BUILDINGS

A. KAREEM*, J. E. CERMAK** and J. A. PETERKA**

*University of Houston, Houston, Texas, USA
**Colorado State University, Fort Collins, Colorado, USA

ABSTRACT

The complex nature of wake-excited crosswind response, that is, in a plane perpendicular to the wind has been investigated. Multivariate probabilistic and spectral characteristics of the fluctuating pressure field around a tall building model have been derived from measurements in a wind tunnel. Through the use of the spatiotemporal characteristics of the fluctuating pressure field around a square cross-section building model, expressions for the crosswind forcing function are developed in time and frequency domains based on statistical integration. A system identification approach is used to evaluate the crosswind forcing function induced due to the building motion. Methods of random vibration theory are used to estimate the dynamic response. The results are compared with the response of an aeroelastically modeled building in a boundary-layer wind tunnel. The predicted results show good agreement with the experimental response. The procedure developed in this study permits the crosswind acceleration levels of square cross-section tall buildings to be estimated at the preliminary design stages.

NOMENCLATURE

A	Area of building face	$Co_{p1p2}(g_1,g_2,n)$	Co-spectra between location g_1,g_2 as a function of frequency
B, b	Building width		
BL1	Boundary layer 1	F_t	Total crosswind force
BL3	Boundary layer 3	$G_F(n)$	Cross spectral density function
[C]	Damping matrix	H	Building height

H_i	Building segment height
$\|H(ik)\|^2$	Mechanical admittance function
$J(k)$	Aerodynamic coefficient
k	Reduced frequency
$[K]$	Stiffness matrix
m	Building mass
$[M]$	Mass matrix
n	Frequency in hertz
$[R]$	Normalized co-spectral density function matrix
$S_F(n)$	Normalized power spectral density function of forcing function
V_h	Longitudinal mean velocity at the building top
y	Crosswind building response
ρ	Air density
ρ_{ij}	Normalized cross-correlation
σ	RMS value
$[\Phi]$	Modal matrix
ξ	Ratio of the critical damping
γ_b	Building density
γ	Net damping
ϕ_n	Natural frequency in Rad/sec

Superscripts

$\overline{(\quad)}$	Time average

INTRODUCTION

The assurance of structural safety and functional reliability of high-rise buildings during strong winds, based on strength and serviceability, is an important design objective. The required strength of the major structural system of a tall building usually depends on wind induced stresses, whereas, serviceability is affected by sway acceleration at the top floors. Therefore, a comprehensive dynamic analysis of buildings subjected to high winds is essential for the evaluation of response parameters to ensure an efficient and economically viable structural system.

Although the knowledge of wind loading on structures has significantly improved over the past decade, an understanding of the mechanism that relates the atmospheric flow field to various wind induced effects on structures has not been developed sufficiently for the functional relationship to be formulated. Not only is the wind approaching a building complex but the flow pattern generated around a building is complicated by the distortion of the mean flow field, flow separation, vortex formation and the wake development. Consequently, large wind pressure fluctuations are observed on the surface of a structure. As a result of these pressure fluctuations, large aerodynamic forces and moments are imposed upon the structural system and intense localized fluctuating forces act on the facade of such structures. Under the collective influence of these fluctuating forces a building vibrates in rectilinear and torsional modes. The amplitude of such oscillations is dependent on the nature of aerodynamic forces and the dynamic characteristics of the structure.

The dynamic response of buildings in the alongwind direction can be predicted with reasonable accuracy (1). The crosswind response, that is, motion in a plane perpendicular to the wind dominates over the alongwind response for most tall buildings with equal mechanical admittance in both rectilinear directions. The complex nature of wake-excited crosswind response which results from a nonlinear interaction of incident turbulence, wake dynamics and building motion has inhibited analytical prediction with sufficient degree of reliability. A torsional motion in a building results from a possible aerodynamic and/or inertial coupling in various degrees of freedom.

Spatiotemporal measurements of the fluctuating pressure field around a square cross-section building model has resulted in a semi-analytical methodology for predicting the crosswind dynamic response of square cross-section tall buildings in urban and suburban environments. The procedure presented in this paper permits the crosswind acceleration levels and associated bending moments at the base of square cross-section tall buildings to be estimated at the preliminary design stage. These estimates allow the designer to assess the need for more exact investigation by an aeroelastic model in a boundary-layer wind tunnel.

DEVELOPMENT OF CROSSWIND FORCING FUNCTION

A number of techniques relying on experimental measurements have been developed to estimate the crosswind forcing function (2,3,4,5). The overall crosswind loading can be derived by integrating simultaneously monitored pressure fluctuations on a building model surface. In the past, limitations on the availability of a large number of transducers, and the subsequent data-acquisition problem rendered this procedure impractical. This procedure can, however, be modified by making simultaneous measurements of pressure at various levels throughout the building model height. Subsequently, by monitoring the cross-correlation between pressures at various levels, an expression for the integral wind loading function on the building can be developed through statistical integration. This approach not only determines the fluctuating integral forces on the building but also provides good insight of the local spatiotemporal variations of dynamic pressure fluctuations. This procedure is presented in the following paragraphs.

PRESSURE MEASUREMENTS

All measurements were made in the industrial aerodynamic wind tunnel at Colorado State University (5). A square cross-section building model was exposed to two approach flows, boundary layers 1 and 3 (BL1, BL3), representing suburban and urban flow conditions respectively. The boundary layer 1 was simulated with flow approaching the model over smooth floor of the test section. Whereas, for the boundary layer 3 the floor was covered with 2-in. randomized roughness blocks. A set of spires and a horizontal barrier at the entrance of the test section were used to stimulate the boundary layer, in addition to the surface roughness. A detailed description of the characteristics of boundary layers 1 and 3 is presented in reference 5.

A 5-in. square and 20-in. tall prism made of "Lucite" was modeled for this study. Flexible "Tygon" tubing connected a pressure tap on the model prism to the electronic pressure transducer. An on-line digital data-acquisition system was used to monitor simultaneously eight channels of pressure data. The important components of the system are an analog-to-digital converter, a minicomputer and a digital tape recorder. The output of the transducers was fed through a signal conditioning unit for amplification. The data was measured and recorded on tapes for the duration of five minutes for each configuration at a sampling rate of 500 samples per second. The data were stored on digital magnetic tapes (6).

A total of 14 configurations of pressure fluctuations were measured in two approach flows. The square prism was divided into six levels at various heights, and at each level, pressure was monitored on both side faces simultaneously to develop chordwise pressure distributions. Similarly, pressure was monitored along four vertical axes on the side faces to establish spanwise pressure distribution.

The data were reduced to evaluate univariate and multivariate probabilistic and spectral characteristics of spatiotemporal pressure fluctuations on the side faces of the scale model (5,6). This statistical information provided a vital input to development of the crosswind forcing function.

TIME DOMAIN MODEL

For the flow normal to the model face the mean square value of total fluctuating force is given by

$$\overline{F_t^2} = 2\left[\int_0^B\int_0^B\int_0^H\int_0^H \rho_{12}(x_1, z_1, x_2, z_2)\sigma_1\sigma_2 dz_1 dz_2 dx_1 dx_2 \right. \tag{1}$$
$$\left. - \int_0^B\int_0^B\int_0^H\int_0^H \rho_{56}(x_5, z_5, x_6, z_6)\sigma_5\sigma_6 dz_5 dz_6 dx_5 dx_6 \right],$$

where $\rho_{12}(x_1, z_1, x_2, z_2)$ is a cross-correlation function between locations (x_1, z_1) and (x_2, z_2) on the same face. Whereas, $\sigma_{56}(x_5, z_5, x_6, z_6)$ represents cross-correlation between locations (x_5, z_5) and (x_6, z_6) on opposite faces. The root-mean-square value of pressure fluctuations is given by σ_i. The above expression can be represented in discrete matrix representation by

$$\overline{F_t^2} = [\sigma_i A_i]^T [\rho_{ij}] [\sigma_j A_j] , \tag{2}$$

where A_i is the area vector. The fluctuating crosswind force coefficients for various segments along the building height and for the overall building are summarized in Table 1.

Table 1. Crosswind Fluctuating Force Coefficients

Level	$(\overline{F_t^2})^{1/2}/(1/2\rho V_h^2 BH_i)$
1	0.2214
2	0.3525
3	0.4201
4	0.4622
5	0.4748
overall for the building	0.3422

FREQUENCY DOMAIN MODEL

The cross power spectral density of the forcing function can be obtained from Eq. (1) using Wiener-Khintchine relationship,

$$S_F(n) = 2\left[\int_0^B\int_0^B\int_0^H\int_0^H Co_{p_1p_2}(x_1, z_1, x_2, z_2, n)dx_1dz_1dx_2dz_2 - \int_0^B\int_0^B\int_0^H\int_0^H Co_{p_5p_6}(x_5, z_5, x_6, z_6, n)dx_5dz_5dx_6dz_6\right], \quad (3)$$

where $Co_{p_1p_2}(x_1, z_1, x_2, z_2, n)$ represents the normalized co-spectral density function of pressure fluctuations between locations (x_1, z_1) and (x_2, z_2) on the same side face. Whereas, $Co_{p_5p_6}(x_5, z_5, x_6, z_6, n)$ represents normalized co-spectral density between locations (x_5, z_5) and (x_6, z_6) on the opposite faces.

The normalized co-spectral density matrices of intraface and interface pressure fluctuations at five levels were used to generate a normalized cross-spectral density matrix of the crosswind forcing function on a square cross-section tall building. A multilevel forcing function represents the load at various levels and describes the inter-level relationships. This is a useful description of wind loads in order to calculate the response components of a multi-degree-of-freedom system. The discrete matrix representation of Eq. (3) is given by

$$G_{F_{ij}}\left(\frac{nb}{V_h}\right) = 2[(A_i)^T (G_{P_i})^T[-R-]_{ij}(G_{P_j})(A_j) - (A_i)^T(G_{P_i}^T)[-R-]_{ij}(G_{P_j})(A_j)],$$

for $i = j$ — intraface intra-level; interface intra-level (4)

$$G_{F_{ij}}\left(\frac{nb}{V_h}\right) = 2[(A_i)(G_{P_i})[-R-]_{ij}(G_{P_j})(A_j) - (A_i)(G_{P_i})[-R-]_{ij}(G_{P_j})(A_j)],$$

for $i \neq j$ — intraface inter-level; interface inter-level (5)

where the matrix $[R]$ represents the normalized co-spectral density function, (Gp_i) is the power spectral density vector and (A_i) describes the area under consideration.

The main diagonal elements of the crosswind forcing function matrix were evaluated from Eq. (4), and the off diagonal terms were formulated from Eq. (5) (5). The cross power spectral density matrix of the crosswind forcing function is a Hermitian, positive definite matrix. In Fig. 1 each element of the upper triangle of the matrix in nondimensional form as a function of reduced frequency for two approach flows, boundary layers 1 and 3, representing urban and suburban environments, is presented.

MOTION INDUCED FORCING FUNCTION

A system identification technique is developed based on the approach used by Scanlan (7) for suspension bridges, to evaluate the forcing function induced due to the motion of the structure. This portion of the forcing function is in addition to the loading given by equations derived from the pressure fluctuations induced by the incident turbulence and the wake excitation on a stationary building model. At low reduced velocities the building motion is normally too small to induce any significant loading (8). However, at higher reduced velocities, the contribution of motion induced loading becomes more significant.

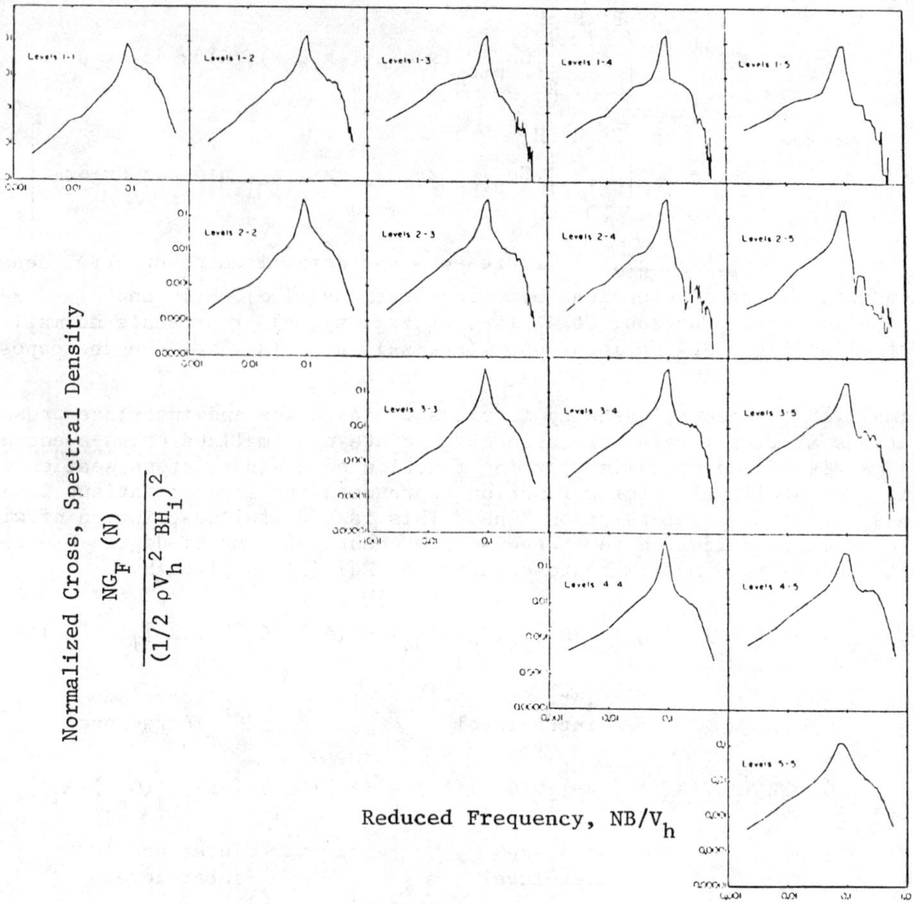

Fig. 1a. Cross power spectral density function of crosswind forcing function at five levels for BL1.

Consider the dynamic equation of motion in the crosswind direction.

$$\ddot{y} + 2\ \xi\omega_n\ \dot{y} + \omega_n^2 y = \frac{F(t)}{m}, \text{ and} \tag{6}$$

$$F(t) = 1/2\ \rho V_h^2\ BH[J(k)\ \frac{\dot{y}}{V_h} + C_F(t)], \tag{7}$$

where $J(k)$ is an aerodynamic force coefficient associated with building motion induced forces, and $C_F(t)$ is the coefficient of forces resulting from pressure fluctuations (Eqs. (4) and (5)). By rearranging Eqs. (6) and (7),

$$\ddot{y} + 2\ \omega_n\ \gamma\dot{y} + \omega_n^2 y = \frac{\rho V_h BH}{2m}\ C_F(t), \tag{8}$$

$$\gamma = \xi - \frac{\rho Bh\ V_h\ J(k)}{4m\ \omega_n}, \text{ and} \tag{9}$$

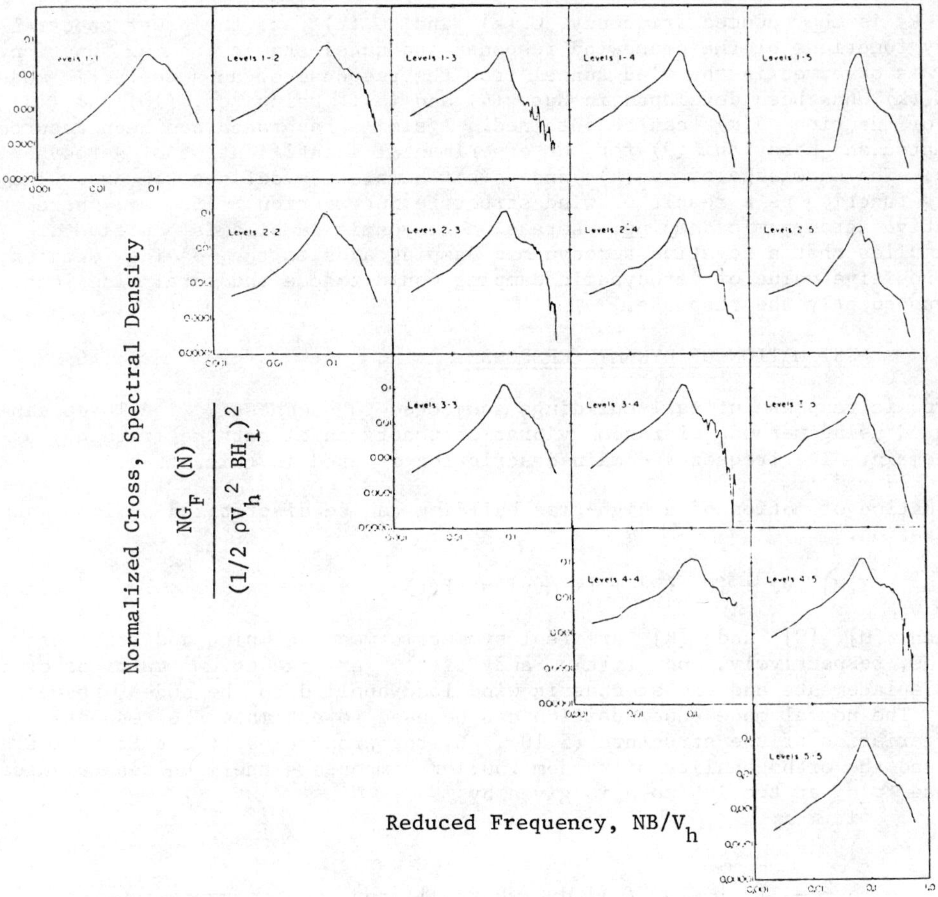

Fig. 1b. Cross power spectral density function of crosswind forcing function at five levels for BL3.

$$J(k) = \frac{4m\,\omega_n\,(\xi-\gamma)}{\rho BH\,V_h}\,. \tag{10}$$

In the frequency domain Eq. (8) is transformed to,

$$G_y(k) = g_c^2\,|H(ik)|^2\,G_F(k), \tag{11}$$

$$|H(ik)|^2 = \frac{G_y(k)}{g_c^2\,G_F(k)}, \tag{12}$$

$$= \frac{1}{\omega_n^4[(1-(\frac{\omega}{\omega_n})^2)^2 + 4\gamma^2\,(\frac{\omega}{\omega_n})^2]}, \text{ and} \tag{13}$$

$$g_c = \frac{\rho\,V_h\,BH}{2m}\,,$$

where k is the reduced frequency, $G_y(k)$ and $G_F(k)$ are the power spectral density functions of the crosswind response and quasi-static loading function. $G_y(k)$ is observed in the wind tunnel from the response of an aeroelastic model and $G_F(k)$ has been developed in Eqs. (4) and (5). Using Eqs. (10) and (12) the value of function J(k) can be obtained. A similar approach has been reported by Komatsu and Kobayashi (9) for the experimental identification of aerodynamic forces. The knowledge of $G_F(k)$ and J(k) completely defines the crosswind loading function as a result of wind-structure interaction. If $\xi > \gamma$, there exists a negative aerodynamic damping, whereas, aerodynamic damping is positive if $\gamma > \xi$. This implies that a negative aerodynamic damping adds to the loading function, and a positive value of aerodynamic damping would reduce the total wind loading and consequently the response.

ESTIMATION OF DYNAMIC RESPONSE

The dynamic response of tall buildings subjected to stochastic wind loads can be estimated using methods of random vibration theory in either the frequency or the time domain. The frequency domain description is used in this study.

The equation of motion of a high-rise building can be discretized by lump mass idealization.

$$[M]\{\ddot{y}\} + [C]\{\dot{y}\} + [K]\{y\} = \{f(t)\}, \tag{14}$$

in which [M], [C] and [K] are real symmetric mass, damping and stiffness matrices, respectively, and {y(t)} and {f(t)} are the column matrices of the node displacements and the stochastic wind load applied to the nodes, respectively. The normal mode superposition can be used to estimate the response characteristics of the structure (5,10). By the properties of the Fourier transforms and the orthogonality of random Fourier components the mean square value of response $\sigma_{y_i}^2$ at the ith node is given by

$$\sigma_{y_i}^2 = \sum_{j=1}^{n} \int_0^{\infty} [G_y(\omega)]\, d\omega, \text{ where i and j designate rows and columns of matrix}$$

$$[G_y(\omega)] = [\phi]\left[\frac{1}{\omega_n^4}\left(\frac{1}{(1-(\omega/\omega_n)^2)^2 + 4\xi_n^2(\omega/\omega_n)^2}\right)\right][G_F(\omega)][\Phi]^T . \tag{15}$$

A solution of the equations of motion can also be obtained in terms of the convolution integral (Duhamel integral) in solving through the time domain.

Using equations of motion and forcing function developed in Eqs. (4) and (5), the dynamic response of a building model was predicted in the fundamental mode only. The dynamic properties of this model are given below:

n = 15 hertz,

B = 5 in.,

H = 20 in.,

ξ = 0.05, and

γ_b = 8.5 lb/ft^3.

EXPERIMENTAL PREDICTION

The response of a 5 x 5 in. cross section and 20 in. tall aeroelastic model of a tall building was experimentally determined for comparison in the industrial wind tunnel for BL1 and BL3 flow conditions. Similitude requirements for aeroelastic modeling are described by the authors in references (5,11). The dynamic characteristics of the model have been reported in the last section. An on-line digital data acquisition system was used to monitor output from a hot-wire anemometer and strain gages mounted on the flexural spring system. Output was recorded on digital tapes and the following response characteristics were computed:

a) mean square value of displacement at the top,

b) probability density and distribution functions,

c) joint probability density histograms of alongwind and crosswind response,

d) auto and cross-correlation functions, and

e) power and cross power spectral density functions.

The computed and experimental rms crosswind displacements for various reduced velocities at the top of the aeroelastic model are compared and plotted in Fig. 2. At high reduced velocities there exists a greater disparity between the predicted and experimental values. This disparity could result from the errors accumulated in development of the crosswind forcing function by statistical integration based on pressure measurements at discrete locations. However, for reduced velocities greater than 5 the predicted values are smaller than the experimental values. This suggests that a contribution to the forcing function is also resulting from motion induced mechanism. At the same time this contribution is included in the response measured from the aeroelastic wind-tunnel model. In order to ensure that the contribution is due to the building motion the variation of net damping in the aeroelastic model response at various reduced velocities was examined. A number of system identification techniques (12,13) were tried to evaluate the damping value from the time history of the aeroelastic response, however, the decay of autocorrelation function of response yielded the most consistent results (14). It was observed that the net damping of the aeroelastic model was continuously decreasing above the reduced velocity of 5. Therefore, the modified estimates of response were computed based on revised damping value,

$$[G_y(k)] = [|H(ik)|^2]_{\text{modified}} \; G_F(k),$$

$$|H(ik)|^2 = \frac{1}{\omega_n^4[(1-(\omega/\omega_n)^2)^2 + 4\gamma^2(\omega/\omega_n)^2]} ,$$

where γ is the net damping given by Eq. (9) and γ is ξ - aerodynamic damping.

The modified prediction of response is compared with the dynamic response obtained from the wind-tunnel study. It is apparent that the inclusion of negative aerodynamic damping has provided a better reconciliation at higher reduced velocities (Fig. 2).

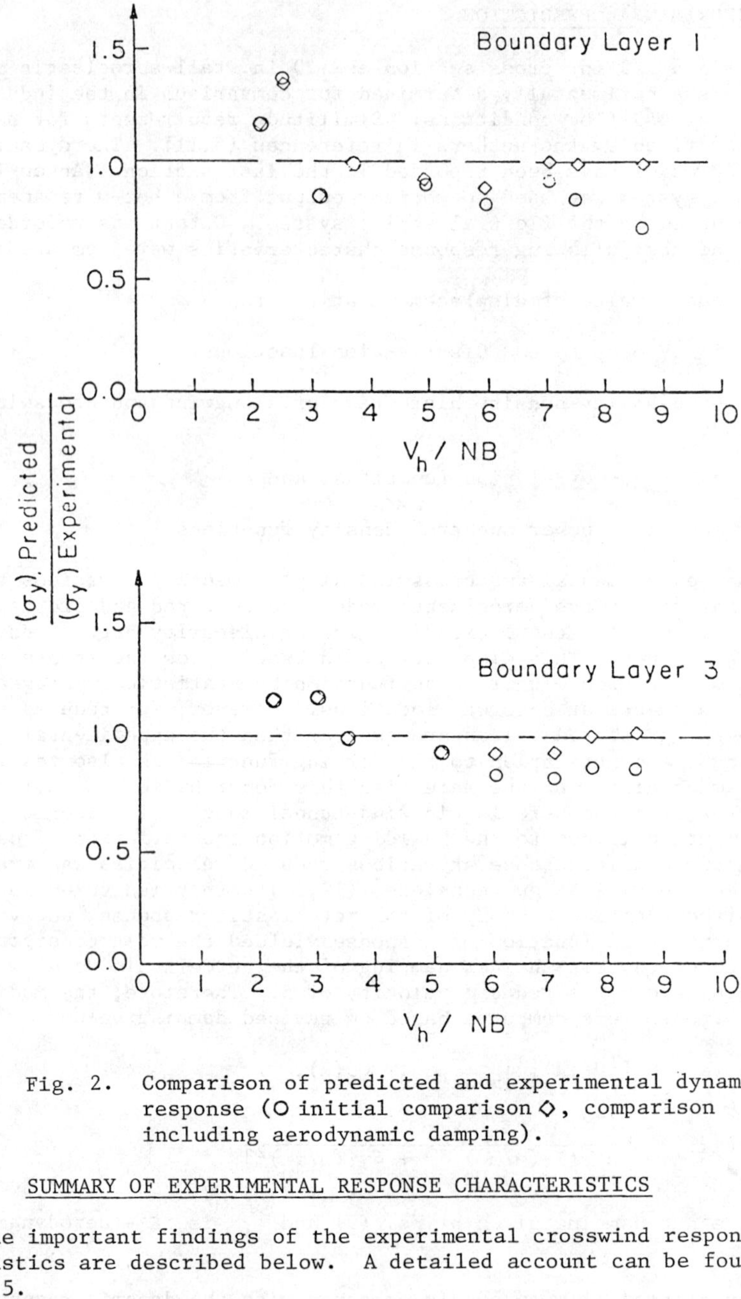

Fig. 2. Comparison of predicted and experimental dynamic response (○ initial comparison ◇, comparison including aerodynamic damping).

SUMMARY OF EXPERIMENTAL RESPONSE CHARACTERISTICS

Some of the important findings of the experimental crosswind response characteristics are described below. A detailed account can be found in reference 5.

1. rms and maximum dynamic response can be expressed as

$$y/B \propto \left(\frac{V_h}{nB}\right)^k ,$$

where k is a constant and varies from 2.79 to 3.

2. Simultaneously monitored incident velocity fluctuations, alongwind and crosswind response reveals that the crosswind response at low reduced velocity is more of intermittent nature, whereas, for higher reduced velocities a well-defined periodic modulation or "beat" phenomenon is exhibited. This phenomenon is more pronounced in boundary layer 1 flow suggesting a transfer of energy between the two translatory directions of motion (Fig. 3).

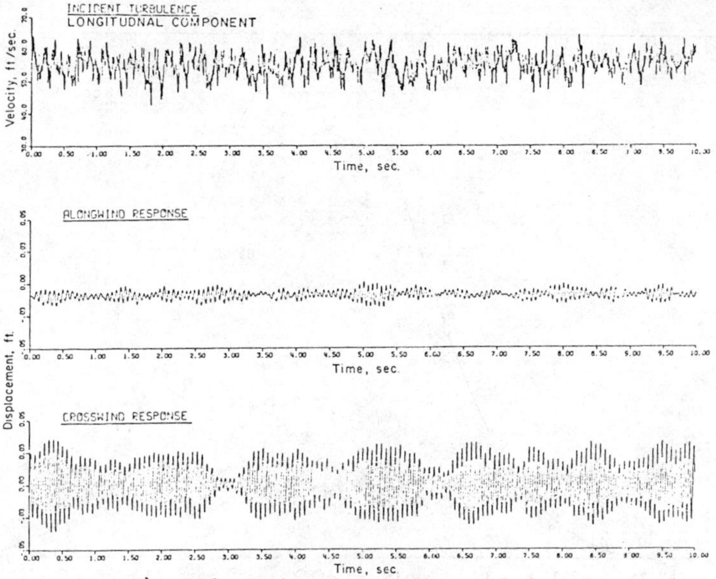

a) BL1, Reduced Velocity = 8.6

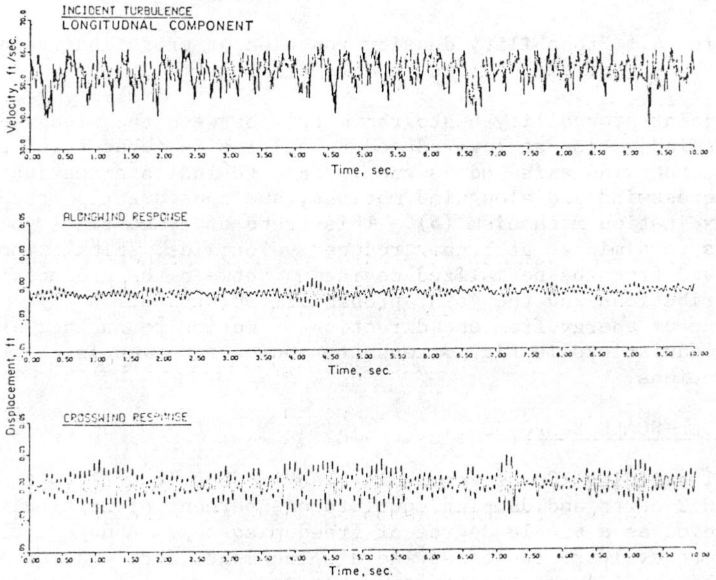

b) BL3, Reduced Velocity = 8.45

Fig. 3. Time histories of incident turbulence, alongwind and crosswind response at the top of a tall building model measured simultaneously.

3. At low reduced velocities the crosswind response follows the Gaussian distribution. However, as the reduced velocity increases beyond 6, the probability density function exhibits departure from the normal distribution and tends to follow the distribution of a sine wave plus a random noise (Fig. 4). This feature of the probability density function is more pronounced for boundary layer 1 than for boundary layer 3, which is also supported by the time history of the crosswind response in Fig. 3.

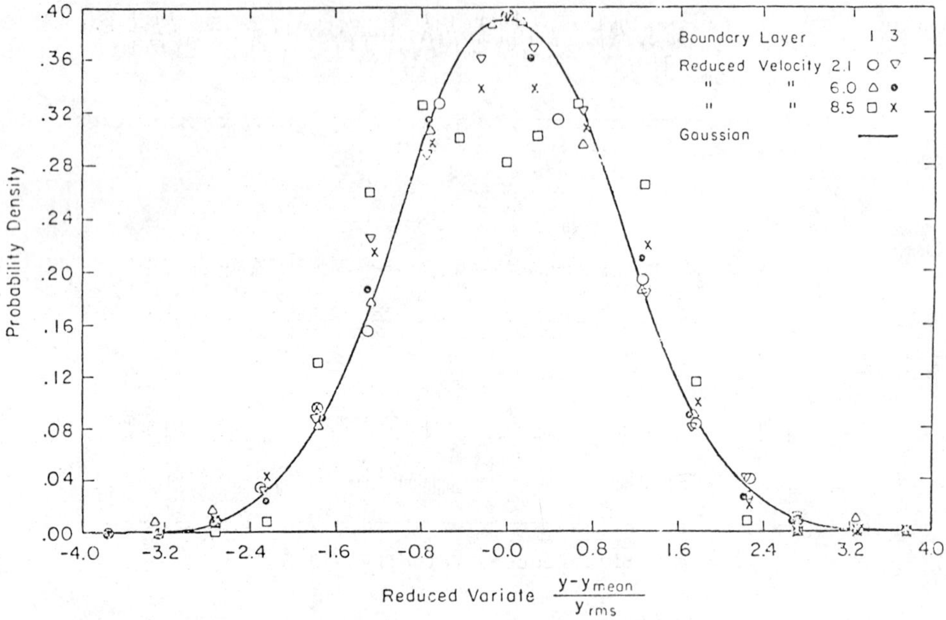

Fig. 4. Probability density function of crosswind response of the building model.

4. The joint probability histogram matrix between the alongwind and crosswind motion at low reduced velocities is found to be symmetrical about the wind axis and as such, tends to indicate the independence of the crosswind and alongwind motions, and consequently, the independence of excitation mechanism (5). This trend of symmetrical joint probability tends to diminish at higher reduced velocities. Similar conclusions were derived from the normalized deviation between the product of marginal distributions and the joint probability distribution (5). The interchange of energy from one direction of motion to another as evidenced by the modulations in Fig. 3 tends to increase motion in the respective directions.

FULL-SCALE RESPONSE

A hypothetical building 100 ft square in plan and 600 ft tall, with a natural frequency of 0.2 hertz and damping equal to one percent of critical was mathematically modeled as a single degree of freedom system. A numerical scheme (5) was developed to evaluate the crosswind displacement, acceleration and jerk response of this building subjected to urban and suburban flow conditions, based on the forcing function model presented in this paper. The results show good agreement with the predictions evaluated by using Vickery's procedure for the

Australian wind loading code (15) and National Building Code of Canada (16). The results are presented in Table 2.

Table 2. Response of 100-ft square section, 600-ft tall building*.

		Boundary Layer 1			RMS Acceleration from Building Codes		Boundary Layer 3		
Reduced Velocity	Velocity ft/sec	RMS Displacement Response ft	RMS Acceleration Response ft/sec^2	RMS Jerk Response ft/sec^3	Canadian Code ft/sec^2	Australian Code ft/sec^2	RMS Displacement Response ft	RMS Acceleration Response ft/sec^2	RMS Jerk Response ft/sec^3
2.75	55	.041	.049	.061	.049	.042	.034	.041	.051
3.00	60	.058	.078	.097	.066	.057	.049	.064	.081
3.25	65	.074	.093	.117	.091	.079	.063	.085	.107
3.50	70	.092	.130	.164	.109	.098	.075	.102	.129
3.75	75	.108	.154	.193	.143	.129	.092	.129	.162
4.00	80	.133	.171	.215	.169	.157	.109	.157	.196
4.25	85	.154	.228	.288	.199	.199	.144	.209	.263
4.50	90	.178	.229	.289	.251	.237	.153	.222	.281
4.75	95	.193	.250	.316	.297	.279	.169	.245	.309
5.00	100	.209	.289	.365	.355	.342	.194	.281	.355
5.25	105	.239	.325	.409	.395	.4	.223	.324	.409
5.50	110	.286	.377	.471	.485	.477	.262	.386	.488
5.75	115	.331	.412	.521	.546	.557	.265	.405	.492
6.00	120	.383	.493	.624	.647	.647	.399	.488	.62

*specific weight = 12 lb/ft^3, natural period = 5 sec and critical damping ratio = 1.0%.

CONTRIBUTION OF HIGHER MODES

Kareem (5) has illustrated from a dimensional argument that the crosswind response of a uniform high-rise building is dominated by the fundamental mode. A multi-mode dynamic response analysis presently in progress would enable to assess the precise contribution of higher modes of vibration to the overall dynamic response of a tall building. The results of this study will be reported in forthcoming reports.

CONCLUSIONS

The procedure presented in this paper permits the crosswind response of square cross-section tall buildings to be estimated at the preliminary design stage. These estimates allow early assessment of the structural requirements to restrict oscillatory response to ensure adequate occupancy comfort, and also to assess the need for wind-tunnel tests.

The reliability and likelihood of unserviceability or functional failure can be derived from the conditional probability of response integrated with the meteorological statistics of the building site to provide predictions of the response expected for certain levels of probability expressed in terms of return periods.

ACKNOWLEDGMENT

The research presented in this paper was sponsored by NSF Grant ENG-76-03135.

REFERENCES

1. Davenport, A.G., Gust loading factors, ASCE Journal of Structural Division, Vol. 93, No. ST3, pp. 11-34, June (1967).

2. Ellis, N., A new technique for evaluating the fluctuating lift and drag force distributions on building structures, Proceedings of the 4th International Conference on Wind Effects on Buildings and Structures, Cambridge University Press, pp. 526-530 (1975).

3. Saunders, J.W., Wind excitation of tall buildings, Ph.D. Thesis, Mechanical Engineering Department, Monash University, Australia (1974).

4. Whitbread, R.E., The measurement of non-steady wind forces on small scale building models, Proceedings of the 4th International Conference on Wind Effects on Buildings and Structures, Cambridge University Press, pp. 507-574 (1975).

5. Kareem, A., Wind excited motion of buildings, Ph.D. Dissertation, Fluid Mechanics and Wind Engineering Program, Department of Civil Engineering, Colorado State University, Fort Collins, Colorado (1978).

6. Kareem, A. and Cermak, J.E., Wind tunnel simulation of wind-structure interactions, Proceedings of 24th International Instrumentation Symposium, Albuquerque, New Mexico, pp. 343-362, May 1-5 (1978).

7. Scanlan, R.H., Theory of the wind analysis of long-span bridges based on data obtainable from section model tests, Proceedings of the 4th International Conference on Wind Effects on Buildings and Structures, Cambridge University Press, pp. 297-308 (1975).

8. Melbourne, W.H., Cross-wind response of structures to wind action, Proceedings of the 4th International Conference on Wind Effects on Buildings and Structures, Cambridge University Press, pp. 343-358 (1975).

9. Komatsu, S. and Kobayashi, H., Experimental identification of aerodynamic forces, ASCE Journal of Structural Division, Vol. 104, No. EM4, pp. 921-938, August (1978).

10. Clough, R.W. and Penzien, J., Dynamics of structures, McGraw-Hill, Inc., pp. 177-207 (1975).

11. Kareem, A., Cermak, J.E., and Peterka, J.A., Wind-induced response of high-rise buildings, Proceedings of the 3rd U.S. National Conference on Wind Engineering Research, Gainesville, Florida, Preprints (1978).

12. VanMarke, E., Properties of the spectral moments with applications to random vibrations, ASCE Journal of Engineering Mechanics Division, Vol. 98, No. EM2, pp. 425-446, April (1972).

13. Pilkey, W.D. and Cohen, R., System identification of vibrating structures, Proceedings ASME Winter Annual Meeting (1972).

14. Cherry, S. and Brady, A.G., Determination of structural dynamic properties of statistical analysis of random vibrations, Proceedings of the 3rd World Conference on Earthquake Engineering, New Zealand National Commission on Earthquake Engineering, Wellington, New Zealand, Vol. II, pp. 50-67 (1965).

15. Annex of the Australian Wind Loading Code, AS 1170-pt. II(1) (1973).

16. Davenport, A.G. and Dalgleish, W.A., Treatment of wind loads in the National Building Code of Canada, ASCE Annual Convention, Building Code Requirements for Wind Loading, Denver, Colorado, November (1975).

COMBINED EFFECTS OF GALLOPING INSTABILITY AND VORTEX RESONANCE

M. A. WAWZONEK and G. V. PARKINSON
The University of British Columbia, Vancouver, British Columbia, Canada

SUMMARY

Some towers and most tall buildings have cross-sectional shapes that are aerodynamically unstable to transverse disturbances in a wind. The oscillation resulting from this form of instability is called galloping, and for systems in which the critical wind speed for the initiation of galloping is considerably higher than the wind speed for wake vortex resonance with the elastic system, a quasi-steady theory can successfully predict galloping behaviour. However, the parameters for towers and tall buildings typically place the critical galloping wind speed close to, or even lower than, the vortex-resonance wind speed, thus introducing mutual effects of the two phenomena, and making the quasi-steady theory inapplicable. In the present experimental investigation, these mutual effects are studied in a wind tunnel for a two-dimensional prism of square section in smooth flow, as a first step in seeking understanding of the phenomena. Interesting results are obtained, and some trends, dependent on the relative values of the two critical wind speeds, are indicated.

GLOSSARY

h	section dimension	$S = \frac{f_v h}{V}$	Strouhal number
ℓ	model length	$n = \frac{\rho h^2 \ell}{2m}$	mass ratio
m	oscillating mass	C_{F_y}	galloping force coefficient
c	system damping constant	α	angle of attack
ρ	air density	$A = \left.\frac{dC_{F_y}}{d\alpha}\right\vert_{\alpha=0}$	galloping criterion
V	wind speed	$\beta = \frac{c}{2m\omega_n}$	fraction of critical damping
$\omega_n = 2\pi f_n$	natural frequency in rad/sec or hz	$U = \frac{V}{\omega_n h}$	dimensionless wind speed
f_v	wake vortex frequency	$U_o = \frac{2\beta}{nA}$	critical galloping speed

GLOSSARY (cont'd.)

$\bar{y}$	oscillation amplitude	$U_r = \frac{1}{2\pi S}$	vortex resonance speed
$\bar{Y} = \bar{y}/h$	dimensionless amplitude	$(\ )_\ell$	values at top of tower

INTRODUCTION

Phenomena of wind-induced oscillation of buildings and structures, and their component elements, have been studied in the authors' laboratory in a continuing program, and progress in understanding the phenomena has been reported at each of the 4 previous International Conferences(1,2,3,4). The program has concentrated on two organized forms of wind-induced oscillation, vortex-induced and galloping. The bodies studied have generally been long prisms of simple cross-section, mounted elastically to permit transverse oscillations in a plane normal to the incident wind.

In addition to sharing a transverse plane of oscillation, the two forms have in common that they are both nearly harmonic and occur at a frequency very close to a natural frequency of the elastic system, so that from an observation at a single wind speed it would be difficult to distinguish vortex-induced oscillation from galloping. However, their dependence on wind speed is quite different. Vortex-induced oscillation, a form of nonlinear resonance, is confined to a discrete range of wind speed containing that at which the vortex-shedding frequency for the stationary body coincides with the system natural frequency, while galloping, resulting from aerodynamic instability to transverse disturbances, begins at a critical wind speed proportional to the system damping, and the amplitude grows with wind speed, asymptotically approaching a linear increase.

If the critical wind speed for galloping is considerably higher than the vortex-resonance wind speed the two forms of oscillation occur independently, and under these conditions a quasi-steady theory can successfully predict the galloping behaviour. Figure 1, taken from Ref. 1, shows a comparison of the theoretical and experimental variations of dimensionless galloping amplitude $\bar{Y}$ with dimensionless wind speed U for a prism of square section in smooth two-dimensional flow. Both have been divided by U_o, the theoretical value of the dimensionless critical wind speed, and the theory then predicts that data points representing different values of system parameters will collapse onto a universal curve. The prediction is seen to be verified in Figure 1, where the branches of the solid curve represent stable limit cycles of oscillation, the dotted branch represents unstable limit cycles and the dashed line is the asymptote. For the 3 experimental runs shown in the Figure, the vortex-resonance wind speed U_r was only 1/3 or less of U_o.

Subsequent investigations showed that the quasi-steady theory of galloping continued to give good predictions for prisms of rectangular section in homogeneous turbulent flow, under mean two-dimensional conditions (Refs. 3 and 4), and for towers of square section in turbulent shear flow(5), again provided that U_r was considerably less than U_o. The wind-tunnel results of Ref. 5 were intended to simulate possible galloping behaviour of full-scale towers, or even tall buildings, in the natural wind, whose mean profile and turbulence properties were reproduced to model scale in the tests.

However, in actual full-scale towers and tall buildings of modern construction, U_o will not be much greater than U_r, and may even be lower. Thus, one of the model towers of Ref. 5 would correspond to a full-scale tower 400 m (1312 ft.) high, and Figure 2, adapted from Ref. 5, shows the variation of 3 critical values of V_ℓ, the wind speed at the top of the tower (approximately the gradient wind speed for the assumed exposure), with lateral dimension h for a family of 400 m towers of square

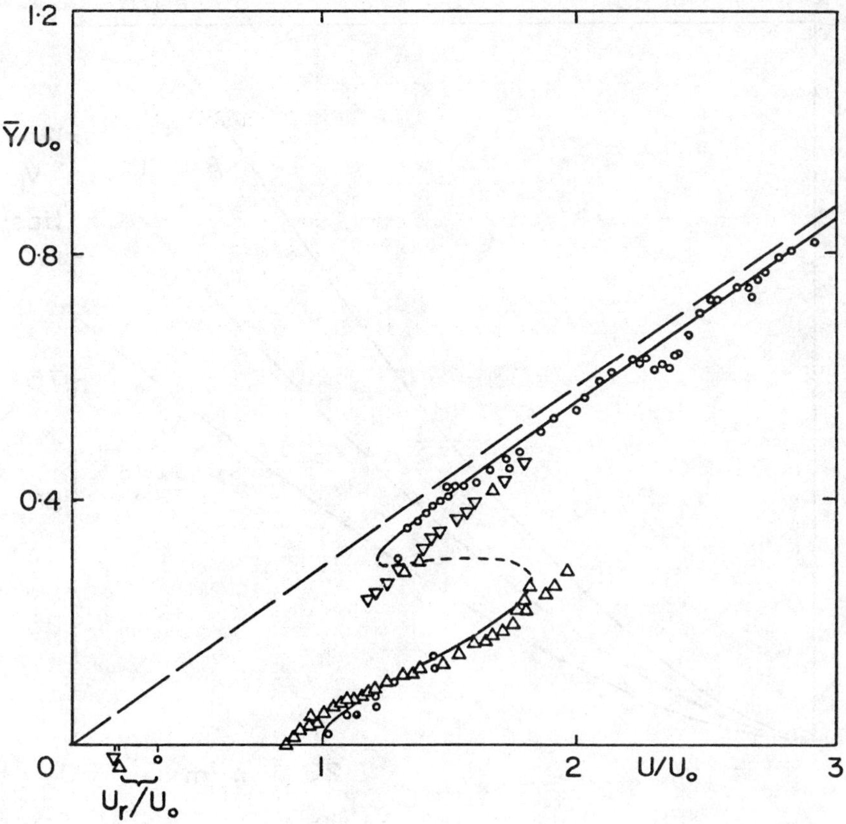

Fig. 1. Galloping behavior of square prism (from Ref. 1)

section. The 3 curves of V_ℓ are for V_{ℓ_r}, representing wake-vortex resonance with the fundamental tower natural frequency, and 2 curves of initial galloping speed $V_{\ell o}$, for 2 assumed values of system damping, β=0.005 and 0.01. The lower value is possible for modern welded-steel structures, and the higher value is typical. The curves imply that galloping is a real danger to towers of square section and, for the lower damping, even to tall buildings. However, the curves were based on the model test results, for which V_{ℓ_o} was always about twice V_{ℓ_r}, whereas the full-scale values are much closer, and V_{ℓ_o} is less than V_{ℓ_r} for the lower damping.

Under such conditions, the quasi-steady theory of galloping cannot be assumed applicable, and it is important to discover how the galloping mechanism is modified by the influence of the nearly resonant wake vortex system. The experimental investigation reported here follows procedures used successfully in the earlier studies, by dealing initially with two-dimensional bodies of square section in smooth flow, to minimize the number of parameters to be controlled. The elastic system is designed with U_o and U_r close, and capable of variation so that U_o can be greater or less than U_r, and the resulting dependence of flow-induced transverse oscillations on wind speed is studied.

EXPERIMENTS

The air film journal bearing model support system designed and used by J.D. Smith[6]

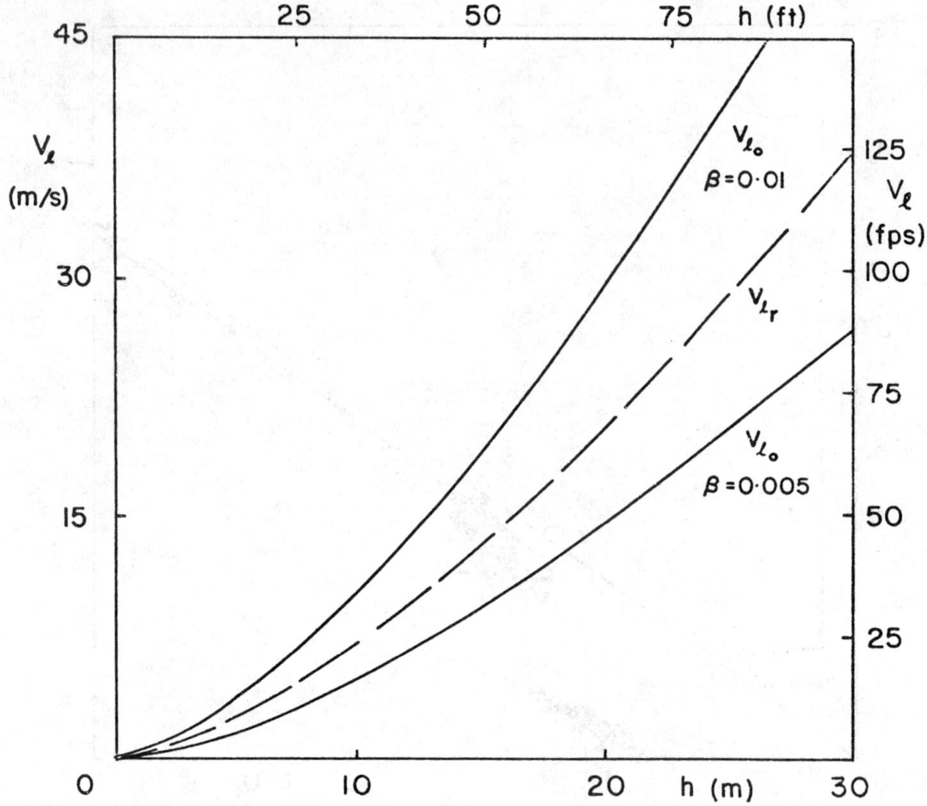

Fig. 2. Critical wind speeds for 400 m. (1312 ft.) towers of square section

was employed in this study, and is shown schematically in Figure 3. It allows simple oscillation in one degree of freedom with adjustable viscous damping.

The model is rigidly attached to two hollow aluminum shafts that are supported by the air bearings. Four linear tension springs provide restoring force. Both shafts pass through electromagnetic eddy current dampers. Since the only solid contact between the oscillating model and the support frame is at the spring attachments, the adjustable system damping coefficient c can go as low as 0.036 N·s/m (0.0025 lb·s/ft.). The value of c was determined from the logarithmic decrement of amplitude for free oscillation (wind off) of a streamlined bar in place of the model. One end of the top shaft oscillates within the annular gap of a coaxial air core transformer, forming a linear displacement transducer. The wind tunnel is a very low turbulence intensity, return flow type, with a test section 0.91 m (36 in.) wide by 0.69 m (27 in.) high.

The displacement was determined from the calibrated output of the displacement transducer as a voltage read from a digital RMS voltmeter, or as the actual voltage observed on a storage oscilloscope. All amplitudes are reported as mean peak values. Steady state displacement amplitudes and flow velocity were recorded for models of square section and 0.69 m (27 in.) length for different damping levels and oscillation frequencies. Two sizes of cross section were used, 0.051 m (2.0 in.) and 0.033 m (1.3 in.). In some cases it was desirable to investigate

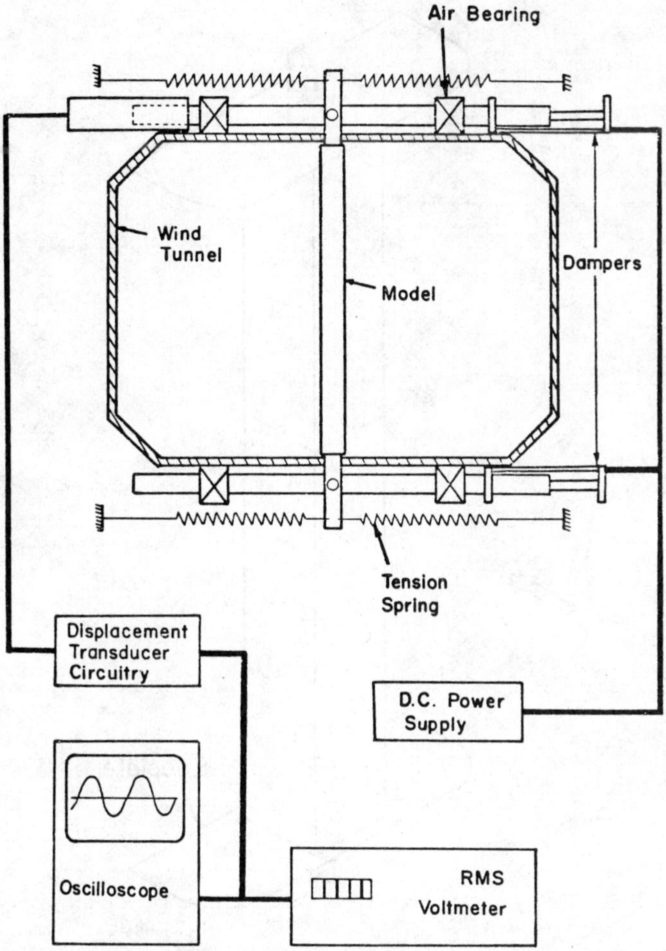

Fig. 3. Schematic diagram of experimental arrangement

whether the model would begin oscillation from rest at a certain flow velocity, so the oscillation was manually stopped, and the model was released from rest in the neutral position and observed. A sketch of the type of model used is shown in Figure 4.

RESULTS

One of the important conditions of the experiments was that the flow be uniform and two dimensional over the length of the model. This was found to be a problem for simple models without endplates, since the wind tunnel wall slots that were necessary to allow model oscillation were found to be admitting outside air into the low pressure region near the models and increasing the three-dimensional end effects. The flow disturbance was sufficient to prevent galloping from rest of the model of Figure 5, which oscillated only near U_r. Narrowing the wall slots helped alleviate the problem, but fixing endplates to the models seemed the better choice for the existing apparatus. The endplates reduce the tendency of outside

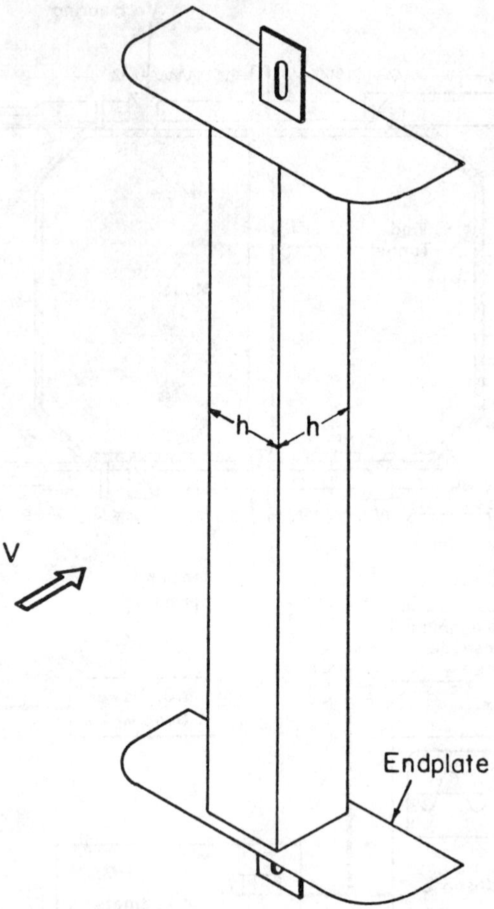

Fig. 4. Typical test model

air to be drawn in, and also deflect any inflow in the streamwise direction, reducing the effects on the flow field near the model. The change in behaviour with endplates is shown by Figure 8, for the same model and nearly the same U_o as in Figure 5. With endplates the model oscillates from rest at all velocities above U_r. All experiments mentioned in the remainder of this section are for models fitted with endplates.

Oscillation behaviour typical of all experiments where $U_o < U_r$ is shown in Figure 6, with oscillation from rest beginning near U_r rather than at U_o. This experiment represents the response for the lowest value of U_o tested. Amplitudes approach the theoretical galloping amplitudes only at higher flow velocities.

In some cases the oscillation amplitude varied in time at a constant flow velocity, and since this was not apparent from the RMS-derived amplitudes, the actual amplitude - time behaviour was determined for the model of Figure 7. The oscillation which began near U_r was shown to be modulated, with the beat frequency correlating well with the difference between the oscillation frequency and assumed Strouhal frequency. At higher velocities near $U = 1.5$, the oscillation was characterised by a periodic buildup - dropoff behaviour. At still higher velocities, the amplitude variation decreased and became more random. This figure shows a waviness in the experimental average amplitude - velocity profile with a periodicity near integer multiples of U_r. Similar waviness is sometimes observed for other models as well.

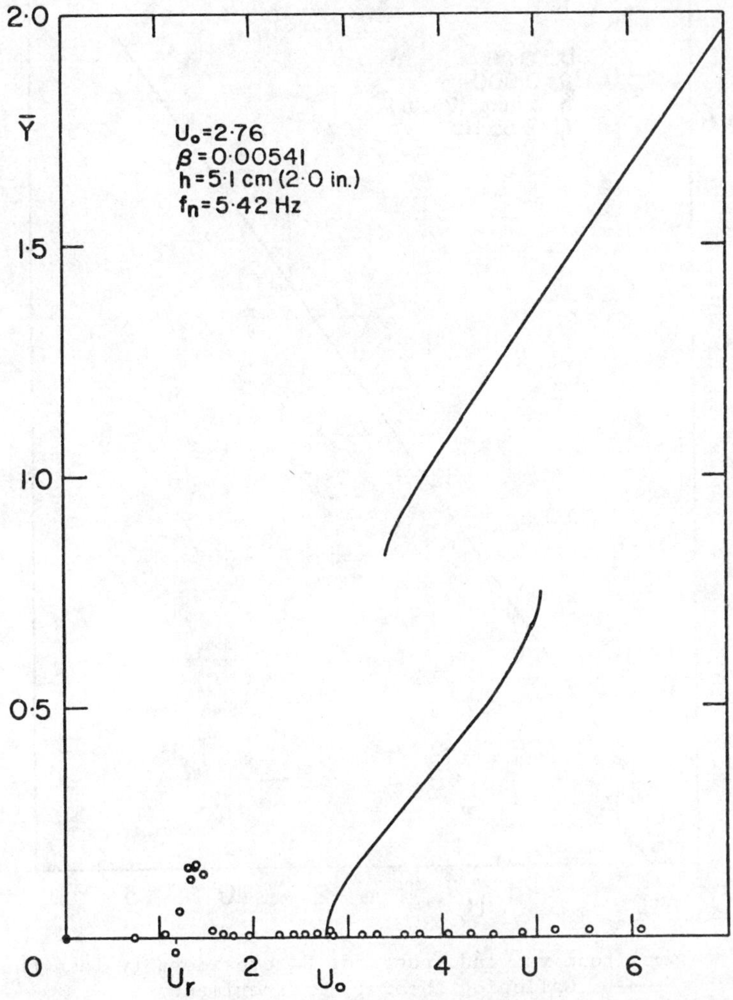

Fig. 5. Behavior of model without endplates.
——— Galloping theory; o Experiments

For values of $U_o > U_r$, oscillation again began near U_r, but the amplitude characteristics then depended on the separation between U_o and U_r. For $U_o > U_r$ but close, as in Figure 8, slowly increasing the velocity from below U_r would result in the model oscillation following the upper set of experimental points, approaching the upper theoretical amplitudes at high velocities. For the range of $1.4 \leq U \leq 1.9$, however, bringing the model to rest and then releasing it would result in the lower amplitude oscillation range shown by the lower set of experimental points. In this region, the experimentally determined unstable limit cycle amplitudes are also shown. These points were determined by observing the amplitudes for which slight perturbation would cause an amplitude change in the direction of perturbation, and eventual convergence on a higher or lower limit cycle.

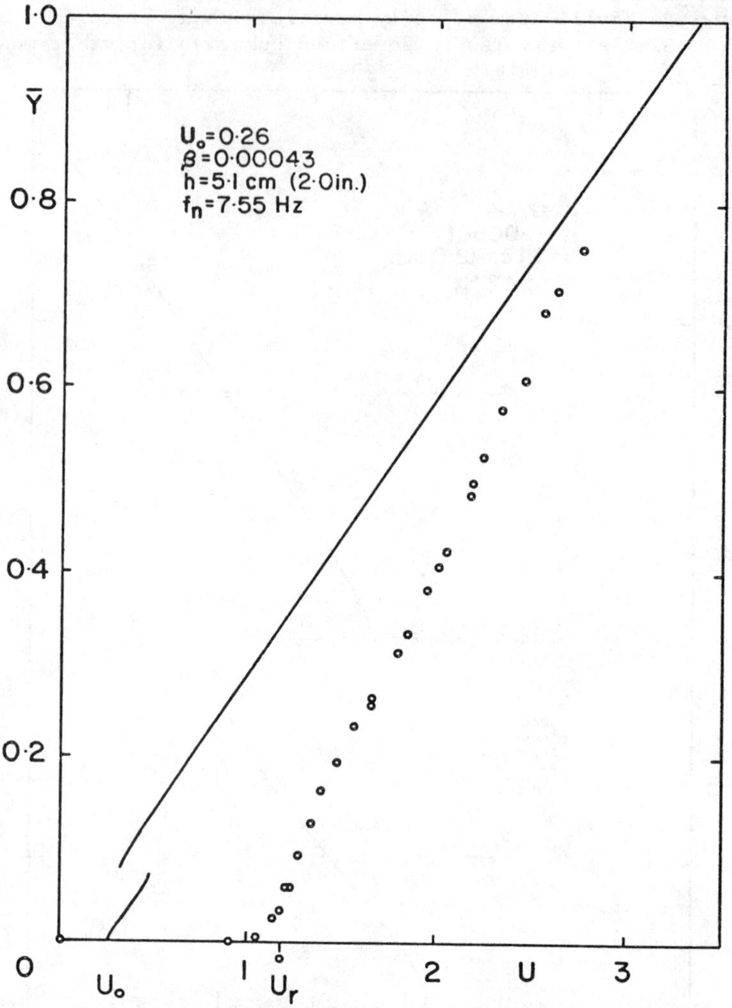

Fig. 6. Amplitude vs wind speed for U_o considerably less than U_r. ——— Galloping theory; o Experiments

Further increase in U_o relative to U_r resulted in separation of vortex resonance and galloping behaviour as shown in Figure 9. Here, the amplitude and velocity are normalised by U_o as in Figure 1, showing fairly good collapse of the data in the galloping regime on a single curve. However, this fitted curve indicates an apparent value of U_o that is 0.62 times the expected value, disagreeing with the data of Figure 1. Note also that the resonant peak near U_r for the 5.1 cm model is larger than the peak for the 3.3 cm model, even though the damping levels for the smaller model are lower.

DISCUSSION

Once satisfactorily two-dimensional flow conditions were obtained, by adding endplates to the models, a consistent progression of flow-induced oscillatory behaviour was observed as U_o was varied from well below to well above U_r, in Figures 6 to 9. In Figures 6 and 7, with $U_o < U_r$, oscillation begins at a wind speed slightly below U_r, and $\overline{Y}$ increases approximately linearly with U, approaching the theoretical galloping amplitudes at higher values. Thus, it appears that at the lower wind speeds the vortex mechanism completely suppresses the galloping mechanism for $U_o < U < U_r$, and partially suppresses it for $U > U_r$. However, in Figure 7, where U_o is closer to U_r, the galloping mechanism does seem to exert more influence than in Figure 6. Thus, the sawtooth amplitude modulation in one of the inset photographs seems likely to be related to the proximity to the knee of the theoretical galloping curve; i.e., the experimental amplitude periodically shifts from a lower value close to that of the theoretical knee to a higher value approaching the upper theoretical branch.

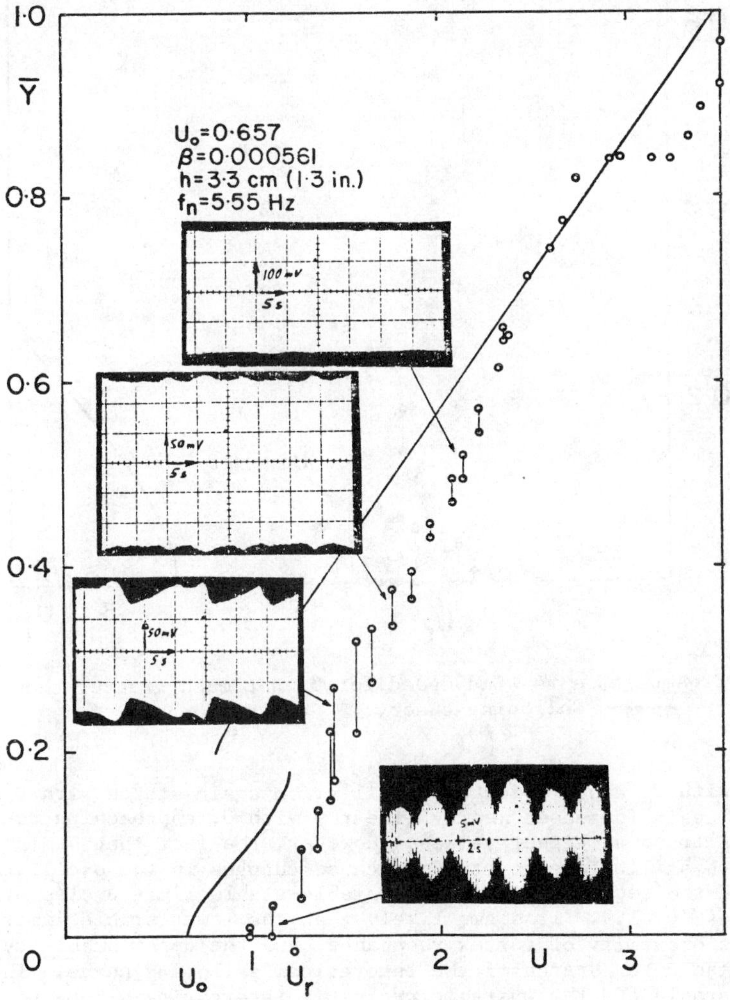

Fig. 7. Amplitude vs wind speed for U_o less than U_r, with inset photographs of amplitude-time records at constant wind speeds. ——— Galloping theory; o Experiments

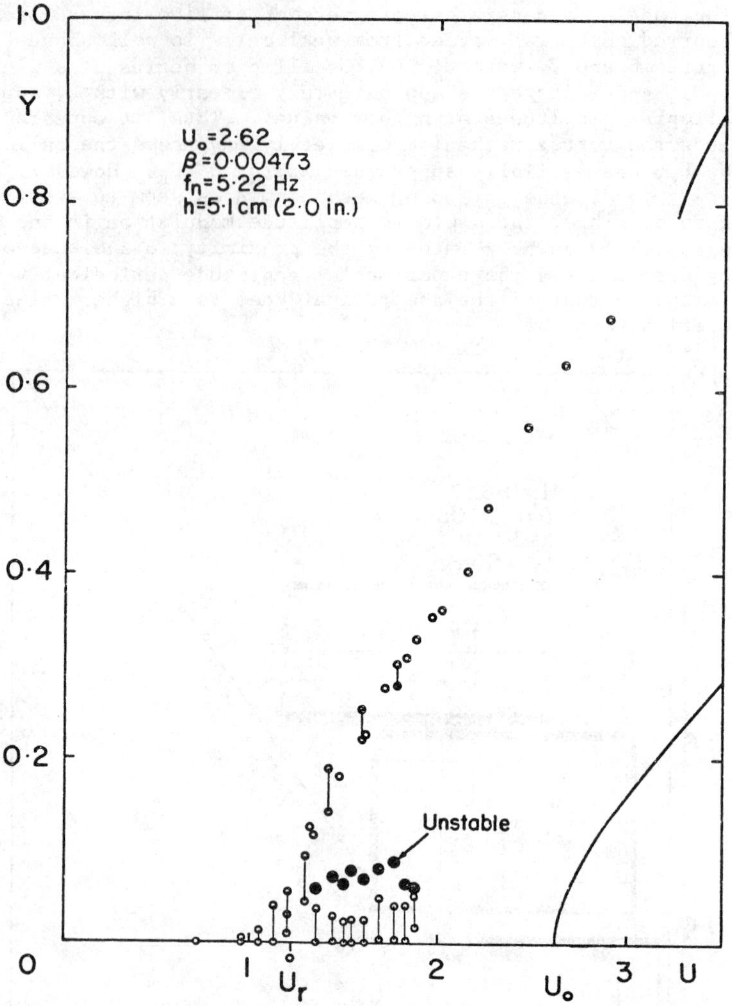

Fig. 8 Amplitude vs wind speed for U_o not much greater than U_r.
——— Galloping theory; o Experiments

In Figure 8, with U_o about twice U_r, oscillation again starts with U slightly below U_r, and $\bar{Y}$ again increases nearly linearly with U, approaching the theoretical galloping amplitudes at higher values. However, the fact that U_o is now greater than U_r, though still close to it, does cause changes in the oscillating behaviour in the form of the sequence of stable-unstable-stable limit cycles over the wind speed range $1.4 \lessapprox U \lessapprox 1.9$. It seems likely that the lower stable limit cycle is related to the proximity of vortex resonance, and the upper stable cycle to the proximity of the upper branch of the theoretical galloping curve. The closed loop of the lower stable and the unstable cycles is interesting. Some nonlinear oscillator models for vortex-induced oscillation predict closed loops of this general form[7].

Figure 9, for the highest values of U_o tested ($2.87 \leqslant \frac{U_o}{U_r} \leqslant 5.35$) shows separation of the effects of vortex resonance and galloping, and again provides confirmation of the quasi-steady galloping theory, in that the collapsed data have reasonably small scatter, particularly in the upper branch, that the two predicted branches of stable limit cycles appear, and that the upper branch has the predicted asymptotic trend at higher wind speeds. What is puzzling is that the data correlates with a theoretical curve for which U_o is only 0.62 of the expected value, based on the measured parameters. Since most of the physical constants that determine U_o are simple to measure (the expression can be put in the form $U_o = \frac{2c}{\rho\omega_n h^2 \ell A}$), the results of Figure 9 suggest that either the effective system damping constant c during galloping was lower than the measured value, or the effective aerodynamic coefficient A was higher than the assumed value (which was the value measured in the experiments of Ref. 1). Neither of these possibilities seems very likely.

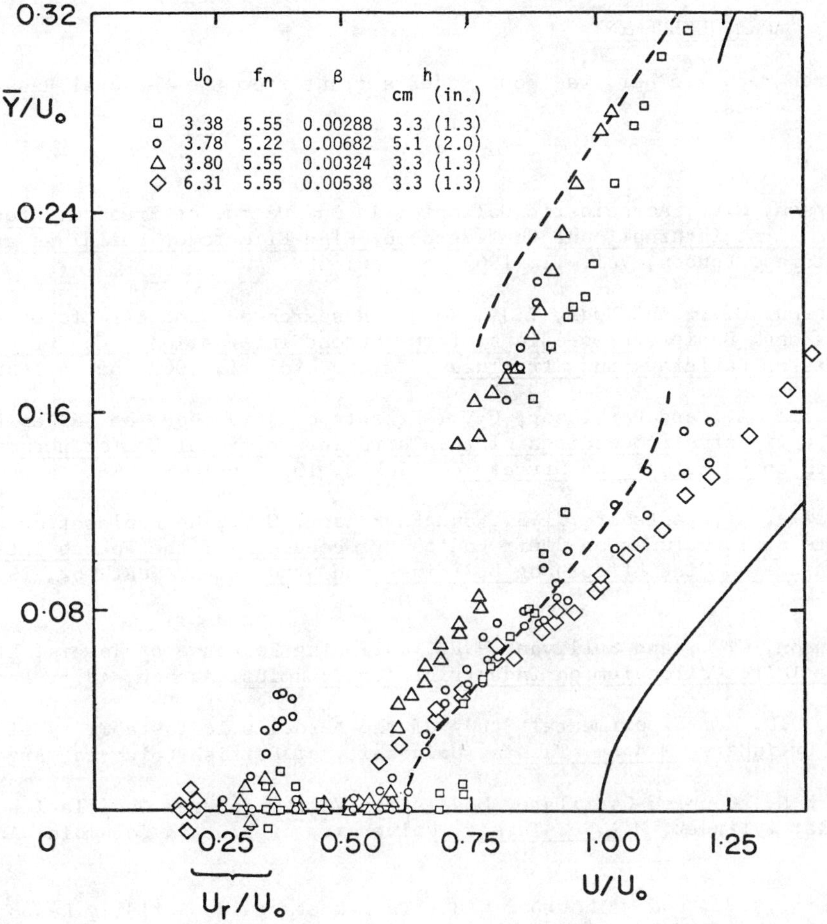

Fig. 9. Collapsed values of amplitude vs wind speed for U_o considerably greater than U_r. ——— Galloping theory; - - - - Adjusted curves; □, ○, △, ◇ Experiments

The principal motivation for the present study was an interest in determining the mutual effects of vortex resonance and galloping on the wind-induced oscillations of towers and tall buildings. Others have considered this problem, and Saunders and Melbourne[8] concluded that for tall buildings with ℓ/h up to 9, and in winds with U_ℓ up to 1.6, the vortex resonance mechanism dominated the aeroelastic behaviour, and the galloping mechanism was not significant.

The present results suggest that it would be dangerous to neglect the possibility of the galloping mechanism combining with the vortex resonance mechanism to produce oscillation amplitudes considerably greater than those from the vortex mechanism alone, even for U_ℓ values as low as 1.6 or less. Of course, one cannot directly translate results obtained in two-dimensional smooth flow into predictions for towers or tall buildings in the natural wind, but the results of Ref. 5, obtained for tower models in simulated natural winds, lend support to the present results as a preliminary basis for prediction.

ACKNOWLEDGEMENT

The research reported here was done under a grant from the National Research Council of Canada.

REFERENCES:

1. Parkinson, G.V., Aeroelastic Galloping in One Degree of Freedom, Proceedings of the First International Conference on Wind Effects on Buildings and Structures, London, Vol. II, 1965, pp. 581-609.

2. Parkinson, G.V., and Modi, V.J., Recent Research on Wind Effects on Bluff Two-Dimensional Bodies, Proceedings of the Second International Conference on Wind Effects on Buildings and Structures, Ottawa, Vol. 1, 1967, pp. 485-513.

3. Laneville, A., and Parkinson, G.V., Effects of Turbulence on Galloping of Bluff Cylinders, Proceedings of the Third International Conference on Wind Effects on Buildings and Structures, Tokyo, 1971, pp. 787-798.

4. Laneville, A., Gartshore, I.S., and Parkinson, G.V., An Explanation of Some Effects of Turbulence on Bluff Bodies, Proceedings of the Fourth International Conference on Wind Effects on Buildings and Structures, Heathrow, 1975, pp. 333-342.

5. Parkinson, G.V., and Sullivan, P.P., Galloping Response of Towers, Proceedings of the Third Colloquium on Industrial Aerodynamics, Aachen, 1978, Paper #24.

6. Smith, J.D., An Experimental Study of the Aeroelastic Instability of Rectangular Cylinders, M.A.Sc. Thesis, University of British Columbia, August, 1962.

7. Wood, K.N., Coupled-Oscillator Models for Vortex-Induced Oscillation of a Circular Cylinder, M.A.Sc. Thesis, University of British Columbia, August, 1976.

8. Saunders, J.W., and Melbourne, W.H., Tall Rectangular Building Response to Cross-Wind Excitation, Proceedings of the Fourth International Conference on Wind Effects on Buildings and Structures, Heathrow, 1975, pp. 369-379.

THE INFLUENCE OF WIND DIRECTION ON THE RESPONSE OF A SQUARE-SECTION TALL BUILDING

T. A. REINHOLD* and P. R. SPARKS**

*National Bureau of Standards, Washington, D.C., USA
**Virginia Polytechnic Institute and State University, Blacksburg, Virginia, USA

SUMMARY

This paper describes a wind-tunnel study to determine the effect of wind-direction on the response of a square-section tall building with an 8.33 to 1 aspect ratio. Rigid models instrumented with pressure transducers were used to determine the modal forces for the fundamental translational and torsional modes of the building. Using these forces and typical stiffness and damping values, the influence of wind-direction on the response of a 500:1 scale prototype building was estimated. The maximum response was generally found to occur when the wind was blowing onto a face of the building. Comparisons are made between predicted response levels of the prototype building, based on experimental results on the one hand, and on current design procedures on the other hand.

NOTATION

$a_{x,y}$ - peak acceleration at top of building in x or y direction.
$C_{xi,yi}$ - force coefficient in x or y direction at ith level.
C_{mi} - twisting moment coefficient in ith level.
$C_{x,y}$ - mean force coefficient in x or y direction.
$C^*_{x,y}$ - RMS modal force coefficient in x or y direction.
C^*_m - RMS modal twisting moment coefficient.
$F_{x,y}$ - mean modal force in x or y directions.
$F_{xi,yi}$ - mean force at ith level in x or y direction.
$F^*_{x,y}$ - RMS modal force in x or y direction.
$G_{x,y,\theta}(n)$ - single sided power spectral density (p.s.d.) response spectrum, x and y translation or θ torsional.
$G_{Fx,Fy,M}(n)$ - single sided p.s.d. force or moment spectrum.
H - height of model or building.
H(n) - complex frequency response function.
K - modal stiffness.
M - mean modal twisting moment.
M_i - mean twisting moment at level i.
M* - RMS modal twisting moment.
m - mass per unit height of building.
n - frequency.
n_o - fundamental natural frequency of building.
U_H - mean wind velocity at top of model or building.
U_g - mean gradient wind velocity.
X - mean response in x direction.
X^P - peak response x direction.
X^P_θ - peak response in x direction due to rotation.
Y^P - peak response in y direction.
z_i - height of ith level.
α - angle of wind approach.
γ - critical damping ratio.
$\sigma_{x,y,\theta}$ - RMS response in x, y or θ directions.
σ_B - RMS background response.
σ_R - RMS resonant response.
$\sigma_{\theta x}$ - RMS response in x direction due to rotation.
ρ - air density.
ρ_b - average building density.
μ_1 - fundamental mode shape of building.

INTRODUCTION

The tendency towards the construction of tall, slender, and flexible buildings has forced designers to recognize the unsteady nature of wind-loading and to modify their designs in an effort to ensure satisfactory performance of their buildings under these fluctuating loads.

Unfortunately, analytical techniques for the prediction of these wind loads and the response of a building to these loads are not well developed. In many cases a wind study has to be performed using an aeroelastic model. Such models are expensive and the stiffness characteristics of the building have to be known before the test can be performed. Yet observations of actual stiffnesses of buildings have shown them to have considerably different characteristics from those predicted by the designer.(1) Measurements made on a model with one particular set of stiffness characteristics cannot readily be extrapolated to other likely stiffnesses and damping values. Consequently, a series of tests must be performed to assess the effects of changes in stiffness and damping. The inclusion of the correct torsional behavior in a model of this type is virtually impossible. Aeroelastic model tests are required only if the movement of the building is likely to influence the flow around it. Yet in relation to their height, movements in buildings are extremely small and it will be shown later that within the range of expected wind speeds, the movement of the building is unlikely to affect the aerodynamic forces on the structure.

The response of a tall building to wind loading consists essentially of a mean response plus resonant responses in each of its fundamental modes. Computation of a mean response is fairly straightforward but computation of resonant responses has been hampered by the lack of suitable data concerning modal forces. This paper describes a wind-tunnel investigation into the effect of wind-direction on the response of a square-section tall building utilizing a rigid model. Pressure measurements were made in such a way that a modal force spectrum could be produced for each of the fundamental modes of the building. Using this technique, only the geometrical shape of the prototype building need be known at the time of the wind tunnel investigation and the influence of changes in stiffness, mass and damping on the expected response of the prototype can then be easily investigated by analytical means.

MODELS AND TEST PROGRAM

A series of tests employing square prisms with sharp edges as models of buildings were carried out at Virginia Polytechnic Institute and State University. These tests were conducted by placing models in a turbulent shear layer which was generated in a 1.8 m square by 8.0 m long test section of a low-speed wind tunnel. Fluctuating pressures on the model's surfaces were measured using numerous transducers. Resultant fluctuating loads on the building models were obtained by summation of the local instantaneous pressures, weighted by appropriate area and moment arm factors.

Flow Environment

The turbulent shear layer was developed over the test section floor using a combination of spires at the test section entrance and a 6.1 m fetch of 102 mm cubic roughness elements placed at 305 mm centers. A detailed study of the shear layer mean flow and three turbulence components was conducted at the model location, 7.5 m downstream from the spires. The results of the study and a comparison of the flow properties with those of the atmospheric boundary layer over urban terrain are contained in references (2) and (3). It is sufficient for purposes of this paper to summarize only the major properties of the flow. The mean velocity profile followed the power law with a power index of 0.37 (corresponding to flow over urban terrain). The turbulence properties, including variances, covariances, power spectra and length scales, were found to compare reasonably well with full scale estimates at a geometric scale ratio ranging from 1/400 to 1/600. The measured spectra of the three turbulence components were in excellent agreement with the empirical von Karman spectrum functions. The longitudinal length scale of the turbulence was fairly constant along the upper half of the model and approximately equal to three times the model width. The lateral and vertical length scales were reasonably constant throughout the model height and about equal to the model width.

Models and Location of Pressure Taps

Two models of identical geometry (76.2 mm x 76.2 mm x 635.0 mm) were built for the test program. One model was fitted with 96 pressure taps, connected by flexible tubing to two scanivalve systems and was used for the mean pressure measurements. The second model was fitted with the flush mounting pressure transducers (exposed diaphram of 2.16 mm diameter) and was used to measure fluctuating pressures at the same locations as the pressure taps. This arrangement was used because the flush mounting pressure gauges originally intended to measure both mean and fluctuating pressures, were found to exhibit a slow voltage drift. The pressure taps and transducers were located in groups of 16 about the model perimeter, with 4 transducers or taps on a side, at 6 elevations throughout the model height. Fifty three transducers were available so it was possible, at most, to instrument three levels simultaneously.

Selection of the locations for pressure taps and transducers was based on results of experiments conducted using a section-model of square cross-section (101.6 mm x 101.6 mm) placed in a grid generated two-dimensional uniform turbulent flow. The flow had a turbulence intensity of 12 percent and the integral length scale in the streamwise direction was 160 mm. Tests were conducted for four different orientations of transducers on a single model face, ranging from 7 to 3 transducers. The signals from the transducers were summed to obtain the resultant normal force on the model face and the resultant twisting moment about the center of the model face. The objective was to determine the positioning across the face of no more than 4 transducers on a side which would closely reproduce the loads measured using 7 transducers on a side. A comparison was made of both RMS values and autocorrelation functions for the normal forces and twisting moments measured with the four configurations of transducers. Measurements were made with the model face rotated to a number of angles with respect to the mean flow direction.

The positions of the transducers and taps adopted for the boundary layer model tests are shown in Figure 1. This configuration reflects the observation that the outer transducers had to be as close to the edge as possible (it was necessary to offset the outer transducers to accomplish this). The maximum estimated error obtained by using the 4 transducers on a side as opposed to 7 is less than 5% for the normal force and less than 15% for the twisting moment (for any of the wind directions studied). More detailed information concerning the placement of transducers can be found in reference (2).

DETERMINATION OF RESULTANT LOADS

Forces and Moments at Particular Elevations

Resultant normal forces and twisting moments were determined at each instrumented level (0.16, 0.32, 0.48, 0.64, 0.80 and 0.96 of the model height). To do this each pressure was assumed to act uniformly over its assigned area, see the bottom face in Figure 1. The mean normal forces and twisting moments were obtained by summation of the individual pressure measurements.

Accumulation of a large mass of data representing simultaneous instantaneous pressures on all the transducers was avoided by performing on analog summation of the transducer output signals to obtain single signals representing the resultant fluctuating loads at a particular level. The electronics necessary to perform this task consisted of two units. In the first the transducer output signals were amplified to obtain uniform pressure sensitivity weighted by the area factor associated with the transducer location. In the second these signals were added and subtracted in proper combinations to obtain signals representing the resultant normal forces and were added and subtracted with weighting for moment arm factors to obtain a signal representing the resultant fluctuating moment.

The resultant mean and fluctuating normal forces and twisting moments determined at each instrumented level can be expressed as a force or moment coefficient per unit height. These coefficients are based on the dynamic pressure of the undisturbed mean velocity at an elevation corresponding to the top of the model $\overline{U}_H$. The coefficients for the various instrumented levels can be found in reference (2).

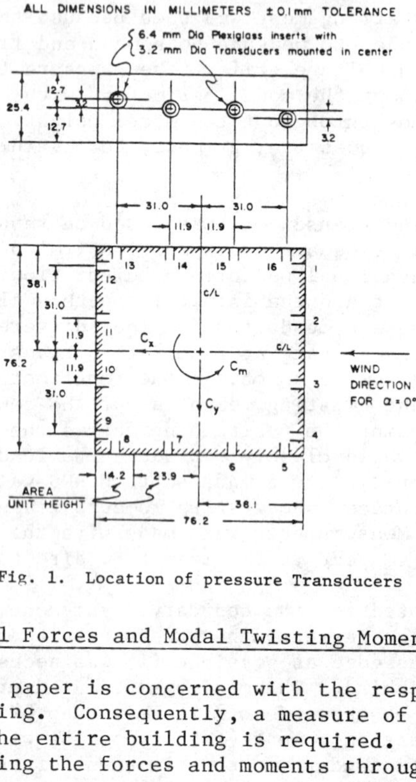

Fig. 1. Location of pressure Transducers

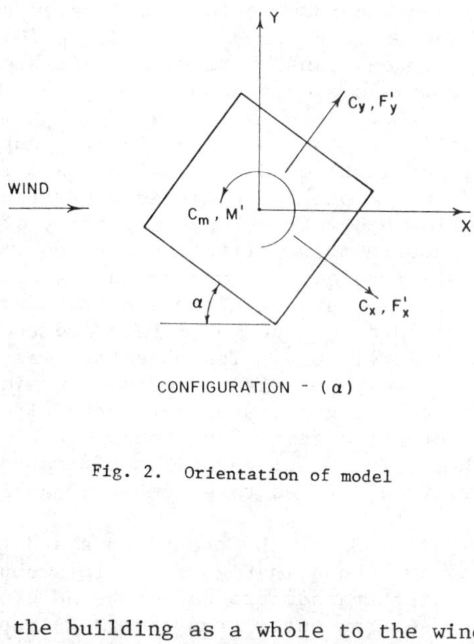

Fig. 2. Orientation of model

Modal Forces and Modal Twisting Moments

This paper is concerned with the response of the building as a whole to the wind loading. Consequently, a measure of resultant wind forces and twisting moments on the entire building is required. These forces and moments are determined by summing the forces and moments throughout the height of the model.

Assuming a linear mode shape, the mean modal forces are given by

$$\overline{F}_x = \sum_{i=1}^{6} F_{xi} \frac{z_i}{H} \Delta z_i = 1/2\ \rho\ \overline{U}_H^2 \left(\frac{B}{H}\right) \sum_{i=1}^{6} C_{xi}\ z_i\ \Delta z_i \tag{1}$$

$$\overline{F}_y = \sum_{i=1}^{6} F_{yi} \frac{z_i}{H} \Delta z_i = 1/2\ \rho\ \overline{U}_H^2 \left(\frac{B}{H}\right) \sum_{i=1}^{6} C_{yi}\ z_i\ \Delta z_i \tag{2}$$

Similarly the mean modal twisting moment is given by,

$$\overline{M} = \sum_{i=1}^{6} M_i \frac{z_i}{H} \Delta z_i = 1/2\ \rho\ \overline{U}_H^2 \left(\frac{B^2}{H}\right) \sum_{i=1}^{6} C_{mi}\ z_i\ \Delta z_i \qquad (3)$$

These mean modal forces and modal twisting moments are presented in Table 1 as coefficients based on the dynamic pressure corresponding to $\overline{U}_H$. The coefficients are given for a number of orientations of the model to the mean wind direction. The coordinate system adopted for expressing the coefficients and for describing the orientation of model is illustrated in Figure 2.

The fluctuating modal forces and twisting moments can be obtained by summing the instantaneous resultant forces at the various levels, weighted by the mode shape. As noted previously, because of the limited number of transducers available, it was only possible to instrument three levels simultaneously. Since forces applied near the top of the building have a greater effect on the modal forces than those applied near the bottom, results from the test carried out with transducers at 0.64, 0.80 and 0.96 of the height i.e. levels 4, 5, 6, were used to compute the fluctuating modal forces and moments. The signals representing forces at these levels were scaled to represent their relative magnitudes, the areas over which they were assumed to act and a linear mode shape factor to represent the first mode of vibration of a tall building. The signals were summed by passing them through analog computer circuits. The resultant signals, representing the modal forces and modal twisting moments, were analyzed to obtain spectra and RMS values. Since these modal force estimates only incorporated force contributions which acted over roughly the top half of the model, it is possible that they could underestimate the actual values by as much as 30%. It will however be shown later, by comparison with results from aeroelastic model tests, that the error can in fact be considerably less. The RMS modal forces and twisting moments are presented in Table 1, as coefficients based on the dynamic pressure corresponding to $\overline{U}_H$, for the model configurations tested. The nondimensional power spectra for the modal forces are shown in Figures 3 and 4 and for the modal twisting moment in Figure 5.

Table 1. Modal force coefficients

Angle α	Mean Coefficients			RMS Coefficients		
	C_x	C_y	C_m	C_x^*	C_y^*	C_m^*
0	.55	.00	.001	.065	.118	.0121
5	.54	-.06	.018	.063	.112	.0126
10	.55	-.09	.038	.059	.085	.0115
15	.55	-.03	.049	.053	.066	.0088
20	.54	.07	.047	.051	.057	.0079
30	.51	.26	.022	.047	.045	.0068
45	.43	.43	.001	.042	.042	.0038

$$C_x = \overline{F}_x / \tfrac{1}{2}\ \rho\ \overline{U}_H^2\ BH$$

$$C_y = \overline{F}_y / \tfrac{1}{2}\ \rho\ \overline{U}_H^2\ BH$$

$$C_m = \overline{M} / \tfrac{1}{2}\ \rho\ \overline{U}_H^2\ B^2H$$

$$C_x^* = F_x^* / \tfrac{1}{2}\ \rho\ \overline{U}_H^2\ BH$$

$$C_y^* = F_y^* / \tfrac{1}{2}\ \rho\ \overline{U}_H^2\ BH$$

$$C_m^* = M^* / \tfrac{1}{2}\ \rho\ \overline{U}_H^2\ B^2H$$

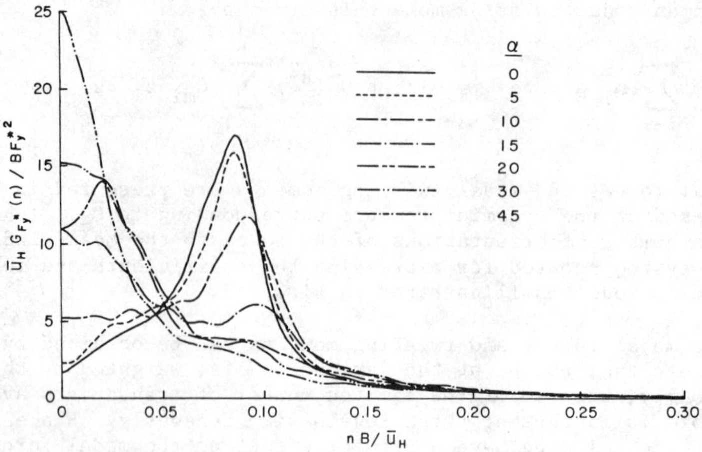

Fig. 3. Modal force spectra for F_y force component

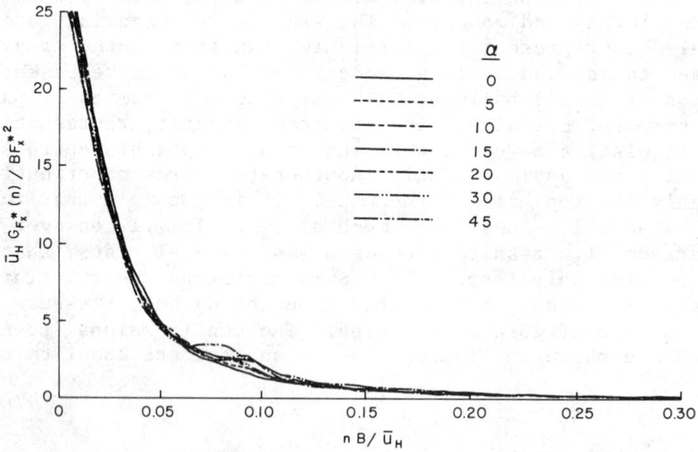

Fig. 4. Modal force spectra for F_x force component

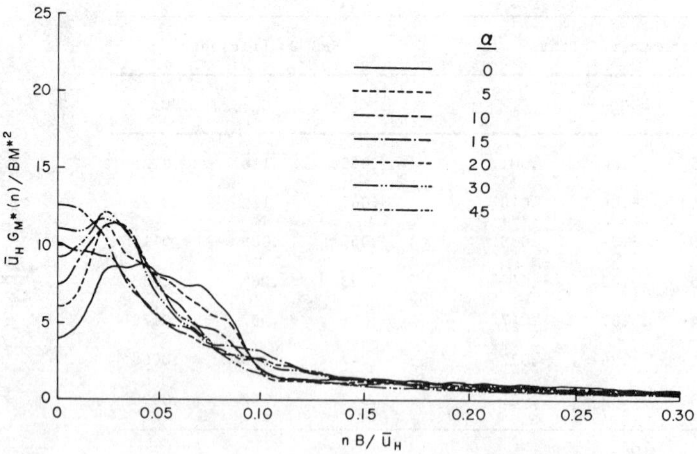

Fig. 5. Modal force spectra for twisting moments

COMPARISON WITH AEROELASTIC TEST RESULTS

Tests on an aeroelastic model with reasonably similar geometry were conducted by Rosati at the University of Western Ontario (4). This model also had a square-cross-section but had an aspect ratio of 7 instead of 8.33. Measurements of the response of the aeroelastic model were made in a flow with similar properties to the one used in this study, for a wide range of velocities.

Using the values of natural frequencies, damping and stiffness provided by Rosati and the modal force spectra given in Figures 3 and 4, the in-wind and cross-wind response of Rosati's model were predicted for the wind blowing onto a face of the model.

The modal response spectrum for each mode is given by

$$G_{x,y,\theta}(n) = \frac{H^2(n)}{K^2}\, G_{F_x, F_y, M}(n) \tag{4}$$

where $$H(n)^2 = \frac{1}{\left[1-\left(\frac{n}{no}\right)^2\right]^2 + \left(2\gamma\frac{n}{n_o}\right)^2},$$

$$K = (2\pi n_o)^2 \int_o^H m\mu_1^2\, dz \qquad \text{for translational response}$$

and $$K = (2\pi n_o)^2 \int_o^H \frac{mB^2}{6}\mu_1^2\, dz \qquad \text{for torsional response}$$

The R.M.S. response $\sigma_{x,y,\theta}$ is given by

$$\sigma^2_{x,y,\theta} = \int_o^\infty G_{x,y,\theta}(n)dn \tag{5}$$

and is numerically equal to the response of the building at the roof level. The computed translational responses are compared with Rosati's measured responses in Figures 6 and 7. The relationship between $\bar{U}_g$ used by Rosati and $\bar{U}_H$ is given by $\bar{U}_g/\bar{U}_H = 1.27$.

Although Rosati tested his models over a wide range of wind velocities and values of damping, the range of practical significance corresponds to values of $\bar{U}_g/n_oB$ less than about 12 and values of damping between 0.5% and 2%. It can be seen that the prediction of the cross-wind response is extremely good, Figure 6, but the rigid model predictions are consistently low for the in-wind response, Figure 7. Differences in test conditions and the omission of the forces at the lower levels may both contribute to the fact that a lower in-wind response was predicted. The differences between the in-wind response levels obtained from the two experimental tests range between 20 and 30 percent. Use of Simiu's analytical technique (5) indicates that neglecting forces at lower levels results in an underestimation of RMS displacements of approximately 15% and an underestimation of RMS accelerations of approximately 12%.

The cross-wind response predictions based on the rigid model tests became considerably lower than the measured response of the aeroelastic model at values of the reduced velocity U_g/n_oB greater than 15 ($\bar{U}_H/n_oB$ greater than 11.8). It is most likely that the differences in response may be attributed to aeroelastic effects. This is consistent with work reported by Saunders and Melbourne (6) where it was noted that, for the prismatic building models they tested, the nondimensional cross-wind force spectra were insensitive to the level of motion up to reduced velocities, $\bar{U}_H/n_oB$, of at least 10.

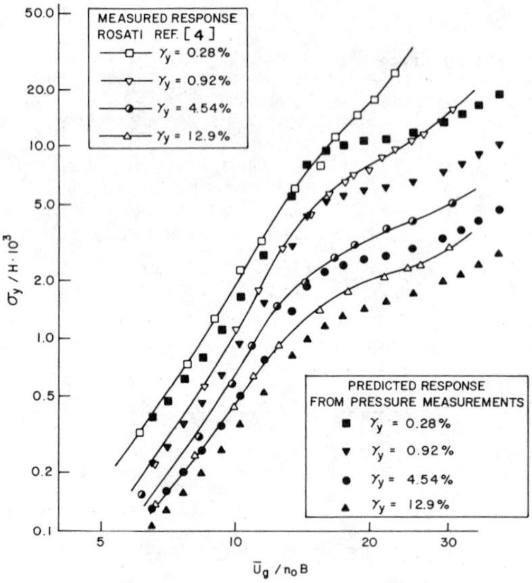

Fig. 6. Comparison of predicted and measured cross-wind response of an aeroelastic model

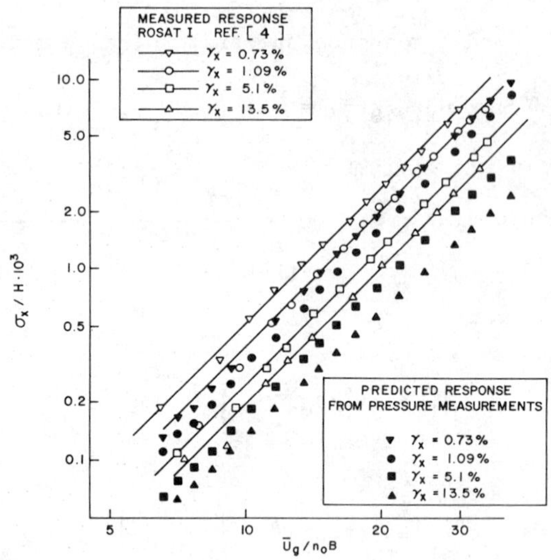

Fig. 7. Comparison of predicted and measured in-wind response of an aeroelastic model

APPLICATION OF THE RESULTS TO THE PREDICTION OF THE RESPONSE OF A PROTOTYPE BUILDING

The modeling technique was used to study the influence of wind direction on the response of a prototype building. The building is considered to respond independently in each of its fundamental modes.

A prototype building 500 times larger than the model was chosen. It was given typical values of natural frequency (0.1 Hz for all fundamental modes), damping (1% of critical for all modes) and average building density (192 kg/m^3). Identical natural frequencies and damping were chosen to facilitate comparisons of the relative effects of the forces.

For the translational and torsional modes RMS responses were computed as described earlier. Mean displacements were found by setting n=0 and replacing $G_{Fx,Fy}$ by $F_{x,y}$ and G_M by M.

RMS translational responses are compared with the RMS in-wind response at 0° in Figures 8 and 9 and the RMS torsional responses are compared with the RMS torsional response at 0° in Figure 10. Peak responses including the mean

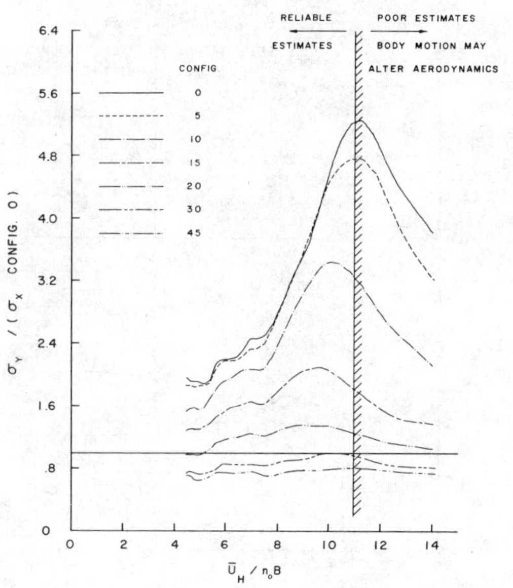

Fig. 8. Ratio of RMS response in y direction to RMS response in x direction at configuration 0°

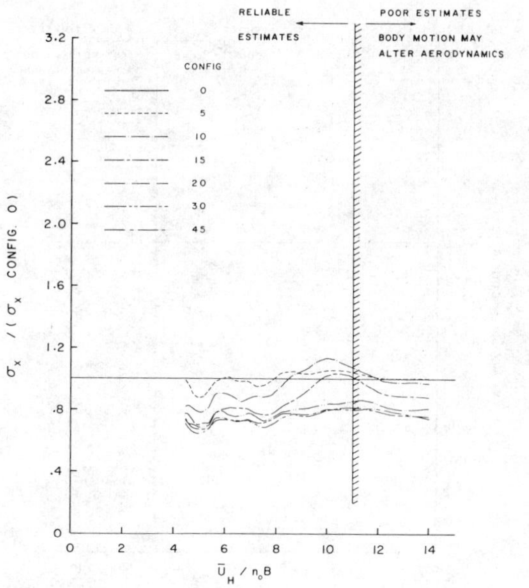

Fig. 9. Ratio of RMS response in x direction to RMS response in x direction at configuration 0°

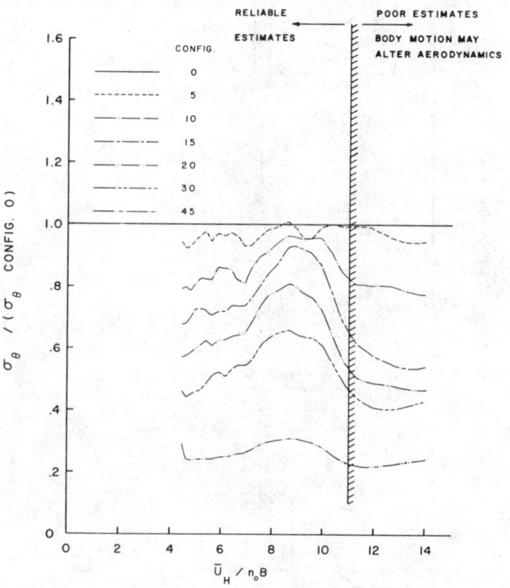

Fig. 10. Ratio of RMS torsional response to RMS torsional response at configuration 0°

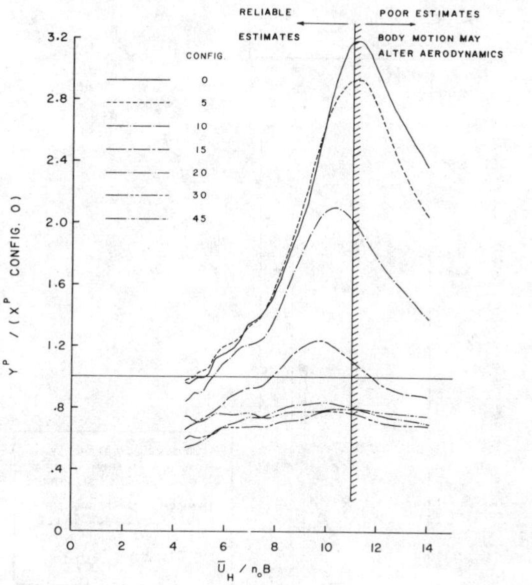

Fig. 11. Ratio of peak response in y direction to peak response in x direction at configuration 0°

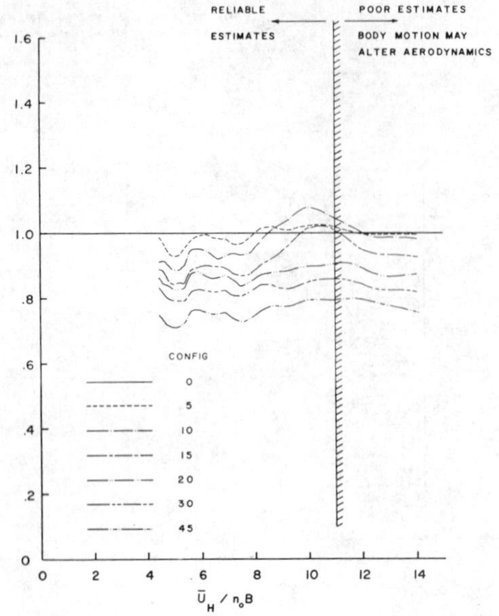

Fig. 12. Ratio of peak response in x direction to peak response in x direction at configuration 0°

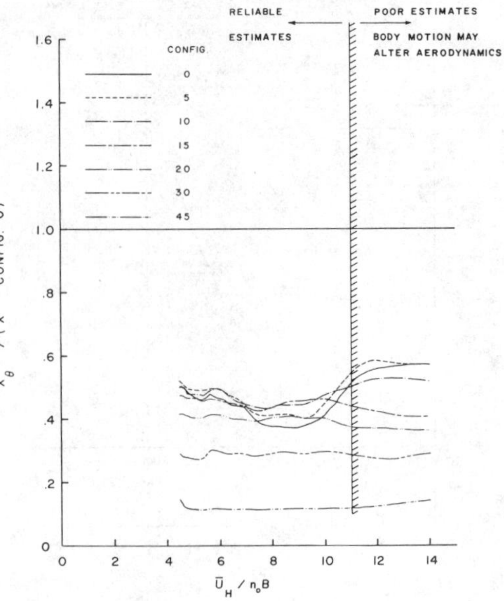

Fig. 13. Ratio of equivalent peak response due to torsion to the peak response in x direction at configuration 0°

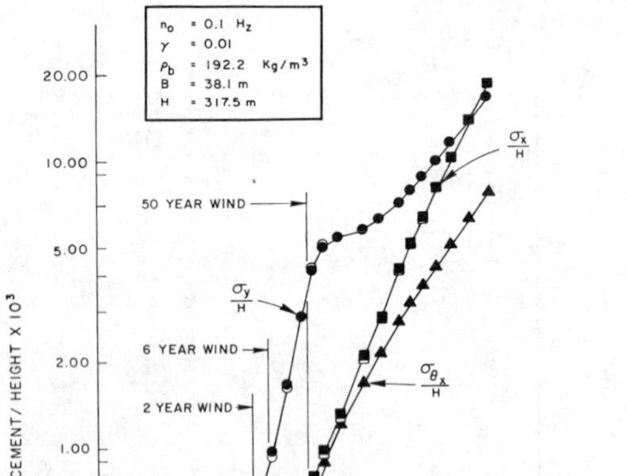

Fig. 14. Predicted RMS displacements of prototype building

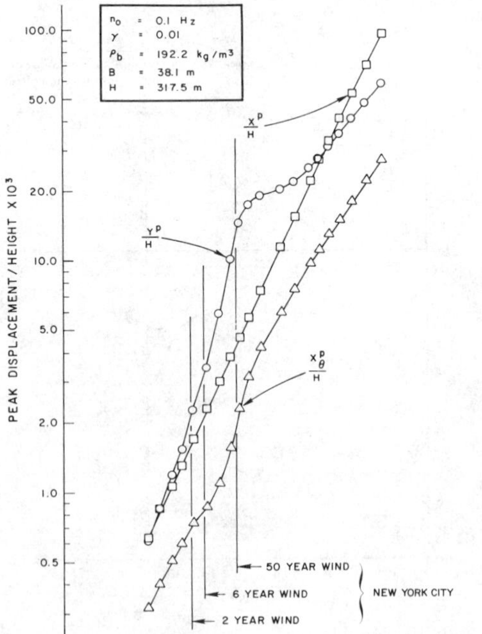

Fig. 15. Predicted peak displacements of prototype building

responses and a peak/RMS factor of 3.5 are compared in a similar way in Figures 11, 12, and 13. In Figure 13 the peak torsional response has been converted to a translation in the center of one side of the building.

For nearly all wind velocities the greatest responses occur when the wind is blowing onto a face of the building. For this wind condition the actual RMS and peak displacements are shown in Figures 14 and 15. In Figure 14 the values predicted by the integration of the full spectrum are compared with values obtained using the approximation that only the value of modal force spectrum at the natural frequency of that mode need be considered. The equation used to determine the approximate response is

$$\sigma^2_{x,y,\theta} = \frac{\pi n_o}{4\gamma} \cdot \frac{G_{Fx,Fy,M}(n_o)}{K^2}, \quad (6)$$

where n_o, γ and K are the values appropriate to the mode under consideration.

It can be seen that this is generally a good approximation, improving as the wind speed increases. Some full-scale investigations of building behavior (7) (8) have made use of this simplified formula to obtain values of modal forces from measured responses. Figure 14 appears to support this approach.

DISCUSSION OF THE RESULTS

It is a commonly held belief by designers of tall buildings that the dynamic response of a building is small in comparison with the steady state drift of the building and need only be considered in exceptional cases where movements of the building may adversely affect the well being of the occupants of the upper floors. Another common assumption is that the worst loading and response will occur when the wind is blowing directly on the face of a building. To a certain extent this latter assumption is justified by the results of these tests. The greatest mean and RMS modal forces do occur when the wind is blowing on or near perpendicular to the face of the building but Table 1 shows that at certain angles, notably between 20° and 45° the mean pressure coefficient in the y direction is very sensitive to the wind direction. Since wind-tunnel tests do not faithfully represent variations of wind direction present in the atmosphere, large responses simply due to variations in wind direction can not be observed although such responses have been measured in an actual building (8).

Most analytical procedures for predicting the response of buildings deal only with in-wind response. These tests show that for a tall slender building the dynamic response in the cross-wind direction can be between 2 and 5 times greater than the in-wind response. These observations were made in the worst case i.e. with the wind on the face of the building, Figure 8.

Furthermore, even when the mean deflection in the in-wind direction is taken into account the peak response in the cross-wind direction is between 1 and 3 times the in-wind peak response. For this shape of building no longer is the cross-wind dynamic response merely computed for serviceability considerations. It represents the direction of the controlling deflection of the building.

The amplitude of motion caused by the torsion of the building can be represented as a displacement if the motion at some point at a distance R from the center of the building is considered. It is convenient to select for this purpose the mid-point of a face of the building. The RMS and peak displacements at this point due to the twisting moments are denoted in Figures 14 and 15 by $\sigma_{\theta x}$ and X_θ respectively. Note that the RMS torsional displacements at a given level, at least for the properties assumed in this building, are as large as the in-wind

RMS response at that level. The motions due to the twisting moments are particularly disturbing to occupants since apparent motion of distant objects is more easily discernible when the building motion is due to torsional rather than to translational motion.

Table 2 shows the predicted response values for the prototype building as given by a number of current design procedures (5)(9)(10). Values have been computed for winds in New York City with return periods of 50 years, 6 years and 2 years.

Table 2. Comparison of response predictions for prototype building

Response to 50 year wind:[11] (U_H = 40 m/s)

Source of Prediction	$\frac{x}{H}$	$\frac{\sigma_x}{H}$	$\frac{x^P}{H}$	$\frac{\sigma_B}{\sigma_R}$; %	$\frac{a_y}{g}$; %	$\frac{a_x}{g}$; %
ESDU[9]	.00178	.00073	.0044	71	-	2.7
NBC[10]	.00175	.00130	.0063	6	15.7	5.6
Simiu[5]	.00179	.00090	.0049	84	-	3.2
Experimental	.00178	.00075	.0044	4	16.8	3.3

Response to 6 year wind:[13] (U_H = 29 m/s)

Source of Prediction	$\frac{x}{H}$	$\frac{\sigma_x}{H}$	$\frac{x^P}{H}$	$\frac{\sigma_B}{\sigma_R}$; %	$\frac{a_y}{g}$; %	$\frac{a_x}{g}$; %
ESDU	.00094	.00037	.00224	76	-	1.3
NBC	.00090	.00052	.00273	11	5.3	2.2
Simiu	.00094	.00044	.00246	95	-	1.5
Experimental	.00094	.00035	.00217	5	4.0	1.6

Response to 2 year wind:[13] (U_H = 26 m/s)

Source of Prediction	$\frac{x}{H}$	$\frac{\sigma_x}{H}$	$\frac{x^P}{H}$	$\frac{\sigma_B}{\sigma_R}$; %	$\frac{a_y}{g}$; %	$\frac{a_x}{g}$; %
ESDU	.00074	.00028	.00173	82	-	1.0
NBC	.00074	.00039	.00207	13	3.7	1.6
Simiu	.00075	.00034	.00193	101	-	1.1
Experimental	.00074	.00025	.00162	7	2.7	1.1

The 50 year return period is the one now normally used for structural design purposes. The associated velocity was obtained from reference (11). Shorter return periods have been suggested for use in determining compliance with serviceability requirements such as occupant comfort (12). Return periods of 6 and 12 years were chosen for illustrative purposes and the velocities were obtained from reference (13).

In discussing these results it must be remembered that they assume a perfect knowledge of the building properties and this, as was pointed out earlier, may be an unwarranted assumption. It should also be borne in mind that the technique used to determine the in-wind dynamic modal forces from the rigid model tests neglects forces at lower levels. Consequently, the RMS in-wind response reported in this paper may be underestimated by about 15% (based on calculations using Simiu's method (5)). A study of the response estimates shown in Table 2 leads to several observations. All of the methods appear to be able to provide reasonably accurate estimates of the mean in-wind response. However, in considering the dynamic response there is a significant difference between the estimates of the response contributions due to the off-resonance response using the ESDU (9) and Simiu (5) methods when compared to estimates using the experimental work and the NBC (10) method.

If the in-wind RMS response obtained from the experimental results is 15% low, then the RMS responses predicted by Simiu's method are quite accurate; those predicted using NBC are high; and those predicted using the ESDU method are low. Peak response estimates follow a similar pattern but estimates of acceleration follow a different one. Assuming that the accelerations predicted from the experimental work are about 12% low, it follows that the NBC prediction is high and predictions using either Simiu or ESDU's methods are low. It has been recommended in the NBC method that the peak acceleration be limited to between 1% and 3% of the gravitational acceleration every ten years. Reference (12) suggests an RMS acceleration of 0.5%g in winds with a return period of 6 years (equivalent to a peak of 1.75%g, assuming a peak to RMS factor of 3.5). The World Trade Center, New York, a similarly shaped building, was designed on the basis that the average number of occurences of 1% accelerations at the top occupied floor should not exceed 12 per year (14).

On the basis of in-wind response this building would meet all the above design recommendations. Yet when the cross-wind response is considered the building's expected performance is very clearly not acceptable and this takes no account of torsional response or coupling between modes.

Only the NBC method gives guidance on computing cross-wind responses. Its method is entirely empirical, based on a large number of wind tunnel tests on aeroelastic models. The results obtained by this method closely match the experimental results reported herein.

Clearly for a building of this shape any designer who ignores the effect of cross-wind excitation in his design omits what is perhaps the most important effect of wind loading.

CONCLUSIONS

This paper has shown that it is possible to use a suitably instrumented rigid model to predict the response of a prototype building. The predictions can be made with the same degree of accuracy as can be obtained using an aeroelastic model provided that wind speeds are lower than those corresponding to $\bar{U}_H/n_oB \simeq$ 10 to 12. This method has an advantage over the conventional aeroelastic method in that the effect of changes in structural characteristics of the prototype can be easily investigated and insight can be gained into the pressure patterns creating the exciting forces.

Using this technique it has been shown that the greatest response of a slender square-section tall building is likely to occur in the cross-wind direction when the wind is blowing directly onto a face. There is however a possibility of significant response at angles of attack of between 20° and 45° if the wind direction is unstable.

Current design procedures appear to provide reasonable predictions of mean and dynamic in-wind response when compared with predictions based on wind-tunnel measurements of modal forces. Only one of the design procedures used gave guidance on the prediction of cross-wind response. Since this type of building is very susceptible to cross-wind excitation it is considered unwise for a designer to neglect the cross-wind response.

ACKNOWLEDGMENTS

The experimental work and some of the data analysis reported in this paper were supported by the National Science Foundation, Grant No. ENG-75-05327 A01. Further data analysis was conducted by one of the authors under a National Research Council/National Bureau of Standards Postdoctoral Research Associateship. The authors also wish to acknowledge the extensive help and guidance of Professors F. J. Maher and H. W. Tieleman of Virginia Polytechnic Institute and State University.

REFERENCES

1. Jeary, A. P. and Sparks, P. R., "Some Observations on the Dynamic Sway Characteristics of Concrete Structures." Pro. ACI Symp. on the Vibr. of Conc. Str., New Orleans 1977. Also BRE Current Paper CP7/78.

2. Reinhold, T. A., "Measurement of Simultaneous Fluctuating Loads at Multiple Levels on a Model of a Tall Building in a Simulated Urban Boundary Layer." Ph.D. Thesis, VPI & SU, Blacksburg, Va., November 1977.

3. Reinhold, T. A., Tieleman, H. W. and Maher, F. J., "Wind Tunnel Simulation of Urban Winds for the Study of Tall Buildings," Report No. VPI-E-77-12, College of Engineering, VPI & SU, Blacksburg, Va., April 1977.

4. Rosati, P. A., "The Response of a Square Prism to Wind Load," Report BLWT-II-68. The Univ. of W. Ontario, London, March 1968.

5. Simiu, E., "Revised Proc. for Estimating Along-Wind Resp.," to be offered for pub. to ASCE; modif. of method in "Equiv. Static Wind Loads for Tall Bldg. Design" by E. Simiu. J. Str. Div. ASCE Vol. 102, No. ST4, April 1970, pp. 719-737.

6. Saunders, J. W. and Melbourne, W. H., "Tall Rectangular Building Response to Cross-Wind Excitation," Proc. Fourth Int. Conf. on "Wind Effects on Bldgs. and Strs.," Heathrow, England, 1975, pp. 369-379.

7. Sparks, P. R., and Crist, R. A., "The Determination of the Response of Tall Buildings to Wind Loading." Proc. ASCE-EMD Conf. on "The Dyn. Resp. of Strs.," Univ. of Cal., Los Angeles, March 1976.

8. Jeary, A. P., Lee, B. E. and Sparks, P. R., "The Determination of Wind Induced Modal Forces." Proc. 3rd U.S. Natl. Conf. on Wind Engr. Res. Univ. of Fla., Gainesville, Feb. 1978.

9. Engineering Sciences Data Unit, "The Response of Flexible Structures to Atmospheric Turbulence," Data Item 76001, London, September 1976.

10. National Research Council of Canada, "Commentaries on Part 4 of the National Building Code of Canada 1975," Ottowa, 1975.

11. ANSI, A58.1. Amer. Natl. Std. on Bldg. Code Req. for Minimum Design Loads in Bldgs. and Other Strs., ANSI, N. Y., 1972.

12. Hansen, R. J., Reed, J. W. and Vanmancke, E. H., "Human Response to Wind-Induced Motion," J. Str. Div., ASCE Vol. 98, No. ST7, July 1973.

13. Simiu, E., Changery, M. J. and Filliben, J. J., "Extreme Wind Speeds at 129 Stations in the Contiguous United States." NBS, BSS 118, Wash., D.C., 1979.

14. Feld, J., "Superstructure of 1350 ft World Trade Center," Civil Engineering, ASCE Vol. 41, No. 6, June 1971, pp. 66-70.

CROSS-WIND RESPONSE OF STRUCTURES DUE TO DISPLACEMENT DEPENDENT LOCK-IN EXCITATION

K. C. S. KWOK* and W. H. MELBOURNE**

*The University of Sydney, Sydney, Australia
**Monash University, Clayton, Victoria, Australia

ABSTRACT

Wind tunnel tests were carried out on a circular and a square tower model in a suburban turbulent boundary layer wind model. It was found that when the towers were operating at close to the critical reduced velocity and particularly at low values of structural damping, displacement dependent lock-in excitation could cause large amplitude cross-wind response. A prediction procedure comprising of a random and a sinusoidal lock-in excitation model is described. The predicted cross-wind response of the towers agreed well with measured response.

INTRODUCTION

Lock-in is a term commonly used to describe the phenomenon in which large amplitude cross-wind response of structures occurs at wind velocities at which the vortex shedding frequency n_s, as defined by the Strouhal number, is close to the natural frequency of the structure n_o. Studies of this phenomenon reported to date, such as those by Bishop and Hassan[1], Bearman and Davis[2], Stansby[3] and others, have been mainly concentrated on the identification of the excitation mechanism of lock-in in smooth flow conditions. Data applicable to modern tall buildings and structures in natural turbulent boundary layer flow are extremely rare. Wind tunnel model studies by Vickery[4] on a reinforced concrete stack and by Melbourne[5] on a slender square tower with a porous shroud on the top half appear to be the only two reported cases in which lock-in has been found to be significant for tall structures under simulated natural wind conditions.

Although the mechanism of lock-in excitation is not yet completely understood, significant clarification has been achieved. In particular, for tall structures in natural turbulent boundary layer flow, lock-in was identified by Melbourne[6] as a cross-wind excitation due to cross-wind displacement in which the cross-wind response of a structure causes an increase in the wake energy which in turn increases the cross-wind response of the structure. It is generally accepted that displacement dependent lock-in is more likely to occur for structures which have relatively low stiffness, are lightly damped and are operating near the peak of the wake energy spectrum.

Wind tunnel model studies were conducted to assess the significance of displacement dependent lock-in in causing large amplitude cross-wind response of towers in natural strong wind conditions. The cross-wind responses of circular and square tower models under simulated turbulent boundary layer flow in a wind tunnel are presented in this paper. A prediction procedure and the relevant data which may be used for the prediction of cross-wind response of structures due to wake excitation, including when displacement dependent lock-in is significant, are also presented.

EXPERIMENTAL PROCEDURES

Wind tunnel model tests were conducted on circular and square towers with a height to diameter or breadth ratio $\frac{h}{b}$ of 9. These models were rigid and pivoted at the base, thus maintaining a straight line deflection mode. The models were restrained by four horizontal helical springs and oil was used to simulate structural damping. A structural density ρ_s of 10 $lbft^{-3}$ (160 kgm^{-3}), was modelled. Strain gauge bridges were used to measure the deflection and overturning moment acting on the models.

The circular and square tower models were tested in the 6.5 ft. x 6.5 ft. (2m x 2m) working section of the Monash University Boundary Layer Wind Tunnel. A wind model of natural wind flow over a suburban area was developed over the 35 ft. (11m) fetch length of the working section. This was done by using the augmented growth method, in which floor-mounted roughness elements and four linearly-tapered vorticity generators spanning the start of the working section were used. The power law exponent of the mean longitudinal wind velocity was 0.23 and the longitudinal turbulence intensity was 0.07 at the top of the tower. A linear length scale of 1/400 was obtained by modelling according to the power spectral density function of the longitudinal velocity, thus giving the towers an equivalent full scale height of about 710 ft. (216m) and a diameter or breadth of about 80 ft. (24m).

The standard deviations of the cross-wind displacement responses at the top of the circular and square tower models were measured and the displacement signals were recorded by an FM tape recorder onto magnetic tapes. These were subsequently digitised and upcrossing probability analysis and spectral analysis were carried out by digital computers.

EXPERIMENTAL RESULTS AND DISCUSSIONS

Cross-Wind Response

The standard deviation cross-wind responses at the top of the circular and square towers σ_y were measured at reduced velocities $\bar{U}(h)/n_o b$ ranging from 4.5 to 24 ($\bar{U}(h)$ is the mean longitudinal wind velocity at the top of the tower), and at values of structural damping ratios ζ_s from 6% to 0.25%. The standard deviation responses were normalised as $\frac{\sigma_y}{b}$ and are presented as a function of reduced velocity in Figure 1, and as a function of structural damping in Figure 2. As the structural damping is reduced, large increase in the cross-wind response of the square tower and particularly the circular tower are evident at reduced velocities close to 10 and 7 respectively. These critical reduced velocities represent the wind conditions at which the vortex shedding frequency n_s is very close to the natural frequency n_o of the tower, that is, the towers are operating near the peak of the wake energy spectrum. These large increases in response, especially that of the circular tower for which there is an increase in response of nearly two orders of magnitude as the structural damping is reduced from 6% to 0.25% of critical, might have been caused by displacement dependent lock-in excitation.

For most tall buildings and structures in turbulent boundary layer flows, the cross-wind excitation associated with the wake has generally been accepted as the most dominant excitation mechanism. Wake excitation is generally random in nature so that the cross-wind response due to this excitation can be adequately described by random excitation theory, and is proportional to $\zeta_s^{-\frac{1}{2}}$. Wind tunnel model studies of a range of structures by Rosati[7], Vickery[4] and others have shown that the increased responses are generally proportional to functions which lie between $\zeta_s^{-1/3}$ and $\zeta_s^{-\frac{1}{2}}$. Such a dependence is evident in the responses shown in Figure 2 of the circular and square towers when they are operating at reduced velocities away from the critical and at high values of structural damping, (above 2% of critical). However, the responses increase much more rapidly with decrease in damping when the

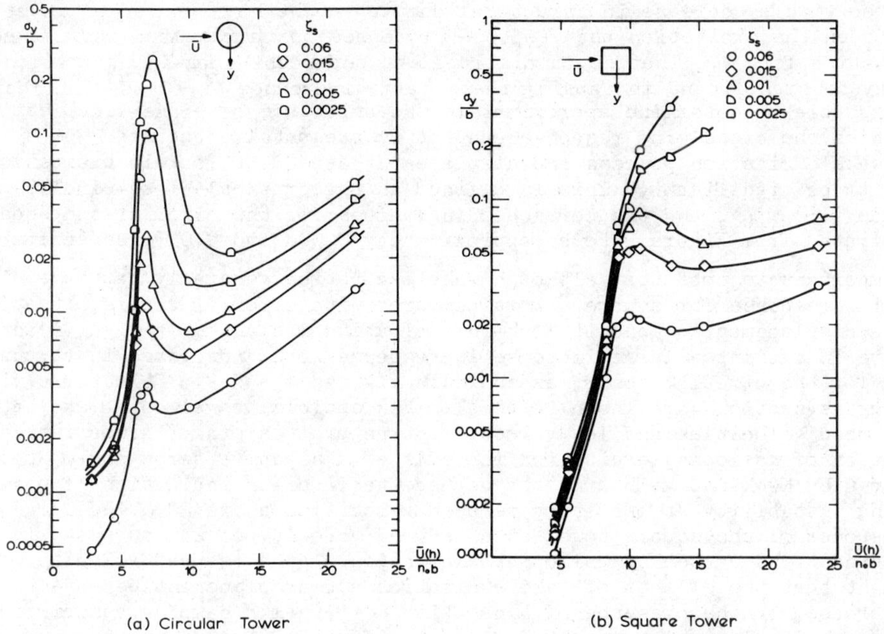

Fig. 1. Cross-wind responses of the circular and square tower models in a simulated turbulent boundary layer flow

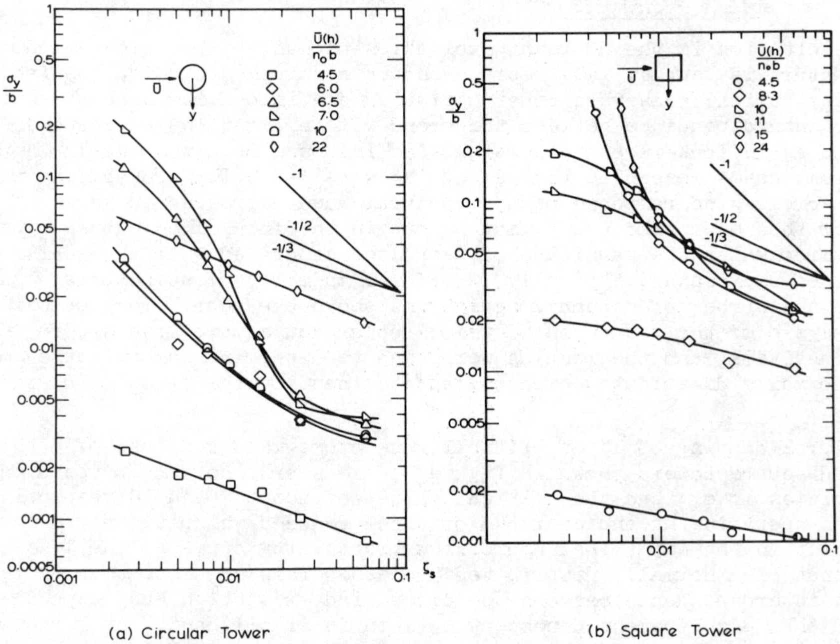

Fig. 2. Cross-wind responses of the circular and square tower models in a simulated turbulent boundary layer flow

towers are operating at close to the critical reduced velocity, especially at low values of damping. For the circular tower, the proportional function is considerably steeper than ζ_s^{-1}. Such a significant departure from the $\zeta_s^{-\frac{1}{2}}$ associated with a random wake excitation process suggests that the increase in cross-wind response associated with a decrease in structural damping might have caused an interdependence between the excitation and response processes which resulted in well-correlated excitation forces. That is, a displacement dependent lock-in excitation mechanism may be significant in causing these large increases in cross-wind response. It is possible to determine approximately the amplitude of cross-wind response above which the cross-wind response can not be adequately accounted for by the random wake excitation process and displacement dependent lock-in excitation is thought to be significant. This is defined as the critical cross-wind response amplitude $(\frac{\sigma_y}{b})_{crit.}$, and the corresponding values for the circular and square towers are estimated from Figure 2 to be approximately 0.006 and 0.025 respectively.

It is important to note that although wake excitation generally causes a higher cross-wind response for square towers compared with circular towers, it appears that the displacement dependent lock-in excitation mechanism is significantly more effective in causing a large increase in response with a decrease in structural damping for the circular tower, as shown in Figures 1 and 2. Structures with square cross-sections are known to be also susceptible to galloping excitation at high reduced velocities and low values of structural damping. A quasi-steady linear formulation of galloping excitation in terms of a negative aerodynamic damping has been shown by Kwok and Melbourne[8] to be acceptable as an indication of possible galloping instability at operating reduced velocities above 20. The large cross-wind response of the square tower at a reduced velocity of 24, as shown in Figure 2, is believed to be caused by galloping excitation. At a reduced velocity of 15, it is thought that the effects of wake excitation and displacement dependent lock-in were increased by the presence of a negative aerodynamic damping associated with galloping excitation which resulted in a significant increase in response, particularly at low values of structural damping.

Upcrossing Probability Analysis

When wake excitation is the dominant excitation mechanism, the cross-wind response y of a structure is usually well represented by a normal, that is Gaussian, probability distribution. However, under operating conditions in which there is a significant interdependence between the cross-wind excitation and response processes, such as in lock-in and galloping, the response is not normally distributed, and in extreme cases resembles that of a sine wave[6,9]. The probability distribution of the cross-wind response of a structure, when expressed in terms of upcrossings, rather than on a time basis, may be put in the form of a Weibull type distribution[9] from which the magnitude of response associated with a certain level of probability of upcrossing, typically 10^{-3} for a majority of structures, can be seen in terms of the number of standard deviations above the mean. The upcrossing probability analysis of the responses of the circular and square towers were performed by digital analysis, and the results were used to determine the variations from that of a normally distributed process, and to identify the cross-wind excitation mechanism.

At a structural damping of 1% of critical, the cross-wind response of both the circular and square towers shown in Figure 3, are nearly normal at operating reduced velocities lower than the critical values of about 7 and 10 respectively. However when operating at and near the critical reduced velocity, there is a significant shift in the normalised upcrossing probability curve to the left of and away from that of a normally distributed process. This suggests that there is a significant interdependence between the cross-wind excitation and response processes, and that displacement dependent lock-in is significant. At a reduced velocity of 22, the response of the circular tower, as shown in Figure 3a, returns to be normally distributed and wake excitation again dominates. At a reduced velocity of 24, galloping excitation becomes significant for the square tower[8]. Although the response process returns to be more normally distributed, the interdependence of the excitation and response processes associated with galloping excitation is evident in Figure 3b. These changes in response characteristics can also be clearly seen in the traces of cross-wind motions and the reduced variates corresp-

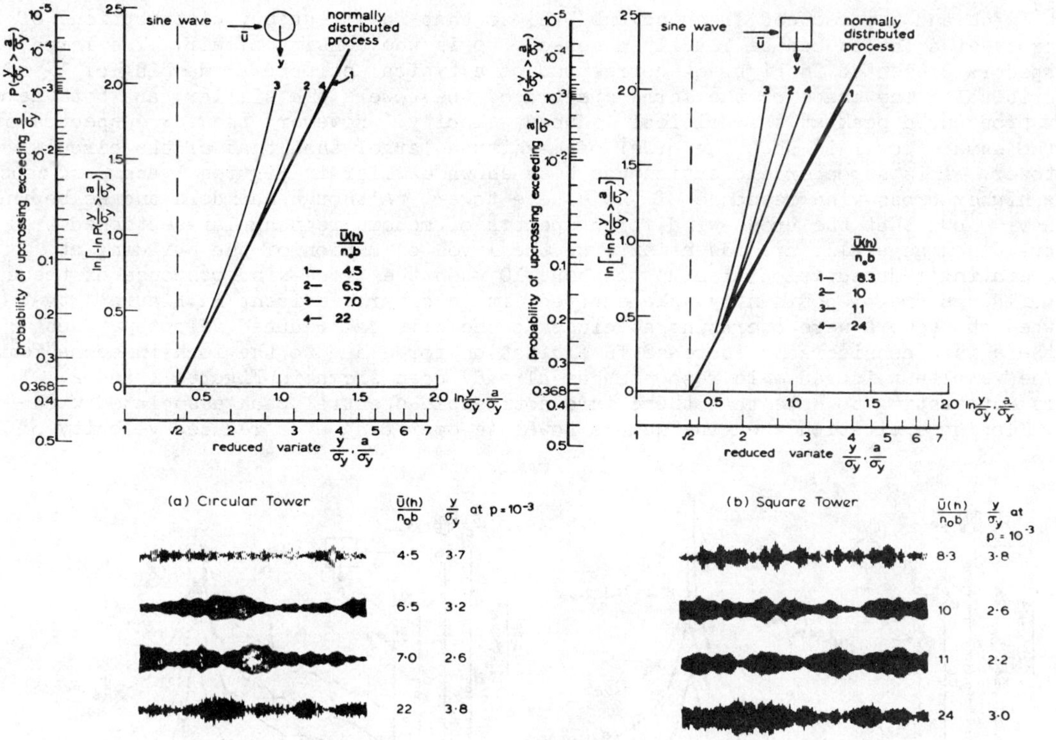

Fig. 3. Upcrossing probability distributions and traces of cross-wind motions of the cross-wind responses of the circular and square tower models

onding to a probability of upcrossing of 10^{-3} given in Figure 3.

Normalised Cross-Wind Force Spectra

Cross-wind displacement spectra $S_y(n)$ were obtained by Fast Fourier Transformation of the recorded displacement signals, using a digital computer. The cross-wind force spectra $S_F(n)$ were determined from the displacement spectra by assuming a linear relationship between the cross-wind excitation and response[10], so that

$$S_F(n) = \frac{(2\pi n_o)^4 m^2 S_y(n)}{|H(n)|^2} \qquad (1)$$

in which m is the modal mass of a mode shape y(z) which is given by

$$m = \int_o^h m(z)\, y^2(z)\, dz \qquad (2)$$

m(z) is the mass per unit height of the structure, and $|H(n)|^2$ is the mechanical admittance function which is given by

$$|H(n)|^2 = \frac{1}{\left[1-\left(\frac{n}{n_o}\right)^2\right] + 4\zeta_s{}^2\left(\frac{n}{n_o}\right)^2} \qquad (3)$$

Each $S_F(n)$ is the equivalent cross-wind force spectrum acting at the top of the tower which represents the integrated effects of all the cross-wind forces acting on the entire structure.

The normalised cross-wind force spectra $nS_F(n)/[\frac{1}{2}\rho\bar{U}^2(h)bh]^2$ of the circular and square towers are presented in Figure 4 as functions of the reduced velocity

$\bar{U}(h)/nb$ and the reduced frequency $nb/\bar{U}(h)$ so that the frequency distributions of cross-wind forces can be readily assessed. ρ is the density of air. The force spectra presented in Figure 4 correspond to a typical structural damping of 1% of critical. The shape of the force spectra of the towers are similar, and both have a pronounced peak at the critical reduced velocity. However, the force spectra of the square tower is about an order of magnitude larger than that of the circular tower. This stronger excitation has been shown earlier in Figures 1 and 2 to cause a higher cross-wind response of the square tower. Although Saunders and Melbourne[11] have shown that the cross-wind force spectra of modern rectangular-sectioned tall buildings generally are insensitive to the level of motion of the building at operating reduced velocities up to about 10 when the cross-wind response of these buildings are predominantly wake excited, it is clearly evident in Figure 4 that when the towers were operating at close to the critical reduced velocity, then there is a considerable increase in excitation force due to the lock-in mechanism. The resultant increase in response has already been shown in Figures 1 and 2. It is interesting to note that there is a noticeable spectral peak associated with galloping excitation when the square tower is operating at a reduced velocity of 24.

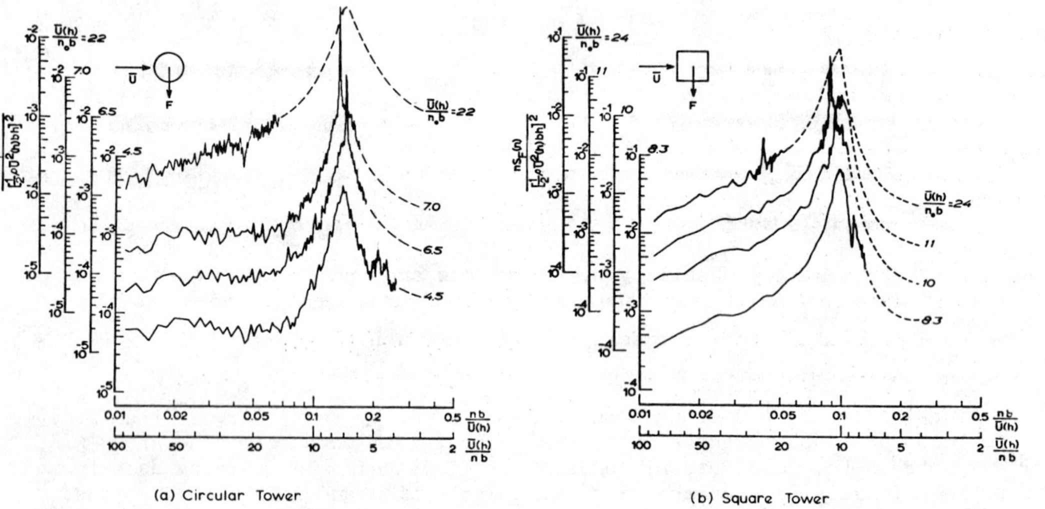

Fig. 4. Cross-wind force spectra of the circular and square tower models

PREDICTION OF CROSS-WIND RESPONSE OF TALL BUILDINGS AND STRUCTURES

A Generalised Design Approach

The cross-wind response of tall buildings and structures can be estimated by using a prediction procedure which consists of a random excitation model and a sinusoidal lock-in excitation model. Each model represents an important mechanism in the cross-wind excitation and response processes. Wake excitation has been shown to be essentially random and can be described by a random excitation model. The sinusoidal lock-in excitation model describes displacement dependent lock-in. The selection of the excitation model to be used in the prediction process is determined by the critical cross-wind response amplitude $(\frac{\sigma_y}{b})_{crit}$ which is the normalised standard deviation cross-wind response at the top of a base-pivoted rigid structure above which displacement dependent lock-in is significant. This critical response amplitude depends on parameters such as cross-sectional shape and turbulence intensity, and has been determined experimentally to be approximately 0.006 of the diameter and 0.025 of the breadth for the circular and square towers respectively in a suburban turbulent boundary layer wind model. In design considerations, when the predicted response computed by using the random excitation model exceeds the critical response amplitude, the sinusoidal lock-in excitation model should be employed thereafter in the prediction process.

Random Excitation Model

By assuming that the excitation is random, the cross-wind response of buildings and structures due to wake excitation can be computed by modal analysis[12,13]. For an effectively rigid structure pivoted at the base, such as the circular and square towers used in this study, the deflection mode shape is adequately represented by a straight line so that there is a considerable simplification in the prediction process[10]. The variance of cross-wind response at the top of the structure can be obtained by a process similar to that of Equation (1), and is given by

$$\sigma_y^{\,2} = \int_0^\infty S_y(n)\,dn = \frac{1}{(2\pi n_o)^4 m^2}\int_0^\infty S_F(n)\,|H(n)|^2\,dn \qquad (4)$$

If excitation by low frequencies is small and the structural damping is less than about 10% of critical so that the excitation bandwidth is large compared with the resonant bandwidth, Equation (4) can be approximated by

$$\sigma_y^{\,2} \approx \frac{\pi n_o S_F(n_o)}{(2\pi n_o)^4 m^2 4\zeta_s} \qquad (5)$$

and the standard deviation cross-wind response is proportional to $\zeta_s^{-\frac{1}{2}}$.

The cross-wind force spectrum used in the prediction process is determined from a normalised cross-wind force spectrum of the structure, which is computed by using Equation 1, for a high value of structural damping at which neither displacement dependent lock-in nor galloping excitation is significant[10]. Either Equation (4) or (5) may be used in the prediction process and the wake-excited cross-wind response can be readily determined as functions of reduced velocity and structural damping. In general, this prediction procedure is adequate when wake excitation is the dominant mechanism. However, at operating conditions at which either displacement dependent lock-in or galloping excitation is significant and particularly at low values of structural damping, the cross-wind response will be significantly underestimated by this procedure.

Sinusoidal Lock-in Excitation Model

For tall buildings and structures, when displacement dependent lock-in becomes significant, the cross-wind forces are well-correlated along the structure. The cross-wind force at any time t may be assumed to be sinusoidal and to take the form

$$F_S(t) = \sqrt{2}\; C_{FS}\,\frac{\rho \bar{U}^2 bh}{2}\,\sin 2\pi n_s t \qquad (6)$$

in which C_{FS} is a sinusoidal lock-in force coefficient which is a measure of cross-wind forces associated with lock-in at a frequency close to the structural natural frequency n_o and the vortex shedding frequency n_s. Therefore for a base-pivoted rigid structure, the cross-wind response at the top of the structure when $n_s \approx n_o$ is

$$\sigma_y = \frac{C_{FS}\rho b^3 h}{2\zeta_s m\left[4\pi\,\dfrac{n_o b}{\bar{U}(h)}\right]} \qquad (7)$$

The cross-wind response predicted by this sinusoidal lock-in excitation model is proportional to ζ_s^{-1}, which may be compared with $\zeta_s^{-\frac{1}{2}}$ by the random excitation model.

In general, the sinusoidal lock-in force coefficient will depend in a complex way on parameters such as cross-sectional shape, response amplitude and operating reduced velocity. The non-linear dependence of this force has been investigated by Hartlen and Currie[14], Iwan and Blevins[15], Blevins and Burton[16] and others. However, as a first approximation, it is assumed that the coefficient is simply a function of the operating reduced velocity. That is, when the cross-wind response exceeds the critical response amplitude, lock-in becomes significant, the excitation force becomes sinusoidal and the response is proportional to ζ_s^{-1}. By using the random excitation model and the appropriate cross-wind force spectrum to compute the cross-wind response up to the critical response amplitude at which there is a change in the dependence of the response from $\zeta_s^{-\frac{1}{2}}$ to ζ_s^{-1}, and by noting the operating reduced velocity and damping, values of the sinusoidal lock-in force coefficient can be estimated from Equation (7). The values of these coefficients for the circular and square towers in a suburban turbulent boundary layer wind model are presented in Figure 5. The coefficients are, as expected, at a maximum at the critical reduced velocity at which $n_o \approx n_s$. Therefore, when the circular and square towers are operating at close to the critical reduced velocity, and the predicted cross-wind response exceeds the critical response amplitude, the sinusoidal lock-in excitation model and the appropriate coefficient given in Figure 5 should be used in the prediction process in place of the random excitation model.

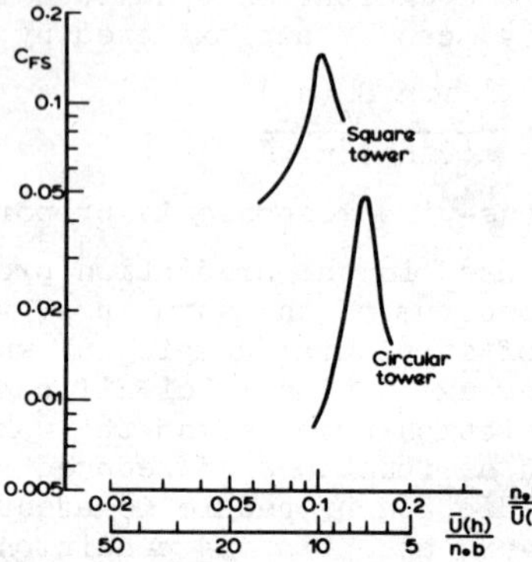

Fig. 5. Sinusoidal lock-in force coefficients of the circular and square tower models

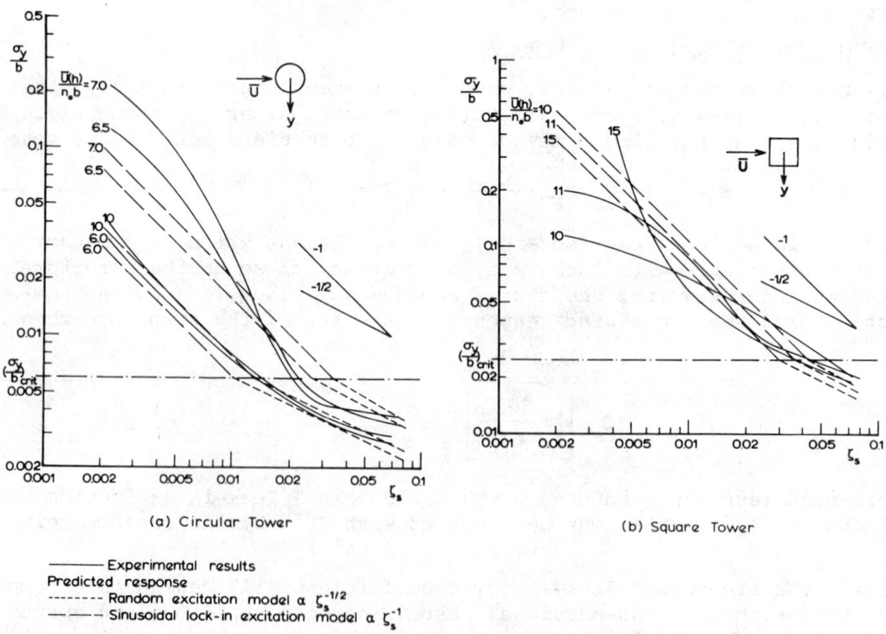

Fig. 6. Predicted cross-wind responses of the circular and square tower models using random and sinusoidal lock-in excitation models

Predicted Cross-Wind Response of the Circular and Square Towers

The predicted cross-wind response of the circular and square towers computed from the random excitation model and the sinusoidal lock-in excitation model are presented in Figure 6. In general, the predicted response obtained from the combined excitation models matches the experimental results more closely than would be achieved by using either model alone. However, at low values of structural damping, the sinusoidal lock-in excitation model becomes quite unsatisfactory. This discrepancy is thought to be largely due to the non-linear effect of lock-in. It should be noted, however, that the deficiency of the sinusoidal lock-in excitation model is more apparent at very large response amplitudes which in terms of design, far exceeds the acceptable level of response. Therefore, under these extreme operating conditions, the magnitude of response becomes purely academic. In practice, the sinusoidal lock-in excitation model should be quite adequate as an assessment which provides useful information of the order of magnitude of cross-wind response that a structure is likely to attain if lock-in becomes significant.

CONCLUSIONS

At close to the critical reduced velocity and particularly at low values of structural damping, displacement dependent lock-in excitation was found to be significant for both the circular and square towers in a suburban turbulent boundary layer flow. This excitation resulted in a large increase in cross-wind response, particularly for the circular tower. When lock-in becomes significant, the increase in response is proportional to ζ_s^{-1} or higher, rather than $\zeta_s^{-\frac{1}{2}}$ as in the case of wake excitation. The increase in cross-wind response associated with a decrease in structural damping causes a significant interdependence between the cross-wind excitation and response processes. This results in an increase in cross-wind excitation forces, and which appears as a peak in the cross-wind force spectrum. The change in response characteristics was also readily identified by the variations from that of a normally distributed process in the results of upcrossing probability analyses.

A prediction procedure, which consists of a random excitation model and a sinusoidal lock-in excitation model, was proposed. The selection of the excitation model to be used in the prediction process is determined by a critical cross-wind response amplitude which is the standard deviation cross-wind response at the top of a structure above which lock-in is significant. This was found to be approximately 0.006 of the diameter and 0.025 of the breadth for the circular and square towers respectively. Cross-wind response predicted by these models compared well with the measured response.

ACKNOWLEDGEMENTS

The authors wish to thank the staff of the Department of Mechanical Engineering and the University Computer Centre of Monash University for their assistance.

REFERENCES:

1. Bishop, R.E.D., and Hassan, A.Y., The Lift and Drag forces on a Circular Cylinder in a Flowing Fluid, Proc. Roy. Soc. London, Series A, 277, 1964, pp. 31-75.

2. Bearman, P.W. and Davis, M.E., The Flow Around Oscillating Bluff Structures, Proc. 4th Int. Conf. Wind Effects on Buildings and Structures, London, 1975.

3. Stansby, P.K., The Locking-on of Vortex Shedding due to the Cross-stream Vibration of Circular Cylinders in Uniform and Shear Flows, J. Fluid. Mech., Vol. 74, 1976, pp. 641-665.

4. Vickery, B.J., Wind Induced Vibrations of Towers, Stacks and Masts, Proc. 3rd. Int. Conf. Wind Effects on Buildings and Structures, Tokyo, 1971, Paper IV-2.

5. Melbourne, W.H., Response of a Slender Tower to Wind Excitation, Proc. 5th Aust. Conf. Hydraulics and Fluid. Mech., Univ. of Canterbury, 1974, pp. 251-257.

6. Melbourne, W.H., Cross-wind Response of Structures to Wind Action, Proc. 4th Int. Conf. Wind Effects on Buildings and Structures, London, 1975.

7. Rosati, P.A., An Experimental Study of the Response of a Square Prism to Wind Load, BLWT-11-68, Univ. of Western Ontario, 1968.

8. Kwok, K.C.S. and Melbourne, W.H., The Effects of Freestream Turbulence on a Galloping Square Tower, Proc. 6th Aust. Conf. Hydraulics and Fluid. Mech., Univ. of Adelaide, 1977, pp. 447-450.

9. Melbourne, W.H., Probability Distributions Associated with the Wind Loading of Structures, Civil Eng. Trans., I.E. Aust., Vol. CE19, No. 1, 1977, pp. 58-67.

10. Kwok, K.C.S., Cross-wind Response of Structures due to Displacement Dependent Excitations, Ph.D. Thesis, Dept. Mech. Eng., Monash Univ., 1977.

11. Saunders, J.W., and Melbourne, W.H., Tall Rectangular Building Response to Cross-wind Excitation, Proc. 4th Int. Conf. Wind Effects on Buildings and Structures, London, 1975.

12. Davenport, A.G., The Response of Slender Line Like Structures to a Gusty Wind, Proc. ICE, Vol. 23, 1962, pp. 389-408.

13. Vickery, B.J. and Clark, A.W., Lift or Across-wind Response of Tapered Stacks, Proc. ASCE, J. Struct. Div., Vol. 98, ST1, 1972, pp. 1-20.

14. Hartlen, R.T. and Currie, I.G., Lift-oscillator model of Vortex Induced Vibration, Proc. ASCE, J. Eng. Mech. Div., EM5, 1970, pp. 577-591.

15. Iwan, W.D. and Blevins, R.D., A Model for Vortex Induced Oscillation of Structures, Trans. ASME, J. Appl. Mech., Vol. 41, 1974, pp. 581-586.

16. Blevins, R.D. and Burton, T.E., Fluid Forces Induced by Vortex Shedding, Trans. ASME, J. Fluid. Eng., 1976, pp. 19-36.

AN ANALYTICAL METHODOLOGY FOR PREDICTING DYNAMIC BUILDING RESPONSE TO WIND

J. F. Y. SIDAROUS and M. D. VANDERBILT

Colorado State University, Fort Collins, Colorado, USA

ABSTRACT

As part of an ongoing research effort in structural and wind engineering, an analytical methodology for computing the response of buildings to wind loading has been developed. This methodology, consisting of three computer implemented models, provides a new tool for simulating the structural properties of a building, modeling wind loadings for rectangular geometries, and computing the dynamic building response. The building is modeled as a multiple degree of freedom (MDOF) structural system, wind loading is treated as a multiple-random process, and response statistics are obtained as the final outcome. The methodology is briefly described and results of limited verification studies are given.

INTRODUCTION

Advances in structural materials, analysis and design, which have led to increasingly taller, lighter, more flexible, and less damped buildings having their lower natural frequencies closer to the frequency range where wind has most of its energy have reduced the ability of traditional design methods to produce buildings which are both safe and serviceable. While traditional design usually produces structures having adequate strength, their in-service performance may be unsatisfactory. For residential or office buildings, an important serviceability factor is tenant comfort during strong wind storms. This depends on nonstructural factors such as whistling noises and the structurally controlled factors of motion, rotation, acceleration and jerk.

To account for dynamic building response, Davenport introduced the gust factor concept and showed the use of spectral analysis in obtaining response spectra (4). He modeled a building as a single DOF (SDOF) system assuming a straight line first vibration mode shape. Vellozzi (13) and Vickery (14) helped refine the gust loading factor concept which has been adopted by many codes. However, it is based on only the along-wind direction with no consideration given to the across-wind and torsional loadings and responses. Simiu (12) reported differences of up to 200% in predicting the along-wind response using different techniques (4, 13). Jayachandran (6) showed that higher vibration modes can contribute significantly to the higher order response factors of acceleration and jerk.

The across-wind response of cylinder-like buildings due to vortex shedding and sometimes galloping, can be more dominant than the along-wind response (8, 10).

Excessive levels of acceleration at top stories due to across-wind vibration can occur. No practical, general, and accurate procedure for estimating across-wind loading and response with the same degree of sophistication of present along-wind procedures presently exists, although considerable related research has recently been reported (8, 10). Saunders (10) concluded that the across-wind response for typical rectangular buildings is primarily due to vortex-shedding, while galloping and negative aerodynamic damping do not have significant effect. Kareem (8) illustrated the independence of the along and across-wind forces and that the response follows a Gaussian distribution.

Most studies have ignored torsional response and assumed the along and across-wind DOF to be uncoupled. However, due to structural asymmetry, center of gravity (c.g.) offsets, and the random nature of wind, a fluctuating torsional moment exists and its effect requires investigation. Occupants are more sensitive to visual effects resulting from rotational than translational response (5), and damage to cladding during strong storms is mostly due to rotation. Patrickson (9), using a single mass system, studied the coupled lateral and torsional response of tall buildings to wind loading and concluded that lateral and torsional displacements are of the same order for realistic values of c.g. offsets.

For this research it is assumed that a proper analysis for the dynamic behavior of buildings under wind loading requires:

1. A good structural model which can faithfully represent static and dynamic building behavior. This model should incorporate a catalog of finite elements, require a minimum of input data, and permit simple reduction of the number of active masses and load levels.

2. An accurate and practical model for simulating wind forces in the frequency domain. The model must include along-wind and across-wind forces, and the torsional moment at selected floor levels, and adequately represent their auto, inter-level, and inter-component correlation structure.

3. An adequate random vibration model for predicting the covariances and spectra of the displacement response and their derivatives for an MDOF structural system under multiple random load input.

A research study to achieve these three goals is outlined in Fig. 1 and 2 (11). The methodology is outlined in Fig. 1 while Fig. 2 shows the input/output and interactions among the computer programs developed for the three models. Given next is a brief description of the methodology.

STRUCTURAL MODEL (WFRAME)

A state-of-the-art sturctural model for performing linearly-elastic, small-deflection structural analysis has been developed and tested. The model embodied in computer program WFRAME provides the core and back storage flexibility found in general purpose programs, while eliminating the limitations and utilizing the data and time saving advantages of special purpose programs. A detailed description of the innovative aspects of this program is the subject of another paper. Space permits listing only the following major features:

1. Arbitrary floor and framing geometry may be considered.

2. Simplified data input is utilized. Similar floors are defined only once and only changes between floors are input.

OBJECTIVE: To develop a methodology for performing dynamic analysis of multidegree of freedom buildings under random wind loads.

Develop state-of-the-art computer program for the structural analysis of buildings having complex geometry modeled as multi-component multilevel systems. Small deflection linear elastic behavior is assumed. Floors are assumed rigid in their own planes.

Computer package WFRAME to perform static analysis and produce the stiffness matrix corresponding to rigid floor DOF at the origin. It can also calculate the modal frequencies and shapes.

Programs LTRM and GERM to reduce the number of DOF used in dynamic analysis and compute the corresponding stiffness matrix, mass matrix and modal frequencies and shapes.

Develop state-of-the-art description of wind forces in the frequency domain. This includes the spectra and cross-spectra of along-wind forces, across-wind forces and torsional moments at the levels of the equivalent masses. Building geometry is rectangular.

Program FREQ to establish the frequency points and intervals based on the natural frequencies of the structure and vortex-shedding frequency.

Program WFSDFM to compute the cross-spectra matrices of wind forces.

Spectral Analysis (random vibration theory) model to perform the required dynamic analysis of building's response in the frequency domain

Input = Wind force spectra + Structural modal frequencies, shapes and damping. Wind forces are assumed to be normally distributed stationary processes.

Output = Structural response covariance matrices (response spectra are optional). The response is normally distributed.

Program DISCOV to compute the spectra and covariance matrices of generalized and geometric coordinate response (displacement, velocity, acceleration and jerk). These matrices can also be calculated for selected points on the equivalent mass levels. Modal analysis is utilized to obtain the uncoupled generalized coordinates.

The final produce is an analytical methodology for computing the response statistics of a building under dynamic wind loading. This methodology has wide application and uses as:

1. A design and analysis tool. It provides better accuracy, speed and economy than current methods and can be used for different types of buildings, exposure and wind conditions. It eliminates the similarity requirements problem involved in wind tunnel investigation and the results can be easily interpreted and used.

2. A research tool to investigate the influence of any individual or combination of wind-related or building-related properties on the dynamic response statistics. Such properties may include wind speed, turbulence intensity, cross-correlation structure, type of structural frame, building height and aspect ratio. The compatability between the different computer packages and the built-in flexibility of the methodology make them an appropriate tool for carrying out several parametric studies in further research.

Fig. 1 Research summary

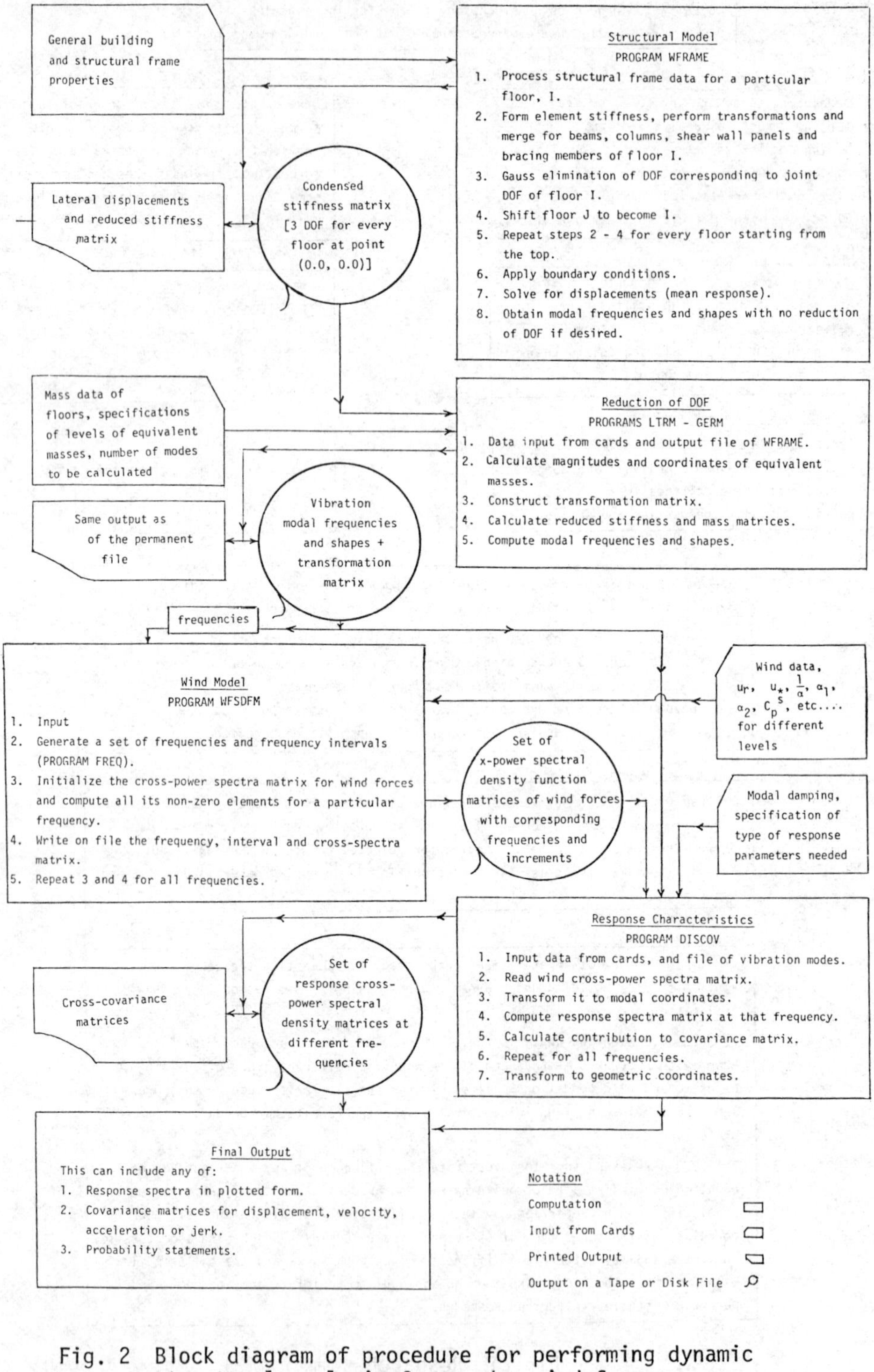

Fig. 2 Block diagram of procedure for performing dynamic structural analysis for random wind forces

3. Floors are treated as rigid in plane, hence the independent DOF for each floor are the three DOF of their c.g. plus three for each joint.

4. The finite element catalog includes beams, columns, diagonal bracing members, shear panels, foundation springs, and tuning masses.

5. Effects of finite sized rigid joints are considered.

6. The user can specify the number and locations of reduced mass floor levels. The program then performs an automatic matrix condensation of mass and stiffness matrices.

7. Using the reduced mass option, virtually any size building can be analyzed. The program automatically adjusts its mode of execution to accomodate the problem size and available computer core.

8. Member self weight can be automatically included and member static loadings can be arbitarily arranged.

9. Temperature effects can be included in the static analysis.

10. Tuning mass, spring, and dash-pot systems can be placed at any level with any orientation in plane.

11. The program performs static analyses for any number of loading vectors and computes the eigenvalues and eigenvectors corresponding to the reduced mass levels.

WIND LOADING MODEL (WFSDFM)

General

The wind loading model consists basically of a selection of the best profiles and spectra available in either analytical or experimental forms for along and across-wind forces and the development of program WFSDFM (Wind Force Spectral Density Function Matrices) to compute the cross-power-spectra of wind forces. The statistical definition of the wind utilized is briefly described followed by a summary of the computations performed by WFSDFM.

Basic Statistical Definitions

For purposes of structural dynamic analysis, wind can be considered as a normally distributed stationary stochastic process (SSP). Let U(t) be a wind related time dependent variable assumed equal to a random fluctuating component u(t) superimposed over the mean value $\overline{U}$, thus

$$U(t) = \overline{U} + u(t), \quad \overline{U(t)} = \overline{U}, \quad \overline{u(t)} = 0 \tag{1}$$

where the overbar indicates a time average.

The autocorrelation function $R_u(\tau)$ quantifies the strength of the linear relationship between pairs of values of u with time separation τ and is given by

$$R_u(\tau) = \overline{u(t)\, u(t+\tau)} = \lim_{T\to\infty} \frac{1}{T} \int_0^T u(t)\, u(t+\tau)dt \tag{2}$$

If v(t) is the fluctuating component of another wind related variable V(t), then the cross-correlation function $R_{uv}(\tau)$ is

$$R_{uv}(\tau) = \overline{u(t)\ v(t+\tau)} = \lim_{T\to\infty} \frac{1}{T}\int_o^T u(t)\ v(t+\tau)dt \tag{3}$$

The Wiener-Khintchine theory shows that for an SSP the correlation function and power spectral density function (PSDF) are a Fourier transform pain (2). Thus the PSDF $S_u(n)$, and the cross PSDF (XPSDF) $S_{uv}(n)$ at frequency n are

$$S_u(n) = 2\int_{-\infty}^{\infty} R_u(\tau)\exp(-i2\pi n\tau)d\tau = 4\int_o^{\infty} R_u(\tau)\exp(-i2\pi n\tau)d\tau,\ n\geq 0 \tag{4}$$

$$S_{uv}(n) = 2\int_{-\infty}^{\infty} R_{uv}(\tau)\exp(-i2\pi n\tau)d\tau,\ n\geq 0 \tag{5}$$

The variance σ_u^2 and covariance C_{uv}^2 are given by

$$\sigma_u^2 = \overline{u^2(t)} = R_u(0) = \int_o^{\infty} S_u(n)dn \tag{6a}$$

$$C_{uv}^2 = \overline{u(t)v(t)} = R_{uv}(0) = \int_o^{\infty} Re(S_{uv}(n))dn \tag{6b}$$

If two multi-component processes $\underline{P1}$ and $\underline{P2}$ are linearly related by a constant transformation matrix $\underline{T}$, then their XPSDF matrices (XPSDFM) $\underline{S}_{P1}(n)$, $\underline{S}_{P2}(n)$ and covariance matrices $\underline{C}_{P1}^2$ and $\underline{C}_{P2}^2$ are related as follows

$$\underline{P1} = \begin{bmatrix} u \\ v \end{bmatrix},\ \underline{S}_{P1}(n) = \begin{bmatrix} S_u(n) & S_{uv}(n) \\ S_{vu}(n) & S_v(n) \end{bmatrix},\ \underline{C}_{P1}^2 = \begin{bmatrix} \sigma_u^2 & C_{uv}^2 \\ C_{vu}^2 & \sigma_v^2 \end{bmatrix}$$
$$\underline{P2} = \begin{bmatrix} \xi \\ \eta \end{bmatrix},\ \underline{S}_{P2}(n) = \begin{bmatrix} S_\xi(n) & S_{\xi\eta}(n) \\ S_{\eta\xi}(n) & S_\eta(n) \end{bmatrix},\ \underline{C}_{P2}^2 = \begin{bmatrix} \sigma_\xi^2 & C_{\xi\eta}^2 \\ C_{\eta\xi}^2 & \sigma_\eta^2 \end{bmatrix} \tag{7}$$

$$\underline{P2} = \underline{T} * \underline{P1} \tag{8}$$

$$\underline{S}_{P2}(n) = \underline{T} * \underline{S}_{P1}(n) * [\underline{T}]^T,\ \underline{C}_{P2}^2 = \underline{T} * \underline{C}_{P1}^2 * [\underline{T}]^T \tag{9}$$

where an underbar indicates a matrix.

For a linear system (structure), if the input (excitation) is a normally distributed SSP, then the output (response) is also a normal SSP (2).

Wind Velocity Profiles and Spectra

The power law and logarithmic profiles (3) are both available for use by WFSDFM. The spectrum for the along-wind velocity u proposed by Kaimal (7), which covers the entire range of frequencies fairly accurately and accounts for its elevation dependence, has been adopted. Thus

$$\frac{n\ S_u(Z,n)}{U_*^2} = \frac{200f(Z,n)}{(1+50f(Z,n))^{5/3}} \tag{10}$$

Similarly, the spectra of the across-wind fluctuating velocity v is (7)

$$\frac{n\,S_v(Z,n)}{U_*^2} = \frac{15\,f(Z,n)}{(1+9.5f(Z,n))^{5/3}} \tag{11}$$

$$\text{where } f(Z,n) = \text{Monin Similarity Coordinate} = n\,\frac{(Z-Z_d)}{\overline{U}(Z)} \tag{12}$$

The XPSDFs S_{uu}, S_{vv}, S_{uv}, correlate the different velocity components at different locations. For two points i, j, in space, Vickery (14) suggested the following form for S_{uu}

$$S_{uu}(i,j,n) = S_u^{1/2}(Z_i,n) * S_u^{1/2}(Z_j,n) * R_u(\Delta Y,\Delta Z,n) * N_u(\Delta X,n) \tag{13}$$

where R_u is an across-wind correlation coefficient function proposed by Davenport as

$$R_u(\Delta Y,\Delta Z,n) = \exp\left(-n\frac{(C_y^2\Delta Y^2 + C_z^2\Delta Z^2)^{1/2}}{\overline{U}_a}\right) \tag{14}$$

and N_u is the along-wind function for correlation between windward and leeward faces of a building given by (13)

$$N_u(\Delta X,n) = \frac{1}{\eta n} - \frac{1}{2\eta^2 n^2}\,(1 - \exp(-2\eta n)) \tag{15}$$

In Eq. 13-15, $\Delta Y = |Y_i - Y_j|$ = horizontal across-wind separation, $\Delta Z = |Z_i - Z_j|$, ΔX = Min(H,B,W of a building, Fig. 3,4), $\overline{U}_a = (\overline{U}(Z_i) + \overline{U}(Z_j))/2$, $\eta = 15.4\Delta X/\overline{U}_a$. The exponential decay factors C_y and C_z, usually taken as $C_y = 16$, $C_z = 10$, can differ (12) and are input to WFSDFM for each wind loading level. An expression similar to Eq. 15 is available for $S_{vv}(i,j,n)$ but with the use of reduced values for C_y and C_z. The form for $S_{uv}(i,j,n)$ assumed (11) for use in conjunction with forming the XPSDFM of wind forces between different levels i, j is:

$$S_{uv}(i,j,n) = \delta_{ij}\,S_u^{1/2}(Z_i,n)S_v^{1/2}(Z_i,n)\exp\left(-n\frac{C_y\Delta}{\overline{U}(Z_i)}\right) \tag{16}$$

where δ_{ij} is the Kronecker delta, and Δ = Max(B,W of the building).

<u>Wind Force XPSDFM</u>

The scope is limited to rectangular buildings with zero angle of wind incidence. To illustrate the procedure, consider a rectangular building modeled by two reduced masses with wind loading applied at the reduced mass levels, Fig. 3. The displacement and action vectors $\underline{DW}$ and $\underline{AW}$ are

$$\underline{DW} = [D_1^x\ D_1^y\ \theta_1\quad D_2^x\ D_2^y\ \theta_2]$$
$$\underline{AW} = [F_1^x\ F_1^y\ M_1\quad F_2^x\ F_2^y\ M_2] \tag{17}$$

For the corresponding wind loading XPSDFM $\underline{SF}_w(n)$, the typical element is $SF_{ij}^{IJ}(n)$ where i, j = level, I, J = direction (x,y for along, across-wind forces, θ for torsional moment). Only one sub- or superscript is used where both are identical. Thus

$$\underline{SF}_w(n) = \begin{bmatrix} SF_1^x(n) & SF_1^{xy}(n) & SF_1^{x\theta}(n) & SF_{12}^x(n) & SF_{12}^{xy}(n) & SF_{12}^{x\theta}(n) \\ & SF_1^y(n) & SF_1^{y\theta}(n) & SF_{12}^{yx}(n) & SF_{12}^y(n) & SF_{12}^{y\theta}(n) \\ & & SF_1^\theta(n) & SF_{12}^{\theta x}(n) & SF_{12}^{\theta y}(n) & SF_{12}^\theta(n) \\ & & & SF_2^x(n) & SF_2^{xy}(n) & SF_2^{x\theta}(n) \\ \text{Symmetric} & & & & SF_2^y(n) & SF_2^{y\theta}(n) \\ & & & & & SF_2^\theta(n) \end{bmatrix} \quad (18)$$

It is assumed that the XPSDF of inter-level actions of different directions are zero, i.e.

$$SF_{12}^{xy} = SF_{12}^{x\theta} = SF_{12}^{yx} = SF_{12}^{y\theta} = SF_{12}^{\theta x} = SF_{12}^{\theta y} = 0$$

Along-wind forces. A practical and flexible approach for obtaining the mean value, the PSDF, and XPSDF of along-wind forces is based on selecting four pressure points per level to represent the loading distribution (Fig. 4). The pressure coefficient CP_i^I of the Ith pressure point on the ith level can be individually defined, which gives the model a potential for use with small angles of incidence and for simulating the effects of surrounding obstructions. The along-wind force vector of the example of Fig. 3, and its XPSDFM are

$$\underline{F}_w^x = \begin{bmatrix} F_1^x \\ F_2^x \end{bmatrix} = \underline{T} * \begin{bmatrix} P_1^1 \\ P_1^2 \\ \vdots \\ P_2^4 \end{bmatrix}, \quad \underline{SF}_w^x(n) = \begin{bmatrix} SF_1^x(n) & SF_{12}^x(n) \\ & SF_2^x(n) \end{bmatrix} = \underline{T} * \underline{SP}(n) * [\underline{T}]^T \quad (19)$$

where

$$\underline{T} = \frac{B}{2} \begin{bmatrix} h_1 & h_1 & h_1 & h_1 & 0 & 0 & 0 & 0 \\ 0 & 0 & 0 & 0 & h_2 & h_2 & h_2 & h_2 \end{bmatrix} \quad (20)$$

$\underline{SP}(n)$ is the XPSDFM of all the pressure points, and h_1, h_2 are the tributary heights of levels 1 and 2.

The pressure P_i^I at pressure point I is

$$P_i^I = 1/2\ \rho * CP_i^I * U^2(Z_i) \quad (21)$$

Linearizing this equation leads to the following equations for the fluctuating components of pressure and along-wind force

$$p_i^I = \rho * CP_i^I * \bar{U}(Z_i) * u \quad (22)$$

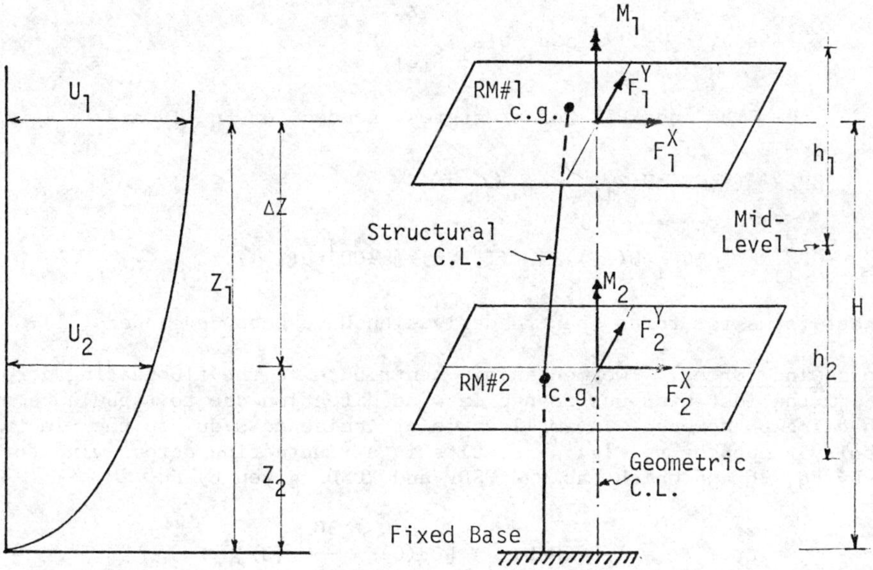

Fig. 3 Example of two reduced mass (RM) system

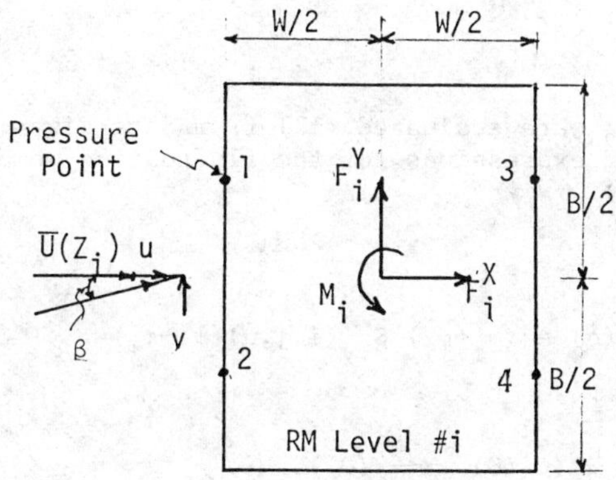

Fig. 4 Typical section, pressure points, and wind forces

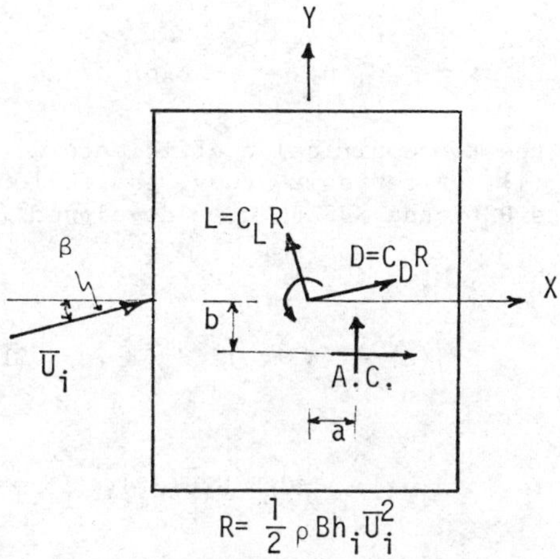

Fig. 5 Drag and lift forces, and aerodynamic center

$$f_i^x = \eta_i u, \quad \eta_i = 1/2\, \rho B h_i\, \overline{U}(Z_i) \sum_{I=1}^{4} CP_i^I \tag{23}$$

From Eq. 22, the PSDF and XPSDF of pressures, needed for Eq. 19 are

$$SP_i^I = (\rho * CP_i^I * \overline{U}(Z_i))^2 * S_u(Z_i, n) \tag{24}$$

$$SP_{ij}^{IJ} = (\rho * CP_i^I * \overline{U}(Z_i)) * S_{uu}(I,J,n) * (\rho * CP_j^J * \overline{U}(Z_j)) \tag{25}$$

More compact forms for force spectra derivation have been developed (11).

Across-wind forces. Two mechanisms contribute to the fluctuating across-wind force of f_i^y, the instantaneous change in wind direction due to turbulence and the vortex shedding phenomenon. A small angle of incidence β due to the fluctuating wind velocity component v, Fig. 4, results in a fluctuating across-wind force f_i^y as given by Eq. 26 and which has the PSDF and XPSDF given by Eq. 27.

$$f_i^y = \xi_i v, \quad \xi_i = 1/2 \rho B h_i \overline{U}(Z_i) \left[C_D(0) + \frac{\partial C_L}{\partial \beta}(0)\right] \tag{26}$$

$$SF_i^y = (\xi_i)^2 S_v(Z_i, n), \quad SF_{ij}^y = \xi_i S_{vv}(i,j,n) \xi_j \tag{27}$$

where C_D and C_L are the drag and lift coefficients, Fig. 5.

For the vortex-shedding, no adequate analytical model could be reached, therefore experimentally determined spectra obtained by Kareem (8) are used.

Torsional moment. For small values of β, the coordinates (a,b) of the aerodynamic center, Fig. 5, are assumed to vary linearly with β(11), i.e.

$$\frac{a}{B} = \alpha_1 |\beta|, \quad \frac{b}{B} = \alpha_2 \beta \tag{28}$$

where the two empirical coefficients α_1, α_2 were evaluated (11) from data given by Akins (1). After some study, the following expressions for the fluctuating moment and its PSDF and XPSDFM were developed (11).

$$M_i = (\psi_i + \phi_i) v \tag{29}$$

$$SF_i^\theta(n) = (\psi_i + \phi_i)^2 S_v(Z_i, n), \quad SF_{ij}^\theta(n) = (\psi_i + \phi_i) S_{vv}(i,j,n)(\psi_j + \phi_j) \tag{30}$$

where,

$$\psi_i = \alpha_2 \frac{\rho}{2} C_D B^2 h_i \overline{U}(Z_i), \quad \phi_i = \alpha_1 \frac{\rho}{2} \left(C_D(0) + \frac{\partial C_L}{\partial \beta}(0)\right) B^2 h_i \sigma_v$$

Expressions giving the XPSDFs for the three wind components at each level have been developed (11).

Computations Performed by WFSDFM

WFSDFM produces an array of frequency values and increments. The frequency increments are set to small values at the natural frequencies of the building and at Strouhal's frequency, and gradually increased between these special frequencies following an exponential interpolation (11). This reduces the number of loading XPSDFMs without significantly affecting the calculated response accuracy. A set of wind loading XPSDFMs are produced corresponding to all frequencies. These XPSDFMs frequencies and increments are stored for later use by DISCOV, Fig. 2.

THE RANDOM VIBRATION MODEL (DISCOV)

The third component of the methodology is implemented by program DISCOV (DISplacement COVariances). The set of XPSDFM, $\underline{SF}_w$ produced by WFSDFM for wind loading at the geometric center of each wind loading level is first transformed to equivalent $\underline{SF}$ corresponding to the reduced mass locations. This transformation eliminates dynamic coupling by obtaining a diagonal lumped mass matrix in the equations of motion. The displacement and action vectors in both wind and reduced mass coordinates are $\underline{DW}$, $\underline{AW}$, $\underline{D}$, and A related by:

$$\underline{DW} = \underline{T}_w \underline{D},\ \underline{A} = [\underline{T}_w]^T \underline{AW} \tag{31}$$

where $\underline{T}_w$ = a transformation matrix.

To obtain uncoupled DOF, modal coordinates are used. The number of modes to be used can be equal to or less than the total DOF (e.g. 1 to 6 for Fig. 3).

The wind loading vector $\underline{A}_Q$ in modal coordinates is

$$\underline{A}_Q = [\underline{\phi}]^T * \underline{A} = [\underline{\phi}]^T * [\underline{T}_w]^T * \underline{AW} \tag{32}$$

where $\underline{\phi}$ = the modal matrix. Therefore the XPSDFM of wind loading in modal coordinates is

$$\underline{SF}_Q(n) = [\underline{\phi}]^T * [\underline{T}_w]^T * \underline{SF}_w(n) * \underline{T}_w * \underline{\phi} \tag{33}$$

and the modal response XPSDFM, $\underline{SD}_Q(n)$, and the corresponding covariance matrix $\underline{C}_Q^2$ are as follows

$$\underline{SD}_Q(n) = \left[H^*(2\pi n)\right] * \underline{SF}_Q(n) * \left[H(2\pi n)\right] \tag{34}$$

$$\underline{C}_Q^2 = \sum_{i=1}^{NF} (\Delta n_i) * \underline{SD}_Q(n_i) \tag{35}$$

where the ith diagonal element of $\underline{H}$ is the complex frequency response or mechanical admittance function of the ith mode, for which ω_i, ξ_i are the angular frequency (rad/sec) and damping ratio

$$H_i(\omega) = \frac{1}{(\omega^2-\omega_i^2)+2i\xi_i\omega_i\omega} \tag{36}$$

and $H_i^*(\omega)$ is its complex conjugate. In Eq. 35, NF is the total number of frequency points, and Δn_i is the ith frequency increment.

The XPSDFM, $\underline{SD}$ and covariance matrix $\underline{C}^2$ of the response of the reduced masses are

$$\underline{SD}(n) = \underline{\phi} * \underline{SD}_Q(n) * [\underline{\phi}]^T,\ \underline{C}^2 = \underline{\phi} * \underline{C}_Q^2 * [\underline{\phi}]^T \tag{37}$$

If the rth derivative of the displacement response (velocity, acceleration, jerk) is needed, then the XPSDFM $\underline{SD}_Q^r$ of that higher order response in modal coordinates is

$$\underline{SD}_Q^r(n) = (2\pi n)^{2r} * \underline{SD}_Q(n) \tag{38}$$

The same procedure of Eq. 35-37 is followed to obtain the higher order response co-variance and XPSDFM, $\underline{C}_Q^{2r}$, $\underline{C}^{2r}$, $\underline{SD}^r(n)$, in modal or reduced mass coordinates.

Program DISCOV has been verified against an independent closed-form analytical method. Is has also the following capabilities and characteristics:

1. Calculation of the response and its derivatives at selected corner points on specific levels.

2. Damping ratios can be set individually for each mode.

3. Modal shapes and frequencies can be read from a data base created by WFRAME (Fig. 2).

4. Response spectra and derivatives, if desired, are extracted for selected responses.

5. The output of response spectra is nondimensionalized and can be printed and/or plotted for each response and its derivatives either individually or combined.

VERIFICATION EXAMPLE

Due to lack of adequately documented data for structural properties and dynamic response characteristics of full scale tall buildings, for which the developed models could be utilized to their fullest potential, it was necessary to use the results of wind tunnel (WT) response measurements on simple models in the verification studies.

Akins (1) and Kareem (8) have performed a series of WT testing of building models using the CSU industrial WT. Akins' work was used (11) in determining the pressure coefficients and α_1 α_2 factors described above. The documentation of boundary layers was used in preparing the input to WFSDFM. Kareem used a 12.7 x 12.7 x 50.8 cm (5 x 5 x 20 in.) rigid model equipped with pressure taps to obtain the XPSDFM of across-wind forces at five different levels for two boundary layers. These spectra were curved-fitted with straight line segments on a log- log scale (11) for use with WFSDFM. For dynamic response measurements, a similar first mode aeroelastic model was used to obtain time histories of roof displacements. The frequencies and damping ratios in the along-wind and across-wind directions are (8)

$$n_x = n_y = 15 \text{ Hertz}, \quad \xi_x = 0.025, \ \xi_y = 0.05 \tag{39}$$

It was not possible to simulate the aeroelastic model with more than one mass located at the top. Hence, equivalent properties for the structure and loading were developed. The equivalent mass and modal shape matrices are

$$\underline{M} = \begin{bmatrix} .0207 & 0 & 0 \\ 0 & .0207 & 0 \\ 0 & 0 & .0006 \end{bmatrix}, \quad \underline{\phi} = \begin{bmatrix} 6.95 & 0 & 0 \\ 0 & 6.95 & 0 \\ 0 & 0 & 40.86 \end{bmatrix} \tag{40}$$

where units are in slugs and ft. The following expression was developed to estimate a tributary height H_D such that, when multiplied by the loading intensity at the top it produces the equivalent along-wind force:

$$H_D = 0.5H/(1+1/\alpha) \quad (41)$$

where $1/\alpha$ is the power law exponent. The derived expression for the equivalent tributary height H_L for the across-wind direction is (11)

$$H_L = \int_o^H \left(\frac{Z}{H}\right)^{1+2/\alpha} * \frac{C_L(Z)}{C_L(H)} dZ \quad (42)$$

In WFSDFM, H_D was used as the tributary height, and since H_L is not equal to H_D, it was necessary to magnify the PSDF of across-wind force per unit height by χ, where

$$\chi = (H_L/H_D)^2 \quad (43)$$

Two boundary layers BL1 and BL3, simulating wind over flat country and urban terrains, and three values of wind velocity at the model top (20, 30, 40, ft/sec) were used in comparing computed versus experimental results. Properties of the boundary layers and tributary heights are given in (11).

Figure 6 shows a comparison of the nondimensional variances of top displacements in the along and across-wind directions. It shows good agreement between analytical and experimental models although the former tends to yield smaller responses. The discrepancy is due to the difference in modeling both wind and structure for full scale in the analytical model and for WT conditions for the experimental one, and to the increased significance of the negative aerodynamic damping in the experimental model (8). However, the discrepancy reduces to a minimum for urban flow (BL3) particularly for the along-wind direction.

Figure 7 shows the computed nondimensional PSDFs of the responses (displacement, velocity, acceleration and jerk) at the top for both directions for BL3 and 30 fps wind velocity. They compare favorably to similar ones given by Kareem for displacements only. Resonant response is dominant in both directions, but an important peak corresponding to Strouhal's frequency appears in the across-wind spectra.

SUMMARY AND CONCLUSIONS

The methodology presented provides a new tool for inexpensive, quick and reliable calculation of structural properties and prediction of dynamic response levels of rectangular buildings under wind loading. It eliminates the difficulties and uncertainties of dimensional analysis involved with testing small models in the laboratory. It can be used as part of the structural design and analysis process. For example, the response variances can be used by the designer to compute the probabilities of exceeding selected response levels, e.g. acceleration at top level exceeds 0.005g during a storm of known return period. Comparison of these probabilities with acceptance criteria will then provide guidance on needed redesign.

The structrual model WFRAME and random vibration model DISCOV incorporate sophisticated and accurate simulation of building behavior for static and dynamic analyses. The wind model WFSDFM, which considers along and across-wind forces and torsional moments at reduced mass levels and accounts for their various correlations, is shown to be adequate for simulating wind forces in the frequency domain for the limited data available.

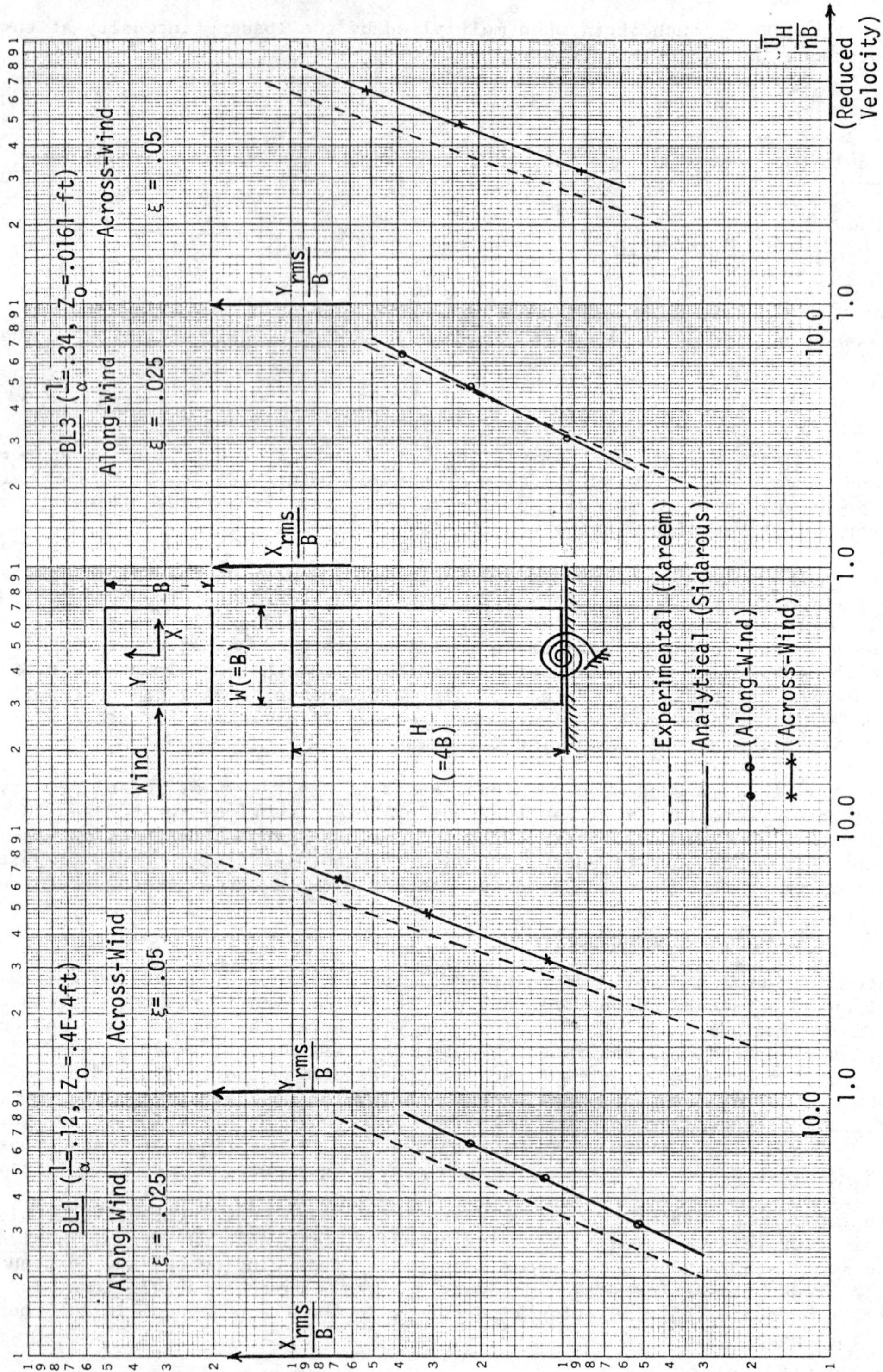

Fig. 6 Comparison of variances of top displacements

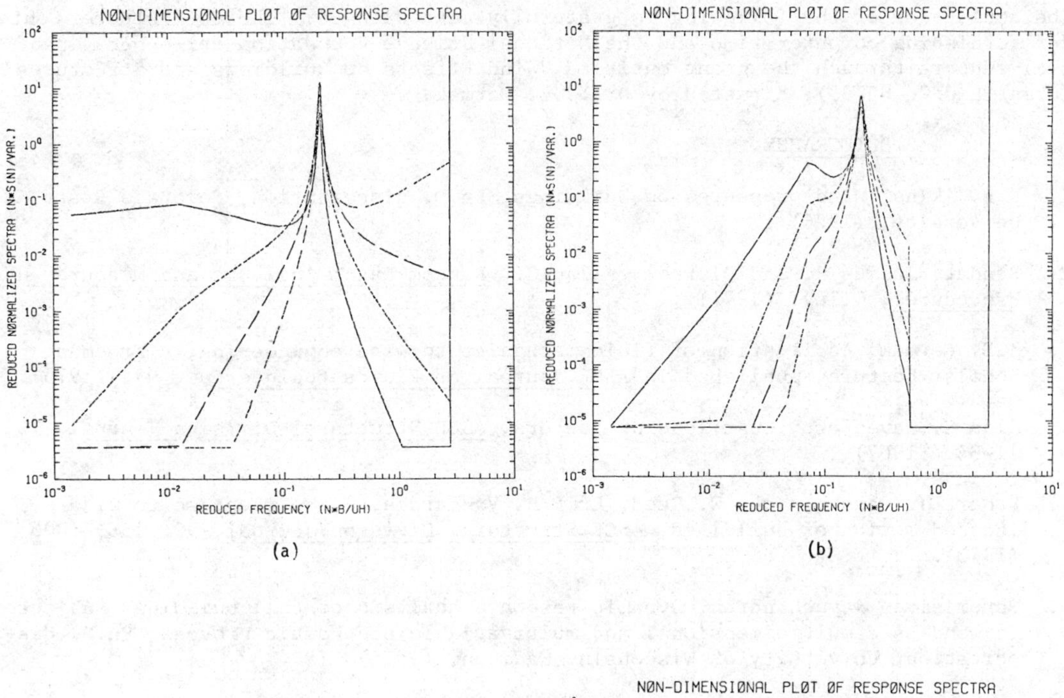

Fig. 7 The response spectra (BL3, reduced velocity = 4.8)

ACKNOWLEDGEMENTS

The assistance of the following is gratefully acknowledged: The CSU Computer Center for furnishing computer time and the National Science Foundation for other financial support through the grant entitled "Wind Effects on Buildings and Structures" (Grant ENG 76-03135), directed by Dr. J.E. Cermak.

BIBLIOGRAPHY

1. R.E. Akins, Wind pressures on buildings, Ph.D. dissertation, Colorado State University, (1976).

2. Bendat, Julius S. and Piersol, Allan G., Random Data: Analysis and Measurement Procedures, Wiley, (1971).

3. J.E. Cermak, Application of fluid mechanics to wind engineering, a Freeman Scholar Lecture, published in ASME Journal of Fluids Engineering, 97, (1975).

4. Alan G. Davenport, Gust loading factors, ASCE Structural Division Journal, 93, 11-34, (1967).

5. Robert J. Hansen, John W. Reed, Erik H. Vanmarcke, Human response to wind-induced motion of buildings, ASCE Structural Division Journal, 99, 1589-1605, (1973).

6. Paramisivan Jayachandran, Dynamic response analysis of tall buildings subjected to wind as a multidimensional and multivariable stochastic process, Ph.D. dissertation, University of Wisconsin, Madison, (1975).

7. J.E. Kaimal et. al. Spectral characteristics of surface-layer turbulence, Quarterly Journal of the Royal Meteorological Society, 98, 563-589, (1972).

8. A. Kareem, Wind excited motion of buildings, Ph.D. dissertation, Colorado State University, (1978).

9. Carlos P. Patrickson and Peredz Friedmann, A study of the coupled lateral and and torsional response of tall buildings to wind loadings, Report for NSF, UCLA, (1976).

10. J.W. Saunders, Wind excitation of tall buildings, Ph.D. dissertation, Monach University, Victoria, Australia, (1974).

11. John F.Y. Sidarous, Dynamic response of buildings under wind loadings, Ph.D. dissertation, Colorado State University, (in progress).

12. Emil Simiu and Daniel W. Lozier, The buffeting of tall buildings by strong winds, National Bureau of Standards, Building Science Series 74, (1975).

13. Joseph Vellozzi and Edward Cohen, Gust response factor, ASCE Structural Division Journal, 94, 1295-1313, (1968).

14. B.J. Vickery, On the reliability of gust loading factors, Civil Engineering Transactions, CE 13, 1-9, (1971).

DYNAMIC RESPONSE OF HIGH RISE BUILDINGS UNDER WIND ACTION WITH INTERFERENCE EFFECTS FROM SURROUNDING BUILDINGS OF SIMILAR SIZE

H. RUSCHEWEYH

Institut für Leichtbau, RWTH Aachen, Federal Republic of Germany

ABSTRACT

In various wind tunnel tests the dynamic response of a downstream high rise building is investigated under interference effects from an upstream building of similar size and different shape. Flow field and pressure measurements on the building walls indicate high increase of turbulence in the wake with a distinct vortex peak at the reduced frequency of 0.1. Auto- and cross-correlation measurements show a phase shift between the fluctuating pressure of the left to the right side of the building of $\Delta\psi = \pi$ and between the front and the rear building wall of $\Delta\psi = \pi/2$ being invariant with building shapes of the investigated models. Systematic variations of the relative distance and position of the two buildings show that the dynamic interference effect is rather stable with increasing distance for bending vibration and less stable for torsional vibration. The investigation showed, too, that the turbulence of the approach flow has only little influence on the intensity of the interference effects. Full scale tests were made on high rise buildings, television towers and chimney stacks to measure the structural damping for different vibration modes.

INTRODUCTION

The trend in the structure design of high rise buildings follows more and more light weight structure principles, i.e. the buildings become more and more sensitive to wind induced vibrations. The turbulence structure of the approach flow and the flow separation with its vortex generation at the building itself on the one hand and the interference effects from other high rise buildings of similar size on the other hand are the dynamic input forces. With regard to the interference problem only little information is published, but shows that the problem may be important. Melbourne (1) shows that the increase of the dynamic response of a tall building in the wake of another is 50 % to 100 %. The NPL report (2) shows for the Quantas-Center an increase of static wind load of 20 % while the dynamic response was found to be equal to the single tower. Worth to mention are the wind tunnel tests for the World Trade Center Towers in New York where an optimal grouping of the two towers was found. The current investigations of Cermak/Peterka show an influence of the wake behind the windward building in a distance downstream up to 80 times the building height (3)

Systematic investigations have been undertaken by the author on a simple elastic building model to look into the dynamic interference mechanisms between two high rise buildings of different shapes, distance and relative position to each other. The main results will be presented in this paper.

WIND SIMULATION TECHNIQUE IN AN AERONAUTICAL WIND TUNNEL

The aeronautical return flow wind tunnel used for these investigations was modified to a boundary layer wind tunnel by using the Counihan simulation technique. A characteristic feature of this tunnel is its "throat-lip" as shown in Fig. 1: The ground plate is lengthened through the throat of the tunnel into the stilling chamber. The vortex generator elements and the barrier wall was set into the throat. With this device the whole length of the free stream working section of 3.2 m (10.5 ft) was made available to the development of the shear stream and the model arrangements. In Fig. 1b one can see the vertical velocity distribution to be ideally smooth for uniform flow with no overshoot due to the displacement of the flow and the shear flow with a power law exponent $\alpha = 0.2$. The turbulence spectrum is similar to that of Harris (4).

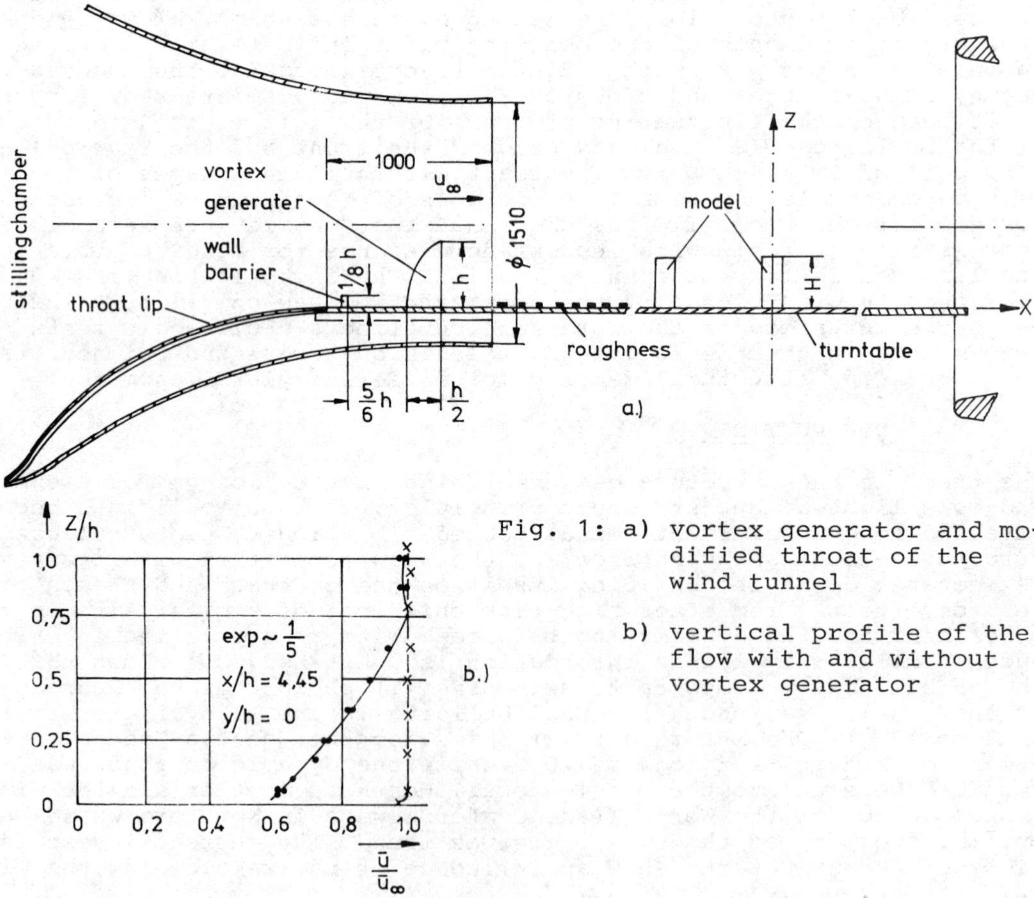

Fig. 1: a) vortex generator and modified throat of the wind tunnel

b) vertical profile of the flow with and without vortex generator

DYNAMIC BUILDING MODELS

Three different building types were modeled for the tests: A rectangular building (type A), a step-house (type B) and a building with Y-shape (type C), s. Fig. 2.

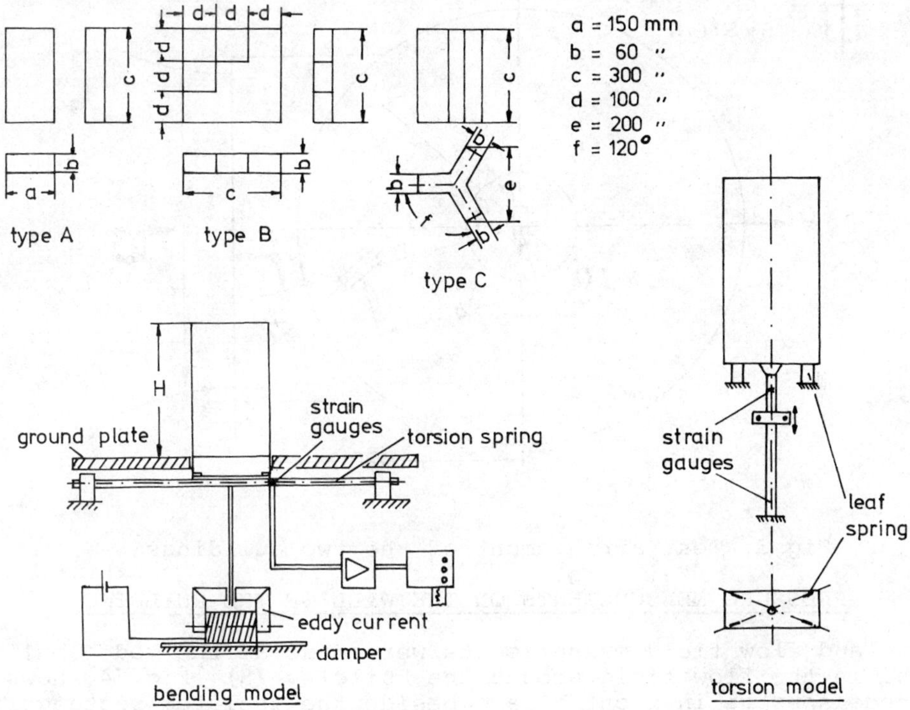

Fig. 2: Types of building models and test arrangement

The extremly light balsa wood models were fitted to a horizontal torsion spring for the bending model which allowed vibrations in a linear bending mode in one direction. With an eddy-current damper at the underfloor of the ground plate the model damping could be modified. The static and dynamic response was measured by a strain gauge bridge at the spring. The torsion model had four leaf springs which allowed a torsional vibration but are rather rigid in the bending mode. With a vertical torsion spring of variable length the torsion frequency could be varied.

The elastic model arrangement was installed in the center of the turntable and was fixed on a separate support to avoid dynamic feed back from the turntable. A rigid building of the three types was set in front of the elastic one (Fig. 3). The distance "a", the angles α, β, and φ could be varied.

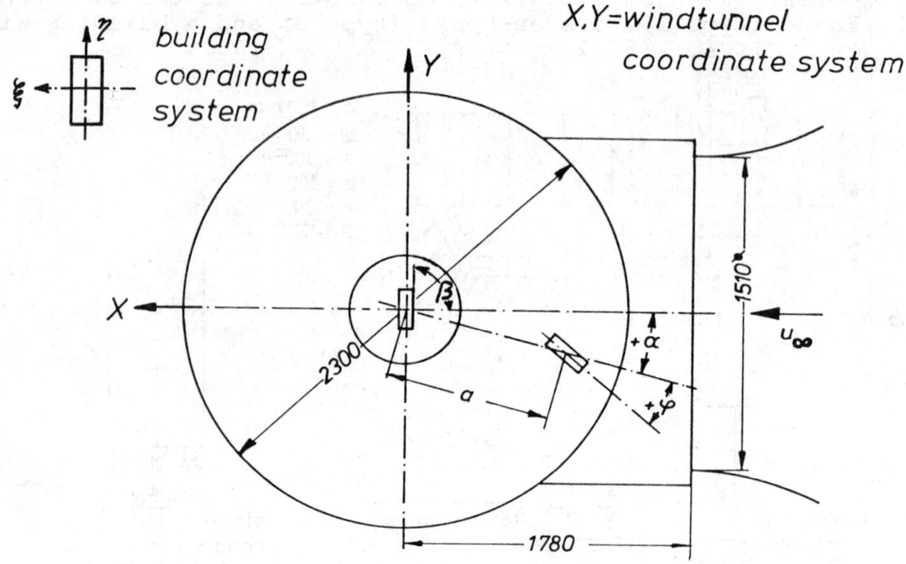

Fig 3: Test arrangement of the two buildings

PRESSURE MEASUREMENTS ON THE WALL OF THE BUILDING

Pressure and flow field measurements were made at the model building wall and in the flow field around the building (5). Fig. 4 shows some turbulence spectra in front of and beside the isolated rectangular building in turbulent flow. For better comparison all curves are normalized with the standard deviation of curve 1 .

Curve 1 represents the flow in a distance of a = 3.83 × H (H = model height) upstream of the building and reflects approximately the turbulence of the approach flow. In the vicinity of the model the turbulence intensity decreases (curve 2 and 3) because of the acceleration effect. At the side of the building the flow separation becomes apparent in the spectrum (curve 4 and 5) with a distinct peak at the reduced frequency of $B \cdot f/U_\infty = 0.093$ (separation at the vertical corner) and less distinct peak at $B \cdot f/U_\infty = 0.6$ (top vortex). The turbulence intensity increases drastically. These vortices are encountered by a downstream building. Measurements of the instantaneous pressure on the windward side show a significant difference between an isolated building in undisturbed turbulent flow and a building in the wake of an upstream building of similar size (Fig. 5 and 6).

The spectra in the first case (Fig. 5) show no distinct peak but represent in general the turbulence spectrum of the approach flow.

In contrast to this, the case with an upstream building shows a very high peak at the reduced frequency of $B \cdot f/U_\infty = 0.09$ especially for the measuring points P 1 and P 5 (Fig. 6). The standard deviation σ increases to 2 ÷ 2,5 times the corresponding value of the isolated building.

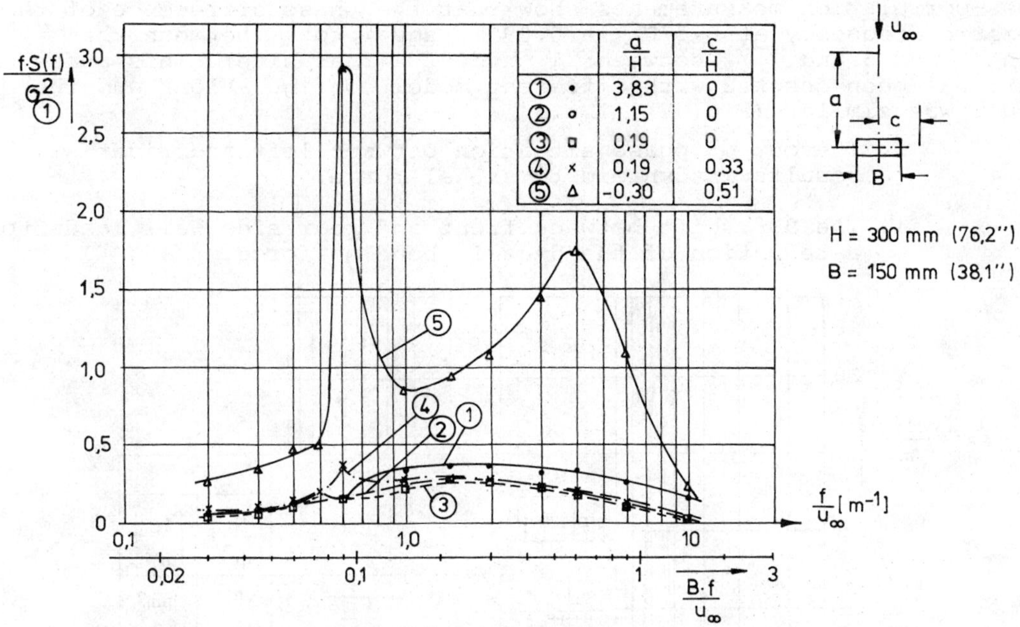

Fig.4: Turbulence spectra of the flow in front and beside the model building, measured at height H/2. All curves are normalized by the rms-value of curve ①

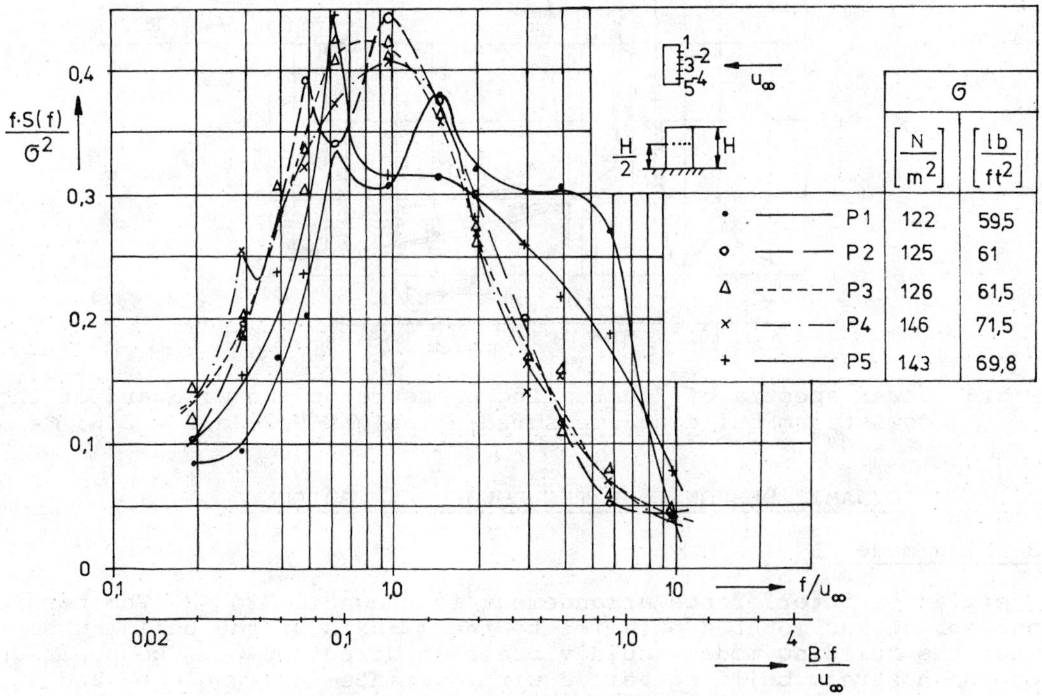

		σ [N/m²]	σ [lb/ft²]
•	P1	122	59,5
∘	P2	125	61
Δ	P3	126	61,5
×	P4	146	71,5
+	P5	143	69,8

Fig.5: Power spectra of fluctuating pressure on the windward wall of an isolated building, measured at height H/2

Cross-correlation measurements show that the phase difference of the pressure frequency at $B \cdot f/U_{\infty} = 0.09$ is equal to π between the point 1 and 5 and $\pi/2$ between 1 and 2 (rear side). This phase shift has been checked with different model configurations and the result was similar (Fig. 7), i.e.:

i. The out of phase situation between left and right side results in dynamic torsional force.

ii. The 90° shift between front and rear side wall leads to a reduction of the dynamic bending force.

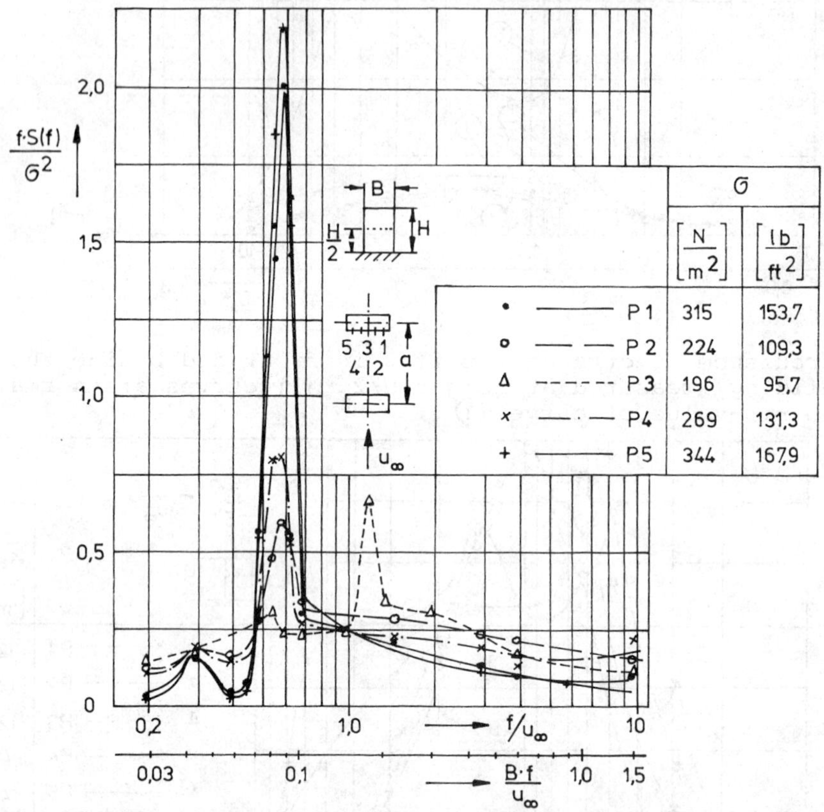

Fig.6: Power spectra of fluctuating pressure on the windward of the downstream building, measured at height H/2; a/H = 2.6; $\varphi = 90°$ $\alpha = 0°$, $\beta = 90°$

DYNAMIC RESPONSE OF THE DOWNSTREAM BUILDING

Bending mode

The plan of interference arrangement is shown in Fig. 3. The torsional spring was mounted parallel to the η-axis of the building so that the building model could vibrate in direction ξ. The position of the upstream building varied with distance "a", angle α and φ.

For $\beta = 90°$ the vibration is in wind direction. Fig 8 shows the static and dynamic reaction versus angle α, for different angles

φ and for the distance ratios a/H = 3.68*). One can see that the dynamic effects increase in the shelter effect range of the upstream building.

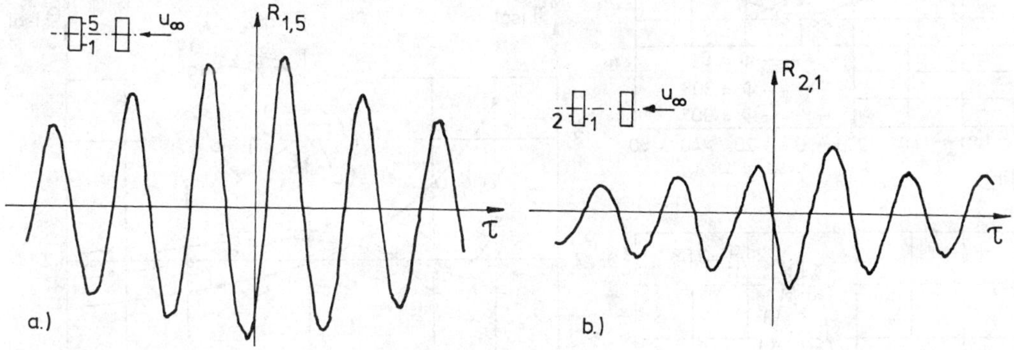

Fig. 7: Cross-correlation function of points 1 ÷ 5 and 1 ÷ 2 on the downstream building wall, measured at height H/2; a/H = 2.6; φ = β = 90°; α = 0°

Fig. 8b shows the effect versus distance ratio a/H. It is evident that the increase of dynamic response from interference of approximately 120 % is rather independent of the relative position of the upstream building (see parameter φ)and the dynamic interference effect decreases only little with increasing distance ratios.

Torsion mode

From the results of the pressure measurements it can be concluded that the torsional vibration of the downstream building may increase due to the wake effects. The dynamic tests with the torsion model (Fig. 2) confirm this presumption. Fig 8b shows the dynamic torsion angle ratio*)versus distance ratio a/H. At small distance ratios the enhancement of the dynamic response grows up to the factor of 3 and more and decreases continiously with increasing distance ratio a/H. In Fig. 8c the influence of the side positions of the upstream building can be seen: Side position of more than ± 1 1/2 widths are necessary for a distinct reduction of the dynamic interference effect. This means in terms of the fluctuating wind angle of attack (B = 1/2 H) a variation of $\Delta\alpha$ = ± 15 degrees. Fluctuating angles of attack of the normal turbulent wind conditions are of the same order and therefore it can be postulated that the turbulence will not reduce this interference effect distinctively.

Furthermore Fig. 8c shows that the upstream building shape is of less importance for dynamic interference effects.

STRUCTURAL DAMPING - FULL SCALE MEASUREMENTS

To apply the model results of full scale buildings the structural damping of the different vibration modes is of dominant importance. Therefore full scale measurements have been undertaken using the auto-correlation method for analyzing the logarithmic decrement of

*) The ordinate is normalized with the static resp. dynamic reaction of the isolated building in turbulent flow

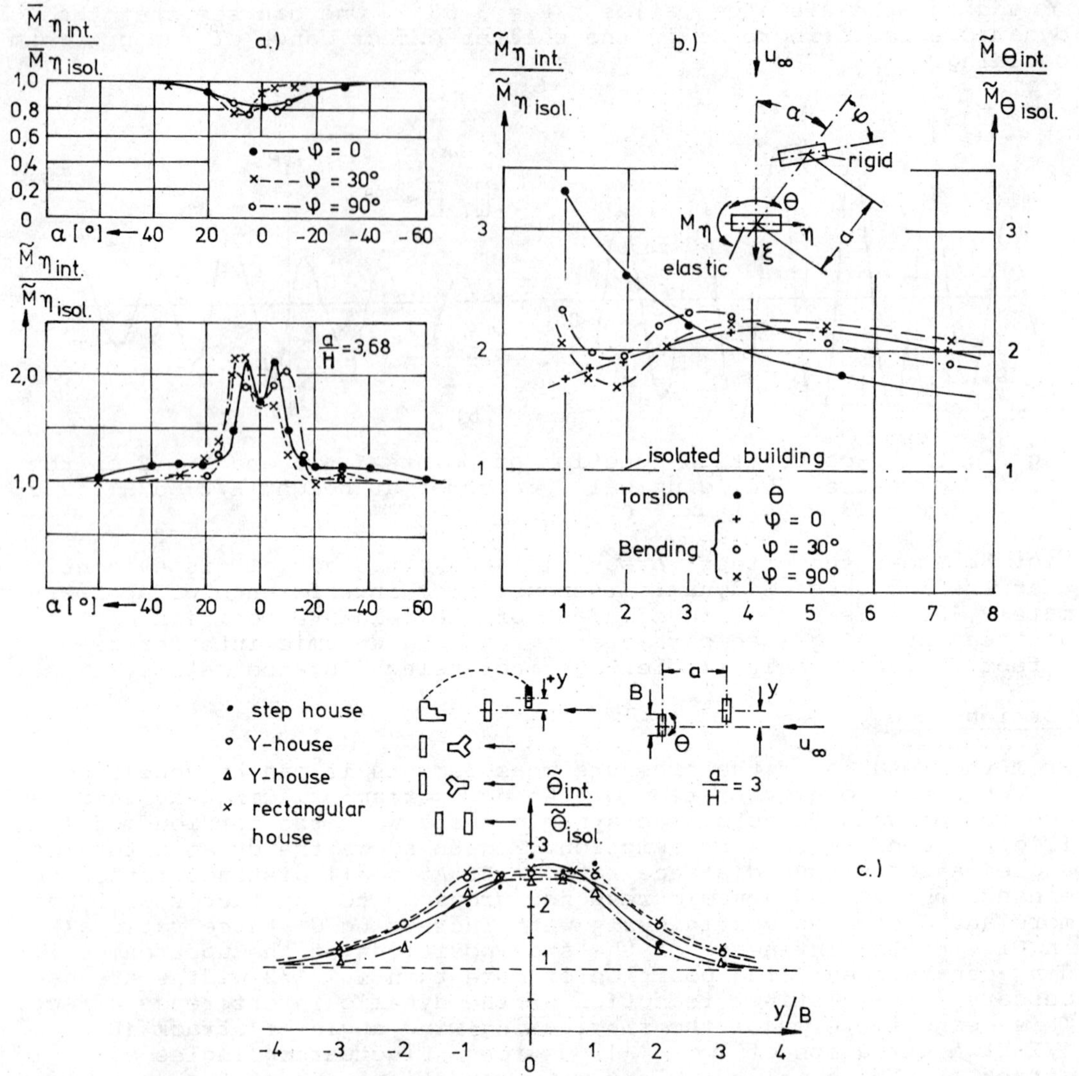

Fig. 8 a), b): Static (Index (-)) and dynamic (Index (~) pitching moment $M_{\eta\, int.}$ of the downstream building with interference effects, normalized by the pitching moment $M_{\eta\, isol.}$ of the isolated building at distance ratio a/H = 3,68 versus side position angle α . c) Dynamic torsional vibrations with interference of an upstream building versus side position ratio y/B

damping. Two buildings and one television tower have been investigated previously by the author in 1975 (6). Four other structures of different shapes are presented here:

<u>Broadcasting station "Deutsche Welle" in Cologne, West Germany</u>:

Fig. 9a shows a sketch of the 141 m (462 ft) high building complex (steel-concrete-combination). The three buildings are linked by concrete beams at floor 24 and 32. The measured frequencies and damping values are summarized in Table I.

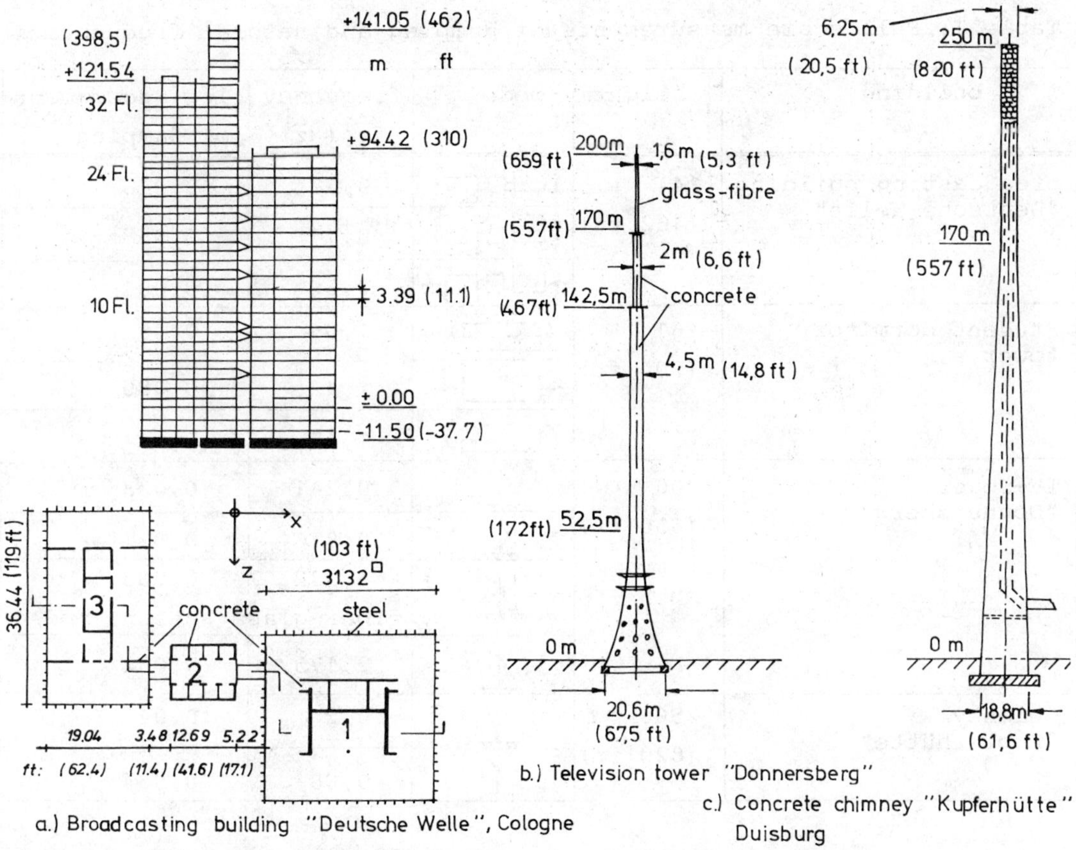

Fig. 9: Full scale structures for damping measurements

Student dormitory tower RWTH Aachen, West Germany: The rectangular concrete building is 61 m (200 ft) high. Bending and twist motion were measured (Table I).

Television tower "Donnersberg", West Germany: Fig. 9 shows the slender concrete television tower "Donnersberg" with a glassfiber reinforced antenna mast. The concrete part reaches up to the altitude of 170 m (557 ft), the antenna top to 200 m (656 ft). Frequencies and damping see Table I.

Concrete chimney "Kupferhütte Duisburg", West Germany: Fig. 9 shows the 250 m (820 ft) high concrete chimney with a long cone at the upper part. This cone gives a relative high stiffness, therefore the fundamental frequency is rather high (Table I).

LITERATURE

(1) Melbourne, W.H., Proc. 4th Intern. Conf. of Wind Effects on Buildings and Structures, London 1975, 343
(2) Whitbread, R.E. et al., NPL Tech. Memo Mar Sci TM 100, Oct. 1972
(3) Zambrano, T.G. et al., Proc. 3rd U.S. Nat. Conf. Wind Engineering Research, Gainesville, Florida 1978
(4) Harris, R.I., Ciria-Publications, London 1971
(5) Ruscheweyh, H., Mitteilungen aus dem Institut für Massivbau der TH Darmstadt, 26, Berlin 1977, 127ff.

Table I: Full scale measurements of damping and natural frequencies

building	height	mode	frequency fe (Hz)	log. decrement of damping
broadcasting building "Deutsche Welle"	141 m (462 ft)	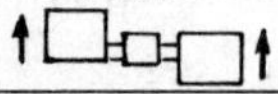	0,52	0,048
		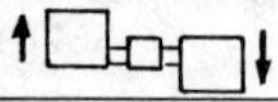	0,74	0,058
		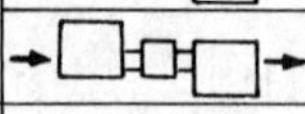	0,50	0,061
Student dormitory tower	61 m (200 ft)	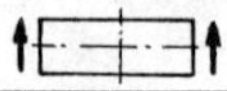	1,06	0,062
		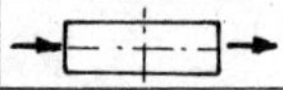	1,42	0,080
		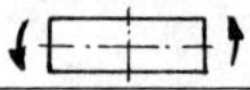	1,45	0,15
TV-Tower "Donnersberg"	200 m (656 ft)		0,241	0,033
			0,976	0,031
			1,532 fiber-glass-antenna mode	0,065
		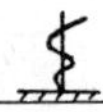	2,175	0,032
Chimney "Kupferhütte"	250 m (820 ft)		0,2	0,04
			0,68	0,034

(6) Ruscheweyh, H. - Hirsch, G., Proc. of the 4th Intern. Conf. on Wind Effects on Buildings and Structures, London 1975, 133ff.

DYNAMIC ALONGWIND RESPONSE OF A STRUCTURAL SYSTEM INCLUDING SOIL FLEXIBILITY

G. SOLARI and D. STURA
Università di Genova, Genova, Italy

ABSTRACT

A general formulation of the problem of the dynamic alongwind response of a structure is given including the dynamic soil-structure interaction phenomenon. The restraint conditions of the physical structure are therefore represented in a more realistic way. A wide parametric study has been carried out in order to investigate the influence of soil flexibility on the structural response, when varying both structural and soil damping.

INTRODUCTION

All the procedures developed for the calculation of the structural response to forces induced by atmospheric turbulence are based on the assumption that the structure is modeled as a linear, homogeneous cantilever beam with no allowed displacements at its base. It should be observed that the hypothesis of a perfectly clamped base, although usual in wind engineering, may be not only unjustified but also detrimental to a correct analysis of the phenomenon.

The total displacement at the top of the structure may be greatly increased by the contribution of small rotations at the base due to the softness of the soil. The natural frequencies of the system are modified when taking into account the soil flexibility, too, and they get closer to the dominant frequencies of the wind gusts. Thus the softer is the soil, the greater is the amplification of the dynamic response. On the other hand, a very soft soil allows for more energy dissipation by the radiation and hysteretic damping mechanisms, leading to a deamplification of the response as partial effect. This fact becomes particularly important for very slender steel structures with welded joints, where the structural damping is very small, in comparison with the damping in the soil.

The soil flexibility will then produce two opposite main effects on the dynamic response: an amplification due to the modification of the natural frequencies of the system and to the rigid motion of the base, and a deamplification due to the damping in the soil. The question is which of these two effects is prevailing?

In this paper the most general problem of the dynamic alongwind response of a soil-foundation-structure system is analyzed. In another paper the Authors (1), starting from this general formulation, derived some simple expressions, particularly suitable for tall and flexible structures, which allow to introduce the concept of "equivalent clamped prismatic structure." In order to evaluate a good approximation of the alongwind response, this technique generalizes the reliability range of the above procedures.

FORMULATION OF THE PROBLEM

Referring to Fig. 1 the symbols are:

H	height of structure
$EJ(z)$, $\gamma(z)$	bending stiffness modulus and linear density of the beam (function of z)
M_j	j-th mass applied to the beam at the z_j $(j=1,2,...s)$ section
D_b	hysteretic damping ratio for the beam
R_o, m_f, I_f	foundation radius, mass and rotatory inertia
G, ρ, ν, c_s, D_s	shear modulus, density, Poisson coefficient, shear wave velocity and hysteretic damping ratio of the soil

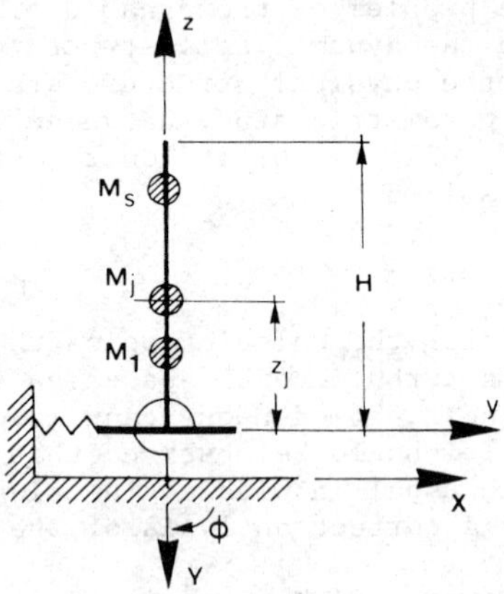

Fig. 1.

The modal superposition method is not rigorously applicable to the study of soil-structure interaction because the foundation stiffness and damping coefficients are frequency-dependent and therefore the soil-foundation-structure system does not possess classical normal modes. But, it has been shown by Jennings and Bielak (2) that modal analysis can still be used; the errors are rather small and relevant simplifications are obtained. Modal superposition has therefore been used in this study also because in this case it allows to utilize the classical formulation of the dynamic alongwind response. Without radical changes in the procedure, soil-structure interaction can also be easily included.

In order to use this technique, natural circular frequencies ω_j and normal modes $\psi_j(z)$ of the soil-foundation-structure system are to be determined. If the structure is considered prismatic, it is possible to obtain an analytical solution for the eigenvalue problem, as in Warburton's analysis (3).

In order to maintain the utmost generality, the Authors (4) studied a non-prismatic beam on which concentrated masses were applied, being a more appropriate model for structures such as arc lamps, antennas for radio and meteorologic devices, and concrete chimneys bearing internal pipes. The eigenvalue problem has been solved through an application of Rayleigh-Ritz's variational direct method.

Dynamic Response to Harmonic Forces

When a horizontal force $Pe^{i\omega t}$ is applied to the structural system of Fig. 1 in a point of ascissa $z=r$, the steady-state response may be expressed in the form

$$y(z,t) = \sum_j \psi_j(z)\, q_j(t) \tag{1}$$

where the Lagrangian coordinates

$$q_j = P(\xi_j + i\eta_j)e^{i\omega t} \tag{2}$$

are the solution for the system

$$(\omega_j^2 - \omega^2)\, q_j + 2i\omega_j \sum_k \omega_k D_{kj} q_k = \frac{P}{m_j}\, e^{i\omega t}\, \psi_j(r) \tag{3}$$

In Eq. (3) m_j is the j-th modal mass defined as

$$m_j = \int_o^H \gamma(z)\, \psi_j^2(z)\, dz + \sum_1^j{}_k M_k \psi_j^2(z_k) + m_f\psi_j^2(o) + I_f\psi_j'^2(o) \tag{4}$$

and

$$D_{kj} = D_b d_{kj} + \frac{1}{2}\, b_{kj} \tag{5}$$

in which

$$d_{kj} = \delta_{kj} - (m_j\omega_j\omega_k)^{-1}\, \{K_{11}\psi_k(o)\psi_j(o) + K_{22}\psi_k'(o)\psi_j'(o) + K_{12}\, [\psi_k(o)\psi_j'(o) + \psi_k'(o)\psi_j(o)]\} \tag{6}$$

$$b_{kj} = (m_j\omega_j\omega_k)^{-1}\, \{C_{11}\psi_k(o)\psi_j(o) + C_{22}\psi_k'(o)\psi_j'(o) + C_{12}\, [\psi_k(o)\psi_j'(o) + \psi_k'(o)\psi_j(o)]\} \tag{7}$$

δ_{kj} being the Kronecker symbol.

In Eqs. (6) and (7) K_{mn} and C_{mn} $(m,n=1,2)$ are respectively the terms of the dynamic stiffness and damping matrix $[K]$ and $[C]$ of the soil-foundation system

$$[K] = \begin{bmatrix} k_{11}K_{xo} & k_{12}K_{xo}R_o \\ k_{12}K_{xo}R_o & k_{22}K_{\Phi o} \end{bmatrix} \tag{8}$$

$$[C] = \begin{bmatrix} (2D_j k_{11} + a_o c_{11})K_{xo} & (2D_j k_{12} + a_o c_{12})K_{xo}R_o \\ (2D_j k_{12} + a_o c_{12})K_{xo}R_o & (2D_j k_{22} + a_o c_{22})K_{\Phi o} \end{bmatrix} \quad (9)$$

$a_o = \omega R_o/c_s$ being the dimensionless frequency parameter, and K_{xo} and $K_{\Phi o}$ being the static stiffnesses. k_{mn} and c_{mn} $(m,n=1,2)$ are the dimensionless dynamic stiffness and damping coefficients respectively (function of a_o and ν) (Veletsos and Wei (5)).

Dynamic Alongwind Response of the System

Considering Eqs. (1) and (2), the frequency influence function $H(z,r;\omega)$ of the soil-foundation-structure system can be expressed in the form

$$H(z,r;\omega) = \sum_j \psi_j(z) \; [\xi_j(r;\omega) + i\eta_j(r;\omega)] \quad (10)$$

Then, knowing the transfer function (10) of the system, it will be possible to correlate the statistical properties of the wind loads, to the statistical properties of the structural response. If $p(r,t)$ is a distributed stationary random load induced by a strong wind and $S_p(M,M';\omega)$ is the cross spectral density of the forces, the complex spectral density function of the stationary response $S_y(z;\omega)$ can be expressed by the frequency-domain relation

$$S_y(z;\omega) = \int_A \int_A S_p(M,M';\omega) \; H(z,r;\omega) \; H^*(z,r';\omega) dA dA' \quad (11)$$

in which $H^*(z,r';\omega)$ is the complex conjugate function of $H(z,r';\omega)$, A is the total exterior area of the structure, and M and M' are the centers of the elemental areas dA and dA' of coordinates r and r'. Assuming

$$S_p(M,M';\omega) = S_p^c(M,M';\omega) + iS_p^q(M,M';\omega) \quad (12)$$

in which the real and imaginary parts in Eq. (12) are the cospectrum and the quadrature spectrum respectively, and substituting Eq. (10) and Eq. (12) in Eq. (11), it follows that

$$S_y(z;\omega) = \sum_j \sum_k \psi_j(z)\psi_k(z) \int_A \int_A \{S_p^c(M,M';\omega) + iS_p^q(M,M';\omega)\}\{[\xi_k(r;\omega)\,\xi_j(r';\omega) + \eta_k(r;\omega)\,\eta_j(r';\omega)] + i[\xi_j(r';\omega)\,\eta_k(r;\omega) - \xi_k(r;\omega)\,\eta_j(r';\omega)]\} dA dA' \quad (13)$$

Considering only the real part of the spectral density function of the response, and assuming that in homogeneous turbulence the quadrature spectrum vanishes, the following expression is obtained

$$S_y(z;\ \omega) = \sum_j \sum_k \psi_j(z)\ \psi_k(z) \int_A \int_A S_p^c(M,M';\ \omega)\ [\xi_k(r;\ \omega)\ \xi_j(r;\ \omega) + \eta_k(r;\ \omega)\ \eta_j(r';\ \omega)]dAdA' \tag{14}$$

In his analysis of clamped structures, Simiu (6) showed that the contribution of the cross-mode product becomes negligible if the damping is small and the resonant peaks are well separated. The Authors (1) confirmed the reliability of this assumption also when the soil-structure interaction is taken into account and observed that the off-diagonal terms of the matrix [D] may be neglected too. Then Eq. (10) may be rewritten in the form

$$H(z,r;\ \omega) = \sum_j \psi_j(z)\ \psi_j(r)\ \frac{H_j(\omega)}{m_j \omega_j^2} \tag{15}$$

in which

$$H_j(\omega) = \frac{1}{(1 - \omega^2/\omega_j^2) + 2iD_{jj}(\omega)} \tag{16}$$

is the frequency response function in the j-th mode.

Defining the cross spectral density of the generalized forces in the j-th mode as

$$I_j(\omega) = \int_A \int_A S_p^c(M,M';\ \omega)\ \psi_j(r)\ \psi_j(r')dAdA' \tag{17}$$

$S_y(z;\ \omega)$ assumes the conclusive expression

$$S_y(z;\ \omega) = \sum_j \frac{\psi_j^2(z)}{m_j^2 \omega_j^4}\ |H_j(\omega)|^2\ I_j(\omega) \tag{18}$$

Expression (18) is more generalized than the formally identical expressions usually found in wind engineering papers, since the effect of soil-structure interaction is also included. In fact ω_j, ψ_j and m_j are frequencies, modes and modal masses of the whole soil-foundation-structure system, and $D_{jj}(\omega)$ includes the above quantities, dynamic stiffness coefficients of the soil-foundation system and foundation mass and rotatory inertia.

The formulation is concluded in the classical way. Calling $\sigma_y(z)$ the standard deviation of the displacement

$$\sigma_y^2(z) = \frac{1}{2\pi} \int_o^\infty S_y(z;\ \omega)d\omega \tag{19}$$

the maximum alongwind displacement may be written in the form

$$Y(z) = \overline{Y}(z) + g_y(z)\ \sigma_y(z) \tag{20}$$

being $\overline{Y}(z)$ the mean alongwind displacement and $g_y(z)$ the peak factor.

PRESENTATION OF THE RESULTS AND COMMENTS

In order to investigate systematically and quantify the various aspects of the problem, a previous numerical program of the Authors on dynamic analysis for alongwind response of tri-dimensional clamped structures has been modified to include soil flexibility.

The program first determines natural frequencies and modes and then solves the dynamic alongwind response problem assuming maximum generalization and with reasonably short computer time. On the IBM 370/168 machine in PISA (CNUCE) the program runs the complete dynamic analysis of free and forced vibration of a slender, line-like structure considering the first three modes of vibration and including soil-structure interaction in about 90 sec. The clamped structure dynamic analysis is performed in only 50 sec computer time, again with the contribution of the three modes. When considering only the first mode, computer time goes down from 90 to 55 sec and from 50 to 10 sec, respectively. Of course, when running the case of a tri-dimensional structure, computer time increases, mainly because the horizontal acrosswind and alongwind cross-correlation functions have to be computerized.

A tapered, unlined, welded steel stack has been chosen as typical slender structure; it is characterized by a height of 100 m, by a bottom and a top diameter of 7 m and 3.80 m respectively and by a hysteretic damping $D_b = 0.2$ percent. The foundation is a circular plate of $R_o = 7$ m radius, laying on the surface of an elastic half-space, being $\nu = 0.33$ the Poisson coefficient. The shear modulus G and the hysteretic damping D_s are the parameters of the analysis and intend to describe the soil. The natural frequencies of the clamped case are respectively $n_{o1} = 0.768$, $n_{o2} = 3.339$ and $n_{o3} = 8.109$. The structure is supposed to be located in a suburban area; the mean velocity profile has been taken according to a surface roughness length $z_o = 0.30$ m and a shear velocity $\overline{V}_* = 2.46$ m/s.

Figure 2 shows the ratio n_i/n_{oi} (i=1,2,3) between the i-th natural frequency including dynamic soil flexibility, and the same frequency in the clamped structure, as a function of G. Figure 3 shows that with increasing the softness of the soil, the second and the third natural frequencies will get closer to the fundamental frequency. But n_3/n_1 and n_2/n_1 have such large values in the case of clamped beam that the natural frequencies are still well separated also for flexible soils.

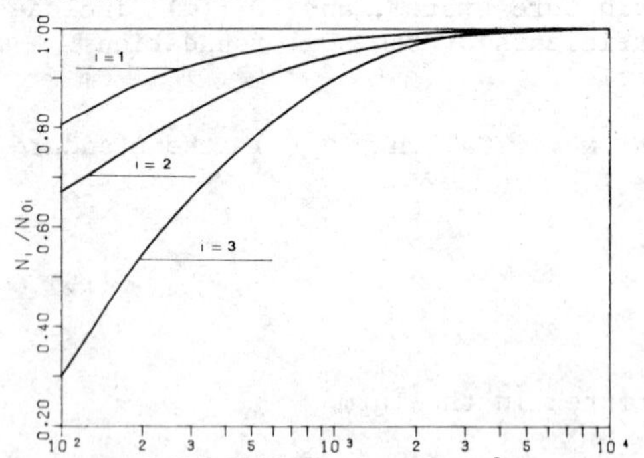

Fig. 2.

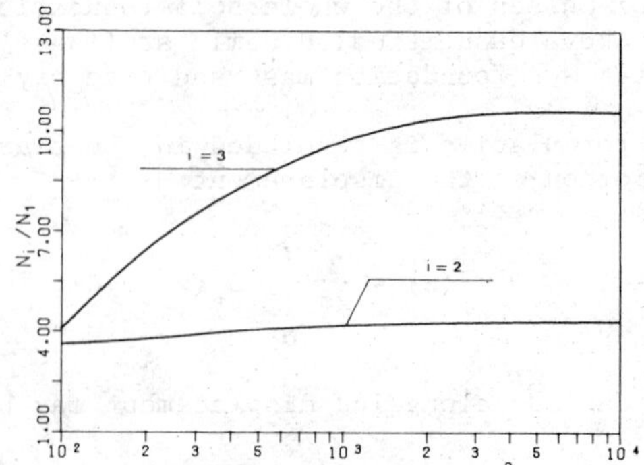

Fig. 3.

Figures 4(a) and 4(b) are reporting the diagrams of the function $|H_1(n)|$ within the range around n_1, for different G, with $D_s = 0$ and $D_s = 5$ percent respectively. The functions $|H_j(n)|$ $(j=2,3)$ have a qualitative behavior like in the figures, so that it can be affirmed that soil softness reduces not only n_3/n_1, n_2/n_1 but also the amplitude of the resonant peaks. The above considerations suggest that it is advisable to neglect higher modes than the first also when the soil-structure interaction is taken into account. A wide numerical analysis carried out by the Authors (1) confirmed the general validity of such procedure, already usually and correctly applied to clamped structures.

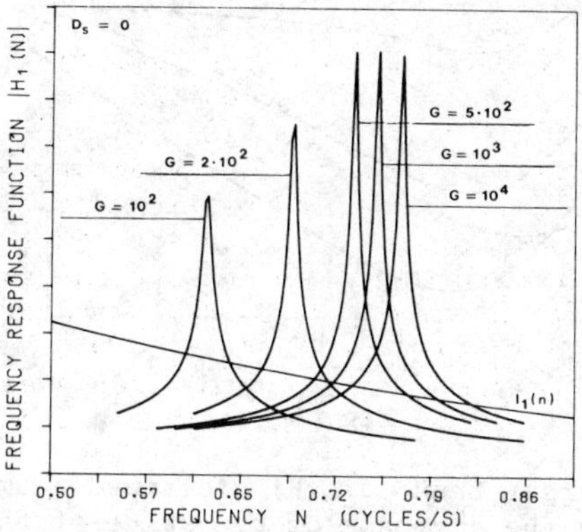

Fig. 4(a).

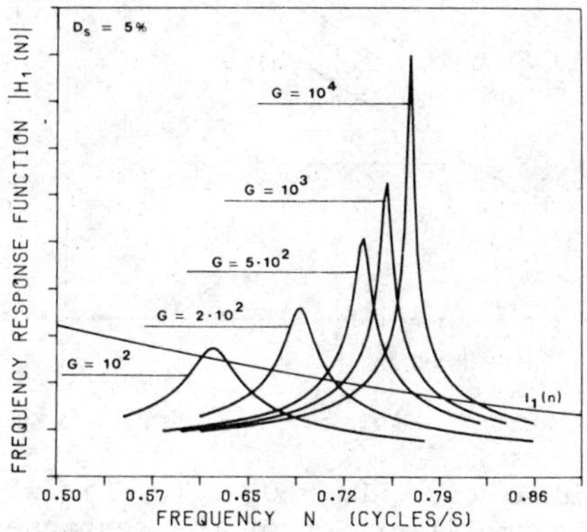

Fig. 4(b).

Therefore, from Eq. (19), when considering only the first mode, the mean square value of the fluctuating alongwind displacement at the top of the structure, can be written in the form

$$\sigma_y^2(H) = \frac{\psi_1^2(H)}{m_1^2(2\pi n_1)^4}(C_1 + D_1) \tag{21}$$

in which

$$C_1 = \int_0^{n_1-\Delta_1} |H_1(n)|^2 I_1(n)dn \qquad D_1 = \int_{n_1-\Delta_1}^{\infty} |H_1(n)|^2 I_1(n)dn \tag{22}$$

being Δ_1/n_1 small (in the numerical analysis the value $\Delta_1 = 0.1\, n_1$ has been chosen, according with Simiu's analysis (6)). The first and the second term of the sum in Eq. (21) are usually referred to as the background and the resonant part of the response, respectively. The latter term is strongly affected by the dynamic properties of the soil-foundation-structure system, while the former is almost independent on them.

Since $I_1(n)$ is a decreasing function of n, and $|H_1(n)|$ has a peak of variable amplitude at variable frequency both increasing with G as shown in Figs. 4(a) and 4(b), it is not possible to predict the law of variation of the product $I_1(n)|H_1(n)|^2$ with soil softness, and then the behavior of D_1. As shown in Fig. 5 the term (C_1+D_1) in Eq. (21) decreases when G decreases if D_s is kept constant and not equal to zero, meaning that energy dissipation by radiation related to a larger softness of the soil is dominant. Moreover, if G decreases, the decrease in n_1^4 is definitely prevailing on the reduction in $\psi_1^2(H)/m_1^2$, so that the coefficient of $(C_1 + D_1)$ increases. Figure 6 shows the behavior of the dynamic response of the structure in terms of standard deviation.

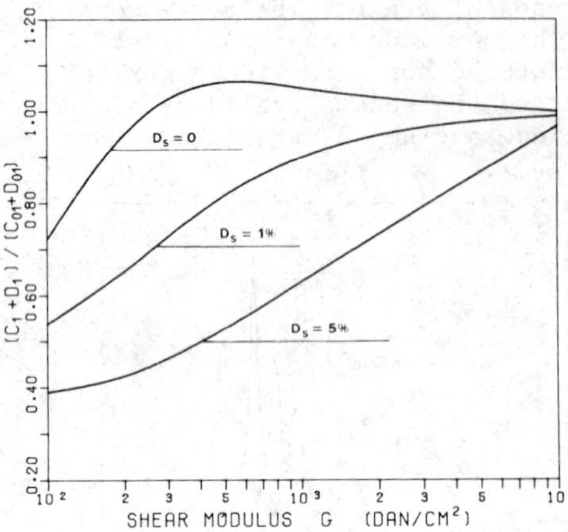

Fig. 5.

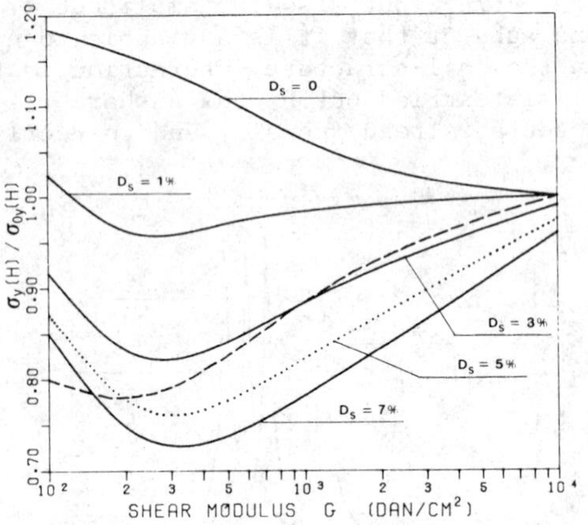

Fig. 6.

In the ideal case $D_s = 0$, $\sigma_y(H)$ is always larger than $\sigma_{oy}(H)$ (clamped beam) since the amplification of the response due to the decrease of the value of the fundamental frequency is prevailing on the attenuation caused by more energy dissipation by radiation. Actually ($D_s \sim 3 \div 7$ percent) the total effect of damping in the soil is always prevailing so that $\sigma_y(H)$ is always less than $\sigma_{oy}(H)$.

All the diagrams have a minimum; this can be explained by the fact that when G is small the attenuation due to damping is partially compensated by the amplification due to a smaller n_1. However, as usually in parametric studies, G, in Fig. 6, is a parameter while the hysteretic damping D_s is constant. Physically it happens that a larger G usually goes together with a larger damping factor. It follows that the physical phenomenon may be better represented by the dashed line, but it cannot be excluded that the dotted line, corresponding to a mean value of D_s = 5 percent may be sometimes appropriate.

On the contrary, the numerical analysis showed that the peak factor g_y is not much sensitive of soil flexibility. It can be taken as constant with good approximation. Figure 7 does not require particular comments. However, it can be noticed that $\bar{Y}(H)$ is the most directly affected quantity by the rigid-type motion of the base.

In conclusion $Y(z)$ is the summation of two terms (Eq. 20)), the first of which only depends on the rigid motion of the base related to the softness of the soil; whereas the second term is a function of all the variables characterizing the dynamic interaction. Which of the two terms is prevailing and how is a question for which a rigorous answer has to be given case by case.

In the stack of this study, for most real cases ($D_s \sim 3 \div 7$ percent), Fig. 8 shows that the attenuation effects due to damping are prevailing for fairly bearing soils; for very soft soils the decrease of n_1 and the rigid motion of the base fully compensate the effect of damping. But, it is rather improbable to find such a high and slender structure on a surface foundation if the soil is so soft. Figures 9 and 10 show the gust factor and the acceleration at the top and their reduction caused by the dominant effect of damping.

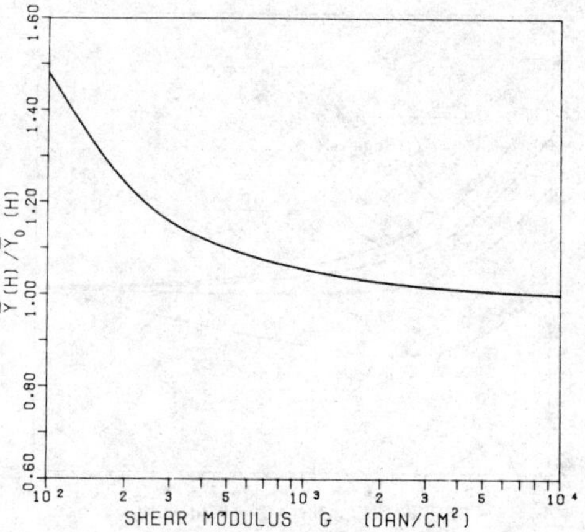

Fig. 7.

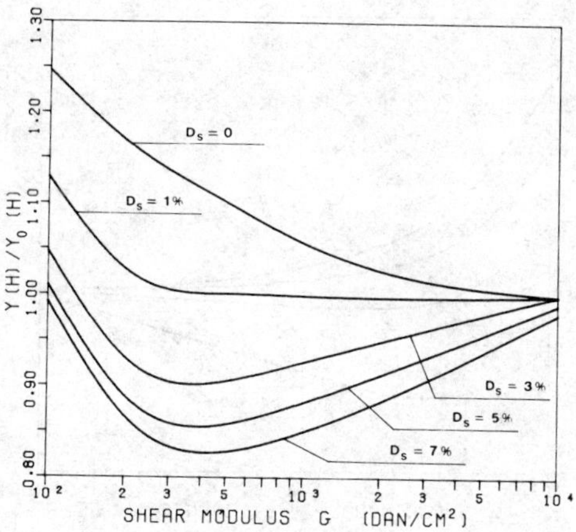

Fig. 8.

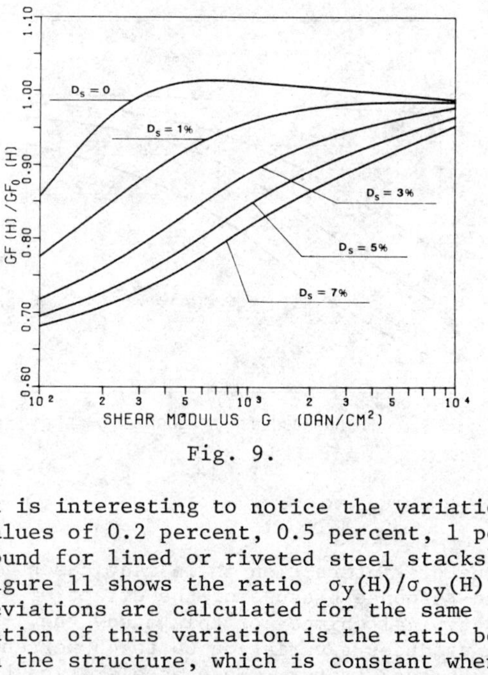

Fig. 9.

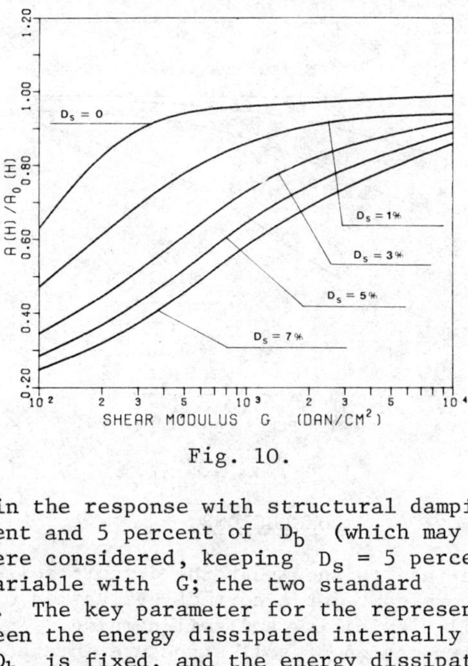

Fig. 10.

It is interesting to notice the variation in the response with structural damping. Values of 0.2 percent, 0.5 percent, 1 percent and 5 percent of D_b (which may be found for lined or riveted steel stacks) were considered, keeping $D_s = 5$ percent. Figure 11 shows the ratio $\sigma_y(H)/\sigma_{oy}(H)$ variable with G; the two standard deviations are calculated for the same D_b. The key parameter for the representation of this variation is the ratio between the energy dissipated internally in the structure, which is constant when D_b is fixed, and the energy dissipated in the soil by both radiation and hysteretic damping, which varies with G. When the former energy is prevailing on the latter energy, as in the case of large D_b values (in Fig. 11 for $D_b \geq 1$ percent), the amplification of the response due to the decrease of n_1, when G decreases, is more significant and larger than the reduction due to damping.

The behavior of the maximum displacement_(Fig. 12) can be easily explained when taking into account Eq. (20), in which $\overline{Y}(z)$ and $g_y(z)$ are still the same as before. Figures 13 and 14 represent the gust factor and the acceleration at the top and show that these two quantities are especially affected by the above ratio of energies. The larger the ratio, the less Gf(H) and a(H) are sensitive of soil flexibility.

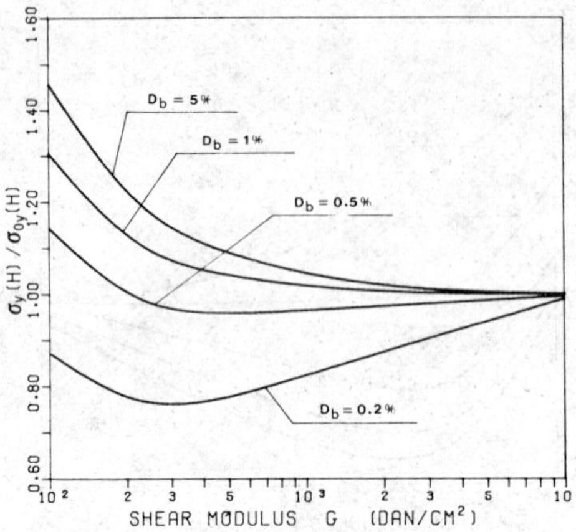

Fig. 11.

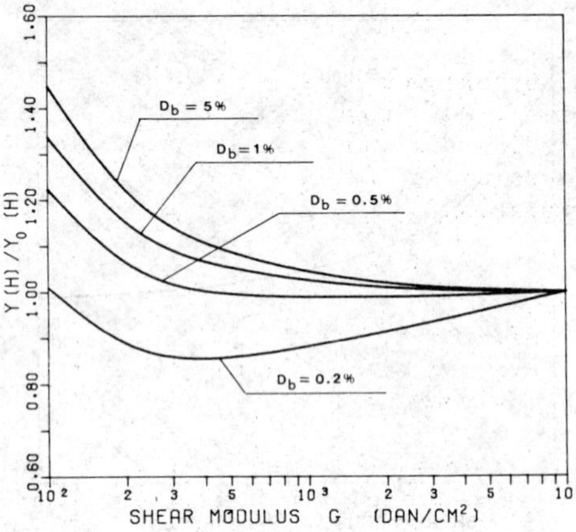

Fig. 12.

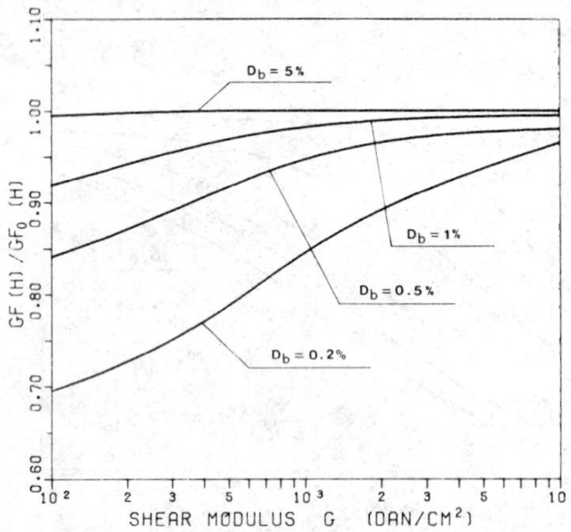

Fig. 13.

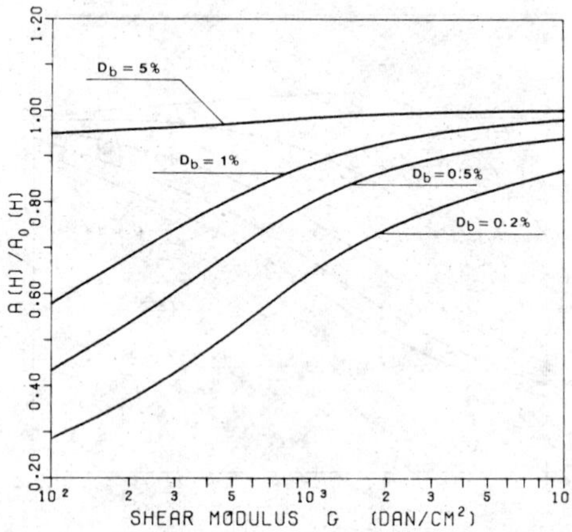

Fig. 14.

CONCLUSIONS

The results show that the dynamic soil-structure interaction is in many cases determinant for a correct evaluation of the response because of the effect of soil flexibility and soil damping. In the typical chimney of this study the introduction of soil-structure interaction produces a variation up to 45 percent in the top displacement and 32 percent in the gust factor. When acceleration has to be considered, variations are even more accentuated and a more general approach is deemed essential for a good representation of the phenomenon.

The conclusions of this study can be extended to the whole class of very slender and flexible structures. The Authors (1) showed that in the case of slender but stiff structures, the soil flexibility will have less influence on the dynamic parameters of the response (as σ_y, Gf, a), so that the static effect of the base motion becomes the determinant aspect of the phenomenon.

ACKNOWLEDGMENTS

The Authors would like to thank their colleagues Prof. A. Corsanego, Prof. A. Del Grosso and Prof. C. Vardanega for their support and helpful advice regarding the manuscript; in particular the capable assistance and the useful suggestions of Prof. A. Del Grosso greatly contributed to the computer program formulation.

REFERENCES

(1) Solari, G. and Stura, D., Effects of soil-structure interaction on the dynamic alongwind response of structures, to appear in Proc. Int. Conf. on Environmental Forces on Engineering Structures, London, England (1979).

(2) Jennings, P.C. and Bielak, J., Dynamics of building-soil interaction, Bull. Seism. Soc. Am., 63, 9 (1973).

(3) Warburton, G.B., Soil-structure interaction for tower structures, Earthq. Engng. Struct. Dyn., 6 (in press) (1978).

(4) Solari, G. and Stura, D., Methods of evaluation of vibration modes of a structural system including soil flexibility, Pubbl. Ist. Sc. Costr. Genova (in press) (1979).

(5) Veletsos, A. and Wei, Y., Lateral and rocking vibration of footings, J. Soil Mech. Found. Div., ASCE, 97, 1227 (1971).

(6) Simiu, E., Equivalent static wind loads for tall building design, J. Struct. Div., ASCE, 102, 719 (1976).

EXPERIMENTAL AND NUMERICAL STUDIES ON THE AERODYNAMIC INSTABILITY OF TOWER-LIKE STRUCTURES

M. IKENOUCHI

Mitsui Engineering and Shipbuilding Co., Ltd., Tokyo, Japan

SUMMARY

This paper deals with the wind tunnel experiments of and numerical analysis of flow around tower-like structure models.

In the experiments, single rigid rectangular cylinders and elastic models of long-span suspension bridge tower were examined by the free oscillation method. The aerodynamic instability, especially vortex-excited oscillation, are discussed in connection with the effects of the ratio of the side lengths and their configuration.

In the numerical analysis, the finite element method based on the Galerkin method was applied. The flow around rectangles were calculated for the analysis of oscillating flow and discussed.

SYMBOLS

d:	side length pararell to wind	$Re=Vh/\nu$:	Reynolds number
h:	side length parpendicular to wind	$St=fh/V$:	Strouhal number
y_o:	bending deflection	$M=\int_0^1 m\phi(\xi)^2 d\xi / \int_0^1 \phi(\xi)^2 d\xi$:	equivalent mass per unit length
f:	frequency		
δa:	aerodynamic logarithmic decrement	$2M/\rho h^2$:	mass parameter
δs:	structural logarithmic decrement	$Ka=2M/\rho h^2 \cdot \delta a$:	aerodynamic damping coefficient
m:	mass		
ρ:	air density	ψ :	stream function
ν:	kinematic viscosity	ω :	vorticity
V:	wind velocity	p :	pressure
$Vr=V/fh$:	reduced velocity	Φ,Ω:	weighting function

INTRODUCTION

Wind-induced oscillation of tower-like structures is of great concern in the practical engineering (1-3). The tower of long-span suspension bridge under construction is a typical important example (1),(2). Among the wind-induced oscillations of these structures, vortex-excited oscillation and galloping are typical. The aerodynamic instabilities of rectangular cylinders and towers have been studied extensively (2-8). It has been noticed that the aerodynamic instabilities depend on the sectional shapes and configuration (4-7). The suspension bridge tower is not such a simple structure as a stack. However, the tower is usually constructed by rectangular cylinders. So, the aeroelastic oscillation phenomena may be considered to be predictable by the knowledge of rectangular cylinders.

In this paper, wind tunnel experiments and numerical flow analysis are discussed for the aerodynamic instability of tower-like structure models. The effects of the sectional shape and configuration are examined. Our primary interest is confined to vortex-excited oscillation.

WIND TUNNEL EXPERIMENTS

Aerodynamic Instability of the Rectangular Cylinder Models

Models and Experimental Procedure Fig. 1 shows a drawing illustrating a rectangular cylinder model used in the experiment. Five kinds of rigid models were made as listed in Table. 1, where Model 3/6 means the model with section size d=3, h=6 (inch). These models were made of plate of balsa wood reinforced by alminium pipe. They were made free to oscillate transversely to the flow direction with the steel spring plate. The deflection was obtained from the measurement of the strain response of the spring plate. The size of the test section of the wind tunnel is 5 feet high and 7 feet wide. Throughout the experiments, the angle of attack was kept zero, where the angle of attack is defined by the angle between the direction of the side d of the model and that of the uniform flow V. The range of Reynolds number Re was $(0.1\sim2.4)\times10^5$.

Using these models, the effects of difference of the side ratio d/h were systematically investigated by means of the free oscillation method.

Experimental Results and Discussion Fig. 2(a),(b) show the aerodynamic responses of the rectangular cylinder models with different side ratio d/h. The ordinate is the non-dimensional bending deflection y_o/h, while the abscissa is the reduced velocity Vr, where d/h appears as a parameter. It is observed that the models of $d/h < 1$ have one instability region around the resonance velocity, which is the flow velocity where the vortex shedding frequency coincides with natural frequency of the model. With the increase of the side ratio d/h, two instability regions appear; one is around resonance velocity, while the other is around half a resonance velocity where the vortex shedding frequency is about half of natural frequency of the model. The oscillation amplitude at half a resonance velocity increases in accordance with the increase of the side ratio d/h for the models with almost same low structural damping. The oscillations with large amplitude for the models of $d/h \geq 1$ persist far beyond the resonance velocity. It may be said that the oscillations around the resonance velocity and half of that are the vortex-excitations, while the oscillations far beyond the resonance velocity are the transverse galloping.

Aerodynamic forces around half of the resonance velocity were examined. Fig. 3 shows the aerodynamic damping coefficient Ka, which is defined in reference (2). The ordinate is Ka, while the abscissa is Vr, where Ka is refered to y_o/h= 0.02. Ka was determined from the measurements of the logarithmic decrement of the oscillation. Negative aerodynamic damping coefficient increases in accordance with the increase of d/h. The tendency of negative aerodynamic damping was observed for the model of $d/h < 1$.

From the experimental results, it may be concluded that the existence of two instability regions is one of the characteristics of the aerodynamic instability of the rectangular cylinder as well as galloping phenomena. Novak (5) has indicated these two instability regions for the model of d/h = 2/1, which is also seen in reference (6),(7). Here these oscillation phenomena were confirmed for the models of $d/h = 0.5 \sim 2$. These oscillation phenomena may be caused by the interaction between the fluid flow and the side d of the rectangular cylinder (4).

Aerodynamic Instability of the Tower Models of Long-Span Suspension Bridge

Models and Experimental Procedure Fig. 4 shows a drawing illustrating tower models used in the experiments. Four kinds of elastic models were made as listed in Table. 2, where Model A is the complete state model and Model B corresponds to the tower under construction. Model AE and BE are the models with additional rigid substructure on their upper ends. These models are made of elastic steel beam covered by formed styrene. These elastic models can oscillate freely in bending and torsional oscillation mode. They were set with the angle of attack 0°, 10° and 90°, where the angle of attack is defined by the angle between the direction along the row of two main structures of the model and that of the uniform flow V. The deflection was measured by the use of the accelermeter placed inside the upper parts. The side of the test section of the wind tunnel is 4 meter high and 2 meter wide. The range of Reynolds number Re was $(0.15\sim1.85)\times10^5$.

Using these models, the effects of the configuration on the aerodynamic instability were investigated by means of the free oscillation method. The experimental results were compared with the results of the rectangular cylinder model.

Experimental Results and Discussion Aerodynamic responses are shown in Fig. 5(a)-(d). Fig. 5(a) shows the aerodynamic response curve of Model A, where the ordinate is y_o, y_o/h and torsion angle θ, while the abscissa is V and Vr. In the figure, bending and torsional oscillations are observed for the elastic tower model. They are indicated as B1 and T1 respectively. Although the cross section ratio d/h of the tower model is less than 1, two instability regions in each oscillation mode are found around resonance velocity and half of that as well as the results of single rectangular cylinder of $d/h \geq 1$. In these oscillation phenomena, non-linear oscillation phenomena such as jumping phenomena are observed. It is noticed that torsional oscillation with large amplitude persists within the wind velocity of the experiment. With the angle of attack 10°, it was damped.

Aerodynamic response of different configuration model, say Model B, is shown in Fig. 5(b). Bending oscillation is larger than that of Model A, but torsional oscillation was damped. This may be resulted in the effect of the structural configuration. The effects of the additional rigid substructure are shown in Fig. 5(b) and 5(c). Additional substructure on the upper part made the response reduce, especially in torsional oscillation mode. The effects of additional damping in bending oscillation mode are shown in Fig. 5(d). The response is reduced in accordance with the increase of structural damping.

Aerodynamic forces around resonance velocity for the bending oscillation mode were examined. Fig. 6 shows the aerodynamic damping coefficient of Model A and Model B at $y_o/h = 0.1$. The ordinate is Ka, while the abscissa is Vr. Negative aerodynamic damping is observed around resonance velocity and half of that.

Vortex shedding frequencies behind the tower models were measured by the use of the hot-wire anemometer. Fig. 7 shows vortex shedding frequency of Model A. The ordinate is vortex shedding frequency f_s, while the abscissa is Vr. Strouhal number St determined from Fig. 7 is 0.10, which is almost same as that of single rectangular cylinder. Around resonance velocity for bending and torsional oscillations, so-called locked-in phenomena are observed. Around half a resonance velocity of bending oscillation, vortex shedding frequency is seemed to be locked-in about half a natural frequency of the model. For the torsional oscillation around half a resonance velocity, the corelation between the model oscillation and vortex shedding was very weak.

From the experimental results of the tower models, it may be concluded that the vortex-excited oscillations in bending and torsional oscillation mode are the characteristics of the aerodynamic instability of the suspension bridge tower with rectangular cross section. These oscillation phenomena may be predictable from the results of single rectangular cylinder. Whole configuration affects the aerodynamic instabilities. Two instability regions in each oscillation mode may be resulted in the effect of the structural configuration. The aerodynamic interaction among the structures may cause these oscillation phenomena.

NUMERICAL FLOW ANALYSIS

Flow analysis of the rectangular section models was carried out using the numerical method. Fromm and Harlow (9) and Okajima et al (10) have developed the finite difference method for the computation of the flow around rectangular sections. Here, we applied the finite element method based on the Galerkin method (11),(12). We confine our problems to two-dimensional imcompressible viscous flow around stationary bodies.

Finite Element Method Two-dimensional Navier-Stokes equations are expressed as vorticity transport equation and Poisson's equation with stream function ψ and vorticity ω. Here, vorticity equation was modified by expressing the unsteady part with ψ. Multiplying proper weighting functions Φ and Ω respectively and integrating the equations in whole domein D, weighted residual equations are obtained (13). Integrating by parts, the following equations are obtained.

$$\iint_D \{ \frac{\partial}{\partial t} \nabla\psi \nabla\Phi + \frac{\partial(\omega,\psi)}{\partial(x,y)} \Phi + \nu \nabla\omega \nabla\Phi \} dxdy = \frac{\partial}{\partial t} \oint_\Gamma \frac{\partial\psi}{\partial n} \Phi \, ds - \nu \oint_\Gamma \frac{\partial\omega}{\partial n} \Phi \, ds \qquad (1)$$

$$\iint_D \{ - \nabla\psi \nabla\Omega + \omega \Omega \} dxdy = - \oint_\Gamma \frac{\partial\psi}{\partial n} \Omega \, ds \qquad (2)$$

We assume that the weighting functions Φ and Ω are zero respectively at the boundary where stream condition ψ and ω are known. Otherwise they are arbitrary.

On the boundary of the bodies, no normal velocity and no slip conditions are given. Besides, another condition that pressure p must be a single-valued is needed. If the body is stationary, the condition is given as follows.

$$\oint_\Gamma \nu \frac{\partial\omega}{\partial n} ds = 0 \qquad (3)$$

Then Φ on the body is set to be an arbitrary constant.
On the downstream boundary, the conditions of no normal derivatives of ψ and ω are given. Taking these boundary conditions into consideration, the righthand side of the equations (1) and (2) vanish.

For the discrete analysis, flow field is divided into small triangular elements. According to the Galerkin method, Φ and Ω are expressed by the same interpolation function as ψ and ω. Finaly, the following discrete equations are obtained.

$$\sum_{\substack{\text{all}\\\text{element}}} \{ \frac{b_i b_j + c_i c_j}{4\Delta_e} \dot{\psi}_j + \frac{b_k c_l - b_l c_k}{12\Delta_e} \omega_k \psi_l I_i + \frac{b_i b_j + c_i c_j}{4\Delta_e} \omega_j \} = 0 \qquad (4)$$

$$\sum_{\substack{\text{all}\\\text{element}}} \{ - \frac{b_i b_j + c_i c_j}{4\Delta_e} \psi_j + \frac{I_{ij} + \delta_{ij}}{12} \Delta_e \omega_j \} = 0 \qquad (5)$$

Where Δ_e = area of element, $\dot{}$ = time derivative

$b_i = y_j - y_k$, $c_i = x_k - x_j$ (i,j,k : cyclic)

$x_i, y_i, \ldots$ = x, y coordinate of the nodal point i,j,k

I_i, I_{ij} = unit vector, tensor ; δ_{ij} = Kronecker's delta

With the proper initial value of ψ, ω is determined from the equation (5). Using these ψ and ω, ψ at $t = t + \Delta t$ is determined from the time difference equation (4). This iteration process is proceeded to get the solution at subsequent time.

Culculated Results and Discussion Flows around stationary cylinders of rectangular section with side ratio d/h = 1/2, 1/1 and 2/1 were calculated at Re = 100. Fig. 8(a)-(c) show the velocity vector distributions at the same time of vortex shedding period. Vortex formation is recognized behind the rectangles. Broad separated flow is observed for the rectangular section of d/h = 1/2, while the section of d/h = 2/1 has a long separated flow along the side d which may interacts with the body. Through the vortex formation process, periodic pressure loading is produced. Fig. 9 shows the variation of lift and drag of the square section. FY and FX are lift and drag coefficient respectively. Strouhal number St, calculated from the period of lift variation, is 0.12. Frequency of drag variation is twice of that of lift. Flow patterns and Strouhal number are reasonble comparing with experimental results (7).

Fig. 10 shows flow around two rectangular sections corresponding to the sectional model of suspension bridge tower at Re = 100. The side ratio d/h is 2/3 and B/d is 8, where B is the distance between two rectangles. Interaction between the rectangular sections and wake is observed. Flows around the downstream section is strongly influenced by the wake of the upstream section. Though the wake interference exists, periodic vortex shedding is formed behind the downstream section. Pressure loading on both rectangles is periodic, as shown in Fig. 11. In this configuration of two rectangles, period and phase of lift variation were almost same. Strouhal number St, calculated from the period of lift variation, is 0.12, which is the same as the result of the single rectangular section.

From the numerical flow analysis, it may be said that the whole structure with two rectangular cylinders, such as suspension bridge tower, oscillate similarly to the single cylinder by the vortex-excitation.

More detailed numerical analysis will be continued.

FINAL REMARKS

In order to investigate the aerodynamic instability of tower-like structures with rectangular cross section, two wind tunnel experiments and numerical analysis were carried out.

From the wind tunnel experiments of the rectangular cylinder models, the effects of the side ratio d/h on the oscillation phenomena were clarified. With the increase of the side ratio d/h from 0.5 to 2, two regions of instability appeared.

From the wind tunnel experiments of the tower models, it is shown that vortex-excited oscillations are the characteristics of the aerodynamic instability of the long-span suspension bridge tower. These oscillation phenomena are considered to be predictable from the results of the single rectangular cylinder. The effects of the structural configuration appeared in two instability regions of each oscillation mode and torsional oscillation. The whole structure is considered to affect the oscillation phenomena.

Through numerical flow analysis by the use of the finite element method, it is shown that the flow patterns vary with the shape and configuration. Vortex shedding was observed for two rectangular sections, simulating the section of tower, as well as that for the single rectangular section.

Needless to say, there remain a number of important problems which are not discussed. Three dimensional effect in the wind tunnel experiments might be important. In the numerical flow analysis, Reynolds number is still low and obstacles are stationary. Further extension of the numerical analysis method to unsteady problems is to be conducted.

ACKNOWLEDGEMENT

The author would like to thank Professor K. Washizu of Tokyo University for his stimulation and useful comments on preparing of the paper.

Professor J. Dugundji of M.I.T. is acknowledged for his suggestion and giving the opportunity of the first wind tunnel experiment in M.I.T.

Professor Y. Nakamura of Kyushu University is also acknowledged for his useful comments and giving the opportunity of the second wind tunnel experiment in Kyushu University.

The author would like to thank Dr. Teshima and Mr. Kimura for their collaborations in the experimental and numerical analysis works.

REFERENCES

1. Anderson, J.K., Forth Road Bridge, Proc. Institution of Civil Engineer, Vol. 32, (1965)

2. Scruton,C., On the Wind-excited Oscillations of Stack, Towers and Masts, Proc. Symp. Wind Effects on Buildings and Structures, NPL, Teddington, Vol. 1, H.M.S.O. (1965)

3. Melbourne, W.H., Cross-wind Response of Structures to Wind Action, Proc. 4th Int. Conf. on Wind Effects on Buildings and Structures, 343-358 (1975)

4. Parkinson, G.V., Mathmatical Models of Flow-Induced Vibrations of Bluff Bodies, Proc. Symp. Flow-Induced Structural Vibrations, Karlsruhe, 81-127 (1972)

5. Novak, M., Galloping and Vortex Induced Oscillations of Structures, Proc. 3rd Int. Conf. Wind Effects on Buildings and Structures, Tokyo, 799-809 (1971)

6. Nakamura, Y. and Mizota, T., Galloping and Vortex Excitation of a Rectangular Block, Rep. Res. Inst. Appl. Mech. Kyushu Univ. Vol. XIX, No. 64, 225-236 (1972)

7. Washizu, K., Ohya, A., Otsuki, Y. and Fujii, K., Aeroelastic Instability of Rectangular Cylinders in a Heaving Mode, Journal of Sound and Vibration, 59(2), 195-210 (1978)

8. Ito, M., Miyata, T. and Morimitsu, T., Fluctuating Drag and Lift Air Forces on a Square Cylinder, Proc. 2nd Japan National Symp. Wind Effects on Structures (1972) (in Japanese)

9. Fromm, J.E. and Harlow, F.H., Numerical Solution of the Problems of Vortex Street Development, The Phisics of Fluids, Vol. 6, No. 7, 82-89 (1963)

10. Okajima, A., Sugitani, K. and Mizota, T., Flows around Rectangular Prisms, Int. Symp. Flow Visualization, Tokyo, 71-76 (1977)

11. Zienkiewicz, O.C., The Finite Element Method in Engineering Science, McGrow-Hill (1971)

12. Ikenouchi, M. and Kimura, N., An Approximate Numerical Solution of the Navier-Stokes Equations by Galerkin Method, Proc. 1st Int. Symp. Finite Element Method in Flow Problems, Swansea, 99-100 (1971)

13. Finlayson, B.A., The Method of Weighted Residuals and Variational Principles, Academic Press (1972)

Table. 1 Characteristics of Rectangular Cylinder Model

Model	$2M/\rho h^2$	f (Hz)	δs
3/6	70	10.1	0.022
3/4	125	15.1	0.022
3/3	199	18.5	0.025
4/3	216	18.0	0.022
6/3	255	12.9	0.022

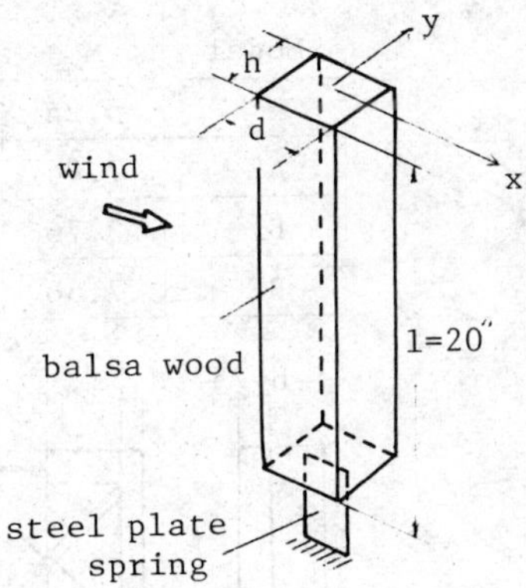

Fig. 1 Rectangular Cylinder Model

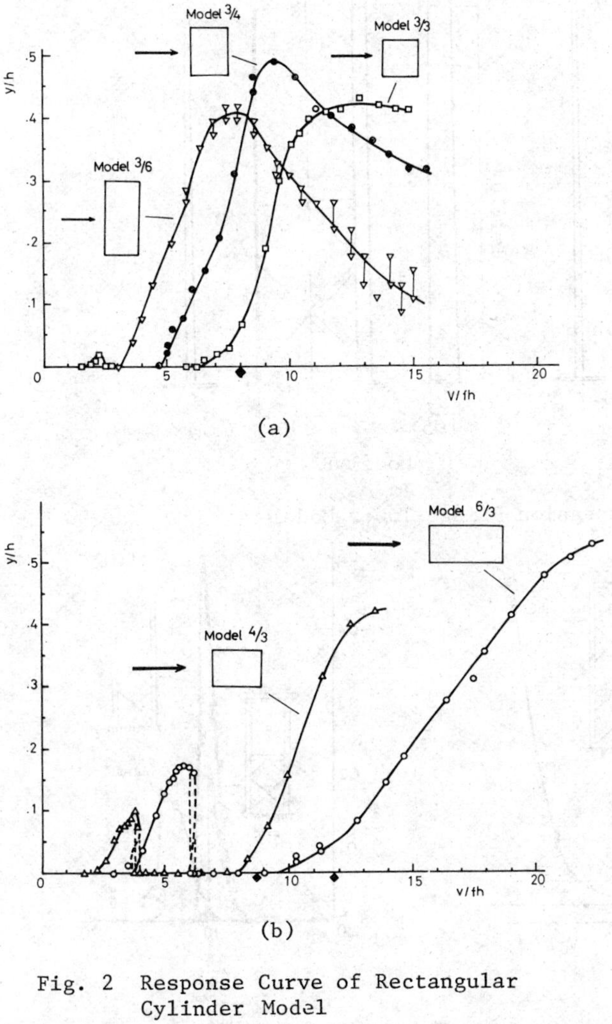

Fig. 2 Response Curve of Rectangular Cylinder Model

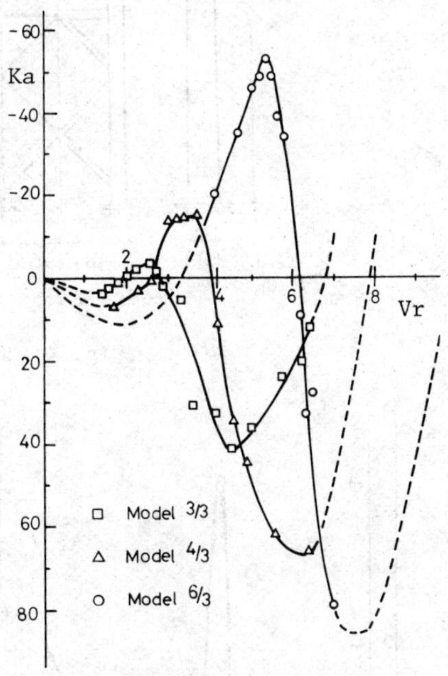

Fig. 3 Aerodynamic Damping Coefficient

Table. 2 Characteristics of Tower Model

Model	M	Bending		Torsion	
		f (Hz)	δs	f (Hz)	δs
A	5.24	4.61	0.0290	13.8	0.0689
AE	7.84	3.76	0.0317	10.2	0.0657
B	3.76	9.65	0.0228	11.6	0.0916
BE	7.56	6.77	0.0259	11.4	0.103

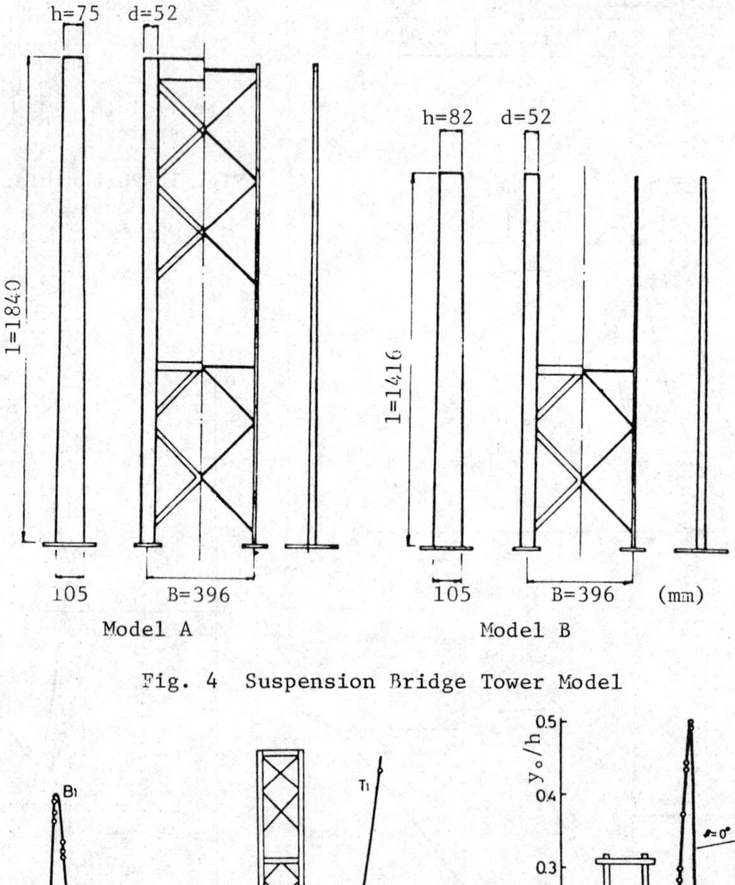

Fig. 4 Suspension Bridge Tower Model

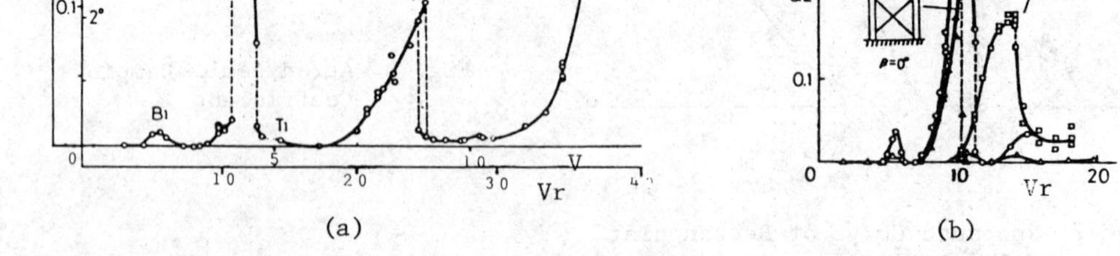

Fig. 5 Response Curve of Tower Model

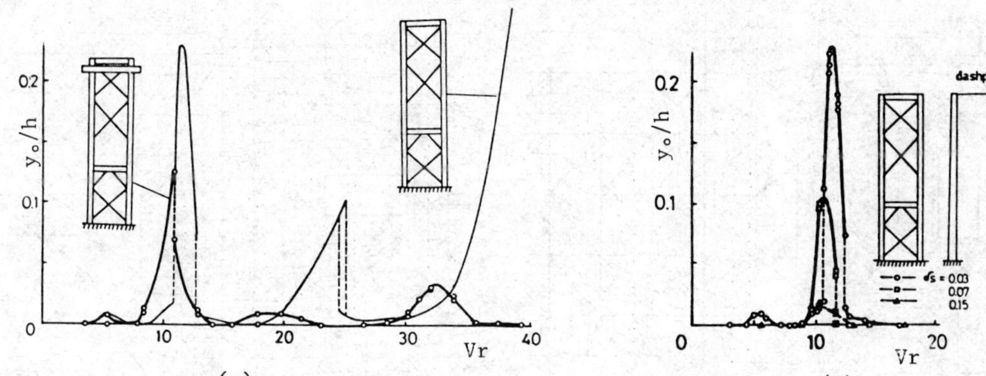

(c) (d)

Fig. 5 Response Curve of Tower Model (continued)

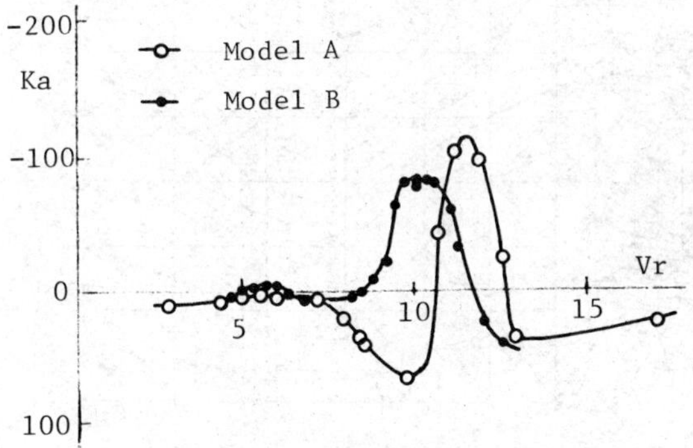

Fig. 6 Aerodynamic Damping Coefficient

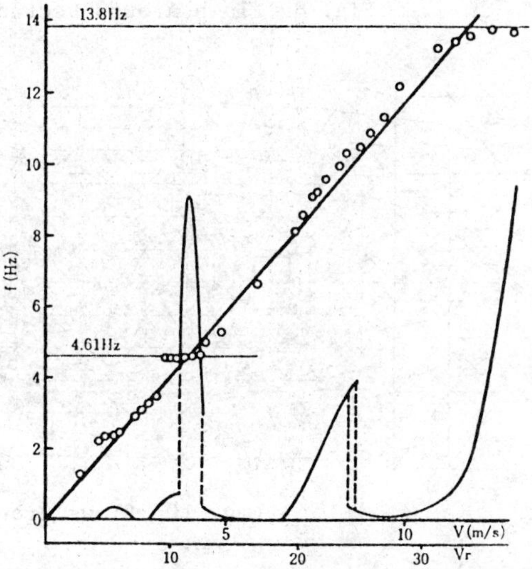

Fig. 7 Vortex Shedding Frequency

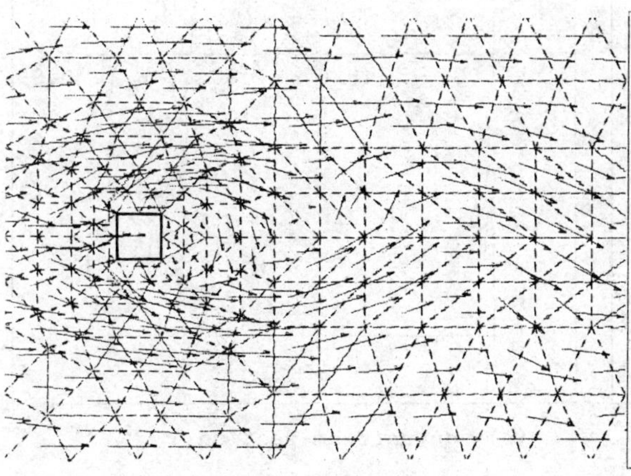

Fig. 8 Flows around Rectangular Section

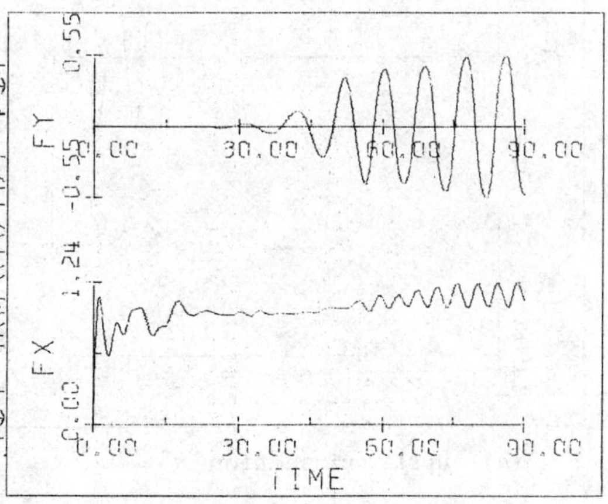

Fig. 9 Lift and Drag Variation (square section)

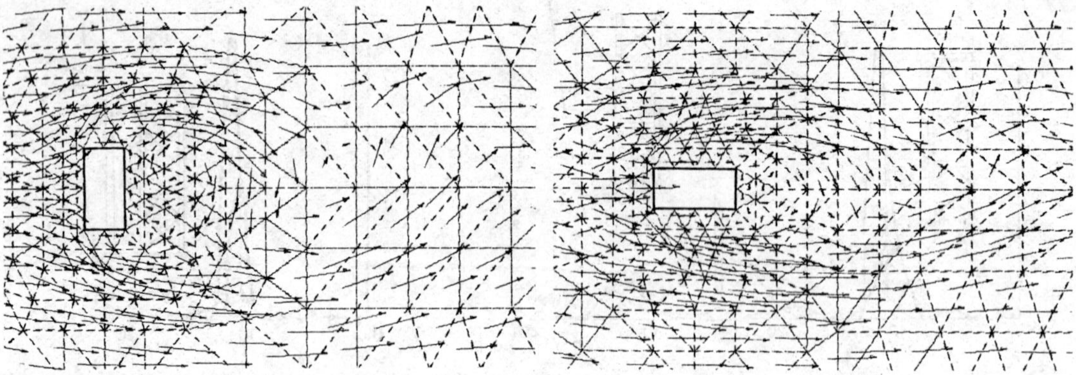

(b) (c)

Fig. 8 Flows around Rectangular Section (continued)

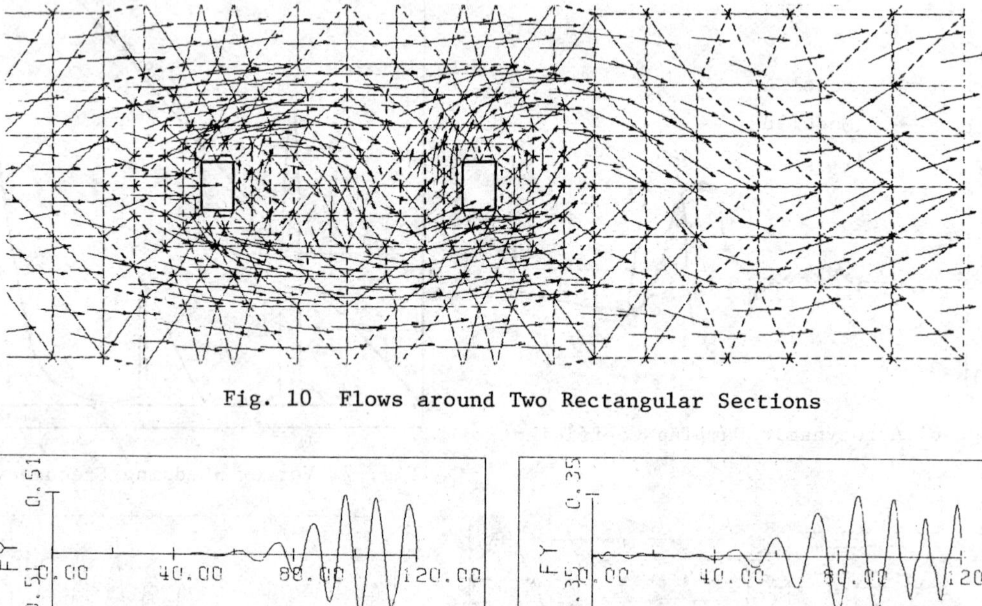

Fig. 10 Flows around Two Rectangular Sections

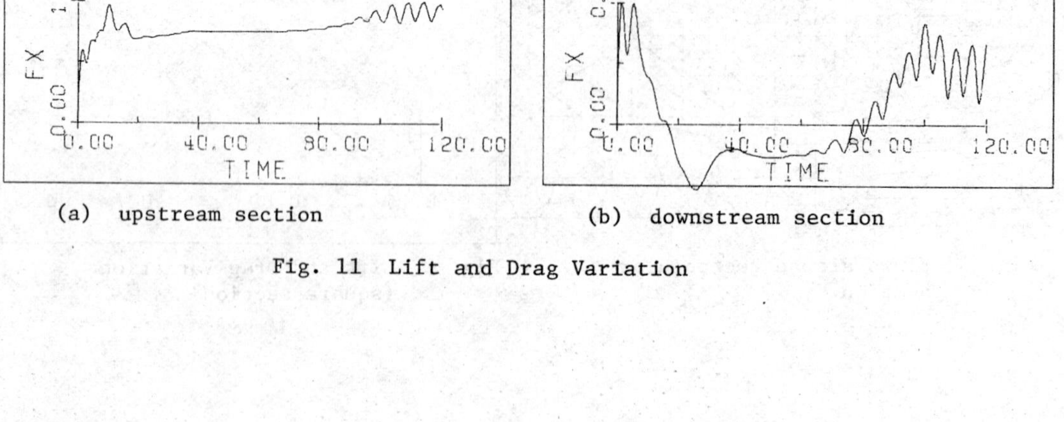

(a) upstream section (b) downstream section

Fig. 11 Lift and Drag Variation

PROTOTYPE AND MODEL EXPERIMENTS ON WIND-INDUCED VIBRATIONS OF A STEEL TOWER

W. SCHNABEL and E. J. PLATE

University of Karlsruhe, Karlsruhe, Federal Republic of Germany

ABSTRACT

To enable a comparison between the forces and dynamic reactions of full-scale structures and wind-tunnel models under ideal conditions, tests have been carried out at the lighthouse "Großer Vogelsand", a steel tower situated off-shore in the North Sea. Similarity criteria are presented and discussed (i.e. velocity profile, intensity and spectra of atmospheric turbulence). The pressure distribution was modelled as well as structural parameters. It is shown that similarity in the response and the energy content of the force acting on the tower agree well in model and prototype.

INTRODUCTION

Early wind-tunnel experiments to determine wind-forces acting on buildings were carried out in smooth, parallel flow with a very low level of turbulence. Jensen (1) was one of the first to successfully compare wind-tunnel data with pressure distributions in the natural environment. His work showed the need to accurately reproduce the atmospheric boundary layer in the wind-tunnel. The recent trend towards higher, slender and more flexible buildings has then precipitated the attempt to obtain relationships between the dynamic response and the turbulent characteristics of the wind,causing the motion. Davenport's (2) research work led to a concept which predicts with statistical methods the dynamic response of a building from terrain, turbulence and building parameters. For nontypical structures further refinement in wind load determination took place in the development of improved wind-tunnel experiments for design purposes. To fulfill the similarity requirements not only the geometrical parameters of the building have to be reproduced, but also the energy distribution of the approaching flow and the dynamic properties of the building.

Up to the present there exists relatively little experimental data to enable an exact comparison between the forces and dynamic reactions measured in full-scale and in wind-tunnel experiments. Hence, an experimental program was initiated to verify simulation parameters and to provide laboratory data for comparison with field measurements from a lighthouse tower in the North Sea.

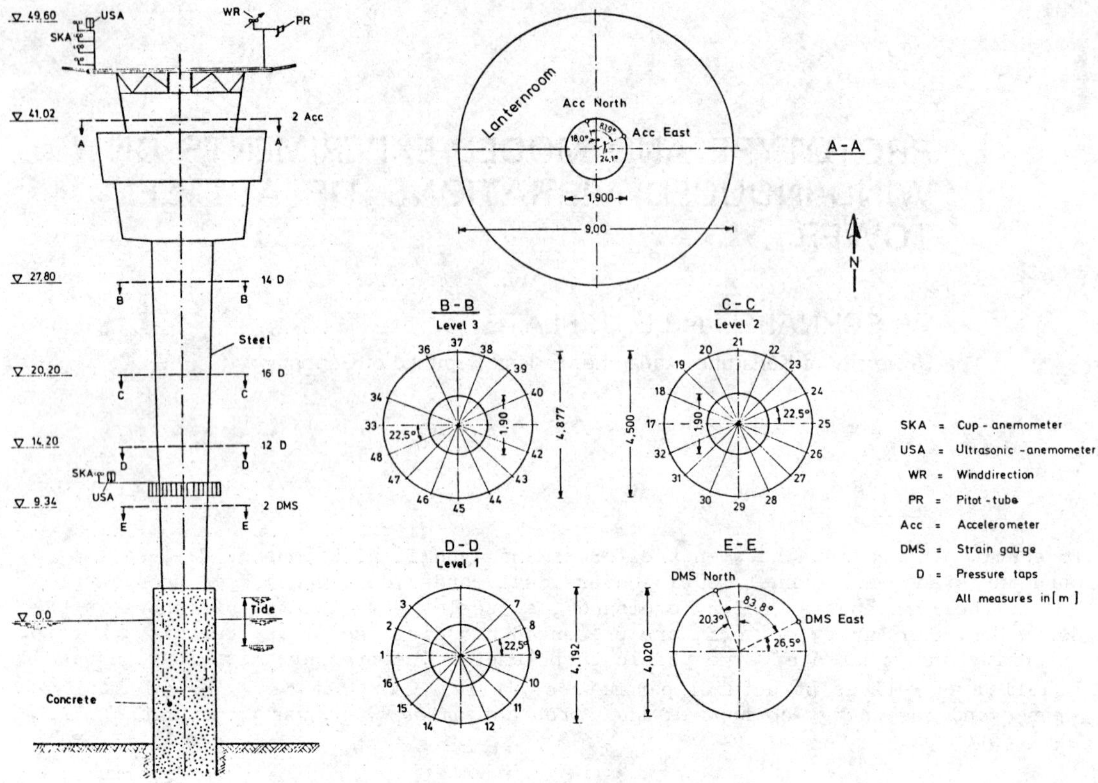

Fig. 1 Experimental configuration at the lighthouse Großer Vogelsand

GEOMETRY OF THE TOWER AND EXPERIMENTAL CONFIGURATION

Prototype

The lighthouse "Großer Vogelsand" (Fig. 1) is situated about 20 km off-shore in the German Bight. It was chosen primarily because it has a simple circular-cylindrical steel construction with a concentrated mass at the top. Furthermore, it was important to ensure a well defined terrain roughness over a long fetch so that a constant velocity profile with a homogeneous structure of turbulence would develop.

Wind measurements were carried out at the top of the tower (50 m above the sea) using both cup- and ultra-sonic-anemometers. A second set of wind measurements was taken at 11 m above water level. Forty-four pressure taps were distributed in three levels along the shaft and connected by tubes and two Scanivalves to difference pressure transducers (Baratron, MKS). In addition 2 difference pressure transducers (Hottinger Baldwin, PD 0,1) were installed at the wall of the tower. This arrangement permits the determination of the mean pressure distribution around the shaft and of cross-correlations of the pressure fluctuations both in the vertical and horizontal direction. The dynamic response of the tower was measured by two accelerometers 41 m above water level as well as by two strain gauges at 9 m which yielded the bending stresses at the base of the tower. All signals were recorded on an eight-channel analog tape recorder and later digitized for the data analysis at a convenient frequency, using well established principles of random data analysis (3).

Model configuration (Fig. 2)

The applicability of aeroelastic modelling and the accuracy of the results depends strongly on the geometrical structure of the building. Because of its concentrated mass at the top the lighthouse sways in its first fundamental mode with no recognizable contribution of higher harmonics. This was confirmed by a dynamic analysis, which showed further that the mode shape could be approximated by a linear mode. Consequently, the dynamic properties of the tower were modelled by representing the tower as a rigid body free to rotate on a set of gimbals. A rigid steel rod attached to the bottom of the model extends downward through the wind-tunnel floor. Below the floor this rod is connected to appropriately selected springs which provide the required stiffness. Strain gauges mounted at thin metal sheets which work as bending beam, yield an electrical signal that is calibrated against the strain. In addition an aluminium disc is attached to the rod. The disc is placed between the jaws of four electro-magnets. The movement of the disc creates an eddy current which provides a damping force proportional to the velocity of the movement. The magnitude of the damping force can be varied over a wide range by adjusting the current through the electro-magnets.

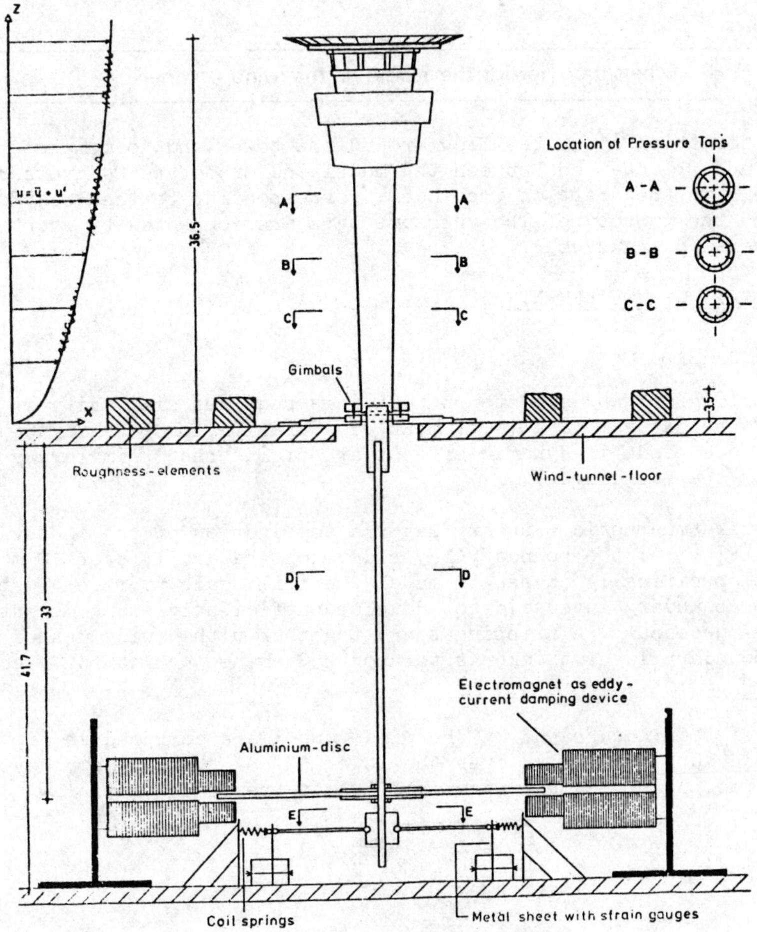

Fig. 2 Construction of the model

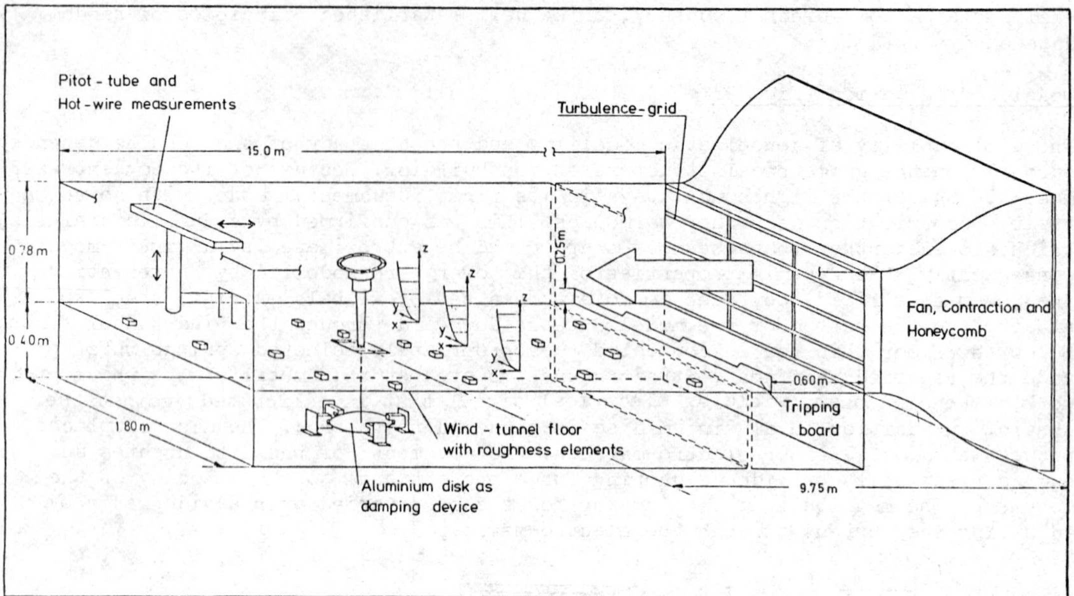

Fig. 3 Schematic view at the model in the wind - tunnel

To provide information about the flow around the tower and to compare the pressure distribution and fluctuation between the model and prototype 24 pressure taps are set at 3 levels in the shaft of the model, corresponding to the locations of the piezometers on the prototype. The pressure taps are connected by short tubes to a pressure difference transducer.

SCALING REQUIREMENTS

Properties of the wind

Physical modelling of wind effects on buildings requires similarity of the salient properties of the natural wind (i.e. mean wind profile, probability density, scales of turbulence) and, if dynamic effects are of interest, similarity of dynamic properties for the building.

To simulate the atmospheric boundary layer a combination of previously used techniques is used (Fig. 3): the boundary layer develops naturally over 10 m of roughness elements specifically chosen to match the full scale profile. Despite the long fetch the boundary layer did not develop sufficiently and some artificial thickening was necessary. A tripping board together with a wide meshed grid increases the boundary layer thickness to about 50 cm, enough to fully cover the model.

Velocity profiles in nature and in the wind-tunnel are compared in Fig. 4. Within the logarithmic sublayer (valid over the lower 15% of the boundary layer) modelling of the atmosphere requires exact scaling of the logarithmic law

$$\frac{\bar{u}}{u_*} = \frac{1}{k} \cdot \ln \frac{z}{z_o} \quad .$$

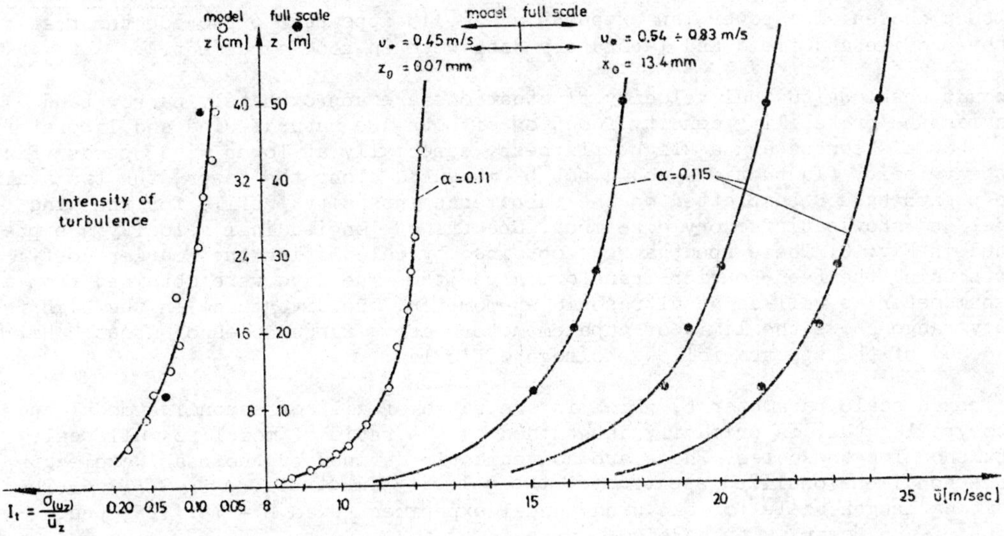

Fig. 4 Profiles of velocity and intensity of turbulence above the sea and in the wind-tunnel

Hence, the shear stress velocity u_* should be the same in field and in laboratory and z_c/z_o should be constant, where z_o is the roughness height and z_c is a scaling length, such as a building dimension (4).

On the other hand, it is difficult to completely simulate the boundary layer in the wind-tunnel using the logarithmic law. Hence, above 8 cm a simulation using the exponential law

$$\frac{\bar{u}_1}{\bar{u}_2} = \left(\frac{z_1}{z_2}\right)^{\alpha}$$

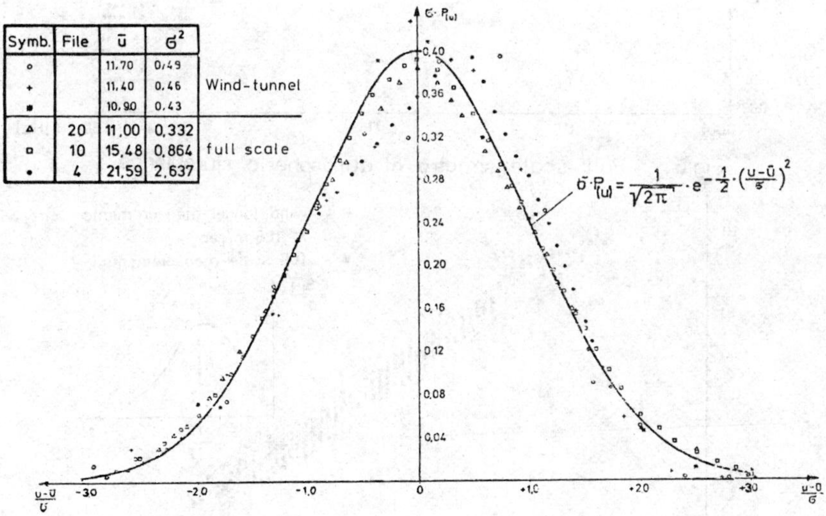

Symb.	File	$\bar{u}$	σ^2
o		11.70	0.49
+		11.40	0.46
▪		10.90	0.43
Δ	20	11.00	0.332
□	10	15.48	0.864
•	4	21.59	2.637

$$\sigma \cdot P_{(u)} = \frac{1}{\sqrt{2\pi}} \cdot e^{-\frac{1}{2}\cdot\left(\frac{u-\bar{u}}{\sigma}\right)^2}$$

Fig. 5 Probability density of longitudinal velocity fluctuations has been chosen. The power law exponent α = 0,115 appeared to provide the best fit of the profile and field and laboratory data were in good agreement.

Although the longitudinal velocity fluctuations are approximately narrow band Gaussian, the probability density function of both the natural wind and laboratory data show a skewness and a slight flatness, especially at lower wind speeds (Fig.5). As the velocity fluctuations could not be measured along the tower, the two available points have been plotted on the turbulence intensity profile for the wind tunnel and show satisfactory agreement. Spectra of longitudinal velocity are presented in Fig. 6. These spectra were obtained by calculating the Fourier coefficients using the Fast-Fourier-Transform algorithm. The data were obtained from a cupanemometer as well as an ultrasonic anemometer. The divergence in the high frequency range shows the limit of cupanemometers above frequencies of about 2 Hz. The decay of the spectra follows Kolmogoroff's law.

The length scale parameter L, which is the ratio of all corresponding model and prototype lengths, is primarily determined by the ratio of model to full scale turbulence length scales. These are conveniently defined by the peak wavelength of the nondimensionalized spectrum of the velocity fluctuations $f \cdot S_{(f)}/\sigma^2$. The resulting length scale for the wind-tunnel experiment was L = 1 : 125, the height of the tower is then 36 cm.

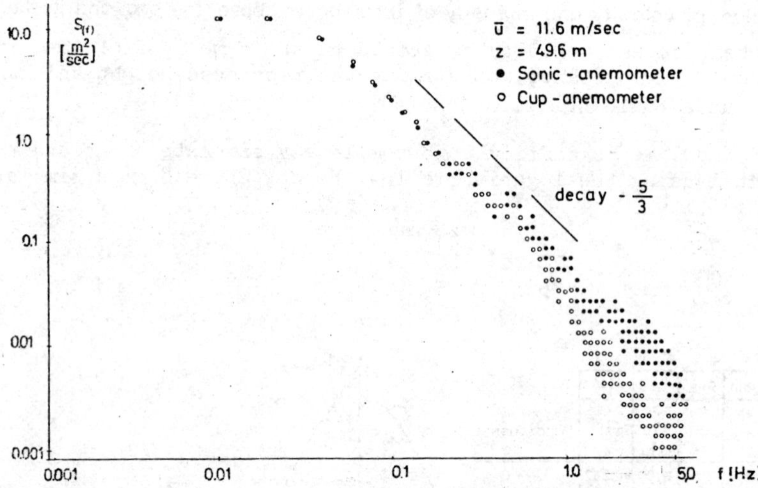

Fig.6 Full-scale spectra of atmospheric turbulence

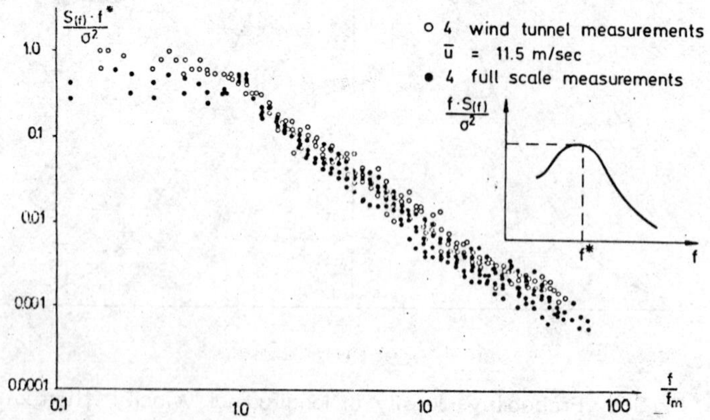

Fig.7 Nondimensionalized spectra of longitudinal turbulence

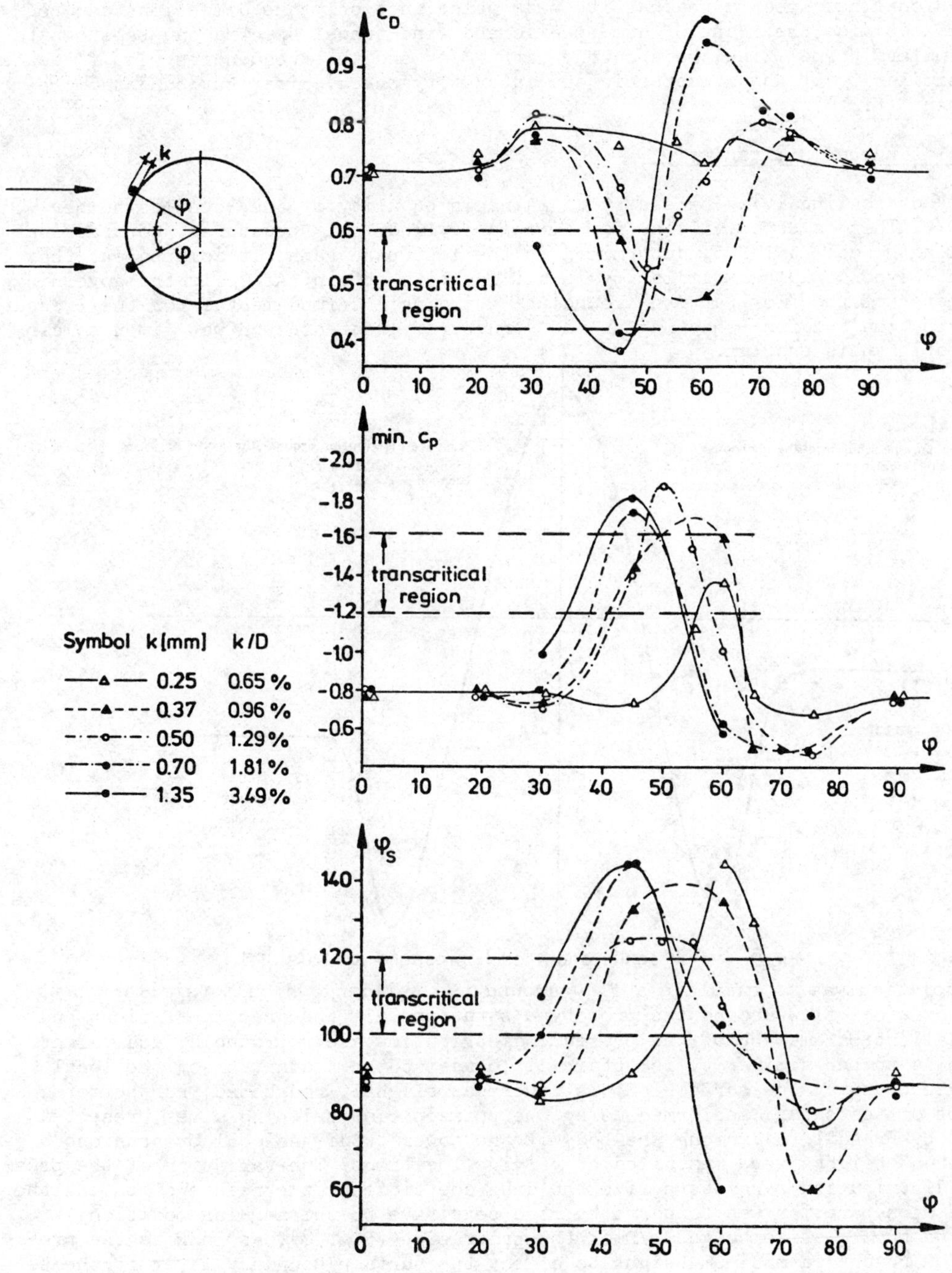

Fig. 8 Variation of drag coefficient, peak suction coefficient and point of separation with location of the protrusion

Taking advantage of the scaling properties of f^*, the frequency at which the non-dimensionalized spectra reaches its peak value, not only the dimensions can be scaled but also the shape of atmospheric and wind-tunnel spectra can be shown to be similar. Normalizing the spectrum with f^*/σ^2 and the frequencies with f^* makes the similarity of the atmospheric and laboratory energy distribution data more apparent (Fig. 7).

Reynolds number similarity

For exact scaling it is necessary to maintain equality of the Reynolds number $\bar{u} \cdot D/\nu$. This is especially so for circular cylindrical structures. Considering the length scale 1 : 125, determined in the last paragraph, it is apparent that strict Reynolds number similarity is difficult to obtain. As Re is the governing parameter of the flow pattern around the building and fundamental for the pressure distribution and forces acting on the structure, great concern was given to circumventing this problem.

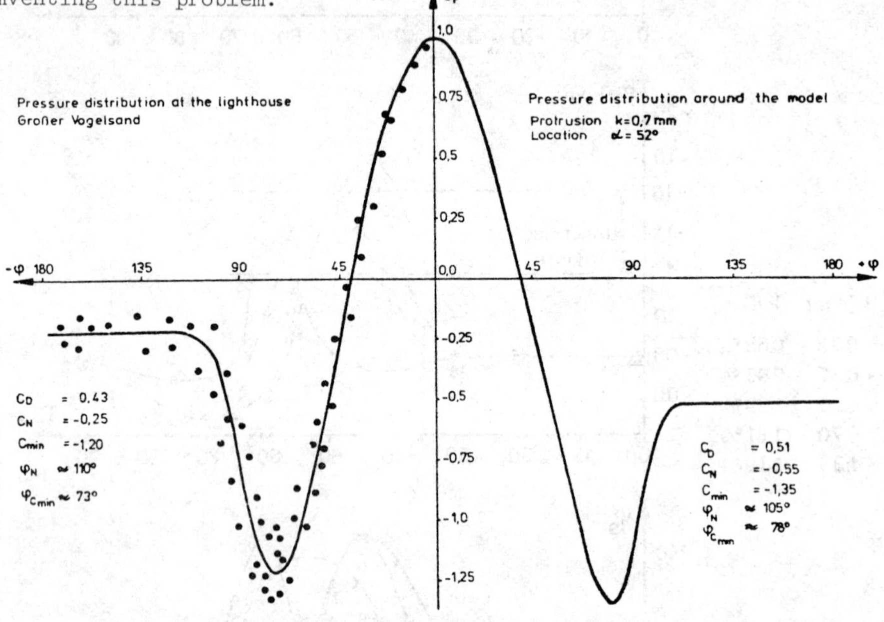

Fig. 9 Full scale and model pressure distribution

The solution was to simulate a flow around the building which is similar to the transcritical flow around the prototype structure. If the drag coefficient and the salient parameters of the pressure distribution are reproduced, equivalent Reynolds number similarity is obtained. To assist the transition of the local surface boundary the surface of the tower was slightly roughened. Furthermore, a series of tests was performed using the principle of tripping wires presented first by Prandtl (5). Along the body of the tower twodimensional protrusions of different heights were installed at varying locations. The variation of the pressure distribution parameters, i.e the drag coefficient, the peak suction and the angle of separation was found to be more sensitive to the angular position than to the relative height k/D of the protrusion (Fig. 8), as long as the protrusion reaches a minimum height to affect the surface boundary layer of the tower. As it was very difficult to accurately evaluate the various parameters of the pressure distribution in turbulent, transcritical shear flows, a range of values in which the parameter was expected to lie was determined by a literature review. However, comparing the results of different authors, there still is a large scatter in the data. The limits used are therefore somewhat in doubt. According to

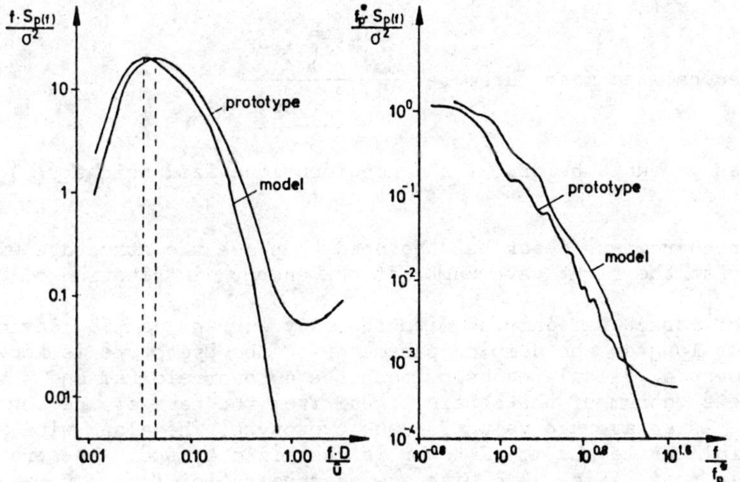

Fig. 10 Normalized pressure spectra

Fig. 8 and the measured full scale pressure distribution (Fig. 9) a protrusion height k = 0,7 mm (k/D = 1,81 %) at φ = 52° from the line of stagnation was chosen. The derived model pressure distribution is compared in Fig. 9 with the average of several pressure measurements at the full scale tower. The difference in the value of the drag coefficient results from the low base pressure coefficient in the wake of the model. Inspecting the full scale pressure distribution one realizes a relatively high base pressure coefficient c_p =-0,25, a trend that recently has been observed elsewhere and that could not be derived in any of the laboratory tests (6,7). As drag coefficients as low as c_D = 0,43 have been reported only in recent times, it appears that they might be a result of modern methods of construction. On smooth, modern towers especially, c_D values below c_D = 0,6 have been observed.

PRESSURES ACTING ON THE TOWER

The response function of the structure is closely related to the fluctuating pressures acting on the surface of the tower. It was shown (9) that buffeting of the tower by gusts is primarily due to the coherency of the pressure fluctuations within the area of positive pressure. This also applies to the model experiment: full scale and model spectra of pressures acting on the surface of the tower are shown to be almost identical if scaled with the frequency at which the nondimensionalized spectra $f \cdot S_p(f)/\sigma^2$ reach their peak value (Fig. 10). Not only do related coherencies of the two pressure signals have the same magnitudes in model and prototype but also their shapes are similar.

RESPONSE OF THE TOWER

Similarity requirements for aeroelastic modelling have been discussed in detail elsewhere (8). The geometric similarity (ratio of all lengths of the building including mode shape) was shown to exist in the previous sections. Flow similarity is achieved and the structural parameters which should be the same in full scale and in the wind tunnel are

the reduced velocity $\frac{f \cdot D}{\bar{u}}$

the damping, expressed by the logarithmic decrement δ

the generalized mass parameter $\frac{\int_0^1 m \cdot y_{(z)}^2 \, dz}{\int_0^1 y_{(z)}^2 \, dz}$

where m is the mass per unit height, z the nondimensionalized height $y_{(z)}$ the deflection at height z.

The first of the requirements describes the need to place the structure with its reaction frequency at the right wavelength of the energy distribution of the wind.

The introduction of structural damping with the eddy current damping device presents no problem as long as the damping parameter of the prototype is known. The logarithmic decrement δ, calculated using both the autocorrelogram and the half power point of the spectrum of the strain signal from the base of the tower, results in $\delta = 0{,}052$ as an average value for the prototype. The along wind vibration of both the full scale and model tower is described by their spectra of deflection and strain resp. (Fig. 11). With a windspeed ratio of $\bar{u}_N : \bar{u}_M = 1 : 6$, where N denotes nature and M model experiment, the shape of the spectrum, drawn over the reduced frequency $f \cdot D/\bar{u}$ are found to be similar. This means that the energy distribution of the response of the tower is fully reproduced over the nondimensionalized frequency range. It is noteworthy that both spectra contain a plateau below the peak of the eigenfrequency. The meaning of this plateau is not fully understood yet. Maybe there exists a coupling of the modes of the longitudinal and lateral vibration.

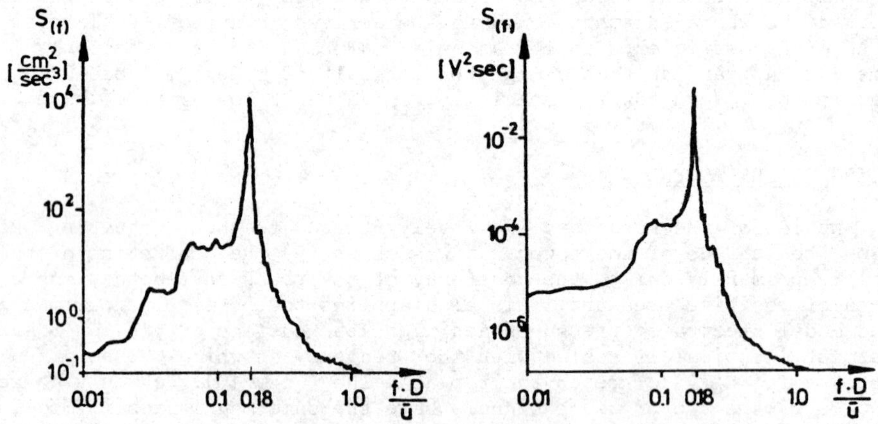

Fig. 11 Response spectra of the prototype and model tower

CONCLUSIONS

Only few experimental data exist on the comparison of full scale and wind-tunnel measurements on the dynamic response of slender structures. Therefore, experiments on a lighthouse tower have been performed and described. Similarity requirements for wind-tunnel experiments demand geometric, dynamic and structural similarity. It has been shown that it was possible to fulfill these requirements and to find satisfying agreement between the field and model results. Therefore

a) the velocity profile of the wind was scaled according to the roughness length

b) the spectra of longitudinal turbulence component was of the same shape if scaled with the frequency f^* of the peak of the nondimensionalized spectra $f \cdot S(f)/\sigma^2$

c) the pressure distribution which implies equivalence to Reynolds number similarity could be reproduced by introducing protrusions into the surface boundary layer of the cylinder

d) the energy content of the pressures acting on the surface within the area of positive pressure is similar in full scale and in model and the coherency is the same

e) the response of the structure under the action of the pressures,described by the spectrum over the nondimensionalized frequency scales well.

The experiments indicate, that by modelling a structure properly in a boundary layer wind tunnel results can be obtained which are dynamically fully comparable to the full scale situation.

ACKNOWLEDGEMENT

The work described herein is supported by the German Science Foundation (Deutsche Forschungsgemeinschaft) within the research program "Building Aerodynamics". The full scale experiments were prepared and conducted by Dr. O.Blohm, the wind measurements with the ultrasonic anemometer were carried out by Dr. N.O. Jensen, Danish AEC, Risø, Denmark.

REFERENCES

1. M. Jensen, The model law for phenomena in natural wind, Ingenioren, Int. Edition (1958)

2. A.G. Davenport, Gust loading factors, Journ. ASCE, Struct. Div.,93 (1967)

3. E.J. Plate, Analyse kontinuierlicher Zufallsfunktionen, Heft 1 der Mitteilungen des Instituts Wasserbau III, Universität Karlsruhe (1971)

4. E.J. Plate, Aerodynamic characteristics of atmospheric boundary layers, AEC Critical Review Series, Springfield, Va. (1971)

5. H. Schlichting, Grenzschichttheorie, Verlag G. Braun, Karlsruhe (1965)

6. O. Blohm and E.J. Plate, Wind wave and windpressure investigation at the lighthouse "Alte Weser", 2nd US Nat. Conf. on Wind Effects on Buildings and Structures, Ft. Collins, Colo. (1975)

7. H. Ruscheweyh, Beitrag zur Windbelastung hoher kreiszylindrischer Bauwerke im natürlichen Wind bei Re = $1.4 \cdot 10^7$, Dissertation,Fakultät für Maschinenbau, TH Aachen (1974)

8. R.E. Whitbread, Model simulation of wind effects on structures, Proc. 1st Int. Conf. on Wind Effects on Buildings and Structures, Teddington (1963)

9. W. Schnabel and E.J. Plate, Experimental study of wind forces and wind induced vibrations at a steel lighthouse tower, Int. Conf. on Environmental Forces on Engineering Structures, London (1979)

A STUDY ON CROSSWIND OSCILLATION OF CHIMNEYS

H. VAN KOTEN

Institute TNO for Building Materials and Building Structures, Delft, The Netherlands

ABSTRACT

Two slender steel chimneys were observed during a number of months to collect information about crosswind oscillations. The results were checked with a calculation method and wind-tunnel tests.

SOME OBSERVATIONS ON STEEL CHIMNEYS

The stresses at the base of two steel chimneys (Fig. 1) have been measured during crosswind vibrations. A part of the records is shown in Fig. 2. The chimneys oscillate in the first resonance frequency almost continuously as if periodically loaded.

If the motion is considered as a random motion, caused by a random load, then the spectrum of the load must have a sharp peak at the resonance frequency. The chimneys were not completely cylindrical as shown in Fig. 1.

There were no provisions to prevent crosswind vibrations. The maximum stresses in the steel during the vibrations derived from the measurements were about 60 N/mm^2 till 80 N/mm^2.

Statistical distributions of the tops of the vibrations are given in Fig. 3. These distributions can be separated in two cases; one for continuous vibrations with almost one amplitude and a more varying one. Both phenomena have been observed at different days.

The statistical distribution of the wind speed in the neighborhood of the chimney is also given in Fig. 3. There was during the time of heavy oscillation an almost constant wind speed. This speed was approximately equal to the critical wind speed with a Strouhal number of 0.2. The critical wind speed for one chimney was 3.5 m/s and for the other chimney it was $V_{crit} = 6.3$ m/s.

From the records the fraction of time, in which crosswind vibrations occur, was estimated. This was for the chimney with $V_{crit} = 3.5$ m/s 0.7 percent and for the chimney with $V_{crit} = 6.3$ m/s 1.2 percent. The number of stress fluctuations in one year can now easily be calculated: $N = 1.5 \times 10^5$ and $N = 3.0 \times 10^5$ respectively.

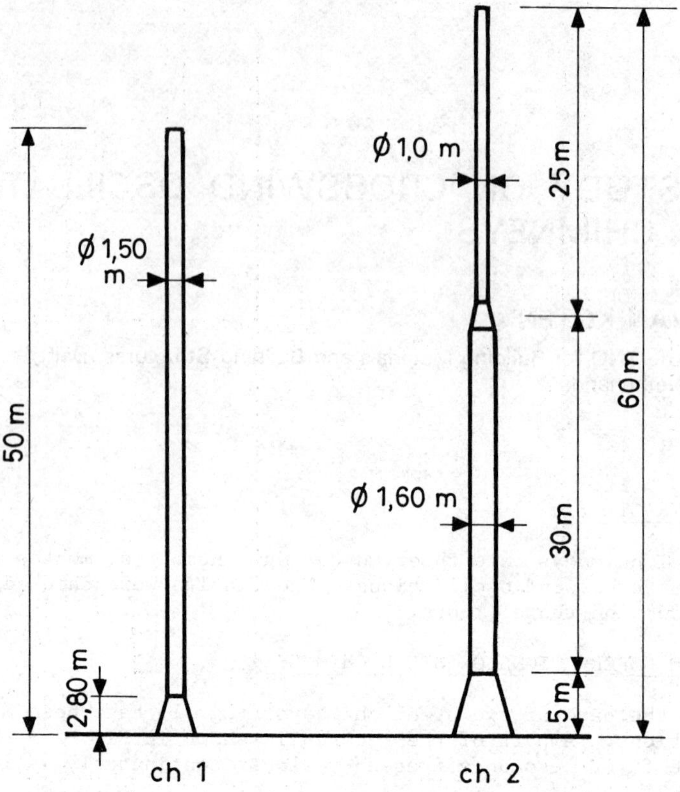

Fig. 1. The sizes of the chimneys.

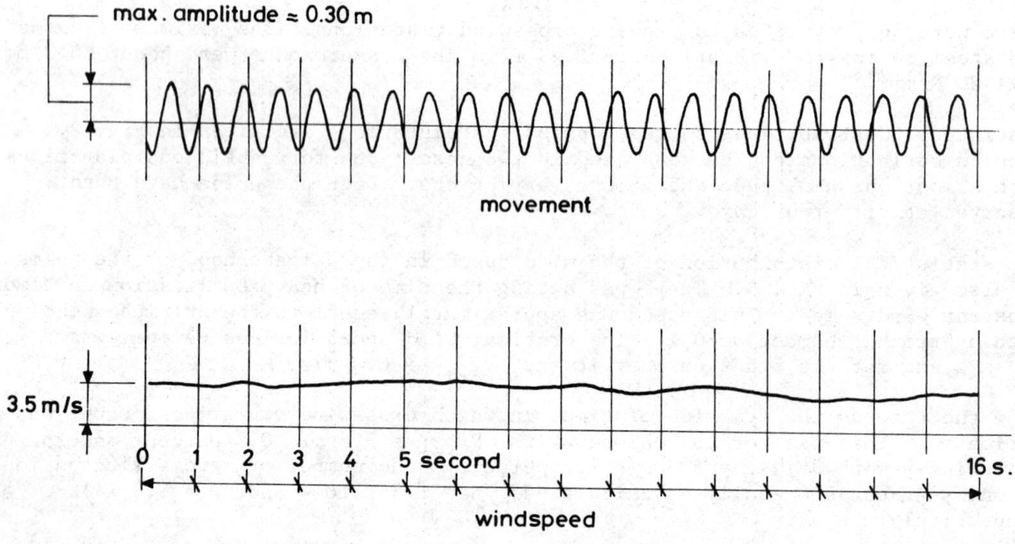

Fig. 2. A part of the measurements.

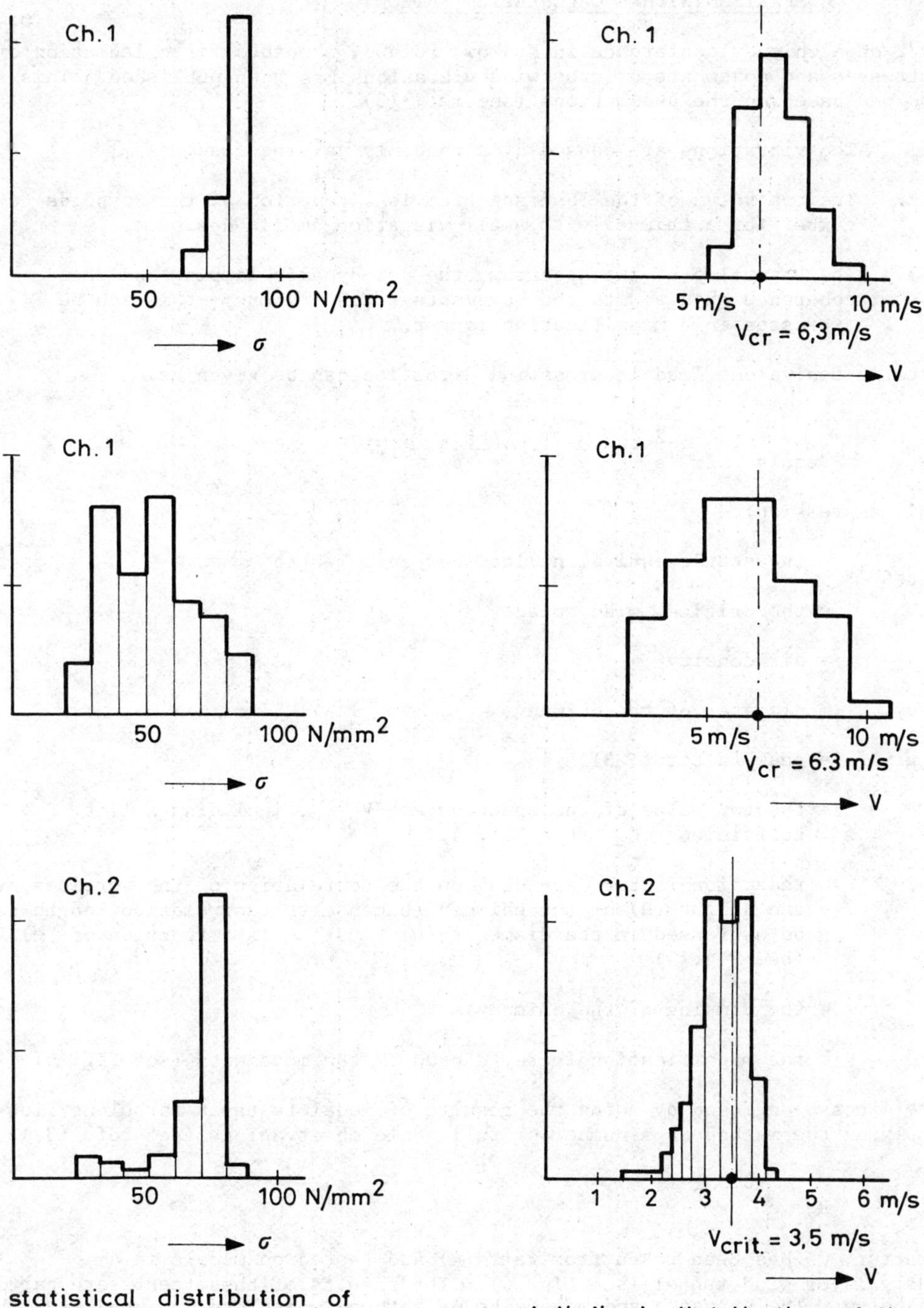
Ch. 1
50
100 N/mm2
σ
Ch. 1
5 m/s
10 m/s
Vcr = 6,3 m/s
V
Ch. 1
50
100 N/mm2
σ
Ch. 1
5 m/s
10 m/s
Vcr = 6.3 m/s
V
Ch. 2
50
100 N/mm2
σ
Ch. 2
1
2
3
4
5
6 m/s
Vcrit. = 3,5 m/s
V
statistical distribution of the stresses
statistical distribution of the windspeed

Fig. 3.

COMPARISON WITH A CALCULATION METHOD

In 1972 on a chimney conference in Krakow, Poland, a method for calculation of the stresses and movements of crosswind vibrations has been published. This method was based on the assumptions (see ref. (3)):

1. The vibrations are caused by a randomly varying load.

2. The top value of the load spectrum is a function of the Reynolds number for a chimney with small vibration amplitudes.

3. The top value of the spectrum, the aerodynamic damping and the coherence change with the movements of the chimney--this can be expressed in a magnification factor.

The static equivalent load in crosswind direction can be given as:

$$q_{equiv} = \frac{1}{2} \rho V_s^2 \cdot D \cdot g \cdot \sqrt{\frac{S_s \cdot R}{\delta}} \quad \emptyset$$

In this expression:

q_{equiv} = the static equivalent load per unit length

V_s = the critical wind speed

ρ = air density

D = diameter of the chimney

g = peak factor (≈ 3)

S_s = the top value of the spectrum at V_{crit}, including a lift coefficient C_c (see Fig. 4)

R = reduction factor depending on the correlation of the vortices over the height (H) of the chimney (because the correlation length can be expressed in the diameter (D) thus R is a function of $\frac{D}{H}$) (see Fig. 5)

δ = the damping of the chimney

$\emptyset$ = the magnification factor caused by the movements (see Fig. 6)

Figure 4 has been found by using the results of model tests in turbulent flow and by applying the method or a number of full-scale observations (see ref. (3)):

$$R = e^{-0.5 \frac{x}{D}}$$

The factor $\emptyset$ has been taken from ref. (4) and tested on models in a boundary-layer wind tunnel (see Fig. 6). The results of these tests are partly given below. The stresses and movements of both chimneys are computed with the given method. In this computation the damping of the chimneys was chosen as 0.3 percent of the critical. This value was calculated from measurement results. The computed values for the stresses were in both cases a factor 1.4 higher as the values found by measurements. The factor of $\emptyset$ as given in Fig. 6 seems to be too high. Some wind-tunnel experiments are carried in a boundary-layer wind

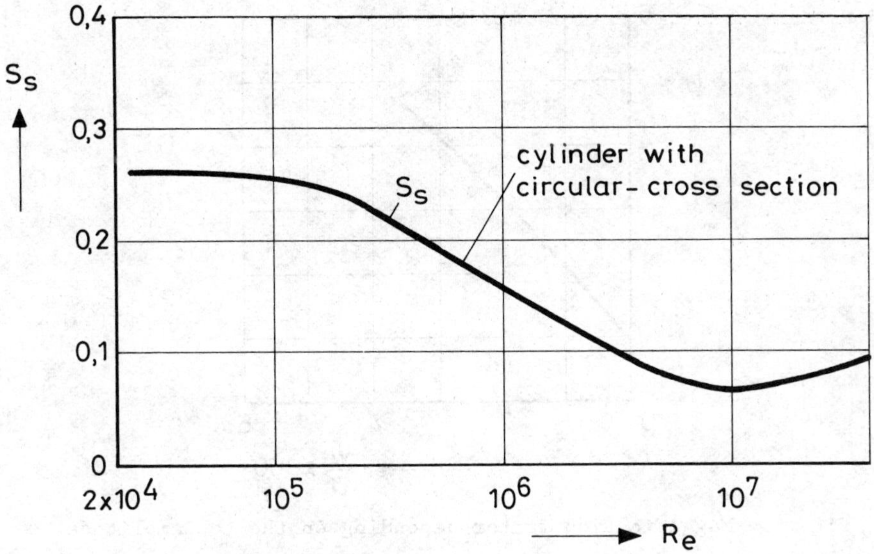

Fig. 4. The top value of the spectrum of crosswind pressure.

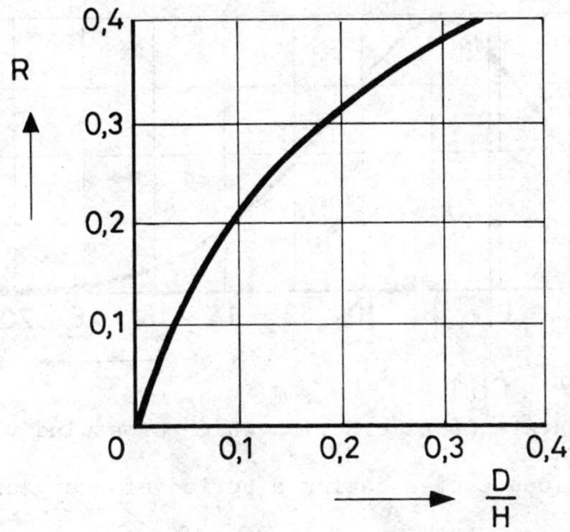

Fig. 5. Reduction factor.

tunnel to test this factor. The results of those tests are reported below. The number of load cycles can be calculated if the probability of occurrence of the critical wind speed can be given.

Supposed is, that vibrations occur if the (10 minutes mean) wind speed is in the range of 0.95 V_{crit} to 1.05 V_{crit}. This occurs roughly during the fraction $0.1 \cdot V_{crit} \cdot P\ (V_{crit})$ of the time. This fraction is far too large compared with the measurements. Probability is another phenomena like stable weather or inverse temperature conditions necessary to cause serious vibrations. To approach the measured value it is suggested to take only one tenth of the fraction, thus the value of $V_{crit} \cdot p\ (V_{crit})$ as percentage of time for vibrations. The curve

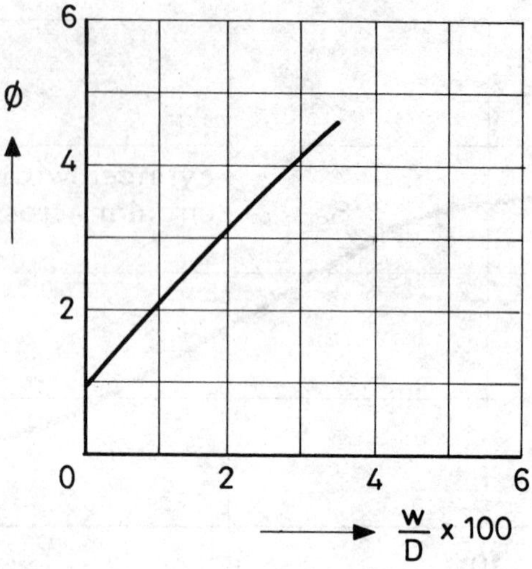

Fig. 6. Magnification factor depending on the top amplitude w.

V • p(V) for a Weibull distribution (at the top of the chimney) is given in Fig. 7. The percentage of occurrence of the critical speeds is now in agreement with the measured values for the two chimneys.

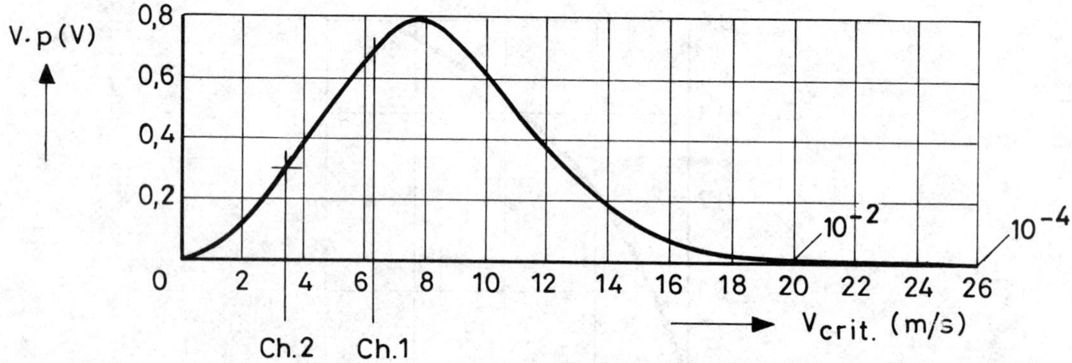

Fig. 7. Percentage of time during which crosswind vibration occurs.

The average number of load cycles during a period of one year can now be given as a function of the ratio:

$$\frac{V_{crit}}{\overline{V}_o}$$

where $\overline{V}_o$ is determined by the statistical distribution of the wind speed at the top of the chimney:

$$p(\overline{V}) = 1 - e^{-\left(\frac{\overline{V}}{\overline{V}_o}\right)^k}$$

$$\overline{V}_o \approx 1.1\ \overline{V}_{mean}$$

The stresses have mainly of the value σ_{max} (see Fig. 3). The number of the excesses of the stress level σ can be estimated with the formula:

$$N = T \cdot p(\sigma) \int_0^\infty \dot{\sigma}\, p(\dot{\sigma})\, d\dot{\sigma}$$

An approximation is:

$$N = T \cdot f_o \cdot \left(\arccos \frac{\sigma}{\sigma_{max}} \right) \cdot \frac{V_{crit}\ p(V_{crit})}{100}$$

This relation between N and σ has been given in Fig. 8. Crosswind vibrations of chimneys with $V_{crit}/V_o > 4$ seems to happen seldom.

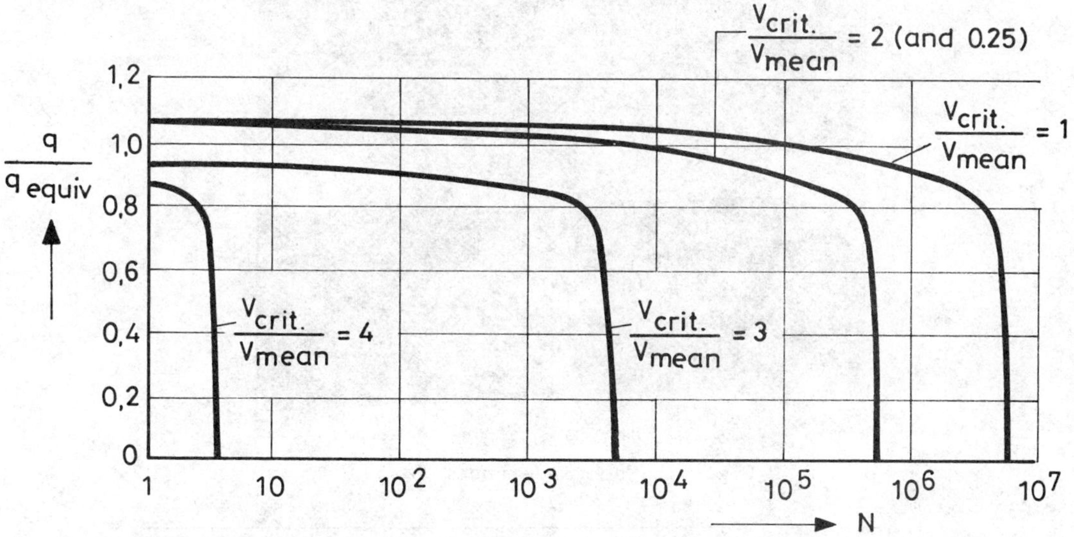

Fig. 8. The number of load cycles in 50 years for different values of V_{crit}/V_{mean}.

WIND-TUNNEL TESTS

Two models were tested in a boundary-layer wind tunnel (see Fig. 9). The turbulence intensity was 0.10. The pressures in crosswind direction were measured over the height at several points as given in Fig. 9. The top amplitude of the vibration was not as constant as found in reality. Only large amplitudes became more constant. The amplitudes were increased by reduction of the damping. The top value of the pressure spectrum for the non-vibrating chimneys was close to the value given in Fig. 4 (see Fig. 10). The top value of the spectrum increased due to aerodynamic damping for the vibrating chimney (see Fig. 11). The correlation and the coherence were not affected by vibrations with top amplitudes of 0.01 D (see Figs. 12 and 13). The coherence of the amplitude of oscillation is clearly noticeable in the different aspects of loading like auto-spectrum and coherence (or better cospectrum and quad-spectrum). The factor of Ø from these tests is given in Fig. 14. The factor Ø seems to depend on the turbulence intensity or possibly on the Reynolds number.

Fig. 9. The wind-tunnel test.

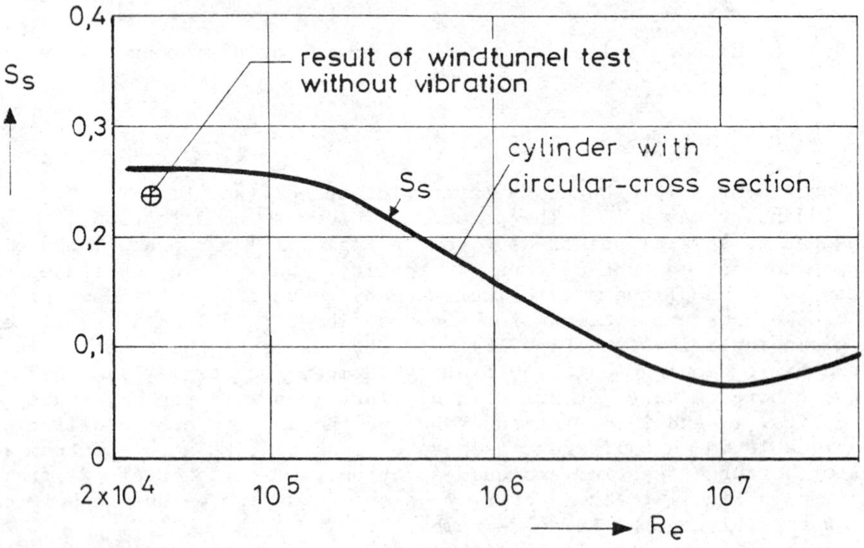

Fig. 10. The top value of the spectrum of crosswind pressure.

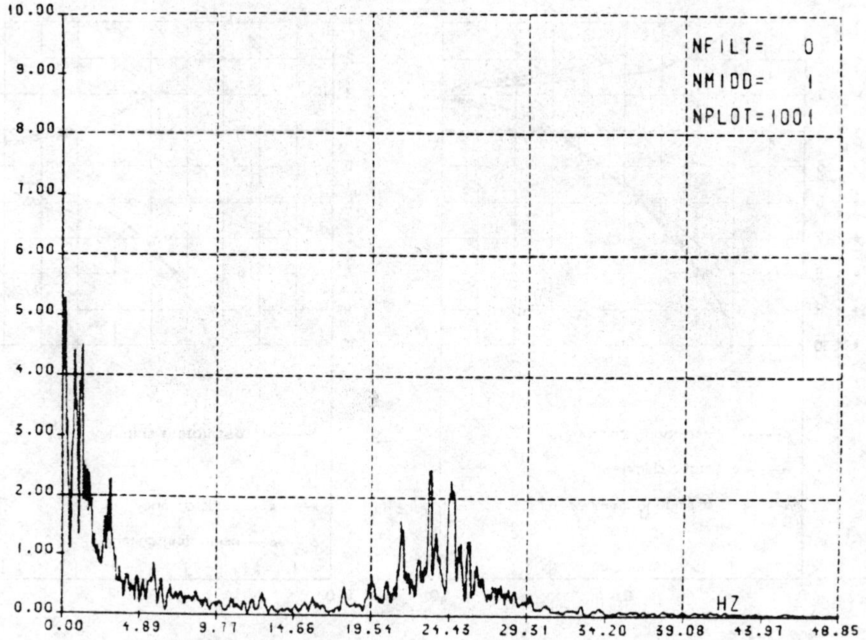

Fig. 11a. Spectral density of the crosswind pressures of a static chimney.

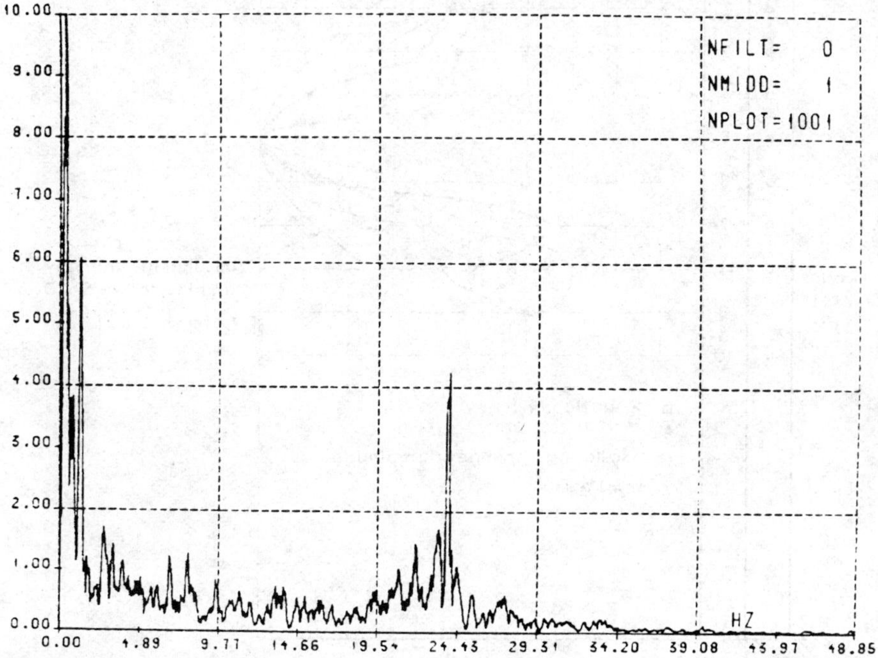

Fig. 11b. Spectral density of the crosswind pressures of an oscillating chimney.

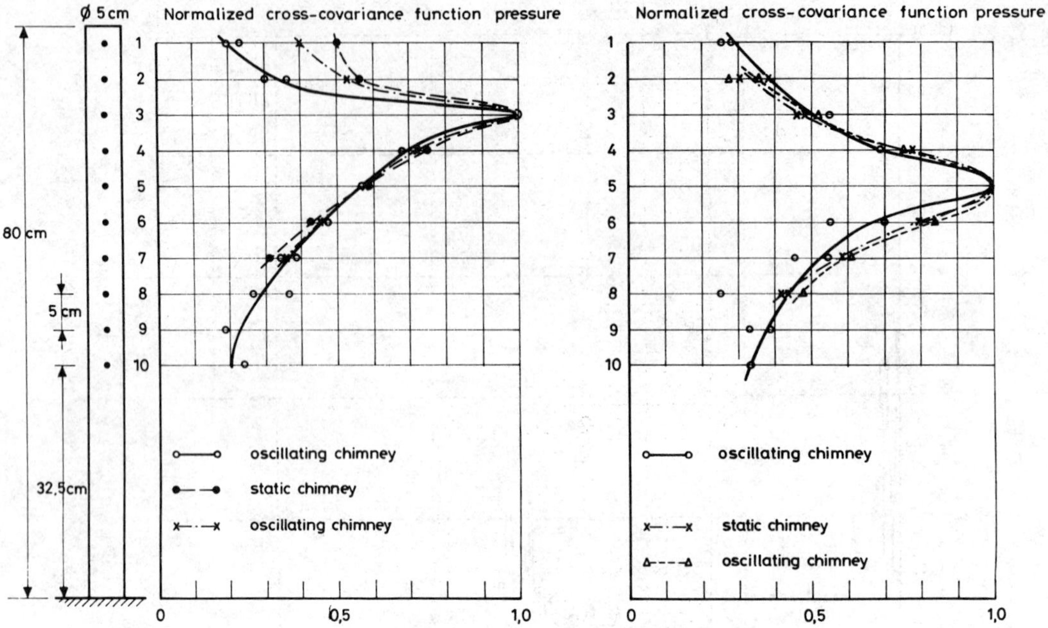
Ø 5cm
80 cm
5 cm
32,5cm
Normalized cross-covariance function pressure
1
2
3
4
5
6
7
8
9
10
oscillating chimney
static chimney
oscillating chimney
0
0,5
1,0
Normalized cross-covariance function pressure
oscillating chimney
static chimney
oscillating chimney

Fig. 12.

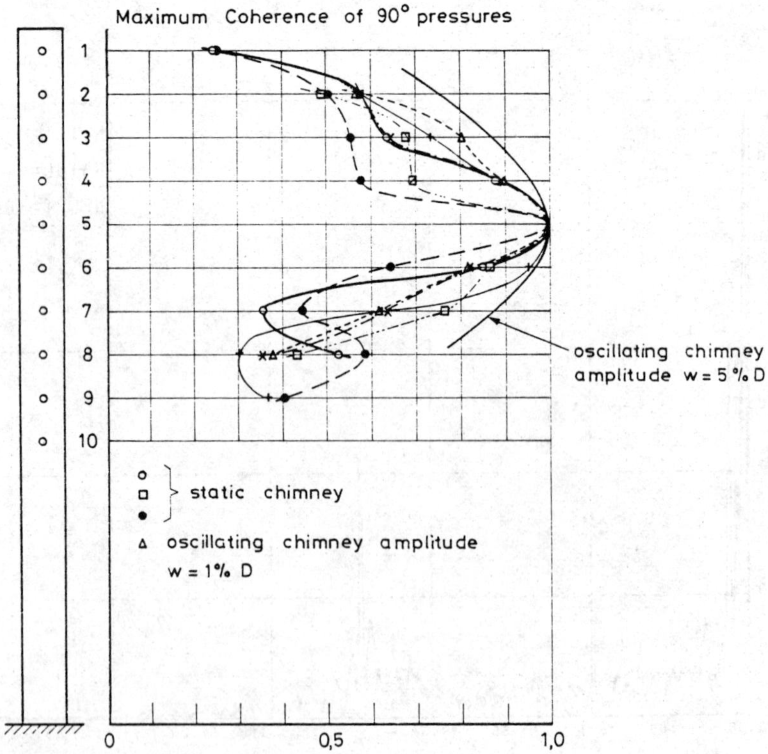
Maximum Coherence of 90° pressures
1
2
3
4
5
6
7
8
9
10
oscillating chimney
amplitude w = 5 % D
static chimney
oscillating chimney amplitude
w = 1% D
0
0,5
1,0

Fig. 13.

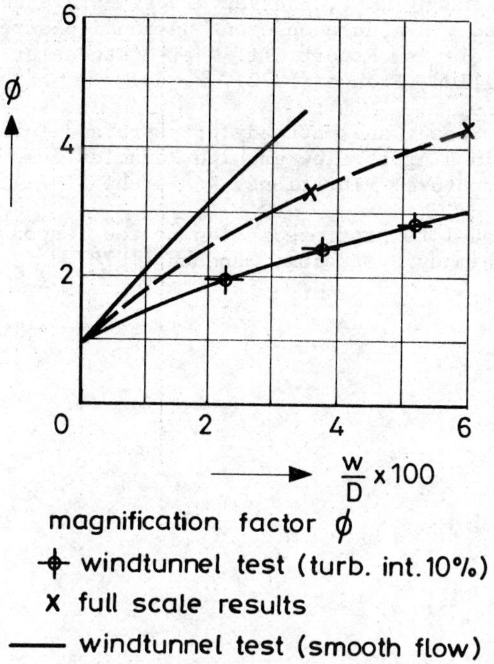

Fig. 14.

CONCLUDING REMARKS

While the number and magnitude of the stress fluctuations in crosswind movements are necessary for chimney design, an attempt has been made to find a simple calculation method. An earlier published method based on smooth flow experiments for the calculation of the stresses had to be corrected. More extended experiments on vibrating chimneys are necessary for a better understanding of the phenomena. The number of load fluctuations must be known for fatigue calculations. It seems that this number is proportional to $\overline{V} \cdot p(\overline{V})$ (where $p(\overline{V})$ is the probability density of $\overline{V}$). In that case the number of crosswind fluctuations of chimneys with a critical speed of about 25 m/s is very limited.

REFERENCES

(1) Vickery, B.J. and Clark, A.W., Lift or across-wind response of tapered stacks, Proc. ASCE, Journal of the Structural Division, January (1972).

(2) Bendat, J.S. and Piersol, A.G., Measurement and analysis of random data, J. Wiley and Sons, New York (1966).

(3) Koten, H. van, Wind load on chimneys, Proc. of the IASS Symposium on Industrial Chimneys, Krakow, Poland, September (1973).

(4) Jones, G.W., Cincotta, J.J., and Walker, R.W., Aerodynamic forces on a stationary and oscillating circular cylinder at high Reynolds numbers, NASA report (1969).

(5) Loisseay, H. and Szechenyi, E., Analyse experimentale des portances sur un cylindre immobile soumis a un ecoulement perpendiculaire a son axe a des mombre de Reynolds eleves, Report Off. Nat. d'Etudes et de Rechersches Aerospatiales Chatillon, France (1972).

(6) Nunen, J.W.G. van, Steady and unsteady pressure and force measurements on a circular cylinder in a cross flow at high Reynolds numbers, Proc. Semaniar on flow-induced Structural Vibrations, Karlsruhe (1972).

(7) Vickery, B.J., A model for the prediction of the response of chimneys to vortex shedding, Chimney Symposium, München (1978).

STATIC AND DYNAMIC BEHAVIOUR OF A COOLING TOWER MADE OF A STEEL STRUCTURE

J. A. HERTIG

Institut d'Economie et de Technologie Energitique, Lausanne, Switzerland

ABSTRACT

The present paper relates the static and dynamic experimentations done on models of cooling towers made of a steel structure. The object of the static load experiment is to determine the influence of the specific form of the steel cooling tower, and of the nearby buildings, on the local shape factors.

A preliminary study of dynamic loads is also presented. This study is done on two models so as to examine the specific dynamic issue of cooling towers, under the aspects of similarity, model technology, and the particular dynamic behaviour of such a structure.

NOTATIONS

a	acceleration of the structure	m/s^2
D	dimension	m
E	elastic modulus	N/m^2
F	force	N
g	acceleration of gravity	m/s^2
I	moment of inertia	m^4
m	mass	kg
t	time	s
δ	deflection	m
ω	frequency	Hz (1/s)

SUBSCRIPTS

o	reference value
M	value for model
R	value for prototype

1. INTRODUCTION

This paper describes a study made as part of a German/Swiss research programme to evaluate a new design of an helium-cooled, nuclear power plant*.

**HHT Programm. Study for Alusuisse Engineering SA and Swiss Federal Institut for Reactor Research.*

The basic question considered here is whether or not a limit exists for the size of cooling towers. In order to minimize problems associated with the structural material, a steel structure has been chosen.

There is no obvious reason to believe that the size of structures could be limited. The question is better formulated by asking how the risks increase with the size of the tower. This problem has not been investigated much, so it seems reasonable to start with the size of existing towers and increase progressively.

The present study investigates the static and dynamic behaviour of a large cooling tower placed in a turbulent shear flow. From an aerodynamic point of view, the size effect is dependent on the ratio of the height of the tower to the thickness of the planetary boundary layer.

2. STATIC LOADS

2.1. Static load measurements

The purpose of these measurements is to provide an answer to the following questions :

- What is the influence of the particular shape of the usual cooling tower (polygonal cylinder) on the associated shape factors ? (Fig.1)
- What would be the influence of protruding beams, added at the edges along the polygonal contour, on the associated shape factors ?
- What is the influence of nearby buildings, such as other cooling towers, machinery buildings, etc., on the associated shape factors ?

Fig. 1 Experimental arrangement
Foreground: cooling tower model, existing cooling tower and buildings.
Background: turbulence generator cylinder ϕ 113 mm.

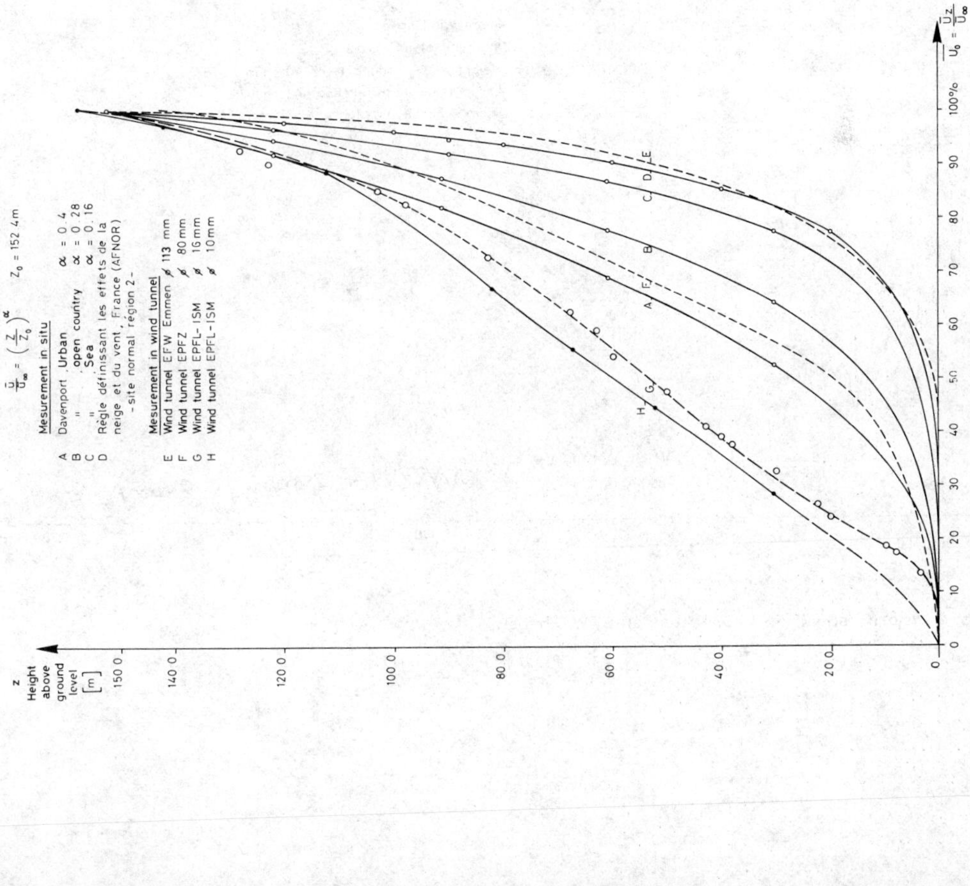

Fig. 2 Mean velocity profiles in function of height.

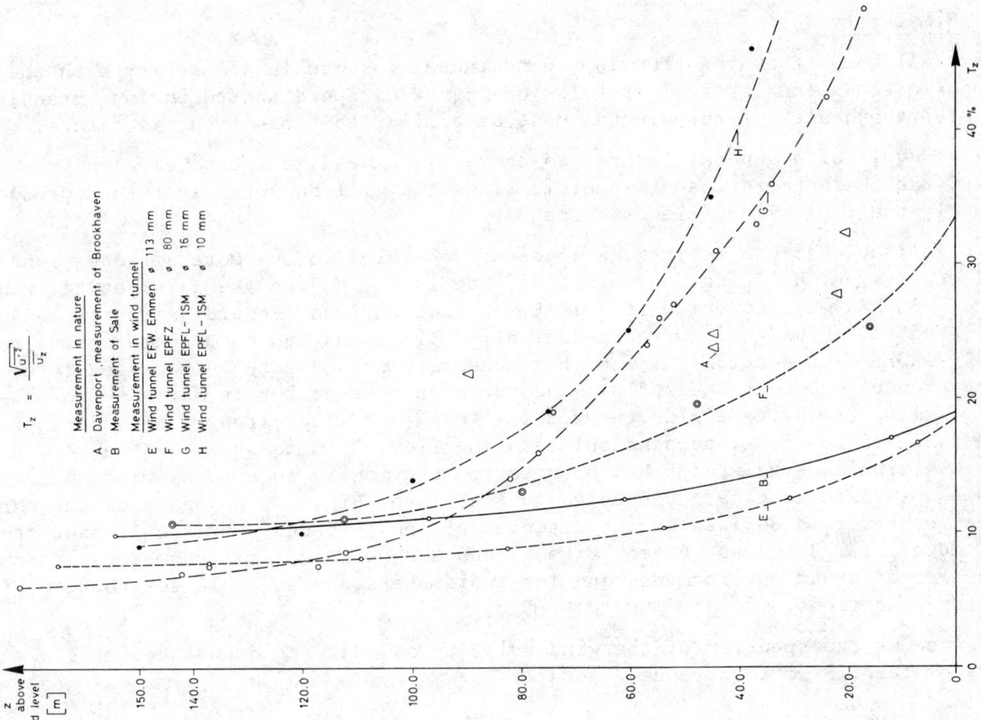

Fig. 3 Turbulent intensity profiles as a function of height.

2.2. Similarity

It is well known that the flow in a wind tunnel must be in similarity with the atmospheric flow. The vertical profiles of mean wind speed and turbulent intensity, and also the scales of turbulence, must be similar to those found in nature.

This study is of a general nature, so there is no need to reproduce with great accuracy the characteristics of specific sites. It will be sufficient to reproduce properly the typical profiles and scales.

It is advisable to use as large a model as possible. Furthermore, if one wishes to include adjacent buildings in the study, one is obliged to use a largesized wind tunnel. Such a wind tunnel exists in Emmen, Switzerland (Federal Aircraft Industry), but it is designed for aeronautical applications and has, therefore, a relatively short test section. It was thus necessary to add a turbulence generator to get a proper turbulent shear flow. The turbulence generator is derived from that of J. Armitt (1) by replacing the vertical elements with circular cylinders placed into two parallel rows, perpendicular to the flow. This is shown in Fig. 1. This type of turbulence generator has been tested previously in several wind tunnels.(2) Results of these tests are presented in Fig. 2 and Fig. 3. The scale of the turbulence generator is defined by the diameter of one cylinder. For the present study a diameter of 113 mm was chosen. Figs. 2 and 3 show that small diameters are suitable for an urban environment, and large diameters are suitable for shore or flat country conditions.

Fig. 4 shows the spectrum of the wind velocity as measured downwind of the elements of the 113 mm turbulence generator.

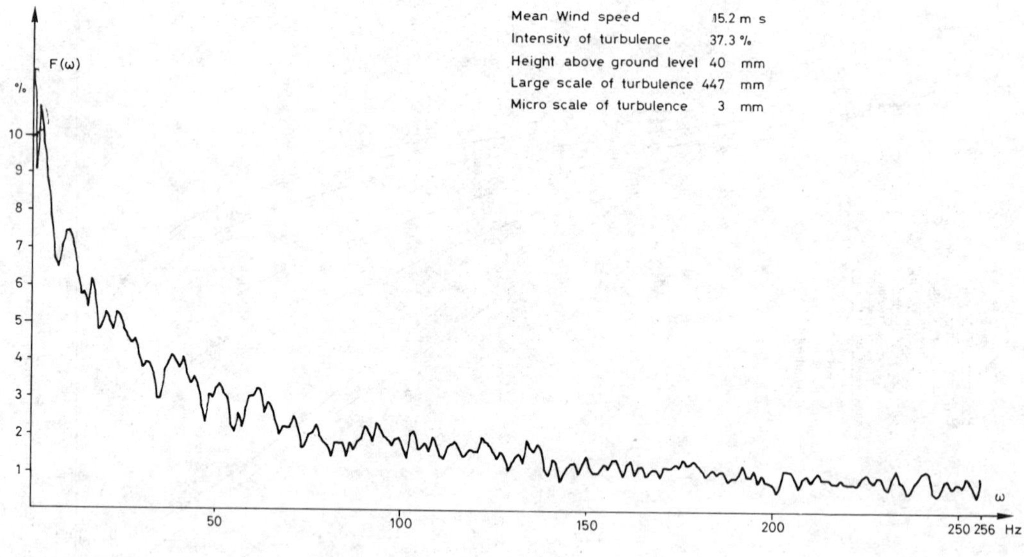

Fig. 4 Velocity spectrum in wind tunnel EFW Emmen.

2.3. Experimental procedure

A rigid model (scale 1 : 150) of the cooling tower has been constructed. It has been placed on a plate inside the 7 x 5 m test section of the large wind tunnel in Emmen. The 113 mm turbulence generator has been put on the same plate, upwind of the model. Between plate and model, a weight indicator has been inserted. This arrangement was first used to determine the shape factors and the effect of protruding beams. Subsequently, the other buildings were added on the plate.

2.4. Results

The distribution of the local shape factors on the model without influence of nearby buildings is shown in Fig. 5.

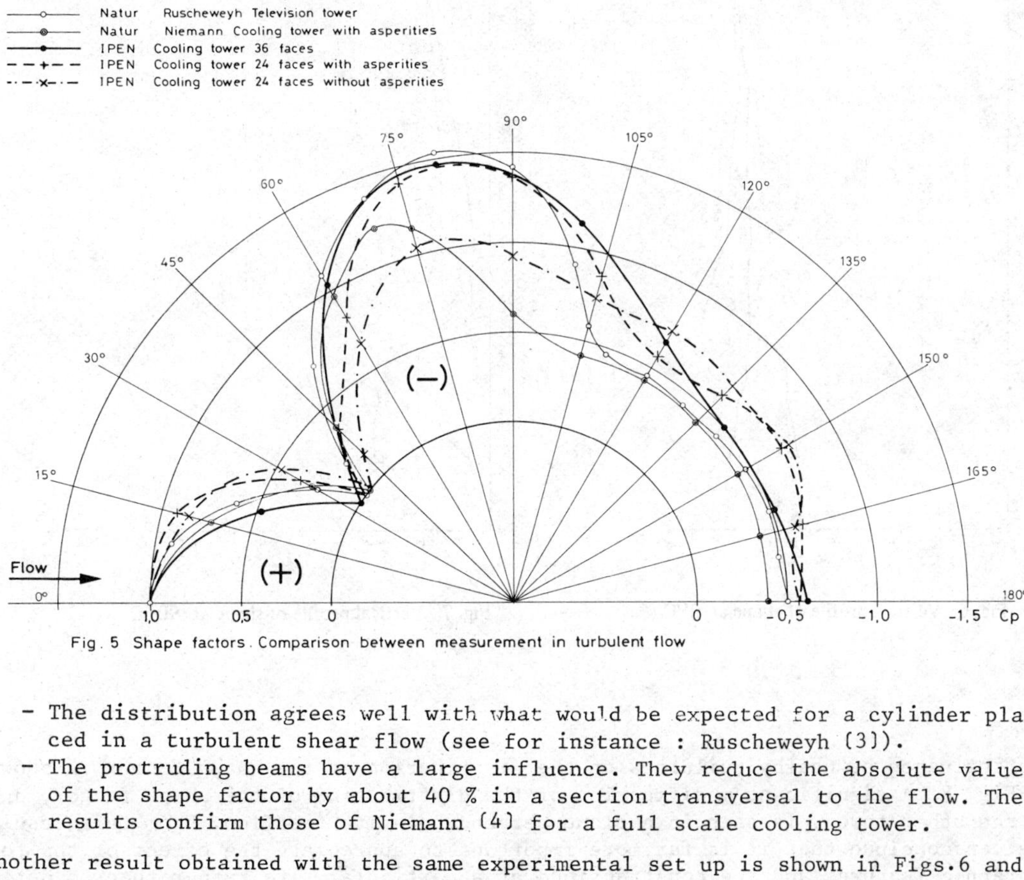

Fig. 5 Shape factors. Comparison between measurement in turbulent flow

- The distribution agrees well with what would be expected for a cylinder placed in a turbulent shear flow (see for instance : Ruscheweyh (3)).
- The protruding beams have a large influence. They reduce the absolute value of the shape factor by about 40 % in a section transversal to the flow. These results confirm those of Niemann (4) for a full scale cooling tower.

Another result obtained with the same experimental set up is shown in Figs.6 and 7. The absolute value of the local shape factor, in a transverse plane, increase in the lower part of the tower if calculations are made by using the dynamic pressure as derived from the wind profile. This is done in almost all codes of practice. If, on the other hand, the calculations are made using the dynamic pressure at the top of the tower, the resulting shape factor is nearly constant for the vertical profile of the tower (Fig. 7). Opposite results are obtained at the upwind edge of the tower (i.e. the shape factor is nearly constant if a variable dynamic pressure is used, but decreases close to the ground when calculations are made with a constant dynamic pressure) (Fig. 6). One can therefore conclude that the flow is three-dimensional. Similar results have been obtained by J. F. Sageau (1978) on a full-scale cooling tower [5].

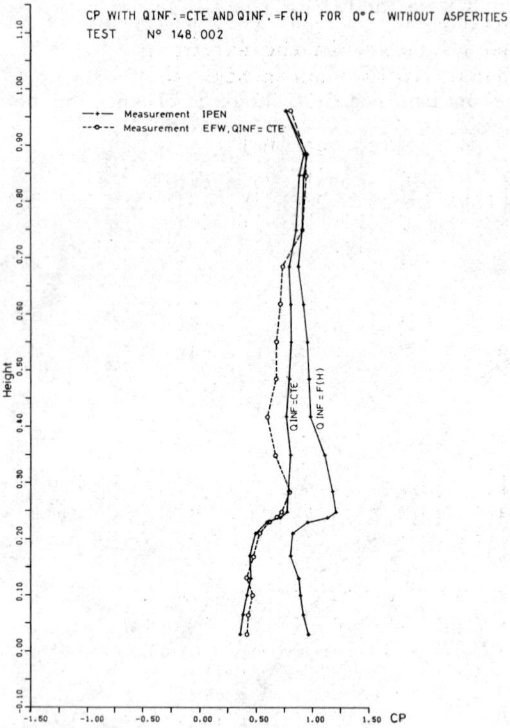

Fig. 6 Vertical profile of shape at 0°C

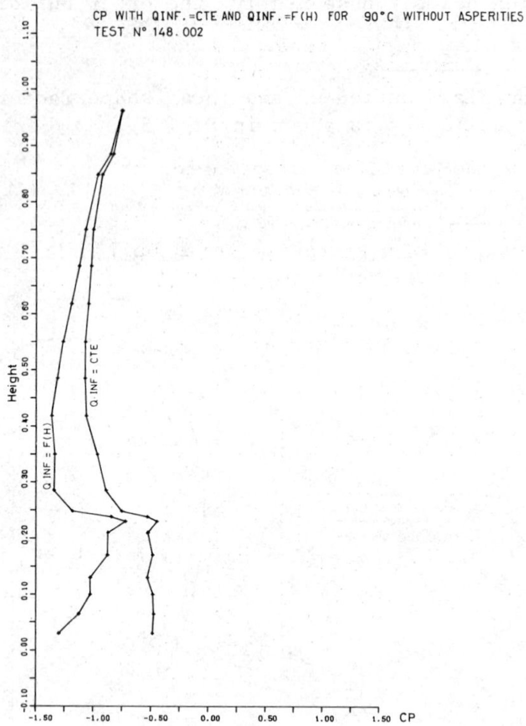

Fig. 7 Vertical profile of shape at 90°C

The influence of nearby buildings on the distribution of the shape factor is shown in Fig. 8. A significant modification of the distribution of the shape factor, due to the other tower, is seen in a plane perpendicular to the flow. From this result one can conclude that it is far more important to appreciate the effect of the topographic features and the constructions at a particular site rather than to determine, in great detail, the shape factors of a tower in an undisturbed environment.

3. DYNAMIC LOADS

3.1. Generalities

Dynamic calculations of the considered cooling towers show general modes with almost identical natural frequency modes. A fundamental question is whether the wind can excite one or sevreal modes. This question is very important in view of the slow damping to be expected with steel structures. It is important, on the other hand, to have an idea of the breakdown of the safety factor with respect to wind speed. (Ratio of the critical wind speed to the maximum wind speed).

In conventional dynamic tests, a high-pressure wind tunnel is normally requested in order to have the correct similarity for the structural aerodynamics and the stability behaviour.

The experimental programme was divided into two parts. The first part consists of conventional experimentation in a large wind tunnel of relatively high velocity turbulent flow. The second part was designed to ascertain the extent to which it is possible to test dynamic and stability behaviour without a high-pressure wind tunnel.

The first part of the experimentation is to be carried out in the wind tunnel of the Institute for Aerodynamics at the ETHZ. This tunnel has a section of 2 x 3 m. Use of an 80 mm turbulence generator gives the velocity profile of Fig. 2 and the turbulent intensity profile of Fig. 3. A 1 : 200 model, similar to the full-scale prototype, is presently being built. The proper experimentation is scheduled to start at the beginning of 1979.

Some preliminary results of the experimentation with two 1 : 500 models are presented here as contribution at the second part of the experimentation.

3.2. Similarity

3.2.1. Structural dynamic similarity

To get rigorous structural dynamic similarity, the following conditions must be met.

a) geometrical conditions : the scale ratio must be applied to all parameters having a length dimension. This determines the scales of displacements, cross sections and moments of inertia. (δ_M/δ_R ; I_M/I_R)

b) Kinematic conditions : since velocity and acceleration are, respectively, the first and the second derivatives of space with respect to time, one must satisfy the conditions of homochrony and equal acceleration for the structure.

$$\frac{D_o}{t_o^2 a_o} = cte$$

$$\frac{V_o^2}{a_o D_o} = cte$$

This corresponds for flow to the conservation of the Strouhal number if $\frac{1}{t_o} = \omega_o$

the velocity of deflection of the structure is given as :

$$\frac{D_o}{t_o V_o} = cte$$

This corresponds to the conservation of the Froude number in the fluid if the scale of the acceleration is homogeneous to the scale of the gravity g.

It is generally accepted that the acceleration of gravity, or the weight of the structure itself, has no influence on the motion of the structure.

c) dynamic conditions : once the kinematic conditions are set, the Newtonian law gives the ratio between the scales of force and mass. The previous remark concerning gravity can be made here also. Precise similarity thus requires that the scale of the force is equal to the scale of mass. Consequently, one is left with only one degree of freedom for satisfying the aerodynamic conditions. One possibility is to choose the scale of force, from which all the other scales :
mass, specific mass, strain and elastic modulus will be derived. One might also change the scale of acceleration, if necessary.

3.2.2. Aerodynamic similarity

The similarity conditions to be satisfied by the turbulent shear flow in the tunnel have been discussed in § 2.2. In addition to these conditions, the Strouhal number must be conserved. This determines the scale of force. In order to have accurate similarity simultaneously for the dynamic and the aerodynamic conditions, it is required that the Froude Number be conserved. This leads to rather low wind velocities; and therefore, the turbulence values are too small.

Consequently, it was decided to make two models, one for Froude similarity with very low speeds and one for higher flow speeds.

3.3. Description of the models

In order to obtain a good similarity of behaviour of the models, in regard to damping, to stress and to strain, steel is chosen for the structural material of the model. The similarity of rigidity is then required, not the similarity of inertia. It is assumed also that the limit of stability is attained with the horizontal deflection of a ring or of a vertical post. Taking these assumptions into consideration, it becomes possible to realize a steel wire model similar to a steel tube cooling tower.

For both models, the scale is 1 : 500. The first one (Model No. 3) is calculated for a velocity scale of 1 : 22.36; that means, a wind speed of 2.24 m/s in the wind tunnel simulates a 50 m/s wind speed in nature. The mean diameter corresponding to the rigidity of the real tubes is 0.2 mm (Fig. 9).

The second model is calculated at the 1 : 3.33 velocity scale, to match the maximum turbulent shear flow of 15 m/s, which corresponds to a 50 m/s wind speed in nature (Model No. 4).

In the present study, we only observed the horizontal deflection spectrum of the top of the cooling tower.

3.4. Results

Measurements of the natural frequencies of the structure show a correct relation between both models (Fig. 10). Fixing a very thin polythene sheet as a wall on the wire structure introduces an important damping. This damping is greater in the case of lower velocity model.

Comparison of the dynamic behaviour of the models is done by dynamic and aerodynamic admittances. An attempt to measure the dynamic admittances of the models has shown a very delicate behaviour.

The deflection spectrum at one point of the structure, submitted to a sinusoidal oscillation at another location, presents several aspects which are dependent upon the excitation frequencies. For low frequencies particularly, the displacement spectrum shows higher values than for high frequencies.

It is almost impossible to find only one mechanism for all oscillations spectrum. This becomes even more difficult when taking other points of oscillations into account. A combined spectrum is determined which is the product of the mechanical admittance and the aerodynamic admittance.

This product is obtained by dividing the displacement by the wind spectrum (Figs 11 and 12).

This combined spectrum shows an important amplification due to wind of the 48 Hz frequency for the high velocity model. This corresponds approximately to the first natural frequency. The amplification of the 75 Hz frequency is very interesting. This does not correspond to a natural frequency mode, but seems to be an aerodynamic excitation. Such amplification can be on the other spectrum, too (Model No 4).

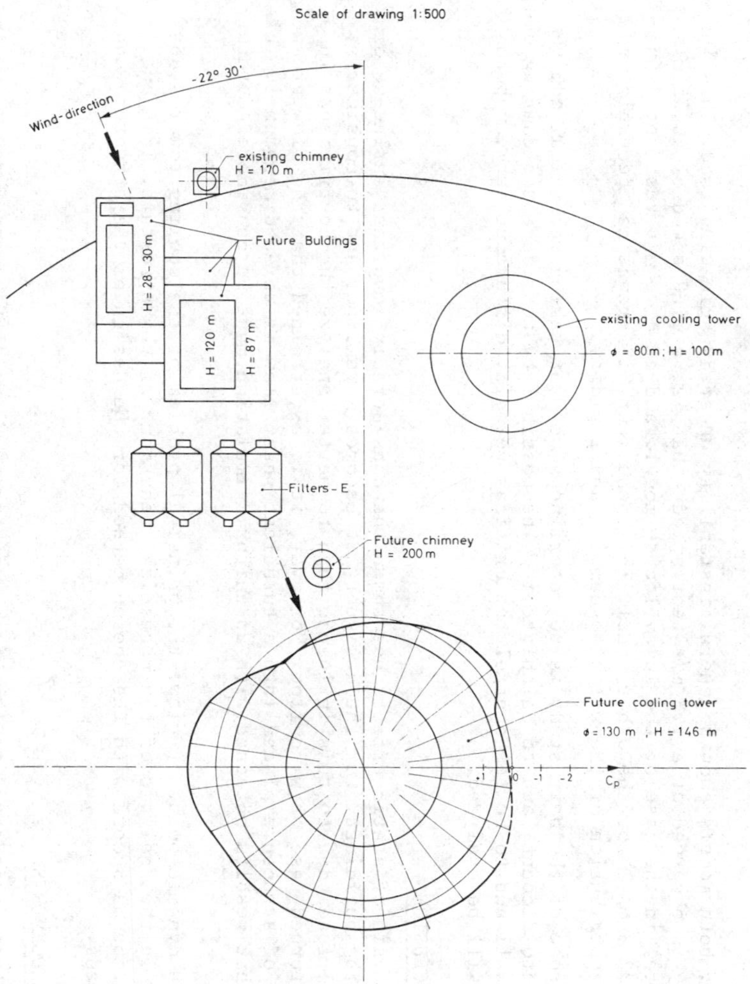

Fig. 8 Shape factor Cp with the influence of near by buldings

Fig. 9 Dynamic model number 3 Measurement of mechanical properties.

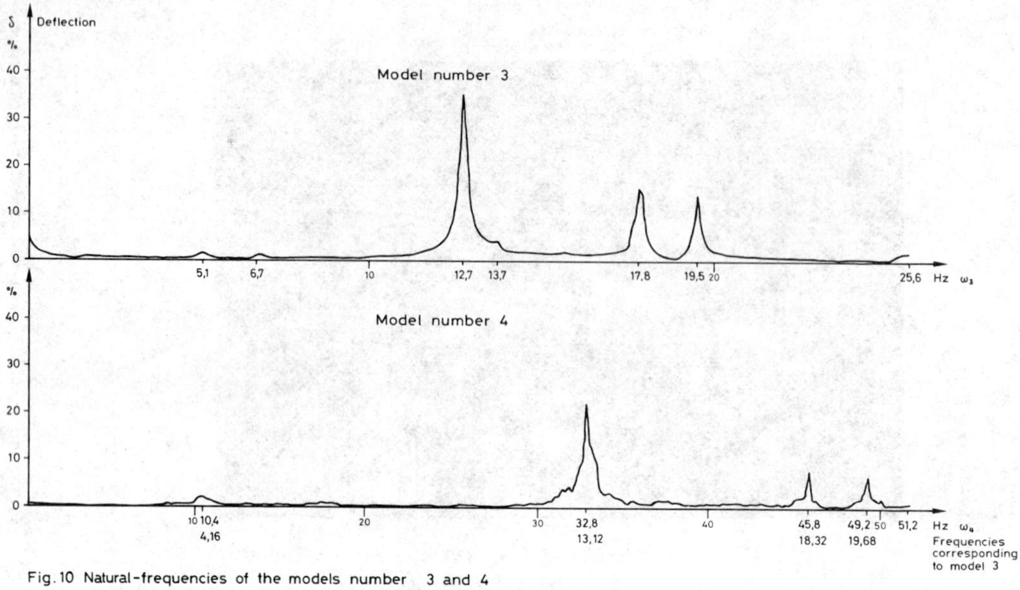

Fig. 10 Natural-frequencies of the models number 3 and 4
Ratio between the frequencies of the two models $\omega_4 = 2{,}5\ \omega_3$

Finally, both models show a general instability under dynamic wind load. This instability is observed at a 1.8 m/s test speed (that is, a wind speed of 40.25 m/s in nature) in the case of the low velocity model, and at a 10 m/s test speed (nature : 33.3 m/s) for the other model. This behaviour difference can be explained by the contribution of the walls to damping and rigidity.

The purpose of the next study is to determine a way for taking mechanical admittance into account, and to calculate it. The possibility to reduce the damping due to the wall and to better control the critical velocity difference between both models will be analyzed.

4. CONCLUSIONS

The analysis of static effects has shown the considerably important influence of the nearby buildings. For the designer, it becomes more important to consider asymetrical loads due to this influence than to better analyze the coefficients in a non-disturbed case of wind-flow. Moreover, it is ascertained that the flow around a piece of structure placed into a turbulent shear flow is three-dimensional. This means that results obtained with an uniform simulation must be considered carefully.

From the dynamic point of view, the principal problem of the cooling tower study resides in the analysis of the mechanical behaviour of the structure. This analysis must be in accord with the chosen method for the design of the structure under wind loads.

ACKNOWLEDGEMENTS

The writter is indebted to Alusuisse Engineering S.A. (ALESA), and Federal Institute for Nuclear Reactor Research (EIR), who support this research.

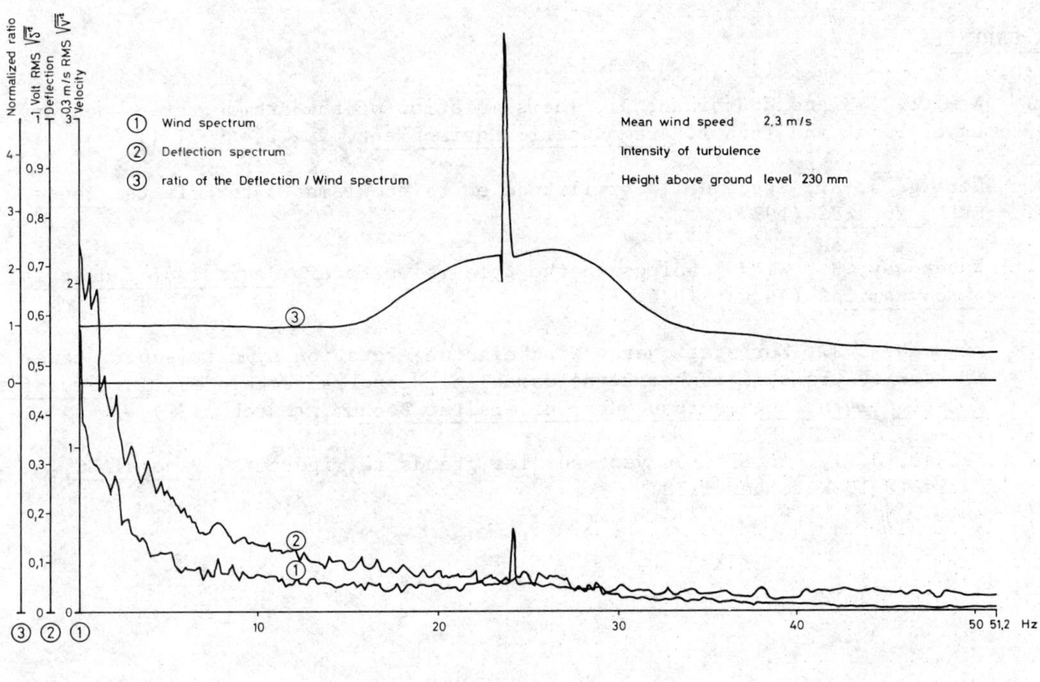

Fig. 11 Comparison between velocity, deflection spectrum and their ratio for model N° 3. Wind tunnel LHYDREP

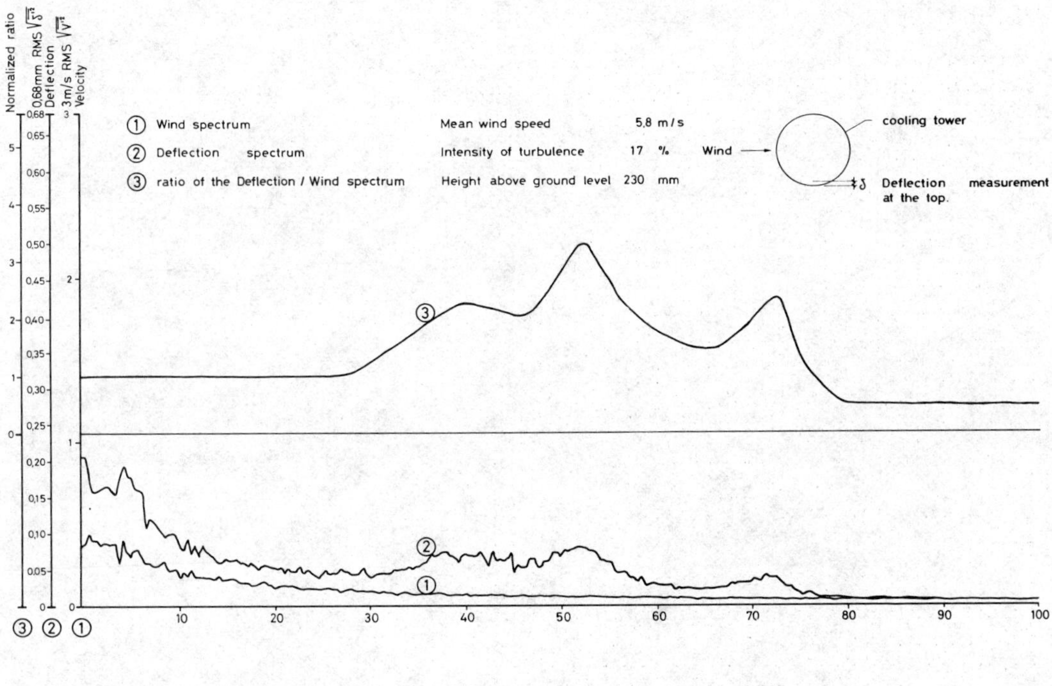

Fig. 12 Comparison between velocity, deflection spectrum and their ratio for model N° 4. Wind tunnel LHYDREP

REFERENCES

1. Armitt, J., and Cournihan, J., The simulation of the Athmospheric Boundary Layer in a Wind Tunnel, Athmospheric Environment, Vol.2, 49-71 (1968).

2. Hertig, J.-A., Etude de la similitude entre écoulements turbulents, Thèse EPFL, No 288, (1977).

3. Ruschewey, H., Wind loadings on the television tower, Journal of industrial aerodynamics, (August 1976)

4. Niemann, J.J., Zur stationären Windbelastung rotationssymmetrischer Bauwerke im Bereich transkritischer Reynoldszahlen, Rapports techniques, Institut für Konstructiven Ingenieurbau Ruhr-Universität Bochum, (March 1971)

5. Sageau, J.F., L'action du vent sur les grands réfrigérants, Paper IASS, Paris (8-10 November 1978)

DISCUSSION ON SESSION VI

Dynamic Response--Tall Buildings and Towers

Rapporteur: A. N. L. Chiu
University of Hawaii at Manoa
Hawaii

Paper by T. A. Reinhold, P. R. Sparks--THE INFLUENCE OF WIND DIRECTION ON THE RESPONSE OF A SQUARE-SECTION TALL BUILDING

Question: N. Isyumov. This is a general question to the authors of several papers which describe the use of pressure data to predict overall building responses. There are basically two difficulties: (1) the significance of aeroelastic effects and the difficulty of their evaluation without actual aeroelastic data; and (2) the extrapolation of single building results to a real environment. The effects of adjacent buildings presented by Dr. Saunders would be difficult to apply in such cases. In view of these difficulties do the authors feel that their methods are a practical alternative to actual aeroelastic models? The latter can be quite simple as usually only the fundamental modes are important.

Reply: T. A. Reinhold, P. R. Sparks. With regard to the first difficulty, we have tried to point out that the rigid models clearly cannot be used when aeroelastic effects are important. However, our comparison with Rosati's work and the conclusions from Saunders' and Melbourne's work indicates that aeroelastic effects are not important for reduced velocities below about 10. Furthermore, if the building is designed for a serviceability limit state rather than an ultimate limit state, the acceptable levels of motion of the building will be so small that aeroelastic effects will be small. Regarding the second difficulty, the particular results presented are for an isolated building, but there is nothing that precludes the testing of a model of this type in a more realistic environment where other structures are adjacent. Since it is possible to use the pressure measurements to estimate response levels for a particular structure, it is difficult for us to understand why the effects observed by Dr. Saunders could not be applied, to provide estimated interference effects, with equal utility to the results of either method. We feel that the two difficulties suggested by Dr. Isyumov are not the most important drawbacks. The single greatest problem is the initial cost of the transducers; but we feel that the direct measurement of dynamic forces is particularly advantageous for the modification of dynamic forces on structures of simple geometric shape because universal force spectra can be determined. Furthermore, we feel that the use of the direct pressure measurements on an aeroelastic model could provide a particularly useful research tool since it would be possible to understand the pressure field more completely.

Reply: A. Kareem. I have proposed an approach to estimate the crosswind response which divides wind loads into two parts. One results from the fluctuation in the approach flow and the excitations caused by the wake, whereas, the other is caused by the motion of the building at high reduced velocities. The first part can be evaluated using pressure fluctuations or a force balance with a high frequency response. The motion induced loading can be derived using the proposed system identification technique.

The forcing function obtained using the force balance will provide an average wind loading and without considerable modifications to the system multilevel forcing function cannot be obtained. However, the pressure fluctuations approach provides the multilevel forcing function and also furnishes information regarding the local characteristics of fluctuations for the cladding design.

With regards to interaction of surrounding buildings, the estimate of any adverse effects can be made using a number of results available. In concluding these remarks, I feel this approach although in its early stages has a promising future.

Paper by T. A. Reinhold, P. R. Sparks--THE INFLUENCE OF WIND DIRECTION ON THE RESPONSE OF A SQUARE-SECTION TALL BUILDING

Question: A. Kareem. In your paper Figure 11 shows peak response based on peak/rms factor of 3.5. This is true for a Gaussian process. However, at high reduced velocities the crosswind response exhibits departure from Gaussian, therefore 3.5 might not be a correct number to use.

Reply: T. A. Reinhold, P. R. Sparks. A peak/rms factor of 3.5 was chosen as being typical of a peak factor which may range from about 3 to about 4. The values quoted were therefore only estimates of expected peaks. While the method presented in the paper is applicable for reduced velocities likely to be encountered by a prototype structure, it is not applicable for very high reduced velocities, i.e., in excess of 10. One of the points we emphasized in the paper was that the peak crosswind response was much more important than the peak alongwind response. The peak factor associated with the non-Gaussian process would likely be larger than 3.5 and would further strengthen the conclusions.

Question: N. Isyumov. Using the technique of the authors, which neglects aeroelastic effects, one would expect better agreement for the dynamic along-the-wind than the lateral response. This is not apparent from Figures 6 and 7. In fact, σ_x/H underestimates Rosati's data by some 100 percent. On a further note, the difficulties of aeroelastic modeling and the extrapolation of the results to other structural properties are over stated. Could the authors comment on both points?

Reply: T. A. Reinhold, P. R. Sparks. The authors accept that there may be points in Figures 6 and 7 in which our results and Rosati's differ by as much as 100% but these points are associated with levels of damping which are far higher than one would normally encounter in a structure or with very low reduced velocities. There are points on Figures 6 and 7 where there is almost perfect agreement between the two sets of results. Nevertheless, the general trend for inwind response is a difference of 20%-40% as stated in the paper. Unfortunately, Rosati did not document his flow conditions in detail and so it was difficult to compare the two flow regimes.
The authors believe that whilst aeroelastic model tests are appropriate for specialized buildings the use of the method described in the paper would be advantageous in terms of cost and time for general modification of dynamic forces on simple rectangular structures.

Paper by K. C. S. Kwok, W. H. Melbourne--CROSS-WIND RESPONSE OF STRUCTURES DUE TO DISPLACEMENT DEPENDENT LOCK-IN EXCITATION

Question: J. C. R. Hunt. In the experiments reported here the ratio of diameter to integral scale is about 24 m/100 m ≃ 0.24, so that the integral scale is about equal to the wavelength of shed vortices. Britter, Hunt and Mumford (JFM, 1979) have found that at this ratio of a/Lx the turbulence amplifies the vortex shedding part and the velocity spectrum, whereas at a higher a/Lx (i.e., for large buildings or buildings in wakes) the effects of turbulence on vortex shedding may be less.

Reply: K. C. S. Kwok, W. H. Melbourne. The authors were not aware of this observation and its effect may indeed be hidden in a part of the face spectra measured on this size of tower. In all our measurements on models of habitable buildings (i.e., stiffer than towers) we have never seen any evidence of amplitude dependent effects.

Question: C. Kramer. You spoke about towers. But, the Reynolds numbers in your experiments were far below the range of practical interest. Wouldn't you expect a change in the crosswind response for the cylindrical towers for a pressure distribution corresponding to a transcritical Reynolds number?

Reply: K. C. S. Kwok, W. H. Melbourne. Yes, we would expect a change in the force spectra for the circular tower at higher Reynolds number. We are in the middle of a program at the moment which is making measurements up to 2×10^6 in turbulent flow, and these measurements are being compared with full-scale data, with a view to validating their use at higher Reynolds number. In fact, when the incident flow turbulence is about 10% or more we are getting base pressures at $Re = 10^5$ within 10% of full-scale measurements at $Re = 10^7$ without using trip or surface roughness on the cylinder.

Paper by J. F. Y. Sidarous, M. D. Vanderbilt--AN ANALYTICAL METHODOLOGY FOR PREDICTING DYNAMIC BUILDING RESPONSE TO WIND

Question: A. P. Jeary. Do the authors have any indication of the accuracy of prediction of natural frequencies and modes of vibration using their method? In particular, have they correlated predictions with results from actual structures, and do they have any indication that the method is either more accurate or cheaper than simpler estimators?

Reply: J. F. Y. Sidarous, M. D. Vanderbilt. Several comparison studies were made on buildings having simple geometries which could be adequately modeled using the analytical and computer models (ETABS-SAP4-EFRAME) and also on a 7-story experimental frame. Excellent agreement was obtained for displacements, vibration frequencies and modal shapes for all cases. No actual structure study was performed due to the lack of the needed types of full-scale data. The methodology is expected to be more accurate and provide more response information than simpler estimators and to be less expensive than experimental investigations, but no comparative study on that aspect has yet been performed.

Question: B. Etkin. The generalized force driving a mode depends strongly on mode shape. The usual approach to mode shape can lead to significant error if the damping is "usual." Then, the various points in the structure move with various phase at the same frequency. The analytical formulation that allows for this is not complicated and is easily carried out on a computer, both calculation of nodes and of forced motion.

Reply: J. F. Y. Sidarous, M. D. Vanderbilt. The suggested approach is interesting, but there is no literature on it. However, the traditional modal analysis has shown to be very adequate.

Question: A. Kareem. It seems that the torsional forcing function you have proposed only accounts for the effect of fluctuations in the approach flow. However, a major portion of the torsional force is induced by the wake. How do you account for that or I assume you are only presenting this model to use in your dynamic analysis model?

Reply: J. F. Y. Sidarous, M. D. Vanderbilt. We agree that the torsional contribution of the wake excitation is not accounted for completely. However, it is partially implemented in an indirect way by considering the spectra of torsional moments created by the across wind forces (including vortex shedding) due to the shifts in the elastic center.

Paper by H. Ruscheweyh--DYNAMIC RESPONSE OF HIGH RISE BUILDINGS UNDER WIND ACTION WITH INTERFERENCE EFFECTS FROM SURROUNDING BUILDINGS OF SIMILAR SIZE

Question: T. Tschanz. I was pleased to see that you measure torsional response from Figure 2. The model mode shape however does not appear to represent the full-scale mode shape. Without modeling over a simple torsional mode shape, how do you think your results apply?

Reply: H. Ruscheweyh. We looked for the torsional mode only to show the principal tendency of the interface effect and therefore we used this simple dynamic model. But nevertheless, dynamic acting wind force is rather small at the building base because of the effect of the ground (dynamic torsional force is coming from the vortex separation) and is small cross-correlated at the top because of the three-dimensional flow. Only the part in the region of approximately 2/3 of height H of the building model will give the input energy to torsion and can represent the generalized input force. I believe that a model with exact simulated torsional mode will react in a similar manner as our simple model. I agree with you, that there is much more work to do in this field.

Question: R. C. F. Dye. Has the effect of using a flexible upstream model been investigated as tests on twin distillation column models at Manchester University have shown increased response of the downstream column when the upstream model is flexible over the similar case of a dummy (rigid) upstream model for conditions of high reduced velocity?

Reply: H. Ruscheweyh. Yes, we have undertaken series of measurements on the dynamic response of two flexible structures, especially on circular cylindrical stacks and have found the well-known instabilities of interference galloping and vortex resources. The results will be presented on the IUTAM-Symposium on September 1979 in Karlsruhe, West Germany. The aim of the paper here was only to show the buffeting interference effect on a rather rigid building.

Paper by G. Solari, D. Stura--DYNAMIC ALONGWIND RESPONSE OF A STRUCTURAL SYSTEM INCLUDING SOIL FLEXIBILITY

Question: J. F. Howell. I would like to draw the authors attention to the work of Tajimi, Kobori, Minai, Baba from Japan and Novak et al. from Canada. Their work develops frequency dependent soil stiffness and damping for all modes of vibration and for all foundation types. Under wind loading the main effect of soil-structure interaction is to moderately reduce alongwind response but to significantly reduce response to vortex shedding and galloping. Higher modes are also much more affected due to the mobilization of soil damping from the larger base rotations and displacements. Could you comment please?

Reply: G. Solari, D. Stura. I agree with Mr. Howell about the relevant importance of soil-structure interaction on the structural response to vortex shedding and galloping, but confirm, as our paper clearly shows , that the dynamic alongwind response of very tall, slender and flexible structures on surface foundations, may be strongly influenced by dynamic soil-structure interaction. More precisely we observed (see also Reference 1 of the paper), that for such structures and normal soils, both the frequency dependence of soil stiffness and the radiation damping in the soil are of negligible importance, while the rocking static motion of the base of the structure and the hysteretic dissipation in the soil become really essential. For this reason the rise of the foundation and the properties of the soil become determinant parameters for the structural response.

Question: A. P. Jeary. Have the authors measured significant soil-structure interaction in the real case? Practical studies conducted on real buildings and chimneys by the Building Research Establishment and Central Electricity Research Laboratories show that soil-structure interaction effects are insignificant for dominant first mode activity (with only one exception).

Reply: G. Solari, D. Stura. The Building Research Establishment and Central Electricity Research Laboratories conducted practical studies on buildings and multiple chimneys with deep foundations, for which the soil-structure interaction effects are surely insignificant. On the contrary when tall steel stacks, or concrete slender chimneys, or radio and television towers or lighting poles on surface foundations are of concern, the above effects may become really determinant for a correct evaluation of the dynamic response of the structure.

Question: J. L. Wood. Street lighting poles are often set in soils of widely varying properties which are not constant. Can the authors suggest a way of adapting their treatment so as to predict the damping requirements at ground level for a weakly damped structure subject to lateral and torsional oscillations?

Reply: G. Solari, D. Stura. In the future, it will be necessary to investigate a greater number of structural types, kinds of foundations and soils characterized by different properties to know in a more detailed way the influence of the dynamic soil structure intersection on the alongwind, crosswind and torsional oscillations. Anyway, I think that both the general treatment presented in our paper, and the simplified technique of the "equivalent clamped structure" (see Reference 1 in the same paper) can be adopted for across wind vibrations in order to obtain a pattern of reliable results. On the contrary, the analysis of torsional oscillations taking the soil-structure interaction into account needs some more detailed studies and researches. This is because torsional motion of footings on soil is very different from rocking and horizontal motions.

Paper by M. Ikenouchi--EXPERIMENTAL AND NUMERICAL STUDIES ON THE AERODYNAMIC INSTABILITY OF TOWER-LIKE STRUCTURES

Question: I. S. Gartshore. The effects of turbulence in the shear layers and wake of the bodies under study are important in describing the flow field correctly. Do you see a possibility of including these effects in your calculations of the flow field and if so, how? Also, have you calculations of oscillating force on your prisms reached at a time at which amplitudes are constant or is the flow pattern still developing from initial condition?

Reply: M. Ikenouchi. The numerical analysis of turbulent flow is a big problem in the field of the numerical flow analysis. Up to date, there seemed to be no effective method to include turbulence in the finite element method analysis except for the analysis of turbulent boundary layer on a flat plate. We are now searching for the new effective turbulence model. To the second question: We stopped our calculation when the amplitude is almost constant. In this case, the flow speed is unity and the length of side facing to the fluid flow is unity. Reynolds number was changed using the different kinematic viscosity.

Question: H. Ruscheweyh. In your very interesting diagrams one can see that your model vibrates with nondimensional amplitude of approximately 0.5. Have you tested the structural damping of your model in dependency of amplitude, in other words: is the damping constant?

Reply: M. Ikenouchi. The tower models were made of elastic thin steel beams covered by formed styrene. They were linear models. The structural damping was almost constant within the experiments.

Paper by W. Schnabel, E. J. Plate--PROTOTYPE AND MODEL EXPERIMENTS ON WIND INDUCED VIBRATIONS OF A STEEL TOWER

Question: J. Durgin. To what would you attribute the apparent failure of the wind tunnel simulation in reproducing the base pressure of the full-scale lighthouse?

Reply: W. Schnabel, E. J. Plate. The divergence of the base pressure coefficient Cpb between the full-scale and model experiment, I would not call that a failure, in my opinion does not result from the model experiments. In those we were able to reach Cpb = -0.45, a value that certainly corresponds to expected values. The divergence is caused by the high base pressure coefficient Cpb = -.25, that was observed at the prototype, and that leads to a drag coefficient as low as C_D = 0.43. Besides the fact, that C_D values as low have been observed by others in recent times, I cannot give you a satisfying answer why the base pressure coefficient in nature is as high.

Paper by H. van Koten--A STUDY ON CROSSWIND OSCILLATION OF CHIMNEYS

Question: K. Handa. The peak factor g is a function of the zero crossing frequency, which is again a function of the spectral moments. Moreover the distribution of the response may not follow a Gaussian distribution. Is the value of $g \simeq 3$ based on the evaluation of spectral moments? If not, then how has it been derived?

Reply: H. van Koten. The value of the peak factor $g \simeq 3$ is based on a Gaussian distribution of the response movements. If the movements are not Gaussian (as in the case of an almost sinusoidal response) the value of the peak factor has to be revised. This point will be studied.

Question: F. H. Durgin. My own on-site measurements as well as the paper by Tieleman point up the variability of the earth's boundary layer due to non-thermal equilibrium, even up to gradient wind speeds of 45 mph (20 m/sec). The variability of your results in Figure 3 point out the importance of this effect.

Reply: H. van Koten. I fully agree with this comment.

Question: K. Handa. Is the curve given in Figure 4 based on experimental data or on some analytical model? In case it is based on the experimental data, then Figure 10 shows only one value which seems rather inadequate for describing the behavior over the whole Reynold's number range.

Reply: H. van Koten. The curve of Figure 4 is based on a number of experimental data from full-scale and boundary layer wind tunnel tests (see Table 3).

Question: A. Kareem. In Figure 14 the smooth approach flow results in maximum values of M.F. and 10% turbulence intensity flow gives the minimum values, whereas, the full-scale values are in between the two. Did you try to match the full-scale values by applying some roughness on the model surface?

Reply: H. van Koten. No, but the influence of the Reynolds number, as far as known, was taken in account as the comparison of the results. It seems a good suggestion to make a few tests with a model with a rough surface.

Question: K. Handa. The reduction factor R given in Figure 5 is not clear. For a constant height, the parameter R becomes a function of D. Now for example, H = 20, D = 1, R = 0.1, H = 20, D = 2, and R = 0.2. It means that when the diameter decreases, the reduction factor also decreases. Is it true in the case of slender structures?

Reply: H. van Koten. The parameter R is derived from the coherence function. This function is $e^{-x/20}$. In this paper it is roughly shown that $R = e^{-x/20}$. R is derived by integrating the coherence function over the height of the structure.

Question: K. Handa. Does the analytical expression given on page 10-4 take into account the influence of higher modes of vibration?

Reply: H. van Koten. The same expression can be used for the vibrations in higher modes. The influence of the different modes can be considered separately. The function R must be recalculated for other mode shapes.

Question: B. E. Lee. Small imperfections which are inevitable in the construction of a normally symmetrical circular chimney will result in the existence of a set of principal axes whose planes may vary with building height. The result of this is that the response at a given value of V_1rD will be shown that the cross-wind response $\eta c = f(\text{Cos}2\theta)$ and the alongwind response $\eta a = f(\sin 4\theta)$, see B.E. Proc. I.C.E. Part II, September, 1974. Did you allow for this effect in your experimental procedure?

Reply: H. van Koten. The influence of imperfection in the form was reduced by measuring in the principal directions mainly. Especially the wind tunnel models were loaded and measured in these directions.

Paper by M. A. Wawzonek, G. V. Parkinson--COMBINED EFFECTS OF GALLOPING INSTABILITY AND VORTEX RESONANCE

Question: Y. Nakamura. I would have gotten an impression that you are trying to combine galloping with vortex resonance rather straight forwardly. However, because vortex resonance occurs at a relatively low wind speed, I would like to remind you that galloping at such a low wind speed must be modified considerably by the unsteady flow effect, or, what I call by the fluid memory effect. So, my scheme is that galloping may be modified by the fluid memory effect initially and then interact with vortex shedding.

Reply: M. A. Wawzonek, G. V. Parkinson. Our paper represents a purely experimental study of an idealized case. We realize that any theory describing the observed effects would be one of unsteady flow. In the paper the existing quasi-steady theory serves merely as a useful reference with which to compare the observations, and, as would be expected, we find agreement between that theory and the observations only at relatively high reduced wind speeds, or, in the usual terms of unsteady flow theory, at low reduced frequencies.

Question: C. M. Sheih. Figures 1 and 9 should not be normalized by the same variable (U_o) in both ordinate and abscissa, i.e., $\overline{Y}/U_o$ vs. U/U_o. In doing so, one would introduce an artificial dependency between these two normalized variables as reported in a paper of Boundary Layer Meteorology by B. B. Hicks in 1978.

Reply: M. A. Wawzonek, G. V. Parkinson. The questioner is mistaken, perhaps because he has misunderstood the reason for plotting $\overline{Y}/U_o$ vs. U/U_o in Figures 1 and 9. In these figures the validity of the quasi-steady theory is being tested. The theory, not given in the paper, but derived in Reference 1 of the paper, predicts that experimental points $(\overline{Y},U)$, if normalized by the theoretical critical galloping speed $U_o = 2\beta/nA$, should collapse to a universal curve, for all values of damping and mass parameters (β,n). This is seen to be true in the paper for Figure 1, and in Mr. Wawzonek's presentation for the modified Figure 9, using corrected values of A obtained after the paper was submitted.

Question: J. Saunders. I must challenge the implication of your conclusion regarding extrapolating your work in two-dimensional flow to tall buildings despite your qualifications. At the last of these conferences, I presented work on buildings which had been tested at levels of motion up to equivalent of 30% of gravity, full-scale and up to reduced velocities of 10 (U_1 = 1.6), in turbulent flow for which aerodynamic damping and galloping was not significant. At 30% of gravity a building would be quite uninhabitable. Work presented at this conference by Reinhold and Sparks (VI-3-1) where they tested model buildings with damping of 0.28%, Kareem (VI-1-1) and Kwok (VI-4-1) have also all supported this conclusion. Above a reduced velocity of about 10 and at levels of amplitude unsatisfactory for buildings, they have found galloping effects to become significant. If a building exhibited galloping combined with vortex resonance, the building would have long since become uninhabitable. The only exceptions to the above are most unusual buildings with low stiffness which may exhibit a reduced velocity greater than 10. These buildings require special damping treatment, which is not normally an economical concept. I commend your work for industrial flows where such phenomena can exist or if someone builds a very slender industrial tower without guying it. However, for buildings galloping is not the design criterion, but rather vortex shedding.

Reply: M. A. Wawzonek, G. V. Parkinson. The questioner has missed the point of the paper, that if the system parameters are such that the critical galloping speed is slightly above the vortex resonance speed, possible for towers and some tall buildings, the aeroelastic effects will bridge the gap and produce much larger amplitudes than would be expected from vortex resonance along at wind speeds slightly above the vortex resonance value. Evidence of this for model towers in a simulated suburban wind exposure is given in Reference 5 of the paper, soon to appear in Jour. Ind. Aero., July 14.

SESSION VII

Dynamic Response - Bridges, Transmission Lines and Roof Membranes

CHAIRMAN

M. Ito
Professor
University of Tokyo
Tokyo, Japan

CO-CHAIRMAN

R. L. Wardlaw
National Research Council of Canada
Ontario, Canada

RAPPORTEUR

R. H. Scanlan
Professor of Civil Engineering
Princeton University
Princeton, New Jersey, USA

RECENT DEVELOPMENTS AND TRENDS: DYNAMIC RESPONSE - BRIDGES, TRANSMISSION LINES AND ROOF MEMBRANES

M. ITO

University of Tokyo, Japan

The dynamic response of horizontally extended structures to wind is dealt with in this session. Design of these very flexible structures exposed in natural wind is not only required to consider gust response, but also governed often by the aerodynamic instability such as vortex-induced oscillation and self-excited motion. The flutter of suspension bridges have been among these the oldest topic of research on structural aerodynamics since the accident of the Tacoma Narrows Bridge in 1940. But there are still various problems to be resolved in this area.

The difficulty in bridge aerodynamics arises from a great deal of complexity and variety of the bridge deck configuration, the cross-section of which is not generally streamlined, and from the temporal and spatial randomness of natural wind. The situation is similar in the case of the cable-stayed bridges recently developed.

In determining the cross-sectional shape of a suspended-span bridge deck that is sensitive to wind excitation, the first step is to select the basic cross-sectional type from the viewpoints of structural and functional requirements, followed usually by wind tunnel experiments. When the results of the model experiment do not satisfy the specified provisions for aerodynamics stability, one must adopt such countermeasures as the increase of stiffness and damping, the modification of the cross-sectional shape by adding fairings, flaps and so forth, or the substantial change of the basic cross-sectional type. At this stage, attention should be paid to the possibility of increasing buffeting amplitude by the improvement of aerodynamic characteristics.

Throughout the above-mentioned procedures, the minor difference in sectional configuration such as the size and shape of curb stones and handrails on road deck may sensitively affect the stability of overall structure, and then the systematic treatment of bridge aerodynamics is not easy. Therefore, the wind-resistant design of suspended-span bridges have been made respectively.

Under these circumstances, the wind tunnel test is inevitable for these structures. As well known, the following three approaches are used in this case: a) The sectional model, b) The taut-strip model, c) The full model.

Although appraisal of the advantages and shortcomings of these three approaches has been done, it appears still impossible to draw up the balance sheet in favour of one of these, as pointed out elsewhere by Wardlaw.

The most concerned for us is the correlation between the results of wind tunnel model test and the prototype behaviour. While the full-scale field measurements of large bridges are costly and laborious, it is gratifying to find that they have recently been conducted in several countries including Canada, Japan, Norway, the United Kingdom and the United States. However, because the modern bridges are carefully designed against wind action and further the attack of very strong wind at the site is infrequent, these measurements have been confined to the buffeting and at best limited-amplitude vortex excitation. In this respect, the large scale bridge model experiment in natural wind, which was conducted by the Honshu-Shikoku Bridge Authority, Japan, is of interest, though it was regrettably only with the sectional model.

Theoretical attacks on the dynamic response of suspended-span bridges to wind continue also in an effort to provide useful method to aid in design. The examples are the study on the initiation mechanism of bridge flutter due to separated flow, the analytical prediction of prototype behaviour from the experimental information with a sectional model, and the application of simulation technique to wind-induced vibration. The verification from actual phenomena is hoped to support the analytical considerations.

As far as the roof-membranes including hanging and pneumatic space structures are concerned, their history is rather new compared with other engineering structures. However, intensive research works on the wind effects have been carried out in Canada, the Federal Republic of Germany, Japan and other countries.

Because the roof-membranes are generally much lighter and have two-dimensional extension, some novel considerations are necessitated. Previour theoretical works are mostly based on various assumptions, so that it is somehow doubtful how accurate they can predict the prototype behaviour. The wind tunnel testing techniques for these structures seem to require the additional considerations to those for bridge model; namely, according to Irwin and Wardlaw, required to allow for the effect of virtual mass of the air that is replaced with movement of the roof, dissipation of energy by acoustic radiation from the roof, and the pneumatic stiffness term relating to the internal volume of the structure below the roof.

Finally, the personal statement of anticipated trends during the coming years is assigned to the author. Concerned with the suspended-span bridges, the following items seem to be pursued;

1. More reliable prediction of the prototype behaviour through experimental and analytical procedures, as already mentioned.

2. Development of aerodynamically stable cross-sections for much longer spanned structures; the conventional practice of bridge deck configuration may have limitation to apply to further long spans in the future, thus some novel considerations will be needed.

3. Establishment of wind-resistant design criteria to ensure the aerodynamic stability; the sample topics will be

a) the definition of critical wind velocity for initiation of flutter in turbulent wind (refer to the figure indicating the results of model test)

b) the quantitative prediction of vortex-excited response; it is widely recognized that the effects of turbulence are conservative for vortex-shedding response. Therefore, if the wind tunnel test is conducted only in smooth flow, unnecessary and uneconomical design may be enforced, while the quantitative information of the turbulence effect on the response is not yet satisfactory to aid in design.

c) the design provisions relating to buffeting; two papers on the large scale gantry cranes and transmission line are presented in this session. This kind of structure has a resemblance to girder bridges. The gust factor approach will be appropriate in these cases. The wind-resistant design code for suspension bridges in Japan provides also the gust response factor approach, but there is the discussion elsewhere that this approach may not be satisfactory to estimating loads for some sort of structures, and the dynamic buffeting response shall be obtained by an alternative.

As to the roof-membranes, it is desired to have more realistic method of analysis and more information from the full-scale measurements. It will be the future subject of study to establish their wind-resistant design criteria.

In preparing the present review, informative suggestions were given by R. L. Wardlaw, the Co-Chairman of this session, to whom the author is much indebted.

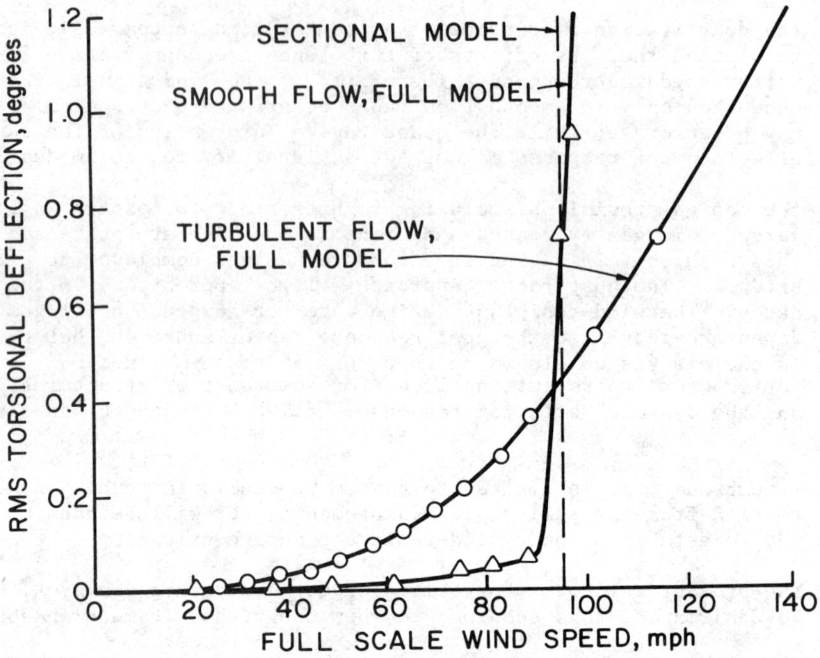

THE LIONS' GATE BRIDGE TORSIONAL RESPONSE AT THE
1/10TH SPAN POSITION IN SMOOTH AND TURBULENT FLOW
(by Irwin & Wardlaw)

WIND INDUCED RESPONSE OF SUSPENSION BRIDGES - WIND TUNNEL MODEL AND FULL SCALE OBSERVATIONS

A. G. DAVENPORT*, N. ISYUMOV*, H. ROTHMAN** and H. TANAKA***

*The University of Western Ontario, London, Ontario, Canada
**Paul Weidlinger Consulting Engineers, New York, New York, USA
***University of Ottawa, Ontario, Canada

ABSTRACT

The wind-induced response of an existing long span suspension bridge was examined by various wind tunnel testing procedures. The wind tunnel findings are compared with observations of the full scale bridge behaviour. Turbulent characteristics of the natural wind were found to have a significant effect on the behaviour of the bridge. Consequently the representative simulation of turbulent flow is an important requirement of wind tunnel model testing. Whereas the use of a taut strip model in turbulent boundary layer flow provided best agreement with full scale observations, the extension of conventional section model tests to include turbulence was found to provide a potential new avenue of study. Also discussed in the paper are the effects of wind yaw angle, which combined with the directional non-homogeneity of the wind at the site significantly influences the predictions of bridge response.

BACKGROUND

A persistent question in the design of major suspension bridges against wind is: How accurate are wind tunnel test predictions of response? Wind effects on long span bridges have been extensively studied over the years and a variety of testing approaches has developed. Each approach attempted to explore or reduce perceived shortcomings of the previous methods. Some provide no more than a conservative bound of the response so that an inherently safe design results. The acid test of any approach is its ability to forecast the behaviour of the actual bridge. Unfortunately, opportunities to observe full scale behaviour are infrequent and often occur years after the bridge is finished.

This paper summarizes a study of a particularly interesting bridge in which both different wind tunnel testing procedures and full scale observations were compared. The bridge is a 700 m center span suspension bridge across the East River in New York City. The siting and overall dimensions are shown in Figure 1. The bridge was built as a girder stiffened suspended structure carrying four roadway lanes and two sidewalks. Because the bridge vibrated in moderately strong winds during construction, modifications were made to the design before it was opened to traffic. These comprised the addition of inclined stay cables from tower tops to stiffening girders and the installation of a frictional device in the tower. From its opening in 1939 until its traffic capacity was expanded in 1946, the bridge functioned satisfactorily, although it underwent occasional episodes of minor oscillation. The oscillations which occurred with winds between 18 and 27 m/sec lasted as long as the winds blew strongly and ranged in amplitude up to about 40 cm.

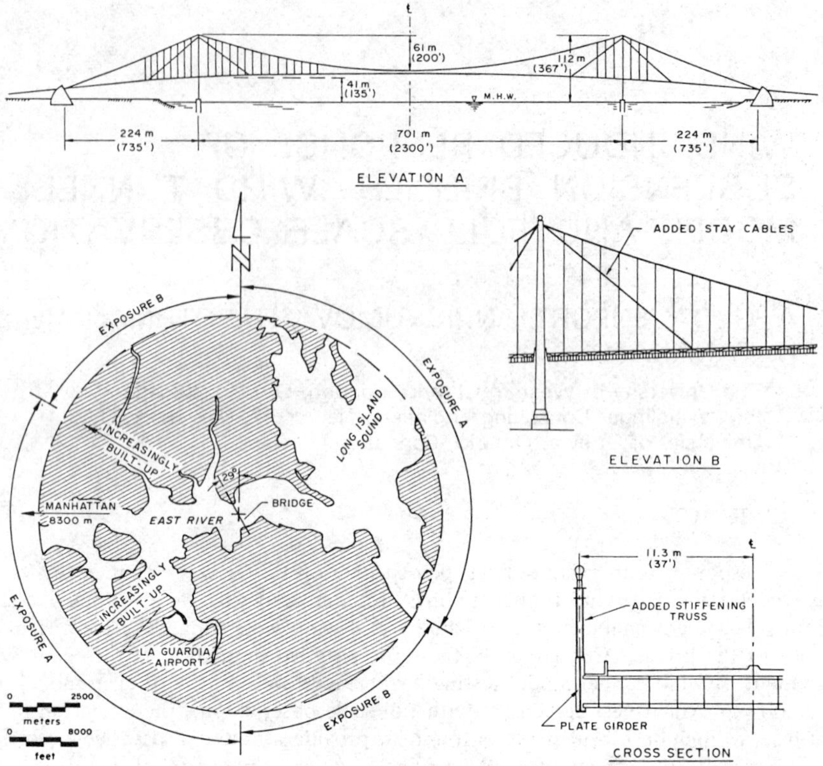

Fig. 1 Location and details of bridge

During this period, which followed the failure of the first Tacoma Narrows Bridge, the bridge was kept under close surveillance by its designer, O. H. Ammann, who was also a member of the board investigating the collapse of the Tacoma Narrows Bridge. In view of the similarity of the sections of these two bridges, section model wind tunnel tests were made at the University of Washington(1). Typical test results indicated that the bridge could not survive in winds over about 20 m/s, whereas the bridge in fact had often withstood far greater winds with minor vibration and no damage. As a result, when it became necessary to modify the structure in order to increase its traffic capacity, Ammann gave little weight to these wind tunnel tests. He instead based the alterations on a Stiffness Index which he developed and which correlated well with observed bridge performance.

In modifying the bridge, the sidewalks were removed, the roadway was widened to six lanes and a stiffening truss was built on top of the existing girders, see cross-section in Figure 1. This is the present configuration of the bridge. The bridge has functioned well for the 32 years since it was modified, although it still has had episodes of minor vibration during strong winds.

When in the 1960's, more powerful analytic tools and reliable motion measuring instrumentation became available, the Triborough Bridge and Tunnel Authority, owners of the bridge, decided to restudy the motion to determine:

i) The potential for fatigue damage to the floor system since the oscillation episodes can accumulate over 100,000 cycles during the life of the bridge.

ii) The risk of local damage due to large amplitude motion in extreme winds.

iii) Cost effectiveness of ameliorating the above risks by bridge modification, i.e., is it cheaper to prevent damage or make repairs after it occurs?

Subsequent publications will summarize the overall study. This paper is limited to a review of wind tunnel modelling techniques that were used(9) and some comparisons with available full scale observations.

CHOICE OF TEST PROCEDURES

The extensive service experience and the full scale measurements available for this bridge placed stringent demands on the selection of the wind tunnel test program. Procedures which fail to corroborate observed behaviour at lower operational wind speeds must be suspect when applied to higher, more critical wind speeds. Pessimistic predictions may be acceptable for design but they are counter productive if the bridge is already known to withstand such winds!

The wind tunnel test program was, therefore, selected to meet the following objectives:

1) Predict the aerodynamic behaviour of the bridge over a full range of wind speeds and over a range of parameters that may vary in the full scale bridge. The parameters were frequency, damping, center of rotation and turbulence in the wind flow.

2) Analyse the full scale measurements in order to determine these parameters and to confirm the model's ability to predict full scale behaviour.

3) Predict the performance of the bridge at long return periods by combining model results with meteorological measurement.

The aeroelastic testing procedures available for bridge decks (in roughly chronological order of development) are:

1) Aeroelastic section models in uniform flow
2) Aeroelastic models of the full bridge in uniform flow
3) Aeroelastic models of the full bridge tested in turbulent boundary layer flow
4) Analytic solutions using aerodynamic derivatives obtained from section models
5) "Taut strip" aeroelastic models tested in turbulent flow
6) Aeroelastic section models tested in turbulent flow

A list of the contributors to the development of the various methods and their virtues is given elsewhere(2).

The main considerations leading to the test approach adopted were as follows: Previous experience reported by Davenport, Isyumov and Miyata(3) suggests that turbulence can have a dramatic influence on aerodynamic stability, either postponing the instability or even changing the mode of the instability. Turbulence can also excite significant motions at sub-critical speeds. These aspects can not be reflected by tests in smooth flow. Further, the behaviour of the three dimensional structure can deviate from that of the two-dimensional section models. These general conclusions appear to have been supported in the studies carried out by Irwin and Wardlaw on the section models and full bridge models of the Lion's Gate Bridge(4).

While these conclusions favoured testing of full bridge models in turbulent boundary layer flow, there are significant drawbacks to such tests. These are: cost, Froude number dependence with consequent low range of model wind speeds, and their small scale. The 'taut strip model' approach suggested by Davenport(5) represents a compromise. In this, the three dimensional character of the mode shapes is retained using a simplified model of the deck supported on taut wires. The separation of the wires is chosen to provide the appropriate torsional/flexural frequency and the deck inertias are correctly scaled. In cases where the mode shapes contain more than two nodes, several lengths of the model can be tested bracketing the half waves of the mode shapes. The model responses are then reassembled from the responses of the half waves (in contrast to the response of strip sections). This approach correctly introduces the three dimensional attributes of turbulence into the modelling. Further advantages of the taut strip model are its comparative ease of construction, the freedom in the choice of velocity scaling (it is not Froude number dependent) and, importantly, the fact that it permits the directional sensitivity of the response and the influence of surrounding topography to be determined. An apparent disadvantage of the taut strip model is its reduced geometric scale. On balance this is considered of less importance than the simulation of the strong three dimensional influences of turbulence.

The wind tunnel test program of this study was based on taut strip model testing in turbulent flow and on section model tests both in smooth and turbulent flow generated by a coarse grid. The taut strip model testing was expected, to provide realistic estimates of bridge motion(6,7) while the larger scale section model would allow comparisons with earlier work on such models, as well as an opportunity to estimate aerodynamic coefficients and the influence of the centre of rotation. Additional testing in a turbulent flow was thought to be a new avenue of exploration which might check the sensitivity of results to turbulence and, if necessary, assist in reconciling taut strip and section model results.

WIND TUNNEL MODEL STUDY

General

The section model, shown in Figure 2, was spring mounted to permit vertical and torsional motion and tested between fixed end plates. Damping was provided with oil dashpots. The taut strip model is shown in Figure 3. The design of the experimental program relied on dynamic properties of the bridge determined in a study of ambient vibration at low wind speeds(8) and summarized in Table 1.

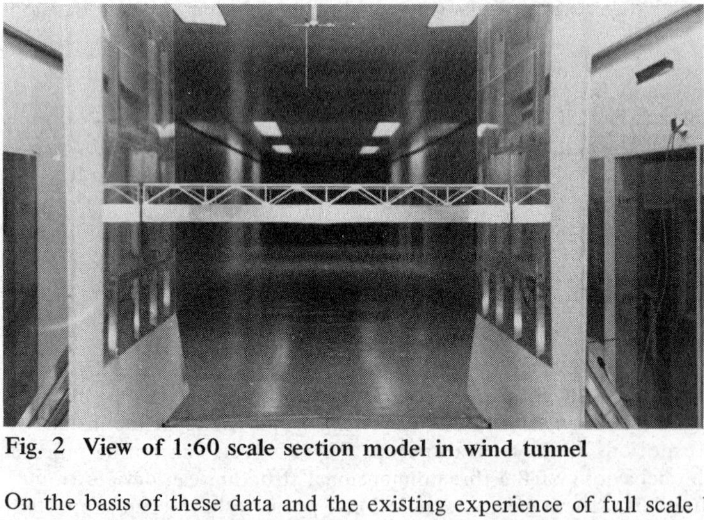

Fig. 2 View of 1:60 scale section model in wind tunnel

On the basis of these data and the existing experience of full scale behaviour during strong winds, both the section and taut strip models were designed to simulate the 1st symmetric vertical and torsional modes of vibration. The frequency of the 1st antisymmetric torsional mode is significantly higher in view of the centre-ties. The lateral dynamic properties of the bridge were not modelled. The details of the 1:400 scale taut strip model are shown in Figure 4. The effective span of this model, taken as 536 m in full scale, was selected so that the sinusoidal mode shape matched closely the first symmetric vertical and torsional modes of the prototype structure The procedure for this matching has been described elsewhere(5,6). The deck was modelled by 11 rigid sections attached to the two piano wires tensioned between the cable anchorages.

Fig. 3 View of 1:400 scale taut strip model in wind tunnel

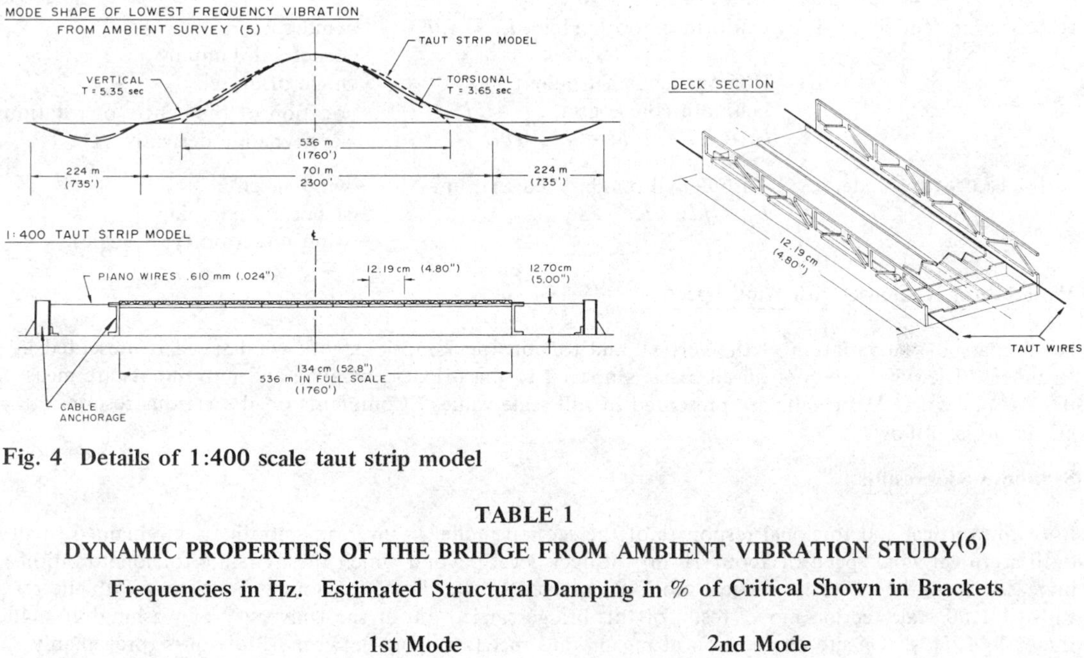

Fig. 4 Details of 1:400 scale taut strip model

TABLE 1
DYNAMIC PROPERTIES OF THE BRIDGE FROM AMBIENT VIBRATION STUDY[6]

Frequencies in Hz. Estimated Structural Damping in % of Critical Shown in Brackets

		1st Mode	2nd Mode
Lateral:	Sym.	.114 (2.9%)	.322 (3.2%)
Vertical:	Sym.	.187 (1.3%)	.300 (0.9%)
	Asym.	.201 (1.9%)	.351 (0.9%)
Torsional:	Sym.	.274 (1.1%)	.422 (0.8%)
	Asym.	.362 (0.9%)	

A summary of the entire wind tunnel test program is presented in Table 2. The wind induced response of the bridge was in all cases measured over a range of wind speed up to and beyond the onset of large amplitude oscillatory motion. In the case of the taut strip model study, most measurements were carried out in turbulent boundary layer flow conditions representative of natural wind over an open water fetch. The intensity of the longitudinal component of turbulence at deck level, defined as $I_u = u_{rms}/\bar{V}$, was approximately 11% in this simulation. Although, these flow conditions were considered to be representative of easterly to north easterly winds coming across Long Island Sound, see Figure 1, winds from the westerly to south westerly directions were expected to be more turbulent. The structure of the flow for this direction is expected to be influenced by the highly built-up Manhattan region, located some 11 km upstream, and by the northern part of Long Island. Also the change in surface roughness from the East River to the northern fringe of Long Island, which is aligned with this wind direction, is expected to produce some lateral shear. Although, full scale measurements of turbulence at deck level were not available, the turbulence intensity for westerly to southwesterly directions was likely to be higher than over an open water fetch. To examine the sensitivity of the bridge response to increased turbulence levels, taut strip model measurements were also carried out in turbulent boundary layer flow conditions with I_u values of 15 and 20 percent.

TABLE 2

SUMMARY OF WIND TUNNEL TEST PROGRAM

Type of Model		Flow Conditions	Parameters Studied
1:60 Section Model	i)	Uniform Smooth Flow, $I_u < 1.0\%$	– wind speed – structural damping
	ii)	Homogeneous Turbulent Flow behind a coarse grid, $I_u \approx 11\%$ ($B = 8.9$ cm, $M = 43.2$ cm)	– angle of attack – location of the centre of rotation – aerodynamic derivatives
1:400 Taut Strip Model		Turbulent Boundary Layer Flow $I_u \approx 11\%$, 15% and 20%	– wind speed – structural damping – wind direction (yaw angle)

Variation of Response With Wind Speed

A summary of the variation of the vertical and torsional deck motions with wind speed is presented in Figure 5. The wind direction in all cases is normal to the bridge centreline and movements are measured at mid-span. All results are presented in full scale values. Comments on the various response curves are as follows:

Section model results

Both the vertical and torsional responses of the section model in uniform smooth flow remained small until a critical wind speed of about 18 m/s at deck level beyond which the dynamic torsional motion increased abruptly while the vertical motion remained small. This behaviour is consistent with the results of 1:50 scale section model tests for this bridge carried out at the University of Washington(1) in the early 1950's. On-site observation at mean wind speeds ranging between 13–18 m/s (presumably at the deck level) reported only vertical oscillations of the bridge with a maximum single amplitude of 46 cm for the original section and 23 cm after the reinforcement. In constrast, the University of Washington wind tunnel study predicted violent torsional oscillations at wind speeds of 13 to 22 m/s. The study explained this "apparent" disagreement between wind tunnel test and on-site measurement, namely that no torsional oscillations had been observed in full scale, to be due to "non-uniformities and unsteadiness in the natural wind". Apparently the other reason of disagreement was an under-estimation of the frequency and damping. The torsional frequency observed in the current study is some 26% higher than the calculated one. The assumed structural damping of $\zeta_T = .64\%$ was also rather low. The full scale vertical oscillation was not explained by the findings of that study.

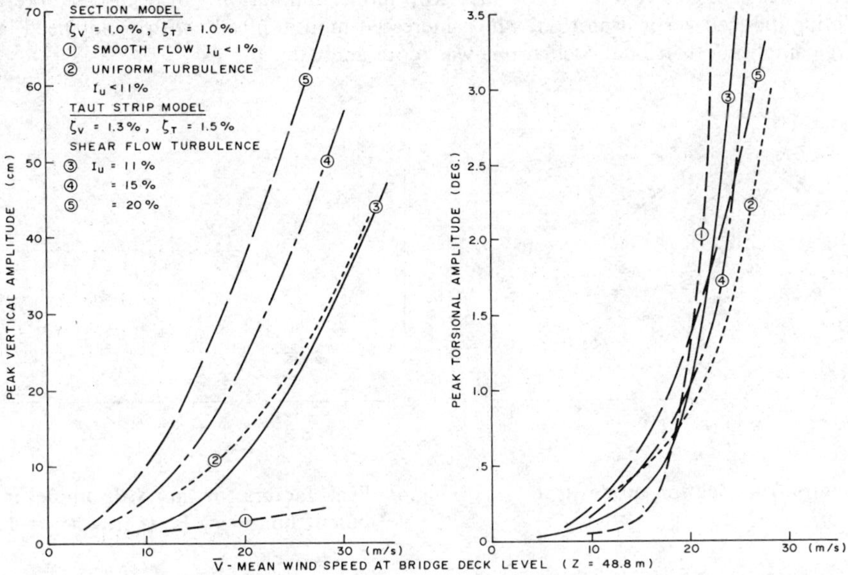

Fig. 5 Summary of wind induced response from section model and taut strip wind tunnel model measurements for wind normal to bridge centreline

The introduction of grid generated turbulence in the approaching flow, attempted to provide an indication of the importance of simulating turbulent flow conditions, produced a significant change in the character of the dynamic response. As seen from curve 2, the critical wind speed in uniform turbulent flow was some 20% higher than that in smooth flow. Furthermore, the random buffeting excitation by turbulence produced higher torsional movements at low wind speed and thus a much more gradual transition to "catastrophic" oscillations. This gradual transition to organized large amplitude motions can also be seen from the peak factors presented in Figure 6. The peak factor $\overline{g}$, for smooth flow case, rapidly approaches $\sqrt{2}$ at critical wind speeds, while $\overline{g}$ values for the turbulent flow remain somewhat higher even at large amplitudes, suggesting the response to be less regular. As seen from Figure 5 there is a significant buffeting induced vertical motion.

Taut strip model results

Taut strip model results in turbulent boundary layer flows with longitudinal turbulence intensities of 11, 15 and 20 percent,measured at deck level,are depicted in Figure 5 by curves 3, 4 and 5 respectively. The observed dynamic vertical motion was characteristic of buffeting excitation by turbulence. The peak factors of the vertical response, as typically shown for the $I_u = 11$ percent case in Figure 7, remained constant with wind speed and were consistent with theoretical predictions for a narrow-band Gaussian process. The torsional response was found to be random in character at low wind speeds but increased more rapidly beyond some critical wind speed. The different character of the vertical and torsional motions at different wind speed is clearly shown by the variation of peak factors presented in Figure 7. Increasing the turbulence level resulted in larger torsional motions at low wind speeds and a postponement of the onset of organized large amplitude motion.

Taut Strip and Section Model Comparisons

Whereas the results of the conventional section model tests in smooth uniform flow differed from those of the taut strip model in turbulent boundary layer flow, the behaviour of the section model in turbulent flow bore a marked resemblence to that of the taut strip model simulation. In both cases there was a significant buffeting induced vertical motion, which increased monotonically with wind speed, and the transition to large amplitude torsional oscillations was more gradual.

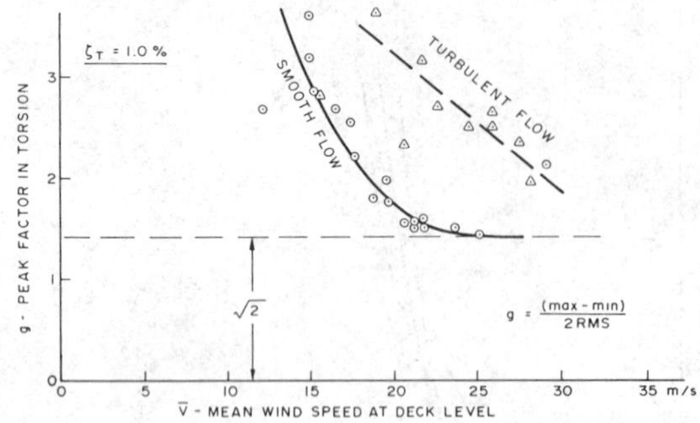

Fig. 6 Peak factors from section model tests

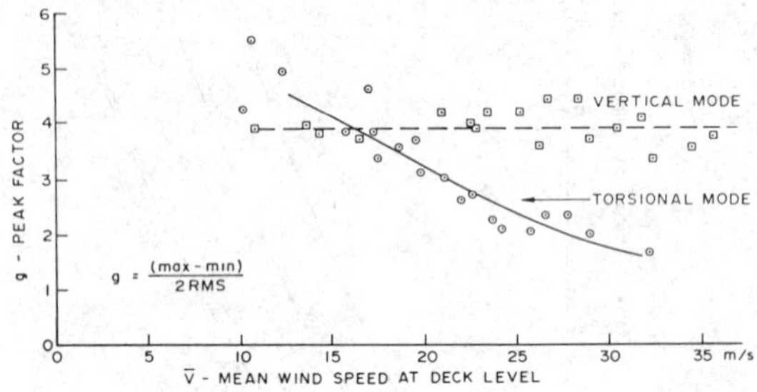

Fig. 7 Peak factors for taut strip model in turbulent boundary layer flow (I_u = 11%)

This similarity was interesting, considering that the section model tests in grid generated turbulence were carried out primarily to examine the sensitivity of this modelling technique to the presence of turbulence in the airstream. Although, the turbulence intensity at deck level was the same as in the taut strip model tests, the scale of turbulence was about one order of magnitude smaller. Spectra of the longitudinal component of turbulence for the two flows are presented in Figure 8. Although the peaks of the spectra are mismatched, the energy contents in the range of reduce frequencies used in these two simulations are similar. At the low end of this range, corresponding to wind speeds near the onset of large amplitude torsional motions, the turbulent energy in the grid generated flow is comparable to that in the turbulent boundary layer simulation with $I_u = 15\%$. Examining the data presented in Figure 5, the torsional response of the section model in turbulent flow, given by curve 2, is quite similar to that of the taut strip model simulation with $I_u = 15\%$. The following comments may be appropriate in view of this similarity:

i) The main effect of turbulent velocity fluctuations on the aerodynamic forcing mechanisms for this bridge section appears to come from wavelengths comparable to the cross-sectional dimensions.

ii) As a result, section models may produce good estimates of the full scale response in real wind provided that the energy content at wavelengths comparable to the bridge deck, rather than the turbulence intensity, is correctly modelled.

iii) The uniform mode shape of the two-dimensional section model leads to dissimilar forcing functions at wavelengths comparable to the scale of turbulence; however this dissimilarity is less significant at wavelengths comparable to the deck width, and small in comparison to bridge span. This is illustrated below:

If a narrow band correlation is assumed to be of the form,

$$R(x, x_1; n) = exp\left[- C \frac{|x - x_1| n}{\overline{V}}\right]$$

the effective gust size, l, at a particular frequency $\frac{n B}{\overline{V}}$ becomes

$$\frac{l}{B} = \frac{2}{C}\ \frac{\overline{V}}{nB} \quad \text{where } C \text{ is typically around } 8.$$

Therefore, at values of $\frac{n_T B}{\overline{V}} \approx 0.25,\ l \approx B$. A correct simulation of the aerodynamic forcing effects at this wavelength is not sensitive to dissimilarities of the mode shape providing that the span is significantly larger than B. The span of the section model used was approximately 4B.

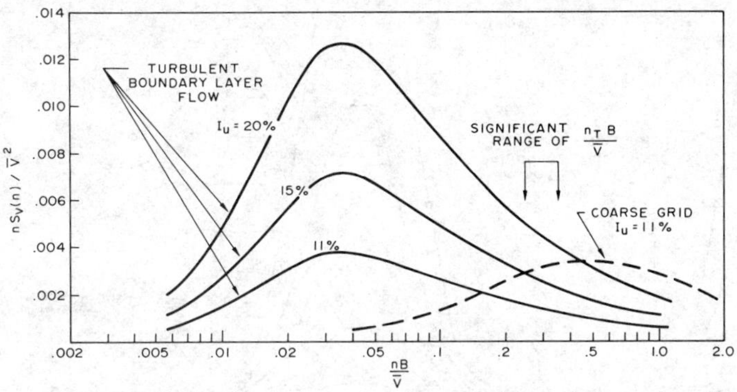

Fig. 8 Spectra of the longitudinal component of turbulence at deck level for the flow simulations used in study

Although the above comments may indicate a rationale for two-dimensional section model tests in small scale turbulent flow generated by coarse grids, further quantitative supporting data are necessary.

Effect of Structural Damping

The onset speed for large amplitude torsional oscillations was sensitive to the structural damping. The critical wind speed was approximately proportional to $(\zeta_T)^m$, where m ranged between 1/4 to 1/5 in section model test and was approximately 1/2 in the taut strip model tests. The aerodynamic damping for the torsional motion is generally negative and the aerodynamic instability is thus in the form of a pure torsional motion. Variations of the structural damping have only a modest effect on the vertical motion as the aerodynamic damping in this case is positive.

The marked sensitivity of torsional motion to structural damping emphasises the importance of this parameter in the prediction of the full scale bridge behaviour. The value of $\zeta_T = 1.1\%$, obtained from the ambient vibration study, may well underestimate the structural damping at larger amplitudes. The refinement of structural damping estimates thus became a major objective in the analysis of the full scale response data.

Effect of Wind Direction

The sensitivity of the torsional response to changes in wind direction, as presented in Figure 9, is pronounced and the wind speed required for the onset of large amplitude motion increases markedly for yaw angles greater than $\pm$ 30 degrees. In contrast to the section model,which only provides information for a zero yaw angle, a taut strip simulation permits the evaluation of the directional sensitivity of the response and providesmore representative estimates of full scale behaviour. An analysis of winds aloft data for the New York area indicated that, while the prevailing winds in the area are from the west, the strong winds are more likely to occur from the south but do not induce significant motion. This directional non-homogeneity of strong winds in the project area is illustrated by the shape of the 10^{-6} probability contour shown in Figure 10. Since the bridge is most sensitive to wind directions within yaw angles of approximately ± 30 degrees, its orientation is a favourable one.

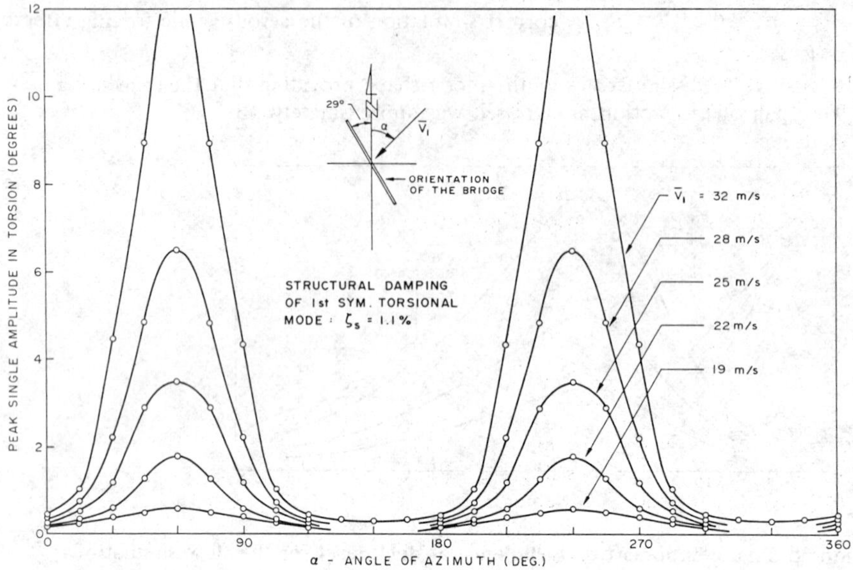

Fig. 9 Influence of wind direction on the torsional response of bridge

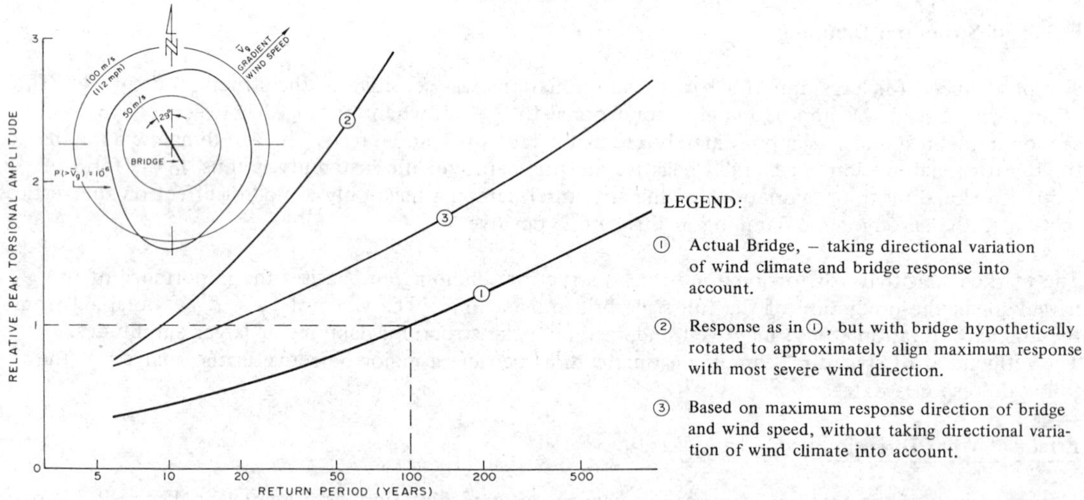

Fig. 10 Influence of the directional sensitivity of the bridge response and the directionality of the local wind climate on predicted bridge motions

The modifying effect of orientation and the directionally varying statistical properties of the wind climate is illustrated in Figure 10. The curves presented are predictions of annual extreme torsional motions of the bridge for various return periods. The procedures followed in combining the wind tunnel derived response data with the statistics of the local wind climate have been described elsewhere[10,11]. Curve 1 depicts predicted motions of the bridge using the taut strip model data and the developed probability distribution of wind speed and direction for the bridge. Curve 2 provides predictions for a hypothetical situation in which the bridge is rotated to approximately align the normal to the bridge centreline with the most severe wind direction. The 100 year return period response, predicted for this least favourable orientation, is some 3 times higher than for the actual orientation. The method used for these response predictions is more refined than that followed in conventional design. It is common practice to base predictions of the response on the maximum wind speed from any azimuth acting along the most sensitive response direction. Predictions based on this approach are given by Curve 3. At the 100 year recurrence interval, the predicted motion is some 60 percent higher than that indicated by Curve 1. Clearly this is a conservative procedure and more realistic estimates of the bridge motion are obtained by accounting for the fact that strong winds do not always occur from the critical response direction.

FULL SCALE MEASUREMENTS

General

Full scale observations were made of the vertical acceleration at the mid and quarter points of the main span and of the wind speed and direction near the middle of the main span. At both locations the accelerometers are positioned at each of the two stiffening trusses. Their output is passed through integrating circuits and the resulting vertical displacements are recorded on a strip chart recorder. Summing and differencing the total displacement at the upstream and downstream stiffening trusses provides measurements of the vertical and torsion bridge motions.

Continuous strip chart records for six periods with significant wind speeds and wind directions approximately normal to the bridge centreline were selected for detailed analysis. The objectives were,

i) to provide comparisons of full scale behaviour with the findings of the wind tunnel studies for this bridge; and

ii) to fine tune existing estimates of the dynamic properties, in particular the structural damping.

Wind speeds are given as 10 min. average at a height of 10 m above the bridge deck.
A summary of the selected records is givenin Figure 12. The wind direction was near normal to the bridge for every cases. Records of the bridge motion at the mid-point of the main span for these storms were dizitized at a sampling rate of 2 Hz. For the first symmetric vertical and torsional modes this corresponds to approximately 11 and 7 equally spaced samples per cycle of oscillation respectively. Typical traces of the bridge motion, reconstructed from these digitized records, are presented in Figure 11. The subsequent analysis of the digitized records included evaluations of various statistical properties of the response, calculations of auto-correlation functions and spectral analysis.

Response Estimates

The dynamic amplitudes of the bridge motion during the various storms, summarized in Figure 12, were computed using both 5 and 10 minute averaging times. Also shown are the intervals of the records during which the mean wind speed remained approximately constant. The variation of the RMS response approximately follows changes in recorded wind speed and the RMS estimates based on 5 and 10 minute averaging times are in most cases similar. This suggests that the response over most parts of the records can be treated as locally stationary over a 10 minute period.

Probability distributions of the dynamic response for all storms are presented in Figure 13. The parent distributions of both vertical and torsional motions are seen to be approximately Gaussian. Also shown in Figure 13 are the distributions of observed 10 minute extreme values. The average values of these extreme value distributions are approximately 3.4 and 3.1 standard deviations above the parent mean for vertical and torsional motions respectively. These observed average peak factors agree well with theoretical predictions. As shown elsewhere by Davenport[12], the average factor of extreme values over a sampling interval T for a narrow band Gaussian distribution is,

$$\bar{g} = \mu + \frac{\gamma \mu}{\mu^2 - 1}, \quad \text{where} \quad \mu = \sqrt{2 \ln \nu T} + \frac{\ln \sqrt{2 \ln \nu T}}{\sqrt{2 \ln \nu T}}$$

is the mode of extreme values, ν is the effective cycling rate and $\gamma = .5772$.

Taking $\nu \approx n_o \sqrt{8\zeta}$, where n_o is the natural frequency and ζ is the total damping, theoretical estimates of the $\bar{g}$ for $T = 10$ minutes are

$\bar{g} = 3.5$ (vertical)
$\bar{g} = 3.2$ (torsional)

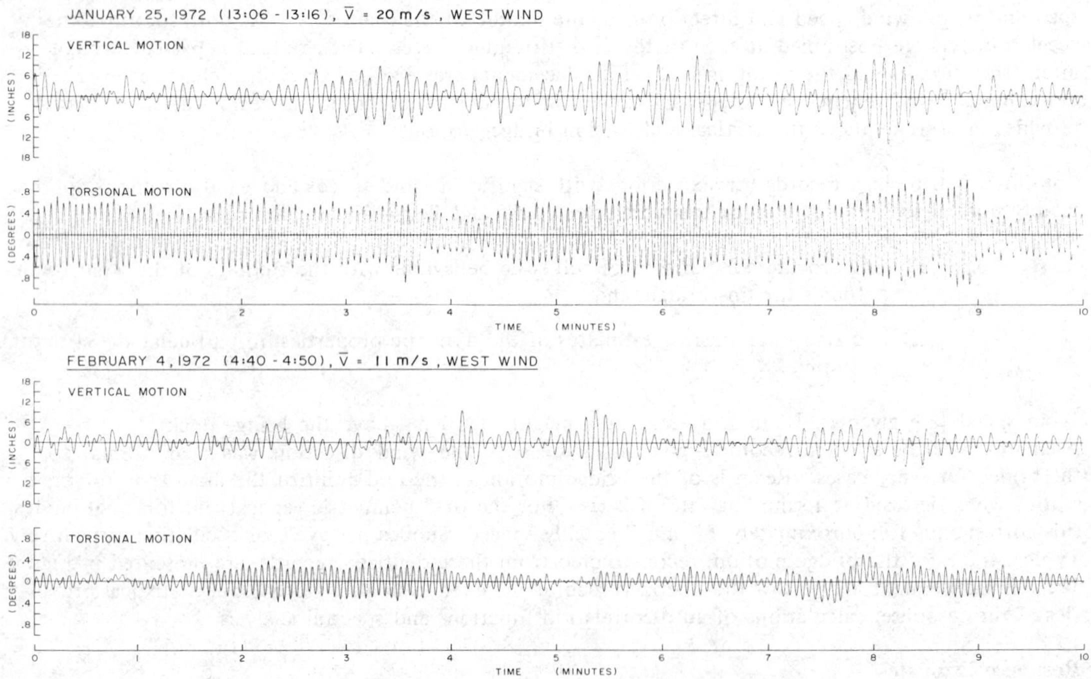

Fig. 11 Typical time records of wind induced motion at mid-span of bridge

This relatively close agreement with observed values suggests that both the vertical and torsional response of the bridge over the observed range of wind speed remain narrow-band Gaussian. It is also noteworthy that the observed full scale values of $\bar{g}$ are consistent with those observed in the taut strip model study, see Figure 7.

Typical power spectra of the full scale motion are presented in Figure 14. Most of the energy is concentrated at the 1st symmetric vertical and torsional modes of vibration with negligible contributions from higher harmonics. The natural frequencies estimated for the 1st symmetric vertical and torsional modes agree very well with those obtained in the earlier ambient vibration measurements, see Table 1, and assumed in the wind tunnel study. The agreement is also excellent for the higher modes determined from these spectral estimates. For example, the 2nd symmetric torsional and the 2nd and 4th symmetric vertical frequencies obtained in the ambient vibration study were 0.422, 0.300 and 0.558 Hz respectively.

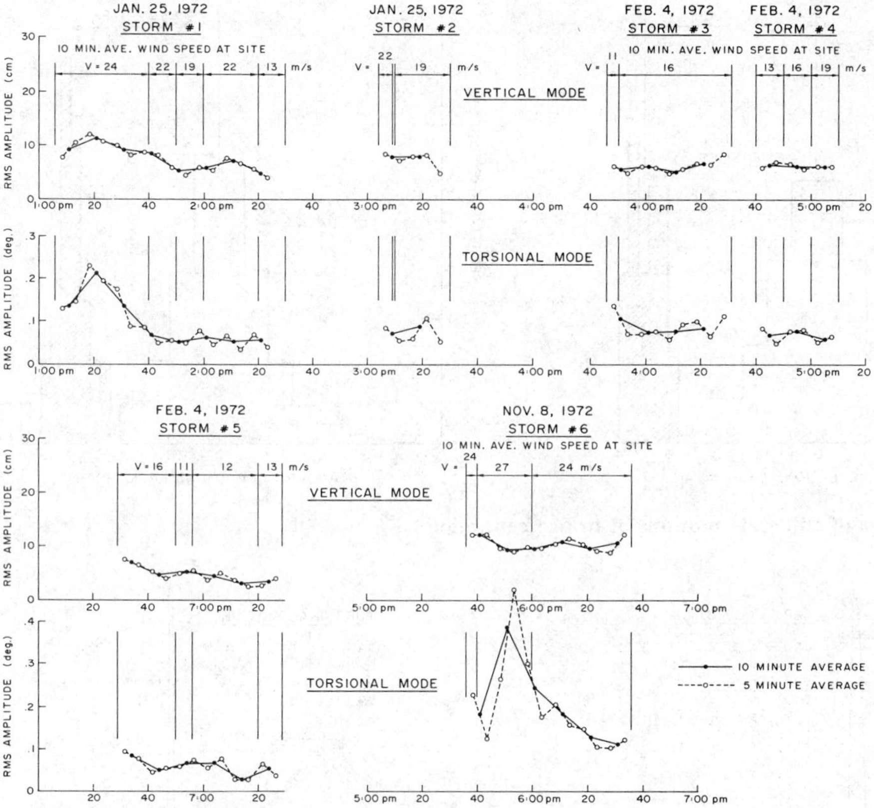

Fig. 12 Summary of storm data

Structural damping estimates

Estimates of the total damping at particular wind speeds were obtained from the decay of the auto-correlation functions of the motion. This technique provides reliable results as both vertical and torsional responses in this case almost entirely comprised motion in one particular mode of vibration. Auto-correlation functions were computed for various maximum time lags up to τ_{max} = 125 seconds. Typical auto-correlation functions are shown in Figure 15. The total damping in both modes for these particular data are estimated as 5.1 and 0.35 percent of critical respectively.

The evaluation of the structural damping from these data requires the estimation of the aerodynamic damping which is a function of wind speed. Estimates of the aerodynamic damping for the 1st symmetric vertical and torsional modes of vibration were obtained from the section model study. These are shown as a dotted line and curve in Figure 16. The aerodynamic damping in the torsional model is negative, while that for the vertical motion remains positive. For the vertical motion the aerodynamic damping is based on quasi-steady theory and increases linearly with wind speed. The measured total damping values, obtained for each of the six storms, are presented in Figure 16. The wind speed in each case is taken as the time weighted average speed for the entire storm. Also plotted in Figure 16 is the computed total damping for various wind speeds.

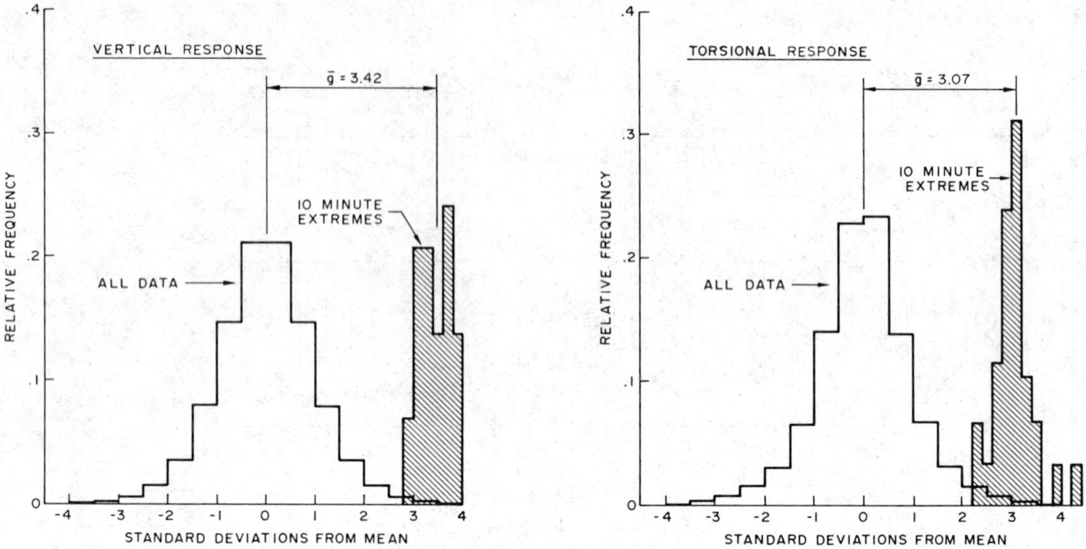

Fig. 13 Histograms of full scale motions at bridge centreline

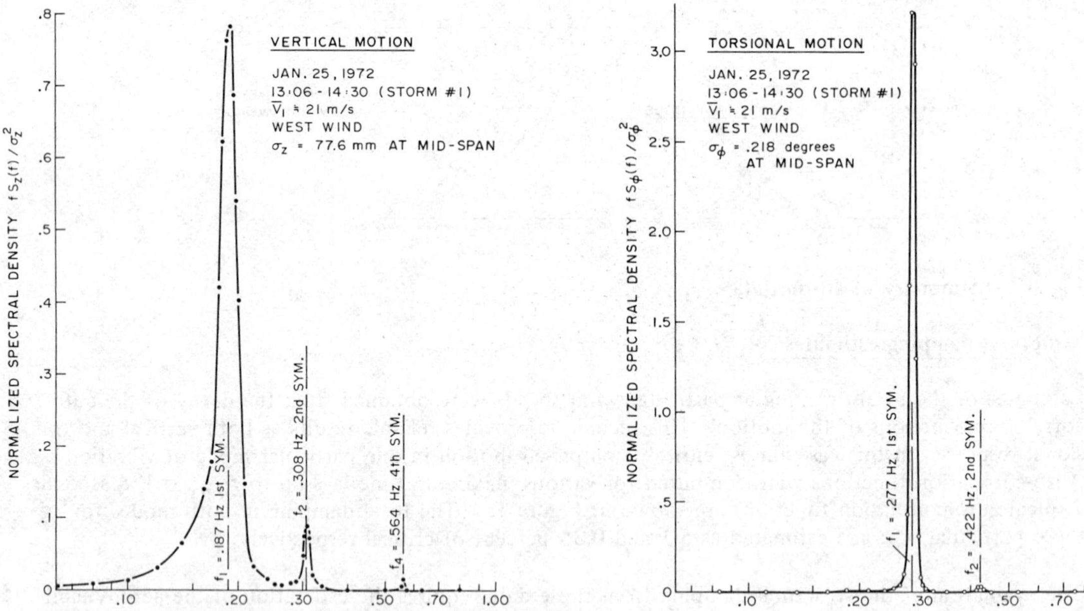

Fig. 14 Typical full scale response spectra at mid-span

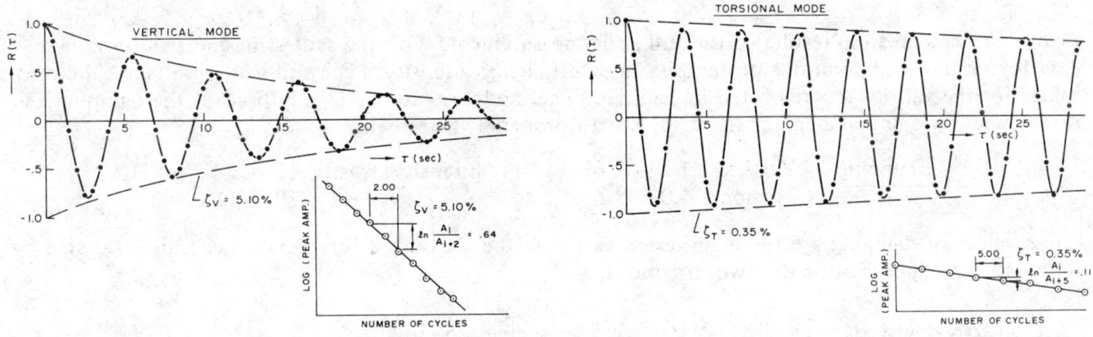

Fig. 15 Typical auto-correlation functions of full scale bridge response

In the case of the torsional motion, it appears that the structural damping is in the range of 1.0 to 1.5 percent with a definite trend towards higher values with increasing wind speed and consequently higher amplitudes of motion. A value of $\zeta_{st} \approx 1.5\%$ or even somewhat higher would appear to be appropriate for estimates of large amplitude torsional motion. Damping values for the vertical mode are scattered and it is thus more difficult to identify a trend. Based on the damping estimates for the higher range of wind speed, the value of ζ_{st} = 1.3% obtained in the ambient vibration study appears to be appropriate. A precise evaluation of ζ_{st} is of lesser practical significance for vertical motions as the aerodynamic damping is large, around + 4% at the upper end of the range of observed wind speeds, and the dynamic response is proportional to $(\zeta_{st} + \zeta_{aero})^{-1/2}$.

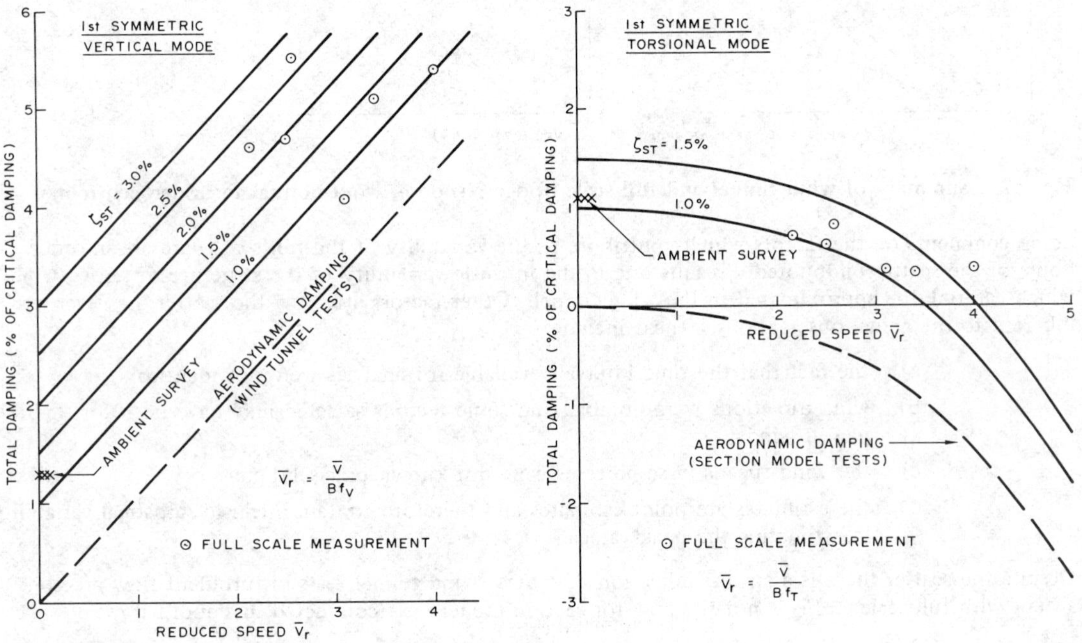

Fig. 16 Summary of full scale damping estimates

COMPARISON WITH WIND TUNNEL FINDINGS

Figure 17 compares the results of the full scale measurements with the taut strip and section model data for several configurations of damping and turbulence intensity. The full-scale results are the peak values for the various records listed in Table 3. The model results effectively bracket the damping and turbulence levels anticipated for the bridge, with probable values being:

Damping	Vertical	1.2 – 2.0%	Intensity	West	15 – 20%
	Torsional	1.0 – 1.6%		East	10 – 15%

These values of damping are those suggested earlier in the study; the turbulence levels are suggested by the upstream exposures for the two directions.

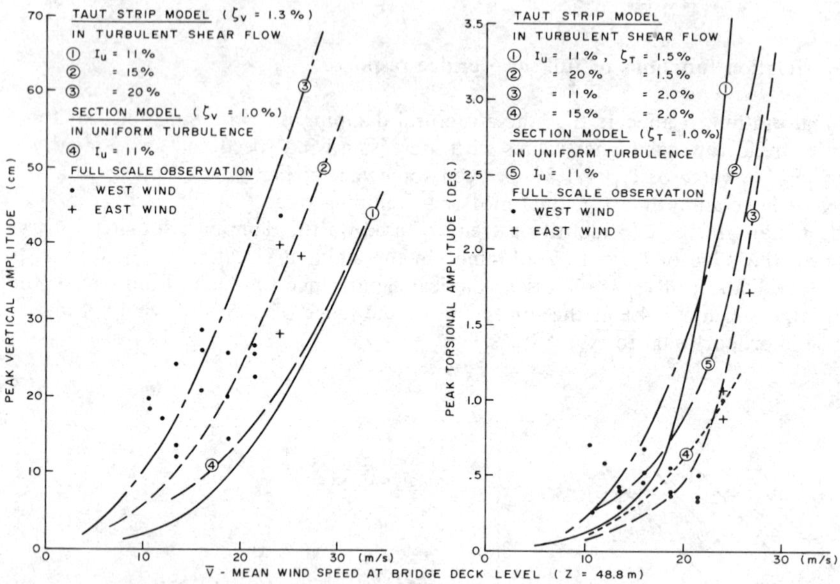

Fig. 17 Summary of wind tunnel and full scale bridge response, wind normal to bridge centreline

Some comments on the factors which contribute to the variability of the full scale data are in order. Some of the scatter undoubtedly results due to the intrinsic variability of the structure of strong winds which can only be approximated in the wind tunnel. Other factors affecting the scatter, however, are inherent to the actual observations. These include,

a) the fact that the time intervals available for analysis were not identical;

b) wind directions were nominal and some records selected may have been for slightly yawed winds;

c) the wind speeds in some cases were not known precisely; and

d) the estimates are point estimates and therefore contain intrinsic statistical variability in estimating the peak values.

Despite the scatter there is a strong indication that these wind tunnel tests in turbulent flow are consistent with full scale. This is not the case for section model tests conducted in smooth flow.

CONCLUDING REMARKS

The experimental approach and the main findings of an extensive wind tunnel study for a long span suspension bridge in New York City have been presented. Existing records of the wind induced motion of this bridge provided a rare opportunity to compare the wind tunnel findings with actual observations. Also, an analysis of the full scale data provided a refined estimate of the dynamic properties of the bridge.

The more important conclusions emerging from this are as follows:

i) The results of this study indicate that conventional section model tests in uniform smooth flow provide a rather pessimistic estimate of wind induced motion. Existing records indicate that this bridge has, without distress, experienced wind speeds in excess of the critical speed for the onset of catastrophic oscillations as predicted from conventional section model tests. This shortcoming of section model tests in smooth uniform flow is consistent with the findings of previously reported studies of other bridge sections[3,4,13].

ii) The presence of turbulence was found to have a pronounced effect on the wind induced response of the bridge. First, buffeting by turbulence produces significant vertical and torsional motions of the bridge even at low wind speeds. Second, the presence of buffeting induced motion results in a more gradual transition to large amplitude torsional oscillations. Both of these trends are consistent with available full scale observations. Increases in turbulence level tend to increase the magnitude of the buffeting induced motion and to further postpone the onset of large amplitude torsional motion.

iii) The taut strip model tests in turbulent flow were found to provide good agreement with observed full scale motions. Advantages of this technique are the correct simulation of the turbulent flow field and the three-dimensional flow-structure interaction, and its ability to provide information on the sensitivity of the response to wind direction. The latter was important for the large amplitude torsional oscillations. These were limited to a relatively narrow range of yaw angle, which leads to a reduction in the predictions of the full scale response.

iv) The rather good agreement observed between the taut strip model and the section model tests in uniform turbulent flow generated by a coarse grid is noteworthy. This suggests that the effects of turbulence may be approximated by correctly reproducing the energy of turbulent fluctuations at wavelengths of the order of the cross-sectional dimensions of bridge deck. This is a promising avenue for study.

v) The estimates of the effective full scale structural damping during strong winds were valuable in comparing model and prototype results and reconfirmed the important role it plays in controlling oscillations. For the case of this bridge, the structural damping in the 1st symmetric torsional mode was found to increase with the amplitude of the motion.

REFERENCES

1. Farquharson, F. B. and G. S. Vincent, Aerodynamic Stability of Suspension Bridges with Special Reference to the Tacoma Narrows Bridge, Part V, Univ. Washington, Eng. Expt. Station, Bulletin 116, 1954.

2. Scanlan, R. H., Theory of the Wind Analysis of Long–Span Bridges Based on Data Obtainable From Section Model Tests, Proc. 4th Int. Conf. on Wind Effects on Bldgs. and Strcts., London, 1975.

3. Davenport, A. G., N. Isyumov and T. Miyata, The Experimental Determination of the Response of Suspension Bridges to Turbulent Flow, Proc. 3rd Int. Conf. on Wind Effects on Bldgs. and Struct., Tokyo, 1971.

4. Irwin, H. P. and G. D. Schuyler, Experiments on the Full Aeroelastic Model of Lion's Gate Bridge in Smooth and Turbulent Flow, N.R.C. Report LTR–LA–206, 1977.

5. Davenport, A. G., The Use of Taut Strip Models in the Prediction of the Response of Long Span Bridges to Turbulent Wind, Proc. Symp. on Flow Induced Struct. Vibrations, Karlsruhe, 1972.

6. Davenport, A. G. and H. Tanaka, The Aerodynamic Stability of the Golden Gate Bridge: Comparison of the Full Scale Bridge Motion with Taut Strip Wind Tunnel Model, presented at Symp. of Full Scale Measurements of Wind Effects on Tall Bldgs. and Other Structures, London (Ont.), 1974.

7. Davenport, A. G. and H. Tanaka, Aerodynamic Instability of Taut Strip Models in Turbulent Boundary Layer Flow, U.W.O., Eng. Sc. Res. Rept., BLWT–1–74, 1974.

8. McLamore, V. R., Ambient Vibration Survey of Bronx-Whitestone Bridge, Teledyne Geotronics, Report 1070–2157 prepared for the Triborough Bridge and Tunnel Authority, New York, 1970.

9. Davenport, A. G., N. Isyumov and H. Tanaka, A Study of Wind Action for the Bronx-Whitestone Suspension Bridge, New York, N.Y., U.W.O., Eng. Sc. Res. Rept., BLWT–SS3–76, 1976.

10. Davenport, A. G., On the Statistical Prediction of Structural Performance in the Wind Environment, A.S.C.E., Natl. Struc. Eng. Meeting, Baltimore, Maryland, April 1971.

11. Davenport, A. G., The Dependence of Wind Action on Wind Direction, ASCE–EMD Conference, Univ. of Waterloo, May, 1976.

12. Davenport, A. G., M. Hogan and B. J. Vickery, An Analysis of Records of Wind Induced Building Movements and Column Strain Taken at the John Hancock Centre, Chicago, U.W.O., Eng. Sc. Res. Rept., BLWT–10–70, 1970.

13. Davenport, A. G., N. Isyumov, D. J. Fader and C. F. P. Bowen, A Study of Wind Action on a Suspension Bridge During Erection and on Completion, U.W.O., Eng. Sc. Res. Rept., BLWT–3–69, 1969.

FIELD OBSERVATION OF AERODYNAMIC BEHAVIOUR OF LONG-SPAN BRIDGES

T. OKUBO, N. NARITA, K. YOKOYAMA and H. SATO

Public Works Research Institute, Ibaraki, Japan

ABSTRACT

A considerable amount of literature in the past has been devoted to the aerodynamics of bridge structures by either analytical or experimental way. The main purpose of the paper is to introduce the wind characteristics and the aerodynamic response of structure observed at the specific bridges and to compare them with the theoretical results. The Kanmon Bridge and the Suehiro Bridge span 712m and 250m respectively and they are the longest in each type of bridge in Japan.

The long-term observation systems were set up and put in operation at the early stage of completion of the bridges.

After introducing the present status of the wind resistant design of long span bridges in Japan, the authors describe the wind characteristics, the lateral vibration of suspended structure and torsional and vertical bending vibration of bridge girder against gusty wind.

SYMBOLS

A_n= exposed area for unit length of structure

A_w=aerodynamic admittance

B=distance between the servo-type accelerometers

B_m, B_f=width of the bridge of model and full-scale, respectively

C_D, C_L, C_M=steady wind force coefficients

C=unsteady aerodynamic coefficient

ex. $C_{L\dot{\eta}}$; lift-vertical bending-in phase with velocity

$C_{M\varphi}$; Moment-torsional-in phase with displacement

D=drag

f=frequency

I=polar moment of inertia per unit length of span

L=lift

M=moment

L_{se}, M_{se}=self-excited aerodynamic force

L_{fo}, M_{fo}=forced aerodynamic force

l=lift due to turbulence of wind

m=mass per unit length of span

q=generalized coordinate

r=order of vibration mode

R(x, x',ω) =cross-correlation coefficient of wind

S=power or cross spectral density

t=time

U=mean wind speed

V_{10}=basic wind speed

w=vertical component of wind speed

x=spanwise coordinate of bridge

α=angle of attack

ζ=structural damping constant

η=vertical bending displacement

φ=torsional displacement

η_0, φ_0=amplitude of forced vibration during the measurement of C.

μ=contribution factor of lift to moment

ρ=air density

Φ=vibration mode

ω=circular frequency

$\omega_{\eta r}$, $\omega_{\varphi r}$=natural circular frequency

$\wedge$=expression of Fourier transform

INTRODUCTION

Wind effects on a structure is generally classified into two groups — static effect and dynamic one. In the usual design only the mean drag component is taken in the design wind load. The design wind load, though it is just a subordinate load for a shorter span bridge, holds an important position in a longer span bridge, that is to say, the cross-sectional area of principal members in a long-span bridge is governed by wind load.

Moreover, there is a fair chance for either suspension bridges or cable-stayed girder bridges to show aerodynamic instability such as flutter or vortex-excited oscillation as a result of dynamic effect of wind. So, it calls for necessary countermeasures to these phenomena in the design procedure.

Then a comprehensive survey including wind tunnel test is conducted from the early stage of preliminary design of long-span bridges in Japan. Design wind speed is determined by wind data gathered at meteorological stations near the construction site, taking into account of the local geographical condition. The stability of the bridge against wind-induced oscillation is confirmed by sectional model tests under laminar flow.

In the last decade, several long-span suspension bridges and cable-stayed girder bridges were constructed in Japan. The representative bridges are;

Kanmon Bridge	1974	suspension bridge	178 + 712 + 178m
Hirado Bridge	1977	suspension bridge	465.4m
Suehiro Bridge	1976	cable-stayed girder bridge	110 + 250 + 110m
Suigo Bridge	1977	cable-stayed girder bridge	178.9 + 111.6m

The construction of these bridges owed much on the investigation results which had been carried out in relation to the Honshu-Shikoku Bridge Project.

The first wind resistant design specifications for the Proposed Honshu-Shikoku Bridges was come into force in 1967. The revision work of the design criteria started in 1972 in running parallel with the field inspection, and the analytical method based on statistical concept was adopted firstly for buffeting problem of suspension bridges.

The design specifications for the Proposed Honshu-Shikoku Bridge (1975) provides that the design wind speed (V_D) and the design wind load (P_D) shall be computed by the following equations:

for horizontal line-like structure: $V_D = \nu_1 \cdot \nu_2 \cdot V_{10}$

$$P_D = \frac{1}{2} \rho V_D^2 \cdot C_D \cdot A_n \cdot \nu_4$$

for vertical line-like structure: $V_D = \nu_1 \cdot \nu_3 \cdot V_{10}$

$$P_D = \frac{1}{2} \rho V_D^2 \cdot C_D \cdot A_n \cdot \nu_5$$

where ν_1, ν_2 and ν_3 are wind modification factor for the elevation of the structure, one for horizontal length of the structure and one for vertical length of the structure, respectively, ν_4 and ν_5 are wind load modification factor for horizontal length of the structure and one for vertical length of the structure, respectively(1).

The fundamental concept used in the design criteria is to determine the wind speed modification factors and the wind load modification factors so that the magnitude of lateral bending moment of the stiffening trusses by natural wind is equal to that induced by uniform wind. The analytical method is based mainly on the statistical concept proposed by A. G. Davenport (2) in 1962. However only a few suspension bridges have reportedly shown lateral vibration in stormy wind, and some suppositions are introduced in the analysis.

Moreover, vortex-induced oscillation should be considered in case of cable-stayed girder bridges. The critical wind speed of it is usually lower than the design wind speed. Then the bridge engineers have to ensure that the oscillation is small enough to be harmless or have to search for suitable countermeasures to reduce the amplitude, if necessary. The vibrational amplitude of a full-scale structure can not directly derived from a sectional model test. The effects of both non-linearity in aerodynamic forces and mode of vibration should be considered.

Furthermore, wind at the site is not smooth but turbulent. It is well known that unsteady aerodynamic forces in smooth flow are different from those in turbulent flow. The bridge structures constructed in recent years are more flexible and have smaller damping capacities, then buffeting problem of structures has become important. Theoretical approaches and three-dimensional wind tunnel tests are conducted, but full-scale measurement seems necessary to ensure the assumptions adopted in the analysis and in the wind tunnel test. The long-term observation systems are installed at four bridges in accordance with the thought. The observation of response at stormy wind is expected to comply with the useful data which confirm the adequency of the bridge design and enhance design philosophy of the long-span bridges.

HORIZONTAL BENDING GUST RESPONSE OF THE KANMON BRIDGE

The Kanmon Bridge and its long-term observation system

The Kanmon Bridge constructed over the Kanmon Straits is the largest suspension bridge in Japan, and it has the span length of 178 + 712 + 178m. The suspended structure carrying six-lane highway traffic is stiffened by trusses with 9m high and with distance of 29m apart. The general view is given in Fig. 1.

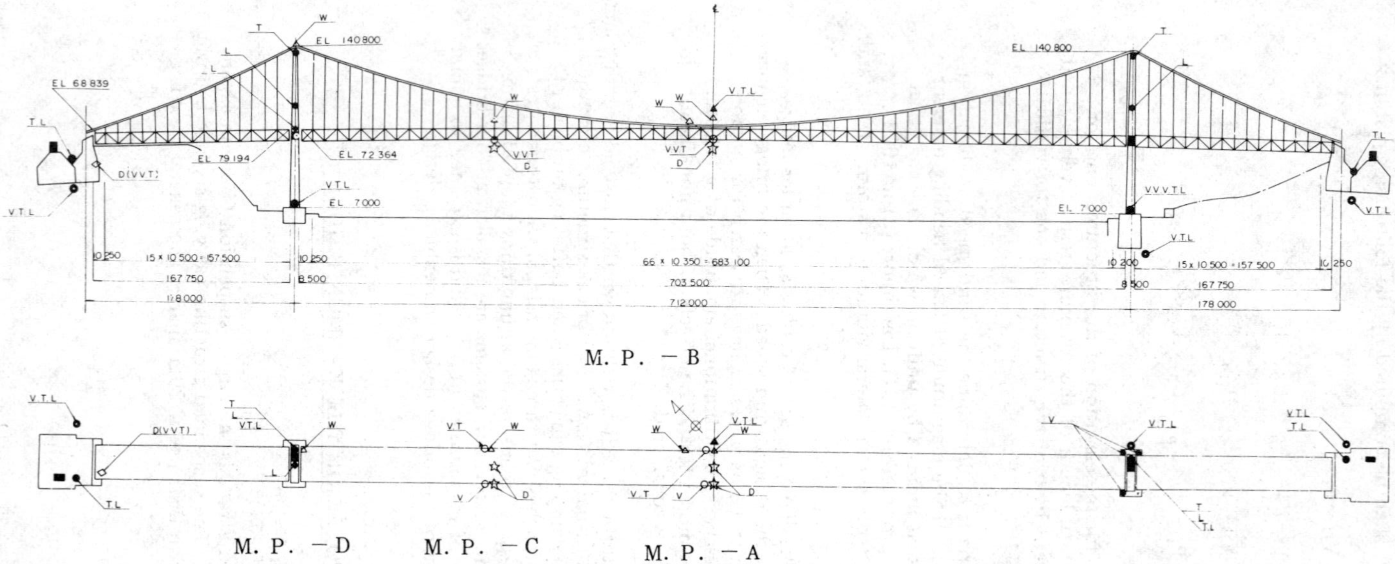

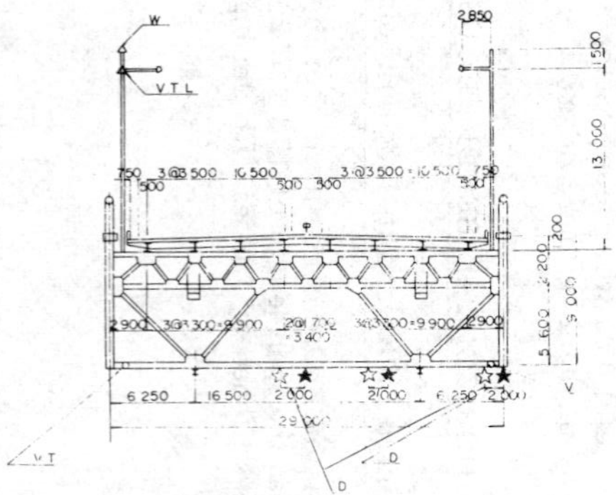

Mark	Instrument
●	electro-magnetic accelerometer
o	submerged seismometer
○	servo accelerometer
▲	supersonic anemometer
△	propellar type anemometer
☆	target for displacement meter
◇	target for calibration
▬ / □ / ◆	signal transmission system

Mark	Direction
V	vertical
T	transverse
L	longitudinal

Mark	Item
W	wind speed and direction

Note

M.P.—A the center of center span

M.P.—B 20m apart from M.P. — A to Shimonoseki

M.P. — C the quarter point of Shimonoseki side

M.P. — D the top of tower of Shimonoseki side

Fig. 1 Long term observation system of the Kanmon Bridge

After completion of the bridge a vibration test was carried out by the use of a twin exciter and heavy trucks to obtain the natural frequencies and the damping capacities in both vertical bending and torsional vibrations.

An observation system for monitoring and recording the speed and the direction of approaching wind and the vibration of the structure was installed by the Japan Highway Corporation. The location of instruments are shown in Fig. 1 and Table 2. The characteristic features of the system are (i) the elements are large in number, (ii) the system is always kept ready to record and monitor, and (iii) the signal is transmitted in digital form to keep good signal-noise condition.

Table 1 Instruments for long term observation system at Kanmon Bridge

Observation item	Location	Measuring range	Instruments	Number of tranducers	Number of elements	Sampling rate (freq./sec)
Acceleration	Ground	±500 gal	Electromagnetic accelerometer	3	9	100
	Abutment			4	4	50
	Pier			8	8	
	Tower			4	4	
	Stiffening truss	±200 gal	Servo-type accelerometer	6	6	25
Displacement	Stiffening truss	Horizontal ±2m Vertical ±2m	Two-dimensional optical analyser	1	3	25
		Longitudinal ±25cm	Differential transducer	1	1	25
Wind speed and wind direction	Stiffening truss	Horizontal ±60m/sec Vertical ±10 m/sec	Ultrasonic anemometer	1	3	25
		Horizontal 60m/sec 540°	Propeller type anemometer	3	6	25
	Top of Shimonoseki tower			1	2	25

Transducers

(a) Propeller type anemometer: Four anemometers are set up at the top of poles spread on the bridge deck for observing horizontal distribution of wind.
One anemometer is installed at the top of main tower.

(b) Ultrasonic anemometer: One set of three-component ultrasonic anemometer is set at the mid-point of main span for observing wind inclination angle.

(c) Servo-type accelerometer: Servo-type accelerometers with high fidelity are set up at the mid-point and the quarter point of the center span. One element for horizontal component and two elements for vertical component are set up at each point.

(d) Displacementmeter: Both horizontal and vertical motion of the stiffening structure are detected by this optical facility.

(e) Others: Vibration transducers for earthquake measurement also installed at the bridge.

Data acquisition system

The output data are digitized at every 40 milli-second by analogue-to-digital convertors set-up in the transducers and transmitted for about 1.5 km as far as the data acquisition system in the administration office. A data processing system equipped with a mini-computer ordinally performs monitoring and recording the data. When the wind speed exceeds a preset level, the data acquisition system starts automatically and the data are stored in the magnetic tape.

Observed wind and response of stiffening girder

Wind characteristics

The observation system was put in operation in April, 1974. Since then, a few typhoons have attacked the area, but strong wind normal to the bridge axis has not been observed. Wind statistics have been obtained in order to compare the real wind characteristics with those of the expected data.

Power spectral density and coherency of wind speed are given in Fig. 2 and 3, respectively. From the data the characteristics of approaching wind are:

(1) The turbulence intensity, gust factor and power spectral density of wind agree fairly well with those obtained in the past, and no clear effect of wind convergence due to topographical condition has been observed.

(2) The longitudinal distribution of horizontal mean wind speed observed along the deck is almost uniform, the deviation of which is less than 10%. The mean wind direction at each point is nearly the same. Instantaneous wind speed and direction, however, fluctuate so violently that cross-correlation of wind speed rapidly decreases with the increase of distance. The cross-correlation coefficient of wind speed shows wind speed has strong correlation between the anemometers set 20m apart, but weak correlation 180m apart.
The decay factor k of spatial correlation obtained from the coherency of wind speed is about 3 or 4 in lower frequency range, which is smaller than that presented in the past.

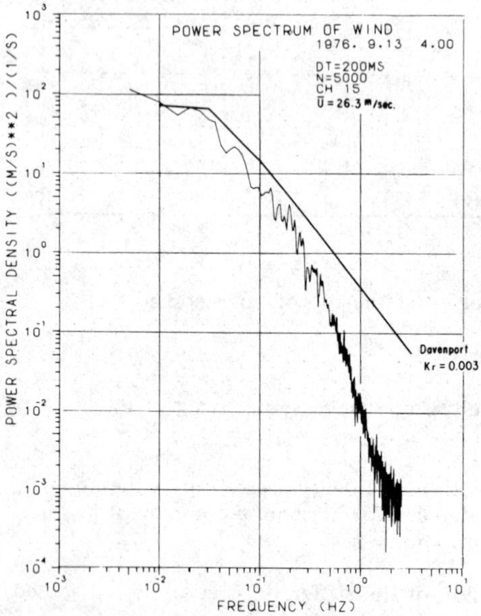

Fig. 2 Power spectral density of wind

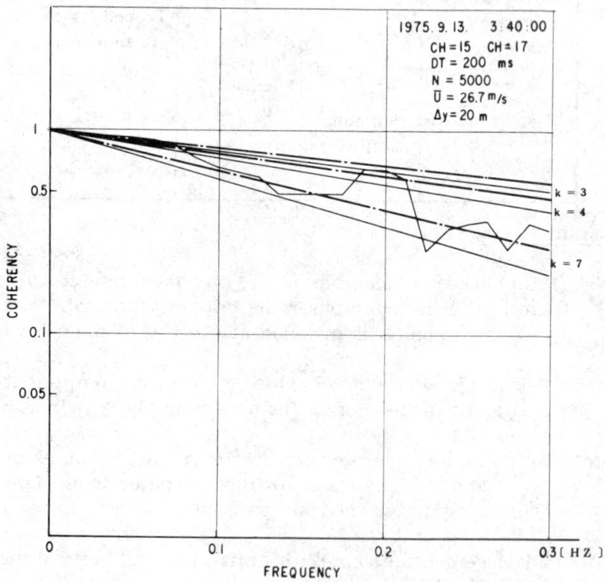

Fig. 3 Coherency of wind

Response of stiffening girder

In order to grasp optically the behavior of suspended structure against wind, a motion picture was made by computer out put microfilming system, and it was found that the bridge vibrated randomly in vertical bending mode, torsional one, and lateral bending one. Fig. 4, Fig. 5, and Fig. 6 show the power spectral density of the vertical bending acceleration, the torsional one, and the lateral one, respectively. The prominent powers appear at 1.21Hz, 0.38Hz, and 0.10Hz in vertical bending mode, torsional one, and lateral bending one, respectively. These frequencies coincide with the lowest natural frequencies obtained from the calculation. Considering the relationship,

$$S_{displacement}(f) = (2\pi f)^{-4} S_{acceleration}(f),$$

the contribution of higher modes to the displacement responses seems to be far smaller than of the lowest mode. It may be said that the vibration of the lowest mode is important not only for flutter phenomena but for gust responses.

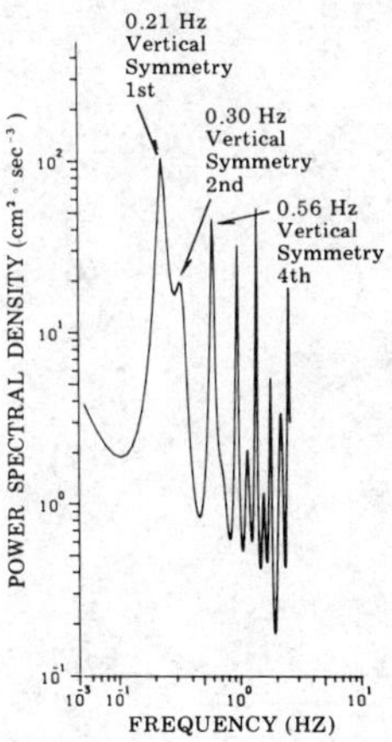

Fig. 4 Power spectral density of vertical bending acceleration at mid-span

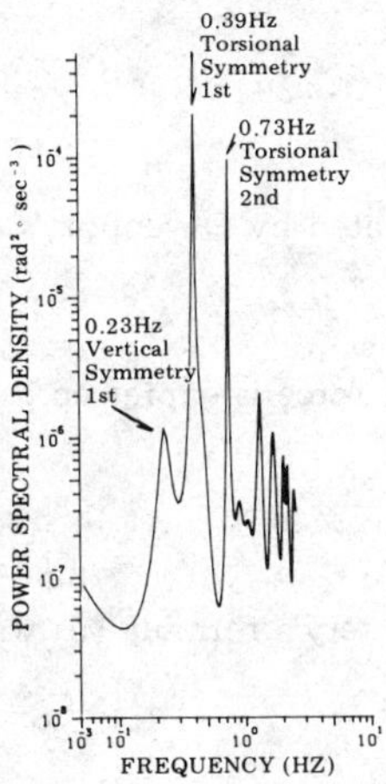

Fig. 5 Power spectral density of torsional acceleration at mid-span

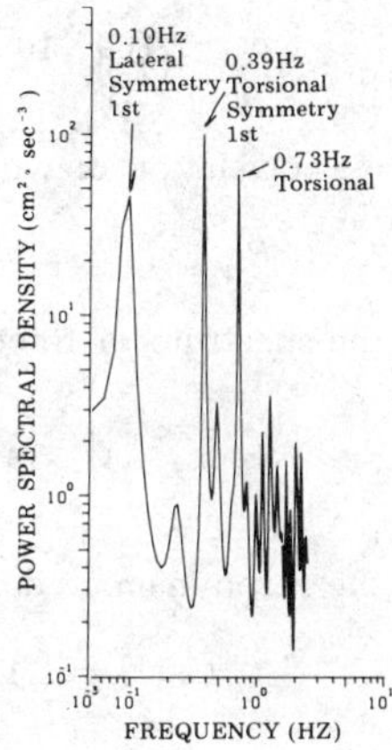

Fig. 6 Power spectral density of lateral acceleration at mid-span

Horizontal gust response

Structural analysis of the system

Lateral motion of the system was analysed by the use of what is called Finite Element Method. The fundamental equation of the system is

$$\begin{cases} \dfrac{d^2}{dx^2}\left(EI\dfrac{d^2v}{dx^2}\right) = q_t + \dfrac{p_s}{h}(u-v) \\ H_d\dfrac{d^2v}{dx^2} = -q_c + \dfrac{p_s}{h}(u-v) \end{cases} \quad (1)$$

and the mechanical admittance for r-th mode of vibration is

$$|H_r(n)|^2 = 1/P_r^4\left[\left\{1-\left(\frac{p}{p_r}\right)^2\right\}^2 + 4\zeta_r^2\left(\frac{p}{p_r}\right)^2\right] \quad (2)$$

The fraction of damping was assumed as

$$\zeta_r = \zeta_{ro} + \zeta_{raero}, \quad \zeta_{ro} = 0.03/(2\pi), \quad \zeta_{raero} = \frac{C_D\, B\, V_Z}{32\pi\, m\, n_r} \tag{3}$$

Spectral density of wind force

The mean wind force $\overline{P}$ corresponding to the mean wind speed V_Z is

$$\overline{P} = \tfrac{1}{2}\rho\, V_Z \;\; C_D\, A_n \tag{4}$$

The spectral density of u-component was evaluated by Hino's formula (3), namely

$$S_u(n) = 0.476\, \frac{6kV_{10}^2}{\beta}\, \left(1 + \frac{n^2}{\beta^2}\right)^{-\frac{5}{6}}$$

$$\beta = 1.169 \times 10^{-3}\, \frac{V_{10}\,\alpha}{\sqrt{k}} \left(\frac{z}{10}\right)^{(2m_e\alpha - 1)} \tag{5}$$

The cross-correlation of wind was evaluated by Davenport's expression (4),

$$R_u(x_1, x_2, n) = exp\left(-\frac{kn}{V_z}\,|\,x_1 - x_2\,|\right) \tag{6}$$

Then, the spectrum of fluctuating wind force is expressed as

$$S_p(n) = 4\overline{P}^2\, |X_u(n)|^2\, \frac{S_u(n)}{V_z^2} \tag{7}$$

as for the aerodynamic admittance, Vickery's formula (5) was used

$$|X_u(n)|^2 = \left\{1 + 2\left(\frac{nD}{V_z}\right)^{\frac{4}{3}}\right\}^{-1} \tag{8}$$

Aerodynamic response

Response spectrum of lateral bending moment of the stiffening structure is

$$S_m(x, n) = \sum_r M_r^2(x)\, |H_r(n)|^2\, |J_r(n)|^2\, S_p(n) \tag{9}$$

where the joint acceptance Jr (n) is given in the following form,

$$|J_r(n)|^2 = \iint v_r(x_1)\, v_r(x_2)\, R_u(x_1, x_2, n)\, dx_1\, dx_2 \tag{10}$$

Variance for bending moment is given

$$\sigma_m^2(x) = \int_0^\infty S_m(x,n)\, dn = \sum_r M_r^2(x) \int_0^\infty |H_r(n)|^2\, |J_r(n)|^2\, S_p(n)\, dn \tag{11}$$

In making computer program, the special attention was paid to the accuracy of numerical calculation for the following terms and the expressions.

(i) Terms $v_r(x)$, $u_r(x)$, $M_r(x)$, $Q_r(x)$, $M(x)$ and $Q(x)$

(ii) Double integral in the joint acceptance

(iii) $\int_0^\infty (\cdots\cdots\cdots)\, dx$ and $\sum_r (\cdots\cdots\cdots)$

The numerical calculation was carried out to make sure of the effect of wind characteristics on the gust response. The power index $m_e\alpha$, the surface drag coefficient K and the decay factor k were chosen as variables. The effect of K and k is so remarkable that they seem to govern the response.

The comparison between the observed lateral gust response and the theoretical values.

The relation between the mean wind speed perpendicular to the bridge axis and the horizontal displacement at mid-span is shown in Fig. 7 and 8. The mean displacement increases with the increase of the mean wind speed. The root mean square value of displacement relates not only to mean wind speed but to intensity of turbulence. It increases with the increase of the mean wind speed and the intensity of turbulence. The observed mean displacement agrees fairly well with the calculated one. In the calculation the intensity of turbulence of wind was assumed to be constant, but the observed rms displacement seems to nearly agree with the calculated one. The main direction of the wind used did not coincide with the transverse axis of bridge so that the effect of it should be considered. Moreover, the discrepancy of statistical characteristics, i.e., power spectral density, spatial correlation of wind should also be considered.

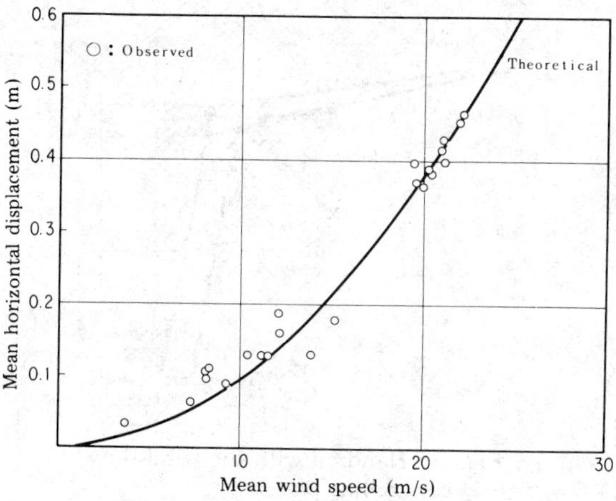

Fig. 7 Wind speed vs. mean horizontal displacement

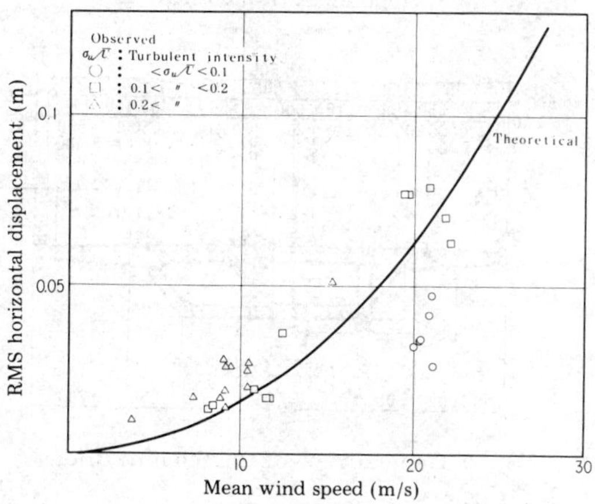

Fig. 8 Wind speed vs. RMS. horizontal displacement

VERTICAL BENDING AND TORSIONAL GUST RESPONSES OF THE SUEHIRO BRIDGE

Outline of the Suehiro Bridge

The Suehiro Bridge, constructed in 1976, is a cable-stayed bridge with the span of 110 + 250 + 110m. The general side-view and cross section of the bridge are illustrated in Fig. 9.

The engineering feature of its superstructure is the wind flow controlling plates attached to the top of the handrails. They restrict the separation of the wind flow in the deck to reduce the vertical bending vibration of the superstructure. Their shape is illustrated in Fig. 10.

Outline of the field observation system

The speed and the incidental angle of approaching wind and the structural response have been observed since the early stage of erection in order to confirm the theoretical and experimental analysis. The observation system consists of a propeller type anemometer, a three-component ultrasonic anemometer, four servo-type accelerometers, and a data recorder. The arrangement of the instruments are illustrated in Fig. 11.

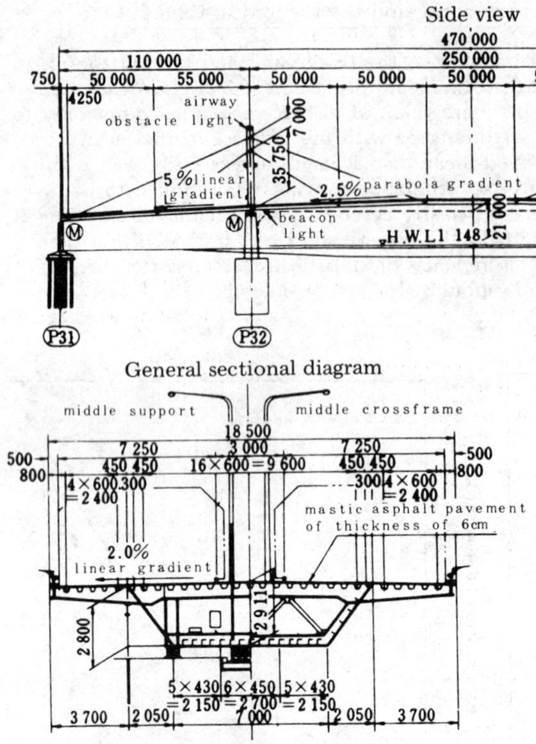

Fig. 9 General view of the Suehiro Bridge

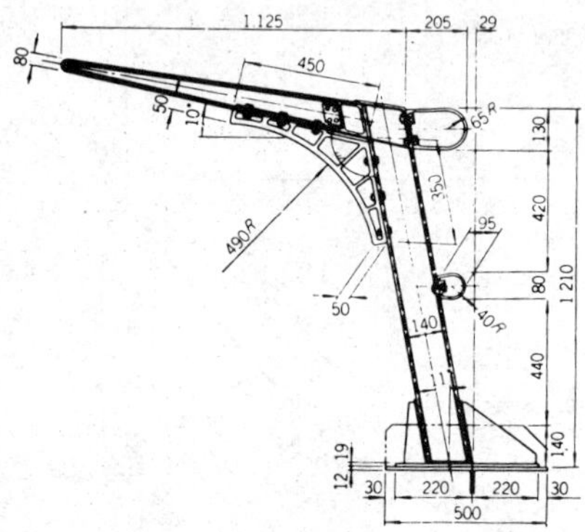

Fig. 10 Handrail with the wind flow controlling plates

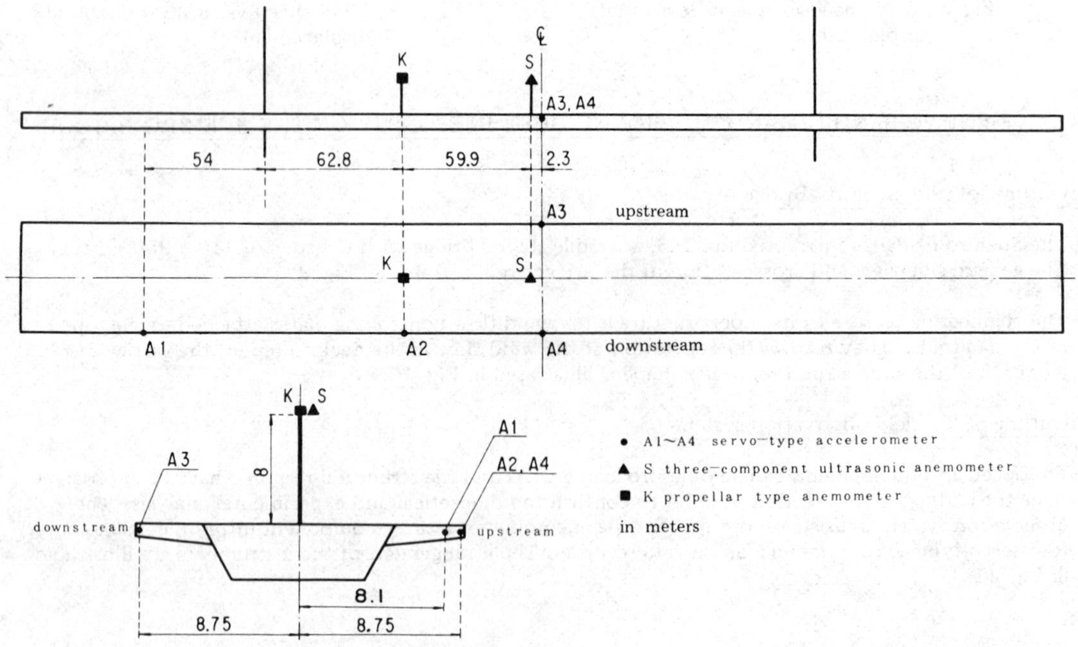

Fig. 11 Arrangement of instruments for long-term observation system of the Suehiro Bridge

The data recorder starts to record wind and response data when the wind speed exeeds preset level. The tape is brought to the Institute, and the data are converted from analogue to digital with the sampling time of 100ms and the sampling number of 6000 for IRUN, then they are processed with a digital computer.

Characteristics of the natural winds at the Suehiro Bridge

The outline of characteristics of the natural winds at the Suehiro Bridge is as follows.

Mean wind speed; The largest mean wind speed is 19m/s in the data obtained in 1976 and 1977.
Mean wind direction in a horizontal plane; The bridge axis stretches almost from north to south (inclines 4.2deg. clockwise in a horizontal plane), and the prevailing wind direction is north-west or south-east. It almost coincides with the direction of the Shinmachi River which the Suehiro Bridge crosses over.
Mean angle of attack; It ranges between −8deg. and 8 deg.
Intensity of turbulence of the horizontal component; It ranges between 0.15 and 0.25.
Power spectral density; Its shape and its value are almost the same as those measured before.

Characteristics of the responses of the Suehiro Bridge.

The responses at the center of the main girder are analysed. From the accelerations (a_3, a_4) gained with two servo-type accelerometers (A_3, A_4), the vertical bending acceleration ($a\eta$) and the torsional one ($a\varphi$) are calculated as follows.

$$a\eta = (a_3 + a_4)/2 \qquad a\varphi = (a_3 - a_4)/B \tag{12}$$

The power spectral densities of $a\eta$ and $a\varphi$ are shown in Fig. 12. The prominent powers appear at 0.48Hz and 1.52Hz in vertical bending acceleration and torsional one, respectively, and these frequencies coincide with the lowest natural frequencies obtained from the calculation.

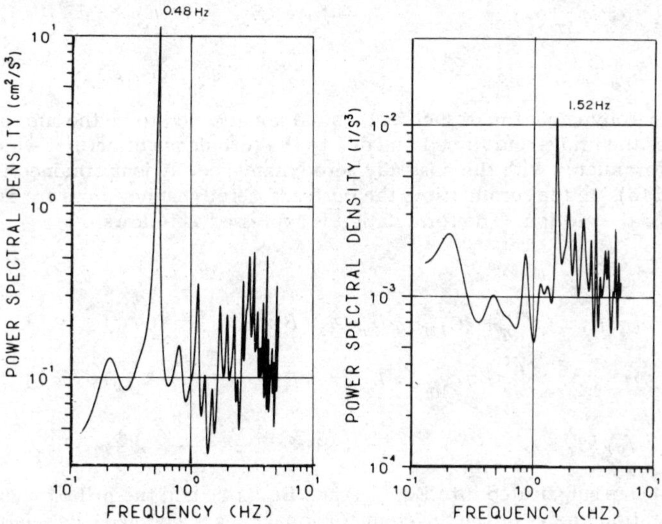

Fig. 12 Power spectral density (vertical bending and torsional acceleration)

Assuming that the responses of the girder consist of the simple harmonic vibration of the first mode only, the relation between the mean wind speed and the amplitude of the vertical bending vibration can be shown in Fig. 13. From the relation, the amplitude is found to be less than 1.3cm under the mean wind speed up to 19m/s.

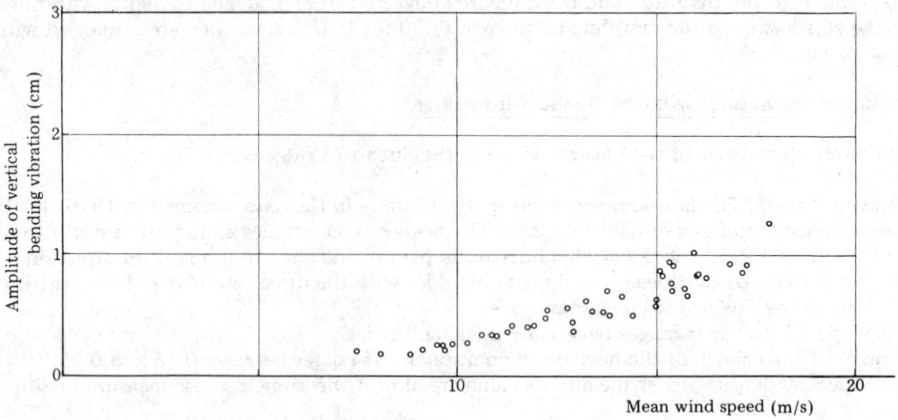

Fig. 13 Relation between mean wind speed and vertical bending response

Calculation method of the gust responses

Equations of motion can be expressed as follows in a frequency domain, when vertical bending and torsion of three-dimensional structure are considered.

$$[(i\omega)^2+2\zeta_{\eta r}\omega_{\eta r}(i\omega)+\omega_{\eta r}^2]\,\hat{q}_{\eta r}(\omega)=\hat{Q}_{\eta r}(x,\omega)/\int_0^{\ell} m\Phi_{\eta r}^2(x)\,dx \tag{13}$$

$$[(i\omega)^2+2\zeta_{\varphi r}\omega_{\varphi r}(i\omega)+\omega_{\varphi r}^2]\,\hat{q}_{\varphi r}(\omega)=\hat{Q}_{\varphi r}(x,\omega)/\int_0^{\ell} I\Phi_{\varphi r}^2(x)\,dx \tag{14}$$

$$\hat{\eta}(x,\omega)=\sum_r \hat{q}_{\eta r}(\omega)\,\Phi_{\eta r}(x) \qquad \hat{\varphi}(x,\omega)=\sum_r \hat{q}_{\varphi r}(\omega)\,\Phi_{\varphi r}(x) \qquad (15),(16)$$

$$\hat{Q}_{\eta r}(x,\omega)=\int_0^{\ell}\Phi_{\eta r}\hat{L}(x,\omega)\,dx \qquad \hat{Q}_{\varphi r}(x,\omega)=\int_0^{\ell}\Phi_{\varphi r}\hat{M}(x,\omega)\,dx \qquad (17),(18)$$

In above equations, the aerodynamic forces such as L and M are assumed to be the sum of self-excited one due to the motion of the bridge and forced one due to the turbulence of natural wind. The self-excited one can be formulated with the unsteady aerodynamic coefficient obtained from the forced vibration method (6). In the formulation, the unsteady aerodynamic forces are assumed to be linear to the amplitude of the motion. The formulation is expressed as follows.

$$\hat{L}(x,\omega)=\hat{L}_{se}(x,\omega)+\hat{L}_{fo}(x,\omega) \qquad \hat{M}(x,\omega)=\hat{M}_{se}(x,\omega)+\hat{M}_{fo}(x,\omega) \qquad (19),(20)$$

$$\hat{L}_{se}(x,\omega)=\tfrac{1}{2}\rho U^2 B_f\,[(C_{L\eta}^{R}+iC_{L\eta}^{I})\hat{\eta}B_m/\eta_0/B_f+(C_{L\varphi}^{R}+iC_{L\varphi}^{I})\hat{\varphi}/\varphi_0] \tag{21}$$

$$\hat{M}_{se}(x,\omega)=\tfrac{1}{2}\rho U^2 B_f^2\,((C_{M\eta}^{R}+iC_{M\eta}^{I})\hat{\eta}B_m/\eta_0/B_f+(C_{M\varphi}^{R}+iC_{M\varphi}^{I})\hat{\varphi}/\varphi_0] \tag{22}$$

$$\hat{L}_{fo}(x,\omega)=\sum_r \hat{L}_{for}(\omega)\,\Phi_{\eta r}(x) \qquad \hat{M}_{fo}(x,\omega)=\sum_r \hat{M}_{for}(\omega)\,\Phi_{\varphi r}(x) \qquad (23),(24)$$

When Eq. (15) — Eq. (24) are substituted into Eq. (13) and Eq. (14), and the orthogonality between the different vertical vibration modes or the different torsional ones is assumed, the relation between $\hat{q}_{\eta r}$, $\hat{q}_{\varphi r}$ (generalized coordinates) and L_{for}, M_{for} (self-excited aerodynamic forces) can be obtained in a matrix form. Then the similar relation between $S_{q\eta r}$, $S_{q\varphi r}$ (power spectral density of $q_{\eta r}$, $q_{\varphi r}$, L_{fot} and M_{fos}, M_{fot} and M_{fos}, respectively) can be obtained.
In order to derive S_{LtLs}, S_{LtMs}, S_{MtMs} from S_{Lo} (power spectral density of lift at a reference point of a bridge), two assumptions (7) are used and shown below.

$$\hat{M}_{fo}(x,\omega)=\mu B_f\hat{L}_{fo} \tag{25}$$

$$S_L(x, x', \omega) = S_{Lo}(\omega) R(x, x', \omega) \tag{26}$$

Lift due to the turbulence of the wind can be related with the vertical component of the wind as follows.

$$\ell(t) = \tfrac{1}{2} \rho B_f U (dC_L/d\alpha + C_D) A_w (f B_f/U) w(t) \tag{27}$$

The gust responses of vertical bending mode and torsional one can be calculated in a frequency domain through the above process, if the wind data are obtained and the values of parameters in the calculation process are determined properly.

Effects of parameters in the calculation

In the calculation of gust responses, the effects of parameters such as aerodynamic admittance, life-curve slope, unsteady aerodynamic coefficient, structural damping constant, cross-correlation coefficient, vibration mode, and contribution factor of lift to moment were investigated with regard to the standard deviation of vertical bending gust responses and torsional ones at the center of the main girder. The result is shown in Table 2. The upper part of the table shows the result calculated by using one of the observed wind data and by setting the values of parameters as those in the table. The lower part of the table shows the result calculated by exchanging each value of parameter in the lower part of the table for the one in the upper part. From Table 2, it is found that the effect of cross-correlation coefficient and higher vibration modes are very small.

Table 2. Effect of Parameters

Parameter		Standard deviation of responses	
		Vertical bending (cm)	Torsion ($\times 10^{-2}$ deg)
Aerodynamic admittance; quasi-steady theory (=1) Lift-curve slope: measured in smooth flow (=2.86) Decay factor of cross-correlation coefficient; 7 Unsteady aerodynamic coefficient; of flat plate Structural damping constant; 0.00318 Vibration mode; vertical bending; 6 modes torsion; 4 modes Contribution factor; 1		0.395	2.422
Aerodynamic admittance;	Davenport	0.207	1.155
	Sears (1)	0.127	1.021
	Sears (2)	0.126	1.020
	Holmes	0.192	1.125
Lift-curve slope; of flat plate		0.868	5.320
Decay factor of cross-correlation coefficient;			
10		0.393	2.417
15		0.392	2.415
20		0.391	2.414
25		0.391	2.414
30		0.391	2.414
Unsteady aerodynamic coefficient; measured in smooth flow		0.461	2.466
Structural damping constant;			
0.00159		0.438	2.787
0.00636		0.340	2.036
0.0127		0.283	1.700
Vibration mode; vertical bending; 1 mode (lowest) torsion; 1 mode (lowest)		0.387	2.395
Contribution factor; $C_M(0)/C_L(0)$ (=0.14)		0.393	0.339

Comparison of calculated responses with observed ones

By considering aerostatic and aerodynamic characteristics obtained from wind tunnel test in smooth flow, the values of parameters were determined as follows.

Aerodynamic admittance; Sears' function
Lift-curve slope; measured in smooth flow (=2.86)
Decay factor of cross-correlation coefficient; 7
Unsteady aerodynamic coefficient; measured in smooth flow
Structural damping constant; 0.00318
Vibration mode; vertical bending (6 modes)
torsion (4 modes)
Contribution factor; measured in smooth flow ($C_M(0)/C_L(0) = 0.14$)

The gust responses were calculated with regard to the standard deviation at the center of the main girder by using observed wind data and compared with observed gust responses. 12 of observed data in which the mean angle of attack ranges from —1 deg. to 1 deg. were chosen and are shown in Table 3. On the other hand, the calculated gust responses are shown in Table 4 and compared with observed ones in Fig. 14. From the results, it is concluded that the gust responses of vertical bending mode can be predicted well, but as for torsional mode, the parameters such as the contribution factor of lift to moment should be evaluated more precisely.

Table. 3 Observed winds and gust responses (mean angle of attack; —1 deg.~ 1 deg.)

Case No.	U (m/s)	σ_u (m/s)	$I_u=\sigma_u/U$	σ_w (m/s)	β (deg)	σ_η (cm)	σ_φ ($\times10^{-3}$ deg)
1	7.6	1.55	0.20	2.21	292.1	0.14	0.87
2	9.4	1.89	0.20	3.17	289.6	0.22	1.22
3	10.8	2.41	0.22	3.13	293.2	0.28	1.30
4	13.5	3.01	0.22	3.74	281.4	0.39	1.51
5	10.3	1.63	0.16	2.69	278.8	0.40	1.76
6	11.6	1.99	0.17	3.48	281.4	0.46	3.26
7	11.5	2.36	0.21	3.57	284.7	0.49	2.87
8	10.9	2.58	0.24	3.62	278.5	0.50	2.46
9	9.4	1.83	0.19	1.84	122.2	0.29	1.87
10	8.8	2.12	0.24	2.68	304.0	0.18	0.77
11	7.8	1.94	0.25	3.10	312.6	0.18	0.99
12	10.7	2.16	0.20	3.34	279.9	0.18	0.95

where u ; horizontal component of wind speed perpendicular to bridge axis
w ; vertical component of wind speed
U ; mean of u
σ ; standard deviation
I ; intensity of turbulence
η ; vertical bending response at the center of the main girder
φ ; torsional response at the center of the main girder
β ; angular inclination in a horizontal plane

Table 4. Calculated gust responses

Case No.	σ_η (cm)	σ_φ ($\times10^{-3}$ deg)
1	0.13	1.43
2	0.22	2.01
3	0.29	2.56
4	0.45	4.55
5	0.33	3.71
6	0.49	5.63
7	0.37	3.61
8	0.48	4.66
9	0.19	1.55
10	0.26	2.46
11	0.26	2.40
12	0.50	5.45

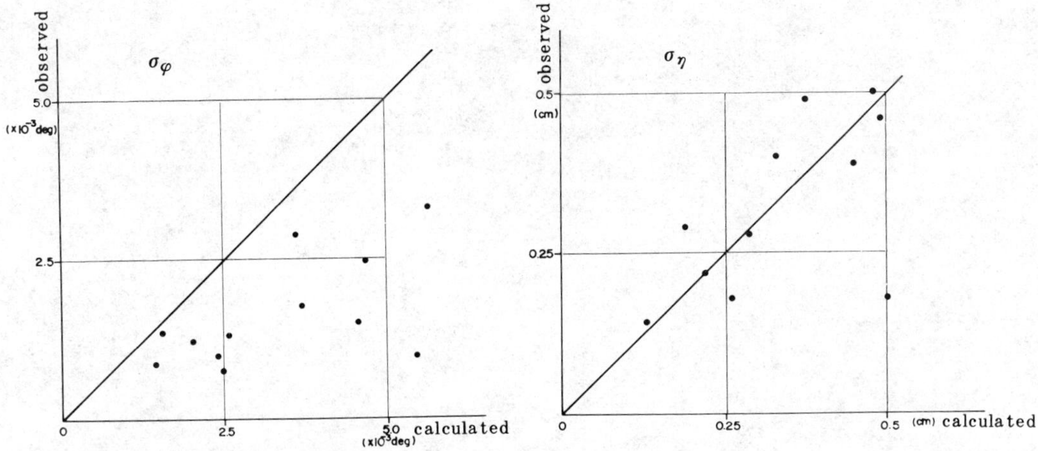

Fig. 14 Calculated responses and observed ones

ACKNOWLEDGEMENT

The authors are indebted to Japan Highway Corporation and Tokushima Prefecture for obtaining the valuable data of wind and response at the Kanmon Bridge and the Suehiro Bridge, respectively. The authors also thank the other members of the Structure Division, P.W.R.I., for a good work in processing the data.

REFERENCES

(1) T. OKUBO and N. NARITA, On the wind resistant design specifications for the proposed Honshu-Shikoku Bridges (1975), *8th Joint Meeting, UJNR, Washington, D.C.,* 1976

(2) A. G. DAVENPORT, Buffeting of a suspension bridge by storm winds, *ST 3, ASCE,* June 1962

(3) M. HINO, Relationships between the instantaneous peak values and the evaluation time, No. 117, *Trans. of JSCE,* May 1965 [in Japanese]

(4) A. G. DAVENPORT, The response of slender, line-like structures to a gusty wind, *Proc. ICE,* Vol. 23, 1962

(5) B. J. VICKERY, On the flow behind a course grid and its use as a model of atmospheric turbulence in studies related to wind loads on buildings, *NPL Aero Report* No. 1296, 1969

(6) T. OKUBO, N. NARITA, and K. YOKOYAMA, Some approrches for improving wind stability of cable-stayed girder bridges, *Proc. 4th Int. Conf. on Wind Effects on Bldgs. and Sts.,* 1975

(7) I. KONISHI, N. SHIRAISHI, and M. MATSUMOTO, Studies on wind resistant design on long span suspension bridges, *Fac. Eng., Univ. Kyoto,* March 1973 [in Japanese]

RESPONSE OF THE LARGE SCALE BRIDGE MODEL TO NATURAL WIND

I. OKAUCHI*, J. TAJIMA** and H. AKIYAMA**

*Chuo University, Tokyo, Japan
**Honshu-Shikoku Bridge Authority, Tokyo, Japan

ABSTRACT

In this paper, the results of analysis related to the observation on the response of the large scale bridge model to natural wind are introduced. The prototype of the model is the proposed Honshu-Shikoku Bridge.

More than 800 records of the wind characteristics and the response of model were collected in three years. As the results of analysis on these data, the reliability of wind tunnel test was assured, and the turbulence effects on the bridge structure under natural wind were clarified qualitatively.

GLOSSARY OF TERMS

$\|A(f)\|^2$	:	Aerodynamic admittance
f	:	Frequency
$\|H(f)\|^2$	:	Mechanical admittance
$\underline{k}$	:	Decay factor
$\bar{L}$	:	Mean lift force
$\bar{P}$	:	Mean drag force
$\underline{S}(f)$	:	Power spectrum
$\bar{U}$	:	Mean wind velocity
u, w	:	Fluctuating wind velocity components in longitudinal and vertical directions respectively
x	:	Displacement of lateral (horizontal) vibration
y	:	Displacement of heaving (vertical) vibration
α	:	Angle of the model in the vertical plane (attack angle)
δ	:	Logarithmic damping ratio
ζ	:	Damping ratio
ϕ	:	Displacement of pitching (torsional) vibration

1. PURPOSE OF THE LARGE SCALE MODEL TESTS

"The Honshu-Shikoku Bridge Project", one of the largest construction projects is now in progress in Japan. This project involves the construction of three long chains of bridges including several long span suspension bridges across the Seto Inland Sea. Since the construction sites for these bridges are situated along the route of typhoons, it is vitally necessary to provide adequate resistance to strong wind.

For that reason, from very early in the project, investigations have been carried out to confirm the aerodynamic stability of bridges. Those investigations have been planned by the Technical Committee for wind resistant design of the Honshu-Shikoku Bridges. Based on these research activities, a wind resistant design code has been developed for the proposed Honshu-Shikoku Bridges. The first edition of that code, issued in 1964, has been revised several times.

The reliability of a design code should be confirmed through the detailed observation of response of a real structure against natural wind of high velocity, that is, the design wind velocity. However, the confirmation of the design code through the way above mentioned seems to be impossible at the existing state of things, and therefore the large scale bridge model test was planned. (1),(2)

The purpose of the test consists in the study of the aerodynamic effects of turbulence on the response of bridge structure, which seems to be one of the most important problems concerning to the wind resistant design of structures, by the use of a far large scale model compared with the ordinary wind tunnel model. It was also intended in the project to confirm the reliability of wind tunnel test results.

Field observation was started in February 1973, and more than 800 pieces of records of the wind characteristics and response of the model were collected in three years.

The project had been carried out by the Honshu-Shikoku Bridge Authority under the guidance of the Technical Committee on the Wind Resistant Design of the Honshu-Shikoku Bridges in Japan Society of Civil Engineers.

2. OUTLINE OF LARGE SCALE MODEL TEST

2.1 Observation Site

As a result of careful research, the site for observation was selected on the beach near Nojimazaki, at the south apex of Boso peninsular about 100 km south of Tokyo. As shown in Fig.1, the site was nearly flat and open to the Pacific Ocean in the west, the westerly wind was expected to be unaffected by the topographical conditions.

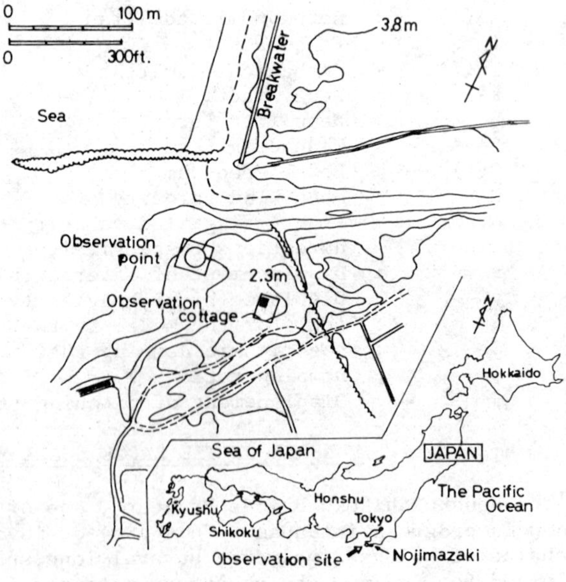

Fig. 1 Observation site

2.2 Test Facilities

(1) Model

The scale of model was chosen as large as possible in order to ease assembling the model and making the observation. For this tests, two models were prepared. The first one termed the MA model imitated in detail the prototype structure and the scale factor was 1/10.52. The second termed the MB model was a more simplified version of the prototype structure. Each model had the original type with slotted deck, which was named type 1, and the modified type with closed deck named type 2. The prototype structure was one of the Honshu-Shikoku Bridges, that had a center span length of 1,500 m and a truss type stiffening girder, 14 m in height and 36 m in width.

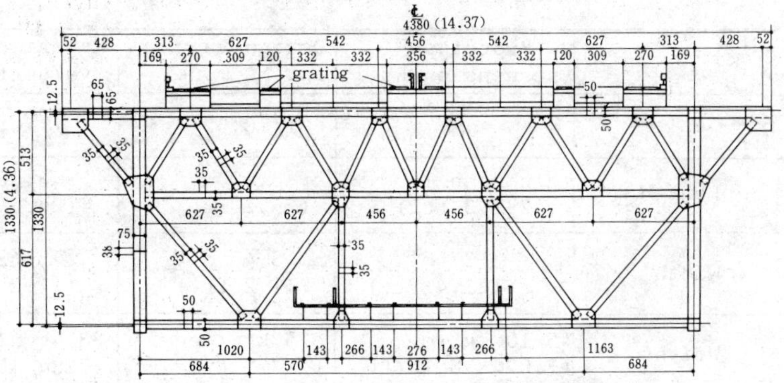

(1) MA-1 Model

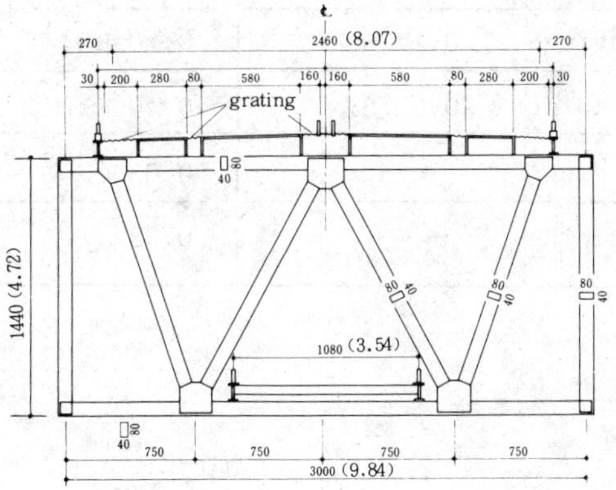

(2) MB-1 Model

Fig. 2 Cross-section of model Unit: mm (ft.)

As shown in Fig. 2, the aluminium sectional model with a width of 3.42 m, a height of 1.33 m and a length of 8 m (MA model), was made with high rigidity. The end plate was attached to both ends of the model to force two dimensional flow of the wind. The governing similitude law is the same as that commonly used for the sectional model tests conducted in wind tunnels, and the test conditions are given in Table 1.

(2) Model Mounting

The models were mounted on the coil spring-balance system shown in Fig.3, in which the dynamic balancing conditions as well as static ones were satisfied.
The elaborate supplementary devices were attached to the system so as to keep the heaving and pitching vibrations and their coupled one. An oil-pressure equipment was installed to the system. It was used for an excitor to put the model in forced oscillation and for a damper to give the additional damping capacity to the model. Also a shock absorber to the system was equipped in order to restrict the vibration with such excessive amplitude that might harm the model and the bearing.

(3) Frame

An overall view of the model and the test frame is shown in Fig. 4. A streamlined steel cover was attached to the supporting frame for the model to prevent wind effects on the coil-balance system and to ensure two dimensional flow of the wind on the model. The whole frame was mounted on a large turntable so that it could be faced to the prevailing wind direction.

Table 1. Similarity Condition

		Prototype (suspension Br.)	Design value for MA model	Experimental value (MA)	Experimental value (MB)
Scale		——	1/10.52	1/10.52	——
Length		main span 1500 m (4291.3 ft.)	8.00 m (26.2 ft.)	8.00 m (26.2 ft.)	8.03 m (26.3 ft.)
Width		37 m (121.4 ft.)	3.42 m (11.2 ft.)	3.42 m (11.2 ft.)	3.00 m (9.8 ft.)
Weight		45×10^3 kgf/m (3.024×10^4 lb/ft)	3253kgf/model (7172 lb/model)	3304kgf/model (7284 lb/model)	3238kgf/model (7138 lb/model)
Polar moment of inertia		1242 $tf\cdot m\cdot s^2$/model (2.738×10^6 $lb\cdot ft\cdot s^2$/ft)	811 $kgf\cdot m\cdot s^2$/model (5866 $lb\cdot ft\cdot s^2$/model)	785 $kgf\cdot m\cdot s^2$/model (5678 $lb\cdot ft\cdot s^2$/model)	589 $kgf\cdot m^2\cdot s^2$/model (4206 $lb\cdot ft\cdot s^2$/model)
Frequency	Heaving	0.11 Hz	0.20 Hz	0.23 Hz	0.30 Hz
	Pitching	0.22 Hz	0.40 Hz	0.43 Hz	0.67 Hz
Logarithmic Damping ratio	Heaving	——	0.03	0.07 - 0.13	0.04 - 0.07
	Pitching	——	0.03	0.01 - 0.11	0.01 - 0.05

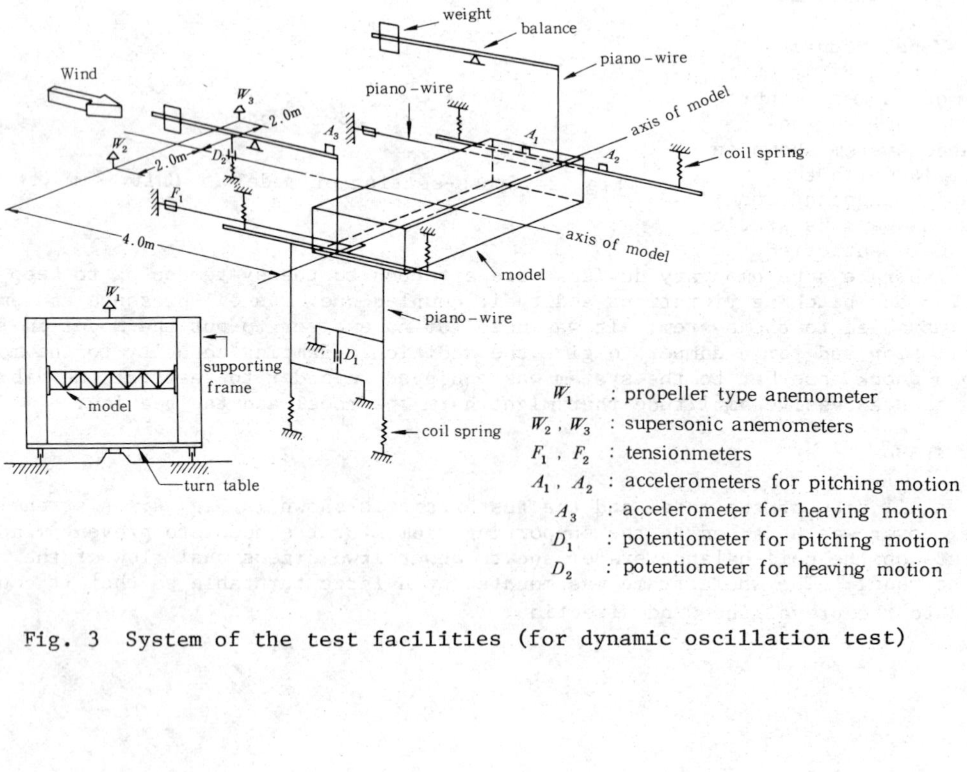

Fig. 3 System of the test facilities (for dynamic oscillation test)

(4) Observation and Instrumentation

Fig. 4 The view of the whole system

Observation was made for the velocities of the approaching wind, the wind forces on the model and the motions of the model. The three components of wind velocity approaching to the model were measured at two points 4 m in front of the model by two three-component supersonic anemometers. The wind forces acting on the model were measured by electronic tensionmeters attached to the piano-wires connecting the model to the test frame. The lateral,heaving and pitching motions were measured by potentio meters and servo-type accelerometers. The output signals were recorded on both a magnetic tape-recorder and a pen-recorder.

2.3 Wind Tunnel Tests

It is clear that the significance of the large scale bridge model test is increased,if its results are compared with the results obtained in ordinary wind tunnel tests. And it was necessary to fix the test conditions in detail beforehand because the large scale models were too big to change the observation conditions frequently. In order to comply with those requirements, wind tunnel tests were carried out both before and after observation of the large scale model tests.

To conduct wind tunnel test two kinds of wind tunnel models were prepared, whose prototypes were models MA and MB, respectively. Like the large scale field test models,each wind tunnel model with a closed deck as well as with a slotted deck was tested. The scale of these models were chosen to be about 1/10.

Measurements were made on the wind forces and the vibrational behavior of wind tunnel models mounted on spring supports not only in laminar flow but also in turbulent flow.

3. TEST RESULTS

More than 800 pieces of data were obtained from the field observations. These data could be divided into 3 groups, which concerned with wind, forces and vibrations. Thus at first the test results had been analysed by checking the wind characteristics and the relation between the wind forces and the wind. Then the relationships among the vibration, the wind and the wind forces were examined by the power spectral density method and so on. Reported in the following are some of the results obtained from that analysis.

3.1 Wind Characteristics

In order to investigate the response characteristics of models, it was necessary to know the characteristics of the approaching wind. Thus, such values as mean velocity, gust intensity, attack angle, auto and cross correlation, power and cross spectra were calculated for almost series of the data. Some examples of the results are given in Fig. 5. In general, values such as the intensity of turbulence and the gust factor were almost the same as the commonly reported ones. The power spectra of fluctuating wind velocity u and w were slightly different from the commonly reported ones, but fortunately such peculiarities were not important for the analysis of the relationship between the wind and the response of the model.

The influence of the walls and stage on the uniformity of wind characteristics perpendicular to wind direction is more important. Although many kinds of effort was made to ensure uniformity at every point of transversal plane to wind direction, that condition was not fully satisfied. Careful attention must be paid to that fact in interpreting the test results.

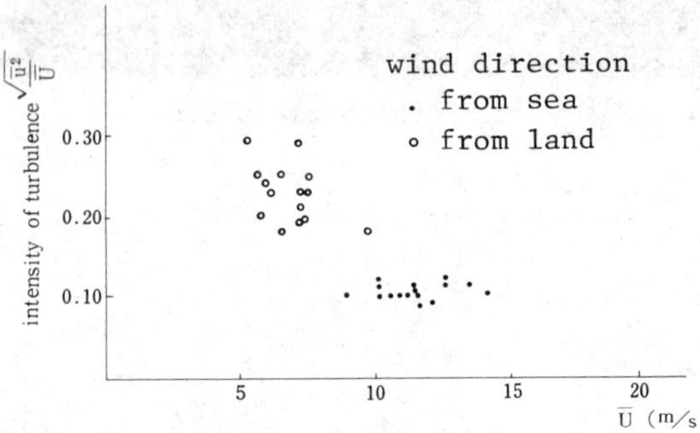

(1) Intensity of turbulence (longitudinal) (sampling time : 300 sec.)

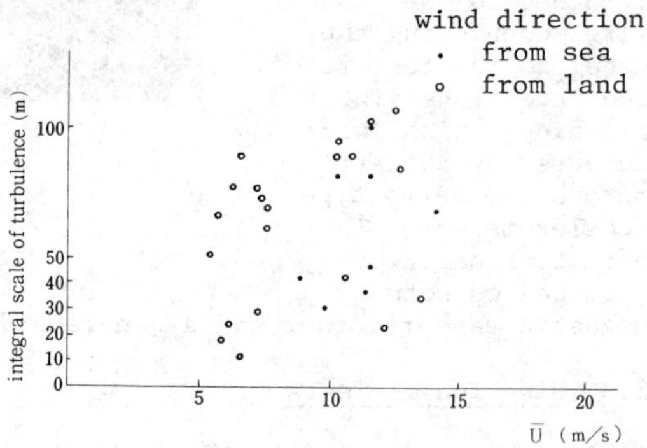

(2) Integral scale of turbulence (longitudinal)

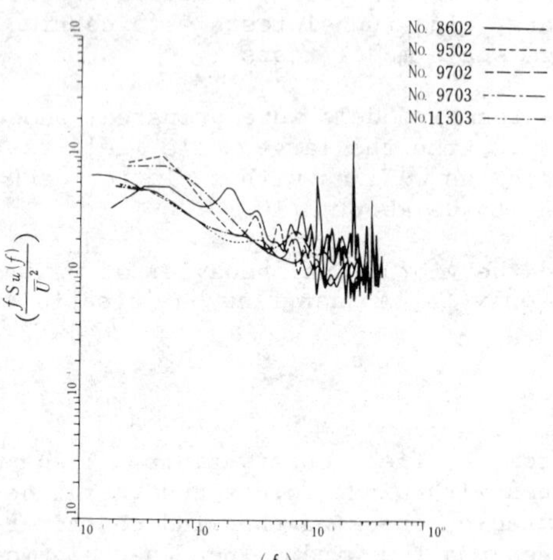

(a) u

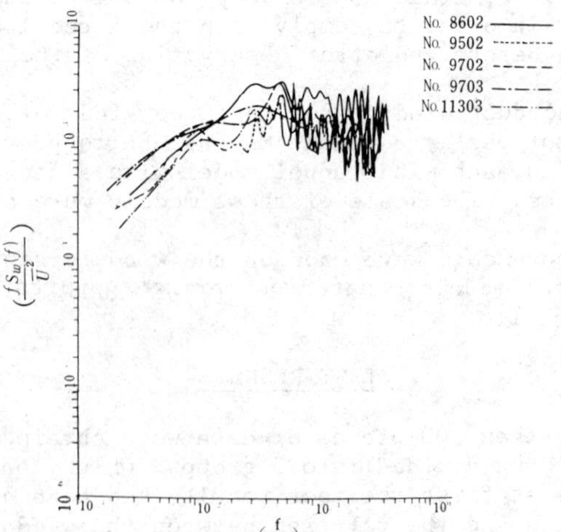

(b) w

(3) Power spectra of fluctuation wind

Fig. 5 Wind characteristics

Wind forces were measured by load-cells attached to the piano-wires supporting the model. Of the three components of wind force, only the drag forces could be measured precisely. The drag coefficient, the power spectrum and aerodynamic admittance were calculated by relating the drag force to the wind velocity. The aerodynamic admittance is the transfer function from wind velocity to aerodynamic force.

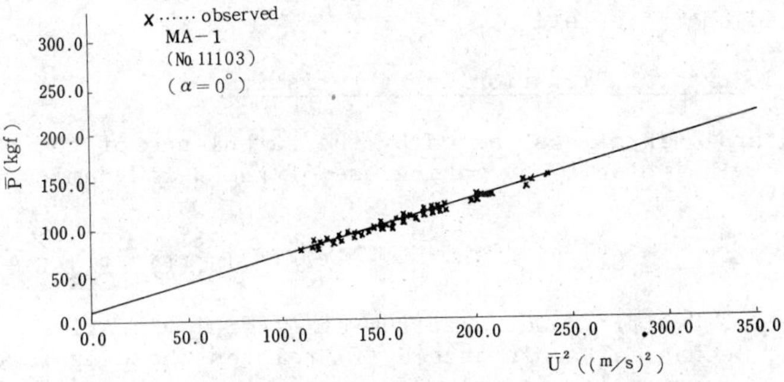

Fig. 6 The relation between drag force and the square of mean wind velocity

(1) Drag Coefficient

As shown in Fig. 6, it is apparent that there is a linear relation ship between the mean drag force and square of the mean wind velocity. From the tangent of the corresponding regression line, the drag coefficient was calculated and compared with the results obtained in ordinary wind tunnel tests. Table 2. shows the results of that comparison. The results of the large scale model tests closely agree with those of the wind tunnel tests.

Table 2. Drag coefficients (α=0°)

	MA-1	MA-2	MB-1	MB-2
(1) Observation	2.45	2.44	2.11	2.66
(2) Wind tunnel tests	2.27	2.31	2.13	2.67
(1) / (2)	1.08	1.06	0.99	0.99

(2) Aerodynamic Admittance

The typical aerodynamic admittance is shown in Fig. 7. Aerodynamic admittance values were obtained from the power spectra of drag and wind velocity by the following equation.

$$|A_p(f)|^2 = \left(\frac{\bar{U}}{2\bar{P}}\right)^2 \frac{S_{pp}(f)}{S_{uu}(f)} \qquad (1)$$

where, $S_{pp}(f)$: power spectrum of fluctuating drag force p
$S_{uu}(f)$: power spectrum of fluctuating wind velocity u

As shown in Fig. 7, admittances' values decrease rapidly from 1.0 in the high frequency region. Therefore the contribution of high frequency part of the fluctuating wind velocity must have decreased. Also shown in Fig. 7 with solid lines are the approximate relationship calculated by the following equation recommended by Devenport (3):

$$|A_p(f)|^2 = \frac{2}{\gamma_y^2}(\gamma_y - 1 + e^{-\gamma_y}) \frac{2}{\gamma_z^2}(\gamma_z - 1 + e^{-\gamma_z}) \qquad (2)$$

where, $\gamma_y = \frac{khf}{U}$, $\gamma_z = \frac{k\ell f}{U}$, h: truss height, ℓ: truss length

In Eq.(2) k is a parameter called as the decay factor and it relates the degree of the decreasing rate of coherence to frequency.

From Fig.7 it can be seen that appropriate value k is assumed to be about 15. That value coincided closely with the result obtained from the analysis of the wind characteristics.

3.3 Lateral Vibration Characteristics

Lateral vibrations caused by the turbulence of the natural wind may be related to the wind velocity by making use of the aerodynamic admittance as indicated in Eq. (3).

$$S_{xx}(f) = \left(\frac{2\bar{P}}{\bar{U}}\right)^2 \cdot |A(f)|^2 \cdot |H_x(f)|^2 \cdot S_{uu}(f) \qquad (3)$$

where, $S_{xx}(f)$: power spectrum of lateral vibration
The mechanical admittance $|H_x(f)|^2$ relates the drag force to the lateral vibration and is generally described as indicated in Eq. (4).

$$|H_x(f)|^2 = \frac{1}{m_x^2 (2\pi f_x)^4 \cdot \{(1 - \frac{f^2}{f_x^2})^2 + 4\zeta_x^2 \frac{f^2}{f_x^2}\}} \qquad (4)$$

where, m_x: mass of the system, f_x: natural frequency of the system,
ζ_x: damping ratio of the system.

An example of comparison of the calculated spectrum with the observed power spectra on lateral vibrations is shown in Fig. 8. The lateral vibration spectra were calculated from the observed wind velocity spectrum by making use of Eq. (3) and are shown in Fig. 8 by broken lines. The lateral vibration spectrum shown in

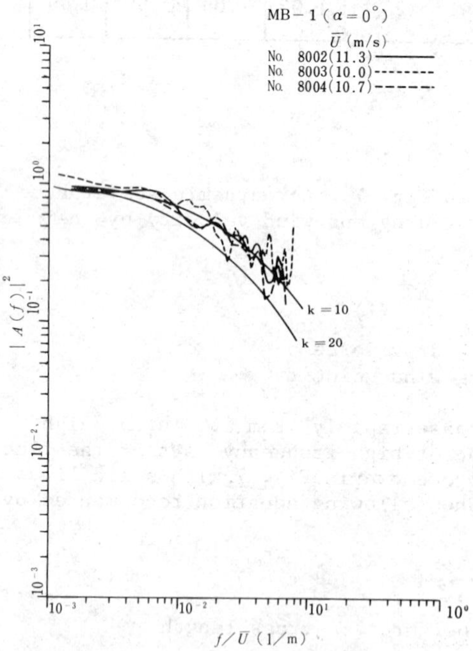

Fig. 7 Aerodynamic admittance of drag force (MB-1)

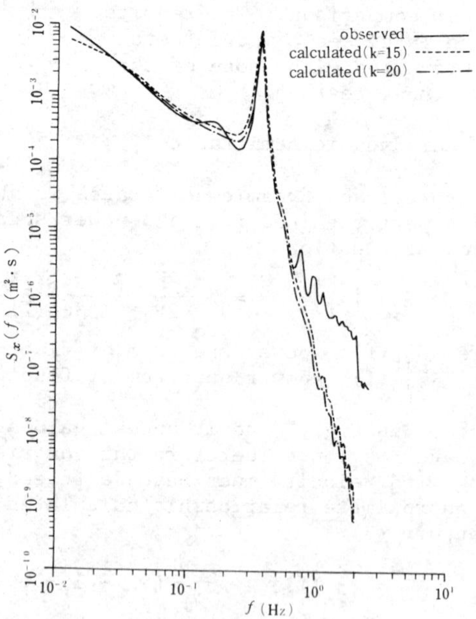

Fig. 8 Comparison of the calculated spectra with observed one on lateral vibration (No.7335, MB-1)

Fig.8 by solid line was obtained directly from the observed vibration data. Each spectrum has a sharp peak at a frequency coinciding with the natural frequency of the system. By comparing these spectra it is apparent as follows:

1. Both of the spectra coincide well in the region of $f \lesssim f_x$.
2. The observed spectrum is larger than those calculated by Eq. (3) in the region of $f \geqq f_x$.
3. The k value has little influence on the lateral vibration spectrum.

3.4 Heaving Vibration Characteristics

The same procedure as described in 3.3 was used in the analysis of the random heaving vibrations. However, as the lifting force is caused not only by the vertical wind component, but also by the longitudinal wind component and others, these components must be considered for the analysis of the heaving vibrations. Thus, the spectrum of the heaving vibration $S_{yy}(f)$ was considered to be represented as the following equations.

$$S_{yy}(f) = S_{yyww}(f) + S_{yyuu}(f) + \text{------------} \qquad (5)$$

where,

$$S_{yyww}(f) = \left(\frac{1}{\bar{U}}\right)^2 \cdot \left\{\frac{d\bar{L}}{d\alpha}(0)\right\}^2 \cdot |A_w(f)|^2 \cdot |H_y(f)|^2 \cdot S_{ww}(f)$$

$$S_{yyuu}(f) = \left(\frac{2}{\bar{U}}\right)^2 \cdot \left\{\frac{d\bar{L}}{d\alpha}(0)\alpha + \bar{L}(0)\right\}^2 \cdot |A_u(f)|^2 \cdot |H_y(f)|^2 \cdot S_{uu}(f)$$

$$|H_y(f)|^2 = \frac{1}{m_y^2 (2\pi f_y)^4 \left\{\left(1 - \frac{f^2}{f_y^2}\right)^2 + 4\zeta_y^2 \frac{f^2}{f_y^2}\right\}}$$

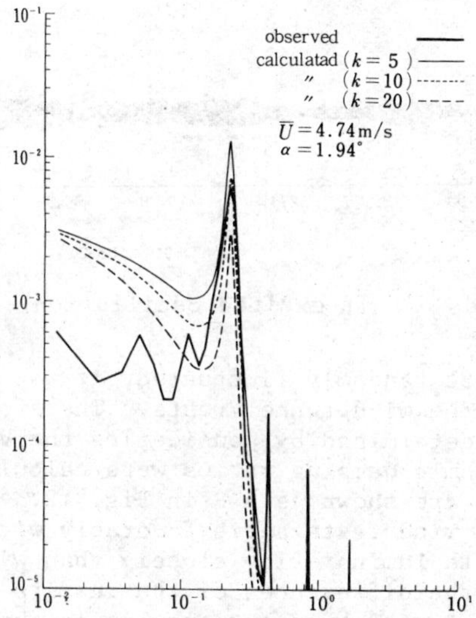

Fig. 9 Comparison of the calculated spectra with observed one on heaving vibration (No.9507, MA-2)

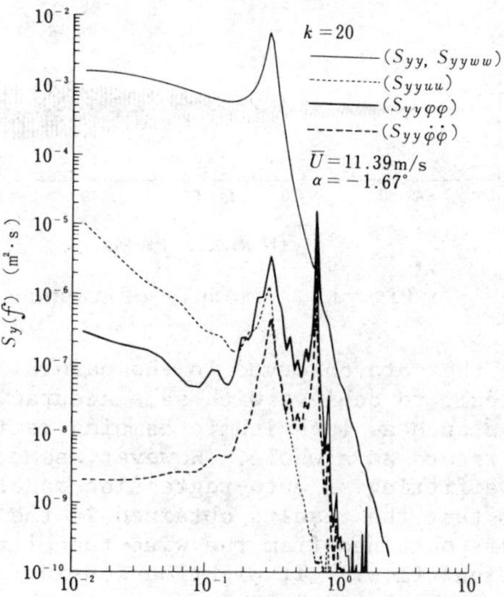

Fig. 10 The ratio of contribution of components to the total spectrum on heaving vibration (No.8903, MB-2)

The spectrum of the heaving vibrations calculated from Eq. (5) by the use of the observed wind spectrum was compared with the spectrum obtained directly from the observed vibration data.

One of the results is shown in Fig. 9 where the influence of the k value is checked again. From these results, it was apparent that, for heaving vibrations, the calculated and observed power spectra coincided but not so precisely as for the lateral vibrations. Particularly there were considerable differences in the peak values of the spectra. In order to clarify the differences, an analysis of the data was tried not only in the frequency domain but in time domain. And it was found that the main differences were caused by the assumed aerodynamic admittance.

The ratios of contribution of the lateral and vertical wind components and others to the heaving vibration are shown in Fig. 10. An attention must be paid for that the broken line of S_{yyww} are almost dropped on the solid line of S_{yy} in Fig. 10. From the figure the heaving vibration forced by wind turbulence may be induced primarily by the vertical component of the wind velocity.

3.5 Self-Excited Vibration Characteristics

The type 1 model with the slotted deck showed random oscillations but no self-excited oscillations. In contrast the type 2 model with the closed deck showed significant flutter phenomenon for certain wind velocities and attack angle. The typical self-excited vibration records are shown in Fig. 11. Since the cross sections of the bridge girders were too complicated to analyse the self-excited vibrations of the bridges theoretically, a check on vibration characteristics is made usually by wind tunnel tests. Accordingly, the results of the large scale model test were compared with those obtained by the wind tunnel tests. This comparison was made by checking the characteristics of logarithmic damping ratios to the mean wind velocities and the values of critical wind velocities to attack angle, as shown in Fig. 12 and 13, respectively.

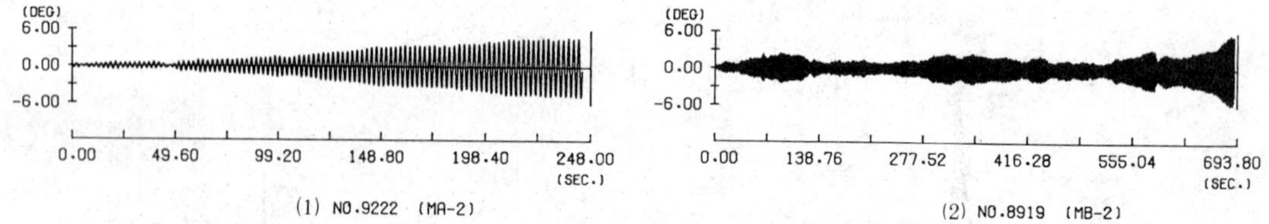

Fig. 11 Example of response records (self-excited oscillation)

Since the data observed in the natural wind were randomly fluctuated, it was very difficult to achieve with same accuracy as in the wind tunnel tests. Therefore, values such as logarithmic damping ratio were determined by considering the vibration record as a whole. However, some logarithmic damping ratios were calculated by the fitting of auto-regression model, which are shown as A R in Fig. 12. It seems that the results obtained in the natural wind tests agree favorably with the results obtained from the wind tunnel tests with laminar flow closely than with turbulent flow. It might be considered that the differences of the results between the natural wind tests and the turbulent wind tunnel tests lay in the wind characteristics, for example, the integral scales of turbulence of the tunnel wind were far less than those of natural wind.

From those results, and since the type 1 models did not show self-excited oscillations in both field tests and wind tunnel tests, it could be said that ordinary wind tunnel tests were reliable for self-excited vibrations.

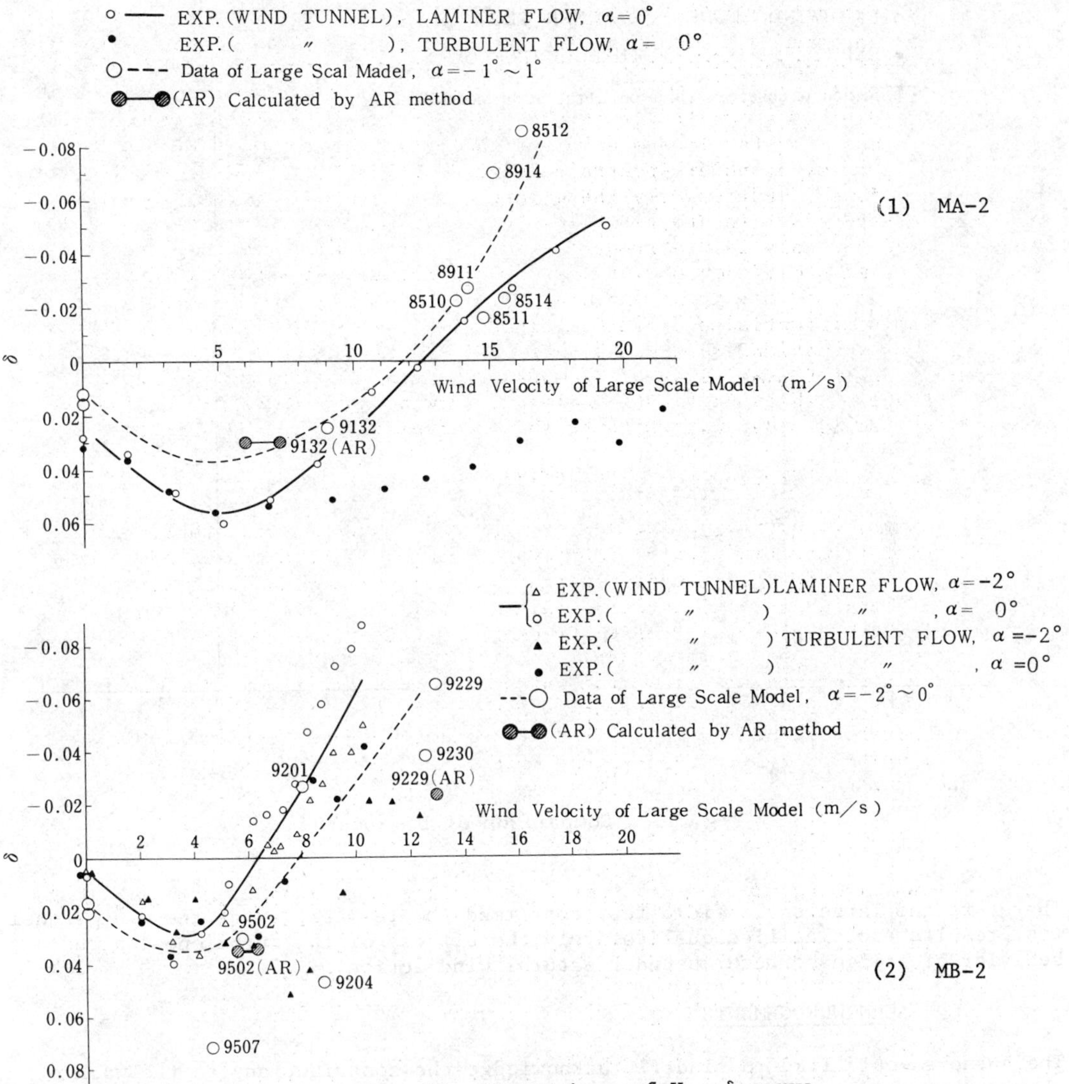

Fig. 12 Comparison of U − δ curve

4. CONCLUSIONS

The significant results of the large scale bridge model test were as follows:

1. The observed drag coefficients agreed closely with those obtained in wind tunnel tests with laminar flow.
2. The power spectral densities of the lateral vibration agreed with the ones calculated from the power spectra of the wind by multiplying the aerodynamic and mechanical admittance.
3. The power spectral densities for observed heaving vibrations agreed qualitatively with the calculated ones and were determined primarily by the vertical component of the fluctuating wind.
4. The observed characteristics of self-excited vibrations agreed reasonably with those obtained in ordinary wind tunnel tests.

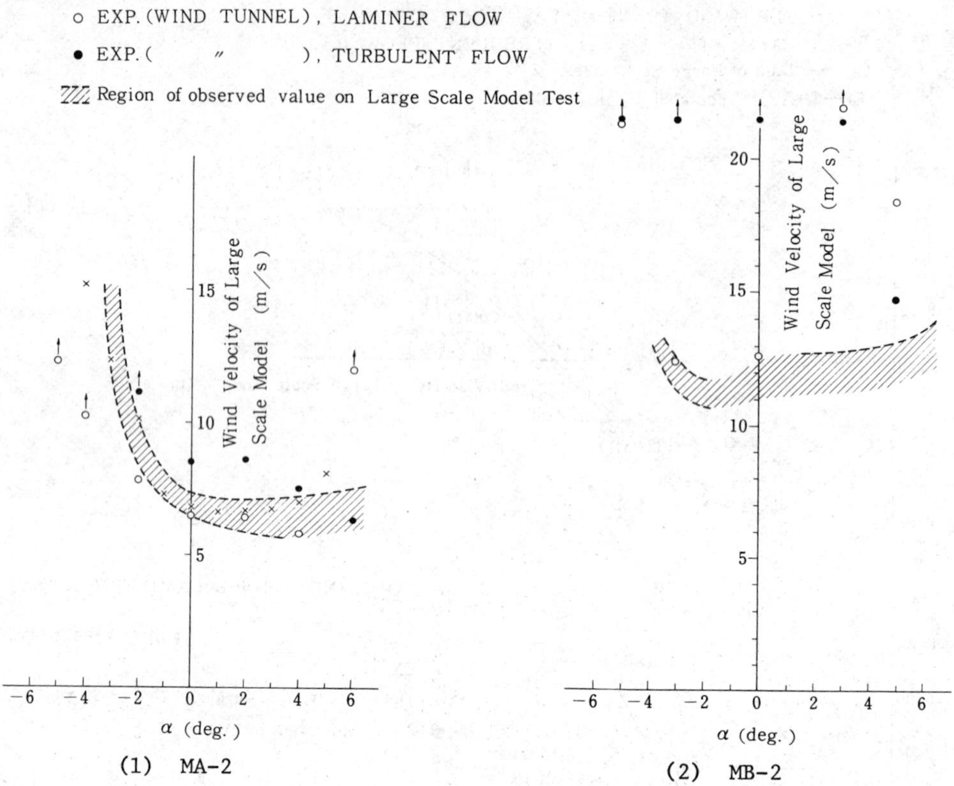

Fig. 13 Comparison of U -α curve

Therefore the large scale field test confirmed the reliability of the wind tunnel test results and clarified qualitatively the effects of the turbulence on the behavior of bridge structures under natural wind action.

ACKNOWLEDGEMENTS

The authors would like to sincerly acknowledge the contributions to the work of this paper by the members of the Technical Committee on the Wind Resistant Design of the Honsh-Shikoku Bridge, JSCE and the members of Honshu-Shikoku Bridge Authority.

REFERENCES

1. Technical Comittee on the Wind Resistant Design of the Honshu-Shikoku Bridge, Final Reports on the Observation of Large Scale Bridge Model (I), Japan Society of Civil Engineers, Mar. 1977. (in Japanese)

2. Technical Comittee on the Wind Resistant Design of the Honshu-Shikoku Bridge, Final Reports on the Observation of Large Scale Bridge Model (II), JSCE, Oct.1977. (in Japanese)

3. Davenport, A. G., Buffeting of a Suspension Bridge by Storm Winds, Proceedings of ASCE, Vol. 88, ST3, June 1962, pp. 233-268.

AN INVESTIGATION OF WIND INSTABILITY OF LONG-SPAN BRIDGES BASED ON THE CONCEPT OF AEROELASTIC MODES

N. SHIRAISHI* and K. OGAWA**

*Kyoto University, Kyoto, Japan
**Kawasaki Heavy Industries, Ltd., Kakogawa, Japan

SUMMARY

Because of their advantages - ease of testing, good control of simulated conditions, etc. - wind tunnel tests using spring-suspended sectional models are generally employed in the examination of aerodynamic instability of long-span bridges. As the spring-suspended sectional model is two-dimensional and rigid, several questions remain in order to accurately estimate aerodynamic instability of bridges based on tests of sectional models. This is because the full bridge is a three-dimensional structure affected by aerodynamic forces - wind velocity, angle of attack, amplitude, sectional shape of bridge decks, etc. - and these forces generally vary across the span.
This paper discusses investigation of wind-induced instability of full bridges from sectional model tests, using the concept of aeroelastic modes. It is intended to clarify the significance of sectional model tests in the wind resistant design of bridges, especially with reference to the relation between two-dimensional models and three-dimensional structures.

LIST OF SYMBOLS

m	- mass per unit span
I_α	- mass moment of inertia per unit span
ρ	- air density
V	- wind velocity
$2b$	- width of bridge between stiffening trusses
h	- vertical displacement
$\bar{h} = h/b$	- non-dimensional displacement
α	- angular torsional displacement
ζ_h	- structural damping ratio of vertical vibration
ζ_α	- structural damping ratio of torsional vibration
ω_h	- natural frequency of vertical vibration in a vacuum
ω_α	- natural frequency of torsional vibration in a vacuum
ω_z	- frequency of vertical vibration in air stream
ω_θ	- frequency of torsional vibration in air stream
ω	- frequency in coupled vibration
δ_z	- logarithmic decrement of vertical vibration
δ_θ	- logarithmic decrement of torsional vibration
δ	- logarithmic decrement in coupled vibration
θ	- phase lag of torsional motion behind vertical motion

$\mu = m/\pi\rho b^2$ - mass ratio
$\nu = I_\alpha/\pi\rho b^4$ - ratio of mass moment
$k = b\omega/V$ - reduced frequency of oscillation
$\bar{V} = \pi V/b\omega$ - reduced velocity
h_0 - amplitude in vertical vibration
α_0 - amplitude in torsional vibration
β - angle of attack
γ - section shape index of bridge deck (*)
$\mu_j = \int_0^1 m\eta_j^2 d\bar{x}/(\pi\rho b^2)$ - mass ratio in j-th vertical mode
$\nu_j = \int_0^1 I_\alpha\varphi_j^2 d\bar{x}/(\pi\rho b^4)$ - ratio of mass moment in j-th torsional mode
x - distance of bridge section from end support
$\bar{x} = x/l$ - non-dimensional distance
l - span length
q_j - generalized coordinate of vertical vibration
p_j - generalized coordinate of torsional vibration

INTRODUCTION

Wind tunnel tests to estimate aerodynamic instability of long-span bridges are generally classified into two types, namely the two-dimensional model test and the three-dimensional one as shown in Fig. 1. A two-dimensional rigid model is used for the two-dimensional model test, while the three-dimensional model test employs a complete reduced scale model of the full bridge.

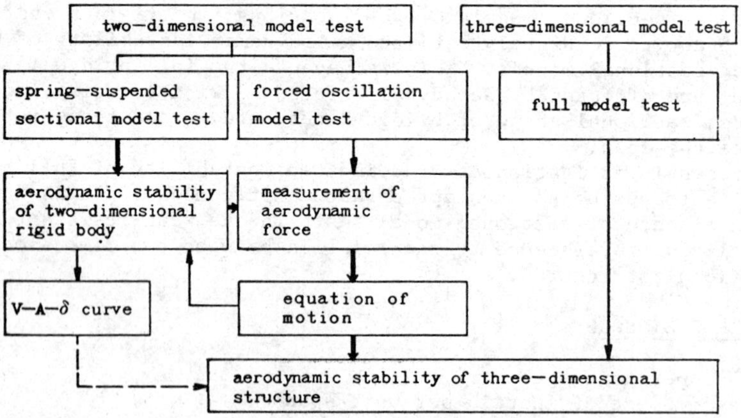

Fig. 1. Wind tunnel test for investigating aerodynamic stability of long-span bridge

The two-dimensional model test is used in many cases. For instance it is used for the Honshu-Shikoku Bridge Projects in Japan, because of the limitation of wind tunnel sections to model scales and convenience in test procedures, though the three-dimensional model test is theoretically desirable for predicting wind-excited oscillation of structures. Based on Wind Tunnel Test Specification for Honshu-Shikoku Bridges (1), to understand aerodynamic stability of suspension bridges, the result of the spring-suspended sectional model test is reduced to the relation between logarithmic decrement δ, wind velocity V and amplitude A (V-A-δ curve). Example of V-A-δ curve is shown in Fig. 2, which plots the torsional logarithmic decrement in connection with wind velocity in test and torsional amplitude, as the

(*) γ corresponds to section shape characteristics, representing the relation between shapes and aerodynamic force coefficients; for example, $\gamma = 0$ in the single deck section, and $\gamma = 1$ in the double deck one.

V-A- δ curve appearing to correspond to the contour line for δ with co-ordinates given in terms of V and A. Considering that the spring-suspended sectional model test depends on the two-dimensional rigid body confined to one-vertical and one-torsional mode, the spring-suspended sectional model test cannot always indicate

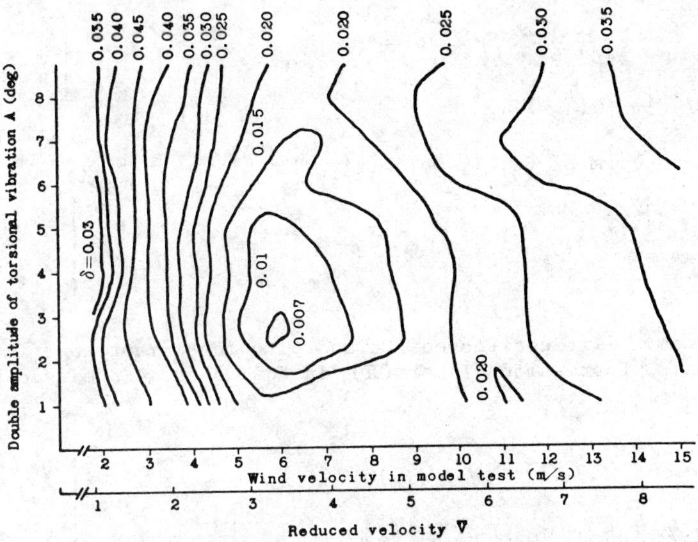

Fig. 2. V-A- δ curve

the motion of the full bridge, i.e., the three-dimensional structure which has an infinite number of natural modes of vibration. As one way to solve the problem, it is attempted to study the aerodynamic instability of the long-span bridge using the strip theory, where the aerodynamic forces measured by the experimental method are used for the external forces in the equation of motion (2). This paper proposes how to measure aerodynamic forces accurately from a practical point of view with the spring-suspended sectional model test from the free oscillation method (3), instead of the forced oscillation method (for example, (4)). Then the wind-induced instability of the three-dimensional structure affected by the aerodynamic forces is investigated with the complex coupled aeroelastic modes.

In this analysis the following assumptions apply:

(1) Aerodynamic force coefficients are approximately represented by polynomial equations as functions of reduced velocity, amplitude, angle of attack and bridge section shape.

(2) The influence of damping on aerodynamic force coefficients is so small that general aerodynamic force coefficients in decay motion are approximately represented by critical aerodynamic force coefficients in which the damping is not essential.

(3) The Reynolds number scarcely effects on aerodynamic force coefficients acting on bridge decks with bluff sections.

(4) It can be said that the behaviour of the long-span bridge associated with non-linear aerodynamic forces depending on amplitude becomes non-linear oscillation. In this paper we assume linearized oscillation equivalent with actual motion in energy dissipation, in which constant logarithmic decrement is used during every one cycle of oscillation.

(5) Aerodynamic force coefficients on a strip of unit length of the bridge deck are determined by reduced velocity, amplitude, angle of attack and aerodynamic shape at each section of the bridge in the span direction (the strip assumption).

(6) The air flow around the model is smooth.

NON-LINEAR AERODYNAMIC FORCE COEFFICIENT

Let us consider a strip of unit length of centrally symmetric section of the bridge model in a smooth flow. We have the following set of equations of motion with two degrees of freedom, h and α :

$$\left.\begin{aligned} m(\ddot{h}+2\zeta_h\omega_h\dot{h}+\omega_h^2 h) &= \tfrac{1}{2}\rho V^2(2b)C_L \\ I_\alpha(\ddot{\alpha}+2\zeta_\alpha\omega_\alpha\dot{\alpha}+\omega_\alpha^2\alpha) &= \tfrac{1}{2}\rho V^2(2b^2)C_M \end{aligned}\right\} \qquad (1)$$

As the non-dimensional form of Eq. 1, we obtain

$$\left.\begin{aligned} \ddot{\bar{h}}+2\zeta_h\omega_h\dot{\bar{h}}+\omega_h^2\bar{h} &= \frac{\omega^2}{\mu}\left(\frac{C_L}{\pi k^2}\right) \\ \ddot{\alpha}+2\zeta_\alpha\omega_\alpha\dot{\alpha}+\omega_\alpha^2\alpha &= \frac{\omega^2}{\nu}\left(\frac{C_M}{\pi k^2}\right) \end{aligned}\right\} \qquad (2)$$

Representing aerodynamic lift coefficient C_L and pitching moment coefficient C_M about the elastic axis, Expression (I) or (II) in Eqs. 3 and 4 are used in general:

Expression (I)

$$\left.\begin{aligned} C_L/(\pi k^2) &= L_h\bar{h}+L_\alpha\alpha = (L_{hR}+iL_{hI})\bar{h}+(L_{\alpha R}+iL_{\alpha I})\alpha \\ C_M/(\pi k^2) &= M_h\bar{h}+M_\alpha\alpha = (M_{hR}+iM_{hI})\bar{h}+(M_{\alpha R}+iM_{\alpha I})\alpha \end{aligned}\right\} \qquad (3)$$

Expression (II)

$$\left.\begin{aligned} C_L &= kH_1^*\frac{\dot{h}}{V}+kH_2^*\frac{b\dot{\alpha}}{V}+k^2H_3^*\alpha+k^2H_4^*\frac{h}{b} \\ C_M &= kA_1^*\frac{\dot{h}}{V}+kA_2^*\frac{b\dot{\alpha}}{V}+k^2A_3^*\alpha+k^2A_4^*\frac{h}{b} \end{aligned}\right\} \qquad (4)$$

Expression (I) separates aerodynamic force coefficients into real and imaginary spheres of displacement, but Expression (II) separates aerodynamic force coefficients into components of displacement and velocity. The relations between Expressions (I) and (II) are given as follows:

$$\left.\begin{aligned} H_1^* &= \pi L_{hI} & A_1^* &= \pi M_{hI} \\ H_2^* &= \pi L_{\alpha I} & A_2^* &= \pi M_{\alpha I} \\ H_3^* &= \pi L_{\alpha R}+\frac{\delta}{2}L_{\alpha I} & A_3^* &= \pi M_{\alpha R}+\frac{\delta}{2}M_{\alpha I} \\ H_4^* &= \pi L_{h\alpha}+\frac{\delta}{2}L_{hI} & A_4^* &= \pi M_{hR}+\frac{\delta}{2}M_{hI} \end{aligned}\right\} \qquad (5)$$

where δ is the logarithmic decrement and aerodynamic coefficients $H_i{}^*$, $A_i{}^*$ (i=1,2, 3,4) in Expression (II) mix real and imaginary parts, excepting stationary oscillation (when damping δ is zero). Therefore, in this paper Expression (I) is used in which aerodynamic force coefficients are divided into real and imaginary parts. In flutter analysis, aerodynamic force coefficents are considered as functions of reduced velocity, but we represent the aerodynamic force coefficients as the following polynamial expressions to estimate the non-linearity dependent on amplitude $\bar{h}_0$, α_0, angle of attack β and section shape index of bridge deck γ, in addition to reduced velocity $\overline{V}$:

$$\left.\begin{aligned}
L_{hR}+i\,L_{hI} &= \sum_j \sum_l \sum_m \sum_n (a_{Rjlmn}+i\,a_{Ijlmn})\gamma^{n-1}\beta^{m-1}\overline{V}^{l-1}\overline{h}_0^{j-1}\\
L_{\alpha R}+i\,L_{\alpha I} &= \sum_j \sum_l \sum_m \sum_n (b_{Rjlmn}+i\,b_{Ijlmn})\gamma^{n-1}\beta^{m-1}\overline{V}^{l-1}\alpha_0^{j-1}\\
M_{hR}+i\,M_{hI} &= \sum_j \sum_l \sum_m \sum_n (c_{Rjlmn}+i\,c_{Ijlmn})\gamma^{n-1}\beta^{m-1}\overline{V}^{l-1}\overline{h}_0^{j-1}\\
M_{\alpha R}+i\,M_{\alpha I} &= \sum_j \sum_l \sum_m \sum_n (d_{Rjlmn}+i\,d_{Ijlmn})\gamma^{n-1}\beta^{m-1}\overline{V}^{l-1}\alpha_0^{j-1}
\end{aligned}\right\} \tag{6}$$

Let us derive the equations that deduce the aerodynamic force coefficients from the sectional model test according to Expression (I). Considering the sectional model to be so restrained in Eq. 2 that only torsional degree of freedom α is permitted, the equation of pure torsional motion becomes as following:

$$\ddot{\alpha}+2\,\zeta_\alpha\,\omega_\alpha\,\dot{\alpha}+\omega_\alpha^2\,\alpha=\frac{\omega_\theta^2}{\nu}(\,M_{\alpha R}+iM_{\alpha I})\,\alpha \tag{7}$$

We shall assume the mode of torsional oscillation:

$$\alpha=\alpha_0 e^{(-\delta_\theta/2\pi+i\,)\,\omega_\theta t} \tag{8}$$

Substituting Eq. 8 for Eq. 7, we obtain

$$\left.\begin{aligned}
M_{\alpha R} &= \nu\left[\left(\frac{\omega_\alpha}{\omega_\theta}\right)^2-\frac{\zeta_\alpha\,\delta_\theta}{\pi}\left(\frac{\omega_\alpha}{\omega_\theta}\right)-1+\frac{\delta_\theta^{\,2}}{4\pi^2}\right]=\sum_j d_{Rj}\,\alpha_0^{\,j-1}\\
M_{\alpha I} &= \nu\left[\,2\,\zeta_\alpha\left(\frac{\omega_\alpha}{\omega_\theta}\right)-\frac{\delta_\theta}{\pi}\right] = \sum_j d_{Ij}\,\alpha_0^{j-1}
\end{aligned}\right\} \tag{9}$$

Analogously with pure torsional motion, the equation of pure vertical motion h gives

$$\left.\begin{aligned}
L_{hR} &= \mu\left[\left(\frac{\omega_h}{\omega_z}\right)^2-\frac{\zeta_h\,\delta_z}{\pi}\left(\frac{\omega_h}{\omega_z}\right)-1+\frac{\delta_z^{\,2}}{4\pi^2}\right]=\sum_j a_{Rj}\,\overline{h}_0^{\,j-1}\\
L_{hI} &= \mu\left[\,2\,\zeta_h\left(\frac{\omega_h}{\omega_z}\right)-\frac{\delta_z}{\pi}\right] = \sum_j a_{Ij}\,\overline{h}_0^{\,j-1}
\end{aligned}\right\} \tag{10}$$

When the model oscillates in coupled mode of two degrees of freedom h and α, the motion is described:

$$\left.\begin{aligned}
\overline{h} &= \overline{h}_0\,e^{(-\delta/2\pi+i)\,wt}\\
\alpha &= \alpha_0\,e^{(-\delta/2\pi+i)\,wt\,-i\theta}
\end{aligned}\right\} \tag{11}$$

Substitution of Eq. 11 into Eq. 2 gives

$$
\left.\begin{aligned}
L_{\alpha R}&=\mu\left(\frac{\overline{h}_0}{\alpha_0}\right)\left[\left\{\left(\frac{\omega_h}{\omega}\right)^2-\frac{\zeta_h\delta}{\pi}\left(\frac{\omega_h}{\omega}\right)-1+\frac{\delta^2}{4\pi^2}-\frac{L_{hR}}{\mu}\right\}\cos\theta-\left\{2\zeta_h\left(\frac{\omega_h}{\omega}\right)-\frac{\delta}{\pi}-\frac{L_{hI}}{\mu}\right\}\sin\theta\right]\\
&=\sum_j b_{Rj}\,\alpha_0^{\,j-1}\\
L_{\alpha I}&=\mu\left(\frac{\overline{h}_0}{\alpha_0}\right)\left[\left\{\left(\frac{\omega_h}{\omega}\right)^2-\frac{\zeta_h\delta}{\pi}\left(\frac{\omega_h}{\omega}\right)-1+\frac{\delta^2}{4\pi^2}-\frac{L_{hR}}{\mu}\right\}\sin\theta+\left\{2\zeta_h\left(\frac{\omega_h}{\omega}\right)-\frac{\delta}{\pi}-\frac{L_{hI}}{\mu}\right\}\cos\theta\right]\\
&=\sum_j b_{Ij}\,\alpha_0^{\,j-1}\\
M_{hR}&=\nu\left(\frac{\alpha_0}{\overline{h}_0}\right)\left[\left\{\left(\frac{\omega_\alpha}{\omega}\right)^2-\frac{\zeta_\alpha\delta}{\pi}\left(\frac{\omega_\alpha}{\omega}\right)-1+\frac{\delta^2}{4\pi^2}-\frac{M_{\alpha R}}{\nu}\right\}\cos\theta+\left\{2\zeta_\alpha\left(\frac{\omega_\alpha}{\omega}\right)-\frac{\delta}{\pi}-\frac{M_{\alpha I}}{\nu}\right\}\sin\theta\right]\\
&=\sum_j c_{Rj}\,\overline{h}_0^{\,j-1}\\
M_{hI}&=\nu\left(\frac{\alpha_0}{\overline{h}_0}\right)\left[-\left\{\left(\frac{\omega_\alpha}{\omega}\right)^2-\frac{\zeta_\alpha\delta}{\pi}\left(\frac{\omega_\alpha}{\omega}\right)-1+\frac{\delta^2}{4\pi^2}-\frac{M_{\alpha R}}{\nu}\right\}\sin\theta+\left\{2\zeta_\alpha\left(\frac{\omega_\alpha}{\omega}\right)-\frac{\delta}{\pi}-\frac{M_{\alpha I}}{\nu}\right\}\cos\theta\right]\\
&=\sum_j c_{Ij}\,\overline{h}_0^{\,j-1}
\end{aligned}\right\}\qquad(12)
$$

Let us assume that the model having two degrees of freedom h and α oscillates in simple harmonic motion in which the damping ratio of the oscillation vanishes, and that the following approximate expressions are applied

$$
\frac{m}{\rho b^2\omega_z}=\frac{m}{\rho b^2\omega},\qquad \frac{I_\alpha}{\rho b^4\omega_\theta}=\frac{I_\alpha}{\rho b^4\omega},\qquad \frac{I_\alpha}{\rho b^4\omega_\theta^2}=\frac{I_\alpha}{\rho b^4\omega^2}\,. \qquad (13)
$$

Taking the preceding assumptions into account, substituting of Eqs. 9, 10 and 12 into Eq. 5, we obtain the expressions of aerodynamic force coefficients given by R.H. Scanlan and Ali Sabzevari (3).

Using Eqs. 9, 10 and 12, we can deduce the aerodynamic force coefficients from experimental data on the spring-suspended sectional model test, such as logarithmic decrement δ, circular frequency ω, amplitude ratio ($\alpha_0/\overline{h}_0$) and phase lag θ. Experimental data, namely, the aerodynamic force coefficients on the spring-suspended sectional model test are affected by reduced velocity $\overline{V}$, amplitude $\overline{h}_0$, α_0, angle of attack β and section shape index of the bridge deck γ. If, by calculation of aerodynamic force coefficients step by step under various conditions of the sectional model test (combinations of $\overline{V}$, h_0, α_0, β and γ), we are given enough combinations of both aerodynamic force coefficients and sectional model test conditions, we can determine aerodynamic derivatives a_{Rjlmn}, a_{Ijlmn}, b_{Rjlmn}, b_{Ijlmn}, c_{Rjlmn}, c_{Ijlmn}, d_{Rjlmn} and d_{Ijlmn} by the method of least squares, taking Eq. 6 into an account.

AERODYNAMIC INSTABILITY OF THE LONG-SPAN BRIDGE

Considering two modes of displacements, h and α, the equations of motion of the long-span bridge situated in a smooth flow can be expressed as follows:

$$
\left.\begin{aligned}
\ddot{q}_j+2\zeta_{hj}\omega_{hj}\dot{q}_j+\omega_{hj}^2q_j&=\frac{\omega^2}{\mu_j}\Big[\sum_k\int_0^1(L_{hR}+iL_{hI})\eta_j\eta_k\,d\overline{x}\cdot q_k\\
&\quad+\sum_k\int_0^1(L_{\alpha R}+iL_{\alpha I})\eta_j\varphi_k\,d\overline{x}\cdot p_k\Big]\\
\ddot{p}_j+2\zeta_{\alpha j}\omega_{\alpha j}\dot{p}_j+\omega_{\alpha j}^2p_j&=\frac{\omega^2}{\nu_j}\Big[\sum\int_0^1(M_{hR}+iM_{hI})\varphi_j\eta_k\,d\overline{x}\cdot q_k\\
&\quad+\sum_k\int_0^1(M_{\alpha R}+iM_{\alpha I})\varphi_j\varphi_k\,d\overline{x}\cdot p_k
\end{aligned}\right\}\qquad(14)
$$

where

$$
\left.\begin{aligned}
\mu_j&=\int_0^1 m\eta_j^2\,d\overline{x}/(\pi\rho b^2)\\
\nu_j&=\int_0^1 I_\alpha\varphi_j^2\,d\overline{x}/(\pi\rho b^4)\\
\overline{x}&=x/l
\end{aligned}\right\}\qquad(15)
$$

and subscripts j and k denote properties of j-th and k-th mode, respectively. At the general oscillatory condition, the bridge motion is specified:

$$q_j = \overline{q}_j e^{zt}, \quad p_j = \overline{p}_j e^{zt}, \quad z = (-\delta/2\pi + i)\omega \tag{16}$$

Substituting Eq. 16 into Eq. 14, we obtain

$$\begin{pmatrix} A_{11} & A_{12} & \cdots & A_{1n} & B_{11} & B_{12} & \cdots & B_{1n} \\ A_{21} & A_{22} & \cdots & A_{2n} & B_{21} & B_{22} & \cdots & B_{2n} \\ \vdots & \vdots & & \vdots & \vdots & \vdots & & \vdots \\ C_{11} & C_{12} & \cdots & C_{1n} & D_{11} & D_{12} & \cdots & D_{1n} \\ C_{21} & C_{22} & \cdots & C_{2n} & D_{21} & D_{22} & \cdots & D_{2n} \\ \vdots & \vdots & & \vdots & \vdots & \vdots & & \vdots \end{pmatrix} \begin{pmatrix} \overline{q}_1 \\ \overline{q}_2 \\ \vdots \\ \overline{p}_1 \\ \overline{p}_2 \\ \vdots \end{pmatrix} = \begin{pmatrix} 0 \\ 0 \\ \vdots \\ 0 \\ 0 \\ \vdots \end{pmatrix} \tag{17}$$

where

$$\left.\begin{aligned} A_{jk} &= (z^2 + 2\zeta_{hj}\omega_{hj} z + \omega^2_{hj})\,\delta_{jk} - \frac{\omega^2}{\mu_j}\int_0^1 (L_{hR} + iL_{hI})\,\eta_j \eta_k \, d\overline{x} \\ B_{jk} &= -\frac{\omega^2}{\mu_j}\int_0^1 (L_{\alpha R} + iL_{\alpha I})\,\eta_j \varphi_k \, d\overline{x} \\ C_{jk} &= -\frac{\omega^2}{\nu_j}\int_0^1 (M_{hR} + iM_{hI})\,\varphi_j \eta_k \, d\overline{x} \\ D_{jk} &= (z^2 + 2\zeta_{\alpha j}\omega_{\alpha j} z + \omega^2_{\alpha j})\,\delta_{jk} - \frac{\omega^2}{\nu_j}\int_0^1 (M_{\alpha R} + iM_{\alpha I})\,\varphi_j \varphi_k \, d\overline{x} \end{aligned}\right\} \tag{18}$$

and δ_{jk} is the Kronecker delta.
Since all the right-hand sides of Eq. 17 equal zero, there can be non-trivial solutions only when the determinant of the coefficients vanishes. This fact leads to a polynomial with complex coefficients of 4n-th degree in the unknown z. Considering imaginary parts of z are equivalent to circular frequencies ω of the bridge and only positive imaginary parts of z are effective in this analysis, we obtain 2n roots of z which determine 2n aeroelastic modes in one-to-one correspondence with the free vibration modes of the bridge with no wind action.
Since generalized aerodynamic forces in the right side of Eq. 14 are considered as generalized external forces in the j-th equation of motion and include coupled j-th and k-th terms, coupled vibration between j-th and all k-th modes can be expected to occur. Then there is the question of investigating the aerodynamic instability of long-span bridges by a sectional model test, such as the spring-suspended sectional model test, in which the vibration modes are confined to only one-vertical and one-torsional modes. Nevertheless the sectional model test is used in general to estimate wind-induced instability of long-span bridges. Let us introduce the conditions in which the two-dimensional model test on the spring-suspended sectional model is useful for investigating the stability of long-span bridges. If we assume the orthogonality relationships among the vibration modes weighted by aerodynamic forces in the right side of Eq. 14, the equations of aerodynamic stability can be expressed as follows:

$$\left.\begin{aligned} &\ddot{q}_j + 2\zeta_{hj}\omega_{hj}\dot{q}_j + \omega^2_{hj} q_j = \frac{\omega^2}{\mu_j}\Big[\int_0^1 (L_{hR} + iL_{hI})\,\eta_j^2 \, d\overline{x} \cdot q_j \\ &\quad + \int_0^1 (L_{\alpha R} + iL_{\alpha I})\,\eta_j \varphi_j \, d\overline{x} \cdot p_j\Big] \end{aligned}\right\} \tag{19}$$

$$\ddot{p}_j + 2\zeta_{aj}\omega_{aj}\dot{p}_j + \omega_{aj}^2 p_j = \frac{\omega^2}{\nu_j}\Big[\int_0^1 (M_{hR} + iM_{hI})\varphi_j\eta_j d\bar{x}\cdot q_j + \int_0^1 (M_{\alpha R} + iM_{\alpha I})\varphi_j^2 d\bar{x}\cdot p_j\Big]$$

In this case where coupling terms among different-degree modes vanish, the spring-suspended sectional model test becomes effective in aerodynamic analysis, because the equations of motion of long-span bridges reduce to one of two dimensional rigid bodies. According to Eq. 19, the vibration modes which should be considered in investigating the aerodynamic stability of two-dimensional rigid bodies are modes with the same degree between the vertical and torsional vibrations, so that the stability can be examined, neglecting the coupled vibration of modes with different degrees.

Concrete conditions for the orthogonality relationships among the vibration modes weighted by aerodynamic forces are given as follows:

(1) The vertical and torsional mode shapes are orthogonal.

(2) Aerodynamic properties of an air-stream, such as wind velocity and angle of attack are uniformly distributed along the span.

(3) The aerodynamic shape of the bridge is constant across the span.

(4) Aerodynamic force coefficients are linear in the amplitude.

In general conditions where the orthogonality relationships among generalized aerodynamic forces given by conditions (1) - (4) do not exist, we cannot estimate the stability of long-span bridges without considering the coupled vibration affected by a number of vibration modes. Then it is necessary to apply the three dimensional analysis for study of aerodynamic instability. As examples of the three-dimensional analysis, let us investigate the response characteristics of the bridge where the aerodynamic properties of an air-stream in one case and aerodynamic shapes of the bridge deck in another case vary across the span.

Influence of the Aerodynamic Properties Varying Across the Span

In many cases the long-span bridges are subjected to wind actions varying in the direction of the span, due to on-site geographical features, such as the presence of hills. Let us investigate the aerodynamic stability of long-span suspension bridges where the distribution of wind velocities varies across the span as an illustration of aerodynamic properties. Regarding the distribution of wind velocities as one of the aerodynamic forces, we can obtain a measure to estimate how the distribution of attack angles and section shapes affects the stability of structures. Let us consider the suspension bridge of one suspended span where vertical and torsional natural modes are $\eta_j(x) = \sin(j\pi x/l)$, $\varphi_j(x) = \sin(j\pi x/l)$, structural dampings are $\delta_{hj} = \delta_{aj} = 0$ and mass parameters are $\mu = 33.33$, $\nu = 20.83$. Natural frequencies of vertical vibration are $N_{h1} = 0.234$ Hz, $N_{h2} = 0.159$ Hz, $N_{h3} = 0.318$ Hz and natural frequencies of torsional vibration are $N_{\alpha 1} = 0.443$ Hz, $N_{\alpha 2} = 0.620$ Hz, $N_{\alpha 3} = 0.951$ Hz where N_{hj} and N_{aj} are related to natural modes $\eta_j(x)$ and $\varphi_j(x)$, respectively. The aerodynamic shape of the bridge section is an airfoil and the angle of attack at the bridge section is $\beta = 0$ (Aerodynamic forces are determined from Theodorsen's function).

We shall consider the aerodynamic stability at the following distribution of wind velocity:

(A-1) Uniform distribution of wind velocity $\bar{V}$ from 0 to l across the span.

(A-2) Wind velocity becomes zero from 3/8 l to 5/8 l.

(A-3) Wind velocity becomes zero from 1/4 l to 3/4 l.

(A-4) Wind velocity becomes zero from 1/8 l to 7/8 l.

where the wind with uniform distribution of velocity $\bar{V}$ acts along other parts of the span in cases of (A-2), (A-3) and (A-4).

To measure the aerodynamic stability, let us calculate the logarithmic decrement δ as a function of reduced velocity $\bar{V}$. Fig. 3 illustrates the logarithmic decrement δ of the first torsional branch (that is, the fundamental aeroelastic mode

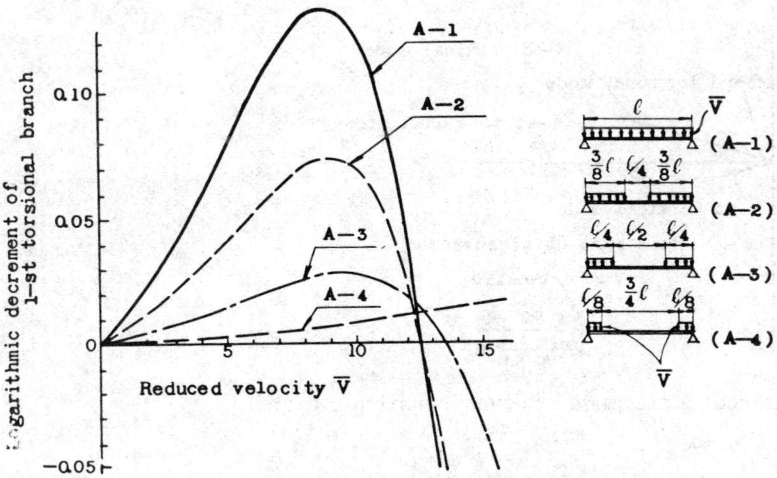

Fig. 3. Relation between wind distribution and aerodynamic stability

in correspondence with the first torsional vibration at rest). Then we investigate how vibration modes considered in the analysis influence the accuracy of calculated results for aerodynamic stability (Fig. 4). We assume the distribution

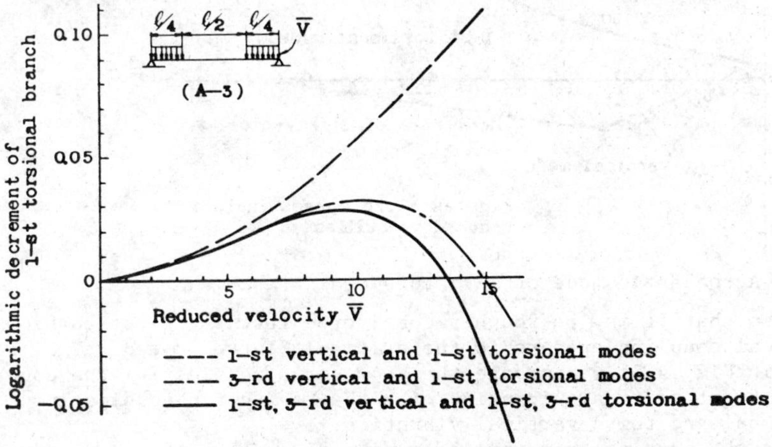

Fig. 4. Effect of considering vibration modes

of wind velocity as (A-3), i.e, the wind acting along the span from 0 to $1/4\ l$ and from $3/4\ l$ to l. Since generalized aerodynamic forces corresponding to even modes vanish in this analysis, taking into account the symmetric distribution of wind velocity, only generalized forces corresponding to odd modes are significant. Fig. 4 indicates that vibration modes considered in the analysis greatly effect the flutter speed. Fig.5 shows how the aeroelastic mode of the first torsional branch is influenced by other vibration modes. It is suggested in the first torsional branch that the first torsional vibration mode prominently couples with the third and first vertical vibration modes and the mutual coupling among the torsional vibration modes is weak. Fig. 6 illustrates how the aeroelastic mode of the first torsional branch varies with the wind velocity, considering the coupling between the first torsional vibration, the first and third vertical vibration. Increasing the wind velocity, the third vertical vibration mode in the first torsional branch

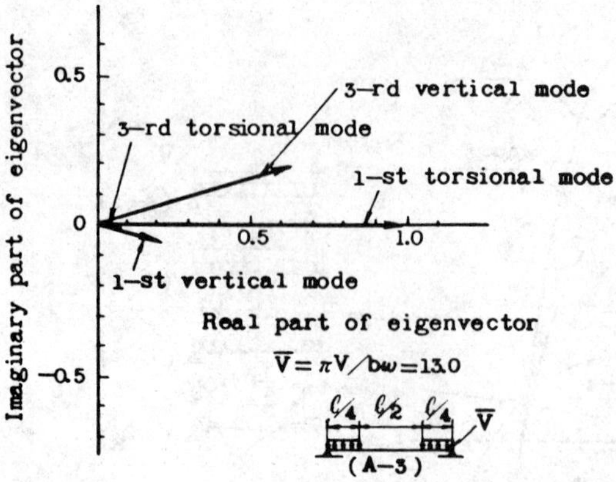

Fig. 5. Aeroelastic mode of 1-st torsional branch

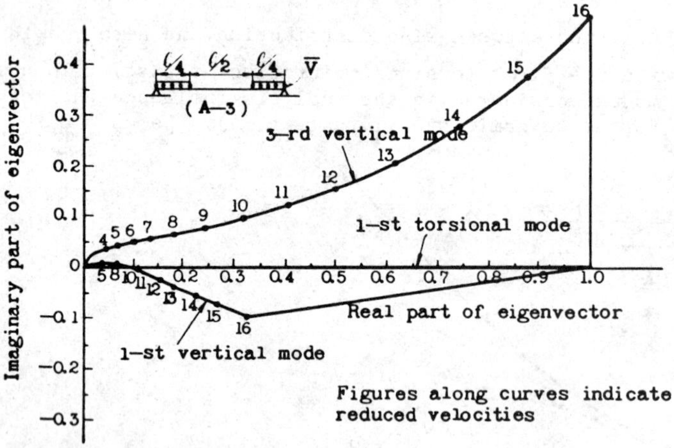

Fig. 6. Aeroelastic mode of 1-st torsional branch

remarkably arises, so that it may be roughly considered that the first torsional and the third vertical coupling appears in the critical flutter speed. This fact is also recognized in Fig. 4 if we compare the V- δ curve considering the coupled first torsional and first vertical vibration with the V- δ curve considering the coupled first torsional and third vertical vibration.

The coupling between the vertical and torsional vibration with different mode shapes indicates that the amplitude ratio $\overline{h}_0 / \alpha_0$ and the phase lag θ of the torsional motion behind the vertical motion varies across the span. Taking into account that the mean work per unit time done by the aerodynamic forces per unit span in a coupled oscillation is closely related with the amplitude ratio $\overline{h}_0/\alpha_0$ and phase lag θ , one may say that the work done by the aerodynamic forces per unit span, namely, the aerodynamic damping, varies along the span in the coupled vertical and torsional vibration with different mode shapes. The three dimensional characteristics of the complex coupled aeroelastic modes can be estimated by the three dimensional analysis in which the long-span bridge is considered as the three-dimensional structure.

Influence of Varying Aerodynamic Shapes Across the Span

Although most bridge cross sections form uniform and constant across the span from an aerodynamic point of view, in some cases aerodynamic characteristics associated

with cross section may vary along the span due to such structural requirements for usage and site conditions (for example, a suspension bridge of three spans with walls for reducing traffic noises on side spans). In this section we shall consider the motion of the three-dimensional structure affected by varying aerodynamic characteristics across the span in a numerical analysis, assuming a single-deck bridge with double-decks in part. Let us assume the structural properties of the bridge is the same as the model considered in the preceding section. For the sake of convenience, we shall use the aerodynamic forces given by Theodorsen on single-deck parts and the double aerodynamic forces on double-deck parts.
We shall consider an aerodynamic stability of the single and double-deck bridges and the single-deck bridges with double-decks in part at the following cases of the variances in the aerodynamic shapes of suspension bridges as

(B-1) Double-deck sections along the full span
(B-2) Single-deck sections from 3/8 l to 5/8 l and double-deck sections at the other parts along the span
(B-3) Single-deck sections from 1/4 l to 3/4 l and double-deck sections at the other parts along the span
(B-4) Single-deck sections from 1/8 l to 7/8 l and double-deck sections at the other parts along the span
(B-5) Single-deck sections along the full span.

Fig. 7 shows the relation between the reduced velocity $\overline{V}$ and the logarithmic decrement δ of the first torsional branch. As is suggested in Fig.7, the wind velocity

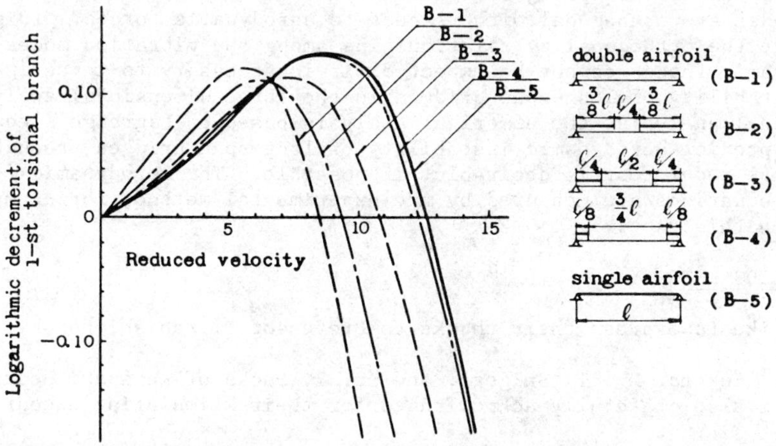

Fig. 7. Relation between aerodynamic shape and aerodynamic stability

in correspondence with each other in V- δ curves is transferred in parallel in proportion to the ratio between the single and double deck parts of suspension bridges. Naturally the increase of double-deck parts reduces the critical flutter speed. The aerodynamic stability is remarkably decayed according to the extension of double-deck parts more than 1/4 l from both end-supports such as (B-3), though the only a little aerodynamic influence of double-deck parts on the suspension bridge is recognized in the case of (B-4) with double deck parts of 1/8 l from end-supports. Fig. 8 illustrates the coupling among vibration modes when the single-deck bridge has double deck parts of 1/4 l from both end-supports, i.e., (B-3). It is shown in Fig. 8 that the first torsional branch is prominently affected by the first and third vertical vibration modes, though the coupling among the torsional vibration modes is insignificant.

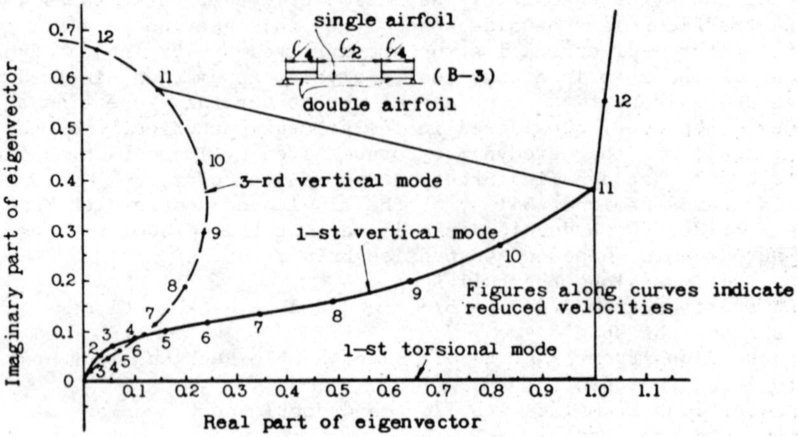

Fig. 8. Aeroelastic mode of 1-st torsional branch

CONCLUSIONS

Two-dimensional analysis of the spring-suspended sectional model test is effective, if the vibration modes are orthogonal with respect to aerodynamic forces. In general conditions where the orthogonality relationships among the vibration modes weighted by aerodynamic forces cannot be expected, it is necessary to investigate the aerodynamic instability of long-span bridges in the three dimensional analysis. As a three-dimensional analysis, the numerical analysis based on Lagrange's equations is useful to predict aerodynamic instability of long-span bridges, though the three-dimensional model test is desirable, if possible. The aerodynamic forces used in Lagrange's equations are measured by the experimental method, for example, Eqs. 9, 10 and 12 in this paper.

ACKNOWLEDGMENTS

The authors would like to express their thanks to Professor I. Konishi for his valuable advice.
Mr. K. Omiya, Mr. T. Fukata, Mr. T. Shigeto, and Mr. Y. Saoka of Kawasaki Heavy Industries, Ltd. are also gratefully acknowledged for their stimulating encouragement.

REFERENCES :

1. Japan Soc. Civil Engineers, Technical Committee Report on Studies on Wind Resistance Design of Honshu-Shikoku Bridges : 117-142 (1975)
2. Shiraichi, N. and Ogawa, K., An Investigation on Aeroelastic Responses of Suspension Bridges due to the Non-Linear Aerodynamic Forces, Proc. JSCE, 244 : 23-35 (1975)
3. Scanlan, R.H. and Sabzevari, A., Suspension Bridge Flutter Revisited, ASCE Structural Engineering Conf., Washington : 11-14 (1967)
4. Tanaka, H., Vibrations of Bluff-Sectional Structures under Wind Action, Proc. 3rd Int. Conf. on Wind Effects on Buildings and Structures : 899-909 (1971)

AEROELASTIC STABILITY OF SUSPENSION BRIDGES

A. PETRE and R. SOFRONIE
Bucharest, Romania

ABSTRACT

An advanced theory on the aeroelastic stability of suspension bridges has been developed by taking into account the drag effects of the aerodynamically shaped decks. The theory includes both structural systems of suspension bridges, the classical system of hanging bridges and the new system of prestressed bridges. Practical methods to solve the stability problems of suspension bridges in a more reliable manner have also been suggested.

1. INTRODUCTION

As it is now well-known, the theory of aeroelasticity deals with the interaction between the elastic and aerodynamic forces of lifting surfaces exposed to an airstream. Such developed aeroelastic phenomena are dynamic or static whether inertia forces do or do not play a significant role. The stability of aeroelastic phenomena is traditionally studied in the limit state of indifferent equilibrium when the induced energy equals the lost one. For a long time the only components of aerodynamic forces included in stability analyses were those of a lifting type. But recently, by taking into account drag effects, in addition to the lifting forces certain aeroelastic phenomena were discovered which cannot be longer disregarded. Consequently, the whole theory of aeroelasticity has been reformularized (1,2,3,4,5).

The suspension bridges are elastic structures of large spans composed of cables and decks. In the last years in order to improve their stability to the wind action the decks were aerodynamically shaped. Thus they became true lifting surfaces of large aspect-ratio. For the same stability reasons the shape of the cables was also modified and the structural system of prestressed bridges with closed cross-sections has been patended (6).

The idea to study aeroelastically the stability of suspension bridges to wind occured soon after the Tacoma failure. Certainly, the first attempts although encouraging had only a general character (7,8). However, in a relatively short period, by an interdisciplinary cooperation, remarkable studies appeared (9,1o,11,12). The most important theoretical and test results have been recently presented by R.H. Scanlan in a synthetical report (13). All these available data on aeroelastic stability are useful either for further studies or for the design of new suspension bridges. However, they remain valid only between the restricted limits of operators adopted for elastic and especially for aerodynamic forces.

The aim of this paper is to develop a more advanced theory on the aeroelastic stability of suspension bridges by taking into account the drag effects of the aerodynamically shaped decks. The theory includes both structural systems of suspension bridges, the classical system of hanging bridges and the new system of prestressed bridges. Practical methods to solve the stability problems of suspension bridges in a more reliable manner are also suggested.

The paper presents first the analysis bases consisting in the assumption adopted to define all the occuring forces as well as the mechanism in which they are balanced. Then it describes the governing equations of the aeroelastic phenomena and the parameters on which they depend. These general equations are further detailed for each structural system of suspension bridges. In order to solve the governing equations they are transformed into algebraic systems. Then two solution methods are presented. Finally, there are discussed the drag effects on critical velocities of divergence and flutter, the differences between the two structural systems of suspension bridges and the efficiency of the analysis technique used in the paper.

2. ANALYSIS BASES

Since suspension bridges have by far larger spans than the dimensions of their cross-sections, in the following they will be considered as one-dimensional structures with continuous and straight elastic lines. As a consequence of elastic deformations any cross-section of such a structure could develop three displacements and three rotations. But as the principal stiffnesses of the bridge decks are highly different from one another only two deformations are significant, namely the vertical deflection of bending and the transversal rotation of torsion. Therefore it is assumed that suspension bridges develop binary phenomena of bending and torsion. The variation of these deformations along the bridge spans is also taken into account in the analysis. Accordingly, the binary phenomena of elastic deformation will be studied in three-dimensions.

The elastic forces corresponding to the above defined deformations will be expressed by differential operators. Since the elastic lines are reference axes, the expressions of elastic forces are independent of one another, i.e. elastic coupling terms do not appear.

The forces of structural damping are considered according to Voigt's model. They will be written with the aid of a complex modulus of elasticity and are included in the expressions of elastic forces. The imaginary component of these expressions shows the loss by friction.

The inertia forces are generated by the mass of decks, cables and hangers spanwisely assumed as a uniformly distributed mass per unit. Due to the vertical symmetry of bridge cross-sections, the mass centres coincide with the elastic centres.

The aerodynamic forces are considered according to the unsteady, incompressible flow theory, and the aerodynamic operators of strip type. Lift and drag are assumed to have fixed directions relating to the wind stream, and the running drag will be taken as a constant across the deck span. This assumption on drag implies, at the very least, that the elastic deformations in bending and torsion must always remain small.

In order to calculate the equilibrium stability between the above defined forces the perturbation method will be used as follows. It is assumed that in the initial state of equilibrium, in absence of wind, the bridge deck has zero incidence and no external forces act on it. Then it is subjected to an arbitrary but small perturbation consisting in a vertical displacement w and/or a transversal rotation φ. As a consequence of this perturbation, under the wind action, aerodynamic forces of lift and drag develop. From the interaction between these aerodynamic forces and the elastic restoring forces there results a motion or a deformation, as the influence of inertia forces are significant or not. However, in any way if the components of the initial perturbation decrease and vanish the resulting aeroelastic phenomenon is stable. On the contrary, if the initial perturbation increases, the phenomenon is unstable. It is easy to imagine that between the two equilibrium states there is a limit state in which the initial perturbation neither decreases nor increases. The aerodynamic forces are then perfectly equilibrated by the elastic forces, although the aeroelastic phenomenon remains non-conservative. In this state any additional perturbation, as small as possible, does not encounter any opposition from the suspension bridge structure and consequently the components of the initial perturbation can dangerously increase. At this instant the suspension bridge behaves as a structure which has completely lost its structural damping of stiffness. It is just that limit state of indifferent equilibrium in which the aeroelastic phenomena of flutter or divergence occur. Only in this limit state the system of balancing forces can be considered quasi-conservative, the resulting motion periodical and the deformation independent of the way in which it is produced.

3. GOVERNING EQUATIONS

In the most general case, to the above assumed initial perturbation of bending and torsion there corresponds the motion

$$\frac{\partial^2}{\partial x^2}\left[(1 + ig_w)EI\frac{\partial^2 w}{\partial x^2}\right] + m\frac{\partial^2 w}{\partial t^2} - L_a(w,\varphi) + \frac{\partial^2}{\partial x^2}(M_z\varphi) = 0 \qquad (1a)$$

$$-\frac{\partial}{\partial x}\left[(1+ig_\varphi)GI_d\frac{\partial\varphi}{\partial x}\right] - J\frac{\partial^2\varphi}{\partial t^2} + M_a(w,\varphi) - M_z\frac{\partial^2 w}{\partial x^2} = 0 \qquad (1b)$$

where in each equation the first three terms are well known and represent the elastic forces with structural damping, inertia forces and lifting airloads. The last terms introduce the drag effects and for a simply supported deck the bending moment M_z associated with the uniform running drag has the expression

$$M_z = \frac{\rho b V^2 C_D}{2} x(l-x) \qquad (2)$$

where V, ρ, b, l and C_D are airspeed, density, deck semichord, bridge span and drag coefficient, respectively. For simple harmonic motion of circular frequency ω,

$$w(x,t) = W_o w(x) e^{i\omega t}$$
$$\varphi(x,t) = \emptyset_o \varphi(x) e^{i\omega t} \tag{3}$$

where

$$w(x) = \bar{w}(x) + i\bar{\bar{w}}(x)$$
$$\varphi(x) = \bar{\varphi}(x) + i\bar{\bar{\varphi}}(x), \tag{4}$$

to the aerodynamic force and moment there correspond the implicite functions

$$L_a(x,t) = L_a\left[w(x),\varphi(x)\right] e^{i\omega t}$$
$$M_a(x,t) = M_a\left[w(x),\varphi(x)\right] e^{i\omega t} \tag{5}$$

and the pair of equations (1) becomes

$$W_o \frac{d^2}{dx^2}\left[(1+ig_w)EI\frac{d^2w}{dx^2}\right] - \omega^2 m W_o w(x) - L_a(w,\varphi)\ \frac{d^2}{dx^2}(M_z\varphi) = 0 \tag{6a}$$

$$\emptyset_o \frac{d}{dx}\left[(1+ig_\varphi)GI_d\frac{d\varphi}{dx}\right] + \omega^2 J\emptyset_o\varphi(x) + M_a(w,\varphi) - M_z\frac{d^2w}{dx^2} = 0. \tag{6b}$$

If in equations (1) the inertia forces are neglected initial perturbation reduces to the bending-torsional deformation

$$W_o\ \frac{d^2}{dx^2}(EI\frac{d^2w}{dx^2}) - L_a(w,\varphi)\ \frac{d^2}{dx^2}(M_z\varphi) = 0 \tag{7a}$$

$$\emptyset_o \frac{d}{dx}(GI_d\frac{d\varphi}{dx}) + M_a(w,\varphi) - M_z\frac{d^2w}{dx^2} = 0. \tag{7b}$$

It is to observe that the two components of the motion (1) or deformation (7) are always aerodynamically coupled. But as the pair of equations (6) and (7) associated with the boundary conditions

$$w(o) = w(1) = 0$$
$$\varphi(o) = \varphi(1) = 0 \tag{8}$$

form Sturm-Liouville systems, they can be put under integral form. Indeed, cf. p. 325 of Ref. (4), equations (6) corresponding to the periodical motion become

$$w(x) = \int_0^x G_w(x,\xi)\left\{\omega^2 m w(\xi) + \frac{1}{W_o} L_a(w,\varphi) - \frac{1}{W_o}\frac{d^2}{d\xi^2}\left[M_z(\xi)\varphi(\xi)\right]\right\} d\xi\ +$$

$$+\int_x^1 G_w(x,\xi)\left\{\omega^2 m w(\xi) + \frac{1}{W_o} L_a(w,\varphi) - \frac{1}{W_o}\frac{d^2}{d\xi^2}\left[M_z(\xi)\varphi(\xi)\right]\right\} d\xi\ , \tag{9a}$$

$$\varphi(x) = \int_0^x G_\varphi(x,\xi)\left[\omega^2 J\varphi(\xi) + \frac{1}{\emptyset_o} M_a(w,\varphi) - \frac{1}{\emptyset_o} M_z(\xi)\frac{d^2w(\xi)}{d\xi^2}\right] d\xi +$$

$$+\int_x^1 G_\varphi(x,\xi)\left[\omega^2 J\varphi(\xi) + \frac{1}{\emptyset_o} M_a(w,\varphi) - \frac{1}{\emptyset_o} M_z(\xi)\frac{d^2w(\xi)}{d\xi^2}\right] d\xi\ , \tag{9b}$$

while equations (7) of the deformation

$$w(x)=\int_0^x G_w(x,\xi)\left\{L_a(w,\varphi)-\frac{d^2}{d\xi^2}\left[M_z(\xi)\varphi(\xi)\right]\right\}d\xi+\int_x^1 G_w(x,\xi)\left\{L_a(w,\varphi)-\right.$$
$$\left.-\frac{d^2}{d\xi^2}\left[M_z(\xi)\varphi(\xi)\right]\right\}d\xi\ , \tag{10a}$$

$$\varphi(x)=\int_0^x G_\varphi(x,\xi)\left[M_a(w,\varphi)-M_z(\xi)\frac{d^2w(\xi)}{d\xi^2}\right]d\xi+\int_x^1 G_\varphi(x,\xi)\left[M_a(w,\varphi)-\right.$$
$$\left.-M_z(\xi)\frac{d^2w(\xi)}{d\xi^2}\right]d\xi\ . \tag{10b}$$

In both pairs of integral equations $G_w(x,\xi)$ and $G_\varphi(x,\xi)$ are Green's influence functions for bending and tŏrsion, respectively, of suspension bridges.

As concerns the aerodynamic force and its moment with respect to the elastic axis, in the adopted strip-theory,cf. (8.100-103) of Ref. (4), they are

$$L_a(w,\varphi) = \pi\varrho b^3\omega^2\left\{L_w\frac{w}{b}+\left[L_\varphi-(\tfrac{1}{2}+a)L_w\right]\varphi\right\} \tag{11a}$$

$$M_a(w,\varphi)= \pi\varrho b^4\omega^2\left\{\left[M_w-(\tfrac{1}{2}+a)L_w\right]\frac{w}{b}+\left[M_\varphi-(\tfrac{1}{2}+a)(L_\varphi+M_w)+(\tfrac{1}{2}+a)^2L_w\right]\varphi\right\} \tag{11b}$$

where

$$L_w = 1 - 2i\frac{V}{b\omega}(F_k+iG_k) \tag{12a}$$

$$L_\varphi = \frac{1}{2} - i\frac{V}{b\omega}\left[1+2(F_k+iG_k)\right] - 2\left(\frac{V}{b\omega}\right)^2(F_k+iG_k) \tag{12b}$$

$$M_w = \frac{1}{2},\qquad M_\varphi = \frac{3}{8} - i\frac{V}{b\omega}, \tag{12c}$$

and

$$F_k = \frac{J_1(J_1+J_o)+Y_1(Y_1-J_o)}{(J_1+Y_o)^2+(Y_1-J_o)^2},\qquad G_k = -\frac{Y_1Y_o+J_oJ_1}{(J_1+Y_o)^2+(Y_1-J_o)^2} \tag{13}$$

are the familiar Theodorsen functions of reduced frequency

$$k' = \frac{b\omega}{V}\ . \tag{14}$$

In order to calculate the stability boundaries the pairs of integral equations (9) and (10) are to be first undimensionalized. Since it is a question of calculating complex eigenvalues as unknowns are chosen the velocity and frequency parameters

$$X = \frac{V}{b\omega}\ \text{and}\ Z = (1+ig)\frac{GI_d}{\omega^2 J1^2},\ \text{where}\ g = g_w = g_\varphi\ , \tag{15}$$

for dynamic stability and

$$V_D = \frac{\pi\varrho V^2 1^2}{EI} \tag{16}$$

for static stability. They are to be determined as functions of the following four dimensionless quantities

$$p=\frac{GI_d 1^2}{EI\ b^2}\ ,\quad r^2=\frac{J}{mb^2}\ ,\quad m_o=\frac{m}{\pi\varrho b^2}\ ,\quad C=\frac{C_D}{2\pi}\ , \tag{17}$$

combined in the ten symbols

$$A_1 = \frac{p}{r^2}\left\{1 + \frac{1}{m_o}\,\frac{Re\left[L_a(w)\right]}{\pi\rho b^2\omega^2}\right\} \qquad A_2 = \frac{p}{r^2 m_o}\,\frac{Im\left[L_a(w)\right]}{\pi\rho b^2\omega^2} \tag{18a}$$

$$A_3 = \frac{p}{r^2 m_o}\,\frac{Re\left[L_a(\varphi)\right]}{\pi\rho b^2\omega^2} \qquad A_4 = \frac{p}{r^2 m_o}\,\frac{Im\left[L_a(\varphi)\right]}{\pi\rho b^2\omega^2} \tag{18b}$$

$$A_5 = \frac{1}{r^2 m_o}\,\frac{Re\left[M_a(w)\right]}{\pi\rho b^4\omega^2} \qquad A_6 = \frac{1}{r^2 m_o}\,\frac{Im\left[M_a(w)\right]}{\pi\rho b^4\omega^2} \tag{18c}$$

$$A_7 = \frac{1}{r^2 m_o}\,\frac{Re\left[M_a(\varphi)\right]}{\pi\rho b^4\omega^2} \qquad A_8 = \frac{1}{r^2 m_o}\,\frac{Im\left[M_a(\varphi)\right]}{\pi\rho b^4\omega^2} \tag{18d}$$

$$A_9 = \frac{p}{r^2 m_o} CX^2 \qquad A_{10} = \frac{1}{r^2 m_o} CX^2. \tag{18e}$$

In the following x and ξ are also rendered dimensionless through division by l, and w and e by b so that

$$\frac{x}{l} \rightarrow x, \quad \frac{\xi}{l} \rightarrow \xi \quad \text{and} \quad \frac{w}{b} \rightarrow w, \quad \frac{e}{b} = 0.5, \tag{19a}$$

while

$$G_w(x,\xi) \rightarrow \bar{G}_w(x,\xi) \quad \text{and} \quad G_\varphi(x,\xi) \rightarrow \bar{G}_\varphi(x,\xi). \tag{19b}$$

4. HANGING BRIDGE

The special feature of this structure consists in the fact that the cables hang freely in vertical plans following the funicular lines of the loads applied on the deck, and its cross-sections are open (Fig.1).

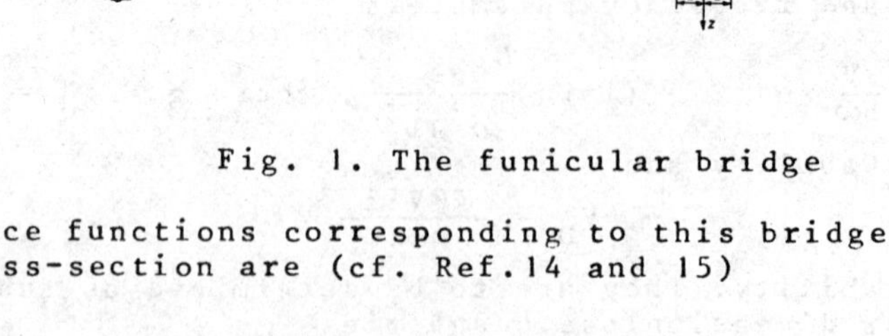

Fig. 1. The funicular bridge

Green's influence functions corresponding to this bridge with a deck of constant cross-section are (cf. Ref.14 and 15)

$$G_w(x,\xi)=\begin{cases}\dfrac{(1-\xi)x}{EI\mu^2 l}-\dfrac{sh\,\mu(1-\xi)\,sh\,\mu x}{EI\mu^3\,sh\,\mu l} & \text{for } x\leqslant\xi\\[2ex] \dfrac{(1-x)\xi}{EI\mu^2 l}-\dfrac{sh\,\mu(1-x)\,sh\,\mu\xi}{EI\mu^3\,sh\,\mu l} & \text{for } x\geqslant\xi\end{cases} \tag{20}$$

where

$$\mu^2=\frac{H+\Delta H}{EI} \quad \text{and} \quad \mu l=\bar{\mu}\,, \tag{21}$$

and

$$G_\varphi(x,\xi)=\begin{cases}\dfrac{(1-\xi)x}{(GI_d+Hb^2)l} & \text{for } x\leqslant\xi\\[2ex] \dfrac{(1-x)\xi}{(GI_d+Hb^2)l} & \text{for } x\geqslant\xi\end{cases} \tag{22}$$

Now, for hanging bridges the dimensionless form of integral equations of dynamic stability (9) and of static stability (10) will be

$$w(x)=\frac{1}{Z}\left(\int_o^x \bar{G}_w\left[(A_1+iA_2)w(\xi)+\frac{\emptyset_o}{W_o}(A_3+iA_4)\varphi(\xi)\right]d\xi+\right.$$
$$+\int_x^1 \bar{G}_w\left[(A_1+iA_2)w(\xi)+\frac{\emptyset_o}{W_o}(A_3+iA_4)\varphi(\xi)\right]d\xi+$$
$$\left.+\frac{X^2 pC}{2r^2 m_o\bar{\mu}}\,\frac{\emptyset_o}{W_o}\left\{\frac{sh\,\bar{\mu}(1-x)}{sh\,\bar{\mu}}\int_o^x (sh\,\bar{\mu}\xi)\,\xi(1-\xi)\varphi(\xi)\,d\xi+\frac{sh\,\bar{\mu}x}{sh\,\bar{\mu}}\int_x^1\left[sh\,\bar{\mu}(1-\xi)\right]\xi(1-\xi)\varphi(\xi)\,d\xi\right\}\right), \tag{23a}$$

$$\varphi(x)=\frac{1}{Z}\left\{\int_o^x \bar{G}_\varphi\left[\frac{W_o}{\emptyset_o}(A_5+iA_6)w(\xi)+(A_7+iA_8)\varphi(\xi)\right]d\xi+\right.$$
$$+\int_x^1 \bar{G}_\varphi\left[\frac{W_o}{\emptyset_o}(A_5+iA_6)w(\xi)+(A_7+iA_8)\varphi(\xi)\right]d\xi-$$
$$\left.-\frac{X^2 C}{r^2 m_o}\,\frac{W_o}{\emptyset_o}\left[(1-x)\int_o^x(1-3\xi)w(\xi)\,d\xi+x\int_x^1(3\xi-2)w(\xi)\,d\xi-\frac{x(1-x)}{2}w(x)\right]\right\}, \tag{23b}$$

and

$$w(x)=V_D\left(\frac{1}{2}\int_o^x \bar{G}_w(x,\xi)\varphi(\xi)\,d\xi+\frac{1}{2}\int_x^1 \bar{G}_w(x,\xi)\varphi(\xi)\,d\xi-\right.$$
$$\left.-\frac{C}{\bar{\mu}}\left\{\frac{sh\,\bar{\mu}(1-x)}{sh\,\bar{\mu}}\int_o^x(sh\,\bar{\mu}\xi)\,\xi(1-\xi)\varphi(\xi)\,d\xi+\frac{sh\,\bar{\mu}x}{sh\,\bar{\mu}}\int_x^1\left[sh\,\bar{\mu}(1-\xi)\right]\xi(1-\xi)\varphi(\xi)\,d\xi\right\}\right)\frac{\emptyset_o}{W_o}, \tag{24a}$$

$$\varphi(x)=V_D\left\{\frac{1}{p}\int_o^x \bar{G}_\varphi(x,\xi)\varphi(\xi)\,d\xi+\frac{1}{p}\int_x^1 \bar{G}_\varphi(x,\xi)\varphi(\xi)\,d\xi+\right.$$

$$+ \frac{2C}{p}\frac{W_o}{\phi_o}\left[(1-x)\int_0^x(1-3\xi)w(\xi)d\xi + x\int_x^1(3\xi-2)w(\xi)d\xi - \frac{x(1-x)}{2}w(x)\right]\Bigg\} \quad , \qquad (24b)$$

respectively.

5. PRESTRESSED BRIDGE

The special features of this structure consist in the fact that the cables follow the geodetic lines of the funicular surface of vertical loads applied on the deck, and its cross-sections are closed (Fig.2).

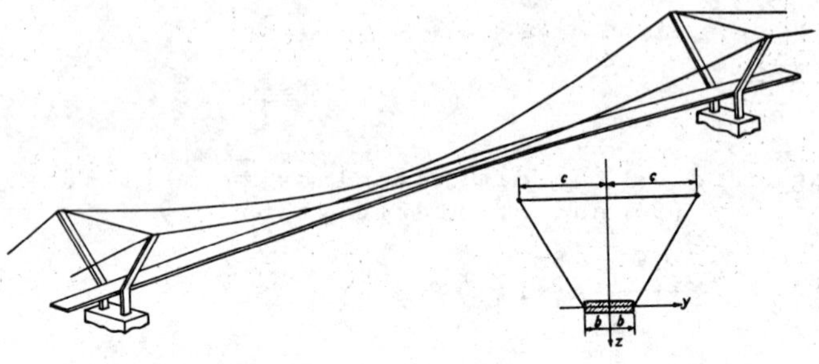

Fig. 2. The geodetic bridge

Green's influence functions corresponding to this bridge with a deck of constant cross-section are (cf. Ref. 14 and 15)

$$G_w(x,\xi) = \begin{cases} \dfrac{(1-\xi)x}{EI\mu^2 l} - \dfrac{sh\mu(1-\xi)sh\mu x}{EI\mu^3 sh\mu l} & \text{for } x \leqslant \xi \\ \dfrac{(1-x)\xi}{EI\mu^2 l} - \dfrac{sh\mu(1-x)sh\mu\xi}{EI\mu^3 sh\mu l} & \text{for } x \geqslant \xi \end{cases} \qquad (25)$$

and

$$G_\varphi(x,\xi) = \begin{cases} \dfrac{sh\lambda(1-\xi)sh\lambda x}{(GI_d+Hb^2)\lambda sh\lambda l} & \text{for } x \leqslant \xi \\ \dfrac{sh\lambda(1-x)sh\lambda\xi}{(GI_d+Hb^2)\lambda sh\lambda l} & \text{for } x \geqslant \xi \end{cases} \qquad (26)$$

where

$$\lambda^2 = \frac{2b^2\frac{P_o}{c}}{GI_d+Hb^2} \quad \text{and} \quad \lambda l = \bar{\lambda}, \qquad (27)$$

p_o being cable pretension and c cable semichord. Now, for prestressed bridges the dimensionless form of integral equations of dynamical stability (9) and of static stability (10) will be

$$w(x) = \frac{1}{Z}\Bigg(\int_0^x \bar{G}_w(x,\xi)\left[(A_1+iA_2)w(\xi) + \frac{\phi_o}{W_o}(A_3+iA_4)\varphi(\xi)\right]d\xi \ +$$

$$+\int_x^1 \bar{G}_w(x,\xi)\left[(A_1+iA_2)w(\xi) + \frac{\phi_o}{W_o}(A_3+iA_4)\varphi(\xi)\right]d\xi \ +$$

$$+\frac{X^2 pC}{2r^2 m_o \bar{\mu}}\frac{\phi_o}{W_o}\left\{\frac{sh\bar{\mu}(1-x)}{sh\bar{\mu}}\int_0^x (sh\bar{\mu}\xi)\xi(1-\xi)\varphi(\xi)d\xi + \right.$$

$$\left. + \frac{sh\bar{\mu}x}{sh\bar{\mu}}\int_x^1 \left[sh\bar{\mu}(1-\xi)\right]\xi(1-\xi)\varphi(\xi)d\xi\right\}\Bigg) , \qquad (28a)$$

$$\varphi(x) = \frac{1}{Z}\left\{\int_0^x \bar{G}_\varphi(x,\xi)\left[\frac{W_o}{\phi_o}(A_5+iA_6)w(\xi)+(A_7+iA_8)\varphi(\xi)\right] d\xi + \right.$$

$$+\int_x^1 \bar{G}_\varphi(x,\xi)\left[\frac{W_o}{\phi_o}(A_5+iA_6)w(\xi)+(A_7+iA_8)\varphi(\xi)\right] d\xi -$$

$$-\frac{X^2 C}{r^2 m_o}\frac{W_o}{\phi_o}\left(\frac{sh\bar{\lambda}(1-x)}{2\bar{\lambda}sh\bar{\lambda}}\int_0^x\left[\bar{\lambda}^{2}(sh\bar{\lambda}\xi)\xi(1-\xi)+2\bar{\lambda}(ch\bar{\lambda}\xi)(1-2\xi)-2sh\bar{\lambda}\xi\right]w(\xi)d\xi + \right.$$

$$\left.\left. + \frac{sh\bar{\lambda}x}{2\bar{\lambda}sh\bar{\lambda}}\int_x^1\left\{\bar{\lambda}^2\left[sh\bar{\lambda}(1-\xi)\right]\xi(1-\xi)-2\bar{\lambda}\left[ch\bar{\lambda}(1-\xi)\right](1-2\xi)-2sh\bar{\lambda}(1-\xi)\right\}w(\xi)d\xi-\frac{x(1-x)}{2}w(x)\right)\right\}$$

$$(28b)$$

and

$$w(x) = V_D\left(\frac{1}{2}\int_0^x \bar{G}_w(x,\xi)\varphi(\xi)d\xi + \frac{1}{2}\int_x^1 \bar{G}_w(x,\xi)\varphi(\xi)d\xi - \right.$$

$$\left. -\frac{C}{\bar{\mu}}\left\{\frac{sh\bar{\mu}(1-x)}{sh\bar{\mu}}\int_0^x (sh\bar{\mu}\xi)\xi(1-\xi)\varphi(\xi)d\xi + \frac{sh\bar{\mu}x}{sh\bar{\mu}}\int_x^1\left[sh\bar{\mu}(1-\xi)\right]\xi(1-\xi)\varphi(\xi)d\xi\right\}\right)\frac{\phi_o}{W_o},$$

$$(29a)$$

$$\varphi(x) = V_D\left\{\frac{1}{p}\int_0^x \bar{G}_\varphi(x,\xi)\varphi(\xi)d\xi + \frac{1}{p}\int_x^1 \bar{G}_\varphi(x,\xi)\varphi(\xi)d\xi + \right.$$

$$+\frac{2C}{p}\frac{W_o}{\phi_o}\left(\frac{sh\bar{\lambda}(1-x)}{2\bar{\lambda}\; sh\;\bar{\lambda}}\int_0^x\left[\bar{\lambda}^2(sh\bar{\lambda}\xi)\xi(1-\xi)+2\bar{\lambda}(ch\bar{\lambda}\xi)(1-2\xi)-2sh\bar{\lambda}\xi\right]w(\xi)d\xi + \right.$$

$$\left.\left. + \frac{sh\bar{\lambda}x}{2\bar{\lambda}sh\bar{\lambda}}\int_x^1\left\{\bar{\lambda}^2\left[sh\bar{\lambda}(1-\xi)\right]\xi(1-\xi)-2\bar{\lambda}\left[ch\bar{\lambda}(1-\xi)\right](1-2\xi)-2sh\bar{\lambda}(1-\xi)\right\}w(\xi)d\xi-\frac{x(1-x)}{2}w(x)\right)\right\},$$

$$(29b)$$

respectively.

6. SOLUTION METHODS

The two problems of aeroelastic stability described above by their governing equations are in fact eigenvalue problems. Consequently, they can be solved by the specific methods of this kind of problems. Certainly, all will be aproximate methods. One of them is the functional collocation. But, as by using Green's influence functions the governing equations were put into integral form, it is more practical and of higher reliability to transform them into algebraic equations. The operation can be done by punctual collocation, with the aid of the tangent method. Thus, the two pairs of integral equations, (9) for flutter and (10) for divergence, take the form

$$\{w\} = \frac{1}{Z}\left((A_1+iA_2)[G_w]\{w\} + \frac{\emptyset_o}{W_o}(A_3+iA_4)[G_w]\{\varphi\} + \frac{\emptyset_o}{W_o}A_9[D_w]\{\varphi\}\right) \tag{30a}$$

$$\{\varphi\} = \frac{1}{Z}\left(\frac{W_o}{\emptyset_o}(A_5+iA_6)[G_\varphi]\{w\} + (A_7+iA_8)[G_\varphi]\{\varphi\} + \frac{W_o}{\emptyset_o}A_{10}[D_\varphi]\{w\}\right) \tag{30b}$$

and

$$\{w\} = V_D\left(\frac{1}{2}[G_w]\{\varphi\} - C[D_w]\{\varphi\}\frac{\emptyset_o}{W_o}\right. \tag{31a}$$

$$\{\varphi\} = V_D\left(\frac{1}{p}[G_\varphi]\{\varphi\} + \frac{2C}{p}\frac{W_o}{\emptyset_o}[D_\varphi]\{w\}\right., \tag{31b}$$

respectively, where by the operators D_w and D_φ the effects of drag were introduced, the ratio $W_o/\emptyset_o$ contains the phase difference between the two components of motion or deformation, and obviously, the normalized vectors $\{w\}$ and $\{\varphi\}$ replace the complex mode shapes $w(x)$ and $\varphi(x)$. Such algebraic systems could be also directly obtained by a finite element technique. But for the problems of aeroelastic stability the above adopted way is much simpler and especially safer. In order to solve these linear systems two methods can be used, the direct method and the iterative one. The well known method of assumed modes is not adequate in the case of suspension bridges because it is difficult to predict the common deformations of the cables and decks. In the direct method the matrix of linear system is transformed into the Hessenberg matrix and the eigenvectors and eigenvalues of this matrix are then determined. The method is useful when the explored field is completely unknown, although it requires more computing time and for plotting the V-g curves each eigenvalue has to be pursued. The iterative method is in fact the inverse power method. It is a much faster method but not always convergent enough, especially when the critical velocities of divergence and flutter, affected by drag, assume close values.

Both methods were used by authors to investigate the aeroelastic stability of suspension bridges, and the obtained numerical results provide lots of informations. For exemple, by assuming for dimensionless quantities (17), (21) and (27) the values $\bar{p} = 0.001$, $r^2 = 0.25$, $m_o = 1$, $C = 0$ (hence without drag), $\bar{\mu} = 1$ and $\bar{\lambda} = 10$, according to (15), the dimensionless critical velocity of flutter

$$V_F = \frac{X}{\sqrt{Re(Z)}} \tag{32}$$

takes for the funicular bridge the value $V_F = 4.3447$, and for the geodetic bridge $\bar{V}_F = 13.673$, i.e. 3.147 times larger. If for the same parameters $\bar{\mu}$ increases from 1 to 10 the critical velocity of the funicular bridge becomes $V_F = 13.937$, i.e. 3.208 times larger. But, as a consequence of aerodynamic drag, the critical velocity of the same funicular bridge decreases to $V_F = 7.244$ for $C = 0.02$ and to $V_F = 1.7552$ for $C = 0.04$, i.e. it becomes 1.924 and 7.940 times respectively, smaller.

7. CONCLUSIONS

1. The aeroelastic stability of suspension bridges is qualitatively and quantitatively influenced by the aerodynamic drag of span decks. By taking into account the effects of drag in stability analyses, the two components of deformation or motion, bending and torsion, are always coupled. Only when the drag effect is neglected an one-degree divergence is possible, namely a torsional divergence. Another consequence of drag is the reduction of the absolute values of divergence and flutter critical velocities, calculated only for lifting forces. In addition, the ratio between the two critical velocities can also be modified. Thus, if without drag the critical velocity of divergence is higher than the corresponding critical velocity of flutter, as a consequence of the drag effect the inequality sign can be reversed.

2. The above determined governing equations allow to draw a comparison between the two structural systems of suspension bridges. Thus, while their bending components are formally identic, the torsional components are essentially different. The hanging bridges are more sensitive to torsion than prestressed bridges. For this reason the effect of drag is greater on the first structural system than on the last. More important, however, is that, while hanging bridges behave by far more differently in bending than in torsion, prestressed bridges have about the same behaviour in both degrees of freedom.

3. The adopted technique of analysis based on Green's influence functions has proved efficient for the study of bridge aeroelastic stability. It allows to control at any instant the developed phenomena and the modify conveniently the aerodynamic and/or structural parameters in order to improve the stability criteria. This technique can be used for further studies on aeroelastic stability, and the results obtained in this way for a realistic design of suspension bridges.

REFERENCES :

1. Petre, A., Teoria aeroelasticitatii, Statica. Ed.Academiei R.S.R., Bucuresti 1966.

2. Petre, A., Divergence et répartition aéroélastic des charges aérodynamiques, par suite de la portance et de la résistance à l'avancement, en tant qu' effets combinés du vent. Wind Effects on Buildings and Structures. Proc. of the Int.Research Seminar, Ottawa, Canada 11-15 Sept. 1967, Vol.II Paper 34, pp.191-241

3. Petre, A., L'éffet de la résistance à l'avancement sur la vitesse critique du flutter. The Sixth Cong. of Int. Council of the Aeronaut. Sci., München, Sept.1968

4. Petre, A., Teoria aeroelasticitatii. Fenomene dinamice periodice. Editura Academiei R.S.R., Bucuresti 1973

5. Petre, A., Ashley, H., Drag Effects on Wing Flutter. Journal of Aircraft, Vol. 13, No. 10, Oct. 1976, pp.755-763

6. Sofronie, R., Suspension Bridge. United States Patent No.3,857,130.

7. Farquharson, F.B., Model Verification of the Classical Flutter Theory as Adapted to the Suspension Bridge. IABSE Publ. 1952, Vol.12

8. Rocard, Y., L'instabilité en Mécanique. Automobiles-Avions-Ponts Suspendus. Masson et C-ie. Editeurs, Paris 1954

9. Ukeguchi, N. et al., An Investigation of Aeroelastic Instability of Suspension Bridges. Proceedings Symposium on Suspension Bridges, Lisbon, Nov. 1966, Paper No.11, pp.11.1-11.12.

10. Sabzevari, A., Scanlan, R.H., Aerodynamic Instability of Suspension Bridges. J. Engrg. Mech. Div. Proc. ASCE, Apr.1968, Vol.94, EM2, pp.489-519.

11. Dicker, D., Aeroelastic Stability of Unstiffened Suspension Bridges. J. Engrg. Mech. Div. Proc., ASCE, Dec. 1971, Vol.97, EM6, pp.1677-1701.

12. Scanlan, R.H., Tomko, J.J., Airfoil and Bridge Deck Flutter Derivatives, J. Engrg. Mech. Div. Proc. ASCE, Dec. 1971, Vol.97, EM6, pp.1717-1737.

13. Scanlan, R.H., Recent Methods in the Application of Test Results to the Wind Design of Long, Suspended-Span Bridges. Report No. FHWA-RD-75-115. Washington D.C., Oct. 1975.

14. Sofronie, R., Vertical Deflection of Suspension Bridges. Rev. Roum. Sci. Techn.-Méc. Appl., Vol.24, No,3, Bucarest, 1979.

15. Sofronie, R., Torsional Deflection of Suspension Bridges. Rev. Roum. Sci. Techn.-Méc. Appl., Vol.24, No.4, Bucarest, 1979.

Bucharest, 5 April 1979.

ON THE INDICIAL AERODYNAMIC MOMENT RESPONSES OF BRIDGE DECK SECTIONS

T. YOSHIMURA* and Y. NAKAMURA**

*Kyushu Sangyo University, Kukuoka, Japan
**Kyushu University, Fukuoka, Japan

ABSTRACT

This paper is concerned with theoretical and experimental investigations on indicial aerodynamic moment responses of bluff prismatic sections for the purpose of elucidating the mechanism of the onset of torsional flutter of bridge deck sections.

1. INTRODUCTION

It has been suggested in a previous paper (1) that the fluid memory effect may be dominant for the onset of torsional flutter of bluff prismatic bodies having an axis of rotation near the center of the cross section. By fluid memory is meant here that the instantaneous aerodynamic forces and moment acting on a body may be dependent on the entire past history of the body motion, or in short, the former are functionals of the latter. While flutter instability is associated with frequency aerodynamic responses, the fluid memory effect may be examined more directly with indicial aerodynamic responses. In the investigations of torsional flutter of bridge deck sections, Raggett (2) and Scanlan (3,4) obtained the characteristic indicial aerodynamic moment responses by the Fourier transform of the measured frequency aerodynamic responses.

This paper is concerned with theoretical and experimental investigations on indicial aerodynamic moment responses for geometrically simple bluff prismatic sections for the purpose of getting a clear insight into the mechanism of the onset of torsional flutter of bridge deck sections. The results of flow visualization and surface pressure measurement are also reported. Since transient motions of a model are more conveniently realized in a towing water tank rather than in a wind tunnel, the experiments were conducted with models moving in a towing water tank.

2. A THEORETICAL ANALYSIS OF TORSIONAL FLUTTER

(a) The condition for the onset of torsional flutter expressed in the frequency and in the time domain

The aerodynamic torsional moment per unit span $M_\theta(t)$ acting on a bluff prismatic section in a uniform flow undergoing a unit sinusoidal torsional motion $\theta(t) = \exp(i\omega t)$ with an axis near the center of the cross section is expressed in a non-dimensional form as $2M_\theta(\tau)/(\pi\rho c^2V^2) = Z_\theta(ik)\theta(\tau)$. Here, ρ, c, V, ω, $\tau = 2Vt/c$, $k = \omega c/(2V)$ and $Z_\theta(ik)$ denote the air density, the deck width, the flow velocity, the circular frequency, the reduced time, the reduced frequency and the transfer function of the aerodynamic moment, respectively.

$Z_\theta(ik)\theta(\tau)$ may be expressed as

$$Z_\theta(ik)\theta(\tau) = C_{M\theta}(k)\theta(\tau) + C_{M\dot{\theta}}(k)\dot{\theta}(\tau), \quad (1)$$

in which $\dot{\theta}(\tau) = d\theta(\tau)/d\tau$. Assuming the system mechanical damping to be zero, an expression for the condition for the onset of torsional flutter is given by

$$C_{M\dot{\theta}}(k) \quad > \quad 0. \quad (2)$$

On the other hand, an indicial aerodynamic moment response is defined as the response for $\theta(\tau) = E(\tau)$ where $E(\tau)$ is the unit step function, and denoted by $\Phi_\theta(\tau)$. Assuming a linear aerodynamic system, the Fourier transform of $d\Phi_\theta(\tau)/d\tau$ gives $Z_\theta(ik)$. Carrying out the Fourier transform, one finds

$$C_{M\dot{\theta}}(k) = \int_0^\infty\{ \Phi_\theta(\tau) - \Phi_\theta(\infty) \}\cos k\tau \, d\tau > 0. \quad (3)$$

In the case of the limiting condition $k \to 0$, Eq. 3 is reduced to

$$C_{M\dot{\theta}}(0) = \int_0^\infty\{ \Phi_\theta(\tau) - \Phi_\theta(\infty) \} \, d\tau > 0. \quad (4)$$

The graphical expression of Eq. 4 is shown in Fig. 1. Thus, torsional flutter can occur if the indicial aerodynamic moment response has a positive area as shown with hatching in Fig. 1. ($\Phi_\theta(\tau)$ in Eq. 3 may contain $\delta(\tau)$ as the so-called added mass effect. However, in view of Fig. 7, its effect on $C_{M\dot{\theta}}(k)$ was found to be negligibly small.)

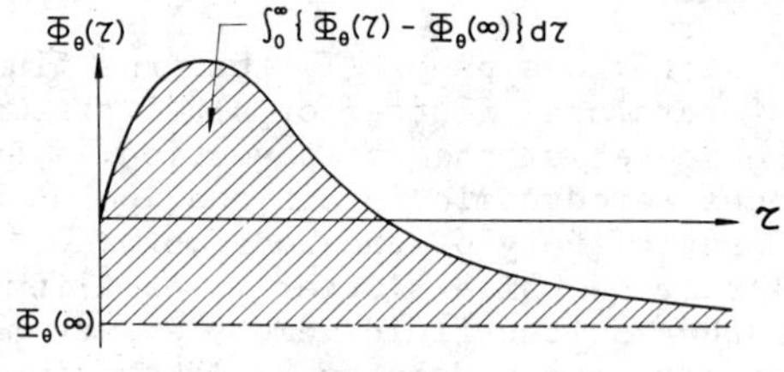

Fig. 1. Graphical expression of Eq. 4

(b) Decomposition of aerodynamic moment

For the purpose of getting an insight into the mechanism of the onset of torsional flutter of bridge deck sections, the decomposition of the aerodynamic moment is useful (5), and will be discussed in this section. Consider for convenience purposes a bluff prismatic section moving in a still air, as shown in Fig. 2,

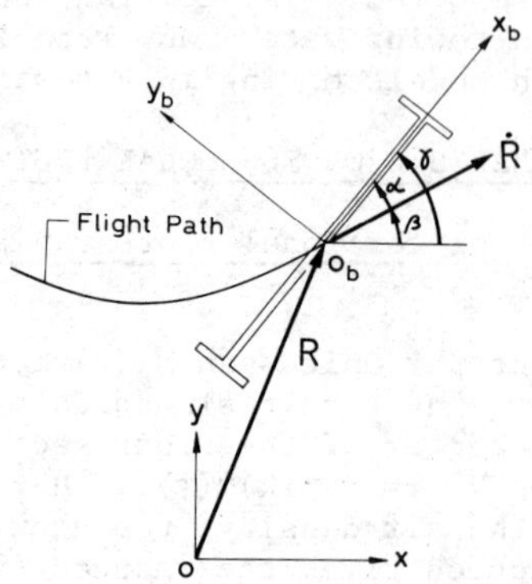

Fig. 2. The motion of a prismatic section in a still air

instead of a section held stationary in a flowing fluid. It is well known (see, for example, Ref. 6) that the aerodynamic moment (or force) at any given instant is dependent not only on the instantaneous values of the state variables of the section, $R(\tau)$ and $\gamma(\tau)$, but also on their entire past history, where $R(\tau)$ and $\gamma(\tau)$ denote respectively the displacement vector and the angle of the body axis O_bX_b relative to the fixed coordinate axis OX. This important phenomenon which is characteristic of unsteady flow is referred to as " fluid memory " in this paper. The aerodynamic moment is thus a functional of the state variables, $R(\tau)$ and $\gamma(\tau)$. Since the aerodynamic moment arises from the relative motion between the fluid and the body, it may be expressed more conveniently in terms of the time derivatives of the state variables, $\dot{R}(\tau)$ and $\dot{\gamma}(\tau)$. The velocity vector may be represented by the magnitude $V(\tau)$ and the angle $\beta(\tau)$ which is relative to the axis OX. We assume that $V(\tau)$ = const for simplicity, and note also that $\beta(\tau) = \gamma(\tau) - \alpha(\tau)$, where $\alpha(\tau)$ denotes the angle of the body axis relative to the velocity vector, that is, the angle of attack. Therefore, the aerodynamic moment is a functional of $\alpha(\tau)$ and $q(\tau)$, where $q(\tau) = \dot{\gamma}(\tau)$, and denoted by

$$M_\theta(\tau) = M[\ \alpha(\tau),\ q(\tau)\ ]. \tag{5}$$

Assuming the applicability of the principle of superposition of small disturbances to a linear aerodynamic system, one finds

$$M[\ \alpha(\tau),\ q(\tau)\ ] = M[\ \alpha(\tau)\ ]_{q=\mathrm{const}} + M[\ q(\tau)\ ]_{\alpha=\mathrm{const}} \tag{6}$$

The first and the second term on the right hand side of Eq. 6 are respectively associated with two special motions. The motion in which $\alpha(\tau)$ is changed while $q(\tau)$ is held constant is referred to as the angle of attack motion. On the other hand, the motion in which $q(\tau)$ is changed while $\alpha(\tau)$ is held constant is referred to as the angular velocity motion.

Equation 6 can be applied to a section moving steadily along a straight line with a perturbed torsional motion about an axis. This is a special motion in which $\alpha(\tau) = \theta(\tau)$ and $q(\tau) = \dot{\theta}(\tau)$. That is, the aerodynamic moment acting on a bluff prismatic section oscillating torsionally in a uniform flow is decomposed into two terms having physically different origins.

The frequency aerodynamic moment response due to the torsional motion which is given by Eq. 1 is now decomposed into two terms, that is,

$$Z_\theta(ik)\theta(\tau) = \{\ C_{M\alpha}(k)\theta(\tau) + C_{M\dot{\alpha}}(k)\dot{\theta}(\tau)\ \} + \{\ C_{Mq}(k)\dot{\theta}(\tau) + C_{M\dot{q}}(k)\ddot{\theta}(\tau)\ \}, \tag{7}$$

in which $C_{M\alpha}(k)$, $C_{M\dot{\alpha}}(k)$ and $C_{Mq}(k)$, $C_{M\dot{q}}(k)$ denote the aerodynamic moment derivatives due to the angle of attack motion, and those due to the angular velocity motion, respectively. Correspondingly, the indicial aerodynamic moment response is decomposed into two terms as

$$\Phi_\theta(\tau) = \Phi_\alpha(\tau) + d\Phi_q(\tau)/d\tau, \tag{8}$$

since $\alpha(\tau) = \theta(\tau) = E(\tau)$ and $q(\tau) = \dot{\theta}(\tau) = \delta(\tau)$. Here, $\Phi_\alpha(\tau)$ and $\Phi_q(\tau)$ denote the indicial aerodynamic moment response for the angle of attack motion and that for

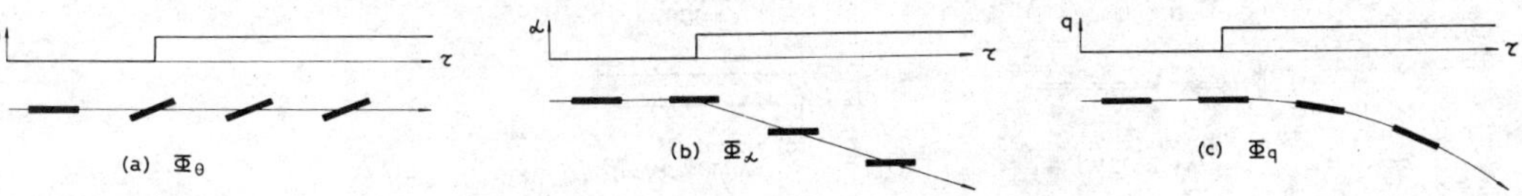

Fig. 3. Three types of indicial motion

the angular velocity motion, respectively. The three types of indicial motion are illustrated in Fig. 3. By the use of Eq. 7, the condition for the onset of torsional flutter is expressed as

$$C_{M\dot{\theta}}(k) = C_{M\dot{\alpha}}(k) + C_{Mq}(k) > 0. \tag{9}$$

Substituting Eq. 8 into Eq. 3, Eq. 9 is now expressed in terms of the indicial aerodynamic moment responses as

$$C_{M\dot{\theta}}(k) = \int_0^{\infty}\{ \Phi_\alpha(\tau) - \Phi_\alpha(\infty) \}\cos k\tau \, d\tau + \int_0^{\infty} d\Phi_q(\tau)/d\tau \cos k\tau \, d\tau > 0. \tag{10}$$

In the case of the limiting condition $k \rightarrow 0$, Eq. 10 is reduced to

$$C_{M\dot{\theta}}(0) = \int_0^{\infty}\{ \Phi_\alpha(\tau) - \Phi_\alpha(\infty) \}d\tau + \Phi_q(\infty) > 0. \tag{11}$$

(c) Mechanism of the onset of torsional flutter

It is clear from Eq. 9 that the aerodynamic derivative $C_{M\dot{\theta}}(k)$ has two constituents of physically different origins, that is, one due to the angle of attack motion, and the other due to the angular velocity motion. Obviously, the major part of the first term $C_{M\dot{\alpha}}(k)$,which is the out-of-phase component arising from the angle of attack motion, comes from the fluid memory. On the other hand, the second term $C_{Mq}(k)$, which is related to the angular velocity motion, represents the quasi-steady aerodynamic moment, since one can approximate it in terms of the steady aerodynamic moment when $k \rightarrow 0$. The situation may be more easily seen in the equivalent expression in the time domain, that is, in Eq. 10 or Eq. 11. One can see from Eq. 11 that the first term on the right hand side is related to the entire history of $\Phi_\alpha(\tau)$, so that the fluid memory can play vital role in it.

The question as to which term in Eq. 11 is dominant is essential for the mechanism of the onset of torsional flutter. If the first term in Eq. 11 is dominant, one would say that the fluid memory effect is responsible for the onset of torsional flutter. Therefore, a measurement of $\Phi_\theta(\tau)$ and $\Phi_\alpha(\tau)$ for some bluff prismatic sections was attempted in the present work to make a comparison between these two. A measurement of $Z_\theta(ik)$ and $Z_\alpha(ik)$ was also made for supplement purposes. Flow visualization and surface pressure measurement were also made for sections following an indicial change in incidence in order to investigate the flow characteristics associated with $\Phi_\theta(\tau)$.

3. EXPERIMENTAL APPARATUS AND MODEL

The experiment was conducted with geometrically simple bluff prismatic sections moving in a towing water tank 600cm in length and 40cm in width and depth. The models used in the present experiment are shown in Fig. 4a to Fig. 4e, and they are referred to as Model H, Model A, Model B, Model B1 and Model B2, respectively. All these models had $c \times \ell = 10 \times 37$cm where ℓ is the span length. The running gear, with which the transient or the sinusoidal motions of the model were realized, was attached to the towing carriage. In the experiment where measurement of the

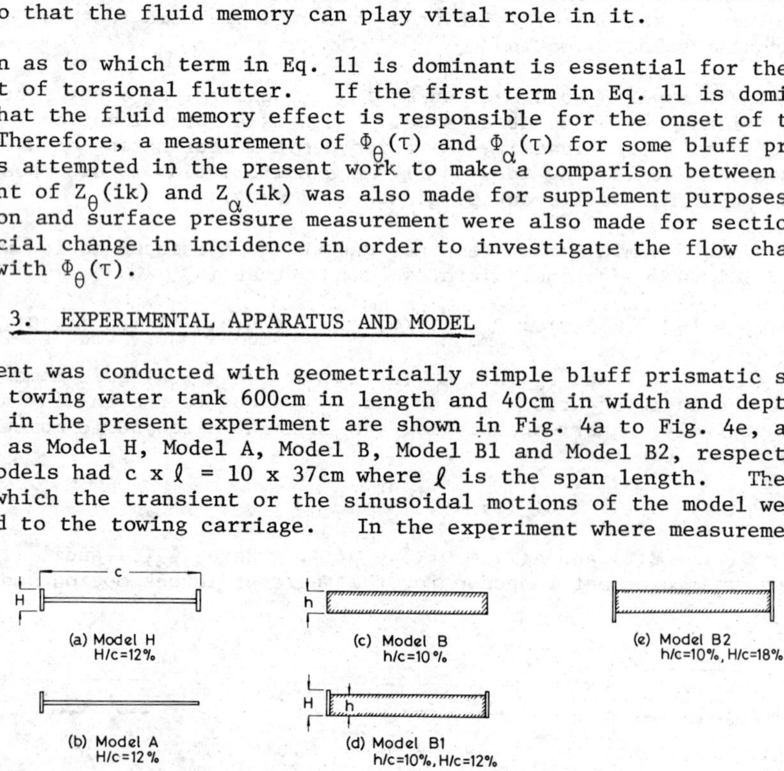

Fig. 4. Cross sections of models used in the present experiment

aerodynamic moment responses was made, the gear was equipped with a relatively hard cross spring pivot. One end of the pivot was fixed to the gear while the other end was connected to the model at the center of the cross section. The model was suspended vertically. The aerodynamic moment was measured by the use of the wire strain gages which were cemented on the flexures of the cross spring pivot. In the experiment where the surface pressure measurement was made, the pressure taps which were distributed around the midspan of the model were connected through tubings to the pressure transducers which were mounted outside the model. The hydrogen bubble method and the aluminium powder method were employed for flow visualization. The flow speed for the moment and pressure measurement was 20cm/s while that for flow visualization was 3cm/s. The corresponding Reynolds number based on c was 2×10^4 for the former experiment while it was 3×10^3 for the latter experiment.

4. MEASUREMENT OF INDICIAL AERODYNAMIC MOMENT RESPONSES

Figure 5 shows the measured indicial aerodynamic moment responses for the present models together with the theoretical one for an airfoil (7). The experimental responses correspond to an initial angle of attack of $\theta_0 = 0$ deg and a step change of $\Delta\theta = 1$ deg; they were approximately linear for $\Delta\theta \leq 1$ deg. Observation indicated

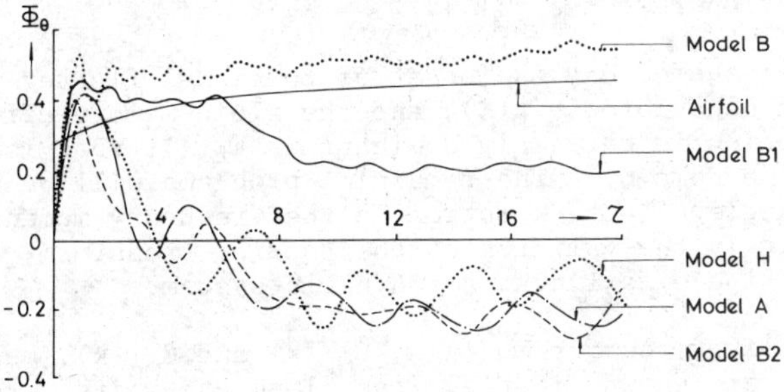

Fig. 5. Indicial aerodynamic moment responses for various prismatic sections

that all the measured responses contained sharp but very transient portions at $\tau = 0$ which were probably due to the virtual mass effect. However, they were very difficult to be reproduced so that they are not shown in the curves of Fig. 5.

Among the responses in Fig. 5, those for Model H, Model A, Model B1 and Model B2 are characterized by an initial abrupt overshoot followed by a rather slow settling down to the steady state condition. Raggett (2) and Scanlan (3,4) also obtained similar characteristic responses. These characteristic responses obviously satisfy Eq. 4 so that the onset of torsional flutter can be anticipated for these models. The free oscillation experiment, not shown here, indicated that torsional flutter did commence at a relatively low reduced flow velocity $\bar{V} = 2\pi/k$ for all these models. The experiment for Model B also indicated that it was torsionally stable, which is another evidence for the validity of Eq. 4.

Figure 6 shows a comparison between $\Phi_\theta(\tau)$ and $\Phi_\alpha(\tau)$ for Model A. Obviously, little difference is seen between the two. That is, the contribution of the angle of attack motion is dominant while that of the angular velocity motion is negligibly small. The experimental results, not shown here, also indicate that the same was true with all the other models tested. In summary, the fluid memory effect arising from the angle of attack motion is suggested to be responsible for the onset of torsional flutter of bridge deck sections at least with an axis of rotation near the center of the cross section.

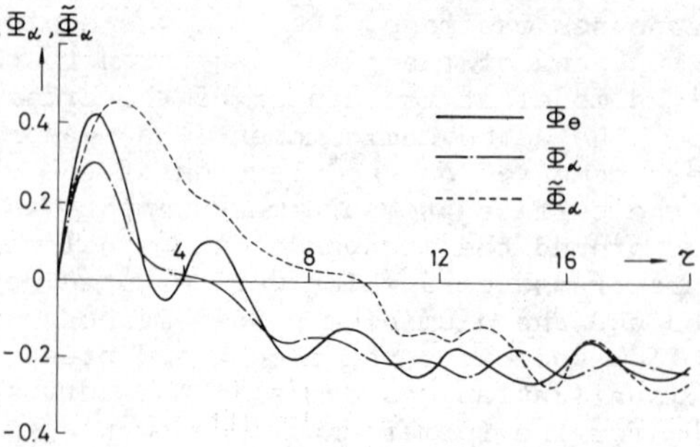

Fig. 6. Three transient aerodynamic moment responses for Model A

It is expected that an impulsive start of a body at an angle of attack from rest would hold the essential fluid mechanical property of the indicial angle of attack motion. Therefore, measurement of the corresponding response $\widetilde{\Phi}_\alpha(\tau)$ was added to get further insight into the mechanism of the onset of torsional flutter. The curve indicated with dotted line in Fig. 6 represents $\widetilde{\Phi}_\alpha(\tau)$ for Model A at $\alpha = 1$ deg. As was expected, $\widetilde{\Phi}_\alpha(\tau)$ is very similar in trend to the corresponding indicial angle of attack motion response $\Phi_\alpha(\tau)$.

So far, it has been shown that the onset of torsional flutter is closely related to the characteristic behavior of $\Phi_\theta(\tau)$, and the fluid memory effect arising from the angle of attack motion is certainly dominant in $\Phi_\theta(\tau)$, that is, in the moment response in the time domain. The remaining problems will be 1) to show the predominance of the angle of attack motion in the frequency moment response, and 2) to examine quantitatively the validity of the Fourier transform, or equivalently, to examine the assumption of a linear aerodynamic system.

The aerodynamic derivatives for Model A, $C_{M\dot{\theta}}(k)$ and $C_{M\dot{\alpha}}(k)$, measured in the same towing water tank, are shown in Fig. 7 as black and white circles, respectively. It is clear from Fig. 7 that the angle of attack motion is also dominant in the frequency domain. The curves with soild and chained lines in Fig. 7 denote respectively the aerodynamic derivatives, $C_{M\dot{\theta}}(k)$ and $C_{M\dot{\alpha}}(k)$, obtained by the Fourier transformation of $\Phi_\theta(\tau)$ and $\Phi_\alpha(\tau)$. Good agreement is seen between the directly measured values and the Fourier transformed ones. Therefore, the assumption of a linear aerodynamic system is valid provided the displacement of a model is small enough.

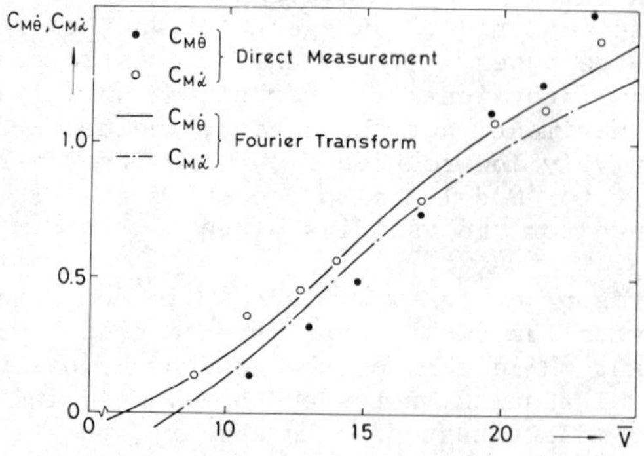

Fig. 7. Aerodynamic derivatives for Model A obtained by direct measurement and by Fourier transformation

5. FLOW VISUALIZATION AND SURFACE PRESSURE MEASUREMENT

The experiment was conducted with Model B2 and Model B which were found torsionally unstable and torsionally stable, respectively. In the experiment, a relatively large step change in incidence of $\Delta\theta$= 4 deg was given to the model. Although this was done at the sacrifice of linearity in the moment response, important trend in the linear region was still retained in the responses of $\Delta\theta$= 4 deg.

Photo 1 shows successive changes in the flow pattern around Model B2 following an indicial change in incidence. Immediately after the incidence change, a sheet of hydrogen bubbles was discharged into the flow as a tracer from a tungsten wire stretched near the leading edge of the model. In the photographs, the shear layers which have been newly formed as a result of an indicial change in incidence are colored white by the hydrogen bubbles, and distinguished from those formed before the incidence change. The flow pattern at the initial stage (photo 1a and photo 1b) is characterized by the formation and growth of the separation bubbles as they travel downstream. Similar phenomenon had been observed for an airfoil during dynamic stall (8). After about τ = 4.64 (photo 1c) when the separation bubble on the upper surface was shed from the trailing edge, the deformation of the shear layers become relatively slow and the flow around the model tends to be nearly steady at about τ = 7.

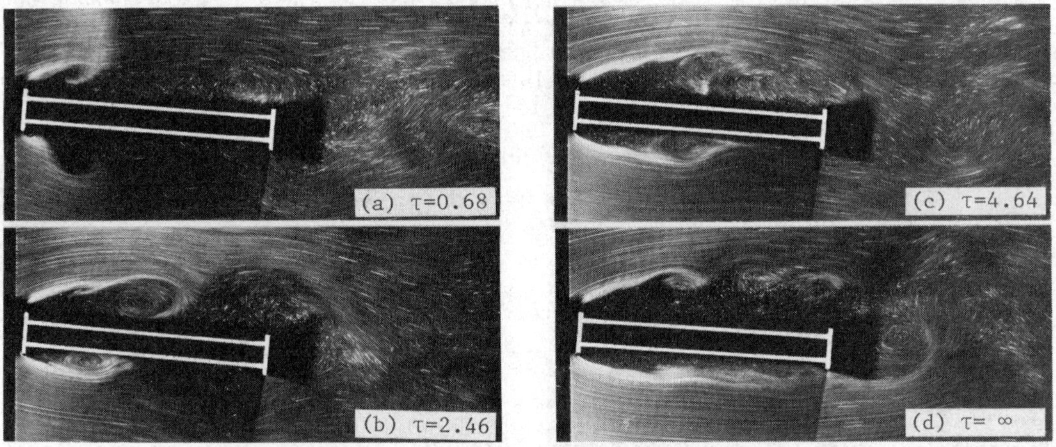

Photo. 1. Temporal development of the flow pattern around Model B2 following an indicial change in incidence

Figure 8 shows the variations of the upper surface pressures with the time for Model B2 following an indicial change in incidence, in which $C_{pu} = 2(P_u - P_o)/(\rho V^2)$ is the upper surface pressure coefficient, LE and TE denote the leading and the trailing edge, respectively, and P_u and P_o denote the upper surface pressure and the hydrostatic pressure at the position of the pressure tap, respectively. The curves with bold broken lines and bold solid lines in Fig. 8 represent the steady-state pressure distributions before and after the indicial change in incidence. The pressure distribution is characterized by the presence and the rearward movement of peak and trough in pressure (τ = 0.6 ∿ 3.0) and also by the subsequent relatively slow variation of pressure (τ = 4 ∿ 12). It is obvious that the peak and the trough in pressure distribution correspond to the presence of a separation bubble. The indicial aerodynamic moment response which was estimated from the differential pressure distribution was found to correspond qualitatively to the directly measured one.

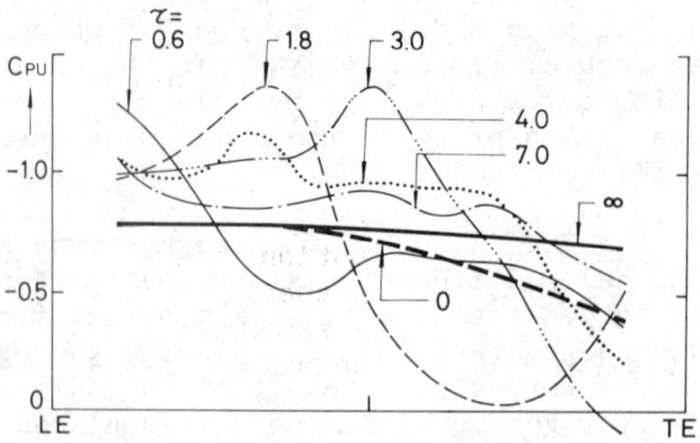

Fig. 8. Temporal development of the upper surface pressures for Model B2 following an indicial change in incidence

It was found that the separation bubble on the upper surface for Model B was much smaller than that for Model B2, and that such a separation bubble could hardly be observed on the lower surface. Again, the indicial aerodynamic moment response which was estimated from the differential pressure distribution was found to correspond qualitatively to the directly measured one. Photo 2 shows successive changes of the flow pattern around Model B2 starting impulsively from rest at an angle of attack of $\alpha = 4$ deg. It should be noted that the formation and the subsequent growth of the separation bubble are very similar to those in Photo 1 in spite of the difference in basic flow.

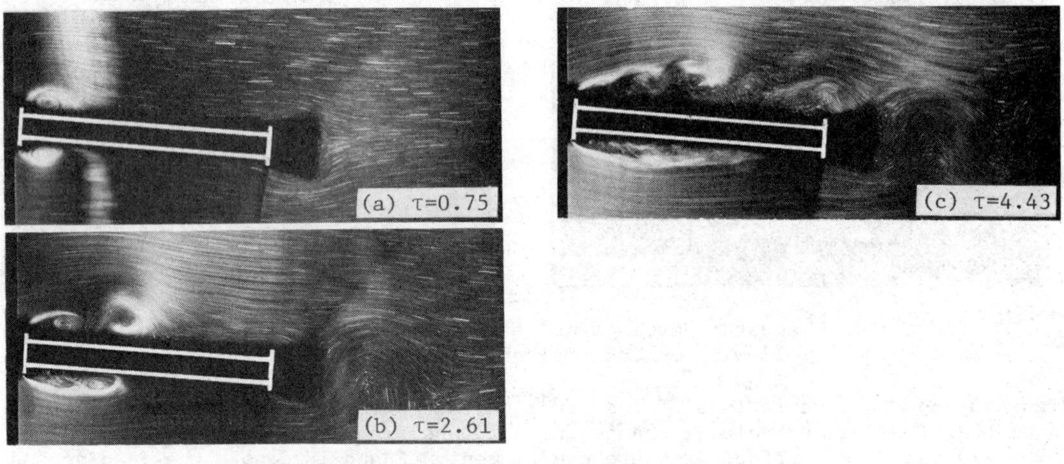

Photo. 2. Temporal development of the flow pattern around Model B2 starting impulsively from rest at an angle of attack of 4 deg

6. CONCLUSIONS

The main results of this paper are summarized in the followings.

1) A theoretical expression for the condition for the onset of torsional flutter in terms of an indicial aerodynamic moment response is obtained on an assumption of a linear aerodynamic system.

2) The aerodynamic torsional moment acting on a bluff prismatic section oscillating torsionally in a uniform flow is decomposed into two terms having physically different origins, that is, one due to the angle of attack motion, and the other due to the angular velocity motion.
3) The measured indicial aerodynamic moment response for a torsionally unstable section with an axis of rotation at the center of the cross section is characterized by an initial abrupt overshoot followed by a rather slow settling down to the steady state condition. This characteristic response satisfies the theoretical condition for the onset of torsional flutter.
4) The term due to the angle of attack motion constitutes a major part of the aerodynamic moment response. Therefore, the fluid memory effect arising from the angle of attack motion is suggested to be responsible for the onset of torsional flutter of bluff prismatic section with an axis of rotation near the center of the cross section.
5) The formation and the subsequent growth of separation bubbles as a result of the change in angle of attack are associated with the unstable aerodynamic moment response that leads to the onset of torsional flutter of bluff prismatic sections.
6) The response for a bluff prismatic section starting impulsively from rest at an angle of attack is similar in trend to the corresponding angle of attack motion response.

REFERENCES:

1. Nakamura, Y. and Mizota, T., Torsional flutter of rectangular prisms, Journal of the Engineering Mechanics Division, Proceedings of the American Society of Civil Engineers 101, EM2: 125-142 (1975)

2. Raggett, J.D., Modeling problems in the study of bridge flutter, thesis presented to Princeton University: (1970)

3. Scanlan, R.H. et al., Indicial aerodynamic functions for bridge decks, Journal of the Engineering Mechanics Division, Proceedings of the American Society of Civil Engineers 100, EM4: 657-672 (1974)

4. Scanlan, R.H., The action of flexible bridges under wind, II--Buffeting theory, Journal of Sound and Vibration 60(2): 201-211 (1978)

5. Nakamura, Y., On the aerodynamic mechanism of torsional flutter of bluff structures, to be published

6. Etkin, B. (1972) Dynamics of Atmospheric Flight, John Wiley & Sons, Inc.

7. Bisplinghoff, R.L. et al. (1957) Aeroelasticity, Addison-Wesley, Inc.

8. Ham, N.D., Aerodynamic loading on a two-dimensional airfoil during dynamic stall, AIAA Journal, the American Institute of Aeronautics and Astronautics 6(10): 1927-1934 (1968)

STOCHASTIC ANALYSIS OF BRIDGE MOTION IN LARGE-SCALE TURBULENT WINDS

Y. K. LIN

University of Illinois at Urbana-Champaign, Urbana, Illinois, USA

ABSTRACT

Using a linear formulation similar to one commonly used for the analysis of the motion of an airfoil, the presence of turbulence in the wind flow is shown to change the self-excited loads and it may cause instability of an otherwise stable motion of a suspension bridge. Furthermore, the self-excited loads and the buffeting loads are statistically correlated since both are related to the same turbulence fields. Therefore, even when the bridge motion remains stable, this correlation must be taken into account in the computation of stochastic moments of the structural response. Numerical examples are given to illustrate the application of this new theory.

INTRODUCTION

Long span bridges are among the more flexible ground structures for which the wind excited motion can be a major cause of failure to warrant a careful analysis at the design state. Since turbulence is always present in natural wind flows, the problem is one of random vibration, and it has been so treated by many investigators. See, for example, Refs. (1), (2), (3). Scanlan and his co-workers (4), (5), (6) have shown that a linear formulation, analogous to one commonly used in the study of airfoil motion in atmospheric turbulence, is applicable to bridges with a lattice or open-truss type deck, and have conducted wind tunnel experiments on bridge deck models to obtain the required aerodynamic constants and functions for such an analysis. In this linear theory the wind loads on a bridge are separated into two groups: the buffeting loads and the self-excited loads. The buffeting loads are independent of the structural motion, whereas the self-excited loads would disappear if the structure were held motionless at all time. Mathematically speaking, the self-excited loads appear in the coefficients of the unknown quantities in the governing equation; therefore, they can cause motion instability. On the otherhand, the buffeting loads appear in the inhomogeneous terms in the equation which do not multiply the unknown quantities. The statistical properties of the buffeting load terms, while affecting the level of structural response, do not change a stable solution to an unstable solution.

In a turbulent wind, the wind velocity is a random function of time. Then the self-excited loads appear in the governing equation as random time-varying coefficients of the unknowns. For such a case, the usual spectral analysis of random vibration is not applicable. To avoid this difficulty, only the mean wind velocity was used in the representation of the self-excited loads in some recent publications (e.g. Refs. 2, 3); thus, the coefficients in the governing equation became independent of time and the convenient spectral procedure could then be carried through.

It may be surmised that turbulence in the wind plays some role in changing the self-excited loads, and consequently changing the stability condition of the structural motion. Furthermore, if this is true, then the self-excited loads and buffeting loads must be statistically correlated since the two types of loads depend on the same turbulence field. These considerations have motivated a series of theoretical investigations at the University of Illinois at Urbana-Champaign (7), (8). In the present paper, these new theoretical developments will be briefly discussed and the major results summarized.

ANALYSIS

Typical scale of turbulence in a natural wind is at least ten times larger than the lateral dimensions of a suspension bridge. Therefore, the wind flow is spatially smooth as "viewed" by the structures at any given instant of time. However, when the turbulence is "carried" by a high convection velocity, the time variation in the total velocity can become fast at any given point on the bridge, resulting in a short correlation time or a broad spectral density. The lowest natural frequency in either torsion or bending of a long-span bridge is in the order of 0.1 - 0.5 Hz, and the structural damping is usually quite low. As a dynamical system such a bridge has a relaxation time (a measure of "memory" of the system) in the order of several minutes. If the correlation time of the random excitation is much shorter than the relaxation time of the system, then the system response, represented in a phase space, is approximately a Markov vector (9).

The above physical reasoning leads to the following simplified assumptions:

1. The aerodynamic constants and indicial functions obtained experimentally for smooth (non-turbulent) two-dimensional flow are applicable to spatially smooth turbulent flow.

2. The effects of turbulence on the self-excited loads can be adequately accounted for in the random variation to the dynamic pressure $1/2\ \rho\ U^2$ where U is the horizontal wind velocity perpendicular to the bridge span (mean horizontal velocity plus random perturbation).

3. The turbulent components in the wind velocity are weakly stationary wide-band random processes, and the bridge deflection and velocity are approximately components of a Markov vector.

4. The buffeting loads are dependent on both the horizontal and vertical velocity components of the turbulence.

Let $U = u[1+\bar{\xi}(t,x)]$ and the vertical component of the wind velocity be $u\bar{\eta}(t,x)$, in which $\bar{\xi}(t,x)$ and $\eta(t,x)$ are weakly stationary wide-band random processes with zero means. This implies that u is the mean wind velocity which is directed perpendicular to the bridge span and in the horizontal direction and that $u\bar{\xi}(t,x)$ and $u\bar{\eta}(t,x)$ are random (turbulence) perturbations from the mean. To concentrate on the main ideas, consider one uncoupled torsional mode alone. The equation of motion may be written as follows:

$$I_s(\ddot{\alpha} + 2\zeta_\alpha\,\omega_\alpha\dot{\alpha} + \omega_\alpha{}^2\alpha) = \frac{1}{2}\,\rho B^2\ [C_M U^2(t,x) + \frac{dC_M}{d\alpha}\,u^2\,\bar{\eta}(t,x)]$$

$$+ \frac{1}{2} \left(\frac{dC_M}{d\alpha}\right) \rho(2B^2) \{X_\alpha U^2(t,x)\dot{\alpha} + U^2(t,x)\alpha + \int_{-\infty}^{t} [X_{M\alpha}(t-\tau)-1]U^2(\tau-x)\frac{\partial\alpha}{\partial\tau}\, d\tau\} \quad (1)$$

in which α = torsional angle; I_s = torsional moment of inertia per unit length; ω_α = natural frequency of a dominant torsional mode; ζ_α = modal damping ratio; ρ = air density; B = bridge deck width; C_M , $\frac{dC_M}{d\alpha}$, X_α are aerodynamic constants; and $X_{M\alpha}(t-\tau)$ is the aerodynamic indicial function which gives the self-excited aerodynamic torsional moment at time t due to an incremental angle of attack at an earlier time τ. In this equation the self-excited loads differ slightly from those used in Refs. (2) and (6) in that the "quasi-steady" term, $U^2\alpha$, has been taken out of the Duhamel type integral. In the approximate theory of quasi-steady aerodynamics, the indicial function $X_{M\alpha}$ takes the constant value of unity, and the integral in Eq. 1 vanishes. Physically, the quasi-steady approximation implies that the self-excited loads depend only on the instantaneous displacements and velocities, not on their past history, of a structure relative to the surrounding flow. Although the two representations of the self-excited loads are equivalent in a deterministic analysis when the wind flow is independent of time, they are not exactly the same in a stochastic analysis. However, the representation here is considered to be physically more meaningful in the stochastic case, and its adoption will also simplify the ensuing analysis.

The experimentally determined aerodynamic indicial function $X_{M\alpha}$ has the form of (6)

$$X_{M\alpha}(t) = 1 + c_1 e^{-\gamma_1 t} + c_3 e^{-\gamma_3 t} \quad (2)$$

in which $\gamma_1 = a_2/B$, $\gamma_3 = a_4/B$, and c_1 , c_3 , a_2 , a_4 are constants. Furthermore, for natural wind flows, it is reasonable to assume that the turbulence perturbations are small compared with the mean velocity. Then $U^2 \approx u^2[1+2\overline{\xi}(t,x)]$. Now let $\phi(x)$ be the dominant torsional mode; that is,

$$\alpha = \overline{\beta}(t)\, \phi(x) \quad (3)$$

and let

$$I = \text{generalized mass} = \int_0^\ell I_s\, \phi^2(x)\, dx$$

$$a_1 = \frac{\rho B^2}{I} \frac{dC_M}{d\alpha} \int_0^\ell \phi^2(x)\, dx \ , \quad a_6 = a_1 u^2$$

$$b_1 = \frac{\rho B^2}{I} C_M \int_0^\ell \phi(x)\, dx \ , \quad b_6 = b_1 u^2$$

$$\omega_o = (\omega_\alpha^{\ 2} - a_6)^{1/2} \ , \quad \zeta_o = (2\zeta_o\omega_\alpha - a_6 X_\alpha)/2\omega_o$$

$$a_7 = a_1 u^2 X_\alpha/\omega_o \ , \quad a_8 = a_6 b_1/\omega_o^{\ 2}$$

$$\xi(t) = \int_0^\ell \bar{\xi}(x,t)\, \phi^2(x)\, dx \Big/ \int_0^\ell \phi^2(x)\, dx$$

$$\eta(t) = \int_0^\ell \bar{\eta}(x,t)\, \phi(x)\, dx \Big/ \int_0^\ell \phi^2(x)\, dx$$

$$\zeta(t) = \int_0^\ell \bar{\xi}(x,t)\, \phi(x)\, dx \Big/ \int_0^\ell \phi(x)\, dx$$

$$\beta = \bar{\beta} - b_6/(2\omega_o^2)$$

Eq. 1 can be cast in the form of

$$\ddot{\beta} + 2\zeta_o\omega_o\dot{\beta} + \omega_o^2\beta = 2a_6\xi(t)\beta + 2\omega_o a_7\xi(t)\dot{\beta}$$
$$+ a_6 \int_{-\infty}^{t} [1 + 2\xi(\tau)]\, [c_1 e^{-\gamma_1(t-\tau)} + c_3 e^{-\gamma_3(t-\tau)}]\, \frac{d\beta(\tau)}{d\tau}\, d\tau$$
$$+ a_8\xi(t) + \frac{1}{2} a_6\eta(t) + b_6\zeta(t) \qquad (4)$$

Denoting $Y_1 = \beta$ and $Y_2 = \dot{\beta}$, Eq. 7 is equivalent to four first order equations:

$$\dot{Y}_1 = Y_2$$
$$\dot{Y}_2 = -\omega_o^2 Y_1 - 2\zeta_o\omega_o Y_2 + a_6(Y_3 + Y_4)$$
$$+ 2a_6\xi(t)Y_1 + 2\omega_o a_7\xi(t)Y_2 + a_8\xi(t) + \frac{1}{2} a_6\eta(t) + b_6\zeta(t)$$
$$\dot{Y}_3 = -\gamma_1 Y_3 + c_1\, [1 + 2\xi(t)]\, Y_2$$
$$\dot{Y}_4 = -\gamma_3 Y_4 + c_3\, [1 + 2\xi(t)]\, Y_2 \qquad (5)$$

These are of the form

$$\dot{Y}_j = f_j + g_{jk}\, \xi_k(t) \qquad (6)$$

in which the usual summation convention of repeated indices in a product is implied and the ξ_k are either $\xi(t)$, $\eta(t)$ or $\zeta(t)$. Since ξ_k are wide-band random processes, Eq. 6 may be converted to the following Itô stochastic differential equations (see e.g. Ref. 10)

$$dY_j = m_j\, dt + \sigma_{jk}\, dW_k \qquad (7)$$

where m_j and σ_{jk} are known as the drift and diffusion coefficients, respectively; W_k are independent unit Wiener processes (also called Brownian motion processes).

We shall assume that the $\xi_k(t)$ in Eqs. 6 are white noise processes. Then the conversion from Eqs. 6 to 7 can be done very simply using the following Stratonovich or Wong and Zakai conversion rule (11), (12)

$$m_j = f_j + \frac{1}{2} D_{kr} \left(\frac{\partial}{\partial Y_\ell} g_{jk}\right) g_{\ell r} \tag{8}$$

$$\sigma_{j\ell}\,\sigma_{k\ell} = D_{rs}\, g_{jr}\, g_{ks} \tag{9}$$

in which D_{kr} is the cross-spectral density of ξ_k and ξ_r per Hz (cycle per second). When k = r the cross-spectral density becomes a spectral density, which is a constant for a white noise process.

A straight forward application of Eq. 8 to the present physical problem results in

$$\begin{Bmatrix} m_1 \\ m_2 \\ m_3 \\ m_4 \end{Bmatrix} = \begin{bmatrix} 0 & 1 & 0 & 0 \\ -\omega_o^2 + 2\omega_o a_6 a_7 D_{\xi\xi} & -2\zeta_o\omega_o + 2(\omega_o a_7)^2 D_{\xi\xi} & a_6 & a_6 \\ 2c_1 a_6 D_{\xi\xi} & c_1(1+2\omega_o a_7 D_{\xi\xi}) & -\gamma_1 & 0 \\ 2c_3 a_6 D_{\xi\xi} & c_3(1+2\omega_o a_7 D_{\xi\xi}) & 0 & -\gamma_3 \end{bmatrix} \begin{Bmatrix} Y_1 \\ Y_2 \\ Y_3 \\ Y_4 \end{Bmatrix} + D \begin{Bmatrix} 0 \\ \omega_o a_7 \\ c_1 \\ c_3 \end{Bmatrix}$$

$$= [F]\,\{Y\} + \{R\} \tag{10}$$

where $D = a_8 D_{\xi\xi} + \frac{1}{2} a_6 D_{\xi\eta} + b_6 D_{\xi\zeta}$. Using Eq. 9, we obtain the elements of the product matrix $[\sigma]\,[\sigma]'$ as follows:

$$([\sigma]\,[\sigma]')_{1k} = 0 \ , \quad 1 \le k \le 4,$$

$$\begin{aligned}([\sigma]\,[\sigma]')_{22} &= 4D_{\xi\xi}(a_6^2 Y_1^2 + 2\omega_o a_6 a_7 Y_1 Y_2 + \omega_o^2 a_7^2 Y_2^2) \\ &\quad + 4a_6 D Y_1 + 4\omega_o a_7 D Y_2 + (a_8^2 D_{\xi\xi} + \frac{1}{4} a_6^2 D_{\eta\eta} \\ &\quad + b_6^2 D_{\zeta\zeta} + a_6 a_8 D_{\xi\eta} + 2b_6 a_8 D_{\xi\zeta} + a_6 b_6 D_{\eta\zeta})\end{aligned}$$

$$([\sigma]\,[\sigma]')_{23} = 4D_{\xi\xi}(c_1 a_6 Y_1 Y_2 + \omega_o c_1 a_7 Y_2^2) + 2c_1 D Y_2$$

$$([\sigma]\,[\sigma]')_{24} = 4D_{\xi\xi}(c_3 a_6 Y_1 Y_2 + \omega_o c_3 a_7 Y_2^2) + 2c_3 D Y_2$$

$$([\sigma]\,[\sigma]')_{33} = 4c_1^2 Y_2^2 D_{\xi\xi} \ , \quad (\sigma\,\sigma')_{34} = 4c_1 c_3 Y_2^2 D_{\xi\xi} \ ,$$

$$([\sigma]\,[\sigma]')_{44} = 4c_3^2 Y_2^2 D_{\xi\xi}. \tag{11}$$

It is only necessary to show the ten elements $([\sigma]\,[\sigma]')_{jk}$ for $1 \le j \le k \le 4$ since this matrix is symmetrical.

Equations for the first moments, $\mu_j = E[Y_j]$, are obtained by taking the ensemble average of Eqs. 7

$$\dot{\mu}_j = E[m_j] \tag{12}$$

Note that the ensemble average of $\sigma_{jk}\ dW_k$ is zero, a consequence of Markov properties. Therefore, the diffusion terms in Eqs. 7 do not contribute to the first moments. Substituting Eq. 10 into Eq. 12, we obtain

$$\{\dot{\mu}\} = [F]\ \{\mu\} + \{R\} \tag{13}$$

This is solved readily

$$\{\mu\} = e^{[F]t}\ \{\mu_o\} + \int_0^t e^{[F](t-\tau)}\ \{R\}\ d\tau \tag{14}$$

in which $\{\mu_o\}$ is a column matrix of initial conditions. It is clear that the first moments are unstable if the real part of any eigenvalue of [F] is positive.

If the motion is stable, the response tends to be stationary as the time of exposure to stationary excitations increases, and $\{\mu\}$ becomes independent of time. Thus, at the state of stationarity,

$$\{\mu\} = -[F]^{-1}\ \{R\} \tag{15}$$

Of course, only the first and the second elements in μ; i.e. $\mu_1 = E[\beta]$ and $\mu_2 = E[\dot{\beta}]$ are of practical interest. The ensemble average of the total modal deflection

$$E[\bar{\beta}] = \mu_1 + b_6/(2\omega_o^2) \tag{16}$$

To obtain a set of equations for the second moments, we apply Itô's differential rule (13) which states that if ϕ is a scalar function of a Markov vector $\{Y\}$ and is twice differentiable with respect to Y_j, then

$$d\phi = [\ \frac{\partial\phi}{\partial t} + m_j\ \frac{\partial\phi}{\partial Y_j} + \frac{1}{2}\sigma_{j\ell}\ \sigma_{k\ell}\ \frac{\partial^2\phi}{\partial Y_j \partial Y_k}\ ]\ dt + \sigma_{jk}\ \frac{\partial\phi}{\partial Y_j}\ dW_k \tag{17}$$

In particular, when $\phi = Y_j\ Y_k$,

$$d(Y_j Y_k) = (m_j Y_k + m_k Y_j + \sigma_{j\ell}\ \sigma_{k\ell})\ dt + (\sigma_{j\ell} Y_k + \sigma_{k\ell} Y_j)\ dW_\ell \tag{18}$$

Equations for the second moments, $\mu_{jk} = E[Y_j\ Y_k]$, are obtained by taking the ensemble average of Eqs. 18. Again, the ensemble average involving dW_ℓ is zero; and m_j, m_k and $\sigma_{j\ell}\ \sigma_{k\ell}$ are found from Eqs. 10 and 11.

Since the second moments are symmetrical; i.e., $\mu_{jk} = \mu_{kj}$, it is only necessary to investigate the ten different combinations $1 \leq j \leq k \leq 4$. Now, if μ_{jk} are arranged in the order $\mu_{11} = \nu_1$, $\mu_{22} = \nu_2$, $\mu_{33} = \nu_3$, $\mu_{44} = \nu_4$, $\mu_{12} = \nu_5$, $\mu_{13} = \nu_6$, $\mu_{14} = \nu_7$, $\mu_{23} = \nu_8$, $\mu_{24} = \nu_9$, $\mu_{34} = \nu_{10}$, the equations for the second moments can be cast in the form of

$$\{\dot{\nu}\} = [H]\{\nu\} + D[L]\{\mu\} + \{S\} \tag{19}$$

in which the elements of the 10 x 10 matrix [H], 10 x 4 matrix [L] and 10 x 1 matrix {S} are given by

$$H_{1,5} = 2 \ , \ H_{2,1} = 4a_6^2 D_{\xi\xi} \ , \ H_{2,2} = 2(-2\zeta_o\omega_o + 4\omega_o^2 a_7^2 D_{\xi\xi}) \ ,$$

$$H_{2,5} = 2(-\omega_o^2 + 6\omega_o a_6 a_7 D_{\xi\xi}) \ , \ H_{2,8} = H_{2,9} = 2a_6 \ , \ H_{3,2} = 4c_1^2 D_{\xi\xi}) \ ,$$

$$H_{3,3} = -2\gamma_1 \ , \ H_{3,6} = 4c_1 a_6 D_{\xi\xi} \ , \ H_{3,8} = 2c_1(1+2\omega_o a_7 D_{\xi\xi}) \ , \ H_{4,2} = 4c_3^2 D_{\xi\xi} \ ,$$

$$H_{4,4} = -2\gamma_3 \ , \ H_{4,7} = 4c_3 a_6 D_{\xi\xi} \ , \ H_{4,9} = 2c_3(1+2\omega_o a_7 D_{\xi\xi}) \ ,$$

$$H_{5,1} = -\omega_o^2 + 2\omega_o a_6 a_7 D_{\xi\xi} \ , \ H_{5,2} = 1 \ , \ H_{5,5} = -2\zeta_o\omega_o + 2\omega_o^2 a_7^2 D_{\xi\xi} \ ,$$

$$H_{5,6} = H_{5,7} = a_6 \ , \ H_{6,1} = 2c_1 a_6 D_{\xi\xi} \ , \ H_{6,5} = c_1(1+2\omega_o a_7 D_{\xi\xi}) \ , \ H_{6,6} = -\gamma_1 \ ,$$

$$H_{6,8} = 1 \ , \ H_{7,1} = 2c_3 a_6 D_{\xi\xi} \ , \ H_{7,5} = c_3(1+2\omega_o a_7 D_{\xi\xi}) \ , \ H_{7,7} = -\gamma_3 \ , \ H_{7,9} = 1 \ ,$$

$$H_{8,2} = c_1(1+6\omega_o a_7 D_{\xi\xi}) \ , \ H_{8,3} = a_6 \ , \ H_{8,5} = 6c_1 a_6 D_{\xi\xi} \ ,$$

$$H_{8,6} = -\omega_o^2 + 2\omega_o a_6 a_7 D_{\xi\xi} \ , \ H_{8,8} = -2\zeta_o\omega_o + 2\omega_o^2 a_7^2 D_{\xi\xi} - \gamma_1 \ ,$$

$$H_{8,10} = a_6 \ , \ H_{9,2} = c_3(1+6\omega_o a_7 D_{\xi\xi}) \ , \ H_{9,4} = a_6 \ , \ H_{9,5} = 6c_3 a_6 D_{\xi\xi} \ ,$$

$$H_{9,7} = -\omega_o^2 + 2\omega_o a_6 a_7 D_{\xi\xi} \ , \ H_{9,9} = -2\zeta_o\omega_o + 2\omega_o^2 a_7^2 D_{\xi\xi} - \gamma_3 \ , \ H_{9,10} = a_6 \ ,$$

$$H_{10,2} = 4c_1 c_3 D_{\xi\xi} \ , \ H_{10,6} = 2c_3 a_6 D_{\xi\xi} \ , \ H_{10,7} = 2c_1 a_6 D_{\xi\xi} \ ,$$

$$H_{10,8} = c_3(1+2\omega_o a_7 D_{\xi\xi}) \ , \ H_{10,9} = c_1(1+2\omega_o a_7 D_{\xi\xi}) \ ; \ H_{10,10} = -(\gamma_1 + \gamma_3) \ ,$$

$$L_{2,1} = 4a_6 \ , \ L_{2,2} = 6\omega_o a_7 \ , \ L_{3,3} = 2c_1 \ , \ L_{4,4} = 2c_3 \ , \ L_{5,1} = \omega_o a_7 \ ,$$

$$L_{6,1} = c_1 \ , \ L_{7,1} = c_3 \ , \ L_{8,2} = 3c_1 \ , \ L_{8,3} = \omega_o a_7 \ , \ L_{9,2} = 3c_3 \ ,$$

$$L_{9,4} = \omega_o a_7 \ , \ L_{10,3} = c_3 \ , \ L_{10,4} = c_1 \ ,$$

$$S_{2,1} = a_8^2 D_{\xi\xi} + \frac{1}{4} a_6^2 D_{\eta\eta} + b_6^2 D_{\zeta\zeta} + a_6 a_8 D_{\xi\eta} + 2b_6 a_8 D_{\xi\zeta} + a_6 b_6 D_{\eta\zeta}$$

All the other elements of [H], [L] and {S} are zero. The second moments are stable if the real part of every eigenvalue of [H] is negative, and the stability boundary is reached when the real part of one or several eigenvalues become zero.

It is seen from Eq. 19 that the values of the second moments are related to those of the first moments. If the second moments are stable (which implies that the first moments are also stable), they will approach to constants as the time of exposure to the stationary excitations increases. At the state of stationarity

$$\{\nu\} = -[H]^{-1}(D[L]\{\mu\} + \{S\}) \tag{20}$$

NUMERICAL EXAMPLE

To illustrate the application of the theory, numerical calculations has been carried out using the following physical data:

Structural Parameters:

$$B = 16.4\text{m},\quad I = 8.98 \times 10^5\ \text{kg-m}^2/\text{m},\quad \ell = 490\ \text{m},\quad \zeta_\alpha = 0.01,$$

$$\omega_o = 0.74\ \pi\ \text{rad/sec.},\quad \phi(x) = \sin\frac{\pi x}{\ell}.$$

Aerodynamic Parameters:

$$\rho = 1.226\ \text{kg/m}^3,\quad X_\alpha = 1.52,\quad C_M = 0.05,\quad \frac{\partial C_M}{\partial \alpha} = 0.93,$$

$$c_1 = 1.64,\quad c_3 = -51.61,\quad \gamma_1 = \frac{0.38u}{B},\quad \gamma_3 = \frac{19.74u}{B}$$

These were taken from Ref. (2), presumably corresponding to an existing bridge. In addition, the horizontal and vertical components of the wind turbulence were assumed to be totally correlated in the span-wise direction individually, but totally uncorrelated of one another; i.e. $\bar{\xi}(x,t)$ and $\bar{\eta}(x,t)$ were independent of x but $\bar{\xi}(x,t)$ was uncorrelated with $\bar{\eta}(x,t)$. These assumptions were made arbitrarily to facilitate the calculation. The theory as presented herein is, of course, applicable to any torsional mode, as well as accounting for any type of correlation for $\bar{\xi}$ and $\bar{\eta}$ processes. Having chosen a particular bridge, a particular mode of vibration, and particular statistical relationships for the turbulence field, the structural response then depends on the mean flow velocity u and the spectral levels $D_{\xi\xi}$ and $D_{\eta\eta}$ of the normalized random excitations. For the particular turbulence field assumed, we also have $D_{\xi\zeta} = D_{\zeta\zeta} = D_{\xi\xi}$ and $D_{\xi\eta} = D_{\eta\zeta} = 0$.

The stability boundaries for the first and second moments of the structural response are shown in Fig. 1 as functions of the spectral density $D_{\xi\xi}$ of the normalized horizontal turbulence and the ratio of the mean wind velocity u to u_{cr}; the latter is the critical wind velocity under an ideal non-turbulence condition. For the particular bridge and the excitation field investigated, turbulence has little effect on the first moment stability. However, the first moment stability only assures that the mean response is bounded while excursion from the mean can still be unbounded. Therefore, the second moment stability condition is much more useful in engineering applications.

The computed first moment $\mu_1 = E[\beta]$ is plotted in Fig. 2 for several u/u_{cr} ratios and a spectral density ratio $D_{\eta\eta}/D_{\xi\xi} = \frac{1}{2}$. It is of interest to note that $E[\beta] \neq 0$ although the ξ and η processes have zero means. The non-zero first moment for the response is due to the choice of a non-zero C_M and the correlation between the self-excited loads and the buffeting loads. If the effect of turbulence on the self-excited loads were neglected, the average response would be zero.

Fig. 3 shows the plots for the second moment $\nu_1 = E[\beta^2]$. Within the range $0 \leq D_{\xi\xi} \leq 0.1$ shown here, the second moment is stable for lower u/u_{cr} ratios (e.g. 0.6 and 0.8). It becomes unstable at $D_{\xi\xi} = 0.040$ for $u/u_{cr} = 0.95$ and at

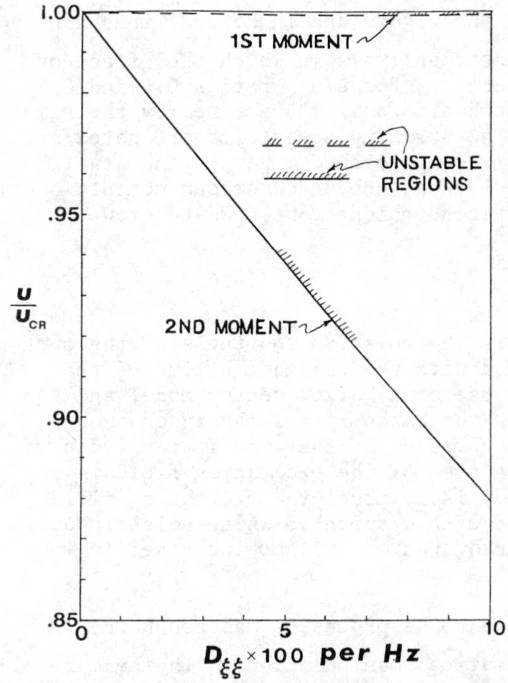

Fig. 1. Stochastic stability boundaries for bridge torsional motion in large-scale turbulent winds.

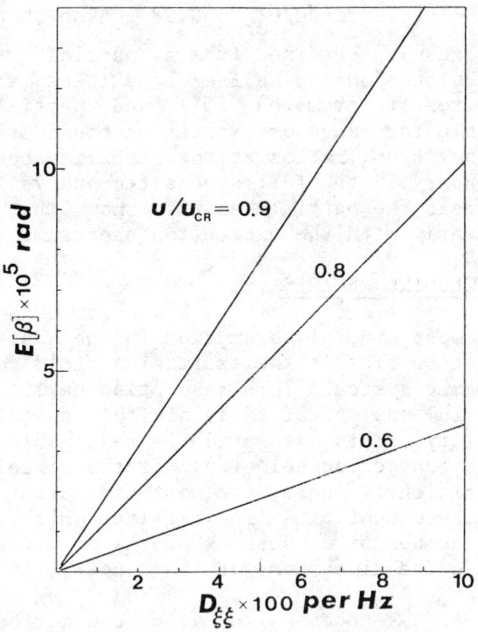

Fig. 2. First moments of bridge torsional motion in large-scale turbulent wind computed for $D_{\eta\eta}/D_{\xi\xi} = 1/2$ and $D_{\xi\eta} = 0$.

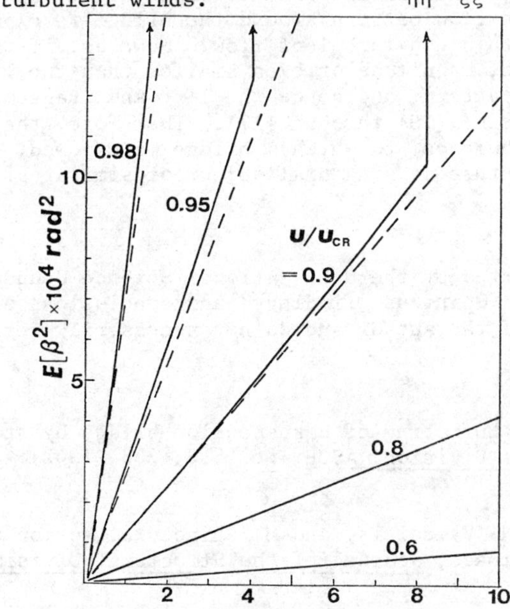

Fig. 3. Second moments of bridge torsional motion in large-scale turbulent wind computed for $D_{\eta\eta}/D_{\xi\xi} = 1/2$ and $D_{\xi\eta} = 0$.

——— Full stochastic analysis

- - - Partial stochastic analysis

$D_{\xi\xi} = 0.015$ for $u/u_{cr} = 0.98$. Also shown in the figure and in dotted lines are the results obtained from a "partial" stochastic analysis in which the effect of turbulence on the self-excited loads is neglected. For u/u_{cr} ratios 0.6 and 0.8 the results from the "full" and "partial" stochastic analyses are nearly the same within the range of excitation considered. The striking departures are noted for higher u/u_{cr} ratios as the stability boundary is approached. Beyond the stability boundary, the full stochastic analysis predicts an unbounded second moment whereas the partial analysis shows that the second moment continues to grow linearly with the excitation spectrum.

CONCLUDING REMARKS

The most crucial assumption in the analysis is the relative shortness of the correlation time in the excitation field compared with the relaxation time of the dynamic system. This assumption permits the use of a Markov vector model and the special analytical tools of Itô's calculus in the mathematical theory of probability. This assumption is reasonable if turbulence in the wind is carried by a high convection velocity. If the correlation time of the excitation field is not sufficiently short, the most convenient remedy is to model the excitation field as the output of a linear filter which is fed with a suitable white noise input. The number of dimensions of the resulting Markov vector will be increased to include the dimensions of the linear filter.

The $\xi_k(t)$ processes in Eq. 6 need not be white noise processes, although the white noise assumption simplifies some mathematical manipulations. In the more general case, the method of stochastic averaging (9), (14) must be used.

When the scale of turbulence in the wind is much larger than the lateral dimensions of a bridge, the flow pattern around the bridge is expected to be similar to that associated with a non-turbulent flow. However, if the turbulence scale is of the same order of magnitude or even smaller than the lateral dimensions of the bridge, the flow pattern, and hence the lift and drag curves, can change greatly as shown in Refs. (15) through (17). Therefore, the present theory is not applicable if the turbulence, to which a bridge is exposed, is generated from the wakes of another structure or a natural object of similar size.

ACKNOWLEDGMENT

The research is supported by the U.S. National Science Foundation under Grant No. NSF PFR 76-84171 A01. Opinions, findings, and conclusions or recommendations expressed are those of the author and do not necessarily reflect the views of NSF.

REFERENCES

(1) A. G. Davenport, Buffeting of a suspension bridge by stormy winds, Journal of the Structural Division, ASCE, 88, ST3, 233 (1962).

(2) J. G. Beliveau, R. Vaicaitis, and M. Shinozuka, Motion of suspension bridge subject to wind loads, Journal of the Structural Division, ASCE, 103, ST6, 1189 (1977).

(3) R. H. Scanlan, and R. H. Gade, Motion of suspended bridge spans under gusty wind, Journal of the Structural Division, ASCE, 103, ST9, 1867 (1977).

(4) R. H. Scanlan, and A. Sabzevari, Experimental aerodynamic coefficients in

the analytical study of suspension bridge flutter, Journal of Mech. Eng. Science, Institution of Mech. Engrs., London, 11, 234 (1969).

(5) R. H. Scanlan, and J. J. Tomko, Airfoil and bridge deck flutter derivatives, Journal of the Engineering Mechanics Division, ASCE, 97, EM6, 1717 (1971).

(6) R. H. Scanlan, J. G. Beliveau, and K. S. Budlong, Indicial aerodynamic functions for bridge decks, Journal of the Engineering Mechanics Division, ASCE, 100, 657 (1974).

(7) Y. K. Lin, and S. T. Ariaratnam, Stability of bridge motion in turbulent wind, to appear.

(8) Y. K. Lin, Response of suspension bridges to turbulent winds, to appear.

(9) Stratonovich, R. L., Topics in the Theory of Random Noise, Vol. 2, Translated by R. A. Silverman, Gordon and Breach, New York, NY, 1967.

(10) Arnold, L., Stochastic Differential Equations: Theory and Applications, John Wiley and Sons, Inc., New York, NY, 1974.

(11) R. L. Stratonovich, A new representation for stochastic integral and equations, Journal of SIAM Control, 4, 362 (1966).

(12) E. Wong, and M. Zakai, On the relation between ordinary and stochastic equations, International Journal of Engineering Science, 3, 213 (1965).

(13) K. Itô, On stochastic differential equations, Memoirs of the American Mathematical Society, 4, 51, (1951).

(14) R. Z. Khasminskii, A limit theorem for the solution of differential equations with random right hand sides, Theory of Probability and Applications, 11, 390 (1966).

(15) P. W. Bearman, An investigation of the forces on flat plates normal to a turbulent flow, Journal of Fluid Mechanics, 46, Part I, 177 (1971).

(16) J. C. R. Hunt, Turbulent velocities near and fluctuating surface pressures on structures in turbulent winds, Proceedings of the 4th International Conference on Wind Effects on Buildings and Structures, Edited by K. J. Eaton, Cambridge University Press, London 309 (1977).

(17) A. Laneville, I. S. Gartshore, and G. V. Parkinson, An explanation of some effects of turbulence on bluff bodies, Proceedings of the 4th International Conference on Wind Effects on Buildings and Structures, Edited by K. J. Eaton, Cambridge University Press, London 333 (1977).

GUST RESPONSE FACTORS FOR TRANSMISSION LINE LOADING

A. G. DAVENPORT
The University of Western Ontario, London, Ontario, Canada

SUMMARY

The objective of the paper is to derive simple factors for estimating the response of a transmission line system, consisting of towers, conductors and ground wires, to gusty wind. The approach is based on the statistical methods which takes account of the spatial correlation and energy spectrum of windspeed and the dynamic response of the transmission line system. The main focus is on the transverse loading. The influence of the variation in drag coefficient is discussed.

INTRODUCTION

The importance of wind loading (as well as the combination of wind and ice) in the design of transmission line systems renders the accurate estimation of these loads economically significant. Historically, one of the imponderables has been the action of gusts. It was recognized that the extreme gusts do not envelope the entire span, and that some reduction reflecting the spatial extent of gusts might be anticipated; but at the same time the possibility of some resonant response and dynamic amplification would tend to aggravate the loads and offset the spatial reductions. The seriousness of both these trends would depend on the characteristics of the turbulence.

A framework for investigating this problem of the response of structures to turbulent wind was provided by the statistical methods suggested by the writer[1,2]. These methods are ideally suited to transmission line structures and have been expanded on in Europe, Japan and elsewhere[3,4,5,6,7]. They also found application in codes for buildings in the form of so-called "gust response factors"[8].

The purpose of this note is to derive simplified gust response factors for a tower and line system. These consider the frequency response characteristics of the conductors as well as the tower and the spatial characteristics of gusts. The influence of terrain roughness is taken into account. Only transverse loading is considered herein; longitudinal loading is however amenable to the same treatment.

DEFINITIONS

Although it is the ultimate objective to consider the response of a total transmission line system consisting of towers, conductors and overhead ground wires we shall initially consider the simpler system shown in Figure 1 consisting of only a single conductor and towers. The following defines properties of the towers, conductor and the wind required in the analysis:

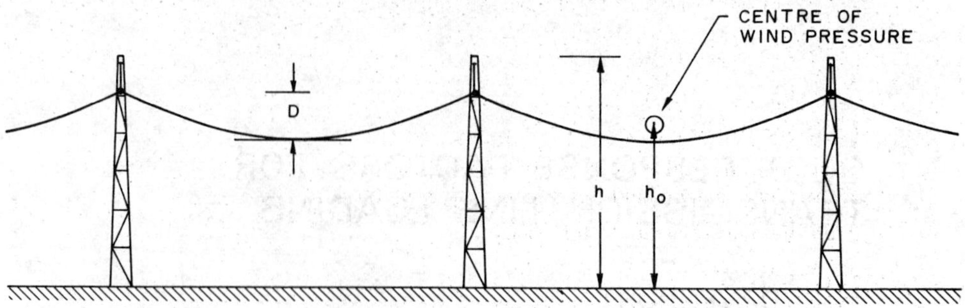

Fig. 1 Definition of transmission line system

Towers	Height above ground – h_t	(T1)
	Fundamental transverse frequency of free-standing tower – f_t	(T2)
	Structural damping – ζ_t	(T3)
	Drag coefficient – C_{D_t}	(T4)
Conductor	Typical span – L;	(C1)
	Sag – D;	(C2)
	Mass/unit length – M_c;	(C3)
	Fundamental frequency for horizontal sway – $f_c = \frac{1}{2\pi}\sqrt{\frac{3g}{2D}}$;	(C4)
	Cable diameter – d;	(C5)
	Average cable density – ρ_c;	(C6)
	Drag coefficient – C_{D_c} – generally a function of Reynolds number	(C7)
Wind[9]	Effective height (at approximate centre of pressure of structure) – h_o	(W1)
	Mean wind speed at effective height – U_o	(W2)
	Mean wind speed at reference height, z_{ref} – U_{ref}	(W3)
	Velocity variation with height, z, $\frac{U(z)}{U_{ref}} = \left(\frac{z}{Z_{ref}}\right)^a$ or $\frac{U(z)}{U_{ref}} = \frac{\sqrt{K}}{4}\ ln\,(z/z_o)$	(W4)
	where U_{ref} is the mean velocity at a reference height z_{ref}.	
	Mean square longitudinal turbulent wind speed fluctuation $\sigma_u^2 = 6\,K\,U_{ref}^2$	
	where K is surface drag coefficient	(W5)
	Representative terrain roughness factors:	(W6)

	Gradient Height Z_G (m)	Power Law Exponent a	Surface Drag Coefft. K (z_{ref} = 10 m)
Open country, flat shorelines	250	.10	.0015
Farmland, scattered trees and buildings	300	.16	.005
Woodland, Suburbs	400	.28	.015

Spectrum of horizontal turbulence at frequency, f,

$$\frac{f\,S_u(f)}{K\,U_{ref}^2} = .28\left(\frac{fz}{\bar{U}_z}\right)^{-2/3} \qquad (W7)$$

(valid for higher frequencies $f > \bar{U}_z/z$)

Transverse integral scale of turbulence L_S; typically $L_S \approx 65$ m. (increasing slightly with height and decreasing with roughness) (W8)

Transverse correlation of turbulence:

$$R_u(x, x') = exp - |x - x'|/L_S \qquad (W9)$$

Transverse narrow band correlation of turbulence:

$$R_u(x, x'; f) = exp - cf|x - x'|/\bar{U}_o \qquad (W10)$$

where typically $c = 8$

MEAN RESPONSE

Consider some response R of the tower (base bending moment for example). Suppose that the mean wind forces on the tower and conductor at the reference height h_o are respectively

$$\bar{p}_t = \tfrac{1}{2}\rho\,\bar{U}_o^2\,C_{D_t}$$
$$\bar{p}_c = \tfrac{1}{2}\rho\,\bar{U}_o^2\,C_{D_c} \qquad (1)$$

Let us define influence functions θ_{tt} and θ_{tc} such that the mean response R of the tower in this wind is given by

$$\bar{R} = \theta_{tt}\,\bar{p}_t + \theta_{tc}\,\bar{p}_c \qquad (2)$$

The values of θ_{tt} and θ_{tc} clearly depend on the geometry of the tower, its structural properties, its aerodynamic properties and the form of the velocity variation with height. Since the effects are static they may be found straight forwardly and are not further elaborated on here.

DYNAMIC RESPONSE

To simply matters we will, in considering the dynamic response, assume that the properties of the wind field are homogeneous over the height of the line and characterized by the actual properties at the reference height h_o. This height is likely to be close to the centre of mass of the conductor i.e. 2/3 D below the suspension point. This simplification will not involve significant errors.

In estimating the fluctuating response we make an important observation and assumption concerning the dynamic properties of the tower and conductors – namely that they are only lightly coupled dynamically and that the free standing tower frequency f_T is much greater than the fundamental frequency of the conductor swing. This assumption means that the tower motion has negligible effect on the conductor behaviour and that the tower "sees" the wind forces transmitted at the cable suspension point as quasi-static forces, at frequencies well below f_T.

With these assumptions the spectra of the responses are as indicated in Figure 2. We can represent the mean sqaure fluctuating response of the tower in the form

$$\sigma_R^2 = B_c + R_c + B_t + R_t \qquad (3)$$

in which the four components of response are as follows:

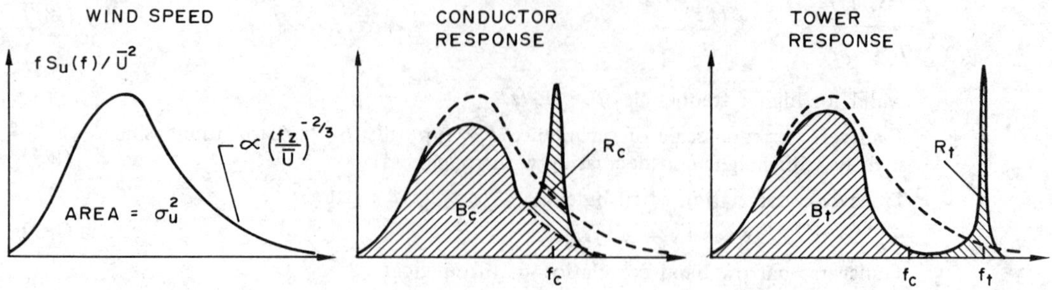

Fig. 2 Spectra of wind speed and conductor and tower responses

B_c – the mean square response to the background, quasi-static, wind loads acting on the conductors at frequencies $\ll f_c$. These loads are only partially correlated over the spans.

R_c – the mean square resonant response of the conductor at frequencies close to f_c.

B_t – the mean square response to the background, quasi-static wind loads acting on the tower, at frequencies $\ll f_t$. These are only partially correlated over the height h.

R_t – the mean square resonant response of the tower at frequenices at or near f_t.

This expression is not exact inso far as it neglects the cross-coupled terms due to the correlated wind loads on the tower and conductor. This is not considered a significant omission and the point is referred to later. Because of their proximity loads on conductors are assumed fully correlated.

With this description of the mean square response σ_R^2, the peak response R can then be expressed as:

$$R = \bar{R} + g\,\sigma_R \tag{4}$$

where g is a statistical factor, weakly dependent on the frequency characteristic of the response and the sampling interval. For present purposes, assuming sampling times of the mean wind of the order of 10 min., appropriate values of g are in the range $3.5 < g < 4.0$.

We now need to evaluate the component factors B_c, R_c, B_t, R_t from the external parameters. The derivation in detail would follow the procedures described in references 4, 5 and 6. Following these results we find to a close approximation:

$$B_c = \theta_{tc}^2 \quad 4\,\bar{p}_c^2 \,\frac{\sigma_u^2}{U_o^2}\, \frac{1}{1 + \frac{4}{\pi^2}\frac{2L}{L_S}}$$

This simplifies, using statement W5 above, to:

$$B_c = \quad 24\,K\left(\frac{z_{ref}}{h_o}\right)^{2a} \bar{p}_c^{2}\;\theta_{tc}^{2} \quad \frac{1}{1 + \frac{8L}{L_S}} \tag{5}$$

$$R_c = \theta_{tc}^2 \quad 4\,\overline{p}_c^{\,2} \quad \frac{f_c\, S_u(f_c)}{U_o^2} \quad \frac{\pi}{4\zeta} \quad \frac{U}{c f_c L} \quad \frac{\pi^2}{4}$$

Using statements W5 and W7 this simplifies to:

$$R_c = \quad 24\,K\left(\frac{z_{ref}}{h_o}\right)^{2a} \overline{p}_c^{\,2} \quad \theta_{tc}^2 \quad .0113\,\left(\frac{f_c\, h_o}{\overline{U}_o}\right)^{-5/3} \frac{h_o}{L} \quad \frac{1}{\zeta} \tag{6}$$

In this ζ = total damping

= structural damping + aerodynamic damping.

For the conductor, aerodynamic damping is likely to be significantly greater than the structural damping which is also difficult to ascertain. For these reasons, we will take

$$\zeta \approx \zeta_{aero}$$

$$= \frac{\overline{p}_c d}{2\,\pi f_c\, m_c\, \overline{U}_o} \quad \approx \quad \frac{\tfrac{1}{2}\,\rho_{air}\, \overline{U}_o^{\,2}\, C_{D_c}\, d}{2\pi\, f_c\, \rho_c\, \frac{\pi\, d^2}{4}\, U_o}$$

$$= \frac{1}{\pi^2} \quad \frac{\rho_{air}}{\rho_c} \quad \frac{\overline{U}_o}{f_c d} \quad C_{D_c} \tag{7}$$

We note here that typically

$\rho_{air} \approx .0024\ slugs/ft^3$ and $\rho_c \approx 5\ slugs/ft^3$ so $\rho_{air}/\rho_c \approx .00048$ and for these values

$$\zeta \approx \frac{.00048}{\pi^2} \quad \frac{U_o}{f_c d} \quad C_{D_c}$$

$$= 4.8 \times 10^{-5}\ U_o/f_c d\ \ C_{D_c} \tag{8}$$

Turning to the tower,

$$B_t = \theta_{tt}^2 \quad 4\,p_c^2 \quad \frac{\sigma_u^2}{U_o^2} \quad \frac{1}{1 + .375\,\frac{h}{L_S}}$$

Substituting from (W5)

$$B_t = \quad 24\,K\left(\frac{z_{ref}}{h_o}\right)^{2a} \quad p_t^2 \quad \theta_{tt}^2 \quad \frac{1}{1 + .375\,\frac{h}{L_S}} \tag{9}$$

Finally

$$R_t = \theta_{tt}^2 \; 4\, p_c^2 \; \frac{f_t\, S_u(f_t)}{U_o^2} \; \frac{\pi}{4\,\zeta_t} \; \frac{\bar{U}_o}{.375\, c\, f_t\, h_o}$$

$$= 24\, K \left(\frac{z_{ref}}{h_o}\right)^{2a} p_t^2\, \theta_{tt}^2 \; .0123 \left(\frac{f_t\, h_o}{\bar{U}_o}\right)^{-5/3} \frac{1}{\zeta_t} \tag{10}$$

Generally ζ_t 1 – 3% and aerodynamic damping is less important.

To clean up the algebra we will write:

$$B_c^* = \frac{1}{1 + .8\, L/L_S} \tag{11}$$

$$R_c^* = .0113 \left(\frac{f_c\, h_o}{\bar{U}_o}\right)^{-5/3} \frac{h_o}{L} \frac{1}{\zeta_c} \tag{12}$$

$$B_t^* = \frac{1}{1 + .375\, h/L_S} \tag{13}$$

$$R_t^* = .0123 \left(\frac{f_c\, h_o}{\bar{U}_o}\right)^{-5/3} \frac{1}{\zeta_t} \tag{14}$$

and $E^2 = 24\, K \left(\frac{z_{ref}}{h_o}\right)^{2a}$ (an exposure factor) (15)

A similar expression may be given if the logarithmic profile is used.

Then

$$\sigma_R^2 = E^2 \; \theta_{tc}^2\, p_c^2\, (B_t^* + R_c^*) \tag{16}$$

The peak response follows as before from equation (4). In this expression all the conductors have been "lumped" together. They could equally well have been treated separately in which case the term for the conductor would be replaced by a sum. The addtional complexity would not seem justified in a simple approach and the improvement in accuracy could hardly be justified. In effect, for purpose of the simplified analysis, the conductors have been constrained to more in phase with one another.

In the above analysis the possibility of tower-conductor correlation has been ignored. This only arises when considering the low frequency, quasi-steady loads B_c and B_t. Since the bulk of the cable excitation is at mid span and remote from the tower this is not likely to be a significant component and, except for very short spans ($\approx L_S$), can be reasonably ignored. Also, the conductors are constrained so their motion is fully correlated which may slightly exaggerates the real situation. These two effects as well as being small should also compensate one another. This simplifies the resulting equation.

GUST RESPONSE FACTOR

The peak response is given in equation (4) and can be written as

$$R = R + g\, \sigma_R$$

$$= \theta_{tc}\,\overline{p}_c + \theta_{tt}\,\overline{p}_c + g\,E\,[\theta_{tc}^2\,\overline{p}_c^2\,(B_c^* + R_c^*) + \theta_{tt}^2\,\overline{p}_t^2\,(B_t^* + R_t^*)]^{1/2} \qquad (17)$$

If the terms for the conductors and tower in the square root are of similar magnitude we may use the approximation

$$(A^2 + B^2)^{1/2} \approx \epsilon\,(A + B) \text{ where } \epsilon \approx .75$$

and rewrite

$$R = \theta_{tc}\,\overline{p}_c\,[1 + g\,\epsilon\,E\sqrt{B_c^* + R_c^*}] + \theta_{tt}\,\overline{p}_t\,[1 + g\,\epsilon\,E\sqrt{B_t^* + R_t^*}]$$

$$= \theta_{tc}\,\overline{p}_c\,G_c + \theta_{tt}\,\overline{p}_t\,G_t \qquad (18)$$

where G_c and G_t are "gust response factors" given by

$$G_c = 1 + g\,\epsilon\,E\sqrt{B_c^* + R_c^*} \qquad (19)$$

$$G_t = 1 + g\,\epsilon\,E\sqrt{B_t^* + R_t^*} \qquad (20)$$

SUSPENDER FORCES

The forces R_{SA} in the suspender arms may also be determined from the above. The detailed development will give

$$R_{SA} = p_c\,d\,L\,(1 + g\,E\,\sqrt{B_c^* + R_c^*}) \qquad (21)$$

"SPAN FACTOR" APPROACH

Armitt[3] has described an approach based on a 2-second gust speed. Denoting this speed by U_o we can show

$$U_o = \overline{U}_o + g\,\sigma_u$$

$$= \overline{U}_o\,(1 + g\,\sqrt{6\,K}\,(\frac{10}{h_o})^a) \qquad (22)$$

Hence, ignoring higher order terms,

$$U_o^2 = \overline{U}_o^2\,(1 + 2\,g\,\sqrt{6\,K}\,(\frac{10}{h_o})^a + \ldots)$$

$$= \overline{U}_o^2\,(1 + g\,E) \qquad (23)$$

Following this Armitt give the maximum suspender arm load as

$$R_{SA} = \tfrac{1}{2}\,\rho\,U_o^2\,C_{D_c}\,d\,L \text{ x span factor}$$

$$= p_c\,d\,L\,(1 + g\,E)\,(\text{span factor}) \qquad (24)$$

Comparing (24) and (21) gives

$$\text{"Span factor"} = \frac{1 + g\,E\sqrt{B_c^* + R_c^*}}{1 + g\,E} \qquad (25)$$

Evaluation of this expression for a typical mean gradient wind speed of say 100 ft/sec should lead to span factors similar to those suggested by Armitt[3]. It would also be possible to incorporate in this expression the appropriate reductions for terrain. This would require the numerator to be computed for the particular terrain category and the denominator for the reference 'open country' condition.

It might also be convenient to include the effect of the changes of mean wind speed and turbulence in terrains differing in roughness from the standard terrain. Denoting the standard terrain by the subscript "st", the modified span factor taking into account the exposure of the line is then

$$\text{“span–exposure factor”} = \left(\frac{\bar{U}}{\bar{U}_{st}}\right)^2 \frac{1 + g\,E\sqrt{B_c^* + R_c^*}}{1 + g\,E_{st}} \tag{26}$$

The factor (U/U_{st}) can be found through the gradient velocity and can be written $(h_o/z_g)^a/(h_o/z_g)^a_{st}$. An alternative expression using the logarithmic profile is $(z_o^{.09}\,ln\,h_o/z_o)/(z_o^{.09}\,ln\,h_o/z_o)_{st}$. In the subscripted bracket the terms affected by terrain, a, z_g and z_o are subscripted.

INFLUENCE OF DRAG COEFFICIENT VARIATIONS

The theory for gust response described above is based on the conventional assumption of a constant drag coefficient. For conductors it is common for there to be a dramatic decrease in the drag coefficient C_{D_c} at Reynolds numbers in the range 1 to 5 x 10^4. Examples of this are shown in Figure 3. These Reynolds numbers frequently coincide with design wind speeds. This change in drag coefficient affects both the appropriate drag coefficient to be used in determining the mean pressure, $\bar{p}_c$, as well as the fluctuating component expressed through the gust response factor.

For the mean pressure, the drag coefficient corresponds to that for the Reynolds number, $\bar{R} = \frac{\bar{U}_o d}{\nu}$ corresponding to the mean wind speed U_o. The fluctuating component is however augmented by an additional term dependent on the slope of the drag coefficient at that Reynolds number. This modified drag coefficient, $C_{D_c}^1$ for the fluctuating component is

$$C_{D_c}^1 = [\,C_{D_c} + \tfrac{1}{2}\,(d\,C_{D_c}/d\,Re)\,\overline{Re}\,]$$

This is derived from the derivative of the force, $p_c = \tfrac{1}{2}\,\rho\,U^2 C_{D_c}(U)$ with respect to wind speed U.

Since the slope at the 'drag crisis' is negative, this has the effect of reducing the effective drag coefficient influencing the dynamic components of conductor sway. This modified drag coefficient should be used in estimating both B_c and R_c above. In R_c the drag coefficient also modifies the aerodynamic damping. As a result the values of R_c are multiplied by the factor $(C_{D_c}^1/C_{D_c})$ and B_c by the square of this factor. If this factor is less than unity, the overall fluctuating response is reduced while the relative importance of resonant fluctuations is increased.

To illustrate this point consider the average slopes of curves 4 and 5 in Figure 3 at $Re = 3 \times 10^{-4}$ (corresponding to 36 ft/sec and 52 ft/sec for the 1.6" and 1.1" diam. cables respectively). At this Re, $C_{D_c} \approx .75$. The average slope of these curves, $(d\,C_{D_c}/d\,Re)$, is approximately -2×10^{-5}. So that

$$C_{D_c}^1 = [.75 - \tfrac{1}{2} \times 2 \times 10^{-5} \times 3 \times 10^4] = .45$$

and $\quad C_{D_c}^1/C_{D_c} = .60$

A reduction of this magnitude, of course, has a significant effect in reducing the dynamic loads seen by the suspender arms and the tower. The apparent differences between the full scale and wind tunnel results is worthy of further study.

Source	Curve	Conductor Diam. in.	No. of outside strands	Wire/Cable diam.
Wind	1	1.125	24	9
tunnel	2	.770	18	7
	3	1.695	42	15
Full scale	4	1.108	16	6.3
estimates	5	1.60	27	10

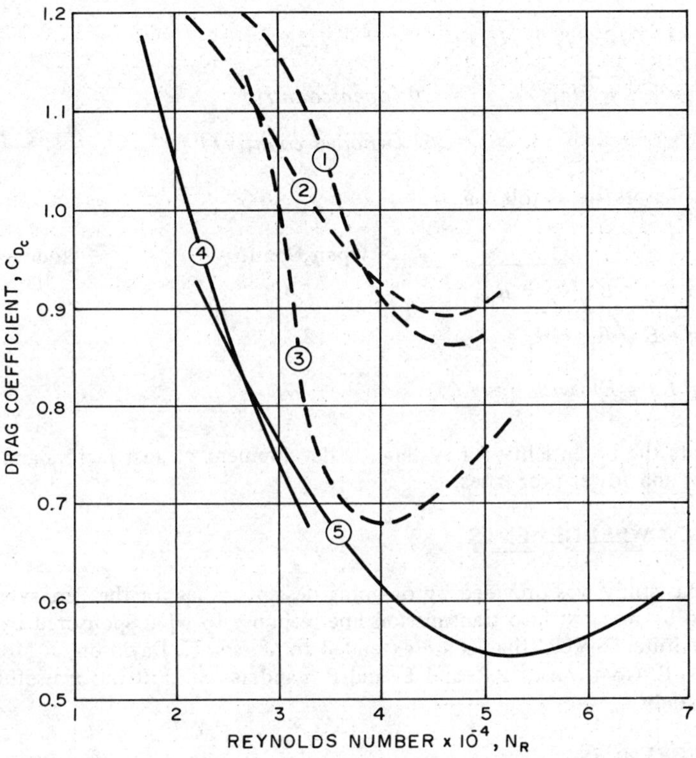

Fig. 3 Conductor drag coefficients C_{D_c}

EXAMPLE

To indicate the general character of the gust factors consider the following examples:

Conductor $L = 1000$ ft; $D = 32$ ft; $d = 0.1$ ft; $f_c = .2$ Hz; $C_{D_c} = 1.0$

Tower $h = 150$ ft; $h_o = 100$ ft; $f_t = 1$ Hz; $\zeta_t = .01$

Wind $U_o = 100$ ft/sec; $L_S = 210$ ft ($\sim$ 65 m)

Open country $\alpha = .16$ $K = .005$
Wooded terrain $\alpha = .28$ $K = .015$

Referring to equations 8 and 11 through 15;

$$B_c^* = \frac{1}{1 + .8\ L/L_S} = .21$$

$$\zeta_c = 4.8 \times 10^{-5}\ \bar{U}_o/f_c d = .24$$

$$R_c^* = .0113\,(f_c h_o/\bar{U}_o)^{-5/3}\,(h_o/L)\,\zeta_c^{-1} = .07$$

$$B_t^* = \frac{1}{1 + .375\ h/L_S} = .79$$

$$R_t^* = .0123\,(f_t h_o/\bar{U}_o)^{-5/3}\,\zeta_t^{-1} = 1.23$$

$$E = \sqrt{24\,K}\,(z_{ref}/h_o)^a = .30 \text{ (open country)}$$

$$= .61 \text{ (wooded country)}$$

Thus gust response factors are as follows:

	Open Country	Wooded Country
Conductor: $G_c = 1 + g\,\epsilon\,E\sqrt{B_c^* + R_c^*}$	1.42	1.61
Tower: $G_t = 1 + g\,\epsilon\,E\sqrt{B_t^* + R_t^*}$	2.12	2.64
"Span factor": $G_c/(1 + g\,E)$ (with $\epsilon = 1.0$)	.76	.71

These results indicate the potentiality for systematic development of gust factors and span factors for ranges of conductor and tower properties.

ACKNOWLEDGEMENTS

The incentive for this study was provided by on-going design studies for the Pennsylvania Power and Light Company and by research into transmission line response to wind sponsored by the Electric Power Research Institute. Special thanks are extended to Messrs. H. Davidson, J. Mozer and T. Di Gioia of GAI Consultants, E. Goodwin of P.P. and L. and P. Landers of E.P.R.I., for useful suggestions and comments on this study.

REFERENCES

1. Davenport, A. G., The Application of Statistical Concepts of the Wind Loading of Structures, Proc. Inst. Civ. Eng., Vol. 19, 1961.

2. Davenport, A. G., The Response of Slender Line-Like Structures to a Gusty Wind, Proc. Inst. Civil Eng., Vol. 23, 1962.]

3. Armitt, J., Cojan, M., Manuzio, C. and Nicolini, P., Calculation of Wind Loadings on Components of Overhead Lines, Proceedings IEE, Vol. 122, No. 11, November, 1975, pp. 1242–1252.

4. Manuzio, C., and Paris, L., Statistical Determination of Wind Loading Effects on Overhead Line Conductors, CIGRE Report 231, Paris, France, 1964.

5. Castanheta, M., Dynamic Behaviour of Overhead Power Lines Subject to the Action of Wind, CIGRE Report 22–08, Paris, France, 1970.

6. Ohtsuki, A., Studies on Wind Pressure Upon Overhead Transmission Lines and Their Movement, Paper Presented to Fujikura Electric Wire Manufacturing Company, Ltd., Japan, January 30, 1967.

7. Fujikura Electric Wire Manufacturing Company, Ltd., Movement of Overhead Transmission Line and Statistical Approach, Japan, January 30, 1967.

8. Davenport, A. G., Gust Loading Factors, J. Str. Div. Proc., ASCE, Vol. 93, 1967.

9. Davenport, A. G., Wind Structure and Wind Climate, Proc. Int. Res. Seminar, Safety of Structures Under Dynamic Loading, June 1977, Norwegian Inst. of Tech. Trondheim.

10. Davenport, A. G., The Prediction of the Response of Structures to Gusty Wind, Proc. Int. Res. Seminar, Safety of Structures Under Dynamic Loading, June 1977, Norwegian Inst. of Tech, Trondheim.

WIND LOADING OF SHIPYARD GANTRY CRANES - A COMPARISON OF FULL-SCALE MEASUREMENT, WIND TUNNEL TEST AND GUST FACTOR APPROACH

G. KÖNIG, K. ZILCH and G. LAPPAS

Institut für Massivbau, Technische Hochschule Darmstadt, Darmstadt, Federal Republic of Germany

SUMMARY

Shipyard gantry cranes are nowadays built up to about 100 m hook height and 200 m span. Wind loading is becoming a major factor in design. Therefore, detailed and modern approaches to wind loading are required. They have to be verified by full-scale measurement. In this paper some results of full-scale measurement are reported and compared to results of a boundary layer wind tunnel test and a gust factor approach. For the general dynamic behavior of the crane a good correspondence can be observed. The gust factors for open terrain are typically in the order of 2. The size of the structure results in a low level of wind correlation and an essential amount of aerodynamic damping which effectively counterbalances the effects of the low first natural frequency of about 0.1 Hz.

INTRODUCTION

The size of shipyard gantry cranes has increased quite fast. Nowadays, these structures are built up to more than 100 m hook height, up to about 200 m span, and with capacities up to 1600 metric ton. An isometric view of such a crane is given in Fig.1. The dynamic behavior of this type of structure is characterized by its mass distribution (heavy equipment in the girder), by the slenderness of the A-frames, and by the low damping (low structural damping and no energy dissipation into the soil). The lowest natural frequency of such cranes is about 0.1 Hz.

Wind loading is becoming a major factor in design. Areas for a detailed consideration may be the gust loading response of the entire structure and, possibly, a galloping-type exitation of parts. The gust factor approach - developed by Davenport for tall buildings (1) - has been modified for this type of structure (2). Furtheron, testing in a boundary layer wind tunnel was carried out (3). These concepts of modern design approaches - gust factors and boundary layer wind tunnel tests - have to be verified by full-scale measurement.

THE STRUCTURE

The structural system of the investigated shipyard gantry crane is a welded steel portal frame. Its girder consists of a trapezoidal box section of approximately 14 m by 16 m. The columns are A-frames, one is fixed and the other is hinged at the girder. They have rectangular box sections. The total height of the structure is 136 m, the span is 174 m (see Fig.1).

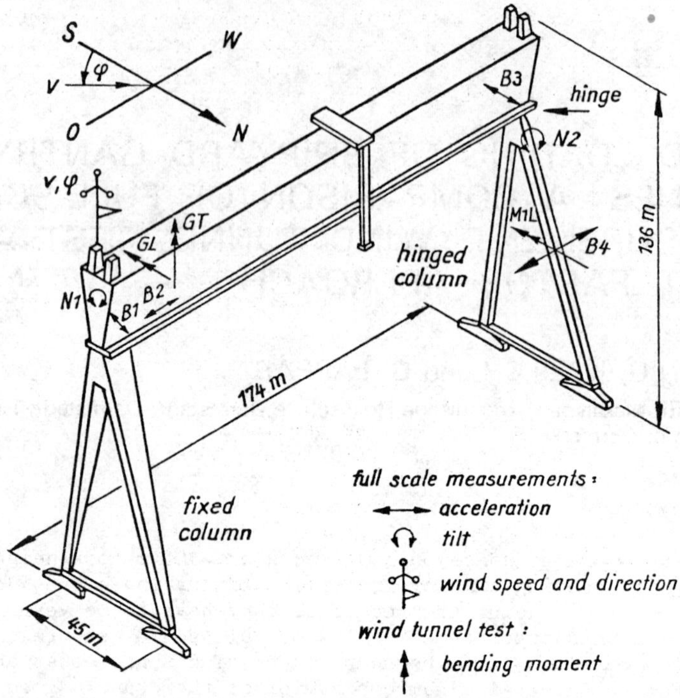

Fig.1. View of the crane and location of instruments

The first mode shapes of the structure have been computed by a finite element program: The first mode corresponds to a movement in girder direction, the second mode essentially to a movement transverse to the girder, the third and fourth modes are symmetric resp. antimetric movements of the hinged columns.

THE PROGRAM OF THE STUDY

Full-Scale measurement

The overall response of the structure is monitored using special seismic instruments with very high sensitivity (threshold less than 10^{-6} g). The natural frequency and damping of this instrument can easily be adjusted to specific requirements by an external electronic feedback circuit. The output of the instrument contains contributions of base acceleration and base tilt (4), (5). By filtering it can be decomposed into a low frequency part proportional to the tilt (mean response) and a high frequency part basically proportional to the acceleration (dynamic response). As a suitable corner frequency 0.005 Hz is used.

The seismic instruments have been located to give representative information of the response in the first modes (compare Fig. 1). Reference wind speed and direction are measured about 7 m above the girder at the fixed support.

The output from all instruments is amplified and transmitted in analogue form to a central unit which is placed in the girder. The signal is conditioned (filtered) and recorded on a magnetic tape using PCM. Additional information - date, hour, minute, and operating conditions of the crane - are also written on the second track of the tape. The data is automatically recorded for wind speeds exceeding a preset threshold. A single magnetic tape contains six hours of recording. The measuring and recording system is able to transmit information up to 12 Hz. The data spacing is 1/60 s which gives 198.000 data points per five minutes.

To recover the data a seperate tape unit is used which is connected via an interface to a small digital computer. Some data processing is done and then the data is written on a computer tape. A multi channel strip chart recorder is also connected to the replay unit. It is used for a first data screening. The final analysis of the data is done digitally on a medium size computer. To check the entire recording - replay - data transfer operations specified test signals have been generated, recorded, and processed. It should be noted that the full-scale measurement is still in progress.

Wind Tunnel Test

A wind tunnel study was carried out for the partiular crane at the Boundary Layer Wind Tunnel Laboratory of the University of Western Ontario (3). An aeroelastic model to a geometric scale of 1:500 was constructed. The main components of the crane; namely the girder and the fixed and pinned A-frame supports; were instrumented with electrical resistance strain gauges (compare Fig.1). The crane was tested in the wind tunnel to provide estimates of full-scale bending moments at all strain gauge locations in flow regimes representative of "open" and "built-up" terrains.

Gust Factor Approach

The along-wind response of structures to turbulent wind can be estimated theoretically by a gust factor approach proposed e.g. by Davenport (1), (6). The total response may be viewed as a sum of the mean response $\bar{R}$, induced by the mean wind, and the fluctuating response induced by gusts. The maximum response R_{max} may then be written as

$$R_{max} = \bar{R} + g\,\sigma_R \tag{1}$$

where the fluctuating part is characterized by a peak factor g and the rms-value σ_R. It is convenient to define an equivalent static wind load

$$R_{max} = G\,\bar{R} \tag{2}$$

where the gust factor G may be written in the form given by Davenport (6)

$$G = 1 + g\,\frac{\sigma_R}{\bar{R}} = 1 + 2\,g\,I_v\sqrt{\frac{s\,F}{\beta} + E} \tag{3}$$

where
I_v = turbulence intensity
s = size reduction factor
F = gust energy ratio
E = background excitation factor
β = damping ratio

The turbulence intensity I_v and the gust energy ratio F characterize the gustiness of the wind. The size reduction factor s and the background excitation factor E define the interaction between the turbulence structure and the structural modes of vibration. They are evaluated by integrating the fluctuating effects over the structure (compare e.g. (2), (6)). The damping ratio β comprises a structural part β_s and possibly an aerodynamic part β_a.

Due to the stiffness distribution dynamic along-wind effects are essential in the first mode. Relevant excitation is limited to an inclination less than 30° between wind and girder. Then the response may be viewed as a sum of contributions from the front and rear leg. Adding the mean values $\overline{R}_1$ arithmetically and the rms-values σ_{R_1} statistically results in:

$$R_{max} = 2\,\overline{R}_1 + g\,\sqrt{2}\,\sigma_{R_1}\sqrt{1+\rho} \tag{4}$$

where ρ is a correlation coefficient between front and rear leg (a somewhat more detailed approach is given in (3)).

The gust factor G_1 of the first mode follows as

$$G_1 = 1 + \sqrt{2}\,g\,I_v\,\sqrt{(1+\rho)\left[\frac{s_1 F}{\beta} + E_1\right]} \tag{5}$$

The size reduction factor s_1 and the background excitation factor E_1 are related to the loading of a single A-frame, they are given in Fig. 2.

For wind directions parallel to girder (maximum inclination approximately 5°) the along-wind response may be effected by the convection of ambient velocity fluctuations or eddies by the mean flow (frozen turbulence). Particular eddies having excited the front A-frame are convected downstream and may arrive - depending upon the mean convective wind speed $\overline{v}_c$, first mode frequency n_1 and distance Δ - in phase ($\rho = 1$) or out of phase ($\rho = -1$) with the motion of the crane. The phenomenology of this effect can be reasonably described by a correlation coefficient of the form

$$\rho = e^{-C\frac{n\Delta}{\overline{v}_c}} \cos 2\pi\,\frac{n_1\Delta}{\overline{v}_c} \tag{6}$$

where C is a empirical constant ($C \simeq 1$ (3)), as approximations $n \simeq n_1$ and $\overline{v}_c \simeq \overline{v}_h$ (speed at girder centre line) may be used. .

For wind directions inclined more than 5° to the girder the correlation coefficient is $\rho = 0$.

The aerodynamic damping - resulting from forces proportional to the velocity of the structure - may be theoretically estimated on the basis of quasi-steady aerodynamics (6). For the first mode the damping ratio β_a can be expressed in terms of a generalized modal damping c_a^* and a generalized mass m_1^* (5):

$$\beta_a = \frac{c_a^*}{4\pi\,n_1\,m_1^*} \tag{7}$$

where for vertical line-like structures and using linearized theory:

$$c_a^* = \rho_a \int_0^H c_D\,D\,\overline{v}_z\,\Phi_1^2\,dz \tag{8}$$

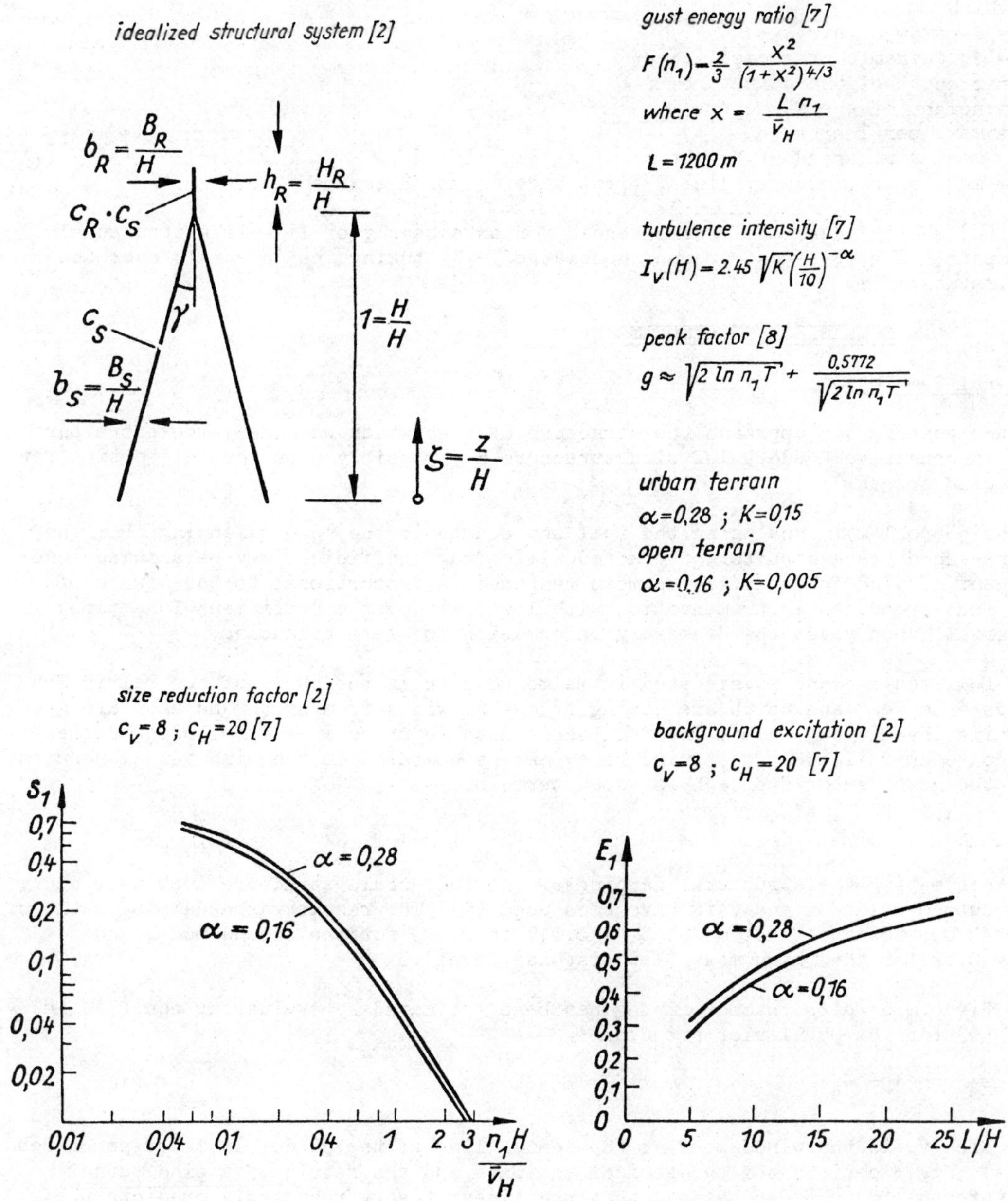

Fig.2. Gust factor approach for the first mode

$$m_1^{*} = \int_0^H m\, \Phi_1^{\,2}\, dz \qquad (9)$$

in which
ρ_a = density of air
c_D = aerodynamic drag coefficient
D = breadth of the structure
H = height of the structure
m = mass per length
$\bar{v}_z$ = mean wind speed at height z
Φ_1 = mode shape, usually linear first mode Φ_1 = z/H assumed

Eqn (8) and (9) can easily be extended for an assembly of line-like structural elements - e.g. the gantry crane considered - by taking the integrals over the entire assembly.

COMPARISON OF RESULTS

General Remarks

In the gust factor approach the structure of high winds was used. Therefore parameters sensitive to detailed wind structure may possibly show some distortion for low wind speeds.

A basic problem in analysing the full-scale data is the zero balancing. For the sharp-edged crane structure the aerodynamic drag coefficient may be assumed independent of wind speed, i.e. the mean response is proportional to the square of the mean speed. An instrumentation with low drift over a sufficient long time intervall then gives the necessary information for zero balancing.

The full-scale crane is situated in Malmo, Sweden at the shore of the Baltic Sea. Winds from west and north are coming from sea; winds from south and east are passing the city of Malmo. The full-scale data reported in the following corresponds to approximately west wind (sea) and is compared to the wind tunnel results and the gust factor approach for open terrain.

Damping

For estimating the structural damping of the full-scale structure amplitude decay and autocorrelation analysis have been used (9). The result was a damping ratio of β_s = 0.005 for the first mode, β_s = 0.017 to 0.029 for the second mode, and β_s = 0.03 for the third mode (low response level).

The first mode aerodynamic damping has been estimated by evaluating eqn (7), (8) and (9) for the particular structure:

$$\beta_a = 0.00015\ \bar{v}_H \qquad (10)$$

in which $\bar{v}_H$ is the wind speed at the center line of the girder in meter per second [m/s]. Fig.3 depicts the theoretical estimate and the result of a wind tunnel measurement (3). It can be seen that the linear theory reasonably predicts the level of aerodynamic damping.

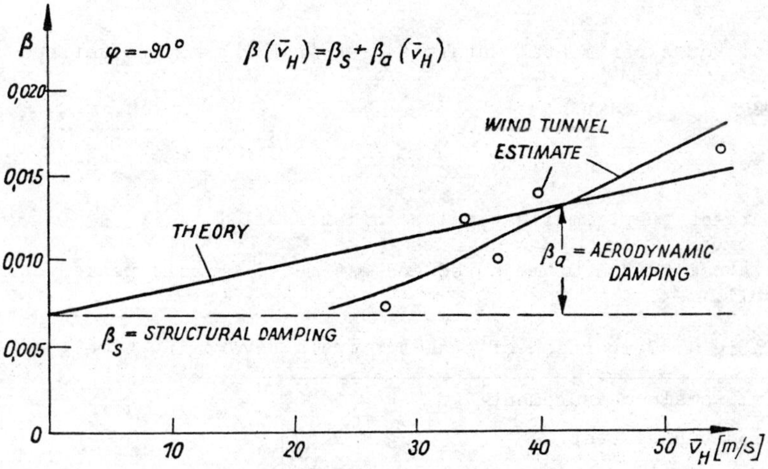

Fig.3. Comparison of aerodynamic damping in the first mode (open terrain)

RMS-values

Estimates for normalized rms-values are given in Fig. 4. The results agree reasonably. It can be seen that the aerodynamic damping effects the response significantly. The simple approach used for the convective effects provides a good phenomenological explanation.

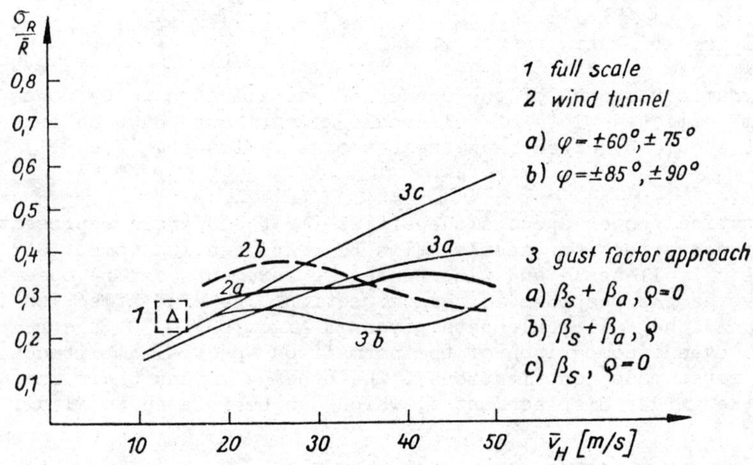

Fig.4. Comparison of rms-values for the first mode

Peak factors

For the analysis of experimental data the peak factor g may be defined as

$$g = \frac{R_{max} - R_{min}}{2\,\sigma_R} \qquad (11)$$

where
R_{max}, R_{min} = largest resp. smallest value in intervall T

Table 1 gives a comparison of measured and estimated (Fig.2) peak factors. They agree quite well.

Table 1. Comparison of peak factors g (averaging time T = 200s)

full-scale measurement	g = 2.56
wind tunnel test	g = 2.62
theory	g = 2.71

Gust factors

Estimates of gust factors G_1 for the first mode assuming independency between front and rear A-frame are given in Fig.5. The overall behavior is reasonably predicted by the theoretical values taking into account the aerodynamic damping.

A more detailed evaluation is given in Fig.6. Here directions with possible convective effects have been separated. Again the simple approach used for the convective effects provides a good phenomenological explanation. It should be noted, that the differences are practically irrelevant as for design a variation of the wind angle of attack of more than the range of possible convective effects would have to be considered. This means, that the maximum of the uncorrelated or convective gust factor must be taken.

Spectral densities

The general dynamic response of the crane is best illustrated by power spectral densities. Some selected plots of full-scale measurement and wind tunnel test are given here. It should be noted that the response varies significantly with wind direction.

Fig. 7 shows typical power spectral densities for the signals representing a movement in girder direction (acceleration B2 resp. bending moment GL). A good correspondence of full-scale and wind tunnel measurements can be observed. The first mode governs the response in this direction. The wind direction of the wind tunnel $\Phi = 0^o$ and the full-scale data given in Fig. 7b $\Phi = 30^o$ are relatively close, so that even a comparison of the normalized spectral amplitudes may be tried (mainly first mode contributions). The bending moment GL is basically proportional to the girder displacement d, which can be related to girder acceleration a by

$$a = \omega^2 d \qquad (12)$$

For the first mode peak this gives a = 0.45 d, which results in a factor of 2 between the peaks of the two estimates of the spectral density.

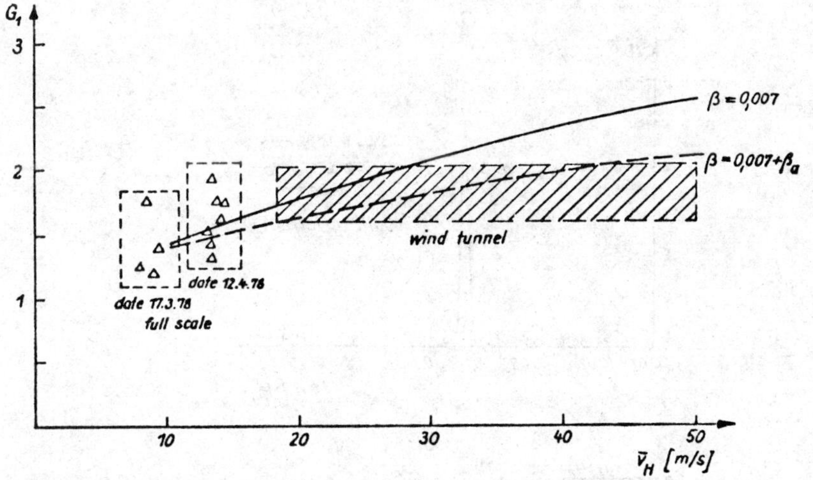

Fig.5. Comparison of overall gust factor G_1 (front and rear leg uncorrelated: ρ=0, averaging time T = 200 s)

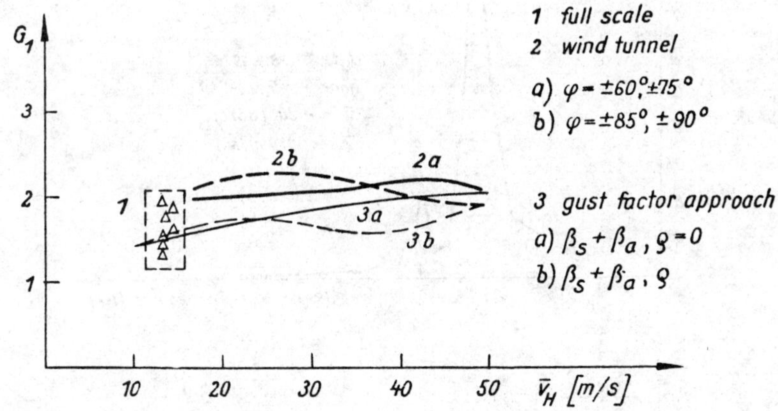

Fig.6. Comparison of detailed gust factor G_1 (averaging time T = 200 s)

Fig.8 shows typical power spectral densities for the reaction of the hinged columns (acceleration B4 resp. bending moment M1L). This reaction is essentially in the third and fourth mode. Considering that a movement in the first mode results in accelerations B4 but neglectible bending moments M1L the general characteristics are similar. The autocorrelation function in Fig.9 also illustrates the combination of these two modes.

Fig. 10 depicts spectral densities for the reaction perpendicular to the girder (acceleration B3 resp. bending moment GT), which show a good agreement in the frequency content.

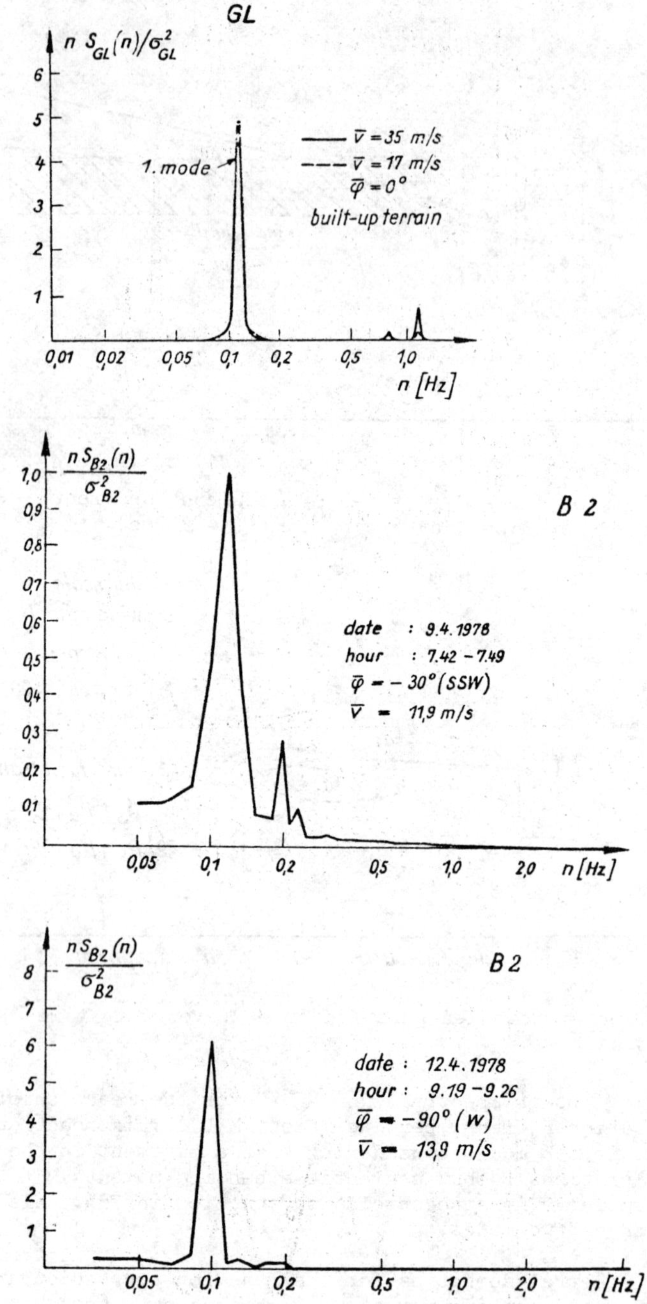

Fig.7. Comparison of spectral densities
a) wind tunnel GL, $\phi = 0^{\circ}$
b) full-scale B2, $\phi = -30^{\circ}$
c) full-scale B2, $\phi = -90^{\circ}$

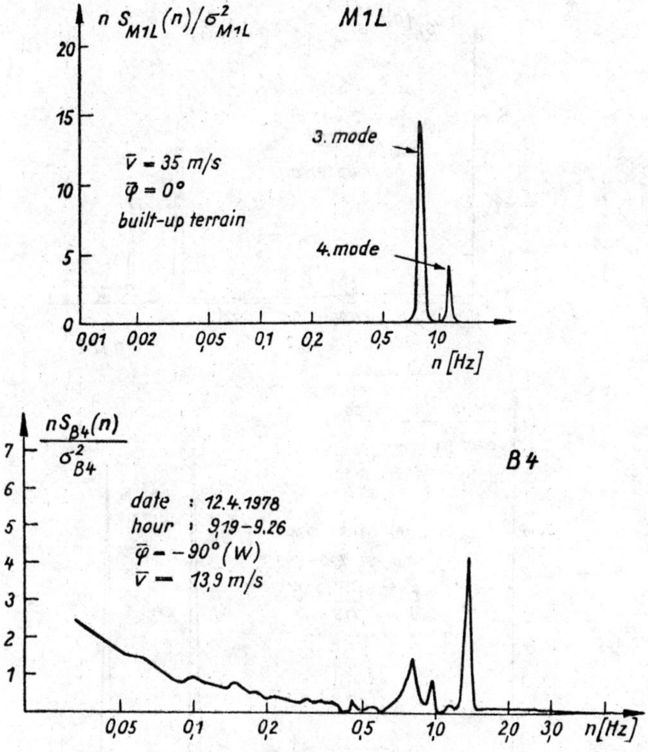

Fig.8. Comparison of spectral densities
a) wind tunnel M1L, ϕ = 0°
b) full-scale B4, ϕ = -90°

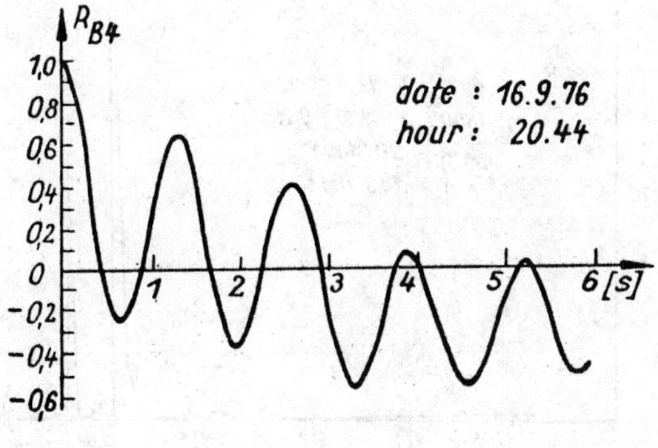

Fig.9. Typical plot of auto-correlation function R_{B4} of full-scale measurements

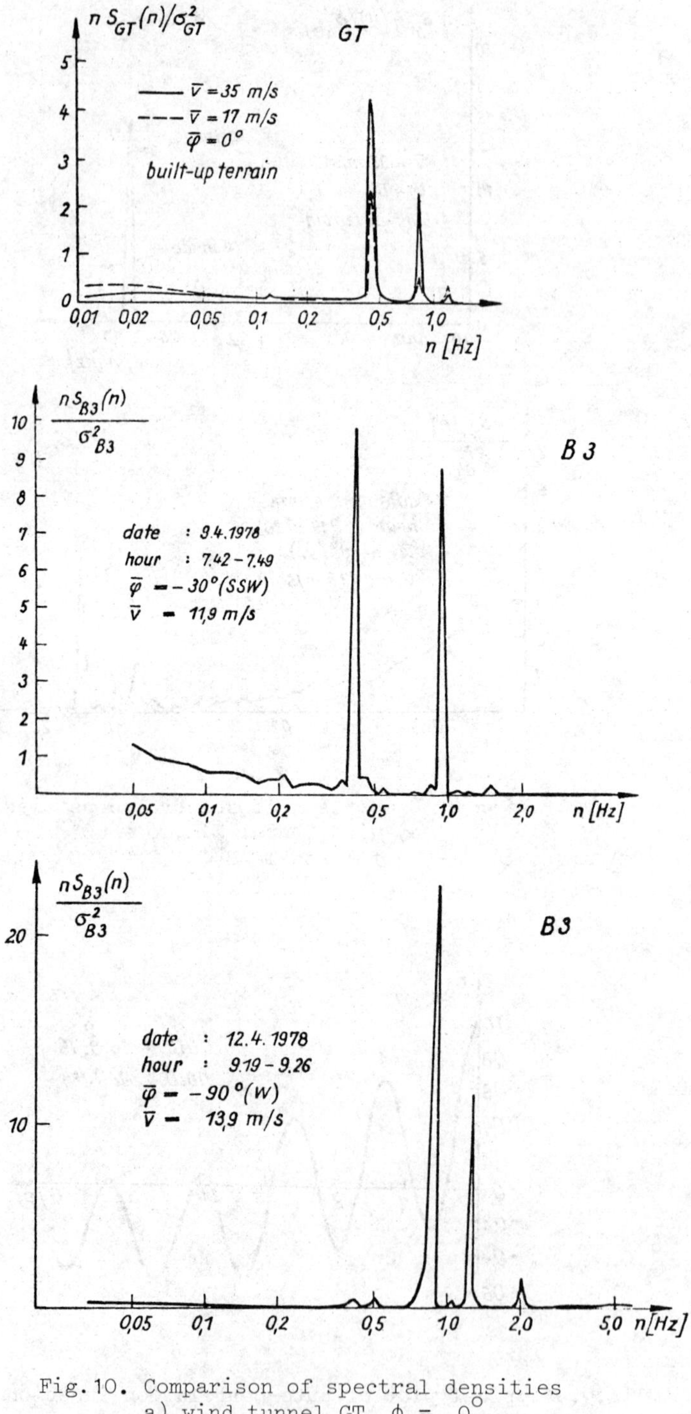

Fig.10. Comparison of spectral densities
a) wind tunnel GT, $\phi = 0^\circ$
b) full-scale B3, $\phi = -30^\circ$
c) full-scale B3, $\phi = -90^\circ$

REFERENCES

1. Davenport,A.G./Dalgliesh,B.A.: Wind Loads. In: Canadian Structural Design Manual. Ottawa: National Research Council of Canada 1970.

2. König,G./Zilch,K.: Böigkeitsfaktoren für Portalkräne. In: 1. Berichtskolloquium des DFG-Schwerpunktprogramms "Bauwerke unter aerodynamischer Belastung". Technische Universität München, 4.Oktober 1976.

3. Isyumov,N./Davenport,A.G.: A Study of Wind Action on a Shipyard Portal Crane. BLWT - SS 6 - 1975. The Boundary Layer Wind Tunnel Laboratory, The University of Western Ontario, London, Ontario, Canada 1975.

4. König,G./Zilch,K./Lappas,G.: Wind Loading of Shipyard Gantry Cranes - Full-Scale Measurements. Journal of Industrial Aerodynamics, to appear.

5. Clough,R.W./Penzien,J.: Dynamics of Structures. New York: McGraw-Hill 1975.

6. Davenport,A.G.: The Prediction of the Response of Structures to Gusty Wind. In: Holand, I. et al. (ed.): Safety of Structures under Dynamic Loading, Vol. 1. Trondheim: Tapir Publishers 1978.

7. Davenport,A.G.: The Dependence of Wind Loads on Meteorological Parameters. In: Wind Effects on Buildings and Structures, Proceedings. Toronto: University of Toronto Press 1968.

8. Davenport,A.G.: Note on the Distribution of the Largest Value of a Random Function with Application to Gust Loading. Proceedings, Institution of Civil Engineers, Vol. 28 (1964), p. 187.

9. Zilch,K./Lappas,G.: Windbeanspruchung eines Werftportalkranes - Meßergebnisse. In: Bauwerke unter aerodynamischer Belastung. Mitteilungen aus dem Institut für Massivbau der Technischen Hochschule Darmstadt, Heft 26. Berlin: Ernst & Sohn 1977.

Acknowledgements

The full-scale measurements presented in this report are supported by a research grant from the Deutsche Forschungsgemeinschaft. This support is gratefully acknowledged. The authors also acknowledge the continued interest and support of the crane owner Kockums Varv at Malmo/Sweden and of the crane manufacturer Pohlig-Heckel-Bleichert at Cologne/Germany. Highly appreciated is the cooperation in the wind tunnel study with Professor Dr. Davenport and Dr. Isyumov of the Boundary Layer Wind Tunnel Laboratory, The University of Western Ontario at London/Canada.

A WIND TUNNEL INVESTIGATION OF A RETRACTABLE FABRIC ROOF FOR THE MONTREAL OLYMPIC STADIUM

H. P. A. H. IRWIN and R. L. WARDLAW
National Research Council of Canada, Ottawa, Canada

ABSTRACT

The Stadium playing field is to be covered by a retractable fabric roof that is attached to the fixed part of the roof covering the stands. The fabric is tensioned by cables rising up to a tower above the Stadium. The purpose of the investigation has been to determine the effect of wind on the forces imparted to the fixed roof structure by the deployed fabric roof. A 1:100 scale model has been studied in the National Research Council 9m × 9m (30ft × 30ft) Low Speed Wind Tunnel. The fabric roof and supporting cables have been modelled elastically.

NOTATION

A	area of roof
A_C	cross-sectional area of a cable
b	typical length of roof (e.g. its major axis)
c	speed of sound
D	aerodynamic drag of cable per unit length
E	Young's modulus
g	gravitational acceleration
K_T	slope of tension versus strain curve for warp or weft direction
K_S	slope of shear loading versus shear strain curve for warp or weft direction
L	integral scale of the turbulence
M_A	added mass
M_C	mass of cable per unit length
M_F	mass of roof fabric per unit area
M_V	virtual mass
n	frequency
P_I	absolute internal pressure
Δp	excess of internal over external pressure in zero wind
q_{60}	$\frac{1}{2}\rho U_{60}^2$
$S(n)$	power spectrum of longitudinal velocity fluctuations
U	mean wind speed
U_e	velocity above boundary layer
U_{60}	value of U at representative height of roof, 60m (200ft)
u'	root-mean-square of longitudinal velocity fluctuations
V_I	internal volume covered by roof

X,Y	horizontal forces at perimeter attachment point, respectively parallel to and normal to the console axis (Fig. 3)
Z	vertical force at perimeter attachment point
$\overline{\Delta X}$, $\overline{\Delta Y}$, $\overline{\Delta Z}$	mean change in X, Y and Z due to wind
γ	ratio of specific heats for air
ζ	damping ratio in vacuo (mechanical damping)
ζ_A	damping ratio due to acoustic damping
μ	viscosity of air
ρ	density of air
ϕ	wind angle, see Fig. 3

INTRODUCTION

An unusual feature of this investigation is that it focuses on the dynamic behaviour under wind action of a membrane-like structure — a retractable roof that will be installed over the playing field of the 70,000 seat Montreal Olympic Stadium (Fig. 1). While there have been some analyses of wind action on such structures (1), the only other similar wind tunnel investigations, to the authors' knowledge, are References 2, 3 and 4. Interesting scaling requirements, acoustic considerations and virtual mass effects have been encountered in the course of the present study.

The 168m (552ft) high tower that will support the roof is, itself, a major structure and will visually dominate the Stadium as can be seen in Figure 2, a photograph of the 1:100 scale model that has been studied in the National Research Council 9m × 9m (30ft × 30ft) Low Speed Wind Tunnel. The base of the tower already houses the Olympic swimming pools and when completed, the tower will accommodate extensive additional athletic facilities, administrative offices and a high level restaurant.

The retractable fabric roof is attached to the fixed roof that covers the stands and is tensioned upwards by cables leading into the tower as illustrated in Fig. 3. The roof is retracted by the paying out of cables at the rigid roof while drawing it upwards and into the tower where the collapsed roof is stored. The open elliptical area to be covered is about 180m (600ft) × 120m (400ft) and the fabric has an area of 18,000m^2 (190,000ft^2).

The purpose of the investigation has been to determine the effect of wind on the forces imparted to the attachment points on the fixed roof structure by the retractable roof while in the deployed position. These forces create bending moments in the long, curved cantilever beams (subsequently referred to as consoles) that support the shell of the fixed structure and there had been some concern that the effect of wind on the forces was not known with adequate precision. A secondary purpose, that will not be discussed in this paper, was the measurement of surface pressures on the tower for cladding design. In 1973, at a much earlier stage in the design of the Stadium, wind tunnel investigations were undertaken by Romani (5) at the Laboratoire Eiffel, Paris, France. Measurements were made of the wind pressure on the roof using a 1:150 scale rigid model. These data were then used by the French roof designers, Technique Avancée de l'Architecture et de l'Amenagement (TAAA), of Paris to compute the roof loading using quasi-static assumptions. In the present investigation, the removable roof was modelled elastically and tested in a simulated turbulent wind layer so that the forces, including dynamic behaviour, could be measured directly using three-component strain gauge balances at each of the seventeen attachment points on the fixed roof. In addition, deflections were measured at six points. It was not considered necessary to model the tower or the fixed shell of the Stadium elastically. Two

Fig. 1. Aerial view of the Olympic site.

configurations were tested — one for winter usage where the gap between the fabric roof and the fixed roof is sealed and the other for summer usage during rainstorms where the gap is left open.

A novel feature of the model design has been the fabrication of a model scale material having the appropriate mass and elastic properties to simulate the full-scale material which is a woven Kevlar mat coated on both sides with plasticized polyvinylchloride. The development of a satisfactory technique for fabricating the material in the laboratory lagged behind the rest of the program. As a result, a preliminary investigation (6) was done using Dacron sailcloth weighing 85 gm/m^2 (0.017 lb/ft^2), about four times the correctly scaled weight of 22 gm/m^2 (0.0045 lb/ft^2). Subsequently, when the lighter roof material was available, a final test was done that is reported here. Some disparity was found between the measured forces and the earlier TAAA force predictions used in the Stadium design but it has been concluded that under wind action, the forces will remain within acceptable limits. Response peaks suggestive of wind-induced instability were observed but it has been determined that they were excited by the characteristic acoustic frequencies of the wind tunnel and no purely aerodynamic instability was detected. Mean deflections of the roof occurred which would have an effect on the measured forces, thus supporting the use of an aeroelastic as opposed to a rigid model.

As well as describing the investigation, the paper considers novel scaling requirements, added mass effects, acoustic damping and the importance of the elastic and mass properties of the roof material.

SIMILARITY REQUIREMENTS

For a cable-supported membrane roof system that is attached to a rigid structure, the following physical quantities have potential roles in determining its behaviour under wind action: M_F, M_c, ρ, b, K_T, K_S, U, μ, EA_c, D, g, Δp, ζ, V_I, γ, P_I.

Since woven fabric membranes have non-linear stress-strain relationships, the stiffnesses K_T and K_S were taken to be representative values in the range of stress of most importance. There are a total of sixteen physical quantities in the preceding list and they are made up from three fundamental units. Thus, by the Pi Theorem, a minimum of thirteen non-dimensional parameters are required to describe the behaviour of the complete roof system. An appropriate set is:

$$\frac{\rho Ub}{\mu}, \frac{U^2}{bg}, \frac{M_F}{\rho b}, \frac{K_T}{\rho U^2 b}, \frac{K_S}{K_T}, \frac{M_c}{\rho b^2}, \frac{EA_c}{\rho U^2 b^2}, \frac{D}{\rho U^2 b}, \zeta, \frac{\Delta p}{\rho U^2}, \gamma, \frac{P_I b^3}{\rho U^2 V_I}, \frac{\gamma P_I}{\rho U^2}$$

The non-dimensional frequency, nb/U, is a dependent parameter that is a function of the preceding independent parameters.

The Reynolds number, $\rho Ub/\mu$, was, as with most tests, ignored. After some analysis it was also considered unnecessary to achieve similarity of the Froude number, U^2/bg, because the gravitational forces at full scale tend to be much smaller, owing to the light weight and high tension of the roof system, than the aerodynamic forces in high winds or those due to cable tensioning. An exceptional situation, where gravitational forces would play a role, would be when a cable went slack at high wind speeds due to the tendency of the wind to pull the roof upwards in some areas. The gravitational force would then be of importance for determining the excitation of the slack cable itself but not for the loads at the perimeter of the roof which were under study here. If a cable were sufficiently slack for gravity effects to influence its stiffness, it would no longer contribute significantly to the loads at the roof edge.

Some pains were taken to obtain the correct value of $M_F/\rho b$ but at a late stage in the work, it was realized that for very lightweight membranes, the added mass effect of the surrounding air is important. Therefore, there was room for some relaxation of the accuracy of the membrane mass simulation. A discussion of the added mass effect is given in the Appendix.

For a completely flat membrane, the natural frequencies and the small deflections under static load are independent of the elastic properties of the membrane material. The elastic properties do, however, affect the behaviour of curved membranes so in the present investigation it was considered worthwhile that they be simulated. Accordingly, the velocity scaling for the experiments was derived from the condition that the stiffness parameter $K_T/\rho U^2 b$ be the same at model and full scale. Since the full-scale fabric had, essentially, negligible shear stiffness, a very low value of K_S/K_T was aimed for on the model. There is some evidence (2,3) that the behaviour of tensioned membranes in wind is not very sensitive to their elastic properties, the membrane deflections being such as to require very little strain, but more work is required in this area to establish the range of applicability of such an assumption and the degree to which elastic similarity can be

Fig. 2. 1:100 scale model in 9m × 9m (30ft × 30ft) Low Speed Wind Tunnel

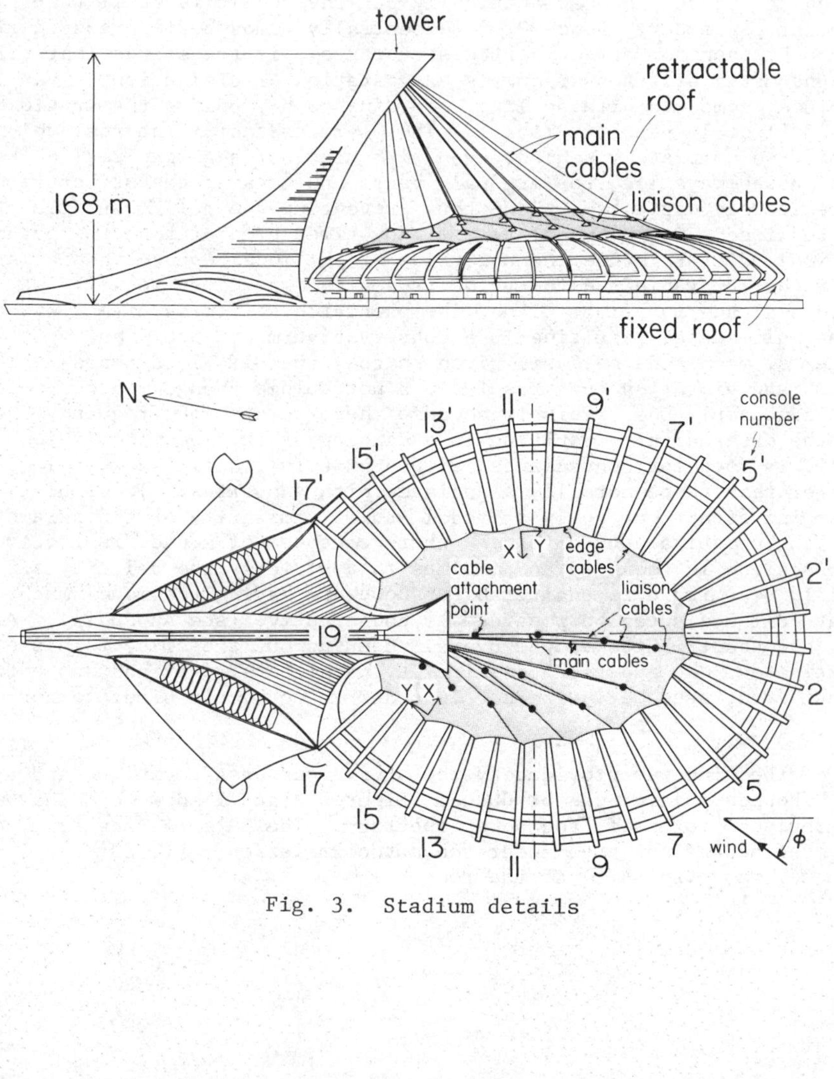

Fig. 3. Stadium details

relaxed. When a cable supported membrane roof is initially tensioned, the resulting movement and the final shape are affected by the elastic properties and this must be borne in mind, and the necessary precautions taken to obtain the correct final shape, if elastic similarity conditions would be relaxed.

It was considered desirable to simulate the cable parameters $M_c/\rho b^2$, $EA_c/\rho U^2 b^2$ and $D/\rho U^2 b$ and this was accomplished using steel piano wire. The mechanical damping ratio, ζ, of the full-scale roof system was unknown but it was thought unlikely that the model mechanical damping ratio would be higher than at full scale because the model cables were mono-filaments compared with the full-scale stranded cables, which have high damping.

The parameter $\Delta p/\rho U^2$ is of importance for air-supported structures but for the present cable-supported roof, the only form of internal pressurization came from the air-conditioning and heating systems and it was estimated to have only a minor effect compared with the wind pressure at high speeds. For a sealed structure, the internal volume contributes a pneumatic stiffness if the membrane roof deflections produce a net change of internal volume. To model this effect correctly, Tryggvason and Isyumov (2) assumed isothermal conditions in the internal volume and deduced that $P_I b^3/\rho U^2 V_I$ should be the same at model and full scale. A better approximation, at least for rapid movements of the roof, would appear to be that the volume change takes place adiabatically but the result is still that $P_I b^3/\rho U^2 V_I$ is the relevant parameter. Since P_I is essentially atmospheric pressure (in practice $\Delta p \ll P_I$ even for air-supported structures), it transpires that correct scaling of pneumatic stiffness requires exaggeration or diminution of the model internal volume, compared with scaling according to b^3, unless that ratio of model velocity to full-scale velocity is 1:1. The Olympic Stadium internal volume could not be considered completely sealed, even with the roof-edge gap sealed, because underground passageways lead to large air vents and because the structure has many points where leakage would take place thus preventing the full pneumatic stiffness from being realized. On the model V_I was many times greater than the parameter $P_I b^3/\rho U^2 V_I$ would dictate since the underside of the model opened into a large volume under the turntable and the velocity scaling was close to 1:1. No attempts were made to seal the interior. Thus, the pneumatic stiffness effect was certainly less than at full scale, resulting in a conservative model behaviour. In fact, from an order of magnitude analysis given in the Appendix, it appears unlikely that modes of membrane vibration that result in a net volume change would ever be excited to large amplitude. This is because they radiate low frequency acoustic waves into the external air resulting in a damping of the membrane motion. The term acoustic is used for convenience, although at full scale the wavelength would be much larger than those normally associated with acoustics. To simulate the acoustic damping effect correctly at model scale, similarity of the parameter $\gamma P_I/\rho U^2$, which is equivalent to $(U/c)^{-2}$ where c = speed of sound, is required. Thus Mach number scaling is required and implies that strictly the velocity scaling should be 1:1, assuming c is unalterable. However, having the model Mach number less than at full scale can be regarded as conservative (see Appendix). Aerodynamic damping also contributes to the overall damping but the size of this effect has not been estimated. No additional similarity parameters, over and above those already considered, need be introduced in order to correctly simulate aerodynamic damping.

Similarity of the wind was provided by the spire/roughness technique and a 3.7m (12ft) thick boundary layer was produced. Figures 4 and 5 show the mean velocity profile, turbulence intensity and power spectrum. The integral scale, L, of the turbulence, defined as the area under the autocorrelation curve for the longitudinal turbulence, was 1.2m (4ft) at the roof height.

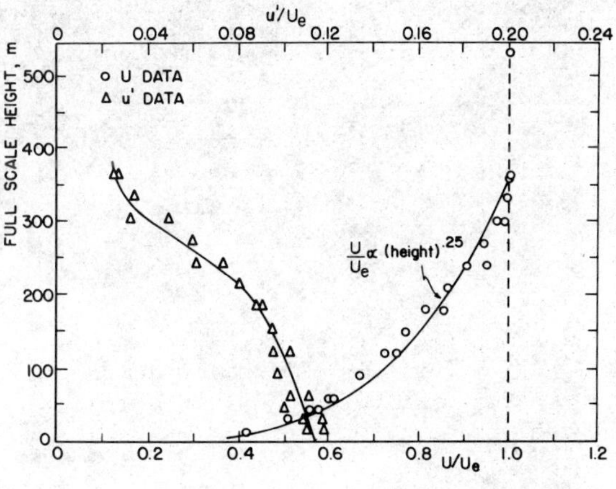

Fig. 4. Mean velocity and turbulence intensity in wind simulation.

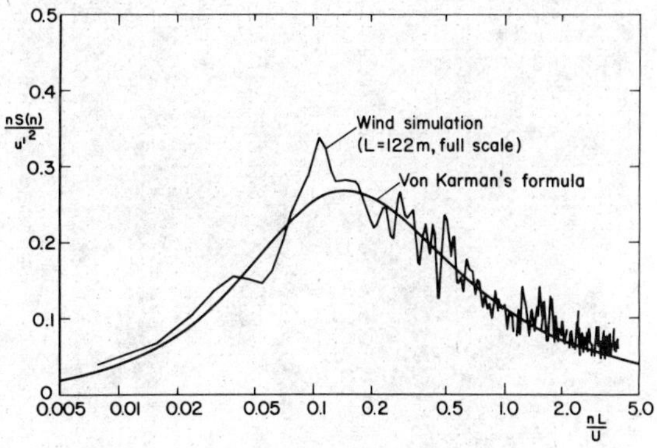

Fig. 5. Power spectrum of velocity at roof height.

MODEL FABRICATION

The details of the full scale and model roof materials are summarized in Table 1 while Figure 6 shows the model roof. The values of K_T and K_S in the table were measured in the laboratory on some samples of the full scale and model fabric.

TABLE 1. Roof Material

	Full Scale	Model
Fabric	1420 denier Kevlar 49 closely woven	100 denier Kevlar 29 woven as an open net
Airtight Coating	PVC on both sides	High density polyethylene sheet on underside
Mass	Kevlar 1.1 kg/m^2 Coating 1.1 kg/m^2 Total 2.2 kg/m^2	Kevlar 0.009 kg/m^2 Coating 0.006 kg/m^2 Adhesive 0.005 kg/m^2 Total 0.020 kg/m^2
Approximate K_T	16 MN/m	0.14 MN/m
Approximate $\frac{K_S}{K_T}$	6.2×10^{-4}	20×10^{-4}

The Kevlar net, from which the model roof was made, was specially woven as a 32 cm (12.6in) wide strip with a mesh size of 0.5cm × 0.5cm (0.20in × 0.20in). Segments of the Kevlar net were stretched over a rectangular frame to maintain stability and the polyethylene sheet was bonded to the Kevlar using diluted contact cement that had been sprayed onto the polyethylene. The roof was then assembled on a mould from the segments of composite material. Joints in the material were made using reinforcing strips of Dacron fabric and epoxy resin, the total weight added from all joints being equivalent to approximately 15% of the membrane weight. The joints in the full-scale roof are also made in a manner that results in significant added weight. At each of the 25 cable attachment points on the membrane surface, the attachment fitting and local reinforcement corresponded to a full-scale

Fig. 6. Model roof.

concentrated mass of 3300 kg (230 slugs). Since the design of full-scale joints and cable attachments had not been settled, and thus the masses were not known accurately, the policy was to keep the model masses as low as possible. Edge cables were installed around the perimeter by folding the material back on itself and reinforcing it with Dacron.

The external shape of the fixed Stadium structure was modelled in wood but inside was a rigid steel frame upon which the seventeen, three-component strain-gauge balances were mounted in order to measure the forces at the perimeter attachment points of the roof. After attaching the roof to the balances, the tension was applied to the main cables from the tower by a system of weights and pulleys. This system circumvented some of the iterative process that would have been required to obtain the correct distribution of cable tensions. However, a considerable time was still spent in obtaining the desired distribution of tension amongst 'liaison' cables (see Fig. 3) and edge cables by means of screw adjustments. The target distribution of tensions had been computed by the roof designers and the tensions on the model were either measured directly using a strain-gauge tension meter or indirectly by measuring the fundamental frequency of vibration with the aid of a microphone. The edge-cable tensions were adjusted by monitoring the force balance readings and comparing them with the computed forces.

Attempts were made to excite the roof in zero wind and measure its natural frequencies which had not been previously computed. Figure 7 gives an example of a signal from a proximeter close to the centre of the roof, after the roof had been tapped at the far end from the tower. This figure also shows the motion power spectrum in which several modes of vibration are evident in the full-scale frequency range 0.1 to 0.3 Hz. No attempts were made to determine mode shapes since this would have been an elaborate experiment in itself because of the complexity of the structure. The roof vibration can be seen to be very highly damped. It is worth noting that on membrane structures the natural frequencies will in general depend on the mean wind speed because the wind changes the mean membrane tension which in turn affects the frequency. On the Stadium, the roof tension was increased by the wind so the natural frequencies would have increased to values above those measured in zero wind.

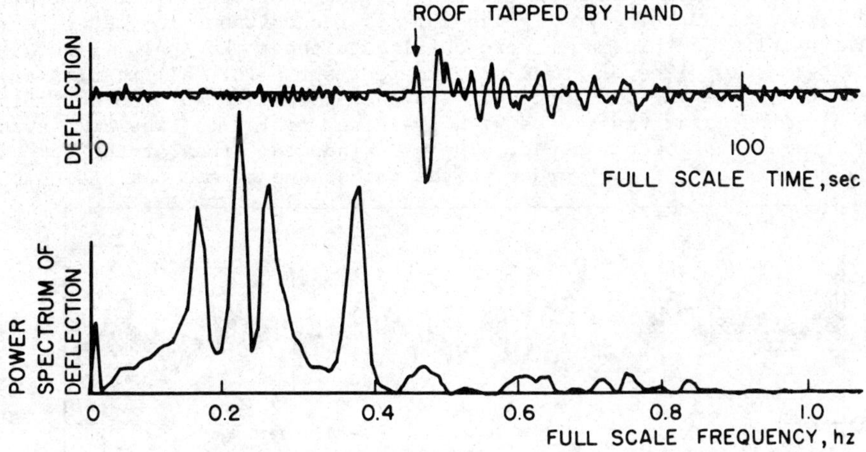

Fig. 7. Example of damping trace and power spectrum of deflection in zero wind.

TESTING METHODS

The strain gauge balances were connected to the tunnel data system which was used to digitally analyze the balance signals. Up to six channels were sampled simultaneously. For some conditions, complete time histories were obtained and for others only statistical quantities such as mean, root-mean-square and peak forces were measured. The averaging period for the statistical quantities was equivalent to ½ hour and in some cases 1 hour at full scale and the sample rate on each channel was 200 points per second at model scale. Six proximeters were used to measure the membrane deflections, again the data being acquired and analyzed on the tunnel data system.

Tunnel noise and vibration were monitored using a microphone, mounted on the tunnel wall above the simulated boundary layer, and accelerometers mounted on the model structure and tunnel structure. A digital Fourier Analyzer was used to compute power spectra.

DISCUSSION OF RESULTS

One of the main reasons for conducting tests of this kind is to bring to light any unforeseen aerodynamic instabilities. Thus, large amplitude vibrations are looked for. It was found that the model roof, being a taut membrane, was sensitive to wind tunnel noise and it was important not to mistake the resulting large amplitude vibration for an aerodynamic excitation. Wind tunnel noise takes three principal forms, fan noise, standing waves and noise originating from vibration of the tunnel structure. In the present experiments, standing waves, whose frequency is independent of fan speed, caused considerable roof excitation at certain resonant speeds and significant contributions from the other two sources were also evident. Therefore, the discrete frequencies of the tunnel noise and vibration were measured over a range of speeds with the roof absent and the results were then used to remove the worst of the acoustic contamination from the force and deflection measurements. After this had been done, no significant aerodynamic instability could be detected over the mean wind speed range tested, equivalent to 0 to 30 m/s (0 to 67 mph) at full scale. The vibratory forces reported in Reference 6, which describes the preliminary investigation on the roof, are now known to have been acoustic in origin.

The roof was tested with and without the edge of the roof sealed but most of the data were obtained without the seal since this last condition produced higher wind loading. All results in this paper were obtained without the seal. Figure 8 shows typical time histories of force at an attachment point (all results are expressed at full scale in this section). In the example shown, the acoustic contamination of the data, visible as high frequency vibration, was at a relatively low level and has not been removed. The wind induced forces at the upper end of the speed range were of the same order as the forces due to the cable tensioning.

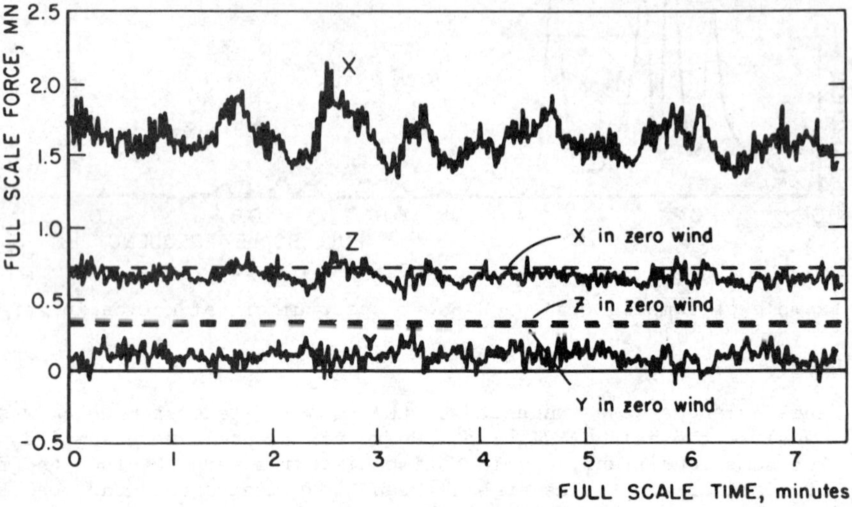

Fig. 8. Time history of forces on console 9, U_{60} = 26.7 m/s, ϕ = 90°

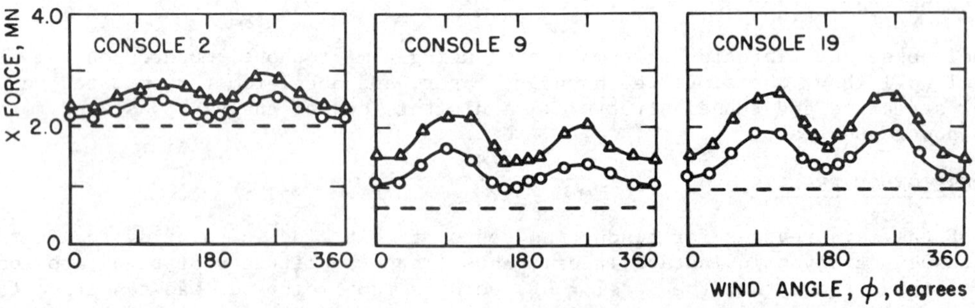

Fig. 9. Variation with wind angle of mean and peak X-component forces on three consoles, U_{60} = 26.7 m/s, ⊙-mean, △-peak, --- zero wind force (due to initial roof tension).

Figure 9 gives examples of the variation with wind angle of the mean and peak X-component forces. The angles $\phi = 90°$ and $\phi = 270°$ produced the greatest wind forces, which may be explained in terms of these being the directions where the highest flow curvature was generated over the roof. It was expected that at $\phi = 180°$, at which angle the tower was upwind of the roof, there might be some excitation due to the tower wake but none was observed. Figure 10 shows the variation with wind speed of the root-mean-square (rms) of fluctuations in the X-component force on console 9 after being corrected for noise. The corrected rms forces varied smoothly with wind speed indicating no aerodynamic instability.

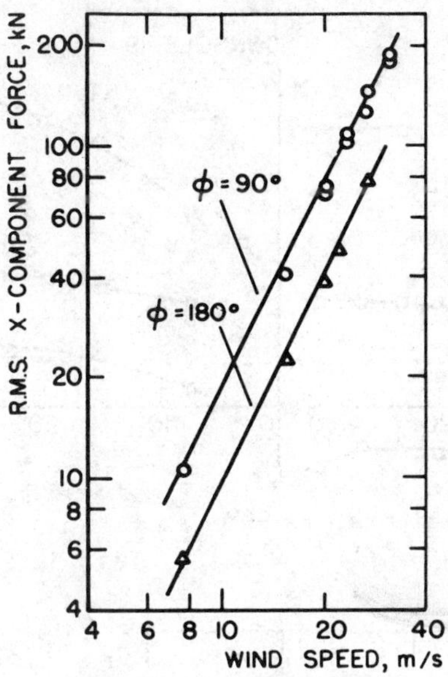

Fig. 10. Rms X-component force console 9.

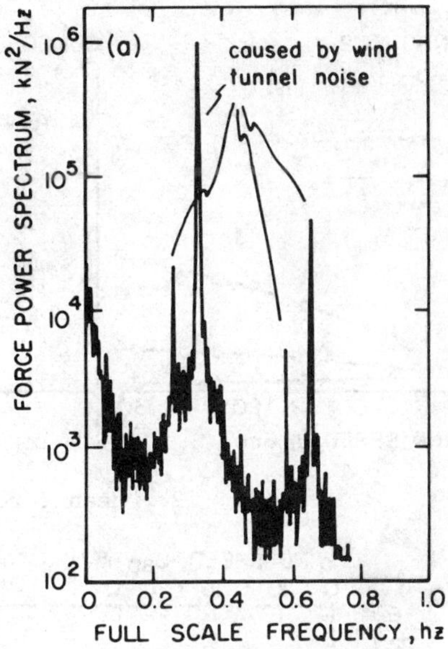

Fig. 11a. Example of data badly affected by wind tunnel noise, U_{60} = 30 m/s, $\phi = 90^{\circ}$.

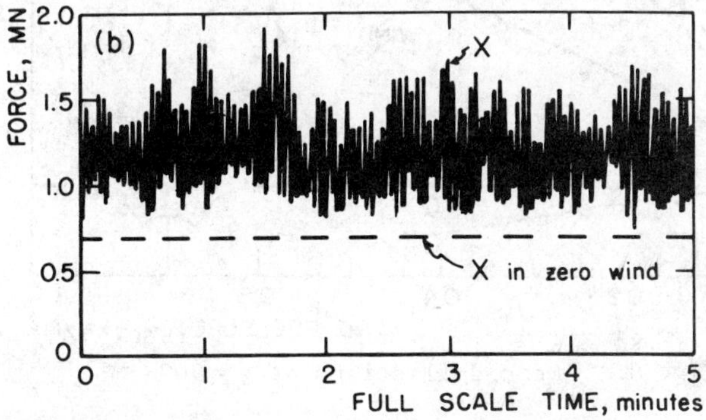

Fig. 11b. Example of data badly affected by wind tunnel noise, U_{60} = 30 m/s, $\phi = 90^{\circ}$.

The rms forces measured in the preliminary tests (see Introduction), where the roof mass was too high by a factor of approximately four, were very similar to the present values, thus verifying that the membrane mass need not be accurately simulated because of the large added mass effect. Figures 11(a),(b) show one of the worst cases of noise contamination and illustrate the degree to which wind-tunnel noise affected some of the measurements. Figure 12 shows examples of the mean force coefficients $\overline{\Delta X}/q_{60}^2A$, $\overline{\Delta Y}/q_{60}^2A$, $\overline{\Delta Z}/q_{60}^2A$ as functions of wind speed. It can be seen that significant changes occurred in some of the force coefficients as the wind speed increased. These changes are attributed mainly to deflections of the roof altering the mean pressure distribution and to the non-linear stiffness of the roof. Examples of mean deflections, some of which exceeded 10% of the local roof depth are given in Figure 13. In view of these results, the use of rigid pressure-tapped models for membrane-like structures should be approached with caution because of the neglect of deflection effects and the consequent danger of underestimating wind forces.

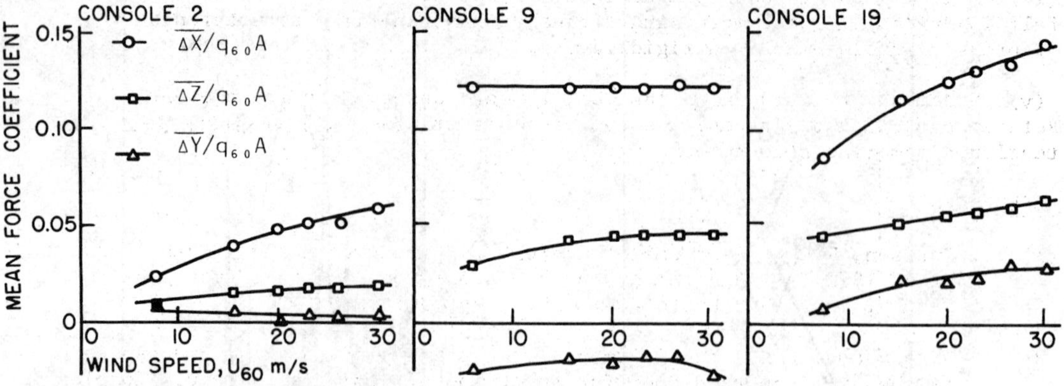

Fig. 12. Mean force coefficients, $\phi = 90^o$

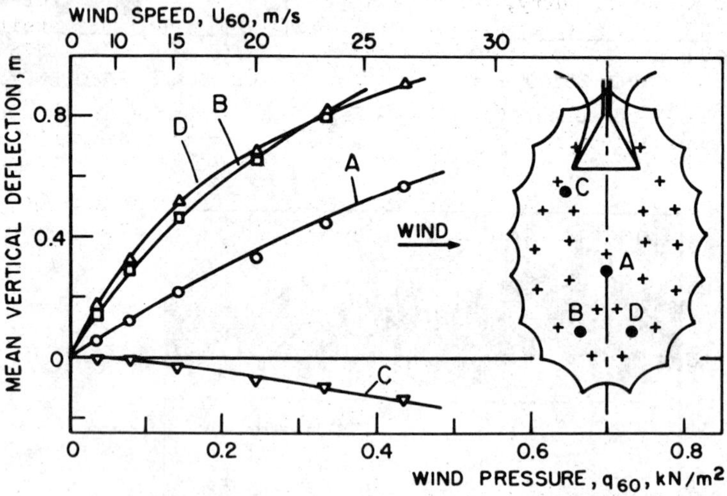

Fig. 13. Mean deflections at $\phi = 90^o$

CONCLUSIONS

From this particular investigation, the following conclusions can be drawn which bear on wind effects on membrane-like structures in general:

(i) Added mass effects are significant and may in some instances be dominant for lightweight membranes. Thus, great precision is not always required in modelling the membrane mass.

(ii) Acoustic damping is likely to prevent volume displacing modes of vibration from being excited. To model this effect correctly requires Mach number scaling, i.e. a velocity scaling of 1:1 in normal wind tunnels, but velocity scaling of less than 1:1 is conservative. A velocity scaling of 1:1 has an additional advantage for sealed models in that no distortion of the internal volume is then required when modelling pneumatic stiffness.

(iii) Membranes are sensitive to wind tunnel noise and a thorough knowledge of the acoustic environment in the wind tunnel is essential in interpreting the data.

(iv) On the present roof the deflections due to wind were significant. It is felt inadvisable to estimate wind forces on membrane-like structures on the basis of pressure measurements on rigid models.

(v) No signs of aerodynamic instability were observed on the present model. References 2, 3 and 4 report the same absence of serious instabilities on other tensioned membrane structures.

REFERENCES

(1) Knudsen, W.C., "Response of Cable-Net Structures under Dynamic Loads", Proc. 1971 IASS Pacific Symposium Part II on Tension Structures and Space Frames, Tokyo and Kyoto, Oct. 1971, pp. 365 - 376.

(2) Tryggvason, B.V., and Isyumov, N., "A Study of the Wind Induced Response of the Air Supported Roof for the Dalhousie University Sports Complex", University of Western Ontario Engineering Research Report BLWT-SS7-1977.

(3) Tryggvason, B.V., Surry, D. and Davenport, A.G., "A Study of Wind Effects for the Haj Terminal at Jedda International Airport", University of Western Ontario Engineering Research Report, BLWT-SS9-78, 1978.

(4) Grillaud, G., and Gandemer, J.,"Étude de la Reponse Dynamique d'une Structure Gonflable", Symposium International sur les Structures Gonfables, C.I.B. Venise, June 1977.

(5) Romani, L., "Stade Olympique de Montréal: Études Aérodynamiques sur Maquette", Report to Architect Roger Taillibert from Laboratoire Eiffel, July 16, 1973.

(6) Irwin, H.P.A.H., and Proulx, J., "A Wind Tunnel Investigation of the Montreal Olympic Stadium", The Second Canadian Workshop on Wind Engineering, Varennes, Quebec, Sept. 1978. Proceedings in press.

(7) Campbell, W.F., "Additional Mass and Inertia Effects", National Research Council of Canada, NAE Report LTR-LA-223, 1979.

(8) Irwin, H.P.A.H., Wardlaw, R.L., Wood, K., and Bateman, K., "A Wind Tunnel Investigation of the Montreal Olympic Stadium", National Research Council of Canada, NAE Report LTR-LA-228, 1979.

APPENDIX

Added mass and acoustic damping

To estimate the order of magnitude of the added mass, consider a sphere, of surface area A, whose surface is somehow made to pulsate with a radial velocity uniform over the surface. Neglecting compressibility, the added mass of the air external to the sphere, per unit surface area, is

$$M_A = \rho\sqrt{\frac{A}{4\pi}} = 0.282\ \rho\ \sqrt{A}$$

Thus, if the surface is a fabric membrane of mass per unit area, M_F, the virtual mass per unit area, M_V, is

$$M_V = M_F + M_A = M_F + 0.282\ \rho\ \sqrt{A}$$

Now suppose A is 18,000 m^2, equal to the total area of the Stadium roof, $M_F = 2.2$ kg/m^2 and $\rho = 1.23$ kg/m^3. Then $M_V = 2.2 + 46.6 = 48.8$ kg/m^2. Thus, the added mass is concluded to be large, an order of magnitude greater than the membrane mass. Campbell (7) and Irwin (8) have computed results of similar magnitude for flat membranes resembling the roof. Clearly, from the preceding analysis, the size of the effect depends on the membrane area and its mass.

In Reference 8 it is shown that for volume displacing modes of vibration of a nearly flat roof covering a sealed volume the radiated acoustic energy gives rise to an increment, ζ_A, in the damping ratio given by

$$\zeta_A = \frac{k}{2} \frac{\rho n A}{c M_v}$$

where c = speed of sound and k is a constant of order unity depending on the mode shape. In arriving at this expression, it was assumed that for typical frequencies of roof motion the wavelength of sound was much larger than the roof. Thus, for a sealed roof, with the same size and mass as the Stadium roof, and with frequencies in the range 0.1 to 1.0 Hz, it is deduced that the order of magnitude of ζ_A is in the range 0.06 to 0.6. This is a very large contribution to the damping. For an unsealed roof, the value of ζ_A would be reduced. The expression for ζ_A can be rearranged into the form

$$\zeta_A = \frac{k}{2} \left(\frac{M_V}{\rho b}\right)^{-1} \left(\frac{nb}{U}\right) \left(\frac{A}{b^2}\right) \left(\frac{U}{c}\right)$$

from which it is seen that for correct simulation of ζ_A it is necessary for the Mach number, U/c, to be the same on the model as at full scale. In many wind tunnels, the full-scale Mach number cannot be achieved but if the model Mach number is less than the full-scale value ζ_A will be less on the model than at full scale, and this is conservative.

AERODYNAMIC STABILITY CRITERIA OF ONE-WAY TYPES OF HANGING ROOFS IN SMOOTH UNIFORM FLOW

S. KAWAMURA and E. KIMOTO

Osaka City University, Osaka, Japan

SUMMARY

With regard to oscillating phenomena of one-way types of hanging roofs due to wind, theoretical equations are derived by the good use of modified thin airfoil theory, improved in Kutta-Joukowski's condition and pressure on roofs by assuming oncoming flow to be smooth and uniform. In favour of qualitative results in wind tunnel tests, two critical reduced velocities are deduced from these equations based on potential energy. Further, these critical velocities are made clear to accord substantially with two stability criteria which are distinguishing characteristics of phenomena. Probably these criteria would be useful for use to design this kind of hanging roofs.

INTRODUCTION

Tension structures such as hanging roofs and pneumatic structure have recently come into wider use. These structures are likely to flutter (1) in the wind because of their small natural frequency and structural damping. This fluttering phenomenon is one of the main subjects of hanging roofs. The response of roofs has been statistically derived from the pressure disturbance (2) which has not interacted with structural movement, measured on rigid flat roofs (3) by making a modified application of thin airfoil theory to the flag fluttering in uniform flow, a critical velocity has been theoretically made clear to have a great effect on mass ratio and reduced velocity. Similarly, a critical velocity has been theoretically and experimentally suggested by making a good choice of roofs pretension as restoring force (4). Further, theoretical equations have been presented for one-way types of hanging roofs with walls by the use of improved pressure in the absence of air stream below roofs (5).

On the other hand, a few experiments (6) on pressure with rigid model had already shown that on the windward, there was a negative falling-off of mean component and its spectra of disturbance made a slight peak. A few qualitative facts (7,8), that roofs frequency was proportional to wind speed, had been presented by the use of wind tunnel tests with flexible roofs. The relations among Strouhal number, Reynolds number, and mass ratio had been revealed by wind tunnel test of flag (9). The effects of pretension, structural damping, and turbulence of stream on roofs response had been studied (10, 11).

Little is known of the all-out relations among these quantities although this is a problem of practical significance. Therefore, the present paper is devoted to the aerodynamic stability problem concerning one-way types of hanging roofs with and without walls.

PHENOMENA

Restricting our consideration to the phenomena in which roofs amplitude has rapidly increased in proportion to the increment of wind velocity, the wind tunnel tests were made on models of one-way types of hanging roofs (12) with and without walls, which were settled alone without cable-nets for pretension, in smooth uniform flow.

Model

The scale of model was 45 cm in span (L), 30 cm in breadth (B), and 11.4 cm in height (H). The sag (b) of roofs was nearly either 2 cm or 4 cm. The roof was covered with a vinyl sheet of 4×10^{-3}cm thickness (t_0), whose weight per unit area ($\rho_s g$) was 6.61×10^{-3}gr/cm^2. Where, ρ_s is roofs density and g is acceleration of gravity.

Experimental technique

The wind velocity of oncoming flow and the velocity around roofs were measured by anemometer set on the windward point of model and the hot wire anemometer respectively. The pressure disturbance in close vicinity to the center of roofs was measured through pressure transducer. Both edge beams were supported by phosphor bronze plates, where strain guages were set to indicate the disturbance of tensile force in roofs. The pictures of roofs behaviour were taken by 8 mm cinecamera. Projecting these pictures on screen, static displacement, amplitude, and its period were counted by rule and stop watch. The arrangements of model and experimental apparatuses are shown in Fig. 1.

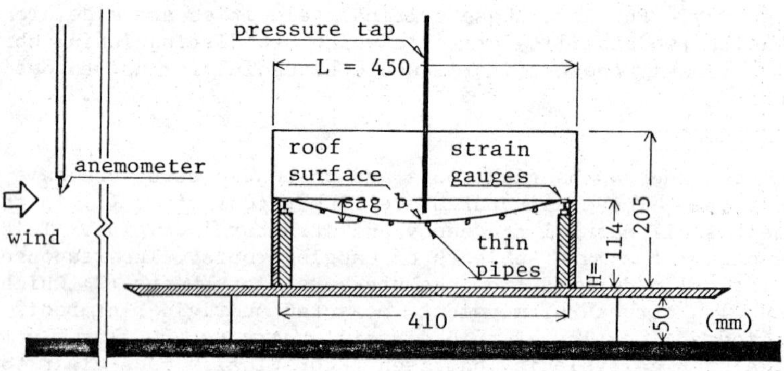

Fig.1 Model and experimental apparatuses

Outline of Phenomena

The roofs began to oscillate under wind action around the neutral state, where the mean wind pressure lifted up surface. The displacements were taken from initial state to this neutral state, being regarded as static because of mean lift force. From this neutral surface, the amplitude and its dominant frequency were read out. The relations among displacement, amplitude, period, and wind speed are shown in Fig. 2 - 4 as illustrative examples.

These results indicate that the displacement increases rapidly in the immediate neighborhood of wind speed where the amplitude increases remarkably and moreover the period of disturbance pressure near the roof center coincident with that of roofs.

On the otherhand, the flowpattern around forced oscillations of roofs were examined with a smoke tunnel.

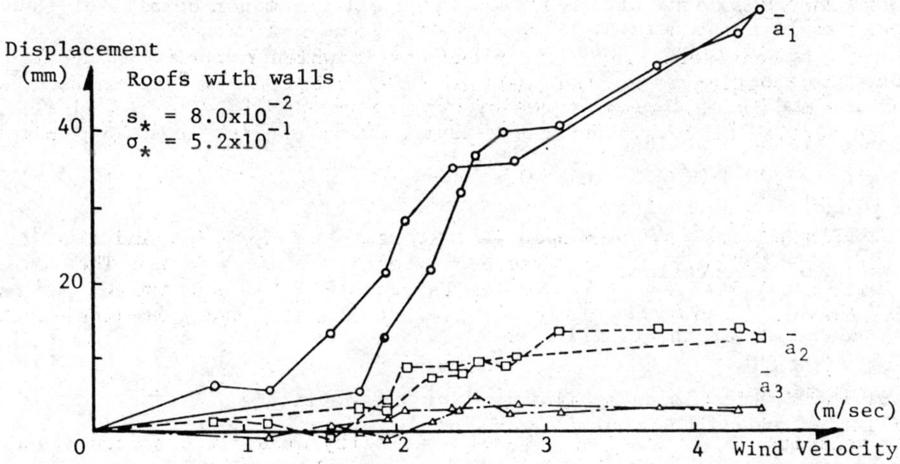

Fig.2 Displacement

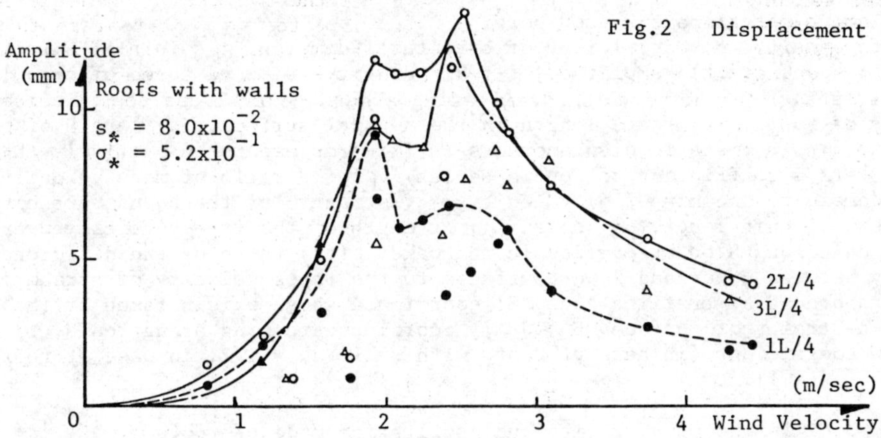

Fig.3 Amplitude

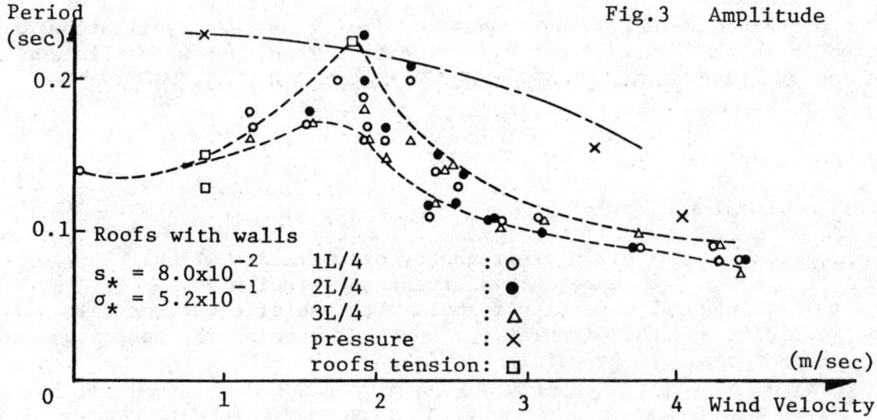

Fig.4 Period

THEORETICAL TREATMENT

Taking account of the interaction between disturbance of flow and oscillation of roofs, fundamental equations are theoretically developed for the movement of neutral roofs in the smooth uniform flow. Then, all the minor details of these equations are left out here.

Equation of motion;

$$\rho_s \frac{\partial^2 a(S,t)}{\partial t^2} + c\frac{\partial a(S,t)}{\partial t} = N\frac{\partial^2 a(S,t)}{\partial S^2} + P_r(S,t) \tag{1}$$

Disturbance pressure acting on roofs;

$$P_r(S,t) = -a_0 \rho_a \frac{\partial}{\partial t}\int_0^S \gamma_s(S',t)dS' - a_0 \rho_a \beta U \gamma_s(S,t) \tag{2}$$

Boundary condition on roofs;

$$-\frac{1}{2\pi}\oint_0^{S_n}\frac{\gamma_s(S',t)}{(S-S')}dS' - \frac{1}{2\pi}\oint_{S_n}^{+\infty}\frac{\gamma_w(S',t)}{(S-S')}dS' = \frac{\partial a(S,t)}{\partial t} + \beta U\frac{\partial a(S,t)}{\partial S} \tag{3}$$

Conservation of circulation;

$$\frac{d}{dt}\int_0^{S_n}\gamma_s(S',t)dS' + \beta U \gamma_w\left\{S, t + \frac{S-S_n}{\beta U}\right\} = 0 \tag{4}$$

Velocity condition at the trailing edge;

$$\gamma_w(S_{n+0}, t) = \alpha(t)\beta U \tag{5}$$

Strain of roofs which is made allowance for the amplitude of oscillation around the neutral surface;

$$\varepsilon = \frac{N}{E t_0} \tag{6}$$

The following notations are used here.
ρ_s: roofs density, c: coefficient of structural damping, ε_0: initial strain of roofs, L: span, a(s,t): amplitude, ρ_a: air density, U: wind speed of oncoming flow, N: roofs tension per unit width, oscillating around neutral surface , Et_0: extension rigidity of roofs, S_n: roofs length in the neutral surface, S_b: length of oscillating roofs, a_1, a_2: static displacements (suffix represents respectively the 1st and the 2nd term's coefficient of Fourie series), β: the ratio of mean velocity in separated domain to oncoming flow velocity, γ_s: strength of the bound vorticity over an element of unit length on roofs, γ_w: strength of the free vorticity over an element of unit length on streamline in the wake, $\alpha(t)$: ratio of the difference in velocity between upper and lower surfaces to the upper velocity of streamline in the neighborhood of trailing edge, S: coordinate whose origin taken at the leading edge of neutral roof surface, S': dummy coordinate running along roofs, t: time, a_0: flow coefficient (in case of roofs with walls a_0 = 1/2, in case of roofs without walls a_0 = 1).

Harmonic oscillation of vorticity and oscillation mode of roofs

Both the disturbance velocity near roofs and its transverse oscillation are assumed to be harmonic in the limit of certain velocity. Then, roofs oscillation may take place in one of many modes.

$$\begin{aligned} \gamma_s(S,t) &= \gamma_s(\eta)e^{i\omega t} \\ \gamma_w(S,t) &= \gamma_w(\eta)e^{i(\omega t+\Delta\theta)} \\ a(S,t) &= a_m(t)\sin\left\{\frac{m\pi}{2}(\eta+1)\right\} \quad \text{(m: the order of m-th mode)} \end{aligned} \tag{7}$$

where, $\eta = \frac{S - S_n/2}{S_n/2}$, ω: circular frequency of vorticity which represents disturbance velocity

According to the integral inversion formula with use of equations (4), (5) and (7), the bound vorticity and the circulation, which is summing the contribution of bound vorticity along roofs, can be written as follows.

$$\begin{aligned} \gamma_s(\eta) = i\frac{2(\beta U)^2}{\pi\omega S_n\sqrt{1-\eta^2}}&\left[1 + i\frac{\omega S_n}{2\beta U}e^{i\frac{\omega S_n}{2\beta U}}\int_1^{+\infty} e^{-i\frac{\omega S_n}{2\beta U}\lambda}\left\{\frac{\sqrt{\lambda^2-1}}{(\lambda-\eta)} - 1\right\}d\lambda\right]\alpha(t)e^{-i\omega t} \\ &+ \frac{2}{\pi}\oint_{-1}^{1}\sqrt{\frac{1-\xi^2}{1-\eta^2}}\frac{w_a(\xi,t)}{(\eta-\xi)}d\xi\, e^{-i\omega t} \end{aligned} \tag{8}$$

$$\Gamma(\eta)=i\frac{2(\beta U)^2}{\pi\omega S_n\sqrt{1-\eta^2}}\left[(\pi-\cos^{-1}\eta)+i\frac{\omega S_n}{2\beta U}e^{i\frac{\omega S_n}{2\beta U}}\int_1^{+\infty}e^{-i\frac{\omega S_n}{2\beta U}\lambda}\left\{\cos^{-1}\eta-\sin^{-1}\left(\frac{\sqrt{\lambda^2-1}\sqrt{1-\eta^2}}{\lambda-\eta}\right)\right\}d\lambda\right]\alpha(t)e^{-i\omega t}$$

$$-\frac{2}{\pi}\oint_{-1}^{1}w_a(\xi,t)\log\left|\frac{1-\eta\xi+\sqrt{(1-\eta^2)(1-\xi^2)}}{\eta-\xi}\right|d\xi\, e^{-i\omega t} \qquad (9)$$

where, $w_a(\eta,t)=\frac{\partial a(\eta,t)}{\partial t}+\frac{2\beta U}{S_n}\frac{\partial a(\eta,t)}{\partial\eta}$

Substituting these equations into Eq. (2) yields a equation of motion which has complex function. Cancelling out $\alpha(t)$ from real and imaginary parts of this equation reproduce a nonlinear equation as follows.

$$\frac{d^2x_m}{d\tau^2}+2\delta_{m*}\frac{dx_m}{d\tau}+4\pi^2\nu_{m*}^2x_m+e_mx_m^3=0 \qquad (10)$$

where, $\delta_{m*}=\delta_m-\delta_{ma}$, $\nu_{m*}^2=f_s^2-f_{ma}$, $e_m=(m^2\pi^4)/(S_*^4\varepsilon_0)$

$$f_s^2=1-\frac{\pi^2S_*}{2\varepsilon_0}X_1+\frac{\pi^2}{4\varepsilon_0}(1-2\pi^2s_*^2)X_1^2+\frac{\pi^2}{\varepsilon_0}X_2^2$$

$$\delta_{ma}=\frac{\omega_{m*}}{\sigma_*}f_{m,21}F_m(\omega_{m*},U_{m*},S_*,\eta_{mr})+2\frac{\beta U_{m*}}{\sigma_*S_*}f_{m,11}$$

$$f_{ma}=\frac{m\beta\omega_{m*}U_{m*}}{\sigma_*S_*}f_{m,22}F_m(\omega_{m*},U_{m*},S_*,\eta_{mr})-2\frac{m(\beta U_{m*})^2}{\sigma_*S_*^2}f_{m,12}$$

$$F_m(\omega_{m*},U_{m*},S_*,\eta_{mr})=\frac{2\beta U_{m*}}{S_*\omega_{m*}}\left[f_{01}^2-2n\pi^2f_{01}\left\{f_{02}+\frac{(7-\eta_{mr}^2)}{6n\pi f_{01}}\right\}\frac{\psi}{\beta U_{m*}}\right]$$

$$\psi=\frac{A_{42}\omega_{m*}^2+A_{41}\beta U_{m*}\omega_{m*}}{\sqrt{A_{31}\beta U_{m*}\omega_{m*}+A_{32}(\beta U_{m*})^2}},\quad f_{01}=\frac{1}{\sqrt{1-\eta_{mr}^2}},\quad f_{02}=1-\frac{1}{\pi}\cos^{-1}\eta_{mr}$$

$$A_{42}=1-\eta_{mr},\quad A_{41}=\frac{4n\pi}{S_*},\quad A_{31}=\frac{32n\pi}{S_*^3},\quad A_{32}=\frac{64n^2\pi^2}{S_*^4},\quad n=1$$

$$f_{m,11}=\frac{1}{\sin\{\frac{m\pi}{2}(\eta_{mr}+1)\}}\oint_{-1}^{1}\sqrt{\frac{1-\xi^2}{1-\eta_{mr}^2}}\frac{\sin\{\frac{m\pi}{2}(\xi+1)\}}{(\xi-\eta_{mr})}d\xi$$

$$f_{m,12}=\frac{-1}{\sin\{\frac{m\pi}{2}(\eta_{mr}+1)\}}\oint_{-1}^{1}\sqrt{\frac{1-\xi^2}{1-\eta_{mr}^2}}\frac{\cos\{\frac{m\pi}{2}(\xi+1)\}}{(\xi-\eta_{mr})}d\xi$$

$$f_{m,21}=\frac{1}{\sin\{\frac{m\pi}{2}(\eta_{mr}+1)\}}\oint_{-1}^{1}\sin\left\{\frac{m\pi}{2}(\xi+1)\right\}\log\left|\frac{1-\eta_{mr}\xi+\sqrt{(1-\xi^2)(1-\eta_{mr}^2)}}{\eta_{mr}-\xi}\right|d\xi$$

$$f_{m,22}=\frac{1}{\sin\{\frac{m\pi}{2}(\eta_{mr}+1)\}}\oint_{-1}^{1}\cos\left(\frac{m\pi}{2}(\xi+1)\right)\log\left|\frac{1-\eta_{mr}\xi+\sqrt{(1-\xi^2)(1-\eta_{mr}^2)}}{\eta_{mr}-\xi}\right|d\xi$$

S_0: roofs length in initial state, ω_{m0}: the m-th natural circular frequency of roofs in initial state, $S_*=S_n/S_0$: ratio of roofs length, $\tau=\omega_{m0}t/(2\pi)$: nondimensional time, $\omega_{m*}=\omega/\omega_{m0}$: the m-th reduced circular frequency, $u_{m*}=U/(\omega_{m0}S_0)$: the m-th reduced velocity, $x_m=a_m/S_0$: the m-th reduced amplitude, $\delta_m=\pi c/(\rho_s\omega_{m0})$: the m-th logarithmic decrement of structural damping, η_{mr} : the representative point of m-th mode, $\sigma_*=\rho_s/(a_0S_n\rho_a)$: nondimensional density

Periodic solution and frequency equation

Eq. (10) can be solved for periodic solution by the asymtotic method.

$$x_m(\tau)=c_0e^{-\delta_{m*}\tau}a\cos\phi\{(1-3c_1)+4c_1\cos^2\phi\}$$

$$\frac{d\phi}{d\tau}=2\pi\nu_{m0}+\varepsilon\frac{3e_{m0}}{16\pi\nu_{m0}}a^2 \qquad (11)$$

where, $\nu_{m0}^2=\nu_{m*}^2-\frac{1}{4\pi^2}\delta_{m*}^2>0$, $e_{m0}=\frac{e_m}{4\pi^2\nu_{m0}^2}$, $c_1=\frac{\varepsilon e_m a^2}{128\pi^2\nu_{m0}^2}$, $c_0=\text{const.}$

The following condition can be deduced from uniqueness theorem of this solution.

$$\varepsilon<1.44\times10^{-1}\frac{S_*^4\varepsilon_0}{m^2a^2}\nu_{m0}^2 \qquad (12)$$

Considering the interaction between roofs oscillation and disturbance flow, the frequency equation is derived as follows.

$$\omega_{m*}^2 = \omega_{km}^2 - 4\pi^2 f_{ma} + 2\delta_m \delta_{ma} - \delta_{ma}^2, \quad (\omega_{m*} = \frac{d\phi}{d\tau}) \tag{13}$$

where, $\omega_{km}^2 = 4\pi^2 f_s^2 + \varepsilon \frac{3}{4} e_m a^2$

f_s means the ratio of structural frequency in neutral state to that in initial state. Eq. (13) is a polynomial equation of ω_{m*} and its coefficient of n-th degree is reciprocal to u_{m*}. We shall confine our attention to the 1st, the 2nd and the 3rd mode respectively considering the effect of damping and the representative point shall be taken as follows, corresponding each mode.

$$\eta_{1r} = 0, \ \eta_{2r} = -0.5, \ \eta_{3r} = 0 \tag{14}$$

After performing numerical calculations of the coefficients which depend on σ_*, $f_{m,ij}$ and others, the minimum circular frequency ω_{m*0} is obtained. Moreover, the investigation is made into the existence of ω_{m*0}. Then, the solution ω_{m*0} is a function of reduced velocity u_{m*}.

The 1st and the 2nd critical reduced velocities

The total potential energy of system oscillating in ω_{m*0} can be obtained as follows.

$$E(\tau) = E(0) - C_0^2 a^2 \delta_{m*0}(\omega_{m*0}^2 + 2\delta_{m*0}^2)\tau \tag{15}$$

Where, E(0): initial value of total potential energy, δ_{m*0} : total damping of system oscillating in ω_{m*0}.

By use of ω_{m*0}, the total potential energy can be rewritten as a function of u_{m*}.

On the basis of aerodynamic knowledge, the 1st critical reduced velocity can be deduced from the condition that the total damping is zero.

$$\delta_{m*0}(u_{m*c1}) = 0 \tag{16}$$

When the total damping is positive, the stability of system can be studied with the characteristic of total potential energy $E(u_{m*})$.

As far as all of three modes treated here, both ω_{m*0} and δ_{m*0} are uniformly increasing and/or decreasing functions of u_{m*} all over the useful range of ω_{m*0}.

The extreme values of $E(u_{m*})$ appear on the boundary of ω_{m*0}'s useful range.

One mode disappears beyond this boundary and another mode becomes to appear in larger velocity.

The 2nd critical reduced velocity is deduced from this boundary where jumping phenomena take place in total potential energy. Rearranging these critical reduced velocities, the following representations are obtained now.

The 1st critical reduced velocity,

$$u_{c11} = \beta^{-1}\lambda_{c11} S_* \omega_{k1} \cdot \theta_{c11}, \quad \theta_{c11} = 1 - \lambda_{c13}\delta_1/\omega_{k1}, \ (10^{-1} \leq \sigma_* < 2.48\times10^{-1})$$

$$u_{c12} = \beta^{-1}\lambda_{c12} S_* \omega_{k1} \cdot \theta_{c12}, \quad \theta_{c12} = 1 - \lambda_{c14}\delta_1/\omega_{k1}, \ (2.48\times10^{-1} \leq \sigma_* < 1.79)$$

There doesn't exist this type of critical reduced velocity when the nondimensional density σ_* is not less than 1.79. (17)

The 2nd critical reduced velocity,

$$u_{c2} = \beta^{-1}\lambda_{c2} S_* \omega_{k1} \cdot \theta_{c2}, \quad \theta_{c2} = 1 - \lambda_{c2}\delta_1/\omega_{k1}, \ (10^{-1} \leq \sigma_* \leq 2.00)$$

$\beta^{-1}\lambda S_*\omega_{k1}$ and Θ, which are functions of σ_* and $f_{m,ij}$, are named here the restoring term and the damping term respectively after their properties. The coefficient of restoring term and the damping term are shown in Fig. 5, which are applicable to the region of σ_* from 0.1 to 2.0.

As far as the relative error which is the ratio of absolute error to calculated value, the theoretical deduction of ω_{1*0} and the numerical calculation of Uc are respectively estimated at 3.47 and 1.73 at most. Therefore, the relative error of critical reduced velocity may be suggested to be roughly on the order of the sum of these values.

Considerations

It seems worthwhile to consider and list up the current information about critical reduced velocity.

1) Both the roofs density and the periodic change of deformed roofs due to mean pressure have remarkable effect on each critical reduced velocity.

2) The weight of roofs with walls has twice effect on critical velocity as compared with that of roofs without walls.

3) There is not in existence the 1st critical velocity which begins self-excited motion in case of very heavy roofs.

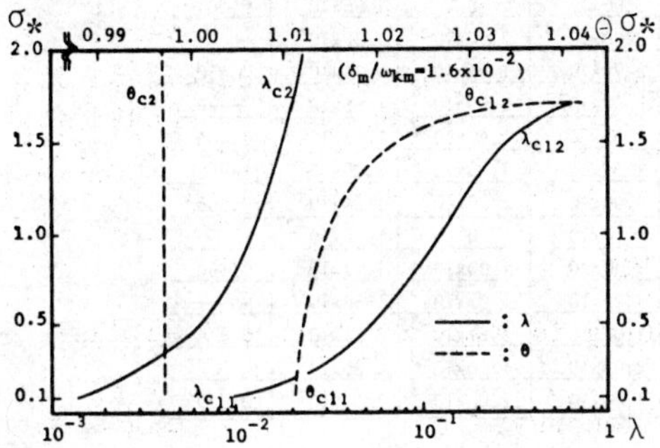

Fig.5 The coefficients λ of restoring terms and the damping terms θ

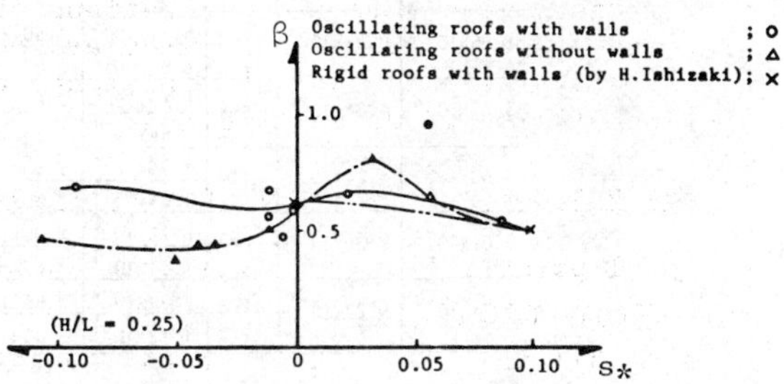

Fig.6 The ratio of mean velocity in separated domain to oncoming flow

EXPERIMENTAL STUDY

Although the detail of model and the experimental technique are much the same as chapter mentioned above, 16 models differing in sag and density are examined here. This chapter will be devoted to the experimental results only because of the lack of enough space to describe in detail.

Starting Velocty(u_{1*s}) and Peak Velocity(u_{1*p})

From the behavior of roofs amplitude according to the increase of oncoming flow, it will be noticed that the amplitude is gradually increasing in the range of small velocity and rapidly increasing beyond a certain velocity and then shows a few peaks. On the basis of these characteristics, the starting velocity is defined to be a intersecting point of initial tangent with second tangent to rapidly increas-

Table 1. Starting velocity and peak velocity of roofs model with walls

Outline of Models		u_{1*s} & u_{1*p}		Displacements, Amplitude and Circular Frequency				
s_*	σ_*			$\bar{a}_1 / L$	$\bar{a}_2 / L$	$\bar{a}_3 / L$	a / S_0	ω / ω_{10}
3.98×10^{-2}	4.50×10^{-1}	u_{1*s}	6.76×10^{-2}	3.30×10^{-3}	3.00×10^{-3}	5.00×10^{-4}	5.44×10^{-3}	
		u_{1*p}	1.10×10^{-1}	3.20×10^{-2}	1.45×10^{-2}	9.95×10^{-3}	2.36×10^{-2}	0.670
9.29×10^{-2}	4.38×10^{-1}	u_{1*s}	8.10×10^{-2}	3.50×10^{-3}	5.00×10^{-4}	2.50×10^{-3}	7.79×10^{-3}	
		u_{1*p}	1.45×10^{-1}	6.50×10^{-2}	3.13×10^{-2}	5.63×10^{-3}	6.49×10^{-2}	0.678
5.97×10^{-2}	6.58×10^{-1}	u_{1*s}	1.08×10^{-1}	3.50×10^{-3}	2.10×10^{-3}	1.60×10^{-3}	8.85×10^{-3}	0.645
		u_{1*p}	2.01×10^{-1}	6.45×10^{-2}	1.10×10^{-2}	7.25×10^{-3}	4.85×10^{-2}	0.818
8.85×10^{-2}	6.44×10^{-1}	u_{1*s}	9.42×10^{-2}	3.50×10^{-3}	1.50×10^{-3}	0	7.42×10^{-3}	0.542
		u_{1*p}	1.95×10^{-1}	6.43×10^{-2}	2.37×10^{-2}	2.60×10^{-3}	7.61×10^{-2}	0.909
5.31×10^{-2}	8.64×10^{-1}	u_{1*s}	7.92×10^{-2}	4.50×10^{-3}	1.40×10^{-3}	0	5.88×10^{-3}	
		u_{1*p}	1.89×10^{-1}	4.75×10^{-2}	2.04×10^{-2}	3.50×10^{-3}	3.27×10^{-2}	0.902
8.85×10^{-2}	8.42×10^{-1}	u_{1*s}	1.48×10^{-1}	1.04×10^{-3}	5.00×10^{-4}	-3.80×10^{-4}	9.38×10^{-3}	0.731
		u_{1*p}	2.44×10^{-1}	6.50×10^{-2}	2.05×10^{-2}	-2.00×10^{-3}	5.61×10^{-2}	1.05
5.75×10^{-2}	1.07×10^{0}	u_{1*s}	1.11×10^{-1}	4.50×10^{-3}	3.20×10^{-3}	-2.80×10^{-3}	8.53×10^{-3}	0.657
		u_{1*p}	2.46×10^{-1}	6.04×10^{-2}	2.14×10^{-2}	5.00×10^{-4}	4.36×10^{-2}	0.917
9.96×10^{-2}	1.05×10^{0}	u_{1*s}	1.51×10^{-1}	4.50×10^{-3}	3.00×10^{-3}	-2.00×10^{-3}	1.18×10^{-2}	0.694
		u_{1*p}	2.41×10^{-1}	7.88×10^{-2}	3.33×10^{-2}	4.04×10^{-2}	7.35×10^{-2}	1.16

Table 2. Starting velocity and peak velocity of roofs model without walls

Outline of Model		u_{1*s} & u_{1*p}		Displacements, Amplitude and Circular Frequency				
s_*	σ_*			$\bar{a}_1 / L$	$\bar{a}_2 / L$	$\bar{a}_3 / L$	a / S_0	ω / ω_{10}
4.20×10^{-2}	2.25×10^{-1}	u_{1*s}	2.78×10^{-2}	2.00×10^{-3}	2.00×10^{-3}	5.00×10^{-4}	6.64×10^{-4}	——
		u_{1*p}	3.72×10^{-2}	1.50×10^{-3}	1.50×10^{-3}	5.00×10^{-4}	6.65×10^{-3}	——
7.52×10^{-2}	2.20×10^{-1}	u_{1*s}	4.94×10^{-2}	1.00×10^{-3}	2.50×10^{-3}	2.00×10^{-3}	9.84×10^{-4}	——
		u_{1*p}	7.41×10^{-2}	4.00×10^{-3}	5.00×10^{-3}	0	1.96×10^{-2}	0.381
3.98×10^{-2}	3.29×10^{-1}	u_{1*s}	5.85×10^{-2}	5.00×10^{-4}	2.00×10^{-3}	2.00×10^{-3}	1.32×10^{-3}	——
		u_{1*p}	9.93×10^{-2}	2.70×10^{-3}	2.70×10^{-3}	1.75×10^{-3}	9.24×10^{-3}	0.460
8.85×10^{-2}	3.21×10^{-1}	u_{1*s}	5.11×10^{-2}	1.50×10^{-3}	5.00×10^{-3}	5.00×10^{-4}	1.95×10^{-3}	——
		u_{1*p}	8.07×10^{-2}	2.00×10^{-4}	1.35×10^{-2}	1.40×10^{-3}	2.14×10^{-2}	0.404
5.31×10^{-2}	4.34×10^{-1}	u_{1*s}	5.83×10^{-2}	2.67×10^{-3}	1.33×10^{-3}	1.33×10^{-3}	3.29×10^{-3}	——
		u_{1*p}	1.64×10^{-1}	7.33×10^{-3}	9.00×10^{-3}	2.00×10^{-3}	1.64×10^{-2}	0.605
8.63×10^{-2}	4.24×10^{-1}	u_{1*s}	5.33×10^{-2}	7.50×10^{-4}	2.10×10^{-3}	0	1.96×10^{-3}	——
		u_{1*p}	1.08×10^{-1}	2.00×10^{-3}	1.10×10^{-2}	1.00×10^{-3}	1.94×10^{-2}	0.460
5.09×10^{-2}	5.38×10^{-1}	u_{1*s}	8.71×10^{-2}	2.30×10^{-3}	4.20×10^{-3}	1.87×10^{-3}	1.64×10^{-3}	——
		u_{1*p}	1.66×10^{-1}	4.67×10^{-3}	1.07×10^{-2}	2.67×10^{-3}	8.21×10^{-3}	0.616
9.74×10^{-2}	5.22×10^{-1}	u_{1*s}	7.35×10^{-2}	2.50×10^{-3}	2.00×10^{-3}	-1.50×10^{-3}	4.86×10^{-3}	——
		u_{1*p}	1.13×10^{-1}	5.00×10^{-3}	1.15×10^{-2}	1.00×10^{-3}	2.33×10^{-2}	0.329

ing curve. The peak velocity is defined to be a point of peak.
Together with displacement, amplitude, and circular frequency, both the starting velocity and the peak velocity are reduced by the use of span L, roofs length S_0 and the 1st natural circular frequency ω_{10} and shown in Table 1 and Table 2. When there are a few peak velocities, the smallest one is listed up in these Tables.

The ratio of mean velocity in separated domain to oncoming flow velocity, β

According to the definition in this paper, the ratio β is experimentally studied by surveying the flow arround oscillating roofs due to wind with hot-wire. Ascertaining the separated flow domain according to the intensity of turbulence, the ratio β is determined by averaging the mean velocity distribution over the separated flow domain near roofs and the results are shown in Fig. 6.

CONSIDERATION

By applying this theory to the models used in this experiment, the 1st and the 2nd critical reduced velocities are obtained in Table 3. Three kinds of wind velocity i.e. zero (U_{n1}), $1/\sqrt{2}$(starting velocity) (U_{n2}), and starting velocity (U_{n3}) are

Table 3. Theoretical values of models

Outline of Model			U_{n1}		U_{n2}		U_{n3}	
Wall	σ_*	s_*	$U_{c2}/(\omega_{10}S_0)$	$U_{c1}/(\omega_{10}S_0)$	$U_{c2}/(\omega_{10}S_0)$	$U_{c1}/(\omega_{10}S_0)$	$U_{c2}/(\omega_{10}S_0)$	$U_{c1}/(\omega_{10}S_0)$
with Walls	4.50×10^{-1}	3.98×10^{-2}	5.52×10^{-2}	4.53×10^{-1}	5.39×10^{-2}	4.42×10^{-1}	5.39×10^{-2}	4.42×10^{-1}
	4.38×10^{-1}	9.29×10^{-2}	7.12×10^{-2}	5.75×10^{-1}	5.87×10^{-2}	4.73×10^{-1}	5.58×10^{-2}	4.48×10^{-1}
	6.58×10^{-1}	5.97×10^{-2}	8.68×10^{-2}	7.60×10^{-1}	8.02×10^{-2}	7.00×10^{-1}	7.97×10^{-2}	6.97×10^{-1}
	6.44×10^{-1}	8.55×10^{-2}	9.94×10^{-2}	8.64×10^{-1}	8.39×10^{-2}	7.30×10^{-1}	8.00×10^{-2}	6.96×10^{-1}
	8.64×10^{-1}	5.31×10^{-2}	1.03×10^{-1}	1.01	1.01×10^{-1}	9.92×10^{-1}	1.01×10^{-1}	9.88×10^{-1}
	8.42×10^{-1}	8.85×10^{-2}	1.31×10^{-1}	1.24	1.25×10^{-1}	1.19	1.25×10^{-1}	1.19
	1.07	5.75×10^{-2}	1.36×10^{-1}	1.44	1.28×10^{-1}	1.36	1.28×10^{-1}	1.36
	1.05	9.96×10^{-2}	1.74×10^{-1}	1.82	1.32×10^{-1}	1.37	8.68×10^{-2}	9.06×10^{-1}
without Walls	2.25×10^{-1}	4.20×10^{-2}	2.55×10^{-2}	2.03×10^{-1}	2.49×10^{-2}	1.97×10^{-1}	2.47×10^{-2}	1.97×10^{-1}
	2.20×10^{-1}	7.52×10^{-2}	3.31×10^{-2}	2.65×10^{-1}	3.16×10^{-2}	2.55×10^{-1}	3.11×10^{-2}	2.51×10^{-1}
	3.29×10^{-1}	3.98×10^{-2}	3.63×10^{-2}	2.84×10^{-1}	3.57×10^{-2}	2.79×10^{-1}	3.53×10^{-2}	2.76×10^{-1}
	3.21×10^{-1}	8.85×10^{-2}	5.29×10^{-2}	4.20×10^{-1}	5.04×10^{-2}	4.00×10^{-1}	4.98×10^{-2}	3.96×10^{-1}
	4.34×10^{-1}	5.31×10^{-2}	5.16×10^{-2}	4.26×10^{-1}	4.88×10^{-2}	4.03×10^{-1}	4.81×10^{-2}	3.97×10^{-1}
	4.24×10^{-1}	8.63×10^{-2}	6.38×10^{-2}	5.38×10^{-1}	6.21×10^{-2}	5.23×10^{-1}	6.23×10^{-2}	5.25×10^{-1}
	5.38×10^{-1}	5.09×10^{-2}	6.45×10^{-2}	5.35×10^{-1}	6.19×10^{-2}	5.13×10^{-1}	6.04×10^{-2}	5.01×10^{-1}
	5.22×10^{-1}	9.74×10^{-2}	8.63×10^{-2}	7.27×10^{-1}	7.67×10^{-2}	6.45×10^{-1}	7.26×10^{-2}	6.12×10^{-1}

set up here corresponding to the wind velocity in neutral state of theory. The static displacements in each state of experiment and the ratio β in Fig. 6 are used for the numerical calculation of theoretical values.

Theoretical values and experimental values

As the lift force increases gradually according to the increase of wind velocity in neutral state, the roofs deformation increases and its natural frequency decreases on one-way types of hanging roofs with single layer. Then, two critical velocities decrease slightly. However, the numerical calculation of theoretical value is very troublesome in neutral state where non-zero wind velocity is set up so that the initial state is very useful and efficient for this calculation.
Of course, the wind velocity is zero in initial state.
The 2nd critical velocity isn't always coincident with the starting velocity in nature. The same holds good of the relation between the 1st velocity and the peak velocity. In the case of the 2nd critical velocity, a higher mode of oscillation appears and the total potential energy discontinuously jumps off to the lower value according to the increase of wind velocity. The increase of energy due to the increase of wind velocity seems to contribute powerfully to the increase of static deformation rather than oscillation.
Further, the 1st critical velocity which begins self-excited oscillation seems to make contact with the peak velocity. As the neutral state is remarkably varied from the starting velocity to the peak velocity, this relation doesn't seem to be close.
Therefore, the ratios such as U_s/U_{c2} and U_p/U_{c1} mean the constants which bear mutual relation between experimental values and theoretical values. With these constants, theoretical values can be transformed to experimental values which represent the phenomena of oscillation. Let's use the constant values from 0.6 to 1.2 in the case of one-way types of hanging roofs with walls which has a sag ratio of about 0.05.
Our consideration will be restricted to the constant of mutual relations between experimental values and theoretical values, which are listed up in Table 4. The other theoretical values are left out of this Table because their theories differ much from ours in the theoretical standing point and some suitable assumption and modification are always necessary for numerical calculations. The constant U_s/U_{c2} is larger than the constant U_p/U_{c1}.

Table 4. The constants U_s/U_{c2}, U_p/U_{c1}

with Walls	σ_*	4.50×10^{-1}	4.38×10^{-1}	6.58×10^{-1}	6.44×10^{-1}	8.64×10^{-1}	8.42×10^{-1}	10.74×10^{-1}	10.48×10^{-1}
	s_*	3.98×10^{-2}	9.29×10^{-2}	5.97×10^{-2}	8.55×10^{-2}	5.31×10^{-2}	8.85×10^{-2}	5.75×10^{-2}	9.96×10^{-2}
the Constants	U_s/U_{c2}	1.22	1.14	1.25	0.95	0.77	1.13	0.82	0.87
	U_p/U_{c1}	0.24	0.25	0.26	0.23	0.19	0.20	0.17	0.13
without Walls	σ_*	2.25×10^{-1}	2.20×10^{-1}	3.29×10^{-1}	3.21×10^{-1}	4.34×10^{-1}	4.24×10^{-1}	5.38×10^{-1}	5.22×10^{-1}
	s_*	4.20×10^{-2}	7.52×10^{-2}	3.98×10^{-2}	8.85×10^{-2}	5.31×10^{-2}	8.63×10^{-2}	5.09×10^{-2}	9.74×10^{-2}
the Constants	U_s/U_{c2}	1.10	1.48	1.61	0.97	1.13	0.84	1.35	0.85
	U_p/U_{c1}	0.18	0.29	0.35	0.19	0.36	0.21	0.31	0.16

Discussion

The ratio β is now obtained for one-way types of hanging roofs whose shape factor, which is the ratio of eaves height to span, is 0.25. It is necessary to collect these data of various shape considering scale, deformation and others.
Cartesian coordinate is used here to represent the flow around roofs because the important nature of viscosity has already been taken into consideration for the pressure disturbance. This treatment is suggested to be reasonable from the aforementioned results.
The assumption of uniform flow seems to be admissible and logical today as preliminary study of this kind. However, the velocity gradient and intensity of turbulence seem to have a certain effect on stability of roofs in natural wind.

Then, they are future problems which have received much attention.

CONCLUSION

The present paper has deduced stability criteria of one-way types of hanging roofs in smooth uniform flow from experimental and theoretical studies.

1) The 1st and the 2nd critical nondimensional velocities are numerically calculated in initial state by the use of the ratio β obtained from experiment.

2) These critical velocities can be transformed to the starting and the peak velocities by the use of the constants in Table 4 according to various shape of hanging roofs.

3) The starting and the peak velocities represent characteristics of oscillating phenomena due to wind and suggest stability criteria for making design of hanging roofs. Particularly, the starting velocity will be very useful and efficient.

ACKNOWLEDGEMENT

This study is partly supported by the Education Ministry and the Matsunaga Memorial Foundation in Japan. The authors would like to thank all the parties concerned in this study.

REFERENCES:

(1) L. Zetlin, Elimination of Flutter in suspension roofs, Hanging Roofs, North Holland Pub., 93-103 (1963).

(2) T. Mori, On dynamic response of a flat long-span roof to action of wind, Trans. Archi. Inst. Japan, 241, 91-100 (1976).

(3) T. Mori, On the dynamic wind forces on a flat roof, Trans. Archi. Inst. Japan, 206, 9-16 (1973).

(4) S. Kawamura, Aerodynamic instability of suspension roofs, Colorado State Univ. CEP70-71SK26, 1-23 (1970).

(5) S. Kawamura & E. Kimoto, An aerodynamic stability of one-way types of hanging roofs in the smooth uniform flow, Proc. 3rd ICWEBS Tokyo, 1067-1076 (1971).

(6) F. Vaessen, Windchannel tests to investigate the wind-pressures on a hypershell roof, Hanging Roofs, North Holland Pub., 87-92 (1963).

(7) H. Ishizaki, I. Sung & T. Kumatori, Wind tunnel experiments of a suspended roof, Trans. Archi. Inst. Japan, 126, 9-13 (1966), (in Japanese).

(8) A. Siev, Experimental study of flutter in suspended roofs, Bull. IASS, 23, 3-10 (1965).

(9) S. Taneda, Waving motions of flags, J. Phys. Soc. Japan, 24, 392-401 (1968).

(10) S. Kawamura & J. E. Cermak, Aerodynamic response of hanging roofs, Colorado State Univ. CEP71-72SK-JEC-5, (1970).

(11) S. Kawamura & J. E. Cermak, Aerodynamic behavior of suspension roofs (second report), Colorado State Univ. CEP70-71SK-JEC-26B, (1970).

(12) S. Kawamura & E. Kimoto, Aerodynamic behavior of one-way types of hanging roofs with walls in wind tunnel tests, Memoirs of Fac. Eng. Osaka City Univ. 16, 143-147 (1975).

DISCUSSION ON SESSION VII
Dynamic Response--Bridges, Transmission Lines and Roof Membranes

Rapporteur: R. H. Scanlan
Princeton University
USA

Paper by A. G. Davenport, N. Isyumov, H. Rothman, H. Tanaka--WIND INDUCED RESPONSE OF SUSPENSION BRIDGES--WIND TUNNEL MODEL AND FULL SCALE OBSERVATIONS

Question: H. P. Irwin. In our work on Lions' Gate Bridge at the National Research Council we found that on the original bridge the critical torsional mode involved a significant amount of coupled lateral motion. This tended to alleviate the flutter instability (see Irwin, H. P., J. Industrial Aero., 1978). It appears likely that the same situation may exist on other bridges where the elastic axis is significantly offset vertically from the center of mass. Was there any indication that this was a significant effect on the bridge in your paper?

Reply: A. G. Davenport, N. Isyumov, H. Rothman, H. Tanaka. The comment is of interest since it points out a potential mode of motion frequently precluded from the study. In the study of the Halifax Narrows Bridge (see Ref. 13) horizontal freedom was included but was of no apparent significance. In the Bronx Whitestone Bridge there was (a) a large separation between lateral and torsional frequencies and (b) no indication from full scale any significance of lateral motions. Consequently, it was not considered. The comment by Irwin will remind us to keep the matter under consideration.

Question: J. D. Holmes. From your comment at the end of your paper on the similarity between the response observed on the section model to that on the taut strip model, it would appear that section model rests for bridges may, in fact, have a future. However, I am concerned that the relative intensities and longitudinal and lateral scales of longitudinal and vertical turbulence may be somewhat different for grid turbulence compared with turbulent shear flows representative of actual wind and boundary layer wind tunnels. Can you comment?

Reply: A. G. Davenport, N. Isyumov, H. Rothman, H. Tanaka. The comments on this promising testing approach are still unsubstantiated by any detailed study. There is hope that at the wavelengths of interest the turbulence of interest, the turbulence both in the boundary layer and behind the grid may be isotropic and similar.

Question: J. A. Hertig. Why in your similarity theory you have only considered the neutral atmosphere? In fact, in the most part of the time the atmosphere present thermal stratification who is determinant on the behavior of the plume. It is equally necessary to consider the associated wind profile who is generally not a power law.

Reply: A. G. Davenport, N. Isyumov, H. Rothman, H. Tanaka. Most of the problems which we have examined relate to aerodynamic downwash; typically the entrainment of the plume by a nearby building. This problem becomes significant at high wind speed and the assumption of neutrally stable atmospheric conditions is a good approximation.

Paper by T. Okubo, N. Narita, K. Yokoyama, H. Sato--FIELD OBSERVATION OF AERODYNAMIC BEHAVIOR OF LONG-SPAN BRIDGES

Question: H. Tanaka. (1) Is there any background to estimate the structural damping of Kammon Bridge as given in the paper (1/2% of critical)? (2) What does "mean angle of attack" mean (p. 9)? Should it be considered with the sampling time of vertical fluctuation? (3) Any comments on the effective μ (p. 8-9) considering the effects of drag force component and approaching flow turbulence?

Reply: T. Okubo, N. Narita, K. Yokoyama, H. Sato. (1) The structural damping (0.03 in logarithmic damping) the authors used was obtained by a field test using a large scale twin-excitor. In this opportunity I would like to introduce values required in Japanese Design Specifications for Long Span Bridges:

Structural Damping	Structure
0.03	Completed truss-stiffened suspension bridges
0.02	Completed cable-stayed bridges with girder type deck
0.01	Free standing steel towers

(2) The value (-8 to +8 deg.) gives the whole range of wind inclination angle averaged over ten minutes. The inclination angle decreases rapidly with the increase of wind speed, and such large angle has not any important meaning from the viewpoint of wind resistant design of bridges. (3) Thank you for the comprehensive comment. The effect may be considered in the future analysis for solving the discrepancy between the measured value and the calculated one shown in Figure 14.

Question: H. Sockel. If you make full-scale experiments with a bridge, you measure the vibrations due to the wind and due to the traffic. Is there only a negligible influence due to the traffic?

Reply: T. Okubo, N. Narita, K. Yokoyama, H. Sato. The effect of road traffic on vibration level seems negligible by the following reasons: (1) Signal transducers are set up on bridge members far from the deck, and the traffic induced vibration with high frequency may decrease in its magnitude at a transducer. (2) Traffic volume at high wind speed condition is not so large that the effect of vehicles on the response is neglected in data processing. (3) The bridges are closed to the traffic for traffic safety when wind speed exceeds at the specific level (say 20 m/s).

Question: W. H. Melbourne. Model and full-scale measurements on the West Gate Bridge (which has a slightly slender section than the Suehiro Bridge) showed that the vertical bending mode varied with $(\sigma/\mu U)^{2.2}$. Measurements on the Suehiro Bridge (Table 3) showed variation of turbulence intensity from 0.17 to 0.25. Can the effect of turbulence intensity be identified in your results, in particular can the turbulence intensity associated with the data points in Fig. 13 be identified?

Reply: T. Okubo, N. Narita, K. Yokoyama, H. Sato. The measured results at the Suehiro Bridge cannot directly be identified if they agree with your suggestion. The identification will be checked after we accumulate the data which has wide variation in both wind speed and turbulence intensity.

Paper by T. Yoshimura, Y. Nakamura--ON THE INDICIAL AERODYNAMIC MOMENT RESPONSES OF BRIDGE DECK SECTIONS

Question: R. H. Scanlan. Would the authors kindly make a brief qualitative description of the (time-dependent) evolution of pressures over the bridge section that accompanies their excellent slides?

Reply: T. Yoshimura, Y. Nakamura. Perhaps Figure 8 which is shown on p. 8 of our paper will answer your question. This figure indicates several stages of the upper surface pressure distribution for Model B2 following an indicial change in incidence of 4 degrees. You will see that at each instant, for example $\tau = 1.8$, there is a peak and a trough of pressure and they move downstream as time goes on. This corresponds to the movement of the leading edge vortex which I mentioned in the slides. In particular, the peak suction appearing just downstream of the leading edge (for $\tau = 0 \sim 1.8$) is the source of the overshoot of the indicial moment response.

Paper by Y. K. Lin--STOCHASTIC ANALYSIS OF BRIDGE MOTION IN LARGE-SCALE TURBULENT WINDS

Question: T. Miyata. I would like to make a brief comment. Due to the assumptions you set, the onset of unstable oscillation in turbulent flow is always lower than that in smooth flow. However, our previous wind tunnel experiments and the large-scale bridge model experiment in natural wind reported in this session indicate that the onset cannot be defined in the form of reduction of velocity and appears to be at higher velocity. Unsteady coefficients affected by the presence of turbulence should be taken into account. The onset of galloping in turbulent flow is presented in the form of shifting to higher velocity.

Reply: Y. K. Lin. As I indicated in my response to Professor Scanlan's question, the change of aerodynamic laws due to the presence of turbulence can indeed contribute to discrepancy. However, the galloping motion is a nonlinear "limit cycle" type behavior which is not an instability in the strict sense because its amplitude is always bounded. Experimentally, one can misinterpret a galloping motion as an unstable motion.

Question: B. Etkin. It appears that the inability to separate forcing terms $U(t)$ from response terms $\alpha(t)$ is because of the products of these in Eq. (1). Is it not true that if the equation is linearized for small disturbances, then the conventional separation will result? (This follows if ξ, η, and α are all of the same order.)

Reply: Y. K. Lin. The smallness of turbulence component compared with the average wind velocity does not always justify its omission. Indeed, as it was shown in the paper the results from the full and partial stochastic analyses were markedly different when the stability boundary is approached. In the partial stochastic analysis the turbulence term in the self-excited load was neglected.

Question: R. H. Scanlan. Would the author kindly comment on what might constitute the main reasons for the apparent inconsistency (to date) between his results and global wind tunnel results on 3D models where turbulence delays the onset of instability?

Reply: Y. K. Lin. In addition to possible change in aerodynamic laws due to the presence of turbulence, modal coupling appears to be a reasonable explanation. This coupling provides a mechanism to transmit energy from one mode to another. Thus, the energy overflow from the most vulnerable mode can be absorbed by other modes and the total energy balance in the overall system can be achieved at a higher mean wind velocity.

Paper by A. G. Davenport--GUST RESPONSE FACTORS FOR TRANSMISSION LINE LOADING

Question: E. Simiu. Does the question of the possible nonlinear behavior of the cables require additional research or is it established that the linearization of the problem is acceptable?

Reply: A. G. Davenport. The principle nonlinearity which may be of interest is that when in extremely high winds the conductor swing angle is no longer small. In this case problems arise including only the hammock swinging of the cable. As remarked in the presentation the Electric Power Research Institute has experiments in hand to look at the problem.

Question: D. W. Ebeling. On the first 500 kva DC line from the Columbia River South through Oregon, one area with a prevailing 15 mph wind across the line caused coupled oscillations stripping the insulators and dropping the lines. As construction was underway the insulators were reinforced and alternate spans weighted to correct the problem - the dimensions were comparable to your example. Any other suggestions?

Reply: A. G. Davenport. It is not clear what was the nature of the coupled oscillations. It sounds as if it may have been a higher frequency galloping problem not intended to be covered by the approach of the paper.

Paper by H. P. A. H. Irwin, R. L. Wardlaw--A WIND TUNNEL INVESTIGATION OF A RETRACTABLE FABRIC ROOF FOR THE MONTREAL OLYMPIC STADIUM

Question: B. Tryggavson. The authors show that for a membrane enclosing a volume the acoustic damping is very high. I wonder if you have a feel for the magnitude of the acoustic damping in membranes not enclosing volumes, and whether you have a feel for the magnitude of the aerodynamic damping?

Reply: H. P. A. H. Irwin, R. L. Wardlaw. For a membrane with unsealed edges some local cancellation of pressure waves emanating from the membrane can be expected, resulting in a reduction in the radiated energy. We have not estimated the size of the reduction or what the contribution would be from aerodynamic damping. As stated in the paper no additional similarity parameters need to be introduced, over and above those already in the paper, to correctly model aerodynamic damping.

Comment: B. M. Leadon. The introduction of Mach Number as a similarity parameter for wind tunnel studies of flow over models of buildings having large openings or flexible roofs is questionable. In those studies the major consideration is to maintain the proper phase relationship between the fluctuating pressures imposed upon the interface from the outside by the wind and the responding pressures excited on the inside. The mean wind speed and the interior speed of sound are both important in determining phase shifts, but their ratio is not properly called the Mach Number which is the ratio of mean wind speed to speed of sound in the external atmosphere.

It is perhaps not of practical importance that pressure interactions at the higher frequencies may not be simulated correctly. However, citing exterior Mach Number is misleading for two reasons: it implies that compressibility of the exterior flow is a factor for wind flow about buildings which it is not, and it distracts attention from possible means of achieving phase similarity through control of the speed of sound in the interior of the model by heating or by substituting, say, ethyl ether for the internal air.

Reply: H. P. A. H. Irwin, R. L. Wardlaw. The term, Mach number, may not be the best to use for the acoustic parameter for membrane structures because of the connotations of compressibility effects in the aeronautical sciences that it brings to mind. However, some other name would have to be invented to replace it. One good reason for sticking to Mach number is that ultimately it is the ratio of wind speed to the speed of sound that must be equal at model and full-scale if acoustic damping is to be accurately simulated. That this is so can be seen by pursuing Professor Leadon's line of reasoning a step further. Let ℓ = total linear distance traveled by the internal wave to the reflection point and back. The time elapsed is thus ℓ/C. A representative time scale for the turbulence is L/U (where L = turbulence integral scale). The ratio of time scales is thus $\ell U/LC$. Since ℓ/L is the same at model and full-scale in a correct turbulence simulation such as ours, then the ratio of time scales is also equal at both scales provided U/C is the same on the model as at full-scale.

SESSION VIII

Physical and Mathematical Modeling

CHAIRMAN

J. C. R. Hunt
University of Cambridge
Cambridge, United Kingdom

CO-CHAIRMAN

S. P. S. Arya
Associate Professor of Meteorology
North Carolina State University
Raleigh, North Carolina, USA

RAPPORTEUR

W. Z. Sadeh
Professor of Civil Engineering
Colorado State University
Fort Collins, Colorado, USA

RECENT DEVELOPMENTS AND TRENDS: PHYSICAL AND MATHEMATICAL MODELING

J. C. R. HUNT
University of Cambridge, Cambridge, UK

1. Introduction

The chairman of the conference, Professor J. E. Cermak, asked me to introduce the session at this conference on physical and mathematical models with a brief review of some of the new developments and trends since the last wind engineering conference and to relate these developments to the papers presented here. I should like to say immediately that this session contains some most interesting papers, which I have enjoyed reading; they certainly form an important part of the past 1975 developments. Since this talk is to be delivered in 15 minutes it has to be short, and since it is delivered at 8:30 a.m. it had better be controversial to wake everyone up. I hope this review meets both aims.

2. Physical models

2.1. Different approaches

The 3 main uses of physical models, which need to be borne in mind when assessing them, are as follows:

(i) Demonstrating qualitative features of wind engineering problems; - e.g., explaining air pollution or wind environment problems by flow-visulisation techniques, particularly to laymen. (as in Jack Cermak's T.V. shows!)

(ii) Investigation and measurement of basic fluid mechanic phenomena of wind engineering, and assisting the development of mathematical models. (Here the precise simulation of the total wind engineering problem may be more of a hindrance rather than a help: idealised experiments can be as useful as idealised theoretical models in increasing understanding and improving prediction methods.)

(iii) Simulation of all the relevant aspects of particular field or full-scale conditions (e.g. using boundary-layer wind tunnels) for prediction purposes. (This use requires more precise simulation than the first use. Different prediction problems pose quite different constraints on different aspects of the flow.)

Some of the most important recent advances in wind engineering have come from the second rather than the third kind of modelling. Several examples are to be found in this and other sessions of the conference. For example, the study of tornadoes by idealised experiments such as those to be described by Jischke and Light (q.v.), air flow over idealised hill shapes in simulated atmospheric boundary layers (Wilson, Winkel, Neiman (q.v.), and in stratified salt water towing tanks (Snyder, et al (q.v.), or grid turbulence studies of fluctuating pressures (Robertson, et al, 1977), or diffusion around bluff bodies, (Vincent (1977) & Puttock, 1979). Since the results of any complex modelling are usually understood or explained by the modeller or the 'clients' by referring explicitly or implicitly to simple configurations (whether or not these simpler configurations have been properly studied), one may as well study the simpler configuration properly so that these essential reference phenomena are properly measured and understood.

2.2 Developments in technique

The past few years have seen existing results stemming from the development of new instruments. At the last conference Bradbury and Moss (1975) showed how the pulsed-wire anemometer (P-W.A.) can be used to measure mean and fluctuating flows in regions of high turbulence and reversing flows, which could not be done with convetional hot-wire anemometers. Since then these instruments have been used in Britain to measure flows around model cubes, fences, and a group of 2 buildings in simulated atmospheric boundary layers. (Robin & Castro, 1977; Davis, et al (q.v.), Britter and Hunt (1979)). One has a better idea now about how the turbulence structure in the wakes changes from one structure to another but one is still lacking measurements by pulsed wire anemometers in these regions of the velocity spectrum and swirl.

Laser-Doppler anemometers (LDA) have been used in large wind tunnels, as at NASA Ames, to study small scale flow features such as trailing vortices, but until the investigation by Ogawa et al (q.v.) reported here, they have not been used in large environmental wind tunnels. Perhaps they will be used more widely by 1983. However, they have been used in small scale air and water experiments to measure flows that are of interest in wind engineering, e.g. recirculating wake flows (Pope & Whitelaw, 1975) and reattaching flow downwind of a step (Etheridge & Kemp, 1978). But I understand that there are significant differences between the LDA and P-WA measurements of such flows.

Two other instrumentation developments need mentioning:

(i) devices for measuring fluctuating concentration, which only now are being used to provide extensive profiles of rms concentration C' and pollutant fluxes $\overline{w'C'}$ in simulated boundary layers (Robins & Faskrell 1979). One hopes they will be more widely used in the future.

(ii) Flow visualization techniques in air flows, analogous to hydrogen bubble visualisation in water (Nagib, 1977).

No completely new simulation techniques have been invented since 1975. Of greater importance is that the wide number of different techniques, most of which have been described at this conference, have been extensively studied so that one has a better idea now of the advantages and limitations of different techniques. I particularly welcome the detailed calibration and comparison of the mean and turbulent flow fields that have been performed in different wind tunnels.

Because of the growing need to understand and predict natural and artificial stratified flows (e.g. dense vapour spills, sea breeze fronts, avalanches, Simpson et al 1977; Hopfinger 1975 and topographic flows) stratified salt water tanks are being increasingly used; their great advantage over wind tunnels being that higher Reynolds numbers are obtained typically by a factor of 10 or more for the same Froude number. Such an increase in Reynolds number can overcome the relatively stronger viscous effects in stratified flows; so that the Reynolds number of the turbulence $u'l/\nu$ (which strongly affects fluctuating concentrations) maybe as much as a factor of 10^2 greater in stratified tank experiments than in wind tunnels. (Whether the fact that the diffusion Peclet number $u'l/D$, where D is the density diffusivity, differs from the Reynolds number by 10^3 in salt water experiments, unlike air experiments, is important, we do not know!)

2.3 Modelling principles and problems

The basic problem bedevilling the laboratory modelling of large scale flows is the low value of the Reynolds number. This limitation has and will not be overcome but it is steadily being better understood and corrected for. See Snyder (1973) for a general review and the papers by Isyumov & Tanaka (q.v.) for some remarks on aspects of roughness modelling and by Wilson, Winkel & Neiman (q.v.) on how Reynolds number affects the separated flow behind hills* The doubts expressed by these authors clearly conflicts with the confidence expressed by Meroney et al (q.v.) in their paper on modelling flow over complex terrain. I look forward to some vigorous argument here!

Physical modelling of diffusion in the laboratory may not have improved fundamentally (unlike wind loading modelling which we have seen at this conference, is now very reliable) but at least the wind engineering community knows a bit more what it is doing. I say this because where detailed wind tunnel diffusion measurements have been made the intensities and scales of turbulent components have been measured (e.g. Robins, 1979; Snyder & Britter, 1979). But these measurements, of course, still show a smaller lateral growth of plumes than the roughly linear growth observed in the atmosphere, because, as explained by Scorer, there is not enough lateral low frequency turbulence. For the same reason when building wakes are modelled the model wakes widen less rapidly than Munn & Cole's (1967) full scale observations show. This in turn always makes concentrations in isolated model building wakes too high (Abbey, 1976). Bearing in mind these limitations, wind tunnel studies must be done, and in that spirit I welcome the important contribution of Koga & Way (q.v.) who have shown how sensitive the diffusion of pollutants in the wake are to the location of roof exhausts; obviously a matter of great practical importance.

The physical modelling of buoyant plumes in a turbulent atmosphere combines both the problems of low Reynolds number flow and insufficient lateral large scale turbulence. The best choice of dimension (as groups for this difficult problem is discussed by Isyumov & Tanaka (q.v.) develop a new integral model of buoyant plume, which they verify in the wind tunnel. But, I feel that practical conclusions based on such modelling (which certainly gives us some interesting insights) need careful examination and discussion.

---

* The unrealistically large roughness elements required to model flow over a plain are of the same order as the inner surface shear layer overhills and therefore initiate the modelling of such layers (Britter, Hunt & Richards, 1979).

A fascinating paper is that of Iverson (q.v.) on modelling snow and sand on earth and on various planets. (Prof. Iversen will be our expert guide on the field trips at the first wind engineering conference on Mars!) Apparently out of the Pandora's box of dimensionless groups produced by these new complexities, he will be able to select a very few relevant ones and shut the box on the others. I look forward to hearing how it is done.

3. Mathematical modelling

3.1 Its role

I believe that for the foreseeable future mathematical modelling of wind engineering phenomena will not be able to replace physical modelling and field investigations. Nevertheless there are a growing number of ways and particular problems where mathematical models can be useful; for example in (i) providing estimates which may be required at the design or planning stage in solving a wind engineering problem, or (ii) the design of laboratory or field studies by indicating the regions of flow or parameter ranges which are poorly understood or where the results depend sensitively on the geometry, (e.g. Mathematical models demonstrate the sensitivity of the wake concentrations to the location and momentum of roof emissions which the papers by Isyumov & Tanaka (q.v.) and Koga & Way (q.v.) strongly confirm). Or (iii) the interpreting, extrapolating or comparing of laboratory and field data (I am sure that ad hoc modelling structures should be compared with simple theoretical models more often, to help the investigator and client understand the results better).

3.2 Recent developments

There have been a number of recent developments involving mathematical <u>and</u> physical modelling which have certainly helped my general understanding of wind engineering fluid mechanics. They seem to exemplify the possibility that new models may <u>simplify</u> rather than complicate our knowledge, which should be our goal. Inevitably this is a somewhat subjective list. (i) studies of flow patterns around three-dimensional surface obstacles using flow visualization studies, kinematic principles (Hunt, Abell, Peterka & Woo, 1978), and, most recently, laminar flow computations (e.g. Mason & Sykes, 1979a) have led to the following conclusions:

(a) whereas, in two-dimensional flows separated flows are confined to regions of closed streamlines, in most three-dimensional separated flows there is no regions of closed mean streamlines; so that the mean separation streamline is <u>not</u> the same as the mean streamline that attaches at the downwind end of the separated flow region.

(b) the <u>form</u> of these three-dimensional flow patterns are much more sensitive to Reynolds number than two-dimensional ones.

(c) although conclusion (a) shows that Halitsky's (1968) model of three-dimensional wake streamline patterns is formally incorrect, nevertheless Halitsky was undoubtedly correct in assuming that within the separated region there is a mean recirculation of the flow. Consequently, the fluid particles reside for a considerable period within this region which leads to rapid mixing and low concentration gradients. (Vincent, 1977). Mathematical models based on this fact can give useful estimates of wake diffusion (Puttock & Hunt, 1979).

(ii) A number of different approaches to the computation of flows over obstacles are developing each having limitations and advantages.

(a) Vortex modelling

The large scale vortex shedding behind bluff cylinders can be modelled mathematically by computing the trajectories and agglomeration of large number of small vortices released sequentially from the separation lines. (Clements & Maull, 1975).

These calculations (which are analogous to coherence structure computations in turbulent shear flows) demonstrate the large differences between the instantaneous velocity field created by the vortices and the highly artificial constructions (in this kind of flow) of the mean velocity field. No other calculation method can currently perform such a task, which is necessary for understanding how mean and fluctuating pressures and also concentrations depend on the vortices. The extension of the method to flow down a step, which involves some assumptions about a suitable vorticity sink at the surface, gives some useful insights, but clearly needs further development. Also, the three dimensionality of the vortices needs modelling.

(b) Oscillator modelling

A more economical approach to modelling the forces on an oscillating cylinder is to make use of the observation that the cylinder acts like a non-linear oscillator and then to refine and generalise the model so that it describes as many aspects as possible of the cylinder's behaviour. This may have some practical benefits but for my part I find it difficult to see the future for such an approach which is not based on the equations of motion or the observed flow patterns. However, despite this limitation the 2 papers on this kind of modelling this morning by Poore & Al-Rawi (q.v.) and Tamura & Matsui (q.v.), certainly show the power of the method and illuminate the phenomena in a way other models have not done.

(c) Turbulence closure models

The application to practical wind engineering problems of turbulence closure models has been and continues (quite rightly) to be cautious. Atmospheric flows are very different to the well defined local flows in typical mechanical or aeronautical engineering problems. The limitations we have noted of wind tunnel modelling of atmospheric diffusion or wakes studies also applies to turbulence closures models. However, there is, I think, a recent success story to be reported on the application of these models in the study of wind over hills, where calculations based on simple mixing length assumptions (i.e. the turbulence is approximately in equilibrium) near the surface of the hill and rapid distortion theory well above the hill enable the mean flow and the turbulence to be surprisingly well predicted both in laboratory and field studies (e.g. Mason & Sykes, 1979b; Bradley, 1979; Britter, Hunt & Richards, 1979).

Over flat terrain turbulence closure models have and are being developed which predict (or postdict?) the observations, in unstable and stable conditions. But these exercises are in the nature of theoretical explorations which together with the observations are certainly helping our physical understanding. But, I am doubtful whether they can yet be reliably applied to wind engineering problems say to predict ana- and kata-batic winds over hills. Hopefully they soon will.

(iii) Two examples of insights provided by mathematical modelling of diffusion near buildings and hills are the following:

(a) Impinging plumes When plumes impinge directly onto buildings or onto three-dimensional hills the maximum surface concentration can be approximately equal to that on the plume centre line in the absence of the obstacle. This, of course, may be very much greater than the maximum surface concentration from level groud close to an elevated source. (Over a ridge this argument does not apply.) This result based on some field data, wind tunnel studies and theory enables useful preliminary estimates to be obtained. (Burt & Slater, 1977; Hunt, Britter & Puttock, 1979.)

(b) Downwash effects Pollutants emitted downwind of buildings or hills experience large significant downwash even when they are placed 10 or 20 building or hill heights downwind. The sources at or above the height of the hill are often the worst affected at these distances. In stratified flows when lee waves occur, the downwash may be worse (Eliseev, 1975; Barrett, Hall & Simmonds, 1978; Schiermeier, 1972).

(iv) In many wind engineering problems one wishes to calculate or know something about flow through and around groups of buildings and hills, not just single obstacles. There has been relatively little systematic physical modelling let alone mathematical modellings in this area. There have been some papers at this conference and I hope that by the next conference more will have been done in this area. (For a brief review see minutes of 52nd Industrial Fluid Mechanics Research Meeting; Nat. Phys. Lab.; Teddington.)

Acknowledgement

I am grateful for discussion with Dr. R. E. Britter in preparing this review.

References

Abbey, R. F. 1976 Concentration measurements downwind of buildings: previous and current experiments. 3rd Symp. Atmos. Turb. Diff. & Air Qual. Raleigh, N.C. (Am. Met. Soc.).

Barrett, C. F., Hall, D. J. & Simmonds, A. C. 1978 Dispersion from chimneys downwind of cubical buildings - a wind tunnel study. NATO/CCMS 9th Int. m't'g Air Poll'n Modelling. Toronto.

Bradbury, L.J.S. & Moss, W. D. 1975 Pulsed wire anemometer measurements in the flow past a normal flat plate in a uniform flow and in a sheared flow. Proc. 4th Int. Conf. Wind Eff. Build. & Struc. Cambridge Univ. Press.

Bradley, E. F. 1979 An experimental study of the profiles of wind speed, shearing stress and turbulence at the crest of a large hill. Q.J. Roy. Met. Soc.

Britter, R. E. and Hunt, J.C.R. 1979 Velocity measurements and order of magnitude estimates of the flow between two buildings in a simulated atmospheric boundary layer. J. Inc. Aero. 4, 165.

Britter, R. E., Hunt, J. C. R. and Richards, K. J. 1979 Analysis and wind tunnel studies of speed-up, roughness effects, and turbulence over a two-dimensional hill. Q. J. Roy. Met. Soc.

Burt, E. W. and Slater, H. H. 1977 Evaluation of the valley model. AMS-APCA Conf. on Appl. of Air Poll'n Met. Salt Lake City, Utah, Nov. 1977.

Clements, R. E. and Maull, D. J. 1975 Representation of sheets of vorticity by discrete vortices. Prog. Aerospace Sci. 16, 129.

Davies, M. E., Quincey, V. G., Tindall, S. J. (q.v.)

Eliseev, V.S. 1975 Trajectories of smoke jets from industrial sources. Tr. Geofiz Obs. 373 EPA Trans. TR 77-235.

Etheridge, D. W. & Kemp. P. H. 1978 Measurements of turbulent flow downstream of a rearward facing step. J. Fluid Mech. 86, 545.

Jischke, M. C. and Light, B. D. (q.v.)

Halitsky, J. 1968 Gas diffusion near buildings. Met. & Atomic En. p. 221 U.S.A.E.C.

Hunt, J. C. R., Abell, C. J. Peterka, J. and Wood, H., 1978 Kinematical studies of the flows around free or surface mounted obstacles; applying topology to flow visualisation. J. Fluid Mech., 86, 179.

Hunt, J. C. R., Britter, R. E. and Puttock, J. S. 1979 Mathematical models of dispersion of air pollution around buildings and hills. Proc. IMA Symp. on Math. Modell. of Turb. Diff. in the Envir. Acad. Press.

Isyumov, N. and Tanaka, H. (q.v.)

Iversen, J. D. (q.v.)

Koga, D. J. and Way, J. L. (q.v.)

Mason, P. J. and Sykes, R. I. 1979a Three-dimensional numerical integrations of the Navier-Stokes equations for flow over surface-mounted obstacles. J. Fluid Mech., 91 433.

Mason, P. J. and Sykes, R. I. 1979b Flow over an isolated hill of moderate slope. Q. J. Roy. Met. Soc., April.

Meroney, R. N., Lindley, D., Bowen, A. J. (q.v.)

Munn, R. E. and Cole, A. F. W. 1967 Turbulence and diffusion in the wake of a building. Atmos. Env. 1, 33.

Nagib, H. M. 1977 Visualisation of turbulent and complex flows using controlled sheets of smoke streamlines. Proc. Int. Symp. Flow Visual, Tokyo.

Ogawa, Y., Diosey, P. G., Oikawa, S. (q. v.)

Petersen, R. L. and Cermak, J. E. (q.v.)

Poore, B. and Al-Rawi, A. R. (q.v.)

Pope, S. B. and Whitelaw, J. H. 1976 The calculation of near wake flows. J. Fluid Mech., 73, 9.

Puttock, J. S. 1979 Turbulent diffusion from sources near obstacles with separated wakes. Pt. II Atmos. Enviro. 13, 15.

Robins, A. 1978 Plume dispersion from ground level sources in simulated atmospheric boundary layers. Atmos. Envir. 12, 1033.

Robins, A. and Castro, I. P. 1977 A wind tunnel investigation of plume dispersion in the vicinity of a surface mounted cube. Pt. I Atmos. Envir. 11, 291; Pt. II Atmos. Envir. 11, 294.

Robertson, J. M., Wedding, J. B., Peterka, J. A. and Cermak, J. E. 1977 Wall pressures of separation - reattachment flow on a square prism in uniform flow. J. Ind. Aero.

Schiermeier, F. A. 1972 Large power plant efficient study. E.P.A. APTD-1143.

Simpson, J. E., Mansfield, D. A. and Milford, J. R. 1977 Inland penetration of sea breeze fronts. Q. J. Roy. Met. Soc., 103, 47.

Snyder, W. H., Britter, R. E. and Hunt, J. C. R. (q.v.)

Snyder, W. H., Britter, R. E. 1979 Wind tunnel measurements of diffusion from ground level and elevated sources.

Snyder, W. H. 1973 Similarity criteria for the application of fluid models to the study of air pollution meteorology. Bound. Layer Met. 3, 113.

Tamura, Y. and Matsui G. (q.v.)

Tochan-Danguy, J.C., and Hopfinger, E. J. 1975 Simulation of the dynamics of power avalanches. Proc. Grindelwald Symp. 1974 IAHA-AISH Publ. no. 114.

Vincent, J. H. 1977 Model experiments on the nature of air pollution transport near buildings. Atm. Envr. 11, 765.

Wilson, D. J., Winkel, G., Neiman, O. (q.v.)

REYNOLDS NUMBER EFFECTS ON FLOW RECIRCULATION BEHIND TWO-DIMENSIONAL OBSTACLES IN A TURBULENT BOUNDARY LAYER

D. J. WILSON, G. WINKEL and O. NEIMAN

University of Alberta, Edmonton, Canada

ABSTRACT

The flow recirculation region behind a vertical wall, and a two-dimensional hill model with 3:1 slopes was observed in a water channel and a wind tunnel. The length of the flow recirculation zone downwind of the model hill decreased with increasing Reynolds number until the flow became completely attached at high Reynolds number. The implications of these observations for terrain modelling are discussed.

Reynolds Number Effects In Physical Modelling

Physical modelling has been particularly useful in simulating the effect on complex terrain variations on wind loading, pollution dispersion, and the siting of wind power generators. Because of the large scale reductions required for a terrain modelling wind tunnel, Reynolds numbers are typically three orders of magnitude less than the full scale values.

The mismatch between model and full scale Reynolds number is very large in modelling of pollution dispersion. In this case, in order to simulate the buoyant rise of effluent gases it is necessary to operate the wind tunnel at speeds which are typically twenty times less than their full-scale counterpart. With this additional constraint, the wind tunnel Reynolds numbers for pollution modelling are usually about four orders of magnitude less than full-scale values. Also, because flow visualization systems usually are not effective at high wind tunnel speeds, physical modelling of flow patterns is often done with mismatch of four orders of magnitude in Reynolds number, even when quantitative tests are carried out at much higher tunnel speeds.

To justify the accuracy of their studies, physical modelers depend on the insensitivity to Reynolds number of flow separation points and recirculation patterns. Terrain and structural features are usually characterized as being "sharp edged," where flow separation points are fixed and Reynolds number independent, or "gently curved" where flow separation does not occur in either the model or full-scale. Unfortunately, there are many terrain features which do not fit either of these simple Reynolds number independent categories, and where physical modelling can give misleading results.

In the present study the flow separation and reattachment behind two types of obstacles immersed in a thick turbulent boundary layer were studied. A sharp edge wall was used to determine the effect of Reynolds number on reattachment length of a free shear layer with a fixed separation point. The second obstacle chosen for study was a flat crested symmetrical model hill with relatively steep 3:1 front and back slopes. This configuration is typical of terrain features which cannot be readily described as either sharp edged or gently curved.

Previous Investigations of Recirculation

Most of the previous studies of flow reattachment behind an obstacle in a boundary layer were carried out at a single Reynolds number. However, a comparison of various investigations to determine Reynolds number effects indicates that the flow reattachment length L_c behind sharp-edged obstacles is not sensitive to variations in Reynolds number, but has considerable variation with the shape of the body before separation occurs, and δ/H, the ratio of approach boundary layer thickness to obstacle height. The results of some previous studies are summarized in Table 1, which shows that the recirculation cavity length to obstacle height ratio L_c/H is reduced from a value of about 17 with no approach boundary layer thickness to 12 with a boundary layer about six times the obstacle height. One explanation for this observed effect is that turbulence in the approach boundary layer will stimulate spreading of the separated shear layers and induce an earlier reattachment. This same general trend was observed by Huber et al (1976) on a gently curved model hill with separation fixed at the crest by means of a sharp-edged barrier. In this case they found that the recirculation cavity length changed from $L_c/H = 9$ when $\delta/H = 2$ to a value of $L_c/H = 11$ for $\delta/H = 1$.

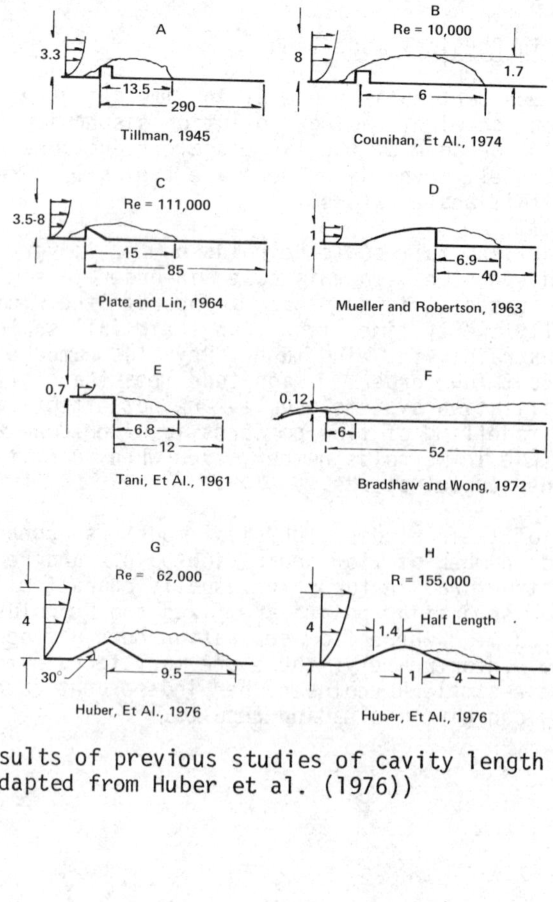

Fig. 1 Results of previous studies of cavity length behind obstacles (adapted from Huber et al. (1976))

The influence of the obstacle shape upstream of the point of sharp-edged separation is illustrated in Figure 1 which summarizes several previous studies.

Full-scale observations of flow separation and reattachment are often difficult to interpret because stable atmospheric conditions can suppress a tendency of flow to separate on the lee side of hills. Scorer (1955) presents a qualitative description of some full-scale observations. Eliseev (1973) carried out a full-scale study over a 100 meter high hill with gradual crest curvature and slopes of 15° to 30°. This study is of particular interest because the hill contour closely resembled the model hill used in the present measurements. Eliseev's observations, using smoke streamers to visualize instantaneous vertical velocity profiles, showed a recirculation region with a length of about $L_c/H \approx 2$ at $R_eH \approx 10^7$. Because the downwind slope of this full-scale hill was about 10° steeper than the model hills which were tested in the present study, it was expected that L_c/H should be even less than 2.0 for the model hills.

Water Channel Flow Visualization

An open channel flume with a test section width of 66 cm, a length of 240 cm and an operating water depth of 30 cm was used for flow visualization studies. A trick turbulent boundary layer was generated using traiangular vortex generators followed by distributed surface roughness. The mean velocity profile is shown in Figure 2, along with the as-built dimensions of the model hills tested.

With the vortex generators and roughness configuration used to generate the profile shown in Figure 2, the lateral uniformity of mean velocity measured at a height Z = 5 cm was within ± 5% of the center-line value, excluding the outer 10 cm at each side of the 66 cm channel width.

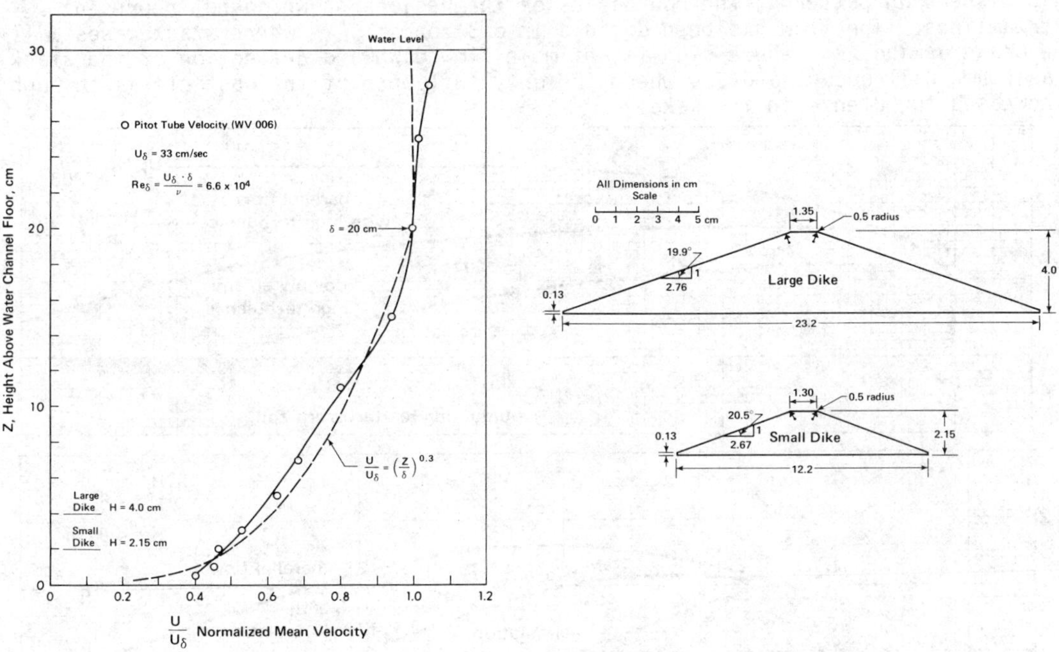

Fig. 2 Velocity profile and model hill dimensions for water channel flow visualization

In both the water channel and wind tunnel studies the velocity profile shape and boundary layer height were characterized by the momentum deficit thickness θ and the displacement thickness δ_*. For the water channel velocity profile these were found to be θ = 2.93 cm and δ_* = 4.87 cm, which give a shape factor δ_*/θ = 1.66. When the velocity profile is represented by a power law

$$\frac{U}{U_{ref}} = \left(\frac{Z}{Z_{ref}}\right)^{\eta}$$

The shape factor is simply

$$\frac{\delta_*}{\theta} = 2\eta + 1$$

The measured shape factor indicates the appropriate power law is η = 0.33. A value of η = 0.30 is shown on Figure 2 based on a best fit to velocity.

Measurements of flow direction and intensity of turbulent mixing were made by releasing a thin stream of dye from a moveable probe positioned at various points in the flow. The position of flow reattachment was determined by releasing small puffs of dye close to the wall and observing their motion. The upstream and downstream limits of the flow reattachment point were taken as the location where all dye puffs moved either upstream or downstream. The mean position of reattachment, where half of the dye puffs moved upstream and half downstream was found to lie approximately midway between these limits, and this midpoint was used in all further studies as the mean position of reattachment. The results of the flow visualization study are shown in Figure 3, where the Reynolds number is characterized by the conditions at the height H of the obstacle. Because the original purpose of these measurements was to determine the effect of obstacles on stack gas dispersion patterns, the boundaries of the regions shown do not represent streamlines. The flow has been divided into 3 zones: A - where stack gases will be drawn upwind, B - where significant mean flow downward deflection of the stack gas plume will occur, and C - where the only influence of the obstacle is through increased turbulence in its wake.

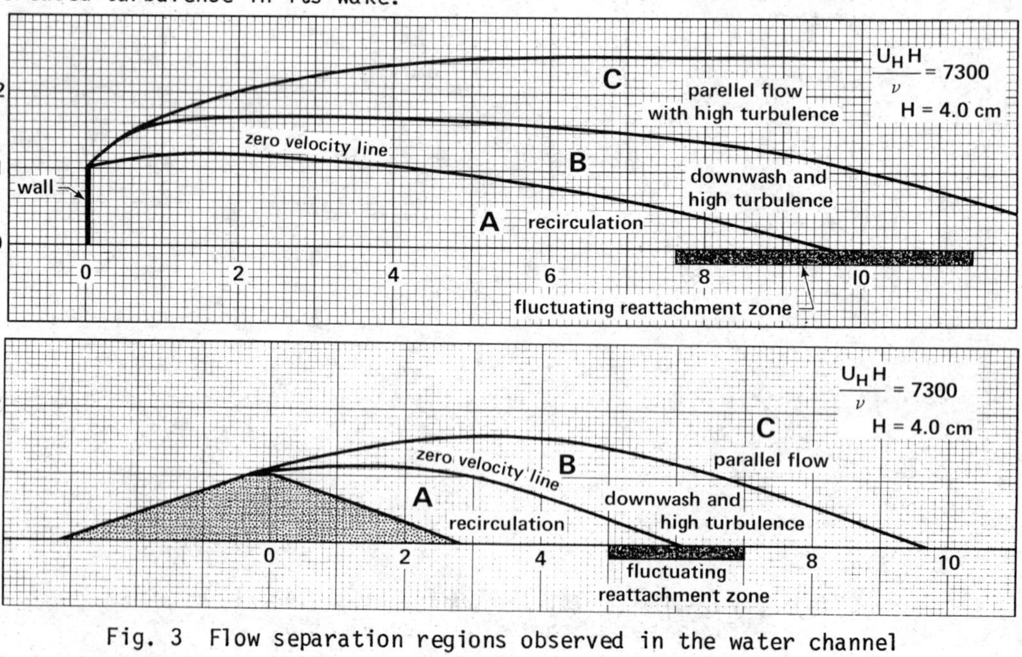

Fig. 3 Flow separation regions observed in the water channel

Wind Tunnel Measurements of Recirculation

The position of flow reattachment downwind of an obstacle in a boundary layer was measured over a wide range of Reynolds numbers in a closed return wind tunnel. The combination of a variable speed motor and closed return design allowed stable tunnel speeds as low as 0.2 m/s to be maintained. The test section was 1.2 m high, 2.4 m wide, and 11 m long, providing sufficient size to accurately simulate atmospheric shear flows over the model terrain. A thick boundary layer is created using a row of triangular spikes and a low barrier on the floor of the tunnel to generate turbulence, with surface roughness to match the wind velocity profile and turbulence to the required terrain type. Mean velocity and turbulence profiles in the simulated atmospheric shear flow are shown in Figure 4. A semi-logarithmic plot of the mean velocity profile indicated a surface roughness of Z_0 = 0.05 cm. Using this value of surface roughness it was found that the turbulent structure in the wind tunnel boundary layer indicated an 800:1 scale factor when compared with Counihan's (1975) compilation of full scale adiabatic boundary layer data. This indicates that the full-scale surface roughness should be about 40 cm, typical of atmospheric boundary layers over vegetated and forested areas. The mean velocity profiles were integrated numerically to determine a momentum deficit thickness of θ = 13.5 cm and a displacement thickness δ_* = 19.15 cm. The shape factor δ_*/θ = 1.46 indicates a velocity power law of η = 0.23. Figure 4 shows that the wall region of the boundary layer has η = 0.18 while the outer region follows η = 0.30.

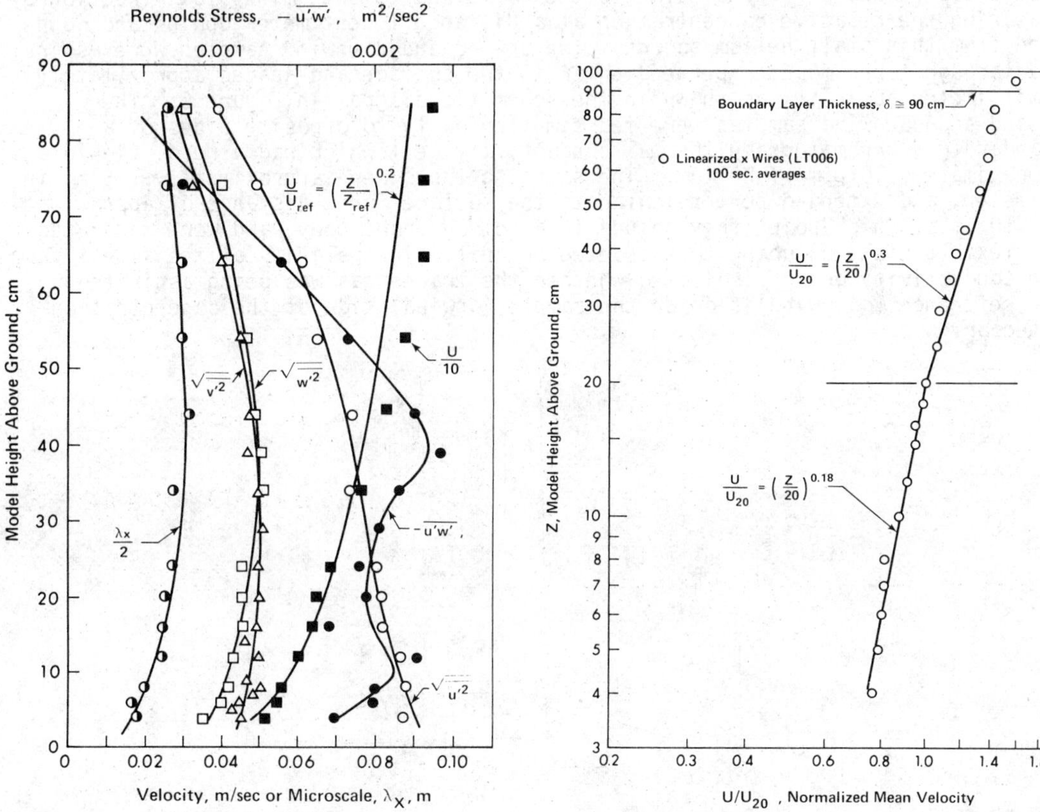

Fig. 4 Mean velocity and turbulence profiles at the obstacle location in the wind tunnel

Three different model hills were tested with crest heights of H = 2.86, 5.72, and 11.75 cm as well as a sharp edged wall with H = 10.4 cm. All the hill models were symmetrical with 3:1 slopes with a 1.9 cm flat across the hill crest.

Flow Reattachment Detector

A few preliminary studies were carried out using titanium tetrachloride ($TiCl_4$) smoke to observe surface flow patterns downwind from one of the model hills in the wind tunnel. At a constant mean tunnel speed, the position of flow reattachment fluctuated over a considerable range, as was also observed in the water channel study (see Figure 3). Because the smoke tended to diffuse rapidly in the highly turbulent wake, it was not possible to measure the mean position of flow reattachment, but only to roughly locate its furthest downwind position where smoke was never entrained into the recirculation zone. When the smoke probe was placed on the wall in the zone of fluctuating reattachment, recirculation of the smoke in the separated flow completely obscured many of the local reattachment processes. From this it was felt that many of the reported values in the literature for the reattachment length L_c probably represent the furthest downwind position of L_c, and not its average value.

In order to determine the mean position of flow reattachment, a tracer gas detector probe was developed to measure instantaneous surface flow direction in the wind tunnel. Its principle of operation was to inject helium tracer gas from a tube mounted perpendicular to the wall, and to determine flow direction by measuring the relative concentration at a distance 1 centimeter upwind and downwind from this small helium source. The upwind and downwind samples were aspirated through tubes mounted perpendicular to the surface and spaced approximately 1 millimeter above it, as shown in the schematic diagram in Figure 5. The upwind and downwind samples were fed simultaneously to opposite sides of a standard gas chromatograph thermal conductivity cell which had a resolution of approximately 10 ppm of helium. The sampling tubes were approximately 1.5 mm in diameter, and extended perpendicular to the surface for a distance of approximately 10 cm at which point they joined in a common probe body, and were conducted by flexible plastic tubing to the detector unit. The polarity of the signal from the conductivity bridge indicated whether the tracer gas was being aspirated by the reference (downwind) side or the sample (upwind) side of the concentration detector.

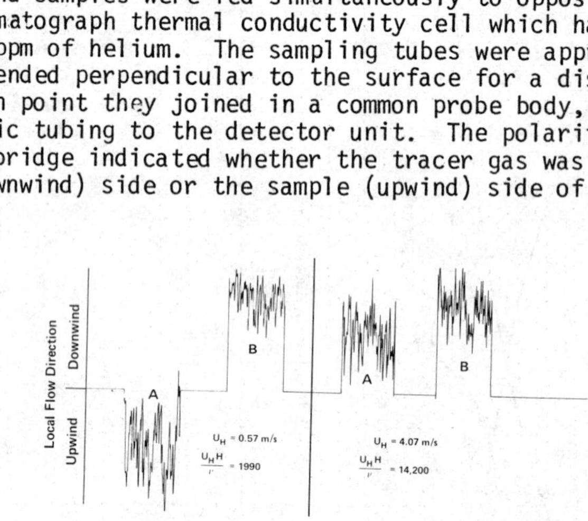

Fig. 5 Measurement of reattachment length with the helium flow direction detector

The length L_c of the flow recirculation zone behind the three model hills was found to be strongly dependent on Reynolds number. This can be seen in Figure 5 where the chart recorder output signals from the flow direction detector is shown at 2 sampling locations for different Reynolds numbers. The flow direction detector signals show that at a Reynolds number based on hill height of 1990 the flow reattachment length L_c lay somewhere between the 2 sampling positions, A at X = 3.6H, and position B at X = 16.0H. In contrast to this, at the higher Reynolds number of 14,200 the detector signals indicate that the position of reattachment lay somewhere upwind of sampling location A, so that the flow was downwind at both A and B locations.

Quantitative measurements of the average position of flow reattachment were made by locating the detector probe at a fixed downwind position from the hill crest, and gradually increasing the wind tunnel speed until the position of flow reattachment coincided with the detector location. When this occurred, the output signal from the detector fluctuated equally above and below its zero value. It was found that with a little experience, the results were readily reproducible using simply a visual observation of the chart recorder signal.

To provide a more sensitive estimate of the position of reattachment, the detector output was amplified and given a DC offset voltage. It was then fed to a voltage to frequency convertor and from there to a counter with a gate time of 100 seconds. With the helium source turned off the counter output for the DC offset was recorded. With the helium source turned on the counter output increased above this DC offset value for downwind flows, and decreased below it for upwind flows. When the position of flow reattachment coincided with the detector location the 100 second average counter output remained at its DC offset level.

The detector system had one serious operating disadvantage. Because Helium was used as the tracer gas (because of its large output signal) in the concentration detector, the gas flow was sensitive to the slope of the surface on which it was placed. In particular, the tracer system gave biased readings when placed on the 1:3 slope of the model hill because the helium gas tended to migrate upward along the hill at low wind tunnel speeds. For this reason, once the position of flow reattachment had reached the base of the hill and began to move upward along the slope the detector could not be used to determine the position of reattachment at low wind tunnel speeds. However, when the wind tunnel was operated at speeds of about 10 meters per second buoyancy effects on the helium tracer gas were negligible, and the detector gave an accurate indication of the position of flow reattachment on the downwind hill slope.

Experimental Results

The reattachment length for the vertical wall for both water channel and wind tunnel studies is given in Table 1 along with the results of other investigators. The slightly shorter reattachment length L_c in the wind tunnel can be attributed to the thicker boundary layer in the wind tunnel approach flow. The flow reattachment results show an almost complete insensitivity of the flow reattachment position to Reynolds number effect over the range tested. The present study supports the assumption that when separation points are fixed by sharp edges, flow reattachment and recirculation patterns will be Reynolds number independent.

TABLE 1. Flow Reattachment Length Behind Two Dimensional Walls Immersed In A Boundary Layer

Investigator	Re_H	$\frac{\delta}{H}$	$\frac{L_c}{H}$
Arie and Rouse (1956)	3×10^4	0	16.8
Good and Joubert (1968)	1.76×10^5	0.75	13.3
Chang (1966)	4.7×10^4	6	12.8
Plate and Lin (1964)	2×10^4	6	12
present study water channel	7.3×10^3	4	9.5
present study wind tunnel	6.3×10^4	8.6	9.05

For the model hill with 3:1 slopes the normalized position of flow reattachment L_c/H is shown in Figure 6, where the sensitivity to Reynolds number is apparent. For the largest hill the reattachment length was consistently about 0.5H longer than the two smaller hills. The reason for this discrepancy may be due to tunnel blockage effects, which amounted to 11% area blockage for the largest hill. Because the wind tunnel test section had a fixed roof with no means of pressure gradient control, the reattachment may have been influenced by pressure gradients induced by blockage. This is consistent with results of Good and Joubert (1968) who found that the measurements of base pressure coefficients on walls in turbulent boundary layers with pressure gradients indicated that the rear separation bubble was affected by pressure gradient.

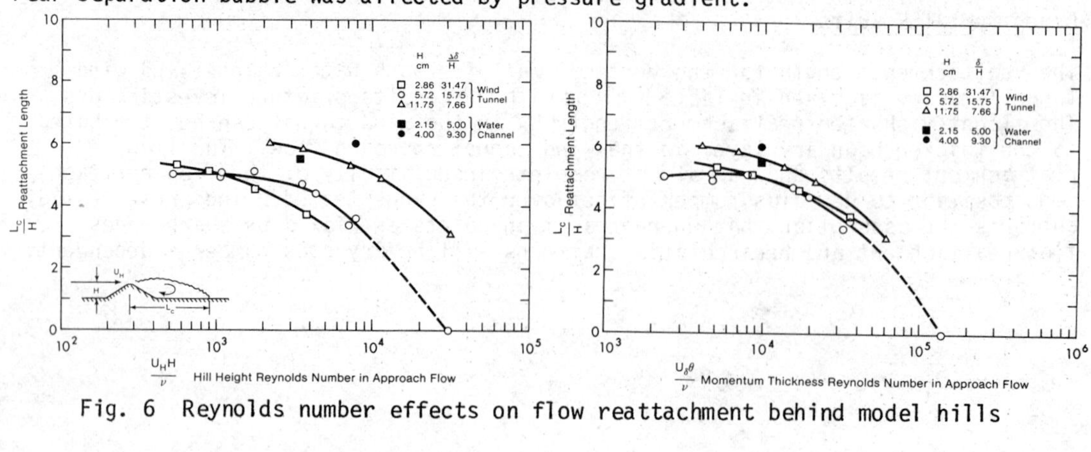

Fig. 6 Reynolds number effects on flow reattachment behind model hills

Two different Reynolds numbers are used for correlating the data in Figure 6; the local hill crest Reynolds number $Re_H = U_H H/\nu$ and the momentum thickness Reynolds number of the approach flow $Re_\theta = \frac{U_\theta \theta}{\nu}$. The momentum thickness Reynolds number, which is independent of hill height gives a better correlation for different hill heights than the hill crest Reynolds number. However, as H increases, δ/H decreases and the observed changes in L_c/H are consistent with the expected effect of decreasing δ/H. In addition, some of the variation in L_c/H may be due to changes in tunnel blockage for the 3 hill heights. Further studies are required using different approach boundary layers for a constant hill height to determine the appropriate Reynolds number for correllating L_c/H.

The most surprising result of the study was that even for 3:1 hill slopes, which at least visually appear quite steep, the flow recirculation zone disappears at a Reynolds number of about $Re_\theta \sim 10^5$. Because full-scale Reynolds numbers for hills are typically 10^7, wind tunnel measurements at low Reynolds number will not give a true picture of the expected full-scale flow field. The full-scale measurements of Eliseev (1973) support this, indicating that even for relatively steep hills, flow recirculation zones will be very small.

Conclusions and Recommendations

The results of this investigation show that there is considerable Reynolds number sensitivity even for relatively steep hills. Wind tunnel tests must be conducted very carefully in order to obtain accurate results which can be extrapolated to full-scale situations.

In contrast, the experiments showed that for separation fixed at a sharp edge, the recirculation and reattachment regions are quite insensitive to Reynolds number effects.

Flow reattachment lengths appear to be only mildly sensitive to the thickness of the approach flow boundary layer. Because only a single boundary layer was used in the wind tunnel experiments, it was not possible to discriminate between any of the thickness parameters H, θ, δ_* or δ as the most logical candiate for forming a Reynolds number. This is not a point of major concern, because physical modelling studies usually accurately simulate the approach flow in the proper scale, thus eliminating any effect on the recirculation and reattachment points.

The Reynolds number sensitivity to physical modelling of flow over hills observed in this study, supports the comment made by Scorer (1955) who in discussing full-scale flow separation noted that, "The influences described here are not readily reproduced experimentally, and I think that further progress in their study is most likely to come from observations made out-of doors. The most likely use of model experiments is to demonstrate the operation of a mechanism revealed by watching nature."

REFERENCES:

1. Arie, M. and Rouse, H. (1956) "Experiments on Two Dimensional Flow Over a Normal Wall" J. Fluid Mech. 1 part 2, pp. 129-141.

2. Bradshaw, P. and Wong, F.Y.F. (1972), "The Reattachment and Relaxation of Turbulent Shear Layer," J. Fluid Mech. 52, 113-135, Part 1.

3. Chang, S.C. (1966) "Velocity Distributions in the Separated Flow Behind a Wedge Shaped Model Hill" Colorado State University Fluid Dynamics and Diffusion Laboratory Tech. Rept. CER 65SCC66.

4. Counihan, J., Hunt, J.C.R. and Jackson, P.S. (1974), "Wakes Behind Two-Dimensional Surface Obstacles in Turbulent Boundary Layer," J. Fluid Mech. 64, 529-563. Part 3.

5. Counihan, J. (1975), "Adiabatic Atmospheric Boundary Layers: A Review and Analysis of Data from the Period 1880-1972," Atmospheric Environment 9, pp. 871-905.

6. Eliseev, V.S. (1973), "Stereophotogrammetric Investigation of the Air Flow in the Boundary Layer of the Atmosphere Above a Hill," from Berlyand, M. (ed.) Air Pollution and Atmospheric Diffusion, John Wiley and Sons, pp. 95-108.

7. Good, M.C. and Joubert, P.N. (1968), "The Form Drag of Two Dimensional Bluff Plates Immersed in Turbulent Boundary Layers" J. Fluid Mech. 31 part 3 pp. 547-582.

8. Huber, A.H., Snyder, W.H., Thompson, R.S. and Lawson, R.E. (1976), "Stack Placement in the Lee of a Mountain Ridge," U.S. EPA Rept.: EPA-600/4-76-047 Research Triangle Park, N.C.

9. Hunt, J.C.R. (1978) "Wind Over Hills" Proc. NOAA-NSF Workshop Boulder Colorado Aug. 1978.

10. Mueller, T.J. and Robertson, J.M. (1963), "Modern Developments in Theoretical Applied Mechanics," 1, 326.

11. Plate, E.J. and Lin, C.M. (1964), "The Velocity Field Downstream from a Two-Dimensional Model Hill," Part 1. Colorado State University, Fort Collins, Colo., Technical Report No. CER65EJP-CWL41.

12. Snyder, W.H. (1972), "Similarity Criteria for the Application of Fluid Models to the Study of Air Pollution Meteorology," Boundary-Layer Met. 3, pp. 113-134.

13. Tani, I., Iuchi, M. and Komoda, H. (1961), Aeronautical Research Institute, Tokyo, Japan, Report No. 364.

14. Tillman, W. (1945), Volkenrode Translation MAP-VG34-45T. British Ministry of Aircraft Production, London, England.

PHYSICAL MODELING OF FLOW OVER COMPLEX TERRAIN

R. N. MERONEY*, D. LINDLEY** and A. J. BOWEN***

*Colorado State University, Fort Collins, Colorado, USA
**Taylor Woodrow Construction Ltd., Southall, Middlesex, UK
***University of Canterbury, Christchurch, New Zealand

ABSTRACT

Physical simulation of wind regimes found over complex terrain offer significant advantages in terms of time, expense, and control of independent variables. The viability of such an approach has been examined by a consideration of inherent physical modeling constraints together with a joint field verification/laboratory simulation study of air flow patterns over complex terrain in the Southern Alps of New Zealand.

INTRODUCTION

Meterologists are increasingly faced with problems requiring quantitative estimates of air flow patterns and turbulence characteristics over complex terrain. Use of the wind information includes air pollution zoning, prediction of smoke movement from forest fires or slash burning, mine tailing dispersal calculations, estimation of the movement of vegetation disease vectors or pests, and the siting of wind powered electrical systems. In view of the extreme difficulties in obtaining practically useful results in this area of meterology over complex terrain, whether by theoretical analysis or field investigation, it is natural to explore the possibilities of simulating the flow over irregular terrain by means of physical model experiments on the laboratory scale. Similitude criteria and previous laboratory case studies have been reviewed by Meroney et al (1). This paper considers the results of a joint field/wind tunnel simulation program which has been completed for a mountain valley/river gorge region in New Zealand (2).

SIMULATION BY PHYSICAL MODELS

The basic tool of laboratory simulation is similitude or similarity, defined as a relation between two mechanical (or flow) systems (often referred to as model and prototype)* such that by proportional alterations of the units of length, mass, and time, measured quantities in the one system go identically (or with a constant multiple of each other) into those in the other. In order that the flow in any laboratory model should be of value in interpreting or predicting the observed flow

*Prototype--actual air flow involving full scale
Model-- air flow involving smaller scale than protorype but usually with geometrically similar boundaries

in the atmosphere, it is essential that the two flow systems should be dynamically, thermally and kinematically similar. This means that it must be possible to describe the flow in the two systems by the same equations after appropriate adjustments of the units of length, time and other variables.

A number of authors including Cermak (3,4), McVehil et al. (5), Bernstein (6) and Snyder (7) have derived the governing parameters for atmospheric heat, mass, or momentum transport by dimensional analysis, similarity theory, and inspectional analysis. Another group justify similitude by considerations of turbulence theory and recent reviews of full scale wind data which present the characteristics of the prototype atmospheric wind on a parametric basis (Nemoto (8), Counihan (9,10), Cook (11), and Melbourne (12)). Although all investigators do not agree concerning details, most would concur that the dominant mechanisms can now be identified and are understandable.

The laboratory method thus consists of obtaining velocity and turbulence measurements over a scale model of selected terrain placed in a simulated atmospheric flow. The wind characteristics of the simulated atmospheric flow are chosen to reproduce the wind profile shape and length scales of the equivalent prototype situations. Since field profiles are rarely available in advance, velocity profiles and turbulence characteristics are chosen to fit an equivalent class of conditions as recorded by earlier investigators over terrain of similar roughness.

The viability of a given simulation scenario is not only a function of the governing flow physics but the availability of a suitable simulation facility and the measurement instrumentation to be employed. It would seem appropriate, therefore, to suggest bounds for the range of field situations which can reasonably be treated by physical modeling. With few exceptions, scientists seem to make rather poor prophets; there is always the danger of what Arthur Clarke (13) calls "failures of nerve" of "failures of imagination." Nevertheless, let us propose limitations of simulation for complex terrain.

A number of boundary layer wind tunnels exist at various laboratories. Generally these tunnels range in size from facilities with cross sections of 0.5 m x 0.5 m to 3 m x 4 m. Several of these facilities are equipped with movable side walls or ceilings to adjust for model blockage. By utilizing a variety of devices such as vortex generators, fences, roughness, grids, screens, or jets a fairly wide range of turbulence integral scales can be introduced into the shear layer. Varying surface roughness permits control of surface turbulence intensity, dimensionless wall shear, and velocity profile shape. Density stratification can be induced by means of heat exchangers, use of different molecular weight gases, or latent heat absorption or release during phase changes.

When one combines various operational constraints into a performance envelope, a clear picture appears of the performance region for wind tunnel facilities. Figure 1 is such a performance envelope prepared for a large facility such as the Environmental Wind Tunnel at Colorado State University. The criteria selected to specify operational ranges are:

Maximum model height $h \leqq 0.5$ m

Minimum convenient model height $h \geqq 0.02$ m

Minimum Reynolds number $Re_h = \frac{U_o h}{\nu} \geqq 10,000$

Maximum model integral scale $L_{u_x} \leq 0.5$ m

Minimum model integral scale $L_{u_x} \geq 0.05$ m

Minimum model measurement resolution $\Delta z \geq 0.1$ mm

Maximum model boundary depth $\delta \leq 2$ m

Minimum model boundary depth $\delta \geq 0.1$ m

Since field values for some parameters are uncertain, the prototype values of δ and L_{u_x} are assumed to range as follows over complex terrain

300 m < δ < 1000 m, and

100 m < L_{u_x} < 1000 m.

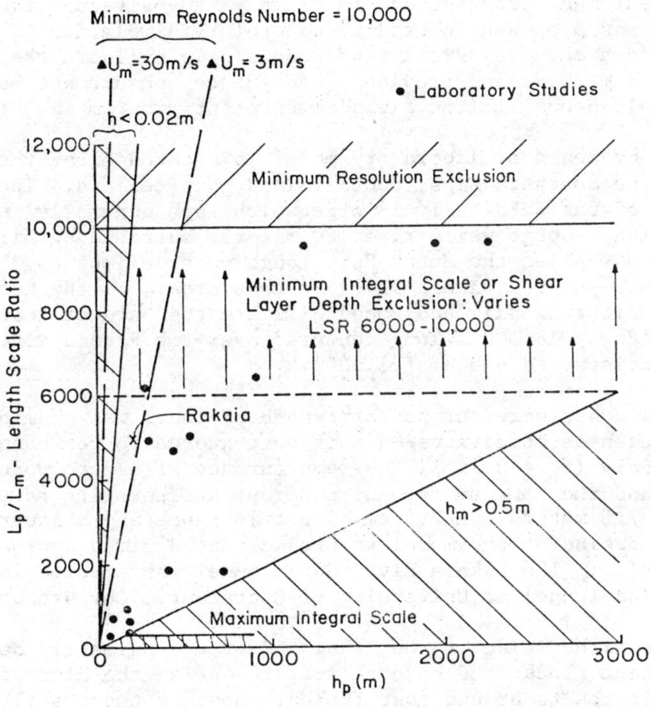

Figure 1. Performance envelope for physical modeling of shear flows over complex terrain

Not all previous laboratory studies meet such similitude restrictions, some experiments were performed to meet other objectives than similitude of turbulence or mean velocity profiles; nevertheless almost all cases noted fall within the indicated operational envelope or just outside the predicted region.

Based on Coriolis force considerations, Snyder (7) suggests a 5 km cut-off point for horizontal length scales for modeling diffusion under neutral or stable conditions in relatively flat terrain. Mery (14) suggests a 15 km limit, Ukejurchi et al. (15) suggest 40 to 50 km, and Cermak et al. (2) and Hidy (16) recommended 150 km. A middle road would be that of Orgill et al. (17) who suggest for rugged terrain in high winds that a length scale of 50 km is not unreasonable.

Assuming an upper value of length scale ratio of 10,000 and a tunnel length of 25 meters, a distance of 50 km is well within the capacity of existing facilities to contain in the windward direction. Assuming a lateral width restriction of 4 m suggests a 40 km lateral maximum for the field area modeled.

LABORATORY MEASUREMENT PROGRAM

New Zealand and the United States are both geographically complex, face similar applied meteorology problems, and yet in many such areas of complex terrain there are "meteorological data" deserts. One such area is the Rakaia River Gorge region on the eastern slope of the Southern Alps in New Zealand. Climatological records obtained from stations somewhat removed from the area suggest moderate to very high wind energy suitable for wind energy conversion system sites. Local farmer and fishermen wisdom and folklore speak of incredible winds in the gorge canyon. Extended field measurement programs are invariably expensive and time consuming; hence a survey program was proposed to utilize laboratory simulation of the relevant wind characteristics in a meteorological wind tunnel. To evaluate the validity of laboratory simulation methods and provide a confidence measurement bound for laboratory data, a simultaneous limited field measurement program was organized.

The area studied by means of laboratory model is located along the Rakaia River as it emerges from the Southern Alps, South Island, New Zealand. The primary terrain features consist of the Rakaia River Gorge which runs generally in a northwest-southeast direction. Gorge walls rise 180 meters, surrounding hills rise to 460 meters. To the south lies the Mount Hutt range which climbs to 2188 m. The range parallels the course of the Rakaia River in this area. To the north lies the Rugged Range but nearby Fighting Hill and Round Hill are the largest features. A model section 6100 m wide by 18,300 m long centered over the Rakaia River Gorge was constructed to an undistorted scale of 1:5000.

A contoured model constructed of polystyrene bead-board was examined for three separate surface roughness conditions--a surface textured to represent typical paddock grass roughness only ($z_o \simeq 0.15$m), the same surface with zero porosity surface shelterbelts added, and the same surface with porous shelterbelts added. The total model length was 7.3 meters. A 2.5 cm high trip fence and a square bar turbulence grid were placed upwind of the model to produce the desired approach wind similitude characteristics. The Rakaia River Gorge model was studied in an Atmospheric Boundary Layer Wind Tunnel at University of Canterbury, Christchurch, New Zealand.

The hill sides to either side of the Rakaia River are primarily devoted to sheep paddock. To protect flocks and paddock surface during the high wind event, farmers have planted shelterbelts around most fields. Most of these shelterbelts are mature coniferous tree stands about 20 m high. The tree stands often consist of several rows and appear quite dense. Aerodynamic studies of flow fields behind shelterbelts in New Zealand have been performed by Sturrock (18). Measurements behind 4 mm high yarn and pipe cleaner wind breaks revealed they simulate the velocity and turbulence field for 1:5000 scale very well.

Figures 2 and 3 compare the velocity defect and turbulence excess measured behind such shelters with full scale behavior. Pipe cleaner shelterbelts were added to the model to simulate the prototype vegetation. Laboratory measurements included horizontal and vertical profiles of mean wind velocity, longitudinal turbulence, wind direction, turbulence spectra and correlation utilizing hot wire anemometry, pitot-static pressure probes, and cobra pressure probes.

FIELD MEASUREMENT PROGRAM

Measurements of wind velocity and directions were desired over the Rakaia River Gorge test region to provide a basis for validation of laboratory methodology and physical modeling. Ideally a network of permanent meteorological instruments would be installed on multiple towers with data recording equipment versatile enough to

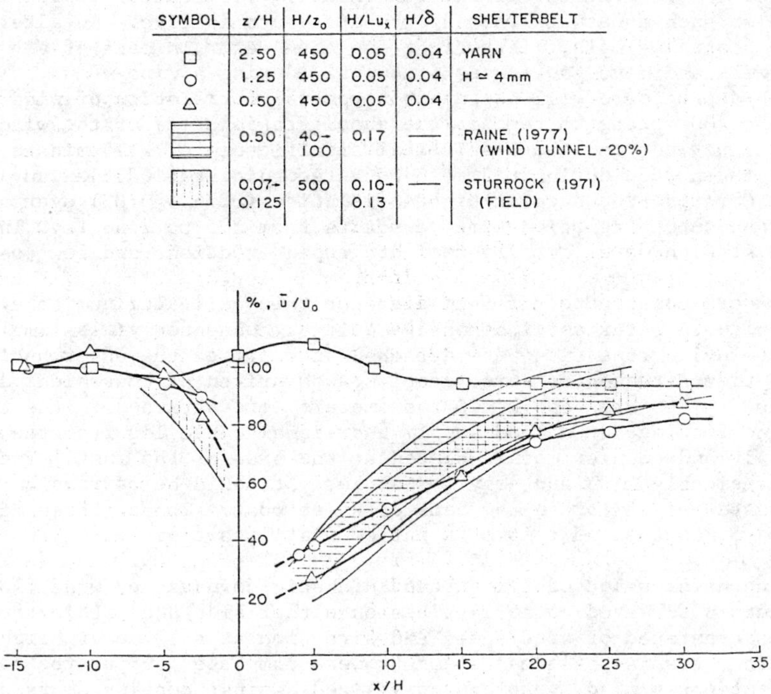

Figure 2. Comparison of wind shelter produced by yarn barriers with typical New Zealand shelterbelt behavior

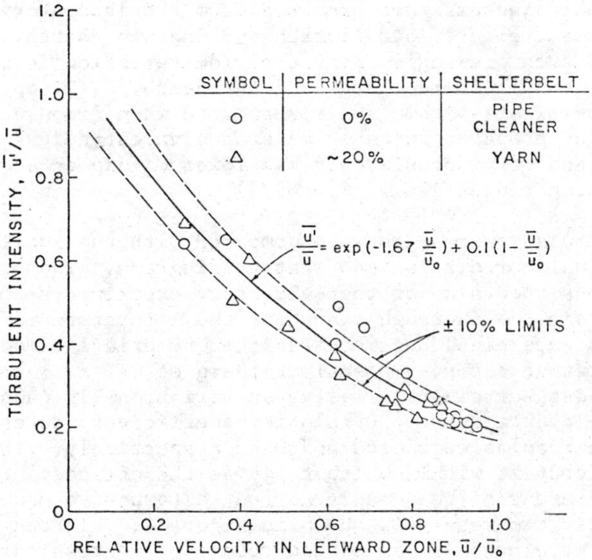

Figure 3. Correlation of turbulent intensity and relative velocity behind model shelterbelts. Solid line-- proposed by Raine and Stevenson (19)

intercept and record a northwesterly wind event. The cost of capitalization and maintenance of such a network was unfortunately prohibitive. An alternative proposed is to place a simple, lightweight cup anemometer on each of several collapsible pole towers and move the towers frequently during a wind event. The effectiveness of such a procedure will depend upon spatial correlation of wind velocities over the same 100 square km region, the quasi-stationarity of the wind event over a 3 to 6 hour period, and the statistical significance of a 15 minute sample at a given point taken once during a 3 to 6 hour recording period. Recent climatological analysis by Corotis (20) suggest high correlation (0.76 - 0.83) over distances less than 22 km and autocorrelation time constants from 3.5 to 7 hours. The criteria for a field station were thus lightweight, rapid erection, and low cost.

Three masts were constructed of 5 cm diameter thin wall aluminum tube. The tubes were made in two 5 meter sections which could be connected via a simple sleeve joint. Three nylon rope ties were attached at 7.5 m to the upper section and when erected the ties attached to three steel stakes driven at convenient distances from the mast base. The three cup RIMCO anemometers were attached to the top of the mast by a threaded fitting. A 3 lead supply and signal cable led from the anemometer to a power supply and counter module placed at the base of the mast. The entire system was conveniently light and easy to handle. It could be carried on the luggage rack of a passenger car or in the back of a jet boat. Two or three men could erect the tower in 5 minutes and remove it in somewhat less time.

On two spring days, selected for strong adiabatic down valley wind flow, three teams of investigators surveyed up to 27 sites on either side and within the river gorge. Measurements consisted of wind speed and direction at a 10 meter height on lightweight portable towers. All measurements were completed during the course of a five hour stationary wind event and normalized against continuous records taken from a fixed anemometer near model terrain center.

RESULTS AND CONCLUSIONS

A series of contour diagrams were prepared from the laboratory velocity and turbulence intensity measurements into isotach and isoturb charts. Figures 4 and 5 show a typical pair of such drawings. Note the wide variation in wind speed near ground level between points within the gorge and the nearby hill top. Simultaneously large relative gustiness exists within the river gorge when compared to the hill crest. Horizontal sections prepared for a 10 meter equivalent height (Figure 6) reveals the river valley and gorge consistenly has lower windspeed and greater gustiness than the surrounding ridges.

The laboratory simulation results were compared with the available field data by means of statistical correlation and scatter diagrams. The model and field results were used to assess the value of the laboratory experiments for predicting wind over complex terrain. A thorough search of the literature reveals that few authors have chosen to compare field and model (either numerical or physical) results in other than qualitative terms. Recently Fozberg et al. (21) compared a numerical model which includes terrain, thermally, and frictionally induced perturbations against seven field data sets. Correlation coefficients determined for the velocity and direction results were 0.60 and 0.62 respectively. These limited results may be used as a context within which to judge the efficacy of the present physical model or as a statement of reasonably current alternative modeling capacity. A typical scatter diagram result is shown in Figure 7. Plotted on the scatter diagram are the co-correlation lines of the result of regressions of abscissa against ordinate and ordinate versus abscissa. When $r = \pm 1$ these lines will be colinear, when $r = 0$ the lines will be perpendicular. The lines thus provide visual evidence of the quality of correlation. Correlation between equivalent field measurements at the same sites taken on independent days were $r = 0.68$. This suggests

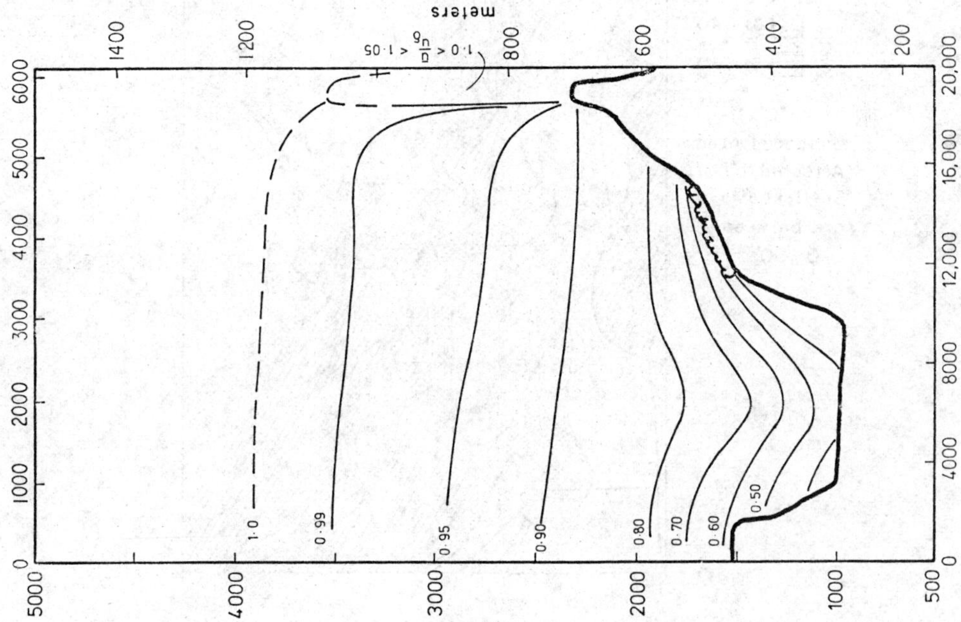

Figure 4. Typical vertical section isotachs

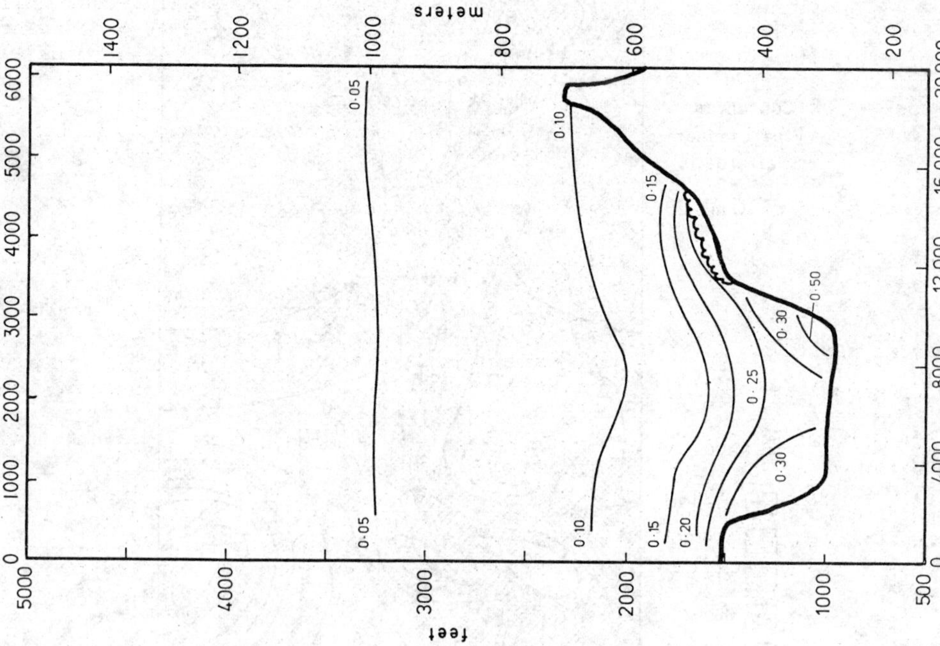

Figure 5. Typical vertical section isoturbs

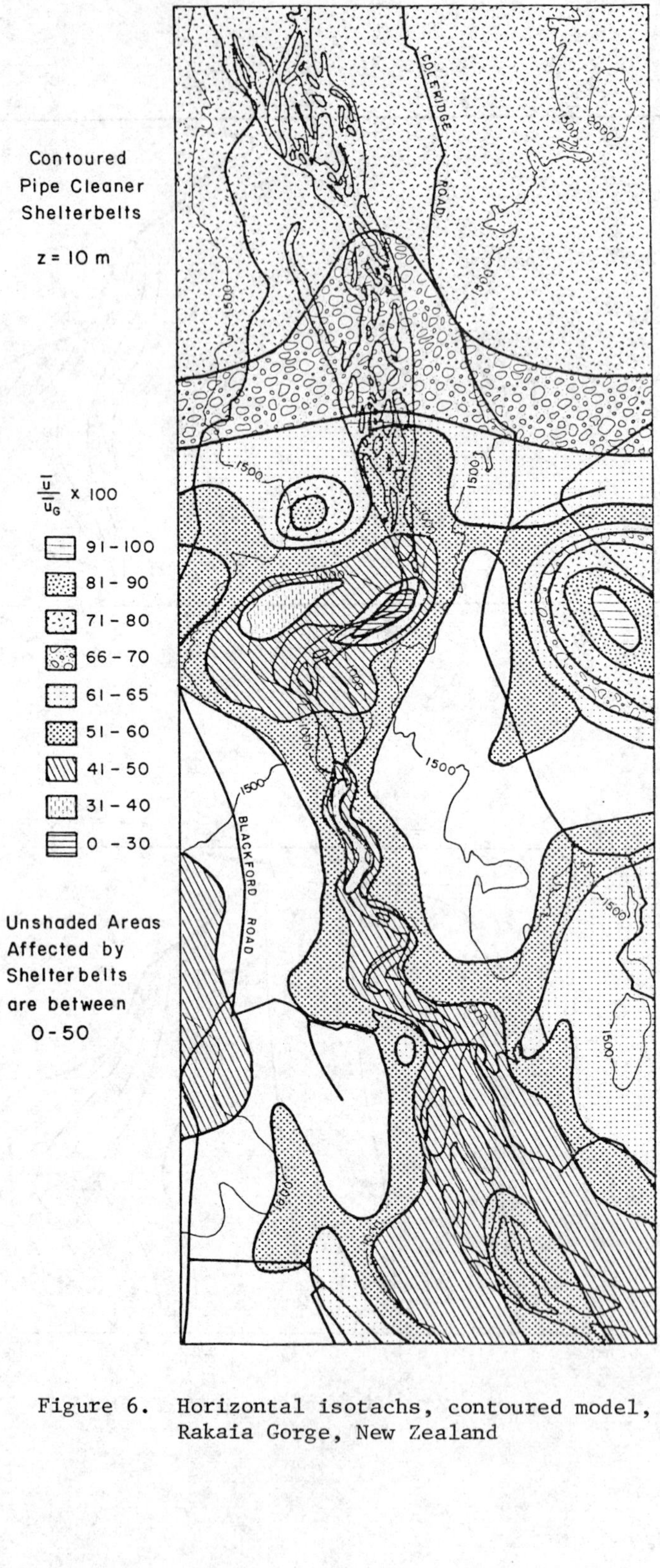

Figure 6. Horizontal isotachs, contoured model, z_p = 10 m.
Rakaia Gorge, New Zealand

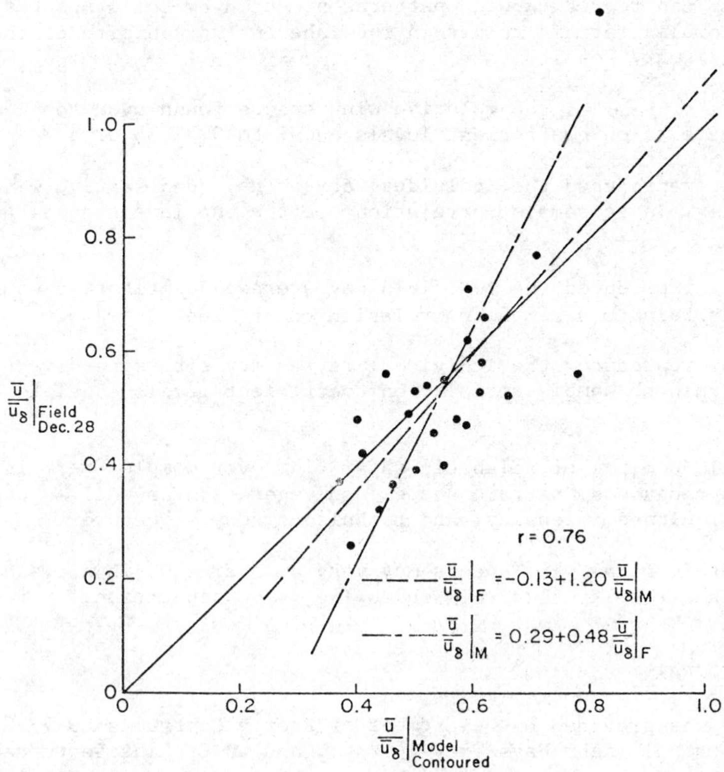

Figure 7. Scatter diagram field test data, December 28 versus contoured model data

that there may be an inherent limitation to the replication of any single realization of a wind flow pattern by any synthetic model whether physical or numerical.

It would appear that the conventional simulation wisdom developed in the past few years is appropriate for physical modeling of flow over complex terrain. Since the flow region of interest is usually in the lowest surface layer ($z < 100$ m) for Wind Energy Conversion Systems siting, great care must be taken that horizontal inhomogenuities in roughness and terrain are faithfully reproduced. Specific conclusions suggest that:

1. A wide range of scales and meteorological conditions may reasonably be simulated in existing boundary layer wind tunnel facilities; (See Performance Envelope).

2. To produce equivalent wind speeds near ground level require accurate reproduction of surface roughness, shape, and accurate reproduction of surface roughness, shape, and vegetation. Hence terraced models, adequate for certain dispersion simulations, are not appropriate for Wind Energy Conversion Systems site analysis; and,

3. Current meteorological data in complex terrain is not yet adequate to stipulate flow conditions to either numerical or physical models with confidence. Hence an adequate approach flow length must be provided to allow the surface layer to come to an equilibrium with underlying terrain undulations.

4. Physical modeling can reproduce wind patterns produced by the atmospheric shear layer flowing over complex terrain to within the inherent variability of the atmosphere to produce stationary results.

5. Physical modeling reproduced the relative wind speeds found over complex terrain by rank to sample correlation coefficient levels equal to 0.78 to 0.95.*

6. Physical Modeling reproduced the individual day to day quantitative wind speeds found over complex terrain to sample correlation coefficient levels equal to 0.70 to 0.76.

7. Physical modeling reproduced the two field day average quantitative wind speeds found over complex terrain to a sample correlation coefficient level equal to 0.81.

8. Physical modeling reproduced the individual day to day site wind directions found on complex terrain to sample correlation coefficient levels equal to 0.65 to 0.67.

9. Adequate physical modeling of adiabatic shear flow over complex terrain requires attention to surface roughness, terrain shape, and vegetation as well as upstream velocity profile, turbulence intensity, and turbulence eddy structure.

10. Over complex terrain local wind speeds may vary by over 100% in a distance of a few hundred meters as a result of terrain shadowing, flow separation, or flow enhancement.

ACKNOWLEDGEMENT

Financial assistance was provided by Department of Energy Contracts EG-77-S-06-1043 and EY-76-S-06-2438, a Fulbright Hays Travel Grant, and an Erskine Lectureship at the University of Canterbury. Messrs. John Pearse and Derrick Neal, University of Canterbury, assisted during the field and laboratory portion of this Project.

REFERENCES

1. R.N. Meroney, V.A. Sanborn, R.J.B Bouwmeester, H.C. Chien and M. Rider, Sites for wind power installations: physical modeling of the influence of hills, ridges, and complex terrain on wind speed and turbulence, Dept. of Energy Report RLO/2438-77/3, (1978).

2. R.N. Meroney, A.J. Bowen, D. Lindley, and J. Pearse, Wind characteristics over complex terrain: laboratory simulation and field measurements at Rakaia Gorge, New Zealand, Dept. of Energy Report RLO/2438-77/2, (1978).

3. J.E. Cermak, et al., Simulation of atmospheric motion by wind-tunnel flows, Tech. Report for DA-AMC-28-043-G20, Fluid Dynamics and Diffusion Laboratory, Colorado State University, Ft. Collins, CO, Report No. CER66JEC-VAS-ESP-GJB-HC-RNM-SI17, (1966).

4. J.E. Cermak, Applications of Fluid Mechanics to Wind Engineering, 1974 Freeman Scholar Lecture, ASME Journal of Fluids Engineering, 97, Colorado State University Report No. CEP74-75JEC7, (1975).

5. G.E. McVehil, G.R. Ludwig, T.R. Sundaram, On the feasibility of modeling small scale atmospheric motions, Cornell AeroLab Report ZB-2328-P-1, Buffalo, NY (1967).

---

*These numbers are somewhat sensitive to the sites chosen to normalize and compare model to field results.

6. A.B. Bernstein, Dimensional analysis applied to the wind distribution in the planetary boundary layer, Monthly Weather Review, 93: 579-585, (1965).

7. W.H. Snyder, Similarity criteria for the application of fluid models to the study of air pollution meteorology, Boundary Layer Meterology, 3: 113-134, (1972).

8. S. Nemoto, Similarity between natural wind in the atmosphere and model wind in a wind tunnel, Papers in Meteorology and Geophysics, Tokyo: 12, 30-52; 12, 117-128; 12, 129-154; 13, 171-195, (1962).

9. J. Counihan, An improved method for simulating an atmospheric boundary layer in a wind tunnel, Atmospheric Environment, 3; 197-214, (1969).

10. J. Counihan, Simulation of an adiabatic urban boundary layer in a wind tunnel, Atmospheric Environment, 7: 673-689, (1973).

11. N.J. Cook, Determination of the model scale factor in wind-tunnel simulations of the adiabatic atmospheric boundary layer, Journal of Industrial Aerodynamics, 2: 311-321, (1977-1978).

12. W.H. Melbourne, Development of natural wind models at Monash University, 6th Australian Hydraulics and Fluid Mechanics Conference, Adelaide, Australia, 190-194. (Dec. 5-9, 1977).

13. Clarke, A.C. Profiles of the Future, Bantam, NY (1958)

14. P. Mery, Reproduction on similitude de la diffusion dans la cauche limite atmospherique, La Houille Blanche, 4: (Trans. Air Poll. Tech. Inf. Cent. No. 1104) (1969).

15. N. Ukejurchi, H. Sakata, H. Okamoto, and Y. Ide, Study on stack gas diffusion, Matsubishi Tech. Bull., 52: 1-13, (1967)

16. G.M. Hidy, Adventures in atmospheric simulation, Bulletin of American Meteorological Society, 48: 143-161, (1967).

17. M.M. Orgill, J.E. Cermak, and L.O. Grant, Laboratory simulation and field estimates of atmospheric transport-dispersion over mountainous terrain, Technical Report, Colorado State University, CER70-71MMO-JEC-LOG40, (1971).

18. J.W. Sturrock, Aerodynamic structure of shelterbelts in New Zealand-2: medium height to tall shelterbelts in mid Canterbury, New Zealand J. of Science, 15: 113-140, (1972).

19. J.L. Raine, and P.C. Stevenson, Wind protection by model fences in a simulated atmospheric boundary layer, Journal of Industrial Aerodynamics, 2: 159-180, (1977).

20. R.B. Corotis, Stochastic modeling of site wind characteristics, Department of Civil Engineering, Northwestern University, DOE Report ERDA/RLO/2342-77/2, (1977).

21. M.S. Fosberg, W.E. Marlatt, and L. Krupnak, Estimating airflow patterns over complex terrain, USDA Forest Service Research Paper RM-162, Rocky Mountain Forest and Range Experiment Station, Ft. Collins, CO, (1976).

WIND TUNNEL MODELLING OF STACK GAS DISPERSION - DIFFICULTIES AND APPROXIMATIONS

N. ISYUMOV* and H. TANAKA**

*The University of Western Ontario, London, Ontario, Canada
**University of Ottawa, Ottawa, Ontario, Canada

ABSTRACT

The similarity requirements for wind tunnel model studies of stack gas dispersion are presented and discussed. Exact similarity of plume behaviour requires explicit scaling of the stack gas density and equality of the model and full scale Froude numbers. For small scale diffusion studies the latter requirement leads to very low wind speeds. This complicates the modelling of the atmospheric boundary layer and makes reliable wind tunnel operation difficult. Approximate modelling methods which relax the requirement of stack density scaling in favour of operating at higher wind tunnel speeds are examined. Experimental data are presented for neutrally stable conditions to evaluate some of the more common approximate techniques. These data are compared with results obtained from an "exact" simulation in which the wind speed and the dynamic characteristics of the efflux are correctly scaled. Such comparisons are presented for an isolated stack and for the case where aerodynamic downwash occurs due to the presence of a nearby building. The approximate simulation technique, which relaxes the scaling of the efflux momentum in favour of a consistent scaling of both ambient and stack exit velocities, produces reasonable results for the examined case of aerodynamic downwash. For isolated stacks, all approximate methods, which reduce the model stack gas density, are found to exaggerate the plume rise and thus potentially underestimate ground level concentration.

INTRODUCTION

While diffusion by turbulent mixing is understood in principle, the estimation of SO_2 concentrations resulting from the operation of chimney stacks is not without practical difficulties. Existing analytical methods, usually based on evaluating the rise of the plume due to buoyancy and momentum and the subsequent treatment of the plume as an elevated point source of non-buoyant pollutant, can provide reasonable estimates under ideal conditions. That is, for stacks, located well outside the aerodynamic influence of their surroundings and situated in homogeneous flat open terrain for which the characteristics of the atmospheric boundary layer can be estimated. On the other hand, no reliable analytical methods for estimating plume dispersion are available for more complex situations. These include:

i) various forms of aerodynamic downwash, where plume behaviour is influenced by the aerodynamic presence of the stack itself or that of surrounding buildings;

ii) terrain downwash in hilly topography; and

iii) situations of unusual flow characteristics and/or temperature stratifications

Wind tunnel model studies provide a practicable means for estimating stack gas concentrations in such situations. The success of such simulations, however, depends upon achieving proper similarity between the model and full scale processes. Strict similarity of atmospheric phenomena is generally not possible and some relaxations of the modelling requirements become unavoidable. While there is consensus on the necessary degree of similarity for simulations of wind induced effects on buildings and structures, such agreement is generally lacking in the area of diffusion modelling. Whilst it is recognized that strict modelling is not practicable, there is disagreement on the degree of relaxation of scaling parameters which is permissible before wind tunnel results become non-representative.

SIMILARITY REQUIREMENTS

Modelling of Natural Wind

Wind tunnel simulations of plume dispersion require representative modelling of both the characteristics of the atmospheric boundary layer and the dynamic and thermal properties of the plume. Similarity requirements for wind tunnel simulations of natural wind have been extensively treated in the literature[1,2,3,4,5,6] and need only brief review at this time. Strict modelling of the atmospheric boundary layer requires:

i) the simulation of boundary conditions, including terrain roughness, topographic relief and surface temperature;
ii) the similarity of the flow field, including the mean and turbulent velocities and the temperature stratification; and
iii) equality of the Reynolds (*Re*), Richardson (*Ri*), Rossby (*Ro*), Prandtl (*Pr*) and Eckert (*Ec*) numbers.

Exact similarity is not possible at reduced geometric scales and practicable models of natural wind are based on relaxing the equality of the above non-dimensional parameters. A common model of the atmospheric boundary layer during strong wind conditions, and $Ri \approx 0$, comprises the simulation of the topographic relief and the terrain roughness and the modelling of natural wind by a turbulent shear flow with zero longitudinal pressure gradient and with the mean and turbulent flow structures matched to known full scale values[5,6,7,8]. Whereas *Pr* similarity is satisfied while air remains the working fluid and *Ec* similarity is of little consequence for incompressible flow, the *Ro* and *Re* numbers remain unscaled. The rotation of the full scale velocity vector with height tends to enhance the lateral plume spread. This effect, however, tends to become significant only at downwind distances of 5 km or more[9]. The absence of cross-shear in the model, $(Ro)_m = 0$, leads to somewhat conservative far field concentrations. *Re* similarity for the entire flow is achieved by maintaining fully developed turbulent model flow conditions. The difficulties of *Re* modelling and the consequences of very small *Re* values on the turbulence structure and local flows around particular objects have been discussed elsewhere[2,4,5].

The above approximate flow simulation is representative of natural wind for neutrally stable and slightly unstable atmospheric conditions. It is generally suitable for dispersion studies of plumes from tall stacks[4], which protrude well above the diurnal inversion layer, and is representative for studies of various forms of plume entrainment which are aerodynamic effects and become pronounced at high wind speeds. Simulations of natural wind during stable conditions, $Ri < 0$, nevertheless are important for some atmospheric diffusion problems. The simulation of temperature stratification effects in such cases are usually based on scaling the bulk Richardson number[1,5] which is of the form of the inverse densimetric Froude number evaluated at a reference elevation. The material presented in this paper only deals with the $Ri = 0$ situation which is well simulated in the wind tunnel by an isothermal boundary layer.

Exact Modelling of Plume Behaviour and Practical Difficulties

Having established a representative simulation of natural wind at a geometric scale of $\lambda_L = L_m/L_p$, where L is a characteristic length and subscripts m and p denote model and full scale quantities respectively, the additional exact modelling requirements for similar plume behaviour can be summarized as follows:

Geometric similarity of stack	$\frac{D_m}{D_p}$	=	λ_L	(1)
Equality of Froude number	$(Dg/U^2)_m$	=	$(Dg/U^2)_p$	(2)
Equality of density ratio	$(\rho_s/\rho_a)_m$	=	$(\rho_s/\rho_a)_p$	(3)
Consistent Scaling of all velocities	$(W_s/U)_m$	=	$(W_s/U)_p$	(4)
Stack Reynolds number	$(DW_s/\nu_s)_m$	=	$(DW_s/\nu_s)_p$	(5)

where D, ρ_s, ρ_a, g, U, W_s and ν_s are the internal stack diameter, the exit stack gas density, the ambient air density, the acceleration of gravity, the average ambient wind velocity, the average stack exit gas velocity and the viscosity of the stack gas. Note that $\rho_s/\rho_a = T_a/T_s$, where T_a and T_s are the ambient air and stack gas temperatures respectively.

The above exact simulation requires that all velocities are scaled to satisfy Froude number similarity. The velocity scale for exact plume modelling thus becomes:

$$\lambda_u = \lambda_w = \sqrt{\lambda_L} \qquad (6)$$

Geometric scales, dictated by modelling requirements of natural wind, are typically in the range of $\lambda_L \approx 1{:}400$ *to* $1{:}900$. Corresponding velocity scales, based on equation *(6)*, are $\lambda_u = 1{:}20$ *to* $1{:}30$. Typical full scale wind speeds which are of interest in diffusion studies are of the order of 5 to 20 m/s. Exact modelling thus requires wind tunnel speeds in the range of 0.15 to 1m/sec. Accurate control of wind speeds towards the lower end of this range is difficult for most wind tunnel facilities. The flow at such speeds is susceptable to small perturbations. Resulting flow instabilities significantly exaggerate the velocity spectra at low frequencies. It is noteworthy that most wind tunnel simulations of wind engineering problems treat the atmospheric boundary layer as locally stationary and determine particular wind related effects for particular values of some reference mean hourly wind speed. Mesoscale and long term variations of the mean wind speed are then allowed for analytically using statistical models of the full scale mean wind speed determined from synoptic data.

Spectral estimates of velocity demonstrating the distortion of spectral shape at very low wind speeds have been presented elsewhere[4]. An exaggeration of energy at low frequencies due to flow non-stationarities, however, is still present even at the high end of wind tunnel speeds normally used in diffusion studies. Fig. 1 depicts the spectrum of the longitudinal velocity component measured in a diffusion study carried out at a geometric scale of $\lambda_L = 1{:}400$. The model mean wind speed at a full scale reference height of about 82 meters for these measurements was $U_R \approx 1.1$ *m/sec.*

Although, the practical difficulties of wind tunnel operation are a major concern, the requirement of very low model wind speeds also complicates the overall simulation of the flow regime. A commonly used criterion for fully developed turbulent flow requires that the roughness Reynolds number $u_* z_o/\nu > 2.5$. This is difficult to satisfy for relatively smooth terrains at small geometric scales without exaggerating the model terrain roughness and consequently the model roughness length z_o. Other modelling problems, arising from basing the velocity scale on *Fr* similarity at very low geometric scales, include the difficulties of achieving flow similarity around particular building shapes[10] and the excessive narrowing of the spectrum of turbulence[4], as the ratio of the peak wavelength to the microscale is proportional to $Re^{3/4}$.

Modelling the ratio of the exit gas density to that of the ambient air, as required by equation *(3)*, in itself is not difficult and can be achieved by using a lighter-than-air gas mixture. Helium gas is commonly used for this purpose. In some experimental techniques[4], helium is used both to satisfy the

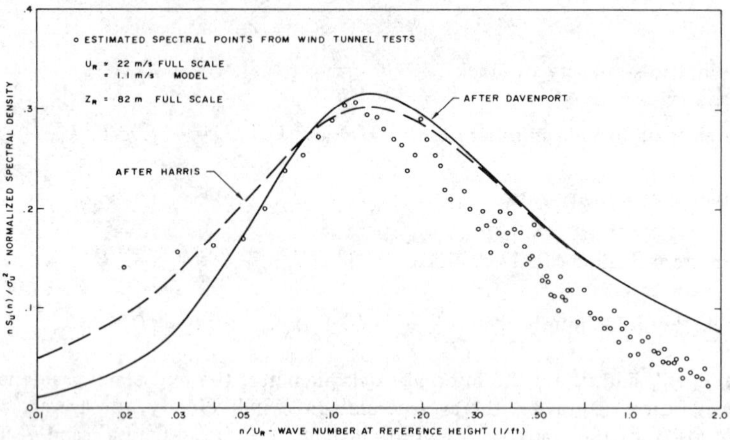

Fig. 1 Typical power spectrum of the longitudinal velocity component in a wind tunnel diffusion simulation (λ_L = 1:400)

stack gas density scaling as well as a tracer gas for detecting pollutant concentrations. As discussed below, the density scaling is sometimes relaxed in favour of achieving higher model wind speeds.

While explicit modelling of the internal stack Reynolds number DW_s/ν_s is not possible, this requirement can be satisfied by achieving approximately similar flow conditions at the stack exit. The flow in full scale stacks is normally turbulent while remaining laminary in the model stack. Without any further treatment, the velocity distributions within the full scale and model stacks are different–approximately logarithmic or power law with an exponent of about 1/7 versus parabolic. With laminar flow in the model stack, the model maximum exit velocity is exaggerated by about 8/5 and the exit momentum by about 4/3 for the same average speed W_s. Two techniques are sometimes used to improve the similarity of the internal stack flow. One of these consists of roughening the inner stack surface(5). This technique is not effective for small diameter stacks and an air-helium mixture, as the Reynolds number of the model internal flow is well below the critical value for pipe flow. The other technique consists of using a somewhat larger diameter stack with a short contraction to the required inside diameter added at the tip of the stack(11).

The practical consequences of improving the modelling of the exhaust flow are of modest significance for stacks well outside the aerodynamic influence of its surroundings. In such cases the total plume rise is usually dominated by buoyancy effects as the component of plume rise due to stack gas momentum is generally small. On the other hand, improved modelling of the exhaust flow becomes more important for aerodynamic downwash effects as the initial plume behaviour is momentum dominated. The effect of improved modelling of the internal stack flow, using a tip contraction, on plume entrainment by an immediately upstream building can be inferred from Fig. 2. This particular simulation was carried out at a geometric scale of λ_L = *1:400* and an exact simulation of U, ρ_s/ρ_a and W_s as required by equations *(2), (3)* and *(4)* respectively. As seen from Fig. 2, the improved exit flow modelling leads to an earlier plume entrainment as the height of the stack is reduced. Conversely, using a simulation in which the velocity distribution within the model stack is near parabolic somewhat underestimates the stack height required to avoid wake entrainment of the plume. Nevertheless, even in this more critical situation, the effects of improved modelling of the stack exhaust remain small.

Approximate simulations of plume behaviour

Difficulties in achieving representative modelling of natural wind and with the operation of wind tunnels, resulting from exact Froude number scaling, have led to the use of various approximate simu-

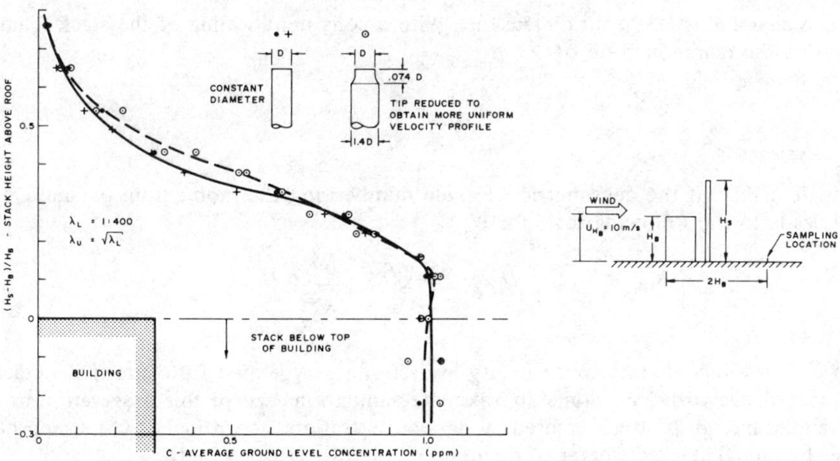

Fig. 2 Influence of changes of the velocity distribution of the exhaust flow on plume dispersion in the presence of wake entrainment

lations of plume behaviour. In all of these, the modelling of the stack gas density ratio as an explicit parameter is relaxed. Instead, equation *(3)* is combined with the Froude number to form the densimetric Froude number

$$(Fr)_d = \frac{(\rho_s - \rho_a)\,Dg}{\rho_a U^2} \tag{7}$$

The velocity scale based on $(Fr)_d$ equality in model and prototype becomes

$$\lambda_u = \sqrt{\lambda_L}\left(\frac{\lambda_{\Delta\rho}}{\lambda_{\rho_a}}\right)^{1/2} \tag{8}$$

where $\Delta\rho = (\rho_s - \rho_a)$ and $\lambda_{\Delta\rho} = (\rho_s - \rho_a)_m/(\rho_s - \rho_a)_p$

The full scale stack gas exit temperature typically ranges around 150 to 250°C. Using pure helium, which is approximately 7.2 times lighter than air, $(\lambda_{\Delta\rho}/\lambda_{\rho_a})^{1/2} \approx 1.5$. Consequently, the use of $(Fr)_d$ rather than *Fr* can result in approximately 50% higher model wind speeds. Using the same ranges of geometric scale and full scale wind speeds, as in the above discussion of exact velocity scaling, required wind tunnel speeds would range between 0.23 to 1.5 m/s. This increase is sufficient to improve wind tunnel operation, which depending on the facility, becomes practicable at speeds of 0.3 to 0.6 m/s.

Although maintaining equality of the densimetric Froude number leads to a correct relative scaling of plume buoyancy forces and fluid inertia forces at the stack exit, a relaxation of the density ratio results in different rates of mixing of the plume and also precludes consistent scaling of other plume properties. For example, modelling the average stack exit velocity W_s to the same scale as the ambient wind speed, namely in accordance with equation (8), leads to an underestimate of the stack gas momentum. The initial phase of the plume is momentum dominated and this approach somewhat underestimates the initial plume rise. On the other hand, maintaining a correct ratio of the plume momentum to that of the ambient flow exaggerates the exit stack gas velocity. Momentum scaling of the plume is based on

$$\frac{\rho_s D^2 W_s^2}{\rho_a L^2 U^2} = constant \tag{9}$$

where L is a characteristic overall dimension. With a consistent scaling of the stack diameter, equation *(9)* reduces to the common form of

$$\frac{\rho_s W_s^2}{\rho_a U^2} = constant \qquad (10)$$

Maintaining equality of the densimetric Froude number and the momentum parameter, given in equation *(10)*, leads to the following scale for W_s;

$$\lambda_{w_s} = \lambda_u (\lambda_{\rho_a}/\lambda_{\rho_s})^{1/2} = \sqrt{\lambda_L} \left(\frac{\lambda \Delta \rho}{\lambda_{\rho_s}}\right)^{1/2} \qquad (11)$$

An increase of the model stack exit velocity leads to an exaggerated flow rate if the stack diameter is scaled in accordance with λ_L. Some approximate simulations accept this exageration of the model source strength and adjust the measured model concentrations accordingly. Other approaches attempt to reduce the model stack diameter to maintain a correct source strength.

Approximate plume modelling techniques reported to have been used at several Laboratories are summarized in Table 1. A summary of the corresponding scaling parameters is presented in Table 2. The various scaling parameters, based on an exact simulation of equation *(1)* to *(4)* and some approximation of the exhaust gas flow, required by equation *(5)*, are given in Tables 1 and 2 along with the procedures followed at the listed Laboratories. The stack gas density is not modelled in any of these approximate methods. As a result, the dynamic and kinematic similarity of the model plume is not exact and the measured model concentrations require interpretation. As seen from Table 2, the scaling of the ambient air speed is based on the densimetric Froude number, except that in the case of Bristol University[11], $(Fr)_d$ is taken as the ratio of the buoyancy and inertia forces of the plume itself. The technique followed at Calspan[12] bases similarity of buoyancy effects on a matching of average inclinations of the model and prototype plumes. The criterion for buoyancy similarity, arrived at on this basis, is given in equation *(12)*. This parameter is of the same functional form as that commonly used for estimating the plume rise for unstable and neutral conditions[15]. The basis for the approach for scaling U, proposed by Monash University[14], is unclear.

$$\frac{(\rho_s - \rho_a) D g W_s}{\rho_a U^3} = constant \qquad (12)$$

The scaling of the exit stack gas velocity at all Laboratories, given in Table 1 and 2, except for C.S.U. [1,5], is supposedly based on exaggerating W_s to maintain a correct scaling of the efflux momentum.

QUANTITY AND SCALING PARAMETERS		EXACT SIMULATION	APPROXIMATE SIMULATIONS				
			B.L.W.T.L. Univ. Western Ont. London, Ontario (4)	C.S.U. Fort Collins Colorado (1,5)	Calspan Corp. Buffalo, N.Y. (10)	Monash Univ. Melbourne, Aust. (11)	Bristol University U.K. (9)
Density Ratio	$\frac{\rho_s - \rho_a}{\rho_a}$	Yes	No	No	No	No	No
Plume Buoyancy	$\frac{(\rho_s - \rho_a) Dg}{\rho_a U^2}$	Yes	Yes	Yes	**	Yes	Yes
Plume Velocity Ratio	W_s / U	Yes	No	Yes	No	No	No
Plume Momentum	$\frac{\rho_s W_s^2}{\rho_a U^2}$	Yes	Yes	No	Yes	Yes*	Yes*
Mass Flow Rate	$\frac{\rho_s D^2 W_s}{\rho_a L^2 U^2}$	Yes	No	No	No	Yes By distorting D	No
Flow Rate	$\frac{D^2 W_s}{L^2 U}$	Yes	No	Yes	No	No	Yes By distorting D

* this is unclear as correct scaling of plume momentum with reduced D should be based on equation (9).

Table 1 Approximate simulations of plume behaviour sometimes used at various wind tunnel laboratories

Quantity	From Exact Simulation	From Various Approximate Simulations				
		B.L.W.T.L. Univ. of Western Ont. London, Ont. (4)	C.S.U. Fort Collins Colorado (1,5)	Calspan Corporation Buffalo, New York (10)	Monash Univ. Melbourne Aust. (11)	Bristol Univ. U.K. (9)
Ambient Air Velocity	$\lambda_u = \sqrt{\lambda_L}$	$\sqrt{\lambda_L}\,(\lambda_{\Delta\rho}/\lambda_{\rho_a})^{1/2}$	$\sqrt{\lambda_L}\,(\lambda_{\Delta\rho}/\lambda_{\rho_a})^{1/2}$	$\sqrt{\lambda_L}\,(\lambda_{\Delta\rho}/\lambda_{\rho_a})^{1/2}(\frac{1}{\lambda_{\rho_s}})^{1/4}$	$\sqrt{\lambda_L}\,(\frac{\lambda_{\Delta\rho}}{\lambda_{\rho_a}^{3/4}\lambda_{\rho_s}^{1/4}})^{1/2}$	$\sqrt{\lambda_L}\,(\lambda_{\Delta\rho}/\lambda_{\rho_s})^{1/2}$
Stack Exit Velocity	$\lambda_w = \lambda_u$	$\lambda_u\,(\lambda_{\rho_a}/\lambda_{\rho_s})^{1/2}$	λ_u	$\lambda_u\,(\lambda_{\rho_a}/\lambda_{\rho_s})^{1/2}$	$\lambda_u\,(\lambda_{\rho_a}/\lambda_{\rho_s})^{1/2}$	$\lambda_u\,(\lambda_{\rho_a}/\lambda_{\rho_s})^{1/2}$
Stack Exit Diameter	$\lambda_D = \lambda_L$	λ_L	λ_L	λ_L	$\lambda_L\,(\frac{\lambda_w}{\lambda_u})^{1/2}$	$\lambda_L\,(\frac{\lambda_w}{\lambda_u})^{1/2}$
Downwind Pollutant Concentration	$\lambda_c = \frac{C_{s_m}}{C_{s_p}} = \lambda_{C_s}$	$\lambda_{C_s}\ \lambda_w/\lambda_u$	λ_{C_s}	$\lambda_{C_s}\ \lambda_w/\lambda_u$	*not known*	λ_{C_s}

Subscripts m and p denote model and prototype properties.

Table 2 Summary of scaling parameters used in approximate simulations

The approach at C.S.U. is to disregard the momentum scaling in favour of a consistent scaling of W_s. The claims of correct plume momentum scaling with a distorted stack diameter scaling are unclear. In such cases the momentum parameter must be based on equation *(9)* and not equation *(10)*, which assumes consistent geometric scaling of all lengths. Furthermore, consistent plume momentum and mass flow rate scaling can only be achieved if the density ratio is correctly modelled and $\lambda_D = \lambda_L$. The unusual form for the ambient air speed scaling given in Table 2 in the case of Monash University results from difficulties with the plume momentum scaling. This has been confirmed by a recent private communication from Prof. Melbourne. Although Barret[11] has reported reasonable success with an exaggerated stack diameter scaling, the advantages of his approach remain unclear to the authors.

Maintaining proper momentum scaling with an undistorted stack diameter unfortunately exaggerates the mass and flow rates. Correspondingly the model downwind concentrations are higher. To allow for this, the resulting downwind concentration scaling becomes

$$\lambda_c = \lambda_{c_s}\,\lambda_{w_s}/\lambda_u \tag{13}$$

where λ_{c_s} is the ratio of model and full scale pollutant concentrations at the stack exit; namely $\lambda_{c_s} = c_{s_m}/c_{s_p}$ and c_s is the stack exit pollutant concentration. As seen from Table 2, this approach is followed both at U.W.O.[4] and at Calspan[12].

The following summary comments follow from the above presented review of approximate plume modelling techniques:

i) In all approximate simulations the density ratio is distorted in order to achieve ambient air speeds which are sufficiently high to permit a reasonable simulation of natural wind and a practicable operation of the wind tunnel facility.

ii) The ambient air speed is scaled in order to achieve similarity of buoyancy effects. The more commonly used buoyancy parameter is the densimetric Froude number.

iii) The plume exit velocity is either scaled as the air speed, while relaxing the momentum scaling of the plume, or on the basis of maintaining correct momentum scaling by exaggerating the stack exit velocity. The advantages of distorting the stack diameter scaling, as reported in the literature, are unclear.

PLUME DISPERSION PREDICTED FROM APPROXIMATE SIMULATIONS

The distortion of the density ratio, inherent to all above reported approximate simulations, results in kinematic dissimilarity between the model and prototype plumes. Some comparisons of the trajectories of stack plumes scaled in accordance to momentum or flow rate similarity, obtained by flow visualization, have been reported in the literature[16]. Quantitative differences of pollutant concentrations obtained by various approximate simulation techniques remain largely undefined. The objective of this part of the paper is to provide some quantitative information on the relative effectiveness of approximate simulation techniques. The approach is to examine downwind stack gas concentrations obtained with various approximate simulation procedures reported by U.W.O., C.S.U. and Calspan Laboratories and to compare these with results obtained with an exact simulation. The approximate methods involving a distortion of the stack diameter are not examined. All measurements are confined to neutrally stable atmospheric conditions.

Experimental Procedure

The experimental procedure used, follows that developed at the Boundary Layer Wind Tunnel Laboratory of the University of Western Ontario. The experimental method, which is described in detail elsewhere[4,17,18], uses either pure helium or an air-helium mixture depending on the required wind speed scaling, to model the stack effluent. In addition to simulating buoyancy effects, the helium is also used as the tracer gas for detecting downwind stack gas concentrations. A schematic view of the entire simulation and gas sampling system is presented in Fig. 3. Only the time average concentrations are measured by this system. Experiments, carried out at a geometric scale of λ_L = 1:400, examined two areas of plume dispersion. Namely the dispersion of a plume from a single stack located in homogeneous, moderately rough terrain and the dispersion from a stack in a similar terrain in the presence of aerodynamic interference from power station buildings. In all cases the internal stack diameter was modelled in accordance with λ_L with no special attempt to improve the simulation of the exhaust flow as required by equation *(5)*. As indicated in Fig. 2, the effects of differences in exhaust flow similarity were found to be modest even in the presence of plume entrainment.

Plume Dispersion in the Presence of Aerodynamic Effects

Close similarity of plume behaviour is of particular importance in situations of aerodynamic downwash. The effect of plume entrainment into the wake of a building is illustrated in Fig. 4. The top diagram shows ground level SO_2 concentration resulting from the operation of one of several relatively short stacks located on top of a power station building. In this case, the stack gas is entrained by the aerodynamic wake formed leeward of the power station building and high levels of SO_2 concentrations result at ground level. As indicated, the entire near wake becomes the pollutant source and relatively high ground level concentrations are experienced not only in the immediate proximity of the plant but also at large downstream distances. Increasing the heights of the stacks by approximately 55 meters (180 feet), as shown in the lower diagram, eliminates wake entrainment. Close similarity of the initial plume rise is particularly important in situations as shown in Fig. 4 as entrainment of the plume is sensitive to small differences in its trajectory.

The results of an experiment, carried out to examine the sensitivity of wake entrainment of a stack plume to various relaxations of the exact modelling requirements, are present in Fig. 5. As indicated in this figure, the stack is located just downstream of a power station building and all concentration measurements are made at ground level, two building heights downstream from the leading edge of the building. The inside diameter of the stack was kept constant at 2.5 meters in full scale and the height of the stack was progressively reduced from a maximum height of 82 meters in full scale. The simulation was carried out in turbulent boundary layer flow conditions representative of natural wind over moderately rough terrain. The mean wind speed at the height of the building was maintained at 10 m/s in full scale.

Curve 1, presented in the Fig. 5, gives the ground level concentration obtained with an "exact" simulation of plume behaviour, namely in accordance with equations *(1)* to *(4)*. Note that this is the same

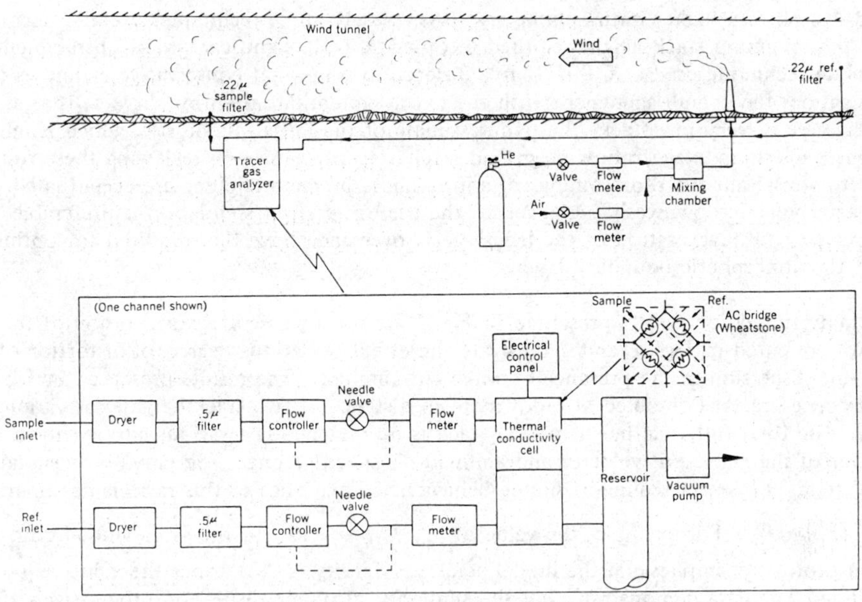

Fig. 3 Schematic of experimental procedures used for plume dispersion studies

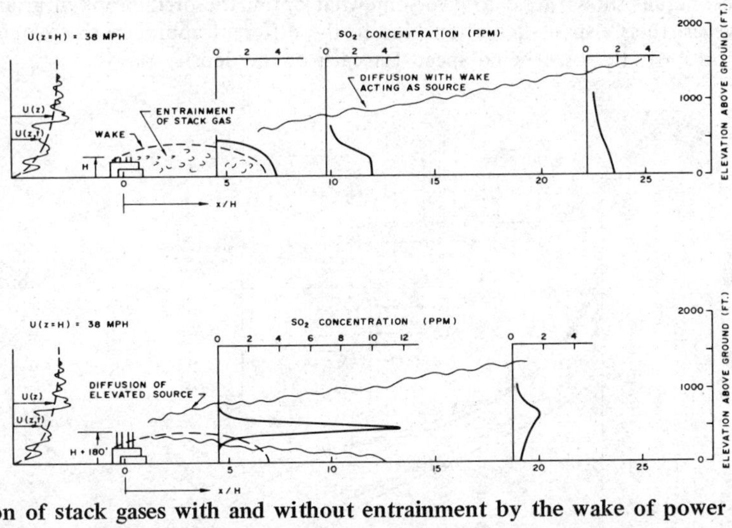

Fig. 4 Diffusion of stack gases with and without entrainment by the wake of power station buildings

curve as that given for the constant diameter case in Fig. 2. The data indicated by curves 2,3 and 4 were obtained from approximate simulations of plume behaviour following approximate procedures reported by U.W.O.[4], C.S.U.[5] and Calspan[12] respectively. The exaggeration of the stack gas density was maintained constant in all three approximate simulations. Using the λ_u scaling presented in Table 2, the ambient air speed is exaggerated by 49% for the U.W.O. and C.S.U. approximate simulations and by approximately 84% for the Calspan simulation in view of the different buoyancy parameter. Again referring to Table 2, there is no exaggeration of the stack exit velocity in the case of the curve 3, whereas the exit plume velocity is exaggerated by some 48% for curves 2 and 4.

From these results, the approximate techniques based on maintaining correct plume momentum scaling, curves 2 and 4, lead to a higher initial plume rise. As a result, these techniques provide unconservative estimates of the minimum stack height required in order to avoid significant wake entrainment. While the approximate technique used at C.S.U., curve 3, leads to somewhat conservative estimates of ground level concentration, the overall agreement with the exact simulation, given by curve 1, is significantly better. In the case of very low stacks,the underestimate of the initial plume rise, which is inherent to this approach, leads to an exaggeration of ground level concentrations. In reviewing these results, it is important to stress that the shortcomings of approximate plume modelling are accentuated in the presence of aerodynamic downwash. As a result, the use of an exact simulation is preferable[18],even if this requires a slight exaggeration of the length scale over and above that required for optimum modelling of the atmospheric boundary layer.

The approximate simulation results presented in Fig. 5 are for a particular exaggeration of the stack gas density. Data presented in Figs. 6 and 7 illustrate the effect of various degrees of distortion of this parameter on plume dispersion in the presence of wake entrainment. The results presented in Fig. 6 are ground level concentrations obtained at a downstream distance of five stack heights for various mean wind speeds. The total full scale height of the stack is 82 meters. The various curves shown are for different ratios of the stack exit velocity and ambient air speed scaling. The curve corresponding to $\lambda_{w_s} / \lambda_u = 1$ represents exact scaling of plume behaviour. For values of this ratio larger than unity, $(\rho_s/\rho_a)_m < (\rho_s/\rho_a)_p$. For example, the value of $\lambda_{w_s}/\lambda_u = 2.17$ for the same ambient air density in model and prototype implies that the model stack gas density is 0.21 times the value required by exact modelling. The data demonstrate that the similarity of plume dispersion progressively deteriorates with an exaggeration of the stack gas density used to achieve higher model wind speeds. Both approximate modelling techniques shown,tend to give somewhat optimistic predicitons of ground level concentrations. However, the Calspan simulation, due to the different approach to scale U, indicates maximum concentrations at a lower wind speed for each of the density ratios.

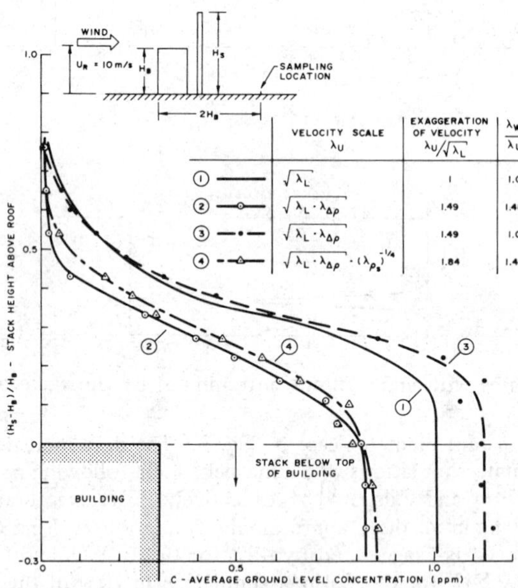

Fig. 5 Plume dispersion in the presence of aerodynamic downwash based on exact and approximate wind tunnel simulations

The far field concentration data presented in Fig. 7 were obtained with the same simulations as described in connection with Fig. 6. Consistent with the data presented in Fig. 6, the exaggeration of the density scaling produces a monotonic departure from ground level concentrations predicted from the exact simulation. The approximate technique suggested by Calspan, as in the case of near field diffusion presented in Fig. 6, tends to give somewhat conservative estimates of concentration at low wind speeds.

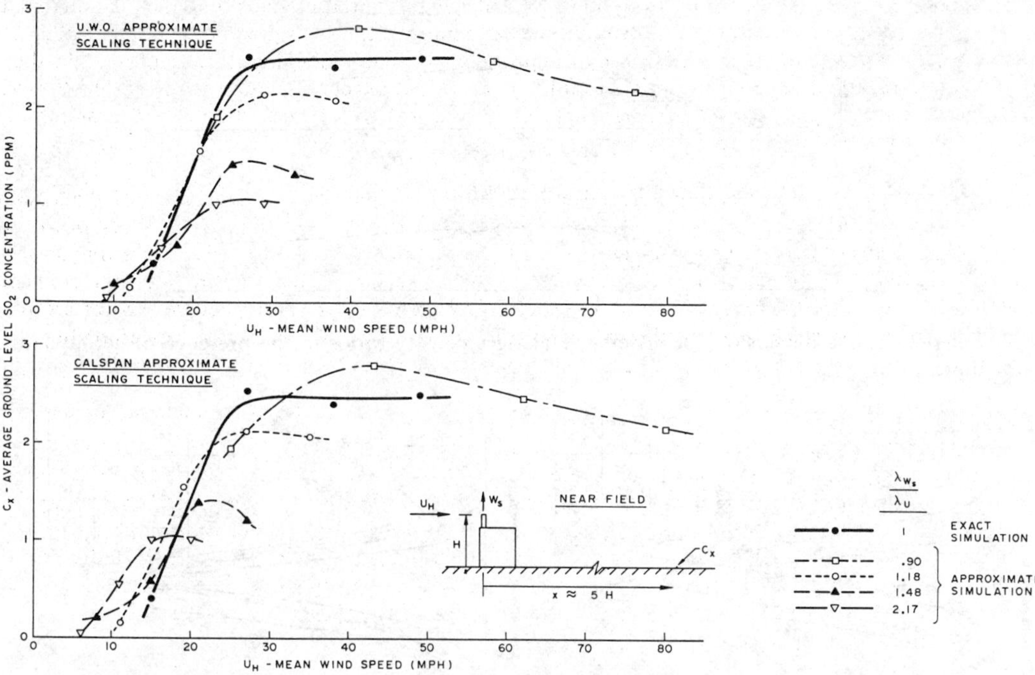

Fig. 6 Near field plume dispersion for different stack gas density ratios in the presence of aerodynamic downwash

Plume Dispersion from Single Stacks

The effectiveness of various approximate plume simulation techniques for different distortions of the stack gas densities were examined for a single stack located in homogeneous,moderately rough terrain. The full scale stack height was 82 meters, with a full scale internal diameter of 2.5 meters. The simulation was carried out at a geometric scale of 1:400. Vertical profiles of average concentration through the centreline of the plume are presented in Fig. 8 for downstream distances of 6 to 12 stack heights. Results depicted by curve 1 are for an exact simulation of the plume, whereas curves 2, 3 and 4 are obtained with approximate simulations following the techniques reported by U.W.O., C.S.U. and Calspan, respectively. Data presented for the downstream location of 6 stack heights are given for several distorted values of the stack gas density scaling. As would be expected, the U.W.O. and Calspan techniques, curves 2 and 4, lead to very similar results. It is noteworthy, that all approximate simulation techniques exaggerate the vertical position of the plume centreline and thus tend to underestimate ground level concentrations.

The effective plume rise, taken as the height of the plume centreline minus the height of the stack, is shown in Fig. 9a. Only the results obtained from the density scaling of $\sqrt{\lambda_{\Delta\rho}} = 1.49$ are presented. Note that λ_{ρ_a} is assumed to be unity. While the C.S.U. simulation,by not modelling the efflux momentum,tends to somewhat underestimate the initial plume rise, the totalplume rise is also significantly above that obtained with exact simulation. The U.W.O. and Calspan approximate simulations result in a similar plume rise at $x/H = 6$.

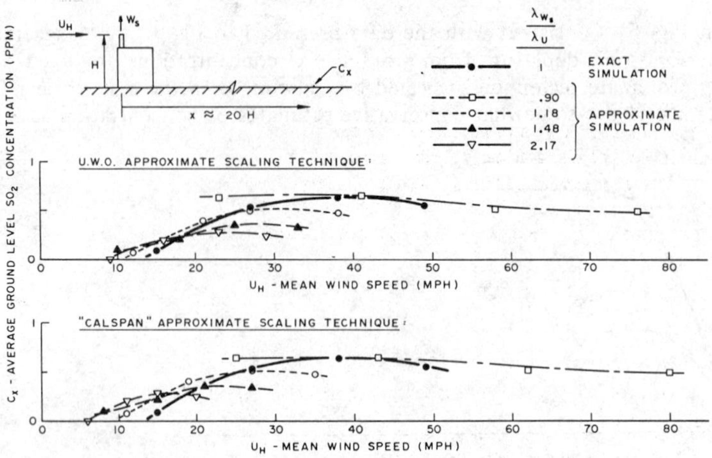

Fig. 7 Far field plume dispersion for different stack gas density ratios in the presence of aerodynamic downwash

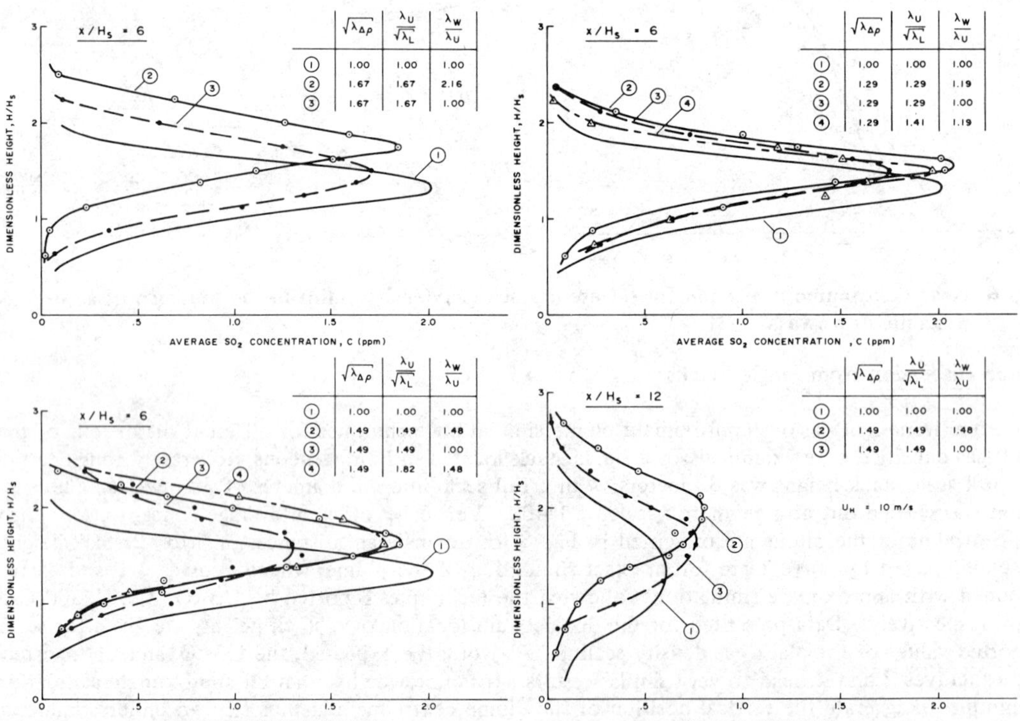

Fig. 8 Vertical profiles of pollutant concentration obtained from exact and approximate plume dispersion simulations

The vertical standard deviation of the plume about its centreline, obtained from the concentration profiles presented in Fig. 8, is given in Fig. 9b. The values of σ_z/H were obtained by fitting each set of data with a Gaussiam distribution. Although the upper parts of all profiles were found to closely follow a Gaussian form, the lower portions in all cases indicated some downward skewness. This contributes appreciably to the experimental scatter. For example, at $x/H = 12$ the exact simulation produces essentially the same σ_z/H value as both the U.W.O. and C.S.U. simulations. At $x/H = 6$ the scatter is larger. The downstream variation of the estimated standard deviations tends to follow the form $\sigma_z/H \propto (x/H)^{1/2}$. This is consistent with the asymptotic solution of the Taylor diffusion equation for x/U much larger than the Lagrangian time scale of turbulence. The concentrations at the plume centreline for the above measurements are summarized in Fig. 9c. It is interesting to note that the maximum concentrations from the approximate techniques generally agree with those of the exact simulation. Unfortunately, no ground level concentrations were measured, however, in view of the additional plume rise, these are expected to be underestimated in the approximate simulations.

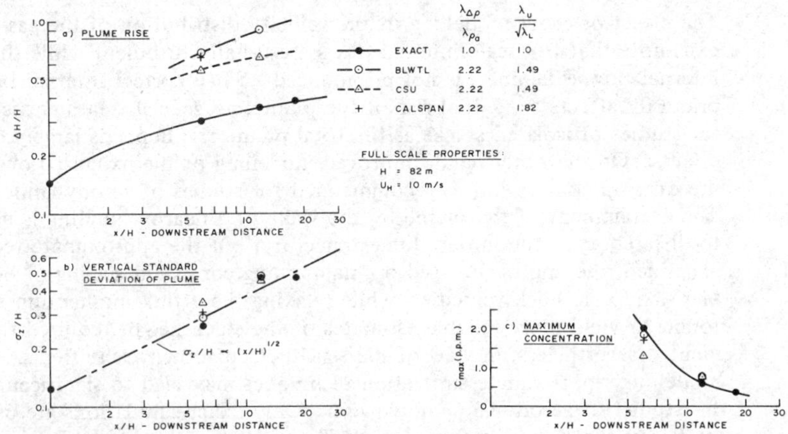

Fig. 9 Comparison of plume characteristics obtained with exact and approximate simulations in homogeneous moderately rough terrain

In summary, although all presented approximate methods lead to a generally similar spreading of the plume, see Figs. 9b and 9c, the use of a lighter model stack gas invariably produces an exaggerated plume rise. Whereas correct buoyancy effects at the stack exit are achieved by $(Fr)_d$ similarity, the rate of mixing of the plume and consequently its buoyant rise depend upon the density differential with respect to ambient air. Thus the presented limited experimental data indicate that the importance of scaling ρ_s/ρ_a may be greater than normally assumed in wind tunnel model studies of plume dispersion. Although an exaggeration of the geometric scale to permit exact modelling of the plume may be justifiable in studies of aerodynamic downwash, approximate simulations are sometimes unavoidable in studies of plume dispersion for stacks located in a complex terrain for which the modelling of the flow remains important. Of the approximate techniques examined, that used by C.S.U. appears to provide somewhat better estimates. Nevertheless, far field concentrations obtained by all approximate techniques remain unconservative due to the exaggerated plume rise. In view of this, further quantitative evaluations of approximate simulation techniques are warranted in order to arrive at systematic methods for evaluating their effectiveness.

CONCLUDING REMARKS

Similarity requirements for wind tunnel model studies of stack plume dispersion were reviewed and a number of approximate modelling techniques reported in the literature were discussed. The need for approximate modelling arises due to difficulties of achieving correct flow similarity and satisfactory wind tunnel operation in small scale diffusion studies if the velocity scaling is based on Froude number similarity. The most satisfactory approach to this problem of course would be the availability of a large wind tunnel facility which would permit the modelling of the atmospheric boundary layers at geometric scales of 1:200 and larger. Current approximate modelling techniques, used to avoid the difficulties resulting from exact Froude number scaling, invariably relax the requirement of correct plume density scaling and attempt to simulate buoyancy effects by modelling the densimetric Froude number or a similar parameter. The effectiveness of some of the reported approximate modelling techniques was examined through comparisons with results from a simulation in which the density of the plume was correctly scaled and all velocities were modelled to maintain Froude number similarity. This comparative evaluation was carried out for a neutrally stable atmospheric boundary layer simulation. A single isolated stack and stacks within the aerodynamic influence of a neighbouring building were examined. Also explored was the importance of correct modelling of the velocity distribution of the internal stack flow.

The following is a summary of the more important findings:

i) The effect of exact modelling of the velocity distribution of the gas at the stack exit; note that full scale internal flows are usually turbulent while the model internal flow is laminar; is not pronounced. Since correct internal flow modelling primarily affects the initial rise of the plume, no special treatment is warranted for studies of isolated stacks as the total plume rise depends largely on buoyancy effects. On the other hand, improved modelling of the exhaust flow, such as the use of a tip contraction, is recommended for studies of aerodynamic downwash.

ii) The entrainment of the plume by the wake of a nearby building is highly sensitive to differences in the initial plume trajectory. Of the approximate techniques examined, the approach based on maintaining consistent scaling of both the ambient and exit stack velocities, while relaxing the efflux momentum scaling, was found to yield more reliable estimates of the stack height required to avoid entrainment. Nevertheless, in view of the sensitivity of downwash effects to correct plume modelling, approximate simulation techniques may lead to significant differences in studies of aerodynamic downwash. Only exact simulations are used for such studies at the Boundary Layer Wind Tunnel Laboratory[18]. The use of an exact simulation may be warranted even at the expense of a somewhat reduced similarity of the overall flow regime.

iii) Correct scaling of the stack gas density, usually considered to be of lesser significance and consequently relaxed in all approximate modelling techniques, was found to be an important parameter for far field diffusion from an isolated stack. All examined approximate methods were found to lead to a significant exaggeration of the plume rise. As a result, approximate techniques are likely to underestimate maximum ground level concentrations. This is an important finding and requires further study.

iv) In situations where approximate modelling techniques are unavoidable, some comparisons with simulations with a correctly scaled density ratio are highly desirable even though this may have to be done at high wind speeds. Such comparisons provide a measure of the effectiveness of the approximate technique used and aid in the interpretation of the findings.

REFERENCES

1. Cermak, J.E., Laboratory Simulation of the Atmospheric Boundary Layer, AIAA Journal, Vol. 6, No. 9, Sept. 1971.

2. Snyder, W.H., Similarity Criteria for the Application of Fluid Models to the Study of Air Pollution Meteorology, Boundary Layer Meteorology, Vol. 3, No. 1, Sept. 1972.

3. Plate, E.J., Aerodynamic Characteristics of Atmospheric Boundary Layers, A.E.C. Critical Review Series, TID25465, 1971.

4. Isyumov, N., Jandali, T. and Davenport, A.G., Model Studies and the Prediction of Full Scale Levels of Stack Gas Concentration, presented at 67th Annual Meeting of APCA, Denver, Colorado, June 1974 and published in APCA Journal, Vol. 26, No. 10, Oct. 1976.

5. Cermak, J. E., Application of Fluid Mechanics to Wind Engineering , Freeman Scholar Lecture, Journal Fluid Engineering, ASME, Vol. 97, No. 1, March 1975.

6. Surry, D. and Isyumov, N., Model Studies of Wind Effects – A Perspective on the Problems of Experimental Technique and Instrumentation , ICIASF 75, Ottawa, Sept. 1975.

7. Davenport, A. G. and Isyumov, N., The Application of the Boundary Layer Wind Tunnel to the Prediction of Wind Loading , Proc. Int. Sem. on Wind Effects on Buildings and Structures, Ottawa, Vol. 1, Sept. 1967.

8. Isyumov, N., Wind Tunnel Methods for Evaluating Wind Effects on Buildings and Structures , Int. Symp. on Exp. Mech. Univ. Waterloo, June 1972.

9. Pasquill, F., The Influence of the Turning of Wind with Height on Crosswind Diffusion , Roy. Soc. London, Phil. Trans. 265 (A), 1969.

10. Halitsky, J., Validation of Scaling Procedures for Wind Tunnel Model Testing of Diffusion Near Buildings , Geophysical Sciences Laboratory, Report No. TR–69–8, New York, Univ., New York, 1969.

11. Barret, R. V., Use of the Wind Tunnel to Investigate the Influence of Topographical Features on Pollution from a Tall Stack , Chimney Design Symposium, Univ. of Edinburgh, April 1973.

12. Skinner, G. T. and Ludwig, G. R., Physical Modelling of Dispersion in the Atmospheric Boundary Layer , Calspan Advanced Technology Center, Calspan Report No. 201, May 1978.

13. Melbourne, W. H., Wind Tunnel Modelling of Buoyant Chimney Plumes , Proc. 3rd Australian Conference on Hydraulics and Fluid Mechanics, Sydney, Nov. 1968.

14. Melbourne, W. H., and Gartshore, I. S., A Wind Tunnel Model Study of Air Conditioning Contamination by Intake of Flue Gas , ICIASF 75, Ottawa, Sept. 1975.

15. Briggs, G. A., Plume Rise , A.E.C. Critical Review Series, TID 25075, 1969.

16. Meroney, R. N. et al, Wind–Tunnel Study of Stack Gas Dispersal at the Avon Lake Power Plant , Fluid Dynamics and Diffusion Laboratory, Techn. Report CER73–74 RNM–JEC–BTY–SKN35, CSU Fort Collins, April 1974.

17. Jandali, T., Isyumov, N. and Davenport, A. G., A Study of Stack Gas Dispersion for the Powerton Plant, Pekin, Illinois , Univ. of Western Ontario, Eng. Science Res. Report, BLWT–SS4–74, 1974.

18. Tanaka, H. and Isyumov, N., A Study of Stack Gas Dispersion for the Tracy Power Plant, Sorel, Quebec , Univ. of Western Ontario, Eng. Science Res. Report, BLWT–SS10–78, 1978.

EFFECTS OF STACK HEIGHT AND POSITION ON POLLUTANT DISPERSION IN BUILDING WAKES

D. J. KOGA and J. L. WAY
Illinois Institute of Technology, Chicago, Illinois, USA

ABSTRACT

Selected smoke visualization and ground level concentration data are presented and discussed for the case of neutrally buoyant plumes influenced by a squat-building wake. The plumes considered were emitted from low stack height sources at various positions atop the building and were studied to ten building heights downstream. The plume smoke visualizations contribute to a method of qualitatively estimating the ground level concentrations with regard to the overall visualized plume behavior. The quantitative ground level concentration data support the qualitative visualization results and show that stack position can affect local ground concentrations by a factor of six and, in addition, the stack height and position parameters, which determine the efflux point of release in the building wake flow, are strongly interrelated in determining the ground level concentrations. In particular, for a building oriented normal to the oncoming flow, roof-top emissions at the trailing building edge produce the highest ground concentrations, whereas for a stack height equal to 1.5 times the building height, the largest ground concentrations occur for the case where the stack is positioned at the leading roof-top edge of the building.

INTRODUCTION

Presently there are many acceptable empirical formulations available for the prediction of ground level concentrations of pollutants emitted from an elevated source [1-4]. Fewer in number, though, are the prediction theories which address themselves to the effects of the building presence in the flowfield [5-8]. The most widely accepted and utilized prediction scheme presently available is that of Briggs [7]. In the Briggs approximation, the building wake effects are significantly generalized so that the ground level concentrations are insensitive to stack position and also to stack height for low effective stack heights. It is the importance and understanding of these effects of the stack height and position parameters to which the following discussion will be addressed, with the eventual goal being an improvement to the present concentration prediction schemes.

EXPERIMENTAL FACILITIES AND TECHNIQUES

The experiments were performed in the I.I.T. Environmental Wind Tunnel, equipped with a counter-jet to simulate atmospheric boundary layers. Further details of these facilities can be found in Tan-atichat and Nagib [9] and Nagib, et al. [10]. The building model employed was 10 cm (4 in) in height, h_b, and 20 cm (8 in) square in plan view. The utilization of a 2 m/sec (6.6 ft/sec) wind tunnel speed insured proper modeling characteristics of neutral atmospheric flows with Reynolds numbers greater than the established similarity criteria values. The building flow Reynolds number, $Re_{h_b} = 1.34 \times 10^4$, was greater than the 11,000 value suggested by Snyder [11] for sharp-edged bluff-body flow independence. The stack flow Reynolds numbers also satisfied respective independence criteria and helium was selected as the stack gas to satisfy the plume momentum and neutral buoyancy modeling criteria as outlined by Isyumov, et al. [12]. The helium stack gas was also selected because of the relative insensitivity of its buoyancy to gas temperature. This property was important in the utilization of the thermal tracer technique.

The smoke visualizations were obtained using a smoke generator which consisted of an electric ribbon heater that was used to vaporize mineral oil from an enclosed wick. Photographs were taken using a 35 mm single lens reflex camera with a shutter speed of two seconds, lens aperature of f3.5 and focal length of 110 mm. The light source used was a high intensity lamp which, through the use of a series of slits, produced a narrow band of light that illuminated the streamwise centerline of the plume.

The ground level concentration data was obtained using a thermal tracer. The helium stack flow was tagged with a constant 100°C temperature at the stack exit and then thermistors were utilized, with the digital acquisition system described by Way [13], to measure the temperature rise at modeled ground level locations downstream of the building. This temperature rise, ΔT, directly yields the corresponding ground level concentrations by $(\Delta C)_m = \Delta T/(100-T_a)$, where T_a is the ambient temperature and $(\Delta C)_m$ represents the ground level concentration, non-dimensionalized using the source concentration for the model case. Conversion to full scale or prototype values can be obtained by $(\Delta C)_p = .373(\Delta C)_m$ for the modeling ratios utilized in this study.

RESULTS AND DISCUSSION

The following selected results [14] of smoke visualization and centerline ground level concentration plots are representative of 5 minute and 1 hour time averages respectively. Although the specific sampling period is important for accurate prototype air monitoring, it is not crucial to the present investigation due to the self-comparison nature of the results presented here. The smoke photographs have square grid lines, superimposed using darkroom techniques, which are representative of the building height. Identical reproduction techniques were used in all photographs to insure consistency. The centerline concentration data $(\Delta C)_m$ is given in units of parts per thousand (p.p.t.), where the "pure" concentration at the stack source is unity, plotted against x/h_b which is the downstream distance (x) non-dimensionalized by the building height (h_b). A small plan-view building symbol is used to the right of each centerline concentration plot to facilitate the identification of the building orientation (square or diamond), stack position and various streamwise stack centerlines considered. The flow is from right to left in all figures and the following non-dimensional variables are used: $HS = h_s/h_b$ is the stack height (h_s) non-dimensionalized by the building height (h_b), $FS = v/u$ is the stack efflux velocity (v) non-dimensionalized by the free-stream mean velocity (u) and PS is the stack position variable defined in Figure 1 for both the square and diamond-oriented buildings.

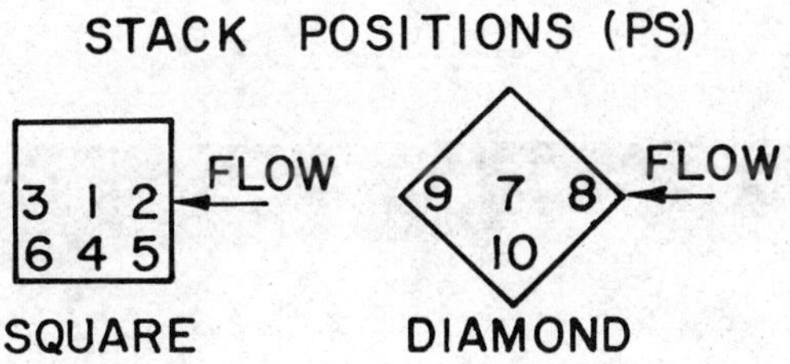

Figure 1. Plan view specifications for building orientation and stack position

Visualization Results

In Figure 2, the effect of stack height at the center of the square-oriented building is examined. In the top photo, with HS=2, the plume is unaffected by the building presence. Lowering the stack height to HS=1.5 causes the vertical thickness of the plume to increase, indicating enhanced vertical mixing of the plume caused by increased turbulence levels in the building wake. At the roof-top emission level, the plume is downwashed and becomes entrapped in the building wake, filling the wake cavity region. Thus from Figure 2, a stack height of HS=1.5 appears to keep the plume elevated for the center stack position (PS=1). However at other stack positions, as evidenced in Figure 3, the plume can be brought much closer to the ground, particularly when the stack is located near the leading edge (PS=2 and 5) of the building. The explanation of this phenomena is that the accelerated flow around the leading building edge causes increased stack downwash, resulting in a lowered effective stack height. At positions away from the leading edge (PS=3,4 and 6) the plumes are quite similar in trajectory and thickness.

Figure 4 shows, for the diamond-oriented building case, that the plumes of the roof-top emissions are downwashed for all stack positions and that increasing the stack height to HS=1.5 in Figure 5 sharply reduces the smoke concentration in the building cavity for leeward stack position (PS=9). The overall trajectory for the plumes at all stack positions are lower for the diamond-oriented building case than the corresponding square building case, with the possible exception of the leading edge stack positions (PS=2 and 8). This indicates that the resulting ground level concentrations should be larger for the diamond-oriented building case. The reason for this effect is that the diamond-oriented building produces a much stronger wake flow with the formation of delta-wing vortices along the two leading roof-top edges of the building. The rotation and line of formation of this strong delta-wing vortex causes, as seen in Figure 4, the "lifting" of the plume from its emission point at PS=10 due to the upward vortex rotation. This same mechanism, as seen in Figure 5, causes the increased stack downwash at PS=7 and 8 due to the downward vortex rotation and increased velocities at these positions, and also the absence of stack downwash at position PS=10 due to the upward vortex rotation. However, the most significant effect of the delta-wing vortices is the overall lowering of the plume trajectories in both figures due to the vortices holding the smoke close to the ground.

Additional qualitative ground level concentration information was obtained from the visualization photographs by determining the plume behavior in terms of its trajectory. The following criteria were used to judge the "amount" of building effect on the plume: D was the level assigned to plumes that were fully downwashed, filling most or all of the building cavity region, D-W was the level given to plumes that appeared to almost touch down or touched the ground only slightly, W was the level assigned to plumes which were slightly

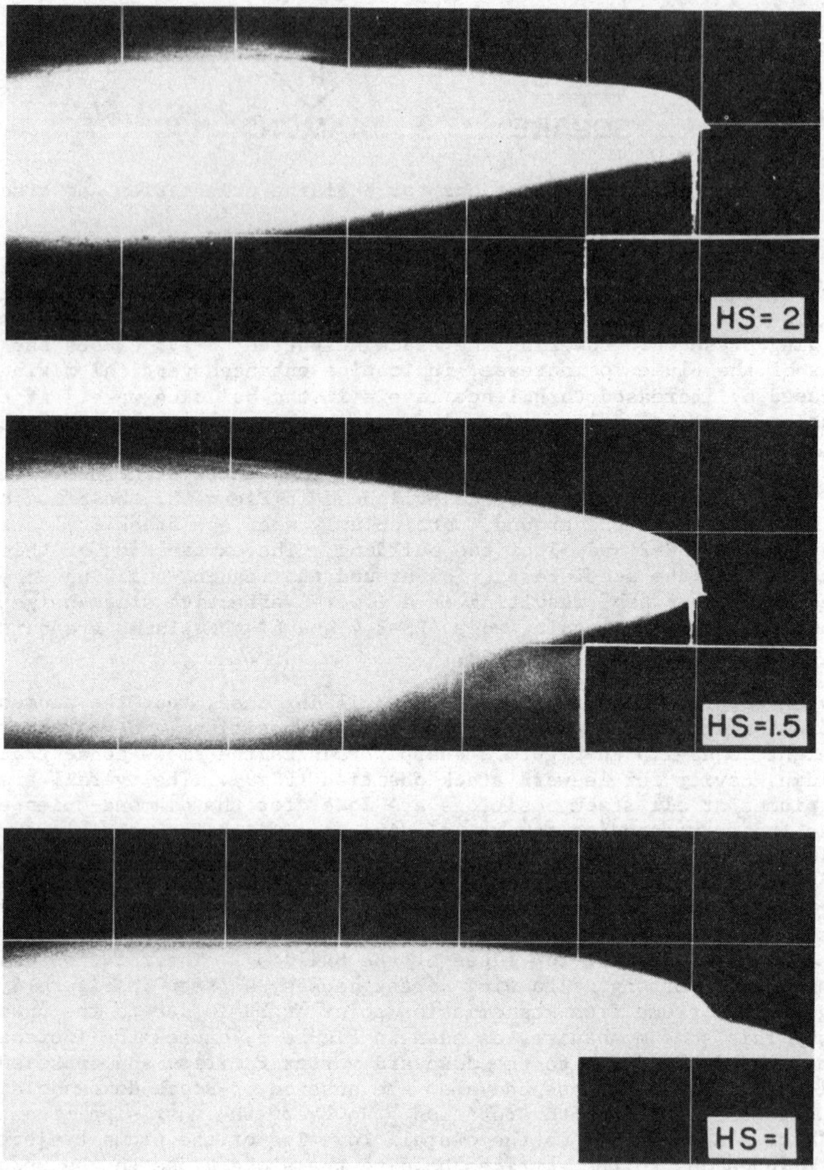

Figure 2. Effect of stack height (HS) for center stack position (PS=1) on the square-oriented building with constant stack flowrate (FS=1.5)

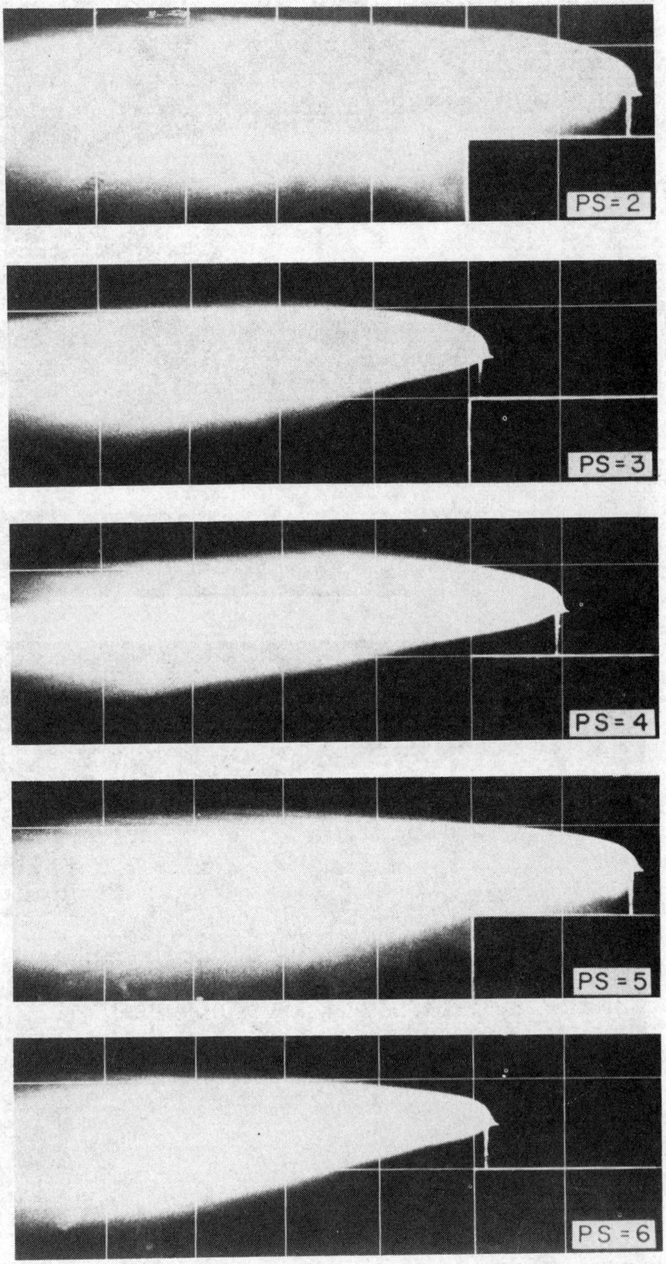

Figure 3. Effect of stack position (PS) for constant stack height (HS=1.5) and flowrate (FS=1) on the square-oriented building

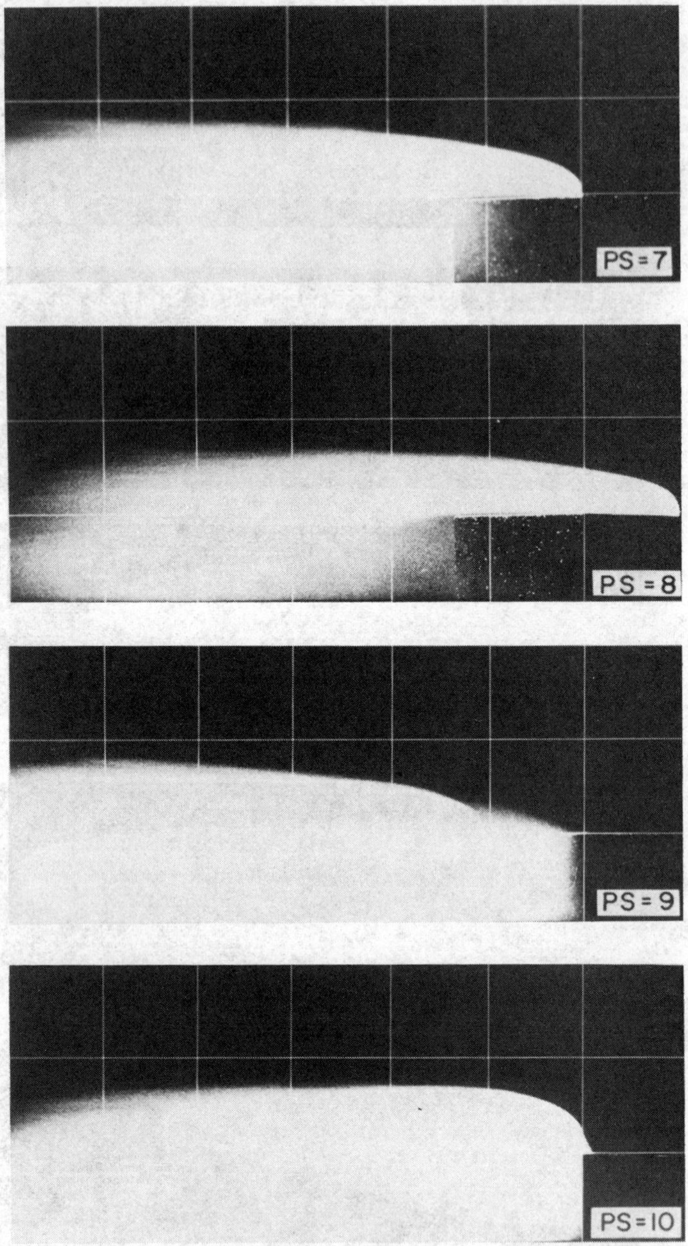

Figure 4. Effect of stack position (PS) for constant stack height (HS=1) and flowrate (FS=1) on the diamond-oriented building

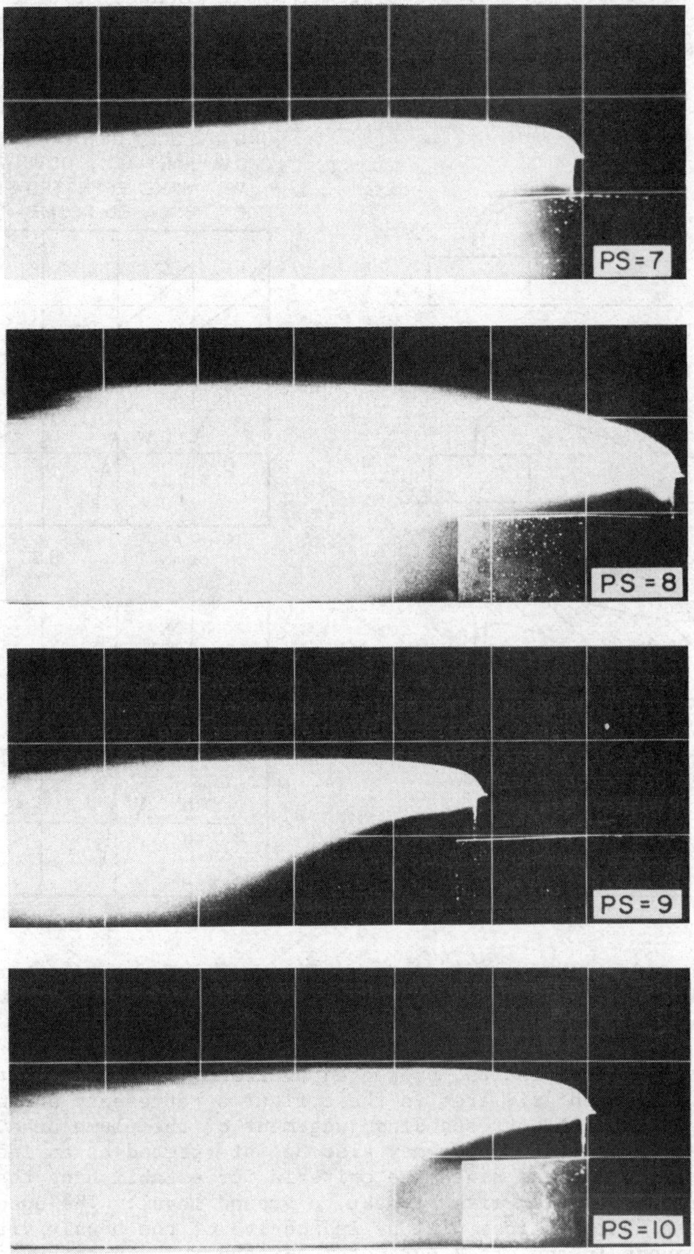

Figure 5. Effect of stack position (PS) for constant stack height (HS=1.5) and flowrate (FS=1) on the diamond-oriented building

turned towards the ground resulting in the overall lowering of the effective stack height, and E was the level for plumes which escaped or were only slightly effected by the building wake. These criteria are illustrated in Figure 6.

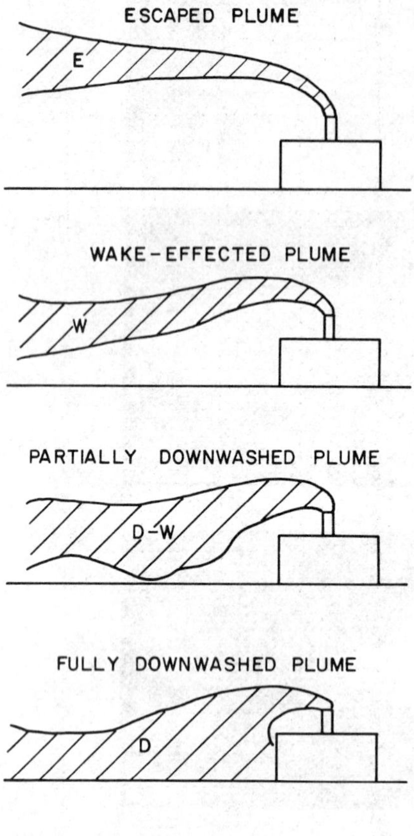

Figure 6. Illustration of categories used to evaluate effects of building wake on plume behavior

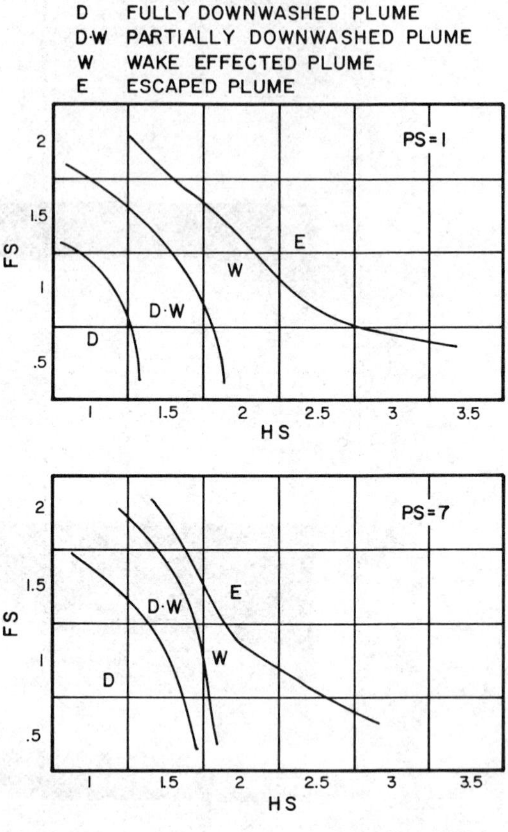

Figure 7. Transition curve comparison of visualization photographs for plume behavior with stack in center position

Figures 7 and 8 are the results, presented by transition curves between the plume behavior regimes. Each grid area in these figures represents a single visualization photograph with its corresponding judgement of the plume behavior. The transition lines in these figures may also be interpreted as an indicator of the ground level concentrations since the criteria for establishing these levels were based on the plume behavior with respect to ground level. The spacing and curvature of the contour lines will be indicative of the sensitivity of the plume behavior to changes in the stack parameters HS and FS. In this manner, it is evident from Figure 7 that for the center stack positions (PS=1 and 7), the diamond-oriented building has the larger downwash (D) region but also exhibits a faster transition to the escaped (E) plume condition. Figure 8 shows that the downwashed (D) or partially downwashed (D-W) regions are much larger for the diamond building cases at all stack positions. Figure 8 also shows that the square-oriented building cases exhibit greater sensitivity to stack flowrate (FS) changes while the diamond building cases exhibit greater sensitivity to stack

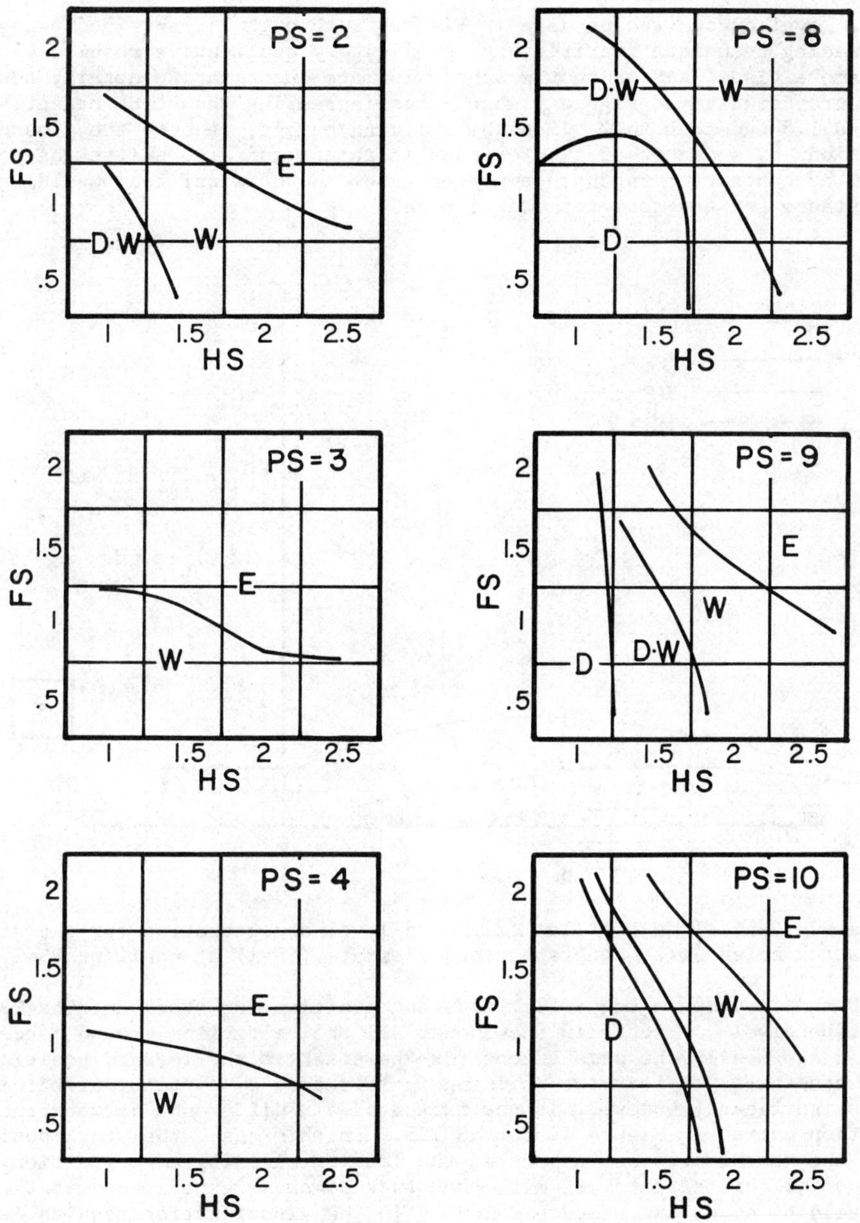

Figure 8. Transition curve comparisons of visualization photographs for plume behavior with stack in non-center positions

height (HS) changes, due primarily to the different type of vortices formed in each building orientation case.

Ground Level Concentration Measurements

The ground level concentration data in Figures 9 through 14, obtained using a thermal tracing technique, verifies the preliminary qualitative results obtained from the smoke visualization photographs. In more quantitative detail, however, the concentration data of Figure 9 shows that increasing the stack height, from roof-top to 1.5 and then to 2 times the building height, reduces the ground concentrations by a factor of two for each increment and also shifts the points of maximum concentration farther downstream from the original four building height distance for the roof-top emission case.

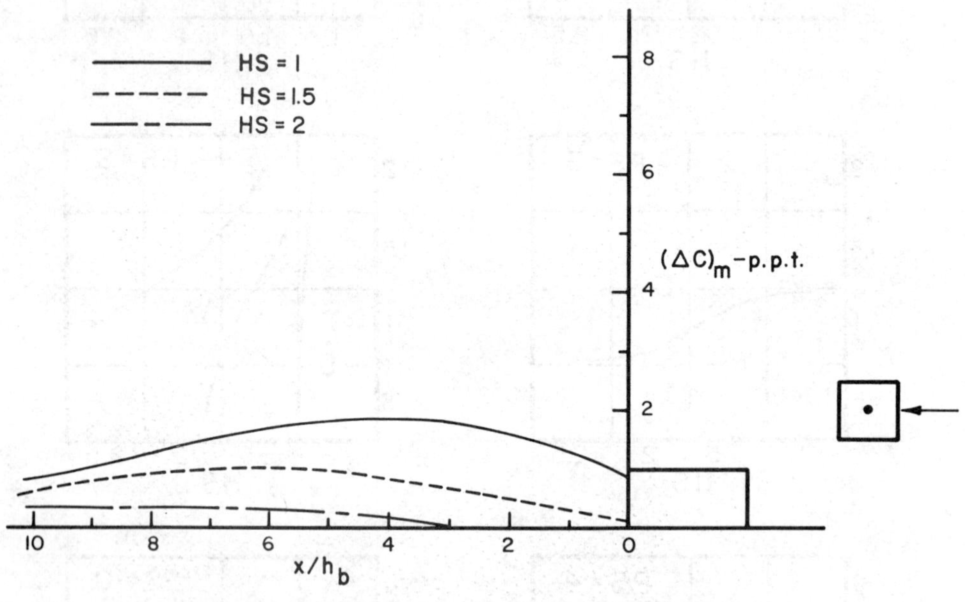

Figure 9. Centerline profiles of ground level concentration for the effect of stack height with constant stack flowrate (FS=1) at position PS=1

Figures 10 and 11 compare the effect of stack position for stack heights HS=1 and 1.5 respectively. Figure 10 shows that all the centerline ground concentration curves are nearly the same except for the stack at the leeward position, where the concentration near the building is increased by a factor of three. Even more significant, however, is the factor of six difference between the concentration curves in Figure 11 for HS=1.5. In this case, the stack position producing the maximum concentrations is the leading building edge position. The explanation for this effect lies with the stack downwash experienced at the leading building edge, discussed previously in the visualization results. The effect of stack position seems to be slightly more significant, in general, for HS=1.5 than the roof-top emission case. However, for both of these stack heights, the effect of stack position on the centerline ground concentrations becomes insignificant at ten building heights downstream of the building. This can be seen by the general collapse of all the curves in Figures 10 and 11 to a value near 1 p.p.t. A similar collapse was experienced with respect to stack height effects as seen in Figures 12 and 13.

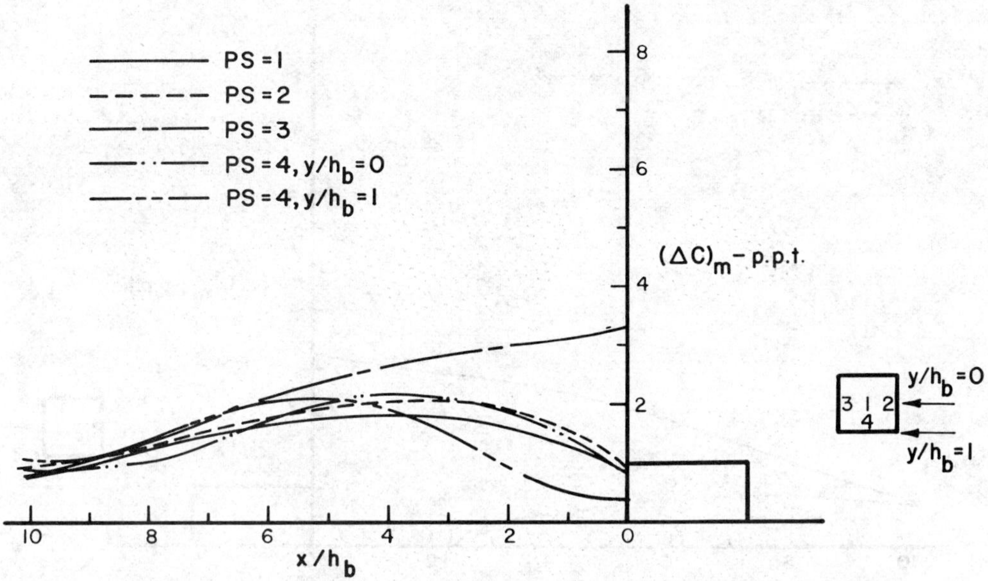

Figure 10. Centerline profiles of ground level concentration for the effect of stack position with constant stack flowrate (FS=1) and height (HS=1)

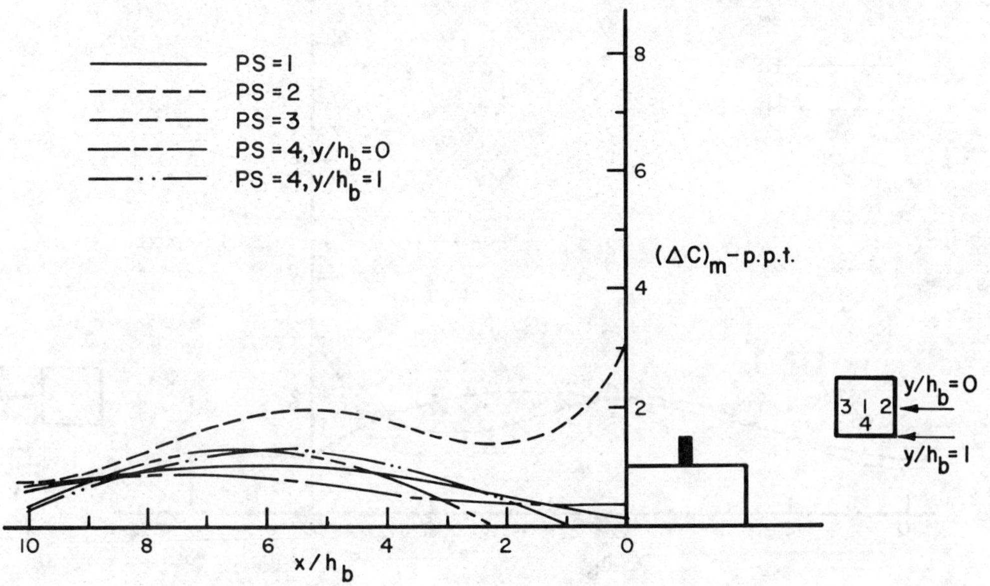

Figure 11. Centerline profiles of ground level concentration for the effect of stack position with constant stack flowrate (FS=1) and height (HS=1.5)

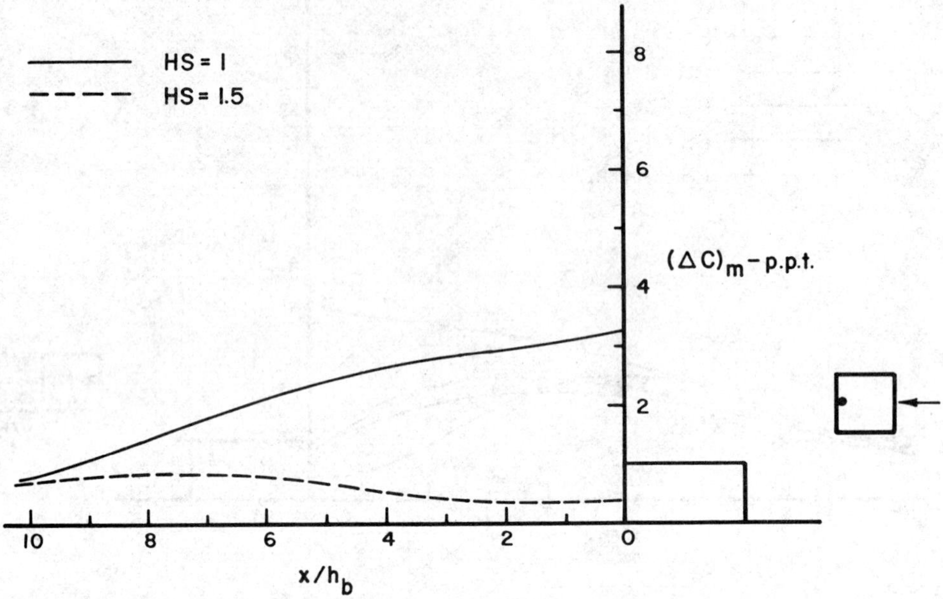

Figure 12. Centerline profiles of ground level concentration for the effect of stack height with constant stack flowrate (FS=1) at position PS=3

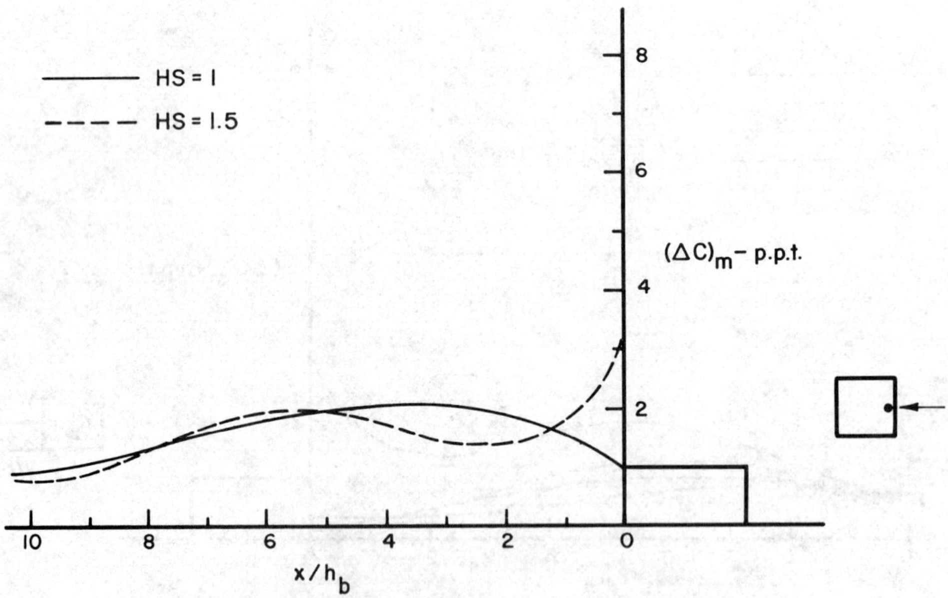

Figure 13. Centerline profiles of ground level concentration for the effect of stack height with constant stack flowrate (FS=1) at position PS=2

The front (PS=2) and leeward (PS=3) stack positions yield the most significant changes with respect to stack height. As previously noted, increasing HS greatly reduces the ground concentration in Figure 12. The apparent reason is that the height increase at this position causes the point of emission to become sufficiently high to avoid immediate plume downwash whereas in Figure 13, where the stack is at the leading edge position, a similar stack height increase serves only to increase the stack downwash phenomenon. As a result, the initial cavity concentration near the building is quite large and decreases to a local minimum value between two to three building heights downstream, the approximate length of the building wake cavity region. Farther downstream, outside of the cavity region, the stack height parameter seemed to have little effect at the leading edge stack position.

Roof-top emissions for the diamond-oriented building case were also examined at all stack positions. Figure 14 shows that stack position has a much more significant effect for the diamond building case than the square-oriented building case, again due to the difference in vortices generated in each building orientation. The delta-wing vortices in the diamond building case produced the highest ground concentrations for the rear (PS=9) and side (PS=10) stack positions. The center and forward stack positions produce significantly lower ground concentrations, although again the effect of stack position seems to be insignificant at ten building heights downstream of the building.

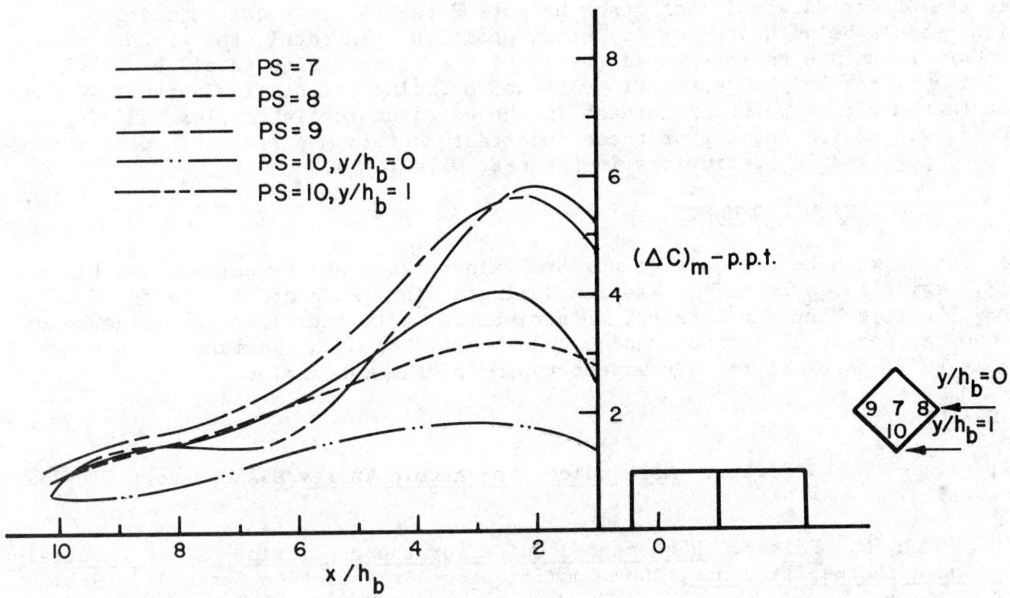

Figure 14. Centerline profiles of ground level concentration for the effect of stack position with constant stack flowrate (FS=1) and height (HS=1)

The results presented here also show that the points of maximum ground concentrations occur near two building heights downstream for the diamond-oriented building case (Figure 14) and from 3 to 4 building heights for the square building case (Figure 10). Also the maximum concentration value for the square building case is $\chi=.24$ compared to the $\chi=.40$ at $x=3h_b$ value obtained by Thompson and Lombardi [15], and the significance of the building orientation agrees well with their factor of six, only for a stack at the building center with HS=1.5. For the other stack positions, the roof-top emissions give maximum ground concentration ratios between the diamond and square building cases of 2, 1.5, 1.8 and 3 for the center, front, rear and side stack positions respectively.

CONCLUSIONS

The effect of the stack height parameter in the near building wake region, for the low stack heights considered, is found to be most significant when the stack is positioned at the leeward building edge. The ground level concentrations at the leeward stack position are the highest for the roof-top emission cases but similar in value to the other stack positions when the stack height is raised to 1.5 times the building height. The stack position is thus an important factor yielding maximum ground level concentration values which can differ by an average factor of three. It is also observed that the leading-edge stack position on the square-oriented building produces the highest ground level concentrations when the stacks are 1.5 times the building height, and it is concluded that this effect is caused by the additional downwashing of the plume due to the accelerated boundary layer flow over the leading edge of the building.

In general, for the design considerations of a stack located atop a squat building, the best overall position regardless of wind direction would be the building center. If the stack needs to be positioned near a building edge then it should be at least 1.5 times the building height at the leeward building edge. If the leeward side is not readily discernable due to varying wind directions, then the Briggs suggestion of stack heights 2 to 2.5 times the building height should be adequate for any stack position. In total, the ground level concentrations are an integrated result of the overall effects of the stack position, stack height, stack flowrate and building orientation parameters which dictate the efflux point of release in the aerodynamically complex building wake flow. Neglecting any of these important factors may result in poor ground level concentration predictions in the near wake field.

ACKNOWLEDGEMENTS

The authors wish to acknowledge the preliminary work and techniques established by H. Nagib, T. Corke and R. Wlezien during a dispersion study done for a General Motors Plant in Lansing, in conjunction with the Illinois Institute of Technology Research Institute under the supervision of R. Norman. Also many thanks to V. Mattioli for her expert typing of the manuscript.

REFERENCES:

[1] Slade, D. H. Editor. Meteorology and Atomic Energy USAEC Report TID-24190, 1968.

[2] Smith, M., Editor. Recommended Guide for the Prediction of the Dispersion of Airborne Effluents, ASME Committee on Air Pollution Controls, 1968.

[3] Turner, D. B., Workbook of Atmospheric Dispersion Estimates U.S. Environmental Protection Agency, 1970.

[4] Briggs, G. A., Plume Rise, US Atomic Energy Commission, TID-25075, 1969.

[5] Robins, A. G. and Castro, I. P., "A Wind Tunnel Investigation of Plume Dispersion in the Vicinity of a Surface Mounted Cube - II. The Concentration Field," Atmospheric Environment, Vol. 11, 1977, p. 299.

[6] Britter, R. E., Hunt, J. C. R. and Puttock, J. S., "Predicting Pollution Concentrations Near Buildings and Hills," Presented at the Systems and Models in Air and Water Pollution Conference at the Institute of Measurement and Control, London, 1976.

[7] Briggs, G. A., "Diffusion Estimation for Small Emissions," Preliminary Draft, NOAA, ATDL Contribution File No. 79, 1973.

[8] Huber, A. H., "Incorporating Building/Terrain Wake Effects on Stack Effluents," Presented at the Joint Conference on Applications of Air Pollution Meteorology, Salt Lake City, Utah, 1977.

[9] Tan-atichat, J. and Nagib, H. M., "Wind-Tunnel Simulation of Neutral Atmospheric Surface Layers by the Counter-Jet Technique," Paper no. 74-214, Proceedings of the 67th Annual Meeting of the Air Pollution Control Association, Denver, Colorado, 1974.

[10] Nagib, H. M., Morkovin, M. V., Yung, J. T. and Tan-atichat, J., "On Modeling of Atmospheric Surface Layers by the Counter-Jet Technique," AIAA Journal, Vol. 14, No. 2, 1976, p. 185.

[11] Snyder, W. H., "Similarity Criteria for the Application of Fluid Models to the Study of Air Pollution Meteorology," Boundary Layer Meteorology, Vol. 3, 1972, p. 113.

[12] Isyumov, N., Jandali, T. and Davenport, A. G., "Model Studies and the Prediction of Full Scale Levels of Stack Gas Concentration," Presented at the 67th APCA Annual Meeting, Denver, CO., 1974.

[13] Way, J. L., "Applications in Fluid Mechanics Research of a Portable Data Acquisition and Processing System," ICIASF '75 Record, 1975, p. 50.

[14] Koga, D. J., "Effects of Stack Height and Position on the Dispersion of Pollutants in Building Wakes," M.S. Thesis, Illinois Institute of Technology, Dec. 1978.

[15] Thompson, R. S. and Lombardi, D. J., "Dispersion of Roof-Top Emissions from Isolated Buildings - A Wind Tunnel Study," U.S. Environmental Protection Agency, EPA-600/4-77-006, 1977.

PLUME RISE FOR VARYING AMBIENT TURBULENCE, THERMAL STRATIFICATION, AND STACK EXIT CONDITIONS—A NUMERICAL AND LABORATORY EVALUATION

R. L. PETERSEN and J. E. CERMAK
Colorado State University, Fort Collins, Colorado, USA

ABSTRACT

Plume rise data from wind-tunnel tests were used to aid in the development and testing of a plume rise model. This model uses the continuity equation and principles of conservation of mass and momentum to predict the plume rise. The system of equations is closed by employing various entrainment relations which are based on theoretical and experimental arguments. The model uses the ambient velocity, temperature and turbulence profiles as inputs. A comparison of the model predictions with wind-tunnel observations showed satisfactory agreement (average error of 8 percent) for all cases studied except the high turbulence tests. The disagreement for these tests was to be expected since upwind ridges produced a two-dimensional flow field.

LIST OF SYMBOLS

a	Radius of curvature
b	Radius of plume
B	Dimensionless radius of plume (b/D)
C_D	Drag coefficient
D	Stack diameter
F_d	Drag force per unit length of jet
F_L	Lagrangian spectral function
Fr	Froude number $\left[\frac{u_{soo}}{\sqrt{g\ \gamma\ D}}\right]$
g	Acceleration due to gravity
h	Stack height
i_x, i_y, i_z	Turbulence intensities in x, y and z direction [u'/u]
i_e	Average turbulence intensity $\left(\frac{i_x + i_z}{2}\right)$
R	Velocity ratio $[u_{soo}/u_{aoo}]$

Ri	Richardson number $\frac{g}{T}\left[\frac{\frac{\partial\theta}{\partial z}}{\left(\frac{\partial u}{\partial z}\right)^2}\right]$	Γ_a	Dimensionless ambient density $\left[\frac{\rho_{ao}}{\rho_{aoo}}\right]$
Re	Reynolds number $\left[\frac{L_o u_o}{\nu_o}\right]$	γ	Density ratio $\left[\frac{\rho_{aoo} - \rho_{soo}}{\rho_{aoo}}\right]$
s,n,ϕ	Plume coordinate system	ν	Kinematic viscosity or angle between plume axis and horizontal plane
V_a	Dimensionless ambient velocity $\left[\frac{u_{ao}}{u_{aoo}}\right]$	ρ	Density
V	Volume flow	θ'	Angle between plume axis and ambient flow
V_e	Dimensionless entrainment rate $[v_e/Du_{soo}]$		
v_e	Entrainment rate		
W	Dimensionless plume velocity $\left[\frac{u_{so} - u_{ao}\cos\theta'}{u_{aoo}}\right]$		
x,y,z	Cartesian coordinates		
X,Y,Z,S	Dimensionless coordinates (x/D, y/D, z/D, s/D)		

Subscripts

a	Pertaining to ambient air as a function of x and z
ao	Pertaining to ambient air as a function of z
aoo	Pertaining to ambient air at stack exit
m	Model
p	Prototype
s	Pertaining to plume as a function of s, n and ϕ
so	Pertaining to plume as a function of s
soo	Pertaining to plume at stack exit

Greek Symbols

$\alpha_{1,2,3}$	Entrainment constants
Γ	Dimensionless density difference $\left[\frac{\rho_{ao} - \rho_{so}}{\rho_{aoo} - \rho_{soo}}\right]$

INTRODUCTION

Methods for estimating the rise of plumes from industrial stacks have received more attention since the promulgation of the Clean Air Act (CAA) of 1970 whereby concentration levels of certain pollutants are regulated. This act requires that prior to the construction of a new power plant the owner must demonstrate that 1) air quality standards (AQS) or significant deterioration limits will not be exceeded and 2) stack heights will not be greater than "good engineering practice" (GEP). GEP is defined in Section 123 of the CAA as

> "...the height necessary to insure that emissions from the stack do not result in excessive concentrations of any air pollutant in the immediate vicinity of the source as a result of atmospheric downwash, eddies and wakes which may be created by the source itself, nearby structures or nearby terrain obstacles..."

Prior to constructing a new power plant or pollutant source studies must be undertaken to determine the GEP stack height and subsequently the maximum ground-level pollutant concentration value. For determining these quantities the plume rise as a function of stack exit conditions, ambient turbulence levels and ambient stratification must be known.

As Briggs (1) reported, at least 50 organizations have developed techniques for estimating plume rise. Some of the methods are site specific such as Montgomery and Carpenter (2) and Carson and Moses (3) and some are more general in nature (1). What is more distressing, the different methods tend to give conflicting results. This leads one to the conclusion that improved and validated plume rise estimates are needed. One of the problems with some of the models is that the turbulence, mean velocity and temperature fields cannot be input and are often neglected. A method for predicting plume rise which allows for the input of these fields was developed and tested over a limited range of ambient conditions by Ooms (4).

The purpose of this research has been to extend Ooms' (4) model and validate it through experiments in a wind tunnel. The laboratory tests were run over a range of ambient turbulence, thermal stratification and stack exit conditions to include tests over a scale model of the topography for an actual power plant site. For the later tests the plume was rising downwind of two ridges and hence was under the influence of a two-dimensional, highly-turbulent flow field.

Included in this paper are a description of the numerical model, the laboratory procedures and a comparison of the numerical and wind-tunnel modeling results. The similarity requirements for the wind-tunnel modeling and detailed description of the experimental procedures are given in Petersen (5).

THE NUMERICAL MODEL

The numerical plume rise estimation technique to be tested in this study is an extension of the work of Fan (6), Abraham (7) and Ooms (4). Their work considered a plume rising in a one-dimensional ambient flow field. While the technique developed here is applicable to one-dimensional flow, with slight modifications cases of two-dimensional ambient flow can also be considered (i.e., flow over a hill). A description of the problem and nomenclature is shown in Fig. 1.

The technique starts with the following equations in cylindrical coordinates for conservation of momentum (along and normal to the plume axis) and mass for an incompressible fluid:

s-momentum:

$$\rho_s\left[\frac{1}{n}\frac{\partial}{\partial n}(n\,u_n u_s) + \frac{\partial}{\partial s}(u_s^2)\right] = -\rho_s\, g \sin\nu - \frac{\partial p}{\partial s} \tag{1}$$

n-momentum:

$$\rho_s\left[\frac{1}{n}\frac{\partial}{\partial n}(n\,u_n^2) + \frac{\partial}{\partial s}(u_s u_n) + \frac{u_s^2}{a}\right] = -\rho_s\, g \cos\nu - \frac{\partial p}{\partial n} \tag{2}$$

continuity:

$$\frac{1}{n}\frac{\partial}{\partial n}(n\,u_n) + \frac{\partial u_s}{\partial s} = 0 \tag{3}$$

mass:

$$\frac{D\rho_s}{Dt} = 0 \tag{4}$$

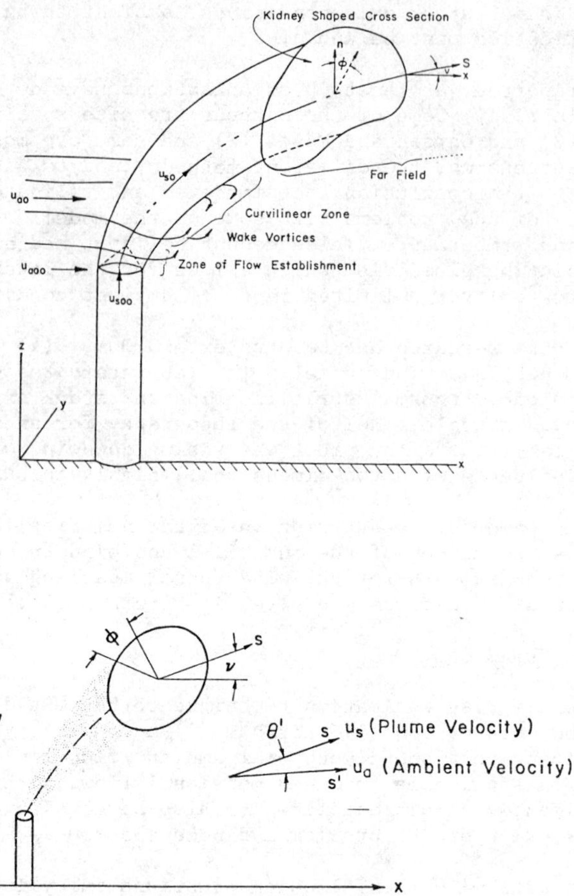

Fig. 1. General description of plume rising in a cross flow and the coordinate system.

The equations result after assuming that the motion is steady and axisymmetric and that the effects of molecular diffusivity or viscosity are negligible. In order to simplify these equations the following additional assumptions are made (4; 6; 7).

1. The plume velocity, density and concentration distributions are normal, or

$$u_s = u^* e^{-n^2/b^2} + u_{ao} \cos\theta' \qquad [u^* = u_{so} - u_{ao}] \tag{5}$$

$$\rho_s = \rho^* e^{-n^2/b^2} + \rho_{ao} \qquad [\rho^* = \rho_{so} - \rho_{ao}] \tag{6}$$

2. The pressure is hydrostatic: $\frac{\partial p}{\partial z} = - \rho_a g$ (7)

3. The ambient velocity and density field can be functions of x and z: $u_{ao} = f_1(x,z)$ and $\rho_{so} = f_2(x,z)$.

With the preceding assumptions the equations can be put in a more manageable form. The simplified equations will now be developed.

Continuity

The continuity equation (2.3) is integrated over the plume cross section as follows

$$v_e = -\int_0^{2\pi}\int_0^{a'} \frac{\partial}{\partial n}(n\,u_n)\,dn\,d\phi = \frac{\partial}{\partial s}\int_0^{2\pi}\int_0^{a'} u_s\,dA$$

where a' goes to infinity for the exponential terms and $\sqrt{2}b$ for the non-exponential term. The limit $\sqrt{2}b$ represents one standard deviation of a normal distribution.

Simplifying and nondimensionalizing

$$\frac{V_e}{\pi} = \frac{\partial}{\partial S} B^2 W . \qquad (8)$$

S-Momentum, N-Momentum, and Mass Conservation

In a similar manner the S-momentum, N-momentum, and mass conservation equations can be simplified. The simplified equations in dimensionless form are

$$\frac{\partial}{\partial S}(B^2W^2) = 2\,\frac{V_e}{\pi R}\,V_a\cos\theta' + \frac{2B^2}{Fr^2}\,\frac{\Gamma}{\Gamma_a}\,S^*\sin\nu \qquad (9)$$

$$B^2W^2\,\frac{\partial\nu}{\partial S} = -\,\frac{2V_e}{\pi R}\,V_a\sin\theta' + \frac{2B^2}{Fr^2}\,\frac{\Gamma}{\Gamma_a}\,S^*\cos\nu \qquad (10)$$

$$\frac{\partial}{\partial S}(B^2W\,\Gamma) = \frac{2B^2W}{\gamma}\,\frac{\partial\Gamma_a}{\partial Z}\,\sin\nu \qquad (11)$$

where $$S^* = \sum_{m=0}^{\infty}\left(\frac{1}{m+1}\right)\left(\frac{\Gamma}{\Gamma_a}\gamma\right)^m .$$

Now in order to solve for the plume height in terms of X the following two equations relating the coordinate systems are needed:

$$\frac{dX}{dS} = \cos\nu \qquad (12)$$

$$\frac{dZ}{dS} = \sin\nu . \qquad (13)$$

Solution of Equations and Closure

Equations (2.8) through (2.13) represent a set of six ordinary differential equations which can be solved numerically once the initial conditions and Fr, γ, V_a, V_e and $\frac{\partial\Gamma_a}{\partial Z}$ are specified. The initial conditions are:

$$B = 0.5 \qquad W = 1.0$$

$$X = Z = S = 0.0$$

The Corrected Euler Method (8) was used to solve the equations. The Froude number (Fr), density ratio (γ), ambient velocity field (V_a and $\theta' - \nu$) and ambient density field $\left(\frac{\partial \Gamma_a}{\partial Z}\right)$ are input parameters for which the solution is desired. The velocity and temperature field may either be specified in empirical form or by the input of actual data. For this study the velocity fields were assumed to be horizontal, hence $\nu = \theta'$. Initially $\theta' = \pi/2$.

The entrainment relations used by Fan (6), Abraham (7), Ooms (4) and Slawson (7) have been incorporated unchanged when appropriate or slightly modified to better agree with the theory.

Consider the definition of entrainment in dimensional form

$$v_e = \frac{\partial}{\partial s}\{\pi b^2[u_{so} + u_{ao}\cos\theta']\}.$$

Close to the source $\theta' \approx \pi/2$ and the relation simplifies to

$$v_e = \frac{\partial}{\partial s}\pi b^2 u_{so}.$$

For this region the free-jet type entrainment relationships should be a good approximation. The equation used here, which appears to fit the observations of Abraham (7) and Ooms (4), is given by

$$v_e = 2\pi b[\alpha_1(u_{so} - u_{ao}\cos\theta') + \alpha_2(u_{ao}\cos\theta'\sin\theta')].$$

Numerical values for the constants α_1 and α_2 are based on the experiments of Albertson (9), Ricou and Spalding (10), Tsang (11) and Richards (12). As derived in Petersen (5), the following relations for α_1 and α_2 based on these earlier studies will be used here

$$\alpha_1 = 0.057\left(\frac{\rho_{soo}}{\rho_{so}}\right)\left(\frac{\rho_{aoo}}{\rho_{soo}}\right)^{1/2}$$

$$\alpha_2 = 0.65 \text{ to } 1.34.$$

In the region where atmospheric turbulence dominates the entrainment $u_{so} \approx u_{ao}$ and the following equation applies:

$$v_e = 2\pi\, u_{ao}\frac{\partial b^2}{\partial s} = 4\pi\, u_{ao}\frac{\partial \sigma_z^2}{\partial s}.$$

From Taylor's (12) theory and a recapitulation by Csanady (13)

$$u_{ao}\frac{\partial \sigma_z^2}{\partial s} = 2\, u'^2_{ao}\int R_L(\tau)\, d\tau .$$

Thus for short travel times

$$v_e = 8\pi\, u'_{ao}\sigma_z = 2\pi b(2\sqrt{2})\, u'_{ao}$$

where $2\sqrt{2}$ is the entrainment constant (α_3) given by Ooms (4) and Briggs (1). They report α_3 is on the order of one. From this analysis the value $\alpha_3 = 2\sqrt{2}$ was derived based on the assumption of normal distributions for u_s and ρ_s. If top-hat distributions were assumed α_3 would be equal to 1.41 in better agreement with the value recommended by Briggs and Ooms. For this study the value of α_3 will be assumed equal to $2\sqrt{2}$.

Fan (6) and Ooms (4) used a similar model as presented here except began the numerical solution after the zone of flow establishment. They felt that the changing velocity distributions within this zone might affect the initial trajectories. Fan used the following relations for the length of the zone and initial plume angle

$$s_o/D = 6.2 - 17.6/R,$$

and

$$\theta_o = 90° - 110°/R$$

Petersen (5) found that neglecting the initial zone of plume rise did not affect the results significantly. Hence, the model as formulated herein should give accurate estimates of the plume trajectory without the added assumptions.

Fan (6) and Ooms (4) also incorporate a drag term, F_d, in their equations. The term only appears in the n-momentum equation in the present theory and is of the form

$$F_d = -\sqrt{2}\, C_d \pi b\, \rho_{ao}(u_{ao}\sin\theta')^2 .$$

Fan found that C_d varied from 0.1 to 1.7 and appeared to be a function of Fr and R. Abraham (7), on the other hand, found that a constant value of $C_d = 0.3$ gave good agreement between theory and observation if the entrainment theory is modified as presented herein. Briggs (1) theoretically argues that C_d should be zero when the ambient fluid is irrotational. The model presented here was run with the drag term and $C_d = 0.3$.

EXPERIMENTAL PROGRAM

Since one of the objectives of this study is to evaluate the effect of ambient turbulence levels due to terrain roughness upon plume rise and dispersion, experimental data were collected in wind tunnels with three different surface roughness configurations. Low turbulence ($\leq$ 5 percent) cases (neutral and stable stratification) were studied in the Meteorological Wind Tunnel (14), whereas moderate ($5 < i_x \leq 10$ percent) and high ($i_x > 10$ percent) turbulence cases (neutral stratification) were studied in the Environmental Wind Tunnel. Since the data were collected as part of two independent research contracts, different scale models were used for each. The first study (15) employed a model stack scaled at 1:300 and stack nozzle attachments to change the exit velocity. The second study (5) used a 1:960 scale model of the Tennessee Valley Authority (TVA), Kingston Steam Plant (KSP) and surrounding topography for three wind directions versus a uniform roughness case. The model and prototype test conditions for each basic study--1) low turbulence and 2) moderate and high turbulence--are given in Tables 1 and 2.

Both tests were conducted in a similar manner. Metered quantities of gas, mixed to appropriate density to simulate prototype exit density, were allowed to flow from the model stacks at the desired speed. Aerial concentration distributions of the resulting plume as indicated with a tracer gas were obtained using gas chromatographic techniques. Sufficient data were collected such that the plume centerline at various downwind distances could be obtained.

The plume was made visible by passing the gas mixture through titanium tetrachloride prior to emission from the model stack. Stills (color and black and white) and motion pictures of the tests for the range of conditions in Tables 1 and 2 were obtained. The black and white photographs were further analyzed to obtain quantitative information on plume rise. The plume center coordinates were digitized off the pictures. Subsequently, these coordinates

Table 1. Model and prototype parameters for the low turbulence wind-tunnel tests.

Parameter	Prototype	Model
1. Stack Parameters		
a) Height (m) - h	60.8	0.204 m
b) Nozzle height (m) - h_1	6.6	0.022
c) Volumetric Emission Rate (m^3/s) - V	320.0	2.14×10^{-4} m^3/s
d) Exit Velocity (m/s) u_{soo}	13.68 to 233.13	0.79 to 13.46
e) Stack Diameter (m) - D	5.59, 3.64, 2.78, 1.80, 1.35	$[1.86, 1.21, 0.93, 0.60, 0.45] \times 10^{-2}$
2. Ambient Parameters		
a) Wind speed (m/s) u_{aoo}	2.0 to 16	0.12 to 0.92
b) Temperature (°K) T_{aoo}	291 to 297	291 to 297
c) Pressure (mb) - P_o	850.0	850.0
d) Reference height (m) z_r	60.8	0.20
e) Viscosity (m^2/s) - ν_a	0.15×10^{-4}	0.15×10^{-4}
3. Dimensionless Parameters		
a) $\gamma = \frac{\rho_{aoo}-\rho_{soo}}{\rho_{aoo}}$	0.20, 0.25, 0.31	0.20, 0.25, 0.31
b) $Fr = \frac{u_{soo}}{\sqrt{gD\gamma}}$	3.70 to 128.19	3.70 to 128.19
c) $R = u_{soo}/u_{aoo}$	0.86 to 112.17	0.86 to 112.17
d) $Ri = \frac{g}{T} \frac{\frac{\Delta\theta}{\Delta z}}{(\frac{\Delta u}{\Delta z})^2}$	0 to 0.57	0 to 0.57
e) $Re_b = \frac{u_{aoo}h}{\nu_a}$	8.2×10^6 to 6.3×10^7	1.6×10^3 to 1.2×10^4
f) $Re_s = \frac{u_{soo}D}{\nu_s}$	5.0×10^6 to 2.0×10^7	954 to 3933

Table 2. Model and prototype parameters for the Kingston Steam Plant moderate and high turbulence tests--Tall Stack #1

Parameter	Prototype	Model
1. Stack Parameters		
a) Height (m) - h	305.0	0.318
b) Diameter (m) - D	7.92	0.0086
c) Exit Temperature (°K) - T_{soo}	408.0	293.0
d) Exit Velocity (m/s) - u_{soo}	27.2	0.88
e) Flow Rate - V	1339 m^3/s	51.12 cm^3/s
f) SO_2 Source Strength (ppm) - χ_o	1470	
g) Exit Density (Kg/m^3)	0.85	0.69
h) SO_2 Emission Rate (g/s) - Q_m	3660	
2. Ambient Parameters		
a) Reference Wind Speed (m/s) - u_r	6.71, 11.18, 15.65	0.22, 0.36, 0.51
b) Temperature (°K) - T_{aoo}	278.0	293.0
c) Density (Kg/m^3) - ρ_{aoo}	1.25	1.02
d) Vegetation Height - h_v	21.0 m	2.19 cm
e) Pressure (mb) - P_a	984.1	850
f) Canopy diameter - D_c	11 m	1.15 cm
g) Wind directions	302, 315, 325	302, 315, 325
h) Reference Height (m) - z_r	110	0.11
i) Viscosity (m^2/s) - ν_a	0.15×10^{-4}	0.15×10^{-4}
3. Dimensionless Parameters		
a) $\gamma = \frac{\rho_{aoo} - \rho_{soo}}{\rho_{aoo}}$	0.32	0.32
b) $Fr = \frac{u_{soo}}{\sqrt{gD\gamma}}$	5.46	5.46
c) $R' = u_{soo}/u_r$	4.1, 2.4, 1.7•	4.1, 2.4, 7.1
d) $Re_b = \frac{u_r z_r}{\nu_a}$	4.90×10^7, 8.20×10^7, 11.5×10^7	1.16×10^3, 2.64×10^3, 3.74×10^3
e) $Re_s = \frac{u_{soo} D}{\nu_s}$	1.44×10^7	505

were corrected for optical distortion (16). The plume rise values obtained photographically and by tracer gas measurements were subsequently used to compare with the numerical model predictions.

Mean and turbulent velocity measurements were performed to 1) quantitatively assess the extent of the disturbed flow due to the surface roughness or terrain, 2) monitor and set wind-tunnel flow conditions, and 3) provide inputs for the numerical plume rise model. The instrumentation used for this study included:

1. Two Thermo-Systems, Inc. (TSI) 1050 Series Anemometers

2. A TSI Model 1287 Split-Film Sensor

3. A TSI Model 1125 Flow Calibrator

4. A Model 800 LV Datametrics Linear Flowmeter and Probe

5. A MKS Baratron Type 77

6. A Systems Development Digital Data Acquisition System

7. A Yellow Springs Precision Thermistor

The techniques used in obtaining the velocity and temperature data with this assortment of equipment and the data processing techniques are described in Petersen (5).

RESULTS AND CONCLUSIONS

General Approach

The numerical model was compared with the laboratory plume rise and concentration measurements. The basic plan for testing the model was to begin with the low turbulence-neutral stratification results. For these tests, the velocity profiles were essentially uniform, the temperature constant, and the turbulence levels low (less than 5 percent). The α_1 entrainment relation was assumed to be valid since it has the least effect on the resulting plume rise and also since more definitive experiments support the relation. This was further supported by running the model for different α_1. It was found that the results were relatively insensitive to the value of α_1 for low R values and became more sensitive with increasing R. The α_2 entrainment constant was varied to find the value which would give the minimum error between model and all experimental data. For this study, the best α_2 for each test case was not determined since it was hoped that one value for α_2 would give sufficiently accurate results and avoid a complex relation for α_2 as a function of stack parameters. The expected value for α_2 is between 0.65 and 1.34.

The next test of the model was under stable stratification, low turbulence, with varying wind and temperature fields with height. The model was run with the actual velocity and temperature field as inputs. The best α_2 determined from the neutral tests was used for these model runs to check its applicability under other stratifications. The final test of the model was to consider moderate and high turbulence and assess whether the α_3 entrainment constant of $2\sqrt{2}$ is valid. For these tests, the upwind velocity and turbulence profiles were used as inputs for the model.

Validation for Neutral Flow--Low Turbulence

Figure 2 shows the model and laboratory plume rise results for the low turbulence, neutral stratification cases. The model best fits the data when $\alpha_2 = 0.9$. In the figures the photo and concentration plume rise results were both plotted along with the rms error between model and photograph plume rise. The rms error is defined by the following equation:

$$\text{error} = \left\{ \frac{1}{N-1} \sum_{i=1}^{N} \left[\frac{[(z_{photo} - z_{model})_i]}{(h + z_{photo})_i} \right]^2 \right\}^{1/2}$$

As can be seen from the figure, the model error values ranged from 0.01 to 0.16 with an average of 0.06. The highest errors were found for the case Fr = 3.7. This may be due to a slight variation of α_2 with Froude number, since the error appeared to generally decrease with increasing Froude number.

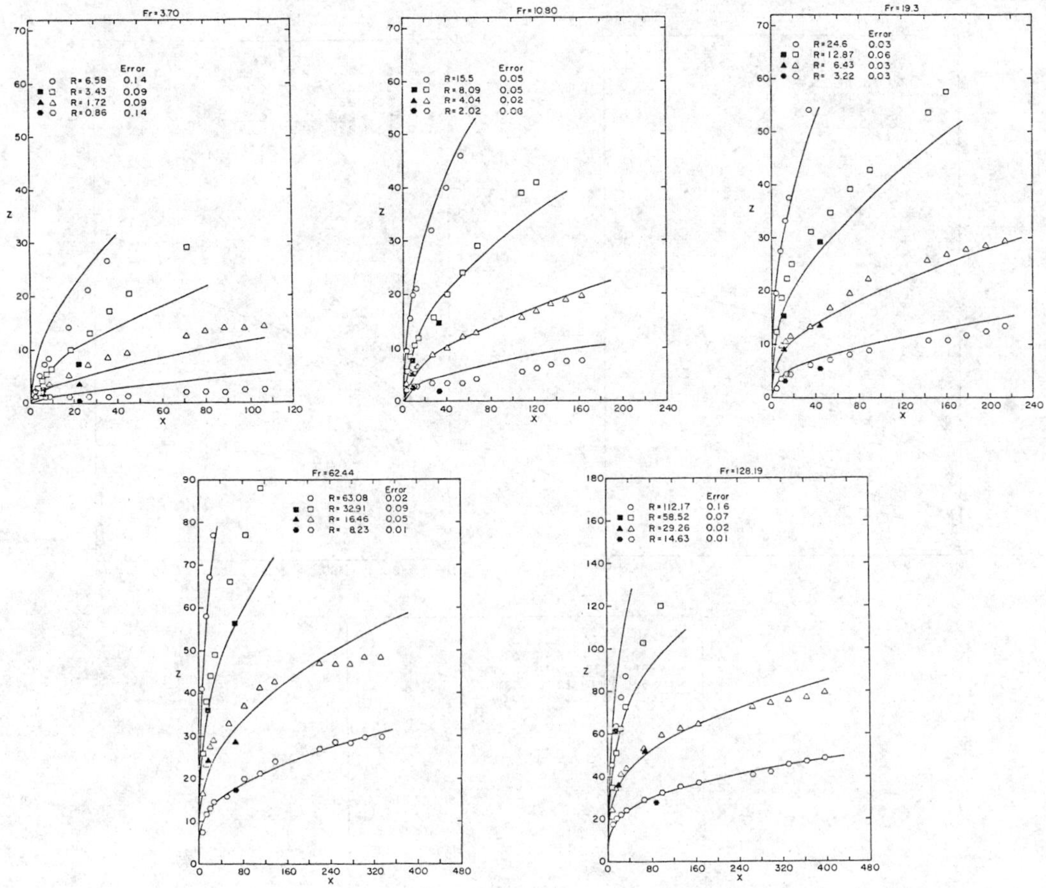

Fig. 2. Comparison of plume rise determined from numerical model and that observed in laboratory (photographs - open symbols; concentration - closed symbols) for various velocity ratios, neutral stratification, negligible turbulence levels and Froude numbers of a) 3.7, b) 10.80, c) 19.3, d) 62.44, and e) 128.19.

Validation of Stable Flow--Low Turbulence

Figure 3 shows the model and laboratory results for $Ri = 0.57$. The model runs were made by using the velocity and temperature profiles as inputs for the appropriate stability as measured in the wind tunnel. Again, the turbulence was assumed to be low enough such as to not contribute to the plume rise entrainment process. Hence, only α_2 and α_2 entrainment parameters were used. The α_2 value of 0.9 determined from the neutral tests was used for both stabilities. Figure 3 shows that the errors for $Ri = 0.57$ range from 0.03 to 0.14 with the error generally decreasing with increasing Fr. The mean error is 0.9. The concentration data plume rise results also compare favorably with the model.

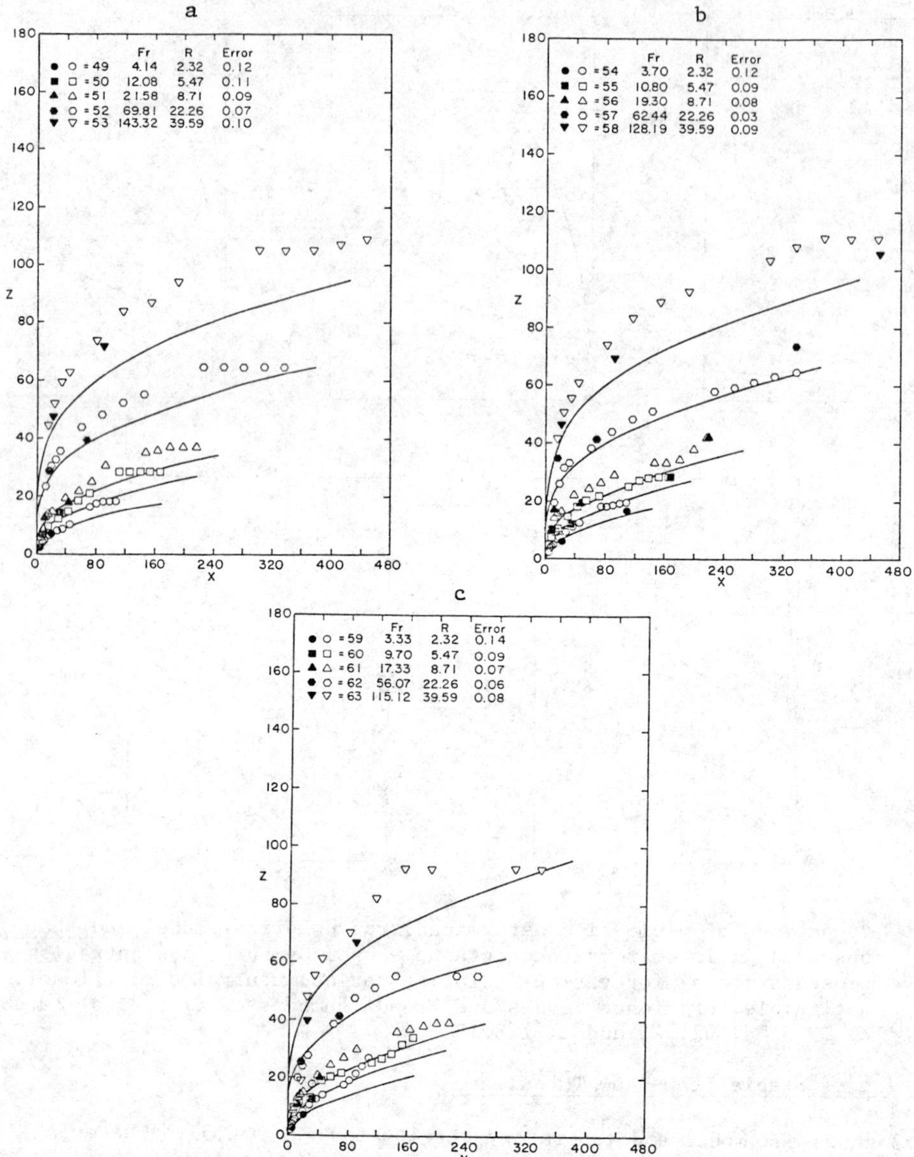

Fig. 3. Comparison of plume rise determined from the numerical model and that observed in the laboratory (photograph - open symbol; concentration - closed symbol) for various velocity ratios, Froude numbers, $Ri = 0.57$ and density ratios (λ) of a) 0.20, b) 0.25, and c) 0.31.

Extension of Model to Turbulent Flow

Up to this point, it is evident that the entrainment relations which have been assumed for α_1 and α_2 are adequate for giving good predictions of plume rise for neutral and stable stratification. The next cases to be considered will include the turbulence term. In considering the third entrainment relation, namely that which takes into account the entrainment due to atmospheric turbulence, the velocity and turbulence profiles taken upwind of the plant site were input into the model. Since the plume has spread in both the vertical and horizontal directions and the longitudinal and lateral turbulence intensities are often nearly equal, the following effective turbulence intensity as measured at various heights was input into the model:

$$i_e = \frac{i_x + i_z}{2} .$$

The results of the moderate and high turbulence tests are shown in Fig. 4. As can be seen, the model predicts very well for the high velocity ratios both under moderate and high turbulence with respective errors of 0.05 and 0.04. For both tests (with and without terrain) the accuracy of the model decreases with decreasing velocity ratio. Overall, the model predicted best for the moderate turbulence tests. This can be explained by the fact that moderate turbulence tests were conducted in essentially a horizontal flow, whereas the high turbulence tests were conducted with simulated terrain roughness. The simulated terrain roughness produced down motions, particularly since the plant was in the lee of several ridges. The effect of this down motion upon the plume rise is particularly evident in these figures which show the lower effective plume altitude for the high turbulence tests and high velocity ratios. The model is not designed to give good predictions under these conditions, but could be modified to handle such cases by using not only the wind speed but also the direction at various heights as a model input. This is an item which is recommended for further research.

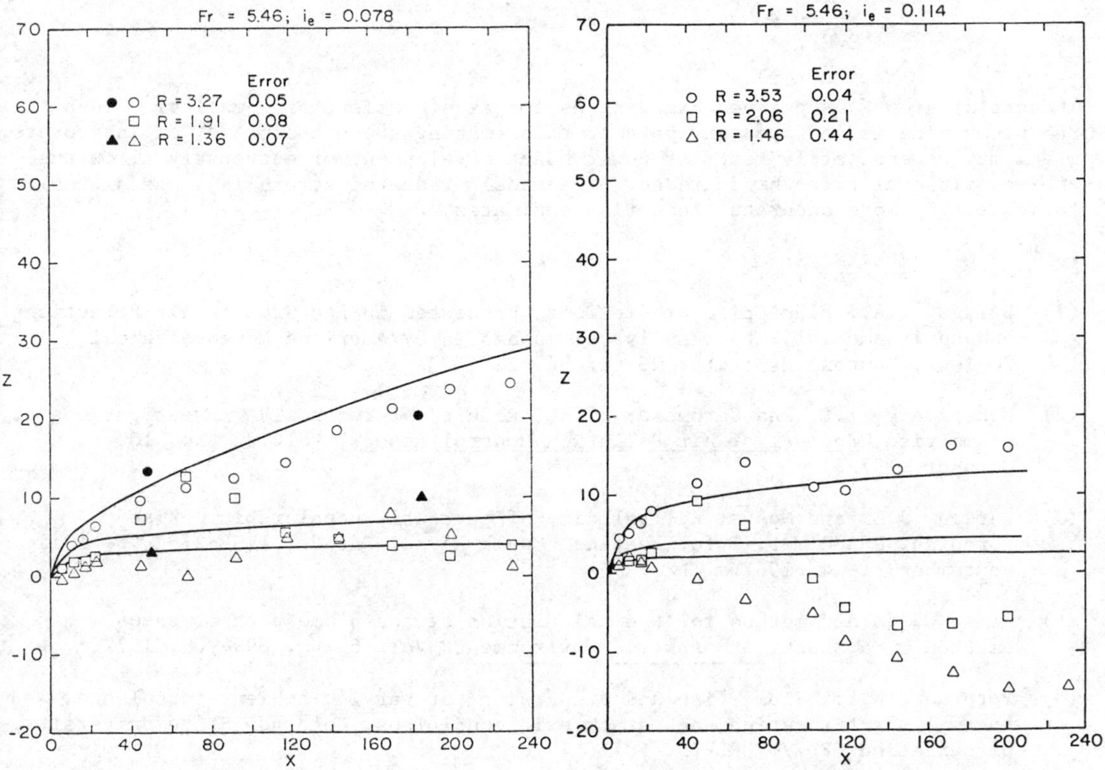

Fig. 4. Comparison of the numerical model and laboratory results (open symbol - photographic data; closed symbol - concentration data) for a Froude number of 5.46, three velocity ratios and stack top turbulence levels of a) 0.078 (without terrain) and b) 0.114 (with terrain).

The results of the moderate and high turbulence tests substantiate the use of $\alpha_3 = 2\sqrt{2}$ as the entrainment constant. Ooms (4) used $\alpha_3 = 1$ and if applied for these tests, the plume rise would be over-predicted.

Conclusions

The numerical plume rise model agreed satisfactorily with the wind-tunnel observations for the low and moderate turbulence tests over the range of thermal stratifications studied. For the high turbulence tests the model compared favorably only when the initial plume rise was sufficient to avoid the down motion in the lee of the simulated ridges. When the plume becomes influenced by the down motion induced by the ridges the model predictions become unacceptable. This was to be expected since only one-dimensional wind fields were input into the model.

The entrainment relations and constants used in this study differ from those used by previous investigators. For the initial plume region good agreement was achieved using the following entrainment constant

$$\alpha_1 = 0.057 \left(\frac{\rho_{soo}}{\rho_{so}}\right)\left(\frac{\rho_{aoo}}{\rho_{soo}}\right)^{1/2} .$$

When ambient-turbulence dominates the mixing processes the following constant was found to give acceptable agreement

$$\alpha_3 = 2\sqrt{2} .$$

Industrial growth is planned based on estimates of ambient air quality. Since the plume rise is an important parameter for making these estimates, an inaccurate model may unnecessarily restrict power plant development or conversely allow the ambient air quality to be degraded. The model presented herein is a useful tool for obtaining more accurate plume rise estimates.

REFERENCES

(1) Briggs, G.A., Plume rise predictions, presented in Lectures on Air Pollution and Environmental Impact Analyses, sponsored by American Meteorological Society, Boston, September 29-October 3, 1975.

(2) Montgomery, T.L. and Carpenter, S.B., Results of recent TVA investigations of plume rise, Journal of Air Pollution Control Assoc., Vol. 22, No. 10, October, 1972.

(3) Carson, J.E. and Moses, H., Validity of currently popular plume rise formulas, USAEC Met. Info. Meeting, Chalk River, Canada, AECL-2787, p. 1015, September 11-14, 1967.

(4) Ooms, G., A new method for the calculation of the plume path of gases emitted by a stack, Atmospheric Environment, Vol. 6, pp. 899-909, 1972.

(5) Petersen, R.L., Plume rise and dispersion for varying ambient turbulence, thermal stratification, and stack exit conditions, Colorado State University, Dissertation CED77-78RLP25, Fall, 1978.

(6) Fan, L.N., Turbulent buoyant jet into stratified or flowing ambient fluids, California Institute of Technology, Report No. KH-R-15, June, 1967.

(7) Abraham, G., The flow of round buoyant jets issuing vertically into an ambient fluid flowing in a horizontal direction, Delft Hydraulics Laboratory, Publ. No. 81, 1971.

(8) Beckett, R. and Hart, J., Numerical Calculations and Algorithms, McGraw-Hill, Inc., 1967.

(9) Albertson, M.L., Dal, F.B., Jensen, R.A., and Rouse, H., Diffusion of submerged jets, American Society of Civil Engineering, Transactions, Vol. 115, Paper No. 2409, pp. 639-697, 1963.

(10) Ricou, F.P. and Spalding, D., Measurements of entrainment by axisymmetrical turbulent jets, Journal of Fluid Mechanics, Vol. 11, Part 1, pp. 21-32, 1961.

(11) Tsang, G., Laboratory study of line thermal, Atmospheric Environment, Vol. 5, pp. 445-471, 1971.

(12) Taylor, G.I., Diffusion by continuous movements, Proceedings of the Meteorological Society, Vol. 20, pp. 196-211, 1931.

(13) Csanady, G.T., Turbulent Diffusion in the Environment, Geophysics and Astrophysics Monograph, 1st Ed., Vol. 3, D. Reidel Publishing Company, Doudrecht, Holland, Chapters 2-3, 1973.

(14) Cermak, J. E., Wind Tunnel for the Study of Turbulence in the Atmospheric Surface Layer, Technical Report CER-58-JEC-42, Fluid Dynamics and Diffusion Laboratory, Colorado State University, Fort Collins, CO, Nov., 1958.

(15) Petersen, R.L., Cermak, J.E., and Meroney, R.N., A wind-tunnel study of plume rise and dispersion under stable stratification, Joint Conference on Applications on Air Pollution Meteorology, pp. 347-352, November 29-December 2, 1977.

(16) Halitsky, J., Single-camera measurement of smoke plumes, International Journal of Air and Water Pollution, Vol. 4, Nos. 3 and 4, pp. 185-198, 1961.

DRIFTING SNOW SIMILITUDE—DRIFT DEPOSIT RATE CORRELATION

J. D. IVERSEN
Iowa State University, Ames, Iowa, USA

ABSTRACT

The similitude requirements for proper modeling of snowdrift phenomena are presented, as well as preliminary results from a series of snowdrift model tests on a 120th scale highway-grade separation. Theoretical means are used to reduce the number of independent dimensionless variables by a combination of terms. A mass rate similitude parameter is shown to successfully correlate model data and is also used to predict full-scale drift deposition as a function of full-scale time and wind speed.

INTRODUCTION

This paper reviews the similitude requirements for proper modeling of snowdrift phenomena in the boundary layer wind tunnel. Although wind tunnel testing to determine snowdrifting characteristics can be very useful if valid similitude relations are used to predict full-scale drift patterns, relatively few studies have been conducted. Experiments that have been conducted related to highways (snow fences, cuts, etc.) include those of Becker [1], Nøkkentved [2], Finney [3], and Norem [4, in water]. Those related to building proximity include Gerdel and Strom [5], Strom et al. [6], Theakston [7, in water], and Calkins [8, in water]. Scale model experiments in the atmosphere using real snow as the modeling material have also been conducted [9]. Preliminary results from a series of experiments conducted by the writer concerning snow-drift control adjacent to grade separations on interstate highways is also presented. This work is supported by the Iowa State Department of Transportation, with the approval and recommendation of the Iowa Highway Research Board, and by the Engineering Research Institute, Iowa State University.

THE SALTATION PHENOMENON

Loose particulate material can move in one of three ways--by creep, saltation, or suspension. Creep is the motion in which large particles do not become airborne but just roll along the surface. Medium size and smaller particles will become airborne as a result of aerodynamic lift or from impact from returning particles. The saltating particle rises from the surface in a nearly vertical direction and then returns via a shallow angle trajectory [10]. Very small particles go into suspension, and if they are dust, can rise to great heights before settling out of the atmosphere. For snow, however, such small particles quickly disappear due to sublimation. Thus, moving snow moves primarily in the saltation mode. In water, more of the particles are suspended and trajectory paths are quite different from those in air because of the much smaller ratio of particle density to fluid density. Thus, use of water as a modeling fluid is not appropriate for modeling snowdrifts.

Bagnold [11] presented the dimensionless threshold friction speed

$$A_1 = u_{*_t} / (\rho_p g D_p / \rho)^{1/2} \tag{1}$$

For water, particle density, ρ_p, is replaced by net density, $\rho_p - \rho$. Bagnold and most later researchers have assumed that A_1 is a unique function of particle friction Reynolds number B

$$A_1 = A_1(B), \; B = u_{*_t} D_p/\nu \tag{2}$$

In addition, however, because of the forces of cohesion that particularly affect small particles, A_1 is also a function of density ratio, ρ/ρ_p. Semiempirical expressions for A_1 have been derived by Iversen et al. [12,13]. The ratio of terminal speed to threshold friction speed, U_F/u_{*_t}, serves as an indicator of the division between suspension and saltation. Since the vertical turbulent fluctuating velocity in the boundary layer is of the same order of magnitude as the friction speed, particles with U_F/u_{*_t} much smaller than one are likely to be suspended, while those with U_F/u_{*_t} much larger than one will move in the saltation mode.

MODELING OF PARTICLE TRANSPORT IN AIR

Jensen [14] and others have observed that the lower portion of the planetary boundary layer follows the logarithmic velocity profile dependent on the surface roughness. Thus, Jensen showed that for simulation of the atmospheric boundary layer, the roughness parameter in the model should be the same as that in the prototype, i.e.,

$$\frac{z_{o_m}}{z_o} = \frac{L_m}{L} \tag{3}$$

where L is a characteristic length. Most investigators have relied on the long test section to simulate the turbulent boundary layer profile as a naturally thick, fully turbulent boundary layer [15].

If simple dimensional analysis is used to group the important variables, the following list of similitude parameters, useful in the description of saltation phenomena, can be written:

(1) D_p/L particle diameter-length ratio

(2) $u(h)/U_f$ reference to particle terminal speed ratio

(3) $[u(h)]^2/gL$ Froude number

(4) e coefficient of restitution

(5)	ℓ/L	topographic geometric similarity
(6)	z_o/L	roughness similitude
(7)	z'_o/L	roughness similitude in saltation
(8)	h/L	reference height ratio
(9)	z_o/L^*	stability parameter
(10)	λ/L	ripple length ratio
(11)	U_F/u_{*_t}	particle property similitude
(12)	$u_{*_t} D_p/\nu$	particle friction Reynolds number
(13)	$u(h)\ L/\nu$	flow Reynolds number
(14)	u_*/u_{*_t}	friction speed ratio
(15)	ρ/ρ_p	density ratio
(16)	$u(h)t/L$	time scale

The first four of these parameters have been suggested by Gerdel and Strom [5] as the scaling parameters of primary importance in the modeling of accumulation of wind-driven snow. In addition, Odar [16] listed $u_{*t}/u(h)$ (a combination of Parameters 2 and 11) and ρ/ρ_p, Parameter 15, as important fundamental parameters. Kind [17] does not agree and lists $u_{*t}/u(h)$, Parameter 2, Parameter 3, and a minimum Reynolds number requirement as the only important parameters.

It is true that some of the above parameters are not independent. The roughness height in saltation, z'_o, for example, has been assumed by different investigators to be proportional to particle diameter [18], to ripple wave length λ[11], or to the square of surface friction speed ($z'_o \sim u_*^2/g$ [19]). Thus Parameter 7 would be proportional to Parameter 10, Parameter 1, or to (3) × $(14)^2/(2)^2$ × $(11)^2$. Discussion of these parameters can be found in work considering simulation of aeolian features observed on the surface of the planet Mars [20,21].

COMBINATION OF TERMS BY THEORETICAL CONSIDERATIONS

It is impossible to satisfy all terms of Parameters 1 to 15 simultaneously (Parameter 16 is a prediction term). However, additional aids to dimensional analysis such as the particle equations of motion and theoretical mass transfer rate can be used to group dimensionless terms and/or determine their importance. The procedure by which this can be done for blowing sand or snow has been outlined earlier by the author [22,23,24]. The particle trajectory equations of motion are employed to obtain the following parameters:

(17) $C_D \dfrac{\rho L}{\rho_p D_p}$

(18) $\dfrac{gL^2}{u_*^{\,2} h}$

The ratio of equivalent roughness height in saltation, z'_o, to characteristic vertical length, h, can be written [24]:

$$(19)\quad A_1^2\,\frac{D_p}{h}\left(\frac{u_*}{u_{*_t}}\right)^2$$

The mass rate of motion of saltating material is used to obtain a combination modeling-prediction equation [24].

$$(20)\quad \frac{d(A/L^2)}{d(u_* t/L)}\Bigg/\left\{\frac{U_F}{u_{*_t}}\left(\frac{\rho}{\rho_p}\right)\left(\frac{u_*^2}{gh}\right)\left(1-\frac{u_{*_t}}{u_*}\right)\right\}$$

In addition to Parameters 17-20, based on the saltation phenomenon, the usual atmospheric boundary layer Parameters, 5, 6, 8, 9, and 13, must of course also be considered.

For modeling of drift features, Parameter 20, from past experience [23], appears to be the most important of the saltation parameters. Parameter 19 is somewhat less important, and generally speaking, Parameters 17 and 18 are important only if small-scale features (such as surface ripples) affect the larger scale drift patterns. The quantity A/L^2 (planform drift area) can be replaced by V/L^2h(drift volume) or A_c/Lh (drift profile area). The coefficient U_F/u_{*t} was empirically derived from erosion rate data [22]. For deposition measurements, and for modeling materials with ratios U_F/u_{*t} near 1, it appears that a slight modification of Parameter 20 concerning the coefficient U_F/u_{*t} may be necessary. To systematically model a snow or sand drifting problem, it is desirable to change variables such as the amount of vertical distortion (to ascertain the effect of Parameter 19), particle material, and wind speed, so that the effect of changes in the values of the dimensionless terms can be evaluated. The use of Parameters 19 and 20 to evaluate these changes is necessary, since not all of the original 15 parameters can be modeled simultaneously.

MODEL PARTICLE SELECTION

The properties of snow and of possible modeling particles are discussed in [24]. Figure 1 illustrates an approach to particle selection. Curves of constant $C_D\rho/\rho_p D_p$ (Parameter 17), constant A^2D_p (Parameter 19) and constant U_F/u_{*t} are shown in Fig 2. First, it is necessary to select materials that lie to the right of the barred curve ($U_F/u_{*t} = 1$), so that the primary transport mode is saltation. (Blowing snow occurs primarily in the saltation mode rather than suspension, perhaps because very small particles disappear quickly due to sublimation.) Conflicting requirements exist for a small scale model. In order to satisfy Parameter 18, it is necessary to pick a particle of less density than snow. To satisfy Parameter 17, a smaller particle must be chosen. Finally to satisfy Parameter 19, a larger density is required.

Five materials were selected for trial for the highway-grade separation experiments (model scale 1/120). Their properties and modeling parameter values are listed in Table 1.

The last three columns of the table list the values of the modeling Parameters 17, 18, and 19, respectively. Particle No. 4 is closest to the terminal speed-threshold speed ratio, Particle No. 2 comes closest to satisfying Parameter 17, Particle No. 1 is nearest the value of Parameter 18 for snow at full scale, and Particle No. 5 comes closest to satisfying the roughness Parameter 19. After testing the five particles, Nos. 1 and 2 were immediately ruled out, No. 1 because the roughness

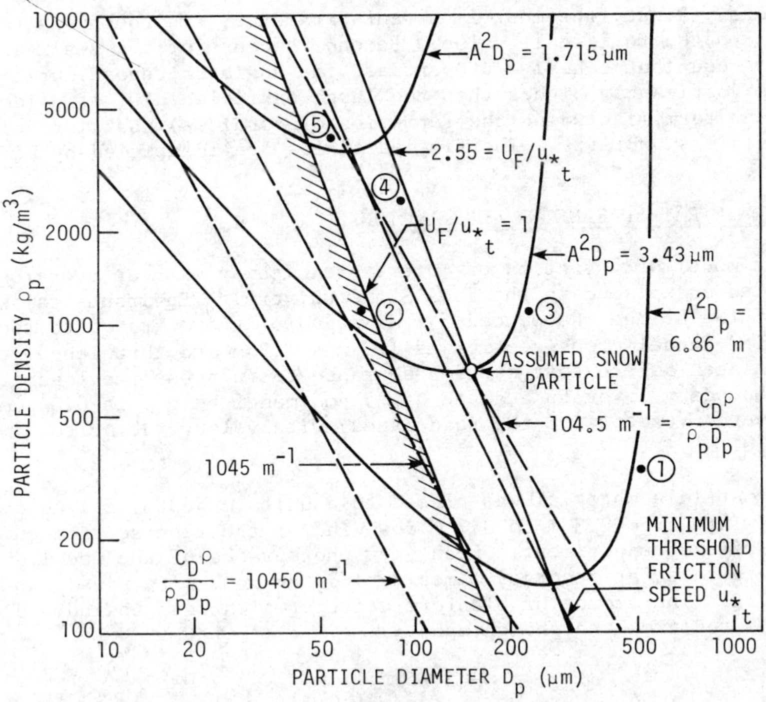

Fig. 1. Variation of similitude values with particle density and diameter. Circled numbers refer to the modeling particles listed in Table 1.

Table 1. Materials selected for trial for highway-grade separation experiments.

Model Particle	Density gm/cm^3	Diameter μm	U_F/u_{*t}	$C_D\rho L/\rho_pD_p$	gL^2/u_{*t}^2h	$A_1^2D_pU^2/U_o^2h$ (typical)
	Particle Properties and Parametric Values					
1. Instant tea	0.2	500	6	3.3	10,000	$3.2(10)^{-3}$
2. Walnut shell	1.1	69	0.9	400	4,000	$8.9(10)^{-4}$
3. Walnut shell	1.1	268	5.96	7.5	3,600	$7.0(10)^{-4}$
4. Glass spheres	2.5	101	2.8	28.0	3,000	$4.4(10)^{-4}$
5. Glass spheres	4.0	49	1.3	125	2,600	$2.5(10)^{-4}$
6. Snow (full scale)	0.7	150	2.55	900	750,000	$7.2(10)^{-5}$

parameter was much too large (in addition to difficulty in working with the material) and No. 2 because most of the material becomes suspended rather than moving in the saltation mode since $U_F/u_{*t} < 1$. The latter three materials are all fairly satisfactory. Number 5, the dense glass sphere, was used for the majority of testing for two reasons. First, the most important parameter, the roughness Parameter 19, comes closest to the full scale value. Second, because of the small size of these particles and resultant cohesive properties, the angle of repose during drift formation can be significantly higher than the large particle value of 34 degrees, and thus drifts can form (including the formation of cornices) that more closely resemble full-scale snowdrifts. The five particles are identified by numbered dots in Fig. 1.

MASS RATE PARAMETER CORRELATION

A 120th scale model of a section of interstate highway with an overcrossing bridge was constructed and placed in the Iowa State University Environmental Wind Tunnel. The center of the bridge was placed 5 m downwind of the entrance to the test section. The wind tunnel cross-section is 1.2 m x 1.2 m and turbulence-generating spires were placed at the test section entrance to increase the logarithmic portion of the boundary layer to a depth of 25 cm (model height is 5.84 cm). The model was covered (except for the road lanes) with velour fabric to simulate a grassy surface.

The modeling particle material was placed to a uniform depth of 1 cm across the test section width from 2.3 m to 3.7 m downwind of the test section entrance prior to the start of each experiment. Plan view photographs of the model during the experiment were taken at recorded times during the run. A typical photograph is shown in Fig. 2. The amount of planform drift area on both upwind and downwind lanes was measured from the photographs.

Fig. 2. Photograph of interstate highway-grade separation model (plan view) during a wind tunnel experiment. Wind direction top to bottom.

The total drift area as a function of time is shown in Fig. 3 for five different wind speeds, with a wind direction parallel to the bridge. The data illustrated in this figure were used to extrapolate the results to zero movement of material, using Parameter 20 as an extrapolation equation to obtain a threshold wind speed. Without the similitude Equation 20, the only time-dependent correlation parameter would be Parameter 16. That this parameter is not satisfactory by itself as a correlation or prediction parameter is illustrated by Fig. 3, which shows drift area as a function of $U_\infty \Delta t/L$.

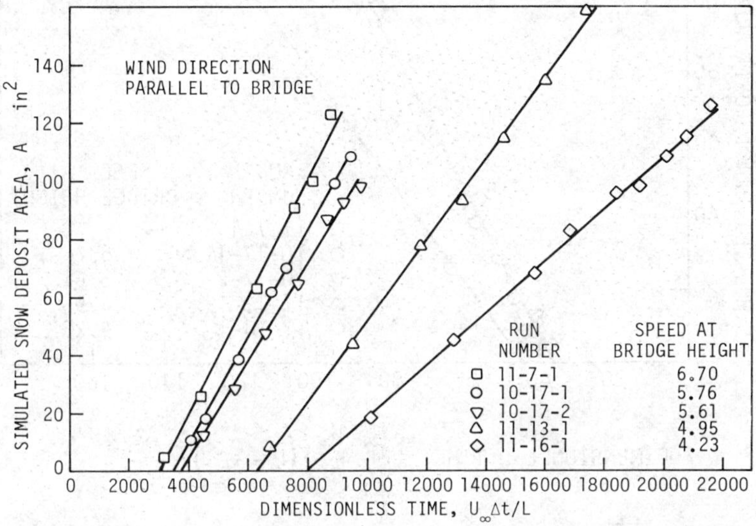

Fig. 3. Accumulated simulated snowdrift plan area as a function of dimensionless time; determined from photographs for five different wind speeds. Wind direction parallel to bridge. Speed in m/s.

Figure 4 illustrates the utility of the mass rate similitude Parameter 20. Data from four of the five runs is well correlated by this parameter. The curve for the highest speed run does not correlate because of the fact that the ratio of U_F/u_{*t} for this material is very close to 1, and for this high wind speed, a significant amount escapes the saltation mode, and goes into suspension, and therefore is not available for drift formation on the highway lanes. Thus the rate of formation is significantly less than it would be if all material stayed in saltation. Succeeding wind tunnel tests using this material were operated at sufficiently low speeds to ensure good correlation.

FULL-SCALE PREDICTION

Most previous investigators who have attempted to model snowdrifts, if similitude has been applied, have assumed that the velocity scale is determined through the Froude number, Parameter 3. The wind speed ratio based on this parameter is the square root of the length scale ratio. Thus, for the data in Figs. 3, 4 the full-scale wind speeds would vary from 46 to 73 m/s (103 to 163 mph) which would be strong winds indeed! It is the author's opinion that Odar was correct in that the density ratio, ρ_p/ρ, is also important. Both the Froude number and density ratio are embedded in the mass rate Parameter 20.

If it is assumed that the deposited drift areas for equivalent scaled times are equal in model and full scale (satisfying geometric similarity),

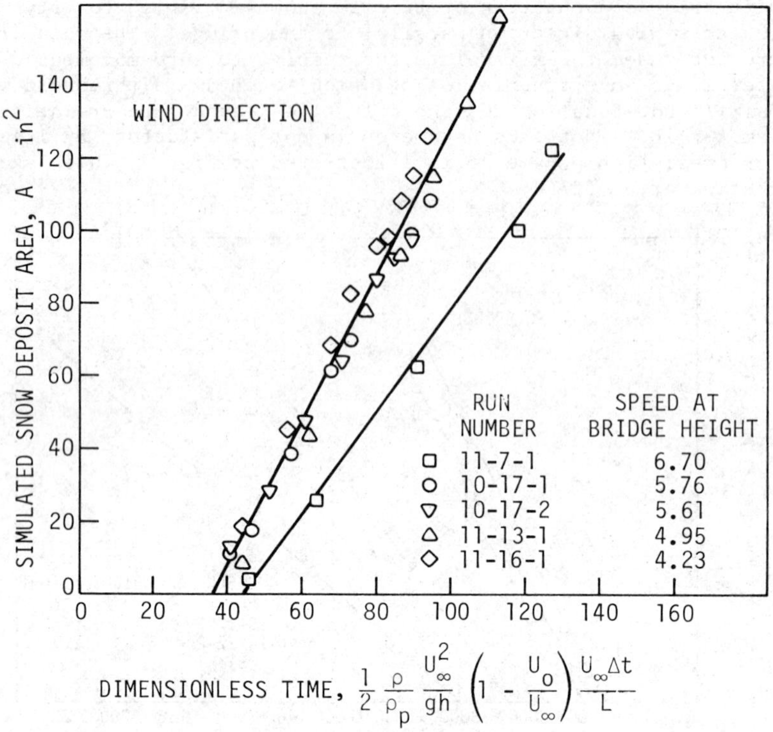

Fig. 4. Accumulated snowdrift area as a function of dimensionless time (mass rate Parameter 20). Same data as Fig. 3.

$$\Delta A/L^2 = \Delta A_m/L_m^2 \tag{4}$$

It is therefore necessary for the rest of the Parameter 20 to be satisfied, i.e.,

$$\left(\frac{\rho}{\rho_p}\right)\frac{U^2}{gh}\left(1 - \frac{U_o}{U}\right)\frac{U\,\Delta t}{L} = \left(\frac{\rho_m}{\rho_{p_m}}\right)\frac{U_m^2}{g_m h_m}\left(1 - \frac{U_{om}}{U_m}\right)\frac{U_m\,\Delta t_m}{L_m} \tag{5}$$

For a geometrically undistorted model, the equation for the elapsed time ratio becomes

$$\frac{\Delta t}{\Delta t_m} = n^2\left(\frac{\rho_m}{\rho}\right)\frac{\rho_p}{\rho_{p_m}}\left(\frac{U_m}{U}\right)^2\left(\frac{U_m - U_{om}}{U - U_o}\right) \tag{6}$$

It is necessary to make a realistic assumption about the velocity ratio, U/U_m, in order to determine both the velocity and time ratios. If the particle and wind speeds in model and full scale are both proportional to wind speed with the same constant of proportionality, then

$$\frac{U}{U_m} = \frac{L}{L_m}\,\frac{\Delta t_m}{\Delta t} = n \Big/ \frac{\Delta t}{\Delta t_m} \tag{7}$$

With this substitution into Equation 6, the equation for time ratio becomes

$$\left(\frac{\Delta t}{\Delta t_m}\right)^2 + \left\{\frac{\frac{\rho}{\rho_m}\left(\frac{\rho_{pm}}{\rho_p}\right)\frac{U_o}{U_m}}{1 - \frac{U_{om}}{U_m}}\right\}\left(\frac{\Delta t}{\Delta t_m}\right) - \left\{\frac{n\left(\frac{\rho}{\rho_m}\right)\frac{\rho_{pm}}{\rho_p}}{1 - \frac{U_{om}}{U_m}}\right\} = 0 \tag{8}$$

The reference speed (at bridge height) for initiation of motion in the wind tunnel was determined from the data of Figs. 3, 4 to be 2.1 m/s (4.7 mph). This value coincidentally is close to the value of critical wind speed at bridge height (4.9 m or 16 ft) in the atmosphere. Using Run No. 10-17-1 as an example, the model reference speed was 5.76 m/s (12.9 mph) and the run time was 17.5 minutes. The model air density was 1.23 kg/m^3 (0.00238 $slugs/ft^3$) and the particle density was 4000 kg/m^3 (7.74 $slugs/ft^3$). If the full-scale air and particle densities are assumed to be 1.23 and 700 kg/m^3 (1.36 $slugs/ft^3$), respectively, the time-scale ratio is calculated to be $\Delta t/\Delta t_m = 31.2$ and the speed ratio $U/U_m = 3.85$. Thus, the equivalent storm speed at full scale would be 22.2 m/s (49.7 mph), a more realistic value than that calculated by using the Froude number. The full-scale equivalent storm would last for 9.1 hours.

SNOW-DRIFT CONTROL

Berms, fences [25], plant arrangements and other kinds of obstacles have been used for drifted snow storage and control. In this series of experiments, simulated rows of bushes were used to collect the simulated snow and protect the highway lanes from becoming drifted over. Figure 5 illustrates the effectiveness of one of the control methods tested. The upper curve shows the lane area covered as a func-

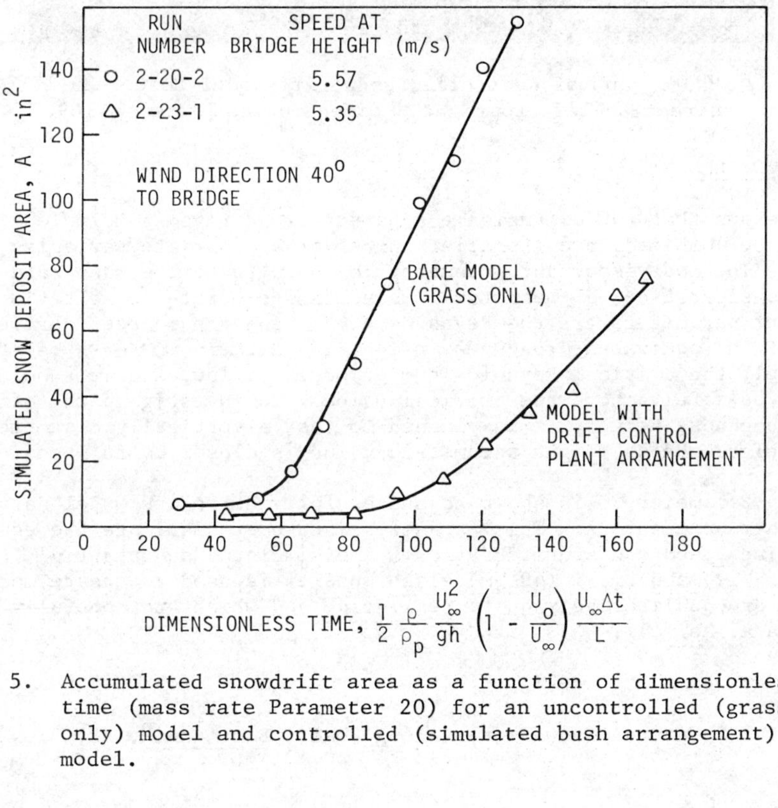

Fig. 5. Accumulated snowdrift area as a function of dimensionless time (mass rate Parameter 20) for an uncontrolled (grass only) model and controlled (simulated bush arrangement) model.

tion of dimensionless time for a bare model, i.e., one covered only with simulated grass. The lower curve shows the same type of data with the plant arrangement control in place. In both cases, for this figure the wind direction is at an angle of 40 degrees to the bridge centerline. The effectiveness of the control is noted by the fact that by the time the bare model has collected 0.065 m^2 (100 in.^{2}), the controlled model, at the same value of dimensionless time, has collected only 0.0077 m^2 (12 in.^{2}). The controlled model test was extended to a dimensionless time of 169, at which time 0.048 m^2 (75 in.^{2}) of lane had been covered. For the length scale of 120, these numbers correspond to 929 m^2 (10000 ft^2), 111 m^2 (1200 ft^2), and 697 m^2 (7500 ft^2). Figure 6 is a photograph of the drift pattern after the test was completed.

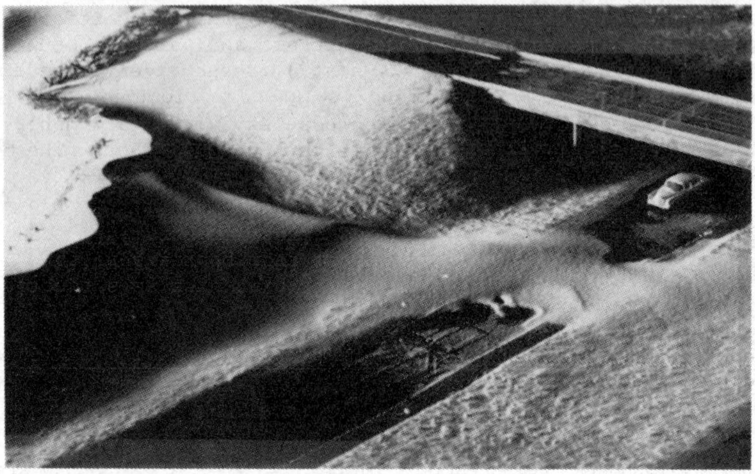

Fig. 6. Photograph of controlled snowdrift model test 2-23-1. Wind direction left to right. Dimensionless time is 169.

CONCLUSIONS

It has been shown that the dimensionless parameters derived by simple dimensional analysis can be combined into a smaller number of appropriate variables by theoretical means. The most important parameter is shown to be the mass rate parameter, derived by consideration of theoretical mass transport rate in saltation. The next most important parameters are the Reynolds number and a parameter derived from consideration of the equivalent roughness height in saltation. For a relatively rough-surfaced model, the critical Reynolds number would be low, and it should not be difficult to be safely above the critical value. To investigate the effect of equivalent roughness height, it is planned to test a vertically distorted model so that the model roughness height parametric value is closer to full scale.

The mass rate Parameter 20 is shown to successfully correlate model data for a range of wind tunnel speeds. This parameter is used to estimate the equivalent full-scale wind speed and storm duration. The simulated snow particle chosen, a dense (4000 kg/m^3) and small (49 μm) glass sphere, is used to create model snowdrifts which are qualitatively quite realistic, and which are hopefully quantitatively accurate, as well.

REFERENCES

(1) Becker, A., Natural Snow Fences Along Roads, Bautechnik, 22, 37-42 (1944).

(2) Nøkkentved, C., Drift Formation at Snow Fences, Stads-og Haveingenøren, 31, 111-114 (1940).

(3) Finney, E. A., Snow Control by Tree Planting, Part VI, Wind Tunnel Experiments on Tree Plantings, Michigan Engineering Bulletin No. 75, Michigan Engineering Experiment Station (1937).

(4) Norem, H., Designing Highways Situated in Areas of Drifting Snow, Draft Translation 503, Cold Regions Research and Engineering Laboratory, Hanover, New Hampshire (1975).

(5) Gerdel, R. W., and G. H. Strom, Scale Simulation of a Blowing Snow Environment, Proc. Inst. Environ. Sci., 53, 53-63 (1961).

(6) Strom, G. H., G. R. Kelly, E. H. Keitz and R. F. Weiss, Scale Model Studies on Snow Drifting, Research Report 73, U.S. Army Snow, Ice and Permafrost Research Establishment, Hanover, New Hampshire (1962).

(7) Theakston, F. A., Model Technique for Controlling Snow on Roads and Runways, Special Report 115, U.S. Army Cold Regions Research and Engineering Laboratory, Hanover, New Hampshire (1970).

(8) Calkins, D. J., Model Studies of Drifting Snow Patterns at Safeguard Facilities in North Dakota, Technical Report 256, Cold Regions Research and Engineering Laboratory, Hanover, New Hampshire (1974).

(9) Tabler, R., Rocky Mountain Forest and Range Experiment Station, Laramie, Wyoming, personal communication, (1979).

(10) White, B. R., and J. C. Schulz, Magnus Effect in Saltation, J. Fluid Mech., 81, 497-512 (1977).

(11) Bagnold, R. A., The Physics of Blown Sand and Desert Dunes, Methuen, London, 1941.

(12) Iversen, J. D., J. B. Pollack, R. Greeley and B. R. White, Saltation Threshold on Mars; the Effect of Interparticle Force, Surface Roughness, and Low Atmospheric Density, Icarus, 29, 381-393 (1976).

(13) Iversen, J. D., R. Greeley, J. B. Pollack, Windblown Dust on Earth, Mars, and Venus, J. Atmos. Sci., 33, 2425-2429 (1976).

(14) Jensen, M., The Model-Law for Phenomena in Natural Wind, Ingeniøren, 2, 121-128 (1958).

(15) Cermak, J. E., Applications of Fluid Mechanics to Wind Engineering — A Freeman Scholar Lecture, J. Fluids Eng., 97, 9-28 (1975).

(16) Odar, F., Simulation of Drifting Snow, Research Report 174, U.S. Army Cold Regions Research and Engineering Laboratory, Hanover, New Hampshire (1965).

(17) Kind, R. J., A Critical Examination of the Requirements for Model Simulation of Wind-Induced Erosion/Deposition Phenomena Such as Snow Drifting, Atmos. Environ., 10, 219-227 (1976).

(18) Zingg, A. W., Wind-Tunnel Studies of the Movement of Sedimentary Material, in Proceedings of the Fifth Hydraulics Conference, Bulletin 34, University of Iowa, Iowa City, 111-135 (1953).

(19) Owen, P. R, Saltation of Uniform Grains in Air, J. Fluid Mech., 20, 225-242 (1964).

(20) Iversen, J. D., R. Greeley, J. B. Pollack and B. R. White, Simulation of Martian Eolian Phenomena in the Atmospheric Wind Tunnel, Space Simulation, NASA Spec. Publ. 336, 191-213 (1973).

(21) Greeley, R., J. D. Iversen, J. B. Pollack, N. Udovich and B. R. White, Wind Tunnel Studies of Martian Eolian Processes, Proc. Roy. Soc. London, A341, 331-360 (1974).

(22) Iversen, J. D., R. Greeley, B. R. White and J. B. Pollack, Eolian Erosion on the Martian Surface, Part 1, Erosion Rate Similitude, Icarus, 26, 321-331 (1975).

(23) Iversen, J. D., R. Greeley, B. R. White and J. B. Pollack, The Effect of Vertical Distortion in the Modeling of Sedimentation Phenomena: Martian Crater Wake Streaks, J. Geophys. Res., 81, 4846-4856 (1976).

(24) Iversen, J. D., Drifting Snow Similitude, J. Hydraulics Division, American Society of Civil Engineers, to be published (1979).

(25) Tabler, R. D., New Engineering Criteria for Snow Fence Systems, Transportation Research Record 506, 65-78 (1974).

LIST OF SYMBOLS

A_1	Bagnold's threshold parameter, Equation (1)
A_c	cross-sectional area of deposition region
A	plan area of deposition region
B	particle friction threshold Reynolds number, $u_{*_t} D_p/\nu$
C_D	drag coefficient
D_p	particle diameter
e	coefficient of restitution
g	gravitational acceleration
h	vertical reference length
ℓ	any horizontal length or saltation grain path length
L	horizontal reference length
L^*	atmospheric stability length (see Cermak [14])
n	length scale, L/L_m
q_s	mass transport rate in saltation, per unit lateral dimension
t	time
u	horizontal air velocity

U reference longitudinal air speed

U_∞ wind tunnel free stream speed

U_o reference or free stream speed at initiation of motion

U_F terminal speed

u_* friction speed, $(\tau/\rho)^{1/2}$

u_{*_t} threshold friction speed

V volume of deposition region

z_o equivalent roughness height

z'_o equivalent roughness height with saltating material

λ ripple wave-length

ν kinematic air viscosity

ρ air density

ρ_p particle density

Subscript

m model

LABORATORY SIMULATION OF TORNADIC WIND LOADS ON A CYLINDRICAL STRUCTURE

M. C. JISCHKE and B. D. LIGHT
University of Oklahoma, Norman, Oklahoma, USA

ABSTRACT

Ward's tornado simulator has been modified to study the interaction of tornadic flowfields with structures. Measurements with and without swirl have been made of the pressure on the surface of a cylindrical model placed in the simulator. The model, similar to a nuclear reactor containment building, is a circular cylinder with a hemispherical roof. Results for the surface pressure coefficient are presented for swirl angles of 0° and 45° with the model at two positions in the simulator: 1) the convergent zone, 2) at the boundary of the convergent and convection zone. The tornado-like vortex is observed to induce a circulation about the cylinder thereby reorienting the net force away from the imposed swirl angle.

INTRODUCTION

Increased concern for structural efficiency and safety has led engineers and architects to devote more attention to the interaction of the wind with structures. Vital structures, such as those housing nuclear reactors, must be able to withstand the forces of the wind and the impact of wind-generated missiles.

The preponderance of research in this growing field of wind engineering has been focused on the interaction of boundary-layer-type winds with structures. These studies have been conducted with scale models in meteorological wind tunnels and in the field with instrumented full-scale buildings . Yet it is well known that the most intense wind conditions in the atmosphere occur with vortical tornado flows. In contrast to boundary-layer-type flows where the vorticity is largely horizontal, vortical flows contain significant vertical as well as horizontal vorticity and thereby exhibit a greater complexity and richness. In spite of their importance, little information is available on the interaction of tornadic flows with structures. Tornadic wind loads, which are the subject of this paper, are currently estimated from boundary-layer-type wind results, completely ignoring the vortical nature of tornadic flows.

In this paper we shall describe the results of a laboratory simulation of the interaction of a tornado and a structure. The structure is a circular cylinder with a hemispherical roof. Pressure measurements have been conducted and are reported in the form of a surface pressure coefficient. In what follows we discuss the relevant characteristics of tornadoes and the tornado simulator used in this study, the scaling analysis of the flow and its simulation, the experimental set-up, and the results.

TORNADO CHARACTERISTICS

Tornadoes represent the most violent of nature's small-scale atmospheric vortices. A typical tornado vortex is believed to extend vertically to the tropopause (about 15 km in midlatitudes) and have a radial extent (e.g. radius of maximum winds) of a fraction of a kilometer -- although the tornadic storm likely draws upon the circulation of a much larger region (15 to 25 km in radius) for its angular momentum and vorticity. Reliable estimates of the maximum wind speed in tornadoes do not exceed 400 kph, although unsubstantiated claims of wind speeds in excess of 800 kph can be found. The pressure drop associated with 400 kph winds is approximately 1.2×10^6 N/m^2 and roughly suggests the magnitude of the wind loadings tornadoes place on structures.

Tornado occurrence in the atmosphere requires an ambient source of rotation and a mechanism to concentrate the associated angular momentum. The rotation source is generally believed to be the rotation of the earth although there are enough examples of anticyclonic tornadoes to indicate that this is not always the case. The concentration mechanism appears to be the low-level convergence accompanying a strong conditional instability that has gone unstable.

The life cycle of tornadoes is usually divided into three distinct stages: genesis, mature stage and decay -- the latter being the least well understood. Typical tornado lifetimes are a quarter to a half of an hour although examples of tornadoes and tornado-spawning storm systems lasting several hours can be cited. Translational velocities of tornadoes are of the order of 40 kph with some cases up to 125 kph. Damage paths range from a few tens of meters wide and a half kilometer long to such extremes as two kilometers wide and 500 kilometers long.

The maximum winds and damage occur during the mature stage of the life cycle. Thus, it is this mature stage that is to be simulated in the laboratory. A schematic sketch of the structure of a mature stage tornado is given in Figure 1 along with estimates of the dimensions of a worst case intense tornado of the Midwest variety. The sketch has been exaggerated for clarity and is not to scale.

The rapidly swirling region is in rough cyclostrophic balance and probably extends to the tropopause. The winds in this region are primarily aximuthal and vary as r^{-n} $(1/2 \leq n \leq 1)$ with radius r. The updraft region is a region of swirling ascending flow in which air from the surface inflow layer rises toward the tropopause, eventually spreading horizontally near the tropopause. Although the net mass flux is upward in this region, the most intense tornadoes show evidence of a downflow along the axis of the tornado suggesting a recirculating core which may or may not extend to the ground. The surface of maximum swirling wind lies close to the axis of the tornado. Assuming axisymmetry, the radius of this surface is typically of the order of a hundred meters for intense, mature stage tornadoes.

The surface inflow layer is a region in which the cyclostrophic balance is upset by the presence of the ground. The inflow layer thickness increases from essentially zero far from the axis to a value of the order of 300 meters in the vicinity of the tornado axis. Most structures are embedded in the surface inflow layer and it is these winds - especially near the core - which must be simulated in the laboratory. A sketch of the lower 1 km of the tornadic storm is given in Figure 2 to illustrate the nature of the flow to be simulated.

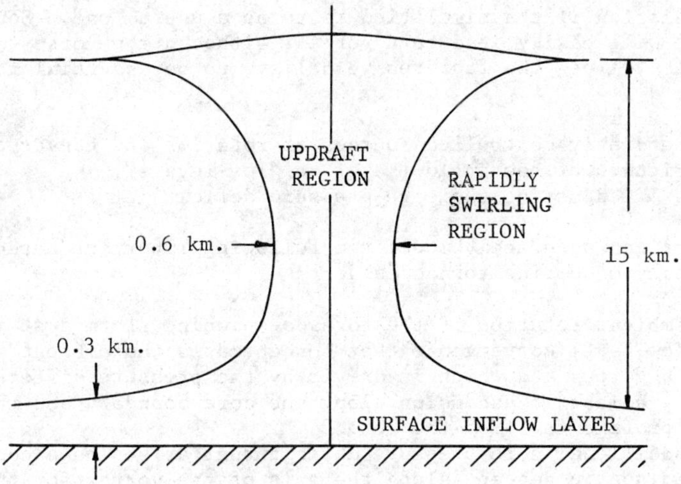

Figure 1. Schematic of mature stage tornadic storm

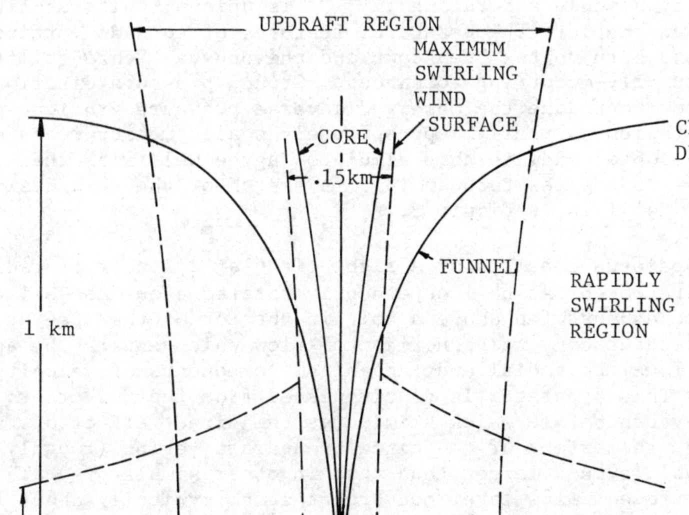

Figure 2. Schematic of lower 1 km. of mature stage tornado flowfield

Given the ease with which one can create a vortex in the laboratory, care must be exercised to ensure that the important features of natural tornadoes are reproduced in a simulation if the simulation is to be a useful one. Sources of rotation and convergence will easily lead to a vortex, although the vortex may not be typical of tornadoes. Among the features we believe to be essential to a meaningful simulation are:

1. Independently controlled sources of rotation and convergence,
2. Restriction of the inflow layer to "low-levels",
3. Means of adjusting the core pressure deficit.

In this way, one can accurately model the following important meteorological variables in naturally-occurring tornadoes:

1. The ambient rotation of the tornado-spawning storm system,
2. The low-level convergence that concentrates the ambient angular momentum into a vortex and which is driven by the pressure difference between the moist, adiabatic ascension along the core boundary and the ambient pressure profile far from the core,
3. The additional core pressure deficit that arises as a consequence of dry-adiabatic descent along the axis of the vortex for the case of the two-celled tornado vortex.

Additional features of a simulation which may have an effect on the modelling of wind loads on structures are surface roughness and the linear translation of the tornado vortex.

Among the tornado simulators that have been constructed, the design by Ward (1), which was used by Jischke and Parang (2), is unique in its ability to model tornadic winds and capture the essential features of tornado vortices. It, for example, exhibits both vortex breakdown and the unusual "core-splitting" phenomenon observed in naturally-occurring tornadoes. Ground pressure distributions obtained in this simulator reproduce the observed adverse pressure gradient near the edge of the updraft region. Finally, appropriately scaled measurements of the radius of maximum winds obtained with this simulator agree well with that observed by Hoecker for the 1957 Dallas tornado (3). A sketch of the Ward design, modified for this study, is given in Figure 3.

Briefly, the apparatus consists of a right circular cylinder of radius 122 cm with sides of meshwire which can be independently rotated about the axis of the cylinder. A variable speed exhaust fan above a hole in the top of the apaparatus creates an updraft and attendant converging horizontal flow which enters the apparatus through the meshwire. Thus the radial and tangential components of velocity can be varied independently. This apparatus is distinguished from that of others in that, first, there is a honeycomb baffle which eliminates the direct effect of fan-induced vorticity. Second, the extent of the circular updraft region (roughly half the radius of the apparatus) is much larger than that in other simulators. Third, the aspect ratio of the surface inflow layer convergent zone (typically about 1/8) is smaller than that in other experiments. Fourth, the large convection zone above the converging surface inflow layer reduces the effect of teacuplike secondary flows observed in other simulators.

SCALING ANALYSIS

Accurate simulation of tornadic wind loadings requires that the important nondimensional parameters characterizing the laboratory simulation be the same as those characterizing naturally-occurring tornadoes. Requiring equal value of these parameters allows the scaling associated with the simulation to be deduced. To proceed,

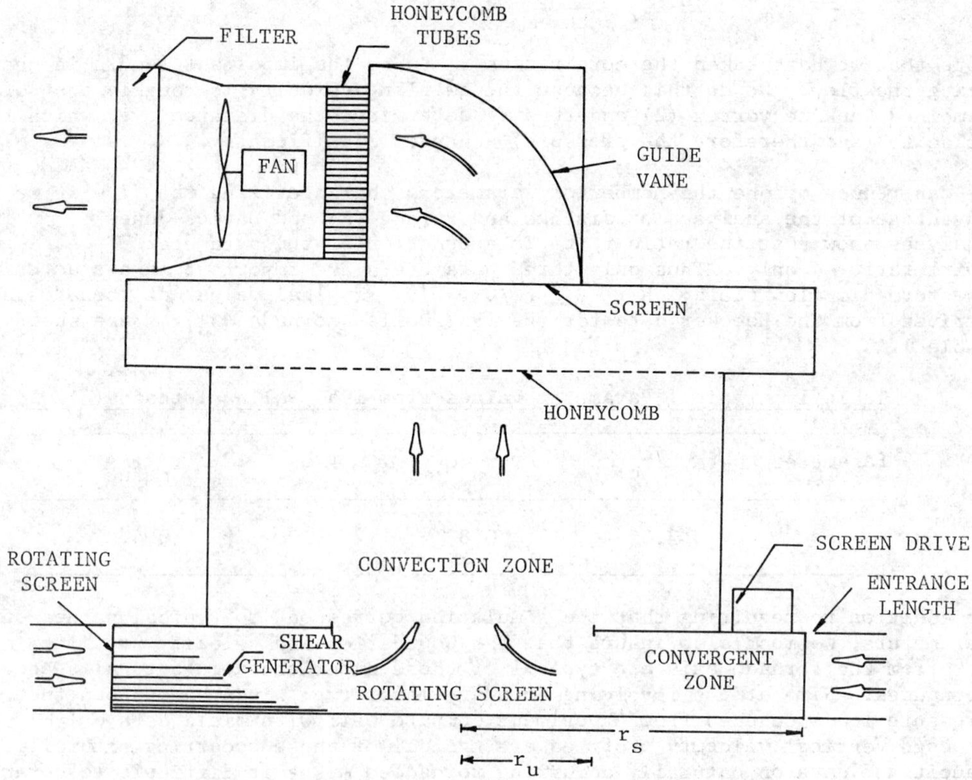

Figure 3. Schematic of tornado simulator

we must first determine the important parameters characterizing the tornado flow field.

Consider the flow due to a tornado vortex near the planar surface (z=0). We shall assume the tornado vortex is axisymmetric with axis along r=0. As suggested by our earlier discussion of the nature of tornadic wind fields, we consider the tornadic flow to consist of a low-level converging swirling inflow region and a central region of updraft characterized by the following seven parameters:

Q	volume flow rate
Γ	imposed circulation far from the axis
h	depth of the inflow layer
r_u	radius of updraft region
r_c	radius of the tornado core
r_m	radius of maximum swirling wind
ν	kinematic viscosity

The additional effects of heat transfer and thermodynamics are believed to be unimportant. We are also ignoring the effects of relative motion between the tornado and the structure. In the laboratory simulation of the intense, mature-stage tornado, $r_c = r_m$. Dimensional analysis then implies that there are four nondimensional parameters that characterize the tornadic flow field. We choose to write these parameters as

$$\frac{h}{r_c}\ ,\quad \frac{r_u}{r_c}\ ,\quad \frac{\Gamma\, r_u}{Q}\ ,\quad \frac{\Gamma}{\nu}\ .$$

Note that we have taken the core radius r_c to be the important length in characterizing the flow. We do this because the swirling velocity is roughly that of a combined Rankine vortex (2) and thus r_c determines the distance over which the velocity (and therefore the pressure) changes significantly.

We can reduce by one the number of parameters characterizing the flow if we take advantage of the analyses of Jischke and Parang (2) and Davies-Jones . These analyses show that the ratio r_u/r_c is a function of the parameter $\Gamma\, r_u/Q$ -- the swirl ratio -- only. Thus only three parameters are important in characterizing the tornadic flow field: h/r_c , $\Gamma\, r_u/Q$, Γ/ν. Typical values of these parameters, derived from the Hoecker data for the 1957 Dallas tornado (3), are shown in Table 1.

Table 1. Typical Parameter Values from 1957 Dallas Tornado (3)

Parameter	h/r_c	$\Gamma\, r_u/Q$	Γ/ν	r_c/r_u
Value	3.0	0.8	2×10^9	0.4

In addition to requiring that the simulation correspond to typical values of these parameters, we must also insure that the boundaries and velocity profiles imposed far from the tornado axis are typical of those experienced in naturally-occurring tornadoes. Simulations involving boundary-layer-type winds have shown that the pressure loads deduced from experiment compare well with field data only if the imposed vertical velocity profiles are typical of those occurring naturally. The paucity of data on naturally-occurring tornadoes makes it difficult to determine which velocity profiles are most appropriate for this case. As a consequence, we shall rely on the data presented by Cermak for naturally-occurring turbulent wind profiles for surface roughnesses typical of grasslands and forest-type topographies (e.g. roughness heights of the order of 1 to 10 meters). The associated turbulent velocity profile is assumed to be of the power law type,

$$\frac{V}{V_{max}} = \left[\frac{Z}{Z_{max}}\right]^{1/\alpha} \tag{1}$$

with α ranging from 3 to 7. In view of the small size of our simulator, it has been necessary to tailor the imposed velocity profile at the rotating screen to match this assumed form. The surface roughness scale has been chosen to represent that of grassland and forest-type topographies. Thus if Z_r denotes the roughness height, we require

$$\left[\frac{Z_r}{r_c}\right]_{model} = \left[\frac{Z_r}{r_c}\right]_{tornado} \tag{2}$$

The roughness height typical of grassland topographies is 1 meter. In the Dallas tornado of 1957, the core radius was deduced by Hoecker to be about 50 meters. Thus (Z_r/r_c) is expected to be approximately 0.02. For forest-type topographies for which Z_r is as large as 10 meters, (Z_r/r_c) approaches 0.2. By tailoring the velocity profile to represent that of a typical naturally-occurring turbulent boundary layer and roughening the simulator floor, we believe that the influence of the Reynolds number Γ/ν on the flow will be small.

Thus, ignoring the Reynolds number Γ/ν as unimportant, there are three essential parameters that characterize the tornadic wind fields: h/r_c, $\Gamma r_u/Q$, Z_r/h. Values for these three parameters typical of naturally-occurring tornadoes can be duplicated in the University of Oklahoma tornado simulation facility. Having evaluated the tornado flow field scaling, the scaling for the structure then follows from

$$\left[\frac{L}{r_c}\right]_{\text{model}} = \left[\frac{L}{r_c}\right]_{\text{tornado}} \tag{3}$$

where L is a typical structure dimension.

EXPERIMENTAL SET-UP

The experimental conditions are chosen to duplicate the parameter values given in Table 1. The imposed swirl angle at the screen, ϕ, is given by

$$\tan\phi = \frac{V_\theta}{V_r} = \frac{\Gamma r_u}{Q}\frac{h}{r_u} \tag{4}$$

Using the parameter values given in Table 1, we find $\phi \approx 45°$. The size of the simulator restricts the depth h of the inflow layer that can be achieved. We have chosen h = 0.508 m so as to obtain model sizes that will allow adequate visualization and resolution without significant flow blockage. With this value of h, we find that the updraft radius r_u is 0.424 m and the vortex core radius r_c is 0.170 m. Taking the relative roughness height for the naturally occurring tornado to be 0.02 corresponding to grassland topographies, the simulator surface roughness height becomes 0.0034 m. These results then define the geometry of the simulator.

The vertical shear of the imposed velocity at the rotating screen has been achieved by varying the porosity of the screen with height, according to the result derived by Cockrell and Lee . The variable porosity is obtained by affixing thin horizontal strips of masking tape to the rotating screen. The vertical spacing between the strips is varied so as to yield various powers $(1/\alpha)$ in the power-law velocity variation given by Eq. (1). Figure 4 shows a typical experimental result for which $\alpha \sim 5.9$. This screen was used in the measurements to be described subsequently.

The model used in these experiments is a circular cylinder with a hemispherical roof. This geometry is typical of containment buildings used to house nuclear reactors. Typical containment buildings are 34 meters in diameter and 65 meters high. With a typical tornado core radius of 50 meters corresponding to the simulated vortex radius, the dimensions of the scaled containment building model then become 0.222 m high and 0.114 m in diameter. A drawing of the model is shown in Figure 5 along with the location of the 63 surface pressure ports on the surface of the model.

The 63 surface pressure ports are connected to a single pressure transducer by means of a Scanivalve switching system. The dynamic pressures to be measured are of the order of a few torr and thus require a sensitive pressure transducer. A Datametrics 570 D 10 T-2A1-V3 capacitance type pressure sensor and 1174-A4A-5A1-A78 electronic manometer with resolution of 10^{-4} torr has been employed for this purpose. Because of the turbulent fluctuations in the flow, the output signal from the manometer is electronically averaged to give the time-averaged surface pressure. The resulting data are presented in the form of a pressure coefficient, C_p, where

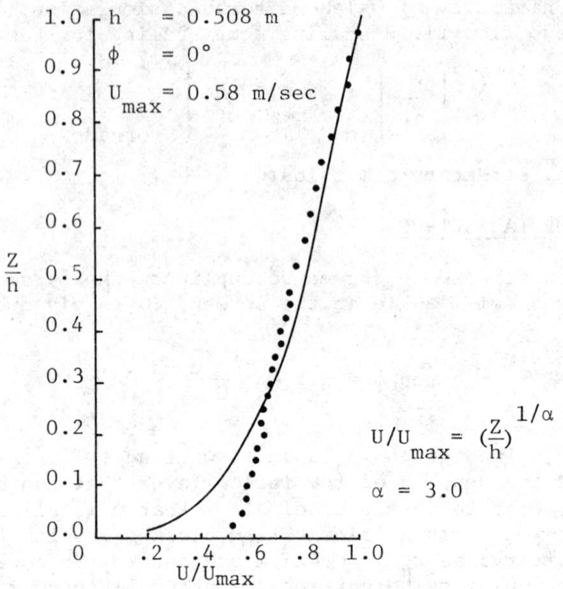

Figure 4. Sheared velocity profile (rough surface)

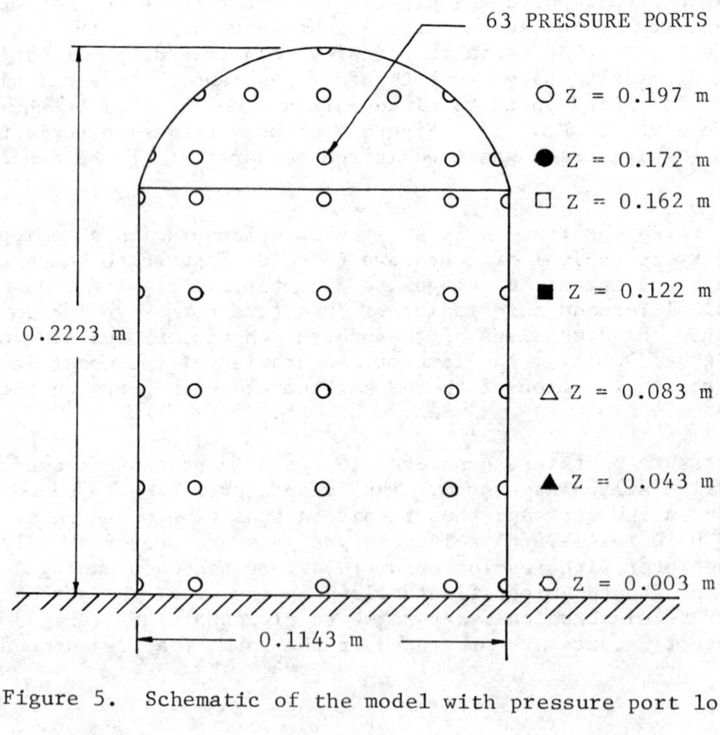

Figure 5. Schematic of the model with pressure port locations

$$C_p = \frac{p - p_{ref}}{\frac{1}{2} \rho V^2_{ref}} \quad (5)$$

Here p_{ref} is the pressure that would be measured on the ground (z=0) at the center of the model, in the absence of the model. The reference velocity V_{ref} is the vertically averaged velocity at the center of the model, in the absence of the model, and is given at any station r in terms of the volume flow rate Q by $V_{ref} = Q/2\pi rh\cos\phi$.

RESULTS

Results have been obtained for two model locations (r = 0.457 m, 0.762 m) and two different swirl angles (ϕ = 0, 45°). The location 0.457 m corresponds to the boundary between the convergent and convection zones. The location r = 0.762 m corresponds to the convergent zone. Figures 6-9 show results for the surface pressure coefficient at various heights z on the model. These data have an uncertainty of ten percent or less.

Figure 6 shows the surface pressure coefficient with zero imposed swirl and the model in the convergent zone where the vertical velocity is zero. The measured pressure coefficient is symmetrical with respect to the flow direction. The variation of C_p with angle θ around the cylinder on the circular cylinder portion of the model are quite like those for an infinite circular cylinder in a uniform flow. The shear flow does cause a slight vertical variation in the pressure coefficient along the line θ = 0° suggesting a stagnation point near z = 0.162 m.

Figure 7 shows the effect of swirl on the surface pressure coefficient when the model is in the convergent zone. The pressure distribution is highly asymmetric with respect to the local flow direction. As a consequence, the force on the model has both drag and side force components, reckoned with respect to the local wind. The maximum pressure coefficient ($C_{p_{max}}$ = 0.68) does not change significantly with

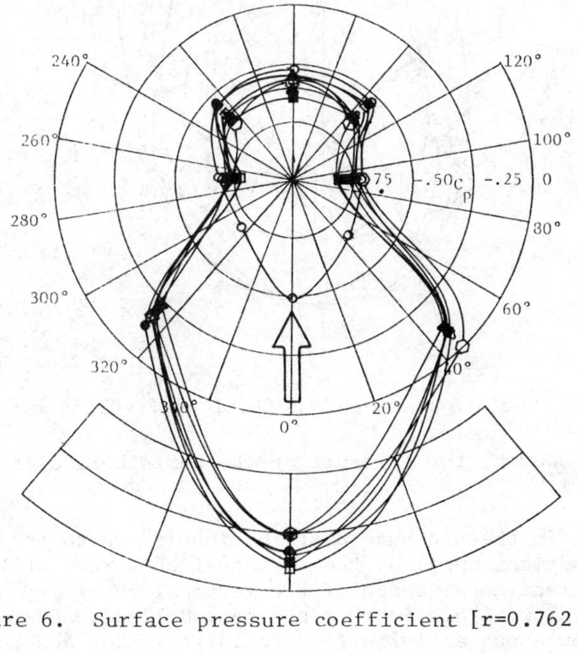

Figure 6. Surface pressure coefficient [r=0.762 m, ϕ=0°]

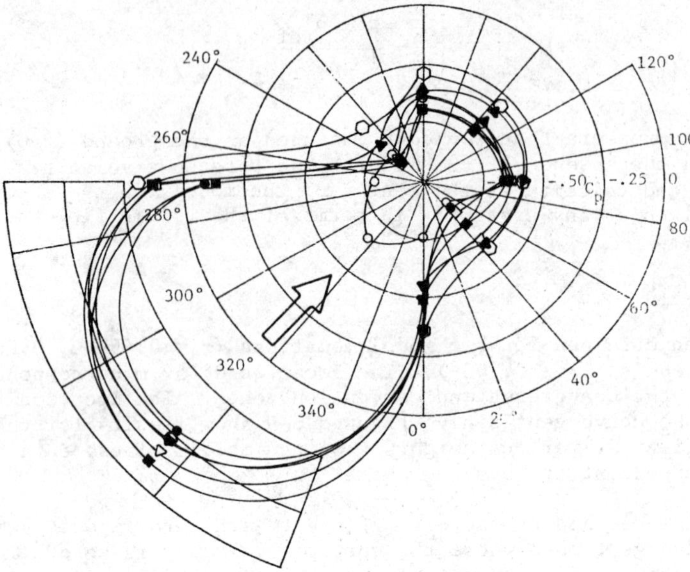

Figure 7. Surface pressure coefficient [r=0.762 m, ϕ=45°]

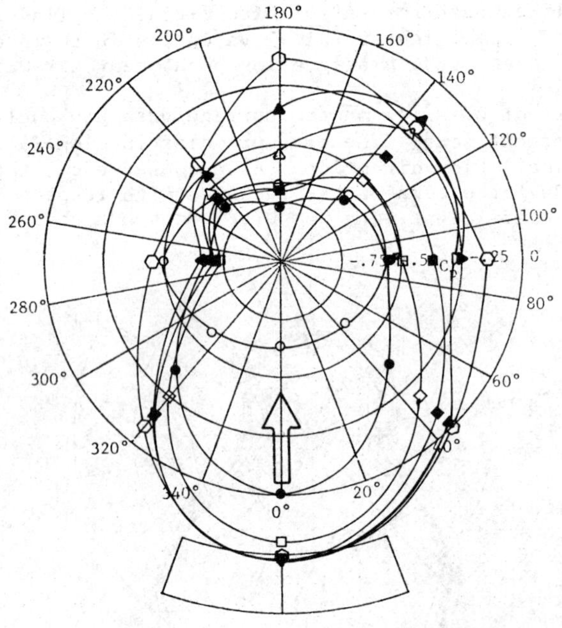

Figure 8. Surface pressure coefficient [r=0.457 m, ϕ=0°]

swirl in this case. However, the pressure on the hemispherical roof is considerably reduced with swirl.

The observed asymmetry in the pressure distribution is due to the circulation induced around the cylindrical model by the tornado-like vortex at the center of the apparatus. It is a direct consequence of the vertical vorticity present in the tornado-like vortical flow. This feature can be modelled theoretically by considering the ideal flow about an infinite circular cylinder due to a combined sink and vortex.

Figures 8 and 9 give the surface pressure coefficient for $\phi=0$ and $\phi=45°$, respectively, when the model is placed at 0.457 m from the center. The vertical velocity in the convection zone has begun to substantially alter the pressure coefficient and its variation with height z and angle θ. The effect of the imposed swirl ϕ remains similar to that at 0.762 m in that a circulation is induced about the cylinder causing an asymmetry (with respect to the undisturbed flow direction) in the distribution of C_p with angle θ.

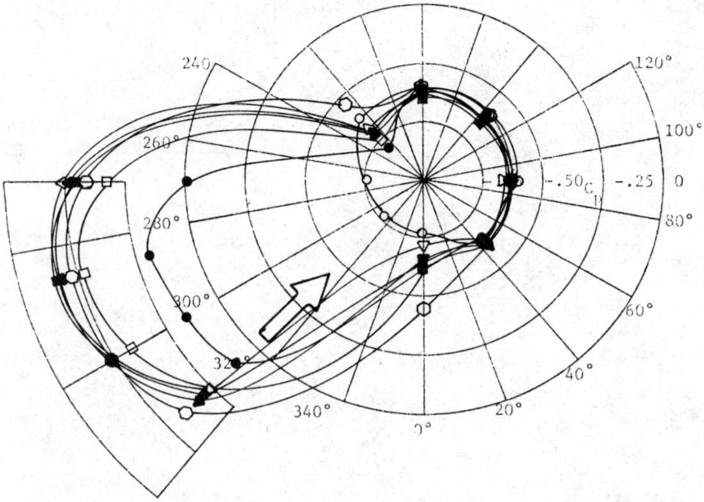

Figure 9. Surface pressure coefficient [r=0.457 m, ϕ=45°]

CONCLUSIONS

The surface pressure coefficient on a model of a cylindrical containment building in the presence of a tornado-like vortex flow has been measured experimentally. When the model is in the convergent flow region where the vertical velocity vanishes, the experimental results are quite like those for an infinite circular cylinder with circulation. Near the boundary of the convection and convergent zones, the results are not as simply interpreted although in both cases, the effect of the vertical vorticity of the tornado-like vortex is to induce a circulation about the model which leads to an asymmetric pressure distribution. As a consequence, the force on the building has both a drag and a side force component.

ACKNOWLEDGEMENTS

This work was supported by the Nuclear Regulatory Commission under Contract NRC-04-78-207. The authors are pleased to acknowledge the encouragement of Dr. Robert Abbey of the Nuclear Regulatory Commission. We also wish to record our appreciation to Mr. Fred Leslie and Mr. Farid Moslehi of the University of Oklahoma for their assistance in using the tornado simulator.

REFERENCES

(1) N.B. Ward, The exploration of certain features of tornado dynamics using a laboratory model, Journal Atmospheric Sciences, 29, 1194 (1972).
(2) M.C. Jischke and M. Parang, Properties of simulated tornado-like vortices, Journal Atmospheric Sciences, 31, 506 (1974).
(3) W.H. Hoecker, Wind speed and air flow patterns in the Dallas tornado of April 2, 1957, Monthly Weather Review, 88, 167 (1960).

AEROELASTIC MODELLING OF PNEUMATIC AND TENSIONED FABRIC STRUCTURES

B. V. TRYGGVASON
The University of Western Ontario, London, Ontario, Canada

SUMMARY

The requirement for modelling the elastic properties of tensioned surfaces is examined and shown to be unnecessary in aeroelastic modelling of most tensioned structures. Simplifications arising from combining the scaling of internal pressurization and Froude number scaling for air-supported structures are illustrated and the importance of scaling the acoustic stiffness of volumes enclosed by such structures is emphasized.

INTRODUCTION

In recent years there has been an increase in the number of light weight tensioned structures such as air-supported roofs and prestressed cable nets. Although tensioned surfaces have actually been used for small scale structures for many centuries, their use is now being extended to very large structures. Examples include air-supported roofs for several sports stadiums in the United States, the tensioned roof for the Olympic Stadium in Montreal, and by far the largest to date, the tensioned roof planned for an airport terminal building in Saudia Arabia which will cover approximately a one square kilometer area.

The engineering design of tensioned structures is quite complex and relies very heavily on computer analysis, not only for determining stress levels, but also for determining the shapes of the surfaces developed. The surfaces are in almost all cases three-dimensional and quite sophisticated techniques are required for their analysis. Furthermore, in comparison to other types of structures, there is a lack of widespread experience and expertise for designing large scale tensioned structures and little guidance is available from codes. One of the problems designers have to consider is the question of the response of these structures to wind loads. Relatively little work has been directed to developing criteria for wind loading for such structures.

Over the past few years the Boundary Layer Wind Tunnel Laboratory (BLWTL) at The University of Western Ontario has been involved in several studies of wind effects on air-supported and tensioned structures. Rigid model studies have been carried out to determine actual wind loads in order to provide designers with appropriate patterns to be used as input loads for computer models. Aeroelastic studies have been carried out to directly determine the response of these structures to wind loading.

The rigid model studies have been relatively straightforward as techniques for these have been developed over many years of testing more conventional structures. However, the aeroelastic studies have presented several problems stemming largely from the difficulties inherent in modelling very light structures. In designing the aeroelastic models it has been found that it is very difficult, and in certain cases seemingly impossible, to satisfy all the usually necessary and adhered to similarity conditions. This has led to exploring theoretically and numerically the importance of some of the similarity conditions. The results of these investigations indicate that some of the scale requirements can be relaxed or modified.

To provide a context from which these simplifications can be viewed an outline of the standard scaling parameters encountered in aeroelastic studies will be given first. Some of the standard modelling parameters are then reformulated to better illustrate how they arise in tensioned structures and the simplifications that are possible. The application of these simplifications is illustrated using several recent wind tunnel studies.

AEROELASTIC SCALING REQUIREMENTS

The scale parameters used in aeroelastic modelling are based essentially on preserving similarity between the various forces that act on and within a structure. As long as the ratios of forces are similar in both full scale and model scale, the model will simulate the response of a full scale structure. In modelling wind effects, the basic driving forces arise through the pressures exerted by the wind, and the resisting forces usually arise from the elastic properties of the structures. Preserving the ratios of the forces leads to relationships between velocity, length, mass and force scaling as summarized in Table 1.

EXAMINATION OF THE NECESSITY FOR SCALING ELASTICITY IN TENSIONED STRUCTURES

When the similarity requirements listed in Table 1 are applied to modelling tensioned membranes or cable nets, reformulation of some of the parameters is advantageous. The mass scaling for structures that are essentially surfaces reduces to scaling the mass per unit area as follows -

$$\lambda_\mu = \frac{\mu_m}{\mu_{fs}} = \lambda_L \tag{1}$$

where μ is the mass per unit area; λ_L is the length scale; and subscripts m and fs refer to model and full scale respectively. This follows from the relation between mass scaling and length scaling and the requirement to scale surface areas according to $\lambda_L{}^2$. Scaling of elastic forces in membranes reduces to scaling the elastic forces per unit length in the membrane as follows -

$$\lambda_{Eh} = \lambda_V{}^2 \lambda_L \tag{2}$$

where h is the thickness of the membrane. For panels, where the restoring forces arise from bending moments, the elastic scaling requirement reduces to scaling the product of the elasticity and the moment of inertia -

$$\lambda_{EI} = \lambda_V{}^2 \lambda_L{}^4 \tag{3}$$

For modelling the dynamic behaviour of panels it is important to satisfy both equations (1) and (3). For modelling membranes, satisfying both (1) and (2) would also be ideal but has, in fact, proved to be quite difficult and sometimes virtually impossible since both the mass per unit area and the elastic stiffness, which is

TABLE 1. Aeroelastic Scale Parameters

Parameter	Scaling	Remarks
Turbulence and Velocity Profile Length Scale	$\lambda_L = \frac{L_{tm}}{L_{tfs}} = \frac{L_{pm}}{L_{pfs}}$	Set by boundary layer developed
Velocity	$\lambda_V = \frac{V_m}{V_{fs}}$	Usually chosen
Density	$\lambda_\rho = \frac{\rho_m}{\rho_{fs}}$	1.0
Model Length Scale	$\lambda_L = \frac{L_m}{L_{fs}}$	λ_L
Mass	$\lambda_M = \frac{(\rho L^3)_m}{(\rho L^3)_{fs}}$	$\lambda_L{}^3$
Pressure	$\lambda_p = \frac{(1/2\rho\ V^2)_m}{(1/2\rho\ V^2)_{fs}}$	$\lambda_V{}^2$
Forces	$\lambda_F = \frac{(1/2\rho\ V^2\ L^2)_m}{(1/2\rho\ V^2\ L^2)_{fs}}$	$\lambda_V{}^2\ \lambda_L{}^2$
Elastic Forces	$\lambda_{EA} = \frac{(EA)_m}{(EA)_{fs}} = \lambda_F$	$\lambda_V{}^2\ \lambda_L{}^2$
Time Scale	$\lambda_t = \frac{t_m}{t_{fs}} = \frac{(L/V)_m}{(L/V)_{fs}}$	λ_L/λ_V
Froude Number (Gravitational Forces)	$\lambda_V{}^2 = \lambda_L$	Usually not satisfied
Reynolds Number (ratio of inertial to frictional forces in air)	$R_m = R_{fs}$	Usually not satisifed

Note: subscripts m and fs refer to model and full scale respectively

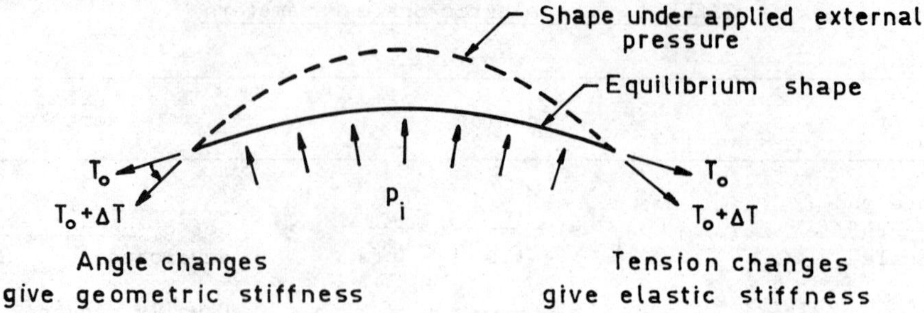

Figure 1. Representation of the mechanisms that give rise to the geometric and elastic stiffnesses in a pneumatic structure.

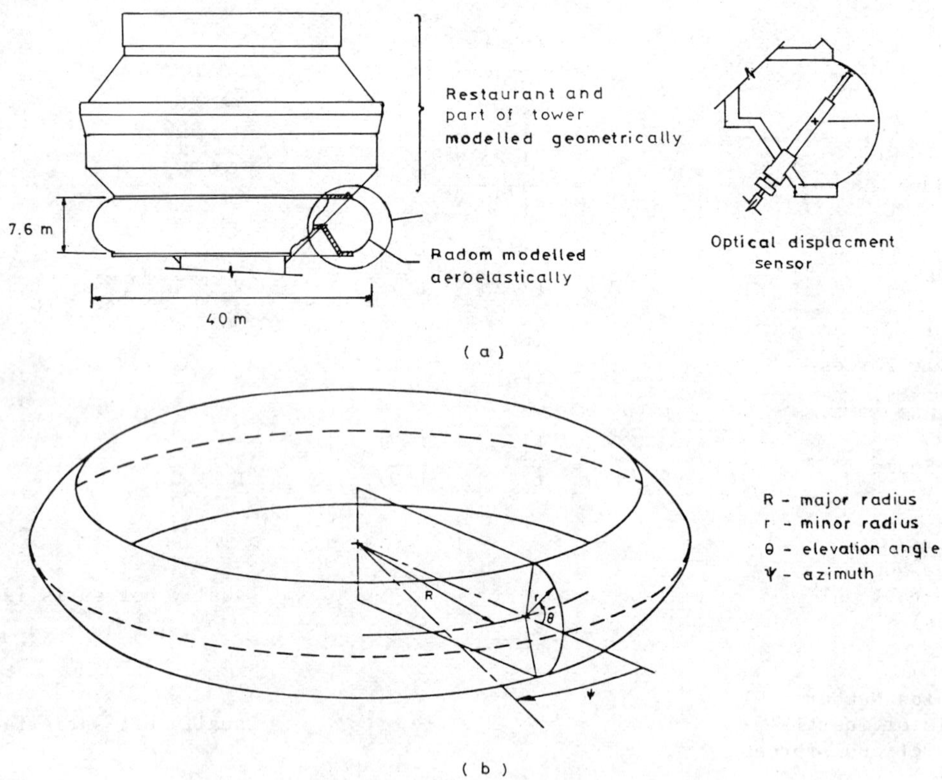

Figure 2. Aeroelastic model of the air-supported microwave radome on the CN Tower.
a) Sketch of the model and instrumentation;
b) View of the membrane surface with co-ordinate definitions.

characterized by the product Eh, are functions of the membrane thickness. As it is quite important to scale the inertial forces correctly, the need for satisfying elastic scaling for membranes has been examined and shown to be unnecessary in all cases investigated.

For curved tensioned surfaces two restoring forces arise when the surface is displaced from its static equilibrium position. The first is due to the change in geometry of the surface and the second is due to an increase in tension in the surface because of additional strain in the material. The corresponding stiffnesses are referred to as geometric and elastic stiffnesses respectively. The mechanisms for these restoring forces are illustrated in Fig. 1 for an air-supported membrane. The illustration also holds for a doubly curved surface where two opposing tensions define the surface.

If elastic properties are scaled according to equation (2) the elastic stiffness will be modelled correctly. To model the geometric stiffness requires modelling the surface shape and tension. The tension scaling in surfaces is related to the velocity and length scale as follows -

$$\lambda_{TS} = \lambda_V^2 \lambda_L \tag{4}$$

and the tensions in cables are scaled as -

$$\lambda_{TC} = \lambda_F = \lambda_V^2 \lambda_L^2 \tag{5}$$

As there are two restoring forces in tensioned structures the relative contribution of each can be examined and the possibility for relaxing or modifying the scale requirement arises. This is examined further through specific examples.

The first study in which difficulty was encountered in scaling elasticity was an aeroelastic study of the radome mounted immediately below the restaurant level of the CN Tower in Toronto (1). The radome is an air inflated toroidal surface with a major radius of approximately 20 m (65 ft.) and a minor radius of 4 m (13 ft.) as illustrated in Fig. 2. As the internal pressure was quite high, Froude number scaling was not required and the pressurization scale was related simply to the velocity scale. With the velocity scale used, correct elastic scaling of the surface was not achieved. The model surface was too stiff by a factor of approximately 2.3. To determine whether this was significant, a numerical model was developed of a strip of the surface and used to examine the significance of the error in elastic scaling.

The results are illustrated in Figs. 3a and 3b. Fig. 3a shows simply the static location of a point on the surface and the initial angle of the surface as functions of the internal pressure and elastic properties. For the range of internal pressures of interest, the error in the elastic scaling had very little effect on the static shape of the membrane and suggests that the error in scaling is insignificant. This is further supported in Fig. 3b which summarizes the results of applying the measured pressure distribution onto the outer surface of the radome and computing the deflected shape as a function of the wind speed for both correctly and incorrectly scaled elasticity. The graph indicates that the effect of the error accepted in the elastic scaling is very small. The non-linear behaviour of the membrane indicated in Fig. 3b was in agreement with the behaviour determined from the aeroelastic tests and is not related to elastic properties.

In another study (2), which examined the wind induced motions of the air-supported roof of the sports complex recently completed at the University of Dalhousie in Halifax, a similar fairly simple analysis indicated that elastic scaling was, again, not required.

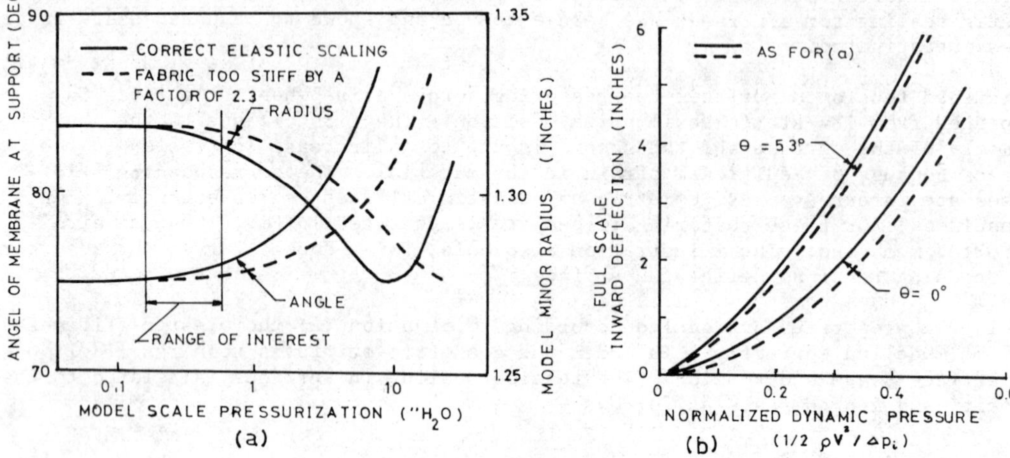

Figure 3. Effect of the elastic properties of the membrane on: a) the static geometry; b) the wind induced deflections.

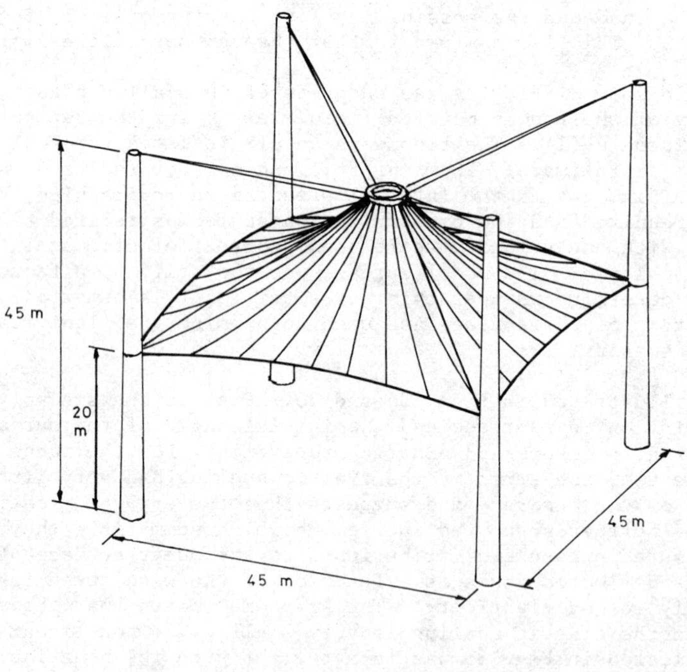

Figure 4. View of a single unit of the tensioned fabric roof planned for the airport terminal at Jeddah airport. The full scale roof will be comprised of 210 similar units.

A more general analysis of the requirement for satisfying elastic scaling of tensioned structures was carried out in conjunction with a recent study of a tensioned fabric roof for an airport terminal building (3).

The size of the structure is rather immense, with plan dimensions of approximately 800 meters square. The lowest level of the roof is 20 meters (66 ft.) above ground and the highest level is 35 meters (115 ft.) above ground. A view of one unit of the roof surface is shown in Fig. 4. The full roof is comprised of 210 similar units. An aeroelastic model was built of part of the structure, including nine units arranged in a 3 x 3 matrix. Again, it was not possible to satisfy elastic scaling of the fabric surface.

An analysis based on energy considerations indicated that the elastic properties of many of the cables and the surface did not require simulation as the geometric stiffnesses were much more significant than the elastic stiffnesses. To determine whether the elastic properties of membranes in general require modelling, an analysis of a non-linear membrane was carried out. Including the non-linear term which arises from the elastic properties of the membrane, to first order, and neglecting damping gave the following equation of motion in the first mode for a unit length of a long narrow membrane.

$$\ddot{Q} + \omega_o^2 Q + \Omega_o^2 \left(\frac{Q}{L}\right)^2 Q = F \qquad (6)$$

where

$$\Omega_o^2 = \frac{Eh}{\mu L^2} \quad \text{and} \quad \omega_o^2 = \frac{\pi^2 T_o}{\mu L^2} \qquad (7)$$

and Q is the generalized co-ordinate for displacement in the first vibration mode; F is the generalized force; ω_o is the frequency for the corresponding linear membrane; E is the elasticity of the membrane; h is its thickness; μ is its mass per unit area; L is its width; and T_o is the initial tension in the membrane.

A solution for the corresponding homogeneous equation was obtained using Lindstedt's perturbation method (4). To second order the expansion obtained for the natural frequency is as follows -

$$\Omega = \omega_o \sqrt{1 + \frac{3}{4}\left(\frac{\Omega_o}{\omega_o}\right)^2 \left(\frac{A}{L}\right)^2} \qquad (8)$$

where A is approximately equal to the amplitude of vibration.

Substituting the expressions for Ω_o and ω_o in the term under the radical then gives

$$\Omega = \omega_o \sqrt{1 + \frac{3}{4}\frac{Eh}{\pi^2 T_o}\left(\frac{A}{L}\right)^2} \qquad (9)$$

indicating that the natural frequency for the non-linear membrane is equal to that of the linear membrane multiplied by a factor that depends on the elastic properties of the membrane, its tension, and the ratio of the amplitude of deflection to the span. For the membrane in the terminal roof, where $Eh = 3.5 \times 10^5$ Nt/m (2,000 lb/in), $T_o = 1.05 \times 10^4$ Nt/m (60 lb/in) and using 0.1 as a conservative estimate for the ratio of deflection to span, the non-linear term increases the natural frequency by only 1.3%. If the elasticity in a model is exaggerated by a factor of 10, the increase in frequency is 12%. Thus, a large error in elastic scaling results in a minimal error in the frequency. Although this error can be eliminated by adjusting the tension in the model membrane, this step is hardly warranted by the small magnitude of the error. In the case of the terminal roof the stiffness of the model membrane was in fact too high by approximately a factor of 5, indicating that the model frequencies would have been in error by only 6% given the above estimate for the deflections. The measured deflections were in fact an order of magnitude less than the above estimate and the error in frequency due to non-linear effects was therefore less than 0.1%. The above indicates that very large errors in elastic scaling of tensioned surfaces can be tolerated, without incurring significant errors in the simulated response.

PRESSURIZATION AND VOLUME SCALING FOR PNEUMATIC STRUCTURES

In pneumatic structures the tension in the surface is determined by geometry and the net pressure across the surface, and achievement of tension scaling therefore requires that the pressurization be scaled. The first natural attempt is to scale this according to the pressure scale listed in Table I, namely

$$\lambda_{\Delta p_i} = \lambda_V^2 \qquad (10)$$

where Δp_i is the pressurization. Scaling the internal pressure in this fashion gives the correct scaling for the tension in the membrane and is sufficient for structures where the pressurization is very much larger than the weight per unit area of the surface. In such cases, Froude number scaling is unnecessary.

However, in air-supported roofs the pressurization is typically comparable to the weight per unit area of the surface, implying that Froude number scaling cannot be ignored. To scale Froude number exactly leads to experimental difficulties as it affects the velocity scaling and dictates that very low speeds be used for testing. Signals are thus very drastically reduced and experimental error increases. This detrimental effect can be overcome by combining the pressurization scaling and Froude number scaling as follows -

$$\frac{(\Delta p_i - \mu g)_m}{(\Delta p_i - \mu g)_{fs}} = \lambda_V^2 \qquad (11)$$

where g is the gravitational acceleration.

For very light structures this reduces to a simple scaling of the pressurization; for structures where the pressurization is very small, which in the limit reduce to hanging structures, this scaling reduces to Froude number scaling. The ability to combine these scale parameters is very fortunate as it allows the selection of reasonable velocity scales. An additional advantage to using reasonably high wind speeds for testing pneumatic structures is outlined below.

For pneumatic structures, and to some extent for those that partially enclose a volume, one further scale requirement arises to complicate matters. This is the requirement to scale the dynamic component of the internal pressure. Deflection of the surface enclosing a volume leads, through the equation of state, to changes in the internal pressure. Fractional changes in the internal volume can lead to pressure changes that are orders of magnitude larger than the static pressurization as volume changes lead to fractional changes in the absolute pressure, which is the sum of the pressurization and the ambient pressure. Volume changes give rise to a third stiffness term for air-supported structures, referred to as acoustic stiffness, which can be very much larger than both the geometric and elastic stiffnesses.

Considering the equation of state leads to a volume scaling which is related to the length and velocity scales as follows -

$$\lambda_{\nu} = \frac{\lambda_L^{\,3}}{\lambda_V^{\,2}} \tag{12}$$

where ν is the volume enclosed by the membrane and supporting structure. This modification to the internal volume scaling arises from the requirement to scale the acoustic stiffness of the enclosed volume and gives correct scaling of the dynamic component of the internal pressure. Equation (12) shows that an exaggeration of internal volumes is required. If Froude number scaling is adhered to this exaggeration is very large and model construction becomes very difficult. Use of reasonable velocity scales as allowed by equation (11) reduces the volume exaggeration required.

The importance of scaling enclosed volumes according to equation (12) is illustrated for the study of the air-supported roof for the sports complex at Dalhousie University. Fig. 5 shows the chamber mounted below the model to achieve correct scaling of the acoustic stiffness, and the very large exaggeration required for the internal volume is evident. In this case a velocity scale of 1:4 was used. If a velocity scale based on the standard scaling of Froude number would have been necessary, the volume would have had to be exaggerated by another factor of 16, which would have made construction of the model practically impossible. Combining the scaling of the internal pressure and membrane mass avoided this prohibiting complication. It should be noted that the mass scaling for the membrane is still based on equation (1), and the mass term in the numerator of equation (11) is negligible. In practice equation (11) reduces to scaling the ratio of model pressurization to the net pressure acting across the full scale membrane.

EFFECT OF ENCLOSED VOLUMES IN PNEUMATIC STRUCTURES

A theoretical analysis was carried out to ascertain the effect of the enclosed volume and details of this analysis are presented in reference 5. Only the results are presented here. An expression was obtained for the natural frequencies of a square membrane with properties similar to those of the Dalhousie sports complex. The expression is as follows -

$$\Omega_{rs} = \omega_{rs} \sqrt{1 + \frac{K'_{rs}}{K_{rs}}} \tag{13}$$

where Ω_{rs} is the natural frequency including the acoustic stiffness; ω_{rs} is the natural frequency for the membrane in a vacuum; K'_{rs} is the generalized acoustic stiffness; K_{rs} is the generalized geometric or membrane stiffness; and the subscripts (r,s) denote the mode shape. For a rectangular membrane r and s indicate the number of half sine waves in each of two orthogonal directions. The acoustic and geometric stiffnesses are given as follows -

$$K'_{rs} = \begin{cases} \dfrac{16}{\pi^4} \dfrac{p_o}{\nu_o} \left(\dfrac{ab}{rs}\right)^2 & \text{for both r,s odd} \\ 0 & \text{for either r,s even} \end{cases} \tag{14}$$

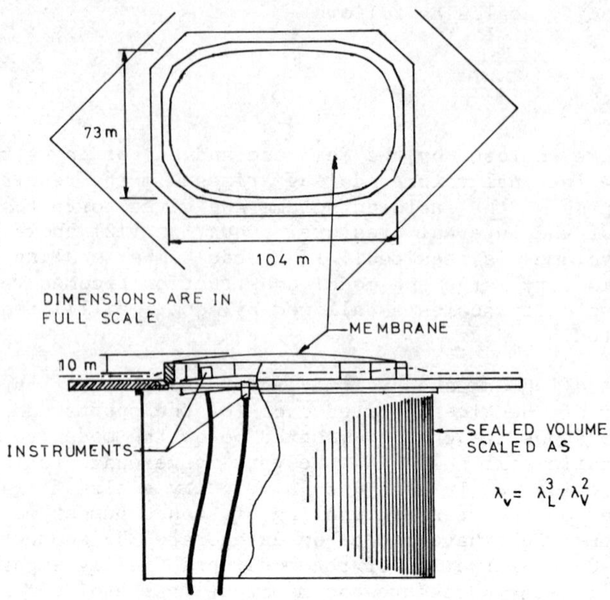

Figure 5. Aeroelastic model of the air-supported roof for the Dalhousie University Sports Complex.

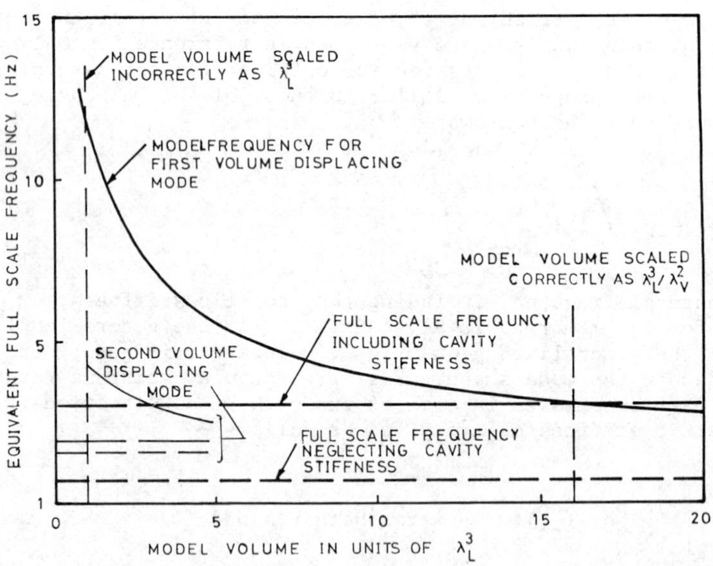

Figure 6. Effect of the enclosed volume on the frequencies of the first two volume displacing modes for a rectangular model membrane simulating the properties of the Dalhousie Sports Complex roof.

TABLE II. Computed Full Scale Frequencies
For An Aeroelastic Model of a Practical Air-Supported Roof

Mode r	Mode s	Frequency in Vacuum (Hz)	Frequency Including Cavity Correctly Scaled as λ_L^3/λ_V^2 (Hz)	Frequency Including Cavity Incorrectly Scaled as λ_L^3 (Hz) for λ_V = 1:4
1	1	.729	3.08	11.8
1	2	1.15	*	*
1	3	1.63	1.93	4.30
1	4	2.13	*	*
1	5	2.63	2.70	3.55
2	2	1.46	*	*

* Indicates no change from the frequency in vacuum

$$K_{rs} = \frac{\pi^2}{4} abT_o \{(\frac{r}{a})^2 + (\frac{s}{b})^2\} \qquad (15)$$

where p_o is the absolute pressure inside the enclosed volume; ν_o is the enclosed volume; a and b are the dimensions of the membrane; and T_o is the tension in the membrane and is assumed to be uniform and isotropic. The volume scaling given by equation (12) can be arrived at by imposing the condition that the ratio of acoustic to geometric stiffness be the same in model and full scale. The vibration modes for which both r and s are odd correspond to the volume displacing modes.

For the Dalhousie roof the effect of the acoustic stiffness term is to very significantly increase the frequencies of the first few volume displacing modes in both full scale and model scale. This is illustrated in Table 2 and Fig. 6. As can be seen the changes in the frequencies for the first few volume displacing modes can be quite marked. Furthermore, if the volume scaling would have been based only on the cube of the length scale, rather than on equation (12), the changes in the model frequencies would have been very much larger than the corresponding changes in the full scale frequencies, and the model would essentially have failed to reproduce a number of the low frequency modes.

CONCLUSIONS

The requirement for scaling the elastic properties of tensioned surfaces has been examined and shown to be unnecessary for aeroelastic modelling of most tensioned structures. For pneumatic structures modifications have been developed for scaling the pressurization which allow reasonable velocity scales to be used without sacrificing Froude number scaling. The ability to relax elastic scaling and modify the pressurization scaling simplifies model construction and test procedures. The importance of scaling acoustic stiffness of volumes enclosed by pneumatic structures has been emphasized, and the modifications required to do so in aeroelastic models have been shown. Equations have been given which can be used to estimate the relative importance of geometric, elastic and acoustic stiffnesses in tensioned structures.

GLOSSARY OF TERMS

A	amplitude of vibration	$p_o, \Delta p_i$	absolute pressure and internal pressurization respectively
a,b	membrane dimensions	Q	generalized co-ordinate
E	tensile modulus	r,s	model co-ordinates
F	generalized force	T_o	membrane tension
fs	subscript indicating full scale	V	velocity
g	gravitational acceleration	ν, ν_o	volumes
h	membrane thickness	λ	general scale parameter, scales for particular variables are indicated by a subscript
I	moment of inertia		
K'_{rs}, K_{rs}	acoustic and geometric stiffnesses respectively		
L	characteristic length; span for a membrane	μ	mass per unit area
		$\omega_o, \omega_{rs}, \Omega, \Omega_{rs}$	frequencies
m	subscript indicating model scale		

REFERENCES

(1) Isyumov, N. and Tryggvason, B. V. A Study of Wind Induced Behaviour of the Air-Supported Microwave Radome on the CN Tower, University of Western Ontario Research Report (in preparation).

(2) Tryggvason, B. V. and Isyumov, N. A Study of the Wind Induced Response of the Air-Supported Roof for the Dalhousie University Sports Complex, University of Western Ontario Engineering Research Report, BLWT-SS7-1977.

(3) Tryggvason, B. V., Surry, D. and Davenport, A. G. A Study of the Wind Induced Response of the Haj Terminal Roof in Jeddah, Saudia Arabia - Phase II - An Aeroelastic Investigation, BLWT, (November 1978).

(4) Minorsky, N. Nonlinear Oscillations, D. Van Nostrand Co. Inc., Princeton, New Jersey, (1962).

(5) Tryggvason, B. V. and Isyumov, N. Similarity Requirements for Inflatable Structures, Proceedings of the Third U.S. Wind Engineering Conference, Gainesville, Florida (February 1978).

THE DYNAMICAL BEHAVIOUR OF THE HARTLEN-CURRIE WAKE OSCILLATOR MODEL

A. B. POORE* and A. R. AL-RAWI**

*Colorado State University, Fort Collins, Colorado, USA
**University of Mustansariyya, Iraq

ABSTRACT

The Hartlen-Currie model for vortex-induced vibration of bluff bodies is investigated to determine the types of dynamical behavior associated with a variable wind speed. By using bifurcation techniques we are able to classify the various types of response diagrams and thus demonstrate the experimentally observed double amplitude responses, hysteresis and several other types of jump phenomena. These response diagrams help explain the dynamic capabilities of the model.

INTRODUCTION

The phenomena of vortex-induced oscillations of bluff bodies has received considerable attention in the literature in the past two decades, principally because these oscillations can cause destructive resonant vibrations when the natural frequency of the body immersed in the flow is near the frequency of the vortex shedding. Examples of these destructive vibrations of structures ranging from large offshore oil terminals to towers, stacks, masts, and heat exchange tubes are given, with references, in the work of Griffin, Skop, and Koopman [4].

The physical situation can be described briefly in the following way. The flow past a bluff body separates into two shear layers which are basically unstable and have the universal tendency to roll up and form the von Karman vortices in the wake of the bluff body. This fluid mechanical phenomenon is known as vortex shedding [3] and results in an oscillatory pressure loading on the bluff body. The response of the bluff body is described mathematically by the dynamic equations of motion for the bluff body coupled with the Navier-Stokes for the flow. Because of the difficulties of solving these general equations, several investigators have sought more tractable models which are capable of explaining the experimentally observed phenomena.

Based on experiments of a flow over a cylinder, Bishop and Hassan [2] suggested that the fluid dynamical behavior of self-excited vortex shedding might be modeled by a simple nonlinear oscillator. This idea was first used by Hartlen and Currie [5] to develop a wake-oscillator model in which a van der Pol type equation is used to simulate the fluctuating pressure on the bluff body. (Actually the equation is one from the class of generalized Lienard's equations of the form $\ddot{x} + f(x, \dot{x})\dot{x} + g(x) = 0$.) Skop and Griffin [10] modified this model by adding a cubic nonlinearity and have presented a method of parameter selection which yields very good agreement with experimental data. A second model is due to Blevins and Iwan [7, 8] who introduced a hidden flow variable to describe the effects of vortex shedding. This allows the model parameters to be determined directly from the physical parameters. Although this model appears to be more promising, it is not quite as tractable as the Hartlen-Currie wake oscillator model.

Most of the investigations into these models have centered on the model development, the parameter determination for experimental simulation, and the calculation of peak amplitude response. As yet no investigation has been made into the dynamic capabilities of these models and it is this aspect of the Hartlen-Currie model that we investigate here. The techniques that we use here are equally applicable to the models of Skop and Griffin [12] and Blevins and Iwan [7].

Based on an investigation of the steady states, bifurcation of periodic solutions and their stability, we are able to classify the response diagrams for a variable wind speed (amplitude of oscillation versus wind speed) in parameter space. In addition to resonant amplitudes, we delineate the experimentally observed double amplitude response, the hysteresis effect, jump phenomena, and regions of stable steady states.

MODELING EQUATIONS

In the Hartlen-Currie model, the bluff body is an elastically supported circular cylinder and the air flow is assumed to be normal to the cylinder. The motion of the cylinder, subjected to the oscillatory pressure loading due to vortex shedding, is assumed to be in a plane perpendicular to the flow. The modeling equations are [5]:

$$\ddot{x}_r + \beta\dot{x}_r + x_r = aW^2c_L \tag{1}$$

$$\ddot{c}_L - \alpha W\dot{c}_L + \frac{\gamma}{W}c_L^3 + W^2c_L = b\dot{x}_r \tag{2}$$

where x_r is a dimensionless displacement of the cylinder and c_L defines the instantaneous lift coefficient. This dimensionless displacement, x_r, is related to the actual displacement, x, by $x_r = x/D$ where D is the diameter of the cylinder. Equation (1) represents a dimensionless dynamic equation of motion for the cylinder in which $-\beta\dot{x}_r$ represents the viscous damping force, $-x_r$, the spring force, and aW^2c_L, the pressure force due to vortex shedding. The second equation (2) for c_L represents a van der Pol type equation in which the lift coefficient is linearly affected by the cylinder velocity. The parameters in (1) and (2) are the damping factor β, the dimensionless wind speed W, the van der Pol coefficients α and γ, and the interaction parameter b. The dimensionless wind speed $W = SV/Df_n$ wherein S is the Strouhal number, V, the wind speed, D, the diameter of the cylinder and f_n is the natural frequency of the cylinder.

The model of Skop and Griffin [12] essentially differs from (1) and (2) in that a cubic term c_L^3 is subtracted from the left side of equation (2). Hartlen, Currie, and Martin [3] have also modified equations (1) and (2) by replacing x_r in equation (1) by $x_r(1 - \varepsilon x_r^2)$ which for $0 < \varepsilon << 1$ represents a slightly nonlinear spring. The model of Blevin and Iwan [7] is substantially different, although it does reduce to the Hartlen-Currie model for certain parameter values.

STABILITY OF THE STEADY STATES AND BIFURCATING PERIODIC SOLUTIONS

The techniques that we use to obtain the classification of the response diagrams are based on an investigation into the stability of the steady states, bifurcation and stability of periodic solutions, and numerical integration of the equations of motion. In this section we discuss the stability of the steady states and the bifurcation of the periodic solutions from some of these steady states.

To explain this general technique, we begin by writing our two second order differential equations as a system of four first order differential equations. Let $x_1 = c_L$, $x_2 = \dot{c}_L$, $x_3 = x_r$, and $x_4 = \dot{x}_r$, so that our modeling equations become $\dot{\underline{x}} = \underline{f}(\underline{x}, \underline{\Gamma})$ where $\underline{\Gamma}$ denotes the vector of system parameters. More explicitly, we have

$$\begin{aligned} \dot{x}_1 &= F^1(\underline{x}, \underline{\Gamma}) = x_2 \\ \dot{x}_2 &= F^2(\underline{x}, \underline{\Gamma}) = -W^2 x_1 + \alpha W x_2 - \frac{\gamma}{W} x_2^3 + b x_4 \\ \dot{x}_3 &= F^3(\underline{x}, \underline{\Gamma}) = x_4 \\ \dot{x}_4 &= F^4(\underline{x}, \underline{\Gamma}) = a W^2 x_1 - \beta x_4 - x_3. \end{aligned} \tag{3}$$

The only steady state, i.e., time-independent solution, of (3) is $\underline{x} = \underline{0}$. The stability of this steady state is based on the exponential growth or decay of solutions of the linearized problem $\frac{d\underline{x}}{dt} = A\underline{x}$ where

$$A = \begin{pmatrix} 0 & 1 & 0 & 0 \\ -W^2 & \alpha W & 0 & b \\ 0 & 0 & 0 & 1 \\ aW^2 & 0 & -1 & -\beta \end{pmatrix}. \tag{4}$$

The growth or decay of these solutions is in turn based on the sign of the real parts of the eigenvalues of this matrix A. If the real parts of the four eigenvalues of A are negative, the zero state is stable whereas if any of the real parts are positive, the zero state is unstable. This stability analysis breaks down whenever there is a zero eigenvalue or whenever there is a pair of complex conjugate purely imaginary eigenvalues, say $\pm i\omega_0$ with $\omega_0 > 0$. Since the determinant of a matrix represents the product of the eigenvalues and since $\det(A)=W^2$ for the matrix in (4), there are no zero eigenvalues for positive wind speeds. However, there are pairs of purely imaginary eigenvalues, $\pm i\omega_0$, for this matrix A and corresponding $2\pi/\omega_0$-periodic solutions of the linear problem $\frac{d\underline{x}}{dt} = A\underline{x}$. Under an additional assumption, to be explained, a bifurcation of periodic solutions from the zero state occurs in the nonlinear problem as one of the system parameters varies.

Our investigations of the oscillations in this model are based on a variable wind speed with the remaining parameters held fixed. As the dimensionless wind speed W varies, so will the eigenvalues of the matrix A. By holding the remaining system parameters fixed and varying the wind speed, the stability of the steady states can be determined by first finding those wind speeds W_0 at which A has a pair of purely imaginary eigenvalues $\pm i\omega_0$. For W close to W_0, the eigenvalues of A can be written as $\alpha(W) \pm i\omega(W)$, $\lambda_3(W)$ and $\lambda_4(W)$ where $\alpha(W_0) = 0$ and $\omega(W_0) = \omega_0$ which is assumed to be positive. The stability of the steady state for all positive wind speeds W can be determined from the sign of $\alpha'(W_0)$ and the sign of the real part of $\lambda_3(W_0)$ and $\lambda_4(W_0)$. For example, when there is only one value of the wind speed W_0 for which A has a pair of purely imaginary eigenvalues, then the steady state is stable for $0 < W < W_0$ and unstable for $W > W_0$ if $\lambda_3(W_0)$ and $\lambda_4(W_0)$ have negative real parts and $\alpha'(W_0) > 0$. We shall only present the results of this analysis; however, the details can be found in the work of A. R. Al-Rawi [1].

As mentioned above the linearized differential equation $\frac{d\underline{x}}{dt} = A\underline{x}$ has $2\pi/\omega_0$-periodic solutions at $W = W_0$. Since our nonlinear differential equation is analytic in all variables and parameters, a periodic solution bifurcates from $\underline{x} = 0$ and $W = W_0$ whenever $\alpha'(W_0) \neq 0$. The mathematical foundation of this fact is the Hopf bifurcation which can be found in several references [1,6,9,10,11]. Derivable from this theory are algebraic expressions which determine the stability of the periodic solution near the bifurcation point and the direction of bifurcation, i.e., whether the orbit grows from zero amplitude at $W = W_0$ as W moves to right of W or as W moves to left of $W = W_0$. To explain how these important algebraic expressions are obtained, we first introduce the notation $W = W_0 + \varepsilon$ and let

$$A^{\varepsilon} = \begin{pmatrix} 0 & 1 & 0 & 0 \\ -(W_0+\varepsilon)^2 & \alpha(W_0+\varepsilon) & 0 & b \\ 0 & 0 & 0 & 1 \\ a(W_0+\varepsilon)^2 & 0 & -1 & -\beta \end{pmatrix} \tag{5}$$

where W_0 is that wind speed for which A^0 has a pair of purely imaginary eigenvalues $\pm i\omega_0$ with $\omega_0 > 0$. Let $\underline{u} = (u_1, u_2, u_3, u_4)$ and $\underline{v} = (v_1, v_2, v_3, v_4)^T$ be left and right eigenvalues corresponding to the eigenvalue $i\omega_0$ of the matrix A^0, respectively, and assume they are normalized so that $\underline{u}\underline{v} = 1$. If $B^0 = \dfrac{dA^0}{d\varepsilon}$, then

$$\alpha'(W_0) + i\omega'(W_0) = \underline{u}B^0\underline{v}$$

from whence one can conclude that the periodic solution bifurcates from $\underline{x} = 0$ and $W = W_0$ if $\alpha'(W_0) = \text{Re}\, uB^0v \neq 0$. Also, there is an algebraic expression $\bar{\delta}'$ which determines the direction of bifurcation in the sense that the periodic orbit bifurcates to $W > W_0$ if $\delta' > 0$ and to $W > W_0$ if $\delta' < 0$. This δ' is determined from

$$8\alpha'(W_0)\cdot\delta' = \text{Re}\{-u_\ell \frac{\partial^3 F^\ell}{\partial x_j \partial x_k\, x_p} v_j v_k \bar{v}_p + 2u_\ell \frac{\partial^2 F^\ell}{\partial x_j \partial x_k} v_j (A^{0^{-1}})_{kr} \frac{\partial^2 F^r}{\partial x_p \partial x_q} v_p \bar{v}_q$$
$$+ u_\ell \frac{\partial^2 F^\ell}{\partial x_j \partial x_k} v_j ((A^0 - 2i\omega_0 I)^{-1})_{kr} \frac{\partial^2 F^r}{\partial x_p \partial x_q}\}, \tag{6}$$

where $(A^{0^{-1}})_{kr}$ denotes the element in the k-th row and the r-th column of $A^{0^{-1}}$ and repeated indices within each term imply a sum from one to n, with n being the number of first order ordinary differential equations. In our problem, n is four. The partial derivatives of F are evaluated at the steady state corresponding to $W = W_0$.

The stability of this bifurcating periodic solution can be determined from the sign of $\alpha'(W_0)\delta'$ and the real parts of the eigenvalues of A^0 other than $\pm i\omega_0$. If $\alpha'(W_0)\delta' > 0$ and the eigenvalues other than $\pm i\omega_0$ have negative real parts, the periodic solution is stable near the bifurcation point. On the other hand, the periodic solution is unstable near the bifurcation point if $\alpha'(W_0)\delta' < 0$ of if there is an eigenvalue A^0 with a positive real part. All of this information has been utilized in the investigation of problems (1) and (2), and the results are given in the next section.

BIFURCATION DIAGRAMS

The key component to finding the various response diagrams is the use of the bifurcation theory discussed above and we now explain how the various types of diagrams are classified. This classification is based on the numerical values $a = 0.002$, $b = 0.4$ and $\gamma = 2/3$ and all values of $\alpha > 0$ and $\beta > 0$. These numerical values of a, b and γ are the same ones used by Hartlen and Currie [5] and correspond to the system parameters used to fit existing experiments. The numerical investigations of A. R. Al-Rawi [1] indicate that there are no additional types of response diagrams other than these reported here. We use the term bifurcation diagram to mean a diagram displaying the steady state, its stability, the bifurcating periodic solution and its stability for all positive wind speeds W.

The classification of the various types of bifurcation diagrams for all parameters α and β begins with the division of the first quadrant, i.e., $\alpha > 0$ and $\beta > 0$, into various regions representing the various types of phenomena. Our classification is presented in Figure 1 with a clarification in Figure 2 for $\alpha > 0.08$. There are eight regions in Figures 1 and 2 labeled I_1, I_2, II_1, II_2, II_3, II_4, III and IV. The curves labeled III, C_1 and C_2 in Figure 1 separate the various regions; the straight line at the bottom of Figure 2 is $\beta = ab$ and separates regions I and II. The curve labeled C_1 separates region IV, where there are not wind speeds for which the matrix A has a pair of purely imaginary eigenvalues, from region II where A will have this pair for some positive wind speed. Next, the curve labeled C_2 denotes a line on which the direction of bifurcation $\delta' = 0$ for the smaller of the two wind speeds at which bifurcation occurs. Note that $\delta' > 0$ in region II_2 and $\delta' < 0$ in II_3 at $W = W_{01}$. Also curve C_2 separates regions II_1 and II_4 with a similar interpretation. Finally, the curve labelled III is defined by $\beta = \frac{ab}{2} + \sqrt{\alpha^2 + (\frac{ab}{2})^2}$ and corresponds to the situation in which A has two pairs of purely imaginary eigenvalues. We believe that the best way to utilize Figure 1 and the resulting eight bifurcation diagrams in Figures 3-11 is to pick an (α, β) from one of the regions and then find the corresponding bifurcation diagram. The bifurcation diagrams are best explained by example.

Figure 3 is intended to convey the information that for $(\alpha, \beta) \in I_1$, the steady state is unstable for all positive wind speeds and an unstable periodic solution bifurcates to the right of $W = W_0$ ($\delta' > 0$). The vertical distance from the W axis to the circle, which indicates an unstable orbit, represents the amplitude of the periodic orbit. In Figure 6 the steady state is unstable for $0 < W < W_{01}$ and for $W > W_{02}$ but is stable for $W_{01} < W < W_{02}$. There are two points of bifurcation, $W = W_{01}$ and $W = W_{02}$ with both bifurcations occuring to the right of $W = W_{01}$ and $W = W_{02}$, i.e., $\delta' > 0$ in both cases. The periodic orbit that bifurcates from $W = W_{01}$ is unstable, while the one that bifurcates to the right of $W = W_{02}$ is stable. Figure 9 shows the bifurcation of two periodic orbits from the zero state at $W = W_0$; one is stable and the other is unstable. This situation is quite different from the usual Hopf bifurcation in which the bifurcation of a stable periodic orbit occurs only when there is an exchange in the stability (from stable to unstable) of the steady states. (The theory and detailed explanations can be found in the work of the authors [1,10,11].) Finally, Figure 10 indicates that there are no bifurcation points whenever α and β are in region IV and that the steady state is always unstable. There are, however, stable oscillatory states for each positive wind speed as we shall see in the next section.

RESPONSE DIAGRAMS

We use the term response diagram in this work to mean a diagram which conveys two types of information: the stability of the steady state and the number and stability of the periodic solution for all positive dimensionless wind speeds W. The stable states on these diagrams are the observable states for the physical problem modeled by the differential equations (1) and (2). These response diagrams displayed in Figures 11-18, allow us to explain the occurrence of the experimentally observed double amplitude response, jump phenomena and hysteresis. Each diagram was obtained for specific numerical values of (α, β) within each region in Figure 1; however, the use of other numerical values within the same region would lead to some quantitative changes but no qualitative ones. The specific numerical values are given in the table.

Table: Parameter Values

Region	Figure	α	β	W_{01} or W_0	W_{02}
I_1	11	0.011	0.0005	1.03706	
I_2	12	0.08	0.0005	1.01	
II_1	13	0.011	0.0083	1.0373	1.0518
II_2	14	0.011	0.0161	1.02637	1.0364
II_3	15	0.011	0.0233	1.0198	1.0348
II_4	16	0.08	0.0015	1.01108	1.45506
III	17	0.011	0.0114	1.037	
IV	18	0.08	0.08		

$a = 0.002$, $b = 0.4$, $\gamma = 2/3$

The process by which one obtains these diagrams starts with the bifurcation diagrams (Figures 3-10), discussed in the previous section. If a stable periodic solution bifurcates from $x_r = c_L = 0$ at $W = W_0$, then one can find this periodic s-lution numerically by using the initial conditions

$$\underline{x}(0) = \begin{pmatrix} c_L(0) \\ \dot{c}_L(0) \\ x_r(0) \\ \dot{x}_r(0) \end{pmatrix} = \left(\frac{W - W_0}{\delta'}\right)^{1/2} b_1\left(\frac{\bar{\underline{v}} - \underline{v}}{2i}\right) + b_2\left(\frac{\bar{\underline{v}} + \underline{v}}{2}\right) \tag{7}$$

where W is close to W_0 with $W - W_0 > 0$ if $\delta' > 0$ and $W - W_0 < 0$ if $\delta' < 0$. Here, b_1 and b_2 are any real numbers satisfying $b_1^2 + b_2^2 = 1$, $\underline{v}$ denotes the right eigenvector corresponding to the eigenvalue $+i\omega_0$ of A^0 when $W = W_0$, and $\bar{\underline{v}}$ denotes the complex conjugate of $\underline{v}$. This formula (7) differs from the true initial conditions by a term of the order magnitude $O(|W - W_0|)$ as $W \to W_0$. However, we found it to be as good an approximation for $|W - W_0| = 0.5$ as for $|W - W_0| = 0.01$. Once a periodic orbit was found numerically, the dimensionless wind speed W was changed and a value of the periodic solution at the old value of W was used as an approximation to a point on the periodic orbit at the new value of W. This technique of numerically tracing the periodic orbit as W varies from W_0 for the stable bifurcating orbits in Figures 14, 15 and 17. To obtain the remaining stable branches, we started with a stable periodic orbit for α and β in one region and varied these α and β with W fixed until we ended up with α and β in another region. For example, to obtain the stable part of that branch that bifurcates from the upper dimensionless wind speed $W = W_{02}$ in Figure 13, we started with the stable periodic solution at $W = 1.0$, illustrated in Figure 15, held $\alpha = 0.011$ fixed and varied β from $\beta = 0.233$ ($(\alpha, \beta) \in II_3$) to $\beta = 0.0083$ so that $(\alpha, \beta) = (0.011, 0.233) \in II_2$. Then the periodic solution was traced numerically as W was varied above and below $W = 1.0$. The unstable part of a branch of periodic solutions near a bifurcation point comes from the bifurcation theory whereas the remaining unstable part is simply conjecture. The in-depth details of how each diagram was obtained is given by A. R. Al-Rawi [1].

Having briefly explained how these diagrams were obtained, we now turn to the dynamical implications of these diagrams. The experimentally observed double amplitude response corresponds to the existence of two stable oscillations. The existence of two stable oscillations occurs for certain wind speeds in Figures 11, 12, 13, 16 and 17, i.e., for α and β in region I_1, I_2, II_1, II_4, and III, respectively. Figures 13 and 16, $(\alpha, \beta) \in II_1$ and $(\alpha, \beta) \in II_4$, respectively, show response diagrams which agree quite well with that obtained experimentally by Griffen and Skop [10]. In fact, the numerical values for the system parameters of their experiment fall into region II_1 in which all response diagrams are qualitatively.

Hysteresis phenomena occurs whenever α and β are in regions II_1 and II_4 as illustrated in Figures 13 and 14. To explain how this occurs, we first suppose that the system is oscillating in a stable mode for a dimensionless wind speed below W = 1.0. (The stable oscillation is the only stable state at this wind speed.) As the wind speed W increases the amplitude increases, reaches a maximum amplitude, and then decreases until the system is oscillating at the point Q. Any further increase in the wind speed causes a jump downward to the point R where the system now oscillates with a much smaller amplitude. As the wind speed increases further, the amplitude decreases very slowly but frequency of oscillation increases. On the other hand, if the wind speed is decreased from that wind speed corresponding to R, the amplitude slowly increases until we reach S at which point a jump of oscillations occurs from S to P. This same hysteresis loop PQRSP occurs as in Figures 16 and 17 when α and β are in region II_4 and III, respectively. Hysteresis involving a stable steady state occurs for α and β in region II_2 as illustrated in Figure 14.

Other types of jump phenomena are illustrated in Figures 11 and 12 with jumps from S to P. If the bluff body is oscillating in a mode corresponding to the stable upper branch in Figures 11 and 12, there will be no jump in amplitude as the wind speed is varied. Also, the amplitude of oscillation appears to grow rapidly as W increases.

CONCLUSIONS

The Hartlen-Currie model of vortex-induced oscillations of bluff bodies is capable of producing a variety of response diagrams which illustrate the experimentally observed double amplitude responses, hysteresis phenomena, and several other types of jump phenomena. More detailed information can be found in the thesis by A. R. Al-Rawi [1]. We have been able to obtain and classify these various types of dynamical behavior by a systematic use of the Hopf-bifurcation theory which, in our opinion, is the most effective and powerful technique presently available for investigating oscillatory solutions of autonomous systems. This method would be equally effective in treating the Skop and Griffin model [10] and the Blevins-Iwan model [7].

ACKNOWLEDGEMENTS

The authors wish to express their appreciation to Professors J. E. Cermak and A. Kareem for motivating this work and for their helpful discussions. This work was supported in part by NSF Grant Eng 76-03-135.

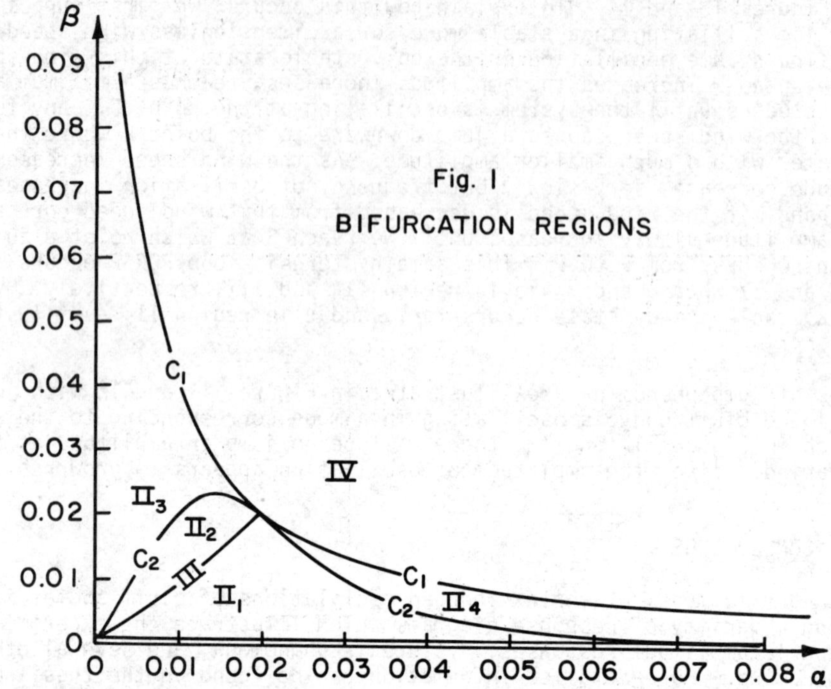

Fig. 1
BIFURCATION REGIONS

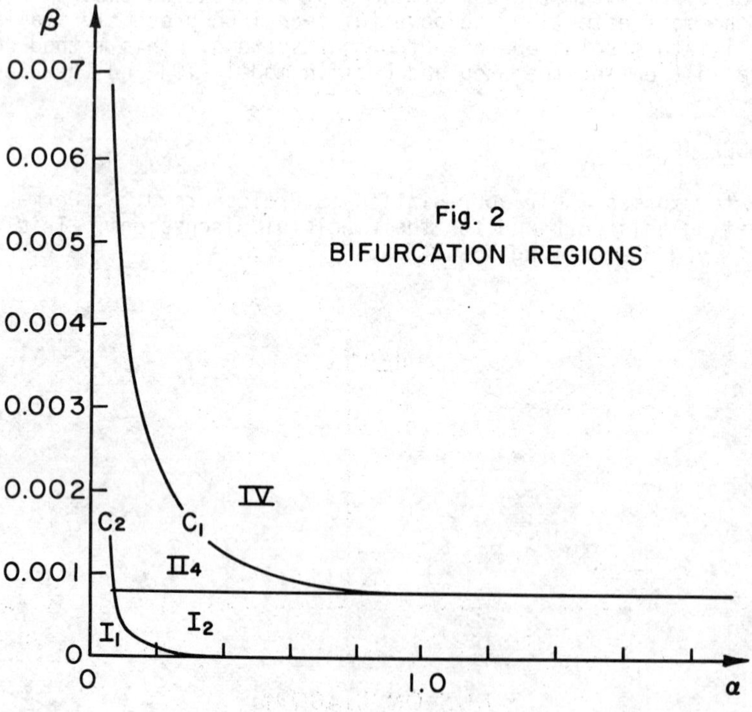

Fig. 2
BIFURCATION REGIONS

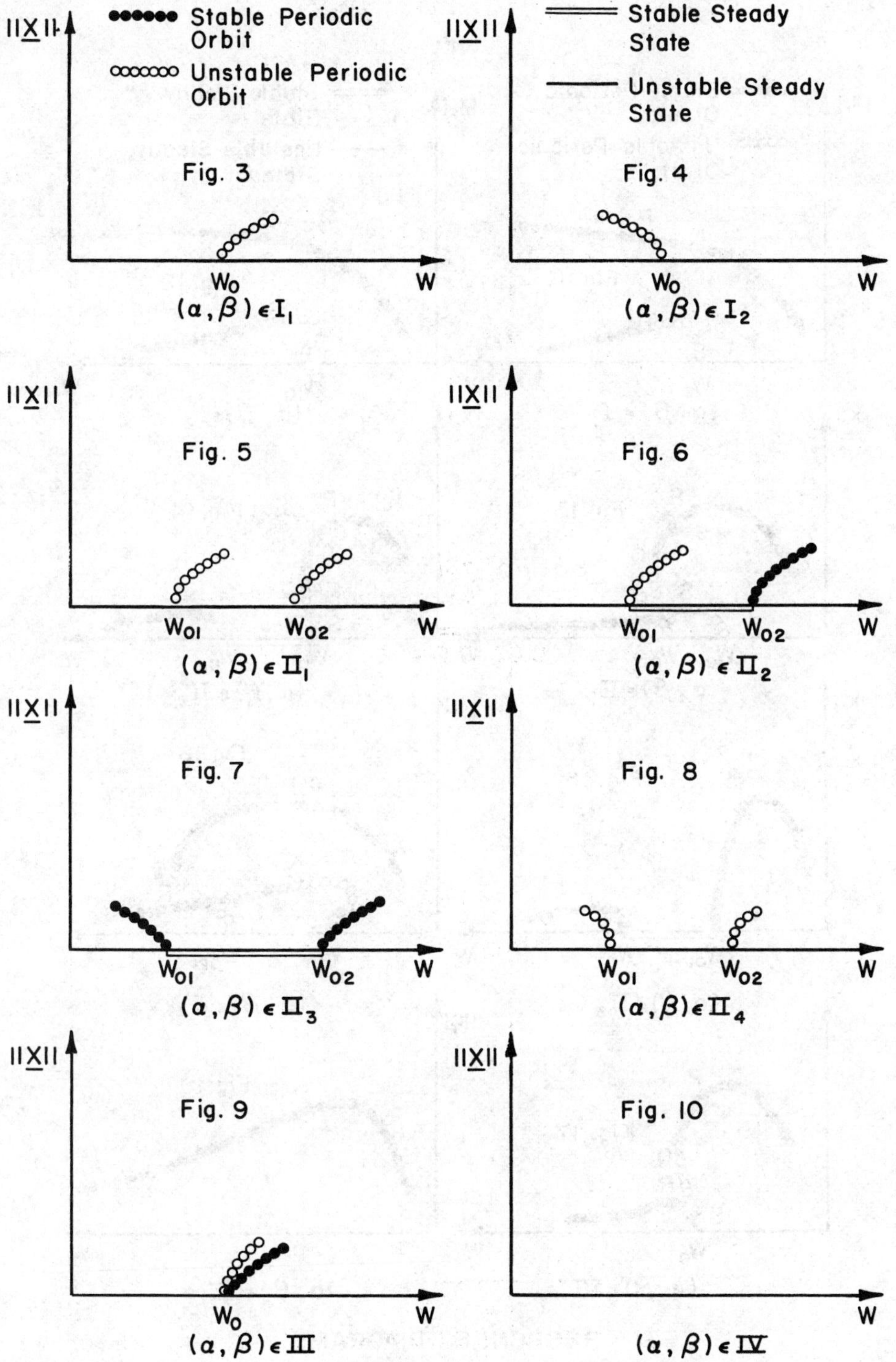

BIFURATION DIAGRAMS

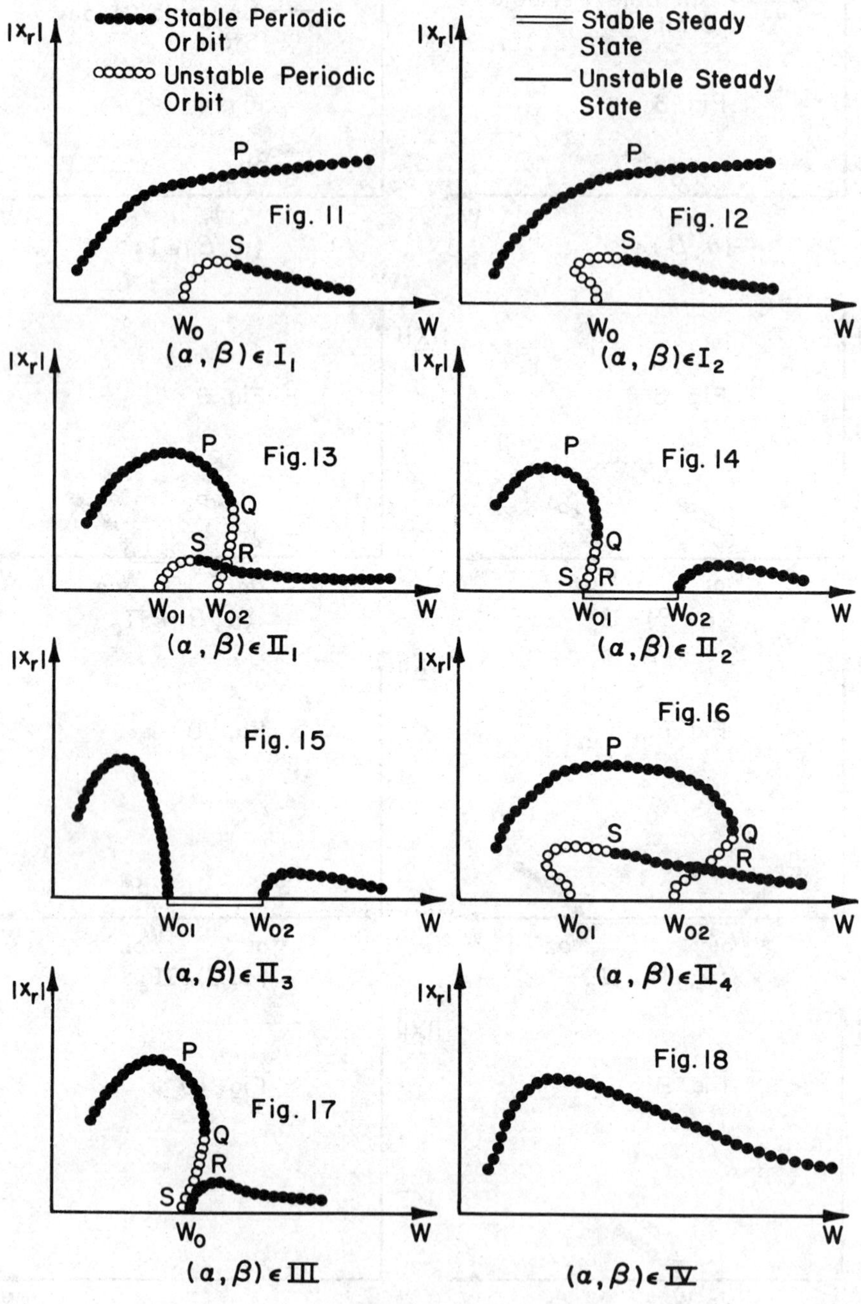

RESPONSE DIAGRAMS

1. Al-Rawi, A. R., Hopf Bifurcation Formulas for Multiple Complex Eigenvalues and Applications in Wind Engineering, Ph.D. Thesis, Colorado State University, 1978.

2. Bishop, R. E. and A. Y. Hassan, "The Lift and Drag Forces on a Circular Cylinder Oscillating in a Flowing Fluid", Proc. of the Royal Society, London, Series A, 277, 51-75 (1964).

3. Currie, I. G., R. T. Hartlen, and W. W. Martin, "The Response of Circular Cylinders to Vortex Shedding", Symposium on Flow-induced Structural Vibration, Karlsruhe, German, 128-142 (1972).

4. Griffin, D. M., R. A. Skop, and G. H. Koopmann, "The Vortex-excited Response of Circular Cylinders", Journal of Sound and Vibration, 31, 235-249 (1973).

5. Hartlen, R. T. and I. G. Currie, "Lift-oscillator Model of Vortex-induced Vibration", Proc. ASCE, J. Engng. Mech., 96, 577-591 (1970).

6. Hopf, E., "Abzweignung einer periodischer Lösung von einer stationären Lösung eines differential system", Ber. Verh. Sächs. Akad. Wiss. Liepzig. Math. Nat. KL95, 3-22 (1942).

7. Iwan, W. D. and R. D. Blevins, "A Model for Vortex-induced Oscillations of Structures", J. Appl. Mech., 41, 581-586 (1974).

8. Iwan, W. D., "The Vortex-induced Oscillation of Elastic Structural Elements", Trans. ASME, J. of Engineering for Industry, Series B, 97, 1378-1382 (1975).

9. Poore, A. B., "On the Theory and Application of the Hopf-Friedrichs Bifurcation Theory", Arch. Rational Mech. and Anal., 60, 371-393 (1976).

10. Poore, A. B. and A. R. Al-Rawi, "Some Applicable Hopf Bifurcation Formulas on an Application in Wind Engineering", to appear in Annals of New York Academy of Sciences.

11. Poore, A. B. and A. R. Al-Rawi, "The Dynamics of a Model of Vortex-Induced Vibration of Bluff Bodies", to appear.

12. Skop, R. A. and O. M. Griffin, "A Model for the Vortex-excited Resonant Response of Bluff Cylinders", J. of Sound and Vibrations, 27, 255-233 (1973).

WAKE-OSCILLATOR MODEL OF VORTEX-INDUCED OSCILLATION OF CIRCULAR CYLINDER

Y. TAMURA and G. MATSUI
Waseda University, Tokyo, Japan

ABSTRACT

The Birkhoff type wake-oscillator whose length varies is proposed for a stationary cylinder. The motion of the wake-oscillator is expressed approximately in a non-linear equation. The system consisting of the oscillator and an elastically mounted cylinder is considered as a binary system in simultaneous equations. The periodic solutions and their rough explanations are obtained and examined for their variation with mechanical damping. The numerical solutions of the model by Runge-Kutta method are calculated and compared with many experimental results. It is found that the proposed model can simulate the vortex-induced oscillation of a two dimensional circular cylinder not only in quality but also in quantity.

NOTATIONS

y = Transverse displacement of circular cylinder, α = Angular displacement of wake-oscillator, D = Diameter of circular cylinder, H = Thickness of wake-oscillator, $2\bar{L}$ = Length of wake-oscillator, $2L$ = Fluctuating length of wake-oscillator, b = Amplitude of fluctuating component of wake-oscillator's length, s = Length of circular cylinder, $f_v = \omega_v/2\pi$ = Natural frequency of wake-oscillator, $f_0 = \omega_0/2\pi$ = Natural frequency of circular cylinder, $\bar{I}$ = Moment of inertia of wake-oscillator about the center of circular cylinder, I = Fluctuating moment of inertia of wake-oscillator about the center of circular cylinder, $\bar{K}$ = Coefficient of fluid dynamical restoring moment, K = Fluctuating coefficient of fluid dynamical restoring moment, S = Strouhal number, V = Fluid velocity, $V_{cr} = f_0 D/S$ = Critical fluid velocity, ρ = Fluid density, M_c = Mass of circular cylinder, K_c = Spring constant of circular cylinder, C_c = Damping coefficient of circular cylinder

INTRODUCTION

Altough vortex-induced oscillation of a two-dimensional circular cylinder immersed in a uniform flow at subcritical Reynolds numbers has been studied by many investigators, the basic mechanisms generating the oscillation are not known. It is difficult to predict the amplitude of the cylinder displacement by some formula. This paper discusses a mathematical model of vortex-induced oscillation based on Birkhoff's oscillator model.

METHODS

WAKE-OSCILLATOR MODEL FOR A STATIONARY CIRCULAR CYLINDER

Funakawa (1) attempted to simulate fluid dynamic effects in vortex-induced oscillation of a circular cylinder by the use of a torsional oscillator based on Birkhoff's oscillator model for the dead air region behind a cylinder (Fig. 1). For a stationary cylinder, his oscillator is governed by the following equation.

$$\bar{I}\,\ddot{\alpha} + \bar{K}\,\alpha = 0 \tag{1}$$

where $\bar{I} = 2\rho\bar{L}H(D/2 + \bar{L})^2,\quad \bar{K} = 2\pi\rho V^2\bar{L}(D/2 + \bar{L})$ (2)

Putting $H = 1.25D$ and $\bar{L} = 1.1D$, the non-dimensional frequency which is numerically equal to the Strouhal number for a stationary circular cylinder at subcritical Reynolds numbers is got as follows.

$$S = f_v D / V = (1/2\pi)\sqrt{\bar{K}/\bar{I}}\cdot D/V \simeq 0.2 \tag{3}$$

Lift coefficient, C_L , is obtained in propotion to the angular displacement.

$$C_L = - f\,\alpha \quad , \quad \text{where} \quad f \simeq 1.16 \tag{4}$$

The value of the constant factor, f , is estimated by the relationship observed experimentally (2)(3) between magnus-effect lift and wake angular displacement for a rotating circular cylinder (1). Nakamura (4) adopted Funakawa's oscillator to his cylinder-fluid model as a binary flutter. Since the model for fluid dynamic effects was a linear one, it could not represent the nonlinear phenomena peculiar to vortex-induced oscillation.

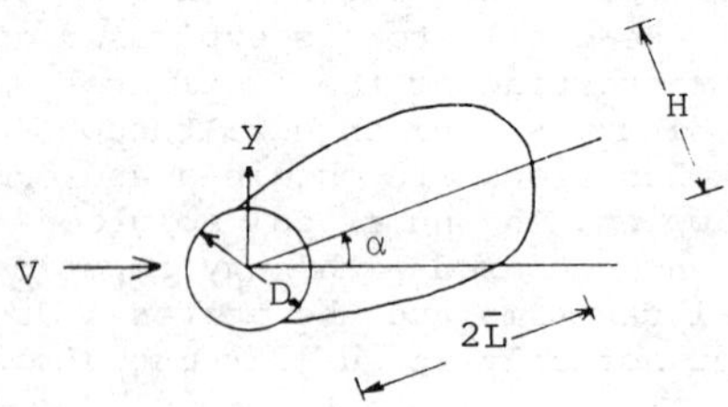

Fig. 1. Birkhoff's oscillator model

The authors suppose a wake-oscillator the length of which varies like Fig. 2. The flow pattern behind a stationary circular cylinder during a half period of Silvio (5) supports the conception of that wake-oscillator (Fig. 3). The oscillator has the minimum length at the instant when a vortex is discharged completely ((a) of Fig. 2 and Fig. 3). The length increases as the opposite vortex-layer curls up, and reaches the maximum just before the next vortex starts to be discharged (near (c) of Fig. 2 and Fig. 3). The fluctuation of the length affects the wake-oscillator system as parametric damping. Assuming the angular displacement of the wake-oscillator to be $\alpha = \alpha_0 \sin\omega_v t$, the wake-oscillator's length is expressed as follows. (Fig. 2)

$$2L = 2\bar{L} - b\,\sin 2\omega_v t = 2\bar{L}(1 - b^*\alpha\,\dot{\alpha}/\,\omega_v) \tag{5}$$

where $b^* = b/\alpha_0^2\bar{L}$. Hence, the circular frequency of the wake-oscillator is given by

$$\begin{aligned}(K/I)^{1/2} &= \{\, 2\pi\rho V^2 L(D/2 + L)\ /\ 2\rho LH(D/2 + L)^2\}^{1/2} \\ &= \{\, \frac{\bar{K}}{\bar{I}}\cdot\frac{1 - D_1\alpha\dot{\alpha}/\omega_v + E_1\alpha^2\dot{\alpha}^2/\omega_v^2}{1 - A_1\alpha\dot{\alpha}/\omega_v + B_1\alpha^2\dot{\alpha}^2/\omega_v^2 - C_1\alpha^3\dot{\alpha}^3/\omega_v^3}\,\}^{1/2} \\ &= \omega_v\{\, 1 + (A_1 - D_1)\alpha\dot{\alpha}/\omega_v + (A_1^2 - A_1D_1 - B_1 + E_1)\alpha^2\dot{\alpha}^2/\omega_v^2 + \cdots\}^{1/2} \\ &\simeq \omega_v\{\, 1 + (A_1 - D_1)\alpha\dot{\alpha}/\omega_v\}^{1/2}\end{aligned} \tag{6}$$

where $A_1 = b^*(0.25 + 2\bar{L}^* + 3\bar{L}^{*2})/(0.5 + \bar{L}^*)^2$, $B_1 = b^{*2}(1 + 3\bar{L}^*)\bar{L}^*/(0.5 + \bar{L}^*)^2$

$C_1 = b^{*3}\bar{L}^{*2}/(0.5 + \bar{L}^*)^2$, $D_1 = b^*(0.5 + 2\bar{L}^*)/(0.5 + \bar{L}^*)$, $E_1 = b^{*2}\bar{L}^*/(0.5 + \bar{L}^*)$, $\bar{L}^* = \bar{L}/D$

Therefore, we assume that the variation of the spring of the wake-oscillator represents the effects of the fluctuation of the length. From eq.(6), we employ

$$\bar{K}\,\{1 + (A_1 - D_1)\alpha\,\dot{\alpha}/\,\omega_v\}$$

for a spring constant of the wake-oscillator.

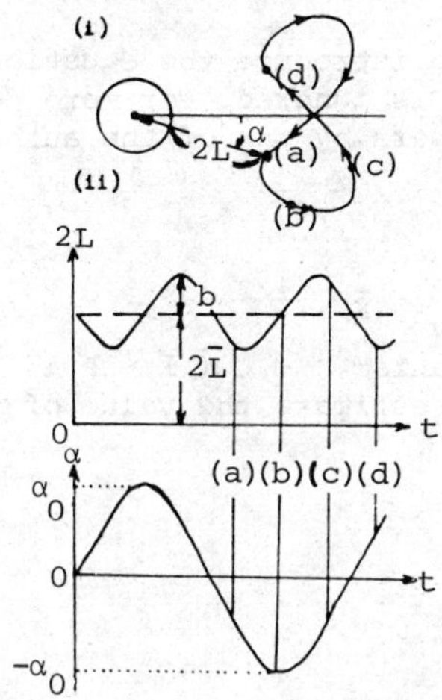

Fig. 2 Fluctuation of the length and the angular displacement of proposed wake-oscillator

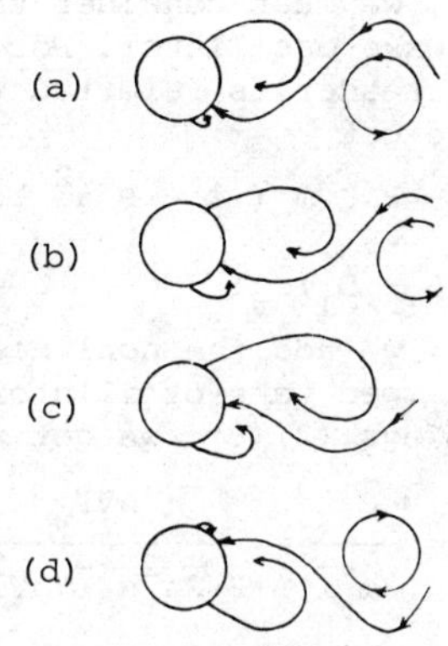

Fig. 3 Velocity pattern behind a stationary circular cylinder during a half period (5)

Next, we consider the effect of the vortex which grows up in the wake-oscillator region and is discharged to its downstream. At the instant (a) of Fig. 2 and Fig. 3 there is a vortex whose intensity is $-\Gamma_v$ behind the wake-oscillator. According to the Kutta-Jukowski's theorem, the lateral force, R, is

$$R = \rho V \Gamma_v$$

during a half period. Assuming the point of action to be the center of gravity of the wake-oscillator, the moment due to R with respect to the center of the cylinder is

$$M_v = \rho V \Gamma_v (D/2 + L)$$

The work done by viscous damping force $C_0\dot{\alpha}$ during a period is

$$-\pi C_0 \alpha_0^2 \omega_v$$

when $\alpha = \alpha_0 \sin\omega_v t$. Then, the equivalent viscous damping coefficient is obtained by the following relation.

$$\begin{aligned} -\pi C_0 \alpha_0^2 \omega_v &= 2\int_{-3\pi/4\omega_v}^{\pi/4\omega_v} M_v \dot{\alpha}\, dt \\ &= \{\, 2\sqrt{2}\alpha_0 (D/2 + \bar{L}) + \sqrt{2}b^*\bar{L}\alpha_0^3/3 \,\}\, \rho V \Gamma_v \\ &\simeq 2\sqrt{2}\alpha_0 (D/2 + \bar{L})\ \rho V \Gamma_v \qquad (7) \end{aligned}$$

The right hand side of the above equation means the work done by the lateral force R during a period. From eq.(7), equivalent damping coefficient is obtained as follows.

$$C_0 \simeq -\frac{2\sqrt{2}\rho V\Gamma_v(D/2+\bar{L})}{\pi\alpha_0\omega_v} \equiv -\bar{C} \tag{8}$$

We assume that the negative damping $-\bar{C}$ represents the effect of the discharged vortex. Then, we obtain the following equation approximately governing the wake-oscillator.

$$\bar{I}\,\ddot{\alpha} - \bar{C}\,\dot{\alpha} + \bar{K}\{1 + (A_1 - D_1)\,\alpha\,\dot{\alpha}/\omega_v\}\,\alpha = 0 \tag{9}$$

More precisely, we must consider the term of $\dot{L}$ to introduce the equation of the motion of the wake-oscillator. For simplicity, it is ignored. For more detailed discussion the reader is referred to the other papers by one of the authors (6)(7). From eq.(9), we get

$$\ddot{\alpha} - 2\,\zeta\,\omega_v(1 - P\,\alpha^2)\,\dot{\alpha} + \omega_v^2\alpha = 0 \tag{10}$$

where $P = (A_1 - D_1)\omega_v\bar{I}/\bar{C}$, $\zeta = \bar{C}/2\omega_v\bar{I}$, $\omega_v = (\bar{K}/\bar{I})^{1/2}$

That is to say, we add the nonlinear damping mechanism, $-\bar{C}(1 - P\alpha^2)\dot{\alpha}$, to the linear undamped wake-oscillator model. We can estimate the value of the damping ratio ζ. From eqs.(2)(8), we get

$$\zeta = \frac{\bar{C}}{2\omega_v\bar{I}} = \frac{\rho V\Gamma_v}{2\sqrt{2}\pi^2\alpha_0\bar{L}\cdot\rho V^2/2} \tag{11}$$

And we can assume roughly,

$$\rho V\Gamma_v \simeq \rho V^2/2\cdot C_{LO}D = \rho V^2/2\cdot f\alpha_0 D \tag{12}$$

where C_{LO} denotes the amplitude of the lift coefficient for a stationary circular cylinder. From eqs.(11)(12) and $\bar{L} = \bar{L}^*D$, we can roughly estimate the value of ζ as follows.

$$\zeta \simeq \frac{1}{2\sqrt{2}\pi^2}\cdot\frac{f}{\bar{L}^*} \simeq \frac{1}{2\sqrt{2}\pi^2}\cdot\frac{1.16}{1.1} \simeq 0.038 \tag{13}$$

Here, the amplitude of its steady state periodic solution is oftained approximately

$$\alpha_0 = 2/\sqrt{P}$$

Therefore, the amplitude of the lift coefficient for a stationary circular cylinder is

$$C_{LO} = f\,\alpha_0 = 2\,f/\sqrt{P} \tag{14}$$

Finally, we can introduce the equation of the wake-oscillator's motion for a stationary circular cylinder, from eqs.(9)(14),

$$\bar{I}\,\ddot{\alpha} - \bar{C}\{1 - (4f^2/C_{LO}^2)\,\alpha^2\}\,\dot{\alpha} + \bar{K}\,\alpha = 0\;,\quad C_L = -f\cdot\alpha \tag{15}$$

CYLINDER-FLUID MODEL FOR AN OSCILLATING CIRCULAR CYLINDER

Let us consider the effects of cylinder motion on the wake-oscillator. The wake-oscillator is exerted by a moment which is propotional to the accelaration of the horizontal movement of the axis of rotation. So, we introduce the following term on the right hand side of eq.(15).

$$-\bar{I}\,\ddot{y}/(D/2+\bar{L})$$

The velocity of the horizontal movement of the axis changes the relative angle of attack of the wake-oscillator. We must transform α into $\alpha + \dot{y}/V$ of the term of the fluid dynamical restoring moment approximately. Hence, we derive the equation of the wake-oscillator motion for a oscillating cylinder as

$$\bar{I}\,\ddot{\alpha} - \bar{C}\{1 - (4f^2/C_{LO}^2)\alpha^2\}\,\dot{\alpha} + \bar{K}(\alpha + \dot{y}/V) = -\,\bar{I}\,\ddot{y}\,/(D/2 + \bar{L}) \tag{16}$$

The equation of the motion of a circular cylinder mounted elastically is

$$M_c\ddot{y} + C_c\dot{y} + K_c y = F_L + F_D \tag{17}$$

where F_L denotes the lift force, and F_D, y-component of the drag force. The values of F_L and F_D are specified as follows.

$$F_L = -\,f\;(\,\alpha + \dot{y}/V\,)\cdot\rho V^2/2\cdot D\;s \tag{18}$$

$$F_D = -\,\rho\;(V^2 + \dot{y}^2)/2\cdot C_D\cdot D\;s\cdot\dot{y}/\sqrt{V^2 + \dot{y}^2} \simeq -\,\rho\;C_D D\;s\;V\;\dot{y}\;/\;2 \tag{19}$$

where $C_D = 1.2$ (1) denotes the drag coefficient. Eqs.(18)(19) are substituted into eq.(17), and we get

$$M_c\ddot{y} + C_c\dot{y} + K_c y = -\,\rho V^2/2\cdot D\;s\;\{\;f(\alpha + \dot{y}/V) + C_D\dot{y}/V\;\} \tag{20}$$

Here, our mathematical model is introduced by eqs.(16)(18)(20). As it is convenient to describe the equations in non-dimensional formation, the following relations are utilized,

$$Y = y/D\;,\;\eta = C_c/2\omega_0 M_c\;,\;n = \rho D^2 s/2M_c\;,\;\nu = \omega_v/\omega_0 = V/V_{cr}\;,\;\tau = \omega_0 t\;,\;S^* = 2\pi S\;,$$
$$m^* = 1/(0.5 + \bar{L}^*)\;,\;\omega_0 = (K_c/M_c)^{1/2}$$

where n denotes a non-dimensional mass ratio, η, mechanical damping ratio of a circular cylinder, and ν, non-dimensional flow velocity. Eqs.(16)(20)(18) now lead to

$$\ddot{\alpha} - 2\zeta\nu\{1 - (4f^2/C_{LO}^2)\alpha^2\}\dot{\alpha} + \nu^2\alpha = -\,m^*\ddot{Y} - \nu S^*\dot{Y} \tag{21a}$$

$$\ddot{Y} + \{2\eta + n(f + C_D)\nu/S^*\}\dot{Y} + Y = -\,f n\nu^2\alpha/S^{*2} \tag{21b}$$

$$C_L = -\,f\cdot(\alpha + S^*\dot{Y}/\nu) \tag{21c}$$

$(\cdot)$ now denotes differentiation with respect to τ. The above equations compose a mathematical model of vortex-induced oscillation of circular cylinder in non-dimensional formation. We can consider that α is an intermediate variable to obtain Y and C_L. ν is a parameter as a velocity. n, η are constants peculiar to a circular cylinder. $\zeta \simeq 0.038$, $m^* \simeq 0.625$, $S^* \simeq 1.26$ ($S = 0.2$), $f \simeq 1.16$ and $C_D \simeq 1.2$ are the constant factors defined already.

APPROXIMATE STEADY STATE PERIODIC SOLUTION

We are most interested in the region where both cylinder motion and lift fluctuation are approximately sinusoidal at a frequency very close to the natural frequency of the cylinder. In order to examine a solution in that region, let us assume it in the form

$$\alpha = \bar{\alpha}\,\cos(\mu\tau + \psi) \tag{22a}$$

$$Y = Y_0\cos\,\mu\tau \tag{22b}$$

$$C_L = \bar{C}_L\cos(\mu\tau + \phi) \tag{22c}$$

Substituting eqs.(22a,b) into eqs.(21a,b) and equating coefficient of cos $\mu\tau$ and sin $\mu\tau$, and eliminating ψ gives

$$\bar{\alpha} = \frac{C_{LO}}{f}\left[1 \pm \frac{1}{2\zeta}\left\{\frac{f^2\nu^2(m^{*2}\mu^2 + S^{*2}\nu^2)}{S^{*2}[(1-\mu^2)^2 S^{*2}/n^2 + \mu^2\{2\delta^* S + (f + C_D)\nu\}^2]} - \frac{(\nu^2-\mu^2)^2}{\nu^2\mu^2}\right\}^{1/2}\right]^{1/2} \tag{23a}$$

$$Y_0 = \frac{C_{L0}}{S^*} \frac{\nu^2}{[(1-\mu^2)^2 S^{*2}/n^2 + \mu^2\{2\delta^* S + (f+C_D)\nu\}^2]^{1/2}}$$
$$\times \left[1 \pm \frac{1}{2\zeta} \left\{ \frac{f^2\nu^2(m^{*2}\mu^2 + S^{*2}\nu^2)}{S^{*2}[(1-\mu^2)^2 S^{*2}/n^2 + \mu^2\{2\delta^* S + (f+C_D)\nu\}^2]} - \frac{(\nu^2-\mu^2)^2}{\nu^2\mu^2} \right\}^{1/2} \right]^{1/2} \tag{23b}$$

where $\delta^* = 2\pi\eta/n$ (damping ratio). Eliminating $\bar{\alpha}$, Y_0 and ψ gives the following frequency equation.

$$(\nu^2-\mu^2)[(1-\mu^2)^2 + \{2\eta + (f+C_D)n\nu/S^*\}^2\mu^2]$$
$$+\nu^2\mu^2(1-\mu^2)fnm^*/S^{*2} - \{2\eta + (f+C_D)n\nu/S^*\}fn\nu^3\mu^2/S^* = 0 \tag{24}$$

Substituting eqs.(22a,b) into eq.(21c), eliminating ψ and considering eq.(22c) gives the following amplitude of the lift coefficient and phase angle.

$$\bar{C}_L = f\left[\bar{\alpha}^2 - \frac{2S^{*2}\{2\delta^* S + (f/2 + C_D)\nu\}\mu^2}{f\nu^3} Y_0^2\right]^{1/2} \tag{25}$$

$$\phi = \tan^{-1}\{(2\eta + C_D n\nu/S^*)\mu/(1-\mu^2)\} \tag{26}$$

As eq.(24) is in the form of a cubic equation for μ^2, we can solve it precisely. Then substituting the positive root of μ^2 into eq.(23a) and eq.(23b) gives $\bar{\alpha}$ and Y_0 respectively. Further, substituting $\bar{\alpha}$ and Y_0 into eq.(25) gives $\bar{C}_L$. Fig. 4 shows a numerical example of eqs.(23a,b)(24)(25)(26) for a soft-damping cylinder.

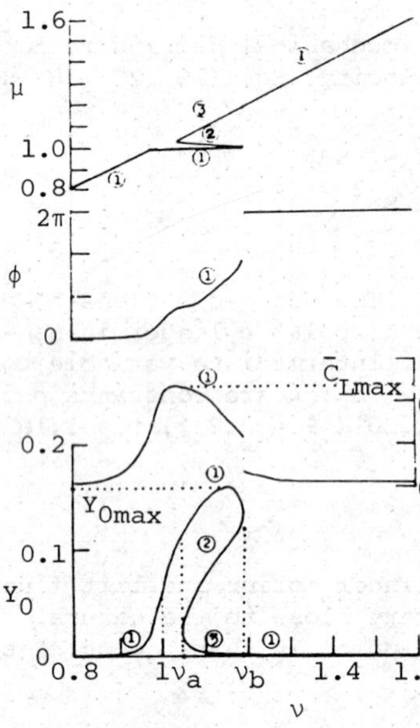

Fig. 4 Numerical example of eqs.(23a,b) (24)(25)(26), n = 0.0034, η = 0.00285

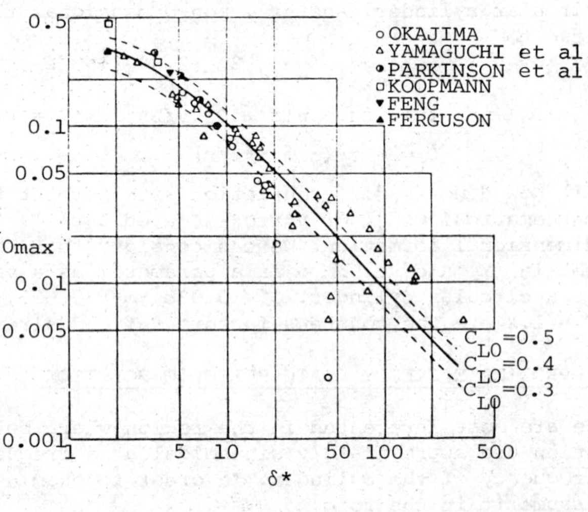

Fig. 5 The maximum displacements of cylinder by rough explanation eq.(27)

So called " synchronization " occurs in the range of velocity $1 \le \nu \le \nu_b$, and many valued response occurs in $\nu_a \le \nu \le \nu_b$. A displacement corresponds to a frequency that has the same number as Fig. 4. Phase angle and lift coefficient are shown only for a frequency ① . The maximum displacement occurs in the range $1 \le \nu \le \nu_b$. At low velocity below unity and high velocity beyond ν_b, the oscillation is of a forced-excited type due to the vortex shedding. But, we must understand that the above discussion is based on the solution assumed to be in the form of eqs.(22a,b,c). The solution that cylinder's frequency is near the natural frequency and lift's frequency is near the vortex shedding frequency for a stationary circular cylinder in the range $\nu > \nu_b$ will be described later (Fig. 6,7,8,9). Fig. 4 shows that the velocity at which the maximum lift occurs is lower than the velocity at which the maximum cylinder displacement occurs (the resonant velocity). For the most part, the results shown in Fig. 4 are in agreement with the experimental results for a soft-damping cylinder (8). The variation of the solution with the increase in mechanical damping is found as follows. Many valued response in the range $\nu_a \leq \nu \leq \nu_b$ cease to exist gradually with the increase in mechanical damping and we imagine that condition as ν_b close on ν_a in Fig. 4. Synchronization with the single valued response occurs. Increasing mechanical damping more, causes synchronization to cease to occur and we imagine that condition as ν_b and ν_a close on unity in Fig. 4. Of course, the displacement of a circular cylinder gradually decreases with the increase in mechanical damping. Finally, the mechanism of oscillation becomes the forced-excited type for a hard-damping cylinder.

ROUGH EXPLANATION OF CYLINDER DISPLACEMENT AND LIFT COEFFICIENT

It is shown that both the maximum displacement of a cylinder and the maximum lift force occur in the range $1 \leq \nu \leq \nu_b$, corresponding to the frequency ①, i.e. $\mu=1$, in Fig. 4. So far as the maximum responses are concerned, we can regard the frequency as unity. Substituting $\mu^2=1$ into eqs.(23a,b) and eq.(25) gives,

$$Y_0 = \frac{C_{L0}}{S^*} \cdot \frac{\nu^2}{2\delta^* S + (f + C_D)\nu} \left[1 + \frac{1}{2\zeta} \left\{ \frac{f^2\nu^2(m^{*2} + S^{*2}\nu^2)}{S^{*2}\{2\delta^* S + (f + C_D)\nu\}^2} - \frac{(\nu^2-1)^2}{\nu^2} \right\}^{1/2} \right]^{1/2} \tag{27}$$

$$\bar{C}_L = C_{L0} \cdot \frac{2\delta^* S + C_D\nu}{2\delta^* S + (f + C_D)\nu} \left[1 + \frac{1}{2\zeta} \left\{ \frac{f^2\nu^2(m^{*2} + S^{*2}\nu^2)}{S^{*2}\{2\delta^* S + (f + C_D)\nu\}^2} - \frac{(\nu^2-1)^2}{\nu^2} \right\}^{1/2} \right]^{1/2} \tag{28}$$

We have only to calculate for ν which makes inside of the $\{ \cdots \}^{1/2}$ positive and $\nu \geq 1$. This condition corresponds to the region of synchronization under vortex-induced oscillation. Calculating Y_0 by eq.(27), and getting the maximum values of them , the variation of the maximum displacement with non-dimensional damping ratio δ^* for some values of C_{L0} is shown in Fig. 5. We see agreement between the results of this model and those of many experiments.

NUMERICAL SOLUTION BY THE RUNGE-KUTTA METHOD

Since the approximate solutions discussed hitherto are in the form of eqs.(22a,b,c) , their meanings and applications are restricted. To compare the solutions of the model with the results of experiments in the wide range of velocity, the numerical solutions of eqs.(21a,b,c) are computed by the Runge-Kutta method. Before computation, we must decide the value of the lift coefficient for a stationary circular cylinder. But it is difficult to decide it uniquely because many different values ($C_{L0} \simeq 0.2 \sim 0.6$, rms value) for the subcritical range of Reynolds number are reported by several investigators. Here, $C_{L0} = 0.4$ is used to calculate, but it is not a definite value. To permit meaningful comparison with experimental data, non-dimensional mass ratio , n , and mechanical damping, η , are made to coincide with those of the experiment. Fig. 6 ∿ Fig. 9 are the comparisons between the experimental results and the solutions of the model. In figures, μ_v denotes the frequency of lift force and μ_c, the frequency of displacement of a cylinder. On the whole, we see agreement between them. As the experimental results of Fig. 7 include the pressure coefficients of an oscillating cylinder, we can qualitatively compare them with the lift coefficient by this model. It shows fairly agreement between them,too. Fig.10 shows the upper limits of velocity at which synchronization and self-excited oscillation occur for the calculated 20 examples, including the above mentioned experimental results.

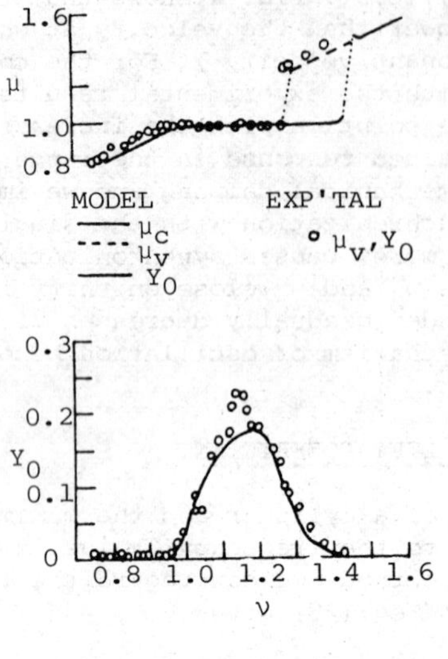

Fig. 6 Comparison of the model with Feng's experiment , n = 0.00257, η = 0.00181, C_{LO} = 0.4

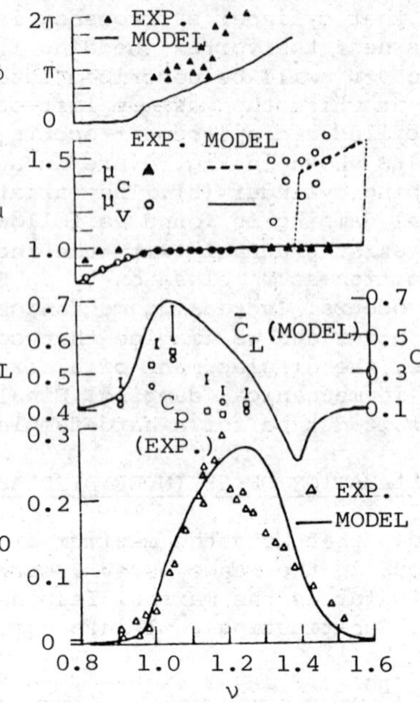

Fig. 7 Comparison of the model with Ferguson's experiment , n = 0.0033, η = 0.001, C_{LO} = 0.4

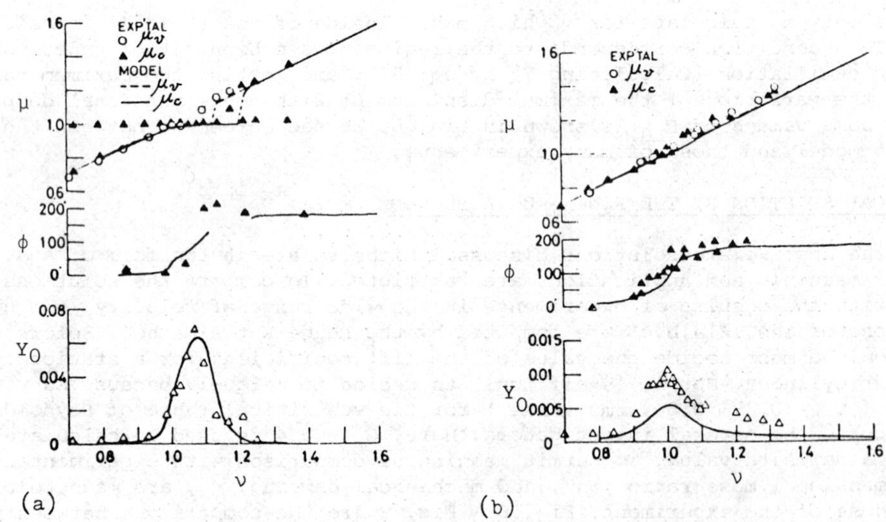

Fig. 8a, 8b Comparison of the model with the experiments of Yamaguchi et al, n = 0.00178, (a) η = 0.004, (b) η = 0.044, C_{LO} = 0.4

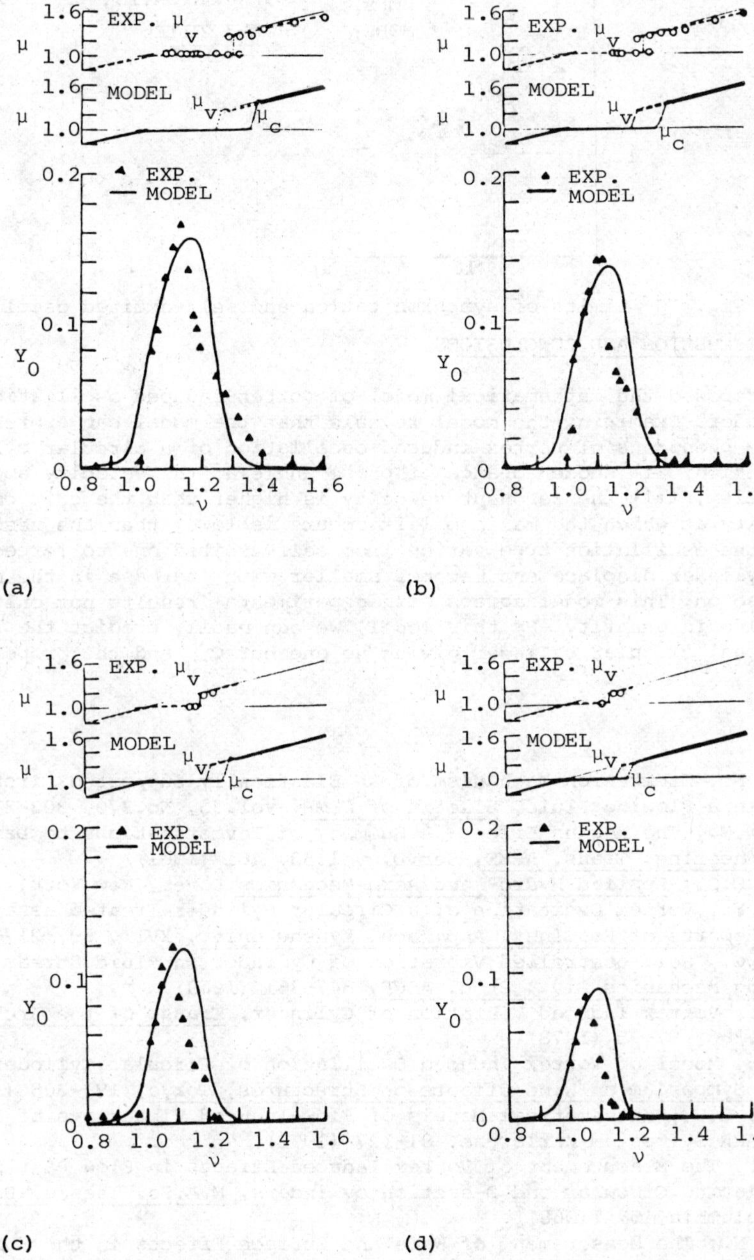

Fig. 9a, 9b, 9c, 9d Comparison of the model with Okajima's experiments, $n = 0.0034$, (a) $\eta = 0.00285$, (b) $\eta = 0.00342$, (c) $\eta = 0.00415$, (d) $\eta = 0.00564$, $C_{L0} = 0.4$

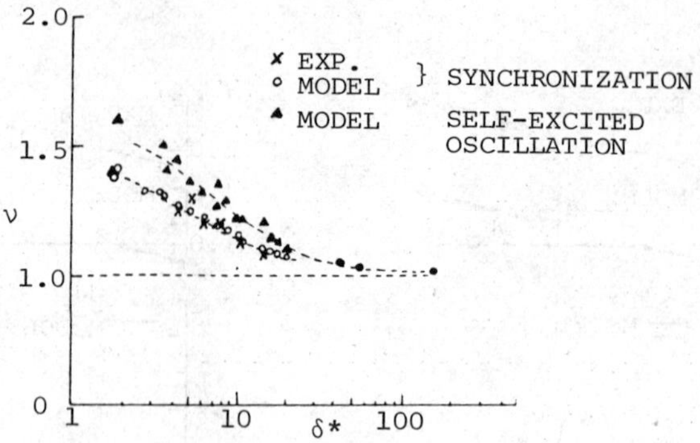

Fig. 10 Limits of synchronization and self-excited oscillation

DISCUSSION AND CONCLUSIONS

The authors proposed the mathematical model of vortex-induced oscillation of a circular cylinder. Examining the model reveals that the model can express the characteristic phenomena of vortex-induced oscillation of a circular cylinder, i.e. i) synchronization, ii) sudden change of phase angle at the velocity about the resonant velocity, iii) the resonant velocity is higher than the critical velocity, iv) the velocity at which the maximum lift occurs is lower than the resonant velocity, v) the oscillation type varies from self-excited one to forced-excited one and the cylinder displacement becomes smaller with increase in the mechanical damping, and so on. This model agrees with experimental results not only in quality but also in quantity. By this model, we can easily predict the response of a two-dimensional circular cylinder giving no one but C_{LO} and the properties of a cylinder.

REFERENCES:

1. Funakawa, M., Excitation Mechanism of an Elastically Supported Circular Cylinder in a Flowing Fluid, Blletin of JSME, vol.35, No.270, 303-312 (1969)
2. Swanson, W.M., The Magnus Effect: A Summary of Investigations to Date, J. of Basic Engineering, Trans. ASME, Ser.D, vol.83, 461 (1961)
3. Tietjens, O.G., Applied Hydro- and Aero-Mechanics,Dover, New York, 1957
4. Nakamura, Y., Vortex Excitation of a Circular Cylinder Treated as a Binary Flutter, Reports of Res.Inst. App.Mech. Kyushu Univ., XVII, 59, 217-234 (1969)
5. Silvio, D.G., Self-controlled Vibration of Cylinder in Fluid Stream, J. of Engineering Mechanics Div., Proc. ASCE, 347-361 (1969)
6. Tamura, Y., Vortex Induced Vibration of Cylinder, Trans. of the Arch. Inst. of Japan, No.266, 87-95 (1978)
7. Tamura, Y., Model of Vortex Induced Oscillation of Circular Cylinder, Proc. of the Fifth Symposium on Wind Effects on Structures, Tokyo, 199-206 (1978)
8. Parkinson, G.V., Mathematical Models of Flow-Induced Vibrations of Bluff Bodies , IUTAM-IAHR Symposium Karlsruhe, 81-127 (1972)
9. Feng, C.C., The Measurement of Vortex Induced Effects in Flow Past Stationary and Oscillating Circular and D-section cylinders, M.A.Sc. Thesis, Univ. of British Columbia, 57 (1968)
10. Ferguson, N., The Measurement of Wake and Surface Effects in the Subcritical Flow Past a Circular Cylinder at Rest and Vortex-Excited Oscillation, M.A.Sc. Thesis, Univ. of British Columbia, 102-110 (1965)
11. Yamaguchi, T., Shiraki, K., Umemura, S., Matsui, Y., On the Vibration of Cylinder by Karman Vortex, Mitsubishi Tech. Rev., vol.7, No.4, 451-457 (1970)
12. Okajima, A., Flows Around an Oscillating Circular Cylinder, Proc. of the 4th Symposium on Wind Effects on Structures, Japan, 147-154 (1976)

DISCUSSION ON SESSION VIII
Physical and Mathematical Modeling

Rapporteur: W. Z. Sadeh
Colorado State University
USA

Paper by J. C. R. Hunt, Chairman--RECENT DEVELOPMENTS AND TRENDS

Comment: J. E. Cermak. Your opening remarks imply that mathematical modeling of diffusion is more reliable than physical modeling. An extremely significant exception exists in the context of wind engineering. Short range dispersion over and around building groups and over complex terrain where advective transport dominates dispersion cannot, at present, be predicted by mathematical modeling but must be treated by physical modeling. Comparisons of concentrations measured in well designed physical models with field measurements are in good agreement.

Reply: J. C. R. Hunt. Abbey (1976) has shown that in the wakes of isolated structures, the wind tunnel concentrations were too large, even quite close to the structure. This is basically because lateral diffusion (σ_y) and the lateral growth at wakes are too small in wind tunnels, unlike σ_z, the vertical growth. But, if a building is located amongst other buildings, then σ_y is probably well modeled. So, there are some cases where I agree with Dr. Cermak, but there are many building and complex terrain diffusion cases where wind tunnel data, field data for σ_y and some theoretical interpretation are, I believe, a more reliable basis for prediction than a wholly physical modeling approach.

Paper by D. J. Wilson, G. Winkel, O. Neiman--REYNOLDS NUMBER EFFECTS ON FLOW RECIRCULATION BEHIND TWO-DIMENSIONAL OBSTACLES IN A TURBULENT BOUNDARY LAYER

Question: C. F. Shieh. Figure 4 showing the Reynolds stresses in dimensional sense gives me an impression that these stresses remain the same in all the cases investigated. If this is the case, the increases of U_H or H reduces the effect of the turbulence mixing process comparing with the effect of inertia force. This can cause the dimensionless reattachment point to move downstream. This contrasts the results shown in Figure 6. Besides the blockage effect, the results of Figure 6 can be caused by the effect of turbulence structure in the approaching wind flow. Has this effect been considered?

Reply: D. J. Wilson, G. Winkel, O. Neiman. The data in Figure 4 are for $U_\delta \approx 1.0$ m/s and are shown in dimensional form to emphasize that normalized values such as $\overline{u'w'}/U_\delta^2$ will change somewhat as Reynolds number changes over the two decade range used in the study. These changes in approach flow structure are lumped with the other Reynolds number effects on the hill.

Question: J. E. Cermak. What was the surface roughness and edge geometry of your wind tunnel models? Have you considered the effects of these variables in attempting to determine Reynolds number effects?

Reply: D. J. Wilson, G. Winkel, O. Neiman. All surfaces were smooth painted metal or wood. Corner radii and edge thickness for water channel models are shown in Figure 2. For wind tunnel models the radii on the largest dike were about 1.0 cm, and 0.2 for the two smaller dikes. There undoubtedly had some effect, but probably these effects are small as all flow separation occurred at the upwind edge of the flat dike top. Only the reattachment point varied with Reynolds number.

Question: W. Z. Sadeh. (1) You imply that for separation fixed at sharp edges the recirculation and reattachment are quite insensitive to Reynolds number effects. The question is if this insensitivity is limited to a certain Reynolds number range or is prevalent at any Reynolds number? (2) What about the effect of surface roughness upon the simulation? This is directly related to the characteristics of the incident flow.

Reply: D. J. Wilson, G. Winkel, O. Neiman. (1) As long as Reynolds number is high enough to assure turbulent approach flow reattachment should be insensitive to Reynolds number. Our tests cover a factor of 20 change in $Re = U_\delta \theta / \nu$ and produce about a 5% change in Lc/H. (2) This study considered only a single approach flow type. The reattachment length will be influenced by interaction with turbulence intensity in a range of turbulence scales. Other turbulent boundary layers may produce somewhat different results by virtue of their different scale and intensity. However, the qualitative effects should be much the same.

Question: C. M. Sheih. In computing Reynolds number for plotting L/H, I feel U* of approaching flow will be more appropriate than U_H used because U* is a better candidate representing the small-scale turbulence with an obstacle. To substantiate the above point, I have done a simple derivation which indicates $L/H \sim \ln \frac{U^*H}{\nu}$. Therefore, if they are plotted in a semi-logarithmic coordinate, the curve will be linear.

Reply: D. J. Wilson, G. Winkel, O. Neiman. The friction velocity U* is certainly a better choice than U for the scaling velocity. Unfortunately, the rough-smooth transition 0.5 m upwind of the model caused a bend in the u'w' profile (Figure 4) near the surface, making estimates for U* uncertain. Also, selected turbulence measurements showed considerable structure changes over the wide velocity range from $U_\delta = 0.2$ to 20.0 m/s. U_δ was chosen as the most accurately measured velocity scale. The term $\ln(U^*H/\nu)$ takes H as the length scale, which does not seem appropriate from the data in Figure 6. Perhaps a scale of turbulence would be more appropriate.

Question: I. P. Castro. To emphasize Dr. Wilson's comments about blockage, we have found that this can successfully 'scramble' the trends in, say, R_L/H as upstream flow parameters are changed (Castro, Fackrell, J. Ind. Assoc. 3, 1, 1978). Regarding flow unsteadiness around reattachment, I do not believe, on the basis of our measurements and those of Eaton and Johnston at Stanford, that this unsteadiness is strongly periodic. Does Dr. Wilson have evidence that his flow has strong periodic components around reattachment, or is it simply large-scale random unsteadiness?

Reply: D. J. Wilson, G. Winkel, O. Neiman. We would prefer to refer to the unsteadiness as "intermittent" rather than periodic. The evidence is based on a few qualitative flow visualization observations.
We make the point at least twice in the paper that some of the δ/H effects observed may be caused by blockage. The δ/H effects are weak compared to the suppression of recirculation at high Reynolds numbers.

Question: R. E. Britter. We (Britter, Hung and Richards, 1979, Subm. to Q.J.R.M.S.) have found that separation and reattachment of flow over hills is strongly influenced by change in roughness near the hill. For example, at a given Reynolds number and a rough wall, turbulent boundary layer, the flow over a rough hill may separate whereas flow over a smooth hill does not; a result of an accelerating inner layer over the hill. Did you authors continue the boundary layer roughness up to and over the hill? Have they observed or studied the above phenomenon?

Reply: D. J. Wilson, G. Winkel, O. Neiman. This is an important point which we have not yet studied.

Paper by R. N. Meroney, D. Lindley, A. J. Bowen--PHYSICAL MODELING OF FLOW OVER COMPLEX TERRAIN

Question: J. C. R. Hunt. Would I be correct in thinking that in your comparison between field and laboratory wind speeds you were referring to location of peak wind speeds such as on hilltops and not in or near recirculating wake regions? Provided the wind speed is not measured very close to the surface (> 10 m) then our theory and wind tunnel modeling work supports your conclusion for hilltops. So, in a sense you are correct in ignoring Reynolds number effects and also Wilson is correct in worrying about them because he is interested in the recirculating region.

Reply: R. N. Meroney, D. Lindley, A. J. Bowen. All measurements were made in well exposed regions; however, one third were made within the gorge, behind, or upwind of topography where separation may have been present.

Question: R. K. Cooper. Topographical modeling has given a 'reasonable' correlation for mean speed ratios. Do the authors have any idea of the correlation for short period extreme gusts?

Reply: R. N. Meroney, D. Lindley, A. J. Bowen. Dr. Holmes, James Cook University, recently completed a study over complex terrain where maximum gusts were compared between field and model. Correlation coefficients were approximately ~ 0.8. See Holmes, J. D., Walker, G. R., and Steen, W. E. (1979), "The Effect of an Isolated Hill on Wind Velocities Near Ground Level-Initial Measurements," Wind Engineering Report 3/79, Dept. of Civil and Systems Engineering, James Cook University of North Queensland, Townsville, Queensland, Australia.

Question: B. Tryggvason. How was U_δ determined in full-scale? What were the averaging times for $\overline{U}$ in both full-scale and model scale? Can you comment on the stationarity of the flow in full-scale?

Reply: R. N. Meroney, D. Lindley, A. J. Bowen. The gradient wind was estimated by assuming the $\overline{U}/\overline{U}g$ values measured over the model was correct for the WETF tower site. Probable gradient wind speeds near noon were then 16.3 m/sec on Nov. 25, 1977, and 23.4 m/sec on Dec. 28, 1977. The equivalent averaging time for model and field measurements was 10 minutes. The wind field over the Rakaia Gorge was associated with a "northwest" wind event typical to the area of high wind speed, constant direction character, on Nov. 25, 1977, the WETF anemometer remained at levels near 8 to 10 m/sec for four hours, on Dec. 28, 1977, the winds remained at levels between 9 and 18 m/sec for ~ six hours.

Paper by N. Isyumov, N. Tanaka--WIND TUNNEL MODELING OF STACK GAS DISPERSION - DIFFICULTIES AND APPROXIMATIONS

Question: R. N. Meroney. Did you apply corrections to your concentration measurements for the volume flux modified model methods?

Comment: R. N. Meroney. Method designated as CSU method was used expecting a conservative plume rise.

Reply: N. Isyumov, N. Tanaka. Yes we did. See Equation 13.

Paper by D. J. Koga, J. L. Way--EFFECTS OF STACK HEIGHT AND POSITION ON POLLUTANT DISPERSION IN BUILDING WAKES

Question: W. Z. Sadeh. I am wondering about the role played by the incident flow and particularly its turbulence characteristics with regard to the stack position and the features of the resulting plume. In particular, the near-wake structure and the plume properties within this region can be strongly affected by the oncoming flow. Could you comment about these points?

Reply: D. J. Koga, J. L. Way. The approaching flow for this study is best represented by Pasquill stability class 'C', with a power-law exponent of .17 and a turbulence intensity of 12% at the building height. We recognize that a study of the sensitivity of the present results to varying approach flow boundary layers would be a suitable topic for future investigation.

Question: L. H. Lamouria. In view of your findings, would you speculate as to the possible effectiveness of plume modifiers such as wind deflector devices which might be more cost and environmentally effective than increased stack height?

Reply: D. J. Koga, J. L. Way. The addition of flow manipulators to control plume behavior on or near buildings is a very complex issue. Since our investigation did not cover the effects of flow manipulators, we choose not to speculate on this important issue.

Question: J. W. Saunders. Can you comment on the reliability and accuracy of your thermistor technique compared with gas chromatograph techniques?

Reply: D. J. Koga, J. L. Way. We performed no direct comparison of the thermal tracing technique with gas chromatography. However, we are confident that the Schmidt number for this problem is close to unity. Resolution of the thermistor circuitry was 0.005°C with no major drift problems.

Comment: R. N. Meroney. Recently NRC incorporated an empirical formulation for entrainment into its regulations based on field measurements over a Nuclear Power Plant where the stack was always on the upwind edge of the structure. Recent field/model study done over Duane Arnold Energy Center, Iowa, revealed a stack on lee building edge entrains far worse just opposite to the effect your study indicates. The stack was fairly short. Can you comment about this discrepancy?

Reply: D. J. Koga, J. L. Way. Figure 10 indicates that our quantitative centerline ground level concentration measurements are indeed consistent with the above results for the rooftop emission case. Our HS = 1 most closely corresponds to Dr. Meroney's "fairly short" stack as indicated by verbal communication. The apparent contradiction with the qualitative results of Figure 8 stems from viewing the integrated smoke density across the finite-width illuminating band of light (that is, when the emission point is at the leading edge of the building, the smoke has a better change of diffusing across the flow to fill the illuminated region).

Paper by R. L. Petersen, J. E. Cermak--PLUME RISE FOR VARYING AMBIENT TURBULENCE, THERMAL STRATIFICATION, AND STACK EXIT CONDITIONS--A NUMERICAL AND LABORATORY EVALUATION

Question: T. B. Morrow. A second paper by Ooms, Mahieu and Zelis showed difficulty in matching the plume rise trajectories photographed by Bodurtha for negatively buoyant plumes in wind tunnel experiments. I have also had difficulty with matching negatively buoyant plume boundaries measured by Brummage in wind tunnel experiments with Ooms model predictions. In both sets of experiments the stack exit Reynolds numbers were less than 1000 and the wind turbulence level was low. In your own work, did you observe any systematic dependence of α_1 and α_2 on stack exit Reynolds number?

Reply: R. L. Petersen, J. E. Cermak. The dependence of α_1 and α_2 was not specifically investigated. However, the results showed that the numerical model did not compare as favorably with the wind tunnel observations for the low Reynolds number cases (i.e., Re = 954). This may suggest a Re dependence for α_1 and α_2 or a Froude number dependence, since the low Re cases also corresponded to the low Froude number cases.

Paper by J. D. Iversen--DRIFTING SNOW SIMILITUDE--DRIFT DEPOSIT RATE CORRELATION

Question: G. T. Skinner. Because small changes in wind angle may dramatically alter the zones of deposition and clearing, we might expect a large influence of meandering of the wind. This is probably more difficult to take into account than in pollution studies. Is anything ever done about this?

Reply: J. D. Iversen. The mean wind in the wind tunnel is of course a constant in direction although lateral turbulent gusts do effect snowdrift shapes to some extent in the wind tunnel. During a typical snowstorm, the mean wind direction may gradually change from a few degrees to a large angle, depending on storm characteristics and location. This is difficult to simulate in the wind tunnel. The mean wind direction, instead was varied in finite jumps of 20°.

Question: J. Templin. You refer to quantitative measurements of snow accumulation based on area of simulated snow cover. It seems that volume or total mass of snow would be more important for most applications including snow removal from highways or snow accumulation on load bearing structures. Do you have any suggestions as to how your area measurements can be related to column or mass accumulation of snow?

Reply: J. D. Iversen. In some experiments, volume of accumulated material was measured by taking drift cross-section measurements as well as plan area. Enough of these measurements were performed to fairly accurately estimate column accumulation for the remainder of the experiments.

Question: R. J. Kind. Our work (Reference 17 of Dr. Iversen's paper) considers the same problem and arrives at somewhat different conclusions. In particular, one of our similarity requirements is concerned with ensuring that the flow over the saltating particle bed is aerodynamically rough as it is in actual snowdrifting. On the other hand, Dr. Iversen's requirement that (z_o'/L) be matched virtually ensures aerodynamically smooth flow in the model since $(z_o'U^*/\nu)$ is near the lower limit for aerodynamically rough flow in actual snowdrifting. Won't this seriously affect similitude?

Reply: J. D. Iversen. The model, without moving material, was aerodynamically rough for the wind speeds which were used. The effect of aerodynamic roughness with moving materials was determined, at model scale, by varying the roughness by variations in particle density, diameter, wind speed, and vertical geometric distortion (roughness was varied by a factor of ten, resulting in a transport rate change by a factor of two). The ultimate proof of ability to extrapolate to full-scale, of course, will have to come from thorough time-dependent full-scale measurements.

Question: W. Z. Sadeh. (1) Could you comment on the role or effect of the underlying terrain roughness upon the snowdrifting? (2) What about the turbulent structure of the wind and its effects upon snowdrifting for a given surface roughness?

Reply: J. D. Iversen. (1) Underlying terrain roughness will cause increase drifting among the roughness elements until the roughness elements are covered. This is a possible effective method of snow control in some situations.
(2) This is a difficult question to answer yet. In our experiments to date, only one turbulence level (caused by upstream turbulence generator and roughness) has existed.

Question: R. L. Wardlaw. In dimensionless considerations of snowdrifting little mention is made of the mechanical properties of the snow. This includes the adhesion properties that lead to the cusped drift foundations that we are accustomed to in full-scale observations. Could Dr. Iversen comment on the importance of the mechanical properties of the snow in predicting extent of drifting, rate of deposition or find depth of snow?

Reply: J. D. Iversen. Certainly the cohesive properties of snow have an effect on the final or equilibrium shapes of snowdrifts associated with simple or complex boundary layer obstruction shapes but the effect is probably greater before equilibrium is achieved because this is the stage when the effects of cohesion are most easily observed in the form of snow cornices, overhangs, etc. It is this property of cohesion which determines the greatest distinction between snowdrift shapes and those in blown sand (ripple, dunes, etc.). The modeling particles chosen approaches snow in its ability to form angles of response greater than 34° (up to 70°) due to electrostatic effects.

Question: J. E. Cermak. Have you attempted to simulate snow deposition rates during a snowstorm having a specified rate of snowfall? If so, what method was used to introduce snow into the flow system?

Reply: J. D. Iversen. This has been attempted by introducing simulated snow particles at a controlled rate from a container located in the tunnel ceiling. Our wind tunnel needs major modification to make this technique successful. Most of the experiments performed to date simulate the situation of a heavy snowfall without wind followed by a strong drift-causing wind, which is not an atypical situation.

Paper by M. C. Jischke, B. D. Light--LABORATORY SIMULATION OF TORNADIC WIND LOADS ON A CYLINDRICAL STRUCTURE

Question: K. Mehta. In many strong tornadoes, there are satellite vortices (often called suction vortices). What effects would these vortices have, if any, on the pressures that were measured? Would the authors care to comment? (Authors are not obliged to comment.)

Reply: M. C. Jischke, B. D. Light. We have not yet made measurements when there are multiple ("suction") vortices. While it is clear that there will be unsteadiness and pressure pealing, we cannot yet estimate their magnitude. We intend to carry out such measurements.

Question: W. Z. Sadeh. The system described does not include simulation of the axial translation and, apparently, of the surface roughness. To what extent can you disregard the foregoing two conditions? Isn't the axial translations significant in ascertaining tornadic wind characteristics and effects?

Reply: M. C. Jischke, B. D. Light. Surface roughness was included in this study. The effects of tornado translation was not included. They are the subject of current on-going work. The significance of relative tornado/structure motion is difficult to assess.

Question: L. Brown. The paper states that this study ignores the effects of relative motion between the tornado and structure. Have you considered these effects in another study? Have you done tests using multiple structures? If so, what were your findings with respect to peak pressure, pressure changes, and rates of pressure changes? If not, what do you think the bounds might be for a "worse case" (using realistic tornado inputs)?

Reply: M. C. Jischke, B. D. Light. We are beginning a study of the effects of relative motion of the tornado and the structure. We have not considered multiple structures. At present, we cannot estimate the effects of the unsteadiness on the pressure loads.

Paper by A. B. Poore, A. R. Al-Rawi--THE DYNAMICAL BEHAVIOR OF THE HARTLEN-CURRIE WAKE OSCILLATOR MODEL

Question: A. Kareem. You have studied the H-C model in great depth, have you tried to come up with some suggestions for the improvement in this model?
Reply: A. B. Poore, A. R. Al-Rawi. No. As we investigate the Iwan-Blevins model and the Tamura model, improvements may suggest themselves so that we can obtain more accurate quantitative agreement with experiments. Our present work is qualitative in nature.

Question: C. V. Parkinson. In bifurcation and limit cycle theory, I am used to finding stable cycles separated by an unstable cycle, and vice versa. This does not seem to be the case in all of the results shown in the paper. Can the authors explain?
Reply: A. B. Poore, A. R. Al-Rawi. For plane autonomous differential equations stable limit cycles are separated by unstable cycles. This is no longer true for higher dimensional (more than two) problems such as the one investigated here.

Comment: I. G. Currie. The principal conclusion of this paper is that the Hartlen-Currie model is capable of exhibiting jumps in amplitude with hysteresis and double-amplitudes (as have been observed physically). It is implied (though not stated explicitly) that the authors have found these phenomena for the first time from this particular model. We reported such phemomena from the Hartlen-Currie model at the previous conference in this series (1975). This work is not referenced on the current paper, hence the implication.
Reply: A. B. Poore, A. R. Al-Rawi. We certainly are not the first to point out the double amplitude response in your model. Thank you for your reference. We do hope that we have shown how prevalent the double amplitude response is in your model.

Paper by Y. Tamura, G. Matsui--WAKE-OSCILLATOR MODEL OF VORTEX-INDUCED OSCILLATION OF CIRCULAR CYLINDER

Question: A. Adenicisa. One important problem in all models of the phenomena of vortex shedding is the choice of parameter. How did you choose your parameters?
Reply: Y. Tamura, G. Matsui. It is an important problem to decice the parameters of the model. In fact, it is difficult for the Hartlen-Currie model. But, those of my model are decided reasonably in my paper, as you can read.

Comment: A. Adenicisa. The most important result is your Equation 21c. In your numerical interpretation was the effect of $\ddot{y}$ very small as expected from other models like Hartlen and Currie since your equations can be transformed to Hartlen and Currie.
Reply: Y. Tamura, G. Matsui. The effect of $\ddot{y}$ is shown as the term of m* in Equations 23b and 23c. It is almost the same order as the effect of $\dot{y}$, as you can see. Therefore, I think you have misunderstood the contents of my paper.

Question: G. V. Parkinson. While applauding the ingenuity of the authors' model, and its good simulation of many observed phenomena of vortex-induced oscillation, I must urge caution in the use of the kind of semi-empirical flow field modeling seen in the paper. In particular, one should not appeal to the Kutta-Joukowsky law, a consequence of the steady-flow Blasius equations, but rather to a solution, for a particular vortex flow model, of the unsteady flow Blasius equations, as for example in my treatment of Etkin's model in Reference 8 of the paper.

Reply: Y. Tamura, G. Matsui. Thank you for your very kind advisory comment. I employed the law and the relation for "steady" phenomena to estimate K and $\overline{C}$. In this paper, I tried to establish my mathematical model in very simplified method and too much complicated consideration was excluded. More precise study and improvement on this model should be taken, of course.

SESSION IX

Wind-Engineering Applications

CHAIRMAN

R. E. Whitbread
National Maritime Institute
Teddington, United Kingdom

CO-CHAIRMAN

F. A. Gifford
Director, Atmospheric Turbulence and Diffusion Laboratory
National Oceanic & Atmospheric Administration
Oak Ridge, Tennessee, USA

RAPPORTEUR

R. N. Meroney
Professor of Civil Engineering
Colorado State University
Fort Collins, Colorado, USA

RECENT DEVELOPMENTS AND TRENDS: WIND-ENGINEERING APPLICATIONS

R. E. WHITBREAD
National Maritime Institute, UK

A history of Wind Engineering would undoubtedly show that it is the practical applications which have stimulated its growth as a subject for research and scientific development. Various separate disciplines have grown up, namely wind structure, wind environment, steady wind loading and dynamic response, and these four themes have been echoed strongly in each of the Conferences held since 1963. However, one has only to scratch the surface of any of these themes to find that the papers presented have particular practical applications; most frequently to stacks, towers, tall buildings and particularly to long-span bridges.

Wind engineering applications have their origins in the problems faced by the practising engineer. He in turn, is faced with social, environmental and economic pressures, and is obliged to seek specialist advice on the likely effects of wind, particularly in those cases where no design precident exists.

It is possible on the other hand, to take the more cynical view that the most significant developments in Wind Engineering can be traced to catastrophic disasters in which the wind either directly or indirectly has been responsible for a failure of a structure or the creation of an environmental hazard. The Tay Bridge disaster in 1879 stimulated a keen interest in wind loading for bridges in the UK which was soon reflected in the classic design by Baker for the Forth Bridge in 1885. In 1940, the Tacoma Narrows Bridge took things a stage further by focussing attention on the dynamic response of long-span suspension bridges; a subject which continues to draw more papers at these Conferences than almost any other single type of structure.

Another major application has been the need to quantify the wind loading and dynamic response of tall buildings. The work of Rathbun published in 1940 was of great significance. He attempted to monitor the wind loading and dynamic response of the Empire State Building for comparison with the earlier wind-tunnel measurements made by Dryden and Hill (1933). Even now, some 40 years on, research workers are still struggling to achieve that same elusive correlation between model and full-scale wind loading.

The majority of the data currently available in the field of Wind Engineering has been determined by wind-tunnel testing. Such measurements are fairly readily made with modern instrumentation, are repeatable, are obtained on a short time scale and are relatively inexpensive. The converse is true of full-scale measurements on virtually all counts.

The number of papers reporting on full-scale measurements has steadily increased since the first Conference in 1963. There are papers relating to full-scale measurements elsewhere and two have been allotted to this particular session. Fortier and Scanlon's paper is an attempt to extend the application of a limited set of full-scale pressure measurements on a cooling tower using a specialised statistical treatment. Fujimoto et al, on the other hand, have investigated the dynamic response of both model and full-scale lattice towers.

The high degree of scatter in their full-scale results has made for difficulties in the quantitative comparison of model and full-scale data.

In future perhaps, we must expect to see treatments of full-scale data being developed which are capable of removing the worst of this scatter so that model and full-scale data can be correlated with greater confidence. Such a correlation is essential if the growing volume of wind-tunnel data is to be made acceptable for use with the next generation of design codes.

Such codes will undoubtedly include treatments of wind loading as a dynamic effect, and the fact that three of the ten sessions of this Conference are concerned directly with unsteady loads and dynamic response bears witness to the continued development in this general area. Methods for predicting the dynamic response of structures to wind are now available to the structural engineer. What he lacks is comprehensive guidance on methods which can be relied on to eliminate or at least minimise these dynamic effects without imposing either unacceptable constraints on design or excessive penalties on cost. The two papers in this session, one by Berlamont and Vanderstappen and the other by Korenev aim to provide such a solution in terms of a parameter which, in the past, has been largely neglected by structural engineers namely, damping. In fact, very few papers in the history of this Conference have dealt explicitly with this topic. Methods for both the prediction and control of the damping properties of structures will assume increasing importance especially as economic constraints dictate lighter and hence more flexible structures.

A topical application of Wind Engineering, which has not appeared in any previous Conference, is that concerned with the so called renewable energy sources of wind and solar power.

Research and development in the field of Wind Energy Conversion Systems (WECS) has progressed rapidly in the last 5 years and meetings have been held elsewhere which have been devoted entirely to this topic. The two papers in this session are not so much concerned with the design of wind power generators but rather with their placement within the wind field to extract the maximum power for a given location. Shieh and Frost have used an analytical model of the flow over a bluff obstacle to determine the optimum size and location of the generator in relation to the local flow field. Meroney, Lindley and Bowen on the other hand, have carried out a wind survey on a topographical wind-tunnel model as a possible means of selecting the optimum site for a WECS.

The future of wind power as a major contribution to the total energy budget remains uncertain. Estimates of the unit cost for wind-generated electricity are put at two or three times that of nuclear power. Nevertheless, the accident at the Three Mile Island power station at Harrisburg has undoubtedly strengthened the case for exploiting alternative energy sources. It should however be remembered that vast numbers of wind generators would be required to produce significant power and this in itself would create problems of interference with the amenity value of mountain or coastal regions which are the sites favoured for such systems. One proposal being actively explored in the UK is for the siting of wind power systems in shallow off-shore areas. Although such areas would seem ideal with respect to minimising their visual impact, it is apparent that relatively little data exists on the wind profile and turbulence structure which occur over large stretches of open water. Pilot studies by NMI are already in progress both to measure wind structure over the sea and to correlate the information with the sea state.

Wind Energy Conversion Systems will no doubt have a part to play in the future and certainly the optimisation of the siting of such systems and possibly even their aerodynamic design offer an application of wind engineering well worth exploiting.

The influence of wind on solar power systems is less direct. The paper by Xerikos, Tang, Cermak and Peterka refers to the effects of wind on an array of some 1800 heliostats. Any distortion or loss of alignment of these heliostats due to wind may seriously reduce the efficiency of the system. Their aerodynamic behaviour is therefore a matter of some importance.

Man-made environmental hazards have been a matter for increasing public and national concern over the past few years. The range of volatile and toxic chemicals which are both stored and transported in large quantities is continuing to grow and a number of horrifying accidents have occurred.

The paper by Morrow reflects a growing concern in this area, where the effects of wind on the dispersion of vapour clouds discharged, either by accident or as part of controlled venting, may be crucial. Work of this type is in progress in the UK, much of it confidential, aimed at identifying the principal parameters which operate in the case of a major spillage of a volatile chemical in the vicinity of a residential or industrial neighbourhood.

Two further papers in this session deal with pollution of the atmosphere by the products of combustion. Nishi and Cermak have developed an analytical model for the prediction of the ground level concentrations of carbon monoxide and the oxides of nitrogen which can be expected to be produced by jet aircraft using approach and take-off flight paths to a proposed new airport off the coast of Japan.

Lethal concentrations of gaseous combustion products combined with smoke can occur in building fires particularly where plastics and other synthetic materials are involved. Most modern buildings are controlled by strict regulations which require that escape routes to ground level be kept smoke free. The degree to which such schemes are successful is not easily verified. The paper by Kandola and Marchant is therefore of particular interest in that they have combined, in a single experimental model, both the pressure differentials induced by the heat of the fire and those caused by the flow of wind around the buildings. Both sets of pressures are then allowed to act on the leakage paths through the building to determine the movement of smoke.

The paper by Cooper is the first to be presented at one of these Conferences dealing specifically with the question of vehicle aerodynamics. There is no doubt that this is a topic of considerable commercial interest to the major vehicle manufacturers with the increasing pressure on them to reduce fuel consumption. Much of the current development work is done with great secrecy and exchange of information is strictly limited. Air dams on the fronts of private cars together with spoilers on the roofs have both reduced the drag and improved road holding in many cases. Even greater benefits are likely to be obtained by the improved aerodynamic performance of heavy freight vehicles, especially those which cruise at relatively high speed over long distances. Opportunities exist for a more rigorous approach to the subject. The necessary facilities and expertise are to be found in most of the centres currently working in the field of Wind Engineering.

Wind Engineering is a continuously evolving subject. A novel application in 1979 may well have become an established discipline by 1983. New challenges are constantly occurring and the innovation of a Session devoted entirely to 'Applications' will, it is hoped, become a feature of future Conferences.

WIND INDUCED MOVEMENT OF SMOKE WITHIN AND AROUND BUILDINGS - A WIND TUNNEL INVESTIGATION

B. S. KANDOLA and E. W. MARCHANT
University of Edinburgh, Edinburgh, UK

ABSTRACT

This paper presents the results of pressure measurements on the external and internal surfaces and in the spaces of a model building. The model had defined barrier leakages and a fire pressure simulation facility. The objective of these measurements was to provide data for the prediction of potential smoke flow paths. The generation of adverse pressures by the natural wind, that is those causing flow into escape routes, need early, pre-construction identification and the partial study presented shows some of the probable flow patterns. It is suggested that appropriate analysis of building designs will lead to the provisions of smoke control systems of greater effectiveness than hitherto.

PURPOSE OF INVESTIGATION

The experimental work presented in this paper is part of a larger research project on smoke control problems in buildings. The main purpose of the total project is to correlate the behaviour of air flow and smoke flow in full size buildings with experimental model and a predictive computer program which is intended for calculating potential smoke movement in buildings.

This paper presents part of the experimental investigation into the wind induced air flows that can be expected in ordinary buildings. The flow potentials generated by wind induced pressures are analysed with respect to the possible movement of smoke within a building.

DESCRIPTION OF MODEL

A 32-cell, 5-storey perspex model scaled to approximately 1/25 full size. Each floor is arranged as shown in Figure 1 (see also Plate 1). The overall dimensions are 425 x 425 x 505 mm high.

D is a sealed vertical duct which contains measuring tubes and air input ducts. These are required for fire pressure simulation and corridor (C) or staircase shaft (S) pressurization. Walls R and T have variable leakage and T contains a sealed doorway. The other two sides are sealed effectively but do contain pressure taps for external pressure movements and sealed removeable panels for the necessary manipulation of internal doors. Spaces 1-6 represent a set of rooms each having the same volume. 'Doors' to the corridor contain holes to represent defined leakage areas and each door is sealed to the corridor wall by a tubular seal (PVC) and kept in place by a turnbuckle arrangement. All doors are removeable. 'Windows' to each room are represented by removeable perspex panels, each with three holes which simulate representative leakage areas. Walls T and R contained the leakage areas; all other wall, floor and roof surfaces were regarded as impermeable.

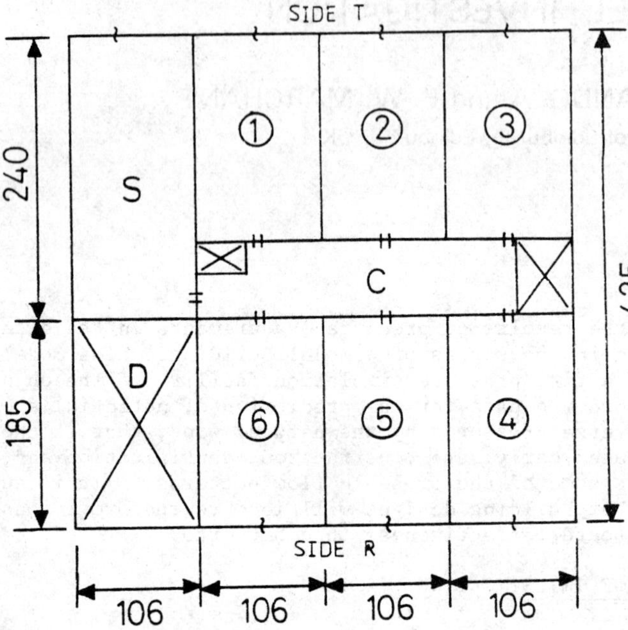

Figure 1: Floor plan of model (dimension:mm).

PRESSURE MEASUREMENT

Pressures inside the model were measured by installing about 200 tubes which ended on surfaces or in the spaces of the model. The tubes were connected to multi-channel pressure scanners then the pressures measured by a micro-manometer and recorded on pen chart recorders. The wind tunnel used was a low velocity open jet tunnel (1070 x 1525 mm), the air being given a turbulent velocity profile by elliptical vortex generators. The increase in velocity with height follows a power law profile having an exponent of 0.4, which is representative of natural wind in urban areas.

LEAKAGES USED

In the results presented, only two air leakage configurations were used. For both configurations the window and door leakages were kept the same, but the leakage available to the stair shaft (S) was changed. The leakage to the shaft represents either that due to windows throughout its height (Leakage (a)) or that due to a single closed entrance door (Leakage (b)).

Table 1

Leakage (a)

T	D	R
1.5	1.5	1.5

Leakage (b)

T	D	R
1.5*	1.5	1.5

The configuration for leakage (a) was that wall T had a leakage equal to 1.5 per cent of the total area of wall T, and wall R the same percentage leakage of wall R but with the outer part of duct (D) excluded. D:1.5 represents the leakage equal to 1.5 per cent of the area of all doors onto the corridor (C).

MEASUREMENTS MADE

Pressure measurements were made on all external surfaces and inside all spaces. These results were normalised to pressure coefficients (Cp) and these were used to calculate differential pressure coefficients (ΔCp) where these are the algebraic difference of the pressure coefficients across a barrier. Cp is defined as the ratio of the actual measured pressure (P) to the dynamic pressure (q) at the building height.

$$Cp = \frac{P}{q} \;; \quad \text{where } q = 0.5\, \rho\, U^2$$

ρ = density of air; U = velocity of air at height of model.

All relevant pressures were measured with and without a simulated fire pressure. This was simulated in room ⑥ on floor 2 or floor 4. In addition the measurements were repeated for wind incidence 0° to 270° at intervals of 45°. For 0° ($\alpha = 0$) the airflow is normal to wall T.

The fire pressure was simulated by introducing air into a room and generating a pressure across the room door $\nless$ 7.5; $\ngtr$ 10 Pa ($\nless$ 11 x 10^{-4}; $\ngtr$ 14.5 x 10^{-4}psi) in still air conditions (1). This range of pressure was chosen because available evidence suggests that fire induced pressure differences do not exceed about 7.5 Pa (11 x 10^{-4} psi) in ordinary buildings (1).

PRESSURE DISTRIBUTION AROUND MODEL BUILDING

Figures 2 to 8 give the Cpo results for the model. Where possible they are presented as contours. These show that the external pressure distribution is not unusual when compared with any other work on bluff bodies.

The variation in stairshaft pressure coefficients with wind incidence for both leakage configurations are presented in Figure 9. With leakage (a) all angles of incidence greater than 60° give a negative pressure whichcould create dangerous conditions for escapers in a real building fire. With reduced leakage (b) the situation is improved but the shaft pressures are not sufficiently positive to overcome adverse pressures elsewhere in the building.

Figures 10-13 are plots of the corridor pressure coefficients for both leakage configurations and all angles of incidence both with and without fire simulation.

With a fire on floor 4 (Figures 10 and 11) the effect of the fire pressure in the corridors is confined to the fire floor, but the pressure is reversed and the ΔCp corridor stair shaft (C/S) is increased which suggests a greater flow potential for smoke to move into the normal escape route.

Measurements for a similar fire on floor 2 show a similar change in flow potential but the pressures differential was less because of the lower values of Cp_o Figures 12 and 13 give the corridor pressure variations for leakage (b). For angles of incidence between 135° and 225° the potential flow of smoke into the stairshaft is reduced (compared with that for leakage (a)) because of the reduced leakage area from the stairshaft. (Compare Figures 11 and 13 with respect to Figure 9 for α = 180°).

DISCUSSION OF RESULTS

Leakage (a) (When T = 1.5; D = 1.5; R = 1.5)

Wind Direction α = 0°

Figure 2 (for side T) shows that the wind pressures vary vertically and horizontally, the maximum pressure occurring at 0.8 of the model height. At this level the highest room and corridor pressures are expected when all doors, throughout the model, have the same leakage areas and are closed. This is confirmed by reference to Figure 10 (α = 0°) for the corridor pressures.

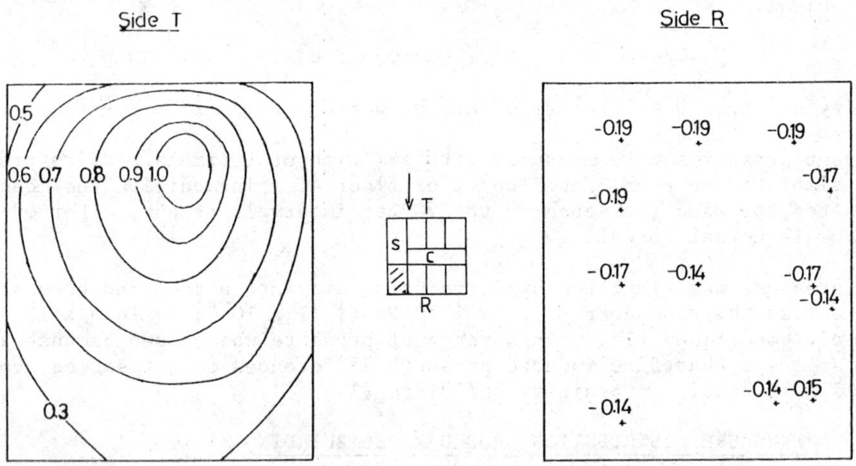

Figure 2: Outside pressure distribution (Cp_o), α = 0°.

Table 2: Room Pressures on Floor 4 (α = 0°)

Room	①	②	③	Shaft	④	⑤	⑥
Cp_i	0.97	0.90	0.65	0.5	−0.19	−0.19	−0.20

The leeward room (④,⑤,⑥) pressures relate directly to the external pressures on side R (Figure 2) both exhibiting small pressure variations only. Measurements on other floors showed similar small variations. Subsequent calculations showed that the use of a single value for Cp on side R (α = 0°) was not inappropriate for smoke flow prediction.

Wind Direction $\alpha = 45^o$

Rotation to 45° gave a marked change in pressure pattern for wall T (Figure 3). There was no definite stagnation point although a pressure gradient is maintained with highest values at 0.8 of the height. The maximum values are somewhat less than when $\alpha = 0^o$.

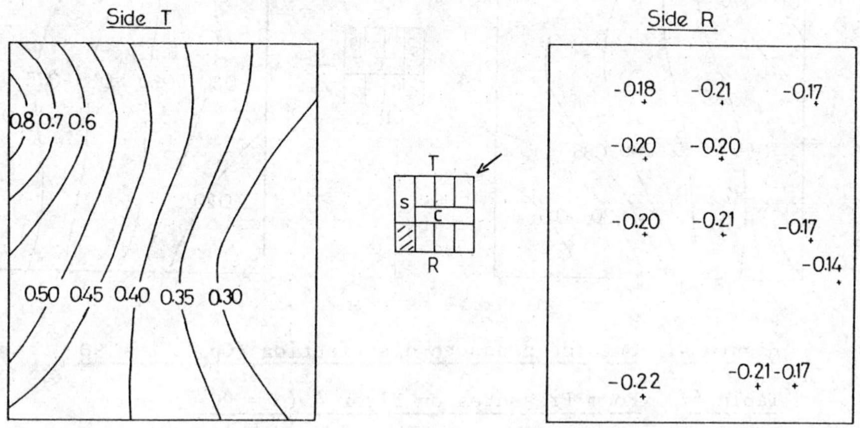

Figure 3: Outside pressure distribution (Cp_o), $\alpha = 45^o$.

Table 3: Room Pressures on Room 4 ($\alpha = 45^o$)

Room	①	②	③	Shaft	④	⑤	⑥
Cp_i	0.44	0.59	0.76	0.15	-0.23	-0.23	-0.29

Table 3 shows that the highest pressure was developed in Room ③ as expected from the skew pressure distribution on wall T. Wall R remained in the wake region but the negative pressure values are higher than when $\alpha = 0^o$, because of the larger eddies, of greater velocity, generated at the windward end of the wall. The average external Cp_o is about -0.2 and can be assumed to be uniform. The room pressures were fairly uniform and this supports the former assumption. Reference to Figure 10 ($\alpha = 45^o$) shows that the corridor pressures are similar to those when $\alpha = 0^o$, although the values are less. Similarly the shaft pressures (Tables 2 and 3) are reduced also.

Wind Direction $\alpha = 90^o$

For this wind angle both leaky walls (T and R) had a negative pressure gradient where the maximum values are near the top of each leading end (Figure 4). Table 4 indicates that the highest negative pressure values were found in Rooms ③ and ④.

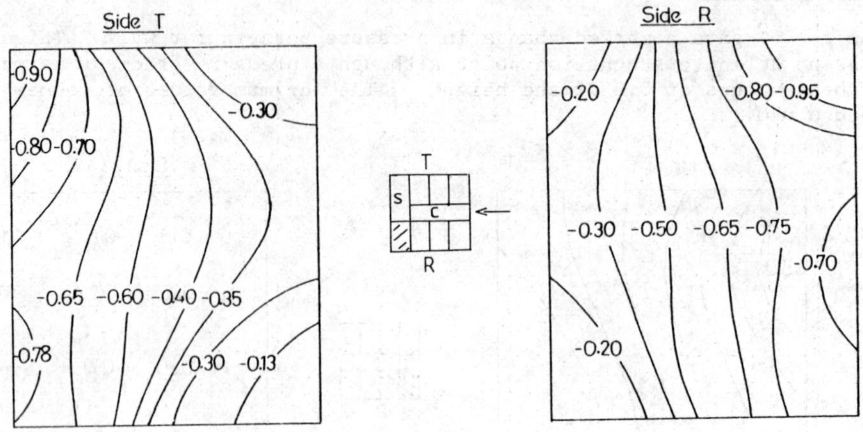

Figure 4: Outside pressure distribution (Cp_o), $\alpha = 90$

Table 4: Room Pressures on Floor 4 ($\alpha = 90^o$)

Room	①	②	③	Shaft	④	⑤	⑥
Cp_i	-0.44	-0.77	-0.81	-0.19	-0.84	-0.73	-0.4

Other measurements showed a similar pressure distribution for other floors which indicated a direct influence of the horizontal pressure gradient.

The corridor pressure for floor 4 (-0.4, Figure 10) is nearly equal to the pressures in rooms ① and ⑥. These volumes are almost in equilibrium so any potential smoke flow would be dictated by other pressure forces. The shaft pressure (-0.19) is greater than those for corridor and room spaces enabling an air flow from the shaft through the corridor and into the various rooms. As it may be assumed that smoke, once disassociated from the fire room, will tend to follow pre-existing air flow paths, this particular configuration would be dangerous for escapers.

Wind Direction $\alpha = 135^o$

The external pressure patterns (Figure 5) should be similar to those for $\alpha = 45^o$. The differences are due to a small experimental error and a comparison of Figures 3 and 5 show that the wind angle is less than 135°. This resulted in greater negative pressures for side T and lower positive values for side R. As a consequence, the corridor pressures (Figure 10) are always negative compared with the values when $\alpha = 45^o$.

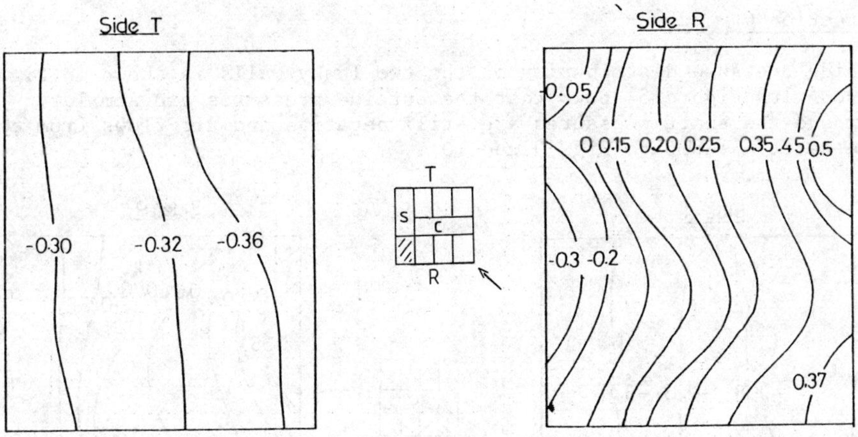

Figure 5: Outside pressure distribution (Cp_o), α = 135°

Table 5: Room Pressures on Floor 4 (α = 135°)

Room	①	②	③	Shaft	④	⑤	⑥
Cp_i	-0.32	-0.32	-0.29	-0.27	0.62	0.44	0.37

Comparing Tables 3 and 5, the rooms have similar pressures but the shaft has a negative pressure when α = 135°. This pressure distribution would allow any smoke produced in Room ④, ⑤ or ⑥ to flow into the corridor, opposite rooms and into the shaft.

Wind Direction α = 180°

For this direction the external pressures are similar to those when α = 0° (Figures 6 and 2). The room pressures were similar also. The major difference is the shaft pressure (Figure 10) which is negative when α = 180° as its leaky surface is on the leeward side of the model.

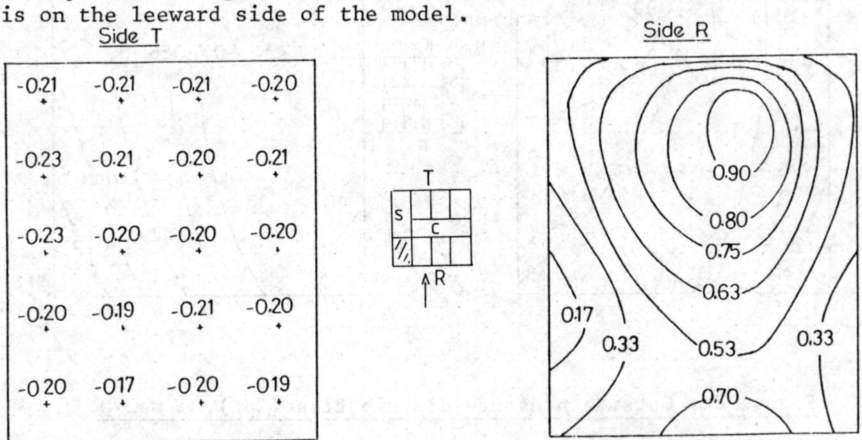

Figure 6: Outside pressure distribution (Cp_o), α = 180°.

Wind direction $\alpha = 225^o$

The outside pressure distribution on the two leaky walls is shown in Figure 7. Comparison with Figure 5 shows that the outside pressures are similar. The corridor and the shaft pressures are still negative and air flows from corridor into the shaft. (Compare Figs 9 and 10).

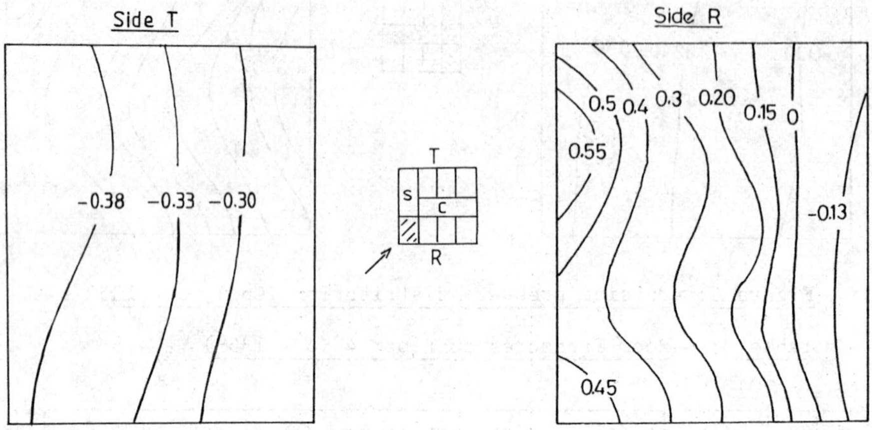

Figure 7: Outside pressure distribution (Cp_o), $\alpha = 225^o$

Wind direction $\alpha = 270^o$

The external pressures (Figure 8) are similar to those for when $\alpha = 90^o$. However the internal pressure distribution is not a reflection of those when $\alpha = 90^o$ because of the asymmetry of the internal spaces. A potential flow from the corridor into the staircase (see Figures 9 and 10) would be generated.

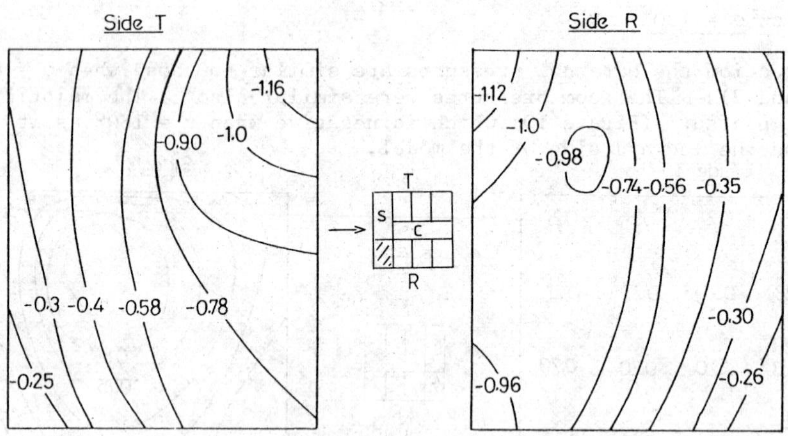

Figure 8: Outside pressure distribution (Cp_o), $\alpha = 270^o$

Leakage (b) (When T=1.5*; D=1.5; R=1.5)

All pressure measurements were repeated for this leakage configuration. For the rooms no significant changes were observed when the results were compared with those from leakage (a). With the reduced leakage to the shaft (staircase) the value of most of the pressure coefficients was reduced (see Figure 9) and this caused the changes in corridor pressures (compare Figures 10 and 12). The implications of such pressure changes is discussed later.

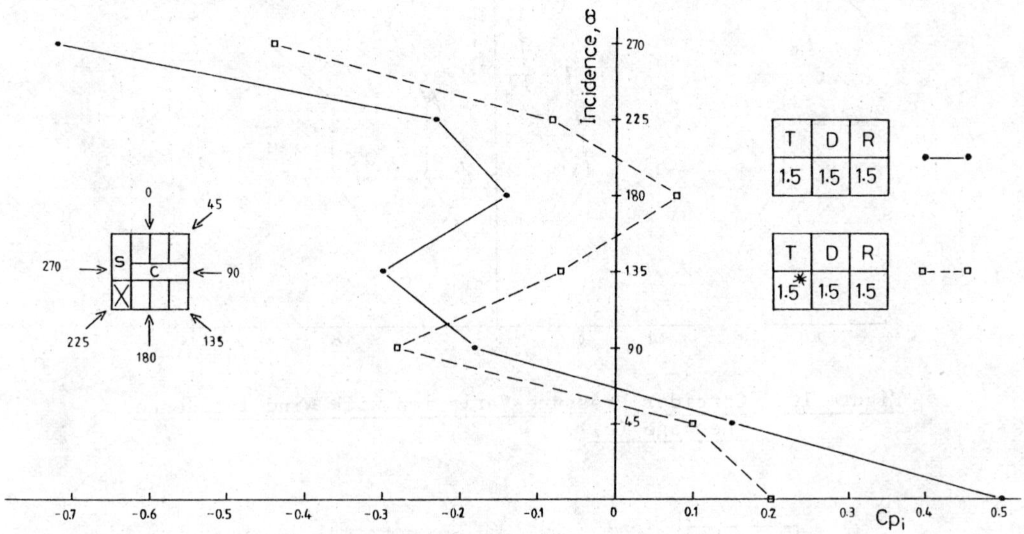

Figure 9: Shaft Pressure Variation with Wind Incidence Leakage (a) and (b)

Changes in Pressure Patterns with Simulated Fire

(i) Leakage (a)

All relevant pressure measurements were repeated with a simulated fire on floor 4 in Room ⑥. It was observed from the detailed results that there was little change from the "no fire" condition anywhere in the model building except on the fire floor. A comparison of Figures 10 and 11 shows the marked changes in measured pressures for the fire floor corridor at all wind angles except when $\alpha = 90^o$. Most measurements showed an increase in pressure which changes the ΔCp values between contiguous spaces.

(ii) Leakage (b)

The corridor pressure results for this reduced leakage configuration (Figures 12 and 13) show that the greatest effect remains on the fire floor but because the reduced staircase pressures (Figure 9) increase the ΔCp value across the corridor to staircase doors, there is a greater flow potential from the fire floor into this shaft. (The detailed implications of the increased flow potential are discussed in the next section).

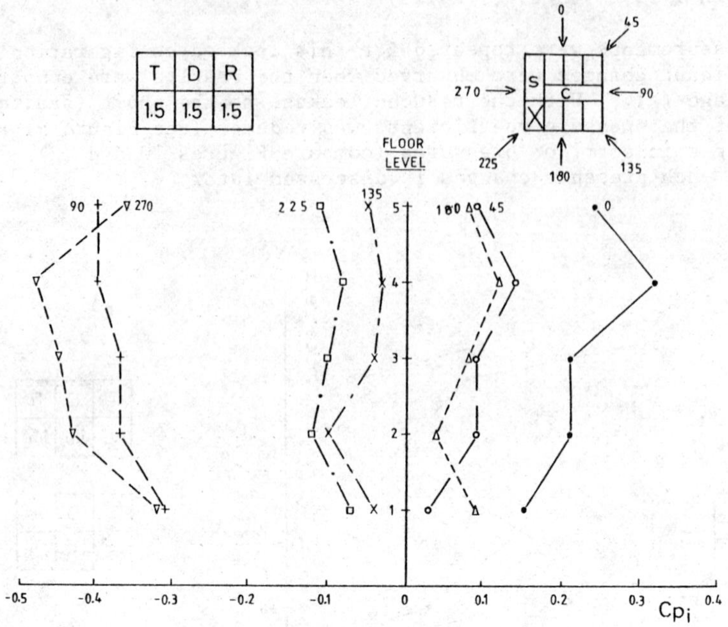

Figure 10: Corridor Pressure Variation with Wind Incidence (Leakage (a)).

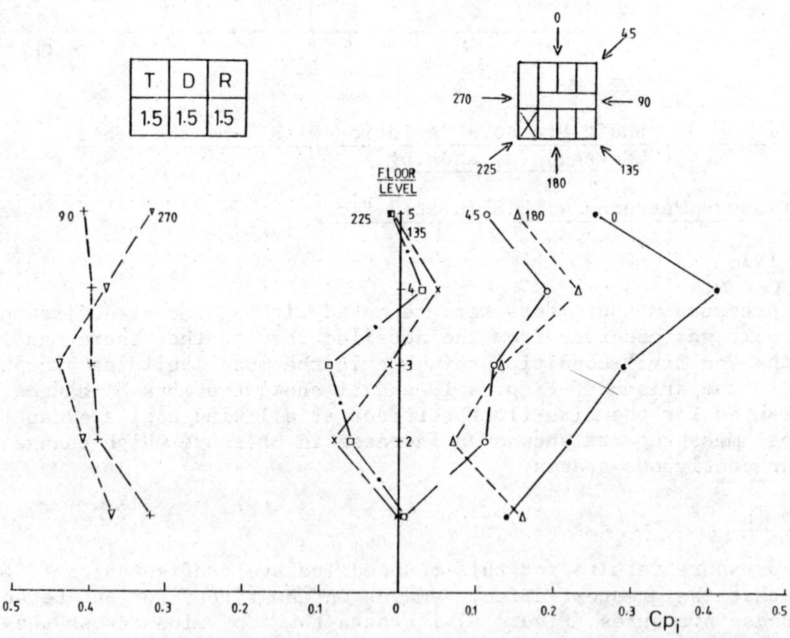

Figure 11: Corridor Pressure Variation with Wind Incidence Leakage (a). Fire Room (6), Floor 4

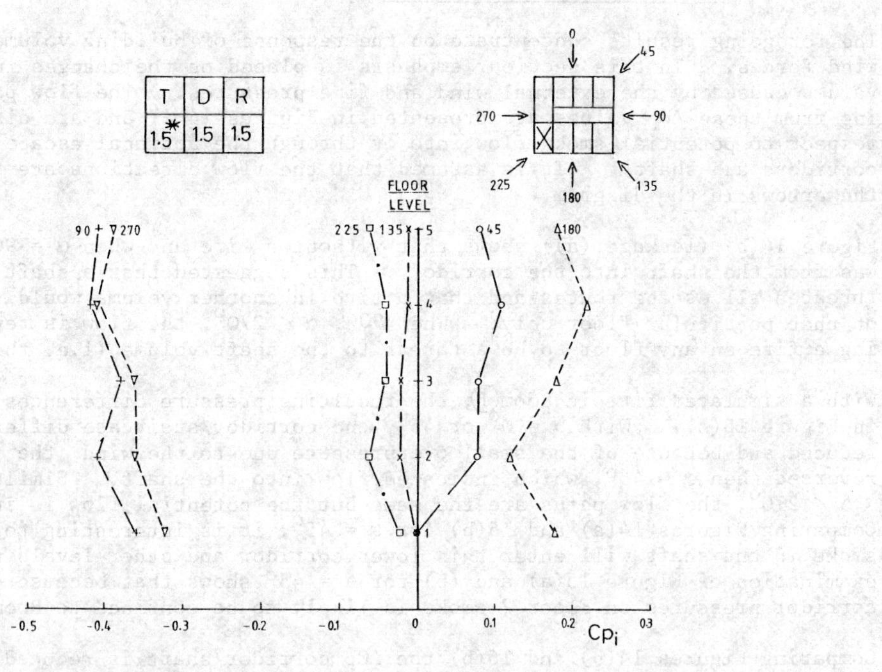

Figure 12: Corridor Pressure Variation with Wind Incidence (Leakage (b)).

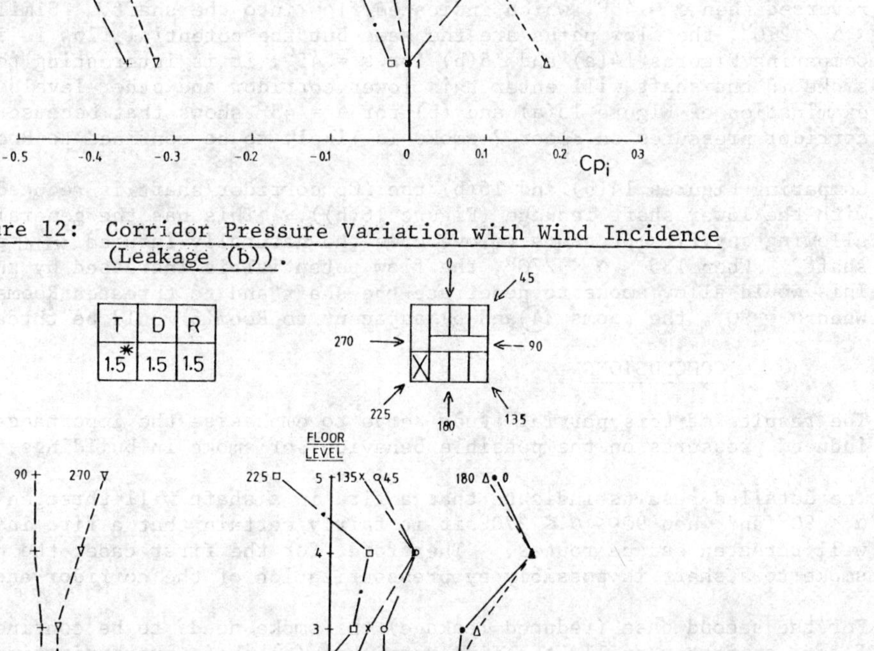

Figure 13: Corridor Pressure Variation with Wind Incidence (Leakage (b)): Fire Room (6), Floor 4.

WIND INDUCED FLOW PATTERNS

The foregoing results concentrate on the response of building volumes to external wind forces. In this section, emphasis is placed on the changes in internal ΔCp values caused by the external wind and fire pressures. The flow patterns resulting from these ΔCp values are presented in Figures 14-17 and are discussed with respect to potential smoke flow into or through the notional escape route (i.e. corridors and shaft). It is assumed that the flow directions are represented by the arrows in the diagram.

Figure 14(b) (leakage (a)) shows that without a fire and when $\alpha = 90^o$, the flow was from the shaft into the corridor. This suggested that a shaft fire would threaten all escape routes and that a fire in another volume would threaten spaces on that particular floor only. When $90^o < \alpha < 270^o$, the flow is reversed, allowing a fire on any floor to be a threat to the shaft volume (i.e. the staircase).

With a simulated fire in Room ⑥, the resulting pressure differences are presented in Figure 15(b). With $\alpha = 0^o$ or 45°, the corridor/staircase differential is reduced and because of the small overpressure due to the wind, the flow is reversed when $\alpha = 45^o$, which indicated flow into the shaft. Similarly when $90^o < \alpha < 270^o$, the flow paths are the same but the potential flow is increased. Comparing Figures 14(a) and 15(b) for $\alpha = 45^o$, it is interesting to note that any smoke in the shaft will enter this lower corridor and other levels also. An examination of Figure 15(a) and (b) for $\alpha = 45^o$ shows that because of the lower corridor pressures on Floor 2 smoke is likely to be confined to Room ⑥.

Comparing Figures 14(b) and 16(b) the ΔCp corridor/shaft is reduced significantly with the lower shaft leakage (Figure 16(b)). This has the general effect of allowing any fire pressure to overcome the naturally induced wind pressures in the shaft. When $135^o < \alpha < 270^o$, the flow potential is increased by the fire pressure. This would allow smoke to penetrate the shaft and to threaten Rooms ①, ② and ③. When $\alpha = 90^o$, the rooms (④ and ⑤) adjacent to Room ⑥ would be threatened also.

CONCLUSIONS

The results of this partial study serve to emphasise the importance of wind induced pressures on the possible behaviour of smoke in buildings.

The detailed results indicate that a fire in a shaft will threaten all levels when $\alpha < 90^o$ and when $90^o < \alpha < 270^o$ it is fairly certain that a fire in a windward space will threaten escape routes. Therefore, for the first case, the confinement of smoke to a shaft is possibly by pressurization of the corridor and/or lobby.

For the second case (reduced leakage) the smoke needs to be confined to the fire floor, or better still, the fire room, by: (a) shaft pressurization; (b) corridor pressurization; or (c) lobby pressurization.

A knowledge of the pressure difference distribution throughout a building for all wind directions is essential input to the definition of design pressure differentials for any pressurization system.

The actual pressures developed in a building will depend on the wind climate at a particular geographical location. An analysis of such a climate with a knowledge of the naturally developed pressure differentials will give the possible limits of problems across such barriers as doors (2). The probable maximum pressures developed will dictate the problems associated with opening and closing doors in escape paths, the selection of control mechanisms for effective pressurization systems and the probable failure potential for smoke control systems, which rely on natural forces for their effective operation.

ACKNOWLEDGEMENT

The authors wish to acknowledge the continuing financial support of the UK Science Research Council on the project "Pressurization as a Smoke Control Device in Buildings", of which this paper forms a part.

REFERENCES

1. Report: Hobson, P.J. and Stewart, L.J., "Pressurization of Escape Routes in Buildings". Fire Research Note No 958. (Fire Research Station, U.K.).

2. Transactions/Proceedings: Morgan, J. and Marchant, E.W., "Some Effects of Natural Wind on Vent Operation in Shopping Malls". Proceedings of the CIB Symposium on the Control of Smoke Movement in Building Fires, pp 171-183.

Plate 1: General view of model.

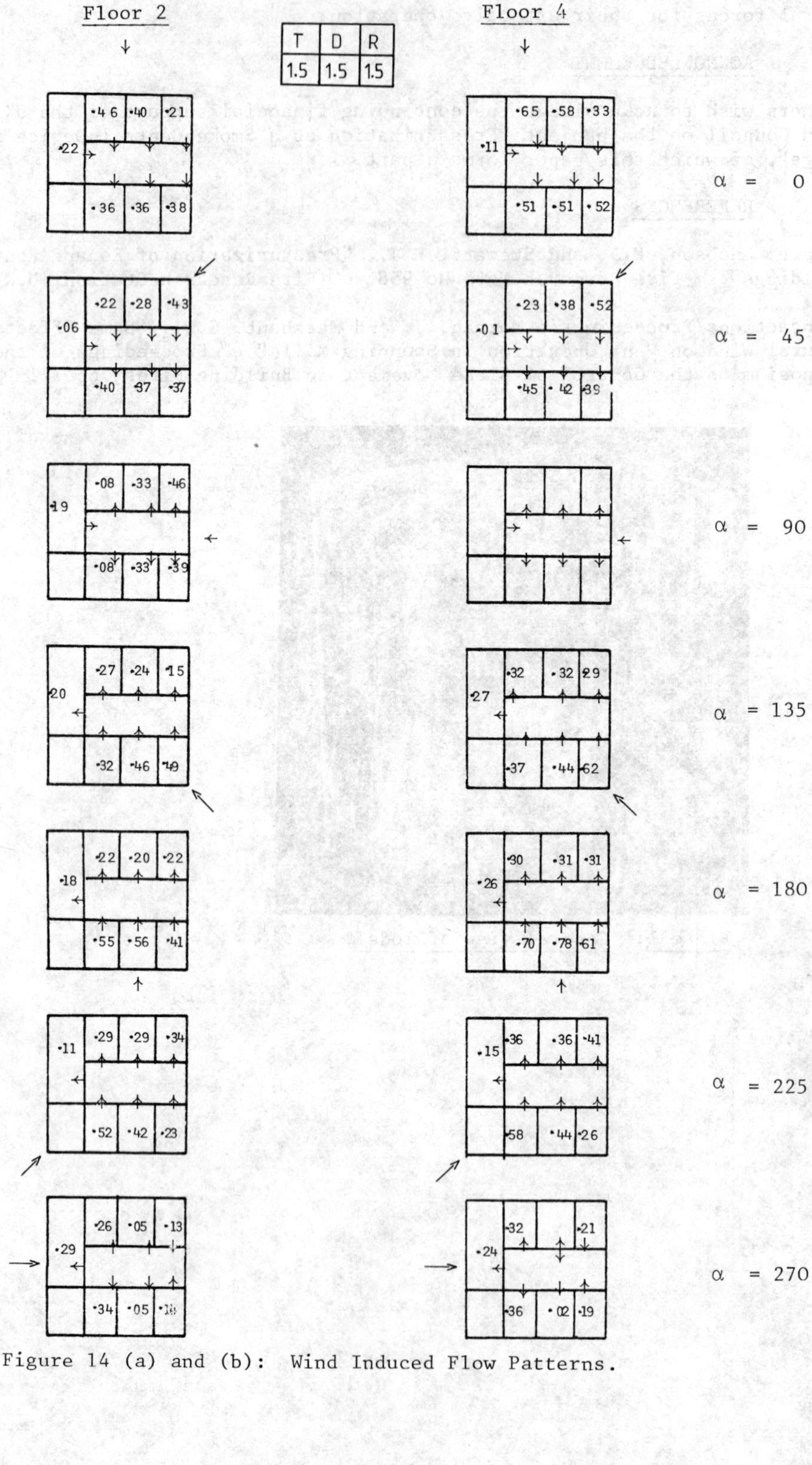

Figure 14 (a) and (b): Wind Induced Flow Patterns.

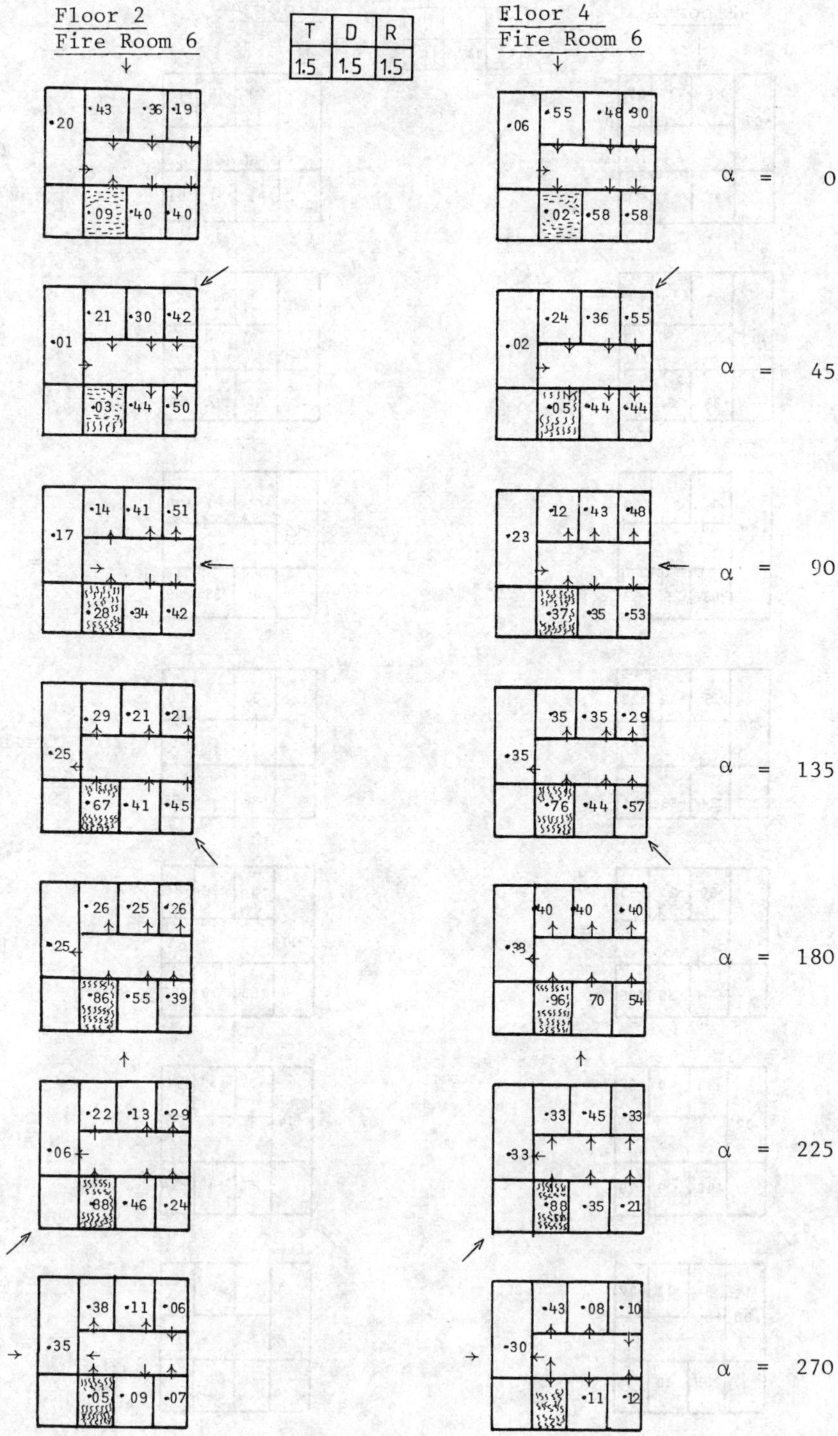

Figure 15 (a) and (b): Wind Induced Flow Patterns (Fire Room 6).

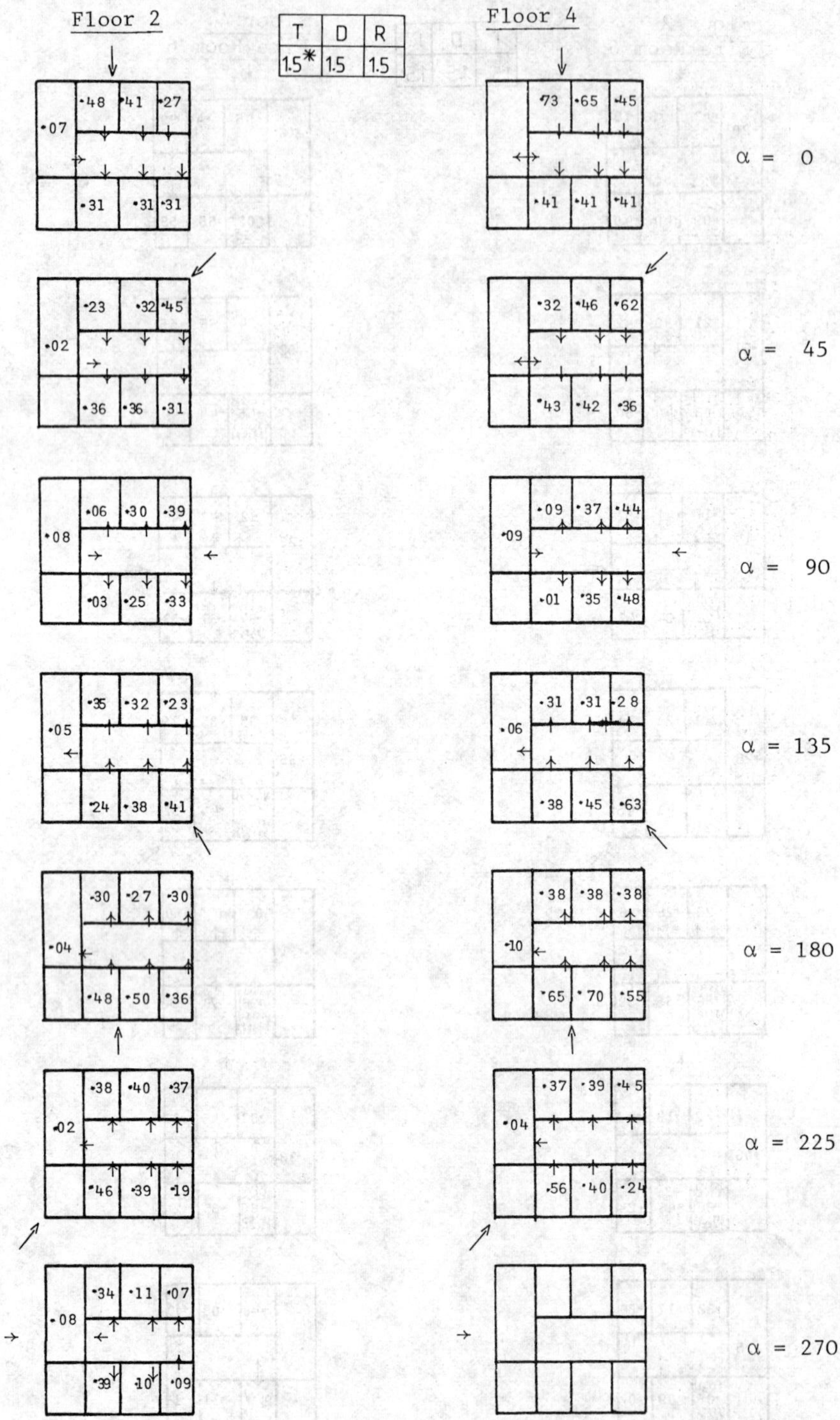

Figure 16 (a) and (b): Wind Induced Flow Patterns.

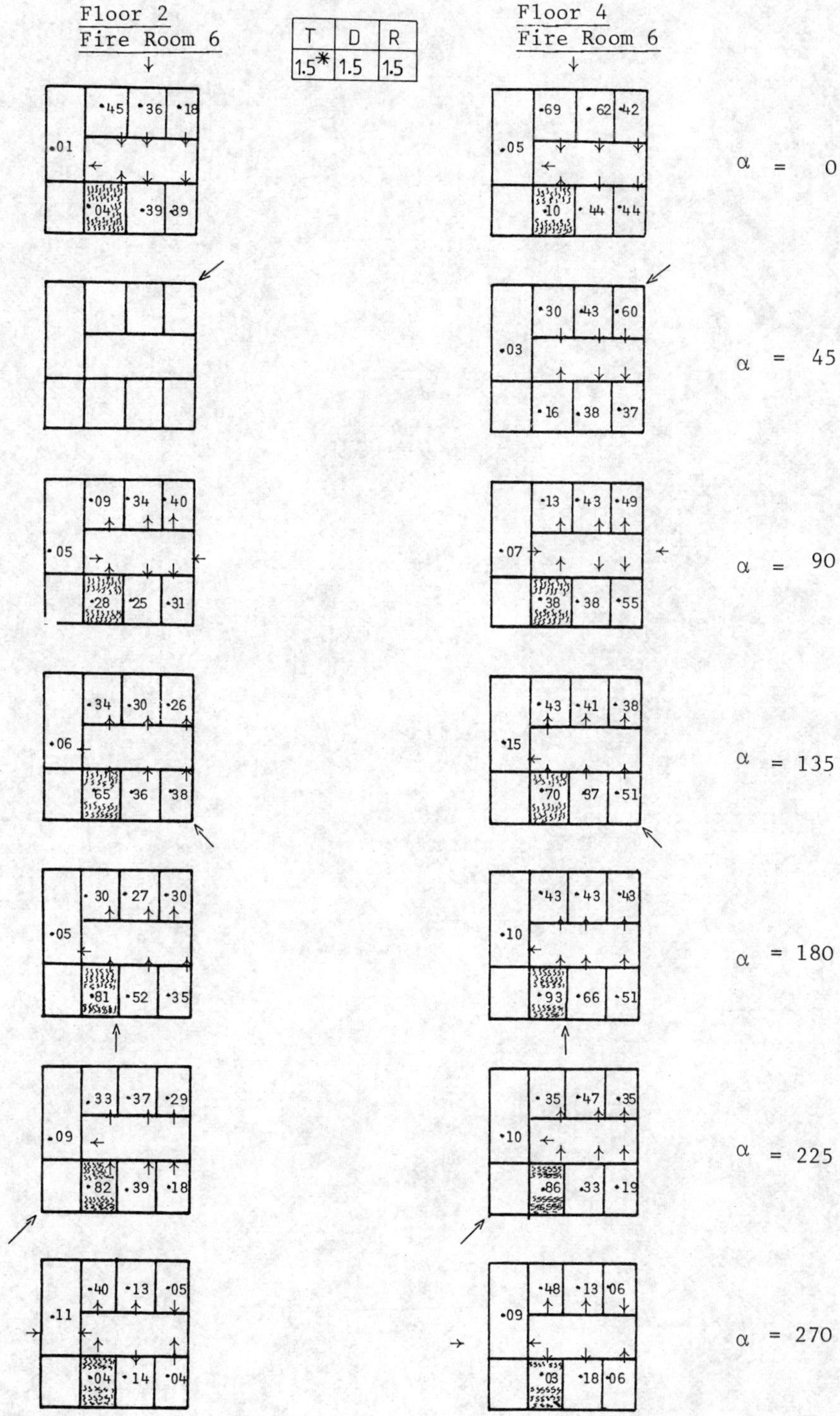

Figure 17 (a) and (b): Wind Induced Flow Patterns (Fire Room 6).

A WIND TUNNEL INVESTIGATION OF THE DISPERSION OF CHEMICAL VAPORS EMITTED FROM MARINE CHEMICAL CARRIERS

T. B. MORROW

Southwest Research Institute, San Antonio, Texas, USA

ABSTRACT

The results of a wind tunnel study of the dispersion of heavier-than-air cargo vapors emitted from vents on a chemical tanker ship during simulated cargo loading conditions are reported. Mixtures of CO_2 and hydrocarbon tracer gases were emitted simultaneously from four simulated tank vents on a 1:150 scale model of a chemical tanker ship while gas atmosphere samples were taken at several locations above the model ship deck and analyzed by a gas chromatograph. Results are presented for the vapor concentration level at man height above the deck for simulated neutral, stable and unstable atmosphere conditions and moderate wind speeds directed from bow to stern.

NOMENCLATURE

d	vent diameter
F	densimetric Froude number
g	acceleration of gravity
J	momentum ratio
Re	Reynolds number
Ri	bulk gradient Richardson number
T	temperature
U_a	wind velocity at deck height upwind of model
U_j	discharge velocity of vented gas
Z	vertical distance above ground
ν	kinematic viscosity
ρ_a	density of wind at deck height
ρ_j	density of vented gas

INTRODUCTION

Gas streams containing chemical vapors mixed with air or inert gas are emitted into the open air around marine chemical carriers (tanker ships and barges) during diurnal breathing of loaded tanks, during gas-freeing (ventilation of empty tanks prior to a cargo change, or as a prerequisite for man entry), and as the tank gas atmosphere is displaced by chemical cargo during loading. The chemical vapor concentration in the gas stream may range from low (a few hundred ppm at the start of cargo loading into a clean, empty tank) to high (approaching the saturation concentration value, several percent by volume for volatile chemicals, near the end of cargo loading). These gas streams are emitted from stack risers high above the ship deck* as well as from P/V valves and open ullage hatches close to deck level.

The cargo vapors emitted from ship tank vents emerge as a plume that is bent over by the wind and eventually dispersed by atmospheric turbulence. However, when cargo vapors are present near the deck and ship superstructure in either flammable or toxic concentrations, they pose a potential occupational health and safety hazard. Brummage (1) carried out extensive low speed wind tunnel experiments using 1/100th scale models of crude oil tanker ships to investigate the size and location of the region of flammable vapor concentration in plumes emitted from ship tank vents during simulated loading conditions. In these tests, the wind tunnel model was placed on a false floor that allowed the tunnel wall boundary layer to pass beneath the model. In this manner, a range of low turbulence, uniform wind velocities were produced equivalent to from 1.1 m/s (2.5 m.p.h.) to 4.5 m/s (10 m.p.h.) in full scale. Propane and butane were emitted at vent locations in concentrations ranging from 25% to 100% by volume at model scale flowrates to simulate the emission of crude oil vapor near to the end of cargo loading. A portable spark igniter was used to identify the combustible region of the plumes. These wind tunnel experiments provided data that was used to formulate guidelines for the minimum height of vent outlets above the deck that will prevent flammable gas concentrations from reaching the deck, and for the minimum separation distance between the vent and the superstructure to eliminate any influence of the superstructure on plume dispersion.

The wind tunnel experiments reported in this paper are part of a larger experimental program (2) aimed at assessing both the flammability and toxicity hazards posed by routine operations (such as cargo transfer) on marine chemical carriers. Model experiments were performed to provide guidance for the design of a vapor sampling plan for use on tanker ships, and to provide concentration distribution data under controlled atmospheric conditions for improving the analytical plume modeling methods used in this program. For this purpose it was necessary to extend the range of concentration measurement below the flammable region investigated by Brummage, and to determine the effect of a realistic simulation of meteorological conditions including atmospheric stability on deck level vapor concentrations. An extensive set of flow visualization experiments was also performed to supplement the vapor concentration measurements.

SCALING CONSIDERATIONS

The similarity of the dispersion of gaseous effluents in an atmospheric boundary layer and in a laboratory scale boundary layer developed in a meteorological wind tunnel has been discussed by Cermak (3). Similarity of velocity and turbulence profiles may be achieved by assuring that the laboratory boundary layer is turbulent, rough and has a long development length. Similarity of atmospheric stability conditions within a layer ΔZ in height is provided by equality of the bulk gradient Richardson number.

$$Ri = \frac{g\ (\Delta T/\Delta Z)}{T\ (\Delta U/\Delta Z)^2}$$

The trajectory of forced plumes emitted normal to the wind direction is known to depend upon the momentum ratio and the densimetric Froude number.

*For certain specified chemicals the vent height is required to be equal to 1/3 the ship beam dimension by U. S. Coast Guard Regulation 46 CFR, Part 153, Safety Rules for Self-Propelled Vessels Carrying Hazardous Liquids.

The momentum ratio,

$$J = \frac{\rho_j U_j^2}{\rho_a U_a^2}$$

influences the initial rise of the plume as it is turned over by the wind. The densimetric Froude number,

$$F = \frac{\rho_a U_a^2}{(\rho_j - \rho_a) g d}$$

affects the rise or fall of the plume due to buoyancy. Also, the jet Reynolds number

$$Re_j = \frac{U_j d}{\nu}$$

and the turbulence of the boundary layer affect the entrainment of ambient air by the plume and the dispersion of gaseous effluents in the plume.

Unfortunately, it was not possible to satisfy both Froude number similarity and Reynolds number similarity in the model experiments. Preference was given to exact modeling of momentum ratio and Froude number since it was believed that correct simulation of plume trajectory was most important in these tests. As a result, the plumes emitted from the vents were laminar, and plume dispersion was affected primarily by boundary layer turbulence.

Model scale and full scale values of wind speed are related by the Froude number, while values of venting rate are related by the momentum ratio. The model experiments were designed to simulate the emission of a plume with a high concentration of cargo vapor. This condition can occur near the end of loading when the tank gas atmosphere becomes saturated with cargo vapor. Many of the liquid cargos transported in marine carriers are volatile and have a molecular weight greater than air. Acetone with a molecular weight of 58 and a partial pressure of 185 mm Hg at 20°C was chosen as the basis for a full scale example. A typical value of ρ_j/ρ_a for a saturated mixture of acetone vapor in air is 1.25. Using Froude number similarity with a gas mixture having the same value of ρ_j/ρ_a in model scale and a length scale factor of 150, a full scale wind speed of 2.2 m/s corresponds to 0.18 m/s in the wind tunnel. However, it was necessary to maintain the wind tunnel air velocity at a speed of 0.3 m/s (1 ft/s) or higher at deck height for both the stable and unstable atmosphere simulations. Consequently, the value of ρ_j/ρ_a in model scale was increased to approximately 1.5 to give an equivalent full scale wind speed of 2.6 m/s (5.8 m.p.h.).

EXPERIMENTAL ARRANGEMENT

A series of wind tunnel experiments was planned to simulate the dispersion of plumes of heavier-than-air chemical vapor and air emitted from vents on a chemical tanker ship into an ambient wind of approximately 2.2 m/s (5 m.p.h.) at deck level. The experiments were performed in the Meteorological Wind Tunnel at Colorado State University. Since current multi-product tanker ships are rather large, typically 183 m (600 ft) and longer, a geometric scale of 1:150 was chosen to permit placing a 1.22 m (4 ft) long model of a 183 m ship at various orientations relative to the wind in the 2 m (6.6 ft) wide tunnel test section. Fig. 1 shows the ship model next to a model of a loading dock.

Fig. 1. 1/150 Scale Model of a Chemical Tanker Ship and Loading Dock.

The tanker ship model was placed directly on the wind tunnel floor. A saw-tooth boundary layer trip at the head of the 22 m long test section insured that the boundary layer ahead of the ship model was turbulent and rough. The wind tunnel floor was cooled and heated to stimulate stable and unstable atmospheric stability conditions, respectively. The value of Richardson number was based upon temperature and velocity measurements in the layer of air around the ship model from 2.5 cm (1 in) to 12.7 cm (5.5 in) above the tunnel floor. The reference wind velocity was measured at deck level, 7.6 cm (3 in) above the tunnel floor.

Four separate gas mixtures consisting of a dilute concentration of a single hydrocarbon tracer gas in carbon dioxide were prepared. Mixture concentrations were 3% methane, 1.5% ethane, 1% propane and 0.75% butane in CO_2 giving values of ρ_j/ρ_a equal to 1.44, 1.51, 1.52, and 1.52, respectively. The tracer gas mixtures were emitted simultaneously from the four vents shown in Fig. 2. The vents were formed from lengths of nominal 0.16 cm (0.063 in) diameter hypodermic tubing that

TABLE 1. Model and Full Scale Values of Test Variables

Vent No.	Dimensionless Parameters		Model Scale Variables			Full Scale Variables		
	J	F	ρ_j/ρ_a	U_a m/s	U_j m/s	ρ_j/ρ_a	U_a m/s	U_j m/s
1	2.82	13.04	1.44	0.3	0.42	1.25	2.77	4.16
2	2.96	11.25	1.51	0.3	0.42	1.25	2.57	3.96
3	2.98	11.03	1.52	0.3	0.42	1.25	2.55	3.93
4	2.98	11.03	1.52	0.3	0.42	1.25	2.55	3.93

extended through the deck and were connected to plastic tubing inside the model. An emission rate of 51 cm^3/min (6.5 ft^3/hr) was selected to give good smoke visibility in the companion set of flow visualization tests. A typical value of full scale emission rate was 11.4 m^3/min (4290 barrels/hr) from a 0.25 m (10 in) diameter vent. Due to the small diameter of the model vents, the jet Reynolds number was 84.

Gas sampling was performed during two runs by drawing gas samples through 26 sampling ports consisting of lengths of 0.16 cm diameter tubing open to the air at scale man height, 1.1 cm (0.44 in) above the ship deck. For the neutral stability tests, both sampling runs were made with the sampling ports arranged in 3 rows spaced 5.1 cm (2 in) apart. The distance between sampling ports along a row was 10.2 cm (4 in). However, the spatial resolution of the vapor concentration contours was marginal. To improve the spatial resolution of vapor concentration for the stable and unstable tests, the second sampling run was made with a modified port arrangement that reduced the spacing between ports to 5.1 cm (2 in) apart in regions downwind of vents. Gas samples were analyzed for the presence of the four tracer gases, corrected for background level. Table 1 summarizes the model scale and equivalent full scale values of the test variables for all experiments.

EXPERIMENTAL RESULTS

Fig. 3 shows the results of measurements of mean velocity profile, one-dimensional turbulence intensity and temperature profile for each stability condition. Note that the shape of the mean velocity profile is different for each stability condition, although the reference velocity was 0.30 m/s (1 ft/s) at ship deck height of 7.6 cm (3 in) in each case. At this low wind tunnel speed, a small temperature difference, $\sim$2.6°C for unstable conditions, was sufficient to enhance the turbulent mixing near the tunnel floor and flatten the mean velocity profile. The value of Richardson number, based upon velocity and temperature readings at 2.5 cm (1 in) and 7.6 cm (3 in) above the tunnel floor was -4.7.

There is a noticeable similarity between the mean velocity and turbulence profiles for the neutral and the unstable atmospheric simulations in the region of the ship deck 7.6 cm (3 in). Note, the velocity at the top of the superstructure 14 cm (5.5 in) is about 10% higher for the neutral simulation than for the unstable simulation.

The stable atmosphere simulation shows a strong suppression of boundary layer turbulence in the 0 to 14 cm (5.5 in) region when compared with either the neutral or unstable atmosphere simulation. The value of Richardson number, based on velocity and temperature readings at 7.6 cm (3 in) and 14 cm (5.5 in) above the tunnel floor was 0.75.

Fig. 4 shows the contours of tracer gas concentration (based upon the concentration measurements at the sampling ports) for gas emitted from vent number 1, a low vent 1.4 cm (0.56 in) above the deck near the bow of the ship. The tracer concentration at the vent was 30,000 ppm of methane. The shaded regions ranging from light to dark correspond to concentration levels of 5-9.9 ppm, 10-19.9 ppm, 20-29.9 ppm and >30 ppm. The equivalent concentration ranges for full scale emission of a saturated mixture of acetone vapor in air are 42-83 ppm, 84-161 ppm, 168-251 ppm and >252 ppm. Even with the closer spacing of sampling ports in the unstable and stable atmospheric simulations, the spatial resolution of concentration was inadequate to resolve the peak concentration region near the vent. The highest measured values of concentration were 139 ppm for the stable simulation and 149 ppm for the unstable simulation at a sampling port placed only 2.5 cm (1 in) from the vent. However, the spatial resolution was sufficient to

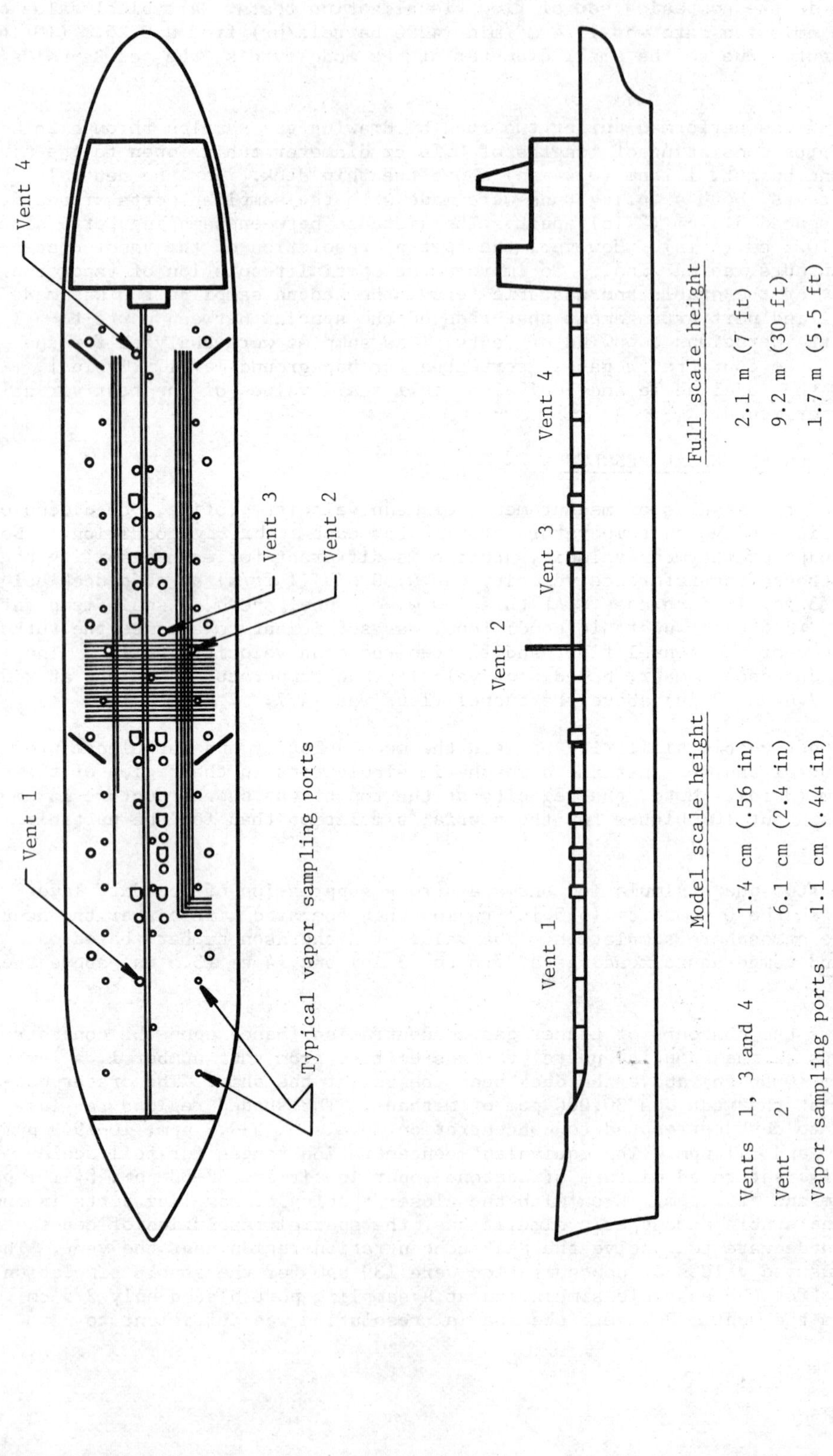

Fig. 2. Vent locations on scale model of tankership.

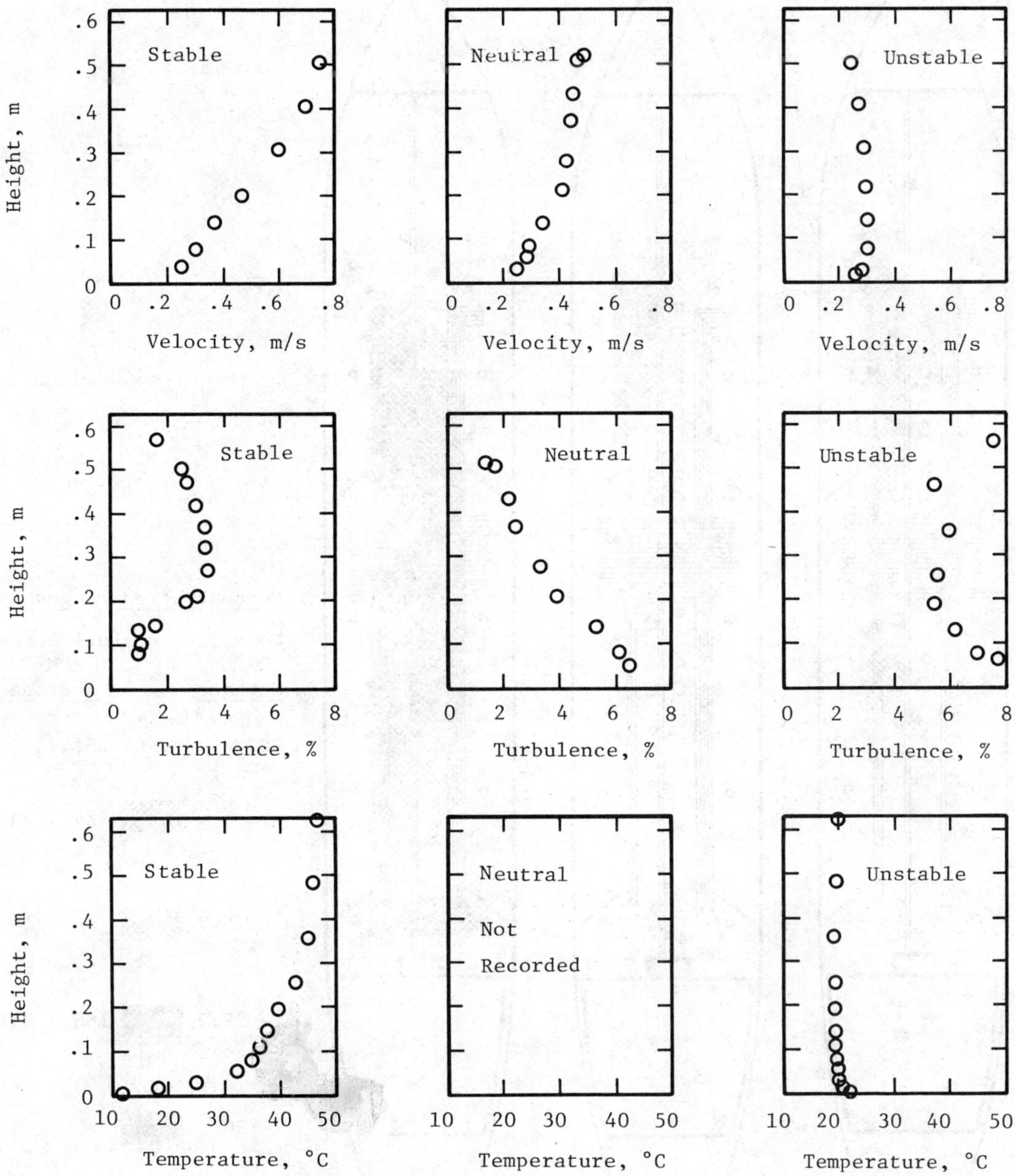

Fig. 3. Velocity, turbulence and temperature profiles for stable, neutral and unstable stability simulations.

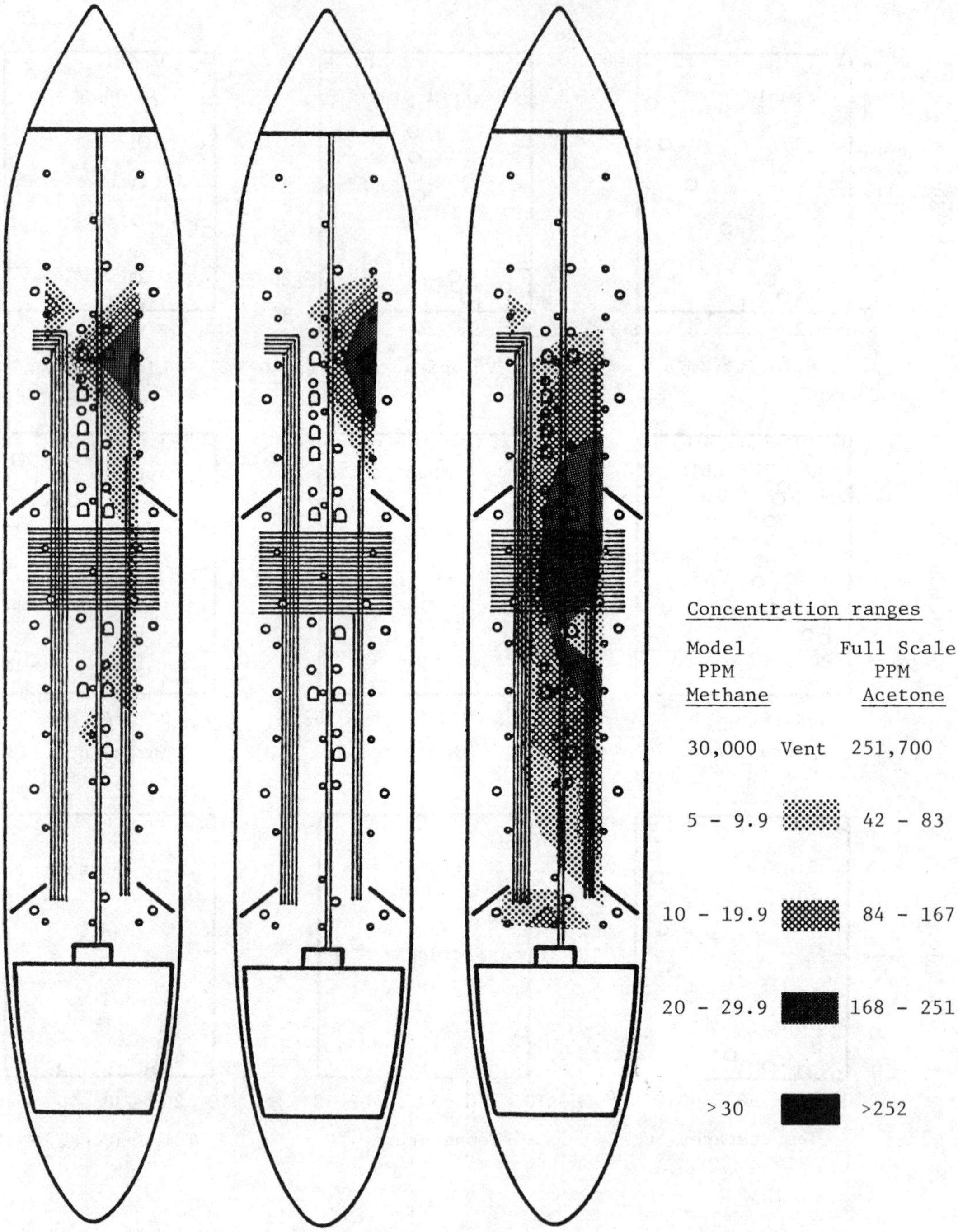

Fig. 4. Concentration contours for vapor emission from vent 1.

distinguish the far field concentration distribution downwind of the vent.

Note that the stable atmospheric simulation gave the greatest area of coverage and the highest values of concentration far away from vent 1. Note also the similarity of concentration contours for the neutral and the unstable simulations.

Vent number 2 was a high vent near the middle of the ship 6.1 cm (2.4 in) above the deck, equivalent to 9.14 m (30 ft) above the deck in full scale. The tracer concentration at the vent was 15,000 ppm of ethane which gave a gas chromatograph (GC) response equivalent to 28,143 ppm of methane. The concentration of ethane was very low near deck level, less than 2 ppm of equivalent methane at all sampling ports for all stability conditions. These results confirm that an elevated vent is an effective method for preventing vented chemical vapors from reaching deck level at wind speeds of 2.2 m/s (5 m.p.h.) and higher.

Fig. 5 shows the tracer gas concentration for gas emitted from vent number 3, a low vent 1.4 cm (0.56 in) above the deck behind the loading manifold piping near the middle of the ship. The tracer concentration at the vent was 10,000 ppm of propane, giving a GC response equivalent to 27,473 ppm of methane. The shaded regions ranging from light to dark correspond to the same concentration levels of equivalent methane as in Fig. 4. The highest measured values of concentration were 120 ppm for stable, 17.1 ppm for neutral, and 54.3 ppm for unstable atmospheric stability simulations. For this vent location, the concentration contours are similar in shape for all three stability simulations.

Fig. 6 shows the tracer gas concentration for gas emitted from vent number 4, a low vent 1.14 cm (0.56 in) above the deck directly in front of the superstructure. The tracer concentration at the vent was 7500 ppm of butane, with a GC response equivalent to 27,174 ppm of methane. The shaded regions have the same interpretation as in Fig. 4 and Fig. 5. The highest measured values of concentration were 35 ppm for stable, 44 ppm for neutral, and 39 ppm for unstable atmosphere conditions. The difference in location of concentration contours for the neutral stability and for the stable and unstable simulations is explained by the difference in sampling port location. The sampling port location closest to vent 4 was used only for the stable and unstable simulations.

DISCUSSION OF RESULTS

These experiments confirm that temperature stratification in the atmospheric boundary layer affects the level of turbulence in the wind, and also the dispersion of chemical vapors emitted from ship tank vents near deck level. The highest levels of vapor concentration at breathing height down wind of vents are expected to occur during atmospheric conditions of low wind speed, 1 m/s or below, and stable temperature stratification. It is note worthy that for the wind speed and venting rate conditions of this study, vapor emitted from the tall (beam width /3) vent was not found near deck level for any of the temperature stratification conditions. This finding reinforces the belief that vapor emission from vents high above deck level is an effective method for dispersing chemical vapors when the wind speed is 1 m/s or above.

The similarity of the concentration contours for neutral and unstable atmospheric simulations is surprising, but it may be explained by the similarity of both the mean velocity and turbulence field upwind of the ship model in the deck and superstructure region. The simulation of unstable atmospheric conditions could have been improved in this instance by heating the deck of the ship model as well as the wind tunnel floor to simulate solar radiant heating of the ship deck.

The distance between gas sampling locations in the matrix of sampling points

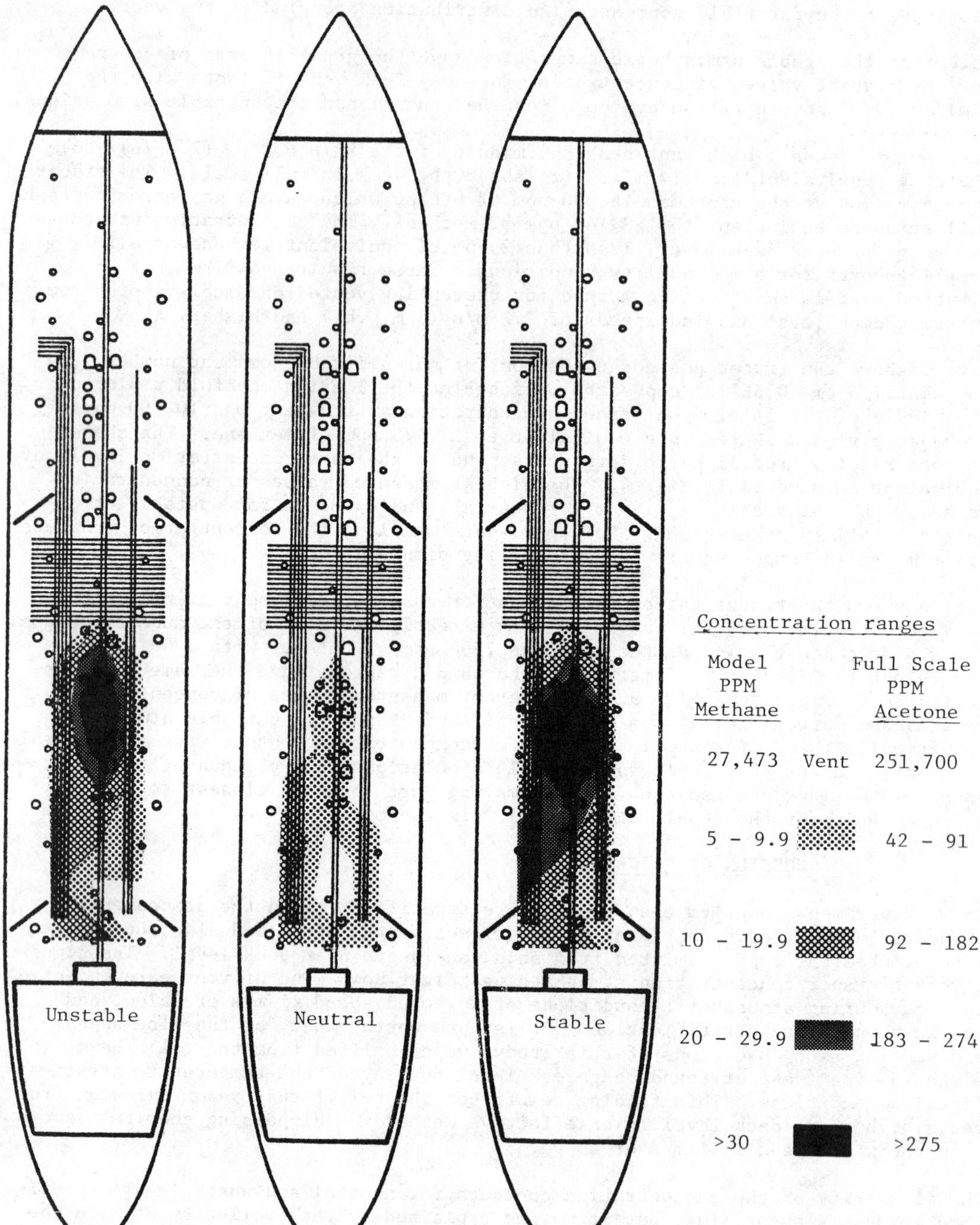

Fig. 5. Concentration contours for vapor emission from vent 3.

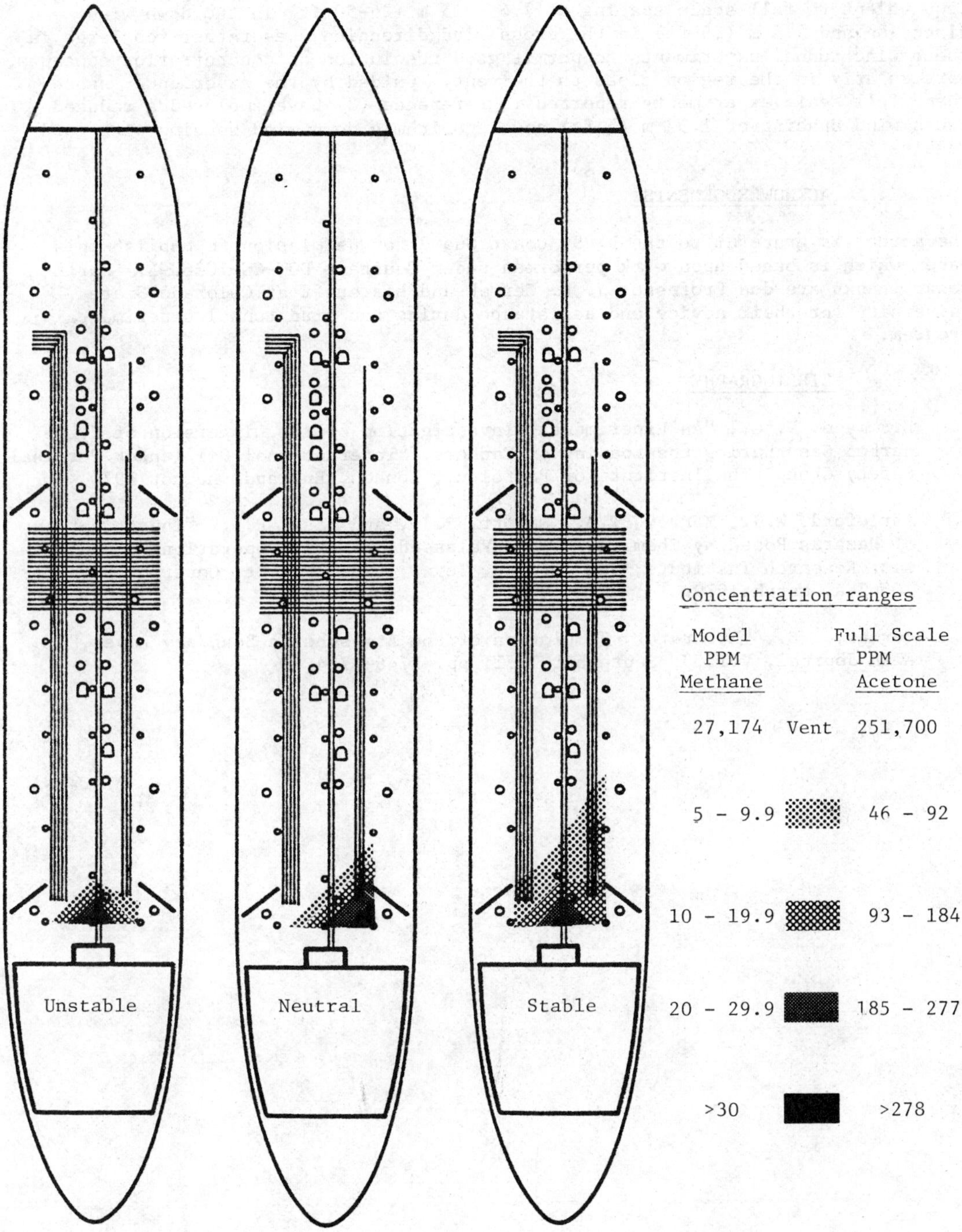

Fig. 6. Concentration contours for vapor emission from vent 4.

[equivalent to full scale spacing of 7.6 - 15 m (25-50 ft) in the down wind direction and 7.6 m (25 ft) in the cross wind direction] was rather too large in these wind tunnel experiments to permit good resolution of concentration contours, particularly in the region close to the vent. Guided by the experience, subsequent full scale experiments reported in reference (2) have employed a reduced cross wind spacing of 1.52 m (5 ft) and logarithmic down wind spacing with good results.

ACKNOWLEDGEMENTS

The author is grateful to the U. S. Coast Guard for permission to publish this paper which is based upon work performed under Contract DOT-CG-70363-A. Particular thanks are due Professor J. E. Cermak and his staff at Colorado State University for their advice and assistance during the wind tunnel experimental program.

BIBLIOGRAPHY

1. Brummage, K. G., "An Experimental Investigation of the Dispersion of Hyrocarbon Gas During the Loading of Tankers," International Oil Tanker Terminal Safety Group, The Institute of Petroleum, London, England, August 1970.

2. Astleford, W.J., Morrow, T.B., Magott, R.J., and Bass, R.L., "Investigation of Hazards Posed by Chemical Vapors Released in Marine Operations", Southwest Research Institute, San Antonio, Texas, Final Report, Contract DOT-CG-70363-A, March 1979.

3. Cermak, J.E., "Laboratory Simulation of the Atmospheric Boundary Layer," AIAA Journal, Vol. 9, September 1971, pp. 1746-1754.

LABORATORY SIMULATION OF LIQUID NATURAL GAS VAPOR DISPERSION OVER LAND OR WATER

R. N. MERONEY and D. E. NEFF
Colorado State University, Fort Collins, Colorado, USA

ABSTRACT

The primary purpose of this paper is to show through basic similarity analysis and comparisons of model and full-scale data that atmospheric transport of dense, cold natural gas clouds can be physically modeled in "meteorological" wind tunnels for a range of real boundary conditions which have great practical importance with respect to liquid natural gas (LNG) spill hazard analysis.

INTRODUCTION

The capability of the wind tunnel to model the gravitational effects associated with negatively bouyant gas discharges in the earth's boundary layer has been established and underlies important potential applications of wind tunnel testing, such as the simulation of the vapor dispersion from large full-scale LNG spills and the generation of gravitationally dominated check cases for theoretical models. Such wind tunnel data can be correlated in a manner that yields an empirical prediction of vapor dispersion resulting from full-scale spills. However, certain facility-specific limitations must be considered in planning and executing an experimental program.

PRIOR EXPERIENCE: LABORATORY SIMULATION OF DENSE GAS AND CRYOGENIC SPILLS

In the event of a liquid natural gas (LNG) release, the dense gas boils from the spilled liquid at the source and is carried downwind in the form of a plume with gas concentration decreasing with distance and height from the source. The negative buoyancy produced by the high density of the cold gas tends to exaggerate lateral movement and inhibit vertical mixing. Hence the spatial distribution of methane dilution is generally found to be quite different from that predicted for neutral density plumes.

A number of controlled laboratory experiments have been prepared previously to evaluate the significance of density on gaseous plume dispersion. Sakagami and Kato (1) measured diffusion and vapor rise from a small 5 x 10 cm LNG well in the floor of a 50 x 50 cm cross-section x 200 cm length wind tunnel. They confirmed a tendency for the gas to remain concentrated at ground level. Boyle and Kreebone (2) released LNG on water, precooled methane, and propane in a specially built 1.5 x 1.2 m cross-section by 5 m long asbestos-wall wind tunnel. No attempt was made to scale the atmospheric surface layer velocity profile or turbulence. They concluded room-temperature propane simulated a LNG spill quite well, but the pre-cooled methane runs lofted suggesting to the authors incorrect release temperature or exaggerated heat transfer from the ground surface. Hoot and Meroney (3) and Hall (4,5) considered point source releases of heavy gases in wind tunnels at ground level. Hoot and Meroney found that releasing gases with specific gravities as great as 3.0 only slightly shifted the decay of maximum concentrations with distance despite significantly different plume cross-sections. Hall considered transient and continuous releases on a rough surface (plume height - roughness height) and on up- and downhill slopes. Hall reported shallow, wide plumes whose shapes were considerably altered by 1 in 12 ground slopes.

Tests were conducted by Neff et al. (6) in wind tunnel facilities to evaluate the rate of dispersion and the extent of downwind hazards associated with the rupture of typical large LNG storage tanks. Concentration and temperature measurements, and photographic records were obtained for different wind speeds, wind direction and constant boiloff rates under both neutral and stable atmospheric stratifications. Subsequent measurements by Meroney et al. (7,8) examined transient releases in similar configurations as well as dense plumes on uphill slopes, and buoyant plume lift off situations. Different model release gasses were used to simulate the behavior of the cold methane plume--heavy isothermal gas mixtures (CO_2, Freon-12 and air, or Argon) or light-cold mixtures (He and N_2).

Currently research on LNG spill behavior over irregular terrain is being performed at Colorado State University for the U.S. Coast Guard Program. Small scale models of the LNG Release Pond and surrounding topography at China Lake, Naval Weapons Testing Center, California, are being examined in a meteorological wind tunnel. Mean and transient concentration contours have been mapped. Overall plume geometry and behavior have been recorded photographically. Results will be used to plan a field experiment and validate tunnel methodologies.

LABORATORY SIMULATION OF CRYOGENIC SPILLS

The reliability of the use of wind tunnel shear layers for modeling atmospheric flows has been demonstrated by several investigators (9). Specific problems associated with the dispersion of cold natural gas plumes have been previously discussed by Meroney et al. (10,11). The Froude number is the primary parameter which governs plume spread rate, trajectory, plume size and entrainment when gases remain negatively buoyant during their entire trajectory. Earlier measurements suggest that heat transfer effects may be small over the significant time scales; hence gas density should be adequately simulated by isothermal high molecular weight gas mixtures (6,7,8). Visualization of similar tests for the range of model scales used (1:130 to 1:666) indicate a similar plume geometry. Concentration results of the different model scales agreed to within the experimental accuracy of approximately ± 20%. Similarly, identical tests also show good agreement; hence the Reynolds number must play a minor part in the dense gas dispersion situations considered.

The major practical limitations of accurate wind tunnel simulation of LNG dispersion are operational constraints, particularly the inability to obtain a steady wind profile or to accurately simulate atmospheric turbulence at the lowest wind speeds of interest, and Reynolds number constraints (as yet somewhat ill-defined) associated with the proper scaling of near-field turbulence. When combined with estimates of the restraint of plume expansion by the tunnel side walls, these considerations permit the development of a performance envelope for a particular wind tunnel facility, examples of which are given in the following sections.

PERFORMANCE ENVELOPES: LAND BASED SPILLS

It is instructive to consider the operational constraints on current large wind tunnels to determine those field situations which may be satisfactorily simulated. Operational limitations include:

1. The inability of most large wind tunnels to function satisfactorily at very low wind speeds (< 0.1 m/sec). At low wind speeds the wind tunnel becomes sensitive to small disturbances, both external and internal, which lead to unrealistic perturbation of the mean flow.

2. The associated inability to maintain large Reynolds number. When the characteristic Reynolds number falls below 3000 wake turbulence no larger remains similar to field conditions.

3. A minimum spatial resolution for concentration measurements of $\pm$0.25 cm. Minimum pertinent resolution in the field may be $\pm$ 1 m.

4. Lateral interference with a spreading dense plume by wind tunnel walls. Current wind tunnel facilities have widths up to about 4 m. One can estimate wind tunnel wall interference by utilizing the spread formulae proposed and tested against field spills by Van Ulden (12). The expression relates spread to boiloff rate, wind speed, and gas density.

The four operational limitations listed above have been incorporated into the performance envelopes shown in Figures 1 and 2. Wind tunnel wall interference lines are conservative for the situation shown since they represent steady boiloff interaction at a distance of 20 diameters downwind of a 0.3 m diameter model source.

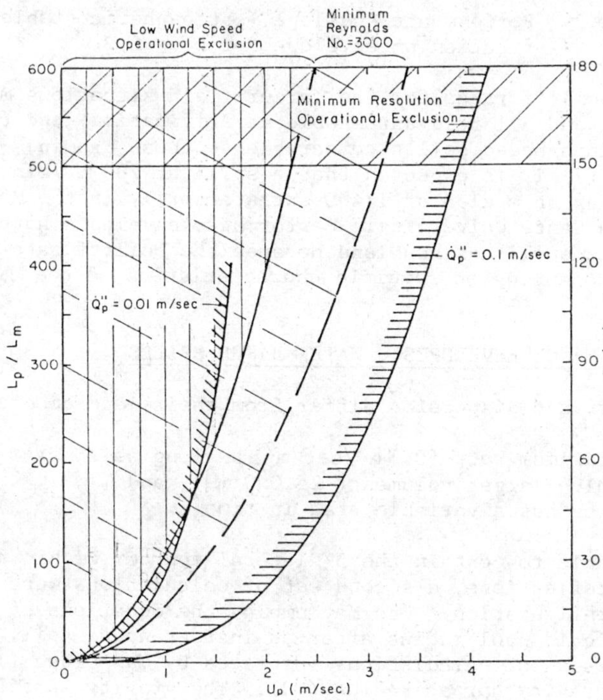

Figure 1. Performance envelope to simulate LNG spills - constant boiloff conditions

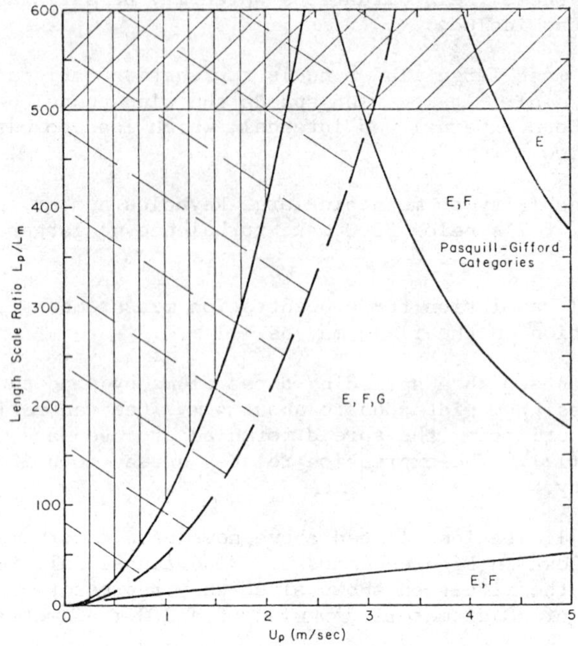

Figure 2. Performance envelope - atmospheric stable stratification on envelope for D_m = 0.3 m

Only the highest boiloff rates for the larger field situations must be eliminated from consideration and all relevant stable stratification conditions can be provided. Assuming land based spills comparable to those examined by Neff et al. (6) or Meroney et al. (7) it is expected that a spill at 200 m^3/min on the floor of an uninsulated dike may be scaled at 1:400 satisfactorily in the Meteorological Wind Tunnel at Colorado State University. Performance envelopes have not been presented here for the finite spill case on land because the boiloff rate tends to vary irregularly with time depending on specific characteristics of the dike or bund shape and its material.

PERFORMANCE ENVELOPES: WATER BASED SPILLS

Water spill boiloff and dispersion differ from their land counterpart because they

1. Boiloff at a maximum rate (0.1 $m^3/sec/m^2$) as long as LNG remains,
2. Generally involve larger volumes (~25,000 m^3), and
3. The spill source has a variable area in time.

Since it is desirable to contain the 5% lateral contour within a test region unaffected by wall reflections, a second set of calculations were prepared assuming a transient spill configuration. One may employ the equations of Raj and Kalelkar to predict maximum liquid pool radius after an instantaneous spill (13). Again following a modified version of calculations suggested by Van Ulden (1974) it is possible to calculate the gravity spread radius (12). The gravity spread is assumed to occur until the frontal velocity equals the mean flow velocity. Subsequently one may emperically allow a 1.5 factor growth in radius before the 5% condition is reached. Since no entrainment is allowed during the gravity spread regime, this calculation is probably reasonable at low wind speeds and conservative at high wind speeds. Figure 3 constructed for a facility similar to the Environmental Wind

Tunnel at Colorado State University suggests that a 20,000 m^3 spill must be modeled at 1:800 to achieve even a 4 m/sec prototype wind speed. A 5000 m^3 spill would be fairly comfortably structured at 1:600; however field wind velocities under 3.0 m/sec would be difficult to reproduce.

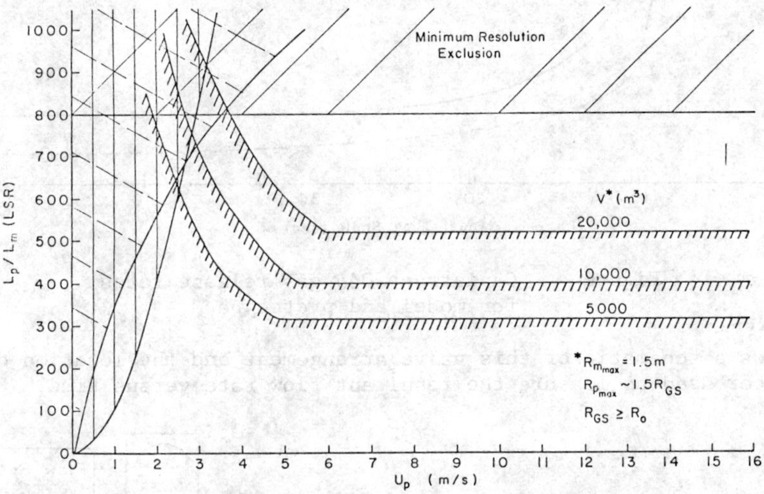

Figure 3. Performance envelope-environmental wind tunnel, water spills LNG, limit to gravity spread when $U_f = \overline{U}$ Final radius $\simeq 1.5\ R_{GS} > 1.5\ R_0$

Hence a 20,000 m^3 spill could be modeled at 7 or 15 m/sec in a facility like the Environmental Wind Tunnel without too much trouble; however a 2.0 m/sec speed would be difficult to achieve. Since the initial cylinder depth of cold natural gas resulting from an LNG spill is $\lesssim$ 10 m then any length scale ratio less than 1/500 has a modeled depth less than 0.02 meters (2 cm). To gain good resolution with some hypothetical probe system it would be preferable not to go to larger length scale ratios.

LABORATORY METHODOLOGY

Simulation of the atmospheric surface layer in Boundary Layer Wind Tunnels has been considered in some detail by previous authors. Special procedures required to inject cold gas plumes are considered by Meroney et al. (11). Representation of transient LNG spills on land or water require additional tools recommended below.

Variable-Boiloff Rate Simulation

To obtain an accurate prediction of the extent of hazard associated with the vaporization of LNG, the model should simulate the variable boiloff rate of the gaseous methane characteristic to that of the spill configuration. Typical boiloff curves for the prototype situation along with the actual model gas release for Capistrano

Test 044 are presented in Figure 4 (7,14). These gas flow rate curves were obtained by use of a programmed cam to close a micrometer needle valve controlling the flow of simulation set at a predetermined rate.

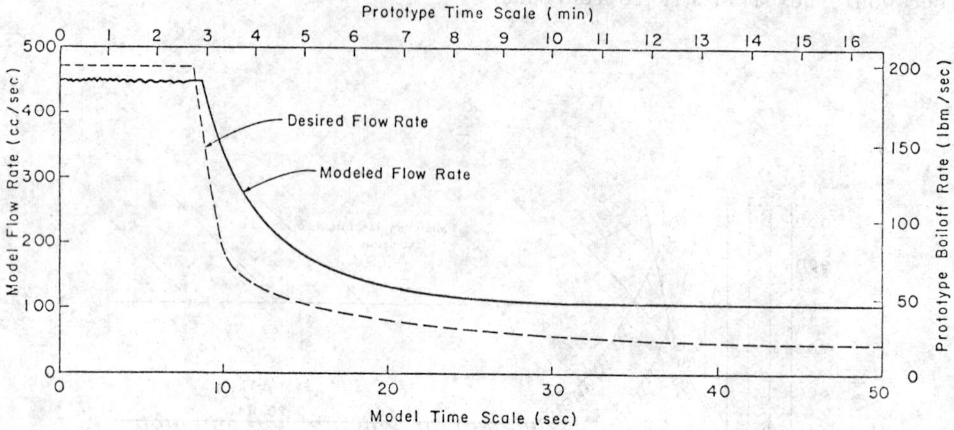

Figure 4. Capistrano 044 gas release rates for model and prototype

Figure 5 shows a schematic of this valve arrangement and the location of the mass flow transducer used to measure the resultant flow rate versus time.

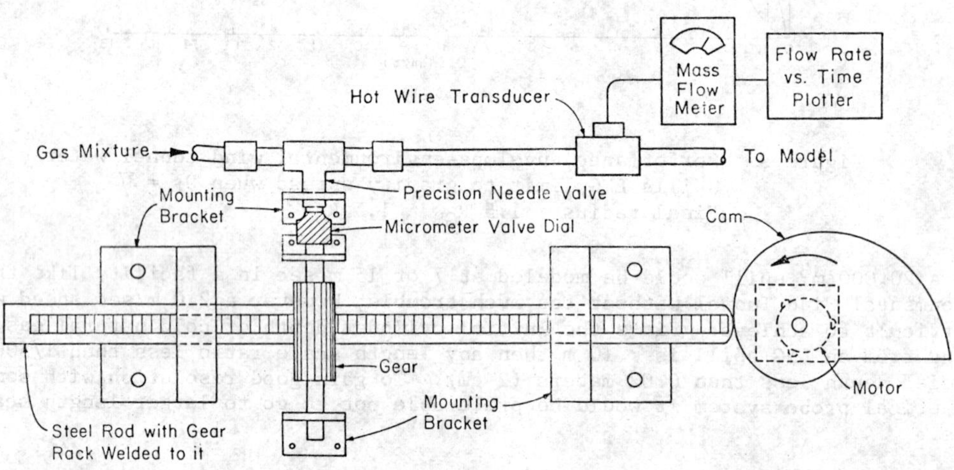

Figure 5. Variable flow rate control valve

Transient-Concentration Instrumentation

The transient nature of the boiloff rates simulated necessitated the use of a fast response, temperature compensated concentration transducer. An aspirated dual-film probe was designed for this project. As noted in Figure 6, dual films operated at different current levels permitted compensation for temperature drift, while a flared inlet reduced the noise of pressure fluctuations. Calibration suggests a noise level of 0.1% by volume CO_2 or Argon and an upper frequency response of 1000 Hz (7).

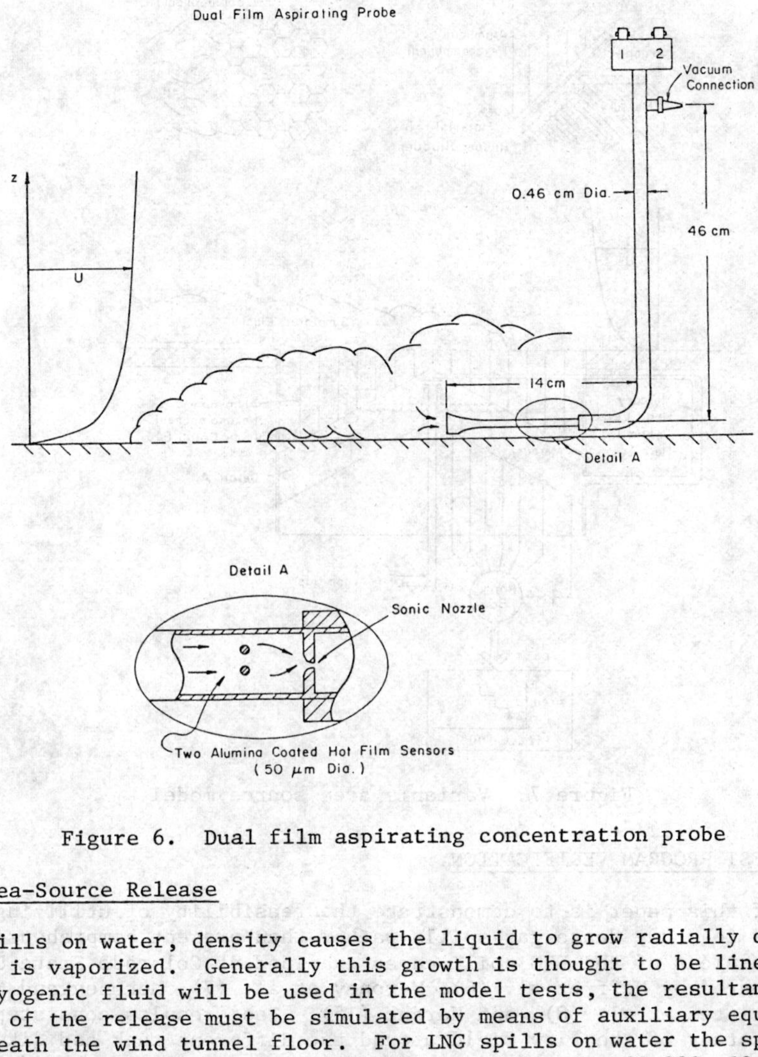

Figure 6. Dual film aspirating concentration probe

Variable-Area-Source Release

When LNG spills on water, density causes the liquid to grow radially outward until all methane is vaporized. Generally this growth is thought to be linear with time. Since no cryogenic fluid will be used in the model tests, the resultant variable area nature of the release must be simulated by means of auxiliary equipment installed beneath the wind tunnel floor. For LNG spills on water the spill radius when all methane is vapor will be ~ 352, 271, or 209 m for 20,000, 10,000 and 5,000 m^3/spills respectively. At a scale of 1:600 one requires a source approximately 1 to 1.5 m diameter.

A variable area source devise has been constructed for the current China Lake spill series. As shown in Figure 7 it consists of a contoured honeycomb surface together with a mercury filled bladder. The device has been constructed at a diameter of 0.25 meters; however a larger version may be possible. A simple alternative for the larger size releases is a set of concentric anuluses metered by flowraters and monitored by cam operated solenoid valves.

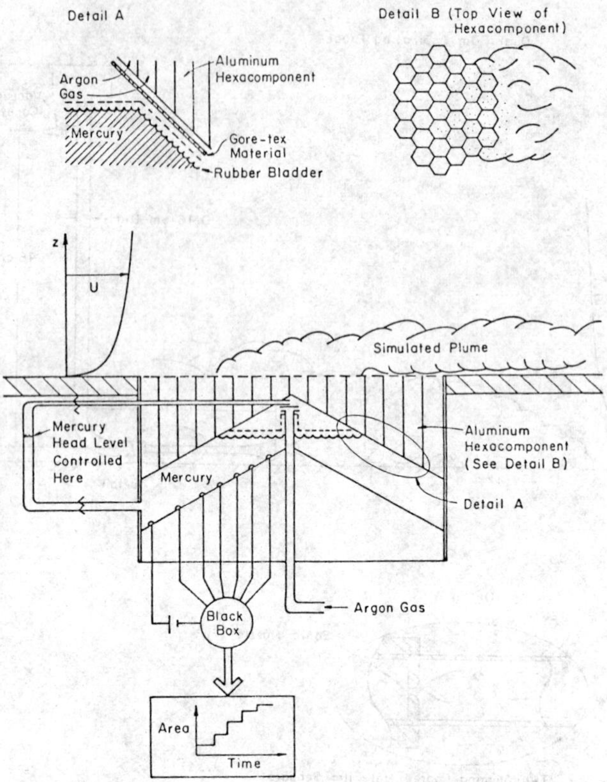

Figure 7. Variable area source model

TEST PROGRAM VERIFICATION

The purpose of this paper is to demonstrate the feasibility of utilizing wind tunnels as a tool to study dense gas spills rather than present comprehensive results. Extended discussions of the LNG spill cases examined at Colorado State University have been prepared by Neff et al. (6), Meroney et al. (7), Meroney and Neff (16), and R & D Associates et al. (8), and Harsha (15). Test results consisted of (a) a qualitative study of the flow field around the different tank and dike localities by visual observation of the plume released from the model area; and (b) a quantitative study of gas concentrations produced by the release of a tracer from the model area. We are now in a position to compare the current wind tunnel data with both additional wind tunnel data taken at the Warren Springs Laboratory and Colorado State University and with field test data taken during the American Gas Association Phase II Program (4,6,7,14).

The Battelle Columbus Laboratories correlation, which represents an upper bound of all concentration resulting from confined LNG land spills during the AGA Phase II program can be arranged as

$$\frac{u\chi \ell_b^{2}}{Q} \simeq 130 \left(\frac{x}{\ell_b}\right)^{-2},$$

where the bouyancy length scale, $\ell_b = \dfrac{g(\rho_g - \rho_a)\, Q}{\rho_a\, \bar{u}^3}$. (13,16)

Figure 8 presents the Colorado State University CO_2 release gas data, the CSU Freon-N_2 release, the CSU neutral stability data, the Battelle Columbus Laboratory correlation, and the Warren Springs data. The data taken under "stable" conditions and with He-N_2 releases in Reference 6 have not been included for the reasons cited there. Thus the data presented in Figure 8 do not reflect the effects of nonadiabatic heat transfer or the effects of atmospheric humidity, both of which would tend to reduce the gravitational forces that are responsible for the initial spreading and miximg of the vapor cloud. With regard to the Warren Springs data, a factor of two reduction in the "advertised" velocity was used to obtain a velocity more typical of that which the vapor plume was actually exposed to. A few Warren Springs and several CSU data points were eliminated because of plume blockage by the wind tunnel walls. In addition, the Warren Springs experiments employed a heavier simulation gas than CSU to allow the wind tunnel to operate at higher speeds. Although the Frounde number was scaled properly, the turbulent mixing processes were probably reduced by the use of the heavy gas. Thus, we would expect the Warren Springs data to exhibit somewhat higher concentration values at a given scaled range. It should also be noted that the Battelle Columbus Laboratory correlation was generated from data occupying a limited region of x/ℓ_b (note for example, the Capistrano 044 data). Thus, the Batelle Columbus Laboratory curve in Figure 8 should be regarded as an extension of the original Battelle Columbus Laboratory correlation. A significant observation from Figure 8 is that the Battelle Columbus Laboratory correlation bounds the current and previous CSU data (with room temeprature gas releases) in addition to the available field test data.

CONCLUSIONS

The wind tunnel can simulate a range of conditions associated with vapor transport and dispersion downwind of LNG spills. Scaling criteria suggest that existing size facilities can simulate land spills boiling from areas up to 150 meters diameter. Wind speeds at a 10 m reference heights may be simulated from lower magnitudes of 0.7 m/s for spills of 15 m diameter (scale ratio = 1/50) or 3.0 m/s for spills of 150 m diameter (scale ratio = 1/500) upwards. A desirable local resolution of 1 m limits model scale ratios to ~ 1/500 or less. Lateral spreading in a typical 4 m width wind tunnel may further limit maximum equivalent volume production of cold vapor at a given wind speed (i.e., for scale ratios = 1/200 a liquid boiloff rate of 25.4 mm/min produces wall interference effects beyond downwind distances of 300 m for velocities less than 3.0 m/sec). For rates less than liquid boiloff rates of 2.5 mm/min there should be no additional constraint. Vapor dispersion downwind of LNG land spills has been reproduced for selected cases of the 1974 AGA landspill program. Water spills may be reproduced if additional consideration is given the characteristics of variable boiloff area. Wind tunnel simulation provides a design tool to pre-scale trajectories and dispersion of cold LNG vapor clouds. This method will provide guidance for instrument placement and numerical model development during the 50 million dollar program planned by Department of Energy to guide LNG hazard analysis.

ACKNOWLEDGEMENTS

Authors gratefully acknowledge support via R & R Associates, Marina del Rey, California, U.S.A., from American Gas Association Project IS-128-1 and Department of Energy Contract EE77C031364 as well as direct support through Department of Transportation Contract No. DOT-CG-75279-A.

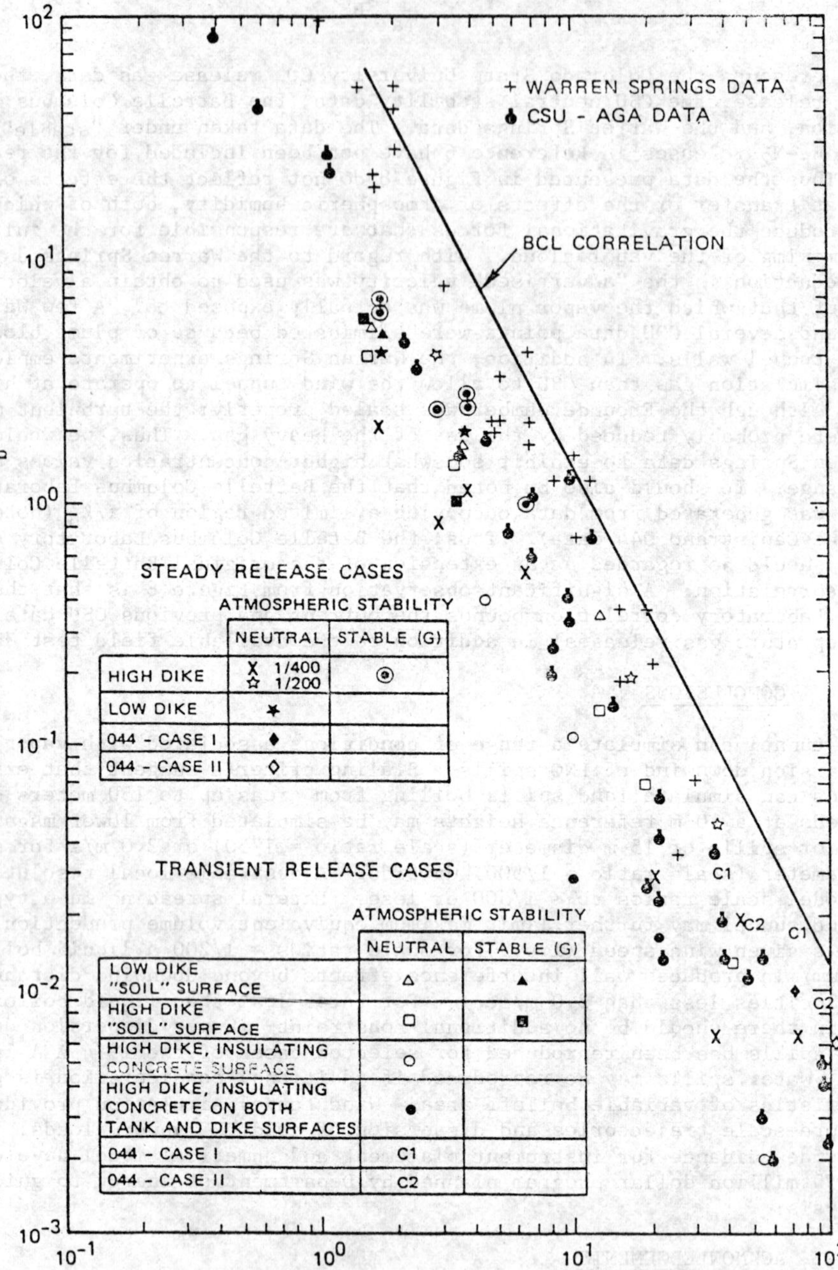

Figure 8. Comparison of selected wind tunnel and field test concentration data

REFERENCES

1. J. Sakagami and M. Kato, Diffusion and vapour rise of methane vapour from a real source in air stream, Natural Science Report of Ochanomizu University, Japan, 19: 59-66, (1968).

2. G.J. Boyle and A. Kneebone, Laboratory investigation into the characteristics of LNG spills on water, evaporation, spreading and vapor dispersion, Shell Research, Ltd. Report to API, (1973).

3. T.G. Hoot and R.N. Meroney, The behavior of negatively buoyant stack gases, 67th Annual Meeting APCA, Denver, CO, 74-210, (June 9-13, 1973).

4. D.J. Hall, C.F. Barrett, and M.O. Ralph, Experiments on a model of an escape of heavy gas, Warren Spring Laboratory Report CR 882 (AP), Department of Trade and Industry, U.K. (1975).

5. D.J. Hall, Further experiments on a model of an escape of heavy gas, Warren Spring Laboratory Report CR 1341 (AP) Department of Trade and Industry, U.K., (1977).

6. D.E. Neff, R.N. Meroney and J.E. Cermak, Wind tunnel study of the negatively buoyant plume due to an LNG spill, Colorado State University, Ft. Collins, CER76-77-22, (1976).

7. R.N. Meroney, D.E. Neff, J.E. Cermak and M. Megahed, Dispersion of vapor from LNG spills - simulation in a meterological wind tunnel, Colorado State University, Ft. Collins, CER76-77RNM-JEC-DEN-MM57, (1977).

8. R&D Associates and Fluid Mechanics and Wind Engineering Program Staff, Colorado State University, LNG wind tunnel simulation and instrumentation assessments, RAD-TR-105700-003 or DOE Report SAN/W1364-01, (1978).

9. J.E. Cermak, Applications of Fluid Mechanics to Wind Engineering--a Freeman Scholar lecture, J. of Fluid Engineering, 97: 9-38, (1975).

10. R.N. Meroney, J.E. Cermak and D.E. Neff, Dispersion of vapour from LNG spills-simulation in a meterological wind tunnel, Proc. of 3rd AMS Symposium on Atmospheric Turbulence, Diffusion and Air Quality, Raleigh, NC, 243-246, (October 19-22, 1976).

11. R.N. Meroney, D.E. Neff and J.E. Cermak, Wind tunnel simulation of LNG spills, Proc. of American Gas Association Transmission Conference, Montreal, Canada, T217-T223, (May 8-10, 1978).

12. A.P. Van Ulden, On the spreading of a heavy gas released near the ground, Loss Prevention and Safety Promotion Seminar, Delft, Netherlands, (1974).

13. A.E. Germeles and E.M. Drake, Gravity spreading and atmospheric dispersion of LNG vapor clouds, Proc. of the 4th International Symposium on Transport of Hazardous Cargoes by Sea and Inland Waterways, Jacksonville, Fla, 519-539, (Oct. 26-30, 1975).

14. American Gas Association, LNG safety program interim report on phase II work, Report on American Gas Association Project IS-3-1, Battelle, Columbus Laboratories, (1974).

15. P.T. Harsha, LNG safety program topical report, wind tunnel tests of vapor dispersion from land facilities, RDA-TR-1100-002, AGA Project IS-128-1 (draft), (1976).

16. R.N. Meroney and D.E. Neff, Behavior of negatively buoyant gas plumes from an LNG spill, 6th Australasian Hydraulics and Fluid Mechanics Conference, University of Adelaide, Australia, (Dec. 5-9, 1977). (Colorado State University, Fort Collins, Paper No. CEP77-78RNM-DEN1).

MODELING OF POLLUTANT DISPERSION FROM A MOVING SOURCE—APPLICATION TO AIRPORT POLLUTION

A. NISHI* and J. E. CERMAK**

*Miyazaki University, Miyazaki, Japan
**Colorado State University, Fort Collins, Colorado, USA

ABSTRACT

The atmospheric dispersion process of pollutants emitted from a moving aircraft is represented by a puff diffusion model to simulate its movements precisely near the airport, under the assumption of a Gaussian distribution of pollutants. Three-dimensional diffusion parameters are estimated by matching calculated concentrations to those measured in the field with tracer experiments. By using these parameters, the time-dependent ground level concentrations of CO and NOx and the isopleths of NOx for quasi-static takeoff operations of aircraft are presented.

INTRODUCTION

Many air-pollution prediction models for airports have been proposed in the past several years.[1,4] The basic ideas of these models depend mostly on the diffusion process of plumes, i.e., plume models for point, line and area sources. The plume model is suitable for stationary sources, aircraft movement, however, is too fast and the spacing between aircrafts is too large to be treated as a stationary line source like for automobiles on the road. Therefore the results will contain a large amount of error if the diffusion process of aircraft emissions is approximated by a plume model, especially when air pollution at the airport and its near vicinity is considered.

In order to develop a more detail and precise model for a moving source, formulation of a diffusion model for instantaneous line sources is necessary. The diffusion process can be simulated by a series of small puffs emitted with small intervals between puffs, i.e., a puff model. As each puff is diffused by the wind turbulence in all directions, three-dimensional diffusion parameters are required to estimate the concentration distribution.

Tracer experiments for moving aircraft are being conducted at Osaka Bay in Japan. Since there is not enough open land space for an onshore international airport at this location, an offshore airport is being planned. Air of the surrounding area is polluted heavily by many industries, therefore, it is very important to predict the impact of additional emissions from airport on air quality. The tracer SF_6 was released by a helicopter along a few flight paths corresponding to those of aircraft movement near the proposed airport, and air samples were collected at onshore receptors about 6-13 km from the proposed offshore-airport site. These data provide a reference for testing prediction by the puff model developed in this paper.

The first step is development of a Gaussian puff model for a moving aircraft. Secondly, some results of the field tracer experiments are analyzed by this method, and the time-dependent concentrations are calculated. By comparing the observed and calculated mean concentrations, three-dimensional diffusion parameters are estimated with the direct search optimization technique.[5,6] That is, both verification of proposed model and the required diffusion parameters are obtained by these calculations. Finally, numerical examples of time-dependent pollutant concentrations and the mean concentration isopleths are shown for single and quasi-static takeoff operations of a B-747 type commercial aircraft respectively.

MODELING OF DIFFUSION PROCESS FOR A MOVING AIRCRAFT

The plume diffusion from a line source is a poor approximation for moving aircraft, since the intervals of aircraft arrival and departure are not as close as the automobiles on the road even for the busiest airport. Moreover the individual source is too large and it moves long distances in both horizontal and vertical directions in a short time.

Because of these considerations, the moving source is represented by an infinite series of puff sources, and each puff is emitted with the infinitesimal intervals, i.e., the puff model is developed in accordance with the following assumptions:

a) Each puff has Gaussian distribution of pollutants.

b) Complete reflection of pollutants occurs at both the sea and ground surfaces and the inversion base.

c) The effect of jet dispersion near the engine is assumed to be negligible compared with the effect of diffusion by atmospheric turbulence.

d) The buoyancy effect can be accounted for by an increase in effective source height.

e) Pollutants are stable chemicals in the air.

The instantaneous pollutant concentration at a receptor (x,y,z) at time t is given as follows for all aircraft operations;

$$c(x,y,z;t) = \frac{q}{(2\pi)^{3/2}} \int_{t_0}^{t_0+t} \frac{1}{\sigma_x \sigma_y \sigma_z}$$

$$\cdot \exp\left\{ -\frac{1}{2}\left[\left(\frac{x-x_0-(V_0t_1+\frac{1}{2}\alpha t_1^2)\cos\gamma\cos\theta-u(t+t_0-t_1)}{\sigma_x}\right)^2\right.\right.$$

$$\left.\left.+\left(\frac{y-y_0-(V_0t_1+\frac{1}{2}\alpha t_1^2)\cos\gamma\sin\theta-v(t+t_0-t_1)}{\sigma_y}\right)^2\right]\right\}$$

$$\cdot\left\{\sum_{n=-\infty}^{\infty}\left[\exp\left\{-\frac{1}{2}\left(\frac{H+z_0+z+(V_0t_1+\frac{1}{2}\alpha t_1^2)\sin\gamma+w(t+t_0-t_1)+2nH_i}{\sigma_z}\right)^2\right\}\right.\right.$$

$$+ \exp\left\{-\frac{1}{2}\left(\frac{H+z_0-z+(V_0t_1+\frac{1}{2}\alpha t_1^2)\sin\gamma+w(t+t_0-t_1)-2nH_i}{\sigma_z}\right)^2\right\}\Bigg]\Bigg\}dt_1 \quad (1)$$

where

- q: emission rate per unit time (g/sec),
- σ's: diffusion parameters of x, y and z directions, standard deviations (m),
- x_0, y_0, z_0: starting point of aircraft at time t_0,
- V_0: initial velocity of aircraft at starting point (m/s),
- α: acceleration of aircraft (m/s^2),
- u, v, w: x, y, z components of wind velocity (m/s),
- θ: angle of runway from N-S direction (deg),
- γ: upward angle of flight path, 3° for landing, 6° for takeoff,
- H: effective source height, depended on the effect of buoyancy and direction of flight path (m),
- t_1: integration time, and
- H_i: inversion base height (m), in practical interest $n = -1, 0, +1$ are important.

There are four major modes of aircraft operations near the airport, they are taxiing, takeoff, climbout and approach. For each mode of operation, the variables in Eq. 1 are given as follows:

a) Taxiing: Aircraft moves on the ground with slow constant speed, then $z_0 = 0$, $\alpha = 0$, $\gamma = 0$. Since the ground-level concentration is considered, $z = 0$, and in general $w = 0$ can be assumed. $H = 10$ m and $t_0 = 0$ are given.

b) Takeoff: Aircraft moves on the ground with acceleration, then $z_0 = 0$, $\gamma = 0$, and $z = 0$, $w = 0$ as above. $V_0 = 0$, if it starts from rest, and $H = 30$ m is given.

c) Climbout: Ascending along an inclined path with acceleration, then $z = 0$, $z_0 = 0$, $w = 0$ and $H = 30$ m and $\gamma = 6°$ are given.

d) Approach: Descending along an inclined path with constant speed, then $z = 0$, $\alpha = 0$, $w = 0$ and $H = 20$ m, $\gamma = 3°$ are given.

For climbout and approach operations, the source height changes quickly, therefore the concern of differences of meteorological conditions along the vertical direction is very important. The changes of wind velocity, direction and diffusion parameters can be taken into account by putting them as the functions of z.

ESTIMATION OF DIFFUSION PARAMETERS FROM THE FIELD TRACER EXPERIMENTS

Tracer Experiment

To verify the mathematical model and estimate the diffusion parameters, the tracer experiments were carried out as mentioned in Sec. 1 for the site shown in Fig. 1. The tracer SF_6 was released around the proposed offshore airport site (about 6 km from the shoreline) for 1.5 hours. It was carried to onshore receptors by the sea breeze. This experiment will continue until the spring of 1979.

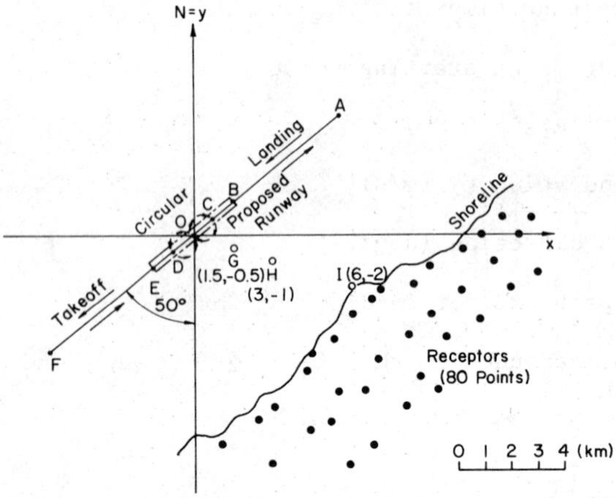

Fig. 1. Proposed offshore airport.

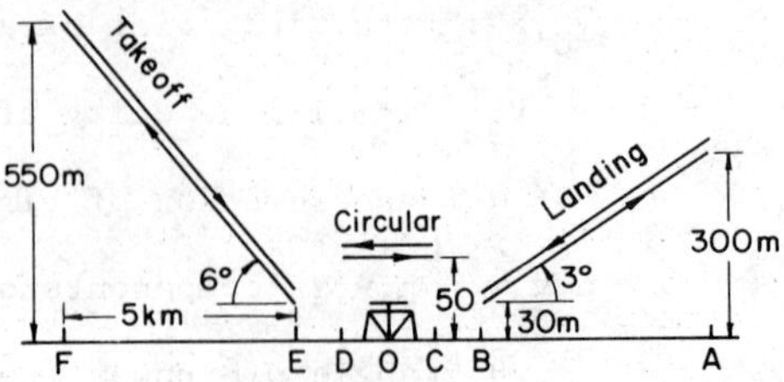

O: meteorological station
C-O-D: circular flight path
E-F: takeoff flight path
A-B: landing flight path
B-E: proposed runway

Fig. 2. Schematics of landing and takeoff flight path.

The tracer was released by a helicopter to simulate the aircraft movements, and its flight path were chosen to duplicate those of aircraft as shown in Fig. 2. A constant height circular flight path at the center of the proposed runway was used to simulate the ground operation of aircraft on the airport. The ascent and descent flights along the paths of 6° and 3° inclination were used to simulate the climbout and approach operations, respectively.

About 80 receptors were arranged near the shoreline. The tracer-air mixture, i.e., contaminated air was collected for 30 minutes into sampling bags at the rate of 8 ℓ/hr by automatic air samplers. Accordingly, the tracer in each bag was the mean concentration over the sampling time, i.e., a time-average dosage. This sampling time was chosen as a characteristic time scale of aircraft operation near the airport.

In these experiments, unsteadiness of the surface concentration resulted from two effects even if the meteorological conditions were in steady state. One effect was related to downwind distance and wind velocity. If the wind velocity was small, each puff took a long time to reach the shoreline, then early sampling before peak values were reached resulting in mean concentrations that were too low. The other effect depended on the flight paths. For inclined paths there were differences of height for each puff source, and larger ground-level concentrations were obtained for the lower source heights.

Estimation of Diffusion Parameters

Since the field tracer experimental data give the mean concentration for sampling time T, the mean value must be calculated to compare the field and model results. It can be obtained by integrating the instantaneous concentration (Eq. 1) for time t

$$\overline{C}\ (x,y,z;T) = \frac{1}{T} \int_{o}^{T} c(x,y,z;t)dt. \qquad (2)$$

The diffusion parameters are estimated by matching the observed results of tracer experiments with the predicted values obtained by Eq. 2 with the direct search optimization technique.[5,6]

At first the diffusion parameters σ_x, σ_y and σ_z are assumed as,

$$\sigma_i = c_i x^{ai} \qquad i = 1,2,3 \qquad (3)$$

where c_i and a_i are the coefficients to be determined, x is downstream distance and $i = 1,2,3$ correspond to x,y,z respectively. This power law is an adequate approximation, since the observed data have many uncertainties. Other than these six coefficients (variables), wind speed and direction are chosen as the variables.

Secondly, as the optimization criterion function, the least-square curve fitting between observed (Y_k) and predicted (calculated) (y_k) values is applied

$$F = \sum_{k=1}^{m} (Y_k - y_k)^2 \qquad (4)$$

where m is the number of given data points.

The procedure of this technique is as follows: Each of these variables is changed a little, and if the criterion function is improved, the new value is adopted, otherwise if it is not improved the initial value is retained. After the check of all variables, which is called "local examination," a "pattern move" follows by the rule.

$$\left.\begin{aligned} \zeta_j^i &= N_j \cdot \delta_j \\ N_j &= n_{1j} - n_{2j} \end{aligned}\right\} \qquad (5)$$

where ζ_j^i: the pattern increment applied to j-th coefficient during i-th pattern move,

δ_j: small increment of coefficient,

n_{1j}: the number of previous successful positive local examinations for j-th coefficient, and

n_{2j}: the number of previous successful negative local examinations for j-th coefficient.

This calculation, local examinations alternating with pattern moves, continues until there is deterioration of the criterion function.

This method is a powerful tool for these multi-variable problems, the difficulties, however, occur when there are many optimums in multidimensional (multi-variable) space, and the combinations of different initial values have to be examined. Fortunately, this problem seems to be unimodal (one optimum), therefore the calculation is relatively easy compared to a multimodal case. It should be noted that the diffusion parameters calculated are for the sampling time T and involve the uncertainties of other factors which cannot be taken into account in the calculations.

NUMERICAL EXAMPLES

Diffusion Parameters

Out of the many tracer experiments two typical cases are analyzed. Measured meteorological conditions of these two cases are presented in Table 1. At first a constant height, circular flight path (No. 1) is analyzed. The tracer was released from 10:30 to 12:00 (1.5 hrs), and air was sampled from 11:30 to 12:00 (30 min). Ten data points (dimensions) and observed concentrations are given, and six coefficients of σ's and wind speed are chosen as the independent variables. The wind direction was determined from the observed concentration pattern, and given as input data to decrease the number of independent variables. The comparison of observed and computed concentrations for each of the points is presented in Fig. 3 with calculated isopleths. A fairly good agreement is obtained and the correlation factor is 0.91 for ten data points.

Table 1. Measured meteorological conditions.

Flight Path	No. 1 Circular			No. 2 Landing		
Height	WD	WS	Temp.	WD	WS	Temp.
50 m	335°	1.7 m/s	7.8°C	299°	10.4 m/s	5.8°C
100 m	331°	2.3 m/s	7.3°C	301°	11.0 m/s	5.2°C
150 m	327°	2.4 m/s	7.0°C	303°	11.4 m/s	4.6°C
200 m	322°	0.3 m/s	7.1°C	304°	11.3 m/s	4.0°C
250 m	305°	0.2 m/s	8.0°C	304°	11.5 m/s	3.5°C
300 m	245°	0.6 m/s	8.6°C	305°	11.7 m/s	3.0°C
350 m	217°	2.1 m/s	9.0°C	304°	11.4 m/s	2.4°C
400 m	210°	4.5 m/s	8.8°C	304°	11.5 m/s	1.9°C
Mean WD	335°			302°		
Approximate WS	1.8 m/s: const			$6.5\ z^{0.103}$		
Inversion base height	180 m			none		

WD: Wind direction
WS: Wind speed

The time-dependent tracer concentration is also calculated, and shown in Fig. 4 for several downwind locations. The calculated wind speed is 1.8 m/s for which more than one hour is required for the first puff to reach the downwind distance of $x = 8$ km. In spite of the fact that sampling began one hour after the tracer release, only a small part of tracer arriving at the receptor was collected during the sampling time at the location of $x = 10$ km. The observed concentration at each location corresponds to a mean value of area under each curve.

Secondly, the landing flight (No. 2 in Table 1) is analyzed. The tracer was released from 15:00 to 16:30, and air was sampled from 15:30 to 16:00. For this case the source height changed from 300 to 30 m as shown in Fig. 2. The comparison of observed and computed results are presented in Fig. 5 with calculated isopleths and the correlation index is 0.73 for 19 points. Measured wind speed along the vertical direction is approximated as

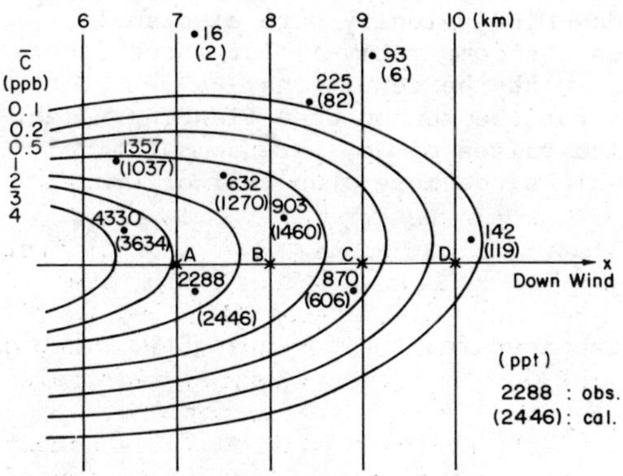

Fig. 3. Comparisons of observed and calculated concentrations, circular flight.

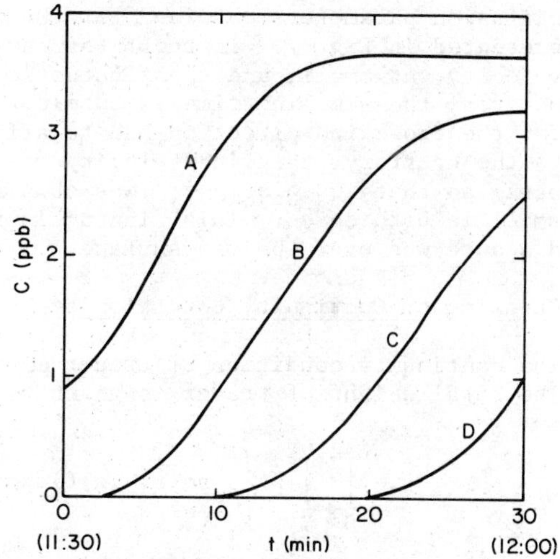

Fig. 4. Time-dependent concentrations at a few downwind locations.

$$\overline{W} = p\ (z)^{1/9.7}$$

where p is a parameter (variable) to be determined.

Seven independent variables are estimated. The wind speed is sufficiently high in this case (about 10 m/s at 30 m), so that enough contaminated air was sampled during the sampling time. The helicopter flight path, however, was inclined, so the time-dependent concentration curves had a few peaks as shown in Fig. 6.

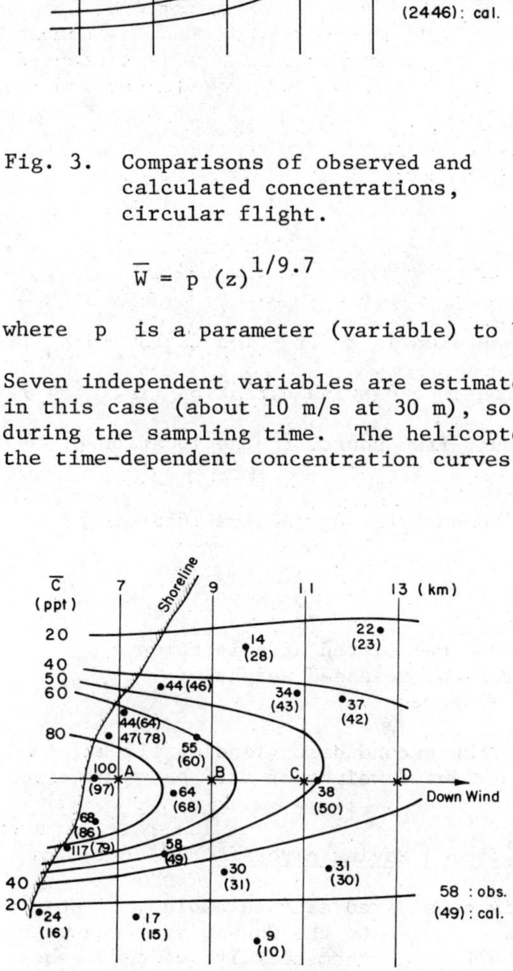

Fig. 5. Comparisons of observed and calculated concentrations, landing flight.

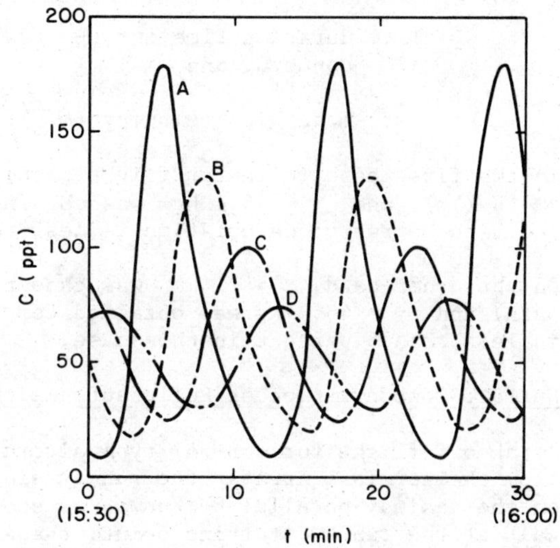

Fig. 6. Time-dependent concentration at a few downwind locations.

Diffusion parameters obtained for the x, y and z directions in both cases are presented in Fig. 7. In the first case (circular flight path), the mean wind velocity was small, and the fluctuations of downwind velocity were also small. However, the sampling time is 30 minutes, then the comparatively large turbulence for the crosswind direction had to affect σ_y with the result that $\sigma_y > \sigma_x$. On the contrary, the wind velocity was larger for the second case (landing flight path) so that $\sigma_x > \sigma_y > \sigma_z$ was obtained. The values of σ_z are comparatively small in both cases. This finding is reasonable since more than half of travel distance was over the sea surface.

Checking of Continuity Condition of Tracer Mass

The continuity condition of tracer mass was examined for the foregoing two cases. The total weight of tracer is given by

$$\begin{aligned} \chi &= \int_0^Z \int_{Y_1}^{Y_2} \int_{X_1}^{X_2} c(x,y,z;t)\,dxdydz \\ &= \int_0^Z \int_{Y_1}^{Y_2} \int_{T_0}^{T_0+T_1} c(x,y,z;t)\overline{W}\,dtdydz \end{aligned} \tag{6}$$

where χ: total weight of tracer (kg),

Z: the upper limit of tracer dispersion in z-direction ($Z \geq (3\sigma_z + H + Z_0)$),

Y_1, Y_2: the limits of tracer dispersion in crosswind direction ($Y_1, Y_2 \geq 3\sigma_y$),

T_0: traveling time of the first puff from source to the location of concern,

T_1: duration time for the tracer to pass through at the location of concern, and

$\overline{W}$: mean wind velocity (m/s).

In the first case (circular flight path) it was calculated at a location of $x = 8$ (km), and $\chi = 45.1$ kg was obtained for the released weight of 40.5 kg, i.e., the error of calculation is less than 12 percent.

On the other hand, $x = 11$ km was chosen for the second case (landing flight path) and $\chi = 45.8$ kg was obtained for the released weight of 43.7 kg. The error is less than 5 percent in this case.

Surface Concentration of Pollutants Emitted from a Moving Aircraft

A takeoff flight for a B-747 type aircraft is considered as an example. It starts from the terminal area at the runway side, and taxies to the runway starting point on the taxiway parallel to runway as shown by Fig. 8 schematically. After a short wait at the runway starting point, the aircraft is accelerated by the takeoff mode of engine operation. Half way down the runway, the aircraft takes off and climbs up along a path inclined at an angle of 6° until it reaches the inversion base height or 900 m.

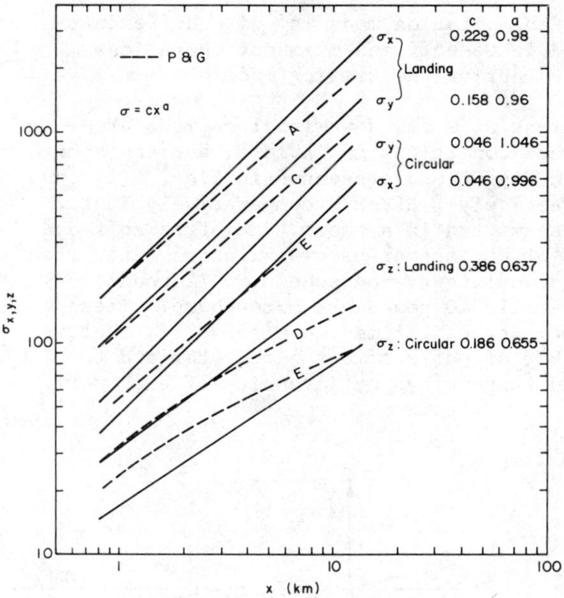

Fig. 7. Estimated diffusion parameters.

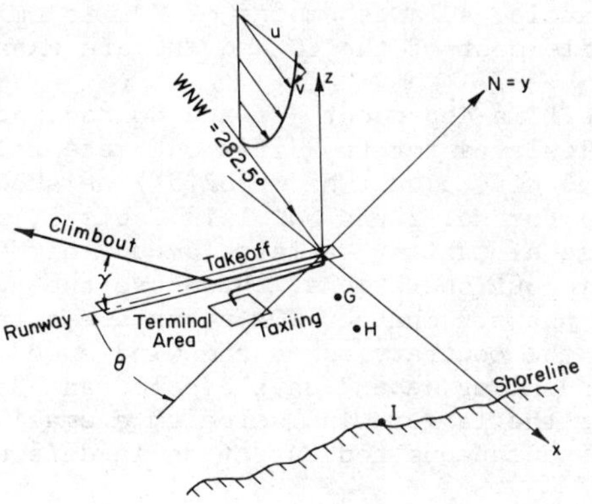

Fig. 8. Schematic view of proposed airport and takeoff operation.

The duration time of each operation has been determined by EPA as a standard takeoff and landing cycle.[7] For this case they are 19, 0.7 and 2.2 minutes for taxiing, takeoff and climbout operations, respectively.

A standard of pollutant emission regulation of the commercial aircraft engine has also been published by EPA, and the corresponding value for the 1981 regulation (given in Table 2) is used in this calculation.

Table 2. Estimated pollutants emission for a B-747 type aircraft (after EPA 1981 regulation).

(a) Emission rate (g/s)

	Idle	App.	Climb.	TO
CO	16.68	5.36	0.88	1.20
THC	2.30	0.12	0.16	0.16
NOx	1.64	13.08	105.56	186.96

(b) Total emission for 1 LTO cycle (kg)

	Idle	App.	Climb.	TO	Total
Time (min)	26	4	2.2	0.7	32.9
CO	26.0	1.29	0.1	0.05	27.5
THC	3.6	0.03	0.02	0.00	3.65
NOx	2.6	3.1	13.9	7.9	27.5

The emission rate of each pollutant in each operation mode is quite different, that is, a large amount of NOx is emitted in takeoff and climbout operations, while most of the CO and THC are exhausted during the taxiing mode.

The time-dependent NOx and CO concentrations at a few downwind locations are calculated for two different meteorological conditions in Table 1, but the same wind direction (WNW = 282.5°) as shown in Fig. 8 and presented in Figs. 9(a) and (b) for No. 2 and No. 1 in Table 1, respectively. After a comparatively long time of taxiing, a large amount of NOx is emitted in a short time of takeoff and climbout operations, therefore the peaks of NOx appear several minutes later than CO peaks. Sharper NOx peaks are obtained in larger wind speed (No. 2 condition). On the contrary, when the wind speed is small, CO peaks are more sharp. These tendencies depend on the pollutant density per unit time and wind speed, that is, for the larger wind velocity a small number of puffs affect a receptor and the instantaneous concentration is determined largely by the intensity of each puff.

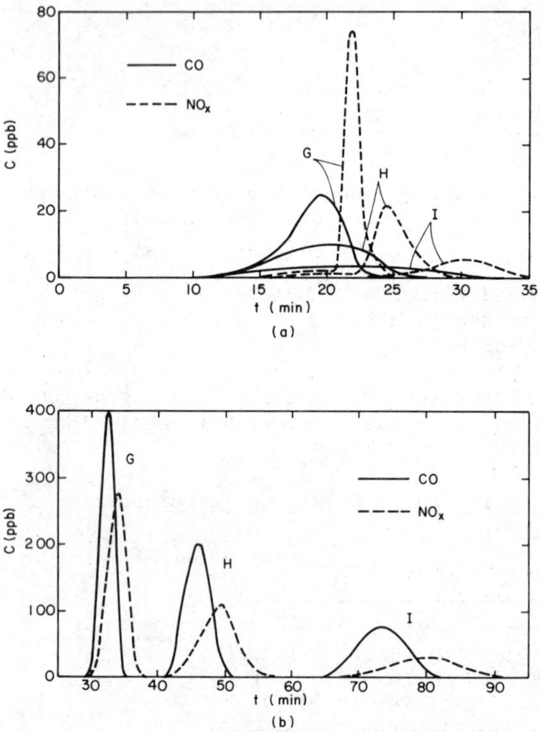

Fig. 9. Time-dependent concentration for a takeoff operation at a few downwind locations.

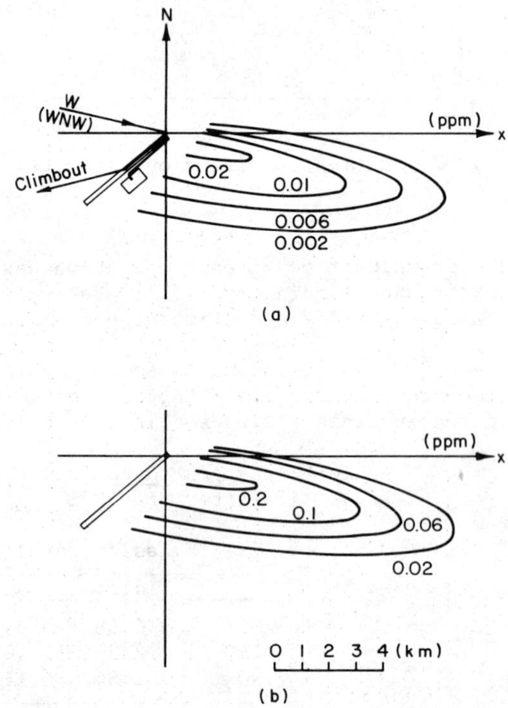

Fig. 10. NOx isopleths for quasi-static takeoff operations.

The NOx isopleths for three minute intervals of quasi-static takeoff operations were calculated, and are shown in Figs. 10(a) and (b) for No. 2 and No. 1 in Table 1, respectively. The concentrations obtained for smaller wind speed (No. 1) are about ten times those for the larger wind speed (No. 2).

CONCLUSIONS AND FUTURE WORK

The atmospheric diffusion process for pollutants emitted from a moving aircraft is given by a puff diffusion model under the assumption of a Gaussian distribution. To apply this equation to the prediction model for airport pollution, the appropriate diffusion parameters must be determined. As the numerical examples, two field tracer experiments are analyzed and three-dimensional diffusion parameters are estimated by comparing the calculated values with the results of tracer experiments. By using these diffusion parameters, the time-dependent surface concentrations of CO and NOx and the mean concentration isopleths of NOx are calculated for single and quasi-static takeoff operations of a B-747 type aircraft, respectively.

The tracer experiments are made for the purpose of predicting offshore airport pollution, however, these results can be applied to other airports directly after a consideration of relations between local meteorological conditions and diffusion parameters.

Two steps of the prediction model must be developed to predict the peak and average pollutant concentrations. They are a short and long-term prediction model. A short-term (1 day) prediction model has already been developed by the same technique as this paper, and in this model, all pollutant sources affecting airport air quality are taken into account.

The next step is a long-term (1 year) prediction model. The local meteorological conditions are classified and the correspondence of each condition with the diffusion parameters needs to be determined. These are available as the basic data for a long-term model. The plume diffusion from point, line and area sources will also be available for many stationary sources in the airport complex.

ACKNOWLEDGEMENT

Authors wish to thank Dr. N. Kouwen, University of Waterloo, Canada, for his recommendation of using the pattern search optimization technique to this problem.

REFERENCES:

(1) Platt, M. et al., The Potential Impact of Aircraft Emission upon Air Quality, NREC Report No. 1167-1, Northern Research and Engineering Co., 1971.

(2) Tammen, H.N. et al., Investigation of Lake Site Alternative Concepts, LERTA Final Report No. 12-1.

(3) Wang, I.T. et al., Airport Vicinity Air Pollution Study-Model Application and Validation and Air Quality Impact Analysis at Washington International Airport, U.S. Dept. of Transportation, FAA Report No. FAA-RD-74-132, 1974.

(4) Rote, D.M. et al., Airport Vicinity Air Pollution Study, U.S. Dept. of Transportation, FAA Report No. FAA-RD-73-113, 1973.

(5) Monro, J.C., Direct Search Optimization in Mathematical Modeling and a Watershed Model Application, NOAA Tech. Memo., NWS HYDRO-12, 1971.

(6) Kuester, J.L. and Mize, J.H., Optimization Techniques with FORTRAN, McGraw-Hill, 1973, pp. 309-319.

(7) EPA Regulations on Control of Air Pollution from Aircraft and Aircraft Engines, 38FR19088, 38FR34734, 38FR35000, 1973.

APPLICATIONS OF A NUMERICAL MODEL TO WECS SITING RELATIVE TO TWO-DIMENSIONAL TERRAIN FEATURES

C. F. SHIEH and W. FROST

University of Tennessee Space Institute, Tullahoma, Tennessee, USA

SUMMARY

A numerical model of wind characteristics about bluff-shaped surface geometries is developed in (1,2,3). The model solves the two-dimensional Navier-Stokes equations of fluid motion coupled with two transport equations, for turbulence kinetic energy and for turbulence length scales, respectively (two-equation model). Verification of the model as demonstrated by comparison with experimental data from both wind tunnel and full-scale field measurements is presented in this paper. The model is shown to give realistic predictions of velocity distributions, turbulence properties and separation flow zones associated with flow fields over bluff objects. Finally, the influence of bluff terrain features on the siting of wind energy conversion systems, WECS, is simulated with the model, and parametric studies are reported.

Nomenclature

h	Height of bluff body
K	Turbulence kinetic energy
ℓ	Turbulence length scale
p	Pressure
P	Potentially available wind power
R	Radius of wind generator rotor
U	Velocity component in X-direction
u_*	Friction velocity
U_h	Undisturbed mean velocity at bluff body height
X	Horizontal coordinate
Z	Vertical coordinate
Z_H	Hub height
z_o	Surface roughness
μ	Laminar viscosity
μ_t	Turbulent viscosity
ρ	Fluid density

Subscript

i, j	Subscripts for Cartesian tensors (i-1,2,3; j=1,2,3)
o	Pertains to the undisturbed flow properties
∞	Notes the undisturbed property at upper boundary (Fig. 1)

INTRODUCTION

Wind fields in the vicinity of surface obstructions have attained major importance in the study of WECS siting techniques. Evidence indicates that the wind over effectively two-dimensional hills or man-made obstacles is generally stronger than that over the surrounding low-lying flat ground (4,5,6). The accelerated wind speed is caused by the flow displacement due to the presence of the obstructions. This accelerated mean wind speed creates a region of potentially high wind energy. However, high turbulence also occurs in these regions of high wind energy and must be carefully considered in the structural design of the wind power generator. The effect of the shear in approaching flow and the strong pressure gradients near sharp corners of the body create zones of recirculation in both the upstream and downstream regions near the obstacles. High turbulence is generated in the shear layers bordering the recirculation regions due to the interaction between turbulent stresses and the deformation of the mean flow field. This generated turbulence diffuses vertically as it is convected downstream with the wind.

Because of the complex nature of these flow fields, determination of the preferred WECS site relative to a bluff body requires either experimental verification or numerical modeling. Full-scale experimental studies of flow phenomena about bluff objects as well as exact simulation in wind tunnels are costly, and therefore analytical methods are needed to predict local atmospheric motion influenced by surface obstructions. Various approaches have been used to theoretically describe the flow properties behind surface obstructions (7-11). However, these theories cannot be used to investigate the characteristics of the wind in the separated flow regions. For these regions it is necessary to solve the complete equations of motion to realistically predict the flow field. Practical methods for solving such equations require the use of numerical modeling techniques. The purpose of this paper is to present the results of a meaningful numerical model for incompressible flow over two-dimensional obstacles and to illustrate the usefulness of these results for the study of WECS siting techniques. The specific flow field considered is neutrally stable atmospheric flow over a bluff rectangular geometry (Figure 1). The upstream, undisturbed atmospheric boundary layer is treated as having a logarithmic velocity profile.

DESCRIPTION OF THE NUMERICAL MODEL

The mathematical model utilized in this paper is described in detail in (1,2,3). The model solves the incompressible two-dimensional Navier-Stokes equations of motion given by:

$$\rho U_j \frac{\partial U_i}{\partial X_j} = - \frac{\partial p}{\partial X_i} + \frac{\partial}{\partial X_j} \mu \frac{\partial U_i}{\partial X_j} - \frac{\partial}{\partial X_j} \overline{\rho u_i u_j}$$

where μ is molecular viscosity and $-\overline{\rho u_i u_j}$ is Reynolds shear stress. The above equations can be integrated if certain assumptions about the Reynolds stresses are made to form a closed set of equations. In the model, the Reynolds shear

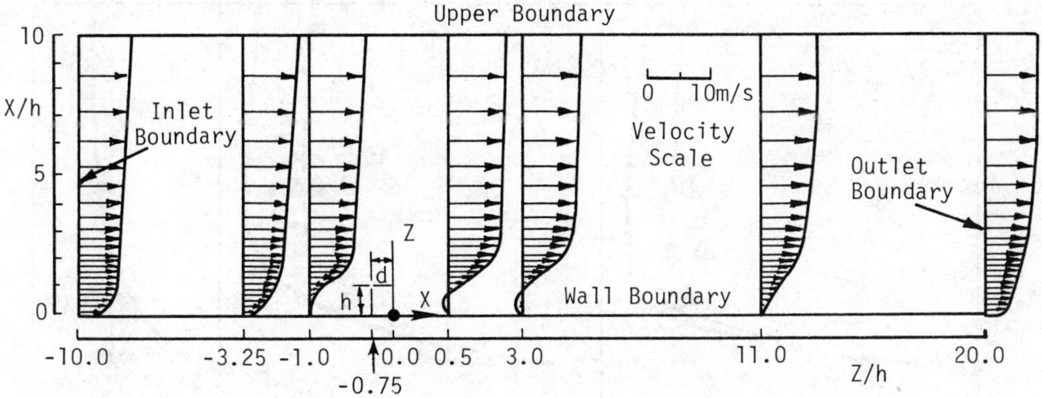

Fig. 1. The flow field considered and typical computed velocity distributions.

stresses are expressed by the product of an eddy viscosity and a velocity gradient,

$$\overline{-\rho u_i u_j} = \mu_t \left(\frac{\partial U_i}{\partial X_j} + \frac{\partial U_j}{\partial X_i}\right)$$

The turbulence viscosity, μ_t, is modeled by the Prandtl-Kolmogorov hypothesis,

$$\mu_t \propto \rho K^{1/2} \ell$$

which calls for simultaneous solution of two additional transport equations (12), one equation for the turbulence kinetic energy, K, and one equation for the turbulence length scale, ℓ.

The solution algorithm is that suggested by Gosman et al (13). This algorithm utilizes stream function and vorticity variables and therefore the solutions are obtained without requiring information about the pressure gradients. The boundary conditions specified in the computer code from which the solutions reported in this paper have been computed are rectangular and therefore the results apply only to building-like blocks, shelter belts, step-like geometry, etc.

VERIFICATION OF RESULTS

A partial comparison of the numerical results with the experimental field program data (14) and with the atmospheric boundary layer wind tunnel data (15) is given in this section. More complete comparison of the numerical results with both field and wind tunnel data plus comparison with other numerical models is given in (3).

The data chosen for comparison are those generated both from a field study and from a wind tunnel study where the exact block geometry and measuring stations as utilized in the field were modeled in the tunnel. The building-like block has the geometric ratio, height/width/depth = h/w/d = 1/8/0.75. Since the analytical results are strictly two-dimensional solutions, only data measured along the center line of the model block structure are used for comparison. Figure 2 compares the computed results with the data measured with wind towers positioned

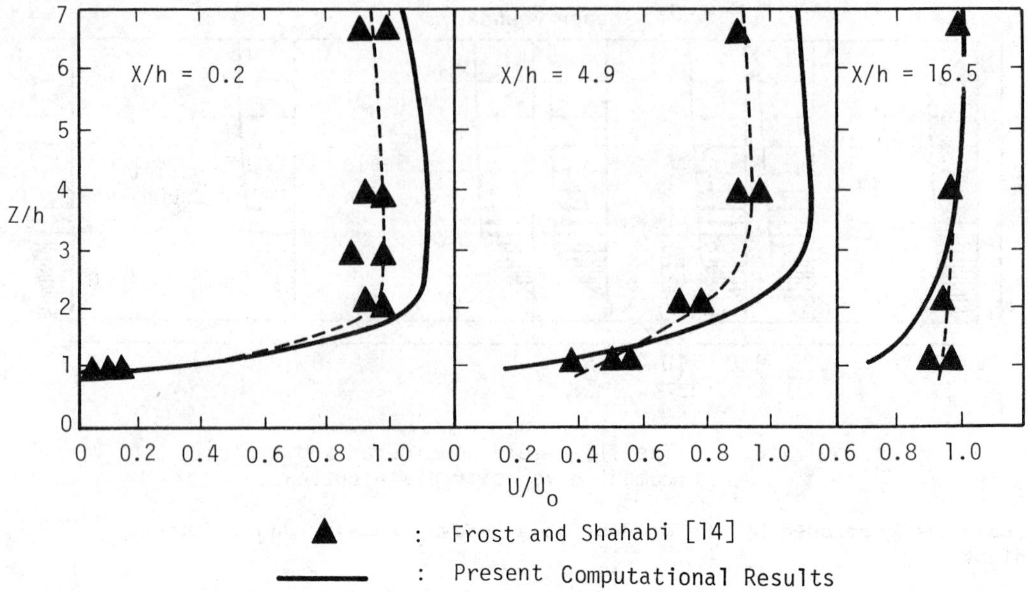

Fig. 2. Nondimensional wind speed.

immediately behind the block-shaped geometry and at the positions X/h = 4.9 and 16.5 downstream. The data are non-dimensionalized with the velocity U_o which is the velocity measured approximately 40 building heights downstream. The agreement is felt to be good considering that the field data were taken when the wind direction was not directly down the array. In over 50 sets of data, no data are available for the wind directly down the tower array. The data shown here are approximately six degrees off center.

In Figure 3 the defect $(U_o-U)/U_\infty$, where U_∞ is the incident mean velocity at the upper boundary (Fig. 1), characterizes the wake generated by the obstacle. Figure 3 shows the agreement between the predicted results and the wind tunnel measurements reported in (15). The agreement between theory and experiment again seems to be very good.

Also of interest to flow over bluff geometries is the extent of the wake. Figure 4 shows a comparison of the flow reattachment distance, X_r, behind a simulated building as a function of upstream flow conditions, where these conditions are expressed as the ratio of u_*/U_h. The parameter u_* designates friction velocity and U_h is the undisturbed mean velocity at the block height.

APPLICATIONS TO WIND TURBINE GENERATOR SITING STUDIES

The wind fields computed from the present model are used to investigate the influence on the potentially available wind power of bluff natural terrain or man-made objects. The barrier geometries defined in the present computational program are highly idealized but can be used to understand the wind phenomenon and to estimate the magnitude of the effect.

In this study, the optimum site for a wind turbine generator is defined as the location where the maximum wind power ratio, P/P_o, is obtained (P_o is the power in the undisturbed wind). The wind power is calculated by the following formula (16):

$$P = 2 \int_{Z_H - R}^{Z_H + R} U^3 \sqrt{R^2 - (Z - Z_H)^2}\, dZ$$

where Z_H is the hub height and R is the radius of the wind generator rotor. (NOTE: Regions of high wind speed may have high turbulence associated with them which is not considered in this study.)

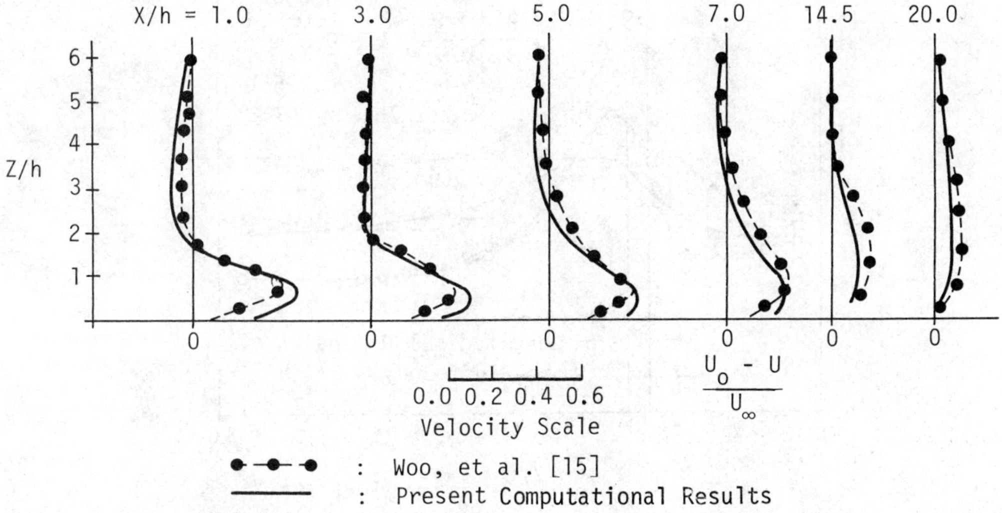

Fig. 3. Decay of mean velocity defect behind the block.

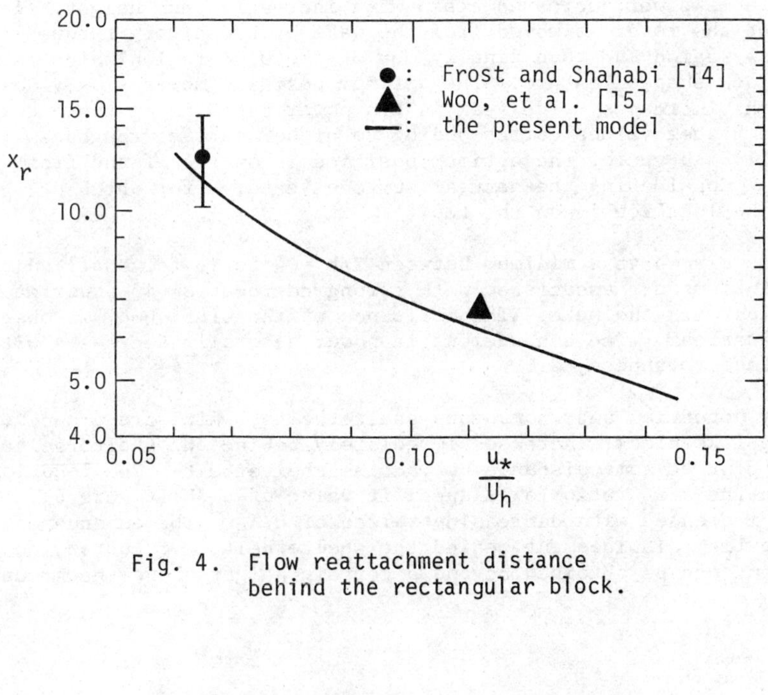

Fig. 4. Flow reattachment distance behind the rectangular block.

Figure 5 illustrates the typical influence of a bluff cliff or building-like block on the power ratio P/P_o. It is observed that a distinct optimum position for maximum wind energy potential exists close to the rear of the bluff body. For $Z_H/h = 3$ this optimum location is approximately h from the rear of the body whereas for $Z_H/h = 2$ it occurs immediately behind the building. The optimum location tends to move forward with increasing surface roughness. The available power increases with increasing hub height to approximately $Z_H/h = 3.5$, beyond

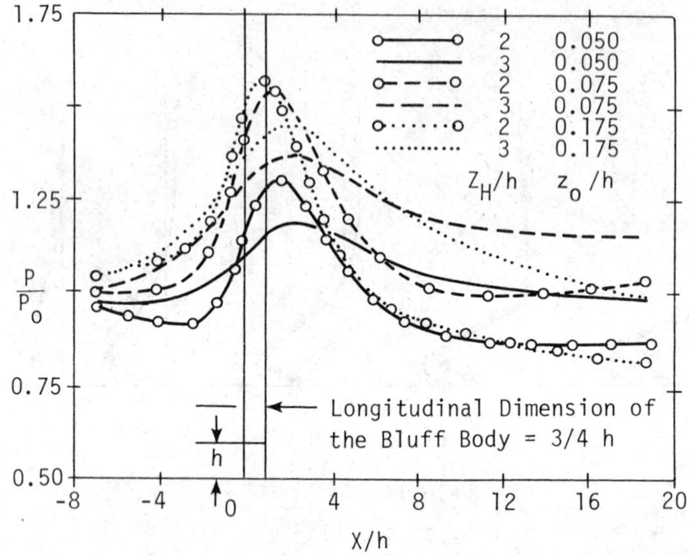

Fig. 5. Variation of wind-power ratio over a bluff, cliff or building-like geometry, R/h = 0.5.

which the available power decreases again with increasing hub height. The reason for this is that the rotor is moved from the wake progressively through the accelerated flow region and then finally out of the wind region which is influenced by the block-shaped feature. The optimum position moves slowly downstream as the hub height increases. The peak in the power ratio is on the order of 1.6 to 1.3 with the higher values corresponding to higher surface roughness values. For a very small Z_H/h ratio, the optimum position is on top of the front edge of the bluff body. Considering the natural surface features for which $h \gg Z_H$, the best wind turbine location is on the top.

The power ratios also have a minimum between X/h = 4 to 16 for small values of Z_H/h. These minimums are associated with strong degradation in power caused by rotor interaction with the wake. The influence of the bluff body is observed to extend well downstream. The enhancement in power generally decreases with decreasing surface roughness, z_o.

The wind energy potential near nonporous shelterbelts, which are a special case of the building-like block, increases immediately behind the shelterbelts, reaching a distinct peak at some distance behind the shelterbelt. The location of the peak depends on the h/Z_H ratio for a specific value of z_o/h (Figure 6). The wind power ratio is increased with decreasing values of h/Z_H. The enhancement of wind power is due to flow displacement behind the shelterbelt. Of course, no wind power enhancement can be obtained beyond a certain height above the ground since the wind field is undisturbed by the shelterbelt above that height. The optimum location is not strongly sensitive to the R/Z_H ratio.

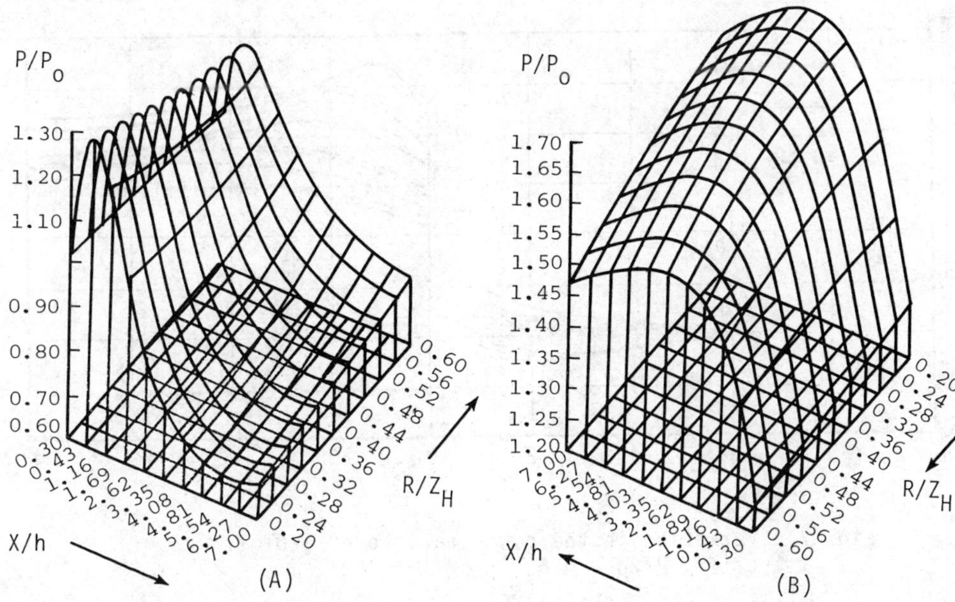

Fig. 6. Variation of wind-power ratio over a solid shelterbelt, $z_o/h = 0.005$. (A) $h/Z_H = 0.5$, (B) $h/Z_H = 0.25$.

Figure 7 illustrates the influence of a plateau or sea coast-type cliff on potential wind energy. Overspeeds in the wind occur above the separated zone at the top of the cliff and provide the optimum wind turbine site, provided the rotor does not interact with the wake. Siting the wind turbine on top of the cliff provides an approximate increase in power of two times that available in the undisturbed flow upstream, and appears to approach a constant value of approximately 1.6 downstream for all cases considered. A factor of 5 increase in the value of z_o causes an approximate 8% difference in the estimated power ratio and an approximate 50% difference in site location. The peaks in power, however, are in general quite broad, thus allowing for considerable tolerance in the site location.

Inspection of Figure 7 illustrates that the optimum site moves closer to the face of the cliff with increasing surface roughness for those conditions where the height of the cliff is greater than approximately one-third of the hub height of the wind power generator. For the case where the hill height is 0.25 Z_H or less, there is no peak in the power profile, but there is an asymptotic approach to higher power, which simply results from the rotor being elevated into the higher wind speed.

It should be noted that the wind turbine generator should not be placed immediately next to the edge of the cliff. Lower power is available there because a large component of the wind is directed upward rather than perpendicular to the plane of rotation of the wind turbine blade. If a wind power generator were designed such that the axis could be tilted out of the horizontal plane, one could possibly achieve equivalent or higher power immediately at the edge of the cliff than farther back. This possibility, however, was not investigated in the current study.

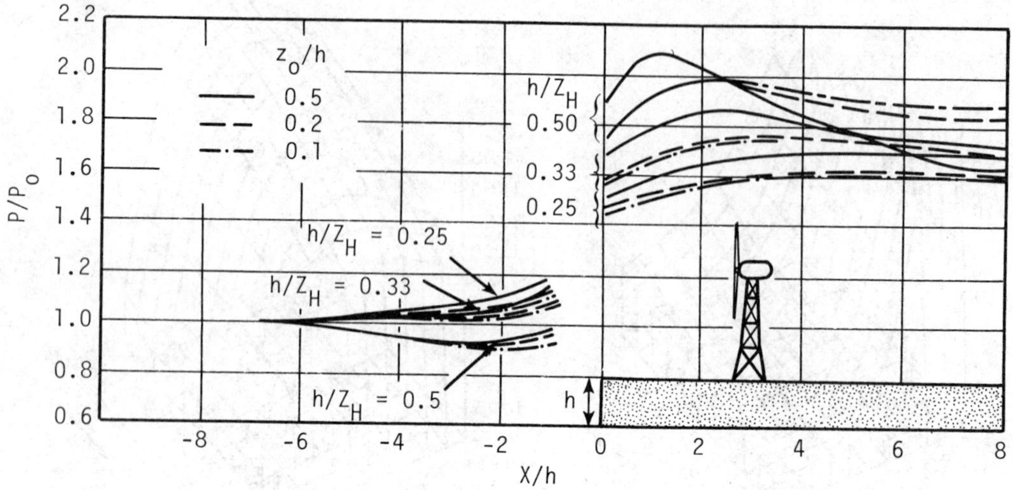

Fig. 7. Variation in the power ratio over a bluff cliff, $R/z_H = 0.6$.

The enhancement of power increases with the increasing value of h/Z_H, the ratio of hill height to hub height. This trend, of course, would be reversed when the value of h/Z_H becomes such that the rotor interacts with the wake, although values of h/Z_H for which this would occur have not been studied. Decreasing the hill height tends to move the optimum site back from the edge of the cliff as is shown in Figure 7. Beyond a certain value of h/Z_H there is only an increase in P/P_o due to tower extension and to contracted flow over the plateau.

Comparing Figures 5 through 7, one observes that the wind potential on a long bluff cliff, e.g., a forward-facing, step-like feature, is larger than on (or near) a rectangular block-shaped terrain feature. Siting a wind turbine near a solid shelterbelt is the worst of the cases investigated in this study.

CONCLUSIONS

The excellent agreement of the computer code with experimental data inspires considerable confidence in the application of the computer code for predicting optimum wind turbine generator site locations and trends in power due to variation of the wind turbine generator location from the optimum site.

Based on the computational results and cautioning the reader to the idealized nature of the terrain features considered, the following rules are available for evaluating the optimum wind turbine generator location relative to bluff terrain features:

1. The highest wind energy input for a wind turbine of fixed tower height is on the top of the bluff features.

2. The potential increase in wind energy induced by the step-like terrain is larger than that induced by the rectangular block-like features.

3. The optimum location off the top of the feature is near the highest point of the separation region originating at the bluff corner and extending downstream several obstacle heights depending on upstream surface conditions.

4. The power ratio P/P_o for a wind turbine on the bluff increases with increasing surface roughness of the terrain. However, on the uniform upstream terrain the wind energy potential decreases with increasing surface roughness due to the wind speed being reduced by the increased friction drag at the surfaces.

5. The wind energy potential increases with increasing hub height Z_H and with rotor diameter, provided the rotor does not enter the separation region on top of the bluff.

* * * * * *

ACKNOWLEDGEMENTS

This research has been supported by three agencies: the National Science Foundation, under Contract CK-42992; the National Aeronautics and Space Administration, NASA/Marshall Space Flight Center, under Contract No. NAS8-29584; and the U.S. Department of Energy, Division of Solar Energy, Wind Energy Branch (prime contractor Battelle Pacific Northwest Laboratories), under Contract No. E(45-1)-2443. The authors are grateful for the support of George Lea, NSF, Dennis Camp, NASA, and Bill Pennel, Battelle DOE.

REFERENCES

(1) J. Bitte and W. Frost, "Atmospheric Flow over Two-Dimensional Bluff Surface Obstructions," NASA Contract Report, NASA CR-2750, October 1976.

(2) U.K. Kaul and W. Frost, "Turbulent Atmospheric Flow over a Backward-Facing Step," NASA Contract Report, NASA CR-2749, October 1976.

(3) C.F. Shieh, W. Frost and J. Bitte, "Neutrally Stable Atmospheric Flow over a Two-Dimensional Rectangular Block," NASA Contract Report, NASA CR-2926, November 1977.

(4) R.N. Meroney, V.A. Sandborn, R.J.B. Bouwmeester, H.C. Chien and M. Rider, "Sites for Wind Power Installations: Physical Modeling of the Influence of Hills, Ridges and Complex Terrain on Wind Speed and Turbulence," Colorado State University, CER 77-78RNM, VAS50a, June 1978.

(5) Golding, E.W., "Wind Flow over Hills," <u>The Generation of Electricity by Wind Power</u>, 2nd ed., John Wiley & Sons, New York, 1976.

(6) C.E. Duchon, "Measurement and Analysis of Wind Flow in Complex Terrain with Application to Siting of WECS," U.S. Dept. of Energy Report, DOE/NSF-00619/75/1, May 1977.

(7) J. Counihan, J.C.R. Hunt and P.S. Jackson, "Wake Behind Two-Dimensional Surface Obstacles in Turbulent Boundary Layers," <u>J. Fluid Mech.</u>, 64, 529-63 (1974).

(8) W. Frost, W.L. Harper and G.H. Fichtl, "Analysis of Atmospheric Flow over a Surface Protrusion Using the Turbulence Kinetic Energy Equation," <u>Boundary Layer Meteor.</u>, 8, 401-17 (1975).

(9) E.J. Plate, "Aerodynamics of Shelterbelts," Agric. Meteor., 8, 203-22 (1971).

(10) N. Tani, "On the Wind Tunnel Test of the Model Shelter Hedge," Nat. Inst. of Agric. Sci., Tokyo Bulletin, Ser. A, 6, 1-80 (1958).

(11) A.A. Towsend, "Self-Preserving Flow Inside a Turbulent Boundary Layer," J. Fluid Mech., 22, 773-97 (1965).

(12) Launder, B.E. and Spalding, D.B., "Two-Equation Models of Turbulence," Mathematical Models of Turbulence, Academic Press, London & New York, 1972.

(13) Gosman, A.D., Pun, W.M., Runchal, A.K., Spalding, D.B. and Wolfshtein, M., Heat and Mass Transfer in Recirculating Flows, Academic Press, London and New York, 1969.

(14) W. Frost and A.M. Shahabi, "A Field Study of Wind over a Simulated Block Building," NASA CR-2804, March 1977.

(15) H.G.C. Woo, J.A. Peterka and J.E. Cermak, "Wind Tunnel Measurements in the Wakes of Structures," CER75-76HGCW-JAP-JEC40, Colorado State University, June 1976.

(16) W. Frost and D.K. Nowak, "Handbook of Wind Turbine Generator Siting Techniques Relative to Two-Dimensional Terrain Features," FWG Associates, Inc., Tullahoma, TN, No. E(45-1)-2443 contract report in preparation, April 1978.

THE AERODYNAMICS OF HELIOSTATS FOR SOLAR POWER PLANT APPLICATIONS

J. XERIKOS*, H. H. TANG*, J. E. CERMAK** and J. A. PETERKA**

*McDonnell Douglas Corporation, Huntington Beach, California, USA
**Colorado State University, Fort Collins, Colorado, USA

ABSTRACT

The heliostat field-array test, conducted under the sponsorship of DOE in the meteorological wind tunnel of Colorado State University, investigated the characteristics of the steady-state wind loads produced by airflow patterns in and around several configurations of heliostat clusters with and without a wind fence. The characteristics of natural boundary-layer winds at the actual site were simulated in the tunnel. A key conclusion of the study was that wind loads can be reduced to a minimum level through the use of properly designed and located fences used in combination with an efficiently distributed field array.

INTRODUCTION

Under research and development contracts supported by the United States Department of Energy, McDonnell Douglas Corporation is currently leading an industrial team in developing an experimental 10-megawatt solar power plant to be built near Barstow, California, that is to be operational in the 1980-1981 period. The power plant consists of a central receiver system for harnessing solar energy. This system consists of a field of two-axis tracking heliostats (approximately 1800), with reflective area utilizing second-surface glass mirror rectangular panels designed for face-down protective storage. The heliostats, mounted on pedestals and computer controlled to follow the sun continuously, reflect sunlight to the central receiver located on top of a high tower. The resulting heat is applied to water pumped up to the receiver. This produces superheated steam which is sent through a turbine and generator on the ground. This concept and design can lead to future commercial systems with 10- to 300-megawatt capacity.

The purpose of the study described in this paper was to investigate characteristics of mean wind loads produced by airflow in and around several configurations of heliostat clusters with or without a wind fence. Twelve model heliostats were constructed and utilized to formulate several combinations of cluster configurations. The configurations represent typical outer field circular and rectilinear heliostat clusters. The 1:22 scale models were made of Plexiglass. One model heliostat was specially constructed for mounting on a force balance to measure two forces and three moments. Another was instrumented with 42 pressure taps for the measurement of wind pressures on the upper and lower mirror-panel surfaces as well as on the support post. The effects of several dominant variables were investigated in this study, namely, wind-azimuth, mirror tilt angle and slot geometry, cluster configuration, and a protective wind fence. All measurements were made in a boundary-layer flow developed in the meteorological wind tunnel of the Fluid Dynamics and Diffusion Laboratory at Colorado State University.

I. SIMULATION OF NATURAL ENVIRONMENT

The primary consideration in modeling wind forces on structures in a wind tunnel is that the wind characteristics in the tunnel simulate natural boundary-layer winds at the actual site. In general, this requires that the vertical distribution of mean velocity and turbulence in the wind-tunnel boundary layer match those at the site and that the Reynolds numbers of the model and the prototype be equal. In addition, the small-scale structure must be geometrically similar to its prototype. A detailed discussion of these requirements and their implementation in the wind-tunnel environment can be found in References 1, 2 and 3.

The construction of a 1:22 scale model of the prototype structure and its immediate surroundings (in this case, a flat, open area), submerged in a turbulent boundary layer of the meteorological wind tunnel shown in Figure 1, satisfies all the above criteria except those of equal Reynolds numbers and similarity of turbulence intensity and scale. In the Reynolds number $(UD)/\nu$, ν is the same for both the tunnel and the full-scale structure. Because of this, the wind tunnel air speed, U, would have to be 22 times the full-scale value if the model and prototype Reynolds numbers are to be equal. Testing at such high wind speeds is not feasible. However, for Reynolds numbers larger than 2×10^4, there is no significant change in the values of aerodynamic coefficients as the Reynolds number increases. Since typical Reynolds number values are 10^6 to 10^7 for full-scale flow and 10^5 to 10^6 for wind tunnel flows, acceptable flow similarity is achieved without equality of the Reynolds numbers.

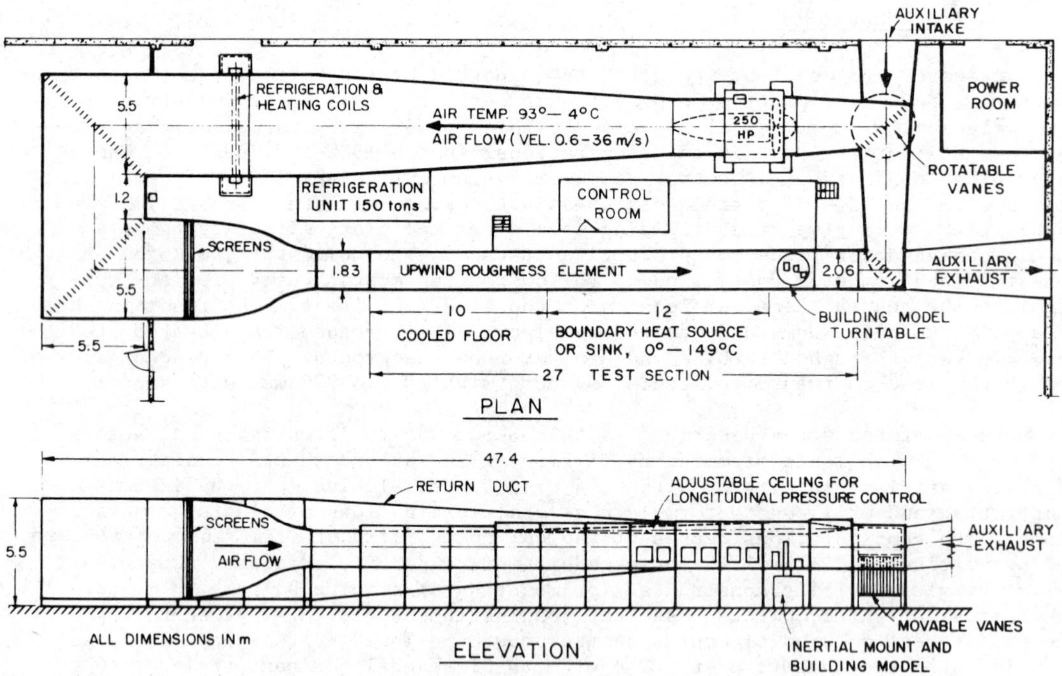

Fig. 1. Meteorological wind tunnel

At a model scale of 1:22, the larger scales of turbulence in the atmospheric boundary layer are not simulated in the wind tunnel flow. However, because the heliostat geometry is that of a flat plate and because the integral scale of the turbulence in the wind tunnel was 2 to 3 times the largest dimension of the heliostat, the influence of the scale of turbulence was not expected to be significant (4). Evidence exists which demonstrates some influence of turbulence intensity on drag of flat plates (4,5,6). Because the turbulence intensity difference between the current simulation and a simulation with complete similarity of turbulent structure is not large, the effects due to turbulence intensity should be small. For cases where an upstream heliostat disturbed the approach flow, turbulence characteristics are dominated by the wake characteristics of the upstream object and differences due to turbulence intensity should decrease.

II. TEST MODELS AND INSTRUMENTATION

The heliostat test model dimensions are shown in Figure 2. The idealized cluster models as shown in Figure 3 include a set of from 7 to 12 heliostats depending upon their position inthe field array. The central heliostat in each cluster was mounted on a six-component strain gage balance.

The study also required that a heliostat be instrumented for measurements of mean pressure on the heliostat surfaces. Physical limitations associated with the installation of pressure tubing limited the number of pressure taps incorporated into the model to 42. These were selected to permit definition of pressure distributions on the surface.

For several heliostat clusters the forces and moments were measured with a porous wind fence located upstream of the cluster. The fence had a porosity of about 33 percent. The fence height selected was 20 feet (full-scale) which approximates the height of a mirror tilted up 30 degrees from the horizontal. The fence extended across the entire width of the tunnel in all cases.

III. TEST PROCEDURES

The study was performed in the meteorological wind tunnel shown in Figure 1. The heliostat field-array test was divided into six phases. In the first phase, forces and moments were measured on a single heliostat (with an angle of tilt of α = 60 degrees) for five wind directions. The second phase was the tilt-angle study. Forces and moments were measured on a single heliostat model at a range of wind directions and tilt angles. The third phase of the test measured the effect of closing the slot between the heliostat reflector panels. An insert was placed between the panels, first to close the top half of the slot and then to close both the top and bottom halves. Forces and moments were measured for the single heliostat model at five wind directions for each of two tilt angles.

The next phase of the test was the cluster study. With each heliostat model in the cluster set to the same tilt angle, configurations 1, 2 and Linear A were tested for four values of α and four wind directions per tilt angle. The instrumented model in each configuration was then used to obtain forces and moments. The fifth phase measured the distribution of mean pressures on the instrumented model in each cluster. Six values of β were tested for a single heliostat model and for configurations 1 and 2, while three wind directions were tested for configuration Linear A.

The sixth and final phase of the field-array test measured the effects of an upwind fence on the heliostat models. Configurations 1, Linear B', and Linear C were tested for two wind directions with the fence at four distances from the cluster for each direction. Configuration Linear B' was also run at a second fence orientation. The instrumented model in each configuration was used to obtain forces and moments on the model.

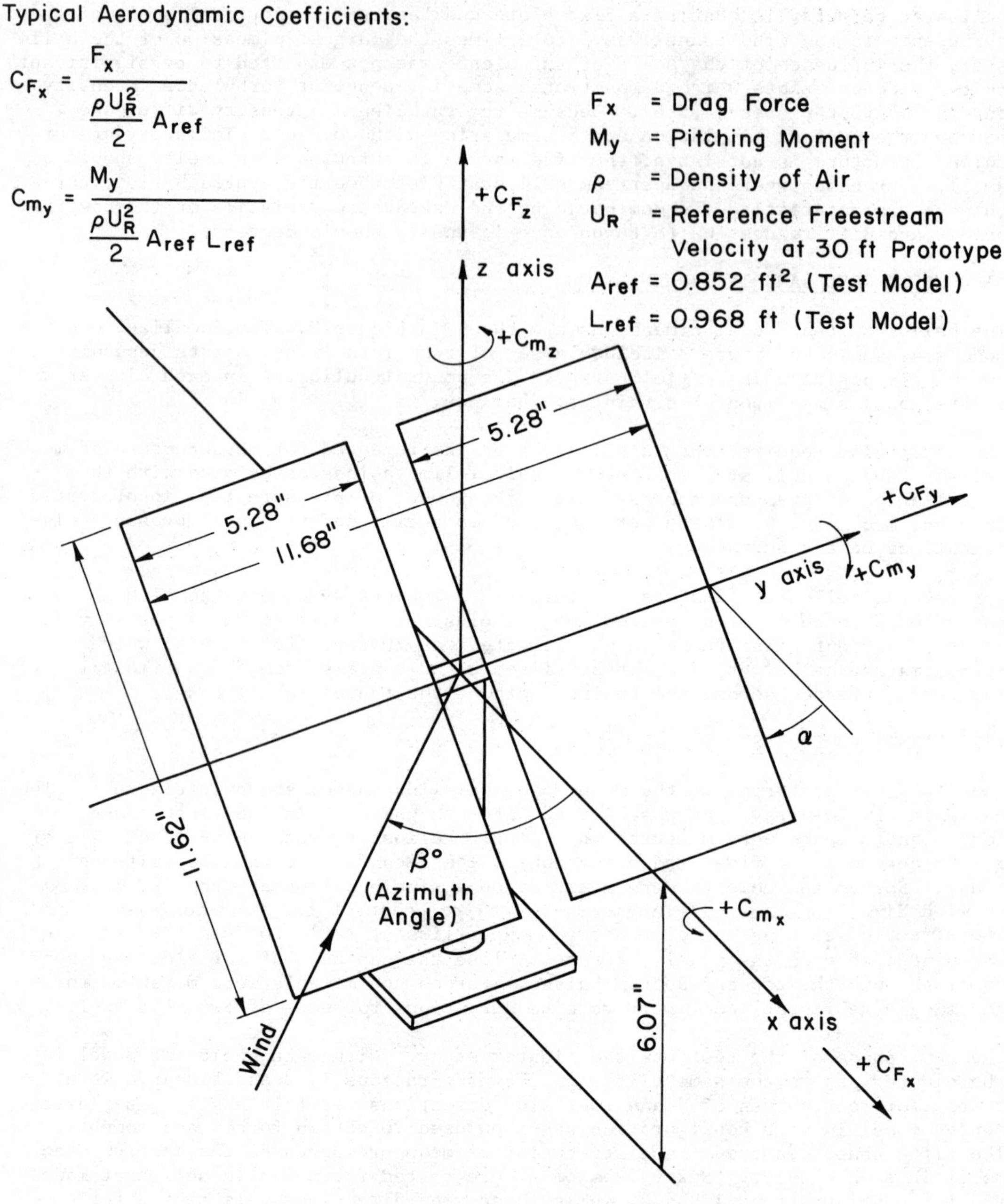

Fig. 2. Test model dimension and definition of angle β and coordinate system for force and moment coefficients

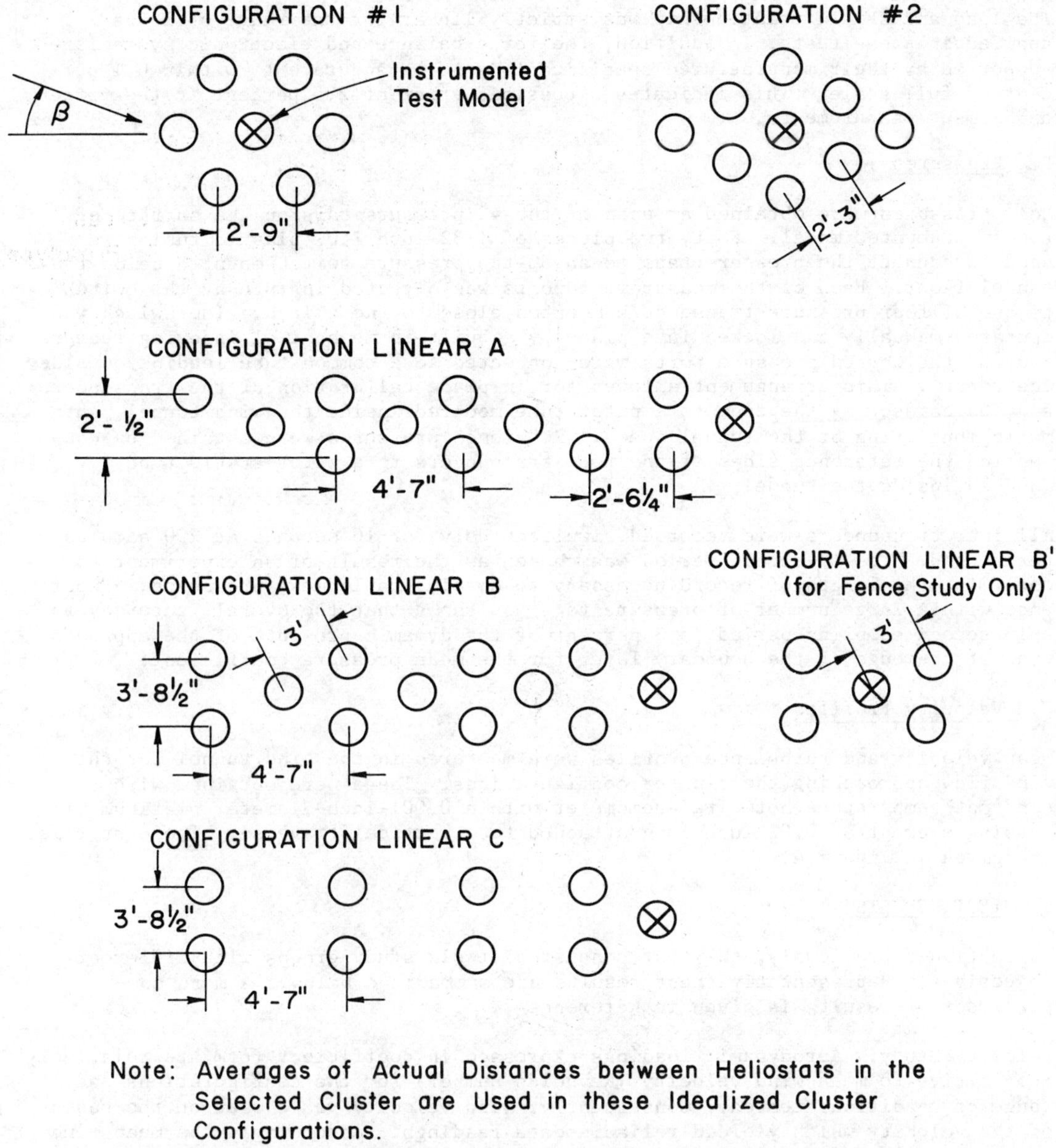

Fig. 3. Idealized cluster configurations

IV. INSTRUMENTATION AND DATA ACQUISITION

A. Forces and Moments

Forces and moments on the instrumented model were measured using an Inca six-component strain-gage balance. Calibration of the balance was performed in a test rig in which known forces and moments could be applied to the balance. A calibration matrix was then developed for reducing the mean output of the strain gages.

The load and strain relationship was strictly linear for the range of loads applied in these tests. In addition, the force balance and electronic system are supported by their manufactured specifications to be accurate to within 0.1 percent of full scale. This indicates a possible error of 2.5 percent in the force and moment measurements.

B. Pressures

Mean pressures were obtained at each of the 42 pressure taps on the surface of the instrumented model. Forty-two pieces of 1/32-inch I.D. plastic tubing were used to connect the pressure taps to an 80-tap pressure switch mounted below the tunnel floor. Each of the measurement ports was directed in turn by the switch to one of four pressure transducers mounted close to the switch. The switch was operated manually and locked into place by a built-in mechanical indexing feature. Four of the unused pressure ports were connected to a common tube leading outside the tunnel. This arrangement allowed for in-place calibration of the transducers and, by connecting the tube to a pitot tube mounted inside the wind tunnel, automatic monitoring of the tunnel speed. Reference pressures were obtained by connecting the reference sides of the four transducers to a pitot-static tube mounted inside the tunnel.

All four transducers were recorded simultaneously for 16 seconds at 250 samples per second. The 16-second period was chosen as the result of an experiment to determine the length of record necessary to obtain stable mean pressures. Experience with a large number of pressure taps has shown that the overall accuracy for a 16 second sampling period is 3 percent of the dynamic pressure of the approach wind at the edge of the boundary layer for the mean pressure coefficient.

C. Velocity Profiles

Mean velocity and turbulence profiles were measured in the wind tunnel for the wind flow approaching the cluster configurations. These were obtained with a constant temperature hot-wire anemometer with a 0.001-inch-diameter platinum film sensing element of 0.02 inch long attached to a vertical traverse. These profiles are given in Figure 4.

V. TEST RESULTS

As mentioned previously, the test consisted of six study groups with different objectives. Representative test results are summarized below. A more complete set of results is given in Reference 7.

Velocity Study: Aerodynamic loadings expressed in coefficient form are relatively insensitive to mean wind velocity (Reynolds number) for the configurations and range of conditions tested. Consequently, wind velocity was chosen on the basis of the velocity which yielded reliable data readings. Accordingly the mean wind velocity referenced to 10 meters (full scale) giving the best instrumentation signal-to-noise ratio was found to be approximately 55 mph and was used for the entire test.

Tilt Angle and Cluster Studies: For a single heliostat, pitching moments C_{m_y} were observed to peak at α = 30 degrees. Typical C_{m_y} variation curves are given in Figure 5. For clustered heliostats, values of C_{m_y} for the center heliostat are usually lower as compared with those for single heliostat and usually peak at α = 30 degrees. Typical curves are given in Figures 6 and 7. The pitching moment for cluster 1 at β = 180 degrees shows that the presence of a cluster does not always guarantee a reduction in the peak moment. Because the mean location of the center of pressure was near the y axis, it should be anticipated that

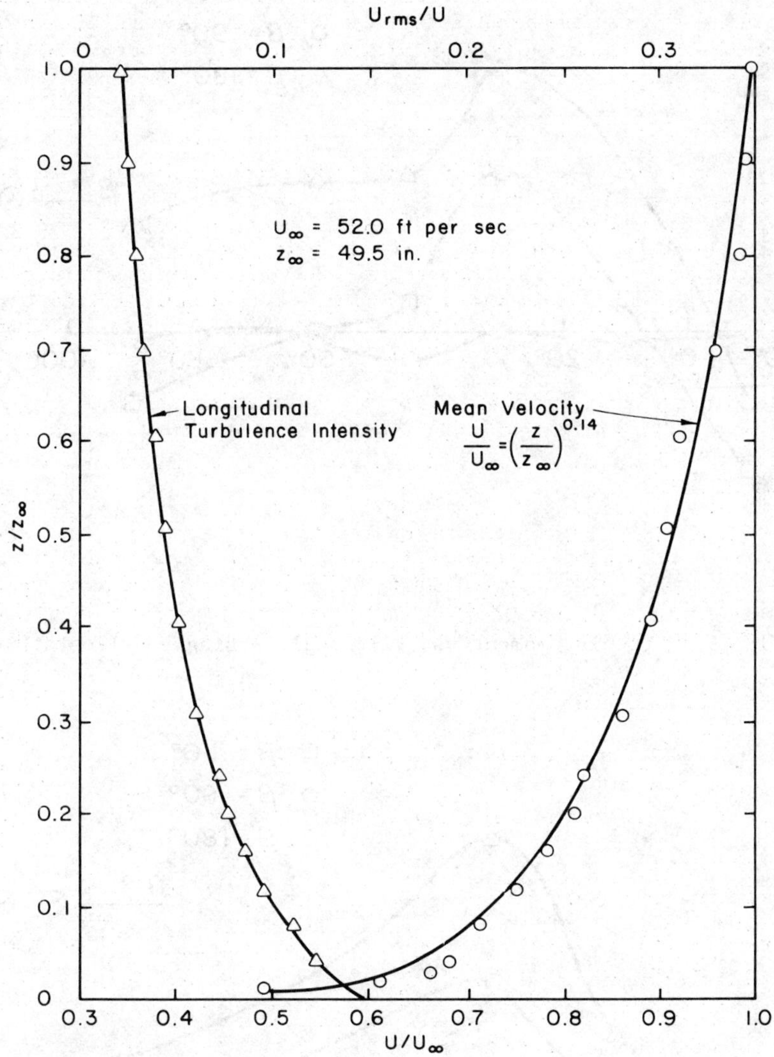

Fig. 4. Wind tunnel velocity and turbulence profile for wind flow approaching model configurations

fluctuating moments will be significantly larger than the mean moments reported here. Drag forces are shown in Figure 8 for a single heliostat and in Figures 9 and 10 for clustered heliostats. These forces usually peak at α = 90 degrees and have somewhat lower values for clustered heliostats. Twist moment, yaw moment and side force were relatively small as compared with pitch moment and drag force. A complete set of data are tabulated in Reference 7.

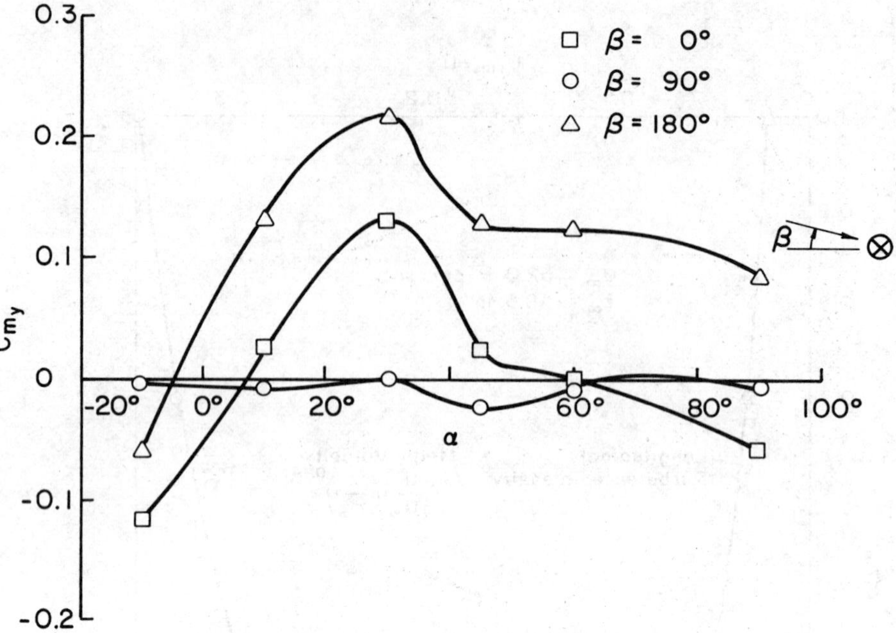

Fig. 5. Pitching moment vs. tilt angle - single heliostat

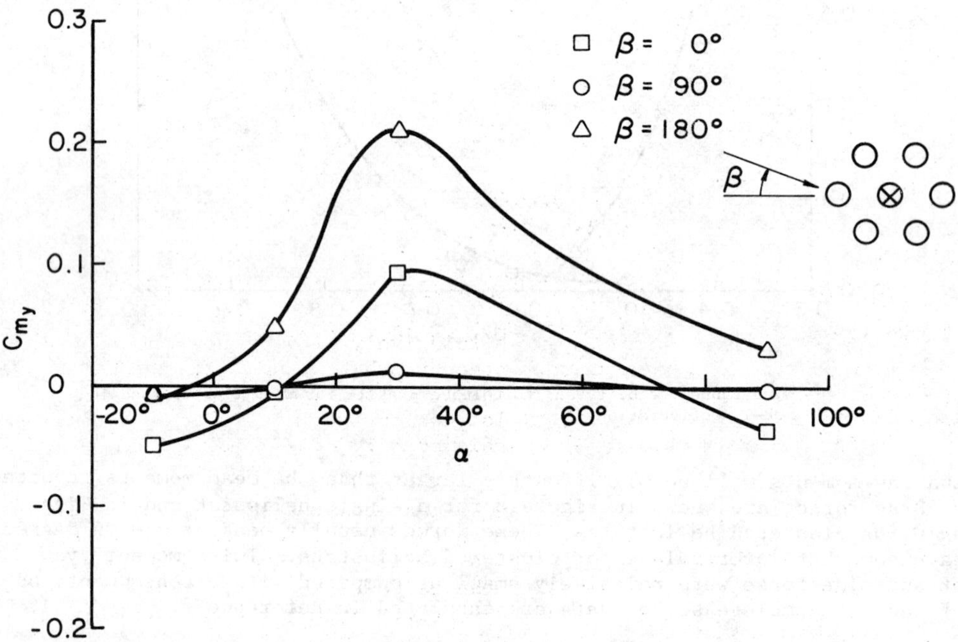

Fig. 6. Pitching moment vs. tilt angle - cluster configuration #1

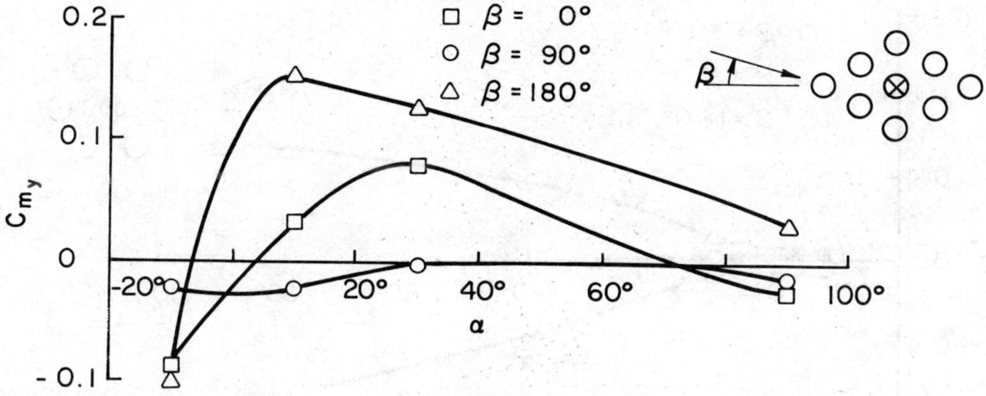

Fig. 7. Pitching moment vs. tilt angle - cluster configuration #2

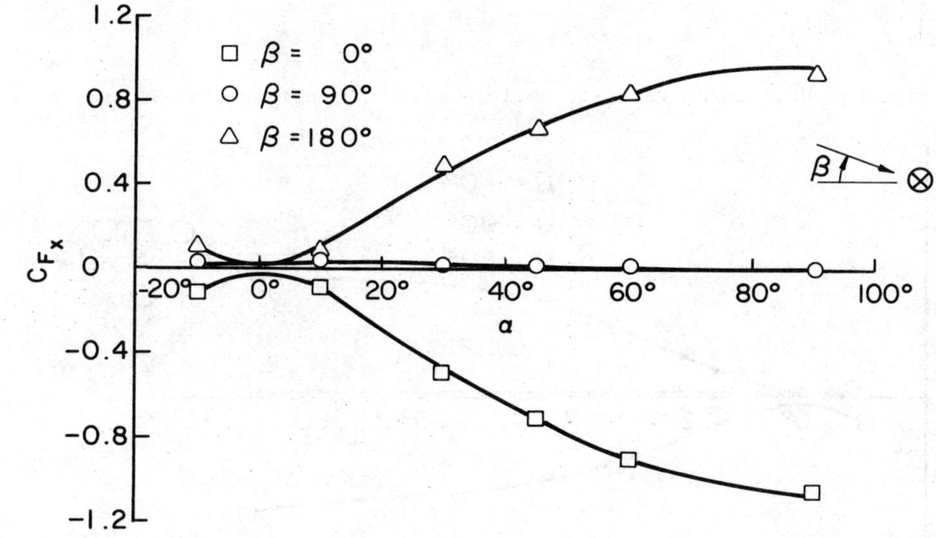

Fig. 8. Drag force vs. tilt angle - single heliostat

Slot Study: Slot effects were found to be relatively small. The selection of slot geometry would consequently be dominated by other design considerations. Owing to the non-monotonic variation of wind load as a function of tilt angle proceeding from an open to fully closed slot, an optimum design from the standpoint of minimizing wind loads could be achieved only over a limited range of α.

Pressure Study: Pressure distributions on heliostat surfaces (front and back) appeared to be relatively nonuniform, both on the planar surfaces and the support post. Nonuniformity of pressures was due to the usual pressure distribution expected on a flat plate, narrow slots in the heliostat surface at mirror intersections, and supporting struts on the back surface of the heliostat. The pressure data provided representative local (as opposed to integrated) load results owing to the limited number of taps available.

Fence Study: The introduction of a fence with 33 percent porosity (provided by 3/8-inch-diameter holes) significantly reduced the aerodynamic loads on heliostats

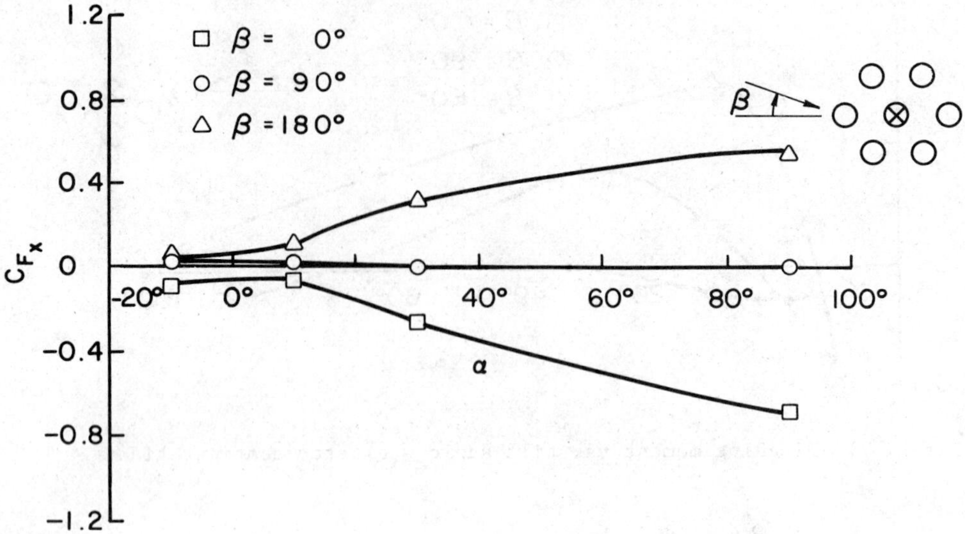

Fig. 9. Drag force vs. tilt angle - cluster configuration #1

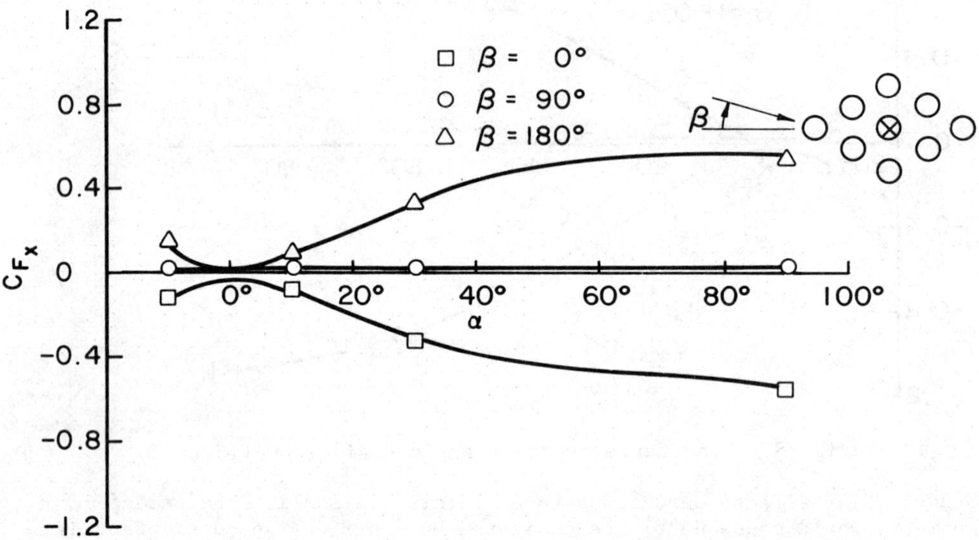

Fig. 10. Drag force vs. tilt angle - cluster configuration #2

as evidenced by the cases given in Figure 11. Pitching moments and drag forces have been reduced to 50 percent or even to one order of magnitude smaller than comparable no-fence cases. The fence height, porosity, and distance to the nearest heliostat were selected such that the free streamline grazing the top of the fence would not impinge directly on the instrumented heliostat.

An overview of the test results is given in Table 1. The table shows that the maximum value of C_{m_y} of 0.13 obtained for a single heliostat at $\beta = 0$ and $\alpha = 30^{\circ}$ is reduced to values of less than 0.1 when in some cluster configurations (configurations 1, 2 and Linear A), but is reduced only slightly for other configurations (Linear B, Linear C) where no adjacent heliostats were directly

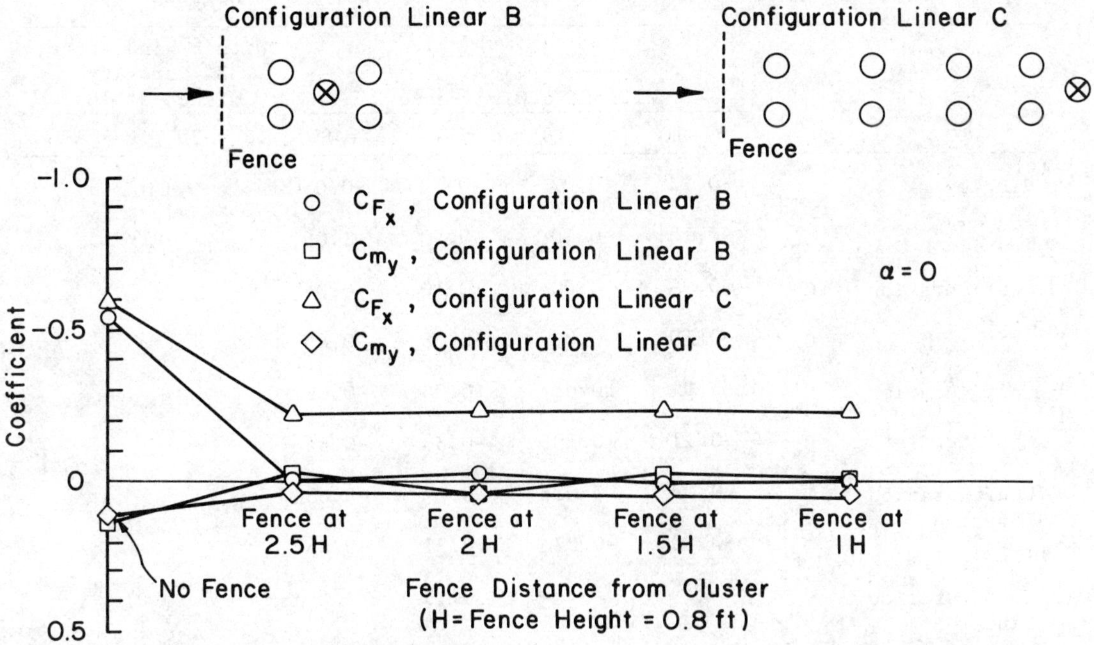

Fig. 11. Plot of force and moment coefficient reduction with varying fence distance

upstream. With a fence included, values of C_{my} were reduced to 0.01 except for configuration Linear C (C_{my} = 0.05) where the instrumented heliostat was at a large distance from the fence with an open channel between the heliostat and fence. Similar trends were observed for the force coefficient C_{FX}.

VI. CONCLUSIONS

The following general trends were observed:

A. Aerodynamic forces and moments on a heliostat embedded in a cluster are, in most cases, smaller than those experienced by an isolated heliostat.

B. Channel-flow effects can be significant.

C. A wind fence can reduce aerodynamic loads significantly.

D. Pitching moments usually peak at α = 30 degrees.

It is concluded that wind loads can be reduced to a minimum level through the use of properly designed and located fences used in combination with an efficiently distributed field array. The optimum design from an aerodynamic viewpoint would tailor the fence upwind of a heliostat array field to lower wind loads on exterior heliostats to the same level as experienced by those in the interior of the array field. Further, more comprehensive studies are needed to achieve this goal.

Additional study is needed to determine the influence of turbulence intensity and scale on aerodynamic coefficients of bluff bodies in a turbulent shear flow in order to define the implications of assessing wind loads in a boundary-layer wind tunnel at scales of 1:20 to 1:50.

Table 1. An overview on experimental results, β = 0 degrees

		Without a Wind Fence				With a Wind Fence, 33% Porosity Fence at 1H
		$\alpha = -10^o$	$\alpha = 10^o$	$\alpha = 30^o$	$\alpha = 90^o$	$\alpha = 30^o$, L = 1H
Single	C_{m_y}	-0.12	0.02	0.13	-0.05	0.01
	C_{F_x}	-0.12	-0.10	-0.50	-1.06	-0.03
Configuration 1	C_{m_y}	-0.05	-0.004	0.10	-0.04	
	C_{F_x}	-0.10	-0.06	-0.28	-0.65	
Configuration 2	C_{m_y}	-0.09	0.03	0.08	-0.02	
	C_{F_x}	-0.11	-0.10	-0.33	-0.54	
Configuration Linear A	C_{m_y}	-0.09	0.02	0.07	-0.03	
	C_{F_x}	-0.10	-0.07	-0.21	-0.55	
Configuration Linear B	C_{m_y}			0.13		0.01
	C_{F_x}			-0.53		-0.01
Configuration Linear C	C_{m_y}			0.12		0.05
	C_{F_x}			-0.58		-0.23

REFERENCES

1. Cermak, J.E., Aerodynamics of buildings, Annual Review of Fluid Mechanics, 8 (1976).

2. Cermak, J. E., Laboratory simulation of the atmospheric boundary layer, AIAA J., 9 (September 1971).

3. Cermak, J. E., Applications of fluid mechanics to wind engineering, A Freeman Scholar Lecture, ASME J. of Fluids Engineering, 97: No. 1 (March 1975).

4. Bearman, P. W., Turbulence effects on bluff body mean flow, Third U.S. National Conference on Wind Engineering Research, 265-272 (1978).

5. Bearman, P. W., An investigation of the forces on flat plates normal to a turbulent flow, J. Fluid Mechanics, 46: 177-198 (1971).

6. Nakamura, Y. and Tomonari, Y., The effect of turbulence on the drag of rectangular prisms, Transactions Japan Society of Aeronautical and Space Science, 19: 81-86 (1976).

7. Cermak, J. E., Peterka, J. A., and Ahsan Kareem, Heliostat field-array wind-tunnel test, Report number CER 78-79 JEC-JAP-AK2 (for McDonnell Douglas Astronautics Company (July 1978).

THE PROBABILITY OF TRAINS OVERTURNING IN HIGH WINDS

R. K. COOPER

Railway Technical Centre, Derby, UK

ABSTRACT

Experience in various parts of the world has shown that trains may overturn in very high winds. The aerodynamic forces on trains in cross-winds and other factors affecting overturning are discussed. A procedure for estimating the probability of overturning per year for fleet operation of trains over a particular route is described. Most of the probability is predicted to come from a small length of track on embankments in coastal locations or in hilly country. The computation is very sensitive to errors in the parameters that describe the exposure of the sites, so the value of the total probability is only approximate.

SYMBOLS

A	reference side area
C_R	coefficient of rolling moment R, about lee rail, $R/\frac{1}{2}\rho AhV_R^2$
h	height of vehicle body
h_c	height of centre of mass above rail level
m	mass of vehicle
r_c	radius of curve
s	semi-track
V_T	train speed, $= V_R \cos\psi$ for wind normal to track
θ_c	track cant angle
ψ	yaw angle between resultant wind and train (0° for zero wind).

INTRODUCTION

Many years of operating experience have shown that train travel is very safe by comparison with other modes of surface transport. However there have been serious accidents caused by trains overturning in strong winds. Garrod (1) and Fukuchi (2) reported incidents in New Zealand and Japan respectively, on narrow gauge (1.067 m) track. The only known incident due to wind on standard gauge track (1.435 m) occurred in England in 1904, when all the carriages of a train, which was stationary on an embankment, were blown over. The low mass of some modern, high speed trains makes it necessary to examine the problem again.

In this paper we examine briefly the aerodynamic forces produced on the train by strong cross-winds, estimate overturning wind speed and give a method for estimating the probability of overturning for fleet operation over a particular route.

2. AERODYNAMIC FORCES DUE TO CROSS-WINDS

A train moving through a cross-wind behaves like a slender body at yaw to a steady flow, but with additional effects due to the ground and the sheared, turbulent wind flow. This is approximated in a wind tunnel by setting the model at a yaw angle to the mean flow, with an appropriate representation of the ground plane and rails. The true effects of wind shear, turbulence and the relative motion between the model and ground can be correctly simulated only by a moving model experiment.

The flow around a train, up to about 50° yaw is dominated by a lee side vortex which grows from a separation line near the top lee corner. The vortex detaches a certain distance down the body, then another begins. The flow on the lower part of the body is strongly affected by local separations around the bogies, which may shed vorticity into the flow in the opposite sense to the main vortex. Depending on the nose shape and yaw angle, there may be a separated flow region on the lee side at the front of the train, and this may extend to ground level (at least in a static wind tunnel simulation).

We will consider the results of two experiments on a 1/5th scale model train consisting of two vehicles (3). The leading vehicle has a low mass body shell suspended by five load cells onto a frame with wheels. The second vehicle, about half the length of an intermediate coach, acts as an aerodynamic fairing to reduce end effects. The model was propelled along a track through a natural cross wind at the Proof and Experimental Establishment, Pendine, Wales (Fig. 1). The forces were averaged over short time intervals then converted to coefficient form using similarly averaged dynamic pressure. The same model was also tested on a ground board in the 7.9 m x 4.4 m wind tunnel at the Motor Industry Research Association Nuneaton. The results for coefficient of rolling moment about the lee rail are shown in Fig. 2. The differences are due to the effects of the relative motion between model and ground, the turbulent onset flow and imperfect modelling of the embankment and rails at Pendine, but it is not possible to judge the relative importance of each at present. The C_R assumed in the examples below is also shown.

Fig. 1. 1/5th Scale model train on test track.

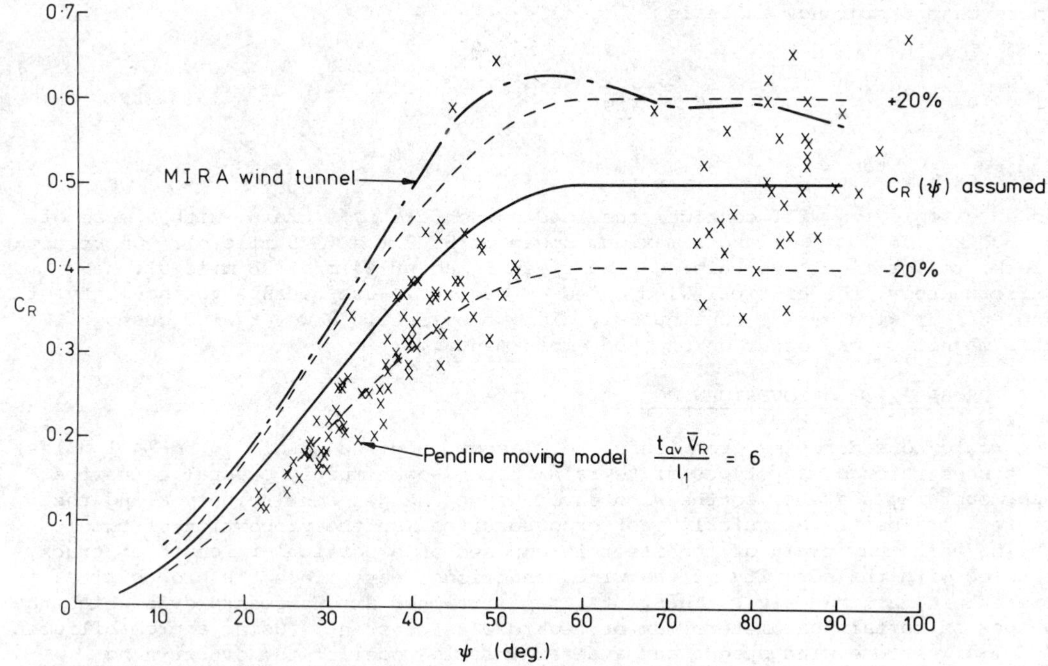

Fig. 2 Rolling moment coefficient v. yaw angle.

The aerodynamic forces on second and subsequent vehicles of the train are lower than those on the leading vehicle, but these vehicles are generally of lower mass, so they must also be considered in an overturning study.

The aerodynamic forces will depend on the boundary layer profile and turbulence intensity of the wind and the gradient of wind velocity in the direction of travel. The flow is strongly affected by embankments, viaducts and topographical features (hills, valleys, etc) and ground roughness (trees, hedges, buildings etc). An attempt is made to allow for these effects on the reference wind speed (to be discussed) but their affect on the aerodynamic behaviour of the train is neglected. In any case it is expected that they would introduce much smaller errors than those associated with the estimation of the reference wind speed.

3. OVERTURNING WIND SPEED

The overturning moment about the lee rail due to cross-wind is opposed by the restoring moment due to gravity minus the moment due to curving. We will ignore the dynamical effect of rough track, although this has been considered by Fukuchi (2). The critical case occurs when the aerodynamic overturning moment and restoring moment are balanced.

The critical wind speed is a function of the angle between the wind vector and the track direction, but for the special case of wind normal to the track and small angle of cant deficiency we may obtain an approximate value of critical overturning wind speed as follows :

$$V_c = \left[\frac{mg\ (s - h_c\ \Theta_d)}{\frac{1}{2}\ \rho\ Ah}\right]^{\frac{1}{2}} \quad \frac{\sin\ \psi}{C_R(\psi)^{\frac{1}{2}}}$$

where cant deficiency angle is

$$\theta_d = \tan^{-1}\left[\frac{V_T^2}{r_c g}\right] - \theta_c \simeq \frac{V_T^2}{r_c g} - \theta_c$$

and $V_c = V_T \tan \psi$.

As an example, we will consider the leading vehicle of a train, with a mass of 34 000 kg (75 000 lbm) and a maximum speed of 55.9 m/s (125 mile/h). On straight track, the critical overturning wind speed is about 44 m/s (98 mile/h). Unfortunately, the critical wind speed is of little use because the severity of exposure to wind varies continuously along the route. However wind gusts of this magnitude may occur on exposed embankments.

4. PROBABILITY OF OVERTURNING

The route considered is from London to Glasgow, a distance of 645 km (401 miles). It passes through many types of terrain; urban, open rural, coastal estuaries and mountainous areas, so the wind environment changes continuously along the route. Trains travel quickly past exposed sites, so the proportion of time during which any train of the fleet is exposed on a particular length of track, coupled with the severity of the wind conditions, determines the probability of overturning at that site. The route is partitioned into segments over which the values of certain parameters may be regarded as constant. Using a probabilistic model of extreme wind speeds and a deterministic model of the overturning situation, the probability of overturning per year for each segment is calculated. These probabilities are then summed for the whole route.

4.1 Reference Wind Speed at Track Side

The track-side reference wind speed at a given location is the speed expected to occur with probability 0.636 in a 50 year period at a height of 3 m above rail level. It is affected by

1) Geographical location of the site, which determines the reference wind speed, V_{10} at a height of 10 m above flat, open countryside with a ground roughness length of $z_o = 0.01$ m, averaged over 1 hour.

2) The topography of the land surrounding the site, (specified by a factor K_L, see later).

3) The ground roughness of the fetch (in the direction from which the wind is blowing), (factor K_S).

4) The height of the embankment or viaduct, (factor K_H).

5) The averaging time, T, over which the velocity is measured (factor K_{TM}).

The track-side reference wind speed is defined to be

$$V_{REF} = V_{10} K_L K_S K_H K_{TM}.$$

All of the factors except K_H are discussed in an ESDU data item (4). The embankment height factor K_H, which is also a function of ground roughness, is given in Fig. 3. It has been determined from a numerical model of wind flow over hills developed by Deaves (5). A field experiment on an 8 m high embankment

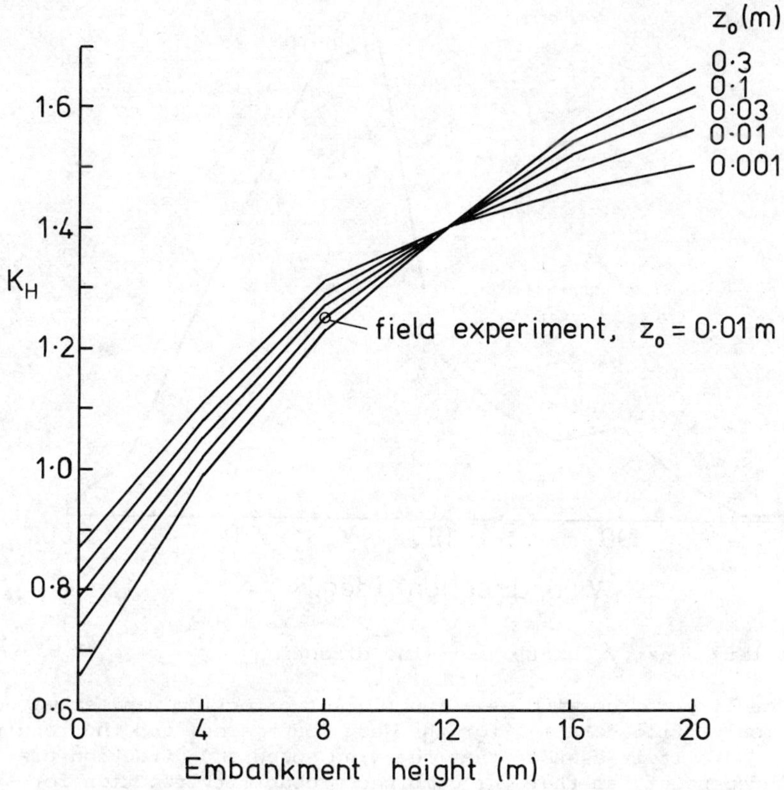

Fig. 3. Embankment height factor

(6) gave a value of K_H slightly lower than the numerical prediction.

As the flow accelerates over an embankment, the turbulence intensity in the mean wind direction is reduced, so we would expect the averaging time factor to be less than the equivalent value over flat ground, K_T. The modified averaging time factor is assumed to be

$$K_{TM} = 1 + \frac{(K_T - 1) \, \mathrm{Log}_{10} \, (3/z_o)}{K_H \, \mathrm{Log}_{10} \, (10/z_o)}$$

The gust averaging time is assumed to be T = 3 s, although the effect of this assumption will be considered later.

4.2 Probability of wind speed and direction

Having obtained the trackside reference wind speed, V_{REF}, the probability of a given wind speed, V, being exceeded in one year is found from the Fisher-Tippet Type 1 distribution function, obtained from ESDU data (4).

$$P\left(\frac{V}{V_{REF}} > k\right) = 1 - \exp\left[-\exp\left(\frac{0.716 - k}{0.07276}\right)\right]$$

This function is assumed to apply to short period gusts as well as mean hourly speeds.

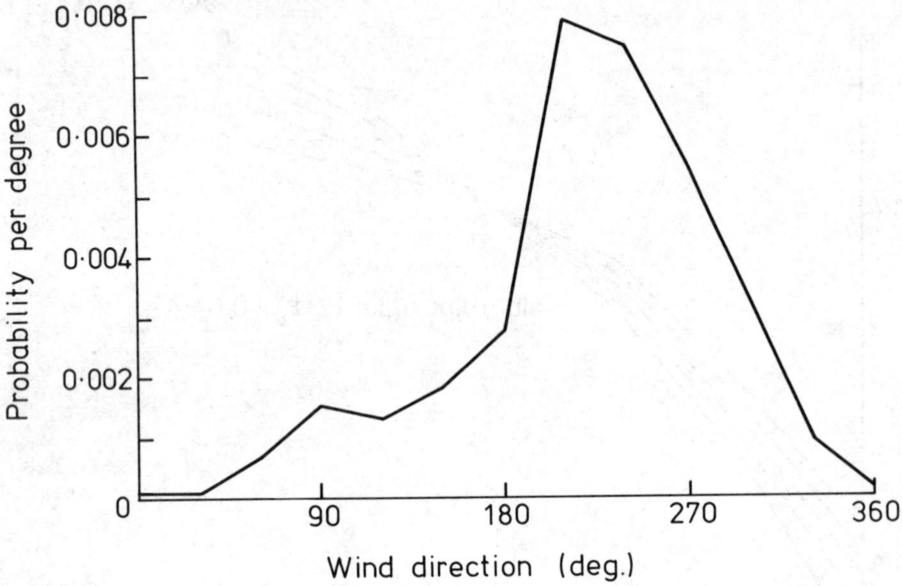

Fig. 4. Probability density function of wind direction

The probability density of the wind as a function of direction was estimated from Meteorological Office data (7) for the West Coast area, and the results are given in Fig. 4. It is assumed that the wind speed and direction are statistically independent, so that the combined probability function for speed and direction is just the product of speed and direction probabilities.

4.3 Data for exposed sections of the route

Only part of the route, a distance of 212 km (132 miles) from Crewe to the Scottish border was considered in detail. The exposed sections had already been surveyed to re-assess the integrity of the overhead electric power system, after several "blow-off" incidents had occurred. (The catenary and contact wires were blown laterally and forced off the end of the pantograph which then tore down the wires). At the exposed sites we had estimated topographical exposure factor (K_L), ground roughness length (z_o) and height of embankment, and so the trackside reference wind speed was derived. The other parameters such as length of exposed segment of track, radius of curvature, track cant angle, track direction, speed of train were then obtained from various sources. In several cases, segments were adjacent to allow for changing track parameters. In this way data was collected for more than 150 segments.

4.4 Computation

For a given yaw angle the computer program (8) finds the critical wind speed (a function of train speed, track curvature etc.) and hence the probability of exceeding this speed. The probability is integrated so as to cover all wind directions. This result is multiplied by the probability of a train being exposed on the segment and then summed over all track segments to give the total probability of overturning per year for operation of the fleet of trains over that part of the route.

4.5 Effect of gust averaging time

It is estimated that a light weight vehicle in a train could overturn in about 2 seconds with an applied moment 10% above the critical value which just starts to lift the windward wheels. Also the aerodynamic moment would develop very rapidly on entry to a gust, so we must choose an averaging time about the same as the duration of the gust as seen by the vehicle, i.e. about 1 second at 200 km/h. With an averaging time of 3 seconds the train would rotate approximately 5° about the lee rail with an applied moment of 1.01 times the critical value. A similar rotation would be produced by a value of 1.1 times critical applied for 1 second.

The probability of reaching critical overturning moment is very sensitive to the averaging time; 65% increase for T = 1 s over the basic result for T = 3 s. However the probability of achieving 1.1 times the critical value for 1 s is only 5% above the basic result. Consequently the critical value of overturning moment with an averaging time of 3 seconds is used to compute the probability of overturning.

5. RESULTS

As an example, the leading vehicle only is considered. For fleet operation, the number of train journeys is assumed to be 18 per day in each direction, over the 212 km (132 miles) section of the route considered. The computed probability of overturning per year in this section is 0.0092, which implies a return period (with probability 0.636) of 109 years.

5.1 Variation of probability with yaw angle

For site 'a' the distribution of probability of $V > V_{critical}$ per degree of yaw angle is plotted against yaw angle in Fig. 5, assuming continuous occupation of the site. Only one train direction (146° from North) is considered, and curves for overturning to both left and right are shown. This example is typical, in that most of the probability is generated in the yaw angle range from 20° to 60°. At the peak, usually between 30° and 35° yaw, the wind angle to the track Ø, is less than 90°. At yaw angles below 25° the low values of rolling moment coefficient cause high critical wind speeds, so the probability of overturning per degree is very low. At yaw angles above 50° the rolling moment coefficient is almost constant, but the wind has a component in the direction of train motion, so the critical wind speed increases with yaw angle, thus producing a low probability.

5.2 Factors which produce a high probability of overturning

The segments have been ranked in order of probability of overturning. Note that the ranking would change if the train parameters, e.g. coefficient of rolling moment, mass or speed were changed. A large part of the probability of overturning comes from a relatively small number of segments, and some segments contribute almost nothing. Of course, the longer segments appear nearer the top of the rank order than shorter ones of the same exposure.

The circumstances which produce a high trackside reference wind speed (greater than 50 m/s say) and hence a high probability of overturning, are usually a combination of at least two of the following :

1) High topographical exposure factor, K_L.

2) Low value of surface roughness length, z_o.

3) Embankment higher than 4 m.

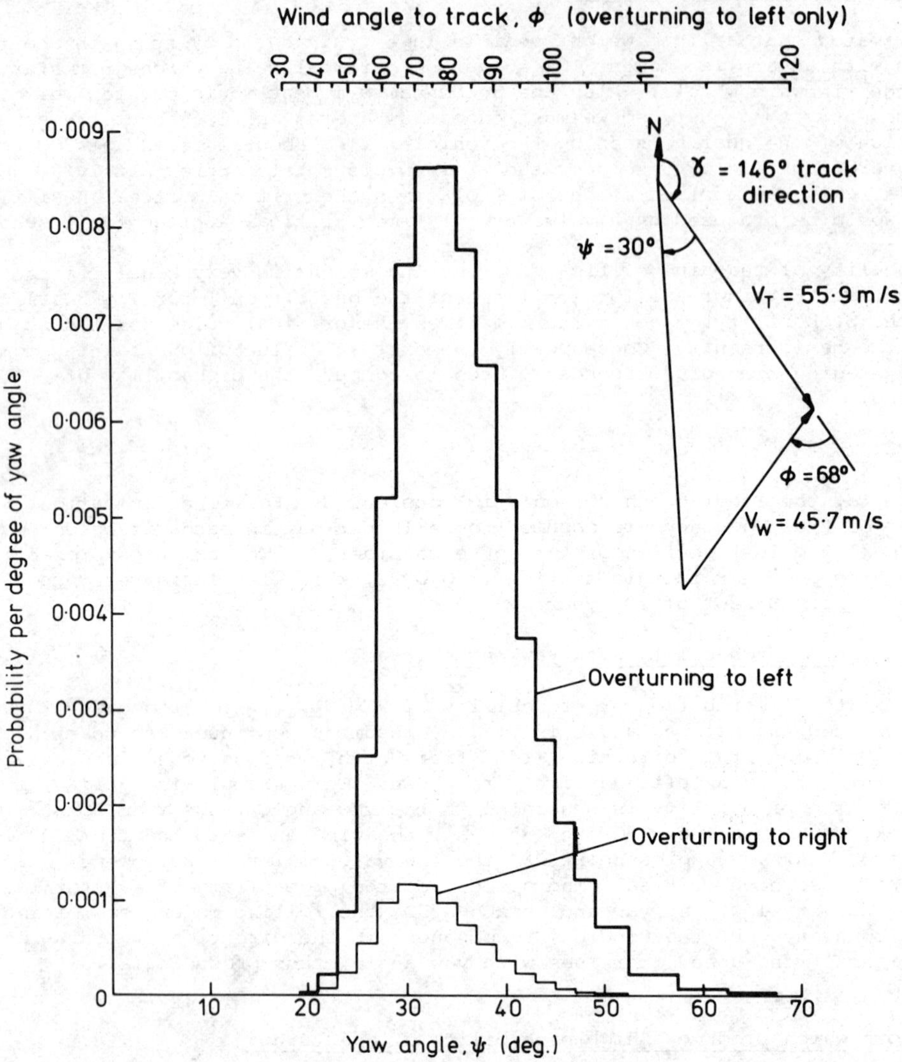

Fig. 5. Probability density of overturning v. yaw angle.

Table 1. Segment parameters for site 'a'

Description		coastal flats, few obstructions	
Topographical exposure	(K_L)	1.1	
Ground roughness z_o	(m)	0.01	
Embankment height	(m)	6	
V_{REF}	(m/s)	58.3	
Segment length	(km)	1.8	
Cant angle	(deg)	0	
Radius of curve	(m)	∞	
Track direction	(deg)	326 (down)	146 (up)
Train speed	(m/s)	55.9	55.9
Probability of overturning per year		0.000864	0.000972

Cant deficiency, caused by a combination of high train speed with fixed values of track curvature and cant angle, is not enough in itself to cause a high probability, but values between 6° and 9° frequently occur on embankments in hilly country, thus making a significant contribution to probability of overturning. (With the cant deficiency set (artificially) to zero the total probability reduces by about 22%). The most exposed segments are in smooth coastal areas with moderate (4 m to 8 m high) embankments, or on curved embankments on hillsides.

The combined length of the top 10 segments in the ranking is only 7 km (4.35 miles), but they contribute 51% of the total probability. The top 30 segments contribute 69% of the total probability in about 15.2 km(9.46 miles). If the total probability is reduced (e.g. by reducing estimates of trackside reference wind speeds) then an even greater percentage of the total occurs in the top ranked segments. The fact that such a large part of the probability is concentrated in a relatively short length of track implies that wind fences may be an economical way of reducing the probability of overturning.

5.3 Sensitivity and Accuracy

The computed probability of overturning per year is very sensitive to changes in some of the most important parameters. For the leading vehicle we obtain the following percentage of changes of probability with the stated changes in parameters:

1) +45% to -36% for ± 5% change in trackside reference wind speed.

2) +26% to -23% for ± 5% change in rolling moment coefficient

3) +12% to -11% for ± 5% change in train speed

4) -21% for 5% increase in mass.

5) +2.3% for 5% increase in height of centre of mass above rail level.

Unfortunately, the computation is most sensitive to errors in the parameters which are likely to have the largest uncertainty.

The uncertainty in estimation of trackside reference wind speed is at least ± 5% in flat open rural sites, which are fairly easy to evaluate, but could be worse than ± 10% for sites in mountainous terrain or with a complex fetch. The topographical exposure factor has been chosen conservatively (1.1 to 1.2) in some cases because of the unknown effect of hilly topography. The modified averaging time factor K_{TM} is only an initial guess, so must be treated with caution.

The other very significant source of error is the coefficient of rolling moment. The conflicting results obtained from the moving model experiment in the natural cross-wind and the conventional wind tunnel test make estimation of the error difficult, however ± 15% is thought to be appropriate. Consequently the estimate of probability could be in error by a factor of 2 either way, although the conservative nature of some of the assumptions should be noted.

The assumptions necessary to make the problem manageable, e.g. (1) the independence of extreme wind gust speed and wind direction, with fixed probability functions for each, (2) rolling moment a determinate function of yaw angle; are obviously sources of errors, which could be reduced by a continuing research program. However, much improved accuracy may not be justifiable when preventive measures may be taken, such as stopping the trains when dangerously high winds are expected.

6. FURTHER WORK

There is a continuing program of wind tunnel work on model trains, to provide basic aerodynamic data. To investigate the true effects of wind shear and turbulence, a simple moving model rig for propelling a model train across a simulation of the earth boundary layer in a wind tunnel, is under construction. The effect of embankments will also be examined.

To provide some information on topographical effects a wind measuring experiment, in conjunction with a wind tunnel investigation of the flows over topographical models, is in progress. It is hoped to correlate the field results with Meteorological Office measurements taken at the same time, and so give more direct prediction of long term statistics for wind speed at the trackside. In turn this will provide data for determining rational limits on safe wind conditions for train operations and enable the Meteorological Office to give warning if the limits are likely to be exceeded.

7. CONCLUSIONS

A method for estimating the probability of overturning of fast, low mass passenger trains has been developed. The results are very sensitive to errors in some of the key parameters, in particular to the trackside reference wind speed and the coefficient of rolling moment. Consequently, further work is needed to improve the aerodynamic data and the wind statistics. This work is necessary to maintain the good safety record of the railways, with the design of train needed for fast economical operation over existing routes.

Acknowledgement

The author wishes to thank the Director of Research, British Railways Board, for permission to publish this paper.

REFERENCES

1. Garrod, S D, "Wind gusts on rolling stock". New Zealand Government Railways Research Report No 136, August 1967.

2. Fukuchi, G, "Statical considerations for railway vehicles overturning due to lateral winds". Japanese National Railways, Quarterly Reports of RTRI Vol 16 (No 2), 1975, 49-53.

3. Cooper, R K, "The effect of cross-winds on trains". ASME Symposium on the Aerodynamics of Transportation, 18-20, June 1979, Niagara NY (to be published).

4. Characteristics of wind speed in the lower layers of the atmosphere near the ground : Strong winds (neutral atmosphere). Engineering Sciences Data Unit Item No 72026, October 1974.

5. Deaves, D M, "Wind over hills : A numerical approach". J Indust. Aerodyn. 1, 1976. 371-391.

6. Cooper, R K, "Edwalton Embankment wind experiment". British Railways Board R & DD, Technical Memorandum TM AERO 27, August 1978.

7. Shellard H C, "Tables of surface wind speed and direction over the United Kingdom". Meteorological Office, Met. O 792, HMSO London 1968.

8. Phillips, R, British Railways Board, R & DD, Unpublished memorandum.

ON THE EFFECT OF BAFFLE WALLS IN A WATER TOWER TANK UNDER WIND ACTION

J. BERLAMONT and N. VANDERSTAPPEN
University of Leuven, Heverlee, Belgium

ABSTRACT

Modern slender water towers have a very low structural damping, which makes them vulnerable to wind-induced oscillations, causing severe wave motion in the tank and high stresses in the shaft. Baffle walls, initially intended for suppressing the wave motion in the liquid which could cause excessive dynamic water pressures against the walls, appear to be very useful in increasing the structural damping considerably.

In situ measurements confirm that a water tower can be modelled mathematically by a two-mass-spring system. A physical model, built after the mathematical model indicates the increased damping, due to the presence of the baffle walls, the influence of the number of baffle walls and their depth.

INTRODUCTION

The damping in modern slender water towers is small (less than 3 % logarithmic decrement (2)) since

- aerodynamic damping is small and may be negative due to vortex formation.
- the damping both in the moving water mass and the shaft is small. Concrete towers have a damping which is 2.5 times greater than steel towers (1,2).
- energy-dissipation through the foundation is small as well. A foundation of piles, however, has a much larger damping than a shallow foundation.

As a consequence, important resonance effects can occur when a light damped water tower is excited by wind gusts (vibrations in the wind-direction) or von Karman vortices (vibrations perpendicular to the wind-direction) when the frequency of the gusts or the vortex shedding is close to one of the natural frequencies of the structure. Large deflections of the tower and the large amplitude wave motion of the water in the tank may increase the stresses in the shaft substantially.
Van Erp (1) reported that specially conceived steel baffles were built into the water tank of a steel water tower for the Eindhoven University of Technology in order to suppress the wave action in the tank caused by the wind-generated vibrations of the water tower. Van Uchelen (2) noticed, however, that the dynamic water pressures against the walls of the tank are less than 5 à 15 % of the hydrostatic pressure, but that the increase of damping added to the structure by the baffle walls could be very useful for stabilizing both the shaft and the structure as a whole.

Since the actual damping is very low, any even small increase of the damping is very welcome and results in a marked decrease of the dynamically magnified deflections and stresses. Model studies show that baffle walls can contribute in an important way to the damping of the structure.

MATHEMATICAL MODEL

A water tower essentially consists in a water reservoir, elastically supported by the shaft. Any dynamical loading (e.g. wind gusts or earth-quakes) can cause a movement of the water content with respect to the walls of the tank, as well as movements of the reservoir with respect to the ground.

Housner (4) and Cloud (5) therefore represented a water tower by a double mass - spring system, as shown in Fig. 1.

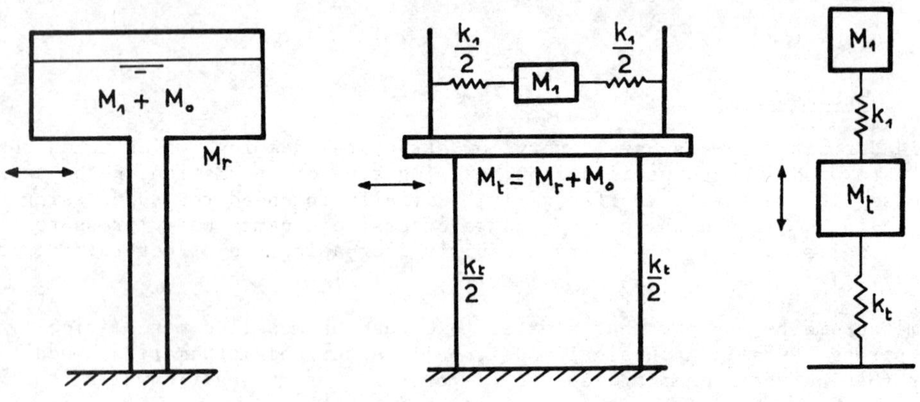

Fig. 1.

Part (M_o) of the water mass (M_1 + M_o), especially the water located in the lower part of the tank, moves solidary with the reservoir (mass M_r). Housner (4) computed this virtual mass M_o, the free mass M_1 and the equivalent stiffness k_1, for a cilindrical and a rectangular reservoir.
The dynamicsof the two-mass-spring system (Fig. 1) under a sinusoidal loading $F_o \sin\omega t$ is well known (8).
The vibration amplitudes are :

$$a_t = a_o \frac{1 - (\frac{\omega}{\omega_1})^2}{\left[1 - (\frac{\omega}{\omega_1})^2\right]\left[1 + \frac{k_1}{k_t} - (\frac{\omega}{\omega_t})^2\right] - \frac{k_1}{k_t}} \quad (1)$$

$$a_1 = a_o \frac{1}{\left[1 - \left(\frac{\omega}{\omega_1}\right)^2\right]\left[1 + \frac{k_1}{k_t} - \left(\frac{\omega}{\omega_t}\right)^2\right] - \frac{k_1}{k_t}} \qquad (2)$$

in which :

$a_o = \frac{F_o}{k_t}$ the static deflection of M_t

$\omega_1 = \sqrt{\frac{k_1}{M_1}}$ the natural frequency of the free mass M_1 (rad/s)

$\omega_t = \sqrt{\frac{k_t}{M_t}}$ the natural frequency of the reservoir (including the virtual mass M_o) (rad/s)

$M_t = M_r + M_o$ reservoir mass plus virtual mass

The two natural frequencies of the system are computed from the frequency equation :

$$\omega^4 - \omega^2 \left(\frac{k_t + k_1}{M_t} + \frac{k_1}{M_1}\right) + \frac{k_1 k_t}{M_1 M_t} = 0 \qquad (3)$$

If $(M_o + M_r) >> M_1$, the natural frequencies of the system of Fig. 1 are close to the natural frequencies ω_1 and ω_t (6).
The fundamental natural frequency of a liquid filled reservoir can easily be computed for rectangular or cilindrical reservoirs (4).

According to Housner (4), the natural frequency of a spherical reservoir can be computed by assimilating it with a cilindric reservoir with the same volume of liquid and a radius equal to the radius of the water surface in the sphere. This is understood by the fact that, during a free vibration, practically only the upper fluid layers are in motion.

Housner's approximation for a spherical reservoir was checked by laboratory measurements. The agreement between theory and measurements is fairly good (table 1)

Table 1 : natural frequencies of a spherical reservoir (diameter .34 m (1 ft))

Waterdepth	natural frequency (Hz)	
(%)	computed	measured
15	1.21	1.28
29	1.32	1.36
44	1.45	1.46
59	1.59	1.61
74	1.74	1.83

The presence of the continuing shaft in the spherical reservoir causes an increase of the vibration period (up to 20 % for small waterdepths).

For a steel water tower at Bierbeek, near Leuven (height of the shaft : 18.5 m (62 ft); diameter of the spherical reservoir 10.6 m (35 ft), water content : 540 m^3) the natural frequencies were computed according to eq. 3 and measured with the aid of an accelerometer, the signal of which was processed by a Fourier analyser :
waterdepth : 8 m

	computed	measured
f_1 =	0.31 Hz	0.3 Hz
f_2 =	0.68 Hz	0.6 Hz
torsion :	6.72 Hz	6.5 Hz

Since $M_1 \simeq 0.24\ M_t$ the natural frequencies are close to the frequencies of the free water mass (0.32 Hz) and the reservoir mass plus the virtual mass (0.66 Hz). A frequency of 0.5 Hz was also present in the response spectrum. This frequency corresponds to a second order vibration mode of the water mass(computed 0.53 Hz). The good agreement between the mathematical model and experiments inspires confidence in the model and allows studying the effect of baffle walls in a laboratory model, constructed according to the two-mass-spring model (according to an idea by Van Uchelen).

PHYSICAL MODEL, INFLUENCE OF THE BAFFLE WALLS

A model of a water tower with a cilindrical tank (Ø 0.40 m) was built in the Hydraulic Laboratory of the University of Leuven (7). Strain-gauges were fixed to the springs k_t. From the frequency-spectrum obtained with the aid of a Fourier analyser, processing the strain-gauges output, the two natural frequencies f_1 and f_2 were obtained and compared with computed values (table 2).

Waterdepth (m)	computed (Hz) f_1	computed (Hz) f_2	measured (Hz) f_1	measured (Hz) f_2
0.05	0.87	1.39	0.84	1.38
0.10	0.89	1.63	0.77	1.62
0.15	0.82	1.72	0.64	1.72
0.20	0.75	1.73	0.50	1.74

The important error between experiments and calculations for the larger waterdepths is caused by the fact that Housner's formulas for M_1, k_1, ... hold good only if the waterdepth in the reservoir is small, as compared with the height of the shaft. This assumption applies to most water towers, but was not valid in the model for large waterdepths (height of the model shaft 0.34 m).

In the absence of baffle walls the dynamically-coupled system (2 masses) has two well pronounced natural frequencies (Fig. 2.) according to eq. (3). The movement of the water tank is hardly damped (Fig. 2.).

By installing two baffle walls (orthogonal to the direction of movement), the lower natural frequency (free movement of the water) becomes less important and even disappears when more baffles are built in (Fig. 2). The upper natural frequency decreases towards the frequency of the single - degree-of-freedom system (all the liquid mass becomes "virtual mass"). At the same time the damping increases considerably (Fig. 2).

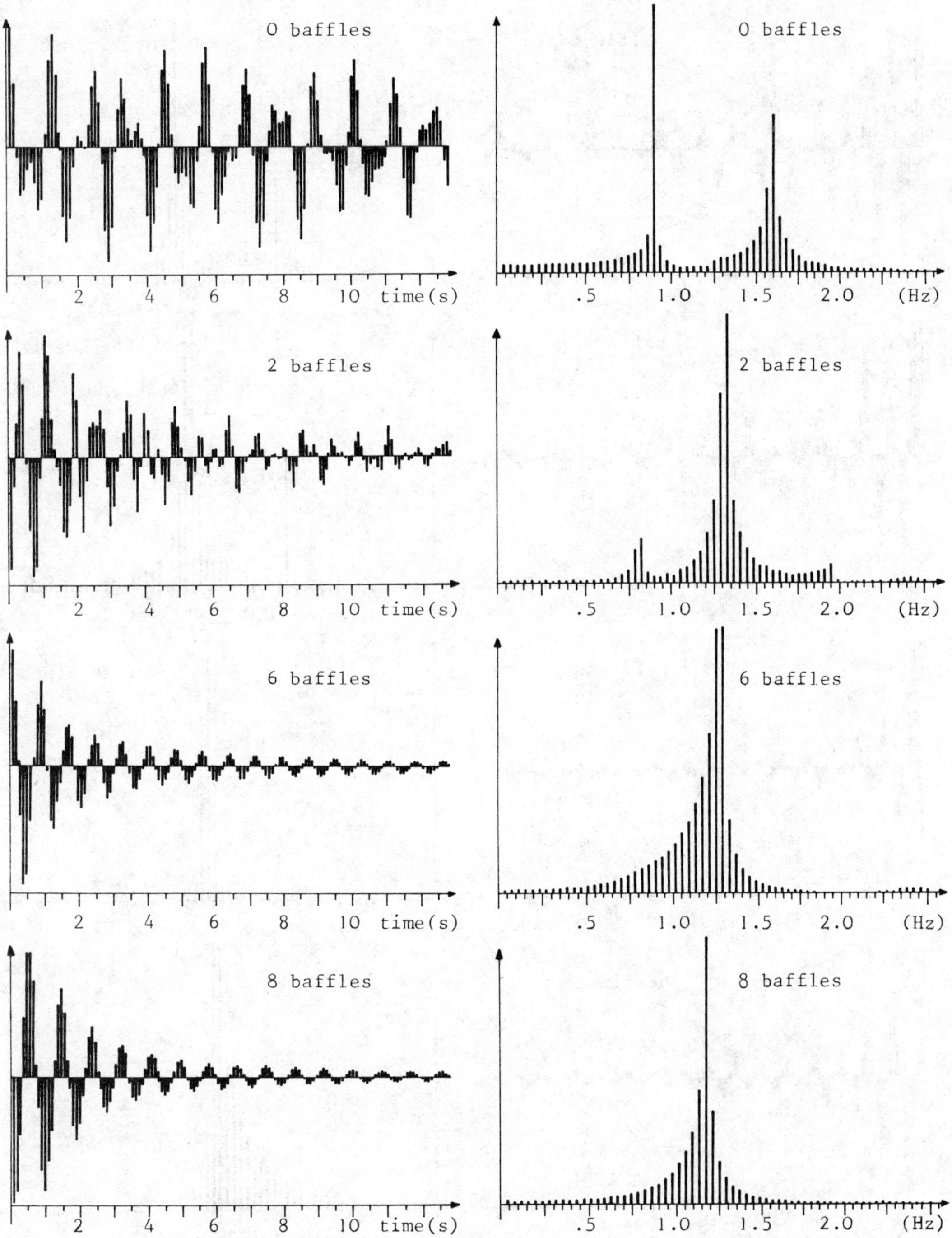

Fig. 2. Model of a water tower with a cylindrical tank : vibration amplitude time history and frequency spectrum (arbitrary units)

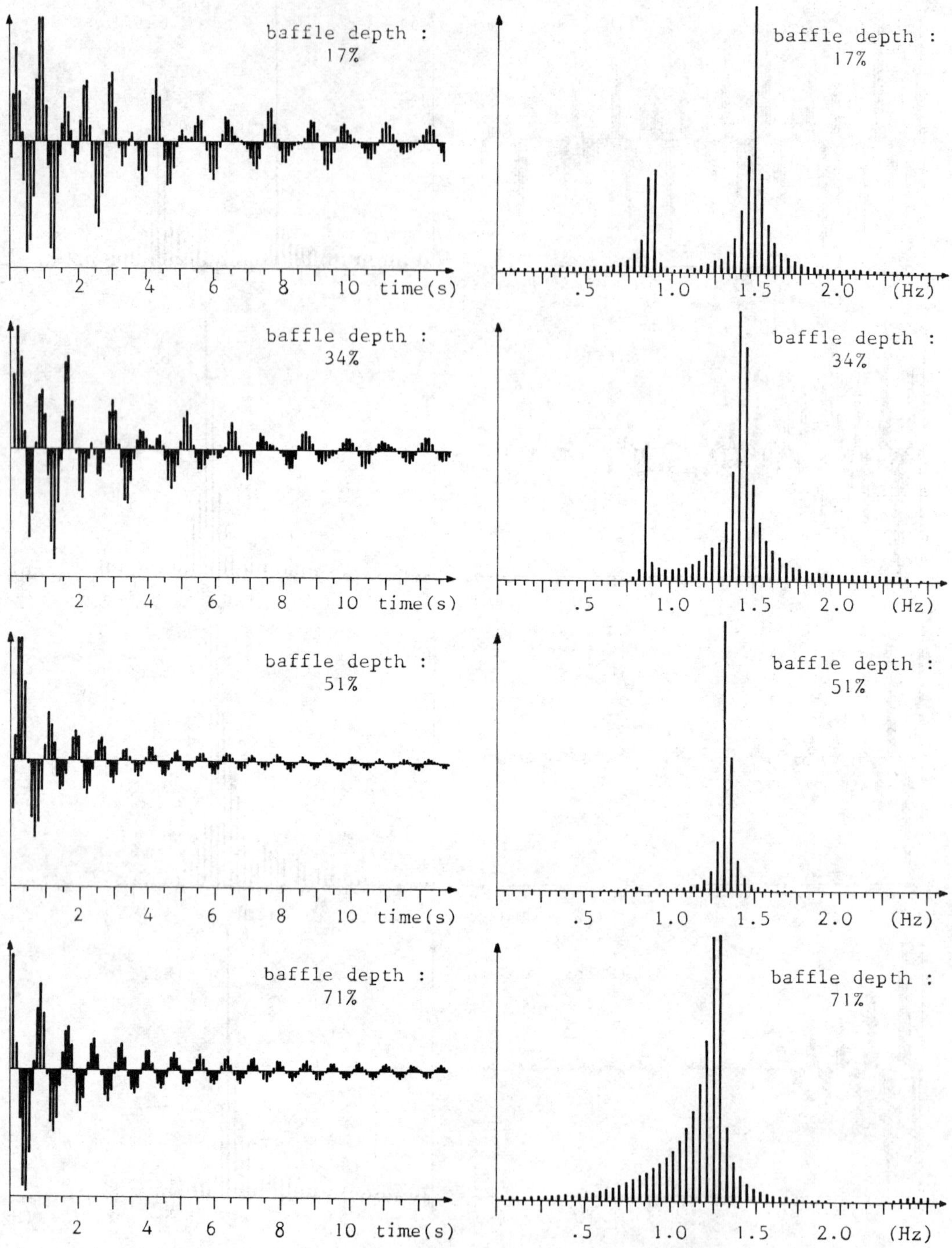

Fig. 3. Model of a water tower with a cilindrical tank : vibration amplitude time history and frequency spectrum (arbitrary units) for different values of the baffle depth

An increased number of baffles (6 or 8), distributed equally over the tank perimeter, emphasize the above mentionned effect. Such a number of baffles is indeed needed since the wind action may be in any direction.

From Fig. 2 it is seen that by increasing the number of baffles from 6 to 8, the logarithmic decrement of damping only increases from 0.18 to 0.23, which however, is considerably more than the value reported by Van Uchelen (2) 3 % or the value used by Davenport (8) 10 %.

The baffle walls divide the reservoir into a large number of smaller reservoirs, communicating with each other through small openings. A vibration of the tower in one particular direction causes a vibration of the different water masses in the compartments, each with its own frequency and mode.
The water movements in the different reservoirs cancell each other, so that the resultant water motion has no definite frequency. The energy added to the system by the wind is dissipated through a disordered and diffuse motion of the water mass. So, apart from preventing wave action against the walls of the tank, the presence of baffle walls damps the wind-induced vibrations of the shaft considerably.

In Fig. 3 the influence of the depth of the baffle walls under the water surface is shown. If the depth equals 50 % of the waterdepth in the tank the system is reduced to a one - degree-of-freedom system. An increase in depth from 50 % to 70 % results in an increase of the damping with 1 % (from $\delta = 0.17$ to $\delta = 0.18$). However, since all waterdepths may occur in a water tower tank the baffles should be as deep as possible. It should be noted that the damping appears to be non - linear so that for large amplitudes values of $\delta = \ln \frac{a_n}{a_{n+1}}$ up to 0.42 were measured. (a_n is the n-th amplitude).
The new concrete water and antenna tower, designed by Prof. F. Mortelmans (9) at Mechelen (30 km (17 miles) from Brussels), which is 140 m (470 ft) heigh with the water reservoir (2500 m^3) at el. 50 m (150 ft), was provided with 16 baffle walls to increase the structural damping. At the same time they serve for prestressing the reservoir (9).

ACKNOWLEDGEMENT

The authors gratefully acknowledge the help of Guido Van Langenhove for processing the experimental data by the Fourrier-analyser and the computer.

REFERENCES

(1) F.C. Van Erp, Trillingsproblemen bij een watertoren (in dutch) : "Oscillation problems with a water tower", De Ingenieur, n° 42, B148-B152, 1969 October 17th
(2) H. Van Uchelen, Discussion on oscillation problems with a water tower by Van Erp, De Ingenieur, n° 9, B13-B17, 1970 February 27th
(3) F.C. Van Erp, Closure of oscillation problems with a water tower, De Ingenieur, n° 24, B69-B70, 1970 June 12th
(4) G.W. Housner, The dynamic behaviour of water tanks, Bulletin of the Seismological Society of America, Vol. 53 n°2, 381-387, 1963
(5) W.K. Cloud, Period measurements of structures in Chile, Bulletin of the Seismological Society of America, Vol. 53 n° 2, 359, 1963
(6) Den Hartog, J.P., Mechnical Vibrations, Mc Graw Hill, New York, 1956
(7) Himpe, J., Vanderstappen, N., Trillingsproblemen bij watertorens (in dutch) : Oscillation problems with water towers, thesis, Department of civil engineering University of Leuven, 1977
(8) Davenport, A.G., The buffeting of structures by gusts, Proc. of the intern. conference on Wind effects on Buildings and Structures, Teddington, Middlesex, June 1963
(9) Mortelmans, F., Tijdschrift der Openbare Werken van België, Water- en Antennetoren te Mechelen, n° 172, 1978, p. 193

ON WIND-PROOF PERFORMANCE OF LATTICED STEEL TOWERS

M. FUJIMOTO*, I. MATSUSHITA**, T. OHKUMA***, T. AMANO**** and H. AKAGI**

*Tokyo Institute of Technology, Tokyo, Japan
**Electrical Communication Lab. NTT, Tokyo, Japan
***Kanagawa University, Kanagawa, Japan
****Ryukyu University, Okinawa, Japan

SYNOPSIS

The wind-proof performance of latticed steel towers has been investigated through full-scale and model tests. The tower tested is fabricated with structural angle steel and the shape of cross section is square, which is most common among various types of microwave telecommunication towers of Nippon Telegraph and Telephone Public Corporation. Together with the experiments presented is an approximate analysis for the buffeting of such structures based on stationary random process, which comes out to agree well with the experimental results. As a consequence of this study, design formulas are proposed from a viewpoint of practical application for wind loads on latticed steel towers considering for dynamic effects of winds.

1. PREFACE

The microwave telecommunication steel towers are one of the most important facilities of Nippon Telegraph and Telephone Public Corporation. There is the wind load among main external forces working at tower structures. The steel towers must be designed so that full functionality and structural safety are obtained against the wind load supposed. At present, the static loads based on maximum instantaneous wind speed are given for steel towers in conventional wind-proof designs [1-4]. The wind load, F is calculated by,

$$F = C \cdot q \cdot A$$

in which C, q and A are wind force coefficient, reference wind pressure based on maximum instantaneous wind speed and projected area. The towers, however, are inclined to become taller and less deformation is required to be allowable due to higher communication density. Therefore, the more rational wind-proof design procedure is now expected to be established in consideration of dynamic effect of winds.

2. OUTLINE OF THE TOWERS

The microwave telecommunication towers constructed so far are numbered about 1,500. Though there are variety of shapes as Fig. 1 shows, they are mostly latticed steel towers using structural angle steel (See Fig. 2).

Factors such as the tower height, the weight, the projected area, the natural period and the damping ratio are thought to give effects on the response beside the shape of the tower. Numerical numbers of towers have been investigated in recent years concerning on these properties [5-6].

The height of towers, H (in meters) is indicated in Fig. 3. Most of them are distributed in the extent of 30 to 50 meters, and percentage of taller ones are increasing in recent years. The reference width, D (in meters) is between 3 to 8 meters at the top of the tower and 7 to 14 meters at the foot. The weight, as seen in Fig. 4, is concentrated at the top at which platforms and antennas are facilitated. As Fig. 5 shows, the projected areas are also concentrated at the top including those of antennas (a projected area of a standard one is approximately 10 square meters). The solidity ratio (ratio of projected area and outside area) is 0.1 to 0.2 in the intermediate portion of the towers and 0.4 to 0.5 at the platform.

The 1st natural period, T_1 is highly correlated to the tower height as seen in Fig. 6, and can be expressed by,

$$T_1\ (=1/n_1) = 0.015\ H$$

in which n_1 denotes 1st natural frequency. The 2nd natural period is in the extent of 1/3.5 to 1/4.5 of that of 1st mode. The typical mode of vibration is shown in Fig. 7. The scatters of the damping ratio of 1st mode, h_1 are shown in Fig. 8, and average value of them is approximately one percent of the critical.

3. EXPERIMENTAL FEATURES

The wind-proof performance of steel towers had been vigorously investigated in the last decade [7-14]. Referring to these researches described in this chapter are the features of dynamic behaviors of microwave telecommunication towers through the model tests recently carried out by the authors in wind tunnel [15], field [16], and full-scale observation at Karagasaki Tower [17].

3.1 Wind Tunnel Test

Two models scaled after typical section of the prototype (Fig. 9, scale ratio is about 1/20) with solidity ratio, φ = 0.26 and 0.42 are tested in uniform flow [15].

The relation of mean wind force coefficient and angle of attack, α is seen in Fig. 10. The relation of maximum dynamic response (expressed in the form of reduced response; $\eta = X_{max}/D$) in cross-wind direction and wind speed (expressed in the form of reduced velocity; $U_r = U/n_1 D$) is shown in Fig. 11. The dynamic response in cross-wind direction increases with the increase of wind velocity. It is said that cross-wind response of latticed tower is very small in uniform flow. By the wind tunnel test for three-dimensional latticed tower with solidity ratio of about 0.3, Ishizaki and others [14] pointed out that there was the case causing large amplitude in cross-wind direction when $U_r = 20 \sim 25$. Such phenomenon, however, was not at all remarkable in this experiment.

3.2 Model Field Test

On the penthouse of five-storied building are set up scaled models (Fig. 12, scale ratio is about 1/50) simulated after the prototype of the microwave telecommunication tower [16]. The solidity ratios of the models are 0.30 and 0.53. Mean wind speed is measured through an propeller-type anemometer and the responses in two orthogonal axes are measured from strains in the plate springs at

Fig. 1 Microwave Telecommunication Towers

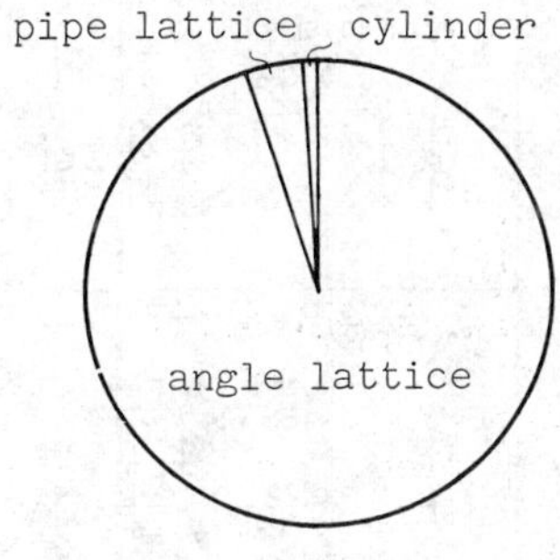

Fig. 2 Ratios of Shapes

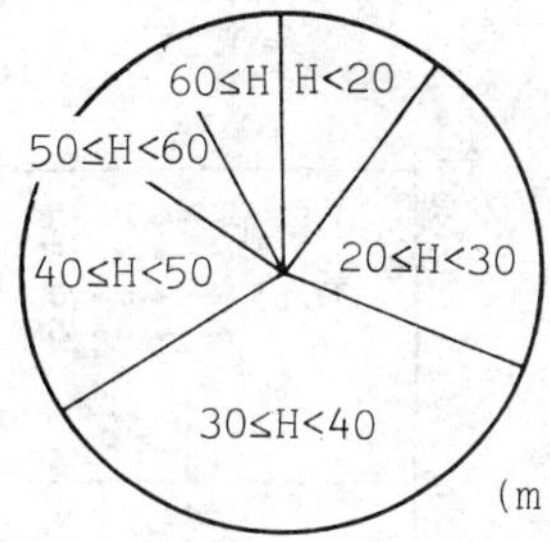

Fig. 3 Ratios of Heights

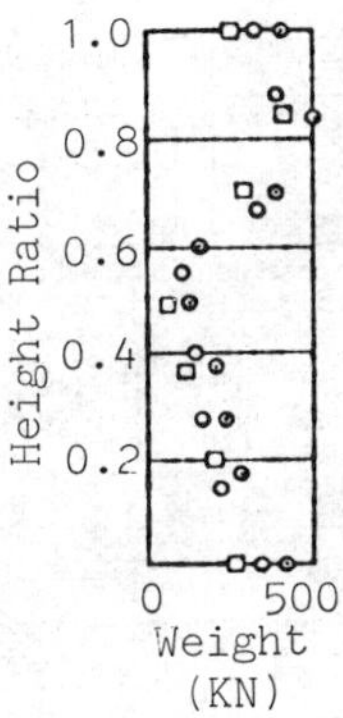

Fig. 4
Distribution of Weights

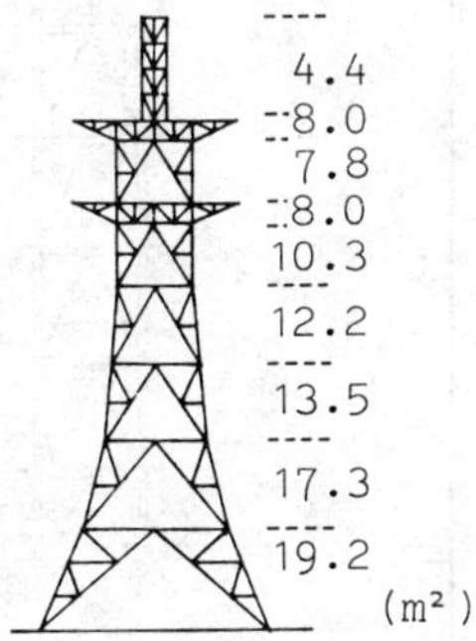

Fig. 5
Typical Pattern of Projected Area

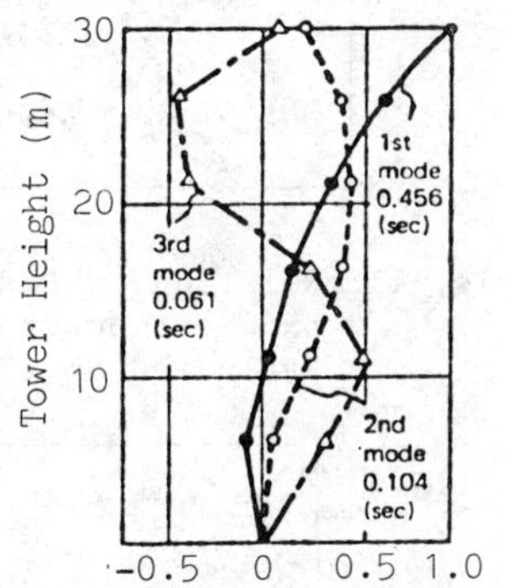

Fig. 7 Modes of Vibration

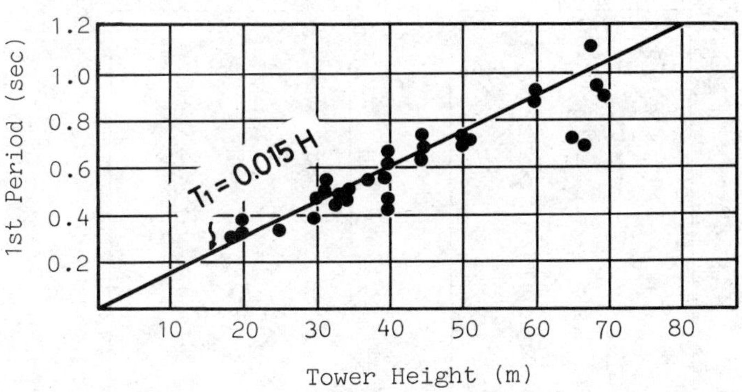

Fig. 6 Relation of the Height and 1st Period

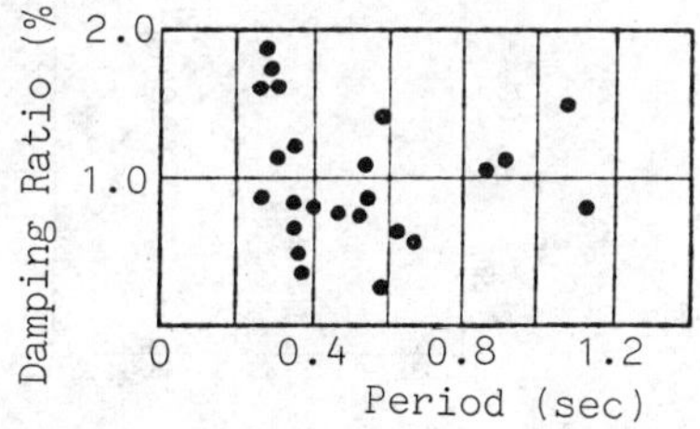

Fig. 8 Damping Ratios

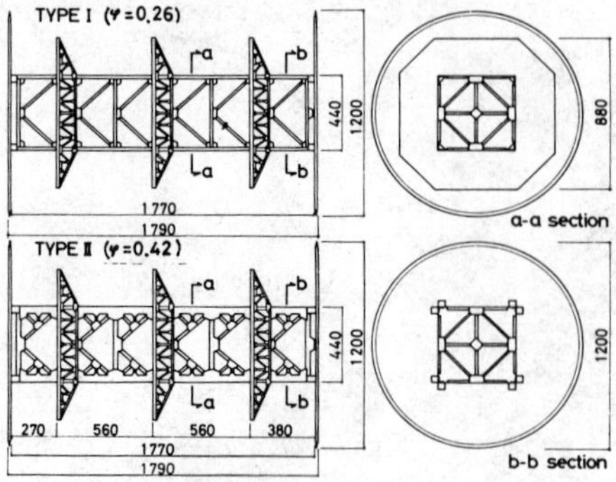

Fig. 9 Models Tested in Wind Tunnel (φ=0.26 & 0.42, n_1=2.2 Hz, h_1=0.3%)

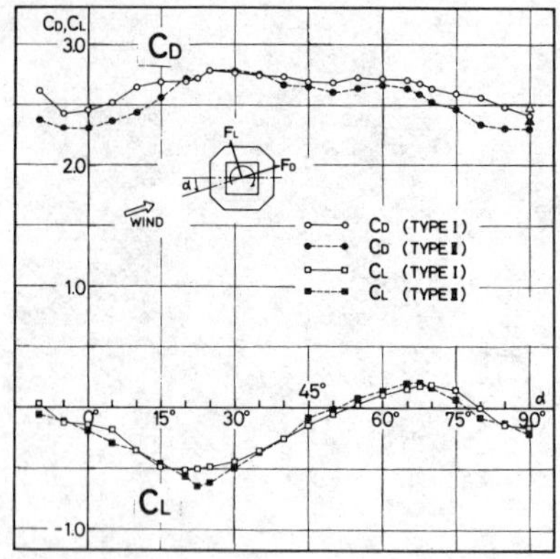

Fig. 10 Drag and Lift Coefficient

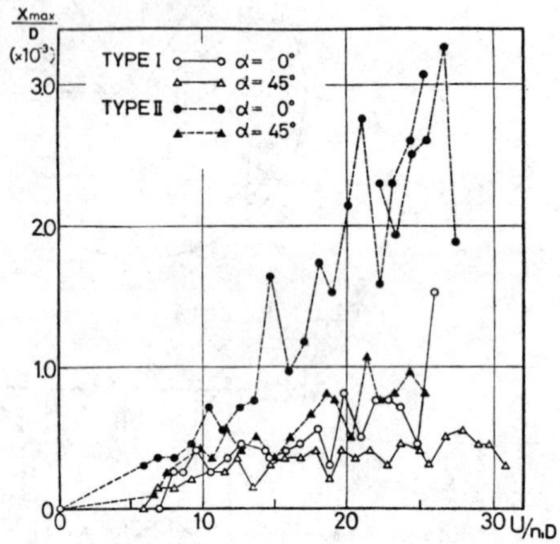

Fig. 11 Relation of Reduced Velocity and Maximum Cross-wind Response Evaluated in 20 seconds

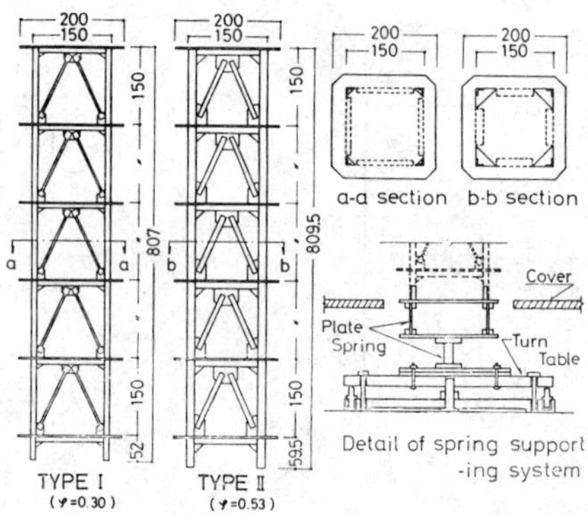

Fig. 12 Models in the Field Test (φ=0.30 & 0.53, n_1=4.5Hz, h_1=0.9%, Iu≃0.15)

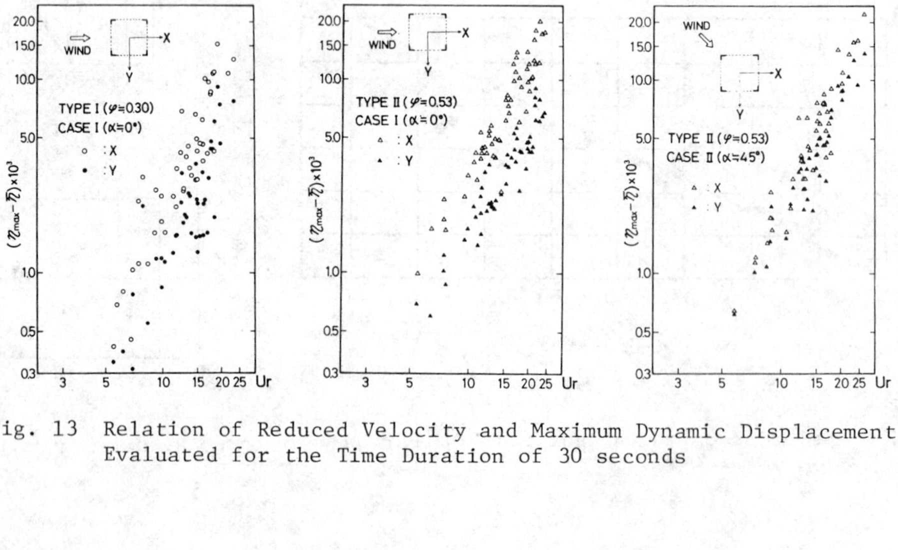

Fig. 13 Relation of Reduced Velocity and Maximum Dynamic Displacement Evaluated for the Time Duration of 30 seconds

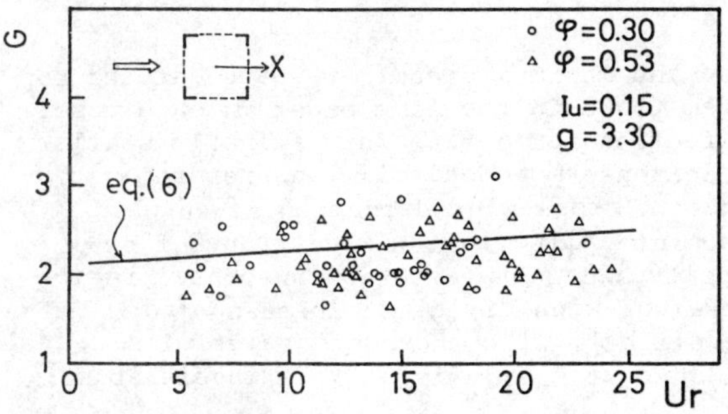

Fig. 14 Relation of Gust Factor of Along-wind Response and Reduced Velocity

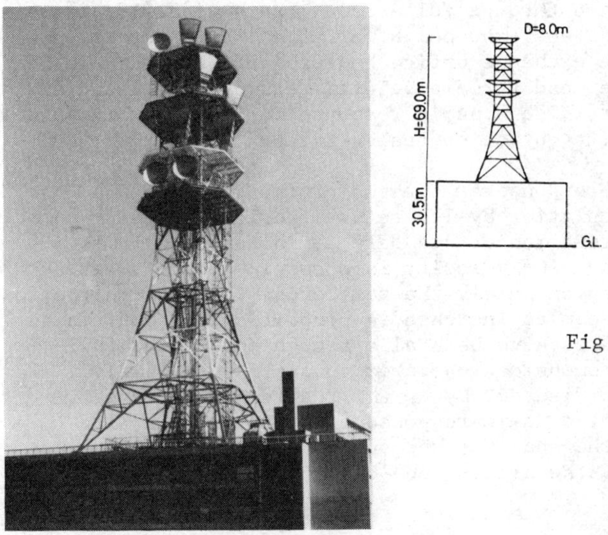

Fig. 15 Karagasaki Tower (n_1=1.1Hz, h_1=1%, α=10°, Iu=0.10-0.15)

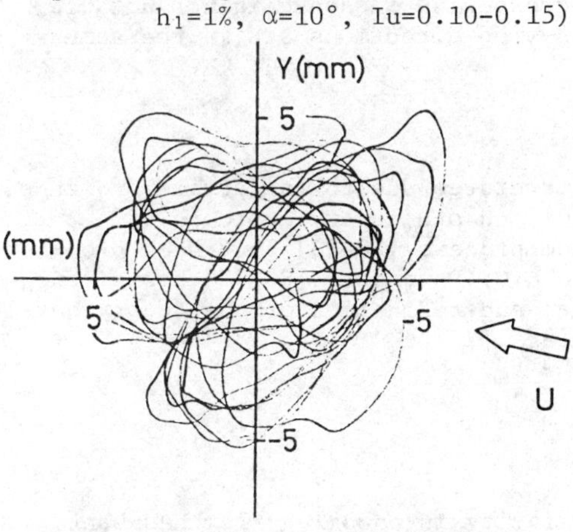

Fig. 16 Locus of the Top of the Tower at U_H=18m/s

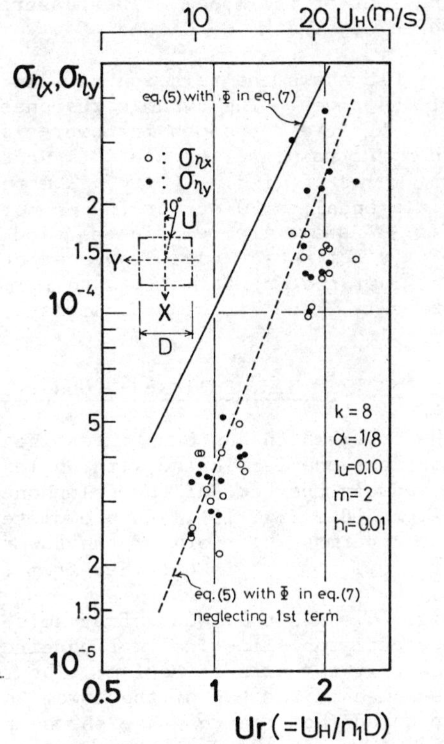

Fig. 17 Relation of Reduced Velocity and Standard Deviation of Dynamic Response Evaluated in 100 seconds

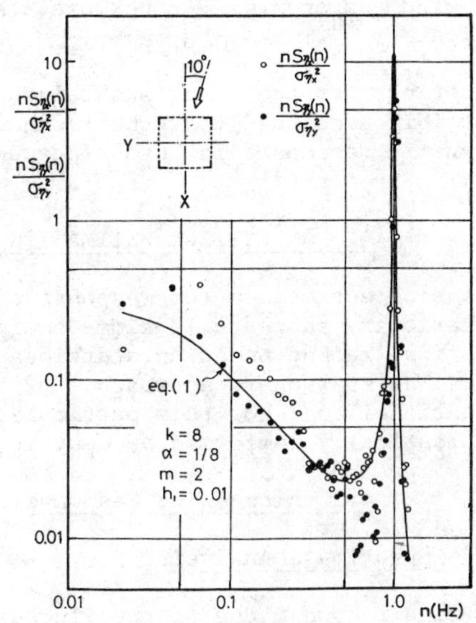

Fig. 18 Power Spectra of Response

the foot of the model. The measurement was done under the condition of the mean angle of attack, $\alpha = 0°$ and 45°.

Fig. 13 shows the relation of wind speed and maximum dynamic response at the top of the tower. The dynamic responses seem to be in the same order in each case, although the responses in tower-axis direction for $\alpha = 45°$ are slightly smaller than the response for $\alpha = 0°$, and they increase monotonically in proportion to the power of wind velocity. The response in cross-wind direction, even when U_r is equal to 20 or more, does not resonante alike in the wind tunnel flow. Fig. 14 shows the relation of wind velocity and gust factor of response (ratio of maximum and average displacement) in along-wind direction in case of $\alpha=0°$. The gust factor is inclined to increase slightly with the increase of wind velocity. The effect of solidity ratio on gust factor is not distinguishable .

3.3 Full-scale Observation

The tower which has been investigated through full-scale observation [17] is a latticed one fabricated with angle steel, the height of which is 69 meters, and stands on the roof of the telephone exchange office building of 30 meters in height (See Fig. 15). An anemometer and three accelerameters are installed on the uppermost platform of the tower. The dynamic response in frequency range from 0.1 to 20 Hz is obtained from the output of analog integration circuit.

Fig. 16 shows the locus of dynamic response at the top of the tower, which indicates an elliptical motion circulating by 1st natural period, shifting succesively the axis of motion. The relation of standard deviation of dynamic response at the top of the tower and wind velocity is shown in Fig. 17. It should be noted that the response shows approximately the same order in either direction of the two axes. Both components seem to increase in proportion to about 2nd power of wind velocity. Such features have been also recognized in the full-scale observation on three-legged communication tower consisted of rigidly connected steel pipes, which was carried out by Machii, Yoshida and others [12-13]. The spectral properties of dynamic response are shown in Fig. 18. Less difference is significant in the shape of the spectrum for both along and cross-wind components. Furthermore, the higher mode of vibration is scarcely observed.

Considering for the results obtained in both model and full-scale tests, it may be possibly concluded that the resonance in cross-wind response is not notable, and dynamic responses in both along and cross-wind directions are in the same order.

4. ANALYTICAL ESTIMATION

As means of estimation for response of the structures due to the action of winds, non-stationary approach in time-domain [18-19] and stationary approach by statistical method based on startionary random process [20-26] have been proposed. The stationary approach attributed to A.G. Davenport [20-22] has given us a rational tool for this particular problem and refinement of the theory has been drastically performed by many investigaters [23-26].

4.1 Along-wind Response

4.1.1 Approximate Analysis

Along-wind response due to the fluctuating velocity in oncoming flow has been formulated by A.G. Davenport [20-22]. It seems that a rational approach for the estimation of along-wind response is now at our hands with not much difficulties. From a viewpoint of practical application, a refined and simplified method is presented by introducing several approximations.

Now, in treating a multi-mass structure as shown in Fig. 19, the following assumptions are introduced;

(1) The vibrational mode of the tower, μ_i is expressed by,

$$\mu_i = Km\ (Z_i/H)^m$$
$$Km = (2m + 1)^{1/2}$$

Where Z_i denotes the height of each mass. And also, the higher order vibration is negligible.

(2) In calculating of crossjoint acceptance [21], projected area and wind force coefficient for each mass are provided constant, and correlation of wind forces within a mass shall be perfect.

(3) The power spectrum, the standard deviation and the intensity of turbulence of velocity fluctuation can be represented by the values at the top of the tower, namely; $Su_H(n)$, σu_H and I_u, respectively. The wind speed profile takes the form of

$$\frac{U_i}{U_H} = \left(\frac{Z_i}{H}\right)^{\alpha}$$

and the coherency of velocity fluctuation, $R_{ij}(n)$ is given by,

$$R_{ij}(n) = \exp\ (-\ k\ n\ |Z_i - Z_j|/U_H\ )$$

Based on the above assumptions, the power spectum, $S\eta(n)$ of non-dimensional dynamic response at the top of the tower, η (=x/D) is obtained straightfowardly, i.e.,

$$S\eta(n) = A_d^2\ U_r^4\ I_u^2\ \frac{Su_H(n)}{\sigma u_H^2}\ |X(n)|^2 J^2(n)\ \ldots\ldots\ldots\ldots\ldots (1)$$

in which

$$U_r = U_H/n_1 D$$
$$A_d = \rho/8\pi^2\rho_s$$

Where U_H, ρ and ρ_s are wind speed at the top of the tower, air density and virtual density of the tower or M_0/A_0D (M_0; total mass of the tower, A_0; total projected area). And furthermore,

$$|X(n)|^2 = [\{1 - (n/n_1)^2\}^2 + 4h_1{}^2(n/n_1)^2]^{-1}\ \ldots\ldots\ldots (2)$$

The crossjoint acceptance, $J^2(n)$ is well fitted as shown in Fig. 20 by eq. (3), i.e.,

$$\left.\begin{aligned} J^2(n) &= \frac{4\ C_0^2\ Km^2}{(m + 1 + \alpha)^2} J_0^2(n) \\ \text{where} \quad J_0^2(n) &= \frac{1}{1 + \dfrac{K\ n\ H}{U_H(m+2)}} \end{aligned}\right\}\ \ldots\ldots\ldots\ldots\ldots\ldots\ldots\ldots (3)$$

in which C_0 denotes reference wind coefficient [27].

Applying a well-known empirical formula [28] for the power spectrum of wind velocity, i.e., as L is a scale length,

$$\frac{nS_u(n)}{\sigma_u^2} = \frac{2}{3} \frac{X^2}{(1 + X^2)^{4/3}}, \quad X = \frac{n\,L}{U} \quad \ldots\ldots\ldots\ldots\ldots \quad (4)$$

the standard deviation of dynamic response at the top of the tower, $\sigma\eta$ and the gust factor of response, G can be written as follows by introducing some approximations.

$$\sigma\eta = [\textstyle\int_0^\infty S\eta(n)\, dn]^{1/2} = A_d\, U_r^2\, I_u \frac{2\, C_0\, K_m}{m + 1 + \alpha}\, \Phi \quad \ldots\ldots \quad (5)$$

$$G = 1 + 2\, g\, I_u\, \Phi \quad \ldots\ldots\ldots\ldots\ldots\ldots\ldots\ldots\ldots\ldots\ldots\ldots \quad (6)$$

in which g and Φ are peak factor and response factor, respectively. Φ is approximately given by function of three parameters λ_1, λ_2 and λ_3, and reduced velocity, U_r.

$$\Phi = [1 + \lambda_1 \frac{(\lambda_2\, U_r)^{2/3}}{1 + \lambda_3/U_r}]^{1/2} \quad \ldots\ldots\ldots\ldots\ldots\ldots\ldots\ldots \quad (7)$$

where

$$\lambda_1 = \frac{1}{2h_1}, \quad \lambda_2 = \frac{D}{L_H} \simeq \frac{D}{1200}, \quad \lambda_3 = \frac{H}{D} \cdot \frac{k}{m + 2}$$

4.1.2 Verification of the Presented Theory

With the analytical formulas obtained in the above section, comparison is made between the analytical and experimental results. The power spectrum of along-wind response based on eq. (1) with eq. (2), (3) and (4) is shown in Fig. 18. Analytical curve seems to be well fitted with the experimental data. The standard deviation of along-wind response based on eq. (5) using Φ in eq. (7) is also shown in Fig. 17. Since the contribution from the components, the frequency of which is less than 0.1 Hz, is neglected in the experiment, the analytical estimation indicated by full line exceeds. On the other hands, the curve indicated by dotted line with the neglection of 1st term in eq. (7) gives a better agreement with the experimental data. It should be noted that $\sigma\eta \propto U_r^2$ when $U_r \ll 1$, and $\sigma\eta \propto U_r^{7/3}$ when $U_r \gg 1$ from eq. (7).

Fig. 21 shows the relation of Φ and U_r based on eq. (7). It is seen that Φ slightly increases with the increase of wind velocity, and the more the tower width and/or the less aspect ratio (H/D), the more the response factor increases. The gust factor of response, G by eq. (6) using Φ of eq. (7) is plotted in Fig. 14. The characteristics of the experimental values are said to be well explained.

4.2 Cross-wind Response

It can be said that theoretical studies for cross-wind response have been mostly dedicated to the one related to resonance due to aeolian vibration [29]. The resonance, however, was not at all recognized in this kind of structures as was mentioned in Chapter 3. The cross-wind response, on the contrary, has much similar features with along-wind response. A quasi-steady method [22] had been suggested to estimate cross-wind response pointing out a possible effect of aerodynamic damping and coupling of the orthogonal motions. According to the calculated results by the authors [17] based on this method, the standard deviation of cross-wind response came out to be very small compared with along-wind response (Fig. 22). In fact, the effects of aerodynamic damping and coupling was found very little, and the theory completely failed in rational explanation of the experimental findings, for both responses being rather comparable as was discussed in Chapter 3.

It is said that a rational interpretation for cross-wind response of the

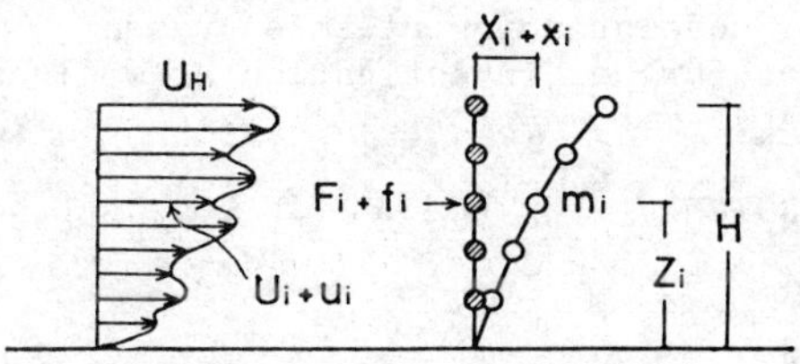

Fig. 19 Analytical Model

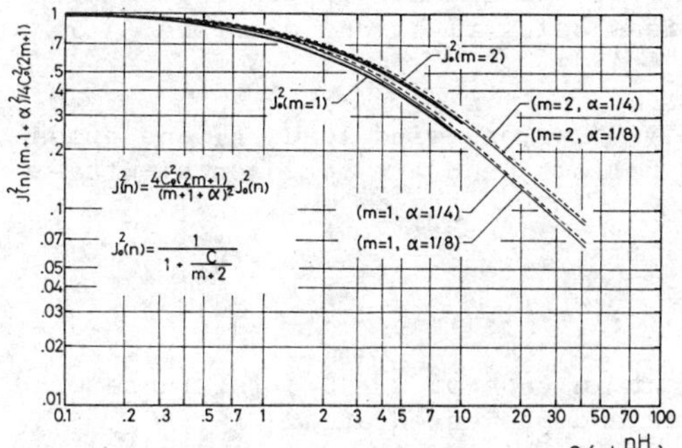

Fig. 20 Joint Acceptance Function

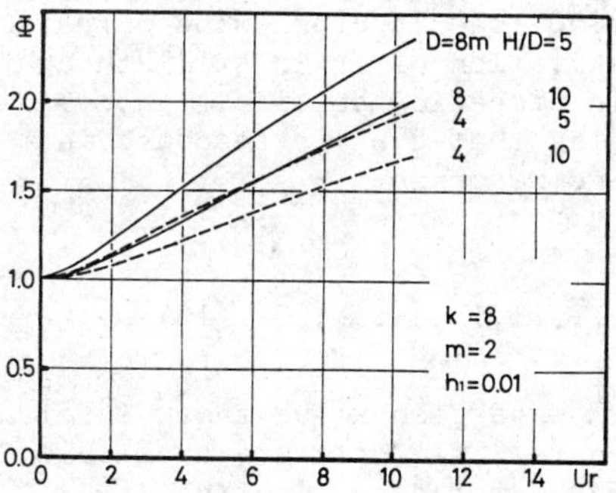

Fig. 21 Relation of Φ and U_r

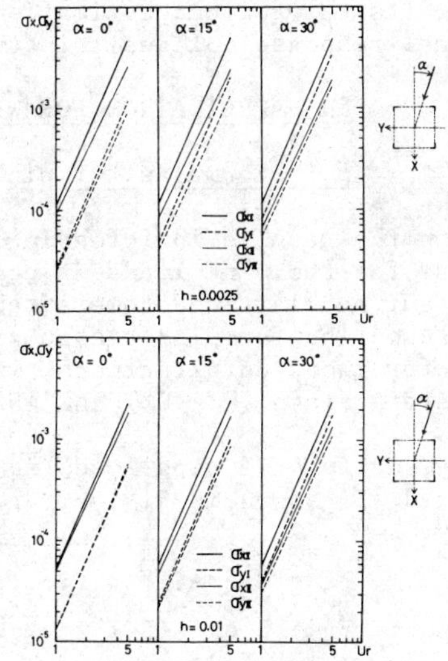

Fig. 22 Calculated Response (Aerodynamic Damping and Coupling are Considered in Theory II and not in Theory I)

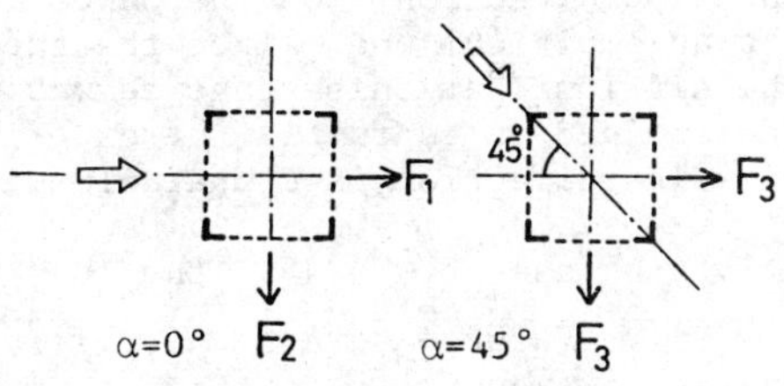

Fig. 23 Wind Loads F_1, F_2 & F_3

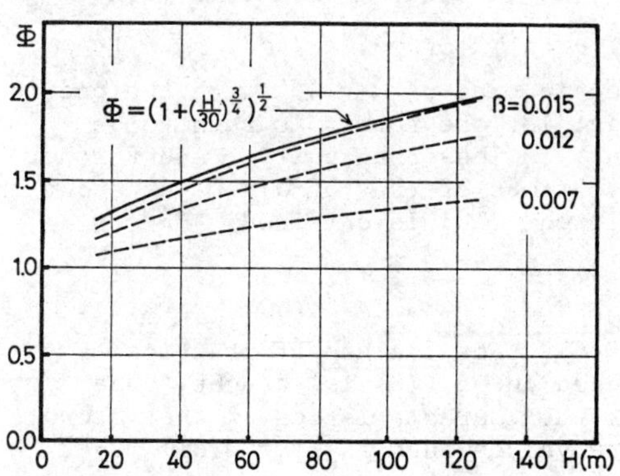

Fig. 24 Approximate Curves for Φ

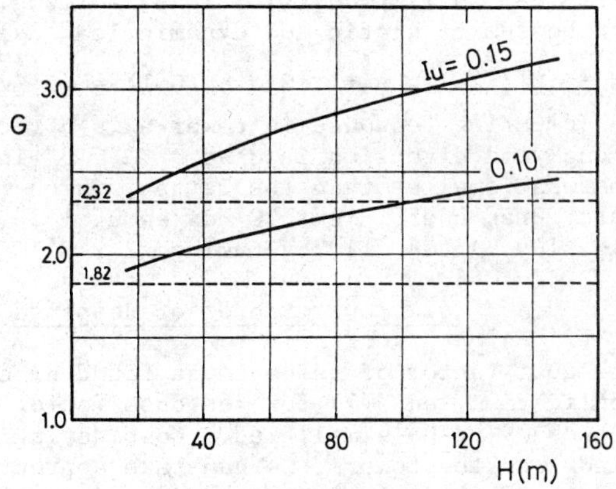

Fig. 25 Proposed Gust Factor

structures in general is still far beyond our understanding at this present stage. Its theoretical estimation is left for further investigation as well as torsional response not dealt in this paper.

5. APPLICATION FOR WIND-PROOF DESIGN

5.1 Formulas for Wind Loads

Based on the results in foregoing chapters presented herein are calculation formulas for the wind loads in consideration of dynamic effects of winds against square surface latticed towers fabricated by structural angle steel without wires and lines. The formulas are so arranged to meet for the conventional procedure [2, 4] for two wind directions; wind directions being right angle to the structure surface ($\alpha = 0°$) and 45 degree ($\alpha = 45°$).

Firstly, as for the wind loads in case of $\alpha = 0°$, along-wind load, F_1 and cross-wind load, F_2 shown in Fig. 23 are taken as follows, and are acted on simultaneously.

$$F_1 = C\,q\,A\,G \qquad (8)$$

$$F_2 = C\,q\,A\,(G-1) \qquad (9)$$

in which C, q and A are wind force coefficient in case of $\alpha = 0°$, reference wind pressure based on expected average wind speed and projected area.
The discussion for the value of gust factor of response, G is given in the following section. The wind load for along-wind direction, F_1 is theoretically given as an equivalent load inducing an equivalent displacement with the maximum one in along-wind direction. On the other hands, the wind load in cross-wind direction, F_2 is an empirical one, since the theoretical estimation of it is still thought to be difficult at this stage as was mentioned in preceding section. Eq. (9) is so determined as to fulfil the experimental finding that dynamic response in cross-wind direction is comparable with along-wind response.

Nextly, in case of $\alpha = 45°$, the load F_3,

$$F_3 = \frac{C'}{\sqrt{2}}\,q\,A\,G \qquad (10)$$

in which C' denotes drag coefficient in case of $\alpha = 45°$, shall be taken in either tower-axis direction as shown in Fig. 23. In general, few data are available in relation to response in case of $\alpha = 45°$. The reason for determination of empirical form of eq. (10) is as follows. Let us consider for F_3, for instance, by the sum of static and dynamic load using γ as an unknown parameter, i.e.,

$$F_3 = (C'/\sqrt{2}\,)\,q\,A + \gamma\,C\,q\,A\,(G-1) = (C'/\sqrt{2}\,)\,q\,A\,\{1 + (\sqrt{2}\,\gamma\,C\,/\,C')\,(G-1)\}$$

In comparing response in tower-axis direction in case of $\alpha = 45°$ with the one in along-wind direction in case of $\alpha = 0°$ in Fig. 13, the former is seen to be somewhat smaller than the latter, which mean γ is less than unity. And furthermore, drag coefficient, C' is about 1.3 or so times of that of $\alpha = 0°$ as shown in Fig. 10. After all, by taking $\sqrt{2}\,\gamma\,C'/C \simeq 1$, eq. (10) is obtained.

5.2 Gust Factor of Response

The gust factor of response is found by eq. (6). For the use of practical application, eq. (7) for response factor Φ used in eq. (6) is thought to be necessary to be simplified. Considering for the characteristics of the natural winds and the towers, we can take appropriate values such as $h_1 = 0.01$, $m = 2$ and $k = 8$, and also the relation,

$$n_1 = 1/\beta H, \ \beta = 0.015$$

By doing so, eq. (7) can be written in the following form as a function of only U_H and H.

$$\Phi = [1 + \frac{0.27\ U_H{}^{5/3} H^{2/3}}{(U_H + 133)}]^{1/2} \quad \text{.......................} \quad (11)$$

Furthermore, taking for instance as U_H [30],

$$U_H = 40\ (H/10)^{1/8}$$

the curves shown in Fig. 24 are obtained, and with some approximation Φ can be rewritten by a rather simpler form;

$$\Phi = [1 + (\frac{H}{30})^{3/4}]^{1/2} \quad \text{................................} \quad (12)$$

The peak factor, g is found by the largest peak value theory of stationary random process [31]. Taking account for the time duration of 240 seconds as a proper time duration for realization of stationality [32] and a typical natural frequency of 1 Hz, an appropriate value for g is suggested by 3.5 for convenience.

Fig. 25 shows the gust factor of response based on eq. (6) using Φ in eq. (12). The gust factor increases gradually together with the increase of height of the tower. In order to show dynamic effects in foregoing formulation, the curves for G_u^2 , as a quasi-static response based on maximum instantaneous wind velocity, are indicated with the dotted line for both $I_u = 0.10$ and 0.15 [33] when the gust factor of wind, G_u is given by,

$$G_u = 1 + 3.5\ I_u \quad \text{..................................} \quad (13)$$

$G_u{}^2$ is found always to be less than G, which suggests that the conventional method for wind load we have been using for the practice has possibly given a somewhat smaller value.

6. CONCLUSION

On latticed steel towers with square section using angle steel, which are the most general microwave telecommunication towers of Nippon Telegraph and Telephone Public Corporation, we practiced a full-scale observation at actual tower and model tests in the field and in wind tunnel. Moreover, we performed the response analysis assuming stationary random process and studied its resonableness in comparison with the experimental results. And, being based on those results, we proposed calculation formulas for wind loads considering for dynamic effect of winds toward this kind of structures.

The conclusions obtained from the study are summarized as follows;

(1) In the experiments, cross-wind response of the towers is in the same magnitude with along-wind response, and the resonance in specific wind speed is not recognized. And also, even if the angle of attack and the solidity ratio are changed, the response characteristics do not vary so much.

(2) The along-wind response of the towers can be explained fully from the response estimation formulas introduced in this paper with some approximation.

(3) From the above two, the design wind loads at each parts of the towers is possibly calculated from suggested formula (8), (9) and (10) using eq. (6) and (12).

We are prescheduled to keep on the study on account of further verification of the findings, an expansion of the application of the formulas and understanding of torsional response of towers.

REFERENCES

(1) Japan Building Standard Law Enforcement Order, Proclaimed in 1950

(2) Architectural Institute of Japan; Standard for Structural Calculation of Steel Tower, 1962

(3) Japan Electrotechnical Committee; Design Standard for Steel Tower for Power Transmission, 1965

(4) Architectural Institute of Japan; Standard for Loads on Buildings (Draft), 1975

(5) Kamei, I., Kimura, E. and Matsushita, I.; Experimental Study of Wind Force Coefficient of SHF Antennas for Telecommunication Tower Design, Proc. 3rd Int. Conf. on Wind Effects on Buildings and Structures, Tokyo 1971 pp. 357-364

(6) Kimura, E. and Konno, T.; Earthquake Effects on Steel Tower for Microwave Antennas atop Building, Review of the Electrical Communication Laboratories, NTT, Vol. 25, No. 3-4, 1975 pp. 301-315

(7) Shichiri, Y.; Wind Withstanding Design for Transmission Line Tower, Proc. 3rd Int. Conf. on Wind Effects on Buildings and Structures, Tokyo 1971 pp. 451-456

(8) Ishizaki, H., Ishida, A. and Kawamura, S.; On the Full Sized Model Test of Transmission Tower Destroyed by a Typhoon (Part 2), Trans. Architectural Inst. of Japan, No. 82, 1963 pp. 29-33

(9) Maeno, Y. and Takano, H.; Dynamic Behavior of a 500kV Transmission Tower under Gusty Winds, Proc. 3rd Int. Conf. on Wind Effects on Buildings and Structures, Tokyo 1971 pp. 1185-1194

(10) Hiroki, M.; Study of the Wind Pressure Fluctuations in the Strong Winds and the Dynamical Effects (Part 1), Trans. Architectural Inst. of Japan, No. 175, 1970 pp. 39-48

(11) Murota, T. and Ishizaki, H.; Deformations and Vibrations of Some Actual Structures due to Wind, Proc. 3rd Int. Conf. on Wind Effects on Buildings and Structures, Tokyo 1971 pp. 1157-1166

(12) Machii, W. and Sanada, H.; Study on Antiseismic and Wind-proof Property of a High-rise tower (Part 5) (in Japanese), Proc. Annual Conf. Architectural Inst. of Japan, 1977 pp. 939-940

(13) Yoshida, M., Nakamura, O. and Yoshida, H.; Ibid (Part 6) (in Japanese), Proc. Annual Conf. of Architectural Inst. of Japan, 1977 pp. 941-942

(14) Ishizaki, H. and Nakamura, S.; Dynamic Test of Steel Tower Model in Wind Tunnel (in Japanese), Trans. Architectural Inst. of Japan, No. 89, 1963 pp. 87

(15) Fujimoto, M., Akagi, H., Ohkuma, T., Amano, T., Makino, T. and Suzuki, T.; Wind Tunnel Test for Wind Effects on Steel Tower for Antennas (in Japanese), Proc. Annual Conf. of Architectural Inst. of Japan, 1976 pp. 813-814

(16) Fujimoto, M., Sugimoto, Y., Amano, T. and Makino, T.; Experiment on the Dynamic Behavior of the Telecommunication Tower due to the Action of the Wind (in Japanese), Proc. Annual Conf. of Architectural Inst. of Japan 1975 pp. 625-626

(17) Fujimoto, M., Ohkuma, T., Matsushita, I., Amano, T. and Makino, T.; Study on the Vibration of Telecommunication Tower due to the Wind (in Japanese), Proc. Annual Conf. of Architectural Inst. of Japan, 1977 pp. 943-944

(18) Chiu, A. N. L., Sawyer, D.N. and Grinter, L.E.; Vibration of Towers as Related to Wind Pulses, J. Structural Div. Proc.ASCE, Vol. 90, No. ST5, 1964 pp. 137-160

(19) Chiu, A. N. L. and Taoka, G. T.; Tower Response of Actual and Simulated Wind Forces, J.Structural Div. Proc. ASCE, Vol. 99. No. ST9, 1973 pp. 1011-1930

(20) Davenport, A. G.; The Application of Statistical Concepts to the Wind Loading of Structures, Proc. Inst. of Civil Engrs. London, Vol. 19, 1961 pp. 449-472
(21) Davenport, A. G.; Gust Loading Factors, J.Structural Div. Proc. ASCE, Vol. 93, No. ST3, 1967 pp. 11-34
(22) Davenport, A.G.; The Treatment of Wind Loading on Tall Building, Proc. Symp. on Tall Buildings, Univ. of Southampton 1967
(23) Vickery, B. J.; Load Fluctuations in Turbulent Flow, J. Engng. Mech. Div. Proc. ASCE, Vol. 94, No. EM1, 1968 pp. 31-46
(24) Ishizaki, H. and Kawai, H.; An Investigation of the Pressure Fluctuations on the Windward Faces of Square Prisms, Disaster Prevention Research Institute Annuals, No. 18B, Kyoto Univ. 1975 pp. 181-188
(25) Bearman, P. W.; Wind Loads on Structure in Turbulent Flow, The Modern Design of Wind-sensitive Structures, Proc. Seminor held at the Inst. of Civil Engrs. London 1971
(26) Ohkuma, T.; Buffeting Analysis óf Multi-mass System, Proc. Annual Meeting of Kanto District of Architectural Inst. of Japan, No. 43, 1972 pp. 301-304
(27) Fujimoto, M. and Ohkuma, T.; Theoretical Study on Oscillatory Tendencies of Cylindrical Structures in Natural Winds (Part 2), Trans. Architectural Inst. of Japan, No. 187, 1971 pp. 59-68
(28) Davenport, A. G.; The Spectrum of Horizontal Gustiness near the Ground in High Wind, Quart. J. Royal Meteorological Soc. London, Vol. 87, 1961 pp. 194-211
(29) Otsuki, Y., Washizu, K., Fujii, K., Ito, T., Tomizawa, H. and Yoshida, M.; Wind-excited Vibration of a Tower of Neary Square Cross Section, Proc. 3rd Int. Conf. on Wind Effects on Buildings and Structures, Tokyo 1971 pp. 853-859
(30) Nakahara,M. and Saito, R.; On the Estimation of the Maximum Mean Wind Speeds Likely to be Exceeded Only Once in 50 years, Trans. Architectural Inst. of Japan, No. 125, 1966 pp. 1-7
(31) Davenport, A. G.; Note on the Distribution of the Largest Value of a Random Function with Application to Gust Loading, Proc. Inst. of Civil Engrs. London, Vol. 28, 1964
(32) Fujimoto, M., Ohkuma, T., Amano, T., Sugimoto, Y. and Akagi, H.; Pressure Measurement at Yokosuka Electrical Communication Laboratory Building (Part 3), Proc. Annual Conf. of Architectural Inst. of Japan, 1977 pp. 935-936
(33) Shiotani, M.; Turbulent Properties of Strong Winds, Proc. 3rd Symp. on Wind Effects on Structures in Japan, 1974 pp. 9-18

A COOLING TOWER WIND LOADING MODEL BASED ON FULL-SCALE DATA

L. J. FORTIER and R. H. SCANLAN
Princeton University, Princeton, New Jersey, USA

Summary

During 1973 and 1974 wind pressure-difference data were taken across the shell at the throat (height 313 ft.) of a new, non-operating hyperboloid cooling tower at Martin's Creek, Pennsylvania. Particular interest in the present paper centers on a digitized data set consisting of the time history at each of 16 pressure-difference sensors recorded during a storm of mean wind velocity 41 mph.

After a brief outline serving both to interpret the data in the context of a tower submerged in the earth's boundary layer and to suggest how to scale such results for other applications, ARIMA statistical methods are used first to fit, and subsequently to model time-varying pressures around the tower. Rather simple AR models quite satisfactorily represent the pressure loadings in an economical 2-parameter manner and, further, suggest insights into the aerodynamic mechanisms operative around the full tower periphery.

Symbols:

ρ	=	pressure
U	=	wind velocity
V	=	local flow velocity
ρ	=	air density
C_p	=	pressure coefficient
$u(t)$	=	fluctuating wind velocity
$v(t)$	=	fluctuating local flow velocity
u_*	=	friction velocity of boundary layer
$k \cong 0.4$	=	Kármán constant
Z	=	height above ground
I	=	turbulence intensity
$\hat{f}, K$	=	dimensionless frequency parameters
S	=	spectrum

n = frequency

C = dimensionless correlation constant

η = deterministic or average pressure

e = stochastic value

z = value of digitized signal

B = backward shift operator

∇ = forward shift operator

$\psi_1, \psi_2, \ldots \psi_p$ = autoregressive coefficeints

$\theta_1, \theta_2, \ldots \theta_q$ = moving average coefficients

N = total number of digitized data items

ρ_k = theoretical value of autocorrelation function at lag k

r_k = actual value of autocorrelation function at lag k

Subscripts and Superscripts

$(\ )_\infty$ = "at infinity"

$(\ )_o$ = "outer surface"

$(\ )_i$ = "inner surface"

$\overline{(\ )}$ = "time averaged" or "steady value"

$(\ )_{max}$ = "maximum value of"

$(\ )_{st}$ = "stagnation value of"

$(\ )_u$ = "upper"

$(\ )_\ell$ = "lower"

$(\ )_{AV}$ = "space average"

$(\ )_j$ = "for position j around tower perimeter"

$(\ )_t$ = "digitized value at time count t"

Introduction

Limited but useful data were reported [1], [2], from the 1973-1974 wind pressure study on the cooling tower at Martin's Creek, Pennsylvania. The data consist of a number of sets of 16 simultaneous pressure-difference time histories obtained by through-shell transducers spaced at about 22.5° around the throat of the tower at 313 ft. elevation. One particular data set (No. 4031), representing 3 minutes of a storm with wind speed averaging 41 mph, is selected here for examination. This data set, originally digitized at 1/2 second intervals, forms the main basis of a study in simulation wherein new time histories are created through use of the methods of parametric time series.

The problem of simulating tower wind loading through the use of such procedures is not fully treated in what follows, but it is considered useful here to present a number of its underlying concepts.

Basic Pressure-Velocity Relationships

The analysis below is based on the idea that a neutrally stratified atmospheric boundary layer flow approaches a free-standing cooling tower. Though this flow is actually turbulent, it is assumed that Bernoulli's relation remains applicable to the average streamlines of the flow.

In this context, the following relations are basic to the analysis:

(1) $$p_\infty + \frac{1}{2}\rho U_\infty^2 = p_o + \frac{1}{2}\rho V^2 = p_{st}$$

(2) $$\bar{p}_o - p_\infty = \frac{1}{2}\rho \bar{U}_\infty^2 C_{po} .$$

Assuming for the present purposes that upstream pressure p_∞ and tower inner pressure p_i are constant, and that

(3) $$U_\infty = \bar{U} + u(t)$$

(4) $$V = \bar{V} + v(t)$$

(5) $$p_o = \bar{p}_o + p(t) ,$$

where the squares of the time-varying portions are negligible with respect to those of the steady portions, leads to some simple deductions; in particular:

(6) $$p_{st}(t) = \rho\bar{U}u(t)$$

(7) $$\bar{V} = \bar{U}\,[1-C_{po}]^{1/2}$$

(8) $$p_o(t) = \rho\bar{U}[u(t) - (1-C_{po})^{1/2}\, v(t)]$$

Steady Pressure Loading

In view of the fact that

(9) $$\bar{p}_o - \bar{p}_i = \frac{1}{2}\,\bar{U}^2(C_{po} - C_{pi}) ,$$

the value of C_{po} can be obtained from experimental $(p_o - p_i)$ data through the formula

(10) $$C_{po} = (1 - C_{pi}) \left[\frac{\bar{p}_o - \bar{p}_i}{(\bar{p}_o - \bar{p}_i)_{max}} \right] + C_{pi} .$$

Assuming further that $C_{pi} \cong -0.4$ permits numerical evaluation of C_{po} around the tower from experimental pressure-difference data (cf. [1], [3]).

Wind-Descriptive Equations

The horizontal mean-wind velocity $U(Z)$ at altitude Z is given [4] by

(11) $$\bar{U}(Z) = \frac{1}{k}\,u_*\,\ell n\,\frac{Z-Z_d}{Z_o} .$$

The value $Z_d \cong 0$ is used in open terrain. Wind velocity at height Z is related to that at the throat height Z_{th} by

$$\bar{U}(Z) = \frac{\ell n(Z/Z_o)}{\ell n(Z_{th}/Z_o)} \bar{U}(Z_{th}) \quad . \tag{12}$$

The turbulence intensity (cf. [4]) is given by

$$I(Z) = \frac{\overline{u^2(t)}^{1/2}}{\bar{U}(Z)} \cong \frac{1}{\ell n(Z/Z_o)} \; . \tag{13}$$

Note that it is independent of wind velocity. As a result of (13) the rms stagnation pressure is:

$$\overline{p_{st}^2(Z,t)}^{1/2} = \frac{\rho \bar{U}^2(Z)}{\ell n(Z/Z_o)} \quad . \tag{14}$$

The value of $\overline{p_{st}^2(Z,t)}^{1/2}$, as calculated on the windward meridian from the experimental data used herein, compares well with the theoretical value given by (14). (The value $Z_o = 1$ ft. was used, with $\bar{U}(Z_{th}) = 60.35$ ft/sec.)

Although by Eq. (8) the external fluctuating pressure depends both on $u(t)$ and $v(t)$, it may be reasonable in further applications to scale experimental rms pressure levels by assuming that

$$\overline{p_j^2(z,t)} = \alpha_j \; \overline{p_{st}^2(Z,t)}^{1/2} \tag{15}$$

with α_j a constant matched to the data, where p_j is a local pressures. A relation of this type is needed to set the level of newly designed time histories $p_j(Z,t)$.

Finally, the question of vertical correlation of fluctuating pressures should be mentioned. If Z_1 and Z_2 are two levels on the tower, the cross co-spectrum between fluctuating velocity components is given [5] by:

$$S_{12}(n) = S^{1/2}(Z_1,n)\, S^{1/2}(Z_2,n)\, e^{-\hat{f}} \tag{16}$$

with

$$\hat{f} = \frac{nC|Z_1 - Z_2|}{\frac{1}{2}\,[\bar{U}(Z_1) + \bar{U}(Z_2)]} \tag{17}$$

and $C \cong 10$. Double integration of (16) over the height H yields approximately

$$S_p(n) = S^{1/2}(Z_u,n)\, S^{1/2}(Z_\ell,n)\, R(n) \tag{18}$$

where $|Z_u - Z_\ell| = H$ and

(19) $$R(n) = \frac{2(K-1+e^{-K})}{K^2}$$

with

(20) $$K = \frac{nCH}{\bar{U}_{AV}} ,$$

$\bar{U}_{AV}$ being the average velocity over height H .

By (6) and (18) - (20) the coherence of fluctuating wind pressures falls off with frequency. Table A lists representative values of K and R .

TABLE A

K	0	0.1	0.2	0.5	1.0	2.0	5.0	10.0	20.0	50.0	100.0
R	1	0.97	0.94	0.85	0.74	0.57	0.32	0.18	0.10	0.04	0.02

Even at frequencies as low as 0.02 Hz, the loss of coherence over a 400 ft. tower at wind velocities of interest can be appreciable. This suggests the eventual need to modify simulated pressure effects with height. The Martin Creek data were insufficient to shed light on this question, however. Space does not permit pursuit of this topic here.

The above basic ideas serve as background to the problem of fluctuating wind pressure simulation over a tower. We now turn to the specifics of the simulation process as derived from real pressure data at one tower level.

The Methodology of Parametric Time Series Analysis

The basic aim of this paper is to show how economical models of time series representing fluctuating pressures around a cooling tower can be constructed for eventual use as dynamic loading models. To this end it is first useful to discuss some general characteristics of the approach used.

Parametric time series analysis has been employed in a variety of digitized applications and forms an integral part of the statistical sciences. Refs. [6] and [7] are borrowed from freely in the discussion that follows. The time series whose properties are in question here are the basic digitized pressure data originally obtained from the Martin's Creek project at the single throat level of that tower.

Broadly speaking, the method considers a given time series such as a pressure fluctuation as the result of random shocks acting upon a dynamic system, the detailed mechanism of which may not yet be known. The problem is to identify the system, through knowledge of its output, in the most economical way. Once this is accomplished, random shocks applied to the identified system reproduce a simulated output, i.e. they "forecast" system behavior.

Consider the output of a pressure sensor as a series of N observations $\tilde{p}$ at times t . The terms of the series are depicted as the sum

(21) $$\tilde{p}_t = \eta_t + e_t \qquad (t=1,2,\ldots,N)$$

where η_t is the response function under study and e_t is a disturbance. Two submodels are postulated for η_t and e_t : η_t is a deterministic, and e_t a stochastic model. It will be assumed that η_t represents the average pressure

(22) $$\eta_t = \frac{1}{N}\sum_{1}^{N} \tilde{p}_t$$

Attention is then directed to the discovery of the model for e_t.

Let z_t be the value at time t of a time series representing any quantity. The ARIMA models for z_t are fully described in the given references. A model may be "autoregressive" (AR) of order p or "moving average" (MA) of order q and may operate on the dth difference of the data. The general "autoregressive, integrated, moving average" (ARIMA) models are then classified by the integers (p,d,q). The "backward shift operator" B is defined by $Bz_t = z_{t-1}$; hence $B^m z_t = z_{t-m}$. The "backward difference operator" ∇ is defined by $\nabla z_t = z_t - z_{t-1}$; hence $\nabla^m z_t = z_t - z_{t-m}$ or, in terms of B: $\nabla^m z_t = (1-B^m)z_t$.

An autoregressive model of order 1 (AR(1) model) is defined as

(23) $$z_t = \psi_1 z_{t-1} + a_t \qquad (t=1,2,\ldots,N)$$

where ψ_1 is the autoregressive coefficient and a_t is a "shock"; a_t is independent for each t and comes randomly from a Gausian distribution with zero mean and fixed variance σ_a^2. An autoregressive model of order 2 (AR(2) model) is defined as

(24) $$z_t = \psi_1 z_{t-1} + \psi_2 z_{t-2} + a_t \qquad (t=1,2,\ldots,N)$$

where ψ_1,ψ_2 are the autoregressive coefficients. An AR(p) model takes the form

(25) $$z_t = \psi_1 z_{t-1} + \psi_2 z_{t-2} + \ldots + \psi_p z_{t-p} + a_t \quad .$$

The first order moving average (MA(1)) model has the form

(26) $$z_t = z_t + \theta_1 a_{t-1}$$

where θ_1 is the first order moving average coefficient. The second order moving average (MA(2)) model has the form

(27) $$z_t = a_t + \theta_1 a_{t-1} + \theta_2 a_{t-2} \quad .$$

The qth order MA(q) model is

(28) $$z_t = a_t + \theta_1 a_{t-1} + \theta_2 a_{t-2} + \ldots + \theta_q a_{t-q} \quad .$$

It is possible to construct combined (ARIMA) models where first, second,...,d^{th} differences constitute the "integrated" portion.

In identifying a model to fit a given time series of data, the properties of it that lend strong insight are its spectrum, autocorrelation and partial autocorrelation functions. The spectrum offers frequency content information, and the autocorrelation (being related as a Fourier transform) reveals this same information in a certain explicit manner. It is the autocorrelation that is prominently used in the present context. It is defined as

$$r_k = \frac{\sum_{t=k}^{N} (z_t \cdot z_{t-k})/(N-k)}{\sigma_z^2} \tag{29}$$

This approaches the continuous theoretical definition ρ_k as $N \to \infty$.

An AR(2) process will exhibit an autocorrelation function dropping off slowly as k increases and approaching zero. An AM(2) process on the other hand, will exhibit an autocorrelation function with two nonzero values. In an AR(2) process, each value of z_t is dependent on all earlier values, while in an MA(2) process, each value depends only on the two previous values.

Figure 5 shows the autocorrelation function calculated for $p_o(t)$ at each of the three sensor locations from the basic data. These plots suggest autoregressive processes and attention is therefore focused on AR models.

The pressure data from Martin's Creek prove to be adequately modeled by AR(1) or AR(2) models. For p=2, the so-called Yule-Walker equations of the ARIMA theory yield

$$\psi_1 = \frac{\rho_1(1-\rho_2)}{1-\rho_1^2} \tag{30}$$

$$\psi_2 = \frac{\rho_2-\rho_1^2}{1-\rho_1^2} \tag{31}$$

Further, the partial autocorrelation functions ψ_{11} and ψ_{22} are

$$\psi_{11} = \rho_1 \tag{32}$$

$$\psi_{22} = \frac{\rho_2-\rho_1^2}{1-\rho_1^2} \tag{33}$$

Insertion of the actual autocorrelation results r_1, r_2 from real data into (30) (31) in place of ρ_1, ρ_2 will yield approximate estimates of the ψ .

The form taken by the theoretical spectrum associated with an AR(2) model is

$$S(n) = \frac{2\sigma_a^2}{|1-\psi_1 e^{-2n\pi i} - \psi_2 e^{-4n\pi i}|^2} \tag{34}$$

and its variance is

$$\sigma_z^2 = \frac{\sigma_a^2}{1 - \rho_1\psi_1 - \rho_2\psi_2} \tag{35}$$

Modeling Martin's Creek Data

Figure 1 is a sketch of the Martins Creek tower geometry. Figure 2 presents original fluctuating pressure data from three representative locations, sensor 7 (essentially windward), sensor 10 (low point of steady pressure distribution), and sensor 15 (wake). These are given to arbitrary scale. It is recalled that the data were originally presented as digitized at 0.5 sec. intervals or a frequency of 2 Hz. Examination of the spectra (suggested in Figure 5) reveals that no strong contributions exist beyond 0.5 Hz, whereas the very important ones occur at much lower frequencies. It was decided therefore to sample only at 1 second intervals or at 1 Hz, implying that at most, 0.5 Hz frequency content will be included. This decision, in fact, influenced the facility with which AR(2) models were later fitted to the data. (There remains no doubt in the authors' minds that in addition to the direct and influential roles played by information garnered from the spectra, autocorrelation, and partial autocorrelation functions, judgment and experience remain key factors in assessing AR models.)

Table B presents the values of ψ_1 and ψ_2 established for the AR(2) models at 16 sensors spaced at 22½° intervals around the tower.

TABLE B

Sensor	ψ_1	ψ_2	Sensor	ψ_1	ψ_2
1	0.394	0.040	9	1.105	-0.213
2	0.684	0.119	10	1.404	-0.462
3	1.508	-0.614	11	1.198	-0.313
4	1.280	-0.345	12*	0.418	0.237
5	1.120	-0.226	13	0.890	-0.170
6	1.270	-0.364	14	0.926	-0.119
† 7	1.280	-0.341	15	0.836	-0.212
8	1.220	-0.285	16	0.284	0.185

† windward *Probably malfunctioning

Plots around the throat periphery of the values from Table B and similar values from 3 other data sets treated analogously are presented in Figure 3, (a) and (b). To emphasize the spread of ψ_1 and ψ_2 values, the same results are plotted to the right of the same figure as bands of varying width at points around the tower circumference. These plots reveal strikingly symmetrical patterns and suggest that there are well-defined and continuously changing mechanisms responsible for the varying states of turbulence as the flow progresses around the tower. These results merit further investigation into the aerodynamic mechanisms which will not be pursued here. It is noted that each pair (ψ_1, ψ_2) in Table B satisfies the stationarity conditions required by the theory.

At the bottom of Figure 2 are graphs of the steady external pressure coefficient (which compare well with those given in Ref. [4]) and of the dimensionless rms of fluctation pressure, fluctuating pressure coefficient vs. angle around the tower from windward. The latter results are compared with some due to Niemann [3] (cited in Ref. [4]). Present results on the windward (0°) meridian differ markedly from other literature values; but on the other hand they coincide with the predictions of the theory cited earlier in the present paper (Eqs. 6, 14), wherein stagnation pressure fluctuations are considered to be directly proportional to wind gust fluctuations. In the wake position the present pressure fluctuations are also greater than those observed by Niemann.

Newly Modeled Time Histories

According to the theory discussed earlier, a new time history for any digitally defined pressure series p_t can be generated as an AR(2) process using the definition

$$p_t = \psi_1 p_{t-1} + \psi_2 p_{t-2} + a_t \quad . \tag{4}$$

This has been done in the present instance, and Figure 4 depicts results, analogous to those in Figure 2, and to the same scale, for sensor locations 7, 10, and 15. These preserve the charactersitic appearance of each original time history.

Finally, Figure 5 presents sketches of the autocorrelation functions (on left) and corresponding spectra (on right) calculated from the basic data, for sensors 7, 10 and 15. Superposed on the sketches of the spectral bar representation on the right are the theoretical spectra (smooth curves) derived from the ARIMA theory presented earlier.

Conclusion

Because of growing interest in the dynamic nature of wind loadings on bluff objects and on cooling towers in particular, the basic pressure-difference data of this study were taken. The first part of the paper discusses simple means for the interpretation of these data in the context of the micrometeorological

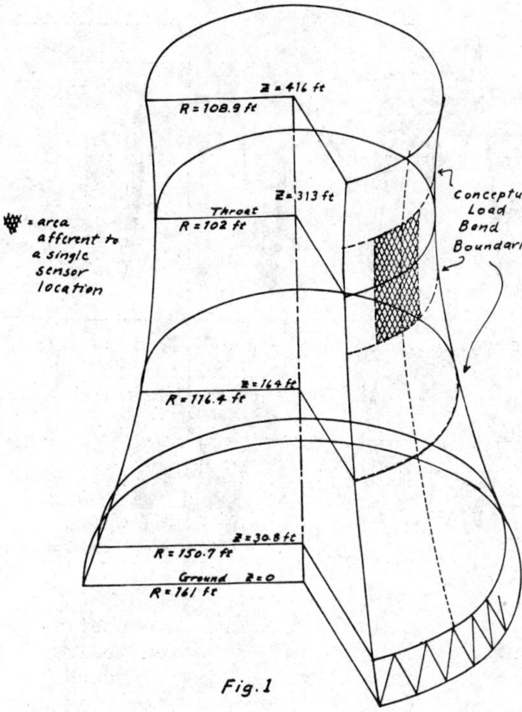

Fig. 1

environment. Further interest in the simulation of wind dynamic loads for design purposes led to the investigation of the applicability of ARIMA methods to this end. The present paper emphasized the suitability of AR2 models to the simulation of fluctuating pressure time histories at all points around the throat of a cooling tower. Incident to the investigation have been results that further suggest that the AR2 models invoked may themselves be indices to the local aerodynamic mechanisms acting at points around the tower.

Acknowledgement

The authors are grateful to Professor J. Stuart Hunter of Princeton University, for the advice and many helpful suggestions offered during the study on which this paper is based.

References

1. Scanlan, R.H. and N.J. Sollenberger, "Pressure Differences Across the Shell of a Hyperbolic Natural-Draft Cooling Tower," *Prod. Fourth Int'l. Conf. on Wind Effects on Buildings and Structures,* Heathrow, England, Sept. 1975, Cambridge University Press, pp. 143-149.

2. Sollengerger, N.J., R.H. Scanlan and D.P. Billington, "Wind Loadings and Response of Cooling Towers," Preprint No. 3108, *ASCE Fall Convention,* San Francisco, Octo. 17-21, 1977.

3. Niemann, H.J., "Zur Stationaeren Windbelastung Rotationssymmetrischer Bauwerke im Bereich Transkritischer Reynoldszahlen," Technischwissenschaftliche Mitteilung, Nr. 71-2, Institut fuer Konstruktiven Ingenieurbau der Ruhr-Universitaet Bochum, 1971.

4. Simiu, E. and R.H. Scanlan, *Wind Effects on Structures,* Wiley, N.Y., 1978.

5. Vickery, B.J., "On the Reliability of Gust Loading Factors," *NBS Bldg. Sci. Series* 30, *Wind Loads on Bldg. & Struct.,* U.S. Dept. of Commerce, 1970, pp. 93-104.

6. Jenkins, G.M. and D.G. Watts, *Spectral Analysis and its Applications,* Holden-Day, San Francisco, 1968.

7. Box, G.E.P. and G.M. Jenkins, *Time Series Analysis,* Holden-Day, San Francisco, 1976.

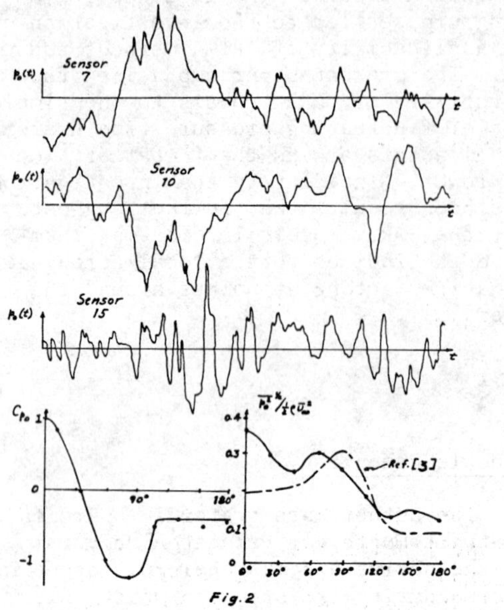

Fig. 2

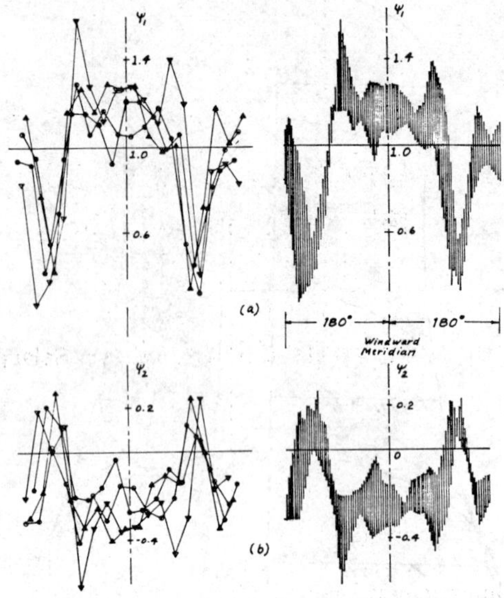

Fig. 3

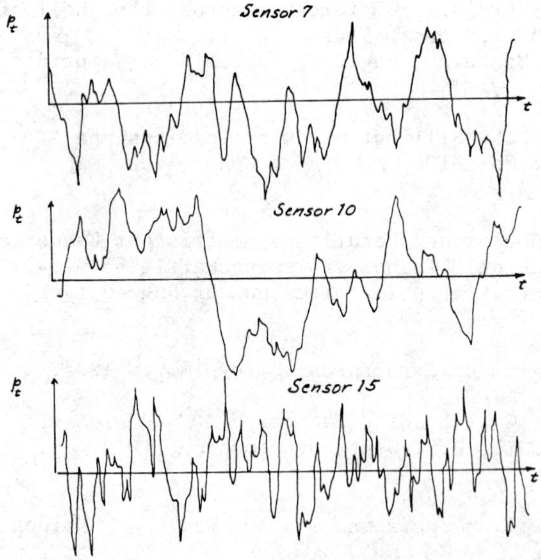

Fig. 4

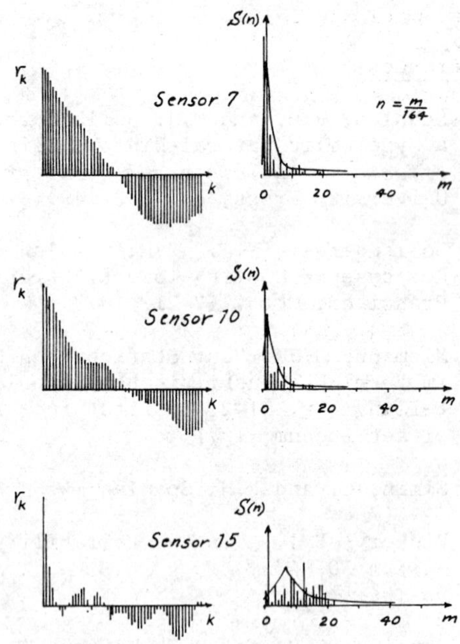

Fig. 5

OSCILLATION RESPONSE OF STRUCTURES IN THE NATURAL WIND

A. J. DUTT

Mara Institute of Technology, Selangor, Malaysia

SYNOPSIS

The paper describes the investigation of the oscillation response of two types of structures under the action of natural gusty wind. The first structure was a vertical cantilever structure having a natural frequency of 0.8 Hz, whose behaviour was accounted for by (a) the motion in the direction of wind (b) the phenomenon of vortex shedding causing aerodynamic instability (c) the dynamic response and (d) by the damping characteristics of the structure. The second structure was a full size Dutch Barn having a natural frequency of 8.00 Hz on which full-scale field experiments were carried out to assess its response to gusty wind. Measurements were made of deflections and of the incident wind; they showed that the maximum deflections of the full scale corresponded closely with the deflections predicted from measurements on a wind tunnel model test.

NOTATION

A	area of body projected in the air stream
C_d	coefficient of drag
C_k	coefficient
D	diameter of the beam
d	logarithmic damping decrement
d_1	dimension of member producing vortex
E	total energy stored in a system
ΔE	work done per cycle against drag (or lift)
F_k	periodic transverse force
F_n	force for an interval
f	frequency of vortex shedding
f_n	natural frequency of vibration of the structure
K	spring constant (stiffness) of the structure
P_1	dynamic sinusoidal load
p	circular frequency of vibration of the structure
m	mass
n_r	natural frequency in the rth mode
R	Reynold's Number
r	damping constant
r_a	aerodynamic damping constant
r_s	structural damping constant
S	strouhal Number

t	time
t_n	time for the beginning of an interval
t_{n+1}	time for the end of the interval
V	wind velocity
W	dead weight of the system
X_n	displacement for the beginning of an interval
x_{n+1}	displacement at the end of the interval
$\dot{x}_n$	velocity for the beginning of an interval
$\dot{x}_{n+1}$	velocity at the end of the interval
ρ	air density
ω	circular frequency of vibration
ω_d	damped circular frequency
Ω	circular frequency of the periodic load P_1

1. INTRODUCTION

All wind forces on structures are dynamic in the sense that they are the effects of a moving and fluctuating fluid. In the majority of cases wind excited oscillation is justifiably regarded as a response for which model tests in the wind tunnel as has been carried out by Peterka and Cermak (1), Ostrowski (2) and Dutt (3) are adequate. There are certain cases, however, where the dynamic character of wind force is of great importance. The analysis of the response of slender structures have been treated by Davenport (4), Melbourne (5), and Dutt (6-10). However, available information on the oscillation response of structures in the natural wind are very limited. The research work described in this paper was therefore, intended to obtain results on the oscillation response of two types of structures in the natural wind.

2. THE VERTICAL CANTILEVER STRUCTURE AND INSTRUMENTATION

The experiments were carried out on a vertical cantilever structure (Fig.1) which consisted of a 100 mm (4 in) diameter vertical steel tube 2.8 mm (9 ft.) high. A steel drum 610 mm (2 ft) diameter by 610 mm (2 ft) high was mounted on the top of the steel tube. The cruciform base was weighted to fix the base in position. The structure was erected on the flat roof of a tower block so that it was exposed to a

Fig.1 The vertical cantilever structure.

natural gusty wind. The drum was filled with weights as to vary the natural period of oscillation and the natural frequency of the structure was calculated using the equation,

$$f_n = \frac{1}{2\pi} \sqrt{\frac{k}{m}} \qquad (1)$$

The calculated value of the natural frequency was 0.8 Hz. Electrical strain gauge of 200ohm nominal resistance were fixed to the surface of the lower end of the structure. The gauges were mounted in pairs at two orthogonal axes. For the correlation of the forces exerted by the wind on the drum, a hot wire anemometer was found most suitable because of its rapid response. The calibration of the anemometer was carried out in the wind tunnel on the ultraviolet Recording galvanometer. Measurements of the dynamic deflections of the drum were performed by means of two differential transformer displacement gauges (6). Each transformer displacement gauge was mounted such that their axes were at right angles to one another and each was in line with one pair of strain gauges. A $\frac{1}{4}$ scale model of the structure was tested in an open jet wind tunnel. The average value of the coefficient of drag, C_d, was found to be 0.67.

3. DAMPING

The presence of damping forces causes the dissipiation of energy that progressively reduces the amplitude of vibration and ultimately stops the motion when all the energy initially stored in the system has been dissipated. The formulation of expressions for damping forces poses an extremely difficult problem that still requires extensive research. The damping of a structure vibrating in wind arises from two sources. First, there is the structural damping occuring at joints of the structure and within the material itself, and second the aerodynamic damping caused by the motion of the object through the strong current of air. The estimate of these damping parameters are extremely important on account of the vital control they exert on the attainable amplitudes of oscillation.

3.1 Structural Damping

The structural damping present in a structure is dependant on such factors as the material it is made of, the foundations, the mode of vibration and the form of construction. The amplitude of a damped vibration decreases and it is assumed that the damping force is proportional to the velocity of vibration and this proportionality factor is denoted by the structural damping constant r_s. This constant was found from the amplitude of free vibration of the structure using the formula for logarithmic damping decrement,

$$d = \frac{r_s \pi}{m\omega_d} = \log e \frac{x_{01}}{x_{02}} \qquad (2)$$

where x_{01} and x_{02} are the successive amplitudes.

3.2 Aerodynamic Damping

A structure vibrating in air will be subject to certain aerodynamic forces tending to damp the vibration. This form of damping has been investigated by Davenport (5). The logarithmic aerodynamic damping decrement is given by the expression

$$d = \frac{\Delta E}{2E} \qquad (3)$$

For a structure such as this, the aerodynamic damping constant, r_a, was evaluated from the formula,

$$d = \frac{\frac{1}{2}\rho c_d V^2 D}{n_r m V} = \frac{r_a \pi}{m\omega_d} \qquad (4)$$

4. DETAILS OF EXPERIMENT

The anemometer, the displacement gauges and the strain gauges were connected to a

12-channel ultraviolet high speed recording galvanometer. The zeros for all devices were fixed when wind was still. Wind velocity record, strain gauge and the deflection gauge records were made for periods of high south-west wind. It was seen that the structure was performing very complex oscillation with component across and along the direction of wind from which it was obvious that the part of the oscillations were due to the phenomenon of vortex excitation and the remaining from the gust action of the wind. Fig. 2 shows a typical portion of galvanometer trace.

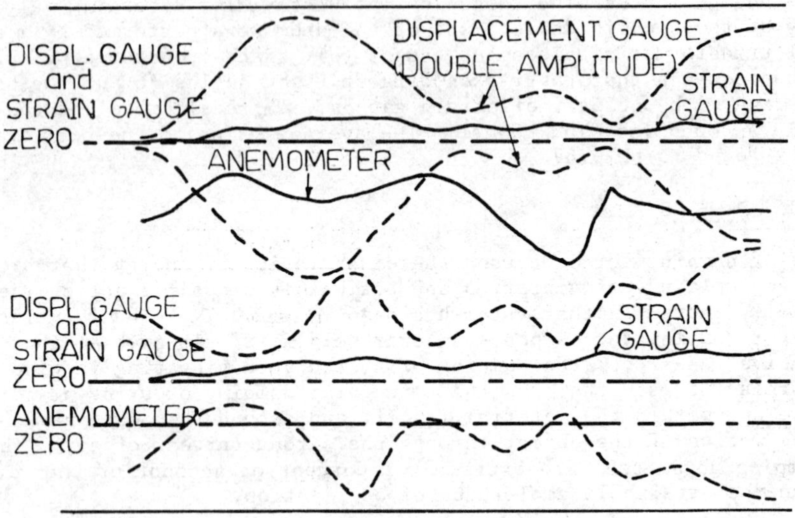

Fig. 2 Typical portion of galvanometer trace
$f_n = 0.8$ Hz; Paper speed: 12.7 mm/s

Karman vortex gives rise to periodic transverse force which is given by equation

$$F_k = C_k\ (\tfrac{1}{2}\rho\ V^2 A)\ \sin 2\pi\ ft \qquad (5)$$

where C_k = unity, $f = \frac{VS}{d}$

If f, d, and V are expressed in Hz, m and m/s respectively, the Strouhal Number for $R < 10^7$ is approximately 0.2 for circular cylindrical bodies. The dynamic displacement and velocity(11) of a one degree system under the action of a sinusoidal dynamic load are given by the equations,

$$x = \frac{P_1 g/W\omega^2}{1-(\frac{\Omega}{\omega})^2}\ (\sin \Omega t - \frac{\Omega}{\omega} \sin \omega t) \qquad (6)$$

$$\frac{dx}{dt} = \frac{P_1 g/W\omega^2}{1-(\frac{\Omega}{\omega})^2}\ (\Omega \cos \Omega t - \Omega \cos \Omega t) \qquad (7)$$

The sinusoidal transverse force due to vortex shedding was given by the equation (5) and the resulting dynamic deflexions and velocity of vibration due to periodic forces corresponding to the different wind velocities were calculated using the equations (6) and (7). When the periodicity of vortex shedding became equal to the natural frequency of the structure, the transverse deflections were very large causing instability which was restricted by the structural and aerodynamic damping. The oscillation of the structure in the direction of the wind was computed as follows: The anemometer trace was replotted in the terms of wind force on the structure using the formula,

$$F = C_d\ (\tfrac{1}{2}\rho\ v^2)A \tag{8}$$

The dynamic deflections and the velocities of vibration of the structure were evaluated by the step by step method(12) making use of the following equations:

$$X_{n+1} = (X_n - \frac{Fn}{k}) \cos p\ (t_{n+1} - t_n) + \frac{\dot{X}_n}{p} \sin p\ (t_{n+1} - t_n) + \frac{Fn}{k} \tag{9}$$

$$\frac{\dot{X}_{n+1}}{p} = -\ (X_n - \frac{Fn}{k}) \sin p\ (t_{n+1} - t_n) + \frac{\dot{X}_n}{p} \cos p\ (t_{n+1} - t_n) \tag{10}$$

From the damping constant and velocity of vibration, the damping force at any particular instant was calculated. The deflections were calculated (Table 1) at intervals of ½s. Dynamic load factor at an instant (defined as the ratio of the dynamic deflection to the static deflection produced by a 'five second gust' applied as a static load) were calculated and plotted in Fig.3. The theoretical and the experimental deflections were in good agreement.

TABLE 1: Dynamic Deflections and D.L.F. Computed From Anemometer Readings:

f_n = 0.8 Hz; k = 35 N/mm

t_n(s)	F_n(N)	X_n(mm)	$\dot{X}_n$|p(mm|s)	D.L.F.
0	8.40	0	0	0
0.5	28.68	0.43	0.14	1.61
1.0	32.21	1.22	0.12	1.74
1.5	21.42	0.75	-0.28	1.07
2.0	10.62	0.33	0.15	0.47
2.5	18.21	0.37	-0.14	0.53
3.0	15.72	0.56	-0.02	0.80
3.5	8.51	0.28	-0.08	0.40
4.0	20.25	0.16	0.04	0.23
4.5	6.54	0.26	-0.28	0.37
5.0	10.67	0.04	0.19	0.06

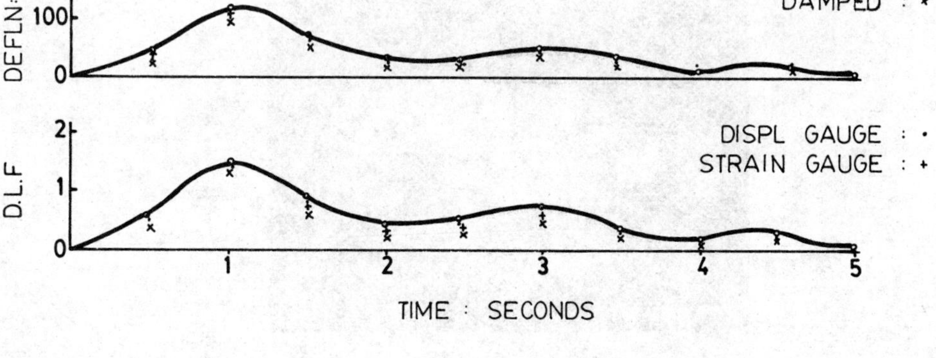

Fig. 3 Force, deflection and DLF of cantilever structure

5. FIELD EXPERIMENT ON A FULL SIZE STRUCTURE

The structure was a steel framed Dutch Barn 13.5 m long x 8.82 m span (45 ft x 29 ft 5 in) having a pitched roof supported on four steel trusses at 4.5 m (15 ft) centres and situated on an exposed site (Fig. 4). The truss members were all 64 x 64 x 6 mm ($2\frac{1}{2}$ x $2\frac{1}{2}$ x $\frac{1}{4}$ in) angles joined by riveted connections and the roof was supported on 178 x 100 mm (7 x 4 in) RSJ columns. The height of the structure to eaves level was 5.1 m (17 ft) and the rise of the truss was 2.25 m (7 ft 6 in). The longitudinal axis of the barn was North-West to South-East.

Fig. 4 The full size structure

5.1 Model Tests

Preliminary model studies were made on a 1/24 scale model (Fig.5) of the full size structure. As far as possible the relative stiffeness of the various members were correctly represented but the precise stiffening effect of the cladding was difficult to assess in advance of the field test. Following this, however, it was possible to determine the stiffeness scale by comparing static deflections to corresponding loads. A diogonal pull of 1.96 N (0.44 lb.) was applied between the eaves of one of the gable columns and the base of the opposite column; the resulting deflection was recorded as 2.11 mm (83 x 10^{-3}in.). This was carried out for comparison with the corresponding measurement on the full size structure.

Fig. 5 The model of full size structure

The model was then placed on the working table of a wind tunnel with the length of the model normal to the wind directions. Four dial gauges were fixed at the eaves level of the downstream face of the leeward columns. Deflections were measured for various wind speeds for different conditions of the model as follows: a) Model empty and with adjacent building; b) Model full and with adjacent building; c) Model empty and without adjacent building; d) Model full and without adjacent building. The model was filled with glass fibre to represent straw in the barn. Graphs were plotted velocity vs deflections of columns for the above conditions. (Fig.6)

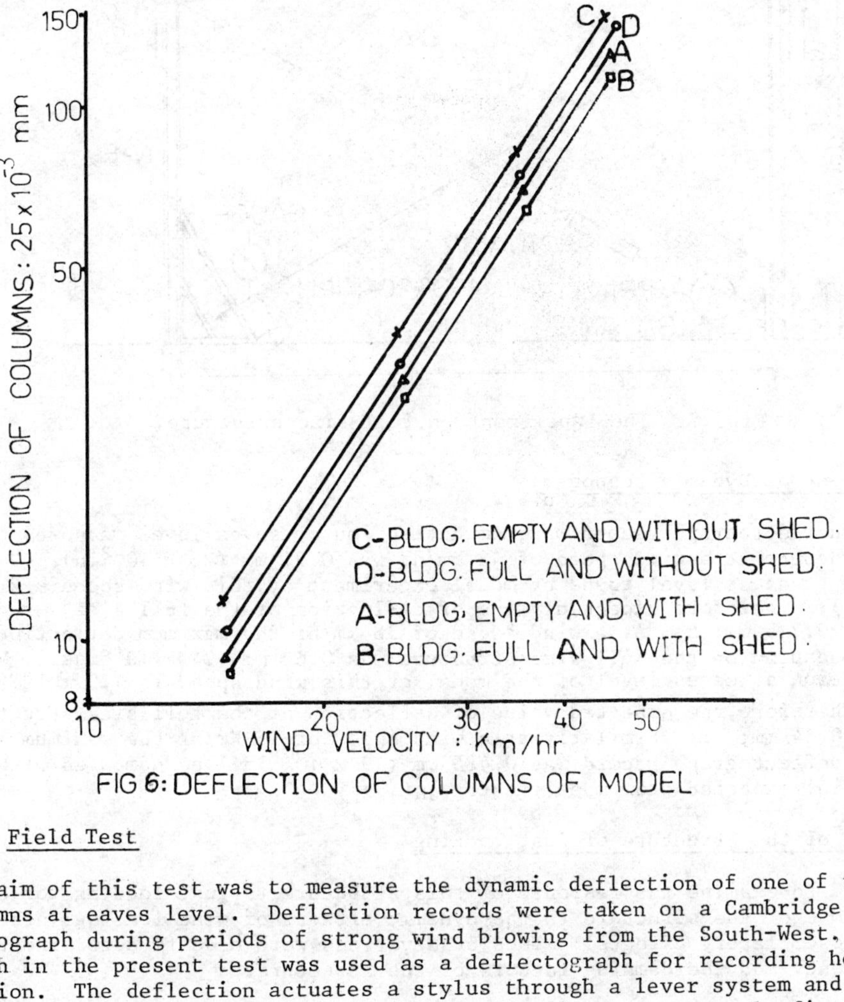

FIG 6: DEFLECTION OF COLUMNS OF MODEL

5.2 Field Test

The aim of this test was to measure the dynamic deflection of one of the outside columns at eaves level. Deflection records were taken on a Cambridge Universal Vibrograph during periods of strong wind blowing from the South-West. The Vibrograph in the present test was used as a deflectograph for recording horizontal deflection. The deflection actuates a stylus through a lever system and records the deflection on a cellulose acetate film. This gives an extremely fine and even trace that is immediately available for reading, when magnified, to 0.01 mm. A marker is provided which gives time intervals of 1 sec. and 1/10 sec. The deflectograph was mounted on a timber bracket fixed to the inner flange of the leeward column. One end of a piano wire was fixed at the eaves level of the windward column whose deflection was to be measured and the other end was connected to the deflectograph. A diagonal pull of 900 N (180 lb) was applied at the eaves level of one of the leeward columns by an actuator and spring balance and the horizontal deflection was noted as 0.5 mm (20 x 10^{-3}in). This was compared with the corresponding model test to obtain the stiffness scale ratio which was found to be

2.98. Fig. 7 shows the arrangement of the field experiment using vibrograph, actuator and spring balance.

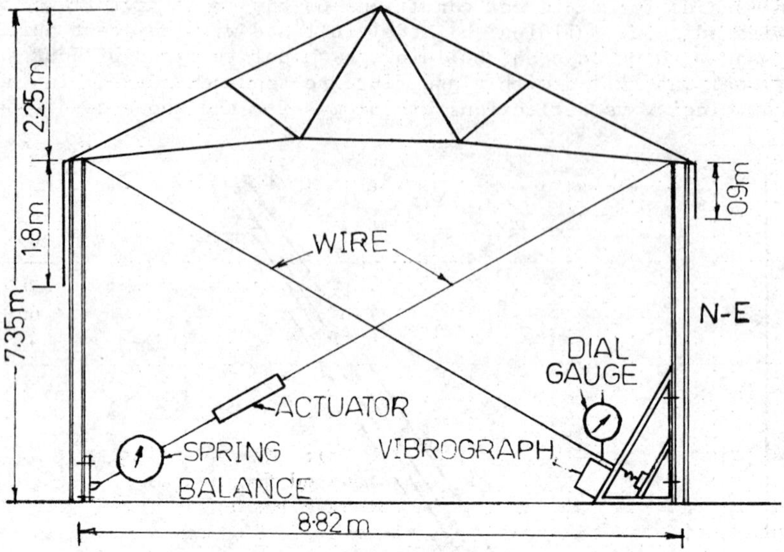

Fig. 7. The experiment on full size structure.

5.3 Measurement of Dynamic Response:

a) From the deflectograph record maximum deflection at eaves level measured on the full size structure at a wind speed of 42 km/hr was 0.71 mm (28 x 10^{-3}in). Value of deflection at eaves level found by model experiment at this wind speed = 2.16 mm (85 x 10^{-3}in.). Therefore expected value of deflection of the full size structure = 2.16/2.98 = 0.725 mm; b) At a wind speed of 38 km/hr the maximum deflection at eaves level measured on the full size structure was 0.62 mm (24 x 10^{-3}ins.). Value of the deflection at eaves level of the model at this wind speed is 1.9 mm (75 x 10^{-3}ins.). Therefore, the expected value of deflection of the full size structure = 1.9/2.98 = 0.64 mm; c) Similarly at a wind speed of 35 Km/hr the maximum deflection from deflectograph record was 0.485 mm (19 x 10^{-3}in.) as compared with 0.5 mm (20 x 10^{-3}in) expected from model test. (Table 2.).

5.4 Response of the Structure of Gust Loading

It was desired to examine the response of this structure to gust loading for assessing the respose of the structure to the dynamic effects of wind loading (other than lateral oscillatory effects). Thus observations were made to determine the natural frequency and the damping factors of the structure.

Natural frequency: It is difficult to calculate the period of vibration of the structure owing to the uncertainties regarding the stiffening effect of the cladding. This quantity was, therefore, found from the record of free vibration of the structure. The structure was shaken and the subsequent movement was recorded by the vibrograph (i.e. free vibration). Examination of this trace showed that the natural frequency was 8.0 Hz.

Damping: Structural and aerodynamic damping constants were found in the same manner as in the case of vertical cantilever structure. The structural damping constant was evaluated from the record of free vibration of the structure and was found to be 30 as compared to 1.54 for the Cantilever Structure.

A calculation of the response of the structure to the gust loading using the step by step method gave a close estimate of the maximum deflection of the structure as shown in Table 2.

TABLE 2. Comparison of deflections obtained by various methods

	Deflection (mm x 10^{-3})		
Wind speed km/hr	From Deflection record	Model and prototype static test	Dynamic deflection with damping for gust loading having average wind speed of
42	710	725	700
38	620	640	600
35	485	500	475

6. CONCLUSIONS

The vertical cantilever structure (f_n = 0.8 Hz): The oscillation response of the vertical cantilever structure is fully accounted for by (a) motion in the direction of the wind (b) vortex shedding (c) the dynamic response of the structure and (d) by the damping characteristics of the structure. The maximum observed dynamic load factor was 1.66 for the vertical cantilever structure having the natural frequency 0.8 Hz.

The Steel Framed Structure (f_n = 8.0 Hz): From wind tunnel test on model of steel framed structure it was found that the maximum deflection of the columns occured when the building was empty and was without the adjoining shed, this deflection being 3.32 mm (131 x 10^{-3}in.) at a wind speed of 45 km/hr. The deflection was minimum when the building was full and the adjoining shed was present, this deflection being 2.8 mm (110 x 10^{-3}in.). This is because the suction effect on the leeward side of the barn was greater when there was no adjacent shed on the leeward side of the barn. From the static deflection test and the measurement of dynamic response it is seen that the dynamic effect on the structure is negligible and the wind tunnel test with steady flow adequately predicted the behaviour of the structure.

Many engineering structures can be represented as one of the two experimental structures described in this paper and it is evident that the oscillation response of all such structures can be predicted by the methods used in this investigation.

REFERENCES:

1. Peterka, J.A. and Cermak, J.E., "Turbulence in Building Wakes", Proc. Fourth International Conference on Wind effects on buildings and structures, London, Cambridge University Press, 1975, pp 447-464.

2. Ostrowski, J.S., Marshall, R.D., and Cermak, J.E., "Vortex information and pressure fluctuations on buildings", Proc. Research Seminar on Wind Effects on buildings and structures, 1967, University of Toronto, pp. 459-484.

3. Dutt, A.J., "Wind pressure distribution on hyperbolic parabloid shell roofs". Civil Eng. Pub. Works Rev. London 1971, 66, pp 65-70.

4. Davenport, A.G. "The application of statistical concepts to the wind loading on structures", Proc. Inst. Civil Engrs, London, 1961, 19, pp 449-460.

5. Melbourne, W.H., "Cross-wind response of structures to wind action", Proc. Fourth Int. Conf. on Wind Effects on Buildings and structures, London, Cambridge University Press, 1975, pp 343-358.

6. Dutt, A.J., "Dynamic behaviour of slender structures in the natural turbulent wind", Proc. International Conference on the Behaviour of slender structures. London City University, 1977, pp 1.6.1 - 22.

7. Dutt, A.J., Discussion Paper "Radio Towers" J. Inst. Structural Eng., April, 1966 pp 151-154.

8. Dutt, A.J., Discussion Paper "Forth Road Bridge", Proc. Inst. Civil Eng. 1967, 36 pp. 117 - 188.

9. Dutt. A. J., Discussion Paper "British Standard Code of Practice for loading: Wind loads". J. Inst. Structural Eng., 1970, 48, pp 243-254.

10. Dutt, A.J., Discussion Paper "Meseum Radio Tower," Proc. Inst. Civil Engrs, 1966, 34, pp 421-446.

11. Norris, C.H., atl. "Structural Design for Dynamic Loads", McGraw Hill Book Co, London 1969. pp 57-58.

12. Rogers, G.L., "Dynamics of Framed Structures," John Wiley & Sons, New York, 1969, pp 61-62.

DYNAMIC SUPPRESSORS TO CONTROL WIND-INDUCED VIBRATIONS OF FLEXIBLE STRUCTURES

B. G. KORENEV
V. V. Kuibyshev Institute for Civil Engineering, Moscow, USSR

ABSTRACT

This paper deals with constructive shapes of a pendulum-type dynamic vibration suppressor meant for absorbing vibrations of tower structures. The problem of practical approximate analysis of a similar structure or a system with one degree of freedom is discussed. Individual results are set out with regard to the theory of vibration absorption in a plate or a shallow spherical shell.

INTRODUCTION

The problem of absorbing vibrations of flexible structures by means of vibration suppressors has long attracted attention of many research workers and it has been studied, in particular, in a series of works fulfilled at the V. A. Kucherenko Central Research Institute for Building Structures (TsNIISK) and the V. V. Kuibyshev Institute for Civil Engineering (MISI).

The present paper is a review, its scope being restricted only to the works which the author personally participated in. Co-authors were A. N. Blekherman, M. Ya. Volotskii, L. M. Reznikov, V. I. Sysoev, V. S. Polyakov, O. M. Fooks. Recommendations (1) were compiled at TsNIISK by B. G. Korenev, M. Ya. Volotskii, V. I. Sysoev, and edited by A. I. Tseitlin and B. G. Korenev.

CONSTRUCTIVE SHAPES OF DYNAMIC VIBRATION SUPPRESSORS

It is typical of dynamic pendulum-type suppressors of vibrations with higher damping, shown in Fig. 1, that near the point of suspension, additional supports are provided which cause local bending stresses that provide the source of dissipative forces. They are described in Ref. 1 and their employment in practice was a success. As a rule, at higher vibration amplitudes damping in pendulums of the mentioned structure is reduced. In this connection a modification in the structure was proposed (see Fig. 2) by introducing an additional fixation at the point of the pendulum load-mass adjoining the cable. The indicated modification has added to increase the damping.

The experimental studies mentioned elsewhere (2) have been further supplemented by some tests and the results are described in Ref. 3. It can be seen that even for rather low frequency pendulums (5s period) at comparatively large amplitudes (ranging from 0.6 m to 3 m) damping created by a supporting device appeared to be

Fig.1

sufficient for practical purposes. It is obvious that at higher frequencies the contribution made the local bending of the terminal portion of the cable, and, consequently, the damping will increase; still more noticeable effect will result from a reduction of design amplitudes of the suppressor. Suppressors of such a type are known for their simple structure and low cost. They possess stable characteristics and can successfully be used under conditions when the suppressor temperature is equal to that of the outdoor air.

ABOUT THE ANALYSIS OF VIBRATION SUPPRESSORS INSTALLED IN TOWERS, CHIMNEYS AND TALL BUILDINGS

Practical problems of calculation of tower structures are discussed in "Recommendations" (1). They are of importance for buildings with a metal skeleton, steel chimneys, latticed and continuous metal towers and reinforced concrete tower structures with a flexible metal top section (e.g., reinforced concrete T.V. towers with a metal top section).

The "Recommendations" (1) deal with linear dynamic suppressors and impact vibration suppressors. The concern of the present paper will be dynamic vibration suppressors only. It should be mentioned that the part relating to impact suppressors was based on the works by V. I. Sysoev (2).

For flexible tower-like structures pendulum suppressors with higher damping described above or with damping devices employing dry friction are recommended (1). In closed heated premises viscous dampers are supposed to be used (4).

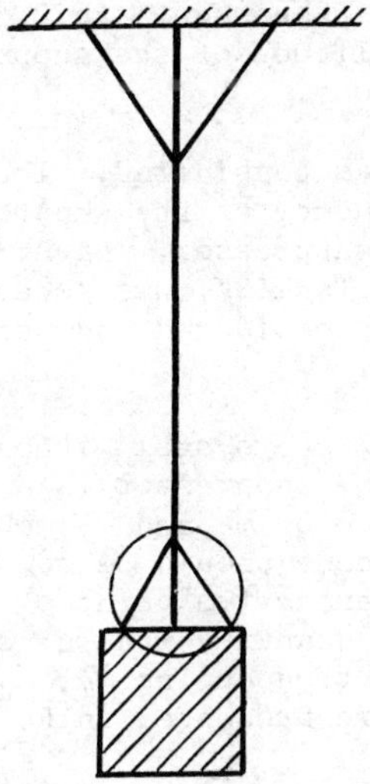

Fig.2

It is recommended to calculate and optimize suppressors for that dynamic effect which contributes most adverse effects in the structure. It is stated that sets of suppressors may be used, being installed at the same elevation and having different tuning. The paper deals in greater detail with calculation of a structure as a system with one degree of freedom for wind effect and loads due to machinery. Along with the selection of optimum parameters of suppressors it is discussed also how the efficiency of vibration suppression is influenced by the deviation of parameter values from the optimum ones.

The problem of determining amplitudes and decrements of vibrations with fixed (non-optimum) parameters of the suppressor is set forth in detail.

For structures that cannot be looked upon as a single mass system the program is given for a design model presenting a cantilever stepped rod with ties possessing elastic and dissipative properties. At different levels one or a few suppressors are attached to the protected system.

Two types of effects are considered: 1) a harmonic one with a steady frequency and amplitude which, as a matter of fact, depends arbitrarily upon the frequency and 2) the stationary random process with the fixed law of load distribution over the height of the construction. The program provides for optimizing suppressor parameters.

Various problems of calculating tall structures were discussed (3) taking as an example one of the versions of design of a tower-tube, 420 m high, described in Ref. 5. The tower-tube has an outer metal stem of constant diameter of 18.5 m.

At the elevation of 150 m an intermediate elastic support is located. Inside the metal stem there are three gas exhaust ducts. Suppressors are placed inside the stem thus making the design more complicated and affecting also the efficiency of vibration insulation since the amplitude of the mass vibration is limited. The total mass of suppressors was taken equal to 0.5% of the structure mass. The finite value of the vibration amplitude of the suppressor mass was taken equal to 3 m for certain construction reasons.

As a design case wind resonance was considered. The task was to choose the optimum value of the partial frequency of the suppressor with the given value of the logarithmic decrement of the suppressor, which due to test data received by the Ukrproertstalkonstruktsia and TsNIISK, was accepted equal to 0.2. Other values of the decrement were also considered and permitted the effect of decay upon the result to be determined.

The partial vibration frequency of the system without suppressor will be denoted by ω_1; the partial frequency of the suppressor is a value near to ω_1. Designating vibration frequencies of the system by $\overline{\omega}_1$ and $\overline{\omega}_2$, where $\overline{\omega}_2 > \omega_1 > \overline{\omega}_1$, resonance phenomena at two critical wind speeds should be considered. The optimum value of tuning is the abscissa of the intersection point of the two curves shown in Fig. 3 where values of the design quantity--the greatest bending moment are plotted (curves 1 and 2 relate to frequencies $\overline{\omega}_1$, $\overline{\omega}_2$). Numerous graphs which help to implement this design scheme are presented in Ref. 1.

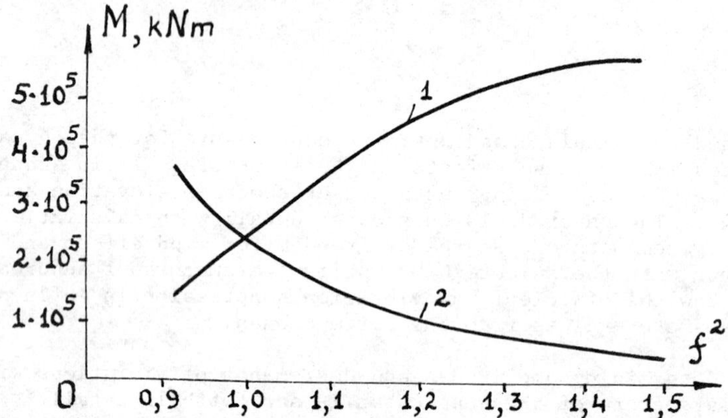

f-ratio of the suppressor partial frequency to the natural frequency of the construction.

Fig. 3

The vibration amplitude of the suppressor equaled 2.6 m; stresses and displacements were reduced by as much as three times. If the system of vibration control consists of a few suppressors having the same frequency, then obviously, one might wish to adjust some of the suppressors to other nearby frequencies. It is, however, clear that with this arrangement the greatest vibration amplitude of the suppressor will increase. The problem of detuning suppressors has been discussed and high efficiency of detuning with small values of decrements in suppressors has been shown (6,7). For suppressors with optimum damping, detuning increases the efficiency insignificantly (8,9).

In Ref. 3 problems of the effect of detuning is regarded in those cases when limitation of the suppressor amplitude is considerable and it is known that the influence of detuning upon the greatest amplitude may be rather large. One of the important merits of dynamic suppressors of vibrations is the multipurpose pattern of their action. Vibration suppressors reduce the level of vibrations not only at wind resonance but also at fluctuations of wind pressure and galloping phenomena. At the same time they reduce vibrations of the structure induced by machinery with non-balanced masses. Vibration suppressors designed for wind effect improve the earthquake resistance of structures. The problem has been dealt with earlier (2). Subsequent work on the design of systems with vibration suppressors for seismic effects has been completed (10).

As can be seen from the above, employment of vibration suppressors is followed in practice by further and deeper examination of traditional problems which had greatly been influenced by the tendency to describe the pattern of the effect and the design structural scheme as accurately as possible. It would, however, be wrong to think that the opportunities for development in this area have been exhausted. The problem of vibrations of structures whose design scheme is a rod of variable cross section with continuously changing rigidity is also of importance in applications (1).

While designing buildings it is sometimes possible to use a design model in the form of a rod working in shear. In this case free vibrations are studied by solving a differential equation of second order, with variable factors, which for some laws of rigidity variation is integrated in Bessel functions. A similar task arises with torsional oscillation of the rod. In this case the differential equation for free vibrations has the form:

$$\frac{d}{d\chi}\left[C\,\frac{d\psi}{d\chi}\right] + B\omega^2\psi = 0 \quad , \qquad (1)$$

where C is the rigidity in torsion, B is the moment of inertia of the rod section mass related to unity, and ψ is the rotation angle.

If $C = C_o\chi^k$, $B = B_o\chi^\gamma$, where k, γ, C_o, B_o are constants, then the solution of (1) has the form:

$$\psi = \chi^{\frac{1-k}{2}}\left[A_1 J_{\frac{1}{2}\nu}\left(\frac{2\beta}{\nu}\, z^{\nu/2}\right) + A_2 Y_{\frac{1}{2}\nu}\left(\frac{2\beta}{\nu}\, z^{\nu/2}\right)\right] \quad , \qquad (2)$$

where

$$\beta = \omega\sqrt{\frac{B_o}{C_o}} \quad , \quad \nu = \gamma - k + 2 \quad .$$

The task of vibrations of a rod with suppressors can be solved by expanding the function $\psi(\chi)$ into a series which is, in fact, the Dinny expansion in the finite interval (11).

Making use of the Lommel transformations other laws of variation of functions $B(\chi)$ and $C(\chi)$ can be found.

VIBRATION ABSORPTION OF PLATES, FOLDS AND SHELLS

In vibration absorption of plates, shells and folds it is hardly allowable to employ comparatively rough simple models of the structure requiring protection. The design of such a structure with attached suppressors for a dynamic load effect is a complicated problem. Yet the problem of vibration absorption gives rise to a few other problems as well. While designing, there arises a problem of local stresses caused by the transfer of load from vibration suppressors, then a problem of how deviations of parameters of vibration suppressors influence the efficiency of vibration suppression, etc.

The notion of optimization includes selection of the place to locate suppressors as well as the extent of their deconcentration, etc. Practical design methods should be devised so that, where possible, computers might be used.

Along with vibration absorption of folds, shallow spherical shells have been studied in detail (12). The desire to draw some general conclusions has led to analyzing continuous vibration suppressors thus allowing for the effect of attachment of a large number of suppressors. At first, there was analyzed the influence of a continuous suppressor without damping on the spectrum of free vibration frequencies, and then the influence of similar suppressors with optimum damping. At the same time similar results were obtained for a circular plate. The task of free vibrations of a shallow spherical shell in the presence of continuous suppressors resolves itself to integration of the equation

$$D\nabla^2\nabla^2 w + \left[\frac{Eh}{R^2} - \rho h\omega^2\left(1 + \frac{\nu\omega_r^2}{\omega_r^2 - \omega^2}\right)\right] w = 0 \quad , \tag{3}$$

where D, h, R, ρ are the cylindric stiffness, radius of the medium surface, thickness and density of the shell; E is the elasticity modulus; ω, ω_r are the frequency of free vibrations of the shell with a suppressor and partial frequency of the suppressor; and ν is the relative mass of the suppressor. The analysis of solutions for a circular shell with hinged and restrained edges has demonstrated that with a comparatively small mass of the suppressor a rather narrow resonantless zone appears to be obtainable. The effect of suppressors upon the logarithmic decrement of vibrations has also been stated. It has been proved that suppressors tuned for a low frequency exert a positive influence with higher frequencies being sufficiently far off. The same phenomenon stated earlier for a suppressor on the rod is accounted for by the fact that in the given situation the suppressor mass is, in fact, immobile and the suppressor works as a damper.

The problem of harmonic vibrations of circular and square plates equipped with a vibration suppressor have been discussed (13). A similar task will be regarded for a plate, extended uniformly by forces distributed over its contour. Such tasks are encountered in the analysis of flexible roofs. Following Ref. 14, an analysis of a circular plate with a suppressor located at an arbitrary point will be described. The differential equation for the elastic plate surface has the form:

$$D\nabla^2\nabla^2 W - P_0\nabla^2 W + \rho \frac{\partial^2 W}{\partial t^2} = q(r,\theta,t) \quad (4)$$

where D is the cylindric rigidity, and ρ is the surface density.

Supposing that the plate makes forced vibrations with a circular frequency k, the solution will be presented in the following form:

$$W(r,\theta,t) = w(r,\theta)\ \sin kt \quad .$$

The Green function for an unrestricted field has the form:

$$K(r) = -\frac{1}{8D}\left[\frac{1}{\lambda_1^2} Y_0(Z_1) + \frac{2}{\pi\lambda_2^2} K_0(Z_2)\right] \quad , \quad (5)$$

where

$$Z_1 = \lambda_1 r, Z_2 = \lambda_2 r, \lambda_{1,2} = \sqrt{\pm b_0 + \sqrt{b_0^2+1}}\ ,\ b_0 = -\frac{P_0}{2D\lambda^2},\ \lambda^4 = \sqrt{\frac{\rho k^2}{D}} \quad .$$

The solution with the load q sinkt uniformly distributed over the circumference of the radius a has the following form:

$$w_0(r) = -\frac{\pi q}{4D}\left[\frac{\alpha_1}{\lambda_1^3} J_0(Z_1)\ Y_0(\alpha_1) + \frac{2\alpha_2}{\pi\lambda_2^3} I_0(Z_2)\ K_0(\alpha_2)\right],\ r \leq a \quad , \quad (6)$$

$$w_0(r) = -\frac{\pi q}{4D}\left[\frac{\alpha_1}{\lambda_1^3} J_0(\alpha_1)\ Y_0(Z_1) + \frac{2\alpha_2}{\pi\lambda_2^3} I_0(\alpha_2)\ K_0(Z_2)\right],\ r \geq a \quad . \quad (7)$$

Having these formulae, the influence of the suppressor can be taken into account which transmits the dynamic response onto the plate over the circumference of the radius a; the influence of the suppressor being presented as:

$$w_r(r,\theta) = w_0 + w_k \quad ,$$

where the function w_k is a compensating solution valid for the homogeneous differential equation (2). If the plate has the form of a circle with the R_1 radius, and the load from the suppressor is imposed over the circumference centered at the b distance off the plate center, then to satisfy boundary conditions, the function w_k with $r > b$ will be presented as follows:

$$w_k(r,\theta) = 2\sum_{n=0}^{\infty}\left\{A_n J_n(\xi_1) + B_n I_n(\xi_2) - \frac{\pi q}{4D}\left[\frac{\alpha_1}{\lambda_1^3} J_0(\alpha_1) Y_n(\beta_1) Y_n(\xi_1) + \frac{2\alpha_2}{\pi\lambda_2^3} I_0(\alpha_2)\ I_n(\beta_2)\ K_n(\xi_2)\right]\right\} \cos n\psi \quad . \quad (8)$$

In this formula the coordinate system begins at the center of the plate. Here the following designations were introduced:

$$\beta_1 = \lambda_1 b,\ \beta_2 = \lambda_2 b,\ \xi_1 = \lambda_1 R,\ \xi_2 = \lambda_2 R\ ;$$

R, ψ are the polar coordinates of a point on the plate.

The constants A_n and B_n are found in terms of boundary conditions on the contour of the plate. The equation of the plate motion caused by the load q* uniformly distributed over the plate surface has the form:

$$w^*(R,\psi,t) = -\frac{q^*}{\rho k^2}\left[1 + B_1 J_0(\zeta_1) + B_2 I_0(\zeta_2)\right], \qquad (9)$$

where

$$B_1 = -\frac{\lambda_2 I_0'(\zeta_2)}{\Delta_1},\quad B_2 = \frac{\lambda_1 J_0'(\zeta_1)}{\Delta_1},$$

$$\Delta_1 = \lambda_2 J_0(\zeta_1)\, I_0'(\zeta_2) - \lambda_1 J_0'(\zeta_1)\, I_0(\zeta_2)\ .$$

Following the supposition that $b \ll R_1$, conditions of the contact of the suppressor and the plate will be written, making equal displacements at the point (b,0). The other part of the analysis is done as in Ref. 13. If the plate has a non-circular shape, the function is determined by the method of compensating loads (11,13,2).

In concluding it should be noted that in determining bearing responses suppressors can be looked upon as point loads. When determining dynamic stresses under the suppressor, the response should be considered as distributed.

REFERENCES:

1. Rekomendatsii po proektirovaniyu gasitelei kolebanii dlya zashchity zdanii i sooruzhenii, podverzhennykh gorizontalnym dinamicheskim vozdeistviyam ot technologicheskogo oborudovaniya i vetra, the V. A. Kucherenko Central Research Institute for Building Structures, Stroiizdat, M., 1978.

2. Korenev, B. G., On damping wind-induced vibrations of flexible structures, Proceedings of the Fourth International Conference on Wind Effects on Buildings and Structures, Cambridge University Press, 1975.

3. Korenev, B. G., Volotskii, M. Ya., and Fooks, O. M., Vibrogashenie vysokikh dymovykh trub pri organichennykh amplitudakh kolebanii mass gassiteley, Stroitelnaya Mekhanika i raschet sooruzhenii No. 5, 1976.

4. Rukovodstvo po proektirovaniyu vibroizolyatsii mashin i oborudovaniya, the V. A. Kucherenko Central Research Institute for Building Structures, Stroiizdat, M., 1972.

5. Arsenyev, L., Promyshlennye truby, Nauka i zhiznj, No. 7, 1973.

6. Pikulev, N. A., Teoriya dinamicheskogo gasitelya kolebanii pri passtroikakh, Collection of articles Borba s shumom i vibratsiyami, Stroiizdat, M., 1966.

7. Pikulev, N. A. and Erdelevskii, A. N., K voprodu proektirovaniya gruppy vibrogasitelei s uchetom rasstroek, Stroitelnaya mekhanika i raschet sooruzhenii, No. 5, 1971.

8. Snowdon, I. C., Vibration and Shock in Damped Mechanical Systems, New York, 1968.

9. Shrinivasan. Issledovanie parallelnykh dinamicheskikh vibropoglotitelei s dempfirovaniem, Konstruirovanie i tekhnologiya mashinostroeniya, No. 1, 1969.

10. Korenev, B. G. and Polyakov, V. S. Reaktsia systemy s odnoi stepeniy svobody i dinamicheskim gasitelem kolebanii pri dvizhenii osnovaniya po zakonu sinusoidy, Collection of articles Sovershenstvovanie metodov raschota i konstruirovaniya zdanii i sooruzhenii, vozvodimykh v seismycheskikh raionakh, TsINIS, Gosstroi of the USSR, Stroiizdat, M., 1976.

11. Korenev, B. G., Vvedenie v teoriyu besselevykh funktsii, Nauka, M., 1971.

12. Korenev, B. G., Volotskii, M. Ya., and Reznikov, L. M., Voprosy gasheniyh kolebanii plastinok, skladok, obolochek. Issledovaniya po teorii sooruzhenii, Issue XXI, Stroiizdat, M., 1975.

13. Korenev, B. G. and Reznikov, L. M., Vynuzhdennye kolebaniya krugloi i kvadratnoi plastinki s dinamicheskim gasitelem pri garmonicheskikh vozdeistviyakh. Collection of articles Dynamika sooruzhenii, Stroiizdat, M., 1971.

14. Korenev, B. G., O metodakh borby s vibratsiyami gibkikh pokryttii. Proceedings of the International IASS Conference on Lightweight Spatious Structures of Roofs to be Constructed in Ordinary and Seismic Regions, Alma-Ata, September 1977, Stroiizdat, M., 1977.

DISCUSSION ON SESSION IX
Wind-Engineering Applications

Rapporteur: R. N. Meroney
Colorado State University
USA

Paper by B. S. Kandola, E. W. Marchant--WIND INDUCED MOVEMENT OF SMOKE WITHIN AND AROUND BUILDINGS - A WIND TUNNEL INVESTIGATION

Question: F. H. Durgin. What is the blockage of your model?

Reply: B. S. Kandola, E. W. Marchant. Area of the model facing the wind is about 14% of the cross-sectional area of the open jet tunnel used herein. Earlier studies have shown that for closed working section wind tunnels no blockage corrections are necessary if the model area is less than about 10% of the tunnel cross-sectional area. For open jet tunnels, this limit is much higher. No blockage correction, therefore, was made in this case.

Question: C. Kramer. Leakage simulation in small-scale models is not only a question or realizing a corresponding open area, it is also a question of simulating the correct volume flow - pressure characteristics during the full-scale leakage. In your case this is surely unlikely. At the low Reynolds number during the model tests a power law with an exponent somewhat between 1 and 2 is to be expected. This would affect the comparability between model and full-scale movement for the same exterior pressure field. Would you please comment on this?

Reply: B. S. Kandola, E. W. Marchant. I agree with Dr. Kramer on this point. It is realized that there are many problems in this field. Further work needs to be done to improve simulation.

Paper by R. N. Meroney, D. E. Neff--LABORATORY SIMULATION OF LIQUID NATURAL GAS VAPOR DISPERSION OVER LAND OR WATER

Question: M. E. Davies. (1) What is the basis for the assertion that a Reynolds number of 3000 is sufficient to generate the correct wake turbulence? Can we be sure that this is appropriate to the flow in the developing cloud? (2) If an isothermal simulation is employed do you correct the measured volume concentration in any way to compensate for the changing volume - mass relation which occurs in practice as the LNG vapor mixture expands?

Reply: R. N. Meroney, D. E. Neff. (1) Halitsky, J. (1963) Geophysical Science Laboratory Report No. 63-2 and 63-3, New York University; states minimum Reynolds numbers for which similarity still exists for the concentration coefficient on and in the wake of a cube. No, we cannot be sure that this is appropriate to the flow in a developing cloud. (2) Yes, we do correct for the differences in modal flux rates admitted for the model.

Question: R. E. Britter. (1) Do the authors consider that any of the full-scale data in Figure 8 (including data that formed the B.C.L. correlation) showed negatively buoyant characteristics? (2) With what confidence would you allow Figure 8 to be used for full-scale hazard analyses?

Reply: R. N. Meroney, D. E. Neff. (1) No, most of the tests were performed at too high of a wind speed to show dominant negatively buoyant characteristics. (2) Crude estimates for reasonably ideal circumstances (flat terrain, no building effects, moderate wind speeds).

Question: R. E. Britter. With reference to Figure 8, the data might be fitted with a (-2) power-law; thus the buoyancy length may (a) be omitted from the correlation or (b) replaced by the geometric scale length. Following the latter alternative we obtain the traditional method of correlating concentration data. However, previous CSU reports contend that their data will not collapse in this form. Could the authors clarify the apparent inconsistency?

Reply: R. N. Meroney, D. E. Neff. The variation of centerline maximum concentration with downwind distance does not appear to be a strong function of gas density for ground level sources. The buoyancy scale length was chosen to correlate results in the near field region where such effects were expected to be dominant; nonetheless, the use of some geometric scale such as spill diameter also collapses data quite well since the power-law exponent is nearly (-2).

Question: R. E. Britter. You only refer to the burning of the LNG cloud. Have you discounted detonation of the cloud?

Reply: R. N. Meroney, D. E. Neff. From the China Lake Tests performed by the Naval Weapons Center of LNG cloud ignition they found that an explosion was unlikely.

Paper by C. F. Shieh, W. Frost--APPLICATIONS OF A NUMERICAL MODEL TO WECS SITING RELATIVE TO TWO-DIMENSIONAL TERRAIN FEATURES

Question: J. A. Peterka. Figure 3 of your paper shows a comparison of your model with wind tunnel data of Reference 15. All data in that reference were for flows about three-dimensional obstacles while the numerical model was for a two-dimensional case. Because the wake structure of a three-dimensional obstacle is fundamentally different in both mean velocity and turbulent structure from a two-dimensional barrier, a two-dimensional simulation should not be compared with that data.

Reply: C. F. Shieh, W. Frost. Only data measured along the centerline of a long block structure have been used for comparison. This model block has the geometric ratio, height/width/depth = 1/8/0.75. Along the centerline of such a long block the flow was assumed in a quasi-two-dimensional state.

Question: I. P. Castro. Wakes of three-dimensional bodies are invariably different from those of two-dimensional bodies. It is therefore, I think, dangerous to make such optimistic claims for numerical prediction techniques based on comparisons between two-dimensional calculations and three-dimensional experiments. In addition, the authors provide no information regarding the adequacy of their numerical techniques. Nearly all current finite difference methods are, arguably, inaccurate in eliptic flow regions. What typical grid/mesh sizes were used, and how was the sharp corner at separation dealt with?

Reply: C. F. Shieh, W. Frost. The data chosen for comparison are those obtained from airflow over a long building-like block with the geometric ratio height/width/depth = 1/8/0.75 (14,15). Along the centerline of a long block the flow structure can be considered a quasi-two-dimensional situation. Thus, only measured along the centerline of the long building have been used for comparison.

The grid/mesh size used in the numerical code is 0.2 h in the region close to the block and is about 1 h in the regions far away from the block. Grid/mesh distribution is given in Reference 3. Details of the numerical accuracy has been discussed in Reference 2 and in a paper by Frost, Bitte, and Shieh recently accepted by AIAA.

Paper by R. K. Cooper--THE PROBABILITY OF TRAINS OVERTURNING IN HIGH WINDS

Question: H. Sockel. If I am right, your critical overturning moment occurs, when the wheels on the windward side of the leading vehicle are unloaded. Isn't it already dangerous, when the leading wheels on the windward side of the leading vehicle are unloaded? This will happen at a lower wind speed, because the side forces on the nose are much higher than those on the rear of the leading vehicle.

Reply: R. K. Cooper. The vehicle structure and suspension are designed so that the bogies (trucks) will remain attached to the body. If the overturning moment exceeds the critical value (defined in the paper), body and bogies will rotate together about the lee rail. The effect of the higher side force near the nose is thus transmitted through the entire vehicle structure.

Question: K. C. Anthony. Having decided on a 3-second averaging interval, why did the author utilize mean hourly data with conversions to 3 seconds rather than use 3-second statistics directly? Presuming that paragraph 5 refers to overall fleet operations and not single journeys, does the quoted 0.0092 annual probability relate to a single overturning event on that section of line under consideration in general or to that when the 50-year return period wind happens to be blowing? Either way, the probability seems high, bearing in mind some 1200 journeys per annum, even allowing for the inevitable calcellations. Should not the quoted probability level be lower than 0.636 since the return period of the reference wind is stated to be 50 years while that of the event is 109 years? Have predictions been made for stationary trains on superelevated track with wind in the adverse direction? This is not an unlikely occurrence and could produce a critical load case.

Reply: R. K. Cooper. Mean hourly wind speeds were used as a basis following the method given by ESDU. This facilitates the use of different averaging times as discussed in Section 4.5. Section 5 refers to the annual probability of one or more overturning incidents for fleet operation over the given route. If this probability is P_1, then the probability of at least one incident in N years is $P_N = 1 - (1-P_1)^N$. We may define the return period as $N = 1/P_1$ ($P_1 << 1$) then P_N is close to 0.636 for all $N > 20$ (say) (e.g., $P_N \simeq 0.634$ at $n = 100$). It is obviously undesirable to stop a train on an exposed segment of track during a gale. No predictions have been made for this case, but it may be avoided by appropriate train management.

Question: K. B. Rojiani. The analysis presented is based on the assumption that wind speed and wind direction are statistically independent. What is the effect of this assumption on the results?

Reply: R. K. Cooper. The probability density function for wind direction was derived from meteorological office data for winds of Beaufort Force 4 and above; hence, it is valid for 'strong winds'. I do not know what the effect of a two-dimensional probability density for wind speed and direction would be. However, I suspect that it would be much less than the uncertainty in the estimates of either of the independent functions assumed in the calculations.

Question: K. B. Rojiani. The only source of uncertainty considered in your study is the estimation of extreme winds. I would think that uncertainties in the modeling of the overturning situation and the estimation of the various parameters (e.g., the wind environment parameters) would be equally important. The inclusion of these uncertainties in the analysis would have a significant effect on the results.

Reply: R. K. Cooper. The deterministic model of the overturning situation is a limitation, but it was necessary to keep the problem simple. The moving model results in Figure 2 suggest that the coefficient of overturning moment may be a statistical variable.
As Dr. Rojiani suggests, a reliability analysis should be done on all parameters. I would expect the uncertainties in the reference wind speed to be the most significant.

Comment: N. J. Cook. I would like to correct Mr.Cooper's statement that "we don't get tornadoes" in the United Kingdom. There are on the average 14 days per year in which conditions are right to spawn tornadoes on any one of these days. There can be severe tornado sightings, tornado damage to buildings and structure runs to about 1/2 million annually.

Paper by J. Berlamont, N. Vanderstappen--ON THE EFFECT OF BAFFLE WALLS IN A WATER TOWER TANK UNDER WIND ACTION

Comment: H. Ruscheweyh. Water tanks may be used as dynamic vibration absorber systems and have been used so very successfully in recent years. Schiffue has developed a theory for the dynamics of cylindrical water tanks which could be applied to your problem.
Question: H. Ruscheweyh. Do you know this fundamental work?
Reply: J. Berlamont, N. Vanderstappen. Unhappily we are not acquainted with this work, maybe because we were mainly interested in spherical and conical reservoirs. Your suggestion may be helpful in our future work.

Paper by L. J. Fortier, R. H. Scanlan--A COOLING TOWER WIND LOADING MODEL BASED ON FULL-SCALE DATA

Question: L. J. Mata. (1) How do the partial autocorrelation functions of the time series look? (2) Was some other model identification technique different from the Box-Jenkins technique used? (3) Did you find any necessity of differencing some of the time series (i.e., those collected from sensor 7 or sensor 10)?
Reply: L. J. Fortier, R. H. Scanlan. (1) In the present context we employed only the autocorrelation function, so that a direct answer to your question cannot be given here. However, clues to a response can be obtained by examination of the B-J text, the methods of which we followed closely. (2) No. In the context of this work, only the B-J techniques were used. In a later paper we examined spectral techniques based on more conventional Fourier transform methods.
(3) Yes. We explored differencing as well as changes in the digitizing interval. The latter changes did influence to some extent the type of model which became appropriate.

Question: A. Kareem. (1) Do you simulated time series satisfy the cross correlation between various locations? (2) Is this approach more economical than other existing approaches based on Monte Carlo schemes? (3) Have you tried simulation based on eigenfunctions of fluctuating pressure field based on K-L (Karhunen-loeve) series?
Reply: L. J. Fortier, R. H. Scanlan. (1) Yes. (2) This approach is indeed economical, but it is not proven that it is more economical than other approaches. (3) No. The authors appreciate the implications of the questions posed, and believe they are indeed important in seeking out really efficient methods for treating the still difficult problems of cooling tower dynamics.

Question: H. Propper. (1) For the numerical evaluation of c_{po} from your data, did you take into account that p_o is height dependent and p_i is independent of height (so that $c_{pi} = -0.47$ for the throat)? (2) Why did you use the logarithmic law instead of the power-law for the wind velocity?

Reply: L. J. Fortier, R. H. Scanlan. (1) The data we have were obtained at a single level. We would have recourse to wind tunnel or other results for assessing a mean loading scheme over a full tower. However, we believe that the character of the fluctuating pressure all the way around the tower at any level can be well simulated by our methods. (2) The full philosophy of why we chose the logarithmic law to represent mean wind velocity variation with height is given in the Simiu-Scanlan test "Wind Effects on Structures." Briefly, the log law represents a viewpoint consistent with that presently held by meteorologists.

Paper by M. Fujimoto, I. Matsushita, T. Ohkuma, T. Amano, H. Akagi--ON WIND-PROOF PERFORMANCE OF LATTICED STEEL TOWERS

Question: B. W. Smoth. The measurements of tower response reported appear to be related solely to response at the tower top. With the heavily perforated section of the Karagasaki tower the bracing forces will be critically dependent on correlation of the gusts on the tower. Were any measurements made of these forces during the investigation? The proposed gust factor (Figure 25), whilst applicable for leg force calculations, would not be adequate for small bracing members.

Reply: M. Fujimoto, I. Matsushita, T. Ohkuma, T. Amano, H. Akagi. During the full-scale measurements, only the displacement at the top of the tower was measured. For such heavily tapered towers the bracing force shauld be treated carefully. Relations between the height and the width of the top and bottom of such towers are limited as for the towers of Nippon Telegraph and Telephone Public Corporation.

Paper by A. J. Dutt--OSCILLATION RESPONSE OF STRUCTURES IN THE NATURAL WIND

Question: J. L. Wood. The work on the vibrating pole does not appear to show the effect of varying the pole top mass on the final results. Was any account taken of the stress induced in the pole by the mass? Was any evidence found that increasing the mass increased the damping effect?

Reply: A. J. Dutt. The result presented in this paper is only for a particular value of mass in the drum on the top of the tube. The natural frequency of the oscillation of the structure was limited to 0.8 Hz. The gravity load stress on the vertical tube was also taken into account. There was an increase in the value of structural damping with the increase in the value of mass, resulting in a decrease of dynamic deflection.

SESSION X

Wind-Engineering Practice

CHAIRMAN

L. E. Robertson
Executive Vice President
Skilling, Helle, Christiansen, Robertson
New York, New York, USA

CO-CHAIRMAN

M. D. Flynn
Senior Associate
I. M. Pei and Partners,
New York, New York, USA

RAPPORTEUR

R. A. Parmelee
Professor of Civil Engineering
Northwestern University,
Evanston, Illinois, USA

RECENT DEVELOPMENTS AND TRENDS: WIND-ENGINEERING PRACTICE

L. E. ROBERTSON* and M. D. FLYNN**

*Skilling, Helle, Christiansen and Robertson, USA

**I. M. Pei and Partners, USA

INTRODUCTION

At the time of the First International Conference on Wind Engineering, the application of wind engineering amongst practitioners was limited almost exclusively to concepts associated with steady-state forces and response; only a few researchers preached the gospel of the turbulent wind and the dynamic characteristic of structure response to that wind. Today, in reviewing the one hundred or so papers submitted to the Fifth International Conference on Wind Engineering, it is obvious that nearly every paper has immediate application to the practice of engineering. These papers, whether prepared by practitioners or by researchers, are useful to those professionals involved in the design and construction of many types of structures.......not just those particularly subject to wind-induced response.

CODES AND STANDARDS

Building codes and standards, such as those under development and in use in the technically-advanced portions of the world, now deal with (or will be dealing with) the broad application of the technology of wind engineering. Gone, are the complete reliances on steady-state concepts — being replaced with probabilistic approaches associated with the dynamic characteristics of the structure, the random and turbulent characteristics of the wind and the dynamic component of structure response. The building codes, of course, will continue to proffer simple, easy-to-use methodologies in analysis and design for those structures not sensitive to the action of the wind.

With the advent of load factor and resistance methods of design, both complex structures (dynamic in character), and simple structures (static in character), are and will be designed with more uniform levels of safety and to more established performance criteria. This will not always result in structures of lower cost, for the evolving field of Wind Engineering has uncovered glaring examples of deficiencies in wind-related design criteria and in wind engineering practice. At the same time, while not necessarily more economical, probabilities of improved protection against failure are taking on a new sense of reliability, not achieved in the past.

GLASS AND CLADDING

The cost of the exterior wall of major buildings continues to dominate the overall cost picture. These components are often more cost-dependent on wind loading than is the basic structural system. Research now underway, particularly that associated with the spacial relationships of wind loading, should result in significant improvements in the reliability of wind load predictions. This work, coupled with strength-related research of glass and other cladding elements, should result in substantial improvements in product reliability. Another important factor here is the trend to increased conservation on the part of designers and manufacturers alike.

PERFORMANCE-BASED DESIGN

We see an ever closing web around the uncertainties in the design for structure safety. The various codes and standards, and the advanced methodologies now available from the literature, while never perfect in their application to a wide variety and configuration of systems, are affording ever more rational, more reliable and more understandable methods of design. These improvements are enormously valuable to the practitioner — but this closing web moves less quickly and with less precision as it attempts to encircle the performance aspects of our structures and to encircle the impact of our structures on the environment. While not completely within the purview of wind engineering, the web closes most slowly on those characteristics of our structure that effect the quality of life of those who live and work in and around these structures.

Performance-based criteria in the field of wind engineering cannot be based on simplistic (or on complex) rules associated only with the wind, with the structure, and with their interaction. Instead, starting with those rules, performance-based criteria must go on to consider other building systems, the surrounding environs and those who will make use of the structure. For example, the acceptable wind-induced lateral deflection (static plus dynamic) between adjacent floors of a building is associated with the characteristics of the exterior wall, partitions, doors, elevators and the like. Further, a clear distinction must be made between the flexural component of response (that associated with axial force in columns, etc.) and the shearing component of response (that associated with bending of columns and girders, axial forces in diagonals, etc.) In kind, the acceptable level of dynamic motion is different from that of steady-state motion because of conditions associated with low-cycle fatigue of materials, the sensitivity of humans and of equipment to dynamic motion and so forth.

It is easy to see that many of these performance-related items are not traditionally within the areas of responsibility of the engineer. It follows that professionals involved in the design, construction and operation of buildings and structures need be educated into many of the fundamentals of wind engineering. The impetus for that educational process will likely come from those in the academic community of Wind Engineers — in communicating on to other disciplines the needs and the state-of-the-art of wind engineering.

THE CLASSIFICATION OF STRUCTURES FOR THE ASSESSMENT AND CODIFICATION OF WIND EFFECTS

R. I. HARRIS
Cranfield Institute of Technology, Cranfield, Bedford, UK

SUMMARY

This paper aims to provide an answer to the problem of how to distinguish between those structures for which wind effects are simple and can be treated by simple, possibly quasi-static, methods, and those for which wind effects are more complex and for which a complex, possibly fully dynamic, treatment is required. A classification method is presented which would be capable of graphical presentation in an introductory section of a Code. It is intended to be used to provide a branching decision to the rest of a Code containing different methods for wind effects of different levels of complexity.

INTRODUCTION

In wind engineering a number of methods for assessing wind effects on structures are available ranging from simple traditional discrete-gust methods to the more modern statistical methods using power spectral or similar techniques. It has often been suggested that the simple methods are adequate for simple structures and that more advanced methods are required only for complex or wind-sensitive structures. This leaves a designer in the unfortunate position of having to decide whether the structure he is considering will be simple or complex in this sense. There is clearly a need to be able to make this distinction quantitatively, based on those properties of a proposed structure which are known or can be estimated. The object of this paper is to present the basis of such a method.

DEFINITION OF CLASSES OF STRUCTURE

Through the advance of techniques for simulating, in a wind tunnel, the flow in the atmosphere, it is now possible to determine from measurements on models, the wind effects on a large proportion of proposed or existing buildings and structures - indeed in some of the more intractable cases, a wind tunnel simulation may be the only available method. In principle, therefore, all design against wind effects could proceed on a wholly experimental or empirical basis using such simulations. In practice, wind tunnel testing is far too expensive and cumbersome a technique to be used as a general design tool, and consequently there is a need for methods of assessing wind effects by calculation. Codification is a process by which standard calculation procedures are defined. Any calculation method will, of necessity, be semi-empirical in the sense that it will include and operate upon measured full scale data - for instance on winds in the atmosphere - and measured laboratory data for instance pressure or force coefficients measured in a wind tunnel in 'typical' conditions, but <u>not</u> in a full simulation of the particular problem under consideration.

By adopting such a semi-empirical method, a user makes some inherent assumptions about the nature of wind effects which are rarely specifically stated. The most important of these is that it is valid to divide a wind effects problem into three separate (but interacting) stages. These are:-

1) A description of the relevant wind structure.
2) Force or pressure coefficients and the derivation of loads.
3) Structural response to loads and hence the evaluation of wind effects.

All standard codified calculations assume that item 2) depends on item 1) and that item 3) depends only on 2), and not directly on 1). Thus it is assumed that all the effects considered under 3) can be calculated by proceeding in these three discrete stages. In fact, it is well known that there is a class of problems involving such phenomena as vortex induced oscillation of chimneys, galloping of transmission lines etc. - all known by the generic term aeroelastic - in which wind structure, aerodynamics and structural motion all react directly together to produce the resulting wind effects in an inherently inseparable way. Such problems are not at present codified, and the discussion just given implies that the absence of codification is not merely due to lack of knowledge, but because such problems are, by their nature, inherently incapable of conventional codification. In fact they are an obvious example of the 'intractable' type of problem mentioned earlier which can only be tackled by special methods or a full wind tunnel simulation. For the purposes of this paper, such structures will be designated as Class E. All that can be codified is a method for deciding if a particular structure belongs to Class E or not, and methods exist for this purpose(1). Consequently Class E structures will not be considered further in this paper.

For all other structures the investigation started from the premise that, given certain information about the site roughness, structural size, damping etc. - it would then be possible to perform some calculations which place any structure in one of four categories as far as wind effects are concerned. The four categories are:

<u>Class A</u>

Structures and/or elements of structures which are stiff enough for wind effects to be determined by statics, and small enough for the relevant wind information to be specified as a wind speed at a single point in space.

<u>Class B</u>

Structures which are stiff enough for wind effects to be determined by statics, but large enough to require the relevant wind information to be specified as multi-point data.

<u>Class C</u>

As for B, but with the additional complication that the shape of the structural load influence lines has to be considered in conjunction with multi-point wind data.

<u>Class D</u>

Structures which are not stiff enough to be treated by static methods, but instead require a full dynamic treatment.

EVOLUTION OF CLASSIFICATION METHOD

The approach adopted was the calculation of the displacement of four hypothetical idealised structures, chosen so that for certain values of their structural parameters they could be clearly identified as belonging to one of the categories A-D enumerated in the preceeding section. Structural displacement was chosen because it simplifies the calculation, but the conclusions will be immediately applicable to situations where the wind effect of interest is linearly related to displacement - this includes important cases such as shear forces and bending moments. The displacements have (with one exception) been obtained as analytical formulae which allows the effect of the variation of the structural parameters upon the classification of the structure to be studied.

The four idealised structures chosen were as follows:-

i) A structure small enough for the wind to be specified as single-point data, but with flexibility, so that dynamic response is possible. A practical example is a motorway sign or a lamp standard (for which the drag of the lantern dominates that of the column).

ii) A rigid horizontal bar, normal to the mean wind direction, and capable of lateral oscillation in a single rigid body mode (in the wind direction). A practical realisation here is a rigid bridge deck resting on slender piers.

iii) A flexible horizontal bar with pin/pin end conditions, capable of lateral oscillation in the mean wind direction. A practical realisation is a cable or a flexible bridge deck with pin/pin end conditions.

iv) A flexible vertical cantilever. A practical realisation here is a tall building or lattice tower.

For all these structures it has been assumed that 'strip theory' aerodynamics are valid. Thus, the local wind force acting on the ith element of the structure has been taken as proportional to the square of the local wind speed which would have occurred at the position of the element, in the absence of the structure. Or symbolically:-

$$F_i(\bar{r}_i;t) = \tfrac{1}{2}\rho A_i C_{Fi} V_i^2(\bar{r}_i;t)$$

where A_i is the area associated with the ith element.

C_{Fi} is the force coefficient associated with the ith element.

$\bar{r}_i$ is the position (vector) of the ith element relative to some origin.

t is time.

This 'strip theory' assumption is usually, of necessity, made in the codification of wind effects problems. For the structures considered the force coefficients were assumed to be independent of wind speed.

In order that the results from these four structures can be used in a classification procedure it is necessary to define consistently and to calculate a performance parameter for all four structures. The parameter chosen is the ratio, R, defined as:-

$$R = \frac{\text{Maximum mean square displacement anywhere in the structure due to gust action}}{\text{Do. for stiff(quasi-static) structure \& gusts completely spatially correlated}}$$

The reasons for this particular choice of R will become apparent as the individual cases are discussed.

Structure I

This structure is defined as being small enough for the wind speed to be specified at a single point - another way of stating this definition is that the structure is small enough for gusts to be regarded as fully correlated over the entire structure. Consequently it is to be expected that spatial gust variations are irrelevant to the determination of wind effects on such a structure, and the structure will therefore fall into Class D or Class A depending upon whether dynamic response is, or is not, significant.

The obvious definition of the performance ratio is:-

$$R = \frac{\text{mean square displacement}}{\text{quasi-static mean square displacement}}$$

and because of the built-in assumption of full spatial correlation and the fact that the displacement of Structure I is specified by a single quantity (not dependent on a spatial co-ordinate), it is clear that this definition of R is identical for Structure 1, to the general definition already given.

In the general case for Structure I it is found that:-

$$R = 1 + f(n_oT,\zeta)$$

where specifically:-

$$f(n_oT,\zeta) = -0.2085(n_oT)^{-2/3}\operatorname{cosec}(2\phi)\sin(8\phi/3) + O(n_o^{-2}T^{-2})$$

Here:-

n_o is the natural frequency of the structure

ζ is the damping ratio (including any aerodynamic damping)

$\phi = \arccos(\zeta) \quad \pi/4 < \phi < \pi/2$

T is the time scale of the gusts at the same height at which the single point data used for Structure I is specified. This height has also to be specified because, in general, T varies with height above ground level.

For most practical structures n_oT is large enough for the terms $O(n_o^{-2}T^{-2})$ to be neglected, but exact results for these terms are available if required.

Since $1/n_o$ is the period of oscillation of the structure, n_oT is a ratio of a time characteristic of the mechanical properties of the structure - its period - to another time which is characteristic of the wind environment - the time scale of the turbulence.

If the structure is very stiff, $n_o \to \infty$, and therefore $R \to 1$, which clearly corresponds to the quasi-static limit. It follows that, in the general case, Structure I will be Class A unless the contribution from $f(n_oT,\zeta)$ causes R to depart significantly from a value of +1.

If the damping is relatively light, $\zeta \ll 1$, $\cos(\phi) \simeq \pi/2$, and therefore $\sin(8\phi/3)$ is negative, and consequently, overall $f(n_oT,\zeta)$ is positive. It will remain positive unless $\zeta > 0.38$ - which is much larger than values likely to be encountered in practice. For all practical cases, therefore, $f(n_oT,\zeta) > 0$ and Structure I can be classified as follows:-

$$f(n_oT,\zeta) > 0.2 \quad \text{Class D}$$
$$f(n_oT,\zeta) < 0.2 \quad \text{Class A}$$

The choice of 0.2 or 20% as the demarcation level is, of course, arbitrary, but seems reasonable.

Structure II

The larger size (length) of Structure II, means that spatial correlations of gusts become important for the determination of wind effects upon it and consequently the more generalised definition of the performance ratio, R, is required. (The displacement is still, however, specified as a single, spatially independent, quantity because the structure is assumed to be capable only of lateral transational motion as a rigid body).

From the definition of Structure II it follows that the results for the performance ratio must become identical to that for Structure I if either:-

i) the structural size parameter (length) tends to zero

or

ii) the lateral (space) correlation of the turbulence is large compared with the size parameter so that gusts can be regarded as fully correlated over the whole structure.

For Structure II it is found that:-

$$R = 1 + f(n_oT,\zeta)F_r(N,\zeta)$$

where

$f(n_oT,\zeta)$	is the same as for Structure I.
N	denotes the ratio $n_oB/\bar{V}$.
B	is the structural size parameter.
$\bar{V}$	is the mean wind speed at the height of the structure. Again since $\bar{V}$ varies with height, this height must be specified.
$F_r(N,\zeta)$	is a complicated function of N,ζ obtained for a rigid body.

The ratio $\bar{V}/n_o$ is the product of the mean wind speed and the period of oscillation of the structure. Dimensionally, it is therefore a length and represents the size of a turbulent eddy which, if it were swept past the structure at the mean wind speed, would produce a speed variation and hence a load variation at the natural resonant frequency of the structure. The parameter N, is thus the ratio of the size of the structure to the size of this 'resonant eddy' and thus characterises the spatial and temporal relationships of the structure to its turbulent wind environment. From the definition, it would be logical to call N the Natural Strouhal Number or Reduced Frequency of the structure.

The classification of Structure II depends on the size and sign of the product $f(n_oT,\zeta)\ F_r(N,\zeta)$, and thus on the properties of the function $F_r(N,\zeta)$. This has been obtained analytically in the form of an infinite series, which has permitted its limiting properties to be examined. Design charts can, of course, be prepared by evaluating the series by computer for particular values of N and ζ.

For $N \to 0, F_r(N,\zeta) \to +1$, independent of the value of ζ. This confirms the requirement that for small values of the size parameter, the performance ratio of Structure II must become identical to that for Structure I.

As N increases from zero, the magnitude of $F_r(N,\zeta)$ decreases from +1, passes through zero, and then increases towards large negative values. Again, this behaviour occurs regardless of the value of ζ, except that the lower the value of ζ, the slower is the decrease of $F_r(N,\zeta)$ with increase of N. However, this result implies that regardless of the stiffness of the structure and hence of the value of the natural frequency, n_o, the product $f(n_o T,\zeta)\ F_r(N,\zeta)$ will always vanish for a set of pairs of values of N and ζ. In terms of whether the structure responds dynamically or not, there is thus a trade-off between low damping, which tends to increase dynamic response, and large size which reduces dynamic response by averaging or smearing out the resonant eddies. Thus any structure, even if it is flexible and lightly damped can be regarded as quasi-static, if it is big enough.

Classification of Structure II thus depends, in the first instance on whether $F_r(N,\zeta)$ is positive, zero or negative. Figure 1, curve (1), shows the locus in the N,ζ plane of those combinations of N,ζ which lead to $F_r(N,\zeta) = 0$. It follows that

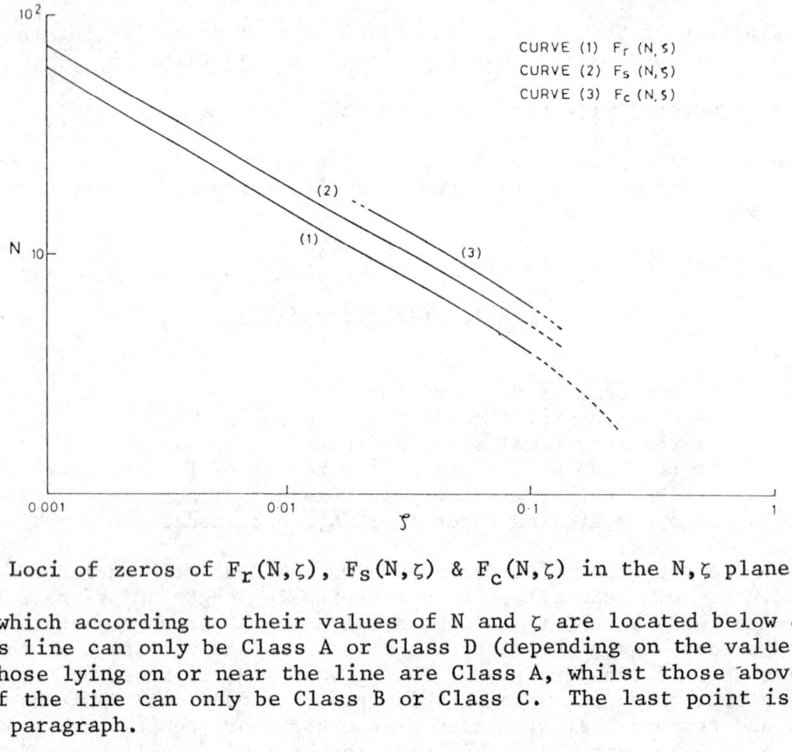

Fig. 1. Loci of zeros of $F_r(N,\zeta)$, $F_s(N,\zeta)$ & $F_c(N,\zeta)$ in the N,ζ plane.

structures which according to their values of N and ζ are located below and to the left of this line can only be Class A or Class D (depending on the value of $f(n_oT,\zeta)$, those lying on or near the line are Class A, whilst those above and to the right of the line can only be Class B or Class C. The last point is discussed in the next paragraph.

As N becomes very large, so that $F_r(N,\zeta)$ is negative:-

$$f(n_oT,\zeta)\ F_r(N,\zeta) \rightarrow g_r(B/\bar{V}T) < 0$$

where, as the notation suggests, the function $g_r(B/\bar{V}T)$ depends only on the ratio $B/\bar{V}T$ and is independent of n_o or ζ. The rate, with increasing N, at which the product fF_r approaches this limiting condition does, however, depend upon n_o and ζ.

Now $\bar{V}T$ is the <u>longitudinal length scale</u> of the turbulence and thus $B/\bar{V}T$ is the ratio of the spatial property of the structure - its size - to the spatial property of the turbulence. Since $g_r(B/\bar{V}T)$ is always negative this limit corresponds to the case when the structure is sufficiently large to behave quasi-statically, and where the quasi-static wind loads (and hence effects) are reduced owing to the incomplete correlation of the gusts. As might be anticipated, the analysis shows that time or frequency relationships are irrelevant (because the behaviour is quasi-static) and the effect of loss of correlation depends only on the spatial relationship of the structure to its environment. This limit thus corresponds to Structure II being in Class B or Class C.

Structure III

Structure III is identical to Structure II except for the fact that it is flexible, and thus consideration of Structure III in comparison with the results for Structure II enables the effect of the flexibility to be evaluated. The main difference is that Structure III has a mode shape or influence line associated with each mode of deflection which varies according to position within the structure - for instance for a bar with pin/pin end conditions the displacement must be a maximum at the centre, and vanish at the points of attachment at the ends. Consequently, the general definition of the performance ratio R must be used. Any flexible structure has an infinity of modes, but for the purposes of classification it is only necessary to examine the mode of lowest frequency.

For Structure III the analysis proceeds along almost identical lines to that for Structure II. The result is:-

$$R = 1 + f(n_oT,\zeta)\ F_s(N,\zeta)$$

Here $F_s(N,\zeta)$ is a complicated function similar in form to $F_r(N,\zeta)$ but as obtained for a string-like body. The function $F_s(N,\zeta)$ can again be obtained analytically as an infinite series and has the following limiting properties:-

$F_s(N,\zeta) \to +1$ as $N \to 0$, independent of ζ, so that again the result for R reduces to that for Structure I as the size of the structure becomes small enough.

As N increases the magnitude of $F_s(N,\zeta)$ decreases from +1, through zero, and then increases to large negative values. The rate of decrease is very similar to that for $F_r(N,\zeta)$. It follows that for $F_s(N,\zeta)$ a locus can be plotted in the N,ζ plane of pairs of values of these quantities for which $F_s(N,\zeta)$ is zero. This is shown in Figure 1 as curve (2). It can be seen that:-

(i) Curve (2) lies very close to curve (1) - the influence of flexibility and mode shape on the classification of the structure is thus not very important.

(ii) Curve (2) lies above and to the right of curve (1). It thus follows that curve (2) dominates curve (1) in the sense that if curve (2) is used for classification of a rigid structure it may result in the structure being classified as Class A when it may really be Class B or C, but can never result in a structure being classified as B or C when it should be D.

When N becomes very large:-

$$f(n_oT,\zeta)\ F_s(N,\zeta) \to g_s(B/\bar{V}T) < 0$$

for the same reasons as in the case of Structure II.

For $B/\bar{V}T$ small, $g_s(B/\bar{V}T)$ for the flexible bar and $g_r(B/\bar{V}T)$ for the rigid bar do not differ appreciably. However as the size ratio $(B/\bar{V}T)$, increases these two results diverge - this is illustrated in Figure 2, where curve R_r is for the rigid

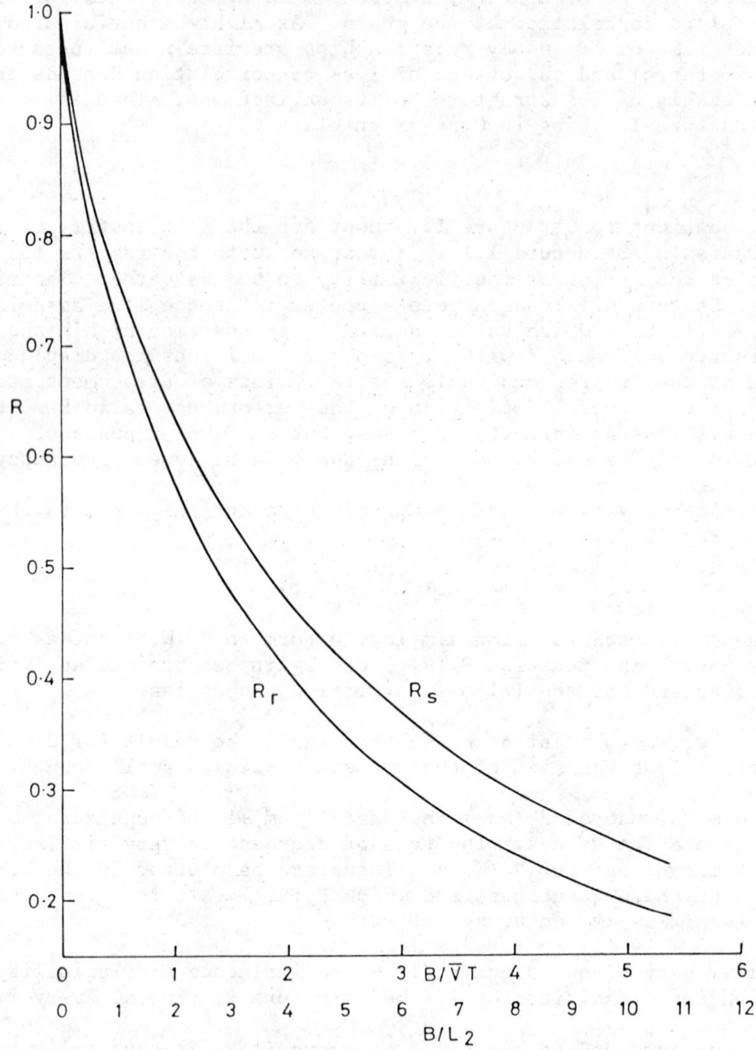

Fig. 2. Quasi-static performance parameters R_r and R_s

bar and curve R_s for the flexible bar. It should be noted that:-

$$| g_r(B/\bar{V}T) | > | g_s(B/\bar{V}T) |$$

which implies, in an obvious notation, that:-

$$R_s > R_r$$

Consequently, in the quasi-static case, the reduction in wind effect which may be assumed for a large flexible body is less than the corresponding reduction for a rigid body of the same size.

The difference between the results for rigid and flexible bodies can be used to distinguish between these structures which are Class A, Class B and Class C.

Because the appropriate modifications are easier to apply in this case a demarcation level of 0.1 or 10% has been taken and gives:-

$$g_r < 0.1 \quad \text{Class A}$$

$$(g_r - g_s)/g_s < 0.1 \quad \text{Class B}$$

$$(g_r - g_s)/g_s > 0.1 \quad \text{Class C}$$

Because g_r and g_s depend only on the ratio $B/\bar{V}T$, it is not necessary to calculate the values of the function g_r and g_s, but merely the ratio $B/\bar{V}T$, since the demarcation between classes based on fixed values of g_r and g_s will occur at fixed values of this ratio. However, on physical grounds, it is also obvious that it is the ratio of the crosswind dimension of the structure to the crosswind or lateral scale of turbulence (and not the longitudinal scale) which is the relevant ratio. In the present work, the result has emerged in terms of the longitudinal scale only because the analysis contains the built-in assumption that:

$$\bar{V}T = 2L_2 \qquad (L_2 = \text{lateral scale})$$

Since this is not always the case and L_2 is dependent on height above ground level, this suggests that the demarcation ratios should preferably be expressed in terms of B/L_2 and in applying these criteria the correct local value of L_2 should be used. With this provision and with the aid of Figure 2, the criteria became:-

$$0 < B/L_2 < 0.25 \quad \text{Class A}$$

$$0.25 < B/L_2 < 2.0 \quad \text{Class B}$$

$$B/L_2 > 2.0 \quad \text{Class C}$$

Structure IV

Structure IV differs from Structure III in two respects. It is a fixed/free cantilever and hence the mode shape will differ from that appropriate to the pin/pin end condition assumed for Structure III (actually the lowest mode shape for Structure IV differs very little from one half of the lowest (symmetrical) mode shape of Structure III). Differences in response arising from this cause will be small.

More important is the fact that the Structure is presumed to be vertical, and thus the variation with height above ground level of the mean wind speed, $\bar{V}(z)$, and of the r.m.s. gust speed, $\sigma(z)$, must be included in the analysis. As before the result is:-

$$R = 1 + f(n_oT, \zeta)\, F_c(N, \zeta)$$

where $F_c(N,\zeta)$ is again a complicated function similar to $F_r(N,\zeta)$ but as obtained for a cantilever body. $F_c(N,\zeta)$ can be obtained analytically in the form of a double integral with respect to the space variable, and includes $\bar{V}(z)$ and $\sigma(z)$ in the integral, but only as their product $\bar{V}(z)\sigma(z)$. Without specifically evaluating this integral it can be shown that $F_c(N,\zeta) \to +1$ as $N \to 0$, independent of ζ, so that once again the result for R reduces to that for Structure I as the size of the structure becomes small enough.

Further investigation of $F_c(N,\zeta)$ is more conveniently carried out by evaluating the integral numerically, and accordingly this was done for a particular choice of site roughness length and height of structure - 0.01m and 100m respectively.

Again as N increases, the magnitude of $F_c(N,\zeta)$ decreases from +1 through zero, and then increases to large negative values, in a very similar way to the behaviour of $F_s(N,\zeta)$. As previously, a locus can be plotted in the N,ζ plane of pairs of values of these parameters for which $F_c(N,\zeta)$ is zero. A portion of this locus is shown in Figure 1 as curve (3). It can be seen that:-

(i) Curve (3) lies very close to curves (1) and (2) - the influence of the vertical variation of $\bar{V}(z)$ and $\sigma(z)$ is thus not very important.

(ii) Curve (3) lies above and to the right of curve (2). It thus dominates curve (2) in the sense that if the entire curve (3) were used for the classification of all structures, it would produce conservative results relative to the use of curve (2).

For most sites, it is known that the product $\bar{V}(z)\sigma(z)$ becomes almost constant above a height which depends on site roughness (about 20m for open country and 60m for suburbia). Consequently, apart from minor differences due to mode shape, it can be confidently asserted that if the size of Structure IV became very large, and hence N became large, Structure IV would approach a quasi-static limit similar to that for Structure III. For practical vertical structures it is found that $N < 2.5$ so that this limit is never attained.

CODIFICATION OF CLASSIFICATION PROCEDURE

From the results of the previous section, a simple classification procedure can be defined which involves a minimum of calculation for the user. Some brief explanatory notes are given in this section. It is not envisaged that they would be included in this form, in any formal code.

The classification procedure starts with the assumption that the first figure in the Code would consist of a map of the 1:50 yr (say) value of the hourly mean friction velocity, u_*, for standard terrain (i.e. of known specified roughness length z_o). It is recognised that this represents a departure from the traditional practice of a map of gust speeds at 10m above ground level for standard terrain, but it is clear that if modern work on wind structure - such as the Deaves/Harris model(2) - is to be utilised, then given a map of $\hat{V}_{10}$ or $\bar{V}_{10}$, any user would first always have to calculate u_* from it. This seems a pointless step and the user might therefore better start with a u_* map.

It is then assumed that the second figure in the Code would consist of a chart identifying values of z_o with categories of terrain roughness (similar or identical to the one in the ESDU Data Sheet(3).

The third figure in the Code would then consist of a graph or nomogram which would enable the user to convert the u_* value from the map to that appropriate to the terrain for which he has selected the z_o. This can be done by using the assumption that the gradient wind speed is the same for both sites and hence by employing the relationship between G and u_* derived from similarity theory.

A user can then select the height above ground level relevant to his problem - obvious for a horizontal structure, and in the case of a vertical structure, the height of the structure itself - and then calculate the values of $\bar{V}$, T and L_2 appropriate to that height by the formulae given in Deaves/Harris(2). It may be remarked that since all these formulae involve only two parameters - u_* and z_o - all the calculations can be replaced by graphs or nomograms.

The user will then assign values of the size parameter (obvious for most structures), the lowest oscillation frequency, n_o, (this may require a table of recommended values for typical structures), and the damping ratio, ζ. This last is much the most difficult, and it is probable that the user will need to estimate upper and lower bound values and observe the results in terms of classification. Having settled for values of B, n_o and ζ, the user will then calculate n_oT and N.

With values of N and ζ available the user will then proceed to a graph in the Code shown as Figure 3 in this report, which is based upon Figure 1. The shaded zone

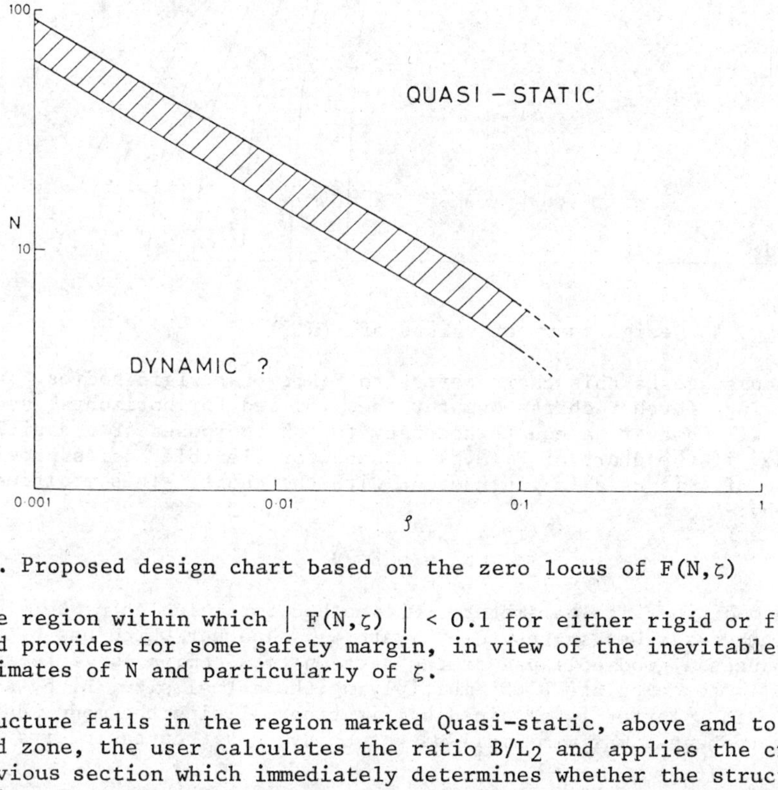

Fig. 3. Proposed design chart based on the zero locus of $F(N,\zeta)$

denotes the region within which $|F(N,\zeta)| < 0.1$ for either rigid or flexible bodies, and provides for some safety margin, in view of the inevitable uncertainties in the estimates of N and particularly of ζ.

If the structure falls in the region marked Quasi-static, above and to the right of this shaded zone, the user calculates the ratio B/L_2 and applies the criteria given in the previous section which immediately determines whether the structure is Class A, Class B or Class C. He can then proceed to the appropriate part of the Code.

If the structure falls within the shaded zone or below and to the left of it, in the region marked Dynamic(?), the user calculates $f(n_oT,\zeta)$ from the formula given in the previous section - a simple task for any electronic pocket calculator having triginometric functions and powers available.

If $f(n_oT,\zeta) < 0.2$, the structure is Class A. Otherwise, the user refers to the next figure, produced as Figure 4 in this report, which is a family of curves for $F_r(N,\zeta)$ shown as a function of N for values of ζ from 0.001 to 0.1 in multiplicative steps of $3\sqrt{10}$. If the required value of ζ does not coincide with one of the curves, interpolation is not necessary; instead the user should refer to the curve for the value of ζ <u>nearest to and below</u> the required value.

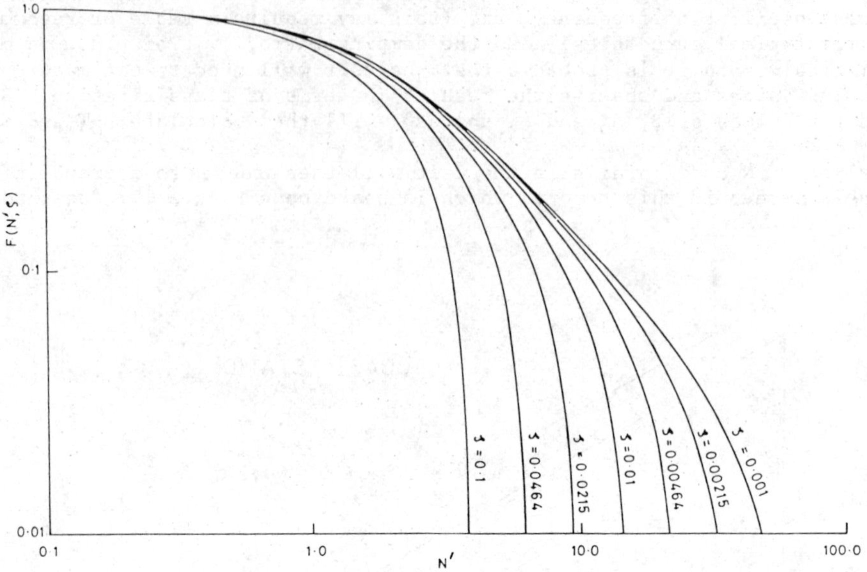

Fig. 4 Design chart of values of $F(N'\zeta)$

As the notation suggests this chart refers to $F(N,\zeta)$ for rigid bodies, and, strictly speaking, further charts ought to be provided for horizontal and vertical flexible bodies. However, adequate accuracy for the purposes of classification can be attained if the chart of $F_r(N,\zeta)$ is used for flexible bodies provided that a modified value of N is used in conjunction with the chart. This modified value, N', is given by:-

$$N' = n_o B'/\bar{V}$$

where B' is a modified size parameter. One method for calculating this parameter for quasi-static cases was suggested by Wyatt(4). Another which has been used here, and produces a good collapse of the data on to a common curve for the cases investigated, is to calculate B' by multiplying the actual size, B, by a mode shape factor. This latter is obtained by first normalising the mode shape of the flexible body to have a maximum amplitude of unity, and then taking its spatial mean.

For example, for a pin/pin beam of length B, the normalised fundamental mode shape is given by:-

$$y = \sin(\frac{\pi x}{B})$$

Mode shape factor

$$= \frac{1}{B}\int_0^B \sin(\frac{\pi x}{B})\, dx = \frac{2}{\pi}$$

Hence $$B' = (2/\pi)B$$

and $$N' = (2/\pi)N$$

Having made this adjustment, the user should then read off the value of $F_r(N',\zeta)$

and calculate the product $f(n_oT,\zeta)F_r(N',\zeta)$. If this product is less than 0.2, the structure is Class A, otherwise it is Class D, and again the user can proceed to the appropriate section of the Code.

CONCLUDING REMARKS

This paper demonstrates that a simple classification process can be devised for practical structures which can be made conservative with respect to likely errors in estimated parameters. It is also satisfying that the analysis shows that whether a structure is 'complex' or not can be shown rigorously to depend upon those quantities which engineering intuition also suggests would be the governing parameters.

ACKNOWLEDGEMENT

This work was carried out for the Building Research Station under Contract F3/2/160. This support is gratefully acknowledged.

REFERENCES

(1) Simiu, E. and Scanlan, R.H., An Introduction to Wind Engineering, Wiley, New York, 1978.

(2) D. M. Deaves and R. I. Harris, A Mathematical Model of the Structure of Strong Winds, CIRIA Report 76, (May 1978).

(3) Engineering Sciences Data Unit, Characteristics of Wind Speed in the Lower Layers of the Atmosphere near the Ground: Strong Winds (Natural Atmosphere), Data Item 72026, (1972).

(4) T. A. Wyatt, The Calculation of Structural Response, Paper 6, Symposium on the modern design of wind-sensitive structures, CIRIA, London, (1971).

DESIGN WIND SPEED AND WIND LOAD FACTOR BASED ON PROBABILISTIC RATIONALE

M. ITO* and Y. FUJINO**

*University of Tokyo, Tokyo, Japan
**University of Tsukuba, Ibaraki, Japan

ABSTRACT

The yearly maximum wind speed data recorded at 136 stations in Japan are statistically analyzed and fitnesses of the theoretical extreme distributions of types I, II and III to the data are discussed. A mixed distribution is found suitable, in some area, to characterize the yearly maximum wind speed. In the next place, the reliability level of the current Japanese specifications for steel highway bridges subject mainly to wind loading are evaluated in terms of the safety index. By reliability-based code calibration, an appropriate value of the wind load factor in the load factor design which will be implemented in lieu of the current allowable stress design is obtained.

INTRODUCTION

For some structures or structural members, wind loading is a significant factor with respect to safety. To ensure the safety of a structure subject to a wind action, wind resistant design should be controlled by many provisions in a structural code and the first step is to give a basic design wind speed at the site.

Although allowable stress design has been used in most of structural codes, load factor design is now widely accepted as suitable for the next generation of structural codes, where safety provisions are expressed in terms of resistance and load factors. In the load factor design, load factors including a wind load factor must provide a comparable level of safety if the code is self-consistent.

Probabilistic rationale is already recognized as appropriate to use in reliability analysis and in determination of design loads and of load factors, i.e., code decision making problems, and is adopted as basic rationale in this study. The objectives of this paper are, from probabilistic and statistical viewpoints, to present some considerations of the yearly maximum wind speeds relevant to the design wind speed and to show, by reliability-based code calibration, how to obtain an appropriate value of the wind load factor with an example of design of steel highway bridges. The wind speed data observed over a number of years at 136 meteorological stations in Japan are used in this study.

STATISTICAL ANALYSIS OF EXTREME WIND SPEED

Distribution of Yearly Maximum Wind Speed and Design Wind Speed

Wind speed varies in a nearly random fashion and can be approximately described as a stationary random process to the exclusion of the seasonal change. This makes it possible to predict a wind speed on the statistical basis. Strong winds which are of interest from a viewpoint of structural safety usually occur only a few times for a year, while the anticipated lifetime of structures is generally more than 30 years. The yearly maximum wind speed data are therefore convenient information form in determining a basic design wind speed which is defined in Japan as the ten-minute averaged wind speed at the altitude of ten meters above ground or water level. These data can be considered as mutually independent samples from the parent distribution.

Accepting probabilistic rationale in design, the basic design wind speed is then defined as the speed v_D which has a specified non-exceedance probability p in the estimated yearly maximum wind speed distribution. Instead of the probability p, an equivalent expression "R-year return period" has been widely used, which is defined by

$$R = 1 / (1 - p) \tag{1}$$

To represent the statistical behavior of the yearly maximum wind speed, two types of extreme distributions with unlimited upper tail have been commonly used. One is the type I distribution, so-called Gumbel distribution, the other the type II, Frechet distribution, of which the cumulative functions take, respectively, the form of

$$F_I(v) = \exp\{-\exp[-\alpha(v - u)]\} \qquad -\infty < v < \infty \tag{2}$$

where α is scale parameter, while u location parameter, and

$$F_{II}(v) = \exp[-(\omega/v)^{\kappa}] \qquad 0 < v < \infty \tag{3}$$

in which ω and κ are scale and tail length parameters, respectively. It should be noted that there is one more type of extreme distribution, called the type III, Weibull distribution, and its cumulative function is given by

$$F_{III}(v) = \exp\{-[(\omega - v)/(\omega - u)]^k\} \qquad v \le \omega \tag{4}$$

where u and k are statistical parameters and ω is the upper limit of the random variable v.

The choice of the underlying distribution is important in determining the design wind speed and in structural reliability analysis. Reliability analysis of a structure subject to a wind loading generally requires the maximum wind speed distribution during its lifetime. Since the yearly maximum wind speeds are assumed mutually independent, the N-year maximum wind speed distribution $F_N(v)$ can be obtained from the yearly maximum distribution F(v) by

$$F_N(v) = [F(v)]^N \tag{5}$$

where the value of N is usually taken as more than 30. The calculated reliability level of the structure from Eq. 5 fairly depends on the tail shape of the distribution F(v), which will be shown later.

The choice of the underlying distribution, usually among the first two (Eqs. 2 and 3), can be made after the analysis of fitting closeness to the data. It was shown that the type II distribution is an appropriate model to employ in most of North America region [1], although recent studies [2,3] have indicated that the type I is more appropriate. In Japan and many other countries, the type I is favored to use on the basis of the statistical data analysis [4]. However, extensive analysis of fitting closeness of the extreme distributions including the type III to the wind speed data in Japan has not been made yet.

Station Data Analysis

The calibrated yearly maximum wind speed records (10 min. averaged)[5] at 136 meteorological stations located in Japan are herein statistically analyzed. Each station has yearly maximum wind speed records over about 45 years. In plotting the station data, Hazen's method is used : the cumulative probability P_i of i-th largest record among m years records is equal to $1 - (2i - 1)/m$. Values of the statistical parameters of these three distributions are obtained by the least square method. It should be noted that value of the upper limit of the type III distribution is tentatively chosen to be 150 m/sec. As examples, the yearly maximum wind speed records and the fitted distributions in Tokyo, Fukuoka and Sapporo are shown in Figs. 1,2 and 3 where the type I probability papers are used. From these figures, it can be said that none of the three types of distributions is given preference and also it is shown that the wind speeds of concerned return periods of 50 ∿ 100 year are not so different for the distribution type. However, in Fig. 4 which presents the 95% fractile values of the N-year maximum wind speed calculated by Eq. 5 for the three types of the fitted distributions, it is shown that the choice of distribution of whether the type I or II makes noticeable differences in the fractile values, leading to the different values of calculated reliability of the structure against the wind loading.

An analysis of fitting closeness of the distributions to the recorded data is herein carried out. The following measure of fitting closeness is adopted:

$$E = \sqrt{\frac{1}{m} \sum_{i=1}^{m} (v_i - v_{i,t})^2 / v_{i,t}^2} \tag{6}$$

where v_i is the i-th largest wind speed while $v_{i,t}$ is the wind speed corresponding to that of the non-exceedance probability P_i in the fitted distribution. The values of E for the three types of the fitted distributions are then calculated.

Since the extreme distributions of types I and II have been commonly used to represent the yearly maximum wind speed, fitness levels among the two are first compared in term of the value E. The criteria for the choice of the underlying distribution is shown in Fig. 5. The results of the choice of the distribution under the criteria are shown in Table I. Among 136 stations, the data at 77 stations are judged to be better fitted by the type I distribution, while 42 stations by the type II. There are approximately 40% of the stations which do not accept the type I and this could be attributed, at some of the stations, to sampling uncertainties due to the limited size of the data[3]. Then conclusively, use of the type I distribution in Japan is statistically reasonable at most of the stations.

Next, the fitness to the distributions of the three types is investigated according to the criteria given in Fig. 6 which are similar to the previous ones given in Fig. 5. The results are shown in Table II, where one can find that the number of the stations of which the data are best fitted by the type III is the largest among the three types. From Table II, it can be pointed out that the type III distribution for the yearly maximum wind speed should be considered to employ in some area, although it has not been used before. Existence of meteorological upper limit of the wind speed may also support to adopt the type III distribution.

Fitting of one type of distribution to the data has been considered so far. However, the case that a mixed distribution is more satisfactorily fitted to data occurs sometimes[1]. The wind speed records in Osaka are shown in Fig. 7, where one can observe that none of the three types is well fitted to the records.

Actually, the records in Osaka were judged to be "Poor Fitness" under both of the criteria presented in Figs. 5 and 6. The top three wind speed records can be attributed to typhoons, and others can be almost to seasonal winds. Accepting the type I distributions to the records due to typhoons and seasonal winds, a refined distribution can be drawn as shown in Fig. 7. It appears to be fitted much closely to the records. This mixed distribution gives significantly larger design wind speed for longer return periods. It should be noticed that such a consideration may produce a remarkable difference in the design wind speed.

RELIABILITY-BASED CALIBRATION OF WIND LOAD FACTOR

Reliability Level of the Current Japanese Specifications for Highway Bridges

The design wind load in JRA Specifications[6)], being applied to designing highway bridges of less than 200m span length, is corresponding to the basic ten-minute averaged wind speed of about 40m/sec. In the commentary of the specifications, it is mentioned that 40m/sec wind speed corresponds to the wind speed of the non-exceedance probability 0.6 for the 50 years lifetime, i.e., approximately 100-year wind speed. Although wind speed characteristics vary from location to location, a constant design wind speed regardless of locations is adopted in the current specifications. Consider this effect on the reliability of steel highway bridge structures mainly subject to wind loading.

In the presence of wind loading, the safety checkings in the JRA specifications which are in the form of allowable stress design are made as follows:

$$\begin{aligned} \sigma_a &\geq D^* + L^* \\ \sigma_a &\geq (D^* + L^* + W^*)/1.25 \\ \sigma_a &\geq (D^* + L^* + W^* + T^*)/1.35 \\ \sigma_a &\geq W^*/1.20 \end{aligned} \tag{7}$$

where σ_a is allowable stress, while D^*, L^*, W^* and T^* are dead, live, wind loads and thermal stress, respectively.

The safety of structures is herein measured, in the sense of the second-moment first-order reliability, by the safety index. Let $\bar{\theta}$ ($\equiv \bar{R}/\bar{S}$) be the central safety factor, and let V_R and V_S be the coefficients of variation of the resistance, R, and the total load effect, S, associated with a design. The safety index is then expressed as

$$\beta = \ln \bar{\theta} / \sqrt{V_R^2 + V_S^2} \tag{8}$$

Design live load L^* and thermal stress T^* are assumed to be zero for the sake of convenience. Statistics of the wind load at the stations are derived from the yearly maximum wind speed distributions. Employing the type I distribution for the yearly maximum wind speed at all the stations, the statistics of the N-year maximum wind speed are given from Eq. 5 by

$$\bar{v}_N = u - (0.577 + \ln N)/\alpha, \quad \sigma_N^2 = 1.645/\alpha^2, \quad V_N = \sigma_N/\bar{v}_N \tag{9}$$

where $\bar{v}_N$, σ_N^2 and V_N are the mean, variance and coefficient of variation of the N-year maximum wind speed while u and α are the statistical parameters of the fitted type I yearly maximum distribution. The value of N is assumed to be 50 years. The wind load can be considered to be proportional to the square of the wind speed. The total uncertainties of the wind load are then given by

$$\bar{W} \propto \bar{v}_N^2 \times \bar{U}, \quad V_W = \sqrt{(2V_N)^2 + V_U^2} \tag{10}$$

where U denotes the uncertainty due to the gust response factor and the drag coefficient etc., and the effect of the direction of winds to the structure are not taken into account. Assume that the total load effect S is expressed as

$$S = E(D + W) \tag{11}$$

where E is the uncertainty denoting error of load analysis. The values of $\bar{E}$, V_E, $D^*/\bar{D}$ and V_D are taken as 1.0, 0.05, 1.0 and 0.07, respectively. Concerning the wind load W, the statistics $\bar{v}_N$, σ_N^2 and V_N are calculated by Eq. 9 from the fitted type I distributions of the yearly maximum wind speed at 136 stations in Japan. The values of $\bar{U}$ and V_U are assumed to be 1.0 and 0.15, respectively from the authors' judgment. On the other hand, concerning the strength of steel structures, yield stress σ_Y is taken as the limit state. Then, the current safety factor which is the ratio of nominal yield stress σ_Y^* to allowable stress σ_a is approximately 1.7. With inclusion of other uncertainties associated with the strength of structures, $\bar{\sigma}_Y/\sigma_Y^*$ and V_{σ_Y} are assumed to be 1.2 and 0.15, respectively.

The values of the safety index β which are calculated with the aforementioned statistical data are different from location to location among 136 stations. The mean $\bar{\beta}$, $\bar{\beta} \pm \sigma_\beta$($\sigma_\beta$: the standard deviation of the β values), and furthermore the β values in Matsumoto and Nagasaki as examples are shown in Fig. 8. It can be observed that large variations of the β values exist.

The average values $\bar{\beta}$ in Fig. 8 are 3.5 ∿ 4.0 and are close to the β values of steel highway bridges subject to traffic live loading[7] and other civil engineering structures[8], and therefore the current design for wind load is relatively consistent with respect to the safety in the average sense.

Next, the safety index values in the case when 100-year wind speed of each station is used as the basic design wind speed are shown in Fig. 9. One can find that the variations of the safety index values are tremendously reduced in comparison with those in Fig. 8. However, it should be noticed with caution that the average values $\bar{\beta}$ are lowered when 100-year wind speed of each location is used instead of the wind speed corresponding to the current JRA specifications.

Calibrated Wind Load Factor

Transition from allowable stress design to load factor design in structural codes are now under consideration in many countries. For a limit state, the load factor design format prescribed by structural codes generally takes the form of

$$\phi R^* \geq \sum_{i=1}^{k} \alpha_i S_i^* \tag{12}$$

where R^* and S_i^* are nominal resistance and nominal load while ϕ and α_i are resistance and load factors, respectively, and k is the number of loads considered. Values of the factors ϕ and α_i must be assigned by the responsible code authorities. Reliability-based code calibration can be defined as selection of values of ϕ and α_i in such a manner that values of these factors in conjunction with the design format provide the structural safety adequately close to the target safety level. A practical method to determine the target safety level is to investigate reliability levels of currently acceptable designs.

Consider the load factor design format in the presence of dead and wind loads only. It takes the form of

$$\phi R^* \geq \alpha_D D^* + \alpha_W W^* \tag{13}$$

When the yield stress is taken as the limit state, R^* can be replaced by σ_Y^*. The values of the safety index β achieved by the current design as shown in Fig. 8 are adopted as the target safety level in this study. The procedure of the calibration to find the appropriate values of factors ϕ, α_D and α_W is now carried out. The values of ϕ and α_D are fixed to be 0.9 and 1.35, respectively, since these values are found suitable from the calibration study of the steel highway bridges subject to dead and traffic loads[7]. The design wind load W^* which is calculated on the basis of the 100-year wind speed of each station is employed. The variable associated with the wind load factor is discretized at interval of 0.05. After successive trial of various values of α_W, the appropriate value is found to be 2.15. Then Eq. 13 yields to

$$0.9\ R^* \geq 1.35\ D^* + 2.15\ W^* \tag{14}$$

The values of the safety index achieved by Eq. 14 are shown in Fig. 10, where one finds that the average safety index values β are close enough to the target values $\bar{\beta}$ shown in Fig. 8. Also, Fig. 10 shows that Eq. 14 maintains the reliability level which is fairly constant over a wide range of the ratio of wind to dead load and which has small variations to wind speed characteristics varying from location to location.

Reliability-based calibration of wind load factor in use for designing structures subject to dead and wind loadings are heretofore carried out. As well known, structures exposed to natural environment are usually subject to other time-varying loadings as well as wind loading. Simultaneous occurrence of such loadings, so-called load combination, should be taken into account in safety provisions in a structural code. Reliability-based code calibration of load factor design in the presence of more than two time-varying loads are given in Ref. 9, in which the load combination of wind and traffic live loadings in highway bridge design is considered in order to show the principle and procedures of the calibration.

CONCLUSIONS

A study has been made of statistical aspects of wind loading in design. Major conclusions obtained in this study are as follows

1) An analysis of fitting closeness of the extreme distributions to the yearly maximum wind speed data at 136 stations in Japan has been carried out. Among the distributions of types I and II, the type I is more appropriate to employ at most of the stations. Among the types I, II and III, the type III has been found to be best fitted to the data at 56 stations and it is suggested to employ the type III at some of the stations. Also a mixed distribution is found, in some area, suitable to represent the maximum wind speed records.
2) Reliability level of the current Japanese specifications for steel highway bridges subject mainly to wind loading has been probabilistically evaluated in terms of the safety index. The calculated values of the safety index vary largely from location to location. It has been found, however, that the current specifications provide relatively consistent level of safety in the average sense. Toward the implementation of load factor design, reliability-based calibration of wind load factor has been carried out. It has been shown that the calibrated wind load factor accompanied with the design wind load which is based on the 100-year wind speed of the site maintains a fairly constant reliability level.

REFERENCES

1. Thom, H.C.S., "Toward a Universal Climatological Extreme Wind Distribution", Proc. Int'l. Res. Seminar on Wind Effects on Building and Structures (Ottawa), Univ. of Toronto Press, 1967

2. Simiu, E., and Filliben, J.J., "Probability Distributions of Extreme Wind Speeds", J. of the Struc. Div., ASCE, 102, ST9, 1976

3. Simiu, E., Bietry, J., and Filliben, J.J., "Sampling Errors in Estimation of Extreme Winds", J. of the Struc. Div., ASCE, 104, ST3, 1978

4. Saito, R., "Probabilistic Estimation of the Upper Limit of Extreme Wind Speed", Res. Report of the Japan Meteorology Agency, 9, 7, 1957

5. The Division of Meteorological Statistics in the Japan Meteorology Agency, "Yearly Maximum Wind Speed Records in Japan", Bulletin of the JMA, 34, 1971

6. Japan Road Association, "Specifications for Highway Bridges", 1973

7. Kinoshita, S., Ito, M., and Fujino, Y., "Reliability-based Code Calibration of Load Factor Design for Steel Highway Bridges", Proc. of the Annual Convention of Japan Society of Civil Engineers, 1978

8. Ravindra, M.K., Cornell, C.A., and Galambos, T.V., "Wind and Snow Load Factors for Use in LRFD", J. of the Struc. Div. , ASCE, 104, ST9, 1978

9. Ito, M. and Fujino, Y., "Some Probabilistic Considerations on Wind Resistant Design", Proc. of the US-Southeast Asia Symp. on Engrg. for Natural Hazard Protection (Manila), Univ. of Illinois Press, 1978

(i) Poor Fitness	(ii) Type I	(iii) Type II	(iv) Type I or II
16 (11.8%)	77 (56.6%)	42 (30.9%)	1 (0.7%)

Table I.- Results of the Choice of the Distribution under the Criteria given by Fig.5, among the Types I and II at 136 Meteorological Stations in Japan (Data: 1929 - 1977)

(i) Poor Fitness	(ii) Type I	(iii) Type II	(iv) Type III
13 (9.6%)	20 (14.7%)	42 (30.9%)	56 (41.2%)
(v) Type I, II or III	(vi) Type I or II	(vii) Type II or III	(viii) Type I or III
0 (0%)	1 (0.7%)	0 (0%)	4 (2.9%)

Table II .- Results of the Choice of the Distribution, under the Criteria Given by Fig.6, among the Types I, II and III at 136 Meteorological Stations in Japan (Data: 1929 - 1977)

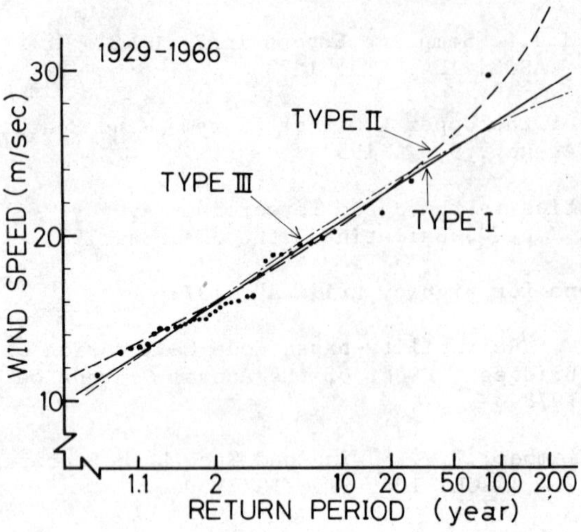

Fig. 1.- Yearly Maximum Wind Speeds in Tokyo

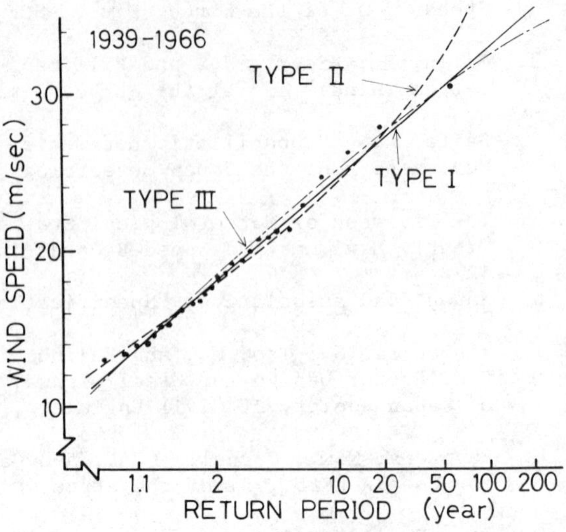

Fig. 2.- Yearly Maximum Wind Speeds in Fukuoka

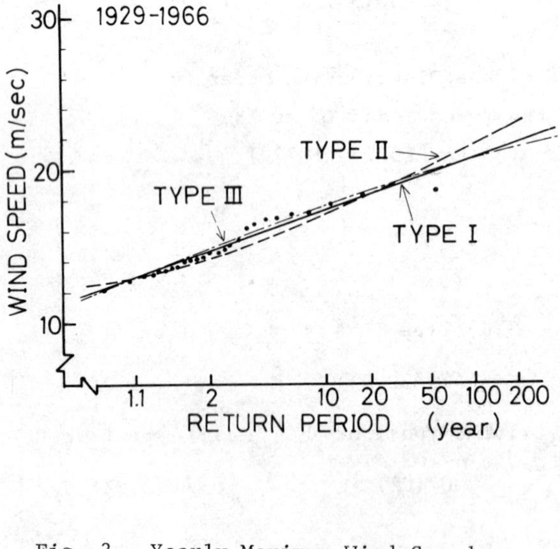

Fig. 3.- Yearly Maximum Wind Speeds in Sapporo

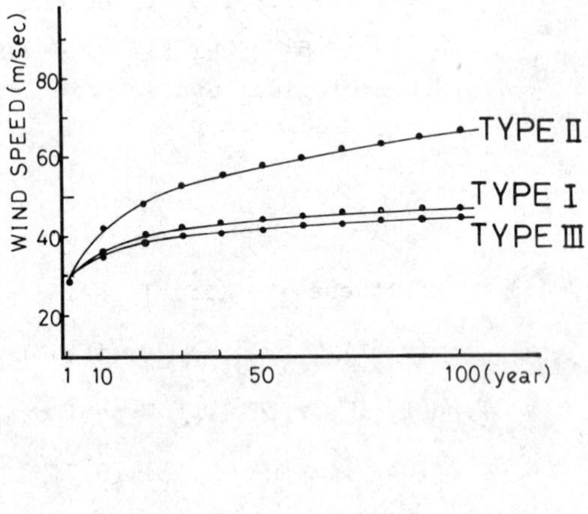

Fig. 4.- 95% Fractile Values of N-year Maximum Wind Speed in Fitted Distributions of Type I, II and III to Records in Nagoya

CRITERIA	JUDGMENT
If $M\ (\equiv \min(E_I, E_{II})) > 0.05$	(i) POOR FITNESS
Otherwise ($M < 0.05$)	
If $E_I/E_{II} \leq 0.99$	(ii) TYPE I
If $E_I/E_{II} \geq 1.01$	(iii) TYPE II
Otherwise	(iv) TYPE I or II

Note : E_I (or E_{II}) = the value of the closeness measure (Eq. 6) of the fitted type I (or II) extreme distribution

Fig. 5.- Criteria for Choice of Underlying Distribution among Extreme Distributions of Types I and II

CRITERIA	JUDGMENT
If $M\ (\equiv \min(E_I, E_{II}, E_{III})) \geq 0.05$	(i) POOR FITNESS
Otherwise ($M < 0.05$)	
If $M = E_I$	(ii) TYPE I
If $M = E_{II}$	(iii) TYPE II
If $M = E_{III}$	(iv) TYPE III
But, the judgment made above is withdrawn in the following cases:	
If $\lvert E_I/E_{II} - 1\rvert < 0.01$ and $\lvert E_{II}/E_{III} - 1\rvert < 0.01$	(v) TYPE I, II or III
If $\lvert E_I/E_{II} - 1\rvert < 0.01$	(vi) TYPE I or II
If $\lvert E_{II}/E_{III} - 1\rvert < 0.01$	(vii) TYPE II or III
If $\lvert E_I/E_{III} - 1\rvert < 0.01$	(viii) TYPE I or III

Note : E_I, E_{II} and E_{III} = the values of the closeness measure (Eq.6) of the fitted extreme distributions of the types I, II and III, respectively

Fig. 6.- Criteria for Choice of Underlying Distribution among Extreme Distributions of Types I, II and III

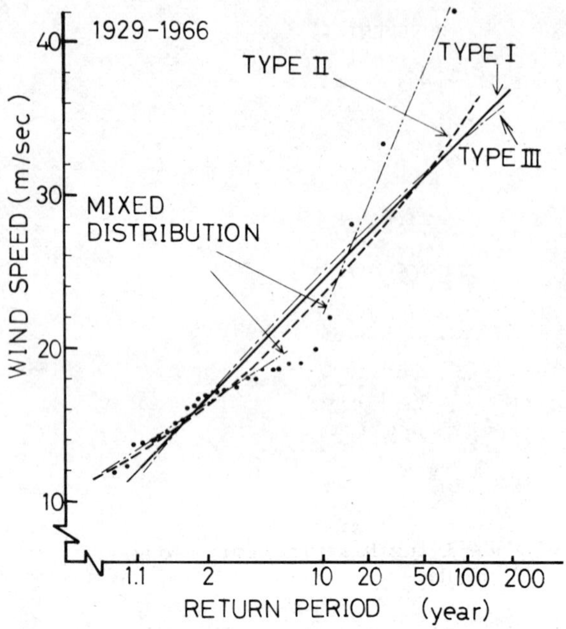

Fig. 7.- Yearly Maximum Wind Speeds in Osaka

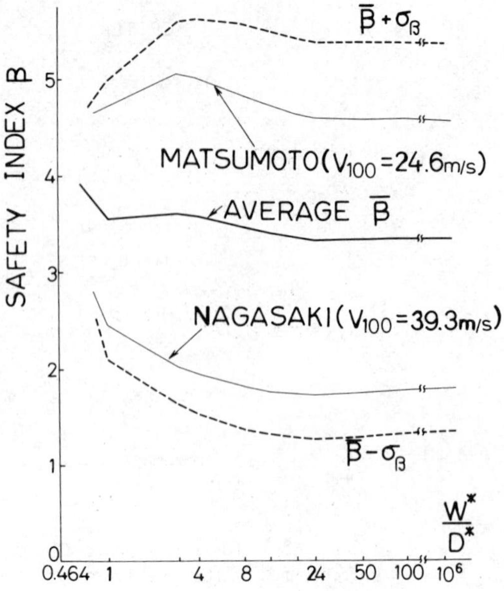

Fig. 8.- Safety Index Values Achieved by the Current Japanese Specifications for Wind Loading

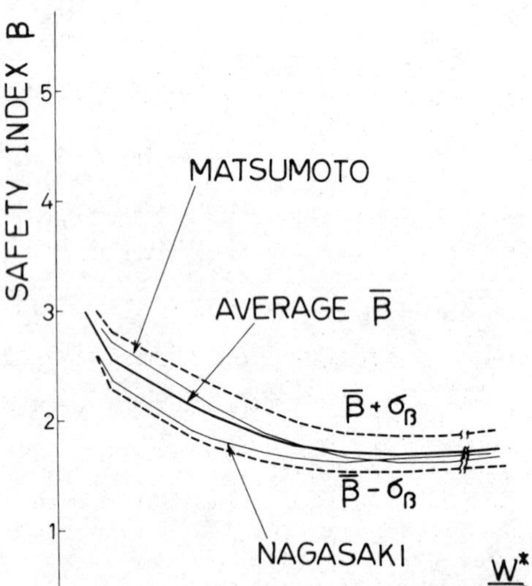

Fig. 9.- Safety Index Values When Wind Speeds of 100-year Return Period Are Used in Design

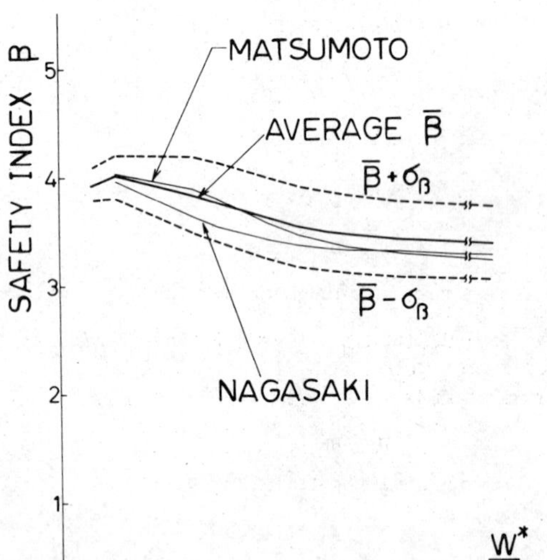

Fig. 10.- Safety index Values Achieved by Load Factor Design
(ϕ = 0.9, α_D = 1.35, α_W = 2.15)

WIND LOADING AND RELIABILITY-BASED DESIGN

E. SIMIU and J. R. SHAVER

National Bureau of Standards, Washington, D. C., USA

ABSTRACT

The implementation of second-moment formats for the design of wind-sensitive structures requires the clarification of a number of questions which are investigated in this paper. These include: the dependence of reliability-based criteria upon type of extreme wind distribution; the influence upon such criteria of sampling and observation errors in the estimation of extreme winds, as well as of errors in the estimation of aerodynamic and structural parameters; the dependence of member safety upon wind climate; and the validity of linear approximations to the expression for the load factor.

INTRODUCTION

In recent years a considerable effort has been devoted to the development of design criteria aimed at ensuring consistent levels of structural safety. In particular, second-moment reliability formats have received wide attention. These rely on the use of a reliability index representing the distance, in standard deviation units, between the failure surface and a point associated with the mean values of the resistance R and of the load Q, the failure surface being defined as the locus of the points (R, Q) such that R = Q. The implementation of second-moment formats for the design of wind-sensitive structures requires the clarification of a number of questions that include: the dependence of reliability-based criteria upon type of extreme wind distribution; the influence upon such criteria of sampling and observation errors in the estimation of extreme winds, as well as of errors in the estimation of aerodynamic and structural parameters; the dependence of the safety of a member upon wind climate, expected life of structure, and mean recurrence interval of the design wind speed; and the validity of certain approximations used in the estimation of load and resistance factors. It will not be attempted in this paper to present an exhaustive investigation into these questions. Rather, simple situations will be dealt with which can provide useful insights into the questions at hand, and may suggest useful directions of further research, particularly with regard to the implementation of reliability-based criteria in design practice and in building codes. Note that basic equations used in this work are included in appendix I. A list of symbols used in this paper is given in appendix II.

ASSUMPTIONS ON LOADS AND RESISTANCE

The stress induced by the wind loads (which will be referred to here as the load effect) will be assumed to be approximately proportional to the square of the wind speeds. (Note that this assumption is not valid in the presence of significant wake or aeroelastic effects.) If gravity loads are present, the stress in the member can be written as

$$Q = cu^2 + GL \tag{1}$$

where GL = stress induced by gravity loads, u = wind speed, and c = coefficient with dimensions [stress]/[velocity]^2, which can be written as

$$c = \frac{F}{u_N^2} \tag{2}$$

In Eq. 2, u_N = wind speed with an N - year mean recurrence interval under the action of which the stress in the member is F + GL.

The response of the structure will be assumed to be independent of wind direction. This assumption applies, for example, to certain members of structures with a circular shape in plan in terrain with uniform roughness around the structure. An approach to the estimation of the reliability of structures whose response depends upon wind direction is suggested in an expanded version of this paper currently in preparation by the writers.

The numerical results will be based on the assumption that the material used is A33 steel (nominal yield stress 227,300kN/m^2 (33 ksi). This, or similar, material has been widely used in the U.S. until about 15 years ago. This assumption is therefore advantageous from the standpoint of calibration against past practice. In addition, a number of statistical data on steel of this type are available in the literature (1,2, p. 237). From these data, and noting that "mill tests give values about 15 percent above the weighted average lower yield point of the rolled shapes from which the specimens were taken" (2), a probability density curve of the yield stress, $f_R\ [(r-\bar{R})/\sigma_R]$ was constructed, which is shown in fig. 1. For A33 steel, $\bar{R}$ = 242,940 kN/m^2 (35.27 ksi) and σ_R = 23,350kN/m^2 (3.39 ksi), where $\bar{R}$ and σ_R = mean value and standard deviation of the yield stress, respectively. Nominal failure is assumed to occur when the yield stress is reached.

DEPENDENCE OF SAFETY INDEX, AND OF LOAD AND RESISTANCE FACTORS, UPON TYPE OF EXTREME WIND DISTRIBUTION

An analysis of extreme wind records at over 100 U.S. locations where hurricane winds are not expected to occur showed that the coefficient of variation of the extreme annual winds, V_u, varied at these locations between 0.085 and 0.226, the average being 0.145 (3). For the calculations presented in this section a station was chosen that was deemed to be fairly representative of average wind climate conditions in the U.S. as expressed by the value of V_u, i.e., Canton, N.Y., where $\bar{u}$ = 23.75 m/s (53.12 mph) and σ_u = 3.68 m/s (8.23 mph) based on the 1912-1948 record of extreme annual speeds converted to fastest-miles (4)]. Calculations were carried out for the case in which the member is subjected to wind loads only (i.e., GL = 0 in Eq. 1). The value $F = F_a$ = 20 ksi was used. The results are listed in table 1. It is seen from table 1 that the safety index β depends strongly upon the type of probability distribution.

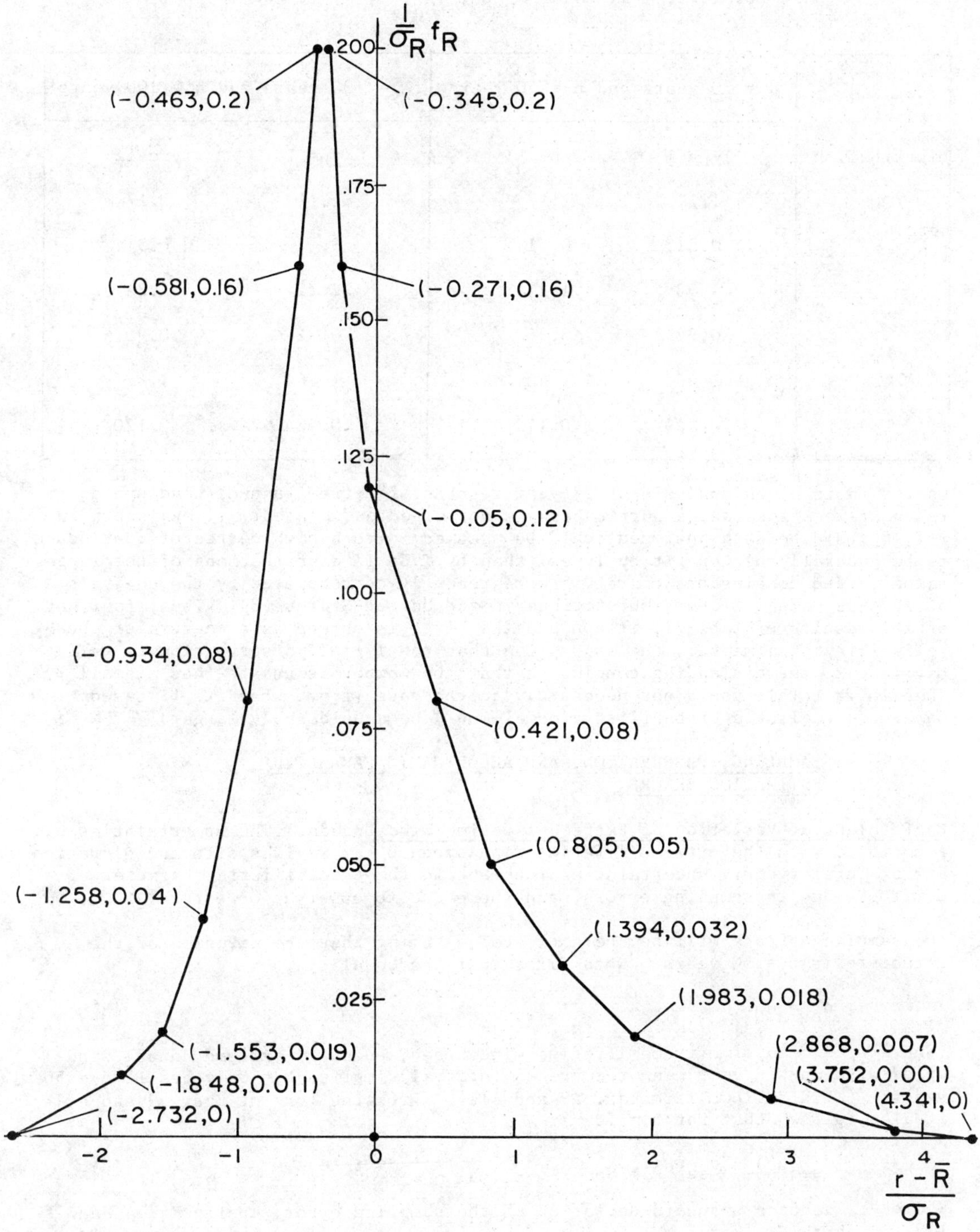

Fig. 1. Probability density curve of the yield stress

Table 1. Values of $\overline{Q}_n/Q_N$, V_{Q_n}, β, Φ, γ, and ϕ_n

Distribution	N = 50 years and n = 50 years		N = 100 years and n = 100 years	
	Type I	Type II	Type I	Type II
$\overline{Q}_n/Q_N$	1.11	1.17	1.11	1.17
V_{Q_n}	0.21	0.31	0.21	0.31
β	1.98	1.21	2.11	1.21
Φ	0.92	0.96	0.92	0.96
γ	1.46	1.42	1.46	1.42
ϕ_n	0.029	0.122	0.024	0.120

On the basis of an analysis of 37-year records of extreme annual wind speeds taken at locations where hurricanes are not expected to occur, it was shown in ref. 5 that the data analyzed could be assumed, with a high degree of confidence, to be generally better fit by Type I than by Type II distributions of the largest values. The conclusions of ref. 5 were recently corroborated by the analysis of records taken at over 100 stations in the United States (3). Thus, in light of the results of table 1, it appears that if β is chosen as a measure of the reliability of a member, the assumption that the Type II distribution is valid may lead to the misleading conclusion that the member is unsafe (has a small β) when in fact this would not necesarily be the case (i.e., when its β based on a more realistic probabilistic model would be considerably larger).

SAMPLING, OBSERVATION, AND AERODYNAMIC PARAMETER ESTIMATION ERRORS

Coefficient of Variation of Extreme Lifetime Wind Speeds. The uncertainties associated with the extreme lifetime wind speed u_n at a given site and elevation are of three types: uncertainties inherent in the probabilistic character of u_n, those due to sampling errors, and those due to observation errors.

The sampling errors will now be estimated assuming that the lifetime of the structure is n = 50 years. Note first that the equality

$$\overline{u}_n = u_N \tag{3}$$

(where $\overline{u}_n$ = mean of extreme lifetime wind speed, and u_N = extreme annual wind speed with an N - year mean recurrence interval) implies N = 86 years if n = 50 years. (This follows from Eqs. A9 and A11). From the form of Eqs. A9 and A11 it also follows that for n = 50

$$\sigma_{se}(\overline{u}_n) = \sigma_{se}(u_N = 86) \tag{4}$$

where σ_{se} denotes standard deviation of the sampling error. Using Eq. 4 and A13 (in which N = 86)

$$\sigma_{se}(u_n) = 3.8 \frac{\bar{u} V_u}{\sqrt{m}} \tag{5}$$

where $\bar{u}$ = mean, V_u = coefficient of variation, of extreme annual wind speeds, and m = sample size.

Let now the standard deviation of the observation error be denoted by σ_{obs}. Following the approach of, e.g., refs. 6, 7 and 8 which, though not rigorous from a theoretical point of view, is useful in practice, the coefficient of variation of u_n will be written as

$$V^*_{u_n} = \frac{[\bar{u}^2 V_u^2 \pm \sigma^2_{se}(\bar{u}_n) \pm \sigma^2_{obs}]^{1/2}}{\bar{u}_n} \tag{6}$$

(Note that $\bar{u}\, V_u = \bar{u}_n V_{u_n}$ - see Eqs. A9 and A10)

Using Eq. 5 and A10, for n = 50 it then follows

$$V^*_{u_n} = [V_u^2 \frac{1 + \frac{14.5}{m}}{(1 + 3.05\, V_u)^2} + V_{obs}]^{1/2} \tag{7}$$

where V_{obs} = coefficient of variation of observation error.

Coefficient of Variation of Extreme Lifetime Wind Pressures. The extreme lifetime wind pressure at elevation z at a given site will be written in the form

$$W_n(z) = \frac{1}{2} \rho\, C_p\, G\, u^2_{sn}\,(z) \tag{8}$$

where ρ = air density, C_p = pressure coefficient, G = gust factor, and u_{sn} (z) = extreme lifetime wind speed at that site, averaged over one hour.

The extreme lifetime hourly wind speed at elevation z at the site under consideration can be written as follows [9]:

$$u_{sn}\,(z) = p \ln \frac{z}{(z_o)_s} \frac{1}{\ln \frac{10}{(z_o)_{ref}}} u_{ref,n} \tag{9}$$

where $u_{ref,n}$ = extreme lifetime value of the reference wind speed at 10 m above ground in open terrain (e.g., at airport sites), averaged over one hour; $(z_o)_{ref}$ = roughness length (i.e., a measure of the terrain roughness [9]) in open terrain, in meters; $(z_o)_s$ = roughness length at, and upwind of, the site under consideration; p = retardation factor, reflecting the fact that the surface friction which retards the wind flow near the ground is higher over rougher than over smoother terrain [9]. Values of the roughness length and of the retardation factor corresponding to various categories of terrain are suggested in table 2 [9]. If, as in the case of most U.S. primary weather stations, extreme annual winds are recorded in terms of fastest-miles, it is necessary to convert the fastest-mile winds into hourly wind speeds. Let the conversion factor--a random variable--be denoted by b, and let

$$p \ln \frac{z}{(z_o)_s} = f_1(z) \tag{10}$$

$$\frac{1}{\ln \frac{10}{(z_o)_{ref}}} = f_2 \tag{11}$$

The product $p \ln [z/(z_o)_s]$ is treated as one variable because p and $\ln [z/(z_o)_s]$ are correlated, being both functions of $(z_o)_s$. [Actually p also depends on $(z_o)_{ref}$. However this dependence is relatively weak and will be neglected in this analysis.] Following the type of analysis used in Refs. 6 and 10, the mean and the coefficient of variation of the extreme lifetime wind pressure can now be written as

$$\overline{W_n} \simeq \frac{1}{2} \bar{\rho}\, \bar{C_p}\, \bar{G}\, \overline{f_1^2(z)}\, \bar{f_2^2}\, \bar{b^2}\, \overline{u'^2_{ref\ n}} \tag{12}$$

$$V_{W_n} \simeq [V_\rho^2 + V_{C_p}^2 + V_G^2 + 4V_{f1}^2 + 4V_{f2}^2 + 4V_b^2 + 4V^{*2}_{u'_{ref\ n}}]^{1/2} \tag{13}$$

where $u'_{ref\ n}$ = fastest mile wind speed at 10 m above ground in open terrain.

Table 2 - Parameters z_o and p for Various Terrain Categories

Terrain Category	I	II	III	IV	V
z_o(m)	0.005	0.07	0.30	1.00	2.50
p	0.83	1.00	1.15	1.33	1.46

Note. Terrain categories are defined as follows: 1 = open water; II = open terrain; III = suburbs at considerable distance from towns (sites in rural zones sparsely built-up, with trees, hedges, etc.); IV = towns, densely built-up suburbs, wooded terrain; V = centers of large cities

The air density depends upon altitude and air temperature. It will be assumed here that the variability of ρ is negligible and, following ref. 10, that $V_{Cp} \simeq 0.1$. It appears reasonable to assume $V_G \simeq 0.1$. Using the data of table 2, it can be verified that from one terrain category to the next the differences between the corresponding values of $f_1(z)$ are about 25% at z = 10 m, 10% at z =100 m, and 6% at z = 400 m. Mistaking one terrain category for the next will be assumed to be a very rare event corresponding, say, to an error of 2.5 σ_{f1}, where σ_{f1}= standard deviation of error in the estimation of f_1. It then follows $V_{f1}(10) \simeq 0.1$, $V_{f1}(100) \simeq 0.04$, $V_{f1}(400) \simeq 0.02$.

As far as the factor f_2 is concerned, it is reasonable to assume that $(z_o)_{ref}$ may vary between, say, 0.03m and 0.15m, so that $\ln (10/0.03) < 1/f_2 < \ln(10./0.15)$

or $5.8 < 1/f_2 < 4.2$, with $1/\bar{f}_2 \simeq 5.0$ corresponding to $(z_o)_{ref} \simeq 0.07$m. Assuming that the distance from $\bar{f}_2$ to the lower and upper values in this inequality is roughly $2\sigma_{f2}$, $V_{f2} \simeq 7.5\%$, say.

Data on V_b can be obtained for example, from ref. 11; it is reasonable to assume $V_b \simeq 0.1$.

In the United States the number of weather stations with records that can be used in the estimation of extreme wind speeds is relatively small (about 120). For this reason, records taken at one station are used to make inferences on the wind climate over areas of tens of thousands of square kilometers, even though these areas might be quite inhomogeneous from a wind climatological point of view. This is a source of errors which are not taken into account in Eq. 13. However, the possibility that these errors could be very significant at certain locations should not be overlooked.

Table 3 shows calculated values of V_{Wn} for n = 50, corresponding to various parameters V_u, z, and m, and to two sets coefficients of variation V_{Cp}, V_G, V_{f1}, V_{f2}, V_b. The first set [columns (a) of Table 3] corresponds to the values $V_{Cp} = 0.1$, $V_G = 0.1$, $V_{f1}(10) = 0.1$, $V_{f1}(100) = 0.04$, $V_{f1}(400) = 0.02$, $V_{f2} = 0.075$, $V_b = 0.1$. In the second set $V_{Cp} = V_G = V_{f1} = V_{f2} = V_G = 0$ [columns (b) of table 3]. The values V_{Wn} of columns (b) thus reflect only the effect of sampling and observation errors.

Table 3. Values of V_{W_n} for Various Values of V_u, z, m

V_{obs}	z meters	V_u	m = 10		m = 30		m = β	
			(a)*	(b)**	(a)*	(b)**	(a)*	(b)**
		0.10	0.45	0.28	0.43	0.24	0.41	0.21
		0.14	0.49	0.34	0.45	0.28	0.43	0.25
	10	0.18	0.53	0.39	0.47	0.32	0.45	0.28
		0.22	0.56	0.44	0.50	0.35	0.46	0.30
		0.10	0.41	0.28	0.39	0.24	0.37	0.21
		0.14	0.46	0.14	0.41	0.28	0.39	0.25
0.075	100	0.18	0.50	0.39	0.44	0.32	0.41	0.28
		0.22	0.53	0.44	0.46	0.35	0.43	0.30
		0.10	0.41	0.28	0.38	0.24	0.36	0.21
		0.14	0.45	0.34	0.40	0.28	0.38	0.25
	400	0.18	0.49	0.39	0.43	0.32	0.40	0.28
		0.22	0.53	0.44	0.46	0.35	0.42	0.30

*$V_{Cp} = 0.1$, $V_G = 0.1$, $V_{f_1}(10) = 0.1$, $V_{f_1}(100) = 0.04$, $V_{f_1}(400) = 0.02$, $V_{f_2} = 0.075$, $V_b = 0.1$

**$V_{Cp} = V_G = V_{f_1} = V_{f_2} = V_b = 0$

Example 1. To illustrate the use of table 3, let z = 100m, the sample size m = 30 years, and $V_u = 0.14$. For the set of values $V_{Cp} = 0.1$, $V_G = 0.1$, $V_{f1}(100) = 0.04$, $V_{f2} = 0.075$, $V_b = 0.1$, it is seen from table 3 that $V_{Wn} = 0.41$. If $V_{Cp} = V_G = V_{f1} = V_{f2} = V_b = 0$ (i.e., in the case in which only the sampling and observation errors are taken into account, in addition to the inherent variability of the wind load), it is seen from table 3 [column (b)] that $V_{Wn} = 0.28$. If, in addition, it is assumed $V_{obs} = 0$ and $m = \infty$ (i.e., if only the inherent variability of the wind load is taken into account), then $V_{Wn} \simeq 0.20$.

Example 2. It will now be assumed that $V_{Qn} \simeq V_{Wn}$. This assumption implies that the member is subjected to wind loads only, and that the variability of the influence coefficient which transforms wind loads into wind stresses is neglected. The magnitude of β inherent in the provisions of ref. 12 will be sought assuming that the structure is located in Canton, N.Y. It was seen previously that if all the errors dealt with in this section are neglected, then $V_{Qn} = 0.21$ and $\beta = 1.98$ (table 1). If N = 50 years, n = 50 years, and $F = F_a = 137{,}800$ kN/m^2 (20 ksi), assuming $V_{Cp} = 0.1$, $V_G = 0.1$, $V_{f1} = 0.04$, $V_{f2} = 0.075$, $V_b = 0.1$, $V_u = 0.155$, $V_{obs} = 0$, $m = \infty$ (i.e., neglecting sampling and observation errors), it follows from table 3 that $V_{Wn} \simeq 0.36$, so that $\beta = 1.22$. If, in addition, it is assumed m = 20 and $V_{obs} = 0.075$, then $V_{W_n} \simeq 0.45$ and $\beta = 0.99$.

Two interesting observations can be made by inspecting table 3. First, the coefficients of variation V_{Wn} are practically invariant with height for z > 100 m or so. Second, the influence upon V_{Wn} of increasing the sample size beyond m = 30 is small, particularly if in the estimation of V_{Wn} the aerodynamic parameter errors are taken into account. Thus, the accuracy of load and resistance factors is only negligibly improved if m = 50, say, instead of m = 30.

We will now calculate load and resistance factors for various values V_u. This is done assuming $V_R = 0.096$, m = 30, $V_{obs} = 0.075$, z = 10 m, $\beta = 3.00$, and $V_{GL} = 0$. Eqs. A19-A22 are used; however, γ was in addition calculated, also for $\beta = 3.00$[10], using the well-known linear approximation to Eq. A20:

$$\gamma = 1 + k\beta\, V_{W_n} \tag{14}$$

with k = 0.75 [14]. The results of the calculations are shown in table 4. It is noted, first, that the load factors calculated in accordance with Eq. A20 are very large, even allowing for possible subsequent reductions due to member ductility or wind direction effects. Thus, the value $\beta = 3.00$ proposed in ref. 10 may be too large for certain types of structures. Second, the differences between results corresponding to various values of V_u and z (i.e., to different geographical locations and to different elevations) are not negligible and might have to be taken into account in certain situations. Third, the differences between results yielded by Eq. A20 and Eq. 14 are quite large in the case investigated here ($\overline{GL}/\overline{W}_{Qn} = 0$, $V_R = 0.096$). It can be verified that these differences decrease little if it is assumed $V_R = 0.2$, say (i.e., if V_R is increased to account for variations in member dimensions and uncertainties in structural theory). It thus appears that the range of ratios $\overline{GL}/\overline{W}_{Qn}$ and V_{Qn}/V_R for which Eq. 14 is applicable should be determined carefully, and that, accordingly, clear restrictions should be set on the use of Eq. 14.

Table 4. Load and Resistance Factors for Various Values of V_u

z (meters)	V_u	V_{W_n}	Φ	γ (Eq. A20)	γ (Eq. 14, k = 0.7)
10	0.10	0.43	0.94	3.54	1.97
	0.14	0.45	0.94	3.75	2.02
	0.18	0.47	0.94	3.98	2.06
	0.22	0.50	0.95	4.35	2.13

SAFETY INDEX, PROBABILITY OF FAILURE, AND STRUCTURAL DESIGN

Relation Between Safety Index and Probability of Failure. Numerical calculations of the probability of failure during the lifetime of the structure, ϕ_n, (Eq. A14) and of the index β (Eq. A17) have been carried out, on the basis of which the following approximate relation was found to hold for $V_u \geq 0.1$ and $\beta < 4.0$ ($\phi_n > 7.8 \times 10^{-4}$):

$$\phi_n \simeq \exp\left[-\left(\frac{\beta+0.45}{0.835}\right)^{1/0.85}\right] \tag{15}$$

$$\beta \simeq 0.835\,(-\ln \phi_n)^{0.85} - 0.45 \tag{16}$$

Eqs. 15 and 16 are valid under the assumption that Eq. 1 holds, that the variability of the gravity loads is negligible (i.e., $GL = \overline{GL}$), and that the probability density function of the reduced yield stress is given by fig. 1.

For example, for $\beta = 1.0$, $\phi_n \simeq 0.15$; for $\beta = 2.0$, $\phi_n \simeq 0.029$; for $\beta = 3.0$, $\phi_n \simeq 5 \times 10^{-3}$; for $\beta = 4.0$, $\phi_n \simeq 7.8 \times 10^{-4}$.

APPENDIX I. BASIC EQUATIONS

The extreme lifetime wind speed has the following distribution function:

$$F^*(u) = [F(u)]^n \tag{A1}$$

where $F(u)$ = cumulative distribution function of extreme annual wind speed, and n = life of structure in years. If the Type I distribution of the largest values is assumed to hold [5]

$$F(u) = \exp\left[-\exp\left(-\frac{u-\mu_I}{\sigma_I}\right)\right] \tag{A2}$$

From Eqs. A1 and A2

$$F^*(v) = \exp\left[-\exp\left(-\frac{u-\mu_I^*}{\sigma_I}\right)\right] \tag{A3}$$

$$\mu_I^* = \mu_I + \sigma_I \ln n \tag{A4}$$

It is recalled that [5]

$$\mu_I \simeq \bar{u} - 0.5772\,\sigma_I \tag{A5}$$

$$\sigma_I \simeq \frac{\sqrt{6}}{\pi}\,\bar{u}\,V_u \tag{A6}$$

where $\bar{u}$ and V_u denote the mean and the coefficient of variation of the extreme annual speeds, respectively.

The mean and the coefficient of variation of the extreme lifetime wind loads corresponding to Eq. 1 can be written as

$$\overline{Q_n} = \frac{F}{u_N^2}\,\bar{u}_n^2\,(1 + V_{u_n}^2) + \overline{GL} \tag{A7}$$

$$V_{Q_n} = \frac{2\,V_{u_n}\,(1 + V_{u_n}^2/2)}{1 + \dfrac{\overline{GL}}{\dfrac{F}{u_N^2}\,u_n^2\,(1 + V_{u_n}^2)}} \tag{A8}$$

where $\bar{u}_n$ = mean, and V_{un} = coefficient of variation of extreme lifetime wind speeds. If the Type I distribution holds:

$$u_n = \bar{u}\,(1 + \frac{\sqrt{6}}{\pi}\,V_u \ln n) \tag{A9}$$

$$V_{V_u} = V_u\,\frac{1}{1 + \dfrac{\sqrt{6}}{\pi}\,V_u \ln n} \tag{A10}$$

$$u_N = \bar{u}\,\{1 - \frac{\sqrt{6}}{\pi}\,V_u\,[0.5772 + y]\} \tag{A11}$$

$$y = \ln\,[-\ln(1 - \frac{1}{N})] \tag{A12}$$

If m = sample size, the error in the estimation of u_N for the Type I distribution can be estimated as [5]

$$\sigma_{se(u_N)} \simeq [\frac{\pi^2}{6} - 1.1396(y+0.5772)\frac{\pi}{\sqrt{6}} + 1.1(y+0.5772)^2]^{1/2}\,\frac{\sqrt{6}}{\pi}\,\frac{\bar{u}V_u}{\sqrt{m}} \tag{A13}$$

If the wind speed is described probabilistically by a Type I distribution, and if the variability of the gravity loads is negligible, the probability of failure of a member during its lifetime n is

$$\phi_n = \int_0^\infty d\left(\frac{u}{\bar{u}}\right) \frac{\pi}{\sqrt{6}V_u} \exp\left[-\frac{\frac{u}{\bar{u}} - M_n}{\frac{\sqrt{6}}{\pi}V_u} - \exp\left(-\frac{\frac{u}{\bar{u}} - M_n}{\frac{\sqrt{6}}{\pi}V_u}\right)\right] \int_0^A f_R\left(\frac{r-\bar{R}}{\sigma_R}\right) d\left(\frac{r-\bar{R}}{\sigma_R}\right) \quad (A14)$$

where

$$M_n = 1 + \frac{\sqrt{6}}{\pi} V_u (\ln n - 0.5772) \quad (A15)$$

$$A = \frac{1}{V_R}\left[\frac{F}{\bar{R}} \frac{(u/\bar{u})^2}{(u_N/\bar{u})^2} + \frac{GL}{\bar{R}} - 1\right] \quad (A16)$$

The safety index of a member is defined as [13,14]

$$\beta = \frac{\ln \bar{R} - \ln \bar{Q}_n}{(V_R^2 + V_{Q_n}^2)^{1/2}} \quad (A17)$$

From Eq. A19 there follows

$$\Phi \bar{R} = \gamma \bar{Q}_n \quad (A18)$$

where the resistance factor ϕ and the load factor γ have the expressions

$$\Phi = \exp(-\alpha_R \beta V_R) \quad (A19)$$

$$\gamma = \exp \alpha_{Q_n} \beta V_{Q_n} \quad (A20)$$

$$\alpha_R = \cos[\tan^{-1}(V_Q / V_R)] \quad (A21)$$

$$\alpha_{Q_n} = \sin[\tan^{-1}(V_{Q_n}/V_R)] \quad (A22)$$

APPENDIX II. NOTATIONS

C_p = pressure coefficient
f_1, f_2 = quantities defined by Eqs. 10, 11, respectively
$f_n(u)$ = probability density function of maximum lifetime wind speed
F, F_a, F_y = stress, allowable stress, yield stress, respectively
$F(u)$, $F_n(u)$ = cumulative distribution function of extreme annual wind speed, of maximum lifetime wind speed, respectively
G = gust factor
GL = gravity load
m = sample size
R = resistance
n = life of structure, in years
N = mean recurrence interval of wind speed, in years
p = retardation factor
Q = load effect
Q_n = maximum lifetime load effect

u, u_n, u_N, u_N , $u_{ref,\ n}$, u_{sn} = extreme annual wind speeds, extreme lifetime wind speeds, wind speed with mean recurrence interval N, wind speed with mean recurrence interval N that induces stress F_a, extreme lifetime wind speed at reference site, extreme lifetime wind speed at building site.

V = symbol denoting coefficient of variation

W_n = extreme lifetime wind pressure

β = safety index

γ = load factor

σ = standard deviation

ϕ_n = nominal probability of failure during the lifetime of the structure

Φ = resistance factor

REFERENCES

1. Galambos, T. V., Ravindra, M. K., "Properties of Steel for Use in LRFD," Journal of the Structural Division, ASCE, pp. 1459-1468 (1978).

2. McGuire, W., Steel Structures, Prentice-Hall, Inc., Englewood Cliffs, 1968.

3. Simiu, E., Changery, M. J., Filliben, J. J., Extreme Wind Speeds at 129 Stations in the Contiguous United States, Building Science Series, 118, National Bureau of Standards, Washington, D.C., 1979.

4. Simiu, E., Filliben, J. J., Statistical Analysis of Extreme Winds, Technical Note 868, National Bureau of Standards, Washington, D.C., 1975.

5. Simiu, E., Biétry, J., Filliben, J. J., "Sampling Errors in Estimation of Extreme Winds," Journal of the Structural Division, ASCE, pp. 491-501 (1978).

6. Cornell, C.A., "A Probability-Based Structural Code," Journal, American Concrete Institute, Vol. 66, No. 12, Dec., 1969.

7. Ellingwood, B. R., Reliability Basis of Load and Resistance Factors for Reinforced Concrete Design, BSS 110, NBS, Washington, D.C., 1978.

8. Ang, A. H.-S., Cornell, C. A., "Reliability Bases of Structural Safety and Design," Journal of the Structural Division, ASCE, 755-1769 (1974).

9. Biétry, J., Sacré, C., Simiu, E., "Mean Wind Profiles and Change of Terrain Roughness," Journal of the Structural Division, pp. 1585-1593 (1978).

10. Ravindra, M. K., Cornell, C. A., Galambos, T. V., "Wind and Snow Load Factors for Use in LRFD," Journal of the Structural Division, ASCE, 1443-1458 (1978).

11. Durst, C. S., "Wind Speeds over Short Periods of Time," Meteorological Magazine, 181-186 (1960).

12. American Institute of Steel Construction, Part I, in Manual of Steel Construction, Seventh Edition, American Institute of Steel Construction, New York, N.Y., 1970.

13. Hasofer, A. M., Lind, N. C., "Exact and Invariant Second-Moment Code Format," Journal of the Engineering Mechanics Division, ASCE, 111-121 (1974).

14. Ravindra, M. ,. Galambos, T. V., "Load and Resistance Factor Design for Steel," Journal of the Structural Division, ASCE, 1337-1353 (1978).

THE DEVELOPMENT OF THE BRITISH DRAFT CODE OF PRACTICE FOR THE LOADING OF LATTICE TOWERS

A. R. FLINT and B. W. SMITH
Flint and Neill Partnership, London, UK

ABSTRACT

The British Standard Code of Practice: Lattice Towers: Loading, has been issued as a draft for public comment. It contains design guidance on wind and ice conditions, wind resistance, structural response and safety. The paper outlines the basis of the draft.

1. INTRODUCTION

A British Standard Code of Practice for Lattice Towers: Loading has been drafted (1) to provide rational design values of load effects due to wind and/or ice on free standing lattice towers, with due allowance for their dynamic response. It is suited to worldwide application with guidance on the use of available meteorological information. The Code also provides guidance on safety factors on wind, ice and dead load effects appropriate for different types of tower, based on the application of reliability theory and calibration against previous practice.

The background to the various provisions of the Draft Code is discussed in this paper. A detailed commentary is appended to the Draft for further reference.

2. METEOROLOGICAL DATA

2.1 Reference Wind Speeds

The Code defines reference values of 'mean' wind speed, $\overline{V}_R$, at 10m. above ground. The averaging period for this speed, taken as one hour, must be long enough to encompass the full spectrum of gusts but less than the total duration of a major storm. The reference value for design is conveniently taken as an annual extreme 'mean' value having the commonly chosen probability of occurrence of 0.02. Isopleths of 50 year return wind speeds $\overline{V}_R$ for sites in open level country are provided.

'Gradient' speeds at heights above the earth's boundary layer are ideally used for reference, since these are directly determined by large scale meteorological factors and are unaffected by local topography. However, there are relatively few direct records of speeds at such levels, and whilst a formulation for the use of gradient wind speeds has been incorporated in the Code, it is expected that this will only be used for designs for regions for which few local meteorological data are available.

The Code makes allowance for the general roughness of the site of a tower by application of roughness coefficient, K_R, to the reference speed. Values are given for five categories of terrain based on the relationship shown in Figure 1, between K_R and the average ratio between the maximum 3 second gust speed in a given hour and the mean speed for the same period.

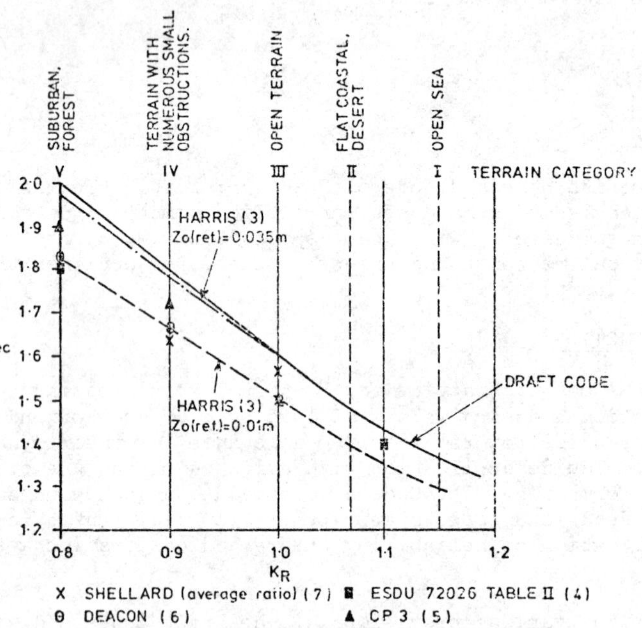

Fig.1 Gust ratios (3 second/hourly) for various terrain categories

The value of 1.0 corresponds to flat open terrain having a surface drag coefficient k = 0.005 for which the gust ratio is approximately 1.6 (2). The relationship is compared with that developed by Harris (3) with those presented in ESDU Data Item 72026 (4) and BS:CP3 (5) and with observations by Deacon (6) and Shellard (7).

For towers which are designed for a fixed orientation and with significant direction-dependent loads, such as those supporting cables or dishes, the designer may wish to consider the variation in extreme wind speeds with direction. Data for eleven stations in the U.K., published by the Meteorological Office, were analysed by using the highest recorded wind speed for each 30° sector. The resulting envelope of the results was used to provide a factor, K_D, to be applied to the reference speed, to give a directional value, $\bar{V}_{RD}$.

2.2 Variation of Mean Wind Speed with Height

A power law has been adopted for variation of mean speed with height. Figure 2 shows the relationship between the power law index α for level terrain and K_R adopted in the Code, based on proposals by the Meteorological Office, compared with that proposed by others for the standard terrain categories.

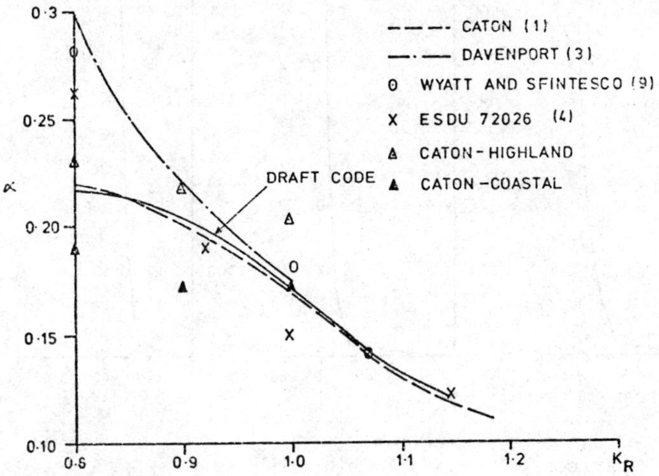

Fig.2 Variation of index of mean velocity with terrain roughness

It will be seen that for the rougher terrains the values of α recommended are significantly less than those previously in use and those evaluated by Davenport (8) in a survey of a number of observations. The power law index approximates closely to the logarithmic relationship suggested by Harris (10) within the range of structural heights covered by the Code.

The velocity profiles for towers situated on hills have also been defined in the Code. Theoretical studies of flow over hills have been made by Deaves (11).

An empirical relationship has been derived which considers the effect of the hill to change the normal velocity profile and to modify the wind speed at 10m. Thus the velocity at height z metres is given by

$$V_z = V_{RD} K_R \bar{K}_\mu \left(\frac{z}{10}\right)^{\mu}$$

$\bar{K}_\mu$ and μ are functions of α, the effective hill height above level terrain, and its slope and were derived by equating the volumetric flow of air at any instant away from the hill to that over the brow of the hill, assuming the flow to be laminar and incompressible with further simplifying assumptions as to the limits to the effects of the hill both above it and downstream. The relationship was compared with the results of wind tunnel tests undertaken by CERL on a series of hills and comparison with field observations, such as those shown in Figure 3 for a site in Scotland showed reasonable agreement for a range of speeds and directions.

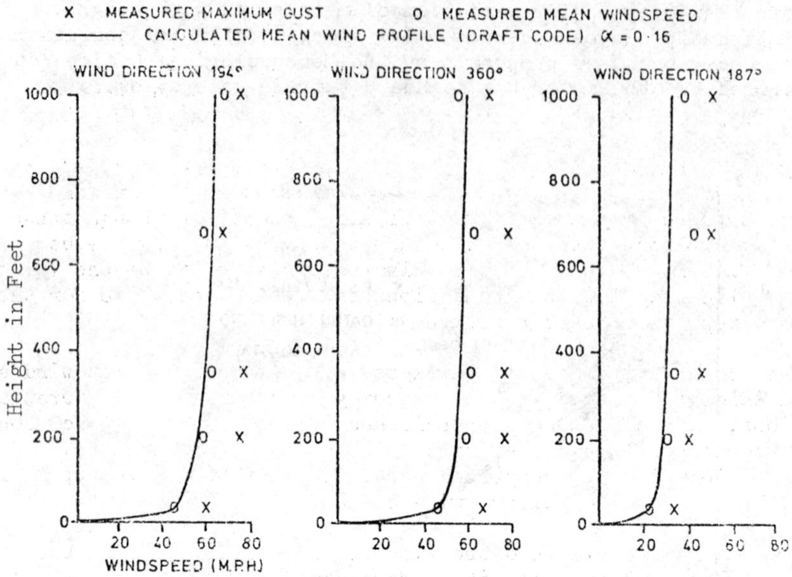

Fig.3 Comparison of proposals for wind over hills with records

No means of assessing the change in turbulence intensity have been proposed in the Code other than to assume that generally the sites on hills are likely to be one category rougher than those in the surrounding terrain.

2.3 Distribution of Wind Speeds

When structures may become unserviceable in windy conditions due to excessive deflection or vibration or may require to be appraised for fatigue life, a statistical distribution of mean wind speeds at the site is required. The draft Code contains guidance, based on observations at five meteorological stations, as to the total numbers of hours per annum during which the mean wind speed will, on average, exceed a given value. When the deflection limitation is confined to a certain direction, the total annual periods may be modified to give the annual periods during which the mean speed within $\pm 15^{\circ}$ of a given direction exceeds that corresponding to unserviceability.

A similar approach may be used for assessing periods of crosswind vibration due to vortex excitation. Again, the draft Code contains wind 'roses' from which may be derived the annual periods during which the wind speed may be assumed to be constant at a given value within $\pm 15^{\circ}$ of any assumed wind direction. Such simplified presentation enables rapid appraisal to be undertaken without data analysis by the designer.

2.4 Ice Loading

Ice loading can govern the self-supporting towers carrying cable aerials or conductors, and ice combined with wind may require consideration in the design of any tower in the U.K. Design guidance is therefore given in the Code.

The Meteorological Office has analysed records at 23 of their stations of the average annual duration of occurrence of relative humidity exceeding 97% in combination with minimum daily air temperatures within the range -4 to 0^oC. This analysis, combined with qualitative information obtained from various sites suggested that the design reference ice radial thickness, r_i, could be best defined by a regional zoning map which allows for the influence of latitudes on icing intensity. Separate allowance is made in the rules for altitude.

The frequency distributions of ice weight, independent of wind speed, have been studied in Czechoslovakia at 3 stations over a period of 28 years and were found to approximate to an extremal Type 1 distribution (12). Although these results are not directly applicable to the U.K. they appear to justify the use of decreasing coefficients of variation for increasing severity of the icing zone. The reference values for low wind conditions selected in the Code for the U.K. are intended to represent 50 year return values.

The appropriate combinations of ice thickness r_{iw} and density with wind speed to be assumed in design depend on the frequency of occurrence and duration of icing conditions, interaction between thickness and speed during such conditions and on the meteorological characteristics of the region of the site. In the absence of sufficient data, assumptions had to be made to derive appropriate reference thicknesses and coincident wind speeds.

The Meteorological Office examined the wind speeds coindicent with the conditions deemed to permit icing, from which it appears that the extreme coincident speeds did not exceed 70% of the 50 year return values. This conclusion has led to the criterion in the Code. The coincident ice thicknesses specified for design are intended to represent the mean annual maxima.

3. WIND RESISTANCE

3.1 General

The term 'wind resistance' has been adopted in the Code to encompass the combinations of area, shielding effects and drag coefficients. Resistance has been related to a panel height to allow for horizontal members and for variations of structure and ancillaries within a panel and to clarify the definition of solidity.

Overall drag coefficients, C_{FN}, are given for square and triangular towers consisting of either flat-sided or round members, being derived from wind tunnel test results. Identical coefficients have been applied to calculate response to mean and fluctuating components of wind speed, although there is limited evidence to support this.

Comparisons were also made with coefficients in current Codes of Practice. There is wide scatter between various standards, resulting in loads which could differ by up to 30% for identical configurations. Any sophistication in other aspects of a Code should be viewed in the context of this order of possible differences in resistance calculations.

3.2 Lattice Towers with Flat-sided Members

Figure 4 shows a typical curve from the Code for a square lattice tower composed of flat-sided members, together with a series of experimental results.

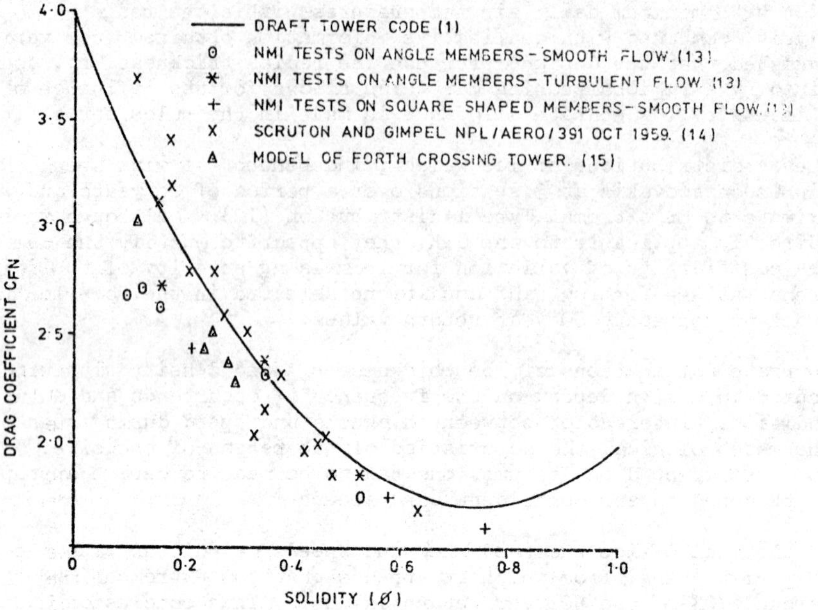

Fig.4 Variation of drag coefficient with solidity for a square lattice tower composed of flat-sided members

There are few experimental data available on tests of lattice sections at high solidity ratios, and it has been assumed in the Code that for all towers the drag coefficient approaches 2.0 as the solidity ratio tends to 1.0. At low values of solidity the proposed curve would again appear to overestimate the drag according to recent tests carried out at the National Maritime Institute (13). However, these tests showed a significant increase in drag at low angles of incidence as shown for example in Figure 5 in which Kθ is the factor to be applied to the resistance for wind normal to a face.

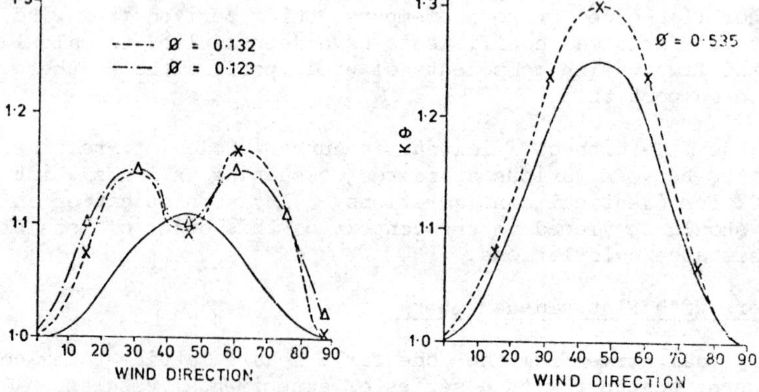

Fig.5 Variation of drag with wind direction for a square lattice tower composed of flat-sided members

It may be seen that for low solidity ratios maximum drag occurs at about $22\frac{1}{2}^{\circ}$ (and $67\frac{1}{2}^{\circ}$) incidence, but at 45° for higher solidity. The Code provides values of $C_{fN}\,K\theta$ which closely predict the maximum resistance over all ranges of incidence and solidity.

3.3 Lattice Towers with Circular Members

For towers composed of circular sections few tests have been undertaken on models in flows providing Reynold's Numbers of the order attained in the field. Gould and Raymer (16) undertook a parametric study into the drag of lattice sections with tubular members in the compressed air tunnel at the National Physical Laboratory. Both ESDU Item 75011 (17) and a Constrado Monograph (18) are based on this work. However, whereas the ESDU Item suggests that there is no systematic variation of drag coefficients with solidity, the Monograph gives detailed curves for varying solidity ratios.

In any normal tower construction each circular member could be in a different flow regime, and it is not practicable to assess the relevant Reynold's Number for the overall frame. Consequently, in codification it was decided to adopt single sets of drag coefficients, over the range of solidity ratio, for both sub-critical and super-critical flow. For sub-critical flow a lower limit to Reynold's Number of about 10^5 was assumed, which at a factored extreme mean wind speed of 45m/sec, with gusts to, say, 67m/sec, implies an average member diameter of about 21mm - a reasonable lower bound for practical sizes. For super-critical flow the drag coefficients correspond to Gould and Raymer's results for a Reynold's Number of about 2×10^6 - implying member sizes up to about 400mm.

Figure 6 shows the Code proposals for a square lattice section composed of circular members, together with test results.

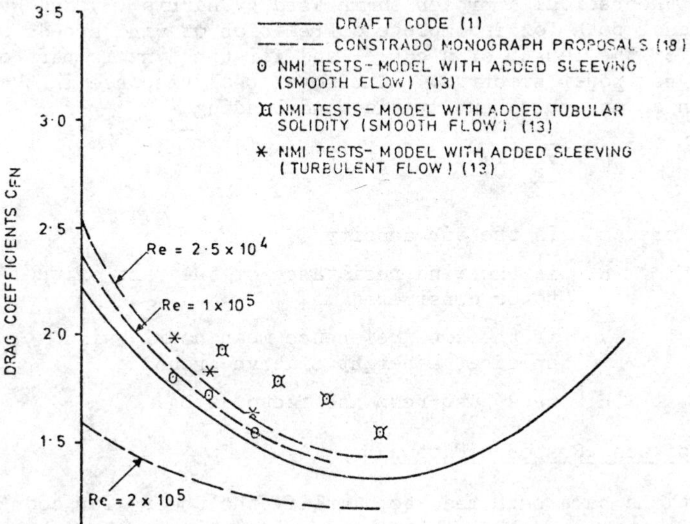

Fig.6 Variation of drag coefficient with solidity for a square lattice tower composed of circular-section members(in sub-critical flow)

The experimental results from N.M.I. are sensitive to Reynold's Numbers. The solidity of the test models was varied by adding identical members to the original frame, thus not affecting the Reynold's Number. By adding sleeving to increase the Reynold's Number the drag coefficients were reduced. The Constrado Monograph proposals (18) are plotted in the Figure showing these to agree with the N.M.I. results for a Reynold's Number of about 2.5×10^4 - approximately the value appropriate to the N.M.I. tests. The Code value would then be compatible with the Reynold's Number of about 10^5 noted above.

3.4 Towers Mounted with Ancillary Equipment

The majority of towers do not consist of simple frameworks but contain a variety of ancillary items such as ladders, feeders, aerials, platforms, etc. and often are constructed of combinations of flat and circular members. To deal with the latter case, simple empirical formulae were developed which obviate the necessity of interpolation for varying solidity ratios and flow regimes. In regions where ancillary items occur previous practice has generally necessitated making an assessment of each panel separately, the solidity ratio being calculated on the basis of the 'total' shadow area. From the results of the recent N.M.I. tests (13) it has been found possible to simplify this procedure, with no undue loss of accuracy, by using drag coefficients appropriate to the basic tower and to add the drag of the ancillary items and allow for their shielding effects separately.

4. STRUCTURAL RESPONSE

4.1 General

In common with other recent wind loading directives the draft Code contains rules for evaluating the response of a structure to turbulent wind. These have been based on the theoretical approach summarised by Harris (3) and Wyatt (19) which makes allowance both for incomplete correlation of wind speeds over the height of a tower or the length of a cable, and for the dynamic part of the response under a gust power spectrum. The total peak response in the downwind direction is given as an equivalent static wind loading

$$\rho \frac{\overline{V}_z R(1+G)}{2}$$

where ρ is the air density

R is the wind resistance of the part of the tower considered

$\overline{V}_z$ is the site reference mean hourly wind speed at a height z above ground

G is a gust response factor = Bj.

4.2 Derivation of Gust Response Factor

Several approximations have been made to simplify the theory for codification. The aerodynamic admittances relevant to the quasi-static response have been combined to give relationships between excitation by background turbulence, B, and the height from the top to a given level of the tower which depend on the terrain roughness coefficient, K_R. These are based on the assumption that a tower will have wind resistance per unit height varying inversely to the wind speed profile. Parametric studies have shown this assumption to be satisfactory for a wide range of towers. The r.m.s. value of the fluctuating component of wind has been taken as independent of height and equal to

$$\frac{\overline{V}_{10}}{3.3}\left(\frac{1.6}{K_R}-1\right)$$ based on observations by Caton (2).

Field measurements have shown the difference between maximum gust speed and mean speed to remain reasonably constant over heights of towers within the scope of the Code. B contains the 'peak' factor and an allowance for dynamic response which, combined, have been shown to produce an effective peak factor of 4.1 for a wide range of natural frequencies typical for lattice towers. This constant value, which represents the ratio of peak to r.m.s. response, results from the allowance made for aerodynamic damping (19) which is inversely proportional to the natural frequency. Such damping represents the major part of the total damping and has been ignored in certain other Codes. As a result, not only has the simplification been justifiable but the resulting gust factors have been lower than adopted when only structural damping has been accounted for as has been shown by the application of several Codes to two heights of tower, having differing frequencies.

The total bending response to turbulence is obtained by multiplying B by a height factor, j, dependent on the velocity profile and the height to the tower top from the level considered. It has been shown that design can be further simplified by assuming the product Bj to be constant for all levels. The shear loading response factor is complicated by the influence of the tower geometry on bracing member forces. Rules have therefore been devised to allow for the effect of leg inclination on such forces. A modifying coefficient is applied to the bending response coefficient to allow for the most adverse correlation of gust speeds as affecting bracings. This overcomes a major defect in most previous loading directives.

Field observations have shown significant response of towers to crosswind turbulence. Allowance has been made for this in the Code by assuming no correlation between longitudinal and lateral turbulence and, conservatively, that the turbulence intensities are identical in the two directions. Thus the gust response factors are the same for both. Crosswind turbulence will not influence the design of a simple square tower but may be of importance when there are large ancillaries or when fatigue life is to be estimated.

Gust factors applicable to the calculation of loads on long cables have been derived by use of the spectral analytical techniques used for the towers. It has been shown that the factors may be simply obtained as the product of two coefficients, one of which is related only to span and the other to span and terrain roughness. The codified rules have provided a close estimate of the observed peak response of 200m. span conductors at the Vada Station of Electricité de France. For cables with chord inclination to the horizontal of less than 45°, the total load effect on the supporting towers has been based on zero correlation between fluctuating tower and cable loads. For steeper inclinations the gust loadings on cables have conservatively been taken to be fully correlated with those on the towers.

5. PERFORMANCE REQUIREMENTS

5.1 General

The aim of the Code is to provide designs which have an acceptable probability of performing satisfactorily during their intended lives. There should be a reasonable certainty that they will not fail nor become unserviceable for unacceptable periods. To this end it has been considered essential to incorporate safety factors on loading and strength commensurate with appropriate risk levels.

5.2 Social Consequences of Failure

Safety margins should reflect the potential risk to life in the event of failure. It has been suggested in a study undertaken by C.I.R.I.A.(20) that the target annual failure probability for towers should be inversely proportional to the number of people at risk in the event of a failure. Thus, the target value should depend on the environment in which a tower is to be built.

Performance requirements in the Code, represented by factors of safety γ on design wind speed, dead load, and material strength, are related to the environment of the tower ranging from unmanned towers in open country to towers in urban environments. The target annual failure probabilities lie between about 10^{-2} to 10^{-4} which corresponds to the range recommended for transmission towers by the I.E.C. (21). The lower value has been considered to be appropriate to structures of inhabited buildings (22).

5.3 Economic Consequences of Failure

When a tower is to be located at a site for which the risk of loss of life in the event of a failure is slight but when the function of the structure is of great economic importance, the safety margins used in design should reflect the financial consequences of collapse. A method used to determine the optimum target levels of notional reliability in these circumstances has been developed by C.I.R.I.A. (20) following the work of Bea (23) and has been used to relate safety factors to estimated consequential costs.

5.4 Classification of Quality

The Code is drafted to apply to towers designed and constructed in accordance with best practice in the U.K., to standards of materials and workmanship comparable with those specified in relevant British Standards, and the safety factors given have been derived by reference to results of tests on towers of such quality. It is, however, recognised that the reliability of a given design may be influenced by errors in the design and detailing. In consequence lower safety margins are suggested in instances in which such errors are controlled, either by appraisal of the design by an experienced engineer or by full-scale testing. Higher margins are proposed where errors in fabrication and erection are not to be controlled and for towers which are not subject to regular inspection and maintenance in service.

Statistical models have been derived to represent the strengths of towers, and these models have been based on the results of full-scale tests on transmission towers together with experience gained in testing of typical elements used in lattice tower construction.

5.5 Safety Effect Assessment

To provide a direct design method applicable to the various quality classes of towers and the different performance requirements of the structures, reliability theory has been used to define factors on loads and strength to be used to achieve the required reliability.

5.6 Load Distribution

It has been shown that a Gumbel Type 1 distribution applies closely to the extreme annual mean hourly wind speeds with return periods up to 50 years. Application of reliability theory showed that the maximum probability density of failure for high reliability towers corresponds to a wind speed, the return period of which is between 1,000 and 3,000 years based on extrapolation from the Gumbel distribution to wind speeds well beyond the range of experience.

From recorded wind data at some 144 stations in the U.K. over a period of 2,500 site years, only 19 sites have experienced speeds greater than or equal to the 50 year return gust speeds predicted by the Meteorological Office for open level terrains. For various reasons it was found expedient to modify the Gumbel distribution for speeds in excess of 50 year return values.

5.7 Calibration

Reliability theory can only be used as a comparative tool in code drafting. There is therefore a need for calibration of any new such code against accepted practice. When dealing with wind loading such calibration should, ideally, be against designs which have in sufficient numbers withstood the range of factored wind speeds to be encompassed by the code. Since no towers in the U.K. have been subjected to wind speeds in the region of the highest failure probability density, it has been found expedient to calibrate the new Code so that it produces designs having design strengths on average equal to those for similar structures which have been in service for a reasonable period of time. It has been shown that this is broadly achieved by the Code provisions, although further calibration is being undertaken before they are finalised.

REFERENCES:

1. Draft Code of Practice - Lattice Towers: Loading. British Standards Institution. Sept. 1978 (With Commentary)

2. Caton, P.G.F. Standardised Maps of Hourly Mean Wind Speed over the United Kingdom and some Implications regarding Wind Speed Profiles. 4th Int. Conf. on Wind Effects on Buildings and Structures. London, Sept.1975

3. Harris, R.I. The Nature of the Wind. C.I.R.I.A. Seminar on the Modern Design of Wind-sensitive Structures. June 1970

4. E.S.D.U. Data Item 72026. Characteristics of Wind Speed in the Lower Layers of the Atmosphere near the ground. 1972

5. British Standard Code of Practice CP:3 Chapter V Part 2. 1972. Wind Loads

6. Hardman, C.E., Heliwell, N.C. and Hopkins, J.S. Extreme Winds over the United Kingdom for periods ending 1971. Climatological Memorandum No. 50A. July 1973

7. Shellard, H.C. Extreme Wind Speeds over the United Kingdom for periods ending 1963. Climatological Memorandum No: 50

8. Davenport, A.G. The application of Statistical Concepts of the Wind Loading of Structures. Proc. I.C.E. Vol. 19, 1961

9. Sfintesco D. and Wyatt, T.A. A proposed European Code of Practice. 4th Int. Conf. on Wind Effects on Buildings and Structures. London. Sept. 1975

10. Deaves, D.M. and Harris, R.I. A Mathematical Model of the structure of strong winds. C.I.R.I.A. Report 76. May 1978

11. Deaves, D.M. Wind over Hills: A numerical approach. Jour. Industrial Aerodynamics. Vol. 1, 1976

12. Fecko, Š., Kružik, J. et al - Investigations of combined stress by Icing and Wind in Czechoslovakia. C.I.G.R.E. Int. Conf. on High Voltage Electric Systems. Aug. 1976

13. Wind Loading on Lattice Towers. N.M.I. Report on Project P/352003 (To be published)

14. Scruton and Gimpel. Memorardum on Wind Forces and pressure exerted on Structures and Buildings. NPL/Aero/39. Oct 1959

15. Cowdrey, C.E. Aerodynamic Forces and Moments on Models of two sections of a Forth Crossing Tower. NPL/Aero Special Report 027. June 1969

16. Gould, R.W.F. and Raymer, W.G. Measurements over a wide range of Reynold's Numbers of the Wind Forces on Models of Lattice Frameworks with Tubular Members. NPL Scie. Report No. 5 - 72

17. Fluid Forces on Lattice Structures. E.S.D.U. - Item No. 75011. Aug. 1975

18. Constrado Publication 1/75. Wind Forces on unclad Tubular Structures. Jan. 1975

19. Wyatt, T.A. The calculation of Structural Response. Paper 6. C.I.R.I.A. Seminar on the modern design of Wind-sensitive Structures. London June 1970

20. C.I.R.I.A. Report 63. Rationalisation of Safety and Serviceability Factors in Structural Codes. July 1977

21. International Electrotechnical Commission Technical Committee No: 11 - Recommendations for Overhead Lines. Nov. 1972

22. Draft Safety Code for Load-carrying Structures. Nordic Committee for Building Regulations. 1974

23. Bea, R.G. Development of Safe Environmental Criteria for Offshore Structures. Oceanology International. 1975

WIND LOAD STANDARDS AND CODES

K. C. MEHTA
Texas Tech University, Lubbock, Texas, USA

ABSTRACT

Wind load standards and codes govern design of buildings and structures to resist wind induced loads. A major portion of wind engineering research effort ultimately translates into development of wind loading standards. This paper compares wind loading standards of four nations, namely the United States, Australia, Britain and Canada. In addition, wind loading requirements of the four model building codes of the United States (Uniform, Standard, Basic, and National Building Code) are discussed. Comparisons between the standards and codes are substantiated by determining design wind pressures for a low-rise building.

INTRODUCTION

Wind load standards and codes govern design of buildings and structures to resist wind induced loads. In a vast majority of cases, prescriptive requirements of the standards and codes are used in design of buildings and structures. The use of prescriptive requirements is the most efficient method of accomplishing the design. In addition, adherence to the requirements of the standards and codes provides a guarantee of meeting a certain minimum requirement of loading. Only in an unusual structure of considerable importance are specific wind loads for that structure developed. Since prescriptive requirements given in the standards and codes control design of the vast majority of structures, it is desirable for practicing engineers and researchers to understand these requirements.

Professional practice of engineering is reflected in standards and building codes. It is sometimes thought that building codes serve to establish standards of practice; actually, the character of code provisions is a good indicator of accepted practice in countries or regions of countries where they apply. Wind resistant design regulations in the form of standards and codes vary considerably (1) in their format and application around the world. In the United States, the American National Standards Institute (ANSI) Standard A58.1-1972 (2) establishes load requirements on a consensus basis. The four U.S. model building codes establish wind load requirements based on ANSI provisions and on data from other sources. Local communities or regions in the United States adopt as regulations the provisions of one of the model building codes.

This paper discusses development of a wind load standard and compares standards of four nations, namely Australia (3), Britain (4), Canada (5) and the United States(2).

In addition, wind load provisons of the four major United States model building codes, these being Uniform (6), Standard (7), Basic (8), and National (9), are discussed. Comparisons of the standards and the codes are substantiated by determining design wind pressures for a low-rise building.

DEFINITION OF STANDARD AND CODE

A standard is a documentation of the state-of-knowledge. It should include current practices, past experiences, and synthesized research knowledge. Most of the standards are developed by consensus groups, professional societies, or governmental agencies for use by building code bodies or other governmental agencies. A consensus standard carries more weight than other types of standards because it is supported by extensive professional review and agreement. The consensus process demands that every individual reviewing the standard is satisfied, and that every question has been answered prior to the adoption of the standard.

A building code is a document embracing a systematic collection of laws, rules and regulations which, if enforced, provides for safe use and maintenance of buildings and structures. The purpose of a building code is to provide legal standards for buildings to protect life, health, and public welfare. Code regulations represent minimum requirements that will assure adequate safety under most conditions.

DEVELOPMENT OF STANDARDS

A standard which is applicable to design of buildings and structures should be formulated carefully. The wind load provisions of a standard are generally based on existing professional practice, past experiences, and on latest research information. A group of people forming a committee that represents various disciplines formulates the provisions of a standard. A wind load standard, in particular, utilizes input from disciplines of atmospheric sciences, fluid dynamics and structural mechanics. In addition, input from practicing engineers in the form of good judgement is most valuable to the development of a good standard. There are three different, but interdependent components in establishing wind loads. These components and associated disciplines are: (1) windspeed--meteorologists, (2) wind structure interaction--structural engineers, and (3) pressure coefficients-fluid dynamicists.

Windspeeds

Geographical distribution of windspeeds are established based on the statistical analysis of the windspeed recordings of many years. Since only extreme values of windspeed are of interest in design of a building, maximum windspeeds recorded in each year at a particular location are used in the analysis. Windspeed recording method generally controls time-averaging of reference windspeed, such as fastest mile or two-second gust speed. It is important to recognize that a value of windspeed would depend on the averaging time; the smaller the averaging time the higher the windspeed value. The wind also fluctuates in real time; this fluctuating component over the average windspeed is termed turbulence.

It has been determined that terrain roughness and height above ground significantly affect reference windspeed and turbulence content of the free-field wind. Terrain roughness is a difficul parameter to quantify. Current standards make a good attempt in providing reasonable quanitification of the effects of terrain roughness by describing terrain in terms of flat, suburban, hilly or urban.

Based on an acceptable level of risk of exceeding a given value, reference windspeeds are established by standards. Meteorologists and engineers need to work very closely to establish dependable values of reference windspeeds that are consistent with an acceptable level of risk.

Wind-Structure Interaction

When free-field wind encompasses a bluff body in its path, it transfers some of its energy to the body. The amount of energy transferred is termed "gust response factor". If the body is small, a total gust may envelope the body and transfer more energy, comparatively, than to a body which is much larger than the gust size. In this latter case the effect of gust on the body as a whole is reduced.

As mentioned previously, turbulence in wind (gustiness) is affected by terrain roughness and height above ground. Hence, the gust response factor would be dependent on the terrain roughness and the height above ground.

Gustiness in wind buffets the body. A tall, slender, and flexible structure could have significant dynamic amplification in its response to wind because of this buffeting. This dynamic amplification of response would depend on gust frequency correlation with the natural frequency of the structure. Thus the structural mechanics discipline, with its knowledge of structural dynamics, has to be involved in establishing gust response factors.

Pressure Coefficients

When wind attempts to negotiate around a body, a considerable amount of turbulence is created near the surfaces of the body. This turbulence creates fluctuating pressures on the surfaces. Non-dimensional ratios of wind-induced pressures on a surface to dynamic free-field wind pressures are termed pressure coefficients. Of necessity, pressure coefficient values are generally obtained through wind tunnel experimentation. Full-scale measurements are used only to spot check the values obtained in the wind tunnel because of the expense involved in full-scale measurements. Since wind tunnel measurements play an important role in establishing wind loads, a brief discussion of wind tunnel technique is warranted.

A good wind tunnel experiment should be able to simulate wind characteristics of turbulence intensity and vertical profile. Data obtained from a smooth flow wind tunnel without turbulence could be very misleading. In addition, the measurement techniques in a wind tunnel should be monitored carefully. The scale in wind tunnels are in the range of 1:100 to 1:500. This range of scale permits development of a complete vertical boundary layer in a reasonably sized wind tunnel. However, the size of the building model for a low rise building becomes very small in the wind tunnel. For example, a building with height of 7 m (22 ft) would require a wind tunnel model with height of 28 mm (1 in) for a scale of 1:250. This reduction in geometric scale indicates that it is essential to construct building models for wind tunnel use with due precision.

Dimensional characteristics of fluid flow require that time scale be compressed in wind tunnel testing. Thus, a test run of 30 seconds in a wind tunnel is equivalent to a real time of approximately one hour. If windspeed in a wind tunnel is averaged over one second, that average is equivalent to an average of 4-5 minutes in real time. This scaling of time indicates that it is difficult, although not impossible, to measure gust speed in a wind tunnel equivalent to a two-second gust in real time. It is essential to measure pressures at 1/30 second interval in the wind tunnel to obtain peak pressures averaged over 2-3 seconds in real time. These numerical values emphasize the importance of good wind tunnel testing techniques.

In recent years, several excellent boundary layer wind tunnels have been developed that have the capability of simulating wind characteristics closely and of measuring pressures accurately. However, most of the pressure coefficients available in the literature and provided in current standards were obtained in the smooth flow wind tunnel. Hence, good judgement should be used in utilizing the pressure coefficient values available in the literature.

DEVELOPMENT OF BUILDING CODES

Building code requirements are developed by committees established specifically to recommend changes in the code provisions. In the case of one of the U.S. model building codes, the Basic Building Code (8), and Ad Hoc committee was charged to recommend changes in the Wind Loading article (10). The committee of fourteen members evaluated requirements given in the ANSI A58.1-1972 (2) standard, and adopted an edited and simpler version of the requirements given in the standard. The committee in its deliberation used considerable engineering judgement and was concerned more about the human safety aspects of wind loads than the economics of damage to buildings. In general, building code requirements are adopted both from recognized standards and past experiences.

NATIONAL STANDARDS

The wind load standards of the four nations considered, these being the United States (2), Australia (3), England (4), and Canada (5), contain similar approaches to wind loading requirements. Each one accomplishes manipulation of windspeed, establishment of gust factors, and introduction of pressure coefficients to obtain design pressures. The factors considered in these standards to obtain wind loads are presented in Table I. In general, all standards consider the same factors. Methods of consideration of these factors in each standard, though, are quite different. These differences and their implications are discussed below.

Reference Windspeeds

An important parameter in obtaining wind loads is the duration of reference windspeed. In the United States, the reference windspeed in standards is the fastest mile wind. The Canadian standard uses mean hourly average windspeed, while British and Australian standards use two-second gust speed. These differences in averaging time of reference windspeed have a dramtic influence on basic dynamic wind pressures. As shown in Table 2 , the Canadian reference windspeed would be 22 percent lower and British and Australian reference windspeed would be 18 percent higher relative to the U.S. reference windspeed. Since wind pressure is a function of the square of the windspeed, the basic dynamic pressure differences are even more dramatic.

Implications of these differences in basic reference windspeed are that gust response factors and pressure coefficient values have to be adjusted relative to each reference windspeed. For example, the gust response factors in the Canadian standard should be higher than the ones in the U.S. standard because gusts in wind above mean hourly value would be much higher than the gusts above fastest mile wind. The British and Australian Standard would have a gust factor of 1.0 because they are already using gust wind.

Levels of Approaches

The flexibility which the standards provide to the designers is reflected in the levels of approaches permitted by each standard. Various levels of approaches to the design are permitted by the four national standards as shown in Table 3. Four possible levels of approaches are as follows:

1. Tables or Simple Procedure: Buildings and structures in which wind loading does not govern the economics of structural design could be designed using simple procedure and tabularized loads. Most low-rise structures come under this category of design.

TABLE 1. Factors considered in national standards

Factors Considered	ANSI A58.1-1972 (2)	Australian (3)	British (4)	Canadian (5)
WINDSPEED				
·Reference Speed	Fastest mile	Two second gust	Two-second gust	Mean hourly
·Variation with height	Yes	Yes	Yes	Yes
·Terrain roughness	Three	Four	Four	Three
·Local terrain	None	Yes	Yes	None
GUSTS				
·Magnitude	Gust response factor	Gust speed	Gust speed	Gust effect factor
·Spatial correlation	Parts and portions	Reduction for large area	None	Gust effect factor
·Gust frequency	Dynamic consideration for $h/b > 5$	Dynamic consideration for $h/b > 5$	Dynamic consideration not included	Dynamic consideration for $h/b > 4$ or for $h > 400$ ft
WIND PRESSURE				
·Pressure coefficients	Tables, text	Tables in appendix, includes sketches	Tables, includes sketches	Figures and tables in commentaries

2. Detailed Procedure: Buildings and structures in which wind loading governs the economics of structural design should use detailed procedure. Mass produced buildings or tall slender structures lend themselves to the detailed procedure, to take advantage of more accurately predicted loads and savings of material.

3. Wind Tunnel: Monumental building in which wind loads control economical design of the building, and in which the cost of the project is high enough to justify additional design expense, should resort to wind tunnel testing. Standardized methods are conservative. Relatively accurate loads and response of structure could be predicted in a boundary layer wind tunnel. A wind tunnel should also be used for a structure with unusual architectural details, particularly when that type of shape has not been tested previously in the wind tunnel.

4. References: Design using referenced material could be advantageous in cases where type and shape of structure is not covered by the standard, and the project is not big enough to justify wind tunnel testing. Only the Australian standard specifically permits load design using acceptable references.

TABLE 2. Comparison of reference windspeed

Reference Windspeed	ANSI A58.1-1972 (2)	Australian (3)	British (4)	Canadian (5)
Averaging time	Fastest Mile	Two-Three Second Gust	Two Second Gust	Mean Hourly Average
Equivalent Reference Windspeed to Fastest Mile 100 mph (161 km/h)	100 mph (161 km/h)	118 mph (190 km/h)	118 mph (190 km/h)	78 mph (126 km/h)

TABLE 3. Levels of approaches permitted in standards

Levels of Approaches	ANSI A58.1-1972 (2)	Australian (3)	British (4)	Canadian (5)
Tables or Simple Procedures	Yes	Yes	Yes	Yes
Detailed Procedures	Yes	No	No	Yes
Wind Tunnel	No	Yes	Yes	Yes
References	No	Yes	No	No

As indicated in Table 3, Australian and Canadian standards are more flexible than the British and U.S. standards.

Special Items

Sophistication in a standard is reflected in consideration of special items, such as building motion and partition walls. Table 4 shows some of the special items considered in each of the standards. It should be noted that most of the special items are covered in the Canadian Code Commentary (11) and British Handbook (12) rather in the texts of the respective standards.

TABLE 4. Special items considered in each standard

Special Items	ANSI A58.1-1972 (2)	Australian (3)	British (4)	Canadian (5)
Partition Walls	No	No	Yes	Yes
Vortex Shedding	No	Yes	Yes	Yes
Building Motion	No	No	No	Yes
Shielding	No	No	No	Yes
Torsional Effects	Signs	No	No	Yes

Because of different ways of specifying basic windspeed in various standards, it is difficult to provide a direct comparison of their provisions. The requirements of standards are compared by obtaining design wind pressures for a typical low-rise building. This comparison is presented later in the paper.

UNITED STATES MODEL BUILDING CODES

Activities of building code organizations are not the same in the United States as in other countries. Canada, England, and Australia have national codes. In the United States, there are four model codes, Uniform (6), Standard (7), Basic (8), and National (9). These four model building codes are established by independent non-profit organizations. Individuals representing various facets of the building industry make-up the committees of these codes organizations. The committees recommend changes to the provisions of the codes, which in turn are approved by the membership at large by a majority vote. The code bodies accept changes annually, though major revisions and publications are on a three to five year cycle. Status of wind load provisions in each of these codes is given below.

Uniform Building Code (6)

The 1978 version of the code has the same wind load provisions as the previous version. Wind load provisions are based on American Standards Association (now ANSI) A58.1-1955 (13). Wind pressures (not windspeed) obtained from a map are used to enter a table which provides wind pressures for various heights above ground. Only one ground roughness category and one value of gust factor are incorporated into the wind pressure values. Pressure coefficients are provided for simple shapes and for overall surfaces. Internal pressures are not considered.

Standard Building Code (7)

This code is based on requirements given in the ANSI A58.1-1972 (2) standard, though the requirements are modified. Windspeed variations with height are based on only one terrain roughness category, namely smooth terrain. Gust response factor is taken as 1.0 for all cases. Pressure coefficients are provided only for the most common shapes; internal pressures are not considered.

BOCA Basic Building Code (8)

The 1975 edition of the code contained specific uniform design pressures irrespective of the location in the country. For external wall design, 2/3 of the total pressure is applied to the windward wall and 1/3 is applied to the leeward wall.

Internal pressure are treated by imposing 10 psf (479 Pa) of pressure if openings equal or exceed 1/3 of total wall area. These requirements of the 1975 edition are revised in the 1978 edition as described in the article by Kominek (10). According to the article, the revised wind loads are based on ANSI A58.1-1972 (2) standard. Variation of windspeed and gust factor with height are based on only one terrain roughness, namely partially built suburban terrain.

National Building Code (9)

This code has adopted ANSI A58.1-1972 (2) standard in its entirety.

As evidenced from the discussion of the codes above, it is difficult to compare the wind load provisons of the codes directly. A comparison is provided through obtaining design wind loads for a typical low-rise building using the provisions of each code. This comparison is presented in the next section.

DESIGN PRESSURES FOR A LOW-RISE BUILDING

Provisions of the National Standards and the U.S. Building Codes are utilized to obtain design pressures on a typical low-rise building. This determination of design pressures serves two purposes: (1) it permits comparison of the requirements in the standards and codes, and (2) it indicates areas of ambiguity that are present in each of the standards.

The typical building chosen for the example is shown in Figure 1. The building is 61 m (200 ft) square in plan with eave height of 6.1 m (20 ft). The roof slope is 1:12. The structural frames are spaced 7.6 m (25 ft) apart. Purlins are secondary roof members, spanning between the frames. The purlins are spaced 1.5 m (5 ft) apart. Metal decking would be attached to the purlins to form the roof. Girts are secondary wall members; they are spaced 2.4 m (8 ft) apart. Metal siding would be attached to the girts. These structural components which would resist wind loads are identified in Figure 1. The windows and doors in the walls are assumed to be uniformly distributed. The openings add up to about 5 percent of the wall surface area.

The building is assumed to be located in the Omaha, Nebraska area. The fastest mile wind for that area at a 0.02 probability of exceedance is between 129 km/h (80 mph) and 145 km/h (90 mph) as specified in ANSI A58.1-1972 (2). The reference fastest mile wind of 145 km/h (90 mph) is used for the example problem. The terrain surrounding the building is assumed to be smooth terrain (airport surroundings). Geographical location for the building must be specified because some of the building codes provide reference pressures rather than reference windspeed. Design pressures for various components of the building are determined.

Design Pressures Using National Standards

Requirements of the four national standards, U.S., Australian, British, and Canadian, are utilized to obtain design pressures. The reference windspeeds in each of the standards are different. These reference windspeeds and design pressures for various building components are shown in Table 5. The values shown in the table are pounds per square foot; metric units are not shown for the purpose of clarity.

A comparison of the design pressures shown in Table 5 indicates that wind loadings under these standards differ by up to 50 percent, and sometimes by even more. For example, roof purlins would be designed for uplift of 21 psf (1006 Pa) under the Canadian standard while for uplift of 32 psf (1522 Pa) under the British standard. These variations in design pressures signify a lack of ability to translate research results into professional practice on a consistent basis.

Influence Areas

Purlin: 125 sq ft (11.6 sq m)
Girt: 200 sq ft (18.6 sq m)
Roof Deck: 48 sq ft (4.5 sq m)
Wall Siding: 48 sq ft (4.5 sq m)
Door: 180 sq ft (16.7 sq m)
Window: 36 sq ft (3.3 sq m)

Location: Omaha, Nebraska
Fastest Mile Wind: 90 mph (145 km/h)
Frame Spacing: 25 ft (7.6 m)
Exposure: Flat Open (Airport)
Openings: Uniformly Distributed

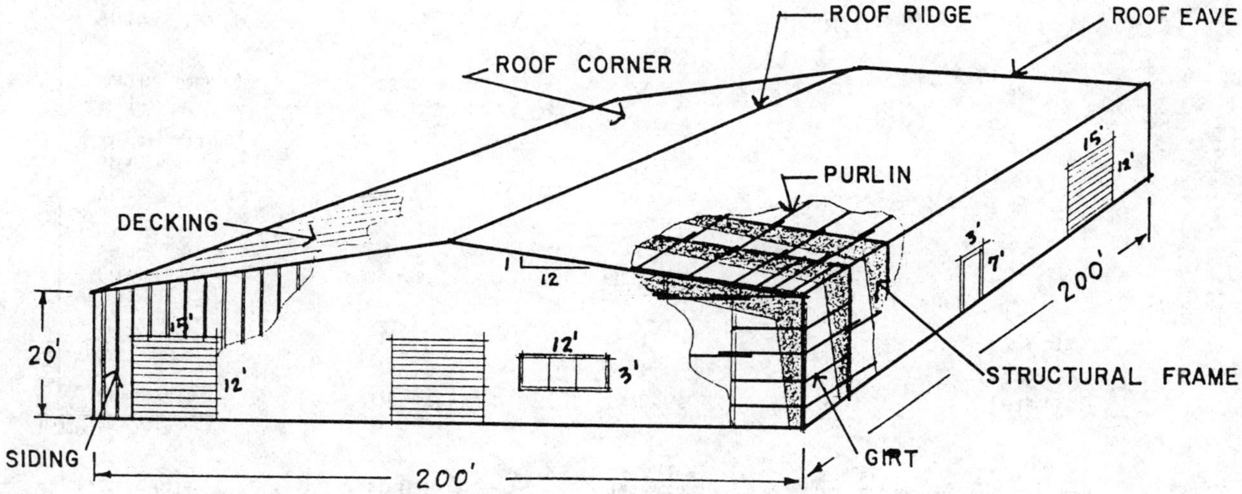

Fig. 1. Building for design pressures

TABLE 5. Design pressures determined using national standards

	Reference Speed	ANSI A58.1-1972 (2)	Australian (3)	British (4)	Canadian (5)
	90 mph Fastest Mile	90 mph[a] Fastest Mile	106 mph Two Second Gust	106 mph Two Second Gust	70 mph Hourly Mean
	Building Component				
FRAME	Windward	+16 psf[b]	+22 psf	+13 psf	+18 psf
FRAME	Leeward	-10[c]	-11	-9	-13
FRAME	Uplift	-14	-24	-20 & -9[d]	-25 & -13[d]
WALL	General: Girt, Door, Window, Siding	+31, -28[e]	+32, -17	+24, -17	+21, -14
WALL	Corner: Width	20 ft[f]	20 ft	50 ft	None
WALL	Girt, Door Window, Siding	-46	-43	-24	--
ROOF	General: Purlin, Decking	-28	-26	-32	-21
ROOF	Eave: Width	20 ft	20 ft	20 ft	20 ft
ROOF	Purlin, Decking	-54	-43	-46	-46[g]
ROOF	Corner: Area	20 ft x 20 ft	20 ft x 40 ft	none	20 ft x 40 ft
ROOF	Purlin, Decking	-96	-58	--	-71[g]

(a) 1 mph = 1.6. km/h; (b) 1 psf = 47.9 Pa

(c) + Sign means inward acting, - sign means outward acting or uplift

(d) Windward and leeward parts of roof

(e) Member to be designed to resist both pressures

(f) 1 ft = 0.305 m; (g) Pressure for decking only

In addition to the variations of design pressures, the standards also necessitate interpretation of the provisions by structural engineers. The areas of ambiguity, where interpretations in each standard are essential, are enumerated below in terms of questions.

ANSI A58.1-1972 (2):
- (a) What constitutes openings?
- (b) What components are parts and portions?
- (c) Are the gust factors included in the local peak pressure coefficients?

Australian (3) :
- (a) What constitutes openings and permeability? Are they different or the same?
- (b) Which wind angle of attack controls?
- (c) Are the reduction factors applicable to purlin loading or only to the frame loading?

British (4) :
- (a) Cladding and immediate fixings include purlins or not?
- (b) Does permeability mean opening for internal pressure?
- (c) Which wind angle of attack controls?

Canadian (5) :
- (a) Should the internal pressure be included for both outward and inward acting pressures?
- (b) What are the dimensions for wall corner and roof corner areas?
- (c) Are the local pressures applicable to cladding only or also applicable to purlins?

The above mentioned areas of ambiguity were encountered during determination of design pressures for this building. There may be additional areas of ambiguity which would be discovered if a different type of building had been used as the example.

The main point of the discussion is that refinement of research results to provide more accurate design pressures could easily be lost in the interpretation of the standards. More effort should be expended in developing standard requirements which are consistent and easy to use by practicing professionals.

Design Pressures Using U.S. Building Codes

Design pressures for the building components are obtained following the requirements of the four model U.S. Building Codes; Uniform Building Code (6), Standard Building Code (7), BOCA Basic Building Code (8), and National Building Code (9). The design pressure values for building components are shown in Table 6. The values shown are in pounds per square foot; equivalent metric values are not shown for purposes of clarity. For comparison, table 6 also presents design pressures obtained using the ANSI A58.1-1972 (2) requirements.

A comparison of values in Table 6 indicates wide variations in design pressures. For example, a roof purlin could be designed for uplift of 8 psf (383 Pa) under the BOCA Code, but would have an uplift of 28 psf (1341 Pa) under the National Building Code requirements. This extent of design pressure variation is disconcerting to practicing engineers. Reasonable consistency in wind load requirements of various building codes is an objective that code bodies should pursue.

FUTURE TRENDS

The discussion of the standards and codes presented in this paper illustrates the

TABLE 6. Design pressures determined using U.S. model codes

	Building Components	ANSI[2] A-58.1-1972	Uniform Bldg.[6] Code 1978	Standard Bldg.[7] Code 1978	BOCA Basic[8] Bldg. Code 1975	National Bldg.[9] Code 1976
FRAME	Windward	+16 psf[a]	+18 psf	+13 psf	+10 psf	+16 psf
	Leeward	-10[b]	-12	-8	-8	-10
	Uplift	-14	-22	-16 & -12[c]	-1 & -8[d]	-14
WALL	General					
	Girt, Door, Windows, Siding	+31, -28	+18, -12	+18, -18	+15, -8	+31, -28
	Corner					
	Width	20 ft.	N.A.[e]	N.A.	N.A.	20 ft.
	Girt, Door, Window, Siding	-46	+18, -12	+18, -18	+15, -8	-46
ROOF	General					
	Purlin, Decking	-28	-22	-16	-8	-28
	Eave					
	Width	20 ft.	N.A.	Unspecified	N.A.	20 ft.
	Purlin, Decking	-54	-22	-24	-8	-54
	Corner					
	Area	20 ft x 20 ft	-22	N.A.	N.A.	20 ft x 20 ft
	Purlin, Decking	-96	-22	-24	-8	-96

(a) 1 psf=47.9 Pa; (b) + means inward acting, - means outward acting
(c) 1/3 & 2/3 surface areas; (d) Windward and leeward surface areas; (e) not applicable

disparities that exist in the wind load requirements. The differences are in the form of windspeed averaging time, gust factors, pressure coefficients, and in interpretation of the provisions. In view of these differences, future standards and codes should attempt to make wind load requirements as simple and consistent as possible. Some specific suggestions to achieve these objectives are as follows:

1. Wind Climate: An attempt should be made to use consistent windspeed averaging time. Whether this be hourly mean or two-second gust can be decided through due consideration of other factors.

2. Simplicity and Flexibility: The provisions in the standards and codes should be written for practicing engineers, not for researchers. Refinements in wind load provisions are not justified if they have a high probability of being misinterpreted. It is difficult to make provisions which are simple in format, and yet applicable to all structures. More time and effort should be expended in writing the provisions.

3. Consistent Terminology: A fair amount of misinterpretation of the provisions is due to inconsistency in terminology. Interpretation of permeability and openings could make a big difference in internal design pressures for building components. Although it is impossible to write a standard which would be void of interpretive judgement on the part of practicing engineers, consistent terminology would help in reducing misinterpretation.

A trend toward an International Standard through a committee of the International Organization for Standardization (ISO) may help to reduce disparities between the national standards.

ACKNOWLEDGEMENTS

The work presented in this paper was accomplished under National Science Foundation (NSF) Grant PFR-17244. Encouragement and advice of Dr. Michael Gaus of the NSF is gratefully acknowledged. Also, the counsel of the author's colleagues, Dr. James R. McDonald and Dr. Joseph E. Minor, was most helpful in the assembly of the paper.

REFERENCES

1. Wind Resistant Design Regulations - A World List, Gakujutsu Bunken Fukju-kai, Oh-Okayama, Megurokv, Tokyo 152, 1975, 817 p.

2. Building Code Requirements for Minimum Design Loads in Buildings and Other Structures: ANSI A58.1-1972, American National Standards Institute, Inc., New York, 1972, pp. 11-25.

3. SAA Loading Code, Part 2 - Wind Forces (Australian Standard 1170, Part 2-1973), Standards Association of Australia, North Sydney, 1973, 52 p.

4. Code of Basic Data for the Design of Buildings, Chapter V. Loading; Part 2, Wind Loads, British Standards Institution, London, 1972, 49 p.

5. National Building Code of Canada, National Research Council of Canada, Ottawa, 1975, pp. 156-157.

6. Uniform Building Code, International Conference of Building Officials, Whittier, Calif., 1976, pp. 130-132.

7. Standard Building Code, Southern Building Code Congress International, Inc., Birmingham, 1976, pp. 12-11 through 12-17.

8. The BOCA Basic Building Code/1975, Building Officials and Code Administrators International, Inc., Chicago, 1975, pp. 199-202.

9. National Building Code, American Insurance Association, New York, 1976, pp. 319-342.

10. Kominek, P.E., "Windloads," The Building Official and Code Administrator Sept. 1978, pp. 12-13, and Oct. 1978, pp. 22-25.

11. Supplement No. 4 to the National Building Code of Canada, National Research Council of Canada, Ottawa, 1975, pp. 53-82.

12. Newberry, C.W., and Eaton, K.J., "Wind Loading Handbook," Her Majesty's Stationery Office, London, 1974, 74 p.

13. American Standard Building Code Requirements for Minimum Design Loads on Buildings and Other Structures, A58.1-1955, American Standards Assoc, Inc. NY, 1955, 32 p.

ECCS-RECOMMENDATIONS FOR THE CALCULATION OF WIND EFFECTS ON BUILDINGS AND STRUCTURES

D. SFINTESCO and C. SOIZE
C.T.I.C.M., Puteaux, France

ABSTRACT

The "Recommendations" prepared by the European Convention for Constructional Steelwork in a code-format are intended as a possible basis for a unified European code on wind effects on buildings and structures.

The up-to-date character of this document and its convenient presentation for practical use are briefly described in this paper.

1. INTRODUCTORY REMARKS

The European Convention for Constructional Steelwork (ECCS) has undertaken, some years ago, on proposal of the first named author, the preparation of a model code of practice for the calculation of wind effects on buildings and structures. This task has been assigned to a specially formed Committee on "Wind effects".

The main reasons of this action were, on the one hand, the need to harmonize the current national regulations at the European level and eventually to provide a basis for a unified European code and, on the other hand, to seize this opportunity to take advantage of the latest findings of the theoretical and experimental research and of field observation, in order to produce an up-to-date document.

There were indeed, and there still are, great discrepancies between the national specifications on this matter in Europe, which result in an illogical and most awkward situation for practical applications and for the implementation of the European Common Market in the building sector.

The need was also felt to have an international document established in consistency with the probabilistic concepts of structural safety adopted in the ECCS-Recommendations for the design of steel structures (1) and in all modern national specifications for structural design. Therefore the calculation methods proposed by the ECCS are based on a probabilistic definition of the design wind loads.

The drafting of these Recommendations was preceded by a fundamental theoretical study carried out on behalf of the ECCS at the Imperial College of Science and Technology, London, UK, under the responsibility of Dr. T.A. Wyatt. This study was an essential contribution to the scientific and practical value of the Recommendations. It has been described in a paper presented at the Fourth International Conference on Wind Engineering (2).

The ECCS document is now completed and already published (3)*. It is written in a code format, in order to facilitate its practical use and its official consideration within the current action for unified building codes for Western Europe ("Eurocodes"). It is in fact a very modern document representing a synthesis of findings of research carried out in the world during the past two decades, put in simple form for direct use in design offices.

But before commenting its contents, we shall briefly describe the state of things which has led to the drafting of such a document.

2. BACKGROUND OF THE ECCS-RECOMMENDATIONS

The various theoretical and experimental studies carried out for many years on national and international level for the knowledge of the wind structure and of its effects on buildings and structures have allowed to define a mathematical model for the representation of the wind speed field in the planetary boundary layer near the ground level for the design of structures.

This mathematical model is based on a probabilistic approach and has been established by means of a statistical treatment of the turbulence. It allows to show up, on the one hand, the spatial effect of the turbulence which results in an "effect of dimensions" when pressures are integrated on surfaces (this operation being needed for determining the forces) and, on the other hand, the effect in time of the turbulence, which reflects the swift fluctuations of the wind speed field and which can produce, in certain cases, dynamic effects on the structures.

This model is principally valid for determining the wind effects in a direction parallel to that of the wind. Of course, other dynamic effects can appear in a structure without being predominantly due to the turbulence. This is so e.g. for the effects of von Karman vortices and of aerodynamic instabilities.

The very great interest of this model consists in giving a good "global" approach of a rather complex physical phenomenon. It takes correctly into account the energy input due to wind as a function of the frequency bands, this allowing to tackle in a realistic way the study of the dynamic behavior of structures.

Experience has shown that rigid structures are generally not sensitive to dynamic wind effects. The above mentioned model is therefore particularly interesting for the determination of the dynamic effects on flexible structures in a direction parallel to that of the wind. It allows however also to determine the effects which we call "quasi-static" on rigid structures ; in this case the spatial effect of the turbulence is the only one taken into account.

One must well realize the essential differences introduced by such a model with regard to models of the type used in "old" specifications, where the fundamental parameter for design was the basic dynamic pressure associated to a gust peak speed. Such specifications concerning wind effects fix the value of the basic dynamic pressure for the design wind depending on the location of the building or structure and prescribe the sequence of calculations to be applied in order to determine the wind effects on the considered building or structure once the basic dynamic pressure is known.

The use of the "new" wind model changes this procedure radically even for low and rigid buildings, by introducing two basic parameters in the calculation instead of one :

----

* It will be available at the time of the Conference.

- The first parameter is still the basic dynamic wind pressure depending on the location of the building or structure, but this basic pressure is no more associated to a gust peak speed, but to a mean speed (e.g. on 10 minutes);

- the second basic parameter, which was not used in the "old" specifications, is a parameter of global roughness of the site where the building or structure is located. The roughness is directly linked to the intensity of the turbulence and conditions the whole model. One does not need to know the laws of turbulence to realize that the greater the site roughness, the greater the turbulence. So the turbulence will be greater leeward of a big city than for the wind coming from the sea.

With regard to "old" methods and once the two basic parameters are fixed, the sequence of calculations which allows to determine the wind effects in a direction parallel to that of the wind is very different and leads in its general formulation to random static and dynamic calculations. The complexity of these calculations is greater or lesser, depending on the approximations used (linear or non-linear calculations, etc...).

In brief, in spite of being still imperfect in certain circumstances, this "new" model can be considered on the international level as sufficiently valid and definitely better than the previous models for the study of the dynamic response of flexible structures in a direction parallel to that of the mean wind.

The formulation of this type of calculation is of course relatively heavy. It implies necessarily the use of computers. However it is possible, by means of certain simplifications, to introduce these concepts in a simple manner into a specification. In this way the designer will have a simple tool of calculation which does not require systematically the use of computers. As an example, a similar thing has been already done for the Canadian code of 1970.

The ECCS has now produced its "Recommendations" on the basis of these new methods. Their contents will be briefly described hereafter.

But we wish to emphasize here the following fundamental point : the choice of the basic dynamic pressure of the design wind to be adopted in a given site is linked to the general treatment of the structural safety. In the present terminology of "safety", this basic pressure would be called "characteristic action" and would be associated with a given "return period". The choice of the value for the characteristic action or for the corresponding return period -which comes to the same thing- supposes the adoption of a probabilistic model of this characteristic action which is then considered as a random value, since this model results, of course, from the statistical treatment of wind speed measurements over a great number of years. This choice does not change anything to the above mentioned sequence of calculations, which is based on a probabilistic model of the turbulence. Each of the two fundamental parameters has then a given value.

It must be understood that the value of the characteristic wind action must be adjusted in connection with the safety factors, by referring to the new wind model and not to the old ones.

The ECCS-Recommendations being an international document, the obvious thing to do was to take realistically into account the still existing diversity among the single countries in matters of structural safety. Therefore they have been established in a general way on this particular point, by recommending for the "characteristic" wind action the value corresponding to the mean wind speed on 10 minutes with a return period of t years.

The value of t and the values of the load factors as well as the load combinations have been left open, in order to let them be defined within the general framework of the treatment of the structural safety in each country or on international level.

3. GENERAL REMARKS ON THE CONTENTS OF THE ECCS-RECOMMENDATIONS

The purpose of the Recommendations is to "give methods of calculation of the wind effects on a whole building or structure, on its various parts and on the comfort of its users".

In order to make things easier and not to have to use systematically a complicated method for all types of buildings and structures, certain criteria are given for the distinction between so-called "rigid" and so-called "flexible" buildings and structures.

The general arrangement of the contents of the ECCS-Recommendations is not made in the logical sequence a research worker would expect (theoretical background, then definition of actions and treatment of their effects, and finally practical design rules), but in a different order which, according to an inquiry we have carried out, has appeared to be more convenient for the practical use of this document.

So its first part allows to find the value of the parameters related to the design wind effects on rigid buildings and structures. It is noteworthy that this part of the document (printed on white paper) is extremely simple and is practically used for the large majority of buildings and structures.

The second part (printed on pink paper) allows to determine the wind effects on flexible buildings and structures. The given method is very elaborate but still perfectly fit for use. Practically it does not concern but a very small percentage of buildings and structures.

The third part is composed of 4 appendices :

Appendix A : a very complete catalogue of pressure and force coefficients (on white paper)

Appendix B : a formulary for the calculation of oscillation frequencies for simple structures and a reminder of some methods of calculation (on pink paper)

Appendix C : a general description and justification of the proposed methods of calculation of the wind effects on buildings and structures, followed by

Appendix D : bibliographic references (both on blue paper).

So for the study of rigid structures the user (or reader) can limit himself to the white pages. When dealing with flexible structures he must refer to the pink pages also. The blue pages can be ignored in both cases. They are intended for those who wish to have an insight into the background of the Recommendations, but their reading is not necessary for their practical use.

In the following sub-chapters some fundamental points of the first and second parts of the Recommendations will be described more in detail.

3.1 Values and definition of the parameters concerning the design wind

As already mentioned in the preceding chapter 2, the first basic parameter is the "characteristic action" of the wind, i.e. the basic speed of the design wind (or the corresponding basic dynamic pressure), which is the mean wind speed on 10 minutes, measured at the height of 10 meters above ground level, on flat terrain without obstacles (roughness degree II as defined below), and of a return period t (i.e. which occurs on an average of every t years).

As mentioned above in chapter 2, the value of the return period t is not fixed in the Recommendations. These contain however a specially prepared wind map of Western Europe which shows the basic design wind speeds for a return period t of 50 years, and a graph giving the conversion factor for return periods different from 50 years.

The second basic parameter is the parameter of global roughness of the site where the considered building or structure is located. The choice of this parameter consists in determining the degree of roughness at the construction site. A graph gives the variation factor of the design wind speed (or pressure) depending on the degree of roughness and on the height above ground level.

The Recommendations envisage five degrees of terrain roughness, defined as follows :

Degree I : Sea shore or broad stretch of water of at least 5 km width

Degree II : Flat terrain with a few isolated obstacles (trees, buildings, etc...)

Degree III : Rural zone with low buildings, trees, hedges, etc...

Degree IV : Urban, industrial or forest-clad zone

Degree V : Center of big cities.

9 aerial photographs illustrating the above mentioned degrees of roughness are given in order to help in the choice of the appropriate degree.

The Recommendations specify also :

- the calculation method to be used if the site has a variable degree of roughness (as for instance by sudden change of roughness)
- the factor of overspeed to be used depending on the relief if the terrain is not flat (building or structure located on a hill, on the slope of a plateau, in a valley, etc...)
- the wind direction to be considered.

3.2 Definition of rigid and flexible buildings and structures

The definition applies to the whole structure as well as to structural parts, such as isolated columns, chimneys, etc...

The vertical or horizontal slenderness λ of the building or structure has first to be defined. This parameter is the ratio of the greatest dimension of the windward face of the building or structure to the smallest dimension of the same face.

A building or structure has to be considered as rigid if λ is less than 5 and its height is smaller than 50 meters. In this case no dynamic analysis has to be made and the part of the document related to rigid buildings and structures can be applied for the calculation of the wind forces.

If the building or structure does not comply with this criterion, this does not necessarily mean that it cannot be considered as rigid. But in this case it is necessary to apply a more sophisticated criterion which implies the determination of the first mode oscillation frequency n_1. This criterion refers to a limit value n_{1lim} of the frequency which is given by a graph as a function of λ and of the damping. If $n_1 \geqslant n_{1lim}$ the structure is rigid ; if not, it has to be considered as flexible.

It is obvious that, by taking into account the first criterion, the great majority of buildings and structures can be considered as rigid.

3.3 Calculation of the wind effects on rigid buildings and structures

The problem is purely static. The wind forces applied on the single faces of the building or structure are obtained by simple integration of the pressure calculated in each point. The latter is equal to the product of the dynamic pressure of the design wind by the pressure coefficient at the considered point (this coefficient being given in Appendix A). This dynamic pressure is equal to the product of the basic dynamic pressure by a coefficient determined from a graph depending on the height of the point above ground level and on the degree of roughness of the site and -if the terrain is not flat- by the factor of overspeed.

3.4 Calculation of the wind effects on flexible buildings and structures

The details of the method will not be described here. The treated points will be listed only.

Buildings and structures called "flexible" are those for which the oscillations due to wind cannot be neglected. They have in general two principal types of oscillations : flexural and torsional. The method consists in calculating separately :

- wind effects for flexural oscillations in the wind direction
- wind effects for torsional oscillations with the axis perpendicular to the wind direction
- effects of von Karman vortices :
  The Recommendations deal with flexural oscillations in the across wind direction and with the ovalization of cylindrical shells
- effects of aerodynamic instabilities :
  The Recommendations deal with flexural oscillations in the across wind direction, with torsional oscillations with the axis perpendicular to the wind direction and with the possible coupling of the two.

For vertical flexible structures the calculations must be carried out successively for the wind blowing in the two principal directions of flexural oscillation, whereas for horizontal flexible structures the calculations are limited to the wind blowing in the principal direction of flexural oscillation.

3.5 Other criteria

The ECCS-Recommendations deal with two other aspects:

- influence of oscillations of flexible structures on the comfort of the occupants, and

- effects of repetitive wind action in view of possible fatigue calculations.

4. CONCLUSIONS

The ECCS-Recommendations for the calculation of wind effects on buildings and structures are an up-to-date, i.e. most modern, document produced by a committee of experts from European and non-European countries.

It has benefited by the progress of knowledge of the relevant phenomena, acquired through research and observation carried out in the various parts of the world and has been drafted in compliance with the currently prevailing probabilistic concepts of safety.

5. BIBLIOGRAPHY

(1) ECCS : Recommendations for Steel Structures, 1978.

(2) Sfintesco D. and Wyatt T.A., A proposed European Code of Practice, Fourth International Conference on Buildings and Structures, London, 1975.

(3) ECCS : Recommendations for the Calculation of Wind Effects on Buildings and Structures, 1978.

CYCLONE TRACY IN RETROSPECT: A REVIEW OF ITS IMPACT ON THE AUSTRALIAN COMMUNITY

G. R. WALKER* and J. E. MINOR**

*James Cook University of North Queensland, Townsville, Australia
**Texas Tech University, Lubbock, Texas, USA

SUMMARY

Nearly five years have passed since Cyclone "Tracy" devastated the City of Darwin on Christmas Day, 1974. The impact of this disaster can be measured, in retrospect, by outlining the nature of changes made in building practices and codes, by reviewing adjustments made by government and other institutions, and by examining the character of the rebuilt city. The authors have remained in close contact with the processes of change brought about by Tracy. They report major adjustments in the manner in which housing is constructed in cyclone prone regions of Australia and find that some of the building code modifications made in the emotional months following Tracy may not have been justified.

INTRODUCTION

On Christmas morning, 1974, the north Australian city of Darwin was devastated by a severe tropical cyclone named "Tracy". In terms of its impact on the community it was the most disastrous single event to have occurred in Australian history.

At the time of Cyclone Tracy, there were between 8000 and 9000 homes in Darwin. Following Tracy it was estimated that approximately 5000 of these had been destroyed or damaged beyond repair as far as their main structure was concerned, and only about 500 remained intact and continuously habitable.

Consequently, about three-quarters of the population of 46,000 had to be evacuated and the life of the community was totally disrupted. Thus, the primary factor in the disaster was the inability of the housing to withstand the severe winds of Cyclone Tracy (1,2). This observation led the technical community to make major adjustments in the manner in which housing is designed and constructed in cyclone prone regions of Australia.

The total economic cost of the disaster was of the order of 500-700 million dollars. The social cost arising from the terrifying experience, the evacuation, and the difficult living conditions for those who remained, will probably never be properly assessed. Further, the impact of Cyclone Tracy was not restricted to Darwin alone. Because of the magnitude of the disaster a large section of the Australian community became involved in the relief and restoration activity, creating a national awareness of the problems of tropical cyclones. This awareness sustained the large amount of technical activity and produced a significant social response which seems destined to have a permanent influence on the Australian community.

HISTORICAL BACKGROUND

On average approximately five tropical cyclones cross the Australian coastline each year (3). The entire coastline above 30°N, approximately 10,000 km (6,250 mi) in length and representing about sixty percent of the total length of the Australian coastline, is at risk, with the most frequent crossings occurring on the north-west and north-east coasts between 15° and 25°S. However, outside the polar regions, this coastline is probably the most sparsely populated for its length of any in the world. Only one major city, Brisbane (latitude 27.5°S), with a population of 900,000 is located within this region, and it is outside the area of maximum risk. There are 10 other small cities, ranging in population from 20,000 to 100,000 of which all but 2 (Darwin and Geraldton) are on a 2,000 km (1250 mi) length of the Queensland coast. Most of these have only developed to a significant size during the last 20 years of so. There are approximately 20 other smaller communities in the same area.

Consequently, many of the tropical cyclones cross the coast well away from communities of any size and direct hits by severe cyclones on sizeable communities are relatively rare. European settlement in this region is little more than 100 years old. The experience of one or two serious tropical cyclones is part of the history of most of the communities, but in most cases they happened many years ago when the communities were much smaller and before most of the current population were either born or had settled in that region.

As a result, although the risk of tropical cyclones occurring was appreciated, public awareness of the dangers that could arise from them was generally low in most of the communities at the beginning of the 1970's; the only exceptions were some small communities in north-west Australia where the frequency of occurrence is greatest. In the space of 4 years, 2 of the small cities and one of the larger towns were hit by tropical cyclones of moderate intensity or greater: Townsville (population 80,000) by Cyclone Althea (952 mb) in December 1971 (4), Darwin (population 45,000) by Cyclone Tracy (945 mb) in December 1974 (1), and Port Hedland (population 11,000) by Cyclone Joan (934 mb) in December 1975 (5). Neither Townsville nor Port Hedland were in the direct paths of the cyclones but both experienced significant damage from winds which produced maximum gust velocities [3 second gusts at 10 m (33 ft.) height in open terrain] of approximately 55 m/s (123 mph). Cyclone Tracy passed across the centre of Darwin and produced maximum gust velocities estimated to be of the order of 65-70 m/s (145-157 mph) which caused massive damage. It was nearly 40 years after the previous tropical cyclone which caused any significant damage to Darwin and nearly 80 years after the previous most damaging tropical cyclone.

IMPACT ON HOUSING CONSTRUCTION

The most dramatic effect of Cyclone Tracy was the widespread failure of housing. This is highlighted in Table 1 which compares the performance of housing with that of engineered buildings, the latter being defined as those for which an assessment of wind loads and strength was made at the design stage in accordance with the basic principles of structural mechanics and engineering codes of practice, or more sophisticated studies.

A second feature of the damage was that newer housing seemed more susceptible to damage than older housing, in contrast to local technical opinion which prior to Tracy has been voicing its concern about the inadequacy of the older housing to resist tropical cyclones (6).

TABLE 1. Damage Pattern in Darwin from Cyclone Tracy (1)

	Housing	Engineered Buildings
Serious Structural Failure	70%	< 5%
Serious Cladding Failure	20%	20%
Little or no Damage	10%	75%

Both these features had been noted in Townsville 3 years earlier following Cyclone Althea, as a result of which recommendations were made for increased engineering input into housing design (4). These recommendations were mainly directed at providing strength against uplift, the most dominant failure mode during Althea. The recommendations resulted in more attention being given to this aspect of housing construction in North Queensland. In Darwin, the recommendations were applied quite rigorously by the Federal Government, the authority responsible for the design of the majority of new houses.

Of the houses in Darwin less than 2 years old at the time of Cyclone Tracy, the majority were probably tied down better than any other comparative group of houses in Australia. Despite this, houses in this age group were the most intensively damaged, with almost 100 percent of them suffering major damage. Instead of failing in uplift they failed under horizontal loading as a result of weaknesses which had not been exposed in Townsville during Althea and had therefore been assumed not to exist.

The same weaknesses existed in housing throughout Australia but it required a major event like Tracy to expose them. The explanation for the damage was therefore not to be found in "scapegoats" but in the inherent weaknesses of the traditional approach to housing construction (7). In Australia, as indeed throughout most of the world, the construction of low rise dwellings has been based on tradition rather than on the application of scientific knowledge of mechanics and material behaviour. This traditional approach is a combination of intuition and experience. The approach is successful if changes in construction practice are slow and experiences of relatively severe conditions are frequent enough for inadequacies to be highlighted and corrected before being incorporated on a large scale, thus preventing a disaster prone situation. In Darwin, changes in construction practice had not been slow, nor had severe wind conditions occurred frequently enough to highlight inadequacies; hence, Cyclone Tracy revealed the weakness in the traditional approach.

The justification for adopting the traditional approach, particularly from engineers, has been that an individual house has small value economically and accommodates few people compared with larger buildings, hence they are of lesser importance and do not warrant detailed engineering design. Cyclone Tracy also showed the fallacy of this argument (8).

IMPACT ON THE COMMUNITY

Considering the community as a unit, housing generally accounts for approximately half of the total value of building construction. Thus, economically, housing represents an economic risk similar to other buildings in the event of a tropical cyclone. In addition, from a social point of view, houses are probably of much greater value since they provide the basic shelter and facilities required for living. If, in Darwin, the relative performances of houses and engineered buildings had been reversed, the scale of the disaster would have been significantly less and no evacuation would have been necessary. Working conditions would have been difficult, but the social cost of difficult working conditions would have been far less than the social cost of difficult, or as in the case of Darwin, impossible living conditions. It must also be borne in mind that in the event of tropical cyclones the population is more likely to be in their homes than in other buildings, as experience shows that schools and most places of work are closed when such a hazard threatens.

CHANGES IN APPROACH TO HOUSING CONSTRUCTION

The most significant consequence that followed from the investigations of damage caused by Cyclone Tracy was recognition of the importance of incorporating structural engineering principles into the design of housing likely to be exposed to extreme winds such as those accompanying severe tropical cyclones. This observation constituted the major recommendation following the initial investigation of the damage (1). This recommendation was incorporated into the regulations governing the reconstruction of Darwin (9) and was vigorously applied during the design and construction of new and reconstructed houses (10,11). Further, this recommendation is incorporated in Queensland's new Building Act (12) and is providing the basis for detailed "deemed to comply" rules for house construction currently being prepared to accompany these regulations. It also forms the basis of a manual on construction of houses in high wind areas currently being prepared by the Standards Association of Australia (13), and underlies the recommendations on performance testing prepared at a recent workshop at the Experimental Building Station by a widely represented group from public, private and commercial interests (14).

This activity in the technical community and building industry represents a radical change in the approach to housing construction. Implementation of the changes is not taking place without considerable pain. There is very little engineering experience available and few design guidelines are applicable in the design of low rise housing. The materials and methods of construction most commonly employed are not those with which the structural engineering profession has been most concerned. Consequently existing design codes, handbooks and methods of structural analysis are often of little value. In Darwin, many of the initial solutions were grossly conservative because of this lack of knowledge and the need for urgent solutions, and because designers were obliged to settle for answers which were certain to be safe. This situation led, in due course, to extensive research and testing related to the wind resistance of houses within private industry, Government organisations and Universities (6,15,16), and some of the initial conservatism was relaxed.

IMPACT ON WIND DESIGN CRITERIA

The studies surrounding Cyclone Tracy have also led to a re-assessment of the criteria used in the design of buildings to resist wind loads. The most immediate reaction to Cyclone Tracy was to increase the design wind velocities in tropical cyclone prone regions. The estimated maximum wind speeds in Tracy were significantly above those specified for design purposes in the Australian wind loading code (17) prior to Tracy, and new studies of extreme wind velocities around the north Australian coastline being undertaken at the time (18) tended to justify such a move. Strong emotional forces supporting this change, generated by the widespread damage, were also present. At the same time, load factors were examined, and some significant increases were made for critical wind loading situations, particularly in relation to the design of steel structures.

These changes have produced significant increases in the structural cost of steel industrial buildings in tropical cyclone prone areas. This result has subsequently raised major questions about the relevance of the new criteria in view of the good performance of steel structures, designed to the earlier standards, during Cyclone

Tracy (19). On the basis of this performance it seems difficult to justify the increased costs arising from the code revisions. One cannot help but feel that if the non-engineered housing had not failed so catastrophically, the increases in design windspeeds and load factors would not have been approved despite the quite logical and rational arguments used to justify them.

One change that was justified by performance is related to the design of cladding and its connection to a building. Cyclone Althea revealed weaknesses in roof cladding connections. As a result, suppliers were required to justify their systems on the basis of load testing in accordance with the Australian wind loading code. Cladding and fixings satisfying these requirements were used widely in Darwin in the two years prior to Tracy. Yet, cladding failures were widespread, contributing greatly to the housing damage, despite having been tested to ultimate loads in excess of those predicted from code coefficients and the estimated maximum wind velocity.

For design purposes, wind loads have generally been regarded as static loads except when dealing with tall flexible structures. It soon became evident that in the case of the cladding this assumption of a static loading was very much in error. Inspection of the damage and laboratory testing revealed that widespread failure of light metal cladding had occurred as a result of low cycle fatigue arising from the fluctuating nature of wind loading (20). This observation introduced an aspect of wind loading which had not previously been considered relevant in the normal engineering design of buildings. Repeated loading tests were immediately introduced and, with subsequent modifications based on detailed studies of the nature of the fluctuations (21), these have now become standard requirements for the specification of roof cladding in tropical cyclone prone areas (14).

Recognition of the importance of considering the fluctuating nature of wind loads in cladding design is probably the greatest impact Cyclone Tracy has had on wind design criteria. It has also caused questions to be asked about fatigue behaviour of other components and connections in buildings under fluctuating wind loads, and standardised criteria have also been developed for these investigations (14).

The application of engineering criteria to the design of housing, and subsequent reflections on the performance of the engineered structures in Darwin, have led to a strong feeling that established engineering criteria in relation to design for wind loads are conservative, particularly for low rise buildings (including housing) located in tropical cyclone prone areas. This feeling has produced a critical examination of the underlying premises and assumptions of design criteria and their applicability to tropical cyclone prone areas (7). An item of concern in this examination is the relevance of code specified pressure coefficients for the design of the structure of low rise buildings. A major study of wind loads on housing has been undertaken by Holmes (16), following a preliminary investigation by Vickery (22). The results of these studies are showing clearly that the total loads on a low rise building tend to be significantly less than those predicted by the current Australian wind loading code (17), although code values seem reasonable with respect to cladding loads. These observations may well explain the good performance of engineered structures in Darwin, the great majority of which suffered only cladding damage, if damaged at all. Another major concern relates to the relevance of current criteria for deriving design loads, assuming the pressure coefficients are satisfactory (6,8,23).

NEW APPROACH TO DESIGNS FOR TROPICAL CYCLONES

It is being recognised that in designing for extreme winds from tropical cyclones: (a) the primary concern is the protection of the community as a whole and not individual buildings or persons, and (b) the aim is to provide an appropriate degree of protection (which may vary as a function of the anticipated frequency and the level of the event), and not to provide guaranteed protection. The degree of protection provided will also vary from building to building, and community to community depending on: (a) the importance of the building in the immediate post-disaster situation, which may not be the same as under normal conditions (8,24); (b) the size of the community, larger communities requiring relatively more protection than smaller communities (8); and (c) variation of vulnerability of similar classes of structure within a community, those with greater variation being less susceptible to catastrophic damage from a severe event (8). These considerations also suggest a need for a multi-level design approach akin to limit state design, with different performance levels being specified for different levels of event (6,8,24). In recognition of this it is now being accepted in Australia that design loads for strength and design loads for serviceability may not be directly related to each other, and may be independently derived from different assumed conditions (14).

Exact definition of what constitutes satisfactory design for events such as tropical cyclones is difficult. Nevertheless, since Cyclone Tracy, Australian engineers have made significant progress towards the solution of this problem with respect to design for extreme winds associated with them.

IMPACT ON STORM SURGE INVESTIGATIONS

In addition to the extreme winds, tropical cyclones also pose a threat of coastal flooding from the associated storm surge. Historically, it is the storm surge which has been the greatest cause of loss of life from tropical cyclones. The 1970 Bay of Bengal cyclone, for instance, is reported to have caused about 300,000 deaths as well as other massive losses, practically all due to the surge.

The Australian technical community was awakened to the threat posed by storm surges by Cyclone Althea, which would have threatened many lives had it crossed the coast at high tide rather than low tide. A considerable amount of investigation was initiated as a result (4). These efforts received a significant boost from Cyclone Tracy. Although the surge effects were small in the case of Tracy (1), the possibilities of what could happen with a different combination of circumstances became a matter of considerable public debate and concern, and led to the imposition in Darwin, for the first time in Australia, of land use controls in potential surge zones (25). As a result of public pressures the controls were subsequently lifted, but the technical community and emergency services officials remain acutely aware of the problem. Sophisticated computer models are now available and being used for predicting storm surge levels and wave heights as a function of return periods for various locations around the North Queensland coast (26,27).

IMPACT ON INSTITUTIONS

The impact of Cyclone Tracy on the Australian community has not been restricted to those inhabitants who suffered loss, the building industry and the engineering profession. Darwin at the time of Tracy was a somewhat unique community in that it was essentially a Federal Government community. The majority of its citizens were Federal Government employees, and the Federal Government was responsible for most of the construction and maintenance of the community and its services. As a result, the Government itself took a major share of the direct brunt of Tracy, and became directly responsible for the reconstruction activities.

This direct involvement has had a major impact on Federal Government policy in relation to mitigation of the effects of tropical cyclones. Most of the responsibility for Federal Government action has been undertaken by its Department of Construction which has taken the lead in the establishment and implementation of new design criteria and structurally engineered housing construction. This Department (then the Department of Housing and Construction) was responsible for commissioning the major investigation of damage (1), making policy recommendations and determining design criteria for the reconstruction (9), providing the technical input to the Darwin Reconstruction Commission (10), educating professional engineers, architects, building tradesmen and the general public involved in the reconstruction on cyclone resistant construction (11) and conducting extensive testing and research into the structural behaviour of cladding and housing construction to provide the required data for implementation of the lessons of Cyclone Tracy.

Although initially directed at the reconstruction of Darwin, these efforts have subsequently been extended to all tropical cyclone prone areas of Australia in respect of Federal Government construction, and they have had a significant impact on the total building industry, through the demands placed on it by the Government. The impetus given by this activity has flowed on to other bodies responsible for establishing standards and building regulations, notably the Standards Association of Australia and State Government authorities. Not only design codes, such as the wind loading code (17), but many of the performance codes relating to roof cladding, windows, masonry construction etc. have been revised or are under revision as a consequence of Cyclone Tracy and the results of subsequent studies. In addition, the Standards Association of Australia and the Queensland Government have separately undertaken the task of preparing "deemed to comply" specifications for single dwelling construction which will satisfy the requirements for structurally engineered construction (13,28).

The other major institution which suffered direct impact from Cyclone Tracy was the insurance industry. The provision of insurance against natural hazards such as tropical cyclones is itself a risky process which requires insurance companies to take out insurance themselves through international re-insurance companies. Cyclone Tracy followed a series of disasters which had led to significant claims on re-insurers by Australian insurance companies. The large claims generated by these disasters led to difficulties in obtaining continuing re-insurance at economic levels. In addition, the plight of the uninsured placed moral pressures on the Government to provide assistance to them placing further burdens on the Federal Government treasury. The withdrawal of insurance against natural hazards threatened to increase the latter problem in future disasters.

In the light of this, the Federal Government, in March, 1976, decided in principle to introduce a natural disaster insurance scheme in Australia. In December 1976, proposed guidelines for such a scheme were tabled in Parliament for public discussion (29). Subsequently, two committees were formed to undertake more detailed studies of likely premiums and mitigation measures to provide maximum protection to the scheme. (The Federal Government eventually decided not to proceed with the scheme but to support mitigation activities and assist the insurance industry in making provision for disaster insurance.)

One important consequence of this activity has been a considerable interaction between the technical community and the insurance industry which is likely to have a long term impact both on the insurance industry and on technical research in the area of risk assessment. The structural engineering profession can make a significant contribution in this area but there are many difficult questions to be answered which will provide a major challenge to researchers.

The impact on the Bureau of Meteorology was probably not as great as on some of the other institutions, mainly because the Bureau was already well equipped for detecting and tracking tropical cyclones. Improvements in forecasting which have taken place since Cyclone Tracy would probably have still taken place if Tracy had not occurred. The Bureau did not anticipate the actual severity of wind which occurred, but without aircraft reconnaissance accurate predictions would have been difficult. Strong recommendations were made for aircraft reconnaissance of tropical cyclones following Cyclone Althea and again following Cyclone Tracy but to date the Government has not been convinced that the benefits of the more accurate forecasts outweigh the costs.

From a wind engineering viewpoint the failure of the Bureau's anemometer in Darwin before the winds reached their maximum intensity was a major loss. A programme of upgrading anemometer installations in tropical cyclone prone areas to avoid repetition of this failure was embarked upon as a consequence. The need for more wind measuring devices in tropical cyclone prone areas has also been recognised and a programme has been initiated to erect a network of inexpensive robust maximum gust recording instruments around the northern part of Australia. The Bureau has also responded to requests for better estimates of cyclone risk and is making a significant contribution in this area (3).

The other major impact at institutional level has been in the area of counter disaster planning and organisation. Australia has a national system of counter disaster arrangements involving both Federal and State organisations which was established in its present form as a result of Federal Government action in 1974 (30). The Natural Disasters Organisation, which is a federal organisation, had been established only a few months at the time of Tracy, which provided its first major test. Because Darwin was a federal government controlled city, this organisation played the major role in relief and rescue operations, a good description of which has been given by its Director General at the time (31).

Cyclone Tracy gave a great boost to the development of the Natural Disasters Organisations and the related State Emergency Services, the latter of which depend a great deal on voluntary assistance for their effective operation. Cyclone Tracy provided the necessary public interest. It also provided an example of the type of extreme emergency with which the counter disaster organisations may have to cope so stimulating interest in developing management procedures for implementation in the event of a disaster. Because much of the detailed planning is done at the individual community level this has been a significant factor in maintaining community awareness of the dangers of tropical cyclones. This interest has also led to studies of human behaviour in disaster and their implications for disaster planning (32,33).

CONTINUING IMPACT

Despite predictions to the contrary, community awareness of the potential dangers from tropical cyclones remains very high nearly five years after Cyclone Tracy, and there are many indications that it will remain so for a considerable length of time, even if no further disasters of this magnitude should occur. During 1978, for instance, there were community based seminars on disaster and community welfare (34), intense discussion on proposed details to provide wind resistance in housing construction and on matters related to the proposed Natural Disasters Insurance Scheme, a major 'state of the art' symposium on design for tropical cyclones (35) a seminar on natural hazards management (36), and a national study of the state of natural hazards research in Australia (37), in addition to continuing extensive technical research into problems related to design for tropical cyclones. Little of this activity would probably have taken place if Cyclone Tracy had not occurred, indicative of its continuing impact on the Australian community.

In 1906, San Francisco was hit by a severe earthquake which, in combination with the fire which followed, produced the United States' greatest disaster. It created an awareness of the need for earthquake resistant construction and fire protection which has had a continuing impact to this day. Although not in the same category of disaster as the San Francisco earthquake, there is considerable reason to believe that the Cyclone Tracy disaster will have a comparable long term impact on tropical cyclone resistant construction in Australia.

The disaster made an impact on the whole Australian community. The residents of Darwin had come from all over Australia and, after the event, were evacuated to all parts. The whole nation viewed the damage and the distress of the inhabitants at close range through television. The disaster occurred on Christmas Day, which evoked a particularly strongemotional response, and the damage was catastrophic. Cyclone Tracy will be remembered for a long time and seems destined to become part of the Australian legend.

CONCLUSION

The reconstruction of Darwin is now complete and the city has now expanded beyond its size at the time of Tracy, although much evidence of the disaster still remains. It is a different city than it was in several respects, chief among them being the structural strength of its houses. It can probably justly claim to be one of the most cyclone resistant cities in the world, with over 70 percent of its dwellings meeting the demanding standards of Australian structural engineering design practice. Also, many of the surviving 30 percent have been strengthened above their pre-Tracy standards. This has been a remarkable achievement which should serve as a model for future reconstruction efforts following disasters.

The repercussions from Cyclone Tracy are continuing to have impacts on the Australian community, particularly in relation to housing construction and design criteria. This impact continues to generate tremendous amounts of wind engineering activity at all levels of the building industry and seems certain to ensure that the lessons from the Cyclone Tracy disaster will not be lost.

ACKNOWLEDGEMENTS

The authors acknowledge their gratitude to the Australian Government's Department of Housing and Construction (now Department of Construction) at whose request they took part in the investigation of the damage from Cyclone Tracy, and from whose officers they have continued to receive much assistance and encouragement. In particular they would like to thank Mr. N. Sneath and Mr. C. Bubb of the Department for their help. Acknowledgement is also due to the Australian-American Educational Foundation which sponsored a visit to Australia by Dr. Minor in 1978, and to the James Cook University of North Queensland for its constant support of these studies.

REFERENCES

(1) Walker, G.R. "Report on Cyclone Tracy - Effect on Buildings - December 1974", 3 Vol., Australian Department of Housing and Construction, Melbourne, 1975.

(2) Marshall, R.D. "Engineering Aspects of Cyclone Tracy, Darwin, Australia, 1974", N.B.S. Building Science Series 86, U.S. Dept. of Commerce, 1976.

(3) Southern, R.L. "The Nature of Tropical Cyclones with Emphasis on their Occurrence in the Australian Region", Design for Tropical Cyclones, Vol. 1, Vacation School Proc., James Cook University, Townsville, 1978, Paper B.

(4) Trollope, D.H. (Ed.) "Cyclone Althea: Part 1 - Buildings, Part 2 - Storm Surges and Coastal Effects", 2 Vols., James Cook University, Townsville, 1972.

(5) Clarke, A.G. "Cyclone Joan - Effects on Structures in Port Hedland, 8th December, 1975", Australian Department of Construction, Perth, 1976.

(6) Walker, G.R. "The Design of Buildings and Their Components for Cyclonic Conditions", Queensland Division Technical Papers, Institution of Engineers, Australia, Vol. 18, No. 31, October 1977, pp. 1-8.

(7) Walker, G.R. "The Rational Design of Low Rise Housing in Tropical Cyclone Prone Areas", Proc., 1976 Annual Engineering Conference, Institution of Engineers, Australia, pp. 248-253.

(8) Walker, G.R. and Stark, K.P. "The Development of Design Criteria for Extreme Events arising from Natural Hazards", Proc. U.S.-S.E. Asia Joint Symposium on Engineering for Natural Hazards Protection, Manila, 1977.

(9) "Darwin Area Building Manual" Darwin Reconstruction Commission, 1975.

(10) "Darwin Area Deemed to Comply Standards", Darwin Reconstruction Commission, 197(

(11) "Structural Design of Cyclone Resistant Dwellings: Part A - Design; Part B - Construction", 2 Vol., Australian Department of Construction, Melbourne 1976.

(12) "Building Act, 1975", Queensland State Government, 1975.

(13) "Draft Manual for Guidance on Construction of Houses in High Wind Areas". DR77109, Standards Association of Australia, 1977.

(14) "Guidelines for Cyclone Product Testing and Evaluation", Technical Record 440, Experimental Building Station, Sydney, 1977.

(15) Walker, G.R. "The Design of Walls in Domestic Housing to Resist Wind", Design for Tropical Cyclones, Vol. 2, Proc. Vacation School, James Cook University, Townsville, 1978, Paper O.

(16) Holmes, J.D. "Wind Loads on Buildings and Other Structures", Design for Tropical Cyclones, Vol. 1, Proc. Vacation School, James Cook University, Townsville, 1978, Paper I.

(17) "SAA Loading Code: Part 2 - Wind Forces", AS1170, Part 2 - 1975, Standards Association of Australia.

(18) Gomes, L. and Vickery, B.J. "Tropical Cyclone Gust Speeds along the Northern Australian Coast", Civil Engineering Transactions, Institution of Engineers, Australia, Vol. CE18, No. 2, 1976, pp. 40-48.

(19) Baker, K.N. and Walker, G.R. "The Behaviour of Steel Structures in Cyclone Tracy and the Implications for Design", Proc. Metal Structures Conference Institution of Engineers, Australia, Adelaide, 1976, pp. 29-34.

(20) Morgan, J.W. and Beck, V.R. "Failure of Sheet-Metal Roofing under Repeated Wind Loading", Civil Engineering Transactions, Institution of Engineers, Australia, Vol. CE19, No. 2, 1977, pp. 1-5.

(21) Beck, V.R. "Loads - The Basis of Repeated-Loading and Impact-Loading Criteria', Design for Tropical Cyclones, Vol. 2, Proc. Vacation School, James Cook University, 1978, Paper L.

(22) Vickery, B.J. "Wind Loads on Low Rise Buildings", Darwin Reconstruction Commission Seminar, Darwin, 1976.

(23) Walker, G.R. and Stark, K.P. "Engineering Design Philosophy in Relation to Tropical Cyclones", Design for Tropical Cyclones, Vol. 1, Proc. Vacation School, James Cook University, 1978, Paper G.

(24) Bubb, C.J. "Survival and Serviceability of Major Facilities in Tropical Cyclones", Design for Tropical Cyclones, Vol. 2, Proc. Vacation School, James Cook University, 1978, Paper Z.

(25) D'Rozario, J. "Land Use Management as a Disaster Mitigation Measure: The Successes and Failures in the Case of Post-Tracy Construction in Darwin", Proc. Seminar on Natural Hazards Management in Northern Australia, Darwin, Australian National University, 1978.

(26) Sobey, R.J., Harper, B.A. and Stark, K.P. "Numerical Simulation of Tropical Cyclone Storm Surges", Research Bulletin No. CS-14, Dept. of Civil and Systems Engineering, James Cook University, Townsville, 1977.

(27) Stark, K.P. "Cyclonic Storm Surges and Sea Characteristics Associated with Tropical Cyclones", Design for Tropical Cyclones, Vol. 1, Proc., Vacation School, James Cook University, Townsville, 1978, Ppaer E.

(28) "Appendix 4", Building Act 1975, Queensland State Government (to be published).

(29) "A Natural Disaster Insurance Scheme for Australia", Australian Government, 1976.

(30) Swan, R.C. "The Role of the Natural Disasters Organisation", Proc. Seminar on Natural Hazards Management in Northern Australia, Darwin, Australian National University, 1978.

(31) Stretton, A.B. "The Furious Days", William Collins, Sydney, 1976.

(32) Oliver, J. "Socio-Economic Effect of Tropical Cyclones", Design for Tropical Cyclones, Vol. 1, Proc. Vacation School, James Cook University, Townsville, 1978, Paper C.

(33) Oliver, J. "An Assessment of Problems and Practice: Natural Hazard Management in Northern Australia", Proc. Seminar on Natural Hazards Management in Northern Australia, Darwin, Australian National University, 1978.

(34) Innes-Reid, J. (Ed.) "Natural Disaster and Community Welfare", Seminar Proc. James Cook University, Townsville 1977.

(35) "Design for Tropical Cyclones", Vacation School Proc., James Cook University, Townsville 1978.

(36) Pickup, G. (Ed.) "Natural Hazards Management in Northern Australia", Seminar Proceedings, North Australia Research Unit, Australian National University, Darwin, 1978.

(37) Pickup, G. and Minor, J.E. "Assessment of Research on Natural Hazards in Australia", Australian National University, 1978.

ACQUISITION, ANALYSIS AND APPLICATION OF WIND LOADING DATA

J. R. MAYNE and N. J. COOK
Building Research Establishment, Garston, Watford, UK

ABSTRACT

This paper describes a complete package of techniques for the acquisition, analysis and processing of wind speed and wind pressure data in order to arrive at the best possible values for use in the design of static structures against wind loads. For the purposes of this paper, static structures are defined as those in which the dynamic response is insignificant and can be ignored in the design calculations. The techniques described comprise:

(1) The definition of speed and loading coefficient parameters so that, for practical purposes, they are statistically independent.

(2) The treatment of analog full-scale data records by trend removal, so as to render such data stationary and compatible with model-scale data, and thus amenable to the application of standard statistical techniques.

(3) Extreme value analysis of wind speed and wind pressure data.

(4) The use of the extreme value parameters so obtained to make a fully probabilistic assessment of wind loads on structures.

All these ingredients are essential to the accurate assessment of wind loads. The fact that such statistical methods have not previously been fully implemented in the analysis of both full-scale and model-scale data is one reason for the lack of consistency between full-scale and model scale on the one hand, and between different model-scale tests on the other. The adoption of these methods should help to resolve some of the difficulties, enable simulation methods at model scale to be better assessed, compared and developed, and also provide more accurate data for design.

The treatment of dynamic structures is not covered by this paper.

INTRODUCTION

Structures can conveniently be divided into three categories:

Static structures are those structures where the dynamic response caused by wind loads is negligible and does not enter into the design considerations. *Dynamic structures* are those in which the stiffness and damping are such that movement will occur under wind excitation which demands consideration in design, but which is not sufficiently large to affect the aerodynamic flow around the structure and the forces acting on it. *Aeroelastic structures* exhibit dynamic response which is sufficiently large to modify the flow around them and the forces acting on them, sometimes but not always increasing the motion. When there is positive feedback between the flow and the structural response, the motion will increase in amplitude until either a limit is reached or failure occurs. Structures that reach a limit are particularly prone to fatigue failure.

In this paper we shall be concerned with the assessment of wind loads on static structures. Although these comprise only one class of structure, it is by far the most common class, and the one covered (in many cases the only one covered) by most wind loading codes of practice throughout the world.

Thus, whereas the production of general design methods for dynamic and aeroelastic structures raises many fundamental problems at the frontiers of wind engineering experience and rightly induces considerable research activity, there are many more data available on the loading of static structures which could be incorporated into regulations. Unfortunately, much of this is defective in various ways.

There has long been a tendency on the part of structural engineers and many research workers to regard the pressures and forces on structures as deterministic quantities. Variability has been recognized, but not the need to apply the same probabilistic techniques that have become generally established for the assessment of wind speeds. One of the reasons for this is the historical development of the field. The complexity of the problem (the turbulence of the atmospheric boundary layer, the nature of the flow around the structures and the effect of one structure on near neighbors) renders a deterministic evaluation impossible and should direct us towards a probabilistic approach.

As far as the engineer is concerned, the misunderstanding is deepened by the mixing of deterministic and probabilistic methods, as in the UK Code of Practice (1). Here, the designer chooses a wind speed of known return period from data prepared statistically and he is led to conclude, in the absence of a denial, that the pressures he calculates from this speed, by means of the pressure and force coefficients provided, also have the same risk of occurrence. A brief consideration shows that this is not so. The coefficients are derived from data samples and are usually based upon one specific parameter such as the highest recorded value, or an upper fractile, or the mean plus so many standard deviations. This value is then introduced into the code as a deterministic value, in spite of the fact that it is actually a statistical variate whose variability depends upon the design of the experiment in which it was measured. Thus the code user is led to believe that the stated pressure coefficient will apply in the design wind condition with probability $p = 1$; that is, it will neither be exceeded nor will it fail to be reached.

In this paper we shall discuss various statistical and data processing techniques which lead to a better assessment of wind loads. We show how the wind speed and pressure coefficient extreme value data can be used together in order to arrive at design loads with a quantified risk of exceedance. A first order method is presented and reference made to the possibility of refinement by higher order methods. These techniques need not complicate the format of codes to any marked degree, but they should produce values of design loads that are more accurate and which have a more consistent risk of exceedance.

DEFINITION OF LOADING COEFFICIENT

The historical definition of loading coefficient is the "shape factor" of early smooth flow wind-tunnel tests as given by

$$P = q \cdot C_P \quad \text{or} \quad F = q \cdot A \cdot C_F \tag{1}$$

where P and F are the actual pressures and loads, respectively, on the model, A is a scaling area for the model, q is the reference dynamic pressure and C is the loading (pressure or force) coefficient. The dynamic pressure is in turn defined by

$$q = \frac{1}{2} \cdot \rho \cdot V^2 \tag{2}$$

where ρ is the density of air and V is the reference wind speed. As long as wind-tunnel simulation techniques gave only simple representation of a few atmospheric boundary layer characteristics, for example mean velocity profile, this simple static and deterministic definition was adequate.

In the measurement of full-scale meteorological conditions it has long been recognized that a deterministic approach is inappropriate. Wind speed is treated with the other parameters as a statistical variate and is assessed on a fully probabilistic basis. This approach is also extended to contemporary model-scale simulation techniques in which the fluctuating velocity components of atmospheric turbulence are represented.

Although loading coefficients are recognized as random variables there has been, until lately, considerable reluctance to treat them in the same probabilistic manner as wind speed. The aerodynamic admittance approach to fluctuating pressures and loads attempted to assign all the variability to the turbulence of the incident wind, ignoring all structure-generated gust loads. The current UK Code of Practice CP 3 Ch. V Pt. 2(1) operates in the same manner, assigning all variability to the design maximum 3-second duration gust and treating the loading coefficients as static and deterministic.

This paper seeks to promote the use of a probabilistic approach to the measurement of loading coefficients and their subsequent assessment for design. Both q and C in equation (1) should be considered as statistical variates. Consider first the dynamic pressure which, at any one site, will contain components covering a wide range of periodicity. A spectrum of wind speed compiled by Van der Hoven (2) is shown in Fig. 1. The following features (3) of the spectrum are applicable to temperate latitudes not prone to tropical cyclones.

(1) The major peak at a center period of about four days, which corresponds to the typical transit time of a fully developed weather system, usually referred to as the Macrometeorological peak.

(2) The second major peak comprising a wide band of much higher frequencies associated with the turbulence of the boundary layer, which range from periodicities of the order of minutes to the order of tenths of a second. This is usually referred to as the Micrometeorological peak.

(3) The well-defined gap between the two peaks, often called the Spectral Gap, in which there is little wind fluctuation of a range of periodicity of about one order of magnitude.

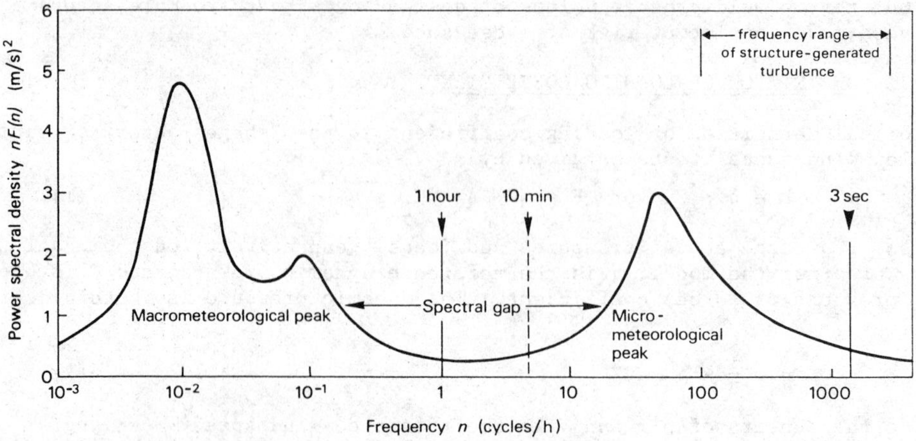

Fig. 1. Spectrum of horizontal wind speed after Van der Hoven.

The fluctuating pressures and forces on a building will contain variations over this whole range of frequency derived directly from the incident wind. In addition, however, neighboring buildings and the areas of separated flow around the building itself generate additional fluctuations over the range marked on Fig. 1 in the Micrometeorological peak, and these fluctuations interact with the turbulence of the incident wind. The conventional definition of dynamic pressure as representing all the wind speed characteristics, and of loading coefficient as representing only the building-generated characteristics, results in an interdependence of the two parameters.

It is possible to achieve statistically independent dynamic pressure and loading coefficient parameters by a fundamental redefinition. This redefinition makes use of the way that the Spectral Gap divides the frequency range into Macro- and Micrometeorological ranges. A period T is defined between 10 minutes and 1 hour. All fluctuations in the incident wind longer than this period V(T) are used to define the dynamic pressure q(T) thus:

$$q(T) = \frac{1}{2} \cdot \rho \cdot V(T)^2 \tag{3}$$

The incident wind turbulence of periodicities shorter than T are absorbed into the definition of the loading coefficient, together with the building generated fluctuations, thus

$$P(T,t) = q(T) \cdot C_P(t) \text{ and } F(T,t) = q(T) \cdot C_F(t) \cdot A \tag{4}$$

where t denotes periods shorter than T. The dynamic pressure, being a function of the Macro-range only, and the loading coefficient, now a function of the Micro-range only, can be considered as being statistically independent since all cross-spectral terms are identically zero. The disadvantage of this redefinition is that, by including the incident wind turbulence of the Micro-range, the loading coefficient becomes terrain dependent as well as dependent on the building shape. This disadvantage is far outweighed by the advantages that statistical independence brings in measurement and analysis.

ACQUISITION OF FULL-SCALE LOADING COEFFICIENT DATA

The current BRE program of pressure measurements on an array of farm silos is a typical example of full-scale data acquisition. The term "full-scale" is slightly misleading in that it applies only to the linear scale. It is unlikely that data will ever be collected in wind speeds approaching the UK Code of Practice's design once-in-50-year values. The majority of full-scale measurements are taken at much lower, more common wind speeds and require the application of a velocity scale factor to bring the data to design conditions. The BRE silos are 3.5 m in diameter, 4 m high cylinders with conical roofs. Two of them have 3 rings of 24 equi-spaced pressure tappings at 1/2, 2/3 and 5/6 the height of the cylinder. Twenty-four Validyne type DP 104 pressure transducers give a continuous signal of the external surface pressure at each selected tapping relative to the internal pressure of the silo. In addition, a static pressure probe and an orthogonal pair of propeller anemometers are mounted on a 10 m high mast upstream of the silos. The time histories of the pressure transducer and anemometer signals are recorded on analog magnetic tape. Each recording is bracketed on the tape by reference calibration and transducer-zero records taken immediately before and after each data run, as shown in Fig. 2.

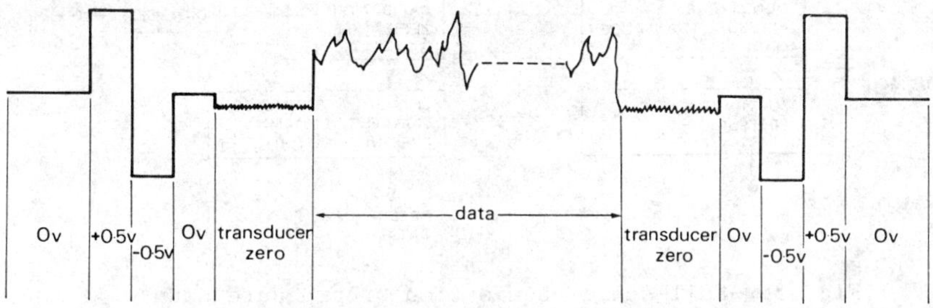

Fig. 2. Typical full-scale analog data record.

The recordings are subsequently digitized, taking the usual precautions against aliasing, at a frequency in excess of the range of interest. Each record is calibrated for tape recorder gain, transducer initial zero offset and for the nonlinear gain characteristic of each individual transducer. The pressure records, initially relative to the internal silo pressure, are made relative to static pressure by subtracting the static probe signal. Finally, the signals from the orthogonal anemometer pair are decoded to form reference dynamic pressure and direction records. This process is summarized in Fig. 3.

The use of a data sample length of at least 160 minutes (16 periods of T = 10 minutes for use in the subsequent analysis, see later) results in a set of pressure records $P(T,t)$, a reference dynamic pressure record $q(T,t)$ and direction record $\Theta(T,t)$ which contain contributions from the whole Micro-range and from part of the Macro-range. In addition, the acquisition process is not perfect and will contain a time-dependent error component. This error will be partly long-term drift $e_1(T)$, caused principally by the temperature drift of the Validyne transducers (6 Pa/°C (0.125 lb/ft^2/°C)), and partly high-frequency noise $e_2(t)$ with zero mean from the recording electronics. The data records are represented from equation (4) by

$$P(T,t) + e_1(T) + e_2(t) = q(T) \cdot C_P(t) + e_1(T) + e_2(t) \quad (5)$$

The propeller anemometers are not subject to temperature drift, thus the dynamic pressure record is represented by

$$q(T,t) + e_2(t) = q(T) \cdot C_q(t) + e_2(t) \quad (6)$$

Equation (6) effectively defines a Micro-range dynamic pressure coefficient C_q in a similar fashion to the silo pressure coefficients but with a mean value of unity.

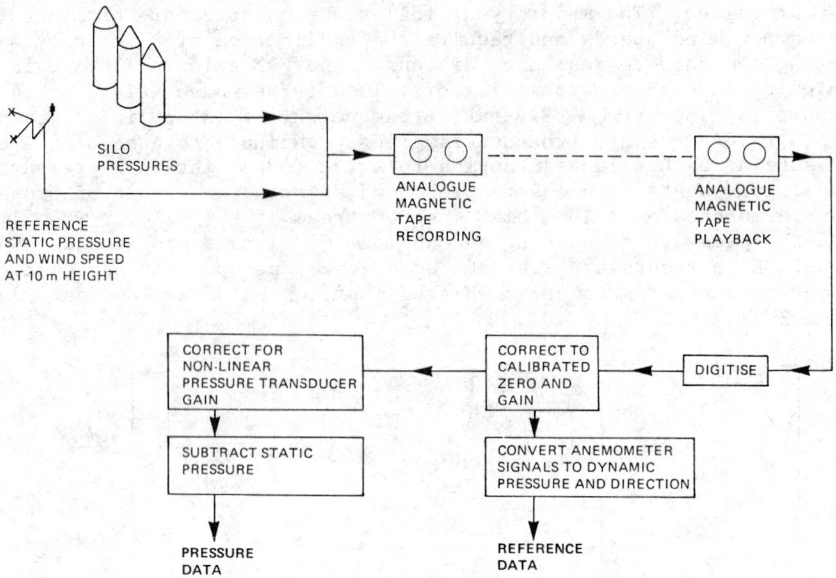

Fig. 3. Full-scale acquisition of pressure data.

Recovery of the loading coefficient data from the calibrated pressure records will be illustrated using two actual data runs. The dynamic pressure record from a data run in very squally conditions is shown in Fig. 4a, and from a run in more typical strong wind conditions in Fig. 5a. In both records, low-frequency fluctuations of the Macro-range appear as trends, rendering the records nonstationary. In addition, the continuously variable scaling velocity implied by the trends produces both amplitude and frequency modulation of the Micro-range fluctuations. This is most noticeable in the "squally" run in Fig. 4a.

Each dynamic pressure record is smoothed by a rectangular recursive digital low-pass filter set at the cut-off period T = 10 minutes (equal to the observation period of the subsequent analysis). The smoothed records (Figs. 4b and 5b) represent the Macro-range trend q(T) of equation (6), i.e.,

$$\frac{1}{T} \cdot \int_{t_1 - T/2}^{t_1 + T/2} (q(T,t) + e_2(t))\, dt = q(T) \tag{7}$$

since $C_q = 1$ and $e_2 = 0$. The amplitude modulation of the raw surface pressure data records is removed by dividing each record by the smoothed dynamic pressure record q(T) to give the statistically stationary pressure coefficient plus superimposed errors; thus,

$$\frac{P(T,t) + e_1(T) + e_2(t)}{q(T)} = C_p(t) + \frac{e_1(T) + e_2(t)}{q(T)} \tag{8}$$

For the special case of the reference dynamic pressure record itself, the dynamic pressure coefficient C_q, Figs. 4c and 5c, is obtained; thus,

$$\frac{q(T,t) + e_2(t)}{q(T)} = C_q(t) + \frac{e_2(t)}{q(T)} \tag{9}$$

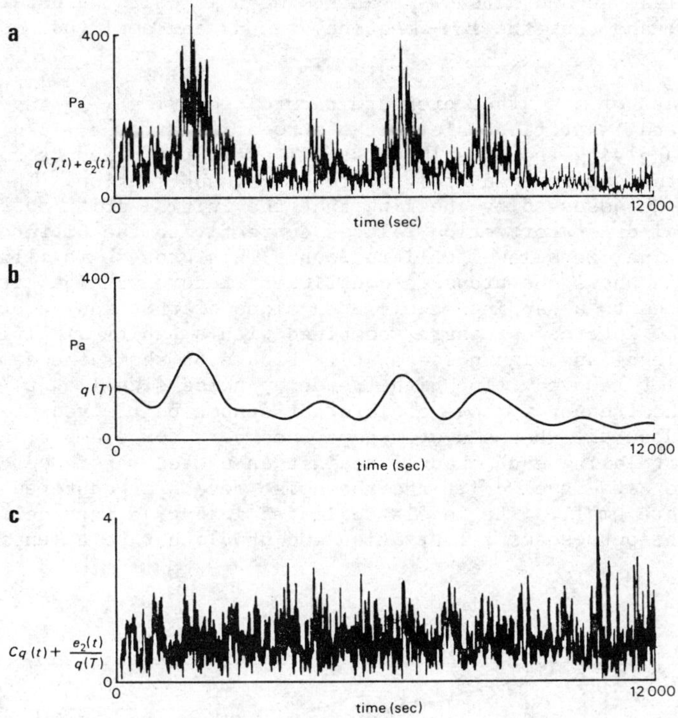

Fig. 4. "Squally" data set.

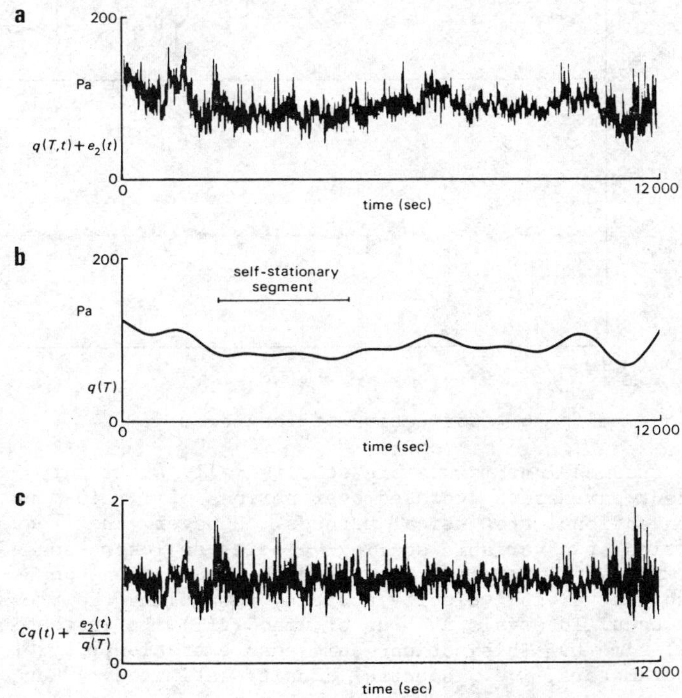

Fig. 5. Typical data set.

Examination of Figs. 4c and 5c shows that the mean is unity, according to the definition of C_q; and that the low-frequency trends and amplitude modulation have been removed.

In the general case of a surface pressure record (equation (8)) any remaining low-frequency trend is attributable to the zero-drift error $e_1(T)/q(T)$. This may be recovered by applying the same low-pass filter to the pressure coefficient record, and may then be removed from the record by subtraction. The validity of this measure can be assessed by checking that the initial drift correction is zero and that the final drift correction relates correctly to the difference between the initial and final zero-level calibrations. This process is illustrated by Fig. 6. Figure 6a shows the pressure coefficient record obtained from a transducer not connected to a tapping, i.e., both sides of the transducer were connected to the silo internal pressure, combined with a faulty amplifier giving step changes of zero-level and high noise $e_2(t)$. Figure 6b shows the same record with the long-term drift removed. The high-frequency noise signal $e_2(t)/q(T)$ which remains within the frequency range of interest cannot be removed if it is random. (Noise introduced on all channels of a tape recorder, for example, can sometimes be removed by correlating each channel against an unused reference channel.) However, both the zero-level drift and the noise levels encountered in the BRE silo experiment are small. The levels indicated in Fig. 6 were deliberately introduced for the purpose of illustration and should not be taken as typical.

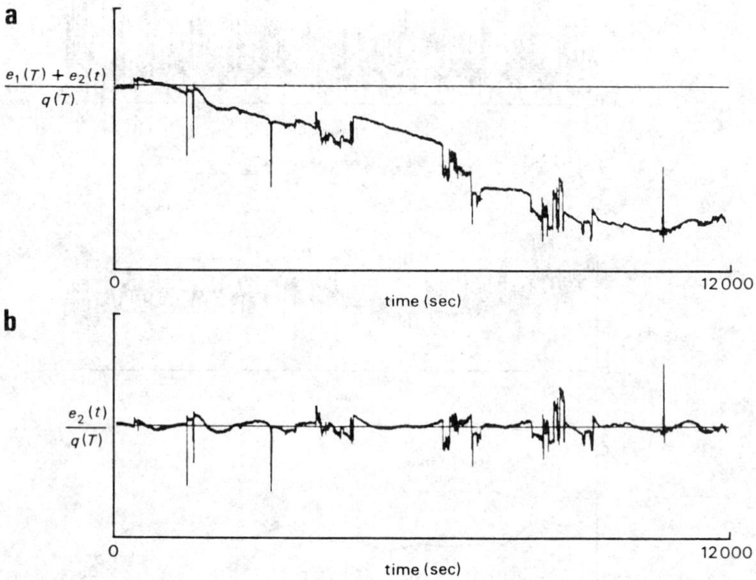

Fig. 6. Correction of long-term drift.

The data records treated in this way are statistically stationary in terms of amplitude-dependent parameters averaged over periods of T = 10 minutes. This may be checked by conventional statistical methods. However, the frequency modulation caused by the action of a variable scaling velocity on length-scale dependent flow characteristics will still remain. The frequency modulation can be removed if time is scaled to the varying velocity scale by resampling the pressure coefficient records at equal intervals of "run of wind" $V(T) \cdot t$ instead of equal intervals of real time t. This action would ensure stationarity of frequency-dependent flow parameters, i.e., spectral density estimates. "Run of wind resampling" is not performed on the BRE silo data for extreme value analysis because an alternative correction is applied during the subsequent analysis.

In the case of response measurements of dynamic structures, the time scale is fixed by the invariant model frequencies of the structure and no variable form of velocity scaling is appropriate. Instead, it is necessary to select self-stationary segments of data, by examination of the reference dynamic pressure record, during which the reference wind velocity remains within defined bounds, and to discard the remainder of the signal. If this process is applied to the examples of Figs. 4 and 5, no self-stationary sections exist in the "squally" data and only one, as marked, in the "typical" data. Experience with BRE full-scale response measurements (4) has shown that only about a quarter of the data obtained in nominally steady wind can be used.

ACQUISITION OF MODEL-SCALE LOADING COEFFICIENT DATA

Contemporary atmospheric boundary layer simulation techniques (5,6) attempt to represent only the Micrometeorological peak of the incident wind spectrum in the wind tunnel, i.e., the mechanically generated turbulence. Typical linear scales of 1/250 and velocity scales of 1/2.5 produce a typical model time scale of 1/100. The lower limit for the definition of T within the spectral gap of T = 10 minutes scales typically to 6 seconds in the wind tunnel. As the Macrometeorological peak of the incident wind spectrum is not represented, i.e., the "mean" wind speed is held constant, the upper limit for T is no longer significant and statistical stationarity is effectively guaranteed without implementing any trend removal techniques. The majority of model-scale data, collected in adequate simulations, automatically comply with the definition of loading coefficient.

EXTREME VALUE ANALYSIS

Wind engineering applications are often concerned with the largest value. In order to assess the data for design against extreme winds in the manner implied by the UK Code, it is necessary to transfer attention from the parent loading coefficients to their extremes. The current UK Code attempts to design for a single application of the maximum gust load experienced by a structure or part of that structure within its design lifetime. (Whether or not it succeeds in this aim has been discussed elsewhere (7).) The present well-established methods of estimation of extreme winds are based upon theoretical methods first presented by Fisher and Tippett (8) and later developed by Gumbel (9). The application of the methods has been described recently by Mayne (10), so only the essentials will be described here.

Fisher and Tippett (8) showed that the probability distribution functions of extreme values, the density function (PDF) and the cumulative function (CDF), converge to one of only three forms as the sample from which each extreme is selected increases in size. These are called Type I, Type II and Type III. They also showed that the form of the upper tail of the parent probability distribution functions determines which of the three forms results. The parent probability distributions of Macrometeorological wind data in temperate climates are a good fit to the Weibull form and the extremes consequently converge to the Type I distribution. Gumbel (9) proposed an analysis method for Type I extremes which is now used as a matter of routine in the analysis of wind data (Shellard (11)).

Over recent years a large amount of data describing the total parent population of pressure and force coefficients has been collected. Peterka and Cermak (12) suggest that the form of the PDF's falls into two classes, one for "attached" and the other for "separated" flows. It has been demonstrated (7) that both classes meet the necessary requirements for the distributions of the positive-going and the negative-going extremes to be of the Type I form. The standard form of extreme value analysis used for the Macrometeorological wind data is therefore applicable to loading coefficient data.

This standard method consists of ranking the extreme values taken from independent observation periods by magnitude. The magnitude of the extreme is then plotted against Gumbel's reduced variate y, estimated from the rank m of the extreme by

$$y = -\ln\left(-\ln\left(\frac{m}{n+1}\right)\right) \tag{10}$$

where n is the number of extremes. This has the effect of transforming the Type I CDF given by

$$P(x) = e^{-e^{-a(x - U)}} \tag{11}$$

where the term

$$a(x - U) = y \tag{12}$$

is Gumbel's reduced variate, to a straight line. Values of the distribution parameters U, the mode, and 1/a, the dispersion, are obtained by fitting a straight line through the plotted points. There are a number of ways of doing this, the most obvious being a "least-squares" fit. However, the Gumbel ranking formula, equation (10), introduces a bias into the estimates of y because there is unequal confidence in the ranking positions of the data. Clearly the least confidence attaches to the largest value; as it is not exceeded in the data sample there is no information on the probability of its exceedance. Lieblein (13) has developed methods of fitting which correctly weigh each extreme to give unbiased estimates of the parameters. Here, "unbiased" is used in the statistical sense to imply that, although the estimates obtained will differ from data sample to data sample, the errors will be minimized and be random, not systematic. Lieblein's method requires that each ranked extreme is multiplied by a corresponding "Best Linear Unbiased Estimator" (BLUE); then summation of the products gives each of the parameters. Lieblein tabulates the values of the BLUE for sample sizes of 2 to 16 extremes, and gives a simplified method for larger sample sizes which is almost as accurate as the basic method.

Figure 7 shows this form of analysis performed on a typical set of data from the BRE Boundary Layer Wind Tunnel, where pressure coefficient minima in a region of high suction were acquired over an observation period equivalent to 10 minutes full-scale, and averaged over periods equivalent to 1, 4 and 16 seconds full-scale. The use of extremes from 16 observation periods gives an efficiency of 98% in the Lieblein BLUE. Acquisition of these data at a model time scale of 1/100 requires 96 seconds of wind-tunnel time, which compares favorably with conventional analysis in contemporary boundary layer simulations. In full-scale, the acquisition time corresponds to the 160-minute length of the data recordings for the BRE silo experiment, described earlier.

The variable scaling velocity of the full-scale data makes the time scale variable. The observation period T is corrected by dividing the 160-minute runs into 16 equal lengths of "run of wind" $V(T) \cdot t$. Each corrected observation period corresponds to 10 minutes at the overall mean wind speed $V(T)$. The results of this form of analysis are values of mode U, the most likely extreme value in the observation period, and corresponding values of dispersion 1/a, a measure of "spread." In terms of conventional analysis of the parent loading coefficients the mode is analogous to the mean and the dispersion is analogous to the standard deviation or root-mean-square value.

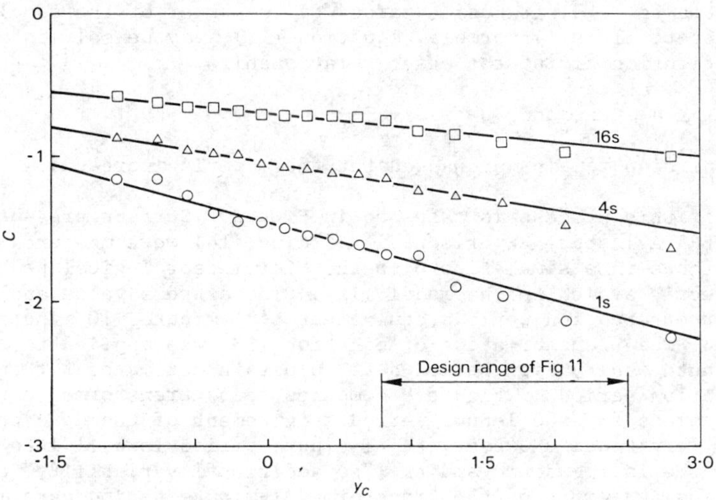

Fig. 7. Gumbel extreme-value plot of model loading coefficient data.

APPLICATION OF EXTREME VALUE RESULTS TO DESIGN

Comparability of Micro-range loading coefficient data with Macro-range wind data

The removal of the Macro-range trends from the full-scale pressure records ensures that the variations already accounted for in the analysis of the Macrometeorological wind are not included twice. However, it is important that no contribution to the variability be lost between the Macro-range wind data and the Micro-range loading coefficients. Therefore, the averaging period for the extreme wind data and the observation period for the extreme loading coefficients should take the same value T within the Spectral Gap. For reasons of economy, we have chosen to use the shortest practicable period equivalent to T = 10 minutes full-scale for both full- and model-scale measurements of loading coefficient data. The majority of European wind data are averaged over a period of 10 minutes and are thus automatically matched. UK wind data are averaged over 1 hour so transformation of the loading coefficient data to the corresponding longer observation periods is required.

Providing that no additional variability is introduced into the parent data by extending the observation period and that the original T minute extremes are independent, the distribution of extreme values over longer observation periods can be exactly related to the original extreme-value distribution. These conditions are effectively met for the loading coefficient data as long as both the original observation period T = 10 minutes and the new, longer observation period T_r remain within the Spectral Gap. The number of independent observations of the original extreme population in each of the new longer periods is the ratio of their lengths $r = T_r/T$. The probability that an extreme value is not exceeded in the new longer period T_r is the probability that all of r original extreme values are not exceeded

$$P_r(C) = (P(C))^r \tag{13}$$

Thus, the CDF of the new extreme value distribution is merely the CDF of the original extreme value distribution raised to the power of the sample-size ratio. The Fisher-Tippett (Type I) formula, equation (11), may be solved for its parameters by substitution giving the exact relationships

$$a_r = a \quad \text{and} \quad U_r = U + \frac{\ln r}{a} \tag{14}$$

i.e., the dispersion remains unchanged but the mode is increased.

The accuracy of this process is assessed in Fig. 8. Surface pressures on a 1/100 model of the BRE Aylesbury Experimental Building (14) were measured at four different roof pitches in a simulation with the Micrometeorological peak of the wind spectrum correctly scaled to the model (15). The extreme value analysis was performed over observation periods equivalent to 1 minute, 10 minutes and 1 hour in full-scale. The transformation of equation (13) was applied to the 1-minute and the 10-minute observation period data to obtain estimates for the mode over 1-hour observation periods. Figure 8 compares these transformed estimates with the measured values for the 1-hour period. Agreement of the data transformed from 10 minutes is very good. However, the 1-minute period is well into the Micro-range so both the independence and the "no additional variability" requirements are broken. The agreement of the transformed 1-minute period data is not good, but it is significant for design purposes that the error is conservative, over-estimating by between 10% and 20%.

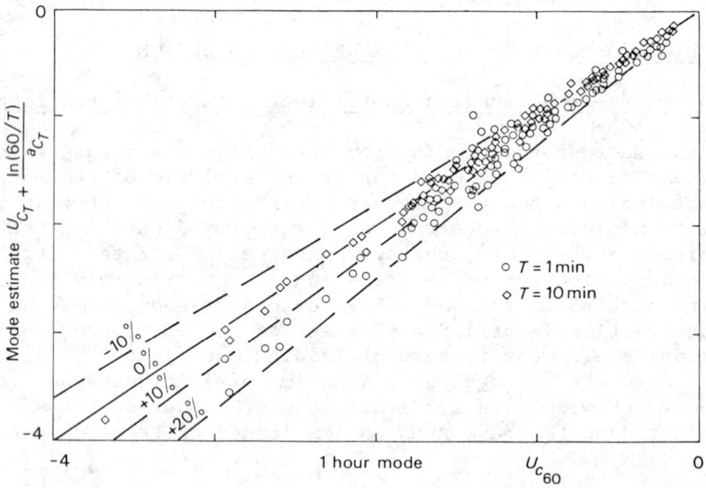

Fig. 8. Aylesbury model, roof pitches 5, 10, 15 and 22.5 degrees.

It is instructive to examine how the error on the transformation from 10 minutes to 1 hour is distributed. Figure 9 shows the 1-hour observation period mode of the 16-second C_p minima measured on the roof of the 1/100 Aylesbury Experimental Building model at a roof slope of 5° and a skew wind direction of 233° (a simulation of full-scale run A38-C (16)). The measured C_p data are plotted as contours on the plan of the roof. Superimposed on the figure are contours of the percentage error in the estimates transformed from the 10-minute observation period. It is seen that significant errors are confined to regions where the C_p mode is low and that the errors are small in the regions of significance for design.

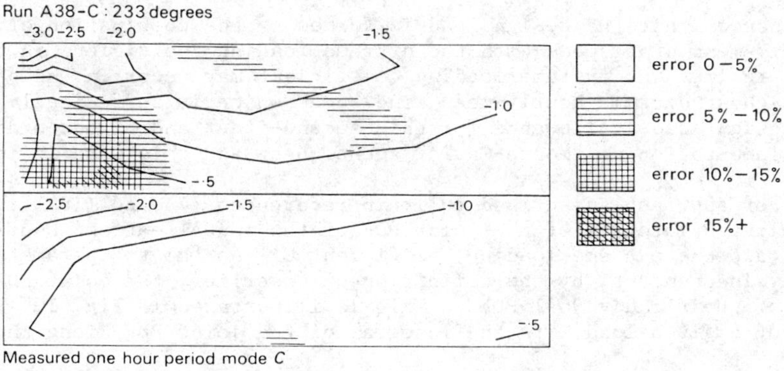

Fig. 9. Aylesbury model, 5 degree roof, 16 sec minima.

Application of the "Gust Factor Approach"

A parameter which was suggested (17) to describe extreme design loading coefficient values in terms of the parent mean and rms values is the Gust Factor, defined by

$$g = (\hat{C}_p - \overline{C}_p)/\sigma_{C_p} \tag{15}$$

where $\hat{C}_p$ is the extreme value, $\overline{C}_p$ is the mean and σ_{C_p} is the rms value of the loading coefficient. This approach has been adopted by other researchers and was tested in the BRE full-scale experiment at Aylesbury (16). The problem with the Gust Factor is that it appears to take on a wide range of values and great difficulty is experienced in finding a coherent pattern to the variation (16). In retrospect, two major reasons for this variability are apparent. Firstly, the assessment of the extreme values was made from the single highest extreme in the data sample. Since there is no assessment of its probability of exceedance, the same doubt attaches to this value as to the largest extreme in the Gumbel analysis. Secondly, the process by which the random variations of the parent conspire to select this extreme is dependent on the frequency spectrum of the parent, which itself varies considerably with position on a structure.

The advantage of extreme value analysis over the selection of a single extreme is that the probability of the extreme is correctly assessed and is described by the values of mode U and dispersion 1/a obtained.

The definition of g in equation (15) is analogous to that for Gumbel's reduced variate y in equation (12). Extreme "Gust Factor" type of approach is possible using y in place of g. The reduced variate y expresses the risk of exceedance of the loading coefficient in the observation period. The risk of the design wind speed is also known from a similar analysis of the wind data. Thus, the actual risk of wind loads on a structure can be assessed from the joint risk of wind speed and loading coefficient. Several approaches to this assessment are under development and a first-order method, developed at BRE, has been proposed (7). This is now described.

A first-order method for design

The method assumes that the design load is formed by the combination of the extreme hourly-mean wind speed with the extreme loading coefficient in that hour of wind. In reality, an extreme loading coefficient may occur in the period of the second highest extreme hourly-mean wind speed which is sufficiently great to exceed the design load. Allowance for this second-order and higher-order effects requires implementation of "Monte-Carlo" techniques, and this work is in progress.

The first-order approach, described fully in references (7) and (18), invokes a simple convolution integral of the joint PDF of the annual extreme hourly-mean wind speed V and the extreme loading coefficient C. As the two parameters are statistically independent, by the definition of C earlier, the joint PDF is simply the product of their individual PDFs. This is illustrated in Fig. 10. The probability of a given load X is the integral of the joint PDF along the curved line

$$X = \frac{1}{2} \cdot \rho \cdot V^2 \cdot C \qquad (16)$$

As the shape of the two individual PDFs is the Fisher-Tippett (Type I) form, the shape of the joint PDF is always known and geometric similarity reduces the four parameters U_V and $1/a_V$, for the wind speed, and U_C and $1/a_C$, for the loading coefficient, to the two independent parameters aU_V and aU_C, the mode/dispersion ratios, and the scaling function of equation (16). The design load probability distribution functions have been computed (18) over the range of aU = 5 to aU = 20 of both parameters.

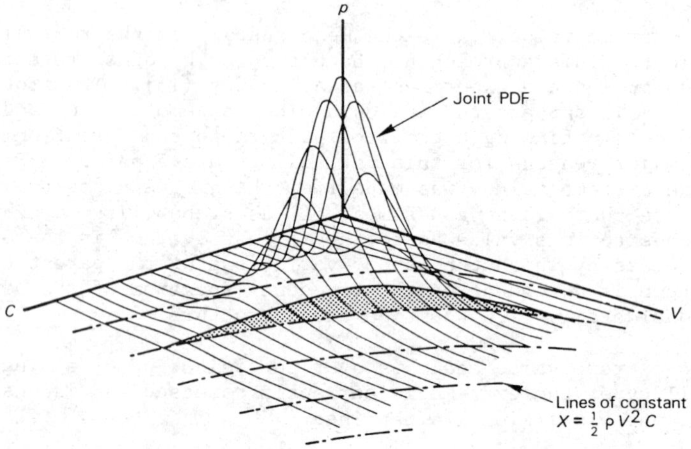

Fig. 10. Joint PDF of C and V.

Consider again the "Extreme Gust Factor Approach" suggested in the previous section. It is now possible to work backwards from the results of the first-order design method to determine, for a given return period, the equivalent static loading coefficient which gives the design result when applied in equation (16) with the extreme mean-hourly wind speed of the same return period. This effectively calibrates the "Extreme Gust Factor Approach" resulting in the design values of y_C shown for return periods of 10, 20, 50 and 100 years in Fig. 11. The range of y_C appearing in Fig. 11 is from 0.8 to 2.2, depending systematically on the parameters aU_V and aU_C. This type of calibration was not possible with the original gust factor approach from the parent, when the gust factor g was observed (16) to take values up to 13 with no coherent pattern to the variation.

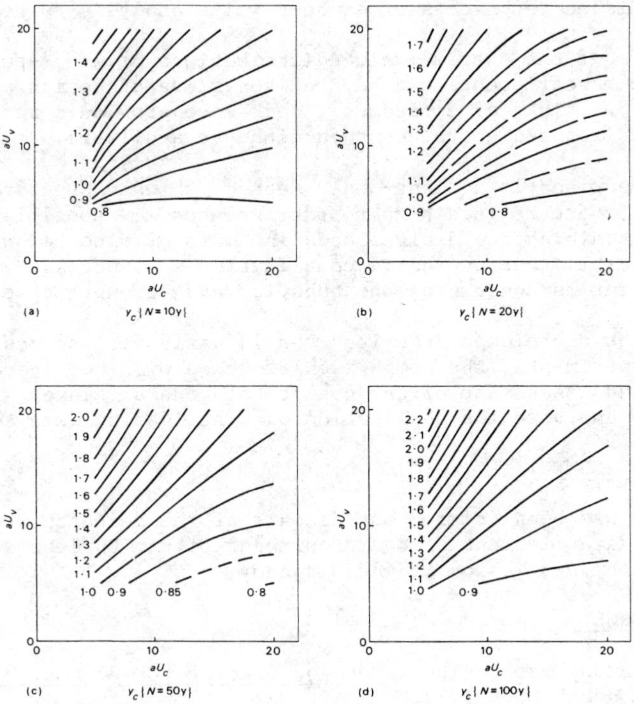

Fig. 11. Design values of y_C for N = 10, 20, 50 and 100 years.

Some simple assessments have been made which indicate that each successive order correction to the first-order method has substantially less effect on the resulting extreme load PDF, especially to the tail of the distribution and hence on the design values. The error in the first-order method is greatest for high aU_V combined with low aU_C, when it will underestimate by up to 10%. A second-order correction would reduce this error to about 1%. Nevertheless, the simple first-order approach gives a much more reliable estimate of design load than was available previously.

CONCLUSIONS

1. The wind speed and pressure/force coefficients should be defined in such a way that they are statistically independent. The Van der Hoven spectrum makes this possible in well-behaved wind climates.

2. Pressure/force coefficient data as well as wind speed data, are best described by an extreme value distribution. This distribution, as in the case of wind speed data, is a Fisher-Tippett (Type I). One of the prine advantages of using extreme values, additional to its convenience, is that it avoids detailed knowledge of the spectral content of the parent distribution necessary for the estimation of extreme values from the parent itself.

3. To achieve this, data must be rendered stationary by certain trend removal procedures which are carefully chosen to match the underlying physical processes. Transducer drifts can be removed by similar methods at the same time.

4. Data samples should contain a sufficient number of observation periods to enable established techniques of extreme value analysis to be employed.

5. The parameters of the extreme value distributions of wind speed and pressure/force coefficients can be used to arrive at design pressures and loads with known risks of exceedance, and hence provide a rational and more accurate basis for design values than hitherto available.

6. The method described is a first-order method, which can underestimate by the order of 10%. More refined higher-order methods are possible, which include the risk of attaining the design loads in hours of wind below the maximum, but it is possible that the extra complication is unnecessary because of the other uncertainties concerning the behavior and strength of structures.

7. Adoption of the techniques described should assist in the design of better full-scale experiments, the comparison of these with model-scale tests, the development and assessment of the quality of boundary layer simulations in wind tunnels, and also the application of wind loading data to codification.

ACKNOWLEDGMENT

The work described has been carried out as part of the research program of the Building Research Establishment of the Department of the Environment and this paper is published by permission of the Director.

REFERENCES

1. British Standards Institution. Code of Practice CP 3: Ch. V: Part 2: Wind Loads, London, BSI, 1972.

2. Van der Hoven, I., Power spectrum of horizontal wind speed in the frequency range from 0.0007 to 900 cycles per hour. Journal of Meteorology, 14, 1957, pp. 160-164.

3. Davenport, A.G., The dependence of wind loads on meteorological parameters. Proceedings, International Conference on Wind Effects on Buildings and Structures, Vol. 1, Ottawa, Sept. 1967, pp. 19-82.

4. Jeary, A.P., Private communication.

5. Hunt, J.C.R. and Fernholz, H., Wind tunnel simulation of the atmospheric boundary layer: a report on Euromech 50, J. Fluid Mech., 70, 1975, pp. 543-559.

6. Cook, N.J., Wind tunnel simulation of the adiabatic atmospheric boundary layer by roughness, barrier and mixing device methods, J. Industrial Aerodynamics, 3, 1978, pp. 157-176.

7. Mayne, J.R. and Cook, N.J., On design procedures for wind loading, Current Paper 25/78, Building Research Establishment, Garston, 1978.

8. Fisher, R.A. and Tippett, L.H.C., The frequency distribution of the largest or smallest member of a sample, Proc. Cambridge Philosophical Society, 24, 1928, pp. 180-190.

9. Gumbel, E.J., Statistics of Extremes, Columbia University Press, 1958.

10. Mayne, J.R., The estimation of extreme winds, J. Industrial Aerodynamics (in press).

11. Shellard, H.C., The estimation of design wind speed, Proceedings, Conference on Wind Effects on Buildings and Structures, 1963, Vol. 1, HMSO, London, 1965, pp. 29-51.

12. Peterka, J.A. and Cermak, J.E., Wind pressures on buildings - probability densities, Journal of Struct. Div., ASCE, 101, 1975, pp. 1255-1267.

13. Lieblein, J., Efficient methods of extreme value methodology, Report 74-602, National Bureau of Standards, Washington, October, 1974.

14. Eaton, K.J. and Mayne, J.R., The measurement of wind pressures on two-story houses at Aylesbury, J. Industrial Aerodynamics, 1, 1975, pp. 67-109.

15. Cook, N.J., Determination of the model scale factor in wind-tunnel simulations of the adiabatic atmospheric boundary layer, J. Industrial Aerodynamics, 2, 1978, pp. 311-321.

16. Eaton, K.J., Mayne, J.R., and Cook, N.J., Wind loads on low-rise buildings - effects of roof geometry, Current Paper 1/76, Building Research Establishment, Garston, 1976.

17. Dalgliesh, W.A., Statistical treatment of peak gusts on cladding, Journal of Struct. Div., ASCE, 97, 1971, pp. 2173-2187.

18. Cook, N.J. and Mayne, J.R., A novel approach to the assessment of wind loads for equivalent static design, J. Industrial Aerodynamics, 4, 1979, pp. 149-164.

THE RESPONSE OF A 190 METRE TALL BUILDING AND THE RAMIFICATIONS FOR THE PREDICTION OF BEHAVIOR CAUSED BY WIND LOADING

A. P. JEARY and B. R. ELLIS
Building Research Establishment, Garston, Watford, UK

ABSTRACT

The behavior of a tall building has been monitored in response to controlled oscillation induced by a vibrator system, and in response to naturally occurring winds.

The oscillation tests allowed the "calibration" of the dynamic properties of the structure, and these results were subsequently used as input for design guide predictions in order to see how they perform.

Spectral analysis of the response of the structure to the naturally occurring wind allowed a full comparison to be made with the design guide predictions.

It has been found that in this case the design guides do not perform well, even when the input parameters have been measured directly.

The implications for the methods of predicting dynamic structural response are considered and a field for further investigation is suggested.

INTRODUCTION

At the invitation of the consultants for the project, the Building Research Establishment has recently conducted a series of tests on a 190 m tall building. The basic purpose of the tests was to provide assurances about the projected dynamic behavior of the structure.

A series of oscillation tests were conducted, using the BRE vibrator system, and the major dynamic structural properties were measured or inferred directly from the tests.

Subsequently the response of the building to the natural wind environment was monitored and analyzed.

These measurements were compared with the predictions obtained from the use of two design guides, and these predictions were found to be unsatisfactory. This was particularly surprising as the structural properties were measured directly and pressure data were obtained from a wind tunnel study.

In this paper these findings are discussed, a possible cause for the shortcomings of the design guides is proposed, and an area for urgent research effort is suggested.

DESCRIPTION OF THE BUILDING

Figure 1 shows an elevation of the 46-story tower and Fig. 2 shows a typical plan. The building consists of a massive, heavily reinforced core, with three leaves attached and supported by large concrete corbels at their bases. Services and lifts are provided inside the core, and the leaves are mainly for office usage. The three leaves are each of different height. Leaf "A" extends from the 1st to 43rd levels, leaf "B" extends from the 3rd to the 41st, and leaf "C" extends from the 5th to the 39th levels. The exterior of the leaves consists of steel columns, onto which are attached the cladding panels. The concrete floors are cantilevered from the core, and are attached to the steel columns at the periphery. Expansion joints are provided at the interfaces between leaves.

The top of the core is approximately 187 m above ground level and at the time of the forced oscillation tests the building was structurally complete, with the sole exception of the 46th floor. Additionally, the upper three floors were open to the atmosphere, and several minor items at or near ground level were not complete.

TEST PROCEDURE

A vibrator was installed at the 43rd floor level in leaf "A" (see Fig. 2). This vibrator was constructed for the BRE, under contract, by the staff of the Civil and Mechanical Engineering Departments at the University of Bristol. The vibrator is capable of delivering a peak force of 1070 Newtons at 1 Hz. This force is developed by two sets of horizontally contra-rotating masses.

Servo-type accelerometers were used to transduce the motion of the building to a varying voltage proportional of the acceleration response. The accelerometers used are very sensitive devices, and when used in conjunction with the appropriate signal conditioning (Fig. 3), are capable of resolving a motion of 0.001 mm (1000A°) in the region of 1 Hz.

The vibrator was set in motion and the response of the structure was monitored. The frequency of the applied force was incremented and the response/frequency curve was noted. In this way the lowest frequency normal modes of vibration were found.

The next step was to set the vibrator to a resonance frequency of the building and then incrementally rotate an accelerometer. The maximum response corresponds with the modal direction. (In practice it is easier to find a minimum response orthogonal to the modal direction.) The modal directions were established for the second translational modes as the separation in frequency was greater than that for the first translational ones, and the interference caused by the wind was less. It has been assumed that the modal directions for the first translational normal modes are the same, although no direct measurements of these were made.

The mode shapes, associated with each normal mode, were investigated by setting the vibrator to the resonance frequency and then taking a traveling accelerometer to various locations throughout the structure. The amplitudes and phases measured by the use of the traveling accelerometer were compared with those measured by the use of the reference accelerometers on the 43rd floor.

The response of the structure to wind excitation was monitored through two nights. These data were then returned to the laboratory for spectral analysis.

For each mode of vibration an estimate of the modal damping value was obtained by a sudden cessation of excitation. The resulting decay envelopes obtained from the accelerometer signals allowed estimates of damping to be made.

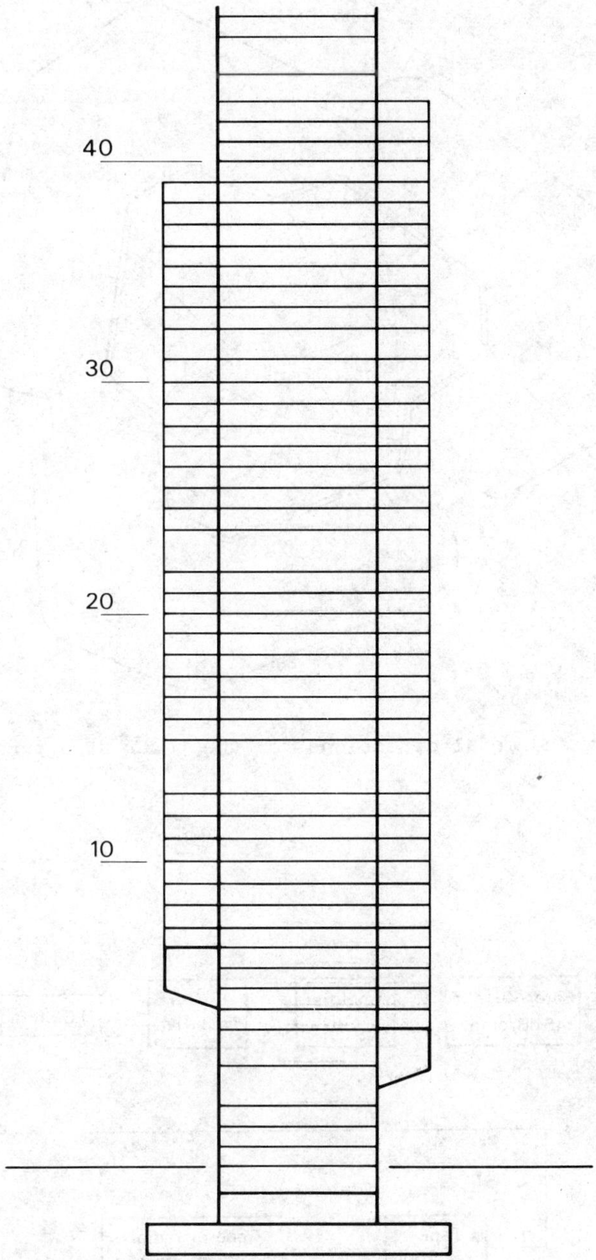

Fig. 1. Elevational cross section of the tower.

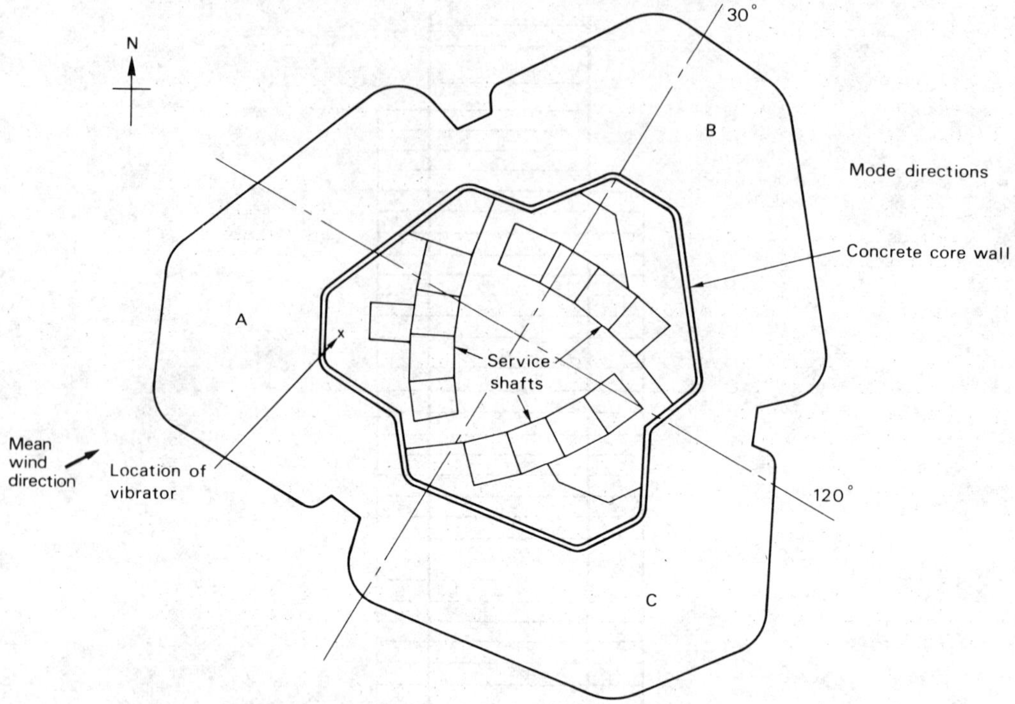

Fig. 2. Plan view and modal directions for the tower at a typical floor level.

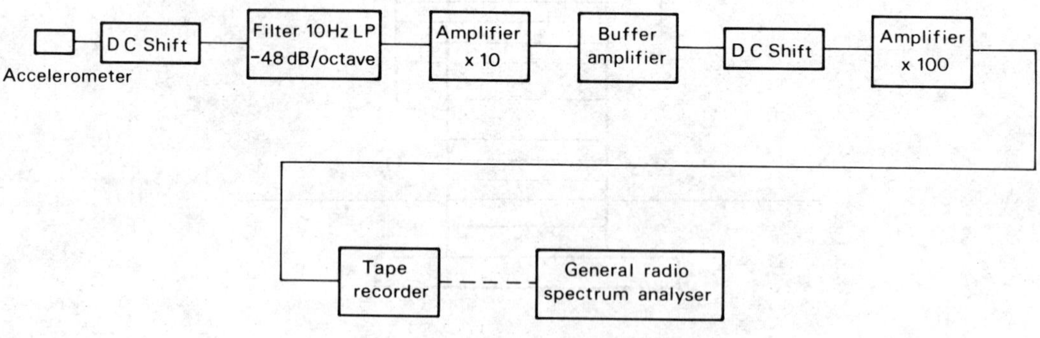

Fig. 3. Signal conditioning system.

SPECTRAL ANALYSIS

The recorded data of the building's response to wind-induced excitation were returned to the laboratory for analysis on a "Genrad" minicomputer.

For the assessment of natural frequencies, Power Spectral Density and Enhanced Incoherent Output Spectra (1) were used. Figures 4 and 5 are spectra taken from the recording made on the night of December 27-28, 1977.

For the case of well-separated modes it is possible to obtain an estimate of the damping associated with individual modes by use of the autocorrelation function (2). Figure 6 is a representation of the autocorrelation function for the first north-south translational mode.

It is usual to treat random quantities in terms of square root of the mean square (rms) or mean square values. However, it is felt that a greater insight to the physical meaning will be given by converting these values to peak-to-peak measurements. This is justified only if the motion is sinusoidal. Since the responses being considered are very narrow band the sinusoidal assumption represents a reasonable approximation. However, it must be remembered that rms and mean square are averaged measurements and the mean peak-to-peak measurements also represent averaged quantities. These measurements have been used throughout this note.

INTERPRETATION OF RESULTS

The fundamental modes and the second translation modes as measured during the oscillation test are depicted in Fig. 7. The natural frequencies and associated damping ratios and stiffness "K" values are noted in Table 1.

TABLE 1. Structural Properties.

Mass (measured)		$72.6 \cdot 10^6$ kg	$\pm 10 \cdot 10^6$ kg
	Frequency	Stiffness	Damping (% of critical)
$E\text{-}W_1$	0.440	185.0 MN/m	0.5
$N\text{-}S_1$	0.440	122.0 MN/m	0.5
θ	1.553	$9.56 \cdot 10^{11}$ Nm/rad	0.7
$E\text{-}W_2$	1.670	799.3 MN/m	2.8
$N\text{-}S_2$	1.690	818.6 MN/m	2.1

Fundamental Time Constant $\tau_1 = \dfrac{1}{\zeta_1 W_1}$

$$\tau_1 = \left(\frac{0.5}{100} \cdot 2 \cdot \pi \cdot 0.44\right)^{-1} = 72 \text{ seconds}$$

In practice, seldom are more than three modes of significance for buildings, and in this case only the first translational orthogonal pair (both resonance frequencies being close to 0.440 Hz) play a significant part in the response. This can be appreciated from Figs. 4 and 5, where it can be seen that the acceleration responses of the first translational modes are two orders of magnitude larger than that caused by torsion at 1.553 Hz. (This is equivalent to more than three orders of magnitude in terms of displacement.)

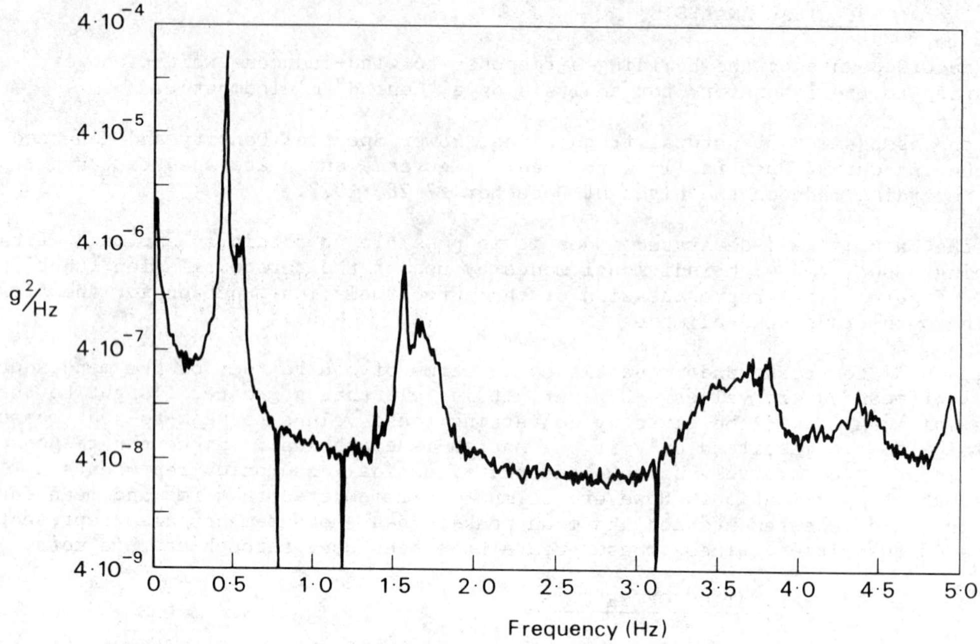

Fig. 4. Enhanced incoherent output spectrum taken from the north-south accelerometer.

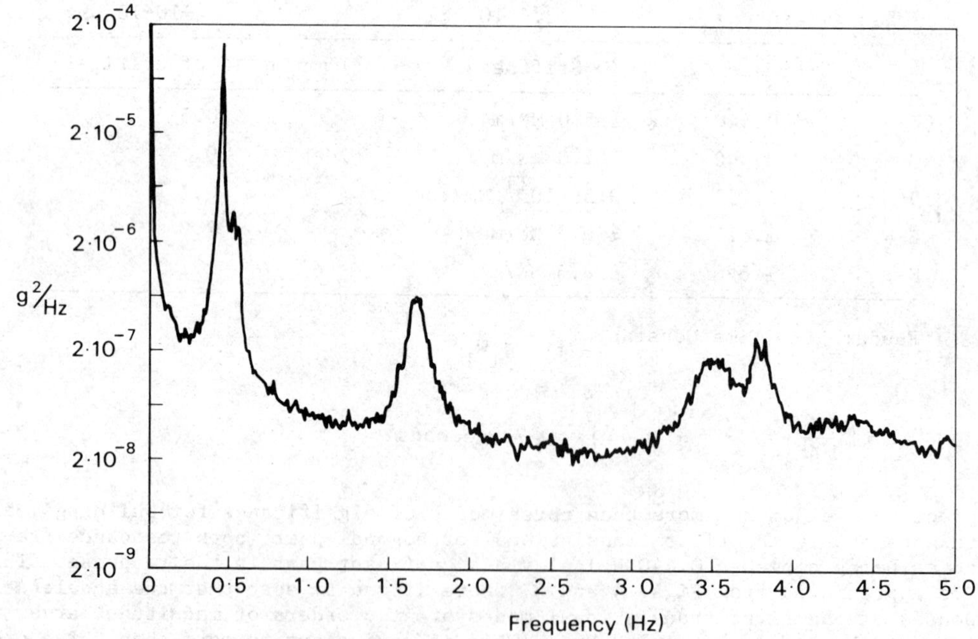

Fig. 5. Enhanced incoherent output spectrum taken from the east-west accelerometer.

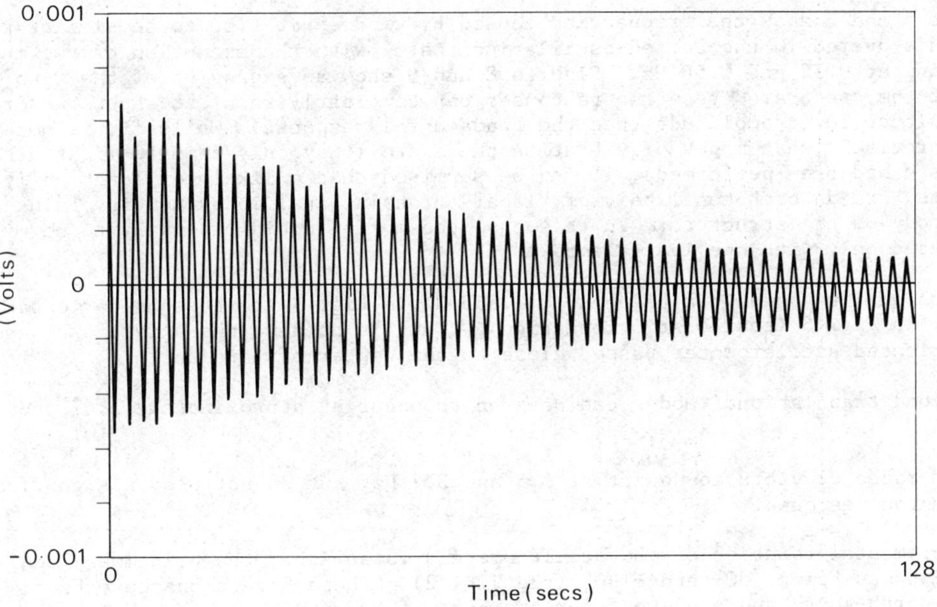

$$\xi_{N-S_1} = \frac{100}{2\pi} \times \frac{1}{40} \ln \frac{19 \cdot 5}{7 \cdot 0} = 0 \cdot 51\%$$

Fig. 6. Autocorrelation taken from north-south accelerometer.

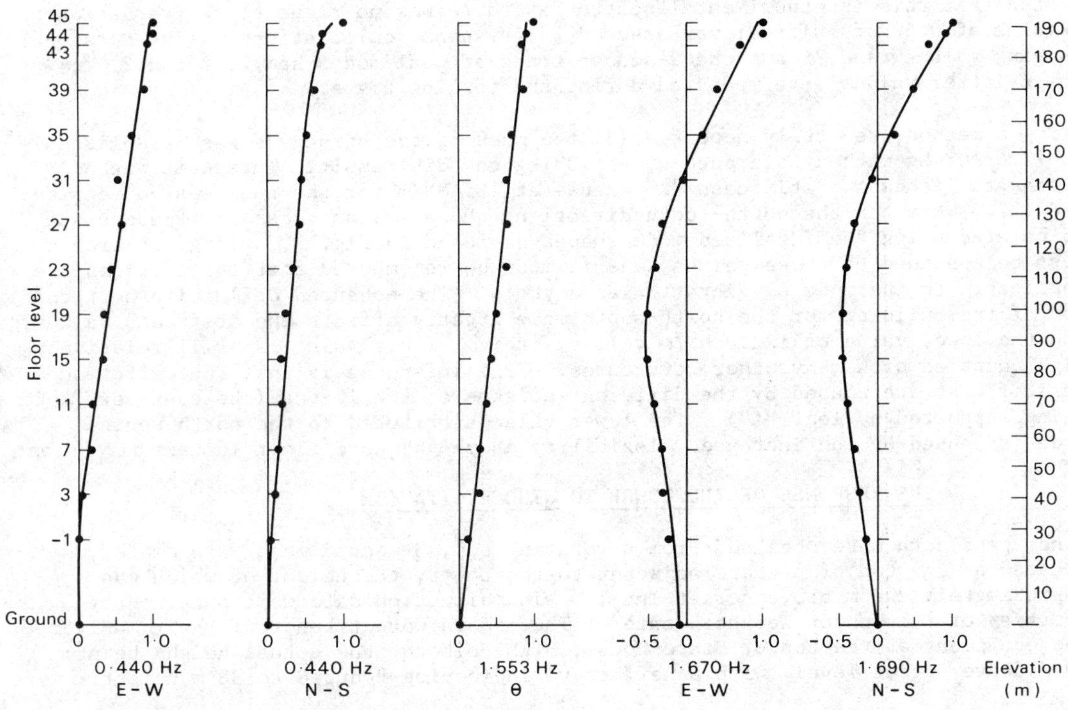

Fig. 7. Normalized mode shapes for the fundamental translation, torsion and second translation modes.

Figures 5 and 6 show the frequencies caused by wind excitation to be similar to those discovered in the forced-oscillation tests, with the exception of two modes occurring at 0.52 and 0.56 Hz. Figures 8 and 9 show an expansion of the 0 to 1 Hz part of the spectra. These two responses can be clearly seen at about one order of magnitude lower amplitude than the fundamental response. Whilst it is not known precisely what modes of vibration these are (they only came to light after the tests had been performed), it can be surmised that a likely explanation is that they are an orthogonal pair of "local" translational modes not involving the whole body of the structure. In this case the motion would be likely to be associated only with one leaf (leaf "A").

The torsional response can be seen at 1.553 Hz in Fig. 5 (north-south orientation), but not in Fig. 6 (east-west orientation). This is because the axis of the east-west oriented accelerometer passed close to the center of torsion.

The second translational modes can be seen to occur at approximately 1.67 and 1.69 Hz.

Further modes of vibration exist at 3.5 and 3.7 Hz, but do not play a significant part in the response.

The incremental rotation of the accelerometers established the principal modal directions as being +30° and +120° (see Fig. 2). These directions have been termed north-south and east-west for the sake of brevity.

The damping values measured for first translational mode activity were approximately 0.5% of critical for both north-south and east-west. These values were backed up by measurements made from autocorrelation plots (e.g., Fig. 6). The values obtained varied between 0.46% and 0.55% of critical, and it is suggested that this is experimental scatter as there was no trend to the variation. A value of 0.5% of critical was assumed for response calculations. (The torsional damping value was 0.7% and the 2-second translational modes had 2.1% and 2.8%.) These latter values were measured during the forcing tests.

Using a method previously reported (3) the mass of the structure was established as $72.6 \cdot 10^6$ kg with a tolerance of $\pm 9 \cdot 10^6$ kg on this result. This same method then establishes the stiffness "K" values at 185 MN/m for the east-west direction and 122.0 MN/m for the north-south direction. However, as these stiffnesses are calculated using the idealized mode shapes depicted in Fig. 7, a large tolerance must be assigned to these values, and it must be remembered that these values are applicable to the mode of vibration as a whole. The enhanced deflection near the top of the building for the north-south mode greatly affects the stiffness values, and the lower value of 122.0 MN/m reflects this. (This value is still relatively high compared with many other structures.) The inference is that the differences in stiffness are caused by the different heights of the leaves (the east-west mode being supported by leaf "A"). The lower value attributed to the north-south mode is caused by the increased flexibility above the 43rd floor in this direction.

THE RESPONSE OF THE TOWER TO WIND EXCITATION

Wind speed data were obtained from a rotating cup, 3-second mean anemometer, located at the top of a crane adjacent to the tower, the height of which was approximately 200 m above street level. Wind direction data were obtained by courtesy of the London Weather Center. These data were taken from their monitoring equipment at the top of State House, High Holborn, the actual height being 70 m above street level, with an effective indication reduced to 38 m height.

During both recordings, the peak, 3-second average, gusts were in the region of 15-17 m/sec, and the mean wind direction was in the region of 250° (west-south-west). On the night of the 27th/28th the mean wind speed varied between 7 and 11 m/sec and the direction from 250° to 290°, and on the 29th/30th the mean wind varied between 4 and 10 m/sec and the direction from 230° to 250°.

Since the response of the tower consists mainly of motion in the two fundamental translational modes, the total motion was taken to be the resolved component of these two modes.

In the following computations, emphasis has been placed on the maximum response, and the prognostications assume that the maximum measured is the result of a similar gust factor to that which would be encountered when a much stronger wind was blowing (i.e., that it is the gust factor applicable to a duration of one hour).

The maximum gust of 17 m/sec occurred when the hourly mean wind speed was approximately 11 m/sec and the resolved maximum peak-to-peak motion at this time was 0.13 mm.

From Figs. 8 and 9 it is ascertained that the response was approximately 0.007 mm rms in both the north-south and east-west modes. The mean wind speed, for the time represented by these spectra, was approximately 11 m/sec.

Whilst it may be a crude approximation in some cases, it is assumed in the following that the force of the wind is proportional to ρAV^2, where ρ is the air density, A is the cross-sectional area of the structure presented to the wind, and V is the velocity of the wind at some reference height. It is further assumed that the response of the structure is proportional to the value of the forcing. Thus,

$$\left(\frac{V_1}{V_2}\right)^2 = \frac{R_1}{R_2} \tag{1}$$

where V_1 is a wind velocity at which a response amplitude R_1 is measured and V_2 is a wind velocity at which a response amplitude R_2 is measured.

The basic wind parameters for the city of London can be calculated by using ESDU document 72026 (7), and gives the following results for a return period of 50 years:

$\bar{V}_{10}$ = 13.26 m/sec (mean wind speed at height 10 m)

$\bar{V}_{183}$ = 30.50 m/sec (mean wind speed at height 183 m)

V_{183} = 43.92 m/sec (maximum, 3-second mean, wind speed at height 183 m)

Thus, using equation (1) and extrapolating from the gust response gives:

$$\left(\frac{43.92}{17}\right)^2 \cdot 0.13 = 0.868 \text{ mm peak-peak (total maximum response)}$$

From the spectra, the total resolved mean response is 0.028 mm peak-peak. The mean hourly response for the "once in 50-year return wind" is 30.50 m/sec at 183 m. Thus, the mean peak-peak response for this wind is calculated, using equation (1), as:

$$\left(\frac{30.5}{11}\right)^2 \cdot 0.028 = 0.215 \text{ mm peak-peak}$$

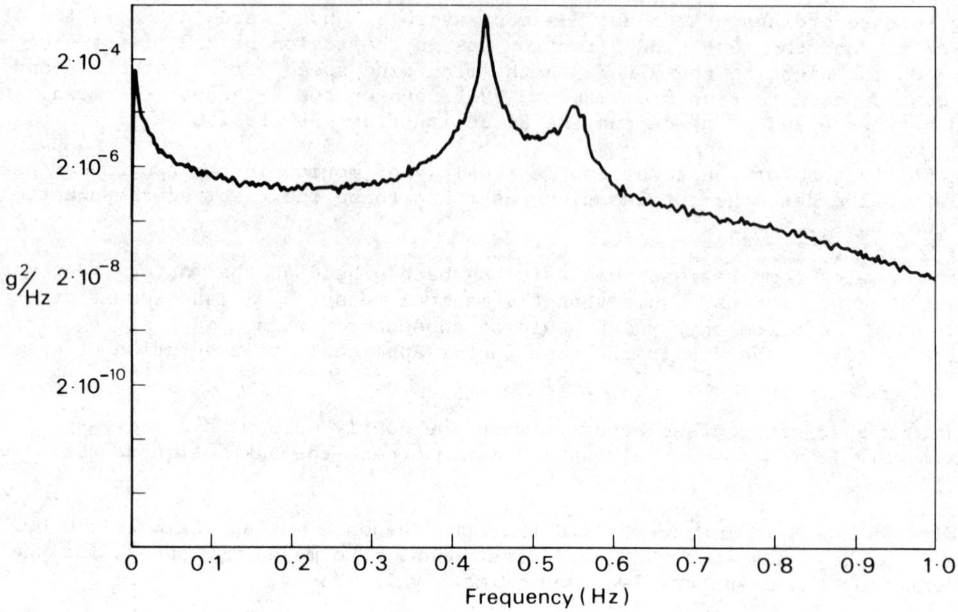

Fig. 8. Power spectral density of north-south accelerometer (0-1 Hz).

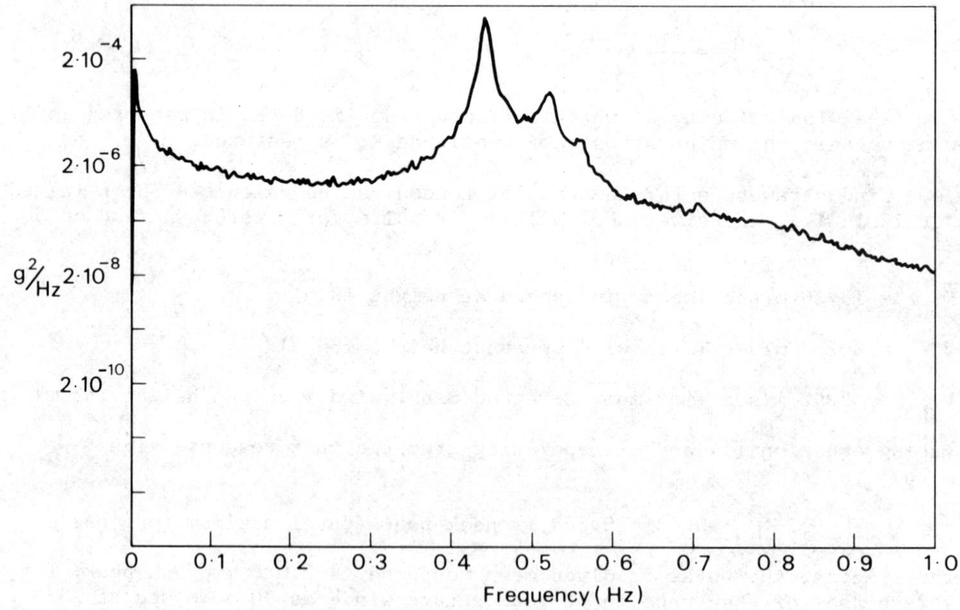

Fig. 9. Power spectral density of east-west accelerometer (0-1 Hz).

The maximum dynamic response factor g_D is obtained from Engineering Sciences Data Unit, item 76001, and is given as:

$$g_D = \sqrt{2 \log_e f\, T_o} + 0.577 \sqrt{2 \log_e f\, T_o}^{\,-1}$$

where f is the natural frequency and T_o is the period to be considered. Here T_o is taken as 3600 seconds.

$$g_D = 3.99$$

Therefore, the likely maximum response is

$$3.99 \cdot 0.215 = 0.859 \text{ mm peak-peak}$$

It is seen, therefore, that consideration of the measured maximum response and the measured mean response have led to two extrapolations for the maximum total resolved response for the "once in 50-year wind," both of which are near to 0.86 mm peak-to-peak response. This response relates to movement at the 43rd floor. The response at the top of the structure would be 0.94 mm peak to peak.

COMPARISON OF RESPONSE WITH ESDU 76001 (5)

In order to calculate the mean maximum acceleration the design guide ESDU 76001 uses the following estimator:

$$x_{max} = g_D \left\{ \sum_{j=1}^{l} \left[\sqrt{\frac{\pi}{4\zeta_j} f_j\, S_{FF}(f_j)}\; \frac{\mu_j}{M_j} \right]^2 \right\}^{\frac{1}{2}}$$

where

l = the total number of modes considered

g_D = the maximum dynamic response factor, for T_o = 3600 seconds

ζ_j = the damping associated with mode j

f_j = the natural frequency of mode j

$S_{FF}(f_j)$ = the generalized modal force spectral density at frequency n

μ_j = a mode shape function describing mode j

M_j = the generalized modal mass associated with mode j

In this case we assume that only the orthogonal fundamental translational modes are significant, and

$n_1 S_{FF}(f_1) = 0.475 \cdot 10^9$

ζ_1 = 0.005 (or 0.5% of critical)

μ_1 = 1.0 for motion at the top of the structure

$M = 72.6 \cdot 10^6$ kg and $M_1 = (72.6 \cdot 10^6)/3$

$g_D = 3.99$

$x_{max} = 0.054\ \text{m/sec}^2 \simeq 0.55\%\ g$

which represents approximately 12 mm peak-peak, at 0.44 Hz. This value is significantly different from the values extrapolated from measurements, but gives a result which is similar to that obtained using the method of Wyatt (6).

DISCUSSION

It can be seen that these design guides do not provide reasonable predictions of the structures response even when the basic input parameters (e.g., resonance frequencies and damping) are known. It is perhaps lucky that the prediction falls on the pessimistic side, although instances like this do not offer any reassurance to the designer.

A consideration for this particular structure is that the openings at the top may have been acting as an aerodynamic spoiler and accordingly the response may have been reduced somewhat. However, it is difficult to believe that this could be responsible for the 1200% difference between measurement and theory.

The predictions used are based on the spectral approach, as suggested by Davenport. This approach states that

$$X_r(f) = H(f) \cdot F_r(f)$$

where

$X_r(f)$ = the response

$H(f)$ = the complex frequency response function for the structure

$F_r(f)$ = the forcing function

An implicit assumption when dealing with random processes in the context of this method is that all the processes must be stationary (i.e., that all statistical properties remain invariant with time).

There are two possible causes of non-stationarity in response data. The first is that the excitation does not form a stationary sample and this can be caused by trends in the wind force or the wind direction data. The second is by the structure having a nonlinear response (i.e., one in which there is a change of response caused by changes of resonance frequency or damping) with amplitude of motion. These effects have been noted by previous workers in several large structures (9,11,12,13), although at normal service loads measurable nonlinearities may not be significant (10).

For the quantities of data necessary to achieve reasonable statistical estimates of structural response to random forces it becomes unlikely that a sequence of stationary data will be found, and consequently it is possible to obtain misleading information from spectra if the data are non-stationary. (For 10% confidence interval and 9% bias error a record length of $\frac{200}{\zeta \cdot f_r}$ seconds is necessary, or approximately 5½ hours for a structure with a natural frequency of 1 Hz and damping of 1% of critical.) It has become obvious, from BRE tests of large structures, that an answer is to use ensemble averaging techniques, in which data occurring when similar conditions exist are averaged. This means that the quantities of data necessary are much larger than often used at present.

It can be seen, therefore, that it is possible to obtain some results by spectral analysis of sequential data, but that these results may be subject to large errors. At present ensemble averaging is the only practicable method of assessing response in terms of spectra for wind-excited structures, and to take shortcuts in assessing response may be a perilous process.

It is also possible that a further factor is significant: If a structure is excited by a force of constant magnitude at its' natural frequency it will take three time constants (3τ) to reach 95% of the steady state response and 4τ to reach 98%. At one time constant, the value is approximately 37%. If spectra were to be taken from the forcing and response times occurring before about 5τ, then the response would be less than that predicted theoretically by the spectral method.

It is probable that this type of situation is occurring for the case of large structures being buffeted by the atmosphere. It may be that time constants are becoming so long that it is no longer possible for large structures to respond in a steady state fashion to a naturally occurring stationary forcing function.

At present this can be considered as a possible explanation for an effect which otherwise would not be predictable, and it may be that a correction factor, based on time constants, may be necessary when using spectral methods. The form of this correction is by no means certain, and will need to be determined by experimentation.

There may be several factors which would affect this correction, all of which would need investigation (for instance the shape of the structure may be important).

It can be seen that there is a large gap between theory and practice and that there is an urgent need for more research effort into the prediction of response. This particular case has been demonstrated with measured values, but there is a tendency for predictions of response to be made from computer-based solutions. Many of these behave well for very simple structures, but as the complexity of the structure increases the accuracy of the prognostications diminishes. Experience gained by BRE of predictions of natural frequencies alone, for complex structures, has shown that a value within 50% of the real value is a reasonable result, at present. More accurate comparisons are often as a result of hindsight.

It is only by making more full-scale measurements and comparisons with theory and predictions that these gaps can be closed.

REFERENCES

1. Jeary, A.P., The estimation of reliable spectral information when recording low intensity data, J. of Sound and Vibration, 3, BRE N 69/76, 1978 (60).

2. Jeary, A.P. and Winney, P.E., Determination of structural damping of a large multi-flue chimney from the response to wind excitation, Proc. Inst. Civil Engineers, Pt. 2, 53, Dec. 1972, pp. 569-572.

3. Jeary, A.P. and Sparks, P.R., Some observations on the dynamic sway characteristics of concrete structures, ACI Symposium on Vibrations of Concrete Structures, New Orleans, BRE CP 7/78, Oct. 1977.

4. Mayne, J.R., On the estimation of extreme winds, J. Industrial Aerodynamics, to be published, BRE N 134/76.

5. Engineering Sciences Data Unit, Item 76001, ESDU 251-259, Regent Street, London, Sept. 1976, amended June 1977.

6. Wyatt, T.A., The calculation of structural response, Paper 6, Proc. Seminar on the Modern Design of Wind Sensitive Structures, Pub CIRIA, 1971.

7. Engineering Sciences Data Unit, Item 72026, ESDU 251-259, Regent Street, London, Nov. 1972.

8. Whitbread, R.E., Private communication.

9. Jennings, P.C. and Kuroiwa, J.H., Vibration and soil-structure interaction tests on a 9 strong reinforced concrete building, Bulletin of Seismological Soc. of America, Vol. 58, No. 3, June 1968.

10. Burrough, H.L., Jeary, A.P., and Wilson, J.M., Structural dynamics of multi-flue chimneys, Proc. 4th International Conference on Wind Effects on Buildings and Structures, Cambridge University Press, 1972, pp. 497-514.

11. Nielson, N.N., Vibration tests of a 9-story steel frame building, J. of Eng. Mech. Div., ASCE, EMI, Feb. 1966.

12. Rea, D., Bouwkamp, J.G., and Clough, R.W., The dynamic properties of McKinley School Building, Earthquake Eng. Res. Center, Report 68-4, California, 1968.

13. Foutch, D.A., The vibrational characteristics of a 12-story steel frame building, Earthquake Eng. and Struct. Dynamics, Vol. 6, 1978.

THE EFFECTIVENESS OF VORTEX SPOILERS ON A CIRCULAR CYLINDER IN SMOOTH AND TURBULENT FLOW

I. S. GARTSHORE*, J. KHANNA** and S. LACCINOLE***

*University of British Columbia, Vancouver, British Columbia, Canada
**Swan Wooster Engineering Co. Ltd., Vancouver, British Columbia, Canada
***Swan Wooster Engineering Co. Ltd., Portland, Oregon, USA

ABSTRACT

As part of the aerodynamic design for a 12' diameter tubular gallery conveyor bridge, wind tunnel investigations were conducted to determine the effectiveness of various vortex spoiler devices. The investigations were carried out on sectional models of a bare circular cylinder, and on cylinders fitted with different configurations of fins. All cases were tested in smooth flow and in grid turbulence having a relative intensity of 14%.

The results of the tests indicate that a helical system of projecting fins, more commonly known as "strakes", was the most effective device for suppressing vortex induced excitations of the cylinder. The effectiveness of this arrangement however, was markedly influenced by the presence of the turbulence in the windstream. In the case of resonant vortex shedding with smooth windstream conditions, the strakes reduced the maximum amplitude of oscillation by a factor of about 20. In the case of resonant vortex shedding coupled with turbulent windstream conditions, the strakes reduced the maximum amplitude of oscillation only by a factor of about 2. In both cases, the strakes caused an upward shift of about 50% in the wind velocity necessary to attain resonance.

1. INTRODUCTION

Analytical and model test investigations were recently made in connection with the design of a 12 foot diameter steel tubular gallery conveyor bridge with spans varying from 147 to 243 feet. The tests were required because the natural frequency calculations for the spans indicated that vortex shedding could result in serious resonant amplification of wind induced lift forces to the cylindrical cross section. The model tests were directed at an assessment of the effectiveness of various vortex spoiler devices.

The most common device used for suppressing vortex excitation of circular tubes is a set of strakes (ref. 1, 2) consisting of a triple start set of helical fins projecting perpendicular to the surface of the cylinder. The projection is usually 10% of the diameter and each helix has a pitch of 5 diameters. This conventional configuration was tested in the present measurements. Strakes are commonly used for circular stacks which are subject to winds from all directions. The present conveyor bridge, being horizontal, will be subjected to winds primarily from two directions, and it was thought that some device simpler than conventional strakes might be as effective at less cost for this application. Sawtooth fins have been used in a similar application (ref. 3) and versions of these were tested, as were straight fins.

Because of the relatively large length-to-diameter ratio of the present conveyor bridge spans (>12) and because the mean velocity approaching the conveyor centerline is essentially uniform along each span, sectional models were used. In addition, since tubulence is known to affect the flow around circular cylinders significantly (ref. 4), these effects were created in the wind tunnel through the introduction of a grid to the inlet stream.

The experimental arrangements and data analysis for the tests are described in section 2, followed by a presentation of results in Section 3 and conclusions in section 4.

2. EXPERIMENTAL ARRANGEMENTS AND DATA ANALYSIS

For each type of spoiler, sectional models were mounted vertically to span the 27 inch (0.69 m) dimension of the 27 inch by 36 inch test section of the UBC small wind tunnel. The basic model tube diameter was 3 inches (76 mm) which produced a blockage ratio of approximately 1:12. Corrections for blockage were not made since most of the results are presented in comparative form.

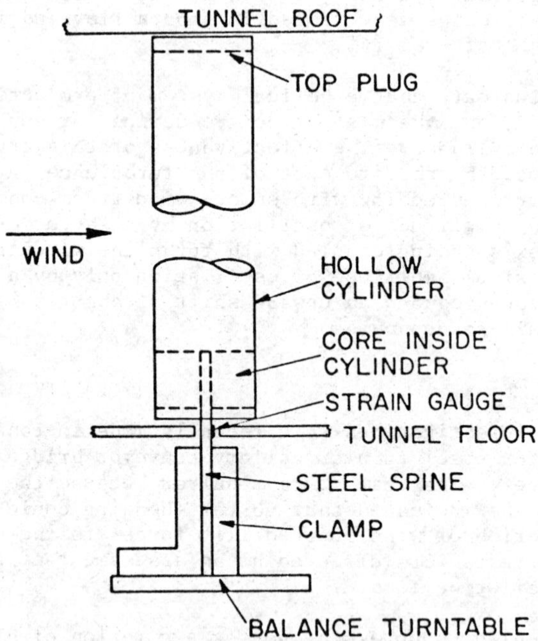

Fig. 1. Mounting arrangements

Each model was mounted as a cantilever beam with a steel spine providing stiffness (Figure 1). A strain gauge on the steel spine was used to measure instantaneous deflection in one direction. The model and spine were then mounted on a balance turntable so the model mean drag could be measured. By rotating the model such that the strain gauge sensed deflections in the along-wind direction, the measured drag was used to determine the strain gauge output for a known uniform wind load applied to the cylinder. With the strain gauge thus calibrated, the model was then rotated 90° about its own axis and oscillation amplitudes in the cross wind direction were measured by the gauge.

The output of the strain gauge was measured in terms of RMS amplitude and an oscillograph trace was used to determine peak factors. Traces representing several minutes of wind, and several hundred oscillations at the various wind speeds, were used for this purpose.

Turbulence was introduced by placing a square mesh grid (9 inch mesh = 3 times the the diameter of the model) about three feet (1 M) upstream from the model. The longitudinal intensity of grid turbulence thus produced was 14%.

Natural frequencies and structural damping values varied slightly in the model tests depending on the exact clamping arrangements used. For the majority of the tests, including those for the bare cylinder and conventional strake arrangements, natural frequencies between 20 and 22½ HZ and damping values between 0.6 and 0.89% of critical were measured. For purposes of data reduction, the damping was assumed constant at 0.7%.

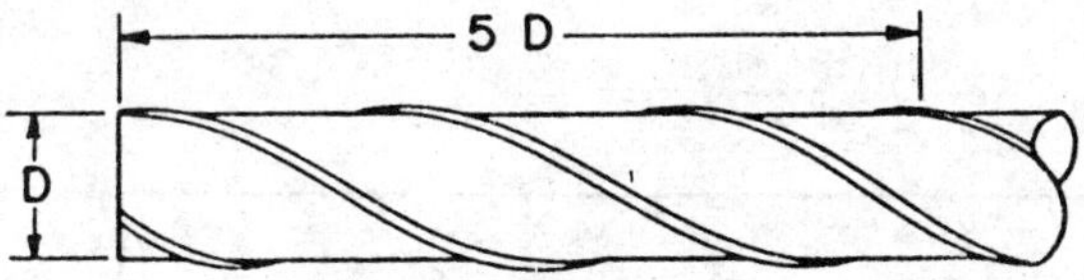

Fig. 2A. Strakes

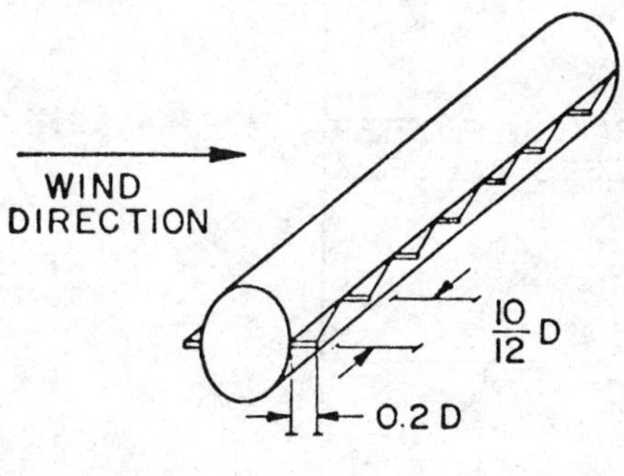

Fig. 2B. Sawtooth fins placed front and back

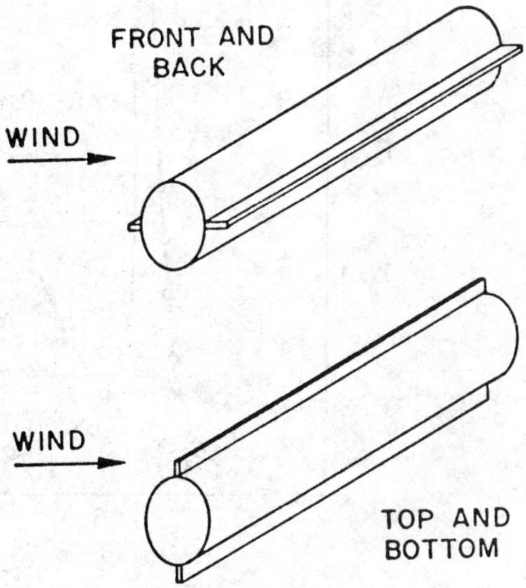

Fig. 2C. Straight fins

The conventional strake arrangements tested are sketched in Figure 2A, the saw-tooth fins in Figure 2B and the straight fins in Figure 2C.

The results are presented in terms of the ratio of the maximum RMS deflection of a model fitted with spoilers to the maximum RMS deflection of a model without spoilers in turbulent flow. This ratio is given the symbol R. In addition, some numerical values of the non-dimensional force on the cylinder are also presented through C_L' as defined from the equation:

$$\gamma(\text{RMS}) = \frac{C_L' \times \frac{1}{2}\rho(\text{Vcrit})^2 D}{K} \times \frac{1}{2\beta} \qquad (1)$$

where K is the stiffness of the spine (the force per unit length of cylinder required to produce a unit deflection at the free end of the cylinder); γ(RMS) is the maximum RMS deflection of the cylinder's free end at the critical velocity Vcrit; β is the fraction of critical damping; ρ is the air density; and D is the cylinder diameter. This form of definition was chosen since oscillation amplitudes were clearly more than 1% of the diameter indicating that lock-in was probably occurring over much of the cylinder span (ref. 5). The coefficient C_L' is a measure of the RMS force acting on the cylinder at the peak of resonance. The symbol R_{max} is used in section 4 to denote the maximum value of R, which corresponds to a peak response of the cylinder at its critical velocity. This number R_{max}, is a rough measure of the overall effectiveness of the spoiler configuration over a wide range of wind speeds.

3. RESULTS

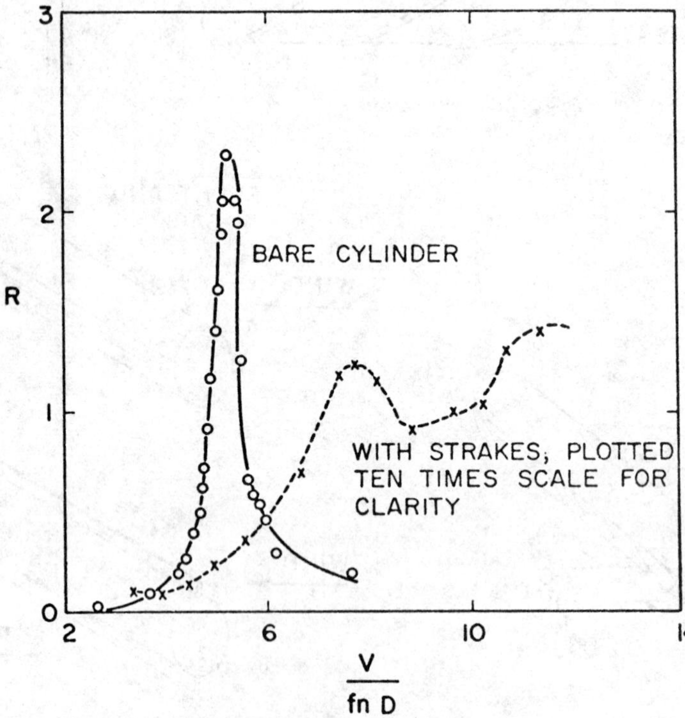

Fig. 3. The effect of strakes in smooth flow

(a) Conventional Strakes

Figure 3 shows the results of strakes applied to a bare cylinder in smooth flow (turbulence intensity less than 0.1%). In this case, the strakes reduce the maximum response at an elevated critical velocity by a factor of more than 18. In contrast, by making a comparison of the straked cylinder response at the critical velocity of a bare cylinder, the maximum response is reduced by a factor of about 100. The elevation of the critical velocity through the addition of strakes can be a significant additional benefit of this spoiler device. The present results are comparable to those obtained by Scruton et al (ref. 2) in his investigations of strakes. The effectiveness of strakes is very evident here.

In Figure 3, the measured response amplitudes of a bare cylinder in smooth flow have been normalized by the observed maximum RMS amplitude of the bare cylinder in turbulent flow. This normalization shows that the reduction in maximum response of a bare cylinder due to turbulent flow alone is in excess of 2. The reduced velocity of figure 3, and subsequent plots, is based on fn, the natural frequency of the section under test.

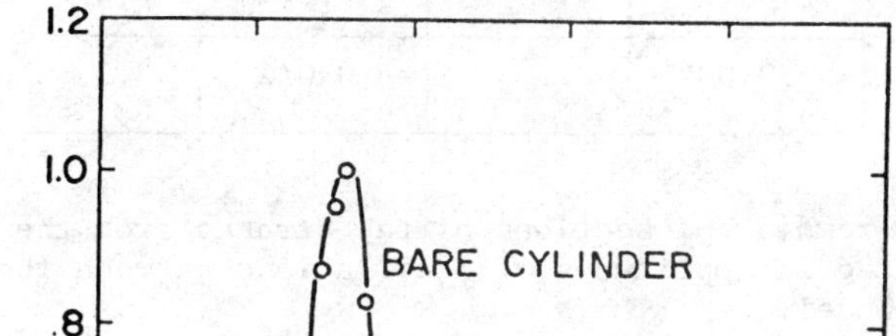

Fig. 4. The effect of strakes in turbulent flow

Figure 4 shows the results obtained with strakes in turbulent flow. These are surprising results since they show that the strakes reduce the response amplitudes of the cylinder by only a factor of about 2 in this flow, while simultaneously raising the critical speed by about 50%. It was anticipated that strakes would be less effective in turbulent flow than in smooth flow but such a large reduction in their effectiveness was not expected. When evaluated at the critical velocity of the bare cylinder, strakes reduce the response amplitudes by a factor of about 5.

Some of these results have been presented in a slightly different way for comparison with previous data. The values of C_L' as defined by equation (1) at maximum response, have been listed in Table 1 together with the peak factors observed in each case. The values of C_L' for smooth and turbulent flows on the bare cylinder are 0.13 and 0.062 respectively. The latter value is close to the figure of 0.05 deduced by Kwok (ref. 5) from model stacks oscillating with large amplitudes (above 0.6% D) in a simulated turbulent wind. When observed peak factors are applied to these two values, the peak values of C_L' become more nearly equal (0.20 and 0.14 for smooth and turbulent flow respectively) and they add support for the peak design value of 0.2 suggested in the Canadian Building Code (ref. 6) which is based on earlier studies by Davenport (ref. 7).

TABLE 1. C_L' and Peak Factor Evaluated at Maximum Oscillation Amplitude

	SMOOTH FLOW		TURBULENT FLOW	
	C_L'	Peak Factor	C_L'	Peak Factor
BARE CYLINDER	0.13	1.5	0.062	2.3
STRAKES	0.0026	2.5	0.012	3.5

As might be expected, removal of sections of the strakes from the front and rear of the cylinder resulted in amplitudes of oscillations between those for the fully straked and the bare model.

(b) Sawtooth Fins

The sawtooth fins tested (Figure 2B) are similar in many respects to those used by Baird (ref. 3) except for their size. The fins tested extend out from the cylinder for a distance of 0.2 D and are 10/12 D in length, whereas Baird's fins projected 1½ D from the cylinder and were 12 D in length.

The results, obtained in turbulent flow and plotted in Figure 5 as relative RMS amplitude R, show that the fins reduce the response of the cylinder by about 35%. These same fins placed on the top and bottom of the cylinder (rather than at the front and back) produced values of R considerably greater than one and increased the critical velocity considerably over that of the bare cylinder.

(c) Straight Fins

The straight fins tested (Figure 2C) project radially by 0.2 D at the front and rear of the cylinder. The results are shown in Figure 5. The measured values are similar to those found using the sawtooth fins with maximum amplitudes being reduced by about 30%. In smooth flow the maximum amplitude is increased over that for the bare cylinder, as shown in Figure 6.

When these straight fins were placed on the top and bottom of the cylinder, rather than at the front and back, their effect was to increase the maximum amplitudes of response. Further, larger amplitudes occurred in turbulent flow than in smooth

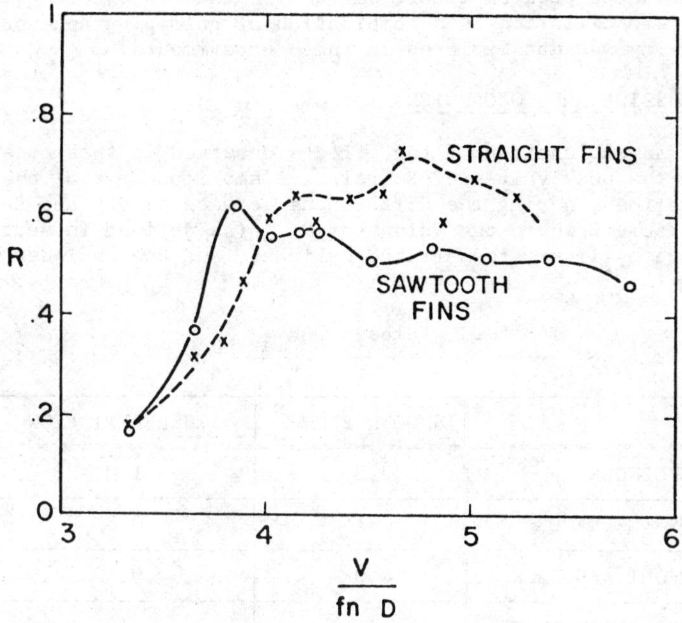

Fig. 5. The effect of fins (front and back) in turbulent flow

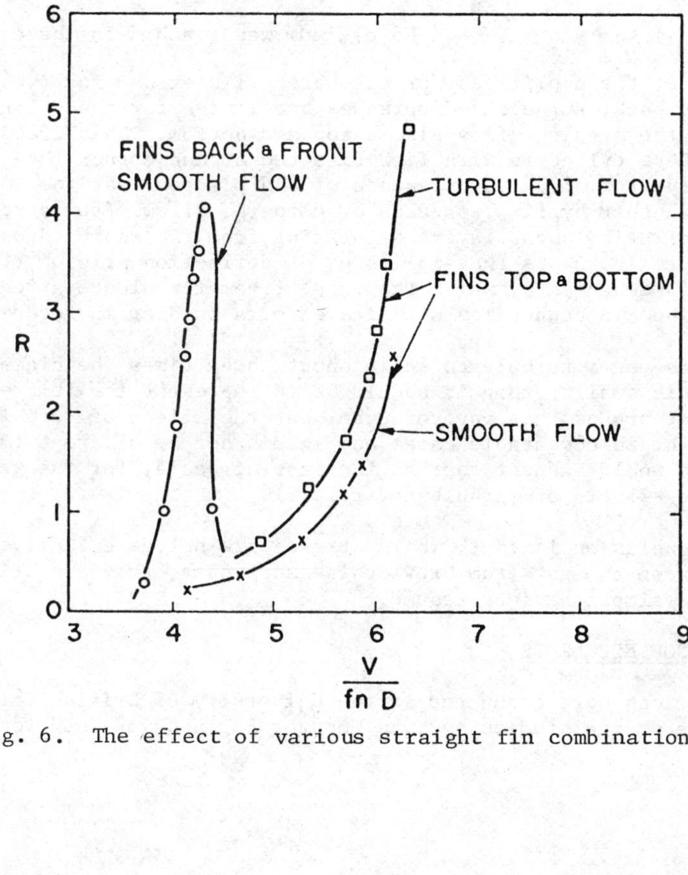

Fig. 6. The effect of various straight fin combinations

flow. The critical velocity increased considerably in both cases as shown in Figure 6, and in fact a peak amplitude was not reached in the tests. It is possible that this oscillation is a combination of galloping and vortex excitation but this possibility was not explored in these experiments.

4. DISCUSSION AND CONCLUSION

In searching for an explanation of the effects observed in these tests, it is useful to summarize the peak values of R (relative amplitude ratio) observed for the various configurations, noting the differences between smooth and turbulent flows. This is done in Table 2 where the values of R_{max} (as defined in section 2), are shown for the bare cylinder, the straked cylinder, and the cylinder with straight fins.

TABLE 2. Values of R_{max}

	SMOOTH FLOW	TURBULENT FLOW
BARE CYLINDER	2.3	1.0
STRAKES IN PLACE	0.14	0.5
FINS FRONT AND BACK	4.0	0.6
FINS TOP AND BOTTOM	2*	3*

* Values taken at $\frac{V}{fnD} = 6.1$; No maximum was reached in these tests.

Turbulence reduces the amplitudes for the bare cylinder and for the straight fins placed front and back; turbulence increases amplitudes for the cylinder with strakes and for the straight fins placed top and bottom. This trend was also observed for square cylinders with flow directed at one corner ($\alpha = 45°$) as reported by Vickery (ref. 8). The presence of sharp edges near the top and bottom of a cylinder, whether by fins, strakes or corners, allows free stream turbulence to move the oscillating shear layers separating from the sharp edges closer to the body and therefore increases the amplitudes of deflection arising from vortex excitation. The mechanism by which the free stream turbulence affects the shear layers is described in connection with its effects on drag in reference 9.

The present tests use a turbulence scale about three times the diameter of the cylinder. This is smaller than it should be to represent the full scale atmospheric turbulence around the conveyor tube, but the effect of turbulence scale is known to be slight in comparable cases and should not be of great importance here. If anything, one would suspect that an increase of scale, for the same intensity, would reduce the effects of turbulence (ref. 9).

The important conclusion is that, while strakes are not as effective a vortex spoiler as had been thought from previous measurements, they are still the most effective of the simple devices tested.

ACKNOWLEDGEMENTS

These investigations were conducted at the University of British Columbia in 1978 as a part of the design studies by Swan Wooster Engineering Inc. for a tubular conveyor bridge for Georgia Pacific Corporation's Paper Division at Toledo, Oregon. The assistance of Mr. M. Wawzonek with the experimental work is gratefully acknowledged.

REFERENCES

1. Scruton, C. and Walshe, D.E.J., A means for avoiding wind-excited oscillations of structures with circular or nearly circular cross-section, NPL Aero Report 335, October 1957.

2. Scruton, C. and Flint, A.R., Wind-exited oscillations of structures, Paper No. 6758 Institute of Civil Engineers (United Kingdom), pp. 673-702, May 1964.

3. Baird, R.C., Wind induced vibration of pipeline suspension bridge and its cure, Trans. A.S.M.E. Vol. 77, pp.797-804, 1954.

4. Surry, D., Some effects of intense turbulence on the aerodynamics of a circular cylinder at subcritical Reynolds Number, Journal of Fluid Mechanics, Vol. 52, pp. 543-563 (1972).

5. Kwok, K.C.S., Cross-wind response of structures due to displacement dependent excitations, Ph.D. Thesis at Dept. of Mechanical Engineering, Monash University, Australia, Dec. 1977.

6. National Building Code of Canada, Commentary on Part 4, Supplement No. 4, Commentary B, "Wind Loads" (1977).

7. Davenport, A.G., The wind-induced vibration of guyed and self supporting cylindrical columns, Trans. E.I.C. Vol. 3, No. 4, pp. 119-141, Dec. 1959.

8. Vickery, B.J., Fluctuating lift and drag on a long cylinder of square cross section in a smooth and in a turbulent stream, Journal of Fluid Mechanics, Vol. 25, pp. 481-494 (1966).

9. Laneville, A., Gartshore, I.S. and Parkinson, G.V., An explanation of some effects of turbulence on bluff bodies, Proc. 4th International Conference on Wind Effects on Structures, Heathrow 1975, pp. 333-341.

CABLES WIND EXCITED VIBRATIONS OF CABLE-STAYED BRIDGE

J. WIANECKI

Centre Experimental de Recherches et d'Etudes, Paris, France

SUMMARY

This paper deals with wind-induced vibrations of the stays of a large bridge spanning over the Seine River and located at about 100 km, North of Paris.

When the foundations of this bridge were near completion and during a storm in October 1976, we could observe that a 15 m/sec wind blowing athwart at 20° - 30° with regard to the longitudinal axis of the bridge was sufficient to excite and sustain the vibration of the stays up to the 3rd mode.

This wind tunnel investigation made it possible to reproduce this phenomenon and ascertain in which zone the occurence of this phenomenon was possible. The hypotheses linked to the cause of this phenomenon were laid down. This paper is concluded by the description of the provisions taken to make sure the safety of the work.

1. - SCOPE OF THE INVESTIGATION

This investigation stems from the observation of the vibration amplitudes of a cable-stayed bridge under a storm in October 1976.

The geometrical characteristics of this bridge, known as Brotonne bridge (fig. 1) spanning the Seine River half-way up Rouen and Tancarville, are as follows :

This linear structure with an overall length of 1278,40 m comprises a main central structure and two access viaducts (fig. 1). The main structure 697,50 m long is a cable stayed prestress concrete bridge, comprising a 320 m central span, two side spans 143,60 long and two partial spans for balance. The stays are displayed fanlike in the median plane of the structure and on the mast rising 70 m above the deck bridge. The left-hand bank (Rive Gauche) access viaduct, 464,40 m long, comprises one 38,90 m span, seven 58,50 m spans and one 16 m hinge console. The right-hand bank (Rive Droite) access viaduct comprises but two spans and one console for an overall length of 116,50 m.

Fig. 1 : Comprehensive view of the bridge.

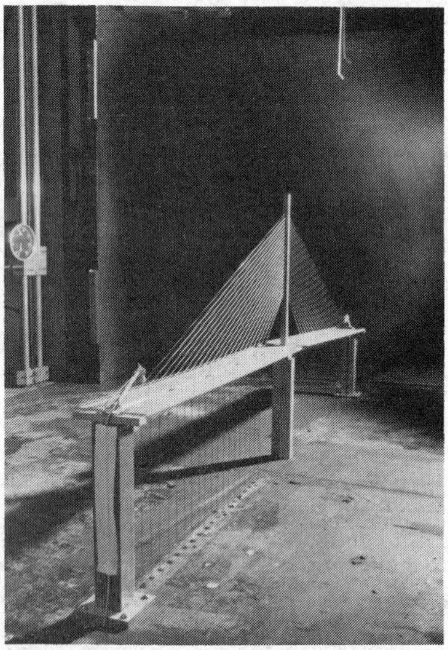

Fig. 2 : General view of the model disposed in the test section of the wind tunnel .

While the completion of the foundation work of the Brotonne bridge was in progress, we ascertained in the course of a storm in October 1976 that a 15 m/sec wind, blowing athwart at 20°- 30° with regard to the longitudinal axis of the bridge was sufficient to excite and sustain the vibration of the stays up to the 3rd mode.

According to several eye-witnesses located at about 100 m off the bridge, the vibration amplitudes were comprised between ± 30 cm for the 1st mode. For higher velocities, the vibrations decreased down to zero.

When the storm abated, the roccurence to 15 m/sec wind velocity caused vibrations again with important amplitudes decreasing in ratio with the subsiding of the wind. Each time, these vibrations lasted for about 30 min. Then the scope of the study was to examine the vibratory behaviour of the bridge stays by means of a partial 1/200 scale model in wind-tunnel to reproduce the phenomenon observed on the site and detect :

1/ the incidences of wind responsible of the stays vibrations
2/ the critical velocities
3/ the vibration modes of the various stays.

2. DESCRIPTION OF THE MODEL AND OF THE INVESTIGATION PROGRAMME

About half the length of the deck of the bridge was carried out and used as a supporting beam for the return axis of the wires representing the stays. The outside shape of the deck was respected in order to realize correctly the aerodynamic form constituted by this deck.

Conversely, the connection deck to mast foot was neglected since it was impossible to reproduce correctly both flexural and torsion rigidity of the bridge at the 1/200 scale.

The elastic components of the model were restricted to the stays represented by wires together with the mast made out of steel so as to resist a 400 daN global resultant force. Thus, the rigidity of the mast was considerably improved which nevertheless did not impede the correct reproduction of the vibration phenomenon of the stays as observed on the site.

The figures n° 2 and 3 show the ready-for test model, attached to the revolving platen in the wind tunnel. The yoke used for the application of the forces on the stays with the help of a large pneumatic jack is shown on figure n° 4.

The model was tested according to four different cases :

1/ without any scaffolding around the mast, the stays being under tension to represent a certain stage of construction (fig. 2)
2/ without any scaffolding around the mast the stays being under tension to represent the last stage of construction
3/ with scaffolding erected around the mast and tensions on the stays to represent a certain stage of construction (fig. 3)
4/ with scaffolding erected around the mast and tensions on the stays to represent the last stage of construction.
   In each case, the model was tested under seven incidences, namely : 0°, 30°, 60°, 90°, 120°, 150° and 180° for wind velocities from 1,3 m/sec up to 12 m/sec (fig. 5) and according to different cases 2/ and 3/ above mentioned.

3. SIMILITUDE SCALES

If we choose the geometrical scale of the model $K_L = 1/200$ and the material for the realization of the stays (either cables or steel wires) we get the following scales :

Fig.3 : Close up view on the scallolding erected around the mast.

Fig.4 : General view on the yoke beneath the revolving platen of the wind-tunnel.

REFERENCE MARK OF THE INCIDENCES

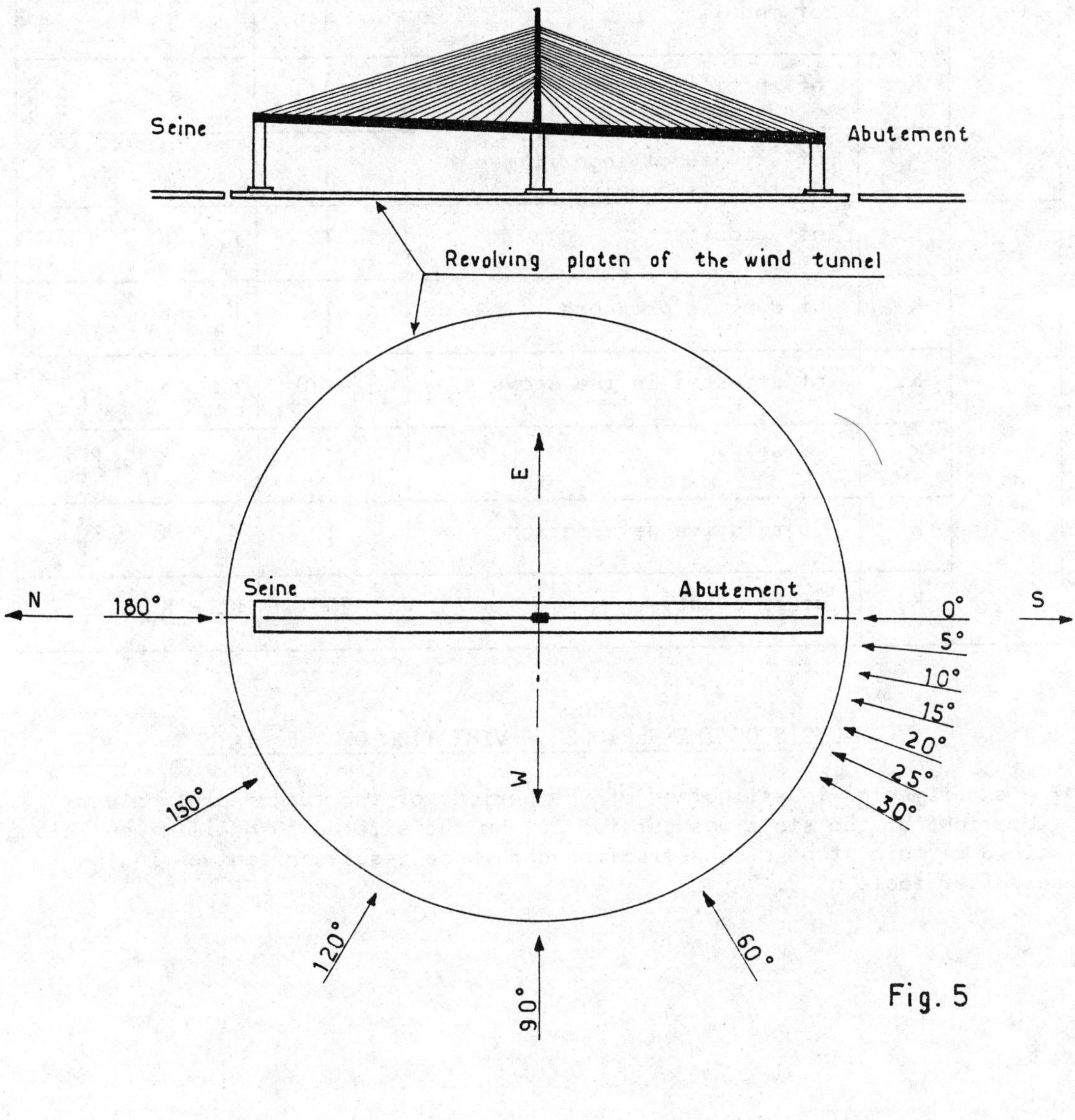

Fig. 5

Table 1

Scales	K_i	Designation	Value	Relation
selected	K_L	geometrical	1/200	-
	K_E	of moduli	1	-
	K_γ	of specific weight	1	-
	K_δ	of structural logarithmic decrements damping	1	-
derived	K_V	of velocity	1	$K_V = K_E^{1/2}$
	K_q	of dynamic pressure	1	$K_q = K_V^2 = K_E$
	K_T	of stresses in the stays	$1/4.10^4$	$K_T = K_q \cdot K_L^2$
	K_σ	of stress	1	$K_\sigma = K_q = K_E$
	K_ε	of relative deformation	1	$K_\varepsilon = K_\sigma \cdot K_E^{-1}$
	K_f	of frequency	200	$K_f = K_V \cdot K_L^{-1}$

4. ANALYSIS OF THE PERIODS OF VIBRATION OF THE STAYS

The experimental investigation of the periods of the fundamental natural vibrations of the stays was carried out on the site on 50 oscillations along with a certain stage of construction and whose results are given in the hereafter table n° 2.

Table 2

Ref. of the stay	Fundamental periods T in sec		Marking of the stays grouted and not grouted
	Abutment side	Seine side	
1	0,492	0,510	grouted
2	0,528	0,544	
3	0,514	0,516	not grouted
4	0,578	0,578	
5	0,620	0,638	
6	0,696	0,748	
7	0,798	0,764	
8	0 806	0,810	
9	0,988	1,040	grouted
10	0,930	0,900	not grouted
11	1,010	0,936	
12	1,074	1,020	
13	1,148	1,094	
14	1,336	1,27	grouted
15	1,254	1,172	not grouted
16	1,306	1,244	
17	1,336	1,288	
18	1,412	1,364	
19	1,508	1,452	
20	1,574	1,584	
21	1,602	1,646	

On the other hand, these fundamental periods were calculated from the formula in relation with the vibration of a fixed end cable :

$$T_i = \frac{2L}{i}\sqrt{\frac{m}{N}} \quad \left[\text{sec}\right]$$

where : T_i = period of vibration related to mode i
L = cable length
m = linear mass of the cable
N = traction force
i = 1,2,......n coefficient respectively corresponding to the

the 1st, 2nd, ... n^{th} mode of vibration.

The calculation of the fundamental periods e.g. related to the 1st mode of vibration was carried out for the grouted stays n° 1, 2, 9 and 14 *).

The comparative values are given in table n°3 :

Table n°3

Ref. of the stay	Fundamental period in "abutment side"		μ %
	Theoritical value	Experimental value	
1	0,505	0,492	+ 2,6
2	0,559	0,528	+ 5,9
9	0,966	0,988	- 2,2
14	1,257	1,336	- 5,9

The cross-checking of both theoritical and experimental values was quite satisfactory and made it possible to calculate the fundamental periods for the other stays after grouting.

Afterwards but the grouted stays were retained for the tests since the related damping is the weaker and represents the most unfavourable case.

5. ANALYSIS OF THE CRITICAL VELOCITIES OF WIND

The results of the calculation of the periods and the mean theoritical frequencies of the first three modes of vibrations linked to the actual stays n° 1, 9 and 21 which are respectively the shorter, the mean and the longer ones, are given in the table n° 4.

Table n°4

Ref. of the stay	Periods T_1 in sec		Mean periods in sec			Mean frenquencies in Hz		
	Abutment side	Seine side	T_1	T_2	T_3	f_1	f_2	f_3
1	0,505	0,482	0,494	0,247	0,165	2,024	4,049	6,073
9	0,966	0,957	0,962	0,481	0,321	1,040	2,079	3,115
21	1,634	1,623	1,623	0,812	0,541	0,616	1,232	1,848

*) Each stay is composed of :
- a cable with parallel strands
- a metal sheath 165 mm in diameter and 4,5 mm thick
- a grout filling the surrounding duct.

Considering a cylinder parallel with the stream, the critical velocities of wind calculated from the formula :

$$V_{cr_i} = \frac{d.f_i}{Sh}$$

where V_{cr_i} = critical velocity in m/sec corresponding to the vibration mode i

d = diameter of the cylinder equal to 0,165 m

f_i = frequency linked to the mode i

Sh = Strouhal's number equal to 0,2 for cylinders

would give the results given in table n°5.

Table n°5

Ref. of the stay	Critical velocity in m/sec		
	V_{cr_1}	V_{cr_2}	V_{cr_3}
1	1,68	3,36	5,04
9	0,86	1,72	2,58
21	0,51	1,02	1,53

This analysis was restricted to the first three modes of vibration observed "in situ".
The table n°5 points out that all the possible critical velocities for everyone of the stays are without exception comprised within the range :

$$0{,}51 \text{ m/sec} \leqslant V_{cr_i} \leqslant 5{,}04 \text{ m/sec}$$

Yet, the observed phenomenon of vibrations on the site is well beyond the maximum velocity of 5,04 m/sec since it occurs for velocities equal or superior to 15 m/sec. It means that the Strouhal's formula applied in a classical way to determine the critical velocities does not give a valuable explanation of the phenomenon observed.

6. CHECK OF THE TUNING IN FREQUENCES OF THE MODEL STAYS

The final check of a satisfactory simulation of the actual bridge through the model was obtained by checking the tuning in frequency of the model stays. The analysis of those results shows that the tuning in frequency of the model stays is satisfying. The differences arise to 7 % f r the stays 14 and 21 and 12 % for those ref. 1 to 13.

7. EXPERIMENTATION METHOD

The rule of conduct agreed upon before testing consisted to detect both amplitude of vibration and their mode for everyone of the stays in compliance with the programme defined in chapter 2, by means of optics making use of a Wild theodolit and stroboscope.

This approach proved to be a failure since the stays constituted by cables did not allow to detect the least vibration amplitudes by means of optical device while running through all the incidences and velocity levels planned in the programme. This may be explained by the shape of the 7 strand cables whose manufacturing made them antivibratil damping the vibrations together with the small geometrical scale K_L = 1/ 200.

These verifications lead to the conclusion that the cable must be replaced by a smooth steel wire which was realized for the n°21 stay only. The first tests carried out on the model thus modified led to a new trend of investigation, namely acoustics since the stay n°21 gave a sound, the proof of vibration. The original investigation method had but to be adjusted to quantify these vibrations. The analogy with the well known vibrating wire method used to measure local deformations and calculate the stresses, was the obvious thing to do. In this method, the wires (gauges and indicators) are sounded by means of an electromagnet with initially was used to actuate the wires by the flow of a capacitor discharge : this is the "damping mode". Since a decade, we actually prefer to keep the wires in vibration by providing them the small quantity of energy, with the help of a second electromagnet, necessary to balance the natural damping : this is the "forced vibration mode". In our special case, the stay is already submitted to wind induced forced vibrations. Suffice to pick them. Thanks to electronics, it is now possible to select the frequency alone, either by comparison with a steady one or by measurement. In our case, we are concerned with direct measurement of the frequency. This was possible thanks to the frequency meter "TELEMAC" whose portable sounding unit equipped with an electronic counting device, type PC 2 is shown in fig. 6. The measurement of the higher frequency is performed by means of a two-way switch-counting lapse 1 or 10 sec making it possible to respectively obtain two precisions of reading : 1 Hz or 1/10 Hz. For our measurements, we used but the first position giving a 1 Hz precision.

As concerning the location of the sounding electromagnet, they were placed on stream lined rigid brackets 90 mm long designed in such a way that the stream disturbance around the stays may be as small as possible and the measurements sufficiently accurate (fig. n°7).

Taking into account that the stay n°21 presents the lowest natural frequencies of all the stays thus the examination of this sole stay allows first to detect the smallest excitation velocities and answer the questions set in chapter 1.

For the other stays, these velocities may be slightly shifted towards higher values hence an extra information.

8. ANALYSIS OF RESULTS AND CONCLUSIONS

The analysis of the results made it possible to ascertain that the sole basic incidences for which the first three modes occur for wind velocities higher than the standard critical velocities (0,51; 1,02; 1,53 of table n°5) are 0° and 180°. Yet, the 30° and 150° incidences present a few values corresponding to the first three modes blurred by harmonics of higher order.

The 60°, 90° and 120° incidences are characterized by vibration modes higher than three. Thus, the analysis made it possible to ascertain that the stays vibration phenomenon induced by high winds, observed on the site has to be looked for within the zone determined by the ± 30°angle with regard to the axis of the bridge.

Hence the reason why the sequence of the tests pointed out by anticipation in chapter 2 consisted in examining the additional incidences : 0°, 5°, 10°, 15°, 20°, 25° and 30° for wind velocities ranging from 7,7 m/sec up to 31,0 m/sec.

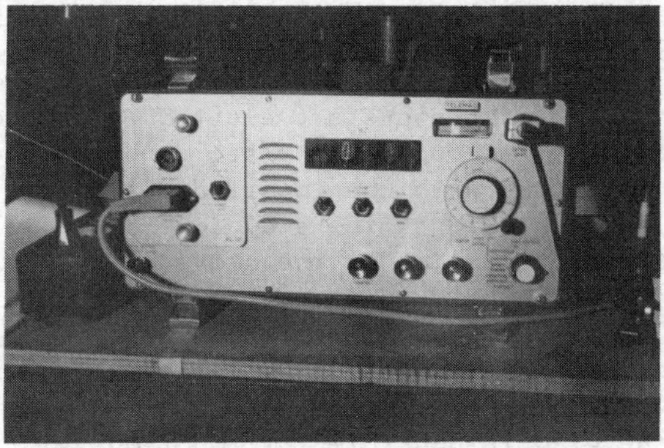

Fig. 6 : Frenquency meter "TELEMAC"

Fig. 7 : Close up view of an electromagnet sounding the stay n° 21.

Thus, the analysis of this investigation allowed to conclude that :
1°) the configuration in relation with the forces applied to the stays and linked to a certain stage of construction together with scaffoldings is more

unf:vourable than the one with final forces and without any scaffolding, since the probability to see the three modes occur is higher.
2°) the 0° incidence is dramatically perilous since its first three modes may occur for very important wind velocities up to 31/sec.
3°) the first three modes of wind are less likely to occur in proportion with the increase of the incidence angle but it is not excluded for so much.
4°) for the same incidence, the probability for these mode to appear is lessening in ratio with the increase of wind velocity.

The general conclusions are defined into two parts :
The first one is linked to the reproductibility in wind-tunnel model of the vibratory phenomenon of the stays generated by high winds observed "in situ".
The tests pointed out that the stays are excited in accordance with the first three modes by 1,3 m/sec up to 31 m/sec winds in the zone delimited by the angle ± 30° in regard with the longitudinal axis of the bridge or, in the same manner, with regard to the stay layers.
This eventuality of the excitation of the stays by high winds, according to the first three modes and linked to a very low structural damping of the stays may lead to very important vibration amplitudes as was observed on actual structures. This explains why all the stays were in vibration for 15 m/sec wind velocity.
The directions of the wind responsible for important vibration amplitudes are located within the well defined zone for our investigation.
The second conclusion is linked to the pursuit of an aerodynamic explanation of the observed phenomenon :
Two hypotheses may be formulated.
The first one assumes a relation between certain modes of vibration of the mast and the stays. In that case, it implies the vibrations of the mast perpendicular to the axis of the bridge and caused by Benard-Karman's whirlwinds; these vibrations will excite the stays according to various modes.
In the framework of our next program of research and in order to get a better approach of the problem, we intend to shield the mast of the actual model so that it may be excited by wind. The cessation of any vibration of the stays will confirm the hypothesis.
Conversely, one may formulate the second hypothesis which explains these vibrations by the geometry of the layer of stays causing interferences in the stream flow within the zone delimited by the angle ± 30° in relation with this layer. These interferences will disappear beyond this zone. It is worth noticing the smoothness of the stay sheats which may force these vibrations.
The increased disturbance caused by the scaffoldings around the mast extends the range of velocities together with the zone for which this phenomenon may take place.
Assuming the validity of this hypothesis, a spiral helix-shaped antivibratil tubing device with both a correctly defined diameter and pitch may solve this problem. The tests carried out on 1/5 and 1 scale stay models allowed to choose a helix constituted by a tubing 16,5 mm

in diameter with a pitch of 330 mm.

Actually, the solution retained for this bridge put into service on July 9, 1977 consisted in improving the damping of the stays by means of hydraulic dampers supported by steel brackets fixed on the deckbridge in the neighbouring of the stay anchorage. The principle of this device is to cause the dampers to absorb the energy generated by the vibrations without knowing their true origin. These dampers, 2 or 4 in number, are placed V-like in order to stop both horizontal and vertical vibrations. A full test carried out on one of the stays of the actual bridge proved to be reliable. This antivibratil device was placed on the whole structure for safety, pending a more aesthetic and as much reliable solution based on the knowledge of the exact cause of these vibrations.

DISCUSSION ON SESSION X
Wind-Engineering Practice

Rapporteur: R. A. Parmelee
Northwestern University
USA

Paper by R. I. Harris--THE CLASSIFICATION OF STRUCTURES FOR THE ASSESSMENT AND CODIFICATION OF WIND EFFECTS

Question: H. Sockel. You give formulas for the performance parameter R for special structures. For structure I, for instance, it is found $R = 1 + f$ and there is a formula for f. Where can I find details? What is the corresponding reference?

Reply: R. I. Harris. Details of the power series and integral functions representing f have not been published, but should be sought from the author.

Comment: H. Söckel. N should not properly be called Strouhal number, it is a nondimensional frequency, whereas the Strouhal number is always based on the cortex frequency.

Reply: R. I. Harris. There is a body of opinion that the parameter referred to should be properly called a type of Strouhal number, but the point is unimportant and a designation of reduced frequency is satisfactory.

Paper by M. Ito, Y. Fujino--DESIGN WIND SPEED AND WIND LOAD FACTOR BASED ON PROBABILISTIC RATIONALE

Question: R. A. Hazelwood. Referring to Figure 7, the highest point of 42 m/s occurs at a 90-year return period. How is this computed from data over only 37 years?

Reply: M. Ito, Y. Fujino. In plotting the station data of the annual maximum wind speed on a Type I distribution paper, various methods have been proposed by Hazen, Gumbel, Gringorten, etc. Among these, the authors used Hazen's method according to the suggestion by H. Kikuchihara, JMA; that is, the probability value for ith data will be $p^i = (i - \frac{1}{2})/N$. Accordingly, if the total number of data, N, is taken 37, the first point on the figure corresponds to the return period of 74 years, because $p_1 = 1/74$.

Question: K. C. Anthony. In those cases of identified mixed distributions, was any trend found which would confirm the mixed distribution expression advocated by Thom in Reference 1, i.e., parameters of 4.5 and 9?

Reply: M. Ito, Y. Fujino. When the annual maximum wind speed data at Osaka (Figure 7) were fitted to the mixed Type II distribution proposed by Thom, namely

$$F(\nu) = P \exp\{-(U_1/\nu)^{k_1}\} + (1-P) \exp\{-(U_2/\nu)^{k_2}\},$$

the values of the parameters became as follows:

$$P \doteqdot 0.9,\ U_1 \doteqdot U_2 \doteqdot 14,\ k_1 \doteqdot 6 \sim 6.5,\ k_2 \doteqdot 2 \sim 3.$$

The second term of the right-hand side of the above equation corresponds to the winds caused by severe typhoons. The values of k_1 and k_2 shown above seem discrepant from those in Reference 1. Since the authors are not meteorologists, it is hard to say whether these discrepancies are due to the difference of local wind conditions or to other causes.

Paper by E. Simiu, J. R. Shaver--WIND LOADING AND RELIABILITY-BASED DESIGN

Question: T. Tschanz. To determine a safety factor one would have to consider the building in the nonlinear range. You mentioned that some adjustments could be made, but that it would be very difficult, e.g., to make a nonlinear model for a wind tunnel test. I would like to ask what suggestions the authors have to obtain estimates for the building response in the large displacement range.

Reply: E. Simiu, J. R. Shaver. Mr. Tschanz is asking an important but almost impossibly difficult question. There is much to be learned in this area from earthquake engineers. Possible approaches include: stepwise nonlinear analyses, strength reserve techniques, and inferences based on testing and on the analysis of damage. Some results are already available and consist of ratios of e.g., failure load to load inducing yield. Wind tunnel tests could perhaps be based on various assumptions concerning such ratios.

Question: V. R. Cartelli. Although buildings have been designed adequately and safely against failure, there have been instances where flexibility and vibrations have affected the occupants, as well as interior partitions. Will you please comment on this aspect.

Reply: E. Simiu, J. R. Shaver. Mr. Cartelli's observation is correct. Generally speaking, structural reliability theory is concerned with the probability of the event that certain performance criteria will fail to be satisfied. Among these performance criteria are those dealing with serviceability. The techniques used in this case are similar to those used in the analysis of structural failure.

Paper by E. Simiu, J. R. Shaver--WIND LOADING AND RELIABILITY-BASED DESIGN
Paper by M. Ito, Y. Fujino--DESIGN WIND SPEED AND WIND LOAD FACTOR BASED ON PROBABILISTIC RATIONALE

Comment: K. Rojiani. We performed a reliability analysis of a 25-story steel frame building subjected of wind loading. One of our conclusions was that the reliability was quite sensitive to the form of the distribution, being typically much higher for the Type I extreme value distribution than the Type II extreme value distribution. As regards uncertainties in the estimation of the distribution parameters our results indicate that the effect of sample size (number of years of record) on the reliability is quite significant when the sample size is small ($n \leq 10$). The results presented here seem to confirm our results. We also found that failure probabilities obtained when uncertainties in the wind environment parameters and structural properties were included were several orders of magnitude higher than when these uncertainties were neglected indicating that these uncertainties do indeed have a significant effect and should be considered in the reliability analysis.

Paper by J. R. Mayne, N. J. Cook--ACQUISITION, ANALYSIS AND APPLICATION OF WIND LOADING DATA

Question: W. Z. Sadeh. I am wondering about how you suggest to assess the statistical independence. Two random processes are statistically independent if their JPDF is given by the product of their individual PDFs. We have not been able to check this independence - it would require a data length that includes all the marco-peak - i.e., seven weeks long. Our rational visualizes the full-scale as a large wind tunnel in which the speed is varying at least one order of magnitude slower in frequency than the largest micro-range eddy. Because this variation is so slow the micro-range effects are effectively quasi-static.

Paper by K. C. Mehta--WIND LOAD STANDARDS AND CODES

Question: A. H. Dutt. I very much welcome your proposal for the formulation of an International Standard through ISO in order to reduce disparities between the national standards. Perhaps ISO can also help to stimulate the updating of the wind standards of the various countries in Asia and Africa, where wind hazard is a very serious problem causing several billion dollar disasters every year. I would be glad to receive your comments on this matter for implementation.

Reply: K. C. Mehta. Process of standardization is a long process because provisions of the standards should be prepared with due care. There is a continuing trend to be aware of the work accomplished in other parts of the world by the standard making bodies. The trend is likely to result into more consistent standards in various countries around the world. The ultimate goal is to prepare one international standard applicable throughout the world; this goal may be a long way in the future.

Question: V. R. Cartelli. In Table 4, ANSI Code neglects special items such as partition walls. Where can a structural engineer get some direction regarding such items. Please comment.

Reply: K. C. Mehta. Some information on special items such as this can be obtained from the literature such as commentary on Canadian Building Code or Wind Loading Handbook published by Newberry and Eaton. There is a move to publish a Design Guideline for Wind Loads in the United States. It may be a year before it is likely to be published. This document should be very helpful to the practicing engineers.

Paper by G. R. Walker, J. E. Minor--CYCLONE TRACY IN RETROSPECT: A REVIEW OF ITS IMPACT ON THE AUSTRALIAN COMMUNITY

Question: C. T. G. Bubb. The authors distill the Austalian experience into the rule adopted in some areas after TRACY namely "In tropical cyclone prone areas housing should be structurally designed yet - traditional housing is not structurally designed at all." It is potentially disastrous to exclude a profession, in this case structural engineering from a certain field in this case housing. The Queensland coastal situation would still largely fall into this situation. What is the situation in USA in the major hurricane risk areas for housing?

Reply: G. R. Walker, J. E. Minor. The projections of catastrophe by Dr. White and of hundreds of billions of dollars of future building damage by Dr. Wiggins are realistic. A large part of the basis for these projections is related to housing. In the USA housing in hurricane prone regions is poorly postured to resist hurricane winds because of the general absence of structural engineering input. While I believe this statement to be true, many would argue that coastal areas are adequately postured for hurricane effects in an insurance context, i.e., the U.S. economy would absorb the costs of a disaster such as occurred in Darwin through insurance mechanisms. The fundamental question of imposing structural engineering input to housing in the USA is thus seen to relate directly to considerations of insurance. In my opinion we cannot continue to absorb wind damage costs in our economy in this fashion; therefore, the structural engineer must become involved in housing design.

Paper by A. P. Jeary, B. R. Ellis--THE RESPONSE OF A 190 METRE TALL BUILDING AND THE RAMIFICATIONS FOR THE PREDICTION OF BEHAVIOR CAUSED BY WIND LOADING

Question: T. Tschanz. For Figure 6, we can see that the damping from a wind excited record was .51% which is close to damping values from forced excitations. Would you conclude that aerodynamic damping is negligible for this building?

Reply: A. P. Jeary, B. R. Ellis. Yes I would, but these tests were all conducted at low amplitude and I would not like to state categorically what this situation would obtain at a very high amplitude. However, the separation of aerodynamic damping from increasing structural damping would be difficult were measurements to be made.

Question: K. Handa. How much was the contribution from the second mode to the total response?

Reply: A. P. Jeary, B. R. Ellis. The second mode in frequency content was a torsional one, and the contribution in terms of displacement was never greater than 10% of the total, and, as an average, was two orders of magnitude less for the fundamental modes. In any one direction then it is seldom the case that this amounts to greater than 2% of the total response if it is averaged. However, on a short-term basis for some structures it is possible to see a short period (from time to time) of the second translational mode being dominant (reference an 11-story office block in Sutton, England).

Question: J. W. Saunders. Your results show the very conservative design of tall buildings using current design techniques (for most buildings). This study highlights the need for parametric studies to fill the gap between code information for framing design for crosswind, torsional and buffeting response to make proper design allowance at an early stage compared with using a wind tunnel. This conference has shown that the scientific tools are now available to do this.

Reply: A. P. Jeary, B. R. Ellis. Yes, I would agree with your basic tenet (although in some cases the code predictors give very good answers), but I would make a plea for a greater effort in full-scale measurement. I think that our knowledge of the manner in which buildings behave can only be increased accurately in this manner.

Paper by I. S. Gartshore, J. Khanna, S. Laccinole--THE EFFECTIVENESS OF VORTEX SPOILERS ON A CIRCULAR CYLINDER IN SMOOTH AND TURBULENT FLOW

Question: H. Ruscheweyh. I am very surprised about your results on strakes in turbulent flow. Your results are opposite to full-scale observations, where the strakes work well in any case. Have you measured the turbulence spectra in respect to frequency peaks to be sure that there is no resonance buffeting? What is the width of the lattices of your turbulence grid?

Reply: I. S. Gartshore, J. Khanna, S. Laccinole. The spectra of turbulence fluctuations was not measured in this case but since the turbulence was grid generated and the grid was about 5 mesh widths upstream of the model, it seems unlikely that the spectrum would have unusual spikes. The grid mesh was three times the basic tube diameter.

Question: W. Z. Sadeh. I am wondering as to whether the characteristics of the incident turbulence were accounted for in determining the response for the different configurations. I suspect that the results presented depend upon the incident turbulence features.

Reply: I. S. Gartshore, J. Khanna, S. Laccinole. More accurately modeled turbulence would have been ideal but more difficult to create. Grid turbulence is frequently used as a means to test bridge sections (see recent papers in this conference) and is at least representative of turbulent atmospheric conditions.

Question: J. L. Wood. The authors may be interested to know that practical trials on high aspect ratio tapered octagonal section street lighting poles showed no significant improvement due to fitting of helical strakes. Column heights 8 m and 10 m cross section dimension 75 mm to 100 mm. Is there some threshold of structure size below which spoilers do not work?

Reply: I. S. Gartshore, J. Khanna, S. Laccinole. We do not know of a threshold below which the spoilers will not work since this has not been rigorously pursued in the laboratory. We suggest that in the case of the octagonal cross section, the geometry may have a far more significant effect on the strake effectiveness than the size.

Appendix 1

Minutes of the IAWE Steering Committee

Fort Collins, Colorado

5:30 - 7:30 p.m., 9 July 1979

Chairman: J. E. CERMAK

Recording Secretary: R. H. SCANLAN

In keeping with its responsibility to provide continuity of organization and planning for the International Conferences on Wind Engineering, the IAWE Steering Committee assembled for a meeting on the evening of 9 July 1979 at Colorado State University. The meeting was opened by Dr. Cermak with a welcome to those present (Exhibit A) and an expression of gratitude for assistance and guidance provided by the Steering Committee during four years of planning and preparation for the Fifth Conference. Discussions and actions by the Steering Committee were as follows:

1. Minutes of the 1975 IAWE Steering Committee meeting at Heathrow Hotel on 10 September 1975 were reviewed and approved.

2. The working agenda proposed by the Chairman was adopted.

3. The following brief reports by Regional Secretaries and other organizational representatives were presented:

   van Koten (Regional Secretary - Europe) commented that IAWE had supported the 1977 meeting at Aachen and is supporting the IUTAM meeting at Karlsruhe in September 1979. The need for international coordination of the dates of wind-engineering-related meetings was emphasized. As an example, the recent four-day meeting on Environmental Forces on Structures, held in London was cited.

   Melbourne (Regional Secretary - Australasia) pointed out that at this time no national organizations exist in Australasia with which IAWE can interact.

   Sfintesco (UNESCO contact) suggested that IAWE-UNESCO contacts be strengthened. The lack of formal organization within IAWE as an official medium of international communication have been barriers to UNESCO recognition.

   > *It was moved, seconded, and carried that Dr. Sfintesco be authorized by the IAWE Steering Committee to pursue the question of further and closer contact with UNESCO.*

   Teunissen (CWEA Chairman) reported on the activities of the Canadian Wind Engineering Association. An organizational difficulty is identification of active workers willing to devote substantial amounts of time to CWEA projects. For example, more assistance is needed for the newsletter "CWEA News", which reports activity on Canadian wind-engineering projects, including those on wind power. CWEA sponsored a workshop on Wind Engineering in September 1978 in Varennes, Quebec.

4. The Chairman reported as follows: Attendance at the Fifth Conference exceeded 275 at the date of reporting (later this exceeded 340).

   The financial support of the Fifth Conference, based on registration fees, and support by NSF appears adequate to cover all costs and allowed CSU to avoid any losses on the week's activities. The Conference expenses were covered approximately in the following fractions: 3/7 from NSF, 3/7 from the fees paid by conferees, and 1/7 from private and industrial sources.

   The Conference Proceedings, in edited form with discussions and papers by Session Chairman, will be published in two volumes by Pergamon Press. The anticipated price will be approximately $75; however, Conference registrants will receive a copy without additional charge.

   Commentaries that had been returned by Steering Committee members in response to a solicitation of the Chairman were briefly reviewed and acknowledged as contributions to the success of the Conference organization and format.

   The Conference Organizing Committee with the assistance of external reviewers selected some 100 papers out of 250 complete papers submitted for presentation. The Organizing Committee decided to not include a Technical Session devoted to wind-power generation at the present conference. This decision resulted from the desire, after having added a new Technical Session on Social and Economic Impact of Wind Storms, to avoid parallel sessions. However, the desirability of including the topic of wind power in future conferences was emphasized.

   The need to consider some legal form for the organization and/or the establishment of a set of by-laws was proposed for discussion. It was emphasized that the principal activity of the IAWE, however, is to keep the International Conference alive at four-year intervals.

5. The location for the 1983 meeting of the Sixth International Conference on Wind Engineering was discussed. The letter of Melbourne proposing that the conference be held in Australia, was given consideration.

   It was moved, seconded, and carried that the Sixth International Conference on Wind Engineering will be held in Australia during 1983 under the general chairmanship of Dr. W. H. Melbourne.

   A poll of the Fifth Conference conferees was instituted to inquire as to the most desired calendar period for the Sixth Conference. This was later established at July-August 1983, with exact dates to be set by the Conference Organizing Committee.

   Preliminary suggestions for the Australiasian conference travel plan included one week of pre-conference meetings or visits, one week of conference, and one week of post-conference activity. The general location suggested for the Sixth Conference is in the near neighborhood of Brisbane, Australia.

6. The following Regional Secretaries were approved: Dr. H. van Koten for Europe, Professor R. A. Parmelee for the Americas, with other Regional Secretaries to remain the same as in the past (except as noted subsequently).

7. The question of formulating by-laws for the IAWE was discussed.

Ito noted that a first draft of proposed by-laws was made in 1971 at Tokyo during the Third Conference. These were offered to, and accepted by, the Chairman.

Holmes commented that by-laws can be minimal if the principal purpose of IAWE is to keep the International Wind Conferences alive.

Eaton emphasized that organization for the Fourth Conference in 1975 had been largely informal.

Mayne remarked that the United Kingdom was somewhat over-represented on the Steering Committee because of the carryover of tenureship of committee places by the previous host country.

Cermak suggested that the Conference Organizing Committee of the next host country should be *ex officio* on the Steering Committee.

Dalgliesh emphasized the need for an "overlap" or carryover provision in Steering Committee membership to provide continuity.

Melbourne spoke of the four-year lead time for preparation of the next conference and inquired whether a provision related to this question should be included in the conference by-laws.

Cermak mentioned that the question had been posed on numerous occasions as to what exactly is considered to constitute membership in the IAWE.

Wardlaw returned to the point that formalization of the IAWE gains importance only if it undertakes other activity than the continuance of the International Conferences.

Walker suggested that a listing of fields of interest of IAWE and the organization's relation to them should be considered in any future organizational format.

Robertson emphasized the need for some organizational formality.

Panofsky suggested that procedure toward organizational form, while important, is not of the highest urgency. He therefore suggested that a mail vote of members could be taken on matters proposed for their consideration.

A motion was made, seconded and carried to the effect that:

a) The Chairman and Secretary shall prepare guidelines, of the type of by-laws, for operation of the IAWE, and that these shall be circulated to the Steering Committee membership for discussion and eventual approval.

b) Membership in the IAWE shall be open and without fee to institutions and individuals.

8. Foreseeable activities of IAWE during the next four years, 1979-1983, were discussed.

Dutt proposed that the following list of possible activities be considered:

1. Establishment of a Wind Engineering Research Federation.
2. Coordination on an international scale of research in progress.
3. The design of a bachelor of science degree in wind engineering.
4. Development of an international code of practice in wind engineering.

5. Organization of bi-annual symposia.
6. Publishing of a monthly news bulletin.
7. Establishment of a formal organization for IAWE.
8. Establishment of links with other societies.

General reactions to this proposal expressed by comments of the Steering Committee were as follows:

a) it would take considerable financial support to accomplish many of the listed objectives, and that sources for this support are not available at this time,
b) that the group generally was reluctant to materially broaden its scope of activities, and
c) that the problem of an official journal for IAWE might be solved by approaching an existing journal. *Cermak* stated that the Journal of Industrial Aerodynamics would be an appropriate international journal and that he thought the publisher and editor would look with favor upon a proposal to become identified with IAWE.

> *A motion was offered, seconded and carried to the effect that the editor of the Journal of Industrial Aerodynamics be approached by the Chairman, Professor Cermak, with a view toward its addition of an appropriate masthead statement representing it as an "official" journal of the IAWE, the exact form of the relationship to be worked out by the new Chairman, Dr. Melbourne, and the Editor.*

9. A number of other business items were presented and discussed.

Ito presented a copy of a draft letter (Exhibit B) from Professor Hirai, of the Committee on Wind Loading of the Japan Society of Steel Construction. This letter solicited new or revised wind-code information, from any source, to be considered for inclusion in a World List of Wind Resistant Design Regulations which will be published an updated edition of a version already published in Japan in 1975.

Several committee members endorsed the value and need of such a document worldwide.

> *It was moved, seconded and carried that Professors Ito's and Hirai's efforts and initiative regarding the establishment of an up-to-date document on worldwide wind codes be endorsed by the IAWE.*

Mayne pointed out that there exists extensive and useful meteorological wind data on the United Kingdom available through R. H. Collingbourne, Meteorological Office, Bracknell, Berkshire, United Kingdom. *Melbourne* agreed to approach Dr. Collingbourne as to the general availability of these data.

> *It was moved, seconded and carried that Dr. G. R. Walker be named Regional Secretary, representing Australasia, of the IAWE.*

A request was made that minutes of the present Steering Committee Meeting be mailed to all conferees. The Chairman noted that the minutes will appear as an Appendix in the Proceedings of the Fifth Conference.

Bubb suggested that IAWE should come to a decision whether or not the topic of wind energy is to be included within the purview of IAWE and be included in future meetings.

The general and emerging importance of the wind power area was again emphasized by several committee members.

> *It was moved, seconded and carried that the topic of wind energy be included on the program of the Sixth Conference in Australia.*
>
> *It was moved, seconded, and carried that a plaque congratulating Dr. D. Sfintesco on the occasion of his 70th birthday, which occurred during the Fifth Conference at Colorado State University, be offered by the Chairman to Dr. Sfintesco, the plaque to read as follows:*
>
> > *Presented to Dr. Deliu Sfintesco on his 70th birthday in recognition of his contributions to development of the European Recommendations on Wind Effects of Buildings and Structures.*

10. The meeting was adjourned at 7:30 p.m., 9 July 1979.

EXHIBIT A

Attendance List -- IAWE Steering Committee

9 July 1979

Colorado State University

Chairman	
Dr. J. E. Cermak	Colorado State University, USA
Meeting Secretary	
Dr. R. H. Scanlan	Princeton University, USA
Mr. K. C. Anthony	Ove Arup and Partners, United Kingdom
Mr. A. J. Bowen	University of Canterbury, New Zealand
Mr. C. T. J. Bubb	Australian Department of Housing and Construction, Australia
Dr. W. A. Dalgliesh	National Research Council, Canada
Dr. A. J. Dutt	Mara Institute of Technology, Malaysia
Dr. K. J. Eaton	Building Research Establishment, United Kingdom
Dr. J. D. Holmes	James Cook University, Australia
Dr. H. Ishizaki	Kyoto University, Japan
Professor M. Ito	University of Tokyo, Japan
Mr. J. R. Mayne	Building Research Establishment, United Kingdom
Dr. W. H. Melbourne	Monash University, Australia
Dr. H. Panofsky	Pennsylvania State University, USA
Professor G. V. Parkinson	University of British Columbia, Canada
Dr. R. A. Parmelee	Northwestern University, USA
Mr. L. E. Robertson	Skilling, Helle, Christiansen, Robertson, USA
Dr. H. Ruscheweyh	RWTH Aachen, West Germany
Dr. D. Sfintesco	CTICM, France
Dr. H. Teunissen	Atmospheric Environment Service, Canada
Mr. H. van Koten	TNO, Netherlands
Dr. B. J. Vickery	University of Western Ontario, Canada
Dr. G. R. Walker	James Cook University, Australia
Mr. R. L. Wardlaw	National Research Council, Canada
Mr. R. E. Whitbread	National Maritime Institute, United Kingdom
Mr. B. G. Wiren	National Swedish Institute for Building Research, Sweden

EXHIBIT B

Dear

This letter is a respectful request for your contribution. A World List of Wind Resistant Design Regulations 1975 was published from the Association for Science Documents Information, Japan, in compliance with the resolution adopted at the steering meeting of the Third International Conference on Wind Effects on Buildings and Structures held in Tokyo, 1971.

On the resolution it was also requested that the World List would be revised in due course of time when any newly revised versions of wind resistant design regulations were issued in member's countries.

It would be highly appreciated you send copies of new or newly revised code of practice to the Committee of Wind Loading, Society of Steel Construction of Japan, Shin-Tokyo Bldg. 848, 3-3-1, Marunouchi, Chiyoda-ku, Tokyo, Japan. Particularly we wish to add the codes which were not listed in the previous edition, such as the wind-resistant regulations for transmission towers and other special structures, and also wish to have the revised regulations, the old one of which was printed in the previous edition.

The collected documents will be published as a supplement of the above-mentioned World List by the Japanese Group, which will enhance the mutual understanding among the peoples who study wind effects on buildings and structures. Your cooperation will be most gratefully acknowledged in its published version.

Yours sincerely,

Atsushi Hirai
Chairman
The Committee of Wind Loading
JSSC

Appendix 2

Conference Organization

Organizing Committee

Dr. Jack E. Cermak, Chairman[1]
Dr. Robert N. Meroney[1]
Dr. Jon A. Peterka[1]
Dr. Ronald L. Petersen[1]
Dr. Willy Z. Sadeh[1]
Professor Virgil A. Sandborn[1]
Dr. M. D. Vanderbilt[2]

Guest Program Committee

Mrs. V. A. Sandborn, Chairwoman
Mrs. J. E. Cermak
Mrs. R. N. Meroney
Mrs. J. A. Peterka
Mrs. W. Z. Sadeh

Support Staff

Mrs. Edith L. Kruse, Conference Secretary
Mrs. Helen V. Sitzman, Secretary[1]
Mr. J. A. Garrison, Coordinator[3]
Mr. D. M. Joseph, Administrative Assistant[1]
Mr. W. B. Bartholomew, Assistant Director[4]

Paper Selection

The Organizing Committee acknowledges with gratitude vital contributions to the paper selection process by a host of reviewers.

---

[1] Fluid Mechanics and Wind Engineering Program, Colorado State University
[2] Structures Program, Colorado State University
[3] Fluid Dynamics and Diffusion Laboratory, Colorado State University
[4] Conference Services, Colorado State University

Appendix 3

Conference Delegates

ARGENTINA

Dr. V. Barros
Department of Meteorology
Centro Nacional Patagonico
Puerto Madryn 912O

AUSTRALIA

Dr. R. M. Aynsley
Department of Architectural Science
University of Sydney
Sydney, N.S.W. 2006

C. T. J. Bubb
Australian Department of Housing
and Construction
Canberra 2611

Dr. J. D. Holmes
Department of Engineering
James Cook University
Townsville, Q4814

Dr. K. T. Kavanagh
Department of Civil Engineering
University of Western Australia
Nedlands 6009

Dr. K. C. S. Kwok
Department of Civil Engineering
University of Sydney
Sydney, N.S.W. 2006

Dr. W. H. Melbourne
Department of Mechanical Engineering
Monash University
Clayton, Victoria 3168

Dr. J. W. Saunders
Mechanical and Production Engineering
Royal Melbourne Institute of Technology
Box 2476 V
Melbourne, Victoria 3001

Dr. G. R. Walker
Department of Civil Engineering
James Cook University
Townsville, Q4814

AUSTRIA

Dr. G. Kiesselbach
Technical University of Vienna
Vienna, A-1040

Dr. H. Sockel
Institut für Strömungslehre
Der Technischen Universität Wien
Karlsplatz 13
Vienna, A-1040

BELGIUM

Professor J. E. Berlamont
Department of Civil Engineering
University of Leuven
Park van Arenbert de Croylaan 2
3030 Heverlee

P. Spehl
S.E.C.O.
1050 Brussels

CANADA

K. R. Cooper
National Research Council
Montreal Road
Ottawa, Ontario K1A OR6

Dr. I. G. Currie
University of Toronto
Toronto, Ontario

W. A. Dalgliesh
Division for Building Research
National Research Council
Montreal Road
Ottawa, Ontario K1A OR6

Dr. A. G. Davenport
Boundary Layer Wind Tunnel Laboratory
University of Western Ontario
London, Ontario N6A 5B9

Dr. A. Davies
Morrison Hershfield Theakston and Rowan Ltd.
Guelph, Ontario N1H 4S5

Professor B. Etkin
Faculty of Applied Science and Engineering
University of Toronto
Toronto, Ontario M5S 1A4

Dr. R. G. J. Flay
Atmospheric Environment Service
4905 Dufferin Street
Downsview, Ontario M3H ST4

Professor J. Gardner
University of Ottawa
Ottawa, Ontario

Dr. I. S. Gartshore
Mechanical Engineering Department
University of British Columbia
Vancouver, B.C. V6T 1W5

P. N. Georgiou
Boundary Layer Wind Tunnel Laboratory
University of Western Ontario
London, Ontario N6A 5B9

Dr. H. P. A. H. Irwin
M2, Room 215
National Research Council
Montreal Road
Ottawa, Ontario K1A OR6

Dr. N. Isyumov
Boundary Layer Wind Tunnel Laboratory
University of Western Ontario
London, Ontario N6A 5B9

Dr. R. J. Kind
Carleton University
Ottawa, Ontario K1V 7S1

Professor A. Laneville
Department of Mechanical Engineering
University of Sherbrooke
Sherbrooke, Quebec J1K 2R1

O. Neiman
Department of Mechanical Engineering
University of Alberta
Edmonton, Alberta

Professor G. V. Parkinson
Department of Mechanical Engineering
University of British Columbia
Vancouver, B. C. V6T 1W5

Dr. T. Stathopoulos
Center for Building Studies
Concordia University
Montreal, Quebec H3G 1M8

Dr. D. Surry
Boundary Layer Wind Tunnel Laboratory
University of Western Ontario
London, Ontario N6A 5B9

Dr. H. Tanaka
Faculty of Science and Engineering
University of Ottawa
Ottawa, Ontario

J. T. Templin
Division of Building Research
National Research Council
Montreal Road
Ottawa, Ontario K1A OR6

Dr. H. Teunissen
Atmospheric Environment Service
4905 Dufferin Street
Downsview, Ontario M3H 5T4

Dr. B. V. Tryggvason
Boundary Layer Wind Tunnel Laboratory
University of Western Ontario
London, Ontario N6A 5B9

T. Tschanz
Boundary Layer Wind Tunnel Laboratory
University of Western Ontario
London, Ontario N6A 5B9

CANADA (continued)

Dr. B. J. Vickery
Boundary Layer Wind Tunnel Laboratory
University of Western Ontario
London, Ontario N6A 5B9

R. L. Wardlaw
National Aeronautical Laboratory
National Research Council
Montreal Road
Ottawa, Ontario K1A OR6

M. A. Wawzonek
297 Erie Avenue
Brantford, Ontario N3S 2H7

Dr. C. D. Williams
National Research Council
Montreal Road
Ottawa, Ontario K1A OR6

Professor D. J. Wilson
Department of Mechanical Engineering
University of Alberta
Edmonton 7, Alberta

G. Winkel
Department of Mechanical Engineering
University of Alberta
Edmonton 7, Alberta

DENMARK

E. T. D. Bjerregaard
Danish Ship Research Laboratory
2800 Lyngby

Dr. S. Frandsen
Research Establishment Risoe
4000 Roskilde

FRANCE

P. Duchêne-Marullaz
Centre Scientifique et Technique
du Batiment
11 rue Henri Picherit
Nantes 43000

Dr. J. Gandemer
Centre Scientifique et Technique
du Batiment
11 rue Henri Picherit
Nantes 43000

J. F. Sageau
Electricite de France
6, Quai Watier
78400 Chatou

Dr. D. Sfintesco
C.T.I.C.M.
20, rue Jean Jaures
92807 Puteaux

P. Sonneville
Electricite de France
6, Quai Watier
78400 Chatou

GERMANY (WEST)

Dr. G. Berz
Munich Reinsurance
Munich 40, D-8000

Dr. W. Hanenkamp
Ruhr University
Bochum

Dr.-Ing. C. Kramer
Strömungslabor der FH Aachen
Goethestr. 1,
D-5100 Aachen

Dr.-Ing. H. Pröpper
in Hochtief AG
Huyssenalle 22-30
4300 Essen 1

Dr.-Ing. H. Ruscheweyh
Institut für Leichtbau
RWTH Aachen
D-5100 Aachen

W. Schnabel
Institut Wasserbau III
Universität Karlsruhe
Karlsruhe

J. W. H. Welsch
Ruhr-Universität
4630 Bochum

Dr. K. Zilch
Institut für Massivbau
Technische Hochschule
6100 Darmstadt

HONG KONG

Dr. L. C. H. Lam
Department of Civil Engineering
University of Hong Kong

Dr. R. Lam
Department of Civil Engineering
University of Hong Kong

ISRAEL

Dr. M. Poreh
Environmental Engineering
Technion City
32000, Haifa

Professor M. Reiss
Technion
Israel Institute of Technology
Technion City, Haifa 32000

ITALY

Dr. G. Solari
Instituto di Scienza delle Costruzioni
Universita di Genova
Via Montallegro 1
16145 Genova

JAPAN

H. Akiyama
Honshu-Shikoku Bridge Authority
Tokyo

H. ChangKoo
Naigai Engineering Consul Company
Kyoto

H. Deno
Hitachi Shipbuilding and Engineering Company, Ltd.
Hiroshima 722

Dr. S. Gotoh
Department of Architecture and Building Sciences
Yokohama National University
Yokohama

Y. Higami
Ishikawagima Harima Heavy Industries, Ltd.
Tokyo

K. Higuchi
Waseda University
Tokyo

M. Ikenouchi
Akishima Laboratory
Mitsui Engineering and Shipbuilding Company, Ltd.
926 Ogami
Akishima, Tokyo 196

Dr. H. Ishizaki
Disaster Prevention Research Institute
Kyoto University
Kyoto

Dr. M. Ito
University of Tokyo
Tokyo

Y. Iwasa
Takenaka Technical Research Laboratory
Tokyo

Professor H. Kawai
Tokyo Denki University
Hatoyama, Hikigun
Saitama 350-03

Dr. Eng. S. Kawamura
Osaka City University
Sumiyoshi, Osaka 1558

Dr. T. Kitahara
Sumitomo Heavy Industries, Ltd.
Tokyo

A. Kitatima
Kawada Industries Inc.
Tokyo

N. Kobayashi
Tokyo Institute of Polytechnics
Kanagawa

H. Komine
Institute of Industrial Science
Tokyo University
Tokyo

K. Makino
Makino Institute of Applied Instruments
Tokyo

I. Matsushita
Nippon Telegraph and Telephone Public Corporation
Musashino-Shi

JAPAN (continued)

Dr. T. Miyata
Department of Civil Engineering
University of Tokyo
Tokyo 113

Dr. M. Miyazaki
Sumitomo Heavy Industries, Ltd.
Tokyo

S. Murakami
Institute of Industrial Science
Tokyo University
Tokyo

Professor Y. Nakamura
Research Institute for Applied Mechanics
Kyushu University
Fukuoka

N. Narita
Public Works Research Institute
Ibaraki

Dr. A. Nishi
Faculty of Engineering
Miyazaki University
Miyazaki 880

Dr. K. Ogawa
Kawasaki Heavy Industries, Ltd.
1224, Shinzaike
Hiraoka-cho, Kakogawa-shi
Hyogo-ken

Dr. Y. Ogawa
National Institute for Environmental Studies
P.O. Yatabe
Ibaraki 300-21

T. Ohkuma
Kanagawa University
Kanagawa

S. Oikawa
Shimizu Construction Company
4-17 Etchujima 3-Chome Koto-ku
Tokyo 135

T. Okamura
Sakurada Iron Works Company, Ltd.
Chiba

Professor I. Okauchi
Chuo University
Tokyo

H. Onchi
Matsuo Bridge Company, Ltd.
Tokyo

Dr. T. Saito
Mitsubishi Heavy Industries Ltd.
Nagasaki

T. Sato
Mitsui Engineering and Shipbuilding Company, Ltd.
Tokyo

Dr. K. Shimoto
Tokyo Steel Rib and Construction Company, Ltd.
Tokyo

Dr. M. Shiotani
Physical Science Laboratories
Nihon University at Narashino
7-24-1 Narashino-Dai
Funabashi, Chiba

Y. Tamura
Waseda University
Tokyo

Y. Taniike
Disaster Prevention Research Institute
Kyoto University
Kyoto

Y. Uchiumi
The Technical Laboratory of Fujita
Tokyo

Professor S. Usuki
University of Akita
Akita-shi 010

N. Watanabe
Nihon Kokan Company, Ltd.
Yokohama

T. Yoshimura
Department of Civil Engineering
Kyushu Sangyo University
Fukuoka

MALAYSIA

Dr. A. J. Dutt
Faculty of Engineering
MARA Institute of Technology
Shah Alam, Selangor

NETHERLANDS

W. Alberts
Research Centre for Physical Planning-TNO
Delft

W. J. Beranek
Institute TNO-IBBC
P.O. Box 49
2600 AA Delft

Ir. N. Feis
Central Technical Institute TNO
Apeldoorn

H. van Koten
Institute TNO-IBBC
P.O. Box 49
2600 AA Delft

NEW ZEALAND

A. J. Bowen
Mechanical Engineering Department
University of Canterbury
Christchurch

Dr. N. Cherry
Agricultural Meteorology
Lincoln College
Canterburry

NORWAY

E. Hjorth-Hansen
Division of Structural Mechanics
Norwegian Institute of Technology
N-7034 Trondheim-NTH

PORTUGAL

Dr.-Eng A. R. J. Borges
Universidade Nova de Lisbon
De Engenharai Civil
Avenida do Brasil
Lisbon 1800

SRI LANKA

K. De Silva
State Engineering Corporation
130, Alston Place
Colombo

Dr. P. Sivaprakasapillai
Department of Buildings
Colombo 1

SWEDEN

Dr. K. Handa
Structural Design
University of Technology
Göteborg

B. G. Wiren
Wind Tunnel Laboratory
National Swedish Institute for Building Research
Box 785
80129 Gävle

SWITZERLAND

Dr. J. A. Hertig
Institut d'Economie et de Technologic Energitique (IETE)
EPFL - Ecublens
CH 1015 Lausanne

Dr. U. H. Oelhafen
Ingenieurschule HTL
8640 Rapperswil

M. Schwarz
Brown Boveri Ltd.
DEP. TX-2
Baden

UNION OF SOVIET SOCIALISTS REPUBLIC

Dr. L. Khurshudyan
U.S.-Soviet Air Pollution Program
Main Geophysics Observatory
Leningrad

Dr. I. Nekrasov
U.S.-Soviet Air Pollution Program
Moscow State University
Moscow

UNITED KINGDOM

K. Anthony
Ove Arup and Partners
London

Dr. R. E. Britter
D.A.M.T.P.
Silver Street
Cambridge CB3 9EW

Dr. I. P. Castro
Mechanical Engineering Department
University of Surry
Guildford GU2 5XH

R. H. Collingbourne
Meteorological Office
Bracknell RG12 2SZ

Dr. N. J. Cook
Building Research Establishment
Garston, Watford

Dr. R. K. Cooper
British Railways Board
London Road
Derby DE2 8UP

Dr. M. E. Davies
National Maritime Institute
Feltham, Middlesex TW14 OLQ

Dr. R. C. F. Dye
Simon Engineering Laboratories
University of Manchester
Manchester M13 9PL

Dr. K. J. Eaton
Building Research Establishment
Garston, Watford WD2 7JR

Dr. W. Ewers
Atkins R and D
Epsom

Dr. R. A. Hazelwood
Redland Technology Ltd.
Horsham

Dr. J. F. Howell
Department of Building Engineering
University of Bath
Bath BA2 7AY

Dr. J. C. R. Hunt
Applied Mathematics and Theoretical Physics Department
University of Cambridge
Cambridge CB3 9EW

A. P. Jeary
Building 14, Room 008
Building Research Establishment
Garston ND7 2TR

B. S. Kandola
Fire Safety Engineering Department
School of Built Environment
University of Edinburgh
Edinburgh EH9 3JL

Dr. B. E. Lee
Sheffield University
Sheffield

J. R. Mayne
Department of the Environment
Building Research Establishment
Building Research Station
Garston, Watford WD2 7JR

B. W. Smith
Flint and Neill Partnership
Portland House, Stag Place
London

D. M. Sykes
Department of Aeronautics
The City University
Northampton Square
London EC1V OHB

N. Thompson
Engineering Sciences Data Unit
251/259 Regent Street
London W1R 7AD

UNITED KINGDOM (continued)

R. E. Whitbread
National Maritime Institute
Teddington, Middlesex TW11 OLW

J. L. Wood
The British Aluminum Company, Ltd.
Chalfont Park, Gerrards Cross
Bucks

UNITED STATES OF AMERICA

Dr. R. F. Abbey, Jr.
Meteorology/Hydrology Research Program
U.S. Nuclear Regulatory Commission
Washington, D.C. 20555

A. B. Adenigba
Fluid Mechanics and Wind Engineering Program
Department of Civil Engineering
Colorado State University
Fort Collins, CO 80523

Dr. R. E. Akins
Department of Engineering Science/Mechanics
Virginia Polytechnic Institute and State University
Blacksburg, VA 24061

Professor S. P. S. Arya
Department of Meteorology
North Carolina State University
Raleigh, NC 27607

S. S. Ayad
Fluid Mechanics and Wind Engineering Program
Department of Civil Engineering
Colorado State University
Fort Collins, CO 80523

R. J. Baer
Ruben J. Baer and Associates
Skokie, IL 60077

Dr. L. V. Baldwin
College of Engineering
Colorado State University
Fort Collins, CO 80523

Y. Baron
Fluid Mechanics and Wind Engineering Program
Department of Civil Engineering
Colorado State University
Fort Collins, CO 80523

Dr. W. R. Barchet
Pacific Northwest Laboratory
Richland, WA 99352

Dr. M. Batts
National Bureau of Standards
Washington, D.C. 20234

A. Bedard
National Oceanic and Atmospheric Administration
Boulder, CO 80302

B. Bienkiewicz
Fluid Mechanics and Wind Engineering Program
Department of Civil Engineering
Colorado State University
Fort Collins, CO 80523

J. M. Bitner
Arizona State University
Tempe, AZ 85281

T. Boland
Stearns-Roger, Inc.
P.O. Box 5888
Denver, CO 80217

H. R. Bosch
Federal Highway Administration
400 7th Street, S.W.
Washington, D.C. 20024

A. G. Brady
U.S. Geological Survey
Menlo Park, CA 94025

H. J. Brauer
Fluid Mechanics and Wind Engineering Program
Department of Civil Engineering
Colorado State University
Fort Collins, CO 80523

L. Brown
U.S. Nuclear Regulatory Commission
Washington, D.C. 20555

UNITED STATES OF AMERICA (continued)

R. A. Bryer
Detroit Edison Company
20900 Oakwood Boulevard
Dearborn, MI 48124

M. Button
University of California
Berkeley, CA 94720

V. R. Cartelli
Rose, Chulkoff & Rose
White Plains, NY

Dr. J. E. Cermak
Fluid Mechanics and Wind Engineering Program
Department of Civil Engineering
Colorado State University
Fort Collins, CO 80523

Dr. A. R. Chamberlain
Colorado State University
Fort Collins, CO 80523

H. Chien
Fluid Mechanics and Wind Engineering Program
Department of Civil Engineering
Colorado State University
Fort Collins, CO 80523

Dr. A. N. L. Chiu
Department of Civil Engineering
University of Hawaii at Manoa
Honolulu, HI 96822

Dr. D. Chock
General Motors Research Labs
Warren, MI 48089

Professor K. H. Christensen
Department of Architecture
Kansas State University
Manhattan, KS 66506

L. W. Crow
3064 South Monroe
Denver, CO 80210

A. Y. Dabholkar
Department of Civil Engineering
Colorado State University
Fort Collins, CO 80523

M. W. Davis
Sverdrup/ARO, Inc.
Tullahoma, TN 37388

R. G. Derickson
Applied Science and Engineering
3244 South Platte River Drive
Englewood, CO 80110

D. Durbin
University of Cincinnati
Cincinnati, OH 45221

Dr. F. H. Durgin
Department of Aeronautics and Astronautics
Massachusetts Institute of Technology
Cambridge, MA 02139

J. R. Durgin
Fluid Mechanics and Wind Engineering Program
Department of Civil Engineering
Colorado State University
Fort Collins, CO 80523

D. Ebeling
Dick W. Ebeling, Inc.
Portland, OR 97223

Dr. J. A. Edwards
North Carolina State University
Raleigh, NC 27650

Dr. C. Etter
419-3A Strafford Avenue
Wayne, PA 19087

R. L. Ewald
Fluid Mechanics and Wind Engineering Program
Department of Civil Engineering
Colorado State University
Fort Collins, CO 80523

M. D. Flynn
I. M. Pei and Partners
600 Madison Avenue
New York, NY 10022

M. Folse
Department of Civil Engineering
Colorado State University
Fort Collins, CO 80523

UNITED STATES OF AMERICA (continued)

J. Fuh
Fluid Mechanics and Wind Engineering Program
Department of Civil Engineering
Colorado State University
Fort Collins, CO 80523

Dr. M. P. Gaus
National Science Foundation
Washington, D.C. 20550

H. Ghavam
Fluid Mechanics and Wind Engineering Program
Department of Civil Engineering
Colorado State University
Fort Collins, CO 80523

D. H. George
NOAA, PROFS, R45x2
Boulder, CO 80303

Dr. F. Gifford
NOAA, ATDL
Oak Ridge, TN 37830

Dr. J. H. Golden
Environmental Research Laboratories
NOAA
Boulder, CO 80302

Dr. O. M. Griffin
Code 8441, Ocean Technology Division
Naval Research Laboratory
Washington, D.C. 20375

Dr. A. C. Hansen
Wind Systems Program
Rockwell International, Box 464
Golden, CO 80401

R. Hayden
Aero-Sciences
Bolt, Beranek and Newman
Cambridge, MA 02138

N. I. Hazan
Fluid Mechanics and Wind Engineering Program
Department of Civil Engineering
Colorado State University
Fort Collins, CO 80523

L. E. Hebb
Defense Civil Preparedness Agency
Lakewood, CO 80226

M. Henderson
Department of Civil Engineering
Colorado State University
Fort Collins, CO 80523

Professor H. R. Holbo
School of Forestry
Oregon State University
Corvallis, OR 97331

N. Hosoya
Fluid Mechanics and Wind Engineering Program
Department of Civil Engineering
Colorado State University
Fort Collins, CO 80523

A. H. Huber
Mail Drop 14
U.S. Environmental Protection Agency
Research Triangle Park, NC 27711

J. Hutson
C. F. Braun and Company
Alhambra, CA 91802

C. Igawa
Technology Transfer Institute
Los Angeles, CA 90017

Dr. J. D. Iversen
Iowa State University
Ames, IA 50010

R. C. Jackson
Stearns-Roger, Inc.
Denver, CO 80217

Dr. P. C. Jennings
Mail Code 104-44
California Institute of Technology
Pasadena, CA 91125

Dr. M. C. Jischke
University of Oklahoma
Norman, OK 73019

Dr. A. Kareem
Department of Civil Engineering
University of Houston
Houston, TX 77004

UNITED STATES OF AMERICA (continued)

D. L. Kelly
National Severe Storms Forecast Center
NOAA
Kansas City, MO 64106

L. Kempner, Jr.
Bonneville Power Administration
Portland, OR 97212

G. K. Kiely
Fluid Mechanics and Wind Engineering Program
Department of Civil Engineering
Colorado State University
Fort Collins, CO 80523

C. S. Kirkpatrick
Haag Engineering Company
Dallas, TX

D. J. Koga
Illinois Institute of Technology
Chicago, IL 60616

K. M. Kothari
Fluid Mechanics and Wind Engineering Program
Department of Civil Engineering
Colorado State University
Fort Collins, CO 80523

S. E. Laccinole
Swan Wooster Engineering
Portland, OR 97206

Professor L. H. Lamouria
Department of Agricultural Engineering
California Polytechnic Institute
San Luis Obispo, CA 93401

Dr. P. Landers
Electric Power Research Institute
Palo Alto, CA 94303

Dr. B. M. Leadon
231 Aerospace Engineering Building
University of Florida
Gainesville, FL 32611

Dr. W. S. Lewellen
Aeronautical Research Associates
Princeton, NJ 08540

W. Li
Fluid Mechanics and Wind Engineering Program
Department of Civil Engineering
Colorado State University
Fort Collins, CO 80523

B. D. Light
University of Oklahoma
Norman, OK 73019

Professor Y. K. Lin
College of Engineering
University of Illinois
Urbana, IL 61801

A. W. Lingnell
Libbey-Owens-Ford Company
Toledo, OH 43695

D. F. Lockard
Johns-Manville Corporation
R. and D. Center
Denver, CO 80217

J. Lou
Fluid Mechanics and Wind Engineering Program
Department of Civil Engineering
Colorado State University
Fort Collins, CO 80523

Professor H. Lundgren
Department of Chemical Engineering
Arizona State University
Tempe, AZ 85281

Dr. R. D. Marshall
Center for Building Technology
National Bureau of Standards
Washington, D.C. 20234

L. J. Mata
Fluid Mechanics and Wind Engineering Program
Department of Civil Engineering
Colorado State University
Fort Collins, CO 80523

Dr. J. R. McDonald
Department of Civil Engineering
Texas Tech University
Lubbock, TX 79409

UNITED STATES OF AMERICA (continued)

Dr. K. Mehta
Department of Civil Engineering
Texas Tech University
Lubbock, TX 79409

Dr. R. N. Meroney
Fluid Mechanics and Wind Engineering Program
Department of Civil Engineering
Colorado State University
Fort Collins, CO 80523

R. D. Miller
Boeing Engineering and Construction
Seattle, WA 98121

Dr. P. H. Miller
Department of Mechanical Engineering
Louisiana State University
Baton Rouge, LA 70803

Professor W. M. Miller
Department of Civil Engineering
University of Washington
Seattle, WA 98195

Dr. J. E. Minor
Institute for Disaster Research
Texas Tech University
Lubbock, TX 79413

Dr. L. Morino
College of Engineering
Boston University
Boston, MA 02215

A. Morris
Ambient Analysis Inc.
Boulder, CO 80302

Dr. T. B. Morrow
Southwest Research Institute
San Antonio, TX 78284

D. G. Mutter
Department of Civil Engineering
Colorado State University
Fort Collins, CO 80523

E. Nazario
PSC 2, Box 697
Sheppard Air Force Base
Wichita Falls, TX 76311

D. E. Neff
Fluid Mechanics and Wind Engineering Program
Department of Civil Engineering
Colorado State University
Fort Collins, CO 80523

Dr. J. L. Noland
Atkinson-Noland and Associates
Boulder, CO 80307

Dr. D. J. Norton
Texas A&M University
College Station, TX 77840

Dr. J. P. Olienyk
Department of Economics
Colorado State University
Fort Collins, CO 80523

Dr. A. O'Mara
1235 St. Albans Road
San Marino, CA 91108

Dr. H. A. Panofsky
Department of Meteorology
Pennsylvania State University
University Park, PA 16802

Dr. R. A. Parmelee
Department of Civil Engineering
Northwestern University
Evanston, IL 60201

R. Peace
2050 New York Drive
Altadena, CA 91001

Dr. J. A. Peterka
Fluid Mechanics and Wind Engineering Program
Department of Civil Engineering
Colorado State University
Fort Collins, CO 80523

Dr. R. L. Petersen
Fluid Mechanics and Wind Engineering Program
Department of Civil Engineering
Colorado State University
Fort Collins, CO 80523

Dr. A. B. Poore
Department of Mathematics
Colorado State University
Fort Collins, CO 80523

UNITED STATES OF AMERICA (continued)

M. D. Powell
National Hurricane and Experimental Meteorology Laboratory
NOAA
Coral Gables, FL 33134

J. V. Ramsdell
Battelle Pacific Northwest Laboratory
Richland, WA 99352

M. A. Ratcliff
Fluid Mechanics and Wind Engineering Program
Department of Civil Engineering
Colorado State University
Fort Collins, CO 80523

D. A. Reed
Princeton University
Princeton, NJ 08544

Dr. T. A. Reinhold
U.S. Department of Commerce
National Bureau of Standards
Washington, D.C. 20234

J. H. Ribbe
Wind Tunnel Services
Texas A&M University
College Station, TX 77840

M. C. Richmond
Meteorology Research, Inc.
Altadena, CA 91001

W. Ridinger
A D & C Group, Inc.
Denver, CO

L. E. Robertson
Skilling, Helle, Christiansen, Robertson
New York, NY 10017

Dr. K. B. Rojiani
Department of Civil Engineering
Virginia Polytechnic Institute and State University
Blacksburg, VA 24061

Dr. A. Roshko
California Institute of Technology
Pasadena, CA 91125

H. J. Royston
1422 Kearney Street
Laramie, WY 82070

Dr. W. Z. Sadeh
Fluid Mechanics and Wind Engineering Program
Department of Civil Engineering
Colorado State University
Fort Collins, CO 80523

D. Saharon
Fluid Mechanics and Wind Engineering Program
Department of Civil Engineering
Colorado State University
Fort Collins, CO 80523

Dr. R. H. Scanlan
Department of Civil Engineering
Princeton University
Princeton, NJ 08544

Dr. J. T. Schaefer
National Severe Storms Forecast Center
NOAA
Kansas City, MO 64106

Dr. A. Schiff
School of Mechanical Engineering
Purdue University
West Lafayette, IN 47907

Dr. C. Sheih
Argonne National Laboratory
Argonne, IL 60440

M. Sherman
Lawrence Berkeley Laboratory
Berkeley, CA 94720

C. F. Shieh
Space Institute
University of Tennessee
Tullahoma, TN 37388

J. F. Y. Sidarous
Department of Civil Engineering
Colorado State University
Fort Collins, CO 80523

Dr. E. Simiu
Center for Building Technology
U.S. Department of Commerce
National Bureau of Standards
Washington, D.C. 20234

UNITED STATES OF AMERICA (continued)

W. L. Simons
W. L. Simons and Associates
Denver, CO 80235

J. Sinou
Fluid Mechanics and Wind Engineering Program
Department of Civil Engineering
Colorado State University
Fort Collins, CO 80523

Dr. G. T. Skinner
Calspan Corporation
Advanced Technology Center
Buffalo, NY 14225

Dr. M. D. Skold
Department of Economics
Colorado State University
Fort Collins, CO 80523

J. R. Snyder
Department of Economics
Colorado State University
Fort Collins, CO 80523

Dr. W. H. Snyder
Meteorology and Assessment Division
Environmental Sciences Research Laboratory
U.S. Environmental Protection Agency
Research Triangle Park, NC 27711

Dr. P. R. Sparks
Engineering Science and Mechanics
Virginia Polytechnic Institute and State University
Blacksburg, VA 24061

M. J. Sramek
Omaha Public Power District
Omaha, NB 68126

S. L. Steely
FWG Associates, Inc.
Tullahoma, TN 37388

Dr. G. R. Stegen
Science Applications Inc.
Bellevue, WA 98005

Dr. J. H. Strickland
Texas Tech University
Lubbock, TX 79409

J. H. Struthers
General Dynamics Convair
San Diego, CA

P. P. Sullivan
Fluid Mechanics and Wind Engineering Program
Department of Civil Engineering
Colorado State University
Fort Collins, CO 80523

Dr. H. R. Swift
Technical Center
Libbey-Owens-Ford Company
Toledo, OH 43605

K. Takeda
Fluid Mechanics and Wind Engineering Program
Department of Civil Engineering
Colorado State University
Fort Collins, CO 80523

Dr. H. W. Tieleman
Virginia Polytechnic Institute and State University
Blacksburg, VA 24061

M. Ui
Fluid Mechanics and Wind Engineering Program
Department of Civil Engineering
Colorado State University
Fort Collins, CO 80523

Dr. M. D. Vanderbilt
Department of Civil Engineering
Colorado State University
Fort Collins, CO 80523

Dr. A. W. Viney
Department of Psychology
Colorado State University
Fort Collins, CO 80523

R. A. Vognild
American Iron and Steel Institute
Dallas, TX 75234

Dr. J. L. Way
Illinois Institute of Technology
Chicago, IL 60616

R. Weaver
1534 E. Morada Place
Altadena, CA

UNITED STATES OF AMERICA (continued)

Professor G. F. White
Natural Hazards Research and Information Center
University of Colorado
Boulder, CO 80302

F. Wilkie
1 Olde Coach Road
Hudson, NH 03051

C. Y. Wong
Koppers Company, Inc.
Pittsburgh, PA 15220

H. G. C. Woo
Fluid Mechanics and Wind Engineering Program
Department of Civil Engineering
Colorado State University
Fort Collins, CO 80523

Professor F. L. Worley
Department of Chemical Engineering
University of Houston
Houston, TX 77004

Dr. J. Xerikos
McDonnell Douglas Astronautics Company
Department 203, Mail Station 13-3
Huntington Beach, CA 92647

Dr. S. Zagajeski
Department of Civil Engineering
Colorado State University
Fort Collins CO 80523

T. G. Zambrano
AeroVironment, Inc.
Pasadena, CA 91107

Appendix 4

Authors Not Present

Mr. H. Akagi, Nippon Telegraph & Telephone Public Corporation, Japan
Dr. A. R. Al-Rawi, University of Mustansariyya, Iraq
Mr. T. Amano, Ryukyu University, Japan
Mr. F. D. Beresford, Commonwealth Scientific & Industrial Research, Australia
Professor J. Blessmann, UFRGS, Brazil
Dr. A. P. Burns, BHRA Fluid Engineering, United Kingdom
Mr. K. Deguchi, University of Hosei, Japan
Ms. P. G. Diosey, National Institute for Environmental Studies, Japan
Mr. C. Dorman, Australian Department of Housing & Construction, Australia
Mr. P. A. Durbin, University of Cambridge, United Kingdom
Mr. B. R. Ellis, Building Research Establishment, United Kingdom
Dr. A. R. Flint, Flint & Neill Partnership, United Kingdom
Mr. L. J. Fortier, Bolt Beranek and Newman Inc., USA
Mr. D. G. Friedman, Travelers Insurance Company, USA
Dr. W. Frost, University of Tennessee Space Institute, USA
Professor M. Fujimoto, Tokyo Institute of Technology, Japan
Mr. Y. Fujino, University of Tsukuba, Japan
Professor R. I. Harris, Cranfield Institute of Technology, United Kingdom
Dr. G. C. Hart, University of California, USA
Mr. Y. Iwatani, Nihon University at Narashino, Japan
Professor J. Katsura, Kyoto University, Japan
Dr. J. Khanna, Swan Wooster Engineering Company, Canada
Professor S. Kim, Yonsei University, Korea
Dr. G. König, Institut für Massivbau, West Germany
Dr. B. G. Korenev, CNIISK, USSR
Ms. G. Lappas, Institut für Massivbau, West Germany
Dr. R. H. Leicester, Commonwealth Scientific & Industrial Research, Australia
Dr. D. Lindley, Taylor Woodrow Construction Ltd., United Kingdom
Mr. R. Lipschutz, Pennsylvania State University, USA
Professor S. Mackey, University of Hong Kong, Hong Kong
Dr. E. W. Marchant, University of Edinburgh, United Kingdom
Professor G. Matsui, Waseda University, Japan
Professor Y. Mitsuta, Kyoto University, Japan
Mr. S. E. Mullins, Virginia Polytechnic Institute & State University, USA
Dr. M. Novak, University of Western Ontario, Canada
Dr. T. Okubo, Public Works Research Institute, Japan
Ms. M. Paciuk, Israel Institute of Technology, Israel
Dr. W. Petak, J. H. Wiggins Company, USA
Dr. A. Petre, Department of Aeronautical Engineering, Romania

AUTHORS NOT PRESENT (continued)

Dr. M. Pirner, Technical & Research Testing Institute, Czechoslovakia
Professor E. J. Plate, University of Karlsruhe, West Germany
Mr. V. G. Quincey, National Maritime Institute, United Kingdom
Professor J. Riera, UFRGS, Brazil
Mr. H. Rothman, Paul Weidlinger Consulting Engineers, USA
Mr. J. A. Saraiva, Laboratorio Nacional de Engenharia Civil, Portugal
Mr. H. Sato, Public Works Research Institute, Japan
Dr. J. R. Shaver, National Bureau of Standards, USA
Mr. Y. P. Sheng, Aeronautical Research Associates, USA
Dr. N. Shiraishi, Kyoto University, Japan
Dr. R. Sofronie, Civil Engineering Department, Romania
Mr. C. Soize, C.T.I.C.M., France
Professor D. Stura, University of Genoa, Italy
Mr. J. Tajima, Hanshu-Shikoku Bridge Authority, Japan
Dr. H. H. Tang, McDonnell Douglas Corporation, USA
Mr. M. E. Teske, Aeronautical Research Associates, USA
Mr. R. L. Thorpe, Aeronautical Research Associates, USA
Mr. S. J. Tindall, National Maritime Institute, United Kingdom
Mr. K. Uehara, National Institute for Environmental Study, Japan
Ms. N. Vanderstappen, University of Leuven, Belgium
Mr. M. Vilardo, Pennsylvania State University, USA
Dr. J. Wianecki, Centre Experimental de Recherches et d'Etudes, France
Mr. W. O. Willis, USDA-AR, USA
Mr. K. Yokoyama, Public Works Research Institute, Japan

FIFTH INTERNATIONAL CONFERENCE ON WIND ENGINEERING

Colorado State University, Fort Collins, Colorado

8-14 July 1979

Left-Hand Portion of Group Photograph

1. S.E. Laccinole
2. I.S. Gartshore
3. H.R. Bosch
4. D. Sfintesco
5. A.G. Davenport
6. A.N.L. Chiu
7. A.P. Jeary
8. M. Poreh
9. T. Miyata
10. J. Sidarous
11. Y. Nakamura
12. J.E. Cermak
13. M.P. Gaus
14. H.R. Holbo
15. H.P. Irwin
16. R.H. Scanlan
17. H. Kawai
18. A. Laneville
19. J. Hutson
20. N.J. Cook
21. B.E. Lee
22. L. Lee
23. R.E. Whitbread
24. T. Yoshimura
25. R.A. Parmelee
26. P.C. Jennings
27. H. Ishizaki
28. H. Tanaka
29. H. van Koten
30. C. Kramer
31. G. Kiesselbach
32. J.E. Minor
33. R. Aynsley
34. N. Thompson
35. G. Mears
36. R. Hazelwood
37. W.A. Ewers
38. J.R. Mayne
39. R.L. Wardlaw
40. L.A. Brown
41. K. Mehta
42. P. Spehl
43. B.W. Smith
44. M.C. Richmond
45. C.S. Kirkpatrick
46. L.E. Hebb
47. T. Reinhold
48. A. Dalgliesh
49. P. Sparks
50. T. Ohkuma
51. G.V. Parkinson
52. Y. Tamura
53. D.W. Ebeling
54. M.A. Wawzonek
55. M. Ito
56. Y.K. Lin
57. R.C.F. Dye
58. C.F. Shieh

Right-Hand Portion of Group Photograph

59. H. Ruscheweyh
60. V.R. Cartelli
61. K.C.S. Kwok
62. T. Tschanz
63. J. Saunders
64. D. Surry
65. C.D. Williams
66. J.V. Ramsdell
67. R.J. Kind
68. M.D. Powell
69. W. Alberts
70. J. Way
71. K. Makino
72. H. Sockel
73. Y. Taniike
74. B. Tryggvason
75. S. Usuki
76. N. Isyumov
77. Ph. Duchene-Marullaz
78. B. Etkin
79. W.M. Miller
80. C.G. Etter
81. J.A. Hertig
82. H. Teunissen
83. K.R. Cooper
84. B. Bienkiewicz
85. M. Miyazaki
86. M.C. Jischke
87. B.D. Light
88. J.H. Strickland
89. J. Xerikos
90. R. Britter
91. S. Gotoh
92. I.G. Currie
93. I. Okauchi
94. T. Okamura
95. T. Sato
96. G.A. Berz
97. P.H. Miller
98. P.N. Georgiou
99. A. Roshko
100. J.F. Howell
101. J.D. Holmes
102. J. Gardner
103. K.M. Kothari
104. J.A. Peterka
105. G.T. Skinner
106. M.W. Davis
107. A.R.J. Borges
108. A.J. Schiff
109. W.H. Melbourne
110. N. Narita
111. K. Takeda
112. K. Ogawa
113. J.S. Fuh
114. M. Ui
115. A.J. Dutt
116. F.H. Durgin
117. W.Z. Sadeh
118. R.L. Petersen
119. J.H. Struthers
120. T.B. Morrow
121. B.M. Leadon
122. J.A. Edwards
123. W. Hanenkamp
124. K. Zilch
125. J. Iversen
126. E. Bjerregaard
127. W. Schnabel
128. P. Landers
129. M. Shiotani
130. M. Ikenouchi
131. S. Kawamura
132. T. Saito
133. K. Higuchi
134. Y. Akiyama
135. H. Akiyama
136.
137. M.D. Flynn
138. K.B. Rojiani
139. S. Frandsen
140. C.Y. Wong
141. H. Panofsky
142. A. Kareem
143. J. Berlamont
144. A.J. Bowen
145. M. Button
146. R.G.J. Flay
147. L.H. Lamouria
148. D.M. Sykes
149. D. Lockard
150. L. Kempner, Jr.
151. J.L. Wood
152. E. Hjorth-Hansen
153. D. Kelly
154. K. Handa
155. N. Hosoya

TECHNICAL SESSION

PRESENTATION OF PLAQUE-OF-RECOGNITION TO DR. DELIU SFINTESCO (left)

RELAXATION AT CONFERENCE PICNIC

OUTING IN ROCKY MOUNTAIN NATIONAL PARK

CONFERENCE DINNER

CONFERENCE DINNER

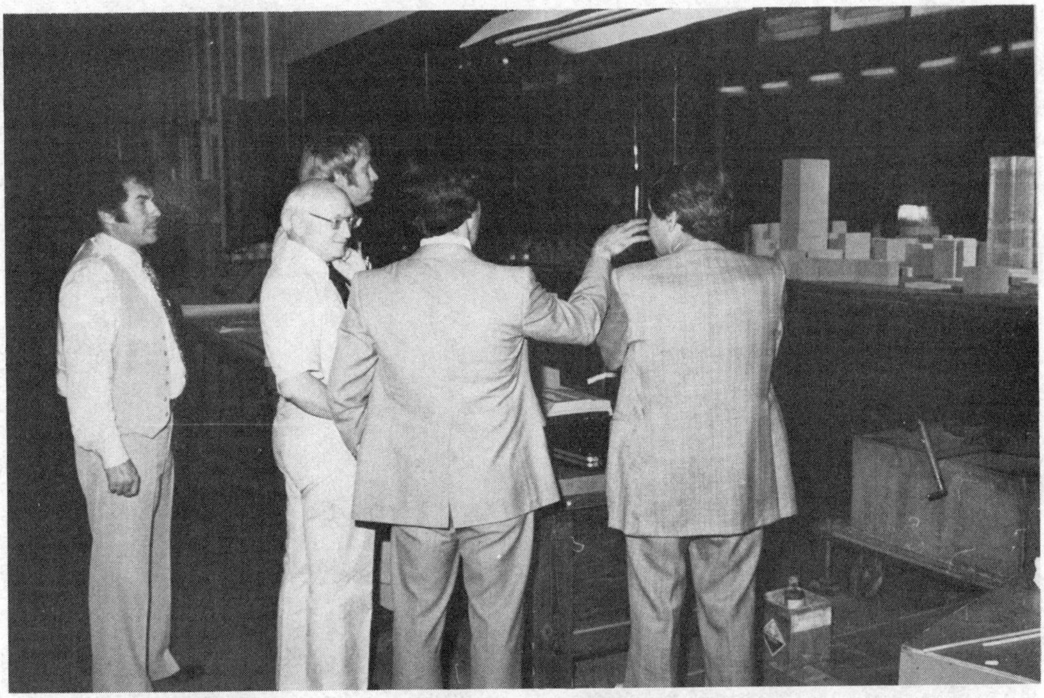

VISIT TO FLUID DYNAMICS AND DIFFUSION LABORATORY

REGISTRATION

CONFERENCE SECRETARIAT

COMMUNICATION

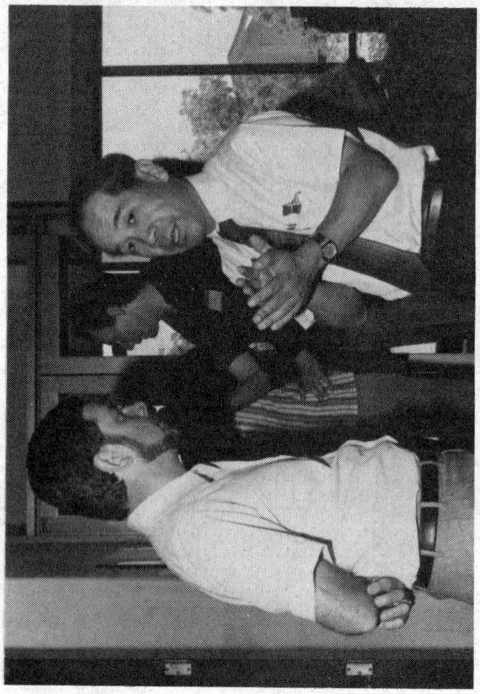

COMMUNICATION

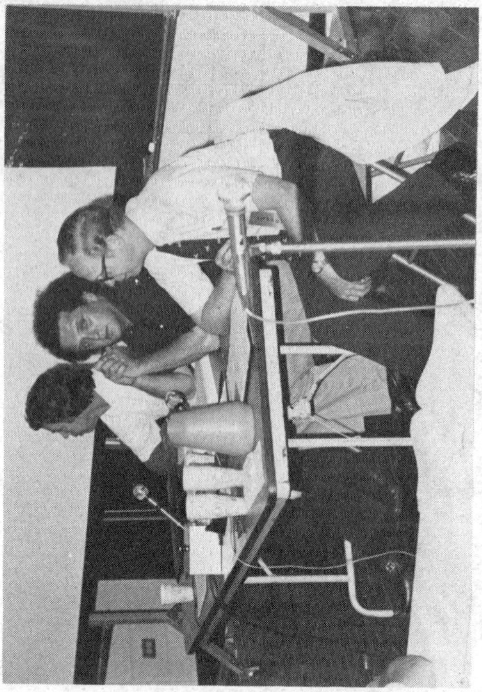

SESSION CHAIRMAN, CO-CHAIRMAN AND RAPPORTEUR

Indices*

Type of Paper

Subject Index

*Page numbers refer to the first page of a paper.

Author Index